SHELTON STATE COMMUNITY
COLLEGE
JUNIOR COLLEGE DIVISION
LIBRARY

D1037447

REF.   Brown, Roland Wil-
PE   bur, 1893-1961.
1175
.B76   Composition of
1956   scientific words

DATE DUE			
REF.	REF.	REF.	REF.
	REF.	REF.	
	REF.	REF.	
	REF.	REF.	

COMPOSITION OF SCIENTIFIC WORDS

A manual of methods and a lexicon of materials
for the practice of logotechnics

ROLAND WILBUR BROWN

Published by Smithsonian Institution Press
Washington, D.C.

Library of Congress Cataloging in Publication Data
Brown, Roland Wilbur, 1893–1961.
  Composition of scientific words.

  Reprint of the 1956 ed., which was first published in
1927 under title: Materials for word-study.
  Bibliography: p.
  1. English language—Word formation. 2. English
language—Foreign elements—Greek. 3. English language
—Foreign elements—Latin. 4. Science—Terminology.
I. Title.
PE1175.B76 1978   422'.4   78-14717
ISBN 0-87474-286-2

Copyright © 1954, 1956 by ROLAND W. BROWN

Revised Edition, 1956

Reprinted 1978, 1985

CONTENTS

FOREWORD

Composition of scientific words is a revised and greatly enlarged version of my *Materials for word-study,* published in 1927, but retains the same general features. Because the lexicon of that book proved inadequate for the widest application in synthesizing words, I decided in 1941, after insistent urging from friendly critics among my scientific colleagues, to undertake the preparation of a larger reservoir of usable material preceded by a concise, introductory discussion of the methods for creating words, especially terms used in the sciences. As the fulfillment of this purpose unfolded I considered the invention of a more descriptive title for the new book.

The Greeks had several distinct terms for words and names, as *onoma,* noun, and *rhema,* verb, the sources of *onomatology* and *rhematology,* which are resounding creations for phases of semantics. *Onomatopoeia,* from *onomatopoieo,* to make names, now has the special sense of 'words made in imitation of natural sounds.' *Epos* and *logos,* used concretely and abstractly, denoted the spoken or written word; and a *logotechnes* was an artificer in words. From this word I derive *logotechnics,* the art of composing words. This inventive process could be implied by the term *wordcraft,* but the latter refers more particularly to the skillful use of words. The Greeks also had a figurative, perhaps slightly satirical synonym for *logotechnes,* namely, *logomageiros,* wordcook. This word conveys a subtle suggestion that not all cooks are artists,—an observation not inapplicable to some post-classical wordmakers.

The Romans apparently had no single term for 'a maker of words', but the Roman wordcreator could well have been called a *verbifex* and his art *verbifactio* or *verbicultura.* The *nomenclator* was essentially a name-caller or identifier, and a *nomenclatura* was a list of names. Other words from the same source (*nomen, -inis,* name), but with accessory or derived meanings, are *nominator,* namer, and *nominatio,* designation by name, especially for public office.

Although the materials just reviewed were suggestive, I concluded that a one-word title, such as *Logotechnics,* however appropriate, might prove to be too forbidding because of its novelty. After selecting that now used, and while cogitating about the menu offered under the new designation, I was reminded of a U. S. Department of Agriculture bulletin, entitled *Turnips, beets, and other succulent roots, and their use as food.* My book also deals with roots, more or less succulent, and the approved methods for cultivating them. It will, I trust, promote skill and pleasure in composing words, and thus, although not intended to be a panacea for all the ills of nomenclature, may, by offsetting some of the bad effects resulting from decreased instruction in classical languages, diminish the area and amount of verbicultural wrongdoing. My fervent hope, however, is that the book will not encourage the making of unnecessary words, the proliferation of jargon. It is intended to be a useful tool, and as such, whether or not it is considered sharp, should be handled carefully and wisely.

The LEXICON of this book is arranged on the cross-reference plan, the English key-words receiving their appropriate Latin, Greek, and other synonyms, with derivative examples. To save space, the etymologies of the entry-words are given as simply as possible without intermediate steps, except in special instances. Sometimes cognate or related words are also cited because they may have pertinence in providing useful derivatives. Many entries have several parts, indicated by semicolons and separate cross-references, as **gum** < AS. *goma,* palate, jaws, see **mouth;** < L. *gummi* (Gr. *kommi*), see **resin.** Separate entries for such homonyms would have greatly increased the size of the book and would not have contributed much to clarity of treatment. At the end of the key-word list of synonyms is a line beginning with See:, followed by one or more words to which the reader is directed for further material suggested by the key-word. The direction Derive:, encountered here and there, is intended to arouse the reader's curiosity and to encourage resort to the unabridged dictionaries. In most instances the purpose is to emphasize caution that words or parts of words spelled like the keyword are not necessarily derived from the same source.

As clear example is potently contagious, this book illustrates generously how words have been and can be constructed. Most examples are good, many are indifferent, and some are very bad. The bad examples should serve as instructive deterrents. In many but by no means all faulty instances I have indicated editorial corrections in brackets, as *Menticirrhus[Menticirrus] saxatilis* (kingfish). These corrections, it should be emphasized, must in no wise be considered by cataloguers as published taxonomic emendations. Examples from the biologic sciences receive preponderant treatment because those sciences have levied most heavily upon Greek and Latin for their terminologies. The scientific names of plants and animals cited are, however, not necessarily the most recent or those thought to be currently valid. In numerous instances, such as Gr. *apoknisma* and L. *pisinnus* under **little,** authentic examples are missing because I had no luck in finding any. When searching for a word to express a given idea, the user of the book should remember that, like English words, Greek and Latin terms may be used both literally and figuratively. Under **dull,** for example, will be found synonyms denoting both concrete and abstract dullness. Is there any need to say that imagination and ingenuity are prime requisites not only for searching out word materials but also for compounding them skillfully?

As this book is not a manual, textbook, or dictionary of etymology, the user must not expect to find all terms listed and analyzed, for many words have no authentic extant etymology, or they are New Latin words whose etymology can be inferred from their components. Some that appear to be missing from the list may be found lurking behind others with nearly the same spelling, as *furnarius,* baker, not listed at the main entry but under **furnus,** oven. So *salubris* occurs under **salus,** and *saxatilis* under **saxum.** Only the most commonly used geographic, personal, and mythological names have been entered. I have deemed it convenient and helpful, where warranted, to give the entry-item in its presently used, transliterated, combining form, as **crocido-** instead of **krokis.** To have listed all the compounds having such prefixes as *amphi-,* *circum-, in-, peri-, pro-, syn-,* and the like, would have been an unnecessarily wasteful procedure. After creating a word, the creator should check the Greek, Latin, and English dictionaries for the possibility that the word may already be in existence.

The ideal aimed at in the discussion of classical grammar was to provide the irreducible minimum necessary for wordmakers to know. Some of these grammatical details and their treatment need explanation here. The genitives of nouns are given only when they help to clarify the spelling of the root-stem or combining base. For this reason the genitives in *-ae* of Latin first and in *-i* of second declension nouns, and those in *-ou* of Greek nouns, are omitted. In some instances, of several possible Greek genitives, I have given that which seems euphonious and has current preferable usage. Genitives characteristic of the several declensions will be found exemplified in the section on NATURE OF GREEK AND LATIN. The genders of nouns are indicated, so far as determinable, by the customary abbreviations, m., f., n., c., and that of the more unusual adjectives by their nominative singular terminations. The perfect participle, but not always the present participle of a Latin verb is given. When these are omitted, assume them. To find discussions of special topics, consult the INDEX.

Many nouns are represented in the classical dictionaries in their plural forms only, but on the analogy of *oat* and *oats,* I have taken the liberty of entering such nouns in the singular, as *clitella,* saddle, for *clitellae; cuna,* cradle, for *cunae; geminus,* twin, for *gemini; hilla,* small intestine, for *hillae; insidia,* ambush, for *insidiae; nuga,* trifle, for *nugae; trica,* nonsense, for *tricae; krossos,* fringe, for *krossoi.*

Finally, the user should not conclude that possession of or exposure to this book, or hasty, erratic flights through it, will make him an expert wordsmith. Only persistent, systematic study, practice, and critical appraisal will produce the desired facility and expertness.

Acknowledgments. Study of the practical application of Latin and Greek in the formation of literary and scientific terms has been my favorite hobby since those days when I found that it sweetened the bitter classroom pill of grammatical routine. Fortunately, I also had several word-conscious teachers in other subjects who made a practice of analyzing the terminology employed. To them I owe a linguistic debt which the preparation of this book may, in part, liquidate. The vicarious sacrifice to this project of many holidays and innumerable leisure moments may also expiate my own verbal follies and nomenclatural sins. During the gathering of the raw materials for this book I was doubtless a considerable nuisance to my colleagues and to the librarians at the United States National Museum. To all these patient, solicitous, and helpful friends I express heartfelt gratitude. My applause and tribute also go to the often unsung but indefatigable compilers of dictionaries, lexicons, catalogues, and indexes, on which, deliberately or unwittingly, I have drawn very heavily. May their products become increasingly more accurate and effective!

ROLAND W. BROWN

U. S. National Museum,
  Washington 25, D. C.

ABBREVIATIONS AND SYMBOLS

Ab.Am.	Aboriginal American
Abyss.	Abyssinian
Ar.	Arabic
AS.	Anglo-Saxon
C.	Celtic
Chin.	Chinese
D.	Dutch
Dan.	Danish
E.	English
Egypt.	Egyptian
F.	French
G.	German
Goth.	Gothic
Gr.	Greek
Heb.	Hebrew
Hind.	Hindi
Ir.	Irish
It.	Italian
Jap.	Japanese
L.	Latin
LL.	Late Latin
ME.	Middle English
MHG.	Middle High German
ML.	Medieval Latin
N.	Norse
NL.	New Latin
OE.	Old English
OF.	Old French
OHG.	Old High German
ON.	Old Norse
Per.	Persian
Pg.	Portuguese
Russ.	Russian
Skt.	Sanskrit
Scan.	Scandinavian
Slav.	Slavic
Sp.	Spanish
Sw.	Swedish
Teut.	Teutonic
Turk.	Turkish

VL.	Vulgar Latin
W.	Welsh
ant.	antonym
c.	common
cf.	compare
chem.	chemistry
cog.	cognate
comp.	comparative
cont.	continued
der.	derived
dim.	diminutive
e.g.	for example
etc.	et cetera
etym.	etymology
ex.	example
exc.	exception
f.	feminine
gen.	genitive
lit.	literally
m.	masculine
n.	neuter
nom.	nominative
part.	participle
perf.	perfect
pl.	plural
pref.	prefix
prob.	probably
q.v.	which see
sing.	singular
suf.	suffix
super.	superlative
syn.	synonym
var.	variant, variety
<	derived from
?	doubtful, probably
[ ]	brackets enclose editorial assumptions, corrections, and opinions
=	equals

INTRODUCTION

HISTORY AND NATURE OF THE ENGLISH LANGUAGE

The prehistoric inhabitants of the British Isles probably left little of their language to English speech. Their relics, which excite our wonder and speculation, consist chiefly of stone, clay, and bronze implements, kitchen middens, burial mounds (cairns, barrows, tumuli), and megalithic monuments comprising isolated stones (monoliths, menhirs), simple tombs (dolmens), and stone circles (cromlechs) such as Stonehenge, in Wiltshire, England, and the Ring of Brogar, near Stromness, in the Orkney Islands. These aborigines were either absorbed or destroyed by Celts or Gaels, the nucleus of the present Irish, Scotch, and Welsh stocks, who came from the Continent during the first millennium B.C., and, seizing the ports, spread out over the good agricultural lands. The Celts, besides practicing agriculture, are said to have brewed beer, mined tin, and introduced the use of iron. To the English vocabulary, through contact with their Roman and later contemporaries, they contributed *bard, bin, crag,* and many geographic names, such as *Cornwall, Dover, Kent, London, Thames, York; avon,* river, in *Stratford-on-Avon, Avondale; bryn,* hill, in *Bryn Mawr; cumb,* valley, in *Duncombe;* and *dun,* hill, town, in *Dundee, Dumbarton, Doncaster.* Our May Day and Halloween revelries are reminiscent of less respectable Celtic ceremonies. Likewise our custom of kissing under the mistletoe is said to date from those barbaric days when that plant was held sacred and druid priests "with voices sad and prophetic" offered up human sacrifices and chanted about the transmigration of the soul. One tribe of southern Celts was called Brythons or Britons, a name perpetuated in the words *Britain, British, Brittany, Briton,* and *Breton.*

In 55 B.C. Julius Caesar and his victorious legions, desirous of extending the sway of Roman power, disembarked on British shores but received scant welcome from the outraged Celts. Rome, however, after gaining a foothold, maintained this Latin-Celtic connection until early in the fifth century. The invasion resulted in the importation of many typical Roman customs, the institution of organized government, the appearance of the inevitable tax-collector, the construction of roads, stone walls, fortifications, and public buildings, and the introduction of such Latin words or roots as *port (portus,* harbor) in *Portsmouth, Portsea, Portchester; coln (colonia,* settlement) in *Lincoln; wick* or *wich (vicus,* village) in *Warwick, Greenwich; cester* or *chester (castra,* camp, fort) in *Chester, Gloucester, Lancaster, Manchester, Winchester,*—places that were former strategic military posts or protected habitations. These words now testify as eloquently to the Roman occupation of Britain as the crumbling walls across northern England from Newcastle on the Tyne to Solway Firth, or of Watling Street, an improvement of the old Roman road leading from London to Chester. The use of Latin, however, seems to have been confined to officialdom and townsfolk, whereas the tribes in the rural districts held somewhat aloof and continued to speak their own tongue.

The Roman legions had no sooner been withdrawn from Britain to protect Rome against Alaric and his Goths, than tall, robust, blue-eyed, fair-faced Angles, Saxons, and Jutes from the region on the Continent now known as Jutland and Schleswig-Holstein began about 450 A.D., under various impelling motives, to enter the country. Despite the valiant resistance of the legendary King Arthur and his Knights of the Round Table, the invasion continued in more or less desultory fashion, but with complete success, the Angles settling in Northumbria, the Saxons in Wessex, Sussex, and Essex, and the Jutes in Kent. This conquest swept away most of the Roman innovations, and gradually, with multiplication in numbers, the invaders overshadowed and displaced the Celtic and aboriginal population and became the dominant race in that part of Britain which presently by common consent was called Angle-land or England, after the Angles, at first the strongest tribe. No unity existed among the Anglo-Saxons except that they spoke dialects of the same language, a kind of Low German. The elements of this speech, which for the earlier developmental period in Britain is sometimes called Early English or Old English, comprise the basic, common words in Modern English. The Anglo-Saxons also reputedly brought with them a small residue of modified Latin, illustrated in English as follows, showing that, while still on the Continent, they had made commercial and cultural contacts with Roman civilization: *bishop, butter, camp, cheese, chest, church, copper, dish, gem, inch, kettle, kitchen, linen, mile, mint, mule, pea, pepper, pile, pillow, plum, poppy, pound, Saturday, seal, spelt, street, table, wall, wine.* Roughly, the stages in the development of the English language from the arrival of the Anglo-Saxons in Britain to the present may be designated by the following periods:

450 to 1100 A.D.   Anglo-Saxon, Early English, Old English.

1100 to 1500 A.D.   Middle or Midland English, characterized by the loss of many Anglo-Saxon inflections and the establishment of a new homogeneity that led to widespread authorship.

1500 to the present.   Modern English, characterized by its heterogeneity and ability to absorb word materials from all sources.

Near the close of the sixth century (597 A.D.) Augustine, the missionary, arrived from Rome and, aided by native Celts from Ireland, preached Christianity throughout the land for the purpose of converting the pagan Anglo-Saxons. This movement, besides giving a strong impetus toward unification, introduced to the native vocabulary many new domestic and ecclesiastical terms derived from Latin and Greek: *alms, altar, balsam, beet, belt, candle, clerk, creed, cross, deacon, devil, fig, hymn, lentil, lily, mass, millet, minster, monk, myrrh, nun, place, plant, pope, priest, psalm, relic, rose, saint, school, shrine, stole, temple, turn.* Christianization, however, did not uproot some long-established pagan customs, a few of which persist in modified form to this day. For example, the winter solstice, falling or or about December twenty-first, being a turning in the astronomical year, was the inspiration for the annual festival of Yule, in which the holly, Yule-log, and wassail-bowl played conspicuous parts. Our Christmas has inherited much of the spirit and outward trappings of the Yule celebration. The word *jolly,* said by some authorities to be from *Yule,* may be regarded as a reminder of that cheerful season. In the spring of the year, at the time of the vernal equinox, the goddess Eastre was worshipped. From those rites another Christian festival has derived much of its outward form and a name, *Easter.* Finally, in the names of the days of the week, as *Monday (mona,* moon), *Tuesday (Tiw,* god of war), *Wednesday (Woden,* chief of the gods), *Thursday (Thor,* god of thunder), *Friday (Frigg,* goddess of marriage), *Saturday (Saeter* [L. *Saturnus*], Saturn), *Sunday (sunne,* sun), orthodox Christians, as well as unbelievers, harmlessly perpetuate religious traditions of Anglo-Saxon heathenism.

The first Anglo-Saxon literary works in England include the unique *Beowulf* and versions of ballads and battle-songs recited by scops. The poets Caedmon and Cynewulf wrote in Anglo-Saxon and with it Alfred is said to have made a translation of the Venerable Bede's Latin *Ecclesiastical history.* The *Anglo-Saxon chronicle* continued that history.

In the eighth century Anglo-Saxon tranquillity was greatly disturbed by marauding invasions of Scandinavians or Vikings. Against this foe, first Egbert, then the renowned Alfred, both Saxons, offered stout resistance and achieved a semblance of Anglo-Saxon unity, but the invaders remained in that part of northeastern England called the Danelaw. In 1016, after repeated exactions of tribute, called Danegeld, the Danes succeeded in placing Canute (Cnut or Knut), a powerful king, upon the English throne. Then followed the gradual commingling of Scandinavians and Anglo-Saxons with an effect on the English language that is hard to estimate because the two tongues resembled each other closely at that time. Clear survivals of the Danish influence are the terminations of the names of many towns and geographical features, as *by,* town, in *Corby, Derby, Kirby, Rugby, Selby, Whitby; ness,* promontory, in *Dungeness, Furness, Langness, Skegness; thorp,* village, in *Claythorpe, Mabelthorpe; thwaite,* piece of cleared land, in *Applethwaite, Thornthwaite.*

The successors of Canute were weak and the line came to an end. In the meantime England again found invaders on her shores. These were the Normans, originally Scandinavians or Northmen, now from Normandy in France where they had come under Roman influence. At first the relations between the two peoples were peaceful, but then came a contest for the throne that culminated in the Battle of Hastings (1066), in which William the Conqueror defeated Harold and became king of England. Shortly after the battle William erected Battle Abbey on the site of the conflict and supposedly placed therein a roll containing the names of his followers from Normandy. The authenticity of this Battle Abbey Roll has been questioned in recent years, but is, nevertheless, considered a source of information concerning the surnames current among the Normans and introduced into England at that time. Later in his reign William ordered a complete census of all England to be made. The result was the Domesday Book, which also is a treasury of information about personal names and in many instances has been the ultimate resort in settling property disputes.

The keeping of more accurate civil records necessitated better distinction and identification of persons, and in consequence, surnames were invented and assumed on a large scale in Britain. Such names were in part derived from (1) personal names of sires and ancestors, as *Johnson, Smithson, Watson, Fitzhugh, MacDonald, O'Keefe;* (2) locality or place of residence, as *Atwell, Bradford, Brook, Hill, Lindley, Wood;* (3) occupations, as *Baker, Brewer, Carpenter, Chandler, Clark, Cooper, Smith, Taylor;* (4) personal peculiarities, which resulted in nicknames, as *Armstrong, Birdseye, Broadhead, Doolittle, Gray, Lightfoot, Longfellow, Short;* and (5) names of natural objects, as *Berry, Birch, Buck, Cherry, Fish, Fox, Oakes, Onions, Stone.*

One temporary effect of the Conquest was the interruption of literary production in Anglo-Saxon. Four native dialects were then spoken in as many parts of Britain: Northumbrian, in the north of England and south of Scotland; Saxon, in the south of England; Mercian, in the Midland area; and Kentish in Kent. A dialect of French was the mode at court, in the schools, and in society. At Oxford it was requested that the students converse, if not in Latin, at least in French. Because polite society used synonymous but frenchified Latin words for common things and ideas, many Anglo-Saxon words, including terms we now regard as obscene, profane, or vulgar, such as those for the facts of life, began to be restricted to colloquial speech, a process illustrated briefly in the opening chapter of Sir Walter Scott's *Ivanhoe.* However, as the Normans were outnumbered, French did not vanquish the Anglo-Saxon dialects and become the national language. Nonetheless, from the eleventh to the fifteenth century the English vocabulary was greatly enriched by a steady influx of French words, such as: *argent, arms, assault, azure, bacon, banner, baron, battle, beef, butler, buttery, castle, chair, countess, court, curfew, dandelion, dinner, duke, fortress, gentle, grace, gules, harness, hauberk, honest, justice, lance, lieutenant, marquis, master, mercy, miracle, mutton, pansy, parlor, peace, peer, pennon, pork, prison, private, privilege, rent, siege, standard, supper, tower, traitor, treason, treasure, veal, victory, viscount, war.* Many of these words were transferred by mouth and ear, in court, church, business, war, and social life, but in the later stages, by numerous translations of French works into English. Scarcely a sentence can be spoken now without employing a few words showing French origin or influence. In the last sentence such words are: *scarcely, sentence, employing, origin, influence.*

A notable change that began in the highly inflected Anglo-Saxon dialects before the Norman Conquest was now accelerated. As the despised speech of the conquered race, although spoken, was not much written, and as there was considerable competition between dialects, these lost most of their superfluous machinery. Inflections softened or sloughed away and auxiliary words took their places, so that eventually the speech of the common people, like Mondamin after wrestling with Hiawatha, was stripped of its excessive garments and adornments but not of its latent usefulness.

In the latter part of the eleventh century the Crusades began and, by undermining feudalism, contributed directly to the rise of the middle classes. This movement was further hastened by the extortion from King John of the celebrated Magna Charta in 1215 at Runnymede.

The times and circumstances were now ripe for the emergence of a dialect sufficiently centralized and prominent to claim and hold precedence as a national language. The Midland, a descendant of Mercian, used in central England, was that dialect, for it was spoken in London, in court, and at Oxford when any English at all was spoken. Wycliff used it for translating the *Bible* and Chaucer for writing the *Canterbury tales*. Stirring movements and events, both in and out of England, as the beginnings of the Reformation, the Revival of Learning, and the subtle but powerful influence exerted by the invention of printing from movable types, stimulated the use of the dialect of London. Thus, by 1450 Middle English had reached its climax and was already ceasing to be Middle English.

Middle English by losses and gains gradually developed into the English of Shakespear and the King James translation of the *Bible*. The losses (about three-fourths of the original Anglo-Saxon vocabulary left no survivors) include such words as: *gelome* (often), *here* (army), *micel* (large), *pomely* (dappled), *sote* (sweet), *wlitig* (beautiful). The gains (about three-fifths of the present vocabulary is derived directly or indirectly from Greek and Latin) may be summarized by saying that in keeping pace with the rapid advances in scientific discovery, the development of international relationships, and the widespread diffusion of knowledge, the English vocabulary has been omnivorous in absorbing words from every language under the sun, so that the modern unabridged dictionary contains a list of more than 600,000 entries. Besides these there are many slang and colloquial terms, which, because they are transitory or objectionable, rarely find their way into good usage and the common dictionary. Recently, out of this huge vocabulary, 850 words, called Basic English, were chosen to serve the purpose of teaching a usable form of English quickly. This, however, is not pidgin (Chinese pronunciation of *business*) English, a verbal jargon used chiefly in trade in many parts of the world.

The almost homogeneous Early English has now become heterogeneous Modern English, having derived materials from many diverse sources and adopted them with or without change, as illustrated by the following examples. The sources indicated, it should be remembered, are largely immediate and not ultimate, trail's end often lying in the mists behind the beyond:

African: *chimpanzee, goober, gorilla, guinea, gumbo, oasis, okra, simba, voodoo, yam, zebra, zombie.*

American (aboriginal): *alpaca, Apache, Aztec, caoutchouc, canoe, caribou, catalpa, cayuse, chautauqua, chicle, chili, chinquapin, chipmunk, chocolate, cougar, coyote, Eskimo, guano, hammock, hickory, hogan, hominy, igloo, Inca, kayak, llama, mackinaw, mahogany, maize, manatee, Massachusetts, menhaden, mescal, mesquite, metate, moccasin, moose, mugwump, muskeg, nunatak, ocelot, opossum, papoose, pawpaw, pecan, pemmican, persimmon, pipsissewa, pocosin, pone, potato, potlatch, powwow, puccoon, quahog, quinine, raccoon, sachem, sagamore, savannah, sequoia, Sioux, skookum, skunk, squash, squaw, succotash, Susquehanna, tamale, tamarack, tapioca, tepee, terrapin, tobacco, toboggan, tomahawk, tomato, totem, tupelo, Utah, wahoo, wampum, woodchuck, Yosemite.*

Anglo-Saxon: *This, is, an, Anglo-Saxon, word.*

Arabic: *admiral, alchemy, alcohol, alfalfa, algebra, Algol, alkali, Allah, amber, arsenal, assassin, attar, azimuth, caliph, candy, cipher, coffee, cotton, crimson, elixir, gazelle, ghoul, giraffe, hashish, hazard, hegira, jar, khamsin, Koran, lute, magazine, mattress, Mecca, minaret, Mohammed, morrocco, nabob, nadir, sahib, salaam, sheik, simoon, sugar, tariff, zenith, zero.*

Australian: *boomerang, dingo, kangaroo, wombat.*

Celtic (Gaelic, Irish, Scotch, Welsh): *avon, bard, blarney, bog, brat, Bridget, brogue, cairn, carpenter, clan, coracle, cromlech, druid, drumlin, esker, flannel, galore, glen, kame, Kennedy, Llewellyn, loch, Lomond, London, McKay, O'Connor, penguin, plaid, ptarmigan, shamrock, slogan, Tory, truant, varlet, vassal, whisky.*

Chinese: *China, chop-suey, chow* (dog), *fan-tan, ginseng, junk, kaolin, kowtow, kumquat, litchi, loquat, nankeen, oolong, pekoe, sampan, satin, silk, tea, tycoon, typhoon.*

Dutch: *ahoy, aloof, ballast, Barnegat, Boer, boom, boss, bowery, brandy, bulwark, coleslaw, cooky, cruise, cruller, deck, easel, freebooter, freight, frolic, fumble, furlough, horst, Kaaterskill, Kinderhook, knapsack, landscape, loiter, luck, morass, mumps, pickle, rover, Santa Claus, scow, skate, skipper, sled, sleigh, sloop, smuggle, spook, spool, staple, stoker, tattoo* (a sound), *trek, waffle, wagon, yacht, Yankee, yawl.*

French: *adroit, belle, blonde, bouquet, brunette, burlesque, cache, camouflage, chauffeur, chiffonier, coquette, crevasse, debris, debut, depot, etiquette, fete, garage, hotel, impasse, jargon, levee, matinee, mayonnaise, moraine, negligee, parole, prairie, reconnaissance, rendezvous, silhouette, trousseau, vaudeville.*

German: *bismuth, cobalt, delicatessen, dollar, feldspar, Fahrenheit, frankfurter, gneiss, hamburger, haversack, hex, howitzer, kibitzer, kindergarten, lager, loess, meerschaum, nickel, noodle, ouch, plunder, poodle, pretzel, quartz, sauerkraut, swindler, wanderlust, wolframite, yodel, zinc.*

Greek: *abyss, acacia, acme, Alexander, analysis, Andrew, asbestos, basilica, basis, biology, bison, calyx, camera, cholera, church, cone, decalogue, dilemma, diphtheria, electricity, epic, geometry, geranium, Helen, idiom, iodine, kilogram, lamp, leopard, lyceum, lyric, mechanic, music, myth, nemesis, neuralgia, ocean, orchid, organ, orthodox, pathos, philosopher, platonic, rheumatism, rhododendron, socratic, squirrel, surgeon, synonym, telephone, thermometer, tyrant, Ulysses, Uranus, urethra, xanthophyll, xenon, xylophone, zealot, zodiac, zone.*

Hawaiian: *aloha, Honolulu, hula, Kilauea, ukulele.*

Hebrew: *Adam, amen, cabal, cherub, cider, Elizabeth, Esther, hallelujah, hosanna, Jehovah, Joseph, jubilee, kosher, manna, rabbi, Ruth, Sabbath, Satan, seraph, shekel, shibboleth.*

Hungarian: *coach, hussar, saber, shako, tokay.*

Italian: *alarm, balcony, bandit, bankrupt, battalion, cameo, canto, carnival, carton, cartoon, citadel, concert, ditto, duet, finale, gondola, granite, imbroglio, influenza, lava, macaroni, madonna, madrigal, malaria, motto, opera, piano, piazza, quarto, regatta, sirocco, solo, sonata, soprano, spaghetti, stanza, stiletto, stucco, studio, umbrella, vista, volcano.*

Japanese: *banzai, Fujiyama, geisha, ginkgo, hara-kiri, Japan, jinrikisha, jujitsu, kimono, Mikado, moxa, sake, soy, tsunami, yen.*

Latin: *alibi, animal, bacillus, calculus, census, doctor, extra, fiat, formula, galena, hernia, hiatus, index, junior, liquor, locus, major, medium, nebula, nominator, nucleus, October, operator, ovum, pauper, pollen, quorum, quota, radius, rostrum, September, splendor, terminus, tuber, tutor, ulterior, umbo, vacuum, veto, victor.*

Malay: *amuck, atoll, bamboo, bantam, batik, cassowary, cockatoo, gong, guttapercha, mandarin, mango, orangutan, rattan, sago.*

Persian: *azure, bazaar, borax, caravan, check, chess, dervish, divan, Iran, jackal, jasmine, kiosk, lemon, lilac, lime* (a fruit), *mogul, musk, orange, pagoda, pistachio, rice, rook* (a chessman), *scimitar, shah, shawl.*

Polish: *mazurka, polka.*

Polynesian: *taboo, tattoo* (impregnation of the skin with color).

Portuguese: *albatross, caste, cobra, commodore, marmalade, molasses, tank.*

Russian: *balalaika, Bolshevik, chernozem, czar, drosky, knout, mammoth, podsol, ruble, samovar, Soviet, steppe, tundra, ukase.*

Sanskrit (Hindi, etc.): *Aryan, bungalow, calico, chaulmoogra, chintz, guru, Himalaya, jungle, khaki, loot, pajama, pundit, shampoo, swami, swastika, thug, toddy, veranda.*

Scandinavian (Norse, Icelandic, Norwegian, Danish, Swedish): *axle, blend, both, boulder, cake, club, dahlia, dirt, egg, fellow, fiord, flake, flat, frog, freckle, gantlet, get, geyser, happen, husband, ill, keel, kid, kilt, law, leg, loan, oaf, odd, rutabaga, saga, scalp, score, ski, skill, sky, slag, they, thrive, tungsten, ugly, varve, want, window, wing, wrong.*

Spanish: *adobe, alligator, anchovy, armada, barbecue, bonanza, bravado, bronco, calaboose, canyon, chaparral, cargo, cigar, cork, corral, creole, desperado, flotilla, lasso, mesa, mosquito, mustang, Negro, palmetto, pickaninny, placer, plaza, poncho, quadroon, quixotic, ranch, rodeo, sherry, sierra, siesta, stampede, tornado.*

Turkish: *bey, caracal, fez, harem, horde, khedive, Ottoman, shagreen, sherbet, sofa, sultan, tulip, turban.*

Five languages contributed elements to the word *remacadamizing:* *re* (Latin), *mac* (Celtic), *adam* (Hebrew), *iz* (Greek), *ing* (Anglo-Saxon). This and many other words in the preceding list exemplify the assimilating process called anglicizing when a foreign word or word-element is adopted into the English language. Although English, as illustrated, has received much from other languages, borrowing from English has also taken place on a relatively large scale. Since the end of the seventeenth century such loans include among many others the following: *baseball, beefsteak, bill, budget, committee, Derby, gin, handicap, jury, mackintosh, speech, strike, Yankee.*

The invasion in force of the English vocabulary by words used in the sciences began in the seventeenth century following the stimulation of scientific activity that resulted from the invention of the telescope and microscope, accompanied by the development of new mathematical, chemical, and other precise techniques for observing, investigating, and explaining natural phenomena and their relationships. Prior to that, mankind had through thousands of years gradually accumulated considerable empirical knowledge. Many penetrating observations and calculations had been recorded and some sporadic, planned, experimental attacks had been made on the nature of things and the mysteries of the universe. The perception that particular effects follow given discernible natural causes presaged the twilight of the gods and the dawn of science. Superstition began to fade from the skies and the earth, but it has resisted obstinately and many persons still regard some naturally explainable phenomena as miracles and prodigies. Not long ago, on the sidewalk in front of the National Museum, I saw a man, obviously undecided about something, solemnly spit into his left palm, then strike the moist spot smartly with his right index finger to determine by the spurt in which direction to go. The Age of Credulity has not yet ended.

In 1620 Francis Bacon published *Novum organum*, or new instrument, urging controlled experiments based on postulated hypotheses, careful observation and collection of facts, and induction of generalizations from those facts—in short, the scientific approach. Whether or not Bacon's statement of procedures inspired scientific activity in other men, the fact is patent that, like ships, many branches of knowledge in the seventeenth and eighteenth centuries slipped away from their static Medieval moorings and departed on lengthy voyages of discovery. Influential scientific societies were founded in centers of learning, as the Accademia dei Lincei in Rome (1603), of which Galileo became a member; Royal Society of London (1662), of which Newton became a member; Academie des Sciences of Paris (1666); Royal Academy of Sciences at Berlin (1700); Royal Academy of Sciences at Stockholm

(1739), of which Linnaeus became a member; American Philosophical Society (1743), of which Benjamin Franklin was the founder; Geological Society of London (1807); and the Academy of Natural Sciences of Philadelphia (1812). The following words with the approximate dates of their appearance in English connote this new movement: *magnetism* (1616), *telescope* (1619), *gravity* (1622), *gravitation* (1645), *electricity* (1646), *microscope* (1656), *fossil* (1661), *geology* (1661), *capillary* (1664), *cell* (1665), *hydraulics* (1671), *botany* (1696), *stratum* (1699), *fluxion* (1704), *zoology* (1726), *oxygen* (1789), *oxidation* (1791), *nitrogen* (1794), *atom* (1801), *biology* (1802), *evolution* (1832), *paleontology* (1838), *bacterium* (1847). Because most scientists of the time had had extensive classical training, Greek and Latin became almost universal media for the exchange of information and those languages supplied the raw materials for coining words—a fertile source that continues to be tapped to this day.

The development of the English language has been governed by factors and processes observable today, especially in the United States, in such phenomena as the rise and fall of slang terms with the passage now and then of one of these pariahs to a higher caste; the coinage and adoption of terms used in trade, advertising, politics, and the learned professions; the absorption of words from foreign languages; the perversion and distortion of terms by faulty spelling and odd pronunciation; and the rise of provincial dialects. A child learning to talk also illustrates unique and striking language-making processes.

What are the basic features of the English vocabulary? Every schoolboy knows, perhaps painfully, that words are classified as nouns, pronouns, adjectives, verbs, adverbs, prepositions, conjunctions, and interjections. Some of these parts of speech may be inflected to show different shades of meaning, and, by a variety of combinations, they may be grouped in phrases, clauses, and sentences in a more or less invariable order, to express every thought. Punctuation marks and numerals, whose evolution is itself an interesting story, are also integral parts of the written and printed language.

Words are composed of one or more syllables. Sometimes but not always coinciding with the syllabic structure are the radical elements— prefixes, suffixes, and primary roots, whose original meanings may still be clear, or may be vague, or perhaps may be entirely lost. Such formatives of Indo-European derivation are relics of a long ancestry traceable backward through one or another branch of the Indo-European family of languages to the original stock by pertinacious students characterized facetiously by Cowper (*Retirement,* line 691) as

> "... Philologists who chase
> A panting syllable through time and space;
> Start it at home and hunt it in the dark,
> To Gaul, to Greece, and into Noah's Ark."

One of these students was Jakob Grimm (1785-1863) probably better known popularly for his (and his brother Wilhelm's) *Fairy tales* than for the formulation of the Law of Consonantal Shifts in the evolution of Indo-European words. This law is based on a tedious study of cognate words like *brotherly* and *fraternal, cow* and *beef, wheel* and *cycle,* and many others that originated in the same cradle but parted company early in their careers, the first words of the couplets becoming English by the northern or Teutonic route through Anglo-Saxon, and the second by the southern or Romance route through Latin and French. Further examples illustrating consonantal shifts follow:

	Greek	Latin	German	Anglo-Saxon	English
b-p	kybos	cubitus	hüfte	hype	hip
c(k)-h	keras	cornu	horn	horn	horn
	kephalos	caput	haupt	heafod	head
d-t	deka	decem	zehn	ten	ten
f(ph)-b	phrater	frater	bruder	brothor	brother
	phero	fero	gebären	beran	bear
g-k	gony	genu	knie	cneo	knee
h-g	——	hostis	gast	gaest	guest
p-f	pous	pes	fusz	fot	foot
t-th	treis	tres	drei	threo	three
th-d	thyra	——	tür	dor	door
v-w	——	vallum	wall	weall	wall
	——	vespa	wespe	waesp	wasp

So great have been the consonant and vowel changes as words were repeated by tongue and pen down the centuries that recognition of ancestry is often difficult, and one must be wary of being misled by coincidental similarities as well as by apparent insuperable differences between words under examination as possible cognates. Thus, Greek *kaleo* and English *call,* though of the same meaning, are unrelated; and Greek *rhythmos* is not ancestral to English *rhyme,* which is from Anglo-Saxon *rim,* number, and allegedly got its *h* from analogy with *rhythm.* On the other hand, Greek *ego* and English *I,* in which little similarity remains, are genetically related. Two words, now of different meaning and spelling but derived from the same ultimate source like cognates, are called doublets, as *dainty* and *dignity* from Latin *dignitas; shirt* and *skirt* from Old Norse *skyrta;* and *whit* and *wight* from Anglo-Saxon *wiht.* There is considerable point to Voltaire's reputed definition of etymology as "a science in which vowels do not count at all and consonants very little."

Changes in the appearance of words by erosion and otherwise, such as *adown* to *down,* an *ewt* to *newt, Bethlehem* to *bedlam, chirurgeon* to *surgeon, esquire* to *squire, history* to *story, hydropsy* to *dropsy, influenza* to *flu, juniper* to *gin, omnibus* to *bus, puliall royal* to *pennyroyal, Purgatoire* to *Picketwire, rime* to *rhyme, Rivoire* to *Revere,* and *Saint Nicholas* to *Santa Claus,* are matched if not surpassed by the changes in their meaning. A *villain* now is not necessarily a dweller in a villa; cultured people of cities may be *pagans,* a term applied originally to countrymen; *reduce* only figuratively means to lead back; *language* is a metonymy for *lingua,* the tongue; and *person* has now only a remote ideological link with *persona,* a mask. With words, as elsewhere, things are not always what they seem. Regarded superficially in spelling and sound, some components of words may belie their true origin and thus breed deceptive derivations—a process called folk or popular etymology. For example, *belfry* has no connection with *bell; crayfish* with *fish; cutlass* with either *cut* or *lass; dropsy* with *drop; horehound* with *hound; Jerusalem artichoke* with *Jerusalem; pantry* with *pan; penthouse* with *house; primrose* and *tuberose* with *rose; serviceberry* with *serve;* sham-

rock with either *sham* or *rock; starboard* with *star; uproar* with *roar;* and *wormwood* with *worm.* The derivation of *realtor* from *real,* genuine, and *tor,* bull, although sly, is admittedly jocose, as is also that implied in the definition of *mugwump*—a voter who sits on the political fence with his mug on one side and his wump on the other. Deeper and more exacting etymological study requires the finding of correct answers to the questions Whence? When? and Who? Here the tangibles of word derivation fade into the intangibles of semantics, which is the philosophical study of words, signs, and gestures considered as symbols and vehicles for communication of significant meaning.

The word *chloroform* will serve as an example of etymological inquiry. It consists of two parts, *chloro* from Greek *chloros,* green, and *form* from Latin *formica,* ant. Literally translated, the word might mean 'green ant' or 'ant green', neither version being satisfactory because chloroform is not a green substance nor is it made from green ants. Chloroform is a compound of carbon, hydrogen, and chlorine. The carbon and hydrogen are associated in a group called the formyl radical because it occurs in formaldehyde and formic acid. The latter was obtained in the seventeenth century by distilling red ants with water and received its name from Latin *formica,* ant. Chlorine is a green gas, whence its name, given to it in 1810 by Sir Humphrey Davy, who discovered its true nature as that of an element, not a compound. Chloroform itself was prepared simultaneously in 1831 by Justus Liebig in Germany, E. Soubeiran in France, and Samuel Guthrie in New York, but its chemical constitution was not fully determined until 1834 when Jean Dumas announced the correct analysis and suggested the present name, which, like *automobile, speedometer,* and many others, is a hybrid, being composed of elements from two distinct languages. Instances like that just given might be multiplied indefinitely. The interested reader will find a wealth of additional material in some of the books cited in the BIBLIOGRAPHY.

The English language, complex and ever-changing, has thus evolved through the vicissitudes of time and circumstance. What of its future? My Delphic oracle intimates that, because it has great versatility and is linked by long usage with the love of liberty, it will some day possess the earth, for those traits have a powerful, universal, human appeal. Those who have inherited or adopted the language can show their appreciation of the noble instrument by thoughtful treatment of it. Generally, full consciousness of its significance and power is not manifest until attempts are made to express original thoughts by voice or pen, or when, in rare, enlightening moments, perception reveals that another has used it in a charming and effective manner.

Inasmuch as the English vocabulary, whose history and nature have just been summarized, is so largely derived from Greek and Latin, a brief study of the essential features of those languages is now indicated.

NATURE OF GREEK AND LATIN

Although ultimately derived from a common Indo-European stock, Greek and Latin aboriginally had become different languages. Ethnologically, the Romans were not Greeks, but, however much they despised other contemporaries, they respected the Greeks and the Greek language and in time borrowed and assimilated a considerable part of the Greek vocabulary.

In early times three principal dialects were spoken among the Greeks or Hellenes—Aeolic(Arcadian), Doric, and Ionic(Attic). Of these, Attic, a form of Ionic, was the speech in Attica, whose capital was Athens. It was the medium used by Aeschylus, Aristophanes, Aristotle, Plato, Plutarch, Thucydides, and Xenophon, thus emerging as the standard literary language and eventually the common language of all the Greek peoples. This was the language into which the Septuagint version of the Old Testament was translated at Alexandria and in which the New Testament was written. By 850 A.D., after conservative colloquial changes in pronunciation and syntax, it had developed the characteristics of Modern Greek.

Similarly, among a number of dialects existing in central Italy, that spoken in Latium, whose capital was Rome, was used by Caesar, Cicero, Horace, Livy, Lucretius, Ovid, Pliny, and Vergil, and became the universal Latin of the Romans. An outstanding example of the colloquial form of this language is Jerome's Vulgate version of the *Bible* in the fourth century. Gradually, classical Latin developed into Late Latin, which by 600 A.D. evolved into the Medieval Latin of the Middle Ages. Beginning about 1500 A.D. the ancient tongue of Caesar and Pliny, now called a dead language, became, with more or less tinkering, the New Latin used by scientists. Concurrently, during this evolution of literary Latin, colloquial Latin with all its variations differentiated into the Romance languages—Italian, French, Spanish, Portuguese, and Rumanian.

A cursory inspection of the contents of the conventional, unabridged Greek and Latin dictionaries shows that, within limitations, the Greeks and Romans "had a word for it". Thus, the Greeks had *hemikakos*, half bad, and the Romans *edormio*, sleep off, that is, the effects of the night's conviviality. The Greek skinflint or pennypincher could be called quite a mouthful—a *kyminopristokardamoglyphos*—, that is to say, a cumin-sawing cress-carver, the seeds of those plants being relatively small. Is there more poetry than truth hidden in the word *damar*, the Greek for wife, meaning one that has been tamed, an unwedded maiden being *adamastos*, untamed? The total vocabularies, however, as in English, doubtless included many slang and colloquial terms that did not survive, except perhaps in some instances as Romance words.

Many examples in the dictionaries show that the Greeks and Romans adopted words, usually with appropriate modification, from their foreign contemporaries. A word, probably of Asiatic origin, became *sapon* in Greek, *sapo* in Latin, and by the Anglo-Saxon route, *soap* in English. Other exotic words in Latin were: *Brito*, Briton, an inhabitant of Britain, from a Celtic word; *celia*, a Spanish beer; *cerevisia* (*cervisia*), a Gallic term for beer; *gabalus*, gallows, related to German *gabel*, fork; *ganta*, goose, related to German *gans*, goose; *minium*, a Spanish term for native cinnabar; *petoritum*, Gallic name for an open, four-wheeled carriage;

rasta, the German mile; *sacal,* Egyptian amber; *sapenos,* a kind of Indian amethyst; *sarabarum,* Persian wide trousers; *urus,* a Celtic name for the wild ox.

Inscriptions and papyri indicate that both languages were originally written with capital letters. Lower case or small letters evolved much later from the cursive letters used in rapid writing. The classic Greek alphabet, said to be derived from Phoenician characters, contained twenty-four letters, many of which, with alterations, were borrowed directly or indirectly by the Romans. About 600 A.D. the classical Latin letters began displacing the Anglo-Saxon runic futhorc, and eventually, with the addition of the further differentiated J, U, and W, which are modified forms of I, V, and Y, became the present English alphabet. This alphabet, in the simplest style of print and script, is here considered to be the ultimate and unmodifiable base to which Greek in particular and other alphabets in general should be transliterated as closely as possible when devising ordinary terms for use in the English language and for binomials in universal biological nomenclature. The busy scientist, however, may take one look at the Greek dictionary and exclaim, "It is all Greek to me!" and do nothing further about alpha, beta, gamma, and omega. Nevertheless, it would seem that the little time necessary to learn the Greek alphabet is well worth spending because this is the preliminary step in transliteration and in finding one's way around in the Greek dictionary, an accomplishment that, at a moment's notice, may be of great importance.

TRANSLITERATION

Transliteration is the somewhat arbitrary process of rendering the letters and sounds of one language into those of another. For the method of transliterating other languages than Greek and Latin into English the reader must consult references dealing with those languages. The recommended transliteration of the classic Greek alphabet of twenty-four letters and diphthongs into Latin and ultimately into English is exemplified as follows. See also GREEK LETTERS under **letter.**

Vowels and Consonants

Alpha	A, α	= A, a
Beta	B, β	= B, b
Gamma	Γ, γ	= G, g

Gamma becomes *n* before γ, κ, ξ, χ:

γγ = *ng*, as in αγγελος (*angelus*), φθογγος (*phthongus*), στρογγυλος (*strongylus*), συγγενης (*syngenes*).

γκ = *nc*, as in αγκιστρον (*ancistrum*), πλαγκτος (*planctus*), ῥεγκω (*rhenco*), σφιγκτηρ (*sphincter*).

γξ = *nx*, as in ελεγξις (*elenxis*), φαλαγξ (*phalanx*), φαρυγξ (*pharynx*), σαλπιγξ (*salpinx*).

γχ = *nch*, as in κογχη (*concha*), εγχεω (*encheo*), λαγχανω (*lanchano*), ῥυγχος (*rhynchus*).

Delta	Δ, δ	= D, d
Epsilon	E, ε	= E, e
Zeta	Z, ζ	= Z, z
Eta	H, η	= E, e

Epsilon and eta are not distinguished in transliteration, nor is there any way of telling from a transliterated word which letter is meant. Eta at the end of most nouns becomes *a*, as in Ανδρομεδη (*Andromeda*), κομη (*coma*), επιστολη (*epistola*), θηκη (*theca*), but there are exceptions, as ακμη (*acme*), Καλλιοπη (*Calliope*), ψυχη (*psyche*), συνεκδοχη (*synecdoche*).

Theta   Θ, θ = Th, th
Iota   I, ι = I, i, or sometimes J, j

In the Middle Ages *i* and *j* were interchangeable, but by the seventeenth century they were distinguished as vowel and consonant, respectively. Widespread usage now favors an *i* before a consonant, as in *imberbis, insignis,* and in nearly all instances (exceptions: *major, majestas*) before a vowel in a post-initial syllable, as in *socialis, vaccinium;* and a *j* before a vowel at the beginning of a word, as in *janitor, justitia,* unless the word is still essentially Greek with an *i* sounded as an initial syllable, as in *iambus, ion, iynx.*

Kappa   K, κ = K, k, C, c

Kappa is retained in many English words, as *karyokinesis, Katharine, keloid, kenotoxin, keratin, ketone, kinetic, kleptomaniac, krypton, kymograph, plankton, skeleton, skeptic.*

Lambda   Λ, λ = L, l
Mu   M, μ = M, m
Nu   N, ν = N, n
Xi   Ξ, ξ = X, x
Omicron   O, o = O, o
Pi   Π, π = P, p
Rho   P, ρ = R, r

In compounds, initial ρ of the second term is doubled after a vowel but remains single after a diphthong, as in διαρρυτος (*diarrhytus*), γλυκυρριζα (*glycyrrhiza*), ευρηκτος (*eurhectus*), χαμαιριφης (*chamaerhiphes*). The double ρρ in words like αρρητος (*arrhetus*), αρρυπτος (*arrhyptus*), συρραπτος (*syrrhaptus*), and συρριζος (*syrrhizus*), results from euphonic assimilation of ν to ρ in the prefixes αν- (*an-*) and συν- (*syn-*). In word-making, doubling of ρ after vowels, unless required by euphony, might well be declared pedantic and unnecessary. *Cyrtorhynchus* is simpler by one letter but surely as good etymologically, mnemonically, and otherwise, as *Cyrtorrhynchus.*

Sigma   Σ, σ, ς = S, s

Initial σ, medial σ, and final ς are not distinguished in transliteration.

Tau   T, τ = T, t
Upsilon   Υ, υ = U, u, Y, y

Upsilon (ypsilon or hypsilon) is generally transliterated *y*, but the Romans kept it in *guberno* (κυβερναω), govern; *muraena* (μυραινα), moray; *tumba* (τυμβος), tomb; and in the diphthongs *au, eu, ou.*

Phi   Φ, φ = Ph, ph
Chi   X, χ = Ch, ch
Psi   Ψ, ψ = Ps, ps
Omega   Ω, ω = O, o

Omicron and omega are not distinguished in transliteration, nor is there any way of telling from a transliterated word which letter is meant, except sometimes in final ον, ος, ων, ως:

ον = *on* = *um*, as in χρυσανθεμον (*chrysanthemum*), αρον (*arum*). Usage has these exceptions: *criterion, Dodecatheon, ganglion, micron, neuron, Rhododendron.*

ος = *os* = *us*, as in ιππος (*hippus*), καρπος (*carpus*), ποταμος (*potamus*), ραμνος (*rhamnus*). Usage has these exceptions: *Apios, cosmos, Diospyros, logos*. Greek genitives in *-os* generally become *-is* in Latin, as in *chlamys*, *-ydos*, (L. *-ydis*), *lampas*, *-ados* (L. *-adis*), *narthex*, *-ekos* (L. *-ecis*), *nema*, *-tos* (L. *-tis*), *onyx*, *-ychos* (L. *-ychis*), *tigris*, *-ios* or *-idos* (L. *-is*, or *-idis*).

ων = on, as in γειτων (*giton*), κανων (*canon*), σιφων (*siphon*), χιτων (*chiton*). Exception: βραχιων, m. (*brachium*, n.).

ως = os, as in αιδως (*aedos*), ερως (*eros*), ρινοκερως (*rhinoceros*).

Diphthongs

αι = ai, ae, e

Ex. αρχαιος (*archaeus*, *archeus*), καινος (*caenus*, *cenus*), αιματο- (*haemato-*, *hemato-*), παλαιοπλουτος (*palaeoplutus*, *paleoplutus*), σκαιος (*scaeus*, *sceus*).

This diphthong is reduced to *e* in many modern English words, as *celestial*, *cerulean*, *demon*, *equal*, *ether*, *heresy*, *pedagogue*, *preface*, *question*, *sphere*. The Latin prefix *prae-*, before, is preferably shortened to *pre-*, as in *precede*, *precept*, *precipice*, *predator*, *predict*, *prefer*, *pregnant*, *prejudice*, *prelate*, *premature*, *prenatal*, *preponderant*, *prerogative*, *president*, *pretend*.

αυ = au

Ex. δαυλος (*daulus*), σαυρος (*saurus*), ταυρος (*taurus*).

ει, ηι = ei, i

Ex. εικονο- (*icono-*), κλειστος (*clistus*), πλειων (*plion*), χειρο- (*chiro-*), εμπειρικος (*empiricus*), νηιστος (*nistus*).

This diphthong is sometimes erroneously transliterated *e*, as in *Potamogeton* and *Anemia*. Many words retain *ei*, as *cleistogamous*, *kaleidoscope*, *Pleistocene*, *Steironema*.

ευ, ηυ = eu

Ex. ευγενης (*eugenes*), γραφευς (*grapheus*), λευκος (*leucus*), ψευδης (*pseudes*), τηυσιος (*teusius*).

Before vowels this diphthong is sometimes transliterated *ev*, as in *evangelist*, *Evernia*.

οι, ωι = oi, oe, e

Ex. κοιλος (*coelus*), Φοιβη (*Phoebe*), στοιχος (*stoechus*), πρωιοτης (*proeotes*).

The *oe* of Latin words may be transliterated *e* in English, as in *cenobium* (< Gr. *koinobion*, L. *coenobium*), *economics* (< Gr. *oikonomikos*, L. *oeconomicus*), *federation* (< L. *foedero*), *penal* (< L. *poenalis*).

ου = ou, u

Ex. ακουστικος (*acousticus*), βουνος (*bunus*), πους (*pus*), ουρα (*ura*), ξουθος (*xuthus*).

υι = yi

Ex. αγυια (*agyia*), γυιος (*gyius*), ἁρπυια (*harpyia*), υιος (*hyius*), οργυια (*orgyia*).

ᾳ, ῃ, ῳ = a, e, o

Ex. ᾳσις (*asis*), ῃσθημενως (*esthemenos*), ῳδικος (*odicus*), ῳον (*oum*), but κωμῳδια (*comoedia*, comedy) is a transitional exception.

The second vowel of these so-called improper diphthongs is iota subscript, generally ignored in transliteration.

Concerning the supposed original sounds of the classical Greek and Latin letters and their subsequent evolution, consult the discussions in the grammars and English dictionaries. Present pronunciation of most letters, especially the consonants, is perhaps not unlike that of the Greeks and Romans, but the tendencies with respect to a few need to be mentioned here: epsilon, ε (*e* as in method); eta, η (*e* as in

Pompeii); iota, ι (*i* variously, as in chloric, chloride, and chlorine); xi, ξ (like *z* at the beginning, but like *ks* elsewhere in words, as in Xanthippe and taxonomy); omicron, o (*o* as in thermostat, that is, a nice mean between the *o* of hot and that of cold); chi, χ (usually like *ch* in chorus, but sometimes like *ch* in church); psi, ψ (like *s* at the beginning, but like *ps* elsewhere in words, as in pseudonym and Cyclops); omega, ω (*o* as in Trojan).

The diphthongs were pronounced as fluid combinations of the sounds of both letters but today are sounded as single letters: αι = *ae* (originally like *ai* in aisle but now like *e* in he, as in algae); αυ = *au* (originally like *ou* in house but now more like *aw* in hawk, as in saurian and thesaurus); ει = *ei* (*ei* as in eight but now like *ei* in height); ευ = *eu* (*eu* as in feud); οι = *oe* (*oi* as in toil but the *oe* tending toward *e* in me, as in coelome); ου = *u* (*ou* as in you); υι = *yi* (*ui* as in suite). See further remarks under SPELLING AND PRONUNCIATION.

DIACRITICAL MARKS

The mark (ʽ) is the symbol indicating aspiration or the rough breathing (*spiritus asper*) denoted by the letter *h*, in contrast with (ʼ) the sign for the smooth breathing (*spiritus lenis*). These signs originated from the halves of eta (H), one half (Ⱶ) for the rough and the other (Ⱶ) for the smooth breathing. One or the other of these marks is shown in the dictionary entry over the vowel when a vowel begins a word. When placed over a vowel or the second vowel of a diphthong, (ʽ) means that in transliteration *h* precedes the vowel or diphthong. Over ρ it means that *h* follows the ρ; but *h* always follows double ρρ even though the mark be not shown.

Ex. ὁ (*ho*), ἡ (*he*), αἱ (*hae*), the; ῥοδον (*rhodum*), πυρρος (*pyrrhus*).

If the initial letter of the second term of a compound requires an *h* when the term stands by itself, it should take an *h* in the compound, except when omission for euphony is indicated.

Ex. ἑδρα (*hedra*), euhedral, rhombohedral; ῥυθμος (*rhythmus*), rhythm, arrhyth-mic, eurhythmic; ανιδρυτος (*anhidrytus*, < αν, ἱδρυω); αννδρος (*anhydrus*, < αν, ὑδωρ), anhydrous; αφηλιξ (*aphelix*, < απο, ἡλιξ); διοδος (*diodus*, < δια, ὁδος), παροδος (*parodus*, < παρα, ὁδος); αἱρεσις (*haeresis*, but *diaeresis* or *dieresis*, not *diahaeresis* or *diaheresis*).

Thus, although the dictionary does not indicate a rough breathing (ʽ) over the second ρ in χρυσορυγχος, the latter should be transliterated *chrysorhynchus*, be-cause, as *chrysorynchus*, the normal identity and spelling of *rhynchus*, beak, nose, is lost. In brief, the *h* serves as a mnemonic signal and becomes a distinctive, identifying mark of many Greek words.

The initial aspirate of numerous Greek words is represented in Latin cognates by *s*, as ἁλς (*hals*; L. *sal*, salt); ἑδος (*hedus*; L. *sedes*, seat); ἑνος (*henus*; L. *senex*, old); ἑπτα (*hepta*; L. *septem*, seven); ἑρπω (*herpo*; L. *serpo*, creep); ἑξ (*hex*; L. *sex*, six); ἡμι- (*hemi-*; L. *semi-*, half); ὑδωρ (*hydor*; L. *sudor*, sweat); ὑπερ (*hyper*; L. *super*, above); ὑς (*hys*; L. *sus*, hog).

No other diacritical marks, except the dieresis (¨), as in *Haliaëtos, Leucothoë, Proërna*, and *coöperatus*, are necessary in transliteration. This mark is a survival from the Greek usage of placing it over the second of two adjacent vowels in a word to show separation in sound and hence that the two vowels did not constitute a diphthong. The umlauted German letters *ä, ö, ü*, however, use the dieresis in a different sense and should be transliterated *ae, oe, ue*.

About 200 B.C. Aristophanes of Byzantium invented accent marks to help foreigners learn to pronounce Greek words. These marks were: the acute ('), which meant that the syllable over which it stood should receive a raised tone or pitch; the grave (`), a somewhat lower tone; and the circumflex (∧ or ⌢), first a raised then a lowered tone. As these marks do not affect transliteration their further consideration will be deferred to the discussion of pronunciation.

The standard method of transliteration having been illustrated, some examples of its violation may prove helpful: *Agkistrodon* for *Ancistrodon*, *Aleyrodes* for *Aleurodes*, *Coïlodera* for *Coelodera*, *Diospyros* for *Diospyrus*, *Evtactus* for *Eutactus*, *Macrosiphum* for *Macrosiphon*, *Oxydendron* for *Oxydendrum*, *Phaiomys* for *Phaeomys*, *Potamogeton* for *Potamogiton*, *Stringocephalus* for *Strigocephalus*, *Sylvilagus* for *Sylvilagos*, *Symphoricarpos* for *Symphoricarpus*, *Yphantes* for *Hyphantes*, and *Ypophaea* for *Hypophaea*.

Having examined the letters of the Greek and Latin alphabets, let us now turn to an inspection of words to determine their properties and behavior.

GREEK AND LATIN WORDS

The items of the Latin and Greek vocabularies most useful to the wordmaker are nouns, adjectives, verbs, adverbs, pronouns, and prepositions. A summary of their structure and behavior follows. The word *stem*, which occurs constantly in this discussion, is the root-stem or operating base in word-formation. This base sometimes coincides with the root and sometimes with the stem as ordinarily explained in the grammars. Only the essential framework of grammar is indicated here but this presumes that the reader is familiar with a number of technical terms.

Nouns

The stem or operating base of a noun can always be determined by removing the case-ending of the genitive singular, that is, for Latin, the -*ae* of the first, -*i* of the second, -*is* of the third, -*us* of the fourth, and -*ei* of the fifth declension; and for Greek, the -*as*, -*es*, or -*ou* of the first, -*ou* of the second, and -*os* of the third declension. The base and the connective vowel, usually *i* in Latin and *o* in Greek, make the combining form for compound words.

Latin nouns

First declension or a-nouns end with -*ae* in the genitive singular; -*ae* in the nominative and -*arum* in the genitive plural, as *hasta*, spear, *hastae*, of a spear, *hastae*, spears, *hastarum*, of spears. The root-stem is *hast-*, and the combining form is *hasti-*. First declension nouns in *a* must not be confused with Greek neuter nouns in *a*, whose genitive singulars end with -*atos*, transliterated -*atis*. Other examples of a-nouns are:

Singular	Plural
aqua, -ae, water, *aqui-*	*aquae, -arum*
causa, -ae, reason, *causi-*	*causae, -arum*
libra, -ae, balance, *libri-*	*librae, -arum*

Singular	Plural
mensa, -ae, table, *mensi-*	*mensae, -arum*
nauta, -ae, sailor, *nauti-*	*nautae, -arum*
rosa, -ae, rose, *rosi-*	*rosae, -arum*
rubia, -ae, madder, *rubii-*	*rubiae, -arum*
trabea, -ae, robe, *trabei-*	*trabeae, -arum*
via, -ae, way, *vii-*	*viae, -arum*

Second declension or o-nouns end with *-i* in the genitive singular; *-i,* or if neuter, *-a* in the nominative, and *-orum* in the genitive plural, as *hortus,* garden, *horti,* of a garden, *horti,* gardens, *hortorum,* of gardens; *ovum,* egg, *ovi,* of an egg, *ova,* eggs, *ovorum,* of eggs. The stems of these words are *hort- ov-,* and the combining forms are *horti- ovi-,* respectively. Second declension nouns in *us* must not be confused with third declension nouns in *us,* whose genitive singulars end with *-is,* nor with fourth declension nouns in *us,* whose genitive singulars end with *-us.* Other examples are:

Singular	Plural
ager, agri, field, *agri-*	*agri, -orum*
amicus, -i, friend, *amici-*	*amici, -orum*
bellum, -i, war, *belli-*	*bella, -orum*
filius, -ii or *-i,* son, *filii-* or *fili-*	*filii, -orum*
folium, -ii or *-i,* leaf, *folii-* or *foli-*	*folia, -orum*
genius, -ii or *-i,* talent, *genii-* or *geni-*	*genii, -orum*
hilum, -i, trifle, scar, *hili-*	*hila, -orum*
liber, libri, bark, book, *libri-*	*libri, -orum*
servus, -i, slave, *servi-*	*servi, -orum*
vir, -i, man, *viri-*	*viri, -orum*

Third declension or i-nouns end with *-is* in the genitive singular; *-es,* or if neuter, *-a* or *-ia* in the nominative, and *-ium* or *-um* in the genitive plural, as *filix,* fern, *filicis,* of a fern, *filices,* ferns, *filicum,* of ferns; *tempus,* time, *temporis,* of time, *tempora,* times, *temporum,* of times; *animal,* beast, *animalis,* of a beast, *animalia,* beasts, *animalium,* of beasts. The stems of these words are *filic-, tempor-,* and *animal-,* and the combining forms are *filici-, tempori-,* and *animali-,* respectively. For convenience in finding a given kind of example the words in the following list are arranged alphabetically by the last two letters of the nominative singular:

Singular	Plural
lac, lactis, milk, *lacti-*	*lactes, -ium*
calcar, -is, spur, *calcari-*	*calcaria, -ium*
aetas, -atis, age, *aetati-*	*aetates, -um*
mas, maris, male, *mari-*	*mares, -ium*
urbs, urbis, city, *urbi-*	*urbes, -ium*
flumen, fluminis, stream, *flumini-*	*flumina, -um*
leo, leonis, lion, *leoni-*	*leones, -um*
imber, imbris, storm, *imbri-*	*imbres, -ium*
iter, itineris, journey, *itineri-*	*itinera, -um*
pater, patris, father, *patri-*	*patres, -um*
venter, ventris, belly, *ventri-*	*ventres, -ium*
abies, abietis, fir, *abieti-*	*abietes, -um*
comes, comitis, companion, *comiti-*	*comites, -um*
eques, equitis, rider, *equiti-*	*equites, -um*
gurges, gurgitis, whirlpool, *gurgiti-*	*gurgites, -ium*
limes, limitis, end, *limiti-*	*limites, -um*
nubes, nubis, cloud, *nubi-*	*nubes, -ium*
pes, pedis, foot, *pedi-*	*pedes, -ium*
termes, termitis, termite, *termiti-*	*termites, -um*
rex, regis, king, *regi-*	*reges, -um*
vertex, verticis, whirl, *vertici-*	*vertices, -um*
virgo, virginis, maid, *virgini-*	*virgines, -um*

Singular	Plural
fusio, fusionis, a melting, *fusioni-*	*fusiones, -um*
bellis, bellidis, daisy, *bellidi-*	*bellides, -um*
civis, civis, citizen, *civi-*	*cives, -um*
glis, gliris, dormouse, *gliri-*	*glires, -ium*
ignis, ignis, fire, *igni-*	*ignes, -ium*
lapis, lapidis, stone, *lapidi-*	*lapides, -um*
navis, navis, ship, *navi-*	*naves, -ium*
turris, turris, tower, *turri-*	*turres, -ium*
xyris, xyridis, wild iris, *xyridi-*	*xyrides, -um*
nix, nivis, snow, *nivi-*	*nives, -ium*
radix, radicis, root, *radici-*	*radices, -um*
cubile, cubilis, couch, *cubili-*	*cubiles, -ium*
calx, calcis, pebble, *calci-*	*calces, -ium*
poema, poematis, poem, *poemati-*	*poemata, -um*
homo, hominis, man, *homini-*	*homines, -um*
cliens, clientis, customer, *clienti-*	*clientes, -ium*
mons, montis, mountain, *monti-*	*montes, -ium*
sol, solis, sun, *soli-*	*soles, -ium*
arbor, arboris, tree, *arbori-*	*arbores, -um*
color, coloris, color, *colori-*	*colores, -um*
cor, cordis, heart, *cordi-*	*cordia, -ium* or *-um*
bos, bovis, ox, cow, *bovi-*	*boves, boum*
custos, custodis, guard, *custodi-*	*custodes, -um*
flos, floris, flower, *flori-*	*flores, -um*
os, ossis, bone, *ossi-*	*ossa, -ium*
nox, noctis, night, *nocti-*	*noctes, -ium*
princeps, principis, chief, *principi-*	*principes, -um*
mare, maris, sea, *mari-*	*maria, -ium*
caro, carnis, flesh, *carni-*	*carnes, -ium*
corpus, corporis, body, *corpori-*	*corpora, -um*
fraus, fraudis, deceit, *fraudi-*	*fraudes, -ium*
genus, generis, race, *generi-*	*genera, -um*
jus, juris, right, *juri-*	*jura, -ium*
mus, muris, mouse, *muri-*	*mures, -ium*
opus, operis, work, *operi-*	*opera, -um*
sus, suis, hog, *sui-*	*sues, -um*
virtus, virtutis, merit, *virtuti-*	*virtutes, -um*
caput, capitis, head, *capiti-*	*capita, -um*
dux, ducis, leader, *duci-*	*duces, -um*
nux, nucis, nut, *nuci-*	*nuces, -um*

Fourth declension or u-nouns end with *-us* in the genitive singular; *-us*, or if neuter, *-ua* in the nominative, and *-uum* in the genitive plural, as *porticus*, porch, *porticus*, of a porch, *porticua*, porches, *porticuum*, of porches; *pecu*, herd of cattle; *pecus*, of a herd, *pecua*, cattle, *pecuum*, of cattle. The stems of these words are *portic-*, *pec-*, and the combining forms are *porticu-* or *portici-*, *pecu-* or *peci*, respectively. Other examples are:

Singular	Plural
cornu, -us, horn, *cornu-* or *corni-*	*cornua, -uum*
domus, -us, house, *domu-* or *domi-*	*domus, -uum*
fructus, -us, fruit, *fructu-* or *fructi-*	*fructus, -uum*
genu, -us, knee, *genu-* or *geni-*	*genua, -uum*
lacus, -us, lake, *lacu-* or *laci-*	*lacus, -uum*
manus, -us, hand, *manu-* or *mani-*	*manus, -uum*
quercus, -us, oak, *quercu-* or *querci-*	*quercus, -uum*
senatus, -us, senate, *senatu-* or *senati-*	*senatus, -uum*
tumultus, -us, uproar, *tumultu-* or *tumulti-*	*tumultus, -uum*

Fifth declension or e-nouns end with *-ei* in the genitive singular; *-es* in the nominative, and *-erum* in the genitive plural, as *effigies*, image, *effigiei*, of an image, *effigies*, images, *effigierum*, of images. The stem is *effig-*, and the combining form is *effigi-* or *effigie-*. Other examples are:

Singular	Plural
acies, -ei, sharpness, *aci-*	*acies, -erum*
dies, -ei, day, *di-* or *die-*	*dies, -erum*
facies, -ei, face, *faci-*	*facies, -erum*
glacies, -ei, ice, *glaci-*	*glacies, -erum*
res, -ei, thing, *re-*	*res, -erum*
species, -ei, appearance, *speci-*	*species, -erum*
spes, -ei, hope, *spe-*	*spes, -erum*

Greek Nouns

First declension or a-nouns end with *-as, -es,* or *-ou* in the genitive singular; *-ai* in the nominative, and *-on* in the genitive plural, as *peira,* attempt, *peiras,* of an attempt, *peirai,* attempts, *peiron,* of attempts; *thalassa,* sea, *thalasses,* of a sea, *thalassai,* seas, *thalasson,* of seas; *hoplites,* armed soldier, *hoplitou,* of an armed soldier, *hoplitai,* armed soldiers, *hopliton,* of armed soldiers. The stems of these words are *peir-, thalass-,* and *hoplit-,* and the combining forms, obtained by transliterating when necessary, and adding, or exceptionally omitting, the connective vowel *o,* are *piro- thalasso-,* and *hoplito-.* Other examples are:

Singular	Plural
chora, -as, country, *choro-*	*chorai, -on*
gephyra, -as, bridge, *gephyro-*	*gephyrai, -on*
petra, -as, rock, *petro-*	*petrai, -on*
skia, -as, shadow, *scio-*	*skiai, -on*
pelte, -es, shield, *pelto-*	*peltai, -on*
skene, -es, tent, *sceno-*	*skenai, -on*
time, -es, honor, *timo-*	*timai, -on*
zone, -es, girdle, *zono-*	*zonai, -on*
boreas, -ou, north wind, *boreo-*	*boreai, -on*
hippotes, -ou, horseman, *hippoto-*	*hippotai, -on*
polites, -ou, citizen, *polito-*	*politai, -on*
siderites, -ou, an iron ore, *siderito-*	*sideritai, -on*
tamias, -ou, steward, *tamio-*	*tamiai, -on*

Second declension or o-nouns end with *-ou* in the genitive singular; *-oi,* or if neuter, *-a* in the nominative, and *-on* in the genitive plural, as in the following examples:

Singular	Plural
anthropos, -ou, man, *anthropo-*	*anthropoi, -on*
astron, -ou, star, *astro-*	*astra, -on*
bios, -ou, life, *bio-*	*bioi, -on*
doron, -ou, gift, *doro-*	*dora, -on*
logos, -ou, word, *logo-*	*logoi, -on*
osteon, -ou, bone, *osteo-*	*ostea, -on*
potamos, -ou, river, *potamo-*	*potamoi, -on*
rhodon, -ou, rose, *rhodo-*	*rhoda, -on*
therion, -ou, beast, *therio-*	*theria, -on*

Third declension or consonant-nouns end with *-os* in the genitive singular; *-eis, -es,* or if neuter, *-a* or *-e* in the nominative, and *-on* in the genitive plural. For convenience in finding a given kind of example the words in the following list are arranged alphabetically by the last two letters of the nominative singular:

Singular	Plural
damar, damartos, wife, *damarto-*	*damartes, -on*
hepar, hepatos, liver, *hepato-*	*hepata, -on*
nektar, nektaros, drink of the gods, *nectaro-*	*nektara, -on*
phrear, phreatos, spring, *phreato-*	*phreata, -on*
stear, steatos, fat, *steato-*	*steata, -on*
gigas, gigantos, giant, *giganto-*	*gigantes, -on*

Singular	Plural
himas, himantos, thong, *himanto-*	*himantes, -on*
keras, keratos, horn, *cerato-*	*kerata, -on*
kreas, kreos, flesh, *creo-*	*krea, -on*
lampas, lampados, torch, *lampado-*	*lampades, -on*
peras, peratos, end, *perato-*	*perata, -on*
phylax, phylakos, guard, *phylaco-*	*phylakes, -on*
chen, chenos, goose, *cheno-*	*chenes, -on*
men, menos, month, *meno-*	*menes, -on*
sphen, sphenos, wedge, *spheno-*	*sphenes, -on*
aner, andros, man, *andro-*	*andres, -on*
aster, asteros, star, *astero-*	*asteres, -on*
gaster, -os or *-tros,* stomach, *gastero-* or *-tro-*	*gasteres, -on*
pater, patros, father, *patro-*	*patres, -on*
ther, theros, beast, *thero-*	*theres, -on*
hippotes, -etos, horsiness, *hippoteto-*	*hippoteis, -on*
trieres, triereos or *-ous,* trireme, *triero-*	*triereis, -on*
alopex, alopekos, fox, *alopeco-*	*alopekes, -on*
myrmex, myrmekos, ant, *myrmeco-*	*myrmekes, -on*
aktis, aktinos, ray, *actino-*	*aktines, -on*
arthritis, arthritidos, gout, *arthritido-*	*arthritides, -on*
aspis, aspidos, shield, *aspido-*	*aspides, -on*
charis, charitos, grace, *charito-*	*charites, -on*
delphis, delphinos, dolphin, *delphino-*	*delphines, -on*
kelis, kelidos, spot, *celido-*	*kelides, -on*
kleis, kleidos, key, *clido-*	*kleides, -on*
kteis, ktenos, comb, *cteno-*	*ktenes, -on*
lophis, lophidos, case, *lophido-*	*lophides, -on*
ophis, ophios, snake, *ophio-*	*ophies, -on*
opsis, opsios, face, *opsi-*	*opsies, -on*
ornis, ornithos, bird, *ornitho-*	*ornithes, -on*
polis, polios, city, *poli-*	*polies, -on*
rhis, rhinos, nose, *rhino-*	*rhines, -on*
this, thinos, shore, *thino-*	*thines, -on*
aix, aigos, goat, *aego-*	*aiges, -on*
mastix, mastigos, whip, *mastigo-*	*mastiges, -on*
rhadix, rhadikos, branch, *rhadico-*	*rhadikes, -on*
thrix, trichos, hair, *tricho-*	*triches, -on*
gala, galaktos, milk, *galacto-*	*galakta, -on*
meli, melitos, honey, *melito-*	*melita, -on*
hals, halos, sea, salt, *halo-*	*hales, -on*
derma, dermatos, skin, *dermato-*	*dermata, -on*
gramma, grammatos, line, *grammato-*	*grammata, -on*
kalymma, kalymmatos, hood, *calymmato-*	*kalymmata, -on*
stoma, stomatos, mouth, *stomato-*	*stomata, -on*
gyne, gynaikos, woman, *gynaeco-* or *-eco-*	*gynaikes, -on*
helmins, helminthos, worm, *helmintho-*	*helminthes, -on*
phalanx, phalangos, phalanx, *phalango-*	*phalanges, -on*
salpinx, salpingos, trumpet, *salpingo-*	*salpinges, -on*
gony, gonatos, knee, *gonato-*	*gonata, -on*
agon, agonos, contest, *agono-*	*agones, -on*
chion, chionos, snow, *chiono-*	*chiones, -on*
daimon, daimonos, god, *daemono-* or *demono-*	*daimones, -on*
geron, gerontos, old man, *geronto-*	*gerontes, -on*
klon, klonos, shoot, *clono-*	*klones, -on*
kyon, kynos, dog, *cyno-*	*kynes, -on*
leon, leontos, lion, *leonto-*	*leontes, -on*
hydor, hydatos, water, *hydato-* or *hydro-*	*hydata, -on*
rhetor, rhetoros, orator, *rhetoro-*	*rhetores, -on*
anthos, antheos, flower, *antho-*	*anthe* or *anthea, -on*
belos, beleos, dart, *belo-*	*bele* or *belea, -on*
euros, eureos, breadth, *euro-*	*eure* or *eurea, -on*
genos, geneos, race, *geno-*	*gene* or *genea, -on*
oros, oreos, mountain, *oro-* or *oreo-*	*ore* or *orea, -on*
phos, photos, light, *photo-*	*photes, -on*
xiphos, xipheos, sword, *xipho-*	*xiphe* or *xiphea, -on*
klops, klopos, thief, *clopo-*	*klopes, -on*
ops, opos, eye, face, *opo-*	*opes, -on*
phleps, phlebos, vein, *phlebo-*	*phlebes, -on*

Singular	Plural
rhips, rhipos, wickerwork, *rhipo-*	*rhipes, -on*
dory, doratos, spear, *dorato-*	*dorata, -on*
asty, asteos, town, *asty-* or *astyo-*	*aste* or *astea, -on*
basileus, basileos, king, *basilo-* or *basileo-*	*basilees* or *basileis, -on*
bous, boös, ox, cow, *boö-* or *bu-*	*boes, -on*
graus, graos, old woman, *grao-*	*graes, -on*
naus, neos, ship, *nau-, nausi-,* or *nauti-*	*nees, -on*
odous, odontos, tooth, *odonto-*	*odontes, -on*
ous, otos, ear, *oto-*	*ota, -on*
pous, podos, foot, *podo-*	*podes, -on*
pyr, pyros, fire, *pyro-*	*pyra, -on*
chlamys, chlamydos, cloak, *chlamydo-*	*chlamydes, -on*
drys, dryos, oak, *dryo-*	*dryes, -on*
ichthys, ichthyos, fish, *ichthyo-*	*ichthyes, -on*
korys, korythos, helmet, *corytho-*	*korythes, -on*
nekys, nekyos, corpse, *necyo-*	*nekyes, -on*
pechys, pecheos, forearm, *pechyo-*	*pechees* or *pecheis, -on*
pitys, pityos, pine, *pityo-*	*pityes, -on*
dioryx, diorychos, canal, *diorycho-*	*dioryches, -on*
nyx, nyktos, night, *nycto-*	*nyktes, -on*
onyx, onychos, claw, *onycho-*	*onyches, -on*

Many Greek nouns were absorbed by Latin and received latinized endings, as *Helene, -es,* Helen, changed to *Helena, -ae; theke, -es,* case, to *theca, -ae; prophetes, -ou,* foreteller, to *propheta, -ae; paidagogus, -ou,* slave who had charge of children, to *paedagogus, -i; Phoibos, -ou,* Apollo, to *Phoebus, -i; psalterion, -ou,* a musical instrument, to *psalterium, -ii; dogma, -atos,* doctrine, to *dogma, -atis; dropax, -akos,* a depilatory, to *dropax, -acis; dryas, -ados,* a tree-nymph, to *dryas, -adis; erysipelas, -atos,* a skin eruption, to *erysipelas, -atis; rhous, -ou,* sumac, to *rhus, rhois; typhon, -os,* typhoon, to *typhon, -is; Xenophon, -ontos,* Xenophon, to *Xenophon, -ontis.*

Plurals of latinized nouns or endings are used in the terms signifying collective classificatory divisions of the plant and animal kingdoms, as *Charophyta, Filicales, Rosaceae, Hominidae, Teleostei, Selachii, Osteichthyes, Pelecaniformes, Testudinata.* To determine the plural of a classical word, observe its gender and genitive singular ending. Then apply the principle of analogy by comparison with similar nouns in the foregoing lists. If this proves futile or unsatisfactory, consult the grammars and dictionaries.

For noun diminutives and their creation see DIMINUTIVES under **little.**

Adjectives

The chief details about Latin and Greek adjectives, exclusive of numerals, necessary for the wordsmith to know, are the forms of the nominative singular in the masculine, feminine, and neuter genders, and the combining forms as determined by adding the connective vowel *i* or *o,* for Latin and Greek, respectively, to the masculine genitive singular stems. Exceptions to this general rule are the *-ys* Greek adjectives like *barys,* heavy, and *polys,* many, whose combining form is the masculine nominative singular without the *s,* as *bary-* and *poly-.* For procedure involving numerals see the discussion of adjectival prefixes in the section FORMATION OF SCIENTIFIC TERMS and also under the required entry in the LEXICON.

Latin Adjectives

Latin adjectives may be classified, in respect to gender, as those having in the nominative singular (1) three different terminations for the masculine, feminine, and neuter genders; (2) two different terminations, the masculine and feminine being the same, the neuter different; and (3) one termination, the same in all genders. In the following lists illustrating these successive groups the arrangement is masculine, feminine, neuter, and combining form. The genitive singulars are like those of analogous nouns of the several declensions.

Three terminations (-us, -a, -um)

aureus, aurea, aureum, golden, *aurei-*
bonus, bona, bonum, good, *boni-*
formosus, formosa, formosum, beautiful, *formosi-*
medius, media, medium, middle, *medi-*
multiflorus, multiflora, multiflorum, many-flowered, *multiflori-*
perditus, perdita, perditum, lost, *perditi-*
unus, una, unum, one, *uni-*

Three terminations (-r, -ra, -rum)

asper, aspera, asperum, sharp, *asperi-*
bigener, bigenera, bigenerum, hybrid, *bigeneri-*
gibber, gibbera, gibberum, humped, *gibberi-*
glaber, glabra, glabrum, smooth, *glabri-*
liber, libera, liberum, free, *liberi-*
macer, macra, macrum, lean, *macri-*
niger, nigra, nigrum, black, *nigri-*
pulcher, pulchra, pulchrum, beautiful, *pulchri-*
satur, satura, saturum, full, *saturi-*
sinister, sinistra, sinistrum, left, *sinistri-*

Three terminations (-r, -ris, -re)

acer, acris, acre, sharp, *acri-*
celeber, celebris, celebre, famous, *celebri-*
paluster, palustris, palustre, marshy, *palustri-*
saluber, salubris, salubre, healthful, *salubri-*
volucer, volucris, volucre, swift, *volucri-*

Two terminations (-is, -is, -e)

cornis, cornis, corne, horned, *corni-*
finalis, finalis, finale, of the end, *finali-*
formis, formis, forme, shape, *formi-*
fragilis, fragilis, fragile, breakable, *fragili-*
hortensis, hortensis, hortense, in or of gardens, *hortensi-*
levis, levis, leve, light, *levi-*
similis, similis, simile, resembling, *simili-*
talaris, talaris, talare, of the ankle, *talari-*
tristis, tristis, triste, sad, *tristi-*

One termination (diverse)

anceps, anceps, anceps, two-headed, *ancipiti-*
atrox, atrox, atrox, fierce, *atroci-*
bicolor, bicolor, bicolor, two-colored, *bicolori-*
cantans, cantans, cantans, singing, *cantanti-*
felix, felix, felix, happy, *felici-*
longipes, longipes, longipes, long-footed, *longipedi-*
par, par, par, equal, *pari-*
repens, repens, repens, creeping, *repenti-*
tenax, tenax, tenax, holding fast, *tenaci-*
tridens, tridens, tridens, three-forked, *tridenti-*
uber, uber, uber, fertile, *uberi-*
vetus, vetus, vetus, old, *veteri-*
vivax, vivax, vivax, lively, *vivaci-*

Some Latin and New Latin adjectives of only one termination may be used as or mistaken for nouns, as *bipes*, two-footed or a biped; *particeps*, sharing or a sharer; *tridens*, three-forked or a trident.

Greek Adjectives

Simple, uncompounded, Greek adjectives are seldom latinized and used as single words, such as the specific terms of binomials. The compounded adjectives, however, have been freely used as specific terms, as *erythroleukos*, reddish-white. In many instances, the compounded adjectives may also serve as nouns, as *erythrokephalos*, red-headed for redhead. Greek adjectives are classed, in respect to gender, as having (1) three terminations, (2) two terminations, and (3) one termination, in the nominative singular. In the following lists of these successive groups, the arrangement is masculine, feminine, neuter, and combining form. The genitive singulars are like those of analogous nouns of the several declensions.

Three terminations (diverse)

agathos, agathe, agathon, good, *agatho-*
barys, baryeia, bary, heavy, *bary-*
charieis, chariessa, charien, graceful, *chariento-*
chrysous, chryse, chrysoun, golden, *chryso-*
glykys, glykeia, glyky, sweet, *glycy-*
haplos, haple, haplon, single, *haplo-*
heteros, hetera, heteron, different, *hetero-*
leucos, leuke, leukon, white, *leuco-*
lyon, lyousa, lyon, loosing, *lyonto-*
makros, makra, makron, long, *macro-*
megas, megale, mega, large, *megalo-*
melas, melaina, melan, black, *melano-*
orthos, orthe, orthon, straight, *ortho-*
oxys, oxeia, oxy, acid, sharp, *oxy-*
pas, pasa, pan, all, *panto-*
polys, polle, poly, many, *poly-*
protos, prote, proton, first, *proto-*
skiotos, skiote, skioton, shaded, *skioto-*
stereos, sterea, stereon, solid, *stereo-*
teren, tereina, teren, soft, *tereno-*
thelys, theleia, thely, female, *thely-*
timon, timosa, timon, honoring, *timonto-*

Two terminations (diverse)

alethes, alethes, alethes, true, *aletho-*
arrhen, arrhen, arrhen, male, *arrheno-*
bathyskios, bathyskios, bathyskion, deep-shaded, *bathyscio-*
eudaimon, eudaimon, eudaimon, happy, *eudaemono-, eudemono-*
eumathes, eumathes, eumathes, learning easily, *eumatho-*
hygies, hygies, hygies, healthy, *hygio-*
megalokeros, megalokeros, megalokeron, large-horned, *megalocero-*
mikrostomos, mikrostomos, mikrostomon, small-mouthed, *microstomo-*
onychoeides, onychoeides, onychoeides, naillike, *onychoido-*
pepon, pepon, pepon, ripe, *pepono-*
pleres, pleres, pleres, full, *plero-*
pseudes, pseudes, pseudes, false, *pseudo-*

In this list of two-termination adjectives, *alethes, arrhen, eudaimon,* and others appear to have only one termination, but in untransliterated Greek the masculines and feminines end with -ης, -ην, -ων, and the neuters with -ες, -εν, and -ον, respectively. The -os masculine ending of two-termination adjectives is, as elsewhere, transliterated -us, but the feminine -os is transliterated -a, like the -e in a three-termination adjective such as *agathos, -e, -on.* These adjectives, therefore, become three-termination (-us, -a, -um) Latin adjectives.

One termination (diverse)

apais, apais, childless, *apaedo-, apedo-*
bathykrepis, bathykrepis, with deep foundations, *bathycrepido-*

> *diadochos, diadochos,* succeeding, *diadocho-*
> *hierochthon, hierochthon,* of sacred soil, *hierochthono-*
> *kenelpis, kenelpis,* with empty hope, *cenelpido-*
> *lagopous, lagopous,* hare-footed, *lagopodo-*
> *makrocheir, makrocheir,* long-handed, *macrochiro-*
> *nomas, nomas,* roaming, *nomado-*
> *phygas, phygas,* fugitive, *phygado-*
> *sosipolis, sosipolis,* city-saving, *sosipoli-*

These one-termination adjectives are masculines and feminines, but they could presumably, for practical purposes, serve as neuters also.

Comparison of Latin Adjectives

Comparatives are formed by attaching *-ior*, and superlatives some form of *-imus* (*-illimus, -irrimus, -issimus*), to the stem of the positive, as *altus*, high, *altior*, higher, *altissimus*, highest. The *-ior* comparatives have two terminations, *-ior* for masculines and feminines, as *formosior, major, melior,* and *-ius* for neuters, as *formosius, majus, melius*. Other neuters are *plus* and *minus*. Diversified examples of regular and irregular comparison are:

> *bonus,* good, *melior, optimus*
> *carus,* dear, *carior, carissimus*
> *formosus,* beautiful, *formosior, formosissimus*
> *inferus,* low, *inferior, infimus*
> *magnus,* large, *major, maximus*
> *malus,* bad, *pejor, pessimus*
> *multus,* much, *plus, plurimus*
> *parvus,* small, *minor, minimus*
> *rarus,* scarce, *rarior, rarissimus*
> *superbus,* excellent, *superbior, superbissimus*
> *tepidus,* lukewarm, *tepidior, tepidissimus*
>
> *acer,* sharp, *acerior, acerrimus*
> *concors,* harmonious, *concordior, concordissimus*
> *deter,* poor, *deterior, deterrimus*
> *facilis,* easy, *facilior, facillimus*
> *felix,* happy, *felicior, felicissimus*
> *fragilis,* breakable, *fragilior, fragilissimus*
> *frugalis,* thrifty, *frugalior, frugalissimus*
> *gracilis,* slender, *gracilior, gracillimus*
> *hebes,* dull, *hebetior, hebetissimus*
> *levis,* light, *levior, levissimus*
> *miser,* wretched, *miserior, miserrimus*
> *vetus,* old, *veterior, veterrimus*

Comparison of Greek Adjectives

Most Greek comparatives are formed by attaching *-teros* or *-ion*, and superlatives *-tatos* or *-istos*, to the stem of the positive:

> *barys,* heavy, *baryteros, barytatos*
> *habros,* pretty, *habroteros, habrotatos*
> *hygies,* healthy, *hygiesteros, hygiestatos*
> *makros,* long, *makroteros, makrotatos*
> *melas,* black, *melanteros, melantatos*
> *orthos,* straight, *orthoteros, orthotatos*
> *oxys,* sharp, sour, *oxyteros, oxytatos*
> *penes,* poor, *penesteros, penestatos*
> *pleres,* full, *pleresteros, plerestatos*
> *teren,* soft, *terenoteros, terenotatos*
>
> *echthros,* hated, *echthion, echthistos*
> *glykys,* sweet, *glykion, glykistos*
> *hedys,* sweet, *hedion, hedistos*
> *kakos,* bad, *kakion, kakistos*

megas, large, *meizon (megion), megistos*
polys, many, *pleion, pleistos*
tachys, swift, *tachion, tachistos*

For adjective diminutives and their creation see DIMINUTIVES under little.

Verbs

Latin Verbs

Latin verbs end with *-o* (*laudo,* praise; *deleo,* destroy; *salio,* leap; *fluo,* flow), and *-or* (*sequor,* follow). Removal of these endings leaves the base to which appropriate suffixes may be attached to give the chief parts of the verb that supply nounal and adjectival material for the wordmaker—present active participle, perfect passive participle, gerundive, and supine. These are illustrated respectively by *creans,* making; *creatus,* made; *creandus,* producible; and *creatum,* product, of the verb *creo,* make; *movens,* moving; *motus,* moved; *movendus,* movable; and *motum,* thing moved, of *moveo,* move; *plectens,* braiding; *plexus,* braided; *plectendus,* braidable; *plexum,* braid, of *plecto,* braid; *sentiens,* feeling; *sensus,* felt; *sentiendus,* feelable; *sensum,* thing felt, of *sentio,* feel; *struens,* building; *structus,* built; *struendus,* buildable; *structum,* thing built, of *struo,* build; *mirans,* wondering; *miratus,* wondered at; *mirandus,* to be wondered at; *miratum,* thing admired, of *miror,* wonder at, admire; *verens,* respecting; *veritus,* respected; *verendus,* respectable; *veritum,* thing respected, of *vereor,* respect; *loquens,* speaking; *locutus,* spoken; *loquendus,* speakable; *locutum,* thing spoken, of *loquor,* speak; *ordiens,* beginning; *orsus,* begun; *ordiendus,* to be begun; *orsum,* thing begun, of *ordior,* begin. These parts are the nuclei of new operating bases to which other suffixes may be attached.

The dictionary entries usually give the verb and the supine in *-um.* If the supine is not indicated the perfect participle can nearly always be inferred by substituting *s* for *m* in the assumed supine. The present participles are adjectives of one termination and the perfect participles are adjectives of three terminations (*-us, -a, -um*). Sometimes the present participle may be used as a masculine noun of the third declension, as *secans, -antis,* one who or that cuts, secant; *praesidens, -entis,* one who or that sits before or presides, president; and the perfect participle may be used as a masculine noun of the fourth declension, as *ductus, -us,* a leading; *status, -us,* a standing. The gerundive is an adjective of three terminations (*-us, -a, -um*), and the supine in form is the accusative of a noun of the fourth declension but in practice is used as a neuter noun of the second declension.

Although Latin verbs are classed in four conjugations, three groups suffice for the present purpose. These may be distinguished by the differing endings of the several parts, as follows:

CONJUGATION	PRESENT PARTICIPLE	PERFECT PARTICIPLE	GERUNDIVE	SUPINE
First	-ans	-atus, -itus, -tus	-andus	-um
amo, love	*amans*	*amatus*	*amandus*	*amatum*
canto, sing	*cantans*	*cantatus*	*cantandus*	*cantatum*
conor, endeavor	*conans*	*conatus*	*conandus*	*conatum*

CONJUGATION	PRESENT PARTICIPLE	PERFECT PARTICIPLE	GERUNDIVE	SUPINE
First	**-ans**	**-atus, -itus, -tus**	**-andus**	**-um**
domo, tame	*domans*	*domitus*	*domandus*	*domitum*
frico, rub	*fricans*	*frictus (fricatus)*	*fricandus*	*frictum*
levo, raise	*levans*	*levatus*	*levandus*	*levatum*
miror, admire	*mirans*	*miratus*	*mirandus*	*miratum*
radico, root	*radicans*	*radicatus*	*radicandus*	*radicatum*
salto, leap	*saltans*	*saltatus*	*saltandus*	*saltatum*
sono, sound	*sonans*	*sonitus*	*sonandus*	*sonitum*
sto, stand	*stans*	*status*	*standus*	*statum*
Second and Third	**-ens**	**-etus, -itus, -sus, -tus, -xus**	**-endus**	**-um**
abdo, hide	*abdens*	*abditus*	*abdendus*	*abditum*
augeo, increase	*augens*	*auctus*	*augendus*	*auctum*
carpo, pluck	*carpens*	*carptus*	*carpendus*	*carptum*
caveo, beware	*cavens*	*cautus*	*cavendus*	*cautum*
cingo, blind	*cingens*	*cinctus*	*cingendus*	*cinctum*
coquo, cook	*coquens*	*coctus*	*coquendus*	*coctum*
credo, believe	*credens*	*creditus*	*credendus*	*creditum*
debeo, owe	*debens*	*debitus*	*debendus*	*debitum*
deleo, destroy	*delens*	*deletus*	*delendus*	*deletum*
duco, lead	*ducens*	*ductus*	*ducendus*	*ductum*
fleo, weep	*flens*	*fletus*	*flendus*	*fletum*
gaudeo, rejoice	*gaudens*	*gavisus*	*gaudendus*	*gavisum*
ludo, play	*ludens*	*lusus*	*ludendus*	*lusum*
medeor, heal	*medens*	*meditus*	*medendus*	*meditum*
misceo, mix	*miscens*	*mixtus*	*miscendus*	*mixtum*
mitto, send	*mittens*	*missus*	*mittendus*	*missum*
moneo, warn	*monens*	*monitus*	*monendus*	*monitum*
moveo, move	*movens*	*motus*	*movendus*	*motum*
neo, spin	*nens*	*netus*	*nendus*	*netum*
pendeo, hang	*pendens*	*pensus*	*pendendus*	*pensum*
placeo, please	*placens*	*placitus*	*placendus*	*placitum*
plecto, braid	*plectens*	*plexus*	*plectendus*	*plexum*
rego, rule	*regens*	*rectus*	*regendus*	*rectum*
rideo, laugh	*ridens*	*risus*	*ridendus*	*risum*
sequor, follow	*sequens*	*secutus*	*sequendus*	*secutum*
suo, sew	*suens*	*sutus*	*suendus*	*sutum*
taceo, be silent	*tacens*	*tacitus*	*tacendus*	*tacitum*
vinco, conquer	*vincens*	*victus*	*vincendus*	*victum*
vivo, live	*vivens*	*victus*	*vivendus*	*victum*
Fourth	**-iens**	**-itus, -sus, -tus**	**-iendus**	**-um**
aperio, open	*aperiens*	*apertus*	*aperiendus*	*apertum*
audio, hear	*audiens*	*auditus*	*audiendus*	*auditum*
blandior, flatter	*blandiens*	*blanditus*	*blandiendus*	*blanditum*
capio, take	*capiens*	*captus*	*capiendus*	*captum*
finio, end	*finiens*	*finitus*	*finiendus*	*finitum*
gannio, yelp	*ganniens*	*gannitus*	*ganniendus*	*gannitum*
haurio, drain	*hauriens*	*haustus*	*hauriendus*	*haustum*
hinnio, neigh	*hinniens*	*hinnitus*	*hinniendus*	*hinnitum*
operio, cover	*operiens*	*opertus*	*operiendus*	*opertum*
orior, rise	*oriens*	*ortus*	*oriendus*	*ortum*
salio, leap	*saliens*	*saltus*	*saliendus*	*saltum*
sentio, feel	*sentiens*	*sensus*	*sentiendus*	*sensum*
vincio, bind	*vinciens*	*vinctus*	*vinciendus*	*vinctum*

Greek Verbs

The commonest endings of Greek verbs are *i* and *o*, or combinations of other letters with them: *-ai* (*mainomai*, rage); *-aino* (*tetraino*, bore); *-ao* (*gelao*, laugh); *-azo* (*damazo*, tame); *-eo* (*kineo*, move); *-euo* (*therapeuo*, serve); *-izo* (*thesaurizo*, store up); *-mi* (*kerannymi*, mix); *-o* (*herpo*, creep); *-oo* (*sphenoo*, use a wedge); *-sko* (*gerasko*, grow old); *-so* (*oneirosso*, dream); *-yno* (*trachyno*, roughen). For these, as for

Latin verbs, the operating bases are obtained by removing the designated endings. No simple rule-of-thumb, however, can be stated to explain the many variations shown by Greek verb stems. The following examples illustrate the variety of combining forms that may be had, but such formatives, if not already in existence, should not be invented by the novice without great care and discrimination: *archo*, lead, govern, *arche-, archi-, archo-; leipo*, leave, *lip- liph- lipo-; lyo*, loose, *lysi-; phaino*, shine, *phaeno-, pheno-; phero*, bear, *pher-, phere-, phero-; phersi-; rhipto*, throw, *rhipto-, rhips-; sozo*, save, *sozo-, sosi-*.

Adverbs, Prepositions, Pronouns

Latin adverbs, *alibi, ibi, modo, ne, non, quippe,* and *ubi,* appear in English alibi, ibidem, modern, negative, nonsense, quip, and ubiquitous. The Greek adverb *mala,* very, much, occurs in its comparative form in malloseismic.

Latin prepositions, *ad, ante, circum, cum, de, ex, in, per, prae, pro, se, sub, super, trans, ultra,* and others; and Greek prepositions, *amphi, ana, anti, apo, dia, en, epi, hyper, hypo, kata, meta, para, peri, pro, syn,* and others, enter into many words. For examples, see these entries in the LEXICON.

Latin pronouns, *alius,* another, *ego* (Gr. *ego*), I, *idem,* the same, *ipse,* self, *noster,* our, *qui, quis, quod,* who, *sui,* of oneself, and *tu,* you, appear in aliquot, egotism, idem, ipsedixitism, nostrum, quibble, quiddity, *qui vive, Quisqualis,* quodlibet, suicide, tuism; and the Greek *autos,* self, occurs in automatic.

FORMATION OF SCIENTIFIC TERMS

According to some sixteenth century chronologists the world was created by fiat from aboriginal, chaotic nothingness during the week of October 18 to 24, 4004 B.C. This awesome event, however, has now been predated many millions of years as a result of a succession of astronomical, chemical, geological, and archeological discoveries whose facts are explainable only by stretching time and space beyond anything heretofore thought probable, and by applying the common-sense principle that physical and chemical forces or processes such as gravitation, heat, light, electricity, circulation of air and water, oxidation, solution, erosion, and sedimentation operated as dependably, invariably, and inexorably in the distant past as they do today. Indeed, an orderly universe without relatively undeviating behavior by at least its inanimate constituents is inconceivable. The universe is wonderful but it is not an infantile, anything-can-happen-at-any-time Fairyland, as all thoughtful minds discover when they escape from the fancies of childhood into the realities of maturity.

Although the Biblical cosmogony was reported as colossal and inclusive some nominal chaos, nevertheless, remained. Things and processes were nameless, and in consequence man, on emerging as a conscious being, aware of his environment and gifted with the power of speech, began to identify persons, things, and phenomena by name, thus ordering his own thoughts and facilitating discussion of matters with his fellows. The animals, so the story runs, were brought before Adam to be named, but there is no record that he was commanded to name the plants. Botanists have been diligently attending to this project. Astronomers, geologists, chemists, physicists, mathematicians, zoologists, physiologists, anatomists, psychologists, and the like, have been busy with other phases of the creation. Practical considerations are customarily given or alleged as the motives for these investigations in pure science, but the basic spur is the desire to discover, invent, or do something new, activated fundamentally by the spirit of inquiry or, to use a less dignified term, curiosity, defined pithily by William James as "the impulse toward better cognition". Whether applied to shallow backyard gossip or to profound study of the materials and mechanism of the universe, this mental stimulus, like body appetite, focuses interest and sharpens attention. When Eve and Newton dally beneath apple-trees sin and science are imminent. As a result of the scientists' curiosity the world is gradually being analyzed, dissected, named, classified, and the how, if not the why of it, understood.

At this stage of the inventory the making and giving of names is recognized as something of an art, perhaps better practiced a century ago than today; and the realization that some words are more happily constructed and applied than others arouses perennial interest in word-coinage. Fundamental to a review of the methods of word-formation attention must be directed to the qualities that should be expected in words when they make their debut in scientific or literary society. Such words should be simple, euphonious, pure, and mnemonically attractive, to say nothing of being descriptively appropriate.

Simplicity is a desirable quality, for it satisfies a primary psychological urge. Because simplicity promotes clarity and definiteness we expect others to exemplify it in their creations and communications, although we excuse the motes and opacity in our own eyes. It is almost a truism to say that short words are usually simple. However, not all words, particularly those of scientific terminology, can be short and easily spelled, but the most easily spelled version of a word, consistent with its sound and etymology, is most likely to be popularly preferred. Thus, the tendency to transliterate some diphthongs to single letters is in the direction of simplicity, and hence *cerulean, prepare, primeval, Cenozoic, paleontology, cenobite,* and *penalty,* are improvements on *caerulean, praepare, primaeval, Caenozoic, palaeontology, coenobite,* and *poenalty. Sulfur* is preferable to *sulphur,* if for no other reason than that the *ph* is a false suggestion of derivation from a Greek source.

The nature of the material to be named and its specificity may be the controlling factors in the construction of names. Short, catchy trade names often identify chemicals for which chemists, in order to express the components of their preparations, feel obliged to invent what seem to the layman to be sesquipedalian jawbreakers, like some notorious German words. Such, for example, are *sulfanilamide, dihydroergotamine, 2.4 dichlorophenoxyacetic acid,* and *p-tertiary-octylphenoxyethyoxy-ethylodimethylbenzyl-ammonium chloride.* Other sciences have also developed formidable terminologies and jargons peculiar to themselves as can be seen by consulting the special dictionaries devoted to their vocabularies.

The comparative ease and speed with which words such as *cyclotron, penicillin, hippopotamus,* and *chrysanthemum* are absorbed into common speech should put one on guard against arguments for too extreme simplicity. Some scientific terms, however, definitely exceed the limits for easy handling. Consider the names given to two amphipods from Lake Baikal: *Leucophthalmoechinogammarus crassus* and *Siemienkiewicziechinogammarus siemienkiewiczi.* The specific term of the name of a Jurassic pelecypod, *Pecten ntlakapamuxanus,* has perhaps little more justification for existence.

Words that sound well appeal to the esthetic appetite of the mind as savory food does to the stomach. Many persons, however, are tone deaf, and to them the line from Keats' *Ode on a Grecian urn,* "Heard melodies are sweet, but those unheard are sweeter", must be doubly unintelligible. The fact remains that, regardless of unappreciative ears, euphonious words are desirable above those that are not, and word-coiners should govern themselves accordingly, discounting the ill-considered assertion sometimes heard that nomenclature is not poetry and music but only a strictly practical matter. Are the roseate color and delicate fragrance not integrated with the practical purpose for which apple blossoms exist?

Such words as *automobile, colorimetry, subway, television,* and *tonsillectomy,* compounded from two different languages, are hybrids like the offspring from biologically different species or sometimes varieties. With the sturdy mule and the productive hybrid corn in mind the realist can hardly be in a receptive mood for listening to arguments against hybridization. On all sides, in trade and the professions, hybrid words are constantly being coined and many of these soon become satisfactory operating parts of the general vocabulary. The objection to them, therefore, is not that they are impractical, but that they too frequently are unesthetic substitutes for possible genuine articles. Elements of the same language usually combine more easily with one another than do those from different languages, and the results are likely to be more attractive and euphonious. The wordmaker should exhaust every resource to obtain a purebred before resorting to mongrelization.

Sometimes speakers and writers are purveyors of words but not conveyors of ideas. The difference lies not only in the choice of words but also in their grouping. Single or grouped words are most effective if, in addition to having the desirable qualities already discussed, they are also mnemonically adhesive, that is, are easily remembered. This is the best of reasons for constructing words from known roots or stems that have been used repeatedly in other combinations. Such words carry mnemonic signals of descriptive value, whereas words of pure invention without legitimate antecedents lack this quality and may have to be in existence for some time before they acquire mnemonic connotations. In biology, if a plant or animal species is worth describing it surely merits an appropriate, uncriticizable name, and the sacrifice of a little time and energy by the creator in devising such a name. He should find this not only a challenge to his ingenuity, but, in subsequent years, a recurring pleasure like that brought by all significant, creative work. Conversely, use of nomenclature as a vehicle for frivolous purposes and the expression of personal opinions derogatory of others, shows poor taste.

In contrast with the use of common, usually single, words for identifying plants and animals—scientifically, a somewhat unsatisfactory practice well discussed by Gordon[1]—there has been in existence, since its establishment by Carolus Linnaeus (1707-1778), a system of biological nomenclature wherein two Latin or latinized words constitute the scientific name of a plant or animal species, comparable to the surname and first name, respectively, of persons. Thus, a binomial, such as *Passer domesticus* (house-sparrow), consists of a first or generic term, which is a noun or a word considered as a noun, in the nominative singular; and a second or specific term that may be (1) an adjective agreeing with the generic term in gender, as *Rosa stellata* (a rose); (2) a present participle, as *Arenaria repens* (creeping sandwort); (3) a noun or a word considered as a noun, in the nominative singular, requiring no agreement in gender, as *Felis leo* (lion); or (4) a noun in the genitive singular or plural, as *Picea engelmanni* (Engelmann spruce), *Ceratostomella ulmi* (a fungus), *Aster scopulorum* (an aster). These binomials are in effect abbreviations of the longer Latin phrases formerly used for designating species, and for scientific purposes are more accurate and distinctive than the more or less indefinite common-name binomials, such as African violet, bottle brush, choke cherry, dwarf rose, Easter lily, fairy bell, ground squirrel, horned toad, ice plant, laughing jackass, monkey flower, nut pine, Oregon grape, poison oak, rain bird, star grass, timber beetle, velvet ant, water scorpion. Physicians employ a kind of binomial nomenclature to identify diseases, as *angina pectoris, chorea nutans, delirium tremens, dementia praecox, diabetes mellitus, scarlatina maligna, tinea tonsurans, verruca vulgaris.* Sometimes others than biologists invent binomials for humorous effect, as *Boobus americanus, Goofus gregarius,* and *Homo frustratus,* for the well-known John Q. Public (*Homo sapiens,* the 'wise guy').

Trinomials have found their way into some of the biological sciences, the third term being a variety designation. Most of these are acceptable, but seeing *Bison bison bison* (bison), *Cardinalis cardinalis cardinalis* (cardinal), and *Rattus rattus rattus* (black rat) arouses suspicions concerning the quality of the scientist's imagination, or perhaps pity for his helplessness in the clutch of a ruling that makes no provision for avoiding such tautonomous reiterations. Nomenclature need not compete with the Hallelujah Chorus.

Scientific terms originate in three ways: (1) adoption directly with appropriate modifications in spelling, from Greek, Latin, and other languages; (2) composition by compounding and affixation; and (3) outright or arbitrary creation without use of evident, antecedent root or stem material. A consideration of these methods follows.

---

[1] Gordon, Myron. *How animals get their names.* Natural History, vol. 60, no. 8, pp. 370-375, 1951.

ADOPTION OF WORDS

As already illustrated, the English vocabulary has taken many words directly from foreign languages, both dead and living. The more specialized scientific vocabulary has also ransacked every corner of the world for word-making material, but the chief source has been Greek and Latin. This immense reservoir, however, has now been practically emptied of common words like *animal, analysis, atlas, carbon, crisis, diabetes, electron, genius, item, nucleus, parallel, premium, senator, stimulus,* and *synopsis,* that could be taken bodily or with little change in spelling. Most of the Greek and Latin names for plants and animals have been used in biological nomenclature, although, as reference to the LEXICON will reveal, a few such words are still available. One difficulty now, however, is the uncertainty in many instances of the identity of the plant or animal species the ancients had in mind. Many Greek and Latin mythological, historical, and geographical names have been utilized by scientists. As this book could include only part of that large list, the interested wordcoiner may satisfy his further needs by consulting the excellent *Oxford classical dictionary.* After exhausting the ancient classical vocabularies for usable terms the taxonomist resorts to new coinage for new concepts. This may require recombination of basic elements in Greek and Latin or adoption of words from modern languages.

If the adopted word is to be used in common parlance like any other English word it may have to be naturalized or anglicized so that it will harmonize with the phonetic pattern peculiar to the language. This process, besides sometimes shifting accents in exotic words, has heretofore usually involved an operation, more or less drastic, on their endings, such as (1) the elision of letters, as in baptism (Gr. *baptismos*), brutal (L. *brutalis*), chasm (Gr. *chasma*), digit (L. *digitus*), element (L. *elementum*), firm (L. *firmus*), grand (L. *grandis*), human (L. *humanus*), hymn (Gr. *hymnos*), nymph (Gr. *nymphe*), petal (Gr. *petalon*), prophet (L. *propheta*, Gr. *prophetes*), similar (L. *similaris*), verb (L. *verbum*); or (2) changes in spelling involving single letters or groups of letters with the development of a new suffix, as in apiary (L. *apiarium*), barbule (L. *barbula*), canine (L. *caninus*), copious (L. *copiosus*), fable (L. *fabula*), hostile (L. *hostilis*), image (L. *imago*), injury (L. *injuria*), lily (L. *lilium*), member (L. *membrum*), movie, moving, moved (L. *moveo*), mundane (L. *mundanus*), particle (L. *particula*), partition (L. *partitio*), pause (Gr. *pausis*), raceme (L. *racemus*), rose (L. *rosa*), scheme (Gr. *schema*), sphere (Gr. *sphaira*), space (L. *spatium*), verbose (L. *verbosus*). In particular, special care should be taken in rendering the sounds of a native, foreign word by the nearest representative letters of the English alphabet. Remarkable ectoplasmic transformations have been accomplished in deriving such words as *Asimina, hickory, hominy, Massachusetts, opossum, quinine, raccoon, skunk, Yosemite,* and many others, from Aboriginal American and other grunts, squeaks, and mumbles.

When words from any language other than Latin are taken for use in biological nomenclature, one of the first rules in the several codes requires that they be latinized in their letters and endings. Considerable latitude, however, is given to the term 'latinized', for, in practice, many vernacular words are employed with little or no significant change in spelling. Thus, already transmuted into English, Amharic *zebra* occurs in *Equus zebra*, Chinese *ginseng* in *Panax schinseng*, Tamil *betle* in *Piper betle,* Aboriginal American *catalpa* in *Catalpa speciosa* and *powwow* in *Triticites powwowensis* (a Permian foraminifer from Powwow Canyon, Texas). Other examples are: *Litchi chinensis* (litchi-nut), *Ferula sumbul* (an umbellifer), *Macrosporium tomato* (a fungus), and *Myrica gale* (sweetgale). A double adoption from Aboriginal American is the name of an Oligocene rodent, *Manitscha tanka,* literally 'big gopher'. Latinized English words occur in *Alwisia bombarda* (a mycetozoan), *Cyclopterus lumpus* (lumpfish), *Khaya ivorensis[ivoryensis]* (African mahogany from the Ivory Coast), *Nilssonia steamboatea* (a fossil cycadophyte), and *Trigla gurnardus* (gray gurnard). Camouflaged English words masquerading as Greek terms occur in *Conacodon cophater* (a fossil mammal), the *cophater* said to be from 'Cope hater'; and in the generic terms for some bugs: *Ochisme, Dolichisme, Elachisme, Peggichisme,* and *Polychisme.* A near miss is *Palustrimus lewisi* (a Miocene rodent named for O. C. Marsh and G. E. Lewis) ; and *Sinclairoceras haha* (an Ordovician cephalopod) is only indirectly a laughing matter, the fossil being from the Ha! Ha! River region of Quebec, Canada. These examples are cited to illustrate the borrowing and not to question the methods or the propriety of the adaptations.

Apropos of the latinization of English and other modern terms, particularly personal and geographic names, the following question, in lieu of further examples, is pertinent. Shall the specific term of a binomial be *neomexicana* or *newmexicana; noveboracensis* or *newyorkensis?* The answer is left to individual judgment, but should correlate, it would seem, with the customary literary advice, Do whatever is necessary to enlighten and please the reader.

In latinization the method of analogy with authoritative precedents is recommended, but pitfalls should be anticipated. For example, the Latin words *ibex, -icis, latex, -icis,* and *vertex, -icis,* not derived from Greek, have the combining forms *ibici-, latici-,* and *vertici-.* The similar Greek words *myrmex, -ekos, narthex, -ekos,* and *Threx, -ekos,* have the combining forms *myrmeco-, nartheco-,* and *Threco-.* Latinized, the genitive ending becomes *-ecis* and the combining forms become *myrmeci-, nartheci-,* and *Threci-,* not *myrmici-, narthici-,* and *Thrici-,* as might be expected by analogy with *ibex, latex,* and *vertex.*

COMPOSITION OF WORDS

The coinage of words by composition may be accomplished in three ways: (1) compounding by joining entire, independent words without using a connective vowel; (2) compounding by joining stems with a connective vowel; and (3) modification of existing roots and stems by appropriate attachment of prefixes and suffixes.

Compounding by joining entire words without a connective vowel. This simple kind of composition is common in German and

English, as illustrated by *fernsprecher,* far-speaker or telephone; *luft-waffe,* air force, *Baumgarten, Schwarzwald, Helmholtz, barnyard, brambleberry, hatrack, Jameson, kingdom, limelight, machineman, northwest, packrat, scofflaw, shipworm, tinsmith, votemonger, whitefish, zebrawood.* The Greeks and Romans used simple composition infrequently, as in *kynosoura,* dog's-tail, cynosure: *myosotis,* mouse-ear; *aquaeductus,* aqueduct; *cornucopia,* horn of plenty; *maledico,* revile; *nomenclator,* namer; *paterfamilias,* head of a family; *respublica,* republic; *satisfacio,* satisfy; *septendecim,* seventeen; *solstitium,* solstice. In general, when the first component of a compound is a preposition (*diaphoresis,* perspiration; *circumspectus,* looking · around cautiously), numeral (*pentameres,* five-parted; *quinquefolius,* five-leaved), Greek adjective in *-ys* (*barypalamos,* heavy-handed), adverb (*eutektos,* easily dissolved; *sempervivus,* ever-living), inseparable particle (*zakalles,* very beautiful; *vecordia,* folly), or if it ends with a diphthong (*naukrates,* pilot; *praedictio,* a foretelling), it is regularly joined directly to the second component without a connective. Except for such instances as those just cited, where euphony obviously dictates the nature of the union, composition without the use of a connective is not recommended wholeheartedly for compounding material from classical sources because a better product can generally be had by using procedures more customary for those languages. In biological nomenclature, however, this method is sometimes applied, more or less felicitously, in connection with the use of proper names, as in *Nucula austinclarki* (an Eocene pelecypod, named for Austin Clark), *Hughmilleria socialis* (a eurypterid, named for Hugh Miller), *Ofcookogona alia* (a milleped, named for O. F. Cook), and *Roystonea regia* (a palm, named for Roy Stone).

These observations lead directly to a discussion of hyphenation. Most dictionaries preface their vocabularies by statements of 'principles' or 'rules' for compounding and hyphenation, and try to adhere to them. Diligent searchers soon discover, however, that practice does not always coincide with preaching and that many exceptions occur,— in brief, that there are no fixed 'rules', except perhaps one, namely, Let usage take its course. The evolution through usage of single words, expressing separate concepts, to hyphenated and finally compound words, expressing united concepts, is illustrated by *water lily, water-lily, waterlily,* and many others. Consistency may never be attained. We see *apple tree* and *apple juice* but also *applejack* and *applesauce.* Hyphenated terms, like the following, should be banned: *Anabaena flos-aquae* (an alga), *Euphorbia caput-medusae* (a spurge), *Hibiscus rosa-sinensis* (Chinese hibiscus), *Rorippa nasturtium-aquaticum* (watercress).

Sometimes words, phrases, and sentences, instead of being combined as such, may be syncopated or telescoped into new terms by fracturing and acrosticizing, as *ACTH* (adrenocorticotropic hormone), *brunch* (breakfast, lunch), *cattalo* (cattle, buffalo), *chortle* (chuckle, snort), *darso* (dwarf, red, sorghum), *fanglomerate* (fan, conglomerate), *GOPolitician* (G.O.P., politician), *motel* (motor, hotel), *smog* (smoke, fog), *snark* (snake, shark), *thob* (think, opine, believe), *thon* (that one), *Socony* (Standard Oil Company of New York), *Uneeda* (you need a),

Nabisco (National Biscuit Company). The reverse of this process is exemplified by *Veep* (V.P., Vice-President).

Compounding by joining stems with a connective vowel. Many Greek and Latin words are compositions made from two or more modified roots, stems, or words, and dissection of them reveals the recipe for most modern compounding. Thus, the Greek *ichthyopoles*, fishmonger, may be separated into *ichthy-o-poles*, *ichthy* being the stem of *ichthys*, *ichthyos*, fish, *o* the connective, and *poles*, dealer. So also *phyll-o-phoros*, bearing leaves, and *somat-o-phylax*, bodyguard. In Latin, *plumipes*, feather-footed, may be separated into *plum-i-pes*, *plum* being the stem of *pluma*, *plumae*, feather, *i* the connective, and *pes*, foot. So also *bell-i-potens*, mighty in war, *carn-i-vorus*, flesh-eating, *corn-i-ger*, bearing horns. Inspection of the foregoing examples of the normal or regular method of compounding shows in each a first component, obtained by removing the genitive case-ending, a connective, *o* in Greek, *i* in Latin, and a final component.

First components of compounds may be the stems of nouns, as in *Potamogale*, *potamophilous*, *Avihospita*, *aviculture;* adjectives, as in *Leucobryum*, *leucocyte*, *Breviceps*, *brevipennate;* numerals, as in *hexagon*, *monologue*, *bisect*, *unison;* verbs, as in *Trechomys*, *trechometer*, *dictation*, *stationary;* adverbs, prepositions, and pronouns, as in *Eucalyptus*, *euphony*, *Peridermium*, *periscope*, *Sempervivum*, *internode*, *egocentric*. Latin is perhaps a little less amenable than Greek to this method and it has employed principally prepositions as first components of compounds. Verb stems, as such, are infrequent as first components, and the wordmaker is advised to proceed with caution in using them. Custom seems to confirm the principle that the first component of a compound should be a qualifying and the final component a governing word, either a nounal or verbal element, as in *melanokomes*, black-haired, and *melanoeides*, black-looking. Thus, the word *hippopotamus* has been criticized as not meaning what it says, horse-river. It should be *potamohippus*, river-horse. On the other hand, *anthropophilus*, man-loving, and *philanthropus*, loving man, have essentially the same meaning and choice of usage would depend upon the context to be served.

Some first components may be sanctioned in two forms cited in the classical dictionaries, thus forcing the wordcoiner to make a difficult choice. Such, for example, are the Greek neuter nouns ending with -*a*, as *derma*, *dermatos*, skin, *haima*, *haimatos*, blood, and *stoma*, *stomatos*, mouth. These normally yield the long combining forms *dermato-*, *haimato-*, and *stomato-*, which appear in many compounds, but the short forms *dermo-*, *haimo-*, and *stomo-* occur in *dermopteros*, skin-winged, *haimorrhagia*, hemorrhage, *stomodokos*, talkative, and others. Further, such derivatives of the short forms as *dermoblast*, *hemoglobin*, and *stomodeum*, seem to be doing well in scientific English. Therefore, the question arises, Which practice shall be recommended to the wordsmith? A satisfactory solution, perhaps, would be to adopt the long forms for all binomials and the short forms only where simplicity, clearness, and euphony permit for words outside of latinized binomial nomenclature.

Final components of compounds may be nouns in the nominative singular, as in *dicotyledon, Geomys, Hesperornis, Leptothrix, Mesohippus, Trifolium, viaduct,* or, exceptionally, in the genitive, as in *Auricula aurismidae* (a snail); adjectives, as in *hypermegas, longivivax, omnipotens, polychloros, seminudus, uniformis;* and transmuted verbal derivatives, as in *dystaktos, fructifer, halogen, monticola, peripheres, solstice, trichotomy.*

Although most neoclassical compounds consist of only two components joined by a connective vowel, there is, theoretically, no limit, except that set by cumbersomeness, to the number of items that may be combined. Examples are: *Adelopteromyia, Baeopleuroceras, Callodontomys, Epiphragmophora, Lepidocephalichthys, Megagnathotermes, Spilochaetosoma, Xanthospilopteryx, apicitripunctella, dipterocarpifolia.* The citation of these polysyllabic terms suggests that one of the commonest causes of bafflement to the novice in word-creation is the predilection to crowd too much descriptive characterization into a single word. In general, it is better to leave some of the less relevant details unexpressed than to create an unwieldy word.

Connective vowels join the components of most compounds smoothly and euphoniously. In Greek the connective *o* is almost universal; but sometimes *e* is used, as in *staphylephoros* for *staphylophoros,* grape-bearing. This double usage should be avoided because it leads to the coinage of duplicates, identical in meaning and differing by only one letter. In Latin the connective is *i,* but for euphony may sometimes be *o,* as in *Urceolabrum* (a mollusk). Fourth declension nouns in *-us* sometimes retain *u,* as in *manumitto,* release, but most change to *i,* as in *domiporta,* a snail, and *fructifer,* fruit-bearing. The genitive ending *-ae* of Latin first declension nouns, in simulation of *aquaeductus,* aqueduct, has been inadvisedly retained as a diphthongal connective in such New Latin words as *rosaeflora, acaciaefolia, algaecide,* and *tabulaemontana.* These would be simpler as *rosiflora, acaciifolia, algicide,* and *tabulimontana,* and use of the connective *ae* is, therefore, not recommended. In hybrid compounds, if the first component is Latin and final component is Greek, the best connective usually is *i,* as in *colorimetry,* but *o* may be better in *aureomycin, cellophane,* and *Merrilliopeltis.* If the first component is Greek, *o* is the preferable connective, as in *electromotive* and *thermocouple.* If the first component is of unknown ancestry and gender, with consequent doubt as to its combining form, nice judgment will have to be used in selecting the best connective.

Omission of the connective in compounding is sometimes justified by the requirements of euphony, particularly in those awkward instances caused by the juxtaposition of vowels, necessitating the elision of one, usually that at the end of the first component, as in *kephalalgia* (*kephale, algia*), headache; *choragos* (*choros, agos*), chorus-leader; *Streptelasma* (*streptos, elasma*), a coral; and *canadensis, coloradensis, iowensis.* For other conditions and examples, see page 39. Sometimes, however, both vowels are retained, as in *perianthes,* surrounded by flowers; and in compounds whose first components end with *-ius, -ia,* or *-ium,* giving the genitive endings *-ii* and *-iae,* as in *Aphidiinae* (*Aphidius,* a wasp), *Tiphiidae* (*Tiphia,* a wasp).

Elision, addition, or assimilation of letters for the sake of euphony may be indicated under other circumstances, as in *ampheliktos* (*amphi, heliktos*), coiled round; *duplex* (*duo, plico*), twofold; *ephippion* (*epi, hippos*), saddle; *kynosoura* (*kyno-, oura*), dog's-tail; *myosotis* (*myo-, otis*), mouse-ear; *quadrangulus* (*quadri-, angulus*), four-cornered; *tredecim* (*tres, decem*), thirteen; and in the combining forms of neuter nouns like *anthos, antheos*, flower (*antho-*, not *antheo-*); *belos, beleos*, dart (*belo-*, not *beleo-*); and *oros, oreos*, mountain (*oro-*, not *oreo-*).

Modification of existing stems by attachment of prefixes and suffixes. This process, sometimes called derivation, though perhaps as spontaneous as compounding, is more complex in practice because the joining of affixes to root-stems without using a connective vowel necessitates not only a choice of desirable word-elements but also a sensitive ear for their skillful amalgamation according to recognized phonetic principles. By this method nouns, adjectives, and verbs may be derived from one another, resulting in new words having new stems. Thus, L. *arena*, sand, with stem *aren-*, plus *-ula*, gives the diminutive *arenula*, small grain of sand; with *-aceus, -arius, -atus, -osus*, the adjectives *arenaceus, arenarius, arenatus*, and *arenosus*, sandy; and with *-arium*, the noun *arenarium*, sandpit. Gr. *morphe*, form, with stem *morph-*, plus *-idion* and *-ion*, gives the diminutives *morphidion* and *morphion*, little form; with *-eus*, *Morpheus*, god of dreams, that is, fanciful forms; with *-eides, morpheides*, formlike; with *-sis, morphosis*, a shaping or forming; with *-ikos, morphotikos*, pertaining to forming; with *-tron, morphotron*, a tool for shaping; with *ana-, anamorphosis*, a forming anew; and with *meta-, metamorphosis*, a transformation. In English, by attaching the verbal suffix *-ize* to the stem of *economy*, we get *economize;* with *-ical, economical;* and with *-er, economizer*. One of the signs of an evolving language and enlarging vocabulary under new or rapidly changing environmental conditions, is the widespread and diversified practice of this method of word-composition.

The coining of commercial and other terms connected with the hurly-burly of daily living is fundamentally the same in principle as the creation of scientific terms, but the technique is generally more naive and nonchalant, greater liberties being taken with the verbal proprieties. Mimicry and imagery play conspicuous parts in the process, as in the invention of *slanguage, scanties, squawkies*, and *telebrities*. Old names may be transferred to new ideas, as *angstrom*, from A. J. Angstrom; *diesel*, from the inventor, Rudolf Diesel; *volt*, from Alessandro Volta; *calico*, from Calicut, India; *denim*, from serge de Nimes, France; *hackney*, from Hackney, England; *frankfurter*, from Frankfurt, Germany; and *sandwich*, from John Montagu, Earl of Sandwich, who ate meat placed between slices of bread to save time at gambling. The suffix *-ician*, meaning agent or practitioner, as in *physician* and *technician*, may, by analogy, be wedded to all sorts of stems, as *beautician, bootician, mortician*, etc. In a restaurant for a snack you may be offered a choice of *-burger's*, such as *beefburger, cheeseburger, fishburger*, and *porkburger*, whose ancestor was *hamburger*, originally made in Hamburg, Germany. *Cafeteria* has spawned many other *-teria's*, including *groceteria*, a grocery, and *sudsateria*, a laundry. The list of *-ade's* as *lemonade, orangeade;* *-eer's*, as *profiteer, racketeer;* *-ee's*, as *draftee, standee;* *-icle's*, as *icicle, popsicle, sapsicle;* *-ism's*, as *agnosticism, Bolshevism, bowdlerism, conservatism, Fletcherism, liberalism, pragmatism;* *-phobia's*, as *acrophobia, bacteriophobia, claustrophobia, hydrophobia, rhypophobia, Russophobia;* *-thon's*, as *marathon, talkathon, telethon, walkathon;* *near-'s*, as *near-beer, near-miss, near-silk;* and *super-'s*, as *super-colossal, super-duper, super-market*, grows daily. The patent medicine, grocery, hardware, and other shelves display a wide variety of plain and fancy word-coinage, of which *Cuticura, Freezone, Pepsodent, Cocomalt, Crisco, Jell-well, Ry-Krisp, Calavo, Linoleum, Nylon, Dacron, Kleenex, Kotex*, and *Waxtex* are samples.

The following list of Greek and Latin affixes used in composition is intended only as a sample and few anglicized affixes are cited. For full lists with many illustrations see the entries under the appropriate keywords in the LEXICON.

Prefixes

Greek and Latin adjectival, adverbial, prepositional, pronominal, and intensive prefixes, with more or less definite meanings, combine with other word elements to make new words. These comprise a large proportion of the classical vocabularies. In a number of instances prefixes have been modified by elision and assimilation of letters, according to euphonic requirements. For example, L. *ad*, to, becomes *a* in *ascendo* and *aspiro*, and the *d* becomes the corresponding consonant before *b, c, f, g, l, n, n, p, r, s*, and *t;* L. *cum*, with, becomes *co* in *coerceo* and *cooperator*, and the *m* becomes the corresponding consonant before *l, n*, and *r;* and the *n* of Gr. *syn*, with, becomes *l, m, r*, and *s* before those consonants. Other examples will be found among the prefixes listed in the LEXICON.

Adjectival prefixes can be used in their combining forms, as illustrated in the discussion of adjectives in the section on NATURE OF GREEK AND LATIN. Most numerals, however, are combined in their entirety or after euphonic elision or substitution of some letters, as *decemplex*, tenfold; *duodecim*, twelve; *November*, ninth Roman month; *novendecim*, nineteen; *octipes*, eight-footed; *octojugis*, eight-yoked; *quadrimanus*, four-handed; *quadrupes*, four-footed; *quattuordecim*, fourteen; *quindecim*, fifteen; *quinquefolius*, five-leaved; *septemplex*, sevenfold; *septendecim*, seventeen; *sexangulus*, hexagonal; *tredecim*, thirteen; *trifolium*, trefoil; *dekamphoros*, holding ten amphorae; *dekastylos*, with ten columns; *dyotokeo*, produce two; *enneapsychos*, nine-lived; *heptakephalos*, seven-headed; *hexapleuros*, six-sided; *hexetes*, of six years; *octohedros*, eight-sided; *octopous*, eight-footed; *pentagrammos*, five-lined; *pentepous*, five-footed; *tessaragonos;* *tetragonos*, four-angled; *triphyllos*, three-leaved.

Adverbial and prepositional prefixes denoting such space and time relations as are implied in the words after, against, apart, around, back, before, between, down, from, in, near, not, out, over, through, to, under, up, with, and without, include the following, which are exemplified under the appropriate references in the LEXICON: *ab-, ad-, ambi-, amphi-*, *an-, ana-, anchi-, ante-, anti-, apo-, circum-, cis-, contra-, cum-, de-, dia-, dis-, ek-, en-, eso-, ex-, hama-, hyper-, hypo-, in-, infra-, inter-, intra-, juxta-, kata-, meta-, ne-, non-, ob-, palin-, para-, pene-, per-, pera-, peri-, post-, pre-, preter-, pro-, re-, retro-, se-, sine-, sub-, super-, syn-, trans-.*

Pronominal prefixes include the following: *auto-, ego-, ipse-, qui-, sui-.*

Intensive prefixes meaning very and exceedingly, include the following: *aexi-, aga-, ari-, bary-, bathy-, bu-, con-, da-, de-, dia-, dis-, dys-, ek-, em-, en-, eri-, eu-, ex-, in-, kata-, la-, mala-, mega-, per-, peri-, poly-, pre-, re-, red-, ve-, za-.*

Suffixes

Almost any variation or play upon an idea can be incorporated in a word by attaching the proper suffix to its base. This process is exemplified in the derivation of terms denoting the following classified categories. See the indicated references in the Lexicon for examples of the method of joining a given suffix to an operating word base.

Action and agency: *-arius, -cida, -eia, -eus, -fex, -fid, -ize, -lyze, -or, -sis, -sor, -sy, -teira, -tes, -ter, -tis, -tor, -tria, -tris, -trix.*

Classification and order: *-aceae, -ales, -ata, -idae, -ina, -oideae.*

Commemoratives and dedicatives: *-ae, -i, -ia, -ius, -ium.*

Diminutives: *-arion, -aster, -culus, -ellus, -essus, -idion, -illus, -ina, -ion, -iskos, -ister, -kin, -ling, -olus, -ulus, -unculus, -ydrion, -yllion.*

Diseases: *-algia, -iasis, -itis, -oma, -osis, -pathy.*

Fullness and abundance: *-bundus, -cundus, -lens, -lentus, -osus, -undus.*

Incipiency: *-escens.*

Intensives and magnifiers: *-ior, -issimus, -istos, -ius, -teros, -tatos.* Although Latin and Greek employ numerous prefixes to make intensives, augmentatives, and the opposite of diminutives, there are no comparable, useful suffixes that can be added to noun bases; but comparative and superlative suffixes are available for adjectives and adverbs. If, therefore, intensive suffixes are needed to make new nouns, they should be invented. Herewith are some suggestions.

LATIN. Dismember *magnus*, large, leaving *-gus*. This suffix, in the appropriate *-us, -a, -um* gender ending, according with that of the parent noun, could be attached as follows to the operating bases of nouns by means of the regular connective *i*:

macula, f. spot, *maculiga*, large spot; *porta*, f. gate, *portiga; ventricola*, m. glutton, *ventricoliga; vespa*, f. wasp, *vespiga.*

amicus, m. friend, *amicigus; bellum*, n. war, *belligum; filum*, n. thread, *filigum; liber*, m. book, *librigus; vir*, m. man, *virigus.*

corpus, n. body, *corporigum; flos*, m. flower, *florigus; navis*, f. ship, *naviga; pes*, m. *pedigus.*

cornus, f. horn, *corniga; genu*, n. knee, *genuga; lacus*, m. lake, *lacuga; quercus*, f. oak, *querciga, quercuga.*

facies, f. kind, *faciga; res,* f. thing, *rega; spes,* f. hope, *spega.*

GREEK.   Dismember *megas,* large, change *a* to *e,* leaving *-ges.* This suffix would be spelled the same in all genders and could be attached to any stem by means of the regular connective *o:*

chora, f. country, *choroges,* large country; *pelte,* f. shield, *peltoges; polites,* m. citizen, *politoges; tamias,* m. steward, *tamioges.*

anthropos, m. man, *anthropoges; dendron,* n. tree, *dendroges; osteon,* n. bone, *osteoges; therion,* n. beast, *therioges.*

aner, m. man, *androges; gigas,* m. giant, *gigantoges; keras,* n. horn, *keratoges; lampas,* f. torch, *lampadoges; pyr,* n. fire, *pyroges; stoma,* n. mouth, *stomatoges; xiphos,* n. sword, *xiphoges.*

Likeness: *-esque, -ish, -ite, -odes, -oid, -oides, -opsis.*

Patronymics: *-ades, -as, -ides, -is.*

Place for or where: *-arium, -orium, -ensis, -ester, -etum, -ile, -terion, -tre, -tris.*

Possession, belonging to, pertaining to, having the nature of, made of, quality of, state or condition of: *-aceus, -acus, -ados, -ae, -ago, -aleos, -alis, -ans, -antia, -anus, -arius, -arum, -aster, -atus, -bilis, -bus, -clis, -ens, -enus, -erum, -etus, -eus, -i, -ia, -icus, -idus, -igo, -ilis, -imus, -inus, -is, -ismus, -ister, -ite, -itus, -ium, -ius, -ivus, -mone, -ness, -orius, -orum, -otes, -otus, -ship, -syna, -tudo, -tus, -ty, -ugo, -us.*

Result of action: *-ion, -ismus, -ment, -monia, -sion, -sura, -tion, -tura, -ura.*

Time: *-ernus, -urnus.*

Tools and means: *-brum, -bulum, -culum, -crum, -mentum, -ter, -tra, -tron, -trum.*

Many of the foregoing suffixes were recombined in transforming the parts of speech, as shown by the following examples:

Adjectives from adjectives.

acerbus, sour, < *acer,* sharp; *-bus*
altissimus, highest < *altus,* high; *-issimus*
breviculus, somewhat short, < *brevis,* short; *-culus*
flaccidus, languid, < *flaccus,* flabby; *-idus*
sextus, sixth, < *sex,* six; *-tus*

melantatos, blackest, < *melas, melanos,* black; *-tatos (-tatus)*
melanteros, blacker, < *melas, melanos,* black; *-teros (-terus)*
sophikos, of wisdom, < *sophos,* wise; *-ikos (-icus)*
trachodes, of rough nature, < *trachys,* rough; *-odes*

Adjectives from nouns.

argenteus, of silver, < *argentum,* silver; *-eus*
Romanus, Roman, < *Roma,* Rome; *-anus*
barbatus, bearded, < *barba,* beard; *-atus*
campester, of fields, < *campus,* plain; *-ester*
censorius, critical, < *censor,* critic; *-orius*
-collis, necked (as in *crassicollis*), < *collum,* neck; *-is, -e*
-cornis, horned (as in *unicornis*), < *cornu,* horn; *-is, -e*
corpulentus, fleshy, < *corpus,* body; *-lentus*
cretaceus, chalky, < *creta,* chalk; *-aceus*
femininus, of women, < *femina,* woman; *-inus*
festivus, joyful, < *festum,* feast; *-ivus*
-florus, flowered (as in *multiflorus*), < *flos, floris,* flower; *-us, -a, -um*

-folius, leaved (as in *latifolius*), < *folium,* leaf; *-us, -a, -um*
-formis, shaped (as in *oviformis*), < *forma,* shape; *-is, -e*
fullonicus, of fullers, < *fullo, -onis,* fuller; *-icus*
generalis, general, < *genus, generis,* race; *-alis*
hibernus, of winter, < *hibernum,* winter; *-ernus*
nefarius, wicked, < *nefas,* something wicked; *-arius*
pastorius, of shepherds, < *pastor,* shepherd; *-ius*
pestilens, unwholesome, < *pestis,* plague; *-lens*
pratensis, of meadows, < *pratum,* meadow; *-ensis*
silvanus, of the woods, < *silva,* forest; *-anus*
-vius, wayed (as in *multivius*), < *via,* way; *-us, -a, -um*

dendroeides, treelike, dendroid, < *dendron,* tree; *-eides (-oides)*
erythrogrammos, red-lined, < *erythros,* red, *gramma,* line; *-os (-us)*
glossos, chattering, < *glossa,* tongue; *-os*
hydrodes, watery, < *hydor, hydro-,* water; *-odes*
kardiakos, cardiac, < *kardia,* heart; *-akos (-acus)*
kephalotos, headed, < *kephale,* head; *-tos (-tus)*
lithikos, of stone, < *lithos,* stone; *-ikos (-icus)*
lithinos, of stone, < *lithos,* stone; *-inos (-inus)*
melanommatos, black-eyed, < *melas, melanos,* black, *omma,* eye; *-os (-us)*

Adjectives from verbs.

creatus, made, < *creo,* make; *-atus*
dubius, doubtful, < *duhibeo,* hold as two; *-ius*
finitus, finite, < *finio,* limit; *-itus*
flexilis, pliant, < *flecto,* bend; *-ilis*
nigricans, blackish, < *nigrico,* be black; *-ans*
relativus, referred, < *refero,* refer; *-ivus*
rigidus, stiff, < *rigeo,* stiffen; *-idus*
rotundus, round, < *roto,* revolve; *-undus*
variabilis, variable, < *vario,* change; *-bilis*
virescens, becoming green, < *vireo,* be green; *-escens*

gnorimos, well-known, < *gnorizo,* make known; *-imos*
lethodes, forgetful, < *lanthano,* forget; *-odes*
praktikos, practical, < *prasso,* accomplish; *-ikos*
tlemon, patient, < *tlao,* bear; *-mon*

Nouns from adjectives.

avaritia, avarice, < *avarus,* greedy; *-tia*
gratitudo, thankfulness, < *gratus,* agreeable; *-tudo*
loquacitas, volubility, < *loquax,* talkative; *-tas*
rubor, redness, < *ruber,* red; *-or*

glykismos, sweetness, < *glykys,* sweet; *-ismos*
hiereus, priest, < *hieros,* holy; *-eus*
melanotes, blackness, < *melas, melanos,* black; *-otes*
sklerosis, a hardening, < *skleros,* hard; *-osis*
sophia, wisdom, < *sophos,* wise; *-ia*
sophrosyne, moderation, < *sophron,* sensible; *-syne*

Nouns from nouns.

balteolus, little girdle, < *balteus,* girdle; *-olus*
candelabrum, candlestick, < *candela,* candle; *-brum*
carnifex, killer, < *caro, carnis,* flesh; *-fex*
codicillus, little note, < *codex, codicis,* a writing; *-illus*
collegium, college, < *collega,* colleague; *-ium*
cornuarius, horn-maker, < *cornu,* horn; *-arius*
hastula, little spear, < *hasta,* spear; *-ula*
herbarium, place for plants, < *herba,* plant; *-arium*
lanugo, down, < *lana,* wool; *-ugo*
libellus, little book, < *liber,* book; *-ellus*
lumbago, pain in the loin, < *lumbus,* loin; *-ago*
matrimonium, marriage, < *mater,* mother; *-monium*
militia, military service, < *miles, militis,* soldier; *-ia*
offella, little bit, < *offa,* bit; *-ella*
oleaster, wild olive, < *olea,* olive; *-aster*
pinetum, place for pines, < *pinus,* pine; *-etum*
portiuncula, little part, < *portio,* share; *-uncula*

vasculum, little vessel, < vas, vessel; -culum
viculus, little village, < vicus, village; -ulus

anemone, windflower, < anemos, wind; -mone
Boreades, son of Boreas, < Boreas, Boreas; -ades
dendrosis, a becoming a tree, < dendron, tree; -osis
Dryas, tree-nymph, < drys, tree; -as
elephantiasis, a disease, < elephas, elephant; -iasis
gnomidion, little opinion, < gnome, opinion; -idion
hippeus, horseman, < hippos, horse; -eus
hoplites, armed soldier, < hoplon, armor; -ites
Kekropides, son of Kekrops, < Kekrops, Cecrops; -ides
kometes, villager, < kome, village; -tes
obeliskos, little pillar, < obelos, pointed pillar; -iskos
odontalgia, toothache, < odous, odontos, tooth; -algia
peiraterion, pirates' nest, < peira, attack; -terion
pissarion, a little pitch, < pissa, pitch; -arion
rhizion, rootlet, < rhiza, root; -ion
sarkoma, a tumor, < sarx, sarkos, flesh; -oma
xenosyne, hospitality, < xenos, guest; -syne

Nouns from verbs.

alimonia, support, < alo, feed; -monia
aratrum, plow, < aro, plow; -trum
auditorium, place for hearing, < audio, hear; -orium
differentia, difference, < differo, separate; -entia
instrumentum, tool, < instruo, equip; -mentum
orator, male speaker, < oro, speak; -tor
oratrix, female speaker, < oro, speak; -trix
pictura, painting, < pingo, pictus, paint; -ura
revisio, revision, < revideo, see again; -sio
sepulcrum, tomb, < sepelio, bury; -crum
stabulum, stall, < sto, stand; -bulum
terebra, auger, < tero, bore; -bra
timor, fear, < timeo, fear; -or
tonsor, barber, < tondeo, tonsus, shear; -or
vehiculum, wagon, < veho, carry; -culum
vertigo, giddiness, < verto, turn; -igo
vocatio, vocation, < voco, call; -tio

bombetes, hummer, < bombeo, hum; -tes
dynastis, female ruler, < dynasteuo, rule; -tis
dynastor, male ruler, < dynasteuo, rule; -tor
ichneuteira, huntress, < ichneuo, track out; -teira
kathartria, cleanser, < kathairo, cleanse; -tria
kentron, spur, < kenteo, sting; -tron (-trum)
logistes, calculator, < logizomai, compute; -istes
louter, bathtub, < louo, wash; -ter
rhyposis, pollution, < rhypoo, befoul; -osis
sympatheia, sympathy, < sympatheuo, feel with; -ia
teretismos, a whistling, < teretizo, whistle; -ismos (-ismus)
thymiatris, censer, < thymiao, burn incense; -tris

Verbs from adjectives.

albo, whiten, < albus, white
clareo, be clear, < clarus, clear
laeto, delight, < laetus, joyful
rubeo, be red, < ruber, red

aletheuo, speak the truth, < alethes, true
dynateo, be mighty, < dynatos, strong
hedyno, sweeten, < hedys, sweet
kathairo, purge, < katharos, pure
melaino, blacken, < melas, melanos, black

Verbs from nouns.

bullio, bubble, boil, < bulla, bubble
finio, limit, < finis, end
metuo, fear, < metus, fear
stimulo, incite, < stimulus, goad

antheo, blossom, < *anthos,* flower
ostrakizo, banish, < *ostrakon,* shell
semaino, indicate, < *sema,* sign
timao, honor, < *time,* honor

Verbs from verbs.

cantillo, hum, < *canto,* sing
facesso, do eagerly, < *facio,* do
iacto, hurl, < *iacio,* throw
rubesco, turn red, < *rubeo,* be red

didasko, teach, < *dao,* teach
gelaseio, desire to laugh, < *gelao,* laugh
klausiao, wish to weep, < *klaio,* weep

As may be inferred from the examples cited, no satisfactory formula can be given for the derivation of verbs from other parts of speech. Concerning the derivation of terms from other parts of speech not thus far considered but of some interest to wordcoiners, a few of the scarce items may suffice as illustrations. The adjective *modernus,* modern, is from the adverb *modo,* just now; and the adjective *superbus,* excellent, is from the preposition *super,* above. The English noun *nostrum,* our own, is from the Latin pronoun *noster,* our; and the English noun *autocrat* and verb *autograph* are combinations using the Greek pronoun *autos,* self. For examples showing how English words are derived from one another by use of affixes, see the appropriate entries in the LEXICON.

Having illustrated the chief materials and shown the methods whereby they have been used previously in compounding by composition, I shall complete this discussion by giving deductive applications in the invention of new words not now found in the dictionaries. Let it be assumed that a New Latin word for 'little sea' is desired. Reference to the dictionary shows no diminutive for *mare, -is,* n. sea, but in the list of nouns with diminutives under **little** occurs the similar *rete,* net, with its diminutive *reticulum.* Analogy, therefore, indicates *mariculum* for 'little sea'. Similarly, as there is no listed diminutive for the Gr. *thalassa,* sea, after the analogy with *glossa,* tongue, *thalassarion, thalassidion,* and *thalassion* are indicated; and for *kardia,* heart, *kardidion, kardion,* and *kardiskos,* analogous to the diminutives of *oikia,* house.

Suppose an adjective meaning 'of a lake or pertaining to lakes' is wanted. In the list of suffixes *-alis* meets the requirements. Attachment of this suffix to the stem of *lacus, -us,* m. lake, gives *lacualis,* of a lake, similar to *manualis,* an already existing example.

The Latin first declension nouns in *-cola,* dweller in, < *colo,* cultivate, like *agricola, limicola, monticola,* and *ruricola,* have instigated the New Latin adjectival endings *-colus, -a, -um,* dwelling in; and the neuter *folium,* leaf, like the feminine *flora* in *multiflorus,* many-flowered, has given *-folius, -a, -um,* leaved. Similarly, the suffix *-cida,* killer, develops *-cidus, -a, -um,* killing. These are natural evolutionary derivations that confirm the modern wordsmith's right to invent words or their parts, such as affixes, to meet special needs, so long as he observes the basic principles and processes.

ARBITRARY CREATION OF WORDS

This procedure might be called creation or invention by fancy, as the materials it uses are entirely, or almost entirely, imaginary, with little or no obvious dependence on antecedent syllables identifiable as definite roots or stems. Here belong the Lilliputian and Brobdingnagian terms of Swift's *Gulliver's travels*, the whimsical names for some items of the flora and fauna encountered by Alice in Wonderland, some trade and brand names like *Kodak* and *Pyrex*, onomatopoeic words like *buzz* and *chewink*, and slang terms like *heebie jeebies* and *palooka*. The generic term in the name of the pintail-duck (*Dafila acuta*) is said to be without ancestry. Similarly, many others, some undoubtedly legitimate and some fanciful, are irritating puzzles because their creators did not see fit to supply derivations and clarify origins.

Metathesis, or the transposition of the letters of an existing word, results in an anagram. Many of these have been invented by biologists to serve as generic terms for plant and animal species when the supply of antecedent material seemed low or insufficient. Such are: *Anogra* < *Onagra; Dacelo* < *Alcedo; Delichon* < *Chelidon; Gifola* < *Filago; Ixerba* < *Brexia; Muilla* < *Allium; Parosela* < *Psoralea; Tellima* < *Mitella; Teonoma* < *Neotoma; Trilisa* < *Liatris*. Spelled backward a word may take on a new look as in the slogan, "Don't be a guberif. Keep Idaho green".

Objection to the creation of words without benefit of orthodox derivation has greatly diminished in recent years with the decline in classical instruction. Apparently the trend is parallel with the pragmatic Shakespearean conclusion that "a rose by any other name would smell as sweet". Thus, also, it is commonly understood that the technical names of plants, animals, minerals, and other objects of scientific study, are merely convenient, verbal handles for labeling purposes. Why, then, be fussy about their creation? Why be concerned as to whether or not they are apt, legitimately derived, and appropriately spelled? To ask these questions is to answer them by putting another: Does the Shakespearean realist not demand that handles, wherever attached, be of good material, well constructed, and harmonious with their best use?

STOP! LOOK! LISTEN! SAFETY FIRST!

Having coined a new word or adapted an old one to a new purpose, ask yourself these questions before rushing your brain-child to a baptism of printer's ink:

1. Is the word really necessary? "Therefore doth Job open his mouth in vain; he multiplieth words without knowledge." (Job 35:16)

2. Are the combined elements pure and euphonious, not hybrid and awkward?

3. Is the connective vowel necessary, proper, and appropriate?

4. Is the transliteration, if necessary, correct?

5. How about the spelling and gender, especially in the case of binomials?

6. Is the word already in the dictionaries or preoccupied in the taxonomic catalogues? Examination of taxonomic papers and catalogues discloses the sad fact that many well-intentioned names are now buried in the nomenclatural graveyard under the epitaph 'Once a synonym always a synonym'. Although some of these words are misconstructed and deserve their fate, others are correct and distinctive. These arouse feelings of loss and indispensability. Is there no way of resurrecting them to a new incarnation of liveliness and usefulness without their former implications?

7. Has the etymology of the word been stated clearly?

GENDER

Gender is a particular concern of the biologist who is expected to abide by International Rules of Nomenclature that prescribe the requirements in creating valid binomials as scientific names for plants and animals. A fundamental statement pertaining to such names is that they must be Latin or latinized. This requirement automatically prompts a corollary, namely, that the two terms of the binomial be adjusted to one another according to the precedents of classical grammar. Furthermore, as the generic term of a binomial is a substantive it must have gender. The question, Can binomial nomenclature get along without gender or the strict observance of gender?, is not deemed pertinent at this juncture, but a review of the history and applications of gender may be enlightening.

Very early in the development of Greek, Latin, and related languages, nouns were inflected to signify gender, but these inflections did not necessarily indicate differences in sex in all instances, as is illustrated by the fact that Latin *verpa*, penis, is feminine; *cunnus*, vulva, is masculine; and *veretrum*, private parts, is neuter. Similar examples occur in Greek. Two chief theories have been elaborated to explain the origin of gender.[2] The first suggests that primitive man personified everything, attributing action, strength, large size, and sometimes frightfulness, to masculinity; and passiveness, delicacy, fertility, and attractiveness to femininity. The second rejects this poetic view and considers that early man, influenced by the hard and prosaic facts of his life, was dominantly extroverted and concerned with concrete realities, and only in his reflective moments recognized the existence of abstract qualities, which he named by inflection of the words for the associated concrete objects. Thus, many nouns of masculine or feminine gender, but having no suggestion of sex, may represent adaptations and abstractions according to analogies and comparisons with other nouns actually denoting sex. For example, many Latin -*us* names for plants are feminine, perhaps because they were formed by analogy with an original -*us* name for some object of the female sex, or because many plants, especially trees, were supposed to incorporate dryads or wood-nymphs.

The origin of the neuter gender for designating sexless objects is also obscure. As in the masculine and feminine genders, inconsistencies

---

[2] Brugmann, Karl (translation by Edmund Y. Robbins). *The nature and origin of the noun genders in the Indo-European languages.* 32 pp. Charles Scribners Sons, 1897.

occur, for many neuter nouns, such as the Greek diminutives, *paidion*, little child; *gynaion*, little woman; and *tauridion*, little bull, represent objects with sex. Probably, once the Greek neuter ending *-ion*, signifying a diminutive, was invented, it was thereafter applied to most diminutives regardless of sex.

Latin dictionaries indicate genders of nouns by the abbreviations m., f., n., and c.; and of adjectives by the terminations of their nominative singular forms. Thus, adjectives of three terminations, all different, are *bonus*, *-a*, *-um*, good; *niger*, *-gra*, *-grum*, black; *celer*, *-is*, *-e*, swift; *celeber*, *-bris*, *-bre*, famous; with two terminations, the masculine and feminine being the same, the neuter different, *brevis*, *-e*, short; *levis*, *-e*, light; and with one termination, all genders the same, *atrox*, fierce; *ferox*, wild.

Greek dictionaries indicate genders of nouns by the article ὁ (ho), m.; ἡ (he), f.; το (to), n.; and ὁ, ἡ, c. The plural forms of the article are respectively οἱ, αἱ and τα. The gender of adjectives, as in Latin, is indicated by the diagnostic terminations of their nominative singular forms. Thus, adjectives of three terminations, all different, are *agathos*, *-e*, *-on*, good; *lampros*, *-a*, *-on*, bright; *pleos*, *-a*, *-on*, full; *glykys*, *-eia*, *-y*, sweet; *melas*, *-aina*, *-an*, black; *teren*, *-eina*, *-en*, tender; *charieis*, *-essa*, *-en*, graceful; with two terminations, the masculine and feminine being the same, the neuter different, *polylogos*, *-on*, talkative; *alethes*, *-es*, true; *odontoeides*, *-e*, toothlike; and with one termination for masculines and feminines, the neuters missing, *makrocheir*, long-handed; *megalenor*, self-confident. The ὁ and ἡ following the dictionary entry of such adjectives indicates this fact. Two-termination adjectives having *-os* endings in the masculine and feminine, and *-on* in the neuter, may, as explained on page 29, become three-termination Latin adjectives, as *boukeros* (L. *bucerus*, *-a*, *-um*), ox-horned, and *leukokomos* (L. *leucocomus*, *-a*, *-um*), white-haired. New Latinists sometimes use the several gender forms of these adjectives as nouns with such results as *kyklostomos* (*cyclostomus*), m., *kyklostoma* (*cyclostoma*), f., *kyklostomon* (*cyclostomum*), n., round-mouthed one; *lithophagos* (*lithophagus*), m., *lithophaga* (*lithophaga*), f. *lithophagon* (*lithophagum*), n. stone-eater. This practice causes an inconsistency in respect to gender when the last term of a compounded adjective is from a neuter noun ending with *-a*. *Cyclostoma* may thus pass as both a neuter noun and a feminine adjective. Sometimes the Greeks avoided this dilemma by using the long combining form of the noun in making the adjective, as *mikrommata*, instead of *mikromma*, small-eyed.

The classical grammars discuss the distribution of gender and record some general rules, but none without exceptions, as shown by the following:

Masculines include names of males, months, mountains, rivers, winds, and Greek diminutives in *-iskos*. Many Latin nouns in *-us* (Gr. *-os*) are masculine, but exceptions are *alvus*, f. belly; *acus*, *-eris*, n. chaff, husk; *acus*, *-us*, f. needle; *Aegyptus*, f. Egypt; *euonymus* (Gr. *euonymos*), f. spindle-tree; *manus*, f. hand; *Venus*, f. Venus; *virus*, n. poison; *vulgus*, n. people; Gr. *anthos*, n. flower; *gnathos*, f. jaw; *ichnos*, n. track.

Feminines include names of females, cities, countries, islands, gems, plants, many birds, conditions, qualities. Exceptions are *acanthus* (Gr. *akanthos*), m. acanthus; *burdunculus*, m. probably a kind of borage; *paliurus* (Gr. *paliouros*), m. a kind of buckthorn. Most Latin nouns in -*a* are feminine but exceptions are *accola*, c. neighbor; *advena*, m., f., n., visitor; *agricola*, m. farmer; *artopa*, m. baker; *auriga*, c. driver; *bubsequa*, m. shepherd; *cacula*, m. servant; *collega*, m. partner; *confuga*, c. refugee; *dumicola*, m. dweller in thickets; *hamiota*, m. angler; *idiota*, m. idiot; *indigena*, m. native; *Jura*, m. a mountain range; *larifuga*, m. vagabond; *legirupa*, m. lawbreaker; *madulsa*, m. a drunken man; *nauta; navita*, m. sailor; *sculna*, m. mediator; *ventricola*, m. glutton; *verna*, c. slave. Some, but not all Greek nouns in -*a* are feminine. The exceptions are many neuter nouns, as *bema*, step; *gramma*, letter; *nema*, thread; *soma*, body; *stigma*, mark; *stoma*, mouth. Most Greek nouns in -*e* are feminine, but the -*e* is generally transliterated -*a* in Latin. Many nouns in -*otes*, denoting quality, are feminine.

Neuters include most indeclinable nouns, Latin nouns in -*um*, Greek diminutives in -*ion*, many Greek nouns in -*a* and -*on*, but exceptions are the -ων nouns, as *chiton*, m. tunic; *kyon*, c. dog; *lakedon*, f. doctrine; *odon*, m. tooth; *pogon*, m. beard.

Some nouns may be masculine and feminine in gender, that is, have common gender, if the name includes both males and females, as L. *bos* (Gr. *bous*), ox, cow; L. *dux*, leader; L. *parens*, parent; and Gr. *phygas*, fugitive; but such nouns are generally considered masculine if no distinction in sex is intended. When nouns include both sexes but have only one gender they are said to be epicene, as L. *animal*, n. beast; L. *lepus*, m. hare; L. *vulpes*, f. fox; Gr. *alopex*, f. fox.

If only the masculine form of a word is given in the dictionary, what procedure can the wordcoiner follow to invent the feminine form? The feminine can be formed by attaching one of the designated suffixes to the stem of the masculine form according to analogy with precedents illustrated in the following diversified examples:

Latin. Attach -*a*, -*trix*: *amicus*, friend, *amica*; *buculus*, bullock, *bucula; deus*, god, *dea; hinnus*, mule, *hinna; leo*, lion, *lea; puer*, boy, *puera; ursus*, bear, *ursa; vir*, man, *vira; dictator*, ruler, *dictatrix; executor*, doer, *executrix; laudator*, praiser, *laudatrix; necator*, slayer, *necatrix*.

Greek. Attach -*aina*, -*eia*, -*iske*, -*issa*, -*teira*, -*tis*, -*tria*, -*tris*: *hys*, hog, *hyaina; lykos*, wolf, *lykaina; sys*, hog, *syaina; tragos*, goat, *tragaina; basileus*, king, *basileia, basilis, basilissa; hiereus*, priest, *hiereia; kyniskos*, puppy, *kyniske; paidiskos*, boy, *paidiske; doter*, giver, *doiteira; geneter*, begetter, *geneteira; lobeter*, slanderer, *lobeteira; soter*, savior, *soteira, sostria; demotes*, citizen, *demotis; dynastes*, ruler, *dynastis; eranistes*, contributor, *eranistis; oiketes*, servant, *oiketis; prophetes*, prophet, *prophetis; erastes*, lover, *erastria; mathetes*, pupil, *mathetria, mathetris; orchester*, dancer, *orchestria, orchestris; poietes*, poet, *poietria; rhetor*, speaker, *rheteira, rhetria*.

From the foregoing analysis it should be clear that for some nouns and particularly for some adjectives inspection of their terminations alone is not a reliable guide to their genders. Thus, the Latin two-termination adjective *fragilis* in the name of the willow, *Salix fragilis*, does not indicate the gender of *Salix* except that it may be masculine or feminine but not neuter. Similarly, the one-termination *ferox*, used as a specific term, may indicate any of the three genders. Greek two-termination and one-termination adjectives are likewise confusing, with an additional complicating feature, namely, that some terminations, as -ης and -ες, although different and distinctive in Greek, are transliterated -es in Latin.

In compound words, according to ancient and modern usage, the final component, if a noun, determines the gender. If the first component is a noun and the final component an adjective, the compound is an adjective but may be intended for use as a noun. In such instances the gender of the compound is that of the governing noun. Here, for example, belong the many substantives having *-oides* and *-opsis* as adjectival terminations. If both components of a compound are adjectival and the compound is the generic term of a binomial, its gender may remain dubious unless clearly indicated by the author. Such words need not and should not be created for use as generic terms. They may be avoided by transformation of the adjectival terminations into noun forms whose gender is definite. In making noun compounds having adjectival terminations in *-oideus, -oidea, -oideum,* and the like, the wordcoiner should be careful to adjust the adjectival ending to the gender of the governing noun. If a new generic term is an outright invention it takes the gender indicated by the author, and the author is, therefore, under moral obligation to so indicate. Analogy and precedent may help to determine the gender of such words but difficult choices are common.

Other pitfalls involving gender, aside from boners such as "The epistles were the wives of the apostles", include the possibility of supposing that a new word like *habryntis* is a misspelling because the Greek dictionary gives only *habryntes*, m. fop. The former, however, is the feminine form. The termination *-on* of Greek nouns is so commonly indicative of the neuter gender that many other nouns in *-on* (-ων), which are masculine or feminine, are mistakenly used as neuters, as *kodon*, m., bell, in *Platycodon grandiflorum* (balloon-flower). The Greek *ops*, eye, is listed first as feminine then as masculine, but *Kekrops*, Cecrops, mythical king of Athens; *Pelops*, son of Tantalus; and *konops*, gnat, are masculine. Such instances of common gender place the coiner of generic terms in dilemmas from which there is no escape except by arbitrary choice according to preponderance of usage.

Because of the obvious diversities and apparent difficulties in regard to gender, some students have considered methods for simplifying and standardizing procedures.[3] The basic suggestion in tentative proposals

---

[3] Blackwelder, R. E. *The gender of scientific names in zoology.* Washington Acad. Sci. Jour., vol. 31, pp. 135-140, 1941.

Emiliani, Cesare. *Nomenclature and grammar.* Washington Acad. Sci. Jour., vol. 42, pp. 137-144, 1952.

Grensted, L. W. *The formation and gender of generic names.* Entomologist's Monthly Mag., vol. 80, pp. 229-233, 1944.

McAtee, W. L. *Automatic nomenclature.* Entom. Soc. Washington Proc., vol. 30, pp. 72-76, 1928.

Richter, Rudolph. *Ist eine unveränderliche Form des Art-Namens möglich?* Senckenbergiana, Bd. 25, pp. 340-356, 1942.

is the arbitrary assignment of gender according to the endings of words, with a minimum deviation from classical precedent. Although such proposals are actuated by good intentions, reluctance to accept them is engendered by inability to reconcile them with well-established, classical practice, as well as by doubts that they would result in the simplicity and uniformity desired. If latinized binomial nomenclature is to apply in botany and zoology, and if the old proverb, What is sauce for the goose is also sauce for the gander, is still apt, should not the same grammatical rules prevail? Were zoologists, for example, to adopt the suggestion that all -us nouns be considered masculine, botanists would be very unhappy because most plant names in -us are primordially feminine. What complicates this particular phase of the problem is the fact that botanists and zoologists have used many of the same names for both plants and animals—an unnecessary and regrettable practice. Thus, *Ficus* is not only the generic term for the fig but also the generic term for a fig-shaped gastropod; *Corydalis* is a fumewort and a hellgrammite; *Crucibulum* is a puffball and a gastropod; *Hystrix* is a grass and a porcupine; *Passerina* is a thymeleacead and a finch; *Pieris* is a heath and a butterfly; and *Scleroderma* is both a puffball and a wasp.

Perhaps two of the underlying causes for some of the uncertainties in practice with respect to gender can be removed gradually without doing violence to classical rules. These are (1) the presence in nomenclature of barbaric or fanciful generic terms whose gender as well as ancestry is unknown or ambiguous, and (2) the lack of a systematic, succinct, and well-exemplified compilation of the Greek and Latin procedures in one authoritative publication. The first can be eliminated if there be incorporated in the International Rules a requirement to the effect that henceforth no newly-coined or adopted generic term shall be considered valid unless the author cites its etymology, if any, and indicates its gender. The second may be met by some such compilation of materials and methods as is attempted in this book, and by convincing non-classically trained biologists that study of classical practice, mixed with judgment, has its rewards.

Many persons are properly disturbed by the inconsistencies that descend from author to author and from catalogue to catalogue without being corrected. Thus, L. *euonymus* (Gr. *euonymos*), spindle-tree, is feminine, but Linnaeus instituted the genus *Euonymus* as masculine with specific terms in agreement. The Greek *anthos*, flower, is neuter, but in such combinations as *Helianthus*, also instituted by Linnaeus, and *Cyclanthus*, by Poiteau, it is treated as masculine. If gender is to be retained in binomial nomenclature, such departures from clear classical usage through an author's whim or inadvertence, must not be countenanced. Consistency is a jewel that can adorn nomenclature as well as other procedures.

Spelling and Pronunciation

Spelling and pronunciation are closely interwoven aspects of word-formation and discussion of them is a notoriously fragile and touchy business. Some damage, however, may be avoided by remembering that the essential problem is not What is right or wrong? but What is the best, consistent usage? Quandaries are frequent on such questions as, Shall it be spelled *labor* or *labour; Romania, Roumania,* or *Rumania; sizable* or *sizeable; skillful, skilful,* or *skilfull; whisky* or *whiskey?* Shall plurals of Latin nouns be anglicized by attaching *-s* or *-es* to the singulars, as *memorandums* for *memoranda, indexes* for *indices?* [4] The variants and inconsistencies in English spelling and pronunciation are so obtrusive that, at least since the sixteenth century, they have caused comment and instigated perennial efforts at corrective action. Shakespear, for example, is said to have spelled his name at least 30 different ways. In 1755 Samuel Johnson's *Dictionary* began to have a strong influence toward fixing the spelling that was then current among the printers. Unfortunately, according to Brander Matthews,[5] Johnson permitted many incongruities and "accepted *comptroller* ignoring the older *controller; sovereign* and *foreign* instead of the older *soverain* and *forain.* He kept a Latin *p* in *receipt* though he left it out of *deceit.* He spelt *deign* one way and *disdain* another. He kept up the accidental and useless distinction in spelling of the final syllables of *accede, exceed, precede,* and *proceed.*"

Proposals to found an English academy, comparable to the Italian, French, and Spanish academies, for overhauling the language periodically and suggesting the most acceptable spelling and usage, have been without issue. In the United States, however, in order to obtain uniformity throughout the government in regard to geographic names, there was established in 1890 the Board on Geographic Names, which publishes lists of approved spellings for such names. Further, to facilitate printing of government matter the Government Printing Office has a *Style manual* that regulates spelling, punctuation, capitalization, abbreviation, compounding, hyphenation, etc. The practical advantages of this setup are clear, and, although there may seem to be in it a tendency to freeze procedures into a Mede and Persian finality, the door to amendment, I understand, is always open.

The simplified spelling movement in the early years of the twentieth century began with modest success but failed to enlist continuing public interest when it expected too much in too short a time. A single symbol for a single sound always the same is the ideal goal sought by the thoroughgoing spelling reformers. This would eliminate the confusion of letters and sounds illustrated by the fact that the same vowel sound is represented by *e* in *let, ei* in *heifer, eo* in *leopard, ay* in *says, ai* in *said,* and *a* in *many.* Ten different symbols may represent the same consonant sound: *s* in *sure, sh* in *ship, sci* in *conscience, ci* in *suspicion, ce* in *ocean, ti* in *motion, se* in *nauseous, ch* in *champagne, sch* in *schist,* and *xi* in *anxious.* Conversely, the symbol *-ough* may stand for the sound of *o* in *borough, ock* in *lough, off* in *cough, oo* in *slough, ow* in *bough, uf* in *tough,* and *up* in *hiccough.* All the vowels and many of the consonants may represent two or more sounds. Some letters, or combinations of letters, duplicate one another, and in many words one or more letters are silent.

---

[4] Ball, Carleton R. *English or Latin for anglicized Latin nouns?* American Speech, vol. 3, pp. 291-325, 1928.

[5] Matthews, Brander. *The spelling of yesterday and the spelling of tomorrow.* Simplified Spelling Board Circular 4, May 7, 1906.

These and other alphabetic discordances, however bewildering, are, nevertheless, not entirely the results of illiterate invention, but, in many instances, are readily explainable accidents of etymology and need be no more disconcerting than that a consignment of freight on a railroad car is a *shipment,* whereas on a ship it is a *cargo.* One serious objection to radical reform in spelling is the possibility of losing letters that serve as clues to derivation and aids to memory. Thus, the printed words of the couplets *ate* and *eight, dying* and *dyeing, fair* and *fare, sea* and *see, marten* and *martin, thyme* and *time,* although sounding the same, have different origins and are distinguished by one or more significant letters. However, by omission of a letter, whose presence is not justified by ultimate etymology, the frenchified *amour, colour, endeavour, humour, labour,* and *saviour,* can be simplified in accord with their Latin antecedents, to *amor, color, endeavor, humor, labor,* and *savior.* How can the retention of the *u* in these words be reconciled with its consistent omission from *debtor, factor,* and *tailor?* So also *glamor,* not derived from or through French but from Scotch *glamer,* is better than *glamour,* the form popularized by Sir Walter Scott.

So-called scientific phonetic alphabets have been devised to meet the condition 'a single symbol for a single sound always the same', but, although useful as keys to pronunciation in dictionaries, they seem to falter so far as general acceptance is concerned, in a morass of unfamiliar, forbidding, printed characters and diacritical marks. Under the name Anglic a simply-spelled English has been proposed as an international language and has received some support. This does not enlarge the present alphabet of twenty-six letters but reallocates some sounds and letters to achieve uniformity. Judging by what has happened in the past, unless our customs become too deeply set, reform in English spelling, in pace with popular approval, will come slowly but surely. The influence of shorthand systems in this direction must not be overlooked. In the meantime, are children being encouraged to work at spelling intelligently and earnestly?

The Greek and Latin vocabularies, like English, contain many instances of variant spellings, thus posing difficulties for the biological scientist. For example, shall a generic term derived from the Latin for swan be spelled *cycnus* (Gr. *kyknos*) or *cygnus?* Shall the Latin for harrow be *irpex* or *hirpex?* For lamp, *lanterna* or *laterna?* For sapling, *orpex* or *horpex?* Shall the Greek for strong be *karteros* or *krateros?* For circle, *kirkos* or *krikos?* For little, *mikros* or *smikros?* For duck, *nessa* or *netta?* For mole, *spalax* or *aspalax?* For sea, *thalassa* or *thalatta?* In these matters must we continue muddling along until somehow by usage a consistent, uniform standard emerges?

The Latin dictionary entry for sulfur is first *sulfur,* then the variants *sulphur* and *sulpur,* but there is no evidence that the word was derived from Greek as might be supposed by the presence of the *ph* in the variant. The Latin *cirrus,* curl, and its derivative *cirratus,* curled, have been given a similar unwarranted Greek appearance by insertion of an *h,* thus becoming *cirrhus* and *cirrhatus* and simulating the derivatives of Greek *kirrhos,* tawny. The English *sepulchre,* from Latin *sepulcrum,* has fared in like manner. However, simplification at the expense of concealing or eliminating the helpful mnemonic signs inherent in etymology is hardly desirable. The *ph* and *rh*, for example, are revealing clues to the Greek origin of many words.

The variants in the Greek and Latin vocabularies lead to apparent errors in the spelling of binomial terms, but biologists themselves have so-called lapses that account for most orthographic errors. The establisher of binomial nomenclature himself was not infallible, as the following Linnaean lapses indicate: *cerris* for *cerrus* in *Quercus cerris; Cinchona* for *Chinchonia; Cypripedium* for *Cypripedilum; Didelphis* for *Didelphys; Diospyros virginiana* for *Diospyrus virginianus*, the final *o* in *diospyros*, m. being from o not ω; *Gleditsia* for *Gleditschia; Laternaria* for *Lanternaria*, although the former could be justified as a variant; *lathyrus* for *lathyris* in *Euphorbia lathyrus; Samyda* for *Semyda; sipedon* for *sepedon* in *Natrix sipedon;* and *Splachnum* for *Splanchnum*. Linnaeus attempted corrections of his 1753 and 1758 spellings in later editions of *Species plantarum* and *Systema naturae*, but these have been frowned upon by many modern naturalists. Thus, David Starr Jordan, in *Guide to the study of fishes*, vol. 1, p. 392, 1905, says: "In taxonomy it is not nearly so important that a name be pertinent or even well chosen as that it be stable. In changing his own established names, the father of classification set a bad example to his successors, one which they did not fail to follow."

One of those who did not fail to follow, at least in respect to correction of misspellings and misconstructions, was Louis Agassiz, whose *Nomenclator zoologicus*, 1842-1846, is liberally sprinkled with emendations. The rejection of these proposals by later workers is largely the result of invoking the Law of Priority, which is the reputed God of Stability, or if you please, the God to Prevent Confusion, supposed to be revered by all biologists. The nomenclatural codes provide only for the correction of unintentional orthographic errors, misprints, and errors of transcription. This, according to some interpreters of the provision, leaves little or no sanction for correcting mistakes like *Didelphis*, *Gleditsia*, *Aleyrodes*, *Pachystima*, *Ypophaemyia*, and similar misspellings, faulty transliterations, and misconstructions. Such a limited interpretation exposes those, who insist upon liberal corrections, to the sneers implied by the terms 'purist' and 'perfectionist', and sentences them to the inescapable propagation of manifest error, as well as to acceptance of the lazy dodge of being satisfied with something less than the best. If stability is to be expected, consistency must be the watchword, for the two are complementary; but harmonious consistency cannot be achieved without temporary confusion and inconvenience. Perpetual confusion, however, often predicted·as the inevitable sequel to controlled emendations, is here adjudged a wholly imaginary bogey conjured up to scare timid souls, rather than an honest objection to sensible corrective measures. To deny that a workable plan can be devised for purifying biological nomenclature according to an acceptable standard without revoking the principle of priority, is to impugn the intelligence of biologists. As scientific terminology is an artificial development it is amenable to control and, therefore, is somewhat like a garden. If a garden shall produce expected flowers and fruits in due season it cannot be left to haphazard evolution but must be cultivated assiduously and the weeds must be extirpated ruthlessly.

Consistency in spelling suggests consideration of uniformity in the typographic treatment of binomials. The trend today is toward capitalization of the generic term and decapitalization of the specific term no matter what its origin, the reason being that this practice enables ready distinction of the rank of the term in printed or written discussions and is therefore an aid to clearness and simplicity. Some biologists, particularly botanists, however, insist upon capitalizing the specific term if it is derived from a personal name, or if it was formerly a generic term. This deviation, as has been pointed out many times, is unnecessary, unsightly, and confusing. Compare, for example, the ugly, mixed, and confusing typography of the species entries on pages 584-585 of the eighth edition of Gray's *Manual of botany*, 1950, with the attractive, uniform, and clear treatment of the same kind of material on pages 288-289 of Jepson's *Manual of the flowering plants of California*, 1925, and pages 690-699 of the Clapham, Tutin, and Warburg *Flora of the British Isles*, 1952.

Pronunciation is first of all a personal matter because no two sets of vocal apparatus are exactly alike. Persons are thus enabled to identify one another, even in darkness or over the telephone, by the timbre or overtone quality peculiar to their voices. Superimposed on these fundamental peculiarities are the enunciation habits of the group acquired by the individual in infancy largely by imitation. These habits are often so characteristic that they serve readily to identify the locale of the group. Finally, in the course of their evolution, languages themselves develop distinctive, ingrained traits inheritable by all who speak those languages. Outstanding among these traits in English is the explosive stress or accent given habitually to a part or parts of polysyllabic words.

According to the grammars, the accent of the Greeks and Romans was not explosive but musical. How closely this modulation of accent coincided with the root elements of words is unknown. Presumably, however, if the grammars and dictionaries can be believed, there should be little trouble in pronouncing scientific or other terms derived from Greek and Latin, provided it be recognized which syllables of a word are long or short, which is penult and antepenult, and which receives the acute, grave, or circumflex accent. Thus, words of two syllables are accented on the first. Those of more than two syllables are accented on the penult (the next to the last syllable), if that is long, as indicated by a macron (ˉ) over the vowel; and on the antepenult, if the penult is short, as indicated by the breve (˘) over the vowel. As these signs are seldom given on the usual page of printed scientific matter, the pronouncer must either learn the classical rules and constantly consult dictionaries, or else adopt a simpler method. For the consolation of those who may be worried about the adjustment of modern speech to the pronunciation of the Greeks and Romans, let them ponder the words of the noted scholar, Goodwin[6] concerning Greek: "While the sounds of most of the letters are well established, on many points our knowledge

---

[6] Goodwin, W. W. *Greek grammar,* p. vii. Ginn & Co., 1899.

is still very unsatisfactory. With our doubts about the sounds of θ, ϕ, χ, and ζ, of the double $\varepsilon\iota$ and $o\upsilon$, not to speak of ξ and ψ, and with our helplessness in expressing anything like the ancient force of the three accents or the full distinction of quantity, it is safe to say that no one could now pronounce a sentence of Greek so that it would have been intelligible to Demosthenes or Plato."

The tendency to adopt or naturalize Greek and Latin words, or new latinized compounds, in modern English has evolved departures from Roman pronunciation so that the sounds of some of the vowels and consonants now differ materially from those of classical Latin.[7] For example, c, originally with a g or k sound, is now pronounced like s in many words, such as *Caesar, civilization, acerbity, procession,* and for most botanists, *Acer,* although some hold out for the k sound here, as well as for the broad rather than the long a. Another difference is in the sound of i. The English *pine,* with the long i, is from the Latin *pinus,* also with a long i but sounded by the Romans like e in *me.* The latter is now pronounced like the English word. Thus the quantity of the vowel i remains the same but the quality changes. In the English *fig,* with a short i, both the quantity and quality of the i have changed from the long i of Latin *ficus.* For the latter, as a generic term, most English dictionaries indicate the pronunciation *feye-cus;* but why not *fick-us* or *fee-cus,* which would harmonize better with the short i in *fig?* In Latin *erica,* heath, and *formica,* ant, the i is long and the syllables containing it receive the accent, to which some English dictionaries give approval; but the natural impulse is to shorten the i and shift the accent forward to the first syllable. Both words have yielded English derivatives, as *ericaceous* and *formicary,* that illustrate these tendencies. Does it not seem as though the pronunciation of their modern derivatives is having a salutary boomerang effect on that of the originals? In English the o of *rose* is long and the generic term *Rosa* is so pronounced, but the Latin dictionary indicates a short o. On the contrary, there is a long o in the similar *Roma,* Rome. Likewise, although the a in *mater,* mother, is long, that in *pater,* father, is short, but the English dictionaries indicate the same sound for both, as seems appropriate and consistent with the spelling of the words, regardless of their original sounds.

The question can now be put baldly. When scientific terms derived from Greek, Latin, or other languages, occur in an English sentence, shall they be pronounced according to the known or assumed methods in those languages, or shall they be pronounced as English? Chandler[7], like Goodwin about Greek, admits that no one knows definitely how the Romans pronounced Latin, and that "in very many syllables it is quite impossible to say whether the vowel is long or short (p. 166)." He, therefore, argues strongly for pronouncing classical terms in English as English. Because much scientific terminology, especially binomial nomenclature, is now in universal use, non-English speaking scientists, accustomed to pronouncing Latin according to either the Roman or

---

[7] Chandler, Charles. *Pronunciation of Latin and quasi-Latin scientific terms.* Denison University Sci. Lab. Bull., vol. 4, pt. 2, pp. 161-176, 1889.

Drewitt, F. Dawltrey. *Latin names of common plants: Their pronunciation and history.* 68 pp. H. H. F. & G. Witherby, 1927.

Continental method, will find Chandler's solution somewhat provincial and in need of modification.

The uncritical application of alleged classical rules to the pronunciation of technical words in English sometimes results in monstrosities that undoubtedly would have shocked the musical ears of the Greeks and Romans. The basic cause for this cacophony is the slavish adherence to the acquired habit of shifting the accent in a polysyllabic word from a root element to a new syllable that is etymologically meaningless. Thus *aragonite* becomes *a-rag-on-ite* instead of *ar-a-gon-ite* (from Aragon, Spain); *atacamite* becomes *a-tac-am-ite* instead of *at-a-cam-ite* (from Atacama, Chile); *labradorite* becomes *la-brad-or-ite* instead of *lab-ra-dor-ite* (from Labrador). *Centimeter* remains *centi-meter*, but *perimeter* becomes *per-im-e-ter* instead of *peri-meter;* and *thermometer* becomes *ther-mom-e-ter* instead of *thermo-meter.* Would any Greek have said *ki-lom-e-ter* for *kilo-meter?* If *biloba* is *bi-loba,* is there any justification for *tril-o-ba* in *Hepatica triloba?* *Oligocene* is sometimes heard as *O-lig-o-cene* instead of *Ol-i-go-cene;* and *Paleocene* as *Pal-eocene* instead of *Paleo-cene,* although there is no connection whatever with the word *Eocene.* *Polygonum* circulates as *Po-lyg-o-num* instead of *Poly-gonum,* meaning many knees. Although the *n* is sounded in Latin *columna,* it is silent in *column* and many persons pronounce *columnist* accordingly. The Greek for fern is *pteris.* How silent was the *p* in this word? As it is not now sounded in *Pteris longifolia,* is there a convincing reason for sounding it in *Dryopteris spinulosa?* *Dryo-pteris, Neuro-pteris,* and *Spheno-pteris* should be compared for expressiveness with *Dry-op-ter-is, Neur-op-ter-is,* and *Sphen-op-ter-is.* Similarly, *Cedro-xylum, Querco-xylum,* and *Xantho-xylum* should be compared with *Ced-rox-y-lum, Quer-cox-y-lum,* and *Xan-thox-y-lum.* If *Fuchsia* were pronounced *Fooks-ia* and *Halesia, Hales-ia,* both Fuchs and Hales would be more intelligently commemorated. By what nuance in expression did the Greeks distinguish *anamiktos,* which, if pronounced *ana-miktos,* means mixed up, but if *an-amiktos,* means unmixed?

The purpose of pronunciation is not to be sonorous but to be understood. Examples like the foregoing illustrate how clarity may be promoted by enunciating the simple components of technical words so that they will be heard as distinct, etymological entities.[8] The single rule, 'Pronounce the root elements clearly', may well be a declaration of independence from too strict application of classical practice. However, the classical rules for accentuation cannot be disregarded entirely because they indicate a natural impulse in dealing with some words of many syllables, such as *biology, geography, philosophy,* and *sympathy,* whose syllabic quantities can be most easily pronounced according to that impulse. In brief, divergence from the classical rules of pronunciation presupposes the exercise of good judgment.

As this book deals primarily with the making of words and not their pronunciation, the latter phase, with all its possibilities for difference of

[8] Gleason, H. A. *The pronunciation of botanical terms.* Torreya, vol. 32, pp. 53-58, 1932.

opinion and argument, is minimized. Accordingly, the words listed in the LEXICON are left unblemished by accent marks and other devices to suggest their sounds.

CROSS-REFERENCE LEXICON

A

a- < AS. *an, on,* at, in, upon, see **on**; < AS. *a, af, of,* of, off, from, out, see **from**

a-; ab- < L. *ab,* from, off, away, see **from**; < L. *ad,* to, see **to**

a-; an- < Gr. *a, an,* not, without, negative, privative, see **not**; < Gr. *a-,* with, union, sameness, copulative, see **with**

aages, Gr. unbroken, hard, strong; see **hard**

aaptos, Gr. invincible; see **strong**

aatos, Gr. insatiable; see **hunger**

ab- < L. *ab,* from, away, see **from**; < L. *ad,* to, see **to**

abaco- < Gr. *abakes,* speechless, childlike, innocent; see **silent**

abactus, L. driven away, aborted; *abactor,* driver off; see *ago* under **drive**

abacus; abax, -acis, L. (Gr. *-akos*), board, counting-board, gaming-board divided into square compartments; *abaculus,* dim. small, colored tile for inlay and mosaic work; see **number**

abandon < OF. *abandoner.*
  L. *abdico, -atus,* disown, renounce, resign: abdicate, abdication.
  L. *abjicio, -jectus,* cast off, give up; see **poor**
  L. *abjuro, -atus; ejuro,* forswear, reject, abandon: abjure, ejurate.
  Gr. *ametor, -os,* motherless; see **alone**
  Gr. *apostasis,* f. defection, departure from; *apostates,* m. runaway, deserter, renegade; *apostatikos,* of desertion, rebellious: apostasy, apostate, *Apostates albiclathratus* (a moth), *Psilodacus apostata* (a fly).
  Gr. *automolos,* m. deserter:
  Gr. *brephotropheion,* foundling or orphan asylum; see **house**
  L. *cedo, cessus,* go, yield, abandon, withdraw; *concedo,* give up; *discedo,* part, leave, forsake; see **move**
  Gr. *cheroö,* bereave, desolate; *cherosis; cherosyne,* bereavement; see *chera* under **woman**
  L. *deficio, -fectus,* desert, leave: defection.
  L. *desero, -ertus,* abandon, forsake, leave: deserted, *Deserta raptoria* (a bug).
  L. *desisto, -stitus,* cease, leave off: desist.
  L. *desolo, -atus,* leave alone, abandon, forsake: desolate, desolation.
  L. *destituo, -utus,* desert, forsake: destitute, destitution.
  L. *deviator, -is,* m. one who leaves the way, forsaker: deviator.
  L. *digredior, -essus,* abandon, deviate from: digress, digression.
  L. *dimitto, -issus,* abandon, forsake: dimission, dimissary.
  Gr. *ekdysis,* a getting out, escape, molt, < *ekdyo,* take off, strip, lay bare; see **bare**
  L. *exul,* outcast, exile; *exilium,* outcast; see **banish**
  Gr. *leipo,* leave, be wanting or without; *leimma, -tos; leipsanon,* n. remnant, relic; *loipos,* remaining; *ekleipsis,* f. a forsaking; *eklipes,* failing, deficient; L. *ellipsis* (Gr. *elleipsis*), f. omission, < *elleipo,* leave out, omit: lipopod, lipogram, Lipalian, eclipse, ecliptic, ellipse, ellipsis, *Lipodectes penetrans* (a Paleocene mammal), *Liposarcus multiradiatus* (a fish), *Lipura hudsonia* (a marmot).
  L. *linquo, lictus,* leave, abandon, lack; *derelictus,* abandoned, deserted, disregarded, neglected; *relictus,* abandoned, forsaken; *reliquia,* f. leaving, remnant; *reliquum,* n. remainder, balance: delinquent, derelict, relic, relinquish, reliquary, *Dorcus derelictus* (a beetle).
  Gr. *loisthos,* left behind; see **after**
  L. *orbo, -atus,* bereave; *orbus,* bereft of parents; *orba,* orphan; see **alone**
  L. *orphanus* (Gr. *orphanos*), bereft; see **alone**
  L. *procrastino, -atus,* put off till tomorrow, defer; see **delay**
  Gr. *prodotes,* betrayer, traitor; *prodotos,* abandoned; see *prodosia* under **treason**
  L. *residuus,* remaining, left over: residue, residuary, residuum.
  L. *resigno, -atus,* cancel, give up, surrender, relinquish: resign, resignation.
  L. *subduco, -uctus,* evacuate, remove, withdraw: subduce, subduction.
  L. *vidua,* widow; see **woman**
  See: **depart, let, run, shun, free, give, alone, end, empty, delay, from, bad, lose, cancel**

abate < OF. *abatre;* see **lessen**

abatos, Gr. untrodden, inviolate, pure, chaste; see **pure**

abba, Heb. father; see **father**

abbreviatus, L. shortened; see *brevis* under **short**

abdico, *-atus,* L. disown, renounce; see **abandon**

abditivus, L. removed from, separated from; see **from**

abditus, L. hidden, concealed; see **secret**

abdomen, L. belly; see **belly**

abductio, L. forcible carrying off, kidnapping; see **steal**

abebaeo- < Gr. *abebaios,* uncertain, unsteady, wavering, fickle; see **change**

abecedarius, L. of the alphabet, alphabetical; see **letter**

abelteros, Gr. asinine, silly, stupid, foolish; see **fool**

aberrans, L. wandering, < *aberro, -atus,* go astray; see *erro* under **wander**

abet < OF. *abeter;* see **help, arouse**

abhor < L. *abhorreo,* dislike, shrink from; see **hate, shun**

abide < AS. *abidan;* see **stand, life, house, sit**

abies, L. fir; *abiegnus; abietinus,* of fir; *abietarius,* joiner; see **fir**

abiga, L. a kind of mint; see **mint**

abigeus, L. cattle-stealer, rustler; see **steal**

abitus, L. departure, exit, < *abeo,* go away; see *eo* under **move**

abjectus, L. cast away, low, mean; see **poor**

ablabes, Gr. harmless, innocent; see **safe**

ablatio, L. a taking away; see **from**

able < L. *habilis,* apt, fit, expert; see **strong, know, make**

-able; -ible; -uble; -ble < L. *-bilis,* tending to be, capable of, worthy of; see **make**

ablechros, Gr. weak, feeble; see **weak**

ablemo- < Gr. *ablemes,* feeble; see **weak**

ablepsia, Gr. blindness; see **blind**

abletos, Gr. unhit; see **safe**

abligurio, L. spendthrift; see **waste**

ablusus, L. different; see **different**

abnormal < L. *abnormis,* deviating from the rule; see **different, strange, wonder, monster**

aboethetos, Gr. hopeless, incurable; see **hopeless**

abolish < L. *aboleo, -litus,* terminate, destroy, cancel; see **destroy, end, cancel, lessen, abandon**

abolla, L. cloak of thick, woolen cloth; see **garment**

abolos, Gr. uncast, unshed; see **hold**

abominor, *-atus,* L. abhor, detest; see **hate**

aborigineus, L. ancestral, native, original; see **native**

abortion < L. *abortio, -onis,* f. premature delivery; *abortus,* m. miscarriage; *abortivus,* prematurely born or terminated; *aborior, -ortus,* set, disappear, perish: abort, abortive, *Brucella abortus* (bacterium causing abortion in cattle). *Ranunculus abortivus* (a buttercup), *Mesochloa abortiva* (a grasshopper).
    L. *abagmentum,* n. means of procuring abortion:
    L. *abigo, abactus,* drive away, abort; see *ago* under **drive**
    Gr. *ambloma, -tos,* n.; *amblosis,* f. abortion, < *amblisko,* cause abortion; *amblothridion,* n. an abortive child: *Ambloma brachyptera* (a moth).
    Gr. *ektroma, -tos,* n. abortion; *ektrosis,* f. miscarriage; *ektrotikos.* of abortion:

about < AS. *abutan,* on the outside, around; see **around, near**

above < AS. *abufan;* see **over**

abramis, Gr. a kind of fish; see **bream**

abricto- < Gr. *abriktos,* wakeful; see **awake**

abrochos, Gr. dry, waterless, unwetted; see **dry**

abrogatus, L. annulled, repealed; see **cancel**

abronia, NL. a genus of the fouro'clock family; see **fouro'clock**

abrotonum, L. (Gr. *abrotonon*), an aromatic plant, southernwood; see **composite**

abrotos, Gr. not eating, fasting; see **hunger**

abrupt < L. *abruptus,* broken off, precipitous, steep; see **break, short, cut, swift, high, cliff**

abscess < L. *abscessus,* purulent tumor; see **sore**

abscissus, L. cut off; see *scindo* under **cut**

absconditus, L. concealed, secret; see *condo* under **cover**

absent < L. *absens, -entis,* away from; see **from, far, empty, poor, neglect**

absidatus, L. arched; see *apsis* under **arc**

absinthium, L. (Gr. *apsinthion*), wormwood; see **wormwood**

absolutus, L. finished, perfect, complete; see **all**

absonus, L. inharmonious, disagreeing with; see **different**

abstain < L. *abstineo, -tentus;* see **hold, from, govern**

abstemius, L. refrain from drinking liquor, moderate, temperate; see **govern**

abstractus, L. drawn away, separated; see **from**

abstrusus, L. hidden, concealed; see **secret**

absumedo, L. consumption; see *edo* under **eat**

absurdus, L. that which offends the ear, foolish, silly; see **fool**

abundance < L. *abundantia,* f. an overflowing fullness, plenty, richness, < *abundo, -atus,* overflow, be copious, luxuriate; *abundus,* copious: abound, abundance, abundant.

Gr. *adaios,* abundant; *aden,* enough, abundantly; *chydaios,* poured out in streams, abundant, common, vulgar: *Chydaeopsis fragilis* (a beetle), *Chydaeus obscurus* (a beetle).

Gr. *adapanetos,* inexhaustible:

Gr. *adinos,* thick, dense, crowded; see **thick**

L. *affluens,* abundant, copious, rich, profuse, < *affluo, -fluxus,* flow abundantly, overflow, have in abundance: affluent, affluence.

Gr. *apheides,* lavish, unsparing: aphid (?).

Gr. *aphthonos,* without envy, ungrudging, plentiful, abundant: *Aphthonomorpha minuta* (a beetle).

Gr. *arketos,* enough:

Gr. *charistikos,* giving freely, bounteous; see *charisma* under **give**

L. *consuo, -utus,* stuff, fill up: *Pachymerus consutus* (a bug).

L. *contentus,* satisfied, having enough; *distentus,* filled up, full: content, contentment, contented, distended.

L. *copia,* f. abundance, plenty, wealth; *copiosus,* abundant; *cornucopia,* f. horn of plenty: copious, cornucopia, copy, copyright, *Hipponyx cornucopia* (a gastropod), *Chrysotus copiosus* (a fly).

L. *crumino, -atus,* fill like a purse; see *crumena* under **bag**

L. *cumulatus,* heaped up, filled; see *cumulus* under **heap**

L. *dapsilis* (Gr. *dapsiles*), abundant, sumptuous, plentiful: *Dapsilotoma testaceipes* (a wasp).

Gr. *elitha,* enough:

Gr. *epeetanos,* sufficient, plentiful, abundant:

Gr. *epidemos,* prevalent, common to a large number; see **common**

Gr. *epirrhytos,* sufficient, plentiful, abundant:

Gr. *euthenia,* f. abundance, < *eutheneo,* be well off, thrive, flourish: euthenics.

L. *fecundus,* fruitful, rich, abounding: fecund.

L. *fetosus,* prolific; see *fetus* under **fertile**

L. *fructuosus,* abounding in fruit; see **fruit**

AS. *fyllan,* make full: fulfill, fill.

Gr. *habrotes,* splendor, luxury; see *habros* under **beauty**

Gr. *hales; halis,* in abundance, plenty, heaps, crowds, swarms:

Gr. *himaleos,* abundant:

Gr. *iladon,* in troops, in abundance:

Gr. *koros,* m. satiety:

L. *largus,* copious, abundant; see **large**

L. *luxuria,* f. profusion, rankness, excess, extravagance; *luxurio, -atus,* abound to excess, grow rapidly; *luxuriosus,* rank, exuberant; *luxus,* m. excess, extravagance: luxury, luxurious, luxuriant.

Gr. *mala,* very, much, exceedingly; see **very**

L. *mammon,* n. riches: mammon.

Gr. *mestos,* full, filled; *mestotes, -tetos,* f. fullness; *emmestos,* filled full: *Mestobregma platti* (a fossil grasshopper), *Perimestocrinus noduliferus* (a fossil crinoid)

AS. *most,* < *maest,* superlative of *more:* utmost, innermost, upmost, topmost, 'devil take the hindmost'.

L. *multus,* much; *plus, -uris,* more; *plurimus,* most; *pluralis,* pertaining to more than one; *plusculus,* dim. a little more; *multifarius,* manifold, various: multitude, multiplication, multiparous, multimillionaire, multifarious, nonplus, surplus, plural, plurality, *E pluribus unum, Phlox multiflora* (flowery phlox).

L. *munificus,* bountiful, generous: munificent.

L. *nepotalis; nepotinus,* extravagant, prodigal, profuse, < *nepotor, -atus,* be prodigal:

L. *nimius,* too much, excessive, superfluous: nimiety, *Steropus nimius* (a beetle).

L. *oneratus,* filled, full, loaded; see *onus* under **weight**

L. *opimus,* rich, abundant, copious; see *ops* under **wealth**

L. *pecorosus,* rich in cattle; see *pecus* under **cattle**

Gr. *periousia,* f. abundance, plenty, surplus; *periousios,* more than enough, abundant:

Gr. *perisseia,* f. abundance, surplus; *perissotes, -etos,* f. excess, superfluity:

Gr. *phoras, -ados,* fruitful; see *phero* under **carry**

L. *plenitas, -atis,* f. fullness, repletion, < *plenus,* full: plenty, plentiful, plenitude, replenish, plenary, *Rostricellula plena* (a fossil brachiopod), *Arabellites plenidens* (a fossil worm), *Himerocrinus plenissimus* (a Devonian crinoid).

L. *pleo, -etus,* fill; *pletura,* f. fullness; *compleo, -etus,* fill up, fulfill; *expletivus,* serving to fill out; *implebilis,* filling up; *repletus,* filled, full: complete, complementary, complimentary, supplementary, accomplish, comply, implement, repletion, expletive, supply, deplete, *Cristellaria impleta* (a foraminifer).

Gr. *pleonasmos,* m. more than enough, excess: pleonasm.

Gr. *pleos; pleres,* full, full of, filled, satisfied, complete; *pleresteros,* fuller; *plerestatos,* fullest; *plethore,* f. fullness, satiety; *plethos,* n.; *plethys,* f. a throng, multitude, fullness; *plethysmos,* m. increase, enlargement; *apeiroplethes,* innumerable; *myrioplethes,* countless, infinite: pleroma, plerosis, plerotic, anaplerosis, plethora, plethomeria, plethysmograph, *Plethodon cinereus* (a salamander)

L. *plerus; plerusque,* very many, most:

Gr. *plesmone,* f. satiety, fullness, plenty:

Gr. *polys,* many; *pleion; pleon,* more; *pleistos,* most; see **number**

L. *prodigus,* wasteful, lavish, extravagant; see **waste**

L. *profusus,* abundant, extravagant, lavish: profuse, profusion, *Syrphus profusus* (a fly).

L. *quantus,* how many, how much; see **number**

L. *redundans,* overflowing, superfluous; see **over**

L. *refertus,* crammed, filled, stuffed; see *farcio* under **press**

Gr. *rhyden; rhydon,* abundantly:

L. *satis,* enough; *satisfacio,* content, pay, make amends; *satio, -atus,* fill, satisfy: satisfactory, satiate, satiety, insatiable, dissatisfaction, assets.

L. *satur, -a, -um,* full, rich; *saturo, -atus,* fill, glut; *satullus,* filled, satisfied: saturate, saturation, satire, *Putorius saturatus* (a weasel).

L. *scateo,* be plentiful, gush, abound; see **spring**

L. *superfluus,* overflowing, unnecessary: superfluous, superfluity.

L. *suppedito, -atus,* have in abundance, supply abundantly:

L. *tantus,* so much: tantamount, tantivity.

Gr. *thaleia,* f. a blooming, luxuriance; *thalia,* f. abundance, good cheer, wealth; *thaleros,* vigorous, fresh, luxuriant, abundant: *Thaleichthys pacificus* (candlefish, eulachon), *Thalia divaricata* (a marantacead).

Gr. *tosos,* so much, so great, so many, so very: *Tosotarsus velutinus* (a beetle).

L. *turnatim,* in troops, in abundance:

L. *uber, -is,* full, abundant, fruitful; *exubero, -atus,* come forth in abundance, abound in: exuberance, exuberant, uberous.

L. *voluminosus,* full of folds, of many volumes: voluminous.

* * * * * * *

L. *-bundus; -cundus; -undus,* denoting continuance or augmentation; see **very**

AS. *-ful, full,* abounding in, filled: full, fulsome, fullness, fulfill, awful, cheerful, pailful, tearful, spoonfuls, wilful, youthful.

L. *-lens; -lentus,* full of, prone to, as *pestilens, -entis,* unhealthy, unhealthful; *vinolentus,* drunk with wine: corpulent, pestilent, pestilence, redolent, truculent, *Echeveria pulverulenta* (a crassulacead).

Gr. *-odes,* denoting fullness, as *bunodes,* hilly; *sarcodes,* fleshy; see *-oid* under **like**

L. *-osus* having the nature or quality of, usually in fullness or abundance; see *-ous* under **nature**

See: **number, very, large, wealth**

abuse < L. *abusus,* misuse; *abusivus,* misapplied, wrongly used; see **hurt, blame**

abutilon, NL. a genus of the mallow family; see **mallow**

abyrtaco- < Gr. *abyrtake,* salad with sour sauce; see **salad**

abyss < L. *abyssus* (Gr. *abyssos*), bottomless pit, the deep sea; bottomless; see **deep**

ac- < L. *ad,* to; see **to**

acacalido- < Gr. *akakalis, -idos,* white tamarisk; see **tamarisk**

acacia, L. (Gr. *akakia*), a thorny plant of the bean family; see **bean**

academy < L. *academia* (Gr. *akademia*), an association of learned men; see **school**

acaeno- < Gr. *akaina,* thorn; see **thorn**

acalepha, L. (Gr. *akalephe*), nettle; see **nettle**

acallo- < Gr. *akalles,* ugly; see **ugly**

acalo- < Gr. *akalos,* peaceful, still; see *hekelos* under **rest**

acalypho- < Gr. *akalyphos,* uncovered; see *akalyptos* under **open**

acamanto- < Gr. *akamas, -antos,* untiring, unresting; see **strong**

acano- < Gr. *akanos,* a kind of thistle; see **thistle**

acanthias, L. (Gr. *akanthias*), a prickly thing, a kind of shark; see **shark**

acanthis, *-idis,* L. (Gr. *akanthis, -idos*), goldfinch; see **finch**

acanthium, L. (Gr. *akanthion*), a species of thistle; see **thistle**

acanthus, L. (Gr. *akanthos*), a prickly plant, < *akantha,* thorn; see **thorn**

acanthyllis, L. (Gr. *akanthyllis*), a little bird, probably a titmouse; see **titmouse**

acarno- < Gr. *akarna,* a kind of thistle; see **thistle**

acaro- < Gr. *akares,* short, small, momentary; see **little**

acarus, L. (Gr. *akari*), mite; see **tick**

acatus, L. (Gr. *akatos*), a light boat; *akation,* dim. sail, a woman's shoe; see **ship**

accensus, L. set on fire, kindled, incited, < *accendo,* light, kindle, inflame, see **burn**; attendant, see **servant**

accentus, L. tone, signal; see **sound**

acceptus, L. welcome, agreeable; see **agreeable**

accessible < L. *accessibilis,* approachable; *accessus, -us,* m. approach, entrance, < *accedo, -essus,* approach, enter upon: access, accession, accessible, accessory, accede.
    L. *adibilis,* accessible, passable:
    Gr. *basimos,* passable, accessible; *batos,* passable; *diabatos,* passable, fordable; *embatos; eubatos; prosbatos,* passable, accessible; see *baino* under **walk**
    Gr. *ephiktos,* accessible, within reach:
    Gr. *hodotos,* passable, feasible; see *hodos* under **way**
    L. *meabilis,* passable; *commeabilis,* permeable; see *meatus* under **way**
    Gr. *perasimos,* passable:
    Gr. *prositos,* approachable, accessible; *dysprositos,* hard to get at: dysprosium.
    See: **near, way, enter**

accident < L. *accidentia,* an unforeseen, unexpected event, chance; see **lot**

accinctus, L. well-girded, armed, equipped; see *cingo* under **bind**

accipiter, L. hawk; see **hawk**

acclivis, L. ascending, steep, up hill; see *clivis* under **slope**

accola, L. neighbor; see **neighbor**

accubitum, L. couch on which guests reclined at table, < *accubo; accumbo, -ubitus,* recline at table; see **bed,** *cubo* under **lie**

accurate < L. *accuro, -atus,* take care, be exact; see **right**

accuse < L. *accuso, -atus,* complain against, reproach, blame; see **blame**

-aceae, L. suffix indicating a family of plants; see **class**

aceano- < Gr. *akeanos,* a kind of plant; see **plant**

acedo- < Gr. *akedes,* careless, heedless, negligent; see **neglect**

aceo- < Gr. *akos, akeos,* cure, remedy, < *akeomai,* cure, heal; see **heal**

-aceous < L. *-aceus,* pertaining to, having the nature of; see **nature**

acer, L. maple; *acernus,* of maple; see **maple**

acer, *acris, acre,* L. pointed, pungent, stinging, sharp, sour; see *acuo* under **point**

aceratus, L. mingled with chaff; *acerosus,* full of chaff, mixed with chaff, < *acus, -eris,* husk of grain, chaff; see **scale**

acerbus, L. sour, sharp, harsh; see **sour**

acerra, L. incense-box; see **box**

acervus, L. heap; see **heap**

acesto- < Gr. *akestos,* curable; *akesis,* cure, salve, plaster, < *akeomai,* heal, cure; see **heal**

acetabulum, L. a vinegar cup, any cup-shaped vessel, socket of the hip-bone; see **cup**

acetum, L. vinegar, < *aceo,* be sour; *acetosus,* sour; see *acidus* under **sour**

acetyl-, denoting the CH₃CO radical of acetic acid; see *acidus* under **sour**

-aceus, L. pertaining to, having the nature of; see **nature**

achanes, Gr. mute, yawning, immense; see **silent, deep**

acharne, L. (Gr. *acharna*), a kind of fish; see **bream**

achates, L. (Gr. *Achates,* a river in Sicily), agate; see **stone**

ache < AS. *acan,* pain; see **pain**

achen, Gr. needy, poor; see **poor**

achenium, NL. a kind of indehiscent fruit; see **fruit**

acherdos, Gr. a prickly shrub used for hedges; see **bush**

Acheron, Gr. a river in the nether world; see **stream**

acheta, L. (Gr. *achetas; echetes*), male cicada; see **cicada**

achillea, L. yarrow; see **yarrow**

achlis, L. a northern wild beast, probably the elk; see **deer**

achlyo- < Gr. *achlys,* mist, darkness, < *achlyoo,* grow dark; see **black**

achne, Gr. any light substance, foam, froth, chaff, down; see **scale**

achoreutes, Gr. joyless, melancholy, banished from the dance; see *choros* under **dance**

achoro- < Gr. *achor, -os,* scurf, dandruff, see **scale**; *achoros,* homeless, see **wander**

achos, Gr. pain, distress, grief, sorrow; see **pain**

achras, *-ados,* Gr. wild pear; see **pear**

achrestos, Gr. useless; see **worthless**

achrostos, Gr. untouched, uncolored; see **color**

achtheros, Gr. annoying, distressing; see **trouble**

achthos, Gr. weight, burden, distress; *achtheinos,* burdensome, oppressive; *achtheros,* annoying, distressing; *achthophoros,* bearing burdens, porter; see **weight**

acia, L. thread for sewing; see **thread**

acicula, L. small pin, dim. of *acus,* needle, pin; see **needle**

acicyo- < Gr. *akikys, -yos,* powerless; see **weak**

acid < L. *acidus,* sour, sharp; see **sour**

acidno- < Gr. *akidnos,* weak, feeble; see **weak**

acido- < Gr. *akis, -idos,* point, barb, dart, < *ake,* point; see **point**

acidotum, NL. (Gr. *akidoton*), a kind of plant; see **plant**

acies, L. sharp edge or point, keenness of mind; see *acuo* under **point**

acinaces, L. (Gr. *akinakes*), short sword, scimitar; see **sword**

acineto- < Gr. *akinetos,* not moving, idle, steadfast; see **rest**

acino- < Gr. *akinos,* basil-thyme; see **mint**

acinus, L. berry, grape; see **grape**

acio- < Gr. *akios,* not worm-eaten; see **all**

acipenser, L. (Gr. *akkipesios*), sturgeon; see **sturgeon**

aclys, *-ydis,* L. short javelin; see **spear**

aclysto- < Gr. *aklystos,* sheltered; see **safe**

acmaea, NL. a genus of gastropods; see **mollusk**

acmeto- < Gr. *akmes, -etos,* untiring, unwearied; see **strong**

acmo- < Gr. *akme,* highest point; *akmaios,* in prime, vigorous; see **top**

acmono- < Gr. *akmon, -os,* anvil; see **anvil**

acniso- < Gr. *aknisos,* without fat, lean, meager, spare; see **thin**

acoce- < Gr. *akoke,* point, edge; see *akis* under **point**

acolo- < Gr. *akolos,* bit, morsel; see **little**

acolutho- < Gr. *akolouthos,* following after, attending on, succeeding; see **follow**

aconio- < Gr. *akonias,* a kind of fish; see **fish**

aconite < L. *aconitum* (Gr. *akoniton*), n. a poisonous plant of the buttercup family: *Aconitum paniculatum* (panicled monkshood), *Delphinium aconiti* (a larkspur), *Begonia aconitifolia* (a begonia), *Geranium aconitifolium* (a geranium).

Gr. *kammaron*, n. a kind of aconite: *Aconitum cammarum* (Hungarian monks-hood).

Gr. *lykoktonon*, n. wolfbane, aconite: *Aconitum lycoctonum* (wolfbane).

Gr. *myoktonos*, m. a kind of aconite said to kill mice:

Gr. *pardalianches*, n. leopardbane:

Gr. *thelyphonon*, n. aconite:

acono- < Gr. *akone*, whetstone, hone; see **stone**

aconto- < Gr. *akon, -tos*, spear, javelin, dart; *akontion*, dim.; see **spear**

acopo- < Gr. *akopos*, without weariness; see **strong**

acoresto-; acoreto- < Gr. *akorestos; akoretos*, unsatisfied, insatiable; see **hunger**

acorn < AS. *aecern*, fruit of the field.

L. *acylus* (Gr. *akylos*), m. acorn of the holm-oak: *Acylocrinus tumidus* (a Missis-sippian crinoid).

Gr. *balanos*, f. acorn, barnacle, bar, peg, glans penis; *balanion*, n. dim.; *balaninos*, of an acorn; *balanites*, like an acorn: balanoid, balanite, myrobalan, balanitis, balanism, valonia, *Balanus eburneus* (acorn-barnacle), *Balaninus rectus* (chestnut-weevil), *Balanites aegyptiaca* (bito), *Balanion comatum* (a protozoan), *Balano-phora elongata* (a parasitic plant), *Chrysobalanus icaco* (coco-plum), *Balano-glossus clavigerus* (a wormlike chordate), *Dryobalanops aromatica* (Borneo camphor-tree).

L. *glans, glandis*, acorn, any acorn-shaped object; see **gland**

acorna, L. (Gr. *akorna*), a kind of thistle; see **thistle**

acoro- < Gr. *akoros*, unceasing; see **always**

acorus, L. (Gr. *akoros*), sweetflag; a genus of the arum family; see **arum**

acosto- < Gr. *akoste*, barley; see **grain**

acoustic < Gr. *akoustikos*, pertaining to hearing, < *akouo*, hear; see **hear**

acranto- < Gr. *akrantos*, futile, idle, unaccomplished; see **worthless**

acrato- < Gr. *akratos*, unmixed, pure, see **pure**; *akrates*, powerless, weak, see **weak**

acredula, L. an unknown bird; see **bird**

acremono- < Gr. *akremon, -os*, bough, branch, spray; see **branch**

acribo- < Gr. *akribes*, exact, precise; see **right**

acrido- < Gr. *akris, -idos*, locust, grasshopper; *akridion*, dim.; see **grasshopper**

acrito- < Gr. *akritos*, confused, disorderly, mixed, doubtful; see **confusion**

acro- < Gr. *akron*, top, summit, peak; *akros*, at the end, tip, top, first, highest; *akris, -ios*, hilltop, peak; see **top**

acroaticus, L. (Gr. *akroatikos*), designed for hearing only; *akroama*, anything heard; see **hear**

acrochordo- < Gr. *akrochordon*, a kind of wart; see **wart**

acrolopho- < Gr. *akrolophos*, mountain crest, ridge; see *lophos* under **crest**

across < AS. *a*, on, L. *crux*, cross; see **over**

acrostichum, NL. a genus of ferns; see **fern**

acrulo- < Gr. *akroulos*, curled at the end; see *oulos* under **curl**

acrylo-, a chem. pref. referring to derivatives of the aldehyde acrolein; see **aldehyde**

act < L. *actum*, deed, transaction, record; see *ago* under **make**

acta, L. (Gr. *akte*), seashore, strand, promontory; *aktaios*, coastal; see **shore**

actaea, L. (Gr. *aktaia; aktea*), elder; now applied to baneberry; see **buttercup**

actaeo- < Gr. *aktaios*, on the coast or shore; see **shore**

actemono- < Gr. *aktemon, -os (akten, -os)*, poor; *aktemosyne*, poverty; see **poor**

actinea, NL. a genus of the composite family; see **composite**

actino- < Gr. *aktis, -inos*, ray, beam; see **rod**

actito- < Gr. *aktites*, shore dweller, < *akte*, seashore, beach, strand; see **shore**

activus, L. engaged in action; *actor; actrix*, doer, performer; see *ago* under **make**

actuariolum, L. small rowboat, barge, dim. of *actuarium*, a swift sailer; see **ship**

actuarius, L. copyist, clerk, shorthand writer, see **write**; swift, agile, see **swift**

actus, L. transacted, done, < *ago*, do, perform, drive, move; *actum*, deed, record; *actuosus*, very active; see *ago* under **make**

aculeus, L. sting; *aculeatus*, sharp-pointed, stinging; see *acuo* under **point**

acuminatus, L. pointed, sharpened, < *acumen, -inis*, sharp point of anything; see *acuo* under **point**

acus, L. needle, pin; *acicula; aculeolus; aculeus,* dim., small pin, sting, point; *aculeatus,* prickly, pointed, stinging, see **needle, point;** *acus, -eris,* husk of grain, chaff, see **scale**

-acus, L. belonging to, of, pertaining to; see **nature**

acutus, L. sharp, pointed, < *acuo,* sharpen, whet; see *acuo* under **point**

acylo- < Gr. *akylos,* acorn of the holm-oak; see **acorn**

acyro- < Gr. *akyros,* without authority, incorrect; *akyrotos,* unconfirmed; see **bad**

ad- < L. *ad,* to, direction toward; see **to**

-ad; -ade < F. *-ade,* < Gr. *-as, -ados;* < F. *-ade,* Sp. *-ada,* < L. *-ata,* pertaining to; see **nature**

adactus, L. < *adigo,* drive, bring, take, urge; see *ago* under **drive**

adaeo- < Gr. *adaios,* abundant; see **abundance**

adaetos, Gr. unknown, unknowing, see **secret;** *adaitos,* inedible, see *dais* under **eat**

adage < L. *adagium,* n. proverb, saying.
   L. *dictum,* n. saying, word; see *dico* under **speak**
   Gr. *epigramma, -atos,* n. inscription, witty or pithy saying: epigram, epigrammatic.
   Gr. *logos,* word, saying; see **word**
   L. *proverbium,* n. saying, saw, maxim, adage: proverb, proverbial.

adamant < Gr. *adamas, -antos,* unconquerable, stubborn, unyielding, hard; diamond; see **diamond**

adamastos; adamatos, Gr. untamed, unbroken, unconquered; see **nature**

adapanetos, Gr. inexhaustible; see **abundance**

adarca, L. (Gr. *adarke*), salt coating on marsh vegetation; *adarkion,* dim.; see **salt**

add < L. *addo, -itus,* give to, put to; see **and, grow**

addax, L. an African animal with twisted horns; see **antelope**

adder < AS. *naedre,* snake; see **snake**

addictus, L. given up to, delivered to, devoted to, habituated; see **hold**

additus, L. given to, put to, joined to; see *addo* under **and**

adductus, L. stretched, strained, contracted; see **draw**

adecasto- < Gr. *adekastos,* unbribed, impartial; see **justice**

adecto- < Gr. *adektos,* not received, incredible, see **doubt;** unbitten, ungnawed, see **all**

adeletos, Gr. unhurt; see **safe**

adelos, Gr. unseen, unknown, obscure; see **secret**

adelphos, Gr. brother, twin, near kinsman; *adelphe,* sister; *adelphikos,* brotherly, sisterly; *adelphixis,* brotherhood, kinship; see **brother**

ademono- < Gr. *ademon, -os,* troubled, distressed; see **trouble**

adeno- < Gr. *aden, -os,* gland, see **gland;** < Gr. *aden,* enough, abundantly, see **wealth**

adephagos, Gr. gluttonous, greedy; see *phagein* under **eat**

adeps, -ipis, L. fat, grease, lard; see **fat**

adeptus, L. proficient; see **know**

aderco- < Gr. *aderkes,* unseen, invisible, unexpected; see **secret**

adercto- < Gr. *aderktos,* not seeing, sightless; see **blind**

aderitos, Gr. without strife; see **rest**

-ades, Gr. a patronymic suffix denoting son of, descendant of; see *-ides* under **son**

adetos, Gr. unbound, loose, free; see *adesmos* under **free**

adexios, Gr. left-handed, awkward; see **hand**

adhaesus, L. clinging to, holding to; see *haereo* under **hold**

adia, NL. (Gr. *adeia*), freedom from fear; see **safe**

adiantos, Gr. unwetted; see **dry**

adiantum, L. (Gr. *adianton*), maidenhair-fern; see **fern**

adiaphoros, Gr. indifferent, neutral; see **middle**

adiastaltos, Gr. somewhat indefinite; see **dim**

adiastolos, Gr. not separated, confused; see **confusion**

adibilis, L. accessible; see **accessible**

adica, L. (Gr. *adike*), nettle; see **nettle**

adico- < Gr. *adikos,* wrongdoing, injurious, unjust; see **hurt**

adicto- < Gr. *adeiktos,* not shown, unknown; see **secret**

adilo- < Gr. *adeilos,* fearless; see **bold**

adimo- < Gr. *adeimos,* fearless, see **bold**; L. *adimo, ademptus,* take away, see *demo* under **take**

adinos, Gr. close, thick, crowded, loud; see **thick**

adipatus; adipalis, L. fatty, greasy, < *adeps, -ipis,* fat, grease, lard; *adipeus,* of fat; see **fat**

aditus, L. entrance, approach; see **door**

adjectivus, L. added to; see *adjicio* under **and**

adjutor; adjutrix, L. helper; see *juvo* under **help**

admetos, Gr. unbroken (cattle), untamed; unwedded (maidens); see *adamastos* under **nature**

adminiculum, L. support, prop; see **pillar**

admodum, L. very, fully, completely; see **very**

admolie, Gr. uncertainty; see **doubt**

admones, Gr. a sea-fish; see **fish**

adnatus, L. joined to, united with; see **bind**

adoceto- < Gr. *adoketos,* unexpected; see **lot**

adocimo- < Gr. *adokimos,* spurious, false, base; see **lie**

adolescens, L. young; see **young**

adoleschos, Gr. prating, garrulous; *adolesches,* idle talker; see *lesche* under **speak**

adolos, Gr. without fraud, honest; see **true**

adonetos, Gr. unshaken; see **stand**

Adonis, L., Gr. a beautiful youth beloved by Venus; see **beauty**

ador, L. a kind of grain, spelt; see **grain**

adoratio, L. honor, worship; see *adoro* under **honor**

adoretos, Gr. not giving or receiving gifts; see *do* under **give**

adorno, -atus, L. decorate, embellish; see **ornament**

ados, Gr. satiety, loathing, see **hate**; decree, see **law**

adoxos, Gr. disreputable, disgraceful, obscure; see **bad**

adrachne, Gr. strawberry-tree; see **arbutus**

adrano- < Gr. *adranes,* inactive, powerless, feeble; see **weak**

adrastos, Gr. not running away; see **stand**

adulatus, L. < *adulor,* fawn, flatter, admire; see **flatter**

adultero, -atus, L. commit adultery, falsify, corrupt; see **lie**

adultus, L. grown up; see **ripe**

adumbratus, L. vaguely outlined, hazy; see *umbra* under **shade**

adunatus, L. united, made one; see *unus* under **one**

aduncus, L. bent, hooked; see *uncus* under **hook**

adustus, L. burned by the sun, tanned, brown, swarthy; see **brown**

advena, L. visitor, stranger; *adventor,* guest, visitor, customer; see **guest**

adventitius, L. additional, extraordinary; see **and**

adversus, L. against, opposed to; *adversarius,* enemy, rival; see **against, enemy**

adynatos, Gr. without strength, weak; see **weak**

adytum, L. (Gr. *adyton*), inmost portion of a temple, shrine; see **temple**

-ae, L. forms plurals; genitives of specific terms derived from feminine personal names and other nouns of the first Latin declension: algae, alumnae, antennae, costae, fossae, larvae, nebulae, striae, vertebrae, Compositae, Cruciferae, Leguminosae, *Rosa helenae* (a rose), *Begonia urticae* (a begonia). See *-ae* under **nature**

aecalo- < Gr. *aikalos,* flatterer, < *aikallo,* flatter; see **flatter**

aechmo- < Gr. *aichme,* spear; see **spear**

aecio-; aecidio- < NL. *aecium* (Gr. *aikia,* injury), a spore fruit or cluster cup of fungal spores; see **fruit**

aecismato-; aecismo- < Gr. *aikisma, -tos; aikismos,* injury, outrage, < *aikizo,* injure, torment; see **hurt**

aedemono- < Gr. *aidemon, -os,* bashful, modest; *aidemosyne,* modesty; see **modest**

aedes, Gr. distasteful, disagreeable, odious, unpleasant; see **bad**

aedicula, L. dim. of *aedis,* building, house, temple; *aedificium,* building of any kind; see **house**

aedo- < Gr. *aidos,* feeling of awe, respect, shame, modesty, see **modest;** *aidoin,* private part, see **sex organs**

aedoeo- < Gr. *aidoios,* august, venerable; see **honor**

aedon, Gr. nightingale; see **nightingale**

aeger, *-gra, -grum,* L. sick, troubled; *aegrotus,* sick, ill; see **disease**

aegialo- < Gr. *aigialos,* seashore; see **shore**

aegilops, L. (Gr. *aigilops*), wild oats, ulcer of the eye; see **grain, sore**

aegiro- < Gr. *aigeiros,* black poplar; see **poplar**

aegis, L. (Gr. *aigis*), shield of Zeus; see **shield**

aegithalo- < (Gr. *aigithalos*), titmouse; see **titmouse**

aegithus, L. (Gr. *aigithos*), hedge-sparrow; see **finch**

aeglo- < Gr. *aigle,* radiance, brightness; see **light**

aego- < Gr. *aix, aigos,* goat; *aigagros,* wild goat, see **goat;** a water-bird, see **duck**

aegolio- < Gr. *aigolios,* probably an owl; see **owl**

aegothelo- < Gr. *aigothelas,* a goatsucker; see **goatsucker**

aegrotus, L. ill, sick; *aegritudo,* sickness, sorrow; see *aeger* under **disease**

aegypio- < Gr. *aigypios,* vulture; see **vulture**

aei, Gr. ever, always; see **always**

aelino- < Gr. *ailinos,* mournful, plaintive; see **weep**

aella, Gr. stormy wind, whirlwind; *aellodes,* stormy; see **wind**

aellos, Gr. a bird; see **bird**

aeluro- < Gr. *ailouros,* cat; see **cat**

aema, Gr. blast, wind, < *aemi,* blow; see **wind, blow**

aeneto- < Gr. *ainetos,* praiseworthy; see **worth**

aeneus, L. brazen, of copper; *aenum,* brazen or copper vessel, kettle, caldron; see **copper, kettle**

aenicto- < Gr. *ainiktos,* puzzling; see *aenigma* under **secret**

aeno- < Gr. *ainos,* tale, see **story;** dreadful, dire, horrible, terrible, see **fear**

Aeolus, L. (Gr. *Aiolos*), god of the winds; *aiolos,* shifting, changeable, variable; see **wind**

aeonio- < Gr. *aionios,* lasting for an age, everlasting, < *aion,* age; see *aeon* under **age**

aeoreto- < Gr. *aioretos,* suspended, hovering, < *aireo,* hang, hover; see **hang**

aephnidio- < Gr. *aiphnidios,* sudden, quick, unforeseen; see **swift**

aepsero- < Gr. *aipseros,* quick, sudden; see **swift**

aepy- < Gr. *aipys,* high, lofty, steep, tall; *aipos,* height; see **high**

aequoreus, L. of any smooth even surface, < *aequor, -is,* an even, level surface, as of the calm sea; see **level**

aera, L. (Gr. *aira*), darnel, a kind of grass; hammer, see **grass, hammer;** L. *aera,* epoch, see **age**

aergos, Gr. not working, idle; see **rest**

aero- < L. *aer, -is* (Gr. *-os*), the lower atmosphere; *aerinos,* aerial, like air; L. *aerius* (Gr. *aerios*), of the air, airy, high; see **wind**

aersi- < Gr. *airo,* lift up, raise; in the air, high; see **high**

aerugo, *-inis,* L. verdigris; *aeruginosus,* greenish; see *aes* under **copper**

aerumna, L. trouble, distress, need; see **trouble**

aes, *aeris,* L. at first an alloy of copper and tin (bronze), later of copper and zinc (brass); *aereus,* of copper, bronze, or brass; *aerarius,* coppersmith; see **copper**

aesalon, L. (Gr. *aisalon*), a kind of hawk; see **hawk**

aeschro- < Gr. *aischros,* deformed, foul, shameful, obscene, ugly; see **bad**

Aeschylus, L. (Gr. *Aischylos*), Greek dramatist; see **play**

aeschyno- < Gr. *aischyne,* disgrace, shame; *aischyntos,* shameful; see **shame**

aeschynomene, L. (Gr. *aischynomene*), a sensitive plant; see **bean**

aeschyntelo- < Gr. *aischyntelos,* bashful, modest; see **modest**

Aesculapius, L. (Gr. *Asklepios*), god of medicine; see **heal**

aesculus, L. ancient name of an oak; horsechestnut; see **oak**

aesio- < Gr. *aisios*, lucky, auspicious; see **lot**

aestivus, L. pertaining to summer, < *aestas*, summer; see **year**

aestus, L. heat, fervor, unrest, passion; see **heat**

aesylo- < Gr. *aisylos*, evil, godless; see **bad**

aesyro- < Gr. *aisyros*, light as air, agile; see **light**

aethalo- < Gr. *aithalos*, thick smoke, soot; *aithalion*, dim.; *aithaleos*, smoky; *aithalodes*, sooty, black; see **dirt**

aethero- < Gr. *aither, -os*, the upper atmosphere; see **wind**

aethes, Gr. unusual, strange; see **strange**

aethio- < Gr. *aithiops* (*aithops*), *-opos*, sunburnt, fiery, black, scorched, swart, < *aitho*, burn; see **burn**

aetholico- < Gr. *aitholix, -ikos*, pustule, pimple; see **pimple**

aethrio- < Gr. *aithrios*, clear, fair; see **clear**

aethusa, NL. (Gr. *aithousa*), sun-porch; see **porch**

aethyctero- < Gr. *aithykter, -os*, darter; see **swift**

aethyia, NL. (Gr. *aithyia*), probably a shearwater; see **albatross**

aetio- < Gr. *aitios*, causing, responsible for, < *aitia*, cause, charge, blame, fault; see **begin**

aetnaeo- < Gr. *aitnaios*, a sea-fish; see **fish**

aetoma, Gr. gable; see **gable**

aëtos, Gr. eagle; *aëtideus*, eaglet; see **eagle**

aex < Gr. *aix, aigos*, goat; a water-bird; see **goat, duck**

aexi- Gr. augmentative, < *aexo*, old form of *auxo*, increase, enlarge, foster, strengthen, grow; see **grow**

aezeno- < Gr. *aizenos*, strong, active, vigorous; see **strong**

af- < L. *ad*, to; see **to**

afer, afra, afrum, L. African; see **Africa**

affectus, L. acted upon, influenced; see **make**

affinis, L. related to, neighboring; see **kin**

afflatus, L. breath, blast; see *flo* under **blow**

afflictus, L. struck down, distressed; see **hurt**

affluens, L. abundant, copious, rich; see **abundance**

afraid < OF. *esfrei*, < L. *ex*, out, OHG. *fridu* (G. *friede*), peace; see **fear**

Africa, L., f. Africa; *afer, afra, afrum; africus*, African: Africanus, Aframerican, *Cinnyris afer* (a sunbird), *Sphex afra* (a wasp), *Lycium afrum* (a spiny shrub), *Afropavo congensis* (Congo peacock), *Agapanthus africanus* (African lily).

L. *Aegyptus* (Gr. *Aigyptos*), m. Egypt: Egyptian, gypsy, gitano, ghetto, *Balanites aegyptiaca* (bito).

L. *Aethiopia* (Gr. *Aithiopia*), f. Ethiopia: Ethiopian, *Limeum aethiopicum* (a phytolaccacead), *Amblyospiza aethiopica* (a weaver).

L. *Libya* (Gr. *Libye*), f. North Africa; *Libys, -yos*, m. a Libyan; *Libykos*, Libyan, hence foreign, strange: Libyan, *Libyostrongylus douglassi* (a nematode), *Libyhipparion ethiopicum* (a fossil horse), *Libycosuchus brevirostris* (a fossil crocodile).

L. *Maurus*, m. a native of North Africa: Moor, Mauretania, blackamoor, Morocco.

L. *Numidia*, f. North Africa: Numidian, *Abies numidica* (Algerian fir).

after < AS. *aefter*, behind: afternoon, afterward, afterglow, aftermath, aft, abaft, hereafter.

Gr. *apo*, from, away, after; see **from**

L. *aversus*, turned away, turned backward, behind; *aversum*, n. the back part: averse, aversion, avert, aversation.

AS. *baec*, to the rear, return: back, backward, backslide, background, backfire, setback.

Gr. *bradys*, slow, late; see **slow**

Gr. *chronios*, after a long time, late, long; see *chronos* under **time**

Gr. *epi*, on, after; see **on**

Gr. *eschatos*, extreme, last: eschatology, *Eschatocerus acaciae* (a wasp).

L. *futurus*, about to be: future, futurity.

Gr. *hysteros*, after, later; *hystatos*, last: hysteresis, hysterogenic, *Hysterocrates greshoffi* (a spider), *Hystatomyia lamellata* (a fly), *Hysteroptissus conspergulus* (a bug).

Gr. *loisthos*, left behind, last:

Gr. *meta,* near, between, after, over; see **over**

Gr. *opisthen,* behind; *opisthios,* hinder: opisthograph, opisthural, opisthodetic, *Opisthocomus cristatus* (hoatzin), *Rana opisthodon* (a frog), *Oxyopisthen funebre* (a beetle), *Puffinus opisthomelas* (a shearwater), *Chaetopisthes sulciger* (a beetle).

Gr. *opsios,* late; *opsiaiteros,* later; *opsiaitatos,* latest; *opsigonos,* late-born; *opsikarpos,* fruiting late; *opsiotes, -etos,* f. lateness; *opsimos,* too late: *Opsiomyia palpalis* (a fly), *Opsigonus kruperi* (a beetle), *Opsitycha squalidella* (a moth).

Gr. *ouraios,* of the tail, hindmost; see *oura* under **tail**

L. *post,* after, behind: *posterus,* following; *posterior,* hinder; *postremus; postumus,* last, hindmost; *posterganeus,* behind one's back; *posticus,* that is behind: postpone, postscript, 8 p.m., post-graduate, post-mortem, posthumous, postil, puny, preposterous, posterity, postern, posterior, *a posteriori, Phytonomus posticus* (alfalfa-weevil), *Sphenocorynus posthumus* (a beetle).

Gr. *prymnos,* hindmost: *Kobus ellipsiprymnus* (waterbuck), *Prymnopteryx glaucina* (a beetle), *Hypsiprymnodon moschatus* (musk-kangaroo).

Gr. *pymatos,* hindmost, last:

L. *retro,* back, backward; see **return**

L. *secundus,* following, second; see *sequor* under **follow**

L. *serus,* late; *serior,* later; *serotinus,* happening late: serotine, serotinous, serotinal, *Prunus serotina* (wild black-cherry), *Amelanchier sera* (a serviceberry), *Salix serissima* (autumn-willow).

Gr. *teleutaios,* last; see *telos* under **end**

L. *ultimus,* last; see *ulter* under **far**

See: **follow, tail, stern, return**

afternoon, postmeridian, between noon and evening; see **day**

ag- < L. *ad,* to; see **to**

aga- < Gr. *agan,* very, much; see **very**

agaeo- < Gr. *agaios,* admirable, enviable; see **honor**

again < AS. *agen;* see **return**

against < AS. *ongean,* toward: gainsay.

L. *adversus,* against, opposed to: adverse, adversity.

Gr. *anarsios,* not fitting together, hostile, strange:

Gr. *antenopios; enopios,* face to face:

Gr. *anti,* against, opposed to, return; *antios,* opposite, facing; *antithetos,* opposed; *antestis,* f. a confronting; *anteres,* opposite; *enantios,* opposed, contrary; *enantiotes,* f. opposition; *enantioma, -tos,* n. obstacle, hindrance; *enantiosis,* f. contradiction, discrepancy; *katantion,* facing, opposite: antithesis, antitoxin, antidote, antiseptic, anticlimax, anti-American, anti-slavery, Antarctic, antagonist, anthem, Antares, antipodes, enantiomorphism, *Enantiocephalus cornutus* (a bug).

Gr. *axenos; apoxenos,* inhospitable:

L. *contra,* against, opposite: *contrarius,* against, opposite: counter, contrary, contrast, contradict, contraband, contralto, pro and con, counteract, controversy, encounter, counterfeit, counterpoint, comptroller, *Allomorphina contraria* (a foraminifer).   Derive: along, answer.

L. *conversus,* turned around, reversed: converse, conversely.

L. *erga,* against, toward:

L. *hypotenusa* (Gr. *hypoteinousa*), f. side opposite the right angle of a triangle: hypotenuse.

L. *inhospitalis; inhospitus,* unfavorable, unreceptive: inhospitable.

L. *interdictum,* a prohibition; see **forbid**

L. *invitus,* unwilling, reluctant: *Capsus invitus* (a bug).

Gr. *kata,* down, against; see **down**

L. *nolens,* unwilling: *Gammarus nolens* (an amphipod).

L. *oppositus,* on the other side, contrary, < *oppono, -positus,* set against: opposite, opposition, opposable, *Chrysosplenium oppositifolium* (a golden-saxifrage).

Gr. *para,* beside, near, against; see **near**

Gr. *peratos,* on the opposite side; see **end**

L. *repugnans,* contrary, opposed, distasteful: repugnant.

L. *rivalis,* competitor; see *rivus* under **stream**

L. *rursus,* turned back, reverse; see **return**

Gr. *zele,* a female rival; see *zelos* under **busy**

See: **near, not, turn, enemy, stand, forbidden, bar, hinder, slow, rebellion**

agalena, NL. a kind of spider; see **spider**

agalliasis, Gr. great joy; see **joy**

agallis, -idos, Gr. a plant; see **plant**

agallochon, Gr. an East Indian wood yielding a dark aromatic resin; see **resin**

agalma, -*atos,* Gr. glory, delight, honor, statue, < *agallo,* glorify, exalt, rejoice; see **form**

agamos, Gr. unmarried, single, asexual; see **alone**

agan, Gr. very, much; see **very**

aganactico- < Gr. *aganaktikos,* fretful, irritable, peevish; see **pain**

aganos, Gr. mild, gentle, kindly; see **mild**

agapetos, Gr. beloved, < *agape,* love, charity; *agapesis,* affection; see **love**

agar, Malay for a gelatinous substance from seaweeds; see **jelly**

agaricum, L. (Gr. *agarikon*), tinder-fungus; see **fungus**

agaso, L. driver, hostler, groom, lackey; see **servant**

agastache, NL. a genus of the mint family, < Gr. *agastachys,* with many ears or spikes; see **mint,** *stachys* under **point**

agastor, Gr. near kinsman, twin; see **brother**

agastos, Gr. admirable; *agasma,* object of adoration; see *agauma* under **honor**

agasyllis, Gr. a plant of the carrot family; see **carrot**

agate < L. *achates* (Gr. *Achates,* river in Sicily); see **stone**

agathis, Gr. ball of thread; *agathidion,* dim., see **ball;** a genus of conifers, see **gymnosperm**

agathos, Gr. good; *areion; ameinon,* better; *aristos,* best; see **good**

agave < L. *Agave* (Gr. *Agaue*), a mythological woman, < *agauos,* noble, admirable; a genus of the amaryllis family; see **amaryllis, honor**

age < L. *aevum,* n. lifetime, period; *aetas,* -*atis; aevitas,* -*atis,* f. lifetime; *coaevus,* of the same age; *longaevus,* of great age: primeval, coeval, longevity, medieval, aged, eternal, nonage, dotage.
L. *aeon,* -*is* (Gr. *aion,* -*os*), m. lifetime, age; *aionios,* lasting an age, everlasting: eon, *Aeonium sedifolium* (a crassulacead).
L. *aera,* f. epoch: era.
Gr. *helikia,* f. time of life, age, manhood, womanhood, maturity; *homelix,* -*ikos,* of the same age: *Homelix cribratipennis* (a beetle).
L. *seculum,* n. age, lifetime, spirit of the times, world: secular, *in secula seculorum.*
L. *senectus,* -*utis,* old age, senility; see *senex* under **old**
See: **time, old, young, always**

-age < L. -*aticum,* collection of, condition, state of being; see **nature**

agelaeo-; agelo- < Gr. *agelaios,* gregarious, < *agele,* herd; see **herd**

agellus; agellulus, L. dim. of *ager,* field; see **field**

agenes, Gr. of low family, mean; see **poor**

agenio- < Gr. *ageneios,* beardless; see **bare**

agent < L. *agens,* -*entis,* doing, strong, < *ago, actus,* do; see **make, strong, tool**

ager, L. field; *agellus; agellulus,* dim.; see **field**

ageratum, L. (Gr. *ageraton*), a genus of the composite family, < *ageratos,* not growing old; see **composite, always**

agerochos, Gr. haughty, arrogant; see **proud**

agetos, Gr. admirable; see *agauma* under **honor**

agger, L. heap, mound; see **heap**

agglomeratus, L. balled, gathered into a mass; see *glomus* under **ball**

aggregatus, L. collected, clustered, united; see *grex* under **herd**

agilis, L. quick, light, nimble; see **light**

agitatus, L. stirred up, aroused, in motion, < *agito,* set in motion, arouse; see **arouse**

aglaos, Gr. splendid, shining, bright, beautiful, noble; *aglaia,* splendor, beauty; see **light**

agleuco- < Gr. *agleukes,* not sweet, sour; see **sour**

aglis, Gr. head of garlic; see **onion**

agma, Gr. fragment, splinter; see *agmos* under **break**

agminalis, L. of any army, < *agmen,* -*inis,* army on the march, multitude, train; see **army**

agnos, Gr. a chastetree; see **vervain**

agnostos, Gr. unknown, unknowing, ignorant of; see **dull, secret**

agnus, L. lamb; *agnicellus; agniculus,* dim.; *agninus,* of a lamb; see **sheep**

ago < AS. *agan*, go away; see **before**

ago, *actus*, L. do, make, drive, see **make**; < Gr. *ago*, lead, guide, train, see **lead**

-ago; **-igo**; **-ugo**, L. having the characteristics of; see **nature**

-agogue, Gr. that which stimulates, causes to function, induces, promotes, < *agogos*, leading, guiding, attracting; *agoge*, a carrying away, movement; see **arouse**

agolum, L. shepherd's staff or crook; see **rod**

agon, *-os*, Gr. gathering, assembly to see games, arena, contest; see **fight**

agonos, Gr. unfruitful, barren, childless; see **sterile**

agony < Gr. *agonia*, contest, pain, anguish; see **pain**

agora, Gr. assembly, market; see **gather, trade**

agos, Gr. leader, chief, see **lead;** curse or pollution that requires expiation or sacrifice, see **curse**

agostos, Gr. the bent arm; see **arm**

agra, Gr. catch, booty, prey, seizure; *agraios*, of the chase, < *agreuo*, take by hunting or fishing; see *agreuo* under **hunt, disease**

agrarius, L. of the land; see *ager* under **field**

agraulos, Gr. dwelling in the field, living outdoors; see *ager* under **field**

agree < L. *ad*, to, *gratus*, pleasing; see **harmony, yes**

agreeable
L. *acceptus*, welcome, agreeable: acceptable.
L. *affabilis*, easy to speak to: affable.
L. *ambrosius* (Gr. *ambrosios*), divine, lovely, pleasant; see *ambrosia* under **food**
L. *amoenus*, pleasant, delightful: amenity, *Passerina amoena* (lazuli bunting).
L. *arcadius*, ideally rustic; see **country**
Gr. *arestos*, pleasing, acceptable; *areskos*, desirous to please, obsequious, flattering, < *aresko*, make amends, placate, please, flatter:
Gr. *armenos*, fitting, acceptable; see **fit**
L. *arrisor*, one who smiles on another, fawner, flatterer; see **flatter**
Gr. *asmenistos*, acceptable, welcome: *Asmenistis cucullata* (a moth).
Gr. *aspasios; aspastos*, gladly welcomed, acceptable: Aspasia.
L. *assentor*, *-tatus*, agree constantly, flatter; see **yes**
Gr. *astikos* (*astykos*); *asteios*, urbane, refined, elegant, nice; see *asty* under **town**
L. *blanditia*, f. caress, flattery, cajolery; *blanditus*, agreeable, charming < *blandus*, of a smooth tongue, flattering, pleasant, mild: bland, blandly, blandness, blandishment, blandiloquent, *Anemone blanda* (Greek anemone).
Gr. *charma*, source of joy, delight; see **joy**
L. *comis*, courteous, kind, friendly, obliging, amiable; *comitas*, *-atis*, f. affability, friendliness: comity.
L. *commodus*, fitting, agreeable, proper; see **fit**
L. *congruus*, suitable, agreeable, fitting; see **fit**
L. *consensus*, agreement, concord; see *consentio* under **harmony**
L. *contentus*, satisfied: content, contentment.
Gr. *dektos; dektikos*, acceptable; see *dechomai* under **take**
L. *delecto*, *-atus*, amuse, delight, please; *delectabilis*, delightful, agreeable; see **joy**
L. *delicatus*, delightful, soft, tender; see **soft**
Gr. *ephimeros*, desired, delightful, agreeable: *Ephimeropus geniculatus* (a beetle).
Gr. *epicharis*, agreeable, pleasing, charming; see *charis* under **beauty**
Gr. *epieros*, pleasant, acceptable: *Epieropsis geminata* (a beetle).
Gr. *ethelo-*, willing, voluntary; see *ethelo* under **wish**
Gr. *eucharistos*, agreeable, pleasant, grateful; see *charis* under **beauty**
L. *favorabilis*, agreeable, auspicious: favorable.
L. *festivus*, pleasant, agreeable, cheerful; see **joy**
L. *genialis*, delightful, jovial, pleasant: genial, congenial, geniality.
L. *gratus*, beloved, pleasing, agreeable, thankful; *gratiosus*, full of favor, popular; see *gratia* under **give**
Gr. *hekon*, *-tos*, willing, agreeable:
Gr. *hestikos*, agreeable, pleasing: *Hesticus pictus* (a bug).
L. *jocundus* (*jucundus*), pleasant, agreeable, delightful, merry, sportive; see *jocus* under **laugh**
Gr. *keletikos; keleterios*, charming, delightful; *kelema*, *-atos; keletron*, *-os*, n. charm, spell; *kelesis*, f. enchantment; *keletes* (*-or*), m. charmer: *Celetothrips breviceps* (a thrips), *Celetes binotatus* (a beetle), *Celetor strigipennis* (a fly).
Gr. *laros*, agreeable, pleasant, lovely: *Larodryas haplocala* (a moth).
L. *lepidus*, pleasant, agreeable, neat, elegant, witty, < *lepor*, *-is*, m. attractiveness, charm: *Rhynchotreta lepida* (a fossil brachiopod).

L. *libitus*, pleasing, agreeable; *libens*, doing willingly, with pleasure; *collibitus*, agreeable: *ad libitum*, libido, quodlibetarian.

L. *morigeror*, *-atus*, comply with, endeavor to please, adapt oneself to: morigerate, morigerous.

Gr. *nedymos*, delightful: *Nedymoserica flavida* (a beetle).

Gr. *ourios*, fair, prospering, favoring, happy: *Uriolelaps argenticoxae* (a wasp).

L. *placeo*, *-citus*, please, be agreeable, satisfy; *complacitus*, pleased, favorable: placid, complacent, plea, displease, pleasure, plead, unpleasant, placebo.

L. *propitius*, favorable, gracious, kind: propitious, propitiate, unpropitious.

L. *prosperus*, agreeable, favorable: prosperous, prosperity.

Gr. *terpnos*, delightful, agreeable; see **joy**

L. *vinnulus*, delightful: *Dicranura vinula*[*vinnula*] (puss-moth).

See: **agree, joy, harmony, fit, beauty, love, sweet, laugh, good, yes**

agremon, Gr. net; see **net**

agrestis, L. of the land, rural, wild; see *ager* under **field**

agreutes, Gr. hunter; see *agreuo* under **hunt**

agri-; agro- < L. *ager, agri* (Gr. *agros*), field, open country, land; see **field**

agricola, L. farmer; *agricolaris*, of farmers; see **farmer**

agrimonia, L. a genus of the rose family; see **rose**

agrios, Gr. wild; see **nature**

agriphe, Gr. harrow, rake; see **rake**

agrippos, Gr. wild olive; see **olive**

agro- < Gr. *agros*, field, country; *agroikos*, of the country; see *ager* under **field**

agronomos, Gr. rural, wild; see *ager* under **field**

agropyron, NL. a genus of grasses; see **grass**

agrostis, Gr. a forage grass; see **grass**

agrotes, Gr. countryman; *agrotikos*, rustic, rural; see *ager* under **field**

agrycto- < Gr. *agryktos*, not to be spoken of; see **ineffable**

agrypnos, Gr. sleepless, wakeful, watchful; see **awake**

aguro- < Gr. *agouros*, a youth; see **young**

agyia, Gr. street, highway; *agyiates*, neighbor; see **way, neighbor**

agyniax, Gr. wifeless; see **alone**

agyrto- < Gr. *agyrtos*, got by begging; *agyrtes*, beggar, collector; *agyrma*, anything collected; see **beggar**, *agora* under **gather**

aid < OF. *aide*, < L. *adjuvo*, help; see **help**

ailanthus, NL. a genus of the quassia family; see **quassia**

aim < OF. *esmer*, < L. *aestimo*, judge.

Gr. *boulema*, *-tos*, n. design, purpose; *bouleutos*, devised, plotted; *bouleutes*, m.; *bouleutis*, *-idos*, f. adviser, planner, schemer; *epiboule*, f. plot, scheme; *epiboulos*, plotting, designing, scheming; *aboulos*, thoughtless, < *boule*, f. will, counsel; *euboulia*, f. good counsel; *euboulos*, well-advised, prudent: abulia, *Eubulopsis edentatus* (a beetle).

L. *collineo*, *-atus*, aim, direct:

L. *consilium*, n. consultation, counsel, plan; see *consulo* under **think**

L. *conspiratio*, *-onis*, f. plot, agreement in plan: conspire, conspiracy.

L. *destino*, *-atus*, determine, aim at; *destinatio*, *-onis*, aim, end, goal, mark; see **end**

L. *dirigo*, *-rectus*, aim, steer: direct, direction, indirect.

L. *institutum*, n. design, purpose, plan: institute, institution.

L. *intentio*, *-onis*, f. aim, purpose, < *intendo*, stretch towards: intention, intend.

Gr. *medos*, *-eos;* pl. *medea*, n. counsel, plan; *mestor*, *-os*, m.; *mestris*, f. adviser, counselor: Archimedes, *Diomedea nigripes* (an albatross), *Apomestris westwoodi* (a beetle).

L. *meta*, goal; specifically, the columns at the extremities of the Roman circus; see **pillar**

Gr. *programma*, *-tos*, n. public notice, prospectus, outline, or plan to be followed: program.

L. *propositum*, plan, design, purpose; see *propositio* under **hypothesis**

Gr. *schema*, *-atos*, n. form, shape, plan; *schemation*, n. dim.: scheme, schemer, schematic, *Gnorimoschema gallaesolidaginis* (a moth).

L. *scopus*, m. a mark at which to shoot:

Gr. *stochos*, m. aim, shot, guess; *stochasmos*, m. guess, conjecture; *stochasma*, *-tos*, n. the thing aimed, arrow, spear: *Stochomys longicaudatus* (a mouse), *Stochasmus exilis* (a crustacean), stochastic.

See: **hypothesis, mark, end**

air < L. *aer, -is,* atmosphere; see **wind**

aïstos, Gr. unseen; see **secret**

aïthales (aeithales), Gr. evergreen; see **green**

aizoon, L. an evergreen plant; see *zoe* under **life**

ajuga, L. a kind of mint; see **mint**

al, Ar. the : alchemy, algebra, alfalfa, admiral, elixir, Algol, apricot, artichoke, lute, Harun al Raschid.

al- < L. *ad,* to; see **to**

ala, L. wing; see **wing**

alabaster, L. a kind of gypsum, usually white, used in carving; also a kind of calcite; see **gypsum**

alabe, Gr. a kind of ink; see **ink**

alabeta, L. (Gr. *alabes*), a Nile fish; see **fish**

alacer, L. lively, active, brisk, eager; *alacritas, -atis,* briskness, ardor, eagerness; see **life**

alalos, Gr. speechless, dumb; see **silent**

alaos, Gr. blind; see **blind**

alapa, L. box on the ear; see **strike**

alapadnos, Gr. feeble, weak; see **weak**

alaris; alarius, L. of the wing; see *ala* under **wing**

alastos, Gr. unforgettable, unceasing, avenging; *alastor,* avenger; see **punish**

alaternus, L. a buckthorn; see **buckthorn**

alatus, L. winged, < *ala,* wing; *alicula; alula,* dim.; see **wing**

alauda, L. lark; see **lark**

alazon, Gr. vagabond, impostor, quack, braggart; see **lie**

albarius, L. plasterer; see **plaster**

albatross < Sp. *alcatraz,* a seafowl, pelican, frigate-bird: *Diomedea albatrus* (short-tailed albatross).
  Gr. *aithyia,* f. probably a shearwater: *Aethyia islandica* (a bird).
  NL. *diomedea,* f. a genus of albatrosses, < Diomedes, a Trojan hero: *Diomedea nigripes* (black-footed albatross).
  It. *fregata,* f. a ship; a genus of sea birds: frigate, *Fregata aquila* (a frigate-bird).
  NL. *fulmarus,* m. a genus of the family Procellariidae: *Fulmarus glacialis* (fulmar).
  NL. *oceanodroma,* f. a genus of petrels: *Oceanodroma furcata* (fork-tailed petrel).
  NL. *puffinus,* m. a shearwater: puffin, *Puffinus griseus* (sooty shearwater).

albizzia, NL. a genus of the bean family; see **bean**

albugo, L. a white spot, a disease of the eye; see *albus* under **white**

album, L. list, register; see **list**

albumen, L. white of an egg; see **protein**

alburnum, L. sapwood of trees; see **wood**

albus, L. white; *albidus; albulus,* whitish; *albico,* whiten; see **white**

alca, NL. a genus of birds including the auks; see **auk**

alcaeo- < Gr. *alkaios,* strong; see *alke* under **strong**

alcea, L. (Gr. *alkea*), a kind of mallow; see **mallow**

alcedo, L. kingfisher; see **kingfisher**

alces, L. (Gr. *alke*), elk; see **deer**

Alcestis, L. (Gr. *Alkestis*), wife of Admetus, whose life she saved by offering her own; see **give**

alcesto- < Gr. *alkestes,* a fish; see **fish**

alchemilla, NL. a genus of the rose family; see **rose**

alcibius, L. (Gr. *alkibios*), a kind of bugloss; see **borage**

alcimo- < Gr. *alkimos,* strong, stout, brave; see *alke* under **strong**

alcohol < Ar. *alkuhl;* the suffix *-ol* signifies an alcohol: alcoholic, alcoholism, acetol, cholesterol, ketol, sterol, glycerol, carbinol, santalol, tribromethanol (avertin), aldehyde, alkyl, alkylamine.
  Pref. *ethyl-,* < Gr. *aither,* ether, *hyle,* wood, material; pertaining to the hydrocarbon radical in grain alcohol: ethyl alcohol (ethanol), ethylamine, ethylate, ethylene, diethylstilbestrol, *Amylobacter ethylicus* (a bacterium).

Suf. *-hyl, -yl,* < Gr. *hyle,* wood, stuff, material; generally an alcohol: methyl, ethyl, benzyl, ceryl, cetyl, cinnamyl, glyceryl, isopropyl, acetylsalicylate.

Pref. *methyl-,* < Gr. *methy,* wine, *hyle,* wood, material; pertaining to the hydrocarbon radical in wood alcohol: methyl alcohol (methanol), methylamine, methylate, methylene, methane, diazomethane.

alcyon, L. (Gr. *alkyon*), kingfisher; see **kingfisher**

alcyonium, L. (Gr. *alkyonion*), polyp; see **coral**

aldehyde < alcohol *dehy*drogenatum, that is, an alcohol deprived of some of its hydrogen; a compound having the CHO group: formaldehyde, benzaldehyde, salicylaldehyde.

Pref. *acrylo-,* pertaining to derivatives of the aldehyde acrolein: acrylic acid, acrylate, acrylyl, polymethyl methacrylate.

alder < AS. *aler,* akin to G. *erle,* L. *alnus.*

L. *alnus,* f. alder; *alneus,* of alder: alnein, *Alnus rugosa* (an alder), *Viburnum alnifolium* (hobblebush), *Alniphyllum fortunei* (China bells), *Microsphaera alni* (a fungus).

Gr. *klethra,* f. alder; now a genus of the heath family; see **heath**

ale < AS. *alu,* see **beer;** < Gr. *ale,* a wandering, see **wander**

alea, L. a game of chance with dice; *aleator,* dice-player, gamester, see **play;** Gr. *alea,* warmth, heat, see **heat**

alec, L. sauce prepared from small fish; see **sauce**

alecto- < Gr. *alektos,* unceasing, see **always;** not to be told, see **secret**

alector, L. (Gr. *alektor; alektryon*), cock; see **chicken**

alectro- < Gr. *alektros,* unmarried; see **alone**

alema, Gr. flour, meal; see *aleuron* under **flour**

alemon; aletes, Gr. wanderer, rover; see *ale* under **wander**

aleos (eleos), Gr. astray, distraught, crazed; see *eleos* under **mad**

alert < F. *a l'erte,* on watch; see **awake**

-ales, L. suffix used to form names of plant orders; see **class**

aletho- < Gr. *alethes,* true, real; *aletheia,* truth; see **true**

aletos, Gr. a grinding; *aletes,* grinder, wanderer; *aletris,* female slave who ground grain, < *aleo,* grind; see **grind, wander**

aleuron, Gr. flour; see **flour**

alexeter, Gr. guard, defender; *alexesis,* defense; *alexis,* help, < *alexo,* avert, defend, ward off; see **guard**

alga, L. seaweed: algae, algoid, algology, algicide, algin, algic, algivorous.

L. *chara,* f. a kind of plant; name now applied to a genus of algae: Characeae, *Chara crinita* (a stonewort).

L. *conferva,* f. an aquatic plant; a genus of algae: Confervoideae, *Conferva bombycina* (a seaweed), *Nitella confervacea* (a characead), *Potamogeton confervoides* (a pondweed).

L. *fucus,* m. rock-lichen; a genus of seaweeds; fucoid, fucivorous, *Fucus vesiculosus* (a rockweed), *Roccella fuciformis* (a lichen yielding orchil), fucose, Fucales, *Didymosphaeria fucicola* (a fungus).

NL. *laminaria,* f. a genus of seaweeds: *Laminaria digitata* (fan-kelp).

Gr. *phykos, -eos,* n. alga, seaweed; *phykarion; phykion,* n. dim.: phycology, phycocyanin, Rhodophyceae, Chlorophyceae, *Calamophycus septum* (a fossil plant).

Pg. *sargasso,* m. gulfweed, < *sarga,* f. grape: Sargasso Sea, *Sargassum natans* (a seaweed).

L. *ulva,* f. sedge; now applied to a genus of seaweeds: *Ulva latissima* (sea-lettuce), *Hydrobia ulvae* (a snail).

-algia < Gr. *algos,* pain, < *algeo,* feel pain, suffer; see **pain**

algidus, L. cold, < *algeo,* feel cold, be cold; see **cold**

alibas, -antos, Gr. corpse; see **body**

alibi, L. elsewhere, at another place, with some other person; *alias,* another (time, person, occasion, name); see **other**

alibilis, L. nourishing; see *alo* under **eat**

alica, L. spelt (probably emmer); *alicarius,* of spelt; see **grain**

alicula, L. a light, upper garment; see **garment**

alienus, L. foreign, strange, not related; *alieno, -atus,* estrange, drive insane; see **strange, change**

aliger, L. winged; see *ala* under **wing**

alimentum, L. food; *alimentarius,* of food; see *alo* under **eat**

alimmato- < Gr. *aleimma, -tos,* fat, oil, unguent; see *aleiphar* under **fat**

alimonium, L. sustenance, support; see **receive**

aliphato- < Gr. *aleiphar, -atos,* fat, oil, unguent; *aleiphò,* anoint with oil; see **fat**

alipta, L. (Gr. *aleiptes*), trainer and teacher in gymnastic schools; see **teach**

aliquot, L. some, few, an indefinite number; see **few**

-alis, L. pertaining to, having the nature or quality of; see **nature**

alisma, Gr. an aquatic plant; see **waterweed**

aliso- < Gr. *aleison,* cup, goblet; see **cup**

alitema, Gr. sin, offense; *alitros,* sinful; see **fault**

alitus; altus < L. *alo,* nourish, feed; see **eat**

alius, L. another, other; see **other**

alkali < OF. *alcali,* < Ar. *al qaliy,* ashes of saltwort (*Salsola kali*): alkaline, alkaloid, alkalosis, kalium (K), *Alcaligenes*(*Brucella*) *melitensis* (bacterium causing Malta fever).

   L. *lix, -icis,* m. ashes, lye; *lixivia,* f. lye; *lixivius,* made into lye: lixiviate.

all < AS. *eall,* the entire quantity or extent: altogether, withal, wherewithal, all-embracing, heal-all, almighty, All-Hallows, alone.

   L. *absolutus,* finished, perfect, complete, unfettered, < *absolvo,* loosen, free: absolute, absolutely, absolution, absolve.

   Gr. *adektos,* unbitten, ungnawed, unmolested:

   L. *admodum,* completely, fully, wholly, very; see **very**

   Gr. *akios,* not worm-eaten: *Aciodoris lutescens* (a gastropod).

   Gr. *artios,* complete, perfect, even (as applied to numbers): artiodactyl, *Artiocotylus speciosus* (a turbellarian), *Artionyx gaudryi* (an Oligocene mammal), *Artiopyteryx elegans* (a lacewing), *Artioposthia triangulata* (a worm).

   Gr. *athroos,* assembled in crowds, collected all at once, once for all, the whole mass; see **number**

   Gr. *autokrates,* absolute, having full power: autocrat, autocracy.

   L. *catholicus* (Gr. *katholikos*), universal, general: catholic, Catholicism.

   L. *compleo, -etus,* fulfill, finish: *expletus,* complete, perfect: completion, incomplete, expletive.

   L. *cunctus,* all, the whole; *cunctalis,* general: *Non sibi, sed cunctis.*

   Gr. *despotikos,* absolute, inclined to tyranny: despot, despotic, despotism.

   Gr. *ekpleos,* complete, entire: *Ecpleopus gaudichaudi* (a lizard).

   L. *funditus,* entirely, totally, completely:

   Gr. *holos,* whole, complete, all; *holikos,* universal, general: holoblastic, holograph, holohedral, holomorphic, catholic, *Holodiscus discolor* (a rosacead), *Holosteum umbellatum* (jagged chickweed).

   L. *idealis,* existing in idea, perfect: ideal, idealistic.

   L. *illabefactus,* unbroken, unimpaired, unshaken:

   L. *illaesus,* unhurt, unimpaired: *Brachydere illaesus* (a beetle).

   L. *illibatus,* undiminished, unimpaired; *illibilis,* incapable of diminution: *Phalaena illibata* (a moth).

   L. *incolumis,* unimpaired, safe, sound, entire, whole; see *columis* under **safe**

   L. *incomminutus,* not broken, entire:

   L. *indiscissus,* undivided, entire:

   L. *indiscretus,* unseparated, undivided, closely connected; see **one**

   L. *indispertibilis,* indivisible:

   L. *indivisus,* undivided; *individuus; indivisibilis,* not to be divided: indivisible, individual, *Psilepyris indivisus* (a wasp).

   L. *infractus,* unbroken: *Cicindela infracta* (a beetle).

   L. *intactus,* untouched, untried, chaste, whole: intact.

   L. *integer, -gra, -grum,* whole, entire, complete, perfect; *integritas, -atis,* f. completeness, soundness: *integro, -atus,* make whole, restore: integer, integrity, entire, integrate, disintegration, integral, *Ceanothus integerrimus* (deerbrush), *Rhus integrifolia* (lemonade-sumac).

   Gr. *oikoumenikos,* worldwide, universal, general; see *oikoumene* under **world**

   L. *omnis,* all: omnipotent, omniscience, omnibus, omnipresent, omnivorous, *Phymatotrichum omnivorum* (a root-rot fungus).

   Gr. *pal-, pam-, pan-, pasi-* < *pas, pantos,* all, the whole, every, very; *pampan,* quite whole; *hapas,* quite all; *sympas,* all together, all at once; *palleukos,* all white: panacea, Pantheon, panorama, panegyric, pandemonium, Pandora, pasigraphy, panoply, Pan-American, diapason, pamphagous, pamphlet, pantaloon, pantomime, Palermo, pantograph, pantology, *Panax trifolium* (dwarf ginseng), *Pasiphila bilineolata* (a moth), *Pamphila linea* (a butterfly), *Pamborus elongatus* (a beetle), *Pamphagus cucullatus* (a grasshopper), *Pantolestes longicaudus* (an Eocene mammal), *Pantolambda bathmodon* (a Paleocene mammal), *Pancratium canariense* (a sea-daffodil).

L. *perfectus,* complete, finished: perfect, perfection, imperfect.

L. *plenus,* full, complete; see **abundance**

L. *sollus,* f. entire, complete: solemn, solicitude.

L. *summa,* f. the entire quantity; *summalis,* of a sum or whole; *consummo, -atus,* sum up, complete: sum, summary, summarily, summit, consummate, *summum bonum.*

Gr. *teleos; teleios,* having reached its end, finished, complete, perfect, < *teleo,* complete, fulfill, accomplish; see *telos* under **end**

L. *totus,* all, whole; *totietas, -atis,* f. the whole: total, teetotaler, factotum, *in toto,* tutsan, totitive, totiety.

See: **one, number, abundance, and, health, safe, very**

allago- < Gr. *allage,* change, exchange; *allagma,* that which is given in exchange, price, < *allasso,* change, alter; see **change, trade**

allanto- < Gr. *allas, -antos,* sausage; see **sausage**

allatus < L. *affero,* bring, carry; see *fero* under **carry**

allegatio, L. assertion of proof, excuse; see **excuse**

allegory < Gr. *allegoria,* a metaphorical story; see **story,** FIGURES OF SPEECH under **form**

allelo- < Gr. *allelon,* one another, reciprocal; see **alternate**

allergy < NL. *allergia,* sensitivity to inoculation with foreign substances; see **disease**

alley < OF. *aler,* go, < L. *ambulo,* go, walk; see **walk, way**

alligator < Sp. *el lagarto,* < L. *ille,* that, *lacerta,* lizard; see **reptile**

allistos, Gr. inexorable; see **merciless**

allisus, L. struck against; see *laedo* under **strike**

allium, L. garlic; see **onion**

allo- < Gr. *allos,* other; *allodapos; allokotos; allophylos; allotrios,* of another kind or race, foreign, strange; see **other, strange**

alloeo- < Gr. *alloios,* of another kind, different; see **different**

allow < OF. *alouer,* < L. *ad,* to, *locus,* place; see **let**

alloy < F. *aloyer,* < L. *ad,* to, *ligo,* bind; see **mix**

allusio, L. a direct or indirect reference; see FIGURES OF SPEECH under **form**

alluvium, L. detritus deposited by running water; see *luo* under **wash**

ally < F. *allier,* < L. *ad,* to, *ligo,* bind; see **companion**

allyl- < L. *allium,* garlic; pertaining to C_3H_5 compounds found in garlic and onions; see **onion**

almond < OF. *almande,* < L. *amygdala* (Gr. *amygdale*); see **peach**

almost < AS. *ealmaest,* nearly; see **near**

almus, L. nourishing, cherishing, fostering, kind, bountiful; see **guard**

alnus, L. alder; *alneus,* of alder; see **alder**

alochos, Gr. bedfellow, spouse; see **companion**

aloco- < Gr. *alox, -okos,* furrow; see **furrow**

aloe, Gr. aloe; see **lily**

alogos, Gr. speechless, irrational, absurd; *alogistos,* silly; *alogia,* absurdity; see **fool**

alone < AS. *eall,* all, *an,* one: loneliness, lonesome.

L. *agamus* (Gr. *agamos*), unmarried (bachelor), single, asexual: agamic.

Gr. *ageiton, -os,* neighborless, isolated, solitary: *Ageiton ater* (a fly), *Agiton idioptila* (a moth).

Gr. *agyniax; agynos,* wifeless:

Gr. *alektros,* unbedded, unmarried: *Alectrosaurus olseni* (a dinosaur).

Gr. *ametor, -os,* motherless: *Ametor rudisculptus* (a beetle).

Gr. *anachoretes,* m. one who has retired from the world, hermit, recluse; *anachoresis,* f. retirement: anchorite, anchoret.

Gr. *anandros,* without men, husbandless, single: *Anandra capriciosa* (a beetle).

Gr. *anomiletos,* unsociable:

Gr. *anymphos,* without bride, unwedded: *Anymphochaeta fuscinervis* (a fly).

Gr. *aphilos,* friendless: *Aphiloctenus virginiensis* (a wasp).

Gr. *apozyx,* separated, single; see *apozeuxis* under **free**

Gr. *aprosmiktos,* isolated, solitary: *Aprosmictus scapulatus* (a bird).

Gr. *aprosmilos,* unsociable:

Gr. *asketikos,* practicing extreme self-denial, solitary devotion: ascetic, asceticism.

Gr. *astibos,* untrodden, desert; *apostibes,* off the road, solitary: *Apostibes griseolineata* (a moth).

Gr. *ataurotos,* unwedded, virgin:

Gr. *autos,* self; *automatos,* self-moving, spontaneous; *philautos,* loving oneself, selfish: automatic, autograph, autobiography, autosuggestion, automobile, autocrat, autonomy, autointoxication, autopsy, autotomy, *Autoplectus torticornis* (a beetle).

L. *avius,* out of the way. unfrequented, remote, solitary; see *via* under **way**

Gr. *azyx, -ygos,* unyoked, unwedded, solitary: *Azygophleps scalaris* (a moth).

ML. *baccalaureus.* m. apprentice, single man who has never married: bachelor, A.B., baccalaureate, *Baccalaureus verrucosus* (a barnacle).

L. *caelebs, -ibis,* unmarried (man), bachelor, single; *caelibatus,* pertaining to single life: celibate, celibacy, *Fringilla caelebs* (European chaffinch).

L. *cellulanus,* m. hermit, recluse; solitary: *Meta cellulana* (a spider).

Gr. *chera,* widow; see **woman**

L. *dissocialis,* selfish, unsocial, < *dissocio, -atus,* separate from fellowship: dissocial, dissociate.

L., Gr. *ego,* I, myself: egoist, egotistical, Ego, egoism.

Gr. *eidikos,* specific, special:

Gr. *ekositos,* feeding at home, living at one's own cost:

Gr. *eremos,* solitary, lonely; *eremia,* f. desert, solitude, wilderness; *eremikos,* of solitude; *eremites,* m. hermit: eremite, hermit, eremitical, *Eremocitrus glauca* (desert-lime), *Cordylanthus eremicus* (Death Valley birdbeak), *Gilia eremica* (desert-gilia), *Eremurus robustus* (a lily), *Eremobates cinerea* (a solpugid).

Gr. *hekastos,* every, each; *hekateros,* singly, either: *Hecastophyllum monetarium* (a legume).

Gr. *hesychastes,* m. hermit:

AS. *ic,* I: I.

Gr. *idios,* one's own, personal, individual, distinct, peculiar; *idikos,* special, peculiar, one's own; *idiotikos,* private, unskillful; *idiogenes,* peculiar; *idiastes,* m. recluse, hermit; *idiazo,* live in retirement; *idioma, -atos,* n. a peculiarity of language; *idiomatikos,* peculiar: idiom, idiot, idiotic, idiosyncrasy, hypidiomorphic, idioblast, *Idiopogon uranopola* (a moth), *Idiogenes otides* (a cestode), *Idiastes costatus* (a beetle), *Pseudidiops camelus* (a spider), *Idiurus macrotis* (a mouse-squirrel).

L. *incomitatus,* unaccompanied, alone: *Orthoceras incomitatum* (a fossil cephalopod).

L. *incomparabilis,* that cannot be equalled, matchless, peerless: incomparable.

L. *increatus,* uncreated, self-existent:

L. *infletus,* unwept, unlamented:

L. *innubus; innuptus,* unmarried: *Buprestis innuba* (a beetle).

L. *inuxorus,* unmarried (man):

L. *ipse,* self: *ipse dixit,* neipsead, ipsedixitism.

NL. *isolatus,* detached, separated, < It. *isola,* < L. *insula,* island: isolated.

Gr. *lipernes,* homeless, outcast: *Lipernes perspectus* (a beetle).

Gr. *monachos,* single, solitary; m. monk; *monache,* f. nun: *Monachocrinus sexradiatus* (a crinoid), *Monachus tropicalis* (West Indian seal), *Myopsittacus monachus* (green parakeet), Munich, Des Moines, *Porthetria monacha* (a moth).

Gr. *moneres,* single, solitary; *monios,* solitary, ferocious; see *monos* under **one**

L. *nonnus,* m. monk; *nonna,* f. nun: nun.

Gr. *oios,* alone, singular, unique, peculiar; *oiobios,* living alone: *Oeophronistus australicus* (a beetle).

L. *orbus,* bereft of parents; *orba,* f. orphan; *orbo, -atus,* bereave: *Euarthrus orbatus* (a beetle).

L. *orphanus* (Gr. *orphanos*), bereft; *orphanistes,* m. one who cares for orphans: *Tulipa orphanidea* (Spartan tulip), Huerfano County (Colo.)

L. *peculiaris,* one's own, singular: peculiar, *Actinotrypa peculiaris* (a fossil bryozoan).

L. *privus,* alone, each, single; *privatus,* individual; *privilegium,* n. a law in favor of an individual, special right: private, privilege, privacy, privily, privy, privateer, deprive, deprivation.

L. *proprius,* one's own, special, particular, distinctive: proper, appropriate, property, propriety.

L. *reclusus; seclusus,* shut up, separated, removed; see **separate**

L. *se,* apart, aside; see **separate**

L. *segrex, -egis,* apart, separate; see **separate**

AS. *self,* self: selfish, selvage, himself, themselves.

L. *seorsus,* apart, separate, severed; see **separate**

L. *singularis,* alone, solitary; *singulus,* one: single, singular, singularity.

L. *solus,* alone; *solitarius,* alone; *solitudo, -inis,* f. lonely place, desert; Sp. *soledad,* solitude, desert; *solivagus,* wandering alone, solitary: sole, solo, solitary, solitaire, solitude, soliloquy, solipsism, sullen, desolation, *Buteo solitarius* (a Hawaiian hawk), *Stilbocoris solivagus* (a bug).

L. *specialis,* individual, particular: special, specialty, especially.

L. *sui, sibi, se,* oneself, himself, herself, itself; *suus,* his, her, its: suicide, suicism, *sui generis.*

L. *umbratilis,* in retirement, private; see *umbra* under **shade**

L. *unicus,* only, sole, singular: unique.

See: **one, separate, different, selfish, free**

alopeco- < Gr. *alopex, -ekos,* fox, see **fox**; *alopekia,* mange in foxes, baldness, see **bare**

alopecurus, NL. foxtail-grass; see **grass**

alosa, L. shad; see **shad**

alpestris, L. of high mountains; see *alpinus* under **mountain**

alpha, Gr. first letter of the Greek alphabet; see **letter**

alphema, Gr. contract sum; *alphesis,* gain; see **gain**

alphestes, Gr. a wrasse; see **wrasse**

alphito- < Gr. *alphiton,* barley meal, see **flour**; < *alphito,* bugbear, see **fear**

alphos, Gr. a white spot on the skin, a kind of leprosy; see **disease**

alpine < L. *alpinus,* of or like high mountains, < *Alpis,* Alp; see **mountain**

alsidena, L. a kind of onion; see **onion**

alsine, Gr. perhaps a kind of chickweed; see **pink**

alsius; alsus, L. chilly, cold, cool; see *algeo* under **cold**

alsos, Gr. grove; see **forest**

altar < L. *altare, -is,* n. place of sacrifice: altarage, altarpiece.
   L. *ara,* f. altar, pyre; *arula,* f. dim.: Ara.
   Gr. *bomos,* m. base, stand, altar:
   Gr. *thymele,* f.; *thysiasterion,* n. place of sacrifice, altar: thymele.

alter, L. the other; see **other, change, alternate**

altercum, L. henbane; see **nightshade**

alternate < L. *alternus,* first one then the other successively, reciprocal: alternation, alternative, alternating, *Alternanthera amoena* (an amaranth), *Cornus alternifolia* (a dogwood), *Alternaria brassicae* (a mold).
   Gr. *allax; enallax; parallax,* crosswise, alternate, in turn: parallax.
   Gr. *allelon,* one another, reciprocal: allelomorph, parallel, allelotaxy.
   Gr. *amoibaios,* alternate, reciprocal; see *ameibo* under **change**
   Gr. *andokaden,* alternately:
   L. *mutuus,* exchanging like for like, reciprocal: mutual, mutuality.
   ME. *or,* < AS. *awther,* alternative: or.
   L. *reciprocus,* returning, back and forth, alternating; *reciproco, -atus,* return, move back and forth, alternate: reciprocal, reciprocity, reciprocate, reciprocornous.
   See: **change, cross,** *zigzag* under **slope**

althaea, L. (Gr. *althaia*), a mallow; see **mallow**

althos, Gr. a healing, medicine; see **heal**

altilis, L. fattened; rich, nourishing; see *alo* under **eat**

altrix, L. nourisher, wet-nurse; see **nurse**

altus, L. high; *altiusculus,* dim.; *altitudo,* height; see **high**

alucita, L. gnat; see **fly**

aluminum < L. *alumen, -inis,* n. alum: alum, alumina, aluminate, aluminiferous, duralumin, aluminosis, aluminite, aluminosilicate, aluminoferric. See ELEMENTS under **thing**

alumnus; alumna, L. foster-child, pupil, < *alumno,* nourish, bring up, educate; see **child**

aluta, L. a kind of soft leather; *alutaceus,* of soft leather; see **leather**

alveus, L. cavity, trough, pit, hollow, channel; *alveolus,* dim; *alveatus,* hollowed out; *alvearium,* beehive, kneading-trough; see **hollow, hive**

alvus, L. belly, womb, hold of a ship; see **belly**

always < AS. *eall,* all, *weg,* way.
   AS. *a,* always; ON. *ei,* ever: aye, each, either, every, ever, no, never.

Gr. *aei*, ever, always: *aeichronios; aeidios; aeigenes; aeizoos*, ever-living, ever-lasting; *aeichloros; aeiphyllos; aeithales*, evergreen: aeipathy, *Draba aizoon* (a whitlowwort).

L. *aeturnus*, < *aeviturnus*, immortal, everlasting: eternal, eternity.

Gr. *ageratos*, not growing old, unwithering: *Achillea ageratum* (sweet yarrow).

Gr. *aionios*, lasting for an age, perpetual; see *aeon* under **age**

Gr. *akalektos; alektos*, incessant, unceasing: acatalectic.

Gr. *akoros*, unceasing: *Acorostoma medicatum* (a moth).

L. *amarantus* (Gr. *amarantos*), unfading, imperishable; m. a never-fading flower: amaranthine, *Amaranthus spinosus* (spiny amaranth).

Gr. *ambrosia*, f. food of the gods that gave immortality; *ambrosios; ambrotos*, immortal, divine, excellent: ambrosia, Ambrose, ambrosial, ambrophyte, *Ambrosia pumila* (dwarf ragweed).

Gr. *anekleiptos*, endless, incessant:

Gr. *anostos*, unreturning, irrevocable: *Anostostoma opacum* (a katydid).

Gr. *apalaiotos*, not growing old:

Gr. *apaustos*, unceasing: *Apaustus menes* (a butterfly).

Gr. *aphthartos*, undecaying, incorruptible, immortal: *Aphthartus ornatus* (a fossil shrimp).

Gr. *aphthitos*, imperishable, immortal, unchanging:

Gr. *asapes*, not likely to decay:

L. *assiduus*, constant, steady, busy; see **busy**

Gr. *ateleutos*, endless, eternal: *Ateleute linearis* (a wasp).

Gr. *athanatos*, undying, immortal; *athanasia*, f. immortality: tansy, Athanasius.

Gr. *atropos*, unchangeable, eternal, inflexible; f. one of the three Fates: Atropos, atropine, *Atropos pulsatorium* (a psocid), *Atropa belladonna* (belladonna).

L. *constans*, steady, firm, unchanging: constant, constancy, Constantinople.

L. *continuus*, unbroken, uninterrupted, successive; see **one**

Gr. *diateles*, continuous, incessant:

Gr. *dienekes*, continuous, unbroken: *Dieneces clarkiae* (a moth).

Gr. *emmenes*, abiding, persistent, steadfast; see **stand**

Gr. *empedos*, firm, steadfast, lasting; see **stand**

L. *immarcescibilis*, unfading:

L. *immortalis*, undying: immortal, immortality, immortalize, immortelle.

L. *impausibilis*, unceasing, incessant:

L. *imputribilis*, incorruptible, not liable to decay:

L. *incessabilis; incessans*, unceasing: incessant.

L. *incorruptibilis*, imperishable: incorruptible, incorruption.

L. *indeficiens, -entis*, unfailing, continuous:

L. *indelebilis*, imperishable; see **indelible**

L. *indesinens, -entis*, incessant:

L. *indissolubilis*, that cannot be loosed, indestructible, imperishable: indissoluble, indissolubly.

L. *inexterminabilis*, immortal: inexterminable.

L. *insedabilis*, incessant:

L. *irrevocabilis*, that cannot be recalled, unchangeable: irrevocable.

L. *jugis*, continual, perpetual, perennial:

Gr. *katamonos*, permanent: *Catamonus cribrarius* (a beetle).

L. *perennis*, through the years, everlasting: perennial, *Lupinus perennis* (a lupine).

L. *permanens, -tis*, remaining forever, enduring: permanent, permafrost.

L. *perpetuus*, continuous, lasting; *perpes, -etis*, continuous, unending; *perpetualis*, constant, permanent: perpetual, perpetuity.

Gr. *phoinix, -ikos*, a fabulous bird symbolic of resurrection and immortality: phoenix, Phoenix (Ariz.).

L. *semper*, always; *sempiternus*, everlasting, eternal: sempiternal, *Sic semper tyrannis*, *Sempervivum arenarium* (sand-houseleek), *Sedum sempervivoides* (a houseleek), *Sequoia sempervirens* (redwood), *Begonia semperflorens* (a begonia).

Gr. *synochos*, holding together, unintermitted, incessant: *Synochodaeus modestus* (a beetle).

See: **indelible, stand, age, long, green, one**

alycto- < Gr. *alyktos*, to be shunned; see **shun**

alypos, Gr. without pain, causing no pain or grief, harmless; *alypon*, a plant with pain-killing properties; see **painless**

alysmos, Gr. disquiet, restlessness, < *alysko*, wander uneasily; *alyo*, be at a loss, be ill at ease; see **move**

alyssum, L. (Gr. *alysson*), a genus of the mustard family; see **mustard**

alytos, Gr. continuous, unbroken; see **one**

alyxis, Gr. an escape, < *alysko*, escape, flee from; see *alyktos* under **shun**

ama, L. (Gr. *ame*), water-bucket, pail; see **bucket**

amabilis, L. lovely, < *amo, -atus,* love; see **love**

amalla, Gr. sheaf; see **bundle**

amalos, Gr. soft, tender, weak; see **soft**

amanita, Gr. a kind of fungus; see **fungus**

amara, Gr. trench, conduit, irrigation ditch; see **ditch**

amaracum, L. (Gr. *amarakon*), marjoram; *amaracinus,* of marjoram; see **mint**

amaranth < L. *amarantus* (Gr. *amarantos*), m. a never-fading flower; unfading, imperishable: *Amaranthus[Amarantus] retroflexus* (redroot-amaranth).
NL. *celosia,* f. a genus of the amaranth family: *Celosia floribunda* (a cockscomb).
L. *gromphaena,* f. a kind of amaranth: *Gomphrena[Gromphaena] globosa* (globe-amaranth).

amarus, L. bitter; *amarulentus,* very bitter; *amaritudo,* bitterness; see **bitter**

amarygma, Gr. a sparkle, twinkle, < *amarysso,* sparkle, twinkle; see **light**

amaryllis, -idis, L. (Gr. *Amaryllis, -idos,* a shepherdess), f. a genus of monocotyledonous plants: amaryllid, Amaryllidaceae, *Amaryllis vittata* (an amaryllis).
L. *agave* (Gr. *Agaue,* a mythological woman), f. a genus of the amaryllis family: agavose, *Agave americana* (century-plant), *Eryngium agavifolium* (an umbellifer).
L. *hypoxis,* f. a genus of the amaryllis family: *Hypoxis stellata* (a stargrass).
L. *narcissus* (Gr. *narkissos*), m. a genus of the amaryllis family; *narcissinus,* of narcissus: narcissus (*Narcissus poeticus*), *Narcissus jonquilla* (jonquil), *Narcissus pseudonarcissus* (daffodil), *Anemone narcissiflora* (an anemone).
L. *pancratium* (Gr. *pankration*), n. a genus of the amaryllis family: *Pancratium maritimum* (sea-daffodil).

amathos, Gr. sand, sandy soil; see *ammos* under **sand**

amator; amatrix, L. lover; *amatorculus,* dim.; *amatorius,* loving; see *amor* under **love**

amauros, Gr. dark, dim, obscure; see **dim**

amaze < AS. *amasian,* stupefy, confound; see **wonder**

Amazon, Gr. one of a race of warrior women, a strong, masculine woman; see **woman**

ambactus, L. vassal, dependent; see **servant**

ambagiosus, L. full of digressions; see *ambages* under **around**

amber < OF. *ambre,* < ML. *ambra,* < Ar. *anbar,* a fossilized resin: ambergris, pomander, *Liquidambar orientalis* (a sweetgum), *Electrocoris fuscus* (a fossil bug).
L. *aromatites,* m. a kind of amber:
L. *electrum* (Gr. *elektron*), n. amber; electrical terms: electron, electricity, *Electromyrmex klebsi* (a fossil ant in Baltic amber).
L. *glaesum* (*glessum*), n. amber; *glaesarius,* of amber: glessite.
L. *langurium,* a kind of amber or perhaps some amber-colored stone, said to have been formed from lizard urine; see *languria* under **lizard**
L. *lyncurium* (Gr. *lynkourion*), a kind of amber or perhaps some amber-colored stone, said to have been formed from lynx urine; see **stone**
L. *sacal,* n. Egyptian amber:
L. *sacrium,* n. Scythian amber:
L. *succinum,* n. amber; *succinus,* of amber; *succinacius,* amber-colored: succinic, succinate, succinite, succinoxidase, succinylsulfathiazole, *Succinea aurea* (a gastropod), *Pinus succinifera* (a fossil pine from Baltic amber).
See: **resin, electricity, brown**

ambi-, L. around; *ambio, -itus,* go around; see **around,** *eo* under **move**

ambico- < Gr. *ambix, -ikos,* cup, beaker; see **cup**

ambiga, L. cap of a still; see **cap**

ambiguus, L. changeable, uncertain, doubtful, of double meaning; see **doubt**

ambitiosus, L. desiring excessively; see *ambitio* under **wish**

ambitus, L. encircling, surrounding; orbit, course; see *ambi-* under **around**

amblosis, Gr. abortion, < *amblisko,* cause abortion; see **abortion**

amblys, Gr. blunt, obtuse; see **dull**

ambon, Gr. ridge, crest, rim, pulpit; see **border, table**

ambrosia, Gr. food of the gods that gave immortality; *ambrosios; ambrotos,* immortal, divine, excellent, see **always;** a composite, see **composite**

ambulo, *-atus,* L. walk, go; *ambulacrum,* walk; *ambulator; ambulatrix,* walker; see **walk**

ambulyx, NL. a kind of moth; see **moth**

ambush < OF. *embuschier* (G. *busch*), go into the woods; see **trap**

ambyco- < Gr. *ambyx, -ykos,* cup; see *ambix* under **cup**

ameba (amoeba), NL. a protozoan, < Gr. *amoibe,* change, exchange, alternation; *amoibaios,* interchanging, alternate, reciprocal; see **protozoan, change**

amelanchier, NL. serviceberry; a genus of the rose family; see **rose**

ameletos, Gr. unworthy of care; see **neglect**

amellus, L. a purple aster; see **composite**

amenenos, Gr. feeble, weak; see **weak**

amentum, L. thong, strap, catkin; see **catkin**

amerimnos, Gr. free from care, unconcerned, driving care away; see **rest**

amersis, Gr. deprivation; see *amerdo* under **take**

ames, *-itis,* L. pole; see **pillar**

amethyst < L. *amethystus* (Gr. *amethystos*), a preventive of drunkenness, a precious stone; *amethystinus,* of amethyst, like amethyst; see **stone**

ametor, Gr. motherless; see **alone**

ametos, Gr. harvest; *ameter, -os; ametris, -idos,* reaper; *ameterion,* sickle; see **reap, sickle**

ametros, Gr. beyond measure, immense; see **large**

amia, Gr. a kind of tunny; see **tunny**

amiantos, Gr. unspotted, pure, see **pure**; asbestus, see **asbestus**

amicarius, L. procurer of a mistress, *amica;* see **prostitute**

amicto- < Gr. *amiktos,* immiscible, unmixed, pure; see **pure**

amiculum, L. mantle, cloak; *amicio, -ictus,* wrap about, clothe, cover; see **garment**

amicus, L. friend, friendly; *amiculus,* dim., dear friend; see **friend**

amido-; amino-; ammon- < *ammoniakos,* pertaining to the temple of Jupiter Ammon at Siwa oasis in Libya, < Egypt. *Amun* (*Amen*), the sun god, represented with ram's horns; compounds having the NH_2 radical, derivatives of ammonia; *imido-, imino-,* variants of *amido-* and *amino-* and denoting NH compounds: amide, carbamide, aminol, trimethylamine, amidase, amphetamine, ammonia, ammonium, ammoniacal, ammonite, amidonaphthol, aminobenzene- (aniline), sulfanilamide, N-trichloromethylthio-tetrahydrophthalimide, cevitamic acid, vitamine, imide, imine, imidazolyl, acetoxime, acetaldoxime, *Ovis ammon* (argali), *Pharaonaster ammon* (a fossil echinoid). Why is the head of Alexander the Great on the coins of Lysimachus adorned with ram's horns?

amis-, *-idos,* Gr. chamber-pot; *amidion,* dim.; see **pot**

amissus, L. dismissed, lost; *amissibilis,* that may be lost; see **lose**

amita, L. aunt; *amitinus,* cousin; see **kin**

ammi; ammium, L. (Gr. *ammi*), a genus of the carrot family; see **carrot**

ammonia < Gr. *ammoniakos,* pertaining to the temple of Jupiter Ammon at Siwa oasis in Libya; see **amido-**

ammos, Gr. sand; see **sand**

amnacum, L. pellitory; see **pellitory**

amnesia, Gr. forgetfulness; see **forget**

amnis, L. stream of water, river; *amniculus,* dim. brook; *amnicus,* of a stream; see **stream**

amnos, Gr. lamb; *amnion,* dim., the membrane around the fetus; see **sheep, membrane**

amoeba, L. (Gr. *amoibe*), change; a protozoan; see **change, protozoan**

amoenus, L. pleasant, delightful; *amoenitas,* pleasantness; see **agreeable**

amolgaeo- < Gr. *amolgaios,* of milk; see *amelktos* under **milk**

amolgeus, Gr. milk-pail; see **bucket**

amolgos, Gr. dead of night; see **night**

amomos, Gr. blameless; see **innocent**

amomum, L. (Gr. *amomon*), an aromatic plant; a genus of the ginger family; see **ginger**

among < AS. *on,* in, *gemange,* crowd; see **in, with, middle, herd, number**

amor, L. love, < *amo, -atus,* love; *amorificus,* awakening or causing love; see **love**

amorbaeo- < Gr. *amorbaios,* rustic, pastoral; see **country**

amorgeus; amorgos, Gr. squeezer, drainer; see **press**

amorpha, NL. a genus of the bean family; see **bean**

amorphos, Gr. formless, misshapen; see **different**

amotos, Gr. furious, savage; see **rough**

ampelion, Gr. a songbird; see **bird**

ampelocarpo- < Gr. *ampelokarpon,* a kind of bedstraw; see **madder**

ampelos, Gr. vine; *ampelinos,* of the vine; *ampelion,* dim.; see **vine**

amphi-, Gr. around, on both sides, double; see **around, two**

amphibian < Gr. *amphibios,* living a double life, adapted to water and land; see **frog, newt, toad**

amphibolos, Gr. attacked on both sides, ambiguous, doubtful; see **doubt**

amphictyono- < Gr. *amphiktyon, -os,* neighbor; see **neighbor**

amphidymos, Gr. double; see **two**

amphidysis, Gr. double cup; see **cup**

amphigyos, Gr. double-pointed, pointed at both ends; see **point**

amphisbaena, L. (Gr. *amphisbaina*), a fabulous serpent having a head at each end; now applied to a genus of lizards; see **lizard**

Amphitrite, Gr. wife of Poseidon; see **sea**

amphora, L. vessel, flagon, pitcher, flask, bottle, jar; Gr. *amphoreus,* jar, cinerary urn; *amphoriskos,* dim.; see **pot**

amphoteros, Gr. both, each; see **two**

amplector, -exus, L. wind, twine, embrace; see **embrace**

amplifiers. See **large, very**

amplus, L. large; *amplior; amplius,* larger, more; *amplissimus,* largest; *amplifico, -atus,* enlarge, increase, widen, extend; see **large**

ampotis, Gr. ebb of the tide; see **return**

ampulla, L. flask, bottle; *ampullaceus,* of a flask; *ampullarius,* bottle-maker; see **bottle**

amputo, -atus, L. cut away, lop off; shorten; see **cut, short**

ampyco- < Gr. *ampyx, -ykos,* headdress, fillet; see **ribbon**

amuletum, L. a magic object worn as a charm against evil and disease; see **magic**

amurca, L. (Gr. *amorge*), dregs or lees from pressing olives; see **dirt**

amuse < F. *amuser,* < *a,* at, *muser,* stare, loiter, trifle; see **play, joy, Musa**

amussis, L. rule, level; see **level, measure**

amycticus, L. (Gr. *amyktikos*), scratching, biting, sharp, pungent, strong; see **point**

amydros, Gr. indistinct, dim, obscure; see **dim**

amygdalus, L. (Gr. *amygdalos*), almond-tree; *amygdala* (Gr. *amygdale*), almond; see **peach**

amylum, L. (Gr. *amylon*), fine meal, starch; see **flour**

amyna, Gr. defense; see **guard**

amyris, NL. a genus of the rue family; see **rue**

amystis, Gr. large draught, large cup; see **cup**

amystos, Gr. profane; see **profane**

amyxis, Gr. a tearing, < *amysso,* tear, rend; see **tear**

an- < AS. *an,* at, in, see **on**; < Gr. *an-,* not, without, see **not**; < L. *ad,* to, see **to**

ana- Gr. up, back, again; see **return**

anacampseros, L. (Gr. *anakampseros*), a sedumlike plant; see **purslane**

anacardium, NL. a genus of the sumac family; see **sumac**

anachoma, Gr. mound, dam; see *choma* under **wall**

anachoretes, Gr. one who has retired from the world, hermit, recluse; *anachoresis,* retirement; see **alone**

anachyma, Gr. an expanse; see **space**

anaclinterium, L. cushion; see **pillow**

anacto- < Gr. *anax, -aktos,* king, lord, chief, master; see **govern**

anaedo- < Gr. *anaides,* shameless, ruthless; see **merciless**

anaereto- < Gr. *anairetes,* destroyer, murderer; see **destroy**

anagallis, Gr. pimpernel; see **primrose**

anagyris; anagyros, Gr. a bean-trefoil; see **bean**

analcido- < Gr. *analkis, -idos,* weak, feeble; see **weak**

analecto- < Gr. *analektos,* choice, select; see *lego* under **choose**

analemma, Gr. support, frame, sling; see **frame**

analepter, Gr. bucket; see **bucket**

anallacto- < Gr. *anallaktos,* unchangeable; see **stand**

analogos, Gr. conformable, proportionate; see **like**

analoma, Gr. cost, outlay; see **pay**

analtos, Gr. insatiate; see **hunger**

analyze < Gr. *ana,* back, *lyo,* loose; see **separate, judge**

anambatos, Gr. unmounted, unbroken; see **nature**

anamestos, Gr. filled full; see *mestos* under **abundance**

ananas, NL. pineapple; see **pineapple**

anancaeum, L. a large cup that must be drained on a wager; see **cup**

ananco- < Gr. *ananke,* force, necessity, see **strong**; *anankaion,* prison, see **pen**

anandros, Gr. unmarried, husbandless; see **alone**

ananepsis, Gr. recovery; see **return**

anantes, Gr. uphill, steep; see **slope**

anapetes, Gr. wide open, expanded; see **open**

anaphalis, NL. a genus of the composite family; see **composite**

anaphes, Gr. tasteless, insipid, not to be touched; see **tasteless**

anaphrygmos, Gr. confusion; see **confusion**

anarhyster, Gr. bucket; see **bucket**

anarsios, Gr. hostile, strange; see **against**

anas, L. duck; *anaticula,* dim., duckling; *anatinus,* of ducks; see **duck**

anasa, NL. a genus of bugs; see **bug**

anaschetos, Gr. endurable; see **let**

anasobe, Gr. disturbance, tumult; see **confusion**

anassa, Gr. queen, lady; see *anax* under **govern**

anastater, Gr. destroyer; see **destroy**

anastomosis, Gr. a new outlet; formation of a network; see **net**

anathema, Gr. curse; see **curse**

anatinus, L. of ducks; see *anas* under **duck**

anatole, Gr. a rising, sunrise, east; see **east**

anatomy < L. *anatomia* (Gr. *anatome*), a cutting up, dissection; see **separate**

anatos, Gr. harmless; see **safe**

anax, Gr. lord, master, king; *anaxios,* royal; see **govern**

anaxios, Gr. unworthy, worthless; see **worthless**

anaxyris, Gr. trousers; see **trousers**

ancala, L. (Gr. *ankale*), the bent arm, see **arm**; *ankalos,* armful, bundle, see **bundle**

-ance; -ancy; -ence; -ency < L. *-antia, -entia,* pertaining to, quality, state; see **nature**

anceps, L. two-headed, two-sided, double, ambiguous, uncertain, dangerous; see **two, head**

ancestor < L. *antecessor,* forerunner; see **kin, before**

anchauros, Gr. near the morning; see **dawn**

anchi-, Gr. near; *anchion; anchoteros; asson,* nearer; *anchistos,* nearest; see **near**

anchialos, Gr. near the sea; see *hals* under **sea**

anchistinos, Gr. close, crowded, in heaps; see **thick**

ancho, Gr. press tight, choke, throttle; *anchone,* a throttling, strangling, cord for hanging; see **choke**

anchor < L. *ancora* (Gr. *ankyra*); *ancoralis,* of an anchor; see **hook**

anchovy < Sp. *anchova;* see **herring**

anchusa, L. (Gr. *anchousa*), bugloss; see **borage**

ancient < OF. *ancien;* see *antiquus* under **old**

ancile, L. the sacred shield that fell from heaven; see **shield**

ancilla, L. maidservant, handmaid, female slave; *ancillaris,* pertaining to maid-servants, subservient; see *ancus* under **servant**

ancistro- < Gr. *ankistron,* fishhook; *ankistrion,* dim.; see **hook**

ancisus, L. cut around or away; see *caedo* under **cut**

anco- < Gr. *ankos,* a bend, curve, hollow, valley; see **bend**

ancon, L. (Gr. *ankon*), bend of the arm, arm of a workman's square, elbow; see **arm**

anctero- < Gr. *ankter, -os,* binder, clasp; see *ango* under **choke**

ancus, L. servant; *ancula,* maidservant; *anculus,* manservant; see **servant**

ancylio- < Gr. *ankylion,* ring of a chain; see **chain**

ancylo- < Gr. *ankylos,* bent, hooked, crooked; see **bend**

ancyro- < Gr. *ankyra,* anchor; see *ancora* under **hook**

and, AS. in addition: ampersand.
  L. *addo, -itus,* give to, add to; *additivus,* added, annexed: add, addition, additive, addendum, *Valgus addendus* (a beetle), *Aphis addita* (an aphid).
  L. *adjicio, -ectus,* add to, increase; *adjectivus,* added to: adjective.
  L. *adjungo, -unctus,* add to, join to: adjunct.
  L. *adventitius* (*adventicius*), additional, extraordinary: adventitious.
  L. *affigo, -ixus,* annex, attach: affix.
  L. *amplifico, -atus,* enlarge, extend, increase: amplify.
  L. *annecto, -exus,* bind to, add to; see *necto* under **bind**
  L. *appendix, -icis,* f. appendage, addition; *appendicula,* f. dim.: appendix, appen-dicitis, penthouse, *Uromyces appendiculatus* (bean-rust).
  L. *auctarium,* an addition; see *augeo* under **grow**
  Gr. *epholkion,* small boat towed after a ship, an appendage; see *holkas* under **ship**
  Gr. *epitithemi,* put to, add; see *tithemi* under **place**
  L. *et,* and : et cetera (etc.).
  L. *extra,* on the outside, beyond, over, in addition to; see *exterus* under **out**
  L. *-farius,* suffix meaning multiplication in numbers or parts; see **fold**
  L. *mantissa,* a worthless addition; see **trifle**
  Gr. *prostheke,* addition, supplement; *prosbletos,* added; see *pros* under **near**
  L. *residuus,* remaining, left over; see **abandon**
  L. *subscivus,* that remains over and above, extra, odd, spare:
  L. *subsidiarius,* belonging to a reserve: subsidiary.
  L. *superamentum,* n. remnant, remainder:
  L. *supplementum,* n. a filling up, addition of what is necessary to complete: sup-plement, supplementary.
  Gr. *symballo,* throw together, add:
  See: **over, grow**

-and; -end; -und, L. having the quality of, full of; see **nature**

anderon, Gr. any raised bank, dike; see **wall**

andinus, NL. pertaining to the Andes Mountains; see **mountain**

andrachle, Gr. a kind of pan; see **pan**

andrachne, Gr. purslane; see **purslane**

andrapodon, Gr. slave; see **servant**

andrianto- < Gr. *andrias, -antos,* image of a man, statue; see **form**

andro- < Gr. *aner, andros,* man, male; see **man**

androgynos, Gr. hermaphrodite; see **mix**

Andromeda, L. (Gr. *Andromede*), daughter of Cepheus and Cassiope; a genus of the heath family; see **heath**

andropogon, NL. a genus of grasses; see **grass**

androsace < L. *androsaces* (Gr. *androsakes*), a genus of the primrose family; see **primrose**

androsaemon, L. (Gr. *androsaimon*), a kind of St. Johnswort; see **St. Johnswort**

-ane ; see *-anus* under **nature**

anebos, Gr. beardless, young; see *hebe* under **young**

aneca- < Gr. *anekas,* upward; see **over**

anecbato- < Gr. *anekbatos,* without outlet; see **close**

anecdote < Gr. *anekdotos,* unpublished, secret; see **story, secret**

aneclipto- < Gr. *anekleiptos,* endless, incessant; see **always**

anecto- < Gr. *anektos,* bearable, tolerable; *anektikos,* patient, enduring; see **let**

anelpsis, Gr. hopeless, desperate; see **hopeless**

anemone, Gr. windflower; see **buttercup**

anemoros, Gr. wild, savage; see **nature**

anemos, Gr. wind; see **wind**

anepaphos, Gr. unharmed, untouched; see **safe**

anerestos, Gr. displeasing; see **bad**

aneristos, Gr. undisputed; see **true**

anesthetic < Gr. *anaisthetos,* insensible, without feeling; *anaisthesia,* insensibility; see **painless, dull**

anethum (anisum), L. anise, dill; see **carrot**

aneticus, L. (Gr. *anetikos*), relaxing; *anetos,* relaxed, slack, set free; see **free**

aneuretos, Gr. undiscovered; see **secret**

aneuros, Gr. sinewless, weak; see **weak**

aneurysma, Gr. permanent, diseased dilatation of a bloodvessel; see **disease**

anfractus, L. bending, winding, crooked, circuitous, prolix; see **bend**

angarius, L. (Gr. *angaros*), courier, messenger; see **ride**

angelos, Gr. messenger, envoy; *angeleia,* message, tidings, news; *angelikos,* of a messenger, < *angello,* bear a message, bring news, report; see **send**

anger < ON. *angr,* affliction, sorrow; a temporary strong emotion aroused by wrong or injury.
  L. *bilis,* gall, anger; see **bile**
  Gr. *brimosis,* f. indignation; *brimodes,* grim, stern; *brimazo,* snort with anger; *barybrimetos,* very indignant: *Brimus spinipennis* (a beetle), *Brimoda pagana* (a beetle).
  Gr. *chalepos,* difficult, testy, angry; see **difficult**
  Gr. *cholos,* m. anger, wrath, < *chole,* bile; *akracholos,* quick to anger, passionate: *cholotos,* angry; *zacholos,* very angry: *Zacholus zonatus* (a reptile).
  Gr. *embrimma, -tos,* n.; *embrimesis,* f. indignation:
  Gr. *erethizo; eretho,* anger, provoke: see **arouse**
  Gr. *eridmaino,* irritate, anger, provoke to strife; see *eris* under **fight**
  L. *indignans,* angry, < *indignor, -gnatus,* be angry; *indignabundus,* full of anger, enraged; *indignativus,* passionate, irascible: indignant, indignation.
  L. *ira,* f. anger, fury, rage, wrath; *iracundus,* inclined to anger, wrathful; *irascibilis,* becoming angry easily, < *irascor, iratus,* be angry: ire, irate, irascible, *dies irae, Crataegus iracunda* (a hawthorn).
  L. *irrito, -atus,* exasperate, vex, inflame; see **arouse**
  Gr. *kotos,* m. grudge, hate, wrath; *zakotos,* very angry: *Cotocripus caridei* (a fly), *Zacotus matthewsi* (a beetle).
  Gr. *menis, -ios (-idos),* f.; *menithmos,* m. wrath; *dysmenis,* wrathful: *Meniomorpha inconstans* (a beetle).
  Gr. *odysis,* f. anger, wrath; *odyssomai,* hate; *Odysseus,* m. Ulysses (Ulixes), famous king of Ithaca; *Penelope,* f. faithful wife of Ulysses: odyssey, *Penelope cristata* (guan), *Motacilla penelope* (gray wagtail).
  Gr. *orge,* f. passion, anger, temper, wrath; *orgilos,* easily angered, irritable; *orgetes,* m. a passionate man; *baryorgetos,* very angry: *Orgilus obscurator* (a wasp), *Orgetorixa australica* (a bug).
  Gr. *oxyno,* sharpen, goad, provoke; see **arouse**
  Gr. *oxythymos,* quick to anger:
  Gr. *skydmainos,* angry: *Scydmaenus capillosulus* (a beetle).
  Gr. *skythros,* angry, sullen: *Scythrophanes stenoptera* (a moth).
  L. *stomachosus,* angry, irritable, choleric, < *stomachor, -atus,* be irritated, vexed, fume, fret: stomachous, *Omophorus stomachosus* (fig-weevil).
  Gr. *thymos,* mind, spirit, passion, anger; see **mind**
  See: **mad, hate, spleen, bitter, choke, arouse, fight, swell, red**

angina, L. quinsy; *angor, -is,* quinsy; see **disease**

angio- < Gr. *angeion,* dim. of *angos,* vessel or container of any kind, capsule, seedcase of plants; see **bag**

angle < L. *angulus,* m. corner, bend; *angellus,* m. dim.; *angularis; angulatus,* with angles; *angulosus,* full of corners; *quadrangulus,* four-cornered; *quinquangulus,* five-cornered; *triangulus,* three-cornered: angular, rectangle, triangle, quadrangular, *Pleurosigma angulatum* (a diatom), *Odontomachus angulatus* (an ant), *Phaseolus angularis* (adsuki-bean), *Hepatica angulosa* (Hungarian hepatica), *Passiflora quadrangularis* (granadilla), *Stereocrinus triangularis* (a Devonian crinoid), *Chalcophora angulicollis* (a beetle).

L. *axilla,* f. armpit, crotch; *axillaris,* of an axil: axil, axillary, *Leucothoë axillaris* (coast-leucothoë), *Delphinium axilliflorum* (a larkspur).

L. *canthus* (Gr. *kanthos*), tire of a wheel, edge, corner; see **border**

Gr. *delta,* the Greek capital letter Δ, shaped like a triangle; see **letter**

Gr. *gonia,* f. angle, corner, prob. < *gony,* bent knee; *gonidion,* n. dim.; *goniodes,* angular: gonial, gonion, goniometry, goniotropous, trigonometry, polygon, diagonal, goniatite, amblygonite, *Goniopholis simus* (a fossil crocodile), *Trigonella caerulea* (a legume), *Exogonium bracteatum* (jicana).   See *gony* under **knee.** Derive: paragon.

L. *hamus,* hook, angle; see **hook**

Gr. *kampter, -os,* bend, angle; see *kamptos* under **bend**

Gr. *koleps, -epos,* hollow or bend of the knee; see **knee**

Gr. *maschale,* f. armpit, hollow, bay: maschalephidrosis.

Gr. *onkos,* hook, barb, angle; see *uncus* under **hook**

L. *-quetrus, -a, -um,* suffix meaning angled, cornered, sided; *triquetrus,* three-cornered, triangular: triquetrous, triquetral, *Lactophrys triqueter* (a trunk-fish), *Prismapora triquetra* (a fossil bryozoan), *Arenaria tetraquetra* (a sandwort), *Allium triquetrum* (an onion).

F. *zigzag,* alternately changing direction by sharp angles; see **slope**

See: **bend, elbow, break, hook, slope**

ango, *anctus; anxus,* L. press together, choke, vex, distress; see **choke**

angos, Gr. vessel or container of any kind, see *angeion* under **bag**; a Frankish javelin, see **spear**

anguilla, L. eel; see **eel**

anguis, L. snake; *anguiculus,* dim.; *anguineus; anguinus,* snaky; see **snake**

anguish < OF. *anguisse;* see **pain, trouble, sad**

angulus, L. corner, angle; *angellus,* dim.; *angularis,* having angles; see **angle**

angurium, L. (Gr. *angourion*), watermelon; see **melon**

angustus, L. narrow, see **narrow;** *angustia,* defile, strait, see *angiportus* under **way**

anhelus, L. out of breath, panting, puffing, < *anhelo, -atus,* pant, puff, gasp; see **breathe**

anhydro- < Gr. *anhydros,* waterless, dry; see **dry**

aniaros, Gr. troublesome, annoying, grievous; *ania,* trouble, distress, grief; see **trouble**

aniatos, Gr. incurable; see **hopeless**

anicano- < *anikanos,* incapable, insufficient; see **weak**

aniceto- < Gr. *aniketos,* unconquered, invincible; see **stand**

anicmo- < Gr. *anikmos,* without moisture, dry; see **dry**

anicula, L. little old woman, dim. of *anus,* old woman; see **woman**

anilastos, Gr. merciless, pitiless; see **merciless**

aniline < Skt. *nila,* dark blue; see **blue**

anilis, L. pertaining to an old woman; see *anus* under **woman**

anima, L. air, breath of life, soul; *animula,* dim.; *animalis,* lively; see **life**

animal, -is, L. n. a living being; *animalculum,* n. dim.: animality, animalism, animalcule.

Heb. *behemah,* beast, a large animal, probably the hippopotamus: behemoth.

L. *belua,* f. beast, monster; *beluatus,* covered with animal figures; *beluilis; beluinus,* beastial, brutal; *beluosus,* full of monsters:

L. *bestia,* f. animal; *bestiola,* f. dim.; *bestialis,* like a beast: beastly, bestial, bestiality, wild beast, *bete noir.*

Gr. *boskema; boton,* fed or pastured animal, cattle, < *bosko,* feed, pasture; see **cattle**

L. *burius,* m. an unknown animal:

L. *campus* (Gr. *kampos*), m. a sea-animal: *Hippocampus ingens* (a sea-horse).

L. *catoblepas* (Gr. *katobleps, -epos*), m. an African animal, perhaps the gnu: *Catobleps blattoides* (a beetle), *Catoblepas gnu* (former name of the gnu).

L. *Centaurus* (Gr. *Kentauros*), m. a monster, half man and half horse: centaur, centaury (*Centaurium umbellatum*), *Centaurea pulcherrima* (a dusty-miller).

L. *Cerberus* (Gr. *Kerberos*), three-headed, doglike monster that guarded the gates of Hades, hellhound; see **dog**

L. *Chimaera* (Gr. *Chimaira*), f. a fabulous fire-spouting monster with goat's body, lion's head, and serpent's tail: chimerical, *Chimaera mediterranea* (a sharklike fish).

L. *Cyclops* (Gr. *Kyklops*), mythical, one-eyed giant; see **large**

L. *draco* (Gr. *drakon*), a fabulous, lizardlike animal; see **lizard**

L. *Fauna*, f. sister of *Faunus*, m. god of agriculture and cattle-raising: fauna, faunal, *The marble faun* (Nathaniel Hawthorne), *Phanaeus faunus* (a beetle).

L. *ferus*, m.; *fera*, f. wild beast; *ferinus*, of wild animals; see **nature**

L., Gr. *Gorgo*, Medusa, the grim one with the snaky locks; see *gorgos* under **fear**

L. *gryphus; gryps, -phis* (Gr. *gryps, -ypos*), m. a fabulous creature, half lion and half eagle, prob. < *grypos*, hooknosed, curved: griffin, griffon, gryphon, hippogrif, *Sarcorhamphus gryphus* (condor), *Gryphus vitreus* (a brachiopod), *Anomia gryphorhynchus* (a fossil pelecypod), *Gryphaea arcuata* (a fossil pelecypod). Derive: epimachus.

L., Gr. *Harpyia*, f. a monster, half woman and half bird, prob. < *harpazo*, snatch: harpy, *Harpia harpyia* (harpy-eagle), *Harpyrhynchus brevis* (a mite).

L. *jumentum*, n. beast of burden:

Gr. *kartazonon*, n. an Indian animal:

Gr. *knodalon*, n. any dangerous, wild animal, monster: *Cnodalum viride* (a beetle), *Cnodalium nodosum* (a beetle).

Gr. *ktenos*, beast, livestock (ox, sheep, horse, etc.); see **cattle**

Gr. *ktisis*, a creation, creature; see *ktizo* under **make**

Gr. *lamia*, f. a fabulous monster said to suck blood and eat human flesh; a bugbear to frighten children: *Lamia textor* (a beetle).

Heb. *liwyathan*, a large aquatic animal: leviathan.

L. *mantichora* (Gr. *mantichoras; martichoras*, m.), f. a fabulous beast with human face, lion's body, and scorpion's tail: manticore, *Mantichora tuberculata* (a beetle), *Mantichorula semenowi* (a beetle).

L. *Medusa* (Gr. *Medousa*), f. a gorgon with snaky locks and capable of turning beholders to stone; see **fear**

L. *Minotaurus* (Gr. *Minotauros*), m. a monster with the head of a bull and body of a man: Minotaur.

L. *monstrum*, n. a supernatural or fabulous wonder; an animal that in form or structure departs widely from normality; *monstrosus*, strange, grotesque, unnatural: monster, *Chimaera monstrosa* (a sharklike fish), *Monstera deliciosa* (ceriman).

Gr. *Mormo* (*Mormon*), bugbear, hobgoblin, used to frighten children into good behavior; see **fear**

L. *pecus, -oris*, domestic animals collectively; *pecus, -udis*, a single head; see **cattle**

Gr. *pelor*, n. monster, prodigy; *peloros*, monstrous, prodigious, huge; see **large**

Gr. *pleumon, -os*, m. probably a kind of jellyfish:

L. *Polyphemus* (Gr. *Polyphemos*), m. a Cyclops in Sicily blinded by Ulysses: *Telea polyphemus* (a moth).

L. *pullus*, young animal, particularly a chicken; see **chicken**

L. *quadrupes, -edis*, c. a four-footed animal: quadruped, quadrupedal.

Gr. *saperion*, n. an unknown animal:

L. *Scylla* (Gr. *Skylla*), f. a sea-monster; a famous rock between Italy and Sicily, opposite the dangerous whirlpool, Charybdis: Scylla and Charybdis.

Gr. *Sphinx*, female monster of Thebes who propounded riddles; see **ask**

Gr. *teras*, monster, wonder; see **wonder**

Gr. *tethyon*, n. a tunicate; *tethynakion*, n. dim.: *Tethyum papillosum* (a tunicate).

Gr. *thambena*, monster; see **wonder**

Gr. *therion; theridion; theraphion*, n. dim. of *ther, theros*, m. wild animal, beast; *theriakos*, of wild beasts; *therodes*, like a wild beast, brutal, savage; *entheros*, full of wild beasts: theromorphic, theriodont, theralite, treacle, *Megatherium americanum* (a sloth), *Theridionexus cavernicolus* (a spider), *Theriosuchus pusillus* (a fossil crocodile), *Theridium bellicosum* (a spider), *Anthracotherium magnum* (a fossil ungulate), *Theriomorphus miasensis* (a beetle), *Theraphosa blondi* (a spider).

Gr. *thremma, -atos*, n. creature, tamed animal; *thremmation*, n. dim.: *Thremmatophilus roseus* (a bird).

L. *veterinus*, pertaining to beasts of burden: veterinary.

Gr. *xandaros*, m. a fabulous sea-monster:

Gr. *zoön*, n. animal; *zoarion; zodarion; zodion; zoyllion; zoyphion*, n. dim.; *zodiakos; zoikos*, of or pertaining to animals: zoology, zoophilous, zoolith, zoophyte, zodiac, metazoon, azote, azotemia, Paleozoic, Mesozoic, Cenozoic, *Ptychozoon homalocephalum* (flying gecko), *Collozoum serpentinum* (a radiolarian), *Sphaerozoum punctatum* (a radiolarian), *Zoyphium* (formerly *Zuphium*) *fasciolatum* (a beetle), *Zodion notatum* (a fly), *Passalus zodiacus* (a beetle), *Zodarion punicum* (a spider), *Eperythrozoon parvum* (a blood-parasite), *Zodiomyces vorticellarius* (a fungus).

See: **cattle, life**

animatus, L. alive, lively, < *animo, -atus*, fill with breath, quicken, give life to; see *anima* under **life**

animo- < Gr. *aneimon,* naked, unclad; see **bare**

animus, L. life, spirit, mind; *animulus,* dim.; *animosus,* full of courage, bold; see *anima* under **life**

anipsalos, Gr. unhurt; see **safe**

aniptos, Gr. unwashed; see **dirt**

aniro- < Gr. *aneiro,* take away, destroy; *anairesis,* destruction; *anairetes,* destroyer; see **destroy**

anise < L. *anisum* (Gr. *anison*); see *anethum* under **carrot**

anisos, Gr. unequal; *anisotes,* inequality; see **different**

ankle < AS. *ancleow;* see **foot**

annexus, L. < *annecto,* tie, bind; *annectens,* linking, joining; see *necto* under **bind**

annona, L. yearly produce, crop, grain, see **reap**; NL. < Ab.Am. *anon,* custard-apple, see **pawpaw**

annosus, L. full of years, aged, old; see **old**

annotinus, L. of last year, a year old; see **year**

annoy < L. *in odium,* in hatred; see **tease, arouse, trouble, fretful, pain**

annul < OF. *anuller,* < L. *ad,* to, *nullus,* none; see **cancel**

annulus, L. ring; *annellus,* dim.; *annularis,* of a ring; *annulatus,* ringed; see **circle**

annus, L. year; *annalis; anniculus; annualis,* of the year; *annosus,* of many years; see **year**

ano-, Gr. up, upward, above; see **over**

anodynos, Gr. allaying pain; *anodynia,* freedom from pain; see **painless**

anoecto- < Gr. *anoiktos,* opened; *anoigma; anoixis,* an opening, door, see **open**; pitiless, ruthless, see **merciless**

anolbos, Gr. poor, wretched; see **poor**

anomalos, Gr. uneven, irregular, inconsistent, abnormal, unusual, deviating from the regular rule; see **different**

anomoeo- < Gr. *anomoios,* dissimilar; see **different**

anomos, Gr. without law; see **different**

anonymos, Gr. nameless, unknown; see **secret**

anoos; anous, Gr. without understanding, silly; see **fool**

anopheles, Gr. useless, hurtful; see **hurt**

anoplos, Gr. without armor, unarmed; see **bare**

anoptos, Gr. unseen; see **secret**

anorecto- < Gr. *anorektos,* without appetite for, undesired; see *orexis* under **wish**

anosios, Gr. unholy, profane; see **profane**

anosos, Gr. healthy, sound; see **health**

anostos, Gr. unreturning, irrevocable; see **always**

anothen, Gr. above, on high; see *ano-* under **over**

another < AS. *an,* a, one, *other;* see **other, different, and**

anozos, Gr. with few or no branches; see **bare**

anquina, L. rope binding sailyard to the mast; see **rope**

ansatus, L. having a handle, < *ansa,* handle, haft; *ansula,* dim.; see **handle**

anser, L. goose; *anserculus,* dim. gosling; *anserinus,* of geese; see **goose**

answer < AS. *andswaru.*
    Gr. *amoibe,* exchange, return, answer; see *ameibo* under **change**
    Gr. *antigraphe,* f. reply in writing, transcription: antigraph.
    Gr. *apokrisis; hypokrisis,* f. answer, decision: apocrisiary.
    Gr. *chresmos,* oracular reply; see **prophecy**
    L. *responsum,* n. reply, answer; *responsivus,* answering: response, responsive, R.S.V.P.
    See: **return**

ant < AS. *aemete,* emmet, pismire: anthill, anteater, ant-cow, ant-lion.
    NL. *anomma,* n. a genus of ants: *Anomma nigricans* (a driver-ant).
    L. *atta,* f. a genus of ants, < *Atta,* name of persons who walk on their shoetips: *Atta laevigata* (a leaf-cutting ant), *Lentinus atticolus* (a fungus), *Attaphila fungicola* (a cockroach).
    NL. *eciton* (prob. an anagram of F. *notice*), n. a genus of ants: *Eciton hamatum* (a legionary ant), *Mimeciton antennatum* (a beetle).

L. *formica,* f. ant; *formicula,* f. dim.; *formicinus,* of ants: formicivorous, formalin, formaldehyde, chloroform, formicine, formicarium, formicide, iodoform, *Formica sanguinea* (red ant), *Formicarius longipes* (an ant-thrush), *Formicivora ferruginea* (an ant-wren), *Eremobates formicaria* (a solpugid), *Endospermum formicarum* (a euphorbiacead), *Poliocephalus formicetorum* (a beetle).
Gr. *laertes,* m. a kind of ant:
Gr. *myrmex, -ekos,* m. ant; *myrmekion,* n. dim.; *myrmedon,* m. ant-nest: myrmecology, myrmecophily, myrmecochorous, *Myrmecophaga jubata* (great anteater), *Myrmecobius fasciatus* (banded anteater), *Myrmecodia tuberosa* (a madder), *Myrmedonobia rufoscutellata* (a bug), *Myrmarachne vehemens* (a spider), *Pogonomyrmex badius* (a harvester-ant), *Myrmeleon obsoletus* (ant-lion).

anta, L. pilaster; see *anteris* under **pillar**

Antaeus, L. (Gr. *Antaios*), a giant Libyan wrestler whose strength was renewed when he touched the earth; see **strong**

antagonistes, Gr. competitor, opponent, rival; see **enemy**

antarcticus, L. (Gr. *antarktikos*), southern; see **south**

antauges, Gr. reflecting light, sparkling; see **light**

ante-; anti- < L. *ante,* before; *anterior,* before, former; *anticus,* foremost; see **before**

anteater
Gr. *phattages,* m. pangolin: *Manis(Phatages[Phattages]) laticauda* (a pangolin).
NL. *tachyglossus,* m. spiny anteater: *Tachyglossus setosus* (a spiny anteater).

antellogos, Gr. compensation; see **pay**

antelope < Gr. *antholops, -opos,* m. a horned animal: *Antilope cervicapra* (sasin), *Antilocapra americana* (pronghorn).
L. *addax, -acis,* m. an African animal with twisted horns: *Addax nasomaculata* an antelope).
Gr. *boubalis, -idos,* f. a kind of antelope, gazelle: *Bubalis caama* (hartebeest), *Bubalis lunata* (sassaby).
Gr. *dorkas, -ados,* f. gazelle; *dorkadion,* n. dim.: *Gazella dorcas* (African gazelle), *Antidorcas euchore* (springbok), *Dorcadotherium aquaticum* (water-chevrotain), *Dorcadium emeritum* (a fossil beetle).
F. *gazelle,* < Ar. *ghazal,* a kind of antelope: *Gazella subgutturosa* (Persian gazelle), *Arctocephalus gazella* (a fur-seal).
Gr. *onelaphos,* m. a kind of antelope:
L. *oryx, -ygis* (Gr. *-ygos*), m. gazelle, antelope: *Oryx gazella* (gemsbok).
L. *pygargus* (Gr. *pygargos*), m. a kind of antelope, probably *addax*: *Gerbillus pygargus* (a rodent).
L. *rupicapra,* f. chamois: *Rupicapra rupicapra* (chamois).
Gr. *strepsikeros,* m. an African antelope: *Strepsiceros kudu* (kudu).
Gr. *tragelaphos,* m. an African antelope: *Tragelaphus sylvaticus* (boschbok).

antenna, L. sailyard; a sensory appendage on the head of an insect; see **feel**

antennaria, NL. a genus of the composite family; see **composite**

anterastes, Gr. rival; see **enemy**

anteres, Gr. opposite; see *anti* under **against**

anterior, L. before, former; see *ante* under **before**

anteris, Gr. prop, stay, support, buttress; see **pillar**

anteros, Gr. avenger of slighted love; see **punish**

antestis, Gr. a confronting; see *anti* under **against**

anthalium, L. a kind of sedge; see **sedge**

anthamillos, Gr. rivaling; see *hamilla* under **fight**

anthedon, Gr. bee; see **bee**

anthelion, Gr. dim. of *anthele,* tuft or plume of a reed; see **bush**

anthemis, Gr. chamomile; see **composite**

anthemon, Gr. flower; see *anthos* under **flower**

anther < L. *anthera,* f. a medicine made from flowers; the pollen-bearing part of a stamen, < Gr. *antheros,* blooming: antheridium, antheriform, *Acidanthera bicolor* (an iridacead), *Heteranthera limosa* (a mud-plantain), *Erica melanthera* (a heath).

anthereon, Gr. chin; see **chin**

antherico- < Gr. *antherix, -ikos,* awn, beard, ear; see **point**

anthericum, L. (Gr. *antherikon*), asphodel; see **lily**

anthias, Gr. a sea-fish; see **fish**

antholco- < Gr. *antholkos*, balancing, equaling; see *antholke* under **balance**

antholops, Gr. a horned animal; see **antelope**

anthos, Gr. flower; *anthion; anthyllion*, dim.; *anthikos; anthinos*, of flowers; *antheros*, flowery; see **flower**

anthraco- < Gr. *anthrax, -akos*, charcoal, coal, precious stone, carbuncle, ulcer; *anthrakinos*, pertaining to coal, coal-black; *anthrakites*, a kind of coal; see **coal, sore, stone**

anthrene, Gr. wild bee, wasp, hornet, see **wasp**; *anthrenion*, wasp-nest, see **nest**

anthriscus, L. (Gr. *anthriskos*), an umbellifer; see **carrot**

anthropos, Gr. man; see **man**

anthurium, NL. a genus of the arum family; see **arum**

anthus, L. (Gr. *anthos*), probably a wagtail; see **wagtail**

anthyllis, Gr. a kind of plant; see **bean**

anti-; anta- < Gr. *anti*, against, opposed to, like; see **against, like**

antia, L. forelock; see **hair**

antiar < Javanese *antjar*, a poisonous resin; see **resin**

antias, *-ados*, Gr. tonsil; see **tonsil**

anticus, L. foremost; see *ante* under **before**

antidote < L. *antidotum* (Gr. *antidoton*), counterremedy; see **drug**

antimony < uncertain origin.   See ELEMENTS under **thing**.
   Gr. *larbason*, n. antimony:
   L. *stibium* (Gr. *stibi; stimmi*), n. antimony; *stibinus*, of antimony: stibine, stibial, stibnite.

antipex, Gr. a kind of cradle; see **cradle**

antiquus, L. old; *antiquarius*, of antiquity; see **old**

antirrhinum, L. (Gr. *antirrhinon*), snapdragon; see **figwort**

antistes; antistita, L. overseer, master; see **govern**

antithesis, Gr. opposition, figurative contrast; see FIGURES OF SPEECH under **form**

antler < OF. *antoillier;* see **horn**

antlia, Gr. bilge-water in the hold of a ship, pump; see **pump**

antlion, Gr. bucket; see **bucket**

antrostomus, NL. a genus of goatsuckers; see **goatsucker**

antrum, L. (Gr. *antron*), hollow, hole, cave, cavity, sinus; *antraios*, of caves; see **hollow**

antyx, *-ygos*, Gr. edge, rim, frame; see **border**

Anubis, L. (Gr. *Anoubis*), Egyptian god of the chase; see **hunt**

anus, L. rectal opening, fundament; see **intestine**

-anus, L. pertaining to, belonging to; see **nature**

anutato- < Gr. *anoutatos*, unwounded; see **safe**

anvil < AS. *anfilt.*
   Gr. *akmon, -os*, m. anvil; *notakmon, -os*, with mailed back: *Acmonorhynchus vincens* (a flowerpecker), *Notacmon fimbriatus* (a fish).
   L. *incus, -udis*, f. anvil; *incudo, -usus*, forge with a hammer, < *cudo, cusus*, strike, beat, pound: incus, incuse, incudal, incudate.

anxius, L. uneasy, troubled, solicitous; *anxiosus*, full of uneasiness; see **fear**

any < AS. *aenig*, < *an*, one; see **one**

anysimos, Gr. effectual; *anystos*, practicable, able, < *anyo*, effect, accomplish; see **make**

aochletos, Gr. calm, still, undisturbed; see **rest**

aocno- < Gr. *aoknos*, without hesitation, brave; see **bold**

aoratos, Gr. invisible, unseeing, blind; see **blind**

aoristos, Gr. indefinite, indeterminate; see **doubt**

aoros, Gr. pendulous, something hanging or waving, see **hang**; untimely, unripe, see **unfit**; *aor, -os*, any weapon, see **tool**

aorta < Gr. *aorte*, the great artery; see **pipe**

ap- < L. *ad*, to, see **to**; < Gr. *apo*, from, see **from**

apages, Gr. not firm, flabby, soft; see **soft**

apanastes, Gr. emigrant; see *astatos* under **move**

apape, Gr. a dandelionlike plant; see **composite**
apargia, Gr. hawkbit; see **composite**
aparine, Gr. bedstraw; see **madder**
apartment < It. *appartamento;* see **room, house**
apate, Gr. deceit, fraud; *apatelos,* deceitful, deceptive, illusory; see **lie**
apathy < Gr. *apatheia,* insensibility, unfeelingness, indifference; see **dull**
apaustos, Gr. unceasing; see **always**
ape < AS. *apa,* ape; see **monkey**
apechema, Gr. echo; see **return**
apeches, Gr. discordant, dissonant; see **different**
apechthes, Gr. hateful, hated; see *echthos* under **hate**
apecto- < Gr. *apektos; apektetos,* uncombed, unkempt; see **confusion**
apedilos, Gr. barefooted, unshod; see **bare**
apemon, Gr. harmless; see **safe**
apene, Gr. a four-wheeled wagon; see **vehicle**
apenes, Gr. rough, harsh, cruel; see **rough**
apeoros, Gr. hanging on high, soaring; see **fly**
aper, *apri,* L. wild boar; *apriculus,* dim.; see **hog**
apertura, L. an opening; *aperio, -ertus,* open, uncover, lay bare; see **hole, open**
apestys, *-yos,* Gr. absence; see **far**
apeucto- < Gr. *apeuktos,* to be deprecated, abominable; see **bad**
apeutho- < Gr. *apeuthes,* unknown, unknowing; see **secret**
apex, *-icis,* L. tip, top; *apiculus,* dim. point; see **top**
apexabo, L. a kind of sausage; see **sausage**
aphaca, L. (Gr. *aphake*), a kind of legume; see **bean**
aphados, Gr. displeasing, hateful; see **bad**
aphanistes, Gr. destroyer; see **destroy**
aphano- < Gr. *aphanes; aphantos,* unseen, invisible, obscure, secret; see **secret**
apharco- < Gr. *apharke,* an evergreen tree; see **arbutus**
apharotos, Gr. unplowed, untilled; see **rough**
aphatos, Gr. ineffable, unutterable; see **ineffable**
aphauros, Gr. feeble, powerless; see **weak**
apheles, Gr. even, smooth; see **smooth**
aphelico- < Gr. *aphelix, -ikos,* past youth, elderly; see **old**
aphenos, Gr. riches, wealth; see **wealth**
aphesmos, Gr. swarm of bees; see *hesmos* under **herd**
aphetos, Gr. at large, let loose, ranging at will; see **free**
aphid < NL. *aphis, -idis,* plant-louse; see **bug**
aphido- < Gr. *apheides,* lavish, abundant; see **abundance**
aphixis, Gr. arrival; see **come**
aphobetos, Gr. fearless; see **bold**
aphodos, Gr. departure; excrement; see **dung**
aphonos, Gr. dumb, silent; see **silent**
aphrades, Gr. senseless, reckless; *aphradia,* folly; see **fool**
aphritis, Gr. a fish; see **fish**
Aphrodite, Gr. goddess of love; *aphrodisiakos,* arousing sexual desire; see **love**
aphron, Gr. silly, foolish, senseless; *aphradia; aphrosyne,* foolishness; see **fool**
aphros, Gr. foam; *aphrodes,* foamy; see **foam**
aphtha, Gr. eruption, thrush; see **disease**
aphthartos, Gr. undecaying, immortal; see **always**
aphthitos, Gr. imperishable, immortal; see **always**
aphthonos, Gr. without envy, ungrudging, plentiful, abundant; see **abundance**
aphya, L. (Gr. *aphye*), a small fish; see **herring**
apiastra, L. bee-eater; see **bird**
apiastrum, L. wild parsley; see *apium* under **carrot**
apiculatus, L. small-pointed, < *apiculus,* dim. of *apex, -icis,* tip, top; see *apex* under **top**

apilo- < Gr. *apeile,* boast, threat; see **brag**

apines, Gr. clean; see **pure**

apios, Gr. pear-tree; *apion,* pear; see **pear**

apiro- < Gr. *apeiros,* inexperienced, ignorant, see **dull**; boundless, infinite, see **infinite**

apis, L. honeybee; *apicula,* dim.; *apiarius,* of bees, see **bee**; Gr. *Apis, -idos,* Egyptian sacred bull, see **cattle**

apistos, Gr. faithless, untrustworthy, doubting; see **doubt**

apithynter, Gr. director, guide; see **lead**

apium, L. celery, parsley; *apiacus,* of parsley; see **carrot**

aplaeoto- < Gr. *aplaiotos,* not growing old; see **always**

aplatos, Gr. monstrous, terrible; see **fear**

aplestos, Gr. insatiate; see **hunger**

apletos, Gr. immense; see **large**

apluda, L. chaff, scale; see **scale**

aplustre, L. stern of a ship; see **stern**

aplysia, Gr. filth; *aplytos,* unwashed; see **dirt**

apo- < Gr. *apo,* from, off, away, after, without, separate; see **from**

apobletos, Gr. rejectable, worthless; see **worthless**

apocha, L. (Gr. *apoche*), receipt; see **pay**

apocnismato- < Gr. *apoknisma, -atos,* snip, bit; see **little**

apocope, L. (Gr. *apokope*), a cutting off, dropping, stoppage; see **short**

apocryphus, L. (Gr. *apokryphos*), concealed, obscure, spurious; see *krypto* under **cover**

apocynum, L. (Gr. *apokynon*), dogbane; see **dogbane**

apodesmos, Gr. away from home, abroad; see **far**

apodicto- < Gr. *apodeiktos,* demonstrable, demonstrated; see *deigma* under **form**

apodocimo- < Gr. *apodokimos,* worthless; see **worthless**

apodosis, Gr. restitution, return; see *do* under **give**

apoeo- < Gr. *apoios,* without quality, neutral; see **sterile**

apolecto- < Gr. *apolektos,* chosen, picked; see *lego* under **choose**

apolegma, Gr. hem of a robe; see **border**

apollinaria; apollinaris, L. a kind of nightshade; see **nightshade**

Apollo, L. (Gr. *Apollon*), god of youth, music, healing; see **young**

apollyon, Gr. destroying; see *olethros* under **destroy**

apolos, Gr. immovable; see **stand**

apomaxis, Gr. a cleansing, < *apomasso,* wipe off, wipe clean; see **wipe**

aponos, Gr. idle, lazy; see **rest**

apopatos, Gr. dung, ordure; see **dung**

apophysis, Gr. offshoot, prominence, process; see **projection**

apoplexy < Gr. *apoplexia,* a stroke; see **disease**

apopyris, Gr. a small fish; see **fish**

aporema, Gr. doubt, question, perplexity; see **doubt**

aporrhaïs, Gr. a kind of mollusk; see **mollusk**

aporrhegma, Gr. fragment; *aporrhox, -ogos,* broken off; see *rhegnymi* under **break**

apositos, Gr. hungry; see **hunger**

aposmileuma, Gr. chip, shaving; see **part**

apospasma, Gr. part torn off, piece; see **part**

aposphax, Gr. abrupt, broken off; see **break**

apostates, Gr. deserter, renegade; see *apostasis* under **abandon**

apostema, Gr. abscess; see **sore**

apostibes, Gr. off the road, solitary; see *astibos* under **alone**

apostle < Gr. *apostolos,* messenger, ambassador; see *stello* under **send**

apostos, Gr. driven away; see **drive**

apostrophe, Gr. figurative turning aside; see FIGURES OF SPEECH under **form**

apotheca, L. (Gr. *apotheka*), storehouse, barn, magazine; *apothetos,* stored up; see **store**

apotmetos, Gr. cut off; see *temno* under **cut**

apotmos, Gr. unlucky; see *potmos* under **lot**

apotomos, Gr. cut off sharp, sheer; see *temno* under **cut**

apoxyros, Gr. sharp, sheer; see **break**

apoxys, Gr. tapering off; see **point**

apparatus, L. equipment, machine; see **tool**

appear < L. *appareo, -ritus,* be visible, manifest, clear; *apparitio,* appearance, ghost; see **come, see, clear, spirit**

appease < OF. *apaisier,* pacify; see **soothe, rest, soft, mild, smooth, atone, agreeable, harmony, painless, dull**

appendix, *-icis,* L. appendage, addition; *appendicula,* dim.; see **and**

appetite < L. *appetitus,* desire, passion; *appetibilis,* desirable; see **wish**

appha, Gr. dear one; *apphidion,* dim.; see **love**

applanatus, L. flattened; see *planus* under **level**

apple < AS. *aeppel:* applebutter, mayapple, Sodom apple, pineapple.
Gr. *epimelis, -idos,* n. a kind of medlar:
L. *malum* (Gr. *melon*), n. apple; *malus* (Gr. *melea; melis, -idos*), f. apple-tree; *malinus; melinus* (Gr. *melinos*), of apples or quinces, apple-green, quince-yellow; *melimelum* (Gr. *melimelon*), n. honey-apple, quince; Gr. *hamamelis, -idos,* f. a kind of medlar, now applied to witchhazel; Pg. *marmelos,* quince: malaceous, malic acid, malate, chamomile, marmalade, *Malus pumila* (apple, formerly *Pyrus malus*), *Passiflora maliformis* (a passion-flower), *Psylla mali* (apple-psylla), *Aegle marmelos* (bael-fruit), *Chaenomeles lagenaria* (flowering quince), *Thecla melinus* (hairstreak-butterfly), *Hamamelis japonica* (Japanese witchhazel), *Hormaphis hamamelidis* (cone-gall aphid), *Sphaeropsis malorum* (a fungus).
Sp. *manzana,* f. apple; *manzanilla; manzanita,* dim.: *Arctostaphylos manzanita* (common manzanita), manzanilla.
L. *mespilum* (Gr. *mespilon*), n. medlar; *mespilus,* m. medlar-tree: *Mespilus germanica* (medlar), *Myrtus mespiloides* (a myrtle), *Sorbus chamaemespilus* (a mountain-ash).
L. *pomum,* n. fruit of any kind, apple; *pomulum,* n. dim.; *pomarius,* of applelike fruits; *pomosus,* abounding in fruit; *Pomona,* f. goddess of fruit and fruit trees: pome, pomace, pomade, pomiferous, pomology, Pomona, *pomum Adami,* Pomeroy, *Aphis pomi* (apple-aphid), *Carpocapsa pomonella* (codling-moth), *Typhlocyba pomaria* (apple-leafhopper), *Fomes pomaceus* (a bracket-fungus), *Bartramia pomiformis* (apple-moss), *Dentalina pomuligera* (a foraminifer), *Turpinia pomifera* (a staphyleacead), *Gloeodes pomigena* (a fungus).
L. *setania,* f. a kind of medlar:

applicatus, L. attached to, connected with; see **bind**

appositus, L. fit, proper, appropriate; see **fit**

approach < L. *ad,* to, *propio,* draw near; see **accessible, near**

appropriate < L. *ad,* to, *proprius,* peculiar; see **fit**

approve < L. *approbo, -atus,* assent to; see **yes**

apricot < Pg. *albricoque;* see **plum**

apriculus, L. dim. of *aper,* wild boar; *aprinus; aprugnus,* of the wild boar; see **hog**

apricus, L. lying open, uncovered, exposed to the sun; *apricarius,* of the open; see **open**

apron < OF. *naperon,* dim. of *nape,* < L. *mappa,* napkin, tablecloth; see **napkin, garment**

apronia, L. a plant, probably bryony; see **melon**

aprosmicto- < Gr. *aprosmiktos,* isolated, solitary; see **alone**

aps, Gr. backward, back again; see **return**

apsis, L. (Gr. *hapsis, -idos*), arch, vault, orbit, rim, loop, mesh, network; see **arc, net**

apteno- < Gr. *apten, -os,* unfledged, unable to fly; see **fly**

aptus; aptatus, L. appropriate, fit, suitable, < *apo,* bind, fasten; see **fit, bind, know**

apud, L. at, near, in, with, among; see **with**

apuro- < Gr. *apouros,* far away, distant; see **far**

apus, *-podis,* L. a kind of swallow, swift; see **swallow**

aqua, L. water; *aquarius,* of water, water-carrier; *aquaticus; aquatilis,* living in or near water; *aquosus,* very moist; see **water**

aquifolium, L. holly, scarlet holm, < *aquifolius,* having pointed leaves; see **holly**

aquila, L. eagle; *aquilinus,* of the eagle; see **eagle**

aquilegia, NL. a genus of the buttercup family, columbine; see **buttercup**

aquilonaris, L. north, northern; see **north**

aquilus, L. dark-colored, blackish, dun, swarthy; see **black**

ar- < L. *ad,* to, see **to**; < Gr. *a-, an-,* not, without, see **not**

ara, L. altar, pyre; *arula,* dim., see **altar**; Gr. *ara,* prayer, curse; *aratos,* accursed; *katara,* imprecation, curse, see **ask**

arabicus, L. Arabian, < L. *Arabs, -bis* (Gr. *Araps*), m. Arab, a native of Arabia: Arab, Arabia, Arabian, Arabic, gum arabic (from *Acacia arabica,* the babul acacia), arabic acid, arabinose, ribose, riboflavin, *Arabis stricta* (a rockcress).

arabilis, L. tillable; see *aratrum* under **plow**

arabis, NL. a genus of the mustard family; see **mustard**

arabos, Gr. chattering, rattling, or gnashing of the teeth; see **rattle**

arachis, NL. (Gr. *arakos; arachos*), a genus of the bean family; see **bean**

arachno- < Gr. *arachne,* spider; *arachnion; arachnidion,* n. dim. cobweb; see **spider, weave**

arados, Gr. a rumbling; see **thunder**

araeo- < Gr. *araios,* thin, narrow, porous, few; see **thin**

aragmos, Gr. rattle, clatter, clash; see **rattle**

aralia, NL. a genus of the ginseng family; see **ginseng**

aranea, L. spider; *araneola,* dim.; *araneus,* of spiders; see **spider**

arasso, Gr. strike hard, dash to pieces; see **strike**

arator, L. plowman, husbandman; *aratrum,* plow; *aro, -atus,* plow; see **plow**

araucaria, NL. a genus of conifers; see **gymnosperm**

arbelos, Gr. shoemaker's rounded knife; see **knife**

arbiter; arbitrator, L. judge, umpire; see **judge**

arbor, L. tree; *arbuscula,* dim. shrub; *arborarius; arboreus,* of trees; *arboretum; arbustum,* place with trees, orchard; *arbustivus; arbustus,* planted with trees; see **tree**

arbutus, L., f. strawberry-tree, madrone; a genus of the heath family: *Arbutus xalapensis* (Mexican madrone), *Rhododendron arbutifolium* (a rhododendron). Gr. *adrachne,* f. strawberry-tree: *Arbutus andrachne[adrachne]* (a madrone). Gr. *apharke,* f. an evergreen tree:
L. *comarum* (Gr. *komaron*), n. fruit of the strawberry-tree, *komaros,* f. arbutus: *Comarum palustre* (bog-cinquefoil), *Comarostaphylis diversifolia* (toothed manzanita).
L. *memecylon* (Gr. *mimaikylon*), n. fruit of the strawberry-tree: *Memecylon ovatum* (a melastomacead).
L. *unedo, -onis,* m. strawberry-tree: *Arbutus unedo* (strawberry-madrone).

arbyle; arbylis, Gr. half-boot; see **shoe**

arc; arch < L. *arcus, -us,* m. bow; *arcuarius,* relating to bows; m. bow-maker; *arcuatilis,* shaped like a bow; *arcuatus* (*arquatus*), bent like a bow: arcade, archery, archivolt, arcuate, arbalest, Ozark, *Corythucha arcuata* (a lacebug), *Numenius arquatus* (European curlew).
L. *apsis* (*absis*), *-idis* (Gr. *hapsis, -idos*), f. arch, vault, juncture, hoop of a wheel, orbit, loop; *absidatus,* arched, vaulted: apse, apsis, synapsis, Parapsida, *Hapsiphyllum indicum* (a Permian coral).
L. *camera* (Gr. *kamara*), chamber or vault with arched roof; see **room**
L. *circumflexus,* arched: circumflex.
L. *concavus,* hollowed or arched inward: concave, concavity.
L. *convexus,* arched outward; *convexa,* f. arch, vault: convex, convexity, *Gryphaea convexa* (a fossil pelecypod), *Baeckmanniolus convexicollis* (a beetle), *Scymnus convexus* (a beetle), *Alloniscus perconvexus* (a pillbug).
L. *crescens, -entis,* shape of the moon in first or last quarter; see **crescent**
L. *curvo, -atus,* bend, bow, crook; see **bend**
L. *falcatus,* sickle-shaped; *falcipedius,* bowlegged; see *falx* under **sickle**
L. *fornix, -icis,* m. arch, vault, brothel in an underground vault; *fornicatus,* arched, vaulted: fornicate, fornication, *Crepidula fornicata* (a slipper-limpet).
Gr. *iaphetes,* m. archer:
L., Gr. *iris,* rainbow; see **color**

L. *luno, -atus,* bend into a crescent or half-moon; see **crescent**

L. *porticus,* arcade, gallery, colonnade; see **porch**

Gr. *psalis, -idos,* low building with a vaulted roof; see **house**

L. *recavus,* hollowed or arched inward, concave:

L. *sagittarius,* archer; see *sagitta* under **arrow**

Gr. *talaor, -os,* m. bow:

L. *testudinatus,* arched or vaulted like a tortoise shell; see **turtle**

L. *tholus* (Gr. *tholos*), m. dome, cupola, rotunda; *tholion,* n. dim.: tholus, *Tholospira spinosa* (a radiolarian), *Tholospyris fenestrata* (a radiolarian), *Staurotholus tetrastylus* (a radiolarian), *Rubus tholiformis* (a blackberry), *Zonidium octotholium* (a radiolarian).

Gr. *toxon,* n. bow; *toxarion,* n. dim.; *toxeutes; toxotes,* m. archer, bowman; *toxikos,* of the bow; *toxeres,* furnished with a bow: toxophily, *Toxostoma redivivum* (a thrasher), *Toxarium costatum* (a radiolarian), *Toxeutes arcuatus* (a beetle), *Toxeutotaenius* (a beetle), *Toxotes jaculator* (archer-fish), *Eutoxeres aquila* (a hummingbird), *Hyalonema toxeres* (a sponge), *Toxylon pomiferum* (osage-orange).

See: **circle, bend, crescent, curl, cup**

arca, L. box, chest; *arcella; arcula,* dim.; *arcarius,* treasurer; see **box, store**

arcadius, L. ideally rustic, < Arcadia, the Switzerland of Greece; see **country**

arcanus, L. shut up, private, secret, silent, mysterious; see **secret**

arcebion-, L. a kind of borage; see **borage**

arcelo- < Gr. *arkelos,* young panther; see **cat**

arcera, L. covered carriage for the sick; see **vehicle**

arcesso, *-itus,* L. call, summon; see **call**

arceto- < Gr. *arketos,* sufficient; *arkesis,* help, service; see **help**

arceutho- < Gr. *arkeuthos,* juniper; see **juniper**

archae-; arche-; archi-; archo- < Gr. *arche,* beginning, first cause, chief; *archaios,* from the beginning, old; *archon,* magistrate, ruler; see **begin, lead, govern**

archezostis, L. bryony; see **melon**

archi-, Gr. chief, leader; see **govern**

archidion, Gr. a petty office, dim. of *arche,* first place; see *arche* under **begin**

Archimedes, Gr. famous Greek physicist and inventor; see **find**

architectus, L. (Gr. *architekton*), master-builder; see *tekton* under **carpenter**

archive < L. *archivum* (Gr. *archeion*); see **store**

archon, Gr. chief, ruler, magistrate; see **govern**

archos, Gr. rectum, anus; see **intestine**

arcio- < Gr. *arkios,* sure, certain, enough; see **sure**

arctatus, L. compressed, confined; see *arcto* under **press**

arctium, L. (Gr. *arktion*), burdock; see **composite**

arcto- < Gr. *arktos,* bear, north; *arktylos,* cub; *arktikos; arktios,* northern; see **bear, north**

arctostaphylos, NL. a genus of the heath family; see **heath**

arculatum, L. sacrificial cake; see **cake**

arcus, L. bow; *arcuarius,* of bows, bow-maker; *arcuatilis,* bowlike; *arcuatus,* bent like a bow; see **arc**

arcyo- < Gr. *arkys, -yos,* net, toils; see **net**

-ard; -art < F. *-ard* (G. *hart*), hard; quality in the highest degree; habitual; one who, that which; see **hard**

ardalion, Gr. watertrough; see **basin**

ardalos, Gr. dirty, < *arda,* dirt; see **dirt**

ardea, L. heron; *ardeola,* dim.; see **heron**

ardelio, L. zealous person, busybody, meddler; see **busy**

ardens, L. burning, glowing, < *ardeo, arsus,* burn; *ardor,* flame, heat, desire; see **burn**

ardeutos, Gr. watered, < *ardo,* water; see **wet**

ardis, Gr. point; see **point**

ardmos, Gr. watering place; see **drink**

arduus, L. difficult to attain, hard, laborious; see **difficult**

area, L. open space, court, threshing floor; *areola,* dim.; *areolatus,* with small spaces; see **space**

areca, NL. a kind of palm; see **palm**

aregon, Gr. helper; see *arexis* under **help**

aremenos, Gr. distressed, harassed; see **trouble**

arena, L. sand; *arenula,* dim.; *arenaceus; arenarius,* of sand, sandy; *arenosus,* full of sand; see **sand**

arenga, NL. a genus of palms; see **palm**

arens, L. dry, parched, < *areo,* dry up; see *aridus* under **dry**

areolatus, L. with small spaces; see *area* under **space**

Ares, Gr. god of war; see **fight**

arestos, Gr. pleasing, acceptable; *areskos,* desirous to please, obsequious, flattering; see **agreeable**

arete, Gr. excellence, goodness, virtue; see **good**

Arethusa, L. (Gr. *Arethousa*), famous spring at Syracuse, named for a wood-nymph; see **spring**

argaleos, Gr. troublesome, vexatious; see **trouble**

argas, NL. a genus of ticks; see **tick**

argemone, Gr. a kind of poppy; see **poppy**

argentum, L. silver; *argenteus; argentinus,* of silver, silvery; see **silver**

argestes, Gr. whitener; see *argos* under **white**

argetes, Gr. a kind of snake; see **snake**

argilla, L. (Gr. *argillos*), white clay, potter's clay; *argillaceus,* clayey; see **earth**

argiope, NL. a genus of spiders; see **spider**

argolas, Gr. a kind of serpent; see **snake**

argon < Gr. *argon,* inactive; see ELEMENTS under **thing**

argos, Gr. white, bright, see **white**; swift, see **swift**; idle, lazy; *argikos,* indolent, see *aergos* under **rest**

argue < OF. *arguer,* < L. *arguo, -utus,* debate, contend; *argumentum,* contention; see **speak, fight**

Argus, L. (Gr. *Argos*), hundred-eyed guardian of Io; see **peacock**

argutus, L. clear, bright, sharp; *argutulus,* dim.; see **clear**

argyros, Gr. silver; see **silver**

ari-, Gr. intensive, very; see **very**

aria, Gr. a kind of oak; see **oak**

aridelos, Gr. quite clear; see *delos* under **clear**

aridus, L. dry; *arefactus,* dried up; *ariditas,* dryness; see **dry**

ariena, L. banana; see **banana**

aries, L. ram; *arietinus,* of a ram; *arieto, -atus,* butt like a ram; see **sheep, push**

arillator, L. haggler; see **trade**

arillus, NL. a fleshy growth around some seeds; see **flesh**

arilus, NL. a genus of bugs; see **bug**

arinca, L. a kind of grain; see **grain**

-arion, Gr. diminutive suffix; see *-idium* under **little**

ariono- < Gr. *areion, -os,* better; see *agathos* under **good**

aris, Gr. auger; see **bore**

arisaron, Gr. a kind of arum; see **arum**

arise < AS. *a,* away, *risan,* rise; see **rise**

arisemos, Gr. very notable, very clear; see **clear**

arista, L. beard of grain, awn, ear; *aristosus,* full of awns; see **point**

aristera, Gr. left hand; *aristeros,* left, on the left; see **hand**

aristida, NL. a genus of grasses; see **grass**

aristolochia, Gr. birthwort; see **birthwort**

aristos, Gr. best; see *agathos* under **good**

Aristoteles, Gr. Aristotle, distinguished natural scientist; see **know**

arithmos, Gr. number; see **number**

-arium, L. place where, place for; see **place**

-arius, L. one who, agent; pertaining to; see **make, nature**

ark < L. *arca,* chest, box; see **box**

arm < AS. *arm*, forelimb.
  Gr. *agostos*, m. the bent arm; flat of the hand: *Agostopus medius* (fossil am-
    phibian tracks).
  L. *ala*, upper part of arm that joins the shoulder; see **wing**
  L. *ancala* (Gr. *ankale*), f. the bent arm: *Ancalochernes mexicanus* (a pseudo-
    scorpion).
  L. *ancon, -is* (Gr. *ankon, -os*), m. bend of the arm, elbow: *Anconocrinus beachleri*
    (a Mississippian crinoid).
  L. *axilla*, armpit; see **angle**
  L. *brachium*, n. (Gr. *brachion, -os*, m.), arm, upper arm, forearm; *brachialis*, of
    the arm; *brachiatus*, with arms or branches: brachial, brachiopod, brachiotomy,
    brachiferous, brachiate, brace, bracelet, embrace, pretzel, *Nannobrachium nigrum*
    (a fish), *Brachiaria plantaginea* (a signal-grass), *Cactocrinus multibrachiatus* (a
    fossil crinoid).
  Gr. *Briareos*, a hundred-armed giant; see **strong**
  L. *carpus* (Gr. *karpos*), wrist; see **hand**
  L. *cubitum*, n. elbow, ell; *cubitalis*, of the elbow: cubit.
  Gr. *epikrion*, n. yard arm, sailyard: *Epicrionops bicolor* (a caecilian).
  L. *humerus*, upper arm, shoulder, bone of the upper arm; see **shoulder**
  L. *lacertus*, m. upper arm from elbow to shoulder; *lacertosus*, muscular, strong:
    Plexippus lacertosus (a spider).
  Gr. *olene*, f. elbow, ulna, forearm: olecranon.
  Gr. *omos*, shoulder, upper arm; see **shoulder**
  Gr. *pechys, -eos*, m. forearm: *Pechyptila rhodocharis* (a moth).
  Gr. *rhethos, -eos*, n. limb (arm, leg):
  L. *ulna*, f. elbow, forearm: ulna, ulnar, ulnocarpal.
  See: **leg, branch, beam, rod, wing**
armarium, L. chest, box, safe; *armariolum*, dim.; see **box**
armeniacum, L. apricot; see **plum**
armenos, Gr. fitting, acceptable; see **fit**
armentum, L. herd; see **herd**
armeria, NL. a genus of the leadwort family; see **leadwort**
armilla, L. bracelet; see **bracelet**
armoracia, L. horseradish; see **mustard**
arms < L. *arma*, weapons; *armamentarium*, arsenal; *armatura*, equipment; see **tool**
armus, L. (Gr. *harmos*), shoulder; see **shoulder**
army < OF. *armee*, < L. *arma*, n. weapons: army-worm (*Heliophila unipuncta*).
  L. *agmen, -inis*, n. army on the march, multitude, crowd, train; *agminalis*, of a
    march, train: agminate.
  L. *bellator; bellatrix*, soldier, warrior; see *bellum* under **fight**
  L. *caligatus*, common soldier, private; see **shoe**
  L. *castrensis, -is*, m. a soldier in camp: see *castrum* under **camp**
  L. *classis*, army, navy; see **ship**
  L. *clibanarius*, m. soldier clad in mail, cuirassier: *Clibanarius vulgaris* (a crus-
    tacean).
  L. *equites*, cavalry; see *equus* under **horse**
  L. *exercitus, -us*, m. army:
  AS. *here*, army; *sciphere*, navy: herald, hership, harbor, harbinger, harry, Walter,
    Hereford, harangue.
  Gr. *hoplites*, armed soldier; see *hoplon* under **tool**
  Gr. *lochos*, body of troops lying in ambush; *lochites*, ambusher, comrade; see **trap**
  L. *miles, -itis*, c. soldier; *militaris*, of soldiers and war; *militia*, f. soldiery; *milito,
    -atus*, perform military service: militia, military, militant, militate, *Orchis
    militaris* (an orchid), *Pheidole militicida* (an ant).
  Gr. *oulamos*, m. crowd of warriors:
  L. *pedes, -itis*, foot-soldier; see *pes* under **foot**
  Gr. *peltastes*, soldier armed with a small shield (*pelte*); see **shield**
  Gr. *stolos*, expedition, army; see **move**
  Gr. *stratos*, m.; *stratia*, f. army; *stratios*, of any army; *stratiotes*, m. soldier;
    strateia, f. campaign; *strategos*, m. general: strategist, strategical, strategy,
    Stratonice marioni (a polychaete), *Stratiomyia nubeculosa* (a fly), *Stratiotes
    aloides* (water-soldier).
  L. *tiro*, young soldier, cadet, beginner; see **begin**
  L. *turma*, f. troop of cavalry; *turmalis*, of a troop: *Dusona turmalis* (a wasp).
  L. *veles, -itis*, m. skirmisher; *velitor, -atus*, skirmish; see **fight**
  See: **fight, number, line**
arnacis, L. sheepskin; see **skin**

arnesis, Gr. denial; *arnetikos,* of denial, negative; see **no**

arneuter, Gr. diver, plunger; see **dip**

arnica, NL. a genus of the composite family; see **composite**

arnos, Gr. sheep, lamb; *arnion,* dim.; see **sheep**

arogos, Gr. helping, < *aroge,* aid; see *arexis* under **help**

aroma, Gr. spice, smell; *aromatikos,* fragrant, spicy; see **smell**

aronia, NL. of genus of the rose family; see **rose**

arotos, Gr. arable, < *aroo,* plow, till; *arotron,* plow; see *aratrum* under **plow**

around < AS. *a,* on, L. *rotundus,* round.

L. *ambages, -is,* f. a roundabout way, evasion, digression; *ambagiosus,* full of windings: ambage, ambagious, ambagitory.

L. *ambi-,* around; *ambifarius,* with two sides, double meaning; *ambitus,* encircling, surrounding; *-us,* m. course, orbit; *ambio, -itus,* go around: ambient, ambiguous, ambitus, amputate, circumambient, ambigenous.

Gr. *amphi,* around, on both sides, double; *amphitermos,* bounded on all sides: amphibrach, amphitheater, amphioxus, amphibious, amphigenous, amphibole, *Amphicarpaea monoica* (hog-peanut), *Amphisbaena fuliginosa* (a lizard), *Amphithoë pygmaea* (an amphipod).

L. *circum,* around, about, on all sides: circumference, circumstance, circumlocution, circumvent, circumscribe, circumcision, circumflex, circumspect, circa, *Begonia circumlobata* (a begonia).

Gr. *peri,* around, near: periscope, perianth, period, pericardium, perinium, perimeter, perihelion, peristalsis, periphery, *Periophthalmus barbarus* (mudskipper), *Peribrotus pustulosus* (a beetle), *Perilachna ixota* (a moth), *Pericoptus punctatus* (a beetle).

See: **circle, near**

arouse < uncertain origin.

L. *accendo, -ensus,* set on fire, kindle, inflame, incite; see *incendo* under **burn**

L. *acerbo, -atus,* aggravate, anything disagreeable; *exacerbo,* irritate, provoke: *exacerbatrix, -icis,* f. exasperator: acerbation, exacerbate.

L. *aculeus,* sting; see **point**

L. *agito, -atus,* set in motion, arouse, stir up, vex; *agitator, -is,* m.; *agitatrix, icis,* f. one who stirs up action: agitate, agitation, agitator, *Trachys agitosa* (a beetle).

Gr. *-agogue,* that which stimulates, causes to function, induces, promotes, < *agoge,* f. a carrying away, movement; *agogos,* leading, guiding, attracting: pedagogue, sialogogue, galactogogue, cholagogue, hypnagogue, isagogic, emmenagogue, synagogue, demagogue, *Haemogogus capricorni* (a mosquito).

Gr. *anasobeo,* rouse, flush:

L. *animatus,* alive, vivacious; *animo, -atus,* give life to, quicken; see *anima* under **life**

Gr. *anistemi,* awake, rouse, stir:

L. *calcar,* spur; see **point**

L. *cieo, citus,* actuate, move, excite, call; *cito, -atus,* put into quick motion, drive rapidly, rouse, incite, stimulate; *concito, -atus,* stir up, arouse, excite; *concitor, -is,* m. exciter; *excio, -itus,* call forth, arouse, animate; *incito, -atus,* urge, spur; *percitus,* greatly moved, stimulated, excited; *suscito, -atus,* arouse, rekindle, stir up: cite, excite, citation, recitation, incite, solicit, suscitate, resuscitate, oscitate, excitement, *Drosophila excita* (a fruitfly), *Anthrax incitus* (a fly).

L. *commotus,* moved, excited, aroused: commotion.

Gr. *egertikos,* waking, stirring; *egersis,* f. a waking, stirring; *egersimos,* awaking; *agersigelos,* causing laughter:

Gr. *ekkineo,* rouse:

Gr. *entatikos,* stimulating, aphrodisiac; *entatikon,* n. a plant with stimulating properties:

Gr. *epeigo,* press upon, drive on, urge forward, hurry, speed:

Gr. *eretho; erethizo,* provoke, arouse, irritate; *erethismos,* m. irritation; *erethistikos,* irritating, provocative: erethism, erethic, erethisia, *Erethizon dorsatus* (Canada porcupine).

Gr. *eridmaino,* provoke to strife; see *eris* under **fight**

L. *exaspero, -atus,* stir up, provoke, roughen: exasperate, exasperation.

L. *expergo, -itus,* awaken, arouse; *expergiscor, -perrectus,* be awakened, rouse and bestir oneself:

L. *flexanimus,* affecting, touching, moving:

L. *focillo, -atus,* resuscitate, revive, refresh: *Bulimus focillatus* (a gastropod).

Gr. *hormao,* excite, set in motion, start, spur; *horme,* f. attack, start, impulse, instinct; *aphorme,* f. starting-point, origin, cause; *hormeterion,* n. stimulant;

hormetikos, exciting, stimulating: hormone, *Aphorme horrida* (a sponge).

L. *hortor, -atus,* urge; *hortativus,* serving to arouse and encourage: exhort, exhortation.

L. *incentivus,* that strikes up, sets the tune, provokes, incites: incentive.

L. *inquietus,* restless; see **move**

L. *inspiro, -atus,* breathe into, excite, inflame: inspire, inspiration.

L. *instigo, -atus,* incite, stimulate: instigate, instigation.

L. *instillo, -atus,* drop in, inspire: instill.

L. *instinctus,* instigated, incited; see **nature**

L. *irrito, -atus,* excite, provoke; *irritabilis,* easily excited: irritating, irritation, *Trombicula irritans* (a chigger).

Gr. *kenteo,* prick, goad, sting; *kentor,* goader, driver; see **point**

L. *lacesso, -itus,* provoke, stimulate, excite, irritate:

Gr. *nysso,* prick, spur; see **bore**

Gr. *ochleros,* troublesome; see **trouble**

L. *oestrus* (Gr. *oistros*), gadfly, sting, fury, frenzy, sexual heat, desire; *oistrao,* sting, incite to madness; see **fly**

Gr. *orino; ornymi,* excite, move, stir up; *orintes,* m. exciter; *orsiktypos,* making noise: *Orinophilus oligopus* (a centiped), *Orinophorus hypsinomus* (a beetle), *Orsilochus cornutus* (a beetle).

Gr. *otryno,* rouse, stir up, spur: *otrynter, -os,* m. instigator: *Otrynter caprinus* (a porgy).

Gr. *oxyno,* sharpen, goad, provoke; see *oxys* under **sharp**

L. *provoco, -atus,* call forth, arouse, challenge; *provocabilis,* easily aroused, excitable: provoke, provocation.

L. *pungo, -unctus,* sting, prick; see **point**

L. *sollicito, -atus,* stir up, disturb, agitate, incite; see *sollicitus* under **fear**

Gr. *speudo,* urge on, press on, hasten, quicken, exert oneself, strive eagerly; see *speustikos* under **swift**

L. *stimulus,* m. goad, incentive; *stimuleus,* having prickles or goads; *stimulosus,* full of incentives: stimulant, stimulation, stimulate, stimulose, *Sibine stimulea* (saddle-back moth), *Cnidoscolus stimulosus* (treadsoftly), *Stomoxys stimulans* (a fly).

L. *vegeto, -atus,* quicken, arouse; see **life**

See: **awake, trouble, whip, press, itch, burn, nettle, tease, anger, begin**

arquatus, L. pertaining to the rainbow, see **arc;** one suffering from jaundice, see **disease**

arrange < OF. *arangier,* put into a rank; see **class**

arrectus, L. upright, steep; *arrectarium,* upright post of a wall; see **upright, pillar**

arrha, L. down-payment, pledge; see **pay**

arrhecto- < Gr. *arrhektos,* unbroken, see **one;** unfinished, see **fault**

arrhemon, Gr. silent, speechless; see **silent**

arrhen, Gr. male; *arrhenodes,* brave; see **man, bold**

arrhenes, Gr. fierce, savage; see **rough**

arrhetos, Gr. inexpressible, unutterable; see **ineffable**

arrhichos, Gr. wicker-basket; see **basket**

arrhogo- < Gr. *arrhox, -gos,* unbroken; see **one**

arrhostos, Gr. weak, sickly; see **weak**

arrogant < L. *arrogantia,* presumption, conceit; see **proud**

arrow < AS. *arewe.*

Gr. *atraktos,* spindle, arrow; see **spindle**

Gr. *belos, -eos,* n. arrow, dart, bolt, sting; *acrobeles,* pointed at the end; *oxybeles,* sharp-pointed; *tribeles; tribolos,* three-pointed: belodont, belomancy, *Belostoma flumineum* (a water-bug), *Oxybelus quadrinotatus* (a digger-wasp), *Acrobeles ciliatus* (a nematode), *Beliophorus viduus* (a beetle), *Mastacembelus armatus* (a spiny-eel), *Platybelodon barnumbrowni* (a mastodon), *Tribelesodon longobardicus* (a Triassic reptile), *Tribelocephala lignea* (a bug), *Tribolocephalus laevis* (a crab), *Tribolium confusum* (a flour-beetle).

F. *fleche,* arrow: Fletcher.

Gr. *glochis, -inos,* point of an arrow; see **point**

Gr. *ios,* m. arrow, poison; see **poison**

Gr. *kelon,* n. shaft of an arrow, arrow: *Celonoptera mirificaria* (a butterfly).

Gr. *keraunos,* thunderbolt; *keraunios,* of a thunderbolt; see **thunder**

Gr. *oistos,* m. arrow; *oisteutes,* m. archer; *oistotheke,* f. quiver: *Oestocephalus* [*Oistocephalus*] *pectinatus* (a fossil amphibian).

L. *penna; pinna,* feather, arrow; see **feather**

L. *sagitta,* f. arrow; *sagittula,* f. dim.; *sagittarius,* of arrows; m. archer: sagittal, sagittate, Sagittaria, *Sagitta elegans* (an arrow-worm), *Sagittaria latifolia* (an arrowleaf), *Lactuca sagittifolia* (a lettuce), *Polygonum sagittatum* (arrow-leaved tearthumb), *Tyntlastes sagitta* (a fish), *Textularia sagittula* (a foraminifer), *Sagittarius serpentarius* (secretary-bird), *Crotalaria sagittalis* (a rattle-box), *Xanthosoma sagittifolium* (yahutia), *Sisyrinchium sagittiferum* (a blue-eyed grass).

Gr. *skeptos,* m. thunderbolt: *Sceptobius dispar* (a beetle), *Sceptonidium mozambicanum* (a coral).

Gr. *stochasma,* the thing aimed, arrow, spear; see *stochos* under **aim**

Gr. *toxeuma, -atos,* n. that which is shot, arrow, bolt: *Toxeumella albipes* (a wasp).

See: **point, spear**

arrugia, L. shaft, pit; see **hole**

arsella, L. probably a kind of poppy; see **poppy**

arsen, Gr. male, masculine; see *arrhen* under **man**

arsenic < L. *arsenicum* (Gr. *arsenikon; arrhenikon),* n. orpiment: arsenical, arsenate, arsonate, arsinic, arsonic, arsenolite, arsenotherapy, arsanilic, trimethyarsine, arrhenal, arrhenite. See ELEMENTS under **thing**
L. *sandaraca* (Gr. *sandarake),* f. realgar: sandarac-tree (*Tetraclinis articulata*).

arsios, Gr. fitting, agreeing, friendly; see **fit**

arsis, Gr. a raising or lifting, < *airo,* raise, lift; see **raise**

art < L. *ars, artis,* f. craft, trade, profession; *artifex, -icis,* m. master, artist, actor; *artificialis,* of art: artist, artisan, artless, artifice, artificial, artificer, artifact, artistical, inertia.

L. *Apollo* (Gr. *Apollon),* god of youth, music, healing; see **young**
Gr. *asketos,* curiously wrought, ornamented; see *askeo* under **make**
L. *astutia,* adroitness, craft; see *astutus* under **know**
Gr. *banausia,* f. handicraft; *banausos,* mechanical:
L. *calliditas,* skill, cunning, artfulness; see *callidus* under **know**
L. *commentus,* devised, fabricated: comment, commentary.
AS. *craeft,* strength, skill, art: crafty, artcraft, handicraft, woodcraft, witchcraft, wordcraft, smithcraft, priestcraft, Roycroft.
L. *daedalus* (Gr. *daidalos; daidaleos,* dappled, spotted), cunningly or skillfully made in the manner of Daedalus, the Athenian artificer; *daidalma, -tos,* n. work of art: daedal, daedalian, *Daedalea cinnabarina* (a fungus), *Daedalocrinus kirki* (a fossil crinoid), *Daedalma inconspicuum* (a butterfly).
Gr. *demosyne,* f. skill, knowledge:
Gr. *dexiotes, -etos,* f. handiness, cleverness:
L. *dexteritas, -atis,* f. skill, facility, aptitude: dexterity.
L. *dolosus,* crafty, cunning; see *dolus* under **lie**
Gr. *eustochos,* well-aimed, making good shots, skillful, successful:
L. *faber, -bra, -brum,* workmanlike, skillful; *affaber,* ingenious; see **make**
L. *factitius; fictitius,* artificial, not natural: factitious, fictitious, fetish, featous.
L. *graphicus* (Gr. *graphikos),* of painting or drawing; see **write**
Gr. *hemosyne,* skill in throwing or shooting; see *hemon* under **throw**
Gr. *Hephaistos,* god of fire used in the arts, identified with Vulcan; see **fire**
L. *Musa* (Gr. *Mousa),* f. Muse, one of the goddesses who presided over the fine arts; *museum* (Gr. *mouseion),* n. temple of the Muses, art school, academy: museum, mosaic, *Ophimusium pulchellum* (a brittle-star), *Anthrenus museorum* (museum-pest), *Chiridium museorum* (a pseudoscorpion). Identify: Calliope, Clio, Erato, Euterpe, Melpomene, Polyhymnia(Polymnia), Terpsichore, Thalia, Urania.
Gr. *palame,* handiwork, work of art; see *palma* under **hand**
L. *peritia,* practical knowledge, skill; see *peritus* under **know**
L. *Pieris, -idis* (Gr. *-idos),* f. one of the nine daughters of Pierus called the Pierides: *Pieris oleracea* (cabbage-butterfly), *Pieris floribunda* (an ericacead).
Gr. *Pygmalion,* m. a sculptor to whose statue of a woman Aphrodite gave life: Pygmalion.
Gr. *skeuastos,* artificial; see *skeuos* under **tool**
Gr. *techne,* f. art, craft; *technydrion; technyphion,* n. dim.; *technikos,* artistic, skillful; *technetes (technites),* m. craftsman; *technao,* contrive, execute, or make cunningly or skillfully; *cheirotechnes,* m. artisan, handicraftsman; *entechnos,* artistic, skilled: technology, technocracy, technique, technical, technetium, logotechnics, *Technocrinus sculptus* (a Devonian crinoid), *Entechnia (Melitoma) taurea* (a bee).
L. *vafer,* artful, crafty, cunning, sly; see **know**

L. *versutus,* adroit, crafty, sly; see **know**
See: **form, know, make, play, write, cut**

artamos, Gr. butcher; see **butcher**

artane, Gr. rope, noose; *artao,* hang, suspend; see **rope, hang**

artatus, L. contracted; see *arcto* under **press**

artema, Gr. pendant, earring; see **earring**

Artemis, Gr. goddess of the chase; see **hunt**

artemisia, L. mugwort, sagebrush; see **wormwood**

artemon, Gr. foresail, pulley; see **sail**

artery < Gr. *arteria,* windpipe, aorta; see **pipe**

arthmos, Gr. bond, league, union; see **bind**

arthritis, Gr. inflammation of the joints, pain in the joints; see **disease**

arthro- < Gr. *arthron,* joint; see **joint**

arti, Gr. now, at this time, just; see **now**

artichoke < It. *articiocco;* see *cynara* under **thistle,** *sol* under **sun**

articulus, L. dim. of *artus,* joint; *articulatus,* jointed; see **joint, part**

artios, Gr. complete, perfect, even (as applied to numbers); see **all**

artocarpus, NL. a genus of the mulberry family; see **mulberry**

artos, Gr. bread; see **bread**

artuatus, L. torn to pieces; see **tear**

artus, L. joint; *articulus,* dim., see **joint;** *artus,* confined, close, see *arcto* under **press**

artytos, Gr. flavored, seasoned; see **spice**

arula, L. dim. of *ara,* altar; see **temple**

arum, L. (Gr. *aron*), n. wakerobin; Gr. *arisaron,* n. a kind of arum: Araceae, aroid, *Arum pictum* (an arum), *Arisaema speciosum* (a jack-in-the-pulpit), *Polygonum arifolium* (a tearthumb).
L. *acorus* (Gr. *akoros*), f. sweetflag; a genus of the arum family: *Acorus calamus* (sweetflag), *Iris pseudacorus* (yellowflag), *Enalus acoroides* (a marine plant).
NL. *alocasia,* f. said to be from *colocasia: Alocasia macrorrhiza* (an arum).
NL. *anthurium,* n. a genus of the arum family: *Anthurium ornatum* (an anthurium).
NL. *arisaema,* n. a genus of the arum family: *Arisaema serratum* (Japanese jack-in-the-pulpit).
NL. *caladium,* n. a genus of the arum family, < Malay *kaladi: Caladium bicolor* (caladium).
NL. *calla,* f. a genus of the arum family: *Calla palustris* (water-arum).
L. *colocasia* (Gr. *kolokasia*), f. an aracead: *Colocasia esculenta* (taro).
L. *dracontium* (Gr. *drakontion*), n. dragonwort: *Dracontium asperum* (Brazil dragonwort), *Arisaema dracontium* (dragonroot).
NL. *lysichitum* (Gr. *lysichiton,* loose tunic), n. a genus of the arum family: *Lysichitum camtschatcense* (a yellow-skunkcabbage).
NL. *orontium,* n. a genus of the arum family: *Orontium aquaticum* (golden-club)
NL. *pistia,* f. a genus of the arum family: *Pistia stratiotes* (water-lettuce).
NL. *pothos,* m. Singhalese name for an aroid: *Pothos loureiri* (an aroid), *Pothomorphe peltata* (caapeba).
Gr. *skindapsos,* m. a plant; now a genus of the arum family: *Scindapsus pictus* (painted ivy-arum).
NL. *symplocarpus,* m. a genus of the arum family: *Symplocarpus foetidus* (skunk-cabbage).

aruncus, L. goatsbeard; see **rose**

arundo, L. reed, cane; see **grass**

arvina, L. fat, suet, lard, grease; *arvinula,* dim.; see **fat**

arvum, L. field; *arvalis,* of a cultivated field; see **field**

arx, *arcis,* L. stronghold, fortress, citadel, castle, height, peak; see **fort**

aryballos, Gr. bag, purse; see **bag**

arytaena, L. (Gr. *arytaina*), ladle, dipper; see **spoon**

as, *assis,* L. unity, unit, a copper coin; see **one, money**

as- < L. *ad,* to; see **to**

asaminthos, Gr. bathtub; see **basin**

asapes, Gr. not likely to decay; see **always**

asaphes, Gr. dim, indistinct, obscure, baffling; see **dim**

asarco- < Gr. *asarkos,* lean; see **thin**

asarotum, L. (Gr. *asaroton*), floor laid in mosaic; see **mosaic**

asarum, L. (Gr. *asaron*), hazelwort, wild-ginger; see **birthwort**

asbestus < Gr. *asbestos,* m. a fibrous mineral; L. *asbestinum,* n. fireproof cloth made of asbestus: asbestus, asbestine, asbestoid.
  Gr. *amiantos,* m. a green, fibrous mineral: amianthus.

asbolos, Gr. soot; see **dirt**

ascalabo- < Gr. *askalabos; askalabotes,* a kind of lizard; see **lizard**

ascalapho- < Gr. *askalaphos,* an unknown bird, probably an owl; see **owl**

ascalo- < Gr. *askalos,* unhoed, unweeded; see **rough**

ascaris < Gr. *askaris, -idos,* an intestinal worm; see **worm**

ascaulo- < Gr. *askaules,* bagpiper; see *aulos* under **pipe**

ascelo- < Gr. *askeles,* dried up, withered; see **lessen**

ascend < L. *ascendo, -ensus,* mount, climb, rise, grow; see **raise**

ascetico- < Gr. *asketikos,* rigorous practice, austere; *asketes,* practician, athlete; *asketos,* practiced, expert, < *askeo,* exercise, practice, fashion; see **make**

asceto- < Gr. *asketos,* curiously wrought, ornamented; see *askeo* under **make**

aschemon, Gr. misshapen, ugly; see **ugly**

aschetos, Gr. ungovernable, resistless; see **strong**

aschium, NL. (Gr. *aschion*), puffball, truffle; see **fungus**

ascholos, Gr. without leisure, busy; see **busy**

ascia, L. ax, carpenter's adze; see **ax**

ascidio- < Gr. *askidion,* dim. of *askos,* bag; see **bag**

ascitus, L. alien, foreign; see **strange**

asclepias, L. (Gr. *asklepias*), a genus of the milkweed family; see **milkweed**

asco- < Gr. *askos,* bag, bladder; *askidion; askion,* dim.; see **bag**

ascyrum, L. (Gr. *askyron*), a kind of St. Johnswort; see **St. Johnswort**

ase, Gr. nausea, distress; *asodes,* attended with nausea; see **vomit**

-ase < Gr. *-asis, -esis,* denoting ferments, enzymes, yeasts; see **yeast**

asebema, Gr. sacrilege; see **profane**

aselgo- < Gr. *aselges,* dissolute, lewd; see **lewd**

asellus, L. dim. of *asinus,* ass, see **horse;** a sea-fish, see **cod**

aseros, Gr. disdainful, irksome; see **hate**

ash < AS. *easc* (G. *esche*), a tree.
  L. *bumelia* (Gr. *boumelia*), a kind of ash; now a genus of the sapodilla family; see **sapodilla**
  L. *farnus,* f. ash-tree; *farneus,* of the ash:
  L. *fraxinus,* f. ash; *fraxineus,* of ash: *Fraxinus americana* (white ash), fraxinella (*Dictamnus albus*), *Pterocarya fraxinifolia* (Caucasian wingnut), *Mycosphaerella fraxinicola* (a fungus).
  Gr. *melia,* ash; now a genus of the mahogany family; see **mahogany**
  Gr. *oa (oia),* mountain-ash, equivalent to *sorbus;* see **rose**
  L. *ornus,* mountain-ash, equivalent to *sorbus;* see **rose**

ashes < AS. *asce,* incombustible residue: ashen, ashery, ashy, Ashland, Ashton, potash.
  L. *cinis, -eris,* m. ashes; *cinisculus,* m. dim.; *cineraceus; cinerarius; cinereus,* of ashes, ash-colored, ashen; *cinericius,* like ashes; *cinerosus,* full of ashes; *cinefactus,* turned to ashes: cineraceous, cinereous, cinereal, cineritious, Mt. Cenis, cinerary, incinerator, *Juglans cinerea* (butternut or white walnut), *Centaurea cineraria* (dusty miller), *Striglina cinereola* (a moth), *Myiarchus cinerascens* (a flycatcher).
  L. *favilla,* f. embers, ashes; *favillaceus,* of ashes; *favillaticus,* like ashes: favilla, favillous.
  L. *lix, -icis,* ashes, lye; see **alkali**
  AS. *sinder,* slag, dross, scoria, ash: cinder, Cinderella, sinter.
  Gr. *spodos,* f. ashes, embers; *spodios,* of ashes, ash-colored, gray: spodogenous, spodomancy, spodumene, *Spodochlamys caesarea* (a beetle), *Spodiornis jelski* (a bird).
  Gr. *tephra,* f. ashes; *tephros; tephrinos,* ash-colored; *tephrodes,* like ashes, ashy: tephrite, tephroite, *Tephrophilus wetmorei* (a bird), *Tephrodornis pondicerianus* (a shrike), *Tephrosia cinerea* (a goat's-rue), *Tephraeops zebra* (a zebra-fish),

Tephrocorys cinerea (a bird), *Tephrinopsis munda* (a moth).
See: **dust, slag, gray**

Asia, L. f. country east of Europe: Asiatic, *Berberis asiatica* (Asiatic barberry).

asilla, Gr. yoke for carrying pails, baskets, etc.; see **bind**

asilus, L. gadfly; see **fly**

asimina, NL. pawpaw; see **pawpaw**

asinus, L. ass; *asellus,* dim.; *asininus,* having the qualities of an ass, stupid, obstinate; see **horse, fool**

asinusca, L. a kind of worthless grape; see **grape**

asio, L. a horned owl; see **owl**

asios, Gr. miry, muddy, < *asis,* alluvium, mud, slime; see **earth**

asiraco- < Gr. *asirakos,* a wingless locust; see **grasshopper**

ask < AS. *ascian,* question, request.
Gr. *aitema, -atos,* n. request, demand; *aitetikos,* fond of asking:
Gr. *anaxia,* f. command, behest, charge:
Gr. *ara,* f. prayer, prayer for evil, curse; *aratikos,* of prayer; *areter, -os,* f. one who prays; *aratos,* accursed, < *araomai,* pray, invoke a curse; *eparatos,* under a curse; *katara,* f. imprecation:
AS. *biddan,* pray, ask; *beodan,* command: bid, bade, forbidden, bode, foreboding.
L. *chartula,* bill; see **pay**
Gr. *elenktikos,* fond of cross-questioning or examining; see *elenchos* under **try**
Gr. *entole,* f. command, order, behest; *entolikos,* of a command, < *entello,* command:
Gr. *erotema, -tos,* n. question; *erotematikos,* interrogative; *erotesis,* f. questioning; *erotetikos,* skilled in questioning, < *erotao,* ask, inquire: croteme, erotesis, erotetic.
Gr. *euche; epeuche,* f. prayer, wish; *euchole,* f.; *euchos,* n. the thing prayed for:
L. *griphus* (Gr. *griphos; gripos*), m. riddle, puzzle: griphite, *Griphopithecus suessi* (a fossil primate).
Gr. *hikesia,* f. prayer, entreaty; *hiketes; hikter, -os,* m. one who prays or entreats; *hikesios,* suppliant; *hiketerios,* of suppliants: *Hicetes innexus* (a fossil worm).
L. *invito, -atus,* ask, bid: invite, invitation, inviting.
L. *jubeo, jussus,* bid, command, order, designate, appoint; *jussum,* n. order, command, law:
Gr. *keleusma, -tos,* n.; *keleusmos,* m. order, command, call; *keleustes,* m. signaler:
L. *litania* (Gr. *litaneia*), f. prayer, entreaty: litany.
L. *mando, -atus,* ask, charge, command, order; *mandatela,* f. charge; *mandativus; mandatorius,* of an order: demand, mandatory, command, commend, recommendation, maundy, commander, mandate, mandative.
L. *mendico, -atus,* beg, ask for alms, solicit; see **beggar**
L. *Oedipus* (Gr. *Oidipous*), m. reputed solver of the riddle of the Sphinx: *Leontocebus oedipus* (pinche, a tamarin).
L. *oro, -atus,* speak, plead, pray; see **speak**
L. *percunctor, -atus,* question strictly, interrogate:
L. *peto, -itus,* seek, ask, desire, strive after; *altipeta,* aspiring: petition, competitor, impetus, appetite, propitious, petulant, repeat, incompetent, centripetal.
Gr. *peuthen,* inquirer, spy; *peusis,* inquiry, question; see **try**
L. *postulo, -atus,* ask, request, < *posco,* beg, demand: postulate, postulant, expostulation.
L. *precor, -catus,* pray, entreat; *prex, -ecis,* f. prayer, request; *precator, -is,* m. one who prays; *deprecabilis,* that may be entreated; *deprecativus,* apologetic: prayer, precative, imprecation, precarious, deprecatory, *Abrus[Habrus] precatorius* (jequirity-bean).
Gr. *problema, -tos,* n. a question propounded for solution, puzzle, riddle; *problematikos,* of a problem: problem, problematical, *Pterygotus problematicus* (a fossil crustacean), *Jivaromyia problematica* (a tipulid fly).
Gr. *pysma, -tos,* n. question; *pysmatikos,* interrogative:
L. *qui, quod, quis, quid,* who, which, what: qui vive, quorum, quiddity, quoddity, quodlibet, Q.E.D., *Quisqualis indica* (Rangoon creeper), *Quisque gilberti* (a Miocene herring).
L. *rogo, -atus,* ask, request, solicit; *rogatio, -onis,* f. question, bill; *interrogo, -atus,* question, inquire: interrogate, derogatory, abrogate, arrogance, prerogative. surrogate, supererogation.
L. *Socrates* (Gr. *Sokrates*), m. a celebrated Greek philosopher whose method of teaching was characterized by the asking of searching questions: socratic. *Socraticum firmum* (a fossil brachiopod).

Gr. *Sphinx, -ingos,* f. female monster of Thebes who propounded riddles: *Sphinx gordius* (a moth), *Pachysphinx modesta* (a moth), *Papio sphinx* (a baboon).

L. *supplico, -atus,* beseech humbly; *supplex, -icis,* suppliant, humble: supplication, supplicant, suppliant.

L. *vindex, -icis,* c. claimant: *Pyrgus vindex* (a butterfly).

See: **beggar, try, call, doubt, secret**

asma, Gr. song; see *aeido* under **sing**

asmenos, Gr. pleased, glad; see **joy**

asodes, Gr. muddy, slimy; see *asis* under **earth**

asotos, Gr. abandoned, without hope, lost, profligate; see **hopeless**

aspalathus, L. (Gr. *aspalathos*), a thorny shrub yielding a fragrant oil; see **bean**

aspalieus, Gr. angler; see **fisherman**

asparagus, L. (Gr. *asparagos*), m.: *Asparagus plumosus* (fern-asparagus), *Crioceris asparagi* (asparagus-beetle).
L. *acanthillis, -idis,* f. wild asparagus:
L. *corruda,* f. wild asparagus:
Gr. *hormenos,* m. wild asparagus:
Gr. *myacanthos,* m. wild asparagus:

aspasmos, Gr. greeting, embrace; see **embrace**

aspect < L. *aspectus,* look, appearance; see *specio* under **see**

aspendios, L. a kind of vine; see **vine**

asper, L. rough, harsh; *asperitas,* harshness, roughness; see **rough**

aspergillus, NL. a genus of fungi whose conidiophores resemble a brush; see **bush**

aspernatus, L. despised, rejected, spurned; see *sperno* under **hate**

aspersus, L. scattered, sprinkled, < *aspergo, -ersus,* scatter, strew, sprinkle; see *spargo* under **spread**

asperugo, L. a plant with rough leaves; see **borage**

asperula, NL. a genus of the madder family; see **madder**

asphales, Gr. steadfast, firm; see **stand**

asphaltes, Gr. bitumen; see **resin**

asphodel < L. *asphodelus* (Gr. *asphodelos*); see **lily**

asphyxia, Gr. stopping of the pulse and respiration; see **choke**

aspido- < Gr. *aspis, -idos,* shield; *aspidion; aspidiskos,* dim.; *aspidiotes,* shield-bearer, warrior, see **shield**; viper, see **snake**

aspiro, *-atus,* L. desire to reach or obtain; see **wish**

asplenium, NL. < L. *asplenum* (Gr. *asplenon*), spleenwort; see **fern**

aspratilis, L. rough, scaly; *aspretum,* rough place; see *asper* under **rough**

aspris, Gr. a kind of oak; see **oak**

aspros, Gr. white; see **white**

ass < AS. *assa,* < L. *asinus;* see **horse**

assa, L. dry-nurse; see **nurse**

assail < L. *ad,* to, *salio,* leap, spring; see **strike**

assarius, L. roasted; *assatura,* roasted meat; see *asso* under **cook**

assassin < Ar. *hashashin,* hashish-eater; see **kill**

assectator, L. attendant, follower; see **servant**

assemble < OF. *assembler,* < L. *ad,* to, *simul,* at the same time, together; see **gather**

assentor, *-atus,* L. agree constantly, flatter; see **yes**

asser, L. beam, stake, pole; *asserculus,* dim.; see **beam**

assert < L. *assero, -sertus,* affirm, declare; see **speak**

assessor, L. assistant of a judge; see **help**

assevero, *-atus,* L. assert firmly, declare positively; see **swear**

assiculus, L. dim. of *axis;* see **axis**

assiduus, L. constant, steady, busy; *assiduitas,* constant attendance; see **busy**

assignatus, L. marked out, allotted, appointed; see *signum* under **mark**

assis, L. plank, board; *assula,* dim., chip, shaving, splinter; see **board, part**

assist < L. *assisto,* stand by, help; see **help**

assitus, L. near; see **near**

associate < L. *associatus,* joined with; see *socius* under **companion**

assula, L. shaving, chip, splinter; see **part**

assume < L. *assumo, -umptus,* take for granted, suppose; see **hypothesis**

assumentum, L. patch; see **patch**

assus, L. roasted; *assum,* roast meat; see *asso* under **cook**

astacus, L. (Gr. *astakos*), lobster; see **crab**

astandes, Gr. courier; see **run**

astaphis, Gr. dried grape, raisin; see **grape**

Astarte, Gr. Phoenician goddess of love and fertility; see **love**

astatos, Gr. unstable, restless; see **move**

aster; astrum L. (Gr. *aster; astron*), star; a plant; *asteriskos; astrion,* dim.; *asterias; asterios,* starred, starry; see **star**

-aster; -astrum, L. dim. suffix with derogatory implication; wild; resemblance; see DIMINUTIVES under **little**

astericto- < Gr. *asteriktos,* unstable; see **shake**

asterion, Gr. a kind of spider; see **spider**

astes, Gr. singer; see *aeido* under **sing**

-astes, Gr. one who, agent; see *-ter* under **make**

astheno- < Gr. *asthenes,* weak; see **weak**

asthma, Gr. hard-drawn breath, panting, gasping; see **disease**

astibos, Gr. untrodden, pathless; see **alone**

astico- (astyco-); astio- < Gr. *astikos (astykos); asteios,* urbane, refined, elegant, nice; see *asty* under **town**

astilbe, NL. a genus of the saxifrage family; see **saxifrage**

astochos, Gr. missing the mark; see **fault**

astonishing < OF. *estoner,* < L. *ex,* out, *tono,* thunder; see **wonder**

astorgos, Gr. heartless, cruel; see **merciless**

astrabe, Gr. saddle; see **saddle**

astragalinos, Gr. goldfinch; see **finch**

astragalus, L. (Gr. *astragalos*), vertebra, anklebone, die, legume; see **foot, bean**

astrape, Gr. flash of lightning; see **light**

astreptos, Gr. inflexible; see **stiff**

astrifer, L. starry; see *aster* under **star**

astringent < L. *astringens,* shrinking, binding; see *stringo* under **bind, lessen**

astro- < Gr. *astron,* star; see *aster* under **star**

astronium, NL. a genus of the sumac family; see **sumac**

-astrum, L. dim. suffix with derogatory implications; wild; resemblance; see *-aster* under **little**

astur, L. a kind of hawk; see **hawk**

astutus, L. adroit, clever, cunning, < *astus,* cleverness, cunning; see **know**

asty, Gr. town, city; *asteios; astikos (astykos),* of town; see **town**

astytis, Gr. a kind of lettuce; see **lettuce**

astytos, Gr. incapable of erection; *astysia,* impotence; see **weak**

asylum < Gr. *asylon,* sanctuary, refuge; see **safe**

asymmetros, Gr. without symmetry, disproportionate; see **different**

asymphoros, Gr. useless, inconvenient; see **worthless**

asymphylos, Gr. not of the same race, unlike; see **different**

asyndctos, Gr. unconnected, loose; see *adesmos* under **free**

asynetos, Gr. stupid, witless; see **dull**

asynteles, Gr. useless; see **worthless**

asyphe, Gr. a kind of spice; see **spice**

asyphelos, Gr. mean, paltry, worthless; see **worthless**

asyres, Gr. filthy, lewd; see **lewd**

asystatos, Gr. having no cohesion, loose; see **free**

at- < L. *ad,* to; see **to**

atacto- < Gr. *ataktos,* not arranged, disordered; see *ataxia* under **confusion**

Atalanta, L. (Gr. *Atalante*), a legendary maiden noted for her fleetness of foot; see **run**

atalos, Gr. tender, delicate; see **soft**

atalymnos, Gr. a plum-tree; see **plum**

atasthalos, Gr. wicked; see **bad**

atavus, L. forefather, ancestor; see *avus* under **kin**

ataxia, Gr. confusion, disorder, incoordination; see **confusion**

-ate, suffix denoting a salt of an *-ic* acid; see *-atus* under **nature**

atechnos, Gr. unskillful; see **awkward**

atecmarto- < Gr. *atekmartos,* baffling, obscure, uncertain; see **doubt**

ateles, Gr. imperfect, unfilled, ineffectual, exempt; *ateleia,* incompleteness, imperfection, exemption from tax and public burdens; see **fault, free**

ateleutos, Gr. endless, eternal; see **always**

atemeles, Gr. neglected, neglectful; see **neglect**

ater, *atra, atrum,* L. black; see **black**

ateros, Gr. baneful, ruinous; see *atarteros* under **hurt**

aterpes, Gr. joyless; see **sad**

ateucho- < Gr. *ateuches,* unarmed; see **bare**

athales, Gr. withered; see **lessen**

athanatos, Gr. undying, immortal; see **always**

Athena, L. (Gr. *Athene*), tutelary of Athens and goddess of wisdom; see **know**

ather, Gr. ear of wheat, spike; see **point**

athere (athare), Gr. goats, grits, porridge; see **flour**

atherine, Gr. smelt; see **smelt**

athesphatos, Gr. inexpressible, enormous, marvelously great; see **large**

athetetes, Gr. violator; see **hurt**

athetos, Gr. useless, unfit; see **unfit**

athicto- < Gr. *athiktos,* untouched, chaste; see **pure**

athlete < Gr. *athletes,* contestant; *athlos; athlesis,* contest, struggle; see **fight**

athroos, Gr. crowded, assembled, collected; see **gather**

athymos, Gr. fainthearted, spiritless, hopeless; see **hopeless**

athyrma, Gr. plaything, top; see **play**

-aticum, L. collection of, condition, state of being; see *-age* under **nature**

atimetos, Gr. despised; see **hate**

atimos, Gr. without honor, dishonorable; *atimastos,* dishonored; see **bad**

Atlas, *-antis,* L. (Gr. *Atlas, -antos*), mythical god who held up the heavens, < *tlao,* bear, carry; see **carry**

atmen, Gr. slave, servant; see **servant**

atmetos, Gr. uncut, undivided; see **one**

atmos, Gr. vapor; *atmis, -idos,* steam; see **cloud**

atolmos, Gr. cowardly, spiritless; see **coward**

atom < L. *atomus* (Gr. *atomos,* uncut, indivisible), a small particle; see **little**

atone < AS. *at,* at, *an,* one.
 Gr. *aresko,* make amends, atone, satisfy, please; see **agreeable**
 L. *expio, -atus,* make atonement, purge by sacrifice: expiate, expiation.
 Gr. *hilaskomai,* propitiate, atone; *hilasmos,* m. atonement; *hilasterios,* propitiatory; *hilastes,* m. propitiator; *anhilastos,* unappeased:
 L. *reparo, -atus,* restore, revive: repair, reparation.
 See: **pay, sad**

atonos, Gr. not taut, relaxed, slack, languid; see **weak**

atopos, Gr. out of place, odd, strange, absurd; *atopia,* absurdity; see **strange**

atracto- < Gr. *atraktos,* spindle, arrow; see **spindle**

atramentum, L. any black fluid, ink; see **ink**

atraphyaxys, Gr. saltbush; see **goosefoot**

atrapos, Gr. path, road, way; see **way**

atratus, L. dressed in black; see *ater* under **black**

atreco- < Gr. *atrekes,* exact, sure, real, true; see **right**

atrestos, Gr. fearless; see **bold**

atri-; atro- < L. *ater,* black; see **black**

atrina, NL. a genus of mollusks; see **mollusk**

atriplex, L. saltbush; see **goosefoot**

atrium, L. vestibule, hall, entry; see **room**

atropos, Gr. unchangeable, eternal, inflexible; one of the three Fates; see **always**

atrotos, Gr. invulnerable; see **safe**

atrox, *-ocis,* cruel, harsh, horrible, severe; see **merciless**

atrygos, Gr. without lees, clear, pure; see **pure**

atrytos, Gr. not worn, unwearied; see **strong**

atta, L. a genus of ants, < *Atta,* name for persons who walk on their shoetips; see **ant**

attack < F. *attaquer;* see **strike**

attacus, L. (Gr. *attakos*), a kind of locust; see **grasshopper**

attagas, Gr. a kind of partridge, francolin; see **quail**

attagen, Gr. probably a kind of grouse; see **grouse**

attagino- < Gr. *attageinos,* a kind of tunny; see **tunny**

attalea, NL. a genus of palms, < L. *Attalus* (Gr. *Attalos*), king of Pergamum; see **palm**

attegia, L. tent; see **tent**

attelabus, L. a small wingless locust; see **grasshopper**

attempt < L. *attento, -atus,* attack, try; see *tento* under **try**

attend < OF. *attendre;* see **heed, care, busy**

attenuatus, L. drawn out, tapered, weakened, thin; see *tenuis* under **thin**

Atticus, L. (Gr. *Attikos*), pertaining to the Greek province of Attica and its capital, Athens; see **Greek**

attignus, L. touching, bordering; see **near**

attilus, L. a sturgeon; see **sturgeon**

attina, L. a stonewall boundary; see **wall**

attributus, L. that is ascribed to a person or thing; see **nature**

-atus, L. provided with, having the nature of, pertaining to; see **nature**

atycto- < Gr. *atyktos,* undone; see **cancel**

atyphos, Gr. not puffed up, modest; see **modest**

atyzelos, Gr. frightful; see **fear**

aucella, L. a little bird; see *avis* under **bird**

auceps, *-upis,* L. bird-catcher; see *avis* under **bird**

auchema, Gr. boast, pride; *auchetes,* boaster; see **brag**

aucheno- < Gr. *auchen, -os,* neck; see **neck**

auchmos, Gr. drought; *auchmeros,* dry; see **dry**

auchos, Gr. a kind of pulse; see **bean**

auctioneer < L. *auctio, -onis,* public sale; see **trade**

auctor, L. originator, causer, doer; see **make**

auctus, L. enlarged, increased, < *augeo,* increase, grow; *augmentum,* increase; see **grow**

aucupatus, L. fowling, bird-catching; *auceps,* fowler; see *avis* under **bird**

audacia, L. boldness, courage; *audax, -acis,* bold, daring; *audaculus,* dim., somewhat bold; see **bold**

auditorium, L. a room for hearing, < *audio, -itus,* hear, listen; see **room, hear**

auge, Gr. luster, shine; see **light**

auger < AS. *nafugar;* see **bore**

auginus, L. henbane; see **nightshade**

augmentatives.   See **very, large, grow**

augur, L. diviner, soothsayer; *auguralis,* of augurs; see **prophecy**

augustus, L. venerable; see **honor**

auk < uncertain origin, but cf. Sw. *alka:* great auk (*Plautus impennis*).
    NL. *alca,* f. a genus of birds including the auks: Alcidae, *Alca torda* (razor-billed auk).
    Scan. *lunde,* puffin: *Lunda cirrhata* (tufted puffin).

aula, L. (Gr. *aule*), court, courtyard, hall; see **space**

auletes, Gr. flute-player; see *aulos* under **pipe**

auleum, L. curtain, tapestry, canopy; see **canopy**

aulido- < Gr. *aulis, -idos,* tent, stall; see **tent**

aulon, Gr. channel, pass, pipe; *aulos,* flute, pipe; *auliskos,* dim.; *auletes,* flutist; see **pipe**

aulula, L. small pipkin or pot; see **pot**

aunt < OF. *ante,* < L. *amita,* father's sister; see **kin**

aura, L., Gr. air, wind; *aurula,* dim.; see **wind**

aurantium, NL. orange; see **citrus**

aurea, L. bridle; see **hold**

auriga, L. charioteer, driver; *aurigo, -atus,* drive a chariot; see **drive**

auris, L. ear; *auricula,* dim., lobe of the ear, little ear; *auritus,* eared, attentive; see **ear**

aurora, L. dawn, morning; see **dawn**

aurugo, L. jaundice; *aurugineus,* jaundiced; see **disease**

aurula, L. dim. of *aura,* air, wind; see **wind**

aurum, L. gold; *aurarius; auratus; aureolus; aureus,* golden, gilt, pertaining to gold, made of gold, color of gold; *aurifer; auriger,* gold-bearing; *aurifex,* goldsmith; see **gold**

ausculto, -atus, L. listen; see **hear**

auspex, -icis, L. bird inspector, diviner, augur; *auspicatus,* fortunate, favorable, lucky; see **prophecy**

auster, L. a dry, hot, south wind; see **wind**

austerus, L. harsh, rough, stern, gloomy, sad, hard; see **serious**

australis; austrinus, L. southern; see **south**

ausus, L. attempt, hazard, < *audeo,* venture, dare, undertake, hazard; see **bold**

authades, Gr. stubborn; see **stubborn**

authentic < Gr. *authentikos,* genuine, true, original; see **true**

authepsa, L. (Gr. *authepses*), self-cooker, teapot; see **pot**

author < L. *auctor,* originator, causer, doer; see **make**

automatos, Gr. self-moving, spontaneous; see *autos* under **alone**

autopsy < Gr. *autopsia,* a making certain by seeing for oneself, post-mortem examination; see **try**

autos, Gr. self; see **alone**

autumnus, L. season of increase and abundance, fall; see **year**

auxesis, Gr. growth; *auximos,* promoting growth, < *auxo,* increase, grow; see **grow**

auxilium, L. help, aid; *auxiliaris,* aiding, helping; see **help**

auxilla, L. small pot; see **pot**

auxis, Gr. a young tunny; see **tunny**

avarus, L. eagerly desirous, covetous, greedy; *avaritia,* greed; see **greed**

avellana, It. filbert, hazelnut, < L. *Abella (Avella),* a city of Campania; see **hazel**

avena, L. oat, oats; *avenaceus,* of oats; see **grain**

avenge < OF. *avengier;* see **punish**

Avernus, L. nether world, infernal regions; see **hell**

averruncus, L. averter, protector; see **guard**

aversus, L. turned away, backward; behind; *aversum,* the back part; see **after**

averta, L. saddlebag; see **bag**

avicennia, NL. a genus of the vervain family; see **vervain**

avidus, L. eagerly desirous, greedy, < *aveo,* crave, desire; see **wish**

avis, L. bird; *avicella; avicula,* dim.; *aviarius,* of birds; see **bird**

avitus, L. of a grandfather, ancestral; see *avus* under **kin**

avium, L. byway, desert, wilderness; *avius,* out of the way; see *via* under **way**

avocet < It. *avocetta,* a shore bird; see **snipe**

avoid < OF. *esvuidier,* < L. *ex,* out, *viduus,* bereft; see **shun, run**

avus, L. grandfather; *avia,* grandmother; *avunculus,* dim., uncle; see **kin**

awake < AS. *awacian.*
  Gr. *abriktos,* wakeful: *Tibicen(Abricta) ferruginosus* (a cicada).
  Gr. *agrypnia,* f. sleeplessness; *agrypnetikos,* wakeful; *agrypnos,* sleepless, wakeful,
    watchful; *agrypneter, -os (-es),* m. watcher: agrypnia, agrypnotic, *Agrypnetes
    crassicornis* (a trichopterid).
  L. *alacer, -cris, -cre,* active, brisk, eager, lively; see **life**
  Gr. *egertikos,* waking, stirring; see **arouse**
  Gr. *egregorikos,* wakeful, watchful; *egregorsios,* keeping awake: Gregory.
  L. *inconnivus,* that does not close the eyes, sleepless:
  L. *indormis,* sleepless:
  L. *insomnia,* f. sleeplessness; *insomniosus; insomnis,* sleepless; *desomnis,* sleepless:
    insomnia.
  L. *insopibilis,* that cannot be lulled to sleep; *insopitus,* sleepless, wakeful:
  Gr. *phylaktikos,* observant, vigilant, alert; see *phylax* under **guard**
  L. *vigilis,* alert, awake, watchful; see **guard**
  See: **arouse, life**

award < OF. *esgarder,* bestow, confer, prize; see **give, pay**

away < AS. *onweg;* see **from, far, separate, alone**

awe < ME. *aw, agh,* < ON. *agi,* fear; see **wonder, fear, honor**

awkard < ON. *ofugr,* turning the wrong way, AS. *weard,* in the direction of.
  Gr. *adexios,* left-handed, awkward; see *dexia* under **hand**
  L. *agrestis,* of the fields, uncultivated, rustic; see *ager* under **field**
  Gr. *atechnos,* unskillful:
  L. *inconcinnus,* inelegant, awkward, absurd: *Conotrachelus inconcinnus* (a beetle).
  L. *inscitus,* ignorant, awkward; see **dull**
  L. *rusticus,* of the country, rough, awkward; see *rus* under **country**
  Gr. *skaios,* left, awkward, ill; see **hand**

awl < AS. *awel.*
  Sp. *alesna,* f. awl: Alesna (a volcanic plug near San Mateo, N. Mex.), Coeur
    d'Alene (Idaho).
  Gr. *kenteron,* n. awl:
  Gr. *opeas, -atos,* n. awl; *opetidion, opetion,* dim.: *Opeatocerus purpuratus* (a fly).
  L. *subula,* f. awl: subulate, *Aster subulatus* (an aster), *Phlox subulata* (moss-
    phlox), *Subularia aquatica* (awlwort), *Caulolepis subulidens* (a fish), *Festuca
    subuliflora* (a fescue).
  See: **needle**

ax < AS. *eax; adesa,* tool with blade at right angles to handle: axman, battleax,
    Saxon, adze.
  L. *ascia,* f. carpenter's ax or adze; *asciculus (acisculus),* m. dim.:
  Gr. *axine,* f. ax, wedge; *axinarion; axinidion,* n. dim.: axinite, *Axinomimus para-
    doxus* (a sponge).
  L. *dolabra,* f. ax, hatchet; *dolabella,* f. dim.; *dolabratus,* shaped like an ax:
    Thujopsis dolabrata (a conifer), *Begonia dolabrifera* (a begonia), *Mesembryan-
    themum dolabriforme* (a fig-marigold), *Dolabella scapula* (a gastropod).
  Gr. *genys, -yos,* jaw, ax; see **jaw**
  Gr. *kopis, -idos,* chopper, cleaver; see **knife**
  Gr. *kybelis, -idos,* f. ax, cleaver:
  Gr. *labrys, -yos,* f. two-bitted ax: labrys.
  Gr. *pelekys, -eos,* m. ax, hatchet; *pelekion,* n. dim.: pelecoid, Pelecypoda, *Pele-
    cyphora aselliformis* (a cactus), *Argyropelecus[Argyropelecys] heathi* (a fish).
  L. *securis, -is,* f. ax or hatchet with a broad edge; *securicula,* f. dim.; *securifer;
    securiger,* ax-bearing: *Securicula gora* (a fish), *Tritonia securigera* (an iridacead),
    Securigera coronilla (hatchet-vetch).
  Gr. *skeparnon,* n. ax: *Sceparnodon ramsayi* (a fossil mammal).
  See: **pick**

axicia, L. shears; see **scissors**

axilla, L. armpit, crotch; *axillaris,* of an axil; see **angle**

axino- < Gr. *axine,* ax; *axinidion,* dim.; see **ax**

axiom < Gr. *axioma,* self-evident principle; see **law**

axios, Gr. worthy, fit; see **worth**

axis, L. (Gr. *axon, -os*), m. axle, pole, hinge, valve; *axiculus; assiculus,* m. dim.;
  Gr. *axonion,* n.; *axoniskos,* m. dim.: axial, abaxial, uniaxial, coaxial, axiferous,
  axiolite, axilema, axopodium, axometer, axonost, *Assiculus punctatus* (a fish),
  Axonopus furcatus (carpet-grass), *Arthraxon hispidus* (a grass), *Petalaxis
  sibiricus* (a fossil coral), *Hexacontium axotrias* (a radiolarian), *Trissocyclus*

triaxonius (a radiolarian).

L. *cnodax, -acis* (Gr. *knodax, -akos*), m. pin, pivot: *Arthrocnodax occidentalis* (a gall-gnat).

Gr. *penion,* n. bobbin, spool; see *penos* under **weave**

L. *polus,* m. pivot, axis: pole, polar, polarity, Polaris, circumpolar, *Lioplax polaris* (a fossil snail).

Gr. *rhachis,* axis, stem or stalk, spine; see **back**

Gr. *strophinx, -ngos,* m. axle, pivot; *strophingion,* n. dim.: *Strophingia ericae* (a bug).

L. *turbo, -inis,* top, spindle; see **cone**

See: **spindle, hinge, center, pillar, stem, turn, back, ridge**

axitiosus, L. acting together, in combination; see **harmony**

axle < ON. *axultre,* axletree; see **axis**

axon, Gr. axle; see **axis**

azalea, NL. a genus of the heath family, < *azaleos,* dry, parched; see **heath, dry**

azo- < Gr. *azoos,* lifeless; pertaining to nitrogen compounds, see **nitrogen**; < *azo,* dry up, see *azaleos* under **dry**

azolla, NL. a genus of ferns; see **fern**

azorella, NL. a genus of the carrot family; see **carrot**

Aztec, Ab.Am. an ancient tribe in Mexico: Aztecan, *Theridium aztecum* (a spider).

azura < F. *azur,* < Per. *lazhuward,* a blue color; see **blue**

azyx, -ygos, Gr. unyoked, unwedded, solitary; see **alone**

B

babax, Gr. chatterer, loud talker; see *baxis* under **speak**

baboon < OF. *babuin;* see **monkey**

babulus, L. babbler, fool; see **speak**

baburrus, L. foolish, silly; see **fool**

baby, dim. of ME. *babe,* < uncertain origin; see **young**

babyas, Gr. mud; see **earth**

bacalusia, L. sweetmeat, confection; see **food**

bacca, L. berry; *bacula,* dim.; *baccalis,* with berries; see **berry**

baccharis, L. (Gr. *bakcharis*), a plant of the composite family; see **composite**

Bacchus, L. (Gr. *Bakchos*), god of wine, shouting, revelry; see **wine**

baccina, L. henbane; see **nightshade**

bacelo- < Gr. *bakelos,* a eunuch; see **sterile**

bachelor < ML. *baccalaureus,* apprentice, single man who has never married; see **alone**

bacillum; bacillus, L. little stick, dim. of *baculum,* rod; see **rod**

back < AS. *baec,* the rear or dorsal part of anything, to the rear: backache, backbone, hunchback.

Gr. *aetoma,* gable; see **gable**

Gr. *astragalos,* a vertebra, anklebone; see **foot**

L. *carina,* keel; see **ridge**

Gr. *choiras, -ados,* like a hog's back; see *choiros* under **hog**

L. *dorsum,* n. back, ridge of a hill; *dorsualis (dorsalis),* of the back: dorsal, endorse, indorsement, addorsed, dorsiferous, dorsiventral, *Chalepus dorsalis* (locust leaf-miner), *Apis dorsata* (dingar), *Didelphys dorsigera* (an opossum), *Typhlochrestus dorsuosus* (a spider).

L. *fastigium,* slope, gable-end; see **gable**

Gr. *hyptios,* supine, lying on the back; see **lie**

Gr. *kyphos,* bent, humpbacked; see **bend**

Gr. *metaphrenon*, n. broad of the back, back:

Gr. *noton*, n.; *notos*, m. back, ridge: notochord, notopodium, *Notonecta undulata* (back-swimmer), *Notropis hudsonius* (white chub), *Prionotus scitulus* (a gurnard), *Arsinotus albipes* (a bug), *Pycnonotus barbatus* (a bulbul), *Callonotacris lophophora* (a grasshopper), *Callistonotus nigroruber* (a bug), *Lacconotus punctatus* (a beetle).

Gr. *osphys*, *-yos*, loin, lower part of the back; see **side**

Gr. *rhachis*, *-ios*, f. spine, backbone, ridge, midrib or axis of a leaf; *rhachites* (*rhachitis*), of the spine: rachis, rachitis, rickets, rachitomous, rachialgia, rachiglossate, rachilla, *Rhachiocephalus prognathus* (a fossil reptile), *Polyrhachis simplex* (an ant).

Gr. *sphondylos* (*spondylos*), m. vertebra, joint; *sphondylion*, n. dim.: spondylitis, spondylium, *Paleospondylus gunni* (a fossil fish).

L. *spina*, f. backbone; *spinalis*, of the backbone: spine, spinal.

Gr. *stropheus*, a vertebra, socket; see *streptos* under **turn**

L. *supinus*, lying on the back; *resupinus*, bent backward with the face upward; see **lie**

L. *tergum*, n. back: tergite, tergiversation, tergal, tergiferous.

Gr. *trophis*, *-idos*, keel; see **ridge**

L. *vertebra*, f. one of the bones of the backbone: vertebral, vertebrate, Vertebrata. See: **ridge, after, return**

bacrio, L. vessel with a long handle, ladle; see **spoon**

bacterium < Gr. *bakterion*, dim. of *baktron*, stick; see **rod**

bactris, NL. a genus of palms; see **palm**

bactro- < Gr. *baktron*, stick, cane, staff; *bakterion*, dim.; see **rod**

baculum, L. stick, rod, staff; *bacillum; bacillus*, dim.; see **rod**

bad < uncertain origin.

Gr. *acharistos*, ungracious, thankless, unpleasant:

Gr. *adoxos*, disreputable, disgraceful, obscure: *Adoxus obscurus* (a grape-root worm).

Gr. *aedes*, unpleasant, disagreeable, odious; *aedismos*, m. disgust: *Aëdes aegypti* (formerly *Stegomyia fasciata*, yellow-fever mosquito).

Gr. *aischros*, deformed, ugly, foul, obscene, shameful: *Aeschrostoma marmoratum* (a moth).

Gr. *aisylos*, evil, godless: *Aesylacris villosula* (a beetle).

Gr. *alastos*, unforgettable, insufferable, abominable; see **punish**

Gr. *alitros*, sinful, wicked; see **fault**

Gr. *anenor*, *-os*, unmanly, dastardly:

Gr. *anerestos*, displeasing:

Gr. *aparestos*, unpleasing:

Gr. *apeuktos*, to be deprecated, abominable:

Gr. *aphados*, displeasing, hateful, odious; *aphadia*, f. displeasure:

L. *aspernabilis*, worthy of contempt, despicable: *Coelosternus aspernabilis* (a beetle).

Gr. *asyphelos*, vile, mean, low, paltry:

Gr. *atasthalos*, wicked: *Atasthalus spectrum* (a beetle), *Atasthalistis pyrocosma* (a moth).

Gr. *Ate*, f. goddess of mischief and cause of rash acts; *ateros*, mischievous, baneful, ruinous: *Ate jovianus* (a butterfly), *Ateropogon cyrtopogonoides* (a fly).

Gr. *atimos*, without honor, dishonorable; *atimastos*, dishonored: *Atimoblatta reniformis* (a fossil cockroach), *Atimus crenatostriatus* (a beetle), *Atimastillas flavicollis* (a bird).

L. *aversabilis*, causing a turning away, abominable:

Gr. *bdelyktos; bdelyros*, abominable, disgusting, loathsome, shameless; *bdelygma*, *-atos*, n. abomination; *bdelyros*, f. nausea, disgust: *Harpyia(Bdelygma) major* (a bat).

L. *contemptus*, despised, despicable, < *contemno*, esteem lightly, despise; see *temno* under **hate**

L. *corruptus*, spoiled, damaged; *corruptibilis*, perishable: corrupt, corruption, incorruptible, *Epilachna corrupta* (Mexican bean-beetle).

L. *despicabilis*, contemptible; see *despicus* under **hate**

L. *deter*, poor, bad; *deterior; deterius*, poorer, worse; *deterrimus*, poorest, worst: deteriorate, deterioration.

L. *detestabilis*, abominable, execrable; see *detestor* under **hate**

L. *dispendiosus*, hurtful, prejudiced:

L. *dissipatio*, *-onis*, wasteful squandering, dissolute indulgence: dissipation.

L. *dissolutus*, lax, loose in morals, dissipated: dissolute.

Gr. *dys*, bad, ill, with difficulty, hard, unlucky, very; *dysalgetos*, hard to bear; *dysbolos*, throwing badly; *dysgnostos*, hard to recognize or understand; *dyseklytos*, hard to undo; *dyserastos*, unhappy in love; *dysippos*, hard to ride; *dyspeptos*, hard to digest; *dysprositos*, hard to get at; *dystektos*, hard to melt; *dysphoros*, hard to endure: dysentery, dyspnea, dysuria, dyspepsia, dyschronous, dysmennorrhea, dyspareunia, *Dysoxylum spectabile* (pencilwood), *Dysichthys coracoideus* (a fish).

Gr. *dyskleia*, f. disgrace, infamy:

L. *Elagabalus* (*Heliogabalus*), m. Roman emperor, whose reign was notorious for debauchery and profligacy:

L. *flagitiosus*, base, dissolute, profligate; see **shame**

L. *foedus*, loathsome, foul, filthy, vile; see **dirt**

L. *ignobilis*, unknown, obscure, low, mean: ignoble.

L. *illegitimus*, unlawful; see *illex* under **forbid**

L. *illiberalis*, unworthy of a freeman, ignoble, sordid, mean: *Tholeria illiberalis* (a butterfly).

L. *improbus*, not according to standard, bad, poor, inferior: *Phygadeuon improbus* (a wasp).

L. *improprius*, unsuitable, inappropriate: improper, impropriety.

L. *impuratus*, infamous, vile: *Geometra impura* (a butterfly).

L. *incomis*, unpleasant:

L. *incommodus*, inconvienient, unsuitable, unfit, troublesome: *Piazurus incommodus* (a beetle).

L. *incorrectus*, wrong: incorrect.

L. *indignus*, unworthy, undeserving; *indignitas*, f. unworthiness, vileness: indignity.

L. *infamis*, of bad repute, disreputable, notorious; *infamo*, *-atus*, brand with infamy, dishonor: infamy, infamous.

L. *infandus*, unspeakable, unnatural, abominable: *Odontomachus infandus* (an ant).

L. *infaustus*, unfortunate; see *faustus* under **lot**

L. *infestivus*, disagreeable, unpleasant:

L. *ingustabilis*, not fit to be eaten or drunk:

L. *iniquus*, unfavorable, unequal: iniquity.

L. *injucundus*, unpleasant: *Naupactus injucundus* (a beetle).

L. *injustus*, wrong: unjust, injustice.

L. *insuavis*, unpleasant, disagreeable:

Gr. *kakos*, bad, harmful; *kakion*; *hesson*; *cheiron*, worse; *kakistos*; *hekistos*; *cheiristos*, worst; *kakodoxos*, infamous; *hemikakos*, baddish; *pankakos*; *perikakos*, very bad: cacophony, cachexia, cacodoxy, cacodyl, kakistocracy, *Cacomorpha aberrans* (a phasmid), *Cacodacnus hebridanus* (a beetle).

L. *malus*, bad; *pejor*, m., f.; *pejus*, n. worse; *pessimus*, worst; *malefactor*, *-is*, m. evildoer; *maleficus*, criminal, evil, wicked; *malevolens; malevolus*, ill-disposed toward, inimical; *malignus*, of evil nature; *malitia*, f. ill will, spite: malefactor, malefic, malodorous, malevolence, malfeasance, malingerer, malediction, malcontent, malapert, malice, malpractice, malign, malignant, malady, malaria, malapropism, maladroit, dismal, pejoration, pejorative, impair, pessimist, *Cratosomus maleficus* (a beetle).

L. *mictilis*, worthless, bad:

AS. *mis-;* OF. *mes-*, < L. *minus*, less; wrong, bad, ill, evil: mistake, misunderstanding, misuse, misjudge, mischief, misplace, misdeed, misfortune, mischance, miscarriage, misfeasance, misspell, mispronounce, mesalliance.

Gr. *mysos*, n. anything disgusting, abomination, defilement; *mysaros*, foul, loathsome, abominable; see **dirt**

L. *nefarius; nefandus; nefastus*, heinous, abominable, criminal, wicked, profane, unlawful, < *nefas*, n. sin, crime: nefarious.

L. *nequam*, worthless, bad, wicked:

L. *nequitia*, f. worthlessness, badness:

L. *obscenus*, indecent, offensive, filthy: obscene, obscenity, *Dolichopus obscenus* (a fly).

Gr. *paraphoros*, bad, inferior: *Paraphorocera tincta* (a fly).

L. *perditus; deperditus*, destroyed, lost, abandoned, corrupt; see **destroy**

L. *perversus*, turned the wrong way, wrong, evil, bad: perverse, perversity.

Gr. *phaulos*, trivial, bad, abject; see **trifle**

Gr. *poneros*, bad, worthless, useless: ponerology, Poneridae, *Ponera trigona* (an ant), *Dinoponera gigantea* (an ant).

L. *pravus*, crooked, deformed, perverse, wrong, bad; *depravo*, *-atus*, pervert, spoil, corrupt: depravity, depraved, *Acacia pravissima* (an acacia).

Gr. *prodotos*, abandoned; see *prodosia* under **treason**

L. *profligatus*, dissolute, corrupt, vile: profligate, profligacy.

L. *scelus, -eris,* n. wickedness, evil deed, crime; *scelestus,* wicked; *scelerosus,* full of wickedness; *scelerator, -is; scelio, -onis,* m. evildoer; *scelero, -atus,* desecrate; *conscelero, -atus,* stain, dishonor, disgrace by wicked conduct: scelestic, scelerate, scelerous, *Ranunculus sceleratus* (a buttercup), *Scelio rugulosus* (a procto-trypid).

Gr. *sindron, -os,* mischievous:

L. *sputatilicus,* that deserves to be spit at, detestable: *Episus sputatilius*[*sputatilicus*] (a beetle).

Gr. *telchin, -os,* m. a person with malicious or spiteful disposition: *Telchina serena* (a butterfly).

L. *teter, -tra, -trum,* foul, hideous, offensive, shameful; *tetricus,* forbidding, harsh, gloomy, stern: tetric, *Gymnetron teter* (a beetle).

L. *turpis,* ugly, foul, base; see **dirt**

L. *vilis,* cheap, base, worthless; see **worthless**

L. *vitio, -atus,* injure, corrupt, spoil; see **hurt**

See: **dirt, fault, hell, hurt, pain, poor, shame, unfit, worthless**

badistes, Gr. goer, walker; *badistikos,* good at walking; see *badizo* under **walk**

badius, L. brown, chestnut-colored; see **brown**

baeno- < Gr. *baino,* walk, go, pass; see **walk**

baeo- < Gr. *baios,* little, small, scanty, few; see **little**

baeticus, L. pertaining to the river Baetis (now Guadalquiver) in southern Spain; see **stream**

baeto- < Gr. *baite,* coat of skin, tent of skin; see **garment**

baetulus; baetylus, L. (Gr. *baitylos*), a meteorite, sacred stone; see **meteor**

baetygo- < Gr. *baityx, -ygos,* leech; see **leech**

bag < ON. *baggi.*   Derive: bagatelle.

L. *aluta,* purse or pouch made of soft leather; see **leather**

Gr. *angeion,* n. dim. of *angos, -eos,* n. vessel or container of any kind, seedcase, jar, box; *angeidion,* n. dim.: angiosperm, angioblast, angioma, angiitis, angioid, gonangium, synangium, hydrangea, *Angiopteris crassipes* (a fern), *Angophora lanceolata* (rusty gum-myrtle).

Gr. *aryballos,* m. bag, purse:

Gr. *askos,* m. bag, bladder, sac; *askidion; askion,* n. dim.: ascus, asciferous, ascidian, ascidium, ascites, ascocarp, ascospore, *Ascophyllum nodosum* (a brown alga), *Exoascus cerasi* (cherry witches-broom), *Gymnoascus uncinatus* (a fungus), *Ascodipteron phyllorhinae* (a fly).

L. *averta,* f. saddlebag:

Gr. *balantion,* n. bag, pouch, purse; *balantidion,* n. dim.: *Balantium inflatum* (a mollusk), *Balantionella elegans* (a sponge), *Balantidium coli* (a protozoan).

AS. *belg,* bag, belly: bellows.

AS. *blaeddre,* bag, sac: bladder, bladdernut, bladderseed, bladderwort, gall-bladder.

L. *bulga,* f. leather knapsack, bag: bulge, budget.

L. *bursa,* f. purse: purse, bursar, disburse, reimburse, bourse, purser, bursitis, bursicle, *Capsella bursa-pastoris* (shepherds-purse), *Geomys bursarius* (pocket-gopher), *Phormosoma bursarium* (an echinoid), *Grammostomum bursigerum* (a foraminifer), *Gymnophallus bursicola* (a trematode).

L. *cassidile, -is,* n. small bag or wallet:

Gr. *chiloter, -os,* m. nosebag:

AS. *codd,* small bag:

L. *coleus* (Gr. *koleos*), sheath, scabbard, scrotum; see **sheath**

L. *corycus* (Gr. *korykos*), m. leather punching-bag used by athletes; *korykion,* n. dim.: *Corycus prolifer* (an alga), *Corycium enigmaticum* (a problematic Archaean fossil).

L. *crumena (crumina),* f. leather moneybag; *crumilla,* f. dim.; *crumino, -atus,* fill like a purse: *Entosolenia crumenata* (a fossil foraminifer), *Leptoptilus crumeniferus* (African stork, marabou), *Crumenaria erecta* (a rhamnacead), *Pelecanichthys crumenalis* (a fish).

Gr. *doros,* m. leather bag, wallet: *Chydorus sphericus* (a cladoceran), *Echinodorus cordifolius* (a burhead), *Dorocidaris granulostriata* (a fossil echinoid), *Dorosoma cepedianum* (gizzard-shad).

Gr. *ellobos,* in a pod:

L. *fel,* gall-bladder, gall; see **bile**

L. *fiscus,* basket, moneybag; see **basket**

L. *follis, -is,* m. bellows, windbag, moneybag; *folliculus,* m. dim.; *folleatus,* loose, baggy: fool, foolish, follicle, *Folliculina ampulla* (a protozoan), *Cephalotus follicularia* (a pitcher-plant), *Folliculus lubricus* (now *Cochlicopa lubrica,* a gastropod), *Carex folliculata* (a sedge).

L., Gr. *gaster*, stomach, belly, paunch, womb; see **stomach, belly, womb**
Gr. *grymea*, f. bag, chest: *Didelphys(Grymaeomys) murina* (an opossum).
Gr. *gylios*, m. an elongate wallet: *Gyliotrachela hungerfordiana* (a gastropod).
Gr. *kelyphos, -eos*, n. pod, shell, case; *kelyphion*, n. dim.: *Celyphomima chrysomelina* (a fly), *Cromyatractus tetracelyphus* (a radiolarian), *Celyphus scutatus* (a fly).
Gr. *kemos*, muzzle, nosebag; see **cover**
Gr. *kibisis*, f. pouch, wallet:
Gr. *kystis, -eos*, f. bladder, sac, cell; *kystinx, -ingos*, f.; *kystion*, n. dim.: cyst, nematocyst, cystoid, cystolith, cystine, cystinuria, statocyst, *Cystopteris fragilis* (a bladder-fern), *Macrocystis angustifolia* (a giant kelp), *Dermocystidium marinum* (a protistan), *Cystidium inerme* (a radiolarian).
Gr. *lobos*, capsule, pod; see **projection**
L. *lomentum*, n. an indehiscent bean pod, bean meal: loment.
L. *lura*, f. mouth of a sack, sack:
OF. *male*, bag: mailbag.
L. *mantica*, f. handbag, wallet; *manticula*, f. dim.: *Mantica horni* (a beetle).
L. *marsupium* (Gr. *marsipion*), n. pouch, bag, purse: marsupial, marsupioid, *Perna marsupium* (purse-perna), *Nototrema marsupiatum* (a frog), *Adenophora marsupiiflora* (a campanulacead), *Didelphys marsupialis* (an opossum).
Gr. *oïstodegmon; oïstodoke; oïstotheke*, quiver; see *oïstos* under **arrow**
Gr. *osche*, f. scrotum: oscheal, oscheoma, oschitis.
Gr. *pera*, f. pouch, wallet; *peridion*, n. dim.; L. *perula*, f. dim.; *askopera*, f. knapsack; *hippopera*, f. saddlebag: perula, perulate, peridium, peridiolum(peridiole), exoperidium, *Pera pellucida* (a tunicate), *Perameles fasciata* (barred bandicoot), *Perognathus fasciatus* (pocket-mouse), *Ascopera gigantea* (a tunicate), *Astropera sabulosa* (a tunicate), *Stichopera pectinata* (a radiolarian).
Gr. *perikarpion*, case of a fruit; see *karpos* under **fruit**
Gr. *perioche*, f. case, pod, husk, shell:
Gr. *pharetra*, f. quiver or case for arrows: *Pharetronema zingiberis* (a sponge).
Gr. *phaskolos*, m. leather bag: *Phascolarctos cinereus* (koala), *Phascogale calura* (a marsupial-rat), *Phascolomys ursinus* (wombat).
Gr. *physa*, f. bellows, bubble; *physallis, -idos*, f. bladder, bubble, plant with bladderlike fruits: *Physaria didymocarpa* (twinpod), *Physalis pubescens* (downy ground-cherry), *Physalia arethusa* (Portuguese man-of-war), *Physocarpus malvaceus* (mallow-ninebark), *Physostegia virginiana* (a mint), *Nicandra physalodes* (apple-of-Peru).
OF. *poche*, bag: pouch, poke, pocket, pucker.
L. *pustula*, pimple, blister, bubble; see **bubble**
L. *reticulum*, a network bag; see *rete* under **net**
L. *saccus* (Gr. *sakkos*), m. bag; *sakkion*, n. dim.; *sakkinos*, of sackcloth; *sacciperium*, n. pocket for a purse: sac, sack, saccate, sacciform, saccule, sackcloth, satchel, knapsack, *Saccolabium ampullaceum* (an orchid), *Sacculina purpurea* (a parasitic crustacean), *Saccinobaculus doroaxostylus* (a protozoan), *Sacciolepis striata* (a grass), *Saccocoma pectinata* (a Jurassic crinoid), *Globigerina sacculifera* (a foraminifer), *Delphinium saccatum* (a larkspur), *Dizygocrinus sacculus* (a fossil crinoid).
Gr. *sagis, -idos*, f. wallet, pouch:
L. *sarcina*, bundle, pack, package; see **bundle**
L. *scrotum*, n. pouch containing the testicles, cod: scrotum, scrotal.
L. *segestre, -is*, n. covering, wrapper:
L. *siliqua*, f. pod; *silicula*, f. dim.: silique, silicle, *Siliqua costata* (a pelecypod), *Ceratonia siliqua* (carob), *Cercis siliquastrum* (a redbud), *Frondicularia silicula* (a foraminifer), *Lysiloma latisiliqua* (sabicu).
L. *theca* (Gr. *theke*), f. case, container, envelope, sheath; *thekion*, n. dim.: thecodont, thecoid, endothecium, gonatheca, tick (mattress), *Thecocyathus tintinnabulum* (a coelenterate), *Dizygotheca elegantissima* (an araliacead), *Brachythecium rivulare* (a moss), *Trichothecium roseum* (a fungus), *Loxoscaphe theciferum[thecifera]* (a fern).
Gr. *thylakos; thylax; thylak, -akos*, m. bag, sack, pouch; *thylakion*, n. dim.: thylacitis, *Thylacocrinus gracilis* (a fossil crinoid), *Thylacium aggregatum* (a tunicate), *Thylacynus cynocephalus* (Tasmanian wolf), *Artiocotylus macrothylax* (a flatworm), *Tetrathylacium macrophyllum* (a flacourtiacead).
L. *utriculus*, m. dim. of *uter, utris*, m. leather bag or bottle; *utricularius*, m. bagpiper: utricle, *Utricularia gibba* (a bladderwort), *Ionopsis utricularioides* (an orchid), *Lomatium utriculatum* (an umbellifer).
L. *valvola*, pod, shell, dim. of *valva*, leaf of a folding door; see **door**
L. *vesica*, f. bladder, purse, blister, pustule; *vesicula*, f. dim.; *vesicarius*, of bladders; *vesiculosus*, full of blisters: vesicle, vesicate, vesicular, vesiculose, vesicotomy, *Fucus vesiculosus* (rockweed), *Caesalpinia vesicaria* (brasiletto).

L. *vidulus,* m. wicker trunk covered with leather, portmanteau, knapsack:
See: **basket, box, cover, cup, garment, sausage, sheath, skin, womb**

baggage < OF. *bague,* bundle; see **pack, weight**

bagoas, Gr. eunuch; see **sterile**

bail < OF. *bailler;* see **sure**

bait < ON. *beita,* food, cause to bite; see **trap, draw**

bajulus, L. porter, carrier; see **carry**

bake < AS. *bacan;* see **cook**

balaena, L. (Gr. *phalaina*), whale; see **whale**

balagros, Gr. a fresh-water fish; see **carp**

balance < L. *lanx, lancis,* f. plate, dish, pan, scale of a balance; *lancicula; lan-cula,* f. dim.: balances, balance-wheel.
Gr. *antholke,* f. counterpoise, resistance: *Antholcus varinervis* (a wasp).
Gr. *antirrhopos,* counterbalancing:
L. *exagium,* n. weight, balance:
L. *halter, -is* (Gr. *-os*), weight used in jumping, balancer; see **weight**
L. *libra,* f. balance, scales; the Roman pound; *libella; libellula,* f. dim.; *libratus,* well-poised, horizontal: libration, equilibrium, level, Libra, deliberate, £, lb., *Libella azurea* (a dragonfly), *Libellula quadrimaculata* (a dragonfly), *Gorgosaurus libratus* (a dinosaur).
Gr. *plastinx, -ingos,* f. scale of a balance, balance:
L. *sacoma* (Gr. *sekoma*), f. a counterpoise; *sekoter, -os,* m. beam of a balance:
Gr. *stachane,* f. balance:
L. *statera,* f. steelyard, balance:
Gr. *stathmos,* m. balance, weight; *stathmion,* n. dim.:
Gr. *talanton,* n. balance, scales; the thing weighed; a piece of money; *atalantos,* equal in weight: *Pleurotomaria(Talantodiscus) mirabilis* (a gastropod).
L. *trutina* (Gr. *trytane*). f. balance, pair of scales, < *trutinor,* weigh: *Trutina solenoidea* (a pelecypod).
Gr. *zygon,* beam of a balance, yoke; *zygikos,* of a balance; see **bind**
See: **equal, weight**

balanos, Gr. acorn, barnacle, peg, bar, glans penis; *balanion,* dim.; see **acorn, penis**

balantion, Gr. bag, pouch, purse; *balantidion,* dim.; see **bag**

balatro, L. babbler, fool; see *blatero* under **speak**

balatus, L. bleating of sheep, < *balo,* bleat; see **bleat**

balaustium, L. (Gr. *balaustion*), flower of the wild pomegranate; see **pomegranate**

balbido- < Gr. *balbis, -idos,* starting point, goal; see **begin**

balbus, L. stammering, stuttering; see **falter**

balcony < It. *balcone,* < OHG. *balcho* (G. *balken*), beam; see **projection**

bald < uncertain origin; see **bare, smooth**

baleros, Gr. a kind of carp; see **carp**

balios, Gr. spotted, dappled, piebald; see **spot**

ball < OHG. *bal, balla,* a spherical body: balloon, ballot, highball, bale, puffball.
Gr. *agathis, -idos,* f. ball of thread; *agathidion,* n. dim.: *Agathis australis* (Kauri pine), *Agathidium globulum* (a beetle), *Phyllagathis rotundifolia* (a melastomacead).
Sp. *bola,* f. ball, globe: bola.
F. *boule,* ball: bullion, bullet, bulletin.
L. *catapotium* (Gr. *katapotion*), n. pill: *Catapotia laevissima* (a beetle).
L. *folliculus,* an inflated ball, football; see *follis* under **bag**
L. *globus,* m. ball, sphere; *globulus,* m. dim.; *globosus,* round as a ball, spherical; *globo, -atus,* make into a ball: globe, globe-trotter, globical, globule, globigerina, hemoglobin, *Eucalyptus globulus* (blue gum), *Porpites globulatus* (a fossil coral), *Physa globosa* (a fossil gastropod), *Globularia cordifolia* (globe-daisy), *Lituola globata* (a foraminifer), *Volvox globator* (a flagellate), *Sporotrichum globuliferum* (chinch-bug fungus), *Dentalina globuligera* (a foraminifer).
L. *glomus, -eris,* n. ball; *glomerabilis,* round; *glomerosus,* like a ball, round; *glomero, -atus,* form into a ball, gather into a rounded heap; *agglomeratus,* balled, gathered into a mass: glomerule, conglomerate, agglomeration, *Glomerella cingulata* (a fungus), *Juncus conglomeratus* (a rush), *Dactylis glomerata* (orchard-grass), *Clasterosporium glomerulosum* (a fungus), *Scleria triglomerata* (a sedge).
Gr. *gongylos,* m. ball; round, spherical: gongylus, *Gongylocormus bivirgatus* (a snake), *Cissus gongylodes* (a treebine), *Rozites gongylophora* (a fungus cultivated by ants), *Gongylidium nigricans* (a spider).

L. *harpastum* (Gr. *harpaston*), n. handball:

Gr. *palla*, f. ball: *Palla decia* (a butterfly), *Biopalla palmata* (a Mississippian geode in a cavity supposed to have been occupied by a sponge).

L. *pastillus*, little roll of bread, little round ball; see *panis* under **bread**

L. *pila*, f. (Gr. *pilos*, m.), ball; *pilula*, f. dim.; *pilaris*, of a ball; *pilicrepus*, m. ball-player: pill, pilule, pellet, aegagropila, *Pilolobus crystallinus* (a fungus), *Pilularia globulifera* (pillwort), *Urtica pilulifera* (Roman nettle), *Geotrypes pilularis* (a dung-beetle), *Stenopilus latus* (a trilobite).

L. *sphaera* (Gr. *sphaira*), f. ball; *sphaerula*, f.; *sphairidion; sphairion*, n. dim.; *sphairikos*, globular: sphere, spheroid, atmosphere, hemisphere, spherule, coccosphere, spheridium, *Sphaerella rosigena* (a fungus), *Spheroides maculatus* (globe-fish), *Sphaeralcea umbellata* (Mexican globe-mallow), *Sphaerococcus coronopifolius* (a red alga), *Sphaerocystites globularis* (a cystid), *Aposphaerion longicolle* (a beetle), *Microsphaera alni* (lilac-mildew), *Drymosphaera dendrophora* (a radiolarian), *Ecnomoneura sphaerotropha* (a moth), *Melasphaerula graminea* (an iridacead), *Dacus sphaeriticus* (a fruitfly).

Gr. *systremma, -atos*, n.; *systrophe*, f. anything aggregated, consolidated, or twisted together, generally a ball or round object: *Systremma crassicornis* (a butterfly), *Systrophoceras arietinum* (a fossil cephalopod).

Gr. *tolype*, f. ball of yarn: *Tolypeutes tricinctus* (apar, an armadillo), *Tolypella prolifera* (a characead), *Tolypeceras marcousanum* (a fossil cephalopod), *Tolypothrix lanata* (an alga).

Gr. *trochos*, m. wheel, ball; *trochiskos*, m. dim. pill: troche, trochiscus, *Trochophora duckei* (a wasp).

See: **circle, berry, bubble, lump, sore**

ballator, L. dancer; *ballematicus*, accompanying the dance; see **dance**

ballen, Gr. king; see **govern**

ballis, Gr. a kind of plant; see **plant**

ballista, L. a machine for hurling missiles, < Gr. *ballo*, throw, cast; see **throw**

ballota, L. (Gr. *ballote*), horehound; see **mint**

ballux, -ucis; balluca, L. gold-sand, gold-dust; see **gold**

balneum, L. (Gr. *balneion*), bath; *balnearis; balnearius*, of baths; see **wash**

balsamum, L. (Gr. *balsamon*), a fragrant gum; *balsameus; balsaminus* (Gr. *balsaminos*), of balsam; see **resin**

balte, Gr. swamp; *baltodes*, swampy; see **marsh**

balteus, L. girdle, belt, border, edge; *balteolus*, dim.; see **belt**

bambax, Gr. cotton; see *bombax* under **cotton**

bambusa, NL. < Malay *bambu*, a kind of grass; see **grass**

bamma, Gr. dye; see *bapheus* under **color**

ban, OHG. proclamation, edict; LL. *bannum*, summons; see **call**

banana < Sp. *banana*, < an Ab.Am. word.

L. *ariena*, f. fruit of pala, plantain:

Amharic *anset*, the Abyssinian banana: *Musa ensete* (Abyssinian banana), *Musa enseteformis* (supposed fossil banana seeds from Colombia).

NL. *heliconia*, f. a genus of the banana family: *Heliconia bihai* (Carib heliconia).

NL. *musa*, f. a genus of the banana family: *Musa paradisiaca* (plantain-banana), *Phoma musae* (banana freckle-fungus), *Gloeosporium musarum* (a fungus), *Pratylenchus musicola* (banana-nematode).

L. *pala*, f. plantain:

NL. *ravenala*, f. a genus of the banana family: *Ravenala madagascariensis* (travelers-tree).

NL. *strelitzia*, f. a genus of the banana family, < Charlotte Sophia, of the Mecklenburg-Strelitz family: *Strelitzia angusta* (a bird-of-paradise flower).

banausia, Gr. handicraft; *banausos*, mechanical; see **art**

banchus (bancus), L. an unknown fish; see **fish**

band < the root of AS. *bindan;* see **bind, belt, bracelet, circle, collar, garter, ribbon**

bane < AS. *bana*, killer, murderer, that which destroys; see **kill**

banish < OF. *banir*, drive away, expel.

L. *abdo, -itus*, put away, remove, exile:

Gr. *apatria*, f. exile; *apatridos*, without country:

L. *deporto, -atus*, carry off, banish: deport, deportation.

L. *ejicio, -ectus*, drive out, expel: eject.

Gr. *elasis*, f. banishment, < *elauno*, drive away:

Gr. *exhoristos*, banished, expelled, exiled:

L. *expello, -pulsus,* drive out, remove, eject: expel, expulsion.

L. *extermino, -atus,* banish, expel, exile: exterminate, extermination, exterminator.

L. *extorris,* banished, exiled:

L. *exul* (*exsul*), *-is,* c. a banished person, exile; *exulatus,* banished, exiled; *exilium* (*exsilium*), n. banishment, a place of exile: exulate, exile, *Sabal exul* (Victoria palm), *Cheirurus exsul* (a trilobite), *Diomedea exulans* (wandering albatross), *Enallagma exsulans* (a dragonfly).

Gr. *ostrakizo,* banish, cast out: ostracize, ostracism.

L. *relego, -atus,* send away, banish: relegate.

L. *sepono, -situs,* separate, banish, exile:

See: **drive, wean**

bank < ME. *bank,* see **shore, wall;** < It. *banca,* counter, see **store**

banker < F. *banquier,* < F. *banque,* < It. *banca,* bench, table, counter; see **trade**

banner < LL. *bandum;* see **flag**

bapheus, Gr. dyer; *baptos,* dipped, dyed; *bapsis,* a dipping, dyeing; *baphikos,* of dyeing, < *bapto,* dip, dye; see **color, dip**

baptisia, NL. a genus of the bean family; see **bean**

bar < LL. *barra,* f.: barrier, embarrass, barracks, embargo, debar, disbar, crowbar.

L. *cancellus,* lattice, grating, bar of a tribunal; *cancellosus,* covered with bars; see **screen**

L. *clathrum* (Gr. *kleithron*), bolt, bar, lattice; see **screen**

L. *claustrum* (*clostrum*), n. bolt, bar, lock, door, gate, dam, dike, wall, fortress; *clausura* (*clusura*), f. lock, bar, bolt; *clostellum,* n. dim.; *claustrarius,* m. locksmith: claustrum, claustration, cloister, claustral, clausula, clausure.

L. *clavis,* key, bar: see **key**

Gr. *embolon,* bolt, bar, peg, wedge; *embolimos,* inserted; see **insert**

Gr. *enchoma, -atos,* n. bar of sediment in a river:

Gr. *epibles, -etos,* m. bar, bolt:

Gr. *harmos,* bolt, peg; see **joint**

Gr. *mandalos,* m. bolt, bar; *mandalotos,* with the bolt shot: *Mandaloceras bohemicum* (a fossil cephalopod), *Mandalotus vetulus* (a beetle).

L. *obex, obicis,* c. bar, bolt, obstacle: *Obex mulveyana* (a gastropod).

L. *obstructio, -onis,* barrier; see *obstruo* under **hinder**

L. *pessulus,* m. bolt; *oppessulatus,* bolted:

L. *repagulum,* n. barrier, bar, bolt:

L. *replum,* n. bolt for closing folding doors:

L. *septum* (*saeptum*), inclosure, barrier, partition; see **wall**

L. *sera,* f. bar for fastening doors; *obsero, -atus,* bar, bolt, fasten:

L. *vectis,* lever, bar; see **lever**

Gr. *zygoma, -atos,* n. bolt, bar, yoke; see *zygon* under **bind**

See: **close, key, hold, against, rod, bank, wall**

barathrum, L. (Gr. *barathron*), pit, gulf, abyss; see **hole**

barba, L. beard; *barbula,* dim.; *barbatus,* bearded; *barbatulus,* dim.; see **hair**

barbarus, L. (Gr. *barbaros*), foreign, therefore outlandish, crude, rude; see **strange**

barber < L. *barba,* beard.

Gr. *korseus; korsoter, -os; korsoteus,* m. barber; *korsoterion,* n. barbershop; *korsotos,* shaved, shorn:

Gr. *koureus; koureutes,* m. barber; *koureutikos,* of a barber; *koureion,* n. barbershop; *kourimos,* shorn; *keiro,* clip, cut short: *Curimus circassicus* (a beetle).

L. *tonsor, -is,* m. barber, hairdresser; *tonsorius,* of barbering; *tonstrina,* f. barbershop; *tonsura,* f. clipping, shearing; *tondeo, tonsus,* shave, shear, clip; *detonsus,* clipped, shorn; *intonsus,* unshaved, bearded, rough: tonsorial, tonsure, *tinea tonsurans* (barber's itch, caused by the fungus, *Trichophyton tonsurans*).

barberry < ML. *berberis,* f.: berberine, Berberidaceae, *Berberis ilicifolia* (a barberry), *Bougainvillea berberidifolia* (a nyctaginacead), *Berberidopsis corallina* (a flacourtiacead).

L. *epimedium* (Gr. *epimedion*), n. a genus of the barberry family: *Epimedium alpinum* (alpine epimedium).

NL. *jeffersonia,* f. a genus of the barberry family, < Thomas Jefferson, author of the Declaration of Independence and third president of the United States: *Jeffersonia diphylla* (twinleaf).

L. *leontice* (Gr. *leontike*), f. a genus of the barberry family: *Leontice altaica* (a lionsleaf).

NL. *mahonia,* f. a genus of the barberry family, < Bernard McMahon, horticulturist: *Mahonia pinnata* (a mahonia).

barbilos, Gr. wild peach-tree; see **peach**

barbitos, Gr. a kind of lyre; see **harp**

barbus, L. a fresh-water fish; see **carp**

barca, L. small boat; see **ship**

bardistos, Gr. variant of *bradistos;* see *bradys* under **slow**

bardus, L. dull, stupid, see **dull**; minstrel, singer, see **sing**

bare < AS. *baer,* naked.

Gr. *ablautos,* unslippered:

Gr. *abotanos,* without plants or vegetation:

Gr. *ageneios,* beardless: *Agenius limbatus* (a beetle)

Gr. *akalyptos; akalyphos,* uncovered, unveiled:

Gr. *akentros,* stingless: *Acentropelma spinulosum* (a spider), *Acentron albitarsis* (a wasp).

Gr. *akomos,* without hair, bald: *Acomus inornatus* (a bird).

Gr. *alopekia,* f. loss of hair, baldness, mange in foxes: alopecia.

Gr. *anamphiestos,* undressed, naked:

Gr. *anaphalantos,* with bald forehead: *Anaphalantus pennatus* (a fly).

Gr. *anarbylos,* without shoes, unshod:

Gr. *anebos,* beardless, young:

Gr. *aneimon,* naked, unclad: *Anemia[Animia] phyllitidis* (a fern), *Animomyia morta* (a fly).

Gr. *anioulos,* downless, beardless:

Gr. *anoplos,* unarmed: *Anoplostoma vivipara* (a nematode), *Anoplotherium posterogenium* (a fossil mammal).

Gr. *anozos,* with few or no branches: *Anozus siphonophorae* (a wasp).

Gr. *anypenos,* beardless: *Anypenus obscurus* (a fly).

Gr. *anypodetos,* barefooted:

Gr. *apampischo,* undress, bare:

Gr. *apedilos,* barefooted, unshod:

Gr. *apeplos,* unrobed:

Gr. *apeskes,* without skin, uncovered:

Gr. *aphares,* unclad, naked: *Apharus muelleri* (a beetle).

Gr. *aphyllos,* without leaves: *Calligonum aphyllum* (a polygonacead), *Buxbaumia aphylla* (a moss).

Gr. *apoglyphe,* f. a bare spot or space:

Gr. *aptilos,* unfeathered:

Gr. *astales,* unarmed, unclad:

Gr. *ateuches,* unarmed: *Ateuchopus armenius* (a wasp).

Gr. *athrix, atrichos,* without hair: *Atrichosema aceris* (a fly).

L. *calvatus,* made bare, bald, < *calvus,* bald; *calvities,* f. baldness: Calvin, *Gymnocephalus calvus* (a fruit-crow), *Calvatia gigantea* (a puffball).

L. *defloccatus,* without locks, bald:

L. *depilis, -e,* without hair: depilatory.

L. *deplumis, -e,* without feathers: deplume, deplumation.

Gr. *dero,* skin, flay; *dartos,* skinned, flayed; see *derma* under **skin**

L. *detego, -ectus,* unroof, disclose, reveal; *detector, -is,* m. uncoverer, revealer; *distectus,* uncovered: detective, detector, detection.

L. *devestivus,* undressed: devest, divest, divestiture, divestment.

L. *discalceatus,* barefooted, unshod: *Elater discalceatus* (a beetle).

Gr. *ekdysis,* f. a getting out, escape, molt, < *ekdyo,* take off, strip, lay bare; *ekdyma, -tos,* n. that which is taken off, skin, garment; *apodyo,* take off, strip; *apodysis,* f. a stripping, undressing: ecdysis, ecdysiast, *Ecdysanthera utilis* (an apocynacead).

Gr. *ekthesis,* f. a putting out, exposure, exhibition:

L. *exutus,* bared, stripped, < *exuo, -utus,* divest, strip: *Chesterella exuta* (an ostracode), exuviate.

L. *glaber, -bra, -brum,* hairless, bald, smooth; *glabellus,* dim.: glabrous, glabella, *Ulmus glabra* (Scotch elm), *Acer glabrum* (Rocky Mountain maple), *Cornus glabrata* (brown dogwood), *Zygadenus glaberrimus* (a camas), *Aster glabriusculus* (alkali-aster), *Bursera glabrifolia* (linaloa tree), *Senecio glabella* (butterweed), *Paepalanthus glabrifolius* (a pipewort).

Gr. *gymnos,* bare, naked: gymnasium, gymnosperm, gymnotus, *Gymnosoma fuliginosum* (a fly), *Gymnadenia odoratissima* (an orchid), *Gymnarchus niloticus* (an African fish), *Rhegmatorhina gymnops* (a bird), *Gymnocladus chinensis* (a bean), *Gymnodinium sanguineum* (a dinoflagellate), *Gymnopleurus flagellatus* (a beetle).

L. *imberbis; inberbus,* beardless; see *barba* under **hair**

L. *implumis, -e,* unfledged, featherless, bald:

L. *inermis, -e,* unarmed, defenseless: *Castanopsis inermis* (a chinquapin), *Agropyron inerme* (beardless wheatgrass).

L. *infrons*, without foliage:

Gr. *kenodontis, -idos*, toothless; see *odous* under **tooth**

Gr. *kranioleios*, bald-headed, bald:

Gr. *lipothrix, -trichos*, hairless: *Lipothrix lubbocki* (a springtail).

Gr. *madaros*, bald; *madesis*, f. loss of hair: *Madarus binotatus* (a beetle), *Madaropsis sulcipectus* (a beetle).

L. *minus*, bare, smooth:

Gr. *nelipos (anelipos)*, barefooted, unshod: *Nelipophygus ramsdeni* (a blattid).

L. *nudus*, bare, naked; *nudulus*, dim.; *nudo, -atus*, make bare, expose, strip; *nudipes*, barefooted; *connudatus*, wholly bare; *denudo, -atus*, lay bare, uncover, strip, plunder: nude, nudity, denudation, nudibranchiate. *Viburnum nudum* (possum-haw), *Hibiscus denudatus* (a rose-mallow), *Conopodium denudatum* (an umbellifer), *Asilus nudipes* (a fly).

Gr. *phalakros*, bald-headed, bald, smooth: *Phalacrocorax carbo* (a cormorant), *Phalacrophorus pictus* (an annulate).

Gr. *phalanthos*, bald in front: *phalantins*, m. a bald man:

Gr. *psednos*, thin, scanty, bald: *Psednothrix belgicaria* (a moth), *Psednoserica amoena* (a beetle).

Gr. *psenos*, smooth, bald: *Psenocerus supernatatus* (a beetle).

Gr. *psilos*, bare, smooth, bald, naked; *psilosis*, f. a stripping; *psilotes, -etos*, f. baldness, nakedness; *psilotikos*, of baring; *psilokorrhes*; *psilokranos*, bald-headed; *psilothron*, n. a depilatory; *hypopsilos*, somewhat bald: psilophyte, psilosis, psilology, psilosopher, psilomelane, psilothrum, epsilon, upsilon, uropsile, psilotum (*Psilotum nudum*), *Psilopa petrolei* (a fly), *Lampsilis luteolus* (a mussel), *Psilomastax pyramidalis* (a wasp), *Psilocephalus barbatus* (a fish).

L. *revelo, -atus*, unveil, uncover, lay bare, disclose: reveal, revelation.

Gr. *steiros*, barren, sterile; see **sterile**

L. *vulsus*, plucked, shorn, hairless; see *vello* under **tear**

See: **smooth, open, abandon, sterile, treason, simple, lewd**

barema, Gr. burden, load; see *barys* under **heavy**

baris, *-idos*, Gr. a flat-bottomed boat; see **ship**

barium < Gr. *barys*, heavy; see ELEMENTS under **thing**

bark of a dog, < AS *beorcan*, bark: bark, barker, barking.

L. *baubor, -atus*, bark gently:

Gr. *bayzo*, bark:

L. *fremo, -itus*, growl, howl, roar, snort; see **roar**

L. *gannio, -itus*, yelp, bark, whine, snarl, growl; *ogganitus*, yelped or barked at:

L. *hirrio, -itus*, snarl: hirrient.

Gr. *hylagma, -tos*, n. bark of a dog; *hylagmos*, m. a barking, baying; *hylake*, f. a barking, howling; *hylaktes (-or)*, m. barker; L. *hylax, -acis*, m. barker; *hylaktikos*, given to barking: *Hylactes megapodius* (a bird), *Hylacobdella grisea* (a leech).

L. *latro, -atus*, bark, howl; *latrans*, barking; *latrator, -is*, m. barker; *latrabilis*, barking: *Canis latrans* (coyote).

Gr. *orythmos*, m. howling of dogs:

Gr. *rhyzeo*, growl, snarl, show the teeth:

Gr. *thoysso*, bark, bay; *thoysmos*, m. bark:

L. *ululo, -atus*, howl, make a mournful outcry; see **call**

bark of a tree, prob. < Sw. *bark*, rind.

Sp. *cascara*, bark; *cascarilla*, dim.: cascara, cascarilla (from the bark of *Croton eleuteria*), cascarin.

L. *cortex, -icis*, m. bark; *corticulus*, m. dim.; *corticatus*, covered with bark; *corticeus*, of bark; *corticosus*, abounding in bark: cortex, cortical, corticate, corticiferous, corticoline, cortin, cortisone, desoxycorticosterone, *Corticium vagum* (a fungus), *Corticivora clarki* (a moth), *Thoracophorus corticinus* (a beetle), *Fuchsia excorticata* (a tree-fuchsia), *Adenocarpus decorticans* (a legume), *Knema corticosa* (a myristicacead)

Gr. *phellos*, m. cork, bark: phellem, phelloderm, phellogen, *Phellodendron amurense* (cork-tree), *Quercus phellos* (willow-oak).

Gr. *phloios*, m. bark: phloem, phloeophagous, phlorhizin, *Juniperus pachyphloea* (alligator-juniper), *Trogophloeus fossulatus* (a beetle), *Phloeosinus cristatus* (a beetle).

It. *scorza*, bark, peel: scorza.

L. *suber, -is*, n. cork-oak, cork; *subereus; suberinus*, of cork; *suberosus*, corky: suberin, suberose, suberization, *Quercus suber* (cork-oak), *Euplectella suberea* (a sponge), *Asperula suberosa* (a woodruff).

See: **skin, cover, scale**

barley < AS. *baerlic;* see **grain**

barn < AS. *bern;* see **store, house**

barnacle < F. *bernicle;* see *balanos* under **acorn,** *lepas* under **mollusk**

baros, Gr. weight; *baryllion,* dim.; see *barys* under **heavy**

barrel < OF. *baril.*
> L. *cupa,* tub, cask, tun, vat; *cupella; cupula,* dim.; see **cup**
> L. *dolium,* n. large earthern jar or wooden cask for wine; *doliolum,* n. dim.; *doliaris; doliarius,* of a cask; *doliarium,* n. place for casks: dolioform, *Dolium perdix* (tun-shell), *Doliolum denticulatum* (a tunicate).
> Gr. *kados,* m. pail, jar, cask: *Cadocrinus amarassicus* (a Permian crinoid).
> L. *maganum,* n. wine-vessel made of wood:
> Gr. *pithos,* large earthen wine-jar; *pithogastros,* potbellied; see **pot**
> See: **pot**

barren < OF. *brehaing;* see **sterile**

barrus, L. elephant; *barrinus,* of elephants; *barritus,* trumpeting of the elephant; see **elephant, call**

barys, Gr. heavy, impressive, grave, very; see **heavy, very**

basalt < L. *basaltes,* a dark, African rock; see **stone**

basanistes, Gr. investigator; *basanos,* touchstone, < *basanizo,* examine, test; see **try**

bascano- < Gr. *baskanos,* practicing sorcery, bewitching, slandering; *baskanion,* amulet, charm; see **magic**

base < L., Gr. *basis,* f. that on which one steps, foundation, bottom, pedestal; *basella,* f. dim.; *basilaris,* at the base: basal, basic, basis, baseless, baselevel, basement, basidium, debase, basommatophorous, *Xylobiops basilaris* (vine-borer), *Sinoxylon basilare* (a beetle), *Goniobasis impressa* (a snail).
> L. *area,* threshing-floor; see **space**
> L. *asarotum* (Gr. *asaroton*), floor laid in mosaic; see **mosaic**
> Gr. *bathron,* n. base, pedestal: *Bathrodon annectens* (a fossil mammal).
> Gr. *bomos,* base, stand, altar; see **altar**
> L. *coaxatio* (*coassatio*), *-onis,* f. boarded floor:
> L. *crepido, -onis,* basis, foundation, < Gr. *krepis,* boot, sandal; see **shoe**
> Gr. *edaphos,* bottom, foundation, base, soil; see **earth**
> L. *elementum,* first principle, rudiment; see **thing**
> L. *fundus,* m. base, bottom, estate; *fundulus,* m. dim.; *fundamentum,* n. basis; *funditus,* from the bottom; *fundo, -atus,* establish, lay a foundation: fund, founder, foundation, fundamental, profound, Fond du Lac (Wis.), *Fundulus heteroclitus* (a killifish).
> Gr. *halos,* a circular threshing floor; see **circle**
> Gr. *hedos,* seat, base, pedestal; see **sit**
> Gr. *hypostasis,* f. substance, foundation, support:
> Gr. *hypostema, -atos,* n. base, support:
> L. *infimum; imum,* n. lowest part, bottom:
> L. *pavimentum,* n. floor of stones: pavement.
> L. *principium,* foundation, beginning; see **begin**
> Gr. *pternis, -idos,* f. bottom of a dish:
> Gr. *pyndax, -akos,* m. bottom of a cup or other vessel:
> Gr. *pythmen, -os,* m. bottom, foundation, stock:
> Gr. *selma, -atos,* n. deck, floor, seat: *Diselma archeri* (a conifer).
> L. *solum,* bottom, floor, ground, soil: see **earth**
> Gr. *sphelas, -atos,* n. footstool, pedestal: *Sphelatus lehoni* (an Eocene mammal).
> L. *subex, -icis,* f. basal layer, substratum, support, underlayer; *subicula,* f. dim.: subiculum.
> L. *tabulatum,* n. boarding, floor:
> Gr. *themethlon,* n. foundation, bottom, root; *themelios,* of a foundation:
> See: **sit, sole, low, deep, begin, hypothesis, shoe**

bashful < OF. *esbahir,* < L. *ex,* out, *bah,* exclamation of astonishment; see **modest, fear**

basidium, NL. a kind of conidiophore, dim. of L., Gr. *basis,* pedestal; see **fruit**

basilaris, NL. at the base; see **base**

basileus, Gr. king; *basiliskos,* dim.; see **govern**

basilica, L. (Gr. *basilike*), a colonnaded public building; see **house**

basimos, Gr. accessible, passable; see *baino* under **walk**

basin < OF. *bacin,* a water vessel: abacinate.
L. *alveus,* hollow, excavation, trough, bathtub, bed of a stream; *alveolous,* dim.;
  see **hollow**
L. *aqualis,* c. washbasin, ewer; *aqualiculus,* m. dim.; *aquiminarium,* n. washbasin:
Gr. *ardalion,* n. watertrough:
Gr. *asaminthos,* f. bathtub:
Gr. *chernibeion* (*chernibon*), n. basin for washing hands:
L. *cisterna,* f. underground reservoir for water:
L. *crater* (Gr. *krater*), mixing vessel, bowl, basin; see **cup**
L. *cupa,* tub, vat; see **cup**
Gr. *dexamene,* f. tank, cistern, reservoir: *Dexaminella aegyptiaca* (an amphipod).
Gr. *droite,* f. bathtub:
Gr. *ekdochion,* n. reservoir, tank:
Gr. *holkeion* (*holkion*), n. large bowl or basin for washing cups:
Gr. *kape,* f. crib, manger, trough:
Gr. *kardopos,* f. kneading-trough: *Cardopostethus annulosus* (a bug).
L. *labrum,* n. basin, tub:
Gr. *lakkos,* cistern, pond, reservoir; see **lake**
L. *lavabrum,* n. basin, bathtub:
Gr. *lekos,* dish, pan, basin; see **plate**
Gr. *lenos,* f. trough, vat, anything with such a cavity: *Helicolenus rufescens* (a
  fish).
L. *linter, -tris,* trough, tub, vat; see **ship**
L. *luter* (Gr. *louter*), bathtub, basin; *louterion,* dim.; see *loutron* under **wash**
Gr. *maktra,* f. kneading-trough: *Mactra ovata* (a surf-clam), *Mactrodesma pon-*
  derosum (a fossil pelecypod).
L. *malluvium,* n. washbasin:
L. *mixtarius,* a mixing-vessel; see *misceo* under **mix**
L. *mortarium,* basin or trough for mixing plastic building material; see **plaster**
Gr. *nipter, -os,* m. basin, laver: *Nipterocrinus arboreus* (a Mississippian crinoid).
Gr. *pelika,* bowl, basin, cup; see *pella* under **cup**
L. *pelluvium,* n. foot-tub for washing the feet:
L. *pelvis,* f. basin; *pelvicula,* f. dim.: pelvis, pelvic, *Lipodectes pelvidens* (a Paleo-
  cene mammal).
Gr. *pelyx, -ykos,* m. bowl, basin, pelvis: pelycosaur, *Pelycodus frugivorus* (a fossil
  mammal).
Gr. *phatne,* f. manger, crib, trough; *phatnion,* n. dim., tooth-socket: *Phatnaspis*
  fenestrata (a radiolarian), *Leptophatnus ruficeps* (a wasp).
Gr. *pistra,* f. watertrough:
Gr. *plynos; plynter, -os,* washtub, trough; see **wash**
Gr. *pollubrum,* n. washbasin, laver:
Gr. *potistra,* f. watertrough:
Gr. *pyelos,* f. trough, tub, pelvis; *pyelis, -idos,* f. setting or socket of a ring;
  pyelion, n. dim.; *pyelodes,* hollow: pyelitis, pyelitic, pyelogram, pyelocystitis,
  Pyelorhamphus molothroides (a bird).
Gr. *pyndax, -akos,* bottom of a vessel; see **base**
Gr. *skaphe,* anything hollowed out, basin, trough, bowl, boat; see **cup, ship**
L. *trulleum,* n. basin:
See: **hollow, lake, vessel, valley, cradle**

basis, Gr. pedestal, foundation, that on which one steps; see **base,** *baino* under
  walk

basium, L. kiss; *basiolum,* dim.; *basio, -atus,* kiss; see **kiss**

basket < uncertain origin, but cf. L. *bascauda,* f. a dish-holder of woven work.
L. *aero, -onis,* m. wicker-basket, hamper:
Gr. *arrhichos,* f. wicker-basket:
L. *calathus* (Gr. *kalathos*), m. basket shaped like a lily or vase, bowl, cup;
  kalathion, n.; *kalathis, -idos,* f.; *kalathiskos,* m. dim.: *Calathocrinus digitatus*
  (a crinoid), *Calathea eximia* (a calathea).
L. *canipa,* f. fruit-basket:
L. *canistrum; canistellum* (Gr. *kanistron; kaneion*), n. bread, fruit, or flower-
  basket; *kanetion,* n. dim.; *kanthelion,* n. large basket, pannier: canister,
  Canistrum aurantiacum (a bromeliad), *Lychnocanium fenestratum* (a radio-
  larian).
L. *cartallus* (Gr. *kartallos; kartalos*), m. basket with small bottom; *kartalamion,*
  n. dim.:
L. *cophinus* (Gr. *kophinos*), m. basket, hamper: coffin.
L. *corbis,* f. basket; *corbicula; corbula,* f. dim.: corbel, corbeil, corbicula, corbicu-
  late, *Corbis ventricosa* (a pelecypod), *Ichthyocrinus corbis* (a Silurian crinoid),

Corbula carinata (a fossil pelecypod), *Corbicula fluminalis* (a fossil pelecypod), *Batophyllum corbicula* (a coral), *Nucula corbuliformis* (a fossil pelecypod), *Discorbina perforata* (a foraminifer).

L. *cratis*, something made of wickerwork; see **screen**

L. *fiscus*, m. wicker-basket, purse; *fiscella*, f.; *fiscellus*, m.; *fiscina*, f. dim.: fiscal, confiscate, confiscation, *Ellopia fiscellaria* (a beetle).

Gr. *gerrhon*, wickerwork; see **screen**

L. *griphus* (Gr. *griphos*), fishing net or basket; see **ask**

Gr. *gyrgathos*, m. wicker-basket for catching fish:

Gr. *helene*, torch, wicker-basket; see **light**

Gr. *hyrichos*, m. wicker-basket:

Gr. *kemos*, wicker-vessel like an eel-basket, weel:

Gr. *kremathra*, f. basket for hanging things up in:

Gr. *kyrte*, m.; *kyrtis, -idos*, f. fish-basket, weel; *kyrtidion*, n. dim.:

Gr. *larkos*, m. charcoal-basket; *larkidion*, n. dim.: *Larcospira quadrangula* (a radiolarian), *Larcidium dodecanthum* (a radiolarian).

Gr. *liknon*, n.; *likmos*, m. basket of wickerwork used in winnowing; *liknarion*, n. dim.: *Licnophora macfarlandi* (a protozoan).

Gr. *lygistes*, basket-maker; see *lygos* under **willow**

L. *metella*, f. basket or crate filled with stones for hurling at besiegers: *Metella breuli* (a spider).

Gr. *myrsos*, m. basket: *Myrsus corrugata* (a pelecypod).

L. *nassa; naxa*, f. wicker-basket with a narrow neck, fish-trap, weel: *Nassa reticulata* (a gastropod), *Cylichnostomum nassatum* (a nematode).

L. *panarium*, breadbasket; see **bread**

Gr. *phernion*, n. fish-basket:

Gr. *phormos*, m.; *phormis, -idos*, f. basket; *phormion*, n.; *phormiskos*, m. dim.: *Sethophormis rotula* (a radiolarian), *Phormosoma uranus* (a sea-urchin).

L. *qualum*, n. wicker-basket, hamper; *quasillum*, n. dim.:

Gr. *sargane*, braid, plait, basket; see **weave**

L. *scirpea*, basket made of rushes; see *scirpus* under **sedge**

L. *soracum*, n. (Gr. *sorakos*, m.), basket, hamper, pannier:

L. *sporta*, f. basket, hamper; *sportella; sportula*, f. dim.:

Gr. *spyris, -idos*, f. basket, creel; *spyrichnion; spyridion*, n. dim.: *Spyridobotrys trinacria* (a radiolarian), *Spyridium globosum* (a rhamnacead), *Spyridiocrinus cheuxi* (a Devonian crinoid), *Ceratospyris radicata* (a radiolarian).

Gr. *talaros*, m. basket; *talarion*, n.; *talariskos*, m. dim.: *Talarocrinus sexlobatus* (a Mississippian crinoid).

Gr. *tarpe*, f. large wicker-basket: *Tarpa hispanica* (a wasp).

Gr. *tarrhos*, m. basket, crate, wickerwork, mat, hurdle; *tarrhion*, n. dim.: *Tarrus[Tarrhus] reticulatus* (a sponge), *Tarroma[Tarrhoma] rubrum* (a sponge).

Gr. *thibe*, f. basket, ark:

L. *vietor*, basket-maker; see *vieo* under **weave**

L. *vimen, -inis*, basket of wickerwork; see **willow**

See: **box, weave, screen**

bass < AS. *baers*, perch; see **perch**

bassaris, Gr. fox; see **fox**

bassus, LL. low, deep; see **low**

bastagma, Gr. burden; *bastaktes*, porter, < *bastazo*, lift, carry; see **weight**

bastard < OF. *bastard* (*fils de bast*, son of a packsaddle), illegitimate, spurious; see **lie**

basterna, L. sedan chair, litter; see **sit**

bat < uncertain origin.

L. *Alcithoë* (Gr. *Alkithoë*), f. mythical woman changed into a bat for ridiculing the orgies of Bacchus: *Voluta(Alcithoë) pacifica* (a gastropod).

L. *molossus* (Gr. *Molossos*, a region in Epirus famous for its hounds), m. a genus of bats: *Molossus fuliginosus* (a bulldog-bat).

Malay *mops*, m. a kind of bat: *Mops indicus* (a bat), *Eumops nanus* (a bat).

Gr. *nykteris, -idos*, f. bat: *Glauconycteris beatrix* (a bat), *Nycteridium mosquense* (a wasp), *Paleunycteris pusilla* (an Eocene bat).

Gr. *phalke*, f. a kind of bat: *Phalcobaenus carunculatus* (a bird).

NL. *vampyrus*, m. a genus of bats, < Slav. *wampir*, a blood-sucking ghost or demon: vampire, *Vampyrus* (now *Trachops*) *cirrhosus* (a bat).

L. *vespertilio, -onis*, m. bat; It. *pipistrello*, dim.: *Vespertilio murinus* (a bat), *Ogcocephalus[Oncocephalus] vespertilio* (batfish), *Vesperugo pipistrellus* (pygmy-bat), *Pipistrellus hesperus* (a bat).

L. *vesperugo, -inis*, f. bat: *Vesperugo serotinus* (a bat).

batalaria, L. a kind of warship; see **ship**

batalos, Gr. anus, hinder parts; see **rump**

bater, Gr. threshold; see **door**

bates, Gr. one that walks, treads, haunts; see *baino* under **walk**

bathe < AS. *baeth,* bath; see **wash**

bathmos (basmos), Gr. step; see **step**

bathron, Gr. base, pedestal; see **base**

bathys, Gr. deep; *bathos,* depth; see **deep**

batia, Gr. bush, thicket; see *batus* under **blackberry**

batillum, L. shovel, fire-pan, chafing-dish; see **shovel**

batiola, L. goblet; see *batiaca* under **cup**

batis, L., Gr. samphire, see **goosefoot**; a flatfish, skate, ray, see **shark**

bato- < Gr. *batos,* bramble, blackberry, thorn-bush, see **blackberry**; passable, see *baino* under **walk**

batrachos, Gr. frog; *batrachion,* dim.; see **frog**

battle < OF. *bataille,* < L. *battuo,* strike; see **fight, strike**

batus, L. (Gr. *batos*), bramble, blackberry, thorn-bush; see **blackberry**

batyle, Gr. female dwarf; see **little**

baubor, L. bark gently; see **bark**

baucalio- < Gr. *baukalion,* a narrow-necked vessel; see **bottle**

bauco- < Gr. *baukos,* affected, prudish; see **modest**

baunos, Gr. furnace, forge; see **oven**

baxa, L. kind of woven shoe; see **shoe**

baxis, Gr. saying, report, rumor, < *bazo,* speak, say; see **speak**

bay < diverse origins; see **sea, hollow, brown, laurel, berry, bark**

bayberry
NL. *comptonia,* f. a genus of the Myricaceae, < Henry Compton, Bishop of London: *Comptonia peregrina* (sweetfern).
AS. *gagel,* a marsh plant with aromatic odor: gale, sweetgale (*Myrica gale*).
L. *myrica* (Gr. *myrike*), f. tamarisk, bayberry: *Myrica cerifera* (bayberry, wax-myrtle).

bdalsis, Gr. a milking; see *bdallo* under **milk**

bdella, Gr. leech; see **leech**

bdellium, L. (Gr. *bdellion*), a fragrant gum; see **resin**

bdelycto- < Gr. *bdelyktos,* abominable, disgusting; *bdelygma,* abomination; *bdelyros,* loathsome, shameless; see **bad**

bdolos, Gr. stench, < *bdeo,* break wind; *bdesma,* stench; see **smell**

be < AS. *beon,* be, exist; see **life**

be- < AS. *be, bi, big,* by, on, about, see **near**; intensifier, see **make**

beach < uncertain origin; see **shore**

bead < AS. *bed* (*gebed*), prayer; see *globulus* under **ball, drop, chain, necklace**

beak < OF. *bec;* see **nose, projection**

beam < AS. *beam,* post, tree, ray of light: sunbeam.
L. *asser, -is,* m. beam, pole; *asserculus,* m. dim.: *Scydmaenus asserculatus* (a beetle).
AS. *balca* (G. *balken*), beam, ridge: balk.
L. *buris,* m. beam of a plow: *Buris brunneus* (a beetle).
L. *dentale, -is,* n. share-beam of a plow:
Gr. *diapegma, -tos,* n. crossbeam:
Gr. *dokos,* m. main beam, balk; *dokidion; dokion,* n.; *dokis, -idos,* f. dim.; *diadokis, -idos,* f. crossbeam: docoglossate, *Diplodocus carnegiei* (a dinosaur), *Docidium baculum* (a desmid), *Docidophryne aqua* (a toad).
Gr. *elyma, -atos,* n. share-beam of a plow:
Gr. *epistylion,* n. crossbeam on the columns: *Helix epistylium* (a snail).
Gr. *gya,* f. beam or tree of a plow:
Gr. *keleon, -ontos,* m. beam of a loom; *kelon, -os,* m.; *keloneion,* n. swing-beam:
Gr. *kmelethron,* n. beam:
L. *jugum,* crossbeam; *jugumentum,* crossbeam, lintel; see *jungo* under **bind**
Gr. *phalkes,* m. beam, rib of a shop: *Phalces coccyx* (a phasmid).
L. *radius,* rod, ray; see **rod**
Gr. *stomix, -ikos,* f. wooden beam:

Gr. *stroter, -os,* m. crossbeam; *stroteridion; stroterion,* n. dim.: *Stroter comatus* (a moth).

L. *temo, -onis,* m. beam, pole, tongue:

L. *tignum,* n. beam, timber; *tigillum,* n. dim.; *tignarius,* m. carpenter; *contigno, -atus,* join with beams, joists, or rafters:

L. *tolleno, -onis,* m. swing-beam:

L. *trabs, -bis,* f. beam, timber; *trabecula (trabicula),* f. dim.; *trabalis,* of beams: trabal, trabecule, trabeculate, architrave, *Sphaerocrinus trabeculatus* (a Devonian crinoid).

L. *transtrum,* n. crossbeam: transom.

Gr. *traphex, -ekos,* m. beam, piece of timber: *Traphecocorynus anxius* (a beetle).

Gr. *zygon,* crossbeam of a balance or loom, yoke; see **bind**

See: **rod, pillar, light**

bean < AS. *bean.*

NL. *abrus,* m. (< Gr. *habros,* pretty, dainty), a genus of the bean family: *Abrus[Habrus] laevigatus* (a rosary-pea), abrin.

L. *acacia* (Gr. *akakia),* f. a thorny plant of the bean family: acaciin, acacetin, *Acacia horrida* (allthorn-acacia), *Robinia pseudoacacia* (black locust), *Loranthus acaciae* (a mistletoe).

L. *aeschynomene* (Gr. *aischynomene),* f. a sensitive plant: *Aeschynomene virginiana* (sensitive jointvetch).

NL. *albizzia,* f. a genus of the bean family, < Albizzi, an Italian family: *Albizzia julibrissin* (silk-tree albizzia).

NL. *amorpha,* f. a genus of the bean family: *Amorpha fruticosa* (indigo-bush).

Gr. *anagyris,* f.; *anagyros,* m. bean-trefoil: *Anagyris foetida* (stinkbush), *Laburnum anagyroides* (golden-chain).

Gr. *anthyllis, -idos,* f. a plant, now a genus of the bean family: *Anthyllis montana* (an anthyllis), *Osteomeles anthyllidifolia* (a rosacead).

L. *aphaca* (Gr. *aphake),* f. a kind of legume: *Lathyrus aphaca* (a pea).

Gr. *arachidna,* f. a vetch:

NL. *arachis,* f. (Gr. *arachos; arakos,* m.), a kind of legume; *arakis, -idos,* f.; *arakiskos,* m. dim.: *Arachis hypogaea* (peanut).

L. *aspalathus* (Gr. *aspalathos),* m. a kind of broom: *Aspalathus virgata* (a legume), *Hypericum aspalathoides* (a St. Johnswort).

L. *astragalus* (Gr. *astragalos),* m. a legume: *Astragalus lentiginosus* (speckled locoweed).

Gr. *auchos,* m. a kind of pulse:

NL. *baptisia,* f. a genus of the bean family: *Baptisia bracteata* (a wild-indigo).

NL. *bauhinia,* f. a genus of the bean family, < Gaspard and Jean Bauhin, Swiss botanists: *Bauhinia tomentosa* (a bauhinia).

Gr. *belekos,* m. a kind of pulse:

L. *brya,* f. a shrub, now a genus of the bean family: *Brya ebenus* (ebony cocoswood).

L. *buceras, -atis* (Gr. *boukeras, -atos),* n. fenugreek: *Bucida buceras* (oxhornbucida).

NL. *caesalpinia,* f. a genus of the bean family, < Andrea Cesalpino, Italian botanist: *Caesalpinia spinosa* (spiny caesalpinia).

NL. *canavalia,* f. a genus of the bean family: *Canavalia gladiata* (jackbean), canavalin.

L. *caprago, -inis,* f. a kind of legume:

NL. *caragana,* f. a Mongolian name for a kind of bean: *Caragana sinica* (Chinese pea-shrub).

L. *cassia* (Gr. *kassia),* f. a genus of the bean family: *Cassia acutifolia* (a senna), *Cinnamomum cassia* (Chinese cinnamon).

L. *ceratonia* (Gr. *keratonia),* f. a genus of the bean family: *Ceratonia siliqua* (carob).

NL. *cercis, -idis* (Gr. *kerkis, -idos),* f. a tree of the bean family; *kerkidion,* n. dim.: *Cercis canadensis* (redbud), *Cercidium texanum* (Texas paloverde), *Cercidiphyllum japonicum* (katsura), *Cercospora cercidicola* (a fungus).

Gr. *chedropon,* n. legume:

L. *cicer, -is,* n. chickpea; *cicercula,* f. dim: *Cicer arietinum* (chickpea), Cicero, *Astragalus cicer* (a vetch).

NL. *cladrastis,* f. a genus of the bean family: *Cladrastis lutea* (yellowwood).

NL. *clitoria,* f. (Asa Gray said modestly, "derivation recondite"); a genus of the bean family: *Clitoria ternatea* (a butterfly-pea).

L. *colutea* (Gr. *koloutea),* f. a pod-bearing tree, now a genus of the bean family: *Colutea persica* (Persian colutea).

L. *conchis,* f. beans or lentils boiled with the pods; *conchicla; conchicula,* f. dim.:

L. *cracca,* f. a kind of pulse: *Cracca spicata* (a goatsrue).

NL. *crotalaria*, f. a genus of the bean family: *Crotalaria lanceolata* (a rattlebox).
L. *cytisus* (Gr. *kytisos*), m. a shrubby legume: *Cytisus hirsutus* (a broom).
NL. *dalbergia*, f. a genus of the bean family, < Nils Dalberg, Swedish botanist: *Dalbergia nigra* (Brazilian rosewood).
Gr. *derris*, f. a genus of the bean family: *Derris elliptica* (source of rotenone).
NL. *desmodium*, n. a genus of the bean family: *Desmodium rhombifolium* (a ticktrefoil).
Gr. *dolichos*, m. kidney-bean: *Dolichos lignosus* (Australian pea).
L. *dorycnium* (Gr. *doryknion*), n. a plant, now applied to a legume: *Dorycnium hirsutum* (a canary-clover), *Cytisopsis dorycnifolia* (a legume).
NL. *entada*, f. a genus of the bean family: *Entada phaseoloides* (a sea-bean).
L. *ervum*, n.; *ervilia*, f. a kind of legume: *Ervum* [now *Vicia*] *hirsutum* (a legume).
NL. *erythrina*, f. a genus of the bean family: *Erythrina tomentosa* (a coral-bean).
L. *faba*, f. bean; *fabaceus; fabaginus; fabalis; fabarius; fabatus*, of beans; *fabalium*, n. beanstalk: fabaceous, fabiculture. Fabaceae, Fabian, *Vicia faba* (broad bean), *Goniobasis fabalis* (a snail), *Echinocystis fabacea* (manroot), *Zygophyllum fabago* (Syrian bean-caper).
L. *genista* (*genesta*), f. broom: *Genista florida* (a broom), *Melaleuca genistifolia* (ridge-myrtle), Plantagenet.
Gr. *gizi*, f. a kind of cassia:
NL. *gleditschia*, f. a genus of the bean family, < Gottlieb Gleditsch, German botanist: *Gleditschia aquatica* (swamp-locust).
NL. *glycine*, f. a genus of the bean family: *Glycine max* (or *soja*) (soybean).
L. *glycyrrhiza* (Gr. *glykyrrhiza*), f. a plant with a sweet root: licorice (*Glycyrrhiza glabra*).
NL. *gymnocladus*, m. a genus of the bean family: *Gymnocladus dioicus* (Kentucky coffee-tree).
Gr. *halex, -ekos*, m. a kind of pulse:
L. *hedysarum* (Gr. *hedysaron*), n. vetch: *Hedysarum occidentale* (western sweetvetch).
Ab.Am. *inga*, f. a kind of legume: *Inga edulis* (an inga).
Gr. *kyamos*, m. bean; *kyamion*, n. dim.: *Cyamocephalus loganensis* (a Silurian crustacean), *Cyamiomactra problematica* (a pelecypod), *Hyoscyamus aureus* (yellow henbane).
L. *laburnum*, n. a legume: *Laburnum alpinum* (Scotch laburnum), *Crotalaria laburnifolia* (a rattlebox).
L. *lathyrus* (Gr. *lathyros*), m. a legume: *Lathyrus odoratus* (sweetpea), *Lathyrophthalmus vitrescens* (a fly).
L. *legumen, -inis*, n. bean, pulse: legume, leguminous. Leguminosae, *Vaginulina legumen* (a foraminifer), *Aspidiotus leguminosum* (a bug), *Dasyneura leguminicola* (clover-seed midge), *Leguminocythereis corrugata* (an ostracode).
L. *lens*. lentil; see **lens**
NL. *lespedeza*, f. a genus of the bean family, < Lespedes, a misspelling of Cespedes, a Spanish governor of Florida: *Lespedeza striata* (bush-clover or lespedeza).
L. *lotus* (Gr. *lotos*), m. name for several different plants, now applied to a genus of the bean family: *Lotus pinnatus* (meadow deer-vetch), *Melilotus alba[albus]* (white sweet-clover), *Diospyros lotus* (a persimmon).
L. *lupinus*, m. a legume: *Lupinus pusillus* (rusty lupine), *Osmia lupinicola* (a bee).
NL. *medicago*, alfalfa; see *medica* under **clover**
NL. *mimosa*, f. a genus of the bean family: *Mimosa argentea* (a mimosa).
NL. *mucuna*, f. a genus of the bean family, < Ab.Am.: *Mucuna urens* (cowhage).
Gr. *norye*, f. a kind of pulse:
Gr. *onobrychis, -idos*, f. a legume: *Onobrychis pulchella* (a sainfoin).
Gr. *ononis, -idos*, f. restharrow: *Ononis serrata* (yellow restharrow).
L. *orobus* (Gr. *orobos*), m. vetch: *Vicia oroboides* (a vetch), *Orobanche ramosa* (a broomrape).
Gr. *osprion*, n. pulse of any kind:
Gr. *pelekinos*, m. hatchet-vetch:
Gr. *phakos*, lentil, seed of the lentil; see **lens**
L. *phaselus* (Gr. *phaselos*), m. kidney-bean; *phaseolus*, m. dim.: phaseolin, frijole, *Phaseolus coccineus* (scarlet runner), *Macrophoma phaseoli* (a beanfungus).
L. *pisum* (Gr. *pison*), n. pea; *pisarion; pisidium*, n. dim.; *pisinos*, of peas: pease, pisiform, pisolitic, *Pisum sativum* (pea), *Cynips pisum* (a gall-insect), *Pisidium pusillum* (a pelecypod), *Sactogaster pisi* (a wasp), *Bruchus pisorum* (pea-weevil), *Pisolithus tinctorius* (a fungus), *Pisocrinus sphericus* (a Silurian crinoid).

Lecanocrinus pisiformis (a Silurian crinoid), *Clinodiplosis pisicola* (a gall-insect), *Chamaecyparis pisifera* (sawara chamaecyparis).

NL. *poinciana,* f. a genus of the bean family, < M. de Poinci, a governor in the West Indies: *Poinciana pulcherrima* (Barbados poinciana).

Gr. *prosopis, -idos,* f. a plant; now a genus of the bean family: *Prosopis glandulosa* (honey-mesquite).

NL. *psoralea,* f. a genus of the bean family: *Psoralea esculenta* (breadroot-scurfpea).

NL. *pterocarpus,* m. a genus of the bean family: *Pterocarpus indicus* (a padauk).

NL. *pueraria,* f. a genus of the bean family, < M. N. Puerari, Swiss botanist: *Pueraria hirsuta* (kudzu-vine).

Gr. *pyanos,* m. a kind of legume; *pyanion,* n. dim.: *Pyanomya gibbosa* (a fossil pelecypod).

NL. *robinia,* f. a genus of the bean family, < Jean and Vespasien Robin, French naturalists: *Robinia viscosa* (clammy locust).

NL. *sesbania,* f. a genus of the bean family: *Sesbania aegyptiaca* (sesban).

NL. *sophora,* f. a genus of the bean family, < Ar. *sufayra: Sophora japonica* (pagoda-tree), *Caragana sophoraefolia* (a legume).

L. *spartum* (Gr. *sparton*), n.; *spartos,* m. Spanish broom; *spartion,* n. dim.; *sparteus; spartinos,* of broom: *Spartium nanum* (dwarf broom), *Spartopteryx serrularia* (a moth), *Spartiomyia martinsi* (a fly), *Brachyspartus costatus* (a beetle), *Stipa spartea* (porcupine-grass).

NL. *tamarindus,* f. a genus of the bean family, < Ar. *tamr hindi: Tamarindus indica* (tamarind).

L., Gr. *telis,* f. fenugreek:

Gr. *thermos,* m. lupine: *Thermopsis fabacea* (a legume).

L. *ulex, -icis,* m. a kind of shrub; a genus of the bean family: *Ulex europaeus* (furze, gorse), *Hakea ulicina* (a proteacead).

NL. *vigna,* f. a genus of the bean family, < Domenica Vigna, Italian botanist: *Vigna sinensis* (cowpea).

NL. *wistaria,* f. a genus of the bean family, < Caspar Wistar, American anatomist: *Wistaria floribunda* (Japanese wistaria).

bear < AS. *bera* (G. *bär*).

Gr. *arktos,* f. bear, north; *arktylos,* m. cub: arctic, antarctic, Arcturus, *Arctostaphylos densiflora* (Sonoma bearberry), *Arctomys pruinosus* (hoary marmot), *Arctocebus aureus* (a lemur), *Arctotis acaulis* (a composite), *Arctium nemorosum* (a burdock), *Arctosa littoralis* (a spider), *Arctonoë pulchra* (an annelid).

L. *Callisto* (Gr. *Kallisto*), a nymph changed into a she-bear; see *kallos* under **beauty**

Gr. *dasyllis, -idos,* f. bear: *Dasyllis thoracica* (a fly).

Gr. *knopeus,* m. bear:

L. *ursus,* m.; *ursa,* f. bear; *ursula,* f. female cub; *ursinus,* of bears: ursine, Ursula, ursiform, ursal, Ursa Major, *Ursus americanus* (black bear), *Anacodon ursidens* (a creodont), *Arctostaphylos uva-ursi* (bearberry), *Rubus ursinus* (a dewberry).

bear < AS. *beran,* carry, suffer; see **carry, birth, pain**

beard < AS. *beard;* see **hair**

beast < L. *bestia,* animal; see **animal**

beat < AS. *beatan;* see **strike**

beatus, L. happy, blessed; *beatificus,* making happy; *beatrix,* gladdener; see **joy**

beauty < OF. *beaute,* < L. *bellus,* pretty, lovely, fine; *bellatulus; bellulus,* dim.: beautiful, beau, beaux, beautify, belle, embellish, Beau Brummel, Beaumont, Arabella, Beauchamp, Bellevue, belles-lettres, beldame, bellflower, belladonna (*Atropa belladonna*), *Calathea bella* (a calathea), *Mesembryanthemum bellum* (a fig-marigold), *Elasmonema bellatulum* (a fossil gastropod), *Cryptolithus bellulus* (a trilobite), *Drosophila bellula* (a fruitfly).

L., Gr. *Adonis,* m. a beautiful youth beloved by Venus, favorite, darling: adonite, *Adonis vernalis* (spring adonis).

Gr. *aglaos,* splendid, shining, bright, beautiful, noble; *aglaia,* splendor, beauty; see **light**

Gr. *aisthetikos,* sensitive or perceptive, especially of beauty; see **feel**

L. *amabilis,* lovely; see *amo* under **love**

L. *amoenus,* delightful, pleasant, lovely; see **agreeable**

Gr. *asteios,* urbane, refined, nice, pretty, see **agreeable**

Gr. *charis, -itos,* f. loveliness, grace, favor, thankfulness; *epicharis; eucharis; eucharistos,* pleasing, charming, gracious, thankful; *charieis, -entos,* beautiful, graceful: *Ammocharis coccinea* (scarlet sandnymph), *Limnocharis aquatica* (a water-mite), *Charitonetta albeola* (bufflehead), *Eucharis candida* (an amaryllid),

Eucharist, *Vaginulinopsis epicharis* (a Jurassic foraminifer), *Charientocrinus ithacensis* (a Devonian crinoid), *Chariessa vestita* (a beetle).

L. *comptus*, ornamented, adorned; see **ornament**

L. *concinnus*, skillfully put together, beautiful, appropriate: *Globigerina concinna* (a foraminifer), *Adiantum concinnum* (a maidenhair-fern), *Compsilura concinnata* (a fly).

L. *cracens, -entis*, neat, graceful:

L. *decoris*, ornamented, elegant, beautiful; see **ornament**

Gr. *eidalimos*, comely, shapely; see *eidos* under **form**

L. *elegans, -antis; electilis*, tasteful, choice, fine, select; *elegantulus*, very fine, < *eligo, electus*, pick, choose: *Cyathea elegans* (a tree-fern), *Aquilegia elegantula* (a columbine), *Chalepus elegantulus* (a beetle), *Mesochorus electilis* (a wasp), *Stipa elegantissima* (a grass).

Gr. *epaphroditos*, lovely, charming: *Epaphroditus placens* (a fly).

Gr. *eratos*, lovely, beloved; see *eros* under **love**

Gr. *eu-*, good, well, beautiful; see **good**

Gr. *eueides*, shapely, beautiful, comely: *Eueides harmonica* (a butterfly).

Gr. *eumorphos*, shapely, comely: *Eumorpha capensis* (a butterfly), *Eumorphocerus laticornis* (a beetle).

Gr. *euphues*, shapely, graceful, clever: Lyly's *Euphues*, euphuism, euphuistic.

Gr. *euprepes*, becoming, comely; *euprepeia*, f. comeliness; *prepon, -ontos*, seemly, fit, proper: *Eupreponotus inflatus* (a grasshopper), *Euprepocnemis scitulus* (a grasshopper), *Euprepia pudica* (a butterfly).

Gr. *euthemon*, well arranged, neat; see **class**

L. *exquisitus*, choice, excellent, fine: exquisite, *Fissurella exquisita* (a gastropod).

L. *facetus*, fine, elegant, well made: *Dizygocrinus facetus* (a Mississippian crinoid).

L. *formosus*, beautifully formed, comely, handsome; *formosulus*, pretty; see **form**

Gr. *glaphyros*, hollowed out, polished, pretty; see **hollow**

L. *gratus*, beloved, pleasing, thankful; *gratia*, kindness, mercy, pleasure, beauty; see *gratia* under **give**

Gr. *habros*, pretty, graceful, dainty, tender, delicate; *habroteros*, prettier; *habrotatos*, prettiest; *habrotes*, f. splendor, luxury; *habrosyne*, f. delicacy; *habrotimos*, delicate, costly; *habropenos*, of delicate texture: *abrocome*, abrin, habronema, habromania, *Abronia latifolia* (sand-verbena), *Abrus[Habrus] pulchellus* (a rosary-pea), *Strigops habroptilus* (kakapo), *Habrosyne[Habrosyna] derasa* (a butterfly), *Anabrus simplex* (Mormon cricket), *Habrophlebiodes brunneipennis* (a mayfly).

Sp. *hermosa*, beautiful: Hermosa Beach (Calif.).

Gr. *himeros*, yearning for, desiring, desirable; *himertos*, desired, longed for, lovely: *Himeropteryx miraculosa* (a butterfly).

Gr. *horaios*, seasonable, timely, ripe, beautiful; see *hora* under **year**

Gr. *idanos*, fair, comely:

Gr. *kallos*, n. beauty; *kalos; kallimos*, beautiful; *kallion*, more beautiful; *kallistos*, most beautiful; *kallosyne*, f. beauty; *pankalos; perikalles*, very beautiful: calligraphy, callisthenics, calliope, callithumpian, calomel, kaleidoscope, Callista, Callisto, *Callorhynchus antarcticus* (a chimaera), *Callotillus eburneocinctus* (a beetle), *Callostoma fascipennis* (a fly), *Calostoma cinnabarinum* (a puffball), *Calosoma frigidum* (a beetle), *Calophyllum angustifolium* (a guttiferad), *Calliocrinus bifurcatus* (a Silurian crinoid), *Calliphora erythrocephala* (bluebottle fly), *Callitriche heterophylla* (a water-starwort), *Callithamnium gracillimum* (a red alga), *Sambucus callicarpa* (an elderberry), *Hymenocallis speciosa* (spider-lily), *Kallima inachis* (leaf-butterfly of India), *Callistochiton palmulatus* (a chiton), *Callidium antennatum* (a beetle).

Gr. *keleterios*, charming, delighting; see **agreeable**

Gr. *kommosis*, beautification; *kommotes*, beautifier; *kommotikos*, of embellishment; see **ornament**

Gr. *kompsos*, elegant, pretty; *kompseia; kompsotes, -etos*, f. elegance; *kompseutikos*, of elegance; *kompsodes*, vainglorious; *mikrokompsos*, bedecked with ornaments: *Compsothlypis americana* (parula warbler), *Compsognathus longipes* (a fossil reptile), *Zacompsus clarki* (a Cambrian trilobite), *Pericompsus sellatus* (a beetle).

Gr. *kosmetikos*, skilled in arranging, decorative, beautifying; see *kosmos* under **ornament**

L. *lautus*, washed, neat, elegant; see *lavo* under **wash**

L. *lepidus*, pleasant, elegant, fine, < *lepor, -is*, attractiveness, charm; see **agreeable**

Gr. *lichnos*, greedy, dainty; see **greed**

L. *limatus*, polished, refined, elegant, fine, < *limo, -atus*, file off, polish; see **scrape**

L. *mundus*, clean, neat, nice; *mundulus*, dim.; *mundo, -atus*, cleanse: *Gongrocnemis munda* (a katydid), *Aclistochara mundula* (a fossil chara).

L. *nitidus*, shining, neat, elegant; see **light**

L. *ornatus,* bedecked, handsome, splendid; see **ornament**
L. *preclarus; perclarus,* very beautiful, splendid: *Cetonia preclara* (a beetle).
L. *pulcher, -chra, -chrum,* beautiful, pretty, fine, lovely; *pulchrior,* prettier; *pulcherrimus,* prettiest; *pulchellus,* dim.; *pulchritudo, -inis,* f. beauty: pulchritude, *Potamogeton pulcher* (a pondweed), *Percarina pulchra* (an Oligocene fish), *Cirrhopetalum pulchrum* (an orchid), *Phrixocoma pulchricornis* (a trichopterid), *Phyllium pulchrifolium* (a phasmid), *Euphorbia pulcherrima* (poinsettia), *Calopogon pulchellus* (an orchid), *Triodia pulchella* (fluffgrass).
Gr. *saukros,* graceful, pretty: *Saucropus rubella* (a fly).
L. *scitulus,* beautiful, handsome, pretty, elegant, neat: *Trochilus scitulus* (a hummingbird).
L. *speciosus,* beautiful, handsome, splendid, showy; *spectabilis,* showy, notable, remarkable: *Lilium speciosum* (showy lily), *Rudbeckia speciosa* (showy coneflower), *Orchis spectabilis* (showy orchid), *Odontoglossum spectabile* (an orchid).
L. *subtilis,* fine, delicate, nice; see **know**
L. *tornatilis,* turned, beautifully rounded; see *torno* under **turn**
L. *venustus,* like Venus, lovely, beautiful, elegant, graceful; *venustulus,* dim.: *Abies venusta* (bristlecone-fir), *Simulium venustum* (a blackfly), *Ulmus venustula* (a Pliocene elm).
See: **agreeable, ornament, fit, form, love, joy, good**

beaver < AS. *beofer.*
L. *castor, -is* (Gr. *kastor, -os*), m. beaver; *castorinus,* of beavers: castoreum, *Castor canadensis* (American beaver), *Myocastor coypus* (coypu), *Platypsyllus castoris* (a beetle).
L. *fiber, -bri,* m. beaver; *fibrinus,* of beavers: *Castor fiber* (European beaver), *Neofiber alleni* (a vole).
Gr. *pyktis (piktis), -idos,* f. probably a beaver: *Pyctis aureolaria* (a butterfly).
Gr. *satherion,* n. probably a beaver: *Satherium ingens* (a Pliocene otter).

bebaeo- < Gr. *bebaios,* firm, steady; see **stand**

bebelos, Gr. profane; see **profane**

bebros, Gr. stupid; see **dull**

because < AS. *be,* by, L. *causa;* see **begin, follow, think**

beccus, L. beak, bill; see **nose**

becho- < Gr. *bex, bechos,* cough; *bechikos,* of a cough; *besso,* cough; see **cough**

beco- < Gr. *bekos,* bread; see **bread**

bed < AS. *bed.*
L. *accubitum,* n. couch on which guests reclined at meals:
Sp. *cama,* f. bed; *camilla,* f. dim.:
L. *coactum,* n. mattress:
L. *cubile, -is,* n. couch, bed, marriage-bed, den, lair, nest, kennel, hole:
Gr. *demnion,* n. bed, mattress:
Gr. *enaulos,* m. bed of a stream:
Gr. *eune,* f. bed; *chameune,* f. bed on the ground: pareunia, dyspareunia, eunuch.
L. *grabatus* (Gr. *krabatos*), m. couch, bed; *grabatulus,* m. dim.: *Trygon grabatus* (a fish).
Gr. *katakliton,* n. couch:
Gr. *kline,* f. bed, couch; *klinidion,* n. dim.; *klinter, -os,* m. couch; *klinterion,* n. dim.; *klismos,* m. couch; *klisia,* f. bed, couch, hut; *klinochares,* fond of bed; L. *clinicus* (Gr. *klinikos*), of a bed, a physician attending a patient in bed: clinic, diclinous, *Clinopodium coccineum* (basil), *Auloclisia mutata* (a fossil coral).
Gr. *koitos,* m. bed; *koitaios,* lying in bed: *Exocoetus callopterus* (a flyingfish), *Hylecoetus flabellicornis* (a beetle).
Gr. *lechos; lektron,* n. couch, bed, marriage-bed, nest; *lechaios,* of a couch; *lecheres,* bedridden; *lechoios,* of childbed:
L. *lectus,* m. couch, bed; *lectulus,* m. dim.; *lectualis; lectuarius,* of a bed: lectual, coverlet, lectisternum, *Cimex lectularius[lectuarius]* (bedbug).
Gr. *phyllas, -ados,* f. bed of leaves:
L. *puerperium,* childbed, childbirth; see *puer* under **child**
L. *pulvinar, -is,* n. cushioned couch, seat: *Pulvinaria tesselata* (a coccid).
L. *sandapila,* f. bier for the lower classes:
L. *scimpodium* (Gr. *skimpodion,* n. dim. of *skimpous, -odos,* m.) n. low bed, couch, litter:
L. *sponda,* f. bedstead, bed, couch:
Gr. *stibas, -ados,* f. bed of straw, mattress, nest; *stibadion,* n. dim.: *Stibadoderus murinus* (a beetle).

L. *stramen, -inis,* straw, litter, any material for bedding down; see **grass**
L. *stratum,* n. cover, blanket, bed, layer: stratum (pl. strata), substrate.
Gr. *stroma, -tos,* n. bed, mattress; *stromation,* n. dim.; *stromne,* f. bed, couch,
   mattress: stromatic, *Stromatium barbatum* (a beetle), *Stromatopora tuberculata* (a hydrozoan).
See: **nest, cradle, pillow, lie**

bee < AS. *beo.*
Gr. *anthedon, -os,* f. bee: *Anthedon compta* (a bee).
Gr. *anthrene,* f. wild bee, wasp, hornet; see **wasp**
L. *apis,* f. honeybee; *apicula,* f. dim.; Sp. *abieja,* f. bee; *apianus,* of bees; *apiarium,*
   n. beehive; *apiarius,* m. beekeeper; *apicius,* sought by bees: apiary, apiculture,
   apivorous, *Apis mellifera* (honeybee), *Ophrys apifera* (a bee-orchid), *Merops
   apiaster* (bee-eater), *Trochilium apiforme* (a hornet-moth), *Trichodes apiarius*
   (bee-beetle), *Trypanea apivora* (a fly), *Nosema apis* (a protozoan).
Gr. *bombylios,* m. bumblebee, < *bombos,* humming, buzzing: *Bombus pennsylvanicus* (a bumblebee), *Bombylius fraudulentus* (a beefly).
Gr. *esmos,* swarm of bees; see **hive**
L. *examen,* swarm, crowd; see **number**
L. *fucus,* m. drone: *Osmia fuciformis* (a bee).
NL. *halictus,* m. a genus of bees: *Halictus inconspicuus* (a bee).
Gr. *kephen, -os,* m. drone; *kephenion,* n. dim.; *kephenodes,* dronelike: *Cephenomyia pratti* (a deerfly).
Gr. *melissa (melitta),* f. honeybee; *melittourgos,* beekeeper: melittology.
Gr. *smenos,* beehive, swarm of bees; see **hive**
Gr. *thronax, -akos,* m. drone:
See: **wasp, honey, wax, hive**

beech < AS. *bece (boc);* G. *buche:* book, beechnut, buckwheat (*Fagopyrum
   esculentum*).
L. *fagus* (Gr. *phegos,* oak, according to some dictionaries), f. beech; *fageus;
   fagineus; faginus* (Gr. *pheginos*), of beech: fagine, Fagaceae, *Fagus sylvatica*
   (European beech), *Nothofagus fusca* (a falsebeech), *Epifagus virginiana* (beech-drops), *Fagopyrum tataricum* (Tartary buckwheat), *Phegopteris polypodioides*
   (a beech-fern), *Cryptococcus fagi* (a scale-insect), *Inocarpus fagiferus* (a
   legume), *Gaultheria fagifolia* (beechleaf-wintergreen), *Curtisia faginea* (assegai-tree).   Derive: fagot.
Gr. *oxya,* f. beech:

beef < OF. *boef,* < L. *bos, bovis,* ox; see **cattle**

beer < AS. *beor.*
Gr. *brytos,* m. beer; *brytikos,* drunk with beer:
L. *camum,* n. a kind of beer:
L. *celia,* f. a Spanish beer:
L. *cervisia (cervisia),* f. a Gallic word for beer; Sp. *cerveza,* f. beer: *Saccharomyces cerevisiae* (beer or bakers' yeast).
L. *Gambrinus,* m. mythical inventor of beer:
Gr. *kourmi,* n. a kind of beer:
Gr. *pinon,* n. beer:
L. *sabaia,* f. drink made of barley; *sabaiarius,* m. beer-drinker:
L. *zythum,* n. (Gr. *zythos,* m.), a kind of Egyptian beer; *zythion,* n. dim.:
See: **drink**

beet < AS. *bete,* < L. *beta,* f.: *Beta vulgaris* (beet), *Cyphomandra betacea* (tree-tomato), *Cercospora beticola* (a fungus), *Phoma betae* (a fungus).
Sp. *acelga,* f. beet:
Gr. *smeche,* f. a kind of beet:
Gr. *teutlon (seutlon),* n. beet; *teutlion,* n. dim.:

beetle < AS. *bitel; betel,* a biting animal.
L. *bruchus* (Gr. *brouchos*), m. a kind of wingless locust, now a genus of beetles:
   Bruchidae, *Bruchus pisorum* (pea-weevil).
L. *buprestis* (Gr. *bouprestis, -idos*), f. a kind of beetle: *Buprestis lineata* (a
   beetle), *Anisomorpha buprestoides* (a phasmid), *Cerceris bupresticida* (a wasp).
L. *carabus* (Gr. *karabos*), m. a kind of beetle: *Carabus serratus* (a ground-beetle),
   Callistocarabus marginatus (a beetle), *Pachylomerus carabivorus* (a spider).
AS. *ceafor* (G. *käfer*), beetle: chafer, cockchafer (*Melolontha vulgaris*).
NL. *cetonia,* f. a genus of beetles: cetonian, *Cetonia aurata* (rose-chafer).
L. *cicindela,* f. glowworm; a genus of beetles: *Cincindela argentata* (a tiger-beetle).
L. *curculio, -onis,* m. weevil; *curculiunculus,* m. dim.: *Curculio proboscideus*
   (chestnut-weevil), *Curculigo latifolia* (an amaryllidacead), *Attelabus curculionoides* (a weevil).

NL. *dorcus* (Gr. *dorkos*, gazelle), m. a genus of beetles: *Dorcus bubalus* (a beetle).
Gr. *kantharis, -idos,* f. a blister-beetle; *kantharos,* m. a kind of beetle: cantharides, cantharis, *Cantharoctonus stramineus* (a wasp).
Gr. *kerambyx, -ykos,* m. a horned beetle: Cerambycidae, *Cerambyx cerdo* (a beetle).
Gr. *kis, kios,* m. a weevil: Ciidae, *Cis lineatocribratus* (a beetle), *Dimerocis aurasiacus* (a beetle), *Pityocis coarctatus* (a beetle).
Gr. *koprion,* dung-beetle; see *kopros* under **dung**
L. *lucanus,* m. a kind of beetle: *Lucanus elaphus* (a stag-beetle).
NL. *lyctus,* m. a genus of beetles. < L. *Lyctus* (Gr. *Lyktos*), a city of Crete: *Lyctus striatus* (powder-post beetle).
L. *melolontha* (Gr. *melolonthe*), f. cockchafer: *Melolontha clypeata* (a beetle).
NL. *popillia,* f. a genus of beetles, < Popillia, a Roman clan: *Popillia japonica* (Japanese beetle).
NL. *ptinus,* m. a genus of beetles: *Ptinus fur* (museum-pest), *Acalles ptinoides* (a beetle).
L. *scarabaeus,* m. a kind of beetle: scarab(*Scarabaeus sacer*), *Leguminocythereis scarabaeus* (an ostracode).
Gr. *silphe* (*tilphe*). f. a kind of insect; now a genus of beetles: *Silpha opaca* (a carrion-beetle), *Silphomorpha quadrisignata* (a beetle), *Silphidium priscum* (a fossil beetle), *Acritotilpha siliginella* (a moth).
Gr. *skorobylos,* m. a beetle:
Gr. *staphylinos,* m. an insect; now applied to the rove-beetles: *Staphylinus maculosus* (a rove-beetle).
Gr. *trox,* a weevil; see *trogo* under **bite**
before < AS. *beforan,* < *be,* by, *fore,* preceding in time and place: fore, aforetime, forearm, forecast, forefather, foregone, foreman, foremost, forenoon, foresight, foreskin, forestall, foretaste, foretell, forewarn, forward.
L. *abhinc,* ago:
L. *abortivus,* prematurely born; see **abortion**
AS. *aer,* before: ere, erstwhile, early.
L. *ante,* before; *anterior,* before, former, previous; *antecedens,* going before; *antecessor; antecursor, -is,* m. forerunner; *anticus,* in front, foremost; *anticipo, -atus,* foresee, forestall, foretaste: antecedent, antedate, antediluvian, antechamber, 3:00 a.m., anterior, antique, anticipate, ancestor, antrorse, antler, van, vanguard, advance, avaunt, disadvantage, ancient.
L. *immaturus; praematurus,* too early, untimely, unripe: immaturity, premature.
L. *omen,* a foreshadowing, portent, sign, prophecy; see **prophecy**
Gr. *periallos,* before all others:
Gr. *phthano,* come before, anticipate, do first: *Phthanocoris occidentalis* (a fossil insect-wing).
L. *pre-,* < *prae,* before, as in *praecox, -ocis,* too early ripe. premature; *praefatio, -onis,* f. introduction, foreword; *praerogativus,* asked before others, privileged; *praevius,* going before, leading: precede, precocious, preface, prefix, prefer, prerogative, prehensile. prescribe, preposterous, pretext, prelude, premature, precaution, prepare, prevention, preamble, presuppose, prejudice, precipitate, preposition, previous, pre-Cambrian, preglacial, praenomen, praetor, *Stachyurus praecox* (a Japanese tree), *Chamaebatia prefoliolosa* (a fossil bearmat).
L. *primaevus,* early, young: primeval.
L. *princeps,* chief, first; *principalis,* first, original; see **govern**
L. *prior; prius,* earlier. former; see *primus* under **first**
L. *pristinus,* early, original, primitive: pristine, *Hybocrinus pristinus* (an Ordovician crinoid).
Gr. *pro,* before; *proteros,* earlier; *protos,* first; *proso* (*porrho*), forward, onward, in front; *prosthe* (*prosthen*), before, in front; *prosthios,* foremost; *prodromos,* running before, preceding, early; m. precursor: program, prophecy, prophylaxis, prognosis, prologue, Procyon, propolis, proglottid, prostate, prosodetic, prosobranch, prosogyrate, proterobase, Proterozoic, protoplasm, prototype, protagonist, prodrome, prodromous, *Proterotherium cervoides* (a fossil mammal), *Prosactogaster venusta* (a proctotrypid), *Porrhothele antipodiana* (a spider), *Prosthogonimus pellucida* (poultry-fluke), *Prosthiostomum elongatum* (a worm).
L. *pro-, pol-, por-.* < *pro,* before, forward, in front of, forth, for; *protenus*(*protinus*), before, forward; *polliceor,* promise; *portendo,* foretell: procession, promotion, procrastinate, protest, prospect, progress, prohibition, projection, provide, pro-German, pro-English, reproduce, portent, portray, prudent, purpose, purchase, pollicitation, porrect, pursue, purvey, pronoun, proglacial, *Didymictis protenus* (a creodont).
See: **begin, face, first, old, prophecy, near, dawn**
beget < AS. *begitan,* engender, generate; see **begin**

beggar < OF. *begard*, mendicant: beggarly, beggarweed.

Gr. *agyrtes*, m. beggar, impostor, vagabond, collector; *agyrtikos*, of beggars; *agyrtos*, got by begging: *Agyrtes primoticus* (a fossil beetle).

Gr. *bioplanes*, m. beggar: *Bioplanes carinatus* (a beetle).

Gr. *bomax*, *-akos*, c. beggar; *bomakeuma*, *-tos*, n. beggary:

Gr. *dekter; dektes*, receiver, beggar; see **receive**

Gr. *epaites; prosaites*, m. beggar: *Prosaites compactus* (a Permian insect).

L. *Irus* (Gr. *Iros*), m. proverbial name for a beggar, a poor man: *Irus crispatus* (a pelecypod).

L. *mendicus*, beggarly, poor; *mendiculus*, dim.; *mendico*, *-atus*, beg, ask alms, solicit: mendicant, maunder, mendicity, *Spheniscus mendiculus* (Galapagos penguin), *Diaphora mendica* (a moth).

Gr. *ptochos*, m. beggar; beggarly; *ptochikos*, of beggars; *ptocheion*, n. poorhouse: ptochocracy, *Ptochostola dimidiella* (a moth), *Ptochomyia afra* (a fly), *Paraptochus californicus* (a beetle).

See: **ask, poor**

begin < AS. *beginnan*, commence.

Gr. *aitios*, causing, responsible for, < *aitia*, f. charge, cause, blame: etiology.

Gr. *alpha*, first letter of the Greek alphabet; see **letter**

Gr. *anekathen*, from the first or beginning:

Gr. *aphorme*, starting-point, origin, cause; see *hormao* under **arouse**

Gr. *arche*, f. beginning, chief, first cause, first place; *archidion*, n. dim.; *archaios*, from the beginning, old, primitive; *archegos*, beginning, originating; *archegenes*, causing the beginning of a thing; *archegonos*, original, primal; *archetypon*, n. original, model: archeology, archegonium, architect, archaic, archangel, archcriminal, Archean, patriarch, *Archidium ohioense* (a moss), *Archaeomyrmex cacabau* (an ant).

L. *auctor*, *-is*, c. founder, progenitor, creator, originator: author.

Gr. *balbis*, *-idos*, f. starting-point: *Balbis assumptana* (a moth).

Gr. *bibastes*, mounter, stallion; see **raise**

L. *causa*, f. agent, motive, reason; *causalis*, of a cause: causation, causal, cause, because, accusation, excuse, recusant, ruse.   Derive: causeway.

L. *coepio*, *-ptus*, begin, commence, undertake; *coeptum*, n. undertaking: *annuit coeptis*.

L. *concipio*, *-ceptus*, receive, become pregnant; *inceptus*, beginning; *inceptor*, *-is*, m. beginner: conception, conceive, inception, incipient, *Polypora incepta* (a fossil bryozoan).

L. *creo*, *-atus*, produce, beget; see **make**

Gr. *eispompe*, f. introduction:

L. *elementum*, first principle; see **thing**

Gr. *embryon*, fetus, the unborn young; see **young**

Gr. *enthoros*, impregnated: *Enthorodera atricolor* (a spider).

L. *-escens*, participial ending denoting beginning of, becoming, incipiency, nascency. To form inceptives, add *-esco* to the stem of a verb, as *nigro*, blacken; *nigresco*, become black: deliquescent, effloresce, quiescent, incandescent, recrudescence, *Lycopodium erubescens* (a clubmoss), *Dendroica nigrescens* (black-throated gray warbler), *Triturus viridescens* (red eft).

Gr. *eu-*, original, primitive: eusporangiate, eumeromorph.

L. *genero*, *-atus*, beget, create; *ingenero*, implant, engender, < *gigno*, *genitus*, beget, produce; see *gigno* under **birth**

Gr. *genesis*, beginning, birth, origin; see **birth**

L. *germen*, bud, seed, sprout; see **bud**

Gr. *gonopoios*, impregnating, fertilizing:

Gr. *hidryo*, establish, found, settle; see **make**

Gr. *horme*, first impulse, stir, start, beginning; *aphorme*, starting-point, origin, cause; see **arouse**

L. *impraegno*, *-atus*, cause to conceive, fertilize, fecundate, engender: impregnate.

L. *incohatus* (*inchoatus*), only begun, incipient, incomplete; *incohamentum*, n. first principle, rudiment, element: inchoate.

L. *incunabulum*, n. cradle, birthplace, origin, beginning: incunabula.

L. *ineo*, *-itus*, go into, enter, begin; *initialis*, first, original, beginning, < *initio*, *-atus*, begin, originate: initial, initiation.

L. *inoculo*, *-atus*, ingraft, implant: inoculate, inoculation.

L. *instituo*, *-utus*, set up, establish, found, organize: institute, institution.

L. *introduco*, *-uctus*, lead into, bring in or forward, institute, originate: introduce, introduction.

L. *isagogicus* (Gr. *eisagogikos*), introductory, elementary: isagogic.

Gr. *kyesis*, conception, pregnancy; *kyeros*, pregnant; see *kyo* under **fertile**

Gr. *mello,* be about to do; *mellesis,* threatening to do; *melletes,* delayer, loiterer; see **delay**

L. *nascens,* arising, beginning: nascent.

Gr. *ocheutos,* impregnated; see *ocheuo* under **coitus**

L. *ordior, orsus,* begin to weave, undertake, commence; *exorsus,* begun, commenced; *exordium,* n. a beginning: exordium.

L. *origo, -inis,* f. source, beginning, birth; *originalis,* pertaining to the beginning, < *orior, ortus,* rise, appear, spring from; *oriundus,* born in, descended or risen from: origin, original, aborigines.

L. *parens, -entis,* c. procreator, progenitor; *parentalis,* of parents: parent, parental.

L. *pario, -itus; partus,* beget, create, produce; see **birth**

Gr. *phitys, -yos,* m. father, begetter:

Gr. *phthano,* anticipate, do first; see **before**

Gr. *phyo; phyteuo,* plant, engender, beget; see *phyton* under **plant**

L. *principium,* n. foundation, beginning: principle.

L. *propago, -atus,* set, fasten, generate, increase: propagate, propaganda.

Gr. *prophasis,* f. cause, pretext:

Gr. *prototypos,* original: prototype.

L. *rudimentum,* n. first principle, beginning: rudiment, rudimentary.

L. *semino, -atus,* sow, seed, engender, beget; *insemino,* implant, impregnate; see *semen* under **seed**

L. *sero, satus,* sow, beget, produce; see **sow**

Gr. *syllepsis,* f. conception, pregnancy: syllepsis, sylleptic.

Gr. *teknosis,* begetting, bearing children, < *tikto,* beget, bear; see *teknon* under **child**

L. *teleta* (Gr. *telete*), f. initiation:

L. *tiro, -onis,* m. recruit, beginner; *tirunculus,* m. dim.; *tirocinium,* n. first military experience, rawness: tiro (tyro), tirocinium.

See: **arouse, enter, birth, new, young, before, fertile, base, hypothesis**

begonia, NL. f. a genus of the begonia family, < Michel Begon, French patron of botany: Begoniaceae, *Begonia coccinea* (scarlet begonia), *Begoniella angustifolia* (a begoniacead), *Symbegonia fulvovillosa* (a begoniacead).

behind < AS. *behindan;* see **after**

being < AS. *beon,* be, exist; see **life, man, woman, animal, spirit, nature**

belch < AS. *bealcian.*

Gr. *apereuxis,* f. a belching forth:

Gr. *eryge,* f.; *erygma, -tos,* n.; *erygmos,* m. belching; *ereuxis,* f. belching; *erygmatodes,* causing belching:

L. *ructo, -atus,* belch; *ructus, -us,* m. a belching; *eructo,* belch, vomit: ructation, eruct.

See: **vomit, spit**

beleco- < Gr. *belekos,* a kind of legume; see **bean**

belemnon, Gr. dart, javelin; see **spear**

belenium, NL. (Gr. *belenion*), a kind of plant; see **plant**

believe < AS. *belifan:* belief, believer, unbelief.

L. *credo, -itus,* believe; *credibilis,* worthy of belief; *credulus,* believing too easily on little or no evidence; *incredibilis,* that cannot be believed; *incredulus,* unbelieving; *incredundus,* not to be believed: credible, creed, credulous, credential, creditable, incredible, incredulity, discredit, grant, miscreant, recreant, *Pardosa credula* (a spider).

Gr. *eikotos,* probably, perhaps, reasonably; see **lot**

Gr. *elpis, -idos; elpore,* f. hope, expectation, < *elpizo,* hope for; *elpisma, -tos,* n. thing hoped for; *elpistos,* hoped; *elpistikos,* arousing hope; *elpidodotes,* m. giver of hope: *Elpidophorus minor* (a fossil mammal), *Elpidia purpurea* (a holothurian).

L. *fido, fisus,* trust, have faith in, rely upon; *fides, -ei,* f. faith, reliance, trust; *fidelis; fidus,* faithful, sure, trustworthy; *fidens,* trusting, confident, *fiducia,* f. confidence, trust, reliance; *confido,* trust implicitly; *confidens, -entis,* assured, trusting, bold; *confisio, -onis,* f. assurance, confidence: fidelity, confidential, fiduciary, fidejussor, affidavit, bona fide, infidel, diffidence, perfidy, defy, defiant, fealty, Santa Fe, faith, fiance, affiance, *semper fidelis, Monadenia fidelis* (a snail). Distinguish between faith and credulity.

L. *fretus,* relying on, trusting in:

L. *opinio, -onis,* belief, fancy, view, conjecture; see *opinor* under **think**

Gr. *pistis,* f. faith, trust; *pistos,* faithful, trustworthy, genuine; *pistikos,* of faith or trust; *pistoma, -tos,* n. assurance, guarantee; *pisynos,* trusting, confiding: pistiology, pistomesite, *Pistosaurus longaevus* (a fossil reptile).

Gr. *pithanos,* persuasive, plausible, probable; see *peitho* under **victory**

L. *praestolor, -atus,* await, expect:

L. *probabilis,* likely, credible: probable, probability, *Rubus probabilis* (a blackberry).

L. *spero, -atus,* hope; *spes, -ei,* f. hope; *specula,* f. dim.; Sp. *esperanza,* f. hope: desperation, despair, desperado, prosper, Esperanto, *Abrochocis esperanza* (a moth).

L. *superstitio, -onis,* f. credulous belief without rational foundation: superstition, superstitious.

See: **think, fancy, doubt, hopeless,** *Penelope* under **odysis**

belittle. See **little**

bell < AS. *belle.* Derive: belfry, beldame.

L. *campana,* f. bell; *campanula;* Sp. *campanilla,* f. dim.: campanile, campanology, elecampane, *Campanula rotundifolia* (bluebell, harebell), *Clematis campaniflora* (a clematis), *Omphalia campanella* (a mushroom), *Sieversia campanulata* (a rosacead).

Gr. *kodon, -os,* m. bell; *kodonion,* n. dim.: codonostome, *Codonopsis ovata* (a campanulacead), *Codonellopsis lusitanica* (an infusorian), *Codaster acutus* (a blastoid), *Platycodon grandiflorus* (Chinese bellflower), *Limnocodium sowerbyi* (a jellyfish), *Hybocodon pendulus* (a medusa).

L. *nola,* f. a little bell: *Nolana prostrata* (a nolanacead).

L. *tintinnabulum,* n. bell: tintinnabulation, tintinnid, *Tintinnus lusus-undae* (an infusorian), *Tintinnopsis campanula* (an infusorian), *Balanus tintinnabulum* (a barnacle). See **ring**

bellarium, L. dessert; see **food**

bellator, L. soldier, warrior; *bellatrix,* female warrior; *bellax, -acis,* warlike; see *bellum* under **fight**

bellis, -idis, L. daisy; *bellio, -onis,* f. corn-marigold; see **composite**

bellow < AS. *bylgan,* sound like a bull; see **roar**

bellows < AS. *belg,* bag, belly; see **bag, blow**

bellum, L. war; *bellicus,* of war, military; *bellicosus,* warlike; *bellifer; belliger,* making war; see **fight**

bellus, L. beautiful; *bellulus,* dim.; see **beauty**

belly < AS. *belg, belig.*

L. *abdomen, -inis,* n. belly: abdomen, abdominal, *Tipula abdominalis* (a cranefly).

L. *alvus,* f. belly, womb, stomach, hold of a ship, beehive:

L. *ampullaceus,* like a flask, big-bellied; see *ampulla* under **bottle**

L. *aqualicus,* m. belly, paunch:

Gr. *araia,* f. belly: *Araea attenuata* (a moth).

L. *coeliacus* (Gr. *koiliakos*), pertaining to the abdomen, bowels, < Gr. *koilia,* f. large cavity of the body, belly, abdomen, intestines; *koilidion,* n. dim.; *mikrokoilos,* with small belly: coeliac. See *koilos* under **hollow**

Gr. *etron,* n. abdomen, belly: cephaletron, *Etroplus maculatus* (a fish), *Etropus crossotus* (a fish), *Stiretrus anchorago* (a bug).

Gr. *gaster,* stomach, belly, paunch, womb; `gastrodes,` potbellied; *hypogastrion,* abdomen; see **stomach**

Gr. *hypochondrion,* n. part of body between ribs and groin, abdomen, belly: hypochondria, hypochondriac.

Gr. *nedys, -yos,* f. belly, stomach, paunch, womb: *Nedyopus cingulatus* (a milleped).

L. *pantex, -icis,* m. paunch, bowels: paunch.

L. *venter, -tris,* m. belly; *ventriculus,* m. dim.; *ventralis,* of the belly; *ventricosus,* (*ventriosus*), potbellied, bulging: ventral, ventricle, ventriloquist, Gross Ventre, *Microtribonyx ventralis* (a rail), *Diplolepis longiventris* (a gall-insect), *Fusulina ventricosa* (a fossil foraminifer), *Valonia(Halicystis) ventricosa* (sea-bottle), *Alopecurus ventricosus* (a grass).

See: **stomach, bag, womb, hollow, swell, groin**

belone, Gr. needle; *belonis, -idos; belonion,* dim.; see **needle**

belong < ME. *belongen* (D. *belangen;* AS. *gelangen*); see **nature, hold, kin**

belos, Gr. dart, arrow, bolt, sting, see **arrow**; threshold, see **door**

below < AS. *be,* by, ON. *lagr,* low; see **under**

belt < AS. *belt,* < L. *balteus,* m. girdle, belt, border, edge: belted, belting, baldric, *Emphytus balteatus* (a wasp).

L. *cestus* (Gr. *kestos*), m. girdle: cestoid, *Cestus veneris* (a ctenophore), *Habrocestum pulex* (a spider), *Cestocrinus striatus* (a Mississippian crinoid).

L. *cinctum; cingulum,* n. girdle, belt, zone; *cincticulum; cingillum,* n. dim.; *cinctorium,* n. sword-belt; *cinctura,* f. girdle; *cinctutus,* girded, < *cingo, cinctus,* surround, gird, bind; *accingo; incingo,* gird about, surround, equip; *discingo,* ungird; *praecinctorium,* n. girdle, apron: cincture, precinct, succinct, surcingle, shingles, enceinte, *Cephus cinctus* (a sawfly), *Calyptocephalus cincticollis* (a beetle), *Metablax cinctiger* (a beetle), *Sebastichthys nigrocinctus* (a rockfish), *Hystatomyia circumcincta* (a fly), *Hydrophis cyanocincta* (chital, a sea-snake), *Scolopendra cingulata* (a centiped), *Heliodiscus cingillum* (a radiolarian), *Rhinocricus cingendus* (a milleped), *Marginula accincta* (a foraminifer), *Cryptodrilus semicinctus* (an oligochaete), *Petrolisthes cinctipes* (a crab).

L. *fascia* (Gr. *phaskia*), f. band, bandage, girdle, zone, stripe; *fasciola,* f. dim.; *phaskidion,* n. dim.; *fascio, -atus,* envelope with bands, swathe: *Nomadacris septemfasciata* (red locust), *Oncopeltus fasciatus* (milkweed-bug), *Fasciola hepatica* (liver-fluke), *Idiacanthus fasciola* (a fish).

Gr. *kombos,* m. roll; band, girth; *kombion,* n. dim.: *Combophora maculata* (a leafhopper).

Gr. *mitra,* girdle, headband, turban; see **cap**

Gr. *prostethion,* n. girdle:

L. *redimiculum,* n. band, fillet, girdle:

Gr. *sparganon,* swaddling band; see **ribbon**

Gr. *strophos,* cord, belt, band, chaplet; *strophion.* dim.; see **rope**

Gr. *telamon,* strap, belt; see **strap**

L. *zona* (Gr. *zone*), f. girdle, belt; *zonula,* f.; *zonarion; zonion,* n. dim.; *zonalis; zonarius,* of a belt or girdle; *zoma; diazoma; perizoma, -tos,* n. that which is girded, girdle: zone, zonation, zonesthesia, *Zonotrichia coronata* (a sparrow), *Zoniolaemus setifera* (a nematode), *Zonarium quadrispinum* (a radiolarian), *Zonidium octostylium* (a radiolarian), *Zoniscus rectangulus* (a radiolarian), *Perizona scutella* (a radiolarian), *Brachygobius xanthozonus* (a fish), *Pomacanthus zonipectus* (a fish), *Pelargonium zonale* (a storksbill), *Erilepsis zonifer* (a fish), *Hemizonia congesta* (a tarweed), *Streptoprocne zonaris* (white-collared swift), *Crocus zonatus* (Cilician crocus), *Rhipozonium lacinulatum* (a coelenterate), *Zomaria interruptolineana* (a moth), *Diazoma mediterranea* (a tunicate), *Perizoma albulata* (a butterfly).

Gr. *zoster, -os,* m. belt, girdle: *Zosterothrix nasutus* (a beetle), *Zosterops halmaturina* (a bird), *Zostera marina* (eelgrass), *Acanthozostera gemmata* (an amphineuran), *Potamogeton zosteriformis* (a pondweed).

See: **bind, collar, garter, ribbon, circle, strap**

beltion, Gr. better; *beltistos,* best; see *agathos* under **good**

belua, L. large animal, beast, monster; *beluinus,* bestial; see **animal**

bema, Gr. pace, step; see *bathmos* under **step**

bembico- < Gr. *bembix, -ikos,* top, whirlpool, buzzing insect; see **cone, wasp**

bembras, Gr. a kind of anchovy; see **herring**

bembros, Gr. stupid; see *bebros* under **dull**

benacus, NL. a genus of bugs; see **bug**

bench < AS. *benc;* see **base, shelf, table**

bend < AS. *bendan.*

L. *aduncus,* bent inward, hooked, aquiline; *reduncus,* bent backward; see *uncus* under **hook**

Gr. *anaklastos,* bent back, reflected: *Anaclastus apicistrigellus* (a moth).

L. *anfractus,* bending, winding, crooked, circuitous, prolix; *anfractuosus,* very winding, sinuous: anfractuous, *Serpula anfracta* (a worm).

Gr. *ankylos,* bent, crooked, hooked, < *ankos,* n. a bend, curve, hollow, glen, valley; *ankyle,* f.; *ankon, -os,* m. bend of the arm, elbow; *ankylosis,* f. a bending, stiffening of the joints: ancylopod, ancylite, ankylosis, *Ancylostoma duodenale* (hookworm), *Ancylus rivularis* (a snail), *Leptocerus ancylus* (a caddisfly), *Ancognatha sellata* (a beetle).

L. *arcuo, -atus,* bend like a bow; see **arc**

Gr. *blaisos,* bandylegged, bent; *blaisotes,* f. crookedness: *Blaesospira echinus* (a gastropod), *Blaesodactylus sakalava* (a lizard).

Gr. *brizo,* nod, be sleepy: *Briza maxima* (a grass).

AS. *bugan,* bend: bow, elbow, oxbow, bowfin.

L. *camur, -a, -um,* crooked, turned inward: *Etheostoma camurum* (a darter).

L. *cernuus,* drooping, stooping, facing earthward, nodding, < *cernuo, -atus,* throw or fall headforemost; *cernulus,* somersaulting: cernuous, *Spiranthes cernua* (an orchid), *Lycopodium cernuum* (staghorn-clubmoss).

L. *clino* (Gr. *klino*), *-atus,* bend, slant, slope, tend; *klisis,* a bending, reclining; see **slope**

L. *compernis*, knockkneed: *Cryptorhynchus compernis* (a beetle).
L. *concavus*, hollowed or arched inward, sunken; see **arc**
L. *convexus*, arched outward, protuberant; see **arc**
L. *cubitum*, elbow, ell; see **arm**
L. *curvus*, bent; *curvo, -atus*, bend, bow, crook; *curvabilis*, that may be bent, flexible: curve, curvature, incurve, recurved, *Salterella curvata* (a fossil cephalopod), *Esacus recurvirostris* (stone-plover), *Hypseloconus recurvus* (a fossil gastropod), *Thysanocarpus curvipes* (a mustard), *Trillium recurvatum* (prairie-trillium), *Eragrostis curvula* (a grass).
Gr. *eiktikos*, yielding readily, pliable, < *eiko*, give way, yield, concede: *Icticus ischanus* (a fish).
L. *falcatus*, curved like a sickle; *falcipedius*, bowlegged; see **sickle**
L. *flecto, flexus*, bend; *flexura*, f.; *flexus*, m. a bending, turning; *flexibilis; flexilis*, pliant; *flexuosus*, full of bends, tortuous, crooked, winding; *reflexus*, bent or turned back: flexure, flexibility, reflect, deflection, genuflection, *Pinus flexilis* (limber pine), *Panicum flexile* (a grass), *Cymbopogon flexuosus* (ginger-grass), *Pycnanthemum flexuosum* (a mountain-mint), *Asarum reflexum* (a wild ginger), *Lonicera deflexicalyx* (a honeysuckle), *Pithecellobium flexicaule* (Texas ebony).
Gr. *gausos*, crooked, bent: *Gausocentrus gyrini* (a wasp).
L. *geniculatus*, like the bent knee: see *genu* under **knee**
L. *gibber; gibbus*, humpbacked, humped, crooked, bent; see **projection**
Gr. *gnamptos*, curved, bent, pliant: *Gnamptogenys rimulosa* (an ant).
Gr. *gnyx; prochny*, with bent knee, kneeling, *gonykrotos*, knockkneed; see *gony* under **knee**
Gr. *grypos*, hooknosed, curved; *grypanios*, bowed by age: gryposis, grypanian, *Grypotherium domesticum* (a sloth), *Grypoceras cancellatum* (a fossil nautiloid).
L. *incoxo, -atus*, bend down, cower:
Gr. *kamptos; kampylos; kampaleos; kamphos; kamptikos*, bent, curved, flexible; *kampe; kampsis*, f. a bend, curve, turn; *kampter, -os*, m. bend, angle: camptosaurus, campylotropal, *Camptosorus sibiricus* (Asiatic walking-fern), *Camphotherium elegans* (a fossil mammal), *Camponotus inflatus* (Australian honey-ant), *Campogaster exigua* (a fly), *Campogramma vadigo* (a fish), *Campeloma multistriatum* (a fossil gastropod), *Campsis radicans* (trumpet-creeper), *Campsicnemus fumipennis* (a fly), *Campteroneura reticulata* (a fossil insect).
Gr. *kirsos*, m. dilated vein, varicocele: cirsocele, cirsoid, cirsotomy.
Gr. *koronos*, curved, crooked; *koronis, -idos*, f. a curved line, flourish, mark of apostrophe: coronis.
Gr. *kybebos*, stooping, with head bent: *Cybebus dimidiatus* (a beetle).
Gr. *kyllos*, crooked, crippled, lame; see **hurt**
Gr. *kyphos*, bent, humped, humpbacked; *kyphos, -eos*, n. hump, hunch; *kyphoma, -tos*, n. hump; *kyphon, -os*, m. crooked piece of wood, bent yoke, pillory, knave; *kypto*, bend forward, stoop: cyphonautes, cyphonism, *Cyphomycter tuberosus* (a fish), *Cyphoderia ampulla* (a protozoan), *Cyphonocephalus smaragdulus* (a beetle), *Cyptocephala cogitabunda* (a bug), *Cypturus aenescens* (a beetle), *Cuphea miniata* (cinnabar-cuphea).
Gr. *kyrtos*, curved, humped; *kyrtoma, -tos*, n. curve, swelling; *kyrton, -os*, m. hunchback; *epikyrtos*, humpbacked: cyrtocone, cyrtodont, crytolite, cyrtosis, *Cyrtophyllum concavum* (broad-winged katydid), *Cyrtomium falcatum* (holly-fern), *Dicyrtomellus pectinatus* (a wasp), *Dicyrta candida* (a gesneriacead), *Encyrtocephalus simplicipes* (a wasp).
L. *lentus*, flexible, pliant, tenacious, tough, slow; *lentulus*, dim.; *lentipes*, slow-footed: relent, relentless, *Betula lenta* (sweet birch), *Viburnum lentago* (nanny-berry-viburnum).
L. *licinus*, bent or turned upward; *relicinus*, bent backward:
Gr. *lordos*, bent backward: lordosis, *Lordophleps curvinervis* (a fly).
L. *loripes, -edis*, bowlegged: *Loripes lacteus* (a pelecypod).
Gr. *lygizo*, bend like a willow, twist; *lygeros*, pliant, flexible; *lygismos*, m. a bending or twisting; *lygistos*, bent, pliant; *lygistes*, m. basket-maker: *Lygistopterus sanguineus* (a beetle).
Gr. *myllos*, crooked, bent: *Myllocerus pici* (a beetle), *Myllocerops psittacinus* (a beetle).
Gr. *neuo*, nod; see **yes**
L. *nuto, -atus*, nod, droop; *numen, -inis*, n.; *nutus, -us*, m. nod of the head; *nutabilis; nutabundus*, tottering, swaying; *annuo, -utus*, nod to, assert, approve; *innuo*, nod to, give a sign: nutant, nutate, nutation, innuendo, *Carduus nutans* (musk-thistle).
L. *pandus*, bent, crooked, curved; *repandus*, bent backward, turned up, undulate: repand, *Cyclamen repandum* (spring cyclamen).
Gr. *parabole*, f. a conic section: parabola, parabolic.

Gr. *pholkos,* bowlegged: *Pholcus phalangioides* (a spider), *Pholcophora americana* (a spider).

L. *pravus,* crooked, deformed, perverse, bad; see **bad**

Gr. *prenes,* hanging forward, drooping, face downward, prone: *Prenanthes alba* (white rattlesnake-root).

L. *resimus,* turned up, bent back:

L. *resupinus,* bent backward, with the face upward; see *supinus* under **lie**

L. *retrorsus,* turned or bent backward; see *torno* under **turn**

Gr. *rhaibos,* bent, crooked, bowlegged: rhaebocrania, rhaeboscelia, *Rhaebothorax piscator* (a spider).

Gr. *rhoikos,* crooked: *Rhoecocoris australasiae* (a bug).

L. *serpentinus,* serpentlike, twisting, winding; see *serpens* under **snake**

Gr. *simoo,* turn up the nose; *simoma, -tos,* n. anything turned up:

L. *sinuo, -atus,* bend, curve, wind; *sinuosus,* full of bendings, winding; *insinuo, -atus,* put or bring in by devious or oblique means: sinuous, sinuosity, sinus, insinuate, *Alnus sinuata* (Sitka alder).

Gr. *skambos,* bent, crooked, bowlegged: *Scambus echinatus* (a beetle), *Scambocarabus bifoveicollis* (a beetle).

Gr. *skellos,* bowlegged: *Scellus monstrosus* (a fly), *Scelloides spinosus* (a fly).

Gr. *skolios,* curved, bent, oblique: scoliosis, *Scoliocystis pumila* (a cystid).

Gr. *strangos,* twisted, crooked, tortuous:

Gr. *streblos,* twisted, crooked, wrinkled: *Streblus asper* (a moracead), *Strebloceras subannulatum* (a gastropod), *Streblosoma magnum* (an annelid), *Streblomastix strix* (a protozoan).

L. *suppes,* with twisted feet; see *pes* under **foot**

Gr. *synkryptes,* m. one who leans forward:

L. *valgus,* bowlegged, bent: *Valgus plumatus* (a beetle), *Valgipes deformis* (a fossil mammal), *Acanthovalgus marquardi* (a beetle).

L. *varus,* bent, distorted, spread; *varix, -icis,* f. a dilated, twisted vein: varicose, varicocele, prevaricate, *Oncidium varicosum* (an orchid), *Phlox divaricata* (blue phlox), *Divaricella dentata* (a pelecypod), *Cratosomus varicosus* (a beetle).

L. *vatius,* bent outward, bowlegged:

See: **arc, angle, hook, fold, form, willow, turn, slope**

Bendis, Gr. the Thracian equivalent of Artemis; see **hunt**

bene, L. well, good, < *bonus,* good; *beneficus,* favorable; see *bonus* under **good**

benetes, Gr. blue; see **blue**

benignus, L. kind, good, friendly, favorable; see **friend**

benna, L. carriage or wagon made of wickerwork; see **vehicle**

benthos, Gr. depth of the sea; see **deep**

benzene; benzo-; benzyl- < NL. *benzoin,* an aromatic resin, < Ar. *luban jawi,* incense of Java: benzene(benzine), benzol, benzoate of soda, benzoic, benzaldehyde, benzidine, benzyl, benzoin, olibanum (from *Boswellia carteri,* a burseracead), *Styrax benzoin* (Sumatra snowbell).

Suf. *-ol,* signifies a benzene derivative: phenol, pyrogallol, resorcinol, salol.

NL. *pheno-; phenyl-,* pertaining to benzene and its compounds: phenol, phenolphthalein, phenothiazine, phenobarbital, phenazine, phenanthine, phenyl salicylate, 2,4-D (dichlorophenoxyacetic acid), thiodiphenylamine.

berberion, Gr. a shabby garment; see **garment**

berberis, ML. barberry; see **barberry**

bereschethos, Gr. booby; see **dull**

bericocco- < Gr. *berikokkon,* apricot; see **plum**

berry < AS. *berige.*

L. *acinus* (*acinum*), berry, grape; *duracinus,* hard-berried; see **grape**

L. *bacca,* f. berry; *baccula,* f. dim.; *baccalis,* with berries: baccate, bacciform, bacciferous, baccivorous, asarabacca, bay (*Laurus nobilis*), *Taxus baccata* (European yew), *Sargassum bacciferum* (a seaweed), *Baccaurea bracteata* (tampoy).

L. *coccum,* n. (Gr. *kokkos,* m.) grain, seed, berry; *kokkion,* n.; *kokkis, -idos,* f. dim.; *kokkalos,* m. seed from a cone, berry, a round insect called kermes (*Coccus ilicis*) on the oak, *Quercus coccifera,* a source of scarlet dye; seed, grain, spherical, red: coccolith, cocciferous, diplococcus, staphylococcus, streptococci, micrococcus, coccidium, coccidiosis, cochineal (from the cochineal insect, *Dactylopius coccus*), *Coccinella novemnotata* (a ladybird-beetle), *Coccolobis grandifolia* (a sea-grape), *Coccus cryptus* (a scale-insect), *Cocculus carolinus* (Carolina moonseed), *Coccidoxenus ugandensis* (a wasp), *Anamirta cocculus* (fishberry), *Rhodococcus roseus* (a bacterium), *Triticum dicoccum* (emmer).

L. *cuscolium,* n. a scarlet berry [*sic*] of the holm-oak, perhaps the kermes insect:
Gr. *rhax, rhagos,* grape, berry; see **grape**
See: **fruit, barberry, blackberry, crowberry, currant, elder, holly, huckle-
berry, mulberry, pokeberry, strawberry, juniper, ball, grape**

berula, L. a kind of watercress; see **mustard**

beryl < L. *beryllus* (Gr. *beryllos*), a precious stone of sea-green color; see **stone**

beryx, Gr. a perchlike fish; see **perch**

besalon, Gr. brick; see **brick**

bessa, Gr. a wooded glen; see **valley**

bessalis, L. pertaining to eight, < *bes, bessis,* eight-twelfths or two-thirds of a unit;
see **eight**

bestia, L. beast, animal; *bestiola,* dim.; *bestialis,* like a beast; see **animal**

bet < OF. *abeter,* bait, arouse, excite; see **lot**

beta, L. beet, see **beet**; Gr. *beta,* second letter of the Greek alphabet, see **letter**

betarmon, Gr. dancer; see **dance**

betonica, L. a genus of the mint family; see **mint**

betray < AS. *be,* and L. *trado,* give up; see **treason**

betula, L. birch; see **birch**

between < AS. *betweonan;* see **middle**

beuthos, Gr. a woman's dress; see **garment**

bex, *bechos,* Gr. cough; see **cough**

beyond < AS. *begeond;* see **over, far**

bi-; bin-; bis-; < L. *bis,* two, twice; *bifarius,* double; *binarius,* of two; see **two**

biaeo- < Gr. *biaios,* forcible, violent; see *biastos* under **rough**

biarco- < Gr. *biarkes,* supplier, provisioner; see **supply**

biastos, Gr. violent; *biastes,* one who uses force; see **rough**

bibastes, Gr. mounter, stallion; see *bibazo* under **raise**

bibio, L. an insect generated in wine, a kind of fly; see **fly**

biblio- < Gr. *biblion,* dim. of *biblos,* paper, book, scroll, < *byblos,* the papyrus-
sedge; see **book**

bibrosco- < Gr. *bibrosko,* eat, gnaw, consume; see **eat**

bibulus; bibax; bibosus, L. drinking much, < *bibo, -itus,* drink; see **drink**

bico- < Gr. *bikos,* wine-jar, drinking-bowl; *bikion; bikidion,* dim.; see **cup**

bidens, L. two-toothed; a genus of the composite family; see **composite**

bier < AS. *baer;* see **carry**

big < ME. *big, bigge;* see **large**

bigener, L. hybrid, mongrel; see **mix**

bignonia, NL., f. a genus of the bignonia family, < Jean Paul Bignon, French
librarian: Bignoniaceae, *Bignonia magnifica* (a bignonia), *Catalpa bignonioides*
(southern catalpa).
NL. *campsis,* f. a genus of the bignonia family: *Campsis grandiflora* (Chinese
trumpet-creeper).
NL. *catalpa,* f. a genus of the bignonia family, < Ab.Am. *kutuhlpa: Catalpa
speciosa* (northern catalpa), *Ceratomia catalpae* (catalpa-sphinx), *Buettneria
catalpifolia* (a sterculiacead).
Ab.Am. *jacaranda,* f. a genus of the bignonia family: *Jacaranda acutifolia* (a jacaranda).
NL. *kigelia,* f. a genus of the bignonia family, < an African word: *Kigelia
pinnata* (sausage-tree).
NL. *tabebuia,* f. a genus of the bignonia family, < Ab.Am. *tabebuya,* antwood:
Tabebuia argentea (a trumpet-tree).
NL. *tecoma,* f. a genus of the bignonia family, < Ab.Am. *tecomaxochitl,* pot-tree
flower: *Tecoma spectabilis* (a trumpet-creeper).

bile < L. *bilis,* f. gall, anger; *biliosus,* full of bile: bilious, biliary, biliation, bilin,
bilicyanin, bilirubin.
Gr. *chole,* f. gall, bile; *cholion,* n. dim.; *cholikos; cholerikos; epicholos,* bilious:
choler, choleric, choline, acetylcholine, cholesterine(cholesterol), cholesterase,
cholemia, cholera, melancholy, choledochoplasty, glycocholate, taurocholic acid,
desoxycholic acid, *Euphoria melancholia* (a sap-chafer), *Gymnophallus chole-
dochus* (a trematode).
L. *fel, fellis,* n. gall, bile; *felleus,* of gall, galllike; *fellitus,* steeped in gall;
fellosus, full of gall: fellic, fellifluous, *Boletus felleus* (a mycorrhiza).

-bilis, L. tending to be, capable of, worthy of, having the quality of; see **nature**

bilix, *-icis,* L. with a double thread; see *licium* under **thread**

bill < AS. *bile,* beak of a bird, see **nose;** < L. *bulla,* edict, see **call, trade**

bimus, L. of two years; see **year**

bin < AS. *binn,* crib, box; see **box**

binarius, L. of two; see *bi-* under **two**

bind < AS. *bindan,* fasten, tie, fix: binder, binding, bindweed, bound, band.

L. *addictus,* bound to, given up to, devoted to, yielding to: addict, addicted, addiction.

L. *adnatus,* joined to, united with, adherent; *connatus,* born together, joined: adnate, connate, *Veronica connata* (a speedwell).

L. *adunatus,* united, made one; *coaduno, -atus,* join together, unite; see *unus* under **one**

L. *amplector, amplexus; complector, -exus,* encircle, embrace; *complexivus,* connecting, copulative: *Hippurites amplexus* (a fossil pelecypod).

Gr. *ankter,* binder, clasp; see *ango* under **choke**

L. *apo, aptus,* bind, fasten, join, fit: apt, aptitude.

L. *applico, -atus,* join, fasten, affix, bring to, devote to; *applicitus,* attached to, connected with: application, apply.

Gr. *aprix,* with closed teeth, fast, tight:

L. *armilla,* bracelet; see **bracelet**

Gr. *arthmos,* m. bond, league, union; *arthmios,* united: *Arthmius globicollis* (a beetle).

Gr. *asilla,* f. yoke for carrying pails, baskets, etc.:

Gr. *bletron,* n. fastener, band, nail, rivet:

L. *camus* (Gr. *kemos*), m. muzzle, gag: *Cemophora coccinea* (a snake).

L. *capistro, -atus,* tie with a halter; *capistrum,* n. halter, muzzle: *Chaetodon capistratus* (a butterfly-fish), *Leioptila capistrata* (black-headed sibia).

L. *cateno, -atus,* chain or bind together; see **chain**

Gr. *chalara,* f. fetter: *Chalara quercina* (a fungus).

L. *cingo, cinctus,* surround, gird, bind; see *cinctum* under **belt**

L. *coagmento, -atus,* join, glue, cement, connect; *coagmentum,* n. joint:

L. *coalesco, coalitus,* grow together, unite; see *alesco* under **grow**

L. *coaxo* (*coasso*), *-atus,* join boards or planks together, floor:

L. *combino, -atus,* unite, join: combine, combination.

L. *commissura,* f. seam, connection, < *committo, -missus,* join, connect, undertake: commissure, commit.

L. *compaginatus,* joined, bordering:

L. *concilio, -atus,* unite separate parts into a whole, connect: conciliate, reconciliation.

L. *confluus,* flowing together, uniting; *confluens, -entis,* m. place where two streams unite: confluence, *Amphibolips confluentus* (a gall-insect).

L. *congregalis,* uniting together; see *grex* under **herd**

L. *conjuro, -atus,* swear together, bind by oath, conspire; *conjuratus,* m. conspirator: conjure, conjuror, conjuration.

L. *consecraneus,* bound by the same oath:

L. *constipo, -atus,* press together, become costive; see *stipo* under **press**

L. *contextus,* interwoven, cohering, connected, < *contexto,* interweave, entwine: context.

L. *continuus,* uninterrupted, joined, connected: continuous.

L. *copula,* f. band, tie, connection; *copulabilis,* that can be connected; *copulativus, copulo, -atus,* join, bind, connect: copulate, couple, couplet, coupler, coupling.

L. *corrigia,* f. shoelace, latchet, rein:

L. *corrivatio, -onis,* f.; *corrivium,* n. confluence of streams, < *corrivo, -atus,* draw together into one stream:

L. *densabilis; densativus,* binding, astringent; see *densus* under **thick**

Gr. *deo,* bind; *dema; desma, -atos,* n. band, bundle; *desmos,* m. bond, fetter, halter, cable, ligament, chain; *demation; desmation,* n. dim.; *deon, -tos,* n. that which is binding, duty; *desis,* f. a binding together; *desmios,* binding; *desmis, -idos,* f. bundle; *detos,* bound; *diadetos,* bound fast; *syndesmikos,* conjunctive; *syndetos,* bound together: desmid, desmology, desmaturgia, desmosis, diadem, anadem, asyndeton, deontology, amphidetic, *Dematochroma picea* (a beetle), *Dematobactron fuscipennis* (a phasmid), *Desmodium nudiflorum* (a tick-trefoil), *Desmodus rufus* (a vampire), *Desmognathus fuscus* (a salamander), *Desmobdella paranensis* (a leech), *Desmiostoma parvulum* (a wasp), *Desmidopora alveolaris* (a fossil coral), *Detodesmus aurantiacus* (a milliped), *Scenedesmus obtusus* (an alga), *Trichodesmium erythraeum* (an alga), *Syndetocrinus nor-*

thrupi (a Silurian crinoid), *Syndesmica homogenes* (a moth), *Diadophis amabilis* (a snake), *Dematium cinereum* (a fungus).

L. *destino*, *-atus*, make fast, bind, establish, determine: destine, destination.

L. *determino*, *-atus*, bound, limit, fix: determine, determination.

Gr. *enete*, pin, brooch; see **needle**

Gr. *epiploke*, f. a connection:

L. *fascia*, bandage, girdle; see **belt**

L. *ferrumino*, *-atus*, cement, solder, glue, bind, join: *conferrumino*, cement together; see **glue**

L. *fibula*, f. buckle, brooch, clasp, clamp, safety pin; *fibulo*, *-atus*, bind together, buckle; *confibula*, f. clamp, clincher; *infibulo*, *-atus*, fasten together with a clasp or buckle: fibula, infibulate. *Dictyocha fibula* (a fossil foraminifer), *Myctophum fibulatum* (a lantern-fish), *Endomyces fibuliger* (a fungus).

L. *figo*, *fixus*, bind, fasten; *affixus*, attached to; *confixus*, fastened together; *confixilis*, that can be joined; *suffixus*, fastened on: fixture, transfix, crucifix, affix, prefix, suffix, infix.

L. *foedero*, *-atus*, form a league, < *foedus*, *-eris*, n. compact, covenant, bargain, league; *confoederatus*, allied: federal, federate, confederation, *Speciospongia confoederata* (a sponge).

L. *frenum*, bridle, rein, bit; see **hold**

AS. *gyrdan*, surround, bind: gird, girdle, girder, Sea Girt (N.J.).

Gr. *hapto*, join, fasten to, fix upon, lay hold of, grasp, touch; *haphe*, f. grip, grasp; *haptos*, touchable; *synaptos*, joined together, united; *synamma*, *-tos*, n.; *synaphe*; *synapsis*, f. bond, connection, union, junction; *aaptos*, untouchable, invincible; *hamma*, *-tos*, n. anything tied or made to tie to, knot, noose, halter, cord, bowstring; *hammation*, n. dim. bandage; *kathamma*, *-tos*, n. anything tied, knot: haptics, hapteron, haptophore, haptogen, synapsis, *Synaptomys fatuus* (a bog-lemming), *Synaphobranchus pinnatus* (a fish), *Hammatoceras insigne* (a fossil cephalopod), *Dichammatus pilosus* (a beetle).

Gr. *harmozo*, fit together, join; *harmotos*, joined, adapted; see **joint**

L. *indiscretus*, unseparated, undivided, closely connected; see **one**

L. *innodo*, *-atus*, fasten with a knot, entangle; see *nodus* under **knot**

L. *jungo*, *junctus*, unite; *jugo*, *-atus*, join, connect; *jugum*, n. yoke, team, pair, crossbeam, ridge; *jugamentum*, n. crossbeam, lintel; *jugalis*, of a yoke; *jugarius*, yoked; *jugis*, *-e*, yoked or joined together; *jugus*, belonging together; *jugabilis*, that may be joined; *adjunctus*; *adjunctivus*, joined to, connected to; *conjugalis*, relating to marriage; *conjugulus*, pertaining to uniting; *conjunctivus*, connective; *conjunctus*, connected, united; *conjux*, *-ugis*, c. spouse: conjugal, subjugate, join, joint, adjunct, rejoinder, joiner, disjointed, adjoining, conjugation, conjunction, conjunctiva, juncture, *Dipteris conjugata* (a fern), *Ajuga reptans* (bugle), *Filaria conjunctivae* (a nematode), *Neoclytus conjunctus* (a beetle). Derive: Yogi, yoke, yokel.

Gr. *krikos*, ring. finger-ring; see **circle**

L. *ligo*, *-atus*, bind, tie, fasten; *ligamentum*, n.; *ligatura*, f. tie, bandage; *alligo*, bind, fetter, obligate; *alligator*, *-is*, m. one who binds; *alligatura*, f. band, tie; *deligo*, bind up, bind fast; *illigo*, tie on, attach, connect; *obligo*, bind, engage, pledge: ligament, ligature, obligation, alliance, ally, rely, reliability, rally, liable, liane, league, lien, liaison.

L. *manica*, sleeve, handcuff; see **sleeve**

F. *mortaise*, < Ar. *murtazz*, fastened in: mortise and tenon.

L. *necto*, *nexus*, knit, tie, bind; *nexibilis*; *nexilis*, tied together, interlaced, knit; *nexuosus*, complicated; *annexus* (*adnexus*); *connexus*; *innexus*, joined; *annectens*, linking, joining: connect, disconnect, annexation, nexus, adnexal, adnexitis, *Diplograptus nexus* (a graptolite), *Psocus nexus* (a psocopterid), *Pelmatochromis annectens* (a fish), *Stephanauge nexilis* (a sea-anemone).

L. *numella*, f. fetter, shackle, pillory:

Gr. *ochma*, *-tos*, hold, tie, band, < *ochmazo*, grip fast; see *ischo* under **hold**

Gr. *opheilo*; *ophello*, owe, be indebted or bound to; *opheiletes*, debtor; see **pay**

L. *pango*, *panctus*; *pactus*, fasten, fix, set; *pactilis*, entwined, plaited; *pactum*, n. agreement, covenant; *compago*, *-inis*, f. a joining together, connection; *compactilis*, joined together: pact, compact, compactly, impinge, propagate, *Modiolaria impacta* (a pelecypod).

L. *patibulum*, n. fork-shaped yoke; *patibulatus*; *patibulus*, yoked:

L. *pedica* (Gr. *pede*), f. anklet, fetter, shackle, bangle; *pedion*, n. dim.; *pedao*, shackle; *compes*, *-edis*, f. fetter, shackle; *compedio*, *-itus*, fetter, shackle; *compesco*, *-itus*, fasten together, restrain: impeach, *Chaetonerius compeditus* (a fly).

Gr. *pelekinos*, dovetailing:

Gr. *periskelis*, garter, anklet; see **garter**

Gr. *perone*, buckle, clasp; see **needle**

Gr. *pegma, -tos,* anything congealed, fastened, fixed; *pexis,* fixation, solidification; *paktos; pektos,* fastened, < *pegnymi,* fasten; see *pegma* under **thick**

Gr. *phimos,* muzzle, stopper; *phimosis,* muzzling or stopping of an orifice; see **close**

Gr. *porpe,* f. buckle, brooch, clasp: *Porpita linnaeana* (a siphonophore), *Upsiloporpa haynei* (a mosquito).

Gr. *pyssachos,* m. muzzle to prevent calves from sucking:

L. *redimiculum,* n. band, fillet, girdle; see **belt**

Gr. *saktos,* crammed, stuffed; *satto,* pack, load; see **press**

L. *scytanum,* n. mordant:

L. *securicula,* mortise shaped like a hatchet-head; see **ax**

L. *sero, sertus,* join, knit, plait, connect; *consero,* join, unite: assert, dissertation, exserted, series, exertion, desert, *Sertularia abietina* (a hydroid).

Gr. *sparganon,* swaddling band; see **ribbon**

Gr. *sphingo,* bind tight; *sphinktos,* bound; *sphinxis,* f. constriction; *sphinkter, -os,* m. binder; *sphigma, -tos,* n. that which is bound tight: sphincter, *Sphinctocoris corallinus* (a bug), *Sysphincta cavernicola* (an ant).

L. *splenium* (Gr. *splenion*), patch, bandage; see **spleen**

Gr. *steriktos,* fixed, firmly set; see **stand**

Gr. *strangale,* halter; see **hold**

L. *stringo, strictus,* draw together, hold in check, bind; *astrictus,* drawn together, tight, narrow, close; *astringens,* shrinking, binding, contracting, puckering; *constrictus,* drawn together, contracted, tightened; *restrictus,* bound up, limited, tight: strict, stricture, district, astringent, stringent, constrictor, stress, distressing, strait, restrain, restriction, prestige, Detroit (Mich.), *Oxalis stricta* (wood sorrel), *Dystactocrinus constrictus* (an Ordovician crinoid), *Agave stringens* (an agave).

Gr. *stryphnos,* astringent: *Stryphnodendron rotundifolium* (alum-bark tree).

Gr. *stymma,* an astringent for thickening an unguent; see *stypho* under **contract**

L. *subscus,* m. tongue or tenon of a dovetail:

Gr. *sympodizo,* bind hand and foot, fetter; *sympodistes,* m. fetterer:

Gr. *synairema, -tos,* n. union: *Synairema*[*Synaerema*] *alpina* (a wasp).

Gr. *synankeia,* f. confluence of glens or valleys: *Synancia verrucosa* (a fish).

Gr. *synarmoge,* f. combination: *Synarmoge ferrari* (a fossil insect-wing).

Gr. *tropos,* twisted leather thong; see **strap**

L. *unio, -itus,* join together, make one; see *unus* under **one**

L. *vador, -atus,* bind by requiring bail or surety; *vas, vadis,* m.; *vadimonium,* n. bail: vadimonium.

L. *vincio, vinctus,* bind; *vinculum,* n. anything used for binding: vinculum, periwinkle (*Vinca minor*), *Embrithes vinculatus* (a beetle), *Limnephilus vinculum* (a trichopterid).

Gr. *zygon,* n.; *zygos,* m. yoke; *zygion,* n. dim.; *zygios,* of a yoke, < *zeugnymi,* yoke, join; *zeugle,* f. loop of a yoke; *zeugos,* m.; *zygas, -ados,* f. team, pair; *zeugma, -tos,* n. band, bond; *zeuktos,* yoked; *zeuxis,* f. a joining, yoking; *homozygos,* yoked together, paired; *syzygos,* united, yoked: zygospore, zygosis, zygote, zygomatic, zeuxite, syzygy, *Zygophyllum coccineum* (bean-caper), *Zygnema cruciatum* (an alga), *Zeuglodon cetoides* (a fossil whale), *Zygadenus gramineus* (a death-camas), *Syzygium operculatum* (a myrtacead), *Zeuxippus histrio* (a spider), *Syzygops prasina* (a beetle), *Zeugmatothrips hispidus* (a thysanopterid), *Hemizyga inflata* (a fossil gastropod), *Tetrazygia bicolor* (a melastomacead).

See: **close, pen, hold, marry, press, knot, chain, trap, thick, belt, collar, ribbon, hook, glue, bar, nail, magic, key, guard, bridge, sew**

bindweed

L. *clymenus,* m. (Gr. *klymenon,* n.), a kind of bindweed: *Lathyrus clymenum* (a pea).

L. *convolvulus,* m. bindweed, morning-glory: *Convolvulus arvensis* (European bindweed), *Convolvulus aureus* (golden bindweed).

Gr. *iasione,* f. probably a bindweed, but now applied to a campanula: *Jasione montana* (sheepsbit).

NL. *ipomoea,* f. a genus of the morning-glory family: *Ipomoea purpurea* (a morning-glory).

NL. *porana,* f. a genus of the morning-glory family: *Porana paniculata* (a porana).

Ab.Am. *quamochitl,* a twining plant: *Quamoclit coccinea* (starglory).

NL. *rivea,* f. a genus of the morning-glory family: *Rivea corymbosa* (source of ololiuqui).

L. *scammonia* (Gr. *skammonia*), f. bindweed: *Convolvulus scammonia* (scammony), *Secamone amygdalina* (an asclepiadacead).

L., Gr. *smilax,* bindweed; now a genus of the lily family; see **lily**
binus, L. two by two; see *bi-* under **two**
bios, Gr. life; *biosis,* manner of life; *bioteia,* livelihood; *biotikos,* of life; see **life**
birch < AS. *birce, beorc:* Birch, Berkeley, Barkley.
  L. *betula,* f. birch; *betulinus,* of birch: betulaceous, betulase, betulin, *Betula populifolia* (gray birch), *Polyporus betulinus* (a bracket-fungus), *Barosma betulina* (a buchu), *Rhynchites betulae* (funnel-twister weevil), *Nothofagus betuloides* (a falsebeech), *Carpinus betulus* (European hornbeam).
  Gr. *semyda,* f. probably a birch: *Samyda*[*Semyda*] *serrulata* (a flacourtiacead), Samydaceae.
bird < AS. *bridd.*
  L. *acredula,* f. a kind of bird:
  Gr. *aellos,* m. a kind of bird:
  Gr. *aigokephalos,* m. a kind of bird:
  L. *ales, alitis,* winged; bird; see *ala* under **wing**
  Gr. *ampelion, -os,* m.; *ampelis, -idos,* f. a songbird: *Ampelion rufaxilla* (a bird), *Ampelis cedrorum* (cedar-waxwing).
  L. *apiastra,* f. a bee-eater: *Merops apiaster* (bee-eater).
  L. *avis,* f. bird; *avium,* gen. pl. of *avis; aucella* (*avicella*)*; avicula,* f. dim.; *avicularius; aviarius,* of birds, bird-keeper; *auceps, -cupis,* c. bird-catcher, fowler; *aucupalis,* pertaining to bird-catching or fowling; *aucupor, -atus,* go bird-catching, give chase to: aviator, aviation, aviculture, augur, inaugurate, auspices, auspicious, avicular, aviary, Avernus, ostrich, *Avihospita chrysorrhea* (a fly), *Prunus avium* (sweet cherry), *Pteria(Avicula) contorta* (a fossil pelecypod), *Bugula avicularia* (a bryozoan), *Polygonum aviculare* (knotweed), *Sorbus aucuparia* (European mountain-ash, rowan), *Aucella elongata* (a fossil pelecypod), *Aviculopecten cancellatus* (a fossil pelecypod).
  Gr. *bokkalis, -idos,* f. a kind of bird:
  Gr. *brenthos,* m. a kind of bird:
  NL. *casuarius,* m. < Malay *kasuari,* a large ratite bird: cassowary (*Casuarius casuarius*), *Casuarina suberosa* (a beefwood).
  L. *ciris* (Gr. *keiris*), f. a fabulous, ravenous sea-fowl: *Passerina ciris* (painted bunting).
  NL. *colymbus* (Gr. *kolymbis*), a diving bird, grebe; see **dip**
  L. *cychramus* (Gr. *kychramos*), m. a bird, probably an ortolan or rail: *Cychramus luteus* (a beetle).
  Gr. *dandalos,* m. a kind of bird: *Dandalus pinetorum* (a bird).
  Gr. *dikairon,* n. an Indian bird:
  Gr. *drakontis, -idos,* f. a kind of bird:
  Gr. *edolios,* m. a kind of bird: *Xiphostylus edolius* (a radiolarian).
  Gr. *elasas,* m. a kind of bird: *Elasas furcatus* (a bird).
  Gr. *epolios,* m. a kind of bird:
  L. *galbula,* f. dim. of *galbina,* f. a small bird: *Galbula ruficauda* (red-tailed jacamar), *Icterus galbula* (Baltimore oriole).
  L. *galgulus,* m. a small bird: *Galgulus caudatus* (a bird).
  L. *haematopus, -podis* (Gr. *haimatopous*), m. a red-footed bird: *Haematopus ostralegus* (a bird).
  Gr. *hypothymis, -idos,* f. a kind of bird: *Hypothymis azurea* (a bird).
  Gr. *kalamodytes,* m. a reed-bird: *Calamodyta caligata* (a bird).
  Gr. *kalaris, -idos,* m. a kind of bird:
  Gr. *kaukalias,* m. a kind of bird: *Caucalias leucogastra* (a bird).
  Gr. *keblepyris, -idos,* f. a kind of bird:
  Gr. *kenklos,* m. a sea-bird:
  Gr. *kepphos,* m. a sea-bird: *Cepphus columba* (pigeon-guillemot).
  Gr. *kerkion,* f. an unknown bird:
  Gr. *kerkoronos,* m. an Indian bird: *Cercoronus melanorhynchus* (a bird).
  Gr. *kerthios,* m. a tree-creeper: *Certhia familiaris* (a tree-creeper), Certhidae.
  Gr. *kinnyris, -idos,* m. a small bird: *Cinnyris venusta* (a sunbird), *Cinnyricinclus leucogaster* (a starling).
  Gr. *koraphos,* m. a kind of bird: *Coraphites albifrons* (a bird).
  Gr. *laedos,* m. a kind of bird:
  Gr. *merops, -opos,* m. a bee-eater: *Merops cinereus* (a bee-eater).
  Gr. *neossis, -idos; neossos,* a young bird, nestling, chick; see **young**
  L. *oce, -es,* f. a small bird:
  Gr. *oionos,* large bird of omen, augury; see **prophecy**
  Gr. *ornis, -ithos,* m.; *orneon,* n. bird; *ornithion,* n. dim.; *ornitheutes,* m. fowler: ornithology, ornithivorous, *Ornithorhynchus anatinus* (duckbill), *Ichthyornis victor* (a Cretaceous bird), *Ornithopus scorpioides* (a legume), *Dinornis maxi-*

mus (an extinct bird), *Ornicholax solitarius* (a bird-louse), *Ornithium imberbe* (a flycatcher), *Smicrornis brevirostris* (a bird), *Orneostoma subpulchra* (a moth).
Gr. *ortalis, -idos,* f.; *ortalichos,* m. young bird, chick:
L. *oscen, -inis,* m. a singing bird: oscine, oscinine, Oscines, *Conioscinella araneolicida* (a fly), *Oscinella gregalis* (a fly), *Oscinosoma discretum* (a fly).
Gr. *ourax, -agos,* f. a kind of bird: *Ourax erythrorhynchus* (a bird).
L. *palara,* f. a kind of bird:
L. *parra,* f. a bird of ill omen: *Parra(Jacana) spinosa* (a wading bird).
Gr. *pergoulos,* m. a kind of bird:
Gr. *peteinon,* n. fledgling:
L. *phoenix, -icis* (Gr. *phoinix, -ikos*), m. a fabulous bird, symbolic of resurrection and immortality; *phoeniculus,* m. dim.: phoenix, *Phoeniculus(Irrisor) senegalensis* (a wood-hoopoe).
Gr. *phoinikouros,* m. redstart: *Phoenicurus auroreus* (a bird).
Gr. *phokion, -os,* m. a kind of bird: *Phocion congener* (a wasp).
Gr. *pindalos (spindalos),* m. an Indian bird: *Pindalus ruficapillus* (a bird), *Spindalus bilineatus* (a bird).
Gr. *pipos,* f. young bird, chick:
NL. *pitta,* f. < Hind. *pitta,* bird; a kind of bird: *Pitta brachyura* (ant-thrush).
Gr. *rhyndake,* f. probably a bird-of-paradise:
Per. *rukh,* a large, fabulous bird: roc.
Gr. *sialis, -idos,* f. a kind of bird: *Sialis infumata* (an alder-fly), *Sialia sialis* (bluebird).
Per. *simurgh,* a large, fabulous bird, probably the same as the roc:
Gr. *stymphalis, -idos,* f. a man-eating bird in Arcadia, slain by Hercules: *Stymphalus rubolineatus* (a bug).
L. *subis,* f. a kind of bird: *Progne subis* (purple martin).
Gr. *tetrix, -igos,* f. a kind of bird: *Tetrao tetrix* (a grouse), *Tetrix ornata* (a grouse-locust).
Gr. *titis, -idos,* f. a small bird:
L. *todus,* m. a small bird: green tody (*Todus viridis*), *Todirostrum maculatum* (a bird).
Gr. *tragopan, -os,* m. a fabulous bird: *Tragopan abyssinicus* (a pheasant).
Gr. *trikkos,* m. a small bird: *Triccus crinitus* (a bird), *Perissotriccus ecaudatus* (a bird).
L. *uria* (Gr. *ouria*), f. a water-bird: *Uria lomvia* (thick-billed murre).
L. *vireo, -onis,* m. a kind of bird: *Vireo griseus* (white-eyed vireo).
L. *volucris,* a flying creature, bird; see *volo* under **fly**
See: **chicken, duck, goose, hawk, thrush,** etc., **feather, wing, fly**

birota, L. two-wheeled cab; see **vehicle**

birrus, L. (Gr. *birrhos*), cloak or mantle to keep off rain; see **garment**

birth < AS. *gebyrd,* < *beran,* carry: birthday, rebirth, birthmark, birthright, birthplace, birthwort.
L. *abortivus,* prematurely born or terminated, not well-developed; *abortus, -us,* m. miscarriage; *abortio,* premature delivery, miscarriage; see **abortion**
Gr. *apokyesis,* f. birth; *apokyema, -tos,* n. offspring:
Gr. *aristodin, -os,* bearing the best children:
L. *consatio, -onis,* f. procreation:
L. *cretus,* arisen, sprung, descended from, born of, < *cresco,* come forth, grow; see *cresco* under **grow**
L. *edo, -itus,* bring forth, publish: edit, editor, editorial.
Gr. *ektexis,* f. childbirth:
Gr. *ektroma, -tos,* n. untimely birth, abortion: *Ectroma fulvescens* (a wasp), *Ectromachernes mirabilis* (a solpugid).
Gr. *emphytos,* implanted, inborn, natural; see **nature**
L. *fetus,* embryo, offspring; see **young**
L. *gigno, genitus* (Gr. *gignomai*), be born, give birth to, produce; *genitalis,* of birth, causing birth; *gen-, -gen,* affixes meaning producing, causing, forming; *-genus, -a, -um,* born in, living in, as *aquigenus,* water-born; *fluctigenus,* wave-born; *indigenus,* native; *montigenus,* mountain-born; *paludigenus,* marsh-born; *spumigenus,* foam-born; *genero, -atus,* beget, create; *ingenero; ingigno,* implant, engender; *genesis,* f. beginning, birth, origin; *genetes (-ter; -tor),* m. begetter, father, ancestor; *geneteira,* f. mother; *genethlion,* n. birthday; *genethlios,* of a birthday; *Genetyllis, -idos,* f. goddess of birth; *gonimos,* productive, fruitful; *goneus,* m.; *goneia,* f. begetter; *congenitus,* born with; *archegenes,* causing the beginning of a thing; *autogonos,* self-produced; *eugenes,* well-born; *oligogonos,* producing few at a birth; *progenes,* first-born, ancestral; *epigone,* f. offspring; *apogennema, -tos,* n. offspring: genitive, congenital, gender, genius, genesis, gene, Eugene, hypogene, supergene, antigen, progenitor, progeny, genus, generic,

eugenic, oxygen, hydrogen, terrigenous, generate, degeneration, indigenous, ingenious, ingenuous, engine, malignant, *sui generis*, gentle, jaunty, genuine, *Poa paludigena* (a grass), *Eugenia axillaris* (white stopper).

L. *incubatio, -onis,* f. a sitting upon eggs, brooding: incubation, incubator.

L. *indigenus,* born in a country, belonging to one's own country, native; see **native**

AS. *-isc,* affix denoting origin; see *-ish* under **nature**

Gr. *lechoios,* of child-bed; see **bed**

L. *lochia* (Gr. *locheia*), f. childbirth, delivery; *lochios,* of childbirth: lochia, lochiorrhagia, *Aristolochia odoratissima* (fragrant birthwort), *Smilax aristolochiifolia* (Mexican sarsaparilla).

L. *Lucina,* goddess of childbirth; see *lux* under **light**

L. *nascor, natus,* be born; *nativus,* inherent by birth; *natalis,* of birth; *naturalis,* by birth, innate; *natus, -us,* m. birth, growth: natal, adnate, cognate, connate, nation, nature, nascent, renascence, Renaissance, nativity, innately, Natal, nee, pregnant, impregnate, puny, naive, eigne, *Embrithes agnatus* (a beetle).

L. *obstetrix,* midwife; see **nurse**

Gr. *oizo,* sit on eggs, brood, hatch:

L. *oriundus,* risen from, descended, born; *ortus,* a rising of the heavenly bodies, birth; see *orior* under **raise**

L. *pario, paritus; partus,* beget, create, produce; *parturio,* be pregnant, bring forth; *partus, -us,* m. a birth, embryo; *partualis,* of birth; *viviparus,* bearing active, living young: parturition, parturient, nulliparous, parent, oviparous, gemmiparous, *Viviparus fasciatus* (a snail).

L. *proles,* offspring; see **young**

L. *puerperus,* parturient, giving birth; see *puer* under **child**

L. *pullulo, -atus,* put forth, produce young; see *pullus* under **chicken**

L. *secunda,* f. afterbirth:

Gr. *tokos,* m. birth, delivery, offspring; *tokarion,* n. dim.; *tokas, -ados,* pertaining to birth; *toketos,* m. that which is brought forth, < *tikto,* give birth to; *tiktikos,* of childbirth; *dystokia,* f. hard delivery; *eutokos,* bearing easily; *kyotokia,* f. childbirth: tocogenetic, tocogeny, tocology, dystocia, tocopherol, *Tocotrema lingua* (a flatworm), *Taeniotoca lateralis* (blue perch).

See: **begin, before, mother, father, child, young, womb, cradle, fertile, abortion**

birthwort

Gr. *aristolochia,* f. birthwort: *Aristolochia grandiflora* (pelican-birthwort or Dutchman's-pipe), *Smilax aristolochiifolia* (Mexican sarsaparilla).

L. *asarum* (Gr. *asaron*), n. hazelwort, wild-ginger: *Asarum virginicum* (a wild-ginger), *Parnassia asarifolia* (brook-parnassia), asaron, asarabacca.

NL. *rafflesia,* f. a genus in the order Aristolochiales, < Thomas Raffles, its discoverer: Rafflesiaceae, *Rafflesia arnoldi* (a parasitic plant with the largest-known flower).

bis, L. twice, double; see *bi-* under **two**

biscuit < It. *biscotto,* < L. *bis,* twice, *coctus,* cooked; see **cake**

bismuth < G. *bismuth* (*wismuth*); see ELEMENTS under **thing**

bison, L., Gr. (OHG. *wisunt*), the humpbacked ox, wisent; see **cattle**

bit < AS. *bite;* see **little, trifle, part**

bitch < AS. *bicce,* female dog; see **dog**

bite < AS. *bitan:* bit, biting, backbite, henbit.

Gr. *aprix,* with closed teeth, fast, tight; see **bind**

Gr. *brygma, -tos,* n. bite; *brygmos,* biting, gnashing, < *bryko,* bite, eat greedily, gnash the teeth: *Brycon striatulus* (a fish), *Hyphyssobrycon rosaceus* (a characine fish), *Onobrychis arenaria* (Hungarian sainfoin), *Ixobrychus neoxenus* (a bittern).

AS. *ceowan,* masticate: chew, cud, quid. What is a tertium quid?

Gr. *dakno,* bite; *daknister, -os; dektes,* m.; *daketon,* n.; *dakos,* n. biter; *dagma* (*odagma*), *-tos; degma, -tos,* n.; *dexis,* f. bite, sting; *dakneros; degmos* (*odagmos*), biting, irritating, itching; *dektikos,* able to bite; *odax,* biting of the lips in irritation or vexation; *odaxo,* feel a biting, stinging pain: Odacinae, *Dacnomys millardi* (a mouse), *Tridacna gigas* (a pelecypod), *Dacnister flavescens* (a bug), *Brachydacne rufozonatum* (a beetle), *Trichodectes canis* (dog-louse), *Daceton armigerum* (an ant), *Dacus dorsalis* (a fruitfly), *Psilodacus rufoscutellatus* (a fly), *Carpodacus purpureus* (purple finch), *Brachydegma caelatum* (a fossil fish), *Anisodexis imbricatus* (a labyrinthodont), *Decticus verrucosus* (a katydid), *Odaxosaurus obliquus* (a Cretaceous reptile).

Gr. *grao,* gnaw, eat; see **eat**

L. *mando, mansus; manduco, -atus,* chew; *manducator, -is; manducus,* m. chewer,
glutton: mandible, manger, *Bracon manducator* (a wasp), *Manducus maderensis*
(a fish), *Manduculus limatus* (a spider).

Gr. *maseter,* m. chewer; *masema, -tos,* n. quid, < *masaomai* (*massaomai*), chew:
masseter, *Uintacyon massetericus* (a creodont).

L. *mastico, -atus,* chew: masticate, mastication.

Gr. *merykizo,* chew the cud; *meryx, -ykos,* m. a ruminant: *Merychippus perditus* (a fossil horse), *Nanomeryx caudatus* (a fossil ungulate).

L. *mordeo, morsus,* bite; *mordax, -acis,* biting, corroding, pungent; *mordicus,*
biting, snappish; *morsus, -us,* m. bite; *morsico, -atus,* bite continually: mordant,
mordella, morsel, remorse, margeline, *Microtus mordax* (a vole), *Momordica
charantia* (balsam-pear), *Hydrocharis morsus-ranae* (frogbit), *Glossina morsi-
tans* (a tsetse-fly).

Gr. *ouloboros,* with deadly bite: *Uloboros flavus* (a spider).

L. *pungens,* sharp, acrid, piercing, biting; see **point**

L. *rodo, rosus,* gnaw; *arrosor, -is,* m. nibbler, gnawer; *derosus; erosus,* eaten away,
gnawed, nibbled: rodent, rodenticide, erosion, corrode, corrosive, rosorial, Corro-
dentia, *Trachyrhynchus erosa* (a gastropod), *Pachyrhizus erosus* (a legume).

L. *rumino, -atus,* chew the cud: ruminate, ruminant.

AS. *screawa,* a biter, shrew-mouse: shrew, shrewd, shrewdness.

Gr. *trogo,* gnaw, nibble, munch; *trox, -ogos,* m. gnawer, weevil; *troktes,* m. gnawer,
nibbler, a sea-fish with sharp teeth; *troktos,* gnawed, eatable; *troktikos,* greedy;
trogalion, n. sweetmeat, dessert: troctolite, trout, trogon, *Trogophloeus tarsalis* (a beetle), *Trogoxylon punctatum* (a beetle), *Xylotrogus brunneus* (a
beetle), *Troctes succineus* (a fossil insect), *Platytroctes apus* (a fish), *Trox
oustaleti* (a fossil beetle), *Diglossotrox chinensis* (a beetle), *Trocticus gibbulus*
(a psocid).

See: **eat, cut, sharp, itch, point, tooth, jaw, sour**

bitis, NL. a genus of snakes; see **snake**

bittaco- < Gr. *psittakos,* parrot; see **parrot**

bitter < AS. *biter:* bitterly, bitterness, bittersweet.

Gr. *aloe,* aloe; bitter aloes; see **lily**

L. *amarus,* bitter; *amaritudo, -inis,* f. bitterness; *amarulentus,* very bitter, full of
bitterness; *amarico, -atus,* make bitter, irritate: amarine, convallamarin, morello
cherry, maraschino, cascara amarga (bark of *Picramnia antidesma*), *Cardamine
amara* (a bittercress), *Gentiana amarella* (a gentian), *Solanum dulcamara* (bittersweet), *Rhodeus amarus* (bitterling), *Panicum amarum* (a grass), *Panicum
amarulum* (a grass).

Gr. *myrrha,* myrrh; see **resin**

Gr. *pikros,* bitter: picrolite, picric acid, picrocrocin, pinipicrin, chlorpicrin, picro-
tin, *Picrodendron baccatum* (bittertree), *Picrasma excelsa* (bitter ash), *Picradeniopsis oppositifolia* (a composite), *Picris echioides* (bristly oxtongue),
Xylopia polycarpa (berberine tree), *Picramnia pentandra* (bitterbush).

L. *ruta,* rue; see **rue**

See: **bile, rue, wormwood, anger, arouse**

bittium, NL. a kind of snail; see **mollusk**

bitumen, L. asphalt, pitch; *bituminosus,* full of asphalt or pitch; see **resin**

bixa, NL. a genus of plants: Bixaceae, bixin, *Bixa orellana* (arnatto-tree).

blaberos, Gr. harmful, noxious; see *blaptikos* under **hurt**

black < AS. *blaec:* blacken, blackbird, blackball, blackmail, blacksmith, Blackburn,
Blackwood, bootblack, lampblack, black-eyed Susan (*Rudbeckia hirta*).

Gr. *achlys, -yos,* f. mist, darkness, obscurity, < *achlyo,* grow dark, obscure: *Achlys
triphylla* (vanilla-leaf), *Achlya prolifera* (a fungus).

Gr. *aithiops,* sunburnt, black, swart; *aithalodes,* sooty, black; see *aitho* under
burn; *aithalos* under **dirt**

L. *anthracinus* (Gr. *anthrakinos*), coal-black; see *anthrax* under **coal**

L. *aquilus,* dark-colored, dun, swarthy, blackish: *Elater aquilus* (a beetle), *Sciaena
aquila* (maigre).

L. *ater, atra, atrum,* black; *atratus,* dressed in black: atrament, atrabilious, *Sauromalus ater* (chuckwalla), *Parus atricapillus* (chickadee), *Gminatus atricornis* (a
bug), *Pellaea atropurpurea* (cliffbrake-fern), *Agave atrovirens* (maguey-plant),
Hymenocephalus aterrimus (a fish), *Tabanus atratus* (horsefly).

L. *caccabatus,* black, sooty, like a cooking utensil:

L. *caligo, -inis,* f. fog, mist, darkness; *caligneus,* dark, gloomy; *caliginosus,* foggy,
misty, dark: *Caligo eurylochus* (a butterfly), *Leptodactylus caliginosus* (a
jungle-frog).

L. *coracinus* (Gr. *korakinos*), ravenlike, black as a crow; see *korax* under **crow**

Gr. *dnopheros,* dark, gloomy, murky; see **night**

L. *ebenus* (Gr. *ebenos*), f. a species of persimmon; *ebeneus; ebeninus,* of ebony, black: ebon, Indian ebony (*Diospyros melanoxylon*).

Gr. *eremnos,* black, swarthy, dark: *Eremnoschema ditissima* (a beetle).

L. *fuliginatus,* sooty, black; see *fuligo* under **dirt**

L. *fumidus,* smoky, smoke-colored; see *fumus* under **cloud**

L. *funebris; funereus; funestus,* of a funeral, black; see *funero* under **cover**

L. *furvus,* dark, dusky, gloomy, swarthy, black; *furvescens,* growing dark: *Lyperosomus furvus* (a beetle).

L. *fuscus,* dark, swarthy, dusky; *fuscitas- atis,* f. darkness, obscurity, < *fusco, -atus,* darken, blacken, obscure; *infuscus,* dusky, dark brown, blackish; *offusco (obfusco),* darken, obscure; *suffusculus,* brownish: fuscous, fuscescent, fuscin, obfuscate, *Formica fusca* (an ant), *Buteo rufofuscus* (a hawk). *Serjania fuscifolia* (a sapindacead), *Stiphrosomus fuscicornis* (a wasp), *Hylocichla fuscescens* (veery), *Sterna fuscata* (sooty tern), *Goniobasis infuscata* (a snail).

L. *illuminus,* without light, dark: *Turdus illuminus* (a bird).

Gr. *kapnodes,* like smoke, dusky, sooty; see *kapnos* under **cloud**

Gr. *kelainos,* black, dark, murky, swart: *Celaenomys silaceus* (a rodent).

Gr. *lignyodes,* smoky, sooty, murky; see *lignys* under **cloud**

Gr. *mauros,* dark: *Epicrates maurus* (a snake), *Maurodactylus alutaceus* (a bug).

Gr. *melas, melanos; melaina; melan; melam-,* black, dark; *melasma, -tos,* n. anything black; *melasmos,* m. a blackening; *melanchimos,* black, dark: melancholy, melanin, melanuria, melanite, melasma, calomel, *Aronia melanocarpa* (black chokeberry), *Melaleuca imbricata* (a myrtacead), *Phidippus melanocephalus* (a fossil spider), *Chalinura ctenomelas* (a fish), *Pharopteryx melas* (a fish), *Callithrix melanurus* (a marmoset), *Melampodium perfoliatum* (blackfoot), *Melanthium virginicum* (bunchflower), *Dalbergia melanoxylum* (African blackwood), *Melasmothrix naso* (a mouse), *Melanolestes picicornis* (a bug), *Melanostigma gelatinosum* (a fish), *Disonycha xanthomelaena* (spinach fleabeetle), *Ameiurus melas* (black bullhead).

L. *memnonius,* black, dark, < Memnon, fabled king of the Ethiopians: *Memnon peragrans* (a spider).

L. *morulus,* black, like a mulberry; see *morus* under **mulberry**

L. *nebulosus,* misty, cloudy, dark; see *nebula* under **cloud**

L. *niger, nigra, nigrum,* black, dark, dusky; *nigellus; nigriculus,* dim.; *nigricans,* blackish, swarthy, dark; *nigror, -is,* m. blackness; *nigro, -atus,* be black, blacken: nigrify, nigrite. nigritic, negro, darnel, *Helleborus niger* (black hellebore), *Salix nigra* (black willow), *Solanum nigrum* (black nightshade), *Conanthalictus seminiger* (a bee), *Mustela nigripes* (a ferret), *Melanopteryx nigerrimus* (a weaver), *Nigella damascena* (love-in-a-mist), *Dendroica nigrescens* (a warbler), *Rhizopus nigricans* (bread-mold), *Garrupa nigrita* (black grouper), *Pseudabaris nigrina* (a beetle), *Enneadesmus nigritulus* (a beetle).

L. *nubilosus,* cloudy, dull; see *nubes* under **cloud**

Gr. *orphnos,* dark, dusky; *orphnaios,* murky; *orphne,* f. the darkness of night, night; *orphninos,* brownish-gray: *Orphnophanes productalis* (a moth).

Gr. *pellos,* dusky: *Pellopsyche signata* (a trichopterid), *Pellaea pumila* (a cliffbrake-fern).

Gr. *perknos,* dark-colored: percnosome, perknite, *Percnostola leucostigma* (a bird).

L. *piceus,* pitchy, pitch-black; see *pix* under **resin**

Gr. *psephos,* n. darkness; *psephenos,* dark, obscure: *Psephenus lecontei* (a water-penny), *Psephenops smithi* (a beetle).

L. *pullus,* dark-colored, blackish, grayish-black: *Otiorhynchus pullus* (a beetle), *Campanula pulla* (a bellflower).

L., Gr. *sepia,* squid, ink; see **mollusk**

L. *tenebrosus; tenebricus,* dark, gloomy; *tenebra* (*tenebrae,* shades of night), f. darkness, gloom, obscurity; *tenebrio, -onis,* f. shunner of light, lover of darkness; *tenebresco,* grow dark: tenebrous, tenebrific, *Staphylinus tenebricosus* (a rove-beetle), *Dicerca tenebrosa* (a beetle), *Tenebrio obscurus* (a meal-worm).

See: **night, shade, coal, crow, ink, Africa, sad, blind, hell**

blackberry

L. *batus* (Gr. *batos*), f. bramble, blackberry, raspberry, thorn-bush; *bation,* n. dim.; *batinos,* of a bush or thicket; *batodes,* bushy, thorny; *batia,* f. bush, thicket: *Sarcobatus baileyi* (a greasewood), *Batocrinus turbinatus* (a Mississippian crinoid), *Ribes cynosbati* (pasture-gooseberry), *Chamaebatia foliolosa* (bearmat), *Chamaebatiaria millefolium* (desert-sweet), *Batopora rosula* (a fossil bryozoan).

L. *rubus,* m. bramble, blackberry, raspberry; *rubetum,* n. bramble-thicket; *Rubus odoratus* (purple-flowering raspberry), *Rubus occidentalis* (black raspberry),

Rubus allegheniensis (blackberry), *Cimicifuga rubifolia* (a bugbane), *Phorbia rubivora* (raspberry-cane maggot).
Sp. *zarza,* f. bramble, blackberry: sarsaparilla.

blaco- < Gr. *blax, -akos,* dull, slow, doltish; see **dull**

bladaros, Gr. flaccid; see **weak**

bladder < AS. *blaeddre;* see **bag**

blade < AS. *blaed;* see **sword, leaf, spoon, shoulder, paddle, oar, plate, knife**

blaeso- < Gr. *blaisos,* bandy-legged; *blaisotes,* crookedness; see **bend**

blaesus, L. speaking indistinctly, lisping, stammering; see **falter**

blame < Gr. *blasphemeo,* revile, reproach.
 L. *accuso, -atus,* blame, complain against, reproach; *incuso,* find fault with, blame; *incusabilis,* blameworthy; *incusativus,* blaming: accuse, accusation.
 Gr. *aitia,* charge, blame, fault, cause; see **begin**
 L. *causa,* agent, reason, motive; see **begin**
 L. *criminor, -atus,* accuse of crime, impeach, charge: incriminate.
 L. *culpo, -atus,* hold a person at fault, blame, censure, reproach; see *culpa* under **fault**
 L. *delator, -is,* m. accuser, informer, spy: *Ichneumon delator* (a wasp).
 Gr. *enklema, -tos,* n. charge, accusation, complaint:
 Gr. *epaitios,* blameworthy:
 Gr. *epegoria,* f. accusation:
 L. *improbo, -atus; reprobo,* disapprove, blame, condemn, reject; *opprobrium,* n. reproach, disgrace: opprobrium, reprobate.
 L. *increpito, -atus,* chide, blame, rebuke; *increpitus, -us,* m. a chiding, rebuking:
 Gr. *memptos,* blameworthy; *mempsis; momphe,* f. blame, censure: *Memptus braueri* (a fossil beetle), *Apomempsis bufo* (a beetle).
 Gr. *momos,* m.; *momar; mymar,* n. blame, ridicule, carping criticism, < *momaomai,* find fault with, blame: Momus, Mymaridae, *Mymar pallidus* (a wasp), *Mymaromma anomala* (a wasp), *Mymarothrips ritchianus* (a thysanopterid).
 L. *obloquor, -locutus,* speak against, blame, condemn: obloquy.
 Gr. *oneidos,* n. blame, reproach; *oneidisma, -tos,* n.; *oneidismos,* m. reproach; *oneidistes,* m. reproacher; *oneidistos,* reproachable, disgraceful: *Onidistus odiosus* (a beetle).
 Gr. *onostos,* to be blamed:
 Gr. *psogos,* m. blame, censure; *psogios,* censorious; *psektes,* m. blamer:
 L. *reprehensibilis,* blamable: reprehensible.
 L. *vitupero, -atus,* censure, blame; see **scold**
 See: **fault, scold, curse**

blamma, Gr. damage, hurt; see *blaptikos* under **hurt**

blandus, L. of a smooth tongue, flattering, friendly, kind, alluring, mild; *blanditus,* agreeable, charming; see **agreeable**

blank < OF. *blanc,* white; see **empty**

blanket < OF. *blanchet,* dim. of *blanc,* white.
 Gr. *blema, -tos,* n. something thrown over, coverlet: epiblema, periblem.
 L. *galumma,* f. cover of any kind:
 L. *inductura,* f. covering, coating:
 Gr. *koas; kos,* n. fleece used as a bedcover: *Diplycosia microphylla* (a heath).
 L. *lodix, -icis,* f. blanket, coverlet; *lodicula,* f. dim.: lodicule.
 Gr. *peristroma,* coverlet, carpet; see **rug**
 Gr. *rhegos,* rug, blanket; see **rug**
 L. *stragulum,* n. blanket, pall, spread; *instragulum; instratum,* n. coverlet; *stragulus,* covering:
 Gr. *stromateus,* m. coverlet; *peristroma, -tos,* n. coverlet, carpet:
 See: **cover, garment, bed**

blanos, Gr. blind; see **blind**

blaptico- < Gr. *blaptikos,* harmful, hurtful; *blapsis,* damage, harm, < *blapto,* hurt; see **hurt**

blasphemo, L. (Gr. *blasphemeo*), revile, reproach, defame, see **curse**; *blasphemia,* evil, defamatory speech, profanity, see **profane**

blastos, Gr. bud, sprout, shoot, germ; *blastema,* offshoot, sucker; see **bud, branch**

blatero, L. babbler, buffoon; see **speak**

blatta, L. cockroach, bloodclot; see **coachroach, lump**

blatteus, L. purple; see **purple**

blaute, Gr. slipper; *blaution,* dim.; see **shoe**

blaze < AS. *blaese;* see **fire**

-ble < L. *-bilis,* tending to be, capable of, having the quality of; see **nature**

bleat < AS. *blaetan.*
> L. *balo, -atus,* bleat; *balatus, -us,* m. bleating:
> Gr. *blechas, -ados,* f. bleater; *bleche,* f. bleat:
> Gr. *mekasmos,* m. bleating:
> L. *miceo,* bleat:
> L. *mutio,* mutter, bleat; see **murmur**

blechado- < Gr. *blechas, -ados,* bleater; see **bleat**

blechnum, L. (Gr. *blechnon*), a kind of fern; see **fern**

blechon, Gr. pennyroyal; see *glechon* under **mint**

blechros, Gr. faint, gentle, slight; see **weak**

blema, Gr. something thrown over, coverlet; see **blanket**

blemish < OF. *blesmir;* see **fault, mark, spot, dirt, hurt**

blemma, Gr. glance, look; see *blepsis* under **see**

blennius, L. (Gr. *blennos*), a fish; see **blenny**

blennos, Gr. mucus; see **slime**

blenny < L. *blennius* (Gr. *blennos*), m. a fish: *Blennius ocellaris* (blenny), *Phycis blennioides* (fork-beard hake).
> NL. *brotula,* f. a fish closely related to the blennies: *Brotula barbata* (a fish).

blepharis, Gr. eyelash, see **hair**; *blepharon,* eyelid, see **eyelid**

blepos, Gr. look; see *blepsis* under **see**

blepsias, Gr. a fish; see **fish**

blepsis, Gr. seeing, sight; *bleptos,* worth seeing, < *blepo,* look; see **see**

bletos, Gr. stricken, smitten; see **strike**

bletron, Gr. a fastener; see **bind**

blight < uncertain origin; see **disease, fungus, hurt**

blind < AS. *blind.*
> Gr. *ablepsia,* f. blindness: ablepsia, niphablepsia.
> Gr. *aderktos,* not seeing, sightless; *dysderkes,* hardly seeing; see *derkomai* under **see**
> Gr. *alaos,* blind: nyctalopia, *Alaus myops* (a click-beetle), *Alaopsylla papuensis* (a siphonapterid).
> Gr. *amauros,* dark, dim, obscure, blind; see **dim**
> Gr. *amblyopos* (*amblopos*), dim-sighted, dim, dark; *aops, -opos,* blind; *nops,* purblind: *Amblyopus rusticus* (a beetle), *Nops guanabacoae* (a spider).
> Gr. *anommatos,* eyeless, sightless; *apommatos,* blind: *Anommatophilus tenellus* (a beetle), *Anommatoxenus clypeatus* (a beetle).
> Gr. *aoratos,* invisible, unseeing, blind: *Aoratothrips tenuis* (a thysanopterid).
> Gr. *atheatos,* unseen, unseeing, blind; see **secret**
> Gr. *blanos,* blind: *Blanodium abnormale* (a beetle).
> L. *caecus,* blind; *caecigenus,* born blind; *caeco, -atus; excaeco,* make blind, darken: caecum, caecal, caecilian, Cecil, Cecilia, *Caecidotea stygia* (an isopod), *Geodytes caecus* (a beetle).
> L. *eluminatus,* blinded:
> L. *gramiosus,* blear-eyed, full of pus, watery, dim; see **pus**
> Gr. *lemaleos,* blear-eyed, dim; see **dim**
> Gr. *lipoglenos,* blind:
> L. *lippus,* blear-eyed, dim-sighted, nearly blind; *lippitudo, -inis,* f. inflammation of the eyes; *lipposus,* blear-eyed; see **dim**
> L. *luscus,* one-eyed, with one eye shut, half-blind; see **dim**
> Gr. *maraugia,* f. loss of sight:
> L., Gr. *Phineus,* m. mythical prophet struck blind by Zeus, synonym for a blind man: Phineus.
> Gr. *typhlos; typhlops,* blind; *typhlinos,* m. blindworm, lizard: typhlosis, typhlitis, *Typhlonectes compressicaudus* (a caecilian), *Typhlochrestus digitatus* (a spider), *Typhlomyrmex rogenhoferi* (an ant), *Typhlops punctatus* (a burrowing snake), *Leptotyphlops myopica* (a snake), *Typhlichthys subterraneus* (a cave-fish), *Typhlina lineata* (a snake).
> See: **black, dim**

blisso (blitto), Gr. take honey from the hive, rob, steal; see **steal**

blister < uncertain origin; see **bubble**

blitas, Gr. a worthless woman; see **prostitute**

bliteus, L. tasteless, insipid, silly, foolish, useless; see **fool**

blitum, L. (Gr. *bliton*), a plant of the goosefoot family; see **goosefoot**

blood < AS. *blod:* bloody, bloodless, bloodhound, bloodroot, bleed, bless.

L. *blatta,* bloodclot; see **lump**

L. *cruor, -is,* m. blood, stream of blood; *crudus,* bloody, raw; *cruentus,* bloody, stained or spotted with blood, blood-red; *cruento, -atus,* make bloody: crude, cruorin, chlorocruorin, *Dianthus cruentus* (a pink), *Rhynchophorus cruentatus* (palmetto-weevil), *Euryscopa bicruenta* (a beetle).

L. *cuferion,* n. nosebleed:

Gr. *haima, -tos,* n. blood; *haimas, -ados,* f. stream of blood; *haimaleos; haimateros; haimatinos; haimatodes,* bloody, blood-red; *haimaktos; haimatikos,* of blood; *haimorrhagia,* f. flow of blood; *haimaxis,* f. a blood-letting; *haimasso,* stain with blood; *exaimos,* bloodless; *kathaimos,* bloody: hemal, hematite, hemostat, hemorrhage, hemocyanin, hematin, hematoma, anemia, anemic, hyperemia, androseme, *Haemodorum paniculatum* (a bloodwort), *Hematoxylon campechianum* (logwood), *Bilharzia haematobia* (a blood-fluke), *Staphylococcus hemolyticus* (a bacterium), *Theridium haematostigma* (a spider), *Arisaema serratum* (a jack-in-the-pulpit), *Cathaemia ada* (a butterfly), *Haemanthus coccineus* (a blood-lily).

Gr. *lythron,* n. gore; *lythrodes,* gory: *Lythrum flexuosum* (a loosestrife), *Lythrodes radiatus* (a moth).

L. *sanguis, -inis,* m. blood; *sanguinarius; sanguineus; sanguinalis,* of blood, bloody, blood-red: sanguine, consanguineous, sanies, sang froid, *Sanguinaria canadensis* (bloodroot), *Sanguisorba tenuifolia* (Siberian burnet), *Chalepus sanguicollis* (a beetle), *Benthodytes sanguinolentus* (a holothurian), *Geranium sanguineum* (a geranium), *Cornus sanguinea* (European dogwood), *Conorhinus sanguisuga* (a cone-nose bug).

Gr. *thrombos,* bloodclot; see **lump**

See: **kin, life, red**

blosis, Gr. arrival; see **come**

blosyros, Gr. grim, awful, terrible; see **fear**

blot < OF. *blotte,* clod of earth; see **spot, cancel**

blothros, Gr. tall, stately; see **high**

blow < ME. *blowe,* stroke; see **strike**

blow < AS. *blawan,* move air.

Gr. *aemi; ao,* blow; *aema, -tos,* n. blast, wind: asthma, aula.

Gr. *askoma, -tos,* n. a leather bellows:

Gr. *bdeo,* break wind, stink; see **smell**

L. *flo, flatus,* blow; *flatus, -us,* m. breath, breeze; *flatilis,* blown,. produced by blowing; *inflatus,* puffed, swollen: flatulence, afflatus, inflate, deflation, *Utricularia inflata* (a bladderwort), *Eriogonum inflatum* (bladderstem).

L. *follis,* bellows; see **bag**

L. *mungo, munctus,* blow the nose: emunctory.   Derive: *Phaseolus mungo* (a bean).

Gr. *perdomai,* break wind; see **smell**

Gr. *physa,* f. bellows; *physeter, -os,* m. blower; *physetos,* blown; *physao,* puff up, distend, inflate; *anaphysma, -tos,* n. blast of wind, eruption of a volcano: physagogue, emphysema, *Physa fontinalis* (a snail), *Physocarpus stellatus* (Alabama ninebark), *Physetocrinus ventricosus* (a Mississippian crinoid), *Physeter catodon* (sperm-whale).

Gr. *pneo,* breathe, blow; see **breathe**

Gr. *psycho,* breathe, blow; see *psyche* under **mind**

Gr. *rhipidion,* bellows, fan, < *rhipizo,* blow up, fan; see **fan**

See: **wind, breathe, swell, fan, horn**

blue < OF. *bleu:* bluet, bluebell, blueberry, bluegum, bluebird, bluefish.

L. *aeroides* (Gr. *aeroeides*), like the sky or air, sky-blue; see *aer* under **wind**

Gr. *amethystinos,* of or like amethyst; see *amethystus* under **stone**

F. *azur,* < Per. *lazhuward,* a blue color: azure, azurite, azulene, azulmin, lapis lazuli, lazurite, *Anchusa azurea* (Italian bugloss), *Primavera azul* (blue thrush).

L. *caeruleus,* sky-blue; *subcaeruleus,* pale-blue: cerulean, *Leuciscus caeruleus* (azurine, blue roach), *Houstonia caerulea* (bluet), *Chen caerulescens* (blue goose).

L. *caesius,* bluish-gray; *caesitius,* bluish; see **gray**

L. *callainus* (Gr. *kallainos*), greenish-blue, turquoise-colored:

L. *cymatilis,* sea-colored, blue:

Gr. *glaukos,* bluish-green or gray, sea-colored; see **gray**

L. *indicum,* n. indigo, < L. *Indicus,* of India: indigo, indigotin, indol, indoline, *Indigofera tinctoria* (indigo-plant), *Heliotropium indicum* (a heliotrope).

Gr. *ion,* violet; *ianthinos,* violet-colored; *iodes,* violetlike; see **violet**

Gr. *kyanos; kyaneos,* dark-blue: cyanide, cyanamide, anthocyanin, cyanosis, Cyanophyceae, *Centaurea cyanus* (cornflower), *Chlorion cyaneum* (a wasp), *Cyanospiza cyanea* (a bird), *Cyanotis cristata* (a spiderwort), *Dermatobia cyaniventris* (a warble-fly), *Apomotis cyanellus* (a sunfish), *Poecilomyia cyanogaster* (a fly), *Cyanea arctica* (a jellyfish), *Circus cyaneus* (hen-harrier).

L. *lividus,* bluish, lead-colored, black and blue, envious, spiteful; *livor, -is,* m. bluish or lead-color, envy, malice; *liveo,* turn lead-colored, ashy: livid, *Metriopepla lividula* (a beetle), *Columba livia* (rock-dove).

Gr. *molybros,* lead-colored; see *molybdus* under **lead**

Skt. *nila,* dark-blue; *nilak,* bluish; Ar. *al nil,* indigo-plant: anil, aniline, acetanilid, carbanilide, sulfanilamide, anil indigo (*Indigofera suffruticosa*), lilac (*Syringa vulgaris*), *Bulbophyllum lilacinum* (an orchid).

Gr. *pelidnos; pelios,* livid, black and blue, grayish-blue: *Pelidnota punctata* (a beetle), *Pelidnotites atavus* (a fossil beetle), *Pelidna alpina* (dunlin), *Peliocichla saturata* (a bird), *Chrysopelea ornata* (a snake).

L. *sapphirus* (Gr. *sappheiros*), sapphire; *sapphirinus,* of sapphire, sapphire-blue; see **stone**

L. *sugillatum,* black and blue spot, bruise; see **bruise**

L. *venetus* (Gr. *benetos*), sea-colored, blue:

L. *viola,* violet; *violaceus,* violet-colored; see **violet**

See: **violet, sky, sea, green, color**

blunt < uncertain origin; see **dull**

blush < AS. *blyscan;* see **red**

blysma, Gr. a bubbling; see **bubble**

boa, L. a large serpent; see **snake**

boama, Gr. shriek, cry, < *boao,* cry, shout, roar; *boë,* a cry, shout; see *boö* under **roar**

boar < AS. *bar;* see **hog**

board < AS. *bord.*

L. *abacus* (Gr. *abax, -akos*), board, slab, counting-board, gaming-board divided into square compartments; see **number**

L. *assis,* m. board, plank; *assula,* f. dim.: ashlar.

Gr. *petauron,* n. perch, pole, springboard; see *petauristes* under **leap**

Gr. *pinax, -akos,* m. board, tablet, chart; *pinakion,* n. dim.: pinacoid, pinacocyte, *Pinacosaurus grangeri* (a fossil reptile).

Gr. *plakinos,* made of boards; see *plax* under **plate**

L. *planca,* f. board, slab: plank.

Gr. *sanis, -idos,* f. board, plank, tablet; *sanidion,* n. dim.; *sanidodes,* like a plank, flat: sanidine, *Sanidophyllum davidis* (a fossil coral).

Gr. *selis, -idos,* plank, page, sheet of paper; see **paper**

L. *tabula,* board, plank, table, a flat piece of any material on which to write; *tabella,* dim.; see **table**

L. *vibia,* f. plank, crosspiece:

See: **wood, beam, plate, flat**

boast < ME. *bosten;* see **brag**

boat < AS. *bat;* see **ship**

boatus, L. roar, shout, bellow; see *boo* under **roar**

bobbin < F. *bobine,* from a root meaning to wind up; see **axis**

boccalis, NL. (Gr. *bokkalis*), a kind of bird; see **bird**

body < AS. *bodig.*

Gr. *alibas, -antos,* m. dead body, corpse:

L. *cadaver, -is,* n. dead body, corpse, carcass; *cadaverinus,* of carrion; *cadaverosus,* like a corpse, ghastly: cadaver, cadaverous, cadaverine, *Dermestes cadaverinus* (a beetle).

Gr. *chros, -otos,* surface of the body, skin, flesh, color of the skin; see *chroma* under **color**

L. *corpus, -oris,* n. body, flesh, substance; *corpusculum,* n. dim.; *corporalis,* pertaining to the body; *corporeus,* fleshly; *corpulentus,* fleshy, fat; *incorporo, -atus,* embody: corpse, corps, corporal, corpuscle, corporation, corpulent, corselet, corset, corsage, incorporate, Corpus Christi (Texas), leprechaun, midriff, Sanskrit, *Leucosomus corporalis* (white chub), *Trophocrinus corpulentus* (a Mississippian crinoid), *Thlipsura corpulenta* (a fossil ostracode).

Gr. *demas, -mos,* n. the living body, frame: apodeme, *Demas coryle* (a moth), *Labidodemas sempenianum* (a holothurian).

AS. *lic,* body, corpse: lich gate, licham, hemlock(?).

ML. *mumia* (*mummia*), f. mummy: *Fistulana mumia* (a pelecypod).

Gr. *nekys, -yos,* m. corpse: *Necyomantes xanthaspis* (a beetle).
Gr. *ptoma, -tos,* n. that which has fallen, corpse, carcass: *Ptomatophagus consobrinus* (a beetle).
Gr. *skinar, -os,* n. the body:
Gr. *soma, -tos,* n. body, body substance; *somation,* n. dim; *somatikos,* of the body: somaplasm, somatic, somatology, chromosome, metasomatic, *Somatogyrus depressus* (a gastropod), *Somateria molissima* (eider-duck), *Piarosoma albicinctum* (a moth), *Strongylosoma persicum* (a milleped), *Trypanosoma avium* (a protozoan), *Heligmosomum costellatum* (a nematode), *Stiphrosomus fuscicornis* (a wasp).
See: **flesh, form, frame, thing**

boëtes, Gr. clamorer, clamorous; see *boö* under **roar**
boëthos, Gr. assisting, aiding; *boëthema,* aid, help, remedy; see **help**
bog < C. *bog,* marsh, moor; see **marsh**
boil < AS. *byl,* abscess; see **sore**
boil < L. *bullio, -itus,* bubble up; see *bulla* under **bubble**
Gr. *brazo,* boil; *brasmos,* m. ebullition, agitation; *brastikos,* of boiling, fermenting; *brastos,* boiling up: abrastol.
AS. *breowan,* ferment; see **yeast**
L. *coquo, coctus,* cook, boil; see **cook**
L. *elixo, -atus,* boil thoroughly, seethe; *elixus,* boiled, cooked: elixate, *Pecten elixatus* (a pelecypod).
L. *ferveo,* boil; *fervidus,* boiling, seething, passionate, hot; *fervor, -is,* m. boiling heat, ardor; *effervesco,* boil over, foam up, ferment: fervid, fervent, fervor, fever, ferment, Confervoideae.
Gr. *hepso; kathepso,* boil, seethe; *hepsesis,* f. a boiling; *hepsetes,* m. boiler; *hepsetikos,* of boiling; *hephthos; hepsanos; hepsetos,* boiled, hence languid: *Hephthopelta lugubris* (a crustacean).
Gr. *paphlasma, -tos,* n. a boiling, blustering:
Gr. *pepto,* boil, digest, ripen, soften; see **digest**
L. *scateo,* bubble out, gush forth; see **spring**
Gr. *zeo,* boil; *zema; apozema, -tos,* n. something boiled, a decoction; *zesis,* f. a boiling, seething; *zestos; ekzestos,* boiled: zeolite, apozem, eczema, zeal, *Zestocormus melanopus* (a wasp), *Galeorhinus zyopteris* (soupfin-shark).
See: **bubble, cook, heat, spring**

bolaco- < Gr. *bolax, -akos,* lump, clod; see *bolus* under **lump**
bolbitis, Gr. a kind of cuttlefish; *bolbidion; bolbition,* dim.; see **mollusk**
bolbos, Gr. a swelling, bulb; *bolbarion; bolbidion; bolbion; bolbiskos,* dim.; see **swell**
bold < AS. *bald, beald.*
Gr. *adeilos; adees; adeimos,* fearless; *adeia,* f. freedom from fear: *Adia oralis* (a fly).
Gr. *andristes,* m. a brave man: *androcardios,* manly-hearted, brave:
L. *animosus,* full of courage, bold, spirited; see *animus* under **life**
Gr. *aoknos,* without hesitation, brave: *Aocnus kolenati* (a beetle).
Gr. *aphobos; aphobetos,* fearless: *Aphobus uncinatus* (a fish), *Aphobetus maskelli* (a wasp).
Gr. *arrhenodes,* brave: *Arrhenodes sulcatus* (a beetle).
Gr. *atarbes; atarbetos; atarmyktos,* fearless, undaunted, unflinching, unwincing: *Atarba picticornis* (a fly), *Atarbodes pallidicornis* (a fly).
Gr. *athambos,* imperturbable, fearless:
Gr. *atrestos,* fearless:
Gr. *atromos; atrometos,* dauntless, fearless: *Atrometus insignis* (a wasp).
L. *audax, -acis,* daring; *audaculus,* dim.: audacious, audacity, *Phidippus audax* (a spider).
L. *confidens, -entis,* assured, bold, daring; see *fido* under **believe**
L. *confirmatus,* courageous, resolute: *Aoplus confirmatus* (a wasp).
Gr. *eupsychos,* courageous; see *psyche* under **mind**
L. *extrilidus,* unterrified, dauntless:
L. *ferox, -ocis,* brave, courageous, warlike, fierce; see *ferus* under **nature**
L. *fidentia,* confidence, boldness; see *fido* under **believe**
L. *fortis,* strong, brave; see **strong**
Gr. *heros,* hero; *heroikos,* of heroes; see **honor**
Gr. *hyperoplos,* trusting in force of arms, defiant: *Hyperoplus lanceolatus* (a fish).
L. *impavidus,* fearless, intrepid: *Didelphys impavida* (an opossum).
L. *imperterritus; interritus,* undaunted, unafraid:
L. *impudicus; impudens,* lewd, unashamed; see **lewd**

L. *intrepidus,* unshaken, undaunted, brave: intrepid, intrepidity.

Gr. *lema, -tos,* n. temper of mind, will, courage, audacity:

Gr. *liros,* bold, shameless:

Gr. *meneptolemos,* steadfast, brave, staunch:

Gr. *parabolos,* bold, reckless, venturesome: *Parabolocratus glaucescens* (a bug).

Gr. *phertos,* brave; *pherteros,* braver; *phertatos; pheristos,* bravest: *Pherterus brasiliensis* (a grasshopper).

L. *procax, -acis,* shameless, bold, saucy, impudent, insolent: procacious.

L. *protervus,* bold, impudent: *Tanymeces protervus* (a beetle).

L. *resolutus,* bold, firm: resolute.

L. *temeritas, -atis,* f. venturesome boldness, rashness: temerity, temerous.

Gr. *tharsos (tharrhos),* n. courage, boldness; *tharsaleos (tharrhaleos),* bold, daring; *eutharses,* of good courage; *peritharses,* very bold, confident: *Tharrhaleus koslowi* (a bird).

Gr. *thrasys,* bold, courageous, bragging; *thrasos, -eos,* n. boldness, courage, impudence: thrasonical, *Thrasycephalus guttatus* (a beetle).

Gr. *tlethymos,* stouthearted, staunch; see *thymos* under **mind**

Gr. *tolma, -tos,* n. courage, boldness; *tolmeros,* daring; *eutolmos,* courageous: *Tolmerolestes rubripes* (a fly), *Eutolmaetus pennatus* (a hawk-eagle).

See: **fight, strong, fear, danger**

bole, Gr. throw, cast, stroke, wound; *bolos,* throw, cast, the thing caught; see *ballo* under **throw**

boletus, L. (Gr. *bolites*), mushroom; see **fungus**

bolido- < Gr. *bolis, -idos,* missile; see **missile**

bolitaena, NL. (Gr. *bolitaina*), a kind of cuttlefish; see *bolbitis* under **mollusk**

boliton; bolitos, Gr. cow dung; see **dung**

bolt, AS. projectile, fastening; see **arrow, bar**

bolus, L. (Gr. *bolos*), lump, clod, choice morsel, see **lump**; throw, cast, see *ballo* under **throw**

bomaco- < Gr. *bomax, -akos,* beggar; see **beggar**

bombax, LL. cotton; see **cotton**

bombus, L. (Gr. *bombos*), a booming, humming, buzzing; *bombinator,* buzzer, hummer; see **murmur**

bombycilla, NL. a genus of waxwings; see **waxwing**

bombylios, Gr. bumblebee; see **bee**

bombyx, Gr. silk, silkworm; *bombycinus,* of silk, silken; see **silk**

bomolochos, Gr. lickspittle, toady, beggar, buffoon; see **flatter**

bomos, Gr. base, stand, altar; see **altar**

bonasus, L. (Gr. *bonasos*), wisent, bison; see **cattle**

bond < the root of AS. *bindan,* bind; see **bind**

bone < AS. *ban:* bony, boneset, bonfire, jawbone, backbone, T-bone steak.

L. *os, ossis,* n. bone; *ossiculum,* n. dim.; *osseus,* of bone; *ossosus; ossuosus,* bony; *ossuarius,* of or for bones, urn, vault; *exosso, -atus,* deprive of bones, bone: ossicle, ossify, osseous, ossification, ossuary, Key West (Fla.), osprey (*Pandion haliaëtus*), *Narthecium ossifragum* (a bog-asphodel), *Lepidosteus osseus* (long-nosed gar).

Gr. *osteon,* n. bone; *ostarion,* n. dim.; *ostinos,* of bone; *osteodes,* bonelike, bony: *anosteos,* boneless: osteology, osteitis, osteopath, anostosis, osteotomy, osteoid, osteomyelitis, teleost, *Osteochilus vittatus* (a fish), *Ostinaspis coronata* (a fossil fish), *Ostinops oleagineus* (a bird), *Stellaria holostea* (a chickweed), *Gasterosteus aculeatus* (stickleback).

Gr. *skeleton,* dried body, mummy, bony framework; see **frame**

bonus, L. good; *melior,* better; *optimus,* best; *bonitas,* goodness; see **good**

book < AS. *boc:* bookish, bookbinder, bookkeeper, bookcase, bookworm, notebook, beech, buckwheat.

L. *album,* n. a white tablet: album.

L. *annalis,* m. yearly record: annals.

Gr. *biblion,* n. dim. of *biblos,* f. paper, book, scroll, < *byblos,* the sedge, *Cyperus papyrus; biblis, -idos,* f.; *biblarion; biblidion,* n. dim.; *bibliographos,* m. writer of books; *bibliodetes,* m. bookbinder; *bibliokapelos; bibliopoles,* m. bookseller; *bibliotheke,* f. bookcase, library: bibliography, bibliophile, bibliomaniac, Bible, *Biblomimus minutus* (a beetle).

L. *codex, -icis,* m. tablet, book, ledger; *codicillus,* m. dim.: code, codicil, codify.

L. *conjectaneum,* m. memorandum book, book of miscellanies:

Gr. *deltos,* f. writing; *deltarion; deltion,* n. dim.:

L. *diarium,* n. daybook, journal: diary.

Gr. *encheiridion,* n. handbook, manual:

Gr. *ephemeris, -idos,* daybook, diary, journal; see *hemera* under **day**

L. *liber, -bri,* m. book; *libellus,* m. dim.; *librarius,* pertaining to books: library, librarian, *ex libris,* libretto, libel, libelous.

Gr. *opsartysia,* f. cookbook:

Gr. *pandektes,* m. encyclopedia: pandect.

L. *pittacium* (Gr. *pittakion*), n. tablet, slip of parchment, ticket, label:

Gr. *plax, -akos,* anything flat and broad, plain, stone, tablet; see **plate**

L. *pugillar, -is,* m. writing tablet:

L. *tabula,* board, a flat piece of any material on which to write; *tabella,* dim.; see **table**

Gr. *tessera,* square tablet, token; see **four**

L. *tomus* (Gr. *tomos*), m. part, book, volume; *tomarion,* n. dim.: tome.

L. *volumen, -inis,* n. book, roll, writing: volume, voluminous.

See: **paper**

boot < OF. *bote;* see **shoe**

bora, Gr. meat, food; see *bibrosko* under **eat**

borage < OF. *bourrache,* < ML. *borrago, -inis,* f. prob. from a word meaning rough-hairy: Boraginaceae, *Borago officinalis* (a borage).

L. *alcibius* (Gr. *alkibios*), f. a kind of bugloss:

L. *anchusa* (Gr. *anchousa*), f. a genus of the borage family: *Anchusa officinalis* (bugloss).

L. *arcebion,* n. a kind of borage:

NL. *arnebia,* f. < Ar. *arnabiyah,* a plant of the borage family: *Arnebia cornuta* (Arabian arnebia).

L. *asperugo, -inis,* f. a plant with rough leaves, a borage: *Asperugo procumbens* (madwort).

Gr. *bouglossos,* m. oxtongue, a boraginaceous plant: bugloss (*Anchusa capensis*).

L. *burdunculus,* m. a plant, probably borage:

L. *cerintha* (Gr. *kerinthe*), f. waxwort: *Cerintha retorta* (Greek waxwort).

L. *cynoglossum* (Gr. *kynoglosson*), n. houndstongue: *Cynoglossum officinale* (common houndstongue).

L. *echium* (Gr. *echion*), n. bugloss: *Echium vulgare* (viper's-bugloss).

L. *enchrysa,* f. a kind of borage:

L. *heliotropium* (Gr. *heliotropion*), n. a genus of the borage family: heliotropine, *Heliotropium spathulatum* (a heliotrope).

L. *leucrium,* n. a kind of houndstongue:

L. *lithospermum* (Gr. *lithospermon*), n. gromwell: *Lithospermum officinale* (gromwell).

Gr. *lycopsis,* f. wild bugloss: *Lycopsis arvensis* (wild bugloss).

NL. *mertensia,* f. a genus of the borage family, < Franz C. Mertens, German botanist: *Mertensia virginica* (Virginia bluebells).

L. *mulcetra,* f. heliotrope:

Gr. *myosotis, -idos,* f. mouse-ear, forgetmenot: *Myosotis laxa* (bay-forgetmenot), *Anchusa myosotidiflora* (alkanet).

Gr. *onophyllon,* n. a kind of borage:

Gr. *onosma, -tos, n.* a plant of the borage family: *Onosma stellatum* (golddrop), *Onosmodium occidentale* (western marbleseed).

Gr. *phlonitis, -idos,* f. a kind of borage:

Gr. *salyx,* f. a kind of borage:

L. *solago, -inis,* f. heliotrope:

L. *symphytum* (Gr. *symphyton*), n. comfrey: *Symphytum officinale* (comfrey).

borassos, Gr. palm fruit; see **palm**

boraton, Gr. a kind of juniper; see **juniper**

borax < Per. *burah:* boron, boronic, borocalcite, boric acid, boracic, boracite, borate.

L. *santerna,* f. borax:

borboros, Gr. mud, mire, filth; see **dirt**

borborygmos, Gr. a rumbling in the bowels; see **thunder**

border < OF. *bordeure,* outer edge, margin, brink, rim, verge: borderland.

L. *ambitus,* circuit, circumference, periphery; see **around**

Gr. *ambon, -os,* m. rim, edge, ridge, crest: ambon, *Ambocoelia curta* (a fossil brachiopod).

Gr. *antyx, -ygos,* f. edge, rim, frame: *Salpis(Antygophanes) orbifera* (a moth).

Gr. *apolegma, -tos,* n. hem of a robe:

L. *balteus*, girdle, belt, border, edge; see **belt**
L. *canthus* (Gr. *kanthos*), m. tire of a wheel, rim, edge, corner: cant, decant, decanter, epicanthus, scantle, *Canthigaster cincta* (a fish), *Eurycantha horrida* (a phasmid).
L. *confinis*, bordering; see *finis* under **end**
L. *coronula*, dim. of *corona*, crown, rim, border; see **crown**
AS. *ecg*, point, sharp border or margin; see **point**
AS. *efes*, brim, brink, lower border of a roof: eaves.
Gr. *eschatia*, f. edge, border, outskirt:
L. *fimbria*, fiber, thread, shred, fringe, border, edge; *fimbriatus*, fringed, fibrous; see **fringe**
L. *finis*, boundary, limit, end; *confinis*, bordering; see **end**
Gr. *geisson* (*geison*), n. eaves, cornice, hem, border; *apogeisoma, -tos,* n. cornice, coping: *Geissoloma marginatum* (a thymeleacead), *Geissorhiza pusilla* (an iridacead), *Geisonoceras[Gisoceras]* *timidum* (a Silurian cephalopod), *Anogeissus latifolia* (axlewood).
AS. *hem*, edge, border, margin: hem, hemstitch.
Gr. *horos*, m. limit, boundary; *horios*, of boundaries; *horismos*, m. definition, setting of a boundary; *horistes*, m. definer; *horistikos*, defining; *horistos*, definable: horograph, horopter, horizon, aorist, aphorism, diorite, orismology, *Penthorum humile* (Amur stonecrop).
L. *impages, -is*, f. border or frame around the panel of a door: *Pleurotoma impages* (a gastropod).
L. *instita*, f. border, flounce:
Gr. *itys, -yos*, f. rim or felly of a wheel: *Itys torquatus* (a bird), *Ityocephala falcata* (a grasshopper).
Gr. *kraspedon*, n. edge, border: craspedote, craspedodrome, *Craspedophyllum americanum* (a fossil coral), *Craspedia uniflora* (a composite), *Phyllocraspedum interjectum* (a bug).
Gr. *kosymbos*, m. edge, border; *kosymbotes*, fringed, tasseled: *Cosymbotes platyurus* (a lizard).
Gr. *krossos*, fringe, tassel; see **fringe**
Gr. *kyklas, -ados*, garment with an encircling border; see *kyklos* under **circle**
L. *labrum*, lip, brim; see **lip**
Gr. *legnon*, n. edge, border; *legnotos*, with a colored border: *Legnotomyia cineracea* (a fly), *Legnophyllum primum* (a fossil coral), *Trogolegnum preambulyx* (a moth), *Saprolegnia monilifera* (a fungus).
L. *limbus*, m. border, hem, fringe, selvage; *limbatus*, bordered: limbate, limbo, limbous, limbus, *Phyllobates limbatus* (a frog), *Canna limbata* (Parana canna), *Phonergates(Clopophora) limbativentris* (a frog), *Halictus limbellus* (a bee).
L. *limes, -itis*, m. boundary; *limitaneus; limitaris*, on the border; *limito, -atus,* enclose within limits, fix, determine: limit, limitation, illimitable, unlimited.
Gr. *loma, -tos*, n. fringe, hem, border, coast; *lomation*, n. dim.: loma, lomastone, *Lomatodactylus cepedianus* (a lizard), *Lomatium orientale* (an umbellifer), *Lomatia ferruginea* (a proteacead), *Campeloma nebrascensis* (a fossil snail), *Tricholoma personatum* (a mushroom).
L. *margo, -inis,* c. border, edge; *margino, -atus,* furnish with a border: margin, marginal, emarginate, *Primula marginata* (a primrose), *Dytiscus marginalis* (a diving-beetle), *Leptysma marginicollis* (a grasshopper), *Marginulina vulgata* (a foraminifer), *Marginella limatula* (a fossil gastropod), *Emarginula striatula* (a gastropod), *Upupa marginata* (Madagascar hoopoe), *Perla immarginata* (a stonefly).
Gr. *messoros*, boundary-stone; see **stone**
Gr. *oa*, f. hem, border: *Oabius pylorus* (a centiped).
L. *ora*, f. extremity, edge, margin, coast; *orarius,* of the coast: orle, orlet, oriconic.
L. *paragauda*, f. border or lace on a garment:
Gr. *parorophis, -idos,* f. eaves, cornice:
Gr. *paryphe*, f. border woven along a robe; *euparyphos*, with a fine border: paryphodrome, *Paryphostomum radiatum* (a worm), *Paryphanta superba* (a gastropod), *Paryphodes werneri* (a fly), *Spatalistis paryphoea* (a moth), *Euparyphus bellus* (a soldierfly).
L. *patagium*, n. a gold edging or border; *patagiatus,* bordered: *Patagium brachydelphium* (a trematode), *Patagifer bilobus* (a flatworm), uropatagium.
Gr. *perimetron*, circumference; see **circle**
Gr. *periphereia*, circumference; see **circle**
Gr. *peza; pezis, -idos,* f. border, edge, ribbon; *pezidion,* n. dim.: *Haplopeza violacea* (a beetle).
L. *protectum,* n. eaves:
L. *revimentum,* n. lappet, edging, fringe:

Gr. *sotron,* n. felly of a wheel:

L. *suggrunda,* f. eaves: *Suggrunda porosa* (a Miocene foraminifer).

Gr. *synoria,* borderland; see **country**

L. *terminus,* boundary, limit, end; see **end**

Gr. *thrinkos,* m. eaves, cornice, coping; *thrinkion,* n. dim: *Thrincophora impletana* (a moth), *Thrincostoma malelanum* (a bee).

Gr. *thysanos,* tassel, fringe; see **fringe**

L. *toral, -alis,* n. valance, border drapery:

See: **lip, end, shore, circle, side, door, touch, point**

bore < AS. *borian.*

Gr. *aris, -idos,* f. auger:

Gr. *choinikis, -idos,* f. trepan:

L. *confossus,* pierced through, full of holes; see **hole**

Gr. *diamperes,* piercing:

Gr. *diaprysios,* piercing, thrilling:

L. *foro, -atus,* bore, pierce; *forabilis,* penetrable; *perforo,* bore through; *perforaculum,* n. auger: perforate, foramen, foraminifer, *Hypericum perforatum* (a St. Johnswort), *Xyleborus perforans* (an ambrosia-beetle).

Gr. *kenteo,* goad, prick, spur; *kentor,* goader, driver, piercer; see **point**

L. *modiolus,* trepan; see **box**

Gr. *nysso; nytto,* prick, spur, pierce, puncture; *nygma, -tos,* n.; *nyxis,* f. puncture, prick, sting: pyronyxis, scleronyxis, *Dermatonyssus gallinae* (poultry-mite), *Nyxeophilus bimaculatus* (a wasp), *Philonygmus alaskensis* (a wasp), *Nyxetes bidens* (a beetle), *Nygmatonchus scriptus* (a worm), *Metanysson solani* (a wasp).

Gr. *peiro,* pierce: *Porpita pacifica* (a siphonophore).

L. *pertundo, -tusus,* beat, push, or thrust through, perforate; *Pertunda,* f. goddess who presided over the loss of virginity: pertuse, pertusion, *Pertusaria communis* (a lichen).

Gr. *pholas, -ados,* a rock-boring mollusk; see **mollusk**

L. *pungo, punctus,* pierce, prick, penetrate; see **point**

Gr. *stizo,* prick, puncture, mark; see *stigma* under **mark,** *stiktos* under **spot**

Gr. *teiro; tetraino,* bore, pierce; *teretron,* n. auger, borer, gimlet; *teretrion,* n. dim.; *tresis,* f. a boring; *tretos,* bored, perforated; *anatresis,* f. a boring through: atresia, *Acrotreta attenuata* (a fossil brachiopod), *Tretocalyx polae* (a sponge), *Tretopileus opuntiae* (a fungus), *Teretrum quiescitum* (a fossil beetle), *Heptatretus profundus* (a hagfish).

L. *terebro, -atus,* perforate, pierce; *terebra,* f.; *terebrum,* n. gimlet, auger, borer: *Terebratula grandis* (a fossil brachiopod), *Terebratulina septentrionalis* (a brachiopod), *Nerius terebrans* (a fly), *Cuterebra cuniculi* (rabbit-botfly).

L. *teredo,* woodworm, borer, shipworm; see **mollusk**

Gr. *toreo,* bore, pierce; *toros,* m. borer; *toretos,* bored, pierced: *Torostoma apicale* (a beetle), *Toretocnemus californicus* (an ichthyosaur).

Gr. *trypao,* bore, perforate; *trypanon,* n. auger, gimlet; *trypetes,* m. borer; *trypetos,* bored: trepan, trypanosome, antrycide (for nagana), *Trypoxylon fabricator* (a wasp), *Trypanosoma cruzi* (protozoan causing Chagas disease), *Trypetes rhinoides* (a beetle), *Trypeta corruca* (a fly), *Proctotrupes[Proctotrypes] abruptus* (a mollusk), *Trypanea flavivena* (a fly), trypaneid, Proctotrypidae, *Trypetomima solitaria* (a fly), *Trypanus cossus* (a moth), *Geotrypetes seraphini* (a caecilian).

See: **hole, push, turn, enter, awl**

boreas, Gr. north wind, north; *boreios,* northern; L. *borealis,* northern; see **north**

boreus, Gr. a mullet; *boridion,* dim.; see **mullet**

born < AS. *beran,* carry; see **birth**

boro < AS. *burg,* fort, city; see **town**

boron < Per. *burah,* borax; see ELEMENTS under **thing**

boros, Gr. devouring, greedy, gluttonous; see *bibrosko* under **eat**

borrow < AS. *borgian;* see **trade**

bos, *bovis,* L. (Gr. *bous*), ox, cow; see **cattle**

bosco- < Gr. *boske,* food, fodder; *bosko,* feed, pasture, maintain; see **eat**

bosmorum, NL. (Gr. *bosmoron*), a kind of Indian grain; see **grain**

bosom < AS. *bosm,* breast; see **breast**

bostrychos, Gr. curl, lock of hair, anything twisted, see **curl;** a kind of winged insect, see **insect**

botany < F. *botanique,* < Gr. *botanikos,* of herbs, < *botane,* herb, grass, pasture; see **plant**

botaurus, NL. a genus of bitters; see **heron**

bother < uncertain origin; see **tease, trouble**

bothros, Gr. trench, pit; *bothrion*, dim.; see **ditch**

bothynos, Gr. hole, pit, trench; *bothynotes*, ditcher, digger; see **ditch**

boton, Gr. a fed beast; see **cattle**

botrychium, NL. a genus of ferns; see **fern**

botrys, Gr. bunch or cluster of grapes; *botrydion*, dim.; *botrychos*, grapevine; see **grape**

bottle < OF. *bouteille*.
  Gr. *alabastron*, n. flask or vase for ointments and perfumes: alabastrum.
  L. *ampulla*, f. flask, bottle; *ampullaceus*, flasklike; *ampullarius*, m. flask-maker: *Saccopharynx ampullaceus* (bottlefish), *Nepenthes ampullaria* (a pitcher-plant), *Chalara ampullula* (a fungus).
  Gr. *baukalion*, n. narrow-necked vessel:
  L. *guttus*, m. a narrow-necked flask:
  L. *hirnea*, f. jug; *hirnula*, f. dim.: *Hirneola polytricha* (a fungus).
  Gr. *kyrillion*, n. a jug with a narrow neck: *Cyrillium purpureum* (a gastropod).
  L. *lagena*, f. (Gr. *lagenos; lagynos*, m. flagon, flask; *lagenion*, n. dim.), large jar or bottle with handles and a narrow neck; *laguncula*, f. dim.: lagenoid, *Lagena globosa* (a foraminifer), *Lagenula reticulata* (a foraminifer), *Laguncularia racemosa* (white mangrove), *Lagenidium entophytum* (a fungus), *Lagenaria vulgaris* (calabash), *Mangifera lagenifera* (a mango), *Lagynocystis pyramidalis* (an Ordovician cystid), *Lagynopora lagena* (a Cretaceous brachiopod), *Lagynophora symmetrica* (a Paleocene characead), *Colletotrichum lagenarum* (a fungus).
  L. *lecythus* (Gr. *lekythos*), f. flask, bottle, vase; *lekythion*, n. dim.: *Lecythocrinus briareus* (a Devonian crinoid), *Lecythoplastes fuliginosus* (a bird), *Lecythis laevifolia* (monkeypot), *Lecythiocrinus urnaeformis* (a Pennsylvanian crinoid).
  L. *obba*, f. beaker, decanter: *Obba planulata* (a gastropod), *Obbinula* (a gastropod).
  Gr. *olpe; olpis, -idos*, f. leather flask, vessel; *olpidion*, n. dim.: *Olpidium brassicae* (a fungus), *Pleolpidium marinum* (a fungus).
  Gr. *oxis, -idos*, f. vinegar-jug, cruet:
  Gr. *phlaske*, f. wine flask; *phlaskon*, n. flagon; *phlaskion*, n. dim.:
  Gr. *pytine*, f. flask covered with plaited work: *Sethocapsa pytine* (an Ordovician radiolarian).
  Gr. *stamnos*, m. earthen jar or bottle: *Stamnocnemis* (a fossil sponge).
  See: **pot, bag, cup, vessel**

bottom < AS. *botm;* see **base, deep, under**

botulus, L. sausage; *botellus*, dim.; see **sausage**

boundary < ML. *butina*, limit; see **border, end**

bovinus, L. pertaining to cattle, < *bos, bovis*, ox, cow; see *bos* under **cattle**

bow < AS. *boga;* see **arc**

bowel < L. *botulus*, sausage, intestine; see **intestine**

bowl < AS. *bolla;* see **cup, basin**

box, a container, < AS. *box*, < L. *buxus*, m. boxwood; *buxeus*, of boxwood: box office, box score, bushing, blunderbuss, buxine, *Buxus sempervirens* (boxwood), *Leiophyllum [Liophyllum] buxifolium* (sand-myrtle), *Berberis buxifolia* (Magellan barberry), *Polygala chamaebuxus* (groundbox-polygala).
  L. *acerra*, f. incense-box:
  Gr. *achane*, f. chest, box:
  L. *arca*, f. chest, box, coffer, casket, coffin; *arcella; arcellula; arcula*, f. dim.; *arcarius*, relating to boxes: ark, *Arca transversa* (an arkshell), *Arcella dentata* (a protozoan).
  L. *armarium*, n. closet, cupboard, chest, safe; *armariolum*, n. dim.:
  AS. *binn*, box, crib: corn bin, coal bin.
  L. *capsa*, f. case, box; *capsella; capsula*, f. dim.; *capsus*, m. wagon body: capsule, case, casket, cash, cashier, caisson, chassis, capsicum, *Elymocaris capsella* (a Devonian crustacean), *Histoplasma capsulatum* (a fungus), *Gloeocapsa aurata* (an alga), *Corchorus capsularis* (jute plant).
  L. *capulus*, m. coffin: *Capulus intortus* (a gastropod).
  L. *cella*, chamber; *cellula*, dim.; see **room**
  Gr. *chelos*, f. large chest, coffer:
  Gr. *chnoë*, f. hub or nave of a wheel:
  Gr. *choinix, -ikos*, f. a dry measure: *Choenicosphaera flosculenta* (a radiolarian).

box 160

L. *cista* (Gr. *kiste*), f. box, chest; *cistella; cistellula; cistula*, f.; *kistidion*, n. dim.: cist, cistern, cistella, chest, cistophore, cistophorus, *Cistella neapolitana* (a brachiopod).

L. *cophinus* (Gr. *kophinos*), basket, hamper; see **basket**

L. *cumera*, f. chest, box, basket:

L. *forulus*, m. bookcase:

L. *fritillus*, dice-box; spotted, see **spot**

Gr. *halia*, f. saltcellar:

Gr. *kadiskos*, urn or box to receive ballots, < *kados*, jar, urn, box; see **jar**

Gr. *kampsa*, f. case, casket; *kampsion*, n. dim.: *Campsiomorpha lata* (a beetle).

Gr. *kape*, crib, manger; see **basin**

Gr. *kethis*, *-idos*, f. ballot-box, dice-box; *ketharion; kethion; kethidion*, n. dim.:

Gr. *kibotos*, f. box, chest; *kibotion*, n. dim.: *Cibotium glaucum* (a fern).

Gr. *koite*, f. chest, box, casket; *koitis*, *-idos*, f. dim.:

Gr. *kypsele*, f. box, chest, beehive; *kypselion*, n. dim.: cypsela.

Gr. *kytis*, *-idos*, f. small chest, box, cell: *Cytidium pulcherrimum* (a myxomycete).

Gr. *larnax*, *-akos*, f. box, chest, coffer; *larnakion; larnakidion*, n. dim.: *Larnacospongus tetrapylifer* (a radiolarian), *Larnacantha bicruciata* (a radiolarian), *Spongophortis larnacilla* (a radiolarian).

L. *loculus; locellus*, m. cell, compartment, box, purse, dim. of *locus*, place: locule, loculicidal, *Ctenobolbina loculata* (an ostracode), *Spiroloculina pulchella* (a foraminifer), *Petalostigma quadriloculare* (a euphorbiacead), *Lonicera quinquelocularis* (Himalayan honeysuckle).

Gr. *lophis*, *-idos*, f.; *lopheion*, n. crest-case, any case:

L. *modiolus*, nave or hub of a wheel, box; see *modus* under **measure**

L. *narthecium* (Gr. *narthekion*), n. ointment-box, medicine-chest: *Narthecium americanum* (a bog-asphodel).

Gr. *nekrotheke*, f. coffin:

L. *phimus* (Gr. *phimos*), dice-box; see **cup**

Gr. *plemne*, f. hub or nave of a wheel: *Plemnocrinus bullatus* (a fossil crinoid).

L. *pyxis*, *-idis* (Gr. *-idos*), f. box; *pyxos*, f. boxwood; *pyxidicula*, f.; *pyxidion; pyxion*, n. dim.; *pyxinos*, of boxwood: pyx, pyxis, pyxidate, pyxidium, pyxie (*Pyxidanthera barbulata*), *Acrotreta pyxidicula* (a fossil brachiopod), *Pyxipoma multistriata* (a gastropod), *Pyxidognathus granulosus* (a crustacean), *Pyxidium inclinans* (a protozoan), *Davallia pyxidata* (a fern), *Pyxine sorediata* (a lichen).

L. *riscus* (Gr. *rhiskos*), m. trunk, chest, coffer; *risculus*, m. dim.: *Risculus molvae* (a crustacean), *Rhiscocoma alpestre* (a milleped), *Rhiscosomides meineri* (a milleped).

L. *salinum*, n. saltcellar: *Acrydium salinum* (a grasshopper).

L. *sarcophagus* (Gr. *sarkophagos*), m. coffin, sepulcher: sarcophagus.

ML. *scatula*, f. a kind of pill-box:

L. *scrinium*, n. cylindrical case, book-box, letter-case, portfolio: scrinium.

Gr. *skiraphos*, m. dice-box:

Gr. *soros*, f. coffin, cinerary urn; *coridion*, n. dim.:

L. *theca* (Gr. *theke*), case, container, envelope, sheath; see **bag**

L. *unguilla*, f. ointment-box:

Gr. *zygastron*, n. box, chest, archives; *zygastrion*, n. dim.: *Zygastropyga aurea* (a fly).

See: **bag, basket, room, pen, cradle, basin**

box < uncertain origin, see **strike**; < L. *box, bocis*, a fish, see **fish**

boy < uncertain origin; see **child**

brabeus; brabeutes, Gr. judge, umpire at the games, see **judge**; *brabeion*, prize at the games, see **gain**

brabylos, Gr. wild plum-tree; see **plum**

braca, L. trousers, breeches; see **trousers**

bracelet < OF. *bracel*, armlet, < L. *brachium*, arm.

Gr. *ampholenion*, n. bracelet:

L. *armilla*, f. bracelet; *armillatus*, ornamented with a bracelet: *Armillaria dimidiata* (a mushroom), *Cortinarius armillatus* (bracelet-mushroom).

Gr. *brachionister*, *-os*, m. armlet:

Gr. *chlidon*, *-os*, m. bracelet, anklet, ornament: *Terebratula(Chlidonophora) incerta* (a brachiopod).

Gr. *choinix*, *-ikos*, shackle; see **box**

L. *compes*, *-edis*, fetter, shackle; see *pedica* under **bind**

L. *dextrale*, *-is*, n. bracelet:

L. *galbeus*, m. armband, fillet:

Gr. *klanion*, n. bracelet:

Gr. *maniakes*, m. armlet, bracelet:

Gr. *pselion* (*psellion*), n. armlet, anklet: *Pselliophorus tibialis* (yellow-thighed sparrow), *Pselioceras ophioneum* (a fossil cephalopod).
L. *spatalium* (Gr. *spatalion*), n. bracelet:
Gr. *sphingion*, n. bracelet, necklace:
L. *spinter, -is*, n. bracelet:
L. *viria*, f. armlet, bracelet; *viriola*, f. dim.: ferrule:
See: **garter, ornament**

brachium, L. (Gr. *brachion*), arm; *brachiatus*, with arms or branches; see **arm**

brachys, Gr. short; *brachyteros*, shorter; *brachistos*, shortest; see **short**

bracon, NL. a genus of wasps; see **wasp**

bractea, L. thin metal plate, gold leaf, scale, small leaf; *bracteola*, dim.; see **scale**

bradys, Gr. slow; *bradyteros*, slower; *bradistos*, slowest; see **slow**

brag < OF. *braguer*, flaunt.
Gr. *alazon*, boaster, quack; *alazonikos*, boastful; see **lie**
Gr. *anemolios*, windy, boastful:
Gr. *apeile*, f.; *apeilema, -tos*, n. boast, threat; *apeiletikos*, threatening: *Apeilesis cinerea* (a fly).
Gr. *aretalogos*, m. braggart about virtue:
Gr. *aucheo*, boast, brag; *auchema, -tos*, n. boast; *auchetes*, m. boaster, braggart; *auchematikos; philauchos*, boastful; *megalauchia*, f. great boasting:
Gr. *euchos; euchole*, prayer, boast; see *euche* under **ask**
Gr. *gaurex, -ekos*, m. braggart:
L. *glorior, -atus*, boast, brag; *gloriator, -is*, m. braggart: glory.
L. *jacto, -atus*, boast of oneself, be ostentatious; *jactator, -is*, m. braggart:
Gr. *kompos*, m. boast, noisy boastful words; *kompastes*, m. braggart; *kompastikos*, of boasting; *komperos*, boastful, < *kompeo*, talk big, boast: *Composuchus solus* (a fossil reptile), *Compastes botanicus* (a bug).
Gr. *lapistes* (*lapithes*), m. braggart, swaggerer; *lapisma, -tos*, n. a boasting, swaggering, < *lapizo*, whistle, swagger: *Lapithes pulicarius* (a psocopterid).
Gr. *lekythistes*, m. loud talker, braggart:
L. *magnidicus*, talking big, boastful, bragging:
L. *ostento, -atus*, show off, boast, vaunt: ostentation.
L. *salaco, -onis* (Gr. *salakon, -os*), m. braggart:
Gr. *stomphos*, bombastic; *stomphastes; stomphax, -akos*, m. bombastic talker, braggart, < *stomphazo*, speak mouthfuls: *Stomphosphinctes stomphus* (a fossil cephalopod), *Stomphax crucirostris* (a beetle).
Gr. *thrasys*, bold, bragging; see **bold**
See: **exceed, speak**

braid < AS. *bregdan*, weave; see **weave**

brain < AS. *braegen*.
L. *cerebrum*, n. brain; *cerebellum*, n. dim.: cerebrum, cerebellum, cerebral, cerebration, cerebriform, cerebrin, cerebrose, cerebrospinal, *Pylobotrys cerebralis* (a radiolarian), *Coniophora cerebella* (a fungus), *Cerebratulus lacteus* (a nemertean).
Gr. *enkaros*, m. brain:
Gr. *enkephalos*, m. brain, pith of palm: encephalitis, electroencephalograph, *Encephalartos cycadifolius* (a cycad).
Gr. *enkranion*, n. cerebellum:
Gr. *kottis, -idos*, f. cerebellum:
Gr. *phren, -os*, mind, brain; see **mind**

brake < AS. *breccan;* see **hold**

bramble < AS. *bremel;* see **blackberry, bush**

bran < OF. *bren;* see **scale**

branch < OF. *branche*.
Gr. *akremon, -os*, m. bough, branch, spray: *Acremonium alternatum* (a fungus), *Acremoniella fusca* (a fungus).
L. *baia* (Gr. *bais*), palm branch; see **palm**
Gr. *blastema, -tos,* offshoot, sucker; see *blastos* under **bud**
L. *clavula*, scion, graft; see *clava* under **club**
L. *cyma*, f. young sprout; *cymula*, f. dim.; *cymosus*, full of shoots: cyme, cymose, cymule, cymulose, *Begonia brevicyma* (a begonia), *Dichapetalum cymosum* (an African plant), *Cymaria acuminata* (a mint).
Gr. *ernos*, n.; *ernyx, -ygos*, m. sprout, shoot; *euernes*, sprouting well; *katernes*, with luxuriant branches: *Ernobius mollis* (a beetle), *Euernia furfuracea* (a lichen).
L. *frons, frondis*, leafy branch, foliage; see **leaf**

L. *germen, -inis,* sprig, sprout, bud; *germino, -atus,* sprout forth; see **bud**

Gr. *klados,* m. branch, twig, stem; *kladion,* n.; *kladiskos,* m. dim.: cladose, cladophyll, cladode, cladanthous, cladoceran, *Cladophora sericea* (an alga), *Cladothrix suffruticosa* (an amaranth), *Phyllocladus glaucus* (a celery-pine), *Sparganium androcladum* (a burreed), *Cladium effusum* (a saw-sedge), *Cosmocladium constrictum* (a desmid), *Cladonia pyxidata* (a lichen), *Cladrastis sinensis* (a yellow-wood), *Cladarachnium ramosum* (a radiolarian), *Tetracladium setigerum* (a fungus).

Gr. *klema, -tos,* n. vine-twig, slip, cutting; *klematidion; klemation,* n. dim.; *klematis, -idos,* f. a plant with lithe branches, now a genus of the buttercup family: *Clematograptus multifasciatus* (a graptolite), *Clematis vitalba* (a clematis), *Mitoclema cinctosum* (a fossil bryozoan).

Gr. *klon, -os,* m. young shoot, sprout, twig; *klonion,* n. dim.: clon, clone, cloning, *Clonothrix ferruginea* (an iron-depositing bacterium), *Cloniophora subfasciata* (a fly), *Rhizoclonium fontanum* (an alga).

Gr. *kolon,* member, limb; see **leg**

Gr. *krade,* f.; *krados,* m. branch, spray, twig: *Cradodesmus subspinosus* (a milleped), *Cradeocrinus elongatus* (a Devonian crinoid), *Taeniocrada lesquereuxi* (a fossil alga).

Gr. *orodamnos,* m. bough, branch:

Gr. *oschos,* m. young branch, shoot: Oschophoria.

Gr. *ozos,* m. branch, bough, twig, bud; *ozaleos; ozotos,* branching, branched:

L. *palmes, -itis,* young branch or shoot of a vine, bough; see **palm**

L. *paniculus,* tuft, thatch; see **bush**

Gr. *phryganon,* dry stick, brushwood; see **wood**

Gr. *phyas; paraphyas, -ados,* f. shoot, sucker; *paraphyadion,* n. dim.: *Paraphyas callixema* (a moth).

L. *propago, -inis,* f. sucker, layer, shoot; *propago, -atus,* set in slips, procreate: propagate.

Gr. *ptorthos,* m. young branch, shoot, sucker: *Ptorthodius ramosus* (a beetle).

Gr. *pyrdalon,* firewood, brushwood; see **wood**

L. *ramus,* m. branch; *ramulus; ramusculus,* m. dim.; *ramosus,* branchy: ramus, ramus communicans, ramicorn, ramify, ramification, *Francoa ramosa* (a saxifragacead), *Equisetum ramosissimum* (a horsetail), *Ceanothus ramulosus* (coast ceanothus), *Rotala ramosior* (a lythracead).

Gr. *rhadix, -ikos,* m. branch, frond: *Rhadicoleptus alpestris* (a caddisfly).

L. *sarmentum,* n. twig, light branch, brush, runner; *sarmentosus,* full of little twigs: sarmentose, *Saxifraga sarmentosa* (strawberry-saxifrage).

F. *scion,* branch, shoot, slip: scion (cion).

L. *scopa,* twigs, broom; see **broom**

L. *sertula,* f. dim. of *serta,* f. (*sertum,* n.), garland, wreath: *sertatus,* garlanded, wreathed: *Sertularia pumila* (a hydroid), *Dinobryon sertularia* (a dinoflagellate), *Delphinium sertiferum* (a larkspur).

L. *soboles* (*suboles*), *-is,* f. shoot, sprout, twig: soboliferous, *Caryota sobolifera* (fishtail-palm).

L. *spadix, -icis* (Gr. *-ikos*), f. frond, branch of a palm; spike of flowers enclosed in a spathe: spadix, *Phyllospadix torreyi* (a naiad), *Acacia spadicigera* (an acacia).

L. *stipes, -pitis,* branch, stem, stalk; see **stem**

L. *stolo, -onis,* m. branch, shoot, runner, sucker: stolon, stoloniferous, *Cornus stolonifera* (red osier-dogwood), *Rhizopus stolonifer* (a yeast), *Stoloniclypeus prostratus* (an echinoderm).

L. *surculus,* m. sprout, sucker, graft, dim. of *surus,* m. shoot, twig; *surcularius,* of shoots; *surcularis,* producing shoots; *surculosus,* full of shoots, branchy: surculus, surculous, surculose, surculigerous, *Begonia surculigera* (a begonia), *Mus surifer* (orange-colored rat).

L. *talea,* cutting, set, layer, slip, rod; *taleola,* dim.; see **rod**

Gr. *terchnos,* n. twig:

L. *thallus* (Gr. *thallos*), m. sprout, young shoot, green branch; plant body of algae and fungi; *thallion,* n. dim.: thallus, thalloid, thallome, thallium, thalliferous, thallophyte, prothallium, *Thallograptus succulentus* (a graptolite), *Geothallus tuberosus* (a liverwort).

Gr. *troxanon,* n. twig:

L. *turio, -onis,* m. shoot, sprout, young branch: turion.

L. *vimen, -inis,* pliant twig, withe, osier; *viminalis; vimineus,* made of osiers, pliant; see **willow**

L. *virga,* f. twig, branch, rod, wand; *virgula,* f. dim.; *virgulatus; virgatulus,* striped; *virgatus,* of twigs, rodlike; *virgosus,* full of twigs, bushy; *virgultum,* n. copse, undergrowth: virgal, virgate, virgule, virgulate, *Virgula rigida* (a fossil) sponge), *Virgularia mirabilis* (a sea-pen), *Panicum virgatum* (switchgrass), *Coryphopterus virgatulus* (a fish), *Halichondria virgultosa* (a sponge).

L. *vitulamen, -inis,* n. shoot, twig:

AS. *wice,* a pliant twig or branch: wych elm (*Ulmus glabra*), witchhazel (*Hamamelis virginiana*).

See: **stem, leaf, bush, broom, rod, leg, part**

branchos, Gr. gill, fin; *branchion,* dim., see **gill, fin;** hoarseness, see **hoarse**

brand < AS. *brand;* see **mark**

brasenia, NL. a genus of the waterlily family; see **waterlily**

brass < AS. *braes;* see **copper**

brassica, L. cabbage; see **cabbage**

brastes, Gr. earthquake, < *brasso,* shake violently; see **shake**

brastos; brasmos, Gr. a boiling up, < *brazo,* boil; see **boil**

bratus, L. a cypresslike tree; see **cypress**

brave < L. *barbarus,* foreign, rough; see **bold**

bray < OF. *braire,* sound off like an ass; see **neigh**

bread < AS. *bread:* breadbasket, breadfruit, breadnut, breadroot, breadstuff, breadwinner. Derive: sweetbread.

Gr. *apomagdalia,* f. inside of the loaf:

Gr. *artos,* m. bread; *artidion,* n.; *artiskos,* m. dim., little loaf: artophagous, *Artocarpus heterophyllus* (jackfruit), *Encephalartos villosus* (a cycad).

Gr. *blekos,* n. bread:

L. *bucellatum,* n. soldier's biscuit:

L. *cerinthus* (Gr. *kerinthos*), m. beebread:

L. *crustum,* n. anything baked, bread, cake; *crustulum,* n. dim.; *crustularius,* m. baker, pastryman: crust, *Crustulum gratulans* (an echinoid).

Gr. *daratos,* m. a kind of bread: *Daratus apicalis* (a butterfly).

Gr. *dramis,* f. a kind of loaf:

Gr. *erithake,* f. beebread:

Gr. *gyrites,* bread made of the finest meal; see *gyris* under **flour**

AS. *hlaf,* loaf: loaf, lord, lady, Lammas.

Gr. *kollix, -ikos,* m. roll or loaf of bread: *Collix flavofasciata* (a moth).

L. *mamphula,* f. a kind of Syrian bread:

L. *orinda* (Gr. *orindes*), f. bread made from rice:

L. *panis, -is,* m. bread, loaf; *panicellus; pastillus,* m. dim.; *paniceus,* of bread; *panosus,* like bread; *panifex -icis,* m.; *panifica,* f. breadmaker; *panicium,* n. bread, cake; *panarium,* n. breadbasket: pannier, pantry, panada, pastille, appanage, companion, accompany, *Sitodrepa panicea* (bread-beetle), *Cratosomus pastillarius* (a beetle).

Gr. *physikillos,* m. a kind of bread:

Gr. *pyrnon,* n. bread made of wheat flour:

L. *sandaraca* (Gr. *sandarake*), f. beebread:

See: **cake, cook, flour, food**

break < AS. *brecan,* separate, crack: breaker, breakfast, windbreak, brake, broken, brick, breccia, brittle.

Gr. *agmos,* m. break, fracture; *agma, -tos,* n. fragment, < *agnymi,* break, shiver: *Agmus lyriformis* (a fish).

Gr. *akrotomos,* cut off sharp, abrupt, precipitous; see *temno* under **cut**

Gr. *anapedema, -tos,* n. outburst:

Gr. *aposphax, -agos,* broken off, abrupt, sheer:

Gr. *apoxyros,* abrupt, sharp, sheer:

L. *caesura,* f. pause, break: caesura.

L. *crepatura,* f. fissure, crack:

Gr. *dialeipsis,* f. interval, intermission: *Dialipsis exilis* (a wasp).

Gr. *diaphye,* f. break, joint, suture, cleft: *Diaphyodus* (now *Lubrodon*) *trigonella* (a Tertiary fish).

L. *discutio, -cussus,* break up, disperse, scatter, dissipate:

L. *dissilio,* fly apart, burst; *dissulto,* burst asunder: *Mycena dissiliens* (a fungus).

Gr. *ekphyma,* an eruption of pimples; see **disease**

Gr. *enterokele,* hernia, rupture: enterocele.

L. *explodo* (*explaudo*), *-plosus,* drive off, burst: explode, explosion.

L. *fissura,* f. crack, cleft, chink: fissure, *Fissurella alternata* (a keyhole-limpet), *Styliolina fissurella* (a fossil gastropod), *Fissurina spinigera* (a foraminifer).

L. *frango, fractus,* break; *fragmentum,* n. piece; *fractor, -is,* m. breaker; *fractura,* f. breach; *fragilis,* easily broken; *confractus; confragus,* broken, rough, uneven; *infringo, -fractus,* break, crack, weaken: fraction, fracture, fragile, fragment, frangula, frailty, infraction, refractive, refractory, diffraction, infringe, bire-fringent, ossifrage, osprey, sassafras, suffragist, *Cladophora fracta* (an alga), *Anguis fragilis* (a slowworm, a lizard), *Saxifraga hirsuta* (hairy saxifrage), *Corvus ossifragus* (fish-crow).

L. *frio, -atus,* break or crumble into small pieces; *friabilis,* easily broken; *infrio,* crumble: friable.

L. *hernia,* f. rupture; *herniosus,* ruptured: hernia, herniotomy, *Herniaria glabra* (rupturewort).

L. *intercapedo, -inis,* f. interval, pause, respite:

L. *intercipio, -ceptus,* interrupt, intervene: intercept.

L. *interfor, -fatus,* interrupt in speaking:

L. *intermissio, -onis,* f. interruption, interval: intermission.

L. *interpello, -atus,* interrupt by speaking, disturb: *Pimplopterus interpellatus* (a wasp).

L. *intervallum,* n. space between two palisades, intermission, pause: interval.

Gr. *katagma, -tos,* n. fracture; *kataktes,* m. breaker; *kataktos,* breakable, < *katagnymi,* break in pieces: *Nucifraga caryocatactes* (Himalayan nutcracker).

Gr. *klao,* break; *klastos,* broken in pieces; *klasis,* f. fracture; *klasma, -tos,* fragment; *kladaros,* easily broken, fragile; L. *clades, -is,* f. a breaking to pieces, destruction: clastic, pyroclastic, orthoclase, euclase, clematis, clasmatocyte, clasmatosis, iconoclast, *Clastobasis tryoni* (a fly), *Clastocnemis maculatus* (a beetle), *Clasma parvispinosum* (a grasshopper).

L. *legirupa,* lawbreaker; see **fault**

Gr. *metaxytes,* f. interval: *Metaxytes intermedius* (a Jurassic ammonite).

L. *nucifrangibulum,* n. nutcracker:

L. *quassus,* shaken, broken, shattered; see **shake**

L. *ramex, -icis,* m. hernia, rupture; *ramicosus,* ruptured: ramex.

Gr. *rhegnymi; rhesso; rhaio,* break, smash; *rhektos,* breakable, brittle; *rhaister, -os,* m. smasher; *rhektes,* m. breaker, shaker, earthquake; *rhagas, -ados; rhage,* f. chink, crack, rent; *rhegma, -tos,* n. fracture, tear; *rhegmin, -os,* f. breakers, surf; *rhexis,* f. a breaking, bursting; *rhox, -ogos,* f. break, cleft, fragment; *rhochmos,* m. cleft, gully; *rhogaleos; rhogas, -ados,* broken, rent; *rhogme,* f. fracture; *aporrhox, -ogos,* broken off, abrupt, sheer, precipitous; *aporrhegma, -tos,* n. fragment; *diarrhoge,* f. gap, interstice; *ekrhegma, -tos,* n. a breaking out, eruption: regma [rhegma], rhexis, rhagades, rhagadiform, hemorrhage, *Rhecto-gramma sherborni* (a rockfish), *Rhectophyllum mirabile* (an aracead), *Rhaestes testaceipes* (a wasp), *Rhagophthalmus scutellatus* (a beetle), *Rhagochila inderiensis* (a beetle), *Rhegmatocerus antennatus* (a beetle), *Rhogadopsis miniacea* (a wasp), *Rhogostoma schüssleri* (a protozoan), *Rhochmogaster dimera* (a bug), *Diarrhegma modestum* (a fly), *Diarrogus[Diarrhogus] pubescens* (a beetle), *Trirhogma caerulea* (a wasp).

L. *rima,* f. cleft, fissure; *rimula,* f. dim.; *rimosus,* full of cracks; *rimo, -atus,* lay open, turn over: *Discorbina rimosa* (a foraminifer), *Streptomyces rimosus* (an actinomycete, source of terramycin).

L. *rumpo, ruptus,* break; *abruptus,* broken off, steep; *disruptus,* broken up, separated; *eruptio, -onis,* f. a breaking out, bursting forth; *interruptus,* broken apart, off, or asunder; *irruptio, -onis,* f. a breaking in: rupture, disrupt, interrupt, eruptive, corruption, abruptly, bankrupt, route, rout, routine, rote, rut, *Gomphoceras abruptum* (a fossil cephalopod), *Hemicoptera interrupta* (a fly).

Gr. *syntrips, -ibos,* c. the smasher, a fiend who broke pots in the kitchen:

Gr. *thlao,* crush; see **bruise**

Gr. *thrauo,* break, shatter; *thraulos; thraustos,* easily broken, brittle; *katathraustos,* broken to pieces; *thrausma, -tos,* n. fragment, piece: *Coccothraustes vulgaris* (a hawkfinch), *Cladrastis polycarpa* (Japanese yellowwood), *Thraulodes limbatus* (a mayfly).

Gr. *thrypto,* break; *thryptikos,* easily broken, brittle; *thrypsis,* f. a breaking into small pieces, shattering: thrypsis, *Thrypticomyia aureipennis* (a fly).

F. *zigzag,* alternately changing direction by sharp angles; see **slope**

See: **tear, separate, tame, space, insert, rest, space, part, cut**

bream < OF. *bresme,* a cyprinoid fish, see **carp**; sea-bream, a sparoid fish.

Gr. *abramis, -idos,* f. a kind of fish; *abramidion,* n.: *Abramis brama* (fresh-water bream), *Prionobrama filigera* (a fish), *Abramidopsis leuckarti* (a fish).

L. *acharne* (Gr. *acharna*), f. a sea-fish, gilthead: *Acharnes speciosa* (a fish).

L. *cantharus* (Gr. *kantharos*), m. a bream: *Sparus cantharus* (black bream), *Pomoxis sparoides* (formerly, *Cantharus nigromaculatus,* a fish).

Gr. *charax, -akos,* m. a sparoid fish: *Charax gibbosus* (a fish), *Thrissocharax ocellicauda* (a fish).

Gr. *chrysaphos,* m. probably the gilthead:

Gr. *chrysophrys, -yos,* m. gilthead: *Chrysophrys taurina* (a fish).

L. *dentix, -icis,* m. a sparoid fish: *Dentex*[*Dentix*] *vulgaris* (a fish), *Brycon dentex* (a characoid fish).

Gr. *ioniskos,* m. a gilthead:

L. *maena* (Gr. *maine*), f. a small, sparoid fish: *mainidion,* n. dim.: *Maena jusculum* (a fish), *Maenichthys temmincki* (a fish), *Archaeomaene robustus* (a Jurassic fish).

L. *mormyr, -is,* f. (Gr. *mormyros,* m.), a sparoid fish: *Mormyrus oxyrhynchus* (sacred fish), *Mormyrorhynchus gronovei* (a fish), *Pagellus mormyrus* (a sea-bream).

L. *pagrus* (Gr. *pagros; phagros*), m. sea-bream; *agriophagros; oxyphagros,* m. a fish: *Pagrus major* (Japanese porgy).

L. *salpa* (Gr. *salpe*), f. a kind of gilthead: *Salpa maxima* (a tunicate), *Cyclosalpa pinnata* (a tunicate), *Boops salpa* (a fish).

L. *sargus* (Gr. *sargos*), m. a sparoid fish: *Sargus vulgaris* (sarge), *Archosargus probatocephalus* (sheepshead), *Diplodus sargus* (a fish).

Gr. *sinodon, -tos,* m. a sea-fish:

L. *smaris, -idis* (Gr. *-idos*), f. a sparoid fish: *Spicara smaris* (picarel).

L. *sparus* (Gr. *sparos*), m. gilthead; *sparulus,* m. dim.: *Sparus auratus* (gilthead), Sparidae, *Sparisoma viride* (a parrot-fish), *Dactlyosparus macropterus* (a mor-wong).

Gr. *synagris, -idos,* f. a sparoid fish: *Synagrops argyrea* (a fish), *Neomaenis synagris* (red-tailed snapper), *Nemipterus* (formerly, *Synagris*) *macronemus* (a fish).

L. *synodus* (Gr. *synodos*), m. a sparoid fish: *Synodus varius* (a fish).

breast < AS. *breost,* front of the chest.

Gr. *kitharos,* m. thorax, chest:

Gr. *kolpos,* bosom, womb; see **womb**

L. *pectus, -toris,* n. breast, chest; *pectoralis,* of the breast: pectoral, pectoriloquy, expectorate, parapet, *Nematistius pectoralis* (rooster-fish).

L. *sinus,* curve, fold, bosom; see **hollow**

Gr. *sternon,* n. breast, chest: sternum, sternalgia, sternocostal, *Sternocryptus bitinctus* (a wasp), *Platysternum megacephalum* (an Asiatic turtle).

Gr. *stethos,* n. chest, breast; *stethidion; stethion,* n. dim.: stethoscope, stethograph, *Stethopristes eos* (a fish), *Rhodostethia rosea* (a gull), *Solenostethium furciferum* (a bug), *Symphostethus collaris* (a beetle), *Cannocapsa stethoscopium* (a radiolarian).

Gr. *thorax, -akos,* m. breastplate, cuirass, corselet; the part of the body covered by this armor, chest; *thorakion,* n. dim.: thoracic, thorax, prothorax, thoracograph, thoracoplasty, thoracocyllosis, *Thoracophorus sculptilis* (a beetle), *Gymnothorax mordax* (a moray), *Dasyllis thoracica* (a robber-fly).

See: **udder, front**

breathe < AS. *braethan.*

Gr. *aspairo,* gasp, pant convulsively, as in dying:

Gr. *asthma,* hard-drawn breath, panting, gasping; see **disease**

L. *dasea,* f. rough breathing or spiritus asper:

L. *flatus,* breath, breeze, breath of life; see *flo* under **blow**

L. *halo, -atus,* breathe; *halitus, -us,* m. breath, wind, whiff; *anhelus,* out of breath, panting, puffing, < *anhelo, -atus,* pant, puff, gasp: inhale, exhalation, anhelous, anhelation, halitosis, halituous.

Gr. *pneuma, -tos,* n. wind, air, breath, < *pneo,* breathe, blow; *pneumatikos,* of air, breath, inflation; *pneustikos,* of breathing; *pneumon, -os,* m. lung; *eispneusis; eispnoë,* f. inhalation; *ekpneusis; ekpnoë,* f. exhalation: pneumatic, pneumatolysis, pneumatolytic, holopneustic, apnoea, dyspnoea, dipnoid, pneumonia, pneumococcus, *Pneustocerus nigricornis* (a bug), *Echinomma toxopneustes* (a radiolarian), *Dyspnoetus dignus* (a beetle), *Amphiuma tridactylum* (an amphibian).

Gr. *psycho,* breathe, blow; see *psyche* under **mind**

L. *spiro, -atus,* breathe, blow; *spirabilis,* breathable; *spiraculum; spiramen, -inis,* n. breathing-hole, air-hole, vent; *aspiro,* breathe or blow upon; *suspiro,* draw a deep breath, heave a sigh; *suspirium,* n. sigh: aspire, expire, inspiring, perspiration, respiratory, suspire, conspiracy, aspirate, spirit, spiracle, spirament.

See: **blow, wind, lung, smell**

brecho, Gr. wet, moisten, rain; see **wet**

breeze < F. *brise;* see **wind**

bregma, Gr. front of the head, sinciput; see **head**

brenthos, Gr. haughtiness; a kind of bird; see **proud**

brephos, Gr. fetus, embryo, babe; *brephion,* dim.; see **young**

bretas, Gr. wooden image, blockhead; see **form**

brevis, L. short; *breviarius,* abridged; *brevitas,* shortncss, narrowness; see **short**

bria, L. a wine-vessel; see **pot**

briaros, Gr. strong; *Briareos,* a hundred-armed giant; see **strong**

bribe < OF. *bribe,* gift; see *largior* under **give**

brick < MD. *bricke:* brickyard, brickbat, brickkiln, brickwork, bricklayer.
L. *abaculus* (Gr. *abakiskos,* dim. of *abax,* counting-board), small, colored tile for inlay and mosaic work; see **number**
Gr. *besalon,* n. brick; *besalotos,* of bricks:
L. *later, -is,* m. brick, tile; *laterculus (latericulus),* m. dim.; *latericius (lateritius),* of bricks; *laterarius,* m. brickmaker; *laterina,* f. brickkiln; *Lateranus,* m. god of the hearth: laterite, lateritious, *Callistemon lateritius* (a bottle-brush).
Gr. *plinthos,* f. brick; *plinthion,* n.; *plinthis, -idos,* f. dim.; *plinthinos,* of bricks; *plinthourgos,* m. brickmaker: plinth, plinthite, plinthoid, *Plintholepis retrorsa* (a fossil fish).
L. *tessera,* paving tile, square piece of stone; see **stone**
L. *testa,* piece of baked clay, brick, tile, potsherd; *testaceus,* of bricks, tiles; see **shell**
See: **shingle, stone**

bride < AS. *byrd;* see **marry**

bridge < AS. *brycg:* bridgework, Cambridge, Bridgeton.
Gr. *gephyra,* f. bridge; *gephyrion,* n. dim.: gephyrean, gephyrocercal, *Gephyrocrinus grimaldi* (a crinoid), *Gephyrometra propinqua* (a crinoid), *Gephyrocharax caucanus* (a fish).
Gr. *magas, -ados,* bridge of the cithara; *magadion,* n. dim.:
L. *pons, pontis,* m. bridge; *ponticulus,* m. dim.; *pontilis,* of a bridge: ponticello, pontifex, pontiff, pontage, ponton (pontoon), *pons asinorum,* pontile.
Gr. *schedia,* f. bridge of rafts, boats, or pontoons:

bridle < AS. *bridel;* see **hold**

brier; briar < AS. *brer,* thorn, prickle, bramble; see **thorn, bush, point, blackberry**

bright < AS. *bryht;* see **light, know**

brime, Gr. bulk, strength; see **weight**

brimosis, Gr. indignation; *brimodes,* grim, stern, < *brimazo,* snort with anger; see **anger**

brinco- < Gr. *brinkos,* a sea-fish; see **fish**

bring < AS. *bringan;* see **carry**

brisa, L. grape-skin, refuse after pressing; see **skin**

bristle < AS. *byrst;* see **hair**

brithys, Gr. heavy; *brithos,* weight; see **heavy**

brittle < AS. *breotan,* break.
Gr. *euaxos,* easily broken: *Euaxes filirostris* (an annelid).
L. *fragilis, -e,* easily broken, brittle: fragile, fragility, *Chara fragilis* (a stonewort), *Secale fragile* (a rye), *Fragilaria parasitica* (a diatom).
Gr. *kladaros,* easily broken, frail; see **weak**
Gr. *krauros,* brittle, friable: *Craurothrix leucura* (a rodent).
Gr. *psapharos; psathyros,* easily powdered, crumbling, friable, loose; *psathyrotes,* f. looseness: *Acanthoderes(Psapharochrus) modesta* (a beetle), *Psathyrocerus rufus* (a beetle), *Psathyrotes ramosissima* (a composite).
Gr. *rhektos,* breakable, brittle; see *rhegnymi* under **break**
Gr. *thraulos; thraustos,* easily broken, brittle; see *thrauo* under **break**
Gr. *thryptikos,* easily broken; see *thrypto* under **break**
See: **break, dainty, thin, weak**

briza, Gr. a grain like rye; see **grain**

broad < AS. *brad,* extended, wide: broadcast, broadcloth, broadside, Broadway, Bradstreet, Bradford, Broadmoor, Bradley.
Gr. *eurys,* broad, wide, far-reaching, widespread: eurycerous, eurygnathous, aneurysm, eurystomatous, *Eurycephalus lundi* (a beetle) *Euryscopa vittata* (a beetle), *Eurygnathus latreillei* (a beetle), *Eurystethus nigropunctatus* (a bug), *Eurymetopon ochraceum* (a beetle), *Eurynogaster nitida* (a fly), *Euryocrinus granulosus* (a fossil crinoid), *Eurynotus muricatus* (a beetle).
Gr. *katatonos,* broader than high:
L. *latus,* broad, wide; *latitudo, -inis,* f. breadth, width: latitude, latipennate, *Typha latifolia* (cattail), *Oedaspis latisfasciata* (a fly), *Cytheropteron latissimum* (an ostracode), *Viola latiuscula* (a violet), *Mantichora latipennis* (a beetle), *Corycaeus laticeps* (a cyclops).

L. *patulus,* spread out, broad, wide open; see *pateo* under **open**
Gr. *petalos,* broad, flat, outspread: *Petalophthalmus inermis* (a crustacean).
L. *planus,* even, flat, level; see **level**
Gr. *platys,* broad, wide, flat, level; *platykos,* broad, flat; *platamon, -os,* m. any broad, flat surface; *platamodes,* broad, even; *platytes, -etos,* f. breadth, bulk: platypod, platyrrhine, plat, plate, platitude, plateau, Plato, plaice, *Platycarya strobilacea* (a Chinese hickory), *Platycerus quercus* (a beetle), *Platydactylus muralis* (gecko), *Platyspermum scapigerum* (a mustard), *Platophrys inermis* (a fish), *Epiplatys fasciolatus* (a fish), *Platanus orientalis* (oriental plane-tree), *Platamus pallidulus* (a beetle), *Platypus abbreviatus* (a beetle).
L. *plautus,* broad, flat, flat-footed: *Plautus impennis* (great auk).
L. *simo, -atus,* press flat, flatten; *simus,* flat-nosed, pug-nosed; see **nose**
See: **spread, level, plate, face, board, sit**

broccus; brochus; broncus, L. projecting; see **projection**

brochetos, Gr. rain, < *brecho,* wet, moisten; see **wet**

brochis, Gr. inkhorn; see **ink**

brochos, Gr. noose, halter, snare; *brochotos,* ensnared; see **rope**

brochthos, Gr. throat; see **throat**

broken < AS. *brecan;* see **break**

broma, Gr. food; *bromation,* dim.; *brosis,* food; *brosimos; brotos,* edible; see *bibrosko* under **eat**

bromelia, NL. a genus of the pineapple family; see **pineapple**

brometes, Gr. ass; see **horse**

bromine < Gr. *bromos,* stench; see ELEMENTS under **thing**

bromios, Gr. noisy, boisterous, Bacchic, < *bromos,* any loud noise; see **sound**

bromos, Gr. stench; *bromodes; bromosus,* stinking; see **smell**

bromus, L. (Gr. *bromos*), a kind of oats; see **grain**

bronchos, Gr. windpipe; see **throat**

bronto- < Gr. *bronte,* thunder; *brontema,* clap of thunder; *brontaios,* thundering; *brontodes,* like thunder; see **thunder**

bronze < It. *bronzo;* see **copper**

brooch < OF. *broche,* awl, spit, clasp pin, ornamental clasp; see **bind**

brook < AS. *broc;* see **stream**

broom < AS. *brom,* thorny bush, bramble.   See **bush, bean.**
L. *genista,* broom, a plant of the bean family; see **bean**
Gr. *kallyntron,* n. broom, brush, < *kallyno,* sweep clean, beautify:
Gr. *korema, -tos; korethron,* n. broom, < *koreo,* sweep out: *Corema conradi* broom-crowberry), *Corematocrinus plumosus* (a Devonian crinoid), *Corethron criophilum* (a diatom), *Corethrogyne filaginifolia* (cudweed-cottonaster), coremium.
Gr. *kosmetron,* n. broom:
L. *molucrum,* n. broom for sweeping out a mill:
L. *muscarium,* n. fly-brush, clothes-brush:
Gr. *ophelma, -tos,* n. broom:
L. *peniculus,* brush; see *penicillum* under **bush**
L. *ruscum (bruscum),* broom; *ruscarius,* of broom; see *ruscus* under **lily**
Gr. *saron; sarotron,* n. broom; *sarotes,* m. sweeper; *saroma, -tos,* n. sweepings, < *saroo,* sweep: sarothrum, *Hedysarum obscurum* (a sweetvetch), *Sarocrinus tentaculatus* (a Mississippian crinoid), *Sarophorus tuberculatus* (a beetle), *Sarothra gentianoides* (orange pineweed).
L. *scopa,* f. thin twigs, broom; *scopula,* f. dim.; *scoparius,* m. sweeper: scopula, scopulite, scopiferous, scopuliferous, scopiform, scullion, *Cytisus scoparius* (broom), *Dicranum scoparium* (broom-moss), *Polystichum scopulinum* (a fern).
Gr. *spartos,* broom, a plant of the bean family; see **bean**
L. *syrus,* m. broom:
See: **sweep**

broomrape
NL. *boschniakia,* f. a genus of the broomrape family:*Boschniakia tuberosa* (a broomrape).
NL. *conopholis,* f. a genus of the broomrape family: *Conopholis americana* (squawroot).
NL. *lathraea,* f. a genus of the broomrape family: *Lathraea clandestina* (an orobanchacead).
L., Gr. *orobanche,* f. broomrape: Orobanchaceae, *Orobanche ramosa* (a broomrape).

brosis, Gr. food; see *bibrosko* under **eat**

brother < AS. *brothor:* brothers, brethren, brotherly, brotherhood, half-brother, brother-in-law.

Gr. *adelphos,* m. brother; *adelphe,* f. sister; *adelphidion,* n. dim.; *adelphikos,* brotherly, sisterly; *adelphixis,* f. brotherhood, kinship: monadelphous, diadelphous, Philadelphia, *Philadelphus coronarius* (mockorange), *Oporornis philadelphia* (mourning warbler), *Clitocybe monadelpha* (a mushroom).

Gr. *agastor, -os,* m. near kinsman:

L. *consors, -ortis,* brother, sister, comrade, sharer, clansman; see **companion**

L. *frater, -tris* (Gr. *phrater, -os,* clansman), m. brother; *fraterculus,* m. dim.; *fraternitas, -atis,* f. brotherhood; *fraternus,* brotherly: fraternal, fraternity, friar, fratricide, *Fratercula arctica* (Atlantic puffin), *Diathermus frater* (a beetle) *Tarsius fraterculus* (a tarsier).

L. *germanus,* having the same parents; see **kin**

Gr. *kasignetos; kasis, -ios,* m. brother:

See: **kin, sister, class**

brotico- < Gr. *brotikos,* voracious; see *bibrosko* under **eat**

broton, Gr. food; *brotos,* edible, eaten, swallowed; see *bibrosko* under **eat**

brotos, Gr. man, human, mortal; see **man**

brotula, NL. a kind of fish; see **fish**

brow < AS. *bru;* see **face, projection**

brown < AS. *brun* (ML. *brunneus*), dusky, dark, tawny: brunette, brunneous, burnish, Brown, Bruno, burnet (*Sanguisorba canadensis*), *Blacicus brunneicapillus* (a bird), *Ptinus brunneus* (a bookworm), *Carex brunnescens* (a sedge).

L. *adustus,* tanned, brown, swarthy: *Drosophila adusta* (a fruitfly). See *uro* under **burn**

Gr. *aithiops,* sunburnt, swart; *aithos,* burnt, reddish-brown; see *aitho* under **burn**

L. *badius* (*balius; baliolus*), chestnut-brown, reddish-brown: bay, *Rhizomys badius* (bamboo-rat), *Baeoscelis badiceps* (a bird).

L. *castaneus,* of the color of chestnuts, brown, bay; see **chestnut**

L. *cinnameus,* of cinnamon; see **cinnamon**

L. *ferrugineus; ferruginus,* rust-colored, rusty; see *ferrugo* under **rust**

L. *fulvus,* tawny, reddish-yellow: fulvous, *Ulmus fulva* (slippery elm), *Acipenser fulvescens* (a sturgeon), *Trogophloeus fulvipes* (a beetle).

L. *furfurosus,* like bran, brownish; see *furfur* under **scale**

L. *fuscus,* dusky, tawny; see **black**

Gr. *gaiophanes,* earth-colored; see *ge* under **earth**

Gr. *karykrous,* nut-brown:

Gr. *kirrhos,* tawny; see **yellow**

Gr. *phaios,* dusky, brown: phaeochrous, phaeoplast, phaeophyll, Phaeophyceae, *Phaeopus borealis* (Eskimo curlew), *Phaeochlaena bracteola* (a butterfly), *Phaeogramma vittipennis* (a fly), *Phaeocyclotomus robustus* (a beetle), *Junco phaeonotus* (redbacked junco).

L. *sepia,* cuttlefish; the pigment secreted by the cuttlefish; see **squid**

OF. *sorel,* dim. of *sor,* reddish-yellow, brown: sorrel, surmullet.

AS. *tannian,* tan, prob. < C. *tann,* oak; see **oak**

L. *ustulatus,* scorched, singed, browned; see *uro* under **burn**

Gr. *xouthos,* yellowish-brown; see **yellow**

See: **amber, rust, shade**

bruchus, L. (Gr. *brouchos*), a kind of wingless locust, now a genus of beetles; see **beetle**

bruise < AS. *brysan* and OF. *bruisier.*

L. *contundo, -tusus,* bruise, grind, crush, pound; *contusio, -onis,* f. a bruise: contusion, *Embaphium contusum* (a beetle).

L. *famex, -icis,* m. bruise, contusion:

Gr. *hypopion,* n. black eye, bruise:

Gr. *molops, -opos,* m. bruise, weal: *Molopophorus striatus* (a fossil gastropod).

Gr. *smodix, -ingos,* f. bruise, weal: *Smodingoceramus virgatus* (a fossil cephalopod), *Smodingium argutum* (an anacardiacead), *Pseudosmodingium perniciosum* (an anacardiacead).

L. *sugillatum,* n. black and blue spot, bruise, < *sugillo, -atus,* beat black and blue: sugillation, *Ophisuris sugillator* (a fish), *Stomoxys sugillatrix* (a fly).

Gr. *thlao,* bruise, crush; *thlastos,* bruised, crushed; *thlastes,* m. bruiser, crusher; *thlasma, -tos,* n. a bruise; *thlasis,* f. a bruising, crushing: *Thlastocoris laetus* (a bug), *Thlattodus suchoides* (a Jurassic fish), *Thlasia brunneipennis* (a bug).

AS. *walu,* ridge, stripe, or mark caused by the blow of a whip or stick: weal wale, gunwale.
  See: **hurt, sore, grind, rub, strike**

-**brum,** L. suffix denoting instrument; see **tool**

bruma, L. shortest day of the year, winter solstice; *brumalis,* of the shortest day, wintry; see **day**

brunneus, ML. < AS. *brun,* brown; see **brown**

bruscum, L. excrescence on the maple-tree; see *ruscum* under **bush**

brush < F. *brosse,* brushwood; see **bush, broom**

brutus, L. rough, stupid, heavy; see **rough**

brychema, Gr. roar, bellow; *brychetes,* roarer, bellower; see **roar**

brychios, Gr. deep, from the deep; see **deep**

bryco- < Gr. *bryko; brycho,* eat noisily, greedily; *brygma,* bite; see **bite**

bryo, Gr. swell, see **swell**; < *bryon,* moss, see **moss**

bryonia, Gr. a vine of the gourd family; see **melon**

bryssos, Gr. a kind of sea-urchin; see **echinoderm**

brytos, Gr. beer; *brytikos,* drunk with beer; see **beer**

bryttus, NL. a genus of fishes; see **fish**

bryum, L. (Gr. *bryon*), moss; see **moss**

bu-, L. (Gr. *bou-*), prefix meaning large, huge, great, monstrous, see **large**; < Gr. *bous,* ox, cow, see *bos* under **cattle**

bubalion, Gr. a kind of gourd; see **melon**

bubalus, L. (Gr. *boubalos; boubalis,* gazelle, antelope) buffalo, wild ox; see **cattle, antelope**

bubble < ME. *bobel.*
  Gr. *blysis,* f.; *blysmos,* m.; *blysma, -tos,* n. a bubbling, < *blyzo,* bubble:
  L. *bulla,* f. bubble, knob, boss, stud, swelling, seal, edict, ornament; *bullula,* f. dim.; *bullatus,* inflated; *bullio, -itus,* bubble up, appear boisterously: bull, bulletin, bill, bowl (ball), ebullient, ebullition, bouillon, budge, *Bulla eburnea* (a bubble-shell), *Globigerina bulloides* (a foraminifer), *Lagena bullaeformis* (a foraminifer), *Ocotea bullata* (stinkwood), *Marginulina subbullata* (a foraminifer).
  Gr. *hydatis, -idos,* f. watery vesicle: *Hydatis acephalocystis* (a cestode).
  Gr. *kausalis, -idos,* f. blister, burn; see *kaio* under **burn**
  Gr. *pemphix, -igos* (*pemphis, -idos*), f. something filled with air, bubble; *pemphigodes,* inflated: pemphigus, *Pemphis acidula* (mentigi), *Pemphigus populicaulis* (poplar-petiole gall), *Pemphigonotus mirabilis* (a fly).
  Gr. *phausinx, -ingos,* f. blister from burning:
  L. *phlyctaena* (Gr. *phlyktaina; phlyktis, -idos*), f. blister, pustule; *phlyzakion,* n. dim.: phlyctena, phlyctenoid, phlyctenule, phlyzacium, *Phlyctaena strobilina* (a fungus), *Phlyctis agelaea* (a lichen), *Cardium phlyctaena* (a fossil pelecypod), *Campanula phlyctidocalyx* (a bellflower).
  Gr. *phois, -idos; phos, -odos,* f. blister, burn: *Phodopus bedfordiae* (a rodent).
  Gr. *physema, -tos,* n. bubble; *physinx, -ingos,* f. bladder, bubble; *physallis, -idos,* f. bladder, bubble; *physke,* f. blister, sausage: emphysema, *Physcomitrium turbinatum* (a moss).
  Gr. *pompholyx, -ygos,* f. bubble; *pomphos,* m. blister: *Pompholyx complanata* (a rotifer), *Histopomphus cervicornis* (a Cretaceous foraminifer), *Pomphorhynchus laevis* (a nematode).
  Gr. *psydrax, -akos,* m. blister:
  L. *pustula,* f. pimple, bubble, blister; *pustulosus,* full of blisters: pustule, *Umbilicaria pustulata* (a lichen), *Asterolecanium pustulans* (fringed scale), *Gnathodus pustulosus* (a conodont).
  L. *vesica,* bladder, blister; see **bag**
  See: **boil, ball, bag, pimple, swell, wind**

bubo, L. owl; see **owl**

bubon, Gr. groin; see **groin**

bubulcus, L. plower with oxen, herdsman; see **farmer**

bubulus, L. pertaining to cattle or oxen; *bubulinus,* of cattle; see *bos* under **cattle**

bucca, L. cheek, cavity; *buccula,* dim.; *bucculentus,* with full cheeks; see **cheek**

buccella, L. dim. of *buccea,* morsel, mouthful; see **little**

buccina, L. shepherd's horn, trumpet; *buccinum,* a horn-shaped mollusk; *buccino,* sound the trumpet; see **horn**

bucco, *-onis,* L. babbler, fool; see **speak**
buck < AS. *buc,* he-goat; see **goat**
bucket < OF. *buket,* < AS. *buc,* vessel, pitcher, belly.
   L. *ama* (Gr. *ame*), f. shovel, water-bucket, pail: *Amebelodon grangeri* (a mastodon).
   Gr. *amolgeus,* m.; *amolgion,* n. milk-pail:
   Gr. *amphotis, -idos,* f. a two-handled pail: *Amphotis marginata* (a beetle).
   Gr. *analepter, -os,* m. bucket:
   Gr. *anarhyster, -os,* bucket:
   Gr. *antlema, -tos,* n.; *antlion,* n.; *antlos,* m. bucket; *antleo,* bail out bilgewater: *Antliodus mucronatus* (a fossil bird).
   L. *gaulus* (Gr. *gaulos*), m. milk-pail, water-bucket:
   Gr. *hydria,* water-pot, bucket, pitcher; see **pot**
   Gr. *ibanos,* m. water-bucket:
   L. *mulctra,* f.; *mulgare, -is,* n. milk-pail:
   Gr. *pella; pellis, -idos,* bowl, milk-pail; see **cup**
   L. *situla,* f. bucket or urn for drawing water; *sitella,* f. dim.: *Situlaspis condaliae* (a bug).
   See: **vessel, pot**
buckle < OF. *boucle,* boss, knob, < L. *buccula,* dim. of *bucca,* cheek; see **bind**
buckthorn
   L. *alaternus,* f. a buckthorn: *Rhamnus alaternus* (Italian buckthorn).
   L. *calabrix, -icis,* f. a buckthorn:
   L. *ceanothus* (Gr. *keanothos*), m. a genus of the buckthorn family: *Ceanothus arboreus* (feltleaf-ceanothus), *Croton ceanothifolius* (a spurge).
   L. *colubrina,* f. a genus of the buckthorn family: *Colubrina reclinata* (soldier-wood).
   NL. *condalia,* f. a genus of the buckthorn family, < Antonio Condal, Spanish explorer: *Condalia lycioides* (a condalia).
   NL. *frangula,* f. buckthorn: frangulin, *Rhamnus frangula* (glossy buckthorn), *Aphis frangulae* (an aphid).
   NL. *hovenia,* f. a genus of the buckthorn family, < David Hoven, Dutch senator: *Hovenia dulcis* (Japanese raisin-tree).
   L. *paliurus* (Gr. *paliouros*), m. a genus of the buckthorn family: *Paliurus ramosissimus* (a buckthorn).
   Gr. *philyke,* f. a kind of buckthorn: *Phylica[Philyca] ericoides* (a rhamnacead), *Pomaderris phylicaefolia* (a rhamnacead).
   Gr. *pyxakantha,* f. buckthorn, boxthorn:
   L. *rhamnus* (Gr. *rhamnos*), f. a prickly shrub, buckthorn: rhamnose, *Rhamnus alnifolia* (alder-buckthorn), *Hippophaë rhamnoides* (sea-buckthorn).
   L. *zizyphus* (Gr. *zizyphos*), m. jujube-tree; Gr. *zizyphon* (Per. *zizafun*), jujube: *Zizyphus jujuba* (a jujube-tree), *Aponictus zizyphi* (a beetle), *Trochus zizyphinus* (top-shell).
buckwheat
   L. *carcinethrum* (Gr. *karkinethron*), n. a kind of knotweed or polygonum:
   NL. *coccolobis* (L. *cocolobis* or *cocolubis,* a Spanish grape), f. a genus of the buckwheat family: *Coccolobis diversifolia* (a sea-grape).
   NL. *fagopyrum,* n. buckwheat: fagopyrism, *Fagopyrum sagittatum* (buckwheat).
   L. *polygonum* (Gr. *polygonon*), n. knotweed, smartweed; a genus of the buckwheat family: Polygonaceae, *Polygonum punctatum* (dotted smartweed), *Polygonum aviculare* (knotweed, doorweed), *Cuscuta polygonorum* (a dodder).
   L. *rheum* (Gr. *rheon; rha*), n. a genus of the buckwheat family; NL. *rhaponticum,* n. rhubarb, < *Rha,* Volga River, *Pontus,* Black Sea: *Rheum rhaponticum* (rhubarb), *Centaurea rhaponticum* (a knapweed).
   L. *rumex, -icis,* sorrel, dock; a genus of the buckwheat family; see **sorrel**
   L. *triplaris,* threefold; a genus of the buckwheat family: *Triplaris caracasana* (an ant-tree).
bucolicus, L. (Gr. *boukolikos*), rural, rustic, pastoral; see **field**
bud < ME. *budde.*
   Gr. *blastos,* m. bud, germ, shoot, sprout; *blastema, -tos,* n. offspring, offshoot, sucker; *blastikos,* of sprouting, shooting up: blastoderm, blastoid, blastomycosis, blastula, epiblast, statoblast, *Blastoconus robertsoni* (a fossil mammal), *Blastomeryx borealis* (a fossil mammal), *Blasticorhinus rivulosus* (a moth), *Blastulidium paedophthorum* (a protozoan), *Cryptoblastus granulosus* (a blastoid), *Synechoblastus nigrescens* (a lichen).
   L. *gemma,* f. bud, eye, jewel: *gemmula,* f. dim.; *gemmarius,* of gems; m. jeweler; *gemmeus,* of gems, glittering; *gemmatus,* with buds, eyes, or jewels: gem, gemmule, gemmation, gemmiparous, gemmiferous, gemmuliferous, *Aranea gemmoides* (a spider), *Lycoperdon gemmatum* (a puffball), *Coscinasterias gemmi-*

fera (a starfish), *Bothrylus gemmulatus* (a beetle), *Pisocrinus gemmiformis* (Silurian crinoid), *Asaphus gemmuliferus* (a trilobite), *Anticarsia gemmatilis* (velvet-bean caterpillar).

L. *germen, -inis,* n. bud, seed, sprout; *germino, -atus,* sprout: germ, germicide, germarium, germinal, *Erica regerminans* (a heath).

Gr. *ozos,* branch, twig, bud; see **branch**

L. *papilla,* f. nipple, teat, bud; *papillatus,* budlike, with buds: papilla, papillose, papillate, *Cidaris papillata* (a sea-urchin), *Cephalobellus papilliger* (a nematode), *Nubecularia papillosa* (a foraminifer).

Gr. *thele,* teat, nipple; see **udder**

See: **eye, wart, projection, seed, branch, swell, child**

budyto- < Gr. *budytes,* wagtail; see **wagtail**

buffalo < L. *bufalus;* see *bos* under **cattle**

bufo, L. toad; see **toad**

bug, prob. < W. *bwg,* hobgoblin, specter, sprite; a colloquial word for an insect or similar animal of another class; here used in the strict sense of hemipteron. Derive: chafer, bugbear, bugloss.

NL. *anasa,* f. a genus of bugs: *Anasa tristis* (squash-bug).

NL. *aphis, -idis,* f. plant-louse: aphides (pl.), aphidicolous, aphidicide, *Aphis rosea* (an aphid), *Aphidius exiguus* (a wasp), *Hormaphis spinosa* (a plant-louse), *Lasius aphidicola* (an ant), *Musca aphidivora* (a fly), Aphididae.

NL. *arilus,* m. a genus of bugs: *Arilus cristatus* (wheel-bug).

NL. *benacus,* m. a genus of bugs: *Benacus griseus* (a large water-bug).

NL. *ceresa,* f. a genus of bugs: *Ceresa bubalus* (buffalo tree-hopper).

L. *cicada,* 'tree-cricket', 'locust'; see **cicada**

L. *cimex, -icis,* m. bug; *cimiculus,* m. dim.; *cimico, -atus,* destroy bugs: cimicic, cimicide, chinch, *Cimex lectularius* (bedbug), *Cimicifuga racemosa* (bugbane), *Coridochloa cimicina* (a grass), *Haplophytum cimicidum* (an apocynacead).

NL. *emesa,* f. a genus of bugs, < Emesa, an ancient city: emesid, Emesidae, *Emesa longipes* (a bug).

F. *kermes,* < Ar. *qirmiz,* deep red; a genus of scale-insects; see **red**

Gr. *koris, -ios; -eos,* m. bug, bedbug; L. *coris, -idis,* f. a plant whose seeds resemble small bugs: coriander, *Coris monspeliensis* (a primulacead), *Coreus cornutus* (a bug), *Corispermum hyssopifolium* (a tickseed), *Corixa mercenaria* (a water-boatman), *Cyrtocoris monstrosus* (a bug), *Dicoria canescens* (a composite), *Rhombocoris syriacus* (a bug), *Baccharis coridifolia* (a composite), *Trichocorixa reticulata* (a water-boatman), *Coreopsis lanceolata* (a coreopsis).

L. *perillus,* m. a genus of bugs, < *Perillus,* a worker in metal: *Perillus circumcinctus* (a bug).

NL. *ranatra,* a genus of bugs; see *rana* under **frog**

NL. *sinea,* f. a genus of bugs: *Sinea diadema* (a soldier-bug).

L. *thrips, -ipis* (Gr. *-ipos*), m. a woodworm; a genus of Thysanoptera; *thripedestos,* wormeaten: *Thrips minutissimus* (a thrips), *Thripoctenus russelli* (a wasp), *Cercothrips modestus* (a thrips).

See: **insect, beetle, flea, grasshopper, louse, tick**

bugle < OF. *bugle,* < L. *buculus,* bullock, heifer, see **horn**; < Sp. *bugula,* < L. *bugillo,* a kind of mint, see **mint**

buglossus, L. (Gr. *bouglossos*), oxtongue; see **borage**

build < AS. *byldan;* see **make**

bulbus, L. (Gr. *bolbos*), a swelling; a fleshy, usually underground, bud or tuber; *bulbulus,* dim.; see **swell**

buleuto- < Gr. *bouleutos,* devised, plotted; see *bouleuma* under **aim**

bulga, L. leather knapsack, bag; see **bag**

bulk < ON. *bulki,* heap; see **heap, large, broad**

bulimus, L. (Gr. *boulimos*), great hunger; see *limos* under **hunger**

bull < AS. *bula; bulluc;* see **cattle**

bulla, L. bubble, knob, boss, seal, edict, ornament; *bullatus,* inflated; see **bubble**

-bulum, L. suffix signifying tool, instrument; see **tool**

bumastus, L. (Gr. *boumastos,* having large breasts), a kind of grape with large clusters; see **grape**

bumelia, L. (Gr. *boumelia*), a kind of ash; now a genus of the sapodilla family; see **sapodilla**

bump < uncertain origin; see **swell, strike**

bunch < uncertain origin; see **bundle, cluster, grape, heap, herd, number, press**

bundle < AS. *byndele.*

Gr. *amalla,* f. sheaf: *Diploconus amalla* (a radiolarian).

Gr. *ankalos,* m. armful, bundle:

Gr. *dema,* band, bundle; *desmis, -idos,* bundle, package; see *deo* under **bind**

Gr. *dragma, -tos,* n.; *drax, -akos,* f. handful, sheaf; *dragmis, -idos,* f. dim.: *Dragmatyle lictor* (a sponge), *Dragmastra lactea* (a sponge).

L. *fascis,* m. bundle, packet, sheaf; *fasciculus,* m. dim.; fascicle, fasciole, fascia, Fascisti, fasciculate, *Clematis fasciculiflora* (a clematis), *Vernonia fasciculata* (western ironweed), *Cypripedium fasciculatum* (a moccasin-flower).

Gr. *komys, -ythos,* f. bundle, sheaf: *Comythovalgus villosus* (a beetle).

L. *manipulus,* m. handful, bundle, sheaf: maniple.

L. *merges, -gitis,* f. sheaf:

Gr. *phakelos,* bundle, cluster; see **cluster**

L. *pugillus,* handful; see *pugnus* under **hand**

Gr. *sage,* pack, baggage; see **weight**

L. *sarcina,* f. package, parcel, bundle; *sarcinula,* f. dim.; *sarcinalis; sarcinarius,* of burdens, baggage; *sarcinatus,* laden: *Sarcina maxima* (a bacterium), *Sarcinula costata* (a fossil coral).

Gr. *satto,* pack, load; *saktos,* crammed, stuffed; see *saktos* under **press**

AS. *sceaf,* bundle of grain: sheaf, Sheffield.

Gr. *tropalis, -idos,* f. bundle, bunch:

See: **cluster, weight, pres, bind**

-bundus; -cundus; -undus, L. denoting continuance or augmentation; see **very**

bunium, L. (Gr. *bounion*), a genus of the carrot family; see **carrot**

buno- < Gr. *bounos,* hill, mound, knob; see **mountain**

bupleurum, L. (Gr. *boupleuron*), a genus of the carrot family; see **carrot**

buprestis, L. (Gr. *bouprestis*), a kind of beetle; see **beetle**

bur < ME. *burre,* burdock.

L. *lappa,* f. bur; *lappago,-inis,* f. a plant with burs; *lappaceus,* burlike: lappaceous, *Arctium lappa* (burdock), *Lappula echinata* (a stickweed), *Nephelium lappaceum* (rambutan).

burden < AS. *byrthen,* load; see **weight**

burdo, L. mule; *burdonarius,* mule-driver; see **horse**

burdunculus, L. a plant, perhaps borage; see **borage**

burg < AS. *burg,* < ML. *burgus,* fort, fortified town; see **town**

buris, L. beam of a plow; see **beam**

burn < AS. *beornan;* G. *brennen:* burning, sunburn, sideburn, burnside, brand, brandy, brimstone, brunt.

L. *adolefactus,* set on fire, kindled:

L. *aestus,* heat, fervor; see **heat**

Gr. *aitho,* burn, be clear; *aithos,* burnt, fiery, reddish-brown; *aithiops (aithops), -opos,* sunburnt, swart, fiery, scorched; *aithinos,* burning, flashing, hot; *aithon, -os,* fiery, blazing; *aithalodes,* sooty, black: Ethiopian, *Aethionema pulchellum* (a stonecress), *Microctonus aethiops* (a wasp), *Macrocephalus aethiopicus* (warthog).

Gr. *anamma, -tos,* n. a burning mass:

L. *ardeo, arsus,* burn, blaze, glow, sparkle; *ardesco,* take fire, kindle; *ardens,* burning, glowing; *ardor, -is,* m. flame, fire, heat, great desire: ardor, ardent, arson, *Pelargonium ardens* (a geranium).

Gr. *asbestos,* unquenchable, inextinguishable; see **asbestus**

L. *coctilis,* cooked, burned; see **cook**

L. *cremo, -atus,* burn; *concremo,* burn up; *cremabilis,* combustible: cremate, crematory.

Gr. *daio,* kindle, burn; *daleros,* burning, hot:

Gr. *empresis, -eos,* f.; *empresmos,* m. conflagration; *emprestes,* m. burner: *Empresmothrips combustipes* (a thysanopterid).

L. *flagro, -atus,* burn; *inflagro,* kindle: flagrant, conflagration, deflagrate.

L. *incendo, -ensus,* set on fire, kindle, light, incite; *incendium,* n. fire; *incendialis,* of fire; *incendiarius,* causing fire, setting on fire; *incendiosus; incensus,* burning, hot; *incensor, -is,* m. one who kindles or sets on fire; *accensus,* kindled; *occensus,* burnt: incendiary, incense, censer, frankincense.

L. *inflammo, -atus,* set on fire, kindle; see *flamma* under **fire**

Gr. *kaio,* burn; *kauma, -tos,* n. burning heat; *kaumateros,* hot, glowing; *kausis,* f. a burning; *kauson, -os,* m. burning heat, summer heat; *kauter, -os,* m. burner; *kauterion,* n. dim. branding-iron; *kausimos; kaustikos,* combustible, corrosive; *kaustos,* burnt; *kauthmos,* m. a burning, scorching; *kaualeos,* burnt up, parched; *kausalis, -idos,* f. blister, burn; *epikautos,* burnt at the end; *heliokaustos,* sunburnt; *kynokauma, -tos,* heat of the dog-days: caustic, cauter, cauterize, cauma, calm, causalgia, caeoma, eremacausis, holocaust, ink, *Epicauta vittata* (striped blister-beetle), *Causus rhombeatus* (night-adder), *Heliocausta limbata* (a moth).

Gr. *keleos; kelos,* burning, dry; see **dry**

ON. *kynda,* set on fire: kindle.

Gr. *phlego,* burn; *phlegma, -tos,* fire, flame, heat, inflammation; *phlegmone,* f. fiery heat, passion, boil; *phlegyros,* burning, hot, inflamed; *phlox, -ogos,* f. flame, a plant; *phlogmos,* m. blaze, flame; *phlogistos,* inflammable; *phlogeos; phlogeros; phloginos,* flaming, fiery; *aeiphleges,* ever-burning: phlegm, phlegmatic, phlogiston, phlogopite, Phlegethon, phlox *(Phlox paniculata), Phlogacanthus thyrsiflorus* (an acanthacead), *Zaphleges leucostigmus* (a wasp).

L. *solatus,* sunburned; see *sol* under **sun**

Gr. *statheutos,* scorched, burned; *statheusis,* f. a scorching, < *statheuo,* scorch, roast:

L. *torreo, tostus,* dry by heat, parch, roast; *torridus,* dry, parched, hot; see **dry**

L. *uro, ustus,* burn; *ustulo, -atus,* burn a little, brown, singe, crisp, scorch; *adustus,* browned, tanned, swarthy; *combustus,* burned up; *deustus,* burned, consumed; *exustus,* burned out; *inustus,* not burned; burned up; *perustus,* burned up; *usta,* f. burnt color; *usticius,* brown as a result of burning; *ambustum,* n. a burn; *exustio, -onis,* f. conflagration; *busticetum; bustum,* n. place where corpses were cremated; *bustualis,* pertaining to such places: combustion, combustible, uredo, Urtica *urens* (a nettle), *Cymindis adusta* (a beetle), *Gyrophora deusta* (a fungus), *Dendrolagus inustus* (a tree-kangaroo), *Hylocichla ustulata* (russet-backed thrush), *Mytilus exustus* (a sea-mussel), *Ustilago tritici* (wheat-smut).

See: **fire, cook, heat, light, nettle, brown, itch, lewd, wish, oven, red**

burra, L. nonsense, trifle; see **trifle**

burranicum, L. vessel for milk and must; see **pot**

burricus, L. a small horse; see **horse**

burrow < uncertain origin; see **hole**

bursa, L. (Gr. *byrsa,* hide, skin, leather), purse; see **bag, skin**

bursera, NL. f. type genus of the Burseraceae, < Joachim Burser, German botanist: *Bursera excelsa* (a bursera).
    NL. *canarium,* n. a genus of the Burseracea: *Canarium commune* (Java almond).

burst < AS. *berstan;* see **break**

bury < AS. *byrgan;* see **cover**

-bus, L. having the quality of; see **nature**

bush < ME. *busch, bosch* (G. *busch*); bushwood, bushwacker, Bushman, ambush.

L. *acanthus* (Gr. *akanthos*), a prickly plant; see **thorn**

Gr. *acherdos,* f. a prickly shrub used for hedges: *Acherdocerus fumipennis* (a wasp).

Gr. *anthelion,* n. dim. of *anthele,* f. tuft, plume, or panicle of a reed: *Dissanthelium californicum* (Catalina grass).

L. *arbuscula,* shrub, dim. of *arbor,* tree; see *arbor* under **tree**

NL. *aspergillus,* m. a genus of fungi whose conidiophores resemble a brush: aspergillum, *Aspergillus ochraceus* (a mold).

L. *batus* (Gr. *batos*), bramble, prickly bush, blackberry; see **blackberry**

Gr. *dasos,* thicket, copse; *dasodes,* bushy; see *dasys* under **hair**

Gr. *drios,* n. copse, wood, thicket: *Drioctistes sclateri* (a bird).

L. *dumus,* m. bramble; *dumalis; dumosus,* bushy; *dumetum,* n. thicket; *dumicola,* m. dweller in thickets: dumetose, *Dumetella carolinensis* (catbird), *Gaylussacia dumosa* (dwarf huckleberry), *Polygonum dumetorum* (hedge-buckwheat).

L. *floccus,* tuft or lock of wool or hair; see **wool**

L. *frutex, -icis,* m. shrub, bush; *fruticetum,* n. thicket; *fruticosus,* bushy: frutescent, fruticulose, fruticetum, *Nipa fruticans* (a palm), *Iva frutescens* (shrubby sump-weed); *Veronica fruticulosa* (a speedwell), *Gomphocarpus fruticosus* (a milk-weed), *Calendula suffruticosa* (shrubby calendula).

L. *genista,* broom, a plant of the bean family; see **bean**

Gr. *harpeza,* f. thorn-hedge, thicket: *Harpezoneura multifurcata* (a psocopterid).

Gr. *hylema, -tos,* n. underbrush, shrubbery:

Gr. *krossos,* fringe, tassel; *krossion,* dim.; *krossotos,* fringed; see **fringe**

Gr. *lochme,* f. bush, copse, thicket; *lochmaios; lochmios,* of thickets; *lochmodes,* bushy: *Lochmorhynchus senectus* (a fly), *Lochmaeotrochus oculeus* (a coral).

L. *lumectum,* thorn-thicket; see *luma* under **thorn**

Gr. *mystax, -akos,* hair on the upper lip, mustache; see **hair**

L. *paniculus,* m. tuft, dim. of *panus,* m. ear of millet: panicle, paniculate, *Aster paniculatus* (an aster).

L. *penicillum,* n.; *penicillus; peniculus,* m. little tail, painter's brush, tuft: pencil, penicillate, penicillin, *Penicillium camemberti* and *P. roqueforti* (blue molds giving the flavor to those cheeses), *Linocrinus penicillus* (a Mississippian crinoid), *Phalacrocorax penicillatus* (a cormorant).

Gr. *rhops, -opos,* f. shrub, bush: *Rhopocichla atriceps* (a bird), *Nannorhops ritchieana* (Mazari palm).

L. *rubus,* bramble, blackberry; *rubetum,* bramble-thicket; see **blackberry**

L. *ruscum (bruscum),* broom; *ruscarius,* of broom; see *ruscus* under **lily**

L. *salicetum; salictum,* willow-thicket; see *salix* under **willow**

L. *scopa,* twigs, broom made of twigs; see **broom**

AS. *scrob; scrybb,* a small, woody plant: scrub, scrub-oak, shrub, shrubbery.

L. *senticetum,* a thorn-thicket; see *sentis* under **thorn**

L. *septum,* hedge, fence; see **wall**

Gr. *spartos,* broom; *spartion,* dim.; *spartinos,* made of broom; see *spartium* under **bean**

L. *spinetum,* thorn-hedge; see *spina* under **thorn**

Gr. *sychneon, -os,* m. thicket:

Gr. *tarphos,* thicket; see *tarphys* under **thick**

Gr. *thamnos,* m. shrub, bush; *thamnion,* n.; *thamniskos,* m. dim.: thamnium, thamnophile, *Thamnophis sirtalis* (garter-snake), *Thamnocrinus devonicus* (a Devonian crinoid), *Chrysothamnus paniculatus* (a rabbit-brush), *Lyonothamnus floribundus* (Lyon shrub), *Lithothamnium isthmi* (a fossil alga), *Zoothamnium arbuscula* (a protozoan), *Thamnidium elegans* (a fungus).

Gr. *thysanos,* fringe, tassel; see **fringe**

L. *vepres, -is,* m. and f. brier, bramble; *veprecula,* f. dim.: *Clathurina(Veprecula) reticulosa* (a gastropod).

L. *virgatus,* of twigs, rodlike; *virgosus,* full of bushes, bushy; *virgultum,* copse, thicket, undergrowth, < *virga,* twig, rod; see *virga* under **branch**

L. *viminetum,* willow copse; see *vimen* under **willow**

Gr. *xylochos,* f. copse, thicket: *Xylochus tibialis* (a beetle).

See: **branch, tree, cluster, crest, sod, point, shade, thorn**

bustum, L. place where corpses were cremated; see *uro* under **burn**

busy < AS. *bysig:* business, busybody, pidgin.

L. *activus,* in motion, working, energetic; *actuosus,* full of activity; see *ago* under **make**

L. *admolior, -litus,* exert oneself, strive, struggle, put hands to:

Gr. *akoimetos,* sleepless, unresting:

Gr. *ananetos,* never relaxed:

L. *annitor, -nisus; -nixus,* take pains, exert oneself, strive:

Gr. *aoknos,* restless, tireless: *Aocnus kolenati* (a beetle).

L. *applicatus,* inclined to, devoted to, attached to: apply, application.

L. *ardelio, -onis,* m. a zealous person, busybody, meddler: *Ardelio brevicornis* (a fly).

Gr. *ascholos; periascholos,* without leisure, occupied, busy:

L. *assiduus,* constant, busy, devoted: assiduous, assiduity.

L. *attentus; intentus,* applied to, directed toward, busy: attention, attentive, intent, *Pezomachus intentus* (a wasp).

L. *bullio, -itus,* bubble up; see *bulla* under **bubble**

L. *deditus,* given up to, devoted, assiduous, diligent:

Gr. *dieros,* active: *Dierobia impressa* (a beetle), *Dierobatis eregoodoo* (a ray).

L. *diligens, -entis,* assiduous, attentive, careful: diligent, diligence, *Dictyna diligens* (a spider).

Gr. *ektenes,* earnest, fervent, assiduous:

L. *enixus; enisus,* strenuous, earnest: *Osmia enixa* (a bee).

Gr. *enthousiasmos,* m. ardent interest, zeal: enthusiasm, enthusiastic.

Gr. *ergastikos,* active, industrious; *energos,* active, busy; *periergos,* over-careful, meddlesome; busybody; see *ergon* under **make**

L. *exerceo, -citus,* keep busy, drive on, practice: exercise.

L. *impiger, -gra, -grum,* diligent, active: *Carabus impiger* (a beetle).

L. *industrius,* active, zealous: industry, industrious, industrial.

L. *inotiosus,* not idle, busy:

Gr. *meletao,* attend to, practice, study; *meletema, -tos,* n. practice, exercise, study; *meletikos,* of practice; *meleteros,* diligent: *Meleterus montanus* (a fly).

L. *navus (gnavus),* active, diligent, energetic, assiduous; *navo, -atus,* perform with zeal: *Agalena nava* (a spider).

L. *negotiosus,* full of business, busy; see *negotior* under **trade**

L. *occupatus,* busy, engaged: occupation.

L. *operosus,* active, busy, industrious, < *operor, -atus,* labor, be busy; see *opus* under **make**

L. *persevero, -atus,* continue steadfastly, persist: persevere, perseverance.

Gr. *philoponos,* industrious; see *poieo* under **make**

Gr. *polypragmon, -os,* very busy, meddlesome:
Gr. *rhekterios,* active, busy; see *rhekter* under **make**
L. *sedulus,* busy, assiduous, zealous, diligent: sedulous.
Gr. *sphedanos,* eager, vehement: *Sphedanolestes pulchellus* (a bug).
Gr. *spoude,* f. zeal, exertion; *spoudaios,* active, earnest, zealous; *spoudasma, -tos,* n.
   something done with zeal:
L. *strenuus,* active, brisk, prompt, vigorous: strenuous.
L. *studeo,* be diligent, eager, zealous; *studiosus,* eager, assiduous: study, studious,
   student, studio, etude, *Anelosimus studiosus* (a spider).
L. *vehemens,* eager, ardent, vigorous: vehement, vehemence, *Otiorhynchus*
   vehemens (a beetle).
L. *zelus* (Gr. *zelos*) m.; *zelosyne,* f. ardor, emulation, enthusiasm; *zelotes,* m.
   devoted enthusiast; *zelos,* m.; *zele,* f. rival: zeal, zealous, zealot, *Zela zeus* (a
   butterfly), *Zelogenes newmanni* (a beetle), *Zelosyne poecilosoma* (a moth),
   Zelotes pallipes (a spider), *Zelotichthys alhambrae* (a fish), *Lepthyphantes*
   zelatus (a spider), *Polyzelos crassicaulis* (a beetle).
See: **make, life, go**

butalis, NL. (Gr. *boutalis*), nightingale; see **nightingale**

butcher < OF. *bouchier.*
Gr. *artamos,* m. butcher: *Artamus fuscus* (a swallow-shrike).
Gr. *boutypos,* m. butcher:
L. *carnarius,* meat-dealer, butcher: see *caro* under **flesh**
L. *carnifex,* executioner, hangman; see **kill**
Gr. *daikter,* slayer, butcher; see **kill**
Gr. *kreopoles; kreourgos,* m. butcher; *kreopolikos; kreourgikos,* of a butcher;
   kreopolion, n. butcheteria: *Creurgus lanius* (a bird).
L. *lanius; laniarius; lanio, -onis,* m. butcher; *lanienus,* of a butcher; *lanio, -atus,*
   rend, tear, mangle: laniary, lanner (*Falco laniarius*), *Lanius borealis* (shrike),
   Laniarius ferrugineus (a shrike).
L. *lardarius,* butcher; see *lardum* under **fat**
L. *macellarius,* butcher, meat-seller; see *macellum* under **trade**
See: **kill, destroy**

buteo, L. falcon, hawk; see **hawk**

butio, L. bittern; see **heron**

butomus, L. (Gr. *boutomos*), a water plant; see **plant**

butter < L. *butyrum* (Gr. *boutyron*); see **fat**

buttercup
L. *aconitum* (Gr. *akoniton*), monkshood; a genus of the buttercup family; see
   aconite
L. *actaea* (Gr. *aktaia; aktea*), f. elder; now applied to baneberry: *Actaea alba*
   (white baneberry)
Gr. *anemone,* f. windflower: *Anemone quinquefolia* (American wood-anemone),
   Anemonella thalictroides (rue-anemone), *Geranium anemonifolium* (a geranium).
NL. *aquilegia,* f. a genus of the buttercup family: *Aquilegia caerulea* (Colorado
   columbine), *Dasyneura aquilegiae* (a gall-midge).
L. *caltha,* f. a yellow flower, marigold: *Caltha palustris* (marsh-marigold).
L. *clematis, -idis* (Gr. *klematis, -idos*), f. a genus of the buttercup family, < *klema,*
   vine-twig; *klematitis, -idos,* f. a climbing plant: *Clematis paniculata* (a clematis),
   Codonopsis clematidea (an Asiabell), *Dasyneura clematidis* (a gall-midge).
NL. *coptis,* f. a genus of the buttercup family: *Coptis asplenifolia* (a goldthread).
L. *delphinium* (Gr. *delphinion*), n. larkspur: *Delphinium speciosus* (Caucasian
   larkspur, *Coreopsis delphinifolia* (a coreopsis).
L. *gallicrus, -uris,* n. crowfoot, buttercup:
NL. *hepatica,* f. genus of the buttercup family: *Hepatica acutiloba* (a hepatica).
NL. *hydrastis,* f. a genus of the buttercup family: *Hydrastis canadensis* (golden-
   seal), hydrastine.
NL. *nigella,* f. a genus of the buttercup family, < *nigellus,* dark: *Nigella sativa*
   (fennelflower).
L. *ranunculus,* m. buttercup, crowfoot: Ranunculaceae, *Ranunculus bulbosus* (a
   buttercup), *Bupleurum ranunculoides* (a thorowax).
L. *thalictrum* (Gr. *thaliktron*), n. meadowrue: *Thalictrum polygamum* (tall mead-
   owrue), *Ceratopteris thalictroides* (waterfern), *Corydalis thalictrifolia* (a corydalis).
NL. *trollius,* m. a genus of the buttercup family: *Trollius europaeus* (globeflower).
NL. *viorna,* f. a kind of clematis: *Clematis viorna* (leather-flower).

butterfly
Sp. *mariposa,* f. butterfly: Mariposa, *Arctostaphylos mariposa* (a bearberry).

L. *papilio, -onis,* m. butterfly; *papiliunculus,* m. dim.: papilionaceous, *Papilio polyxenes* (black swallowtail), *Oncidium papilio* (butterfly-orchid), *Viola papilionacea* (butterfly-violet).
Gr. *psyche,* f. butterfly: Psychodidae, *Psychoda alternata* (a mothfly), *Psychomastax psylla* (a grasshopper).
Gr. *setodokis, -idos,* f. a butterfly:
NL. *thecla,* f. a genus of butterflies: *Thecla gibberosa* (a hairstreak).
See: **moth**
buttock < uncertain origin; see **rump**
button < OF. *boton;* see **bind, projection, swell, navel, needle, ornament**
butyl < L. *butyrum,* butter, *hyle,* material; see *butyrum* under **fat**
buxus, L. boxwood; *buxeus,* of boxwood; *buxetum,* boxwood grove; see **box**
buy < AS. *bycgan;* see **trade**
buzz < imitative origin; see **murmur**
buzzard < L. *butio,* hawk; see **hawk**
by < AS. *bi, big,* near, about, see *be* under **near;** < Dan. *by,* town, village, see **town**
byas, Gr. a kind of owl; see **owl**
Byblis, Gr. a nymph changed into a spring; see **spring**
byblos, Gr. papyrus; see *biblion* under **book**
bycano- < Gr. *bykane,* trumpet; *bykanistes,* trumpeter; see **horn**
bycto- < Gr. *byktes,* swelling, blustering; see **wind**
byrsa, Gr. hide, skin, leather; *byrsine,* strap, thong; see **leather**
bysma, Gr. plug; see **close**
byssos, Gr. flax, cotton, thread; see **flax**
bythos, Gr. the deep, the depths of the sea; *bythios,* of the deep; see **deep**

C

caballus, L. horse, nag; see **horse**
cabbage < F. *caboche,* head.
L. *brassica,* f. cabbage: Brassicaceae, *Brassica oleracea* (cabbage), *Hylemia brassicae* (cabbage-maggot).
AS. *cawl,* cabbage, < L. *caulis,* stem: cole, coleslaw, colewort, kohlrabi, kale.
L. *crambe* (Gr. *krambe*), f. cabbage; *krambidion,* n. dim.: *Crambe cordifolia* (a colewort), *Crambidium achrovelfum* (a moth), *Crambodoxa platyaula* (a moth).
L. *halmyridium,* n. a kind of cabbage:
L. *lacuturris,* m. a kind of cabbage:
Gr. *rhaphanos,* cabbage, a term applied to the radish by the Romans; see **radish**
cabomba, Ab.Am. an aquatic plant; see **waterlily**
cacalia, L. (Gr. *kakalia*), a genus of the composite family; see **composite**
caccabido- < Gr. *kakkabis, -idos,* a partridge; see **quail**
caccabus, L. (Gr. *kakkabos*), a cooking pot; see **pot, black**
cachexia, L. (Gr. *kachexia*), general ill-health attended by wasting; see **disease**
cachinno, L. laugh immoderately; see **laugh**
cachla, L. (Gr. *kachla*), oxeye; see **composite**
cachleco- Gr. *kachlex, -ekos,* a pebble on a stream bed, gravel; see **stone**
cachryo- < Gr. *kachrys, -yos,* parched barley, catkin, ament; *kachrydion,* dim.; see **catkin**

cackle < D. *kakelen,* sound off like a hen after laying an egg.
L. *cacabo, -atus* (Gr. *kakkabizo*), cackle:
Gr. *chenizo,* cackle like a goose:
L. *gingritus, -us,* m. cackling of geese:
L. *glocio, -itus,* cluck like a hen:
L. *gracillo, -atus,* cackle, cluck:
Gr. *klosso,* cluck like a hen; *klogmos; klosmos,* m. a clucking sound:
Gr. *poppyzo,* smack or cluck with the tongue and lips:
L. *tetrinnio, -itus,* quack like a duck:

caco, -atus, go to stool; see **dung**

caco- < Gr. *kakos,* bad, harmful; see **bad**

cactus, L. (Gr. *kaktos*), m. a prickly plant, a kind of artichoke: Cactaceae, *Melocactus violaceus* (violet melon-cactus), *Phytophthora cactorum* (a fungus).
L. *cereus,* m. a genus of the cactus family; a wax taper; *Cereus quadricostatus* (a cactus), *Acanthocereus horridus* (a cactus), *Selenicereus inermis* (a cactus).
Sp. *nopal* (Ab.Am. *nopalli*), m. a kind of cactus: *Nopalea dejecta* (a cactus).
L. *opuntia,* f. a genus of the cactus family, < Gr. *Opous, -ountos,* a Grecian town: *Opuntia spinosissima* (a cactus).
NL. *rhipsalis,* f. a genus of the cactus family: *Rhipsalis clavata* (a cactus), *Erythrorhipsalis pilocarpa* (a cactus).

cacula, L. servant; see **servant**

cacumen, L. extreme point or end; see **end**

cadaver, L. dead body, corpse, carcass; see **body**

cadisco- < Gr. *kadiskos,* urn or box to receive ballots of the dicasts in criminal trials, < *kados,* jar, urn, box; see **jar**

cadivus, L. falling of itself; see *cado* under **fall**

cadmia, L. (Gr. *kadmia*), calamine, an ore of zinc; see **zinc,** ELEMENTS under **thing**

Cadmus, L. (Gr. *Kadmos*), fabled introducer of the Greek alphabet; see **letter**

caduceus, L. herald's staff, staff or wand of Mercury, see **rod**; *caduceator,* herald, see **call**

caducus, L. falling, deciduous; see *cado* under **fall**

cadus, L. (Gr. *kados*), jar, jug; *kadion; kadiskos,* dim.; *cadialis,* of a jar; see **pot**

cadytas, L. (Gr. *kadytas*), a parasitic plant; see *cassytha* under **laurel**

caecilia, L. a kind of lizard; see **lizard**

caecus, L. blind; *caeco, -atus,* make blind, darken; see **blind**

caeduus, L. fit for cutting, < *caedo, caesus,* cut, cut down, kill; see **cut**

caelamen, L. embossed work, bas-relief, < *caelo, -atus,* engrave in relief; see **cut**

caelebs, L. unmarried (man), single; bachelor; *caelibatus,* single life; see **alone**

caelum (coelum), L. sky, heaven; *caelestinus (coelestinus); caelestis (coelestis),* of the sky; see **heaven**

caementarius, L. mason; see **mason**

caenis, L. (Gr. *kainis*), knife; see **knife**

caeno- < Gr. *kainos,* new; see **new**

caenum, L. filth, mud, mire; *caenosus,* dirty, filthy, muddy; see **dirt**

caerefolium, L. (Gr. *chairephyllon*), an umbelliferous plant; see **carrot**

caero- < Gr. *kairos,* fit, opportune, seasonable; *kairios,* at the right time or place, see **fit**; thrum or ends of the thread in a loom, see **thread**

caeruleus, L. sky-blue; see **blue**

caesalpinia, NL. a genus of the bean family; see **bean**

Caesar, L. Gaius Julius Caesar; any autocratic monarch, emperor, or ruler; see **govern**

caesariatus, L. covered with hair, long-haired; see **hair**

caesius, L. bluish-gray; see **gray**

caesura, L. pause, break; see **break**

caffer, -fra, -frum, NL. pertaining to the Kaffirs; see *kafir* under **doubt**

cage < L. *cavea,* cavity, pen; see **pen, box, room, house**

cake < ON. *kaka,* akin to G. *kuchen.*
Gr. *ames, -etos,* m. a milk-cake.
Gr. *amitha,* f. a kind of cake:
Gr. *amora,* f. a sweet cake:
Gr. *amylos,* a cake made of fine meal; see *amylum* under **flour**

L. *arculatum*, n. a sacrificial cake:

Gr. *barax, -akos*, m. a kind of cake: *Baracus vittatus* (a butterfly).

Gr. *basynias*, m. a kind of cake:

L. *buccellatum*, n. soldier's biscuit:

Gr. *choirinas*, m. a kind of cake:

L. *copta* (Gr. *kopte*), f. a cake made of materials cut, chopped or pounded small:

L. *crustum*, bread, cake; *crustulum*, dim.; see **bread**

Gr. *dendalis, -idos*, m. a kind of barley-cake:

Gr. *diakonion*, n. a kind of cake:

L. *dulcium*, n. sweet cake, honey-cake; *dulciolum*, n. dim.; *dulciarius*, m. pastry cook:

Gr. *ellytes*, m. a kind of cake:

Gr. *epipaston*, n. a cake sprinkled with confections:

L. *erneum*, n. cake baked in an earthen pot:

L. *farreum*, spelt-cake; *farriculum*, dim.:

L. *fertum*, n. a kind of cake:

Gr. *gouros*, m. a kind of cake:

Gr. *gouttaton*, n. a kind of cake:

L. *gratilla*, f. a kind of cake:

Gr. *gyrine*, f. a kind of cake:

Gr. *itrion*, n. a kind of cake: *Itrium sociale* (a mite).

Gr. *kloustron*, n. a kind of cake:

Gr. *kollabos*, m. a kind of cake or roll:

L. *laganum* (Gr. *laganon*), n. a thin, broad cake made of flour and oil, pancake: *Laganum cingulatum* (an echinoid).

L. *libum*, n. cake, pancake; *libacunculus*, m. dim.; *libarius*, m. pastry cook:

L. *lixula*, f. pancake:

L. *lucuns, -untis*, f. a kind of pastry; *lucunculus*, m. dim.:

Gr. *magis, -idos*, f. cake; *magidion*, n. dim.:

Gr. *maza*, f. barley-cake, barley-bread, placenta; *mazion*, n.; *maziske*, f. dim.; *mazinos*, made of barley meal: mazic, mazolytic, mazopathia, *Mazatrochus* (a fossil echinoderm).

L. *mustaceum*, n. a wedding cake mixed with must and baked on laurel leaves:

Gr. *nastos*, a kind of cake; see *nasso* under **press**

L. *nucunculus*, m. a kind of nut-cake:

Gr. *pemma, -tos*, n, pastry, cake, sweetmeat; *pemmation*, n. dim.: *Pemmatodiscus socialis* (a mesozoan).

Gr. *phthoïs, -ios*, m. a kind of cake; *phthoïskos*, m. dim.:

Gr. *physte*, f. a kind of barley-cake:

L. *placenta*, f. cake, < Gr. *plakous, -ountos*, m. flat cake, pancake; *plakountion*, n. dim.: placenta, placental, *Placenticeras intercalare* (an ammonite), *Phormosoma placenta* (a sea-urchin).

Gr. *plikion*, n. a kind of cake:

Gr. *popanon*, n. a round sacrificial cake: *Popanoceras antiquum* (an ammonite).

Gr. *pyramous, -ountos*, m. a cake made with roasted wheat and honey:

L. *savillum*, n. a cheese-cake:

L. *scriblita*, f. tart:

Gr. *tagenias*, m. pancake:

cakile, NL. a genus of the mustard family; see **mustard**

cala, L. piece of wood; see **wood**

calabrix, L. a shrub, perhaps buckthorn; see **buckthorn**

caladium, NL. a genus of the arum family; see **arum**

calamintha, L. (Gr. *kalaminthe*), a mint; see **mint**

calamistratus, L. crisped, curled; see **curl**

calamitas, L. disaster; see **destroy, hurt**

calamodyto- < Gr. *kalamodytes*, a reed-bird; see **bird**

calamus, L. (Gr. *kalamos*), reed; *calamellus*, dim.; see **grass**

calandro- < Gr. *kalandros*, a kind of lark; see **lark**

calarido- < Gr. *kalaris, -idos*, a bird; see **bird**

calathus, L. (Gr. *kalathos*), a basket shaped like a lily or vase; bowl, cup; see **basket**

calator, L. caller, servant; see *calo* under **call**

calcar, L. spur; see **point**

calcarius, L. of lime, lime-burner; see *calx* under **lime**

calcatorium, L. wine-press; see **press**

calceus, L. shoe; *calceolus,* dim.; see **shoe**

calci- < L. *calx, calcis,* heel; *calcaneum,* heel, see **foot**; < *calx, calcis* (Gr. *chalix, -ikos,* pebble, gravel), stone, lime, see **stone, lime,** ELEMENTS under **thing**

calcitro, L. kick; see **kick**

calculus, L. pebble, dim. of *calx, calcis,* stone, lime; *calculo, -atus,* compute, reckon; see **stone**

caldaria, L. vessel for heating, see **vessel**; *caldarium,* hot bath, see **wash**

calends < AS. *calend,* month, < L. *calendae,* first day of the month; *calendulus,* at the calends; *calendarium,* account-book; see **day, list**

calf < AS. *cealf,* the young of cattle and some other mammals; see **cattle**

calia, NL. (Gr. *kalia*), hut, barn, bird-nest; *kalidion,* dim.; see **house**

calidris, L. (Gr. *kalidris,* var. of *skalidris*), a speckled shore bird; see **snipe**

calidus, L. warm, hot, < *caleo,* be warm, hot, glow; *calendus,* glowing; *calefacio,* make warm, heat; see **heat**

caliendrum, L. a high headdress made of false hair; see **crest**

caliga, L. soldier's shoe, boot; *caligula,* dim.; *caligatus,* booted, shod; see **shoe**

caligo, L. fog, mist, darkness; *caliginosus,* foggy, mist, dark; see **black**

calix, L. cup; *caliculus,* dim.; see **cup**

call < AS. *ceallian:* caller, calling, recall. Derive: tallyho.
 Gr. *alalazo,* shout, give the battle-cry; *alale,* f.; *alaletos,* m. war-cry, victory shout; *alalage,* f. a shouting; *alalagma, -tos,* n.; *alalagmos,* m. a loud noise:
 L. *arcesso, -itus,* call, summon:
 Gr. *auo,* shout, call:
 L. *balo,* bleat; see **bleat**
 OHG. *ban,* proclamation, edict; L. *bannum,* n. summons: ban, banns, banal, banish, bandit, abandon, contraband.
 L. *barritus,* m. the trumpeting of the elephant:
 Gr. *boao,* cry, shout, roar, howl; see **roar**
 L. *caduceator, -is,* m. herald, < *caduceus,* staff or wand of Mercury:
 L. *calo, -atus,* call; *calator, -is,* m. caller, servant: intercalary, council, conciliate, irreconcilable.
 L. *cieo, citus,* call, invoke; *accitus, -us,* m. summons; *praecia; praeco, -onis,* m. public crier; *praeconius,* of an auctioneer: cite, citation, recitation.
 L. *clamo, -atus,* call, cry out, shout; *clamator, -is,* m. bawler, shouter; *clamor, -is,* m. loud call, din; *clamosus,* noisy: clamor, acclaim, disclaim, claimant, exclamatory, proclamation, declaimer, reclamation, *Clamator glandarius* (a cuckoo), *Rana clamitans* (green frog), *Zygogramma exclamationis* (a beetle).
 L. *clueo (cluo)*; *clueor,* to be named, called:
 L. *coaxo,* croak; see **croak**
 L. *convicium,* n. clamor, loud noise, outcry, reviling: convicium, convitiate.
 L. *crocio,* caw like a crow, croak; see **caw**
 L. *cucubo,* hoot like an owl; see **hoot**
 L. *cuculo, -atus,* cry cuckoo:
 L. *cucurio, itus,* crow:
 Gr. *enope,* f. cry, shout, < *enepo,* tell, call:
 Gr. *epyo,* call; *epyta,* m. a calling, crying; *epye,* f. sound, voice:
 L. *gloctoro, -atus,* cry like a stork:
 Gr. *goës, -ëtos; goëtes,* howler, wailer, wizard; see *goëros* under **weep**
 Gr. *iache,* f. cry, shout, shriek, sound, < *iacho,* shout, sound: *Iache latirostris* (a bird).
 L. *indigito, -atus,* call upon, invoke, proclaim:
 Gr. *iygmos,* m. shout, cry, < *iyzo,* cry out, yell:
 L. *jubilo, -atus,* shout for joy; see **joy**
 Gr. *kaleo,* call, summon; *kiklesko,* reduplicate form of *kaleo,* call by name; *kaletor, -os; kletor, -os,* m. caller, crier; *klesis,* f. a calling, summons; *kletos,* called, invited, chosen; *proklesis,* f. challenge: *Cletopsyllus papillifer* (a copepod), *Procletes biangulatus* (a decapod).
 Gr. *keleusma, -tos,* call, summons, command; see **ask**
 Gr. *keryx, -ykos,* m. herald, crier, messenger; *kerykion,* n. dim.; *kerygma, -tos,* n. proclamation; *kerykinos,* of a herald, < *kerysso,* proclaim: kerygma, kerystic, *Ceryx holodiaphana* (a moth), *Hyetoceryx obscurus* (a bird).
 Gr. *klazo,* make a sharp sound, scream, shout; *klageros; klanktos,* screaming; *klange,* f. any sharp sound, scream, cry, bark:
 Gr. *krazo,* cry, scream, shriek; *kragetes; kraktes; kraugasos; kraugetes,* m crier, screamer; *kraugastikos,* screaming; *krauge,* f. a crying or screaming; *kratikos,*

noisy; *diakrazo*, scream continually: *Cractes infaustus* (a bird), *Cracticus fuligi-nosus* (a bird).

Gr. *lakeryza*, f. screamer, < *lasko*, cry, howl, shriek: *Laceryzon nigricollis* (a bird).

Gr. *laryngizo*, shout, bellow, bawl; *laryngismos*, m. a shouting, bawling:

L. *latro*, bark, howl; see **bark**

L. *mugio*, bellow, low; see **roar**

L. *nico, nictus*, beckon:

L. *nuncupo, -atus*, call by name, name, appoint; *nuncupatio, -onis*, f. a naming, name; *nuncupativus*, oral, nominal; *nuncupator, -is*, m. namer: nuncupative.

L. *nuntio, -atus*, call out: announce, enunciate, nuncio, renounce, pronunciation, denunciatory.

Gr. *ololyzo*, cry with a loud voice, shout:

Gr. *ops*, voice; see **speak**

L. *pello, -atus*, call; *appello, -atus*, accost, address, request: appellation, appeal, repeal, appellate, *Cecropia pellata* (trumpet-tree).

Gr. *phoneo*, speak, call, make a sound; see *phone* under **sound**

Gr. *phylopis, -idos*, f. battle cry, din of battle:

L. *quirito, -atus*, cry out, scream, shriek:

Gr. *Stentor, -os*, m. Stentor, a herald with a loud voice: stentorian, *Stentor poly-morpha* (a trumpet-shaped protozoan).

Gr. *trizo*, cry, scream:

L. *ululo, -atus*, howl, shriek, wail; *ululabilis*, howling, wailing: ululant, ululation.

L. *voco, -atus*, call; *vox, vocis*, f. call, sound, voice; *vocula*, f. dim.; *vocalis*, of the voice; *vocativus*, of calling; *vocifero, -atus*, cry out, clamor: vocal, vocabulary, vouch, vociferous, vocation, voice, vowel, avocation, advocacy, equivocate, dis-avow, revoke, irrevocable, provocative, *vox populi vox dei, Hyla avivoca* (a frog), *Haliaëtus vocifer* (African sea-eagle), *Antrostomus vociferus* (whippoor-will), *Tyrannus vociferans* (western kingbird).

See: **ask, speak, roar, bleat, bark, caw, croak, weep, wink**

calla, NL. a genus of the arum family; see **arum**

callaeo- < Gr. *kallaion*, cockscomb; see **crest**

callainus, L. turquoise-colored, greenish-blue; see **blue**

callarias, L. (Gr. *kallarias*), a kind of cod; see **cod**

calli-; callo-; calo- < Gr. *kalos*, beautiful; *kallion*, more beautiful; *kallistos*, most beautiful; *kallos*, beauty; see **beauty**

callidus, L. expert, shrewd, crafty, cunning; see **now**

callionymus, L. (Gr. *kallionymos*), a fish; see **fish**

Calliope, L. (Gr. *Kalliope*), the Muse of epic poetry; see *ops* under **speak, art**

Callirrhoë, L. (Gr. *Kallirrhoë*), a celebrated spring at Athens; see **spring**

callis, L. footpath, trail; see **way**

callisto- < Gr. *kallistos*, most beautiful; see *kallos* under **beauty**

callithrix, L. a monkey; see **monkey**

callitris, NL. a genus of conifers; see **gymnosperm**

callo- < Gr. *kallos*, beauty; see **beauty**

callum; callus, L. hard skin; *callosus*, with a hard skin; see **skin**, *tyle* under **knot**

calluna, NL. a genus of the heath family; see **heath**

callyntro- < Gr. *kallyntron*, broom, brush; see **broom**

calm < Gr. *kauma*, burning heat, heat of the day; see **rest**

calo- < Gr. *kalon*, billet of wood, see **wood**; < *kalos*, beautiful, see **beauty**; < *ka-los*, rope, cable; *kalodion*, dim., see **rope**

calobamo- < Gr. *kalobamon*, on stilts; see **pillar**

calotypo- < Gr. *kalotypos*, woodpecker; see **woodpecker**

calor, L. heat; *caloratus*, heated, hot; see *calidus* under **heat**

calpido- < Gr. *kalpis, -idos*, pitcher, urn, vessel; *kalpion*, dim.; see **pot**

caltha, L. a yellow flower, marigold; see **buttercup**

caltrop < ML. *calcitrapa*, a four-pointed weapon placed on the ground to impede the movements of the enemy; see **point**

calumnior, L. depreciate, misrepresent, accuse falsely; see **lie**

calvaria, L. skull; see **head**

calvatus, L. made bare, bald, < *calvus*, bald; see **bare**

calybo- < Gr. *kalybe*, cabin, hut; see **house**

calyco- < Gr. *kalyx, -ykos*, cup; *kalykion*, dim.; see **cup**

calymma, L. (Gr. *kalymma*), a covering, hood, veil; see *kalypto* under **cover**

calypto- < Gr. *kalyptos*, covered; *Kalypso*, Calypso, nymph who hid Ulysses; see *kalypto* under **cover**

calyx, L. (Gr. *kalyx*), cup, cover, outer envelope of a flower; *calyculus*, dim.; see **cup**

camaco- < Gr. *kamax*, *-akos*, pole, prop, shaft; see **pillar**

camassia, NL. a genus of the lily family; see **lily**

camato- < Gr. *kamatos*, toil, labor, distress, trouble; see *kamno* under **make**

cambarus, NL. a genus of crayfishes; see **crab**

cambio, L. exchange, barter; see **change**

camel < L. *camelus* (Gr. *kamelos*), m.; *camelinus*, pertaining to camels: camelopard, camlet, camleteen, *Camelus bactrianus* (Bactrian camel), *Boselaphus tragocamelus* (nilgai), *Apoderus camelus* (a beetle), *Alticamelus priscus* (a Miocene camel), *Alhagi camelorum* (a camel-thorn).
   L. *dromedarius*, m. camel: dromedary, *Camelus dromedarius* (Arabian camel).
   NL. *lama*, f. < Ab.Am. *llama*, guanaco: *Lama huanaco* (guanaco, alpaca, llama).

camellia, NL. a genus of the tea family; see **tea**

camelopodium, L. (Gr. *kamelopodion*), a kind of horehound; see **mint**

camera, L. (Gr. *kamara*), chamber or vault with arched roof; *camella*, dim., winecup; see **room, cup**

camilla, L. freeborn maiden; *camillus*, freeborn youth; see **child**

camilo- < Gr. *kamilos*, rope; see **rope**

caminus, L. (Gr. *kaminos*), furnace, oven, fireplace; see **oven**

camisia, L. shirt, nightgown; see **garment**

cammaron, L. (Gr. *kammaron*), a kind of aconite; see **aconite**

cammarus, L. (Gr. *kammaros*), sea-crab, lobster; see **crab**

camp < L. *campus*, plain, field; see **fort**

campa, L. (Gr. *kampe*), caterpillar; see **caterpillar**

campagus, L. a military boot; see **shoe**

campana, L. bell; *campanula*, dim.; see **bell**

campester, *-tris*, *-tre*, L. of fields; see *campus* under **field**

camphor < Ar. *kafur*; see **resin**

campo- < Gr. *kampos*, a sea-animal, see **animal**; < *kampe*, a bend, turn, see *kamptos* under **bend**

campsanema, L. (Gr. *kampsanema*), rosemary; see **mint**

campsio- < Gr. *kampsion*, dim. of *kampsa*, case, casket; see **box**

campto-; campylo; campo-; campso-; campho- < Gr. *kampto*, bend; *kampe*; *kampsis*, a bend, curve; *kampter*, *-os*, bend, angle; see **bend**

campus, L. field, plain; *campester*, *-tris*, *-tre*, of fields, see **field**; < Gr. *kampos*, a sea-animal, see **animal**

camur, L. turned inward, crooked; see **bend**

camus, L. (Gr. *kemos*), muzzle, gag; see **bind**

can < AS. *cunnan*, be able, know, see **strong, know**; < AS. *canne*, see **vessel**

canaba, L. hovel, hut; *canabula*, dim.; see **house**

canabino- < Gr. *kanabinos*, lean, slender; see **thin**

canacho- < Gr. *kanachos*, noisy; *kanache*, sharp sound, clang, ring; see **ring**

canal < L. *canalis*, waterpipe, channel; *canaliculus*, dim.; see **ditch, way**

canarium, NL. a genus of the Burseraceae; see **bursera**

canavalia, NL. a genus of the bean family; see **bean**

cancamum, L. (Gr. *kankamon*), an Arabian gum; see **resin**

cancel < L. *cancello*, *-atus*, make like a lattice, strike out, cross out: cancellation.
   L. *aboleo*, *-litus*, terminate, destroy; see **destroy**
   L. *abrogo*, *-atus*, annul, recall, repeal: abrogate, *Ichneumon abrogator* (a wasp).
   L. *annullo*, *-atus*, void, abolish, < *ad*, to, *nullus*, none: annul, annulment.
   Gr. *apaleipho*, wipe off, expunge, cancel:
   Gr. *atyktos*, undone:
   L. *casso*, *-atus*, annul, bring to naught, destroy:
   L. *deleo*, *-etus*, blot, efface, erase; *delebilis*, that may be erased; *deletilis*, that wipes out; *deletrix*, *-icis*, f. destroyer: delete, deletion, delible, indelible.
   Gr. *diagrapho*, draw a line through, cancel:
   L. *elido*, *-isus*, strike out, annul; see *laedo* under **strike**

L. *erado, erasus,* scrape or rub out, expunge, obliterate, abolish, cancel: erase, erasure.

L. *expungo, -unctus,* strike out, blot out, erase: expunge.

Gr. *kathaireo,* put down, cancel, rescind, destroy; see **destroy**

L. *lituro, -atus,* blot out, erase; *litura,* f. erasure, blot, blur: *Teuchnocnemis lituratus* (a fly), *Diaborus lituratorius* (an ichneumonid).

Gr. *lyo,* loose, dissolve, free; see **free**

L. *oblitero, -atus,* blot out, strike out, efface, erase: obliterate, *Zabrotes obliteratus* (a beetle), *Nymphula obliteralis* (a moth).

L. *rescindo, -scissus,* abrogate, repeal: rescind.

L. *resigno, -atus,* annul, cancel, rescind: resign, resignation.

See: **no, rub, scrape, destroy, kill, abandon, pay, free**

cancellus, L. lattice, grill; *cancellatus,* latticed, cross-barred, grilled; see **screen**

cancer, L. crab, see **crab**; a benign or malignant growth, see **sore**

canchasmo- < Gr. *kanchasmos,* loud laughter; see **laugh**

candelabrum, L. candlestick; see *candeo* under **light**

candidate < L. *candidatus,* m. office-seeker clad in white: candidacy, candidature.
  Gr. *agonistes,* contestant, candidate; see *agon* under **fight**
  L. *ambitor, -is,* m. candidate:

candidus, L. shining white, bright, radiant, frank, < *candeo; candico,* shine, glow, be white; see **light**

candle < L. *candela;* see *candidus* under **light**

candor, L. openness, sincerity, clearness; see **true**

candy < Ar. *quand,* cane sugar; see **food**

candytalis < Gr. *kandytalis,* clothes-press; see **press**

cane < Gr. *kanna,* reed; see **grass, rod**

canescens, L. gray, hoary; see *canus* under **gray**

canica, L. a kind of bran; see **scale**

canipa, L. fruit-basket; see **basket**

canis, L. dog; *canicula,* dim.; *caninus; canicularis,* pertaining to dogs; see **dog**

canistrum; canistellum, L. (Gr. *kanistron; kaneon; kaneion*), basket; see **basket**

canna, L. (Gr. *kanna*), reed; *cannula,* dim.; see **reed**

cannabis, L. (Gr. *kannabis*), hemp, tow; see **hemp**

canon, L. (Gr. *kanon*), measuring-rod, rule, standard, model; see **measure**

canopum, NL. (Gr. *kanopon*), elder; see **elder**

canorus, L. singing, melodious; see *cano* under **sing**

canteriolus, L. trellis for supporting plants; see **frame**

cantharis, L. (Gr. *kantharis*), a blister-beetle; see **beetle**

cantharus, L. (Gr. *kantharos*), wide-bellied vessel with handles; *kantharion,* dim., see **pot**; a kind of beetle, see *kantharis* under **beetle**; a bream, see **bream**

canthelio- < Gr. *kanthelion,* pack-saddle, pannier; see **saddle**

cantherius, L. (Gr. *kanthelios; kanthon*), ass, mule, gelding; see **horse**

canthus, L. (Gr. *kanthos*), tire of a wheel, edge, corner; see **border**

canthylo- < Gr. *kanthyle,* swelling, tumor; see **swell**

cantus, L. song; *canticulum,* dim.; *cantor,* singer, minstrel; see *cano* under **sing**

canus; canutus, L. gray, ash-colored, hoary; see **gray**

canyon < Sp. *canon,* < L. *canna,* reed, pipe; see **valley**

caoutchouc < Ab.Am. *cauchu,* rubber; see **resin**

cap < AS. *caeppe,* < LL. *cappa,* f. hood: cape, caparison, nightcap.
  L. *ambiga,* f. cap of a still:
  L. *arsinium,* n. a woman's headdress:
  L. *capidulum,* n. a kind of head-covering:
  L. *capitium,* n. covering for the head:
  L. *cassida; cassis, -idis,* f. helmet; *cassidula,* f. dim.: cassideous, *Cassida bivittata* (a tortoise-beetle), *Cassidula angulifera* (a gastropod), *Cassidulina sicula* (a foraminifer), *Cristellaria cassidata* (a foraminifer), *Crossocassis pilosa* (a beetle).
  L. *causia* (Gr. *kausia*), f. a light, broad-brimmed hat: *Sabal causiarum* (hat-palm).
  L. *cucullus,* m. cap, hood: cucullate, cowl, *Cuculligera flexuosa* (a grasshopper), *Viola cucullata* (a violet), *Dicentra cucullaria* (Dutchman's-breeches).
  L. *cudo, -onis,* m. helmet made of raw skin:
  Gr. *daktylethra,* f. finger-sheath, thimble:

Gr. *epelis, -idos,* f. cover, lid, freckle: *Epelis truncataria* (a moth), *Epelichthys michaelis* (a fish).

Gr. *epikranon,* n. headdress, cap, helmet:

Gr. *epitygma, -tos; epityche,* over-fold, flap, operculum; see *ptyx* under **fold**

Gr. *epithema, -tos,* n. cover, lid: *Epithematus nitidus* (a beetle).

L. *galea,* f. helmet; *galeole,* f. dim.; *galerum,* n.; *galerus,* m. bonnet, cap, hat, helmet; *galericulum,* n. dim.; *galearis,* of a helmet; *galeritus,* wearing a hood; *galeo, -atus,* cover with a helmet: galeate, *Galerucella luteola* (elm-leaf beetle), *Galericulum ovatum* (a gastropod), *Robulina galeata* (a foraminifer), *Scutellaria galericulata* (marsh-skullcap), *Kakatoë galerita* (a cockatoo).

Gr. *hyrtane,* f. potlid: *Hyrtanommatium crassum* (a wasp).

Gr. *korys, -ythos,* f. helmet; *korystes,* m. one armed with a helmet; *korystos,* crested, helmeted; *korysso,* furnish with a helmet: corystosperm, *Corythophanes cristatus* (abbess-lizard), *Corythucha arcuata* (a lacebug), *Corystes punctatus* (a decapod), *Coryssopus hexasticus* (a beetle), *Loxocorys sericea* (a moth), *Calamospiza melanocorys* (lark-bunting), *Aulacoryssus vermiculatus* (a beetle), *Otocoris*[*Otocorys*] *alpestris* (horned lark).

Gr. *kranos,* n. helmet; *perikranon,* n. helmet, cap: *Cranocrinus timoricus* (a Permian crinoid).

Gr. *kynee,* f. cap made of dog skin:

Gr. *kyrbasia,* f. a Persian hat with a peaked crown: *Cyrbasia pupina* (a fossil gastropod), *Cyrbasiodon boycei* (a cynodont).

L., Gr. *mitra,* f. headband, headdress, turban; *mitrula; mitella,* f.; *mitrion,* n. dim.; *mitratus,* wearing a miter: miter, mitriform, mitral, *Mitra vulpecula* (a gastropod), *Mitrula phalloides* (a mushroom), *Mitella nuda* (mitewort), *Mitrocrinus weatherbyi* (an Ordovician crinoid), *Mitriostigma axillare* (a rubiacead), *Gyromitra brunnea* (a fungus), *Numida mitrata* (a guinea-fowl), *Chalcomitra amethystina* (a sunbird), *Physcomitrium turbinatum* (a moss).

L. *operculum,* n. cover, lid; *operculo, -atus,* cover with a lid: operculum, *Megalopyge opercularis* (a puss-caterpillar), *Operculina punctata* (a foraminifer).

Gr. *pelex, -ekos,* f. helmet: *Pelecocrinus aqualis* (a Mississippian crinoid).

Gr. *perikephalon,* n. cap:

L. *petasus* (Gr. *petasos*), m. broad-brimmed hat; *petasunculus,* m.; *petasion,* n. dim.; *petasites,* m. one with a hat; *petasatus,* with hat on: *Petasina fulva* (a gastropod), *Petasata eucope* (a medusa), *Petasites niveus* (snowleaf-butterbur), *Petasodes reflexa* (a cockroach), *Petasometra helianthoides* (a crinoid), *Carex petasata* (a sedge), *Pinna petasunculus* (a pelecypod), *Halipetasus scaber* (a jellyfish).

L. *pileus* (Gr. *pileos*), m. cap; *pileolus,* m. dim.; *pilus* (Gr. *pilos*), m. felt cap; *pilidion; pilion,* n. dim.; *pileatus,* capped: pilidium, pileorhiza, *Pilea pumila* (clearweed), *Ceophloeus pileatus* (pileated woodpecker), *Theophormis callipilium* (a radiolarian).

Gr. *poma, -tos,* n. cover, lid, operculum, gill-cover; *pomation,* n. dim.: pomatic, pomarine, pomatorrhine, *Pomatostomus ruficeps* (chestnut-crowned babbler), *Pomatomus saltatrix* (bluefish), *Pomacentrus fuscus* (a fish), *Helix pomatia* (a snail), *Stercorarius pomarinus* (pomarine jaeger).

Gr. *tholia,* f. comical hat with a broad brim: *Tholia clypeata* (a spider).

L., Gr. *tiara,* f. turban; *tiaratus,* turbaned: tiara, *Tiarophorus elongatus* (a beetle), *Tiarospyris amphora* (a radiolarian), *Tiarella cordifolia* (Allegheny foamflower), *Tiaris olivacea* (Mexican grassquit), *Extatosoma tiaratum* (a phasmatid).

Sp. *toca,* headdress, hood, cowl; *toquilla,* dim.:

See: **crest, top, cover, roof**

capax, *-acis,* L. roomy, spacious, able, fit; see *capio* under **take**

cape < LL. *cappa,* hood, see **cap;** < L. *caput,* head, end, point, see **head**

capedo, L. cup used in sacrifices; *capeduncula,* dim.; see *capis* under **cup**

capella, L. she-goat; *caper, capri,* he-goat; see *caper* under **goat**

capelo- < Gr. *kapelos,* huckster, retailer, peddler; see **trade**

caper < L. *capparis* (Gr. *kapparis*), f.: Capparidaceae, *Capparis spinosa* (common caper).

NL. *cleome,* f. a genus of the caper family: *Cleome hispida* (hairy spiderflower).

caperatus, L. wrinkled, < *capero, -atus,* wrinkle; see **fold**

capillus, L. hair; *capillaceus; capillaris,* of hair; *capillatus,* hairy; see **hair**

capis, *-idis,* L. cup or bowl with one handle; *capula,* dim.; see **cup**

capistrum, L. halter, muzzle; see **bind**

capital < L. *capitalis,* relating to the head, < *caput, -itis,* head; *capitellum; capitulum,* dim.; see **head**

capitaneus, L. large; see **large**

Capitolium, L. temple of Jupiter, statehouse; see temple
capno- < Gr. *kapnos*, smoke; *kapnodes*, like smoke; *kapne*, smoke-hole, chimney; see cloud
capo- < Gr. *kape*, crib, manger, see basin; < *kapon*, a castrated cock, see chicken
cappa, LL. hood, cape; see garment
capparis, L. (Gr. *kapparis*), caper; see caper
capreolatus, L. tendriled, < *capreolus*, tendril, goat; see *caper* under goat
caprifolium, L. honeysuckle; see honeysuckle
caprimulgus, L. goatsucker; see goatsucker
caprinus, L. of goats, < *caper, capri*, goat; *capra*, she-goat; see goat
caprisco- < Gr. *kapriskos*, a fish; see *kapros* under hog
capro- < Gr. *kapros*, boar, wild boar; *kaprios*, like a boar, lustful; see hog
caprunculum, L. an earthen vessel; see pot
capsa, L. case, box; *capsula*, dim.; see box
capsi- < Gr. *kapsis*, a gulping down, < *kapto*, gulp down; see eat
capsicum, NL. chili pepper; see pepper
capstan < F. *cabestan*, < L. *capistrum*, halter; see turn
captio, L. deception, fraud, fallacy; see lie
captivus, L. taken prisoner, a prisoner; see pen
captus, L. a taking or seizing, that which is taken, < *capio*, take; see take
capulus, L. holder, sepulcher, tomb, coffin, handle; see box, grave, handle
caput, L. head, end, point; *capitulum*, dim.; see head
capyro- < Gr. *kapyros*, dry; see dry
cara- < Gr. *kara*, head, top; see head
carabus, L. (Gr. *karabos*), a kind of beetle, a crustacean, a light ship; see beetle
caragana, NL. a genus of the bean family; see bean
carano- < Gr. *karanos*, a chief; see lead
caranx, NL. a genus of fishes; see fish
carassius, NL., < F. *carassin*, carp; see carp
carato- < Gr. *kara, -tos*, top; see head
carapa, NL. a genus of the mahogany family; see mahogany
carbasus, L. (Gr. *karpasos*), a fine Spanish flax; see flax
carbo, -onis, L. charcoal, coal; *carbunculus*, dim., a red, precious stone, a severe boil; see coal, stone, sore, ELEMENTS under thing
carcer, L. (Gr. *karkaron*), jail, prison; see pen
carchaleo- < Gr. *karchaleos*, rough, fierce; see rough
carcharo- < Gr. *karcharos*, sharp-pointed, jagged; *karcharias*, a kind of shark; see point, shark
carchesium, L. (Gr. *karchesion*), cup contracted in the middle and with handles that reached from rim to bottom; see cup
carcino- < Gr. *karkinos*, crab; see crab
carcinoma, L. (Gr. *karkinoma*), cancerous ulcer; see sore
card < L. *charta*, leaf of paper; see paper
cardamum, L. (Gr. *kardamon*), a kind of cress; *kardamomon*, a spice; see mustard
cardio- < Gr. *kardia*, heart; *cardiacus* (Gr. *kardiakos*), of the heart; see heart
cardita, Sp. small vessel; F. *cardite*, a mollusk; see mollusk
cardo, < -inis, L. hinge; *cardinalis*, pertaining to a hinge, chief; see hinge
cardopo- < Gr. *kardopos*, kneading-trough; see basin
carduelis, L. thistlefinch; see finch
carduus, L. thistle; see thistle
care < AS. *carus, cearu;* see guard, heed, feel, think, nurse, give, busy, exact, heal, help, mercy
careless; see neglect, rest
careno- < Gr. *karenon*, head; see *kara* under head
carenum, L. a sweet wine; see wine
caress < F. *carresser*, < L. *carus*, dear; see touch, embrace, kiss
carex, -icis, L. sedge; see sedge
carica, L. a dried fig; see fig

caries, L. decay; *cariosus,* rotten, decayed; see **rot**

carina, L. keel; see **ridge**

caris, L. (Gr. *karis*), shrimp; see **crab**

carisa, L. an artful woman; see **woman**

carissa, NL. a genus of the dogbane family; see **dogbane**

caritas, L. dearness, costliness, high price, regard, esteem, affection, love; see *carus* under **love**

caritus, L. lacking, devoid of, free from; see **poor**

carmen, L. song, poetry, ballad; see **sing**

carminator, L. carder, < *carmino, -atus,* card, comb, cleanse; see **comb**

carmine < Ar. *qirmiz,* deep red, L. *minium,* red lead; see *qirmiz* under **red**

carnabadium, NL. (Gr. *karnabadion*), caraway; see **carrot**

carnalis, L. fleshly, < *caro, carnis,* flesh, pulp; *carnicula; caruncula,* dim.; *carneus,* of flesh; *carnosus,* fleshy; *carnarius,* butcher; see **flesh**

carnifex, L. executioner, hangman; see **kill**

carno- < Gr. *karnon,* a Gallic horn; see **horn**

carota, L. carrot; see **carrot**

caroticus, L. (Gr. *karotikos*), soporific, stupefying, < *karos,* heavy sleep, torpor; see **sleep**

carp < OF. *carpe.*
    L. *alburnus,* m. a whitefish: *Alburnus lucidus* (blay).
    Gr. *baleros,* m. a kind of carp:
    L. *barbus,* m. a fresh-water fish: *Barbus fluviatilis* (barbel), *Stentoropsis barbi* (a protozoan).
    NL. *carassius,* m. < F. *carassin,* carp: *Carassius vulgaris* (crucian carp).
    L. *cyprinus* (Gr. *kyprinos*), m. carp: cyprinoid, Cyprinidae, *Cyprinus carpio* (carp), *Cyprinodon variegatus* (round minnow), *Megastomatobus cyprinella* (buffalofish), *Megalops cyprinoides* (a tarpon).
    Gr. *lepidotos,* m. probably a carp: *Lepidotus maximum* (a fossil fish).
    Gr. *leukiskos,* m. chub: *Leuciscus lineatus* (a chub).
    Gr. *phoxinos,* m. probably a minnow: *Phoxinus flammeus* (a chub), *Phoxinopsis typicus* (a characinid).
    NL. *scardinius,* m. name for a cyprinoid fish: *Scardinius erythrophthalmus* (rudd).
    L. *tinca,* f. tench: *Cyprinus tinca* (tench).

carpalimo- < Gr. *karpalimos,* swift; see **swift**

carpenter < L. *carpentarius,* m. wagon-maker.
    L. *abietarius,* joiner; see *abies* under **fir**
    L. *faber,* craftsman, carpenter, workman; see **make**
    Gr. *hylourgos,* carpenter; see *hyle* under **wood**
    L. *lignarius,* woodworker, carpenter; see *lignum* under **wood**
    Gr. *oikodomos,* m. architect, builder: *Oecodoma hystrix* (a wasp).
    L. *structor,* builder, carpenter; see *struo* under **make**
    Gr. *tekton, -os,* m.; *tektaina,* f. carpenter, joiner, any craftsman; *architekton, -os,* m. master-builder, engineer; *tektonikos,* pertaining to building; *tikto,* bring forth, produce, make: tectonic, eutectic, architect, architecture, *Tectona grandis* (teak), *Architectonica granulata* (a gastropod), *Dystecta mendica* (a bug).
    L. *tignarius,* carpenter; see *tignum* under **beam**
    Gr. *xylourgos,* m. woodworker, carpenter: *Xylurgus subrufinus* (a bird).

carpentum, L. carriage, coach; *carpentarius,* of a wagon; carriage-maker; see **vehicle**

carpesium, L. (Gr. *karpesion,* an aromatic Asiatic wood), a genus of the composite family; see **composite**

carpet < OF. *carpite,* rug; see **rug**

carpho- < Gr. *karpho,* dry up, wither; *karphaleos,* dry; *karphos,* any dry particle, straw, chaff, chip; *karphion,* dim.; see **dry, scale**

carpinus, L. hornbeam, blue beech; *carpineus,* of hornbeam; see **hornbeam**

carpo- < Gr. *karpos,* fruit; see **fruit**

carptus, L. plucker, < *carpo,* pluck, gather, cull, tear, snatch, slander; see **tear**

carpus, L. (Gr. *karpos*), wrist; see **hand**

carpyco- < Gr. *karpyke,* a kind of plant; see **plant**

carrot < L. *carota,* f. (Gr. *karoton,* n.): carotene, carotenoid, *Daucus carota* (carrot).
    Gr. *agasyllis, -idos,* f. a cow-parsnip:
    L. *ammi; ammium* (Gr. *ammi*), n. a genus of the carrot family: Ammiaceae, ammiaceous, *Ammi majus* (ammi).

L. *anethum; anisum* (Gr. *anethon; anison*), n. dill, anise; *anethinos*, made of dill or anise: *Anethum graveolens* (dill), anise (*Pimpinella anisum*), *Illicium anisatum* (Japanese anise-tree), *Chrysanthemum anethifolium* (a chrysanthemum).

NL. *angelica*, f. a genus of the carrot family: *Angelica villosa* (hairy angelica).

L. *anthriscus*, f. (Gr. *anthriskos*, m.), an umbellifer: *Anthriscus sylvestris* (a chervil), *Delphinium anthriscifolium* (a larkspur).

L. *apium*, n. celery, parsley; *apiastrum*, n. wild celery; *apiacus*, pertaining to celery: apiol, apiose, ache, smallage, *Apium ammi* (a wild celery), *Apiastrum angustifolium* (California apiastrum), *Xanthorrhiza apiifolia* (yellow-root).

NL. *arracacia*, f. a genus of the carrot family, < Ab.Am. *aracacha: Arracacia xanthorrhiza* (arracacha).

NL. *azorella*, f. a genus of the carrot family: *Azorella caespitosa* (yareta).

L. *bunium* (Gr. *bounion*), n. an unidentified plant; name now applied to a genus of the carrot family: *Bunium bulbocastanum* (an umbellifer).

L. *bupleurum* (Gr. *boupleuron*), n. a genus of the carrot family: *Bupleurum rotundifolium* (thorowax), *Oxalis bupleurifolia* (a sorrel).

L. *caerefolium* (Gr. *chairephyllon*), n. an umbelliferous plant: chervil, *Anthriscus cerefolium* (salad-chervil).

L. *carum* (Gr. *karon*), n. caraway: *Carum carvi* (caraway).

L. *caucalis, -idis* (Gr. *kaukalis, -idos*), f. an umbelliferous plant: *Caucalis microcarpa* (false carrot).

Gr. *chrysoxylon*, n. a name for *thapsos*, that yields a yellow dye:

L. *cicuta*, f. water-hemlock: *Cicuta maculata* (water-hemlock), *Erodium cicutarium* (alfileria, storksbill).

L. *conium* (Gr. *koneion*), n. poison-hemlock: *Conium maculatum* (poison-hemlock), *Conioselinum chinense* (hemlock-parsley).

L. *coriandrum* (Gr. *koriannon*), n. coriander: *Coriandrum sativum* (common coriander).

L. *crithmum* (Gr. *krithmon*), n. samphire: *Crithmum maritimum* (samphire), *Inula crithmoides* (golden samphire), *Ranunculus crithmifolius* (a buttercup).

L. *cuminum* (Gr. *kyminon*), n. cumin: *Cuminum cyminum* (cumin).

L. *daucus* (Gr. *daukos*), m. carrot: *Daucus pusillus* (a carrot).

L. *eryngium* (Gr. *eryngion*, dim. of *eryngos*, f.), n. a genus of the carrot family: *Eryngium maritimum* (eryngo), *Actinolema eryngioides* (an umbellifer).

L. *feniculum* (*faeniculum; foeniculum*), n. a genus of the carrot family: fennel (*Feniculum vulgare*), *Lomatium feniculaceum* (an umbellifer).

L. *ferula*, f. giant fennel: *Ferula foetida* (an asafetida plant), *Prangos ferulacea* (an umbellifer).

L. *gingidium* (Gr. *gingidion*), n. a kind of carrot: *Daucus gingidium* (French carrot).

Gr. *karnabadion*, n. caraway:

L. *laserpitium* (*laserpicium*), n. a genus of the carrot family: *Laserpitium latifolium* (laserwort).

L. *ligusticum* (*levisticum*), n. an umbelliferous plant indigenous to Liguria: lovage (*Levisticum officinale*), *Ligusticum filicinum* (a ligusticum), *Clematis ligusticifolia* (a clematis).

L. *marathrum* (Gr. *marathron; marathon*), n. fennel: Marathon.

L. *meum* (Gr. *meon*), n. spignel: *Meum athamanticum* (spignel), *Meoma grandis* (sea-urchin).

L. *myophonus* (Gr. *myophonos*), m. an umbellifer said to kill mice:

Gr. *myrrhis, -idos*, f. sweet cicely: *Myrrhis odorata* (sweet cicely).

L. *narthex, -ecis* (Gr. *-ekos*), m. an umbellifer: *Ferula narthex* (a fennel).

L. *pastinaca*, f. parsnip: *Pastinaca sativa* (parsnip), bisnaga.

L. *peucedanum* (Gr. *peukedanon*), n. a genus of the carrot family: *Peucedanum officinale* (hog-fennel).

It. *pimpinella*, f. pimpernel; now a genus of the carrot family: *Pimpinella saxifraga* (a pimpinella), *Lycopersicum pimpinellifolium* (a tomato).

Gr. *sagapenon*, n. a plant of the carrot family: sagapenum.

NL. *sanicula*, f. a genus of the carrot family: *Sanicula tuberosa* (tuber-sanicle), *Delphinium saniculaefolium* (a larkspur).

L. *scandix, -icis* (Gr. *skandix, -ikos*), f. an umbellifer: *Scandix pectenveneris* (ladyscomb).

L. *selinum* (Gr. *selinon*), n. parsley; *hipposelinon*, n. horse-parsley: celery (*Apium graveolens*), *Selinum officinale* (an umbellifer), *Petroselinum crispum* (parsley).

Gr. *seselis*, f. (*seseli*, n.), a plant; now a genus of the carrot family: *Seseli gummiferum* (an umbellifer.)

L. *silaus*, m. a genus of the carrot family: *Silaus flavescens* (sulfur-silaus).

L. *sirpe*, n. an asafetida plant:

L. *sisarum* (Gr. *sisaron*), n. skirret: *Sium sisarum* (skirret water-parsnip).

L. *sisum* (Gr. *sison*), n. a genus of the carrot family: *Sison*[*Sisum*] *amomum* (honewort).

L. *sium* (Gr. *sion*), n. water-parsnip: *Sium suave* (hemlock water-parsnip).

L. *smyruium* (Gr. *smyrnion*), n. a genus of the carrot family: *Smyrnium perfoliatum* (an umbellifer).

L. *sphondylium* (Gr. *sphondylion*), n. cow-parsnip: *Heracleum sphondylium* (a cow-parsnip).

L. *staphylinus* (Gr. *staphylinos*), m. a kind of carrot or parsnip:

L. *thapsia* (Gr. *thapsia; thapsos*), f. a poisonous umbellifer from Thapsos that was the source of a yellow dye; *thapsinos*, yellow: *Thapsia garganica* (deadly carrot), *Thapsimillas affinis* (a bird), *Verbascum thapsus* (woolly mullein).

L. *tordylium* (Gr. *tordylion*), n. an umbelliferous plant: *Tordylium officinale* (hartwort), *Coriandrum tordylioides* (Mediterranean coriander).

NL. *zizia*, f. a genus of the carrot family, < I. B. Ziz, Rhenish botanist: *Zizia aurea* (golden alexander).

Gr. *zomile*, f. a kind of dill:

carrus, L. wagon, cart; *carruca*, four-wheeled coach; see **vehicle**

carry < OF. *carrier*, < L. *carrus*, cart.

L. *angarius* (Gr. *angaros*), mounted courier, messenger; see **ride**

L. *aquarius; aquator*, water-carrier; see *aqua* under **water**

L. *aquilifer*, eagle-bearer, standard-bearer; see *aquila* under **eagle**

L. *Atlas, -antis* (Gr. *-antos*), m. mythical Titan who held up the heavens, < *tlao*, bear, carry, suffer; *tlemon, -os; tletos*, patient, enduring, suffering; *tlemosyne; tlesis*, f. endurance; *tlesikardios*, hard-hearted, miserable; *tlesiphonos*, toiling patiently; *polytlas*, long-suffering: atlas, Atlas Mountains, Atlantic, *Polyancistrus atlas* (a tettigonid), *Atlasaster jeanneti* (a Jurassic echinoid), *Atlantidium secundum* (a pillbug), *Tlaoceras saga* (a beetle), *Tlemon delicatus* (a wasp), *Cedrus atlantica* (Atlas cedar).

L. *bajulus*, m. porter, carrier, < *bajulo, -atus*, carry a burden: bail, bailiff, bailiwick.

Gr. *bastaktes*, porter, < *bastazo*, lift, carry; see *bastagma* under **weight**

L. *basterna*, sedan chair, litter; see **sit**

AS. *beran*, carry: bear, birth, born, forebears, forbearance, bier, bairn, burden.

Gr. *Charon, -os*, m. Charon, the ferryman who, for an obol, carried the souls of the dead over the river Styx: Charon.

Gr. *diadekter, -os*, m. transmitter:

L. *dossuarius*, carrying on the back:

Gr. *epaktos*, imported: *Epactophanes richardi* (a copepod).

L. *fero, latus*, carry, bear; *-fer, -a, -um*, suffix meaning bear, carry, have; *feretrum; ferculum*, n. litter, bier, tray, barrow; *fertorius*, serving to carry; *allatus*, brought, conveyed; *circumlatus*, portable; *illatus*, carried or brought in; *perlatus*, carried through, completed; *lator, -is*, m. bearer, proposer: afferent, efferent, defer, confer, conifer, difference, fertility, offering, interfere, proffer, transfer, preferable, circumference, suffer, Cruciferae, odoriferous, feretory, collate, relate, prelate, relativity, translate, superlative, ablation, oblate, dilatory, fructiferous, Porifera, vociferate, *Mangifera odorata* (Kuwini mango), *Otiorhynchus setifer* (a beetle), *Cidaris spinifera* (a sea-urchin), *Dioscorea bulbifera* (a yam).

L., Gr. *Ganymedes, -is*, m. a beautiful youth who became the cupbearer of the gods: Ganymede, *Ganymedes anaspidis* (a protozoan), *Ganymedebdella cratera* (a leech).

L. *gero, gestus*, bear, carry, perform; *-ger, -a, -um*, suffix meaning bear, carry, have; *gerula*, f.; *gerulus*, m. bearer, carrier; *gestabilis*, portable; *gestatus*, borne, carried, < *gesto, -atus*, bear, carry; *congero*, bring together, heap up: belligerent, armiger, cornigerous, globigerina, gerundive, register, digest, indigestion, gestion, suggest, gesture, gesticulate, jest, congeries, congestion, gestation, *Myrmecocystus melliger* (honey-ant), *Polyancistrus gerulus* (a tettigonid), *Tachina plumigera* (fly), *Probosciger aterrima* (black cockatoo).

L. *gravidus*, laden, pregnant; see *gravis* under **heavy**

Gr. *komizo*, carry off to preserve; see **guard**

L. *lecticarius*, litter-bearer; see *lectica* under **vehicle**

Gr. *ocheo*, bear, carry; *ochesis*, f. a bearing or carrying:

Gr. *omistes*, m. porter, < *omizomai*, take on one's shoulders:

Gr. *phero*, bear, carry, bring; *-pher, -phor*, suffix meaning bear, carry, have; *pheretron; phoreion*, n. litter, bier, sedan chair; *pherma, -tos; phorema, -tos*, n. burden, load; *phortax, -akos*, m. porter; *phortikos*, of bearing, carrying, vulgar; *phoretos*, borne, carried; *phorimos; phoras, -ados*, bearing, fruitful; *ekphero*, carry out; *notophoros*, carrying on the back: Christopher, periphery, semaphore, metaphore, phosphorus, euphoria, diaphoretic, Berenice, *Symphoricarpus occidentalis* (a snowberry), *Sceptrophorus convexus* (a chalcid), *Sphenophorus* *sculptilis* (a beetle), *Pherocera signatifrons* (a fly), *Calliphora vomitoria* (a blue-bottle fly), *Dysphorus torquatus* (a copepod), *Chlorophora tinctoria* (fustic), *Veronica officinalis* (common speedwell), *Ecphora tricostata* (a gastropod), *Ixophorus italicus* (a millet), *Fortax morio* (a beetle).

L. *pincerna* (Gr. *pinkernes*), m. cupbearer, butler: *Pincerna liratula* (a gastropod).

Gr. *porthmeus*, ferryman; see **passage**

L. *porto, -atus,* carry; *portabilis,* that may be carried: portable, porter, deportment, transport, important, importer, portly, portfolio, exportation, disport, support, sportsman.

L. *praegnans,* with child; see **fertile**

Gr. *syrtos,* washed along by a stream, as alluvial material, < *syro,* sweep, drag, drift, or trail along; *syrmas, -ados,* f. drift, snowdrift; *syrma, -tos,* n. robe with a train, anything trailed along, sweepings, litter; *pansyrtos,* swept together, accumulated: *Hyria syrmatophora* (a pelecypod).

L. *tolero, -atus,* bear, endure; *tolerabilis,* bearable, endurable; *intolerabilis,* that cannot be borne, insupportable; *intolerans,* that cannot bear a thing, impatient: tolerate, tolerably, intolerable, intolerance.

L. *-uchus,* < Gr. *echo,* hold, carry: daduchus, *Crenuchus spilurus* (a fish), *Dictyuchus monosporus* (a fungus), *Dragmatucha proaula* (a moth).

L. *veho, vectus,* carry; *vehiculum,* n. wagon, conveyance; *vector, -is,* m. carrier, rider; *conveho,* bring together, gather; *evectus,* carried out, exported, spread; *invectus,* brought in, imported: vector, invective, vehicle, evectant, convex, convection, vexation, vex, vein, inveigh, palfrey.

L. *veredarius,* post-boy, courier; see *veredus* under **horse**

See: **weight, hold, pain, stand**

carsio- < Gr. *karsios,* crosswise; see **cross**

cartallus, L. (Gr. *kartallos*), basket; see **basket**

cartero-; see *krateros* under **strong**

carthamus, NL. safflower; see **composite**

cartilago, L. gristle; *cartilagineus; cartilaginosus,* gristly; see **gristle**

carto- < Gr. *kartos,* shorn close, shortened; see **short**

caruca, ML. plow; see **plow**

carum, L. (Gr. *karon*), caraway; see **carrot**

caruncula, L. dim. of *caro,* flesh; see **flesh**

carus, L. dear; *carior,* dearer; *carissimus,* dearest; see **love**

carve < AS. *ceorfan;* see **cut**

carya, L. (Gr. *karya*), walnut-tree, now applied to hickory; see **walnut**

caryo- < *karyon,* walnut, any nut, kernel; *karydion,* dim.; see **nut**

caryophyllus, NL. (Gr. *karyophyllon*), clove-tree; see **clove**

caryota, L. a kind of palm; see **palm**

casa, L. cottage, cabin, hut, house; *casula,* dim.; see **house**

casabundus, L. ready to fall, tottering, < *caso, -atus,* shake, waver; see **shake**

casalbado- < Gr. *kasalbas, -ados,* strumpet, whore; see **prostitute**

cascus, L. old; see **old**

case < L. *capsa,* box, see **box, bag;** < L. *casus,* event, chance, see **lot**

caseus, L. cheese; *caseolus,* dim.; *casearius,* of cheese; *caseatus,* cheesy; see **cheese**

casio- < Gr. *kasis, -ios,* brother, sister; *kasignetos,* brother; *kasignete,* sister; see **brother, sister**

cask < Sp. *casco,* potsherd, skull, helmet, < L. *quasso,* shake, shatter; see **barrel**

Cassandra, L. (Gr. *kassandra*), prophetess of impending evil but believed by no one; see **prophecy**

cassia L. (Gr. *kassia*), a genus of the bean family; see **bean**

cassiculus, L. small net, cobweb, dim. of *cassis,* hunting-net, snare; see **net**

cassida; cassis, -idis, L. helmet; *cassidula,* dim.; see **cap**

cassidile, L. small bag, wallet; see **bag**

Cassiope, L. (Gr. *kassiope*), a legendary beauty; a genus of the heath family; see **heath**

cassis, L. helmet, see *cassida* under **cap;** hunting-net, snare, see *cassiculus* under **net**

cassita, L. crested lark; see **lark**

cassiterum, L. (Gr. *kassiteros*), tin; see **tin**

casso, -atus, L. annul; see **cancel**

cassus, L. empty; see **empty**

cassymato- < Gr. *kassyma, -atos,* shoe sole; see **sole**

cassytha, NL. a genus of the laurel family; see **laurel**

cast < ON. *kasta;* see **throw**

castanea, L. chestnut; *castaneus,* of the color of chestnuts, brown; see **chestnut**

castigo, *-atus,* L. punish, chastise, censure, correct; see **punish**

castle < AS. *castle,* < L. *castellum,* fort; see *castrum* under **fort**

castor, L. (Gr. *kastor),* beaver; *castorinus,* of beavers; see **beaver**

castro, *-atus,* L. geld; see **geld**

castrum, L. fort, camp, settlement; *castellum,* dim.; see **fort**

castula, L. petticoat; see **garment**

castus, L. pure; *castificus,* purifying; see **pure**

casuarius, NL. a genus of ostrichlike birds; see **ostrich**

casus, L. a falling down, event, accident, chance; *casualis,* fortuitous; see *cado* under **fall**

cat < AS. *cat;* G. *katze;* F. *chat;* L. *catus,* m.; NL. *cattus,* m.; *catulus,* m. dim., puppy: catkin, catbird, catcall, caterwaul, caterpillar, catnip *(Nepeta cataria),* polecat, Catskill (Kaaterskill), kitten, *Felis catus* (European wildcat), *Lemur catta* (ring-tailed lemur), *Catulus brunneus* (a shark), *Pithecellobium unguiscati* (catsclaw), *Murrubium catariaefolium* (a horehound). Derive catsup (ketchup).
   Gr. *ailouros,* m. cat, weasel: *Ictalurus punctatus* (channel-catfish), *Ailurus[Aelurus] fulgens* (panda), *Ailuropus melanoleucus* (giant panda).
   Gr. *arkelos,* m. young panther:
   L. *felis,* f. cat; *felicula,* f. dim.; *felinus,* of cats: feline, Felidae, *Felis domestica* (cat), *Profelis aurata* (golden cat).
   L. *leo, -onis* (Gr. *leon, -tos),* m. lion; *lea; leaena* (Gr. *leaina),* f. lioness; *leunculus,* m.; Gr. *leontarion,* n. dim.; *leoninus,* of lions, lionlike; Bantu *simba,* lion: leonine, Leonard, Lenore, leopard, dandelion, Sierra Leone, Cœur de Leon, Leander, Leona, pantaloon, Napoleon, lion *(Felis leo),* chameleon *(Chameleon vulgaris), Vermileo comstocki* (a snipe-fly), *Mrymeleon formicarius* (an ant-lion), *Rhampholeon spectrum* (a chameleon), *Leontodon nudicaulis* (rough hawkbit), *Leontopodium alpinum* (edelweiss), *Callistoleon erythrocephalum* (a neuropterid), *Ulochaetes leoninus* (a beetle), *Leonotis laxifolia* (a lionsear), *Trachys simba* (a beetle).
   L. *lynx -ncis* (Gr. *lynx, -nkos),* c. wildcat; *lynkion,* n. dim.: *Lynx rufus* (bay lynx, bobcat), ounce *(Felis uncia), Metriophilus lynx* (a beetle), *Solenostethium lynceum* (a bug), Accademia dei Lincei.
   Ab.Am. *ocelotl,* jaguar: ocelot *(Felis pardalis).*
   L. *panthera,* f. (Gr. *panther, -os,* m.), panther: panther, *Lecidea pantherina* (a lichen).
   Gr. *pardos,* m.; *pardalis, -idos,* f. leopard, ounce, panther; *pardalion,* n. dim.; *pardalotos,* spotted like a leopard: leopard *(Felis pardus),* pardalote *(Pardalotus punctatus), Giraffa camelopardalis* (giraffe), *Lynx pardina* (Spanish lynx), *Lilium pardalinum* (leopard-lily), *Polypedates pardalis* (a flying-frog), *Pardosa cursoria* (a spider).
   Ab.Am. *puma,* a cougar: puma *(Felis concolor), Acrotomodes puma* (a moth). Derive: cougar.
   L. *rufius,* m. Gallic name for a lynx:
   L. *tigris, -is; -idis* (Gr. *-ios; -idos),* f. tiger; *tigrinus,* of tigers: tiger *(Felis tigris), Patriofelis tigrina* (a creodont), *Tigridia pavonia* (tiger-flower), *Tigridoptera exul* (a butterfly), *Tigriopus fulvus* (a copepod), *Proales tigrida* (a rotifer), tigrine, *Haemanthus tigrinus* (a blood-lily).

cata-; cat-; cath-; cato- < Gr. *kata,* down, against; very; *kato,* beneath, below, under; see **down**

cataclito- < Gr. *katakliton,* couch; see **bed**

cataclysm < Gr. *kataklysmos,* flood, deluge, disaster; see *klyzo* under **wash**

catacto- < Gr. *kataktos,* breakable; see *katagma* under **break**

catactrio- < Gr. *kataktria,* spinner; see **spin**

catacumba, L. underground burial gallery; see **grave**

cataegis, L. (Gr. *kataigis),* hurricane, whirlwind; see **wind**

catagmato- < Gr. *katagma, -tos,* flock of wool, see **wool**; fracture, see **break**

catalepsis, L. (Gr. *katalepsis),* fit, seizure; see *lepsis* under **disease**

catalogue < F. *catalogue,* < L. *catalogus* (Gr. *katalogos),* register, list; see **list**

catalpa, NL. a genus of the bignonia family; see **bignonia**

catamono- < Gr. *katamonos,* permanent; see **always**

catanto- < Gr. *katantes,* downhill, steep; see *anantes* under **slope**

cataplasma, L. (Gr. *kataplasma*), poultice, plaster, salve; see **plaster**

cataract < Gr. *kataraktes,* waterfall; see **fall**

catarato- < Gr. *kataratos,* cursed; see *ara* under **ask**

catasta, L. scaffold, stage; see **frame**

catastrophe < Gr. *katastrophe,* overturning, calamity, misfortune; see **destroy**

catax, -acis, L. limping, lame; see **hurt**

catch < L. *capio, captus,* take; see **take, trap**

catechetico- < Gr. *katechetikos,* of instruction; *katechetes,* teacher; see **teach**

catechu < Malay *kachu;* see **resin**

category < L. *categoria* (Gr. *kategoria*), class, division; see **class**

catelipho- < Gr. *katelips, -iphos,* ladder, staircase, upper story; see **ladder**

catellus; catulus, L. whelp; see **young**

catena, L. chain; *catella; catenula,* dim.; see **chain**

catepho- < Gr. *katephes,* downcast, mute; *katepheia,* dejection, sorrow; see **sad**

catero- < Gr. *kateres,* furnished, fitted; see **supply**

caterpillar < ME. *catyrpel,* < OF. *chatepelose,* < LL. *cattus,* cat, *pilosus,* hairy; the larval stage of some insects.
> L. *campa* (Gr. *kampe*), f. caterpillar; *pityokampe,* f. a caterpillar on pine trees; *skolekampe,* f. a caterpillar: pityocampa, *Campephilus principalis* (ivory-billed woodpecker, *Campodea staphylinus* (a bristletail), *Hemerocampa vetusta* (a tussock-moth), *Lithocampe ovata* (a radiolarian), *Metriocampa spinigera* (a bristletail).
> L. *eruca,* f. caterpillar, cankerworm: erucivorous, eruciform, *Galeruca notata* (a beetle), *Brachiaria erucaeformis* (a signal-grass), *Machaerocereus eruca* (a creeping-cactus).
> Gr. *trox, -ogos,* a gnawer, weevil, caterpillar; see **bite**
> See: **young**

caterva, L. crowd, troop; see **number**

catfish
> NL. *ameiurus [amiurus],* m. a genus of catfishes: *Ameiurus catus* (a catfish).
> L. *glanis, -idis* (Gr. *-idos*), m. probably a sheatfish: *Silurus glanis* (sheatfish), *Chimarrhoglanis leroyi* (a fish).
> NL. *ictalurus,* m. a genus of catfishes: *Ictalurus punctatus* (channel-cat).

catha, NL. a genus of the staff-tree family; see **staff-tree**

cathammato- < Gr. *kathamma, -tos,* anything tied, knot; see *hapto* under **bind**

catharmato- < Gr. *katharma, -tos,* refuse; see **dirt**

catharo-; catharto- < Gr. *katharos,* clear, pure; *kathartes,* a cleanser, purifier; *katharmos; katharsis,* a cleansing, purification; see **pure**

cathemato- < Gr. *kathema, -tos,* necklace; see **necklace**

catheter, L. (Gr. *katheter*), a pipelike instrument for emptying the bladder; see **pipe**

catheto- < Gr. *kathetos,* perpendicular; see **upright**

cathismato- < Gr. *kathisma, -tos,* seat, buttock; see *hisma* under **sit**

cathodo- < Gr. *kathodos,* a way down, descent; see *hodos* under **way**

catholicus, L. (Gr. *katholikos*), universal, general; see **all**

cathormio- < Gr. *kathormion,* necklace; see *hormos* under **necklace**

catillamen, L. sweetmeat; see **food**

catillo, L. glutton, gourmand; *catillo, -atus,* lick a plate; see **eat**

catinus, L. deep vessel, pot, bowl, dish, cup; *catillus; catinulus,* dim.; see **pot**

catkin
> L. *amentum,* n. thong, strip, catkin: ament, amentiferous, *Amentotaxus argotaenia* (a gymnosperm).
> L. *iulus* (Gr. *ioulos,* down, wool), m. catkin, ament: iulus, Julius, Julia, July, *Prosopis juliflora* (mesquite).
> Gr. *kachrys, -yos,* f. catkin, ament, parched barley; *kachrydion,* n. dim.: *Cachryphora serotinae* (an aphid), *Microcachrys tetragona* (a gymnosperm).
> Gr. *krossos,* fringe, tassel; see **fringe**

catoblepas, L. an African animal; see **animal**

catonium, L. the lower world; see **hell**

catoptero- < Gr. *katopter, -os,* scout, spy; see *optikos* under **see**

catoptro- < Gr. *katroptron,* mirror; see *enoptron* under **mirror**

catorycto- < Gr. *katoryktos,* buried deep; see *orysso* under **dig**

catreo- < Gr. *katreus,* a kind of peacock; see **peacock**

cattle < OF. *catel,* < L. *capitalis,* property, < *caput,* head, stock: cattalo.
  Heb. *aleph,* ox; first letter of the Hebrew alphabet: alif, alpha.
  Gr. *Apis, -idos,* m. Egyptian sacred bull: Apis, Serapis, *Serapis latifolia* (an
  (orchid).
  L. *armentum,* herd; see **herd**
  L. *bison, -ontis* (Gr. *bison, -os*), m., < OHG. *wisunt,* the humpbacked ox, wisent:
  Bison americanus (American bison, 'buffalo'), *Enophrys bison* (a sculpin).
  L. *bonasus* (Gr. *bonasos*), m. bison, wisent: *Bison bonasus* (wisent).
  L. *bos, bovis* (Gr. *bous, boös*), c. ox, cow; *buculus,* m.; *bucula,* f.; *boidarion;*
  boidion, n. dim.; *bubalus; bufalus* (Gr. *boubalos*), m. wild ox; *boukolion,* n.
  herd of cattle; *boutes; boötes,* m. herdsman, ox-driver, plowman; *bovarius;*
  bovillus; bovinus; bubulus; bubulinus, pertaining to shepherds, rural: bovine,.
  bovate, Bosporus, bugle, bucolic, bulimia, LeBoeuf, Boötes, Bucephalus, buffalo,
  bugloss, *Bos bubalus* (Indian buffalo), *Bos pumilus* (an African buffalo), *Bibos
  frontalis* (gayal), *Ictiobus[Ichthyobus] urus* (a buffalo-fish), *Ovibos moschatus*
  (muskox), *Boöphilus bovis* (a tick), *Megaptera boöps* (a humpbacked whale),
  Dielasma bovidens (a brachiopod), *Hypoderma bovis* (heel or oxwarble fly),
  Buphagus erythrorhynchus (oxpecker), *Buceros rhinoceros* (a hornbill), *Buto-
  mus umbellatus* (flowering-rush) *Bubalus brevicornis* (a fossil water-buffalo),
  Tabanus bovinus (gadfly), *Buchloë dactyloides* (buffalo-grass).
  Gr. *boskema, -tos; boton,* n. fed or pastured animal, < *bosko,* feed, pasture:
  AS. *bula; bulluc,* male of cattle: bull, bullock, bulldog, bullfrog.
  Sp. *cibolo* (< Ab.Am.), m. bison: Seven Cities of Cibola.
  AS. *cu* (G. *kuh;* pl. *kühe;* L. *ceua,* f.), pl. *ky; kyen,* female of cattle: cow, cow-
  bird, cowherd, cowpea, cowslip, sea-cow, kine.
  L. *damalio, -onis,* m. (Gr. *damalos,* m.; *damalis,* f.), calf: *Damalichthys argyro-
  somus* (white perch), *Hydrodamalis gigas* (an extinct seacow).
  L. *junix (juvenix), -icis,* f. young cow, heifer; *juvencus,* m. bullock; *juvenca,* f.
  heifer:
  Gr. *killix, -ikos,* m. ox with crooked horns:
  Gr. *ktenos,* n. beast, livestock; *ktenikos,* of beasts; *ktenodes,* like a beast, brutish;
  kteniatros, cattle-doctor, veterinary: *Ctenobium antennatum* (a beetle), *Cteno-
  des zonata* (a beetle).
  Gr. *moschos,* c. calf; *moscharion; moschion,* n. dim.:
  L. *Pales,* m. god of shepherds and cattle; *Palilis,* of Pales: Pales.
  L. *pecus, -oris,* n. cattle collectively; *pecus, -udis,* f. cattle singly; herd, flock,
  animal; *pecusculum,* n. dim.; *pecoralis; pecualis; pecuarius; pecudalis; pecui-
  nus,* of cattle; *pecorosus,* rich in cattle; *peculium,* n. property in cattle: peculate,
  peculiar, pecuniary, impecunious, Pecora, *Eusimulium pecuarum* (buffalo-gnat).
  Gr. *portis, -ios; portax, -akos,* f. calf, heifer; *hypoportis,* with calf:
  L. *taurus* (Gr. *tauros*), m. bull; *taura,* f. freemartin, a barren female of twins, one
  of which is a male; *taurulus,* m.; Gr. *tauridion,* n. dim.; *taureus; taurinus,* of
  bulls: taurine, Taurus, tauromachy, taurian, taurocholic, sodium taurocholate,
  Minotaur, Thoreau, *Bos taurus* (cow), *Onthophagus taurus* (a beetle), *Tau-
  romyia pachyneura* (a fly), *Connochaetes taurinus* (brindled gnu), *Aphilanthops
  taurulus* (a wasp).
  L. *vacca,* F. cow; *vaccula,* f. dim.; *vaccinus,* of cows: vaccine, vaccination, vaquero
  (buckaroo), *Damalichthys vacca* (a white perch), *Saponaria vaccaria* (cow-soap-
  wort), *Boletus vaccinus* (a mushroom).
  L. *vitulus* (Gr. *italos,* bull), m.; *vitula,* f. calf; *vitellus,* m. dim.; *vitulinus,* of
  calves, of veal: vitular, vituline, vitellus, vitelline, veal, vellum, Italy, *Onthophagus vitulus* (a dung-beetle), *Ascaris vitulorum* (a roundworm), *Phoca vitulina* (harbor-seal), *Candelariella vitellina* (a lichen).
  L. *urus,* m. wild ox: *Bos urus(primigenius)* (auroch, urus).
  See: **deer, goat, sheep, animal**

cattyo- < Gr. *kattys, -yos,* piece of leather; see **leather**

catulio, L. desire the male; see **lewd**

catulus, L. puppy, whelp, young animal, dim. of *catus,* cat; *catulinus,* pertaining to
  a young dog; see **dog, cat**

catus, L. intelligent, wise, clever, cunningly sly, see **know**; *catus,* cat, see **cat**

cauca, L. (Gr. *kauke*), a kind of cup; *caucula,* dim.; see **cup**

caucalias, L. (Gr. *kaukalias*), a kind of bird; see **bird**

caucalis, L. (Gr. *kaukalis*), a genus of the carrot family; see **carrot**

cauda, L. tail, appendage; *caudicula,* dim.; see **tail**

caudex, *-icis,* L. stem, trunk; *caudiculus,* dim.; *caudeus,* of wood, wooden; see **stem**

caula, L. opening, hole; see **hole**

caulino- < Gr. *kaulines,* a kind of gudgeon; see **goby**

caulis, L. (Gr. *kaulos*), stalk, stem; *cauliculus,* dim.; see **stem**

cauma, L. (Gr. *kauma, -tos*), burning heat; see *kaio* under **burn**

caunaco- < Gr. *kaunakes,* thick cloak; see **garment**

caupo, L. innkeeper, huckster; *caupona,* landlady, tavern; see **trade, house**

caurinus, L. of the northwest wind, *caurus;* see **wind**

causalido- < Gr. *kausalis, -idos,* blister, burn; see *kaio* under **burn**

cause < L. *causa,* agent, reason, motive; *causalis,* of a cause; see **begin, arouse**

causia, L. (Gr. *kausia*), a light, broad-brimmed hat; see **cap**

causticus, L. (Gr. *kaustikos*), burning, corrosive; *kausos,* burning heat; see *kaio* under **burn**

cauter, L. (Gr. *kauter*), branding-iron, burner; *kausis,* a burning; *kaustikos,* capable of burning; *kauma, -tos,* burning heat; *kausimos,* combustible; see *kaio* under **burn**

cautes, L. a rough, pointed rock; see **stone**

cautus, L. careful, wary; see *caveo* under **guard**

cavannus, L. night-owl; see **owl**

cave < L. *cavus,* hollow, hole; *cavatus,* hollowed out; *cavea,* a hollow place, den, enclosure for animals, cage; *caverna,* cave, grotto, hole; see **hollow, den, pen**

caveo, *cautus,* L. beware, guard against, avoid; see **guard**

cavia, NL. f. < Ab.Am. for cavy, a kind of rodent: *Cavia porcellus* (guinea-pig), *Procavia syriaca* (daman), capybara (*Hydrochoerus capybara*).

cavilla, L. jeering, scoffing, irony, < *cavillor, -atus,* jest, joke, satirize, rail, scoff; see **laugh**

cavus, L. hollow, hole; see **hollow**

caw < imitative origin.
   L. *coracino, -atus,* caw:
   L. *cornicor, -atus,* caw:
   L. *crocio, -itus* (Gr. *krozo*), caw like a crow; *krogmos,* m. a cawing: crocitation. Gr. *klozo,* caw:

ceanothus, L. (Gr. *keanothos,* a kind of thistle), a genus of the buckthorn family; see **buckthorn**

cease < L. *cesso, -atus,* stop; see *cesso* under **end, rest**

ceblepyris, NL. (Gr. *keblepyris*), redpoll; see **finch**

ceblo- < Gr. *keblos,* a baboon; see **monkey**

cebus, L. (Gr. *kebos*), a long-tailed monkey; see **monkey**

cecheno- < Gr. *kechenos,* gaping, yawning; see **open**

cecibalo- < Gr. *kekibalos,* a kind of shellfish; see **mollusk**

cecido- < Gr. *kekis, -idos,* gall; *kekidion,* dim.; see **gall**

Cecropia, L. citadel of Athens, < Gr. *Kekrops,* mythical king of Athens; a genus of the mulberry family; see **mulberry**

cecryphalo- < Gr. *kekryphalos,* hairnet, net; *kekryphalion,* dim.; see **net**

cedar < L. *cedrus* (Gr. *kedros*); *cedrinus,* of cedar; see **gymnosperm**

cede < L. *cedo, cessus,* go, yield; see **give, move**

cedno-; cedo- < Gr. *kednos,* careful, trustworthy; *kedos,* care, concern, anxiety, trouble; *kedemon,* protector, guardian; *kedistos,* most worthy of care; see **guard, trouble, love**

cedrela, NL. a genus of the mahogany family; see **mahogany**

cedrostis < Gr. *kedrostis,* bryony; see **melon**

ceiba, Sp. silk-cotton tree; see **cotton**

ceiling < L. *caelum,* sky; see **roof, heaven**

celado- < Gr. *kelados,* sound of rushing waters, loud noise; see **sound**

celaeno- < Gr. *kelainos,* black, dark, murky, swart; see **black**

celas < Gr. *kelas,* a stork; see **stork**

celastrus, L. (Gr. *kelastros*), a kind of evergreen tree; see **staff-tree**

celatus, L. concealed, hidden, < *celo, -atus,* hide; see **cover**

-cele < Gr. *kele,* tumor, rupture, hernia; see **swell**

celebo- < Gr. *kelebe,* cup, jar, pan; *kelebeion,* dim.; see **cup**

celebro, -atus, L. solemnize, publish, praise; *celeber,* honored, famous; see **honor**

celeo- < Gr. *keleos,* a woodpecker; see **woodpecker**

celeonto- < Gr. *keleon, -tos,* beam of a loom; see **beam**

celepho- < Gr. *kelephos,* leper; *kelephia,* leprosy; see **disease**

celer, L. swift; see **swift**

celery < L. *selinum* (Gr. *selinon*), parsley; see *selinum* and *apium* under **carrot**

celestial < L. *caelestis,* of the heavens, heavenly, < *caelum,* sky, heaven; see **heaven**

celetico- < Gr. *keletikos,* charming, delightful; *keletes,* charmer; see **agreeable**

celeto- < Gr. *keles, -etos,* courser, riding-horse; *keletion,* dim.; *keletistes,* leaper, < *keletizo,* ride two or more horses at the same time; see **ride**

celeusmo- < Gr. *keleusmos,* command, order; *keleustes,* signaler, < *keleuo,* command, order, urge; see **ask**

celeutho- < Gr. *keleuthos,* road, path, track; *keleuthetes,* wayfarer; see **way**

celia, L. a Spanish beer; see **beer**

celibate < L. *caelibatus,* < *caelebs, -ibis,* unmarried, single (man); see **alone**

celido- < Gr. *kelis, -idos,* stain, spot, blemish; see **spot**

cell < L. *cella,* storeroom, chamber; *cellula,* dim.; see **room**

cellar < L. *cellarium,* storeroom; *cellarius,* butler, steward; see *cella* under **room**

cellulanus, L. hermit, recluse; see **alone**

cellulose, F. a carbohydrate constituent of plant-cell walls; see *cella* under **room**

celo- < Gr. *kelon,* shaft of an arrow; see **arrow**

celono- < Gr. *kelon, -os,* swing-beam; see *keleon* under **beam**

celoro- < Gr. *kelor, -os,* son; see **son**

celosia, NL. a genus of the amaranth family; see **amaranth**

celox, -ocis, L. swift, quick; cutter, yacht; see **swift, ship**

celsus, L. raised, high, lofty; see **high**

Celt < L. *Celta,* m. an individual of an ancient race in western Europe and the British Isles, called by the Greeks, *Keltoi,* and the Romans, *Galli:* Celtic.
L. *Brito, -onis,* an individual of a Celtic people in southern Britain; see *Britannicus* under **English**
L. *Gallus,* m. an individual of a Celtic people of central and western Europe; *Gallia,* f. country of the Gauls; *Gallicus,* of the Gauls: Gaul, Gallic, Galatia, *Tamarix gallica* (a tamarisk).
AS. *Iras,* a Celtic people of western Britain, < Ir. *Eire* (L. *Hibernia,* f.), Ireland: Eire, Erin, Ireland, Irish, Hibernian, *Rosa hibernica* (Irish rose).
L. *Scotus,* m. an individual of a Celtic people in northern Britain called Gaels; *Scoticus,* of the Scots: Scot, Scotland, Scotch, Scotic, Gaelic, Caledonia, *Ligusticum scoticum* (Scotch ligusticum).
AS. *Waelisc,* a Celtic people in southwestern Britain, called by themselves, *Cymry:* Welsh, Wales, New South Wales, Cymric.

celtis, L. a genus of the elm family, hackberry; see **elm**

celypho- < Gr. *kelyphos,* pod, shell, case; *kelyphion,* dim.; see **bag**

cemado- < Gr. *kemas, -ados,* young deer, pricket; see **deer**

cembra, NL. < G. *zember,* timber; see **wood**

cement < OF. *ciment,* < L. *caementum,* chips, pebbles, now applied to a substance used as a binder; see *caementarius* under **mason, glue**

cemetery < Gr. *koimeterion,* sleeping-room, burial-place; see *koma* under **sleep, grave**

cemo- < Gr. *kemos,* muzzle, see *camus* under **bind**; wicker-vessel, weel, see **basket**

cena, L. dinner, supper; *cenula,* dim.; *cenaculum,* dining-room; see **eat**

cenchramido- < Gr. *kenchramis, -idos,* fig seed; see **seed**

cenchris, L. (Gr. *kenchris*), a kind of hawk; see **hawk**

cenchro- < Gr. *kenchros,* millet; see **grain**

cenclo- < Gr. *kenklos,* a sea-bird; see **bird**

cenebrio- < Gr. *kenebreion,* carrion; see **flesh**

ceneono- < Gr. *keneon, -os,* flank, any hollow; see **side**

ceno- < Gr. *kainos,* new, recent, see **new;** < *kenos,* empty; *kenotes,* emptiness, see **empty;** < *koinos,* common, see **common**

census, L. register, list, < *censeo, -sus,* judge, appraise, assess; see *censeo* under **judge, list**

centaur < Gr. *Kentauros,* a monster, half man and half horse; see **animal**

centaureum, L. (Gr. *kentaurion*), a plant of the composite family; see **composite**

center < L. *centrum* (Gr. *kentron*), midpoint of a circle, prickle, point, spike, sting, spur; *kentrion,* dim.; see **point, middle**

centipeda, L. a myriapod with one pair of legs to each segment; see **myriapod**

cento, L. patchwork, covering of rags; *centunculus,* dim.; see **rag**

centoro- < Gr. *kentor, -os,* goader, pricker, < *kenteo,* goad, prick, spur; see **point**

centrisco- < Gr. *kentriskos,* a fish; see *centrum* under **point**

centum, L. hundred; *centesimus,* hundredth; see **hundred**

ceo- < Gr. *keio,* cleave, split; see **cut**

ceodo- < Gr. *keodes,* fragrant like incense; see **smell**

cepa, L. onion; *cepula,* dim.; see **onion**

cepaea, L. (Gr. *kepaia*), a succulent salad herb; see **stonecrop**

cepaeo- < Gr. *kepaios,* of a garden; see *kepos* under **field**

cephalo- < Gr. *kephale,* head; *kephalidion; kephalion; kephalis,* dim.; *kephalotos,* with a head, headed, see **head;** *kephalalgia,* headache, see **disease**

cephalus, L. (Gr. *kephalos*), a mullet; see **mullet**

cepheno- < Gr. *kephen, -os,* drone, lazy fellow; *kephenion,* dim.; see **bee**

cephus, NL. a genus of sawflies; see **wasp**

cepo- < Gr. *kepos,* garden, orchard, plantation; see **field**

ceppho- < Gr. *kepphos,* a sea-bird; see **bird**

-ceps < L. *caput,* head, see **head;** < *capio,* take, see **take**

cera, L. (Gr. *keros*), beeswax, wax; *cereus,* waxy; *cerinus* (Gr. *kerinos*), waxen, wax-colored; see **wax**

cerambyco- < Gr. *kerambyx, -ykos,* a horned beetle; see **beetle**

ceramo- < Gr. *keramos,* potter's clay, anything made of clay, pot, vessel, tile; *keramikos,* pertaining to pottery, earthen; *keraminos,* of clay; *keramion,* pot, jar; see **earth, pot**

ceraphido- < Gr. *keraphis, -idos,* a lobster or prawn; see **crab**

cerasto- < Gr. *kerastes,* horned, see *keras* under **horn;** < *kerastos,* mixed, see *kerannymi* under **mix**

cerasus, L. (Gr. *kerasos*), cherry-tree; *cerasinus,* of cherry; see **plum**

cerato- < Gr. *keras, -atos,* horn; *keration,* dim.; *keraos; kerastes,* horned; see **horn**

ceratonia, L. (Gr. *keratonia*), a genus of the bean family; see **bean**

cerauno- < Gr. *keraunos,* thunderbolt, thunder and lightning; *keraunios,* of a thunderbolt; see **thunder**

Cerberus, L. (Gr. *Kerberos*), doglike monster that guarded the gates of Hades; see **dog**

cercaria, NL. tailed, tadpolelike larval stage of trematodes; see **young**

cerceris, L. a kind of bird, now applied to a genus of wasps; see **wasp**

cerchno- < Gr. *kerchne,* a kind of hawk, see **hawk;** < *kerchnos,* hoarseness, see **hoarse**

cercis, -idis, NL. (Gr. *kerkis, -idos*), a tree of the bean family; rod or comb for striking the woof; *kerkidion,* dim.; see **bean, rod**

cerco- < Gr. *kerkos,* tail; *kerkion,* dim.; see **tail**

cercocarpus, NL. a genus of the rose family; see **rose**

cercurus, L. (Gr. *kerkouros*), a light vessel, boat; see **ship**

cercyo- < Gr. *Kerkyon,* a notorious Attic robber; see **steal**

cercyrus, L. a sea-fish; see **fish**

cerdaleo- < Gr. *kerdaleos,* crafty, cunning, shrewd, wily, < *kerdos,* craft, wile; see **know**

cerdo- < Gr. *kerdos,* gain, profit, see **gain;** craft, wile, see *kerdaleos* under **know**

cereal < L. *Ceres,* goddess of grains and agriculture; see **grain**

cerebrum, L. brain; *cerebellum,* dim.; see **brain**

ceremonia, L. rite; see **rite**

ceresa, NL. a genus of bugs; see **bug**

ceresio- < Gr. *keresios*, deadly; see *ker* under **death**

cereus, L. of wax, waxy; wax taper, candle; *cereolus*, dim., see *cera* under **wax**; a genus of the cactus family, see **cactus**

cerevisia (cervisia), L. beer; see **beer**

cerintha, L. (Gr. *kerinthe*), waxwort; see **borage**

cerinthus, L. (Gr. *kerinthos*), beebread; see **bread**

cerinus, L. (Gr. *kerinos*), waxen, wax-colored, yellowish; see *cera* under **wax**

cerio- < Gr. *kerion*, honeycomb; see *cera* under **wax**

cerithium, NL. a genus of gastropods; see **mollusk**

cermato- < Gr. *kerma, -tos*, small piece, mite; *kermation*, dim.; see **little**

cerno- < Gr. *kernos*, earthen dish with wells in the bottom for fruits, see **plate**; < L. *cerno, cretus*, separate, distinguish, see **separate**

cernuus, L. drooping, nodding, stooping, facing earthward; see **bend**

cero- < Gr. *keros*, beeswax, see *cera* under **wax**; < *keros*, horn, see *keras* under **horn**; < *ker, -os*, death, doom, destruction, see **death**

cerritus, L. frantic, mad; *cerritulus*, dim.; see **mad**

cerrus, L. a kind of oak; *cerreus*; *cerrinus*, of the Turkish oak; see **oak**

certamen, L. contest, rivalry, struggle; *certator*, disputant; see *certo* under **fight**

certhio- < Gr. *kerthios*, a tree-creeper; see **bird**

certus, L. definite, fixed, sure, < *cerno, certus*, sift, distinguish; see **sure, separate**

cerulean < L. *caeruleus*, sky-blue; see **blue**

cerussa, L. white lead; see **lead**

cervix, -icis, L. neck; *cervicula*, dim.; *cervicatus*, stiffnecked, stubborn; see **neck**

cervus, L. deer, stag; *cervulus*, dim.; *cervinus*, of deer; see **deer**

cerycium, L. (Gr. *kerykion*), a herald's staff, caduceus; see **rod**

ceryco- < Gr. *keryx, -ykos*, herald, messenger; see **call**

cerylo- < Gr. *kerylos*, kingfisher; see **kingfisher**

cespitose < L. *caespes, -pitis*, turf, sod, mat, tuft; see **sod**

cessator, L. idler, loiterer, < *cesso, -atus*, stop, rest; see **rest**

-cester; -chester, < AS. *ceaster* (L. *castrum*), fort, fortified town, camp; see *castrum* under **fort**

castro- < Gr. *kestra*, hammer, see **hammer**; < *kestron*, a graving tool, see **cut**

cestrum, L. (Gr. *kestron*, a plant), a genus of the nightshade family; see **nightshade**

cestrus, NL. mullet; see **mullet**

cestus, L. (Gr. *kestos*), girdle, belt; see **belt**

cetarius, L. pertaining to fish; fishmonger; see **fish**

ceterus, L. the other, the rest; see **other**

cethario- < Gr. *ketharion*, dice-box, < *kethis, -idos*, ballot-box; see **box**

cethido- < Gr. *kethis, -idos*, ballot-box; see **box**

cetio- < Gr. *keteios*, monstrous, large; see **large**

cetonia, NL. a genus of beetles; see **beetle**

cetra, L. a short Spanish shield; see **shield**

cetus, L. (Gr. *ketos*, any large sea-animal), whale; see **whale**

ceutho- < Gr. *keuthos*, hidden; *keuthmon*, hiding-place, hole, corner, cave; see **cover**

ceyx, -ycis, L. (Gr. *keyx, -ykos*), kingfisher; see **kingfisher**

chaeno- < Gr. *chaino*, gape; see **open**

chaero- < Gr. *chairo*, rejoice; see **joy**

chaeto- < Gr. *chaite*, long hair, mane; see **hair**

chaff < AS. *ceaf*, husks or glumes of grasses; see **scale**

chain < L. *catena*, f.; *catella*; *catenula*, f. dim.; *catenarius*, of a chain; *catenatus*, with chains, chained; *cateno, -atus*, chain or bind together; *concateno*, link together: catenary, catenulate, Concatenated Order of Hoo Hoo, *Gonyaulax catenella* (a dinoflagellate), *Lactophrys concatenatus* (a fish), *Syringopora ca-*

tenoides (a fossil coral).

Gr. *ankylion,* n. ring of a chain:

Gr. *chalaston,* n. chain:

Gr. *desmos,* bond, chain; see *deo* under **bind**

Gr. *halysis, -eos,* f. chain; *halysion,* n. dim.; *halysidotos,* chainlike: *Halysites catenulatus* (a chain-coral), *Halysiocrinus carinatus* (a Devonian crinoid).

Gr. *hormos; hormathos,* chain, string, necklace; see **necklace**

L. *monile,* necklace, string of beads; see **necklace**

Gr. *psalion,* n. curb-chain, part of a bridle:

Gr. *streptos,* a twisted or linked collar or chain; see **turn**

Gr. *sysphigma, -tos,* n. chain:

L. *vinculum,* bond, fetter, chain; see *vincio* under **bind**

chair < Gr. *kathedra,* seat; see **sit**

chalara, Gr. fetter; see **bind**

chalaro- < Gr. *chalaros,* loose, slack, limp; *chalasma,* relaxation; see **free**

chalaston, Gr. chain; see **chain**

chalaza, Gr. hail, sleet, pimple, tubercle; *chalazion,* dim.; see **pimple**

chalcedony < L. *chalcedonius,* of Chalcedon; see **stone**

chalcis, NL. a kind of parasitic wasp; see **wasp**

chalco- < Gr. *chalkos,* copper; *chalkeos,* of copper or bronze; *chalkion,* copper kettle; see **copper, kettle**

chalepo- < Gr. *chalepos,* difficult, severe, harsh; see **difficult**

chalico- < Gr. *chalix, -ikos,* pebble, rubble; see *calx* under **stone**

chalinos, Gr. bridle, bit; *chalinarion,* dim.; see **hold**

chalk < AS. *cealc,* < L. *calx, calcis,* pebble, lime; see **lime, gypsum**

challenge < OF. *chalangier;* see **call**

chalybo- < Gr. *chalyps, -ybos,* iron, steel; see **iron**

chama, L. (Gr. *cheme*), a gaping mollusk, cockle; see **mollusk**

chamadytes, Gr. snail; see **mollusk**

chamae- < Gr. *chamai,* dwarf; *chamelos,* on the ground, creeping; see **low**

chamaecyparis, NL. a genus of conifers; see **gymnosperm**

chamaedaphne, L. (Gr. *chamaidaphne*), a genus of the heath family; see **heath** ·

chamaedora, NL. a genus of palms; see **palm**

chamaedrys, L. (Gr. *chamaidrys*), germander; see **mint**

chamaepitys, L. (Gr. *chamaipitys*), a plant of the mint family; see **mint**

chamaerops, L. (Gr. *chamairops*), a genus of palms; see **palm**

chamelaea, L. olive-daphne; see **mezereon**

chameleon < L. *chamaeleon* (Gr. *chamaileon*), a kind of lizard; see **lizard**

chamelos, Gr. on the ground, creeping; see **low**

chamomile < L. *chamomilla* (Gr. *chamaimelon,* earth-apple), a plant of the composite family; see **composite**

champsos, Gr. crocodile; see **lizard**

chance < L. *cado, casus,* fall; see **lot**

change < OF. *changer,* < L. *cambio, -itus,* exchange, barter: changing, changeless, changeling, cambium, cambist, *Phytophthora cambivora* (a fungus). Explain: Ring the changes.

Gr. *abebaios,* uncertain, unstable, fickle: *Abebaeus dorsalis* (a beetle).

Gr. *aiolos,* shifting, changeable, variable; see *Aeolus* under **wind**

L. *alieno, -atus,* estrange, drive insane; see *alienus* under **strange**

Gr. *allasso,* change, alter; *allage,* f. change, exchange; *allaxis,* f. barter, exchange; *allaktikos,* of exchange; *alloiotos,* changed, changeable; *allophyes,* changeful in nature; *alloprosallos,* fickle; *diallage,* f. interchange; *exallage,* f. complete change; *epallaxis,* f. exchange: allagite, allactite, allasotonic, diallage, trophallaxis, allassotherapy, *Allagecrinus dignatus* (a Pennsylvanian crinoid), *Metallactus ridibundus* (a beetle), *Epallagites avus* (a fossil odonatid).

Gr. *ameibo,* change, exchange; *amoibe,* f. change, alteration, recompense; *amoibaios,* interchanging, reciprocal; *ameipsis,* f. exchange; *epamoibos,* exchanging, bartering; *diameiptos,* changeable: ameba (*Amoeba proteus*), amebocyte, ameboid, amebic dysentery, *Amoebobacter roseum* (a bacterium), *Collozoum ameboides* (a radiolarian), *Endamoeba insolita* (a protozoan).

L. *convertibilis,* changeable, < *converto, -versus,* turn around, wheel about: convertible, convert, conversion.

L. *desultorius*, fickle, inconstant, superficial, skipping about; *desultor*, *-is*, m. leaper, inconstant person: desultory, *Liphistius desultor* (a spider).

Gr. *eutrepsia*, f. changeableness; *eutreptos*, easily changing: *Eutrepsia inconstans* (a moth).

L. *inconstans*, changeable, fickle, capricious: inconstant.

Gr. *keletistes*, leaper, inconstant person; see *keles* under **ride**

Gr. *meta*, preposition meaning between, among, after, but often used in compounds to imply change or exchange; *metabasis*, f. change, shift, transition; *metabolos*, changeable; *metaboulos*, changeful; *metachoresis*, f. change of place; *metadosis*, f. exchange, barter; *metalepsis*, f. substitution, exchange, alternation; *metameipsis*, f. exchange; *metamelos*, m.; *metanoia*, f. repentance; *metamorphosis*; *metaplasis*, f. transformation; *metatopsis*, f. change; *metathesis*, f. transposition, particularly of letters; *metathetos*, changed; *metatropos*, turn about; *metonymia*, f. change of name: metamorphic, metamorphosis, metastasis, metonymy, histometabasis, metaphrase, metabolism, metabletic, metaleptic, metasomatic, *Metabolocerus pilosus* (a beetle), *Metanoea flavipennis* (a trichopterid), *Metaplasia pyxidata* (a Devonian brachiopod).

L. *miscix*, *-icis*, changeable, inconstant:

Gr. *mixaithria*, changeable weather; see *aithros* under **clear**

L. *muto*, *-atus*, change; *mutabilis*, changeable; *mutator*, *-is*, m. changer: mutation, mutant, mutable, mutability, mutual, commute, immutable, *Nephelium mutabile* (pulasan), *Aphrophora permutata* (a spittle-insect), *Bromus commutatus* (a grass).

L. *proselytus* (Gr. *proselytos*), m. convert: proselyte.

L. *Proteus*, m. a sea-god capable of changing his form: protean, Proteaceae, *Proteus anguinus* (an amphibian), *Protea mellifera* (honey-flower), *Proteoceramus callosus* (a fossil pelecypod).

L. *translator*, *-is*, m. transferrer, one who renders from one language to another: translate, translator, translation.

Gr. *treptos*, turned, changed; see *trepo* under **turn**

L. *vario*, *-atus*, change, diversify; *variabilis*, changeable; see *varius* under **different**

L. *versabilis*, changeable, movable; *versatilis*, turning with ease, movable; see *verto* under **turn**

L. *Vertumnus*, m. god of change and trade: Vertumnus, *Vertumnus arabicus* (a beetle).

L. *vicis*, change, reciprocity, substitution; see *vice-* under **equal**

See: **turn, trade, alternate, give, different, other, move, over**

channe; **channos**, Gr. a sea-perch; see **perch**

channel < OF. *chanel*, < L. *canalis*, pipe, groove, conduit, trench, bed; see **ditch, pipe, furrow, bed, passage, valley, way**

chaos, Gr. yawning abyss, condition of utter formlessness, confusion, and disorder; see *chaino* under **open**

chara, L. a kind of plant; now applied to a genus of algae; see **alga**

characo- < Gr. *charax*, *-akos*, pointed stake, prop, pole; *charakion*, dim., see **pillar**; a sparoid fish, see **bream**

character, L. (Gr. *charakter*, an instrument for marking or engraving, < *charasso*, engrave), mark, sign, distinctive nature; see **mark, letter, nature**

charadra; **charadros**, Gr. deep gully, rift, ravine, bed of a mountain stream; see **valley**

charadrius, L. (Gr. *charadrios*), a yellowish bird dwelling in clefts; a plover; see **plover**

charagma; **charagmos**, Gr. any mark, cut, engraved, or stamped; see *character* under **mark**

chariento- < Gr. *charieis*, *-entos*, graceful, beautiful; see *charis* under **beauty**

charis, *-itos*, Gr. loveliness, grace, favor; see **beauty**

charisma, Gr. free gift; *charistikos*, giving freely; see **give**

charitia, Gr. jest, joke; see *charientisma* under **laugh**

charm < L. *carmen*, song; see **sing, draw, beauty, joy, agreeable, manner, magic**

charma, Gr. source of joy, delight; *charmosyne*, joyfulness, < *chairo*, rejoice; see **joy**

Charon, Gr. the ferryman who carried the souls of the dead over the river Styx; see **carry**

charops, *-opos*, Gr. glad-eyed, joyous; see *chairo* under **joy**

chart < L. *charta* (Gr. *charte*), leaf of paper, thin plate, lamina; *chartula*, dim.; see **paper, map**

chascaco- < Gr. *chaskax, -akos*, a gaper; see *chaino* under **open**

chascano- < Gr. *chaskanon*, a cocklebur; see **composite**

chase < OF. *chacier*, < L. *capto*, strive to catch; see **hunt**

chasm < Gr. *chasma, -tos*, yawning hole, gulf, open mouth; see *chaino* under **open**

chaste < L. *castus*, pure; see **modest**

chatter < ME. *chatteren;* of imitative origin; see **speak**

chaulios, Gr. outstanding, prominent; see **projection**

chaunax, Gr. liar; see **lie**

chaunos, Gr. porous, spongy, empty; see **hole**

cheap < AS. *ceap*, trade, sale, price; see **pay**

cheat < L. *ex*, out, *cado*, fall; see **lie**

check < OF. *eschec*, stop, hindrance; see **hold, bind, bar, hinder**

checkered < OF. *eschequier*, chessboard; see **mosaic**

chedropon, Gr. legume; see **bean**

cheek < AS. *ceace*, chap, jowl.
> L. *bucca*, f. cheek, cavity; *buccula*, f. dim.; *bucculentus*, with full cheeks: buccal, buccula, debouch, rebuke, boca, *Bucculatrix pomifoliella* (apple-leaf moth), *Ericymba buccata* (silver-jawed minnow), *Perlissus bucculentus* (an ichneumonid).
> L. *gena*, f. cheek: genal.
> L. *mala*, cheek, cheekbone, jaw; see **jaw**
> Gr. *melon*, n. cheek: meloplasty.
> Gr. *pareion*, n. cheek: *Parioglossus taeniatus* (a fish), Opisthoparia, *Pareiasaurus [Pariosaurus] horridus* (a fossil reptile).

cheer < OF. *chiare;* see **joy, agreeable, laugh, rest**

cheese < AS. *cese* (G. *käse*), < L. *caseus*, cheese; *caseolus*, m. dim.; *casearius*, pertaining to cheese; *caseatus*, mixed with cheese, cheesy: cheesemonger, casease, caseation, casein, caseous, *Piophila casei* (cheese-fly).
> Gr. *hippake*, f. cheese made of mare's milk:
> Gr. *sausax, -akos*, m. a kind of cheese:
> Gr. *trophalis, -idos*, f. fresh cheese; *trophalion*, n. dim.:
> Gr. *tyros*, m. cheese; *tyridion; tyrion*, n.; *tyriskos*, m. dim.: tyrosine, tyrosinase, tyrotoxine, tyroma, butter, *Tyroglyphus siro* (cheese-mite).

chela, L. (Gr. *chele*), claw; *chelion*, dim.; see **nail**

cheleuma, Gr. cord, bond; see **rope**

chelinos; cheleutos, Gr. netted, plaited; see **net**

chelidon, Gr. swallow; *chelidonion*, swallowwort; see **swallow, poppy**

chellaries, Gr. a fish; see **fish**

chellon (chelon), Gr. a mullet; see **mullet**

chelo- < Gr. *chele*, claw, see **nail**; < *chelos*, chest, coffer, see **box**

cheloma, Gr. notch; see **notch**

chelone, Gr. tortoise, turtle; *chelonarion*, dim.; *chelonion*, tortoise-shell; see **turtle**

chelonis, Gr. lyre; see **harp**

chelos, Gr. large chest, coffer; see **box**

chelyne, Gr. lip; *chelynion*, dim.; see **lip**

chelys; chelydros, Gr. tortoise, turtle, water-serpent; see **turtle**

chelyscio- < Gr. *chelyskion*, a slight cough; see **cough**

chemo-, Gr. pertaining to chemistry and chemical terms, < Ar. *alchimia*, alchemy: chemistry, chemical, chemist, chemotaxis, chemotherapy, chemotropism, alchemy.

chennion, Gr. a kind of quail; see **quail**

cheno- < Gr. *chen, -os*, goose; *chenarion; chenion; cheniskos*, dim.; *chenizo*, cackle like a goose; see **goose, cackle**

chenopodium, L. (Gr. *chenopous*), goosefoot, pigweed; see **goosefoot**

chera, Gr. widow; *chereia*, widowhood; *chereios*, widowed; *cherikos*, of a widow; *cheroo*, bereave, desolate; see **woman**

cherado- < Gr. *cheras, -ados*, alluvium, silt, detritus; see **earth**

cherambe, Gr. a kind of pelecypod; see **mollusk**

cheramis, Gr. a kind of mussel; see **mollusk**

cheramos, Gr. hole, cleft, hollow; see **hollow**

chermado- < Gr. *chermas, -ados,* large pebble, stone; *chermadion,* dim.; *chermaster,* slinger; see **stone, throw**

chernes, *-etos,* Gr. laborer; see **make**

chernibion, Gr. chamber-pot; see **pot**

cherry < OF. *cherise,* < L. *cerasus* (Gr. *kerasos*); see **plum**

chersonesos, Gr. peninsula; see **peninsula**

chersos, Gr. land, dry land; *chersaios; chersinos,* living on land; see **earth**

cherub < Heb. *kerubh,* an angel; later, a chubby child; see **child**

chess < OF. *eschec,* check, stop, < Per. *shah,* king; see **play**

chest < AS. *cest,* < L. *cista* (Gr. *kiste*), box; see **box, breast**

chester < AS. *ceaster,* < L. *castrum,* fortified town, camp, fort, settlement; see **fort**

chestnut < L. *castanea* (Gr. *kastanos*), f. chestnut-tree; *castaneus,* of the color of chestnuts, brown, bay: castanet, *Castanea dentata* (chestnut), *Castanella wyvillei* (a radiolarian), *Castanissa challengeri* (a radiolarian), *Castanidium murrayi* (a radiolarian), *Anas castanea* (a duck), *Castanopsis caudata* (Chinese chinquapin), *Castanospermum australe* (a legume), *Aesculus hippocastanum* (horse-chestnut), *Cinclosoma castanotum* (ground-bird), *Taeniopygia castanotis* (a weaver-bird), *Cybelus castaneus* (a beetle).

chetos, Gr. need, want; *chetosyne,* need, destitution; see **poor**

cheuma, Gr. anything poured, stream, flow, < *cheo,* pour; see **pour**

chew < AS. *ceowan,* bite, eat; see **bite**

chi, Gr. twenty-second letter of the Greek alphabet; *chiasmos; chiastos,* arranged diagonally, crosswise; *chiazo,* mark with a cross like the letter *chi;* see **letter**

chickadee, a bird so named from its note; see **titmouse**

chicken < AS. *cicen.*

     L. *alector, -is* (Gr. *alektor, -os*), m. cock; *alektoris, -idos,* f. hen; *alektryon, -os,* m. cock; *alectorius,* pertaining to a cock: alectryomachy, Alectyron, *Alectryon excelsum* (titoki), *Alectoris rufa* (red-legged partridge), *Crax alector* (curassow), *Alectoria sarmentosa* (a lichen), *Alectorurus yedoensis* (a lily), *Hexalectris spicata* (an orchid).

     L. *capo, -onis* (Gr. *kapon, -os*), m. a castrated cock: capon, caponize.

     AS. *cocc,* male fowl, rooster: cock, cockerel, cockade, cockalorum, cockscomb, cockspur. Derive: cockroach, cockney, cockatrice, cockatoo, cocktail, cockeye.

     L. *gallus,* m. cock, fowl; *gallina,* f. hen; *gallinula,* f. dim. pullet; *gallinaceus,* of poultry: gallinaceous, gallinule, *Gallinago delicata* (Wilson snipe), *Gallus domesticus* (domestic fowl, originally from the jungle-fowl, *Gallus ferrugineus*), *Gallinula chloropus* (water-hen), *Gallicrex cristatus* (a rail), *Limnocryptes gallinula* (jacksnipe), *Crataegus crus-galli* (cockspur-thorn), *Nucula gallinacea* (a pelecypod), *Ascaridia galli* (fowl-roundworm), *Erythrina crista-galli* (a coral-bean).

     Gr. *kikkos,* m. cock; *kikka,* f. hen:

     Gr. *nossax, -akos,* m. chick, cockerel:

     Gr. *ortalis; ortalichos,* young bird, chick, fowl; see **bird**

     Gr. *pselex, -ekos,* m. cock without a comb:

     L. *pullus,* m. young animal, particularly a fowl; *pullulus,* m. dim.; *pullarius,* of young animals; *pullulo, -atus,* put forth, produce young: pullet, poultry, pullulation, polecat, pool (game), Punchinello, *Agapornis pullaria* (red-faced love-bird), *Syllis pulligera* (an annelid), *Galeruca semipullata* (a beetle), poulard.

     Gr. *thakothalpas,* f. a sitting hen:

chickweed, some plants of the pink family; see *stellaria* under **pink**

chicory < L. *cichorium* (Gr. *kichorion*); see **lettuce**

chief < L. *caput,* head; see **govern, first, lead**

child < AS. *cild.*

     L. *alumnus,* m.; *alumna,* f. foster-child, pupil, < *alumno,* nourish, bring up, educate: alumnus, alumna, *Nautilus alumnus* (a nautilus).

     Gr. *brephos,* babe, child, young of any animal; see **young**

     L. *camilla,* f. free-born maiden; *camillus,* m. free-born youth: Camilla, Camillus.

     L. *Catamitus* (Gr. *Ganymedes*), m. cupbearer of the gods; youth kept for unnatural purposes: catamite.

     L. *catlaster* (*catulaster*), *-tri,* m. boy, lad, stripling:

     AS. *cnafa,* boy, youth: knave, knavish, knavery.

     L. *conceptum,* the fetus; see **young**

     L. *filiolus,* little son; *filiola,* little daughter; see **son, daughter**

     L. *infans, -antis,* speechless; c. a little child; *infantulus,* m. dim.; *infantilis,* pertaining to children, childish: infant, infancy, infanticide, infantry, *Crator infantulus* (a beetle).

Heb. *kerubh,* an angel; later, a chubby child: cherub, cherubim, cherubic.
Gr. *kore* (L. *cora*), f. girl, pupil of the eye; *korasidion; korasion; koridion; korion,* n.; *koriske,* f. dim.; *koreios,* of a maiden, youthful; *hypokorizomai,* play the child, talk baby-talk: Cora, Corinna, hypocoristic, correctomy, *Coregonus oxyrhynchus* (schnabel), *Callicore rosea* (an amaryllidacead), *Planorbis corinna* (a gastropod).
Gr. *koros* (*kouros*), m. boy, lad; *koriskos,* m. dim.; *kourios,* youthful: *Curius concinnatus* (a beetle), Dioscurus.
L. *liber, -eri,* m. child:
Gr. *meirax, -akos,* f. young girl, lass; *meirakidion; meirakion;* n.; *meirakiske,* f. dim.; *meirakikos,* youthful, juvenile: miracidium, *Oncidium meirax* (an orchid).
L. *nata,* daughter; *natula,* dim.; *natus,* son; *natulus,* dim.; see **daughter, son**
Gr. *neanis, -idos,* f. girl, maiden; *neaniskos,* m. youth; *neanikos,* fresh, youthful: neanic.
Gr. *neophron,* childish in spirit: *Neophron percnopterus* (Egyptian vulture).
Gr. *nepios,* c. infant: nepionic.
Gr. *pais, paidos,* c. child; *paidarion; paidion,* n.; *paidariskos,* m. dim.; *paidiske,* f. young girl; *paidiskos,* m. dim. young boy; *paidnos,* childish; *aeipais,* ever-maiden: pedagogue, pedant, pedagogy, pediatrics, orthopedic, propaedeutic, pederasty, page, *Paedisca saligneana* (a moth), *Arctiopais ambusta* (a moth).
Gr. *pallax, -akos,* c. youth just before puberty: *Pallax prevosti* (a fossil beetle).
L. *parvulus,* m. child:
L. *puella,* f. girl; *puellula,* f. dim.; *puellaris,* girlish: *Irena puella* (fairy-bluebird).
L. *puer, -eri,* m. boy, child; *puera,* f. girl; *puerculus; puerulus,* m. dim.; *puerilis,* youthful, childish, silly; *puerperus,* giving birth, parturient; *puerperium,* n. childbed, childbirth: puerile, puerperal, puerility.
L. *pupus,* m. boy; *pupa,* f. girl, doll, in some insects a stage intermediate between larva and adult; *pupillus,* m.; *pupilla; pupula,* f. dim., minor, learner, pupil of the eye: pupil, puppet, puppy, pupa, pupate, puparium, pupivorous, *Pupa muscorum* (a gastropod), *Vulvulina pupa* (a foraminifer), *Bulimina pupula* (a foraminifer).
L. *pusus,* m. boy; *pusa,* f. girl; *pusio, -onis,* m. little boy; *pusiola,* f. dim. girl; *putillus,* m. boy; *putilla,* f. girl: *Onthophagus pusio* (a dung-beetle), *Climacograptus putillus* (a graptolite).
Gr. *teknon,* n. that which is born, child, young; *teknidion; teknion,* n. dim.; *teknosis,* f. a begetting, bearing of children, < *tikto,* beget, bear: *Tecnophilus nigricollis* (a beetle), *Spanotecnus filicornis* (a wasp).
Gr. *thalos,* n. young (of persons), child, scion: *Spermatothalus* (a protozoan).
Gr. *tryganion,* little turtle dove, a pet name for a girl; see *trygon* under **dove**
See: **young, birth, son, daughter, kin**

chilio- < Gr. *chilios,* thousand; see **thousand**

chilo- < Gr. *cheilos,* lip, rim; *chilarion,* dim.; see **lip**

chilos, Gr. green fodder; see **food**

chiloter, Gr. nosebag; see **bag**

Chimaera, L. (Gr. *Chimaira,* she-goat), a fabulous fire-spouting monster; see **animal**

chimaphila, NL. a genus of the heath family; see **heath**

chimato- < Gr. *cheima, -tos,* winter, cold, frost; see *cheimon* under **year**

chimetlon, Gr. chilblain; see **sore**

chimney < L. *caminus,* furnace, fireplace; see **pipe, hole**

chimo-; cheimono- < Gr. *cheimon, -os,* winter, frost; *cheimerinos,* of winter; see **year**

chin < AS. *cin.*
Gr. *anthereon, -os,* m. chin:
Gr. *geneion,* n. chin; *akrogeneios,* with prominent chin: *Geneion maculatum* (a puffer-fish).
L. *mentus,* n. chin: *Menticirrhus[Menticirrus] saxatilis* (kingfish).

chink < AS. *cinu,* fissure, crack; see **break**

Chione, Gr. mythological woman; a genus of mollusks; see **mollusk**

chiono- < Gr. *chion, -os,* snow; see **ice**

chip < AS. *cippian,* clip, pare; see **cut, part**

chirido- < Gr. *cheiris, -idos,* glove, sleeve; *cheiridion,* dim.; *cheirodotos,* with gloves, sleeves; see **glove**

chiro- < Gr. *cheir, -os,* hand; *cherydrion,* dim.; see **hand**

Chiron, L. (Gr. *Cheiron*), a centaur reputed for his skill in medicine, see **heal**; *cheiron; cheiroteros,* worse; *cheiristos,* wôrst, see *kakos* under **bad**

chironomos, Gr. a pantomimic; see **equal**

chiroto- < Gr. *cheirotos*, tamable; *cheirotikos*, good at taming; see **tame**

chirp < imitative origin.
  L. *frigutio, -itus,* chirp, twitter:
  L. *fritinnio, -itus,* chirp, twitter:
  L. *gryllo, -atus,* chirp like a cricket; see *gryllus* under **grasshopper**
  Gr. *ligypterygos,* chirping with the wings:
  L. *pipilo; pipio; pipo, -atus,* chirp, peep, twitter; *pipulus,* m. a chirping, peeping:
    sandpiper, pigeon, *Pipilo erythrophthalmus* (chewink, towhee), *Pipunculus rufipes*
    (a fly), *Rana pipiens* (leopard-frog).
  Gr. *spizo,* chirp like a sparrow:
  Gr. *strouthizo,* chirp:
  L. *zinzilulo, -atus,* chirp:

chirurgus, L. (Gr. *cheirourgos*), an operating physician; see **heal**

chisel < OF. *chisel.*
  L. *caelum,* n. burin, engraving tool:
  LL. *celtes, -is,* m. chisel: celt, celtiform.
  L. *cilio, -onis,* m. chisel, graver:
  Gr. *glaris, -idos,* f. chisel: *Glaridoglanis andersoni* (a fish).
  Gr. *glyphanos,* m. carving tool, chisel, knife; *glypter, -os,* m. chisel: *Ampeloglypter sesostris* (a beetle).
  Gr. *kolapter, -os,* m. chisel; *kolaptos,* engraved: *Colapteroblatta compsa* (a cockroach).
  Gr. *kopeus, -eos,* m. chisel: *Copeocoris abscissus* (a bug), Copeognatha.
  L. *scalprum,* chisel, knife; see **knife**
  See **cut, knife, tool**

chiton, Gr. tunic; see **garment**

chlaeno- < *chlaina,* cloak, wrapper; *chlainion,* dim.; see **garment**

chlamydo- < Gr. *chlamys, -ydos,* mantle; see **garment**

chlaros, Gr. gay, lively; see **joy**

chledos, Gr. mud, dirt, debris; see **earth**

chleuastes, Gr. mocker, scoffer; see **laugh**

chliaros, Gr. lukewarm, tepid; see **lukewarm**

chlidanos, Gr. delicate, luxurious, < *chlide,* delicacy, luxury; see **soft**

chlidono- < *chlidon, -os,* bracelet, anklet, ornament; see **bracelet**

chloë; chloa, Gr. first green shoot of plants in spring, grass; see **grass**

chlorine < Gr. *chloros,* green; see ELEMENTS under **thing**

chlorion, Gr. probably the golden oriole; see **oriole**

Chloris, Gr. goddess of flowers, see **flower**; a finch, see **finch**

chloro- < Gr. *chloros,* green, < *chloë,* young shoot; compounds of chlorine; see **green**

chlossos, Gr. fish; see **fish**

chnauma, Gr. piece, slice; see **part**

chnauros, Gr. dainty; see **dainty**

chnoë, Gr. hub or nave of a wheel; see **box**

chnu- < Gr. *chnous,* any light, porous substance, foam, fine down, bloom; see **foam**

choano- < Gr. *choane,* funnel; *choanos,* crucible, melting-pot; *chonion,* dim.; see **funnel**

chocolate < Ab.Am. *chocolatl:* cacao, cocoa, *Theobroma cacao* (chocolate-tree).
  NL. *abroma,* f. a genus of the chocolate family: *Abroma sinuosa* (an abroma).
  NL. *cola,* f. a genus of the chocolate family, < African *kola:* Coca-Cola, *Cola acuminata* (kola-nut).
  NL. *sterculia,* f. a genus of the chocolate family, < L. *Sterculius,* god of manuring: Sterculiaceae, *Sterculia lanceolata* (a sterculia).
  NL. *theobroma,* f. a genus of the chocolate family: *Theobroma angustifolia* (monkey-chocolate).

chodanos, Gr. breech, buttocks; see **rump**

chodos, Gr. dung; see **dung**

choë, Gr. a pouring, drink-offering, stream; see **pour**

choenicido- < Gr. *choinikis, -idos,* trepan; see **bore**

choenico- < Gr. *choinix, -ikos,* a dry measure; see **box**

choerado- < Gr. *choeras, -ados,* scrofula; see **disease**

choerino- < Gr. *choirine,* a sea-mussel; see **mollusk**

choero- < Gr. *choiros,* pig; see **hog**

choico- < Gr. *choikos,* of earth or clay, < *chous,* earth heaped up, soil; see **earth**

choke < AS. *ceocian.*
> L. *ango, anctus; anxus* (Gr. *ancho*), press together, throttle, strangle, choke, distress; Gr. *anchone,* f. a throttling, strangling; *anchonios,* of strangling; *ankter, -os,* m. clasp, binder; *deranches,* throttling; *kynanche,* f. quinsy, dog-collar: anger, anguish, anxiety, anxious, angina pectoris, quinsy, cynanche, *Cynanchium nigrum* (black swallowwort), *Orobanche minor* (a broomrape), *Hexanchus griseus* (cow-shark), *Anchonocerus ruficeps* (a beetle), *Anchonidium unguicularis* (a beetle).
> L. *anhelus,* out of breath, panting, puffing; see **breathe**
> Gr. *asphyxia,* f. stopping of the pulse and respiration: asphyxia, asphyxiate.
> Gr. *asthma,* hard drawn breath, shortness of breath, panting, gasping; see **disease**
> Gr. *pnigo,* choke, stifle; *pnigma, -tos; pnigos,* n.; *pnigmos,* m.; *pnixis,* f. a choking, smothering, stifling; *pnigeus,* m. cover, damper; *pniktos,* strangled, suffocated: *Apopnictus longisetis* (a beetle).
> L. *strangulo, -atus,* choke, stifle: strangle, strangulation, *Pseudopolemius substrangulatus* (a beetle).
> L. *suffoco, -atus,* choke, stifle, strangle: suffocate, suffocation.
> See: **bind, press**

cholado- < Gr. *cholas, -ados,* bowels, guts; *cholix, -ikos,* gut; see **intestine**

cholasma, Gr. lameness; see *cholos* under **hurt**

cholera, L., Gr. a disease; see **disease**

cholo- < Gr. *chole,* bile, gall, see **bile**; < *cholos,* anger, wrath, see **anger**; lame, halt, maimed, see **hurt**

chomato- < Gr. *choma, -tos,* bank, mound, dam; *chomation,* dim.; *chostos,* banked up, heaped; see **wall**

chondrilla, L. (Gr. *chondrille*), a kind of endive; see **lettuce**

chondros, Gr. grit, grain of wheat or spelt; cartilage; *chondrion,* dim.; see **grain, gristle**

chone; chonos, Gr. contraction of *choane,* funnel; see **funnel**

choose < AS. *ceosan.*
> Gr. *apoklestos,* chosen, select:
> L. *conscriptus,* chosen, elected, enrolled: conscript, conscription.
> L. *excerpo, -ptus,* pick out, choose, select, extract; *excerptum,* n. extract, selection: excerpt.
> Gr. *hairesis,* choice, election, plan; *hairetikos,* able to choose; *hairetos,* eligible, desirable, chosen; see *haireo* under **take**
> Gr. *kletos,* called, chosen; see *kaleo* under **call**
> L. *lego, lectus* (Gr. *lego*), collect, pick, choose, select; *lektos,* picked out, chosen; *analektos,* choice, select; *apolektos,* chosen, picked; *deligo, -lectus,* choose out, select; *eligo, -lectus,* pick, choose; *eklektos,* selected; *epilektos,* chosen; *seligo, -lectus,* cull, choose out: election, eclectic, eclogue, analects, select, selection, lectotype, *Apolectus stromateus* (a fish).
> Gr. *merikos,* particular:
> L. *opto, -atus,* choose, select, elect; *optio, -onis,* f. choice, selection; assistant of one's own choice; *optabilis,* desirable; *optativus,* of a wish: option, adoption, optative.
> L. *particularis,* of parts, partial: particular.
> OF. *piquer,* pierce, prick, choose: pick, picket, pickpocket, picklock, pickwick.
> See: **lot, take, judge, separate, vote, gather, good**

chora; chorema, Gr. room, space, place; *choraphion,* dim.; *choretos,* containing; see **space**

chord < L. *chorda* (Gr. *chorde*), gut, string of a musical instrument; see **rope**

chordeiles, NL. a genus of goatsuckers; see **goatsucker**

choreutes, Gr. dancer; *choreutikos,* of a dance; see *choros* under **dance**

chorion, Gr. membrane; see **membrane**

choris, Gr. apart, asunder; *choristos,* separated, < *chorizo,* separate, see **separate**

choro- < Gr. *choros,* land, country; *choridion; chorion,* dim., see **country**; dance, see **dance**

chortos, Gr. grass, fodder; see **grass**

chorus, L. (Gr. *choros*), dance; see **dance**

chostos, Gr. made by heaped up earth; see *choma* under **wall**

chowder < F. *chaudiere,* kettle, pot, < L. *calidus,* warm; see **soup**

chremato- < Gr. *chrema, -tos,* goods, money; see **money**

chremetismos, Gr. neigh, whinny; see **neigh**

chremma: chrempton, Gr. spit; see **spit**

chreo- < Gr. *chreos,* debt, obligation; see **pay**

chresis, Gr. use, employment; *chresimos; chrestos,* useful, good, < *chrao,* use; see **use**

chresmo- < Gr. *chresmos,* oracle; *chrestes,* prophet, soothsayer; *chresterios,* prophetic; see **prophecy**

christos, Gr. anointed, < *chrio,* anoint; see **fat**

chroa; chroia, Gr. surface of the body, skin, color of the skin; *chroikos,* colored; see *chroma* under **color**

chromato- < Gr. *chroma, -tos,* color, color of the skin, complexion; *chromation,* dim.; see **color**

chromis, Gr. a sea-fish; see **wrasse**

chromium < Gr. *chroma,* color; see ELEMENTS under **thing**

chromos, Gr. noise, neighing; see **sound**

chrono- < *chronos,* time; *chronikos,* of time; *chronios,* lasting; see **time**

chroster, Gr. dyer; see *chroma* under **color**

chroto- < Gr. *chros, -otos,* skin, color of the skin; *chrotidion,* dim.; see *chroma* under **color**

chrysalis < L., Gr. *chrysallis,* gold-colored pupa of a butterfly; see **young**

chrysanthemum, L. (Gr. *chrysanthemon*) a genus of the composite family; see **composite**

chrysaphos, Gr. probably the gilthead; see **bream**

chryso- < Gr. *chrysos,* gold; *chryseos; chrysous,* golden; see **gold**

chrysocolla, L. (Gr. *chrysokolla*), a silicate of copper; see **copper**

chrysometris, Gr. goldfinch; see **finch**

chrysogonum, L. (Gr. *chrysogonon*), a genus of the composite family; see **composite**

chrysophrys, Gr. gilthead; see **bream**

chthamalos, Gr. on the ground, low; see **low**

chthono- < Gr. *chthon, -os,* earth, soil; see **earth**

chub < uncertain origin; see **carp**

church < AS. *cyrice,* < Gr. *kyriakon,* the Lord's house; see **temple**

chydaeus, L. (Gr. *chydaios*), poured out in streams, abundant, common; see *adaios* under **abundance**

chylos, Gr. juice, moisture; see **juice**

chymos, Gr. juice, see **juice**; *chyma,* anything poured, see *cheo* under **pour**

chytos, Gr. poured, fluid, liquid; see *cheo* under **pour**

chytro- < Gr. *chytra; chytros,* earthen pot, pipkin; *chytridion; chytrion,* dim.; see **pot**

cibarius, L. pertaining to food, < *cibus,* food; see **food**

cibdelus, L. (Gr. *kibdelos*), spurious, base, adulterated, false; see **lie**

ciborium, L. (Gr. *kiborion*), cup shaped like the pericarp of a waterlily; see **cup**

ciboto- < Gr. *kibotos,* box, chest; *kibotion,* dim.; see **box**

cicada, L. 'tree cricket', 'locust': *Cicada orni* (a cicada), *Cicadula sexnotata* (a bug), *Cicadella vittata* (a bug), *Magicicada septendecim* (seventeen-year cicada).
  L. *acheta* (Gr. *achetas; echetes*), m. male cicada: *Acheta domesticus* (a house-cricket).
  Gr. *kixios,* m. cicada: Cixiidae, *Cixius montanus* (a bug).
  Gr. *membrax, -akos,* m. a kind of cicada: membracid, *Membracis foliacea* (a leaf-hopper).
  Gr. *sigion,* n. a kind of cicada:
  Gr. *tephras, -ados,* m. a cicada:
  Gr. *tettix, -igos,* m. cicada; *tettigonion,* n. dim.: *Tettigometra debilis* (a fossil insect), *Circotettix verruculatus* (an orthopterid), *Tettigomyces vulgaris* (a fungus).
  L. *tibicen,* piper, flutist; a genus of cicadas; see **music**
  See: *Tithonus* under **old, grasshopper, bug**

cicatrix, L. scar; *cicatricula,* f. dim.; see **mark**

ciccabo- < Gr. *kikkabe,* screech-owl; see **owl**

ciccus, L. trifle; see **trifle**

cicer, L. chickpea; *cicercula,* dim.; see **bean**

Cicero, L. Roman orator and writer; see **speak**

cichla, L. (Gr. *kichle*), thrush, see **thrush**; a sea-fish, see **wrasse**

cichorium, L. (Gr. *kichorion*), chicory, succory; see **lettuce**

cicilendrum, L. spice; see **spice**

cicindela, L. glowworm; a genus of beetles; see **beetle**

cicinus, L. of the castor-oil plant, *cici* (Gr. *kiki*); see *ricinus* under **spurge**

ciconia, L. stork; see **stork**

cicuma, L. (Gr. *kikymis*), screech-owl; see **owl**

cicur, L. tame; see **tame**

cicuta, L. poison-hemlock; a genus of the carrot family; see **carrot**

cicyo- < *kikys, -yos,* strength, vigor; see **strong**

-cida; -cide < L. *caedo,* cut, kill; see *caedo* under **cut**

cidapho- < Gr. *kidaphos,* wily; see **know**

cidaris, L. (Gr. *kidaris*), a Persian diadem, tiara; see **crown**

cider < L. *sicera* (Gr. *sikera;* Heb. *shekar*), a fermented liquor; see **drink**

cigar < Sp. *cigarro,* prob. < an Ab.Am. word meaning to smoke; see **cloud**

cilicium, L. a garment of haircloth; *ciliciolum,* dim.; see **cloth**

cilio, L. chisel; see **chisel**

cilium, L. eyelash, eyelid; see **hair**

cilliba, L. (Gr. *killibas*), round dining-table; see **table**

cillico- < Gr. *killix, -ikos,* ox with crooked horns; see **cattle**

cillo- < Gr. *killos,* ass; see **horse**

cilluro- < Gr. *killouros,* wagtail; see **wagtail**

cimbex, NL. a kind of wasp; see **wasp**

cimbico- < Gr. *kimbix, -ikos,* niggard, miser; see **stingy**

cimelio- < Gr. *keimelion,* treasure; see **store**

cimex, *-icis,* L. bug; *cimiculus,* dim.; *cimico, -atus,* destroy bugs; see **bug**

cimolia, NL. (Gr. *kimolia*), white clay from the island of Cimolus; see **earth**

cinabra, NL. (Gr. *kinabra*), smell of a he-goat; see **smell**

cinaedus, L. (Gr. *kinaidos*), lewd, wanton, unchaste, shameless; see **lewd**

cinchona, NL. a genus of the madder family, species of which yield quinine and other alkaloids; see **quinine**

cincinnus, L. curl; *cincinnatus,* curly; *cincinnatulus,* dim.; see **curl**

cinclido- < Gr. *kinklis, -idos,* latticed gate or bar; see **screen**

cinclus, L. (Gr. *kinklos*), a wagtail; see **wagtail**

cinctum, L. girdle, belt; *cincticulum,* dim.; *cinctus; cinctutus,* girded; see **belt**

cinder < AS. *sinder,* slag, scoria; see **ash**

cindyno- < Gr. *kindynos,* hazard, risk, danger; see **danger**

cinema < Gr. *kinema, -tos,* movement; *kinesis,* movement, motion; *kinetikos,* setting in motion, exciting; *kinetos,* movable; *kinetes,* mover, author, < *kineo,* move, set in motion; see **move**

cinereus, L. ash-colored, gray, < *cinis, cineris,* ashes; *cinerarius,* pertaining to ashes; see **ashes**

cinetro- < Gr. *kinetron,* ladle, stirring-rod; see **spoon**

cingulum, L. girdle, zone; *cingillum,* dim.; see *cinctum* under **belt**

cinna, L. (Gr. *kinna*), a Cilician grass; see **grass**

cinnabar < OF. *cenobre,* < L. *cinnabaris* (Gr. *kinnabari*), n. the red sulfide of mercury; *kinnabarinos,* red like cinnabar: *Polystictus cinnabarinus* (a fungus).
   Gr. *ammion,* n. cinnabar in its sandy state:
   Gr. *miltos,* red earth, minium, red-lead; see **red**
   L. *minium,* native cinnabar, red-lead; *mineus,* of cinnabar-red color; *miniaceus; minius,* of cinnabar; *minianus,* painted with cinnabar, < *minio, -atus,* paint red with cinnabar; see *miniatus* under **red**

cinnamon < L. *cinnamum* (Gr. *kinnamon; kinnamomon*), n. bark of *Cinnamomum zeylanicum; cinnameus,* of cinnamon: *Cinnamomum camphora* (camphor-tree), *Osmunda cinnamomea* (cinnamon-fern).

L. *comacum,* n. a kind of cinnamon found in Syria:
Gr. *mosylon,* n. a kind of cinnamon:
See: **spice, brown**

cinnyris < Gr. *kinnyris,* a small bird; see **bird**

cinopeto- < Gr. *kinopeton,* a serpent; see **snake**

cinyra, L. Gr. *kinyra*), an instrument with ten strings, a kind of lyre; see **harp**

cinyro- < Gr. *kinyros,* plaintive, wailing, < *kinyrizo,* lament, wail; see **weep**

cio- < Gr. *kis, kios,* a weevil; see **beetle**

ciono- < Gr. *kion, -os,* column, pillar, uvula; see **pillar, uvula**

cipher < OF. *cifre,* < Ar. *sifr,* empty; see **empty, number**

cippus, L. gravestone, palisades; see **pillar**

circaea, L. (Gr. *kirkaia*), a plant; see **evening-primrose**

Circe, L. (Gr. *Kirke*), notorious enchantress; see **magic**

circinatus, L. coiled, curled away from an apex, < *circino,* make round; see **curl, circle**

circle < L. *circulus,* m. dim. of *circus* (Gr. *kirkos*), m. ring; *circellus,* m. dim.; *circularis,* round; *circum,* around, about, on all sides; *circes, -itis,* m.; *circinatio, -onis,* f. circumference, circle; *circuitus,* going round, revolving; *circulator,* m.; *circulatrix,* f. stroller, mountebank, quack; *circino, -atus,* make round; *circular, -atus,* form a circle, go around; *circinus* (Gr. *kirkinos*), m. pair of compasses: encircle, semicircle, circus, circuit, cirque, circulation, circinate, circumference, circuitous, search, research, *Circulifer tenellus* (beet-leafhopper), *Lygodium circinatum* (Malay climbing-fern).
L. *ambitus, -us,* m. circuit, course, orbit, circumference:
L. *amplexus,* embracing, encircling, surrounding; see **embrace**
L. *annulus,* m. ring; *annellus,* m. dim.; *annularis,* of a ring, circular; *annulatus,* ringed, circular: annular, annulate, Annulata, annulet, annelid, *Polistes annularis* (a wasp), *Dictyna annulipes* (a spider).
L. *cesticillus,* m. ring or hoop placed on the head to support a burden:
L. *condalium,* n. a little ring for slaves:
L. *corona,* crown, halo; see **crown**
L. *cyclus* (Gr. *kyklos*), m. circle, ring; *kykliskos,* m. dim.; *kyklas, -ados; kyklikos; kyklios,* round, circular; *artikyklos,* perfectly round; *enkyklios,* rounded, periodical: cycle, bicycle, cyclone, cyclostome, encyclopedia, encyclical, polcyclic, cyclotron, cyclopean, Cyclades, *Cyclorhynchus psittaculus* (parrot-awklet), *Cyclolina carinata* (a foraminifer), *Cyclotella striata* (a diatom), *Cyclina lunulata* (an Eocene pelecypod), *Cyclidium distortum* (a protozoan), *Cyclanthera pedata* (a gourd), *Cyclops bicuspidata* (a copepod), *Cyclicopora praelonga* (a bryozoan), *Cycladophora fenestrata* (a radiolarian), *Cyclopleurus irroratus* (a coelenterate), *Copotocycla bicolor* (tortoise-beetle), *Euryschema tricycla* (a moth), *Chrysichthys cyclurus* (a fish), *Lithocyclia lenticula* (a radiolarian), Ku Klux Klan, *Cyclosa turbinata* (a spider), *Actinocyclus crassus* (a diatom).
Gr. *daktylios,* ring; see *daktylos* under **finger**
Gr. *dinotos,* turned, rounded; see *dine* under **turn**
Gr. *diskos,* quoit, flat circular plate; see **plate**
L. *evexus,* rounded at top; see **dull**
L. *globosus,* round like a ball, spherical; see *globus* under **ball**
L. *glomerosus,* round like a ball; see *glomus* under **ball**
Gr. *gyros,* circle, round; *gyraleos,* rounded, curved; *perigyris,* circumference; see **turn**
Gr. *halos,* f. circle around the sun or moon, corona, a round threshing floor; *halonion,* n. dim.: halo, halation.
L. *horizon, -ontis* (Gr. *-ontos*), m. the circle where earth and sky meet: horizon, horizontal, *Juniperus horizontalis* (creeping-juniper).
AS. *hring,* circle, band: ring, harangue.
Gr. *krikos*(a form of *kirkos*), m. ring, finger-ring; *krikallion; krikion,* n. dim.; *krikoma, -tos,* n. ring, circle; *kriotos,* ringed, made of rings: cricoid, *Cricosaurus elegans* (a fossil reptile), *Cricotus heteroclitus* (a fossil amphibian), *Eurhinocricus fissus* (a milleped).
Gr. *omilla,* f. a circle for playing a game:
L. *orbis, -is,* m. circle, ring, disk, globe; *orbiculus,* m. dim.; *orbicularis; orbiculatus; orbitus,* circular; *orbita,* f. track made by a wheel, path, course, track, rut: orb, orbicular, orbit, exorbitant, *Orbiculus exoletus* (a pelecypod), *Orbiculina nummismalis* (a foraminifer), *Orbitus purpuripennis* (a beetle), *Orbitolites complanata* (a foraminifer), *Spirorbis lucidus* (a tubeworm), *Symphoricarpus orbiculatus* (a coralberry), *Planorbis carinatus* (a snail), *Cornuspira planorbis* (a foraminifer), *Opidnus orbitalis* (a wasp).

Gr. *perimetron,* n. circumference: perimeter.

Gr. *periodos,* f. a going round, cycle: period, periodical, periodicity.

Gr. *periphereia,* f. circumference: periphery.

L. *rotundus,* round, circular, spherical: rotundity, round, around, orotund, *Dizygocrinus rotundus* (a Mississippian crinoid), *Obolus rotundatus* (a fossil brachiopod), *Ctenodonta subrotunda* (a fossil pelecypod), *Viola rotundifolia* (a violet).

Gr. *saros,* m. the Babylonian 18-year cycle for the recurrence of solar and lunar eclipses: saros, saronic.

Gr. *strongylos,* round, rounded: strongyle, strongylate, strongylosis, strongyloid, *Strongylus equinus* (a roundworm), *Strongylocrinus uralicus* (a Permian crinoid), *Libyostrongylus hebrenicutus* (a nematode).

L. *teres, -etis,* rounded, cylindrical; see **cylinder**

L. *tornus* (Gr. *tornos*), lathe, compass; see **turn**

Gr. *trochalos,* running, round; see *trecho* under **run, wheel, ball**

L. *ungulus,* m. finger-ring, ring:

L. *verticillus,* whirl, whorl, circle, dim. of *vertex,* whirl, eddy; see *verto* under **turn**

See: **ball, around, turn, belt, berry, wheel**

circum, L. around, about, on all sides; see **around**

circumspectus, L. cautious, wary; see **guard**

circus, L. (Gr. *kirkos*), circle, see **circle;** < Gr. *kirkos,* a kind of hawk, see **hawk**

ciris, L. (Gr. *keiris*), a fabulous bird; see **bird**

ciro- < Gr. *keiro,* cut (hair), cut up, waste, destroy; see *koureus* under **barber**

cirrhido- < Gr. *kirrhis, -idos,* a sea-fish; see **fish**

cirrho- < Gr. *kirrhos,* orange-colored, yellow, tawny; see **yellow**

cirrus, L. curl, ringlet, tendril; *cirratus,* curly, fringed; *cirritus,* filamentous; *cirrosus,* full of curls; see **curl**

cirsium, L. (Gr. *kirsion*), thistle; see **thistle**

cirso- < Gr. *kirsos,* a dilated vein; see **bend**

cis, NL. (Gr. *kis, kios*), a weevil; see **beetle**

cis-, L. on this side; *citerior,* nearer; *citinus,* nearest; see **near**

cisero- < Gr. *kiseris,* pumice; see **stone**

cisium, L. a two-wheeled cab; *cisiarius,* cab-driver; see **vehicle**

cisorium, L. a cutting instrument; see **knife**

cissus, L. (Gr. *kissos*), ivy; see **vine**

cissybium, L. (Gr. *kissybion*), a wooden drinking-cup; see **cup**

cista, L. (Gr. *kiste*), box, chest; *cistella; cistellula; cistula,* dim.; see **box**

cisterna, L. underground reservoir for water; see **basin**

cistus, L. (Gr. *kistos*), rockrose; see **rockrose**

citatus, L. hastened, urged on, incited, quick, swift, < *cito, -atus,* drive, incite, stimulate; see **arouse, swift**

citellus, NL. a kind of spermophile; see **squirrel**

citer, -tra, -trum, L. lying near, near, close; *citerior,* nearer; *citimus,* nearest; see *cis* under **near**

citeria, L. caricature, effigy, likeness; see **form**

cithara, L. (Gr. *kithara*), harp, lyre; see **harp**

citharoedus, L. (Gr. *kitharoidos*), a Red Sea fish; see **fish**

citharus, L. (Gr. *kitharos*), a kind of turbot, see **flatfish;** thorax, chest, see **breast**

citimus, L. nearest; see *cis* under **near**

citizen < OF. *citeain,* < L. *civis;* see *civitas* under **town**

citocacium, L. olive-daphne; see **mezereon**

citrago, L. a kind of balm; see **mint**

citrullus, NL. a genus of the gourd family; see **melon**

citrus; citrea, L. (Gr. *kitrea*), f. citron-tree, a plant of the rue family; *citreum* (Gr. *kitrion; kitron*), n. citron; *citreus; citrinus,* of citron: citral, citric acid, magnesium citrate, citrul, citron (*Citrus medica*), citronella (from *Cymbopogon nardus*), *Citrus sinensis* (sweet orange), *Citropsis latialata* (a cherry-orange), *Pseudococcus citri* (citrus mealy-bug), *Morinda citrifolia* (a madder), *Eucalyptus citriodora* (lemon-eucalyptus), *Cymbopogon citratus* (lemon-grass), *Taenianotus citrinellus* (a fish), *Hoplichthys citrinus* (a fish), *Atalantia citroides* (a rutacead), *Erythronium citrinum* (lemon fawn-lily), *Ceratina citrinifrons* (a bee), *Xanthomonas citri* (a bacterium).

NL. *aurantium,* n. < OF. *orenge,* < Ar. *naranj,* orange: orangeade, naringin, orange hawkweed (*Hieracium aurantiacum*), *Citrus aurantium* (sour orange), *Rhinonycteris aurantia* (orange-bat).

NL. *fortunella,* f. a genus of the citrus tribe, < Robert Fortune, who introduced the kumquat to Europe: *Fortunella margarita* (oval kumquat).

F. *limon,* lemon: *Citrus limon* (lemon), limoncillo (*Pectis angustifolia*), lime (*Citrus aurantifolia*), *Feronia limonia* (wood-apple), *Pholiota limonella* (a mushroom).

F. *poncire,* a kind of citron: *Poncirus trifoliata* (a Chinese orange).

citta, L. (Gr. *kitta; kissa*), a jay; see **crow**

citus, L. quick, swift, < *cieo, -itus,* arouse, agitate, excite, move, call; *citellus; citillus,* dim.; see **swift, arouse, call**

city < F. *cite,* < L. *civis,* citizen; *civitatula,* dim. small city; see *civitas* under **town**

civet < F. *civette;* ML. *zibethum;* It. *zibetto;* Ar. *zabad;* see **weasel**

civis, L. citizen; *civicus; civilis,* of citizens; *civitas,* citizenship, the body-politic, city; see **people, town**

cixallo- < Gr. *kixalles,* a highway robber; see **steal**

cixio- < Gr. *kixios,* cicada; see **cicada**

clabulare, L. large, open wagon; *clabularius,* of wagons; see **vehicle**

clacendix, L. a kind of shellfish; see **mollusk**

cladaro- < Gr. *kladaros,* easily broken, frail; see **weak**

clades, L. destruction, disaster; see **destroy**

clado- < Gr. *klados,* branch, twig, stem; *kladion,* dim.; see **branch**

cladrastis, NL. a genus of the bean family; see **bean**

clagalopes, L. a kind of eagle; see **eagle**

clagero- < Gr. *klageros,* screaming; see *klazo* under **call**

clam < AS. *clamm;* see **mollusk**

clambo- < Gr. *klambos,* mutilated; see **hurt**

clamo, -atus, L. cry out, shout, call; *clamito,* cry loudly; *clamor,* loud call; see **call**

clan < C. *clann,* offspring, tribe, family; see **kin, class**

clancularius, L. anonymous, secret, unknown; see **secret**

clandestinus, L. secret, hidden; see **secret**

clangor, L. a loud, ringing sound, noise; see **ring**

clanio- < Gr. *klanion,* bracelet; see **bracelet**

clarigo, L. declare war; see **fight**

clarus, L. clear; *claritas; claritudo,* clearness; see **clear**

clasmato- < Gr. *klasma, -tos,* fragment; see *klao* under **break**

class < L. *classis,* f. rank, division; *classicula,* f. dim.; *classicus,* of rank: classify, classic, classmate, classification, outclass.

L. *categoria* (Gr. *kategoria*), f. class, division: category, categorical.

L. *compositus,* orderly, well-arranged: composite, *Dioscorea composita* (a yam).

L. *concinno, -atus,* order, arrange, adjust:

L. *curia,* f. one of the divisions of the Roman people:

Gr. *diathetes,* arranger, placer; see *tithemi* under **place**

L. *digeries, -ei,* f. orderly distribution, disposition, arrangement:

L. *dispono, -ositus,* distribute, arrange, set in order; see *pono* under **place**

Gr. *ethnos,* n. nation, race, people, tribe; *ethnikos,* of a race: ethnology, ethnography, ethnical.

Gr. *euthemon, -os,* well-arranged, ordered, neat; *euthosyne,* good management, tidiness; *diathesis,* arrangement; disposition; *eudiathetos,* well-arranged; *thesis,* a placing, arranging; see *tithemi* under **place**

L. *genus, generis* (Gr. *genos, -eos*), n. kind, race, stock; Gr. *genea,* f. race, stock, family; *genikos,* of the race, generic; L. *gens, -entis,* f. clan; *gentilis,* belonging to the same race or clan; *congener, -a, -um,* of the same race: genus (pl. genera), generic, genotype, genocide, genealogy, congeneric, congener, gentle, miscegenation, *Gentilicamelus wyomingensis* (a fossil camel).

L. *instituto, -utus,* arrange, establish: institute, institution.

Gr. *kosmos,* order, arrangement; *kosmetikos,* skilled in arranging; *kosmios,* well-ordered; see **ornament**

L. *methodus* (Gr. *methodos*), f. mode of procedure; *eumethodos,* well-arranged: method, methodical.

L. *natio, -onis,* f. breed, race, tribe: nation, national, *Doliolum nationalis* (a tunicate).

L. *ordo, -dinis,* m. methodical arrangement, line, row, series; *ordinalis,* of a sequence or succession; *ordinarius,* of regular or usual manner; *ordino, -atus,* set in order, arrange, regulate: order, disorderly, ordinal, ordain, extraordinary, insubordination, *Ceratina ordinaria* (a bee).

Gr. *phrater,* clansman, member of the same tribe or race; see **kin**

Gr. *phyle,* f. tribe, race; *phyletes,* m. tribesman; *phyletikos,* of a tribe: phylum, phylogenesis, phylogenetic, phylogeny, *Phyletobius equestris* (a beetle).

L. *prosapia,* stock, race, family; see **kin**

L. *sors, sortis,* lot, share, kind; see **lot**

L. *species, -ei,* f. look, appearance, form, model, kind; *specialis,* particular, individual: species, specie, special, specific, spice, *Rheum speciforme* (a rhubarb).

Gr. *stello,* place, arrange, send; see **send**

L. *stirps,* stem, stock, family, race; see **stem**

Gr. *systema, -tos,* n. an ordered arrangement; *systematikos,* arranged, orderly: system, systematic, *Systemodon tapirinum* (a fossil mammal).

Gr. *tasso,* arrange, classify, place; *taxis, -eos* or *-ios,* f. order, rank, row, line, arrangement; *tagma, -tos,* n. an arrangement, rank, brigade; *taktikos,* pertaining to arranging or ordering; *taktos,* ordered; *eutaktos,* orderly, disciplined: taxonomy, taxodont, tagma, taxon, taxidermy, homotaxis, syntax, phyllotaxy, biotaxis, ataxia, ataxite, tactics, locomotor ataxia, *Protaxocrinus robustus* (a Silurian crinoid), taxeopodous, *Taxeotis delogramma* (a moth), *Taxoploca ovata* (a fossil sponge), *Taxorchis schistocotyle* (a trematode).

Gr. *thesis,* an arranging, proposition; see **hypothesis**

L. *tribus, -us,* f. stock, group; *tribulis,* of the same tribe: tribal, tribe.

Gr. *typos,* copy, model, pattern; see **form**

* * * * * * *

NL. *-aceae,* suffix denoting a family of plants, < L. *-aceus,* pertaining to: Magnoliaceae, Rosaceae, Salicaceae, Tiliaceae, Urticaceae.

NL. *-ales,* suffix denoting an order of plants: Bryales, Coniferales, Filicales, Geraniales, Malvales, Rosales, Sapindales.

NL. *-ata,* suffix denoting a division of animals, < L. *-atus,* pertaining to: Chordata, Echinodermata, Mcrostomata, Tabulata.

NL. *-idae,* suffix denoting a family of animals, < Gr. *-ides,* a patronymic suffix; *-ida* denotes various classifications: Bovidae, Canidae, Felidae, Pierididae, Veneridae, Arachnida, Eucarida.

NL. *-ina,* suffix denoting an order of animals; *-inae,* denoting subfamilies of animals, < L. *-inus,* pertaining to, like: Acarina, Sarcodina, Tetracladina, Terebratulinae, Scaphitinae, Formicinae, Pieridinae.

NL. *-oideae,* suffix denoting a tribe of plants; *-oidea,* a class of animals, < Gr. *-oides,* like: Asteroideae, Celastroideae, Crinoidea, Molluscoidea, Geometroidea.

NL. *-phyta,* denoting a phylum of plants: Thallophyta, Bryophyta, Pteridophyta, Spermatophyta.

See: **separate, list, govern, kin, nature**

classicum, L. engagement signal, the war-trumpet itself; see **horn**

clasterio- < Gr. *klasterion,* pruning-knife; see *kladeuterion* under **knife**

clasto- < Gr. *klastos,* broken in pieces; *klasis,* a breaking; see *klao* under **break**

clathratus, L. latticed, grated, screened; see **screen**

claudus, L. crippled, limping, lame, defective; see **hurt**

clausto- < Gr. *klaustos,* mournful; *klauster,* weeper; *klausimos,* plaintive; see *klaio* under **weep**

claustrum, L. bar, bolt, gate; *claustrarius,* pertaining to locks, locksmith; see **bar**

clausus, L. closed, shut, < *claudo,* close, shut; *clausula,* close. end; see **close**

clauthmo- < Gr. *klauthmos,* weeping; see *klaio* under **weep**

clava, L. club, graft; *clavula,* dim.; see **club**

clavis, L. key, bar; *clavicula,* dim.; *clavicarius,* key-maker, locksmith, see **key, curl**

clavus, L. nail; *clavulus; claviculus,* dim.; see **nail**

claw < AS. *clawu;* see **nail, tongs, finger**

clay < AS. *claeg;* see **earth**

-cle, diminutive suffix, < L. *-culus;* see DIMINUTIVES under **little**

clean < AS. *claene,* clear; see **pure, sterile, clear, wash, sweep**

clear < L. *clarus,* bright, plain, distinct, famous; *claritudo, -inis,* f. clearness, brightness, fame: clarity, clarion, clarinet, clarain, chanticleer, declaration.

Gr. *aithrios,* clear, fair; *aithria,* f. clear weather; *dysaithrios,* not clear, murky; *mixaithria,* f. changeable weather: *Aethriostoma undulata* (a beetle), *Aethriella conspicua* (a wasp), *Aethria leucaspis* (a butterfly), *Aethriamanta brevipennis* (a dragonfly), *Dysaethria pasteopa* (a moth).

Gr. *anephelos,* cloudless: *Anephilus[Anephelus] longulus* (a beetle).

Gr. *anomiklos,* without mist, clear:

L. *argutus*, clear, bright, shiny, sharp; *argutulus*, dim.: *Dolatocrinus argutus* (a Devonian crinoid).

Gr. *arisemos*, very clear, very notable; *diasemos*, clear, distinct, conspicuous: *Psilopa(Diasemocera) nigrotaeniata* (a fly).

Gr. *arizelos*, clear, conspicuous, distinct: *Arizelocichla nigriceps* (a bird).

Gr. *atholos*, clear: *Atholus bimaculatus* (a beetle).

L. *conspectus*, visible; *conspicuus*, manifest, visible, prominent; *perspicuus*, clear, transparent, evident; *prospicuus*, that may be seen in the distance or future: conspectus, conspicuous, *Cirsium conspicuum* (Mexican thistle), *Ischnochiton conspicuus* (showy chiton), *Delphinium inconspicuum* (a larkspur), perspicuous, *Thalpophila prospicua* (a butterfly).

L. *definitus*, distinct, clear, limited; *definitivus*, explanatory, exact, conclusive: definite, definitive.

Gr. *delos*, evident, visible, clear; *aridelos*; *catadelos*; *eridelos*; *peridelos*, very clear, conspicuous; *epidelos*, clearly seen; *eudelos*, quite clear, manifest; *delotikos*, indicative; *delotos*, demonstrable; *delosis*, f. explanation, < *deloo*, explain, reveal, show: delomorphous, adelite, Urodela, *Spirodela polyrhiza* (duckweed), *Delostoma roseum* (a bignoniacead), *Aridelus bucephalus* (a wasp), *Delonix* [*Delonyx*] *regia* (royal poinciana).

Gr. *dieides*, transparent:

L. *distinctus*, separate, different; see **separate**

Gr. *enarges*, visible, palpable, manifest; *enargeia; enargotes*, f. clearness, distinctness: *Enargopelta obscura* (a wasp).

L. *enodo, -atus*, free from knots, explain, elucidate: *Coryssoglymma enodata* (a beetle).

L. *enubilo, -atus*, free from clouds, make clear; *innubilus*, cloudless:

Gr. *epopsios*, in full view, conspicuous: *Epopsia metreta* (a moth).

Gr. *eudios*, calm, clear, fine; *eudia*, f. fair weather, calm: *Eudiospilus tricolor* (a wasp), *Eudia pavonia* (a moth).

Gr. *eukrines*, clear, distinct, well-separated:

L. *evidens, -entis*, clear, plain, visible: evident, evidence.

Gr. *exegesis, -eos*, f. explanation, interpretation; *exegetes*, m. expounder, interpreter; *exegetikos*, explanatory: exegesis, exegetical, exegete.

L. *explano, -atus*, make clear, elucidate, < *planus*, clear, distinct: explain, explanatory, explanation, plain.

L. *explicatus*, unfolded, arranged, plain, clear; *explicabilis*, that may be explained; *explicitus*, disentangled, apparent, clear, precise: explicate, inexplicable, explicit, *Crepicephalus explicatus* (a trilobite), *Dendropora explicata* (a fossil bryozoan).

L. *expressus*, clear, distinct, evident, plain: expressly.

Gr. *hermeneuo*, interpret, explain, expound; *hermeneus; hermeneutes*. m. interpreter; *hermeneuma, -tos*, n. interpretation, explanation: hermeneutics, hermeneutical.

Gr. *horatos*, visible; see *horao* under **see**

L. *hyalinus* (Gr. *hyalinos*), glassy, transparent; see **glass**

L. *illimis*, without mud, clear, pure:

L. *insignitus*, marked, clear, plain; see *signum* under **mark**

L. *interpretor, -tatus*, explain, translate; *interpres, -etis*, c. agent, intermediary: interpreter, interpretation, *Arenaria interpres* (turnstone).

Gr. *lampros*, bright, clear; see *lanterna* under **light**

Gr. *ligys; ligyros*, clear, sharp, distinct: *Ligypterus heydeni* (a cricket), *Ligyrus ebenus* (a beetle), *Ligyromorphus rufiventris* (a beetle).

L. *limpidus*, clear, transparent, pure: limpid, limpidity.

L. *liquidus*, clear, transparent; *eliquatus*, clear: liquid.

L. *lucidus*, clear, bright; *pellucidus*, transparent: lucid, pellucid, *Geryonia pellucida* (a medusa).

L. *manifestus*, clear, evident, visible: manifest, manifestation.

L. *obvius*, in the way, exposed, easy of access: obvious, obviate.

L. *perceptibilis*, able to be discerned: perceptible.

Gr. *phaneros*, visible, evident; *phantos*, visible; *diaphanes*, transparent, distinct; see *phaino* under **light**

Gr. *saphes*, clear, plain, distinct, certain; *asapheia*, f. indistinctness, obscurity: *Lepidosaphes ulmi* (oyster-shell scale), asaphia, *Asaphus expansus* (an Ordovician trilobite).

L. *screor*, clear the throat, hawk; see **sound**

L. *serenus*, clear, bright, fair: serene, serenity, serenade. Derive: serendipity.

L. *sudus*, without moisture, dry, cloudless, bright, clear: *Noctua suda* (a moth).

Gr. *toros*, piercing, sharp, clear:

Gr. *tranos*, clear, distinct: *Tranopelta gilva* (a wasp).

L. *translator, -is*, one who renders from one language to another; see **change**

See: **pure, open, see, separate, light, display**

cledo- < Gr. *kledos,* enclosure, hedge; see **wall**

cledono- < Gr. *kledon, -os,* omen; *kledonistes,* observer of omens; see **prophecy**

cleft < AS. *cleofan,* cleave, split; see **break, cut, open**

clematis, L. (Gr. *klematis*), a genus of the buttercup family; see **buttercup**

clemato- < Gr. *klema, -tos,* shoot, twig, cutting; see **branch**; *klema,* vine-twig; see **buttercup**

clemens, L. calm, mild, merciful; *clementia,* mercy; see **mild**

clemmato- < Gr. *klemma, -tos,* thing stolen, fraud; *klemmatistes,* thievish person, < *klepto,* steal; see **steal**

clemmyo- < Gr. *klemmys, -yos,* tortoise; see **turtle**

cleo- < Gr. *kleos,* rumor, report, fame, glory; see **honor**

cleome, NL. a genus of the caper family; see **caper**

cleonia, L. (Gr. *kleonia*), elecampane; see **composite**

cleonicum, L. a kind of basil; see **mint**

clepsydra, L. (Gr. *klepsydra*), water-clock; see **time**

clepto-; clepsi- < Gr. *klepto,* steal; *klepsia,* theft; see **steal**

clerk < AS. *clerc,* < OF. *clerc,* < L. *clericus* (Gr. *klerikos*), priest, clergyman; see **servant**

clero- < Gr. *kleros,* chance, lot; *klerion,* dim., see **lot**; clergy, see **servant**

-cles, termination of many Greek proper names, signifying famous; see *kleos* under **honor**

clethra, L. (Gr. *klethra*), alder; now a genus of the heath family; see **heath**

cleto- < Gr. *kletos,* called, invited; see *kaleo* under **call**

clever < uncertain origin; see **art, know**

clibanarius, L. soldier clad in mail, cuirassier; see **army**

clido- < Gr. *kleis, -idos,* key; *kleidion,* dim.; see **key**

client < L. *cliens, -entis,* follower, retainer, dependent, vassal, customer; see **servant**

cliff < AS. *clif,* steep bank or face of rock, precipice: Clifford, Clifton, Clive, Cleveland, Radcliffe.
    L. *abruptum; deruptum,* n. precipice; *abruptus,* precipitous, steep; see *rumpo* under **break**
    Gr. *eripne,* f. cliff, crag: *Eripneura criodes* (a moth)
    Gr. *kremnos,* m. precipice, crag, overhanging wall or bank; *apokremnos,* precipitous, craggy; *kremnobates,* m. frequenter of steep places: cremnophobia, *Cremnomys cutchicus* (a mouse), *Cremnophila auranticiliella* (a moth), *Apocremnus ancorifer* (a bug).
    Gr. *kynouron,* n. sea-cliff:
    L. *praecipitium,* n. a steep place, cliff; *praeceps, -cipitis,* headlong; n. a steep place: precipice, precipitate, precipitous.
    Gr. *rhaktos,* m. cliff; broken: *Rhactorhynchia subtetrahedra* (a fossil brachiopod).
    L. *scopulus* (Gr. *skopelos*), projecting rock, cliff; see **stone**
    See: **slope, high, break, cut, stone, projection, upright**

clima, L. (Gr. *klima*), supposed inclination or slope of the earth from the equator to the poles; region, zone; see **slope**

climax, -acis, L. (Gr. *klimax, -akos*), ladder, staircase; *klimation,* dim.; *klimakter,* rung of a ladder, critical period; see **ladder,** FIGURES OF SPEECH under **form**

climb < AS. *climban.*
    Gr. *anarrhichesis,* f. a clambering up:
    Gr. *epibatos,* climbable, mountable, accessible; *epibates; epibetor, -os,* m. one who mounts, embarks, passenger; *anabatos,* easy to scale or mount; *anambatos,* unmountable (horse), unbroken; *oreibates,* m. mountain-climber:
    L. *scando, scansus,* climb; *scansilis,* that may be climbed; *scansorius,* of climbing: scansorial, scansion, ascend, ascent, descendant, transcendental, *Celastrus scandens* (American bittersweet), *Saxifraga adscendens* (a saxifrage), *Taxodium ascendens* (pond bald-cypress).
    AS. *stigan,* climb: stair, stile, stirrup, sty, pigsty.
    See: **ladder, raise, step**

clinatus, L. sloping, bent, < *clino* (Gr. *klino*), bend, slope, slant, tend; see **slope**

cling < AS. *clingan;* see **hold, glue**

clinicus, L. (Gr. *klinikos*), of a bed, a physician attending a patient in bed; see *kline* under **bed**

clino- < Gr. *kline,* bed, see **bed;** < *kleinos,* famous, see *kleos* under **honor;** < L. *clino* (Gr. *klino*), bend, slant, slope, tend; see **slope**

clinopodium, L. (Gr. *klinopodion*), basil; see **mint**

clintero- < Gr. *klinter, -os,* couch, sofa; *klinterion,* dim.; see *kline* under **bed**

clintonia, NL. a genus of the lily family; see **lily**

Clio, L. (Gr. *Kleio*), the Muse of history; a sea-nymph; see **story**

clisia < Gr. *klisia,* place for reclining or lying down, bed, couch, hut; *klision,* out-house, shed; *klisis,* a bending, inclination, reclining, lying down; see **bed, house,** *clino* under **slope**

clismo- < Gr. *klismos,* couch; see *kline* under **bed**

clisto- < Gr. *kleistos,* shut, closed; see **close**

clitella, L. pack-saddle, pair of panniers; see **saddle**

clithridio- < Gr. *kleithridion,* dim. of *kleithria,* keyhole, cleft, chink; see **hole**

clithro- < Gr. *kleithron,* bolt, bar, lock, lattice; see *clathratus* under **screen**

clito-; clityo- < Gr. *klitos; klitys, -yos,* slope, hillside, see *clino* under **slope;** < *kleitos,* famous, excellent, see *kleos* under **honor**

clitoris, L. (Gr. *kleitoris*), female organ homologous to the male penis; see **vulva**

clivus, L. sloping side of a hill, hill; *clivulus,* dim.; see **mountain**

cloaca, L. sewer, drain, canal; *cloacula,* dim.; see **pipe**

cloak < OF. *cloke,* < ML. *cloca,* bell; see **garment**

clobo- < Gr. *klobos,* bird-cage; see **pen**

clock < D. *klok* (G. *glocke;* OF. *cloche;* ML. *cloca*), bell; see **time**

cloeo- < Gr. *kloios,* dog-collar, pillory; see **collar**

clog < ME. *clogge,* stump, block of wood; see **press, thick, hinder, delay, close**

clogmo- < Gr. *klogmos,* a clucking sound; see *klosso* under **cackle**

clomaco- < Gr. *klomax, -akos,* heap of stones; see **stone**

clono- < Gr. *klon, -os,* twig, slip, clon; *klonion,* dim., see **branch;** < *klonos,* violent, confused motion, see **move**

clopo- < Gr. *klops, -opos,* thief; *klopikos,* thievish; see *klepto* under **steal**

close < OF. *clos,* < L. *claudo, clausus; cludo, clusus,* shut; *clausula; conclusio, -onis,* f. close, end; *clausum,* n. an enclosed place; *clusor, -is,* m. closer; *clusaris; clusilis,* closing or shutting easily; *concludo,* shut up closely, end, finish; *excludo,* shut out; *occludo,* close, shut; *secludo,* shut up, confine, hide: closure, cloture, closet, clause, claustrum, cloister, conclusion, inclosure, occlude, seclusion, recluse, exclusive, sluice, *Pinus clausa* (sand-pine), *Clausilia bulimoides* (a pelecypod), *Planorbis clausulatus* (a gastropod).

L. *ango, anctus; anxus,* choke; see **choke**

L. *arceo, arctus,* shut up, inclose, restrain: *Prothiostomum arctum* (a planarian).

Gr. *bysma, -tos,* no; *bystra,* f. plug, bung; *epibystra,* f. stopper; *parabystos,* stuffed in, < *byo,* stuff, bung up: bysmalith.

L. *coangusto, -atus,* compress, confine, enclose: *Eumenes coangustata* (a wasp).

L. *conniveo,* close the eyes, blink, overlook: connive, *Ninox connivens* (barking owl).

Gr. *embolos,* m.; *embolon,* n. anything pointed for insertion, plug, wedge, bot, bar: embolus, embolism, embolite, Collembola, *Embolichthys mitsukuri* (a fish).

Gr. *emplastikos,* clogging, closing:

Gr. *enekbatos,* without outlet:

Gr. *epistomion,* n. bung, stopper, cock:

L. *epitonium* (Gr. *epitonion*), n. key for tightening the strings of an instrument, stopcock: *Epitonium pretiosum* (wentletrap).

L. *farcio, fartus,* stuff, cram; see **press**

L. *finio, -itus,* bring to an end, close; see *finis* under **end**

L. *fiscella,* little wickerwork basket, muzzle; see *fiscus* under **basket**

L. *impervius,* impenetrable: impervious.

L. *incommeabilis,* impassable, not traversable:

Gr. *kleistos,* shut, closed, < *kleio,* close; *perikleisma, -tos,* an enclosed place: cleistogamous, cleistocarp, *Clistoconcha insignia* (a pelecypod), *Clistomorpha hyalomoides* (a fly), *Clistocoeloma balansae* (a crab), *Clistopsocus dilatus* (a psocopterid).

Gr. *myo,* close, shut; *mysis, -eos,* f. a closing of the eyes, lips, pores, etc.: myopic, myosis, *Mysis stenolepis* (a crustacean).

L. *obturo, -atus,* close up, stop up; *obturaculum,* n. stopper, plug: obturate, obturator, *Leptothorax obturator* (an ant).

L. *oppilo, -atus,* stop up, shut up, < *pilo, -atus,* ram down, thrust home: oppilate, oppilation.

Gr. *perioche,* f. part circumscribed or closed off, case, fence:

L. *pessum* (Gr. *pesson*), plug, tampon; *pessulum,* dim.; see **drug**

Gr. *phimos,* m.; *phimotron,* n. muzzle, stopper; *phimosis, -eos,* f. a muzzling or stopping of an orifice: *Phimus violaceus* (a bird), *Phimophorus spissicornis* (a bug), *Dictyophimus platycephalus* (a radiolarian), phimosis.

Gr. *phrasso,* enclose, fence in; *emphrasso,* bar, block; see *phragma* under **fence**

Gr. *pykazo,* close, shut up, wrap up; *pykasmos,* m. a closing, shutting, covering:

L. *sphincter* (Gr. *sphinkter*), that which binds tight, a closing muscle; see *sphingo* under **bind**

L. *spissamentum,* n. stopper, plug:

L. *stegnus* (Gr. *stegnos*), watertight, closed, constricted; *stegnosis,* f. a making close or costive, stopping natural excretion; *stegnotikos,* making costive: *Stegnolaema montagni* (a bird).

L. *vectis,* bar, bolt, lever; see **lever**

See: **press, cover, bind, pen, end, lessen, narrow, wall, cancel, destroy**

closet < OF. *clos;* see **room**

closmato- < Gr. *klosma, -tos,* thread, line, clue; see **thread**

clostellum, L. small lock; see *claustrum* under **bar**

clostero- < Gr. *kloster, -os,* spindle; *klosterion,* dim.; see **spindle**

clot < AS. *clott,* lump, mass; see **lump**

cloth < AS. *clath.*

L. *cilicium,* n. coarse cloth made of Cilician goat's hair; *ciliciolum,* n. dim.; *cilicinus,* made of hair-cloth: cilicious.

L. *coactilium,* n. thick, fulled cloth; *coactiliarius,* m. maker of thick cloth or felt:

L. *gausapa* (Gr. *gausapes*), f. shaggy, woolen cloth; see **garment**

F. *mousseline,* < Ar. *musili,* of Mosul, where this cloth was made: muslin, Mussolini.

Gr. *othonion,* linen cloth, < *othone,* fine linen; see **linen**

L. *pannus,* piece of cloth, rag; see **rag**

Gr. *sindon, -os,* f. cotton cloth, muslin: *Sindon speciosa* (a fossil cockroach), *Conus sindon* (a gastropod).

Gr. *speiron,* n. piece of cloth, garment; see *speirion* under **garment**

L. *textum,* that which is woven, web, cloth; see *texo* under **weave**

See: **cotton, linen, silk, wool, garment, weave, napkin**

clotho- < Gr. *klotho,* twist, spin; *klostos,* spun; *klostes,* spinner; see **spin**

cloud < AS. *clud:* cloudy, becloud, cloudburst, Cloudmaker.   Explain: Make the welkin ring.

Gr. *achlys,* mist, darkness, obscurity; see **black**

Gr. *aithalos,* thick smoke, soot; see **dirt**

Gr. *atmos,* m.; *atmis, -idos,* f. steam, vapor, gas: atmosphere, atmolysis, atmometer, atmiatry.

L. *caligo, -inis,* fog, mist, vapor, haze, darkness; *caliginosus,* foggy, misty, dark; see **black**

Sp. *cigarro,* m. cigar, prob. < an Ab.Am. word meaning to smoke: cigar, cigarette, *Mesembryanthemum cigarettiferum* (a fig-marigold).

L. *fumus,* m. smoke, steam; *fumeus; fumicus; fumidus; fumosus,* smoky; *fumariolum,* n. smoke-hole; *fumarium,* n. smoke-chamber; *fumigo, -atus,* smoke; *infumo,* smoke, cure in smoke: fume, fumigate, fumarole, femerell, fumagillin, fumagine, fumigatin (from *Aspergillus fumigatus*), fumitory (*Fumaria officinalis*), *Sorex fumeus* (smoky shrew).

Gr. *glamyros,* bleary-eyed; see **dim**

Gr. *homichle* (*omichle*), f. mist, fog, steam: *Homichloda pauli* (a beetle), *Omichlospora incertula* (a moth).

Gr. *kapnos,* m. smoke; *kapne,* f. smoke-hole, chimney; *kapnikos,* smoky; *kapnodes,* like smoke, sooty, dusky: capnomancy, *Capnodium citri* (a mold), *Allocapnia pygmaea* (a stonefly).

Gr. *knisa,* vapor, steam from cooking and sacrifice; *knisotos,* steaming; see **smell**

Gr. *koniortos,* cloud of dust; see **dust**

Gr. *lignys, -yos,* f. thick smoke; *lignyodes,* smoky, sooty: *Lignostola pemphiargyra* (a butterfly), *Lignyodes suturatus* (a beetle).

L. *nebula,* f. mist, fog, cloud, smoke; *nebulosus,* misty, cloudy, dark, indefinite: nebula, nebular, nebulous, nebulite, nebuliferous, *Vorticella nebulifera* (a protozoan), *Ameiurus nebulosus* (a bullhead).

Gr. *nephele* f.; *nephos, -eos*, n. cloud; *nephelion; nephion*, n. dim., cloudlike spot; *nephelodes; synnephos*, cloudy: nephoscope, nephelite, *Nephopteryx subochrella* (a moth), *Nephelium daedalum* (a sapindacead), *Epinephilus striatus* (grouper).

L. *nidor, -is*, m. vapor, steam from cooking; *nidorosus*, steaming: nidor, nidorous.

L. *nimbus*, m. rain-cloud; *nimbosus*, stormy: nimbus, nimbiferous, nimbification, nimbose, *Macrocallista nimbosa* (a pelecypod), *Stethomyia nimbus* (a mosquito),

L. *nubes, -is*, f. cloud; *nubecula*, f. dim. little cloud, dark spot; *nubigosus; nubilosus; nubilus*, cloudy; *nubilo, -atus*, make cloudy, dull; *subnubilus*, overcast, gloomy: nubecula, nubilous, nubiliferous, obnubilation, *Nubecularia lucifuga* (a foraminifer), *Canis nubilus* (a wolf), *Lestricothynnus nubilipennis* (a wasp), *Podocarpus nubigenus* (Chile podocarpus).

Gr. *psolos*, soot, smoke; see **dirt**

AS. *smoca*, smoke: smoky, smoker, smokeless, smog, smokehouse, smokestack.

Gr. *typhos*, m. smoke, vapor, stupor, vanity; *typhodes*, smoky, delirious: typhus, typhoid (caused by *Eberthella typhi*, formerly *Bacillus typhosus*), typhosis, typhomania, *Minotaurus typhoeus* (a dung-beetle).

L. *vapor, -is*, m. steam, smoke; *vaporosus; vaporus*, steaming, reeking, smoking; *vaporalis*, of steam; *vaporarium*, n. steam-pipe in the baths; *vaporo, -atus*, emit steam: vapor, vaporous, vaporize, vaporarium, vaporium, evaporate, *Trialeurodes vaporariorum* (a whitefly).

See: **water, drop, spot, shade**

clove < OF. *clou*, < L. *clavus*, nail: clove (*Syzygium aromaticum*).

Gr. *karyophyllon*, n. clove-tree: Caryophyllaceae, caryophyllin, gillyflower, *Dianthus caryophyllus* (clove-pink, carnation), *Dicypellium caryophyllatum* (a lauracead).

clover < AS. *clafr*.

L. *medica*, f. clover from Media, lucerne, < Gr. *Medike*, Mede, Persian; NL. *medicago, -inis*, f. alfalfa: *Medicago sativa* (alfalfa), *Pyrenopeziza medicaginis* (a fungus), *Chara medicaginula* (a fossil chara).

L. *melilotus* (Gr. *melilotos*), f. sweetclover: *Melilotus alba* (white sweetclover).

L. *trifolium*, n. trefoil, clover: *Trifolium pratense* (red clover), *Peronospora trifoliorum* (a mildew).

Gr. *triphyllon*, n. trefoil, clover: *Arisaema triphyllum* (jack-in-the-pulpit).

clown < uncertain origin; see **fool, laugh**

club < ON. *klubba*.

L. *baculum;* Gr. *baktron*, stick, staff, rod, cudgel; see **rod**

L. *caia*, f. cudgel, < *caio*, beat:

L. *clava*, f. club, cudgel, graft; *clavula*, f. dim.; *claviger, -a, -um*, club-bearing: clavate, clavicorn, clavellate, clavigerous, clavula, *Clavaria flava* (coral-fungus), *Claviceps purpurea* (ergot), *Clavulina caperata* (a foraminifer), *Xanthoxylum clava-herculis* (Hercules-club), *Lycopodium clavatum* (a clubmoss), *Gymnadeniopsis clavellata* (an orchid), *Vaginulina clavula* (a foraminifer), *Glyptodon clavipes* (a fossil mammal), *Cordyceps clavulata* (a fungus), claviformin (from *Penicillium claviforme*).

L. *fustis, -is*, m. club, bludgeon; *fusticulus*, m. dim; *fusterna*, f. knotty part of a tree; *fustibalus*, m. slingstaff: fustigate, *Fustigeropsis peringueyi* (a beetle), *Tergipes fustifer* (a gastropod).

Gr. *hyperos*, pestle, club; see **pestle**

Gr. *kordyle*, f. club, swelling; *kordylinos*, clublike: cordylite, *Cordylophora lacustris* (a hydrozoan), *Cordylocrinus ramulosus* (a Devonian crinoid), *Cordylanthus maritimus* (coast-birdbeak), *Cordyline rubra* (a lily), *Cordyceps militaris* (a fungus).

Gr. *koryne*, f. club, mace; *korynetes*, m. club-bearer; *korynodes*, clublike: corynite, *Corynocarpus laevigatus* (a New Zealand tree), *Corynopoma rüsei* (a fish), *Corynepteris stellata* (a fossil fern), *Corynanthe yohimbe* (a rubiacead), *Brevicoryne brassicae* (cabbage-aphid), *Cryptocoryne ciliata* (an aroid).

L. *pistillum*, a club-shaped pounder used in a mortar; see **pestle**

Gr. *pleganon*, rod, stick; see **rod**

Gr. *rhopalon*, n. club; *rhopalion*, n. dim.; *rhopalikos; rhopalotos*, clublike: rhopalium, rhopalic, Rhopalocera, *Rhopalosiphum prunifoliae* (an aphid), *Rhopalum pedicellatum* (a wasp), *Rhopalocrinus gracilis* (a Devonian crinoid), *Amphirhopalum echinatum* (a radiolarian), *Cryptorhopalum atrum* (a beetle), *Zarhopalus crassus* (a chalcid).

See: **rod, pestle, strike**

clubmoss

NL. *lycopodium*, n. clubmoss: *Lycopodium prostratum* (carpet-clubmoss).

L. *selago, -inis*, f. a kind of clubmoss: *Lycopodium selago* (fir-clubmoss), *Selaginella flabellata* (fan-selaginella), *Athrotaxis selaginoides* (a conifer).

cluck < imitative origin; see **cackle**

clue; **clew** < AS. *cleowen,* ball of thread; see **mark, help, thread**

clump, prob. < ON. *klumba;* see **bush, lump, thick**

clumsy < Scan. *klumsa,* speechless, benumbed; see **awkward, rough**

clunis, L. (Gr. *klonis*), buttocks, rump; *cluniculus,* dim.; see **rump**

clupea, L. a herringlike fish; see **herring**

clura, L. ape; *clurinus,* of apes; see **monkey**

clusaris; **clusilis,** L. closing easily; see **close**

cluster < AS. *cluster; clyster.*
> Gr. *botrys, -yos,* bunch or cluster of grapes; *botrydion,* dim.; see **grape**
> L. *bumammus,* with large clusters; see **breast**
> L. *bumastus* (Gr. *boumastos*), with large clusters; see **breast**
> L. *caespes, -itis,* turf, sod, mat, turf; see **sod**
> L. *corymbus* (Gr. *korymbos*), m. bunch of flowers or fruit, cluster of berries, peak, top; *corymbiatus,* clustered; *korymbodes,* clustered: corymb, corymbose, corymbiate, *Corymbogonium capillare* (a hybrid), *Vaccinium corymbosum* (swamp-blueberry), *Fuchsia corymbiflora* (a fuchsia).
> L. *cymosus,* full of shoots; see *cyma* under **branch**
> L. *fascis, -is,* m. bundle, packet; *fasciculus,* m.; *fasciola,* f. dim.: fascicle.
> L. *gremialis,* growing in a cluster from a stump; see *gremium* under **lap**
> Gr. *hormathos,* cluster, string, chain; see **necklace**
> L. *paniculus,* tuft; see **bush**
> Gr. *phakelos,* m. bundle, cluster; *phakiolion,* n. dim.: *Phacelobranchus braueri* (an ephemerid), *Phacelia sericea* (silky phacelia).
> L. *phoba* (Gr. *phobe*), f. curl, lock, corymb:
> L. *racemus,* bunch or cluster of grapes; see **grape**
> L. *sagmen, -inis,* n. tuft of sacred herbs, rendering the bearer inviolable:
> Gr. *staphyle,* bunch or cluster of grapes; *staphylion,* dim; see **grape**
> L. *thyrsus* (Gr. *thyrsos,* a wand entwined with ivy), m. a close-branched cluster or panicle: thyrse, thyrsus, *Thyrsopelma striginotum* (a simulid), *Ceanothus thyrsiflorus* (blue-blossom ceanothus).
> L. *uva,* grape, bunch, cluster; see **grape**
> See: **bush, branch, heap, thick**

clusus, L. closed, shut; see **close**

clybatis, L. (Gr. *klybatis*), a kind of pellitory; see **pellitory**

clydono- < Gr. *klydon, -os,* billow, wave; *klydonion,* dim. ripple; see **wave**

clymeno- < Gr. *klymenos,* famous; see *klytos* under **honor**

clymenum, NL. (Gr. *klymenon*), a bindweed; see **honeysuckle**

clypeus, L. shield; *clypeolus,* dim.; see **shield**

clyster, L. (Gr. *klyster*), syringe; see **pump**

clyto- < Gr. *klytos,* heard of, famous, renowned; see **honor**

clyzo- < Gr. *klyzo,* wash, rinse, purge; *klysis,* a drenching; see **wash**

cmelethro- < Gr. *kmelethron,* beam; see **beam**

cnapho- < Gr. *knaphos (gnaphos), teasel,* carding-comb; *knapheus,* fuller; see **thistle,** *gnaphalon* under **wool**

cnecium, L. (Gr. *knekion*), marjoram; see **mint**

cneco- < Gr. *knekos,* pale-yellow; see **yellow**

cnemido- < Gr. *knemis, -idos,* greave, legging; see **sleeve**

cnemo- < Gr. *kneme,* leg, shin, see **leg;** *knemos,* shoulder of a mountain, see **shoulder**

cneorum, L. (Gr. *kneoron; knestron*), nettle; see **nettle**

cnephosus, L. (Gr. *knephas, -tos,* darkness, dusk, twilight), dark; see **shade**

cnesto-; **cnetho-** < Gr. *knetho (knao),* scratch, scrape, tickle, itch, sting; *knestis; knestron,* scraper; *knestos,* scraped; see **scrape**

cnicus, L. a kind of thistle; see **thistle**

cnido- < Gr. *knide,* nettle; *knidion,* dim.; see **nettle**

Cnidus, L. (Gr. *Knidos; Gnidos*), Doric city with a Praxitelean statue of Aphrodite; see *Aphrodite* under **love**

cnipo- < Gr. *knipos,* niggardly, stingy, see **stingy;** < *knips, -ipos,* a kind of insect, see *cynips* under **wasp**

cnisa, L. (Gr. *knisa; knissa*), vapor and smell from cooking fat meat; see **smell**

cnismato < Gr. *knisma, -tos,* scratch; *knismos,* itching, tickling; see **itch**

cnodalo- < Gr. *knodalon*, any dangerous wild animal; see **animal**

cnodax, L. (Gr. *knodax*), pin, pivot; see **axis**

cnodo- < Gr. *knodon*, sword; see **sword**

cnopo- < Gr. *knopeus*, bear, see **bear**; < *knops, -opos,* a snake or venomous beast, see **snake**

cnyzemato- < Gr. *knyzema, -tos,* whimper, whine; *knyzethmos,* a whimpering, whining; see **weep**

co- < L. *cum*, together, with; see **with**

coactilis, L. made thick, < *cogo, coactus,* assemble, collect, compress; see **thick, gather,** *ago* under **drive**

coactum, L. mattress; see **bed**

coagmentum, L. joint, < *coagmento, -atus,* join; see **bind**

coagulo, L. curdle; see **thick**

coal < AS. *col:* coaler, coal measures, coalpit, coalsack, coaltit*(Parus ater),* coalfish *(Pollachius virens),* charcoal, Collier, colliery.   What is meant by 'carry coals to Newcastle'?   Derive: coalition.

  Gr. *anthrax, -akos,* m. charcoal, coal, carbuncle, ulcer; *anthrakion,* n. dim.; *anthrakeus,* m. charcoal-burner; *anthrakinos,* pertaining to coal: anthrax, anthracite, anthracnose, anthracene, anthranilic, anthranol, anthraphenone, anthraxylon [anthracoxylon], anthragallol, anthraxolite, *Anthrax laetus* (a fly), *Anthracosaurus russelli* (a labyrinthodont), *Thyridanthrax amoenus* (a fly), *Eumeces anthracinus* (a lizard).

  Pr. *brasa;* Sp. *braza,* live coal, hence red, by transference of idea: Brazil (from a red wood, braziletto, brazil-wood, *Caesalpinia brasiliensis*).

  L. *carbo, -onis,* m. charcoal, coal; *carbunculus,* m. dim.; *carbonarius,* m. charcoal-burner, collier: carbon, carbonate, carbonado, carbohydrate, Carboniferous, carbonize, carbuncle, carbamate, carbamine, carbanilic, carbazide, carbanol, carbolic, carburetor, carborundum, carbonyl, carboxyl, triphenylcarbinol, carbobenzoxyglycylphenylalanine, *Phalacrocorax carbo* (a cormorant), *Baridius carbonarius* (a beetle).

  Gr. *gagates,* m. jet: jet, *Gagaticeras gagateus* (an ammonite), *Milax gagates* (a slug).

  Gr. *marile,* f. embers of charcoal: *Marilochen brevirostris* (a bird).

  L. *pruna,* f. a live coal:

  Gr. *thymalops, -opos,* m. a hot coal:

  See: **fire, black, red, stone**

coalemo < Gr. *koalemos,* dunce; see **dull**

coalitus, L. grown together, united; see *alesco* under **grow**

coarctatus, L. compressed, confined, shortened; see *arcto* under **press**

coarse < L. *cursus,* course; see **rough**

coast < OF. *cote,* < L. *costa,* rib, side; see **shore**

coat < OF. *cote,* a German word for mantle; see **garment, skin, cover**

coax < uncertain origin; see **draw, tease**

coaxo, *-atus,* L. join boards together, floor, see **bird, base;** croak, see **caw**

cobalo- < Gr. *kobalos,* rogue, knave; see **rogue**

cobalt < G. *kobold,* goblin, earth spirit; see ELEMENTS under **thing**

cobbler < uncertain origin.

  L. *diabathrarius,* m. shoemaker: *Diabathrarius variegatus* (a beetle).

  Gr. *pisyngos,* m. shoemaker, cobbler:

  Gr. *rhapheus; rhaptes,* sewer, patcher, cobbler; see *rhapto* under **sew**

  Gr. *skyteus; skytotomos,* m. cobbler, shoemaker; *skyteia,* f. shoemaking:

  L. *sutor, -is,* cobbler, shoemaker; see *suo* under **sew**

cobelo- < Gr. *kobele,* needle; see **needle**

cobidio- < Gr. *kobidion,* dim. of *kobios,* gudgeon; see **goby**

cobium, L. (Gr. *kobion*), a kind of spurge; see **spurge**

coca < Ab.Am. *coca:* Coca-Cola, cocaine, novocaine, *Erythroxylum coca* (coca-plant).

cocalio- < Gr. *kokalion,* a land snail; see **mollusk**

coccalo- < *kokkalos,* seed from a cone; see *coccum* under **berry**

coccido- < Gr. *kokkis, -idos,* dim. of *kokkos,* berry, grain, seed; see *coccum* under **berry**

coccineus, L. red like a berry, scarlet; *coccinatus,* clad in scarlet; see **red**

coccolobis, NL. a genus of the buckwheat family; see **buckwheat**

coccum, L. (Gr. *kokkos,* grain, seed) berry, a round insect on the oak, a source of scarlet dye; see **berry**

coccygia, L. (Gr. *kokkygia*), a kind of sumac; see **sumac**

coccygo- < Gr. *kokkyx, -ygos,* cuckoo; see **cuckoo**

cochlea, L. snail-shell, spiral, < Gr. *kochlias; kochlos,* snail with a spiral shell; see **mollusk**

cochlear, L. spoon; see **spoon**

cocio, L. broker, agent; see **trade**

cockroach < Sp. *cucaracha.*
   L. *blatta,* f. cockroach, moth; *blattarius,* of a moth: *Blatta orientalis* (black cockroach), *Blatella germanica* (croton-bug, German cockroach), *Grylloblatta campodeiformis* (a cricket), *Verbascum blattaria* (moth-mullein).
   Gr. *mylakris,* f. a kind of cockroach: *Mylacris elongata* (a fossil insect-wing).

coconut < Pg. *coco,* of uncertain derivation; see *cocos* under **palm**

coctus; coctilis, L. cooked, well-considered; *coctivus,* easily cooked; see *coquo* under **cook**

coculum, L. vessel for cooking; see **kettle**

cocyto- < Gr. *kokytos,* lamentation; see **weep**

cod < uncertain origin: tomcod *(Microgadus tomcod).*
   L. *asellus,* a valued sea-fish, probably cod; see *asinus* under **horse**
   L. *callarias* (Gr. *kallarias*), m. a kind of cod: *Lotella callarias* (a cod).
   Gr. *delkanos,* m. a fish, probably a burbot:
   Gr. *gados,* m. cod: *Gadus macrocephalus* (Alaska cod), *Microgadus proximus* (a tomcod).
   Gr. *hepatos,* m. a fish, probably a ling: *Hepatus fasciatus* (a fish), *Acanthus hepatus* (tang).
   Gr. *lebias,* m. a fish, probably a kind of cod: *Lebias ellipsoides* (a fish), *Lebistes reticulatus* (guppy).
   NL. *lota,* a kind of fish allied to the cod: *Lota maculosa* (burbot, ling).
   NL. *merluccius,* m. a kind of hake: *Merluccius bilinearis* (silver hake).
   NL. *molva,* f. a fish, probably a kind of cod: *Molva elongata* (a ling).
   NL. *morrhua,* f. cod: morrhuate, morrhuine, *Gadus morrhua* (common cod), *Pitar morrhuana* (a pelecypod).
   NL. *motella,* f. a kind of rockling, < L. *mustela,* weasel, a fish: *Motella mustela* (a rockling).
   Gr. *onos,* ass, a sea-fish, probably a cod or hake; see *onos* under **horse**

code < L. *codex, -icis,* tablet, book, writing; *codicillus,* dim.; see **book**

codio- < Gr. *kodion,* dim. of *koas,* fleece, sheepskin, see **wool**; < *kodeia,* head, head of plants, poppy-head, see **head**

codomo- < Gr. *kodomeus,* roaster of barley; see **cook**

codono- < Gr. *kodon, -os,* bell; *kodonion,* dim.; see **bell**

coelio- < Gr. *koilia,* belly; *koilidion,* dim.; see *coeliacus* under **belly**

coelo- < Gr. *koilos,* hollow; *koiloma,* a cavity, see **hollow**; < L. *coelum,* sky, see **heaven**

coeno- < Gr. *koinos,* common; see **common**

coepio, -eptus, L. begin, commence; *coeptum,* undertaking; see **begin**

coerano- < Gr. *koiranos,* master, ruler; see **govern**

coeto- < Gr. *koitos,* bed; see **bed**

cofanus, L. pelican; see **pelican**

coffee < NL. *coffee,* < Ar. *qahwah;* see **madder**

coffin < L. *cophinus,* basket, hamper; see **box**

cogito, -atus, L. consider, think; *cogitabilis,* imaginable; *cogitabundus,* thoughtful; see **think**

cognatus, L. kindred, related; see **kin**

cognitus, L. know; *cognobilis,* intelligible; see *nosco* under **know**

cohibilis, L. shortened; see **short**

cohibitus, L. confined, limited; see *habeo* under **hold**

cohors, -tis, L. enclosure, company; see **pen**

coil < OF. *coillir;* see **turn, curl, hold**

coin < L. *cuneus,* wedge; see **money**

coitus, L., m. copulation, sexual intercourse, < *coeo,* go or come together, copulate, unite: coitus, coition.

Gr. *aphrodisiasmos,* m. sexual intercourse, lustfulness; see *Aphrodite* under **love**

Gr. *bibazo; epibibasko,* put the female to the male:

Gr. *bineo,* have illicit intercourse:

L. *concubitus, -us,* m. a lying together, intercourse; *concubitalis,* pertaining to copulation; *concubinatus, -us,* m. adulterous intercourse: concubine.

L. *conjunctio, -onis,* f. conjugal connection: conjunction, conjunctive.

L. *consuetio, -onis,* f. carnal intercourse, < *consuesco, -etus,* become intimate with, know:

L. *contrecto, -atus,* touch carnally, have illicit intercourse with:

L. *copulatio, -onis,* f. a coupling, joining, connecting, < *copulo, -atus,* couple, join: copulate, copulation, copulatory.

L. *draucus,* m. a sodomite, one who copulates unnaturally:

L. *futuo, -utus,* copulate; *fututio, -onis,* f. copulation; *fututor, -is,* m.; *fututrix, icis,* f. copulator; *defututus,* exhausted by sensuality:

Gr. *gamoklopeo,* have illicit intercourse; *gamoklopia,* f. adultery:

Gr. *himeroomai,* have sexual intercourse:

L. *ineo, -itus,* enter, go into, have intercourse with, know; see **enter**

Gr. *katapygon, -os,* given to unnatural lust, lecherous, lewd; see **lewd**

Gr. *klinopale,* f. a wrestling in bed, intercourse:

Gr. *lagneia,* f.; *lagneuma, -tos,* n. coitus, lust; *lagnikos,* of lust; *lagnos,* lustful: lagnosis, coprolagnia, osmolagnia.

Gr. *mixis; mixoiphia,* f. sexual intercourse:

Gr. *myllo,* have sexual intercourse; *myllas, -ados,* f. prostitute; *myllos,* m. female pudenda:

L. *notitia,* f. carnal knowledge:

Gr. *ocheuo,* mount, cover, copulate; *ocheia; ocheusis,* f. coitus; *ocheuma, -tos,* n. result of coitus, embryo; *ocheutos,* covered, impregnated; *ocheion,* n.; *ocheutes,* m. stallion, lecher: *Ocheutes scopuliferus* (a beetle).

Gr. *opyo,* marry, have legitimate intercourse:

Gr. *orgasmos,* m. excitement, swelling, kneading, especially the turgidity, pulsation, and discharge accompanying the culmination of sexual intercourse: orgasm, orgastic.

Gr. *splekoma, -tos,* n. sexual intercourse:

L. *stuprum,* n. defilement, adultery, illicit intercourse, < *stupro, -atus,* ravish, debauch, defile: stuprum, stuprate.

Gr. *syneunasis,* f. sexual intercourse; *syngignomai,* have sexual intercourse; *synkatheudesis,* f. sexual intercourse; *synousia,* f. communion, intercourse, copulation; *synousiastikos,* promoting sexual intercourse, lewd, salacious:

See: **bind, begin, sex organs, prostitute**

coix, L. (Gr. *koix*), an Egyptian palm; now applied to a genus of grasses; see **grass**

col- < L. *cum,* together, with; see **with**

cola < African *kola;* a genus of the chocolate family; see **chocolate**

-cola, L. dweller, inhabitant; see *colo* under **life**

colaco- < Gr. *kolax, -akos,* flatterer, fawner; see **flatterer**

colapto- < Gr. *kolapto,* peck, strike, hew, cut, chisel; *kolaphus,* blow, cuff; *kolapter,* chisel; see **strike, chisel**

colasto- < Gr. *kolastes,* punisher; see *kolasis* under **punish**

colatus, L. strained, filtered; see *colo* under **sieve**

colchicum, L. (Gr. *kolchikon*), a kind of crocus; see **crocus**

cold < AS. *cald; ceald:* coldness, cold-chisel, coldfinch, cool, cooler, coolness, coolly, Cold Harbor.

L. *algidus,* cold; *adalgidus,* very cold; *algor, -is,* m. cold, coldness; *alsius; alsus,* chilly, cold, cool; *alsiosus,* easily getting cold, < *algeo,* be cold, feel cold; *algesco,* catch cold: *Natica algida* (a gastropod).

Gr. *astages,* frozen hard: *Astagobius* (a beetle).

AS. *cele; cyle,* cold: chill, chilblain.

Gr. *cheimerios; cheimerinos,* of winter, cold, bleak; see *cheimon* under **year**

L. *frigidus,* cold, inactive; *frigidulus,* dim.; *frigor, -is,* m. *frigus, -goris,* n. cold; *frigusculum,* n. dim.; *frigidarius,* of or for cooling; *frigorificus,* cooling; *frigeo,* be cold, inactive; *refrigero, -atus,* make cold, cool off: frigid, frigidity, frigorific, refrigerator, *Artemisia frigida* (a sagebrush).

L. *gelidus,* cold, frosty, icy, < *gelo, -atus,* freeze, solidify; *congelo,* freeze up: gelid, gelable, gelatin, gelatinize, gelatinous, gel, jelly, gelose, gelosin, congeal, *Gelidium cartilagineum* (a red alga).

L. *gillo, -onis,* m. a cooling-vessel, cooler:

L. *glacialis,* icy, frozen; *conglacio, -atus,* freeze up: glacial, glaciation, *Crymonessa glacialis* (a bird).

Gr. *koryza,* cold, catarrh; see **disease**

Gr. *kryos,* n.; *krymos,* m. icy cold, chill, frost; *krymaleos,* frosty, chilly, icy; *krymodes,* icy, frozen; *kryeros,* icy, cold: cryolite, cryogen, cryometer, cryophyllite, cryoscopy, crymodynia, crymotherapy, *Cryophila lapponica* (a mosquito), *Cryobdella levigata* (a leech), *Cryobius ventricosus* (a beetle), *Crymophilus rufus* (a bird).

Gr. *malkios,* freezing, benumbing:

Gr. *pachnodes,* chilly, cold; *pageros; pagetodes,* frosty, cold; see *pagetos* under **ice**

Gr. *phrike,* a shivering from cold; see *phrix* under **shake**

Gr. *psychros,* cold, frigid; *psyktos,* cool; *psyktikos,* cooled; *psykter, -os,* m. cooler; *psykteridion; psykterion,* n. dim.; *psykterios,* cooling; *psyxis; anapsyxis; katapsyxis,* f. chill: psychrometer, psyctic, psykter, psychrophile, psychrophyte, *Erigone psychrophila* (a spider), *Psychrocoris cuneifera* (a fossil bug).

Gr. *psygmos,* m. chilliness:

Gr. *rhigos,* cold; *rhigion,* colder; *rhigistos,* coldest; see *rhigos* under **ice**

See: **ice, winter**

coleatus, L. pertaining to the penis; see *colis* under **penis**

colecano- < Gr. *kolekanos,* a long, thin person; see **thin**

coleno- < Gr. *kolen, -os,* thigh, leg; see *kolon* under **leg**

coleo- < Gr. *koleos,* sheath, scabbard, scrotum; see *coleus* under **sheath**

colepo- < Gr. *koleps, -epos,* hollow of the knee; see **knee**

colero- < Gr. *koleros,* short-wooled; see **wool**

coleus, NL. a genus of the mint family, see **mint**; < L. *coleus,* scrotum, see **sheath**

colias, L. (Gr. *kolias*), a mackerel; see **mackerel**

colic < L. *colicus* (Gr. *kolikos*), of colic; see **disease,** *colon* under **intestine**

colinus, NL. < Ab.Am. *colin,* partridge; see **quail**

collabo- < Gr. *kollabos,* a cake; see **cake**

collar < L. *collare,* n. iron band or chain for the neck: *Arctonyx collaris* (balisaur).

L. *armilla,* bracelet, dog-collar; see **bracelet**

L. *boja,* f. collar:

L. *columbar, -is,* n. collar:

Gr. *deiropede,* f. necklace, collar; see **necklace**

Gr. *deranche,* f. collar; *kynanche,* f. dog-collar, quinsy:

L. *helcium,* n. horse-collar, yoke:

L. *jugum,* horse-collar, yoke; see **bind**

Gr. *kloios,* m. dog-collar, pillory: *Cloeoascaris spinicollis* (a worm).

L. *occabus,* m. collar, armlet:

L. *mellum,* n. dog-collar:

L. *monile,* necklace, collar; see **necklace**

Gr. *streptos,* a collar of twisted or linked metal; see **turn**

L. *torquatus,* adorned with a necklace or collar, < *torques,* c. twisted neck-chain, necklace, collar: *Saxicola torquata* (stonechat).

See: **belt, garter, bind, ribbon, necklace**

collarbone

L. *clavicula,* f. small key, bar: clavicle.

L. *jugulum,* n. collarbone:

collatus; collativus, L. gathered together, collected; see **gather**

collect < L. *colligo, -lectus,* assemble, gather; see *lectus* under **gather**

college < L. *collegium,* an association of colleagues; see **school**

colleto- < Gr. *kolletes,* gluer, fastener; see *kolla* under **glue**

collido, -isus, L. strike together, come into violent contact; see *laedo* under **strike**

collinus, L. of a hill, hilly, < *collis,* hill; *colliculus,* dim.; see **mountain**

collitus, L. defiled, dirtied, smeared, smudged; see *lino* under **spread**

collo- < Gr. *kolla,* glue; *kollesis,* a glueing, cementing, welding; *kolletos,* glued, joined; see **glue**

collopo- < Gr. *kollops, -opos,* peg, screw; see **turn**

colloquium, L. conference; see **gather**

colluco, -atus, L. let light into the forest, clear, thin; see **thin**

collum, L. neck; *collaris,* of the neck; see **neck**

collybus, L. (Gr. *kollybos*), a small coin, rate of exchange; *kollybistes,* money-changer; see **money**

collyra, L. macaroni; see **food**

collyrium, L. (Gr. *kollyrion*), dough, poultice, eye-salve, suppository, pillar; *collyriolum,* dim., see **drug**; a kind of thrush, see **thrush**

colo-; colobo- < Gr. *kolos; kolobos,* docked, clipped, curtailed, shortened, incomplete, truncated, see **short**; a kind of goat, see **goat**; < *kolon,* colon, see **intestine**

colocasia, L. (Gr. *kolokasia*), an arum; see **arum**

colocynthis, L. (Gr. *kolokynthis*), a kind of gourd, pumpkin, or melon; see **melon**

coloeus, NL. (Gr. *koloios*), jackdaw, grackle; see **crow**

colon, L. (Gr. *kolon*), large intestine; *kolikos,* pertaining to the colon, pain in the colon, see **intestine**; food, meat, fodder, see **food**; limb, member, arm, leg, see **leg**

colono- < Gr. *kolone; kolonos,* hill, mound, barrow; see **mountain**

colonus, L. farmer; see *colo* under **till**

colony < L. *colonia,* estate, farm, settlement; see **country**

colophono- < Gr. *kolophon, -os,* summit, top, end, climax; see **top**

color < L. *color, -is,* m. hue, tint, complexion; *colorabilis; coloratus; coloreus,* colored, variegated; *coloro, -atus,* color, tinge; *concolor,* colored uniformly; *decolor,* discolored, faded; *discolor,* not of the same color, variegated; *versicolor,* of various colors, variegated: coloration, colorable, Colorado, coloratura, colorimetry, color-blind, colorful, colorful, discolor, decolorize, *Poecilopsetta colorata* (a fish), *Solidago bicolor* (silverrod), *Felis concolor* (mountain-lion, cougar, puma), *Salix discolor* (pussywillow), *Iris versicolor* (fleurdelis), *Tarbaleus decoloratus* (a grasshopper), *Geranomyia unicolor* (a fly), *Steganopus tricolor* (a phalarope), *Ameles decolor* (a mantis).
L. *arquatus,* bent like a bow; rainbow; see **arc**
Gr. *bapheus,* m. dyer; *baphe; bapsis,* f. a dipping, dyeing; *baphikos,* of dyeing; *baptos,* dipped, dyed; *bama, -tos,* n. dye; *apobama, -tos,* n. tincture, infusion, < *bapto,* dip, dye: *Baphia nitida* (camwood), *Baptisia tinctoria* (yellow wild-indigo), *Baphothrips tricolor* (a thysanopterid).
Gr. *chroma, -tos,* n. color of the skin, complexion, color, < *chroa; chroia,* f.; *chros, -otos,* m. surface of the body, skin, color of the skin; *chromation; chrotidion,* n. dim.; *chroikos,* colored; *chromatikos,* of color; *chroster, -os,* m. dyer; *achrostos,* uncolored; *anachrosis,* f. discoloration, taint; *anthesichros; heterochros,* of different colors, variegated; *chrozo,* taint, tinge, stain: chromatic, chrome, chromium, chromate, chromatin, chromosome, chroatol, chromatography, polychrome, achromatic, achroite, schizochroal, monochromatic, dichroism, *Chromatocera setigena* (a fly), *Chromatonotus heterus* (a blattid), *Chromocryptus albopictus* (a wasp), *Chromatium okeni* (a bacillus), *Callichroma moschatum* (musk-beetle), *Polychrus marmoratus* (a lizard), *Chroicocephalus atricilla* (laughing-gull), *Microtus chrotorrhinus* (rock-vole), *Chrotopterus auritus* (a bat), *Dyschromus opacus* (a beetle), *Achrostus rufonitens* (a beetle).
Gr. *deuso-,* dye, stain, < *deuo,* wet, drench; *deusopoios,* deeply dyed, fast, ingrained:
L. *fuco, -atus,* color, dye, rouge; *fucinus,* colored with orchil; *fucosus,* painted, counterfeited; *infucatus,* painted; *offucia,* f. paint or wash for the face: fucate, *Trifolium fucatum* (a clover).
L. *immedicatus,* painted:
L. *infectus,* dyed, stained, tainted; *infector, -is,* m. dyer; *infectorius,* of or for dyeing; *inficio, -fectus,* stain, color, dye: infection, infectious, disinfectant, *Ficus infectoria* (dotted fig).
L. *interstinctus,* checkered, variegated: *Asymphoroides interstincta* (a moth).
L. *iris, -idis* (Gr. *-idos*), f. rainbow; a plant; *irinos,* of iris; *iriodes,* like the rainbow: iris, iridescent, iridium, iridosmine, iridectomy, *Iris variegata* (Hungarian iris), *Iridomyrmex humilis* (argentine ant), *Iridio radiata* (doncella), *Trachypterus iris* (ribbon-fish), *Salmo irideus* (rainbow-trout), *Begonia iridescens* (a begonia), *Xyris iridifolia* (a yellowgrass).
Gr. *peridaedalos,* variegated; see *daedalus* under **art**
L. *pingo, pictus,* paint; *pictilis,* colored, painted, embroidered; *pictor, -is,* m. painter; *pictura,* f. a painting; *pigmentum,* n. color, paint; *depingo,* portray, describe, sketch: picture, pictorial, picturesque, depict, pigment, orpiment, orpine, pimento, paint, painter, *Chrysolophus pictus* (golden pheasant), *Scleropogon picticornis* (robber-fly), *Perilissus pictilis* (an ichneumonid), *Elaphrosyron multipictus* (a wasp), *Caladium picturatum* (a caladium), *Streptopharagus pigmentatus* (a nematode), *Onthophagus impictus* (a dung-beetle).
Sp. *pintado,* painted, mottled: pinto.

Gr. *poikilos,* varicolored, pied, mottled, spotted: poikilitic(poecilitic), Poecile, *Poecilobrycon auratus* (a fish), *Platypoecilus maculatus* (platyfish), *Poecilia vivipara* (a fish).

Gr. *rhegeus,* m. dyer; *rhegma, -tos,* n. the thing dyed:

Gr. *skiotos,* shaded by gradation in color; see *skia* under **shade**

L. *tingo, tinctus,* dye, paint; *tinctilis; tinctorius,* of dyeing; *tinctor, -is; tingens, -entis,* m. dyer: tinge, tinctorial, tincture, tint, taint, stain, stainless, Rio Tinto, *Cytisus tinctoria* (a broom), *Dendrobates tinctorius* (a frog), *Uloborus tingens* (a spider), *Rubia tinctorum* (madder).

L. *variegatus,* of different sorts, particularly colors; see *varius* under **different** See: **spot, mark, dip, shade, black, white, red, green, yellow,** etc.

colossus, L. (Gr. *kolossos*), a large statue; see **large**

colpo- < Gr. *kolpos,* bosom, womb; *kolpodes,* full of bays, sinuous, folded; see **womb**

coluber, L. snake; *colubrinus,* snakelike; see **snake**

colubrina, L. a plant; now a genus of the buckthorn family; see **buckthorn**

colum, L. strainer, colander; see **sieve**

columba, L. dove; *columbula,* dim.; *columbarius; columbinus,* of doves; see **dove**

columen, L. top, crown, summit; see *culmen* under **top**

columis, L. safe, unhurt; see **safe**

column < L. *columna,* pillar; *columella,* dim.; see **pillar**

colus, L. distaff; see **spindle**

-colus, -a, -um, NL. dwelling in, inhabiting, living among, < L. *-cola,* inhabitant of; see *colo* under **life**

colutea, L. (Gr. *koloutea*), a pod-bearing tree; see **bean**

coluthium, L. a kind of snail; see **mollusk**

colymato- < Gr. *kolyma, -tos,* hindrance, < *kolyo,* hinder, prevent; see **hinder**

colymbethra, NL. (Gr. *kolymbethra*), swimming-pool; see **swim**

colymbus, L. (Gr. *kolymbis*) a diving bird, grebe; *kolymbetes; kolymbos,* diver; see **dip**

colythro- < Gr. *kolythron,* a ripe fig, see **fig;** *kolythros,* testicle, see **testicle**

colytico- < Gr. *kolytikos,* checking, hindering, preventing; *kolysis,* hindrance; see *kolyma* under **hinder**

com- < L. *cum,* together, with; see **with**

coma, L. (Gr. *kome*), hair of the head, mane, see **hair, crest;** < *koma, -tos,* slumber, see **sleep**

comaco- < Gr. *komax, -akos,* reveller, debauchee; see *komos* under **joy**

comarido < Gr. *komaris, -idos,* a fish; see **fish**

comarum, L. (Gr. *komaron*), fruit of the strawberry-tree; *komaros,* strawberry-tree, arbutus; see **arbutus**

comasto- < Gr. *komastes,* reveller; see *komos* under **joy**

comatus, L. with long hair; shaggy; *comatulus,* dim.; see *coma* under **hair**

comb < AS. *camb,* card, crest, masses of cells built by bees and wasps.   See **crest, wax**

L. *carmino, -atus,* card, comb, cleanse; *carminator, -is,* m. carder: carminative.

L. *como, comptus,* arrange, comb, adorn; see **ornament**

Gr. *knaphos (gnaphos),* carding-comb, teasel; see **thistle**

Gr. *kteis, ktenos,* m. comb; *ktenidion; ktenion,* n. dim.; *ktenistes,* m. comber, hairdresser; *ktenistos,* combed; *ktenodes,* comblike: ctenodont, ctenophore, *Ctenocephalides canis* (dog-flea), *Ctenanthe setosa* (a marantacead), *Ctenium aromaticum* (a grass), *Ctenidium molluscum* (a moss), *Cteniscus annulipes* (a wasp), *Ctenistes consobrinus* (a beetle), *Ctenomyces serratus* (a fungus), *Dactyloctenium aegyptium* (a grass), *Pithecoctenium dolichoides* (a bignoniacead), *Euctenopus novazealandicus* (a wasp), *Euctenodes mirabilis* (a fly).

L. *pecten, -inis,* m. comb, rake, mollusk; *pectinatus,* comblike, toothed; *pexatus,* clothed in a napped garment, < *pecto, pexus (pectitus),* comb: pectinal, pectinate, pectiniform, pectineus, pectinella, *Pecten maximus* (a scallop), *Pectinatella magnifica* (a fresh-water bryozoan), *Potamogeton pectinatus* (sago-pondweed), *Dinorthis pectinella* (a fossil brachiopod), *Discocyrtus pectinifemur* (an opilionid), *Thiasophila pexa* (a beetle), *Ischnopteryx pexata* (a moth), *Cestus pectinalis* (a ctenophore).

L. *praenum,* n. hatchel, an instrument for combing flax:

Gr. *psektra,* scraper, currycomb; see **scrape**

Gr. *xanion,* n. comb for carding wool; *xantes,* m. carder: *Xaniopelma sericans* (a wasp).
See: **crest, tooth, rake, wax**

combine < L. *combino, -atus,* unite; see **bind**

combo- < Gr. *kombos,* roll, band, girth; *kombion,* dim.; see **belt**

combretum, L. a kind of rush; now applied to a genus of the myrobalan family; see **myrobalan**

combustion < L. *combustio,* a burning, < *comburo, -ustus,* burn up, consume; see *uro* under **burn**

come < AS. *cuman.*
L. *accedo, -essus,* approach, draw near: access, accession.
L. *adipiscor, adeptus,* arrive at, reach, attain by effort, acquire: adept.
L. *allapsus,* silent or stealthy approach:
Gr. *aphixis,* f. arrival:
L. *appropinquo, -atus,* draw near:
L. *appulsus,* m. an approach, landing:
Gr. *blosis,* f. arrival, < *blosko,* come or go:
L. *emico, -atus,* appear suddenly, spring forth, become apparent; see **open**
Gr. *ephodos,* f. approach, onset, attack:
Gr. *erchomai,* come or go:
Gr. *hikano,* come to, arrive at, reach:
Gr. *pelazo,* approach, come, draw near; *pelates (pelastes),* approacher, neighbor, hireling; see **neighbor**
Gr. *phthano,* come before, anticipate:
L. *prodeo, -itus,* appear, come forth, spring up:
Gr. *prosodes,* approach, advance, onset, attack: *Prosodoscelis cordicollis* (a beetle).
L. *venio, ventus,* come; *adventus,* m. arrival, approach: venture, venue, advent, avenue, adventurer, adventitious, convene, convenient, conventional, covenant, circumvent, inventive, inventory, parvenu, prevent, intervention, saunter, souvenir, revenue, uneventful, eventually, peradventure.
See: **gather, near, clear, depart**

comedo, L. glutton; see *edo* under **eat**

comedy < L. *comoedia* (Gr. *komoidia*) a humorous drama; see *comicus* under **laugh**

comes, -itis, L. companion, associate, comrade, participant; see **companion**

comestibilis, L. edible; see *edo* under **eat**

comet < L. *cometa* (Gr. *kometes*), see **star**; < *kometes,* countryman, see *kome* under **town**

comfort < L. *conforto, -atus,* strengthen much; see **rest, soothe**

comicus, L. (Gr. *komikos*), pertaining to comedy; see **laugh**

comis, L. friendly, kind, affable, courteous; see **friend**

comisto- < Gr. *komistes,* caretaker, guardian, < *komizo,* care for; see **guard**

comitor, -atus, L. accompany, attend; see **follow**

comma, L. (Gr. *komma*), chip, piece, clause, mark of punctuation; *kommation,* dim.; see **part**

commeabilis, L. permeable; see *meatus* under **passage**

commeator, L. messenger; see **send**

comminus, L. near, at close quarters, hand to hand; see **near**

commissura, L. joint, juncture; see **joint**

committee < L. *committo,* arrange, give to, intrust; see **gather**

commo- < Gr. *kommos,* decoration, embellishment, see **ornament**; lamentation, see **weep**

commodus, L. fit, proper, suitable; see **fit**

common < L. *communis,* general, public, universal: common sense, commonplace, commonwealth, commonalty, commoner, commons, commune, communion, community, communicate, communism, excommunication, uncommonly, Comintern, Cominform, *Juniperus communis* (common juniper), *Polytrichum commune* (haircap-moss).
Gr. *adiaphoros,* not different, neutral:
L. *chydaeus* (Gr. *chydaios*), common, vulgar; see *adaios* under **abundance**
L. *collactaneus; collacteus; collacticius,* nourished at the same breast:
L. *consortium,* community, fellowship, society; *consors,* sharing in common; partner; see **companion**
L. *conspiro, -atus,* breathe together, plot and plan to act in concert; see **companion**

Gr. *epidemos*, prevalent, common to a large number; *demosios*, belonging to the people, common, vulgar; *pandemos*, common, public, general, affecting all the people: epidemic, pandemic.

Gr. *eranos*, m. a common fund, a society of subscribers; *eranion*, n, dim.; *eranistes*, m. contributor to a club; *eranikos*, of a contribution: *Eranistes pandora* (a moth).

Gr. *exoterikos*, outside, common, popular: exoteric.

L. *generalis*, pertaining to all, common: general, generality, jeep.

Gr. *genikos*, generic; see *genus* under **class**

L. *gregarius*, belonging to the herd, common; see *grex* under **herd**

Gr. *koinos*, common; *koinotes*, sharing in common, fellowship; *koinonikos*, social; *koinonia*, f. communion, association; *epikoinos*, promiscuous: cenobite, cenogamy, coenocyte, coenenchyma, coenosarc, biocenose, epicene, *Coenomyia pallida* (a fly), *Coenoptychus pulcher* (a spider), *Cenobita diogenes* (a crab), *Onthophagus coenobita* (a dung-beetle).

L. *mutuus*, borrowed, loaned, exchanged, reciprocal; *mutuarius*, joint, reciprocal: mutual, mutualism, mutuality.

L. *plebeius*, of the people, common; see *plebus* under **people**

L. *proletarius*, m. citizen of the lowest class, low, common: proletarian, proletariat.

L. *publicus*, belonging to the people, < *publico*, *-atus*, make common or known to all, disclose, impart; see **people**

L. *societas*, *-atis*, f. association, community; *socialis*, pertaining to companionship: society, social, *Nasua socialis* (a coati).   See **companion**

L. *tritus*, well-worn, familiar, commonplace; see *tero* under **rub**

L. *trivialis*, belonging to the crossroads, commonplace, vulgar: trivial, *Alisma triviale* (a water-plantain).

L. *usitatus*, common, customary, familiar, usual: *Linum usitatissimum* (common flax).

L. *vulgaris*, common, commonplace; *vulgus*, n. the people, public, rabble; *vulgo*, *-atus*, make common; divulge, publish, spread: vulgar, vulgarism, Vulgate, divulge, *Phaseolus vulgaris* (kidney-bean), *Armadillidium vulgare* (a pillbug), *Ophioglossum vulgatum* (an adderstongue-fern), *Opegrapha vulgata* (a lichen).

Gr. *xynos*, common; *xynon*, *-os*, m. companion, partner: *Xynobius pallipes* (a wasp).

See: **companion, equal, people, profane, other, use**

como- < Gr. *komos*, a jovial festivity, revel; *komastes*, reveller; *komax*, *-akos*, debauchee; see **joy**

comosus, L. hairy; see *coma* under **hair**

compactus, L. thick, firm; see **thick**

compaginatus, L. joined, bordering; see **bind**

companion < L. *cum*, with, *panis*, bread.   Derive: mate, checkmate, stalemate.

Gr. *akoites*, m.; *akoitis*, f. bedfellow, spouse; *parakoites; synkoites*, m. bedfellow: *Acoetes pleei* (an annelid).

Gr. *alochos*, f. bedfellow, spouse: *Neomorpha (Heteralocha) acutirostris* (huia-bird).

L. *coalitus*, m. communion, fellowship: coalition, *Vorticella coalita* (a protozoan).

L. *coarmio*, *-onis*, m. comrade in arms:

L. *collega*, m. partner, comrade: colleague.

L. *collusor*, playmate; see *ludo* under **play**

L. *combibo*, *-onis*, m. pot-companion:

L. *comes*, *-itis*, c. companion, associate, participant, comrade: comitial, comity, concomitant, constable, constabulary, *Erythronema comes* (grape-leafhopper).

L. *compotor (compotator)*, *-is*, m.; *compotrix*, *-icis*, f. drinking-companion:

L. *concellita*, m. cellmate:

L. *concubitor*, *-is*, m. bedfellow:

L. *condiscipulus*, m. schoolmate:

L. *confidelis*, m. fellow-believer:

L. *congregabilis*, social; see *grex* under **herd**

L. *conjunx (conjux)*, spouse, mate; see **spouse**

L. *consacerdos*, *-otis*, c. fellow-priest or priestess:

L. *conscius*, having knowledge in common with an accessory, accomplice or confidant; see *scientia* under **know**

L. *consedo*, *-onis*, m. sitter with one:

L. *consors*, *-tis*, c. partner, sharer, mate; *consortium*, n. community, fellowship, society: consort, consortium.

L. *conspiro*, *-atus*, breathe together, plot and plan to act in concert: conspire, conspirator, conspiracy.

L. *contogatus*, m. law colleague:

L. *contubernalis*, c. tent companion, comrade, mate: *Torymus contubernalis* (a wasp).

L. *convector, -is,* m. fellow-passenger:
L. *conveteranus,* m. fellow-veteran:
L. *convicanus,* m. fellow-villager:
L. *convictio, -onis,* f. companionship, intimacy, < *convivo, -ictus,* live with; *convictor, -is,* m.; *conviva,* c. messmate, table companion: convictor, *Poecilochroa convictrix* (a spider), *Dianthidium plenum convictorium* (a bee).
L. *correus,* m. partaker in guilt, companion in crime:
Gr. *epetes,* sociable:
Gr. *euneter; eunetor, -os; eunetes; pareunetes; syneunetes,* m.; *syneunos; synomeunos,* c.; *eunetis; eunis; pareunetis; syneunetis, -idos,* f. bedfellow, spouse, consort: dyspareunia, *Syneunetis inopella* (a moth).
AS. *gesith,* companion: gesithcund.
Gr. *hetairos,* m.; *hetaira,* f. comrade, companion; *hetaireia,* f. companionship, brotherhood; *philetairos,* fond of companions: *Hetaerius puberulus* (a beetle), *Hetaerodipsas colubrina* (a snake), *Philetaerus socius* (a weaverbird).
Gr. *homelys, -ydos,* c. companion: *Homelys lapponicus* (a wasp).
Gr. *koinonos,* companion, partner, fellow; see *koinos* under **common**
Gr. *metochos,* sharing, partaking; partner; *metoche,* f. communion; *methektos,* participating, sharing, < *metecho,* share in, enjoy, partake: *Metochus abbreviatus* (a bug), *Cepobroticus symmetochus* (an ant).
Gr. *opados,* accompanying, attending; see **servant**
L. *par, -is,* m. companion, comrade, mate, spouse: pair.
Gr. *paredros,* sitting beside, near; associate; see **near**
L. *particeps, -ipis,* m. comrade, partner, sharer; *participalis,* sharing, partaking, < *participo, -atus,* share in: participate, participant, participial, participle.
L. *satelles, -itis,* c. companion, escort, guard, lackey, accomplice: satellite:
L. *socius,* m. companion; *socialis,* companionable; *sociabilis,* disposed to companionship; *societas, -atis,* f. association, fellowship: social, socially, sociability, socialism, sociology, associate, dissociation, unsociable, *Apis socialis* (a bee), *Psammophax consociata* (a foraminifer).
L. *sodalis,* c. companion, comrade, crony: sodality, *Sodaliscala multistriata* (a gastropod).
Gr. *sylleptor, -os,* accomplice, assistant, partner:
Gr. *symbiotes,* m. companion, partner; *symbiotikos,* of companionship; *symbiosis,* f. a living together: symbiote, symbiont, symbiotic, symbiosis, *Symbiotes* (now *Chorioptes*) *bovis* (a mite).
Gr. *symmachos,* m.; *symmachis, -idos,* f. ally: *Symmachis lacteipennis* (a grasshopper).
Gr. *sympaistes,* m. playmate:
Gr. *symphenax, -akos,* m. partner in deceit:
Gr. *symphoitetes,* m. school-fellow:
Gr. *synetheia,* f. intimacy, custom; *synethes,* living together: *Carmeyerius (Synethes) gregarius* (a flatworm).
Gr. *synemon,* united; comrade; *synemosyne,* f. tie of friendship, covenant: *Synemosyna formica* (a spider).
Gr. *synemporos,* c. fellow-traveler, companion:
Gr. *synodia,* journey in company, caravan; see **gather**
Gr. *synoikos; synoiketes,* m. house-fellow: *Synoicus[Synoecus] ypsilophorus* (a quail), *Synoecetes sedulus* (a wasp).
Gr. *synomodites,* m. fellow-traveler:
Gr. *synousia,* f. communion, society, intercourse; *metousia,* f. partnership, communion; *synousiastes,* m. companion; *synousiastikos,* sociable, promoting intercourse:
Gr. *syntrapezos,* m. messmate:
Gr. *syskenos,* m. comrade: *Syscenus infelix* (an isopod).
Gr. *syssitos,* m. messmate: *Syssitos[Syssitus] rostratus* (a beetle).
Gr. *xynon, -os,* companion, partner; see *xynos* under **common**
See: **spouse, friend, neighbor, help**

compare < L. *comparo, -atus,* place side by side to discover resemblances and differences; see *par* under **equal**

comparatives
L. *-ior,* m., f.; *-ius,* n., as in *carior; carius,* dearer; *felicior; felicius,* happier; *major; majus,* greater; *melior; melius,* better: inferior, interior, posterior, superior, ulterior, *Chelidonium majus* (greater celandine).
Gr. *-ion,* as in *beltion,* better; *echthion,* more hateful; *hedion,* sweeter; *kallion,* more beautiful; *kydion,* more glorious:
Gr. *-teros,* as in *axioteros,* worthier; *melanteros,* blacker; *picroteros,* bitterer; *oxyteros,* sharper; *mikroteros,* smaller; *sophoteros,* wiser: *Basileuterus rufifrons* (a warbler).

compass < OF. *compasser*, go round, divide; see **circle**

compeditus, L. fettered, < *compedio*, fetter, shackle; see *pedica* under **bind**

compel < L. *compello, -ulsus*, constrain by force; see *pello* under **push**

compendium, L. abridgement, abstract, summary; see **short**

compernis, L. knockkneed; see **bend**

compertus, L. ascertained, learned; see *comperio* under **learn**

compes, -edis, L. fetter, shackle; see *pedica* under **bind**

competitor, L. rival; see **enemy**

compitum, L. crossroads; *compitalis*, of crossroads; see **way**

complain < OF. *complaindre*; see **fault, scold, trouble, sad**

complanatus, L. flattened; see *planus* under **level**

complete < L. *compleo, -etus*, fulfill, finish; see **all, end**

complex < L. *complector, -exus*, entwine, embrace, enfold; *complico, -atus*, fold together; *perplexus*, tangled, involved, intricate; *plectilis*, complicated, intricate: complexity, complication, perplex, perplexity, accomplice.
L. *contortus*, involved, intricate, complicated; see *torqueo* under **turn**
L. *intricatus*, entangled, complicated, < *intrico*, entangle, perplex, embarass: intricate, intricacy, *Echinopsis intricatissima* (a cactus), *Tolypella intricata* (a characead).
L. *involutus*, intricate, obscure: involved, involute, *Calamagrostis involuta* (a grass).
L. *labyrinthicus*, of a labyrinth, intricate; see *labyrinthus* under **turn**
L. *nexosus*, much intertwined, involved, complicated; see *necto* under **bird**
Gr. *symplokos*, interwoven, involved; see *plecto* under **weave**
See: **mix, confusion, weave**

compo- < Gr. *kompos*, boast, noisy boastful words; *komperos*, boastful; *kompastes*, braggart; see **brag**

composite, a species of the Compositae (sometimes subdivided into Ambrosiaceae, Asteraceae, Carduaceae, and Cichoriaceae), the largest natural group of plants, including the chicories, dandelions, ragweeds, thistles, goldenrods, asters, daisies, chrysanthemums, dahlias, and sunflowers, < L. *compositus*, aggregated, made up of parts, well-arranged. See **class**
L. *abrotonum* (Gr. *abrotonon*), n. an aromatic plant, probably *Artemisia abrotanum* (European wormwood or southernwood): *Gilia abrotanifolia* (a gilia), *Chalcis abrotani* (a chalcid).
L. *achillea*, yarrow; see **yarrow**
NL. *actinea*, f. a genus of the composite family: *Actinea odorata* (Colorado rubberplant).
L. *ageratum* (Gr. *ageraton*), n. a genus of the composite family: *Ageratum houstonianum* (Mexican ageratum), *Achillea ageratum* (sweet yarrow), *Eupatorium ageratoides* (white snakeroot).
Gr. *ambrosia*, f. food of the gods; a genus of the composite family: *Ambrosia artemisiaefolia* (a ragweed), *Artemisia ambrosiifolia* (a sagebrush), *Chenopodium ambrosioides* (a goosefoot).
L. *amellus*, m. a purple aster: *Aster amellus* (Italian aster).
NL. *anaphalis*, f. a genus of the composite family: *Anaphalis cuneifolia* (an everlasting).
NL. *antennaria*, f. a genus of the composite family: *Antennaria lanata* (woolly everlasting).
Gr. *anthemis, -idos*, f. chamomile: *Anthemis nobilis* (Roman chamomile).
Gr. *apape*, f. a dandelionlike plant:
Gr. *apargia*, f. hawkbit: *Apargidium boreale* (a composite).
L. *arctium* (Gr. *arktion*), n. a plant; now a genus of the composite family: *Arctium tomentosum* (a burdock).
NL. *arnica*, f. a genus of the composite family: *Arnica cordifolia* (an arnica), *Trypeta arnicivora* (a fly).
L. *artemisia*, mugwort, sagebrush; see **wormwood**
L., Gr. *aster*, star; a composite; see **star**
L. *baccharis* (Gr. *bakcharis*), f. plant with roots yielding a fragrant oil: *Baccharis salicina* (willow-baccharis).
Gr. *bechion*, n. coltsfoot:
L. *bellis, -idis*, f. daisy; *bellio, -onis*, f. corn-marigold: *Bellis integrifolia* (western daisy), *Aceste bellidifera* (a sea-urchin), *Bellium minutum* (a composite), *Campanula bellidifolia* (a bellflower), *Monoptilon bellidiforme* (a composite).
L. *bidens*, f. a genus of the composite family: *Bidens comosa* (a beggarticks).
L. *brumaria*, f. a kind of edelweiss:

L. *cacalia* (Gr. *kakalia*), f. a plant; now a genus of the composite family: *Cacalia arborea* (a composite), *Cacaliopsis nardosmia* (a composite).

L. *cachla* (Gr. *kachla* = *bouphthalmon*), f. oxeye:

NL. *calendula*, f. a genus of the composite family: *Calendula stellata* (a calendula), calendulin.

L. *carduus*, thistle; see **thistle**

L. *carpesium* (Gr. *karpesion*, an aromatic Asiatic wood), n. a genus of the composite family: *Carpesium cernuum* (a composite).

NL. *carthamus*, m. a composite, < Ar. *qartam*, safflower: *Carthamus tinctorius* (safflower).

L. *centaureum; centaurium* (Gr. *kentaureion; kentaurion*), n. a plant of the composite family: *Centaurea dealbata* (Persian centaurea), *Centaurium pulchellum* (a centaury).

L. *chamomilla* (Gr. *chamaimelon*), f. chamomile: *Chamomilla hookeri* (Arctic chamomile), *Matricaria chamomilla* (German chamomile).

Gr. *chaskanon*, n. cocklebur: *Chascanopsetta lugubris* (a fish).

L. *chrysanthemum* (Gr. *chrysanthemon*), n. a goldflower: *Chrysanthemum segetum* corn-marigold), *Dimorphotheca chryanthemifolia* (a Cape marigold), *Orneoascaris chrysanthemoides* (a nematode).

L. *chrysogonum* (Gr. *chrysogonon*), n. a genus of the composite family: *Chrysogonum virginianum* (gold star), *Leontice chrysogonum* (a lionsleaf).

L. *cirsium*, thistle; see **thistle**

L. *cleonia* (Gr. *kleonia*), f. elecampane: *Cleonia lusitanica* (a Spanish mint).

L. *cnicus*, thistle; see **thistle**

L. *conyza* (Gr. *konyza*), f. fleabane: *Inula conyza* (an elecampane), *Ageratum conyzoides* (tropic ageratum), *Conyza viscidula* (a composite).

NL. *coreopsis* (Gr. *koreopsis*, buglike), f. a genus of the composite family: *Coreopsis rosea* (pink tickseed).

NL. *cosmos* (Gr. *kosmos*, ornament), a genus of the composite family; see **ornament**

L. *crepis*, f. a plant; now a genus of the composite family: *Crepis setosa* (a hawksbeard).

NL. *dahlia*, f. a genus of the composite family, < Andreas Dahl, Swedish botanist: *Dahlia rosea* (a dahlia), *Dahlia maxoni* (a dahlia), *Verticillium dahliae* (a fungus).

NL. *doronicum*, n. a genus of the composite family: *Doronicum cordatum* (a leopardbane).

Gr. *erechthites*, f. groundsel: *Erechtites[Erechthites] hieracifolia* (burnweed).

Gr. *erigeron, -tos*, m. groundsel: *Erigeron strigosus* (a fleabane).

L. *eupatorium* (Gr. *eupatorion*), n. a genus of the composite family: *Eupatorium cannabinum* (hemp-eupatorium).

L. *farfarus*, m. coltsfoot:

NL. *gaillardia*, f. a genus of the composite family, < Gaillard de Marentonneau, French botanist: *Gaillardia aristata* (common gaillardia).

L. *gerontea* (Gr. *geronteia*), f. groundsel:

L. *gnaphalium* (Gr. *gnaphalion*), n. cudweed: *Gnaphalium uliginosum* (low cudweed.)

NL. *grindelia*, f. a genus of the composite family, < Hieronymus Grindel, Russian botanist: *Grindelia robusta* (gumweed).

L. *helenium* (Gr. *helenion*), n. elecampane: *Helenium autumnale* (sneezeweed), *Inula helenium* (elecampane), horseheal.

NL. *helianthus* (Gr. *helianthes*), m. sunflower: *Helianthus annuus* (sunflower).

L. *helichrysum* (Gr. *helichrysos*), n. strawflower, marigold: *Helichrysum angustifolium* (whiteleaf-everlasting).

L. *hieracium* (Gr. *hierakion*), n. hawkweed: *Hieracium floribundum* (a hawkweed).

L. *inula*, f. elecampane: *Inula ensifolia* (swordleaf-inula).

NL. *iva*, f. a genus of the composite family: *Iva ciliata* (a sumpweed), *Ajuga iva* (bugle).

Gr. *krossion*, n. lionsfoot:

L. *lactuca*, lettuce; see **lettuce**

L. *lapsana* (Gr. *lapsane*), f. a plant said to be the crucifer, charlock, but the name is now applied to a genus of the composite family: *Lapsana communis* (nipplewort).

Gr. *leontopodion*, n. lionsfoot: *Leontopodium sibiricum* (Siberian edelweiss).

L. *leucanthemum* (Gr. *leukanthemon*), n. a composite: *Chrysanthemum leucanthemum* (oxeye-daisy).

NL. *liatris*, f. a genus of the composite family: *Liatris spicata* (gayfeather), *Trilisa odoratissma* (wild vanilla).

NL. *madia*, f. a genus of the composite family, < Ab.Am. *madi: Madia sativa* (melosa, a tarweed), *Centromadia pungens* (spikeweed).

NL. *matricaria*, f. a genus of the composite family: *Matricaria aurea* (a mayweed).
L. *melampodium* (Gr. *melampodion*), n. blackfoot: *Melampodium cinereum* (pale blackfoot).
Gr. *melanthemon,* n. a kind of chamomile:
Gr. *orestion,* n. elecampane:
L. *parthenium* (Gr. *parthenion*), n. a composite: *Parthenium argentatum* (guayule), *Chrysanthemum parthenium* (feverfew).
L. *pectis, -idis,* f. a plant; now a genus of the composite family: *Pectis angustifolia* (a limoncillo).
Gr. *petasites,* m. a broadleaved plant; a genus of the composite family: *Petasites palmatus* (a butterbur).
L. *prapedilum,* n. lionsfoot:
L. *pulicaria,* f. a plant, probably fleabane: *Pulicaria dysenterica* (a fleabane).
L. *pyrethrum* (Gr. *pyrethron*), n. a plant of the composite family: pyrethrum *(Chrysanthemum coccineum),* pyrethrin.
NL. *ratibida,* f. a genus of the composite family: *Ratibida pinnata* (a coneflower).
NL. *rudbeckia,* f. a genus of the composite family, < Olaus Rudbeck, Swedish botanist: *Rudbeckia hirta* (blackeyed Susan).
NL. *scorzonera,* f. a genus of the composite family: *Scorzonera hispanica* (black salsify), *Agoseris scorzoneraefolia* (a composite).
L. *senecio, -onis,* m. groundsel: *Senecio aureus* (golden groundsel).
L. *serratula,* f. betony; now a genus of the composite family: *Serratula nudicaulis* (a sawwort).
L. *silphium* (Gr. *silphion*), n. a plant thought to be the same as *laserpitium* of the carrot family, but the name is now used for a genus of the composite family: *Silphium scaberrimum* (rough rosinweed).
NL. *solidago, -inis,* f. goldenrod: *Solidago speciosa* (showy goldenrod), *Solidago bicolor* (silverrod), *Eurosta solidaginis* (a gallfly).
L. *stoebe* (Gr. *stoibe*), f. a genus of the composite family: *Stoebe fusca* (a composite).
NL. *tagetes,* f. a genus of the composite family: *Tagetes patula* (French marigold).
NL. *tanacetum,* n. a genus of the composite family: *Tanacetum vulgare* (common tansy), *Aster tanacetifolius* (an aster), *Tanacetifex tanaceti* (a bug).
NL. *taraxacum,* n. a genus of the composite family: *Taraxacum officinale* (dandelion), *Agoseris taraxacifolia* (a composite).
Gr. *tragopogon, -os,* m. goatsbeard: *Tragopon porrifolius* (oyster-salsify).
L. *tussilago, -inis,* f. coltsfoot: *Tussilago farfara* (coltsfoot).
NL. *vernonia,* f. a genus of the composite family, < William Vernon, English botanist: *Vernonia anthelmintica* (an ironweed).
L. *xanthium* (Gr. *xanthion*), n. a genus of the composite family: *Xanthium spinosum* (spiny cocklebur), *Thalassoxanthium octoceras* (a radiolarian).
NL. *zinnia,* f. a genus of the composite family, < J. G. Zinn, German botanist: *Zinnia elegans* (common zinnia).

compound < OF. *compondre,* < L. *compono, -ositus,* put together, mix; see **mix, make,** *tithemi* under **place**

compso- < Gr. *kompsos,* elegant, pretty; see **beauty**

comptonia, NL. a genus of the bayberry family: see **bayberry**

comptus; comtus, L. ornamented, adorned; see **ornament**

comytho- < Gr. *komys, -ythos,* bundle, sheaf; see **bundle**

con- < L. *cum,* together, with; see **with**

conabilis, L. laborious, difficult; *conamen,* effort, struggle, < *conor, -atus, try,* endeavor; see *conor* under **make**

conabo- < Gr. *konabos,* clashing, din; see **sound**

conaro- < Gr. *konaros,* fat, well-fed; see **fat**

concavus, L. hollowed or arched inward; see **arc**

concentus, L. harmony, symphony; see *concino* under **harmony**

conceptio, L. comprehension, pregnancy; see **learn, begin**

concessio, L. permission; see **let**

concha, L. (Gr. *konche; konchos*), snail, shell; *conchula; konchylion,* dim.; see **mollusk**

conchis, L. lentils boiled with the pods; see **bean**

conchylium, L. (Gr. *conchylion*), dim. of *concha (konche);* see *concha* under **mollusk**

conciliatus, L. devoted to promoting friendliness; see **friend**

concilium, L. assembly, meeting; see **gather**

concinnus, L. well-arranged, skillfully joined, beautiful, striking; see **beauty**

concisus, L. brief, short; see **short**

conclusio, L. close, end; see *claudo* under **close**

concolor, L. colored uniformly; see **color**

concordia, L. harmony, agreement, union, sympathy; see **harmony**

concrete < L. *concretus*, hardened, thick, stiff; see **hard**

concubina, L. kept mistress; *concubinus*, male paramour; see **prostitute**

concussus, L. a shaking, earthquake; see *concutio* under **strike**

condaco- < Gr. *kondax, -akos*, a kind of game; see **play**

condalia, NL. a genus of the buckthorn family; see **buckthorn**

condimentum, L. a pungent or spicy substance added to food, < *condio, -itus*, pickle, preserve, season; see **spice, save**

condition < L. *conditio, -onis*, agreement, compact, circumstance, nature, state; see **nature**

conditus, L. savory, seasoned, see **taste**; founded, made, see **make**; hidden, see **cover**; preserved, see **save**

condylus, L. (Gr. *kondylos*), knuckle, knob, enlarged end of a bone; see **projection**

cone < L. *conus* (Gr. *konos*), m.; *coniculus; conulus*, m.; *konarion; konion*, n. dim.; *konikos*, conelike; *conifer; coniger*, bearing cones: conical, conicle, conarium, conoid, coniferous, condont, conopodium, conoscope, phragmacone, polyconic, *Conocardium hibernicum* (a fossil pelecypod), *Conorhinus megistus* (brabeiro, a bug that transmits the trypanosome causing Chagas disease), *Conularia qilad-risulcata* (a fossil coral), *Primula obconica* (a primrose), *Conicodrilus gracilis* (an annelid), *Aechmea conifera* (a bromeliad), *Oecetes coniferarum* (a moth), *Baioconodon[Baeoconodon]* *denverensis* (a Paleocene mammal), *Palaeocurria (Calloconus) humilis* (a Silurian gastropod), *Paratimia conicola* (a beetle).

Gr. *bembix, -ikos*, f. top, a buzzing insect: *Bembicidium cupreum* (a beetle).

L. *cedris, -idis* (Gr. *kedris, -idos*), f. cedar-cone:

L. *galbulus*, m. cone of the cypress:

L. *meta*, conical or pyramidal figure; see **pillar**

L. *pinea*, f. pine-cone: pineal, *Diplodia pinea* (a fungus).

Gr. *rhombos*, a top; see *rhembo* under **turn**

L. *strobilus* (Gr. *strobilos*), m. anything twisted, cone, top, < *strobos*, m. anything twisted or turned; *strobilion*, n. dim.; *strobilinos*, of a pine-cone: strobilus, strobila, strobiloid, *Strobilocephalus triangularis* (a worm), *Strobilanthes anisophyllus* (an acanthacead), *Stroboceras harti* (an ammonite), *Strobilomyces strobilaceus* (cone-mushroom), *Lepidostrobus ornatus* (cone of a lepidodendron), *Cynips strobilana* (pine-cone oak-gall), *Amanita strobiliformis* (a mushroom), *Actinostrobus acuminatus* (a conifer), *Gomphostemma strobilinum* (a mint), *Chermes strobilobius* (woolly larch-aphid), *Rhabdophaga strobiloides* (pine-cone willow-gall), *Chalara strobilina* (a fungus).

Gr. *strombos*, m. top; *strombeion*, n. dim.; *strombodes*, toplike: strombiform, strombuliform, stromboid, *Strombus gigas* (conch-shell), *Strombidium sulcatum* (a protozoan), *Strombocerus schüppeli* (a beetle).

L. *turbo, -inis*, m. a spinning-top; *turbinatus; turbineus*, top-shaped, conical: turbinate, turbine, *Turbinella pyrum* (chank). *Turbonilla rufa* (a fossil gastropod), *Dipterocarpus turbinatus* (gurjun-tree), *Quercus turbinella* (a scrub-oak).

L. *turris*, tower; see **tower**

confectio, L. something prepared, < *conficio, -ectus*, make, prepare; see *facio* under **make**

confertus, L. pressed together, crowded, thick, dense; see **thick**

conferva, L. an aquatic plant; see **alga**

confess < L. *confiteor, -fessus*, acknowledge, own, avow, admit; see **hold**

confidens, L. trusting, bold; see *fido* under **believe**

confinis, L. neighboring, adjoining; see *affinis* under **kin**

confixilis, L. joinable; see *figo* under **bind**

conflictus, L. contest, fight; see **fight**

confluens, L. place where two streams meet; see *confluus* under **bind**

conformis, L. like, similar; see **like**

confossus, L. pierced through, full of holes; see **hole**

confragus; confractus, L. broken, rough, rugged, uneven; see *fragosus* under **rough**

confusion < L. *confusio, -onis*, f. mixture, disorder, < *confundo, -usus*, pour together, mingle; *confusaneus*, mixed, miscellaneous: confuse, *Dactylopsis confusus* (a cochineal-insect), *Pelastoneurus confusibilis* (a fly).

Gr. *adiastolos,* not separated, confused: *Adiastola americana* (a wasp).
Gr. *akosmos,* disorderly, unarranged, unorganized:
Gr. *akritos,* confused, disorderly, mixed, doubtful: *Acritochaeta pulmonata* (a fly), *Galgupha(Acritophleps) grossa* (a bug).
Gr. *anaphyrmos,* m. confusion:
Gr. *anarchos,* without head or leader, lawless: anarchy, anarchist.
Gr. *anasobe,* f. disturbance, tumult:
Gr. *apektos; apektetos,* uncombed, unkempt: *Apectolophus phalangioides* (a mite).
Gr. *ataxia,* f. disorder, confusion; *ataktos,* not arranged, disordered, irregular: locomotor ataxia, *Atactopora hirsuta* (an Ordovician bryozoan).
Gr. *athemistos,* lawless: *Athemistus rugulosus* (a beetle).
Gr. *atyptos,* shapeless, unformed:
Heb. *Babel,* a city and tower in Assyria; symbolic of noisy, confused speech: babel, babeldom, Babylon.
Gr. *chaos,* condition of formlessness, utter disorder and confusion; see *chaino* under **open**
L. *incompositus,* disarranged, confused:
L. *inconditus,* unformed, rough, rude; see **rough**
L. *inconfectus,* not wrought out, undigested:
L. *inconscriptus,* unarranged:
L. *indigestus,* unarranged, orderless, confused:
L. *indispositus,* orderless, confused: *Omophorus indispositus* (a beetle).
L. *inordinatus,* not arranged, disorderly, irregular: inordinate.
Gr. *kakothemon, -os,* disorderly; *kakothemosyne,* f. disorderliness:
Gr. *klonos,* m. violent, confused motion, tumult, turmoil, rout:
Gr. *synchysis,* commixture, confusion; see **mix**
Gr. *syrphetodes,* jumbled together; see *syrphetos* under **dirt**
Gr. *tarache; taraxis,* f. confusion, disturbance, trouble, < *tarasso,* stir up, confound; *taragmos,* m. disturbance; *taraktos,* disturbed; *tarakitikos,* disturbing; *tarakates; taraktor, -is,* m. disturber: *Tarachomantis rubiginosa* (a mantid), *Tarachodes irrorata* (a mantid), *Taracticus octopunctatus* (a fly), *Taraxineura carbonaria* (a moth), *Taraktogenos[Taractogenus] kurzi* (a flacourtiacead, source of chaulmoogra oil).
L. *tumultus,* m. commotion, confusion, disturbance, uproar: tumult, tumultuous.
L. *turba* (Gr. *tyrbe*), f. turmoil, hubbub, uproar; *turbella; turbula,* f. dim.; *turbidus,* confused, disordered; *turbulentus,* agitated, troubled, stormy; *turbator, -is,* m.; *turbatrix, -icis,* f. disturber, trouble-maker; *turbo, -atus,* agitate, confuse, disorder; *tyrbasma, -tos,* n. confusion, trouble; *tyrbastes,* m. agitator, disturber, trouble-maker; *conturbo; disturbo; perturbo,* throw into disorder, derange, confuse: turbid, turbulent, disturbance, perturbation, *Ammobatrachus turbatans* (a fossil amphibian), *Turbatrix* (proposed new name for *Anguillula aceti,* the vinegar-eel), *Otiorhynchus turbatus* (a beetle), *Graphis turbulenta* (a lichen).
See: **mix, turn, dishevel, complex, move**

congelo, *-atus,* L. freeze, stiffen, harden; see *gelidus* under **cold**
congener, L. of the same race; see *genus* under **class**
conger, L. (Gr. *gongros*), a marine eel; see **eel**
congestus, L. collected, dense, thick, see **thick**; *congeries,* heap, mass, see **heap**
congress < L. *congressus,* assembly, conference, meeting; see **gather**
congruus, L. suitable, agreeable, fitting; see **fit**
conidium, NL. an asexual spore, dim. of Gr. *konis,* dust; see **fruit**
conido- < Gr. *konis, -idos,* nit, egg of a louse; see **egg**
conifer, L. bearing cone-shaped fruit; a kind of gymnosperm; see **gymnosperm**
conio- < Gr. *konis, -ios,* dust; *koniates,* plasterer; see **dust, plaster**
conium, L. (Gr. *koneion*), poison-hemlock; a genus of the carrot family; see **carrot**
conjectaneum, L. memorandum book; see **book**
conjectura, L. guess; see **hypothesis**
conjugalis, L. pertaining to marriage; see **marry**
conjunctus, L. connected, united; see *jungo* under **bind**
connarus, NL. (Gr. *konnaros*), a thorny evergreen tree; see **plant**
connatus, L. born together, joined; see **bind**
connecto, *-exus,* L. bind, fasten, join; see *necto* under **bind**
connisus; connixus, L. labored, struggled; see **make**
conniveo, L. close, shut, blink at, overlook; see **close**
conno- < Gr. *konnos,* beard; see **hair**
cono- < Gr. *konos,* cone; *konikos,* conelike; see **cone**

conopo- < Gr. *konops, -opos,* gnat; see **fly**

conor, *-atus,* L. endeavor, try; see **make**

conquer < L. *conquaero,* search for, procure; see **victory**

conquisitus, L. sought out eagerly; see *quaero* under **try**

consanguineus, L. related by blood, kindred; see **kin**

conscius, L. having knowledge in common with an accomplice or confidant; see *scientia* under **know**

conscriptus, L. chosen, elected, enrolled; see **choose**

consensus, L. agreement, concord; *consentaneus,* agreeing, fitting; see *consentio* under **harmony**

considero, *-atus,* L. contemplate, examine, reflect upon; see **think**

consilium, L. consideration, sense, purpose, plan; see *consilio* under **think**

consimilis, L. like in all respects; see *similis* under **like**

consobrinus, L. cousin; see *sobrinus* under **kin**

consortium, L. community, fellowship, society; *consors,* sharing in common; partner; see **companion**

conspicillum, L. lens, spectacles; see **lens**

conspicuus, L. manifest, visible, prominent; see *conspectus* under **clear**

constant < L. *constans,* standing firm; see *sto* under **stand**

constipatus, L. crowded or pressed together; see *stipo* under **press**

constrictus, L. drawn together, contracted; see *stringo* under **bind, narrow**

consuetus, L. customary, usual; see *sueo* under **use**

consul, L. a chief magistrate of the Roman state; see **govern**

consultus, L. considered, pondered; see *consulo* under **think**

consume < L. *consumo, -umptus,* use up completely, eat, destroy, waste; see **eat**

contaco- < Gr. *kontax, -akos,* shaft, spear; *kontakion,* dim.; see **spear**

contain < OF. *contenir,* < L. *contineo,* hold together; see *teneo* under **hold**

contempt < L. *contemno, -emptus,* esteem lightly, despise; see **hate**

contentus, L. satisfied; see **agreeable**

contest < L. *contestor, -atus,* call a witness, bring an action; see **fight, play**

contextus, L. interwoven, connected, united; see *texo* under **weave, bind**

conticinium, L. evening; see **night**

contiguus, L. bordering, neighboring, near; see **near**

continuous < L. *continuus,* hanging together, uninterrupted, incessant; see **one, always, chain, follow, line**

conto- < Gr. *kontos,* short, see **short**; pole for pushing a boat, see **rod**

contortus, L. twisted, intricate, complex; see *torqueo* under **turn**

contra, L. against, opposite; see **against**

contract < L. *contraho, -actus,* draw together; see **lessen**

contrarius, L. against, opposite; see *contra* under **against**

control < F. *controler,* check, govern, regulate; see **govern**

controversia, L. dispute, fight, quarrel; see **fight**

contubernalis, L. companion, mate; see **companion**

contumax, *-acis,* L. defiant, insolent, obstinate, stubborn; see **stubborn**

contumelia, L. abuse, insult, invective; *contumeliosus,* insolent, abusive, see **hurt**

contus, L. (Gr. *kontos*), long pole for pushing a boat, pike; *kontarion; kontilos,* dim.; see **rod**

contusio, L. bruise; see *contundo* under **bruise**

conula, L. a kind of plant; see **plant**

conus, L. (Gr. *konos*), cone; *coniculus,* dim.; see **cone**

convallaria, NL. a genus of the lily family; see **lily**

conventio, L. assembly, meeting; see **gather**

conversus, L. turned around, reversed; see **against**

convexus, L. arched outward, protuberant; see **arc**

convicium, L. outcry, clamor; see **call**

convictio, L. companionship; *convictor,* messmate; see **companion**

convivium, L. feast, banquet, entertainment; see *convivialis* under **eat**

convolvulus, L. bindweed, morning-glory; see **bindweed**

convulsus 230

convulsus, L. shaking, spasmodic; see **shake**
conyza, L. (Gr. *konyza*), fleabane; see **composite**
coo < L. imitative origin; see **murmur**
cook < AS. *coc*, < L. *cocus (coquus); coctor, -is*, m.; *coqua*, f., < *coquo, coctus*,
    bake, boil, fry, poach, roast, stew; *cocibilis(coquibilis); coctivus*, that can be
    cooked easily; *coctilis,* cooked, baked; *coctus*, cooked, well-considered; *coctio,
    -onis*, f. a cooking; *coquinus*, of cooking; *coquina*, f. kitchen; *coquinarius,* of the
    kitchen, culinary; *concoquo*, boil together, digest; *decoquo*, boil down; *incoctilis,*
    cooked in anything; *incoctus*, uncooked, raw: cook, cooky, concoction, decoction,
    precocious, coctile, kitchen, cuisine, biscuit, apricot.
    L. *asso*, roast, broil; *assus; assarius*, roasted; *assum*, n. a roast; *assatura*, f. roast
       meat: *Bonasa umbella* (ruffed grouse).
    AS. *bacan*, bake: baker, batch, Baxter.
    L. *crustularius*, maker of pastry, baker, confectioner; see *crustum* under **bread**
    L. *dapifex, -icis*, m. servant who prepared food:
    L. *dulciarius*, m. pastry cook:
    L. *focarius*, m. cook:
    L. *frigo, frixus*, roast, parch, fry: fry, fryer, fritter, frixion.   Derive: small fry.
    L. *furnarius*, baker; see *fornax* under **oven**
    Gr. *ipnos*, oven, furnace, kitchen; see **oven**
    Gr. *kodomeus*, m.; *kodome*, f. roaster of barley:
    L. *libarius*, pastry cook; see *libum* under **cake**
    L. *magirus* (Gr. *mageiros*), m. cook; *mageirike*, f. cookery; *mageirikos*, of cooking,
       skilled in cookery; *mageus*, m. kneader, baker: magirics.
    Gr. *maison, -os*, m. cook:
    Gr. *optao*, roast, bake, cook; *optaleos; optos*, roasted, cooked; *opsartytes; opso-
       poios*, m. cook; *opsartytike*, f. cookery; *opsartytikos*, of cooking; L. *artopta*
       (Gr. *artoptes*), m. baker:
    L. *panifex; panifica*, breadmaker; see *panis* under **bread**
    Gr. *pepto (pesso; petto)*, soften, ripen, cook, digest; *peptos*, cooked; see **digest**
    Gr. *phoktos*, roasted, baker, < *phogo*, roast, bake:
    Gr. *phrygo*, roast, fry; *phryktos*, roasted; m. firebrand:
    L. *pistor, -is*, m. grinder, miller, baker; *pistrina*, f. bakery:
    L. *popinarius*, m. cook:
    L. *siliginarius*, baker; see *siligo* under **grain**
    See: **boil, heat, oven, fire**
coot < uncertain origin.
    L. *fulica*, f. coot: *Fulica atra* (European coot), *Achatina fulica* (a gastropod),
       Cataclysta fulicalis (a pyralid moth).
    L. *phalacrocorax, -acis* (Gr. *phalakrokorax, -akos*), m. cormorant: *Phalacrocorax
       perspicillatus* (an extinct cormorant).
    Gr. *phalaris, -idos*, f. coot: *Phalaropus fulicarius* (red phalarope).
copa, L. female tavern-keeper; see *caupo* under **trade**
copaiba, Ab.Am. an aromatic resin; see **resin**
copano- < Gr. *kopanon*, pestle; see **pestle**
cope- < Gr. *kope*, oar, handle, haft, see **oar**; stroke, pounding, cutting, see **cut,
    strike**
copeo- < Gr. *kopeus, -eos*, chisel, see **chisel**; < *kopis, -eos*, prater, wrangler, liar,
    see **lie**
cophias, L. (Gr. *kophias*), an adder; see **snake**
cophinus, L. (Gr. *kophinos*), basket, coffin; see **basket**
copho- < Gr. *kophos*, blunt, dull, deaf, dumb, dim-sighted; see **dull**
copia, L. abundance, plenty, wealth; *copiosus*, abundant; see **abundance**
copido- < Gr. *kopis, -idos*, cleaver, kitchen-knife; *koparion*, dim. surgical knife;
    see **knife**
copo- < Gr. *kopos*, weariness, fatigue; see **weak**
copper < AS. *coper*, < L. *cuprum*, n. < Gr. *Kypros*, Cyprus, ancient source of
    copper; *cupreus; cuprinus*, of copper, coppery: cupreous, cuprous, cupric, cupri-
    ferous, cuprite, cuprammonium, cuproplumbite, *Callithrix cuprea* (a monkey),
    Episcia cupreata (a gesneriacead), *Chalcodermus cupreolus* (a beetle).   See
    ELEMENTS under **thing**
    L. *aes, aeris*, n. at first an alloy of copper and tin (bronze), later of copper and
    zinc (brass); *aeneus; aereus; aeratus*, of copper, bronze, or brass, bronzy, brassy;
    aerosus, full of copper; *aerarius*, m. coppersmith; *aerugo, -inis*, f. copper rust,
    verdigris; *aeruginosus*, full of verdigris, greenish-blue; *aenum(ahenum)*, n. a
    bronze vessel: aeruginous, *Quiscalus aeneus* (bronze grackle), *Chlorosplenium
    aeruginosum* (a fungus), *Hedychridium ahenum* (a wasp), *Gloeocapsa aeruginosa*
    (an alga).

Gr. *chalkos*, m. copper, bronze; *chalkydrion*, n. dim.; *chalkeos*, of copper or bronze; *chalkeus*, m. coppersmith; *chalkion*, n. a copper kettle; *chalkitis, -idos*, f. copper ore; *oreichalkos* (L. *orichalcum*, n.), m. copper ore: chalcocite, chalcopyrite, chalcophyllite, chalcanthite, orichalceous, *Chalcodermus aeneus* (cowpea-curculio), *Chalcophora virginiensis* (a beetle), *Chalcosoma atlas* (a beetle), *Chalcoscirtus insularis* (a spider), *Chalcostigma olivaceum* (a hummingbird).

copro- < Gr. *kopros*, dung; *koprinos*, full of dung, filthy; *koprion*, dung-beetle; see **dung**

copsicho- < Gr. *kopsichos*, a blackbird; see **thrush**

coptis, NL. a genus of the buttercup family; see **buttercup**

copto- < Gr. *koptos*, chopped small, < *kopto*, cut, strike, chop; see **cut**

copula, L. bond, tie, connection, < *copulo, -atus*, couple, join; see **bind, coitus**

copy < L. *copia*, abundance; see **equal**

coquibilis, L. easily cooked; *coquinus*, pertaining to cooking; *coquus*, cook; see **cook**

cor, *cordis*, L. heart; *corculum*, dim.; *cordatus*, heart-shaped; see **heart**

cor- < L. *cum*, together, with; see **with**

coracias, NL. (Gr. *korakias*), daw; see *korax* under **crow**

coracinus, L. (Gr. *korakinos*), a perchlike fish; ravenlike; see **perch, crow**

coral < OF. *coral*, < L. *corallum; corallium* (Gr. *korallion*), n.: coralline, coralloid, *Corallium nobile* (red coral), *Corallorrhiza maculata* (spotted coralroot), *Coralliochama antillarum* (a Cretaceous pelecypod), *Coralliocrangon perrieri* (a crab), *Mactra corallina* (a clam), *Chaetodon corallicola* (a fish), *Erythrina corallodendron* (coral-tree).

L. *alcyonium* (Gr. *alkyonion*), n. polyp: Alcyonaria, *Alcyonium digitatum* (a coral).

Gr. *antipathes*, n. a black kind of coral; a remedy for suffering:

L. *polypus* (Gr. *polypous*), m. the many-footed, the coral animal, octopus, a growth in the nose: polyp, *Polypetta mammillata* (a radiolarian).

corapho- < Gr. *koraphos*, a bird; see **bird**

corax, L. (Gr. *korax*), raven; *korakion*, dim.; *korakinos*, ravenlike; *koraxos*, raven-black; see **crow**

corbis, L. basket; *corbicula; corbula*, dim.; see **basket**

corbita, L. a slow freight ship; see **ship**

corchorus, L. (Gr. *korchoros*), a genus of the linden family; see **linden**

cord < L. *chorda* (Gr. *chorde*), string; see **rope**

cordatus, L. heart-shaped; see *cor* under **heart**

cordax, L. (Gr. *kordax*), a lively, indecent dance of Greek comedy; see **dance**

cordialis, L. hearty, warm; see *cor* under **heart**

cordillera, Sp. a chain of mountains; see **mountain**

cordylo- < Gr. *kordyle*, club, swelling; *kordylinos*, clublike; see **club**

cordylus, NL. (Gr. *kordylos*), a kind of newt; see **newt**

core- < Gr. *kore*, girl, maiden, pupil of the eye; *koridion; koriske*, dim.; see **child**

coremato- < Gr. *korema, -tos*, broom; see **broom**

coreo-; corio- < Gr. *koris, -eos, -ios*, or *-idos*, bedbug, bug; see **bug**

coreopsis, NL. a genus of the composite family; see **composite**

corethro- < Gr. *korethron*, broom; see *korema* under **broom**

coriaceus, L. leathery; *coriarius*, of leather; tanner; see *corium* under **leather**

coriander < L. *coriandrum* (Gr. *koriannon*), a plant of the carrot family; see **carrot**

coris, L. a plant whose seeds resemble small bugs, < Gr. *koris*, bug; see **bug**

corissum, L. a kind of mint; see **mint**

corium, L. leather; see **leather**

cork < Sp. and Ar. *alcorque;* see **bark**

cormo- < Gr. *kormos*, trunk of a tree; *kormion*, dim.; see **stem**

corn < AS. *corn* (G. *korn*), grain; see **grain**

corner < OF. *cornier*, < L. *cornu*, horn, end, point; see **angle**

cornix, *-icis*, L. crow; *cornicula*, dim.; *cornicor*, caw like a crow; see **crow, caw**

cornu, L. horn; *corniculum*, dim.; *cornicen*, horn-blower; *cornuarius*, horn-maker; *cornualis*, of horns; *cornutus*, horned; see **horn**

cornus, L. dogwood; *corneus,* of dogwood; see **dogwood**

coro- < Gr. *koros,* satiety, see **abundance**; < *koros,* boy, lad; *koriskos,* dim., see **child**

corona, L. (Gr. *korone*), crown, wreath, halo; *corolla; coronula,* dim.; *coronarius,* of a wreath or garland; see **crown**

corono- < Gr. *koronos,* curved, crooked, see **bend**; < *korone,* a kind of crow, see **crow**; crown, wreath, see **crown**

coronopus, L. (Gr. *koronopous*), crowfoot; see **crowfoot**

corpus, L. body; *corpusculum,* dim.; *corporalis; corporeus,* of the body; *corpulentus,* fleshy, fat; see **body**

correct < L. *corrigo, -rectus,* make straight, set right, improve; see **right, true, straight**

correus, L. partaker in guilt, companion in crime; see **companion**

corrigia, L. shoelatchet, thong, rein; see **strap**

corroco, L. a fish; see **fish**

corruda, L. wild asparagus; see **asparagus**

corrugatus, L. wrinkled, ridged; see *ruga* under **fold**

corruptus, L. spoiled, damaged; *corruptibilis,* perishable; see **bad**

corso- < Gr. *korse; korrhe,* temple, side of the forehead, hair on the temple; *korseus; korsoter,* barber; see **head, barber**

cortadera, Sp. knife or other cutting instrument; see **knife**

cortex, *-icis,* L. bark; *corticulus,* dim.; see **bark**

corthylo- < Gr. *korthylos,* a little king or chieftain; see **govern**

corthyo- < *korthys, -yos,* heap; see **heap**

corticatus, L. covered with bark; see *cortex* under **bark**

cortina, L. round kettle, caldron with tripod, see **kettle**; curtain, veil, see **curtain**

corusco, *-atus,* L. flash; see **light**

corvus, L. crow, raven; *corvinus,* of ravens; see **crow**

corybantic < Gr. *korybantikos,* pertaining to the frenzied rites of the Corybantes; see **mad**

corycaeo- < *korykaios,* spy; see **see**

corycido- < Gr. *korykis, -idos,* gall on leaves; see **gall**

corycus, L. (Gr. *korykos*), leather punching-bag used by athletes; *korykion,* dim.; see **bag**

corydo- < Gr. *korydos; korydallis; korydalos,* crested lark; see **lark**

corylus, L. hazel, filbert; see **hazel**

corymbus, L. (Gr. *korymbos*), bunch of flowers, top, peak; see **cluster**

coryno- < Gr. *koryne,* club, mace; *korynetes,* club-bearer; see **club**

coryphaena, L. (Gr. *koryphaina*), a dolphinlike fish; see **fish**

corypho- < Gr. *koryphe,* head, top; *coryphaios,* leader, chief; see **head, palm**

corysto- < Gr. *korystos,* crested, helmeted; *korystes,* one armed with a helmet; see *korys* under **cap**

corytho- < Gr. *korys, -ythos,* helmet; *korysso,* furnish with a helmet; see **cap**

coryza, L. (Gr. *koryza*), cold, catarrh; see **disease**

cos, *cotis,* L. any hard stone, flintstone, whetstone, hone, grindstone; *coticula,* dim. touchstone; see **stone**

coscino- < Gr. *koskinon,* sieve; *koskinion,* dim.; see **sieve**

cosmetic < Gr. *kosmetikos,* skilled in arranging, beautifying; see *kosmos* under **ornament**

cosmetro- < Gr. *kosmetron,* broom; see **broom**

cosmo- < Gr. *kosmos,* order, universe, government, ornament; *kosmarion; kosmion,* dim.; *kosmikos,* of the universe; *kosmetos,* trim, neat; *kosmios,* well-ordered; see **world, ornament**

cossus, L. a kind of larva under the bark of trees; see **young**

cossypho- < Gr. *kossyphos,* probably a kind of thrush; see **thrush**

cost < OF. *coster,* < L. *consto, -atus,* stand firm, remain unchangeable; see **pay**

costa, L. rib, side; *costula,* dim.; *costabilis,* riblike; *costatus,* ribbed; see **rib**

costus, L. (Gr. *kostos*), a kind of ginger; see **ginger**

cosymbo- < Gr. *kosymbos,* edge, border; *kosymbotos,* fringed; see **border**

cotalis, NL. (Gr. *kotalis*), pestle; see **pestle**

cothono- < Gr. *kothon, -os,* a drinking-vessel, cup; see **cup**

cothurnus, L. (Gr. *kothornos*), a half-boot; see **shoe**

coticula, L. touchstone, dim. of *cos,* stone; see **stone**

cotilo- < Gr. *kotilos,* babbling, chattering, prattling, twittering, < *kotillo,* chatter, prattle; see **speak**

cotinus, L. (Gr. *kotinos*), the wild olive-tree, oleaster; now a genus of the sumac family; see **sumac**

coto- < Gr. *kotos,* grudge, hate, wrath; see **hate**

cotonea, L. name applied to a number of plants; see **plant**

cottabus, L. (Gr. *kottabos*), a game in which wine was thrown into cups; see **play**

cottido- < Gr. *koltis, -idos,* cerebellum; see **brain**

cotto- < Gr. *kotta,* head; *kottarion,* dim.; see **head**

cotton < Ar. *qutun.*
  LL. *bombax,* n. cotton, < corruption of Gr. *bombyx, -ykos,* silk, and *bambax, -akos,* cotton: Bombacaceae, bombazet, bombast, *Bombax flammeum* (flame-bombax).
  L. *byssus* (Gr. *byssos*), flax, cotton; see **flax**
  Sp. *ceiba,* f. silk-cotton tree, < Ab.Am.: *Ceiba pentandra* (source of kapok), *Bombax ceiba* (Malabar bombax).
  Gr. *erioxylon,* n. cotton:
  L. *gossypium* (Gr. *gossypion*), n. cotton: *Gossypium hirsutum* (upland-cotton), *Jatropha gossypifolia* (bellyache-bush), *Platyedra gossypiella* (pink bollworm), *Chrysopsis gossypina* (a composite), *Bemisia gossypiperda* (a bug), Spermophthora *gossypii* (a fungus), gossypol.
  Skt. *karpasa,* cotton; see *carbasus* under **flax**
  L. *xylinum* (Gr. *xylinon*), n. cotton; *xylinos,* of cotton: *Xylinophorus scobinatus* (a beetle).

cottus, L. (Gr. *kottos*), sculpin; see **sculpin**

coturnix, L. quail; see **quail**

cotyla, L. (Gr. *kotyle*), a cup, cavity; L. *cotyledon* (Gr. *kotyledon*), any cup-shaped cavity; a plant of the stonecrop family; see **cup, stonecrop**

couch < OF. *coucher,* put to bed; see **bed**

cough < uncertain origin.
  Gr. *bex, bechos,* f. cough, < *besso,* cough; *bechion,* n. dim.; *bechia,* f. hoarseness; *bechikos,* of a cough; *anabesso,* cough up: antibechic.
  Gr. *chelyskion,* n. a slight cough:
  L. *screo, -atus,* hawk, hem, cough, clear the throat; *screator, -is,* m. hawker, hemmer; *conscreor,* hawk much:
  L. *tussis,* f. cough; *tussicula,* f. dim.: tussal, tussicular, tussive, tussol, pertussis, *Tussilago farfara* (coltsfoot), *Hemophilus pertussis* (bacterium of whooping-cough).

council < OF. *concile,* < L. *concilium,* meeting; see **gather**

counsel < OF. *conseil,* < L. *consilium,* consideration, plan; see **think, teach, warn**

count < OF. *compter,* reckon; see **number**

counter; see *contra* under **against**

country < OF. *contree.*
  Gr. *amorbaios,* rustic, pastoral: *Amorbaeus infestus* (a beetle).
  Gr. *anakoma, -tos,* n. district:
  Gr. *apoikia,* f. settlement far from home, colony:
  L. *arcadius,* ideally rustic, < Arcadia, the Switzerland of Greece: Arcadia, Arcadian.
  Gr. *boukolikos,* rural, rustic, pastoral; see **field**
  Gr. *choros,* m. place, land, country, district; *choridion; chorion,* n. dim.; *chorites,* m. countryman; *choritikos,* like a countryman, rustic; *enchoros,* rustic, rural; *mesochoros,* midland: chorology, chorography, anchorite, *Mesochoros perniciosus* (an ichneumonid).
  L. *clima,* region, zone; see **slope**
  L. *colonia,* f. estate, farm, settlement: colony, colonial, Cologne, Lincoln.
  Gr. *demos,* country, district, tract of land; see *demos* under **people**
  L. *districtus,* m. political or geographical division: district.
  AS. *dom,* jurisdiction, condition; see **nature**
  L. *indigenus,* belonging to a country: indigenous.
  Gr. *kome,* country town; see **town**
  L. *latifundium,* n. farm, large estate: latifundium.
  Gr. *mesochthon,* midland, interior; see *chthon* under **earth**
  Gr. *mesogaios,* inland; see *ge* under **earth**
  L. *natio, -onis,* race, tribe; see **class**

L. *ora*, edge, border, coast, clime, country; see **border**

L. *pagus*, m. country, district, province; *paganus*, of the country, rustic: pagan, paganism, peasant, *Chenopodium paganum* (pigweed), *Tortula pagorum* (a moss).

L. *pastoralis*, of shepherds, rural; see *pastor* under **guard**

L. *patria*, f. fatherland: patriotic, expatriate, repatriation.

L. *plaga*, f. district, zone, tract, region:

L. *provincia*, f. division of a country or empire: province, provincial.

L. *regio*, *-onis*, f. district, quarter, tract: region, regional.

L. *respublica*, commonwealth, state; see *publicus* under **people**

L. *rus*, *ruris*, n. country, farm, land; *ruralis*, *rurestris*; *rusticus*, of the country, simple; *rusticulus*, dim.; *ruricola*, m. countryman; *rusticor*, *-atus*, live in the country: rural, rustic, rusticity, rusticate, rurigenous, R.F.D., roister, *Gecarcinus ruricola* (a crab), *Nicotiana rustica* (Aztec tobacco), *Cathypna rusticula* (a rotifer), *Buprestis rusticorum* (a beetle), *Scolopax rusticola* (woodcock).

AS. *scire*, county, district, < *sciran*, cut off: shire, sheriff, Yorkshire, Berkshire, New Hampshire.

Gr. *synoria*, f. borderland: *Synoria euglyphella* (a moth).

L. *territorium*, n. district, domain: territory, territorial.

L. *tractus*, m. district, region, territory: tract.

L. *villa*, country place; see **house**

See: **earth, field, forest, space, place**

coup, F. < L. *colaphus*, blow, < Gr. *kolapto*, strike; see **strike**

courage < OF. *corage*, < L. *cor*, heart; see **bold**

course < F. *cours*, < L. *cursus*, way, journey; see **way**

court < L. *cohors*, an enclosed space; see **pen**

cousin < OF. *cosin*, < L. *consobrinus*; see *sobrinus* under **kin**

cover < OF. *couvrir*, < L. *cooperio*, *-pertus*, cover completely; *operio*, *-pertus*, cover hide; *opertaneus*, concealed, secret: coverlet, covert, coverture, curfew, discover, handkerchief, operculum, overt.

L. *abditus*, hidden, concealed, < *abdo*, *-itus*, put away; see **secret**

Gr. *aphanizo*, hide, remove, suppress, cause to disappear; see *aphanes* under **secret**

L. *celo*, *-atus*, hide: conceal, celation, celative, *Clione celata* (a sponge).

L. *clepo*, *-eptus*, steal, conceal oneself, listen covertly, withdraw secretly:

L. *condo*, *-itus*, hide; *conditivus*, laid away, preserved; *absconditus*; *reconditus*, concealed, hidden: abscond, ensconce, recondite, testicond, *Carex abscondita* (a sedge), *Dipetalonema reconditum* (a worm).

L. *dissimulo*, *-atus*, feign, disguise, hide; see **lie**

Gr. *ekleipsis*, a forsaking, failing, eclipse; see *leipo* under **abandon**

Gr. *entaphizo*, prepare for burial, bury; see *taphos* under **grave**

L. *funero*, *-atus*, bury, inter; *funus*, *-eris*, n. burial ceremony; *funebris*; *funerarius*; *funereus*, of a funeral; *funestus*, causing death, fatal, mournful: funeral, funereal, *Cupressus funebris* (mourning cypress), *Cancellaria funerata* (a gastropod), *Ephedra funerea* (Death Valley ephedra), *Anopheles funestus* (a mosquito).

L. *humo*, *-atus*, cover with earth, bury; see **earth**

L. *incrusto*, *-atus*, cover with a coating of some substance: incrustation.

L. *induo*, *-utus*, put on clothes, dress; see *indusium* under **garment**

L. *infero*, *illatus*, bring to a place for burial, bury:

Gr. *kalypto*, cover, conceal, hide; *kalypter*, *-os*, m. covering, sheath; *kalyptra*, f. veil, cover, lid; *kalymma*, *-tos*, n. covering, hood, veil; *kalyptos*, covered; *Kalypso*, f. Calypso, the nymph who hid Ulysses: calypter, calyptra, calyptrate, acalyptrate, Apocalypse, eucalyptus, *Calyptranthes pallens* (a lidflower), *Calyptrophora japonica* (a coral), *Calyptrella bertae* (a sponge), *Calyptodera robusta* (a bug), *Calyptospadix cerulea* (a medusa), *Calymmatotheca obtusiloba* (a fossil fern), *Adenocalymma comosum* (a bignoniacead), *Calypso bulbosa* (an orchid), *Calymene pulchra* (a trilobite), *Eucalyptus salicifolia* (a eucalyptus). See *calyx* under **cup**

Gr. *kataphraktos*, covered, mailed: *Gasterosteus cataphractus* (a stickleback).

Gr. *keuthos*, hidden, < *keutho*, cover, hide; *keuthmon*, *-os*, m. hiding-place, hole, corner, cave: *Ceuthophilus maculatus* (cave-cricket).

Gr. *krypto*, hide, cover, conceal; *kryphaios*; *kryphios*; *kryptos*; *krybelos*, hidden, secret; *krybetes*, m. one hidden; *krypsis*, f. hiding-place, concealment; *apokryphos*, concealed, hidden, obscure, spurious: crypt, cryptogam, cryptogram, cryptonym, anticryptic, grotto, grotesque, Apocrypha, apocryphal, *Cryptorhynchus sceleratus* (a beetle), *Cryptophagus impressus* (a beetle), *Cryptocarya rubra* (a lauracead), *Cryphaea pendula* (a moss), *Eucryphia cordifolia* (muermo), *Cryphiocrinus rotundus* (a Mississippian crinoid), *Crypsis aculeata* (prickle-grass), *Catacrypsis inconspicua* (a moth), *Crybelus mexicanus* (a bird), *Crybelocephalus megalurus* (an amphipod).

L. *lateo,* lie hidden; *latebra,* f. hiding-place, shelter; *latibulum,* n. den, refuge; *latebrosus,* full of hiding-places; *latesco,* hide oneself; *delitesco,* hide, lurk: latent, latescent, *Latibulus tuberculatus* (an ichneumonid).

L. *obrutus,* buried, covered, hidden; see *ruo* under **destroy**

L. *occulto, -tus,* cover, conceal; see **secret**

L. *pelliculo,* cover with skins; see *pellis* under **skin**

Gr. *pykazo,* close, shut, wrap, cover; see **close**

L. *retrusus,* thrust back, removed, concealed; see *trudo* under **push**

L. *sepelio, -pultus,* bury; *sepulcrum,* grave, tomb; see **grave**

Gr. *skepastos,* sheltered, covered; see *skepe* under **safe**

Gr. *stego,* cover; *stege,* f. roof; *stegaster, -os; stegastes,* m. coverer; *stegasma, -tos; stegastron,* n. covering, wrapper; *steganos; stegastos,* covered, enclosed, sheathed: stegosaurus, steganography, steganopod, *Steganocrinus globosus* (a Mississippian crinoid), *Stegotrachelus finlayi* (a Devonian fish), *Stegasmonotus longissimus* (a cranefly), *Stegastopsis babylonica* (a beetle).

L. *tego, tectus,* cover; *tectum; tegulum,* n. roof; *tegmen, -inis; tegmentum; integumentum,* n. cover, covering; *tector, -is,* m. coverer; *tectorius,* of a cover; *obtectus,* covered over: tegmen, tegumentum, integument, tegminal, detect, protection, tectorial, tegulum, protegulum, otga, protege, tectrices, *Pandanus tectorius* (thatch-screwpine), *Sempervivum tectorum* (houseleek), *Acer tegmentosum* (a maple), *Acanthoscelides obtectus* (bean-weevil).

Gr. *thapto,* conduct funeral rites, bury:

L. *velo, -atus,* cover, conceal; *velum,* n. sail, curtain: reveal, revelation, envelope, development, veil, velamen, velarium, *Velella lata* (a siphonophore), *Calyptraphorus velatus* (a fossil gastropod), *Pertusaria velata* (a lichen), *Cristellaria obvelata* (a foraminifer).

See: **bag, blanket, box, cap, crown, curtain, feather, garment, membrane, roof, rug, shade, shell, shield, skin, close, guard, secret, wall**

covinus, L. war chariot of the Britons; see **vehicle**

cow < AS. *cu;* see **cattle**

coward < OF. *couard,* with tail between legs, < L. *cauda,* tail.

Gr. *apokakesis,* f. cowardice:

Gr. *apoknesis,* f. a shrinking from:

Gr. *atolomos,* cowardly, spiritless; *atolmia,* f. cowardice: *Motacilla (Atolomodytes) clara* (a bird).

L. *cussiliris,* cowardly:

Gr. *deilos,* cowardly, craven, wretched, weak; see **poor**

L. *ignavus,* coward, poltroon; see **dull**

L. *murcidus,* cowardly, slothful, < *murcus,* m. coward:

L. *muricidus,* m. coward:

Gr. *Phryx, -ygos,* m. a Phrygian; a coward:

Gr. *phyxelis, -idos; phyzelos,* cowardly, shy: *Phyzelus fasciatus* (a wasp), *Phyxelida makapanensis* (an arachnid).

Gr. *trestes,* m. coward, < *treo,* flee: *Trestis tricincta* (a wasp).

See: **fear, run**

coxa; coxendix, L. hip, hipbone; see **hip**

coy < OF. *coi,* < L. *quietus,* rest; see **modest**

crab < AS. *crabba.*

L. *astacus* (Gr. *astakos*), m. lobster: astacine, *Astacus fluviatilis*(European crayfish).

L. *cammarus; gammarus* (Gr. *kammaros*); NL. *homarus,* m. lobster, crab: *Cambarus pellucidus* (a crayfish), *Gammarus gibbosus* (an amphipod), *Gammarotettix californicus* (a grasshopper), *Homarus americanus* (lobster).

L. *cancer, -cri,* m. crab: cancer, cancriform, Tropic of Cancer, *Cancer scutellatus* (a crab), *Triops cancriformis* (a phyllopod), *Didelphys cancrivora* (an opossum).

L. *carabus* (Gr. *karabos*), a sea-crab; see **beetle**

L. *caris, -idis* (Gr. *karis, -idos*), f. shrimp; *karidion,* n. dim.: carid, caridoid, Phyllocarida, *Echinocaris punctata* (a phyllocarid), *Hymenocaris vermicauda* (a Cambrian crustacean).

Gr. *chlorokyrtis, -idos,* f. a kind of prawn: *Chlorocyrtis miser* (a shrimp).

Gr. *grapsaios,* m. crab: *Grapsus strigosus* (a crab).

Gr. *garkinos,* m. crab; *karkineutes,* m. crab-catcher: carcinology, carcinoma, *Sacculina carcini* (a parasitic cirriped), *Carcinomyia hirta* (a fly), *Carcinognathus sulcifrons* (a beetle), *Asyndetus carcinophilus* (a fly), *Carcineutes amabilis* (a bird).

Gr. *keraphis, -idos,* f. a lobster or prawn:

Gr. *kolybdaina,* f. a kind of crab:

Gr. *krabyzos,* m. a kind of shellfish: *Crabyzos longicaudatus* (an isopod).

Gr. *krangon, -os,* f. shrimp: *Crangon vulgaris* (common shrimp), *Eucrangonyx gracilis* (an amphipod).

L. *locusta,* f. lobster, grasshopper: lobster, *Gammarus locusta* (an amphipod).

Gr. *maia,* f. a kind of crab: *Maia squinado* (a spider-crab), *Maiopsis panamensis* (a crab).

L. *pagurus* (Gr. *pagouros*), m. a kind of crab: *Pagurus longicarpus* (a crab), *Cancer pagurus* (a crab), *Physalozercon paguroxenus* (a mite).

L. *plagusia,* f. a kind of crustacean: *Plagusia tuberculata* (a crab).

L. *scilla; squilla,* f. a shrimp: *Squilla crocea* (a shrimp), *Cancer squilla* (a shrimp), *Oniscus squillarum* (a sowbug).

Gr. *skyllaros (kyllaros),* m. hermit-crab: *Scyllarus carinatus* (a crab), *Cyllarus brevirostris* (a bug).

Ab.Am. *uca,* f. fiddler-crab: *Uca minax* (a fiddler-crab).

crabro, L. hornet; see **wasp**

crabyzo- < Gr. *krabyzos,* a kind of shellfish; see **crab**

cracca, L. a kind of pulse; see **bean**

cracens, -entis, L. neat, graceful, slender; see **beauty, thin**

crack < AS. *cracian,* break; see **break**

cractico- < Gr. *kraktikos,* noisy; see **sound**

cradalo- < Gr. *kradalos,* quivering; see **shake**

cradle < AS. *cradel.*

Gr. *antipex, -egos,* f. cradle on wheels:

L. *cuna,* f. cradle; *cunula,* f. dim.; *cunabulum,* n. cradle, earliest abode, birth, origin; *incunabulum,* n. cradle, birthplace, beginning: cunabular, incunabula, *Cuna gibbosa* (a pelecypod).

Gr. *liknon,* cradle of wickerwork, basket used in winnowing; *liknarion,* dim.; see **basket**

See: **bed, begin, birth, basin**

crado- < Gr. *krade,* branch, twig, fig-branch; *krados,* blight, especially on fig-trees; see **branch**

craepno- < Gr. *kraipnos,* swift, rushing; see **swift**

craft < AS. *craeft,* strength, skill, art, cunning; see **art, ship**

crageto- < Gr. *kragetes,* screamer; see *krazo* under **call**

cram < AS. *crammian,* stuff, crowd, pack, press; see **press**

cramato- < Gr. *krama, -tos,* mixture; see *kerannymi* under **mix**

crambaleo- < Gr. *krambaleos,* dry, parched; *krambos,* dry; see **dry**

crambo- < Gr. *krambe,* cabbage; *krambidion,* dim.; see **cabbage**

cranao- < Gr. *kranaos,* rugged, rocky; see **rough**

crane < AS. *cran,* a long-legged bird: cranefly, cranesbill, Cranford, European cranberry *(Oxycoccus palustris).*

L. *crex, -ecis* (Gr. *krex, -ekos*), f. a bird with a sharp bill and long legs: *Crex pratensis* (corn-crake), *Gallicrex cinerea* (a gallinule), *Megacrex inepta* (a bird), *Creciscus jamaicensis* (black rail).

Gr. *geranos,* f. crane; *geranodes,* cranelike; *geranion,* n. cranesbill: geranomorph, *Geranomyia maculipennis* (a fly), *Geranopus purpuratus* (a soldier-fly), *Geranoides jepseni* (an Eocene bird).

L. *gromphena,* f. a kind of crane:

L. *grus, gruis,* f. crane: gruine, Gruidae, pedigree, gromwell *(Lithospermum officinale),* *Grus canadensis* (a crane), *Erodium gruinum* (crane-heronbill).

Gr. *ibyx, -ykos,* m. crane, < *Ibykos,* Ibycus, a lyric poet, whose murder was revealed by cranes: *Ibycus fissidens* (a snail).

L. *porphyrio, -onis* (Gr. *porphyrion, -os*), m. a gallinule: *Porphyrio caeruleus* (purple gallinule).

NL. *rallus,* m. a genus of wading birds, < OF. *ralle: Rallus aquaticus* (water-rail), *Rallicula leucospila* (red mountain-rail).

Gr. *syrikter, -os,* m. male crane: *Syricter cyanocephalus* (a bird).

L. *vipio, -onis,* m. a small crane: *Grus vipio* (a crane).

crangono- < Gr. *krangon, -os,* shrimp; see **crab**

cranio- < Gr. *kranion,* skull, head; see **head**

crano- < Gr. *kranos,* helmet, head, see **cap, head**; < *kraneia,* dogwood, see **dogwood**

crantero- < Gr. *kranter, -os,* completer, finisher; see *kraino* under **end**

crapulatus; crapulentus, L. drunk, intoxicated; see **drink**

cras, L. tomorrow; *crastinus,* of tomorrow; see **day**

crasis < Gr. *krasis,* a mixing; see *kerannymi* under **mix**

craspedo- < Gr. *kraspedon*, edge, border; see **border**

crassula, NL. a genus of the stonecrop family; see **stonecrop**

crassus, L. thick, fat, stout; *crassitudo*, thickness; *crasso*, thicken; see **thick**

-crat; -cracy < Gr. *krateo*, rule, < *kratos*, strength, might, power; *kratys*, strong; *kreisson*, stronger; *kratistos*, strongest; see **govern, strong**

crataegus, L. (Gr. *krataigos*), a flowering thorn; see **rose**

crate < L. *cratis*, wickerwork, hurdle; *craticula*, dim.; *craticulus*, latticed, wattled; see **screen, basket, box**

crater, L. (Gr. *krater*), a mixing vessel, bowl, basin; see **cup**

cratero-; crato-; craty- < Gr. *krateros; kratys*, strong, sturdy; *kratos*, might, strength; see **strong**

craugo- < Gr. *kraugos*, a woodpecker; see **woodpecker**

crauro- < Gr. *krauros*, brittle, friable; see **brittle**

crave < AS. *crafian*, long for; see **wish**

crawl < ON. *krafla*, paw; see **creep**

crazy < uncertain origin; see **mad**

creagro- < Gr. *kreagra*, a meathook; see **hook**

creak < uncertain origin; see **squeak**

cream < OF. *cresme*, < L., Gr. *chrisma*, unguent; see **fat**

create < L. *creo, -atus*, make, produce; *creator*, maker; *creatura*, the thing made; see **make**

creber, -bra, -brum, L. thick, close, pressed together, numerous; see **thick**

crecto- < Gr. *krektos*, sounded by striking, < *kreko*, sound an instrument; see **sound**

credemno- < Gr. *kredemnon*, veil; see **curtain**

credit < L. *credo, -itus*, believe; *credulus*, believing too easily on little or no evidence; *creditum*, loan; see **believe, trade**

creek < ON. *kriki*, a small stream; see **stream**

creep < AS. *creopan*.
   Gr. *chamai*, on the ground, dwarf; *chamelos*, creeping; see **low**
   Gr. *herpo*, creep; *herpestikos*, creeping; see *herpes* under **snake**
   L. *repo, reptus; repto, -atus*, crawl, creep; *repens; reptatus; reptilis*, creeping, crawling: reptile, repent, reptant, Reptilia, *Agropyron repens* (quack-grass), *Simulium reptans* (pellagra-fly).
   L. *serpo, serptus*, creep; see *serpens* under **snake**
   AS. *snican*, crawl: sneak, snake, snail.
   See: **lie, low**

cregyo- < Gr. *kregyos*, good, useful; see **good**

cremasto- < Gr. *kremastos*, hung, hanging; *kremaster*, suspender; see **hang**

cremathro- < Gr. *kremathra*, basket for hanging things up in; see **basket**

crematus, L. burned; *cremabilis*, combustible; see *cremo* under **burn**

cremno- < Gr. *kremnos*, precipice, crag, overhanging wall or bank; see **cliff**

cremor, L. thick juice, broth, gravy; see **sauce**

crena, L. notch, rounded projection; *crenatus*, notched, toothed; *crenulatus*, minutely crenate; see **notch**

creno- < Gr. *krene*, spring, see **spring**; < *kraino*, accomplish, fulfill, execute, see **end**

creo- < Gr. *kreas, kreos*, flesh, see **flesh**; *kreopoles*, butcher, see **butcher**

creper, L. dark, obscure; *crepusculum*, twilight, dusk; see **shade, night**

crepido- < L. *crepida* (Gr. *krepis, -idos*), boot, sandal, base; *crepidula*, dim.; see **shoe**

crepis, L. a genus of the composite family; see **composite**

crepitans; crepulus, L. rattling; *crepitaculum*, rattle; see *crepo* under **rattle**

crescent < OF. *creissant*, < L. *cresco*, grow; figure of the moon in its first or last quarter; crescentic, Crescent City (La.).
   L. *falcatus*, sickle-shaped, curved; see **sickle**
   L. *lunatus*, shaped like a crescent moon; *lunula*, f. crescent, dim. of *luna*, f. moon; *luno, -atus*, bend like a half-moon, crescent, or sickle: lunate, *Megarhyssa lunator* (thalessa, an ichneumonid).
   Gr. *meniskos*, m. crescent, dim. of *mene*, f. moon: meniscus, menisciform, meniscoid, meniscitis, *Meniscotherium robustum* (an Eocene condylarth), *Meniscium*

reticulatum (a fern), *Meniscoëssus robustus* (a Cretaceous mammal), *Chilomeniscus stramineus* (straw-snake).
Gr. *selenis, -idos,* crescent; see *selene* under **moon**
See: **arc, moon**

cress < AS. *cresse; cerse;* see **mustard**

crest < OF. *creste,* < L. *crista,* f. tuft, comb, or plume on the head of animals; *cristula,* f. dim.; *cristatus,* tufted, crested: cristate, Hillcrest, *Cristellaria vetusta* (a foraminifer), *Sphenostoma cristatum* (wedgebill), *Orpacophora cristulata* (a katydid), *Tutankhamen cristatipes* (a crab), *Corythosaurus bicristatus* (a dinosaur), *Cladonia cristatella* (a lichen), *Onobrychis cristagalli* (cockscomb-sainfoin).
Gr. *akrokomos,* hair on the crown, hair at the tip, leaves at the top or end: *Acrocomia sclerocarpa* (mucaja palm).
L. *caliendrum,* n. a high headdress made of false hair:
L. *coma* (Gr. *kome*), hair on the head, mane; see **hair**
L. *juba,* mane, crest; *jubatus,* maned, crested; see **hair**
Gr. *kallaion,* n. cockscomb, wattles: *Callaionautilus turgidus* (a Triassic cephalopod).
Gr. *korystos,* crested, helmeted; see *korys* under **cap**
Gr. *lophos,* m.; *lophia,* f. mane, crest, comb, tuft, ridge; *lophidion; lophion,* n. dim.; *lophotos,* crested; *akrolophos,* m. mountain crest, ridge; *alectrolophos,* m. cockscomb: lophophore, lophodont, *Lophophora williamsi* (a cactus, source of peyotl or peyote and mescaline), *Lophophytum mirabile* (fel de terra), *Lophodytes cucullatus* (hooded merganser), *Lophiola americana* (an amaryllidacead), *Lophiornis obliquus* (a bird), *Conolophus subcristatus* (Galapagos iguana), *Eulophidium maculatum* (an orchid), *Henricosbornia lophiodonta* (a fossil mammal), *Ocyphaps lophotes* (a pigeon), *Baeolophus atricristatus* (a titmouse), *Lophotocarpus spongiosus* (a waterweed).
L. *tutulus,* m. hair dressed in a high cone over the forehead; *tutulatus,* having a *tutulus: Hypocolobus tutulus* (a beetle).
See: **cap, comb, feather, ridge, top, cover, crown**

creta, L. chalk; *cretula,* dim.; *cretaceus,* chalky, limy; see **lime**

cretus, L. separated, sifted, distinguished, < *cerno,* separate, distinguish, see **separate**; arisen, sprung, grown, < *cresco,* come forth, arise, be born, grow, see **grow**

crevice < OF. *crevace,* < L. *crepo,* break; see **break**

crex, -ecis, L. (Gr. *krex, -ekos*), a bird with long legs; see **crane**

cribano- < Gr. *kribanos (klibanos),* pot or pan wider at bottom than at top; see **pot**

cribrum, L. sieve; *cribellum,* dim., < *cribro, -atus,* sift; see **sieve**

cricetus, NL. hamster, a ratlike European rodent; see **mouse**

cricket < D. *crikel;* see **grasshopper**

crico- < Gr. *krikos,* ring; *krikoma,* ring, circle; *krikotos,* ringed, made of rings; see **circle**

crime < L. *crimen, -inis,* accusation, charge, fault, offense; see **fault, hurt**

crimno- < Gr. *krimnon,* coarse meal; see **flour**

crimson < Sp. *cremesin,* < Ar. *quermez,* kermes; see **red**

crinanthemum, L. (Gr. *krinanthemon*), houseleek; see **stonecrop**

crinitus, L. hairy, with long hair, < *crinis,* hair; see **hair**

crinum, L. (Gr. *krinon*), lily; *crininus,* of lilies; see **lily**

crio- < Gr. *krios,* ram; see **sheep**

crionto- < Gr. *kreion, -tos,* ruler, master; see **govern**

cripple < AS. *crypel;* see **hurt**

crisis, L. (Gr. *krisis*), issue, decision, turning-point; *krisimos,* decisive; see **turn**

crispus, L. curly; *crispulus,* dim.; *crispicans,* curling; see **curl**

crissum, NL. rump of a bird, < *crisso, -atus,* move the haunches; see **rump, shake**

crista, L. tuft, comb, plume, ridge; *cristula,* dim.; see **crest**

crithmo- < Gr. *krithmon,* samphire; see **carrot**

critho- < Gr. *krithe,* barley; see **grain**

criticus, L. (Gr. *kritikos*), capable of judging; *kriterion,* means of judging, standard; < *krino,* distinguish, separate, judge; see **judge**

croak < AS. *cracettan,* call like a frog.
L. *coaxo, -atus,* croak:
Gr. *ololygon, -os,* f. croak of a frog: *Ololygon strigillata* (a frog)

crobylus, L. (Gr. *krobylos*), roll or knot of hair on top of the head; see **hair**

crocalo- < Gr. *krokale*, beach, seashore; see **shore**

crocatus; croceus; crocinus, L. saffronlike; see **crocus**

crocido- < Gr. *krokis, -idos*, nap or downy fibers on woolen cloth; *kroke*, thread, woof in a piece of cloth; see **wool**

crocitus, L. cawing of a crow; see **caw**

crocodile < L. *crocodilus* (Gr. *krokodeilos*), lizard; see **lizard**

crocus, L. (Gr. *krokos*), m. a plant of the iris family, saffron; *crocatus* (Gr. *krokotos*), saffron-yellow; *croceus; crocinus*, of saffron: croceous, crocein, crocin, crocetin, croconic, crocoite, *Crocus sativus* (saffron), *Crocosmia aurea* (an iris), *Oenanthe crocata* (water-dropwort), *Cladodactyla crocea* (a holothurian), *Chlorissa subcroceata* (a moth), *Antholyza crocosmioides* (an iridacead), *Colchicum crociflorum* (an autumn-crocus), *Pericrocotus montanus* (a bird).
  L. *colchicum* (Gr. *kolchikon*), n. a kind of crocus: *Colchicum autumnale* (autumn-crocus), colchicine, N-methyl-colchicamid.
  L. *orsinus*, m. a kind of crocus:

crocuta, L. hyena; see **hyena**

crocydo- < Gr. *krokys, -ydos*, nap on woolen cloth; see *krokis* under **wool**

Croesus, L. (Gr. *Kroisos*), rich Lydian king; see **wealth**

croft, AS. small field, plot of ground; see **field**

crogmo- < Gr. *krogmos*, cawing of a crow, < *krozo*, caw; see *crocio* under **caw**

cromaco- < Gr. *kromax, -akos*, heap of stones; see **heap**

cromyo- < Gr. *kromyon* (*krommyon*), onion; *krommydion*, dim.; see **onion**

Cronus, L. (Gr. *Kronos*), former ruler of heaven and earth, equivalent to Saturn; see **spirit**

crop < AS. crop, top, bunch, see **reap, fruit, grain**; craw, see **stomach**

cropio- < Gr. *kropion*, scythe; see **sickle**

cross < L. *crux, -ucis*, f.; *crucio, -atus*, put to death on a cross, put to the rack, torture; *cruciabilis; cruciabundus*, tormenting, torturing, painful; *cruciarius*, of the cross; *cruciator, -is*, m. torturer, tormentor; *cruciatus, -us*, m. torment, torture: crux, crucial, cruciform, crucifixion, Cruciferae, cruciferous, Rosicrucian, crusade, excruciating, Santa Cruz, across, crisscross, crossword, cruise, *Crucianella stylosa* (crosswort), *Galium cruciatum* (a bedstraw), *Zabrotes cruciger* (a beetle), *Hyla crucifer* (spring peeper), *Cucumaria crucifera* (a holothurian), *Anopheles crucians* (a mosquito).   Derive: crucible.
  Gr. *chiasmos; chiastos*, arranged diagonally, crosswise, < *chiazo*, mark crosswise like the Greek letter *chi;* see *chi* under **letter**
  L. *compitum*, crossroads; see **way**
  L. *decusso, -atus*, form crosswise like the letter X, the Roman numeral ten: decussate, *Actinopteria decussata* (a fossil pelecypod).
  Gr. *enallax*, crosswise, alternate:
  Gr. *gammation*, swastika, dim. of *gamma*, third letter of the Greek alphabet; see *gamma* under **letter**
  Gr. *karsios*, crosswise: *Carsioptychus coarctatus* (a fossil mammal).
  Gr. *lechrios*, slanting, crosswise; see **slope**
  Gr. *stauros*, cross; *stauridion; staurion;* n. dim.; *staurikos*, of a cross; *staurotos*, cruciform: staurolite, staurolatry, stauroscope, *Staurognathus cruciformis* (a conodont), *Stauridium productum* (a hydrozoan), *Staurastrum polymorphum* (a desmid).
  L. *transversus*, crosswise: transverse.
  See: **alternate, slope**

crosso- < Gr. *krossos*, fringe, tassel; *krossion*, dim.; *krossotos*, fringed, tasseled, see **fringe**; sail, pitcher, jar, urn, see **pot**

crotalaria, NL. a genus of the bean family; see **bean**

crotalum, L. (Gr. *krotalon*), rattle; *krotos*, rattling noise; see **rattle**

crotaphido- < Gr. *krotaphis, -idos*, pointed hammer; see **hammer**

crotapho- < Gr. *krotaphos*, side of the forehead, temple; see **head**

crotch < uncertain origin; see **fork, angle, hook**

croteo- < Gr. *kroteo*, beat; *krotema*, a piece of hammered work; see **strike**

croton, L. (Gr. *kroton*), tick, see **tick**; croton-oil plant, see **spurge**

crotono- < Gr. *krotone*, a gall on trees; see **gall**

crow < AS. *crawe*.
  Gr. *bomolochos*, a jackdaw; see **flatter**

L. *citta* (Gr. *kitta; kissa*), f. a chattering, greedy bird, probably a jay: *Cyanocitta cristata* (bluejay), *Urocissa erythrorhyncha* (a magpie).

NL. *coloeus* (Gr. *koloios*), m. jackdaw, grackle: *Coloeus major* (a jackdaw).

L. *corax, -acis* (Gr. *korax, -akos*), m. raven; *korakion*, n.; *korakiskos*, m. dim.; *korakias*, m. daw; *korakinos*, ravenlike; *koraxos*, raven-black; *pyrrhokorax*, m. a kind of crow with a red beak: coracine, coracoid, *Corvus corax* (raven), *Coracias garrulus* (roller), *Phalacrocorax bougainvillei* (a cormorant, guanay), *Pyrrhocorax alpinus* (a bird).

L. *cornix, -icis*, f. crow; *cornicula*, f. dim.: *Corvus cornix* (hooded crow).

L. *corvus*, m. crow, raven; *corviculus*, m. dim.; *corvinus*, of the crow: corvine, corvoid, corviform, cormorant, Corwin, corvina (*Eriscion reticulatus*), *Corvus brachyrhynchus* (crow).

L. *cucus*, m. jackdaw:

L. *garrulus*, babbling; a jay: *Garrulus glandarius* (European jay).

L. *graculus*, m. jackdaw: grackle, *Pyrrhocorax graculus* (Alpine chough), *Strepera graculina* (a crow-shrike).

Gr. *korone*, f. a kind of crow; *koronideus*, m. young crow: *Coronopus didymus* (wartcress), *Corvus corone* (carrion-crow).

L. *monedula*, f. jackdaw: *Corvus monedula* (jackdaw), *Monedula punctata* (a wasp).

L. *pica*, f. magpie: pie, pied, piebald, magpie (*Pica caudata*), *Pica hudsonia* (American magpie), *Chalinolobus picatus* (a bat).

crowberry

NL. *ceratiola*, f. a genus of the crowberry family: *Ceratiola ericoides* (a sand-heath).

NL. *corema* (Gr. *korema*), broom, a genus of the crowberry family; see **broom**

L. *empetrum* (Gr. *empetron*), n. a rock-plant; a genus of the crowberry family: Empetraceae, empetraceous, *Empetrum nigrum* (crowberry), *Hypericum empetrifolium* (a St. Johnswort), *Phyllodoce empetriformis* (red mountain-heath).

crowd < AS. *crudan*, push, press, shove; see **abundance, heap, herd, number, people, press, thick, wealth**

crowfoot. Name applied to many unrelated plants having pedatiform leaves simulating crows' feet.

L. *coronopus* (Gr. *koronopous*), m. crowfoot: *Coronopus procumbens* (creeping wartcress), *Plantago coronopus* (crowfoot-plantain), *Gilia coronopifolia* (a gilia).

crown < L. *corona* (Gr. *korone*), f. crown, wreath, halo, rim, border; *corolla; coronula*, f. dim.; *coronalis; coronarius*, of a crown, wreath, or garland; *corono, -atus*, crown, wreathe: uncrowned, coronation, corolla, corollary, coronary, coronet, coroner, coronium, *Coronilla glauca* (honey-coronilla), *Coronocrinus polydactylus* (a Devonian crinoid), *Coronidium cervicorne* (a radiolarian), *Malus coronaria* (wild crabapple), *Chonetes coronatus* (a fossil brachiopod), *Melongena corona* (crown-shell), *Ophryoscolex bicoronatus* (a protozoan).

L. *cidaris* (Gr. *kidaris*), f. Persian diadem, tiara: *Cidaris coronata* (a fossil sea-urchin), *Goniocidaris canaliculata* (a sea-urchin), *Tretocidaris spinosa* (a sea-urchin).

Gr. *diadema, -tos*, n. royal circlet, headband, crown: diadem, *Ceratotrochus diadema* (a coral), *Cristellaria diademata* (a foraminifer), *Diadophis punctatus* (a ring-neck snake).

Gr. *eiresione*, f. garland or wreath of laurel wound round with wool: *Iresine celosia* (bloodleaf).

Gr. *mesokranon*, n. crown of the head; see *kranion* under **head**

L., Gr. *mitra*, headband, snood, turban; see **cap**

Gr. *ploke; plokos*, wreath, chaplet; see *plecto* under **weave**

L. *sertula*, dim. of *serta* (*sertum*), garland, wreath; see **branch**

Gr. *stemma, -tos*, n. garland, wreath, crown; *stemmation*, n. dim.: *Stemmatocrinus cernuus* (a fossil crinoid), *Sarcostemma acidum* (soma).

Gr. *stephanos*, m. crown, wreath, diadem; *stephanion*, n.; *stephaniskos*, m. dim.; *steptos*, crowned: stephanome, Stephen, *Stephanodiscus niagarae* (a diatom), *Stephanotis floribunda* (waxflower), *Stephanoceras coronatum* (a Jurassic ammonite), *Stephanocrinus angulatus* (a Silurian crinoid), *Circostephanus coronarius* (a radiolarian), *Vermetus* (*Steptopoma*) *roseus* (a gastropod).

See: **cap, cover, crest, top, head**

cruciatus, L. torment, torture; *cruciabilis*, painful, tormenting; *crucio, -atus*, put to death on a cross, put to the rack, torture, < *crux, -ucis*, cross; see **cross**

crucibulum, ML. earthen pot; see **pot**

crudus, L. bloody, raw, rough, rude; *cruditas*, rawness; see **rough**

cruel < OF. *cruel*, < L. *crudelis*, unmerciful, rude, severe; see **merciless**

cruentus, L. stained or spotted with blood, bloody; see *cruor* under **blood**

-crum, L. suffix signifying tool; see -*brum* under tool

crumato- < Gr. *kruma, -tos,* beat, note, sound; see sound

crumb < AS. *cruma,* bit, morsel, small fragment; see little

crumena, L. leather money-bag; *crumilla,* dim.; see bag

cruno- < Gr. *krounos,* spring; see spring

cruor, -*is,* L. blood; see blood

crupper < OF. *croupe,* buttocks.
L. *postilena,* f. crupper:

cruralis, L. of the leg, < *crus, cruris,* leg, shank; *crusculum,* dim.; see leg

crush < OF. *cruisir;* see break, bruise

crusta, L. hard outer surface of a body, rind, shell; *crustula,* dim., see skin; *crustum,* pastry, bread, cake; *crustulum,* dim., see bread

crustico- < Gr. *kroustikos,* striking; *krousis,* a striking; see *krouma* under strike

crux, -*ucis,* L. cross; see cross

cry < F. *crier,* < L. *queror,* complain, lament; see call, weep

crybelo- < Gr. *krybelos,* hidden; see *krypto* under cover

cryo- < Gr. *kryos; krymos,* icy cold, chill, frost; *krymaleos; kryeros,* icy, frozen, cold; see cold

cryphio- < *kryphios,* hidden, secret; *kryphos,* concealment, obscurity; see *crypto* under cover

crypto- < Gr. *krypto,* hide, conceal; *kryptos,* hidden, secret; see cover

cryptomeria, NL. a genus of gymnosperms; see gymnosperm

crystal < L. *crystallum* (Gr. *krystallos*), ice, rock-crystal; *crystallinus* (Gr. *krystallinos*), of crystal; see stone

ctamero- < Gr. *ktameros,* killed; see *kteino* under kill

ctedono- < Gr. *ktedon, -os,* fiber; see thread

ctemato- < Gr. *ktema, -tos,* property, possession; see wealth

cteno- < Gr. *kteis, ktenos,* comb; *ktenidion; ktenion,* dim., see comb; < *ktenos,* livestock; *ktenikos,* of cattle; *kteniatros,* veterinary, see cattle

cteristo- < Gr. *kteristes,* undertaker; < *kterizo,* bury; see undertaker

ctesi- < Gr. *ktesios,* of property; *ktesis,* property; see *ktema* under wealth

ctetico- < Gr. *ktetikos,* acquisitive; *ktetos,* that may be had; see *ktema* under wealth

ctilo- < Gr. *ktilos,* tame, obedient, see tame; ram, see sheep

ctisto- < Gr. *ktistos,* founded, established; *ktisis,* a founding, creation; see *ktizo* under make

ctono- < Gr. *ktonos,* murder; see *kteino* under kill

ctypo- < Gr. *ktypos,* any loud noise; see sound

cubans, L. reclining; see *cubo* under lie

cube < L. *cubus* (Gr. *kybos*), a solid with six equal square sides, die; see three

cubeb < ML. *cubeba,* < Ar. *kababah: Piper cubeba* (cubeb-pepper).

cubiculum, L. sleeping-room, bed chamber; see room

cubile, L. couch, bed, marriage-bed, den, lair, nest, kennel, hole; see bed

cubitum, L. elbow; see arm

cubitus, L. reclined, laid, < *cubo* (*cumbo*), -*itus,* recline, lie down, sleep; see lie

cuckoo < ME. *cuccu;* F. *coucou;* L. *cuculus,* m.: cuckoo clock, cuckold, *Cuculus canorus* (common cuckoo), *Trigla cuculus* (a gurnard), *Lychnis floscuculi* (ragged robin).
Gr. *kokkyx, -ygos,* m. cuckoo; *kokkyzo,* cry cuckoo: *Coccygomimus nudus* (a wasp), *Chrysococcyx cupreus* (emerald cuckoo), *Idiococcyx chlorocephalus* (a bird), *Coccyzus erythrophthalmus* (black-billed cuckoo).

cucubalus, L. a kind of plant, probably a nightshade; see nightshade

cucullus, L. cap, hood; *cucullatus,* hooded; see cap

cuculus, L. cuckoo; see cuckoo

cucuma, L. kettle; see kettle

cucumis, -*eris,* L. cucumber; *cucumeraceus,* of cucumbers; see melon

cucupho- < Gr. *koukouphas,* hoopoe; see hoopoe

cucurbita, L. gourd; *cucurbitula,* dim.; see melon

cucus, L. jackdaw; see crow

cudo, L. helmet made of raw skin; see **cap**

culcita, L. pillow, cushion; *culcitella; culcitula,* dim.; *culcitarius,* pillow-maker; see **pillow**

culeus, L. bag, scrotum, sheath; see **bag**

culex, *-icis,* L. mosquito; *culicellus; culiculus,* dim.; see **fly**

culina, L. kitchen; *culinarius,* pertaining to the kitchen and cooking; see **kitchen**

culmen, *-inis,* L. top, summit, ridge; see **top**

culmus, L. stalk, stem, straw; *culmulus,* dim.; see **stem**

culpa, L. fault; *culpabilis; culpatus,* blamable; see **fault**

culter, *-tri,* L. knife, plowshare; *cultellus,* dim.; see **knife**

cultus, L. cultivated, tilled, civilized; see *colo* under **till**

culullus, L. cup, goblet; see **cup**

-culum, L. suffix denoting diminutives; see *-cle* under **little**

culus, L. fundament, rump; see **rump**

cum, L. together, with; see **with**

cumarin < Ab.Am. *cumaru,* tonka-bean tree *(Dipteryx odorata):* cumarone, cumaric, dicumarol, cumarilic acid.

cumera, L. box, chest; see **box**

cuminum, L. (Gr. *kyminon*), a plant whose seeds are used for seasoning; see **carrot**

cumulus, L. heap; *cumulo, -atus,* heap up; see **heap**

cuna, L. cradle; *cunabulum,* earliest abode, cradle, birth, origin; *cunula,* dim.; see **cradle**

cunctatus, L. delayed, < *cunctor, -atus,* delay, stay, tarry; see **delay**

cunctus, L. all, the whole; see **all**

-cundus, L. denoting continuance or augmentation; see **very**

cuneatus, L. wedge-shaped, < L. *cuneus,* wedge; *cuneolus,* dim.; see **wedge**

cuniculus, L. (Gr. *kyniklos*), rabbit; tunnel, burrow; *cunicularius,* miner; *cuniculum,* rabbit burrow, hole, underground passage; see **hare, mine**

cunila, L. (Gr. *konile*), a mint; see **mint**

cunning < AS. *cunnan,* know, be able; see **know**

cunnus, L. vulva; see **vulva**

cup < AS. *cuppe,* < L. *cupa,* f. cask, tub, vat; *cupella; cupula,* f. dim.; *cuparius,* m. cooper: cupboard, cupping, cupule, cupola, coop, Cupuliferae, Cooper, Cuvier, *Mammillopora cupula* (a fossil bryozoan), *Tholospyris cupola* (a radiolarian), *Ocotea cupularis* (a lauracead), *Potamogeton bicupulatus* (a pondweed), *Cupulita cara* (a siphonophore).

    L. *acetabulum,* n. vinegar cup, any cup-shaped vessel, socket of the hipbone: acetabulum, *Acetabularia mediterranea* (a seaweed), *Dosinia acetabulum* (a Miocene pelecypod).

    Gr. *aleison,* n. cup, goblet, socket: *Alisocrinus laevis* (a Silurian crinoid).

    Gr. *ambix, -ikos (ambyx, -ykos),* m. cup, beaker: alembic, *Ambicocrinus arborescens* (a Devonian crinoid).

    Gr. *amphidysis,* f. double cup:

    Gr. *amystis, -ios* or *-idos,* f. large draught, large cup:

    L. *anancaeum,* n. a large cup that must be drained on a wager:

    Gr. *apothysanion,* n. a drinking-vessel:

    Gr. *arake,* f. cup, bowl:

    Gr. *argyris, -idos,* f. silver cup or vessel:

    L. *atanuvium (athanuvium),* n. earthen bowl used in offering sacrifices:

    L. *bacar,* n. wine-glass:

    L. *batiaca; batiola,* f. cup, goblet:

    Gr. *bessa,* f. a drinking-cup:

    Gr. *bikos,* m. cup, bowl; *bikidion; bikion,* n. dim.: *Bicidium parasiticum* (a medusa), *Bicidiopsis arctica* (a coral).

    L. *calathus* (Gr. *kalathos*), wicker-basket, vase, bowl, cup, calyx; see **basket**

    L. *calix, -icis,* m. cup, goblet; *calicellus; caliculus,* m. dim.: chalice, Calixtine, calicle, caliciform.

    L. *calvariola,* small cup, dim. of *calvaria,* skull; see **head**

    L. *calyx, -ycis* (Gr. *kalyx, -ykos*), m. cup or covering of a flower; *calyculus,* m.; *kalykion,* n. dim.: calyx, calycle, calyculus, calyculate, calycoid, calycine, *Calycophyllum candidissimum* (a rubiacead), *Calycanthus floridus* (sweet-shrub), *Calycularia radiculosa* (a liverwort), *Gentiana calycosa* (a gentian), *Hypericum calycinum* (a St. Johnswort), *Calyptrocalyx spicatus* (a palm), *Gymnocalycium loricatum* (a cactus).

L. *camella*, f. cup, goblet:

L. *capis, -idis*, f. bowl with one handle; *capula*, f. dim.; *capedo, -inis*, f. cup or bowl used in sacrifices; *capeduncula*, f. dim.:

L. *carchesium* (Gr. *karchesion*), n. cup contracted in the middle and with handles that reached from rim to bottom: *Carchesium polypinum* (a protozoan).

L. *catillus; catinulus*, cup, bowl; see *catinus* under **pot**

L. *cauca* (Gr. *kauke*), f. cup; *caucula*, f.; *kaukion*, n. dim.:

Gr. *chonnos*, m. a copper cup:

L. *ciborium* (Gr. *kiborion*), n. cup shaped like the pericarp of a waterlily: ciborium.

L. *cissybium* (Gr. *kissybion*), n. a wooden rustic drinking-cup:

L. *cotyla* (Gr. *kotyle*), f. cup, cavity; *kotyliskos*, m. dim.; *kotyledon, -os*, f. any cup-shaped cavity or hollow, socket: cotyligerous, cotyliscus, cotyledon, dicotyledon, monocotyledonous, hypocotyl, *Cotyledon decussata* (a crassulacead), *Cotylogonimus heterophyes* (a fluke), *Hydrocotyle umbellata* (a marsh-pennywort), *Hexacotyle dissimilis* (a worm), *Diploconus cotyliscus* (a radiolarian), *Lewisia cotyledon* (a bitterroot), *Anthemis cotula* (a mayweed).

L. *crater, -is* (Gr. *krater, -os*), m. mixing vessel, bowl, basin; *kraterion*, n. dim.: crater, crateriform, grail, *Craterostigma pumilum* (a figwort), *Craterellus cornucopioides* (a fungus), *Hymenocrater calycinus* (a mint), *Urnula craterium* (a fungus).

L. *culullus*, m. beaker, cup, bowl:

L. *cyathus* (Gr. *kyathos*), m. cup, ladle; *kyathion*, n. dim.: cyathus, cyathiform, casserole, cyathium, *Cyathus vernicosus* (a birdsnest-fungus), *Cyathea microphylla* (a tree-fern), *Cyathophyllum hexagonum* (a coral), *Cyathiscus actinia* (a sponge), *Cyathodium foetidissimum* (a liverwort), *Cyathula capitata* (an amaranth), *Diploconus cyathiscus* (a radiolarian), *Cyathocrinus scitulus* (a Mississippian crinoid), *Calvatia cyathiformis* (a puffball).

L. *cymba* (Gr. *kymbe*), f. cup, bowl, boat; *cymbium(kymbion)*, n. dim.; *kymbos*, hollow: *Cymbiocrinus tumidus* (a fossil crinoid), *Cymbium melo* (a gastropod), *Cymbidium giganteum* (an orchid), *Pterocymbium tinctorium* (a sterculiacead), cymbal, chime, *Cymbopetalum penduliflorum* (earflower).

L. *cypellum* (Gr. *kypellon*), n. beaker, goblet, cup: *Cypella plumbea* (an iridacead).

L. *cytinus* (Gr. *kytinos*), m. calyx of the pomegranate blossom: *Cytinus sanguineus* (a rafflesiacead), *Dicypellium caryophyllatum* (a lauracead).

Gr. *depas, -aos; depastron*, n. beaker, goblet: *Depastrum cyathiforme* (a jellyfish).

Gr. *dinos*, m. a large, round goblet: *Dinocrinus cornutus* (a Permian crinoid).

Gr. *ekpoma, -tos*, n. beaker, cup:

Gr. *embaphion*, n. small vessel for saucers, cup: *Embaphium muricatum* (a beetle).

L. *futile, -is*, n. vessel broad above and pointed below:

L. *galeola*, a vessel shaped like a helmet; see *galea* under **cap**

Gr. *gyale*, f. a kind of cup: *Gyalecta cupularis* (a lichen), *Gyalostoma jucundum* (a beetle).

Gr. *hemitomos*, m. a kind of cup: *Hemitoma tricostata* (a gastropod).

Gr. *kelebe*, f. cup, jar, pan; *kelebeion*, n. dim.: *Celebomastax curiosa* (a grasshopper).

Gr. *kichetos*, n. an incense vessel:

Gr. *kothon, -os*, m. cup; *kothonion*, n. dim.: *Cothocrinus[Cothonocrinus] verrucosus* (a fossil crinoid).

Gr. *kylix, -ikos*, f. cup; *kylikion; kyliskion*, n. dim.; *kylichne*, f. small cup; *kylichnion*, n. dim.: cylix, *Cylicocrinus spinosus* (a Silurian crinoid), *Cylichnium domitum* (a gastropod), *Cylichnina laevisculpta* (a gastropod), *Cylichnostomum coronatum* (a nematode), *Hyalocylix striata* (a pteropod).

Gr. *labronios*, m. a large, wide cup:

Gr. *lepaste*, f. a limpet-shaped cup:

Gr. *louterion; louteridion*, a kind of cup, dim. of *louter*, bathtub; see *loutron* under **wash**

Gr. *manes*, m. a kind of cup, brazen figure, slave:

Gr. *mathalis, -idos*, f. a kind of cup:

L. *myobarbum*, n. a kind of pointed drinking-vessel:

Gr. *nestoris, -idos*, f. a kind of cup:

Gr. *ollix, -ikos*, f. a wooden drinking-bowl:

Gr. *onos*, beaker, wine-cup; see **horse**

L. *panaca*, f. a drinking-vessel:

Gr. *pella; pellis, -idos*, f. cup, bowl, milk-pail; *pelichne*, f. cup; *pelika (pelike)*, f. cup: pelike, *Pella limbata* (a beetle), *Pelichnobothrium speciosum* (a beetle), *Pelicopsis dubius* (a fossil mammal), *Diadectes sideropelicus* (a fossile reptile).

Gr. *pelyx, -ykos*, wooden bowl, basin, pelvis; see **basin**

Gr. *petachnon*, n. broad, flat cup: *Petachnum triaropsis* (a jellyfish).

L. *phiala* (Gr. *phiale*), f. broad, flat vessel, saucer, bowl; *phialidion; phialon*, n.

dim.: phial, vial, *Phialidium hemisphaericum* (a hydrozoan), *Pachyphiale fagicola* (a lichen).

Gr. *phimos,* m. cup used as a dice-box: *Phimocrinus americanus* (a Devonian crinoid).

L. *poculum,* n. cup, goblet, beaker; *pocillum,* n. dim.: poculiform, pocilliform, poculary, poculation, poculent, *Cecidomyia poculum* (an oak gall-insect), *Hebella pocillum* (a hydrozoan), *Cylichnostomum poculatum* (a nematode), *Begonia poculifera* (a begonia), *Edriocrinus pocilliformis* (a Devonian crinoid).

L. *poterium* (Gr. *poterion*); *potorium,* n. a drinking-cup, dim. of Gr. *poter, -os,* m. wine-cup: *Poteriocrinites nuciformis* (a fossil crinoid).

Gr. *ptomatis, -idos,* f. cup that must be emptied at once because it will not stand upright, tumbler: *Ptomatis forbesi* (a fossil gastropod).

Gr. *sabrias,* m. a kind of cup:

L. *scaphium* (Gr. *skaphion*), n. cup shaped like a boat; *skaphis, -idos,* f. bowl; *skaphidion,* n. dim.: *Scaphiocrinus inordinatus* (a fossil crinoid), *Scaphiophryne marmorata* (a frog), *Scaphidium deletum* (a fossil beetle).

L. *scyphus* (Gr. *skyphos*), m. cup; *scyphulus; skyphion; skyphidion,* n. dim.: scyphiform, scyphistoma, Scyphozoa, scyphulus, scyphose, *Scyphocrinus subornatus* (a Silurian crinoid), *Scyphium rubiginosum* (a myxomycete), *Scyphidium septentrionale* (a sponge), *Dasyscypha willkommi* (a fungus).

L. *simpuvium,* n. a kind of bowl:

L. *sinum,* n. a large cup or bowl: *Sinum perspectivum* (a gastropod).

L. *testum,* crucible for testing ores and metals; see **try**

L. *thuribulum,* n. censer:

L. *tryblium* (Gr. *tryblion*), n. cup, bowl: *Trybliocrinus flatheanus* (a Devonian crinoid).

See: **dipper, basin, funnel, pot, hollow, spoon**

cuparius, L. cooper; see *cupa* under **cup**

cupboard; see **room, shelf**

cuphea, NL. a genus of the loosestrife family; see **loosestrife**

cupho- < Gr. *kouphos,* light, airy, easy; see **light**

cupidus, L. desiring, longing, eager; *cupiditas,* avarice, lust; *Cupido,* god of love < *cupio, -itus,* desire; see **wish, love**

cuppedo, L. titbit, delicacy; *cuppes, -edis,* fond of dainties; see **food**

cupressus, L. (Gr. *kyparissos*), cypress; see **cypress**

cuprum, L. copper; *cupreus; cuprinus,* of copper; see **copper**

cupula, L. dim. of *cupa,* tub, vat; see **cup**

cura, L. care, attention; *curator,* caretaker, custodian, < *curo, -atus,* care for, guard, manage; see **guard, heal, drug**

curb < F. *courbe,* < L. *curvus,* bent; see **hold, bind**

curculio, L. weevil; see **beetle**

curcuma, NL. < Ar. *kurkum,* a genus of the ginger family; see **ginger**

curdle < ME. *curd;* see **thick**

cure < L. *cura,* care; see **heal**

curidio- < Gr. *kouridios,* wedded, nuptial; see **marry**

curido- < Gr. *kouris, -idos,* razor; see **knife**

curimo- < Gr. *kurimos,* shorn off; *koureus,* barber; see *keiro* under **cut**

curiosus, L. inquisitive, odd, strange; see **wish**

curl < ME. *crul,* < MHG. *krol:* curly, curling.

Gr. *bostrychos,* m. curl; *bostrychion,* n. dim.; *bostrychodes,* curly: bostryx, bostrychoid, *Bostrychus angustus* (a beetle), *Bostrychoplites armatus* (a beetle), *Bostrychopsis affinis* (a beetle).

L. *calamistratus,* crisped, curled, < *calamistrum,* n. curling iron: calamistrum.

L. *capreolus,* tendril; *capreolatus,* tendriled; see *caper* under **goat**

L. *cincinnus* (Gr. *kikinnos*), m. curled lock, ringlet; *cincinnatus,* with curled hair: *Cicinnocnemis cornuta* (a moth), *Cincinnurus regius* (a bird-of-paradise), Cincinnatus.

L. *cinerarius; ciniflo, -onis,* m. hair-curler:

L. *circinatus,* coiled, curled away from an apex, < *circino,* make round: circinate, *Acer circinatum* (vine-maple), *Cycas circinalis* (a cycad).

L. *cirrus,* m. curl, ringlet, tendril; *cirratus,* curly, fringed, tufted; *cirritus,* filamentous; *cirrosus,* full of curls: cirrate, cirriform, cirrus clouds, Cirripedia, *Parnassia cirrata* (a saxifragacead), *Andropogon cirratus* (Texas beardgrass), *Cirratulus inhamatus* (an annelid), *Hapalocrinus cirrifer* (a Silurian crinoid), *Mecistocirrus digitatus* (an annelid), *Umbrina cirrosa* (a fish), *Odontoglossum cirrhosum*

[*cirrosum*] (an orchid), *Dactylometra quinquecirra* (a jellyfish), *Smilax ecirrhata* [*ecirrata*] (a catbrier).
L. *clavicula*, f. tendril:
L. *crispus*, curly; *crispulus*, dim., < *crispo, -atus*, curl, crimp: crisp, crispate, crape, crepe de chine, Crispin, *Odontoglossum crispum* (an orchid), *Begonia crispipila* (a begonia).
Gr. *heligma, -tos*, n. curl: *Heligmostrongylus sedecimradiatus* (a nematode).
Gr. *helinos*, m. vine-tendril: *Helinus lanceolatus* (a rhamnacead).
L. *Medusa* (Gr. *Medousa*), a gorgon with snaky locks; see **fear**
Gr. *oulos*, woolly, curly; *oulotes*, woolliness, curliness; see **wool**
Gr. *ostlinx, -ingos*, curled hair, lock of hair, tendril; see **hair**
L. *pampinus*, m. tendril; *pampinatus*, with tendrils; *pampinarius, pampineus*, of tendrils; *pampinosus*, full of tendrils: pampiniform, pampinocele.
Gr. *phobe*, lock, curl; see **cluster**
Gr. *plokos*, curl of hair; *plektane*, anything coiled, twined or wreathed; see *plecto* under **weave**
Gr. *skorpioeides*, scorpionlike, bent like a scorpion's tail: scorpioid.
L. *ustricula*, f. hair-dresser, hair-curler:
L. *viticula*, tendril, dim. of *vitis*, grapevine; see **grape**
See: **bend, turn, circle, arc, hook, vine, hair**

curlew < OF. *corlieu;* see **snipe**

curo- < Gr. *kouros*, boy, youth; see *koros* under **child**

currant < Gr. *Korinthos*, Corinth, ancient Greek city.
NL. *grossularia*, f. gooseberry, < F. *groseille*, currant, gooseberry: grossularite, grossulin, *Grossularia hirtella* (a gooseberry), *Styelopsis grossularia* (a tunicate), *Abraxas grossulariata* (a moth).
NL. *ribes*, n. currant, gooseberry, < Ar. *ribas*, berry with an acid flavor: *Ribes aureum* (golden currant), *Cronartium ribicola* (white-pine blister-rust), *Pseudopeziza ribis* (a fungus).

currax, -acis, L. swift; *curriculum*, race, career; *cursor*, runner, < *curro, cursus*, run; see **run, swift**

currus, L. chariot; *currulis (currilis)*, of a chariot; see **vehicle**

curse < AS. *cursian*, use profane language, imprecate, execrate.
Gr. *agos*, n. curse or pollution that requires expiation or sacrifice:
Gr. *anathema, -tos*, n. curse: anathema.
Gr. *ara*, prayer, prayer for evil, curse; *aratos*, accursed; see **ask**
AS. *ath*, an affirmation by appeal to divinity or other revered persons or things, curse; see **swear**
Gr. *baskanos*, slanderous, malicious; see **magic**
L. *blasphemo* (Gr. *blasphemeo*), revile, reproach, defame: blasphemy, blasphemous.
L. *bovinor*, bellow at, revile; *bovinator, -is*, m. brawler, blusterer, reviler:
L. *detestor, -atus*, curse, execrate, hate: detest, detestation.
L. *execror(exsecror), -atus*, curse; *execrabilis*, detestable: execration.
L. *expletivus*, filling out, exclamatory: expletive.
L. *imprecor, -atus*, call down upon: imprecate, imprecation.
L. *improperium*, n. reproach, taunt:
L. *insector, -atus*, censure, blame, rail at; *insectator, -is*, m. censurer, persecutor: insectation, *Pezomachus insectator* (an ichneumonid).
L. *insulto, -atus*, leap upon, revile, abuse, taunt: insult.
L. *invectivus*, scolding, reproachful, abusive; see **scold**
Gr. *katalalia*, f. slander:
L. *loedoria* (Gr. *loidoria*), f. railing, reviling, abuse; *loidoros*, abusive, railing, < *loidoreo*, revile, abuse:
L. *maledicus*, abusive, foul-mouthed, scurrilous, slanderous, < *maledico, -ictus*, speak ill, slander, abuse, revile: malediction.
L. *objurgator*, chider, blamer, rebuker; see **scold**
L. *profanus*, unholy, irreverent, wicked; see **profane**
L. *vitupero, -atus*, censure, blame; see **scold**
See: **scold, blame, swear, shame**

curtain < OF. *curtine*, < L. *cortina*, f.: *Cortinarius cinnamomeus* (a mushroom).
L. *aulaeum* (Gr. *aulaia*, f.), n. curtain, tapestry, canopy:
L. *conopeum* (Gr. *konopeion*), n. curtain of fine gauze, mosquito net; see **net**
L. *flammeum*, n. bridal veil:
Gr. *kalyptra*, veil; see *kalypto* under **cover**
Gr. *katablema, -tos*, n. anything let down, deposit, curtain:
Gr. *katapetasma; parapetasma, -tos*, n. curtain, veil:
Gr. *kredemnon*, n. veil: *Credemnon sylvellum* (a moth).
L. *plagula*, f. bed-curtain, veil:

L. *rica*, f. veil; *ricula*, f. dim.; *ricinium*, n. a small veil; *ricinus*, veiled:
L. *siparium*, n. a theater curtain:
Gr. *skepastron*, n. veil:
L. *velum*, curtain, sail, veil; see **sail**
See: **cover**

curtus, L. short; see **short**

curvus, L. bent; *curvabilis*, that may be bent; see **bend**

cuscuta, NL. dodder, < Ar. *kushkut;* see **dodder**

cushion, prob. < L. *culcita*, mattress, bolster, pillow; see **pillow**

cusor, L. coiner or stamper of money; see *cudo* under **strike**

cusparia, NL. a genus of the rue family; see **rue**

cuspidatus, L. pointed, < L. *cuspis, -idis*, point, pointed end of anything: see **point**

cussiliris, L. cowardly; see **coward**

custodian < L. *custo, -odis*, keeper, guardian; *custodia*, prison, guardhouse; see **guard, pen**

custom < OF. *costume*, < L. *consuetudo*, habit, use; see **manner,** *sueo* under **use**

cut < uncertain origin: cutter, cutting, cutback, cutoff, cutthroat, cutworm, cutup, shortcut.
L. *abruptus*, broken off, separated, sharp, sheer, steep; see *rumpo* under **break**
L. *amputo, -atus*, cut away, lop off, curtail, shorten; *amputatio, -onis*, f. a pruning: amputate, amputation.
L. *annodo, -atus*, cut off suckers from vines, prune:
L. *arborator, -is*, m. pruner of trees:
L. *bisulcus*, cloven, forked, two-furrowed; see *sulcus* under **furrow**
L. *caedo, caesus*, cut, kill; *caeduus*, fit for cutting; *caesio, -onis*, f. a cutting, pruning; *caesor, -is*, m. cutter, hewer; *-cida*, suffix denoting cutter, killer, killing, as *homicida*, c. man-killer, murderer; *lapicida (lapidicida)*, m. stonecutter; *patricida*, c. father-killer; *ambecisus; ancisus*, cut round; *concisus*, cut away; *incilis*, cut in; *incisus*, cut into; *intercisus*, cut off: caesura, decide, decision, chisel, concise, excision, incision, circumcision, incisor, precise, scissors, herbicide, homicide, insecticide, larvicide, matricide, miticide, infanticide, pesticide, suicide, fratricidal, cement, *Stivalius ancisus* (a flea), *Adiantum excisum* (a maidenhair-fern), *Thoracophorus excisicollis* (a beetle), *Stephanandra incisa* (a rosacead), *Succisa pratensis* (blue scabious), *Coccotorus prunicida* (now *Anthonomus scutellaris*, a beetle).
L. *caelo, -atus*, engrave in relief; *caelator, -is*, m. carver, engraver; *caelamen, -inis*, n. embossed work, bas-relief: celature, *Lycoperdon caelatum* (a puffball).
AS. *ceorfan*, cut, carve; *cyrf*, the channel resulting from cutting or sawing: carving, kerf.
L. *cestrum* (Gr. *kestron*), n. a graving tool: *Cestrophorus paradoxus* (a grasshopper).
Gr. *charasso*, cut, engrave; *charaktos*, graven, cut in; *charagma, -tos*, graven mark; see *character* under **mark**
AS. *cleofan*, split: cleave, cleaver, cleavage, cleft, clove (of garlic), Kaaterskill Clove, cloven.
L. *curto, -atus*, shorten, diminish; see **short**
AS. *daelan*, divide: deal, ordeal, dole.
Gr. *daio; dateomai*, divide, apportion, share; *daithmos*, m. division, boundary; *daitros*, m. carver (especially of meat at table); *daitrosyne*, f. art of carving meat; *daitron*, n. portion; *datetes*, m. divider, distributor; *daterios*, dividing: datolite, geodetic, geodesy.
L. *decollo, -atus*, behead, decapitate:
L. *demeto, -messus*, mow, reap; see *meto* under **reap**
Gr. *dioros*, m. divider:
Gr. *dischides*, cloven, divided: *Dischidia rafflesiana* (an asclepiad).
L. *distribuo, -butus*, divide, apportion: distribute, distribution, distributor.
L. *divido, -visus*, separate, cleave, apportion, cut: divide, dividend, divisor, divisible, subdivision, undivided, devise, device, individual.
L. *divortium*, separation; see **separate**
L. *dolo, -atus*, hew, cut; *dedolo*, hew away, hew smooth; *dolatilis*, easily hewn; *dolamen, -inis*, n.; *dolatus*, m. a hewing; *indolatus*, unhewn: *Dolatocrinus speciosus* (a Devonian crinoid).
L. *feniseca; fenisex, -ecis*, m. mower:
L. *fimbriatus*, fringed, fibrous; see *fimbria* under **thread**
L. *findo, fissus*, cleave, split; *fissilis*, splittable; *fissura*, f. cleft, chink; *-fid*, suffix denoting division into parts, as, *bifidus*, two-cleft, forked; *multifidus*, many-cleft; *infissus*, cut into or through: fission, fissile, fissility, vent, fissure, Fissipedia, bifid,

multifid, pinnatifid, *Viola pedatifida* (prairie-violet), *Philodendron bipinnatifidum* (an aroid), *Pteris multifida* (spider-brake), *Potentilla fissa* (a cinquefoil), *Fissidens taxifolius* (a moss), *Pelargonium fissifolium* (a storksbill).

Gr. *glypho*, carve, engrave; *glyptes*, m. carver; *glypter, -os*, m. carving tool; *glyphikos*, of carving; *glyptos*, carved; *anaglyptos*, wrought in low relief; *glymma, -tos*, n. an engraved figure: anaglyph, triglyph, hieroglyphics, glyptography, *Glyptodon clavipes* (a fossil mammal), *Ampeloglypter sesostris* (grape gall-weevil), *Cicada hieroglyphica* (a cicada), *Glymma candezei* (a beetle), *Glymmatophora submetallica* (a bug), *Tritrichis glymmigera* (a beetle), *Glymmatacanthus rudis* (a fossil fish), *Ophioglypha bullata* (a brittlestar).

AS. *grafan*, dig, cut: grave, groove, graft, engrave.

Gr. *keio*, cleave, split: *Ceocephalus depressus* (a beetle).

Gr. *keiro*, clip, cut short, cut out, destroy; *kourimos*, shorn; see *koureus* under **barber**

Gr. *kolaptos*, engraved; see *kolapter* under **chisel**

Gr. *kopto*, cut, fell, slay; *koptikos*, of cutting; *koptos*, cut, chopped; *kope*, f. a stroke, cutting to pieces; *kopadion*; *kopaion*, n. piece; *komma, -tos*, part cut off; *kommation*, n. dim.; *kommos*, m. stroke; *apokope*, f. abscission, a cutting off; *diakope*, f.; *diakomma, -tos*, n. cut, gash; *ekkopto*, cut out; *perikomma, -tos*, n. trimmings; *perikope*, f. clipping, excerpt: apocope, apocopate, comma, syncope, syncopation, *Coptocycla bicolor* (tortoise-beetle), *Coptis laciniata* (a goldthread), *Xylocopa virginica* (a carpenter-bee), *Dendrocopus major* (a woodpecker), *Melicope ternata* (a rutacead), *Phyllocoptes quadripes* (a maple-gall), *Psalidocoptus scaber* (a beetle), *Eccoptometopus proximus* (a beetle).

Gr. *krossotos*, fringed, tasseled; see *krossos* under **fringe**

L. *lacero, -atus*, tear to pieces, mangle, cut; see **tear**

L. *laciniosus*, full of flaps, cut into thin strips, shredded; see *lacinia* under **fringe**

Gr. *latypos*, m. stone-cutter, mason; see **mason**

Gr. *laxeuo*, cut stone; see **mason**

AS. *mawan*, mow: math, aftermath, meadow, mower, mowing.

Gr. *meristos*, divided, divisible; *meristes*, m. divider; *merismos*, m. division, analysis, < *merizo*, divide, split, distribute; see *meros* under **part**

L. *mutilo, -atus*, main, cut short, lop off; see **hurt**

L. *partio, -itus*, divide, share, distribute; see **part**

Gr. *pelekesis*, f. hewing of wood; *peleketes*, m. hewer; *pelekema, -tos*, n. chips:

Gr. *psalis, -idos*, clipper, scissors; *psalidion*, dim.; *psalistos*, clipped; see **scissors**

Gr. *psolos*, one circumcised, penis; see **penis**

L. *recutitus*, circumcised:

Gr. *rhachistos*, cut up, divided:

L. *sarpo, sarptus*, trim, prune:

L. *scalpo, scalptus*, cut, carve; *scalptor, -is*, cutter, engraver; see *scalprum* under **knife**

AS. *sceran*, cut: shear, share, sharp, shard, potsherd, score, scar (rock cliff), skerry.

Gr. *schizo*, cleave, split; *schisis*, f. cleavage, parting, division; *schisma, -tos*, n. a split, cleft, division; *schistos*, split, divided; *aposchisis*, f. division, branching: schism, schismatic, schist, schistosity, schizoid, schizolite, schizocarp, schizont, schizophyte, schizogenous, Schizomycetes, *Schizophoria striatula* (a fossil brachiopod), *Schizopetalon walkeri* (a crucifer), *Schizostylis coccinea* (an iridacead), *Schizaea pusilla* (curly-grass), *Schistocerca damnifica* (a grasshopper), *Schismatothele lineata* (a spider), *Schismatoglottis pulchra* (an aracead), *Schizandra coccinea* (a magnoliacead), *Schizura concinna* (a moth).

L. *scindo, scissus*, cut, tear, separate, split; *scissilis*, easily split; *scissor, -is*, m. cleaver, splitter; *scissura*, f. rent, cleft; *abscissus*, cut off, separated; *conscissus*, torn to pieces, shredded; *discissus*, separated, cleft; *discidium*, n. separation: scission, scissure, abscissa, abscission, rescind, discide, shingle, circumscissile, *Scissurella costata* (a gastropod), *Scleria scindens* (razor-sedge), *Frondicularia inscissa* (a foraminifer). Derive: scissors.

L. *sculpo, sculptus*, carve, hew, grave, cut, chisel; *sculptor, -is*, m. carver, cutter, engraver; *sculptilis*, produced by carving or graving; *exsculptus*, cut out, chiseled out; *insculptus*, engraved: sculptile, sculptor, sculpture, *Centruroides sculpturatus* (a scorpion), *Steganocrinus sculptus* (a Mississippian crinoid), *Micromitra sculptilis* (a fossil brachiopod), *Onychocrinus exsculptus* (a Mississippian crinoid), *Beryx insculptus* (a fossil fish), *Eurylepis insculpta* (a fossil fish), *Laccoptera sculpturata* (a beetle).

L. *seco, sectus*, cut; *sector, -is*, m. cutter; *sectilis*, cut, cleft, divided; *sectivus*, cuttable; *conseco*, cut to pieces; *dissectus*, cut up; *resigminum*, n. clipping, paring: section, sect, sectarian, sector, secant, segment, sickle, sex, bisect, dissection, vivisection, *Botrychium dissectum* (a grape-fern), *Formica exsectoides* (an ant).

L. *talea*, a cutting, twig, rod; see **rod**

Gr. *temno*, cut, divide, sever; *tmesis; tome*, f. a cutting, separation; *tmema, -tos*, n. portion, piece, section; *tmemation*, n. dim.; *tmetikos; tomikos*, of cutting; *tmetos*, cut, shaped by cutting; *temachos*, n. cut, slice; *temachion*, n. dim.; *tomion*, n. sacrificial victim, usually cut to pieces; *tomos*, m. cut, part (of a book); *tomeus*, m. cutter, knife; *-tomy*, suffix denoting cutting, dissection; *-ectomy*, suffix denoting excision; *anatome*, f. a cutting up, dissection; *apotomos*, cut off, abrupt, precipitous; *entmema, -tos*, n.; *entome*, f. nick, notch, incision; *ektmema, -tos*, n. section, segment: anatomy, tome, microtome, entomion, atom, epitome, tmesis, cardiotomy, lithotomy, autotomy, appendectomy, colpectomy, embolectomy, tonsillectomy, dichotomous, fleam, *Temnocrinus tuberculatus* (a Silurian crinoid), *Tmesiphorus costalis* (a pselaphid), *Tmesipteris tannensis* (a psilotacead), *Tmesisternus trivittatus* (a beetle), *Tmetothrips subapterus* (a thysanopterid), *Tomicoproctus eichhoffi* (a beetle), *Tomopteris smithi* (an annelid), *Acrotomus lucidulus* (a wasp), *Acrotomodes hepaticata* (a moth), *Acrotmetus cetratus* (a bug), *Amphitmetus gibbosus* (a beetle), *Apotmetus montanus* (a beetle), *Apotomus ovatus* (a beetle), *Catatemnus congicus* (a solifugid), *Diplotmema dissectum* (a fossil fern), *Eurytoma prunicola* (a wasp), *Ectemnorhinus viridis* (a beetle), *Macrotoma heros* (a beetle), *Phlebotomus perniciosus* (a sandfly).

Gr. *therizo*, mow, reap; *therister; theristes*, mower, reaper; see **reap**

L. *tondeo, tonsus*, shave, shear, clip; *tonsor*, barber; see **barber**

Gr. *toreuma, -tos*, n. work in relief, embossed work; *toreutos*, turned on a lathe:

L. *trucido, -atus*, cut to pieces, kill cruelly; see **kill**

L. *trunco, -atus*, shorten by cutting off; see **short**

L. *verpus*, m. one circumcised: *Verpulus spumatus* (an arachnid).

Gr. *xyleus*, m. wood-cutter:

Gr. *xyreo*, shave; see *xyron* under **knife**

See: **separate, dig, ax, knife, sword, chisel, geld, short, open, break, sore, barber, strike**

cutis, L. skin; *cuticula*, dim.; see **skin**

cyamo- < Gr. *kyamos*, bean; *kyamion*, dim.; see **bean**

cyano- < Gr. *kyanos; kyaneos*, dark-blue; see **blue**

cyaro- < Gr. *kyar, -os*, hole, orifice, eye of a needle; see **hole**

cyathea, NL. a genus of ferns; see **fern**

cyathus, L. (Gr. *kyathos*), cup, ladle; see **cup**

cyba, L. (Gr. *kybe*), head; see **head**

cybebo- < Gr. *kybebos*, stopping, with head bent; see **bend**

Cybele, L. (Gr. *Kybele*), goddess of nature; see **nature**

cybelido- < Gr. *kybelis, -idos*, ax, cleaver; see **ax**

cybelium, L. (Gr. *kybelion*), the blue violet; see **violet**

cyberneto- < Gr. *kybernetes*, pilot, director; see **govern**

cybistemato- < Gr. *kybistema, -tos*, a somersault; *kybisteter*, tumbler, driver; see **turn**

cybium, L. (Gr. *kybion*), tunny; see **tunny**

cybo- < Gr. *kybos*, a solid with six equal sides, cube, die; see *cubus* under **three**

cycad < Gr. *kykas, -ados*, name for an African plant; see **gymnosperm**

cycethro- < Gr. *kykethron*, ladle, see **spoon**; *kykethra*, mixture, see *kykao* under **mix**

cychramus, L. (Gr. *kychramos*), a kind of bird; see **bird**

cyclamen < Gr. *kyklaminos*, a kind of primrose; see **primrose**

cyclo- < Gr. *kyklos*, circle; *kykliskos*, dim.; *kyklas, -ados; kyklikos; kyklios*, round, circular; see *cyclus* under **circle**

Cyclops, L. (Gr. *Kyklops*), mythical one-eyed giant; see **large**

cydo- < Gr. *kydos*, glory, renown, fame; *kydalimos; kydimos; kydros*, famous; *kydion*, more famous; *kydistos*, most famous; see **honor**

cydonia, L. (Gr. *kydonia*), quince; see **quince**

cydro- < Gr. *kydros*, glorious, noble; see *kydos* under **honor**

cyemato- < Gr. *kyema, -tos*, embryo; *kyesis*, pregnancy; see **young**, *kyo* under **fertile**

cygnus, L. (Gr. *kyknos*), swan; *kyknarion*, dim.; see **goose**

cylico- < Gr. *kylix, -ikos*, cup; *kyliskion*, dim.; *kylichne* (L. *culigna*), small cup; *kylichnion*, dim.; see **cup**

cylinder < L. *cylinder* (Gr. *kylindros*), m. roller, roll of a book; *cylindratus*, in the form of a cylinder: cylindrical, calender, *Cylindrophyma milleporata* (a fossil sponge), *Cylindrophyllum elongatum* (a fossil coral), *Fusulina cylindrica* (a fossil foraminifer), *Ptilopora cylindracea* (a fossil bryozoan), *Cordia cylindristachya* (stringbush).

L. *magdalium*, n. cylindrical figure, pill: magdaleon.

L. *teres, -etis*, rubbed off, rounded, cylindrical: terete, *Catostomus teres* (sucker), *Talinum teretifolium* (flameflower).

See: **turn**

cylisto- < Gr. *kylistos*, rolled, turned; *kylistikos*, expert at rolling; see *kylindo* under **turn**

cylix, L. (Gr. *kylix*), cup; see **cup**

cyllaro- < Gr. *kyllaros*, hermit-crab; see *skyllaros* under **crab**

cyllo- < Gr. *kyllos*, crooked, crippled, lame; *kylloma*, lameness; see **hurt**

cylo- < Gr. *kylon*, part under the eye; see **face**

cymado- < Gr. *kymas, -ados*, a pregnant woman; see **fertile**

cymatilis, L. sea-colored, blue; see **blue**

cymato- < Gr. *kyma, -tos*, wave; *kymation*, dim., see **wave**; a young sprout, see **branch**

cymba, L. (Gr. *kymbe*, bowl, cup; *kymbion*, dim.), cup, boat, skiff; see **cup, ship**

cymbacho- < Gr. *kymbachos*, head foremost, headlong; see **straight**

cymbal < L. *cymbalum* (Gr. *kymbalon*); see **plate**

cymbalaria, NL. < L. *cymbalaris*, a plant; see **figwort**

cymbium, L. (Gr. *kymbion*, dim. of *kymbe*), cup, bowl; see *cymba* under **cup**

cymbo- < Gr. *kymbos*, hollow; see *cymba* under **cup**

cyme < L. *cyma*, young shoot, sprout; *cymula*, dim.; *cymosus*, full of shoots; see **branch**

cymindis, L. (Gr. *kymindis*), probably a hawk; see **hawk**

cynanche, L. (Gr. *kynanche*), quinsy; see **disease**

cynara, L. (Gr. *kynara*), a kind of artichoke; see **thistle**

cyndalo- < Gr. *kyndalos*, wooden peg; see **nail**

cynips, NL. a gall-insect; see **wasp**

cyno- < Gr. *kyon, kynos*, dog, bitch; *kynarion; kynidion; kyniskos*, dim. puppy; *kynikos*, doglike; see **dog**

cynopo < Gr. *kynops, -opos*, a plant; see **plant**

cynosbatus, L. (Gr. *kynosbatos*), dogrose; see **rose**

Cynthia, L. (Gr. *Kynthia*), goddess of the moon; see **moon**

cynuro < Gr. *kynouron*, sea-cliff; see **cliff**

cyparissias, L. (Gr. *kyparissias*), a kind of spurge; see **spurge**

cypasado- < Gr. *kypas, -ados; kypassis, -idos*, a short frock or tunic; see **garment**

cypellum, L. (Gr. *kypellon*), beaker, goblet, cup; see **cup**

cyperus, L. (Gr. *kyperos*), a sedge; see **sedge**

cyphello- < Gr. *kyphellon*, hollow of the ear; see **hollow**

cypho- < Gr. *kyphos*, bent, humped, humpbacked; hump; *kyphon*, bent yoke, pillory; see **bend**

cypo- < Gr. *kype*, hut; see **house**

cypress < L. *cupressus* (Gr. *kyparissos*), f.: cupressineous, *Cupressus arizonica* (a cypress), *Chamaecyparis obtusa* (Japanese chamaecyparis), *Erica cupressina* (a heath), *Athrotaxus cupressoides* (a conifer).

L. *bratus*, f. a cypresslike tree:

cyprinus, L. (Gr. *kyprinos*), carp; see **carp**

cypripedium, NL. a genus of the orchid family; see **orchid**

Cypris, L. Venus, < Gr. *Kypros*, Cyprus, supposed birthplace of Aphrodite; see *Venus* under **love**

cypselido- < Gr. *kypselis, -idos*, earwax; see **wax**

cypselo- < Gr. *kypsele*, chest, box, beehive, hollow vessel; *kypselion*, dim., see **box**; < *kypselos*, swift, swallow, see **swallow**

cypto- < Gr. *kypto*, bend forward, stoop; *see kyphos* under **bend**

cyrbasio- < Gr. *kyrbasia*, a Persian hat with a peaked crown; see **cap**

cyrebio- < Gr. *kyrebion*, bran, husk; see **scale**

cyrilla, NL., f., < Domenico Cyrillo, Italian physician: Cyrillaceae, *Cyrilla racemi-flora* (a cyrilla).
   NL. *cliftonia*, f. a genus of the cyrilla family, < Francis Clifton, English physician: *Cliftonia monophylla* (black titi).

cyrillio- < Gr. *kyrillion*, a jug with a narrow neck; see **bottle**

cyrio-; **cyro-** < Gr. *kyrios; kyros*, authority; *kyriakos*, of a lord or master; see **govern**

cyrmato- < Gr. *kyrma, -tos*, booty, prey, spoil; see **plunder**

cyrto- < Gr. *kyrtos*, curved, see **bend**; < *kyrte*, fish-basket, weel, see **basket**; < *kyrton*, humpback, hunchback, see **humpback**

cystho- < Gr. *kysthos (kysos)*, female pudenda; see **vulva**

cysti-; **cysto-** < Gr. *kystis*, bladder, sac, cell; *kystion*, dim.; see **bag**

Cytherea, L. (Gr. *Kythereia*), a name for Aphrodite; see *Venus* under **love**

cytido- < Gr. *kytis, -idos*, small chest, box; see **box**

cytinus, L. (Gr. *kytinos*), calyx of the pomegranate blossom; see **cup**

cytisus, L. (Gr. *kytisos*), a shrubby legume; see **bean**

cyto- < Gr. *kytos*, hollow place, container, vessel, cell; see **room**

cyttaro- < Gr. *kyttaros*, cell of a honeycomb; *kyttarion*, dim.; see **room**

cyuro- < Gr. *kyoura*, a plant used to produce abortion; see **plant**

D

da-, Gr. intensive prefix; see **very**

dabla, L. a kind of Arabian palm; see **palm**

daceto- < Gr. *daketon*, a biting animal; see *dakno* under **bite**

dacno- < Gr. *dakno*, bite; *daknister*, biter; *dakos*, animal that bites; see **bite**

dacryo- < Gr. *dakryon*, tear, drop; *dakrydion*, dim.; *dakrytos*, tearful; see **tear**

dactylo- < Gr. *daktylos*, finger; *daktylidion*, dim.; see **finger**

dado- < Gr. *das, dados*, firebrand, torch, pine-wood; see **light**

daedalus, L. (Gr. *daidalos; daidaleos*, spotted), cunningly or skillfully made in the manner of Daedalus, the Athenian artificer; see **art**

daemonorops, NL. a genus of palms; see **palm**

daeo- < Gr. *daios*, hostile, dreadful; see **danger**

daer, Gr. brother-in-law; see **kin**

daeto- < Gr. *dais, daitos*, feast, meal; *daitaleus*, banqueter; *daitymon*, guest, see **eat**; *daetos*, cunning, wise, see **know**

daetro- < Gr. *daitros*, carver of meat at table; *daitrosyne*, art of carving meat; see *daio* under **cut**

daffodil < OF. *afrodille*, < L. *asphodelus* (Gr. *asphodelos*); see *narcissus* under **amaryllis**, *asphodelus* under **lily**

dafila, NL. a genus of ducks; see **duck**

dagger < *dague*, < a Celtic word.
   Gr. *dolon, -os*, m. dagger, stiletto, pike: *Dolocerus*[*Dolonocerus*] *reichi* (a beetle).
   Gr. *encheiridion*, n. dagger: *Enchiridium periommatum* (a flatworm).
   L. *pugio, -onis*, m. dagger, dirk, poniard; *pugiunculus*, m. dim.: *Mesembryanthemum pugioniforme* (a fig-marigold), *Cantheliophorus pugiatus* (a fossil lepidophyte).
   L. *sica*, f. dagger, poniard; *sicula*, f. dim.; *sicarius*, m. assassin: sicula, sicarius, *Chalicodoma sicula* (a mason-bee), *Acacia siculiformis* (an acacia), *Cleome*

siculifera (a spider-flower), *Eusmilus sicarius* (a sabertooth), *Cantheliophorus sicatus* (a fossil lepidophyte).
Gr. *xyele,* a kind of dagger, a tool for scraping wood; see *xeo* under **scrape**
See: **point, knife, sword, thorn**

dagma, Gr. bite; see *dakno* under **bite**

dagydo- < Gr. *dagys, -ydos,* doll, puppet; see **doll**

dahlia, NL. a genus of the composite family; see **composite**

daïcto- < Gr. *daïktos,* slain; *daïkter; daïktes; daïktor,* slayer; see **kill**

dainty < OF. *daintie,* delicacy, < L. *dignus,* worthy.
Gr. *chlidanos,* delicate, luxurious; see **soft**
Gr. *chnauros,* dainty: *Chnaura octavialis* (a moth).
L. *cuppes, -edis,* fond of delicacies, dainty; see *cuppedo* under **food**
L. *fastidiosus,* disdainful, dainty, overnice; see *fastidibilis* under **hate**
Gr. *lichnos,* dainty, greedy; see **eat**
L. *ligurio, -itus,* eat daintily; see **eat**
Gr. *trypheros; tryssos,* dainty, delicate; see *tryphe* under **soft**
See: **soft, beauty, thin**

daisy < AS. *daeges eage,* day's eye; see **composite**

dalbergia, NL. a genus of the bean family; see **bean**

dale < AS. *dael* (G. *thal*), valley; see **valley**

daleros, Gr. burning heat, < *daio,* kindle, burn; see **burn**

dally < OF. *dalier,* chat, idle, linger, trifle; see **speak, rest, delay, trifle**

dalos, Gr. firebrand, torch; see *das* under **light**

dam < D. *dam,* stop up; see **wall, heap, hinder**

damalio, L. calf; Gr. *damalis; damalos,* young cow, heifer, calf, see **cattle;** < *dama,* a general name for members of the deer kind, see **deer**

damarto- < Gr. *damar, -tos,* one tamed or yoked, wife, spouse; see **spouse**

damasonium, L. (Gr. *damasonion*), a kind of alisma; see **waterweed**

dammar < Malay *damar,* a resin, kauri gum; see **resin**

damn < L. *damno, -atus,* harm, sentence to punishment; *damnosus,* injurious, pernicious; *damnum,* hurt, injury; see **punish, hurt**

dance < OF. *dancer;* G. *tanzen.*
Gr. *anthema, -tos,* n. a kind of dance:
Gr. *apokinos,* m. a comic dance: *Apocinocera herbacea* (a beetle).
L. *ballator, -is,* m. dancer; Gr. *ballizo,* dance; *ballismos,* m. a dancing, jumping about; *ballematicus,* accompanying the dance: ballet, bal, ball.
Gr. *betarmon, -os,* m. a dancer: *Betarmon bisbimaculatus* (a beetle).
Gr. *bibasis, -eos,* f. a kind of dance: *Bibasis sena* (a butterfly).
Gr. *choros,* m. dance; *choreutes,* m. dancer; *choreutikos,* of a dance; *achoreutos,* banished from the dance; *synchoreutes,* m.; *synchoreutria,* f. partner; *Terpsichore,* f. the Muse of dancing: chorus, choir, carol, Terpsichore, terpsichorean, *Terpsichore delapidans* (an annelid), *Choreutes nemorana* (a butterfly), *Achoreutes armatus* (a springtail).
L. *cordax, -acis* (Gr. *kordax, -akos*), m. a lively, indecent dance of Greek comedy:
Gr. *diapodismos,* m. a kind of dance:
L. *ephalmator, -is,* m. tumbler, dancer:
L. *funambulus,* m. rope-dancer: funambulist, *Funambulus maximus* (a rodent).
Gr. *kallabis, -idos,* f. a Laconian dance:
Gr. *kolabrismos,* m. a wild, Thracian dance:
Gr. *mongas,* n. a wild dance:
Gr. *orcheomai,* dance; *orchesis,* f. dance; *orchestes,* m.; *orchestris, -idos,* f. dancer; *orchestra,* f. place where the chorus danced: orchestra, orchestral, orchestrate, orchestic, orchestrion, *Orchestia agilis* (a beach-flea), *Lagorchestes conspicillatus* (a hare-wallaby).
Gr. *orsites,* m. a Cretan dance:
L. *pantomimus* (Gr. *pantomimos*), mime, actor; see *mimus* under **equal**
L. *praesul, -is,* c. public dancer, presider: presul.
L. *salto, -atus,* hop, dance; *saltator, -is; saltatrix, -icis,* dancer; *saltito,* dance much; see *salio* under **leap**
Gr. *schoinobates,* m. rope-dancer: *Schoenobates walkeri* (a spider).
L. *sicinnis* (Gr. *sikinnis*), f. a satyric dance:
Gr. *skairo,* dance, skip; *skartes,* nimble, quick; *skarthmos,* leap, skip; see **leap**
L. *staticulus,* m. a kind of dance:

L. *thiasus* (Gr. *thiasos*), m. company or troop of dancers in honor of Bacchus: *Thiasophila angulata* (a beetle).
See: **leap, run, joy**

dandruff, prob. < E. *dander,* scurf, and ON. *hrufa,* crust, scab; see **scale**

danger < OF. *danger.* Derive: jeopardy.
Gr. *daios,* hostile, dreadful: *Daeodon shoshonensis* (a fossil mammal), *Daeochaeta harveyi* (a fly).
L. *Damocles* (Gr. *Damokles*), m. a courtier who was punished for incessant flattery by being seated at a banquet table with a sword meanwhile hanging over his head by a single hair; hence, a synonym for being in imminent serious danger: Damoclean.
Gr. *dyschimos,* dangerous, fearful, horrible: *Dyschimus ensifer* (a trichopterid).
L. *funestus,* causing death, calamity, deadly: *Musca funesta* (a fly).
L. *infestus,* hostile, dangerous, unsafe; *infesto, -atus,* attack, injure: infest, infestation, *Pteromalus infestus* (a wasp), *Tabanus infestans* (a fly).
L. *insidiosus,* cunningly deceitful, dangerous; see *insidia* under **trap**
L. *interfectivus,* deadly:
L. *intutus,* unguarded, defenseless, dangerous: *Brachycerus intutus* (a beetle).
Gr. *kindynos,* m. hazard, risk, danger; *kindynodes,* dangerous; *epikindynos,* in danger:
L. *minor, -atus,* threaten, menace; *minator,* threatener; *minax, -acis,* jutting out, threatening; see **projection**
L. *periculum,* n. danger; *periculosus,* dangerous, hazardous: peril, perilous, imperil.
L. *permitialis,* fatal, destructive, ruinous:
L. *Scylla* (Gr. *Skylla*), f. famous rock between Italy and Sicily, opposite the dangerous whirlpool, Charybdis: Scylla and Charybdis.
See: **trap, hurt, fear, death, trouble**

danista, L. (Gr. *daneistes*), money-lender, usurer; see **trade**

danos, Gr. gift, loan, see *do* under **give**; dry, parched, see **dry**

dapanos, Gr. extravagant, lavish, expensive; see **pay**

dapedo- < Gr. *dapedon,* any level surface, floor, ground; see *pedinos* under **level**

daphne, Gr. laurel; now applied to a genus of the Thymeleaceae; see **laurel**

daphoeno- < Gr. *daphoinos,* blood-red, gory; see **red**

dapido- < Gr. *dapis, -idos,* carpet, rug; *dapidion,* dim.; see **rug**

daps, *dapis,* L. solemn religious feast; *dapifer,* waiter; see **eat**

dapsilis, L. (Gr. *dapsiles*), abundant, bountiful; see **abundance**

dapto, Gr. devour, gnaw; *daptes; daptria,* eater, bloodsucker; see **eat**

dare < AS. *dear,* have courage; see **bold**

dark; darkness, < AS. *deorc;* see **dim, brown, black, night, shade,** cloud, blind, sad, hell, Africa

dart < OF. *dard;* see **swift, spear, point, arrow**

dartos, Gr. skinned, flayed; see *derma* under **skin**

dascillo- < Gr. *daskillos,* a kind of fish; see **fish**

dash < ME. *daschen;* see **swift, run, throw**

dasmos, Gr. tribute, impost, tax; see **pay**

dasos, Gr. thicket, copse; *dasodes,* bushy; see *dasys* under **hair**

daspleto- < Gr. *dasples, -etos,* frightful, horrible; see **fear**

dasyllis, Gr. bear; see **bear**

dasys, Gr. thick with hair, hairy, shaggy; *dasytes,* hairiness, roughness; see **hair**

daterios, Gr. dividing, distributing; *datetes,* divider; see *daio* under **cut**

datisca, NL. a genus of plants: Datiscaceae, datiscin, *Datisca cannabina* (a datiscacead).

datum, L. gift, present, fact; *datio,* a giving; *dativus,* of giving; see *do* under **give**

datura, NL. a genus of the nightshade family; see **nightshade**

daucus, L. (Gr. *daukos*), carrot; see **carrot**

daughter < AS. *dohter.*
Gr. *-as; -is,* suffixes denoting daughter of, as *Dryas, -adis,* f. tree-nymph; *Priamis, -idos,* f. daughter of Priam: Thestias, Boreas, Nereis.
L. *filia,* f. daughter; *filiola,* f. dim.; *filialis,* of a son or daughter: filial.
L. *nata,* f. daughter; *natula,* f. dim.:
Gr. *thygater, -os, -tros,* f.; *thygatride,* f. granddaughter; *thygatrion,* n. dim.: *Thygater terminata* (a bee).
See **child**

daulias, Gr. nightingale; see **nightingale**

daulos, Gr. thick, shaggy, dark; see **thick**

dawn < AS. *dagian,* become day.
Gr. *akroria,* f. daybreak:
Gr. *amphilyke,* f. gray of dawn, morning twilight:
Gr. *anchauros,* near the morning:
L. *antelucanum,* n. dawn; *antelucanus,* before daybreak; *sublucanus,* toward morning; *diluculum,* n. daybreak, dawn; *lucesco (lucisco),* day is beginning to break:
L. *antemeridanus,* pertaining to the forenoon, morning: 6:00 a.m.
L. *aurora,* f. dawn: Aurora, auroral, aurora borealis, *Cypraea aurora* (orange cowry).
Gr. *eërios,* early in the morning, early:
Gr. *eos (heos),* f. dawn, morning, east, early; *heothinos; heothen(eothen),* in the morning, early, eastern: eolithic, eophyte, eozoon, eosin, *Eohippus validus* (an Eocene horse), *Eomecon chionantha* (snow-poppy), *Machaeroides eothen* (a creodont), *Eosentomon pusillum* (a proturan), *Eospermatopteris textilis* (a fossil plant).
Gr. *eri,* early, at dawn: *Erigeron argentatus* (a fleabane).
L. *gallicinium,* n. cockcrowing, break of day:
L. *mane,* n. morning, dawn:
L. *matutinus,* of the morning, early, < *Matuta,* f. goddess of the dawn: matutinal, matin, matinal, matinee, *Synbathocrinus matutinus* (a Devonian crinoid), *Obolus matinalis* (a fossil brachiopod).
AS. *morgen,* morning: morn, morning-glory, morrow, tomorrow, morganatic.
Gr. *orthros,* at daybreak, dawn, early morning; *orthrios,* early: *Orthrosanthus multiflorus* (an iridacead), *Orthriomys umbrosus* (a mouse), *Orthriocorisa longipes* (a fossil bug).
L. *primaevus,* early, young; see **before**
L. *primoticus; primotinus,* happening early, blossoming early; see *primus* under **first**
Gr. *proimos; proios,* early morn; *proiotes, -etos,* f. earliness (especially of fruits):
See: **begin, before, first, old**

day < AS. *daeg:* daybreak, daylight, dawn, dayspring, daydream, daymare, daily, daisy, noonday, today, yesterday, birthday, heyday, weekday, payday, dog days, Dayton, Thursday, Domesday Book, sundae.
Days of the week:
Roman: dies solis, d. lunae, d. Martis, d. Mercurii, d. Jovis, d. Veneris, d. Saturni.
Italian: domenica, lunedi, martedi, mercoledi, giovedi, venerdi, sabato.
Spanish: domingo, lunes, martes, miercoles, jueves, viernes, sabado.
French: dimanche, lundi, mardi, mercredi, jeudi, vendredi, samedi.
Modern Greek: Kyriake, Deutera, Trite, Tetarte, Pempte, Paraskeue, Sabbaton.
English: Sunday, Monday, Tuesday, Wednesday, Thursday, Friday, Saturday.
L. *aequinoctium,* n. time when day and night are equal: equinox, equinoctial.
Gr. *aurion,* tomorrow:
L. *bruma,* f. shortest day of the year, winter solstice, winter; *brumalis,* of the winter solstice, wintry: brumal, Brumalia, *Spiraea brumalis* (winter-spiraea), *Tuber brumale* (a truffle), *Cheimatobia brumata* (a moth).
L. *calendae,* f. calends, first day of the month; *calendalis,* of the calends; *calendulus,* at the calends: calendar, calendulin, *Calendula officinalis* (a composite), *Regulus calendula* (ruby-crowned kinglet), *Rhododendron calendulaceum* (flame-azalea).
Gr. *chthes,* yesterday:
L. *cras,* tomorrow; *crastinus,* of tomorrow; *procrastino, -atus,* put off until tomorrow: procrastinate.
Gr. *deile,* afternoon, evening; see **night**
L. *dies, -ei,* m. day; *diecula,* f. dim.; F. *jour,* m. day; *dialis; diurnus,* of the day, daily; *hodiernus,* of today; *interdianus; quotidianus,* daily, recurring; *meridies,* m. midday, noon, south; *meridialis; meridianus,* of midday; *meridionalis,* southern; *antemeridianus,* before noon; *postmeridianus(pomeridianus),* after noon; *biduus,* continuing two days; *perdius,* all day long; *triduum,* n. a three-day period: dial, diary, diurnal, journal, sojourn, journey, adjournment, *sine die,* meridian, 7:45 a.m., 4:15 p.m., quotidian, triduum, biduous, dismal, prandial, Mardi Gras, *Chlorogalum pomeridianum* (soap-plant), F. *samedi*(Saturday), F. *dimanche*(Sunday), *Cestrum diurnum* (day-cestrum).
Gr. *endios,* at midday, at noon, in the open air:
Gr. *epaulion,* n. the day after a wedding:
Gr. *epibda,* f. the day after a festival:

L. *fastus, -us,* m. day on which the praetor could administer justice; the Roman calendar:

L. *feria,* f. holiday, festival; *ferio, -atus,* rest from work, keep holiday: ferial, fair, Mayfair.

L. *festum,* holiday, festival; see *festus* under **joy**

L. *fetalium,* n. birthday:

Gr. *genethlion,* birthday; see *gigno* under **birth**

Gr. *hemera,* f. day; *hemerinos,* of day; *hemerios,* for a day; *ephemeros,* for the day, living only a day, short-lived; *emar, -tos,* n. day; by day, daily; *ephemeris, -idos,* f. journal, diary, calendar; *isemeria,* f. equinox; *mesembria,* f. noon, midday, south; *mesembrinos,* at or of noon, southern: hemeranthous, ephemeral, Nautical Ephemeris, *Hemerocallis fulva* (tawny daylily), *Hemerocampa gulosa* (oak tussock-moth), *Hemerobius speciosus* (a lacewing), *Ephemerella rotunda* (a mayfly), *Pronucula mesembria* (a pelecypod), *Mesembriomys macrurus* (a rabbit-rat), *Mesembryanthemum[Mesembrianthemum] roseum* (a fig-marigold).

L. *heri (hes),* yesterday; *hesternus,* of yesterday, yesterday's: *Camelops hesternus* (a Pleistocene camel).

L. *idus, -us,* f. ides, middle of the month, thirteenth or fifteenth day: ides.

Sp. *manana,* f. tomorrow: manana.

L. *natalis,* m. birthday: natal.

AS. *non* (L. *nona,* ninth hour of the day), originally 3:00 p.m. and then midday: noon, noonday.

L. *nonae,* f. ninth day before the ides, fifth or seventh day of the month: nones.

Gr. *pasteile,* f. last day of the year:

L. *repotium,* the day after a festival; the drinking and carousing on such a day; see *poto* under **drink**

L. *solstitium,* n. longest day of the year, summer solstice, summer; *solstitialis,* of the summer solstice, summery: solstice, solstitial, *Centaurea solstitialis* (yellow centaurea), *Sisyrinchium solstitiale* (a blue-eyed grass).

See: **week, month, year, time, night, dawn**

de, L. down, from, of; intensive; see **down, from, very**

dead < AS. *dead;* see **death**

deaf < AS. *deaf,* unable or unwilling to hear.

Gr. *anekoustos,* not heard, not hearing, not willing to hear, deaf; *baryekoos,* hard of hearing:

Gr. *kophos,* dull, deaf, dumb. dim-sighted: *dyskophos,* stone-deaf; see **dull**

L. *surdus,* deaf. dull, silent, faint: surd, absurd, absurdity, *reductio ad absurdum.*

See: **silent, dull**

deal < AS. *daelan,* apportion, distribute, divide, parcel out; *dael,* part, portion, share; see **separate**

dealbatus, L. whitened; see *albus* under **white**

dear < AS. *deor,* precious; see **love**

death < AS. *death.*

L. *abitio, -onis,* departure, death; *abitus,* departed; see *eo* under **move**

Gr. *akerios,* lifeless:

Gr. *apnoos (apnous); apopnoos,* dead: *Apopnus magniclavus* (a fossil bug).

Gr. *apobiosis,* f. death:

L. *cadaver,* dead body, carcass, corpse; see **body**

L. *defunctus,* dead, deceased, < *defungor,* have done with, finish, die: defunct.

L. *expiro (exspiro), -atus,* breathe out, die, cease, perish: expire, expiration.

L. *exsequiae; obsequiae,* f. funeral procession, funeral rites: obsequies.

L. *extinctus (exstinctus),* dead, put out, destroyed, vanished; see *extinguo* under **destroy**

L. *fatum,* lot, destiny, death; see **lot**

L. *feralis,* of the dead, funereal, deadly: feral.

L. *funestus,* causing death, calamity; see *funero* under **cover**

L. *gangraena* (Gr. *gangraina*), an eating sore, mortification; see **sore**

L. *interitus,* perished, destroyed, lost, < *intereo,* go away and be lost; *interibilis,* perishable, mortal:

Gr. *kedos,* care, trouble, funeral rites; see **trouble**

Gr. *kenebreion,* carrion; see **flesh**

Gr. *Ker, -os,* f. goddess of death, doom; *epikeros,* perishable; *keroulkos,* causing destruction; *keresios; kerotrophos,* deadly, destructive: *Ceresiosaurus calagnii* (a Triassic reptile).

Gr. *lazaros,* m. corpse, < Lazarus; see John, chapter 11:

L. *lethalis (letalis),* deadly, mortal, < *lethum (letum),* n. death; *lethifer (letifer),* death-causing, deadly, fatal: lethal, letisimulate, strychnolethalin. *Strychnos lethalis* (a loganiacead), *Tipula letalis* (a cranefly).

Gr. *moros*, fate, destiny, death; see *moira* under **lot**

L. *mors, mortis*, f. death; *mortalis*, subject to death; c. human being; *moribundus*, dying; *morticinus*, dead; *mortuarius*, of the dead, < *morior, mortuus*, die; *emorior*, die off, perish: mortal, mortify, morgue, mortuary, post mortem, mortgage, amortize, immortality, immortelle, mortmain, mortification, moribund, murder, murrain.

Gr. *nekros*, m. a dead body; *nekrosis*, f. deadness; *nekrikos*, of the dead; *nekrophoros*, burying the dead: necrology, necrolatry, necrosis, necremia, necrectomy, necromancy, necropsy, necropolis, *Necrophorus marginatus* (a carrion-beetle), *Necrobia ruficeps* (a larder-beetle).

L. *nex, necis*, violent death, natural death; Gr. *nekys*, corpse; see *neco* under **kill**

L. *obitus*, a going down, death, destruction; see **destroy**

L. *occido, occasus*, fall down, perish, set; *occidens*, setting; see *occidens* under **west**

L. *occubitus*, m. a going down, setting, death:

Gr. *olethros*, destruction, death; see **destroy**

L. *peritus*, passed away, disappeared, died, < *pereo*, pass away, vanish, disappear: *Murex peritus* (a gastropod), *Saxifraga peritula* (a fossil saxifrage).

Gr. *phthitos*, mortal:

Gr. *potmos*, fate, death: see **lot**

L. *rogus*, funeral pile; *rogalis*, of a funeral pile; see **heap**

L. *sandapila*, bier for the lower classes; see **bed**

Gr. *teleute*, end, death; see *telos* under **end**

Gr. *thanatos*, m. death; *thanasimos; thanatikos; thnetos; katathnetos*, deadly, fatal, mortal; *athanasia*, f. immortality: thanatology, thanatophobia, euthanasia, Athanasius, Bryant's *Thanatopsis*, tansy, *Thanatochlamys tristis* (a bug), *Thnetoschistus revulsus* (a bug), *Thanatus flavidus* (a spider).

L. *vanesco*, pass away, disappear; *evanidus*, passing away; see **depart**

See: **kill, destroy, fall, down, undertaker, depart, lot**

debilis, L. weak; *debilitas*, weakness; see **weak**

debt < OF. *dette*, < L. *debitum*, what is owed, < *debeo, debitus*, owe; see **pay**

decado- < Gr. *dekas, -ados*, the number ten, company of ten; see *decem* under **ten**

decanus, L. chief of ten, dean; see *decem* under **ten**

decay < L. *de*, down, *cado*, fall; see **rot, lessen, destroy**

deceive < OF. *deceivre*, < L. *decipio, -ceptus*, take away, catch, beguile, cheat; see **lie**

decem, L. (Gr. *deka*), ten; *decimus; decumus*, tenth; *decima*, tenth part, tithe; see **ten**

decens, L. seemly, proper, fit; see **fit**

decessus, L. departure; see *cedo* under **move**

decide < L. *de*, off, *caedo*, cut; see **judge, end**

deciduus, L. falling off; see *cado* under **fall**

decimus, L. tenth; see *decem* under **ten**

decipula, L. snare, trap, < *decipio, -ceptus*, ensnare, deceive; see **trap, lie**

deck < D. *dek*, floor, cover; see **base**

decline < L. *declino, -atus*, bend away, deviate, decrease, refuse; see **lessen, no**

declivis, L. downhill, sloping; see *clivus* under **slope**

deco- < Gr. *dex, dekos*, a worm in wood; see **worm**

decoctum, L. drink, potion; see **drink**

decolor, L. discolored, faded; see **color**

decorus, L. becoming, fitting, proper, beautiful; *decus, -oris*, ornament, < *decoro, -atus*, adorn; see **ornament, fit**

decrementatives. See **lessen, little**

decrepitus, L. worn down and feeble with the infirmities of old age; see **weak**

decretum, L. decision; see *cerno* under **separate**

dectes < Gr. *dektes*, biter; *dektikos*, able to bite, biting, pungent; see *dakno* under **bite**

decto-; dectico- < Gr. *dektos; dektikos*, acceptable, < *dechomai*, take, accept, receive, see *dekter* under **receive**; < *dektes*, biter, < *dakno*, bite, see **bite**

decumbens, L. lying down, reclining; see *cubo* under **lie**

decussatus, L. like the letter X, the Roman numeral ten; see **cross**

deditus, L. given up to, addicted to, eager, diligent; see **busy**

deductivus, L. derivative; see **from**

deep < AS. *deop.*

 L. *abyssus* (Gr. *abyssos*), f. a deep pit, the deep sea, hell; bottomless, unfathomed: abyss, abyssal, *Abyssocucumis abyssorum* (a holothurian), *Aspidospira abyssicola* (a holothurian).

 Gr. *achanes,* yawning, immense; *achaneia,* immense width, chasm; see *chaino* under **open**

 L. *bassus,* low, deep; see **low**

 Gr. *bathys,* deep; very; *basson; bathypteros,* deeper; *bathytatos,* deepest; *bathos,* n. depth; *apeirobathes,* unfathomable: bathymetric, bathometer, batholith (bathylith), bathysphere, bathybius, *Bathycrinus gracilis* (a crinoid), *Bathybates ferox* (a fish), *Bathynectes brevispina* (a crab), *Bathylychnus cyaneus* (a fish), *Bathophilus nigerrimus* (a fish), *Bathotrauma lyromma* (a cephalopod).

 Gr. *benthos, -eos,* n. depth of the sea, sea-bottom: benthos, benthonic, *Benthonectes filipes* (a decapod), *Bentheocaris exuens* (a decapod), *Benthoscolex caecus* (an annelid), *Benthodytes abyssicola* (a holothurian).

 Gr. *brychios,* deep, from the deep; *bryx, -ychos,* m. depth of the sea:

 Gr. *bythos,* m. the deep, the depths of the sea; *bythios,* of the deep, in the deep: *Bythocrinus braueri* (a crinoid), *Bythiolophus acanthinus* (a starfish), *Bythinus gallicus* (a beetle), *Bythinogaster simplex* (a beetle).

 Gr. *hyphalos,* under the sea, submerged, in the deep; see *hals* under **sea**

 L. *indespectus,* unfathomable:

 L. *inferus,* low; see **low**

 Gr. *katoryktos,* buried deep; see *orysso* under **dig**

 L. *profundus,* deep, vast: profound, *de profundis, Lophophyllum profundum* (a fossil coral).

 See: **down, low, hell, base, sea**

deer < AS. *deor* (G. *tier*), wild animal: deerlet, deerwort, deerberry, Deerslayer, Deerfield, reindeer.

 Gr. *achaïnes,* m. brocket; *achaïne,* f. deer: *Achaenops dorsalis* (a beetle).

 L. *achlis,* f. a northern wild beast, probably the elk:

 L. *alces, -is* (Gr. *alke*), f. elk: *Alces americana* (moose), *Elaphomyia alcicornis* (elk-horned deerfly), *Platycerium alcicorne* (staghorn-fern).

 L. *capreolus,* roebuck, roe-deer; see *caper* under **goat**

 L. *cervus,* m. deer, buck; *cerva,* f. doe, hind; *cervulus,* m.; *cervula,* f. dim.; *cervarius; cervinus,* of deer: cervine, cervuline, cervoid, Cervidae, cervicorn, *Cervus canadensis* (wapiti, elk), *Lucanus cervus* (a stagbeetle), *Thalassoxanthium cervicorne* (a radiolarian), *Melomys cervinipes* (a naked-tail rat), *Ovis cervina* (bighorn).

 L. *dama,* f. general name for members of the deer kind; *damula,* f. dim.; *Cervus dama* (a deer).

 Gr. *elaphos,* m. deer, stag; *elaphion,* n. dim; *elapheios,* of deer; *elaphodes,* deerlike: elaphine, elaphure, *Elaphoglossum villosum* (a fern), *Elaphomyces cervinus* (a fungus), *Elaphidium mucronatum* (a beetle), *Tragelaphus sylvaticus* (a brushbuck), *Cervus elaphus* (stag).

 Gr. *hellos,* m. young deer, fawn:

 L. *hinnuleus,* m. young stag; *hinnulea,* f. young hind:

 Gr. *kemas, -ados,* f. young deer, pricket:

 Gr. *ladas,* m. young stag: *Ladas planorboides* (a gastropod).

 Gr. *nebros,* m. young deer, fawn; *nebrias,* spotted like a fawn:

 Gr. *prox, -okos,* f. a kind of deer: *Prox moschata* (a deer).

 L. *rangifer,* m. reindeer: *Rangifer caribou* (caribou), *Onthophagus rangifer* (a beetle), *Cladonia rangiferina* (a lichen).

 L. *reno (rheno), -onis,* m. animal of northern Europe, probably the reindeer:

 L. *subulo, -onis,* m. a kind of hart:

 L. *tarandus* (Gr. *tarandos*), m. reindeer: *Rangifer tarandus* (European reindeer), *Tarandichthys cavifrons* (a fish), *Oedemagena tarandi* (a warblefly).

defassus, L. wearied, weakened, exhausted; see *fatigo* under **weak**

defeat < OF. *desfait,* undoing; see **lose**

defect < L. *defectus,* fault, failure, lack; *defectivus,* imperfect; see **fault**

defend < L. *defendo, -fensus,* protect, guard, repel; see **guard**

defer < L. *differo,* put off, delay; see **delay**

definitus, L. distinct, limited; *definitivus,* explanatory; see **clear**

deformis, L. misshapen, ugly; *deformitas,* disfigurement; see **ugly**

defunctus, L. finished, dead; *defunctorius,* quickly despatched; see **death**

degener, L. departing from its kind, debased; see **different**

deglubo, L. peel off, husk, shell; see **skin**

degma, Gr. bite, sting; *degmos,* biting; see *dakno* under **bite**

degree < OF. *degre,* < L. *de,* down, *gradus,* step; see *gradus* under **step**

dehisco, L. split open, gape, yawn; see *hio* under **open**

deios, Gr. enemy; see **enemy**

dejectus, L. cast down, lowered, dispirited; see **sad**

delator, L. accuser, informer, spy; *delatura,* accusation; see **blame**

delay < OF. *delaier.*
L. *amplio, -atus,* delay, defer, enlarge: ampliate, ampliation.
Gr. *anablesis,* f. postponement, delay; *anabole,* f. delay; *anabolikos,* delayed:
Gr. *bradyno,* slow up, delay; see **slow**
L. *cessator, -is,* dilatory one, idler, loiterer; see **rest**
Gr. *chronizo,* pass time, linger, delay; see *chronos* under **time**
L. *cunctor, -atus,* delay, stay, tarry; *cunctabundus,* lingering, loitering; *cunctator, -is,* m. delayer, loiterer: cunctatious, cunctative, cunctator, cunctatory, *Scarabaeus cunctator* (a beetle).
Gr. *dethyno,* delay, tarry:
Gr. *diago,* carry over, past time, tarry:
Gr. *diatribe,* pastime, waste, delay, bitter discussion; see *triba* under **rub**
L. *differo, dilatus,* put off, delay; *dilatio, -onis,* f. delay, deferment; *dilatorius,* delaying: defer, deferment, dilatory.
Gr. *echeneis* (ship-delaying), f. a sucking fish: *Echeneis naucrates* (a remora).
L. *emansor, -is,* m. one who overstays his furlough:
Gr. *epecho,* hold over, stay, delay; see *echo* under **hold**
Gr. *epischesis,* f. check, delay; *epischetikos,* checking, delaying:
L. *Fabianus,* of Fabius, the Roman general surnamed Cunctator for his delaying tactics in fighting Hannibal: Fabian Society, Fabianism.
Gr. *ischanao,* stay, stop, hinder, delay; see *echo* under **hold**
Gr. *melletes,* m. delayer; *mellesis,* f. delay, < *mollo,* intend to do, only think of doing: *Melletes papilio* (a fish).
Gr. *mone,* a tarrying, stay, delay; see *meno* under **stand**
L. *mora,* f. delay; *morula,* f. dim. brief delay; *morator, -is,* m. delayer; *moratorius,* delaying, dilatory; *remora,* f. delay, hindrance; a sucking fish, < *remoror, -atus,* stay, tarry, delay; *remorator, -is,* m.; *remoratrix, -icis,* f. delayer: moratorium, demur, demurrer, *Pinus remorata* (Santa Cruz Island pine), *Echeneis remora* (a remora).
L. *muginor; musinor, -atus,* dally, trifle, delay:
Gr. *oknos,* hesitation, shrinking, delay; see **slow**
L. *opperior, -atus,* wait, wait for, expect:
Gr. *parelkysis,* f. delay, protraction:
L. *piger, -gra, -grum,* reluctant, slow, lazy, dilatory; see **slow**
L. *postpono, -ositus,* put off, defer: postpone, postponement.
L. *procrastino, -atus,* put off till tomorrow; see *cras* under **day**
L. *retardo, -atus,* hinder, delay; see *tardus* under **slow**
L. *suspensus,* waiting in doubtful uncertainty, hestitating, hovering: suspense.
L. *tricor, -atus,* delay, trifle; see *trica* under **trifle**
See: **stand, slow, hinder, sit, bind**

delcano- < Gr. *delkanos,* a kind of fish; see **cod**

deleato- < Gr. *delear, -atos,* bait; *deleastikos,* enticing; see **draw**

delectabilis, L. delightful, agreeable; see *delecto* under **joy**

delemon, Gr. baneful, noxious; *delema,* damage, mischief; see **hurt**

delete < L. *deleo, -etus,* blot, efface, erase; *delebilis,* that may be erased; see **cancel**

deleterios, Gr. harmful, noxious; *deleter,* destroyer; see **hurt**

deletron, Gr. lantern; see **light**

delibo, L. sip, taste; *delibamentum,* libation; see **taste**

delicate < L. *delicatus,* tender, dainty, fastidious; see **soft**

deliculus, L. defective, see *delinquo* under **fault**; *delicus,* weaned, see **wean**

delight < L. *delecto,* intensive of *delicio,* charm, allure; *deliciosus,* delightful; see **joy, agreeable**

delinquentia, L. failure, fault; see *delinquo* under **fault**

deliquia, L. gutter; see *colliquia* under **ditch**

delirus, L. silly, doting, crazy; *delirium,* madness; see **mad**

deliver < OF. *delivrer;* see *liber* under **free, give, save**

dellis, *-ithos,* Gr. a kind of wasp; see **wasp**

delos, Gr. evident, clear, visible; *delotikos,* indicative; see **clear**

delphaco- < Gr. *delphax, -akos,* young pig, porker; see **hog**

delphico- < Gr. *delphix, -ikos,* tripod; see **frame**

delphinium, L. (Gr. *delphinion*), larkspur; see **buttercup**

delphinus, L. (Gr. *delphis, -inos*), dolphin; see **whale**

delphys, Gr. womb; see **womb**

delta, Gr. fourth letter of the Greek alphabet; see **letter**

delubrum, L. temple, shrine; see **temple**

deludo, *-usus*, L. play false, deceive; *delusio, -onis,* deception; see **lie**

demand < OF. *demander;* see **ask**

demas, Gr. the living body, frame; see **body**

demato- < Gr. *dema, -tos,* band, bundle; *demation,* dim.; see *deo* under **bind**

dementia, L. insanity; *demens,* insane; see **mad**

Demeter, Gr. goddess of agriculture; see **grain**

demi, F. < L. *dimidius,* half; see **half**

demissus, L. hanging down, drooping, feeble, weak, low; see **hang**

demnion, Gr. bed, mattress; see **bed**

demo- < Gr. *demos,* the people, district, country; *demios,* of the people, see **people**; *demos,* fat, see **fat**; *demo,* build, see **make**; L. *demo, demptus,* take away, see **take**

demolitus, L. torn down, destroyed, < *demolior, -molitus,* throw down, destroy; see **destroy**

demon < Gr. *daimon,* god, spirit, soul; *daimonion,* dim.; *daimonikos,* of demons; see **spirit**

Demosthenes, Gr. Greek orator; see **speak**

demptus, L. taken away, removed, < *demo, demptus,* take away, remove; see **take**

den < AS. *denn,* lair.
    L. *cavea,* den, enclosure for animals; *caverna,* cave, grotto, den; see **hollow**
    L. *cubile,* den, lair, nest, hole; see **bed**
    Gr. *eilythmos; eilyos,* m. den, lurking-place:
    Gr. *kemma, -tos,* n. lair:
    L. *latibulum,* hiding-place, den, refuge; see *lateo* under **cover**
    Gr. *pholeos,* hole, cave, den; *pholeter,* one who lurks in a hole; see **hole**
    L. *specus,* cave, grotto, den; see **hollow**
    Gr. *thalame,* f. den, hole, lurking-place:
    See: **hole, hollow, nest, room, pen, house**

denaeo- < Gr. *denaios,* long-lived, old; see **old**

denarius, L. a silver coin equivalent to ten asses; see **money**

dendalis, Gr. a kind of barley-cake; see **cake**

dendro- < Gr. *dendron,* tree; *dendrion,* dim.; *dendrikos,* of trees; see **tree**

dendrobium, NL. a genus of the orchid family; see **orchid**

dennos, Gr. reproach, disgrace; see **hurt**

denotatus, L. marked out, conspicuous; see *nota* under **mark**

dens, *-entis*, L. tooth; *denticulus,* dim.; *dentatus,* toothed; see **tooth**

densus, L. thick, close, < *denso, -atus,* thicken; see **thick**

dentale, L. share-beam of a plow; see **plow**

dentaneus, L. threatening; see **fear**

denticulus, L. dim. of *dens,* tooth, see **tooth**; *dentifricium,* tooth-powder, see **dust**; *dentiscalpium,* toothpick, see **pick**

dentix, L. a sparoid fish; see **bream**

deny < L. *denego,* say it is not, refuse; see *nego* under **no**

deodar < Hind. *deodar,* tree of the gods; see *deus* under **spirit**

deonto- < Gr. *deon, -tos,* that which is binding, duty; see *deo* under **bind**

deorsus, L. downward; see *de* under **down**

depart < OF. *departir,* go or pass away.
    L. *abeo; exeo, -itus,* go away; see *eo* under **move**
    L. *aborior, -ortus,* set, disappear, perish; see **abortion**
    Gr. *alyxis,* escape, < *alysko,* flee from, forsake, avoid; see *alyktos* under **shun**
    Gr. *apanastates,* emigrant: see **move**
    Gr. *aparsis,* f. departure:
    Gr. *aperchomai,* go away, depart:
    Gr. *berres,* m. fugitive:
    L. *decessus,* departure; see *cedo* under **move**
    L. *deficio, -fectus,* loosen, set free, desert, leave; see **abandon**

L. *descisco, -citus,* leave, depart, withdraw:

L. *digredior, -gressus,* depart or deviate from; *egressus,* m. departure: digress, digression, egress.

L. *drapeta* (Gr. *drapetes*), runaway slave, fugitive; see *dromos* under **run**

Gr. *ektopizo,* migrate, wander; *ektopismos,* m. migration; *ektopistikos,* migratory; see **move**

L. *exitus,* a going out, a way out; see *eo* under **move**

L. *fuga* (Gr. *phyge*), f. flight, escape, avoidance; *fugela,* f. flight; *fugitivus,* fleeing, a runaway, deserter, < *fugio, -itus* (Gr. *pheugo*), flee, run away, shun, avoid; *confuga,* c. refugee; *defuga; perfuga; transfuga,* m. deserter; *larifuga,* m. vagabond, wanderer; *lucifugus,* light-shunning; Gr. *pheuktos,* to be avoided or shunned; *phyximos; phyxios,* of flight and refuge; *phyxis,* f. flight; *phygas, -ados,* c. fugitive, runaway: fugitive, fugacious, centrifugal, febrifuge, feverfew, vermifuge, refuge, refugee, *Lucifuga subterranea* (blind brotula), *Obisium lucifugum* (a pseudoscorpion), *Parmelia centrifuga* (a lichen), *Pheucticus aurantiacus* (a bird), *Phygadeuon exiguus* (a wasp), *Phyxioschema raddei* (a spider). Derive: hegira.

L. *labilis,* slipping, gliding, fleeting, transient; see **slip**

Gr. *metanastes,* migrant; see **move**

L. *migro, -atus,* move, change habitation; see **move**

L. *pereo, -peritus,* pass away, vanish, disappear; see **death**

Gr. *phroudos,* fled, gone, departed: *Phrudophleps viridis* (a beetle).

L. *transitorius,* evanescent, fleeting, passing, temporary: transitory, *Malus transitoria* (Tibetan crabapple).

Gr. *treo,* flee:

L. *vanesco,* disappear, pass away; *evanidus,* disappearing: evanescent, vanish.

See: **move, walk, wander, death, run**

depas; depastron, Gr. beaker, goblet, bowl; see **cup**

deperditus, L. ruined, lost, abandoned; see *perdo* under **destroy**

depilis, L. without hair; see **bare**

depositus, L. laid, put, or set down; see *pono* under **place**

depressus, L. pressed down, low; see **low**

depstus, L. kneaded, mixed, softened; see *depso* under **mix**

deputy < L. *deputo, -atus,* esteem, consider; see **equal**

deraeo- < Gr. *deraion,* necklace; see **necklace**

derbiosus, L. scabby; see **scale**

dercea, L. a plant; see **nightshade**

derco- < Gr. *derkomai,* see clearly; see **see**

dere, Gr. neck; see *deire* under **neck**

derelictus, L. abandoned, neglected; see *linquo* under **leave**

dergmato- < Gr. *dergma, -tos,* look, glance; *derxis,* sense of sight; see *derkomai* under **see**

deris, Gr. battle, contest, fight; see **fight**

derisus, L. mocked, scorned, laughed at; *derisor,* mocker, scoffer; see *rideo* under **laugh**

derivativus, L. proceeding from; see **from**

derma, -tos; *deros,* Gr. skin, hide, < *dero,* skin, flay, cudgel, thrash, see **skin**; *dermatinos,* leathern, see **leather**

derobios, Gr. long-lived; see *bios* under **life**

deros, Gr. long (time), too long, see **long**; < *deros,* skin, see *derma* under **skin**

derris, Gr. leather coat; *derrion,* dim., see **garment**; a genus of the bean family, see **bean**

dertron, Gr. caul, omentum; see **membrane**

des-; see *de* under **from**

deseps, L. insane; see **mad**

desert < L. *desertum,* n. a waste place, wilderness: Sahara Desert, deserticolous, desert-nut *(Eremocarya micrantha),* *Agave deserti* (an agave), *Pogonomyrmex desertorum* (desert-ant), *Eriogonum deserticola* (a polygonacead).

Gr. *astibos,* untrodden, desert, solitary, off the road; see **alone**

Gr. *astiptos,* untrodden, desert:

L. *desolatio, -onis,* f. desert, waste: desolation.

Gr. *eremia,* desert, solitude, wilderness; see *eremos* under **alone**

L. *inaquosum,* n. desert:

L. *solitudo, -inis,* lonely place, desert; see *solus* under **alone**

L. *tescum,* n. waste place, desert:
L. *vastus,* waste, desert; see **large**
See: **empty**

desertus, L. abandoned, forsaken, solitary, waste; see *desero* under **abandon**

deses, *-idis,* L. idle, lazy, slothful; *desidiosus,* slothful, idle; see **rest**

desideratus, L. desired, longed for; *desiderabilis,* desirable; see **wish**

designatus, L. marked out, appointed; see *signum* under **mark**

desipiens, L. silly, foolish; see **fool**

desire < OF. *desire,* < L. *desidero,* long for, wish for; see **wish**

desis, Gr. a binding together; see *deo* under **bind**

desitus, L. ended, stopped; see *desino* under **end**

desk < L. *discus;* see **table, box, frame**

desmido- < Gr. *desmis, -idos,* bundle, package; see **bundle**

desmo- < Gr. *desmos,* bond, fetter, halter, chain; *desmion,* dim.; see *deo* under **bind**

desmodium, NL. a genus of the bean family; see **bean**

desmoncus, NL. a genus of palms; see **palm**

desmotes, Gr. captive, prisoner; see **pen**

desolator, L. destroyer, waster; see **destroy**

despair < OF. *desperer,* < L. *desperatus,* without hope; see **hopeless**

despicus, L. disdained; *despicabilis,* contemptible; see **hate**

despotes, Gr. master, see **govern**; *despotikos,* absolute, inclined to tyranny, see **all**

dessert < F. *desservir,* clear the table; see **eat**

destiny < L. *destino, -atus,* determine, aim at; see **end, lot**

destitus, L. ceased; see *desisto* under **end**

destitutus, L. lacking, without possessions; see **poor**

destroy < OF. *destruire,* < L. *destruo, -uctus,* destroy; *destructibilis; destructilis,* that can be destroyed; *destructivus,* causing destruction; *destructor, -is,* m. destroyer: destruction, destructive, destructible, destructor, *Phytophaga destructor* (Hessian fly), *Anthostromella destruens* (a fungus).
L. *abolio, -itus,* destroy, terminate; *abolitio, -onis,* f. annulment; *abolitor, -is,* m. abolisher: abolish.
L. *absumo, -umptus,* take away, consume, destroy, ruin:
Gr. *anaireo,* take away, destroy; *anairetes,* m. destroyer, murderer; *anairema, -tos,* n. spoil, booty; *anairesis,* f. destruction; *anairetos,* destructive; *kathaireo,* put down, destroy; *kathairetes,* m. destroyer: *Anaeretus guanajuatensis* (a beetle).
Gr. *analisko,* destroy, waste:
Gr. *anastater, -os; anastates,* m. destroyer:
Gr. *anatropeus,* m. destroyer: *Anatropomyia flavicornis* (a fly).
L. *annihilo, -atus,* bring to nothing, destroy utterly; see *nihil* under **not**
Gr. *aphanistes,* m. destroyer: *Aphanistes bellicosus* (a wasp), *Aphanisticus pusillus* (a beetle).
L. *assolo, -atus,* level to the ground, destroy:
L. *calamitas, -atis,* f. injury, misfortune, disaster; *calamitosus,* destructive, disastrous: calamity, calamitous.
L. *casso, -atus,* annul, bring to naught; see **cancel**
L. *clades, -is,* f. destruction, disaster, injury:
L. *confector,* maker, destroyer; see *facio* under **make**
L. *corruptibilis; corruptivus; corruptorius,* destructible, perishable, likely to decay: see *corruptus* under **bad**
Gr. *deleter,* destroyer; see *deleterios* under **hurt**
L. *demolior, -molitus,* throw down, tear down, destroy: demolish, demolition.
OF. *desastre,* 'evil star', misfortune: disaster, disastrous.
L. *desolator, -is,* m. destroyer, waster: desolator, desolate.
L. *detero, -tritus,* rub off, wear out; see *tero* under **rub**
L. *devasto, -atus,* lay waste; *devastator, -is,* m. destroyer: devastate, *Melanoplus devastator* (a grasshopper), *Phylloxera vastatrix* (a plant-louse).
L. *dilabor, -lapsus,* fall apart, go to pieces, go to ruin, perish: *dilabidus,* that soon goes to pieces:
L. *diripio, -reptus,* tear to pieces, plunder, destroy:
L. *dissolvo, -solutus,* loosen, destroy; see *solvo* under **free**
Gr. *ekkopto,* cut out, knock out, destroy; see *kopto* under **cut**
Gr. *ekrhizotes,* m. uprooter, destroyer:

L. *eradico, -atus,* root out, destroy: eradicate.

L. *erado, erasus,* scratch out, remove, cancel; see **cancel**

Gr. *ereipion,* n. ruin, wreck; *ereipsimos,* thrown down, ruined; *ereipsis,* f. a ruining, < *ereipo,* thrown down, fall in ruins:

Gr. *exaleipho,* obliterate, wipe out; *exaleipsis, -eos,* f. destruction:

L. *everto, eversus,* overturn, destroy, ruin: *Cardium eversum* (a fossil pelecypod).

L. *exitiosus; exitiabilis; exitialis,* destructive, fatal, deadly: *Nasutitermes exitiosus* (a termite), *Conopia exitiosa* (peach-tree borer).

L. *exscindo, -scissus,* extirpate, destroy:

L. *extermino, -atus,* banish, abolish, destroy: exterminate.

L. *extinguo (exstinguo), -inctus,* put out quench, destroy, annihilate: extinguish, extinct, extinction.

L. *extirpo (exstirpo), -atus,* uproot, eradicate, destroy: extirpate, extirpation.

Gr. *kataklysmos,* flood, deluge, disaster; see *klyzo* under **wash**

Gr. *katastrophe,* f. overturning, calamity, misfortune: catastrophe.

Gr. *kathaireo,* put down, rescind, cancel, destroy; *kathairesis,* f. a destroying:

Gr. *kathypago,* destroy utterly:

Gr. *keiro,* cut off, destroy; see *koureus* under **barber**

Gr. *ker,* death, doom, destruction; see **death**

Gr. *loigas,* m. ruin, havoc; *loigios,* ruinous, destructive; *loigistria,* f. destroyer:

Gr. *lyme,* f. outrage, ruin; *lymanter, -os; lymantes,* m. destroyer, spoiler; *lymanterios,* destructive; *lymainomai,* outrage, injure, ruin: *Lymantes scrobicollis* (a beetle), *Lymantor sepicola* (a beetle), *Lymexylon nivale* (a beetle).

Gr. *nauagia,* f. shipwreck; *nauagion,* n. piece of a wreck; L. *naufragus,* shipwrecked; *naufragiosus,* full of wrecks, dangerous:

L. *obitus,* m. a going down, downfall, destruction, death: obit, obituary.

L. *occisus,* ruined, lost, undone:

Gr. *olethros,* m. ruin, destruction; *olethrios,* destructive, deadly; *oletor, -os,* m. destroyer; *olesiptolis,* city-destroying; *oloos,* destroying; *apollymi,* destroy utterly; *exoles,* utterly destroyed, ruined, abandoned, < *allyo (ollymi),* destroy, ruin: Apollyon, *Octopus apollyon* (an octopus), *Olethreutes arcuana* (a moth), *Olethroblatta americana* (a fossil cockroach), *Olesicampa longipes* (a wasp).

L. *perdo, -ditus,* destroy, ruin, waste, kill; *perditor, -is,* m.; *perditrix, -icis,* f. destroyer; *deperditus,* utterly lost, abandoned: perdition, perdue, *Formica perditor* (an ant), *Anthomyia perdita* (a fly), *Dendroctonus piceaperda* (a bark-beetle), properdine.

L. *perimo, -emptus,* abolish, destroy; *peremptalis,* of destruction; *peremptor, -is,* m. destroyer: peremptory.

L. *perniciosus; pernicialis,* baneful, injurious, destructive; *pernicies, -ei,* f. destruction, ruin, calamity: *Aspidiotus perniciosus* (San Jose scale), anemia perniciosa.

Gr. *pertho; portheo,* destroy, plunder, ravage, waste; *porthesis,* f. sack, destruction; *portheon, -os; porthetes; porthetor, -os,* m. destroyer, ravager: *Porthetes zamiae* (a beetle), *Porthetria dispar* (gypsy-moth), *Diaporthe perniciosa* (a fungus), *Portheus molossus* (a Cretaceous fish).

Gr. *phthora,* f.; *phthoros,* m. destruction, corruption, decay, ruin; *phtharsis,* f. corruption; *phthartos,* perishable; *phthartikos; phthorimos; phthorios,* destructive; *phthersigenes,* race-destroying; *thymophthoros,* soul-destroying, < *phtheiro,* corrupt, destroy, ruin, spoil, waste: *Phytophthora infestans* (potato-blight), *Phthorophloeus frontalis* (a beetle), *Phthorimaea operculella* (a moth), *Phthartomicrus externus* (a beetle).

L. *populor, -atus,* devastate, lay waste, ravage, plunder, pillage: depopulate.

L. *ruo, rutus,* fall down, tumble down; *ruina,* f. downfall; *ruinosus,* going to ruin; *corruo,* fall together; *diruo,* demolish, destroy, overthrow; *eruo,* root out, destroy; *obruo,* overwhelm, overthrow, bury: ruin, ruinous, *Pteromalus dirutor* (a wasp), *Turritella obruta* (a fossil gastropod).

Gr. *sinis, -idos,* plunderer, destroyer; see *sinos* under **hurt**

L. *strages, -is,* f. overthrow, ruin, destruction:

Gr. *syncheo,* pour together, confuse, destroy; see *synchysis* under **mix**

See: **kill, break, mix, cut, waste, hurt, cancel, lessen, disease, eat, end, fall**

desuetus, L. discontinued, disused; see *sueo* under **use**

desultorius, L. fickle, inconstant, superficial, skipping about; *desultor,* leaper; see **change,** *assilio* under **leap**

dete, Gr. fagot; see **wood**

detector, L. uncoverer, revealer; see *detego* under **bare**

deter, L. poor, bad; *deterior,* poorer, worse, less; *deterrimus,* poorest, worst; see **bad**

detersus, L. cleansed; see *tergeo* under **wash**

detest < L. *detestor, -atus,* curse, execrate; *detestabilis,* execrable, abominable; see **hate**

detha, Gr. for a long time; see **long**

detis, -*idos,* Gr. head of garlic; see **onion**

detos, Gr. bound; see *deo* under **bind**

detrimentum, L. damage, loss; see **hurt**

detritus, L. worn away; pertaining to alluvial material; see *tero* under **rub, earth**

deus, L. god; *dea,* goddess; Skt. *deva,* god; see **spirit**

deuso-, Gr. dye, stain, < *deuo,* wet, drench; see **color**

deustus, L. burned up; see *uro* under **burn**

deuteros, Gr. second; *deuterios,* secondary; see **two**

Deverra, L. goddess of brooms and sweeping; see *verro* under **sweep**

deversor, L. guest, inmate; *deversorium,* inn, lodging-place; see **guest**

devexus, L. sloping, descending; see **slope**

devil < AS. *deofol,* < L. *diabolus* (Gr. *diabolos*); see **spirit**

devius, L. off the highway, out of the way; *deviator,* one who leaves the way; see **wander**

devotus, L. attached, faithful; see **true**

dew < AS. *deaw;* see **drop**

dex, *dekos,* Gr. a worm in wood; see **worm,** *dakno* under **bite**

dexios, Gr. on the right, see **hand;** *dexiotes,* handiness, cleverness, see **art**

dexis, Gr. bite; see *dakno* under **bite**

dexter, -*tra, -trum,* L. right, on the right hand; see **hand**

di-; dif-; dir-; dis- < L. *dis,* in two, apart, asunder, away from, without, not, see **from:** < Gr. *dis,* twice, double, see **two**

dia, Gr. through, between, during; see **through**

diabathra, Gr. ladder, see **ladder;** *diabathron,* slipper, see **shoe, cobbler**

diabetes, Gr. a disease, see **disease;** siphon, see **pipe**

diabolus, L. (Gr. *diabolos*), devil, evil spirit; see **spirit**

diabrotico- < Gr. *diabrotikos,* eating through, corrosive; see *bibrosko* under **eat**

diacheton, L. a plant; see **plant**

diaconus, L. (Gr. *diakonos*), servant, menial, minister of the church; see **servant**

diadem < L., Gr. *diadem,* headband, fillet, crown; see **crown**

diadexios, Gr. presaging good luck; see **lot**

diadochos, Gr. succeeding, receiving; *diadektor,* inheritor; see **receive**

diadysis, Gr. passage; see **way**

diaereto- < Gr. *diairetos,* divided, divisible; *diairesis,* division, separation; *diairetes,* divider; see *dia* under **separate**

diagnosis, Gr. conclusion from understanding of appearances or symptoms; see **know**

diagonal < L. *diagonalis,* < Gr. *diagonios,* from angle to angle, oblique; see **slope**

diagram < Gr. *diagramma,* figure, form, plan; see **map, form**

dial < L. *dialis,* daily, < *dies,* day.
  Gr. *analemma, -tos,* n. sundial:
  Gr. *chronos,* time, *meter,* measure: chronometer, *Amphicyclia chronometra* (a radiolarian).
  L. *horarium,* n. dial, clock:
  Gr. *skiatheras,* m. sundial: *Sciatheras trichotus* (a wasp)
  L. *solarium,* n. sundial, balcony: *Solarium perspectivum* (a sundial-shell).

dialect < Gr. *dialektos,* common language, conversation; see *lego* under **speak**

Dialis, L. pertaining to Dis(Jupiter), see *Dis* under **spirit;** *dialis,* daily, see *dies* under **day**

diallactero- < Gr. *diallakter,* mediator; see **judge**

diallagma, Gr. substitute; see **equal**

dialysis, Gr. dissolution, separation; see *lyo* under **free**

diamond < L. *adamas, -antis* (Gr. *-antos*), unconquerable, unyielding, hard, durable; lozenge; *adamanteus,* of diamond, like a diamond or lozenge; *adamantinus,* very hard: adamant, adamantine, *Crotalus adamanteus* (diamond-rattlesnake), *Didymograptus adamantinus* (a graptolite).

L. *anancites,* m. a name for the diamond when used as a remedy for distress of mind:

F. *losange,* a diamond-shaped or rhomboid figure: lozenge.

L. *rhombus* (Gr. *rhombos*), m. an equilateral parallelogram with unequal pairs of angles: rhomb, rhomboid, rhombohedron, *Rhombodictyon reniforme* (a fossil sponge), *Fasciolaria rhomboidea* (a fossil gastropod), *Phyllocladus rhomboidalis* (a taxacead), *Microcentrum rhombifolium* (a katydid), *Eugenia rhomba* (spice-berry), *Lagodon rhomboides* (pinfish-bream).

L. *scutula,* f. diamond or lozenge-shaped figure; see *scutella* under **plate**

diamperes, Gr. piercing; see **bore**

diamphidios, Gr. utterly different; see **different**

Diana, L. goddess of the chase; see **hunt**

dianthus, NL. carnation; a genus of the pink family; see **pink**

diantos, Gr. that may be wetted; see *diaino* under **wet**

diapegma, Gr. crossbeam; see **beam**

diaperama, Gr. strait, ferry; see **way**

diaphano- < Gr. *diaphanes,* transparent, distinct; see *phaino* under **light**

diaphoresis, Gr. perspiration; *diaphoretikos,* promoting perspiration; see **drop**

diaphoros, Gr. different; see **different**

diaphragm < Gr. *diaphragma,* partition wall, the muscle separating the chest and abdominal cavities; see *phragmos* under **wall**

diaphyge, Gr. refuge; see **safe**

diaptyxis, Gr. evolution, explication; see **turn**

diarrhea < L. *diarrhoea* (Gr. *diarrhoia*), a flowing through; see **disease**

diary < L. *diarium,* daybook, journal; see **book**

diasphax, Gr. rocky gorge; see **valley**

diastatos, Gr. divided; *diastatikos,* separable; see *dia* under **separate**

diastema, Gr. space between, interval; see **space**

diastole, Gr. dilatation; see **spread**

diateles, Gr. continuous, incessant; see **always**

diatribe, Gr. a rubbing away, waste, delay, bitter discussion; see *tribo* under **rub**

dibolia, Gr. a two-edged lance; *dibolos,* two-pointed; see **spear**

dicabula, L. chatter, idle talk; see *dico* under **speak**

dicaeo- < Gr. *dikaios,* observant of the right, decent, just, civilized; *dikaiosyne,* justice; see *dike* under **law**

dicaero- < Gr. *dikairon,* an Indian bird; see **bird**

dicastico- < Gr. *dikastikos,* of law and trials; *dikastes,* judge; see *dike* under **law**

dicatus, L. dedicated, devoted, proclaimed; see *dico* under **speak**

dicax, -acis, L. witty, satirical; *dicaculus,* facetious; see **laugh**

dice, plural of die, < OF. *de,* < L. *datus,* given, thrown; see **play**

dicello- < Gr. *dikella,* two-edged hoe, mattock; see **hoe**

dicho- < Gr. *dicha,* in two; see *di-* under **two**

diclido- < Gr. *diklis, -idos,* double-folding, as two doors or valves; see **fold**

dico- < Gr. *dike,* right, custom, law, justice; *dikaios,* right, decent, just; see **law**

dicrano- < Gr. *dikranon,* pitchfork; see **fork**

dicro- < Gr. *dikros,* forked, cloven; see **fork**

dictamnus, L. (Gr. *diktamnos*), dittany, a mint; name now applied to a genus of the rue family; see **rue**

dictate < L. *dicto, -atus,* assert, declare, designate, prescribe; see *dico* under **speak**

dictator, L. emergency ruler with absolute authority; see **govern**

dicto- < Gr. *deiktos,* capable of proof; *deitikos,* able to prove; *deiktes,* an exhibitor; see **try**

dictyo- < Gr. *diktyon,* net; *diktydion,* dim.; see **net**

didacto- < Gr. *didaktos,* instructed, taught; *didagma, -tos,* lesson; see *didasko* under **teach**

didus, NL. genus of the dodo; see **dove**

didymos, Gr. double, twin, testicle; see *di-* under **two, testicle**

die < ON. *deyja,* see **death**; < OF. *de,* < L. *datus,* given, thrown, see **play**

diecula, L. dim. of *dies,* day; see **day**

diedros, Gr. sitting apart; see *dia* under **separate**

dielo- < Gr. *deielos,* of the evening; see *deile* under **night**

dieneco- < Gr. *dienekes,* continuous, unbroken; see **always**

dierama, Gr. strainer, sieve; see **sieve**

diero- < Gr. *dieres,* double, see *di-* under **two**; < *dieros,* active, see **busy**

dies, L. day; *diecula,* dim.; see **day**

diesthio, Gr. eat through, consume; see *edo* under **eat**

diet < L. *diaeta* (Gr. *diaita*), a prescribed regimen of food; see **eat**

dif- < L. *dis,* away from; see *di-* under **from**

different < L. *differens, -entis,* dissimilar.

L. *aberrans,* wandering, deviating, abnormal; see *erro* under **wander**

L. *abludo, -lusus,* be out of tune with, be unlike, differ:

L. *abnormalis,* departing from the rule; *enormis,* irregular, unusual: abnormal, abnormality, *Eurymetopon abnorme* (a beetle).

L. *absonus; dissonus,* discordant, different, inharmonious: dissonant, *Murex absonus* (a gastropod).

Gr. *akatallelos,* not fitting together, heterogeneous:

Gr. *alloios,* of another kind, different: *Alloeomimus unifasciatus* (a bug).

Gr. *amorphos,* formless, misshapen: amorphous, *Amorpha glabra* (a legume).

Gr. *anartios,* uneven, odd: *Anartioschiza major* (a beetle).

Gr. *anisos,* unequal; *anisotes, -etos,* f. inequality: *Anisolabis annulipes* (an earwig), *Anisota rubicunda* (maple-worm), *Anisophyllea disticha* (leechwood), *Anisoptera thurifera* (palosapis-tree).

Gr. *anomalos, uneven,* irregular, inconsistent, abnormal, unusual, deviating from the general rule; *anomos,* without law, lawless, different: anomalous, *Anomala orientalis* (Asiatic beetle), *Anomalopteryx parva* (a moa), *Anomalina calymene* (a foraminifer), *Fraxinus anomala* (simple-leaved ash), *Campostoma anomalum* (a stone-roller), *Anomia aculeata* (a pelecypod).

Gr. *anomoios,* dissimilar: *Anomoeocerus hispidus* (a fly).

Gr. *apeches,* discordant, dissonant: *Apochoneura longicauda* (a wasp).

Gr. *apochordos,* discordant, inharmonious:

Gr. *apodos,* out of tune:

Gr. *asymbatos,* irreconcilable; *asymbasia,* f. inconsistency: *Asymbata roseiventris* (a moth).

Gr. *asymmetros,* without symmetry, disproportionate: asymmetry, asymmetrical, *Asymmetrocrinus poteriocrinoides* (a Permian crinoid).

Gr. *asymphylus,* not of the same race, unlike: *Asymphylomyrmex balticus* (a fossil ant).

Gr. *asynkritos,* unlike: *Asyncritus reticulatus* (a fossil insect).

Gr. *ataktos,* not arranged, disordered, irregular; see **confusion**

L. *curiosus,* odd, strange, inquisitive; see **wish**

L. *degener,* departing from its kind, debased: degenerate.

Gr. *diadelos,* distinguishable, distinctive; see *delos* under **clear**

Gr. *dialeptos,* distinguishable:

Gr. *diamphidios,* utterly different: *Diamphidia femoralis* (a beetle).

Gr. *diaphonos,* dissonant:

Gr. *diaphoros,* different: *Diaphoromys fiegi* (a fossil mammal).

Gr. *diatropos,* various: *Diatropornis ellioti* (a bird).

L. *discors, -ordis,* different, unlike: *Querquedula discors* (blue-winged teal).

L. *discrepo,* disagree, differ, sound differently: discrepancy.

L. *disgregus,* unlike, different:

L. *disjunctus,* separate, distinct, different, remote: disjunct.

L. *dispar, -is; disparilis,* unlike, dissimilar; *impar, -is; imparilis,* unequal, odd, different: disparity, *Trichocephalus dispar* (a nematode), *Marginalis disparilis* (a foraminifer), imparity, imparipinnate, *Polynoë impar* (an annelid), *Megaselia impariseta* (a fly).

L. *dissensus,* m. disagreement, discord; *dissentaneus,* contrary, disagreeing; *dissentio,* differ, disagree: dissent, dissension.

L. *dissidens,* differing, disagreeing: dissident, *Hippelates dissidens* (a gnat).

L. *dissimilis; absimilis,* unlike: dissimilar, *Rhipsalis dissimilis* (a cactus).

L. *distinctus,* different, separate; see *distinguo* under **separate**

L. *disto,* stand apart, be separate, differ; see **separate**

L. *diversus,* different, separate: diverse, diversity, diversification, *Arctostaphylos diversifolia* (a manzanita).

L. *dividia,* f. dissension, discord:

L. *eccentricus* (Gr. *ekkentros*), not having the same center, different, odd: eccentric, eccentricity.

Gr. *ekmeles,* dissonant, irregular: *Ecmeles fuscescens* (a bird).

Gr. *ektonos,* out of tune:

Gr. *exallos,* quite different; *enallos,* changed, contrary: *Enallocrinus scriptus* (a Silurian crinoid), *Exallus semirugosus* (a beetle).

L. *exceptio, -onis,* f. something taken out or excluded: exception.

Gr. *hairetikos,* holding views opposed to accepted doctrine: heretic, heretical, heresy.

Gr. *heteros,* different; *heteroklitos,* deviating, abnormal: heterodox, heterogeneous, heterocercal, heterogamous, heterandrous, heterogynous, heterodyne, heteroclite, heterocyst, heterodont, heteromorphic, heteronym, heteroplasm, heterosporous, heterosis, heterothallic, heterozygous, *Heteropogon contortus* (tanglehead), *Heterotermes aureus* (a termite), *Heteranthera dubia* (a mud-plantain), *Tsuga heterophylla* (western hemlock), *Fundulus heteroclitus* (mummychog, a killifish).

Gr. *idiogenes,* distinctive, peculiar; see *idios* under **alone**

L. *impertinens,* irrelevant, inappropriate: impertinent, impertinence.

L. *inaequalis,* uneven, rough, unlike, different: unequal, inequality, *Phytobius inaequalis* (a beetle).

L. *insuetus,* unaccustomed to, inexperienced in, unusual; see **strange**

L. *intempestus,* unseasonable, inopportune:

L. *multifarius,* manifold, various: multifarious.

Gr. *perissos (perittos),* beyond the regular number or size, extraordinary, odd or uneven (as applied to numbers): perissodactyl, *Perissonota protexta* (a fossil pelecypod), *Perittocrinus radiatus* (an Ordovician crinoid).

Gr. *plemmeles,* discordant, erroneous; see **fault**

L. *praecipuus,* peculiar, special, extraordinary: *Sciara praecipua* (a fly).

L. *scalenus* (Gr. *skalenos*), uneven, unequal, odd, said of a triangle with unequal sides: scalene, scalenohedron.

L. *secus,* otherwise, different: extrinsic, sequester.

L. *seorsus,* especially, particularly, apart, separate; see **separate**

L. *separatus,* apart, sundered, distinct; see **separate**

L. *singularis,* solitary, different; see **alone**

L. *specialis,* individual, particular: special, especially.

AS. *steop,* pertaining to relatives by marriage: stepmother, stepfather, stepson, stepdaughter, stepchild.

L. *unicus,* only, sole, singular; see **alone**

L. *varius,* different; *variabilis,* changeable, < *vario, -atus,* change, diversify; *variegatus,* of different sorts, particularly colors: vary, various, variety, variegated, invariably, *Ceratopogon varius* (punkie, no-see-um), *Dermacentor variabilis* (dog-tick), *Stapelia variegata* (an asclepiadacead), *Psathyrocerus variegatus* (a beetle), *Equisetum variegatum* (a horsetail).

L. *versicolor,* of different colors, changeable in color; see **color**

See: **separate, other, strange, alone, mix, against**

difficult < L. *difficultas, -atis,* f. distress, perplexity, trouble; *difficilis, -e,* not easy, hard, troublesome: difficulty, *Agrilus difficilis* (a beetle).

L. *angustia,* narrowness, difficulty; see *angustus* under **narrow**

Gr. *aporema,* doubt, question, perplexity; *aporia,* difficulty; see **doubt**

L. *arduus,* difficult to attain, laborious: arduous.

Gr. *chalepos,* difficult, painful, severe, harsh, troublesome: *Chalepus trilineatus* (a beetle), *Chalepoderus hiaticollis* (a beetle), *Xenochalepus platymerus* (a beetle).

L. *conabilis,* laborious, difficult; *conamen,* effort, struggle; see *conor* under **make**

Gr. *dilemma,* double proposition; see *lemma* under **think**

Gr. *dyscheres,* hard to manage, annoying, vexatious, difficult; see **trouble**

Gr. *dyschrestos,* difficult to use, inconvenient:

Gr. *dyskolos,* difficult to please, peevish: *Dyscolorhinus squalinus* (a grasshopper).

Gr. *dysprositos,* hard to get at, difficult of access; see *aprositos* under **impassable**

L. *gravabilis,* oppressive, troublesome; see *gravis* under **heavy**

L. *ingestabilis,* that cannot be borne, insupportable:

L. *laboriosus,* toilsome, wearisome, difficult: laborious.

L. *operosus,* laborious, difficult; see *opus* under **make**

L. *praedicamentum,* n. unpleasant condition: predicament.

L. *vix,* with difficulty, hardly, scarcely, barely:

See: **hard,** *dys* under **bad, hinder, trouble**

diffusus, L. spread out, extended, dispersed; see **spread**

dig < AS. *dician.*

L. *cavator,* excavator; see *cavus* under **hollow**

L. *copiates* (Gr. *kopiates*), m. grave-digger, sexton:

L. *cunicularius; cuniculator,* miner; see **mine**

Gr. *dikellites,* m. digger:

L. *fodio, fossus; fosso, -atus,* dig; *fossa,* f. ditch, canal; *fossula,* f. dim.; *fossilis,* digging, dug up; *fossor, -is,* m. digger; *fossorius,* adapted to digging; *confodio,* dig thoroughly; *defodio,* dig deep, dig up; *effodio,* dig out, dig up: fossil, fossil-ization, fossor, fossorial, fossa, fossette, fossula, fodient, *Cobitis fossilis* (a loach), *Paleocastor fossor* (a fossil beaver found in demonelix burrows in Nebraska), *Tylosurus fodiator* (a needlefish), *Chalepus fossulatus* (a beetle), *Clostridium fossicularum* (a bacterium), *Neomys fodiens* (a shrew).

L. *funero, -atus,* bury, inter; see **cover**

AS. *grafan,* dig, cut; see **cut**

Gr. *lachaino,* dig:

L. *lacuno, -atus,* hollow out; *lacunarius,* grave-digger; see **hollow**

Gr. *orysso,* dig; *oryktes,* m. digger; *oryktos,* dug, mined; *oryxis,* f. a digging; *orygma, -tos,* n. pit, trench, tunnel, mine, moat; *aporyx; dioryx; katoryx, -ygos,* f. canal, trench, conduit, pit; *georychos,* throwing up earth: oryctology, *Orycteropus capensis* (aardvark), *Notoryctes typhlops* (pouched mole), *Oryctes prolixus* (a beetle), *Dermatoryctes fossor* (a mite), *Orygmatobothrium versatile* (a worm), *Catorygma curvipes* (a beetle), *Catoryctis subnexella* (a moth).

L. *pastino, -atus,* dig and trench ground in preparation for planting vines: pastinate.

Gr. *skapto,* dig; *skaptos,* dug; *skapter, -os; skapheus; skapheutes,* m. digger; *askaphos,* undug, unhoed: *Scapheus ancylochelis* (a Jurassic crustacean), *Scapteromys tumidus* (a rodent), *Scapteriscus abbreviatus* (southern mole-cricket), *Scapheutes mocsaryi* (a wasp), *Ascaphus truei* (an amphibian).

See: **cut, furrow, till, mine, grave, hollow, shovel, pick**

digest < L. *digero, -estus,* separate, dispose, classify, consider, render food assimil-able: digestion, digestible, indigestion.

L. *concoquo, -coctus,* boil together, digest, consider; *concoctio,* digestion; see **cook**

Gr. *pepto (pesso; petto),* soften, ripen, cook, digest; *pepsis, -eos,* f. a softening, digestion; *peptikos,* of digestion; *peptos,* cooked: peptone, peptic, pepsin, dyspepsia, carboxypeptidase, *Pepsis elegans* (a spider-wasp), *Pepsonema pellucidum* (a worm), Pepsi-Cola.

digitabulum, L. a kind of glove; see **glove**

digitellum, L. houseleek; see **stonecrop**

digitus, L. finger; *digitulus,* dim.; *digitalis,* of the finger; see **finger**

digmato- < Gr. *deigma, -tos,* sample, pattern, specimen, proof; see *deigma* under **form**

dignus, L. worthy, deserving, honorable; *dignitas,* merit, rank; see **honor**

digressus, L. departure, deviation; see *digredior* under **depart**

dilatorius, L. delaying; see *differo* under **delay**

dilatus, L. spread, expanded; *dilatio,* delay; see *differo* under **spread, delay**

dilectus, L. beloved, dear; see **love**

dilemma, Gr. double proposition; see *lemma* under **think, trap**

diligent < L. *diligens,* careful, industrious; see **busy**

dill < AS. *dile;* see **carrot**

dilo- < Gr. *deile,* afternoon, evening, see **night**; < *deilos,* cowardly, paltry, see **poor**

diluculum, L. daybreak, dawn; see *antelucanum* under **dawn**

dilutus, L. mixed, weak, thin; see **mix**

diluvium, L. deluge, flood; *diluvialis,* of a flood; see *luo* under **wash**

dim < AS. *dim.*

Gr. *achlys, -yos,* mist, darkness, obscurity; *epachlyo,* to dim, darken; see **black**

Gr. *adelos,* unseen, unknown, obscure, dim.; see **secret**

Gr. *adiastaltos,* not clearly defined, indefinite, ambiguous: *Adiastaltus habilis* (a fossil monotreme).

Gr. *agnostos,* unknown, obscure; see *nosco* under **know**

Gr. *amauros,* dark, dim, faint, obscure, blind: amaurosis, *Amaurobius socialis* (a spider), *Amaurocichla kempi* (a babbler), *Rhamphalcyon amauroptera* (a king-fisher).

Gr. *amblyopos,* dim-sighted; see **blind**

Gr. *amydros,* indistinct, dim, obscure: *Amydrus morio* (red-winged starling), *Amydrosoma discedens* (a fly), *Amydraulax pulchra* (a wasp).

Gr. *aphanes; aphantos; apophanes; asymphanes,* unseen, invisible, obscure; see *aphanes* under **secret**

Gr. *apoptos,* dimly seen, far away:

Gr. *asaphes,* dim, indistinct, obscure, baffling: *Asaphus vetustus* (a trilobite), *Asaphocrinus excavatus* (a Silurian crinoid), asaphia.

Gr. *dysaithrios,* not clear, murky; see *aithrios* under **clear**

Gr. *dysderkes,* seeing badly; see *derkomai* under **see**

Gr. *glamyros,* blear-eyed, watery; see **pus**
L. *gramiosus,* full of pus, blear-eyed, watery, dim; see **pus**
L. *inconspicuus,* not readily seen, not prominent: inconspicuous, *Xylocopa incons-
picua* (a carpenter-bee).
L. *indefinitus,* without fixed limit, vague: indefinite, indefiniteness.
L. *indeterminatus,* undefined, unlimited; *indeterminabilis,* that cannot be defined:
indeterminate, indeterminable.
L. *indistinctus,* obscure, dim: indistinct, *Petrographus indistinctus* (a graptolite),
Xylophanes indistincta (a moth).
Gr. *lemaleos,* blear-eyed, dim:
L. *lippus,* blear-eyed, dim-sighted; *lippitudo, -inis,* f. inflammation of the eyes;
lipposus, blear-eyed:
L. *luscinus; luscus,* one-eyed, half-blind, blinded; *luscitiosus,* dim-sighted: *Gulo
luscus* (wolverine), *Mutilla lusca* (a velvet-ant).
L. *nebulosus,* misty, hazy, indefinite, obscure; see *nebula* under **cloud**
L. *obscurus,* indistinct, dim, dark: obscure, obscurity, chiaroscuro, *Lycopodium
obscurum* (a clubmoss), *Bibio obscuripennis* (a fly).
L. *opacus,* shady, obscure, dim; see **shade**
L. *septuosus,* obscure:
See: **shade, cover, secret, dull, blind**

dimidius, L. half; *dimidiatus,* halved; see *demi* under **half**

diminutives.   See **little**

dimo- < Gr. *deimos,* fear, terror; *deima, -tos,* a fearful thing; see *deinos* under **fear**

dimple < uncertain origin; see **hollow**

dino- < Gr. *deinos,* terrible, fearful, see **fear**; < *dinos,* a whirling, rotation, eddy,
see **turn**; < *dinos,* a large, round goblet, see **cup**

dinotos, Gr. turned, rounded; *dine,* whirl, eddy; see **turn**

dinuptila, L. a kind of bryony; see **melon**

Dio- < Gr. *Dis, Dios,* Zeus, chief of the Greek gods, see **spirit**; < *Dione,* Aphrodite,
see **love**

dioche, Gr. distance; see **far**

diocto- < Gr. *dioktos,* pursued; *diokter,* pursuer; see **hunt**

diodos, Gr. thorofare; see *hodos* under **way**

Diogenes, Gr. a Cynic who went about in daytime with a lighted lantern looking
for an honest man; see **doubt**

diogma, Gr. chase, pursuit; see *dioktos* under **hunt**

diomedea, NL. a genus of albatrosses; see **albatross**

Dionysus, L. (Gr. *Dionysos*), god of wine; see *Bacchus* under **wine**

diopter, Gr. scout, spy, an optical instrument; see *optikos* under **see**

diorygma, Gr. canal, channel; *dioryx,* canal, trench; see *orysso* under **dig**

dioscorea, NL. a genus of yams; see **yam**

diospyros, Gr. a kind of plant; see **persimmon**

diota, L. a two-handled jar; see **pot**

dioxis, Gr. chase, pursuit; see *dioktos* under **hunt**

dip < AS. *dyppan.*
Gr. *arneuter; arneutes,* m. diver, plunger:
Gr. *bapto,* dip, dye; *bapsis; baptisis,* f. a dipping, dyeing; *baptos,* dipped, dyed;
baptizo, immerse: baptize, baptism, John the Baptist, *Baptornis advenus* (a fossil
bird), *Baptisia perfoliata* (Georgia wild-indigo).
Gr. *dyo; dypto,* plunge in, dive, duck, enter; *dysis,* f. a dipping, setting; *dytes;
dyptes,* m. diver; *dytikos,* able to dive, diving: troglodyte, *Telmatodytes palus-
tris* (marsh-wren), *Eudyptes chrysocoma* (a crested penguin or rock-hopper),
Palaeeudyptes antarcticus (an Eocene penguin), *Ammodytes tobianus* (sand-eel),
Dytiscus latissimus (a diving-beetle), *Benthodytes abyssicola* (a holothurian).
L. *inficio, -fectus,* put or dip into anything, dye: see **color**
Gr. *kolymbao,* dive, swim; *kolymbetes; kolymbos,* m. diver, swimmer; *kolymbis,
-idos,* f. a diving bird: *Colymbetes signatus* (a beetle), *Colymbus auritus* (horned
grebe), *Tetraposthia colymbetes* (a worm).
L. *mergo, mersus,* dip, plunge in, sink; *mergus,* m. diver; *mergulus,* m. dim.;
demergo, sink, plunge in, dip; *immergo,* dip under, plunge in; *submergo,* plunge
under, sink, overwhelm: merge, emergency, immersion, submerge, merger, merganser, *Mergus americanus* (American merganser), *Hydrophilus mergus* (a
beetle), *Aptenodytes demersus* (a king-penguin).
Gr. *skinthos,* diving:

L. *tingo, tinctus,* bathe, dip, soak; *intingo,* dip in, pickle, preserve; see **color**

L. *urinor, -atus,* dive; *urinator, -is,* m. diver: *Pelecanoides urinatrix* (a diving-petrel).

See: **color, wash, swim, wet**

diphas, Gr. a kind of serpent; see **snake**

diphetor, Gr. searcher; see **try**

diphros, Gr. chariot, seat; *diphrion,* dim.; see **vehicle**

diphthera, Gr. leather; see **leather**

diphy- < Gr. *diphyes,* of double nature; see *di-* under **two**

diplo- < Gr. *diploos,* twofold; see *di-* under **two**

diploma, Gr. folded letter, passport, certificate; see **letter**

dipno- < Gr. *deipnon,* a meal; see **eat**

dipper < AS. *dyppan,* dip; see **spoon**

dipsa, Gr. thirst; see **dry**

dipsacus, L. (Gr. *dipsakos*), teasel; see **thistle**

dipsas, Gr. a venomous snake; see **snake**

dirado- < Gr. *deiras, -ados,* ridge of a hill; see **ridge**

Dirce, L. (Gr. *Dirke*), a spring in Thebes; see **spring**

direction < L. *directio, -onis,* a making straight; bearing of a point or object; see **north, east, south, west, wind, hand**

directus, L. straight, direct, set in order; see *dirigo* under **govern,** *rectus* under **right**

diremptus, L. separation; see *dirimo* under **separate**

direptus, L. plundered, ravaged, raped; see *rapio* under **take**

diribitor, L. separator, sorter; see *dis* under **separate**

diro- < Gr. *deire,* neck, throat; see **neck**

dirt < ON. *drit,* excrement.

Gr. *aithalos,* m. thick smoke, soot; *aithalion,* dim.; *aithaleos,* smoky, sooty; *aithalodes,* sooty, black: *Aethalium septicum* (a fungus), *Dictydiaethalium plumbeus* (a beetle).

Gr. *akathartos,* impure, foul: *Acathartus insignis* (a beetle).

Gr. *aloutos,* unwashed, filthy:

L. *amurca; amurga* (Gr. *amorge*), f. dregs or lees from pressing olives:

Gr. *anagnos,* impure, unclean:

Gr. *aniptos,* unwashed:

Gr. *aplytos,* unwashed; *aplysia,* f. filth; *apoplyma, -tos,* n. filth: *Apoplymus pectoralis* (a bug).

Gr. *aprobrasma, -tos,* n. scum:

Gr. *arda,* f. dirt; *ardalos,* dirty: *Ardalus aciculatus* (a wasp).

Gr. *asbolos; asbole,* f. soot; *asbolodes,* sooty: *Asbolus laevis* (a beetle).

Gr. *asyres,* filthy, lewd; see **lewd**

Gr. *borboros,* m. mud, mire, filth; *borborodes,* filthy, miry: *Borborocoris pallescens* (a bug), *Psiloniscus borborurus* (a beetle).

L. *caenum,* n. filth, mud, mire; *caenosus,* muddy, dirty, foul; *caenulentus,* covered with mud: obscene.

L. *colluvium,* collection of washings, sweepings, dregs; see *luo* under **wash**

L. *commingo, -inctus,* defile, pollute; *commictilis,* despicable, vile, that deserves to be defiled:

L. *conforio,* pollute, defile:

L. *contamino, -atus,* corrupt, defile, stain: contaminate, contamination.

L. *crassamentum,* n. dregs, grounds:

Gr. *deisa,* f. filth; *deisaleos,* filthy:

L. *faex, faecis,* f. dirt, *dregs,* lees, grounds; *faecula,* f. dim.; *faecarius,* of dregs; *faecatus,* made of dregs; *faecinus,* leaving dregs; *faecosus; faeculentes,* abounding in dregs, sediment; *faecus,* impure; *defaeco, -atus,* cleanse, purify, void excrement: feces, fecal, defecate, feculent.

L. *foedus,* foul, filthy, detestable: *Tachina foeda* (a fly), *Scarabaeus foedatus* (a beetle).

L. *frax, -acis,* m. dregs, grounds, refuse from oil-making:

L. *fuligo, -inis,* f. soot, smut; *fuliginatus,* sooty, painted black; *fuligineus,* like soot; *fuliginosus,* sooty: fuliginous, *Fuligo septica* (a myxomycete), *Fuligula rufina* (a duck), *Tipula fuliginosa* (a cranefly), *Xylocopa fuliginata* (a carpenter-bee).

L. *illuvies, -ei,* f. dirt, filth, mud, inundation; *illutilis,* that cannot be washed out; *illuviosus; illotus (illutus),* unwashed, unclean, dirty: *Drosophila illota* (a fruit-fly).

Gr. *ilys, -yos,* mud; see **earth**

L. *imbubino; imbulbito,* befoul, defile:

L. *immundus,* unclean, impure, foul; *immundita,* f. filth: *Chironomus immundus* (a midge).

L. *impurus,* unclean, filthy, defiled: impure, impurity.

Gr. *infectus,* tainted, dyed; see **color**

L. *inquino, -atus,* befoul, pollute, defile; *inquinatum,* n. filth: *Pelargonium inquinans* (sticky storksbill).

Gr. *katharma, -tos,* n. offscouring, refuse: *Catharma orthura* (a bird).

Gr. *korema, -tos,* refuse, sweepings, broom, < *koreo,* sweep out; see **broom**

L. *lino, litus,* daub, anoint, spread over; *collino,* besmear, defile, pollute; see **spread**

L. *lutum,* mud, mire, dirt, loam, clay; *lutulentus,* muddy, dirty, filthy, turbid, impure; *luto, -atus,* debaub with mud or clay; see **earth**

Gr. *lyma, -tos,* n. filth, dirt:

Gr. *miaros,* stained with blood, polluted, foul, coarse, disgusting; *miantos,* dyed, stained, defiled; *miasma, -tos,* stain, defilement; *miastor, -os,* m. polluter, < *miaino,* stain, defile, sully: miasma, amianthus, *Miaroblatta ella* (a fossil blattid), *Amianthium muscaetoxicum* (fly-poison), *Miastor metraloas* (a fly).

Gr. *molyno,* stain, defile; *molynsis,* f.; *molysmos,* m. pollution: *Molynoptera multiguttata* (a beetle).

Gr. *mysos,* n. dirt, uncleanness, anything disgusting; *mysaros,* foul, impure, loathsome: mysophobia, *Mysaromima liquescens* (a moth).

Gr. *pelos,* clay, earth, mud, dirt; see **earth**

Gr. *peripsema, -tos,* n. offscouring, refuse:

Gr. *phoryktos,* stained, defiled: *Phoryctus mucoreus* (a beetle).

Gr. *phorytos,* m. sweepings, chips, rubbish:

Gr. *phtharsis,* corruption; see *phtheiro* under **destroy**

Gr. *pinos,* m. dirt, filth, squalor; *pinaros,* dirty, squalid; *pinarotes, -etos,* f. filthiness; *pinodia,* f. dirt; *pinodes,* dirty: *Pinarolestes dissimilis* (a bird).

L. *polluo, -utus,* befoul, defile: pollute, pollution, *Asymphorodes polluta* (a moth).

Gr. *psegma,* that which is rubbed off, scrapings, dust; see **scrape**

Gr. *psolos,* m. soot, smoke: *Psoloessa ferruginea* (a grasshopper).

L. *purgamentum,* n. garbage, rubbish, filth:

L. *quisquilium,* n. refuse, trash, rubbish: *Anthriscus quisquiliarius* (a beetle).

Gr. *rhypos, m.* dirt, filth; *rhyparos,* dirty, filthy; *rhypax, -akos,* m. dirty person; *rhypasma, -tos,* n. dirt, filth; *rhyposis, -eos,* f. pollution; *rhyptikos,* cleansing from dirt; *arrhyptos,* unwashed: rhypophagy, rhyparography, rhypophobia, rhyptic, rupia, *Rhyparochromus irroratus* (a bug), *Rhypticus bistrispinus* (a soapfish).

Gr. *saroma,* sweepings; see *saron* under **broom**

L. *scelero, -atus,* pollute, defile, desecrate; see *scelus* under **fault**

L. *sordidus,* dirty, filthy; *sordidulus,* dim., < *sordeo,* be filthy, squalid; *sordes, -is,* f. dirt, filth: *sordicula,* f. dim.; *sordidatus,* dressed dirtily or shabbily: sordes, sordid, sordidness, sordor, *Sordaria fimiseda* (a fungus), *Chamaesarache sordida* (a nightshade).

L. *squalidus,* rough because of neglect, dirty; *squalor, -is,* m. roughness from dirt and filth; *squales, -is,* f. dirt, filth: squalid, squalor, *Leptops squalidus* (root-weevil).

Gr. *stemphylon,* n. pomace of pressed olives or grapes:

L. *stupro, -atus,* debauch, defile, dishonor; *constupro,* violate, ravish; see **coitus**

Gr. *syrphetos,* m. sweepings, refuse, litter; *syrphetodes,* jumbled together: *Syrphetodes marginatus* (a beetle).

L. *teter,* foul, hideous, offensive, shameful; see **bad**

Gr. *tholos,* mud, dirt; *tholeros,* muddy, impure, turbid; see **earth**

L. *turpis,* ugly, foul, base, shameful, filthy; *turpiculus,* dim.; *turpitudo, -inis,* f. ugliness, unsightliness, < *turpo, -atus,* soil, pollute: turpitude, *Scarabaeus turpis* (a beetle).

Gr. *tryx, -ygos,* f. new wine, lees of wine, dregs:

See: **earth, dung, lewd, rot, slag, worthless, bad**

dirus, L. fearful, ominous; see **fear**

dis, L. in two, asunder, away from, without, see **separate, not,** *di-* under **from;** < Gr. *dis,* twice, double, see *di-* under **two;** < L. *dis, ditis,* rich, abundant, see **wealth**

disagreeable < F. *desagreable;* see **bad, rough, unfit**

disaleo- < Gr. *deisaleos,* filthy; see *deisa* under **dirt**

disappear < L. *dis,* without, *appareo,* become visible; see **depart, death, move, cover**

disaster < OF. *desastre,* 'evil star', misfortune; see **destroy**

dischides, Gr. cloven, divided; see **cut**

discidium, L. discord, divorce; see *dis* under **separate**

discipulus, L. learner, pupil, follower; *disciplina,* instruction; see *disco* under **learn**

disco- < Gr. *diskos,* flat, circular plate; see **plate**

discolor, L. of different colors, variegated; see **color**

discontent < L. *dis,* not, *contentus,* satisfied; see **trouble**

discord < L. *discordia,* disagreement, disunion, dissension; *discors,* different, unlike; see **fight, different**

discover < OF. *descovrir,* uncover; see **find**

discrepo, L. disagree, differ; *discrepantia,* discordancy; see **different**

discretus, L. separated; *discretivus,* serving to distinguish; see *cerno* under **separate**

discrimino, *-atus,* L. distinguish, separate; see *dis* under **separate**

discus, L. (Gr. *diskos*), quoit, flat, circular plate; see **plate**

disease < L. *dis,* without, F. *aise,* ease.

L. *aeger, -gra, -grum,* sick, ill, troubled; *aegrotus,* sick; *aegritudo, -inis,* f. sickness, sorrow; *aegritatio, -onis,* f.; *aegror, -is,* m. sickness; *aegrimonia,* f. grief, trouble; *aegroto, -atus,* be sick: aegrotat, aegritude, egrimony.

Gr. *agra,* f. seizure: pellagra, chiragra, podagra.

NL. *allergia,* f. sensitivity and susceptibility to inoculation with foreign substances: allergy, allergic, allergen, allergin.

L. *alopecia* (Gr. *alopekia*), mange in foxes, baldness; see **bare**

L. *alphus* (Gr. *alphos*), m. a white spot on the skin, a kind of leprosy: alphosis, *Alphopteryx maculata* (a grasshopper).

Gr. *aneurysma, -tos,* n. permanent diseased dilatation of a bloodvessel: aneurysm (aneurism). See *A study in scarlet* (Conan Doyle).

L. *angina,* f. quinsy; distress: angina pectoris.

Gr. *ankylosis,* f. a bending, stiffening, and fixing of the joints: ankylosis (ancylosis).

Gr. *anthrax, -akos,* m. a kind of tumor: anthrax (caused by *Bacillus anthracis*), anthracnose, anthracosis.

Gr. *aphtha,* f. thrush: aphtha, aphthoid, aphthous, *Aphthomonas infestans* (organism causing foot and mouth disease), *Peltigera aphthosa* (a lichen).

Gr. *apoplexia,* f. stroke; *apoplektos,* crippled by a stroke, paralyzed: apoplexy, apoplectic.

L. *arquatus,* m. one who has jaundice:

Gr. *arrhostos,* sickly, weak; *arrhostia,* lingering illness; see **weak**

Gr. *arthritis,* f. inflammation of the joints: arthritis, arthritis deformans, arthritic.

Gr. *ase,* f. loathing, nausea, distress; *aseros,* causing nausea:

Gr. *askites,* m. a kind of dropsy: ascites.

Gr. *asthma, -tos,* n. hard-drawn, difficult breath, panting, gasping: asthma, asthmatic, *Typhlophora asthmatica* (ipecac-tree).

L. *aurugo, -inis,* f. jaundice; *aurugineus,* jaundiced, yellow:

Gr. *bdelygmia,* f. nausea, disgust; see **bad**

Gr. *bletos,* stricken by disease:

L. *cachexia* (Gr. *kachexia*), f. general ill-health attended by wasting; *kachektes; kachektikos,* consumptive, disaffected: cachexia, cachectic.

L. *cancer,* crab; a benign or malignant growth; see **sore**

L. *cardimoma,* f. pain in the stomach:

L. *catarrhus* (Gr. *katarrhous*), m. a morbid discharge, particularly from the nose: catarrh, catarrhal, catarrhous.

L. *catalepsis* (Gr. *katalepsis*), f. fit, seizure, sudden attack of illness: catalepsy, cataleptic.

L. *catocha* (Gr. *katoche*), f. possession, catalepsy:

I. *causarius,* sick, diseased, ill:

L. *cephalaea* (Gr. *kephalaia*); *cephalalgia* (Gr. *kephalalgia*), f. headache: cephalalgia.

L. *cerium* (Gr. *kerion*), n. a kind of skin eruption:

L. *chiragra* (Gr. *cheiragra*), f. gout in the hand: chiragra.

L. *choeras, -adis* (Gr. *choiras, -ados*), f. scrofula:

L., Gr. *cholera,* f. a bacterial disease: cholera.

Gr. *chordapsos,* m. disease in the large intestine:

L. *colicus* (Gr. *kolikos*), m. spasmodic pain in the colon or digestive tract: colic.

L. *contagio, -onis,* f. a touching of something unclean, infection: contagion, contagious.

L. *coriago, -inis,* f. a skin disease of animals:

L. *coryza* (Gr. *koryza*), f. a running from the nose, catarrh, cold: coryza.

L. *cruditas, -atis,* f. indigestion:

L. *destillatio, -onis,* f. a dripping down, running, catarrh:

Gr. *diabetes,* m. a name for two kinds of disease denoted by excessive passage of urine or of urine containing sugar: diabetes insipidus, diabetes mellitus.

L. *diarrhoea* (Gr. *diarrhoia),* f. a flowing through: diarrhea.

Gr. *dysenteria,* f. an intestinal disease: dysentery, *Pulicaria dysenterica* (a composite), *Holarrhena antidysenterica* (an apocynacead).

Gr. *dyspepsia,* f. indigestion: dyspepsia, dyspeptic.

L. *dyspnoea* (Gr. *dyspnoia),* f. difficulty in breathing, asthma: dyspnoea.

L. *dysuria* (Gr. *dysouria),* f. retention of urine, difficult urination: dysuria.

Gr. *ekphyma,* an eruption of pimples; see **pimple**

Gr. *ekzema, -tos,* n. tetter, a skin eruption: eczema, eczematous.

Gr. *empyema, -tos,* n. suppuration, abscess: empyema.

Gr. *enterokele,* rupture, hernia; see **break**

Gr. *epialos,* m. fever with shivering; *epialodes,* aguish:

Gr. *epidemos,* said of diseases that spread through a large number of people: epidemic.

Gr. *epiphora,* f. afflux, excessive watering of the eyes: epiphora.

L. *erysipelas, -atis* (Gr. *-atos),* n. a reddish skin eruption, St. Anthony's fire: *Streptococcus erysipelatis* (germ causing erysipelas), *Erysipelothrix porci* (cause of swine erysipelas).

Gr. *exanthema, -tos,* n. eruption: exanthematous.

L. *exochadium,* n.; *exochas, -adis* (Gr. *-ados),* f. external hemorrhoid; *esochas,* internal hemorrhoid:

L. *farciminum,* n. a disease of horses and cattle; *farciminosus,* of farcy: farcy, *Actinomyces farcinicus* (cause of cattle farcy).

L. *febris,* f. fever; *febricula,* f. dim.; *febrilis,* of fever; *febriculosus,* feverish: febrile, febricity, febricant, febrific, febrifuge, fever, feverfew, feverish, *Dichroa febrifuga* (a saxifragacead).

L. *foria,* f. flux, diarrhea:

Gr. *glaukoma, -tos,* n. a disease of the eye: glaucoma, glaucomatous.

Gr. *gomphiasis,* toothache; see *gomphios* under **tooth**

L. *gonorrhoea* (Gr. *gonorrhoia),* f. clap: gonorrhea.

L. *gravedo, -inis,* f. cold in the head, catarrh; *gravedinosus,* subject to colds: gravedo.

L. *gutturosus,* with a tumor in the throat, goitered:

Gr. *haimorrhagia,* f. violent bleeding: hemorrhage.

L. *haemorrhoida* (Gr. *haimorrhois, -idos),* f. pile: hemorrhoids, *Sarcophaga hemorrhoidalis* (a flesh-fly), *Gonophora hemorrhoidalis* (a beetle).

L. *hernia,* rupture; *herniosus,* ruptured; see **break**

L. *herpes, -tis* (Gr. *-etos),* m. a creeping skin eruption: herpes, herpetic, herpes zoster (shingles).

Gr. *hydrophobia,* f. fear of water, rabies: hydrophobia.

Gr. *hydrops, -opos,* m. edema: dropsy, dropsical.

NL. *hysteria,* uncontrolled emotional excitement; see *hystera* under **womb**

Gr. *ikteros,* m. jaundice; *ikterikos,* jaundiced, yellow: icteric, icteroid, icteritious, icterine, *Sphenarium ictericum* (a grasshopper).

L. *ileus* (Gr. *eileos; ileos),* m. severe colic; *iliacus,* of colic; *iliosus,* m. sufferer from colic: ileus, iliac.

L. *impalpebratio, -onis,* f. loss of motion in the eyelids:

L. *impetigo, -inis,* f. a skin eruption: impetigo, *Tabebuia impetiginosa* (a trumpet-tree).

L. *impetus,* attack, fit, spasm; see **strike**

L. *indigestio, -onis,* f. failure of food to undergo normal change in the alimentary canal: indigestion, indigestible.

L. *infectio, -onis,* f. contamination by disease-producing agents, < *inficio, -ectus,* taint, spoil: see **color**

L. *inflammatio, -onis,* f. congestion attended by heat, redness, swelling, and exudation: inflammation.

L. *intercus, -utis,* f. dropsy:

L. *jecorosus,* having the liver complaint; see *jecur* under **liver**

Gr. *karebareia,* f. headache:

Gr. *kardialgia,* f.; *kardiogmos,* m. heartburn, bellyache:

Gr. *karkinos,* m. crab, cancer; *karkinoma, -tos,* n. cancerous sore: carcinoma, carcinogen.

Gr. *kelephia; kelephiasis,* f. leprosy; *kelephos,* m. leper:

Gr. *kerioma, -tos,* n. a disease of the eyes:

Gr. *kynanche,* f. an inflammation of the throat: cynanche, quinsy, *Cynanchum acuminatifolium* (a swallowwort).

L., Gr. *lepra,* f. leprosy; *leprosus,* scaly, scabby, leprous: leper, leprosy, leprous, leprosarium.

Gr. *lepsis*, f. seizure: catalepsy, cataleptic, diabolepsy, epilepsy, epileptic.

Gr. *leukoma*, *-tos*, n. a white spot in the eye, cataract: leucoma.

Gr. *loimos*, m. pestilence, plague: loimia, loimology, *Loemopsylla nilotica* (a siphonapterid).

L. *lues*, *-is*, f. plague, pestilence: luetic.

L. *lumbago*, *-inis*, f. pain in the small of the back: lumbago.

F. *malaria*, < It. *mala*, bad, *aria*, air: malaria, malarial, *Plasmodium malariae* (the protozoan causing quartan malaria).

Gr. *melancholia*, having black bile, depression, sadness; see **sad**

L. *meliceris*, *-idis* (Gr. *melikeris*, *-idos*), f. a kind of skin eruption:

L. *mentigo*, *-inis*, f. an eruption on lambs:

L. *morbus*, m. sickness; *morbidus; morbosus*, sickly, diseased; ML. *morbillus*, m. dim. measle: morbid, morbidity, morbific, cholera morbus, morbillus, morbilliform, *Pamborus morbillosus* (a beetle), *Plowrightia morbosa* (plum black-knot).

Gr. *mydriasis*, f. prolonged dilatation of the pupil of the eye: mydriasis, mydriatic.

L. *nausea* (Gr. *nausia*), f. seasickness, retching; *nauseabilis; nauseosus*, causing vomiting: nausea, ad nauseam, *Chrysothamnus nauseosus* (a rabbit-brush).

Gr. *nosos*, f.; *nosema*, *-tos*, n. disease, sickness; *noseros*, diseased; *nosematikos*, sickly; *nosokomeion*, n. hospital, infirmary: nosology, nosomycosis, nosohemia, *Nosopsyllus fasciatus* (rat-flea), *Nosema bombycis* (protozoan causing pebrine).

L. *ostigo*, *-inis*, f. an eruption on lambs:

Gr. *paralysis*, f. palsy; *paralytikos*, struck with palsy: paralysis, palsy, paralytic.

Gr. *paroxysmos*, m. attack, fit, spasm, throe: paroxysm.

L. *pernio*, *-onis*, m. chilblain; *perniunculus*, m. dim.: pernio.

L. *pestis*, f. plague, contagion; *pestibilis; pestilentus; pestilis*, unhealthful, noxious: pest, pestilence, pestiferous, pesticide, pestilential, *Pasteurella pestis* (bacterium of plague, Black Death).

Gr. *phlegma; phlogmos*, inflammation; see *phlego* under **burn**

Gr. *phthisis*, f. consumption, decline, decay: phthisis, phthisical.

Gr. *podagra*, f. gout in the feet; *podagrikos*, gouty: podagra, podagrous, antipodagric, *Aegopodium podagraria* (goutweed), *Jatropha podagrica* (a euphorbiacead), *Rhyparochromus podagricus* (a bug).

Gr. *psoriasis*, a skin disease; see *psora* under **itch**

Gr. *pyretos*, m. burning heat, fever; *pyretion*, n. dim.; *lexipyretos*, allaying fever: pyretic, antipyretic, *Pyretophorus* (a group of mosquitoes).

L. *ramex*, *-icis*, hernia, rupture; see **break**

Gr. *rheuma*, *-tos*, n. discharge, catarrh; *rheumatikos*, subject to a flux; *rheumatismos*, m. rheum, flux: rheumatism, rheumatic.

Gr. *sarkoma*, *-tos*, n. tumor, cancer: sarcoma, sarcomatous.

L. *scabies*, mange, itch; see *scaber* under **rough**

ML. *scorbutus*, m. scurvy: scorbutic, antiscorbutic, *Pringlea antiscorbutica* (a crucifer).

L. *scrofula*, f. dim. of *scrofa*, sow; a swelling, like little pigs, of the glands in the neck: scrofula, scrofulous, crewels.

L. *serniosus*, covered with an eruption, scabby:

L. *sideratio*, *-onis*, f. stroke, apoplexy, blight, paralysis, thought to have been caused by an unlucky configuration of stars:

Gr. *spasma*, *-tos*, n.; *spasmos*, m. convulsion, cramp, fit, throe; *spasmation*, n. dim.; *spasmodes; spastikos*, convulsive, fitful, < *spao*, draw, pluck, pull, rend, tense: spasm, spasmatic, spasmodic, spastic, spasmotoxin, *Spasmostoma viride* (a ciliate protozoan).

Gr. *strophos*, m. a twisting of the bowels, colic:

L. *struma*, scrofulous tumor; see **swell**

L. *suppuro*, *-atus*, form pus, be ulcerous: suppurate.

NL. *syphilis*, a venereal disease; the name said to be from Syphilus, a swineherd in a Latin poem by Fracastoro: syphilis, syphilitic, *Ephedra antisyphilitica* (an ephedra), *Lobelia syphilitica* (blue lobelia).

L. *tabes*, a wasting disease; see *tabeo* under **lessen**

L. *tetanus* (Gr. *tetanos*), lockjaw; see **stiff**

L. *tormen*, *-inis*, m.; *torminum*, n. gripes, colic; *torminalis*, of or for colic; *torminosus*, subject to colic: *Sorbus torminalis* (a mountain-ash), *Lactarius torminosus* (a mushroom).

NL. *torticollis*, wryneck: torticollis.

Gr. *trismos*, a gnashing, grinding; lockjaw; see **grind**

NL. *tuberculosis*, < L. *tuberculum*, n. small swelling: tuberculin, *Mycobacterium tuberculosis* (tuberculosis bacillus).

L. *tumor*, a swelling; see **swell**

LL. *variola*, f. smallpox: variola, variolate.

L. *verminum*, n. bellyache:

L. *vitiligo, -inis,* f. a skin disease manifested by milk-white spots: vitiligo, vitili-
ginous.
L. *vomica,* f. ulcer, sore, boil, plague; *vomicus,* ulcerous; *vomicosus,* full of sores:
nux vomica (seed of *Strychnos nuxvomica*).

*   *   *   *   *   *   *

Gr. *-algia,* pertaining to pain, < *algeo,* feel pain, suffer; see **pain**
NL. *-iasis,* denoting disease, < Gr. *iasis,* f. cure, process of healing: acariasis,
elephantiasis, filariasis, psoriasis.
Gr. *-itis,* denoting inflammation, disease, pain, < *-ites (-itis),* belonging to, pertaining to: appendicitis, arthritis, endocarditis, laryngitis, meningitis, nephritis,
neuritis, osteomyelitis, peritonitis, phrenitis, pleuritis, poliomyelitis, rhinitis,
tonsillitis, uteritis, uvulitis, vaginitis.
Gr. *-oma,* denoting a tumor or morbid condition, < *-oma, -omatos,* a termination
suggesting condition or result: adenoma, adipoma, carcinoma, chloroma, chon-
droma, glaucoma, lipoma, mycetoma, papilloma, sarcoma, tyroma, xyloma.
Gr. *-osis,* a suffix denoting condition, usually but not always morbid: ankylosis,
arteriosclerosis, berylliosis, chlorosis, cirrhosis, diabrosis, metamorphosis, neurosis,
osmosis, psychosis, silicosis, thrombosis, trichinosis, tuberculosis.
Gr. *-pathy,* denoting suffering from disease; see *pascho* under **pain**
See: **pain, sore, hurt, trouble, sad, weak, bad, lessen, pus, swell, weak**

disgrace < It. *disgrazia;* see **shame**

dish < AS. *disc,* < L. *discus* (Gr. *diskos*), quoit; see **plate**

dishevel < OF. *deschevler,* < L. *dis,* apart, *capillus,* hair; let the hair remain
uncombed.
Gr. *aktenistos,* uncombed:
Gr. *apektos,* uncombed: *Apectolophus phalangioides* (a mite).
See: **confusion**

disi- < Gr. *deisi-,* < *deido,* fear; see *deimos* under **fear**

disjunctus, L. unyoked, disunited; see *dis* under **separate**

disk < L. *discus* (Gr. *diskos*), quoit, flat, circular plate; see **plate**

disorder; see **confusion, dishevel**

dispalatus, L. wandered, strayed, scattered; see *palor* under **wander**

dispar, L. different, unequal; *disparilis,* dissimilar; see **different**

dispendium, L. cost, see *pendo* under **pay;** *dispensator,* manager, steward, see
govern

dispersio, -onis, L. a scattering; see *spargo* under **spread**

display < OF. *despleier,* < L. *displico,* unfold.
Gr. *deiknymi,* bring to light, make known, inform; see *deiktos* under **try**
L. *deoperio, -ertus,* uncover, disclose, reveal:
L. *detego, -tectus,* uncover, reveal, disclose: detect, detective.
Gr. *ekphantor, -os,* m. revealer; *ekphantos,* revealed:
Gr. *enargma, -tos,* n. a phenomenon:
L. *manifesto,* show clearly, reveal; see **clear**
Gr. *menyo,* disclose, reveal, betray; see **treason**
L. *monstro, -atus,* show, point out; see *moneo* under **warn**
L. *ostendo, -entus; -ensus,* show, display, exhibit; *ostentatio, -onis,* f. display,
parade: ostensible, ostentation, *Thyropygus ostentatus* (a myriapod).
L. *phaenomenon* (Gr. *phainomenon*), n. appearance, happening, event: phenome-
non, phenology, phenomenal.   See *phaino* under **light**
L. *pompa* (Gr. *pompe*), f. procession, parade; *pompalis; pompaticus; pomposus;
pompikos,* showy, splendid: pomp, pomposity, pompous.
L. *promulgo, -atus,* make known, publish: promulgate, promulgation.
L. *revelo, -atus,* unveil, disclose: reveal, revelation, revelatory.
L. *speciosus,* showy, splendid; see **beauty**
See: **bare, open, teach, light, clear, rite, honor, manner, beauty, treason**

displicatus, L. scattered; see **spread**

disposition < < L. *dispositio,* arrangement; see **class, nature, manner, mind**

dispute < L. *disputo, -atus,* contend, discuss; see **fight**

disruptus, L. broken off, separated; see *rumpo* under **break**

dissectus, L. cut up; see *seco* under **cut**

dissepimentum, L. partition; see *septum* under **wall**

dissertatio, L. discourse; see *assero* under **speak**

dissidens, L. differing, disagreeing; see **different**

dissimilis, L. unlike; see **different**

dissipo, -*atus,* squander wastefully, indulge dissolutely; see **spread**

dissitus, L. apart, remote; see *absitus* under **far**

dissolutus, L. lax, dissipated, loose in morals; see **bad**

dissolve < L. *dissolvo,* disunite, loosen, separate; see **free,** *dis* under **separate**

dissonus, L. discordant, different; see **different**

dissos (dittos), Gr. double; see *di-* under **two**

dissutus, L. unstitched, ripped, opened; see **open**

distaff < AS. *distaef;* see **axis**

distagmos, Gr. doubt; see *distazo* under **doubt**

distance < L. *distantia,* remoteness; see **far**

distentus, L. filled up, full, swollen; see *distendo* under **swell**

distichlis, NL. a genus of grasses; see **grass**

distichus, L. (Gr. *distichos*), of two rows; see *stichos* under **line**

distinctus, L. separate, different; see *dis* under **separate**

distolos, Gr. in pairs; see *di-* under **two**

distortus, L. misshapen, deformed, twisted; see **hurt**

distress < OF. *destrece;* see **pain, trouble, disease, fretful, sad**

distribute < L. *distribuo,* -*utus,* allot, divide, give; see **give, separate, spread**

ditch < AS. *dic,* dike, trench.

    Gr. *amara,* f. trench, irrigation ditch: *Amarodytes duponti* (a beetle).

    Gr. *bothros,* m. trench, pit, trough; *bothrion,* n. dim.: bothrenchyma, bothrium, *Bothrodendron punctatum* (a lepidodendron), *Bothrops insularis* (island-viper), *Bothriolepis tuberculata* (a Devonian fish), *Chrysobothris femorata* (apple-tree borer), *Embothrium coccineum* (a proteacead), *Stenobothrus maculipennis* (a grasshopper), *Dibothriocephalus latus* (a tapeworm), *Bothriocyrtum californicum* (a trapdoor-spider).

    Gr. *bothynos,* m. trench, pit; *bothynotes,* m. ditcher, digger:

    L. *canalis,* m. channel, gutter, water-pipe; *caniculus,* m. dim.: canal, channel, caniculate, Panama Canal, *Busycon caniculata* (a snail), *Mytilus caniculus* (a sea-mussel).

    L. *cloaca,* sewer, drain; see **pipe**

    L. *colliquia* (*collicia*), f. drain, gutter, channel; *deliquia,* f. gutter:

    L. *corrugus,* m. canal, conduit:

    Sp. *cuneta,* f. gutter, drain, ditch:

    L. *elix,* -*icis,* m. drain, trench:

    L. *fossa,* ditch, canal; *fossula,* dim.; see *fodio* under **dig**

    L. *fusorium,* n. sink, drain:

    L. *incile,* -*is,* n. ditch, trench; *incilis,* pertaining to ditches:

    Gr. *kapetos* (*skapetos*), f. ditch, groove:

    Gr. *ochetos,* aqueduct; see **pipe**

    Gr. *orygma,* -*tos,* excavation, ditch, tunnel, mine, pit, trench, moat; see *orysso* under **dig**

    L. *scrobis,* m. ditch, trench; *scrobiculus,* m. dim.: scrobe, scrobiculate, *Cupania scrobiculata* (a sapindacead).

    Gr. *skamma,* -*tos,* n. trench, pit: *Scammatonotum herero* (a wasp).

    Gr. *taphros,* f.; *taphreuma,* -*tos,* n., trench, ditch: taphrenchyma, taphrophilous, taphrinose, *Taphrospira convallata* (a gastropod), *Taphrina aurea* (a fungus), *Deretaphrus alveolatus* (a beetle).

    See: **dig, furrow, grave, lacuna, hollow, passage, valley, stream**

ditissimus, L. very rich, abundant; see *dis* under **wealth**

diurnus, L. of the day; see *dies* under **day**

diutinus; diuturnus, L. lasting long; *diutius,* longer; see **long**

divaricatus, L. spread apart; see **spread**

dive < AS. *dyfan,* sink; see **dip**

diversus, L. different; *diversitas,* variety, difference; see **different**

diverticulum, L. bypath; see *devortium* under **way**

dives, -*itis,* L. rich; *divitior,* richer; *divitissimus,* richest; see **wealth**

divide < L. *divido,* -*visus,* cut; see **cut, separate**

divine < L. *divinus; divus* (*dius*), belonging to a god, see *deus* under **spirit;** *divino,* -*atus,* foretell, see **prophecy**

divisus, L. cut, separated, distributed; see *divido* under **cut**

divorce < L. *divortium,* separation; see *dis* under **separate**

divum, L. sky; see **heaven**

dixo- < Gr. *dixos,* a variant of *dissos,* double, see *di-* under **two**; < *deixis,* proof, specimen, see *deiktos* under **try**

dizzy < AS. *dysig,* giddy.
Gr. *ilingos,* m. dizziness, vertigo; see *ilinx* under **turn**
Gr. *skotodinia,* f. dizziness; *skotodes,* giddy: *Scotodes annulatus* (a beetle).
L. *vertigo, -inis,* f. dizziness; *vertiginosus,* dizzy: vertigo, vertiginous.
See: **turn, shake, mad, feel**

dmetos, Gr. tamed; *dmeter,* tamer; *dmesis,* a taming; *dmoë; dmos,* slave taken in war; *dmoios,* servile; see **tame**

dnopheros, Gr. dark, gloomy, murky, < *dnophos,* darkness, gloom; see **night**

do < AS. *don;* see **make**

doche; docheion, Gr. holder, receptacle; see *dochos* under **hold**

dochlea, L. a plant; see **mint**

dochmos; dochmios, Gr. slanting, oblique; see **slope**

dochos, Gr. able to hold; see **hold**

docido- < Gr. *dokis, -idos; dokidion,* dim. of *dokos,* beam; see **beam**

docilis, L. easily taught; *docibilis,* learning easily, teachable; see *doceo* under **teach**

docimo- < Gr. *dokimos,* assayed, tested, approved; *dokimastes,* assayer, examiner; see **try**

doco- < Gr. *dokos,* main beam; *dokidion,* dim.; see **beam**

doctor, L. teacher; *doctus,* educated, skilled; see *doceo* under **teach, heal, know**

documentum, L. pattern, specimen, example; see **form**

dodder < ME. *doder.*
L. *cadytas* (Gr. *kadytas, kasytas*), a parasitic plant, probably dodder; see *cassytha* under **laurel**
NL. *cuscuta,* f. dodder, < Ar. *kushkut: Cuscuta indecora* (a dodder), *Diastrophus cuscutaeformis* (a blackberry gall).
Gr. *epithymon,* n. dodder:

dodecatheon, L. (Gr. *dodekatheon*), a genus of the primrose family; see **primrose**

doedyco- < Gr. *doidyx, -ykos,* pestle; see **pestle**

dog < AS. *docga:* dogged, dogbane, dogfish, dog-days, bulldog.
L. *canis,* c. dog, bitch; *canicula,* f. dim.; *canarius; caninus,* of dogs; *canicularis,* of the dog-days: canine, canary, *cave canem!,* canicular, canaille, chenille, kennel, *Canis familiaris* (dog), *Canis aureus* (jackal), *Rosa canina* (dogrose), *Cynias canis* (smooth dogfish), *Cynocephalus caninus* (pink-capped stinkhorn), *Serinus canarius* (a finch), *Junco caniceps* (gray-headed junco), *Homalomyia canicularis* (a fly).
L. *catulus,* m. puppy, whelp, young animal; *catula,* f. little bitch; *catulinus,* pertaining to a young dog: *Catulus stellaris* (spotted dogfish).
L. *Cerberus* (Gr. *Kerberos*), m. three-headed doglike monster with serpent's tail that guarded the gates of Hades, hell-hound: *Cerbera fruticosa* (Cerberus tree), *Cerberus rhynchops* (an aquatic snake).
AS. *hund,* dog: hound, bloodhound, houndstongue, dachshund, greyhound. Derive: horehound.
L. *hylax, -acis,* the barker; see *hylagma* under **bark**
Gr. *kyon, kynos,* m. dog; f. bitch; *kynarion; kynidion,* n.; *kynideus; kyniskos,* m. dim., puppy, whelp; *kyneios,* of dogs; *kynikos,* doglike: cynosure, cynic, quinsy, *Cyon javanicus* (adjag, wild dog of Java), *Cynoglossum officinale* (houndstongue), *Cynomys ludovicianus* (prairie-dog), *Cynoscion regalis* (weakfish), *Cynictis penicillata* (meerkat), *Hydrocyon vittatus* (a tigerfish), *Procyon pygmaeus* (a raccoon), *Apocynum pumilum* (a dogbane), *Otodectes cynotis* (a mite).
L. *Maera* (Gr. *Maira*), f. a woman changed into a dog; name of a dog: *Aglaoglypta maera* (a fossil gastropod).
Sp. *perro,* m. dog; *perra,* f. bitch; *perrico; perrillo; perrito,* m. dim.:
Gr. *pholys, -yos,* m. a kind of dog:
Gr. *skylax, -akos,* c. young dog, puppy, whelp; *skylakion,* n. dim.; *skylakinos,* of puppies; *skylakodes,* like a pup; *skylakeutes,* m. dog-trainer: *Scylacosaurus sclateri* (a fossil reptile).
L. *vertagus,* m. greyhound: *Isotoma(Vertagopus) cinerea* (a collembolid).
Derive: beagle, collie, cur, dane, malemute, mastiff, mongrel, mutt, poodle, pug, pup, spitz, terrier, whippet.

dogbane
NL. *acocanthera,* f. a genus of the dogbane family: *Acocanthera abyssinica* (a bushman's poison).

L. *apocynum* (Gr. *apokynon*), n. a genus of the dogbane family: *Apocynum scopulorum* (cliff-dogbane).
NL. *carissa*, f. a genus of the dogbane family: *Carissa grandiflora* (Natal plum).
L. *vinca* (*pervinca*), f. a genus of the dogbane family: *Vinca herbacea* (a periwinkle).

dogma, Gr. opinion, doctrine; *dogmatikos,* of doctrine; see *dokeo* under **think**

dogwood < uncertain origin.
NL. *aucuba*, f. a genus of the dogwood family, < a Jap. word: *Aucuba japonica* (a cornacead).
L. *cornus*, f. cornel, dogwood; *cornum*, n. fruit of the dogwood: cornel, *Cornus florida* (flowering dogwood), *Pittosporum cornifolium* (a pittosporum), cornelian, Cornelia, *Elsinoë corni* (a fungus).
Gr. *kraneia*, f. dogwood:
NL. *nyssa*, f. once a genus of the dogwood family: *Nyssa aquatica* (a tupelo)'.

dolabra, L. ax, hatchet; *dolabella,* f. dim.; see **ax**

dolatus, L. hewn, cut; *dolatilis,* easily hewn; see *dolo* under **cut**

doleros, Gr. deceitful; see *dolus* under **lie**

dolichos, Gr. long, see **long**; kidney-bean, see **bean**

dolium, L. large jar or cask for wine; *doliolum,* dim.; see **barrel**

doll, contraction of *Dorothy.*
Gr. *dagys, -yos,* f. doll, puppet:
Gr. *glene,* eyeball, doll (from the small image in the eyeball); see **eye**
L. *planguncula,* f. (Gr. *plangon, -os,* m.) wax doll, puppet:
L. *pupa,* girl, doll; see **child**

doloma, Gr. trick, deceit; *dolomedes,* crafty, wily; see *dolus* under **lie**

dolon, Gr. dagger, stiletto, pike; see **dagger**

dolopo- < Gr. *dolops, -opos,* ambusher; see *dolus* under **lie**

dolor, L. ache, pain, distress, grief, smart; *dolorosus,* painful, sad; see **pain, sad**

dolosus, L. crafty, cunning, deceitful; see *dolus* under **lie**

dolphin < OF. *daulphin,* < L. *delphinus* (Gr. *delphis*); see **whale**

dolus, L. (Gr. *dolos*), artifice, deceit, guile; see **lie**

dom, AS. state, condition, jurisdiction; see **nature**

domabilis, L. tamable; see *domo* under **tame**

dome < F. *dome,* < L. *domus,* roof, house; see **arc**

domesticus, L. of a house or household; *domicilium,* abode, dwelling; see *domus* under **house**

dominus, L. master; *domina,* mistress; *dominulus,* dim.; see **govern**

domitus, L. tamed; see *domo* under **tame**

domus; domicilium, L. house, home, abode; *domucula; domuncula,* dim.; *domitius,* pertaining to the house; see **house**

donax, L., Gr. reed, cane, see **grass**; mollusk, see **mollusk**

donum, L. gift; *donator; donatrix,* giver; see *do* under **give**

door < AS. *dor.*
L. *abitus,* m. outlet, exit, egress:
L. *accessus,* m. entrance; see **enter**
L. *aditus,* m. entrance; *aditiculus,* m. dim.; *aditialis,* of an entrance: adit.
Gr. *anapiesma, -tos,* trap-door on a stage:
Gr. *anoigma, -tos; anoixis,* an opening, door; see **open**
Gr. *bater, -os,* m. threshold:
Gr. *belos* (*balos*), m. threshold:
L. *claustrum,* bar, door, gate, wall; see **bar**
Gr. *diklis, -idos,* double-folding (doors); see **fold**
Gr. *eisbasis; eisodos,* f. entrance, entry:
L. *exitus,* a way out; see *eo* under **move**
L. *foris,* f. door, entrance; *fornicula,* f. dim., little door, shutter:
L. *janua,* f. door, entrance; *Janus,* m. god of gates and doors, represented with two faces; *janitor, -is,* m.; *janitrix, -icis,* f. doorkeeper: janitor, Janus, January, janiceps, *Janusia californica* (a malpighiacead), *Galerita janus* (a beetle).
Gr. *klisias, -ados,* f. folding door or gate:
L. *limen, -inis,* n. threshold, entrance; *liminaris,* of a threshold or lintel: limen, eliminate, preliminary, subliminal.
L. *ostium,* n. entrance, door, opening; *ostiolum,* n. dim.; *ostiarius,* of a door: ostium, ostiary, ostiolate, ostiolum, usher.

Gr. *oudos,* m. threshold:

L. *porta,* f. gate, door; *portella; portula,* f. dim.: portal, portcullis, porthole,
Sublime Porte.

L. *postica,* f.; *posticum,* n. backdoor, backhouse, privy; *posterula; posticula,* f.
dim.: *Psorophora posticatus* (a mosquito).

L. *praefurnium,* n. door of a furnace:

Gr. *pyle,* f.; *pylos,* m. gate, orifice; *pylis, -idos,* f. dim.; *pyloma, -tos,* n.; *pylon, -os,*
m. gateway; *pyloros,* m. gatekeeper; *parapylion,* n. side-door or gate; *propylaion,*
n. gateway, entrance: pylon, propylon, propylaeum, pylorus, micropyle, Thermopylae, propylite, propylitization, *Pylodiscus triangularis* (a radiolarian), *Dipy-
lidium caninum* (dog-tapeworm).

Gr. *thyra,* f. door; *thyrion,* n; *thyris, -idos,* f. dim., window; *thyrotos,* with a door
or aperture; *thyridotos,* having windows; *dithyros,* two-doored, bivalved; *para-
thyros,* f. side-door; *pseudothyron,* n. secret door: thyridium, *Thyropygos luxuriosus* (a myriapod), *Thyrioclostera trespunctata* (a moth), *Thyridocrinus patulus*
(a fossil crinoid), *Athyrium esculentum* (a fern), *Athyris polita* (a fossil brachiopod), *Synthyris dissecta* (a figwort).

L. *valva,* f. leaf of a folding door; *valvola* (*valvula*), f. dim.; *valvatus,* with folding
doors: valve, valvate, bivalve, valvular, *Valvulina tribullata* (a foraminifer),
Nothoscordum bivalve (yellow false-garlic), *Cochlophora*(*Apterona*) *valvata*
(a moth), *Nama quadrivalvis* (hairy nama).

See: **hole, harbor, window, enter, way**

dora, Gr. hide; see *derma* under **skin**

doracino- < Gr. *dorakinon,* apricot; see **plum**

dorato- < Gr. *dory, -ratos,* spear; *doration; dorydion,* dim.; see **spear**

dorcado- < Gr. *dorkas, -ados,* gazelle; see **antelope**

dorcus, L. (Gr. *dorkos,* gazelle), a genus of beetles; see **beetle**

dorido- < Gr. *doris, -idos,* knife; see **knife**

dorea; dorema, Gr. gift, present; see *do* under **give**

dormio, *-itus,* L. sleep; *dormitator,* sleeper; *dormitorium,* sleeping-room; see **sleep,
room**

doron, Gr. gift; see *do* under **give**

doronicum, NL. a genus of the composite family; see **composite**

doros, Gr. leather bag, wallet; see **bag**

dorpon, Gr. supper, meal; see **eat**

dorsum, L. back, ridge of a hill; see **back**

dory < F. *doree,* gilded, < L. *deauro,* gild.
L. *faber, -bri,* m. dory: *Zeus faber* (a dory), *Chaetodipterus faber* (spadefish).
L. *zeus* (Gr. *zaios*), m. dory: *Zeus australis* (a dory), Zeidae.

dory, *doratos,* Gr. spear; *doration; dorydion,* dim.; *dorybolos,* hurling spears; see
spear

dorycnium, L. (Gr. *doryknion*), a plant; name now applied to a legume; see **bean**

dosinia, NL. a genus of mollusks; see **mollusk**

dosis, Gr. gift; see *do* under **give**

dossenus, L. clown, jester; see **laugh**

dot < AS. *dott,* speck, head of a boil; see **spot, mark**

dotalis, L. pertaining to a dowry, < *dos, dotis,* a marriage portion; *dotatus,* gifted,
endowed; see *do* under **give**

dothien, Gr. abscess, boil; see **sore**

dotos, Gr. gift; *doter,* giver; see *do* under **give**

double < L. *duplus,* twofold; see **two**

doubt < L. *dubito, -atus,* hesitate, waver, be uncertain; *dubiosus; dubitativus;
dubius,* wavering, uncertain, doubtful; *dubitalis,* to be doubted: dubious, dubiety,
indubitable, *Leptoptilus dubius* (adjutant), *Baeomorpha dubitata* (a fossil
insect).

Gr. *adektos,* not received, incredible:

Gr. *admolie,* f. uncertainty:

L. *ambiguus,* changeable, uncertain, doubtful, of doubtful meaning, < *ambigo,* go
about or around, wander, waver, hesitate, debate: ambiguous, ambiguity, *Phoebe
ambigens* (a lauracea).

Gr. *amphibolos,* encompassed, between two fires, ambiguous, doubtful: amphibole,
Amphibolips ilicifoliae (an oak-gall maker).

Gr. *amphideritos,* disputed, doubtful: *Amphideritus formosus* (a beetle).

Gr. *amphidoxos,* dubious, doubtful: *Amphidoxotherium cayluxi* (an Eocene insectivore).

Gr. *amphignoia,* f. doubt; *dysgnoia,* f. ignorance, doubt:

Gr. *amphilektos; antilektos,* doubtful, disputed: *Amphilectus papillatus* (a sponge).

Gr. *amphilogos,* questionable, doubtful:

Gr. *antonimia,* f. ambiguity or contradiction in law: antinomy, antinomian.

Gr. *aoristos,* indefinite, indeterminate; *adioristos,* undefined: aorist.

Gr. *apistos,* incredulous, doubting, distrustful, untrustworthy; *apistosyne,* f. unbelief, distrust; *dyspistos,* hard to believe, incredible: *Apistus evolans* (a fish), *Apistocerus wasmanni* (a beetle).

Gr. *apithanos,* incredible, unlikely, improbable: *Apithanus jocularis* (a fossil insect).

Gr. *aporema, -tos,* n. doubt, question, perplexity; *aporia,* f. difficulty: *Aporemodon tomlini* (a gastropod).

Gr. *atekmartos,* baffling, obscure, uncertain:

L. *Cynicus* (Gr. *kynikos,* doglike), m. a Cynic philosopher; a sneeringly distrustful and disbelieving person: cynic, cynical, cynicism.

L. *diffidentia,* f. lack of confidence, doubt, mistrust: diffident.

Gr. *Diogenes,* m. a Cynic who went about in daytime with a lighted lantern looking for an honest man: *Cenobita diogenes* (a crab).

Gr. *distazo,* doubt, hesitate; *distaktikos,* expressive of doubt, doubtful; *distagmos; distasmos,* m. doubt, uncertainty:

Gr. *doie,* f. doubt, uncertainty:

Gr. *dyskritos,* hard to determine, doubtful: *Dyscritomyia fulgens* (a fly).

L. *forsan; forsit; forsitan; fortasse,* perhaps:

Gr. *hypopsios,* viewed with suspicion; *hypopsia,* f. suspicion:

L. *incertus,* doubtful, unsettled: uncertain, incertae sedis, *Rhagodinus incertus* (a spider), *Lepthyphantes incertissimus* (a spider).

L. *infidelis,* unbelieving, faithless, false; see **lie**

Ar. *kafir,* unbeliever, infidel: Kaffir, *Sisyphus caffer* (a beetle), *Erythrina caffra* (a coral-bean).

Gr. *mermerizo,* be in doubt:

L. *precarius,* obtained by prayer, doubtful, uncertain, transient: precarious.

Gr. *skeptikos,* thoughtful, critically doubtful: skeptic, skepticism, skeptical.

L. *suspicio, -pectus,* regard with mistrust; *suspicax, -acis,* distrustful; *suspiciosus,* full of mistrust; *suspicor, -atus,* mistrust, surmise, conjecture: suspect, suspicion, suspicious, *Heloderma suspectum* (Gila monster).

See: **lot, two, think, lie**

dough < AS. *dag;* see **mix**

dove < AS. *dufe.*

L. *columba,* f.; *columbus,* m. dove, pigeon; *columbula,* f. (*-lus,* m.), dim.; *columbarius; columbinaceus; columbinus,* pertaining to doves, dovelike: Columbus, columbite, Sitka columbine(*Aquilegia formosa*), *Columba inornata* (blue pigeon), *Columbus picui* (picui dove), *Columbicola extincta* (a bird-louse), *Columbella lunata* (a gastropod), *Tremex columba* (a horntail), *Exogyra columbella* (a fossil pelecypod), *Geranium columbinus* (a geranium) *Scabiosa columbaria* (a scabious), *Columbia longipetiolata* (a tiliacead).

NL. *didus,* m. the genus including the dodo: *Didus cucullatus* (a dodo), *Didunculus strigirostris* (tooth-billed pigeon).

Gr. *oinas, -ados,* f. a wild pigeon: *Chrysoenas viridis* (a pigeon), *Alectroenas pulcherrima* (a wart-pigeon).

L. *palumbus,* m.; *palumba,* f. ringdove, wood-pigeon; *palumbulus,* m. dim.; Sp. *paloma,* f.; *palomica; palomita,* f. dim.: La Paloma, Mt. Palomar, *Columba palumbus* (ringdove).

Gr. *peleia,* f. wild pigeon: *Melopelia asiatica* (a bird).

Gr. *peristera,* f. pigeon, dove; *peristerion,* n. dim.: peristerite, *Peristeria elata* (an orchid).

Gr. *phaps, phabos,* f. a wild pigeon: *Phaps chalcoptera* (bronzewing), *Geophaps scripta* (partridge-pigeon), *Phabotreron*(for *Phapitreron*) *amethystina* (a bird), *Phabotypus palumbarius* (a bird).

Gr. *phassa (phatta),* f. ringdove; *phattion,* n. dim.: *Petrophassa albipennis* (a bird).

L. *teta,* f. a kind of dove:

Gr. *treron, -os,* f. wild dove; shy, timorous: *Treron pompadora* (a dove), *Lamprotreron superba* (a fruit-pigeon).

Gr. *trygon, -os,* f. turtledove; *trygonion,* n. dim. pet name for a girl:

L. *turtur, -is,* m. turtledove; *turturilla,* f. dim.: *Streptopelia turtur* (turtledove).

dovetail; see **bind**

down < AS. *adun,* off the hill.

L. *aborior, -ortus,* set, disappear, perish; see **abortion**

L. *de*, down, as in *declivis*, sloping downward; *decurro*, run down; *dejectus*, downcast; *deorsus*, downward; *dependeo*, hang down; *descendo*, come down; *destruo*, pull or tear down; *devolvo*, roll down: declivity, decurrent, dejected, depend, descend, destruction, devolution, *Euryproctus depressus* (a wasp).

Gr. *dysis; dysme*, f. setting of the sun; *dytikos*, setting, < *dyo*, sink, set, enter:

Gr. *hizema*, a setting down, sinking; see *hisma* under **sit**

Gr. *kata*, down, away, against, very; *kato*, beneath, below; *katelys*, going down; *katotatos*, lowest: cataclysm, catacomb, catastrophe, catalysis, catalog, category, catabolism, cataract, catodont, catogenic, cathedral, cathode, catheter, catholic, *Catadyptes chrysolophus* (a penguin), *Catascopus auratus* (a beetle), *Catasetum tridentatum* (an orchid), *Catocala fraxini* (a moth), *Cathypna sulcata* (a rotifer).

Gr. *klisis*, inclination, decline, sinking; see *clino* under **slope**

L. *obitus*, a going down, destruction; see **destroy**

L. *occido, -casus*, go down, set; see **west**

L. *sido*, sit down, settle, sink; see **sit**

See: **low, deep, under, from, hell, slope**

doxa, Gr. opinion, glory, praise; *doxarion*, dim.; *doxastos*, conjectural; see *dokeo* under **think**

dozen < F. *douze*, < L. *duodecim*, twelve; see **twelve**

draba, L. (Gr. *drabe*), a plant of the mustard family; see **mustard**

draco, L. (Gr. *drakon*), a fabulous lizardlike animal; L. *dracaena* (Gr. *drakaina*), she-dragon; *dracunculus*, dim.; see **lizard**

dracontido- < Gr. *drakontis, -idos*, a kind of bird; see **bird**

dracontium, L. (Gr. *drakontion*), dragonwort; see **arum**

dragma, Gr. handful, sheaf; see **bundle**

dragmos, Gr. grasping, < *drassomai*, grasp; see **take**

dragon < L. *draco* (Gr. *drakon*), a fabulous lizardlike animal; see **lizard**

drain < AS. *dreahnian;* see **flow, suck, run, stream, ditch**

drake. Cf. G. *enterich*, male duck; see **duck**

dram < OF. *drame*, < L. *drachma* (Gr. *drachme*), a weight and coin; see **weight, money**

drama, L., Gr. act, deed, play; see **play**

drapeta, L. (Gr. *drapetes*), runaway slave, fugitive; see *dromos* under **run**

drasmos, Gr. flight; see **run**

drastic < Gr. *drastikos*, powerful, efficacious; see *drao* under **make**

draw < AS. *dragan:* drag, draggle, drawl, draft, dray.

L. *adductus*, stretched, strained, contracted: adductor, adduction.

Gr. *anapastos*, drawn up, drawn back:

L. *aquilegus*, water-drawing:

L. *blandior, -ditus*, flatter, caress, coax, cajole; see *blanditia* under **agreeable**

L. *contentus*, stretched, strained, tense, tight; see *tendo* under **spread**

Gr. *delear, -atos*, n. bait, incitement; *deletion*, n, dim.; *delastikos*, enticing, seductive; *deleama; deleasma, -tos*, n. bait; *deleastron*, n. baited trap; *deleazo*, bait, entice: *Deleastes daector* (a fish).

L. *delenio, -itus*, cajole, charm, win, captivate, entice:

Gr. *dolos*, bait for fish; see **trap**

Gr. *eidar, -atos*, bait; see *edo* under **eat**

Gr. *epagogus*, attractive, alluring, seductive:

Gr. *eryo; eiryo*, draw, drag, pluck; *erysis*, f. a drawing, plucking, dragging: *Eryops megacephala* (a Permian stegocephalian), *Erysimum asperum* (western wallflower).

L. *esca*, bait, food; *inesco, -atus*, allure with bait, entice; see **food**

L. *fascino, -atus*, bewitch, charm, enchant; see **magic**

L. *harpax*, drawing, seizing; see **take**

L. *haurio, haustus*, draw up, drain, drink in, consume; see **suck**

Gr. *helko*, draw, drag, tow; *helxis*, f. a drawing, dragging; *helketer, -os*, m. dragger, drawer; *helkyster, -os*, m. instrument for drawing, midwife's forceps; *helkysma, -tos*, n. that which is drawn; *helkystos*, drawn; *helktos*, that can be drawn; *holke*, f. a drawing, towing; *holkos*, attractive; *holkimos*, ductile; *epholkos*, alluring: *Holcocneme lucidus* (a wasp), *Notholca longispina* (a rotifer), *Helxine soleroli* (babytears).

L. *incanto, -atus*, charm, spell; see *cantamen* under **magic**

L. *inductus*, m. a persuasion:

L. *invitamentum*, n. allurement, inducement, < *invito, -atus*, attract, entertain: invitation.

L. *lacio*, entice, allure, charm; *lacto, -atus*, cajole, flatter, wheedle; *allicio, -ectus*, draw to oneself, attract; *allector, -is*, m. an enticer; *delicio*, delight, entice; *elicio, -itus*, draw out, lure forth; *elecebra*, f. allurer, golddigger, sponger, wheedler; *illicio, -ectus*, allure, entice, inveigle, decoy, seduce; *illecto, -atus*, allure, attract, invite; *illecebra*, f. enticement, allurement, bait; *illecebrosus; illex, -icis; illicibilis*, attractive, seductive; *illicitator, -is*, m. a sham bidder at an auction to raise others' bids; *illicium*, n. attraction, allurement; *oblecto, -atus*, delight, divert; *pellicio, -ectus*, allure, coax, decoy, inveigle, wheedle; *pellax, -acis*, deceitful, seductive: lace, latchet, delicious, delight, delectable, elicit, dilletante, illecebrous, oblectate, *Pellax huttoni* (a gastropod), *Gynaecoserica pellecta* (a beetle), *Illicium parviflorum* (yellow anise-tree), *Illecebrum verticillatum* (whitewood), *Rubus illecebrosus* (a raspberry), *Citrus deliciosa* (tangerine).

L. *leno, -onis*, m. procurer, seducer; see **prostitute**

L. *magnes, -etis*, m. lodestone, < Magnesia, a district of Thessaly: magnesium, magnet, magnetic, magnetism, manganese, manganic, manganous, permanganate.

Gr. *manganon*, charm, spell; see **magic**

Gr. *paleutes*, m.; *paleutria; paleutris, -idos*, f. decoy-bird; *paleuma, -tos*, n. allurement; *paleuo*, decoy:

L. *Siren* (Gr. *Seiren*), f. a sea-nymph who lured mariners to destruction: siren, *Siren lacertina* (mud-eel), *Lepdiosiren paradoxa* (an eellike dipnoan fish).

Gr. *spao*, draw, pluck, pull, tense; see *spasma* under **disease**

Gr. *stergethron*, love-charm; see **love**

Gr. *syro*, draw, drag, trail; see *syrtos* under **carry**

Gr. *tonos*, m. a stretching, tightening, bracing, strain, tension, rope, cord; *tonikos*, pertaining to tension; *syntonia*, f. tension: tonic, *Tonoscolex ferinus* (a worm).

L. *tormentum*, torture, rack; see **pain**

L. *traho, tractus*, draw, haul, pull; *traha*, f. drag, sled; *contractus*, draw together, drawn up; *distractus*, drawn away: tractor, contraction, detractor, intractable, subtract, retraction, extract, attractive, abstract, trail, trait, train, treat, treaty, entreat, treatise, portrayal, distraught, *Camarotoechia contracta* (a fossil brachiopod), *Volutoderma protracta* (a fossil gastropod).

L. *vello, vulsus*, pull, pluck, tweak; see **tear**

See: **suck, carry, tear, bind, lessen, pump, wick, form, magic, flatter**

dread < AS. *draeda*, fear; see **fear**

dream < AS. *dream*; G. *traum*.

Gr. *enypnion*, n. dream; *enypniastes*, m. dreamer: *Enypnium quadripunctatum* (a fly), *Enypniastes eximia* (a holothurian).

Gr. *ephialtes*, m. nightmare: *Ephialtes cristata* (a bird).

L. *hallucinor, -atus*, wander in mind, dream; *hallucinatio, -onis*, f. vision, fantasy: hallucination.

L. *incubus*, nightmare, male demon; see **spirit**

Gr. *Morpheus*, god of dreams; see *morphe* under **form**

Gr. *oneiros*, m.; *onar, -os*, n. dream, vision; *oneiration*, n. dim.: oneiromancy, oneirology, oneirodynia, onirotic, *Oniromyia pachycerata* (a fly), *Onar nebulosum* (a fish), *Oneirophanta[Onirophanta] mutabilis* (a holothurian).

L., Gr. *phantasia; phantasma*, appearance, image, illusion, dream; see **fancy, spirit**

Gr. *pnigalion*, n. nightmare: *Pnigalion oweni* (a fossil reptile).

L. *somnium*, dream; see *somnus* under **sleep**

Gr. *tiphys, -yos*, m. nightmare: *Tiphyocetus temblorensis* (a fossil cetacean), *Tiphys podagricus* (a crustacean).

See: **fancy, form, see, wander, spirit**

dreg; dregs < ON. *dregg*; see **dirt**

drepanis, Gr. a bird, perhaps the swift; see **swallow**

drepano- < Gr. *drepane; drepanon*, sickle; see **sickle**

dreptos, Gr. plucked; see **tear**

dress < OF. *drecier*, make straight, prepare, arrange; see **garment**

drester, Gr. laborer; see *drao* under **make**

drift, something driven, < AS. *drifan*, drive; see **heap, carry, mine**

drill < D. *drillen*, bore, pierce; see **bore, teach**

drilos, Gr. penis; see **penis**

drimys, Gr. sharp, acid, pungent, piercing; see **point**

drink < AS. *drincan*.

L. *amystis, -idis* (Gr. *-idos*), f. deep draught, large cup, the emptying of a cup at one draught:

Gr. *ardmos*, m. watering place:

L. *bibo, -itus*, drink; *bibilis*, drinkable, potable; *bibax, -acis; bibosus; bibulus*, given to drink, drinking much, fond of drink; *bibitor, -is; bibo, -onis; bibonius*,

m. hard drinker, tippler, drunkard; *ebibo*, drink up, drain: imbibe, bibber, bibacious, bibulous, imbrue, beverage.

Gr. *brytikos*, drunk with beer; see *brytos* under **beer**

L. *calidum*, n. a hot drink:

L. *cinnus*, m. a mixed drink of spelt and wine: *Papilio cinnus* (a butterfly).

L. *crapulatus*, drunk, inebriated, < *crapula* (Gr. *kraipale*), f. result of a debauch, drunken headache, intoxication; *crapulentus*, very drunk; *crapulosus*, inclined to drunkenness: crapulence, crapulent, crapulous.

L. *cyceon, -is* (Gr. *kykeon, -os*), m. a mixed drink:

L. *decalicator, -is*, m. a hard drinker:

L. *decoctum*, n. medicinal drink, potion: decoction.

L. *ebrius*, drunken; *ebriolus*, dim.; *ebriolatus*, tipsy; *ebriosus*, addicted to drink, boozy; drunkard, sot; *inebrio*, make drunk: ebrious, inebriety, inebriated, *Ribes inebrians* (squaw-currant).

L. *haustus*, drink, draught; see *haurio* under **suck**

Gr. *kothonizo*, drink hard; *kothonistes*, m. tippler; *kothon, -os*, m. drinking-cup, drinking bout, carousal:

Gr. *lapto*, lap with the tongue, drink, suck:

L. *libatio, -onis*, f. a drink offering, < *libo, -atus* (Gr. *leibo*), taste, pour out: libation, delibate.

L. *liquor*, fluid; see *liqueo* under **flow**

L. *madidus*, sodden, drunk; *madulsa*, a drunken man; see **wet**

L. *melicratum* (Gr. *melikraton*), n. a kind of mead:

L. *melina*, f. mead:

L. *merulentus*, drunk; *merulator, -is*, m. wine-drinker:

Gr. *methysos*, drunk with wine; see *methy* under **wine**

Sp. *mezcal* (Ab.Am. *mexcalli*), a drink derived from maguey: mescal (chiefly from tequila, *Agave tequilana*).

L. *nectar, -is* (Gr. *nektar, -os*), n. drink of the gods; *nectareus* (Gr. *nektareos*); *nectarius*, of nectar: nectar, nectary, nectarine, *Nectarina scutellaria* (a wasp), *Nectandra(Ocotea) coriacea* (a lauracead).

Gr. *oinophlyx, -ygos*, drunk; *oinosis*, drunkenness; see *oinos* under **wine**

L. *posca*, f. a drink composed of vinegar and water:

Gr. *posis*, f. a drinking; *poma, -tos*, n.; *potos*, m. drink; *potes*, m.; *potis, -idos*, f. drinker; *potimos*, drinkable; *pino*, drink; *potizo*, give to drink; *metrioposia*, f. moderation in drinking; *propoma, -tos*, n. a drink before meals; *symposion*, n. a drinking party; *symposiarchos*, m. toastmaster; *zapotes*, m. hard drinker: potomania, symposium, *Hydropotes inermis* (river-deer).

L. *poto, -atus; potus*, drink; *potio, -onis*, f. drink, draught; *potiuncula*, f. dim. snort, snifter; *potabilis; poculentus; potulentus*, drinkable; *potilis; potorius; potatorius*, of drinking; *potax, -acis*, fond of drink; *potor, -is*, m.; *potrix, -icis*, f. drinker; *Potua*, f. goddess of drinking; *appotus*, drunk, intoxicated; *repotium*, n. the drinking on the day after a festive occasion: potable, potation, potion, poison, *Cosmotricha potatoria* (drinker-moth), *Conidiophrys guttipotor* (a suctorial protozoan), *Strychnos potatorum* (a strychnine-plant), *Haematopinus eurysternus* (cattle-louse).

Gr. *rhopheo*, sip, sup; see **soup**

L. *sicera* (Gr. *sikera;* Heb. *shekar*), n. a fermented liquor: cider, *Lagenaria siceraria* (a gourd).

L. *sitis*, thirst, dryness; see **dry**

L. *sorbitio, -onis*, f. drink, potion:

Gr. *sponde*, f. drink-offering, libation; *spondix*, m. one who pours a libation; *spondeios* (L. *spondeus*), of libations; *spondeion* (L. *spondeum*), n. cup or vessel used in libations: spondee, spondaic.

L. *temulentus*, drunken; *temetum*, n. any intoxicating drink: temulence, temulentive, *Lolium temulentus* (darnel).

L. *vinolentus; vinosus*, drunk; see **wine**

Gr. *zema, -tos*, n. decoction: *Chorizema ilicifolium* (flame-pea).

Derive: absinthe, ale, beer, brandy, chocolate, cocoa, coffee, cognac, gin, mead, metheglin, porter, rum, vodka, water, whisky, wine; bridal, cocktail, highball, swizzle, toast.

See: **water, beer, wine, alcohol, drop, cup**

drios, Gr. copse, wood, thicket; see **bush**

drive < AS. *drifan*.

L. *abortus*, driven away, prematurely born or terminated; see **abortion**

L. *agaso, -onis*, driver, hostler, groom, lackey; see **servant**

L. *ago, -actus*, do, drive; *abactor, -is*, m. one who drives off, < *abigo, abactus*, drive away, remove, abort; *adactio, -onis*, f. a driving to, forcing to, < *adigo*, drive, bring, urge; *coactus, -us*, m. a driving together, constraint, compulsion;

exactor, -is, m. demander, enforcer, tax collector; *inigo, -actus,* drive into, thrust: abigeus, abaction, exactor, exaction, exigent, exigency, exigeant, exiguous, redaction, *Andrena abacta* (a bee), *Tabanus abactor* (a horsefly).

Gr. *apostes,* m. one who drives away, rejector; *apostos,* driven away; *aposis,* f. repulsion:

L. *auriga,* c. charioteer, driver; *aurigo, -atus,* drive a chariot: Auriga.

L. *carrucarius,* m. coachman:

Gr. *diphreutes,* m. charioteer:

L. *eques, -itis,* horseman, rider; see *equus* under **horse**

Gr. *elater, -os,* m. driver; *elasis,* f. a driving out; *elaterios,* driving; *elatikos,* of driving, < *elauno,* drive: elater, *Ecballium elaterium* (squirting cucumber), *Hippelates stramineus* (a gnat).

Gr. *henioche,* f.; *heniochos,* m. driver: *Heniochus macrolepidotus* (a fish).

Gr. *hippeuo,* ride, drive; *hipposyne,* horsemanship; see **ride,** *hippos* under **horse**

Gr. *keleuo,* command, order, urge; *keleuma,* command; see **ask**

Gr. *kentor,* goader, driver; see *kenteo* under **point**

L. *mulionicus,* m. mule-driver:

L. *pello, pulsus,* beat, drive, push; see **push**

AS. *wrecan,* drive out, banish, impel: wreck, wreckage, wretch.

See: **arouse, push, press, banish, busy, make, move**

dromos, Gr. a running, race; *dromikos,* fleet, swift; *dromeus,* runner; see **run**

drone < AS. *dran;* see **bee, rest**

droop < ON. *drupa; see* **hang, low, nod**

drop < AS. *dropa:* droplet, dropper, dropwort, eavesdrop, eardrop.

Gr. *brochetos,* rain; see *brecho* under **wet**

Gr. *dakryon,* n. tear; *dakrydion,* n. dim.; *dakryodes,* like tears; *dakrytos,* tearful; *dakryo,* shed tears, weep: dacryoma, dacryops, dacryolith, dacryoadenitis, *Dacrydium cupressinum* (rimu), *Podocarpus dacrydioides* (kahika).

Gr. *diaphoresis,* f. perspiration: diaphoresis, diaphoretic.

Gr. *drosos,* f. dew; *droseros,* dewy: drosometer, *Drosophila melanogaster* (a fruit-fly), *Drosera rotundifolia* (a sundew), *Bythocrates drosocycla* (a moth), *Drosophyllum lusitanicum* (a droseracead).

Gr. *enormis,* m. drop of an earring:

L. *globulus,* droplet, bead; see *globus* under **ball**

L. *gutta,* f. drop, spot; *guttula,* f. dim.; *guttatus,* dappled, speckled, spoted: gutta, guttule, guttate, guttiferous, gutter, gout, *Cattleya guttata* (an orchid), *Globulina guttula* (a foraminifer), *Guttulina gibbosa* (a foraminifer), *Hypsopsetta guttulata* (a flounder), *Zuphium* [*Zoyphium*] *biguttatum* (a beetle). Derive: guttapercha.

Gr. *herse,* f. dew; *herseeis,* dewy:

Gr. *hidros, -tos,* m. sweat; *hidrothion,* n. dim.; *hidrotikos,* sudorific; *kathidros,* sweating much; *exidrosis,* f. violent sweat: hidrosis, hidrotic, *Hidrosis incostata* (a beetle), *Hidroticus rufus* (a medusa).

Gr. *hyetos,* m; *hysma, -tos,* n. rain; *hyetios; hyetodes,* rainy, < *hyo,* rain: hyetograph, hyetology, *Hyetornis pluvialis* (a cuckoo).

L. *imber, -bris,* m. rain, shower, storm; *imbrialis; imbrilis,* of rain; *imbricus; imbridus; imbrifer,* rainy:

Gr. *kremasterion,* n. drop in a necklace, bead:

L. *lacrima* (*lacryma*), f. tear; *lacrimula,* f. dim.; *lacrimosus,* tearful, < *lacrimo, -atus,* shed tears, weep: lacrimal, lacrimose, lacrimae Christi (a Neapolitan wine), *Merulius lacrymans* (dry-rot fungus), *Guttulina lacryma* (a foraminifer), *Productella lachrymosa* (a fossil brachiopod), *Lachrymaria olor* (a protozoan).

Gr. *libos,* n. tear, drop: *Libocedrus plumosa* (an incense-cedar).

Gr. *ombros,* m. rainstorm; *ombresis,* f. a raining; *ombrema, -tos,* n. rain-water; *ombrios,* rainy, < *ombreo,* rain; *dysombros,* stormy: ombrology, ombrophyte, ombrophilous, ombrophobous, ombrifuge.

L. *pluo,* rain; *pluvialis; pluvius,* of rain, rainy: pluvial, pluvious, Jupiter Pluvius, plover, *Haematococcus pluvialis* (a flagellate), *Pluvialis dominica* (golden plover).

Gr. *prox, -okos,* f. dewdrop:

Gr. *psakas* (*psekas*), *-ados,* f. small drop, droplet, bit, grain, morsel; *psakadion* (*psekadion*); *psakion,* n. dim.; *psakastos,* dripping: *Psacadoptera simulatrix* (a beetle), *Psacadonotus irregularis* (a katydid), *Psacadium bilimbatum* (a psocopterid), *Psecadium ovatum* (a foraminifer).

Gr. *psias, -ados,* f. drop: *Psiadosporus caenosus* (a trichopterid).

Gr. *rhanis, -idos,* f. drop, spot; *rhasma, -tos,* n. shower, < *rhaino,* sprinkle: *Rhanidopsis neophantes* (a moth).

Gr. *rhathaminx, -ingos,* f. drop, grain, bit:

L. *ros, roris,* m. dew; *roridus; rorulentus; roscidus,* dewy, bedewed; *irroratus,* bedewed, sprinkled, covered with granules; *roro, -atus*, bedew, drizzle, sprinkle: roscid, roric, rorulent, rosemary, *Lorius roratus* (a parrot), *Adelocera rorulenta* (a click-beetle), *Roridula muscipula* (a sundewlike plant), *Roripa amphibia* (marsh-cress), *Idothea irrorata* (an isopod), *Cancer irroratus* (a crab), *Rhus rosmarinifolia* (a sumac).

Gr. *stagma, -tos,* n.; *stagon, -os,* f. drop; *staxis,* f. a dropping, dripping; *stagonias; staktos,* dripping, trickling, < *stazo,* let fall, drip; *epistaxis,* f. nosebleed: apostaxis, epistaxis, *Stagmatoptera biocellata* (a mantid), *Stagonospora carpathica* (a leafspot-fungus), *Stactolaema anchietae* (a bird).

Gr. *stalagma, -tos,* n. drop; *stalagmion,* n. pendant, eardrop; *stalagmos,* m. a dropping, dripping; *stalaktikos; stalaktos,* dropping, dripping, trickling, < *stalasso,* fall in drops, drip, trickle: stalactite, stalagmite, *Holospira(Stalactella) rosei* (a gastropod).

L. *stilla,* f. drop; *stillativus,* dropping; *substillus,* dribbling, sprinkling, < *stillo, -atus,* drip, drop, trickle: still, distill, distillery, distillation, instill, instillation, *Nepenthes distillatoria* (a pitcher-plant).

Gr. *stranx, -angos,* f. drop:

L. *sudo, -atus,* sweat; *sudor, -is,* m. sweat; *sudatilis,* flowing like sweat; *sudatorius,* of sweat, promoting sweating; *sudorus,* dripping with sweat; *sudabundus,* sweating; *sudatorium,* n. sweating-bath; *consudo; desudo,* sweat profusely; *exsudo (exudo),* discharge by sweating: sudatory, exude, exudate, exudation, sudorific, sudarium, sudatorium, *Clitocybe sudorifica* (a mushroom).

See: **water, wet, weep, drink, stream, pour, fall, flow, ball**

dropax, Gr. a pitch plaster used as a depilatory; see **plaster**

drosallis, Gr. a kind of vine; see **vine**

drosos, Gr. dew; *droseros*, dewy; see **drop**

dross < AS. *dros*, filth, dirt, lees, refuse; see **dirt, slag**

drug < F. *drogue.*

Gr. *akos, -eos,* cure, relief, remedy, < *akeomai,* cure, heal; see **heal**

L. *antidotum* (Gr. *antidoton*), n. something given against, counterremedy: antidote, antidotal.

Gr. *antipathes,* n. remedy; a black coral: *Antipathes flabellum* (a coral).

L. *apothecarius,* storekeeper, now a druggist; see *apotheca* under **store**

Sp. *bezoar,* antidote, < Per. *badzahr,* against poison: bezoar, phytobezoar, trichobezoar (hairball, aegagropila).

L. *chlora,* f. medicament:

Gr. *chraismenion,* n. remedy:

L. *collyrium* (Gr. *kollyrion*), n. dough, poultice, eye-salve, suppository; *collyriolum,* n. dim.: collyrium, *Dyschimus collyrifer* (a trichopterid).

L. *decoctio, -onis,* f.; *decoctum,* n. a medicinal potion: decoction.

L. *electuarium* (*electarium*); *ecligma* (Gr. *ekleigma; ekleiton*), n. a medicine compounded with syrup: electuary, eclegma.

Gr. *embregma, -tos,* n.; *embroche,* f. lotion: embrocation.

Gr. *hiera,* f. medicine, antidote:

Gr. *iama, -tos,* n. medicine, remedy; see *iaomai* under **heal**

Suf. *-in, -ine,* denoting chemical terms, drugs, enzymes, < L. *-inus* (Gr. *-inos*), pertaining to, having the nature of: adenine, allethrine, anabasine, arbutin, aureomycin, berberine, clavatin (from *Aspergillus clavatus*), deguelin, digitalin, emulsin, fluorine, gasoline, insulin, lignin, luciferin, medicine, morphine, nicotine, oestrin, penicillin, pepsin, ptyalin, rennin, santalin, santonin, streptomycin, strychnine, theelin, thyroxin, trypsin, viridin (from *Tricoderma viride*).

Ab.Am., Pg. *ipecacuanha,* a South American plant that contains emetine and cephaëline: ipecac (*Cephaëlis ipecacuanha*), *Euphorbia ipecacuanhae* (ipecacspurge).

Gr. *katapoton; katapotion,* n. pill:

L. *linimentum,* n. stuff for smearing or rubbing on the skin: liniment.

L. *medicina,* f. remedy; *medicamen, -inis; medicamentum,* n. drug, remedy; *medicinalis,* of drugs: medicine, medicinal, medicament, *Tegenaria medicinalis* (a spider).

L. *nostrum,* n. our own, a quack remedy: nostrum.

L. *officinalis,* pertaining to drugs and medicines, < *officina,* f. laboratory: officinal, *Euphrasia officinalis* (eyebright), *Delphinium officinale* (a larkspur), *Ceterach officinarum* (scale-fern).

L. *panacea* (Gr. *panakeia*), a universal remedy, heal-all: panacea.

L. *pessum* (Gr. *pesson*), n. plug, tampon; *pessulum,* n. dim. suppository: pessary.

Gr. *pharmakon,* n. drug, medicine, poison; *pharmakion,* n. dim.; *pharmakeutikos,* of drugs; *pharmakeus; pharmakeutes; pharmakopoles; pharmakos,* m. druggist, poisoner, sorcerer; *pharmaktos,* drugged, poisoned; *alexipharmakon,* n. antidote, remedy: pharmaceutical, pharmacy, pharmacist, pharmacopeia, pharmacognosy, pharmacology.

L. *pilula,* pill; see *pila* under **ball**

L. *remedium,* n. medicine, cure: remedy, remedial.

Gr. *therapidion,* n. means of cure, remedy:

L. *venenum,* poisonous drug; see **poison**

Gr. *xerion,* n. a drying powder for wounds:

See: **heal, dull, poison**

druid < L. *druis, -idis,* a Celtic priest or wiseman; see **priest**

drum < uncertain origin.

Gr. *echeion,* n. kettle-drum or gong:

L. *sucula,* windlass, winch, capstan: see **turn**

L. *tympanum* (Gr. *tympanon*), n. drum; *tympanion,* n. dim.: tympanum, tympanitis, tympany, tympanic, timbrel, *Tympanuchus pallidicinctus* (a prairie chicken), *Tympanidium binoctonum* (a radiolarian), *Tympanophorus canaliculatus* (a beetle).

See: **membrane, music, sound**

drunk < AS. *druncan;* see **drink**

drupe < L. *drupa* (Gr. *dryppa*), an overripe, wrinkled olive; any pulpy fruit with a stone or pit that encloses the seed; see **fruit**

dry < AS. *dryge.*

Gr. *abrochos,* dry, waterless, unwetted; *abrochia,* f. drought: *Abrochia zethes* (a butterfly).

Gr. *achylos,* without juice, insipid; see **tasteless**

Gr. *adiantos,* unwetted: *Adiantum pedatum* (maidenhair-fern).

Gr. *anamatos,* without water, dry:

Gr. *anhydros,* without water, dry, arid: anhydrous, anhydrite.

Gr. *anikmos,* without moisture: *dysikmos,* hard to moisten:

Gr. *anombros,* without rain, dry:

Gr. *apeirodrosos,* unbedewed, parched:

L. *aridus,* dry, dim.; *arens, -entis,* dry, parched; *arefactus,* dried up; *aritudo. -inis,* f. dryness, drought, < *areo; aresco,* dry up, wither: arid, aridity, semiarid, *Anthidium aridum* (a bee). Derive: Arizona.

Gr. *arrhantos,* unwatered:

L. *assus,* dried, roasted; see *asso* under **cook**

Gr. *auchmeros,* dry; *auchmos,* m. drought; *auchmodes,* looking dry: *Auchmobius sublaevis* (a beetle).

Gr. *auos* (*ahyos*), dry; *auotes* (*ahyotes*), *-etos; auone,* f. dryness; *aualeos,* dry, parched; *ahyetos,* without rain; *austaleos,* dried or burned by the sun, squalid; *austeros,* drying the tongue, harsh, severe: austere.

Gr. *azaleos,* dry, parched, < *azo,* parch: azalea, *Azorella glacialis* (an umbellifer).

Gr. *danos,* dry, parched, burned: *Danosoma conspersa* (a beetle).

Gr. *dipsa,* f. thirst; *dipsios; dipsodes; dipsaleos,* thirsty; *ekdipsos,* very thirsty: dipsomaniac, *Dipsalidictis platypus* (a creodont), *Haemadipsus ventricosus* (a louse).

L. *exuccus* (*exsuccus*), juiceless, sapless; *exsuccidus,* sapless, dry:

L. *inaquosus,* lacking water, dry; *inaquosum,* n. desert:

L. *inhumectus,* not moist, dry:

Gr. *ischnos,* dry, withered, thin, lean, weak; see **thin**

Gr. *kankanos,* dry: *Cancanodes orthometalla* (a moth).

Gr. *kapyros; kapyrodes,* dry:

Gr. *karphaleos,* dry, < *karpho,* dry up, wither: *Carphalea madagascariensis* (a madder).

Gr. *keleos* (*kelos*), burning, dry: *Celosia cristata* (cockscomb).

Gr. *kerchaleos,* dry, hoarse, rough; see **hoarse**

Gr. *krambaleos; krambos,* dry, parched:

L. *passarius,* dried in the sun:

Gr. *phrygios,* dry:

Gr. *skeliphros* (*sklephros*), dry-looking, lean, thin; see **thin**

L. *siccus; siccanus,* dry; *siccitas, -atis,* f. dryness; *siccativus,* drying; *persiccatus,* thoroly dry, < *sicco, -atus,* dry up; *assico,* dry up: siccate, siccative, siccimeter, desiccator, exsiccate, *Carex siccata* (a sedge), *Phyllium siccifolium* (a leaf-insect), *Labidesthes sicculus* (a fish).

L. *sitis,* f. thirst, dryness; *sititor, -is,* m. one who thirsts; *sitiens; sitibundus; siticulosus,* thirsty, < *siteo,* be thirsty: *Eumastia sitiens* (a sponge).

L. *sudus,* dry, clear; see **clear**

L. *torridus,* dry, parched, hot, scorched; *retorridus,* dried up, < *torreo, tostus,* dry by heat, parch, roast: torrid, torrefy, torrent, toast.

Gr. *trasia,* grate, drying kiln; see **screen**

Gr. *xeros,* dry: xerophyte, xerophilous, xerography, elixir, *Xerus rutilans* (an African ground-squirrel), *Philoxerus vermicularis* (an amaranth), *Xerophyllum asphodeloides* (a beargrass), *Phylloxera vitifolia* (grape-aphid), *Alternanthera philoxeroides* (an amarantacead), *Xeranthemum annuum* (immortelle).

See: **empty, heat, drink**

Dryas, L., Gr. a wood or tree-nymph; see *drys* under **tree**

drymos, Gr. oak-coppice, forest, wood; see **forest**

dryo- < Gr. *drys, -yos,* tree, oak; see **tree**

dryops, Gr. a kind of woodpecker; see **woodpecker**

dryopteris, Gr. a kind of fern; see **fern**

drypepes; drypetes, Gr. quite ripe, ready to fall; see *pepon* under **ripe**

dryphacto- < Gr. *dryphaktos,* railing, latticed partition, bar; see **wall**

drypis, Gr. a kind of thorn; a genus of the pink family; see **pink**

drypto, Gr. tear; *drypsos,* torn, worn; see **tear**

dualis, L. of two; see *duo* under **two**

dubius, L. doubtful, wavering, < *dubito, -atus,* be uncertain, doubt; see **doubt**

ducalis, L. pertaining to a leader, < *dux, ducis,* leader; see *duco* under **lead**

duck < AS. *duce.*

Gr. *aix, aigos,* c. a water-bird: *Aix[Aex] galericulata* (mandarin-duck).

L. *anas, -atis,* f. duck, drake; *anaticula,* f. dim.; *anatinus; anatarius,* relating to ducks: anatine, *Anas domestica* (duck), *Anatina canaliculata* (a pelecypod), *Lepas anatifera* (a barnacle), *Anseranas semipalmata* (a goose).

Gr. *baskas (phaskas),* f. a kind of duck:

NL. *dafila,* f. a genus of ducks: *Dafila acuta* (pintail-duck).

NL. *fulgula,* f. a genus of ducks: *Fulgula marilla* (scaup-duck).

Gr. *glaukion,* n. a kind of duck: *Glaucionetta islandica* (a goldeneye).

L. *mergus,* a kind of waterfowl; see *mergo* under **dip**

Gr. *nessa (netta),* f. duck, drake; *nessarion (nettarion)*; *nettion,* n. dim.: *Netta rufina* (red-crested pochard), *Nettion carolinense* (green-winged teal), *Nettapus auritus* (a pygmy-goose), *Arctonetta fischeri* (spectacled eider), *Rhodonessa caryophyllacea* (pink-headed duck).

NL. *nyroca,* f. a genus of ducks, < Russ. *nyrok,* diver: *Nyroca americana* (red-head duck).

Gr. *penelops, -opos,* m. a kind of duck: *Mareca penelope[penelops]* (European widgeon).

L. *querquedula,* f. a kind of duck: *Querquedula cyanoptera* (cinnamon teal).

duckweed

Gr. *lemna,* f. an aquatic plant; a genus of the duckweed family: Lemnaceae, *Lemna minor* (a duckweed).

ductilis, L. that may be led, that may be drawn out thin; see *duco* under **lead**

dulario- < Gr. *doularion,* dim. of *doule,* female slave; see *doulos* under **servant**

dulcis, L. sweet; *dulciculus,* dim.; *dulcitas,* sweetness; see **sweet**

dulichium, L. a kind of sedge; see **sedge**

dulio- < Gr. *doulios,* servile; *duleia,* servitude; see *doulos* under **servant**

dull < AS. *dol,* foolish.

Gr. *abakcheutos,* without Bacchic excitement, uninspired, joyless:

Gr. *abelteros,* silly, stupid:

Gr. *adaemon,* unknowing, ignorant:

Gr. *adenes,* ignorant, inexperienced; *adeneia,* f. ignorance:

Gr. *agnomon, -os,* ignorant, unaware, senseless; *agnostos,* unknown, unknowing: agnostic, agnosticism, *Agnostus americanus* (a trilobite).

Gr. *agrammatos,* without learning, unlettered, illiterate:

Gr. *akkistikos,* pretending indifference, coy:

Gr. *amathes,* ignorant, stupid; *amathia,* f. ignorance, stupidity; *dysmathes,* slow at learning, hard to learn: *Dysmathes sahlbergi* (a beetle), *Dysmathia costalis* (a butterfly), *Dysmathosoma picipes* (a beetle).

Gr. *amblys,* blunt, obtuse; *amblyogmos,* m. dim sight; *amblyopos,* dim-sighted; *amblytes, -etos,* f. bluntness, dullness: amblyopia, amblypod, amblygonite, *Amblystoma tigrinum* (axolotl), *Amblycorypha rotundifolia* (round-winged katydid), *Amblyolepis setigera* (a composite), *Amblyopsis spelaeus* (blindfish of Mammoth Cave), *Amblyoproctus torulosus* (a beetle).

Gr. *anaisthetos,* insensible, without feeling; see *anaisthesia* under **painless**

Gr. *aneuthynos,* not accountable, irresponsible:

Gr. *apatheia,* f. indifference, insensibility: apathy, apathetic.

Gr. *apeiros,* inexperienced, ignorant; *apeirobios,* without experience of life; *apeiro-gamos,* unwedded; *apeirokakos,* without experience of evil; *apeirokalos,* without appreciation of beauty, vulgar; *apeiropathes,* free from suffering: *Apeirocalus cornutus* (a beetle).

Gr. *apsychos,* spiritless, lifeless, faint:

Gr. *argos,* idle, lazy, inactive; see **rest**

Gr. *asynetos,* stupid, witless:

L. *attonitus,* thunderstruck, stupefied, stunned:

L. *bardus,* dull, stupid: *Tanyrhynchus bardus* (a beetle).

L. *baro, -onis,* m. blockhead, simpleton:

Gr. *bebros (bembros),* stupid: *Bebrornis seychellensis* (a warbler), *Bembrops caudimacula* (a fish).

Gr. *bereschethos,* m. booby:

Gr. *blax, -akos,* dull, slow, doltish; c. dolt; *blakikos,* slow, stupid: *Blacops gym-nophthalmus* (a bird), *Metablax acutipennis* (a beetle), *Blacicus flaviventris* (a bird).

L. *blennus,* m. blockhead, dolt, simpleton:

L. *Boeotius* (Gr. *Boiotios*), of Boeotia, a district of Greece, whose inhabitants were dull and obtuse; hence, a synonym for stupidity: Boeotian.

L. *brutus,* stupid, rough; see **rough**

L. *caroticus* (Gr. *karotikos*), stupefying, soporific, < *karos,* deep sleep, stupor; see **sleep**

L. *catocha* (Gr. *katoche*), f. complete stupor, catalepsy: *Catocha latipes* (a fly).

L. *crassedo, -inis,* f. thickness, stupidity:

Gr. *dysgnoia,* ignorance, doubt; see *amphignoia* under **doubt**

L. *elucus,* m. a drowsy or dreaming person:

L. *evexus,* rounded at the top:

L. *gurdus,* m. dolt, numbskull; *gurdonicus,* stupid:

L. *hebetatus,* blunted, dulled, weakened; *hebes, -etis,* blunt, dull, sluggish, stupid; *hebetudo, -inis,* f. bluntness, < *hebeo, -etus,* be blunt, dull; *hebeto, -atus,* dull, deaden, dim, weaken: hebetate, hebetude, *Torpedo hebetans* (a ray), *Macrourus [Macrurus] hebetatus* (a fish).

Gr. *heolos,* stale: *Heolus providentiae* (a fossil insect).

L. *hiberno, -atus,* spend the winter, keep in winter quarters, be inactive; see *hibernum* under **year**

L. *idiota* (Gr. *idiotes*), m. ignorant person: idiot, idiotic.

L. *ignarus,* ignorant, unacquainted, inexperienced; *ignavus,* inactive, lazy, idle, sluggish, listless; *ignavia,* f. inactivity, listlessness, sloth; *ignorantia,* f. lack of knowledge or information, < *ignoro, -atus,* have no knowledge, be unaware: *Buprestis ignara* (a beetle), *Staphylinus ignavus* (a beetle), ignore, ignorance, ignorant, ignoramus.

L. *illiteratus,* unlettered, uneducated: illiterate.

L. *imbecillus (imbecillis),* weak, feeble (especially in mind): imbecile, imbecility, *Dentalina imbecilla* (a foraminifer).

L. *impassibilis,* incapable of passion: impassible, impassive.

L. *imperitus,* inexperienced, unskilled, ignorant: *Colaspis imperitus* (a beetle).

L. *inanimis, -e; inanimus,* lifeless, spiritless, dull: inanimate.

L. *inanis, -e,* empty, lifeless, silly; see **empty**

L. *incallidus,* unskillful, simple, stupid: *Sphadasmus incallidus* (a beetle).

L. *inconscius,* unaware: unconscious.

L. *indiscretus,* undiscerning, undiscriminating: indiscretion.

L. *indocibilis; indocilis,* unteachable: indocility.

L. *indoctus,* untaught, ignorant:

L. *ineptus,* unsuitable, absurd, silly; see **fool**

L. *iners, -ertis,* inactive, lazy, sluggish: inert, inertia.

L. *ineruditus,* uninstructed, illiterate, ignorant:

L. *infrunitus,* unfit for enjoyment, tasteless, senseless, silly:

L. *inscitus,* ignorant, absurd, silly: *Occisor inscitus* (a fly).

L. *insensibilis,* incapable of feeling: insensible, insensibility.

L. *insipidus,* tasteless; see **tasteless**

L. *invalidus,* infirm, weak; see **weak**

Gr. *kenokranos,* empty-headed; see *kranos* under **head**

Gr. *koalemos,* m. dunce:

Gr. *kophos,* blunt, dull, deaf, dumb, dim-sighted; *kophosis, -eos; kophotes, -etos,* f. dullness, deafness; *dyskophos,* stone-deaf: *Cophus thoracicus* (a cricket), *Cophocetus oregonensis* (a fossil whale).

Gr. *lethargikos,* drowsy; see *lethe* under **forget**

L. *mutus,* speechless, dumb; see **silent**

Gr. *narke,* f. numbness, torpor; *narkotikos,* benumbing, dulling: narcotic, narcosis, narcolepsy, narcissus.

Gr. *neis, -idos,* unknowing, unpracticed, feeble:

L. *nescius,* unknowing, ignorant, unaware, stupid: *Bruchus nescius* (a beetle).

Gr. *nokar, -os,* sleep, sloth; see **sleep**

Gr. *nothes; nothros,* lazy, dull, stupid, sluggish, torpid; see **slow**

L. *obscurus,* dark, indistinct; see **dim**

L. *obtusus,* blunt, dull; *retusus,* rounded and notched: obtuse, retuse, obtund, obtundent, *Gnaphalium obtusifolium* (a cudweed), *Eumerus obtusiceps* (a fly).

L. *oscitans,* yawning, listless, drowsy: oscitant.

L. *phlegmaticus* (Gr. *phlegmatikos*), full of phlegm, like phlegm, sluggish: phlegmatic.

L. *piger, -gra, -grum,* reluctant, slow, lazy, dilatory; see **slow**

L. *rupex, -icis,* m. uncultivated man, boor, rustic: *Pompilus rupex* (a wasp).

L. *socors, -ordis,* weak-minded, stupid, silly; see *excors* under **fool**

L. *sopor,* deep sleep, stupor; *sopio,* lull to sleep; see **sleep**

L. *stagno, -atus,* form a pool of standing water, hence, become foul or dull from lack of motion, < *stagnum,* pool: stagnate, stagnation.

L. *stolidus,* dull, stupid, unmovable: stolid, stolidity.

L. *stupidus,* senseless, dull; *stupor, -is,* m. dullness, insensibility, < *stupeo,* be stunned: stupid, stupor, stupefaction, stupidity, stupendous.

L. *surdus,* deaf, dull, silent, faint; see **deaf**

L. *torpidus,* benumbed, stupefied; *torpor, -is,* m. numbness, sluggishness; *torpedo, -inis,* f. numbness, stiffness: torpid, torpor, torpedo.

Gr. *typhedanos,* m. one with cloudy wits, dullard:

L. *vapidus,* dull, stale: vapid.

L. *vervex, -ecis,* wether, mutton-head, a stupid fellow; see **sheep**

L. *veternosus,* lethargic, sleepy, dreamy, dull:

See: **dim, slow, sleep, weak, forget, fool, shade, sad, rest, empty, short, painless**

dulo- < Gr. *doulos,* bondman, slave; *doulios,* servile; see **servant**

dumb < AS. *dumb,* mute; see **silent**

dumus, L. thorn-bush, bramble; *dumosus,* covered with thorn-bushes; *dumetum,* thicket; see **bush**

dung < AS. *dung,* excrement, manure.

Gr. *aphodos,* m. excrement, evacuation: *Aphodobius misellus* (a beetle), *Aphodiopsis latipes* (a beetle), *Aphodius abditus* (a beetle).

Gr. *apopatos,* m. dung, ordure, privy:

L. *assellor, -atus,* go to stool, void:

Gr. *boliton,* n; *bolitos,* m. cow dung: *Bolitophagus costulatus* (a beetle).

L. *caco, -atus,* go to stool; *concaco,* defile with dung; Gr. *kakke,* f. human dung:

Gr. *chodos,* m. dung; *chezo,* pass excrement, evacuate: allochezia.

L. *editus,* m. excrement, evacuation:

L. *egeries,* f. excrement, dung: *Trochocyathus egerius* (a fossil coral).

L. *excrementum,* n. dung, ordure, refuse: excrement.

L. *faex, -ecis,* dregs, grounds, excrement; see **dirt**

L. *fimus,* m. manure, dung; *fimetum,* n. dunghill: fimicolous, *Sordaria fimicola* (a fungus), *Coprinus fimetarius* (a mushroom).

Sp. *guano,* < Ab.Am. *huanu,* dung: guano, guanine, guanamine, guanidine, guaniferous, guanase, guanay, guanidopropionic, guanophore, guanazol, guanyl.

Gr. *kanthis, -idos,* f. ass's dung:

Gr. *kopros,* f. dung, ordure; *koprion,* n. dim.; *koprodes,* like dung; *koprinos,* full of dung, filthy; *kopria,* f. dunghill; *kopron, -os,* m. privy: coprolite, coprophagous, coprophilist, coprostasis, koprosterin, copremia, *Coprophilus striatulus* (a beetle), *Coprosma foetidissima* (a rubiacead), *Coprionophagus* (a mite), *Coprinus comatus* (a mushroom), *Copris signatus* (a dung-beetle).

L. *laetamen, -inis,* n. dung, manure:

L. *meconium,* n. first excrement of new-born infants: meconium, meconioid.

L. *merda,* f. dung, ordure; *merdaceus; merdaleus,* defiled with excrement: *Onthophagus merdarius* (a dung-beetle), merdivorous.

Gr. *minthos,* m. human dung: *Minthomyia abdominalis* (a fly).

L. *muscerda,* f. mouse dung:

Gr. *oïspote,* f. sheep dung:

L. *olenticetum,* n. dungheap:

Gr. *onis, -idos,* f. ass's dung:

Gr. *onthos,* m. dung: *Onthophagus politus* (a dung-beetle), *Philonthus fraternus* (a beetle).

Gr. *pelethos,* m. human dung; *hyspelethos,* m. pig dung: *Pelethophila flava* (a fly).

Gr. *skor, skatos,* n. dung, ordure: skatol, scatology, scatomancy, scatophagi, *Scatophagus argus* (a fish), *Scatophaga furcata* (a fly).
Gr. *skybalon,* n. dung; *skybaliktos,* dirty, mean:
Gr. *spatile,* f. thin excrement:
Gr. *spyras (sphyras), -ados,* f. ball of dung:
L. *stercus, -coris,* n. dung; *stercorarius,* relating to dung; *sterculinium (sterquilinium)* n. dungpit; *Sterculius (Stercutius),* m. god of manuring; *stercoro, -atus,* dung, manure: stercoral, stercorary, stercoremia, stercorol, stercovorous, stercoricolous, stercorite, *Stercorarius parasiticus* (a jaeger), *Sterculia foetida* (a sterculia), *Coprinus sterquilinus* (a mushroom), *Anguillula stercoralis* (a nematode), *Geotrypes stercorarius* (a dung-beetle).
L. *sucerda,* f. swine dung:
Gr. *tilos,* m. a thin stool as in diarrhea; *ektilao,* go to stool: tiglic, *Croton tiglium* (purging-croton).
See: **dirt, empty**

duo, L. two; see **two**

duodecim, L. twelve; see **twelve**

duodenum, ML. first part of the small intestine; see **intestine**

duplico, -atus, L. double; *duplex, -icis,* double; *duplus,* double; see *duo* under **two**

dupo- < Gr. *doupos,* dead, heavy sound, thud; *doupema,* crash, peal; *doupetor,* clatterer; see **sound**

durus, L. hard, tough; *durabilis,* lasting, enduring; see **hard**

dusk < ME. *deosc;* see **black, brown, night, shade**

dust < AS. *dust.*
L. *dentifricium,* n. tooth-powder: dentifrice.
L. *fuligo, -inis,* soot, smut; *fuliginosus,* full of soot, sooty; see **black**
Gr. *konis, -ios,* f. dust; *konia,* f. dust, sand, powder; *konios,* duty; *koniortos,* m. cloud of dust; *konistra,* f. a dusty rolling-place: conidium, conidiophore, coniotheca, otoconium, pneumoconiosis, coniosis, *Coniogramma japonica* (bamboofern), *Coniophora cerebella* (a dry-rot), *Coniocybe pallida* (a lichen).
Gr. *odontosmegma, -tos,* n. tooth-powder:
L. *pollen,* fine flour, milldust, dust; see **flour**
Gr. *prisma; apoprisma, -tos,* n. sawdust, < *prizo,* saw:
L. *pulvis, -veris,* m. dust, powder; *pulvisculus,* m. dim.; *pulvereus,* dusty: pulverize, pulverulent, pulveraceous, pulvereous, pulverable, pulverin, powder, Powder River, Cache la Poudre River, *Primula pulverulenta* (a primrose).
L. *scobis, -is,* f. filings, scrapings, sawdust: scobiform, *Scobicia declivis* (short-circuit beetle).
L. *serrago, -inis,* f. sawdust:
Gr. *smyris, -idos,* f. emery powder:
See: **dirt, sand, ashes, earth, flour, cloud, dry**

duty < L. *debeo, -bitus,* owe, be indebted; see **rite, servant**

dux, *ducis,* L. leader; see *duco* under **lead**

dwarf < AS. *dweorg;* see **little, short, low**

dwell < AS. *dwelan;* see **life**

dye < AS. *deag;* see **color**

dynamis, Gr. power, strength, ability; *dynatos,* strong, powerful, able; see **strong**

dynastes, Gr. ruler; *dynastikos,* arbitrary; see **govern**

dyo, Gr. two, double; *dyas, -ados,* two; see **two**

dys, Gr. bad, ill, with difficulty, hard, unlucky, very; see **bad**

dysaxiotos, Gr. inexorable; see **merciless**

dyscheres, Gr. hard to manage, annoying, vexatious, difficult; see **trouble**

dyschimos, Gr. dangerous, fearful, horrible; see **danger**

dyschros, Gr. discolored; see *chroma* under **color**

dysderco- < Gr. *dysderkes,* seeing badly; see *derkomai* under **see**

dysenteria, Gr. an intestinal disease; see **disease**

dysis, Gr. setting of the sun, an entering; see **down**

dysmico- < Gr. *dysmikos,* western; see **west**

dysodes, Gr. stinking, foul; *dysodia,* stench; see *ozo* under **smell**

dyspepsia, Gr. indigestion; see **disease**

dyspetes, Gr. falling out badly, unfortunate; see lot
dyspnoea, L. (Gr. *dyspnoia*), asthma; see disease
dysprositos, Gr. hard to get at, difficult of access; see *aprositos* under impassable
dyspsycto- < Gr. *dyspsyktos*, tolerant of cold, hardy; see strong
dystheratos, Gr. hard to catch; see free
dytes; dyptes, Gr. diver, enterer, < *dyo*, dip, enter, set; see dip

E

e- < L. *ex*, out of, from; see out
eager < L. *acer*, sharp; see /busy, life, burn, wish, move, make, sharp
eagle < OF. *aigle*, < L. *aquila*, f. eagle; *aquilinus*, of eagles; *aquilifer*, m. eagle-
bearer, standard-bearer: aquiline, *Aquila chrysaëtus* (golden eagle), *Aetobius
aquila* (eagle-ray), *Alethopteris aquilina* (a fossil fern).
Gr. *aëtos*, m. eagle; *aëtideus*, m. eaglet; *aëtodes*, eaglelike; *chrysaëtos*, m. golden
eagle; *haliaëtos*, m. sea-eagle, osprey: *Haliaëtus leucocephalus* (bald eagle),
Pandion haliaëtus (osprey).
L. *clagalopes, -is*, f. a kind of eagle:
Gr. *lagotheras*, m. eagle:
Gr. *morphnos*, m. a kind of eagle: *Morphnos capistratus* (a bird).
Gr. *nettoktonos*, m. duck-killer, a kind of eagle:
L. *ossifraga*, f.; *ossifragus*, m. sea-eagle, osprey: ossifrage, osprey, *Corvus ossifragus*
(fish-crow).
Gr. *ostokataktes; ostokorax, -akos*, m. osprey:
Gr. *phene*, f. bearded vulture: *Phene ossifraga* (a bird).
L. *plancus* (Gr. *plangos*), m. a kind of eagle: *Polyborus plancus* (a caracara).
L. *pygargus* (Gr. *pygargos*), m. a kind of eagle: pygarg.
L. *valeria*, f. a kind of eagle:
See: hawk
ear < AS. *eare*, organ of hearing.
L. *auris*, f. ear; *auricula; oricilla*, f. dim.; *auritus*, eared, attentive: aural, aurated,
auriform, auricle, auriculate, auscultate, Pend Oreille Lake, orejon, ormer
(Haliotis tuberculata), Auricula turrita (a snail), *Otogyps auricularis* (a vul-
ture), *Coreopsis auriculata* (eared coreopsis), *Acacia auriculaeformis* (an acacia),
Phalacrocorax auritus (a cormorant), *Ursitaxus inauritus* (a carnivore), *Primula
auricula* (a primrose).
Gr. *ous, otos*, n. ear; *otarion; otion*, n. dim.; *otikos*, of ears; *otaros*, large-eared:
otitis, otidium, otolith, otocyst, otoconium, oticodinia, otogenic, otorrhea, otary,
parotid, *Otolithus aequidens* (a fish), *Otaria californiana* (an eared seal), *Otariono-
mus blattoides* (a beetle), *Otiorhynchus gibbicollis* (a beetle), *Microtus pennsylvanicus* (meadow-mouse), *Plecotus auritus* (a bat), *Pleurotus serotinus* (a
mushroom), *Corynorhinus macrotis* (a bat), *Lepomis megalotis* (a fish), *Myostis
sylvatica* (woodland-forgetmenot), *Arctotis fosteri* (a composite), *Otocyon
megalotis* (a foxlike mammal), *Pholiota praecox* (a mushroom).
Gr. *parouatios*, with hanging ears; see hang
Gr. *stachys*, ear of grain, spike; see point
Gr. *tragos*, m. small front lobe at the ear opening: tragus.
See: point, crest, hear, projection
ear < AS. *ear* (G. *ähre*), fruit-spike of a cereal; see point
earinus, L. (Gr. *earinos*, < *ear, -os*, spring), of spring, green; see *ear* under year
early < AS. *aerlice*, near the beginning, in advance of; see before, begin, dawn
earn < AS. *earnian*; see again, receive
earnest < AS. *eornoste*, serious; see busy, serious

earring

Gr. *artema, -tos,* n. earring: *Artematocis longirostris* (a beetle), *Artematopus caliginosus* (a beetle).

Gr. *ellobion,* n. earring: *Ellobium tumidum* (a gastropod), *Pithecellobium arboreum* (a legume).

Gr. *enklastridion,* n. earring:

Gr. *enope,* f. earring: *Enopa mediella* (a moth).

Gr. *enotion,* n. earring:

Gr. *enstrophos,* m. an earring:

L. *inauris,* f. earring, eardrop, pendant: *Marginella inauris* (a fossil gastropod).

Gr. *krotalia,* ear pendants that rattled against one another; see *krotalon* under **rattle**

Gr. *okis, -idos,* f. earring:

Gr. *siglos,* m. shekel, earring: *Siglophora bella* (a moth).

earth < AS. *eorthe,* the globe, land, ground, soil, dirt, mud, clay: earthen, earthly, unearth, earthling, earthquake, earthnut, earthworm, earthstar, aardwolf.

Sp. *adobe,* earth or clay and the sun-dried bricks made therefrom: adobe.

L. *alluvium,* detritus deposited by running water; *alluvius,* pertaining to alluvium; see *luo* under **wash**

L. *argilla* (Gr. *argillos*), f. white clay; *argillaceus* clayey; *argillosus* full of clay: argillaceous, argil, *Alabama argillacea* (cotton-leaf worm), *Oenothera argillicola* an evening-primrose).

Gr. *aroura,* tillable land; see *arura* under **field**

Gr. *asis, -eos,* f. alluvium, mud, slime; *asios; asodes,* miry, muddy, slimy: *Asiobates rufomarginatus* (a beetle).

Gr. *babyas (babylas),* m. mud: *Babylas bihamatus* (a bug).

L. *bolus* (Gr. *bolos*), lump of earth, clod; *bolinos,* made of clay; see **lump**

Gr. *cheras, -ados,* n. alluvium, silt, detritus: *Cheraphilus[Cheradophilus] bispinosus* (a shrimp).

Gr. *chersos (cherros),* f. land, dry land, waste or barren land; *chersaios; chersinos; chersobios,* living on land: *Chersophilus duponti* (a bird), *Chersobleptes crassus* (an opilionid), *Cherrocrius bruchi* (a beetle), *Chersaeus struthioneus* (a phasmid).

Gr. *chledos,* m. mud, dirt, debris: *Chledophila annularis* (a beetle).

Gr. *chous,* earth heaped up, soil; *choikos,* of earth or clay; *chostos,* made by heaped up earth; see **wall**

Gr. *chthon, -os,* f. earth, ground; *chthonios,* in, of, or from the earth; *mesochthon, -os,* f. midland, interior: chthonian, chthonophagy, chthonography, autochthonous, allochthonous, antichthon, *Chthonius spinosus* (a pseudoscorpion).

L. *cicerculum,* n. an African ocher:

L. *creta,* chalk, white earth or clay; *cretula,* dim.; see **lime**

L. *detritus,* worn away; pertaining to alluvial material: detritus.

Gr. *edaphos,* n. ground, soil, bottom; *edaphion,* n. dim.: edaphic, edaphology, *Edaphosaurus pogonias* (a fossil reptile).

Gr. *epeiros,* f. land, mainland, continent; *epeirotikos,* continental: epirogenic.

L. *fictilis,* earthen, made of clay; *fictiliarius,* m. potter: fictile, *Rhomphaea fictilia* (a spider).

L. *figulinus,* pertaining to pottery; *figulus,* potter, worker in clay; see **pot**

Gr. *ge; gaia,* f. the earth; *gedion,* n. dim.; *gaion, -os,* m. heap of earth; *geïkos,* of land; *geios,* of the earth, indigenous; *geïnos,* of earth; *geodes,* earthlike, earthy; *apogaios,* from the earth; *hypogaios,* underground; *mesogaios,* inland, interior: geology, geography, geode, geometry, geotropism, apogee, amphigean, geophagy, hypogeous, George, *Geomys personatus*(a gopher), *Geococcyx californianus* (road-runner), *Geonoma gracilis* (a palm), *Geaster hygrometricus* (a puffball), *Epigaea asiatica* (a trailing-arbutus), *Psoralea hypogaea* (a scurfpea), *Graptemys geographica* (map-turtle).

L. *gleba,* lump of soil, clod; *glebula,* dim.; see **lump**

Gr. *gounos,* m. fertile land:

AS. *haeth,* waste land; see **heath**

L. *humus,* f. ground, earth, soil; *humilis,* on the earth, low; *humo, -atus,* cover with earth, inter, bury; *inhumatus,* unburied: humus, humble, humility, humiliation, exhume, *Rosa humilis* (a wild rose), *Avicula humata* (a fossil pelecypod), *Arabis humifusa* (Arctic rockcress), *Pholiota humicola* (a mushroom).

Gr. *ilys, -yos,* f. mud, dirt; *ilyodes,* muddy: ilyogenic, *Ilyobius flavicollis* (a neuropterid), *Ilybius oblitus* (a beetle), *Ilyodrilus caspicus* (an annelid), *Ilyphagus hirsutus* (an annelid), *Ilyanassa obsoleta* (a gastropod).

Gr. *keramos,* m. potter's clay, anything made of clay, pot, vessel; *kerameus,* m. potter; *keramikos,* pertaining to pottery; *keraminos,* of clay: ceramics. See **pot**

Gr. *kimolia,* f. white, chalky clay from the island of Cimolus: *Cimolodon nitidus* (a Cretaceous mammal), *Cimolomys gracilis* (a Cretaceous mammal).

AS. *land,* ground, country: land, landscape, landlord, landmark, fairyland, dreamland, folkland, tableland, highland, Roland, Maryland, Holland, England, Greenland, Newfoundland, island, Iceland.

L. *limus,* m. mud, mire, slime; *limarius,* of mud; *limosus,* muddy, miry; *limigenus,* mud-born: limose, limicoline, limicolous, *Limosa lapponica*(a godwit), *Limosella aquatica* (a mudwort), *Oscillatoria limosa* (an alga).

L. *lutum,* n.; *lutus,* m. mud, mire; *lutarius,* of mud; *lutosus,* muddy, miry; *lutulentus,* bedaubed with mud, muddy; *luto, -atus,* smear with mud or clay: lutaceous, lutation, lutite, lutose, lutulent, *Metrodora lutosa* (a grasshopper), *Onthophagus lutulentus* (a dung-beetle).

L. *marga,* f. a kind of limy earth: marl, marlite, marly.

AS. *molde,* earth: mold (mould), moldboard.

Gr. *morochthos; moroxos,* m. a kind of pipe-clay: moroxite.

L. *mundus,* world, earth, heavens; see **world**

L. *murrha,* f. a kind of clay for making porcelain or china; *murrhinus,* of porcelain: murrhine.

ON. *myrr,* bog, swamp, deep mud; see **marsh**

L. *Nerthus,* f. German goddess of the earth: *Nerthus dudgeoni* (a bug).

Gr. *ochra,* f. earthy oxide of iron; see *ochros* under **yellow**

Gr. *oikoumene,* the world; see **world**

Gr. *oudas, -eos,* n. surface of the earth, ground; *oudaios,* on the earth, earthly; *katoudaios,* underground: *Udeocoris nigro-aeneus* (a bug).

Gr. *pedon,* n. ground, earth, soil; *epipedos,* on the ground: pedology, pedograph.

Gr. *pelos,* m. mud, clay, earth; *pelinos,* of clay or mud; *pelodes,* muddy; *pelopoios; peloplathos; pelourgos,* m. worker in clay, potter; *propelokizo,* cover with mud, abuse: pelophilous, pelopathy, *Pelobates fuscus* (a frog), *Pelopoeus lunatus* (a mud-dauber), *Pelurga sagittata* (a butterfly), *Pelinobius muticus* (a spider), *Pelocoetes exul* (a coral).

L. *rubrica,* f. red earth, red ocher; *rubricosus,* full of ruddle or red ocher: rubric, *Scirpus rubricosus* (a sedge).

L. *segutilum(segullum),* n. earth supposed to indicate presence of gold: *Segutilum sydneyanum* (a fish).

L. *sil, -is,* n. a kind of yellowish earth, yellow ocher; *silaceus,* ocherous:

L. *sinopis, -idis* (Gr. *-idos*), f. a red ocher or earth from Sinope on the Black Sea: sinopite, sinopia, sinople.

Gr. *smektis, -idos,* a kind of fuller's earth; see *smecticus* under **wash**

L. *solum,* n. bottom of anything, floor, ground, land: soil, subsoil.

L. *tasconium,* n. a white, clayey earth:

L. *tellus, -uris,* f. the earth, earth, land; *telluster, -tris,* of the earth: tellurian, tellurium, Telluride(Colo.), telluric, tellural.

Gr. *telma,* standing water, pool, marsh, mud of a pool or bank; see **marsh**

Gr. *temenos,* n. piece of land marked off as a reservation: *Temenuchus nemoricolus* (a bird).

L. *terra,* f. earth, ground, soil; *terrula,* f. dim.; *terrenus; terreus,* earthy, earthen; *terrestris; terrulentus,* of the earth, earthy: terra firma, terrestrial, terrace, terrain, subterranean, Mediterranean, interment, disinter, terreous, terrier, terramycin, territory, fumitory, Terre Haute(Ind.), Tierra del Fuego, turmeric, *Tribulus terrestris* (puncture-vine), *Leptolegnia subterranea* (a fungus), *Turdoides terricolor* (a bird), *Aspergillus terreus* (a fungus), *Erica mediterranea* (a heath).

Gr. *tholos (olos),* m. mud, dirt, sepia; *tholeros (oleros),* muddy, impure, turbid: *Tholocrinus spinosus* (a Mississippian crinoid), *Tholemys passmorei* (a fossil turtle), *Tholerostola omphalos* (a moth), *Oloessa minuta* (a beetle).

Gr. *tyntlos,* m. mud; *tyntlodes,* muddy: *Tyntlastes brevis* (a fish).

See: **dirt, stone, field, country, world, plaster**

earthquake; see **shake**

ease; easy < F. *aise,* comfort, without toil; see **rest, accessible**

east < AS. *east.*

Gr. *anatole,* f. a rising, sunrise, east; *anatolikos,* eastern: *Anatole penthea* (a butterfly), *Anatolichthys splendens* (a fish).

Gr. *eos,* dawn, morning, early, east; *heothinos,* in the morning, early, eastern; see **dawn**

L. *oriens, -entis,* m. the rising sun, east; *orientalis,* of the east: orient, oriental, orientation, *Rubus oriens* (eastern dewberry), *Platanus orientalis* (oriental plane-tree).

L. *subsolanus,* eastern, oriental:

See: **dawn**

eat < AS. *etan.*

Gr. *akratisma, -tos,* n.; *akratismos,* m. breakfast:

L. *alo, alitus; altus,* nourish, feed, sustain; *alibilis,* nutritious; *alimentum,* n. food, nutriment; *alimonia,* f. food, sustenance, support; *alimentarius,* pertaining to nourishment; *almus,* nourishing, cherishing, fostering, kind, bountiful; *altilis,* fattened, rich, nourishing; *altor, -is,* m.; *altrix, icis,* f. nourisher, sustainer: alimentary, alimony, aliment, alumnus, alma mater, *Artocarpus altilis* (breadfruit).

Gr. *ariston,* n. breakfast, lunch: *Ariston albicans* (a spider).

Gr. *bibrosko,* eat, gnaw, consume; *brosimos; brotos,* edible, eaten; *brotikos,* voracious; *broter, -os,* m. eater, eating; *bora,* f.; *broma, -tos; brosis, -eos,* n. food, an eating; *bromation,* n. dim.; *boros,* devouring, greedy, gluttonous; *diaboros,* eating through; *molobros,* greedy, greedy fellow; *polyboros,* voracious; *zabros,* gluttonous; *eubrotos,* good to eat; *halibrotos,* swallowed by the sea: anabrotic, diabrosis, theobromine, *Theobroma cacao* (chocolate-tree), *Brosimum paraense* (a breadnut-tree), *Abroma angusta* (devil's-cotton), *Boromys offella* (a spiny rat), *Diabrotica duodecimpunctata* (a cucumber-beetle), *Borophagus diversidens* (a fossil canid), *Uloborus americanus* (a spider), *Xyleborus dispar* (a timber-borer), *Halibrotus chauseicus* (an amphipod), *Pamborus viridus* (a beetle), *Molobrus fuliginosus* (a fly), *Molobrosichthys patagonicus* (a fossil fish), *Zabrus gibbus* (a beetle), *Zabrotes spectabilis* (a beetle), *Zabroscelis ditomoides* (a beetle), *Zabromorphus pachysomus* (a beetle), *Catabrosa aquatica* (a grass).

Gr. *bosko,* feed, pasture; *boske; bosis,* f. food, fodder; *botane,* f. grass, fodder; *boter, -os; botes,* m. herdsman; *eubosia,* f. good pasture, plenty: botany, proboscis, *Hippobosca equina* (horse tickfly).

Gr. *bryko; brycho,* eat nosily, greedily, gnash the teeth; see **bite**

L. *catillo, -onis,* m. plate-licker, glutton, < *catillo, -atus,* lick a plate:

L. *cena,* f. dinner, supper; *cenula,* f. dim.; *cenaticus,* pertaining to dinner; *cenaculum,* n. dining-room, upper story:

L. *cibo, -atus,* feed; see *cibus* under **food**

L. *consumo, -umptus,* use up completely, eat, destroy, waste: consumption, consume.

L. *convivialis,* of a feast, < *convivium,* n. social feast, banquet: convivial.

Gr. *dais, -itos,* f. meal, banquet; *daitymon, -os,* m. guest; *daitaleus,* m. banqueter; *adaitos,* inedible, forbidden: *Daetaleus purpureus* (a fly).

L. *daps, dapis,* f. solemn religious feast, feast, banquet; *dapalis,* of a sacrificial feast; *dapifer, -i,* m. steward; *dapifex, -icis,* m. servant who prepared food: dapifer.

Gr. *dapto,* devour, gnaw; *daptes,* m.; *daptria,* f. eater: *Daptophilus squalidus* (a fossil mammal), *Daptomys venezuelae* (a rat).

Gr. *deipnon,* n. meal, dinner, < *deipno,* dine; *deipnestos,* m. mealtime; *deipnesterion,* n. dining-room; *deipnetes,* m. diner, guest; *deipnetikos,* fond of dining; *epideipnis, -idos,* f. dessert; *deipnizo,* entertain at dinner: deipnosophist, *Deipnopsocus[Dipnopsocus] spheciophilus* (a psocopterid).

L. *diaeta* (Gr. *diaita*), f. manner of living, a prescribed regimen of food: diet, dietary, dietetics.

Gr. *dorpon,* n. supper, dinner, meal; *epidorpis, -idos,* f. dessert:

L. *edo, esus* (Gr. *edo; esthio*), eat; *edax, -acis,* greedy, gluttonous; *edibilis; edulis,* eatable; *esor, -is,* m.; *estrix, -icis,* f. eater; *esurio, -itus,* be hungry; *ambesus,* gnawed around; *comedo, -onis,* m. glutton; *comesor, -is,* m. gourmand; *comesus; exesus,* eaten up; *peresus,* eaten through, wasted; *semesus,* half-eaten; *comestibilis,* edible; *inedax, -acis,* eating little; Gr. *edanos; edestos; edodimos,* eatable; *edesma, -tos,* n.; *edetys, -yos; edode,* f. food, meat; *edestes,* m. eater; *edodes,* given to eating; *eidar, eidatos,* n. food; *diesthio,* eat through, corrode: edible, edaceous, edacity, inedible, comestible, comedo, esurient, obese, dermestid, *Pinus edulis* (pinyon or nut-pine), *Mesembryanthemum edule* (a fig-marigold), *Edestus vorax* (a fossil fish), *Diestogyna chalybeata* (a butterfly), *Diestecopus erodioides* (a beetle), *Myiadestes townsendi* (solitaire).

Gr. *eilapine,* f. feast, banquet; *eilapinastes,* m. feaster, guest: *Ilapinastes davidsoni* (a wasp).

L. *epula,* f. sumptuous food, banquet, feast; *epularis,* pertaining to a banquet; *epulatio, -onis,* f. feasting; *epulator, -is,* m. feaster; *epulo, -onis,* m. guest at a banquet; *epulor, -atus,* feast, eat: epulary, epulation.

AS. *faestan,* abstain from food; see **hunger**

L. *festum,* banquet, feast; see *festus* under **joy**

L. *ganeum* (Gr. *ganeion*), n. restaurant; *ganearius,* of an eating-house; *ganeo, -onis,* m. glutton:

L. *glutio, -itus,* devour, swallow; *gluto* (*glutto*), *-onis,* m. gormandizer: deglutition, glutton, gluttony.

Gr. *grao,* gnaw, eat; *grastis* (*krastis*), *-eos,* f. green fodder, grass; *grastizo,* graze: gangrene.

L. *gulo, -onis,* m. glutton, gormandizer, epicure; *gulosus,* gluttonous, greedy; *degulo, -atus,* devour, consume: *Gulo luteus* (a wolverine), *Lycosa gulosa* (a spider), *Scydmaenus gulosus* (a beetle).

L. *gumia,* c. epicure, gourmand:

L. *heluor, -atus,* guzzle, gormandize; *heluo, -onis,* m. glutton:

Gr. *hestiasis,* f. a feasting, banquet, < *hestiao,* entertain, feast:

Gr. *hydneo,* nourish: *Hydnum repandum* (spreading hydnum), *Hydnora africana* (a parasitic flowering plant), *Hydnocarpus anthelminticus* (a chaulmoogra-tree).

L. *ingluviosus,* voracious, gluttonous; see *ingluvies* under **stomach**

L. *ingurgito, -atus,* pour in, gormandize, guzzle: ingurgitate.

L. *jentaculum,* n. breakfast:

Gr. *kapto,* gulp down; *kapsis; anakapsis,* f. a gulping down; *kapeton,* n. fodder: *Eucapsis alpina* (an alga).

Gr. *labros,* greedy, gluttonous; see **greed**

Gr. *laimargos,* greedy, gluttonous: *Laemargus muricatus* (a fish).

Gr. *lamyros,* greedy, gluttonous; see **greed**

Gr. *laphysso,* eat greedily, devour, gorge; *laphygmos,* m. gluttony; *laphystios,* gluttonous; *laphyktes,* m. glutton, gourmand: *Laphygma exigua* (beet army-worm), *Laphyctes insidiator* (a wasp).

Gr. *laryngikos,* gluttonous:

Gr. *lichnos,* greedy, gluttonous, dainty: *Lichnochromis acuticeps* (a fish).

L. *ligurio, -itus,* eat daintily, be fond of good things, lick; *liguritor, -is,* m. epicure, gourmand: ligurition.

L. *lurco, -onis,* m. gourmand, glutton; *lurcabundus,* voracious, < *lurco,* eat voraciously:

L. *manducus,* chewer, glutton; see *mando* under **bite**

L. *merenda,* f. afternoon snack:

Gr. *nemo,* feed, graze, dispense; *nemos* (L. *nemus*), n.; *nomos,* m. feeding-place, pasture, glade, grove; *pronomos,* grazing forward: *Anthonomos signatus* (straw-berry-weevil), *Pronomotherium altiramum* (a fossil mammal).   See *nemus* under **forest**

L. *nutrio, -itus,* nourish; *nutrico, -atus,* suckle, nourish; *nutribilis; nutritius; nutritorius,* nourishing; *nutrimentum,* n. nourishment: nutritious, nutrition, nurture, nurse, nutriment, nourishment.

L. *pasco, pastus,* feed; *pascuum,* n. pasture; *pasticus,* fed, fattened; *pascalis; pascuus,* of pasturing and grazing; *compasco,* feed together; *compascuus,* of common pasture; *depastus,* consumed: pascual, pascuage, pascuous, pasture, pastor, pastoral, repast, pester, pastern.

Gr. *phagein,* to eat; *phagos,* m. glutton; *adephagos,* gluttonous, greedy; *anthropophagos,* m. cannibal; *microphagos,* eating little; *polyphagos,* eating much: adephagous, bacteriophage, phagocyte, sarcophagus, anthropophagous, geophagy, esophagus, *Geophagus brasiliensis* (a fish), *Onthophagus bubalus* (a dung-beetle), *Pamphagus fuscus* (a grasshopper), *Phytophaga destructor* (Hessian fly), *Sitophagus complanatus* (a beetle), *Xylophaga dorsalis* (a wood-boring mollusk).

Gr. *pherbo,* feed; *phorbe,* f. pasture, food; *phorbas, -ados,* c. a giving food, grazing; *pamphorbos,* all-feeding: *Euphorbia corollata* (flowering-spurge), *Phorbas amaranthus* (a sponge).

L. *popina,* f. cook-shop, eating-house; *popinalis,* of an eating-house; *popinor, -atus,* frequent eating-houses; *popinarius,* m. cook:

L. *prandium,* n. lunch: prandial, postprandial.

Gr. *psizo,* feed pap; *psomizo,* feed on little bits:

L. *rodo, rosus,* gnaw, eat away, consume, waste away; *erodo,* gnaw off, eat away; see **bite**

Gr. *rhopheo,* sup, gulp, bolt; *rhophema, -tos,* n. gruel, porridge; *rhophemation,* n. dim.; *rhophesis,* f. a supping; *rhophetikos,* absorbing; *rhophetos,* that can be supped:

L. *sagina,* f. a feasting, stuffing, fattening; *sagino, -atus,* fatten: saginate, *Sagina subulata* (Corsican pearlwort), *Taenia saginata* (a tapeworm).

AS. *swelgan,* gulp, eat, drink, take in: swallow, groundsel.

Gr. *tenthes,* m. epicure, < *tentho,* eat daintily: *Tenthes citatus* (a fly).

Gr. *thoine,* f. meal, dinner, banquet, feast; *thoinatikos,* of a feast; *thoinater, -os,* m. giver of a feast, feaster:

Gr. *tithene; tithenos; titthe,* nurse, rearer; see **nurse**

Gr. *trepho,* feed, nourish, maintain; *trophe,* f.; *trophema, -tos,* n. food, nourishment; *trophos,* c. feeder, nurse; *trophimos; trophodes,* nutritious; *hapalotrephes; trapheros; trophis, -ios,* well-fed, plump, stout; *trophikos,* nursing; *threpsis,* f. nourishment; *threpterios; threptikos,* feeding, nourishing; *threpter, -os,* m. feeder, rearer; *threptos,* m. slave, servant: trophoblast, trophocyte, trophema.

trophonema, trophothylax, trophallaxis, abiotrophy, atrophy, dystrophy, hypertrophy, threpsology, threptic, athrepsia, *Trophocrinus tumidus* (a Mississipian crinoid), *Trophis americana* (a moracead), *Trophimus aeneipennis* (a beetle), *Trophithauma splendidum* (a fly), *Melithreptus lunatus* (a honey-eater).

Gr. *trogo*, gnaw, nibble, munch; *troktes*, gnawer, nibbler; see **bite**

L. *tuburcinor, -atus*, eat greedily, devour:

L. *ventricola*, m. one who worships his belly, glutton:

L. *vesperna*, f. evening meal:

L. *voro, -atus*, eat greedily, devour; *vorator, -is*, m. devourer; *vorax- -acis*, gluttonous, greedy; *carnivorus*, flesh-eating; *herbivorus*, plant-eating: voracious, voracity, devour, carnivore, herbivorous, insectivorous, omnivorous, *Anthrenus vorax* (buffalo-beetle), *Melanerpes formicivorus* (a woodpecker), *Vermivora celata* (orange-crowned warbler).

See: **food, drink, bite, destroy, greed, nurse**

eaves < AS. *efes*, brim, brink, lower border of a roof; see **border**

ebaeo- < Gr. *ebaios*, little, small, poor; see *baios* under **little**

ebenus, L. (Gr. *ebenos*), the ebony tree, a species of persimmon; *ebeneus*, of ebony, black; see **persimmon, black**

ebrius, L. drunken; *ebriolus*, dim.; *ebriosus*, given to drunkenness; see **drink**

ebulus, L. a dwarf elder; *ebulinus*, of elder; see **elder**

ebur, *-oris*, L. ivory; *eburatus; eburneus; eburnus*, of ivory; see **ivory**

ec- < Gr. *ek*, out of, from; see **out**

ecbleto- < Gr. *ekbletos*, thrown away, rejected; see *ballo* under **throw**

ecchymosis, L. (Gr. *ekchymosis*), extravasation of blood; see *cheo* under **pour**

ecclesia, L. (Gr. *ekklesia*), assembly of citizens, the church; see **gather, temple**

eccopto- < Gr. *ekkopto*, cut out, fell, destroy; see *kopto* under **cut**

eccremo- < Gr. *ekkremes*, hanging, pendent; see *kremastos* under **hang**

ecdemo- < Gr. *ekdemos*, away from home, abroad; see *apodemos* under **far**

ecdysis < Gr. *ekdysis*, a getting out of, escape from, molt; *ekdyma*, that which is taken off, < *ekdyo*, take off, strip, lay bare; see **bare**

-eces < Gr. *-ekes*, a suffix meaning sharp, < *ake*, point, sharp edge; see *akis* under **point**

echeneis, Gr. a sucking fish; see **delay**

echephron, Gr. prudent, sensible; *echephrosyne*, prudence, good sense; see **think**

echetle, Gr. plow-handle; see **handle**

echetrosis, Gr. *bryony;* see **melon**

echeveria, NL. a genus of the stonecrop family; see **stonecrop**

echidna, Gr. adder, viper; see *echis* under **snake**

echinoderm < Gr. *echinos*, sea-urchin; spiny, *derma*, skin: Echinodermata, Echinoidea.

NL. *asterias*, a genus of starfishes; see **star**

Gr. *bryssos*, m. a kind of sea-urchin: *Byssoteres pinniger* (a fish).

L. *cidaris* (Gr. *kidaris*), Persian diadem; a genus of sea-urchins; see **crown**

NL. *crinus* (L. *crinum;* Gr. *krinon*, lily), sea-lily, crinoid; see **lily**

Gr. *holothurion*, n. a sea-cucumber: holothurian, *Holothuridium papillosum* (an echinoderm), *Holothuria tubulosa* (trepang).

L. *spatangius* (Gr. *spatanges*), m. a sea-urchin: spatangoid, *Spatangus purpureus* (a sea-urchin).

echinopus, L. (Gr. *echinopous*), a prickly plant; see **plant**

echinus, L. (Gr. *echinos*), hedgehog, sea-urchin; *echinatus*, prickly; see **hedgehog**

echis, Gr. adder, viper; *echidion*, dim.; see **snake**

echium, L. (Gr. *echion*), bugloss; see **borage**

echma, Gr. hindrance, holdfast, obstacle; see **hinder**

echo < Gr. *echo*, a returned sound; *eche; echos*, sound, noise; *echetes*, shrill, see **sound**; *echo; ischo*, hold, restrain, see **hold**

echtho-; echthro- < Gr. *echthos*, hate; *echthodopos*, detestable; *echthros*, hateful; see **hate**

echyros, Gr. strong, secure; see **strong**

eciton, NL. a genus of ants; see **ant**

eclecto- < Gr. *eklektos*, choice, selected; *eklektikos*, choosing; see *lego* under **choose**

ecligma, L. (Gr. *ekleigma*), a medicine compounded with syrup, electuary; see **drug**

eclipse < Gr. *ekleipsis,* a forsaking, failing, cessation; see *leipo* under **abandon, cover**

ecnomo- < Gr. *eknomos; eknomios,* monstrous, unlawful, unusual, marvelous; see **wonder**

eco- < Gr. *oikos,* house, home; see **house**

ecpaglo- < Gr. *ekpaglos,* wondrous, terrible; see **wonder**

ecphato- < Gr. *ekphatos,* beyond the power of speech, ineffable; see **ineffable**

ecphora, L. (Gr. *ekphora*), a projection in buildings; see **projection**

ecphymato- < Gr. *ekphyma, -tos,* an eruption of pimples; see **pimple**

ecpomato- < Gr. *ekpoma, -tos,* beaker, cup; see **cup**

ecstasy < Gr. *ekstasis,* beside oneself, rapture, trance; see **joy**

ectato- < Gr. *ektatos,* capable of extension or prolongation; see *teino* under **spread**

ectemno- < Gr. *ektemno,* cut out; see *temno* under **cut**

ecteno- < Gr. *ekteino,* stretch out; see *teino* under **spread**

echthymo- < Gr. *ekthymos,* insane, demented; see **mad**

ecto- < Gr. *ek,* out of, from; see **out**

-ectomy, a suffix denoting excision; see *temno* under **cut**

ectopistico- < Gr. *ektopistikos,* migratory, < *ektopizo,* migrate; see **move**

ectopo- < Gr. *ektopos,* away, distant, strange; see *atopos* under **strange**

ectromato- < Gr. *ektroma, -tos,* abortion; see **abortion**

ectyon, NL. a kind of sponge; see **sponge**

ectypho- < Gr. *ektyphos,* puffed up, empty; see **swell**

ectypo- < Gr. *ektypos,* embossed, in relief; see *typus* under **form**

ecumenico- < Gr. *oikoumenikos,* worldwide, universal, general; see *oikoumene* under **world, all**

eczema < Gr. *ekzema,* tetter, a skin eruption; see **disease**

edaphos, Gr. soil, ground, earth; see **earth**

edax, -acis, L. greedy, gluttonous; *edacitas,* voracity; see *edo* under **eat**

edema < Gr. *oidema,* a swelling (with fluid); *oidaleos,* swollen; see **swell**

edeo- < Gr. *aidoion,* private part, genital; see **sex organs**

edesma; edetys, Gr. food, meat; see *edo* under **eat**

edge < AS. *ecg;* see **border, end, sharp, point, angle**

edibilis, L. eatable, < *edo, esus,* eat; see **eat**

edictum, L. ordinance, proclamation; see *dico* under **speak**

edifice < L. *aedificium,* any kind of building, < *aedifico, -atus,* build; see **house, make**

edit < L. *edo, -itus,* prepare for publication, publish: editor, edition, editorial.
    L. *emendo, -atus,* free from errors, improve, correct, < *ex,* from, *menda,* fault: emend, amend, emendation, amendment.
    Gr. *epanorthosis,* revision, correction; see *orthos* under **straight**
    Gr. *euthynter,* straightener, corrector, auditor; see *euthys* under **straight**
    L. *recenseo,* review, revise: recension.
    L. *recognosco, -itus,* examine, inspect, review:
    L. *redactus,* edited, revised, reduced: redact, redactor, redaction.
    L. *revisio, -onis,* f. a seeing again, correction: revise, revision.
    See: **right, straight, smooth**

editus, L. high, lofty; see **high**

edolios, Gr. a kind of bird; see **bird**

educate < L. *educo, -atus,* bring up a child, train; see **teach**

edulis, L. eatable; see *edo* under **eat**

eel < AS. *ael:* eelfare, elver, eelware.
    L. *anguilla,* f. eel: anguilliform, *Anguilla rostrata* (eel), *Anguillula aceti* (vinegar-eel), *Zoarces anguillaris* (eelpout).
    L. *conger, -gri* (Gr. *gongros*), m. a marine eel: *Conger caudilimbatus* (a conger-eel).
    Gr. *enchelys, -yos,* f. eel; *encheleion; enchelydion,* n. dim.: *Enchelyopus americanus* (an eelpout), *Simenchelys parasitica* (pugnosed eel).
    L. *fluta,* f. a large eel: *Fluta javaensis* (an eel).
    ML. *lampetra,* f. an eellike fish: lamprey, *Lampetra fluviatilis* (lampern), *Petromyzon marinus* (sea-lamprey).

L. *muraena* (Gr. *myraina*), f. moray: *Muraena helena* (a moray).

L. *smyrus* (Gr. *smyros; myros*), m. a kind of eel: *Myrus vulgaris* (an eel), *Myrophis longicollis* (an eel).

ef- < L. *ex*, out of, from; see *e-* under **out**

effect < L. *efficio, -fectus,* execute, accomplish, make; see *facio* under **make**

effeminatus, L. womanish; see *femina* under **woman**

efferatus, L. fierce, savage, wild; see *ferus* under **nature**

effetus, L. exhausted, worn out; see **weak**

efficax, -acis, L. effectual, powerful; see **strong**

effigies, L. image, likeness; see **form**

effrenus, L. free from the bridle, unrestrained; see **free**

effusus < L. *effundo,* pour out; see *fundo* under **pour**

egelidus, L. lukewarm, tepid, chilly; see **lukewarm**

egenus, L. poor, needy, indigent; *egens,* needy; see **poor**

egeries, L. excrement; see **dung**

egersis, Gr. a waking, stirring; *egertikos,* waking, stirring; see **arouse**

egestus, L. an emptying, voiding; *egestivus,* purgative; see **empty**

egg < ON. *egg;* AS. *aeg;* G. *ei:* eggnog, eggplant, eggshell, cockney.

AS. *knitu,* egg of a louse: nit.

Gr. *konis, -idos,* f. egg of a louse, nit:

L. *lens, -endis,* f. egg of a louse, nit:

Gr. *oön (obeon; oion),* n. egg; *oarion,* n. dim.: oogenesis, ooblast, ooecium, oocyst, oology, oophrium, oolite, oophyte, oarium, *Dioon edula* (a cycad), *Oidium* (now *Saccharomyces*) *albicans* (a fungus causing thrush), *Stereognathus ooliticus* (a fossil mammal).

L. *ovum,* n. egg; *ovulum,* n. dim.; *ovalis; ovatus,* egg-shaped: ovate, oval, ovule, ovulary, ovary, ovoid, obovate, oviduct, oviparous, ovipositor, synovial, *Ovalipes ocellatus* (a crab), *Ovoides setosus* (a fish), *Viola ovata* (a violet), *Hadrophyllum ovale* (a fossil coral), *Plethomytilus oviformis* (a fossil pelecypod), *Eriogonum ovalifolium* (a polygonacead), *Viburnum ovatifolium* (a viburnum), *Condalia obovata* (a rhamnacead), *Mycale ovulum* (a sponge).

See: **seed, yolk, bird**

ego, L., Gr. I, myself; see **alone**

egregius, L. not of the common herd, distinguished, eminent; see **honor**

egula, L. a kind of sulfur; see **sulfur**

Egypt < L. *Aegyptus* (Gr. *Aigyptos*); see **Africa**

eido- < Gr. *eidos,* form, resemblance; *eidolon,* image, fancy; see **form,** *-oid* under **like**

eight < AS. *eahta.*

L. *bessalis,* of eight, < *bes, bessis,* m. eight-twelfths or two-thirds of a unit:

L. *octo* (Gr. *okto*), eight; *octavus,* eighth; *octonalis; octonarius; octonus,* consisting of eight; *octuplus,* eightfold; *octodecim,* eighteen; *octoginta,* eighty; *octogesimus,* eightieth; Gr. *ogdoos,* eighth; *ogdas (ogdoas; ogtas), -ados,* f. the number eight; *ogdoekonta,* eighty; *ogdoekostos,* eightieth: octave, octuple, octagon, octopus, octogenarian, octaroon, octadecyl, 8 vo., October, ogdoad, *Ogdoconta*[*Ogdoeconta*] *cinereola* (a moth), *Ogdoecosta sagitta* (a beetle), *Dryas octopetala* (a rosacead), *Herpobdella octoculata* (a leech), *Heliosestrum octonum* (a radiolarian), *Buprestis octoguttata* (a beetle), *Aesculus octandra* (buckeye).

ejulo, -atus, L. wail, lament; see **weep**

ejuncidus, L. rushlike, lean, slender; see **thin**

eka, Skt. one; see **one**

-el, OF., L. diminutive suffix; see *-cle* under **little**

elacato- < Gr. *elakate,* distaff, spindle; see **spindle**

elachys, Gr. small, short, low; *elasson (elatton),* smaller, less; *elachistos,* smallest, least; see **little**

elaeagnus, NL. (Gr. *elaiagnos*), a genus of the oleaster family; see **oleaster**

elaeo- < Gr. *elaia,* olive-tree; *elaion,* olive oil; see **olive, fat**

elanos, Gr. kite; see **hawk**

elaphos, Gr. deer, stag; see **deer**

elaphros, Gr. light in weight, nimble; see **light**

elaps, NL. a kind of snake; see **snake**

elasas, Gr. a bird; see **bird**

elasma; elasmos, Gr. metal plate; see **plate**

elasson (elatton), Gr. smaller, less; see *elachys* under **little**

elastic < Gr. *elastikos,* rebounding, springing, < *elauno,* drive, beat out, stretch out; see **leap,** *elater* under **drive**

elate, Gr. fir or spruce, oar; *elatinos,* of fir; *elatine,* a plant with firlike leaves; see **fir, waterweed**

elater, Gr. driver; see **drive**

elatus, L. exalted, high, lofty; *elatior; elatius,* taller; see **high**

elbow < AS. *elboga,* < *eln,* ell, forearm, *boga,* bend; see **arm**

elder < AS. *ellern:* red-berried elder (*Sambucus racemosa*).
    L. *actaea* (Gr. *aktaia; aktea*), elder; now applied to baneberry; see **buttercup**
    L. *ebulus,* m. a dwarf elder; *ebulinus,* of elder: *Sambucus ebulus* (danewort).
    Gr. *kanopon,* n. elder:
    L. *olma,* f. another name for *ebulus:*
    L. *sambucus,* f. elder: sambucene, *Sambucus canadensis* (common elder), *Sorbus sambucifolia* (Siberian mountain-ash), *Aphis sambuci* (an aphid).

elea, Gr. a kind of warbler; see **warbler**

elecebra, L. sponger, wheedler, golddigger; see *lacio* under **draw**

electilis, L. choice, dainty, selected; see *elegans* under **beauty**

elector, L. chooser; *electio, -onis,* choice, selection; see *lego* under **choose**

electoro- < Gr. *elektor, -os,* the beaming sun; see **sun**

electricity < L. *electricus,* pertaining to *electrum* (Gr. *elektron*), n. amber: electron, electrolysis, electroscope, dielectric, *Electrophorus electricus* (electric eel). NL. *galvano-,* pertaining to current electricity, < Luigi Galvani, Italian physiologist: galvanic, galvanize, galvanometer, galvanoscope.

electrum, L. (Gr. *elektron*), amber; see **amber**

electuarium, L. a medicine; see **drug**

eleemosyna, L. (Gr. *eleemosyne*), alms; see **mercy**

elegans, L. tasteful, choice, fine, select; *elegantulus,* very fine; see **beauty**

elegos, Gr. mourning song or poem; see **poem**

elematos, Gr. vain, trifling; see **trifle**

elementum, L. first principle, rudiment, letter of the alphabet; see **thing**

elemi, Sp. < Ar. *al lami,* an oleoresin; see **resin**

elencho- < Gr. *elenchos,* trial, test, examination, reproach, disgrace; see **try**

eleo- < Gr. *eleos,* kitchen table, dresser, see **table**; pity, mercy, see **mercy**; astray, distraught, crazed, see **made**; < *elaion,* olive oil, any oil, see **olive, fat**; a kind of owl, see **owl**

elephant < L. *elephas, -antis* (Gr. *-antos*), m.; *elephantinus,* of the elephant, of ivory; elephantine, elephantiasis, *Elephas indicus* (Asiatic elephant), *Elephantella groenlandica* (elephant-head), *Elephantopus tomentosus* (woolly elephant's-foot), *Elephantomyia angusticellula* (a tipulid), *Phytelephas macrocarpa* (ivory-nut palm), *Megasoma elephas* (elephant-beetle), *Testudinaria elephantipes* (a dioscoreacead).
    L. *barrus,* m. elephant; *barrinus,* elephantine:
    G. *mammuth* (Russ. *mammot*), a kind of elephant: *Mammut ohioticum* (original name of the American mastodon), hairy mammoth *(Elephas primigenius).*
    L. *proboscis* (Gr. *proboskis*), trunk of an elephant; see **nose**
    Gr. *rhizis,* m. a kind of elephant:

eleusine, NL. a genus of grasses; see **grass**

eleutheros, Gr. free; see **free**

elevatus, L. raised; *elevator,* raiser; see *levo* under **raise**

eleven < AS. *endleofan.*
    Gr. *endeka,* eleven; *endekatos,* eleventh: *Endecatomus reticulatus* (a beetle), *Indigofera endecaphylla* (an indigo-plant), *Solaster endeca* (a starfish).
    L. *undecim,* eleven; *undecimus,* eleventh: *Centropomus undecimalis* (snook).

elibatos, Gr. high, steep; see **high**

elido, -isus, L. strike out, annul; see *laedo* under **strike**

elimatus, L. filed, finished, elaborated, adorned; see *lima* under **scrape**

eliquatus, L. clear; see **clear**

elithios, Gr. idle, vain, foolish; see **fool**

elix, *-icis,* L. channel, ditch, trench; see **ditch**

elixatus; elixus, L. boiled, soaked; see **boil**

elk < ON. *elgr* (AS. *eolh*); see **deer**

ellimenistes, Gr. collector of harbor fees; see **pay**

ellipsis, L. (Gr. *elleipsis; ellipes*), omission, lack, < *elleipo,* leave out; see *leipo* under **abandon, form**

ellobion, Gr. earring; see **earring**

ellobos, Gr. in a pod; see *lobus* under **projection**

ellogimos, Gr. esteemed, famous; see **honor**

ellops, *-opos,* Gr. mute; see **silent**

-ellus, *-a, -um,* L. diminutives; see *-cle* under **little**

ellychnium, L. (Gr. *ellychnion*), lampwick; see **wick**

elm < AS. *elm.*
> L. *celtis,* f. a genus of the elm family, hackberry: *Celtis reticulata* (netleaved hackberry), *Morus celtidifolia* (Texas mulberry).
> NL. *planera,* f. a genus of the elm family, < J. J. Planer, German botanist: *Planera aquatica* (water-elm).
> Gr. *ptelea,* f. elm: *ptelein, Ptelea trifoliata* (wafer-ash or hop-tree), *Euptelea polyandra* (a trochodendracead), *Chaetoptelea mexicana* (Mexican elm).
> L. *ulmus,* f. elm: *ulmeus,* of elm: ulmaceous, ulmous, ulmin, ulmic, *Ulmus glabra* (wych or Scotch elm), *Ficus ulmifolia* (a fig), *Ceratostomella ulmi* (Dutch elm-disease, carried by a beetle, *Scolytus multistriatus*), *Colopha ulmicola* (cockscomb gall-insect) *Filipendula ulmaria* (a meadowsweet), *Urvillea ulmacea* (a sapindacead), *Gnomonia ulmea* (a fungus).
> NL. *zelkova,* f. a genus of the elm family, < Caucasian *tselkva: Zelkova serrata* (Japanese zelkova).

elodea, NL. a genus of waterweeds; see **waterweed**

elongatus, L. prolonged; see *longus* under **long**

elops, Gr. a large fish, probably a sturgeon; see **sturgeon**

elpido- < Gr. *elpis, -idos,* hope, expectation; see **believe**

eluctor, *-atus,* L. struggle out, surmount; see *luctor* under **fight**

eludo, *-lusus,* L. avoid, evade, frustrate, baffle; see **shun, secret**

elyma, Gr. share-beam of a plow; see **beam**

elymos, Gr. case, quiver, see *elytron* under **bag;** a kind of grain, see **grain**

Elysium, L. (Gr. *Elysion*), abode of the blessed; see **heaven**

elytron, Gr. sheath, husk; see **sheath**

em-; en- < Gr. *en;* < L. *in,* in, into, within; see **in**

emacio, *-atus,* L. waste away, make lean; see **lessen**

emancipo, *-atus,* L. set free; see **free**

emansor, L. one who overstays his furlough; see **delay**

emarginatus, L. without margin, notched at the apex; see **notch**

emasculo, L. castrate; see **geld**

emato- < Gr. *emar, -atos,* day; see *hemera* under **day**

embaphion, Gr. shallow vessel for sauces; see **cup**

embas, Gr. a felt shoe; see **shoe**

emberiza, NL. < G. *ammer,* bunting; see **finch**

embios, Gr. for life, in life, long-lived; see *bios* under **life**

emblem < Gr. *emblema,* inlaid work, thing put in or on; see **mark, ornament**

embole, Gr. invasion, foray, insertion, pass; *embolimos,* inserted, see **insert;** *embolon; embolos,* bolt, bar, peg, stopper, wedge, see **close;** *embolion,* javelin, see **spear**

embrace < OF. *embracer,* < L. *in,* in, *brachium,* arm: embracement.
> Gr. *amphiptyche; periptyche,* f. embrace, enfolding:
> L. *amplector, -exus,* encircle, enfold, embrace; *implexus,* m. an enfolding, embrace: amplexicaul, *Lamium amplexicaule* (henbit), *Streptopus amplexifolius* (a liliacead).
> Gr. *aspasmos,* m. greeting, embrace: *Aspasmogaster spatula* (a cling-fish).
> Gr. *euankalos,* pleasant to embrace, holding easily:
> Gr. *peribole,* f. embrace:
> See: **kiss, fold, circle, take**

embrithes, Gr. weighty, grave; see *brithys* under **heavy**

embroche; **embregma**, Gr. lotion; see **drug**

embryo < Gr. *embryon*, fetus, the unborn young; see **young**

emendatus, L. corrected, perfect; *emendator*, corrector; see *emendo* under **edit**

emerald < OF. *esmeralde*, < L. *smaragdus* (Gr. *smaragdos*); see *smaragdus* under **stone**

emeritus, L. honorably discharged or retired, veteran; see *mereo* under **honor**

emesa, NL. a genus of bugs; see **bug**

emeticus, L. (Gr. *emetikos*), turning the stomach, causing vomiting; *emesis*, vomiting; see **vomit**

emico, *-atus*, L. appear suddenly, become apparent; see **open**

emigro, *-atus*, L. move out; see *migro* under **move**

eminens, *-entis*, L. projecting, high; *eminulus*, dim.; see *minor* under **projection**

eminus, L. aloof, distant, a spear's throw off; see **far**

emmeles, Gr. harmonious, fit, right; see **harmony**

emmeno- < Gr. *emmenes*, abiding, steadfast, see *meno* under **stand**; < *emmenos*, lasting a month, monthly, see *men* under **month**

emmestos, Gr. filled full; see *mestos* under **abundance**

emmochlion, Gr. socket for a bar; see **hollow**

emmochthos, Gr. toilsome; see *mochtheros* under **trouble**

Emodos, Gr. Himalaya Mountains; see **mountain**

emosyne, Gr. skill in throwing, < *emon*, a thrower; see **throw**

empedos, Gr. firm, steadfast, lasting; see **stand**

empetrum, L. (Gr. *empetron*), a rockplant, a genus of the crowberry family; see **crowberry**

emphabion, Gr. shallow vessel for sauces; see **vessel**

emphasis, Gr. suggested importance, stress; see **strong**

empis, *-idos*, Gr. mosquito, gnat; see **fly**

emplastrum, L. (Gr. *emplastron*), salve, mortar; see **plaster**

emplecto- < Gr. *emplektos*, inwoven; see *plecto* under **weave**

empodios, Gr. at one's feet, in the way, hindering; see **hinder**

emporium, L. (Gr. *emporion*), market; see **trade**

empresis; **empresmos**, Gr. conflagration; *emprestes*, burner; see **burn**

emptus, L. bought, < *emo*, buy; *emax*, *-acis*, fond of buying; *emptor*, buyer; see **trade**

empty < AS. *aemtig*.
   Gr. *apousia*, f. absence, deficiency, want:
   L. *assellor*, *-atus*, go to stool, void; see **dung**
   Gr. *atisia*, f. inability to pay, insolvency:
   L. *avium*, a byway, desert, wilderness; see *via* under **way**
   L. *caco*, *-atus*, go to stool; see **dung**
   L. *careo*, *-ritus*, be without, lack, feel the want of; see **poor**
   L. *cassus*, empty; *incassus*, vain: cassation.
   L. *deliquus*, wanting, lacking:
   L. *desertus*, abandoned, forsaken, solitary, waste; see **abandon, desert**
   L. *desolatio*, *-onis*, desert, waste; see **desert**
   L. *destitutus*, lacking, without possessions; see **poor**
   L. *egero*, *-estus*, discharge, void, empty; *egestivus*, purgative: *Apatura egesta* (a butterfly).
   Gr. *ektyphos*, puffed up, empty; see **swell**
   Gr. *eremia*, desert, solitude, wilderness; *eremikos*, of solitude; see *eremos* under **alone**
   Gr. *hyperinos*, cleared out, purged, exhausted:
   L. *inanis*, empty, void, useless, vain: inane, inanity, inanition, *Amphibolips inanis* (a gall-insect).
   L. *irritus*, invalid, vain, useless, infertile; see **worthless**
   Gr. *kenos*, empty; *kenosis*; *kenotes*, f. emptiness; *kenosimos*, purgative; *kenotikos*, of emptying, purgative; *kenerion*, n. an empty place; *apokenos*; *diakenos*, quite empty, hollow, vacuous: cenotaph, *Cenocephalus thoracicus* (a beetle).
   Gr. *lapatikos*, purgative, < *lapasso*, empty, discharge; *lapaxis*, f. evacuation:
   Gr. *leipo*, be wanting or without, lack; see **abandon**
   L. *nudus*, naked, bare, empty; see **bare**
   L. *purgo*, *-atus*, cleanse, expel; see **wash**
   Ar. *sifr*, empty: cipher, zero.

L. *solitudo, -inis,* a desert place; see *solus* under **alone**

L. *tescum,* rough region, desert, waste; see **desert**

L. *vaco, -atus,* be empty, free from; *vacans; vacivus; vacuus,* empty, void; *vacatio, -onis,* f. freedom from duty; *vacuum,* n. empty space, void; *vacuo, -atus,* empty, clear; *evacuo,* empty, void: vacant, vacation, vacuum, vacuole, evacuate, evacuation.

L. *vanus,* empty, fruitless, idle, proud: vain, vanity, evanescent, vanish, vaunt.

L. *vastus,* empty, desolate; *vasto, -atus,* make empty, desolate: devastate.

See: **hollow, not, desert, abandon, alone, worthless, trifle, vomit, hunger, free, rest**

Empusa, L. (Gr. *Empousa*), a hobgoblin; see **spirit**

empyema, Gr. suppuration, usually in the pleural cavity; see *pyon* under **pus**

emulatus, L. rivaling, < *aemulor, -atus,* strive to equal, rival, envy; see **equal**

emydo- < Gr. *emys, -ydos,* fresh-water tortoise; see **turtle**

en- < Gr. *en,* in; see *em-* under **in**

enalios, Gr. in, on, of the sea; see **sea**

enallos, Gr. changed, contrary; *enallax,* alternate; see *allos* under **other, alternate**

enamma, Gr. a garment; see **garment**

enantios, Gr. opposite, facing, against; see *anti* under **against**

enapo-, Gr. in; see *em-* under **in**

enargma, Gr. a phenomenon; see **display**

enargo- < Gr. *enarges,* visible, manifest; *enargotes,* clearness; see **clear**

encephalo- < Gr. *encephalos,* brain, pith of a palm; see **brain, pith**

enchelys, Gr. eel; see **eel**

enchiridio- < Gr. *encheiridion,* dagger, handbook, handle; see **dagger, book, handle**

encho- < Gr. *enchos,* spear, lance; see **spear**

encomium, L. (Gr. *enkomion*), praise; see **honor**

encrasicholus, NL. (Gr. *enkrasicholos*), a small fish; see **herring**

encyclio- < Gr. *enkyklios,* periodical; see **return**

encyo- < Gr. *enkyos,* pregnant; see *kyo* under **fertile**

end < AS. *ende.*

L. *aboleo, -olitus,* destroy, terminate; see **destroy**

L. *abortivus,* prematurely terminated; see **abortion**

L. *abrogo, -atus,* annul, invalidate, repeal: abrogate.

Gr. *akrokolion,* any extremity of the body; see **projection**

L. *annihilo, -atus,* bring to nothing, destroy utterly; see *nihil* under **not**

Gr. *anysis,* accomplishment, end, < *anyo,* effect, accomplish; see **make**

Gr. *apolego,* desist, cease; *apolexis,* f. cessation:

Gr. *apotelesma, -tos,* n. event, result:

Gr. *balbis, -idos,* starting point, goal; see **begin**

L. *cacumen, -inis,* n. extreme point, end: cacumen, cacuminate.

L. *cesso, -atus,* stop, stand back, delay; *cessator, -is,* loiterer, dawdler; see **rest**

L. *clausula,* close, end, conclusion; see **close**

L. *concludo, -usus,* end, close, infer; see *claudo* under **close**

L. *conficio, -fectus,* make, prepare, finish, destroy; see *facio* under **make**

L. *consummo, -atus,* accomplish, complete, finish, perfect: consummate, consummated, consummation.

L. *crisis* (Gr. *krisis*), issue, decision, turning-point; see **turn**

Gr. *daithmos,* boundary, division; see *daio* under **cut**

L. *defremo, -itus,* cease raging, abate:

L. *defungor, -functus,* have done with, finish, die; see **death**

L. *desino, -itus,* cease, desist, stop, end:

L. *desisto, -stitus,* leave off, cease: desist.

L. *destinatum,* n. mark, aim, end; *destino, -atus,* determine: destine, destiny, destination.

L. *desuetudo, -inis,* f. discontinuance of a practice, disuse, < *desuesco, -suetus,* disaccustom: desuetude.

Gr. *diaprasso,* accomplish, effect:

L. *dissolvo, -solutus,* abolish, abrogate, annul: dissolve, dissolution.

L. *effectum,* n. result: effect.

Gr. *ekphero,* carry out, fulfill:

L. *elimo, -atus,* file up, finish, perfect, polish; see **ornament**

Gr. *eschatos,* extreme, last; see **after**

Gr. *exeko,* run out, expire, come true:

Gr. *exerkomai*, come to an end, expire, come true, reach:

L. *expiro, -atus,* come to an end, cease: expiration, expire.

L. *exsequor, executus,* follow to the end, accomplish: executor, executive, execution.

L. *extinguo, -inctus,* put out, quench, abolish, destroy; see **destroy**

L. *extremum,* last, end; see *exterus* under **out**

L. *fatum,* destiny, lot, end; see **lot**

L. *finis,* m. end, boundary, limit; *finio, -itus,* complete; *finalis,* of the end; *confinis,* bordering; *confinium,* n. boundary: fine, final, finale, finite, financial, finish, confine, definition, indefinite, infinity, ad infinitum, affinity, finesse, infinitesimal, refinement, unfinished, paraffin.

L. *fossatus,* m. boundary:

L. *hactenus,* thus far, so much, indicating a limit:

Gr. *horos,* limit, boundary; *horios,* of boundaries; *horistes,* definer; see **border**

L. *illativus,* inferring, concluding; see **think**

L. *imum,* lowest part, bottom, last, end; *imus,* lowest, last; see **low**

L. *indutia,* cessation of hostilities, truce, armistice; see **rest**

L. *interemo, -emptus,* abolish; *peremo,* annihilate, extinguish: peremptory.

Gr. *karanoo,* accomplish, achieve, complete:

Gr. *katalexis,* f. end, termination: *Catalexis tapinota* (a moth).

Gr. *kraino,* accomplish, fulfill, execute; *kranter, -os,* m. a completer: *Cranterophorus clavicornis* (a beetle).

Gr. *kyrosis,* f. execution, accomplishment, ratification, < *kyroo,* confirm, attain:

L. *limes, -itis,* boundary; *limito, -atus,* enclose within boundaries, fix; see **border**

Gr. *loisthos,* last; see **after**

L. *margo, -inis,* brink; see **border**

L. *meta,* end, goal, column at end of Roman circus; see **pillar**

Gr. *neatos,* last, uttermost; see *neos* under **new**

Gr. *omega,* last letter of the Greek alphabet; see **letter**

L. *pausa* (Gr. *pausis*), temporary stop, halt; *pauso, -atus,* halt, cease, rest; see **rest**

Gr. *peras, -atos,* n. end, extremity, goal; *peratos,* on the opposite side, west: *Peratogonus reversus* (a beetle), *Onychoteuthis peratoptera* (a cephalopod).

L. *rescindo, -scissus,* abolish, annul, repeal; see **cancel**

L. *resulto, -atus,* leap back, rebound, reecho, terminate as a consequence: result, resultant.

L. *-sura, -tura,* suffixes denoting result of action; see **make**

Gr. *tekmar, -tos,* n. goal, end, mark, fixed sign; *tekmarsis,* f. skill in diagnosis of signs; *tekmartos,* determinable:

Gr. *telos, -eos,* n. end, completion, fulfillment; *teleios,* having reached its end, complete; *telesios,* finishing; *telestes,* m. completer, finisher; *teleute,* f. completion, end, finish; *teleutaios,* last; *teleo,* accomplish, complete, finish; *diateles,* permanent; *enteles,* complete, perfect; *epiteles; panteles,* accomplished; *proteles,* perfect before: teloblast, telophase, telome, telolemma, telomitic, Telotremata, Teleostomi, teleology, teleutospore, teliospore, entelechy, ateleosis, *Teleodactylus roscidus* (a beetle), *Isotelus gigas* (a trilobite), *Apanteles congregatus* (a braconid), *Entelostylops completus* (a fossil mammal), *Proteles cristata* (aardwolf), *Syntelestes carbonarius* (a beetle), *Enteles ocellatus* (a beetle).

Gr. *telson,* n. boundary, end: telson.

L. *terminus,* m. (Gr. *terma, -tos,* n.; *termon, -os,* m.), boundary, end, limit, goal; *Terminus,* m. god of boundaries; *terminalis,* of ends or boundaries; *termino, -atus,* bound, limit, end; Gr. *termios,* at the end: term, terminate, terminal, determine, exterminate, interminable, indeterminate, *Terminalia superba* (Afara terminalia), *Ceuthorrhynchus terminatus* (a beetle), *Termatosaurus alberti* (a fossil reptile), *Termioptycha cyanopa* (a moth). Said Henry Thoreau in *Walden* of an old fence: "I sacrificed it to Vulcan, for it was past serving the god Terminus."

Gr. *terthron,* n. end, extremity, crisis: *Terthrothrips sanguinolentus* (a thysanopterid).

L. *trunculus,* tip, end, extremity of the body; see *truncus* under **short**

See: **aim, top, tail, begin, death, lot, cancel, abandon, rest, border, projection, rump, point, destroy, kill, after, shore, gable**

endemos, Gr. living in, native; see **native**

endico- < Gr. *endikos,* right, just, fair; see **right**

endios, Gr. at midday, at noon; see **day**

endo- < Gr. *endon,* within, inside, < *en,* within, in; see *em-* under **in**

endoxos, Gr. esteemed, notable; see *dokeo* under **think**

endure < OF. *endurer,* < L. *induro,* harden; see *durus* under **hard**

endyton, Gr. garment, dress; see **garment**

-ene < L. *-enus;* chemical terms; see *-anus* under **nature**

enelico- < Gr. *enelix, -ikos,* of age, in prime of manhood; see **ripe**

enema, Gr. clyster, injection; see *eneter* under **pump**

enemy < OF. *enemi,* < L. *inimicus,* m. foe; hostile; *inimicalis,* hostile: inimical, *Inimicus japonicus* (okoze), *Stylochus inimicus* (oyster-leech).
    L. *adversarius,* m. antagonist, enemy, rival: adversary.
    Gr. *antagonistes,* m. adversary, competitor, opponent, rival: antagonist, antagonism, antagonistic.
    Gr. *anterastes,* m. rival: *Anterastes serbicus* (a grasshopper).
    Gr. *anthamillos; synamillos,* competing with, rivaling; see *hamilla* under **fight**
    Gr. *antipalos,* m. antagonist, rival, adversary: *Antipalus varipes* (a fly).
    Gr. *antipraktor, -os,* m. adversary:
    Gr. *antistates; enstates,* adversary, enemy, opponent; see *stasis* under **stand**
    Gr. *antizelos,* c. adversary, rival:
    Gr. *aprosphilos,* hostile, unfriendly:
    L. *colluctator, -is,* m. antagonist, adversary, < *luctator,* wrestler:
    L. *competitor, -is,* m. rival; *competo, -itus,* strive in rivalry: competitor, competition, competitive.
    L. *confutator, -is,* m. opponent:
    Gr. *deios,* m. enemy: *Deiphobe yunnanensis* (a mantid).
    Gr. *dysmenikos,* hostile:
    Gr. *echthros,* hateful, hostile; enemy; see *echthos* under **hate**
    Gr. *enantios,* m. enemy:
    AS. *feond,* enemy, foe, devil: fiend, fiendish.
    L. *hostis,* c. stranger, foreigner, enemy; *hosticus,* of an enemy; *hostilis; hostificus,* inimical: hostile, hostility, *Tipula hostifica* (a cranefly).
    L. *infensus,* hostile, unfriendly:
    L. *malevolens,* ill-disposed, inimical; see *malus* under **bad**
    Gr. *polemios,* m. enemy:
    L. *rivalis,* user of the same brook, rival; see *rivus* under **stream**
    L. *subnuba,* f. rival:
    Gr. *zelos; zele,* rival; *antizelos,* adversary, rival; see *zelus* under **busy**
    See: **fight, against, hate, hurt, bad**

eneos, Gr. dumb; see **silent**

energos, Gr. active, busy; see *ergon* under **make**

eneter, Gr. syringe; see **pump**

engine < F. *engin,* skill, machine, < L. *ingenium,* natural capacity, talent, genius; see **tool**

engion, Gr. nearer; *engistos,* nearest; see *engys* under **near**

English < AS. *Englisc,* < *Engle* (*Angle*), < *Angul,* a hook-shaped region in Schleswig-Holstein: Angles, Anglo-Saxon, Anglican, England, *Aster novae-angliae* (New England aster), *Cicerocrinus anglicus* (a Silurian crinoid).
    L. *Britannicus,* of Britain, < AS. *Bryttisc,* < C. *Brython* (*Briton*), an inhabitant of Britain: British, Britannic, *Encyclopaedia Britannica,* Brittany, Breton, Bretagne.

engraulis, Gr. a small fish; see **herring**

engrave < OF. *engraver,* cut, carve; see **cut, dig**

engyos, Gr. giving surety, bail, promise, pledge; see **promise**

engys, Gr. near, at hand, hard by; see **near**

enhydris, Gr. otter; water-snake; see **weasel**

enicmo- < Gr. *enikmos,* humid; see *ikmas* under **wet**

enigma < L. *aenigma* (Gr. *ainigma*), something obscure, inexplicable. a riddle, mystery; see **secret**

enipto; enisso, Gr. reprove, reproach; see **scold**

enixus; enisus, L. strenuous, earnest, zealous, < *enitor,* exert oneself, struggle, strive; see **busy**

ennea, Gr. nine; *enneadikos,* of nine; see **nine**

enochlesis, Gr. annoyance; see *ochleros* under **trouble**

enope, Gr. earring; see **earring**

enopios, Gr. face to face; see *antenopios* under **against**

enoplos, Gr. armed; see *hoplon* under **tool**

enoptron, Gr. mirror; see **mirror**

enormis, L. irregular, unusual, huge, vast, see **large;** < Gr. *enormis,* drop of an earring, see **drop**

enosis, Gr. a shaking, quake; see **shake**

enotion, Gr. earring; see **earring**

enough < AS. *genoh* (G. *genug*); see **abundance, equal**

ens, *entis,* L. being, thing, that which has existence; see *esse* under **life**

ensis, L. sword; *ensiculus,* dim.; *ensifer; ensiger,* sword-bearing; see **sword**

-ensis, L. a suffix denoting place, locality, country; see **place**

enstatico- < Gr. *enstatikos,* stubborn, opposing, resisting; *enstates,* adversary; see *stasis* under **stand**

entada, NL. a genus of the bean family; see **bean**

entatico- < Gr. *entatikos,* stimulating; *entatikon,* a stimulating plant; see **arouse**

entaxis, Gr. insertion; see **insert**

entechema, Gr. echo; see **return**

enteles, Gr. perfect, complete; see *telos* under **end**

enteon, Gr. fighting gear, armor, appliance; see **tool**

enter < OF. *entrer,* < L. *intro, -atus,* go in: entrance, entry, entree.
L. *accessus,* approach, entrance; *accessibilis,* approachable; see **accessible**
L. *admissio, -onis,* f. entrance, a letting in, < *admitto, -missus,* permit access to, grant an audience to, put male to female: admit, admission.
L. *atrium,* vestibule, hall, entry; see **room**
Gr. *dyo,* enter; *dytes,* diver, enterer; see **dip**
Gr. *eisbole,* f. inroad, invasion; *eisodos,* f. entrance; *eisthme,* f. entrance:
Gr. *embole,* insertion, invasion; see *embolon* under **insert**
L. *ineo, -itus,* go into, enter, begin, have intercourse with, know; *initus,* m. approach, advent; *introitus,* m. a going into, entering, entrance: initiate, introitus.
L. *ingressus,* m. a going into, entrance: ingress, ingressive.
L. *intrabilis,* that can be entered:
L. *invado, -vasus,* go into, enter violently; *invasor, -is,* m. intruder: invade, invasion.
L. *irruo, -utus,* invade, enter eagerly:
L. *penetro, -atus,* enter: penetrate, penetralia, impenetrable, *Sarcopsylla penetrans* (chigoe).
See: **bore, dip, insert, begin, door, mouth**

enterione, Gr. pith, inmost part; see **pith**

enteron, Gr. intestine; *enteridion; enterion,* dim.; see **intestine**

enteuxis, Gr. conversation; see **speak**

enthesis, Gr. insertion; see **insert**

enthusiasm < Gr. *enthousiasmos,* ardent interest, zeal; see **busy**

entimos, Gr. honored, prized; see **honor**

entire < OF. *entier,* < L. *integer,* complete, whole; see **all**

ento- < Gr. *entos,* within, inside; see **em-** under **in**

entome, Gr. incision, notch; see *temno* under **cut**

entomon, Gr. insect; see **insect**

entychia, Gr. conversation; see *enteuxis* under **speak**

-enus, L. pertaining to; see *-anus* under **nature**

envelope < OF. *enveloper,* wrap up; see **bag, cover, garment, skin**

envy < OF. *envie,* < L. *invidia,* jealousy; see **jealous**

enystron, Gr. fourth chamber of the ruminant stomach; see **stomach**

enzyme < Gr. *en,* in, *zyme,* yeast; an organic catalyst; see **yeast**

eon < L. *aeon* (Gr. *aion*), lifetime, age; see **age**

eos, Gr. dawn, morning, early, east; *eothen,* early in the morning; see **dawn**

epachthes, Gr. heavy, grievous; see **heavy**

epacmo- < Gr. *epakmos,* in full bloom, mature; see *akme* under **top**

epacro- < Gr. *epakros,* pointed at the end; see *akros* under **top**

epactero- < Gr. *epakter, -os,* hunter, fisher; see **hunt**

epactio- < Gr. *epaktios,* on the shore; see *acta* under **shore**

epacto- < Gr. *epaktos,* imported; see **carry**

epaeto- < Gr. *epaites,* beggar; see **beggar**

epagogion, Gr. foreskin; see **penis**

epallelos, Gr. in sequence, in close order, continuous; see **line**

epalpnos, Gr. cheerful, happy; see **joy**

epanthracido- < Gr. *epanthrakis, -idos,* small fry; see **fish**

epaulos, Gr. house; see *aulion* under **house**

epedanos, Gr. weak, infirm; see **weak**

epeetanos, Gr. sufficient, plentiful, abundant; see **abundance**

epeira, NL. a genus of spiders; see **spider**

epelis, *-idos,* Gr. cover, lid; see **cover**

epelys, Gr. stranger, foreigner; see **strange**

epelyx, *-ygos,* Gr. overshadowing; see **shade**

ependyma, Gr. upper garment; see **garment**

eperatos, Gr. lovely, charming; see *eros* under **love**

eperopeus, Gr. cheat, deceiver, see **lie**

epetes, Gr. mender; see **sew**

epetrion, Gr. needle; see **needle**

epeuchion, Gr. prayer-rug; see **rug**

epharmoge, Gr. accommodation; see **fit**

ephebos, Gr. arrived at puberty; see *hebe* under **young**

ephedra, Gr. horsetail, a plant resembling equisetum; see **gymnosperm**

ephelis, Gr. freckle; see **spot**

ephemeros, Gr. living only a day, short-lived; see *hemera* under **day**

ephestris, Gr. outer garment, mantle; see **garment**

ephicto- < Gr. *ephiktos,* easy to reach, accessible; see **accessible**

ephimeros, Gr. agreeable, delightful, desired; see **agreeable**

ephippium, L. (Gr. *ephippion*), saddle; see **saddle**

ephodos, Gr. approach, attack; see **come**

epholcio- < Gr. *epholkion,* small boat towed after a ship, appendage; *epholkis,* burdensome appendage; see *holkas* under **ship**

ephydros, Gr. wet, moist, rainy; see **wet**

epi-, Gr. upon, on, over; see **on**

epialos, Gr. fever with shivering; see **disease**

epibathra, Gr. ladder, stairs; see **ladder**

epibole, Gr. a throwing or placing upon, penalty, purpose; see **place**

epibulo- < Gr. *epiboulos,* designing, plotting, scheming; see *boulema* under **aim**

epic < L. *epicus* (Gr. *epikos*); see *epos* under **word**

epicaerio- < Gr. *epikairios,* important; see **worth**

epicauto- < Gr. *epikautos,* burnt at the end; see *kaio* under **burn**

epicedium, L. funeral song, dirge; see **sing**

epicharis, Gr. pleasing, agreeable, charming; see *charis* under **beauty**

epicoeno- < Gr. *epikoinos,* common to many, promiscuous; see *koinos* under **common**

epicrato- < Gr. *epikrates,* master; see *krateo* under **govern**

epicrio- < Gr. *epikrion,* yard arm, sailyard; see **arm**

epicrocum, L. (Gr. *epikrokon*), a woman's garment; see **garment**

epicteto- < Gr. *epiktetos,* acquired, gained; see *ktema* under **wealth**

epicuro- < Gr. *epikouros,* assister, ally, helper; see **help**

Epicurus, L. (Gr. *Epikouros*), a Greek philosopher who taught that pleasure (but not sensuality) should be the chief purpose of living; see **joy**

epidemos, Gr. prevalent, common to a large number; see **common**

epidendrum, NL. a genus of the orchid family; see **orchid**

epidermis, Gr. outer skin; see *derma* under **skin**

epidosis, Gr. increase, growth, progress, voluntary gift; see *do* under **give**

epigaea, NL. a genus of the heath family; see **heath**

epigram < Gr. *epigramma,* a pithy expression; see FIGURES OF SPEECH under **form**

epilepsy < Gr. *epilepsis,* seizure; see *lepsis* under **disease**

epileus, L. a hawk; see **hawk**

epilineutes, Gr. one who catches with a net; see *linon* under **net**

epilobium, L. (Gr. *epilobion*), a genus of the evening-primrose family; see **evening-primrose**

epimedium, L. (Gr. *epimedion*), a genus of the barberry family; see **barberry**

epimethes, Gr. afterthinking, careless; see *promethes* under **think**

epiolos, Gr. moth; see **moth**

epios, Gr. gentle, kind, soothing; see **mild**

epipactis, L. (Gr. *epipaktis*), a plant; now a genus of the orchid family; see **orchid**

epipedos, Gr. on the ground, level, flat; see **level**

epipetrum, L. (Gr. *epipetron*), a genus of the Dioscoreaceae; see **yam**

epiploon, Gr. caul, omentum; see **membrane**

epipole, Gr. surface; *epipolaios,* superficial; see **open**

epiptygma; epiptyche, Gr. over-fold, flap, operculum; see **cap**

epiro- < Gr. *epeiros,* land, mainland; see **earth**

epirrhytos, Gr. overflowing, abundant; see **abundance**

epischetico- < Gr. *epischetikos,* checking, delaying; see *epischesis* under **delay**

episcopus, L. (Gr. *episkopos*), overseer, bishop; see **govern**

episio- < Gr. *episeion,* pubic region; see **groin**

episteme, Gr. knowledge, science; *epistemos,* knowing, wise; see **know**

epistola, L. (Gr. *epistole*), letter, see **letter**; *epistolos,* messenger, see *stello* under **send**

epistomium, L. bung, stopper, cock, valve; see **close**

epistylion, Gr. crossbeam on the columns; see **beam**

epithema, Gr. cover, lid; see **cap**

epithetos, Gr. added to, see *tithemi* under **place**; *epitheton,* a term, see **word**

epitole, Gr. rising of anything; see **raise**

epitome, Gr. abstract, brief; see **short**

epitonium, L. (Gr. *epitonion*), key for tightening the strings of an instrument, stopcock, bung, peg; see **close**

epitragias, Gr. a fish; see **fish**

epoche, Gr. stop, pause; see **rest**

epolios, Gr. probably an owl; see **owl**

epomidios, Gr. on the shoulder; see *omos* under **shoulder**

epops, Gr. hoopoe; see **hoopoe**

epos, Gr. word, tale, speech, song; *epyllion,* dim.; see **word**

epsilon, Gr. fifth letter of the Greek alphabet; see **letter**

epula, L. banquet; *epularis,* of a banquet; *epulator,* feaster; see **eat**

epulido- < Gr. *epoulis, -idos,* gumboil; see **sore**

epy- < Gr. *aipys,* tall, high, steep; see *aipos* under **high**

equal < L. *aequalis,* like, same, uniform; *aequabilis,* consistent, uniform, equal; *aequus,* even, flat, level; *aequator, -is,* m. equalizer; *aequo, -atus,* equalize: equality, equable, equanimity, equator, Ecuador, adequate, equity, equitable, equilateral, equinoctial, equipoise, equivalent, unequivocal, unequal, equinox, equilibrium, equidistant, iniquity, *Hipalmus aequatorius* (a beetle), *Porolithon aequinoctiale* (an alga), *Cholomyia inaequipes* (a fly).

L. *aemulor, -atus,* strive to equal: emulate, emulation.

Gr. *antandros,* substitutional:

Gr. *antholkos,* balancing; see **balance**

L. *apocrisarius,* m. deputy:

Gr. *apographon,* n. transcript, copy:

Gr. *atalantos,* equal in weight, equal to; see **balance**

Gr. *axios,* of equal value, worth; *axia,* value, worth; see **worth**

Gr. *chironomos,* m. a pantomimic: *Chironomus plumosus* (a midge).

L. *commensuratus,* equal, proportionate: commensurate.

L. *compenso, -atus,* equalize by weighing, balance, make good: compensate, compensation.

L. *competo, -itus,* agree, coincide, be equal to, rival; see **enemy**

L. *congener,* of the same race; see *genus* under **class**

L. *consector, -atus,* emulate, imitate:

L. *denuo,* anew, again, a second time; see **return**

Gr. *diallagma, -tos,* n. substitute: *Diallagma luteum* (a moth).

Gr. *ephamillos,* equaling, matching, rivaling; see *hamilla* under **fight**

L. *exemplum; exemplar, -aris,* n. copy, pattern: example, exemplary.

L. *geminus,* twin-born; see **two**

L. *gesticulor, -atus,* make expressive motions of the body, pantomime: gesture, gesticulate.

Gr. *homos; homoios,* same, uniform, like, similar; *homalos,* even, level, equal; *homelix, -ikos,* of the same age, equal; *homologos,* agreeing, corresponding; *anchomolos; paromolos,* nearly equal: homocercal, homochromatic, homonym, homotaxis, homoplastic, homogenous, homogeneous, homologous, homosexual, Homoptera, homeopathic, homeomorph, *Homocrinus proboscidalis* (a Devonian crinoid), *Homotelus obtusus* (a trilobite), *Homoeospira sobrina* (a fossil brachiopod), *Homalodisca triquetra* (a leafhopper), *Homalanthus populifolius* (a euphorbiacead), *Ipomoea purpurea* (a morning-glory).

L. *hypocrita* (Gr. *hypokrites*), mime, player, actor, dissembler; see **lie**

L. *idem,* the same: identical, identity, identify.

L. *imitor, -atus,* copy, mimic; *imitabilis,* that may be copied; *imitativus,* inclined to copy; *imitamen, -inis,* n. copy, likeness; *imitator, -is,* m. mimic: imitate, imitation, imitator, imitative, *Phragmidium imitans* (raspberry-rust), *Acronyches imitator* (a fly).

Gr. *isos,* equal, like; *isotes, -etos,* f. equality, impartiality; *exisazo,* make equal; *parisos,* evenly balanced, equal: isobar, isotherm, isomeric, isomorphic, isochronous, isopod, isosceles, isostasy, anisotropic, *Isopogon roseus* (horny conebush), *Isoetes lacustris* (a quillwort), *Parisocrinus radiatus* (a fossil crinoid).

L. *itero, -atus,* repeat; see **return**

L. *mimus* (Gr. *mimos*), in imitator, actor; *mimulus,* m. dim.; *mimicus,* imitative; *mimeomai,* imitate; *mimesis,* f. imitation; *mimeolos; mimetikos,* imitative; *pantomimos,* m. mime, actor: mimicry, mime, mimetic, mimesis, mimeograph, pantomime, *Mimus polyglottus* (mocking-bird), *Mimosa sensitiva* (a sensitive-plant), *Mimulus guttatus* (monkey-flower), *Mimusops emarginata* (wild dilly), *Callistomimus amoenus* (a beetle), *Formicomimus mirabilis* (a beetle), *Vespamima sequoiae* (a pitch-moth), *Mimetes hirtus* (a proteacead).

L. *mutuus,* reciprocal; see **alternate**

L. *par, paris,* equal, like, even; *parilis,* equal, like; *pariator, -is,* m. balancer; *pario, -atus,* make equal, balance; *pararius,* of a pair; *compar; comparilis,* equal, like; *comparo, -atus,* couple, match, pair; *impar,* uneven, odd: par value, parity, disparity, compare, incomparable, peer, pair, nonpareil, umpire, pari passu, *Plaxocrinus parilis* (a Pennsylvanian crinoid), *Nephroma parile* (a lichen).

L. *parallelus* (Gr. *parallelos*), beside one another, side by side equidistantly: parallel, parallelogram, *Parallelodon obsoletus* (a fossil pelecypod), *Orectogyrus parallelus* (a beetle), *Aphrophora parallela* (a spittle-insect), *Dorcus parallelepipedus* (a beetle).

L. *pretendo, -tus (-sus),* allege, simulate: pretend, pretender, pretense.

L. *proportio, -onis,* f. ratio, relative equality: proportion, proportional.

L. *simulo, -atus,* imitate, copy; see *similis* under **like**

L. *substituto, -utus,* put instead or in place of: substitute, substitution.

L. *succedaneus,* following after, substitute; see **follow**

L. *summissum,* n. substitute:

L. *surrogo, -atus,* choose in place of another, substitute: surrogate, surrogation.

Gr. *symmetros,* in measure with, proportional, corresponding part for part: symmetry, symmetrical, *Icriodus symmetricus* (a conodont).

L. *talis,* such, the like, in kind: talion, retaliate.

Gr. *tautos,* same: tautology, tautonym, tautological, tautegorical, tautometric.

L. *vice-,* instead of, in place of, < *vicis,* f. change, reciprocity, substitution; *vicarius,* substituted: vicar, vicarious, viceregent, vicegerent, vice-president, viceroy, viscount, vice versa.

Gr. *zelos,* rivalry, emulation, ardor; see **busy**

See: **like, balance, alternate, return, two, common**

equester, *-tris, -tre,* L. pertaining to riding and cavalry; see *equito* under **ride**

equisetum, L. horsetail; see **horsetail**

equus, L. horse; *equa,* mare; *equulus,* dim.; *equinus,* of horses; see **horse**

-er; -es, L., Gr. suffix signifying agent; see *-ist, -ter* under **make**

era < L. *aera;* see **age**

erannos, Gr. lovely; see *eros* under **love**

eranos, Gr. a common fund; a society of subscribers; see **common**

erase < L. *erado, -asus,* scrape or rub out, expunge, obliterate; see **cancel**

eratos, Gr. lovely, beloved; see *eros* under **love**

erebinthos, Gr. a kind of legume; see **bean**

Erebus, L. (Gr. *Erebos*), a place of darkness in the nether world; see **hell**

erechthites, Gr. groundsel; see **composite**

erectus, L. upright; see **upright**

eremaeo- < Gr. *eremaios,* quiet, still, gentle; see **rest**

eremites, Gr. hermit; see *eremos* under **alone**

eremnos, Gr. black, swarthy, dark; see **black**

eremos, Gr. solitary, lonely; *eremia,* solitude, desert, wilderness; *eremikos,* of solitude; see **alone**

erepsis, Gr. roof; *erepsimos,* of roofing; see **roof**

ereptus, L. snatched, taken; see *rapio* under **take**

ereschelia, Gr. sport, raillery; see **play**

erethizon, NL. porcupine, < *eretho,* provoke; see **porcupine, arouse**

eretmon, Gr. oar; *eretmion,* dim.; see **oar**

ereuna, Gr. inquiry, search, examination; *ereuneter,* inquirer, searcher; see **try**

ereuthedanon, Gr. madder; see **madder**

erga, L. against, toward; see **against**

ergastulum, L. workhouse, penitentiary; see **prison**

ergates, Gr. worker; *ergastikos,* working, active, < *ergon,* work; see **work**

ergo, L. consequently, hence, therefore; see **follow**

ergodes, Gr. irksome, troublesome; see **trouble**

ergot, F. spur; a fungus; see **fungus**

eri-, Gr. intensive particle, very, see *ari-* under **very**; early, see **dawn**

-eria, suffix denoting place where; see *-arium* under **place**

erica, L. (Gr. *erike*), heath, heather; *ericaceus,* of heath; see **heath**

ericinus, L. of a hedgehog; *er; ericius; erinaceus,* hedgehog; see **hedgehog**

ericto- < Gr. *eriktos,* bruised, pounded; *ereixis,* a pounding, grinding; see *ereiko* under **strike**

Eridanus, L. (Gr. *Eridanos*), a mythical river; see **stream**

erigeron, Gr. early old; groundsel; see **old, composite**

Erigone, Gr. daughter of Icarus translated to the sky as the constellation Virgo; see **woman**

erinaceus, L. hedgehog; see *er* under **hedgehog**

erineos, Gr. the wild fig-tree; see **fig**

erinus, L. (Gr. *erinos*), a genus of the figwort family; see **figwort**

Erinys, Gr. an avenging goddess, a Fury; see **punish**

erio- < Gr. *erion,* wool, see **wool**; mound, barrow, tomb, see **heap**

eriole, Gr. hurricane, whirlwind; see **wind**

erioxylon, Gr. cotton; see **cotton**

eriphos, Gr. young goat; see **goat**

eripne, Gr. cliff, crag; see **cliff**

eripsimo- < Gr. *ereipsimos,* thrown down, ruined; *ereipsis,* a ruining; see *ereipion* under **destroy**

eris, -idos, Gr. strife, quarrel, debate, contention; *eristes,* wrangler; *eristikos,* fond of wrangling, given to strife; *eristos,* contested; see **fight**

erisma, L. (Gr. *ereisma*), prop, stay, support, buttress, see **pillar**; < Gr. *erisma,* cause of strife, see *eris* under **fight**

eristalis, L. an unknown precious stone; now applied to a genus of flies; see **fly**

eristico- < Gr. *eristikos,* fond of strife, given to wrangling; see *eris* under **fight**

erithacus, L. (Gr. *erithakos*), probably a robin; see **thrush**

erithalis, Gr. an unknown plant; now a genus of the madder family; see **madder**

erithos, Gr. hired servant; see **servant**

eritimos, Gr. precious; see **worth**

erizoos, Gr. long-lived; see *zoe* under **life**

ermineus, L. white like ermine; see *hearma* under **weasel**

erneum, L. cake baked in an earthen pot; see **cake**

ernos; ernyx, Gr. sprout, shoot; see **branch**

erodios, Gr. heron; see **heron**

eros, -tos, Gr. love; *erotikos,* amorous; see **love**

erosus, L. eaten away, consumed, corroded; see *rodeo* under **bite**

erotema, Gr. question; see **ask**

erratum; error, L. fault, mistake; *erroneus,* straying, < *erro, -atus,* wander, stray, make a mistake; see **fault, wander**

erromenos, Gr. in good health, vigorous; see *menos* under **strong**

eruca, L. caterpillar, see **caterpillar**; a colewort, see **mustard**

eructo, L. belch, vomit; see **belch**

eruditus, L. learned, accomplished, skilled; see **know**

eruption < L. *eruptio, -onis,* < *erumpo, -uptus,* break out or burst forth, pour out, shoot out; see *rumpo* under **break, pour, blow, throw, pimple, disease**

ervum, L. a kind of vetch; see **bean**

eryge; erygma; erygmos, Gr. a belching; see **belch**

erygmelos, Gr. loud-bellowing; see **roar**

eryma, Gr. fence, guard, wall; *erymnos,* fenced; see **wall**

eryngium, L. (Gr. *eryngion*), a genus of the carrot family; see **carrot**

eryo, Gr. draw, drag; see **draw**

erysi- < Gr. *erythros,* see **red**; < *erysis,* a dragging, drawing, see *eryo* under **draw**

erysibe, Gr. mildew; see **fungus**

erysimum, L. (Gr. *erysimon*), hedge-mustard; see **mustard**

erysipelas, Gr. a reddish skin eruption; see **disease**

erythrina, NL. a genus of the bean family; see **bean**

erythrinos, Gr. a kind of mullet; see **mullet**

erythronium, L. (Gr. *erythronion*), a genus of the lily family; see **lily**

erythropus, L. (Gr. *erythropous*), redshank; see **snipe**

erythros, Gr. red; *erythraios,* reddish; *erythrotes,* redness; see **red**

esca, L. food, victuals, bait; *escalis,* of food; *escatilis; esculentus,* edible; see **food**

escape < OF. *escaper;* see **free**

-escent < L. *-escens,* a participial ending signifying beginning of, becoming, inceptive; see **begin**

eschara, Gr. hearth, fireplace, scar, scab; see **oven, mark**

eschatos, Gr. extreme, last; see **after**

esculentus, L. edible; see *esca* under **food**

-ese < OF. *eis,* of, pertaining to, characteristic of, originating in; see **nature**

esmos, Gr. swarm of bees; see **hive**

eso- < Gr. *eso,* within; *esoterikos,* internal; see **in**

esophagus < Gr. *oisophagos,* gullet; see **throat**

esor; estrix, L. eater; see *edo* under **eat**

esox, L. pike; see **pike**

-esque, F. (It. *-esco*), like in manner or style; see **like**

-ess < OF. *-esse,* < L., Gr. *-essa; -issa,* feminines, sometimes with diminutive significance; see **woman**

esseda, L. a two-wheeled war-chariot; see **vehicle**

essence < L. *essentia,* substance of things, < *esse,* to be, *ens,* being; see **life, thing**

-essus, NL. diminutive suffix, < Gr. *hesson,* less; see *mikros* under **little**

essymenos, Gr. hurrying, eager; see **swift**

ester, G. an alkyl salt; see **salt**

esthema; esthes, Gr. dress, clothing, raiment; see **garment**

esthetic < Gr. *aisthetikos,* sensitive, perceptive; see *aisthanomai* under **feel**

esthio, Gr. eat; see **eat**

esthlos, Gr. good; see **good**

estimate < L. *aestimo, -atus,* value, appraise; see **judge**

estrix, L. female eater; see *edo* under **eat**

estuary < L. *aestuarium,* channel subject to tidal action at the mouth of a river; see **mouth**

esurio, -itus, L. desire to eat, be hungry; *esurialis,* of hunger; see **hunger**

-et; -ette; -ot, F. diminutives; feminines; see DIMINUTIVES under **little, woman**

eta, Gr. seventh letter of the Greek alphabet; see **letter**

etelis, Gr. a kind of fish; see **fish**

eteos, Gr. true, real, genuine; see **true**

eternal < L. *aeturnus,* immortal, everlasting; see **always**

etesius, L. (Gr. *etesios*), yearly; see *etos* under **year**

ethas, Gr. customary, ordinary, tame; see **use**

ethelo (thelo), Gr. wish, will; *etheletos,* willing; *ethelonter,* volunteer; see **wish, free**

etheo, Gr. filter, strain; see **sieve**

ether < L. *aether* (Gr. *aither*), the upper atmosphere; see **wind**

ethicus, L. (Gr. *ethikos*), moral, pertaining to custom, habit or use; see *ethas* under **use**

ethiro- < Gr. *etheira,* hair, mane; see **hair**

ethisma, Gr. habit, custom; see *ethas* under **use**

ethmos, Gr. sieve, strainer; see **sieve**

ethnos, Gr. nation, race; *ethnikos,* of a nation or race; see **class**

ethos, Gr. character, custom, habit; see *ethas* under **use**

ethyl, pertaining to the hydrocarbon radical in grain alcohol; see **alcohol**

etio- < Gr. *aitios,* causing, responsible for, < *aitia,* cause, blame, fault; see **begin**

etnos, Gr. pea soup; see **soup**

etor, Gr. heart; see **heart**

etos, Gr. year; see **year**

etosios, Gr. fruitless, useless, in vain; see **worthless**

etrion, Gr. warp; see **weave**

etron, Gr. belly; see **belly**

-ette, F. diminutive; see *-et* under **little**

-etus, L. having the nature of, pertaining to; see *-atus* under **nature**

etymos, Gr. true; *etymon,* the true, original, literal root-meaning of a word; see **true**

eu- < Gr. *eu,* good, well, agreeable, easy, very, true, original, primitive; see **good, begin**

eucalyptus, NL. a genus of the myrtle family; see **myrtle**

euche, Gr. prayer, wish; *euchos,* the thing prayed for; see **ask**

eucrato- < Gr. *eukratos,* well-tempered, temperate, mild; see **mild**

eudios, Gr. calm, clear, fine; *eudia,* fair weather; see **clear**

eudoros, Gr. generous; see *do* under **give**

eugenes, Gr. well-born; see *gigno* under **birth**

euido- < Gr. *eueides,* shapely, beautiful; see **beauty**

eulabes, Gr. taking hold well, holding fast; see *lambano* under **take**

eule, Gr. worm, maggot; see **worm**

eulonchos, Gr. fortunate, propitious; see **lot**

eumaro- < Gr. *eumares,* easy, convenient, without difficulty; see **rest**

eumeco- < Gr. *eumekes,* of good length, tall, great; see *mekos* under **long**

eumeles, Gr. musical, euphonious; see *melos* under **sing**

eumenes, Gr. friendly, gracious, kind, well-disposed; see **friend**

eune, Gr. bed; see **bed**

eunuch < L. *eunuchus* (Gr. *eunouchos*), a castrated male person; see **sterile**

euonymos, Gr. of good name, honored, see **honor**; spindle-tree, see **staff-tree**

eupatorium, L. (Gr. *eupatorion*), a genus of the composite family; see **composite**

eupetes, Gr. easy; see **rest**

euphemia, Gr. good speech, good names for bad things; *euphemos,* uttering auspicious words; see *phemi* under **speak,** FIGURES OF SPEECH under **form**

euphony < Gr. *euphonia,* pleasing sound; see *phone* under **sound**

euphorbia < L. *euphorbea; euphorbeum* (Gr. *euphorbion*), a plant with acrid juice; see **spurge**

euphrosyne, Gr. mirth, merriment; see **joy**

euphues, Gr. shapely, graceful, clever; see **beauty**

euporia, Gr. means, resources, plenty; see **wealth**

euprepes, Gr. becoming, comely; see **beauty**

eurax, Gr. on one side, sideways; see **side**

eurhopos, Gr. inclining or sliding easily; see *rhepo* under **slope**

euripos, Gr. strait, channel; see **way**

euros, Gr. east wind; see **wind**

eurostos, Gr. stout, strong; see *rhonnymi* under **strong**

euroto- < Gr. *euros, -otos,* mold, decay; see **rot**

eurys, Gr. broad, wide, widespread; *euryno,* widen, broaden; see **broad**

-eus; -eutes, Gr. one who, agent, see **make**; *-eus,* L. made of, see **nature**

eusomatos, Gr. sound in limb, able-bodied; see **health**

euthemon, Gr. well-arranged, neat; *euthemosyne,* tidiness; see **class**

euthicto- < Gr. *euthiktos,* to the point, clever; see **fit**

euthys, Gr. straight, direct; *euthynter,* straightener, corrector; see **straight**

eutolmos, Gr. courageous; see *tolma* under **bold**

evanescens; evanidus, L. disappearing; see *vanesco* under **depart**

evectus, L. carried out, exported, spread, raised aloft; see *veho* under **carry**

even < AS. *efen;* see **smooth, all**

evening < AS. *aefnung;* see **night**

evening-primrose

    L. *circaea* (Gr. *kirkaia*), f. a genus of the evening-primrose family: *Circaea latifolia* (an enchanter's-nightshade).

    NL. *clarkia,* f. a genus of the evening-primrose family, < William Clark, of the Lewis and Clark exploring party: *Clarkia pulchella* (an onagracead).

    NL. *epilobium,* n. a genus of the evening-primrose family: *Epilobium luteum* (yellow willow-herb).

    NL. *fuchsia,* f. a genus of the evening-primrose family, < Leonard Fuchs, German botanist: *Fuchsia coccinea* (scarlet fuchsia).

    NL. *godetia,* f. a genus of the evening-primrose family, < C. H. Godet, Swiss botanist: *Godetia amoena* (farewell-to-spring).

    L. *oenothera,* f. (Gr. *oinotheras; onotheras,* m.), a genus of the evening-primrose family: *Oenothera fruticosa* (sundrops).

    Gr. *onagra,* f. an evening-primrose: Onagraceae, *Onagra*(now *Oenothera*) *cruciata* (an evening-primrose), *Anogra coronopifolia* (an onagracead).

    L. *onear, -atis,* n. an evening-primrose:

eventum, L. consequence, issue, result; see **lot**

ever < AS. *aefre;* see **always**

everriculum, L. dragnet; see *verriculum* under **net**

evexus, L. rounded at the top; see **dull**

evidens, L. clear, plain, visible; *evidentia,* clearness; see **clear**

evil < AS. *yful,* bad, wrong, wicked, sinful; see **bad, wrong**

eviratus, L. effeminate, weak; see **weak**

evitabilis, L. avoidable; see *vito* under **shun**

evolution < L. *evolutio, -onis,* an unrolling, < *evolvo, -utus,* unroll, unfold; see *volvo* under **turn**

evum (aevum), L. lifetime, age; *coaevus,* of the same age; see **age**

ex- < L. *ex,* out of, from, see *e-* under **out**; < Gr. *ek,* out of, from, see **out**

exact < L. *exactus* precise, accurate; see **right**

exactor, L. demander, enforcer, tax-collector; see *ago* under **drive**

exacum, L. a genus of the gentian family; see **gentian**

exaesio- < Gr. *exaisios,* extraordinary, beyond demand; see **over**

exaggerate < L. *exaggero, -atus,* heap up, amplify, stretch the truth; see **exceed**

examen, -inis, L. swarm, crowd; see **number**

examine < L. *examino, -atus,* scrutinize, test, consider; see **try**

examma, Gr. handle; see **handle**

exanclo, -atus, L. suffer, endure to the end; see **pain**

exanthema, Gr. eruption; see **disease**

exastis, Gr. fringe; see **fringe**

exauster, Gr. meat-fork; see **fork**

exceed < L. *excedo, -essus,* go beyond: excessive, exceedingly.
　L. *exaggero, -atus,* heap up, amplify, stretch the truth: exaggerate, exaggeration.
　L. *immoderatus,* unrestrained, excessive, extravagant: immoderate.
　L. *immodicus,* excessive, extravagant, immoderate:
　L. *intemperatus,* immoderate, excessive: intemperate, intemperance.
　Gr. *pleonazo,* exceed, claim too much, go beyond bounds; *pleonasmos,* m. **more** than enough, superfluity: pleonasm, pleonastic.

excelsus, L. high, lofty, distinguished; *excelsior,* higher, < *excello,* elevate; *excellens,* rising, lofty, distinguished; see **high, raise**

excerptum, L. extract, selection; see *excerpo* under **choose**

excetra, L. snake, serpent; see **snake**

excipulum, L. receptacle, vessel, basin; see **vessel**

excite < L. *excio, -itus; excito, -atus,* arouse, call forth; see *cieo* under **arouse**

exclamatio, L. a calling out, interjection; see *clamo* under **call,** FIGURES OF SPEECH under **form**

excrementum, L. dung, refuse; see **dung**

excubitor, L. guard, sentinel; see **guard**

excultus, L. adorned, polished, refined; see **ornament**

excuse < OF. *escuser,* < L. *excuso, -atus,* apologize for, seek extenuation: excusable.
　L. *allegatio, -onis,* f. assertion of proof, excuse: allegation.
　L. *praetextum,* n. allegation, excuse, pretense: pretext.
　Gr. *skepsis,* f. plea, excuse, pretense:
　L. *venia,* pardon, forgiveness; *venialis,* pardonable; see **free**

excussus, L. extended, stretched out, stiff; see **lie**

exeches, Gr. prominent; see *exochos* under **projection**

exedros, Gr. away from home, strange, extraordinary; see **far**

exegesis, Gr. explanation, interpretation; *exegetes,* expounder, interpreter; see **clear**

exemplum; exemplar, L. sample, model, pattern, copy; *exemplaris,* patternable; see **form**

exemptus, L. taken out, removed, excused; see **from**

exercise < L. *exerceo, -itus,* keep busy, drive on, practice; see **busy, make, teach**

exetastes, Gr. examiner, auditor, inspector; see **try**

exhomilos, Gr. strange, unfamiliar; see **strange**

exhoristos, Gr. banished, expelled; see **banish**

exiguus, L. small, short, poor, scanty; *exiguum,* trifle; see **little**

exile < L. *exilium,* banishment, < *exul (exsul),* a banished person; see *exul* under **banish**

exilis, L. thin, slender, meager, poor; *exilitas,* thinness, weakness; see **thin**

eximius, L. exceptional, uncommon, extraordinary; see **few**

exist < L. *existo, exstitus,* appear, be manifest, be; see **life**

exitelos, Gr. lessening, fading, weakening; see **lessen**

exitiosus, L. destructive, fatal, deadly; see **destroy**

exitus, L. a going out, a way out; see *eo* under **move**

exo- < Gr. *ex,* out of, without; *exoterikos,* outside, external; see *ek* under **out**

exochadium, L. external hemorrhoid; see **disease**

exochos, Gr. jutting out, projecting, eminent; *exoche,* projection, point, wart; see **projection**

exodus, L. (Gr. *exodos*), departure, exist; see *hodos* under **way**

exoeno- < Gr. *exoinos,* drunken; see *oinos* under **wine**

exoles, Gr. utterly destroyed, ruined, abandoned; see *olethros* under **destroy**

exoletus, L. grown up, mature; see **ripe**

exordium, L. a beginning; see *ordior* under **begin**

exonero, L. free from a burden, discharge; see **free**

exotico- < Gr. *exotikos,* from the outside, alien, foreign; see **strange**

expand < L. *expando, -sus,* spread out; see *pando* under **spread, grow**

expedition < L. *expeditio, -onis,* campaign, enterprise; see **move**

expeditus, L. free, unimpeded; see **free**

expello, *-pulsus,* L. drive out; *expulsio, -onis,* a driving out; see **push**

expensum, L. cost, charge, payment; see *pendo* under **pay**

expergo, *-itus,* L. awaken, arouse; see **arouse**

experiment < L. *experimentum,* test, trial; see **try**

expers, *-ertis,* L. having no part in, not sharing, free from; see **innocent**

expert < L. *expertus,* tried, proved, experienced; see *peritus* under **know**

expio, *-atus,* L. atone; *expiator; expiatrix,* atoner; see **atone**

explain < L. *explano,* make plain, clear; see **clear**

expletus, L. complete, perfect, see *compleo* under **all;** *expletivus,* filling out, exclamatory, see **curse**

explicatus, L. unfolded, arranged, plain, clear; *explicitus,* disentangled, apparent, clear, precise; see **clear**

explodo (explaudo), *-osus,* L. drive off, burst; see **break**

explorator, L. searcher, examiner; see *exploro* under **try**

expositus, L. open, accessible, free; see **open**

expressus, L. clear, manifest, plain; see **clear**

exquisitus, L. choice, excellent, fine; see **beauty**

exsertus, L. projecting, thrust forth; see **projection**

exspes, L. without hope; see *despero* under **hopeless**

extalis, L. rectum; see **intestine**

extensivus, L. spread or stretched out, wide, large; see *tendo* under **spread**

exterus, L. out; *exterior,* outer; *extremus; extimus,* outermost, utmost, farthest, last; see **out**

extinctus (exstinctus), L. dead, destroyed, put out; see *extinguo* under **destroy**

extorris, L. banished, exiled; see **banish**

extra, L. on the outside, besides, beyond; see *exterus* under **out, over**

extraordinarius, L. above or beyond the common or ordinary; see **wonder**

extremus, L. outermost, farthest, last, superlative of *exterus,* out; see **out**

extricatus, L. freed; see *extrico* under **free**

extrilidius, L. dauntless; see **bold**

exuberans, L. abounding, overflowing; see *uber* under **abundance**

exul (exsul), L. banished person; see **banish**

exultatus, L. leaping up, rejoicing; see *assilio* under **leap**

exutus, L. bared, stripped; see **bare**

exuvia, L. cast, skin, slough; see **skin**

eye < AS. *eage.*
   L. *Argus* (Gr. *Argos*), a hundred-eyed monster; see **peacock**
   L. *gemma,* bud, eye, jewel; see **bud**
   Gr. *glene,* f. pupil of the eye, eyeball, eye-socket, doll; *glenion,* n. dim.; *glenos,* n. thing to stare at, show, wonder: glenoid, *Glenotremites paradoxus* (a fossil echinoid), *Glenochrysa typica* (a neuropterid), *Euglena viridis* (a flagellate), *Hypsiglena ochrorhynchus* (a snake), *Trichoglenus complanatus* (a wasp).
   Gr. *kore,* maiden, doll, pupil of the eye; *koridion,* dim.; see **child**
   Gr. *kylon,* part under the eye; see **face**
   L. *lippus,* blear-eyed, having inflamed or watery eyes; see **dim**
   L. *luscus,* one-eyed, half-blind; see **dim**
   L. *oculus,* m. eye, bud; *ocellus,* m. dim.; *ocellatus,* having little eyes, marked with spots; *ocularis,* of the eyes; *oculatus,* having eyes; *oculeus,* full of eyes: desioculus, m. one who has lost an eye: ocellus, oculist, oculate, binocular, inoculate, inveigle, antler, *Oculina varicosa*(a coral), *Euphorbia ocellata* (a spurge), *Cecidomyia ocellaris* (maple-spot gall-insect), *Filaria oculi* (eye threadworm), *Hemigrammus ocellifer* (a fish), *Raia ocellifera* (a skate), *Alaus oculatus* (a click-beetle), *Molgula oculata* (a tunicate), *Ceratopsis oculifera* (a fossil ostracode).
   Sp. *ojo,* m. eye, spring; *ojito,* m. dim.: Ojo Alamo.
   Gr. *okkos,* m. eye:
   Gr. *omma, -tos,* n. eye; *ommation,* n. dim.; *mikrommatos,* small-eyed: ommateum, ommatidium, ommatophore, *Ommatodium volucris* (an orchid), *Ommatothrips gossypii* (a thysanopterid), *Ommastrephes sagittatus* (a squid), *Dysommatus rufulus* (a beetle), *Anomma arcens* (a driver-ant), *Amblyomma maculatum* (Gulf Coast tick), *Megommation insigne* (a bee), *Leptolucania ommata* (a fish), *Psilommiscus sumatranus* (a wasp), *Schismatomma ocellatum* (a lichen).

Gr. *ophrys,* brow, eyebrow; see **face**

Gr. *ophthalmos,* m.; *ops, opos,* f.; *optilos,* m. eye; *ophthalmidion,* n. dim.; L. *Cyclops, -opis* (Gr. *Kyklops, -opos*), m. mythical one-eyed giant; *panops,* all-seeing; *phainops,* bright-eyed, conspicuous; *charopos,* glad-eyed; *megalopos,* large-eyed; *exophthalmos,* with bulging or protruding eyes, eyes popping out, eyes prominent: ophthalmia, ophthalmectomy, exophthalmic, myopia, Cyclops, Cyclopean, *Coccyzus erythrophthalmus* (black-billed cuckoo), *Buphthalmum salicifolium* (an oxeye), *Cyclops coronatus* (a copepod), *Chrysops perpensa* (a fly), *Hypsypops rubicunda* (garibaldi), *Syzygops cyclops* (a beetle), *Anomalops graeffei* (a fish), *Liriope graminifolia* (lilyturf).

L. *pupula,* f. pupil of the eye: pupil.

See: **lens, spot, bud, see, light, hole**

eyelid

L. *cilium,* eyelid, eyelash; see **hair**

Gr. *blepharon,* n. eyelid: blepharal, blepharoblast, blepharitis, *Blepharocysta splendor* (a protozoan), *Blepharocorys caudata* (a protozoan), *Ablepharus cupreus* (a skink), *Arabis blepharophylla* (a rockcress), *Photoblepharon palpebratus* (a fish). See *blepharis* under **hair**

Gr. *epikylion,* n. upper eyelid:

L. *palpebra,* f. eyelid; *palpebralis,* of the eyelids; *palpebro,* wink: palpebral, palpebrate, *Anomalops palpebratus* (a fish), *Zosterops palpebrosa* (spectacle-bird).

Gr. *tarsos,* edge of the eyelid; see **foot**

F

faba, L. bean; *fabaceus; fabalis; fabatus;* see **bean**

faber, L. artisan, workman; workmanlike, skillful, see **make**; dory, see **dory**

Fabius, L. Roman general surnamed Cunctator for his method of fighting Hannibal; see **delay**

fabricatus, L. made, wrought; see *faber* under **make**

fabula, L. tale, story; *fabella,* dim.; *fabulor, -atus,* talk; see **story, speak,** FIGURES OF SPEECH under **form**

fac-; -fact; -fect; -fex; -fic; -fy < L. *facio,* make, do; see **make**

face < L. *facies, -ei,* f. countenance, external form, figure, or surface; *bifax, -acis,* two-faced: facial, facet, facade, deface, effacement, ineffaceably, prima facie, surface, surficial, superficial, *Anopheles culicifacies* (a mosquito).

Gr. *bregma,* front of the head, sinciput; see **head**

Gr. *brikelos,* m. a tragic mask:

Sp. *cara,* f. face; *carilla,* f. dim.:

L. *frons, frontis,* f. brow, forehead, fore part of anything: front, frontal, frontier, frontispiece, affront, effrontery, confrontment, *Megaderma frons* (an African bat), *Vireo flavifrons* (yellow-throated vireo), *Falcunculus frontatus* (shrike-tit).

Gr. *hedra,* seat, base, plane, side; see **sit**

Gr. *kylon,* n. part under the eyes:

L. *larva,* ghost, mask, early stage of some animals; see **young**

Gr. *metopon,* forehead, front; see **head**

Gr. *ophrys, -yos,* f. brow, eyebrow; *ophrydion,* n. dim.; *synophrys, -yos,* with joined eyebrows: ophryon, *Ophrys apifera* (a bee-orchid), *Actinophrys sol* (a protozoan), *Ophryophryne subviolaceus* (a toad), *Synophryostreptus punctatus* (a centiped).

Gr. *ops, opos,* f. eye, face, countenance; *opsis, -ios (-eos),* f. sight, look, appearance, aspect, view; *enope,* f. face, countenance; *enopios,* face to face; *hypopion,* n. part of the face under the eyes; *kelainops, -opos,* black-faced, swarthy; *tauropos,* bull-faced: Ethiopian, *Galeopsis ochroleuca* (a hemp-nettle). See **eye, see**

L. *persona,* f. mask, person; *personella,* f. dim.; *personalis,* of a mask or person; *personatus,* masked: *Gomphus personatus* (a dragonfly), *Cercospora personata* (a fungus).

Gr. *phasis,* appearance, aspect, look; see *phaino* under **light**

L. *pro,* in front of, before, forward; see **before**

Gr. *pronopion,* n. front of a house:

Gr. *prora,* f. prow, bow: *Nematoprora polygonifera* (a fish), *Chlorophthalmus proridens* (a fish), *Chascanopsetta prorigera* (a fish), *Proroblema polystriga* (a moth).

L. *prorsus (prosus),* turned forward, onward, straight, direct: *Triceratops prorsus* (a dinosaur).

Gr. *prosopon, -os,* n. face of a man, countenance, front; *prosopeion,* n. mask; *prosopikos,* of the face: prosopography, prosoposchisis, prosopotocia, atelopro-sopia, megaprosopous, *Engyprosopon xenandrus* (a fish), *Exoprosopa onusta* (a fly), *Prosopothrips vejdovskyi* (a thysanopterid).

Gr. *protome,* f. face of an animal:

L. *sannio, -onis,* one who makes faces, grimacer, buffoon; see **fool**

Gr. *skynion,* n. the skin above the eyes; *episkynion,* n. eyebrows:

L. *supercilium,* n. eyebrow: supercilious, *Ceryle superciliosa* (a kingfisher).

L. *vultus, -us,* m. countenance, face, look, aspect; *vultuosus,* full of expression: *Hemiganus vultuosus* (a taeniodont).

See **cheek, head, jaw, nose, side**

facetus, L. well-made, fine, elegant, jocose; *facetia,* drollery, jest, wit; see **beauty, laugh**

facilis, L. easy, without difficulty; see **rest**

facinus, L. misdeed, crime; see **fault**

fact < L. *factum,* act, certainty, reality; see **thing**

factiosus, L. powerful, eager for power; see **strong**

factitius, L. artificial, not natural; see **art**

factum, L. act, deed; *factor,* doer; *factura,* creation; see *facio* under **make**

facula, L. dim. of *fax,* firebrand, torch; see *fax* under **fire**

facultas, L. capability, means; *facultatula,* dim.; see **know**

facundus, L. eloquent; see *for* under **speak**

fade < OF. *fader;* see **lessen, gray, dry, death**

faex, *faecis,* L. dregs, grounds, dirt, dung; *faecula,* dim.; see **dirt**

fagus, L. beech; *faginus,* of beech; see **beech**

fail < F. *faillir,* < L. *fallo, falsus,* deceive, disappoint. Derive: fiasco.

Gr. *apoteuxis,* f. miscarriage, failure:

L. *improsper,* unfortunate, unsuccessful:

See: **fault, neglect, lessen**

faint < OF. *feindre;* see **dim, weak, shade**

fair < AS. *faeger;* see **beauty, clear, justice, right**

fairy < OF. *faerie,* enchantment; see **spirit**

faith < OF. *feid,* < L. *fides,* belief, trust; see **believe**

fala, L. scaffold; see **frame**

falcatus, L. sickle-shaped, hooked, < *falx,* sickle; *falcicula,* dim.; see **sickle**

falco, L. a kind of hawk; see **hawk**

falisca, L. rack in a manager; see **frame**

fall < AS. *feallan:* fallaway, fallfish, fall line, befall, nightfall, landfall, waterfall, pitfall, windfall, offal, Niagara Falls.

L. *cado, casus,* fall, happen, perish; *cadivus; caducus,* that falls, inclined to fall; *casabundus,* about to fall, tottering; *casus, -us,* m. a falling down, event, accident, chance; *casualis,* fortuitous; *casito, -atus,* fall down repeatedly; *deciduus,* falling off; *incido, -casus,* fall into, light upon; *occidens; occiduus,* setting of the heavenly bodies, western; *occasus, -us,* m. downfall: cadence, caducous, case, casual, chance, accident, incidence, coincide, occidental, decadent, decay, deciduous, cadaver, casuistry, cheat, escheat, recidivist, occasion, chute, *Euphorbia caducifolia* (a spurge), *Larix decidua* (European larch).

L. *cataracta* (Gr. *kataraktes*), f. waterfall: cataract, *Begonia cataractarum* (a begonia).

L. *cernuo, -uatus,* throw or fall head foremost; see *cernuus* under **bend**

Gr. *ereipsimos,* thrown down, ruined; see *ereipion* under **destroy**

L. *labor, lapsus,* slip, glide, fall; *labes, -is,* f. fall, first slip; *labilis,* slipping, prone to slide; *labidus,* slippery; *labina,* f. slippery place; *lapsus, -us,* m. slip, fall, error; *dilabor,* fall asunder, go to ruin; *interlabor,* fall between: lapse, labile, collapsible, elapse, prolapse, relapse, avalanche.

Gr. *pesema, -tos,* n. fall:

Gr. *pipto,* fall; *ptoma, -tos,* n. that which has fallen; *ptosimos; ptotos,* fallen, slain; *ptosis,* f. fall; *paraptoma, -tos,* n. false step, blunder, slip: ptosis, ptomaine, symptom, asymptote, *Piptanthus laburnifolius* (a legume), *Piptadenia peregrina* (cohoba).

L. *ruo, rutus,* fall down; *ruina,* downfall, a tumbling down; see **destroy**

Gr. *sphallo,* cause to fall, overthrow:

Gr. *stalasso,* fall in drops, drip; see **drop**

See: **drop, slip, lot**

fallax, *-acis,* L. deceitful, false; *fallaciosus,* deceptive; see *fallo* under **li***ₑ*

false L. *fallo, falsus,* lie, deceive, cheat; see **lie**

falter < uncertain origin.

L. *atypus* (Gr. *atypos*), stammering:

L. *balbus,* stammering, stuttering, foolish; *balbutio, -itus,* stammer, stutter: balbuties, balbutient, balbutiate, booby (*Sula leucogaster*).

L. *blaesus,* speaking indistinctly, lisping, stammering:

L. *haesito, -atus,* stick fast, halt, be undecided; see **stand**

L. *inarticulatus,* indistinct, dumb, not fluent: inarticulate.

L. *indisertus,* loss for words, not eloquent:

L. *infacundus,* ineloquent:

L. *instabilis,* unsteady, changeable; see **shake**

L. *offendo, -fensus,* stumble, trip, falter, blunder; *offensio, -onis,* a tripping, stumbling-block; see **hurt**

Gr. *psellos,* faltering in speech; *psellismos,* m. stammering: psellism, *Psellogrammus kennedyi* (a fish).

Gr. *ptaio,* stumble, trip, fall, make a mistake; *ptaisma, -tos,* n. false step, stumble, mistake:

Gr. *sphalma, -tos,* n. false step, trip, stumble: *Sphalma quadricollis* (a beetle), *Sphalmatoblattina latinervis* (a fossil insect-wing).

Gr. *traulos,* lisping; *traulotes, -etos,* f. a lisping: *Traulotes ornata* (a beetle).

See: **stand, shake, slow, fault**

falx, *-cis,* L. sickle, scythe; *falcicula; falcula,* dim.; see **sickle**

fame < L. *fama,* reputation, rumor; *famigeratus; famosus,* celebrated; see **honor**

fames, L. hunger; *famelicus,* hungry; see **hunger**

famex, *-icis,* L. bruise, contusion; see **bruise**

family < L. *familia; familiaris,* of a family or household, domestic; see **kin**

famulus, L. servant; see **servant**

fan < AS. *fann,* < L. *vannus,* f. fan for winnowing grain: fanleaf, fantail, fanwort, fanglomerate, *Vanellus cristatus* (lapwing), *Pterotocrinus vannus* (a Mississippian crinoid).

Gr. *anemosyris, -idos,* f. a kind of fan:

L. *flabellum,* n. fan, dim. of *flabrum,* breeze; *conflabello, -atus,* fan violently: flabellation, *Flabellum curvatum* (a fan-coral), *Flabellina spatulata* (a foraminifer), *Gorgonia flabellum* (a sea-fan), *Homotrypa flabellaris* (a fossil bryozoan), *Dictyonema flabelliforme* (a graptolite), *Rhipidura flabellifera* (fan-tailed flycatcher).

Gr. *likmas, -ados,* f.; *likmos,* m. winnowing fan, basket, or shovel; *likmeterion,* n. dim.; *likmao,* winnow: *Licmophora flabellata* (a diatom).

Gr. *liknon,* a winnowing fan, basket, or cradle; *liknarion,* dim.; see **basket**

Gr. *psygma, -tos,* n. fan, a means of cooling; *psygmos,* m. chilliness: *Psygmatocerus wagleri* (a beetle), *Psygmophyllum flabellatum* (a fossil plant).

Gr *ptyon,* n. winnowing fan or shovel; *ptyarion,* n. dim.: *Ptyonodus paucicristatus* (a fossil fish).

Gr. *rhipis, -idos,* f.; *rhipister, -os,* m. fan; *rhipidion,* n. dim.: rhipidate, rhipidoglossal, *Rhipidopteris peltata* (a fern), *Rhipidiomorphus malaccanus* (a beetle), *Rhipidura leucophrys* (a flycatcher), *Rhipipteryx insignis* (a cricket).

L. *ventilo, -atus,* fan, agitate or change the air; see **fan**

fanaticus, L. over-zealous, mad; see **mad**

fancy < OF. *fantasie,* < L., Gr. *phantasia,* f. appearance, imagination, image, illusion: fantasy, fanciful, fantastic.

L. *fabulosus,* celebrated in fable, fictitious; see *fabula* under **story**

L. *fictio, -onis,* f. a making, feigning, imagining; *confictio, -onis,* f. invention, fabrication: fiction, fictitious.

L. *illusio, -onis,* f. deception, false fancy; see **lie**

L. *imaginatio, -onis,* f. mental image, fancy; *imaginor, -atus,* picture mentally; *imaginalis,* figurative; *imaginarius; imagineus,* of an image; *imaginosus,* full of fancies: imagine, imagination, imaginary.
Gr. *indalmatikos,* imaginary; see *eidos* under **form**
L. *opinio,* belief, fancy, view, conjecture; see *opinor* under **think**
L. *putativus,* imaginary; see *puto* under **think**
See: **form, dream, believe, think, spirit**

fandus, L. right, correct, lawful; see **justice**

fanum, L. temple; *fanulum,* dim.; see **temple**

far < AS. *feor,* remote in space and time.
L. *absitus; dissitus,* distant, apart, remote: *Aphis dissita* (an aphid), *Myosotis dissitiflora* (Swiss forgetmenot), *Vicia dissitifolia* (a vetch).
Gr. *aneuthe,* away from, far away, distant:
Gr. *apestys, -yos,* f. absence:
Gr. *apios,* far away:
Gr. *apodemos; ekdemos,* away from home, abroad, foreign: ecdemic, ecdemite, *Apodemus agrarius* (a rodent), *Ecdemus hypoleucus* (a butterfly).
Gr. *apoikos,* away from home, abroad: *Apoecus colonus* (a gastropod).
Gr. *apothen,* at a distance, far off:
Gr. *apouros,* far away, absent: *Apura xanthosoma* (a moth).
L. *avius,* out of the way, remote, solitary; see *via* under **way**
Gr. *dioche,* f. distance:
L. *distantia,* f. remoteness: distance, distal.
Gr. *ektopos,* away, distant, strange; see **strange**
L. *eminus,* aloof, a spear's throw off:
Gr. *epekeina,* beyond, on the far side:
Gr. *eurys,* broad, far-reaching; see **broad**
Gr. *exedros,* away from home, strange, extraordinary: *Exedrus indentatus* (a beetle).
L. *exilium (exsilium),* banishment; see *exul* under **banish**
Gr. *exorios,* beyond the frontier:
Gr. *hekas,* far, far off; *hekatos,* far-shooting:
L. *inaccessibilis,* unapproachable: inaccessible.
L. *inadibilis,* unapproachable, inaccessible:
L. *incongressibilis,* inaccessible:
L. *incontiguus,* that cannot be touched:
L. *longe; longinquus,* far, remote, prolonged:
Gr. *peraios,* on the other side, beyond; see *pera* under **over**
L. *peregrinus,* foreign, exotic, strange, < *pereger,* away from home, abroad: peregrination, *Tropaeolum peregrinum* (canary-bird flower).
L. *porro,* forward, farther, afar off:
L. *procul,* far but within sight:
L. *remotus; semotus,* distant, far off: remote, remoteness, *Lolium remotum* (hardy rye-grass), *Adenophora remotiflora* (a campanulacead), *Senecio remotifolius* (a groundsel).
L. *seductus,* remote, apart; see **separate**
Gr. *sychnos,* far, many; see **number**
Gr. *tele,* far; *teloteros,* farther; *telistos,* farthest: telegraph, telescope, telephone, television, *Telephonus australis* (a bird), *Telephorus bilineatus* (a beetle).
L. *Thule* (Gr. *Thoule*), northernmost part of the world; see **north**
L. *ultra,* < *ulter,* beyond, on the other side, far; *ulterior,* farther; *ultimus,* farthest, last: ultraviolet, ultramarine, ultra-critical, ultramundane, ulterior, ultimatum, ultimately, penult, antepenult, outrage, *Pythium ultimum* (a fungus).
See: **out, banish, over**

far, *farris,* L. spelt (probably emmer); *farraceus; farreus,* of spelt; see **grain**

farcimen, L. sausage, see **sausage**; *farciminum,* a disease in horses, see **disease**

farcio, *fartus (farctus),* L. stuff, cram; *farsilis; fartilis,* stuffed, crammed; see **press**

faredo, L. a kind of abscess; see **sore**

farfarus, L. coltsfoot; see **composite**

farina, L. flour, meal, starch; *farinula,* dim.; *farinaceus,* of meal, mealy; see **flour**

fario, L. a sea-trout; see **trout**

-farius, L. suffix meaning multiplication in numbers or parts; see **fold**

farmer < OF. *fermier,* < ML. *firmarius,* renter, < *firma,* farm.
L. *agricola,* m. farmer, husbandman: Agricola, *Sarcophaga agricola* (a fly).
L. *arator, -is* (Gr. *aroter, -os*), plowman, farmer; see *aro* under **till**
Gr. *boötes,* ox-driver, plowman; see *bos* under **cattle**

L. *bubulcus,* m. plower with oxen, herdsman: *Bubulcus ibis* (a heron).
L. *colonus,* tiller of the soil, farmer; see *colo* under **till**
L. *consitor; -sator,* planter, sower; see *sativus* under **sow**
L. *cultor,* farmer, planter, tiller; see *colo* under **till**
Gr. *ernokomon, -os,* m. tender of young plants:
Gr. *geïtes; geoponos; georgyos,* m. farmer, husbandman; *georgikos,* agricultural:
  georgic, George, Georgia.
L. *hortulanus,* gardener; see *hortus* under **field**
Gr. *kepouros,* gardener; see *kepos* under **field**
L. *olitor,* gardener; see *olus* under **plant**
Gr. *paradeisarios,* m. gardener:
L. *pastor,* shepherd; see **guard**
Gr. *phytourgos,* gardener; see *phyton* under **plant**
See: **field, till, earth, country**

farnus, L. ash-tree; *farneus,* of ash; see **ash**

farrago, L. mixed fodder for cattle, mixture, medley; see **mix**

farreum, L. a spelt-cake; *farriculum,* dim.; see **cake**

fascia, L. band, bandage, girdle, zone, strip, stripe; *fasciola,* dim.; *fascio, -atus,*
  envelop with bands, swathe; see **belt**

fascino, -atus, L. bewitch, charm, enchant; see **magic**

fascinum, L. penis; see **penis**

fascis, L. bundle, packet, sheaf; *fasciculus,* dim.; see **bundle**

fast < AS. *faest,* firm, strong, bound, see **strong, stand, bind, glue**; swift, see
  swift; < *faestan,* abstain from food, see **hunger**

fastidibilis, L. loathsome, disagreeable; *fastidiosus,* overnice, squeamish; see **hate**

fastigium, L. gable, pediment, slope; see **gable**

fastigo, -atus, L. sharpen to a point; see **point**

fastus, L. pride, scorn, arrogance; *fastosus; fastuosus,* proud, haughty, see **proud**;
  calendar, almanac, see **list**

fat < AS. *faett.*
L. *adeps, adipis,* c. fat, grease, lard; *adipalis; adipatus,* fatty, greasy; *adipeus,* of
  fat: adipose, adiposis, adipescent, adipoma, adipocere, adipic acid, *Pholiota
  adiposa* (a mushroom).
Gr. *aleiphar, -phatos,* n. fat, unguent, oil; *aleimma, -tos,* n. fat, oil, unguent;
  aleiptos, anointed, smeared, < *aleipho,* anoint with oil: aliphatic, aliptic.
L. *altilis,* fattened, rich, nourishing; see *alo* under **eat**
L. *arvina,* f. fat, suet, lard, grease:
L. *axungia,* f. axle-grease:
L. *butyrum* (Gr. *boutyron*), n. butter: butyrin, butyrate, butyric, butane, butyl,
  butylene, butter, butterfly, *Butyrospermum parki* (shea-tree), *Pentadesma buty-
  racea* (a butter-tree), *Combretum butyrosum* (a butter-tree), *Clostridium buty-
  ricum* (a bacterium), *Amylobacter butylicus* (a bacterium).
L. *carnatus; carnosus,* fleshy, corpulent, fat; see *caro* under **flesh**
L. *ceroma, -atis* (Gr. *keroma, -tos*), n. an unguent for wrestlers:
Gr. *chrio,* anoint; *christos,* anointed; *chrisma, -tos,* n. ointment, unguent: christen,
  Jesus Christ, Christian, Christman, Christmas, Christopher, Gilchrist, chrism,
  chrismon, cretin, cream, creme de menthe, 399 B.C., *Paliurus spinachristi* (a
  buckthorn), *Christensenia aesculifolia* (a fern).
L. *cnisa* (Gr. *knisa*), vapor and smell from cooking fat meat, fat; see **smell**
L. *corpulentus,* fleshy, fat; see *corpus* under **body**
L. *crassus,* thick, fat, stout; see **thick**
Gr. *demos,* m. fat: *Demodex folliculorum* (follicle-mite).
Gr. *episomos,* bulky, fat:
L. *hedychrum* (Gr. *hedychroun*), n.   a sweet-smelling ointment: *Hedychrum
  cupreum* (a wasp), *Hedychridium roseum* (a wasp).
L. *incrasso, -atus.* thicken, fatten; see *crassus* under **thick**
Malay *kayuputih,* a myrtaceous tree yielding cajuput: cajuputol, cajuputene,
  cajuput-tree (*Melaleuca leucadendron*).
Gr. *konaros,* fat, well-fed:
Gr. *kyprion,* n. oil from a tree growing in Cyprus:
L. *lardum,* n. pork fat, bacon; *lardarius,* m. butcher, pork-seller, pantry: lard,
  lardaceous, lardacein, larder, *Dermestes lardarius* (larder-beetle).
Gr. *larinos,* fatted, fat: *Larinus vitellinus* (a beetle), *Larinorhynchus sturnus* (a
  beetle).
L. *linimen, -inis,* n. grease:
Gr. *lipos,* n. fat; *liparos,* oily, shiny, sleek, bright; *liparotes,* m. fatness, fattiness:
  lipase, lipuria, lipoid, lipogenous, lipolysis, *Lipotropha microspora* (a protozoan),
  Lipoptena cervi (stag-tick), *Neoliparis mucosus* (snailfish), *Chromatium lipo-
  ferum* (a bacillus).

L. *megallium*, n. an ointment:

Gr. *myron*, n. any aromatic oil used in perfumes and unguents, oitment; *myraphion*, n. dim.; *myrisma, -tos*, n. ointment; *myrismos*, m. an anointing; *myrepsos*, m. preparer of unguents: myronic, myrobalan, myrosin, myristin, myroxylin, *Myroxylon peruiferum* (a balm-tree), *Myrothecium roridum* (a fungus), *Myristica argentea* (Macassar nutmeg), *Amyris balsamifera* (Jamaica rosewood).

Gr. *naphtha*, f. a volatile oil: naphthalene, naphthene, naphthol, naphthalylamine, naphthoxyacetic acid, phthallic, phthalazine, phthalein, phthalyl sulfacetimide, phthalocyanine, phenolphthalein, naphthoquinone.

L. *nardus* (Gr. *nardos*), m. an aromatic oil derived from the valerianacead, *Nardostachys jatamansi; nardinus*, of nard: nard, spikenard, American spikenard *(Aralia racemosa), Nardus stricta* (matgrass), *Carex nardina* (a sedge), *Cacaliopsis nardosmia* (a composite).

L. *neopum*, n. olive oil:

Gr. *netopon*, n. almond oil, oil of bitter almonds:

L. *obesus*, fat, corpulent: obese, obesity, *Rosmarus obesus* (walrus), *Isodon obesulus* (short-nosed bandicoot), *Euphorbia obesa* (baseball-plant).

L. *oesypum*, n. (Gr. *oisypos*, m.), grease from sheep's wool; *oisyperos; oisypodes*, containing grease, greasy: *Oesyperus unctulus* (a beetle).

L. *oleum* (Gr. *elaion*, olive oil), n. oil; *oleaceus; oleaginus*, oily; *olearis; olearius*, of oil; *oleatus*, oiled, put up in oil; *oleosus*, full of oil: oleic, olein, oleaginous, oleomargarine, acrolein, petroleum, linoleum, olefin, olefinic, ricinoleic, eucalyptol, ichthyol, menthol, mentholatum, petrolatum, elaeoblast, elaeoplast, *Psilopa petrolei* (petroleum-fly), *Camellia oleosa* (a camellia), *Moringa oleifera* (horseradish-tree), *Ligustrum ibolium* (ibota-privet).

L. *opimus*, well-fed, fat, < *opimo, -atus*, fatten, enrich; see *ops* under **wealth**

Gr. *piar*, n. fat, suet, tallow, cream; *piaros*, fat, rich; *piasmos*, m. fatness; *pion, -os*, oily, rich, fat, plump, sleek; *pialeos; pialos*, fatty; *piantikos*, fattening: piarhemia, pioscope, propionic, *Piaropus crassipes* (water-hyacinth), *Piophila casei* (cheese-fly), *Pionopsitta pyrrhops* (a bird).

Gr. *pimele*, f. fat, lard; *pimeles; pimelodes*, fat, fatty; *zapimelos*, very fat: pimelite, pimelitis, pimelorrhea, *Pimelobdella gracilis* (a fish), *Pimelea ferruginea* (a pimelea).

L. *pinguis*, fat; *pinguiculus*, dim., somewhat fat; *pinguitudo, -inis*, f. fatness: pinguid, pinguitude, pinguite, pinguecula, *Pinguicula vulgaris* (a butterwort), *Salvia pinguifolia* (a sage).

Gr. *psagdan, -os (psagdas; sagdas)*, m. an unguent:

L. *ricinus*, castor-oil plant; see **spurge**

L. *saginatus*, fattened; see *sagina* under **eat**

Gr. *sarkinos*, fleshy, fat; see *sarx* under **flesh**

L. *sebum*, n. fat, grease, suet, tallow; *sebaceus*, fatty, greasy: *sebalis*, of tallow; *sebosus*, full of tallow or grease: sebaceous, sebiferous, sebific, sebolith, seborrhea, *Sapium sebiferum* (Chinese tallow-tree).

Gr. *sialos*, m. fat, fat hog; *sialodes*, fatty: *Sialocyttara erasta* (a moth).

L. *stacta* (Gr. *stakte*), f. myrrh-oil:

Gr. *stear, -atos*, suet, tallow, fat; *steation*, n. dim.; *steatodes,* fatty: steatite, steatopygous, steatoma, stearate, stearin, *Steatomys pratensis* (a rodent), *Steatonyssus spinosus* (a mite).

Gr. *stymma*, chief ingredient of an unguent; an astringent; see *stypho* under **lessen**

L. *taxea*, f. lard:

Gr. *trapheros; trophis*, well-fed, fat, stout; see *trepho* under **eat**

L. *unguen, -inis*, n. fatty substance, fat; *unguentum*, n. ointment; *unguentarius*, of ointment; *unguinosus*, fatty; *unctulus*, anointed, < *unguo, unctus*, smear with oil, anoint: unguent, unguentary, unction, unctious, anoint, ointment, prune (preen).

See: **thick, large, flesh, swell, eat, wax**

fate < L. *fatum*, destiny, lot, death; *fatalis*, of destiny; see **lot**

fateor, *fassus*, L. confess, admit, acknowledge; see *confiteor* under **hold**

father < AS. *faeder:* fatherhood, fatherland, fatherless, fatherly, Father of Waters, godfather.

Heb. *abba*, father: abbot, abbess, abbacy, abbey, Abbotsford.

Gr. *archimandrites*, m. abbot; *archimandritis*, f. abbess:

Gr. *goneus, -eos; genetes*, m. father, parent:

Gr. *papas*, m. father: papacy, papal, paparchy, pope.

L. *pater, -tris* (Gr. *-tros*), m. father; *paterculus*, m. dim.; *paternus; patricus; patrius*, of a father, fatherly, ancestral; *patria*, f. fatherland: paternal, patricide, parricide, patrimony, patrician, Patricia, patron, patronage, pattern, patriotic,

Cleopatra, expatriate, repair, patronymic, *Eupatorium sessilifolium* (upland-boneset).

Gr. *phitys, -yos,* begetter, father; see *phyton* under **plant**

L. *tata* (Gr. *tetta; atta*), m. daddy; *tatula,* m. dim.:

See: **begin, mother, kin**

fatigo, *-atus,* L. tire, plague, vex; *fatigabilis,* that may be wearied; see **weak**

fatus, L. spoken, < *for, fari, fatus,* speak, say, predict; see **speak**

fatuus, L. foolish; see **fool**

fault < OF. *faute,* failure, defect, imperfection: faulty, faultless, fail, failure.

Gr. *ableptema, -tos,* n. mistake, oversight:

Gr. *agrammos,* not on the line (of dice), counting nothing: *Agrammos agrammus* (a fish).

Gr. *aitia,* blame, fault, cause; see *aitios* under **begin**

Gr. *alitema, -tos,* n. sin, offense; *alitros,* sinful: *Alitropus typus* (an isopod).

Gr. *arrhektos,* unfinished:

Gr. *astochema, -tos,* n. failure, fault; *astochia,* f. error, imprudence; *astochos,* missing the mark: *Astochia metatarsata* (a fly), *Astochus aldrichi* (a wasp).

Gr. *asymbletos,* not true to standard: *Asymbletia dispar* (a moth).

Gr. *ateles,* imperfect, unfulfilled, ineffectual, exempt; *atelestos,* without end, to no purpose, imperfect: ateliosis, atelestite, *Ateles ater* (a spider-monkey), *Ateleopus plicatellus* (a fish).

L. *captio, -onis,* f. fallacy, deception; *captiosus,* fallacious: captious.

L. *causa,* agent, motive, cause; see **begin**

L. *claudus; clodus,* limping, halt, lame, crippled, imperfect, defective; see **huit**

L. *corrigendum,* n. error and correction in a manuscript: corrigendum.

L. *crimen, -inis* (Gr. *krima, -tos*), n. charge, fault, offence, decision; *criminalis,* of crime; *criminosus,* full of guilt: crime, criminal, incriminate.

L. *culpa,* f. crime, fault; *culpabilis,* blamable, criminal; *deculpatus,* faulty: culpable, culpability, inculpate, culprit.

L. *defectivus,* imperfect, faulty; *defectus,* m. fault, failure, lack; *deficiens,* lacking: defect, defective, deficient, deficiency, indefectible.

L. *deformis,* misshapen, ugly: deformed, deformity.

L. *degener,* departing from its kind, base, depraved: degenerate.

L. *delinquo, -lictus,* fail, do wrong, offend; *delictum,* n. fault, crime, wrong; *deliculus,* blemished, defective; *delinquio, -onis,* f. failure, want: delinquent, delinquency, *in flagrante delicto, corpus delicti.*

L. *erratum,* n.; *error, -is,* m. fault, mistake: error, erroneous, errata.

L. *facinus, -oris,* n. misdeed, crime: facinorous.

AS. *gylt,* crime: guilt, guilty, guiltless.

Gr. *hamartia,* f. failure, fault, sin; *hamartetikos,* prone to failure; *hamartalos,* sinful; *diamartia,* f. great mistake; *examartia,* f. error, transgression:

L. *imperfectus,* unfinished, defective, faulty: imperfect, imperfection, *Begonia imperfecta* (a begonia).

L. *inconsummatus,* unfinished, incomplete, imperfect:

L. *ineffectus,* incomplete, not carried out: ineffectual, ineffective.

L. *labes, -is,* collapse, ruin, blemish, defect: *lapsus,* fall, error, a slipping; see *labor* under **fall**

L. *legirupa,* m. lawbreaker:

L. *mancus,* maimed, crippled, lame; see **hurt**

L. *menda,* f.; *mendum,* n. fault, error, blemish, bodily defect; *mendosus,* full of faults, faulty: amend, amendment, emendation.

Gr. *menima, -tos,* n. guilt, cause of wrath: *Menimopsis excaecus* (a beetle).

L. *nefastum,* n. wicked deed; *nefastus,* wicked, criminal; see *nefarius* under **bad**

L. *noxia,* fault, offense, crime; see *noceo* under **hurt**

Gr. *odyrmos,* a complaining; *odyrtikos,* querulous; see **sad**

L. *offensa,* fault, violation, transgression; see *offendo* under **hurt**

Gr. *parabasis,* f. trespass:

Gr. *paranoema, -tos,* n. error, folly; *paranomema, -tos,* n.; *paranomesis,* f. illegal act, transgression:

Gr. *paraptoma, -tos,* false step, blunder, slip; see *pipto* under **fall**

L. *pecco, -atus,* err, sin, commit a crime; *peccamen, -inis,* n.; *peccatio, -onis,* f.; *peccatum,* n. fault, mistake, sin; *peccator, -is,* m.; *peccatrix, -icis,* f. sinner: peccancy, peccant, peccadillo, impeccable.

Gr. *plemmeleia,* f. fault, error; *plemmeles,* faulty, erroneous, discordant: *Plemmeles semipunctata* (an isopod).

Gr. *ptaisma, -tos,* false step, mistake, failure; see **falter**

L. *queror, questus,* complain, find fault; *queribundus; querulus,* plaintive, complaining: querulous, querulity, querulent, querent, cry, decry, quarrel, quarrelsome, *Querula purpurata* (piahau), *Zonotrichia querula* (a sparrow).

L. *reus*, m.; *rea*, f. accused person, defendant; *reatus*, m. charge:
Gr. *sathroma*, something unsound, flaw; see **rot**
L. *scelus*, *-eris*, evil deed, crime; *scelero*, *-atus*, pollute, defile, desecrate; see **bad**
Gr. *soloikos*, making a grammatical error, speaking incorrectly like the Soloi: solecism.
Gr. *sphalma*, trip, stumble, error; see **falter**
L. *stribligo*, *-inis*, f. error in language, violation of propriety, solecism:
AS. *syn*, transgression of moral law: sin, sinful.
L. *vitium*, n. fault; *vitiabilis*, corruptible; *vitiosus*, faulty, defective; *vitio*, *-atus*, injure, mar, spoil: vitiate, vicious, vice, vituperation.
See: **fail, blame, scold, bad, unfit, falter, disease, pain, short**

Fauna, L. sister of Faunus, god of agriculture and cattle-raising; see **animal**

faustus, L. favorable, fortunate, lucky; see **lot**

fautor, L. patron; *fautrix*, patroness; see *faveo* under **give**

faux, *faucis*, L. throat, pharynx; pl. *fauces*, throat; see **throat**

favilla, L. ashes, embers; *favillaceus*, of ashes; see **ashes**

favissa, L. cellar or reservoir of a temple; see **room**

favonius, L. west wind; see **wind**

favor, L. goodwill, preference; *favorabilis*, agreeable, auspicious; see **agreeable, give**

favus, L. honeycomb; see **wax**

fawn < AS. *fagnian*, rejoice, flatter, see **flatter**; < F. *faon*, a young deer, see **deer**

fax, *-acis*, L. torch, firebrand; *facula*, dim.; see **light**

fear < AS. *faer*, dread or anxiety in anticipation of danger or misfortune.
Gr. *ainos*, dreadful, dire, horrible, terrible: *Aenocyon dirus* (a Pleistocene wolf).
Gr. *akko*, f. bugbear for frightening children: *Acco bicolora* (a moth).
Gr. *alphito*, f. bugbear for frightening children:
L. *anxius*, uneasy, troubled, solicitous; *anxiosus*, full of uneasiness: anxiety, anxious, *Agrilus anxius* (a beetle).
Gr. *aplatos*, monstrous, terrible: *Dictyopterus(Aplatopterus) rubens* (a beetle).
Gr. *athymos*, fainthearted, spiritless; see *thymos* under **mind**
Gr. *atyzelos*, frightful:
ME. *aw; agh*, < ON. *agi*, fear: awe, awful, awesome.
Gr. *blosyros*, grim, awful, terrible: *Blosyropus spinosus* (a beetle).
Gr. *brimodes*, grim, stern; *brimosis*, indignation; see **anger**
L. *consternatio*, *-onis*, f. amazed and bewildered alarm, terror, < *consterno*, *-atus*, throw up the ground, overcome, overwhelm: consternation.
L. *cussiliris*, cowardly; see **coward**
Gr. *dasples*, *-etos; daspletis*, frightful, horrible:
Gr. *deinos*, terrible, fearful, dread, dire; *dinops*, *-opos*, fierce-eyed; *deimos*, m. fear, terror; *deima*, *-tos*, n. a fearful thing; *deimaleos*, timid; *deimatodes*, frightful; *deisidaimon*, *-os; deisitheos*, fearing the gods, superstitious; *deidemon*, *-os*, fearful, cowardly; *deos*, n. fear, awe, < *deido*, fear, be anxious: dinosaur, dinoceras, dinotherium, *Dinorhynchus dybowskii* (a bug), *Dinornis maximus* (a moa), *Dinichthys intermedius* (an arthrodire), *Deimatostages contumax* (a bug), *Deima validum* (a holothurian).
L. *dentaneus*, showing the teeth, threatening:
L. *diffidens*, *-entis*, lacking confidence, doubtful: diffident, diffidence.
L. *dirus*, fearful, dreadful, ominous: dire, direful.
Gr. *dyschimos*, dangerous, fearful:
AS. *earh*, timid: eerie.
Gr. *ekpaglos*, terrible; see **wonder**
L. *formido*, *-atus*, fear, dread; *formidabilis*, causing fear, terrible: formidable, *Eremobates formidabilis* (a solpugid).
Gr. *gorgo*, grim, terrible, < *Gorgo*, *-onos* (L. *-onis*), f. Medusa with the snaky locks who turned beholders to stone; *gorgopos*, fierce-eyed: gorgon, gorgonid, *Gorgospyris medusa* (a radiolarian), *Gorgonetta mirabilis* (a radiolarian), *Gorgonia verrucosa* (a sea-fan), *Gorgosaurus sternbergi* (a dinosaur).
L. *horreo*, bristle, shrink from, be terrified; *horrendus; horribilis*, frightful, dreadful; *horridus*, bristly, dreadful, rough: horrible, horror, horrify, horrid, horrendous, abhorrent, ordure, *Ursus horribilis* (grizzly bear), *Oplopanax[Hoplopanax] horridus* (devilsclub), *Cirsium horridulum* (yellow thistle).
L. *immanis*, enormous, frightful, fierce: immane.
L. *inausus*, unventured, unattempted:
L. *infandus*, unventured, unattempted:
Gr. *kerambelon*, n. scarecrow:
L. *mania*, f. a bugbear to frighten children; *maniola*, f. dim.:

L. *Medusa* (Gr. *Medousa*), f. a **gorgon** with snaky locks capable of turning beholders to stone: medusa, *Hydromedusa tectifera* (long-necked turtle), *Thalassoxanthium medusinum* (a radiolarian), *Cirrhopetalum medusae* (an orchid), *Cenocrinus caput-medusae* (a crinoid), *Metoecus medusarum* (an amphipod), *Medusagyne oppositifolia* (a thealead).

L. *metus, -us,* m. fear, apprehension; *meticulosus,* fearful, excessively careful; *metuens,* fearing, < *metuo, -utus,* fear, hesitate: meticulous, *Fortax meticulosa* (a beetle).

L. *mineo; minor, -atus,* jut out, hang over, threaten; *minax, -acis,* overhanging, full of threats; see **projection**

Gr. *Mormo* (*Mormon*), *-os,* f. bugbear, hobgoblin, used to frighten children into good behavior: *Mormopterus minutus* (a bat), *Mormops blainvillei* (a bat), *Papio mormon* (mandrill).

Gr. *okneria,* timidity, shrinking, sluggishness; see *oknos* under **slow**

Gr. *opidnos,* awful, terrifying: *Opidnus ferrugineus* (a wasp).

Gr. *orrhodia,* f. dread, terror: *Orrhodia polita* (a butterfly).

L. *pavidus,* trembling, fearful, terrified; *pavor, -is,* m. fear, alarm; *pavibundus,* fearful, anxious, < *paveo,* fear, quake: *Conotrachelus pavidus* (a beetle).

L. *perturbatus,* agitated, troubled; see **trouble**

Gr. *phobos,* m. fear, dread; *phoberos,* fearful, terrible, formidable, frightful; *phobetikos,* timid; *phobetos,* to be feared; *phobetron,* n. scarecrow, bugbear: phobia, hydrophobia, nyctophobia, *Phoberotherium sylvaticum* (a fossil mammal), *Phobetus comatus* (a beetle), *Phobetor tricuspis* (a fish), *Phobetron pithecium* (hag-moth), *Phobetromimus exiguus* (a beetle), *Euphoberia granosa* (a fossil centiped), *Periphobus ferox* (a beetle).

Gr. *phriktos,* horrible, terrible; *phrixos,* standing on end, bristling, shuddering, shivering; *phrixothrix,* with bristling hair; see *phrix* under **shake**

Gr. *phyzelos,* shy, timid: *Phyzelus fasciatus* (a wasp).

Gr. *ptakismos,* m. shyness, timidity:

Gr. *ptexis,* f. terror, < *ptesso (ptosso),* crouch, cower: *Scotiaptex barbata* (Siberian owl).

Gr. *ptyrmos,* m. consternation; *ptyrtikos,* timorous, < *ptyromai,* be frightened:

L. *pudicus,* bashful, modest, chaste; see *pudeo* under **shame**

ON. *skirren,* frighten: scare, scarecrow, scaremonger.

Gr. *smerdaleos,* terrible, fearful: *Smerdalea horrescens* (a bug).

Gr. *sobaros,* frightening, pompous; *sobesis,* f. agitation; *sobetes,* m. frightener, one who scares away, < *sobeo,* scare away: *Sobarocephala ferruginea* (a fly).

L. *sollicitus (solicitus),* agitated, anxious, uneasy, < *sollicito, -atus,* stir up, disturb, agitate, incite; *sollicitudo (solicitudo), -inis,* f. uneasiness: solicitous, solicitude, *Aëdes sollicitans* (a mosquito).

Gr. *tarbos,* n.; *tarbosyne,* f. fear, terror; *tarbaleos,* fearful, frightful, terrible; *tarmysso,* frighten: *Tarbophis fallax* (a snake), *Tarbaleus pilosus* (a grasshopper), *Tarbaleopsis tuberculata* (a grasshopper).

L. *terribilis; terrificus,* dreadful, frightful; *terror, -is,* m. alarm, dread, fear; *terriculum,* n. scarecrow, bugbear, < *terreo, -itus,* frighten, fill with fear: terror, terrible, terrific, terrify, terrorism, deter, deterrent, *Crotalus terrificus* (a rattlesnake).

L. *timidus,* fearful; *timor, -is,* m. anxiety, dread, fear, < *timeo,* fear: timid, timidity, timorous, intimidate, *Larinus timidus* (a beetle), *Chrysomela timidula* (a beetle).

L. *trepidus,* anxious, alarmed: trepidity, trepidation, intrepid.

Gr. *treron,* trembler, wild dove; shy, timorous; see **dove**

L. *vereor, veritus,* fear, honor; *verecundus,* bashful, modest, coy; *verendus,* awesome, venerable; see **honor**

See: **danger, hate, honor, shake, coward, large, modest, wonder**

feast < L. *festus,* joyful; see **eat, joy, agreeable**

feather < AS. *fether.*

Gr. *anthelion,* dim. of *anthele,* plume of a reed; see **bush**

Gr. *chnous (chnoos),* any light porous substance, foam, fine down, bloom; see **foam**

Gr. *mnous (mnoos),* m. down; *mnoudion,* n. dim.: *Mnuphorus flavus* (a beetle).

L. *penna (pinna),* f. feather, wing, arrow, pen, fin; *pennula (pinnula),* f. dim.; *pinnatus,* feathered, plumed, winged; *pinniger,* bearing feathers, wings: pennate, pinnate, pen, pennant, panache, pinnatifid, pinion, pinnule, bipinnately, *Pennisetum glaucum* (pearl-millet), *Pennatula aculeata* (a sea-pen), *Bidens bipinnata* (Spanish needles), *Plautus impennis* (great auk), *Crataegus pinnatifida* (Chinese hawthorn), *Pilocarpus pennatifolius* (a rutacead), *Syringa pinnatifolia* (a lilac), *Stipa pennata* (needle-grass), *Pinnularia viridis* (a diatom), *Dryopteris pennigera* (a fern).

L. *pluma,* f. soft feather, down; *plumella; plumula,* f. dim.; *plumalis; plumosus,* feathered; *plumarius,* of feathers; *plumeus,* downy; *plumo, -atus,* cover with feathers: plume, plumage, plumule, plumose, plumigerous, nom de plume, *Plumatella punctata* (a bryozoan), *Plumularia cetacea* (a hydroid), *Ptilograptus plumosus* (a graptolite), *Meliphaga plumula* (a honey-eater), *Spermatolonchaea plumata* (a fly), *Uloborus plumipes* (a spider), *Corycaeus deplumatus* (a cyclops).

Gr. *ptenos (ptanos),* feathered, winged: ptenoglossate, *Ptenopus garrulus* (a gecko), *Calliptanus italicus* (a grasshopper).

Gr. *pteron,* n. feather *pteriskos,* m. dim.; *pterinos,* made of feathers; *pterotos,* feathered: *Pterinocrinus quinquenodus* (a Devonian crinoid), *Pterotocrinus spatulatus* (a Mississippian crinoid).

Gr. *ptilon,* n. soft feather, down, wing, leaf; *ptilion,* n. dim.; *ptilos,* m. plumage; *ptilotos,* feathered, winged: ptilolite, ptilosis, coleoptile, neossoptile, ptilinum, *Ptilophyllum acutifolium* (a fossil cycad), *Ptilocephalus barbatus* (a fish), *Ptilorhis magnifica* (rifle-bird), *Ptilidium pulchrinum* (a liverwort), *Ptilimnium capillaceum* (a mock bishopweed), *Ptilota serrata* (a red alga), *Ptilium crista-castrensis* (a moss), *Hydroptila hamata* (a caddisfly), *Oxyptilus tenuidactylus* (a plume-moth), *Crossoptilon auritum* (eared pheasant), *Trichoptilium incisum* (a composite), *Leptoptilus javanicus* (an adjutant), *Ptilotoceraeus visendus* (a beetle).

See: **wing, fly, bird, pen**

feature < OF. *faiture,* fashion, make, < L. *factura,* a making, creation, formation; see **form, face**

febris, L. fever; *febricula,* dim.; *febrilis,* of fever; see **disease**

februo, *-atus,* L. purify, expiate; *Februarius,* month of purification; see **pure**

fecula, L. dim. of *faex,* dirt, lees, dregs; see **dirt**

fecundus, L. fruitful, fertile, abundant, prolific, plentiful; see **fertile**

federate < L. *foederatus,* allied, leagued, < L. *foedus, -eris,* compact, covenant, league; see **bind**

feed < AS. *fedan;* see **eat**

feel < AS. *felan.*

Gr. *aisthanomai,* perceive, feel; *aisthema, -tos,* n. the thing perceived by the senses; *aisthesis,* f. feeling, sensation; *aisthetikos,* sensitive, perceptive (especially of beauty); *aisthetos,* sensible, perceptible: esthetics, anesthetic, esthematology, hyperesthesia.

L. *antenna,* f. sailyard; feeler, a sensory appendage on the head of an insect: antenna, *Antennaria pulcherrima* (an everlasting), *Antennarius marmoratus* (a fish), *Trapella antennifer* (a Pliocene pedaliacead).

Gr. *keraia,* f. any hornlike projection, antenna:

L. *passio, -onis,* suffering, feeling; see *patior* under **pain**

Gr. *patheticus* (Gr. *pathetikos*), capable of feeling, sensuous; *pathos,* n. tender feeling; *pascho,* suffer, feel; *ekpathes,* very passionate; *sympatheia,* f. fellow-feeling: pathetic, pathos, empathy, sympathy, sympathetic.

L. *sentio, sensus,* perceive by the senses, feel, experience, think; *sensibilis,* perceptible, capable of perceiving; *sensilis; sensualis,* endowed with feeling; *sensorium,* n. organ of sensation: sentiment, sentient, dissent, assent, sensation, sensible, scent, sensitive, insensibility, nonsense, senses, consent, sensual, sensuous, resentment, presentiment, sentence, sensorium, sensillum, *Onoclea sensibilis* (sensitive fern), *Biophyton sensitivum* (an oxalidacead), *Amphorophora sensoriata* (an aphid).

See: **touch, pain, joy, hurt, beauty, sad, dull, sharp, think**

fel, *fellis,* L. gall, bile; *felleus,* of gall; *fellosus,* full of gall; see **bile**

felicitas, L. happiness, < *felix, -icis,* happy; see **joy**

felis, L. cat; *felinus,* of cats; see **cat**

felix, *-icis,* L. happy; *felicior,* happier; *felicissimus,* happiest; see **joy**

fellator; fellatrix, L. sucker; *fellebris,* sucking; see *fello* under **suck**

fellow < AS. *feolaga,* < ON. *felagi,* comrade; see **companion**

femina, L. female, woman; *femella,* dim.; *femininus,* of females; see **woman**

femur, L. thigh; see **leg**

fence < L. *defendo, -fensus,* protect, guard, repel; see **wall**

fenestra, L. window; *fenestella; fenestrula,* dim.; *fenestratus,* windowed; see **window**

fennel < L. *feniculum;* see **carrot**

fenum, L. hay; *fenarius; feneus,* of hay; *fenilium,* hayloft; see **grass**

fenus, L. interest, gain; see **gain**

-fer, -a, -um; -ferous, < L. *fero, latus,* carry, bear; see *fero* under **carry**

feralis, L. of the dead; see **death**

ferax, -acis, L. fertile, fruitful; see **fertile**

ferculum; feretrum, L. litter, bier, barrow, tray; see *fero* under **carry**

feria, L. holiday, festival; *feriatus,* keeping holiday; see **day**

ferinus, L. pertaining to wildness and wild animals; see *ferus* under **nature**

fermentum, L. leaven, yeast; *fermentarius,* of leavening; see **yeast**

fern < AS. *fearn.*
    NL. *acrostichum,* n. a genus of ferns: *Acrostichum aureum* (a fern), *Polystichum acrostichoides* (Christmas fern).
    L. *adiantum* (Gr. *adianton*), n. maidenhair-fern: *Adiantum bellum* (Bermuda maidenhair-fern), *Polystichum adiantiforme* (a fern).
    NL. *asplenium,* n. < L. *asplenum* (Gr. *asplenon*), n. spleenwort: *Asplenium platyneuron* (ebony spleenwort).
    NL. *azolla,* f. a genus of floating ferns: *Azolla caroliniana* (a floating fern).
    L. *blechnum* (Gr. *blechnon*), a kind of fern: *Blechnum serrulatum* (sawfern).
    NL. *botrychium,* n. a genus of ferns: *Botrychium dissectum* (cutleaved grapefern).
    NL. *cyathea,* f. a genus of ferns: *Cyathea medullaris* (a tree-fern).
    L. *dryophonum,* n. a kind of fern:
    Gr. *dryopteris,* f. a kind of fern: *Dryopteris disjuncta* (oak-fern).
    L. *filix, -icis,* f. fern; *filicula,* f. dim.; *filicatus,* adorned with ferns; *filictum,* n. fernery: filicoid, filicin, Filicales, *Polyscias filicifolia* (an araliacead), *Aspidium filix-mas* (male fern), *Cyperus filicinus* (a sedge), *Quercifilix zeilanica* (a fern), *Azolla filiculoides* (a fern), *Acacia filicioides* (fern-leaved acacia), *Jacaranda filicifolia* (jacaranda).
    Gr. *hemionitis, -idos,* f. a fern: *Hemionitis palmata* (mulewort).
    NL. *lygodium,* n. a genus of ferns: *Lygodium palmatum* (a climbing-fern).
    NL. *marsilia,* f. a genus of aquatic ferns, < Luigi Marsigli, Italian naturalist: *Marsilea quadrifolia* (clover-fern), *Marsilidium speciosum* (a fossil fern).
    L. *onoclea* (Gr. *onokleia,* a plant), f. a genus of ferns: *Onoclea sensibilis* (sensitive fern).
    NL. *ophioglossum,* n. a genus of ferns: *Ophioglossum vulgatum* (adder's-tongue), *Stelis ophioglossoides* (an orchid).
    ML. *osmunda,* f. a kind of fern: *Osmunda regalis* (royal fern), *Polybotrya osmundacea* (a fern).
    NL. *pellaea,* f. a genus of ferns: *Pellaea falcata* (Australian cliffbrake).
    Gr. *phyllitis, -idos,* f. a fern, hart's-tongue: *Phyllitis scolopendrium* (hart's-tongue), *Polypodium phyllitidis* (strap-polypody).
    L. *polypodium* (Gr. *polypodion*), n. a kind of fern: *Polypodium fraxinifolium* ash-leaved polypody).
    NL. *polystichum,* n. a genus of ferns: *Polystichum setiferum* (a holly-fern).
    L. *pteris, -idis* (Gr. *-idos*), f. a kind of fern: pteridophyte, aminopterin, *Pteris longifolia* (rusty brake), *Cystopteris bulbifera* (bladder-fern), *Dryandra pteridifolia* (a euphorbiacead).
    NL. *salvinia,* f. a genus of ferns, < A. M. Salvini, Italian scholar: *Salvinia rotundifolia* (a water-fern).
    NL. *schizaea,* f. a genus of ferns: *Schizaea dichotoma* (a fern).
    L. *scolibrochum,* n. a kind of fern:
    L. *scolopendrium* (Gr. *skolopendrion*), n. a kind of fern: *Phyllitis scolopendrium* (hartstongue-fern).
    Gr. *thelypteris, -idos,* f. a fern: *Dryopteris thelypteris* (marsh-fern), *Athyrium thelypteroides* (silvery ladyfern).
    Gr. *trichomanes,* n. a kind of fern: *Trichomanes lineolatum* (a filmy-fern), *Phyllocladus trichomanoides* (a celery-pine).
    NL. *woodwardia,* f. a genus of ferns, < Thomas J. Woodward, English botanist: *Woodwardia virginica* (a chain-fern).

Feronia, L. a Roman goddess; a genus of the rue family; see **rue**

ferox, -ocis, L. brave, warlike, courageous, fierce; see *ferus* under **nature**

ferrarius, L. pertaining to iron; blacksmith; see *ferrum* under **iron**

ferret < OF. *furet,* < L. *fur,* thief; see **weasel**

ferrugo, L. rust; *ferrugineus; ferruginus,* rust-colored, rusty; see **rust**

ferrum, L. iron; *ferreus,* of iron; *ferratilis; ferratus,* ironed; see **iron**

ferrumen, L. cement, solder; see **glue**

ferry < AS. *ferian,* carry; see **carry**

fertile < L. *fertilis,* fruitful, prolific, capable of producing fruit or young; *fertus,* fertile, productive: fertility, fertilizer.
L. *auctifer,* fruitful, fertile:
Gr. *enteknos,* having children; *euteknos,* blest with children:
Gr. *eponkos,* pregnant:
Gr. *eukarpos,* fertile, fruitful:
Gr. *euochthos,* fertile, rich:
L. *fecundus,* fruitful, fertile, prolific: fecund, fecundity, *Ctenodonta fecunda* (a fossil pelecypod).
L. *ferax, -acis,* fertile, fruitful:
L. *fetus,* pregnant, fruitful, productive; *fetificus,* fructifying, fertilizing; *fetosus,* prolific:
L. *fordus,* pregnant:
L. *fructifer; frugifer,* fruitful, fertile; see **fruit**
L. *genitabilis,* fruitful, productive:
Gr. *gonimos,* productive, fruitful, endowed with generative power; *eugonos,* productive; *polygonia,* f. fertility: gonimoblast, gonimium, *Prosthogonimus pellucidus* (poultry-fluke).
L. *gravidus,* laden, pregnant; see *gravis* under **heavy**
L. *inciens, -entis,* pregnant:
L. *Inuus,* m. god of fecundity in herds, corresponding to Pan: *Macacus innuus* (former name of the Barbary ape, *Simia sylvanus*).
Gr. *kyo (kyeo),* swell, be pregnant; *kymas, -ados,* f. a pregnant woman; *kyesis,* f. conception, pregnancy; *arikymon, -os,* prolific; *enkymon, -os; enkyos; kyeros,* pregnant: cyesis, *Enkianthus[Encyanthus] cernuus* (a heath).
Gr. *metridios,* fruitful, fertile, filled with seed: *Metridiochoerus andrewsi* (a Pliocene pig).
Gr. *panoros,* produced in every season: *Panorosemus frivalskyi* (a beetle).
Gr. *phoras, -ados; phorimos,* fruitful, bearing; see *phero* under **carry**
L. *Picumnus; Pilumnus,* m. brother tutelary deities of wedlock and fertility: *Picumnus obsoletus* (a piculet), *Climacteris picumnus* (a tree-creeper).
Gr. *polychous,* prolific:
L. *praegnans,* with child: pregnant, pregnancy, impregnate.
NL. *prolifer, -a, -um,* fruitful, productive; *prolificus,* producing abundantly or freely, fruitful: proliferate, prolific, *Madrepora prolifera* (a madrepore-coral), *Paleofavosites prolificus* (a fossil coral).
Gr. *syllepsis,* conception, pregnancy; see **begin**
L. *uber, -eris,* full, fertile, fruitful, copious; see **abundance**
See: **fruit, abundance, young, begin, birth**

fertum, L. a kind of cake; see **cake**

ferula, L. giant fennel, see **carrot**; rod, staff, whip, see **rod**

ferus, L. wild, untamed, uncultivated, rough, savage; see **nature**

fervidus, L. boiling, fiery, hot, passionate; *fervor,* ardor; see *ferveo* under **boil**

fessus, L. weary, weak, feeble; see **weak**

festinus, L. hasty, quick, < *festino, -atus,* hasten, hurry; see **swift**

festuca, L. stalk, stem, straw; *festucula,* dim.; see **stem**

festus, L. joyful; *festivus,* gay, joyous, merry; see **joy**

fetalium, L. birthday; see **day**

fetialis, L. pertaining to the priests who sanctioned war and approved treaties; see **priest**

fetidus; fetulentus, L. stinking; *fetutina,* a stinking place; see *feteo* under **smell**

fetus, L. pregnant, fertile, fruitful; *fetosus,* prolific, see **fertile**; that which is brought forth, offspring, fruit, the unborn young, embryo, see **young**

fever < AS. *fefer,* < L. *febris*; see **disease**

few < AS. *feawe.*
L. *aliquantus,* some, moderate, somewhat; *aliquantulus,* little; *aliquatenus,* for some distance, partly, to some extent; *aliqui, -qua, -quod,* some, any; *aliquis, -quid,* somebody, someone, something; *aliquot,* some, few, several: aliquot.
Gr. *araios,* few, intermittent, thin, loose, porous; see **thin**
L. *eximius,* exceptional, uncommon, extraordinary, choice: *Carpinus eximia* (Korean hornbeam), *Limonium eximium* (a sea-lavender).
L. *infrequens, -entis,* seldom, rare: infrequent, *Trioxys infrequens* (a wasp).
L. *inusitatus,* rare, uncommon, unusual: *Robulus inusitatus* (a fossil foraminifer).
Gr. *litos,* plain, simple, frugal; see **simple**
L. *nonnullus,* some, several:

Gr. *oligos,* few, little, scanty; *oligistos,* very few, very little: oligarchy, oligo-chrome, Oligochaeta, oligoclase, oliguria, oligist.

L. *parcus,* frugal, scanty: see **stingy**

L. *paucus,* few, little; *pauculus; pauxillus,* dim.: paucity, pauciloquy, *Polygala paucifolia* (fringed polygala), *Artemisia pauciflora* (santonica, source of san-tonin).

Gr. *psednos,* thin, scanty, bald; see **bare**

L. *rarus,* scarce, scattered, dispersed, thin: rare, rarely, rarefy, rarefaction, rare-ness, *rara avis.*

Gr. *spanios; spanos; sparnos,* rare, scarce; *spanistos,* scanty: spanopnea, spano-menorrhea, *Spaniotherium speciosum* (a fossil mammal), *Spaniotoma senex* (a fly), *Spanophatnus ruficeps* (a wasp), *Comanthus spanoschistum* (a crinoid).

L. *sparsus,* few, rare, scattered; see *spargo* under **spread**

AS. *sum,* an indefinite quantity or number, sometimes a considerable degree: somehow, something, somewhat, sometime, somewhere, awesome, fearsome, burdensome, lonesome, wholesome, foursome.

See: **poor, little, stingy, alone, thin**

-fex, L. suffix denoting maker; see *facio* under **make**

fiber < L. *fibra,* thread, filament, see **thread;** < *fiber, -bri,* beaver; *fibrinus,* of the beaver, see **beaver**

fibula, L. clasp, buckle, safety pin; *fibulatorius,* with clasps; see **bind**

-fic, L. suffix denoting make, cause; see *facio* under **make**

ficedula, L. figpecker; see **warbler**

fickle < AS. *ficol;* see **change**

fictilis, L. earthen, made of clay; *fictiliarius,* potter; see **earth**

fictus, L. feigned, false; *fictitius,* artificial, assumed; see **lie**

ficus, L. fig-tree, fig; *ficulnea,* fig-tree; *ficula,* dim., see **fig;** piles, see **disease**

-fid, divided into many parts, < L. *findo, fissus,* cleave, split; see *findo* under **cut**

fiddle < AS. *fithele;* see *pandura* under **music,** *vitulor* under **joy**

fidelia, L. earthen vessel, pot; see **pot**

fidelis; fidus, L. faithful, sure, true, trustworthy; see *fido* under **believe**

fidicen, L. lyre player, < *fides,* lyre, lute; *fidicula,* dim.; see **harp**

field < AS. *feld:* fielder, fieldfare, feldspar, felwort, veldt, Fielding, Chesterfield, Clearfield, Stubblefield, Sheffield, Wakefield, Roosevelt, Westerveldt.

AS. *aecer,* field: acre, Longacre. Derive: wiseacre.

L. *ager, agri* (Gr. *agros*), m. field, open country, land; *agellus; agellulus,* m. dim.; *agrarius,* of the land; *agrestis, -e,* rural, rustic, wild; *agricola,* m. husbandman; Gr. *agrios,* living in the fields, wild, savage; *agriotes,* f. wildness, savageness; *agroikos,* of the country, rustic, boorish; *agraulos,* dwelling in the field, living out of doors, rural; *agronomos,* rural, wild; *agrotes; agrostes,* wild; country-man; *agrotikos,* rustic, rural: agriculture, agrarian, agrestial, agrestal, Agricola, agronomy, peregrination, pilgrim, *Agropyron cristatum* (created wheatgrass), *Arvicola agrestis* (vole), *Agriocharis ocellata* (Yucatan turkey), *Agrion aequabile* (blackwing), *Agrilus egenus* (a beetle), *Agriotes mancus* (wheat-wireworm), *Agroecomyrmex duisbergi* (an ant), *Agroecodes comata* (a moth), *Mesagroecus obscurus* (a beetle).

L. *aratio, -onis,* f. a plowed field; *aratiuncula,* f. dim.:

L. *areola,* garden; see *area* under **space**

L. *arura* (Gr. *aroura*), f. field, tillable land, earth; *arourion,* n. dim.; *arouraios,* rural, rustic:

L. *arvum,* n. field; *arvalis,* of a cultivated field: arval, arvicolous, arviculture, Auvergne, *Arvicola* (now *Microtus*) *pennsylvanicus* (meadow-mouse), *Equisetum arvense* (a horsetail), *Sonchus arvensis* (a sowthistle).

Gr. *botamion,* n. pasture:

Gr. *boukolikos,* rural, rustic, pastoral: bucolic.

Gr. *bucetum,* n. pasture for cattle:

L. *caespes, -pitis,* grassy field; see **sod**

L. *campus,* m. field, plain; *campester(campestris), -tris, -tre; campanius,* of or pertaining to fields: camp, campaign, Campania, champagne, campion, cham-pignon, champion, encampment, decamp, *Lepus campestris* (jackrabbit), *Acer campestre* (hedge-maple), *Grindelia camporum* (field-gumweed).

L. *cepina,* field or bed of onions; see **onion**

Gr. *chora; chorion; choros* land, country; see **country**

L. *colonia,* farm, estate, settlement; *colonus,* husbandman, tiller of the soil, < *colo, cultus,* till, inhabit; see **country, till**

AS. *croft,* small field or plot of ground: croft, crofter, Bancroft, Moorcroft(Wyo.).

Heb. *eden,* delight, pleasure; see **joy**

L. *floralium,* flower-garden; see **flower**

L. *fundus,* bottom, base, soil, estate; see **base**

OF. *gardin,* an enclosure for cultivating flowers and vegetables; akin to AS. *geard,* an enclosure, court: garden, yard, courtyard, vineyard, *Gardenia latifolia* (a gardenia).

Gr. *georgikos,* agricultural; see *georgyos* under **farmer**

AS. *haeth,* open waste land covered with low herbs and shrubs; see **heath**

Gr. *heiamene,* f. meadow:

L. *hortus,* m. garden; *hortulus,* m. dim.; *hortensis; hortensius; hortualis; hortulanus,* of a garden, gardener: Hortense, Huerta, ortolan *(Emberiza hortulana),* *Carabus hortensis* (a beetle), *Myrmecocystus hortideorum* (a honey-ant), *Prunus hortulana* (a plum), *Gladiolus hortulanus* (gladiolus), *Adoretus horticola* (a beetle).

L. *idyllium* (Gr. *eidyllion*), a poem describing pleasant rustic conditions; see **poem**

ON. *ing,* meadow: Ingham, Ingthorpe, Dorking, Deeping.

L. *jugerum,* a measure of land, acre; see **measure**

Gr. *kepos,* m. garden, orchard, plantation; *kepidion; kepion,* n. dim.; *kepaios; kepeutikos,* of a garden; *kepeutos,* cultivated; *kepouria,* f. gardening; *kepourikos,* of gardening; *kepouros,* m. gardener: *Kepolestes coloradensis* (a fossil mammal).

Gr. *lachania,* f. kitchen-garden:

AS. *leah,* meadow, open field: lea, Ashley, Oakley, Lindley, Bradley, Leigh, Stoneleigh.

Gr. *leimon, -os,* m.; *leimax, -akos,* f. meadow; *leimonarion,* n. dim.; *leimonios,* of a meadow; *leimakodes,* meadowlike, grassy, moist: limonite, *Limodorum multiflorum* (an orchid), *Limacocarenus curtulus* (a bug), *Limacodes rectilinea* (a moth), *Acantholimon glumaceum* (a pricklythrift).

Gr. *libethron,* n. meadow:

Gr. *nomos,* feeding place for cattle, pasture; see *nemo* under **eat**

Gr. *oasis,* f. fertile spot in a desert: oasis, oasitic.

Gr. *orgas, -ados,* f. meadow, field:

L. *pagus* (Gr. *pagos*), district, country; *paganus,* rustic; see **country**

L. *pascuum,* a pasture; see *pasco* under **eat**

Gr. *pedion,* plain, field; see *pedinos* under **level**

L. *pastor,* shepherd; *pastoralis,* of shepherds, rustic, rural; see **guard**

ME. *plot;* plat, a piece of ground: plot, plat.

L. *praedium,* n. plot of ground, estate, manor; *praediolum,* n. dim.:

L. *pratum,* n. meadow; *pratulum,* n. dim. lawn; *pratensis,* found in meadows; *pratens,* meadow-green, grassy: pratal, pratincole, pratincolous, prairie, *Alopecurus pratensis* (meadow-foxtail), *Strix pratincola* (barn-owl), *Eleocharis praticola* (a sedge), *Phleum pratense* (timothy), *Pardosa prativaga* (a spider).

Gr. *rhodonia,* rose-garden; see *rhodon* under **rose**

L. *rosetum,* rose-garden; see *rosa* under **rose**

L. *rus, ruris,* fields, lands, country; *rurestris; ruralis; rusticus,* belonging to the country, of the country; see **country**

L. *seges, -etis,* f. grainfield, crop; *segetalis,* of standing crops: *Petroselinum segetum* (corn-parsley), *Noctua segetum* (an owlet-moth), *Euphorbia segetalis* (a spurge).

L. *seminarium,* n. seed-plot, nursery: seminary.

L. *terrula,* small piece of land, field, dim. of *terra,* earth; see **earth**

ON. *thveit,* a tract of tilled land, clearing, meadow: thwaite, Goldthwaite, Postlethwaite, Stonethwaite, Thistlethwaite.

L. *topia,* ornamental gardening; *topiarius,* pertaining to ornamental gardening; see **ornament**

Russ. *tundra,* f. marshy plain: tundra, *Sorex tundrensis* (tundra-shrew).

Sp. *vega,* f. open country, plain, meadow: Las Vegas(Nev.), vegasite.

L. *vervactum,* n. fallow field; *vervago, -actus,* plow a fallow field: *Vervactor typicalis* (a bug).

L. *veteretum,* n. old, fallow ground:

L. *viridarium,* plantation of trees, pleasure-garden; see **forest**

See: **country, earth, forest, grass, park, marsh, till, flower**

fiend < AS. *feond,* enemy, foe, devil; see **enemy**

fierce < OF. *ferus,* wild, savage, cruel; see **wild, rough, fear**

fifty; see **five**

fig < L. *ficus,* f. fig-tree, fig; *ficula,* f. dim.; *ficulnea,* f. fig-tree; *ficarius,* of figs; *ficaria,* f.; *ficetum,* n. fig plantation; *caprificus,* f. wild fig-tree: fig, ficoid, ficin, caprification, *Ficus palmata* (punjab fig), *Cucurbita ficifolia* (a gourd), *Hylesinus fici* (a beetle), *Opuntia ficus-indica* (prickly pear), *Strepsidura ficulnea* (a fossil gastropod), *Motacilla ficedula* (figpecker), beccafico *(Sylvia hortensis),* *Ranunculus ficaria* (a buttercup).

L. *aracia*, f. a kind of white fig-tree:

L. *carica*, f. a dried fig: caricious, *Carica papaya* (papaya), *Ficus carica* (common fig), *Pucciniopsis caricae* (papaya leaf-ant).

L. *cottanum* (Gr. *kottanon*), n. a small kind of fig:

Gr. *erineos*, m. wild fig-tree; *erineon*, n. fig; *erinasmos*, m. caprification: *Erineophilus schwarzi* (a beetle).

L. *grossus*, m. an unripe fig; *grossulus*, m. dim.: *Blastophaga grossorum*(now *psenes*) (fig-wasp).

Gr. *ischas*, *-ados*, f. dried fig; *ischadion*, n. dim.:

Gr. *kolythron*, n. a ripe fig:

Gr. *krade*, fig branch; see **branch**

L. *marisca*, f. a large kind of fig:

Gr. *olynthos*, m. an untimely fig: *Olynthus hispidus* (a sponge), *Olynthium splendidum* (a sponge).

Gr. *phelex*, *-ekos*, m. a wild fig:

Gr. *phibaleos*, m. a kind of early fig: *Phibalothrips exilis* (a thysanopterid).

Gr. *proknis*, *-idos*, f. a kind of dried fig:

Gr. *sykon*, n. fig; *sykarion; sykidion*, n. dim.; *sykinos*, of figs; *sykodes*, figlike; *syke; sykea*, f. fig-tree; *sykon*, *-os*, m. fig-garden; *bousykon*, n. a large fig: syconium, sycosis, sycophant, sycamore, *Ficus sycomorus* (sycomore-fig), *Sycum ciliatum* (a sponge), *Sycoscapter insignis* (a wasp), *Sycarium villosum* (a sponge), *Sycidium reticulatum* (a fossil bryozoan), *Syconella tubulosa* (a sponge), *Sycopsis sinensis* (fig-hazel), *Dyssycum fistulosum* (a sponge), *Metasycocrinus piriformis* (a Permian crinoid), *Busycon perversa* (a winkle-shell), *Calosoma sycophanta* (a ground-beetle).

Gr. *thrion*, fig leaf, leaf; see **leaf**

fight < AS. *feohtan*.

Gr. *agon*, *-os*, m. contest, struggle; *agonistes*, m. contestant, combatant; *agonios*, of a contest; *agonistikos*, fit for contest: agony, antagonist, antagonize.

Gr. *aichmetes*, spearman, warrior; see *aichme* under **spear**

L. *altercatio*, *-onis*, f. dispute, debate; *altercor*, *-atus*, quarrel, wrangle: altercation.

Gr. *Ares*, *-eos*, m. god of war; *Areios*, devoted to Ares, warlike: Areopagus (Mars Hill), areotectonics, Milton's *Areopagitica*.

Gr. *aspidiotes*, shield-bearer, warrior; see **shield**

Gr. *athletes*, m. contestant; *athletikos*, of an athlete; *athlema*, *-tos*, n.; *athlesis; athlosyne*, f.; *athlos*, m. contest, struggle; *athlios*, struggling; *athlon*, n. prize: athlete, athletic, Pentathlon.

L. *bellum*, n. war; *bellicus*, of war, military; *bellatorius; bellax*, *-acis; bellicosus; belliger; bellosus*, warlike; *bellator*, *-is*, m.; *bellatrix*, *-icis*, f. warrior; *bellicum*, n. signal for attack; *Bellona*, f. sister of Mars and goddess of war: bellicose, belligerent, Bellatrix, *status quo ante bellum*, rebellion, rebel, revelry, duel, *Termes bellicosus* (a termite), *Bellator militaris* (a fish).

L. *certo*, *-atus*, decide by contest, fight, struggle; *certator*, *-is*, m. disputant; *certamen*, *-inis*, n. contest, rivalry, struggle; *certabundus*, contending, disputing; *concertativus*, pertaining to disputation and controversy: concert, disconcert.

L. *clarigo*, *-atus*, declare war: clarigation.

L. *competitor*, *-is*, rival; see **enemy**

L. *conflictus*, m. fight, contest, < *confligo*, strike together, oppose: conflict.

L. *consero*, *-ertus*, join in combat:

L. *contentio*, *-onis*, f. combat, contest, strife; *contentiosus*, given to strife, disputatious: contention, contentious.

L. *controversia*, f. dispute, fight, quarrel: controversy, controversial.

L. *contumax*, *-acis*, stubborn, defiant, insolent; see **stubborn**

Gr. *daiphron*, *-os*, warlike: *Daiphron*[*Daephron*] *ochraceum* (a beetle).

Gr. *deris*, *-ios*, f. battle, contest, fight: *Derichthys serpentinus* (a fish).

L. *dimicatio*, *-onis*, f. fight, struggle, < *dimico*, *-atus*, fight, contend:

L. *disceptatio*, *-onis*, f. debate, controversy:

L. *discordia*, f. disagreement, dissension: discord.

L. *disputo*, *-atus*, contend, discuss; *indisputabilis*, uncontestable: dispute, disputation.

L. *dissensus*, m. disagreement, discord; *dissentaneus*, disagreeing: dissension, dissent.

Gr. *eris*, *-idos*, f. dispute, quarrel, strife; *eridantes; eristes*, m. wrangler; *eristikos*, fond of wrangling, given to strife; *eristos*, contested; *aneristos*, undisputed; *erisma*, *-tos*, n. cause of strife; *dyseris*, *-idos*, contentious; *erizo*, dispute, quarrel, litigate: eristic, *Eridantes erigonoides* (a spider), *Aneristus ceroplastae* (a wasp), *Dyseris albicincta* (a fly).

L. *gladiator*, swordsman, fighter; see *gladius* under **sword**

Sp. *guerra*, f. war; *guerilla*, dim.: guerilla.

Gr. *hamilla*, f. conflict, contest; *hamilleter, -os*, m. competitor; *hamilletikos*, of a contest; *anthamillos*, m.; *anthamillestria*, f. rival; *ephamillos; synamillos*, matching, rivaling, vying with: *Cephaloon(Ephamillus) variabilis* (a beetle).

Gr. *hysmine*, f. battle, fight:

L. *impacatus; impacificus*, not peaceable, warlike:

L. *implacatus*, unappeased, unallayed, unsatisfied: implacable.

L. *jurgium*, n. verbal quarrel, strife, brawl, < *jurgo, -atus*, contend, quarrel, scold, rebuke, chide; see **scold**

L. *lis, litis*, f. strife, quarrel; *litigator, -is*, m. disputant; *litigiosus*, contentious, quarrelsome; *litigo, -atus*, dispute, quarrel, sue at law: litigable, litigator, litigatory, litigation, litigious.

L. *luctor, -atus*, wrestle, struggle, strive, contend, oppose; *luctator, -is*, m. wrestler; *colluctator, -is*, m. adversary; *eluctabilis*, surmountable; *reluctans*, struggling against, resisting: luctation, eluctate, ineluctable, reluctance, reluctant, reluctivity.

Gr. *machetikos; machimos; philomachos*, warlike, quarrelsome; *machetes*, m. fighter, warrior; *epimachos*, equipped for battle; *lamachos*, eager to fight; *machomai*, fight: logomachy, naumachia, *Machetornis rixosa* (a bird), *Machetes* (now *Philomachus*) *pugnax* (ruff), *Machimostola commatias* (a moth), *Neopromachus injucundus* (a grasshopper), *Odontomachus animosus* (an ant), *Epimachus speciosus* (plume-bird), *Lysimachia vulgaris* (European loosestrife), *Lamachus semperi* (a phasmid), *Lamachodes laevis* (a phasmid).

L. *Mars, Martis*, m. god of war; *Martialis*, of Mars: martial, court-martial, mavortial, Martin, March, Mardi Gras.

L. *miles, -litis*, soldier; see **army**

L. *mirmillo, -onis* (Gr. *mermillon; mormillon, -os*), m. gladiator:

Gr. *mothos*, m. battle:

Gr. *neikos*, n. quarrel, strife, wrangle; *neikester, -os*, m. wrangler, brawler; *philoneikos*, contentious; *polyneikes*, wrangling much: Polynices, *Polinices triseriata* [*Polynices triseriatus*] (a moon-shell):

L. *palaestes* (Gr. *palaistes*), m. wrestler; *pale*, f. a wrestling, fight; *palaio*, wrestle; *boupalis*, wrestling hard:

Gr. *polemos* (*ptolemos*), m. war, fight; *polemios*, of war; *polemikos*, warlike, hostile; *polemistes*, m. warrior: polemics, polemical, polemarch, Ptolemy, Ptolemaic, *Polemius mimicus* (a beetle), *Lycopolemius depressus* (a beetle).

L. *procinctus*, girded for battle:

L. *proelium*, n. battle, fight; *proeliaris*, of a battle; *proelior, -atus*, engage in battle:

L. *pugilo, -atus*, fight with the fists, box; *pugil, -is; pugilator, -is*, m. boxer, fighter who uses the cestus: pugilist, *Strombus pugilis* (a gastropod), *Gelasimus pugilator* (a fiddler-crab).

L. *pugno, -atus*, contend, fight; *pugnax, -acis*, combative, contentious: pugnacious, repugnant, impugn, poniard, *Betta pugnax* (fighting-fish).

L. *pycta* (Gr. *pyktes; pygmachos*), m. boxer, pugilist; *pyktikos*, skilled in boxing: *Pyctoderes egenus* (a beetle).

L. *rabulatus*, m. a brawling, wrangling; *rabula*, m. brawler, wrangler, pettifogger: rabulous, rabulistic.

L. *rixa*, f. quarrel, brawl, strife, dispute, fight; *rixosus*, quarrelsome; *rixator, -is*, m. brawler: *Mustela rixosa* (a weasel).

L. *scordalus*, m. wrangler, quarreler, brawler:

Gr. *syrrhagma, -tos*, n. conflict:

Gr. *teuchestes*, m. armed man, warrior: *Teuchestes brunneus* (a beetle).

Gr. *toxeutes*, archer, bowman; see *toxon* under **arc**

L. *velitor, -atus*, skirmish; *veles, -etis*, m. skirmisher: velitation, *Semiphorus velitans* (a Miocene fish).

See: **army, enemy, strike, tool, hurt**

figulinus (figlinus), L. pertaining to pottery; *figulus*, potter, worker in clay; see **pot**

figure < L. *figura*, form, shape; *figuro, -atus*, form, mold, shape; see **form, number**

figwort

L. *antirrhinum* (Gr. *antirrhinon*), n. snapdragon: *Antirrhinum latifolium* (a snapdragon), *Heteropatella antirrhini* (a fungus).

NL. *calceolaria*, f. a genus of the figwort family: *Calceolaria verticillata* (a slipper-flower).

NL. *cymbalaria*, f. a genus of the figwort family, < L. *cymbalaris*, a plant: *Cymbalaria pilosa* (a basket-ivy), *Saxifraga cymbalaria* (a saxifrage).

NL. *digitalis*, f. a genus of the figwort family: digitalis, digitalin, *Digitalis purpurea* (common foxglove).

L. *erinus* (Gr. *erinos*), m. a genus of the figwort family: *Erinus alpinus* (liver-balsam), *Lobelia erinus* (border-lobelia).

NL. *euphrasia*, f. a genus of the figwort family: *Euphrasia arctica* (an eyebright).

NL. *gratiola*, f. a genus of the figwort family: *Gratiola pilosa* (a gratiola).

NL. *linaria*, f. a genus of the figwort family: *Linaria purpurea* (purple toadflax).

L. *mimulus*, m. dim. of *mimus*, actor; a genus of the figwort family: *Mimulus luteus* (golden monkey-flower).

NL. *pedicularis*, f. a genus of the figwort family: *Pedicularis verticillata* (a louse-wort), *Dasystoma pedicularia* (a scrophulariacead).

NL. *scrophularia*, f. a genus of the figwort family, Scrophulariaceae: *Scrophularia aquatica* (water-figwort).

L. *verbascum*, n. mullein: *Verbascum nigrum* (black mullein), *Helicteres verbascifolia* (a screw-tree).

NL. *veronica*, f. a genus of the figwort family: *Veronica spicata* (a speedwell).

file < AS. *fil;* see **scrape**

filipendula, NL. a genus of the rose family; see **rose**

filius, L. son; *filiolus*, dim.; *filia*, daughter; *filiola*, dim.; see **son, daughter**

filix, *-icis*, L. fern; *filicula*, dim.; *filictum*, fernery; see **fern**

fill < AS. *fyllan*, make full; see **abundance**

filter < ML. *filtrum; feltrum*, felt, fulled wool; see **sieve**

filth < AS. *fylth;* see **dirt, dung, pus**

filum, L. thread; see **thread**

fimbria, L. fiber, thread, fringe, shred; *fimbriatus*, fringed; see **fringe**

fimus, L. manure, dung; *fimetum*, dunghill; see **dung**

fin < AS. *finn*.

Gr. *aphareus, -os*, m. ventral fin of the tunny:

Gr. *branchos*, n. fin, gill; *branchion*, n. dim.: *Dibranchus erythrinus* (a fish).

Gr. *notidanos*, with pointed dorsal fin: *Notidanus griseus* (a shark).

L. *pinna* (*penna*), f. fin: Pinnipedia, pinniped, *Aphyocharax rubripinnis* (a fish).

Gr. *pterygion*, fin, dim. of *pteryx*, wing; see *pteron* under **wing**

See: **wing**

finalis, L. pertaining to the end; see *finis* under **end**

finch < AS. *finc*.

L. *acanthis, -idis* (Gr. *-idos*), f. goldfinch: *Acanthis linaria* (redpoll).

L. *aegithus* (Gr. *aigithos*), m. hedge-sparrow: *Aegithocichla terrestris* (a bird).

Gr. *astragalinos*, m. a goldfinch: *Astragalinus columbianus* (a bird).

L. *carduelis, -is*, f. thistlefinch, goldfinch: *Carduelis carduelis* (European goldfinch).

Gr. *chloris, -idos*, f. greenfinch: *Chloris*(or *Fringilla*) *chloris* (greenfinch).

Gr. *chrysometris, -idos*, f. goldfinch: *Chrysometris longirostris* (a goldfinch).

NL. *emberiza*, f. < G. *ammer*, bunting: *Emberiza citrinella* (yellowhammer).

L. *fringilla*, f. chaffinch: *Fringilla caelebs* (chaffinch), Fringillidae.

Gr. *keblepyris*, m. redpoll: *Ceblepyris luctuosus* (a bird).

L. *passer, -eris*, m. sparrow; *passerculus*, m. dim.; *passerinus*, of a sparrow: *Passer domesticus* (house-sparrow), *Passerella iliaca* (fox-sparrow), *Passerina cyanea* (indigo bunting), *Passerculus rostratus* (a sparrow), *Cypripedium passerinum* (a moccasin-flower).

Gr. *phrygilos*, m. a kind of finch: *Phrygilus alaudinus* (a bird).

NL. *pipilo*, m. towhee: *Pipilo erythrophthalmus* (towhee, chewink).

Gr. *poikilis, -idos*, f. a finch:

Gr. *pyrrhoulas*, m. a kind of finch: *Pyrrhula pyrrhula* (bullfinch), *Pyrrhuloxia sinuata* (gray cardinal).

NL. *serinus*, m. a finch: *Serinus flaviventris* (a bird).

L. *spinus* (Gr. *spinos*), m. a finch: *Spinus tristis* (American goldfinch), *Fringilla spinus* (a siskin).

Gr. *spiza*, f. a finch; *orospizos*, m. mountain-finch, brambling: *Spiza americana* (dickcissel), *Melospiza georgiana* (swamp-sparrow), *Spizella pusilla* (field-sparrow), *Thryospiza maritima* (a sparrow).

Gr. *strouthos*, m. sparrow; *stroutharion; strouthion*, n.; *strouthis, -idos*, f. dim.:

Gr. *thlypis, -idos*, f. a kind of finch: *Thlyposis sordida* (a tanager), *Compsothlypis mexicana* (a warbler).

Gr. *thraupis, -idos*, f. a kind of finch: *Thraupis archiepiscopus* (a bird), *Hemithraupis flavicollis* (a bird).

Gr. *zene*, f. a goldfinch:

find < AS. *findan*.

Gr. *Archimedes*, m. famous Greek physicist and inventor, noted particularly for devising the screw to raise water: Archimedean, *Archimedes wortheni* (a fossil bryozoan).

Gr. *heurisko,* find, discover, invent, get; *heurema, -tos,* n.; *heuresis,* f., that which is found, discovery, invention; *heuretor, -os; heuretes,* m. finder, inventor, discoverer; *heuretos,* discovered, found out: eureka!, heuretic, heuristic, *Heuretes picticornis* (a moth).

L. *indicium,* disclosure, discovery; see *dico* under **speak**

Gr. *kyreo,* light upon, find, meet:

L. *nanciscor, nactus; nanctus,* get, obtain, stumble on, find:

L. *reperio, -ertus,* find out, discern, invent; *reperticius,* met or found; *repertor, -is,* m. author, discoverer, inventor: *Decadocrinus repertus* (a fossil crinoid).

See: **learn, see, open, bare**

fine < L. *finis,* end, or *finitus,* finished; see **beauty, good, honor, ornament, pure, little**

finger < AS. *finger.*

Gr. *anticheir, -os,* m. thumb:

Gr. *daktylos,* m. finger, toe; *daktylidion,* n. dim.; *daktylikos,* of the finger; *antidaktylos,* m. thumb; *daktylios,* m. finger-ring: dactyl, dactylic hexameter, dactylogram, dactylology, dactylorhiza, dactylotheca, dactylioglyph, dactyliography, pterodactyl, *Dactylopius tomentosus* (a cochineal-insect), *Dactylotum pictum* (a grasshopper), *Dactylopterus volitans* (flying gurnard), *Dactylis glomerata* (orchard-grass), *Phoenix dactylifera* (date-palm), *Cyclopes didactylus* (two-toed anteater), *Beryx decadactylus* (a fish), *Cynodon dactylon* (Bermuda-grass), *Hemidactylium scutatum* (a salamander), *Lagostomus trichodactylus* (viscacha), *Dactylella minuta* (a fungus).

L. *digitus,* m. finger; *digitulus,* m. dim.; *digitalis,* of the finger; *digitatus,* having fingers: digit, digitate, digitigrade, digitorium, *Digitalis laciniata* (a foxglove), *Digitaria serotina* (a crabgrass), *Carex digitalis* (a sedge), *Adansonia digitata* (baobab-tree), *Calliocrinus digitatus* (a Silurian crinoid), *Epidermophyton interdigitale* (fungus causing athlete's foot).

L. *hallex, -icis,* m. great toe: hallux.

L. *index, -icis,* forefinger, sign, token; see **mark**

Gr. *lichanos,* m. forefinger: *Lichanura roseofusca* (a snake).

L. *phalanx, -angis* (Gr. *-angos*), line, battle-array, bone of the finger or toe; see **line**

L. *pollex, -icis,* m. thumb, great toe; *pollicaris,* of the thumb: pollex, pollical, *Megascelia pollex* (a fly), *Pagurus pollicaris* (a hermit-crab).

See: **hand, foot, nail, projection**

finis, L. end, boundary; *finio, -itus,* determine, define, conclude, complete; see **end**

finish < L. *finio, -itus,* determine, define, end; see **end, polish**

finitimus, L. adjoining, bordering, neighboring; see **near**

finta, Sp. tax; see **pay**

fir < Dan. *fyr.*

L. *abies, -etis,* f. fir; *abiegnus; abietinus,* of fir; *abietarius,* pertaining to fir; m. a joiner: abietin, abietic, abietite, *Abies magnifica* (red fir), *Picea abies* (Norway spruce), *Abietinaria anguina* (a hydroid), *Chermes abietis* (spruce gall-aphid), *Thuidium abietinum* (a fern-moss), *Polyporus abietinus* (a bracket-fungus), *Abiesgraptus[Abietigraptus] multiramosus* (a Silurian graptolite).

Gr. *elate,* f. a kind of fir or spruce; oar; *elatinos,* of fir; *elatine,* f. a plant with firlike leaves: *Elatocladus heterophylla* (a fossil conifer), *Elatine americana* (a waterwort), *Glossostigma elatinoides* (a figwort).

fire < AS. *fyr:* firearm, firebrand, firefly, fireman, fireplace, fireside, fireweed, foxfire, spitfire. Derive: bonfire.

Gr. *aithygma, -tos,* n. spark:

L. *craticula,* gridiron; see **screen**

Gr. *das, dados; dalos,* firebrand, torch; see **light**

Gr. *eschara,* f. hearth, fireplace; *escharion,* n. dim.; see **oven**

L. *fax, facis,* firebrand, torch, flame, light; see **light**

L. *flamma,* f. blaze, fire; *flammula,* f. dim.; *flammeus; flamidus; flamosus,* blazing, fiery, red; *flammo, -atus,* blaze, burn; *inflammo,* set on fire, kindle: flame, flamboyant, flambeau, flamingo, oriflamme, *Euplectes flamiceps* (a bishop-bird., *Ranunculus flammula* (spearwort-buttercup).

L. *focus,* hearth, fireplace, center, central point; *foculus,* dim.; *focarius,* of the hearth; Sp. *fuego,* fire, hearth; see **oven**

L. *fomes,* tinder, punk, kindling wood; see **fungus**

Gr. *Hephaistos,* m. god of fire used in the arts: Hephestian, *Hephaestus tulliensis* (a fish), *Hephaestion ocreatus* (a beetle).

Gr. *hestia,* hearth, fireside; see **house**

L. *ignis,* m. fire; *igniculus,* m. dim., spark; *igneus,* of fire, fiery, ardent; *igniarius,* of fire; *ignicans,* flaming; *ignitus,* glowing; *ignitulus,* dim.; *igninus,* m. fireman;

ignicolor, flame-colored; *ignivagus,* spreading like fire; *ignio, -itus,* set on fire: igneous, ignite, ignis fatuus, ignescent, ignition, *Cuphea ignea* (cigar-flower), *Regulus ignicapillus* (a kinglet), *Fomes igniarius* (a bracket-fungus used as punk).

L. *incendium,* fire, < *incendo, -census,* set on fire; see **burn**

Gr. *phepsalos,* m. spark, ember: *Phepsalostoma electracma* (a moth).

Gr. *phlox, phlogos,* flame; see *phlego* under **burn**

Gr. *Prometheus,* m. fabled Titan who stole fire from heaven and gave it with its derivative arts to man: Promethean, *Callosamia promethea* (a moth).

Gr. *pyr, pyros,* n. fire; *pyridion,* n. dim. spark; *pyrinos; pyrios,* of fire, fiery; *pyrodes; pyropos,* fiery, fiery-red; *pyra,* f. funeral pyre; *pyretos,* m. burning heat, fiery; *pyrsos,* m. torch, fire, beacon; *pyrosis,* f. a burning: *pyrpolema, -tos,* n. signal-fire, beacon; *pyrpolos,* burning; *pyribrotos,* devoured by fire; *pyrikaustos,* inflammatory; *pyriphlogos,* flaming; *pyrsotokos,* fire-producing; *diapyros,* red-hot, fiery, inflamed; *empyrios; empyros,* fiery, burning; *zopyron,* n. spark; *zopyros,* lighting up: pyre, pyrotechnics, pyrite, pyroxene, pyrochlore, pyrolusite, pyromorphite, pyrometer, pyroscope, pyromania, pyrogen, pyridine, pyridoxine, pyrimidine, empyrean, antipyretic, pyrethrum (*Chrysanthemum coccineum*), *Pyrostegia venusta* (a bignoniacead), *Pyracantha crenulata* (a firethorn), *Pyrocephalus rubinus* (a flycatcher), *Pyronema confluens* (a fungus), *Pyrosoma elegans* (a tunicate), *Pyrocystis noctiluca* (a dinoflagellate), *Pyrsonympha vertens* (a protozoan), *Diapyra igniflua* (a moth), *Fontinalis antipyretica* (a moss).

L. *scintilla,* f. spark, glimmer, trace; *scintillula,* f. dim.: scintilla, scintillometer, *Ceratina subscintilla* (a bee).

AS. *spearca,* ember: spark.

Gr. *spinther, -os,* m.; *spintharis, -idos,* f. spark: *Spinther oniscoides* (an annelid), *Spintherobolus papilliferus* (a fish), *Spintharidius cerinus* (a spider).

L. *titio, -onis,* m. firebrand: entice.

L. *torris,* m. firebrand:

L. *Vulcanus* (*Volcanus; Mulciber*), m. Vulcan, god of fire: vulcanize, volcano, volcanic, *Emmydrichthys vulcanus* (a poison-fish), *Fissurella volcano* (a limpet), *Mulciber pullatus* (a beetle).

See: **burn, light, heat, oven, cook, hell, boil, arouse**

firmamentum, L. support, sky; see **heaven**

firmus, L. strong, stout, durable; *firmitudo,* strength; see **strong**

first < AS. *fyrst,* earliest, foremost: first class, firstling, firstly, headfirst.

Gr. *alpha,* first letter of the Greek alphabet; see **letter**

Gr. *arche,* beginning, first cause; *archaios,* from the beginning, see **begin**; *archon,* chief, leader, ruler, see **govern**

L. *causa,* agent, motive, reason; see **begin**

L. *ducator; ductor,* leader, chief; see **lead**

L. *elementum,* first principle; see **thing**

Gr. *etymon,* the true, original, literal root-meaning of a word; see *etymos* under **true**

Gr. *orchamos,* first: *Orchamus zebratus* (a grasshopper).

L. *primus,* first, original, foremost, chief; *primulus,* dim.; *prior; prius,* earlier, former; *primarius; primatus,* of the first rank, chief; *primaevus,* in the first period of life, young; *primigenius,* first of its kind; *primitivus,* early, first; *primordius,* original; *primoris,* first, earliest; *primoticus; primotinus,* happening early; *apprimus,* the very first; *imprimis,* among the first: prime, primer, primeval, primary, primitive, Primates, premier, primordial, prince, prior, *a priori,* priority, priory, primrose (*Primula florida*), *Phenacodus primevus* (a condylarth), *Oryctocephalus primus* (a trilobite), *Elephas primigenius* (woolly mammoth).

L. *principalis,* first, original; see *princeps* under **govern**

L. *priscus,* of former times, ancient; see **old**

Gr. *protos,* first; *proteion,* primary; *protistos,* the very first; protoplasm, prototype, proton, protonema, Protozoa, Protista, protocol, protein, *Protococcus nivalis* (red-snow alga), *Protopterus annectens* (a lungfish).

L. *rudimentum,* first principle; see **begin**

See: **begin, before, dawn, old, top**

fiscalis, L. pertaining to the treasury and money matters; see **money**

fiscus, L. basket, treasury; *fiscina,* small fruit basket; *fiscella; fiscellus,* dim; see **basket**

fish < AS. *fisc:* fishery, fishy, fisherman, fishwife, Fisher, codfish, lungfish, kingfisher.

Son (at dinner with father's boss as guest): "What, beefsteak?"
Mother: "Yes, beefsteak. Why?"
Son: "This morning dad said he would bring a big fish home for dinner!"

L. *acus*, a fish with pointed snout; see **needle**

Gr. *admones*, m. a sea-fish:

Gr. *agnotidion*, n. a kind of fish:

Gr. *aitnaios*, m. a sea-fish:

Gr. *akonias*, m. a kind of fish: *Aconias spinitarsis* (a wasp).

L. *alabeta*, f. (Gr. *alabes*, m.), a Nile fish: *Alabes cuveriae* (a fish).

Gr. *alkestes*, m. a fish:

Gr. *anthias*, m. a sea-fish: *Anthias anthias* (barber-fish), *Hemianthias vivanus* (a fish), *Serranus anthias* (a fish).

Gr. *aphritis*, *-idos*, f. a fish: *Aphritis undulata* (a fish).

Gr. *apopyris*, *-idos*, f. a small fish, fry:

L. *apriculus*, a fish; see *aper* under **hog**

L. *banchus*; *bancus*, m. an unknown fish: *Banchus pictus* (a wasp), *Banchogastra nigra* (a wasp).

Gr. *blepsias*, m. a kind of fish: *Blepsias cirrhosus* (a fish).

Gr. *boupais*, *-aidos*, m. a fish:

L. *box*, *-ocis* (Gr. *boax*, *-akos*; *box*, *-okos*), m. a sea-fish: *Box salpa* (bamboo-fish).

Gr. *brinkos*, m. a sea-fish:

NL. *bryttus*, m. a genus of fishes: *Bryttus*(now *Lepomis*) *punctatus* (a fish), *Chaenobryttus coronarius* (a fish), *Bryttosus kawamebari* (a fish).

L. *callionymus* (Gr. *kallionymos*), m. a fish: *Callionymus dracunculus* (a dragonet).

NL. *caranx*, m. a genus of fishes, < Ab. Am. *caranga:* carangoid, Carangidae, *Caranx equula* (a fish).

L. *cercyrus*, m. a sea-fish: *Cercyra hastata* (a turbellarian).

L. *cetarius*, pertaining to fish; m. fishmonger; *cetarium*, n. fishpond:

Gr. *chellaries*, m. a sea-fish:

Gr. *chlossos*, m. a fish:

L. *citharoedus* (Gr. *kitharoidos*), m. a Red Sea fish: *Citharoedus ornatissimus* (a fish).

L. *corroco*, *-onis*, m. an unknown fish:

Gr. *daskillos*, m. a fish: *Dascyllus*[*Dascillus*] *trimaculatus* (a fish).

Gr. *epanthrakis*, *-idos*, f. small fry:

Gr. *epitragias*, m. a fish:

Gr. *etelis*, m. a fish: *Etelis oculatus* (a fish).

Gr. *galaxias*, m. a fish: *Galaxias punctifer* (a fish).

L. *garus* (Gr. *garos*); *garinos*; *gariskos*, m. a kind of small fish for making sauces:

L. *hippurus* (Gr. *hippouros*), m. a fish: *Coryphaena hippurus* (dorado, a dolphin-like fish).

Gr. *hykes*; *hykos*, m. a fish:

Gr. *ichthys*, *-yos*, m. fish; *ichthydion*, n. dim.; *ichthyeros*, fishy: ichthyology, ichthyosis, ichthyologist, ichthyosaurus, *Ichthyodectes ctenodon* (a Cretaceous fish), *Ichthyornis victor* (a Cretaceous bird), *Ictiobus*[*Ichthyobus*] *bubalus* (white buffalo-fish), *Xanthichthys ringens* (a trigger-fish).
Explain this acrostic: *Iesous CHristos THeou Yios Soter.*

Gr. *iops*, *-opos*, m. a small fish: *Iops sparsus* (a beetle).

Gr. *kapriskos*, a fish; see *kapros* under **hog**

Gr. *kentriskos*, a fish; see *centrum* under **point**

Gr. *kirrhis*, *-idos*, f. a sea-fish:

Gr. *komaris*, *-idos*, f. a fish:

Gr. *koryphaina*, f. a dolphinlike fish: *Coryphaena equisetis* (a fish).

Gr. *larimos* (*larinos*), m. a kind of fish: *Larimus fasciatus* (a banded drum), *Larimichthys rathbunae* (a fish).

Gr. *lebias*, m. a kind of fish: *Lebias perpusillus* (a fish).

Gr. *myllos*, m. a fish:

Gr. *odinolytes*, m. a fish:

Gr. *omis*, f. a fish:

Gr. *pempheris*, *-idos*, f. a fish: *Pempheris schomburghi* (a fish), *Pempherichthys guntheri* (a fish).

L. *piscis*, m. fish; *pisciculus*, m. dim.; *piscarius*, of fish or fishing; *piscator*, *-is*, m. fisherman; *piscatorius*, of fishing; *piscosus*; *pisculentus*, abounding in fish: piscary, piscatorial, Pisces, piscivorous, piscine, pisciculture, grampus, *Lophius piscatorius* (angler), *Agkistrodon*[*Ancistrodon*] *piscivorus* (water-moccasin), *Piscicola geometra* (a leech), *Piscidia piscipula* (fish-poison tree), *Lepidium piscidium* (a pepperweed), *Alampetes pisciformis* (a beetle), *Cymbospondylus piscosus* (a fossil reptile).

L. *pompilus* (Gr. *pompilos*), m. a fish that follows ships; now applied to a genus of wasps: *Pompilus expulsus* (a wasp), *Atopopompilus venans* (a wasp), *Centrolophus pompilus* (now *C. niger*) (a rudder-fish).

Gr. *prepon, -ontos,* m. a sea-fish:
Gr. *probaton,* a sea-fish; see **sheep**
Gr. *psamathis, -idos,* f. a sea-fish:
Gr. *psoropetalos,* m. a fish:
Gr. *psyros,* m. a fish:
Gr. *pytinos,* m. a fish:
Gr. *rhyas, -ados,* m. a school of fish:
Gr. *sakoutos,* m. a fish:
Gr. *salanx, -ngos,* m. a kind of fish: Salangidae, *Salanx hyalocranius* (icefish).
Gr. *seserinos,* m. a sea-fish: *Seserinus xanthurus* (a fish).
Gr. *skiridion,* n. a fish:
Gr. *smerdos,* m. a fish:
Gr. *smylla,* f. a fish:
Gr. *spingos,* m. a fish:
L. *surena,* f. a kind of fish:
Gr. *syax, -akos,* m. a fish; *syakion,* n. dim.: *Syacium ocellatum* (a fish).
Gr. *tathrision,* n. a fish:
Gr. *tilon, -os,* m. a fresh-water fish:
NL. *trachinus,* m. a genus of fishes: *Trachinus vipera* (lesser weever).
Gr. *typhle,* f. a Nile fish: *Tiphle*[*Typhle*] *hexagona* (a fish).

fisherman
Gr. *aspalieus, -eos; aspalieutes,* m. angler:
Gr. *dictybolos,* m. fisherman:
Gr. *gangameutes,* m. oyster-fisher:
Gr. *gripeus; gripon, -os,* m. fisherman, < *gripizo,* fish:
Gr. *halieus; halieutes,* m. fisherman; *halieutikos,* of fishing: *Halieutichthys aculeatus* (a fish), *Halieus forficatus* (a cormorant).
L. *hamiota,* m. angler, < *hamus,* hook:
Gr. *hormiebolos,* m. angler:
Gr. *ichthyoulkos,* m. angler:
Gr. *kalameus, -eos,* m. angler:
L. *piscator, -is,* m. fisherman: piscatorial.
Gr. *sageneutes,* one who fishes with a net; see *sagena* under **net**
Gr. *thalasseus, -eos,* m. fisherman:

fissilis, L. that can be split; see *findo* under **cut**

fissura, L. crack, cleft, chink; see **break**

fist < AS. *fyst;* see **hand**

fistuca, L. beetle, rammer, pile-driver; see **hammer**

fistula, L. pipe; *fistella,* dim.; *fistulosus,* porous, see **pipe**; a kind of ulcer, see **sore**

fit < ME. *fyt,* adapted, suitable, competent, qualified, worthy.
L. *accinctus,* well-girded, armed, equipped, prepared; see *cingo* under **bind**
L. *appositus,* fit, proper, appropriate: apposite, apposition, appositive.
L. *aptatus; aptus,* appropriate, fit, suitable, < *apto, -atus,* fit, adjust, < *apo, aptus,* bind, fasten, join; *adapto, -atus,* fit, adjust: apt, aptitude, attitude, adapt, adaptation, adept, inept, ineptitude.
Gr. *armenos,* fitting, acceptable, suitable, < *ararisko,* join, fit well: *Armenoceras hearsti* (a fossil cephalopod).
Gr. *arsios,* fitting, agreeing, friendly:
Gr. *axios,* worthy, fit; see **worth**
L. *commodus,* fit, proper, suitable: commodius, commodity, incommode, accommodate, *Clerus commodus* (a beetle).
L. *concinnus,* fitting, appropriate; see **beauty**
L. *congruus,* suitable, fitting: congruent, incongruous, *Alampetis incongruus* (a beetle).
L. *consentaneus,* agreeing with, suited to, fit, proper:
L. *conveniens, -entis,* fit, suitable, accordant: convenient, convenience.
L. *decens, -entis,* seemly, proper, fit; *decorum,* n. propriety, fitness; *decorus,* becoming, fitting, proper: decent, decency, indecent, decorous, decorum. See *decoro* under **ornament**
L. *dignus,* worthy, fitting, proper; see **honor**
Gr. *emmetros,* fitting, suitable, proportioned: *Emmetrus auratus* (a beetle).
Gr. *epharmoge,* f. accomodation, agreement, < *epharmozo,* suit, fit, adapt:
Gr. *euthiktos.* to the point, clever; *euthixia,* f. cleverness, tact:
L. *habilis,* apt, fit, suitable: *Dytiscus habilis* (a beetle).
Gr. *harmostos,* adapted, suitable, fit:
Gr. *hetoimos,* ready, feasible: *Hetoemis cinerea* (a beetle).
Gr. *hikanos,* befitting, becoming, competent; see **know**
L. *idoneus,* fit, capable, appropriate, ready: *Arca idonea* (a pelecypod).

Gr. *kairos,* fit, opportune, seasonable; *kairios,* at the right time or place: *Epicaerus imbricatus* (a beetle).

Heb. *kasher,* fit, proper, clean: kosher.

L. *opportunus,* fit, suitable: opportune, opportunity, opportunist, inopportune.

L. *paro, -atus,* prepare, make ready, furnish, equip, order; see **supply**

Gr. *prepon, -ontos,* seemly, fit, proper; see *euprepes* under **beauty**

L. *Procrustes* (Gr. *Prokroustes*), m. a legendary highwayman who conformed his captives to a celebrated bed by stretching or amputation as required: Procrustean, *Procrustes punctatus* (a beetle).

Gr. *skeuastos,* prepared, artificial, < *skeuazo,* prepare, make ready; see *skeuos* under **tool**

Gr. *skopimos,* suitable for a purpose:

Gr. *stello,* arrange, array, prepare, send; see **send**

Gr. *stolizo,* make ready, trim, equip, adorn; see **supply**

L. *tempestivus; temporaneus,* happening at the right time, timely, opportune, seasonable; see *tempus* under **time**

See: **right, supply, harmony, beauty, agreeable, ripe**

fitilla, L. gruel; see **food**

Fitz- < OF. *filz* (F. *fils*), < L. *filius,* son; see *filius* under **son**

five < AS. *fif.*

Gr. *pente,* five; *pentekonta,* fifty; *pentekostos,* fiftieth; *pentas (pempas), -ados,* f. group of five; *pentadikos,* of five; *pemptos,* fifth; *pentameres,* five-parted; *pentaxos,* fivefold; *pentakosioi,* five hundred: pentagon, pentameter, pentamerous, Pentateuch, Pentecost, pentathlon, pentachlorophenol, pinxter, *Pentacrinus decorus* (a Jurassic crinoid), *Pentremites elongatus* (a Mississippian blastoid), *Pentstemon[Pentastemon] hirsutus* (a scrophulariacead), *Acanthopanax pentaphyllum* (an araliacead), *Collettosaurus pentadactylus* (a Permian footprint), *Diapensia lapponica* (Arctic diapensia), *Steganocrinus pentagonus* (a Mississippian crinoid).

L. *quinque; quini,* five; *quintus,* fifth; *quinarius,* consisting of five; *quintanus,* of the fifth; *quintarius,* of five; *quintuplex,* fivefold; *quindecim,* fifteen; *quinquaginta,* fifty; *quinquagesimus,* fiftieth; *quingenti,* five hundred: quinquefid, quiquennial, quinary, quintette, quintuple, quintessence, quintillion, Quinton, Quentin, quincuncial, quinisext, keno, cinquefoil *(Potentilla argentea), Quinqueloculina asperula* (a foraminifer), *Quincula lobata* (a Chinese-lantern), *Quintana atrizona* (a fish), *Parthenocissus quinquefolia* (Virginia-creeper), *Vitex quinata* (a chastetree), *Pisocrinus quinquelobus* (a Silurian crinoid).

fixus, L. fast, immovable; see *figo* under **bind**

flabellum, L. fan; *flabellulum,* dim.; *flabello, -atus,* fan; see **fan**

flabrum, L. blast of wind, breeze; *flabralis; flabilis,* airy, breezy; see **wind**

flaccidus; flaccus, L. weak, drooping; see **weak**

flacourtia, NL. f. a genus of hypericalean plants, < Etienne de Flacourt, a governor of Madagascar: Flacourtiaceae, *Flacourtia indica* (ramontchi).

NL. *azara,* f. a genus of the flacourtia family: *Azara microphylla* (boxleaved azara).

NL. *banara,* f. a genus of the flacourtia family: *Banara dioica* (a shrub).

NL. *idesia,* f. a genus of the flacourtia family: *Idesia polycarpa* (a tree).

NL. *oncoba,* f. a genus of the flacourtia family: *Oncoba echinata* (gorli oncoba).

NL. *pangium,* n. a genus of the flacourtia family: *Pangium edule* (pangium).

NL. *samyda,* f. a genus of the flacourtia family, < Gr. *semyda,* probably a birch: *Samyda mexicana* (a shrub).

flag < uncertain origin.

Sp. *bandera,* f. flag, streamer; *banderilla,* f. dim.: banderilla, banderole.

L. *labarum* (Gr. *labaron*), n. Roman imperial standard:

Gr. *semeion,* signal, standard, ensign, flag, < *sema,* mark, sign, token; see **mark**

L. *signum,* sign, mark, flag; see **mark**

L. *vexillum* (Gr. *bexillon*), n. banner, flag, standard: *Spyridium vexilliferum* (a rhamnacead).

flagellum, L., dim. of *flagrum,* whip, lash; *flagello, -atus,* scourge, whip; see **whip**

flagitium, L. a disgraceful or shameful act; *flagitiosus,* infamous, shameful; see **shame**

flagrans, L. burning, vehement; see *flagro* under **burn**

flake < uncertain origin; see **scale**

flamen, L. priest, see **priest**; blast, wind, see **wind**

flamma, L. blaze, fire; *flammula,* dim.; *flammeus,* fiery, fiery-red; see **fire**

flammeum, L. bridal veil; see **curtain**

flange < OF. *flangier;* see **border**

flank < F. *flanc,* side between hip and ribs; see **side, border**

flap < D. *flappen.*
  L. *auricula (oricilla),* earlap; see *auris* under **ear**
  Gr. *epiptygma,* overflap, operculum; see *ptyx* under **fold**
  L. *lacinia,* lappet, dewlap, fringe; *laciniosus,* full of flaps, cut into thin strips, fringed; see **fringe**
  L. *legula,* f. earlap:
  Gr. *loganion,* n. dewlap of cattle: *Loganiopharynx rarus* (a fossil gastropod).
  L. *palear, -aris,* n. dewlap, wattle:
  See: **tongue, strap, fold, ribbon, border**

flask < F. *flasque,* large bottle; see **bottle**

flat < ON. *flatr,* even, level, smooth; see **level, smooth, broad, low, spread**

flatfish
  L. *citharus* (Gr. *kitharos*), m. a kind of turbot: *Citharus platessoides* (a fish), *Anticitharus debilis* (a fish).
  Gr. *escharos,* m. a fish, probably a sole: *Escharion townleyi* (a fish).
  L. *platessa,* f. flounder: plaice *(Pleuronectes platessa).*
  Gr. *psetta,* f. plaice, sole; *psettarion,* n. dim.: *Psetta maxima* (turbot), *Lophopsetta masculata* (a flounder).
  L. *rhombus,* m. turbot: *Rhombus laevis* (brill).
  L. *solea,* f. a flatfish: sole, *Solea vulgaris* (European sole).

flatilis, L. blown; *flator,* blower; see *flo* under **blow**

flatter < OF. *flater,* smooth, caress.
  L. *adulor, -atus,* fawn, flatter, admire; *adulabilis,* flattering; *adulator, -is,* m.; *adulatrix, -icis,* f. flatter, sycophant: adulation, adulatory.
  Gr. *aikalos,* m. flatterer, < *aikallo,* flatter:
  Gr. *aresko,* placate, flatter; see *arestes* under **agreeable**
  L. *arrisor, -is,* m. one who smiles on another, flatterer, fawner:
  L. *assentor, -atus,* agree constantly, flatter; see **yes**
  L. *blandus,* flattering, kind, mild; see *blanditia* under **agreeable**
  Gr. *bomolochos,* m. lickspittle, toady, beggar, buffoon:
  L. *ceveo,* move the haunches, wag the tail, fawn, flatter; see **shake**
  Gr. *haimylos,* flattering, wheedling, wily:
  Gr. *kolax, -akos,* m.; *kolakis, -idos,* f. flatterer, fawner; *kolakeia,* f. flattery; *kolakikos,* flattering: *Colax apulus* (a butterfly), *Colacium vesiculosum* (a euglenid), *Psomocolax oryzivorus* (a bird).
  Gr. *saino,* fawn upon, beguile; *sainouros,* wagging the tail: *Saenolophus eberthi* (a protozoan), *Saenura flava* (a moth).
  Gr. *sykophantes,* m. slanderer, flatterer: sycophant, *Calosoma sycophanta* (a beetle).
  Gr. *thops, -opos,* m. flatterer, fawner; *thopeutikos,* fawning, flattering, < *thopeuo,* flatter: *Thopomyia dichroa* (a fly).
  L. *vernilis,* cringing, fawning, servile; see *verna* under **servant**
  See: **speak, smooth, agreeable, honor, draw**

flatus, L. breath, breeze, wind; see *flo* under **blow**

flavor < OF. *fleur,* odor; see **taste, smell, spice, sauce**

flavus, L. yellow; *flavidus,* yellowish; see **yellow**

flaw < uncertain origin; see **fault**

flax < AS. *fleax.*
  Gr. *amorgis, -idos,* f. fine flax from Amorgos; *amorgidion,* n. dim.; *amorginos,* of fine flax:
  L. *byssus* (Gr. *byssos*), f. fine flax, cotton, thread; *byssinus,* of fine flax: byssus, byssal, byssinosis, byssiferous, byssolite, *Byssochlamys fulva* (a fungus), *Byssodes politata* (a moth).
  L. *carbasus* (Gr. *karpasos,* < Skt. *karpasa,* cotton), f. a fine Spanish flax, fine linen:
  L. *linum* (Gr. *linon*), n. flax; *lineus,* flaxen: linen, lint, linnet, linseed, linoleum, linsey-woolsey, lingerie, crinoline, gridelin, *Linum flavum* (golden flax), *Linaria vulgaris* (toadflax), *Linanthus dianthiflorus* (fringed pink), *Camelina sativa* (gold-of-pleasure), *Cuscuta epilinum* (a dodder), *Leptilon linifolium* (hairy horseweed), *Byblis liniflora* (a sundewlike plant), *Melampsora lini* (flax-rust).
  L. *stupa; stuppa; stipa* (Gr. *styppe*), f. coarse fiber of flax or hemp, tow, oakum; *stupeus,* of tow: stupe, stupeous, stupose, *Larnacostupa spinosa* (a radiolarian), *Stipa viridula* (green needlegrass).
  See **linen**

flay < AS. *flean,* strip, skin; see **bare, skin**

flea < AS. *flea:* fleabane.

Gr. *psylla,* f.; *psyllos,* m. flea: *Psylla striata* (a jumping plant-louse), *Psyllosphex saltator* (a wasp), *Psyllioides gibbosa* (a beetle), *Xenopsylla cheopis* (a rat-flea), *Odontopsyllus fulgurans* (a siphonapterid).

L. *pulex, -icis,* m. flea; *pulicarius,* of fleas; *pulicosus,* full of fleas: pulicene, pulicifugous, pennyroyal, *Pulex irritans* (human flea), *Pulicaria odora* (a flea-bane), *Daphnia pulex* (a minute crustacean), *Corimelaena pulicaria* (raspberry-bug).

See: **tick, louse**

flebilis, L. tearful, doleful, lamentable; see *fleo* under **weep**

flecto, *-exus,* L. bend; *flexibilis; flexilis,* pliant; *flexura,* a bending; see **bend**

fleminum, L. a kind of boil; see **sore**

flesh < AS. *flaesc.*

L. *aprugma,* flesh of the wild boar; see *aper* under **hog**

NL. *arillus,* a fleshy growth around some seeds: aril, arillode, *Arillaria robusta* (a legume).

Gr. *bora,* meat, flesh; see *bibrosko* under **eat**

L. *bubula,* f. beef:

L. *caro, carnis,* f. flesh, pulp; *carnicula; caruncula,* f. dim.; *carnalis; carnarius; carneus,* of flesh, fleshy; *carnatio, -onis,* f. fleshiness; *carnatus; carnosus, carnu-lentus,* fleshy, corpulent, fat: carnivorous, carnival, carnage, reincarnate, in-carnadine, carnation, carnal, charnel, caruncle, carnassial, *Sarcophaga carnaria* (fleshfly), *Euryscopa carnifex* (a beetle), *Oxalis carnosa*(a wood-soorel), *Coleia carunculata* (wattle-bird), *Lupinus subcarnosus* (Texas lupine), *Trifolium in-carnatum* (crimson clover), *Puffinus carneipes* (a shearwater). Derive: canni-bal, carnallite, carnotite, caribe.

L. *cervina,* f. venison:

Gr. *chros, -otos,* surface of the body, skin, flesh, color of the skin; see *chroma* under **color**

L. *colepium* (Gr. *kolepion*), n. knuckle of beef or pork:

L. *corpus, -oris,* body, flesh, substance; see **body**

L. *ferina,* f. flesh of wild animals, game:

Gr. *histos,* tissue; see **weave**

Gr. *is, inos,* muscle at the back of the neck, sinew, fiber; see **thread**

Gr. *kenebreion,* n. carrion:

Gr. *kole,* f. thigh-bone with its flesh, ham:

Gr. *kreas, kreos,* n. flesh, meat; *kreadion; kreyllion,* n.; *kreiskos,* m. dim.; *kreodes,* fleshy: creatin, creosote, creolin, creophagous, cresol, iodocresol, *Creochiton pudibunda* (a melastomacead), *Puffinus creatopus*[*creopus*] (a shearwater).

L. *musculus,* m. muscle, dim. of *mus,* mouse: muscle, mussel, muscular, muscula-ture, *Neascus musculicola* (a worm), *Unicapsula muscularis* (a protozoan).

Gr. *mys, myos,* m. mouse, muscle; *myon, -os,* m. muscle: myology, myositis, myocardiac, myocyte.

L. *opson,* meat, rich fare; see **food**

L., Gr. *perna,* f. leg of pork, ham:

Gr. *phlogis, -idos,* f. broiled flesh, beefsteak:

L. *porcina,* f. pork:

L. *pulpa,* f. flesh of animals and fruits; *pulposus,* fleshy: pulp, pulpy, pulpify, pulpefaction, pulpalgia, *Collema pulposum* (a lichen).

Gr. *sarx, sarkos,* f. flesh; *sarkion,* n. dim.; *sarkinos; sarkodes,* of flesh, fleshy: sarcocarp, sarcocele, sarcode, sarcolite, sarcosine, sarcology, sarcoma, sarcasm, sarcastic, sarcophagus, anasarcous, *Sarcophilus ursinus* (Tasmanian devil), *Sarco-rhamphus gryphus* (condor), *Sarcodes sanguinea* (snow-plant), *Sarcobatus baileyi* (a greasewood), *Sarcophaga chrysostoma* (a fleshfly), *Callionyma sarcodes* (a moth), *Polysarcodes maestus* (a fly), *Gymnosarca bathybia* (a coral).

Gr. *soma,* body, dead body; see **body**

L. *succidia,* f. flitch or side of bacon:

L. *suilla,* f. pork:

L. *taniaca,* f. long strip of pork:

Gr. *tarichos,* m. mummy, pickled meat; *tarichion,* n. dim.; *taricheros,* pertaining to pickled food:

Gr. *temachos,* cut. slice (of meat); see *temno* under **cut**

L. *torosus,* muscular, fleshy, lusty; see *torus* under **projection**

L. *viscus, -eris,* flesh or organ inside the body; see *viscera* under **intestine**

L. *vitellina; vitulina,* f. calf's flesh: veal, *Nesolechia vitellinaria* (a lichen).

See: **body, skin**

fletus, L. wept, lamented; see *fleo* under **weep**

flexible < L. *flexibilis,* bendable; *flexuosus,* with many bends, winding; see *flecto* under **bend, willow**

flictus, L. struck, beaten; see *fligo* under **strike**

flint < AS. *flint;* see **stone**

float < AS. *flotian;* see **swim**

floccus, L. lock of wool; *floccosus,* full of locks; see **wool**

flock < AS. *flocc;* see **herd**

flood < AS. *flod;* see **flow, wash**

floor < AS. *flor;* see **base, level**

Flora, L. goddess of flowers; *floridus,* full of flowers; see *flos* under **flower**

flosculus, L. dim. of *flos, floris,* flower; see **flower**

flounder < OF. *flondre;* see **flatfish**

flour < OF. *flor,* < L. *flos, floris,* flower.
　Gr. *aleuron,* n. wheat flour, meal, < *aleo,* grind; *alema; alesma, -tos; aleton,* n. flour, meal; *aleurites,* of flour; *aleurodes,* like flour: aleurone, *Aleurites fordi* (tung-oil tree), *Aleurodes pruinosa* (powderwing), *Trialeurodes vaporariorum* (a whitefly).
　Gr. *alphiton,* n. barley meal; *alphiteros,* of meal: *Alphitobius punctatus* (a beetle), *Alphitodes indutus* (a beetle), *Alphitophagus bifasciatus* (a beetle), *Microsphaera alphitoides* (a mildew).
　L. *amylum* (Gr. *amylon*), n. fine meal, starch: amylolytic, amylodextrin, amylase, amyloid, amylaceous, amyl acetate, carbamyl, *Bacillus amylobacter* (a bacterium).
　Gr. *athere* (*athare*), f. groats, meal, gruel: atheroma, atheromatous, atherosclerosis.
　Gr. *autopyros,* m. unbolted wheat flour:
　L. *cibarium,* n. coarse meal, shorts:
　L. *farina,* f. flour, meal, starch; *farinula,* f. dim.; *farinaceus; farinarius; farinosus,* of meal; *farinulentus,* mealy: farina, farinaceous, farinose, farinulent, *Aleurobius farinae* (a mite), *Salvia farinacea* (a sage), *Combretum farinosum* (mealy combretum), *Sitophagus farinarius* (a beetle).
　Gr. *gyris, -eos,* f. the finest meal; *gyrites,* m. bread made of this meal:
　Gr. *krimnon,* n. coarse meal:
　L. *lomentum,* bean meal; see **bag**
　AS. *melu,* coarsely ground grain: meal, mealy, mealywing, oatmeal.
　Gr. *pale; paipale; paspale,* f.; *palema, -tos,* n. finest meal; *palemation,* n. dim.: *Paepalanthus falcifolius* (a pipewort), *Paepalophorus mucoreus* (a beetle).
　L. *polenta,* f. pearl-barley, barleymeal:
　L. *pollen, -inis,* n. fine flour, mill-dust, dust; *pollinarius,* of flour: pollen, pollinize, pollinium, polliniferous, pollinate, cross-pollination, pollinator, pollinosis, *Pollinula hispida* (a sponge), *Pollenia floralis* (a fly).
　Gr. *semidalis,* f. finest wheat flour:
　L. *siligo, -inis,* white wheat, fine wheat flour; see **grain**
　L. *simila; similago, -inis,* f. finest wheat flour:
　Gr. *stais, staitos,* flour of spelt made into dough; see **mix**
　See: **dust, grind**

flow < AS. *flowan:* flowage, inflow, flood.
　Gr. *apektos,* not solidifiable:
　Gr. *blyzo,* bubble or gush forth:
　Gr. *chytlon,* anything that can be poured, fluid, liquid: *chytos,* fluid, liquid; see *cheo* under **pour**
　L. *conflo, -atus,* melt, fuse; *conflatilis,* cast, molten; *conflatile,* n. molten image:
　L. *diluvium,* inundation, flood, deluge; *diluvialis,* of a flood; see *luo* under **wash**
　Gr. *epiklysis,* flood; see *klyzo* under **wash**
　L. *fluo, fluxus,* flow; *fluito, -atus,* float, undulate; *fluctuo, -atus,* flow, waver; *fluidus; fluvidus; fluxilis; fluxuosus,* flowing, liquid; *fluctus, -us,* m. wave, flood, surge: *flucticulus,* m. dim.; *fluvius,* m. river, stream; *affluo,* flow toward; *effluo,* flow out; *confluus,* flowing together; *refluamen, -inis,* n. runoff: fluid, fluent, fluviatile, fluxions, influence, flux, affluent, effluent, effluvium, confluence, Coblentz, influenza, influx, fluctuate, fluorite, fluorine, fluorid, fluosilicate, fluorescence, fluoroscope, superfluous, *Ranunculus fluitans* (eelware), *Difflugia urceolata* (a protozoan), *Dipterocarpus vernicifluus* (balao).　Derive: float, flotsam.
　L. *fusilis,* fluid, molten, pourable; *fusio,* a melting; see *fundo* under **pour**
　L. *inundo, -atus,* overflow, flood; see *unda* under **wave**
　L. *liqueo,* be fluid, melt; *liquidus,* fluid; *liquidum,* n.; *liquor, -is,* m. fluid; *deliquesco,* melt, dissolve: liquid, liquor, liqueur, liquidate, liquefaction, deliquesce, colliquative, *Liquidambar orientalis* (oriental sweetgum).

L. *mano, -atus,* flow, trickle; *manalis,* flowing; *emano,* flow out: emanate, emanation, manation.

Gr. *naros (neros),* flowing, liquid, fluid, < *nao,* flow: *Nerophilus oregonensis* (a trichopterid), aneroid.

Gr. *pelanos,* m. any liquid of thick consistency:

L. *regelo, -atus,* thaw: regelation.

Gr. *rheo,* flow; *rheos,* n.; *rhoë,* f. flow, current, stream; *rhoedion,* n. dim.; *rheuma, -tos,* n. discharge, flux; *rheumation,* n. dim.; *rhysis, -idos,* f. flow, issue; *rheustikos; rheustos; rhyodes; rhytos,* flowing, fluid, liquid; *rheumatismos,* m. rheum, catarrh; *rhyas, -ados,* fluid, flaccid; *aeirrhytos,* ever-flowing; *cheimarrhos,* swollen by melting snow and rain; *epirrhytos,* flowing in, abundant; *halirrhytos,* washed by the sea; *syrrheusis,* f. a flowing together, confluence; *syrrhoos,* confluent: rheophilous, rheostat, rheumatism, catarrh, diarrhea, gonorrhea, rhyolite, hemorrhoid, rhythm, *Callirhoë digitata* (a poppy-mallow), *Xanthorrhoea arborea* (a liliacead), *Syrrheuma cretata* (a moth), *Oceanodroma leucorhoa* (a petrel).

L. *sudatilis,* flowing like sweat; see *sudo* under **drop**

Gr. *teko,* melt, dissolve, waste away; *tektos,* soluble; *tektikos,* pertaining to fluidity; *takeros,* melting, tender, dissolving; *tekedon, -os,* f. a melting away, wasting, decline; *tekedonikos,* wasting away; *tekedanos,* fusible, melting, molten; *texis,* f. a melting, wasting; *anatexis; syntexis,* f. a melting or fusing together; *teximeles,* wasting of the limbs: *Syntegmodus altus* (a Cretaceous fish), *Tacerus upoluensis* (a beetle).

See: **run, drop, stream, pour, free, -rhage**

flower < OF. *flor,* < L. *flos, floris,* m. blossom; *flosculus,* m. dim.; *Flora,* f. goddess of flowers; *floralis,* of Flora or flowers; *floreus,* of flowers; *florifer; floriger; florigenus; floriparus,* bearing flowers; *floribundus; floridus; florosus; florulentus,* abounding in flowers, flowering profusely; *floridulus,* dim. flowering somewhat; *florilegus,* flower-gathering; *infloresco,* begin to blossom: flora, florist, Florence, Florida, floret, inflorescence, florin, florilegium, uniflorous, cauliflorous, ramiflorous, floweret, deflower, Mayflower, flour, fleur de lis (*Iris florentina*), *Clematis florida* (a clematis), *Passiflora mollissima* (a passion-flower), *Cypripedium parviflorus* (a moccasin-flower), *Coreopsis grandiflora* (a tickseed), *Calceolaria biflora* (a slipper-flower), *Malus floribunda* (flowering crabapple), *Florisuga mellivora* (a hummingbird), *Lagenula flosculosa* (a foraminifer), *Floscularia campanulata* (a rotifer), *Tomoplagia deflorata* (a fly), *Cercidium floridum* (blue paloverde), *Leitneria floridana* (corkwood), *Fragaria semperflorens* (alpine strawberry), *Anthodiscus floreatus* (a diatom), *Cetonia floricola* (a beetle).

Gr. *anthos, -eos,* n.; *anthe,* f.; *anthemion; anthemon,* n.; *anthemis, -idos,* f. flower; *anthion; anthyllion,* n. dim. floret; *anthikos; anthinos,* of flowers, with flowers; *antheios; anthemodes; antheros; anthetikos; anthodes,* flowery, blossoming; *anthesis,* f. the full flower, the blossoming; *anthokomos,* bedecked with flowers; *anthonomos,* feeding on flowers; *anthologos,* gathering flowers; *apanthisma, -tos,* n. a plucked or culled flower; *dianthes,* double-flowering; *perianthes,* with flowers all around: anthophilous, anthology, anthodium, anthocyanin, anthoxanthin, antholite, anther, anthesis, Anthesterion, chalcanthite, chrysanthemum, perianth, *Anthoxanthum odoratum* (sweetgrass), *Anthonomus grandis* (cotton-boll weevil), *Anthoceros punctatus* (a liverwort), *Anthoscopus minutus* (a bird), *Anthemis palestina* (a camomile), *Anthidium oblongatum* (a bee), *Anthericum liliago* (St. Bernard lily), *Anthidiellum rhodesianum* (a bee), *Anthurium scherzerianum* (tailflower), *Dianthus barbatus* (sweet-william), *Chrysanthemum segetum* (corn-marigold), *Calyptranthes syzygium* (a lidflower), *Mesembryanthemum spectabile* (a fig-marigold), *Helianthus tuberosus* (Jerusalem artichoke), *Osmanthus serrulatus* (an osmanthus), *Schizanthus pinnatus* (butterfly-flower), *Phyllanthus speciosus* (a euphorbiacead), *Emmenanthe parviflora* (sagebrush-bells), *Spiranthes gracilis* (an orchid), *Crocus chrysanthus* (a spring crocus), *Gymmanthes lucida* (oyster-wood), *Rhus rhodanthema* (a sumac).

AS. *blowan,* be in flower: blow, bloom, blossom.

Gr. *Chloris, -idos,* f. goddess of flowers: *Chloris verticillata* (windmill-grass), *Trichloris pluriflora* (a grass), *Leptochloa chloridiformis* (Argentine sprangletop).

See: **ripe, fruit, seed, good, beauty**

fluctus, L. wave, flood; *flucticulus,* dim.; see *fluo* under **flow**

fluidus, L. flowing, liquid; see *fluo* under **flow**

fluito, -atus, L. float, undulate; see *fluo* under **flow**

flumen, -inis, L. stream; *flumicellum,* dim.; *flumineus,* of a stream; see **stream**

fluo- < fluorine, < L. *fluor,* flow; see **flow,** ELEMENTS under **thing**

flustrum, L. calm, quietness; see **rest**

flute < OF. *flaute;* see **pipe**

fluvius, L. river; *fluvialis; fluviatilis,* of a river; see *flumen* under **stream**

fluxus, L. flowing, loose, slack; *fluxilis,* fluid; see *fluo* under **flow**

fly, an insect, < AS. *fluge; fleoge.*
L. *alucita,* f. gnat: *Alucita acanthodactyla* (a gnat).
NL. *anthrax,* a genus of flies; see **coal**
L. *asilus,* m. gadfly: *Asilus sericeus* (a robberfly).
L. *bibio, -onis,* m. a kind of fly: *Bibio albipennis* (a March fly).
L. *culex, -icis,* m. gnat, midge, mosquito; *culicellus; culiculus,* m. dim.: culicide, Culicidae, *Culex pipiens* (a mosquito), *Culicoides guttipennis* (no-see-um), *Sesia culiciformis* (a clearwing-moth).
Gr. *empis, -idos,* m. mosquito, gnat: *Empis villosa* (a fly), *Empicoris*[*Empidocoris*] *maculatus* (a bug), *Empidonax griseus* (a flycatcher).
L. *eristalis,* f. an unknown precious stone; now applied to a genus of flies: *Eristalis tenax* (dronefly).
Gr. *konops, -opos,* m. gnat; *konopion,* n. dim.: conopid, *Conops tibialis* (a gnat), *Conophaga melanops* (an ant-thrush), *Leptoconops carteri* (a punkie).
L. *mulio, -onis,* m. a kind of gnat: *Mulio americanus* (a fly).
L. *musca,* f. fly; *muscula,* f. dim.; *muscarius,* of flies; *muscarium,* n. fly-trap, fly-brush: *Musca domestica* (housefly), mosquito, musket, Muscidae, muscarine, *Muscivora forficata* (scissor-tailed flycatcher), *Muscari*[*Muscarium*] *botryoides* (grape-hyacinth), *Muscicapa hypoleuca* (pied flycatcher), *Empusa muscae* (a fungus), *Amianthemum muscaetoxicum* (fly-poison), *Amanita muscaria* (fly-agaric), *Ophrys muscifera* (fly-orchid), *Muscina assimilis* (a fly).
Gr. *myia,* f. fly; *myiodes,* like flies: myiasis, stegomyia, *Myiadestes unicolor* (a thrush), *Myioceyx leconti* (a kingfisher), *Myiozetetes similis* (a flycatcher), *Myiarchus crinitus* (a flycatcher), *Nannoceryx myiella* (a moth), *Ornithomyia chiliensis* (a fly).
Gr. *myops, -opos,* m. gadfly, horsefly: *Myopa seminuda* (a fly).
NL. *nerius,* m. a genus of flies < *Nerius,* a Roman clan: *Nerius solitarius* (a fly).
L. *oestrus* (Gr. *oistros*), m. gadfly, horsefly; sting, frenzy, madness, sexual heat, rut, desire; *oistregma, -tos,* n. smart of a gadfly's sting; *oistrodes,* frantic, raging; *oistromanes,* mad from the gadfly's sting; *oistrao,* sting to madness, make frantic: oestrid, Oestridae, oestriasis, oestrual, oestruation, estrogen, estrone, estriol, estradiol, *Oestrus ovis* (sheep-botfly), *Cymothoa oestrum* (a fish-louse).
NL. *pipunculus,* m. a genus of flies: *Pipunculus cingulatus* (a fly), *Pipunculopsis bivittata* (a fly), *Pipunculosyrphus globiceps* (a fly).
Gr. *syrphos,* m. a kind of fly: *Syrphus politus* (a fly), *Syrphoctonus biguttatus* (a wasp).
L. *tabanus,* m. gadfly, horsefly: *Tabanus atratus* (horsefly), *Sesia tabaniformis* (a clearwing-moth), *Leucotabanus pallidus* (a fly).
NL. *tachina,* f. a kind of fly, < Gr. *tachinos,* swift: *Tachina robusta* (a fly) *Tachinosoma corpulentum* (a fly).
L. *tipula,* f. water-spider; now applied to the cranefly: *Tipula bella* (a cranefly), *Tipularia unifolia* (cranefly-orchid), *Bittacus topularis* (a scorpion-fly), *Sesia tipuliformis* (a clearwing-moth).
NL. *volucella,* f. a genus of flies: *Volucella lunulifera* (a fly).
L. *zinzala,* f. a kind of gnat:

fly, move more or less swiftly, < AS. *fleogan.*
Gr. *apeoros,* hanging on high, soaring:
Gr. *apten, -os,* unable to fly, unfledged, wingless: *Aptenodytes patagonica* (king-penguin), *Aptenopedes rufovittata* (a grasshopper).
L. *fugio* (Gr. *pheugo*), fly from, flee; see *fuga* under **depart**
Gr. *petomai,* fly; *ptesis,* f. flight; *ptesimos,* able to fly, winged; *potetos,* flying, winged; *eupetes,* flying well; *okypetes,* swift-flying; *hiptamai,* fly: *Eupetes caerulescens* (a bird), *Hiptage madablota* (a malpighiacead).
L. *volo, -atus,* fly; *volatilis,* flying, fleeting; *volito, -atus,* fly to and fro, flit, flutter; *volucer, -cris, -cre,* flying, winged, swift; *involucer,* unable to fly; Sp. *volador,* flyer; *subvolo,* soar: volant, volatile, volador, volley, volplane, *muscae volitantes* (spots before the eyes), *Glaucomys volans* (flying-squirrel), *Exocoetus volitans* (a flying-fish), *Prionotus evolans* (a sea-robin), *Dictyna volucripes* (a spider), *Ardetta involucris* (a heron).
See: **depart, from, run, shun, swift**

foam < AS. *fam,* froth.
Gr. *achne,* any light substance, foam, froth, chaff, down; see **light**
Gr. *aphros,* m. foam, froth; *aphrioeis; aphrodes,* foamy; *aphrestes,* m. foamer; *aphrismos,* m. a foaming, < *aphrizo,* foam; *epaphros,* covered with foam or froth: aphrite, aphrizite, aphrolite, Aphrodite, *Aphrophora quadrangularis* (a

froghopper), *Aphriza virgata* (surf-bird), *Aphriosphyria communis* (a fly), *Aphropsylla conversa* (a flea), *Aphrodes bicincta* (a bug), *Aphrimyobia simillima* (a fly).

Gr. *chnous* (*chnoos*), m. any light, porous substance, foam, fine down, bloom; *chnodes*, soft, porous, downy; L. *calamochnus*, m. a kind of sea-foam: *Sporochnus pedunculatus* (an alga).

L. *spuma*, f. foam, froth; *spumeus; spumidus; spumosus*, foamy, frothy; *spumo, -atus*, foam, froth: spume, spumescent, *Philaenus spumarius* (a spittle-insect).

focus, L. hearth, fireplace, central point; *foculus*, dim.; see **oven**

fodina, L. pit, mine, < *fodio, fossus*, dig; see **mine, dig**

foedus, L. foul, filthy, detestable; *foeditas*, foulness; see **dirt**

fog, probably < Dan. *fog*, spray, storm; see **cloud**

fold < AS. *fealdan:* twofold, fourfold, manifold, folder, enfold.   Derive: blind-fold.

Gr. *amphelisso*, wrap, fold; *ampheliktos*, coiled round: *Amphelissa isisensis* (a fossil gastropod), *Amphelictus melas* (a beetle).

L. *capero, -atus*, wrinkle: caperin, *Parmelia caperata* (a lichen).

Gr. *diklis, -idos*, double-folding: dicliditis, *Dicliptera brachiata* (an acanthacead).

L. *-farius*, suffix meaning multiplication in numbers or parts: bifarious, multifarious, omnifarious.

Gr. *kolpotos*, formed into folds; *kolpodes*, folded, sinuous; see *kolpos* under **womb**

L. *pannuceus*, wrinkled, shriveled:

Gr. *pharkis, -idos*, f. wrinkle; *pharkidodes*, wrinkled: *Pharkidinotus[Pharcidonotus] tricarinatus* (a fossil gastropod).

Gr. *phrix, -ikos*, ruffling of a smooth surface, ripple; see **rough**

L. *plico, -atus; -itus*. fold; *plicatilis*, foldable, pliable; *complico, -atus*, fold together; *complex, -icis*, folded up, connected, intricate: plicate, ply, pliable, plait, pleat, plier, plight, application, accomplice, complex, complicated, complicity, decemplex, deploy, display, duplex, duplicate, employment, exploit, explicable, explicit, implicit, inexplicable, implication, manyplies, multiple, perplexity, plication, quadruple, quadruplicate, quadriplicate, reply, simple, simpleton, splay, supple, suppliant, solar plexus, triplicate, *Dichocrinus plicatus* (a Mississippian crinoid), *Dictyophora duplicata* (a stinkhorn-fungus), *Lamium amplexicaule* (henbit), *Collema plicatile* (a lichen).

Gr. *ptyx, -ychos; ptyche*, f. fold, leaf, layer; *ptychion; ptyktion*, n. dim.; *ptyxis*, f. a folding, fold; *ptygma, -tos*, n. anything folded; *ptygmation*, n. dim.; *ptychodes*, in folds; *ptychios; ptyktos*, folded, < *ptysso*, fold; *epitychma, -tos*, n.; *epityche*, f. overfold, flap, operculum: ptychodont, ptyxis, ptygmatic, aptychus, *Ptychopyxis costata* (a euphorbiacead), *Ptychozoon homalocephalum* (flying-gecko), *Ptychatractus ligatus* (a gastropod), *Holoptychius nobilissimus* (a Devonian fish), *Leptoptygma virgatum* (a fossil gastropod), *Metazaptyx hachijoensis* (a gastropod), *Sternoptyx diaphana* (a fish), *Actinoptychus undulatus* (a diatom).

Gr. *rhakodes*, ragged, wrinkled (skin): *Neorhacodes enslini* (a wasp).

Gr. *rhiknos*, shriveled, withered, wrinkled; *rhiknosis*, f. a shriveling: *Rhicnoessa cinerea* (a fly).

Gr. *rhytis, -idos*, f. fold, pucker, wrinkle; *rhysema, -tos*, n. wrinkle; *rhysos* (*rhyssos*); *rhytidodes*, wrinkled; rhytidosis, rhytidectomy, rhyssoid, *Rhyssomatus impolitus* (a beetle), *Callirhytis yosemite* (oak gall-wasp), *Megarhyssa superba* (an ichneumon-wasp), *Viburnum rhytidophyllum* (leather-leaved viburnum), *Rhytidoglymma aenescens* (a beetle), *Rutiodon[Rhytidodon] manhattanensis* (a Triassic reptile).

AS. *rifel*, wrinkle, forrow: rivel.

L. *ruga*, f. crease, wrinkle; *rugula*, f. dim.; *rugosus; ruginosus*, wrinkled; *rugo, -atus*, crease, wrinkle; *corrugis; corrugatus*, wrinkled, ridged; *irrugo, -atus*, wrinkle: rugate, rugose, rugulose, corrugated, *Alnus rugosa* (an alder), *Cenchridium rugulosum* (a foraminifer), *Ligyrus rugiceps* (sugarcane-beetle), *Pseudosiopelus rugulifer* (a beetle), *Trypoxylon rugifrons* (a wasp), *Nephalius rugicollis* (a beetle), *Gryphaea corrugata* (a fossil pelecypod).

L. *sinus*, bend, curve, fold; see **hollow**

Gr. *stolidoma*, fold of a robe; *stolidotos*, in folds, folded; see *stola* under **garment**

L. *valva*, leaf of a folding door; see **door**

L. *vietus*, shriveled, shrunken, wrinkled; see **weave**

See: **furrow, ridge, wave, rough, flap, form, bend, embrace, weave, complex**

folium, L. leaf; *foliolum*, dim.; *foliaceus; foliatus; foliosus*, leafy; see **leaf**

folk < AS. *folc*, people; see **people**

follis, L. bellows, windbag; *folliculus*, dim.; see **bag**

follow < AS. *folgian.*

Gr. *akolouthos,* following after, attending on, succeeding: anacoluthon, *Acolutha pictaria* (a moth).

Gr. *alytos,* unbroken, continuous; see **one**

L. *comitor, -atus,* accompany, attend; *comitabilis,* accompanying; *concomitor, -atus,* accompany: concomitant, *Pelurga comitata* (a butterfly).

Gr. *diadochos,* succeeding, following; *diadexis,* f. succession: *Diadochus antigai* (a beetle), *Diadexia parodes* (a moth).

Gr. *dioko,* pursue, follow; *dioktos,* pursued; *diokter, -os; dioktes,* pursuer; see **hunt**

L. *discipulus,* m. learner, follower: disciple.

Gr. *eirmos,* train, series; see **line**

Gr. *epallelos,* one after another, continuous; see **line**

L. *ergo,* consequently, therefore: *Post hoc, ergo propter hoc.*

Gr. *hepomai,* follow:

L. *lixa,* m. sutler, camp-follower:

L. *morigerus,* complying, obedient: morigerous, *Attus morigerus* (a spider).

Gr. *opados,* accompanying, attending; see **servant**

Gr. *peisa,* f. obedience:

Gr. *scheros,* in a line, successive; see **line**

L. *sequor, secutus,* follow; *sequax, -acis,* following, pursuing; m. attendant; *sequela,* f. that which follows; *sector, -atus,* follow eagerly, pursue; *sectator, -is,* m.; *sectatrix, -icis,* f.; *secutor, -is,* m.; *secutrix, -icis,* f. follower; *secundus,* following, ranking below, unilateral; *consector, -atus,* follow, pursue eagerly, strive after; *consectator: consectatrix,* eager follower; *consectarius,* that follows logically; *consectaneus,* following eagerly; *consequus,* following; *exsecutus (executus),* followed to the end; *obsequialis,* complying, yielding; *persequor, -secutus,* follow pertinaciously, chase: sequence, sequel, sequela, sue, suit, lawsuit, suite, suitor, solisequious, second, secund, subsequently, sequester, secondary, consequential, consectary, consecutive, exequy, ensuing, executor, extrinsic, persecute, electrocute, prosecution, pursue, *Micropterna sequax* (a caddisfly), *Kyphosus* [*Cyphosus*] *sectatrix* (a chopa), *Taeniothrips inconsequens* (pear-thrips), *Pyrola secunda* (sidebells), *Combretum secundum* (Trinidad combretum), *Sophora secundiflora* (coral-bean), *Osteoborus secundus* (a fossil canid).

L. *series,* row, succession, train; see **line**

L. *succedaneus,* following after; m. substitute; *successor, -is,* m. follower, < *succedo, -essus,* follow after: succedaneous, successor, succeed, succession, *Rhus succedanea* (wax-tree).

See: **line, after, companion, hunt**

fomentum, L. warm application, poultice; see *foveo* under **heat**

fomes, L. tinder, punk, kindling wood; a genus of fungi; see **fungus**

fontinalis, L. of a spring, < *fons, fontis,* spring; *fonticulus,* dim.; see **spring**

food < AS. *foda; fodor,* fodder.

L. *alimentum,* food; see *alo* under **eat**

Gr. *ambrosia,* food of the gods that gave immortality; see **always**

L. *bacalusia,* f. a kind of sweetmeat, confection:

L. *bellarium,* n. dessert (fruit, nuts, candy, wine, etc.):

Gr. *broma; brosis,* food; see *bibrosko* under **eat**

L. *buccellatum,* n. soldier's biscuit:

L. *caballatio, -onis,* f. fooder for a horse:

L. *capetum* (Gr. *kapeton*), n. fodder for cattle:

L. *catillamen,* sweetmeat; see *catillo* under **eat**

Gr. *chilos,* m. green fodder:

L. *cibus,* m. food; *cibarius,* pertaining to food; *cibatio, -onis,* f. meal, repast; *cibo, -atus,* feed: cibarian, cibation, cibarium, *Cantharella cibarius* (chanterel).

L. *collyra* (Gr. *kollyra,* f. a kind of pastry; *kollyris, -idos,* f. dim.), f. macaroni, vermicelli; *collyricus,* of macaroni: *Collyris postica* (a beetle).

L. *confectio, -onis,* something prepared; a sweetmeat; see *facio* under **make**

L. *cuppedo, -inis,* f.; *cuppedium,* n. titbit, delicacy; *cuppes, -edis,* fond of dainties: *Ochotona cuppes* (tawny pika).

Gr. *edesma,* food; see *edo* under **eat**

L. *esca,* f. food, victuals, bait; *escalis; escarius,* of food; *escatilis; esculentus,* edible; *inesco, -atus,* fill with food, lure with bait, entice: esculent, *Rana esculenta* (a frog), *Cyperus esculentus* (chufa).

L. *farrago,* mixed fodder for cattle, mash; see **mix**

L. *fitilla,* f. gruel:

Gr. *grastis (krastis),* green fodder, grass; see *grao* under **eat**

L. *gustulum,* small dish of food, relish; see *gustus* under **taste**

Gr. *hypotrimma, -tos,* n. a sour dish of grated or rubbed ingredients:

L. *insicium*, n. dish of minced meat:
Gr. *kolon*, n. food, meat, fodder:
Gr. *manna*, morsel, grain, gum of a tree; see **resin**
L. *mattea* (Gr. *mattye*), f. dainty dish, delicacy; *matteola*, f. dim.:
Gr. *melka*, f. a cold preparation from milk:
L. *moretum*, n. a dish composed of garlic, rue, vinegar, and oil:
Gr. *nogalon*, n. sweetmeat, dessert:
L. *nutrimentum*, nourishment; see *nutrio* under **eat**
L. *offa*, bit, morsel, mouthful; see **little**
Gr. *opson*, n. meat, rich fare; *opsaridion; opsarion*, n. dim.; *opsonion*, n. provisions; *paropsis, -idos,* f. dessert; *paropsidion,* n. dim.: opsonin, opsonic, opsomania, opsophagist, *Paropsivora grisea* (a fly).
Gr. *ompne*, f. food:
L. *pabulum*, n. food, fodder; *pabularis*, of fodder; *pabulosus*, abounding in fodder; *pabulor, -atus,* forage: pabulum, pabulation, *Prangos pabularia* (an umbellifer).
Gr. *phorbe*, food, pasture; see *pherbo* under **eat**
Gr. *polphos*, m. a kind of starchy food:
L. *puls, pultis,* f. (Gr. *poltos*, m.), porridge, pottage; *pulticula*, f. dim.: pulse.
L. *scitamentum*, n. dainty food:
Gr. *sitos*, m. food, grain; *sitarion; sition,* n. dim.; *sitikos,* of food; *sitistes,* m. feeder; *sitesis,* f. eating; *sitistos,* fed up, fattened; *micrositos,* eating little: sitology, sitophobia, sitotoxin, parasite, asitosis, *Sitophagus pallidus* (a beetle), *Sitanion hystrix* (squirreltail-grass), *Sitophilus granarius* (a beetle).
Gr. *tragema, -tos,* n. sweetmeat:
Gr. *trogalion*, sweetmeat, dessert; see *trogo* under **bite**
Gr. *trophe; threpsis,* nourishment; see *trepho* under **eat**
L. *turunda*, f. ball of paste for fattening geese:
L. *victus, -us,* m. food, diet; *victualis,* of food; *Victa*, f. goddess of food: victuals.
See: **eat, flesh, plant, sauce, soup**

fool < OF. *fol,* < L. *follis,* bellows, windbag.
Gr. *abelteros*, asinine, silly, stupid, foolish; *abelteria*, f. silliness, stupidity: *Abelterus incarnatus* (a bug).
L. *absurdus*, that which offends the ear, harsh, foolish, silly: absurd, absurdity.
L. *alogus* (Gr. *alogos*), speechless, irrational, absurd; *alogistos,* silly; *alogia,* f. absurdity: *Alogopteron caribbeum* (a grasshopper), *Alogocarabus caerulans* (a beetle), alogia, alogism.
L. *amens*, senseless, mad, foolish; see *demens* under **mad**
Gr. *analistos*, unsalted, silly:
Gr. *anoos; anous; anoemon, -os; anoetos,* without understanding, senseless, silly; m. fool: *Anoüs stolidus* (noddy, a tern).
Gr. *aphrades*, senseless, reckless; *aphradia,* f. folly:
Gr. *aphron, -os,* crazy, silly, foolish, senseless; *aphrosyne,* f. folly: aphronia, *Aphronorus fraudator* (a fossil mammal), *Aphronastes subfasciatus* (a beetle).
L. *asininus*, having the qualities of an ass, stupid, obstinate; *asinalis,* stupid; see *asinus* under **horse**
Gr. *asophos; kenosophos,* frivolous, foolish, silly:
Gr. *atopos*, out of place, odd, strange, absurd, unnatural; *atopia,* absurdity; see **strange**
L. *babulus*, babbler, fool; see **speak**
L. *baburrus*, foolish, silly:
L. *blatero (balatro),* babbler, buffoon; see **speak**
L. *bliteus*, silly, foolish, useless, tasteless, insipid:
L. *bucco*, babbler, fool; see **speak**
L. *delirus*, silly, doting, crazy; see **mad**
L. *desipiens*, silly, foolish: desipience, desipient.
Gr. *elithios*, idle, vain, foolish; *elithiotes, -etos,* f. folly: *Elithiotes hirsuta* (a beetle).
Gr. *euethes*, simple, silly: *Euethogonus hardyi* (a milleped).
L. *excors; socors; vecors,* stupid, silly, foolish, mad; *socordia,* f. stupidity, folly: *Ellimenistes vecors* (a beetle).
L. *fatuus*, foolish; *infatuo, -atus,* make a fool of: fatuous, infatuation, ignis fatuus, *Avena fatua* (wild oats).
L. *frivolus*, silly, trifling: frivolous, frivolity, *Baetis frivolus* (a mayfly).
Gr. *habryntes*, m. coxcomb, fop: *Habryntis scita* (a moth).
L. *ineptus*, unsuitable, improper, silly, absurd; *ineptiola,* f. an absurdity, folly: inept, ineptitude, ineptness, *Didus ineptus* (dodo).
L. *insubidus*, stupid, foolish: *Cryptorhynchus insubidus* (a beetle).
L. *insulsus*, unsalted, insipid, silly, absurd; see **tasteless**
Gr. *kakoboulos*, foolish, unwise:

Gr. *kemphos*, m. simpleton, booby:

Gr. *konnophron*, *-os*, m. foolish one: *Connophron repletus* (a beetle).

Gr. *leros*, silly, nonsensical, foolish; *lerodes*, frivolous, silly; *lerema*, *-tos*, n. silly talk, nonsense; *paraleros*, talking foolishly, delirious: *Lerodes fulgurita* (a moth).

Gr. *lotax*, *-akos*, m. buffoon:

L. *ludibrosus*, ridiculous:

L. *maccus*, m. buffoon, simpleton: *Tephritis maccus* (a fly).

Gr. *mataios*, vain, idle, foolish: *Mataeocephalus acipenserinus* (a fish), *Mataeopsephus nitidipennis* (a beetle).

L. *morus* (Gr. *moros*), foolish, stupid; *morio*, *-onis; morus*, m. fool, simpleton; Gr. *moria*, f. folly: moron, sophomore, oxymoron, morotherium, *Epinephelus morio* (red grouper), *Glymmatophora morio* (a bug).

L. *nesapius*, unwise, foolish:

L. *nugalis*, frivolous, trifling; see *nugor* under **trifle**

L. *opicus*, rude, stupid, foolish:

Gr. *parapaistos*, foolish, mad:

Gr. *paratolmos*, foolhardy:

Gr. *phlyaros*, silly talk, nonsense; *phlyax*, *-akos*, jester, fool, buffoon; see *phledon* under **speak**

L. *praeposterus*, reversed, absurd, distorted: preposterous.

L. *ridiculus*, exciting laughter, funny, absurd; see *rideo* under **laugh**

L. *samardacus* (Gr. *samardakos*), m. buffoon:

L. *sannio*, *-onis* (Gr. *sannion*, *-os*), m. one who makes grimaces, buffoon, fool, harlequin; *sanna*, f. grimace: *Sannio rubrioculus* (a mite), *Diathetes sannio* (a beetle).

L. *stipes*, log, stump, blockhead; see **stem**

L. *stultus*, foolish: stultify, stultification, *Halictus stultus* (sweat-bee).

L. *trossulus*, m. fop, dandy, coxcomb, < Trossulum, an Etrurian town: *Batrachylodes trossulus* (a frog).

L. *truncus*, trunk of a tree, blockhead; see **stem**

See: **dull, dizzy, mad**

foot < AS. *fot:* football, footman, footprint, footpath, footstep, afoot, barefoot, Lightfoot, tenderfoot.

L. *astragalus* (Gr. *astragalos*), m. anklebone, die; *astragaloi*, dice: astragalus.

L. *calx, calcis*, f. heel; *calcaneum*, n. heel; *calco*, *-atus*, tread, trample; *calcator*, *-is*, m. treader; *inculco*, *-atus*, tread down, cram: calcaneum, caltrop, recalcitrant, inculcate, causeway, calcaneus, *Centaurea calcitrapa* (a centaury), *Trapa bispinosa* (singhara-nut).

Gr. *chele*, claw, hoof; see **nail**

Gr. *epibalos*, m. heel:

L. *gamba*, f. hoof:

Gr. *hople*, f. hoof: *Hoplebaea colmanti* (a beetle).

Gr. *ichnos*, footstep, track; see **mark**

Sp. *pata*, f. foot, leg: Patagonia.

Gr. *pelma*, *-tos*, sole of the foot; see **sole**

L. *pes, pedis*, m. foot; *pediculus; pedicellus; pedunculus; petiolus*, m. dim.; *pedalis; pedarius; pedulis*, of the foot; *pedaneus*, of the size of a foot; *pedester*, *-tris*, *-tre*, on foot; *pedes, peditis*, m. foot-soldier; *antepes*, *-pedis*, m. forefoot; *nudipes*, barefooted; *solidipes(solipes)*, uncloven(hoof), m. horse; *suppes, -pedis*, with feet turned under, with twisted feet: biped, cirriped, expedite, expedient, expedition, impediment, impeach, pedal, pedestal, petiole, pedicel, peduncle, pedigree, velocipede, centiped, milleped, pedestrian, piedmont, peon, pioneer, pawn, A.E.F., pedicellaria, trivet, vamp, *Viola pedata* (bird's-foot violet), *Viola pedunculata* (California violet), *Elaeagnus longipes* (goumi), *Avipes dillstedtianus* (fossil footprint), *Artemisia pedatifida* (bird's-foot sagebrush), *Curcuma petiolata* (a turmeric), *Syzygops fuscipes* (a beetle), *Exaesiopus grossipes* (a beetle), *Cucumaria multipes* (a holothurian), *Sisyphus rubripes* (a beetle), *Ipomoea pes-caprae* (a morning-glory), *Scolops sulcipes* (a planthopper).

Gr. *peza*, f. foot; *pezos*, on foot, walking; *apezos*, footless: *Pezophaps solitarius* (an extinct bird).

L. *plancus*, flat-footed:

L. *plautus*, flat-footed; see **broad**

Gr. *pous, podos*, m. foot; *podarion; podion*, n. dim.; *podabros*, tender-footed; *podister*, *-os*, m. one who does anything with the feet; *podizo*, measure with the feet, tie the feet; *antipous*, *-odos*, with the feet opposite; *apeiropous*, many-footed: decapod, megapode, cephalopod, gastropod, pelecypod, diplopod, tripod, octopus, antipodes, polyp, pseudopodium, Arthropoda, dactylopodite, pew, Depew, Dupuy, *Podophyllum peltatum* (mayapple), *Podocarpus spicatus* (matai), *Podocnemus expansa* (arrau), *Podisus spinosus* (a bug), *Podismopsis* *altaica* (a grasshopper), *Macropus giganteus* (a kangaroo), *Macropodus opercularis* (a paradise-fish), *Dipus aegypticus* (jerboa), *Dipospyris forcipata* (a radiolarian), *Hippopus maculatus* (a pelecypod), *Coelopus porcellus* (a rotifer), *Perissopus dentatus* (a copepod), *Melampus bidentatus* (a gastropod), *Barypus speciosus* (a beetle), *Rhegmatopoda leptocerca* (a grasshopper), *Chromocryptus antipodalis* (a wasp), *Apus cancriformis* (a phyllopod), *Apodichthys flavidus* (a blenny), *Isidora antipodea* (a snail), *Dipodomys deserti* (a kangaroo-rat), *Platypus dentatus* (a beetle), *Polypodium vulgare* (common polypody).

Gr. *pterna*, f. heel; *pternion*, n. dim.: *Pternandra capitellata* (a tree), *Micropterna maculata* (a caddisfly), *Vespertilio(Pternopterus) lobipes* (a bat).

L. *ramula*, f. hoof:

Gr. *skauros*, with swollen ankles; see **swell**

Gr. *sphyron*, n. ankle:

L. *talus*, m. ankle, heel, die: *taxillus*, m. dim.; *talaris*, of the ankle; *talipes*, club-foot: talus, talon, talipes, talipomanus, taligrade, *Campsicnemus brevitalus* (a fly).

Gr. *tarsos*, m. ankle, flat of the foot between toes and heel, any flat surface, woven mat or grate: tarsus, tarsal, metatarsus, tarsectomy, *Tarsius saltator* (a tarsier), *Stirotarsus abnormis* (a bug), *Tarsipes rostratus* (a marsupial), *Tarsophlebia longissima* (a dragonfly), *Anopheles albitarsis* (a mosquito), *Culex tarsalis* (a mosquito).

L. *vatrax, -acis;* *vatricosus*, with crooked feet, clubfooted:

L. *vestigium*, footstep, track; see **mark**

See: **hand, sole, shoe, base, measure**

for; fore, AS. preceding in time and place; see **before**

for-, AS. prefix with privative force; see **not**

foramen, L. hole, aperture, opening; *foraminosus*, full of holes; see **hole**

foraminifer < L. *foramen*, hole, *fero*, bear; see **protozoan**

forasticus, L. out of doors, public; see *foras* under **open**

foratus, L. bored; *forabilis*, that may be bored; see *foro* under **bore**

forbidden < AS. *forbeodan*, prohibit, deny, oppose.

Gr. *akthemos*, unlawful, lawless, contraband:

L. *exlex; illex, -egis,* without law, contrary to law, lawless; *illegitimus*, unlawful: illegitimate, illegitimacy, *Cryptorhynchus illex* (a beetle).

L. *illicitus*, forbidden, illegal: illicit.

L. *interdictum*, n. a prohibition: interdict, interdiction.

L. *prohibitio, -onis,* f. a forbidding, preventing: prohibit, prohibition.

L. *veto, -itus,* forbid; *vetitum*, n. forbidden thing: veto.

See: **hold, no**

force < L. *fortis*, strong; see **strong, push, press, make**

forceps, *-cipis,* L. pincers, nippers, tongs, tweezers; see **tongs**

ford < AS. *ford*, shallow passage across a stream; see **way**

fordus, L. pregnant; *forda*, f. cow with calf; see **fertile**

forehead; see **face**

foreign < L. *foras; foris,* out of doors, abroad; see **strange**

forensis, L. of the forum, public; see *foras* under **open**

foreskin, prepuce; see **penis**

forest < OF. *forest*, < ML. *forasta*, < L. *foras; foris,* out of doors: forester, forestry, deforest, reforestation, Black Forest.

Gr. *alsos*, n. grove: *Alsophila excelsa* (a tree-fern), *Alsomyia indica* (a fly).

Gr. *drymos*, m. oak-coppice, forest, wood: *Drymoglossum carnosum* (a fern), *Drymobius margaritiferus* (a snake), *Heliodrymus ramosus* (a radiolarian).

L. *dumentum*, thicket; see **bush**

AS. *holt*, a wooded place, copse: holt.

Gr. *hyle*, wood, forest; *hyloros*, forester; *hylaios*, of the forest; see **wood**

AS. *hyrst*, wood, grove: Hurst, Midhurst, Lyndhurst, Pinehurst.

Gr. *ide*, f. wood, forest, tree: Ida.

L. *lucus*, m. sacred grove; *luculus*, m. dim.; *lucaris*, of a grove: *Labrosaurus lucaris* (a dinosaur).

Gr. *nape*, wooded vale, dell, glen; see **valley**

L. *nemus, -oris* (Gr. *nemos*), n. forest or wood with pasture for cattle, grove, glade; *nemorosus*, full of woods, woodsy, shady; *Nemestrinus*, m. god of groves: nemoral, nemophilous, *Nemophila aurita* (fiesta flower), *Nemestrinus reticulatus* (a fly), *Nemorhedus goral* (a goat-antelope), *Anemone nemorosa* (European wood-anemone), *Solidago nemoralis* (a goldenrod), *Lysimachia nemorum* (a loosestrife).

L. *nucetum*, a nut-orchard; see *nux* under **nut**

L. *pomarium; pometum,* orchard; see **tree**

L. *saltus, -us,* m. woodland, glade; *saltuarius,* pertaining to forests; m. forester, ranger; *saltuosus,* forested, woody: *Juniperus saltuarius* (a Chinese juniper).

AS. *scaga,* thicket, grove, wood: shaw, Shaw, Bradshaw.

L. *silva (sylva),* f. woods, trees, forest; *silvula,* f. dim.; *silvaticus; silvestris,* of woods; *Silvanus,* m. god of woods: silviculture, Silvanus, sylvan, Sylvester, Sylvia, Pennsylvania, sylvestral, sylvanite, savage, *Pinus sylvestris* (Scotch pine), *Cephalophus sylvicultrix* (duiker), *Chrysobothris sylvania* (a beetle), *Lepus sylvaticus* (cotton-tail rabbit), *Prunus pennsylvanica* (fire-cherry).

L. *viridarium,* n. plantation of trees, pleasure garden:

AS. *weald* (G. *wald*), forest, wood, open country: weald, Wealden, Walden, wealdsman, wold, woldsman, waldgrave, waldhorn, waldmeister, Eichwald, Schwarzwald.

Gr. *xylochos,* f. copse, thicket: *Xylochus substriatus* (a beetle).

See: **tree, wood, park, valley**

forever; see **always**

forfex, *-icis,* L. shears, scissors; *forficula,* dim.; *forficatus,* scissors-shaped, forked; see **scissors**

forget < AS. *forgitan.*

Gr. *amnesia,* f. forgetfulness; *amnestos,* forgotten, < *amnesteo,* forget: amnesia, amnesty.

L. *dedisco,* unlearn, forget:

Gr. *eklesis,* f. a forgetting and forgiving:

L. *immemor, -is,* forgetful, unmindful: immemorial.

Gr. *lethe,* f. a forgetting, oblivion; *lethaios; lethios,* oblivious, causing forgetfulness; *lethargos; lethodes; lesmon, -os; lathiphron, -os,* forgetful, unmindful; *lethargikos,* drowsy; *lathiphrosyne; lesmosyne; lethosyne,* f. forgetfulness; *lathetikos,* likely to be forgotten or overlooked; *lethomai (lanthano),* forget, be unseen, escape notice: Lethe, lethargy, lethargic, encephalitis lethargica, lanthanum, lanthopin, *Lethocerus americanus* (giant waterbug), *Lethonymus difformis* (a beetle), *Lathiphronus cupreus* (a beetle), *Monoletes ovatus* (a fossil spore), *Triletes brevispiculus* (a fossil spore).

Gr. *nepenthes,* banishing pain and sorrow; see **painless**

L. *obliviscor, oblitus,* forget; *oblivialis,* causing forgetfulness; *obliviosus,* forgetful, unmindful; *oblivius,* forgotten: oblivious, oblivion, oublette, *Raphidia oblita* (a coddling-moth destroyer).

See: **dull, neglect, painless**

forgive < AS. *forgifan;* see **free, forget, atone**

foris, L. door, gate, see **door**; out of doors, in the open, abroad, see **open**

fork < L. *furca,* f. pitchfork; *furcilla; furcula,* f. dim.; *furcatus; furcillatus,* forked; *furcosus,* full of forks: furcate, furcula, furcellate, bifurcation, carfour, carfax, *Lyramula furcula* (a fossil bryozoan), *Russula furcata* (a mushroom), *Thamniscus furcillatus* (a fossil bryozoan), *Heliomaster furcifer* (a hummingbird), *Trifurcula subnitescens* (a moth), *Platycerium bifurcatum* (a staghorn-fern), *Myallosoma furculigerum* (a milleped).

L. *bifidus,* split into two parts, bifurcated; see *findo* under **cut**

L. *bisulcus,* two-furrowed, cloven, forked; see *sulcus* under **furrow**

Gr. *dikranon,* n. pitchfork: *Dicranum fulvum* (a broom-moss), *Dicranophorus forcipata* (a rotifer), *Dicranostomus nitidus* (a tettigonid).

Gr. *dikros,* forked, cloven: *Dicrostonyx hudsonius* (false lemming), *Dicrurus musicus* (a drongo), *Dicroglossus adolfi* (a frog), *Oniticellus dichrous* (a beetle).

L. *divarico, -atus,* spread apart, separate, fork; see **separate**

Gr. *exauster, -os,* m. fork for taking meat out of a pot:

L. *forficatus,* scissors-shaped, forked; see **scissors**

L. *fuscina,* f. a three-pronged fork, trident; *fuscinula,* f. dim.: foin.

L. *merga,* f. a two-pronged pitchfork: *Onthophagus mergacerus* (a beetle).

L. *pastinum,* two-pronged instrument for digging and trenching ground; see **hoe**

Gr. *pempobolon,* n. five-pronged fork:

Gr. *thrinax, -akos,* m. trident, three-pronged fork: *Thrinax microcarpa* (a thatch-palm), *Thrinacodus albicauda* (a rodent).

Gr. *triaina,* f. trident: *Triaenophorichthys trigonocephalus* (a fish).

L. *tridens, -entis,* m. fork with three tines: trident, *Purshia tridentata* (antelope-brush).

Gr. *triodous, -ontos,* m. trident: *Triodontopyga tridens* (a fly).

See: **spear, cut, scissors, point, branch**

form < L. *forma,* f. shape, figure, model, beauty; *formella; formula,* f. dim.; *formalis,* having a set form; *formosus,* beautifully formed, comely, handsome; *formosulus,* dim. pretty; *formaster, -tri,* m. coxcomb, dandy; *conformo, -atus,* form skillfully, shape symmetrically: formal, formula, cruciform, conformity, information, performance, uniform, malformed, deformity, informality, unconformable, Reformation, Formosa, *Dendrobium formosum* (an orchid), *Sprekelia formosissima* (Aztec lily), *Pentremites pyriformis* (a Mississippian blastoid), *Flabellum cuneiforme* (a fossil fan-coral), *Busycon fusiforme* (a fossil gastropod).

Gr. *agalma, -tos,* n. statue in honor of a god, statue, image, ornament; *agalmation,* n. dim.; *agalmatias,* like a statue, < *agallo,* glorify, honor: agalmatolite, *Agalmatosaurus timoriensis* (a lizard), *Agalmopsis elegans* (a medusa), *Coelagalma mirabile* (a radiolarian).

Gr. *andrias, -antos,* image of a man, statue; see *aner* under **man**

Gr. *bretas, -teos,* n. wooden image, mere image, blockhead:

L. *circus* (Gr. *kirkos*), ring; see **circle**

L. *citeria,* f. caricature, effigy, likeness:

L. *colossus* (Gr. *kolossos*), large statue; see **large**

L. *conflatilis,* cast, molten; *conflatio, -onis,* a casting, *conflatile,* molten image; see *conflo* under **flow**

L. *cubus* (Gr. *kybos*), a solid with six equal sides; see **three**

Gr. *deigma; hypodeigma; paradeigma, -tos,* n. sample, pattern, model, plan, specimen: paradigm, hypodigm, apodictic.

L. *delineatio, -onis,* f. sketch, design, outline: delineator, delineation.

Gr. *diagramma,* outlined figure, form, plan; see **map**

L. *documentum,* n. pattern, specimen, example: document, documentary.

L. *effigies, -ei,* f. image, likeness, copy, bust: effigy.

Gr. *eidos,* n. form, figure, likeness, thing seen, < *eido,* see; *idein,* to see; *eidyllion* (L. *idyllium*), n. dim.; *eidalimos; eueides,* shapely, comely; *eidolon* (L. *idolum*), n. image, form; *eidolikos* (L. *idolicus*), of idols; *idea,* f. mental image; *indalma, -tos,* n. form, appearance; *indalmatikos,* imaginary: idea, ideal, ideology, idol, idolatry, idyl, kaleidoscope, geode, deltoid, ovoid, *Eidalimus annulatus* (a fly), *Idalima maculosa* (a moth), *Idolum diabolicum* (a mantid), *Idolomorpha defoliata* (a mantid), *Cycadeoidea pulcherrima* (a fossil cycadophyte).

Gr. *ekmageion; ekmagma, -tos; ekmaktron,* n. impression, model, mold:

L. *ellipsis* (Gr. *elleipsis*), f. omission, defect; a defective circle: ellipse, *Polypora elliptica* (a fossil bryozoan), *Rubus ellipticus* (yellow Himalayan raspberry), *Arachnocalpis ellipsoides* (a radiolarian), *Saccharomyces ellipsoideus* (wineyeast).

L. *exemplum; exemplar, -is,* n. model, pattern, copy; *exemplaris, -e,* patternable: example, sample, exemplar, exemplary, e.g.

L. *facies,* external form or figure, face, countenance; see **face**

L. *fictio,* a making, forming, < *fingo- fictus,* form, shape; see **make**

L. *figmentum,* n. figure, image: figment.

L. *figura,* f. shape, form: figure, disfigure, configuration, figurative, *Mycosphaerella effigurata* (a fungus), *Siculodes figurata* (a moth).

Gr. *glymma,* an engraved figure; see *glypho* under **cut**

L. *icon, -is* (Gr. *eikon, -os*), f. image, figure; *icuncula,* f.; *eikonion,* n. dim.; *iconicus* (Gr. *eikonikos*), of an image: iconoclast, iconography, *Hygrobates iconicus* (a spider).

L. *imago, -inis,* f. copy, imitation, likeness, phantom; *imaguncula,* f. dim.; *imaginalis,* figurative; *imagineus,* of an image; *imagino, -atus,* represent, fancy, fashion: image, imagine, imaginary, imagination, imago.

L. *instar,* n. form, figure, image, likeness: instar.

L. *modulus,* measure, model, norm; see *modus* under **measure**

Gr. *morphe,* f. form, figure, shape; *morphosis,* f. a shaping, forming; *Morpheus,* m. god of dreams; *Morpho,* f. the shapely one, Aphrodite; *amorphos,* formless: morphology, morphine, Morpheus, amorphous, polymorphic, metamorphosis, rhizomorph, *Morpho achilles* (a butterfly), *Amorpha fruticosa* (a false indigo), *Dimorphotheca sinuata* (a Cape marigold), *Tapinomorphus setosus* (a beetle).

Gr. *opsis,* sight, appearance, face, likeness; see *optikos* under **see**, *ops* under **face**

L. *pictura,* a painting; see *pingo* under **color**

Gr. *plaision,* n. an oblong body, figure, or form: *Plaesius pudicus* (a beetle).

Gr. *plasma,* image, figure, model, substance; *proplasma,* artist's model; see *plasso* under **make**

Gr. *prisma, -tos,* n. anything sawn; a geometrical figure: prism, prismatic, *Prismatocerus auritulus* (a bug), *Mesembryanthemum prismaticum* (a fig-marigold).

Gr. *pyramis, -idos,* pyramid; see **pillar**

L. *rhombus* (Gr. *rhombos*), an equilateral parallelogram with unequal pairs of angles; see **diamond**

AS. *sceap,* form, contour: shape, shapeless, shipshape.

L. *scena* (Gr. *skene*), tent, stage, decorative setting or place; see **play**

Gr. *schema, -tos,* form, shape, plan; *schemation,* dim.; see **aim**

L. *sculptura,* carving in relief; see *sculpo* under **cut**

L. *sigillum,* little figure or image, dim. of *signum,* mark, flag, seal; *sigillatus,* adorned with little figures or images; see *signum* under **mark**

L. *simulacrum,* image, likeness, portrait, effigy; see *similis* under **like**

L. *species,* look, form, kind; see **class**

L. *spectrum,* appearance, form, image; see **spirit**

L. *statua,* f. image in marble, bronze, etc.; *staticulum,* n. dim.: statue, statuary, statuette, statuesque.

L. *topia,* ornamental gardening; see **ornament**

Gr. *trapezion,* n. an irregular four-sided figure: trapezium, trapezoid, *Trapezostigma variegatum* (a neuropterid), *Adiantum trapeziforme* (a maidenhair-fern), *Plectopyramis trapezomma* (a radiolarian), *Pithecellobium trapezifolium* (a legume).

L. *typus* (Gr. *typos*), m. figure, impression, model, shape, < *typto,* strike; *typarion,* n. dim.; *typicalis,* like the type; *typicus* (Gr. *typikos*), conformable; *typoma, -tos,* n. figure, form, outline; *typosis,* f. a forming, molding; *typotos,* formed, molded; *ektypos,* in relief, embossed; *archetypos; prototypos,* original, primitive: typography, typology, typophile, typonym, typothetae, type, typical, genotype, archetype, prototype, linotype, stenotypy, *Rhodotypus kerrioides* (a rosacead), *Ectypodus musculus* (a fossil mammal), *Atypus piccus* (a spider).

Gr. *xoanon,* n. wooden image; *xoanoglyphos,* m. sculptor: xoanon.

Gr. *zodion,* small painted or carved figure; *zodiakos,* of animals; see **animal**

See: **frame, face, thing, spirit, fancy, dream, beauty, make, map, flesh, line, think**

FIGURES OF SPEECH OR RHETORIC

The formal description of scientific materials and processes demands literal, specific language because the prime purpose of the speaker or writer is to be as clearly understood as words will permit, even at the risk of being dry-as-dust. However, at appropriate times and places, scientists, being human, relax from strict exactness by spicing their talk and writing with those delightful deviations from literalness called figures of speech. These arresting, colorful expressions contain a dash, more or less, of distortion and fiction, and, therefore, like gift horses, must not be examined too critically. Their sudden or surprising turns are pleasing because the hearer or reader finds wit where none may have been expected.

Although some figures are achieved by extraordinary arrangement of letters, sounds, and words, most are essentially plays on words by which ideas connected with unrelated objects are transposed. Their effectiveness, which varies as the connotation and suggestion implied, depends largely upon the degree of unlikeness of the objects and upon the experience and background of the hearer or reader. If Jones says, "Smith's house is a bungalow", he makes a plain, literal statement. If, however, he says, "Smith's house is like a barn", he adds an airy, rural touch to the idea; but if he says, "Smith's house is a rat hole", he reaches a libelous climax.

Figures are sometimes manhandled and mixed, as in the quip, "A virgin forest is one in which the hand of man has never set foot"; or they may be strained and overdone, as in much so-called poetry, and by those pathetic persons who cannot or will not say plainly and exactly what they mean but rely on the vague and ambiguous terms of slang, profanity, and other less picturesque forms of indefinite circumlocution. Malapropisms are grotesque or incongruous misusages of words, sometimes accidental, sometimes deliberately posing as blunders, as in the statements: "Corporal Buck is in the veterinary hospital", and "The state flower of Colorado is the concubine". Anachronisms are misdated allusions, as in the preacher's fervent assertion, "If the King James version of the Bible was good enough for St. Paul, it is good enough for me."

Spoonerisms, accidentally or with humor aforethought, transpose letters or sounds in a succession of words, as in "the fly crows", "the cry flows", for "the crow flies"; and in the editor's unsuccessful galley corrections of "the battle-scored veteran" to "the battle-scared veteran" and "the bottle-scarred veteran".

The Greeks and Romans used all the figures familiar to us today. Many English words, such as *convivial, dilapidated, ecstasy, melancholy, symposium, trivial,* and *vacillate,* are more or less faded relics of classical metaphors. The figures common now are:

allegory < Gr. *allegoria,* an extended metaphorical story about an underlying theme thinly disguised.

> Ex. *Pilgrim's progress* (John Bunyan); *Gulliver's travels* (Jonathan Swift); *The termitodoxa, or biology and society* (William M. Wheeler)

alliteration < L. *alliteratio,* a successive repetition of the same sound, represented by the same or different letters, usually to suggest a special effect.

> Ex. Sing a song of sixpence.
>
> How much wood would a woodchuck chuck
> If a woodchuck could chuck wood?

allusion < L. *allusio,* a reference, direct or indirect, generally for figurative effect.

> Ex. They saw the handwriting on the wall.
>
> He was accompanied on the journey by his man Friday.
>
> The adolescent's mustache is the substance of things hoped for, the evidence of things not seen.
>
> Yes, Santa Claus, there is a Virginia. (Heard after the 1952 presidential election)

antithesis < Gr. *antithesis,* figurative contrast or opposition, chiefly epigrammatic.

> Ex. Law without liberty is tyranny, but liberty without law is anarchy. (William Penn)
>
> Never in the field of human conflict was so much owed by so many to so few. (Winston Churchill)
>
> We must all hang together or we shall all hang separately. (Benjamin Franklin)
>
> A cow does not give milk; it must be taken from her.
>
> The man who never thought anything of walking ten miles a day now has a grandson who never thinks of it either.
>
> When you drive, don't drink; when you drink, don't drive.

apostrophe < Gr. *apostrophe,* addressing an absent person as though present, or an inanimate thing as human.

> Ex. Oh Death, where is thy sting? (1 Cor. 15:55)
>
> My country, 'tis of thee,
> Sweet land of liberty,
> Of thee I sing. (Samuel F. Smith)

climax < L. *climax* (Gr. *klimax*), a succession of ideas progressing in force to a culmination. Sometimes the last idea is so inferior to those gone before that the result, whether intentional or not, is an anticlimax.

> Ex. Tax and tax, spend and spend, elect and elect.
>
> Now we learn what patient periods must round themselves before the rock is broken, and the first lichen race has disintegrated the thinnest external plate into soil, and opened the door for the remote Flora, Fauna, Ceres, and Pomona, to come in. How far off yet is the trilobite! How far the quadruped! How inconceivably remote is man! All duly arrive, and then race after race of men. It is a long way from granite to the oyster; farther yet to Plato, and the preaching of the immortality of the soul. (*Nature:* Ralph W. Emerson)

epigram < Gr. *epigramma,* originally an inscription, now a pithy saying. Proverbs are mellowed epigrams.

> Ex. Millions for defense, but not one cent for tribute. (Charles C. Pinckney)
>
> Leave the table with an appetite and you will return with one. (Benjamin Franklin)
>
> How sharper than a serpent's tooth it is
> To have a thankless child! (Lear in *King Lear:* William Shakespear)

euphemism < Gr. *euphemismos,* the use of pleasing words for disagreeable ideas.

Ex. By that time I shall be pushing up daisies.

He deliberately misstated the facts and, therefore, belongs to the Ananias Club.

Young Quaker lady to man who insulted her: "Go home to thy mother and let thy mother bark at thee!"

exclamation < L. *exclamatio,* a sudden elliptical interjection. The Greeks called this ecphonesis.

Ex. A horse! A horse! My kingdom for a horse! (Richard in *King Richard III:* William Shakespear)

fable < L. *fabula,* a short metaphorical tale, usually with a moral, in which animals, plants, or inanimate objects are personified.

Ex. Aesop's *Fables; Uncle Remus and Br'er Rabbit.* (Joel C. Harris)

hyperbole < Gr. *hyperbole,* a throwing beyond, that is, overstatement, exaggeration.

Ex. Perturbed mother to young daughter: "How many million times have I told you not to exaggerate?"

I'll be tickled to death to come to your party.

A GI, losing himself in the Pentagon at Washington, became a general before he emerged.

interrogation < L. *interrogatio,* a figurative question that implies affirmation or denial.

Ex. How shall we escape if we neglect? (Heb. 2:3)

irony < L. *ironia* (Gr. *eironeia*), dissimulation in which the opposite is intended. This grades into sarcasm and satire, which are somewhat more biting. Parody is imitation, sometimes comic or burlesque, for the purpose of ridicule or irony.

Ex. Oath for dentists: I swear to pull the tooth, the whole tooth, and nothing but the tooth.

Sign in National Park: Hospital facilities are provided for visitors who feed bears!

The dry-rot of our academic biology. (William M. Wheeler)

The celestial railroad (in *Mosses from an old manse:* Nathaniel Hawthorne)

It is easy to quit smoking, for I have done so at least a hundred times. (Attributed to Mark Twain)

litotes < Gr. *litotes,* understatement, affirmation of an idea by denying the contrary.

Ex. He was a man of no small attainments.

metaphor < Gr. *metaphora,* a simile with the comparing word omitted.

Ex. He is a wolf in sheep's clothing.

She wrapped herself in the damp blanket of self-pity.

Silently, one by one, in the infinite meadows of heaven,
Blossomed the lovely stars, the forgetmenots of the angels. (*Evangeline:* Henry W. Longfellow)

metonymy < Gr. *metonymia,* change of name, such as the naming of a container for the thing contained, cause for effect, sign for the thing signified, place for its inhabitant, subject for attribute, instrument for agent, part for the whole, whole for a part, etc. Synecdoche is essentially metonymy.

Ex. He was addicted to the bottle.

The scholar burned the midnight oil.

She was born with a silver spoon in her mouth.

May your shadow never grow less!

The attorney read Blackstone.

He defended his fireside.

She: "The new lady next door is a widow."
He: "Grass or sod?"

The court charged the jury.

Washington sweltered in August and shivered in January.

onomatopoeia < Gr. *onomatopoiia,* use of words that imitate or suggest natural sounds.

Ex. Bobwhite, buzz, chickadee, katydid, whippoorwill.

Through the clear, frosty air came the jingling and tinkling of bells.

parable < Gr. *parabole,* analogy by means of a short allegory, the scenes and events of which could be possible.

    Ex. Parable of the sower. (Luke 8: 5-15)

paronomasia < Gr. *paronomasia,* pun, a play on words that sound alike but have different meanings.

    Ex. Here lies Elizabeth Mann

        Who lived an old maid and died an old Mann.

        Apatite is found in Hungary.

        The Benedicts are being heir-conditioned.

        Three sons, having acquired a cattle ranch, asked their mother to suggest a name for it. She replied: "Call it Focus, because that is where the sun's rays meet."

        The Pilgrims first got down on their knees, then got down on the aborigines.

personification < L. *persona,* mask, person. This figure endows animals, plants, and inanimate objects with human attributes.

    Ex. Babbling brooks, angry sores, murmuring pines, stubborn granite.

        And the Lord opened the mouth of the ass, and she said unto Balaam, "What have I done unto thee, that thou hast smitten me these three times?" (Num. 22: 28)

        "How absurd," said the gnat to the gnu,
        "To spell your queer name as you do!"
        "For the matter of that,"
        Said the gnu to the gnat,
        "That's just how I feel about you."

simile < L. *similis,* like, figurative comparison with the comparing word expressed.

    Ex. How far that little candle throws his beams!

        So shines a good deed in a naughty world. (Portia in *The merchant of Venice:* William Shakespear)

        Mosquitoes are like children. When you don't hear them you know they are getting into something.

        Wife to husband: "It will be harder to keep a secret from me than to smuggle daylight past a rooster."

synecdoche < L. *synecdoche* (Gr. *synekdoche*), substitution of a part for the whole, the whole for a part, etc. This is a form of metonymy; see **metonymy.**

vision < L. *visio,* sight, appearance, considering the past as though it were present. This is similar to apostrophe.

    Ex. Lincoln steps forward and begins to read, "Fourscore and seven years ago . . ."

formica, L. ant; *formicula,* dim.; *formicinus,* of ants; see **ant**

formidabilis, L. causing fear, terrible; see *formido* under **fear**

formosus, L. beautiful; *formosulus,* dim. pretty; see **form**

formula, L. rule, method; see **law**

fornax, *-acis,* L. oven, kiln; *fornacula,* dim.; *fornacalis,* of ovens; see **oven**

fornix, *-icis,* L. arch, vault, brothel in an underground vault; *fornicatus,* arched, vaulted; see **arc**

fors, *fortis,* L. chance, luck; see **lot**

forsythia, NL. a genus of the olive family; see **olive**

fort < L. *fortis,* strong.

    L. *arx, arcis,* f. stronghold, fortress, citadel, castle, height, peak:
    AS. *burg,* fortified town; see **town**
    L. *castrum,* n. (AS. *ceaster*), fort, walled town, stronghold, garrison, castle, settlement; *castra,* f. camp; *castellum,* n. dim. citadel, fortress; *castrensis,* of camp; Sp. *alcazar,* fortress, castle: castral, castramentation, castle, castellan, castellated, chalet, chateau, Chester, Worcester, Leicester, Winchester, Gloucester, Lancaster, Exeter, Rochester, Newcastle, Castile, alcazar, Neuchatel, Castilian, *Castilla elastica* (a gum-tree).
    L. *munimentum,* fortification, rampart; see *munio* under **strong**

Gr. *phrourion,* fort, garrison; see *phrouros* under **guard**
See: **strong, guard, town, house**

fortax, L. (Gr. *phortax*), carrier, bearer; see *phero* under **carry**

fortis, L. strong, brave; *forticulus,* dim.; see **strong**

fortune < L. *fortuna,* chance, luck; *fortuitus,* accidental, by chance; see *fors* under **lot**

forum, L. an open, public place, market-place; *forus,* out of doors; see **open**

forus, L. gangway; see **way**

forward < AS. *foreweard;* see **before, face**

fossa, L. ditch; *fossula,* dim.; *fossatus,* dug; see *fodio* under **dig**

fossilis, L. digging, dug up; *fossor,* digger; see *fodio* under **dig**

fotus, L. a warning, < *foveo, fotus,* warm, cherish; see **heat**

foul < AS. *ful,* filthy; see **bad, dirt, rot**

fountain < L. *fons, fontis,* spring; see **spring**

four < AS. *feower:* fourteen, fourth, forty, farthing, firkin. See **diamond**
   L. *abacus; abax, -acis* (Gr. *-akos*), counting-board, gaming-board divided into square compartments; see **number**
   Gr. *gnomon,* carpenter's square, rule; see *nosco* under **know**
   L. *laculatus,* four-cornered, checkered:
   L. *norma,* carpenter's square; see **law**
   L. *quattuor,* four; *quartus,* fourth; *quadrans, -tis; quartarius,* m. a fourth; *quartanus,* of the fourth; *quadrigatus; quadrigeminus; quadruplex; quadruplus; quadruus,* fourfold; *quadrifidus,* split into four parts; *quaternarius,* consisting of four; *quattuordecim,* fourteen; *quadrus,* square; *quadra,* f. a square; *quadrula,* f. dim.; *quadro, -atus,* make four-cornered, square: quarter, quarto, quartet, quart, quaternary, quaternion, quarantine, quarry, quadrant, quadrangle, quadruped, quadrate, quadrature, quadrigeminal, quadrille, quire, cadrans, cater-corner, carillon, casern, square, squad, trocar, *Lysimachia quadrifolia* (a loosestrife), *Chiropsalmus quadrigatus* (a jellyfish), *Delocrinus quadratus* (a Permian crinoid), *Tetradella quadrilirata* (an ostracode), *Acanthostracion quadricornis* (a trunkfish), *Passiflora quadrangularis* (granadilla), *Fraxinus quadrangulata* (blue ash), *Dioscorea quaternata* (a yam).
   Gr. *tetra,* < *tessares (tettares),* four; *tetartos,* fourth; *tetradymos,* fourfold; *tetragonos,* square; *tessarakonta,* forty; *tessarakostos,* fortieth; L. *tessera,* f. small square stone, tile, die; *tessella,* f. dim.; L. *tetradium,* n. quaternion: tessera, tesseratomic, tessellate, tetrad, tetragonal, tetrameter, tetrarchy, tetradynamous, tetrahedron, trapezoid, *Tetrapanax papyriferum* (rice-paper plant), *Tetragonia crystallina* (an aizoacead), *Tetragonops rhamphastinus* (a barbet), *Tessaromerus quadrimaculatus* (a bug), *Tetraodon meleagris* (a globefish), diatessaron, *Episcia tessellata* (a gesneriacead).

fouro'clock
   NL. *abronia,* f. a genus of the fouro'clock family: *Abronia umbellata* (pink sand-verbena).
   NL. *allionia,* f. a genus of the fouro'clock family, < Carlo Allioni, Italian botanist: *Allionia incarnata* (trailing allionia).
   NL. *boerhaavia,* f. a genus of the fouro'clock family, < Herman Boerhaave, Dutch physician: *Boerhaavia coccinea* (scarlet spiderling).
   NL. *bougainvillea,* f. a genus of the fouro'clock family, < L. A. de Bougainville, a French navigator: *Bougainvillea spectabilis* (Brazilian bougainvillea).
   L. *mirabilis,* wonderful, a genus of the fouro'clock family: *Mirabilis longiflora* (sweet fouro'clock).
   NL. *nyctago,* f. a former genus of the fouro'clock family: Nyctaginaceae, *Nyctaginia capitata* (a fouro'clock), *Allionia nyctaginea* (prairie-allionia).
   NL. *pisonia,* f. a genus of the fouro'clock family: *Pisonia aculeata* (devil's-claw-pisonia).

fovea, L. pit, pitfall; *foveola,* dim.; see **hole**

fowl < AS. *fugol,* bird of any kind, an edible bird; see **chicken, bird**

fox < AS. *fox,* male fox; *fyxen,* female fox, vixen.
   Gr. *alopex, -ekos,* f.; *alopos,* m. fox: *alopekion,* n. dim.; *alopekodes,* foxlike, sly: alopecia, alopecoid, *Alopex lagopus* (Arctic fox), *Aloposaurus gracilis* (a fossil reptile), *Alopochelidon fucata* (a bird), *Alopecurus geniculatus* (a foxtail-grass), *Lycopodium alopecuroides* (foxtail-clubmoss), *Falco alopex* (a hawk).
   Gr. *bassaris, -idos,* f. fox; *bassarion,* n. dim.: *Bassariscus astutus* (cacomistle).
   Gr. *kaphore(skaphore),* f. vixen: *Caphora humilis* (a beetle).
   Gr. *kerdo,* the wily one, fox; see *kerdaleos* under **know**
   Gr. *kinados,* n. fox; *kinadion,* n. dim.: *Cinadus spurius* (a fly).

Gr. *lampouris, -idos,* f. fox:

Gr. *skindaphos,* f. vixen:

L. *vulpes, -is,* f. fox; *vulpecula,* f. dim.; *vulpinus,* of a fox: vulpine, vulpecular, vulpinism, *Vulpes fulva* (red fox), *Trichosurus vulpecula* (brushtail-possum), *Albula vulpes* (ladyfish), *Vitis vulpina* (frost-grape), *Hyracotherium vulpiceps* (an eohippus).

fracidus, L. overripe, soft, mellow; see **ripe**

fractus, L. broken; *fractor,* breaker; *fractura,* break; see *frango* under **break**

fragilis, L. easily broken, brittle; see **brittle**

fragmentum, L. piece, bit; see **part**

fragor, L. loud noise, crash; see **sound**

fragosus, L. broken, rough, uneven; see **rough**

fragrans, L. smelling agreeably; see *fragro* under **smell**

fragum, L. strawberry; see **strawberry**

frame < AS. *framian,* fashion, prepare.
  L. *abacus,* counting-board or frame; see **number**
  Gr. *analemma, -tos,* n. support, frame, sling:
  Gr. *antyx, -ygos,* edge, border, rim, frame; see **border**
  L. *canteriolus,* m. trellis for supporting plants:
  L. *catasta,* f. stage, platform, scaffold:
  L. *contrabium,* n. framework of beams:
  L. *crux,* cross; see **cross**
  Gr. *delphix, -ikos,* m. tripod:
  Gr. *demas,* body, frame; see **body**
  L. *equuleus,* m. rack in the shape of a horse:
  L. *fala,* f. scaffold:
  L. *falisca,* f. rack in a manger:
  L. *gabalus,* m. gallows:
  ML. *gibetum,* n. gallows: gibbet.
  Gr. *histos,* loom, frame, web; see **weave**
  Gr. *ikrioma, -tos,* n. scaffold; *ikrion,* n. flooring of a deck, platform, bench: *Icriodus curvatus* (a conodont), *Icriodina irregularis* (a conodont).
  L. *jacea,* f. hayrack:
  Gr. *killibas, -antos,* m. stand, frame, easel:
  L. *lasanum* (Gr. *lasanon*), n. stand, gridiron, nightstool:
  L. *machina,* frame, engine; see **tool**
  Gr. *petauron,* perch, roost, springboard; see **leap**
  L. *proscenium* (Gr. *proskenion*), n. stage: proscenium.
  L. *scena* (Gr. *skene*), tent, stage; see **play**
  Gr. *schedia,* f. frame, scaffold, bridge:
  Gr. *skeleton,* n. dried body, mummy; *skeletos,* dried up, withered, mummified: skeleton, *Sceletophorus biserialis* (a fossil fish).
  L. *sponda,* bedstead, bed; see **bed**
  Gr. *stereoma,* skeleton; see *stereos* under **thick**
  Gr. *triben, -os,* m. tripod:
  L. *tripus, -podis* (Gr. *tripous, -podos*), m. a three-legged stand: tripod, *Cortina tripus* (a radiolarian), *Cortinetta tripodiscus* (a radiolarian).
  L. *vara,* f. trestle, horse:
  See: **form, border, carry, cross, body**

frangula, NL. buckthorn; see **buckthorn**

frank < OF. *franc,* < OHG. *franko,* free; see **free**

franklinia, NL. a genus of the tea family; see **tea**

frater, L. brother; *fraterculus,* dim.; *fraternus,* brotherly; see **brother**

fraud < L. *fraus, fraudis,* deceit; *fraudator; fraudatrix,* cheater; see **lie**

frax, -acis, L. dregs, grounds; see **dirt**

fraxinus, L. ash-tree; *fraxineus,* of ash; see **ash**

freckle < ON. *freknur;* see **spot**

free < AS. *freo;* G. *frei:* freedom, freely, freehold, freestone, Freeman, Freeland, Freemason, Freiburg. Derive: freemartin.
  Gr. *adeia,* freedom from fear, safety, security; see **safe**
  Gr. *adesmos,* unfettered; *adetos,* unbound, loose, free; *asyndetos,* unconnected, loose, free: asyndeton, *Adetococcyx intermedius* (a bird), *Asyndetus inermis* (a fly).
  Gr. *akolastos,* unbridled, undisciplined, intemperate: *Acolastus pictus* (a beetle), *Acolastodes oenotripta* (a moth).

Gr. *alyxis*, an escape; see alyktos under **shun**
Gr. *amochthos*, free from toil and trouble:
Gr. *amomos*, blameless:
Gr. *anepischetos*, not to be stopped, unrestrained: *Anepischetus bipartita* (a moth).
L. *aneticus* (Gr. *anetikos*), relaxing, < *anetos*, relaxed, slack, set free; *anesimos*,
    let loose, having a holiday; *anesis*, f. relaxation, license:
Gr. *apheides*, lavish, unsparing; see **abundance**
Gr. *aphetos*, ranging at will, at large, free, loose; *aphesios*, m. releaser: *Aphetoceras*
    americanum (a fossil cephalopod).
L. *apocha* (Gr. *apoche*), f. receipt, acquittance:
Gr. *apodoulos*, m. a freedman:
Gr. *apozeuxis*, f. an unyoking;*apozyx*, *-ygos*, separated, single; *azyx*, *-ygos*, un-
    yoked, unwedded:
Gr. *asphinktos*, not bound, loose:
L. *assero*, *-atus*, declare freedom, affirm; see **speak**
Gr. *asystatos*, incoherent, inconsistent: *Asystata brevipes* (a phasmid).
Gr. *ateles*, imperfect, exempt; *ateleia*, f. exemption from public burdens, **tax**:
    philately.
L. *avocamentum*, n. alleviation, relaxation, recreation:
L. *candor*, *-is*, m. frankness, openness, integrity, sincerity: candor.
L. *caritus*, free from, devoid of; see **poor**
AS. *ceorl*, freeman: churl, Charles, Carl.
Gr. *chalaros*, loose, slack, limp, languid, supple; *chalasis*, f. a loosening; *chalasma*,
    -tos, n. relaxation; *chalastikos*, laxative, < *chalao* (L. *chalo*, *-atus*), relax, slack-
    en: chalone, chalastic, *Chalaraspis unguiculata* (a mysid).
L. *dissolutus*, loose in morals, lax; see **bad**
Gr. *dystheratos*, hard to catch:
L. *effrenatus; effrenus*, unrestrained, unbridled, < *effreno*, *-atus*, unbridle, let
    loose; *infrenatus*, without a bridle: *Carduelis effrenata* (a bird).
Gr. *eleutheros*, free; *eleutheria*, f. freedom, liberty: eleutherian, *Eleutherodactylus*
    latrans (a frog), *Eleutherocrinus cassedayi* (a crinoid), *Croton eleuteria* (cas-
    carilla).
L. *eludo*, *-lusus*, avoid, evade, frustrate, baffle; see **shun**
L. *emancipo*, *-atus*, set free, deliver: emancipate, emancipation, emancipator.
L. *emendo*, *-atus*, free from faults, correct, improve: emend, emendation.
L. *emergo*, *-mersus*, come forth, rise, free oneself: emerge.
Gr. *ethelonter*, *-os*; *ethelontes*, m. volunteer; *etheletos*; *ethelousios*, voluntary:
L. *exemptus*, taken out, removed, excused; see **from**
L. *exlex*, *-legis*, beyond the law, bound by no law:
L. *exonero*, *-atus*, free from a burden, unload: exonerate.
L. *expeditus*, free, unimpeded, < *expedio*, *-itus*, free the feet from a snare, dis-
    engage, extricate, hasten: expedite, expedition.
L. *extrico*, *-atus*, disentangle, free: *Pipunculus extricatus* (a fly).
OF. *franc*, < OHG. *franko*, free: frank, franchise, franc, France, French, Francis,
    Franklin, Frankfort, San Francisco, frankincense.
Gr. *hapalasso*, free, release, deliver; *hapallage*, f. deliverance, release:
Gr. *hekousios*, voluntary:
L. *ignosco*, *ignotus*, pardon, forgive, excuse, overlook; see *nosco* under **know**
L. *immunis*, tax-free, exempt: immune, immunity, *Apolectus immunis* (a fish),
    Triadocidaris immunita (a sea-urchin).
L. *impunitus*, free from danger, safe, secure: impunity.
L. *incoactus*, voluntary:
L. *incoercitus*, unrestrained:
L. *incohibilis*, that cannot be kept together or restrained:
L. *incompos*, *-otis*, not having control over:
L. *incontinens*, *-entis*, not retaining, uncontrolling: incontinent.
L. *indemnatus*, uncondemned, unsentenced:
L. *ingenuus*, free-born, freeman: ingenuous, ingenuity.
L. *injussus*, unbidden, voluntary:
L. *intemperatus*, immoderate, uncontrolled: intemperate.
Gr. *lagaros*, loose, slack, pliant, thin: *Lagaroceras megalops* (a fly), *Lagarodes*
    jacetus (a katydid).
L. *laxus*, loose, slack, unstrung; *laxativus*, loosening, mitigating, < *laxo*, *-atus*,
    undo, unloose, slacken, lighten; *relaxo*, *-atus*, loosen, slacken, unbend: laxity,
    relax, laxative, release, leash, *Athrotaxis laxifolia* (a conifer), *Spirorbis laxus*
    (a fossil tubeworm), *Myosotis laxa* (a forgetmenot).
Suf. *-less*, < AS. *leas*, free from, without; see **not**
L. *liber*, *-a*, *-um*, free; *liberalis*, relating to freedom, kind, generous; *libertas*,
    -atis, f. freedom; *libertus*, set free, freedman; *libertinus*, pertaining to the con-

dition of a freedman; *libero, -atus,* set free, deliver: liberty, liberal, deliver, libertine, illiberal, livery, Liberia, liberation.

L. *licentia,* f. freedom to do as one pleases, liberty; *licentiosus,* with unrestrained freedom, unbridled, wanton: license, licentious, licentiate.

L. *lustro, -atus,* purify by propitiation; see *lustrum* under **pure**

Gr. *lyo,* loose, dissolve, break up; *lysis,* f. a loosing, freeing, releasing; *lyter, -os,* m. looser, deliverer; *lytos,* soluble; *analyter, -os,* m. deliverer; *apolytos,* loosened, free; *dialysis,* f. dissolution, separation: analysis, catalyzer, histolysis, dialyzer, hydrolysis, proteolytic, catalytic, paralytic, dysluite, dyslysin, hemolysin, tachylyte, *Lysiphlebus testaceipes* (a braconid), *Lysimachia punctata* (spotted loosestrife), *Hippolyte projecta* (a prawn).

Gr. *methiemi,* let go, let loose, give up, relax; *methetikos,* relaxing:

Gr. *phyxis,* refuge, escape, flight; see *fuga* under **depart**

Gr. *psapharos; psathyros,* easily powdered, friable, crumbling, loose; see **brittle**

L. *remitto, -missus,* pardon, forgive: remit, remission.

L. *renodis,* loose, untied, free:

L. *resero, -atus,* unlock, open; see **open**

Gr. *rhysis,* deliverance; see **save**

L. *runco, -atus,* remove weeds; see **till**

Gr. *schazo,* let loose, let go:

L. *securus,* free from care, safe; see **safe**

L. *solvo, solutus,* loosen, set free, explain; *solubilis,* that can be freed, loosed, explained; *dissolvo,* loosen, destroy; *absolutio, -onis,* f. acquittal, forgiveness, remission: solve, dissolve, insoluble, absolute, absolve, dissolute, indissolubly, insolvent, irresolution, solution, hydrosol, aerosol, *Clostridium cellulosolvens* (a bacterium).

L. *spontaneus; spontalis,* voluntary: spontaneous, spontaneity, *Hordeum spontaneum* (wild barley).

Gr. *syngnome,* f. pardon, forgiveness:

L. *ultroneus,* of one's own accord, voluntary:

L. *vacatio, -onis,* freedom from duty; see *vaco* under **empty**

L. *venia,* f. pardon, forgiveness, indulgence, grace; *venialis,* pardonable: venial.

L. *vindico, -atus,* avenge, deliver, justify; see **punish**

L. *voluntarius,* willing, by free will: voluntary, involuntary.

See: **separate, innocent, wash, empty, not, waste, rest, depart, wean, justice, let**

freeze < AS. *freosan;* see **cold, ice, thick, bind**

fremitus, L. a dull sound, < *fremo, -itus,* roar, growl, murmur; see **roar**

frenum, L. bridle, rein, bit; *frenulum,* dim.; *frenator,* controller, tamper; see **hold**

frequent < L. *frequens, -entis,* taking place often, repeatedly; see **number**

fresh < OF. *fresche;* see **new**

fresus, L. ground, crushed, < *frendo, fresus,* gnash the teeth, grind to pieces; see **grind**

fretale, L. frying-pan; see **pan**

fretful < AS. *fretan,* consume, wear away.

Gr. *aganaktikos,* fretful, irritable, peevish; see **pain**

Gr. *dyscheres,* annoying, peevish, captious: *Dyscheres griseus* (a beetle).

Gr. *dyskolos,* hard to please, discontented, fretful:

Gr. *dysthesia,* f. fretfulness, peevishness:

L. *morosus,* fretful, peevish, gloomy: morose, *Xystophora morosa* (a moth).

L. *petulans,* pert, saucy, impudent, peevish, irritable, quick in taking offense: petulant.

L. *pigeo, -itus,* feel annoyance, be irked, displeased:

See: **tease, arouse**

fretum, L. strait, channel, sound; *fretalis,* of a strait; see **way**

fretus, L. trusting to, relying upon; see **believe**

friabilis, L. easily broken or crumbled; see *frio* under **break**

friction < L. *frictio, -onis,* a rubbing; see *frico* under **rub**

friend < AS. *freond:* friendly, friendlily, friendliness, friendship, friendless, befriend.

L. *amicus,* m.; *amica,* f. friend; *amiculus,* m. dim.; *amicabilis; amicalis; amicus,* friendly; *amicitia,* f. friendship; *peramicus,* very friendly: amiable, amicable, amity, inimical, *Sympherobius amiculus* (a neuropterid), *Bolitophagus amicorum* (a beetle), *Metroxylon amicarum* (Tahiti-nut palm).

L. *benevolens, -entis,* c. well-wisher, friend:

L. *benignus,* kind, good, friendly, favorable: benign, benignant.

L. *comis, -e,* friendly, kind, affable, courteous: comity.

L. *conciliatus,* friendly, devoted to promoting agreement and friendliness: con-
ciliatory, reconciliation, conciliator.

Gr. *eumenes,* friendly, gracious, kind, well-disposed; *eumenetes,* m. friend, well-
wisher: Eumenides, Eumenidae, *Eumenes fraternus* (a potter-wasp).

Gr. *hilaos (hileos),* kind, gracious:

Gr. *idioxenos,* m. private friend:

Gr. *oaristes,* m. a familiar friend:

Gr. *philos,* dear one, friend; *philios,* friendly; see *phileo* under **love**

See: **companion, neighbor, love**

fright < AS. *fyrhtu;* see **fear**

frigidus, L. cold, inactive; *frigor,* cold; *frigidarius,* of cooling; see **cold**

frigo, *frixus,* L. fry, roast, parch; see **cook**

fringe < OF. *fringe (frange).*

L. *cirritus,* filamentous; see *cirrus* under **curl**

L. *crinitus,* long-haired, fringed; see *crinis* under **hair**

Gr. *exastis, -ios,* f. rough edge, fringe:

L. *fimbria,* f. fiber, thread, fringe; *fimbriatus,* fibrous, fringed: fimbriate, fimbrilla,
Fimbristylis autumnalis (a sedge), *Stenoglottis fimbriata* (an orchid), *Arisaema
fimbriatum* (a jack-in-the-pulpit).

Gr. *krossos,* m. fringe, tassel; *krossion,* n. dim.; *krossotos,* fringed, tasseled: crosso-
pterygian, Crossorhinidae, *Crossocephalus viviparus* (a nematode), *Crossosoma
peyerimhoffi* (a milleped), *Crossogaster triformis* (a wasp), *Crossaster papposus*
(a starfish).

L. *lacinia,* f. fringe or lappet on the border of a garment; *laciniosus,* full of flaps,
cut into thin strips, fringed: laciniate, lacinose, *Dentaria laciniata* (toothwort),
Carya laciniosa (shellbark-hickory), *Rubus laciniatus* (a blackberry).

Gr. *thysanos,* m. fringe, tassel; *thysanotos,* fringed: Thysanura, *Thysanocarpus
elegans* (a mustard), *Athysanus pusillus* (a mustard), *Thysanoëssa longicaudata*
(a shrimp).

See: **border, thread, cut, flap**

fringilla, L. chaffinch; see **finch**

fritillus, L. dice-box, hence spotted like dice; see **spot**

fritinnio, L. chirp, twitter; see **chirp**

frivolus, L. silly, trifling; see **fool**

frog < AS. *frogga.*

Gr. *batrachos,* m. frog; *batrachion,* n. dim.: batrachian, *Batrachus tau* (toadfish),
Batrachium trichophyllum (a water-crowfoot), *Batrachospermum moniliforme*
(an alga), *Batrachostomus auritus* (frogmouth), *Ammobatrachus montanensis*
(a Paleocene amphibian track), *Trichobatrachus robustus* (a frog with false
hairs), *Nitella batrachosperma* (a characead).

Gr. *gyrinos,* m. tadpole; *gyrinodes,* like a tadpole; *Gyrinophilus porphyriticus* (a
salamander), *Gyrinophorus luteipes* (a wasp).

NL. *hyla,* a genus of frogs, < Gr. *hyle,* wood, forest; see **wood**

Gr. *lalax, -agos,* babbler, croaker; green frog; see *lalia* under **speak**

L. *rana,* f. frog; *ranunculus,* m. dim. tadpole: ranarium, ranid, *Rana sylvatica*
(wood-frog), *Ranina dorsipeda* (a frog-crab), *Ranunculus aquatilis* (a buttercup),
Ranatra americana (water-scorpion), *Phacops rana* (a trilobite), *Basidiobolus
ranarum* (a fungus).

from < AS. *fram.*

L. *a, ab,* from, off, away, as in *avoco,* call away from; *absolvo,* loosen from, free:
avocation, absolve, avaunt, amanuensis, avulsion, advantage, abuse, absent, abject, abduction, absorb, abound, abstract, abhor, abjure, ablution, abolish, abscess,
abstain, abrogate, abnormality, avert, *Colletes aberrans* (a bee).

AS. *a, af, of,* of, off, from, out: arise, arouse, awake, aghast, amaze, abide, adown,
akin, ashamed, athirst, ordeal, of, off, offal, offset, offspring, checkoff, o'clock,
man-of-war.

L. *abditivus,* removed from, separated from:

L. *ablatio, -onis,* f. a taking away; *ablator, -is,* m. a remover: ablation, ablative,
ablator.

L. *absens, -entis,* away from, < *absum,* be away from: absent, absentee, absence,
in absentia.

L. *abstractus,* drawn away, separated: abstract, abstraction.

Gr. *apo,* from, off, away, separate, without, after, as in *apopempsis,* a sending
away: apology, apostate, apostle, apostrophe, apothecary, apotheosis, aphelion,
aphorism, apospory, apogamy, Apocrypha, *Apodytes dimidiata* (an icacinacead),
Apotettix eurycephalus (a grasshopper).

Gr. *ater; aterthe; aterthen,* apart from, away from, aloof:

L. *de,* from, down, of, as in *deflecto,* bend from, turn aside: debar, deciduous, derelict, debut, delicate, debonair, deduction, d'Orbigny, de la Croix, decrepitude, dedicate, definition, descend, dependent, degeneration, deliberate, dejected, deposit, defense, dehiscent, demolish, deny, detritus, delouse, dandelion, debt, devour, aldehyde, *Ctenochira deplanata* (a wasp).

L. *deductivus,* derivative; *dedux,* derived: deductive, deduction.

L. *demo, demptus,* take off, withdraw, subtract, remove; see **take**

L. *desum,* be absent:

L. *derivativus,* arising from, proceeding from, secondary, < *derivo, -atus,* draw off, divert: derivative, derivation, derive, *Chrysis derivata* (a wasp).

L. *di-, dif-, dir-, dis-,* < *dis* (OF. *des*), away from, asunder, not, as in *dimitto,* send away; *diffluo,* flow away, divide; *dirimo,* separate, divide; *dispendo,* weigh out, distribute; *dissimilis,* not like, unlike, different: diminish, digress, dilate, dilute, divert, digestion, divulge, differ, diffusion, difficulty, disagree, disband, disqualify, disclose, discourse, dishonest, disease, dispel, disdainful, disarm, disengage, disintegrate, disjunctive, disyoke, distant, dissect, disuse, discomfiture, dissimilar, diremption, direction, defeat, desquamate, descant, dessert, desoxybenzoin, desiodothyroxine, desoxycholic, *Dasymutilla digressa* (a wasp).

L. *e; ex,* out of, from; see **out**

Gr. *ek,* out of, from; see **out**

Gr. *ektotes, -etos,* f. absence:

L. *exemptus,* taken out, removed, excused; see **free**

Gr. *exhoristos,* banished, expelled; see **banish**

L. *exilium,* banishment, exile; see *exul* under **banish**

D. *van;* G. *von,* of, from: Van Buren, Van Dyke, van Loon, van Beethoven, von Goethe, von Hindenburg, Vanderbilt.

See: **separate, out, not, far, take, banish, far, alone, wean**

frond < L. *frons, frondis,* leafy branch, bough, foliage, leaf; *frondiculus,* dim.; see **leaf**

front < L. *frons, frontis,* brow, fore part of anything; see **face, before**

frost < AS. *frost;* see **ice**

fructus, L. fruit, see **fruit;** < *fruor, fructus; fruitus,* enjoy, see *fruor* under **joy**

frugalis, L. of fruits, economical, sparing, thrifty; see **save, fruit**

frugifer, L. fruit-bearing, fertile, < *frux, frugis,* fruit, produce; see **fruit, fertile**

fruit < L. *fructus, -us,* m. consequence, produce; *frux, frugis,* f. fruit, produce, grain; *frugalis,* of fruits; thrifty; *fructifer; frugifer,* fruitful; *fructuosus,* abounding in fruit; *frugilegus,* fruit-gathering: fructify, frugal, frugality, fruition, fruitful, unsufruct, fructivorous, frugivorous, unfruitfully, *Corvus frugilegus* (rook), *Gloeosporium fructigenum* (a fungus), *Laphygma frugiperda* (fall armyworm), *Sclerotinia fructicola* (brown rot).

NL. *achenium,* n. a kind of indehiscent fruit: achene, achenocarp, *Achyrachaena mollis* (a composite).

NL. *aecium,* n. a spore fruit or cluster of fungal spores; *aecidium,* n. dim.: aecium, aecial, aecidia, aecidiospore, aeciospore, *Aecidium punctatum* (a fungus).

Gr. *akrodryon,* n. a hard-shelled fruit, nut; see **nut**

NL. *basidium,* n. a kind of conidiophore, dim. of Gr. *basis,* pedestal: basidium, basidiospore, Basidiomycetes, *Exobasidium lauri* (a fungus).

Gr. *borassos,* palm fruit; see **palm**

NL. *conidium,* n. an asexual spore, dim. of Gr. *konis,* dust: conidium, conidiophore, conidial.

L. *drupa* (Gr. *dryppa*), f. an overripe, wrinkled olive; any pulpy fruit with a stone or pit enclosing the seed: drupe, drupiferous, drupaceous, *Juniperus drupacea* (Syrian juniper), *Syntomaspis druparum* (a chalcid fly), *Camellia drupifera* (Himalayan camellia), *Drypetes diversifolia* (a euphorbiacead).

L. *genimen, -inis,* n. fruit, product:

NL. *gonidium* (Gr. *gonidion,* dim. of *gone,* seed), n. an asexual cell or spore: gonidium, gonidial, gonidiophore.

Gr. *karpos,* m. fruit; *karpion,* n. dim.; *karpimos; karpophoros,* fruitful; *karpismos,* m. a gathering of fruit; *karpologos,* gathering fruit; *aeikarpos,* ever-bearing: carpel, carpology, acrocarpous, pericarp, schizocarp, parthenocarpy, pilocarpine, *Carpophilus hemipterus* (fruit-beetle), *Cercocarpus alnifolius* (a mountain-mahogany), *Podocarpus ferrugineus* (miro), *Callicarpa americana* (beautyberry), *Symphoricarpus albus* (snowberry), *Quercus macrocarpa* (bur-oak).

L. *palatha* (Gr. *palathe*), f. a cake of dried fruit, usually figs:

L. *pomum,* fruit of any kind, apple; see **apple**

NL. *pycnidium,* n. spore-bearing organ of a rust: pycnidium, pycnium, pycnidial.

L. *samara,* f. seed of the elm, any dry, indehiscent, winged fruit, key: samara, samaroid, *Cardiocarpon samariforme* (a fossil seed).

See: **seed, apple, berry, grain, nut, young, fertile, abundance**

fruitful; see **abundance, fertile, fruit**

fruitless; see **sterile**

frumentum L. grain, corn; *frumentalis; frumentarius,* of grain; see **grain**

frustror, *-atus,* L. disappoint, prevent, circumvent, thwart, deceive; see **hinder**

frustum, L. bit, piece, morsel; *frustillum; frustulum,* dim.; see **part**

fruticosus, L. bushy, shrubby, < *frutex, -icis,* shrub, bush; see **bush**

frux, *frugis,* L. fruit, produce; *frugifer,* fruitful, fertile; see **fruit**

fry < L. *frigo,* parch, roast; see **cook**

fucatus; fucinus; fucosus, L. colored, painted, rouged; see *fuco* under **color**

fuchsia, NL. a genus of the evening-primrose family; see **evening-primrose**

fucus, L. a genus of seaweeds, see **alga**; drone-bee, see **bee**

fugax, *-acis,* L. fleet, fleeting, swift; see **swift**

fugitivus, L. fleeing, flying; see *fuga* under **depart**

-ful < AS. *full,* abounding in; see **abundance**

fulcimen, L. prop, support, pillar; see **pillar**

fulcrum, L. bedpost, prop; see **pillar**

fulgidus, L. shining, gleaming; *fulgor, -is,* lightning, brightness; *fulgur, -is,* lightning, thunderbolt; see *fulgeo* under **light**

fulica, L. coot; see **coot**

fuligo, L. soot, smut; *fuliginatus; fuliginosus,* sooty, painted black; see **dirt**

full < AS. *full;* see **abundance, press**

fullo, L. fuller; *fullonicus; fullonius,* of fullers; see **thick**

fulmarus, NL. a genus of birds; see **albatross**

fulmen, *-inis,* L. flash of lightning, thunderbolt; *fulmino, -atus,* lighten; see **light**

fultura, L. prop, stay, < *fulcio, fultus,* stay, support; see *fulcimen* under **pillar**

fulvus, L. tawny, reddish-yellow; *fulvaster, -tra, -trum,* yellowish; see **brown**

fumidus, L. smoky, smoking, < *fumus,* smoke; *fumeus; fumicus,* smoky; *fumariolum,* smoke-hole; see **cloud**

fun < uncertain origin; see **joy, laugh, play**

funalis; funarius, L. of a rope; see *funis* under **rope**

function < L. *functio,* performance, < *fungor, functus,* perform, do, discharge; see **make**

funda, L. sling; *fundalis,* of a sling; *funditor,* slinger; see **throw**

fundatus, L. firm, grounded, < *fundo, -atus,* establish, lay a foundation; see *fundus* under **base**

fundibulum, L. funnel; see **funnel**

funditus, L. entirely, totally, completely; see **all**

fundula, L. street without outlet, blind alley; see **way**

fundulus, L. piston, see **pestle**; dim. of *fundus,* base, see **base**

fundus, L. bottom, base, estate; *fundamentum,* basis; see **base**

funero, *-atus,* bury, inter; *funus, -eris,* burial ceremony; *funebris; funereus,* pertaining to burial; *funestus,* causing death, mournful; see **cover**

fungus < L. *fungus,* m. mushroom; *fungillus; fungulus,* m. dim.; *funginus,* of mushrooms: fungi(pl.), fungal, fungate, fungoid, fungicide, fungilliform, fungologist, *Fungia symmetrica* (a coral), *Adlumia fungosa* (climbing fumitory), *Nanosella fungi* (a beetle), *Camerospongia fungiformis* (a sponge).   Derive: punk.

L. agaricum (Gr. *agarikon*), n. tinder-fungus: *Agaricus campestris* (a mushroom), *Agaricobia fulvicollis* (a fly), *Coeloptychium agaricoides* (a sponge), *Dragmacidon agariciformis* (a sponge).

Gr. amanita, f. a kind of fungus: *Amanita solitaria* (a fungus), *Amanitopsis vaginata* (a mushroom).

Gr. aschion, n. truffle, puffball:

NL. aspergillus, a genus of fungi whose conidiophores resemble a brush; see **bush**

G. bofist, puffball: *Bovista nigrescens* (a puffball), *Bovistella ohiensis* (a puffball).

L. boletus (Gr. *bolites*), m. mushroom: *Boletus edulis* (an edible mushroom), *Cis boleti* (a beetle).

NL. cronartium, n. a genus of fungi: *Cronartium strobilinum* (a fungus).

F. *ergot*, spur; a fungus: ergot *(Clavipes purpurea)*, ergotism, ergotamine, dihydroergotamine.

Gr. *erysibe*, f. mildew; *erysibodes*, mildewed; *erysibios*, preventing mildew: *Erysibe* (now *Erysiphe) graminis* (a mildew).

L. *fomes, -itis*, m. tinder, punk, kindling-wood; a genus of fungi: *Fomes applanatus* (a shelf-fungus), *Fomes idahoensis* (a Pliocene shelf-fungus).

NL. *fusarium*, n. a genus of fungi: *Fusarium avenaceum* (a fungus).

L. *hydnum* (Gr. *hydnon*), n. an edible mushroom: *Hydnum coralloides* (coral-mushroom), *Hydnocarpus ilicifolius* (a flacourtiacead).

Gr. *iska*, f. a fungus on trees:

Gr. *iton*, n. a kind of mushroom:

Gr. *krados*, blight, especially on fig-trees; see *krade* under **branch**

Gr. *krambos*, m. a grape blight:

NL. *morchella*, f. < G. *morchel*, a mushroom: *Morchella esculenta* (morel).

L. *mucedo, -inis*, f.; *mucor, -is*, m. mold, mildew, < *muceo*, be moldy, musty; *mucidus*, moldy: mucedinous, *Mucor mucedo* (a bread-mold), mucorine, mucormycosis.

Gr. *mykes, -etos*, m. fungus, mushroom; NL. *mycelium*, n. hyphal filaments: mycelium, mycorrhiza, actinomycete, aureomycin, bacillomycin(from *Bacillus subtilis*), chloromycetin, dermatomycosis, mycophenolic acid, streptomycin (from *Streptomyces griseus*), myxomycete, Ascomycetes, Mycetozoa, *Actinomyces scabies* (potato-scab), *Mycetophagus punctatus* (a beetle), *Mycetophilus pusillima* (a fossil fly), *Cyphomyrmex(Mycetophylax) simplex* (an ant).

NL. *nectria*, f. a genus of fungi: *Nectria cucurbitula* (a fungus).

NL. *penicillium*, a genus of fungi, < *penicillus*, painter's brush; see **bush**

L. *pezica* (Gr. *pezis, -ios*), f. a kind of stalkless mushroom: *Peziza[Pezica] coccinea* (a bloodcup), *Pseudopeziza trifolii* (a fungus).

NL. *phoma*, f. a kind of fungus: *Phoma oleracea* (a fungus), phomose, *Phomopsis citri* (a fungus).

Gr. *phoringes*, f. truffle:

Gr. *phthina*, f. mildew:

NL. *polyporus*, m. a genus of fungi: *Polyporus sanguineus* (a shelf-fungus).

L. *rubigo (robigo), -inis*, rust, blight, mildew, smut, mold; see **rust**

NL. *stereum*, n. a genus of fungi: *Stereum hirsutum* (a fungus).

NL. *tremella*, f. a genus of fungi: *Tremella mesenterica* (a fungus), tremelline, tremellose, *Leptogium tremelloides* (a lichen).

L. *uredo, -inis*, f. blight, smut: uredo, uredospore, uredinous, uredinium, *Uredo fici* (fig-rust).

L. *ustilago, -inis*, f. plant name now applied to a fungus: ustilaginous, Ustilaginaceae, *Ustilago nuda* (a smut), *Ustilago hordei* (barley-smut).

NL. *venturia*, f. a genus of fungi, < A. Venturi, Italian botanist: *Venturia inaequalis* (apple-scab).

See: **disease, rust, alga, rot**

funis, L. rope, line, cord; *funiculus*, dim.; see **rope**

funnel < L. *infundibulum*, n.: infundibulum, infundibuliform, *Cinclopyramis infundibulum* (a radiolarian).

Gr. *choane; chone*, f.; *choanos; chonos*, m. funnel-shaped hollow, crucible, melting-pot; *chonion*, n. dim.; choanocyte, chonolith, *Choanephora cucurbitarum* (a fungus), *Chonocephalus dorsalis* (a fly), *Choniangium epistomum* (a nematode), *Chonetes carinatus* (a fossil brachiopod).

L. *infundibulum*, n. funnel: infundibulum, infundibular, infundibulate, *Choanotaenia infundibulum* (a tapeworm).

L. *infurnibulum*, n. funnel for inhaling smoke:

See: **hollow, cup**

funus, *-eris*, L. burial ceremony; see *funero* under **cover**

fur, L. thief; *furunculus*, dim. petty thief, boil; *furax, -acis*, thievish; *furtivus*, stolen, secret, clandestine, see **steal, sore**; < OF. *fuerre*, sheath, case, cover, see **skin, garment**

furca, L. fork; *furcilla; furcula*, dim.; *furcifer*, fork-bearing; see **fork**

furfur, L. branlike scales, scurf, dandruff; *furfuraceus*, scaly; see **scale**

furiosus, L. mad, raging; *furia; furor*, madness, rage; see **mad**

furnish, < OF. *furnir*; see **supply**

furnus, L. oven; *furnaceus*, of ovens; *furnarius*, of ovens, baker; see *fornax* under **oven**

furrow < AS. *furh*, the trench made by a plow.

L. *agina*, f. channel in which the index of a balance moves:

Gr. *alox, -okos*, f. furrow: *Alocorhinus gemmatus* (a beetle).

Gr. *alveus,* bed of a stream; *alveolus,* dim.; see **hollow**

L. *aulax, -acis* (Gr. *-akos*), f. furrow; *aulakion,* n. dim.; *aulakodes,* furrowlike, <
aulakizo, plow; *micraulax, -akos,* with small furrows: aulacode, aulacocarpous,
Aulax cneorifolia (a proteacead), *Aulacomnium heterostichum* (a moss),
Cryptaulax glabella(a trilobite), *Aulacostethus melleus* (a wasp).

L. *canalis,* groove, conduit, channel; see **ditch**

Gr. *holkos,* m. furrow, track: *Disholcaspis globulus* (a gall-insect).

L. *lacunatus; laqueatus,* fluted, paneled, fretted, pitted; see *lacuna* under **hollow**

L. *lira,* earth thrown up by a plow, furrow; see **ridge**

L. *meatus,* passage, path; see **passage**

Gr. *ogmos,* m. furrow, swath, row, path, orbit: *Ogmogaster plicata* (a worm),
Dihogmochilus latimarginata (a fossil ostracode).

L. *stria,* f. furrow, channel, stripe, hollow, flutting, line; *striola,* f. dim.; *strio,*
-atus, hollow out, channel, groove, furrow, flute: striae (pl.), striated, striation,
Corallorhiza striata (an orchid), *Hymenocephalus striatulus* (a fish), *Campeloma*
multistriatum (a fossil gastropod), *Pareora striolata* (a gastropod).

L. *strix, strigis,* f. furrow, channel, groove, flute:

L. *sulcus,* m. furrow, groove; *sulculus,* m. dim.; *sulco, -atus,* furrow, plow; *bisulcus,*
two-furrowed, cloven; *desulco,* plow up, furrow through; *insulco,* make furrows:
sulcate, sulculus, sulciform, bisulcate, sulcal, *Diplocolpus sulcatus* (a radiolarian),
Chrysochus sulcaticeps (a beetle), *Cinyra sulcifera* (a beetle), *Catreus sul-*
caticollis (a grasshopper), *Divaricella quadrisulcata* (a pelecypod), *Alastor sulcifer*
(a wasp), *Lemna trisulca* (a duckweed), *Tabanus sulcifrons* (a gadfly).

L. *versus,* line, furrow; *versiculus,* dim.; see **poem**

See: **plow, dig, ditch, ridge, hollow, valley, fold**

furunculus, L. petty thief, boil, dim. of *fur,* thief; see **steal, sore**

furvus, L. dark, dusky, swarthy, black; *furvescens,* growing dark; see **black**

fury < OF. *furie,* < L. *furia,* violent rage; see **mad**

fusarium, NL. a genus of fungi; see **fungus**

fuscina, L. three-pronged fork, trident; *fuscinula,* dim.; see **fork**

fusus, L. spindle, see **spindle**; < *fundo, fusus,* pour, see **pour**

fuscus, L. dusky, dark, swarthy; see **black**

fusilis, L. fluid, liquid, molten; *fusor,* founder; see *fundo* under **pour**

fustis, L. club, bludgeon; *fusticulus,* dim.; *fustigo, -atus,* cudgel; see **club**

fusulina, NL. a genus of foraminifers; see **protozoan**

futile, L. a vessel broad above and pointed below; see **cup**

futilis, L. useless, vain, worthless; *futilitas,* uselessness; see **worthless**

futuo, *-utus,* L. copulate; *fututor; fututrix,* copulator; see **coitus**

future < L. *futurus,* about to be; see **after**

G

gabalium, L. a kind of shrub; see **plant**

gabalus, L. gallows; see **frame**

gabata, L. dish, platter; see **plate**

gable < F. *gable,* end wall of a building, particularly the triangular portion between the eaves and ridge of the roof, pediment: gable-end, *The house of*
the seven gables (Nathaniel Hawthorne).

Gr. *aetoma, -tos,* n. gable, pediment; *aetosis,* f. a forming of a gable: *Aetomoceras*
scipionianum (an ammonite).

L. *fastigium,* n. a slope up or down to a point, gable, pediment: fastigiate, fastigium, *Hypolaena fastigiata* (a restionacead).

See: **back**

gacino- < Gr. *gakinos,* earthquake; see **shake**

gadus, L. (Gr. *gados*), cod; see **cod**

gaesum, L. (Gr. *gaison*), a kind of javelin; see **spear**

gag < imitative origin; see **bind**

gagates, Gr. jet; see **coal**

gaillardia, NL. a genus of the composite family; see **composite**

gain < F. *gagner.*
  Gr. *alphema, -tos,* n. contract sum; *alphesis,* f. gain:
  Gr. *athlon,* prize of contest; see *athlos* under **fight**
  L. *brabeum* (Gr. *brabeion*), n. a prize in the games:
  L. *captura,* f. that which is taken, gain, profit, wages: capture.
  L. *compendium,* a saving, gain, profit, shortening; see **short**
  L. *fenus, -oris,* n. interest, gain; *fenebris; fenilis,* of interest; *fenerarius,* m. usurer; *fenerator, -is,* m. money-lender, capitalist, < *fenero, -atus,* lend on interest:
  L. *incrementum,* n. growth, increase, addition: increment.
  L. *interest,* it concerns, it claims attention; premium paid for the use of money: interest, interesting.
  L. *interpretium,* n. difference between buying and selling price, profit:
  Gr. *kerdos,* n. gain, profit:
  Gr. *ktaomai,* get; *ktema, -tos,* property, possession; *ktesis,* acquisition; *ktetikos,* acquisitive; *ktetos,* that may be had; *epiktetos,* acquired, gained; see *ktema* under **wealth**
  L. *lucrum,* n. gain; *lucellum,* n. dim.; *lucrativus; lucrosus,* gainful, profitable; *lucrator, -is,* m. gainer, winner; *lucrius,* of gain, < *lucror, -atus,* gain, win, get, make: lucre, lucrative, Lucretia.
  L. *mereo, -ritus,* deserve, earn; see **honor**
  L. *obventio, -onis,* f. income, revenue:
  Gr. *onetos,* profitable; *onesis,* f. profit:
  Gr. *pasis,* f. gain, possession:
  L. *premium,* n. profit, reward: premium.
  L. *quaestus,* m. profit, advantage:
  L. *redactus,* m. proceeds, < *redigo,* bring back:
  L. *reditus,* m. return, proceeds, income:
  Gr. *sostron,* n. reward, fee:
  Gr. *tokos,* m. interest, profit; *tokarion,* n. dim.; *anatokismos,* m. compound interest:
  L. *usuria,* f. use of money lent, interest paid on a loan; *usurula,* f. dim.; *usuarius,* of interest: usuary, usurious, usurer.
  See: **grow, receive, fruit, wealth, plunder**

galacto- < Gr. *gala, -aktos,* milk; *galaktion,* dim.; *galaxaios,* milky; see **milk**

galax, Gr. a genus of the diapensia family; see *gala* under **milk**

galaxias, Gr. milkyway, see *gala* under **milk**; a kind of fish, see **fish**

galbanum, L. (Gr. *chalbanon*), resinous sap of an umbellifer in Syria; see **resin**

galbeum, L. armband, fillet; see **ribbon**

galbula, L. dim. of *galbina,* a small bird; see **bird**

galbulus, L. cone of the cypress; see **cone**

galbus, L. yellow; *galbinus,* yellowish; see **yellow**

gale < AS. *gagel,* a marsh plant with aromatic odor, see **bayberry**; < Gr. *gale,* polecat, weasel, see **weasel**; < uncertain origin; a strong wind, see **wind**

galea, L. helmet; *galeola,* dim.; *galearis,* of a helmet; see **cap**

galena, L. an ore of lead; see **lead**

galenos, Gr. calm; see **rest**

galeopsis, L. (Gr. *galiopsis*), a genus of the mint family; see **mint**

galeos, Gr. shark; see **shark**

galeros, Gr. cheerful; see **joy**

galerum, L. bonnet, cap, hat, helmet; *galericulum,* dim.; see *galea* under **cap**

galgulus, L. a small bird; see **bird**

galium, L. (Gr. *galion*), bedstraw, cleavers; see **madder**

gall < AS. *gealla,* bile; see **bile**

gall < L. *galla,* f. a pathologic swelling or excrescence on plants; *gallula,* f. dim.: gall, gallic acid, gallanol, gallocyanin, galloflavin, pyrogallol, ellagic, *Podapion gallicola* (pine gall-weevil), *Nectria galligena* (a fungus), *Gnorimoschema gallaesolidaginis* (goldenrod gall-moth).

L. *bruscum,* an excrescence on the maple-tree; see *ruscum* under **bush**

Gr. *gongros,* excrescence, swelling, knot; see **swell**

Gr. *kekis, -idos,* f. gall; *kekidion,* n. dim.: cecidium, cecidology, cecidomyian, *Cecidomyia leguminicola* (clover-seed midge), *Cecidobracon asphondyliae* (a wasp).

Gr. *korykis, -idos,* f. gall on leaves:

Gr. *krotone,* f. a gall on trees:

Gallia, L. country of the Gauls; see **Celt**

gallicrus, L. a ranunculacead; see **buttercup**

gallicula, L. dim. of *gallica,* a Gallic shoe; see **shoe**

gallidraga, L. hairy teasel; see **thistle**

gallus, L. cock; *gallina,* hen; *gallinula,* dim. pullet; *gallinaceus,* of poultry; see **chicken**

galumna, L. cover of any kind; see **blanket**

galvano-, pertaining to electricity, < Luigi Galvani, Italian physiologist; see **electricity**

gamba, L. hoof; see **foot**

gamble < AS. *gamen,* amusement, sport; see **lot, play**

Gambrinus, L. mythical inventor of beer; see **beer**

gambros, Gr. akin, related; see **kin**

game < AS. *gamen,* sport; see **play**

gamelios, Gr. bridal, nuptial; see *gamos* under **marry**

gameto- < Gr. *gamete,* wife; *gametes,* husband; see **spouse**

gamma, Gr. third letter of the Greek alphabet; see **letter**

gammarus, L. variant of *cammarus,* lobster; see **crab**

gamos, Gr. marriage, union; see **marry**

gamphela, Gr. jaw; see **jaw**

gampsos, Gr. crooked, curved; see *kamptos* under **bend**

ganeum, L. (Gr. *ganeion*), restaurant; *ganearius,* of a restaurant; see **eat**

gangamo- < Gr. *gangamon,* a small, round net; see **net**

ganglion, Gr. tumor or swelling under the skin, knot or plexus of nerves; see **nerve**

gangrene < L. *gangraena* (Gr. *gangraina*), mortification; see **sore**

gannio, L. yelp, bark, whine, snarl, growl; see **bark**

ganos, Gr. brightness; see **light**

gap < ON. *gapa,* open, yawn, gape; see **open, space, hole, hollow, empty**

gar < AS. *gar,* spear; see **spear, pike**

garden < OF. *gardin;* see **field, park, farmer, till**

gardenia, NL. a genus of the madder family; see **madder**

gargalismos, Gr. a tickling; see **tickle**

gargareon, Gr. uvula; see **uvula**

gargle < F. *gargouille,* throat.

Gr. *gargarizo,* gargle; *gargarisma, -tos,* n. a gargle:

garinos, Gr. a fish; see **fish**

garland < OF. *garlande;* see **crown**

garlic < AS. *garleac;* see **onion**

garment < OF. *garnement.*

L. *abolla,* f. cloak of thick, woolen cloth: *Abolla pallicosta* (a moth).

L. *alicula,* a light, upper garment, coat; see *ala* under **wing**

Gr. *allix, -ikos,* f. a man's upper garment, coat:

L. *amiculum,* n. cloak, mantle; *amictus,* m. garment, vestment, < *amicio, -ictus,* wrap about, clothe, cover, conceal: amice, *Amictus tigrinus* (a fly).

L. *ampechone,* f. shawl, robe:

Gr. *amphiblema, -tos,* n. garment: *Amphiblema eucharis* (a gastropod).

Gr. *amphiesma, -tos; amphion,* n. garment; *amphiestris, -idos,* f. nightgown: *Amphion brennus* (a butterfly), *Amphiesma flavipunctatum* (a snake), *Amphiestris baetica* (a katydid).

L. *armilausa,* f. a military upper garment:

L. *arnacis, -idis* (Gr. *arnakis, -idos*), sheepskin; see **skin**

Gr. *baite,* f. coat of skin, tent of skin:

Gr. *berberion,* n. a shabby garment:

Gr. *beuthos,* n. a woman's dress:

L. *birrus* (Gr. *byrrhos*), m. cloak or mantle to keep off rain: biretta, byrrus, *Byrrhus gigas* (a pill-beetle)

Gr. *brakos*, n. a woman's garment:

L. *calthula*, f. a yellow garment for women, a yellow robe:

L. *camisia*, f. shirt, nightgown: chemise, camisia, camisole.

LL. *cappa*, f. hood, cape: cape, cope, cap, capeador, caparision, capellet, chape, chaplet, chapel, chaperon, chaplin, escape, escapade, handicap, capeline, foolscap, Aix la Chapelle.

L. *caracalla*, f. long tunic or great coat with a hood:

L. *castula*, f. petticoat:

L. *cento, -onis*, m. garment of pieces sewn together, patchwork:

Gr. *cheimas, -ados*, winter; a winter garment; see *cheimon* under **year**

Gr. *chiton, -os*, m. tunic, garment worn next to the skin; *chitonarion; chitonion*, n.; *chitoniskos*, m. dim.: chitin, chitinous, chiton *(Chaetopleura apiculata)*, *Ischnochiton conspicuus* (showy chiton), *Brachychiton populneus* (kurrajong bottletree), *Triplochiton scleroxylon* (a sterculiacead), *Lysichitum americanum* (an aroid).

Gr. *chlaina; laina* (L. *laena*), f. cloak, mantle, wrapper; *chlainion*, n. dim.; *chlanis, -idos*, f. a finer garment than the *chlaina; chlanidion*, n.; *chlaniskos*, m. dim.: *Euchlaena mexicana* (teosinte), *Tricholaena rosea* (a grass), *Thysanolaena maxima* (a grass), *Notholaena aurea* (cloakfern).

Gr. *chlamys, -ydos*, f. mantle, cloak; *chlamydion*, n. dim.: chlamydospore, achlamydeous, Archichlamydeae, *Chlamydophorus truncatus* (pichiciago), *Chlamydomonas nivalis* (red snow-alga), *Callichlamys riparia* (a bignoniacead), *Spodochlamys poultoni* (a beetle), *Stylochlamydium asteriscus* (a radiolarian).

L. *cilicium*, n. a garment of haircloth; *cilicinus*, made of haircloth:

Gr. *derris, -eos*, f. leather coat; *derrion*, n. dim.: *Derris scandens* (Malay jewel-vine), *Scleroderris fuliginosa* (a fungus).

Gr. *enamma, -tos*, n. a garment: *Enamma striatum* (a Jurassic beetle).

Gr. *endyma, -tos; endyton*, n. garment, dress; *ependyma, -tos*, n.; *ependytes*, m. an upper or outer garment, tunic worn over another, cloak; *endysis*, f. a putting on; *endytos*, put on; *ekdyma, -tos*, n. something taken off:

Gr. *ephestris, -idos*, f. outer garment, mantle, cloak: *Ephestris vinosa* (a butterfly).

L. *epicrocum* (Gr. *epikrokon*), n. a transparent woman's garment:

Gr. *esthema, -tos*, n.; *esthes, -etos*, f. dress, clothing, garment, raiment; *esthesis*, f. clothing, < *esthio*, clothe: *Astronesthes nigra* (a fish), *Haploesthes greggi* (a composite), *Penthestes[Penthesthes] rufescens* (chestnut-backed chickadee).

Gr. *exomis, -idos*, f. vest: *Exomis peplopteroides* (a beetle).

L. *gausapa*, f. (Gr. *gausapes*, m.), a shaggy, woolen garment: *Gausapa globifera* (a mite), *Stereum gausapatum* (a fungus).

L. *habitus*, m. dress, attire: habit.

Gr. *hima (heima), -tos*, n. dress, garment, carpet; *himatidarion; himatidion; himation*, n. dim.; *himatiotheke*, f. clothes chest; *dyseimon, -os*, ill clad: *Himatiopetalum ictericum* (a milleped).

L. *indusium*, n. tunic, garment; *inducula*, f. undergarment; *indumentum*, n. garment, clothing; *induvia*, f. garment; *indutor, -is*, m. wearer, < *induo, -dutus*, put on clothes, dress: indusium, indumentum, *Digitaria induta* (a grass), *Buxbaumia indusiata* (a moss).

Gr. *kalasiris*, f. a garment fringed at the bottom:

Gr. *kaunakes*, m. a thick cloak; *kaunakion*, n. dim.: *Caunaca bicingulata* (a moth).

Gr. *kimmerikon*, n. a woman's garment:

Gr. *kolobion*, n. undergarment with short sleeves:

Gr. *kyklas, -ados*, f. a woman's garment with a border:

Gr. *kypas, -ados; kypassis, -idos*, f. a short frock; *kypassion*, n.; *kypassiskos*, m. dim.: *Peliocypas hamata* (a beetle), *Cypassis palliata* (a radiolarian).

L. *lacerna*, f. a kind of cloak; *lacernula*, f. dim.: *Lacerna hosteensis* (a bryozoan).

Gr. *laiphos*, a shabby, tattered garment, canvas sail; see **sail**

Gr. *ledos*, n. a cheap dress; *ledion*, n. dim.:

L. *limus*, m. apron worn by priests and attendants at sacrifices:

Gr. *lope*, f. cloak, mantle. robe; *lopion*, n. dim.: *Lopomorphus carinatus* (a bug).

Gr. *mandya*, f. a woolen cloak:

L. *mantelum (mantellum)*, n. cloth, cloak, garment; Sp. *manta*, f. blanket, cloak; *mantilla*, f. cape, cloak; *mantuelis*, cloaklike: mantle, manteau, mantel, manteltree, mantelet, dismantle, mantilla, *Manta birostris* (devilfish, a ray).

L. *opertorium*, n. cover, garment:

L. *pallium*. n. mantle, robe, cloak, coverlet; *palla*, f. robe, mantle; *palliolum*, n.; *pallula*, f. dim.; *palliatus*. cloaked, protected, mitigated; *palliolatus*. covered with a hood: pallial, pallium, pall, palliate, palliative, tarpaulin, *Pallium con-*

vexum (a pelecypod), *Palliolum fosterianum* (a pelecypod), *Hemotopus palliatus* (oyster-catcher), *Eremobates pallipes* (a solpugid).

L. *paludamentum,* n. a military cloak; *paludatus,* dressed in a military cloak: L. *penula,* f. woolen cloak, mantle: *Penula ocellata* (a turbellarian).

L. *peplum,* n.; *peplus* (Gr. *peplos*), m. robe, tunic: peplum, *Euphorbia peplus* (a spurge), *Callipepla squamata* (scaled quail), *Brochopeplus reticulatus* (a grasshopper), *Pheucticus chrysopeplus* (a bird).

Gr. *peribole,* f. anything thrown around as a covering, garment:
Gr. *perizostra,* f. apron:
L. *pexatus,* covered with a napped garment:
Gr. *pharos,* n. cloak, mantle, shroud: *Pharomacrus mocinno* (quetzal).

L. *sagum,* n.; *sagus* (Gr. *sagos*), m. cloak, mantle, blanket; *sagulum,* n. dim.; *sagarius,* of a cloak; *sagatus; sagulatus,* cloaked, mantled: *Sagoplegma* (a radiolarian).

Gr. *samakion,* n. a woman's garment:
Gr. *sisyra; sisyrna,* f. goatshair cloak, garment of skin; *sisyrnion,* n. dim.; *sisyrnodes,* like a fur garment: *Sisyra fuscata* (a lacewing), *Sisyromyia auratus* (a fly).

Gr. *speirion,* n. light garment, < *speiron,* n. piece of cloth; *speirodes,* with many thin coats:

L. *stola* (Gr. *stole; stolis, -idos*), f. a long garment, robe, fold; *stolidion; stolion,* n. dim.; *stolidoma, -tos,* n. fold of a robe; *stolidotos,* in folds, folded; *stolmos,* m. raiment, equipment; *chryseostolmos,* decked with gold: *Brachystola magna* (a grasshopper), *Stolidosoma cyaneum* (a fly), *Stolidoma crassidens* (a gastropod), *Stolmorhynchia stolidota* (a Jurassic brachiopod), *Stolopsyche libytheoides* (a fossil butterfly).

L. *subligaculum,* n. apron:
L. *subucula,* f. shirt: *Echinopsis subucula* (an echinoid).
Gr. *syrma,* robe with a train; see *syrtos* under **carry**
Gr. *tebenna,* f. a robe of state: *Tebennophorus sallei* (a gastropod), *Dystebenna stephensi* (a moth).

Gr. *theristrion,* n. a light summer garment:
L. *toga,* f. the outer garment of a Roman citizen; *togula,* f. dim.; *togatus,* wearing a toga: toga, *toga candida* (white toga worn by candidates), *Solenomya togata* (a pelecypod).

L. *trabea,* f. a robe of state; *trabealis,* of a trabea; *trabeatus,* wearing a trabea: *Oecomys trabeatus* (a mouse).

Gr. *tribon, -os,* m. a worn cloak; *tribonarion; tribonion,* n. dim.: *Tribonophorus desmaresti* (a bat), *Tribonium spectrum* (a blattid).

L. *tunica,* f. garment; *tunicula,* f. dim.: tunic, tunicle, Tunicata, *Tunica velutina* (a pink), *Halysiocrinus tunicatus* (a Mississippian crinoid), *Zea mays tunicata* (pod-corn).

L. *velamen, -inis,* n. robe, garment, cover: velamen, velamentum.

L. *vestis, -is,* f.; *vestimentum,* n.; *vestitus,* m. garment, clothes; *vesticula,* f. dim.; *vestiarius,* of clothes; *vestitor, -is,* m. tailor: vest, divest, investment, transvestite, vestry, travesty, revetment, *Saperda vestita* (linden-borer), *Bracon vestitor* (a wasp), *Pediculus vestimenti* (body-louse, cootie).

Gr. *zeira,* f. a wide garment, robe: *Zeiraphera argyrana* (a butterfly).
Gr. *zoma,* a girded garment; the position of the girdle; see *zona* under **belt**
See: **cover, trousers, skin**

garrulus, L. talkative; a jay, < *garrio, -itus,* talk, chatter, prate, babble, gossip; see **speak, crow**

garter < OF. *gartier,* band for supporting for supporting a stocking.
 L. *genualium,* n. garter:
Gr. *periskelis, -idos,* f. legband, anklet, garter:

gas < uncertain origin: gaseous, gasoline, gasser, gashouse. See **wind**

gaster, *-os; -tros,* Gr. stomach, belly, paunch, womb; *gastridion; gastrion,* n. dim.; *gastris, -idos,* potbellied; see **stomach**

gate < AS. *gaet;* see **door**

gather < AS. *gaderian.*
 L. *accursus; concursus,* m. a running together, meeting, assembly: concourse.
 Gr. *aboletys, -yos,* f. a meeting:
 Gr. *agarrhis,* f. a meeting:
 Gr. *agora,* f. assembly, market; *agyris, -ios,* f. gathering, crowd; *agyrma, -tos,* n. anything collected; *agyrtes,* m. collector, < *ageiro,* gather, bring together; *panegyris,* f. national or festal assembly; *panegyrikos,* festive, adorned: agora, agoraphobia, allegory, category, categorical, paregoric, panegyric.

Gr. *athroisma, -tos,* n. a gathering, collection; *athroistikos,* of collecting; *athroos,* assembled, collected, crowded, < *athroizo,* gather, collect: *Athrotaxis laxifolia* (a conifer).

L. *borrio,* swarm:

L. *bule* (Gr. *boule*), f. council of elders, senate:

L. *carpo, carptus,* pluck, pick, gather; see **tear**

L. *cogo, coactus,* assemble, collect, compress; *coactor, -is,* m. collector of money; *coactura,* f. a collection:

L. *collatus,* gathered together, collected; *collaticius; collativus,* brought together, collected; *collatio, -onis,* f. collection: collation, collate.

L. *colloquium,* n. conference, conversation: colloquium.

L. *colluvium,* collection of washings, sweepings, dregs; see *luo* under **wash**

L. *comitia,* f. assembly of the Romans to elect magistrates: comitial.

L. *commissio, -onis,* f. a bringing together, undertaking: commission, committee.

L. *compilo, -atus,* gather, plunder; *compilator, -is,* m. gatherer, plunderer: compile, compilation.

L. *compono, -positus,* bring, put, or set together: compose, composition, compositor.

L. *concentratus,* gathered, focused: concentrate, concentration.

L. *concilium,* n. assembly, gathering, meeting: council, councilor.

L. *conditum,* thing laid up, store; see **store**

L. *conglobo, -atus,* gather into a ball, crowd together:

L. *conglomero, -atus,* roll together, crowd, concentrated: conglomerate, conglomeration.

L. *congregatio, -onis,* f. assembly, association: congregation, congregational.

L. *congressus, -us,* m. assembly, conference, meeting: congress, congressional.

L. *congruo,* run together, meet, agree, coincide: congruent.

L. *conquisitor, -is,* m. recruiting officer:

L. *consistorium,* n. place of assembly: consistory.

L. *contio, -onis,* f. assembly, meeting, speech; *contionalis,* of an assembly:

L. *contribuo, -butus,* bring together, unite, collect: contribution, contribute, contributor.

L. *conveho, -vectus,* bring together, gather: convey, conveyance, convection.

L. *convena; conventio, -onis,* f.; *conventus,* m. assembly, meeting: convention, convent, conventicle.

L. *convoco, -atus,* call together, assemble: convocation.

Gr. *ekklesia,* f. assembly of the church: ecclesiastic.

Gr. *ellimenistes,* collector of harbor dues; see **pay**

L. *exactor, -is,* m. tax-collector, demander: exactor, exaction.

L. *foedero, -atus,* form a league; see **bind**

L. *grego, -atus,* gather in a herd, flock, collect, assemble; *aggregatus; congregatus,* collected, gathered; see *grex* under **herd**

Gr. *halia,* f. assembly of the people: *Platygaster halia* (a proctotrypid).

Gr. *homileo,* consort with, have communion with, be in company of: homiletic.

L. *lego, lectus* (Gr. *lego*), collect, pick, read, speak; *legulus,* m. collector, gatherer; *collectio, -onis,* f.; *collectus,* m. gathering, concentration; *collectaneus; collecticius,* collected, gathered together; *collegium,* n. an association of colleagues; *relectus,* gathered, recollected; *analektos,* select, choice; *syllegma, -tos,* n. collection: legion, eligible, recollect, elite, diligent, predilection, election, selection, sacrilege, collective, analects, catalog, gastrilegous, podilegous, *Corvus frugilegus* (rook), *Syllegomydas cinctus* (a fly).

AS. *metan,* come together, gather; *mot; gemot,* meeting: meet, meeting, moot, gemot.

L. *meto, messus,* reap, gather; see **reap**

Gr. *pansyrtos,* swept together, accumulated; see *syrtos* under **carry**

Gr. *pnyx, pyknos,* f. Pnyx, place of public assembly in Athens: Pnyx.

L. *presbyterium* (Gr. *presbyterion*), as assembly or council of elders; see *presbys* under **old**

L. *senatus,* m. council of elders: senate, senator.

Gr. *symphoresis,* f. a bringing together; *symphorema, -tos,* n. an accumulation; *symphoretos,* brought together:

L. *synagoga* (Gr. *synagoge*), f. a bringing together, gathering, place of meeting: synagogue.

Gr. *synakter, -os,* m. collector; *synaktikos,* able to bring together:

Gr. *synkomizo,* gather in, store up; *synkomistes,* m. gatherer; *synkomide,* f. harvest; *synkomistos,* brought together:

L. *synodus* (Gr. *synodos*), f. assembly, meeting; *synodia,* f. journey in company, caravan: synod.

Gr. *systremma, -tos; systrophe,* anything aggregated or twisted together; see **ball**

See: **cluster, heap, number, reap, herd, hunt**

gaudialis, L. glad, joyful; *gaudium,* joy; see *gaudeo* under **joy**

gaultheria, NL. a genus of the heath family; see **heath**

gaulus, L. (Gr. *gaulos*), bucket, pail; see **bucket**

gauros, Gr. haughty, arrogant, proud; *gaurex, -ekos,* braggart; see **proud, brag**

gausapa, L. shaggy, woolen cloth or garment; see **garment**

gausos, Gr. crooked, bent; see **bent**

gavia, L. a bird; now applied to loons; see **loon**

gavisus, L. causing delight; see *gaudeo* under **joy**

gaylussacia, NL. a genus of the heath family; see **heath**

gaza, Gr. treasure, riches; see **wealth**

gazelle, F., < Ar. *ghazal,* a kind of antelope; see **antelope**

gegonos, Gr. loud-sounding, sonorous; see **loud**

gelasimos; gelastos, Gr. laughable, ridiculous; *gelos,* laughter; see *gelao* under **laugh**

gelasinus, L. (Gr. *gelasinos*), dimple; see **hollow**

gelatin < L. *gelatus,* congealed, stiffened; see **jelly**

gelatus, L. frost; congealed, stiffened; see *gelidus* under **cold**

geld < ON. *geldr,* barren.
    Gr. *aorches,* gelded:
    L. *castro, -atus,* remove the testicles, geld, emasculate: castrate, castration.
    Gr. *ektmesis,* f. castration, < *ektemno,* cut out: *Ectmesopus pallidus* (a beetle).
    L. *emasculo, -atus,* geld: emasculate, emasculation.
    OF. *espee,* cut with a sword, < L. *spatha,* sword; remove the ovaries: spay.
    L. *eunuchus* (Gr. *eunouchos*), a castrated male person; see **sterile**
    L. *eviro, -atus,* unman, emasculate:
    L. *exseco, -sectus,* cut out, geld; *sectarius,* gelded, castrated:
    L. *majalis,* gelded boar, barrow, eunuch; see **sterile**
    Gr. *thladias; thlibias,* eunuch; see **sterile**
    See: **sterile**

geleches, Gr. sleeping on the ground; see **sleep**

gelgis, -idos, Gr. head or clove of garlic; see **onion**

gelidus, L. cold, frosty, icy; *gelicidium,* frost; see **cold**

gelsemium, NL. a genus of the logania family; see **logania**

gem < L. *gemma,* bud, precious stone; *gemmarius,* jeweler; see **stone**

gemellar, L. vessel for holding oil; see **pot**

geminus, L. twin; *gemellus,* dim.; *gemino, -atus,* double, pair, join; see **two**

gemitus, L. groan, sigh; *gemebundus; gemulus,* moaning, sighing; see *gemo* under **sad**

gemma, L. bud, eye (of a plant), jewel, precious stone; *gemmula,* dim.; *gemmatus; gemmeus,* pertaining to buds or gems; see **bud, stone**

gemos, Gr. load, freight; *gemistos,* laden; see **weight**

gempylus, NL. a mackerellike fish; see **mackerel**

gemursa, L. a swelling between the toes; see **swell**

gen-; -gen < L. *gigno, genitus* (Gr. *gignomai*), be born, causing, producing, forming; *genea,* race, stock, family; *genero, -atus,* beget, create; see **birth, class, begin**

gena, L. cheek; see **cheek**

general < L. *generalis,* pertaining to all, common; see **common**

genero, -atus, L. beget, create; *generator; generatrix,* begetter, creator; see **begin**

generous, L. noble, kind, liberal, magnanimous; *generositas,* liberality; see **give**

genesis, Gr. beginning, birth, origin; see **birth**

genethlon, Gr. birthday; *genethlios,* of a birthday; see *gigno* under **birth**

geniado- < Gr. *genias, -ados,* beard; *geneiates,* bearded; see **hair**

genialis, L. delightful, jovial, pleasant; see **agreeable**

genico- < Gr. *genikos,* generic; see *genus* under **class**

geniculatus, L. like the bent knee, knotty, jointed; see *genu* under **knee**

genio- < Gr. *geneion,* chin; see **chin**

genista, L. broom, a plant of the bean family; see **bean**

genital < L. *genitalis,* pertaining to procreation, < *gigno, genitus,* beget, produce; see **sex organs, birth**

genius, L. protecting attendant spirit of a person or place, see **spirit;** taste, wit, talents, see **know**

gennaeo- < Gr. *gennaios,* noble, excellent; see **honor**

gens, *gentis,* L. clan; *genticus,* of the clan or race; see *genus* under **class**

gentian < L. *gentiana* (Gr. *gentiane*), f.: Gentianaceae, *Gentiana lutea* (yellow gentian), *Veronica gentianoides* (a speedwell).
L. *exacum,* n. a genus of the gentian family: *Exacum affine* (a gentianacead). NL. *sabbatia,* f. a genus of the gentian family, < Liberatus Sabbati, Italian botanist: *Sabbatia dodecandra* (marsh rose-gentian).

gentle < L. *gentilis,* of the same clan or race; see **mild,** *genus* under **class**

genu, L. knee, joint, knot; *geniculum,* dim.; see **knee**

genuine < L. *genuinus,* native, natural, authentic; see **true**

genus, *-eris,* L. (Gr. *genos*), birth, origin, race, stock, kind; see **class**

genys, *-yos,* Gr. jaw, cheek, chin, edge of an ax; see **jaw**

geo- < Gr. *ge; gaia,* earth; *gedion,* dim.; *geodes,* earthlike, earthy; see **earth**

geonoma, NL. a genus of palms; see **palm**

georgyos, Gr. farmer; *georgikos,* agricultural; see *geïtes* under **farmer**

gephyra, Gr. bridge; see **bridge**

-ger, *-a, -um; -gerous,* < L. *gero, gestus,* bear, carry, perform; *gerula; gerulus,* bearer, carrier; see *gero* under **carry**

geranium, L. (Gr. *geranion*), n. cranesbill: Geraniaceae, geraniol, *Geranium maculatum* (a geranium).
NL. *pelargonium,* n. a genus of the geranium family: *Pelargonium tomentosum* (woolly storksbill).

geranos, Gr. crane; *geranodes,* cranelike; see **crane**

geraros, Gr. dignified, honored; see **honor**

geras, Gr. old age; *geraios,* aged, old; see **old**

geridus, L. weaver; see **weave**

German < L. *Germanus,* m., < a Celtic word; *Germania,* f. Germany; *Germanicus,* of the Germans: Germanic.
L. *Gothus,* m. an individual of a Teutonic people; *Gothicus,* of the Goths: Goth, Gothic, Gotham, Ostrogoths, Visigoths, *Tetrapedia gothica* (an alga).
L. *Teuto, -onis,* m. an individual of a German tribe; *Teutonicus,* of the Teutons: Teuton, Teutonic, Deutschland, Dutch.

germanus, L. having the same parents, genuine, true; see **kin**

germen, *-inis,* L. bud, seed, sprout; *germino, -atus,* sprout; see **bud**

geronto- < Gr. *geron, -tos,* old man; *gerontion,* dim.; *geros,* old age; see *geras* under **old**

gerra, L. trifle, nonsense; see **trifle**

gerres, L. a basslike, marine fish; see **perch**

gerrho- < Gr. *gerrhon,* screen, wickerwork; see **screen**

gerula; gerulus, L. bearer; see *gero* under **carry**

gerys, Gr. speech, voice, sound; *Geryon,* the shouter, a three-bodied monster; *geryma,* voice, sound; *gerygone,* born of sound; see **speak**

gestatus, L. borne, carried; *gestamen,* burden, load, < *gesto, -atus,* frequentative of *gero, gestus,* bear, carry, see **carry, weight;** *gestum,* act, event, see **make**

gesticulor, *-atus,* L. make expressive motions of the body, pantomime; see **equal**

gestor, L. tattler; see **speak**

get < ON. *geta* (AS. *gitan*); see **receive, take, gain, find, make**

gethosynos, Gr. glad, joyful; see *getheo* under **joy**

gethyum, L. (Gr. *gethyon*), a kind of leek; *gethyllis,* dim.; see **onion**

geum, L. avens, bennet; a genus of the rose family; see **rose**

geustes, Gr. taster; *geustos,* to be tasted; *geustikos,* of taste; see **taste**

ghost < AS. *gast,* soul, spirit; see **spirit**

giant < OF. *jaiant,* < Gr. *gigas;* see *gigas* under **large**

gibber; gibbus, L. humped, humpbacked, protuberant, bent; *gibberosus; gibbosus,* very crooked, twisted, protuberant; see **projection**

gift < ON. *gift;* see **give**

gigarton, Gr. grape seed; *agigartos,* seedless; see **seed**

gigas, *-gantos,* Gr. giant; see **large**

gigerum, L. entrail, particularly of birds; see **intestine**

giggle < imitative origin; see **laugh**

gilia, NL. a genus of the phlox family; see **phlox**

gill < uncertain origin.
>   Gr. *branchos,* n. gill, fin; *branchion,* n. dim. fin; *branchia* (pl.), gills: branchial, branchiate, elasmobranch, nudibranch, *Branchiostoma lanceolatum* (lancelet, amphioxus), *Eubranchipus vernalis* (fairy-shrimp), *Melanobranchus micronema* (a fish), *Cryptobranchus alleganiensis* (hellbender).
>   Gr. *epitygma, -tos,* n. flap, gill:

gillo, -onis, L. a cooling-vessel, cooler; see **cold**

gilvus (gilbus), L. pale yellow; see **yellow**

ginger < L. *zingiber,* n. (Gr. *zingiberis,* f.), < Skt. *srngavera:* Zingiberaceae, *Zingiber officinale* (ginger).
>   L. *amomum* (Gr. *amomon*), n. an aromatic plant; a genus of the ginger family: *Amomum cardamon* (a zingiberacead), *Aframomum granum-paradisi* (a zingiberacead), *Cornus amomum* (silky dogwood).
>   L. *costus* (Gr. *kostos*), m. a kind of ginger: costus-root, alecost, costmary *(Chrysanthemum balsamita), Costus igneus* (spiralflag).
>   NL. *curcuma,* f. < Ar. *kurkum;* a genus of the ginger family: *Curcuma longa* (turmeric), curcumin.
>   NL. *hedychium,* n. a genus of the ginger family: *Hedychium coccineum* (scarlet ginger-lily).
>   See: *asarum* under **birthwort**

gingidium, L. (Gr. *gingidion*), a kind of carrot; see **carrot**

gingiva, L. gum; *gingivula,* dim.; see **mouth**

ginglaros, Gr. flute, fife; see **pipe**

ginglismos, Gr. a tickling; see **tickle**

ginglymos, Gr. hinge-joint; see **hinge**

gingras, Gr. a Phoenician flute or fife; see **pipe**

gingritus, L. cackling of geese; see **cackle**

ginkgo, NL. a genus of gymnosperms; see **gymnosperm**

ginseng < Chin. *jenshen:* Asiatic ginseng *(Panax schinseng).*
>   NL. *aralia,* f. a genus of the ginseng family: Araliaceae, *Aralia spinosa* (devil's walking-stick), *Aralidium pinnatifidum* (a Malay tree), *Cecropia araliaefolia* (a tree).
>   L. *panax, -acis* (Gr. *-akos*), m. a plant with cure-all properties; a genus of the ginseng family: *Panax quinquefolius*(American ginseng), *Nothopanax anomalus* (a false-panax), *Pseudopanax crassifolius* (an araliacead), opopanax.
>   L. *pharnaceon,* n. a kind of ginseng: *Eriogonum pharnaceoides* (a polygonacead).

giraffe < Ar. *zarafa;* NL. *giraffa,* f.: *Giraffa camelopardalis* (giraffe), *Giraffomyia solomonensis* (a fly), *Apoderus giraffa* (a beetle), *Acacia giraffae* (an acacia).
>   L. *camelopardalis,* f. giraffe: camelopard.
>   Gr. *hippardion,* n. probably the giraffe: *Hippardium vittatum* (a beetle).
>   L. *nabis,* f. giraffe, camelopard: *Nabis ferus* (a bug).

girdle < AS. *gyrdel,* belt; see **belt**

girl < uncertain origin; see **child**

gisso- (giso-) < Gr. *geisson (geison),* eaves, cornice, hem, border; see **border**

gith (git), L. a kind of plant; see **plant**

gito- < Gr. *geiton,* neighbor; see **neighbor**

give < AS. *gifan:* giver, giving, gift, forgive, Thanksgiving Day, Gifford.
>   L. *addictus,* given up to, delivered to, habitual; see **hold**
>   L. *Alcestis* (Gr. *Alkestis*), wife of Admetus whose life she saved by offering her own: *Alcestis pallescens* (a bug).
>   Gr. *antipsychos,* given for life:
>   Gr. *aphthonos,* bounteous, generous, ungrudging; see **abundance**
>   L. *cedo, cessus,* withdraw, yield, give up; *cessicius,* pertaining to giving up: cede, concede, cessation, cession.
>   Gr. *charisma, -tos,* n. free gift; *charistikos,* giving freely: *Charistica rhodopetala* (a moth).
>   L. *confero, collatus,* give, transfer: confer, collate.
>   L. *coniptum; conitum,* n. an oblation made by sprinkling flour:
>   L. *dico, -atus,* consecrate, devote, proclaim; see **speak**
>   L. *do, datus* (Gr. *doreo; didomai*), give; *dono, -atus,* present; *dedo, -ditus,* give up, yield; *doto, -atus,* endow; *dativus,* of giving; *datum; donum,* n. gift;

donator, -is, m. giver; donabilis; deserving to receive; donaticus, presented; dos, dotis, f. dowry; dotalis, of a dowry; additus, given to, added to; deditio, -onis, f. surrender; dediticius, of surrender or capitulation; traditus, given over, delivered; Gr. danos, n. gift, loan; dorea; dosis, f.; dorema, -tos; doron, n. gift, present; doreter, -os, m. giver; doretikos; doretos, generous, freely given; dotos, granted; diadosis, f. distribution; epidosis, f. voluntary gift, growth; eudoros, generous; apodoma, -tos, n. offering, gift; apodosis, f. restitution; apodoter, -os, m. repayer: date, dative, data (pl.), dowry, dowager, donor, donation, pardon, condone, rendition, surrender, rendezvous, betray, editorial, endow, dotation, command, endue, extradite, mandate, maundy, rent, die, dice (pl.), addition, tradition, traitor, treason, recommendation, reddition, anecdote, epidote, dose, dosage, apodosis, Dorothy, Pandora, antidote, Dorema aureum (a sumbul), Menodora robusta (an oleacead), Chamaedorea graminifolia (a palm), Geodorum purpureum (an orchid), Epidosis incompleta (a fly), Eudoromyia jocosa (a fly), Coloradia pandora (a moth), Adoretus versutus (a beetle).

Gr. eiktikos, yielding readily, pliable; see **bend**

Gr. ektage, f. dole: Ectage lictor (a moth).

L. eleemosyna (Gr. eleemosyne), pity, alms; see **mercy**

Gr. eranistes, contributor to a fund; see eranos under **common**

L. faveo, fautus, be well disposed toward, regard, esteem; favor, -is, m. goodwill, preference; favorabilis, popular, agreeable; fautor, -is, m. patron; fautrix, -icis, f. patroness: favor, favorite, favorable, favoritism, faun, fautor.

L. generosus, noble, kind, liberal, magnanimous: generous, generosity, Cicindela generosa (a beetle).

AS. gieldan, pay, give: guild, yield, Danegeld.

L. gratia, f. favor, kindness, mercy, thanks, pleasure, beauty; gratus, beloved, pleasing, thankful; gratificus, kind, obliging; gratiosus, full of favor, popular; gratulor, -atus, wish one joy, render thanks; ingratus, unthankful: gratitude, gratuity, agree, grace, graceful, gratify, e. g., congratulation, ingratitude, ingrate, ungrateful, disgraceful, ingratiate, Gratiola lutea (golden gratiola), Persea gratissima(americana) (avocado), Callistomimus gratus (a beetle), Rubus pergratus (a blackberry), Stemmatrophora ingrata (a moth).

Gr. hednon, wedding gift; see hednios under **marry**

L. immolo, -atus, sprinkle a sacrificial victim with meal, sacrifice, offer: immolate, immolation.

L. impertio, -titus, share with another, communicate, impart, bestow:

L. indulgeo, -dultus, be forbearing, concede, grant: indulge, indulgent, indult.

L. inferia, f. sacrifice in honor of the dead; inferius, offered or sacrificed:

L. largior, -gitus, give bountifully, bestow freely, lavish; largitio, -onis, f. bribe, bribery: largess, largition.

L. lito, -atus, offer a sacrifice; litabilis, fit for sacrifice; litamen, -inis, n. sacrifice:

L. mancipo, -atus, give up, deliver, transfer; see **trade**

L. munus, -eris, n. gift, duty, service; munusculum, dim.; muneralis, of gifts; munis, obliging; munero, -atus, give, reward: munificence, remuneration.

Gr. nemo, dispense, distribute, allot; aponometes, m. distributor; dianome, f. distribution; nemesis, f. a distribution, retribution: Nemesis.

L. offero, oblatus, present for acceptance, adduce; oblativus freely given; oblatio, -onis, f. offering: oblation, oblate, offer.

L. patrimonium, inheritance; see **receive**

Gr. pherne, f. dowry: Pherne parallelaria (a moth), paraphernalia.

L. prebenda, f. state allowance or support, dole: prebend, prebendary.

L. privilegium, n. law in favor of an individual, special right; see lex under **law**

Gr. proix, -ikos, f. gift, present, dowry:

L. promus, pertaining to giving or distributing:

Gr. prosthetos, applied: Prosthetopteryx barbata (a moth).

L. repedabilis, giving way, yielding, retreating, recoiling:

L. sacrifico, -atus, offer a sacrifice; sacrificium, n. offering to the gods: sacrifice, sacrificial.

L. sportula, f. gift, dole:

L. stips, -ipis, f. gift: stipend.

Gr. syllexis, contribution; see lanchano under **lot**

Gr. thyos, n; thysia, f. sacred rite, sacrifice, offering; thyma, -tos, n. victim; thymele, f. place for sacrifice, altar; thytikos, of sacrifice, < thyo, sacrifice: thymele, idolothyte, Thysiotorus brevipennis (a wasp).

L. tribuo, allot, give, pay; see **pay**

L. xenium (Gr. xenion), n. gift to guest; xeniolum, n. dim.:

See: **pay, return, trade, mercy, change, receive**

gizzard < L. gigerium, entrail, particularly of birds; see **intestine**

glaber, -bra, -brum, L. hairless, bald, smooth; glabellus, dim.; see **bare**

glacialis, L. icy, frozen, < *glacies,* ice; *glacio, -atus,* freeze; see **ice**

glad < AS. *glaed,* bright, happy; see **joy**

gladius, L. sword; *gladiolus, dim.* sword-lily; *gladiator,* swordsman; see **sword, iris**

glaesum (glessum), L. amber; *glaesarius,* of amber; see **amber**

glageros, Gr. full of milk; see **milk**

glamyros, Gr. blear-eyed, watery; see **pus**

gland < L. *glans, glandis,* f. acorn, any acorn-shaped object; *glandula,* f. dim.; *glandarius,* of acorns: glandular, glandiferous, glandulose, glanders, *Juglans regia* (English or Circassian walnut), *Betula glandulosa* (bog-birch), *Garrulus glandarius* (European jay), *Cactocrinus glans* (a Mississippian crinoid), *Quercus glandulifera* (an oak), *Sapium biglandulosum* (Brazil sapium), *Phyllodoce glanduliflora* (yellow mountain-heath).
  Gr. *aden, -os,* c. gland: adenoid, adenectomy, adenology, adenylic, adenitis, aden-osine, adenine, adenomyoma, adenosinephosphatase, *Adenostoma fasciculatum* (chamiso), *Zygadenus paniculatus* (a camas), *Adenanthera microsperma* (bead-tree), *Acacia pentadenia* (an acacia), *Picradeniopsis woodhousei* (a composite), *Calycadenia truncata* (rosinweed).
  Gr. *konarion,* n. pineal gland: conarium.
  Gr. *pankreas,* n. sweetbread: pancreas, pancreatic.
  Gr. *parotis, -idos,* f. gland beside the ear: parotid, parotiditis.
  F. *prostate,* < Gr. *prostates,* m. one who stands first, chief, protector; a gland surrounding the beginning of the male urethra: prostate, prostatitis, pros-tatectomy.
  See: **acorn**

glanis, L., Gr. probably a sheatfish; see **catfish**

glanos, Gr. hyena; see **hyena**

glans, L. acorn, any acorn-shaped object; gland; *glandula,* dim.; see **acorn, gland**

glaphyros, Gr. hollowed out, smoothed, polished, finished, pretty; see **hollow**

glarea, L. gravel; *glareosus,* gravelly; see **stone**

glaris, Gr. chisel; see **chisel**

glass < AS. *glaes.*
  Gr. *hyalos,* f. glass; *hyaleos; hyalinos,* of glass, glassy, transparent: hyaline, hyalite, hyaloid, *Nitella hyalina* (a characead), *Hyalomma aegypticum* (a tick), *Hyalonema lusitanicum* (a sponge), *Hyalosaurus koellikeri* (a lizard), *Hyalo-stola phoenicochyta* (a butterfly).
  L. *vitrum,* n. glass; *vitreus,* glassy: vitreous, vitrification, vitriol, devitrify, vitrain, vitrics, vitrella, vitrotype, *Vitrina major* (glass-snail), *Chiridopsis vitreicollis* (a beetle), *Afropompilus vitripennis* (a wasp), *Corycaeus vitreus* (a cyclops).

glastum, L. woad; see **mustard**

glaucium, L. (Gr. *glaukion*), a genus of the poppy family, see **poppy**; a duck, see **duck**

glaucus, L. (Gr. *glaukos*), bluish-green or gray, sea-colored; see **gray**

glaux, Gr. owl; *glaukidion,* n. dim., see **owl**; a plant, see *gala* under **milk**

glax, Gr. milk vetch; see **primrose**

gleba, L. clod, lump; *glebula,* dim.; *glebalis; glebarius,* of clods; see **lump**

glechon, Gr. pennyroyal; see **mint**

gleditschia, NL. a genus of the bean family; see **bean**

gleme, Gr. pus in the corner of a sore eye; *glemion,* dim.; see **pus**

gleno- < Gr. *glene,* pupil of the eye, eyeball, eye socket, socket of a joint, doll; *glenion,* dim.; *glenos,* thing to stare at, show, wonder; see **eye**

glessum, L. amber; see *glaesum* under **amber**

gleuco- < Gr. *gleukos,* must, sweet new wine; see **wine**

glia, Gr. glue; see **glue**

glide < AS. *glidan;* see **slip**

glinos, Gr. a kind of maple; see **maple**

glis, *gliris,* L. dormouse; *glirarium,* place for dormice; see **mouse**

glischros, Gr. sticky, clammy, viscous; see **glue**

globigerina, NL. a genus of foraminifers; see **protozoan**

globus, L. ball; *globulus,* dim. bead; *globosus,* spherical; see **ball**

glochin, Gr. point of an arrow; see **point**

glocio, L. cluck like a hen; see **cackle**

gloeo- < Gr. *gloios (gloia; glia)*, a sticky substance; see *glia* under **glue**

glomus, *-eris*, L. ball; *glomero, -atus,* form into a ball; see **ball**

gloom < AS. *glom*, twilight; see **sad, shade**

glory < L. *gloria*, fame, honor, renown, praise; *gloriola*, dim.; see **honor**

glos, *-oris*, L. sister-in-law; see **kin**

glossa (glotta), Gr. tongue; *glossidion (glottidion)*, dim.; see **tongue**

glottis, Gr. mouth of the windpipe; see **throat**

glove < AS. *glof.*
  L. *caestus,* m. boxer's glove: cestus.
  Gr. *cheiris, -idos,* f. glove, sleeve; *cheiridion,* n. dim.; *cheiridotos,* gloved, sleeved:
    Chiridopsis signaticollis (a beetle), *Chiridula apicalis* (a beetle), *Chiridium ferum* (a pseudoscorpion), *Chiridota rigida* (a holothurian).
  L. *digitabulum,* n. a kind of glove:
  L. *manica,* long sleeve, glove; *manucium,* glove, muff; see **sleeve**
  ML. *mita,* f. mitten: mitiform.

glox, *-ochos,* Gr. beard of wheat; see *glochin* under **point**

glubo, L. strip of bark, peel, skin; see **skin**

gluco- < Gr. *gleukos,* must, sweet new wine; see **wine**

glue < L. *glus, glutis,* f; *gluten, -inis,* n. a sticky substance in wheat flour; *glutineus; glutinosus,* viscous, sticky; *glutino, -atus,* glue, paste together; *conglutinosus,* viscous: gluten, glutinous, glutamine, agglutination, agglutinin, *Alnus glutinosa* (balck alder), *Xenophora agglutinans* (a gastropod), *Gluta elegans* (an anacardiacead), *Fusarium conglutinans* (a fungus).
  L. *caementum,* chips of stone; now applied to a substance used as a binder; see *caementarius* under **mason**
  L. *coagmento, -atus,* join, glue, connect; see **bind**
  L. *ferrumen, -inis,* n. cement, solder, glue; *ferrumino, -atus,* cement, bind, join: ferruminate.
  Gr. *glia,* f. (*gloia,* f.; *gloios,* m.), glue, any sticky substance: gliadin, ectoglia, glioma, gliosis, neuroglia, *Gliocladium fimbriatum* (a fungus), *Gloeocapsa sanguinea* (an alga), *Gloiopeltis tenax* (an alga), *Gloeosporium concentricum* (a fungus).
  Gr. *glischros,* sticky, clammy, viscous, greedy, niggardly; *glischrasma, -tos,* n. gluten; *glischrotes,* f. stickiness; *glischron, -os,* m. niggard: *Glischrocentrus cucullatus* (a bug), *Vesperugo(Glischropus) nanus* (a bat), *Bacterium glischrogenes* (a bacterium), *Rhododendron glischrum* (a rhododendron).
  L. *haereo, hesus,* stick, hold fast; see **hold**
  Gr. *ixos,* m. mistletoe, birdlime; *ixeuterios; ixeutikos; ixodes,* like birdlime; *ixeutes,* m. catcher with birdlime, fowler: *Ixobrychus minutus* (little bittern), *Ixia lutea* (yellow ixia), *Ixodes ricinus* (dog-tick), *Physalis ixocarpa* (a groundcherry), *Ixeuticus martius* (a spider), *Chironia ixifera* (a starpink).
  Gr. *kolla,* f. glue; *kollesis,* f. a glueing, fastening, welding; *kolletes,* m. gluer, fastener; *kolletikos,* adhesive; *kolletos,* glued, joined; *kollema, -tos,* that which is glued; *echekollos,* glutinous, gluey; *ichthyokolla,* f. fish-glue, isinglass; *lithokolla,* f. cement; *synkollos,* glued together: collagen, chrysocolla, collenchyma, collargol, colloid, collodion, collophore, sarcocolla, protocol, Collembola, *Collembolothrips mediterraneus* (a thysanopterid), *Collospermum hastatum* (a lily), *Colletotrichum lini* (a fungus), *Colletes azteca* (a bee), *Collomia linearis* (a phlox).
  AS. *lim,* a viscous substance: lime, limy, limestone, quicklime, birdlime.
  L. *mucilago, -inis,* f. a sticky, gelatinous juice: mucilage.
  Gr. *propolis,* f. bee-glue: propolis.
  L. *tractuosus,* clammy, gluey:
  L. *viscum,* n. (*viscus,* m.), mistletoe, birdlime; *viscidus; viscosus,* sticky, clammy: viscous, viscoid, viscosity, *Viscum album* (European mistletoe), *Malvaviscus arboreus* (wax-mallow), *Geranium viscosissimum* (a geranium), *Acacia viscidula* (waxy acacia), *Calocera viscosa* (antler-fungus), *Lychnis viscaria* (a campion).
  See: **hold, resin, slime, bind, jelly**

gluma, L. hull, husk, bract; see **scale**

gluten, L. glue; *glutinosus,* viscous, sticky; *glutinator,* gluer; see **glue**

glutio, *-itus,* L. devour, swallow; see **eat**

gluto- < Gr. *glutos,* rump, buttocks; see **rump**

glutton < OF. *glouton,* < L. *gluto (glutto)*, gormandizer; see *glutio* under **eat**

glyceria, NL. a genus of grasses; see **grass**

glycero- < Gr. *glykeros,* sweet; see *glykys* under **sweet**

glycine, NL. a genus of legumes; see **bean**

glyco-; glycy- < Gr. *glykys,* sweet; *glykysma; glykytes,* sweetness; see **sweet**

glycymeris, L. a kind of shellfish; see **mollusk**

glycyrrhia, L. (Gr. *glykyrrhiza*), licorice; see **bean**

glypho, Gr. carve, engrave; *glyptos,* carved; *glyphanos,* a carving tool, chisel; *glyphis,* knife; *glymma,* an engraved figure; see **cut, chisel, knife**

gnamptos, Gr. curved, bent; see **bend**

gnaphalium, L. (Gr. *gnaphalion*), cudweed; see **composite**

gnaphalo- < Gr. *gnaphalon; gnaphallon; knaphallon,* wool; see **wool**

gnaphalos, Gr. probably a waxwing; see **waxwing**

gnaphos, Gr. teasel, carding-comb; see **thistle**

gnarus, L. knowing; see *nosco* under **know**

gnash < ON. *gnastan,* grind the teeth together; see **grind**

gnat < AS. *gnaet;* see **fly**

gnathos, Gr. jaw; *gnathidion; gnathion,* dim.; see **jaw**

gnaw < AS. *gnagan;* see **bite**

gnesios, Gr. genuine, true, legitimate; see **true**

gnetum, NL. n. a genus of the jointfir family: Gnetaceae, *Gnetum gnemon* (gnemon).
    NL. *welwitschia,* f. a genus of plants usually referred to the jointfir family, < Friedrich Welwitsch, Austrian botanist: *Welwitschia mirabilis* (an African plant).

gnome < NL. *gnomus,* diminutive fabled being, dwarf; see **little**

gnomon, Gr. index, judge; *gnome,* mind, opinion, maxim, mark; see *nosco* under **know, mark**

gnophos, Gr. darkness, dusk, gloom; see *dnophos* under **night**

gnorimos, Gr. well-known, see *nosco* under **know;** mark of identification, see *gnoma* under **mark**

gnosis, Gr. wisdom; *gnostos,* known; see *nosco* under **know**

gnypetos, Gr. falling on the knees, weak; see **weak**

gnythos, Gr. cave, pit, hollow; see **hollow**

go < AS. *gan;* see **move, depart, walk**

goal < uncertain origin; see **aim, end**

goat < AS. *gat:* goatherd, goatee, goatsucker, scapegoat, goatish.
    Gr. *aix, aigos,* c. goat; *aigidion,* n.; *aigiskos,* m. dim. kid; *aigagros,* c. wild goat; *aigokeros,* goat-horned; *aigeios,* of goats: aegagropila, egophony, *Aegopodium podagria* (goutweed), *Capra aegagrus* (a wild goat), *Aegopogon tenellus* (a grass), *Egocerus niger* (sable antelope).
    L. *attagus* (Gr. *attegos*), m. he-goat:
    AS. *buc;* ME. *buk;* D. *bok;* G. *bock,* he-goat, ram, male deer: buck, roebuck, butcher, springbok.   Derive: bock beer.
    L. *caper, capri,* m. goat; *capra; caprea,* f. she-goat, nanny; *capella,* f.; *capellus,* m. dim.; *capreolus,* m. dim. roebuck, tendril; *caprilis; caprinus,* of goats; *caprimulgus,* m. a milker of goats; Sp. *cabra,* f. goat; *cabrilla,* f. dim.; *rupicapra,* f. chamois: caper, capriole, caprice(?), capreolate, capric, caproic acid, caprylic, caprification, Capricorn, Caprifoliaceae, cabriolet, chevron, chevrotain, Capella, *Caprimulgus ruficollis* (a goatsucker), *Ancistrocerus capra* (a wasp), *Vulvulina capreolus* (a fossil foraminifer), *Caprella grandimana* (an amphipod), *Bignonia capreolata* (cross-vine), cabrilla *(Epinephelus capreolus), Balistes capriscus* (triggerfish), *Salix caprea* (goat-willow).
    Gr. *chimaira,* she-goat; *chimairos,* he-goat; see *chimaera* under **animal**
    Gr. *eriphos,* m. young goat, kid; *eriphion,* n. dim.; *eripheios,* of a kid:*Eriphoserica camentoides* (a beetle).
    L. *hedus (haedus; hoedús);* m. young goat, kid: *hedillus; hedulus,* m. dim.; *hedinus,* of a kid: *Nemorhedus goral* (a goat-antelope).
    L. *hircus,* m. he-goat, buck; *hirculus,* m. dim.; *hircinus,* of goats; *hircosus,* goatish: hircine, hircosity, hircocervus, *Capra hircus* (domestic goat), *Chenopodium hircinum* (a goosefoot).
    L. *ibex, -icis,* m. a kind of goat, chamois: *Capra ibex* (Alpine ibex).
    Gr. *kolos,* m. a kind of goat:
    Gr. *psinathos,* f. a wild goat:
    L. *tragus* (Gr. *tragos*), m. he-goat; *tragion,* n.; *tragiskos,* m. dim.; *trageios,* of goats; *tragikos,* like a goat, goatish: tragopan, tragacanth, tragule, tragedy,

tragic, *Tragopogon pratensis* (meadow-salsify), *Oreotragus saltator* (klip-springer), *Tragulus kanchil* (chevrotain), *Tragiscus dimidiatus* (a beetle).

goatsucker
Gr. *aigothelas*, m. goatsucker, nightjar: *Aegotheles cristatus* (a bird).
NL. *antrostomus*, m. a genus of goatsuckers: *Antrostomus carolinensis* (a whip-poorwill).
L. *caprimulgus*, m. goatsucker: *Caprimulgus europeus* (a goatsucker).
NL. *chordeiles*, m. a genus of goatsuckers: *Chordeiles virginianus* (nighthawk).

goby < L. *gobio, -onis; gobius* (Gr. *kobios*), m. a fresh-water fish; *kobidion*, n. dim.; *kobitis*, like the gudgeon: gudgeon *(Gobio fluviatilis), Gobiomorus dormitor* (guavina), *Cobitis taenia* (a loach), *Cottus gobio* (miller's-thumb), *Platygobio gracilis* (a minnow), *Typhlogobius californiensis* (blind goby).
Gr. *kaulines*, m. a kind of gudgeon:

god < AS. *god*, a spiritual being, deity; see **spirit**

goëto- < Gr. *goës, -etos; goëtes*, howler, wailer, wizard; *goëros*, mournful; see **weep**

gold < AS. *gold:* goldsmith, goldfish, goldfinch, goldenrod, gulden, golden, goldi-locks *(Chrysocoma coma-aurea)*, marigold, guilder, gilt, giltedge. Derive: Golconda, bonanza, Eureka.
L. *aurum*, n. gold; *auratus; aureus; aurulentus,* golden, made of gold, ornamented with gold; *aurarius,* golden, pertaining to gold, goldsmith; *aureolus,* golden, glittering, splendid; *aurifer; auriger,* gold-bearing; *aurifex, -icis,* m. goldsmith; *aurilegulus,* m. gold-washer: aureate, aureole, auramine, auric acid, ammonium aurate, auriferous, oriole, orpiment, ormolu, orpine, El Dorado, Marcus Aurelius, Orofino(Idaho), Oroville(Calif.), aureomycin(from *Streptomyces aureofaciens*), *Aurelia flavidula* (a jellyfish), *Aureolaria virginica* (a figwort), *Ficus aurea* (a fig), *Carex aurea* (golden sedge), *Matricaria aurea* (a composite), *Carassius auratus* (goldfish), *Drosophila auraria* (a fly), *Blaesoxipha aurulenta* (a fly), *Buprestis aurulenta* (a beetle), *Mitocybe auriportae* (a milleped), *Ploceus sub-aureus* (a weaver), *Xylocopa orpifex* (a carpenter-bee), *Solidago virgaurea* (European goldenrod), *Polypodium aureum* (golden polypody).
L. *ballux, -ucis; balluca,* f. gold-sand, gold-dust:
L. *bractea,* a thin plate of metal, gold-leaf, scale; *bractearius,* a gold-beater; *bracteatus,* gilt; *imbracteo, -atus,* overlay with leaf-metal; see **scale**
Gr. *chrysos*, m. gold; *chrysaphion; chrysidion; chrysion,* n. dim.; *chryseos; chrysikos; chrysinos,* of gold, golden, pertaining to gold; *chrysops,* gold-colored, shining like gold; *chrysotos,* gilt; *chrysis, -idos,* f.; *chrysoma, -tos,* n. a gold object, vessel, plate, dress, etc.; *chrysites,* like gold; *chrysendetos,* set in gold; *chrysostiktos,* gold-spotted; *chryselektes,* m. gold-washer; *chrysoryktes,* m. gold-digger; *chrysotechnes,* m. goldsmith; *enchrysos,* golden: chrysanthemum, chrysalis, chrysoberyl, chrysolite, chrysotile, chrysocolla, Chrysostom, chrys-arobin, chrysin, chrysazin, *Chrysosplenium alternifolium* (a golden-saxifrage), *Chrysolophus amherstiae* (a golden pheasant), *Chrysops vittatus* (deerfly), *Chrysidium antiquum* (a wasp), *Chrysopsis falcata* (a golden aster), *Helichrysum angustifolium* (an everlasting), *Ocyurus chrysurus* (a snapper), *Penicillium chrysogenum* (a mold), *Enchrysa dissectella* (a moth).
Gr. *Midas*, m. mythical king at whose touch everything turned to gold, and whose ears were changed to ass's ears by Apollo for deciding a musical contest in favor of Pan: Midas, *Midas ursulus* (a marmoset), *Auricula aurismidae* (a snail).
L. *palaga*, f. an ingot of gold; Gr. *pala*, f. gold nugget:
Ar. *zarqun*, gold-colored: zircon, zirconium, jargon (a mineral).
See: **yellow, citrus,** ELEMENTS under **thing**

goldenrod; see *solidago* under **composite**

goleos, Gr. hole; see **hole**

gompharion, Gr. a mullet; see **mullet**

gomphios, Gr. molar tooth; *gomphiasis*, toothache; see **tooth**

gomphos, Gr. nail, peg, bolt; *gomphotikos*, fastened with nails; see **nail**

gonad < Gr. *gone*, seed, that which produces seed; see **seed**

gonato- < Gr. *gony, gonatos*, knee, joint, node; *gonation*, dim.; see **knee**

gongros, Gr. excrescence, swelling, knot, see **swell**; eel, see *conger* under **eel**

gongylis, Gr. turnip; see **turnip**

gongylos, Gr. ball; round, spherical; see **ball**

gonidium, NL. an asexual cell or spore, < Gr. *gonidion*, dim. of *gone*, seed, see **fruit, seed**; < *gonidion*, dim. of *gonia*, angle, see **angle**

gonimos, Gr. productive, fruitful, endued with generative power; see **fertile**

gonio- < Gr. *gonia,* angle, corner; *gonidion,* dim.; see **angle,** *gone* under **seed**

gono- < Gr. *gonos; gone,* seed, offspring, product; *gonion; gonidion,* dim., see **seed**; < *goneus,* father, parent, see **father**

gonorrhea < L. *gonorrhoea* (Gr. *gonorrhoia*), clap; see **disease**

gony, *gonatos,* Gr. knee, joint, node; *gonation,* dim.; see **knee**

good < AS. *god:* goods, goodness, gospel, goodwill, Goodnow, Goodrich, Goodyear. Gr. *agathos,* good; *ameinon; areion; belteros; beltion,* better; *aristos; beltistos,* best; *agathosyne,* f. goodness: agathin, agathism, Agathocles, *Agathomerus rubrinotatus* (a beetle), *Beltion peronides* (a Miocene fish), *Beltista rubella* (a beetle), aristocracy, aristarchy, aristol, aristotype, Aristotle, *Aristolochia fimbriata* (a birthwort), *Aristoppus fenestratus* (a bug).
Gr. *aotos,* m. the best, choicest:
Gr. *arete,* f. excellence, goodness, virtue: aretaics.
L. *auspicatus,* fortunate, favorable, lucky; see *auspex* under **prophecy**
L. *bonus,* good; *melior,* m., f.; *melius,* n. better; *optimus,* best; *bene,* well; *beneficus,* favorable; *bonitas, -atis,* f. goodness, excellence; *perbonus,* very good: bonus, bona fide, debonair, bon voyage, summum bonum, bonny, bonanza, bonbon, bounty, bounteous, Pueblo Bonito (N. Mex.), Buena Vista (Mexico), Buenos Aires (Argentina), benevolence, benefactor, benediction, benefit, benign, benison, beneficent, meliorism, ameliorate, optimist, soroptimist, bennet *(Geum urbanum),* Boniface, *Cnicus benedictus* (blessed thistle).
Gr. *chrestos,* good, useful; *chresimos,* useful; see *chrao* under **use**
L. *conducibilis,* advantageous, profitable: conducible, conducive.
Gr. *esthlos,* good; *esthloteros,* better; *esthlotatos,* best: *Esthlodora versicolor* (a moth).
Gr. *eu,* good, well, true, beautiful, very, original, primitive. The preferred spelling is *eu* for all scientific words; but some English compounds have *ev* before words beginning with a vowel, as *evangel, Evander:* eugenics, eulogy, euphony, Eugene, euphoria, Eutopia, Evangeline, Eunice, *Eucalyptus globulus* (a eucalyptus, bluegum), *Eunectes murinus* (anaconda), *Euonymus alatus* (winged spindle-tree).
Gr. *exaitos,* desired, choice, excellent: *Exaetoderes scabripennis* (a beetle).
L. *excellens, -entis,* surpassingly good, choice, superior: excellent, excellency.
L. *eximius,* choice, excellent, exempt; see **few**
L. *expetendus,* desirable, excellent:
L. *exquisitus,* choice, excellent, fine: exquisite.
Gr. *kregyos,* good, useful: *Cregya vetusta* (a beetle).
Gr. *loion,* better; *lostos,* best:
L. *probus,* good, excellent, upright; *approbus,* very good, excellent: probity, *Rubus probus* (Queensland raspberry), approbation, approved.
L. *prosum,* be useful, beneficial, do good:
Gr. *soteriodes,* wholesome; see *soter* under **save**
L. *superbus,* excellent, superior, splendid: superb, *Lilium superbum* (Turkscap lily), *Terminalia superba* (a terminalia), *Adelphocoris superbus* (a bug).
L. *virtus, -utis,* f. manliness, strength, goodness, value: virtue, virtuous.
See: **agreeable, joy, health, true, honor, beauty**

goose < AS. *gos; ges:* geese, gosling, goshawk, goosebone.   Derive: gooseberry.
L. *anser, -eris,* m. goose, gander; *anserculus,* m. dim. gosling; *anserinus,* of geese: anserine, anserated, vulpanser, merganser, *Anser albifrons* (a goose), *Potentilla anserina* (silverweed), *Dileptus anser* (a protozoan).
NL. *bernicla,* f. a kind of goose: *Branta bernicla*(brant), barnacle.
NL. *branta,* f. a genus of geese: *Branta nigricans* (black brant).
Gr. *chen, -os,* m. goose, gander; *chenarion; chenion,* n.; *chenideus; cheniskos,* m. dim. gosling; *cheneios (chenios),* of geese; *chenodes,* like geese; *chenalopex, -ekos,* m. an Egyptian goose; *chenizo,* cackle like a goose: chenocholic, *Chen hyperboreus* (snow-goose), *Chenonetta jubata* (a goose), *Chenopis atrata* (black swan), *Chenopodium album*(pigweed), *Alopochen*(formerly *Chenalopex*) *aegyptiacus* (fox-goose).
L. *cygnus* (Gr. *kyknos*), m. swan; *kyknarion,* n. dim., cygnet; *cycneus; cygneus,* of swans: cygnet, cygneous, Cygnus, *Cygnus gibbus* (a swan), *Cygnophis cygnoides* (Chinese goose), *Cycnoces chlorochilon* (swan-orchid), *Anodonta cygnea* (swan-mussel), *Dendrocygna fulva* (a tree-duck).
L. *ganta* (G. *gans*), f. goose: gander, gannet *(Moris bassana).*
Gr. *helorios,* m. a water-bird, perhaps a swan: *Helorius paludicola* (a fossil bird).
L. *olor, -is,* m. swan; *olorinus,* of swans: *Olor buccinator* (trumpeter-swan), *Cygnus olor* (European white swan).

goosefoot
Gr. *atraphaxys, -yos,* f. saltbush: *Atraphaxis buxifolia* (a polygonacead).

L. *atriplex, -icis,* f. saltbush: *Atriplex rosea* (red orach), orach *(Atriplex hortensis), Cycloloma atriplicifolium* (a goosefoot).

L. *batis, -idis* (Gr. *-idos*), f. samphire, saltwort: *Batis maritima* (seaside-saltwort).

L. *blitum* (Gr. *bliton*), n. a plant of the goosefoot family: blite, *Blitum capitatum* (strawberry-blite), *Blitopertha pallidipennis* (a beetle), *Blitophaga nuda* (a beetle), *Amaranthus blitoides* (prostrate amaranth).

L. *chenopodium,* n. (Gr. *chenopous,* f.), goosefoot, pigweed: *Chenopodium polyspermum* (a goosefoot), Chenopodiaceae.

Gr. *halimon,* n. shrubby seashore plant of the goosefoot family: *Atriplex halimus* (Mediterranean saltbush).

L. *polycnemum* (Gr. *polyknemon*), n. an unknown plant; now a genus of the goosefoot family: *Polycnemum arvense* (a chenopodiacead).

NL. *salicornia,* f. a genus of the goosefoot family: *Salicornia europaea* (marshfire-glasswort), *Rhipsalis salicornioides* (a cactus).

NL. *salsola,* f. a genus of the goosefoot family: *Salsola laricina* (a saltwort).

NL. *sarcobatus,* m. a genus of the goosefoot family: *Sarcobatus vermiculatus* (greasewood).

ML. *spinacia,* f. a genus of the goosefoot family, < Ar. *isbanakh:* spinach (*Spinacia oleracea*).

Gordius, L. (Gr. *Gordios*), mythical Phrygian king noted for the intricate knot on his chariot pole; see **knot**

gordonia, NL. a genus of the tea family; see **tea**

gorge < L. *gurges,* abyss, whirlpool; see **valley**

gorgos, Gr. grim, < *Gorgos,* Medusa, the gorgon; see **fear**

gorgyra, Gr. drain, sewer; see **pipe**

gorytos, Gr. quiver, bow-case; see **sheath**

gossip < AS. *godsib,* sponsor; see **speak**

gossypium, L. (Gr. *gossypion*), cotton; see **cotton**

Goth < L. *Gothus,* an individual of a Teutonic people; see **German**

gourd < F. *gourde,* < L. *cucurbita;* see **melon**

gout < OF. *goute,* < L. *gutta,* drop; see **disease**

govern < OF. *governer,* < L. *guberno, -atus* (Gr. *kybernao*), steer, guide, manage: *gubernator, -is* (Gr. *kybernetes*), m. pilot, director, manager; *kybernetikos,* good at steering; *gubernum,* n. helm, rudder: governor, government, gubernatorial, cybernetics, *Gubernetes*[*Cybernetes*] *yetapa* (a bird).

L. *abstemius,* refrain from drinking liquor, moderate, temperate: abstemious.

L. *abstinens,* holding off from, temperate; see *teneo* under **hold**

L. *arbitrium,* n. mastery, power, will:

L. *aedilis,* m. superintendent of public buildings, works, markets, and police: edile.

Gr. *anax, -aktos; anaktor, -os,* m. king, lord, chief, master; *anassa,* f. queen, lady; *anaxios,* royal, kingly; *astyanax, -akos,* m. mayor, < *anasso,* rule, sway: *Anax junius* (a dragonfly), *Anaxarchus reyi* (a bug), *Hydranassa tricolor* (a heron), *Nyctanassa violacea* (a heron), *Astyanax mexicanus* (a fish), *Empidonax minimus* (a flycatcher), *Ilyanassa irrorata* (a gastropod).

L. *antistes, -itis,* m. overseer, master; *antistita,* f. female overseer:

Gr. *archon, -ontos; archos,* m. chief, ruler; *archi-,* chief, leader; *archikos,* of rule, royal: *monarchos,* m. sovereign: monarch, matriarchy, oligarchy, patriarch, Plutarch, anarchist, archives, archipelago, architect, *Archontophoenix cunninghamiana* (piccabeen king-palm), *Archilestes californica* (a dragonfly), *Myiarchus crinitus* (crested flycatcher).

Gr. *ballen,* m. king:

Gr. *basileus, -eos,* m. king; *basileia; basilis, -idos; basilissa,* f. queen; *basiliskos,* m. dim. a kind of lizard; *basilikos,* royal: basilica, basilisk, basil (*Ocimum basilicum*), *Basilometra boschmani* (a crinoid), *Basilarchia archippus* (viceroy butterfly), *Basiliscus americanus* (a lizard), *Basileocephalus thaumatonotus* (a leafhopper).

L. *Caesar, -is,* m. family name of Gaius Julius Caesar; any autocratic monarch, emperor, ruler: Caesar, Kaiser, Czar, Caesarea, caesarean, Jersey, Cherbourg, *Lucilia caesar* (green-bottle fly), *Amanita caesarea* (a mushroom).

L. *cardinalis,* principal, chief; see *cardo* under **hinge**

L. *compositus,* orderly, regular, well-arranged; see **class**

L. *consul, -is,* m. a chief magistrate of the Roman state; *consularis,* of a consul; *consulatus,* m. office of consul: consul, consular, consulate, proconsul, *Proconsul africanus* (a fossil primate).

L. *coryphaeus* (Gr. *koryphaios*), leader, chief, head; see *coryphe* under **head**

L. *decanus,* chief of ten, dean; see *decem* under **ten**

L. *designator, -is,* m. master of ceremonies: designator.

Gr. *despotes, -ou,* m. master, owner, lord; *despoina; despotis, -idos,* f. mistress, queen; *despotidion*, n. dim.; *despotikos*, absolute, inclined to tyranny; *desposyne*, f. absolute rule: despot, despotic, despotism, *Despoena superba* (a katydid).

L. *dictator, -is*, m.; *dictatrix, -icis*, f. emergency ruler with absolute authority: dictator, dictatorial, dictatrix.

L. *dirigo, -rectus*, arrange, order, guide, set straight: direct, director, direction, dirge, dirigible, directory.

L. *dispensator, -is*, m. superintendent, manager, stewart, treasurer; *dispensativus*, of management: dispensation, dispensatory.

L. *dominus*, m. master; *domina*, f. mistress, lady; *dominulus*, m. dim.; *dominor, -atus*, rule, govern; Sp. *don*, gentleman; *dona; duenna;* It. *donna*, lady: domino, dominate, domain, domineer, dominion, predominant, don, Dom Pedro, dam, dame, damsel, madame, madonna, Notre Dame, demoiselle, belladonna, 79 A.D., F. *dimanche* (Sunday), duenna, prima donna, demesne, Dan Cupid, *Amaryllis belladonna* (an amaryllis), *Anthracothorax dominicus* (a Haitian hummingbird).

AS. *dom*, jurisdiction, condition; see **nature**

L. *domnaedius*, m. landlord; *domnifunda; domnipraedia*, f. landlady:

Gr. *dynastes*, m. lord, master, ruler; *dynastis, -idos*, f. mistress, ruler; *dynastikos*, arbitrary: dynasty, dynastic, *Dynastes hyllus* (a beetle), *Myiodynastes luteiventris* (a flycatcher).

L. *episcopus* (Gr. *episkopos*), m. overseer, bishop: episcopal, bishopric, *Episcopotettix sulcirostris* (a grasshopper), *Mitra episcopalis* (a gastropod), *Episcoposaurus horridus* (a fossil crocodile), San Luis Obispo (Calif.).

Gr. *epistates*, m. overseer, superintendent:

L. *erus (herus)*, m. master; *era (hera)*, f. mistress of a house; *erilis*, of the master or mistress of a household:

L. *formula*, rule, pattern, scheme; see **law**

Gr. *harmostes*, m. governor of a colony: *Harmostes costalis* (a bug).

Gr. *hegemon, -os*, leader; see **lead**

Gr. *hodegos*, guide, teacher; *hodegia*, a guiding, teaching; see **teach**

Gr. *hodouros*, m. conductor, guide: *Hodurus dispar* (a beetle).

Gr. *hyksos*, m. a dynasty of Egyptian nomad kings: Hyksos.

L. *impero, -atus*, command; *imperator, -is*, m. commander-in-chief, general, chief, ruler; *imperialis*, of the empire or emperor; *imperiosus*, possessed of command, mighty, domineering: imperial, imperious, imperative, emperor, empire, *Archidiskodon imperator* (imperial mammoth), *Fritillaria imperialis* (a lily).

Sp. *inca*, < Ab.Am. *ynca*, a former ruler in Peru: Inca, *Incacetus broggi* (a fossil whale), *Scardafella inca* (Inca dove).

Gr. *ithynter, -os*, m. guide, pilot, ruler, < *ithyno*, make straight, guide, rule:

L. *jubeo, jussus*, bid, command, order, designate, appoint; *jussum*, order, command, law decree; see **ask, law**

Gr. *koiranos*, m. ruler; *koiranikos*, of a ruler: *Coeranica isabella* (a moth).

Gr. *korthylos*, m. a little king or chieftain: *Corthylus punctatissimus* (a beetle), *Corthylomimus fasciatus* (a beetle), *Metacorthylus nigripennis* (a beetle).

Gr. *kosmos*, order, arrangement, government; see **world**

Gr. *krateo*, rule, < *kratos*, strength, might; *kratynter, -os; epikrates*, m. master; *naukrates*, m. pilot, master: autocrat, bureaucrat, democracy, plutocracy, theocracy, aristocratic, Hippocrates, *Epicrates angulifer* (a snake), *Naucrates ductor* (pilot-fish), *Zorocrates fusca* (a spider), *Carex gynocrates* (a sedge).

Gr. *kreion, -tos*, m. ruler, master; *kreiousa*, f. queen:

Gr. *kyrios*, m. lord, master; of power and authority, < *kyros*, n. authority; *kyriakos*, of a lord or master: cyriologic, kyrine, church, kirk, *Cyriocrates ruber* (a beetle).

L. *magister, -tri*, m. chief, leader, commander; *magistratus*, m. public administrator; *magistralis*, of a master: master, magistrate, magistral, magisterial, mister (Mr.), mistress (Mrs.), miss.

L. *mando, -atus*, order, ask, command; see **ask**

Gr. *metrios*, within measure, moderate; see *metron* under **measure**

L. *modero, -atus*, keep within bounds, regulate; *moderator, -is*, m. manager, governor: moderate, moderator, moderation, immoderate.

Gr. *nephalios; nephon, -os*, drinking no wine, abstemious, sober: *Nephalius rutilus* (a beetle).

L. *numen, -inis*, nod, will, sway, majesty; see **spirit**

OHG. *Oberon*, king of the fairies: *Oberonia acaulis* (an orchid).

Gr. *oiakizo*, steer, govern:

L. *Palinurus* (Gr. *Palinouros*), m. pilot of Aeneas: *Palinurus vulgaris* (spiny lobster), *Palinurichthys perciformis* (rudder-fish).

Gr. *pedalion*, rudder; see **rudder**

L. *Pharao, -onis,* m. Pharaoh, title of Egyptian kings: *Pharaonaster japonicus* (a Miocene echinoid), *Orbitolites pharaonum* (a foraminifer), *Monomorium pharaonis* (a small, red ant).

Gr. *phrastor,* guide; see *phrasis* under **speak**

L. *potitor, -is,* m. master, possessor, < *potior, -itus,* get, obtain, become master of: Gr. *potnia,* f. mistress, queen: *Potnia venosa* (a bug).

L. *praefectus,* m. overseer, director, chief: prefect, prefecture.

L. *praepositus,* m. president, chief, director:

L. *praesidens, -entis,* m. director, governor, ruler: president, presidency.

L. *praetor, -is,* m. leader, chief, magistrate: pretor, pretorian.

L. *princeps, -ipis,* m. chief, leader; *principalis,* chief, foremost; m. overseer, super-intendent: prince, princess, principal, principle, unprincipled, principality, *Barycrinus princeps* (a Mississippian crinoid).

L. *procurator, -is,* m. manager, keeper: procurator, proctor, proxy.

Gr. *prohedros,* m. presiding officer:

Gr. *prymnetes,* m. steersman: *Prymnetes longiventer* (a fossil fish).

Gr. *prytanis, -eos,* m. chief magistrate; *prytaneion,* n. town hall: prytaneum, prytany.

L. *quaestor, -is,* m. a magistrate of the Roman state: questor, questorian.

L. *rex, regis,* m. king; *regina,* f. queen; *regulus,* m. dim.; *regalis; regillus; regius,* kingly, royal; *regimen, -inis,* n. guidance, rule, government; *rector, -is,* m. leader, director; *regens, -entis,* m. governor, prince, ruler; *rego, rectus,* keep straight, direct, guide, rule; *regno, -atus,* govern, rule: regal, regalia, regale, regent, regicide, royal, realm, regnant, corduroy, regulus, Leroy, pennyroyal, Vercingetorix, Montreal, Mount Royal, rectify, correction, direct, rectum, regulate, regimen, regiment, reign, region, incorrigible, dirigible, rector, rule, ruler, rail, dress, adroit, escort, *Regulus cristatus* (a kinglet), *Victoria regia* (royal waterlily), *Citheronia regalis* (regal moth), *Strelitzia reginae* (bird-of-paradise flower), *Hippocampus regulus* (a fish), *Cibotium regale* (a fern), *Eucalyptus regnans* (giant eucalyptus), *Balearica regulorum* (Kaffir crane). Derive: Regulus (alpha Leonis).

AS. *rice* (G. *reich*), dominion, jurisdiction, empire, kingdom: bishopric, rich, Reichstag, Austria (Österreich), Sverge (Sweden), Henry.

L. *satrapa* (Gr. *satrapes*), m. governor of a province: satrap, *Satrapes sartorii* (a beetle), *Satraparchis bijugata* (a moth).

Gr. *strategos,* m. general, commander; *strategikos,* of a general: strategy, strategist, stratagem.

Gr. *tagos,* m. commander, chief, ruler:

L. *tempero, -atus,* control, moderate: temper, distemper, intemperance, temperature, intemperate, temperamental, *Odynerus tempiferus* (a wasp).

L. *tracto, -atus,* handle, manage, haul; *tractabilis,* manageable, < *traho, tractus,* draw, haul: tractable, intractable.

L. *tribunus,* m. chieftain, representative: tribune, tribunal.

L. *tyrannus* (Gr. *tyrannos*), m. master, despot; *tyrannis, -idos,* f. sovereignty, despotic rule; *tyrannikos,* despotic: tyrant, tyrannize, tyrannical, tyranny, *Tryannus tyrannus* (kingbird), *Tyrannosaurus rex* (a dinosaur), *Tyrannophryne pugnax* (a fish), *Myiarchus tyrannulus* (a flycatcher).

See: **lead, hold, law, strong, rudder, right, sober, victory, country.** Derive: mogul, tycoon.

grabatus, L. (Gr. *krabatos*), small couch, bed; *grabatulus,* dim.; see **bed**

grabion, Gr. torch; see **light**

grace < L. *gratia,* favor, pleasure, beauty; see **beauty**

gracilis, L. slender, thin; *gracilitas,* slenderness; see **thin**

graculus, L. jackdaw; see **crow**

gradus, L. step, degree, stage, pitch; *gradatus,* step by step; see **step, walk**

graft < L. *graphium,* stylus, grafting knife.

Gr. *embolas, -ados,* grafted:

Gr. *enophthalmizo,* graft, inoculate:

L. *inoculatio, -onis,* f. an ingrafting, < *inoculo, -atus,* ingraft an eye or bud:

L. *insitum,* n. graft, scion; *insitivus,* grafted, < *insero, -itus,* graft:

grain < L. *granum,* n. seed, small kernel, pellet; *granulum,* n. dim.; *graneus* of grain; *granatus; granosus,* seedy, full of seeds or grains; *granarium,* n. place for storing grain: granular, granary, granite, ingrain, garner, garnet, grange, pomegranate *(Punica granatum),* granadilla *(Passiflora edulis),* Elaphomyces granulatus (hart-truffle), *Alampetis granulosissima* (a beetle), *Dianthus graniticus* (a pink), *Coeloconus granosus* (a fossil bryozoan), *Batostomella granulifera* (a fossil bryozoan), *Delocrinus granulosus* (a Pennsylvanian crinoid),

Porcellio graniger (a sowbug), *Ascophanus granuliformis* (a fungus), *Aphodius granarius* (a beetle), *Calandra granaria* (a beetle).

L. *ador, -is; adoreum,* n. spelt; *adoreus,* of spelt; *adorea,* f. reward in spelt:

Gr. *agastachys,* rich in grain: *Agastache urticifolia* (a giant-hyssop).

Gr. *aigilops, -opos,* m. wild oats: *Aegilops ovata* (a goat-grass).

Gr. *akoste,* f. barley:

L. *alica,* f. spelt (probably emmer); *alicarius,* of spelt; *alicastrum,* n. summer spelt: *Brosimum alicastrum* (breadnut).

L. *annona,* yearly produce, crop, grain; see **reap**

L. *arenula,* grain of sand; see *arena* under **sand**

L. *arinca,* f. a kind of grain, probably rye:

L. *asia,* f. a Ligurian name for rye:

Gr. *athere (athare),* groats, grits; see **flour**

L. *avena,* f. oat, oats: *Avena sativa* (oats), avenaceous.   Derive: haversack.

Gr. *bosmoron,* n. a kind of Indian grain:

L. *brace,* f. a kind of white grain:

Gr. *briza,* f. a grain like rye: *Briza media* (quaking-grass), *Bromus brizaeformis* (a chess).

L. *bromus* (Gr. *bromos*), m. a kind of oats: *Bromus ciliatus* (fringed brome).

L. *Ceres,* f. goddess of grains and agriculture: cereal, *Sitotroga cerealella* (a grain-moth).

Gr. *chidron,* n. grain from the ear, groat:

Gr. *chondros,* m. grit, grain of wheat or spelt; granular; *chondrion,* n. dim.; *chondrodes,* like grit, granular: chondrite, chondritic, chondrule, chondriosome.

AS. *corn* (G. *korn*), grain (barley, maize, oats, rye, spelt, wheat); *cyrnel,* dim.: corn, kernel, peppercorn, John Barleycorn, einkorn *(Triticum monococcum).* Derive: acorn.

Gr. *Demeter, -tros,* f. goddess of agriculture; *Demetrios,* of Demeter: Demeter, Demetrius, *Demetridula pallidula* (a beetle).

Gr. *elymos,* m. a kind of grain: *Elymus riparius* (a wild rye).

L. *far, farris* (Gr. *phar*), n. spelt (probably emmer); *farraceus; farrarius,* of spelt; *farreatus; farreus,* made of spelt: farrago, *Hemirhamphus far* (a fish), *Gnypetosoma farrea* (a beetle), *Pyrenula farrea* (a lichen).

L. *frumentum,* n. grain; *frumentaceus,* of grain: frumenty, *spiritus frumenti* (whisky), *Echinochloa frumentacea* (Japanese millet).

L. *holcus* (Gr. *holkos*), m. a kind of grain, mouse-barley: *Holcus lanatus* (velvet-grass).

L. *hordeum,* n. barley; *hordeaceus,* of barley: hordeolum, hordein, *Hordeum vulgare* (barley), *Hormodendrum hordei* (a fungus).

L. *irroratus,* bedewed, covered with granules; see *ros* under **drop**

Gr. *kachrys, -yos,* parched barley, catkin, ament; see **catkin**

Gr. *kenchros,* m. millet; *kenchridion,* n. dim.; *kenchrinos,* of millet; *kenchrias; kenchrites; kenchrodes,* like millet: *Cenchrus echinatus* (southern sandbur), *Cenchridium globosum* (a foraminifer), *Aegopogon cenchroides* (a grass).

Gr. *kokkos,* seed, grain, berry; see **berry**

Gr. *krithe,* f. barley; *krithidion,* n. dim.; *krithinos,* of barley: crithomancy, *Crithophaga miliaria* (a bird).

Sp. *mais,* < Ab.Am. *mahiz; mayz,* Indian corn: maize, mayzin, *Zea mays* (maize, corn), *Aphis maidis* (an aphid).

Gr. *manna,* morsel, grain, gum of a tree or exudate of an insect; see **resin**

Gr. *meline,* f. millet; *melinos,* of millet: *Melinus minutiflora* (molasses-grass).

L. *mica,* crumb, grain; see **little**

L. *milium,* n. millet; *miliaceus, miliarius; miliginus,* of millet: millet, miliary, gromwell, *Milium effusum* (millet-grass), *Miliola rostrata* (a foraminifer), *Miliolina venusta* (a foraminifer), *Panicum miliaceum* (broomcorn-millet), *Noctiluca miliaris* (a flagellate).

Gr. *olyra,* f. a kind of grain, perhaps rye:

Gr. *oryza,* f. rice; *oryzion,* n. dim.: *Oryza sativa* (rice), *Oryzopsis miliacea* (a ricegrass), *Oryzomys palustris* (rice-rat), *Sitophilus oryzae* (rice-weevil), *Dolichonyx oryzivorus* (bobolink).

Gr. *oula,* f. barley-corn:

L. *panus,* ear of millet; *paniculus,* dim.; see **bush**

NL. *paspalum,* n. (Gr. *paspalos,* m.), a kind of millet: *Paspalum distichum* (knot-grass).

Gr. *psakas* (*psekas*), droplet, bit, grain; see **drop**

Gr. *ptisane,* f. hulled barley:

Gr. *pyros,* m. wheat, grain; *pyraminos; pyrinos,* of wheat: *Diospyros discolor* (camagon, a persimmon), *Fagopyrum esculentum* (buckwheat), *Melampyrum lineare* (cowwheat), *Isopyrum biternatum* (a ranunculacead), *Agropyron spicatum* (bluebunch-wheatgrass).

Gr. *rhathaminx,* drop, grain, bit; see **drop**

L. *sandala,* f. a very white kind of grain:

L. *secale, -is,* n. rye: *Secale cereale* (rye), *Triticites secalicus* (a foraminifer), *Globulina secale* (a foraminifer), *Bromus secalinus* (cheat, chess).

L. *siligo, -inis,* f. white wheat, winter wheat, fine wheat flour; *siligineus; silignus,* of wheat, wheaten; *siliginarius,* m. baker:

Gr. *sitos,* food, grain; see **food**

L. *spelta,* f. a kind of wheat: spelt(*Triticum spelta*).

L. *triticum,* n. wheat; *triticeus; triticinus,* of wheat, wheaten: *Triticum aestivum* (wheat), *Triticites ventricosus* (a foraminifer), *Harmolita tritici* (joint-worm), *Puccinia triticina* (orange leaf-rust), *Agropyron triticeum* (a grass).

L. *zea* (Gr. *zea; zeia*), f. a kind of grain, probably spelt: *Zea mays* (maize, corn), *Ustilago zeae* (corn-smut), *Zeacrinus peramplus* (a Permian crinoid).

See: **seed, little, sand, grass, flour**

gralla, L. stilt; *grallator,* one who walks on stilts; see **pillar**

gramen, -inis, L. grass; *gramineus; graminosus,* grassy; see **grass**

gramia, L. pus in the corners of the eyes; *gramiosus,* full of pus, watery, bleareyed; see **pus**

gramma, Gr. letter, mark, picture, see **letter**; *gramme,* line; *grammikos,* linear, see **line**

grammar < OF. *gramaire,* < L. *grammatica* (Gr. *grammatike*); see *gramma* under **letter**

grampus, NL. a kind of whale; see **whale**

grand < L. *grandis,* large, great, noble, sublime, magnificent; see **large, honor**

grando, -inis, L. hail, hailstorm; see **ice**

granum, L. grain, seed, kernel, pellet; *granulum,* dim.; *graneus,* of grain; see **grain**

grao- < Gr. *graus, -aos; graia,* old woman, old; see **woman**

grape < OF. *grape.*

L. *acinus,* m.; *acinum,* n. berry, grape; *acinarius,* pertaining to grapes; *acinaticius,* prepared from grapes; *acinosus,* like grapes; *duracinus,* hard-berried: aciniform, acinarious, acinose, duracine.

L. *argitis, -idis,* f. a vine with clusters of white grapes:

L. *asinusca,* f. a kind of grape:

Gr. *astaphis, -idos,* f. dried grape, raisin:

L. *atrusca,* f. a kind of grape:

L. *bannanica,* f. a variety of grape:

Gr. *botrys, -yos,* m. bunch or cluster of grapes; *botrydion,* n. dim.; *botrychos,* m. grapevine: botryoidal, *Botrydium granulatum* (an alga), *Botrychium virginianum* (a grapefern), *Botrymyces equi* (a fungus), *Botryllus smaragdus* (a tunicate), *Artabotrys odoratissimus* (ylang), *Polybotrya cervina* (a fern), *Protosiphon botryoides* (an alga), *Strongylodon macrobotrys* (jade-vine), *Botrytis squamosa* (a fungus).

L. *brisa,* f. grapes after being pressed, refuse:

L. *bumastus* (Gr. *boumastos*), f. a kind of grape with large clusters:

L. *cocolobis,* f. Spanish name for a kind of grape: *Coccolobis floridana* (a seagrape).

Gr. *gigarton,* grape seed; see **seed**

L. *irtiola,* f. a kind of grape in Umbria:

L. *labrusca,* f. a wild grapevine; *labruscum,* n. wild grape: *Vitis labrusca* (foxgrape).

L. *leptorax, -agis,* f. a kind of grapevine with very small berries:

L. *massaris,* f. a grape from a wild vine:

Gr. *oine,* f.; *oinaron,* n. grapevine; *oinanthe,* f. blossom of the vine; an umbellifer; a bird: *Oenanthe fistulosa* (a water-dropwort), *Saxicola oenanthe* (wheatear).

Gr. *omphax, -akos,* f. an unripe grape, sour grape; *omphacodes,* like unripe grapes; *omphakinos,* of unripe grapes; *omphakion,* n. juice of unripe grapes: *Omphax plantaria* (a moth), *Omphacodes directa* (a moth).

It. *passula,* f. dried grape, raisin: passulate, passulation, *Carpoglyphus passularum* (a mite).

L. *precia,* f. a kind of grapevine:

L. *prusinius,* pertaining to a kind of grape:

L. *racemus,* m. bunch or cluster of grapes; *racematus; racemifer,* having or bearing clusters; *racemosus,* full of clusters: raceme, racemic, racemose, raisin, *Cyrilla racemiflora* (titi, leatherwood), *Platanus racemosa* (California plane-tree), *Solanum racemigerum* (a nightshade).

Gr. *rhax, -agos,* f. grape, berry; a kind of spider; *rhagion,* n. dim.; *rhagikos,* of grapes; *rhagodes,* grapelike; *microrrhax, -agos,* small-berried: rhagite, rhagon, rhagose, *Rhagodia hastata* (a goosefoot), *Rhagiosoma madagascariense* (a beetle), *Rhagiomorpha plagiata* (a beetle).

L. *sircula*, f. a kind of grape:

L. *spionia*, f. a kind of grape:

Gr. *staphyle; staphylis, -idos*, f. bunch or cluster of grapes; *staphylion*, n. dim.; *staphylinos*, of a bunch of grapes: *Staphylococcus aureus* (pus-forming bacterium), *Arctostaphylos nummularia* (fire-manzanita), staphylion, staphyloma, *Staphylea trifolia* (bladdernut).

L. *sticula*, f. a kind of grape:

L. *talpona*, f. a kind of grape:

L. *tamnus*, f. wild grapevine: *Smilax tamnifolia* (a greenbrier).

L. *uva*, f. grape; *uvula*, f. dim.: uvate, uvula, uvea, *Uvularia perfoliata* (a bellwort), *Coccolobis uvifera* (sea-grape), *Uvigerina cristata* (a fossil foraminifer), *Pestalozzia uvicola* (a fungus), *Synura uvella* (an alga).

L. *vitis*, f. grapevine; *viticula*, f. dim., tendril: Vitaceae, *Vitis rotundifolia* (muscadine grape), *Vitoxylon opalinum* (a Miocene grapevine), *Peronospora viticola* (a mildew), *Vaccinium vitis-idaea* (cowberry), *Selaginella viticulosa* (a selaginella), *Rubus vitifolius* (California blackberry).

See: **vine, wine, cluster, berry**

grapho, Gr. write; *graphikos*, of writing; *graphis*, pen; *graptos*, written, marked; see **write, pen**

grapsaeo- < Gr. *grapsaios*, a crab; see **crab**

grass < AS. *graes:* grassy, grasshopper, grassland, grassquit, bluegrass, Sweetgrass (Mont.).

L. *aera* (Gr. *aira*), f. darnel, a kind of grass; *aericus; aerinus*, of darnel: *Aira* [*Aera*] *caryophyllea* (silver hairgrass).

NL. *agropyron*, n. a genus of grasses: *Agropyron inerme* (a grass).

Gr. *agrostis, -idos*, f. a forage grass: *Agrostis stolonifera* (a bentgrass), *Eragrostis spicata* (a grass), *Calamagrostis canadensis* (bluejoint reedgrass), *Xiphagrostis floridula* (a sword-grass), *Panicum agrostoides* (redtop).

NL. *alopecurus*, m. a genus of grasses: *Alopecurus saccatus* (a foxtail-grass).

NL. *andropogon*, m. a genus of grasses: *Andropogon maritimus* (a beardgrass).

NL. *aristida*, f. a genus of grasses: *Aristida divaricata* (a grass).

L. *arundo, -inis*, f. reed, cane: arundinaceous, arundineous, *Arundinaria gigantea* (a bamboo), *Bambusa arundinacea* (thorny bamboo), *Tromatobia arundinator* (a wasp).

NL. *bambusa*, f. < Malay *bambu*, a kind of grass: *Bambusa vulgaris* (common bamboo), *Halictus bambusarum* (a bee), *Aedes bambusicolus* (a mosquito).

L. *caespes, -pitis*, turf, sod; see **sod**

L. *calamus* (Gr. *kalamos*), m. reed, stalk, pen, pipe, flute; *calamarius*, of reeds; *calamellus*, m. dim.; *kalamion*, n.; *kalamiskos*, m. dim.; *kalaminos*, reedlike; *kalamodes*, reedy; *kalamites*, m. a reedlike plant: calamary, calumet, *Calamus rotang* (a rattan-palm), *Calamites carinatus* (a fossil plant), *Dendrocalamus membranacea* (a bamboo), *Acorus calamus* (sweet-flag), *Cylindrophyllum calamiforme* (a fossil coral), *Calamogrostis gigantea* (big reedgrass), *Calamodyta arundinacea* (a bird).

L. *canna* (Gr. *kanna*), f. reed, cane, pipe; *cannula*, f. dim.; *cannetum*, n. reed thicket: cannula, cane, cannon, canal, canyon, canister, caramel, *Canna coccinea* (scarlet canna), *Canella winterana* (a plant), *Cannorrhaphis spathillata* (a radiolarian).

Gr. *chilos*, m. grass, forage:

Gr. *chloë, chloa*, f. first green shoot of plants in spring; grass; *chloanos*, greenish: Chloë, chloanthite, *Echinochloa paludigena* (a grass), *Hierochloë occidentalis* (California sweetgrass).

Gr. *chortos*, m. grass, fodder; *chortinos*, of grass: *Chorthippus parallelus* (a grasshopper), *Calochortus luteus* (a mariposa).

L. *cinna* (Gr. *kinna*), f. a Cilician grass: *Cinna arundinacea* (stout woodreed).

L. *coix, -icis* (Gr. *koix, -ikos*), f. an Egyptian palm; now applied to a genus of grasses: *Coix lacrymajobi* (Job's-tears).

NL. *distichlis*, f. a genus of grasses: *Distichlis stricta* (a saltgrass).

L. *donax, -acis* (Gr. *-akos*), m. reed, cane: *Arundo donax* (a reed).

Gr. *Eleusine*, f. Ceres; a genus of grasses: *Eleusine indica* (goosegrass).

L. *fenum* (*faenum*), n. hay; *fenarius; feneus*, of hay; *fenilium*, n. hayloft: fennel, fenugreek, *Carex faenea* (a sedge).

L. *festuca*, stalk, stem, straw; a genus of grasses; see **stem**

NL. *glyceria*, f. a genus of grasses: *Glyceria erecta* (a manna-grass).

L. *gramen, -inis*, n. grass; *gramineus; graminosus*, of grass, grassy: graminivorous, graminaceous, gramineous, *Puccinia graminis* (wheat-rust), *Potamogeton gramineus* (a pondweed), *Stellaria graminea* (a chickweed), *Alisma gramineum* (a water-plantain), *Chrysopsis graminifolia* (a golden aster), *Sisyrinchium graminoides* (a blue-eyed grass), grama (*Bouteloua hirsuta*), *Toxoptera graminum* (an aphid).

Gr. *grastis,* grass, green fodder; see *grao* under **eat**

L. *herba,* green vegetation, grass; see **plant**

Gr. *karphos,* dried grass stem; see **stem**

Gr. *kenchros,* millet; see **grain**

Gr. *lechepoie,* grassy, meadowy, with grass fit to make a bed:

L. *lolium,* n. darnel, cockle, tares; a kind of grass: *Lolium perenne* (ryegrass).

NL. *melica,* f. < It. *melica,* sorghum: *Melica bulbosa* (onion-grass), *Trisetum melicoides* (a grass), *Glyceria melicaria* (a grass).

Gr. *orophos,* the reed used for thatching, < *orophe,* roof; see **roof**

L. *panicum,* n. panic grass, wild millet, < *panus,* ear of millet: *Panicum scoparium* (velvet panicum).

Gr. *pankynion,* n. a kind of grass:

L. *phalaris, -idis* (Gr. *-idos*), f. a kind of grass: *Phalaris arundinacea* (canary-grass).

NL. *phleum,* n. (Gr. *phleos,* m.), a kind of rush or reed: *Phleum subulatum* (Italian timothy), *Phleocryptes melanops* (a bird), *Mycobacterium phlei* (a bacterium).

L. *phragmites,* m. a reed, < Gr. *phragmites,* growing in hedges: *Phragmites communis* (common reed).

Gr. *poa,* f. grass; *poarion,* n. dim.; *poesichroos,* grass-colored: Poaceae, *Poa annua* (annual bluegrass), *Poa pratensis* (Kentucky bluegrass), *Pooecetes gramineus* (vesper-sparrow).

L. *pratulum,* lawn, dim. of *pratum,* meadow; see **field**

NL. *sitanion,* n. a genus of grasses: *Sitanion jubatum* (squirreltail-grass).

NL. *sorghum,* n. (It. *sorgo*), a genus of grasses: *Sorghum vulgare*(sorghum), *Sorghastrum nutans* (Indian grass), *Sphacelotheca sorghi* (a smut).

NL. *spartina* (Gr. *spartine,* cord, rope), f. a genus of grasses: *Spartina pectinata* (prairie-cordgrass).

NL. *sporobolus,* m. a genus of grasses: *Sporobolus ramulosus* (red dropseed).

NL. *stipa,* f. a genus of grasses: *Stipa speciosa* (desert-needlegrass).

L. *stramen, -inis,* n. straw, litter, any material used for bedding down; *stramineus,* of straw: *Stentor stramineus* (a protozoan), *Physarum straminipes* (a fungus), *Panicum stramineum* (a grass), *Carex straminea* (a sedge).

Gr. *thyaros,* m. darnel, a kind of grass:

NL. *tripsacum,* n. a genus of grasses: *Tripsacum lanceolatum* (Mexican gama-grass).

L. *uniola,* f. a plant; a genus of grasses: *Uniola paniculata* (sea-oats).

NL. *vilfa,* f. a genus of grasses: *Vilfa arguta* (now *Sporobolus argutus,* pointed dropseed).

L. *viretum,* greensward, sod, turf; see **sod**

NL. *zizania,* f.; *zizanium* (Gr. *zizanion*), n. a kind of weed; now a genus of grasses: *Zizania aquatica* (wild rice), *Zizaniopsis miliacea* (southern wild rice).

See: **sod, reed, grain**

grassator, L. vagabond, footpad; *grassor, -atus,* go about; see *gradior* under **walk**

grasshopper

Gr. *akris, -idos,* f. locust, grasshopper; *akridion,* n. dim.: Acrididae, *Acridotheres tristis* (myna), *Acridium ornatum* (a grasshopper), *Nomadacris septemfasciata* (red locust), *Coccobacillus(Aerobacter) acridiorum*(a bacterium).

Gr. *asirakos,* m. a kind of wingless locust:

L. *attacus* (Gr. *attakos*), m. a kind of locust: *Attacus atlas* (a moth).

L. *attelabus* (Gr. *attelabos*), m. a small, wingless locust: *Attelabus octomaculatus* (a beetle).

L. *bruchus* (Gr. *brouchos*), a kind of wingless locust; now a genus of beetles; see **beetle**

L. *gryllus,* m. cricket; *gryllo, -atus,* chirp like a cricket: *Gryllus bimaculatus* (a cricket), *Gryllotalpa hexadactyla* (a mole-cricket), *Gryllulus angusticeps* (a cricket) *Entomophthora grylli* (a fungus).

Gr. *herpyllis, -idos,* f. a grasshopper:

Gr. *kalamaia,* f. a kind of grasshopper:

Gr. *kerkope,* f. a long-tailed grasshopper: *Cercopidium mimas* (a fossil insect-wing).

Gr. *kornops, -opos,* m. a kind of locust: *Cornops bivittatus* (a grasshopper).

L. *locusta,* f. lobster, grasshopper: locust, *Locusta viridissima* (green grasshopper), *Locustana pardalina* (brown locust), *Trombidium locustarum* (a mite).

Gr. *molouris, -idos,* f. a kind of locust:

Gr. *olingos,* m. a kind of locust:

Gr. *parnops, -opos,* m. a kind of locust: *Parnops glasunowi* (a beetle), *Gammaro-parnops crassicruris* (a grasshopper).

Gr. *petelis, -idos,* f. a kind of locust:

L. *troxalis, -idis,* f. grasshopper, cricket: *Troxalis brevicornis* (a grasshopper).

grastis, Gr. grass, green fodder; *grastizo,* graze; see *grao* under **eat**

grate < L. *cratis,* hurdle; see **screen**

gratia, L. favor, mercy, pleasure, beauty; *gratiosus,* full of favor, popular; see **give**

gratilla, L. a kind of cake; see **cake**

gratiola, NL. a genus of the figwort family; see **figwort**

gratus, L. beloved, agreeable, pleasing, thankful; *gratulor, -atus,* wish one joy, render thanks; see *gratia* under **give**

grave < AS. *graef,* < *grafan,* dig.

L. *bustum,* place where bodies were burned, grave, sepulcher; *bustualis,* pertaining to such places; see *uro* under **burn**

L. *capulus,* m. sarcophagus, bier, tomb; *capularis,* pertaining to a bier:

L. *catacumba,* f. underground burial gallery: catacomb.

L. *conditivum; conditorium,* n. grave, tomb:

Gr. *ensorion,* n. sarcophagus:

Gr. *hades,* hell, grave; see **hell**

L. *mausoleum* (Gr. *mausoleion*), n. magnificent tomb erected for Mausolus at Halicarnassus: mausoleum.

L. *puticulus,* m. grave for the poor and slaves:

L. *sarcophagus* (Gr. *sarcophagos*), m. grave, sepulcher: sarcophagus.

L. *sepulcrum* (*sepulchrum*), n. grave, tomb; *sepulcralis,* of a tomb; *sepelio, sepultus,* bury: sepulcher, sepulchral, sepulture, *Entomophthora sepulchralis* (a fungus).

Gr. *taphos,* m. grave, tomb; *tapheus,* m. burier; *entaphios,* of burial: taphephobia, epitaph, cenotaph, *Taphonycteris affinis* (a bat).

L. *tumba,* f. (Gr. *tymbos,* m.), sepulchral mound, grave; *tumbula,* f.; *tymbion,* n. dim.: tomb, *Tymbophora peltastis* (a moth).

See: **ditch, hole, furrow, dig**

gravedo, L. heaviness in the head, cold; *gravedinosus,* subject to colds; see **disease**

gravel < OF. *gravelle;* see **stone**

gravis, L. heavy; *gravidus,* laden, pregnant; *gravo, -atus,* load; see **heavy**

gravy < uncertain origin; see **sauce**

gray < AS. *graeg:* graylag, grayling, graywacke, gray drake (*Ephemera vulgata*). Derive: greyhound.

Gr. *anaimos,* bloodless, pale: anemic.

Gr. *apochlorias, -ou,* m. a pale person:

L. *caesius,* bluish-gray; *caesitius,* bluish: cesious, cesium, *Rubus caesius* (European dewberry), *Acer caesium* (Himalayan maple), *Prosthetopteryx caesiata* (a moth), *Phorocera caesifrons* (a fly).

L. *canus; canutus,* gray, ash-colored, hoary; *canitia; canitudo, -inis,* f. grayness, hoariness; *canaster,* grizzled; *canens, -entis; canescens,* becoming gray, gray, grayish, hoary; *incanus,* quite gray: canescent, incanous, *Agastache cana* (a giant-hyssop), *Alnus incana* (an alder), *Calidris canutus* (knot), *Amorpha canescens* (leadplant), *Dictyoloma incanescens* (a rutacead), *Junco caniceps* (a junco).

L. *cineraceus; cinereus; cinericius,* like ashes, ash-colored, gray; see *cinis* under **ashes**

I. *decolor,* discolored, faded; see **color**

Gr. *empelios,* gray:

L. *exsanguis,* bloodless, pale, feeble:

L. *glaucus* (Gr. *glaukos*), bluish-green or gray, sea-colored, hoary, covered with a bloom: glaucous, glauconite, glaucoma, *Glaucus pacificus* (a sea-snail), *Glaucium flavum* (yellow horn-poppy), *Glaucosoma hebraicum* (a jewfish) *Picea glauca* (white spruce), *Elymus glaucus* (a wild rye), *Chamaedorea glaucifolia* (a palm), *Acacia glaucescens* (brigalow acacia).

L. *gravastellus,* m. a gray-headed fellow:

ML. *griseus,* < OF. *gris,* gray; *griseolus,* grayish: grisette, gridelin, ambergris, grison (*Galictis vittata*), *Hapalemur griseus* (gray lemur), *Blaesioxipha cinereo-grisea* (a fly), *Cercopithecus griseoviridis* (grivet).

AS. *har,* gray, old: hoar, hoary, hoarfrost, horehound.

Gr. *killos,* ass-colored, gray; see **horse**

Gr. *knekis, -idos,* pale spot, dim cloud; see **spot**

Gr. *leiros,* like a lily, pale; see *leirion* under **lily**

L. *leucophaeus* (Gr. *leukophaios*), ash-colored, dun:

Gr. *liphaimos*, lacking blood, pale:

L. *lividus*, black and blue, lead-colored, ashy; see **blue**

Gr. *molybros*, lead-colored; see *molybdus* under **lead**

L. *murinus*, mouse-gray; see *mus* under **mouse**

Gr. *ochroma, -tos,* paleness, wanness; *ochros,* pale, wan, sallow, pale-yellow; *ochriasis,* paleness; *enochros,* somewhat pale; *exochros,* very pale; *achromos,* colorless; see *ochros* under **yellow**

L. *pallidus,* ashen, wan; *pallidulus,* dim.; *pallor, -is,* m. paleness; *impallesco,* turn pale: pallid, pallor, pale, appall, *Ditrichum pallidum* (a moss), *Baetis pallidulus* (a mayfly), *Iron subpallidus* (a mayfly), *Fritillaria pallidiflora* (Siberian fritillary), *Calyptranthes pallens* (a lidflower), *Glossina pallidipes* (a tsetse-fly), *Systoechus pallidipilosus* (a fly).

Gr. *pelios,* black and blue, livid, grayish-blue; see **blue**

Gr. *phorkos,* gray, white: *Phorcorrhaphis edwardsi* (an amphipod).

Gr. *polios,* ashy-gray: poliomyelitis, *Polioptilia caerulea* (a gnat-catcher), *Pteropus poliocephalus* (Australian fruit-bat), *Lagonosticta polionota* (a bird).

L. *pruinosus,* frosty, rimy; see *pruina* under **ice**

L. *ravus,* grayish-yellow, tawny, gray; *ravidus,* grayish: *Clavatula rava* (a gastropod), *Epicaerus ravidus* (a beetle).

Gr. *spodios,* ash-colored, gray; see *spodos* under **ashes**

Gr. *tephros; tephrinos,* ash-colored; see *tephra* under **ashes**

See: **ashes, flour**

great < AS. *great;* see **large, honor, abundance**

greed < AS. *graed.*

Gr. *aplestos,* greedy, insatiate; see **hunger**

L. *avarus,* eagerly desirous, covetous, greedy; *avaritia,* f. greediness: avarice, *Baridius avarus* (a beetle).

L. *avidus,* eagerly desirous, greedy; see **wish**

L. *cupiditas,* avarice, greed, lust; see *cupio* under **wish**

L. *edax, -acis,* greedy, gluttonous; see *edo* under **eat**

Gr. *glischros,* sticky, greedy: *Harpalus caliginosus* (a beetle).

Gr. *labros,* furious, hasty, greedy, gluttonous; *labrosyne,* f. greed; *labrosytos,* rushing furiously: *Labrosaurus ferox* (a dinosaur).

Gr. *lamyros,* greedy, wanton: *Lamyrodes philerus* (a moth).

Gr. *molobros,* greedy, greedy fellow; see *bibrosko* under **eat**

L. *petax, -acis,* catching at, greedy for:

L. *rapax, -acis,* grasping, greedy; see *rapio* under **take**

Gr. *sintor, -os,* ravenous, tearing: *Sintor quadrilineatus* (a beetle).

L. *trahax, -acis,* greedy, grasping: *Theridium trahax* (a spider).

See: **eat, hunger, wish, steal, take**

Greek < L. *Graecus* (Gr. *Graikos*), pertaining to Greece: Grecian, Grecism, Greco-Roman, *Amaranthus graecizans* (tumbleweed-amaranth), *Testudo graeca* (a tortoise), *Procrustes graecus* (a beetle), *Trigonella faenumgraecum* (fenugreek).

L. *Atticus* (Gr. *Attikos*), pertaining to the Greek province of Attica and its capital, Athens: attic, Atticism, *Graecophalangium atticum* (an arachnid).

Gr. *Doris, -idos,* f. a district of ancient Greece; *Dorikos,* of Doris or the Dorians: Dorian, Doric, doricism(dorism).

Gr. *Hellas, -ados,* f. Greece; *Hellen, -os,* m. a Greek; *Hellenikos,* Greek: Hellas, Hellenic, Hellenist, *Helladotherium duvernoyi* (a Miocene giraffe).

Gr. *Ionia,* f. a former region comprising islands in the Aegean Sea and part of the coast of Asia Minor; *Ionikos,* of Ionia or the Ionians: Ionian, Ionic, ionicism.

L. *Laconia* (Gr. *Lakonike*), f. Sparta or Lacedaemon: Laconian, laconic, laconicism (laconism).

green < AS. *grene.*

Gr. *aeichloros; aeiphyllos; aeithales,* evergreen:

L. *aerugo, -inis,* verdigris; *aeruginosus,* full of verdigris, greenish-blue; see *aes* under **copper**

L. *callainus,* bluish-green:

F. *celadon,* a sea-green color: celadonite.

Gr. *chloros,* green; *chlorotes,* f. greenness: chloroform, chloromycetin, chlorophyll, chlorine, chloral, chloropicrin (tear gas), chloroxylenol, sodium chloride, pyrochlore, dichlorodiethylsulfide (mustard gas), betachlorovinyldichloroarsine (lewisite), dichlorodiphenyltrichloroethane (DDT), chlorellin (from *Chlorella pyrenoidosa*), Chlorophyceae, *Chlorophytum elatum* (a lily), *Chloroxylon swietenia* (satinwood), *Chrysochloris trevelyani* (a golden mole), *Zizyphus chloroxylon* (cogwood).

L. *earinus* (Gr. *earinos,* < *ear,* spring), of spring, the color of spring, green; see *ear* under **year**

L. *herbaceus,* pertaining to herbs, green; see *herba* under **plant**
L. *malinus,* pertaining to apples, apple-green; see *malum* under **apple**
L. *olivarius,* of olives; see **olive**
L. *prasinus* (Gr. *prasinos; prasios*), leek-green, green; see *prason* under **onion**
L. *pratens, -entis,* meadow-green, grassy; see *pratum* under **field**
L. *salor, -is,* m. color of the sea, sea-green:
L. *smaragdinus* (Gr. *smaragdinos*), emerald-green; see *smaragdus* under **stone**
Gr. *thalassinos,* sea-colored, green: see *thalassa* under **sea**
L. *viridis* (F. *vert;* Sp. *verde*), green; *viror, -is,* m. greenness; *vireo,* be green; *viresco,* become green: viridin, virescent, verdant, verdure, Vermont, verdigris, Cape Verde, Mesa Verde, farthingale, paloverde(*Cercidium floridum), Trillium viride* (green trillium), *Forsythia viridissima* (a forsythia), *Festuca viridula* (green fescue), *Acerates viridiflora* (green milkweed), *Peucetia viridans* (a spider), *Macropodus viridiauratus* (paradise-fish), *Triturus viridescens* (a newt), *Vireo olivaceus* (red-eyed vireo), *Icteria virens* (yellow-breasted chat), *Sminthurus viridis* (lucerne-flea), *Rubus vermontensis* (Vermont blackberry), *Verdithorax prasina* (a fish), *Castanopsis sempervirens* (Sierra chinquapin), *Empidonax virescens* (a flycatcher).
See: **young,** *chloe* under **grass, apple, pea, copper, mallow, sea, onion**

greet < AS *gretan,* address, salute; see **speak**

gregarius, L. pertaining to a flock or herd, common, < *grex, gregis,* flock, herd; *grego, -atus,* gather in a herd, collect, assemble; see **herd**

gremium, L. lap, bosom; *gremialis,* of a lap; see **lap**

grex, *gregis,* L. flock, herd; *gregalis,* of a flock or herd; see **herd**

grieve < OF. *grever,* < L. *gravis,* heavy; see **sad, weep**

grill < F. *grille,* < L. *craticula,* grating, gridiron; see **screen**

grin < AS. *grennian;* see **laugh**

grind < AS. *grindan:* grinder, grindstone, grist.
Gr. *aleo,* grind, crush, pound; *aletes,* m. grinder, upper millstone; *alesis,* f.; *aletos,* m. a grinding; *aletris, -idos,* f. female slave who ground grain: *Aletodes ferrugineus* (a fossil chimaeroid), *Aletocyon multicuspis* (a fossil carnivore), *Aletris farinosa* (colicroot).
Gr. *arabos,* a gnashing or chattering of the teeth; see **rattle**
Gr. *bryche,* f. a gnashing of the teeth:
L. *contundo, -tusus,* beat, bruise, grind; see **bruise**
L. *coticula,* small stone mortar; see *cos* under **stone**
L. *frendo, -esus,* gnash the teeth, grind to pieces, rage; *nefrens, -dis,* that cannot bite, toothless:
L. *molo, -litus,* grind; *mola,* f. millstone, mill; *molaris,* of a millstone; m. grinder; *molina,* f. mill; *molinus,* of a mill; *molitor, -is,* m. miller: moline, molar, Miller, Milner, Moline (Ill.), molendinaceous, immolate, emolument, ormolu, *Mola mola* (a fish, formerly *Cephalus brevis*), *Tenebrio molitor* (a mealworm).
Gr. *myle,* f. mill, millstone; *mylax, -akos,* m. millstone; *mylikos,* of or for a mill; *mylon, -os,* m. mill, grinder; *mylion,* n. dim.; *mylias; mylikos; mylonikos,* of a mill; *mylothros,* m. miller: mylodont, mylonite, amylon, *Mylax batoides* (a fossil fish), *Mylodon robustus* (a fossil sloth), *Myliobatis aquila* (eagle-ray), *Mylacodus quadratus* (a fossil fish).
L. *pinso, pinsus; pistus,* pound, stamp, crush, grind; *pistor, -is; pistrinarius,* m. miller; *pistrinum,* n. mill; *pistrilla,* f. dim.: pestle, piston, *Trimerotropis pistrinaria* (a grasshopper).
L. *tero, tritus,* rub, grind, wear, waste; see **rub**
Gr. *trismos,* m. a gnashing, grating, grinding; lockjaw: trismus.
See: **bruise, break, rub, scrape, pestle, mortar**

grindelia, NL. a genus of the composite family; see **composite**

grip < AS. *gripe;* see **bind, hold, take**

gripeus, Gr. fisherman; see **fisherman**

griphus, L. (Gr. *griphos; gripos*), a fishing-basket, creel, see **basket;** anything intricate, riddle, puzzle, see **ask**

griseus, ML. < OF. *gris,* gray; *griseolus,* grayish; see **gray**

gristle < AS. *gristle,* cartilage.
L. *cartilago, -inis,* f. gristle; *cartilagineus; cartilaginosus,* gristly: cartilage, cartilaginous, *Achillea cartilaginea* (a yarrow).
Gr. *chondros,* m. cartilage; *chondrodes,* cartilaginous: chondroblast, chondrotomy, chondrin, chondritis, chondrosin, chondrosteoma, achondroplasia, hypochondriac, *Chondrus crispus* (Irish moss, a seaweed), *Chondrodendron tomentosum* (pareira-root), *Halichondria panicea* (a sponge).   See **grain**
Gr. *traganon,* n. cartilage, gristle:

groan < AS. *granian;* see **sad**

groin < uncertain origin; the line or region between abdomen and thigh.
>Gr. *boubon, -os,* m. groin; swelling in the groin: bubonalgia, bubo, bubonic, bubonocele.
>Gr. *episeion,* n. pubic region: episiocele, episiotomy.
>L. *ilium,* n. groin, flank: ilium, iliocaudal, iliopubic, jade.
>L. *inguen, -inis,* n.; *inguina,* f. groin; *inguinalis,* of the groin: inguinal, *Epidermophyton inguinale* (a fungus).

gromphaena, L. a kind of amaranth; see **amaranth**

gromphas, *-ados,* Gr. an old sow; see **hog**

gromphena, L. a kind of crane; see **crane**

gronos, Gr. eaten out, hollowed out; *grone,* cavern, any hollow vessel; see **hollow**

groove < MD. *groeve;* see **furrow, ditch, pipe**

grosphus, L. (Gr. *grosphos*), a kind of javelin; see **spear**

grossularia, NL. gooseberry; see **currant**

grossus, L. big, coarse, thick, see **large;** an unripe fig; *grossulus,* dim., see **fig**

ground < AS. *grund;* see **earth**

groundsel < AS. *grundeswelge;* see *senecio* under **composite**

grouse < uncertain origin.
>Gr. *attagen, -os,* m. probably a kind of grouse: *Attagenus piceus* (black carpet-beetle).
>NL. *bonasa,* f. a genus of grouse: *Bonasa umbellus* (ruffed grouse), *Tetrastes bonasia* (hazel grouse).
>C. *tarmachan,* a kind of grouse: ptarmigan *(Lagopus mutus).*
>L. *tetrao, -onis,* m. moor-fowl: *Tetrao lagopus* (a bird).
>Gr. *tetrax, -agos,* m. a kind of grouse: *Otis tetrax* (little bustard).

grow < AS. *growan.*
>L. *addo, -ditus,* bring to, give to, increase; see **and**
>L. *adjectivus,* that is added; *adjectus,* added to, applying to: see *adjicio* under **and**
>Gr. *alsis, -eos,* f. growth, < *aldaino,* grow, multiply; *aldeeis, -essa, -en,* growing, increasing:
>L. *amplifico, -atus; amplio, -atus,* enlarge, increase, widen, extend; see *amplus* under **large**
>L. *augeo, auctus,* increase, enlarge, multiply, grow; *auctificus,* increasing, enlarging; *auctio, -onis,* f. increase; *auctarium,* n. an addition; *augmentum,* n. increase: auction, augment, augmentative, auxiliary, auctary, author, August, autumn.
>Gr. *auxo,* increase, grow; *auxesis,* f. growth; *auximos,* promoting growth; *aexi-,* fostering, increasing, strengthening; *aexikakos,* multiplying evil; *aexitrophos,* fostering growth: auxin, auxochrome, auxospore, auxesis, auxotome.
>Gr. *blothros,* growing up, shooting up; see **high**
>L. *convalesco,* grow strong, regain strength after illness: convalescence.
>L. *cresco, cretus,* grow, increase; *crementum,* n. growth, addition, increase: crescent, increment, concrete, excrescence, decrease, accrue, accrementition, accretion, crescendo, recruit, crew.
>Gr. *epidosis,* increase, growth, progress; see *do* under **give**
>Gr. *epitheke,* f. an addition, increase: *Epitheca bimaculata* (a dragonfly).
>Gr. *euernes,* flourishing; see *ernos* under **branch**
>L. *glisco,* grow up, increase, swell, spread:
>L. *multiplico, -atus,* increase, < *multiplex, -icis,* manifold: multiply, multiplication, multiplex.
>L. *olesco,* grow; *alesco,* grow up, increase; *adolesco, adultus,* grow up; *adolescens,* growing up; *coalesco, -alitus,* grow together, unite; *exolesco, -oletus,* grow out, mature: adolescent, adult, coalition, coalescence, abolish, *Rhantus exoletus* (a beetle).
>Gr. *phyo,* bring forth, produce, cause to grow; *phye,* f. full growth, prime; *physis,* f. growth, stature, nature: paraphysis, symphysis, *Symphysodon discus* (a fish).
>Gr. *plethysmos,* increase, enlargement; see *pleos* under **abundance**
>L. *profectus,* m. advance, progress, growth:
>L. *progressus,* m. a going forward: progress, progression, progressive.
>L. *provectus,* m. progress, increase:
>L. *proventus,* m. growth, increase, crop:
>Gr. *spartos,* grown from seed, cultivated; see *speiro* under **sow**
>Gr. *thaleros,* blooming, fresh, vigorous, active, luxuriant, abundant; *tethelos,* abounding, growing, < *thallo,* bloom, flourish, prosper: tethelin, *Thalerothele fasciata* (a spider).

ON. *thrifa,* grasp, prosper: thrive, thrifty, thriftless, spendthrift.
AS. *weaxan* (G. *wachsen*), grow: wax, wax and wane, waxy.
See: **large, gain, swell, and**

growl < uncertain origin; see **bark**

grudge < OF. *groucher;* see **hate**

grumble < D. *grommelen;* see **murmur**

grumus, L. heap, mound, hillock; *grumulus,* dim.; see **heap**

grunt < AS. *grunian,* produce short, guttural sounds: grunter, disgruntled, yellow
grunt *(Haemulon sciurus).*
L. *grunnio, -itus,* grunt: gurnard, *Bos grunniens* (yak).
Gr. *grylismos,* m. a grunting, < *grylizo,* grunt:
L. *quirrito, -atus,* grunt:

grus, *gruis,* L. crane; see **crane**

grylismos, Gr. a grunting; see **grunt**

gryllus, L. cricket; see **grasshopper**

grylos, Gr. pig; *grylion,* dim.; see **hog**

grymea, Gr. bag, chest; see **bag**

grynos, Gr. fagot; see **wood**

gryphus, L. < *gryps, -yphis* (Gr. *gryps, -ypos*), a fabulous creature half lion and
half eagle; see **animal**

grypos, Gr. curved, hook-nosed; see **bend**

grysmos, Gr. a grumbling; *gryktos,* grumbling; see **murmur**

gryte, Gr. trash; see **worthless**

guaiacum, NL. < Ab.Am. *guayacan,* a kind of tree and the resin derived therefrom; see **resin**

guano, Sp. < Ab.Am. *huanu,* dung; see **dung**

guard < OF. *guarder, warder,* < OHG. *warten,* watch; G. *wehren,* protect; AS.
waer, cautious; *weard,* watchman, keeper; *hlafweard,* bread-keeper: guardian,
guardhouse, vanguard, guarantee, regard, garrison, lord, ward, weir, warrant,
warden, warder, wardrobe, reward, beware, wary, Wareham, Warwick, Ward,
Hayward, Woodward, Howard, Seward, Steward, aware.
L. *aedituus,* m. keeper, guardian of a temple:
Gr. *agrypneter,* watcher; see *agrypnia* under **awake**
Gr. *aipolos,* goatherd; see *aipolion* under **herd**
Gr. *alexo,* ward off; *alexeter, -os,* m. defender, protector; *alexesis,* f. defense:
alexin, alexipyretic, Alexander, alexiteric.
L. *almus,* fostering, nourishing, cherishing, kind: alma mater, *Achthophora alma*
(a beetle).
Gr. *alytes,* m. police at the Olympic games:
Gr. *amorbos,* m. attendant, shepherd: *Amorbus planus* (a bug).
Gr. *amphibater, -os,* m. guardian, defender:
Gr. *amyno,* defend, guard; *amyntikos,* fit for defense: *Amynodon sinensis* (a fossil
mammal).
L. *arceo* (Gr. *arkeo*), shut in, prevent access to, guard, hinder; *arx, arcis,* f.
citadel, fortress, stronghold; *coerceo, -itus,* encompass, restrain: coerce, coercion,
exercise, *Zoarces anguillaris* (eelpout).
L. *Argus* (Gr. *Argos*), hundred-eyed monster, guardian of Io; a synonym of
vigilance; see **peacock**
L. *atriarius,* m. porter, doorkeeper:
L. *averruncus,* m. an averter, protector, < *averrunco, -atus,* avert: *Averruncus
emmelane* (a fish), averruncate.
AS. *beorgan,* protect, save: harbor, harbinger, hauberk, belfry.
Gr. *boter; boutes,* herdsman; see *bosko* under **eat,** *bos* under **cattle**
L. *bubsequa,* m. herdsman, cowboy:
L. *cancellarius,* m. porter, doorkeeper:
L. *carcerarius,* m. jailer, keeper:
L. *carrago, -inis,* f. a barricade made of wagons:
L. *caveo, cautus,* guard against, beware, avoid; *cautela,* f. caution, security: cau-
tion, cautious, precaution, *caveat emptor.*
L. *cellarius,* steward, butler; see *cella* under **room**
L. *Cerberus* (Gr. *Kerberos*), a three-headed doglike monster that guarded the
gates of Hades; see **dog**
L. *chartularius,* m. keeper of archives:
L. *cimeliarcha* (Gr. *keimeliarches*), m. treasurer:

L. *circitor; circuitor, -is,* m. one who goes the rounds, watchman, guard, patrol; *circitorius,* of patrols:

L. *circumspectus,* considered, guarded, cautious: circumspect.

L. *cura,* f. care, attention, management; *curo, -atus,* care for guard, manage; *curator, -is,* m. caretaker; *curans, -antis,* m. physician; *curandus,* m. the patient; *procurator, -is,* m. guardian, manager: curator, cure, curative, accurate, curiosity, procure, procurer, procurator, proctor, proxy, sinecure, insecure, security, assure, sure, surety, insurance, incurable, inaccuracy, manicure, scour.

L. *custos, -odis,* c. guardian, keeper; *custodela; custodia,* f. a watch, guard, guard-house; *custodiarius,* m. jailer, < *custodio, -itus,* watch, protect, guard: custodian, custody, *Plagiolepis custodiens* (an ant).

L. *defendo, -fensus,* keep off, protect, guard: defend, defendant, defensive, indefensible, fend, fender, fence.

L. *diligens,* assiduous, attentive, careful; see **busy**

Gr. *epalxis, -eos,* f. defense, protection: *Epalxis crenulata* (a gastropod), *Epalzeorhynchos kalopterus [Epalxeorhynchus calopterus]* (a fish).

Gr. *epiouros,* m. guardian, watcher:

Gr. *eryo,* drag, draw, protect; see **draw**

L. *excubitor, -is,* m. guard, sentinel. < *excubo, -itus,* sleep out of doors, lie out on guard: *Lanius excubitor* (a shrike).

L. *indemnitas, -atis,* f. protection against or compensation for damage or loss: indemnity.

Gr. *ioros,* m. guard, keeper, watchman: *Ioropus strigulus* (a bird).

L. *janitor,* doorkeeper, porter; see *janua* under **door**

Gr. *kednos,* careful, trustworthy; *kedemon, -os,* m. protector, guardian; *kedos,* n.; *kedemonia; kedosyne,* f. care, concern, trouble; *kedestes,* m. one cared for, kin; *kedistos,* most worthy of care: epicedian, epicedium, *Kedestes[Cedestes] lepenula* (a moth).

Gr. *komistes,* m. caretaker; *komistikos,* of caretaking; *komide,* f. care, provision for, < *komizo,* attend to, preserve: gerocomia.

L. *meddix, -icis,* m. caretaker, curator:

Gr. *medon; medeon, -ontos,* n.; *medeousa,* f. guardian, < *medo,* guard, protect: Medusa, Laomedon.

Gr. *meledonos,* m. caretaker, guardian; *meledone,* f. care, sorrow; *melema, -tos,* n. object of care, < *melo,* care for: *Meledonus latipennis* (a fly).

Gr. *neokoros,* m. custodian of a temple:

Gr. *nomeus,* m. herdsman, shepherd: *Nomeus zonatus* (a fish).

L. *nutrix, -icis,* nurse; see **nurse**

L. *oeconomus* (Gr. *oikonomos*), m. housekeeper, steward, overseer: economy.

Gr. *oikouros,* m. guard of a house:

L. *opilio, -onis,* m. shepherd: opilionine, Opiliones, opiliaceous, *Opilio decoratus* (a harvestman), *Phalangium opilio* (a daddylonglegs), *Opilia amentacea* (a santalacead).

L. *ostiarius,* m. doorkeeper:

Gr. *ouros,* m. guard, watcher, warden: Arcturus.

L. *pastor, -is,* m. shepherd, herdsman; *pastoralis; pastorius,* of shepherds, rustic, rural: pastor, pastoral, pastorate, Pastorius.

L. *patronus,* m. protector, supporter, advocate; *patrocinium,* n. defense, protection; *patrocinor, -atus,* protect, support: patron, patronize, patronage, Patrons of Husbandry, pattern.

Gr. *peripolos,* m. patrol, watchman: *Peripolocetus vexillifer* (a fossil whale).

Gr. *phrouros,* m. guard, watcher; *phrourion,* n. fort, garrison: *Phrurolithus corollatus* (an arachnid), *Calliphruria hartwegiana* (an amaryllidacead).

Gr. *phylax, -akos; phylaktor, -os,* m. guard; *phylake,* f. prison; *phylakistes,* m. jailer; *phylakterion,* n. fort, safeguard, amulet; *phylaktikos,* alert, observant, vigilant; *phylaxis, -eos,* f. a watching, guarding, < *phylasso,* guard; *gazophylax,* m. treasurer; *hierophylax,* m. keeper of a temple; *himatiophylax,* m. keeper of the wardrobe; *nyctophylax,* m. night watchman: phylaxin, phylactery, phylactocarp, prophylaxis, anaphylaxis, *Phylacophora strigifera* (a moth), *Phylacteropoda tarsalis* (a fly), *Paraphylax albiscapus* (a wasp), *Dendrophylax varius* (an orchid), *Ceraticelus phylax* (a spider).

Gr. *poimen, -os,* herdsman; see *poimne* under **herd**

L. *portarius,* m. doorkeeper: porter.

L. *praeses, -idis,* c. guard, protector, chief; *praesidens, -entis,* m. director, ruler; *praesidium,* n. protection, guard: president, preside, presidium.

Gr. *probaskanion,* n. amulet, charm, guard against witchcraft:

Gr. *prostates,* one who stands first, chief, protector; see **gland**

L. *protector, -is,* m. guard; *protectus,* defended, guarded, shielded: protector, protection, *Omalium protectum* (a beetle).

L. *providus; provisus,* foreseeing, prudent; see **think**

Gr. *pyloros,* m. gatekeeper: pylorus.

L. *reus;* m.; *rea,* f. prisoner, defendant:

L. *satelles,* guard, companion, lackey; see **companion**

L. *sequester, -tris,* trustee; see *sequestro* under **separate**

L. *subulcus,* m. swineherd:

Gr. *sybotes; syphorbos,* swineherd; see *sus* under **hog**

Gr. *tamias,* steward, treasurer; see **store**

Gr. *temeleo,* take care of, look after, attend; *temeleia,* f. care, attention; *temeles,* careful:

Gr. *tereo,* watch over, guard; *teresis,* f. a watching, guarding; *teretes,* m. keeper, watcher; *teretikos,* observant: *Tereticus pectinicornis* (a beetle).

L. *tueor, tuitus; tutus,* watch over, defend; *tutela,* f. defense, guard; *tutelina,* f. a guardian deity; *tutaculum; tutamen, -inis; tutamentum,* n. defense; *tutus,* safe, secure; *tutelaris,* of guardianship; *tutelarius,* m. custodian, warden; *tutor, -is,* m. watcher, protector: tutor, tutelary, tutelage, tuition, intuitive, *Tipula tuta* (a cranefly), *Pelycodus tutus* (a fossil primate), *Cymindis tutelina* (a beetle).

L. *vigil, -is,* alert, awake, watchful; m. sentinel, watchman; *vigilans, -antis; vigilax, -acis,* watchful; *vigilo, -atus,* be alert, watchful: vigilant, surveillance, reveille, vigilance, *Xantusia vigilis* (a lizard).

See: **wall, fort, pen, hold, store, save**

gubernator, L. pilot, director, < *guberno, -atus,* steer, guide, govern; see **govern**

gudgeon < OF. *goujon,* < L. *gobio,* a fresh-water fish; see **goby**

guess < uncertain origin; see **hypothesis, prophecy, think**

guest < AS. *gaest.*

L. *advena,* c. visitor, stranger; *adventor, -is,* m. guest, visitor, customer: *Hesperonoë adventor* (a scaleworm), *Nuphar advena* (spatterdock).

Gr. *daitymon,* guest; see *dais* under **eat**

Gr. *deipneustes,* diner, guest; see *deipnon* under **eat**

L. *deversor, -is,* m. guest, inmate; *deversorium,* n. inn, lodge:

Gr. *eilapinastes,* feaster, guest; see *eilapine* under **eat**

Gr. *entrapezites,* m.; *entrapezitis, -idos,* f. parasite:

L. *epulo, -onis,* m. guest at a banquet, feaster, carouser: *Epulotrochus epulus* (a gastropod).

L. *hospes, -pitis,* m.; *hospita,* f. guest, host; *hospitalis,* of a guest or host: hospital, hostler, hostel, hotel, host, table d'hote, hospice, inhospitable, *Hospitalitermes nemorosus* (a termite), *Avihospita azurea* (a fly).

L. *inquilinus,* m. sojourner, tenant, lodger, temporary dweller in another's house: inquiline, *Termes inquilinus* (a termite).

L. *mansor, -is,* m. guest, sojourner; *mansorius,* abiding:

L. *parasitus* (Gr. *parasitos*), m.; *parasita,* f. one who eats at the table of another, guest, sponger; *parasiticus,* sponging: parasite, parasitic, parasitical, parasitism, *Ptomatophagus parasitus* (a beetle), *Endothia parasitica* (chestnut-blight), *Choleva parasita* (a beetle), *Acanthostigma parasiticum* (a fungus).

Gr. *xenos,* m. stranger, foreigner, guest; *xenyllion,* n. dim.; *xenia,* f. hospitality; *xenikos,* foreign, strange; *xenismos,* m. strangeness, novelty; *xenisma, -tos,* n. amazement, surprise; *apoxenos,* inhospitable; *proxenos,* m. public guest: xenial, xenon, xenobiosis, xenotime, Xenophon, pyroxene, Xenia (Ohio), *Xenos vesparum* (a stylops), *Xenophora trochiformis* (a gastropod), *Xenocrinus penicillus* (a Silurian crinoid), *Xenopus laevis* (a toad-frog), *Xenichthys agassizi* (a fish), *Xeniconympha leprea* (a butterfly), *Xenisma catenatum* (a studfish), *Chelyoxenus xerobatis* (a beetle), *Proxenus hospes* (a moth), *Prionotus xenisma* (a gurnard), *Formicoxenus nitidulus* (an ant), *Papilio polyxenes* (celery-moth), *Hirundo neoxena* (welcome-swallow).

See: **eat, strange**

guide < OF. *guider,* watch, observe; see **govern**

guilty < AS. *gyltig,* < *gylt,* crime; see **fault**

gula, L. gullet, weasand, throat; see **throat**

gulf < Gr. *kolpos,* bosom, bay; see **sea**

gull, prob. < C. *gwelan.*

L. *larus* (Gr. *laros*), m. gull, mew: *Larus marinus* (a gull), *Hydrochelidon lariformis* (a black tern).

NL. *rissa,* f. a kind of gull: *Rissa tridactyla* (Atlantic kittiwake).

gullet < OF. *goulet,* < L. *gula,* throat; see **throat**

gully < OF. *goulet,* throat; see **valley**

gulosus, L. greedy, gluttonous; *gulo, -onis,* epicure, glutton; see **eat**

gum < AS. *goma,* palate, jaws, see **mouth**; < L. *gummi* (Gr. *kommi*), see **resin**

gumia, L. epicure, gourmand; see **eat**

gummosus, L. full of *gummi;* see **resin**

guno- < Gr. *gounos,* fertile land; see **earth**

gurdus, L. dolt; see **dull**

gurgito, L. surge, boil, toss about; *gurges, -itis,* whirlpool, abyss; see **turn**

gurgle < uncertain origin; see **sound**

gurgulio, L. gullet, weasand, throat; see *gula* under **throat**

gurgustium, L. hovel, hut, shanty; *gurgustiolum,* dim.; see **house**

gurnard < OF. *gornart.*
   L. *milvago, -inis,* f. a fish, probably a gurnard: *Milvago ochrocephalus* (a bird).
   NL. *prionotus* (Gr. *prionotos,* jagged, serrated), m. a genus of gurnards: *Prionotus strigatus* (a sea-robin).
   Gr. *trigle,* f. red mullet; now the typical genus of gurnards: *Trigla hirundo* (gurnard).

gush < uncertain origin; see **pour**

gustus, L. taste; *gustator,* taster; *gustabilis,* appetizing; see **taste**

gut < AS. *gut;* see **intestine**

gutta, L. drop, spot; *guttula,* dim.; *guttatus,* dappled, speckled, spotted; see **drop**

gutter < OF. *goutiere,* < L. *gutta,* drop; see **ditch, pipe**

guttur, L. throat, see **throat;** *gutturosus,* goitered, see **disease**

guttus, L. a narrow-necked flask; see **bottle**

gya, Gr. beam or tree of a plow; see **beam**

gyale, Gr. a kind of cup; see **cup**

gyalos, Gr. hollow; see **hollow**

Gyas, L. a giant with a hundred arms; see **large**

gygis, NL. < Gr. *gyges,* a water-bird, tern; see **tern**

gyion, Gr. limb; *gyios,* lame; see **leg, hurt**

gylios, Gr. an elongate wallet; see **bag**

gymnasium, L. (Gr. *gymnasion*), school for bodily exercise; see **school**

gymno- < Gr. *gymnos,* bare, naked; see **bare**

gymnocladus, NL. a genus of the bean family; see **bean**

gymnosperm < Gr. *gymnos,* naked, *sperma,* seed; a plant whose seeds are borne naked, that is, not enclosed in ovarian envelopes or pericarps: gymnospermous, Gymnospermae.
   L. *abies, -etis,* fir; see **fir**
   Gr. *agathis, -idos,* f. a genus of conifers: *Agathis robusta* (dammar-pine). See **ball**
   NL. *araucaria,* f. a genus of conifers, < Ab.Am. *Araucanos,* a Chilean Indian tribe: *Araucaria excelsa* (Norfolk Island pine), *Araucarioxylon arizonicum* (a fossil wood).
   NL. *callitris,* f. a genus of conifers: *Callitris robusta* (an Australian conifer).
   L. *cedrus* (Gr. *kedros*), f. cedar; *cedrinus,* of cedar: cedar, cedrine, cedrene, cedron, cedrium, cedrol, *Cedrus brevifolia* (Cyprian cedar), *Cedrus libani* (Lebanon cedar), *Cedrela mexicana* (Mexican cedrela), *Libocedrus uvifera* (an incense-cedar), *Juniperus oxycedrus* (prickly juniper), *Bombycilla cedrorum* (cedar-waxwing).
   NL. *cephalotaxus,* a genus of gymnosperms; see *taxus* under **yew**
   NL. *chamaecyparis* (Gr. *chamaikyparissos*), f. ground-cypress; a genus of conifers: *Chamaecyparis obtusa* (Hinoki cypress).
   L. *conifer,* bearing cone-shaped fruit; a kind of gymnosperm: conifer, Coniferae, *Saxifraga conifera* (a saxifrage).
   NL. *cryptomeria,* f. a genus of conifers: *Cryptomeria japonica* (cryptomeria).
   NL. *cunninghamia,* f. a genus of conifers, < R. Cunningham, English physician: *Cunninghamia lanceolata* (China fir).
   L. *cupressus* (Gr. *kyparissos*), cypress; see **cypress**
   L. *cycas, -adis* (Gr. *kykas, -ados*), f. an African plant, now applied to a gymnosperm: cycad, cycadaceous, Cycadofilicales, *Cycas media* (a cycad), *Cycadeoidea dartoni* (a Cretaceous cycadophyte), *Blechnum cycadifolia* (cycad-fern).
   NL. *dacrydium,* a genus of gymnosperms; see **tear**
   Gr. *ephedra,* f. horsetail, a plant resembling equisetum: *Ephedra altissima* (climbing ephedra), *Colletia ephedra* (a rhamnacead), *Genista ephedroides* (a broom), *Lasioptera ephedricola* (a gall-midge).
   NL. *ginkgo,* f. a genus of gymnosperms, < a Chin. or Jap. word: *Ginkgo biloba* (ginkgo or maidenhair tree).

NL. *glyptostrobus,* m. a genus of conifers: *Glyptostrobus pensilis* (a Chinese conifer).

L. *juniperus,* f. juniper: juniper, juniperic, gin, *Juniperus virginiana* (redcedar-juniper), *Juniperus scopulorum* (Rocky Mountain juniper), *Gnidia juniperifolia* (a thymeleacead).

NL. *keteleeria,* f. a genus of conifers, < Jean Keteleer, French nurseryman: *Keteleeria fortunei* (a conifer).

L. *larix, -icis,* f. larch; *larignus,* of the larch: laricic, laricine, *Larix americana* (tamarack), *Antipathes larix* (a coral), *Pseudolarix amabilis* (golden larch), *Boletus laricinus* (a fungus), *Arenaria laricifolia* (a sandwort), *Coleophora laricella* (a moth).

NL. *libocedrus,* a genus of conifers; see *cedrus* above

NL. *phyllocladus,* m. a genus of gymnosperms: *Phyllocladus alpinus* (a celery-pine).

L. *picea,* f. spruce; *piceus,* of pitch or spruce: *Picea spinulosa* (Sikkim spruce), *Parharmonia picea* (a spruce-moth), *Phthorophloeus piceae* (a beetle).

L. *pinus,* pine; see **pine**

NL. *podocarpus,* m. a genus of gymnosperms: *Podocarpus elatus* (tall podo-carpus).

NL. *pseudotsuga,* a genus of conifers; see *tsuga* below

NL. *saxegothaea,* f. a genus of gymnosperms: *Saxegothaea conspicua* (a conifer).

NL. *sciadopitys,* f. a genus of gymnosperms: *Sciadopitys verticillata* (umbrella-pine).

NL. *sequoia,* f. a genus of conifers, < Sikwayi(Sequoyah), George Guess(Gist), inventor of the Cherokee alphabet: *Sequoia sempervirens* (coast redwood), *Sequoia gigantea* (Sierra redwood; *Sequoiadendron giganteum* has been pro-posed as a new name for this species), *Metasequoia disticha* (a fossil shuihsa), *Vespamima sequoiae* (pitch-moth).

NL. *taiwania,* f. a genus of conifers, < *Taiwan,* Formosa: *Taiwania crypto-merioides* (taiwania).

NL. *taxodium,* bald cypress; see *taxus* under **yew**

L. *taxus,* yew; see **yew**

NL. *tetraclinis,* f. a genus of conifers: *Tetraclinis articulata* (formerly *Callitris quadrivalvis,* sandarac-tree).

NL. *thuja,* < L., Gr. *thya; thyia,* f. a resinous, sweet-scented tree; *thyinos,* of thyia; arborvitae: *Thuja plicata* (western arborvitae), *Chamaecyparis thyoides* (white cedar), *Thujopsis dolabrata* (a conifer), *Thuidium pygmaeum* (a fern-moss).

NL. *torreya,* f. a genus of gymnosperms, < John Torrey, American botanist: *Torreya californica* (California torreya).

Jap. *tsuga,* f. hemlock: *Tsuga heterophylla* (western hemlock), *Pseudotsuga taxi-folia* (Douglas fir), *Hemerocampa pseudotsugata* (a tussock-moth), *Phomopsis pseudotsugae* (a fungus), *Panicum tsugetorum* (a grass), *Opidnus tsugae* (a wasp).

NL. *widdringtonia,* f. a genus of gymnosperms, < a Capt. Widdrington, botanical explorer: *Widdringtonia cupressoides* (a conifer).

NL. *zamia,* f. a genus of cycads: *Zamia furfuracea* (a cycad), *Ceratozamia mexi-cana* (a cycad).

gyneco-; gyneo- < Gr. *gyne, -naikos,* woman, female; *gynaikion; gynaion,* dim.; see **woman**

gynnido- < Gr. *gynnis, -idos,* a womanish man; see **man**

gype, Gr. vulture-nest; *gyparion,* dim.; see **nest**

gyps, *gypos,* Gr. vulture; *gypinos,* of vultures; see **vulture**

gypsum, L., n. (Gr. *gypsos,* f.), white plaster; *gypseus,* of gypsum: gypsiferous, *Gypsophila paniculata* (babys-breath), *Achorion gypseum* (a fungus), *Sedum gypsicola* (a stonecrop).

L. *alabaster, -tri* (Gr. *alabastros*), m. a perfume-box carved out of white gypsum or calcite: alabaster, alabastrine, alabastrum, *Flabellum alabastrum* (a fan-coral).

Gr. *selenites,* m. moonstone, gypsum: selenite.

Gr. *skiros,* m. gypsum, stucco:

Gr. *titanos,* f. a white earth, chalk or gypsum; *titanotos,* whitened:

gyrgathos, Gr. wicker-basket; see **basket**

gyrine, Gr. a kind of cake; see **cake**

gyrinos, Gr. tadpole; see **frog**

gyris, Gr. finest meal; see **flour**

gyro, -atus, L. turn around, < L. *gyrus* (Gr. *gyros*), circle; see **turn**

H

habena, L. thong, strap, rein; *habenula,* dim.; see **strap**

habenaria, NL. a genus of the orchid family; see **orchid**

habilis, L. apt, fit, suitable; see **fit**

habit < L. *habeo, -itus,* have, hold; see **hold, nature**

habitat < L. *habito, -atus,* dwell, have possession of; see **life**

habra, Gr. favorite slave; see **servant**

habros, Gr. pretty, graceful, dainty, tender; *habrotinos,* delicate, costly; see **beauty**

habrynos, Gr. mulberry; see **mulberry**

habryntes, Gr. coxcomb, fop; see **fool**

hackberry; hagberry, a tree of the elm family; see *celtis* under **elm**

hactenus, L. so far, so much; see **end**

Hades < Gr. *Haides,* god of the nether world, hell; see **hell**

hadros, Gr. well-developed, bulky, stout, ripe, large, strong, great; *hadrotes,* thickness, fullness, vigor, abundance; see **large**

hadrynsis, Gr. a coming to maturity, a ripening; *hadryntikos,* of ripening, < *hadryno,* ripen; see **ripe**

haemato- < Gr. *haima, -atos,* blood; *haimatikos,* of blood; *haimatinos,* bloody; see **blood**

haematopus, L. (Gr. *haimatopous*), a red-footed bird; see **bird**

haemylo- < Gr. *haimylos,* flattering, wheedling, wily; see **flatter**

haereto- < Gr. *hairetos,* chosen, desirable, taken; see *haireo* under **take**

haesus, L. held, stuck; see *haereo* under **hold**

hagios, Gr. sacred, holy; *hagiosyne,* holiness; see **holy**

hagnos, Gr. pure, innocent, sacred; see **pure**

hail < AS. *haegel;* see **ice**

hair < AS. *haer:* hairy, hairbreadth, hairpin, haircloth, hairsplitting, hairball.
Gr. *anaphrisso,* bristle up:
Gr. *anasillos,* m. bristling hair:
L. *antia,* f. forelock:
L. *barba,* f. beard; *barbula,* f. dim.; *barbatus,* bearded; *barbatulus,* dim.; *barbiger, -a, -um,* wearing a beard; *imberbis, -e; imberbus,* beardless: barb, barbate, barbule, barbel, barbicel, barbet, barbellate, barber, burbot, Barbarossa, Barbados, barbital, barbituric, *Usnea barbata* (beard-lichen), *Mallotus barbatus* (a euphorbiacead), *Clematis barbellata* (a clematis), *Cryptantha barbigera* (a borage), *Gossypium barbadense* (sea-island cotton), *Nemachilus barbatulum* (loach), *Brotula imberbis* (a fish), *Camptostoma imberbe* (a flycatcher).
Gr. *blepharis, -idos,* f. eyelash: *Blepharipappus glandulosus* (a composite), *Blepharipeza adusta* (a fly), *Blephilia ciliata* (a mint), *Monoblepharis fasciculata* (an aquatic fungus), *Habenaria blephariglottis* (white fringed-orchid). See *blepharon* under **eyelid**
Gr. *bostrychos,* curl, lock of hair; see **curl**
L. *caesariatus,* covered with hair, long-haired, < *caesaries,* f. a beautiful head of hair:
L. *capillus,* m. hair; *capillulus,* m. dim.; *capillamentum,* n. wig; *capillaceus; capillaris,* of hair; *capillatus; capillosus,* hairy: capillose, capillary, capillarity, capillitium, dishevel, *Adiantum capillus-veneris* (a maidenhair-fern), *Cyanea capillata* (a jellyfish), *Ceratostomella capillaris* (a fungus), *Muscicapa atricapilla* (pied flycatcher), *Astur atricapillus* (American goshawk).
L. *caprona,* f. forelock: *Caprona pillaana* (a butterfly).
Gr. *chaite,* f. long hair, mane; *chrysochaites,* golden-haired; *melanchaites,* black-haired: chaetognath, chaetiferous, chaetigerous, chaetophorous, chaetotaxy, *Chaetochloa glauca* (foxtail-grass), *Chaetocalyx parviflora* (a legume), *Chaetura pelagica* (chimney-swift), *Chaetodon ephippium* (a fish), *Chaetetes cylindraceus* (a fossil coral), *Coleochaete pulvinata* (an alga), *Metachaetodus brunneicollis* (a beetle), *Connochaetes taurinus* (a gnu), *Zachaetomyia atriventris* (a fly).
Gr. *chnoos,* fine down, any light substance; see **foam**
L. *cilium,* n. eyelash, eyelid: ciliate, cilary, supercilious, Ciliata, *Habenaria ciliaris* (yellow fringed-orchid), *Cryptocoryne ciliata* (an aracead), *Psorophora ciliata* (gallinipper), *Solidago ciliosa* (a goldenrod), *Begonia ciliifera* (a begonia), *Brachiaria ciliatissima* (a single-grass), *Eumomota superciliosa* (a motmot), *Paspalum ciliatifolium* (a grass), *Erica ciliaris* (a heath).

L. *cirritus,* having filaments; see *cirrus* under **curl**

L. *coma* (Gr. *kome*), f. hair of the head; *comula,* f. dim.; *comans,* covered with hair; *comatus,* with long hair, shaggy; *comatulus,* with hair neatly curled; *comosus,* hairy; *cometes* (Gr. *kometes*), m. one with long hair, a comet; Gr. *komion,* n. dim.; *albicomus,* white-haired; *amphikomos,* covered with hair; *auricomus; chrysokomes,* golden-haired; *erythrokomos; horricomus,* with hair on end, bristly; *prokomion,* n. forelock; *pyrsokomos,* red-haired: comet, comate, comatulid, komeceras, *Comatricha obtusata* (a slime-mold), *Comandra umbellata* (pale comandra), *Comatula cratera* (a crinoid), *Callicoma serratifolia* (a cunoniacead), *Ophiocoma pumila* (a brittlestar), *Aphelocoma californica* (a jay), *Pycnocoma macrophylla* (bomah-nut), *Stipa comata* (a needle-grass), *Eucomis punctata* (a liliacead), *Calligonum comosum* (a polygonacead).

L. *crinis, -is,* m. hair; *crinalis,* of hair; *crinitus,* hairy, long-haired: crinose, crinoline, criniferous, *Gentiana crinita* (fringed gentian), *Trichocladus crinitus* (a hamamelidacead).

Gr. *dasys,* hairy, shaggy, tufted, dense; *dasytes, -etos,* f. hairiness, roughness; *dasos, -eos,* n. thicket, copse; *dasodes,* bushy: dasymeter, dasyure, dasypaedal, *Dasylirion glaucophyllum* (a liliacead, sotol), *Dasypus sexcinctus* (banded armadillo), *Dasyprocta agouti* (agouti).

Gr. *etheira,* f. hair, mane: *Ethirostoma semiacma* (a moth).

Gr. *geneias, -ados,* f. beard; *geneiates,* bearded: *Geneiadolaelaps barbatus* (a mite), *Agenia bombycina* (a wasp).

Gr. *gereion,* n. pappus, down:

Gr. *hebe,* youth, youthful prime; pubescent; see **young**

L. *hirsutus,* hairy, bristly, rough, shaggy: hirsute, hirsutulous, *Helianthus hirsutus* (a sunflower), *Viola hirsutula* (a violet), *Begonia hirsuticaulis* (a begonia), *Carex hirsutella* (a sedge).

L. *hirtus,* hairy, rough, shaggy, uncultivated; *hirticulus,* dim.: hirtic, hirtellous, *Hirtella silicea* (a rosacead), *Potentilla hirta* (a cinquefoil), *Ribes hirtellum* (American gooseberry), *Erica hirtiflora* (a heath), *Rhamphomyia hirticula* (a fly).

L. *hispidus,* bristly, rough, hairy, prickly: hispid, hispidulate, hispidulous, *Strophanthus hispidus* (inee plant), *Adiantum hispidulum* (a maidenhair-fern), *Cercidocerus hispidulus* (a beetle), *Begonia hispidissima* (a begonia).

Gr. *hypene,* f. mustache; *hypenetes,* m. one who is getting a beard: *Hypena proboscidialis* (a moth), *Hypenorhynchus erectilineatus* (a moth).

L. *intonsus,* unshaved, bearded; see **rough**

Gr. *ionthas, -ados,* long-haired, shaggy; *ionthos,* m. young hair: *Ionthadophrys longihirtus* (a fly), *Ionthocerus crematus* (a beetle).

L. *juba,* f. mane; *jubatus,* maned, crested: *Hordeum jubatum* (foxtail-barley), *Connochaetes albojubatus* (a gnu).

Gr. *konnos,* m. beard: *Connochaetes gnu* (wildebeest).

Gr. *krobylos,* m. roll or knot of hair on top of the head: *Crobylus hystrix* (an arachnid), *Crobylocerus megilliformis* (a fly).

Gr. *krokis, -idos,* flock, nap, or down of woolen cloth; see **wool**

Gr. *krossos,* fringe, tassel; see **fringe**

L. *lanugo, -inis,* down on plants and cheeks; see *lana* under **wool**

Gr. *lasios,* hairy, woolly, shaggy; see **wool**

L. *Medusa* (Gr. *Medousa*), a gorgon with snaky locks who turned beholders to stone; see **fear**

Gr. *merinx (smerinx), -ingos,* f. bristle: *Smeringaspis setifera* (a beetle).

Gr. *mystax, -akos,* m. hair on the upper lip; *mystakion,* n. dim: mustache, *Mystacomyia rubriventris* (a fly), *Mystacoleucus padangensis* (a cyprinid), *Mystacops tuberculatus* (a bat), *Mystacidium distichum* (an orchid), *Callomystax gagatus* (a fish), *Inca mystacalis* (Inca tern), *Epinephelus mystacinus* (a grouper).

Gr. *ophrys,* eyebrow, brow; see **front**

Gr. *ostlinx, -ingos,* m. curled hair, lock of hair, tendril: *Ostlingoceras puzosianum* (an ammonite).

L. *pappus* (Gr. *pappos*), m. down, bristles, hair, teeth, awns, or setae on achenes of some Compositae; *pappodes,* downy: pappus, *Isopappus divaricatus* (yellow aster), *Pectis papposa* (a limoncillo).

Gr. *penike,* f. false hair, wig: *Penica peritheta* (a moth).

Gr. *phenake,* f. wig: *Phenacephorus appendiculatus* (a phasmid).

L. *pilus* (Gr. *pilos*), m. hair, felt; *pilulus,* m. dim.; *pilosus,* hairy; *pilo, -atus,* grow hairy: piliferous, pilose, pilosis, pilonidal, caterpillar, horripilation, pilocarpine, *Pilizetes africanus* (a mite), *Pilocarpus jaborandi* (a rutacead), *Triodia pilosa* (hairy triodia), *Paepalanthus pilifer* (a pipewort), *Dioscorea pilosiuscula* (a yam), *Urtica pilulifera* (a nettle), *Polygala rectipilis* (a polygala), *Tethys depilans* (a gastropod).

Gr. *plokos*, lock of hair, braid curl; *plokion*, dim.; *euplokamos*, with beautiful locks; see *plecto* under **weave**

Gr. *pogon*, *-os*, m. beard: *pogonion*, n. dim.; *pogonodes*, beardlike; *pogonias*, bearded: pogoniate, pogoniasis, pogonology, pogonotomy, pogonite, *Pogonopus tubulosus* (a rubiacead), *Pogonia verticillata* (an orchid), *Pogonatum pennsylvanicum* (a moss), *Andropogon furcatus* (big bluestem), *Baeopogon indicator* (a bird), *Ophiopogon japonicus* (lilyturf), *Tragopogon porrifolius* (salsify), *Epipogon aphyllus* (an orchid).

Gr. *ptilon*, down, feather; see **feather**

L. *pubes*, *-is*, f. down of adulthood and maturity; the private parts: *puber*, *-eris*, downy, ripe; *pubens*, arrived at puberty, pubescent; *pubesco*, put on the down of puberty, ripen: pubescent, puberulent, puberty, pubis, pubic, puberulic acid (from *Penicillium puberulum*), *Sambucus pubens* (scarlet elder), *Viola pubescens* (downy yellow violet), *Sericosoma pubipes* (a fly), *Aquilegia pubiflora* (a columbine), *Aposites pubicollis* (a beetle), *Pyrularia pubera* (oil-nut), *Phthirius pubis* (crab-louse).

L. *reburrus*, m. one with bristling hair:

L. *seta*, f. bristle; *setula*, f. dim.; *setosus*, bristly; *setifer; setiger*, bearing bristles: seta, setarious, setaceous, setiferous, setigerous, *Setaria barbata* (a bristle-grass), *Ophiacantha setosa* (a brittlestar), *Juncus setaceus* (a rush), *Carex setifolia* (a sedge), *Chaetodon setifer* (a butterfly-fish), *Aristida longiseta* (a grass), *Equisetum limosum* (a horsetail), *Pennisetum villosum* (feathertop), *Rosa setigera* (prairie-rose), *Sisyphus setosulus* (a beetle).

L. *striga*, swath, bristle; see **ridge**

Gr. *thrix*, *trichos*, f. hair; *trichidion; trichion*, n. dim.; *trichinos*, of hair; *trichosis*, f. hairiness; *trichotos*, hairy; *entrichos*, hairy; *euthythrix*, straight-haired; *leucothrix*, white-haired; *perithrix*, *-trichos*, f. first hair; *phryxothrix*, with bristling hair, making the hair stand on end; *pyrrhothrix*, red-haired; *tanythrix*, long-haired: trichology, trichite, trichome, trichosis, trichophore, trichode, trichoblast, trichocyst, trichogenous, trichinosis(caused by trichina, *Trichinella spiralis*), *Trichostema lanatum* (blue curls), *Trichogaster trichopterus* (a fish), *Trichomanes speciosum* (a filmy-fern), *Trichiocampus aeneus* (a sawfly), *Trichidium octonemus* (a threadfin), *Chrysothrix sciurea* (squirrel-monkey), *Aleurothrixus howardi* (woolly whitefly), *Callitriche palustris* (a water-starwort), *Ithytrichia confusa* (a caddisfly), *Ophiothrix spiculata* (a brittle-star), *Phrixothrix hirta* (a beetle), *Hapalothrix lugubris* (a fly), *Malacothrix glabrata* (a composite), *Trichiurus lepturus* (cutlass-fish), *Polytrichum commune* (haircapmoss), *Zatrichodes thyrsota* (a moth), *Eriophorum callithrix* (a sedge).

Gr. *thysanos*, fringe, tassel; see **fringe**

L. *tomentum*, n. woolly hairs: tomentum, tomentose, tomentulose, *Paulownia tomentosa* (empress-tree).

L. *vesticeps*, *-cipis*, bearded, arrived at puberty:

L. *vibrissa*, f. whisker, stiff tactile hair: vibrissa, vibrissal.

L. *villus*, m. shaggy hair, tuft of hair; *villosus*, hairy; *velutinus*, NL. velvety: villous, villose, villus(pl. villi), villosity, villiform, villiferous, velvet, velvety, *Dioscorea villosa* (a yam), *Dasypus villosus* (an armadillo), *Eristalis brevillosa* (a fly), *Quercus velutina* (black oak), *Melanetta velvetina* (a duck), *Velutina laevigata* (a gastropod), *Psephenus veluticollis* (a water-penny), *Collybia velutipes* (a fungus), *Xestobium rufovillosum* (a beetle), *Rattus villosissimus* (a rat).

See: **wool, curl, silk, thread, fringe, tail, barber**

halatus, L. odor, fragrance, perfume; see **smell**

hales; halis, Gr. in abundance, in plenty, heaps, crowds, swarms; see **abundance**

halex, Gr. a kind of pulse; see **bean**

half < AS. *healf*.

F. *demi*, < L. *dimidius*, half; *dimidiatus*, halved: demigod, demiwolf, demilune, demirelief, demitasse, *Inoceramus dimidius* (a fossil pelecypod), *Discorbina dimidiata* (a foraminifer). Derive: demijohn.

Gr. *dichas*, *-ados*, f. half:

Gr. *hemi-*, < *hemisys*, half; *hemimeres*, halved; *hemiolios; hemisytriton*, one and one-half: hemisphere, hemiplegia, hemicollin, hemicyclic, hemihedral, hemimorphic, Hemiptera, migraine, *Hemigalus hardwicki* (a weasel), *Hemicyon ursinus* (a fossil mammal).

L. *semi-*, < *semis*, *-issis*, half; *semissalis; semissarius*, of a half; *sesqui*, one half more, one and a half; *sesquipes*, one and a half feet; *sequipedalis*, excessively long; *sestertius*, two and a half: semilunar, semicircle, semitone, semiweekly, semipalmate, semibituminous, sesquioxide, sesquicentennial, sesquipedalian, sesterce, *Natica semilunata* (a fossil gastropod), *Didymoprora semipunctata* (a

beetle), *Angraecum sesquipedale* (an orchid), *Vigna sesquipedalis*(yardlong cowpea).
See: **two**

hali- < Gr. *hals, halos,* sea, salt; *halios,* of the sea; see **sea, salt**

halia, Gr. assembly of the people, see **gather**; saltcellar, see **box**

haliaetus, L. (Gr. *haliaetos*), sea-eagle; see *aetos* under **eagle**

halicacabus, L. (Gr. *halikakabon*), a plant with inflated fruits; see **plant**

halictus, NL. a genus of bees; see **bee**

halieutes, Gr. fisherman; see **fisherman**

halimon, Gr. shrubby seashore plant of the goosefoot family; see **goosefoot**

halimos, Gr. of the sea; see *hals* under **sea**

haliotis, NL. a genus of mollusks; see **mollusk**

haliphthoros, Gr. pirate; see **steal**

halitus, L. breath, wind; *halo, -atus,* breathe; see **breathe**

hall < AS. *heall* (G. halle); see **room**

hallex, *-icis,* L. great toe; see **finger**

hallucinor, *-atus,* L. wander in mind, dream; see **dream**

halma, Gr. leap, bound; *haltikos,* good at leaping; see **leap**

halmyridium, L. a kind of cabbage; see **cabbage**

halmyros, Gr. salty, briny; see *hals* under **salt**

halo- < Gr. *hals, halos,* sea, salt, see **sea, salt**; < *halos,* a round threshing-floor, circle around the sun or moon; *halonion,* dim., see **circle**

halosimos, Gr. easy to take, catch, conquer, attain; *halotos,* attainable; see **take**

Halosydne, Gr. a name for Aphrodite; see *hals* under **sea**

halter < AS. *haelfter,* see **hold**; < L. *halter, -is,* weight used in jumping, balancer, see **weight**

halurgo- < Gr. *halourgos,* sea-purple, purple; *halourgis,* purple robe; see **purple**

halysis, Gr. chain; see **chain**

ham < AS. *ham,* back part of thigh; see **flesh, leg**

-ham, AS. home, enclosure; see **house**

hama, Gr. together, at the same time, mutual; see **with**

hamamelis, Gr. a tree with pomelike fruit, a kind of medlar; now applied to witchhazel; see **witchhazel**

hamartia; hamartema, Gr. failure, error; see **fault**

hamatus, L. hooked; see *hama* under **hook**

hamaxa, Gr. wagon; *hamaxis, -idos,* dim., see **vehicle**; *hamaxitos,* road, highway, see **way**

hamilla, Gr. contest; see **fight**

hamma, Gr. anything tied or made to tie to, knot, noose, halter, cord, bowstring; *hammation,* dim. bandage; see *hapto* under **bind**

hammer < AS. *hamer.*
   Gr. *aira,* f. hammer:
   L. *fistuca,* f. beetle, rammer, pile-driver; *fistuco, -atus,* ram, pound: *Nodosaria fistuca* (a foraminifer).
   Gr. *kestra,* f. a kind of hammer, perhaps with one end pointed; a pikelike fish: *Cestradoretus tenuirostris* (a beetle), *Cestracion philippi* (Port Jackson shark).
   Gr. *krotaphis, -idos,* f. pointed hammer:
   Gr. *krotema,* a piece of hammered work; see *kroteo* under **strike**
   L. *malleus,* m. hammer, mallet, maul; *malleolus,* m. dim.; *malleatus,* wrought or beaten with a hammer; *malleator, -is,* m. hammerer: mallet, malleable, malleus, maul, Pall Mall.
   L. *marcus,* m. a large hammer; *marculus,* m. dim.:
   L. *mateola,* f. probably a beetle or mallet:
   L. *portisculus,* m. hammer for beating time for rowers:
   Gr. *rhaister, -os,* m. hammer, smasher:
   Gr. *sphyra,* f. hammer; *sphyrion,* n. dim.; *sphyrelatos,* wrought with a hammer; *xylosphyron,* n. wooden mallet: *Sphyrocoris obliquus* (a bug), *Sphyrapicus varius* (yellow-bellied sapsucker).
   L. *tudes, -is,* m. hammer, mallet; *tudicula,* f. dim.: *Tudicula spinosa* (a gastropod).
   Gr. *tykos,* m. mason's hammer: *Tycodesmus medius* (a milleped).

Gr. *typas, -ados; typis, -idos,* f. hammer, mallet: *Tupidanthus*[*Typidanthus*] *calyptratus* (an araliacead).
See: **pestle, strike**

hamus, L. hook; *hamulus,* dim.; *hamatilis; hamatus,* hooked; see **hook**

hand < AS. *hand:* handy, handle, handsome, handful, handicap, handicraft, handsel, handcuff, handbreadth, handwriting.
Gr. *aristera,* f. left hand; *aristeros,* left, on the left; *eparisteros,* toward the left: *Aristerospira globularia* (a foraminifer).
L. *carpus* (Gr. *karpos*), m. wrist: carpal, carpus, metacarpal.
Gr. *cheir, -os,* f. hand; *cherydion,* n. dim.; *cheironomos,* m. pantomimic; *encheiridios,* in the hand: chiral, chiragra, chiromancy, chiropractor, chiropodist, chirography, Chiroptera, enchiridion, surgeon, *Chirotherium angustum* (fossil footprints), *Chironomus meridionalis* (a midge), *Cheiranthus cheiri*(wallflower), *Metachirus nudicaudatus* (an opossum), *Laemonema rhodochir* (a fish), *Eriocheir sinensis* (mitten-crab).
L. *condylus* (Gr. *kondylos*), knuckle, fist; see **projection**
Gr. *dexia,* f. right hand; *dexios,* on the right, lucky, clever; *adexios,* on the left, left-handed, awkward; *amphidexios,* literally, with two right hands, ambidextrous; *peridexios,* ambidextrous: *Polydexia pusilla* (a foraminifer), *Dexia rustica* (a fly).
L. *dexter, -tra, -trum,* on the right hand, to the right; *dextratus,* lying to the right; *dextrorsus,* to the right; *dextra,* f. the right hand; *dextella,* f. dim.: dexterity, dextrous, ambidextrous, dextrose, dextral, dextrocardia.
Gr. *gronthos,* m. fist:
L. *laevus,* on the left hand, left; *laeva,* f. left hand; *laevorsus,* toward the left: levorotatory, levogyrate, levulose, levolactic, levotartaric, levulin.
Gr. *laios,* left: laeotropic, *Laeogyra bohemica* (a fossil gastropod).
L. *manus, -us,* f. hand; *manciola; manicula,* f. dim.; *manualis; manuarius,* of or for the hand; *manipulus,* m. handful; *quadrimanus,* four-handed; *unimanus,* one-handed: manual, manufacture, manicure, manuscript, maneuver, manage, mandate, mancipation, maintain, mortmain, manifestation, manipulate, manure, manner, amanuensis, command, emancipate, legerdemain, quadrumanous, *Peromyscus maniculatus* (a deermouse), *Autonoë grandimana* (an amphipod).
L. *palma* (Gr. *palame*), f. palm of the hand, hand, palm-tree; *palmula,* f. dim.; *palmaris; palmarius; palmeus; palmularis,* of palms; *palmo, -atus,* make a palm print; *palmatus,* marked or shaped like the palm of the hand, radiate like the veins of a palm leaf; *palmus,* m. palm of the hand: palm, palmate, palmistry, palmetto, *Viola palmata* (a violet), *Cucurbita palmata* (coyote-melon), *Charadrius semipalmatus* (semipalmated plover), *Palmatodella delicatula* (a conodont), *Pachyrhizus palmatilobus* (a legume).
L. *palpus,* soft palm of the hand, feeler; see *palpo* under **touch**
L. *pugnus,* m. first; *pugillus,* m. dim. handful: *Pugnellus fusiformis* (a fossil gastropod), *Pugillaria stowae* (a gastropod).
Gr. *pygme,* f. fist: *Pygmophorus spinosus* (a mite).
L. *scaevus,* on the left, toward the left, awkward, unlucky: *Scaevola attenuata* (a goodeniacead).
L. *sinister, -tra, -trum,* on the left, left, unlucky; *sinistrorsus,* toward the left: sinister, sinistral, sinistrous, sinistrorse, sinistrogyrate, *Mammillaria sinistrohamata* (a cactus).
Gr. *skaios,* on the left, left, awkward, ill; *skaiotes,* f. lefthandedness: *Scaeochlamys livida* (a pelecypod), *Scaeopus torquatus* (maned sloth).
Gr. *thenar, -os,* n. palm of the hand: *Thenarocrinus gracilis* (a Silurian crinoid).
L. *vola,* f. hollow of the hand, palm: *Vola dissecta* (a beetle).
See: **finger, foot, nail**

handle < AS. *hand,* hand.
L. *ansa,* f. handle, haft; *ansula,* f. dim.; *ansatus,* having a handle: ansate.
L. *capulus,* m. handle, hilt of a sword:
Gr. *echetle,* f. plow-handle: *Echetlus peristhenes* (a phasmatid).
Gr. *encheiridion,* n. handle, hilt:
Gr. *examma, -tos,* n. handle:
Gr. *kollops, -opos,* m. handle, peg, screw:
Gr. *kope,* handle, haft, oar; *kopion,* dim.; see **oar**
Gr. *labe,* handle, haft, hilt, hold; *labion,* dim.; see **tool**
L. *manubrium,* n. haft, handle: manubrium.
Gr. *ous,* ear, handle; see **ear**
Gr. *porpax, -akos,* m. handle of a shield: *Porpax asperipes* (an odonatid).
AS. *snaeth,* scythe handle: snath(snathe), snead.
Gr. *steleon* (*steileion*), n. handle of an ax; *steilarion,* n. dim.: *Steleoxiphus catastates* (a phasmatid), *Stelidium cressoni* (a wasp).

L. *stiva*, f. plow-handle:
See: **tool, projection, hold**

hang < AS. *hangian*.
Gr. *aioretos*, hanging, hovering, suspended; *aoros*, pendulous, waving, hanging, < *aioreo*, hang, hover; *kateoros*, hanging down: *Aeorestes*[*Aeoretes*] *villosissimus* (a bat), *Aorocrinus formosus* (a Devonian crinoid).
Gr. *analeptris*, *-idos*, f. a suspensory bandage:
Gr. *anchone*, a strangling, hanging; a cord for hanging, halter; see *ango* under **choke**
Gr. *artao*, hang, suspend; *artema*, *-tos*, n. a pendant: *Artabotrys uncinatus* (tail-grape).
L. *carnifex*, *-icis*, hangman, executioner; see **kill**
L. *demissus*, hanging down, drooping, feeble, weak: *Prunus demissa* (a wild cherry).
Gr. *epholkis*; *epholkion*, burdensome appendage; see *holkas* under **ship**
L. *flaccidus*, flabby, languid, drooping; see **weak**
Gr. *kremastos*, hung, hanging; *kremasis*, f.; *kremasmos*, m. a hanging, suspension; *kremaster*, *-os*, m. suspender, suspensor; *kremastra*, f. pedicel, peduncle; *ekkremes*, hanging, pendent; *epikremes*, suspended: *Alnus cremastogyne* (an alder), *Cremastochilus brunneus* (a beetle), *Eccremocarpus scaber* (a bignoniacead), *Epicremastus concolor* (a wasp).
Gr. *parouatios*, with hanging ears:
L. *pendo* (*pendeo*), *pensus*, hang, weigh; *pendens*; *pendulus*; *pensilis*, hanging, swinging; *appendix*, *-icis*, f. that which hangs to or from, a supplement; *impendens*, overhanging, imminent; *suspendium*, n. a hanging: pendant, pendulum, perpendicular, appendage, appendix, depend, independent, impending, suspenders, *Cymbopetalum penduliflorum* (Aztec car-flower), *Dermophylla pendulina* (a gourd), *Betula pendula* (European white-birch), *Filipendula rubra* (prairie-meadowsweet), *Glyptostrobus pensilis* (a conifer), *Ophioglossum pendulum* (an adderstongue-fern), *Viburnum suspensum* (a viburnum).
L. *stalagmium* (Gr. *stalagmion*), pendant, eardrop; see *stalagma* under **drop**
See: **weight, projection, upright, uvula, weak, frame**

hapalus, L. (Gr. *hapalos*), soft, tender, gentle, delicate; see **soft**

hapax, Gr. once only; see **one**

haphe, Gr. a touch, a kindling; see *haptos* under **touch**

haplo- < Gr. *haploos* (*haplos*), single, simple; *haplotes*, singleness; see **one**

happen < ON. *happ*, good luck; see **lot**

happy < ON. *happ*, good luck; see **joy**

haptos, Gr. touchable; *haphe*; *hapsos*, joint; *aptos*, untouchable, < *hapto*, join, fasten to, fix upon, lay hold of, touch; see **touch, bind**

haptra, Gr. wick; *haptrion*, dim.; see **wick**

hara, L. pen, coop, sty; see **pen**

harbor < G. *hereberge*, shelter for soldiers.
Gr. *hormos*, m. harbor, haven; *hormizo*, bring to harbor, moor, anchor: Palermo.
Gr. *limen*, *-os*, m. harbor, haven, refuge; *limeniskos*, m. dim.; *limenios*, of a harbor:
Gr. *neorion*, dockyard; see *navis* under **ship**
L. *portus*, *-us*, m. harbor, haven, entrance; *portuosus*, full of harbors: port, portico, Porto Rico, Portugal, porch, passport, importune, opportunity.
See: **door, enter, safe**

hard < AS. *hearde*; F. *-ard*, *-art*; G. *hart*, hard; excessiveness; habitual: harden, hardy, hardihood, hardness, hardware, hardwood, hardhead, hardhack, diehard, drunkard, dullard, sluggard, hazard, dastard, laggard, wizard, reynard, dotard, niggard, standard, Gerhart, Gerard, Reinhart.   Derive: mustard, orchard, lizard, leopard, spikenard, steward, coward, poniard, scabbard, buzzard, Bayard.
Gr. *aages*, unbroken, hard, strong: *Aages prior* (a beetle).
Gr. *abiotos*, not to be lived, intolerable:
L., Gr. *adamas*, unconquerable, unyielding, hard; *adamantinus*, very hard; see **diamond**
Gr. *anelatos*, not malleable or ductile, stubborn:
Gr. *apokrotos*, hard:
L. *callosus*, having a hard skin; see *callum* under **skin**
L. *concretus*, hardened, thick, stiff: concrete, concretion.
L. *consisto*, *constitus*, set, become hard: consistency.
L. *durus*, hard, tough; *duriusculus*, dim.; *durabilis*, lasting; *duramen*, *-inis*, n. hardness; *duro*, *-atus*, harden, hold out, last; *edurus*, very hard; *induro*, harden; *obduro*, hold out: durable, durability, durance, duration, dura mater, duress,

durene, duridine, during, endure, obdurate, durain, indurated, *Triticum durum* (a wheat), *Aristolochia durior* (a Dutchmans-pipe).

L. *immitigabilis*, that cannot be softened or allayed: immitigable.

L. *indetribilis*, that cannot be worn out; *indetritus*, not worn out:

L. *intolerabilis*, not bearable, insupportable: intolerable.

L. *ossilago*, *-inis*, f. bony hardness:

L. *petro*, *-onis*, m. something hard or rough as stone; a rustic; an old wether:

Gr. *poroma*, callus; *porosis*, formation of a callus; see **skin**

Gr. *skirrhos*, hard, hardening: scirrhus, scirrhous, scirrhoid, scirrhosity.

Gr. *skleros*, hard, tough; *skleria; sklerotes*, f. hardness; *skleriasis*, f.; *skleroma*, *-tos*, n.; *sklerosis*, m. hardening, induration: sclerotic, sclerenchyma, sclerotium, sclerotin, arteriosclerosis, *Scleroderma vulgare* (a puffball), *Sclerocephalus haeuseri* (an amphibian), *Sclerolepis uniflora* (a composite), *Sclerosomus incommodus* (a beetle), *Sclerostomum equinum* (a nematode), *Sclerotinia fructigena* (brown-rot), *Scleranthus annuus* (knawel), *Scleria ciliata* (fringed razor-sedge).

L. *solidus*, dense, hard, firm, thick, entire, enduring; see **thick**

Gr. *stereos; steriphos*, stiff, firm, hard; see **thick**

Gr. *stomoo*, harden iron; *stomosis*, f. hardening of iron; *stomotes*, m. hardener of iron; *stomotos*, hardened: *Stomotocella atra* (a jellyfish).

Gr. *striphnos*, firm, solid: *Striphnopteryx edulis* (a moth).

See: **strong, thick, stubborn, difficult, slow, diamond, iron, oak, shell, stiff, trouble**

hare < AS. *hara.*

L. *auritus*, having ears; the long-eared animal, the hare; see *auris* under **ear**

L. *cuniculus* (Gr. *koniklos; kyniklos*), m. rabbit; *cunicularis*, of rabbits: cunicular, *Lepus cuniculus* (European rabbit), cony (*Oryctolagus cuniculus*), *Psoroptes cuniculi* (a scab-mite), *Conilurus penicillatus* (a rabbit-rat).

L. *dasypus* (Gr. *dasypous*), m. a kind of rabbit; now applied to a kind of armadillo: *Dasypus fenestratus* (an armadillo).

Gr. *lagos*, m. hare; *lagidion; lagion; lagodarion; lagodion*, n. dim.; *lagideus*, m. leveret: lagopous, lagostoma, *Lagorchestes leporoides* (a hare-wallaby), *Lagopus leucurus* (a ptarmigan), *Lagophylla filipes* (a composite), *Lagotis glauca* (a figwort), *Lagidium peruanum* (a rodent), *Lagodiopsis pinguis* (a bryozoan), *Lagurus ovatus* (a grass), *Sylvilagus floridanus* (a cottontail), *Hydrolagus alberti* (a chimaerid).

L. *lepus*, *-poris*, m. hare; *lepusculus*, m. dim.; *leporarius; leporinus*, of hares: leporine, leveret, *Lepus timidus* (European hare).

NL. *ochotona*, f. a rabbitlike rodent: *Ochotona princeps* (pika).

Gr. *ptox, ptochos*, m. the cowering one, the hare: *Celastrina(Ptox) coalita* (a butterfly).

hariolus, L. soothsayer, prophet; see **prophecy**

harlot < OF. *arlot;* see **prostitute**

harm < AS. *hearm*, insult; see **hurt, bad**

harma, Gr. chariot; see **vehicle**

harmala, Gr. wild rue; see **rue**

harmful; see **bad, destroy, disease, hurt**

harmless; see **innocent, good**

harmony < Gr. *harmonia*, f. a fitting together; *harmonikos*, suitable, musical; *harmonios*, fitting: harmonious, harmonica, harmonics, *Colluricincla harmonica* (a bird).

Gr. *arsios*, fitting, agreeing, friendly; see **fit**

L. *axitiosus*, acting together, in combination:

L. *concino*, *-centus*, sing in chorus, play in concert, join, agree; *cuncticinus*, sounding together, harmonious: concentus, concentual, concentuous.

L. *concordia* f. agreement, union, harmony, sympathy; *concordo*, *-atus*, agree together, harmonize: concord, concordance.

L. *congruo*, meet, coincide, agree: congruous, incongruity.

L. *consentio*, *-sensus*, agree, assent; *consentaneus*, agreeing with, suited to, fitting: consent, consensus, consentaneous, *Combophora consentanea* (a bug).

L. *consonantia*, f. harmony; *consonus*, sounding together, harmonious: consonant.

L. *conspiro*, *-atus*, breathe together, plot and plan to act in harmony and concert: conspire, conspiracy, conspirator.

L. *dispungo*, *-unctus*, balance an account:

Gr. *emmeles*, harmonious, in tune, fit, right: *Emmelichthys vittatus* (boga).

Gr. *homologia*, f. agreement, assent, conformity; *homologos*, accordant, correspondent; *homonoia*, f. unanimity, concord; *homophonos*, in unison, chiming in; *homophrosyne*, f. unanimity, accord: homologous.

L. *paciscor, pactus*, make a bargain, agree, covenant, contract: pact.

Gr. *prosodos,* singing in accord, harmonious: prosody.
Gr. *symphonia,* f. concord in sound; *symphonos,* accordant: symphony, *Symphonia globulifera* (a guttifer).
Gr. *symphylia,* f. accordance, agreement:
Gr. *synchordia,* f. harmony:
Gr. *synodos,* singing in accord, harmonious:
L. *unanimus,* of one mind, accordant: unanimity, unanimous.
See: **fit, agreeable, joy, equal, follow, yes, common, rest**

harmos, Gr. joint, fastening, bolt, peg; see **joint**

harmostos, Gr. adapted, suitable, fit; see **fit**

harness < OF. *harneis.*
Gr. *himas, -mantos,* strap, trace for attaching horses to chariot; see **strap**
L. *lorica,* f. leather cuirass or corselet; *loricula,* f. dim.; *loricatus,* clad in mail; *lorarius,* m. harness-maker; *lorico, -atus,* clothe in mail, harness: loricate, lorica, illoricate, *Loricaria aurea* (a catfish), *Onthophagus loricatus* (a dung-beetle).
See: **bridle** under **hold**

harp < AS. *hearpe,* and LL. *harpa,* f.: harpsichord, arpeggio, *Harpa mutica* (a gastropod), *Flabellina harpa* (a foraminifer), *Marginula harpula* (a foraminifer).
Gr. *barbitos,* m. a kind of lyre:
Gr. *chelonis, -idos,* f. lyre:
L. *cinyra* (Gr. *kinyra*), f. an instrument with ten strings, a kind of lyre: *Cinyra costulata* (a beetle).
L. *cithara* (Gr. *kithara*), f. harp, lyre, lute: cithara, zither, *Citharexylum fruticosum* (Florida fiddle-wood), *Citharomantis falcata* (a mantid), *Citharichthys evermanni* (a fish).
L. *fides, -is* (Gr. *sphide*), f. harp, lyre, lute, gut, string; *fidicula,* f. dim.; *fidicularius,* like a cord, twisted; *fidicen, -inis,* m. harpist, lyrist: *Sphida obliquata* (a moth).
L., Gr. *lyra,* f. lyre, lute; *lyricus,* of the lyre; *lyricen, -inis,* m. harpist: lyric, lyrate, *alpha Lyrae* (Vega), *Lyrurus tetrix* (black grouse), *Quercus lyrata* (overcup-oak) *Chitonastrum lyra* (a radiolarian), *Pardosa lyrifera* (a spider), *Mimetes lyrigerus* (a proteacead).
Gr. *magadis, -idos,* f. a harplike instrument: *Magadiceramus petrachecki* (a fossil pelecypod).
L. *nablium,* n. (Gr. *nabla,* f.), a kind of harp; *nablistes,* m. player of a *nabla:*
Gr. *pektis, -idos,* f. a kind of harp:
Gr. *phorminx, -ingos,* f. a kind of cithara or lyre; *phormiktes,* m. harper: *Phormingochilus tigrinus* (an arachnid), *Phormictopus canceroides* (a tarantula).
L. *psalterium* (Gr. *psalterion*), n. a harplike, stringed instrument; *psaltria,* f. female harpist: psaltery, psalterium, *Psaltriparus minimum* (bush-tit), *Spinus psaltria* (a goldfinch).

harpago, L. grappling hook, drag; Gr. *harpazo,* seize, snatch; *harpagimos,* ravished; *harpax,* robber; see **take**

harpaleos, Gr. grasping, greedy; see **greed**

harpastum, L. (Gr. *harpaston*), handball; see **ball**

harpe, Gr. sickle, hook, kawk; see **sickle**

harpedon, Gr. cord, thread; see **rope**

harpeza, Gr. thicket; see **bush**

harpis, Gr. a kind of shoe; see **shoe**

harpy < Gr. *Harpyia,* a monster, half bird and half woman; see **animal**

harrow < AS. *hearge;* see **rake, till**

haruspex, L. soothsayer; *haruspicalis,* of soothsayers; see **prophecy**

harvest < AS. *haerfest;* see **reap**

hasta, L. spear; *hastula,* dim., see **spear;** *hastile,* shaft of a spear, see **rod**

haste < AS. *haest;* see **swift**

hat < AS. *haet;* see **cap**

hatch < uncertain origin; see **birth**

hatchet < F. *hachette,* dim. of *hache,* ax; see **ax**

hate < AS. *hatian.*
L. *abhorreo,* dislike, shrink from: abhor, abhorrent, abhorrence.
L. *abominor, -atus,* abhor, detest: abominate, abomination.
Gr. *ados, -eos,* n. satiety, loathing: satiety.
Gr. *antipatheia,* f. opposition, aversion: antipathy.
Gr. *aphilos,* unfriendly, hateful: *Aphiloctenus virginiensis* (a wasp).
Gr. *apolaktizo,* kick away, spurn, scorn:

Gr. *apoptystos,* spat out, detested:
Gr. *apothestos,* despised, hated:
Gr. *apothymios,* unpleasant, hateful:
Gr. *aseros,* disdainful, irksome:
Gr. *atimetos,* despised: *Atimeta kirschi* (a beetle).
L. *aversio, -onis,* f. a turning away, loathing: aversion.
L. *consputus,* spit upon contemptuously:
L. *dedignor, -atus,* scorn: disdain.
L. *despicus,* disdained; *despicabilis,* contemptible, < *despicio, -spectus,* look down
  upon, disdain: despicable, despise, despite.
L. *detestor, -estus,* invoke the curse of a god, execrate, abominate; *detestabilis,*
  abominable, execrable: detest, detestation, detestable.
Gr. *echthos, -eos,* n. hate, hatred, enmity, abhorrence; *echthros,* hateful, hated; m.
  enemy; *echthion,* more hated; *echthistos,* most hated; *echthodopos,* detestable;
  apechthes, hateful, hated; *demechthes,* hated by the people: *Echthrogaster*
  lugubris (a beetle).
L. *execrabilis* (*exsecrabilis*), detestable, abominable: execrable.
L. *fastidibilis,* loathsome, disagreeable; *fastidium,* n. loathing; *fastidiosus,* full of
  loathing, squeamish, overnice, dainty, < *fastidio, -itus,* feel disgust, loathe,
  dislike: fastidious, *Hebeloma fastidibile* (a mushroom).
L. *inamabilis,* unlovely, repugnant, revolting, hateful:
L. *invisus,* hated, hateful, detested: *Anthomyia invisa* (a fly).
Gr. *kotos,* m. grudge, hate, wrath; *enkotos,* spiteful; *epikotos,* wrathful, hateful:
L. *malevolus; malignus,* ill-disposed toward, envious, spiteful; *malitia,* spite, bad-
  ness; see *malus* under **bad**
Gr. *misos,* n. hate; *misema, -tos,* n. object of hate; *misetos,* hateful; *misetes,* m.
  hater: misanthropy, misogamy, misogynist, misoneism, misology, *Misocosmus*
  ceyanicus (a fly).
L. *obtrectatio, -onis,* f. detraction, envious disparagement:
L. *odium,* n. hatred, < *odi, osus,* hate; *odibilis,* hateful; *odiosus,* hateful, offensive;
  osor, -is, m. hater; *exosus,* hated exceedingly, detestable: odious, annoy, noisome,
  ennui.
Gr. *odysis,* anger, wrath, < *odyssomai,* hate; see **anger**
L. *offensio, -onis,* f. aversion, disgust, dislike, transgression: offensive, offense.
L. *rancor, -is,* m. grudge, hate: rancor, rancorous.
L. *repugnans, -antis,* offensive, distasteful: repugnant, repugnance.
L. *satietas, -atis,* f. sufficiency, loathing: satiety.
Gr. *sinchaino,* loathe, dislike; *sinchos,* sickening, offensive:
L. *sperno, spretus,* reject, despise, scorn, spurn; *spretor, -is,* m. despiser; *asperna-*
  bilis, worthy of contempt, < *aspernor, -atus,* reject with aversion: *Amanita*
  spreta (a poisonous mushroom), *Panicum spretum* (a grass).
Gr. *stygeo,* hate, abhor, loathe; *styganos; stygeros; stygetos; stygnos,* hated,
  hateful, loathsome: *Anodonta(Styganodon) tenebricosa* (a pelecypod), *Styge-*
  romyia maculosa (a fly), *Stygetoblatta latipennis* (a fossil cockroach wing),
  Stygnoclonius glaberrimus (a beetle).
L. *taedium,* weariness, disgust; see **weak**
L. *temno,* despise, scorn; *temnibilis,* despicable, hateful; *contemno, -emptus,*
  despise, disdain: contemn, contemptible, contemptuous.
See: **anger, enemy, fear, jealous**

haughty < F. *haut,* high; see **proud**

haul < OF. *haler;* see **carry, draw**

haustus, L. a drawing, draught; *haustor,* drawer; see *haurio* under **suck**

have < AS. *habban,* hold, own; see **hold, own, nature**

haven < AS. *haefen;* see **harbor**

hawk < AS. *hafoc:* hawkbit, hawkweed, hawk-nosed, hawk-moth, goshawk, night-
  hawk.
L. *accipiter, -tris,* m. hawk: *Accipiter fuscus* (sharp-shinned hawk), *Deroptyus*
  accipitrinus (hawk-parrot).
L. *aesalon, -is* (Gr. *aisalon, -os*), m. a kind of hawk: *Falco aesalon* (merlin).
L. *astur, -is,* m. a kind of hawk: *Astur gentilis* (European goshawk), *Micrastur*
  plumbeus (a hawk).
L. *buteo, -onis,* m. falcon, hawk: buzzard, *Buteo borealis* (red-tailed hawk).
L. *cenchris, -idis,* f. a kind of hawk: *Cenchris tinnunculus* (a bird).
Gr. *chthamaloptetes,* m. a kind of hawk:
Gr. *elanos,* m. kite: *Elanus leucurus* (white-tailed kite), *Elanoides forficatus*
  (swallow-tailed kite), *Haemoproteus elani* (a protozoan).
L. *epileus,* m. a kind of hawk:
L. *falco, -onis,* m. hawk: falconry, falconer, falcon (*Falco peregrinus*), *Falcunculus*
  flavigulus (a hawk).

Gr. *harpe*, a kind of hawk; see **sickle**

Gr. *hierax*, *-akos*, m. hawk; *hierakiskos*, m. dim.; *hierakion*, n. hawkweed: *Micro-hierax melanileucus* (a falconet), *Melierax musicus* (a hawk), *Hieracium auran-tiacum* (orange hawkweed).

Gr. *iktinos*, m. kite: *Ictinia mississippiensis* (Mississippi kite).

L. *immusulus*, m. a kind of falcon:

Gr. *kerchne*, f. a kind of hawk, probably the kestrel: *Cerchne pennipes* (a bird).

Gr. *kirkos*, m. a kind of hawk: *Circus hudsonius* (marsh-hawk).

Gr. *kymindis*, *-idos*, f. probably a kind of hawk: *Cymindis sabulosa* (a wasp).

Gr. *mermnos*, m. a kind of hawk:

L. *milio*, *-onis*, m. a kind of hawk, kite:

L. *milvus*, m. kite, hawk; *milvinus*, of kites: *Milvus ictinus* (European kite), *Milvulus forficatus* (a flycatcher), *Pardosa milvina* (a spider).

L. *Nisus* (Gr. *Nisos*), m. fabled king said to have been changed into a sparrow-hawk: *Accipiter nisus* (European sparrow-hawk).

Gr. *oinonos*, m. bird of prey, omen; see **prophecy**

Gr. *Pandion*, m. a king of Athens; now a genus of hawks: *Pandion haliaëtus* (osprey).

Gr. *perdikotheras*, m. partridge-catcher, a hawk:

Gr. *perkos*, m. a kind of hawk: *Percus ebenus* (a beetle).

Gr. *phabotypos*, m. dove-striker, a hawk: *Phabotypus plumbarius* (a hawk).

Gr. *phassophonos*, m. dove-killer, a hawk:

Gr. *pternis*, m. a kind of hawk: *Pernis[Pternis] apivorus* (pern, honey-buzzard).

Gr. *spizias*, m. sparrow-hawk:

L. *tinnunculus*, m. kestrel: *Falco tinnunculus* (kestrel).

Gr. *triorches*, m. a kind of hawk: *Triorches fluvialis* (a bird).

See: **eagle, bird**

hawthorn < AS. *haga*, hedge, haw, *thorn*, thorn; see *crataegus* under **rose, thorn**

hay < AS. *heg;* see **grass**

hazel < AS. *haesel*, filbert.

It. *avellana*, f. filbert, hazelnut, < L. *Abella* (*Avella*), city of Campania: *Corylus avellana* (filbert).

L. *corylus*(*colurnus*), f. hazel, filbert: *Corylus americana* (hazelnut), *Corylopsis spicata* (winter-hazel), *Betula corylifolia* (a birch), *Corylus colurna* (Turkish filbert), *Apoderus coryli* (a weevil).

Gr. *hamamelis*, witchhazel; see **witchhazel**

head < AS. *heafod:* heading, headland, headlong, headache, headdress, headquarters, headway, ahead, blockhead, forehead, Whitehead, bullhead.

Gr. *brechmos*, m.; *bregma*, *-tos*, n. front part of the head, sinciput: *Bregmato-thrips venustus* (a thysanopterid), *Brechmotriplax usambarensis* (a beetle), *Eurybregma nigrolineata* (a bug), *Hadrobregmus gibbicollis* (a beetle).

L. *calvaria*, f skull; *calvariola*, f. dim.: calvarium, Calvary.

L. *caput*, *-pitis*, n. head, end, point; *capitellum; capitulum*, n. dim.; *capitalis*, of the head; *capitatus*, having a head; *capito*, *-onis*, m. one with a large head; *capitulatus*, having or ending in a small head; Sp. *cabeza*, f. head; *anceps*, *-pitis*, two-headed; *occiput*, *-pitis*, n. back of the head; *sinciput*, *-pitis*, n. front of the head: capital, capitol, captain, capitulate, cabochon, cape (point of land), Cape of Good Hope, chieftain, chef, chapter, cattle, cadet, cad, cap-a-pie, chattel, achievement, ancipitous, sinciput, centriciput, occiput, decapitate, precipitate, recapitulate, biceps, handkerchief, mischief, precipice, occipital, sincipital, hatchment, Cabezon (N. Mex.), Capitonidae, cabeza de negro, (a yam, *Dioscorea mexicana*, a source of cortisone), *Capito peruvianus* (a barbet), *Pediculus capitis* (head-louse), *Claviceps paspali* (an ergot), *Cordyceps capitatum* (a fungus), *Baculites anceps* (an ammonite), *Opidnus nigriceps* (a wasp), *Pterotocrinus capitalis* (a Mississippian crinoid), *Limeum capense* (a phytolacca-cead), *Rana occipitalis* (a frog), *Begonia capituliformis* (a begonia).

L. *frons*, *frontis*, brow, forehead; see **face**

Gr. *inion*, n. back of the head, occiput: inion, *Inioteuthis japonica* (a squid).

Gr. *kara*, n. head, top; *karenon* (*karanon*), n. head; *akarenos*, headless; *amphi-karenos*, two-headed; *karatomos*, beheaded; *miltokarenos*, red-headed: *Aulono-cara rostrata* (a fish), *Caryocar villosum* (a souari nut).

Gr. *kephale* (*keble*), f. head; *kephalidion; kephalion*, n.; *kephalis*, *-idos*, f. dim.; *kephalaios; kephalikos*, of the head; *kephalotos*, with a head, headed; *amphi-kephalos*, two-headed; *megakephalos*, large-headed; *mikrokephalos*, small-headed: cephalic, cephalopod, encephalitis, hydrocephalus, megacephalous, cephalothorax, *Cephalotaxus oliveri* (a conifer), *Cephalosporium sacchari* (a fungus), *Cephalotus follicularis* (Australian pitcher-plant), *Cephalaria radiata* (a dipsacead), *Cephaëlis emetica* (an ipecac), *Cephalanthus occidentalis* (button-bush), *Cynocephalus volans* (flying lemur), *Didymocephalus ferrugineus* (a bug), *Conocephalum japonicum* (a liverwort), *Encephalartos horridus* (a cycad), *Pachycephala pectoralis* (a bird), *Draconema cephalatum* (a nematode), *Gnaphalium polycephalum* (an everlasting).

Gr. *kodeia* (*kodia*), f. head, head of plants: codeine, *Codium tomentosum* (an alga).

Gr. *korse* (*korrhe*), f. temple, side of the forehead: *Caenocorse depressa* (a beetle), *Lyprocorrhe anceps* (a beetle).

Gr. *koryphe*, f. head, crown, top, highest point; *koryphaios*, m. head, leader, chief: of the head: coryphaeus, coryphodon, *Corypha utan* (a palm), *Conocoryphe elegans* (a trilobite), *Coryphaena equisetis* (a dolphinlike fish), *Coryphanta pallida* (a cactus), *Coryphella pilata* (a nudibranch), *Melanocoryphus punctatiguttatus* (a bug).

Gr. *kotta*, f. head; *kottarion*, n. dim.:

Gr. *kranion*, n. skull, head; *mesokranon*, n. crown of the head; *olekranon*, n. head or point of the elbow; *boukranos*, bullheaded; *kenokranos*, emptyheaded: cranium, cranial, craniometry, olecranon, migraine, *Crania antiqua* (a brachiopod), *Craniolaria annua* (a martyniacead), *Idiocranium russelli* (a caecilian).

Gr. *krotaphos*, m. side of the forehead, temple: *Crotaphopeltis rufescens* (a snake), *Gymnocrotaphus curvidens* (a fish), *Crotaphytus collaris* (collared lizard).

Gr. *kybe*, f. head: *Clitocybe candida* (a mushroom).

Gr. *metopon*, n. brow, forehead; *metopidios*, of the forehead: *Metopoceros cornutus* (horned iguana), *Metopidiothrix lacertosa* (a milleped), *Eurymetopon rufipes* (a beetle), *Pimelometopon pulcher* (fathead), *Glyphometopus ornithodorus* (a beetle).

Gr. *ophrys*, brow; see **face**

See: **top, projection, point, brain, first, lead, crown**

heal < AS. *haelan*, < *hal*, sound, whole.

L. *Aesculapius* (Gr. *Asklepios*), m. god of medicine: Aesculapian, *Asclepias tuberosa* (butterfly-weed).

Gr. *akeomai*, cure, heal; *akesis*, f.; *akesma, -tos; akestron; akos, -eos*, n. remedy, cure; *akesimos; akesios*, healing; *akesmos; akestos*, curable; *akester, -os; akestes*, m. healer; *akestoria*, f. the healing art; *dysakestos*, hard to heal; *panakes*, all-healing: aceology, panacea, *Panax trifolius* (dwarf ginseng), *Oreopanax reticulatus* (an araliacead).

Gr. *althos*, n. a healing, medicine, < *althaino*, heal; *altheeis*, healing; *altheus*, m. healer, physician; *althexis, -eos*, f. a healing, cure; *althesterion*, n. remedy; *dysalthetos*, hard to cure; *polyalthes*, curing many diseases: *Althaea officinalis* (marshmallow), *Proboscidea altheifolia* (a devilsclaw).

Gr. *analeptikos*, curative, restorative:

Gr. *antidotos*, remedy; see **drug**

L. *cataplasma* (Gr. *kataplasma*), plaster, poultice; see **plaster**

L. *chalasticus* (Gr. *chalastikos*), soothing, alleviating; see **soothe**

L. *Chiron, -is* (Gr. *Cheiron, -os*), m. a centaur reputed for his skill in medicine: *Chironia baccifera* (a starpink), *Opopanax chironium* (woundwort), *Hesperochiron pumilus* (a hydrophyllacead).

L. *chirurgus* (Gr. *cheirourgos*), m. an operating physician: surgeon, surgical, surgery, *Hydrophasianus chirurgus* (a water-pheasant).

L. *clinicus* (Gr. *klinikos*), m. physician who attends patients in bed: clinic, clinician, clinical.

L. *curans, -antis*, physician; *curandus*, patient; see *cura* under **guard**

Gr. *Hippocrates*, m. a celebrated physician: Hippocratic.

Gr. *iaomai*, heal; *iasis*, f. a healing; *iama, -tos*, n. remedy; *iasimos*, curable; *iater, -os; iatros*, m. healer, physician; *iatromaia*, f. midwife; *iamatikos*, of healing; *iatikos*, healing; *iatrikos*, of physicians and the medical art; *aniatros*, m. quack; *kteniatros*, m. veterinary: elephantiasis, psychiatry, pediatrics, archiater, *Jatropha multifida* (coral-plant).

Gr. *katoulosis*, f. a healing over, cicatrization; *katoulotikos*, good for healing over:

L. *medicatus*, healing, curing, < *medico; medeor*, heal, cure; *medens, -entis*, m. physician; *medicamentum*, n. drug, remedy; *medicabilis*, curable; *medicinus*, pertaining to the art of healing; *medicus*, healing, curing; m. physician: medical, medication, medicine, medicinal, remedy, irremedial.

Gr. *Paion, -os*, m. Paeon, physician to the gods: paeon, peony (*Paeonia officinalis*).

Gr. *pharmakon*, drug, poison; see **drug**

L. *sano, -atus*, heal, cure; *sanabilis*, curable; *sanator, -is*, healer; *sanatorius*, giving health; see **health**

Gr. *therapeuo*, serve, care for, give medical treatment; *therapeutikos*, serving, healing: therapeutic, hydrotherapy.

L. *veterinarius,* m. animal doctor: veterinary, veterinarian.
See: **health, drug, soothe, guard**

health < AS. *haelth,* < *hal,* whole, sound: hale, heal, healthy, healthful, hail, whole, wholesome, wholly, wassail-bowl.   Derive: hello!
Gr. *anosos* healthy, sound:
L. *consipio,* be of sound mind:
Gr. *euexia,* f. health, vigor:
Gr. *euphoria,* bearing well, feeling good; see *phreo* under **carry**
Gr. *eusomatos,* sound in limb, able-bodied:
Gr. *hygieia (hygeia),* f. health, soundness of body; *hygieimos,* healthful; *hygies,* healthy: hygiene, Hygeia, hygienics, hygeiolatry.
L. *integer,* whole, sound, entire; see **all**
L. *salus, -utis,* f. health, prosperity, safety; *salubris; salutaris,* healthful, wholesome: salubrious, salubrity, salute, salutatory, salutary, salutation.
L. *sanus,* sound, healthy; *sanabilis,* curable; *sanator, -is,* m. healer; *sanatorius,* giving health; *sanitas, -atis,* f. health, soundness of mind: sane, sanitary, sanitarium, insanity, tutsan, *Sanicula marylandica* (sanicle).
Gr. *sophron, -os,* of sound mind, sensible, wise, temperate; see **mind**
Gr. *sos,* safe, sound, well; see *soter* under **save**
AS. *wel,* not ill, in good health, satisfactory: welfare, farewell, well.
See: **heal, all, one**

heap < AS. *heap,* pile, crowd, mass.
L. *acervus,* m. heap; *acervulus,* m. dim.; *coacervo, -atus,* heap together: acerval, acervate, *Acervularia ananas* (a fossil coral), *Acervulina globosa* (a foraminifer), *Myrmecophila acervorum* (a cricket).
L. *agger, -is,* m. heap, mound; *aggero, -atus,* heap up, pile up; *aggestio, -onis,* f. a heaping up; *aggestus,* m. accumulation, collection: exaggerate, exaggeration, aggerose, *Pholiota aggericola* (a mushroom).
Gr. *anabole,* f. something thrown up, heap, mound:
Sp. *apacheta,* f. heap of stones for religious purposes:
Gr. *choma; chosis,* bank, mound, dam; *chomation,* dim.; *chostos,* piled up; see **wall**
L. *congeries, -ei,* f. heap, pile, mass, < *congero, -gestus,* bring together, heap up, accumulate: congeries, congestion.
L. *conjectus,* m. heap, crowd, pile:
L. *cumulus,* m. heap, pile; *cumulo, -atus,* heap up: cumulus clouds, accumulate, cumulative, *Ilex cumulicola* (hummock-holly), *Carex cumulata* (a sedge).
Gr. *erion,* n. barrow, mound, tomb:
Gr. *gaion,* heap of earth; see *ge* under **earth**
L. *grumus,* m. heap, mound, hillock; *grumulus,* m. dim.: grume, grumose, grumous.
Gr. *halis,* in heaps, plenty; see *hales* under **abundance**
Gr. *hermax,* heap of stones; see *herma* under **stone**
Gr. *hyperbole,* exaggeration; see FIGURES OF SPEECH under **form**
Gr. *klomax (kromax), -akos,* heap of stones; see **stone**
Gr. *kolonos,* hill, mound, barrow; see **mountain**
Gr. *kopria,* dunghill; see *kopros* under **dung**
Gr. *korthys, -yos,* f. heap:
L. *moles, -is,* f. heap, mass, pile; *molecula,* f. dim.: molecule, molecular, molar.
Gr. *nagma, -tos,* n. anything piled up:
Gr. *nekas, -ados,* f. heap or row of the dead:
Gr. *netos,* heaped up, < *neo,* pile up:
L. *rogus,* m. funeral pile; *rogalis,* of a funeral pile:
Gr. *sarmos,* m. heap of earth or sand:
L. *sepono,* lay aside, put by, hoard, separate; see **separate**
L. *sorus* (Gr. *soros*), m. heap; *soredion,* n. dim.; *soreuma, -tos,* n. heap, pile; *soreutos,* heaped up, accumulated: sorus, soredium, sorosis, sorites, soritical, *Camptosorus rhizophyllus* (walking-fern), *Ceanothus sorediatus* (jimbrush-ceanothus), *Perisoreus obscurus* (Oregon jay).
L. *strues, -is; struix, -icis,* f. heap, pile, < *struo, -uctus,* build, pile up:
Gr. *syrmas, -ados,* drift, snowdrift; *syrma, -tos,* robe with a train, anything trailed along, sweepings, litter; see *syrtos* under **carry**
Gr. *themon, -os,* m. heap:
Gr. *this, -inos,* beach, sand heap; see **shore**
Gr. *thomos,* m. heap: *Thomomys fuscus* (a pocket-gopher).
L. *tumulus,* m. raised mound of earth, barrow, hillock, tomb: tumular, tumulose, *Megapodius tumulus* (mound-bird).
See: **mountain, wall, herd, lump, cluster, knot, wart, gather, number**

hear < AS. *heran:* hearer, hearsay, hearing, overhear.   Said Saint-Beauve of the charming Madame Recamier: "She listens with seduction."

L. *acroaticus* (Gr. *akroatikos*), designed for hearing only; *akroama, -tos,* n. anything heard; *akroasis,* f. a hearing, < *akroaomai,* listen to: acroama, acroatic,
acroamatic.

Gr. *akouo,* hear; *akousimos; akoustos,* audible; *akoe,* f.; *akousma, -tos,* n. the
thing heard, news, report, sound; *akoustikos,* of hearing; *akoustes,* m. listener;
otakoustes, m. spy, eavesdropper: acoustics, acousticon, acouometer, acousmatic,
Craspedacustes sowerbyi (a medusa).

L. *audio, -itus,* hear; *auditor, -is,* m. hearer: audit, auditor, auditorium, audience,
inaudible, obey, obeisance, oyer, oyez, disobedience.

L. *auritus,* attentive, giving ear to, listening; see *auris* under **ear**

L. *ausculto, -atus,* listen: auscultation, auscultatory, auscultoscope.

L. *cluo; clueo* (Gr. *klyo*), hear, listen to; *cliens, -entis,* m. listener, dependent:
client, clientele.

See: **ear, deaf**

heart < AS. *heorte:* hearty, heartburn, heartily, heartrending, heartwood, dis-
hearten, sweetheart.

L. *cor, cordis* (Gr. *kear; ker*), n. heart; *corcillum; corculum,* n. dim.; *cordialis;
cordicitus,* from the heart, hearty, warm; *cordatus,* heart-shaped, wise, prudent;
kerothi, from the heart: cordate, cordial, cordiality, concord, accord, core,
courage, misericordia, encouragement, discordant, Coeur de Leon, Cordelia,
record, *Isocardia cor* (a heart-shell), *Isocardium corculum* (a pelecypod), *Salix
cordata* (a willow), *Carya cordiformis* (bitternut-hickory), *Aster cordifolius* (an
aster), *Corculum leptopus* (coralvine).

Gr. *etor,* n. heart:

Gr. *kardia,* f. heart; *kardiakos,* of the heart: cardiac, cardioid, carditis, endocarditis,
cardiograph, cardiogram, pericardium, *Cardium costatum* (a pelecypod), *Cardiospermum halicacabum* (heartseed), *Anacardium occidentale* (cashew), *Venericardia purpurata* (a pelecypod), *Benthocardiella obliquata* (a pelecypod),
Leonurus cardiaca (motherwort).

Gr. *metra,* pith or heart of a tree; see **womb**

See: **center, pith, nut**

heat < AS. *haetu.*

L. *aestus, -us,* m. heat, fervor, unrest, passion, surge; *aestuosus,* hot, agitated;
aestifer, producing heat, heated, hot: estuous, estuation, *Erax aestuans* (a
robberfly).

Gr. *alea,* f. warmth, heat; *aleantikos,* fit for warming; *aleeinos,* lying open to the
sun, warm, hot:

Gr. *andromanes,* lusting after men; *hippomanes* (said of a mare in heat), lusting
after horses; see *mania* under **mad**

L. *apricus,* exposed to the sun, sunny; *apricatio,* a basking in the sun; see **open**

L. *apsyktos,* uncooled, incapable of being cooled:

L. *ardor,* flame, heat, desire; see *ardeo* under **burn**

L. *calidus; caldus,* warm, hot; *caldor; calor, -is,* m. warmth, heat; *caldarius,* pertaining to warming; *calendus,* glowing; *caloratus,* hot, heated, excited; *caleo,*
be warm; *calesco,* grow warm, glow; *calefacio, -factus,* make warm, heat: calory,
caldron, calorific, chowder, chafe, chafing-dish, nonchalant, scald, *Calosoma
calidum* (a beetle).

L. *cauma, -atis* (Gr. *kauma, -tos*), burning heat, fever heat; *kaumateros,* hot,
glowing; *kynokauma,* heat of the dogdays; see *kaio* under **burn**

Gr. *chliaros,* lukewarm; *chliodes,* slightly warm; see **lukewarm**

L. *cordialis,* hearty, warm; see *cor* under **heart**

Gr. *daleros,* burning, hot; see *daio* under **burn**

Gr. *diapyros,* red-hot, fiery; *pyretos,* burning heat, fever; see *pyr* under **fire**

L. *fervidus,* hot, glowing; *fervor,* boiling heat, ardor; see *ferveo* under **boil**

L. *formidus,* warm:

L. *foveo, fotus,* warm, cherish; *fotus, -us,* m. a warming; *fomentum,* n. warm
application, poultice: foment, fovilla, *Fomes fomentarius* (a shelf-fungus).

Gr. *hele; heile,* f. heat of the sun:

L. *incendiosus; incensus,* burning, hot; see *incendo* under **burn**

Gr. *mydros,* m. mass of redhot metal: *Mydrothauma semperi* (a butterfly).

L. *oestrus* (Gr. *oistros*), gadfly, sting, madness, sexual heat, rut, desire; see **fly**

Gr. *proseilos,* toward the sun, sunny, warm:

L. *regelo, -atus,* thaw; see **flow**

Gr. *seirios; seiros,* m. scorcher; *seirinos,* hot, scorching: Sirius, *Sirinopteryx
rufivinctata* (a moth).

L. *subo, -atus,* be in heat, have sexual ardor; see **lewd**

Gr. *thalpos,* n. heat, warmth; *thalpsis,* f. a warming; *thalporos,* warm: *Thalpomys
lasiotis* (a mouse).

Gr. *thalykros,* hot, glowing: *Thalycra fervida* (a beetle).

Gr. *therme; thermostes, -etos,* f. heat; *thermole,* f. feverish heat; *thermodes,* lukewarm; *thermos,* hot; *perithermos,* very hot: thermal, thermometer, isotherm, thermite, thermostat, thermos bottle, diathermanous, Thermopylae, *Thermobia domestica*(firebrat, silverfish), *Thermoniphas plurilimbata* (a moth), *Diathermus globosus* (a beetle), *Exosphaeroma thermophilum* (an isopod). Derive: *Thermopsis rhombifolia* (a legume).

Gr. *tinthos,* boiling hot:

L. *torridus,* dry, parched, hot, < *torreo, tostus,* dry by heat, roast; see **dry**

Gr. *trasia,* grate, kiln; see **screen**

L. *urigo, -inis,* lustful heat, desire; see **lewd**

Gr. *zatheres,* scorching; see *thereia* under **year**

See: **burn, fire, cook, boil, fever, lukewarm, lewd,** *mania* under **mad**

heath < AS. *haeth,* open waste land covered with low herbs and shrubs; a plant growing on such land, now particularly an ericacead: heather, heathen.

L. *Andromeda* (Gr. *Andromede*), f. daughter of Cepheus and Cassiope; a genus of the heath family: *Andromeda polifolia* (moorwort), *Pterospora andromedea* (pinedrops).

L. *arbutus,* strawberry tree, madrone; see **arbutus**

NL. *arctostaphylos,* f. a genus of the heath family: *Arctostaphylos manzanita* (common manzanita).

NL. *azalea,* f. a genus of the heath family now generally regarded as synonymous with *Rhododendron:* swamp-azalea *(Rhododendron viscosum), Exobasidium azaleae* (a fungus).

NL. *calluna,* f. a genus of the heath family: *Calluna vulgaris* (Scotch heather).

L. *Cassiope* (Gr. *Kassiope*), f. a legendary beauty; a genus of the heath family: *Cassiope tetragona*(firemoss-cassiope), Cassiopeia.

L. *chamaedaphne* (Gr. *chamaidaphne*), f. a genus of the heath family: *Chamaedaphne calyculata* (leatherleaf).

NL. *chimaphila,* f. a genus of the heath family: *Chimaphila umbellata* (pipsissewa).

NL. *epigaea,* f. a genus of the heath family: *Epigaea repens* (trailing arbutus).

L. *erica* (Gr. *erike*), f. heath, heather: Ericaceae, erical, ericoid, ericolin, ericital, *Erica arborea* (tree-heath), *Aster ericoides* (an aster), *Melaleuca ericifolia* (a myrtacead), *Hypericum pulchrum* (a St. Johnswort).

NL. *gaultheria,* f. a genus of the heath family, < Dr. Gaultier, Canadian physician: *Gaultheria ovatifolia* (Oregon wintergreen).

NL. *gaylussacia,* f. a genus of the heath family, < J. L. Gay-Lussac, French chemist: *Gaylussacia baccata* (black huckleberry).

NL. *kalmia,* f. a genus of the heath family, < Peter Kalm, Swedish botanist: *Kalmia polifolia* (bog-kalmia), *Phyllosticta kalmicola* (a fungus).

Gr. *klethra,* f. alder; now a genus of the heath family: *Clethra alnifolia* (white alder), *Clethropsylla alni* (a bug).

L. *ledum* (Gr. *ledon,* a cistacead yielding ladanum), n. now a genus of the heath family: *Ledum groenlandicum* (Labrador tea), *Cercocarpus ledifolius* (a mountain-mahogany).

L. *Leucothoë,* f. a princess said to have been changed into a sweet-scented shrub: *Leucothoë populifolia* (Florida leucothoë).

NL. *lyonia,* f. a genus of the heath family, < John Lyon, American botanist: *Lyonia lucida* (fetterbush).

NL. *menziesia,* f. a genus of the heath family, < Archibald Menzies, Scotch naturalist: *Menziesia pilosa* (Allegheny menziesia).

NL. *moneses,* f. a genus of the heath family: *Moneses uniflora* (woodnymph),

NL. *monotropa,* f. a genus of the heath family: *Monotropa uniflora* (Indian-pipe).

NL. *oxydendrum,* n. a genus of the heath family: *Oxydendrum arboreum* (sourwood).

L. *Phyllodoce* (Gr. *Phyllodoke*), a sea-nymph; a genus of the heath family; see **sea**

L., Gr. *Pieris,* one of the Muses; a genus of the heath family; see **art**

NL. *pyrola,* f. a genus of the heath family: *Pyrola asarifolia* (Alpine pyrola).

Gr. *rhododendron,* n. rosebay; a genus of the heath family: rhododendron, *Rhododendron maximum* (rosebay-rhododendron), *Exobasidium rhododendri* (a fungus).

L. *rhodora,* f. a species of the heath family: rhodora *(Rhododendron canadense).*

NL. *sarcodes,* f. a genus of the heath family: *Sarcodes sanguinea* (snow-plant).

L. *sisyrus,* m. heath:

L. *tetralix, -icis,* f. a heath: *Erica tetralix* (a heath).

L. *vaccinium,* n. blueberry, whortleberry: *Vaccinium deliciosum* (a blueberry), *Spiraea vacciniifolia* (a spirea), *Exobasidium vaccinii* (a fungus).

heaven < AS. *heofon.*
L. *aether; aethra* (Gr. *aither; aithra*), the upper, pure air, sky, heaven; see **wind**
L. *caelum,* n. heaven, sky; *caelestinus; caelestis; caelicus,* heavenly, divine; *caelitus,* from heaven: celestial, celestite, ceiling, *Eupatorium caelestinum* (mist-flower).
L. *divum,* n. sky:
L. *Elysium* (Gr. *Elysion*), n. abode of the blessed: Elysian, *Elysia chlorotica* (a sea-slug).
L. *firmamentum,* n. support, sky: firmament.
L. *Olympus* (Gr. *Olympos*), m. a celebrated Greek mountain, fabled to be the seat of the gods, hence heaven: Olympian, Olympia (Wash.).
Gr. *ouranos,* m. heaven, sky; *ouraniskos,* m. dim. roof, ceiling: Uranus, uranium, uraniferous, uranography, uranophane, uranothallite, *Uranoscopus bicinctus* (a fish), *Trachicephalus uranoscopus* (a fish).
L. *paradisus* (Gr. *paradeisos*), park, abode of the blessed; see **park**
ON. *sky,* cloud, region of clouds; sky, skylark, skylight, skyscraper.
L. *supernus,* on high, celestial; see *super* under **over**
See: **roof, cover**

heavy < AS. *hefig.*
Gr. *barys,* heavy, impressive, grave; very; *baryntikos,* weighing down; L. *barycus,* topheavy; *barema, -tos,* n. burden, load; *baros,* n. weight; *barydion; baryllion,* n. dim.; *karybares,* topheavy; *katabares,* heavy-laden: barometer, barograph, barite, baritone, isobar, barium, *Barylambda faberi* (a fossil mammal), *Baryproctus barypus* (an ichneumonid), *Barosma pulchella* (buchu), *Barosaurus lentus* (a dinosaur), *Baropus lentus* (fossil amphibian footprints), *Barytettix crassus* (a grasshopper), *Barycnemis claviventris* (a wasp).
Gr. *brithys,* heavy; *brithos,* n. weight; *embrithes,* weighty, grave; *peribrithes,* very heavy: *Embrithes obesus* (a beetle), *Embrithosaurus schwartzi* (a fossil reptile).
Gr. *emmochthos,* toilsome; see **make**
Gr. *epachthes,* heavy, burdensome, annoying: *Epachthes erythropalpus* (a wasp).
L. *gravis,* heavy; *gravidus,* laden, pregnant; *gravamen, -inis,* n. trouble, inconvenience; *gravabilis,* troublesome; *gravedo, -inis,* f. heaviness (in head, limbs, etc.); *gravo, -atus,* load, weigh down: gravid, gravity, grave, gravitation, aggravation, grief, grievous, gravedo, *Astronium graveolens* (an anacardiacead), *Bruclarkia gravida* (a fossil gastropod).
L. *obesus,* fat, corpulent; see **fat**
L. *onerosus,* burdensome, heavy, oppressive; see *onus* under **weight**
L. *ponderosus,* of great weight; see *pondus* under **weight**
See: **weight, large, fat, trouble, serious, difficult**

hebdomas, Gr. seven, week; see *hepta* under **seven, week**

hebe, Gr. youth, youthful prime, time of pubescence; *hebetikos,* youthful; *hebos,* at the threshold of manhood; see **young**

hebetatus, L. dulled, blunted, weakened; *hebes, -etis,* blunt, dull; see **dull**

hecasto- < Gr. *hekastos,* every, each; see **alone**

Hecate, L. (Gr. *Hekate*), goddess of magic and witchcraft; see **magic**

hecato-; hecto- < Gr. *hekaton,* hundred; see **hundred**

hecelo- < Gr. *hekelos (hekalos),* at ease, at will; see **rest**

hecisto- < Gr. *hekistos,* least; see *mikros* under **little**

hectico- < Gr. *hektikos,* habitual; *hektor,* holding fast; see *echo* under **hold**

hecusio- < Gr. *hekousios,* voluntary; see **free**

hecyro- < Gr. *hekyros,* father-in-law; see **kin**

hedeoma, NL. a genus of the mint family; see **mint**

hedera, L. ivy; *hederaceus,* of ivy; *hederatus,* adorned with ivy; see **vine**

hedge < AS. *hecg (haga)*; see **bush, wall**

hedgehog
L. *echinus* (Gr. *echinos*), m. hedgehog, sea-urchin; *echiniskos,* m. dim.; *echinatus,* spiny, prickly; *echinodes,* like a hedgehog, prickly: echinoderm, echinosis, echinoid, *Echinus esculentus* (edible sea-urchin), *Echinops exaltatus* (Russian globe-thistle), *Echinomyia splendidula* (a fly), *Echinacea purpurea* (a composite), *Echinocactus cinereus* (a cactus), *Echinocystis lobata* (mock-cucumber), *Pinus echinata* (shortleaf pine), *Antechinus flavipes* (a marsupial-mouse), *Medicago echinus* (an alfalfa), *Xanthium echinatum* (beach-cocklebur), *Taenia echinococcus* (a tapeworm).

L. *er, -is; ericius; erinaceus,* m. hedgehog; *ericinus,* of hedgehogs: urchin, caprice,
    Erinaceus europaeus (hedgehog), *Ericulus spinosus* (a tenrec), *Raja erinacea*
    (skate), *Hydnum erinaceum* (hedgehog-mushroom).
Gr. *cher, -os,* m. hedgehog: *Cheroscelis oospila* (a moth).
Gr. *schyr, -os,* m. hedgehog:

hedno- < Gr. *hednon,* wedding gift; *hednios,* bridal, nuptial, < *hednoo,* betroth,
    marry; see **marry**

hedolion, Gr. seat, abode; see *hedos* under **seat**

hedone; hedosyne, Gr. delight, enjoyment, pleasure; see *hedomai* under **joy**

hedos, Gr. seat, abode, base, pedestal; see **sit**

hedra, Gr. seat, chair, abode, base, plane; *hedrion,* dim.; *hedraios,* sitting, sedentary,
    sessile; see **sit**

hedus, L. young goat, kid; see **goat**

hedys, Gr. sweet, pleasant; *hedyntos,* sweetened; *hedysmos,* sweetness; see **sweet**

hedysarum, L. (Gr. *hedysaron*), vetch; see **bean**

heed < AS. *hedan,* pay attention to, regard carefully.
Gr. *alego,* care for, mind, heed:
L. *attendo, -tentus,* give heed to, consider, regard carefully: attend, attention.
L. *circumspectus,* cautious, wary; see **guard**
L. *meticulosus,* excessively careful; see *metus* under **fear**
See: **think, guard, follow**

heedless; see **neglect**

heel < AS. *hela;* see **foot**

hegemon, Gr. leader; *hegetes,* leader, guide; *hegemonikos,* of leading; see **lead**

helco- < Gr. *helkos,* wound, sore, ulcer; *helkydrion,* dim.; see **sore**

helcto- < Gr. *helktos,* that can be dragged or drawn; *helkystos,* drawn; see *helko*
    under **draw**

heledone, Gr. a kind of polyplike mollusk; see **mollusk**

helene, Gr. torch, basket; see **light**

helenium, L. (Gr. *helenion*), elecampane; see **composite**

helianthus, L. (Gr. *helianthes*), sunflower; see **composite**

helichrysum, L. (Gr. *helichrysos*), starflower, marigold; see **composite**

helico- < Gr. *helix, -ikos,* spiral, whirl, eddy, tendril, coil, curl; *heliktos,* rolled,
    twisted, spiral; *heligmos,* a winding, see *helisso* under **turn, spiral;** < *helix,*
    -ikos, of the same age, see *helikia* under **age**

heligma, Gr. curl; see **curl**

helinos, Gr. a vine-tendril; see **curl**

helio- < Gr. *helios,* sun, see **sun;** < *heleios,* of a marsh, see *helos* under **marsh**

heliotropium, L. heliotrope; see **borage**

helisco- < Gr. *heliskos,* dim. of *helos,* nail; see **nail**

helium < Gr. *helios,* sun; see ELEMENTS under **thing**

helix, L., Gr. coil, spiral, turn; see **spiral, turn**

hell < AS. *hell,* place of the dead, place for ultimate punishment of the wicked,
    the nether world, abode of evil spirits: hellcat, hellhound, hellish, hellhole, Hell
    Gate.
L. *abyssus* (Gr. *abyssos*), bottomless pit, hell; see **deep**
Gr. *Acheron,* a river in the nether world, the nether world itself; see **stream**
L. *Avernus,* m. nether world, infernal regions; *avernalis,* of Avernus: *Avernus*
    interruptus (a bug).
L. *barathrum* (Gr. *barathron*), pit, gulf, abyss; see **hole**
L. *catonium,* n. the lower world:
Gr. *chthonios,* in or under the earth, pertaining to the nether world: chthonian.
Gr. *enerteros,* lower, pertaining to the nether world; see **low**
L. *Erebus* (Gr. *Erebos*), m. a place of darkness in the nether world; god of
    darkness: *Erebus odora* (a moth), *Peritrochia erebus* (a fossil cephalopod).
L. *gehenna* (Gr. *geenna,* < Heb. *Ge Hinnom*), f. a refuse dump near Jerusalem
    where perpetual fires burned, hell: gehenna.
Gr. *Haides,* m. god of the nether world, nether world, grave; *Persephone,* f. wife
    of Hades: Hades.
L. *infernus,* m. hell; *infernalis,* belonging to the lower regions: inferno, infernal,
    Eurynotus infernalis (a beetle), *Vampyroteuthis infernalis* (a cephalopod).

L. *Orcus,* m. the nether world, abode of the dead, god of the infernal regions; *Orcinus,* of the realms of the dead, of hell: *Orcinus orca* (a cetacean), *Orconectes clypeatus* (a crayfish).

Gr. *Phlegethon,* a river of fire in Hades; see **stream**

L. *Pluto, -onis* (Gr. *Plouton, -os*), m. god of the nether world; *Proserpina,* f. wife of Pluto: Pluto, plutonic, plutonium, *Pluto infernalis* (a fish), *Proserpina nitida* (a gastropod), *Proserpinella berendti* (a gastropod), *Eristalis proserpina* (a fly).

Heb. *Sheol,* underworld, hell: Sheol, sheolic.

Gr. *Stygios,* of the nether world, < *Styx,* a river in the nether world; see **stream**

L. *Tartarus* (Gr. *Tartaros*), m. the infernal regions, nether world, regions of the damned: Tartarean, *Tartarothyas micrommata* (a mite).

See: **deep, down, fire, punish**

hellebore < Gr. *helleboros,* m. a plant of the buttercup family: *Helleborus guttatus* (a hellebore), *Astrantia helleborifolia* (black masterwort).

Gr. *ektomon,* n. black hellebore:

L. *veratrum,* n. hellebore: *Veratrum viride* (swamp false-hellebore), *Calanthe veratrifolia* (an orchid).

Helleno- < Gr. *Hellen, -os,* a Greek; *Hellenikos,* Greek; see *Hellas* under **Greek**

helmet < OF. *helme;* see **cap**

helmintho- < Gr. *helmins, -inthos,* worm; *helminthion,* dim.; see **worm**

helor, Gr. booty, spoil, prey; see **plunder**

helorios, Gr. a water-bird; see **goose**

helos, Gr. marsh, meadow, see **marsh**; nail, stud, see **nail**; wart, callus, see **wart**

help < AS. *helpan.*

L. *advocatus,* m. legal assistant, counselor: advocate.

Gr. *alexis,* help, defense; see **guard**

Gr. *arexis; eparexis,* f. help, aid; *arogos,* aiding, helping; *aregon; eparegon, -os,* c.; *eparogos,* m. helper: *Arexion suavis* (a grasshopper).

Gr. *arkesis; eparkesis,* f. help, service; *arketos,* sufficient: *Eparces quadriceps* (a wasp).

L. *assessor, -is,* m. assistant of a judge, < *assideo, -sessus,* sit beside: assessor, assess.

L. *assisto,* stand by, aid, help: assist, assistance, assistant.

L. *auxilium,* n. help, aid: *auxilio, -atus,* aid, assist: auxiliary.

Gr. *boëthos,* assisting, aiding; *boëtheia,* f.; *boëthema, -tos,* n. help, aid, remedy; *boëthetikos,* ready or able to help, < *boëtheo,* assist, help, succor: Boethius.

Gr. *chraismeo,* help, defend; *chraisme,* f. help; *chraismetor, -os,* m. helper, defender; *chraismenion,* n. means of help, remedy:

Gr. *epikouros,* m. assister, ally, helper; *epikouria,* f. help, aid; *epikourikos,* auxiliary: Epicurus.

L. *fautor, -is,* m.; *fautrix, -icis,* f. favorer, promoter, patron: *Fautor comptus* (a gastropod).

L. *juvo, jutus,* help; *adjutor, -is,* m.; *adjutrix, -icis,* f. helper; *adjutabilis,* helping, serviceable; *juvamen, -inis,* n. help, aid: aid, aide-de-camp, jocund, coadjutor, adjutant, *Athyrma adjutrix* (a butterfly).

L. *auxilium,* n. help, aid; *auxilio, -atus,* aid, assist: auxiliary.

Gr. *koinonos,* companion, partner, associate, accomplice; see **companion**

Gr. *metaitios; paraitios; synaitios,* accessory, sharing:

Gr. *metochos,* partaking: partaker, partner; *metoche,* communion; see **companion**

Gr. *ophelia,* help, aid, use; see **use**

L. *opifer, -a, -um,* aid-bringing, helping; *opitulus,* helper; *opitulor, -atus,* help, aid; see *ops* under **wealth**

L. *optio, -onis,* helper, assistant, adjutant; see *opto* under **choose**

L. *paracletus* (Gr. *parakletos*), m. advocate, helper, comforter: paraclete.

Gr. *paredros,* near; assistant; see **near**

Gr. *prostheke,* f. aid, assistance, addition: *Prosthecobothrium dujardini* (a cestode).

Gr. *spoudastes,* m. partisan, supporter:

L. *subsidium,* n. reserve troops, aid, help: subsidy.

L. *suppetia,* f. aid, help:

L. *sustineo, -tentus,* uphold, support: sustain, sustenance, sustention, sustentor.

Gr. *sylleptor, -os,* m. partner, assistant, accomplice:

Gr. *sympraktor, -os,* helper, assistant; see *prasso* under **make**

Gr. *synepimeletes,* m. helper:

Gr. *synergos,* m. associate, partner, fellow-workman; *synergia,* f. cooperation, collusion; *synergates,* m. partner, colleague: *Synergatus indigens* (a beetle).

Gr. *timoros*, helping, aiding; *timoria*, f. help: *Timorus personatus* (a beetle).
See: **guard, servant, companion, strong, pillar, mercy, give, care**

helpless; see **hopeless, hurt, weak**

heluor, *-atus*, L. guzzle, gormandize; *heluo, -onis*, glutton; see **eat**

helvella, L. a small potherb; see **plant**

helvus, L. bay-yellow, yellow; *helvenacus; helvolus*, pale-yellow; see **yellow**

helxine, Gr. probably a species of pellitory; see **pellitory**

helxis, Gr. a drawing, dragging; see *helko* under **draw**

hem < AS. *hem*, edge, border, margin; see **border**

hema-; **hemato-** < Gr. *haima, -tos*, blood, see **blood**; < *hema, -tos*, wart, javelin, see **spear**

hematite < L. *haematites* (Gr. *haimatites*), an ore of iron; see **stone**

hemedapos, Gr. native; see **native**

hemero- < Gr. *hemera*, see **dry**; *hemeros*, tame, cultivated, see **tame**

hemerocallis, L. (Gr. *hemerokallis*), daylily; see **lily**

hemi- < Gr. *hemisys*, half; *hemiolios*, one and one-half; see **half**

hemionitis, Gr. a fern; see **fern**

hemionos, Gr. half-ass, mule; see *onos* under **horse**

hemitomos, Gr. a kind of cup; see **cup**

hemlock < AS. *hemlic*, name for a poisonous plant of the carrot family; see *conium* and *cicuta* under **carrot**, *tsuga* under **gymnosperm**

hemorrhage < L. *haemorrhagia* (Gr. *haimorrhagia*), a bleeding; see *haima* under **blood**

hemorrhoid < L. *haemorrhoida* (Gr. *haimorrhois*), pile; see **disease**

hemp < AS. *henep*.
L. *cannabis* (Gr. *kannabis*), f. hemp, tow: cannabin, cannabinol, *Cannabis sativa* (hemp, hashish, marijuana, cannabis indica), *Apocynum cannabinum* (hemp-dogbane).
L. *mylasia*, f. a kind of hemp:

hen, *-os*, Gr. neuter of *heis, henos*, one; *henikos*, single; see **one**

henion, Gr. rein, bit, bridle; *heniochos*, driver; see **hold, drive**

henos, Gr. old; see **old**

heolos, Gr. stale; see **dull**

heortios, Gr. festal; *heorte*, festival; see **joy**

heothinos, Gr. in the morning, early, eastern; see *eos* under **dawn**

hepatica, NL. a genus of the buttercup family; see **buttercup**

hepaticus, L. pertaining to the liver, < Gr. *hepar, -atos*, liver; see **liver**

hepatos, Gr. a kind of fish; see **cod**

Hephaestus, L. (Gr. *Hephaistos*), god of fire used in the arts; see **fire**

hephthos, Gr. boiled, hence languid; see *hepso* under **boil**

hepsetos, Gr. a small fish; see **smelt**

hepta, Gr. seven; *heptas, -ados*, the number seven; see **seven**

Hera, Gr. wife of Zeus, Juno of the Romans; see *Dis* under **spirit**

herba, L. green vegetation, plant; *herbula*, dim.; *herbaceus*, of plants; *herbarius*, botanist; see **plant**

herco- < Gr. *herkos*, wall, fence; *herkane*, enclosure; see **wall**

Hercules, L. (Gr. *Herakles*), mythical hero celebrated for his strength and twelve great labors; see **strong**

herd < AS. *heord:* herder, shepherd, goatherd, swineherd.
Gr. *agele*, f. herd; *agelaios*, of a herd, gregarious; *agelasma, -tos*, n. a gathering, crowd; *agelastikos*, disposed to herd together, social: *Agelacrinus cincinnatiensis* (an Ordovician cystid), *Agelaius[Agelaeus] phoeniceus* (red-winged blackbird).
Gr. *aipolion*, n. herd of goats; *aipolos*, m. goatherd:
L. *armentum*, n. herd; *armentalis; armentivus*, pertaining to a herd:
Gr. *boukolion*, herd of cattle; see *bos* under **cattle**
L. *grex, gregis*, m. flock, herd; *gregalis; gregarius*, of the herd, common; *grego, -atus*, gather in a herd, collect, assemble, *aggregatus*, collected, clustered; *congregabilis*, easily brought together, social; *congregalis*, uniting together; *congregus*, united in flocks: gregarious, congregate, segregate, aggregation, egregious, gregal, gregarian, gregaritic, gregarine(*Monocystis agilis*), *Gregarina*

rigida (a sporozoan), *Schistocerca gregaria* (desert-locust), *Coenothele gregalis* (a spider), *Cymatogaster aggregatus* (sparada).

Gr. *hesmos*, m. swarm, flock; *aphesmos*, m. swarm of bees:

Gr. *nomeuma*, *-tos*, n. flock, herd:

Gr. *Pan*, *-os*, m. god of hills, flocks, herds; *Paniskos*, m. dim.; *panikos*, of Pan: panic, tragopan, *Ateles paniscus* (a spider-monkey).

L. *pecu*, *-us*, n. herd, flock; see *pecus* under **cattle**

Gr. *poimne*, f. flock; *poimen*, *-os*, m. shepherd; *poimenikos*, of a shepherd: *Poemenorthrus cinereus* (a beetle), *Corymbites(Poemnites) aeratus* (a beetle).

L. *turba*, crowd, throng, troop; see **turba**

L. *turma*, f. troop, squadron; *turmalis*, of a troop:

See: **number, heap, abundance**

here < AS. *her;* see **now, life**

heredity < L. *hereditas*, inheritance; *heres*, *-edis*, heir; see **receive**

herma, Gr. prop, stay, ballast, sunken rock, reef, mound; *hermax*, heap of stones; see **stone**

hermaphroditos, Gr. a person having both male and female characters; see **mix**

hermeneutes, Gr. interpreter; see *hermeneuo* under **clear**

Hermes, Gr. messenger of the gods; see **mercury**

hermis, Gr. bedpost; see **pillar**

hermosa, Sp. beautiful; see **beauty**

hernia, L. rupture; *herniosus*, ruptured; see **break**

hero < Gr. *heros; heroine,* a female with heroic qualities; see **honor**

heron < OF. *hairon.*

F. *aigrette*, heron: *Egretta thula* (snowy egret), *Bellator egretta* (a fish).

L. *ardea*, f. heron; *ardeola*, f. dim.: *Ardea cinerea* (a heron), *Ardeola comata* (a heron).

NL. *botaurus*, m. a genus of bitterns: *Botaurus lentiginosus* (American bittern).

L. *butio*, *-onis*, m. bittern: bittern (*Botaurus stellaris*).

Gr. *erodios*, m. heron: *Ardea herodias* (blue heron), *Erodium moschatum* (musk-heronbill).

It. *garza*, f. heron; *garzetta*, f. dim.: *Ardea garzetta* (a heron).

L. *ibis*, *-idis* (Gr. *-idos*), f. a wading bird: *Ibis fuscata* (an ibis), *Ibidium immaculipenne* (a beetle), *Carphibis spinicollis* (an ibis), *Ibidorhyncha struthersi* (ibisbill), *Pseudibis papillosa* (an ibis).

Gr. *phoyx; poyx; poynx,* m. a kind of heron: *Mesophoyx intermedius* (a bird).

L. *platalea*, f. spoonbill: *Platalea regia* (an Australian spoonbill).

herpeton; herpestes, Gr. snake, creeping animal, < *herpo*, creep, see **snake**; L. *herpes*, *-etis*, a creeping cutaneous eruption, see **disease**

herpilla, Gr. a marine animal; see *herpeton* under **snake**

herring < AS. *haering*, < ML. *harengus: Pomolobus[Pomatolobus] pseudoharengus* (alewife).

Sp. *anchova*, f. a herringlike fish: anchovy.

L. *aphya* (Gr. *aphye*), f. probably a kind of anchovy; *membraphya*, f. a kind of anchovy: *Aphya meridionalis* (a fish), *Aphyonus mollis* (a fish).

Gr. *bembras* (*membras*), *-ados*, f. a kind of anchovy: *Bembras japonica* (a fish), *Bembradium roseum* (a fish), *Parabembras curta* (a fish).

L. *chalcis*, *-idis* (Gr. *chalkis*, *-idos*), f. a kind of herring:

L. *clupea*, f. a herringlike fish: *Clupea harengus* (common herring), *Coregonus clupeiformis* (a whitefish), clupeoid.

Gr. *engraulis*, *-idos*, f. a small fish: *Engraulis mordax* (an anchovy).

Gr. *enkrasicholos*, m. a small fish: *Engraulis encrasicholus* (anchovy).

Gr. *iktar*, m. a herringlike fish:

Gr. *lykostomos*, m. a kind of anchovy: *Lycostomus coccineus* (a beetle).

L. *saperda* (Gr. *saperdes*), m. a kind of fish from the Black Sea, a herring or sardine; *saperdion*, n. dim.: *Saperda candida* (apple-tree borer), *Saperdirhynchus priscotitillator* (a fossil beetle).

L. *sardina* (*sarda; sardene),* f. a kind of fish caught near Sardinia and pickled, pilchard: sardine, *Sardinia anchovia* (a sardine), *Sardinella sagax* (a fish), *Gymnosarda alliterata* (a tunny), *Hubbsiella sardina* (a grunion).

Gr. *thrassa* (*thratta*), f. a small, herringlike fish; *thrattidion*, n. dim.:

Gr. *thrissa* (*thritta*), f. a kind of anchovy: *Thrissops exigua* (a Cretaceous fish), *Ctenothrissa vexillifera* (a Cretaceous fish).

Gr. *trichis*, *-idos*, f. a kind of anchovy: *Trichis maculata* (a fish).

herse, Gr. dew; see **drop**

hesis, Gr. a sending forth, see *hiemi* under **send**; delight, see *hedomai* under **joy**

hesitate < L. *haesito, -atus,* stick fast, remain fixed, halt, be undecided; see **stand, falter, hold**

hesmos, Gr. swarm, flock; see **herd**

hesperis, L. a genus of the mustard family; see **mustard**

hesperus, L. (Gr. *hesperos*), the evening star, evening, west; *hesperis; hesperius,* western; see **night**

hesson, Gr. less, worse; see **little,** *kakos* under **bad**

hesternus, L. of yesterday; see **day**

hestia, Gr. hearth, fireside, house; see **house**

hestico- < Gr. *hestikos,* agreeable, pleasing; see **agreeable**

hestos, Gr. glad; see *hedomai* under **joy**

hesychastes, Gr. hermit; see **alone**

hesychos, Gr. still, quiet, at ease; see **rest**

hetaero- < Gr. *hetairos,* companion, comrade; see **companion**

heteros, Gr. other, different; *heteroklitos,* deviating, abnormal; see **different**

hetoemo- < Gr. *hetoimos,* ready; see **fit**

heuresi- < Gr. *heuresis,* discovery, invention; *heuretes,* discoverer; see *heurisko* under **find**

hevea, NL. a genus of the spurge family; see **spurge**

hexa- < Gr. *hex,* six; *hexas, -ados,* the number six; see **six**

hexis, Gr. possession, condition, state, habit; see *echo* under **hold**

hiameno- < Gr. *heiamene,* meadow; see **field**

hiatus, L. opening, aperture, gap; see *hio* under **open**

Hibernia, L. Ireland; see *Iras* under **Celt**

hibernum, L. winter; *hibernus,* of winter; *hibernalis,* wintry; see **year**

hibiscus, L. (Gr. *hibiskos*), mallow; see **mallow**

hicano- < Gr. *hikanos,* able, competent, sufficient, fit, see **know**; < *hikano,* arrive at, reach, see **come**

hiccup (hiccough) < uncertain origin; probably imitative.
  Gr. *lynx, -ngos,* f.; *lygmos,* m. hiccup; *alalynx, -ngos,* f. a gulping; *luzo,* gulp:
  L. *singultus, -us,* m. hiccup, < *singulto, -atus,* sob, gulp, hiccup: singultus, singultous, singultient.

hiceto- < Gr. *hiketes,* suppliant, fugitive; *hikesia,* prayer, entreaty; see *hikesia* under **ask**

hickory < Ab.Am. *pawcohiccora;* see *carya* under **walnut**

hide < AS. *hydan,* see **cover, secret, problem**; < *hyd,* skin, see **skin**

hidros, Gr. sweet; see **drop**

hiems, L. winter; *hiemalis,* of winter; *hiemo, -atus,* winter; see **year**

hieracium, L. (Gr. *hierakion*), hawkweed; see **composite**

hierax, -akos, Gr. hawk; *hierakiskos,* dim.; see **hawk**

hieros, Gr. sacred, priestly; *hieratikos,* of sacred matters; see **holy**

high < AS. *heah:* higher, highest, highland, Highgate, highness, highly, highway, heyday, height.
  L. *abruptus,* precipitous, steep; see *rumpo* under **break**
  Gr. *aersi-,* in the air, high, < *airo,* lift up, raise; *aersiporos,* going up; *metarsios,* high up:
  Gr. *aigilips, -ipos,* steep, sheer:
  Gr. *aipys,* tall, high, sheer, steep; *aipos,* n. height: *Aepyornis maximus* (an extinct bird).
  Gr. *akme,* highest point; *akmaios,* in full bloom, prime; see **top**
  Gr. *akrotenes,* stretching high:
  L. *alpinus,* at high altitudes, < *Alpis,* Alp; see **mountain**
  L. *altus,* high; *altitudo, -inis,* f. height; *praealtus,* very high: altitude, alto, Alta-mont, Palo Alto, contralto, exalt, haughty, Terre Haute, *Begonia altissima* (a begonia), *Phenacodon alticuspis* (a fossil mammal), *Delphinium exaltatum* (tall larkspur), *Hieracium praealtum* (a hawkweed), *Alticamelus procerus* (a Miocene camel).
  Gr. *anastema, -tos,* n. height, tallness:
  Gr. *anatonos,* stretching upward:
  Gr. *apopsis,* f. a lofty spot or eminence that gives a commanding view:
  Gr. *aporrhox, -ogos,* broken off, abrupt, sheer, precipitous; see *rhegnymi* under **break**

Gr. *apotomos,* cut off, abrupt, precipitous; see *temno* under **cut**

L. *arrectus,* steep, upright; see *erectus* under **upright**

Gr. *blothros,* tall, stately: *Blothrophyllum decorticatum* (a fossil coral), *Ideoblothrus similis* (an arachnid).

L. *celsus,* high, upright, lofty; *excelsus,* high, lofty, distinguished; *excelsior,* higher: excel, excellent, *Excelsior,* excellency, *Fraxinus excelsior* (European ash), *Bertholletia excelsa* (brazilnut).

L. *editus,* high, lofty: *Turritella edita* (a gastropod).

L. *elatus,* high, lofty, exalted; *elatior,* m., f.; *elatius,* n. taller; *elatio, -onis,* f. exaltation, elevation: elate, elation, elative, *Aralia elata* (Japanese aralia), *Primula elatior* (oxlip-primrose), *Arrhenatherum elatius* (a grass).

Gr. *elibatos,* high, steep; *katabatos,* descending, steep:

L. *eminens, -entis,* projecting, high; see **projection**

Gr. *eumekes,* of good length, tall; see *mekos* under **long**

Gr. *hypsi,* on high; *hypsos,* high; *hypselos,* high; *hypsiteros,* higher; *hypsistos,* highest; hypsometer, hypsodont, hypsophobia, *Hypsignathus monstrosus* (hammer-headed bat), *Hypsilophodon foxi* (a dinosaur), *Hypselocomus recurvus* (a fossil gastropod), *Ticholeptus hypsodus* (an oreodon).

Gr. *kremnos,* precipice, crag, overhanging wall or bank; *apokremnos,* precipitous, craggy; see **cliff**

Gr. *meteoros,* high in the air, raised, suspended; see **meteor**

L. *orthius* (Gr. *orthios*), high, lofty, steep: *Orthiostola lyroda* (a moth).

L. *praeceps, -cipitis,* headlong, steep; a steep place; see **cliff**

L. *procerus,* tall, slender, long: *Rubus procerus* (Himalaya berry), *Ulmus procera* (English elm).

L. *sublimis,* uplifted, exalted, lofty, distinguished; *sublimo, -atus,* elevate, lift up: sublime, sublimity, sublimation.

L. *suggestus, -us,* m. raised place, platform, elevation:

L. *superus,* high; *superior,* higher; *supremus; summus,* highest; *supernus,* on high, celestial; see *super* under **over**

See: **top, heaven, over, upright, cliff, honor, proud**

hilaos (hileos), Gr. kind, gracious; see **friend**

hilaris; hilarus, L. (Gr. *hilaros*), gay, cheerful; see **joy**

hilasmos, Gr. atonement; *hilastes,* propitiator; see *hilaskomai* under **atone**

hill < AS. *hyll;* see **mountain, heap**

hilla, L. small intestine, sausage; see *hira* under **intestine**

hilum, L. bit, scar, trifle; see **mark**

himanto- < Gr. *himas, -antos,* leather strap, thong, leash; *himantidion; himantion,* dim.; see **strap**

himantopus, L. (Gr. *himantopous*), a stilt; see **snipe**

himasthle, Gr. whip; see **whip**

himato- < Gr. *hima* (*heima*), *-tos,* dress, garment, carpet; *himation,* dim.; see **garment**

himeros, Gr. yearning for, desiring; *himertos,* desired, longed for, lovely; see **beauty**

himonia, Gr. rope of a well; see **rope**

hinder < AS. *hindrian.*

Gr. *antikope,* f. resistance:

Gr. *aperomeus,* m. thwarter:

Gr. *ascholema, -tos,* n. hindrance:

L. *capistrum,* muzzle; see **bind**

OF. *combre,* hinder, obstruct: cumber, encumbrance, cumbrous, cumbersome.

Gr. *dysergema, -tos,* n. difficulty, hindrance:

Gr. *echma, -tos,* n. hindrance, holdfast, obstacle: *Echmatophorus pascoei* (a beetle), *Echmatemys septaria* (a fossil reptile).

Gr. *empodios,* at one's feet, in the way, impeding, hindering; *empodisma, -tos,* n. hindrance: *Empodius empodius* (a worm), *Gigantorhynchus empodius* (a worm).

Gr. *enantioma,* obstacle, hindrance; see *anti* under **against**

L. *encomma, -tis* (Gr. *enkomma, -tos*), n. hindrance, incision, mark:

L. *frustror, -atus,* disappoint, prevent, thwart, circumvent, deceive, trick: frustrate, frustration.

L. *impedio, -itus,* hinder, prevent, obstruct: impede, impediment.

Gr. *kateirgo,* hem in, hinder:

Gr. *kolyma, -tos,* n. *kolysis,* f. hindrance, impediment; *kolyter, -os,* m. hinderer; *kolytikos,* checking, hindering, preventing; *kolyterios,* preventive: colytic.

L. *obstaculum,* n. hindrance: *obsto, -atus,* oppose, resist, hinder, obstruct; *obstino, -atus,* be firmly set, inflexible: obstacle, obstinate.

L. *obstrigillo, -atus,* hinder, oppose, impede:

L. *obstruo, -uctus*, barricade, hinder, impede: obstruct, obstruction, obstructionist.
L. *offendiculum*, n. stumbling-block, obstacle, hindrance:
L. *officio, -ectus*, hinder, obstruct, oppose:
Gr. *parapodizo*, entangle the feet, hinder:
Gr. *pyssachos*, muzzle to prevent calves from sucking; see **bind**
L. *retardo, -atus*, hinder, delay; see *tardus* under **slow**
Gr. *skolon*, n. stumbling-block, hindrance:
L. *sufflamen, -inis*, brake, clog, hindrance, impediment; see **hold**
See: **against, slow, delay, bind, bar, close, difficult**

hinge < the root of *hang*.
L. *cardo, -inis*, m. hinge, that on which something turns or depends; *cardinalis*, of a hinge, principal, chief, red (by transfer of meaning); *cardinatus*, hinged: cardinal, incardinate, *Richmondena cardinalis* (cardinal), *Otiorhynchus cardiniger* (a beetle).
Gr. *ginglymos*, m. hinge joint: ginglymoid, *Ginglymostoma cirratum* (gata).
Gr. *stropheus*, vertebra, socket; see *streptos* under **turn**
Gr. *thairos*, m. hinge of a door or gate, axle of a chariot: *Thaerodontis tessellata* (a fish), *Schizothaerus nuttalli* (horse-clam), *Sternotherus odoratus* (musk-turtle).
See: **axis, turn**

hinnio, -itus, L. neigh; see **neigh**

hinnuleus, L. young stag; *hinnulea*, young hind; see **deer**

hinnus, L. mule; *hinnulus*, dim.; see **horse**

hint < AS. *hentan;* see **help**

hip < AS. *hype*.
L. *clunis* (Gr. *klonis*), haunch, buttocks; *klonion*, hip; see **rump**
L. *coxa; coxendix, -icis*, f. hip, hipbone: coxitis, coxalgia, coxopodite, *Canthyloscelis nigricoxa* (a fly), *Odontium*(now *Bembicidium*) *coxendix* (a beetle).
Gr. *ischion*, n. hip.; *ischiadikos*, f. pain in the hip: ischium, ischial, sciatica, *Ischiodontus soleatus* (a beetle), *Trachischium fuscum* (a snake).
See: **side, groin**

hippardion, Gr. probably the giraffe; see **giraffe**

hippeus; hippeutes, Gr. horseman, rider, driver; see *hippeuo* under **ride**

hippo- < Gr. *hippos*, horse; *hipparion; hippidion*, dim. pony; see **horse**

hippophaës, Gr. a spiny plant, now a genus of the oleaster family; see **oleaster**

hippopotamus, L. (Gr. *hippopotamos*), river-horse; see *hippos* under **horse**

hippuris, L. (Gr. *hippouris*), horsetail, a genus of the water-milfoil family; see **water-milfoil**

hira, L. gut, intestine; *hilla*, dim.; see **intestine**

hircto- < Gr. *heirkte*, prison; *heirgmos*, cage, prison; see *ergastulum* under **pen**

hircus, L. goat; *hirculus*, dim.; *hircinus*, of goats; see **goat**

hire < AS. *hyrian*, engage the services of; see **trade**

hirnea, L. jug; *hirnula*, dim.; see **pot**

hirpex, L. rake, harrow; see **rake**

hirritus, L. the snarl of dogs; see **bark**

hirsutus, L. hairy, rough, shaggy; see **hair**

hirtus, L. shaggy, rough, hairy, prickly, uncultivated; see **hair**

hirudo, L. leech; see **leech**

hirundo, L. swallow; *hirundineus*, of swallows; see **swallow**

hisco, L. open, gape; see *hio* under **open**

hisma, Gr. foundation, seat; see **sit**

hispidus, L. hairy, bristly, rough; see **hair**

hiss < imitative origin; see **whistle**

hister, L. actor; see *histrio* under **play**

histio- < Gr. *histion*, sail; see **sail**

historia, L., Gr. information obtained by inquiry, report, narrative, < *histor*, informed, acquainted; see **story, know**

histos, Gr. the upright web-beam of a loom, web, tissue; *histion*, dim.; see **weave**

histrio, L. actor; see **play**

hiulcus, L. gaping, split, open; see *hio* under **open**

hive < AS. *hyf.*
L. *alvearium,* n. beehive, apiary: alveary.
Gr. *anthrenion,* wasp-nest; see **nest**
L. *apiarium,* n. beehive: *Liodora apiarium* (a wasp).
L. *examen,* swarm of bees, throng, crowd; see **number**
Gr. *hesmos,* swarm of bees; see **herd**
Gr. *hyron,* n. beehive:
Gr. *keron, -os,* m. beehive:
Gr. *kypsele,* any hollow vessel, box, chest, beehive; *kypselion,* dim.; see **box**
Gr. *melissophatne,* f. beehive:
L. *mellarium,* n. beehive:
Gr. *simblos,* m.; *simblon,* n. beehive; *simblios,* of a hive: *Simblum sphaerocephalum* (a fungus).
Gr. *smenos,* n. beehive, swarm of bees: *Smenodoca erebenna* (a moth).
Gr. *sphekia,* wasp-nest; *sphekion,* cell in a wasp-nest; see **nest**
See: **house, nest, box, herd, number**

hizema, Gr. a settling down, sinking; see *hisma* under **sit**

hoard < AS. *hord;* see **store**

hoarse < AS. *has.*
Gr. *bechia,* hoarseness; see *bex* under **cough**
Gr. *branchos,* m. hoarseness; *branchaleos,* hoarse:
Gr. *kerchnos,* m. hoarseness; *kerchaleos,* dry, hoarse, rough:
L. *raucus,* hoarse: raucous, *Otiorhynchus raucus* (a beetle).
L. *ravus,* hoarse; *ravis,* f. hoarseness:
See: **rough**

hodegos, Gr. guide, teacher; see **teach**

hodiernus, L. of today; see *dies* under **day**

hodo- < Gr. *hodos,* way, road, journey; *hodios,* pertaining to a way; *hodotos,* passable, feasible; see **way**

hoe < OHG. *houwa* (G. *haue*).
Gr. *dikella,* f. two-pronged hoe, mattock: *Dicellocephalus minnesotensis* (a trilobite), *Dicellocerus vibrans* (a chalcid), *Dicellograptus divaricatus* (a graptolite).
L. *ligo, -onis,* grubbing-hoe, mattock; see **pick**
L. *marra,* f. hoe, weeding-hook:
L. *pastinum,* n. two-pronged instrument for digging and trenching ground; *pastino, -atus,* prepare the ground:
L. *rastrum,* n. hoe, mattock, rake; *rastellum,* n. dim.; *rastrarius,* of a hoe: rastrum, rastellum.
L. *runco, -onis,* hoe; see *runco* under **till**
L. *sarculum,* n. hoe; *sarculo, -atus,* hoe:
L. *sarritor,* hoer, weeder; see **till**
Gr. *skalis, -idos,* f. hoe, mattock; *skalidion,* n. dim.; *skaleus,* m. hoer; *skaleuthron,* n. hoe, oven-rake, poker; *skalsis,* f. a hoeing, < *skallo,* stir up, hoe: *Scalidognathus seticeps* (a spider), *Scalidium hilarei* (a beetle).
Gr. *skapane,* a digging tool, spade, hoe, mattock; see **shovel**
Gr. *sminye,* f. two-pronged hoe, mattock; *sminydion,* n. dim.: *Sminyothrips biuncinata* (a thysanopterid).
See: **till, rake**

hog < AS. *hogg.*
L. *aper, apri,* m. wild boar; *apra,* f. wild sow; *apriculus,* m. dim.; *aprarius; aprinus; aprugnus; aprugineus,* of wild hogs: *Capros[Caprus] aper* (boarfish), *Hyocephalus aprugnus* (a bug), *Haematopinus apri* (boar-louse):
Sp. *cerdo,* m. hog, pig; *cerdillo,* m. dim.:
Gr. *choiros,* m. pig; *choiridion; choirion,* n. dim.; *choireios; choirikos; choirinos,* of swine; *choiras, -ados,* like a hog's back: cherodian, *Choeropus castanotis* (a bandicoot), *Choeradoplana spatulata* (a turbellarian), *Potamochoerus porcus* (red river-hog), *Centrobranchus choerocephalus* (a fish), *Choeridium squalidum* (a beetle).
Gr. *delphax, -akos,* f. sucking pig, young pig, porker; *delphakion,* n. dim.: *Delphax elegantula* (a bug), *Delphacinus mesomelas* (a bug), *Delphaciognathus paucidens* (a fossil reptile), *Dictyophorodelphax usingeri* (a bug).
Gr. *errhaos,* m. wild boar:
AS. *fearh,* little pig: farrow.
Gr. *gromphas, -ados,* f. an old sow: *Gromphadorhina portentosa* (a cockroach).
Gr. *grylos,* m. pig; *grylion,* n. dim.: cherogryl (*Procavia sinaitica*).
Gr. *hapalias,* m. a sucking pig: *Hapalioloemus machaeralis* (a fly).
Gr. *hys, hyos,* m.; *hyaina,* f. hog: *hyidion,* n. dim.; *hyeios,* of swine; *hyenos,* swinish: *Hyaena striata* (striped hyena), *Hyoscyamus niger* (black henbane),

Hyomeryx breviceps (a fossil mammal), *Hyorhynchus platyceps* (a fossil reptile).

Gr. *kapros*, m. boar, wild boar; *kapraina*, f. wild sow; *kapridion*, n.; *kapriskos*, m. dim.; *kaprios*, like a boar: *Capromys prehensilis* (hutia), *Capromimus abbreviatus* (a fish), *Percina caprodes* (a darter).

L. *lardum*, fat of bacon, lard; see **fat**

L. *majalis*, a gelded boar, barrow; see **sterile**

Gr. *molobrion*, n. piglet of wild swine:

L. *orthagoriscus* (Gr. *orthagoriskos*), m. a sucking pig: *Orthagoriscus hispidus* (a sunfish).

Sp. *pecari*, m. a piglike animal, < an Ab.Am. word: *Pecari angulatus* (collared peccary), *Tayassu pecari* (white-lipped peccary).

L. *porcus*, m. hog, pig, swine; *porca*, f. sow; *porcellus; porculus*, m.; *porcula*, f. dim.; Sp. *puerco*, m. hog; *porcarius; porcellinus; porcinus*, of swine; *porcina*, f. pork: pork, porker, porcupine, porcelain, porcellanite, Purcell, Rio Puerco, *Scorpaena porcus* (a hogfish), *Papio porcarius* (chacma-baboon), *Hylastes porculus* (a beetle), *Cavia porcellus* (a guinea-pig).

Gr. *ptelas*, m. wild boar:

L. *scrofa*, f. breeding-sow; *scrofula*, f. dim.; *scrofinus*, of sows: scrofula, scrofulous, *Sus scrofa* (hog), *Scrophularia nodosa* (a figwort), *Scorpaena scrofa* (a hogfish).

Gr. *sialos*, a fat hog, fat; see **fat**

Gr. *sigras*, m. a wild pig:

L. *sus, suis*, c. (Gr. *sys, syos*, m.; *syaina*, f.), hog, pig; *suculus*, m.; *sucula*, f. dim.; *suarius; suillus; suinus*, of swine, swinish; *suilla*, f. pork; Gr. *syagros*, m. hunter of wild boars, wild boar; *sybax, -akos*, swinish; *sybotes*, m. swineherd: suine, swine, suilline, suiform, suoid, *Hemisus marmoratum* (a frog), *Haematopinus suis* (hog-louse), *Sybacodes lutulenta* (a beetle), *Sus jubatus* (a hog).

L. *verres, -is*, m. boar; *verrinus*, of boars: *Verriculus sanguineus* (a fish).

holcado- < Gr. *holkas, -ados*, a towed ship; see **ship**

holcimo- < Gr. *holkimos*, ductile; *holke*, a drawing, towing; see *helko* under **draw**

holcio- < Gr. *holkeion (holkion)*, rudder; large bowl or basin; see **rudder, basin**

holco- < Gr. *holkos*, an instrument for hauling or towing, rein, strap, the effect of hauling, track, attractive; see *helko* under **draw**

holcus, L. a kind of grain; see **grain**

hold < AS. *haldan*: holding, holder, holdfast, household, behold, stronghold, withhold, uphold, upholsterer.

Gr. *abolos*, uncast, unshed (said of a foal with milk teeth): *Abolodoria abdominalis* (a fly).

L. *addictus*, given to, devoted to, habituated: addict, addicted, addiction.

Gr. *agogeus; aktor*, leader, leash; see *ago* under **lead**

Gr. *anteris, -idos*, prop, buttress, support; see **pillar**

L. *attentus*, intent on, applied to, directed toward; see **heed, busy**

Suf. *-atus, -itus, -utus, -tus*, provided with; see **nature**

L. *aurea*, f. bridle:

L. *capax, -acis*, that can hold much; see *capio* under **take**

L. *capistrum*, halter, muzzle; see **bind**

Gr. *chalinos*, m. bridle, bit; *chalinarion*, n. dim.: *Chalinocerus longicornis* (a wasp).

L. *confiteor, -fessus*, acknowledge, own, admit, avow, < *fateor, fassus*, confess, own, grant: confession, confessor.

L. *confuto, -atus*, suppress, restrain, check, disprove: confute, confutation.

Gr. *dochos*, able to contain or hold; *docheion*, n.; *dochos*, m. holder, receptacle: choledoch, elaeodochon, sporodochium, *Dochephra bullata* (a moth).

Gr. *echma*, hindrance, holdfast; see **hinder**

Gr. *echo; ischo*, have, hold, restrain; *anecho*, hold back; *anoche*, f. stop, pause, armistice; *anokoche*, f. stay, hindrance; *antecho*, hold against; *aphexis*, f. abstinence; *echma; ochma, -tos*, n. holdfast, stay, tie, band; *enochos*, held, bound, liable; *hektikos*, habitual; *hektor, -os*, holding fast; *hexis*, f. possession, condition, habit; *kathektos*, held back, restrained; *katoche*, f. possession, catalepsy; *ochos*, holding, securing; *schesis*, f. a holding, checking, retaining; *schetikos*, retentive; *synechos*, holding together, continuous; *synichos*, joined or held together: entelechy, Hector, eunuch, cachexia, ochopetalous, cathexis, hexiology(hexicology), hectic, ephectic, ischemia, ischuria, *Echeneis naucrates* (remora), *Trichechus latirostris* (Florida manatee), *Chrysochus auratus* (a beetle), *Stylochus ellipticus* (a worm), *Ochmacanthus flabellifer* (a fish), *Rhinochetus jubatus* (kagu), *Diochus antennatus* (a beetle), *Synechostictus decoratus* (a beetle), *Tympanuchus americanus* (prairie-chicken).

Gr. *enkrates*, with a firm hold, having the mastery, self-disciplined: *Encrateola subimpressa* (a wasp).

Gr. *enstomisma, -tos,* n. bit, curb:

Gr. *eukratos,* well-tempered, temperate, mild; see **mild**

Gr. *euleron,* n. rein:

L. *frenum,* n. bridle, rein, bit; *frenulum,* n. dim.; *freno, -atus,* bridle, curb, restrain; *infreno, -atus,* put on a bridle; *infrenis; infrenus,* unbridled, untamed: frenulum, refrain, *Apogon frenatus* (coral-fish).

L. *glutinosus;* see **glue**

L. *habeo, -bitus,* have, hold, keep; *cohibeo, -bitus,* hold together, contain, confine, limit; *inhibeo, -bitus,* restrain, curb, check: habit, habitual, habiliment, habitation, habeas corpus, inhabit, exhibition, prohibitive, able, ability, debility, inhibition, habitat, rehabilitate, inhabitant, disable, enable, inability, binnacle, *Habenaria tridentata* (an orchid).

L. *haereo, haesus,* hold fast, stick, cling: adhere, adhesion, coherent, cohesive, incoherent, inherent, hesitate.

Gr. *hapto,* lay hold of, grasp, touch; see **bind**

Gr. *henion,* n. rein, bit, bridle:

Gr. *labe,* something for taking hold of or grasping; see **tool,** *lambano* under **take**

Gr. *metrios,* within measure, moderate; see *metron* under **measure**

L. *modero, -atus,* keep within bounds; see **govern**

L. *noster, -tra, -trum,* our, ours, our own: nostrum, paternoster, Nostradamus.

L. *obses, -idis,* hostage, pledge, surety; *obsidatus,* of a hostage; see **promise**

L. *orea,* f. bit:

L. *possideo, -sessus,* have and hold; *possessor, -is,* m. holder, owner: possessor, possessive, possession, dispossess.

L. *potior, -itus,* get, obtain, become master of; see **govern**

Gr. *psalion,* curb-chain, part of a bridle; see **chain**

L. *reprimo, -ressus,* check, curb, restrain: repress, repression, repressive.

Gr. *rhyter, -os,* m. drawer, rein, trace:

Gr. *strangale,* f. halter: *Strangalomorpha tenuis* (a beetle), *Strangalia sexnotata* (a beetle).

L. *stringo, strictus,* draw together, bind, hold in check; see **bind**

L. *sufflamen, -inis,* n. brake, clog, check, hindrance, impediment:

L. *teneo, tentus,* hold; *tenax, -acis,* holding firmly, steadfast; *tenor, -is,* m. a holding fast, keeping a course; *tentaculum,* n. feeler, holdfast; *abstinens,* refraining from, *abstentus,* held away; *continens,* holding together; *contentus,* held together, satisfied; *detentus,* held back; *retentus,* held back or in; *retinaculum,* n. holdfast, tether, cable, halter: tenure, tenor, tenet, tenacity, tentacle, tentillum, tenon, abstain, abstinence, appurtenance, content, continence, continuous, countenance, detention, entertainment, impertinent, incontinent, malcontent, lieutenant, obtainable, maintenance, pertinacious, retain, sustenance, untenable, *Xerophyllum tenax* (beargrass), *Stipa tenacissima* (esparto-grass).

See: **bind, trap, pen, bar, govern, hinder, handle, glue, vessel, delay, hook**

hole < AS. *hol.*

L. *abyssus* (Gr. *abyssos*), bottomless pit, sea, hell; see **deep**

L. *anus,* rectal opening; see **intestine**

L. *apertura,* f. opening, hole: aperture.

Gr. *araios,* thin, loose, porous; see **thin**

Gr. *archos,* anus; see **intestine**

L. *arrugia,* f. shaft or pit in a mine:

L. *barathrum* (Gr. *barathron; bethron*), n. pit, gulf, abyss: *Barathrobius digitatus* (a coral).

Gr. *bothros,* trench, pit, hole, hollow; see **ditch**

Gr. *bothynos,* hole, pit, trench; see **ditch**

L. *caula,* f. opening, hole, passage, sheepcote:

L. *cavus,* hole, hollow; *multicavus,* porous; see **hollow**

L. *cellulosus,* full of little cells; see *cella* under **room**

Gr. *chasma,* yawning abyss, gulf; see *chaino* under **open**

Gr. *chaunos,* porous, spongy, loose, empty, frivolous; *chaunosis,* f. a slackening, lightening: *Chaunograptus gemmatus* (a graptolite), *Chaunoderus angulosus* (a beetle), *Chauna torquata* (crested screamer, chaja).

Gr. *cheia,* f. hole in the ground:

L. *commeabilis,* permeable; see *meo* under **passage**

L. *confossus,* pierced through, full of holes:

L. *cuniculus,* underground passage, rabbit burrow; see **mine**

L. *demeaculum,* n. underground passage:

Gr. *ektresis,* f. hole:

Gr. *epikoilos,* porous, spongy:

L. *fenestra,* window; see **window**

L. *fissura,* crack, cleft, chink; see **break**

L. *fistulosus,* full of tubes, holes, porous; see *fistula* under **pipe**

L. *fodina,* pit, mine; see **mine**

L. *foramen, -inis,* n. hole, aperture, opening; *foraminosus,* full of holes: foramen, foraminifer, *Collix foraminata* (a moth).

L. *fovea,* f. pit used as a trap, pitfall; *foveola,* f. dim.: fovea, foveolate, *Anomalina foveolata* (a foraminifer).

L. *fumarolium,* n. smoke-hole: fumarole.

L. *fungosus,* full of holes, spongy: *Adlumia fungosa* (climbing fumitory.)

Gr. *goleos,* m. hole:

L. *gurges,* whirlpool, abyss; see **turn**

L. *hiatus,* opening, aperture, cleft, gap; see *hio* under **open**

Gr. *iauthmos,* sleeping-place, den, lair; see *iauo* under **open**

L. *impluvium,* n. the skylight in the roof of the atrium of a Roman house: *Crataeomus impluviatus* (a beetle).

Gr. *keuthmon,* hiding-place, hole, cave; see *keuthos* under **cover**

Gr. *kleithria,* f. keyhole; *kleithridion,* n. dim.: clithridium, clithridiate.

Gr. *kyar, -os,* m. hole, orifice, eye of a needle:

L. *lacuna,* hole, opening, cleft; see **hollow**

Gr. *ope,* f. opening, hole; *metope,* f. space between two triglyphs; *opaios,* with a hole: metope, *Opegrapha saxicola* (a lichen), *Desmometopa sordida* (a fly).

L. *orificium,* opening, < *os, oris,* mouth; see *os* under **mouth**

Gr. *orygma,* tunnel, mine; see *orysso* under **dig**

L. *pala,* socket of a ring for a jewel, bezel; see **hollow**

L. *pertusus,* perforated; see *pertundo* under **bore**

Gr. *pholeos,* m. hole, cave, den; *pholion,* n. dim.; *pholeter, -os,* m. one who lurks in a hole: *Pholeoptynx cunicularia* (a burrowing-owl), *Pholeter gastrophilus* (a worm), *Pholeoteras euthrix* (a gastropod), *Pholeogryllus geertsi* (a cricket).

L. *porta,* gate, door; see **door**

L. *porus* (Gr. *poros*), m. hole, passage, ford, ferry; *porosus,* full of holes; *brachyporos,* m. short passage: pore, porous, porosity, nullipore, madrepore, porodine, Porifera, emporium, Bosporus(Oxford), *Poronotus triacanthus* (harvest-fish), *Porichthys porosissimus* (singing-fish), *Porella pinnata* (a liverwort), *Poria incrassata* (a dryrot-fungus), *Polyporus frondosus* (a fungus), *Batopora scrobiculata* (a fossil bryozoan), *Zaporus dorsalis* (a wasp), *Porodiscus splendidus* (a diatom).

L. *pumicosus,* porous like pumice; see *pumex* under **stone**

L. *punctura,* f. hole, prick: puncture.

L. *puteus,* m. well, cistern, pit, trench; *puteolus; puticulus,* m. dim.; *putealis; puteanus,* of a well: *Puteolus drepanensis* (a gastropod), *Intrapora puteola* (a fossil bryozoan), *Nuphar puteorum* (a pondlily), *Pseudomonas citriputeale* (a bacterium).

Gr. *salambe,* f. vent, chimney, door:

Gr. *sarma, -tos,* n. chasm in the earth: *Sarmatoechinogammarus sarmatus* (an amphipod).

Gr. *serangodes,* porous; see *seranx* under **hollow**

L. *sirus* (Gr. *siros*), m. pit, pitfall, underground granary; Sp. *silo,* m.: silo, *Sironectes anguliferus* (a fossil reptile).

Gr. *somphos,* porous, spongy: *Somphocyathus coralloides* (a fossil sponge).

L. *spiraculum; spiramen,* breathing-hole, airhole, vent; see *spiro* under **breathe**

Gr. *steileie,* f. hole for the handle of an ax:

Gr. *stiktos,* pricked, punctured, spotted; see **spot**

Gr. *stoma,* mouth; see **mouth**

Gr. *tormos,* m. hole, socket; *tormion,* n. dim.: *Tormopsolus osculatus* (a trematode).

Gr. *trema, -tos,* n. hole; *tremation,* n. dim.; *tretos,* perforated; *polytretos,* full of holes, porous: trematode, Monotremata, *Pentremites obesus* (a fossil blastoid), *Tremex columba* (a horntail), *Rhynchotrema peregrina* (a brachiopod), *Trematofusus venustus* (a fossil gastropod), *Anisotremus virginicus* (porkfish), *Trematosaurus brauni* (a fossil amphibian track).

Gr. *trogle,* f. hole; *troglodytes,* m. inhabitant of holes, caves, etc.: troglodyte, *Troglodytes aedon* (house-wren), *Nannus troglodytes* (European wren), *Troglichthys rosae* (a fish).

Gr. *tryma, -tos,* n. hole; *trymation,* n. dim.; *trypa,* f.; *trypema, -tos,* n. hole; *trypemation,* n. dim.; *Diatryma gigantea* (an Eocene bird), *Trymatoderma spongiicollis* (a beetle), *Atrypa oblata* (a fossil brachiopod), *Trypematella papulifera* (a fossil bryozoan).

L. *vorago, -inis,* f. abyss, gulf, chasm, whirlpool; *voraginosus,* full of pits: voraginous.

See: **hollow, mouth, throat, door, sieve, sponge, screen, break, bore, spot, open, window, mine, grave, ditch, cut**

hollow < AS. *holh*.
L. *acetabulum,* vinegar cup, socket of the hipbone; see **cup**
L. *alveus,* m. cavity, pit, channel; *alveolus,* m. dim.; *alveatus,* hollowed out: alveolus, alveolar, alveary, *Alveolina costulata* (a foraminifer), *Sabellaria alveolata* (a worm).
Gr. *antlia (antlos),* hold of a ship, bilge-water; see **water**
L. *antrum* (Gr. *antron*), n. sinus, cavity; *antraios,* of caves; *antrodes,* full of caves: antrum, antrocele, *Antrophyum plantagineum* (a fern), *Hymenocephalus antraeus* (a fish).
L. *axilla,* armpit; see **angle**
Gr. *bothros,* trench, pit, hole, hollow; see **ditch**
L. *bucca,* cheek, cavity; see **cheek**
L. *camera,* vault with an arched roof; see **arc**
L. *caula,* hole, sheepcote; see **hole**
L. *cavus,* m. hollow, hole, < *cavo, -atus,* hollow out; *cavamen, -inis,* n.; *cavatio, -onis,* f.; *cavatura,* f. cavern, hollow, cavity; *cavea,* f. a hollow place, den, enclosure for animals, cage; *caverna,* f. cave, grotto, hole; *cavernula,* f. dim.; *cavaticus,* hollowed out; *cavaticus,* born or living in caves; *cavator, -is,* m. excavator; *cavealis,* kept in a cave or cellar; *cavernosus,* full of hollows or cavities; *cavositas, -atis,* f. hollow, cavity; *concavus,* hollowed or arched inward; *multicavus,* porous: cave, Mammoth Cave, cavern, cavity, cavitation, cage, jail, decoy, concave, excavate, *Megistocrinus concavus* (a Devonian crinoid), *Bactrognathus excavata* (a conodont), *Hymenocephalus cavernosus* (a deep-sea fish), *Obisium cavernarum* (a pseudoscorpion), *Cavernicola pilosa* (a bug), *Ptomatophagus cavernicola* (a beetle).
Gr. *cheramos,* m. hole, cleft, hollow: *Cheramus orientalis* (a crab).
L. *colyx, -ycis,* f. cavern where natron is forming:
Gr. *emmochlion,* n. socket for a bar:
Sp. *ensenada,* bay, cove; see **sea**
L. *fornix,* arch, vault; see **arc**
L. *funda,* f. bezel:
L. *gelasinus* (Gr. *gelasinos*), m. dimple:
Gr. *glaphyros,* hollowed out, carved, smoothed, polished, finished, < *glapho,* hollow out: *Glaphyrocystis compressa* (a cystoid), *Glaphyrophlebia pusilla* (a fossil insect-wing).
Gr. *glene,* eye socket, socket of a joint; see **eye**
Gr. *gnythos,* n. cave, pit, hollow:
Gr. *grone,* f. cavern, any hollow vessel; *gronos,* hollowed out, cavernous: *Gronotherium decrepitum* (a fossil mammal).
It. *grotta,* f. cave, vault: grotto, grotesque.
Gr. *gyalon,* n. a hollow; *gyalos,* hollow: *Gyalocephalus capitatus* (a nematode), *Gyaloceras smithi* (a Cretaceous cephalopod).
Gr. *keneon,* flank, any hollow; see **side**
Gr. *keuthmon,* hiding-place, hole, corner, cave; see *keuthos* under **cover**
Gr. *koilos,* hollow; *koilas, -ados,* f.; *koiloma, -tos,* n.; *koilon,* n. cavity, hollow; *koilia,* f. large cavity of the body, belly; *koilidion,* n. dim.; *koiliakos,* pertaining to the belly; *koilotes, -etos,* f. hollowness; *koilodes,* with cavities: coelome, coelenterate, enterocele, Coelacanthidae, coeliac, *Coelogyne ocellata* (an orchid), *Coelodendron cervicorne* (a radiolarian), *Coelococcus amicarum* (a palm), *Mesocoelopus niger* (a beetle), *Odocoileus hemionus* (mule-deer).
Gr. *kolpos,* bosom, womb, any bosomlike hollow; see **womb**
Gr. *krypte,* f. vault, cavern: crypt.
Gr. *kymbos,* hollow; *kymbe,* bowl; see *cymba* under **cup**
Gr. *kyphellon,* n. hollow of the ear:
Gr. *kypsele,* any hollow vessel, box, chest; see **box**
Gr. *kytos,* any hollow place or container; see **room**
L. *lacuna,* f. cavity, hollow, pool, place where water collects; *lacunosus,* full of hollows, cavities, gaps, sloughs; *lacunar; laquear, -is,* n. a fluted or paneled ceiling, sunken panel; *lacunatus; laqueatus,* fluted, paneled; *lacunarius,* m. gravedigger; *lacuno, -atus,* hollow out, pit: lacuna, lacunose, *Lacuna variegata* (a chink-shell), *Mytilus lacunatus* (a pelecypod), *Pinna laqueata* (a fossil pelecypod), *Nymphoides lacunosum* (a floating-heart), *Pachylocrinus lacunosus* (a Mississippian crinoid).
Gr. *maschale,* armpit, hollow, bay; see **angle**
It. *miarolo,* name for the Baveno granite that contains drusy cavities lined with crystals: miarolite, miarolitic.
Gr. *mychos,* inmost part, recess, closet, bay, alcove, sinus; see **room**
L. *pala,* f. socket of a ring for a jewel, bezel:
Gr. *phatnoma, -tos,* n. lacunar, sunken panel of a framed structure: *Phatnoma laciniata* (a bug), *Phatnomatorhina platycephala* (an amphibian).

Gr. *pholeos,* hole, cave, den; *pholeter,* on who lurks in a hole; see **hole**
Gr. *pyelis, -idos,* setting or socket of a ring; see *pyelos* under **basin**
L. *recessus,* m. corner, nook, retreat: recess.
Gr. *seranx, -angos,* f. cave; *serangion,* n. dim.; *serangodes,* full of caverns, porous: *Serangium mysticum* (a beetle), *Serangodes strongylosomoides* (a milleped).
L. *sinus,* m. curve, fold, hollow, pocket, recess, bay: sinus, sinusoid, sine, sinusitis. See *sinuo* under **bend**
Gr. *skaphe,* f.; *skaphos,* m. anything hollowed out, scooped: scaphoid, scaphopod, *Scaphognathus stejnegeri* (a fish).   See: **cup, ship, basin, shovel, dig.**
L. *specus, -us,* m. cave, den:
L. *speleum* (Gr. *spelaion*), n.; *spelunca* (Gr. *spelynx, -yngos*), f.; *speos,* n. cave, cavern, den, grotto: spelean, speleology, *Ursus spelaeus* (a cave-bear), *Spelerpes ruber* (a salamander), *Speomyia absoloni* (a fly), *Speonomus troglodytes* (a beetle), *Speotyto cunicularia* (a burrowing-owl), *Speoxenus cundalli* (a fossil mammal).
See: **hole, room, belly, pipe, basin, funnel, valley, cup, spoon, vessel, arc, bag, lake**

holly < AS. *holen:* holm.
L. *aquifolium,* n. holly, scarlet holm, < *aquifolius,* having pointed leaves: Aquifoliaceae, *Ilex aquifolium* (European holly), *Mahonia aquifolium* (Oregon grape).
Ab.Am. *cassine,* black drink derived from yaupon *(Ilex vomitoria),* a species of the holly family: *Ilex cassine* (dahoon-holly), *Viburnum cassinoides* (witherod), *Cassinopsis ciliata* (an icacinacead).
L. *ilex, -icis,* f. holm-oak; a genus of the holly family: *Ilex glabra* (inkberry), *Quercus ilex* (holly-oak), *Quercus ilicifolia* (scrub-oak), *Lecanium ilicis* (kermes-insect).

holmos, Gr. mortar; *holmiskos,* dim.; see **mortar**

holo- < Gr. *holos,* whole, entire, all; *holikos,* universal, general; see **all**

holocyrum, L. a kind of mint; see **mint**

holothurion, Gr. a sea-cucumber; see **echinoderm**

holy < AS. *halig:* holiness, unholy, hallow, halibut, halidom, holiday, hollyhock, Halloween.
Gr. *abebelos,* sacred, inviolable:
Gr. *eusebes,* holy, sacred, religious; *eusebeia,* f. piety, reverence: Eusebius.
Gr. *hagios,* sacred, holy; *hagiosyne,* f. holiness: hagiolatry, Hagiographa.
Gr. *hagnos,* pure, innocent, sacred; see **pure**
Gr. *hieros,* holy, sacred, priestly; *hieratikos,* priestly, devoted to sacred use; *hiereus,* m. priest; *hiereia; hieris, -idos,* f. priestess; *hierosynos,* priestly: hieroglyphic, hieratic, hierarchy, hierophant, Jerome(Hieronymus), *Hierochloë odorata* (holygrass), *Hierocrobyla othopyrrha* (a moth), *Hieremys annandalei* (a turtle), *Vriesia hieroglyphica* (a bromeliad).
Gr. *hosios,* holy; *hosia,* f. sacred rite, worship; *hosioma, -tos,* n. a pious act: *Hosia spinulosa* (a starfish).
L. *inviolabilis,* hallowed: inviolable.
L. *pius,* devout; see **rite**
L. *sacer, -cra, -crum,* holy; *sacerdos, -otis,* c. priest, priestess; *sacrum,* n. holy thing; *sacellum,* n. dim.; *sacro, -atus,* set apart as holy; *consecro, -atus,* dedicate to a holy purpose: sacred, sacrifice, sacrilege, sacrament, sacristan, Sacramento, sacrum, sacrosanct, sacellum, sacroiliac, sacerdotal, sexton, consecrate, desecration, execrable, cascara sagrada(bark of *Rhamnus purshiana*), *Harpacticus sacer* (a copepod), *Pleurotoma sacra* (a gastropod), *Artemisia sacrorum* (Russian wormwood).
L. *sancio, sanctus,* make holy; *sanctus,* holy, sacred, inviolable; *sanctum,* n. a holy place: sanctify, sanctuary, sanction, sanctimonious, saint, sanctum sanctorum, samphire, *Halcyon sancta* (sacred kingfisher), *Guaicum sanctum* (lignum-vitae).
Gr. *semnos,* revered, holy, solemn; see **serious**
Gr. *zatheos,* sacred, sainted; *egatheos,* most holy: *Egatheus falcinellus* (a bird).
See: **temple, priest, rite, honor, spirit, serious**

homados, Gr. noise, din; see **sound**

homalos, Gr. even, level, equal; see **level,** *homos* under **equal**

homarus, NL. lobster; see *cammarus* under **crab**

homaulos, Gr. living together; see **life**

home < AS. *ham;* see **house**

homelix, Gr. of the same age; see *helikia* under **age**

homelys, -*ydos,* Gr. companion; see **companion**

homeo- < Gr. *homoios*, like, resembling; see *homos* under **equal**

homeros, Gr. pledge, security, hostage; see **promise**

homichlo- < Gr. *homichle*, mist, fog, steam; *homichlodes*, foggy; see **cloud**

homiletico- < Gr. *homiletikos*, affable, conversable; *homilia*, conversation; *homiletes*, disciple, scholar; see **speak, learn**

homilos, Gr. crowd, throng; see **number**

homo, *-inis*, L. man; *homullus; homunculus; homunicio*, dim.; see **man**

homo- < Gr. *homos*, same, uniform, similar; see **equal**

homologos, Gr. agreeing, corresponding; see *homos* under **equal**

homoros, Gr. having the same borders, neighboring; see **neighbor**

homunculus, L. dim. of *homo*, man; see *homo* under **man**

honest < L. *honestus*, honored, distinguished, noble, upright, truthful; see **true, honor**

honey < AS. *hunig*.
> L. *Hybla*, f. ancient region of Sicily famous for its flowers and honey: Hyblean.
> Gr. *Hymettos*, Mount Hymettus, near Athens, famous for its marble and honey; see **mountain**
> Skt. *madhu*, honey; sweet; *madhuka*, a tree with honey-bearing flowers: *Madhuca butyracea* (illipe or butter-tree).
> L. *mel*, *mellis* (Gr. *meli*, *-tos*), n. honey; *melinus; melitinus; mellarius; melleus; mellitus; mellosus*, of honey, honey-colored; *melliculus; mellitulus*, honeyed, darling; *melliculum*, n. honeykin; *mellifer*, *-a*, *-um*, honey-bearing; *mellificus*, honey-making; *mellifluus*, flowing with honey, honeyed; Gr. *melichros*, honey-sweet; *melichroos*, honey-colored; *melitodes*, like honey: mellifluous, meliceris, melrose, melilot, Malta, melitose, mellite, melilite, diabetes mellitus, melitemia, melituria, marmalade, molasses, melichrous, *Mellivora ratel* (honey-badger, ratel), *Meliosma cuneifolia* (a sabiacead), *Melicocca bijuga* (genip-tree), *Melitodes ochracea* (a gorgonid coral), *Meliphaga penicillata* (a honey-eater), *Armillaria mellea* (honey-agaric), *Salvia mellifera* (a sage), *Myrmecocystus melliger* (honey-ant).
> L. *mulsus; mulseus*, mixed or sweetened with honey; *mulsa*, f. sweetheart, honey:

honeycomb; see **wax**

honeysuckle < AS. *hunigsuce*.
> L. *caprifolium*, n. former generic name of the honeysuckles: Caprifoliaceae, *Lonicera caprifolium* (sweet honeysuckle).
> Gr. *klymenon*, n. a bindweed, honeysuckle; *periklymenon*, n. a honeysuckle: *Lonicera periclymenum* (woodbine).
> NL. *linnaea*, f. a genus of the honeysuckle family, < Carolus Linnaeus, Swedish botanist (1707-1778), establisher of binomial nomenclature: Linnean(Linnaean), linnaeite, *Linnaea borealis* (European twinflower, the favorite of Linnaeus).
> NL. *lonicera*, f. a genus of the honeysuckle family, < Adam Lonicer, German botanist: *Lonicera flava* (yellow honeysuckle), *Alphitoaphis lonicericola* (an aphid).
> NL. *symphoricarpus*, m. a genus of the honeysuckle family: *Symphoricarpus hesperius* (Washington snowberry).
> Gr. *thraupalos*, m. probably a viburnum:
> L. *viburnum*, n. a genus of the honeysuckle family: *Viburnum tomentosum* (a viburnum).

honor < L. *honor*, *-is*, m. esteem, repute, respect; *honorabilis*, estimable; *honorarius*, of honor; *honorus*, worthy; *honestus*, honored, upright, truthful; *honoro*, *-atus*, respect, decorate: honorable, honorary, honorarium, honesty, dishonest.
> L. *admirabilis*, worthy of regard, wonderful: admirable.
> L. *adorea*, f. reward of valor, glory, honor:
> L. *adoro*, *-atus*, honor, esteem, worship; *adoratio*, *-onis*, f. worship: adore, adoration.
> L. *adulor*, *-atus*, fawn upon, flatter, admire; see **flatter**
> Gr. *aeimnestos*, ever to be remembered; *aeiphatos*, ever famed; *aeithryletos*, ever talked of, celebrated:
> Gr. *agaios*, admirable, enviable: *Agaeocera scintillans* (a beetle).
> Gr. *agallo*, glorify, honor, exalt; *agalma*, *-tos*, glory, delight, honor, statue; see **form**
> Gr. *agasma*, *-tos*, n. object of adoration; *agastos (agetos)*, deserving admiration; *agauos*, noble, admirable; *agamenos*, with admiration, < *agamai*, admire: *Agastopsylla boxi* (a siphonapterid), *Agave americana* (century-plant).
> Gr. *agerochos*, honored, noble:
> Gr. *aidoios*, regarded with awe, venerable: *Aedoeophasma anglica* (a beetle).

Gr. *ainesis,* f. praise; *ainetos,* praiseworthy: *Aenetus prasinus* (a butterfly).
Gr. *amymon, -os,* noble, excellent, blameless: *Amymonus elephantinus* (a cephalopod).
L. *augustus,* venerable: august, August, St. Augustine, Augsburg, Augusta.
L. *aulicus,* princely, of a prince's court: *Aulicus viridissimus* (a beetle).
Gr. *axios,* worthy, honorable; see **worth**
L. *celebro, -atus,* honor, praise, publish; *celeber, -bris, -bre,* renowned, famous: celebrate, celebrant, celebrity.
L. *claritudo,* clearness, brightness, fame, splendor; see **clear**
L. *colens, -entis,* honoring, revering:
L. *dignus,* worthy, deserving, fitting, honorable: dignify, dignity, dignitary, indignant, indignation, dainty, deign, disdain, condign.
Gr. *doxa,* glory, praise, opinion; *endoxos,* esteemed, notable; see *dokeo* under **think**
L. *egregius,* not of the common herd, eminent, distinguished: egregious.
Gr. *ellogimos,* esteemed, famous, honored: *Ellogima macroperalis* (a moth).
L. *eminens,* prominent, illustrious; see **projection**
L. *encomium* (Gr. *enkomion*), n. praise, eulogy: encomium.
AS. *eorl,* nobleman: earl, earldom, Earl.
AS. *ethel,* noble: Ethel, Ethelred, Ethelwyn.
Gr. *euonymos,* of good name, honored:
Gr. *euphemia,* praise, worship; see *phemi* under **speak**
L. *excellens, -entis,* lofty, distinguished, noble, remarkable: excellent, excellence.
L. *eximius,* exceptional, distinguished; see **few**
L. *fama,* f. reputation, rumor; *famigerabilis; famigeratus; famosus,* celebrated, renowned: fame, famous, defame, infamy.
L. *faveo,* be well disposed toward, regard, esteem; see **give**
Gr. *gennaios,* noble, high-minded, excellent: *Gennaeus nycthemerus* (silver pheasant), *Gennaeocrinus carinatus* (a Devonian crinoid).
Gr. *geraros,* dignified, honored: *Gerarus vetus* (a fossil insect).
L. *gloria,* f. fame, honor, renown, praise; *gloriosus,* famous, splendid; *glorifico, -atus,* honor: glory, glorious, glorification, vainglory, morning-glory, inglorious, *Gloriosa superba* (Malabar climbing-lily).
L. *hallelujaticus,* devoted to praise, < Heb. *hallelujah,* praise Jehovah!: hallelujah.
Gr. *heros,* m.; *heroine,* f. an exceptionally brave and valorous person; *heroikos,* of heroes: hero, heroine, heroic, heroism, Herod, *Polynices heros* (a gastropod), *Andronymus hero* (a butterfly).
Gr. *hyperochos,* prominent, eminent, distinguished; see *exochos* under **projection**
L. *inclitus; inclutus,* celebrated, famous, renowned:
L. *insignis,* remarkable, notable, distinguished; *praesignis,* excellent, illustrious; see *signum* under **mark**
Gr. *kleos,* n. fame, glory; *kleinos; kleitos,* famous, illustrious; *-kles,* termination of many Greek proper nouns, signifying famous: *agaklytos,* very famous: Cleopatra, Androcles, Empedocles, Heracles, Sophocles, Themistocles, Euclid, *Euclea pseudebenus* (Orange River ebony), *Pterocles personatus* (sand-grouse), *Agaclitus dromedarius* (a bug).
Gr. *klytos; klymenos,* heard of, famous, renowned: *Clytocosmus helmsi* (a tipulid), *Clymenonautilus ehrlichi* (a fossil cephalopod).
Gr. *kydos,* n. glory, renown, fame; *kydalimos; kydimos; kydros,* renowned, famous, glorious, noble; *kydion; kydroteros,* more famous; *kydistos,* most famous: kudos, *Cydista aequinoctialis* (a bignoniacead), *Cydrocrinus coxanus* (a Mississippian crinoid).
L. *laudo, -atus,* praise, extol; *laudabilis,* praiseworthy; *laudativus,* of praise; *laudator, -is,* m.; *laudatrix, -icis,* f. praiser, eulogizer, < *laus, laudis* f. praise: laud, laudable, laudatory, applause, allow, disallow, *summa cum laude.*
L. *lautus,* washed, splendid, distinguished; see *lavo* under **wash**
L. *mactus,* honored, worshipped:
L. *magnificus,* noble, eminent, splendid: magnificent, *Abies magnifica* (red fir).
L. *mereo, -ritus,* deserve, earn; *emeritus,* honorably discharged or retired; *meritorius,* of earning, deserving: merit, meritorious, demerit, emeritus, unmerited, mercy, turmeric, *Emerita rathbunae* (a crab).
L. *nobilis, -e,* well-known, celebrated: noble, nobility, *Amherstia nobilis* (a legume), *Dendrobium nobile* (an orchid).
L. *notabilis,* noteworthy, distinguished, prominent, remarkable: notable.
Gr. *onomastos,* named, noted, famous; see *onoma* under **name**
L. *ovatio,* a lesser exultation or triumph; see *ovo* under **joy**
L. *patricius,* dignified, noble: patrician, Patricia.
Gr. *periboetos,* noised abroad, celebrated, famous; *peridoxos,* very famous; *perisemos,* notable, very famous: *Perisemus minimus* (a wasp).

L. *plaudo, plausus,* strike, clap the hands in praise; *plausibilis,* praiseworthy; see **strike**

L. *praestabilis; praestans,* standing out, preeminent, distinguished, excellent: *Anoplodera praestans* (a beetle).

Gr. *preptos,* distinguished, renowned: *Preptothauma oxydiata* (a moth).

L. *respectus, -us,* m. regard, consideration: respect, respectful.

OHG. *ruod,* fame: Robert, Robin Hood, robin, Roland, Orlando, Rollo, Rudolph, Roger.

Gr. *sebastos,* venerable, august; *sebasmos,* m. reverence: Sebastian, *Sebastodes flavidus* (a rockfish), *Sebastosemus entaxis* (a fish).

L. *sublimis,* uplifted, exalted, lofty, distinguished; see **high**

L. *summas, -atis,* noble, eminent:

Gr. *thapto,* honor with funeral rites: *Thaptomys subterraneus* (a mouse).

Gr. *time,* f. esteem, honor; *timetos,* honorable; *timios,* worthy, costly; *entimos,* honored, prized; *epitimetes,* m. estimator, valuer: timocracy, Timothy, xeno-time, *Timodora chrysochroa* (a moth), *Timarcha obsoleta* (a beetle), *Timiobius brevicornis* (a centiped), *Entimus imperialis* (diamond-beetle).

L. *triumphus,* celebration for a victorious general; see **victory**

L. *veneror, -atus,* honor, respect, worship; *venerabilis,* worthy of respect: venerable, venerate, veneration.

L. *vereor, veritus,* be in awe of, respect, fear; *reverens,* respectful, regardful, filled with awe: revere, reverent, reverend, reverential, irreverent.

See: **love, rite, holy, joy**

hood < AS. *hod,* akin to G. *hut,* hat, cowl; see **cap**

-hood < AS. *had,* condition, state; see **nature**

hoof < AS. *hof;* see **foot**

hook < AS. *hoc.*

L. *ancora* (Gr. *ankyra*), f. anchor: anchor, ankyroid, *Ancyrocoris hastatus* (a bug), *Ancyrobdella biwae* (a leech), *Ancyronyx robusta* (a beetle), *Ankyropteris corrugata* (a fossil fern).

AS. *angel; angul,* hook: angle, angler, Anglo-Saxon, England, English.

Gr. *ankistron,* n. fishhook; *ankistrion,* n. dim.; *ankylis, -idos,* f. hook, barb; *polyankistros,* with many hooks: ancistroid, *Agkistrodon[Ancistrodon] mokasen* (formerly *contortrix*) (copperhead), *Ancistrocactus brevihamatus* (a cactus), *Polyancistrus serrulatus* (a grasshopper).

ML. *croca,* f. shepherd's crook: crozier.

L. *hamus,* m. hook, barb, angle; *hamulus,* m. dim.; *hamatilis; hamatus,* with hooks, hooked: hamate, hamular, hamulate, adhamate, *Idahoia hamulus* (a trilobite), *Wubana hamulifer* (a spider), *Ancistrocladus hamatus* (a Ceylon climbing-shrub), *Leptopsylla hamifer* (a flea), *Dentalina hamulifera* (a foraminifer), *Rostrhamus sociabilis* (Everglade kite).

L. *harpago, -onis,* m. grappling-hook, drag: see **take**

Gr. *kreagra; kreagris, -idos,* f. meathook: *Creagris labrosa* (a beetle), *Creagrophorus bihamatus* (a beetle).

Gr. *rhamphis, -idos,* f. hook: *Rhamphidoiulus bujukderensis* (a milleped).

L. *uncus* (Gr. *onkos*), m. hook, barb, angle; *uncatus; uncinatus; uncinus,* hooked, barbed; *aduncus,* bent inward, hooked, aquiline; *obuncus,* bent in, hooked; *reduncus,* curved back: uncinate, uncinaria, unciform, unciferous, aduncity, *Uncinula clandestina* (a mildew), *Uncaria gambir* (a pedaliacead), *Uncarina peltata* (a pedaliacead), *Aconitum uncinatum* (wild monkshood), *Mimodacne inuncata* (a beetle), *Bacunculus phyllopus* (a phasmid), *Mimosa biuncifera* (catclaw-mimosa), *Sminyothrips biuncata* (a thysanopterid), *Viola adunca* (a violet), *Cervicapra redunca* (reedbuck), *Oncosperma filamentosum* (nibong-palm), *Oncidium tigrinum* (an orchid), *Oncorhynchus tschawytscha* (chinook or quinnat salmon), *Desmoncus oxyacanthus* (a bramble-palm), *Uncaspis micans* (a trilobite).

See: **nail, bend, bur,** *lituus* under **rod**

hoopoe < L. *upupa,* f. the bird, *Upupa epops: Haemoproteus upupae* (a protozoan).

Gr. *apaphos,* m. hoopoe:

L. *epops, -opis* (Gr. *-opos*), m. hoopoe:

Gr. *koukouphas,* m. hoopoe:

NL. *phoeniculus,* the genus of wood-hoopoes; see *phoenix* under **bird**

hoot < imitative origin.

L. *bubulo, -atus,* hoot like an owl:

L. *cucubo, -atus,* hoot:

L. *tutubo, -atus,* hoot:

hop < D. *hoppe*, akin to G. *hopfe*.
  ML. *humulus*, m. < *Slav. qumlix*, hop: humulene, *Humulus americanus* (American hop), *Ampelopsis humulifolia* (hop-ampelopsis), *Delphinium humulinum* (a larkspur), *Sphaerotheca humuli* (a fungus).
  L. *lupulus*, m. dim. of *lupus*, m. hop: lupulin, lupuline, *Humulus lupulus* (hop), *Medicago lupulina* (hop-medic), *Carex lupuliformis* (a sedge).

hope < AS. *hopa;* see **believe**

hopeless
  Gr. *aboethetos*, incurable, hopeless: *Aboetheta pteronoma* (a moth).
  Gr. *anekestos*, incurable, fatal:
  Gr. *anelpis*, hopeless, unexpected; *aelptos*, hopeless, desperate: *Anelpistus americanus* (a beetle).
  Gr. *aniatos*, incurable:
  Gr. *asotos; asostos*, abandoned, without hope, lost, profligate, debauched: *Asotocerus ochraceellus* (a trichopterid).
  Gr. *athymos*, fainthearted, spiritless, hopeless: *Athymodictya parva* (a Pennsylvanian insect).
  L. *despero, -atus*, have no hope, give up; *desperabilis*, hopeless, incurable; *exspes*, without hope: despair, desperate, desperation.
  Gr. *dyskelos*, without remedy:
  L. *immedicabilis*, incurable:
  L. *incorrigibilis*, not to be corrected, incurable: incorrigible.
  L. *incurabilis*, without remedy: incurable.
  L. *insanabilis*, incurable, hopeless:
  L. *perditus*, destroyed, ruined, lost, abandoned, corrupt, profligate; see **destroy**
  L. *Sisyphus* (Gr. *Sisyphos*), m. a mythological king whose perpetual punishment in Hades was to roll to the top of a hill a large stone that always rolled down again; symbolic of endless frustration, futility, and unavailing effort: Sisyphean, *Sisyphomyia pygmaea* (a fly), *Sisyphus calcaratus* (a beetle).

hoplo- < Gr. *hoplon*, any tool or implement, armor, shield; *hoplites*, armed; see **tool**

hora, L. hour; *horalis*, of an hour, see **time**; Gr. *hora*, any time or season of the year; *horaios*, seasonable, timely, ripe, see **year**

horama, Gr. thing seen, object, scene; *horasis*, seeing; *horatos*, visible; see *horao* under **see**

horco- < Gr. *horkos*, oath; *horkios*, by oath; see **swear**

hordeum, L. barley; *hordeolus*, dim. sty in the eye; see **grain, sore**

horia, L. fishing smack; *horiola*, dim.; see **ship**

horimos, Gr. seasonable, timely, ripe; see **ripe**

horistos, Gr. definable; see *horos* under **end**

horizon, L., Gr. the circle where earth and sky meet; see **circle**

hormenos, Gr. wild asparagus; see **asparagus**

hormetico- < Gr. *hormetikos*, exciting, stimulating; *horme*, attack, impulse, start; see *hormao* under **arouse**

horminum, L. (Gr. *horminon*), a kind of sage; see **mint**

hormone < Gr. *hormao*, instigate, start, spur; see **arouse**

hormos, Gr. necklace, chain; *hormiskos*, dim., see **necklace**; harbor, see **harbor**

horn < AS. *horn:* horny, hornbeam, hornblende, greenhorn, stinkhorn, Claghorn. Derive: hornet.
  L. *antenna*, feeler, sensory appendage on the head of an insect; see **feel**
  L. *barrio, -itus*, trumpet like an elephant; see **call**
  Gr. *brochis, -idos*, inkhorn; see **ink**
  L. *buccina*, f. shepherd's horn, trumpet; *buccinator, -is*, m. trumpeter; *buccinum*, n. a horn-shaped mollusk; *buccino, -atus*, sound the trumpet: buccina, buccinal, buccinator, *Buccinum undatum* (whelk), *Buccinulum robustum* (a gastropod), *Cygnus buccinator* (trumpeter-swan).
  Gr. *bykane*, f. trumpet; *bykanistes*, m. trumpeter: *Bycanistes buccinator* (a bird).
  L. *classicum*, n. engagement signal, trumpet itself:
  L. *concha*, trumpet shaped like a snail shell used by Triton to control the waves; see **mollusk**
  L. *cornu, -us*, n. horn; *corniculum; cornulum*, n. dim.; *cornualis*, of horns; *corneus*, of horn; *corneolus*, dim. horny; *corniculans*, horned; *corniculatus*, horned, horn-shaped; *cornuatus*, hornlike, horn-shaped; *corniger; cornis; cornutus*, bearing horns, horned; *cornicen, -inis*, m. hornblower; *cornupeta*, goring with the horns: cornet, cornea, corniferous, cornigerous, cornicle, unicorn, corn, corner, bickern, cornucopia, carnelian, *Cornuspira occlusa* (a foraminifer), *Cornicularia*

aculeata (a lichen), *Cornus racemosa* (a dogwood), *Salicornia herbacea* (a glass-wort), *Ceratophrys cornuta* (horned-frog), *Lithocarpus corneus* (a tan-oak), *Spirobranchus bicornis* (a worm), *Conochilus unicornis* (a rotifer), *Fratercula corniculata* (horned-puffin), *Acacia cornigera* (bullhorn-acacia), *Gelidium corneum* (a red alga).

Gr. *karnon*, n.; *karnyx, -ychos*, m. a Gallic horn:

Gr. *keras, -atos*, n; *keros*, m. horn; *keration*, n. dim.; *keraos; kerastes; keratitis, -idos; kerouchos*, horned; *keratinos*, of horn; *keraia*, f. any hornlike projection, antenna; *keraules*, m. hornblower; *amphikeros*, two-horned; *dikeras, -atos*, n. a double horn; *dikeros, -otos*, two-horned; *nekeros*, hornless: cerargyrite, ceratin (keratin), carat, rhinoceros, keratosis, *Ceratophyllum demersum* (hornwort), *Ceratomeryx prenticei* (a fossil mammal), *Ceratostephanus antennatus* (a fly), *Ceratium tripos* (a dinoflagellate), *Ceratiomyxa fructiculosa* (a slime-mold), *Cerastes cornutus* (a viper), *Cerastium arvense* (field-chickweed), *Cerithium serratum* (a fossil gastropod), *Ceratitis capitata* (Mediterranean fruitfly), *Cerastopsilus rufipes* (a wasp), *Cerastoderma echinatum* (a pelecypod), *Amphicerus bicaudatus* (a beetle), *Ceratiola ericoides* (sand-heath), *Bucida buceras* (a combretacead), *Ancistrocerus unifasciatus* (a wasp), *Acerates pumila* (a milkweed), *Dichoceros bicornis* (homrai, a hornbill), *Cryptocerus atratus* (an ant), *Diceros bicornis* (black rhinoceros), *Rhinoceros unicornis* (Indian rhino), *Platycerium grande* (a fern), *Orthoceras multicameratum* (an Ordovician cephalopod), *Trachyceras austriacum* (a Triassic ammonite).

Gr. *likros*, m. antler: *Licrophycus flabellum* (a fossil plant).

L. *lituus*, curved trumpet, curved staff; see **rod**

L. *ramus*, branch, antler; see **branch**

L. *rhytium* (Gr. *rhyton*, dim. of *rhyton*), n. a drinking-horn: *Rhytiodus microlepis* (a fish).

Gr. *salpinx, -ingos*, f. trumpet, tube; *salpingion*, n. dim.; *salpinktes*, m. trumpeter; *salpistikos*, of a trumpet: salpinx, salpingian, hydrosalpinx, salpingitis, salpingion, salpingotomy, *Salpingomimus deceptor* (a beetle), *Salpiglossis sinuata* (a solanacead), *Salpinctes obsoletus* (rock-wren), *Urosalpinx cinerea* (oyster-drill).

Gr. *Triton*, a sea-god who used a conch to control the waves; see **sea**

L. *tuba*, f. trumpet; *tubula*, f. dim.; *tubicen, -inis*, m. trumpeter: tuba, tuba-phone, tubal, tubicinate, *Hemerocallis longituba* (a daylily).

See: **projection, point, pipe, funnel**

hornbeam < AS. *horn* and *beam*.

L. *carpinus*, f. hornbeam, blue-beech; *carpineus*, of hornbeam: *Carpinus caroliniana* (American hornbeam), *Acer carpinifolium* (hornbeam-maple), *Psylla carpinicola* (a jumping plant-louse).

L. *ostrya* (Gr. *ostrye*), f. hop-hornbeam, ironwood: *Ostrya virginiana* (American hop-hornbeam), *Acalypha ostryaefolia* (a spurge), *Rubus ostryifolius* (a black-berry).

Gr. *zygia*, f. probably the hornbeam:

hornotinus, L. of this year, this year's; see **year**

horos, Gr. boundary, limit, rule, standard, landmark; *horios*, of boundaries; see **border**

horreum, L. storehouse, barn, granary, magazine; see **store**

horridus, L. dreadful, bristly, rough; *horribilis; horrendus*, dreadful, frightful; see **fear**

horse < AS. *hors:* horseman, horseback, horseshoe, horsetail, horsefly, unhorse, horsy.

L. *admissarius*, m. horse or ass used for breeding, stallion:

L. *asinus*, m. ass, jack; *asina*, f. jenny; *asellus*, m. dim.; *asinalis; asinarius; asininus*, of, by, or like an ass, obstinate, stupid: ass, jackass, asinine, asininity, aselline, *Equus asinus* (ass), *Asellus aquaticus* (a sowbug), *Pyralis asinalis* (a butterfly), *Haliotis asininus* (a fossil gastropod), *Haematopinus asini* (horse-louse).

L. *auritulus*, m. the long-eared one, ass:

Gr. *bibastes*, mounter, stallion; see *bibazo* under **raise**

L. *biga*, pair of horses; see *bijugus* under **two**

Gr. *brometes*, m. the brayer, ass:

L. *burdo, -onis*, m. mule; *burdonarius*, m. mule-driver:

L. *burricus*, m. a small horse; Sp. *borrico*, m. ass; *borrica*, f. jenny; *burro*, m. ass; *burra*, f. jenny; *burrito*, m. dim.: burro.

L. *caballus*, m. horse, nag: caballa, f. mare; *caballio, -onis*, m. small horse, pony; Sp. *caballo*, m. horse; *caballarius*, m. rider, horseman; *caballinus*, of horses: cavalry, cavalcade, cavalier, caballine, chivalry, chevalier, caballero, cavally *(Caranx hippos)*, *Equus caballus* (horse).

L. *cantherius* (Gr. *kanthelios; kanthon, -os*), m. pack-ass, mule; *cantherinus*, of mules:

L. *Epona*, f. protecting goddess of horses: *Eponides repandus* (a fossil foraminifer).

L. *equus*, m. horse; *equa*, f. mare; *equulus*, m.; *equula*, f. dim. colt, foal, filly; *equiferus*, m. wild horse; *equinus*, of horses; *eques, -itis*, m. horseman, rider; *equester, -tris, -tre*, of riding and cavalry; *equito, -atus*, ride horseback: equine, Equidae, equestrian, equitation, *Equus pacificus* (a fossil horse), *Equula edentula* (a fish), *Eques lanceolatus* (a ribbon-fish), *Equisetum sylvaticum* (a horsetail), *Casuarina equisetifolia* (a beefwood), *Trypanosoma equinum* (a protozoan).

G. *esel*, ass: easel.

L. *hinnus* (Gr. *hinnos*), m. mule (from a stallion and a female ass); *hinnulus*, m. dim.: hinny, *Onthophagus hinnulus* (dung-beetle).

Gr. *hippos*, m. horse; *hipparion; hippidion; hippiskos*, n. dim.; *hippikos; hippios*, of horses; *hippagros*, m. a wild horse; *hippelates*, m. driver of horses; *hippeus; hippeutes*, m. horseman, rider: hipparch, hippodrome, hippopotamus, Hippocrates, Xanthippe, Philip, philippic, hippuric, eohippus, *Hipposideros larvatus* (a horseshoe-bat), *Hippocampus hudsonius* (a sea-horse), *Hipponyx mitratus* (a gastropod), *Hipparion gratum* (a fossil horse), *Hippidium principale* (a fossil horse), *Hipparionyx proximus* (a Devonian brachiopod), *Gymnohippus marmoratus* (a grasshopper), *Phidippus audax* (a spider), *Onohippidium compressidens* (a fossil horse), *Mesohippus exoletus* (a fossil horse), *Hippeastrum equestre* (Barbados lily), *Metaphidippus nitidus* (a spider), *Hippelates plebeius* (a gnat).

L. *jumentum*, beast of burden, horse, mule; see **animal**

Gr. *keles*, courser, riding-horse; see **ride**

Gr. *killos*, m. ass, ass-colored, gray; *hemikillos*, m. half-ass, mule:

L. *lalisio, -onis*, m. foal of a wild ass:

L. *mannus*, m. a small Gallic horse, coach horse; *mannulus*, m. dim. pony:

OHG. *marah*, horse; AS. *mere*, horse: marshal, mare.

L. *mulus*, m.; *mula*, f.; Sp. *mulita*, f. mule (from a male ass and a mare); *mularis; mulinus*, of mules; *mulionicus*, m. mule-driver: mule, mulatto, muleteer, mulish, muline, mulita, muledeer, mulewort (*Hemionitis palmata*).

Gr. *myklos*, m. stallion ass:

Gr. *ocheion; ocheutes*, stallion; see *ocheuo* under **coitus**

Gr. *onketes*, brayer, ass; *onkema, -tos*, a braying; see **neigh**

Gr. *onos*, m. ass; *onagros* (L. *onager, -gri; onagrus*), m. wild ass; *onarion; onidion*, n.; *oniskos*, m. dim.; *hemionos*, m. half-ass, mule: oniscoid, onager (*Equus onager*), *Oniscus asellus* (a sowbug), *Onus ensis* (a fish), *Onosma rupestre* (a borage), *Ononis arvensis* (a restharrow), *Onopordum arabicum* (a cottonthistle), *Barathronus bicolor* (a fish), *Equus hemionus* (kiang), *Psiloniscus brunneus* (a beetle).

Gr. *oreus, -eos*, m. mule; *orikos*, of a mule:

L. *Pegasus* (Gr. *Pegasos*), m. the winged horse of the Muses: *Pegasus papilio* (a fish).

Gr. *polos*, c. foal, colt, filly, pony; *polion*, n. dim.; *polikos*, of colts; *poleusis*, f. horse-breaking; *poleutikos*, skilled in horse-breaking: *Polocentrus rufus* (a bug), *Poliocnemis ungulatus* (a moth).

Gr. *sagmarion*, n. pack-horse; *Sagmarium spongodictyum* (a radiolarian).

L. *thieldonis*, m. a kind of Spanish horse:

L. *veredus*, m. post-horse; *veredarius*, m. post-boy, courier: palfrey.

Amharic *zebra*, name for the striped equine of Africa: *Equus zebra* (zebra), *Zebrina pendula* (a wanderingjew), *Calathea zebrina* (a calathea), *Galaxias zebratus* (a fish), *Zebrasoma deani* (a fossil fish).

See: **frame**

horsetail.   A name applied to a number of different plants.

L., Gr. *ephedra*, horsetail, a plant resembling equisetum; see **gymnosperm**

L. *equisetum*, n. horsetail: *Equisetum palustre* (marsh-horsetail), *Casuarina equisetifolia* (a beefwood), *Ephedra equisetina* (Mongolian ephedra).

L. *hippuris* (Gr. *hippouris*), horsetail, a genus of the water-milfoil family; see **water-milfoil**

hortativus, L. serving to arouse and encourage; see *hortor* under **arouse**

hortus, L. garden; *hortulus*, dim.; *hortulanus*, gardener; *hortensis*, of gardens; see **field**

hosios, Gr. holy; see **holy**

hospitalis, L. of a guest or host; *hospita*, guest, host; see *hospes* under **guest**

hostis, L. enemy; *hostilis; hostificus*, inimical; see **enemy**

hot < AS. *hat;* see **heat**

hour < L. *hora,* hour; see *hora* under **time**

house < AS. *hus,* abode, dwelling: housekeeper, housewife, White House, husband, hussy, household, housemaid, houseleek, housefly, smokehouse, outhouse.   Derive: penthouse.

L. *aedificium,* n.; *aedes, -is,* f. building, house, temple; *aedicula,* of dim.; *aedilis,* m. superintendent of public buildings and grounds; *aedifico, -atus,* build, erect, establish: edifice, edify, edification, aedicula.

Gr. *anaktoron,* n. royal dwelling, temple, shrine:

Gr. *anapaula,* f. rest, resting-place, inn:

L., Gr. *andron,* men's apartment; see **room**

Gr. *aphedron, -os,* m. privy:

L. *aquarium,* n. watering-place, place for keeping fish and other water animals: aquarium.

Gr. *aulion,* n. country home, cottage, stable; *enaulion,* n. interior of a dwelling, *epaulos,* m. place for passing the night, house, quarters, fold: *Epaulophasia officialis* (a fly).

L. *basilica* (Gr. *basilike),* f. a colonnaded public building; *basilicola,* f. dim.; *basileion,* n. royal palace: basilica.

Sp. *borda,* f. cottage, hut; It. *bordello,* m. dim.: bordel.

Gr. *brephotropheion,* n. foundling or orphan hospital:

Pg. *cabana,* f. cabin, hut: cabin.

L. *caespes, -pitis,* cot, hut, hovel, shed; see **sod**

L. *canaba,* f. hovel, hut; *canabula,* f. dim.:

L. *caprile, -is,* n. goat-stall:

L. *casa,* f. house, hut, cabin; *casellula; casula,* f. dim.; *casarius,* of a cottage: casino, Casablanca.

AS. *castle,* < L. *castellum,* fort; see *castrum* under **fort**

L. *caula,* sheepcote; see **hole**

L. *caupona,* f. tavern, inn; *cauponula,* f. dim.; *caupo, -onis,* m. innkeeper: *Urechis caupo* (innkeeper-worm).

L. *cenaculum,* dining-room, garret; see *cena* under **eat**

L. *columbarium,* n. dovecote, pigeonhouse:

L. *contignatio, -onis,* f. floor, story:

AS. *cote; cot,* pen, fold, shelter for domestic animals: sheepcote, dovecote, cot, cottage, cotter, coterie, cotswold.

L. *deversorium,* inn, hotel, hostelry, lodging-place; see *deversor* under **guest**

L. *domus,* f. (Gr. *doma, -tos,* n.), house, home; *domucula (domuncula),* f. dim.; *domicilium,* n. dwelling, abode; *domesticus; domitius,* of the house: domicile, domestic, domesticate, *Passer domesticus* (house-sparrow), *Thermobia domestica* fire-cricket), Domitian, *Chalicodoma muraria* (a mason-bee), *Saxidomus giganteus* (a pelecypod).

Gr. *ektrope,* f. inn, lodging-place: *Ectropa ancilis* (a butterfly).

Gr. *embioterion,* n. house, dwelling:

L. *equile, -is,* n. horsestable:

L. *forica,* f. outhouse, public privy:

L. *gallinarium,* n. henhouse, coop:

L. *ganea,* f. eating-house:

Gr. *geisoma, -tos,* n. penthouse:

L. *gurgustium,* n. hovel, hut:

L. *habitatio, -onis,* f. dwelling:

AS. *ham,* home, a group of houses, village, town; G. *heim,* home: hamlet, Chatham, Durham, Birmingham, Hampton, Hampstead, Blenheim, Amherst, Waltheim, homely, homesick, homeward, homestead.

Gr. *hedolion; hedos,* seat, abode; see **sit**

Gr. *hedra,* seat, abode, dwelling; see **sit**

Gr. *hestia,* f. hearth, fireside, house, home; *anestios,* homeless; *ephestios,* at home; *homestios,* sharing the same fireside, dwelling with: *Hestia juventa* (a butterfly). *Anestia ombrophanes* (a moth). *Ephestia cinerosella* (a moth).

AS., G. *hof,* home, yard, household: Hofmeister, Hoffman, Freihof.

L. *hospitaculum; hospitium,* n. inn, lodging-place; *hospitiolum,* n. dim.:

Gr. *kalia; kalias, -ados,* f. hut, barn, bird's-nest; *kalidion,* n. dim.: kalidium, *Collocalia fuciphaga* (a swift, constructor of edible nests).

Gr. *kalybe,* f. cabin, hut; *kalybion,* n. dim.; *kalybites,* m. dweller in a hut: *Calybitia picata* (a moth), *Calybium massiei* (a gastropod).

Gr. *kasorion,* brothel; see *kasalbas* under **prostitute**

Gr. *katagogion,* n, hotel, inn:

Gr. *klisia,* f. hut, cabin; *klision,* n. outhouse, shed, brothel: *Clisiocampa silvatica* (a moth).

Gr. *kopron, -os,* outhouse, privy; see *kopros* under **dung**

Gr. *kype,* f. hut:

L. *Lar, -is;* pl. *Lares,* m. household gods; hearth, home: *Laria dulcamarae* (a beetle), *Larifuga weberi* (an opilionid), Lares and Penates.

L. *lupanar, -is,* house of ill repute; see *lupa* under **prostitute**

L. *mansio, -onis,* f. abode, dwelling; *mansiuncula,* f. dim.: mansion, menage, menagerie.

L. *mapalium,* n. an African hut:

Gr. *mossyn, -os,* m. wooden house:

Gr. *Nephthys,* f. an Egyptian household goddess: *Nephthytis afzeli* (an aracead).

L. *nosocomium* (Gr. *nosokomeion*), hospital, infirmary; see *nosos* under **disease**

L. *officina,* workshop, laboratory; see *officialis* under **make**

Gr. *oikos,* m.; *oikia,* f.; *oikema, -tos,* n. house, abode, home; *oikarion; oikidion; oikion,* n.; *oikiskos,* m. dim.; *oikeios,* of a household, domestic; *oikesis,* f. a living; *oiketer, -os; oiketes; oikeus,* m.; *oiketis, -idos,* f. dweller, servant; *oiketikos,* pertaining to the household; *oikonomia,* f. management of a household; *oikonomikos,* having experience in household management, thrifty; *oikouros,* guarding the house; *synoikos,* living in the same house: oecium, monoecious, dioecious, autoecious, heteroecious, ecology, economics, economy, gynoecium, *Oecodoma hystrix* (an ant), *Oecobius texanus* (a spider), *Oecophylla smaragdina* (an ant), *Oecetodella singularis* (a trichopterid), *Oeceticus kirbyi* (a moth), *Oecanthus niveus* (snowy tree-cricket), *Cacoecia argyrospila* (a leaf-roller), *Dendroica tigrina* (Cape May warbler), *Cystioecetes nimbosus* (a moth), *Gymnocladus dioicus* (Kentucky coffee-tree), *Urtica dioica* (a nettle), *Thalictrum dioicum* (meadowrue).

Gr. *ornithon, -os,* m. poultry-house:

Gr. *ostrimon,* n. stable:

L. *ovile, -is,* n. sheepfold:

L. *palatium,* n. palace, < *Palatium,* one of the seven hills of Rome on which stood the royal residence: palace, Palatine, palatial, Palatinate, palatine.

L. *Penas, -atis;* pl. *Penates,* m. guardian deities of the household, hearth, home: Penates.

L. *pluteus,* shed, parapet; see **wall**

L. *popina,* eating-house; see **eat**

Gr. *porneion,* n. brothel, whorehouse:

L. *postica; posticum,* backdoor, backhouse, privy; see **door**

Gr. *prytaneion,* town hall; see *prytanis* under **govern**

Gr. *psalis, -idos,* f. low building with a vaulted roof, crypt:

L. *sedes,* seat, abode, home; *sessorium,* seat, residence; see *sedeo* under **sit**

Gr. *skenema,* abode, nest; see *skene* under **tent**

L. *stabulum,* n. stall, shed: *stabulor, -atus,* house, harbor: stable, *Muscina stabulans* (a fly).

Gr. *stathmos,* m. stall, stable, abode, post; *stathmidion,* n. dim.: *Stathmonyma anceps* (a moth).

L. *statio, -onis,* place of sojourn, post; see *sto* under **stand**

Gr. *stegos,* roof, house, home; see **roof**

L. *structura,* something built, building; see *struo* under **make**

Gr. *sypheos,* m. pigsty:

L. *taberna,* f. hut, stall, shed, shop, inn; *tabernula,* f. dim.; *tabernaculum,* n. tent; *tabernarius,* of shops: tavern, tabernacle, *Amsonia tabernaemontana* (an apocynacead).

L. *trichila,* f. arbor, bower, summer-house:

L. *Vesta,* f. goddess of the hearth and domestic life; *Vestalis,* of Vesta: vestal, *Psithyrus vestalis* (a cuckoo-bee).

L. *tugurium,* n. peasant's hut, cottage: *Xenophora(Tugurium) exuta* (a gastropod).

L. *villa,* farmhouse, country-seat, farm; *villula,* dim.; *villaticus,* of a farmhouse; see *vicus* under **town**

L. *xenodochium* (Gr. *xenodocheion*), n. inn, lodge; *xenodochos,* m. innkeeper, host: xenodochium.

Gr. *xylon, -os,* m.; *xylotheke,* f. woodhouse:

See: **place, hive, nest, den, store, temple, sit, stand, room, tent, pen, guard, grave**

houseleek; see **stonecrop**

houstonia NL. a genus of the madder family; see **madder**

hovenia, NL. a genus of the buckthorn family; see **buckthorn**

howl, akin to G. *heulen,* wail, cry out; see **bark, roar**

hub < uncertain origin; see **axis, box, point**

huckleberry. Popularly, but not technically, includes the blueberry and whortleberry; see *gaylussacia* and *vaccinium* under **heath**

hull < AS. *hulu,* husk, case; see **bag, scale, sheath**

hum < imitative origin; see **murmur**

humanus, L. of man; see *homo* under **man**

humble < OF. *humble,* < L. *humilis,* on the ground, low; see **low**

humectus, L. moist, damp; see *humidus* under **wet**

humerus, L. shoulder, bone of the upper arm; see **shoulder**

humidus, L. moist, damp, wet; *humor,* liquid, moisture; *humorosus,* wet; see **wet**

humifusus, L. low, procumbent; see *humilis* under **low**

humilis, L. on the ground, low; see **low,** *humus* under **earth**

humor, L. fluid, liquid, moisture; fun, comicality; see *humidus* under **wet, laugh**

hump < uncertain origin; see **swell, heap, mountain, projection**

humpback; hunchback
>   L. *dorsennus(dossenus),* m. humpback: *Dossenus marginatus* (a spider).
>   L. *gibber, -eris,* hump on the back; *gibbus,* humped; see **projection**
>   Gr. *kyrton, -os,* m. humpback; *kyrtonidion; kyrtonion,* n. dim.: *Cyrton sanguineus* (a beetle).

humulus, ML. hop; see **hop**

humus, L. earth, ground; see **earth**

hundred < AS. *hundred.*
>   L. *centum,* hundred; *centesimus,* hundredth; *centurialis,* of a hundred; *centuplex,* hundredfold: cent, centennial, centimeter, centiped, centigrade, century, centurion, percentage, centenary, *Rosa centifolia* (cabbage-rose).
>   Gr. *hekaton; hekatonta-,* hundred: hectare, hectograph, hectogram, hecatomb, hectocotylus, hecatophyllous.

hunger < AS. *hungor.*
>   Gr. *aatos,* insatiable: *Aatocrinus robustus* (a Pennsylvanian crinoid).
>   Gr. *aboskes,* unfed, hungry:
>   Gr. *abrotos,* not eating, fasting, not fit to be eaten: *Abrotocrinus cymosus* (a Mississippian crinoid).
>   Gr. *achortastos,* unfed, hungry; *achortasia,* f. hunger:
>   Gr. *akorestos; akoretos,* unsatisfied, insatiate: *Acoretus elevatus* (a serphid).
>   Gr. *analtos,* insatiate:
>   Gr. *apastos,* fasting:
>   Gr. *aplestos,* insatiate:
>   Gr. *apositos; asitos,* fasting, hungry: *Aposites macilentus* (a beetle).
>   L. *appetitus,* desire, hunger, passion; see **wish**
>   L. *esurio, -itus,* desire to eat, be hungry; *esuries, -ei,* f. hunger; *esurialis,* of hunger; *esurio, -onis,* m. hungry person: esurient.
>   AS. *faestan,* abstain from food: fast, fast-day, breakfast.
>   L. *fames, -is,* f. hunger; *famelicus,* hungry: famish, famine, *Apospasma famelicum* (a fly).
>   L. *impastus,* unfed, hungry:
>   L. *impransus,* without breakfast:
>   L. *incenatus,* not having dined, hungry:
>   L. *inedia,* f. abstention from food, fasting:
>   L. *inexpletus,* unfilled, unsatisfied, unsated; *inexplebilis,* insatiable: inexpleble.
>   L. *insatiatus,* unsatisfied, unsatisfiable: *insatiabilis,* that cannot be satisfied: insatiable.
>   L. *insaturatus,* unfilled, insatiate: unsaturated.
>   L. *jejunus,* hungry; *jejunium,* n. fast: jejune, jejunum, dine, dinner.
>   Gr. *kenangia,* f. hunger:
>   Gr. *kissao (kittao),* crave strange food (as by pregnant women):
>   Gr. *limos,* m. hunger, famine; *limeros,* hungry; *limodes,* famished; *boulimos* (L. *bulimus*), m. great or insatiable hunger: limosis, bulimia, bulimy, *Bulimina obesa* (a foraminifer).
>   Gr. *nestis, -ios,* fasting, hungry; *nestimos,* of fasting; *nestis, -idos* (L. *-idis*), f. jejunum: nestitherapy.
>   Gr. *orexis,* appetite, desire, longing; see **wish**
>   Gr. *peina,* f. hunger, famine; *peinaleos,* hungry; *peinetikos,* suffering from hunger: *oxypeinos,* ravenous: pinotherapy, *Pinaleus bivittatus* (a fly).
>   See: **empty, eat, wish, greed**

hunt < AS. *huntian:* hunter, huntsman, Huntington.
>   Gr. *agreuo,* take by hunting or fishing; *agremon, -os; agreus; agreuter, -os, agreutes,* m. hunter; *agreutikos,* skilled in hunting; *agreutos,* caught; *agreuma, -tos,* n. booty, prey; *agra,* f. a catch, prey, seizure; *agraios,* of the chase: pellagra, *Agreuocoris noualhieri* (a bug), *Chlaenius(Agreuter) elegantulus* (a beetle).

L. *Anubis* (Gr. *Anoubis*), m. Egyptian god of the chase: *Megasoma anubis* (a beetle).

L. *Artemis, -idis* (Gr. *-idos*), f. goddess of the chase; *Dictynna* (Gr. *Diktynna*), f. another name for Artemis and Diana: artemisin (from *Artemisia maritima*, a wormwood), *Artemia fertilis* (a brine-shrimp).

Gr. *Bendis, -idos*, f. the Thracian equivalent of Artemis: *Bendis caranea* (a moth).

L. *Diana*, f. goddess of the chase and the moon: *Argynnis diana* (a butterfly).

Gr. *dioktos*, pursued; *diokter, -os; dioktes*, m. pursuer, hunter; *diogma, -tos*, n.; *dioxis*, f. pursuit, chase, < *dioko*, pursue, follow, cause to run; *iochmos*, m.; *ioke*, f. rout; *palioxis*, f. counter-pursuit: *Myiodioctes pusillus* (a warbler).

Gr. *epakter, -os*, m. hunter, fisher: *Epacter guttatus* (a beetle).

Gr. *ichneuo*, trace, seek out; see *ichnos* under **mark**

L. *indagator, -is; indagatrix, -icis*, hunter, searcher; see *indago* under **try**

Gr. *kynagos*, m. hunter:

Gr. *nyktereutes*, m. one who hunts at night: *Nyctereutes procynoides* (a carnivore).

L. *Orion, -is* (Gr. *-os*), m. fabled hunter transformed into a constellation: Orion, *Scoteinus orion* (a bat).

Gr. *therao*, hunt; *therates*, m. hunter; *therama, -tos*, n. prey: theralite, *Acridotheres tristis* (myna), *Conopotheras grayi* (a bird), *Therates cyaneus* (a beetle).

L. *venor, -atus*, hunt; *venator, -is*, m. hunter; *venaticus*, of hunting: venatic, venatory, venery, venison, Grosvenor, *Synageles venator* (a spider), *Heteropoda venatoria* (banana-spider), *Acinonyx venatica* (Asiatic cheetah).

Gr. *zetetes*, searcher, seeker; see **try**

See: **ask, try, follow**

hurricane < Ab.Am. *huracan;* see **wind**

hurry < uncertain origin; see **swift**

hurst < AS. *hyrst*, wood, grove; see **forest**

hurt < OF. *hurter*.

L. *abusivus*, tending to use wrongly, injurious; *abutor, -usus*, use badly, hurt: abuse, abusive.

Gr. *adikos; adiketikos*, doing wrong, injurious, unjust; *adikema, -tos*, n.; *adikia*, f. wrong, injury; *adikesis*, f. wrongdoing; *adiketes*, m. injurer: *Adicocrita araria* (a moth).

L. *afflictus*, struck down, attacked, distressed: affliction, *Apatela afflicta* (a moth).

L. *agonia*, f. victim; *Agonalis*, pertaining to the Agonalia, a festival in honor of Janus:

Gr. *aikizo*, injure, torture, plague; *aikia*, f.; *aikisma, -tos*, n.; *aikismos*, m. injury, outrage; *aikistikos*, prone to torture: aecium, aecidium, aecial, aeciospore.

Gr. *anopheles*, useless, hurtful; *anopheletos*, fruitless, unprofitable: *Anopheles quadrimaculatus* (a malaria mosquito).

Gr. *atarteros; ateros*, baneful, mischievous: *Ateropogon cryptopogonoides* (a fly).

Gr. *athetetes*, m. violator, disregarder, < *atheteo*, disregard, reject, revolt: *Athetetes globicollis* (a beetle).

L. *atrox, -ocis*, cruel, harsh, severe; see **merciless**

AS. *bana*, harmer, slayer; see **kill**

Gr. *blaptikos*, harmful, hurtful; *blapsis*, f. damage, injury, < *blapto*, hurt; *blaberos*, harmful, noxious; *blabe*, f.; *blabos; blamma, -tos*, n. harm, hurt, damage: *Blapticus leucostomus* (a wasp), *Blapsilon scutellatum* (a beetle), *Blapstinus vestitus* (a beetle), *Blaberus giganteus* (drummer-cockroach).

L. *catax, -acis*, limping, lame:

Gr. *cholos*, lame, halt, maimed; *cholasma, -tos*, n.; *cholosis*, f. lameness, < *choleuo*, become lame, maimed: *Cholocrinus obesus* (a silurian crinoid), *Choloepus[Cholopus] didactylus* (two-toed sloth), *Choleva fumata* (a beetle), *Chologaster cornutus* (a fish).

L. *clades*, destruction, damage, loss, defect, disaster; see **destroy**

L. *claudus*, crippled, limping, lame; *claudigo, -inis; clauditas, -atis*, f. lameness: Claude.

L. *colobicus*, mutilated; see *kolos* under **short**

L. *contumelia*, f. abuse, insult, invective; *contumeliosus*, insolent, abusive, contemptuous: contumely, contumelious.

L. *corrumpo, -ruptus*, injure, mar, spoil, adulterate, seduce; *corruptibilis; corruptivus; corruptorius*, perishable, likely to decay: corrupt, corruption, incorruptible.

L. *crucio, -iatus*, put to the rack, torture, torment, crucify; *cruciabilis*, painful, tormenting: excruciating.

L. *damnosus*, injurious, harmful, pernicious; *damnum*, n. hurt, damage, injury, penalty, fine, < *damno, -atus*, harm, sentence to punishment: damage, damnify, *Damnacanthus indicus* (a rubiacead), *Simulium damnosum* (a blackfly).

L. *deformis,* misshapen, crippled, ugly: deformed, *Stephanocrinus deformis* (a Silurian crinoid).

Gr. *deleterios,* harmful, noxious; *delema, -tos,* n.; *delesis,* f. mischief, bane; *deleter, -os,* m. destroyer, < *deleomai,* hurt, harm: deleterious.

L. *delumbis,* lamed in the loins, weak; see *lumbus* under **side**

L. *deminoratio, -onis,* f. injury, degradation:

Gr. *dennos,* m. reproach, disgrace, < *dennazo,* abuse, revile:

L. *detrimentum,* n. damage, injury, loss; *detrimentosus,* harmful: detriment.

L. *devenusto, -atus,* disfigure, deform:

L. *diritas, -atis,* f. misfortune, fatal mischief:

L. *dispendiosus,* hurtful, prejudicial:

L. *distortus,* misshapen, deform, twisted: distorted, *Dentalina distorta* (a foraminifer).

Gr. *epereia,* f. spiteful abuse, insult: *Eperia albicans* (a fly).

Gr. *epesbolos,* scurrilous:

L. *facinus, -oris,* bad deed, outrage, crime; see **fault**

Gr. *gyios,* lame; *paragyios,* with crippled limbs:

Gr. *hybris,* f. wanton violence, outrage, censure, insult, assault and battery; *hybrisma, -tos,* n. wanton act, outrage; *hybrister, -os; hybristes,* m. a violent, wanton, insolent person; *hybristikos; hybristos,* outrageous, insulting, violent, < *hybrizo,* run violent: hubristic [hybristic].

L. *infesto, -atus,* attack, injure; see *infestus* under **danger**

L. *inficio, -fectus,* spoil, corrupt, taint; see **color**

L. *inimicus,* unfriendly, hostile, harmful; see **enemy**

L. *iniquus,* unjust, hurtful, inimical: iniquity, iniquitous.

L. *injuria,* f. injustice, wrong; *injurius,* wrongful, unjust: injury, injurious.

Gr. *iptomai,* oppress, hurt, harm; *ipoterion,* a pressing-place, press; *ipsis,* a pressing; see *ipoo* under **press**

Gr. *kakegoria,* f. slander, calumny, abuse; *kakegoros,* abusive, slanderous; *kakologos,* slanderous, abusive; *kakotikos,* harmful, noxious; *kakizo,* abuse, blame; *kakismos,* m. blame, reproach:

Gr. *kertomesis; kertomia,* f. a jeering, mocking; *kertomios; kertomos,* cutting, stinging, reproachful, < *kertomeo,* taunt, mock, jeer:

Gr. *klambos,* mutilated:

Gr. *kyllos,* crooked, crippled, lame; *kylloma, -tos,* n. lameness: cyllosis, trachelocyllosis, *Cyllorhamphus tuberosus* (a beetle), *Cylloepus aciculus* (a beetle).

L. *laedo, laesus,* hurt by striking, wound; *laesio, -onis,* wound, injury; see **strike**

Gr. *lobe,* f. ill usage, outrage, indignity, mutilation; *lobeter, -os,* mutilator; *lobetos,* outraged, dishonored, < *lobaomai,* insult, maltreat, maim: *Lobetus torticollis* (a beetle).

L. *lutito,* bring into contempt, asperse, < *luto, -atus,* bedaub, besmear; see *lutum* under **earth**

Gr. *lygros,* baneful, harmful: *Lygronoma sporimaea* (a moth).

Gr. *lyme,* outrage, defilement; *lymanter, -os,* destroyer; *lymanterios,* destructive; see **destroy**

L. *maledico, -ictus,* abuse, curse, revile, slander; see **curse**

L. *mancus,* maimed, crippled, lame: *Manculus quadridigitatus* (dwarf salamander), *Agriotes mancus* (wheat-wireworm).

Gr. *mitylos,* hurt, shortened:

L. *mulco, -atus,* beat, handle roughly, injure:

L. *mutilus; muticus,* maimed, curtailed, docked, shortened; *mutilo, -atus,* maim, cut short, lop off: mutilate, mutilation, mutilous, muticous, *Melica mutica* (a grass), *Hypericum mutilum* (a St. Johnswort).

L. *noceo, -citus,* hurt, harm; *nocivus; nocuus; noxius,* harmful; *noxiosus,* very harmful; *noxalis,* of an injury; *noxia,* f. fault, offense, crime; *obnoxius,* offensive, hateful: noxious, obnoxious, innocuous, innocent, ninny, nuisance, *Myllocerinus innocuus* (a beetle), *Antiaris innoxia* (ipoh-tree), *Lissodendoryx noxiosa* (a sponge), *Dermatobia noxialis* (a botfly), *Peckhamia noxicosa* (a spider).

L. *offendo, -fensus,* hit or strike against something, do damage or injury, displease; *offensa,* f. fault, violation, transgression: offend, offensive, offense.

Gr. *outao,* wound; *outatos,* wounded; *outesis,* f. a wounding; *outetes,* m. wounder: *Utetes testaceus* (a wasp).

Gr. *pema,* suffering, woe; see **trouble**

L. *perniciosus,* baneful, injurious, destructive; see **destroy**

Gr. *peros,* disabled, maimed, crippled; *anaperos,* maimed, crippled; *peromeles,* maimed in the limbs: *Peromyscus leucopus* (a deer-mouse), *Anaperus labradoricus* (a holothurian), *Peropus mutilatus* (a lizard).

L. *persecutor, -is,* m. one who pursues with intent to injure: persecutor, persecute, persecution.

L. *praejudicialis,* of judgment beforehand, damaging: prejudice, prejudicial.

Gr. *propelakizo,* cover with mud, treat with indignity, abuse foully; see *pelos* under **earth**

L. *protero, -tritus,* trample under foot; *protervus,* violent, impudent:

L. *saucius,* wounded, hurt, < *saucio, -atus,* wound, hurt; *consaucido,* wound severely: *Noctua saucia* (a butterfly).

L. *scelus, -eris,* evil deed, crime, villain; *scelestus,* wicked; *scelero, -atus,* pollute, profane; see **bad**

Gr. *sinos,* n. hurt, harm, lesion, damage; *sinis, -idos,* m. ravager, plunderer, destroyer; *sinas, -ados,* destructive; *sinotikos,* mischievous; *sinamoros,* hurtful, mischievous: *Sinodendron rugosum* (a stag-beetle), *Sinoxylon basilare* (a beetle), *Sinis brevicollis* (a fossil beetle).

Gr. *siphlos (sipalos),* crippled, maimed: *Siphlopteryx antarcticus* (a fly), *Siphlurus alternatus* (a mayfly), *Sipalocyon gracilis* (a fossil mammal).

Gr. *skerbolos,* scolding, abusive; see **scold**

L. *stupro, -atus,* ravish, debauch, defile; *stuprum,* debauchery, pollution; see **coitus**

L. *sugillo, -atus,* beat black and blue; see **bruise**

Gr. *telchin, -os,* m. a michievous, spiteful person: *Telchinia serena* (a butterfly).

Gr. *thlastos,* crushed, bruised; *thlasma,* a bruise; see *thlao* under **bruise**

L. *tribulo, -atus,* oppress, afflict: tribulation.

Gr. *trotos,* wounded, vulnerable; *troma, -tos,* festering wound; *trosis,* a wounding; see *trauma* under **sore**

L. *violo, -atus,* transgress, abuse, outrage, injure; *violator, -is,* m. injurer, outrager: violate, violence, inviolate.

L. *vitio, -atus,* injure, mar, spoil: vitiate.

L. *vulnero, -atus,* hurt, wound; *vulnerabilis,* subject to injury; see *vulnus* under **sore**

Gr. *zemia,* f. damage, loss, penalty; *zemiodes,* ruinous; *zemiotes,* m. punisher; *epizemios,* hurtful, penal: *Zemiophora scutulata* (a wasp), *Zemiodes coloradensis* (a wasp), *Zemiothes albitarsis* (a wasp), *Zemiagrammus isistius* (a fish).

See: **destroy, kill, sore, pain, cut, scold, curse, fight, strike, plunder, bad, enemy, trouble**

husband < AS. *husbonda,* master of a house; see **spouse**

husk < D. *huus,* house; see **bag, scale, sheath**

hyacinth < L. *hyacinthus* (Gr. *hyakinthos*); see **lily**

hyalinus, L. (Gr. *hyalinos*), of glass; see *hyalos* under **glass**

hybos, Gr. hump; see **projection**

hybrid < L. *hybrida,* mongrel; see **mix**

hybris, Gr. wanton violence, outrage; *hybristos,* wanton, insolent, see **hurt;** *hybris, -idos,* a night bird of prey, see **owl**

hyco- < Gr. *hykes; kykos,* a fish; see **fish**

hydato- < Gr. *hydor, hydatos,* water; *hydation,* dim.; *hydatinos,* watery; *hydatis, -idos,* drop of water, watery vesicle; see **water**

hydnum, L. (Gr. *hydnon*), an edible fungus; see **fungus**

hydra, Gr. a fabulous many-headed serpent; *hydros,* a water-snake; see **snake**

hydrangea, NL. a genus of the saxifrage family; see **saxifrage**

hydrargyros, Gr. quicksilver, mercury; see **mercury**

hydrastis, NL. a genus of the buttercup family; see **buttercup**

hydria, Gr. water-pot, jug, ewer, urn, bucket; *hydrion; hydriske,* dim.; see **pot**

hydro- < Gr. *hydor, -datos,* water; *hydrias,* of water; *hydrotis,* moisture; see **water**

hydrogen < Gr. *hydor,* water; see ELEMENTS under **thing**

hyena < L. *hyaena* (Gr. *hyaina*), f.: *Hyaena striata* (striped hyena).
L. *belbus,* m. hyena:
L. *crocuta,* f. hyena: *Hyaena crocuta* (spotted hyena).
Gr. *glanos,* m. hyena: *Glanosuchus macrops* (a fossil reptile).
Gr. *kynolykos,* m. hyena:

hyetos, Gr. rain; see **drop**

hygiene < Gr. *hygieinos,* healthful, < *hygieia,* health; see **health**

hygro- < Gr. *hygros,* wet; see **wet**

hyios, Gr. son; *hyionos,* grandson; see **son**

hyksos, Gr. a dynasty of Egyptian nomad kings; see **govern**

hylacto- < Gr. *hylaktes; hylaktor,* barker; *hylake,* a barking; see **bark**

hylia, Gr. shoe sole; see **sole**

hylister, Gr. collander, filter, strainer; see **sieve**

hyllos, Gr. ichneumon; see **weasel**

hylo- < Gr. *hyle,* wood, matter; see **wood, alcohol**

hymen, Gr. membrane; *hymenion,* dim., see **membrane;** *Hymen,* god of marriages, see **marry**

Hymettus, L. (Gr. *Hymettos*), a mountain famous for its honey; see **mountain**

hymnos, Gr. song of praise; see **sing**

hynis (hynnis); hynne, Gr. plowshare; see **plow**

hyo- < Gr. *hys (sys), -yos,* hog, see **hog;** pertaining to upsilon, see **letter**

hyoscyamus, L. (Gr. *hyoskyamos*), henbane; see **nightshade**

hyoseris, Gr. a kind of endive; see **lettuce**

hypatos, Gr. uppermost, highest; see *hyper* under **over**

hypelate, Gr. a shrub; see **soapberry**

hypene, Gr. mustache; see **hair**

hyper-; hypero- < Gr. *hyper,* beyond, over, above, very; see **over**

hyperbole, Gr. exaggeration; see FIGURES OF SPEECH under **form**

hypericum, L. (Gr. *hyperikon*), St. Johnswort; see **St. Johnswort**

hyperoche, Gr. projection, prominence, summit; *hyperochos,* prominent, eminent; see *exochos* under **projection**

hyperos, Gr. pestle; see **pestle**

hypha, L. (Gr. *hyphe*), web; *hyphantes,* weaver, < *hyphaino,* weave; see **weave**

hyphalos, Gr. submerged; see *hals* under **sea**

hyphear, Gr. mistletoe; see **mistletoe**

hyphen, Gr. a punctuation mark; see **separate**

hyphesson, Gr. somewhat less, smaller; see *mikros* under **little**

hypnos, Gr. sleep; *hypnikos,* of sleep; *hypnotikos,* sleep-inducing; see **sleep**

hypnum, L. (Gr. *hypnon*), a kind of moss; see **moss**

hypo- < Gr. *hypo,* under, beneath, less than; see **under**

hypochoeris, L. (Gr. *hypochoiris*), a kind of succory; see **lettuce**

hypocrite < Gr. *hypocrites,* pretender, dissembler; see **lie**

hypolais, Gr. a small bird; see **warbler**

hyponome, Gr. underground passage, tunnel, mine; see **mine**

hypopion, Gr. black eye, bruise; see **bruise**

hypopitys, NL. a genus of the heath family; see **heath**

hypotenusa, L. (Gr. *hypoteinousa*), side opposite the right angle of a triangle; see **against**

hypothesis, Gr., f. assumption, condition, foundation, supposition, theory: hypothesis, hypothetical, hypothecate.

    L. *assumptio, -onis,* f. hypothesis, proposition, < *assumo, -umptus,* take for granted, suppose; *praesumptio, -onis,* f. preconception, audacity: assume, assumption, presumption, presumptive.

    L. *conjectura,* f. opinion, guess; *conjecturalis,* of guesswork; *conjector, -is,* m. soothsayer, seer, < *conicio, -jectus,* infer, foretell: conjecture, conjectural.

    Gr. *dogma,* opinion, doctrine; see *dokeo* under **think**

    Gr. *lemma, -tos,* n. assumption, premise, subject; *lemmation,* n. dim.; *dilemma, -tos,* n. double proposition: lemma, dilemma.

    L. *opinio, -onis,* conjecture, view, fancy, belief; see *opinor* under **think**

    L. *propositio, -onis,* f.; *propositum,* n. plan, design, purpose, theme: proposition, proposal, propose, purpose, "jeep".

    L. *ratio, -onis,* doctrine, theory, understanding; see *reor* under **think**

    Gr. *stochos; stochasmos,* aim, conjecture, guess; see **aim**

    L. *suppositio, -onis,* f. a putting under, substitution, postulation, assumption: suppose, supposition.

    Gr. *theoria,* f. something seen in the mind, guess, hypothesis, plan, scheme: theory, theoretical.

    Gr. *thesis,* f. affirmation, proposition, principle: thesis.

    See: **begin, believe, think, base**

hypothymis, Gr. a bird; see **bird**

hypozoma, Gr. diaphragm; see **wall**

hypselo-; hypsi-; hypso- < Gr. *hypselos,* high; *hypsiteros,* higher; *hypsistos,* highest; *hypsi,* on high; *hypsos,* height; see **high**

hyptios, Gr. supine; *hyptiotes,* flatness; see **lie**

hyptis, NL. a genus of the mint family; see **mint**

hyraco- < Gr. *hyrax, -akos,* shrew; see **mole**

hyrcha, Gr. pickle-jar; see **pot**

hyrichos, Gr. basket; see **basket**

hyrtane, Gr. potlid; see **cap**

hys, -*yos,* Gr. hog; see **hog**

hysginon, Gr. crimson or scarlet dye; see **red**

hysma, Gr. rain; see *hyetos* under **drop**

hysmine, Gr. battle; see **fight**

hyssop < OF. *ysope,* < L. *hyssopus* (Gr. *hyssopos*); see **mint**

hyssos, Gr. javelin; see **spear**

hystera, Gr. womb; *hysterikos,* of the womb; see **womb**

hysteros, Gr. after, later; *hystatos,* last; see **after**

hystrix, -*icis,* L. (Gr. *hystrix, -ichos*), porcupine; see **porcupine**

hythlos, Gr. idle talk, gossip, nonsense; see **speak**

I

-i, L. forms second declension genitives, and masculine specific terms with personal names; see *-ae* under **nature**

-i < L. *in,* not; see **not**

-ia, L., Gr. suffix denoting pertaining to; see **nature**

iacho- < Gr. *iache,* shout, cry, shriek; see **call**

iachros, Gr. softened, calm, tranquil; see **rest**

ialemos, Gr. dirge, lament, wail, melancholy; see **weep**

ialtos, Gr. sent forth; see **send**

iama, Gr. medicine, remedy; see *iaomai* under **heal**

iambus, L. (Gr. *iambos*), a metrical foot; see **measure**

ianthinus, L. (Gr. *ianthinos*), violet-blue; see *ion* under **blue**

iaphetes, Gr. archer; see **arc**

iapyx, Gr. northwest wind; south Italian; see **wind**

iasione, Gr. probably a bindweed, but now applied to a campanula; see **bindweed**

-iasis, Gr. a diseased condition; see **disease**

iaspis, Gr. jasper; see **stone**

-iatry < Gr. *iatreia,* a healing, treatment of disease; *iatros,* physician; see **heal**

ibanos, Gr. water-bucket; see **bucket**

iberis, L., Gr. a kind of pepperwort; see **mustard**

ibex, L. a kind of goat, chamois; see **goat**

ibis, L., Gr. a wading bird; see **heron**

-ible < L. *-bilis,* capable of, having the quality of; see **nature**

ibyctero- < Gr. *ibykter, -os,* leader in a war song; see **sing**

ibyx, Gr. crane; see **crane**

-ic, Suf. denoting the greater valence in a chemical compound; see *-acus* under **nature**

icaco, Ab.Am. coco-plum; see **plum**

icano- < Gr. *ikanos,* competent, sufficient; see **strong**

Icarus, L. (Gr. *Ikaros*), protege of Dionysus, who taught him viticulture; see **vine**

ice < AS. *is:* icy, icicle, Iceland, iceberg, *Cetraria islandica* (a lichen).
  Gr. *chalaza,* hail, sleet, pimple; *chalazion,* dim.; *chalazios,* pertaining to hail; see **pimple**
  Gr. *cheima,* winter weather, cold, frost; see **year**
  Gr. *chion, -os,* f. snow; *chioneos; chionodes; chionopos,* like snow, snow-white; *chionotos,* snowy: chiolite, chionoblepsia, *Chionodoxa cretica* (a liliacead), *Chionea nivicola* (snow-gnat), *Chiogenes hispidula* (creeping snowberry), *Chiococca alba* (cahinca-root), *Hedychium flavum* (yellow ginger-lily).
  AS. *frost (forst),* white, congealed dew or vapor: frost, frosty, hoarfrost, frostbite.
  L. *gelatus; gelicidium,* frost, < *gelo, -atus,* freeze; see *gelidus* under **cold**
  L. *glacies, -ei,* f. ice, icy cold, chill; *glacialis,* icy, frozen; *glacio, -atus,* freeze, solidify; *conglacio,* freeze up: glacier, glacial, glaciation, subglacial, Mer de Glace, *Eubalaena glacialis* (right whale), *Theridium glaciale* (a spider).
  L. *grando, -inis,* f. hail, hailstorm, sleet; *grandineus; grandinosus,* full of hail:
  Gr. *kryos; krymos,* icy cold, frost, chill; see **cold**
  Gr. *krystallos,* ice, rock-crystal; see **stone**
  L. *ninguis, -is,* f. snow; *ninguidus,* snowy:
  Gr. *nipha,* f. snow; *niphas, -ados,* f. snowflake; *niphetos,* m. falling snow, snow-storm; *dysniphos,* snowy, wintry: niphablepsia, *Niphopelta imperialis* (a beetle), *Niphadolepis nivata* (a moth), *Niphetophora maculipes* (a beetle).
  L. *nix, nivis,* f. snow; Sp. *nieve,* f. snow; *nivalis; nivarius; niveus,* of snow, snow-white; *nivosus,* snowy: nival, nivation, neve, Nevada, *Boehmeria nivea* (ramie), *Chasmorhynchus niveus* (bell-bird), *Lissomerus niveicomatus* (a fly), *Taeniopteryx nivalis* (a snowfly), *Trillium nivale* (early trillium), *Achoreutes nivicola* (a snowflea), *Charadrius nivosus* (a plover), *Chlamydomonas nivalis* (red-snow flagellate), *Potentilla nivea* (snowy cinquefoil), *Opidnus nivarius* (a wasp).
  Gr. *pachne,* f. frost, rime; *pachnodes,* chilly, cold: pachnolite, *Pachnobia tecta* (a moth).
  Gr. *pagetos; pagos,* m. frost; *pageros; pagetodes,* frosty, cold: *Pagomys foetidus* (a ringed-seal).
  L. *pruina,* f. rime, frost; *pruinosus,* frosty: *Halictus pruinosus* (a bee).
  Gr. *rhigos,* n. frost, cold; *rhigion,* colder; *rhigistos,* coldest; *rhigosis,* f. a shivering; *rhigelos,* chilling: rhigolene.
  AS. *snaw,* snow: snowball, snowbell, snowberry, snowdrop, snowflake, snowshoe.
  Gr. *stibe,* f. frost, rime:
  L. *stiria,* f. frozen drop, icicle; *stiricidium,* n. snowfall:
  See: **cold, year, thick**

icelo- < Gr. *ikelos,* like; see **like**

ichneumon, Gr. the tracker, mongoose; a wasp; see **weasel, wasp**

ichnos, Gr. footprint, track; *ichnion,* dim.; *ichneutes,* tracker; see **mark**

ichor, Gr. serum; see **juice**

ichthys, -yos, Gr. fish; *ichthydion,* dim.; *ichthyeros,* fishy; see **fish**

-ician < F. *-icien,* one who specializes in or practices; see **make**

-icle < L. *-iculus,* diminutive suffix; see *-culus* under **little**

icmado- < Gr. *ikmas, -ados,* moisture; see **wet**

ico- < Gr. *eikos,* likely, probable, reasonable; *eikatos,* reasonably, fairly, see **true**; < *ix, ikos,* a kind of worm, see **worm**

icon, L. (Gr. *eikon*), image, figure; *icuncula,* dim.; see **form**

icosi- (icosa-) < Gr. *eikosi,* twenty; *eikostos,* twentieth; see **ten**

icoto- < Gr. *eikotos,* probably, perhaps, reasonably, naturally; see **lot**

icrio- < Gr. *ikrion,* scaffold, bench, flooring, platform; see **frame**

ictaro- < *iktar,* a herringlike fish; see **herring**

ictero- < Gr. *ikteros,* jaundice; *ikterikos,* jaundiced, yellow, see **disease**; a yellow bird, see **oriole**

ictico- < Gr. *eiktikos,* yielding readily, pliable; see **bend**

ictino- < Gr. *iktinos,* kite; see **hawk**

ictis, L., Gr. marten, weasel; see **weasel**

ictus, L. blow, stroke, stab, sting; see **strike**

-icus, L. (Gr. *-ikos*), belonging to, pertaining to; see *-acus* under **nature**

-id; -ide < L. *-idus,* having the nature of; see **nature**

-idae, NL. suffix denoting a family of animals, < Gr. *-ides,* a patronymic suffix; see **class**

idaeus, L. (Gr. *Idaios*), pertaining to wooded Mount Ida, < *ide,* tree; see **tree**

idalimo- < Gr. *eidalimos,* comely, shapely; see *eidos* under **form**

idanos, Gr. fair, comely; see **beauty**

-ide, Suf. of binary chemical terms whose first part is the name of a nonmetal; see *-id* under **nature**

idea, L. mental image, form, appearance; see *eidos* under **form**

idechtho- < Gr. *eidechthes,* of repulsive appearance, ugly; see **ugly**

idem, L. the same; see **equal**

idemono- < Gr. *eidemon, -os,* expert, skilled; see **know**

-ides, Gr. son of, descendant of; see **son**

idetico- < Gr. *eidetikos,* scientific; see *eidemon* under **know**

idico- < Gr. *eidikos,* specific; see **alone**

idio- < Gr. *idios,* one's own, personal, individual; *idiastes,* recluse; see **alone**

idiot < L. *idiota,* ignorant person, < Gr. *idiotes,* private person; see **dull**

-idium, L. (Gr. *-idion*), a diminutive suffix; see DIMINUTIVES under **little**

idle < AS. *idel,* empty, vain, useless; see **rest**

idmono- < Gr. *idmon, -os,* skilled, expert; *idmosyne,* skill, knowledge; see *eidemon* under **know**

ido- < Gr. *eidos,* form, resemblance; see **form,** *-oid* under **like**

idol < L. *idolum* (Gr. *eidolon*), image, form; see *eidos* under **form**

idoneus, L. fit, capable, appropriate; see **fit**

idris, *-ios,* Gr. experienced, skillful, knowing; see **know**

idyl < L. *idyllium* (Gr. *eidyllion*), a poem describing pleasant rustic conditions; see **poem**

idylo- < Gr. *eidylos,* expert; see *eidemon* under **know**

if < AS. *gif,* on condition that; see **lot**

igdis, Gr. mortar; *igdion,* dim.; see **mortar**

ignarus, L. ignorant; see **dull**

ignavus, L. inactive, lazy, idle, sluggish, coward; see **dull**

ignis, L. fire; *igniculus,* dim. spark; *igneus,* of fire, fiery, ardent; *ignitus,* fiery, glowing; see **fire**

ignobilis, L. unknown, obscure, low, mean; see **bad**

ignominia, L. disgrace, dishonor, shame; see **shame**

ignorant < L. *ignorans, -antis,* < *ignoro, -atus,* have no knowledge, be unaware; see **dull**

ignotus, L. unknown, strange; see **secret, strange**

ignya, Gr. ham; see **leg**

il- < L. *in,* in, not; see **in, not**

ilapinasto- < Gr. *eilapinastes,* feaster, guest; see **guest**

ile, Gr. crowd, band, troop, company; *iladon,* in troops, bodies; see **number**

ilemato- < Gr. *eilema, -tos,* wrapper, envelope; see **sheath**

iletico- < Gr. *eiletikos,* wriggling; see *eilo* under **turn**

ileum, L. last part of the small intestine; see **intestine**

ileus, L. (Gr. *eileos; ileos*), severe colic; *iliacus,* of colic; see **disease**

ilex, *-icis,* L. holm-oak; now a genus of the holly family; *iligneus,* oaken; see **holly**

ilingo- < Gr. *ilinx, -ngos,* whirling, agitation; *ilingos,* dizziness; see **turn**

ilipodo- < Gr. *eilipous, -odos,* with rolling gait; see **walk**

-ilis, L. having the nature or quality of; see **nature**

ilium, L. groin, flank; see **groin**

ill < ON. *illr;* see **bad, disease, fault, pain**

illabefactus, L. unbroken, unimpaired; see **all**

illacrimabilis, L. unmoved by tears, pitiless; see **merciless**

illaesus, L. unhurt, unimpaired; see **all**

illas, *-ados,* Gr. rope, band, see **rope;** thrush, see **thrush**

illativus, L. inferring, concluding; see **think**

illecebra, L. enticement, allurement, bait; *illecebrosus,* attractive, seductive; see *lacio* under **draw**

illegitimus, L. unlawful; *illex, -egis,* contrary to law, lawless; see *exlex* under **forbidden**

illepidus, L. impolite, rude; see **rough**

illex, -egis, L. contrary to law, lawless, see *exlex* under **forbidden;** *illex (illix), -icis,* seductive, see *lacio* under **draw**

illibatus, L. undiminished, unimpaired; see **all**

illiberis, L. childless; see **sterile**

illicitus, L. forbidden, unlawful; see **forbid**

illicium, L. attraction, enticement, see *lacio* under **draw;** a genus of the magnolia family, see **magnolia**

illimitatus, L. boundless; see **infinite**

illisus, L. struck in; see *laedo* under **strike**

illitus, L. a bedaubing, smearing; see *lino* under **spread**

illogical < L. *in,* not, *logicus,* reasonable; see *logicus* under **think**

illos; illis, Gr. squint; *illodes,* looking obliquely, squinting; see **slope**

illotus, L. unwashed, dirty; see **dirt**

illumino, -atus, L. light up, embellish; see *lumen* under **light**

-illus, L. diminutive suffix; see *-cle* under **little**

illusio, L. mocking, deceit, < *illudo, -lusus,* make sport of, jest, mock; see **lie**

illustro, -atus, L. light up, embellish; see *lustro* under **light**

illuvies, L. dirt, filth; see **dirt**

ilymato- < Gr. *eilyma, -tos,* wrapper; see **sheath**

ilys, -yos, Gr. mud; see **earth**

im- < L. *in,* in, not; see **in, not, on**

image < L. *imago, -inis,* copy, likeness, imitation; see **form, fancy**

imbecillus, L. feeble, weak; see **dull**

imber, -bris, L. shower; see **drop**

imberbis; imberbus, L. beardless; see *barba* under **hair**

imbricatus, L. overlapping like roofing-tiles and shingles; see *imbrex* under **shingle**

imbuo, -utus, L. wet, dip, infect, inspire, instruct; see **wet**

imido-; imino-; variants of *amido-* and *amino-* and denoting NH compounds; see **amido-**

imitor, -atus, L. copy, mimic; see **equal**

immanis, L. enormous, monstrous, huge; see **large**

immaturus, L. unripe, premature; see **before**

immensus, L. vast, boundless; see **large**

immitigabilis, L. that cannot be softened or allayed; see **merciless**

immitis, L. harsh, rough; see **rough**

immobils, L. immovable; see **stand**

immodestus, L. unrestrained, indecent; see **lewd**

immortal < L. *immortalis,* deathless, eternal, everlasting; see **always**

immundus, L. unclean, dirty; see **dirt**

immunis, L. tax-free, exempt; see **free**

immutabilis, L. unchangeable; see **stand**

impar; imparilis, L. unequal, odd; see *dispar* under **different**

imparatus, L. unprepared, not ready; see **not**

impassable

Gr. *abatos; dysbatos,* untrodden, impassable, inaccessible: *Dysbatus singularis* (a moth).

Gr. *anexodos,* impassable, with no outlet; *anodos,* without a road; *aprosodos; dysephodos,* impassable, inaccessible:

Gr. *anodeutos,* impassable:

Gr. *aperatos,* impassable:

Gr. *aporos,* impassable, impracticable:

Gr. *aprosiktos,* unattainable: *Westwoodia(Aprosictus) duivendodei* (a beetle).

Gr. *aprositos; dysprositos; dysprosoistos,* unapproachable, hard to get at: *Aprosita ulothrix* (a moth), *Dysprosoestus costatus* (a beetle).
L. *intransibilis,* impassable:
L. *intransmeabilis,* impassable:
L. *invius,* without a road, impassable:
See: **difficult**

impastus, L. unfed, hungry; see **hunger**

impatient < L. *impatiens,* desiring immediate action; *impatientia,* restlessness, want of endurance; see **move**

impavidus, L. fearless, intrepid; see **bold**

impedio, *-itus,* L. hinder, prevent, obstruct; see **hinder**

impendens, L. overhanging, imminent; see *pendo* under **hang**

impendium, L. charge, cost; see *pendo* under **pay**

impensus, L. ample, great, large, strong; see **large**

imperfect < L. *in,* not, *perfectus,* complete; see **fault**

impero, *-atus,* L. command; *imperator, -is,* commander-in-chief, general, ruler; *imperialis,* of the empire or emperor; *imperiosus,* possessed of command, mighty, domineering; see **govern**

imperitus, L. inexperienced, ignorant; see **dull**

impertilis, L. indivisible; see **one**

impervius, L. impassable, impenetrable, closed; see **close**

impetigo, L. a skin eruption; see **disease**

impetrabilis, L. easily obtainable; see **rest**

impetus, L. attack, charge, spasm; *impetuosus,* violent; see **strike**

impexus, L. uncombed; see **rough**

impiger, L. diligent, active; see **busy**

impilium, L. a felt shoe; see **shoe**

impingo, *-actus,* L. push, strike, drive, thrust; see **strike**

impius, L. irreverent, wicked; see **profane**

implacabilis, L. unappeasable; see **merciless**

implexus, L. entwined, interwoven; see *plecto* under **weave**

implicate < L. *implico, -atus,* involve, entangle; see **trap**

impluvium, L. the skylight in the roof of a Roman house; see **hole**

impolitus, L. rough, unpolished; see **rough**

important < L. *importo, -atus,* bring in, bring about, cause; see **worth**

impossible < L. *impossibilis,* not possible.
  Gr. *anenchoretos,* impossible:
  Gr. *anendektos,* impossible, inadmissible:
  Gr. *apoietos,* undone, unfinished, impossible:

impotens, L. powerless, weak; see **weak**

impregnate < L. *impraegno, -atus,* engender; see **begin**

improbo, *-atus,* L. disapprove, blame, condemn, reject; see **blame**

improbus, L. below standard, of bad quality, poor, inferior; see **bad**

improper < L. *improprius,* unfit, unsuitable, unseemly; see **unfit, bad**

improvidus, L. not anticipating or foreseeing; *improvisus,* unexpected, unforeseen; see **lot**

impudicus, L. lewd, unchaste, unashamed; see **lewd**

impunitus, L. free from danger, safe; see **free**

impure < L. *impurus,* unclean, base; see **dirt, mix**

imus, L. lowest; see *inferus* under **low**

-imus, L. (Gr. *-imos*), pertaining to, having the quality of; see **nature**

in < AS. *in:* into, income, inlet, inner, inmost, inward, within.
  L. *centrosus,* inward, internal; see *centrum* under **point**
  Gr. *eis,* in, into, to: eisegesis, isagoge.
  Gr. *em; en,* in; *endon; entos,* within; *enapo-,* in; *endoteros,* interior; *endotatos,* inmost; *endothi,* within: emblem, embryo, emphasis, enthusiasm, endocrine, endophyte, endopodite, endogen, endoderm, entoptic, endocarp, Entozoa, ental, imp, ink. *Endomychus chinensis* (a beetle). *Endothia parasitica* (chestnut-blight), *Encephalartos caffer* (Hottentot breadfruit), errhine.

Gr. *eso,* within; *esoteros,* inner; *esotatos,* inmost; *esoterikos,* internal: esoderm, esoteric, esotropia, esophoria.

L. *il-, im-, in-, ir-,* < *in,* in, as in *illuceo,* shine in, light up; *immigro,* go in; *invado,* move in; *irrumpo,* rush in: illuminator, illustrate, illation, immigrant, imbibe, imply, impact, incident, include, incinerate, indigenous, indicator, infusion, inhale, inquisitive, investigate, inspire, irruption, irrigate, anoint, embrace, embargo, enclose, enfold, enforce, encourage, entitle, engine, encounter, emperor, *Ctenochira inflata* (a wasp).

L. *intra-, intro-,* within, < *inter,* among, amid; *interior,* inner; *intimus,* inmost; *internus,* inside; *interaneus,* inward, interior, internal; *interulus,* inward, inner; *intraneus,* inner; *intrinsecus,* on the inside, inwardly; *introrsus,* toward the inside; *introversus,* directed inward: intracellular, intramural, intravenous, introspection, introduction, intromission, introrse, introvert, interior, intimate, internal, interne, enter, intrinsic, *Juncus interior* (inland-rush).

L. *intus,* within, inside; *intestinus,* internal: intussusception, intestine, denizen.

Gr. *mychos,* inmost part, recess, sinus, closet; see **room**

L. *penitus,* inward, inner, interior: Penates, penetrate, *Pyrausta penitalis* (a moth).

in- < L. *in,* in, into, against, see **in**; not, without, see **not**; at, on, see **on**

ina, L. fiber; see *is* under **thread**

-ina, L. feminine suffix sometimes with diminutive implications, see **woman;** designating an order of animals, see **class**

inaccessible; see **impassable**

-inae, L. suffix denoting subfamilies of animals; see **class**

inaequalis, L. uneven, rough, unequal, different; see **different**

inanis, L. empty, void, useless, vain; see **empty**

inauris, L. earring; see **earring**

incanto, *-atus,* L. charm, spell; see *cantamen* under **magic**

incanus, L. quite gray, hoary; see *canus* under **gray**

incassus, L. vain; see *cassus* under **empty**

incendiarius, L. causing fire; *incendium,* fire; see *incendo* under **burn**

incense < L. *incensum,* resinous material that yields a fragrant odor or smoke when burned; see **resin**

inceptives; see *-escens* under **begin**

incertus, L. doubtful; see **doubt**

incessans, L. unceasing; see **always**

incestus, L. defiled, impure, unchaste; see **lewd**

inch < L. *uncia,* a twelfth part; see *uncia* under **twelve**

inciens, L. pregnant; see **fertile**

incilis, L. cut in; *incile,* ditch; see *caedo* under **cut, ditch**

incipient < L. *incipio, -ceptus,* begin; see *coepio* under **begin**

incisus, L. cut into; see *caedo* under **cut**

inclemens, L. rough, severe, unmerciful; see **rough**

incognitus, L. unknown, strange; see **secret, strange**

incohatus (inchoatus), L. only begun, incomplete; see **begin**

incola, L. resident, native; see *colo* under **life**

incolumis, L. unimpaired, safe, sound, entire, whole; see *columis* under **safe**

incomptus, L. unadorned; see **simple**

inconditus, L. unformed, rough, rude; see **rough**

incongruus, L. unfit, unsuitable; see **unfit**

inconspicuus, L. not readily visible, not prominent; see **dim**

inconvenient; see **unfit**

incredibilis, L. unbelievable, extraordinary; see **wonder**

incrementum, L. growth, increase; see *cresco* under **grow**

incubus, L. nightmare, a male demon supposed to lie upon persons, especially women, in their sleep; see **dream, spirit**

inculco, *-atus,* L. tread down, stuff, cram; see **press**

incultus, L. untilled, rude, unpolished; see **rough**

incunctabilis, L. admitting no delay; see **press**

incus, L. anvil; *incudo, -cusus,* forge with a hammer; see **anvil, strike**

indago, *-atus,* L. hunt for, investigate; *indagator; indagatrix,* hunter, searcher; see **try**

indecens, L. unbecoming, unseemly, unfit, improper; see **unfit**

indefessus, L. unwearied, indefatigable; see **strong**

indefinite < L. *indefinitus,* indeterminate, uncertain, vague; see **doubt**

indelibatus, L. untouched, uninjured; see **safe**

indelible < L. *in,* not, *delebilis,* erasable.
　Gr. *anekniptos,* indelible:
　Gr. *anekplytos,* indelible:
　Gr. *anektriptos,* indelible:
　Gr. *anexaleiptos,* indelible:
　Gr. *anexitelos,* indelible:
　L. *ineluibilis,* that cannot be washed out, indelible:

indemnis, L. unhurt, uninjured; see **safe**

indeprensus, L. unobserved, undiscovered; see **secret**

indeptus, L. attained, reached; see *apiscor* under **take**

index, *-icis,* L. pointer, sign, token, forefinger, catalogue, spy; *indicium,* sign, mark, disclosure, discovery; *indico, -atus,* point out, reveal; see **mark,** *dico* under **speak, list**

indicum, L. indigo, < *Indicus,* of India; see **blue**

indifferent; see **dull**

indigens, L. in want, needy; see *egenus* under **poor**

indigenus, L. belonging to a country, native; see **native**

indignans, L. angry; see **anger**

indigo < L. *indicum;* see **blue**

indigus, L. in want, needy, poor; see *egenus* under **poor**

indiscretus, L. undivided, closely connected; see **one**

indistinctus, L. obscure, dim; see **dim**

inditus, L. put upon, imposed; see **place**

indivisus, L. undivided; *individuus; indivisibilis,* not to be divided; see **all, one**

indoles, L. nature, disposition, talents; see **nature**

indomitabilis, L. that cannot be conquered, see **strong;** *indomitus,* untamed, wild, see **nature**

inductura, L. covering, coating; see **blanket**

indulgeo, *-dultus,* L. be forbearing, concede, grant, favor; see **give**

indusium, L. tunic, garment; *indumentum,* clothing, garment; *induvia,* garment; see **garment**

industrius, L. busy, active, diligent; see **busy**

indutia, L. truce, armistice; see **rest**

indutilis, L. that can be inserted or joined in; see **insert**

indutus, L. clothed, covered, < *induo, -utus,* put on clothes; see *indusium* under **garment**

-ine; -in < L. *-inus,* suffix denoting chemical terms, drugs; see **drug,** *-anus* under **nature**

ineffable < L. *ineffabilis,* inexpressible, unutterable: ineffability.
　Gr. *agryktos,* not to be spoken of:
　Gr. *anaudetos,* unutterable:
　Gr. *aphatos,* ineffable, unutterable: *Aphatum rufulum* (a beetle).
　Gr. *arrhetos,* unutterable, inexpressible: *Arrhetus volxemi* (a beetle), *Arretotherium* [*Arrhetotherium*] *acridens* (an Oligocene mammal).
　Gr. *athesphatos,* inexpressible, ineffable, marvelous; see **large**
　Gr. *ekphatos,* beyond the power of speech, ineffable:
　L. *incompellabilis,* that cannot be named or addressed by name:
　L. *infandus,* unspeakable; see **bad**
　L. *tacendus,* not to be uttered, secret; see *taceo* under **silent**
　See: **secret, wonder, silent**

ineptus, L. unsuitable, silly, absurd; see **fool**

inermis, L. unarmed, without weapons; see **bare**

inert < L. *iners, -ertis,* inactive, idle, lazy; *inertia,* idleness; see **rest**

inesco, *-atus,* L. allure with bait, entice; see *esca* under **food**

inevitable < L. *in,* not, *evitabilis,* avoidable; see **sure**

inexorabilis, L. that cannot be moved by entreaty, inflexible, relentless; see **merciless**

inexplebilis, L. that cannot be filled, insatiate; see **hunger**

inexplicabilis, L. unexplainable; see **secret**

infacetus, L. rude, coarse, blunt; see **rough**

infallible < L. *infallibilis,* not making mistakes, inerrant; see **sure**

infandus, L. unspeakable, abominable; see **bad**

infans, -antis, L. speechless; a little child; *infantilis,* pertaining to children, childish; see **child, silent**

infatigatus, L. unwearied; see **strong**

infaustus, L. unfortunate; see *faustus* under **lot**

infectus, L. dyed, stained, tainted; *infectorius,* serving for dyeing; see **color**

infelix, -icis, L. unhappy; see **sad**

infensus, L. hostile, unfriendly; see **enemy**

infer < L. *infero, illatus,* bring in, conclude; see *illativus* under **think**

infernalis, L. pertaining to the lower regions, underground; see *infernus* under **hell**

inferus, L. low; *inferior,* lower; *infimus; imus,* lowest, last; see **low, end**

infestus, L. hostile, inimical, dangerous, unsafe; *infesto, -atus,* attack, injure; see **danger**

infidelis, L. unreliable, unfaithful, false; see **lie**

infimus, L. lowest; see *inferus* under **low**

infinite < L. *infinitus,* boundless, unlimited: infinity, infinitive, infinitesimal, ad infinitum.
  Gr. *apeiros; aperastos,* boundless, infinite, countless; *apeirometros,* immeasurable: L. *illimitatus,* boundless, illimitable:
  L. *infinibilis,* endless:
  See: **number**

infirmus, L. weak, feeble; see **weak**

inflammo, L. set on fire, kindle; see *flamma* under **fire**

inflatus, L. puffed up, swollen; see *flo* under **blow**

infra, L. underneath, below; see **under**

infractus, L. unbroken; see **all**

infrequens, L. seldom, rare; see **few**

infucatus, L. painted; see *fuco* under **color**

infula, L. band, bandage, fillet; see **ribbon**

infundibulum, L. funnel; see **funnel**

infusio, L. a pouring into, a steeping; see **wet**

infusorium, L. receptacle, vessel, can; see **pot**

-ing < AS. *ende; -ing,* forms present participles and verbal nouns, see **make;** patronymics and diminutives, see **son**

inga, Ab.Am. a kind of legume; see **bean**

ingeniosus, L. superior in intellect, gifted, talented; *ingenium,* genius; see **know**

ingens, L. vast, huge, great, immoderately large; see **large**

ingluvies, L. crop, maw; *ingluviosus,* gluttonous; see **stomach**

inguinalis, L. of the groin; see *inguen* under **groin**

inion, Gr. back of the head, occiput; see **head**

iniquus, L. uneven, unjust, hurtful; see **hurt**

inis, Gr. son, daughter; see **son**

initialis, L. beginning, first, original; see *ineo* under **begin**

injectivus, L. pertaining to insertion; see **insert**

injury < L. *injuria,* harm, wrong, injustice; see **hurt, bruise, pain, sore, cut, break**

injustus, L. wrong, harsh; see **bad**

ink < OF. *enque,* < L. *encaustum* (Gr. *enkauston*), a purple-red ink.
  Gr. *alabe,* f. a kind of ink:
  L. *atramentum,* n. any black fluid, ink; *atramentarium,* n. inkstand: *Coprinus atramentarius* (a mushroom).
  Gr. *brochis, -idos,* f. inkhorn:
  Gr. *kekis, -idos,* gall, ink; *kekidion,* dim.; see **gall**

Gr. *melan, -os,* n.; *melanteria,* f. black pigment, ink:
L., Gr. *sepia,* cuttlefish, squid, ink; see **mollusk**
Gr. *tholos,* mud, dirt, squid-ink; see **dirt**

inlay; see **mosaic**

inn, AS. house, tavern, lodging-place; see **house**

innocent < L. *innocens, -entis,* harmless, guiltless: innocence.
Gr. *adolos,* guileless, honest; see **true**
Gr. *akakos,* guileless, innocent; *akakia,* f. guilelessness: Acacius.
Gr. *amomos,* blameless:
Gr. *amomphos,* blameless:
Gr. *anaitios,* guiltless:
Gr. *apemon,* harmless; see **safe**
L. *expers, -ertis,* having no part in, not sharing, devoid of, free from:
L. *immerens,* undeserving, innocent; *immertus,* undeserved, guiltless: unmerited.
L. *inculpabilis; inculpatus,* blameless, faultless, flawless:
L. *indelictus,* blameless, faultless, innocent:
L. *insons, -ontis,* guiltless, innocent: *Xiphomyrmex spinosus insons* (an ant).
Gr. *nelites,* guiltless, harmless:
See: **pure, free**

innominatus, L. nameless; see **secret**

innocuus; innoxius, safe, harmless; see **safe**

ino- < Gr. *is, inos,* muscle, fiber, sinew, strength; see **thread**

inoculo, -atus, L. ingraft, implant; see **begin**

inopinatus; inopinus, L. unexpected; see **lot**

inops, L. poor, weak, helpless; see **poor**

inornatus, L. unadorned; see **simple**

inquilinus, L. temporary inhabitant, sojourner, tenant, lodger; see **guest,** *colo* under **life**

inquino, -atus, L. befoul, stain, pollute, defile; see **dirt**

inquire < L. *inquiro, -quisitus,* seek after, search for, pry into, examine; see *quaero* under **try, ask**

insanus, L. of unsound mind; see **mad**

insatiable < L. *insatiabilis,* that cannot be satisfied; see *insatiatus* under **hunger**

insaturatus, L. unfilled, insatiate; see **hunger**

inscitus, L. ignorant, awkward, absurd, silly; see **dull**

inscrutable < L. *inscrutabilis,* incomprehensible, mysterious; see **secret**

insculptus, L. engraved; see *sculpo* under **cut**

insect < L. *insectum,* n., < *inseco, -sectus,* cut into: insectarium, insectary, insectean, insectile, insectine, insecticide, insecticidal, insectival, insectivore, *Troxochrus insectus* (a spider).
L. *bestiola,* small animal, insect; see *bestia* under **animal**
Gr. *bostrychos,* a kind of insect; see **curl**
L. *cutio, -onis,* m. a small insect:
Gr. *entomon,* n. insect, < *entomos,* cut into: entomology, entomologist, entomophilous, entomic, entomogenous, entomoid, entomolite, Entomostraca, *Entomolithus paradoxus* (a trilobite), *Entomophthora sphaerosperma* (a fungus), *Eosentomon yosemitense* (a proturan).
Gr. *sphondyle,* f. an insect that lives on the roots of plants: *Sphondylia tomentosa* (a beetle).
See: **ant, bee, beetle, bug, butterfly, flea, fly, grasshopper, moth, wasp**

insert < L. *insero, -ertus,* introduce into, place in or between, mix or mingle with: insection.
Gr. *embolon,* n.; *embolos,* m. anything inserted, bolt, bar, peg, wedge; *embole,* f. insertion, invasion; *embolimos,* inserted; *emballo,* insert, graft; *epembole; parembole,* f. insertion, parenthesis: embole, embolism, embolic, Collembola, *Embolichthys misukurii* (a fish), *Embolimus americanus* (a wasp), *Embola xanthocephala* (a moth), *Parembola litterata* (a pelecypod).
Gr. *enetos,* injected:
Gr. *enophthalmizo,* inoculate, graft; see **graft**
Gr. *entaxis,* f. insertion:
Gr. *enthesis, -eos,* f. insertion; *epenthesis; parenthesis,* f. a putting in extra, insertion; *enthetos,* capable of being put in: parenthesis, parenthetic, enthetic.
L. *indutilis,* that can be inserted or joined in:

L. *injectivus,* pertaining to insertion or injection; *interjectus,* thrown or placed between, interposed: inject, injector, interjection, *Hedulia injectiva* (a moth).

L. *insinuo, -atus,* introduce gently, craftily, or insidiously: insinuate, insinuation.

L. *insiticius,* grafted, inserted, < *insero, -situs,* implant, graft: insititious, *Prunus insititia* (bullace).

L. *intercalaris; intercalarius,* of or for insertion, extra; *intercedo, -cessus,* go between, intervene, arbitrate; *intergerivus,* placed between; *interloquor, -locutus,* speak between, interrupt in speaking; *interludo, -lusus,* play between; *interpolatio, -onis,* f. intercalation of new or modified material; *interpono, -positus,* place between, insert, introduce, interfere; *intersitus,* interposed; *intervenio, -ventus,* come between, interrupt, interpose: intercalate, intercalary, *Gryllus intercalatus* (a cricket), intercede, intercession, interlocutor, interlude, interpolate, *Actidium interpolatum* (a beetle), interpose, *Melania intersita* (a gastropod), intervene.

L. *introduco, -uctus,* lead in, put in, insert: introduce, introduction, introductory.

L. *intromitto, -issus,* put in, introduce: intromittent.

L. *intrudo, -usus,* enter without invitation or permission: intrude, intrusive, intrusion.

See: **graft, middle, space, break, coitus, enter**

insicium, L. mincemeat; see **food**

insidiosus, L. cunningly deceitful; *insidia,* ambush; see **trap**

insignis, L. remarkable, notable, distinguished; *insignitus,* marked, clear, plain; see *signum* under **mark**

insilium; insubulum, L. treadle of a loom; see **step**

insipidus, L. tasteless; see **tasteless, dull**

insist < L. *insisto, institus,* press upon, urge; see **press**

insiticius, L. grafted, inserted; see **insert**

insolens, L. haughty, arrogant; see **proud**

insolitus, L. unusual, uncommon, strange; see **strange**

insomnia, L. sleeplessness; see **awake**

insons, L. innocent; see **innocent**

insopitus, L. wakeful, sleepless; see **awake**

inspector, L. examiner, searcher; see **try**

inspissatus, L. thickened; see *spissus* under **thick**

instabilis, L. unsteady, changeable; see **shake**

instant < L. *instans,* standing by; see **now**

instar, L. form, figure, image, likeness; see **form**

instigo, -atus, L. incite, stimulate; see **arouse**

instinct < L. *instinctus,* natural impulse; see **nature**

instita, L. border, flounce; see **border**

institutor, L. founder, creator; *institutum,* establishment, organization; see **make, plan**

instragulum, L. covering, coverlet; see **cover**

instruct < L. *instruo, -uctus,* build in, equip, teach; see **teach**

instrument < L. *instrumentum;* see **tool, vessel**

insubidus, L. stupid, foolish; see **fool**

insuetus, L. unaccustomed to, inexperienced in, unusual; see **strange**

insula, L. island; *insulanus; insularis,* of islands; see **island**

insulsus, L. unsalted, tasteless, insipid, foolish; see **tasteless**

insult < L. *insultus,* scoffing, reviling; see **curse**

intactus, L. untouched, entire, chaste; see **all**

intaminatus, L. unpolluted, pure; see **pure**

integer, L. whole, entire, sound; *integro, -atus,* make whole, restore; see **all**

integumentum, L. covering; see *tego* under **cover**

intellectus, L. understanding, perception; see *intellego* under **know**

intemeratus, L. undefiled, inviolate, virgin, pure; see **pure**

intemperatus, L. immoderate, uncontrolled; see **free**

intensifiers; intensives; see **very**

inter- < L. *inter,* between, among; see **middle**

interaneum, L. gut, intestine; see **intestine**

intercalaris, L. of or for insertion, extra; see **insert**

interdictum, L. a prohibition; see **forbid**

interest < L. *intersum,* be between, be different, be of concern; premium paid for the use of money; see **gain**

intergerivus, L. placed between; see *intercalaris* under **insert**

interim, L. meantime, meanwhile; see *inter* under **middle**

interior, L. inner; see *intra-* under **in**

interitus, L. perished, destroyed, lost; *interibilis,* perishable, mortal; see **death**

interluvies, L. strait; see **way**

internecinus, L. deadly, destructive; see *neco* under **kill**

internode < L. *internodium,* the space between knots; see *nodus* under **knot**

internus, L. inward, interior; see *intra-* under **in**

interpolo, *-atus,* L. insert, alter; see *intercalaris* under **insert**

interpretor, *-atus,* L. explain, translate; *interpres, -etis,* agent, intermediary; see **clear**

interrogatio, L. question, inquiry; see *rogo* under **ask,** FIGURES OF SPEECH under **form**

interruptus, L. broken apart, between, off, or asunder; see *rumpo* under **break**

interstinctus, L. checkered, variegated; see **color**

interstitium, L. space between, interval; see **space**

interulus, L. inward, inner; see *intra-* under **in**

interval < L. *intervallum,* space between two palisades, intermission, pause; see **space, break**

intestine < L. *intestinum,* n. bowels, gut: intestinal, intestiniform, *Gastrophilus intestinalis* (a botfly).

L. *anus,* m. rectal opening, fundament: anus, anal, *Chaetogyne analis* (a fly).

Gr. *archos,* m. rectum, anus: architis, archocele, *Crossarchus obscurus* (kusimansel).

Gr. *batalos,* anus, hinder parts; see **rump**

Gr. *cholas, -ados,* f. bowels, gut; *cholix, -ikos,* f. gut; *cholikion,* n. dim.:

L. *chorda* (Gr. *chorde*), gut, string made of gut, catgut; *epichordis, -idos,* mesentery; see **rope**

L. *colon* (Gr. *kolon*), n. large intestine; *kolikos,* pertaining to the colon; pain in the colon: colitis, colic, colotomy, *Bacillus coli* (a bacterium).

ML. *duodenum,* n. first part of the small intestine: duodenum, duodenal, *Coelothauma duodenum* (a radiolarian).

Gr. *endinon,* n. entrail:

Gr. *enteron,* n. intestine; *enteridion; enterion,* n. dim.; *enterikos,* intestinal: enteritis, enteroid, enterocele, dysentery, mesentery, lientery, Coelenterata, *Enterolobium cyclocarpum* (a legume), *Enteridium rozeanum* (a myxomycete).

Gr. *entosthion,* n. entrail:

L. *extalis,* m. rectum, < *extum,* n. an internal organ:

L. *gigerium,* n. entrail, particularly of birds: gigerium, gizzard.

L. *hira,* f. gut, intestine; *hilla,* f. dim. small intestine, sausage:

L. *ileum,* n. last part of the small intestine: ileum, ileac, ileocolic.

L. *interaneum,* n. gut, intestine: entrail.

NL. *jejunum,* n. middle part of the small intestine: jejunum, jejunectomy.   See *jejunus* under **hunger**

L. *lactis,* f. small intestine, chitterling:

L. *longano, -onis,* m. straight gut, sausage:

L. *nestis, -idis* (Gr. *-idos*), f. jejunum:

Gr. *proktos,* m. anus, rectum, fundament, hinderparts, tail: proctal, proctitis, proctotrypid, proctoclysis, proctodaeum, proctoplegia, proctoscope, proctoptosis, *Dasyprocta agouti* (agouti), *Euproctis chrysorrhea* (brown-tailed moth), *Proctonotus mucroniferus* (a gastropod), *Thyroproctus townsendi* (a milleped), *Aphonoproctus pentodontinus* (a beetle), *Gymnoproctus abortivus* (a grasshopper).

L. *rectum,* n. straight gut or end of the large intestine: rectum, rectal, rectocele.

Gr. *splanchnon,* n. entrail; *splanchnidion,* n. dim.: splanchnic, asplanchnic. *Splachnum[Splanchnum] ampullaceum* (a moss).

L. *viscera,* n. entrails, pl. of *viscus, -eris,* flesh or organ inside the body: *viscereus.* of flesh: viscera, visceral.

See: **sausage, pipe**

intimus, L. inmost, deepest; see *intra-* under **in**

intitubabilis, L. firm, unwavering; see **stand**

intolerable < L. *intolerabilis,* not bearable, insupportable; see **hard, difficult, trouble**

intonsus, L. unshorn, bearded, rough; see **rough**

intra-; intro-, L. within; see **in**

intrabilis, L. that can be entered; see **enter**

intractabilis, L. unmanageable; see **rough**

intraneus, L. inner; see *intra-* under **in**

intrepidus, L. unshaken, undaunted, brave; see **bold**

intricate < L. *intrico, -atus,* entangle, perplex; see **complex, trap**

intrinsecus, L. on the inside, inward; see *intra-* under **in**

intro-, L. inside, within, in; see *intra-* under **in**

introrsus, L. toward the inside, inward; see *intra-* under **in**

intrude < L. *intrudo, -usus,* enter without invitation or permission; see **insert**

intuition < L. *intueor, -uitus,* look at, contemplate; see **know**

intus, L. within, inside; see **in**

intybus, L. endive, succory; see **lettuce**

inula, L. elecampane; see **composite**

-inus, L. pertaining to; see *-anus* under **nature**

inusitatus, L. rare, unusual; see **few, strange**

inutilis, L. useless, worthless; see **worthless**

Inuus, L. god of fecundity in herds; see **fertile**

invalidus, L. infirm, weak; see **weak**

invasor, L. intruder; *invasio, -onis,* attack; see **enter**

invectivus, L. scolding, abusive; see **scold**

invent < L. *invenio, -ventus,* devise, contrive; see **make**

inventarium, L. list; see **list**

inversus, L. turned upside down; see *verto* under **turn**

investigate < L. *investigo, -atus,* inquire into, search after, trace; *investigator,* inquirer, searcher; see **try**

invictus, L. unconquered, strong; see **strong**

invidus; invidiosus, L. envious, jealous, hostile; see **jealous**

inviolatus, L. unhurt; see **safe**

invisibilis, L. unseen; see **secret**

invisus, L. hated, hateful, detested; see **hate**

invito, -atus, L. bid, ask; see **ask**

involucrum, L. wrapper, case, envelope; see **sheath**

involutus, L. complex, intricate; see **complex**

invulnerabilis, L. unwoundable; see **safe**

Io, Gr. daughter of Inachus, a stream god; see *Inachus* under **stream**

iochmos, Gr. rout; see *dioktos* under **hurt**

iodo- < iodine, < Gr. *iodes,* violetlike in color; see ELEMENTS under **thing**

ion < Gr. *ion, -tos,* participle of *eimi,* go, move, see **move**; < *ion,* violet; *iodes,* violetlike, see **violet**

-ion; -sion; -tion < L. *io, -ionis,* act of, process of, having the nature of, see **nature**; < Gr. *-ion,* comparative ending, see **comparatives**; diminutive suffix, see *-idium* under **little**

ionisco- < Gr. *ioniskos,* a gilthead; see **bream**

ionthas, -ados, Gr. shaggy; *ionthos,* young hair; see **hair**

iops, Gr. a small fish; see **fish**

-ior; see *-or* under **make, comparatives**

ioros, Gr. guard, keeper; see **guard**

ios, Gr. arrow, poison; see **poison**

iota, Gr. ninth letter of the Greek alphabet; anything very small; see **letter**

iotoco- < Gr. *iotokos,* poisonous; see *ios* under **poison**

ipecac < an Ab.Am. word; see **drug**

iphi, Gr. strongly, mightily; *iphthimos,* stout, strong; see **strong**

iphyum, L. (Gr. *iphyon*), a plant; see **plant**

ipne, Gr. a woodpecker; see **woodpecker**

ipnos, Gr. oven, furnace, kitchen, lantern; see **oven**

ipnum, L. (Gr. *ipnon*), a marsh plant; see **plant**

ipomoea, NL. a genus of the morning-glory family; see **bindweed**

ips, -*pos*, Gr. a kind of worm; see **worm**

ipse, L. self; see **alone**

ipsis, Gr. a pressing; see *ipoo* under **press**

ipsus, L. (Gr. *ipsos*), a kind of ivy; see **ivy**

ir- < L. *in*, in, not; see **in, not**

ira, L. anger, fury, rage, wrath; *iracundus,* inclined to anger, wrathful, < *irascor, iratus;* see **anger**

ireno- < Gr. *eirene*, peace; see **rest**

irio, L. a kind of cress; see **mustard**

iris, -*idis*, L. (Gr. *-idos*), f. rainbow, see **color;** a genus of plants: *Iris pallida* (sweet iris), *Haliotis iris* (an abalone), *Canna iridiflora* (a canna), *Oberonia iridifolia* (an orchid), *Moraea iridioides* (an iridacead).
NL. *acidanthera,* f. a genus of the iris family: *Acidanthera bicolor* (a gladixia).
L. *crocus* (Gr. *krokos*), a plant of the iris family; see **crocus**
NL. *cypella,* f. a genus of the iris family: *Cypella peruviana* (an iridacead).
L. *gladiolus,* m. sword-lily: *Gladiolus carmineus* (carmine gladiolus).
Gr. *ixia,* f. a plant; now a genus of the iris family: *Ixia maculata* (African ixia), *Ixiolirion montanum* (an amaryllidacead).
L. *sisyrinchium* (Gr. *sisyrinchion*), n. a genus of the iris family: *Sisyrinchium angustifolium* (blue-eyed grass).
NL. *tritonia,* f. a genus of the iris family: *Tritonia deusta* (a tritonia).
L. *xyris, -idis* (Gr. *-idos*), f. wild iris: *Xyris fimbriata* (a yellow-eyed grass).

Irish < AS. *Iras,* a Celtic people in western Britain; see **Celt**

irmo- < Gr. *eirmos*, series, train; see **line**

iron < AS. *iren; isen* (G. *eisen*).   See ELEMENTS under **thing**
Gr. *chalyps, -ybos,* m. hardened iron, steel; *chalybikos,* of steel: chalybeate, chalybeous, *Chalybion caeruleum* (a wasp), *Cinnyris chalybeus* (a sunbird), *Altica[Haltica] chalybea* (a flea-beetle).
L. *ferrum,* n. iron; *ferreus,* of iron; *ferratilis; ferratus,* ironed, armed; *ferrarius,* m. blacksmith; *ferruginus; ferrugineus,* like iron rust, rust-colored: ferreous, ferrous, ferric, ferromagnesian, ferrocyanide, ferricyanide, farrier, ferruginous, nife, Taliaferro(Tolliver), fer-de-lance(*Trigonocephalus lanceolatus*), *Mesua ferrea* (a clusiacead), *Rhinolophus ferrum-equinum* (a horseshoe-bat), *Colubrina ferruginosa* (snakebark), *Spirophyllum ferrugineum* (an iron-depositing bacterium).
Gr. *sideros,* m. iron; *sidereos; siderikos,* of iron; *siderion,* n. iron implement; *siderites,* m. lodestone: siderographer, siderite, pharmacosiderite, *Sideroxylon quadriloculare* (Trinidad ironwood), *Sideritis syriaca* (Syrian ironwort), *Metrosideros robusta* (rata), *Rhinolophus hipposideros* (a horseshoe-bat).
Gr. *solos,* m. a mass of iron:

irono- < Gr. *eiron, -os,* dissembler; see **lie**

irony < L. *ironia* (Gr. *eironeia*), dissimulation; see FIGURES OF SPEECH under **form, lie**

irpus, L. wolf; see **wolf**

irrationalis, L. without reason; see **mad**

irregular < OF. *irreguler,* < L. *in*, not, *regularis,* according to rule; see **different, confusion**

irretitus, L. caught in a net; see *rete* under **net**

irrigate < L. *irrigo, -atus,* supply with fluid; see *rigo* under **wet**

irrisor, L. derider, mocker, scoffer; see *rideo* under **laugh**

irrito, -*atus*, L. excite, provoke; see **arouse**

irritus, L. invalid, vain, useless, infertile; see **worthless**

irroratus, L. bedewed, moistened, covered with granules; see *ros* under **drop**

irrumator, L. one who practices beastly obscenity; see **lewd**

irtiola, L. a kind of grape in Umbria; see **grape**

Irus, L. (Gr. *Iros*), proverbial name for a beggar, poor man; see **beggar**

is- < Gr. *eis*, in, into, to; see **in**

-is, L. adjectival suffix meaning with, having, nature of; see **nature**

isatis, L., Gr. woad; see **mustard**

isca, L. (Gr. *iska*), a kind of fungus on trees; see **fungus**

ischas, *-ados,* Gr. dried fig; see **fig**

ischion, Gr. hip; see **hip**

ischnos, Gr. withered, thin, lean, weak; *ischnoites,* thinness, leanness; see **thin**

ischo, Gr. form of *echo,* have, hold, restrain; see *echo* under **hold**

ischys, Gr. strength, force, might; *ischyros,* strong, mighty, stubborn, excessive; see **strong**

-iscus, L. (Gr. *-iskos*), diminutives; see DIMINUTIVES under **little**

-ise < L. *-itia,* make, give, act; see **make**

-ish < AS. *-isc,* denoting origin, pertaining to; see **nature**

Isis, L., Gr. an Egyptian goddess of fertility; see **spirit**

-isk < Gr. *-iskos,* diminutives; see *-iscus* under **little**

island < AS. *igland:* islander, island of Langerhans, Rhode Island, Lille. Derive: lisle. Malay *adal,* island with a reef surrounding a central lagoon: atoll, *Atolla wyvillei* (a jelly-fish).
  Sp. *cayo,* m. islet: cay, key, Florida Keys, Key Largo.
  L. *insula* (OF. *isle*; It. *isola*), f. island; *insulanus; insularis,* of an island; *insulatus,* isolated; *insulosus,* full of islands: insular, insulate, isolate, peninsula, insulin, insulinase, islet, British Isles, Isle of Man, Belle Isle, *Bothrops insularis* (a snake), *Mustela peninsulae* (Florida weasel), *Begonia insularum* (a begonia), *Tiphia isolata* (a wasp), *Ferocactus peninsulae* (a cactus).
  Gr. *nesos,* f. island; *nesidion; nesion; nesydrion,* n. dim.; *nesiotes,* insular; m. islander: nesiote, nesogean, Polynesia, chersonese, Peloponnesian, *Nesiophasma spinulosum* (a phasmatid), *Nesydrion fuscum* (a neuropterid), *Nesotragus moschatus* (a fossil antelope), *Nesidobius ramulus* (a beetle), *Speonesiotes hirsutus* (a beetle).

-ism < L. *-ismus* (Gr. *-ismos*), denoting condition, quality, doctrine, sect; see **nature**

iso- < Gr. *isos,* equal, like; see **equal**

isoëtes, L., Gr. name for a kind of houseleek; see **quillwort**

isolatus, NL. detached, separate, < It. *isola,* < L. *insula,* island; see **island**

-issimus, L. adjectival superlatives; see **very**

issue < OF. *eissue,* < L. *exeo, -itus,* go forth; see **end, lot**

-ist < F. *-iste,* < L. *ista* (Gr. *-istes*), agent, one who; see **make**

isthmus, L. (Gr. *isthmos*), neck, narrow passage, neck of land; see **neck**

-istos, Gr. adjectival superlatives; see **very**

-ita, Sp. diminutives; see DIMINUTIVES under **little**

italos, Gr. bull; see *vitulus* under **cattle**

itamos, Gr. hasty, headlong, impetuous, reckless; see **swift**

itch < AS. *gicce.*
  Gr. *knesis; knesmos,* an itching, < *knetho,* scratch, scrape, tickle, sting; see **scrape**
  Gr. *knismos,* m. an itching, tickling; *knisma, -tos,* n. scratch, < *knizo,* annoy, nettle, tease, tickle, scrape: *Cnismatocentrum* (a section of Brachiopoda).
  Gr. *odagmos,* m. itching, irritation: *Odagmia ornata* (a fly), *Discophyria odagmiina* (a fly).
  L. *prurigo, -inis,* f. itch; *pruritus, -us,* m. an itching, itch; *pruriosus; pruritivus,* causing itching; *pruriginosus,* itchy, scabby; *prurio,* itch, be eager for: prurient, pruritic, prurigo, pruriginous, pruritus, *Stizolobium pruriens* (cowhage).
  Gr. *psora,* f. itch, scab, mange; *psoraleos; psoros,* itchy, scabby, mangy, rough; *psorikos,* of the itch: psoriasis, *Psoralea esculenta* (breadroot-scurfpea), *Psoroptes communis* (mange-mite), *Melampsora lini* (flax-rust), *Psorophora ferox* (a mosquito).
  L. *scabies,* f. roughness, scurf, mange, itch: scabies. See *scaber* under **rough**

-ite; -ites < L., Gr. *-ites,* having the nature of, like, descendant, follower; salts of *-ous* acids, see **nature**; mineral terms, see **stone**

itea, Gr. willow; now a genus of the saxifrage family; see **saxifrage**

item, L. so, likewise, in like manner, as, part of a series; see **one**

iter, *itineris,* L. journey; *itinerarius,* of a journey; see *eo* under **move**

itero, *-atus,* L. repeat; *iterabilis,* that may be repeated; see **return**

-ites; see *-ite* under **nature**

ithris, Gr. eunuch; see **sterile**

ithycteano- < Gr. *ithykteanos,* tall, slender; see *eukteanos* under **thin**

ithys, Gr. straight, direct, upright; see **straight**

-itis, Gr. suffix denoting inflammation, disease, pain; see **disease**

iton, Gr. a mushroom; see **fungus**

-itus, L. having the nature of, pertaining to; see *-atus* under **nature**

itys, *-yos,* Gr. rim or felly of a wheel; see **border**

iulis, L. (Gr. *ioulis*), rainbow-wrasse; see **wrasse**

iulus, L. (Gr. *ioulos,* down, wool), catkin, see **catkin;** milleped, see **myriapod**

-ium, L. quality or nature of, see **nature;** diminutives, see *-idium* under **little**

iva, NL. a genus of the composite family; see **composite**

ive < L. *-ivus,* denoting nature, quality, action; see **nature**

ivory < L. *ebur, eboris,* n.; *eboreus; eburneus; eburnus,* of ivory; *eborarius,* m. worker in ivory; *eburatus,* inlaid with ivory: eburneous, eburnine, eburine, ivory-nut palm *(Phytelephas macrocarpa), Cymbidium eburneum* (an orchid), *Yoldia eborea* (a fossil pelecypod), *Eburia quadrigeminata* (a beetle), *Eburna spirata* (a gastropod), *Eburniola leucogaster* (a beetle), *Pinnixa eburna* (a crab), *Ficus eburnea* (ivory-fig).
    Gr. *elephantinos,* of ivory; see **elephant**

ivy < AS. *ifig;* see **vine**

ix, Gr. a kind of grub; see **worm**

ixalos, Gr. bounding, springing; see **leap**

ixeutes, Gr. bird-catcher; see *ixos* under **glue**

ixia, Gr. a plant; now a genus of the iris family; see **iris**

ixine, Gr. a kind of thistle; see **thistle**

ixoboros, Gr. missel-thrush; see **thrush**

ixos, Gr. birdlime, mistletoe; see **glue**

ixys, Gr. waist, loins; see **side**

iygmos, Gr. shout, cry; see **call**

iynx, L., Gr. wryneck; see **woodpecker**

-ize < Gr. *-izein,* make, cause to be; action; see **make**

J

jacaranda, Ab.Am. a genus of the bignonia family; see **bignonia**

jacea, L. hayrack; see **frame**

jaceo, *-citus,* L. lie; see **lie**

Jacob < Heb. *Yaqob* (Gr. *Iakobos;* L. *Jacobus;* OF. *Jaques;* It. *Giacomo;* Sp. *Iago; Diego*), supplanter: Jacobean, Jacobin, Jacobite, jacobsite, Jack, jack, jackknife, blackjack, Jackson, Santiago, San Diego, dago, jockey, James, Jamestown, Jim, jimsonweed, applejack, *Senecio jacobaea* (a groundsel).

jactans, L. boasting, bragging, < *jacto, -tatus,* toss about, make gestures, display ostentatiously, boast; see **brag**

jactus; jaculatus, L. hurled, cast; *jaculus,* thrown; see *jacio* under **throw**

jaculum, L. dart, javelin; see **spear**

jail < ONF. *gaiole,* < L. *cavea,* cage, stall, coop, hive, enclosure; see **pen**

jalapa, Sp. a resin, < Ab.Am. *Xalapa,* a town in Mexico; see **resin**

janua, L. door, entrance; *Janus,* god of gates and doors, represented with two faces; *janitor,* doorkeeper; see **door**

jar < Ar. *jarrah*, earthen water-vessel; see **pot, barrel, vessel**

jasmine < Ar. *yasamin*, a plant of the olive family; see **olive**

jasper < OF. *jaspre*, < Gr. *iaspis*; see **stone**

jatropha, NL. a genus of the spurge family; see **spurge**

jaundice < L. *galbinus*, yellowish; see **disease**

javelin < a Celtic word; see **spear**

jaw < uncertain origin.
> Gr. *chelyne*, lip, jaw; see **lip**
> Gr. *gamphela*, f. jaw of an animal, beak of a bird:
> Gr. *genys, -yos*, f. jaw, cheek, chin, edge of an ax; *amphigenys*, double-edged: genoplasty, *Genyornis newtoni* (a fossil bird), *Dermogenys pusillus* (a fish), *Phoxogenys mülleri* (a beetle), *Hylobates leucogenys* (a gibbon).
> Gr. *gnathos*, f. jaw; *gnathidion; gnathion*, n. dim.: gnathic, gnathism, gnathidium, gnathite, gnathotheca, gnathobase, gnathopod, plectognath, prognathous, *Polygnathus lobata* (a conodont), *Gnathodus pustulosus* (a conodont), *Gnathium francilloni* (a beetle), *Gnathidium cephalotes* (a beetle), *Syngnathus acus* (a pipefish).
> L. *magulum*, n. jaw, mouth:
> L. *mala*, f. cheekbone, jawbone, cheek: malar.
> L. *mandibula*, f. jaw: mandible, *Ptorthodius mandibularis* (a beetle).
> Gr. *mastax, -akos*, f. jaws, mouth: mastax, *Mastacomys fuscus* (broad-toothed rat), *Psychomastax psylla* (a grasshopper).
> L. *maxilla*, f. jaw, jawbone; *maxillaris*, of the jaw: maxillary, maxilliped, maxillodental, *Maxillaria picta* (an orchid).
> Gr. *siagon, -os*, f. jaw; *siagonion*, n. dim.: *Siagonophorus loricatus* (a protozoan), *Siagonium punctatum* (a beetle).
> See: **chin, face, tooth, bite**

jay < uncertain origin; see **crow**

jealous < OF. *jelous*.
> Gr. *dyszelos*, very jealous:
> L. *emulor, -atus*, vie with enviously, be jealous of; see **equal**
> L. *invidus; invidiosus*, envious, jealous; *invidia*, f. envy, jealousy, < *invideo, -visus*, look askance at, cast an evil eye upon: envy, envious, invidious.
> L. *lividus*, envious, spiteful; *livor, -is*, envy, spite; see **blue**
> Gr. *phthonos*, m. ill will, envy, jealousy; *phthoneros*, envious, jealous: *Phthonosema tendinosaria* (a moth), *Phthonerodes scotarcha* (a moth).
> See: **hate**

jecur, -coris, L. liver; see **liver**

jeer < uncertain origin; see **laugh, tease, hurt**

jeffersonia, NL. a genus of the barberry family; see **barberry**

jejunus, L. hungry; NL. *jejunum*, middle part of the small intestine; see **hunger, intestine**

jelly < OF. *gelee*, frost, < L. *gelatus*, congealed, stiffened: jellification, jel(gel), jellyfish, gelatin, gelatinous, gelatigenous.
> Malay *agar*, a gelatinous substance from seaweeds: agar (from *Eucheuma spinosum*), *Agarum cribrosum* (a sea-colander).

jentaculum, L. breakfast; see **eat**

jest < OF. *geste*, < L. *gesta*, acts, deeds; see **laugh**

jewel < L. *jocus*, jest, play, plaything; see **ornament, stone, earring**

jocosus; jocularis, L. facetious, < *jocus*, jest, play; *joculator*, jester, joker; *jocundus*, merry, sportive; see **laugh**

John < Heb. *Yohanan* (Gr. *Ioannes;* L. *Joannes;* OF. *Jean;* It. *Giovanni;* Sp. *Juan*), Jehovah is gracious: John Doe, Johnny, Johnson, Johannine, johannite, Johannesburg, Johnstown, Hans, Hansel, Ivan, San Juan River, Jane, jenny.,

join < L. *jungo, junctus*, unite, connect; see **bind**

joint < OF. *joinct*, < L. *junctus*, a connection.
> Gr. *arthron*, n. joint; *arthridion*, n. dim.; *arthrikos*, of a joint; *arthritikos*, of the joints; *arthritis, -idos*, f. pain in the joints; *arthrodes*, well-knit; *arthrosis*, f. a jointing: arthropod, arthrodire, arthrospore, arthrosis, arthrolite, arthroderm, arthrodesis, arthroplasty, arthritis, arthritic, Condylarthra, *Triarthrus becki* (a trilobite), *Arthropodium candidum* (a lily).
> L. *articulus*, m. joint, member, knuckle, dim. of *artus*, m. joint; *articularis*, of joints; *articulatus*, jointed, distinct; *articulosus*, full of joints: article, articulation, articular, articulatory, inarticulate, *Articulina sagra* (a foraminifer), *Tetraclinis articulata* (arar-tree).

L. *coagmentum*, joint; see *coagmento* under **bind**
L. *colepium* (Gr. *kolepion*), knuckle of beef or pork; see *koleps* under **knee**
L. *commissura*, f. joint, seam, juncture: commissure.
L. *condylus* (Gr. *kondylos*), knob of a joint, knuckle; see projection
Gr. *ginglymos*, hinge, joint; see **hinge**
Gr. *glene*, socket, pupil of the eye; see **eye**
Gr. *gonation*, knot or joint of a reed, dim. of *gony*, knee; see **knee**
Gr. *hapsos; haphe*, juncture, joint; see *hapto* under **bind**
Gr. *harmos*, m. joint, fastening, bolt, peg; *synarmos*, joined together: *Harmosto-
mum leptostomum* (a trematode), *Harmophorus manticoroides* (a beetle),
Harmothoë imbricata (an annelid).
Gr. *rhaphe*, seam, suture; see *rhapto* under **sew**
L. *sutura*, seam; see *suo* under **sew**
Gr. *symblema, -tos*, n. joint, seam:
Gr. *symphysis; diaphysis; diaphye*, f. juncture, break, seam: symphysis.
Gr. *synkampe*, bight, joint; see *kamptos* under **bend**
Gr. *syzygia*, union, coupling; see *zygon* under **bind**
See: **break, hinge, dovetail, mortise, knot, back, knee, angle**

joke < L. *jocus*, jest; see **laugh**

journey < OF. *journee*, a day's travel; see **move**

joy < OF. *joie*, < L. *gaudium*, n. delight, gladness, joy; *gaudialis*, glad, joyful;
gavisus, causing joy: joyful, rejoice, enjoy, gaudy, killjoy.
Gr. *agalliasis*, f. great joy, exultation, < *agalliao*, rejoice exceedingly: *Agallia-
phagus americanus* (a beetle), *Agalliastes pulicarius* (a bug).
Gr. *asmenos*, pleased, glad; *asmenistos*, acceptable, welcome: *Asmenistis cucullata*
(a moth).
L. *Bacchanalis*, pertaining to Bacchus and the festivals in his honor; see *Bacchus*
under **wine**
L. *beatus*, happy, blessed; *beatulus*, dim.; *beatificus*, making happy; *beatrix, -icis*,
f. she that makes happy: beatitude, beatific, Beatrice, Trixy, *Oryx beatrix* (an
antelope).
Gr. *chairo*, rejoice; *charma, -tos*, n. source of joy, joy, delight; *charmosyne*, f.
joyfulness, delight; *chartos*, causing delight; *charops, -opos*, glad-eyed, joyous;
haimatochares, rejoicing in blood: *Charmatometra bakeri* (a bug), *Charmosyna
pulchella* (a parrot), *Anthochaera carunculata* (wattle-bird), *Chaerophyllum
bulbosum* (chervil), *Microchaera albocoronata* (a hummingbird), *Xanthocharops
primus* (a wasp), *Haimatochares tenebrioides* (a beetle).
Gr. *chlaros*, gay, lively:
L. *contentus*, satisfied; see **agreeable**
L. *convivor, -vatus*, feast, banquet and carouse together; *convivium*, feast, ban-
quet; *convivialis*, of a feast; see **eat**
L. *delecto, -atus*, cause pleasure; *delectabilis*, agreeable, delightful; *delicio*, charm,
allure; *deliciosus*, delightful, tasty: delectation, delectable, delight, delicious,
Rubus deliciosus (boulder-raspberry).
L. *ecstasis* (Gr. *ekstasis*), f. state of being beside oneself, rapture, trance: ecstasy,
ecstatic.
Heb. *eden*, delight, pleasure: Garden of Eden.
Gr. *epalpnos*, cheerful, happy:
L. *Epicurus* (Gr. *Epikouros*), m. a Greek philosopher who taught that pleasure
(but not sensuality) should be the chief purpose of living: Epicurean, epicure,
Epicura laetiferana (a moth).
L. *euax* (*evax*), joy be! good!: *Rhynchospira evax* (a fossil brachiopod).
Gr. *eudiagogos*, cheerful: *Eudiagogus episcopalis* (a beetle).
Gr. *euthymos*, cheerful, kind; see *thymos* under **mind**
Gr. *euphrosyne*, f. mirth, merriment; *euphrosynos*, cheery, merry: Euphrosyne,
Euphrosyne cirrata (an annelid).
L. *exulto, -atus*, leap up, rejoice exceedingly; see *assilio* under **leap**
L. *felix, -icis*, happy: Felix, Felicia, felicity, felicitate, felicitous, infelicity.
L. *feria*, holiday, festival; see **day**
L. *festus*, joyful; *festivus*, gay, joyous, merry; *festum*, n. feast, holiday; Sp.
fiesta, f. holiday, festivity: feast, festive, festival, fete, festal, fiesta, festivity,
Amazona festiva (a parrot), *Cichlasoma festivum* (a fish).
L. *fortunatus*, happy, lucky, prosperous; see *fors* under **lot**
L. *fruor, fructus; fruitus*, enjoy, delight in; *fruniscor, -nitus*, enjoy: fruition.
Gr. *galeros*, cheerful: *Galeruclerus[Galeroclerus] crinitus* (a beetle).
L. *genialis*, jovial, pleasant, friendly; see **agreeable**
AS. *geol* (ON. *jol*), December: Yule, Yuletide, Yule log, jolly(?).

Gr. *getheo*, rejoice; *gethos*, n.; *gethosyne*, f. delight, joy; *gethosynos*, glad, joyful; *perigethes*, very joyful: *Gethosyne aequivocalis* (a moth), *Meligethes aeneus* (a beetle).

AS. *glaed*, bright, cheerful, happy, shining: glad, gladly, gladden, gladness, gladsome, Gladstone, glade, Everglades.

L. *grator*, *-atus*, rejoice with one, wish one joy; *gratulor*, *-atus*, wish one joy; see *gratia* under **give**

Gr. *hedomai*, take pleasure, enjoy oneself; *hedone; hedosyne; hesis*, f. delight, enjoyment; *hestos*, glad; *philedonos*, fond of pleasure: hedonism, hedonistic, *Moneses uniflora* (woodnymph).

Gr. *heortios*, festal; *heortasma, -tos*, n.; *heorte*, f. festival, holiday; *heortastikos*, festive; *heortastes*, m. reveller: *Heortia dominalis* (a moth).

L. *hilaris; hilarus* (Gr. *hilaros*), cheerful, gay, merry; *hilarulus*, dim.: hilarity, Hilary, hilarious, exhilaration, *Acrosternum hilaris* (stinkbug), *Synageles hilarulus* (a spider).

L. *jocundus* (*jucundus*), merry, sportive; see *jocus* under **laugh**

L. *jubilo, -atus*, shout for joy; *jubilum*, n. wild shout: jubilant, jubilation.

Gr. *komos*, m. a jovial festivity, revel, carnival; *komastes*, m. reveller; *komax, -akos*, m. debauchee; *komastikos*, of a revel, < *komazo*, revel: *Comastes robustus* (a fly).

L. *laetus; laetabilis*, joyful, glad, pleasant; *laetitia*, f. gladness, joy; *laetificus*, causing joy, making glad; *laetabundus*, full of joy; *laeticulosus*, overjoyful: Letitia, *Acacia laeta* (an acacia), *Sceliphron laetum* (a mud-dauber), *Pericompsus laetulus* (a beetle).

L. *Liburnus*, m. god of lustful enjoyment: *Liburnelenchus heidemanni* (a beetle), *Liburnia campestris* (a bug).

AS. *maesse*, feast-day, festival, < L. *missa*, f. the mass: Mass, Candlemas, Childermas, Christmas, Hallowmas, Hilarymas, Lammas, Michaelmas, Roodmas.

L. *majuma*, f. a festival on the Tiber in the month of May, featuring a mock sea-fight:

Gr. *makarios; makaros; makar*, happy, fortunate; *makarioteros* (*makarteros*), happier; *makariotatos* (*makartatos*), happiest; *makaria*, f. happiness; *makarismos*, m. blessing: *Macarocrinus terfurcatus* (a Devonian crinoid).

AS. *myrige*, gay, happy; *myrth*, joyous merriment: merry, merriment, mirth, mirthfulness.

L. *oblecto, -atus*, amuse, delight, divert, entertain, please; *oblector, -is*, m. charmer, delighter, pleaser:

Gr. *olbios*, blessed, happy, < *olbos*, m. happiness; *olbodotes*, m. giver of happiness: *Olbiorchilus fumigatus* (a bird), *Olbodotes copei* (a Paleocene mammal).

L. *ovo, -atus*, exult, rejoice; *ovatio, -onis*, f. a lesser exultation: ovation.

Gr. *panegyris*, f. festal assembly; *panegyrikos*, festive, showy, pompous: panegyric.

L. *paradisus* (Gr. *paradeisos*), delightful spot, pleasure grounds, park, abode of the blessed; see **park**

L. *Saturnalis*, of Saturn, the god of agriculture; see *Saturnus* under **till**

L. *Sybariticus* (Gr. *Sybaritikos*), pertaining to the luxury and debauchery of Sybaris and the Sybarites: Sybarite, Sybaritic.

Gr. *terpnos*, delightful, agreeable; *terpole; terpsis*, f. delight, enjoyment, pleasure, < *terpo*, delight, cheer, make merry; *euterpes*, delightful, charming: terpsichorean, *Terpnomyia angustifrons* (a fly), *Terpsiphone paradisea* (a flycatcher), *Euterpe oleracea* (assai palm).

Gr. *thiasos*, m. band of street-singers; *thiasodes*, festive: *Thiasophila inquilina* (a beetle).

L. *vitulor*, celebrate, rejoice (probably by sacrificing a calf, *vitulus*): fiddle, viol, violin, violincello, *Amphilonche violina* (a radiolarian).

L. *voluptas, -atis*, f. pleasure, satisfaction; *voluptuosus*, full of pleasure, delightful; *voluptuarius*, of pleasure: voluptuary, voluptuous.

Heb. *yobel*, blast of a trumpet, announcing the beginning of the year of jubilee: jubilee, jubilist.   Derive: jubilant.

See: **laugh, agreeable, play, sing, beauty, rest, heaven, light**

juba, L. mane; *jubatus*, maned, crested; see **hair**

jubar, L. radiance of the heavenly bodies, light, sunshine, splendor, glory; see **light**

jubilo, -atus, L. shout for joy; see **joy**

jucundus, L. agreeable, pleasant, merry; see *jocus* under **laugh**

judge < *judex, -icis*, c.; *judicium*, n. decision, opinion, sentence: judicial, judiciary, judgment, adjudicate, prejudice, injudicious.

L. *adaero, -atus*, estimate, rate, appraise, value:

L. *aestimo, -atus*, appraise, value; *aestimator, -is*, m. appraiser; *aetimabilis*, valuable: estimate, esteem, aim, inestimable.

L. *arbiter, -tri; arbitrator, -is,* m. judge, umpire: arbiter, arbitrator, arbitrary, arbitration, arbitrament.

Gr. *brabeus; brabeutes,* m. judge, umpire at the games:

L. *censeo, -sus,* judge, appraise; *censor, -is,* m. a Roman magistrate; *census, -us,* m. register of citizens and property: censor, censure, censorious, census, recension.

L. *conicio, -jectus,* discuss, draw a conclusion, infer, foretell; see **prophecy**

L. *consulo, -sultus,* reflect, weigh, deliberate, consider; see **think**

L. *criticus* (Gr. *kritikos*), capable of judging; *kriterion,* n. means of judging, standard, < *krino,* judge, estimate, discern, distinguish; *apokritos,* separated, chosen; *diakrisis,* f. decision, judgment: critic, crisis, critical, criticism, criterion, diacritical, hypocritical, hypercritical, endocrine.

L. *damno, -atus,* judge, sentence, doom; see **punish**

L. *decerno, -cretus,* decide, judge; *decretum,* decision; *decretorius,* decisive; see *cerno* under **separate**

Gr. *diallakter, -os; diallaktes; synallaktes,* m. mediator: *Diallactes melanocephalus* (a bird), *Synallaktes aenigma* (a holothurian).

L. *digero, -estus,* separate, dispose, classify, render food assimilable; see **digest**

Gr. *dikastes,* judge; see *dike* under **law**

L. *discepto, -atus,* decide, determine, judge:

L. *examino, -atus,* weigh, consider; see **try**

L. *mediator, -is,* m. go-between, arbitrator, umpire: mediator, mediatorial, mediatrix, mediation.

Gr. *mesites,* m. umpire, mediator: *Mesites variegatus* (a bird), *Mesitoblatta brongniarti* (a fossil insect).

L. *opto, -atus,* choose; see **choose**

Gr. *rhabdouchos,* m. staff-bearer, beadle, judge, umpire: *Rhabduchus denticornis* (a beetle).

L. *sculna (seculna),* m. arbiter, mediator, umpire:

L. *sufes (suffes), -etis,* m. Carthaginian judge:

See: **separate, class, think, choose, worth, try**

jug < uncertain origin; see **bottle, jar, vessel**

jugerum, L. a measure of land; see **field**

juggle < OF. *jogler,* < L. *joculator,* jester; see **play**

jugis, L. perpetual, perennial; see **always**

juglans, L. walnut; see **walnut**

jugulum, L. collarbone, throat; *jugulator,* cutthroat; see **throat, kill**

jugum, L. yoke, team, pair, ridge, lintel; see *jugo* under **bind**

juice < L. *jus, juris,* n. soup, broth, sauce, fluid; *juscellum; jusculum,* n. dim.; *jurulentus; jussulentus,* juicy: juice, juicy, juiciness.

L. *aqua,* water; see **water**

Gr. *chylos,* m. juice, moisture; *diachylos; enchylos,* juicy, succulent: chyle, chyluria, chyliferous, chylocele, chylous, *Chylocladia kaliformis* (an alga).

Gr. *chymos,* m. juice; *enchymos,* juicy: chyme, chymiferous, chymous, *Garcinia xanthochymus* (a gamboge-tree).

L. *cremor,* thick juice, broth; see **sauce**

Gr. *gala,* milk; see **milk**

L. *humor,* fluid of any kind; see *humidus* under **wet**

Gr. *ichor, -os,* m. serum, watery part of the blood: ichorrhemia, ichorous, ichor.

L. *lac,* milk; see **milk**

L. *latex, -icis,* m. fluid, liquid, sap, juice: latex, laticiferous.

L. *liquidum,* n.; *liquor, -is,* m. fluid: liquid, liquor.

L. *meconium* (Gr. *mekonion*), poppy-juice, opium; see *mecon* under **poppy**

Gr. *myron,* any sweet juice from plants used for ointment; see **fat**

Gr. *opos,* m. juice; *opion* (L. *opium*), n. dim., poppy-juice; *opisma, -tos,* n. milky sap of plants: opium, opiane, opianic, opiate, opianyl, opiophagy, opodeldoc, opopanax, opobalsam, opotherapy.

Gr. *oros (orrhos),* m. watery or serous part of anything, serum, whey: orrhotherapy.

AS. *saep,* juice: sappy, sapless, saphead, winesap, sapsucker.

L. *sanies,* f. bloody matter, serum from sores: sanies, sanious.

L. *serum,* n. a thin, watery fluid, whey of curdled milk: serous, serum, serosity, serodermatosis, serotonine.

L. *sicera* (Gr. *sikera*), a fermented liquor; see **drink**

L. *succus,* m. juice, sap; *succidus,* juicy, sappy, fresh; *succosus; succulentus,* full of juice, juicy: succade, succulent, succivorous, *Cinchona succirubra* (redbark quinine-tree), *Astragalus succulentus* (a vetch).

Gr. *zomos,* sauce, soup, broth; see **soup**

See: **water, drink, resin, sauce, soup, wine, pour**

julus < L. *iulus* (Gr. *ioulos*), catkin, multiped; see **catkin, myriapod**

jumentum, L. beast of burden; see **animal**

jump < uncertain origin; see **leap**

junctus, L. joined; see *jugum* under **bind**

juncus, L. rush; *junceus; juncinus,* of rushes; *juncosus,* full of rushes; see **sedge**

junior, L. younger; see *juvenis* under **young**

juniper < L. *juniperus,* f. savin; *junipereus,* of juniper: juniperic, gin, *Juniperus excelsa* (Greek juniper), *Tamarix juniperina* (juniper-tamarisk), *Phoradendron juniperinum* (juniper-mistletoe), *Saxifraga juniperifolia* (a saxifrage).
Gr. *arkeuthos,* f. juniper; *arkeuthis, -idos,* f. juniper berry; *arkeuthinos,* of juniper; *Arceuthobium pusillum* (dwarf mistletoe).
Gr. *boraton,* n. a juniper:
Gr. *brathy, -yos,* n. savin:
L. *sabina* f. the Sabine juniper: savin, *Juniperus sabina* (savin-juniper), *Lycopodium sabinaefolium* (a clubmoss).

junix (juvenix), L. young cow, heifer; see **cattle**

Jupiter, *Jovis,* L. chief of the Roman gods, equivalent to Zeus; *Juno,* wife of Jupiter; see **spirit**

Jura, L. mountain range between France and Switzerland; see **mountain**

jurgo, -atus, L. contend in words, quarrel, scold; *jurgium,* quarrel; see **scold**

juridicus, L. pertaining to the administration of justice; see *jus* under **law**

juro, -atus, L. swear, take an oath; see **swear**

jus, *juris,* L. law, right, see **law**; juice, broth, soup; *juscellum; jusculum,* dim., see **juice**

jussum, L. order, command, law, decree; see *jubeo* under **ask**

justice < L. *justitia,* f. < *justus,* upright, righteous, fair, impartial, equitable, legal: just, justice, adjust, justification, *Justicia secunda* (an acanthacead).
Gr. *adekastos,* unbribed, impartial:
Gr. *dikaiosyne,* justice; see *dike* under **law**
L. *fandus; fas,* right, correct, lawful:
L. *juridicus,* pertaining to the administration of justice; see *jus* under **law**
Gr. *Nemesis,* goddess of retributive justice; see **punish**
L. *vindico, -atus,* avenge, deliver, justify; see **punish**
See: **law, right, punish**

justus, L. right, equitable, fair; see **right, justice**

Juturna, L. a spring in Latium; a nymph; see **spring**

juvenis, L. young; *junior,* younger; *juvencus,* young; *juvenilis,* of youth; see **young**

juxta, L. near, next to; see **near**

K

kalanchoë, NL. a genus of the stonecrop family; see **stonecrop**

kali < Ar. *qaliy,* ashes of saltwort; see **alkali**

kappa, Gr. tenth letter of the Greek alphabet; see **letter**

keel < ON. *kjölr* (Dan. *kjöl*); see **ridge**

keep < AS. *cepan;* see **hold, save, guard**

keg, prob. < Scan. *cag;* see **barrel**

kermes, F. < Ar. *qirmiz,* deep red; a genus of scale insects; see **red**

kernel < AS. *cyrnel,* dim. of *corn,* grain; see **seed, nut, grain**

ketone < G. *ketone,* an aromatic compound having a carbonyl group in the middle of a chain of carbon atoms; the suffix *-one* signifies a ketone: acetone, testosterone, pulegone, monoketone, menthenone, rotenone, ketosis.

kettle < AS. *cetel.*
L. *aenum (ahenum),* brazen vessel, kettle, caldron; see *aes* under **copper**
Gr. *chalkion,* n. copper kettle, caldron; *chalkidion,* n. dim.:
L. *coculum,* n. a vessel for cooking:
L. *cortina,* f. round kettle, caldron with tripod; *cortinula,* f. dim.: *Cortina furcata* (a radiolarian), *Cortinetta cortiniscus* (a radiolarian), *Theopera cortina* (a radiolarian).
L. *cucuma,* f. kettle; *cucumella,* f. dim.:
L. *lebes, -etis* (Gr. *-etos*), m. bronze kettle, caldron; *lebetion,* n. dim.: *Lebetodiscus dicksoni* (a fossil cystid), *Lebetocrinus grandis* (a Mississippian crinoid), *Lebistes reticulatus* (a guppy).
Gr. *thermanter, -os,* m. kettle:

key < AS. *caeg:* keyhole, keyboard, keynote, Keystone State, turnkey.
L. *clavis,* f. key, bar; *clavicula,* f. dim.; *clavicarius,* m. key-maker, locksmith: clavicle, clavichord, clavier, claviger, clavicular, clavecin, claviature, cleft, conclave, autoclave, claviol, *Clavicula polymorpha* (a diatom), *Crassula clavifolia* (a stonecrop), *Vanilla claviculata* (an orchid).
Gr. *kleis, kleidos,* f. key, latch, collarbone; *kleidion,* n. dim.: cleidomastoid, ophicleide, *Hydrocleis nymphoides* (water-poppy), *Clidophleps distanti* (a cicada).
See: **bar, close**

kick < uncertain origin.
L. *calcitro, -atus,* kick; *calcitro, -onis,* m. kicker; *calcitrosus,* apt to kick: calcitrate, recalcitrant, *Stomoxys calcitrans* (stable-fly).
Gr. *laktizo,* kick, trample, spurn; *laktisma, -tos.* n.; *laktismos,* m. a kick; *laktistes,* m. a kicker; *laktikos,* kicking; *analaktizo,* kick out behind: *Lactistomyia insolita* (a fly), *Lactistes rastellus* (a bug).
Sp. *patada,* f. kick:
Gr. *pternizo,* strike with the heel, trip; *pternister, -os;* *pternistes,* m. tripper: *Pternistis capensis* (a bird).

kidney < uncertain origin.
Gr. *nephros,* m. kidney; *nephridios,* of the kidneys: nephrocyte, nephrolith, nephropexy, nephrolysis, nephrosis, nephrology, nephrotoxin, nephritic, nephritis, nephridium, nephrectomy, epinephrin(adrenalin), *Nephrolepis exaltata* (swordfern).
L. *ren, renis,* m. kidney; *reniculus; renunculus,* m. dim.; *renalis,* of the kidneys: renal, reniform, renogastric, adrenal, adrenalin, suprarenal, *Renicola pinguis* (a nematode), *Renilla amethystina* (a sea-kidney), *Pyrola renifolia* (a pyrola), *Heteranthera reniformis* (a mud-plantain), *Trichomanes reniforme* (a filmy-fern), *Nephropyris renilla* (a radiolarian).

kill, prob. < ON. *kolla,* hit on the head.
Gr. *andranchos,* m. executioner:
Gr. *artamos,* butcher; see **butcher**
AS. *bana,* slayer, that which hurts or destroys: bane, baneful, henbane, dogbane, baneberry.
L. *carnifex, -icis,* m. hangman, executioner: carnifex, *Hamachthes carnifex* (a bee).
L. *-cid; -cida; -cide,* pertaining to killing, < *caedo,* cut down, kill; *homicida,* murderer; see *caedo* under **cut**
L. *cultrarius,* m. slayer of the sacrificial victim: *Cephus cultrarius* (a wasp).
Gr. *daikter; daiktes; daiktor, -os,* m. slayer, killer; *daiktos,* slain: *Daictes fukaii* (a wasp).
Gr. *delemon, -os,* baneful, noxious; *delema,* damage, mischief; see **hurt**
Gr. *ekbibastes,* m. executioner:
L. *exstinguo, -inctus,* put out, kill, destroy: extinction, extinguish.
L. *fatifer, -a, -um,* deadly, destructive; see *fatum* under **lot**
L. *funestus,* causing death; see *funero* under **cover**
L. *interficio, -fectus,* destroy, kill, murder; *interfector, -is,* m. slayer, murderer: *Mimetus interfector* (a spider).
L. *jugulo, -atus,* cut the throat, kill, murder; *jugulator, -is,* m. cutthroat, murderer:
Gr. *karanistes,* m. beheader: *Caranistes lineatus* (a beetle).
Gr. *kateino; katakteino,* kill; *ktamenos,* killed; *ktantes,* m. killer; *ktonos,* m. murder: embryoctony, *Dendroctonus frontalis* (a bark-beetle), *Cynoctonum mitreola* (miterwort), *Rhizoctonia violacea* (a fungus), *Helioctamenus lusitanicus* (a beetle).
L. *letho, -atus,* kill, slay; *lethalis; lethifer,* deadly, mortal; see **death**

Gr. *loigios*, deadly, fatal:

L. *macto, -atus*, sacrifice, immolate, kill: mactation, *Latrodectus mactans* (black-widow spider).

L. *neco, -atus*, kill; *necator, -is*, m.; *necatrix, -icis*, f. slayer, killer; *nex, necis*, f. death; *internecinus*, deadly, destructive, pertaining to mutual slaughter; Gr. *nekys, -yos*, m. corpse: internecine, pernicious, *Necator americanus* (hookworm), *Uncinula necator* (a fungus), *Dermatophora necatrix* (a fungus).

L. *occisor, -is*, m. killer, < *occido, -cisus*, strike down, kill; *Occisor versutus* (a fly).

Gr. *oulios*, baneful, deadly: *Uliodon albipunctatus* (a spider), *Uliolepis pilosa* (a moth).

L. *peremptor, -is*, m. destroyer, killer: *Peremptor pavidus* (a fly).

Gr. *pertho*, destroy, kill, plunder; see **destroy**

Gr. *phonos*, m. murder, slaughter; *phonergates; phoneus; phontes*, m. murderer, slayer; *phonikos*, bloody, murderous; *phonax, -akos*, bloodthirsty: *Thelyphonus* (now *Mastigoproctus*) *giganteus* (vinegaroon), *Nesophontes hypomicrus* (an insectivore).

L. *pio, -atus*, sacrifice; see *pietas* under **rite**

Gr. *promysso*, snuff a lamp:

L. *sacrifico, -atus*, offer a sacrifice: sacrifice, sacrificial.

L. *sicarius*, m. assassin, murderer: sicarius, *Sicarius extinctus* (a fossil fish).

Gr. *sphazo*, slaughter, kill; *sphage*, f. slaughter; *sphageus; sphaktes*, m. slayer; *sphagion*, n. victim; *sphaktos*, slain; *sphagios*, slaying, deadly: *Sphagodus pristodontus* (a Silurian fish), *Sphagebranchus flavicaudus* (a fish), *Sphactes politus* (a bug), *Sphagiocrates lusoria* (a moth).

L. *strages, -is*, f. defeat, overthrow, massacre:

Gr. *thymophthoros*, soul-destroying, deadly:

Gr. *thyo*, sacrifice; see **give**

L. *trucido, -atus*, cut to pieces, kill cruelly, butcher, massacre; *contrucido*, put to the sword, slay: *Thamnophilus trucidatus* (a beetle).

See: **destroy, butcher, victim, cancel**

kiln < AS. *cyln*, < L. *culina*, kitchen; see **oven**

kilo- < Gr. *chilioi*, thousand; see **thousand**

kin < AS. *cynn*, kind, race: akin, kind, kindred, kinsfolk, kinsman, kinship, kith and kin.

Gr. *adelphideos*, m. nephew; *adelphide*, f. niece:

L. *affinis*, related to, neighboring, bordering; *confinis*, adjoining: affinity, confine, paraffin, *Sophora affinis* (Texas sophora).

L. *amita*, f. father's sister; *amitinus*, of an aunt; m. cousin:

Gr. *amnamos*, m. descendant, son, grandson:

Gr. *anchisteus*, m. the next of kin: *Anchisteus peregrinus* (a beetle).

Gr. *anepsios*, m. cousin, nephew: *Anepsius delicatulus* (a beetle).

L. *antecessor, -is*, m. forerunner: ancestor, ancestral.

L. *avus*, m. grandfather; *avia (ava)*, f. grandmother; *avunculus*, m. dim. uncle; *avitus*, of a grandfather, ancestral, old; *abavus; atavus; proavus*, m. father of a grandfather successively backward, forefather, ancestor: aval, avuncular, uncle, atavism, atavistic, *Prolimnocyon atavus* (a creodont), *Ischyodus avitus* (a fossil fish), *Serpula avita* (a worm), *Presbytiscus avunculus* (a primate).

L. *cognatus*, kindred, related: cognate.

L. *congener, -eris*, of the same kind or race: congener, congeneric.

L. *consanguineus*, related by blood, kindred: consanguineous, *Panicum consanguineum* (a grass).

Gr. *daer, -os*, m. brother-in-law: *Daer ales* (a beetle).

L. *descendens, -entis*, c. offspring, posterity: descendant.

Gr. *engonos; synengonos*, m. grandson:

Gr. *ethnos*, race, nation; *ethnikos*, of a race or nation; see **class**

L. *familia*, f. household; *familiola*, f. dim.; *familiaris*, of a family or household, knowing intimately: family, familist, familiar, familiarity, familiarize, *Canis familiaris* (dog).

L. *fratruelis; matruelis; patruelis*, m. cousin:

Gr. *galos*, f. sister-in-law:

Gr. *gambros*, m. relative by marriage: *Gambrostola imposita* (a moth).

L. *gener, -eri*, m. son-in-law:

Gr. *genetes*, begetter, father, ancestor; see *gignò* under **birth**

L. *gens, gentis*, clan; see *genus* under **class**

L. *germanus*, having the same parents, genuine, true: germane.

L. *glos, gloris*, f. sister-in-law:

Gr. *hekyra*, f. mother-in-law; *hekyros*, m. father-in-law: *Hecyrida sordida* (a beetle).

Gr. *homophylos; homogenios,* of the same race: *Homophylus castaneus* (a beetle).
Gr. *hyionos,* grandson; see *hyios* under **son**
Gr. *-ides,* patronymic suffix denoting son of, descendant of; see **son**
Gr. *kasis,* brother, sister; see **brother, sister**
Gr. *kedestes,* m.; *kedestria,* f. kin by marriage, in-law; *kedestikos,* of affinity:
L. *levir, -iri,* m. brother-in-law:
L. *majores,* m. adults, forefathers:
L. *matertera,* f. aunt, mother's sister:
Gr. *metryia,* f. stepmother; *metryios,* m. stepfather:
Gr. *nennos,* m. uncle; *nenna,* f. aunt:
L. *nepos, -otis,* m. grandson, nephew; *nepotulus,* m.; *nepotula,* f. dim.: nepotism, nephew.
L. *neptis,* f. granddaughter, niece; *nepticula,* f. dim.: niece, nepticulid, *Nepticula splendissima* (a moth).
L. *noverca,* f. stepmother; *novercalis,* of a stepmother:
L. *nura,* f. daughter-in-law:
Gr. *nyos,* f. daughter-in-law:
L. *pappus* (Gr. *pappos*), m. old man, grandfather: *Papposaurus traquairi* (a fossil reptile), *Pappogeomys albinasus* (a rodent).
L. *parens, -entis,* procreator, progenitor; see **begin**
L. *patraster, -tri,* m. father-in-law:
Gr. *patrios; patrikos,* ancestral, hereditary; see *pater* under **father**
L. *patruus,* m. father's brother, uncle; *patruelis,* of an uncle, cousin: patruity, *Gambusia patruelis* (a fish).
Gr. *penthera,* f. mother-in-law; *pentheros,* m. father-in-law; *sympenthera,* f. stepmother; *sympentheros,* m. stepfather:
Gr. *phrater, -os,* m. clansman; *phratra,* f. clan, brotherhood; *phratrios,* of a clan: *Phratriodes curvisignis* (a moth).
L. *privignus,* m. stepson; *privigna,* f. stepdaughter: *Phaeochlaena privigna* (a butterfly).
L. *progenitor, -is,* m. founder of a family, ancestor: progenitor.
L. *prosator, -is,* m. ancestor, < *prosero, -atus,* beget, produce:
L. *prosapia,* f. stock, race, family:
L. *relativus,* referring to, akin to, near to: relative, relativity.
AS. *sib,* kin: sib, sibling, gossip.
L. *sobrinus,* m.; *sobrina,* f. cousin on the mother's side; *consobrinus,* m. **any** cousin: cousin, *Nodosaria consobrina* (a foraminifer).
L. *socer, -eri,* m. father-in-law; *socra,* f. mother-in-law:
L. *stirps,* stem, stock, family; see **stem**
Gr. *synaimos; synaimon, -os; authaimon, -os,* of the same blood; m. kinsman: *Synemia polygonaria* (a moth), *Authaemon stenonipha* (a moth).
Gr. *tethe,* f. grandmother; *tethis, -idos,* f. aunt; *epitethe,* f. greatgrandmother:
L. *vitricus,* m. stepfather:
See: **birth, class, begin, father, mother, son, daughter, brother, sister**

-kin < D. *-kin,* diminutives; see DIMINUTIVES under **little**

kind < AS. *cynd;* see **mild, mercy, give, class, kin**

kinetico- < Gr. *kinetikos,* pertaining to motion; see *kineo* under **move**

king < AS. *cyng;* see **govern**

kingfisher
L. *alcedo, -inis,* f. kingfisher: Alcedinidae, *Alcedo ispida* (European kingfisher).
L. *alcyon, -is* (Gr. *alkyon, -os; alkyonis, -idos*), f. kingfisher: halcyon, *Alcyonium palmatum* (a coral), *Halcyon pyrrhopygia* (a kingfisher), *Rhamphalcyon amauroptera* (a kingfisher), *Ceryle(Megaceryle) alcyon*(belted kingfisher), *Haemoproteus halcyonis* (a protozoan), *Dacelo gigas* (laughing jackass), *Xiphostylus alcedo* (a radiolarian).
L. *ceyx, -ycis* (Gr. *keyx, -ykos; kex, -ekos: kauax, -akos*), m. kingfisher: *Ceyx tridactyla* (three-toed kingfisher), *Clytoceyx rex* (a kingfisher), *Celastrina ceyx* (a butterfly).
Gr. *kerylos,* m. a sea-bird, halcyon: *Ceryle americana* (a kingfisher), *Metacerylon parallelum* (a beetle).
L. *milesium,* n. a kind of kingfisher:

kirk < AS. *cyrice,* church; see **temple**

kiss < AS. *cyssan; coss,* a kiss.
L. *basio, -atus,* kiss; *basium,* n. kiss; *basiolum,* n. dim.; *basiator, -is,* m. kisser:
L. *blanditia,* caress, flattery; see *blandus* under **mild**
L. *columbor, -atus,* bill or kiss like doves:
Gr. *glottismos,* m. a lascivious kiss, < *glottizo,* kiss lasciviously:
Gr. *kyneo,* kiss; *kynetinda,* f. a kissing-game:

L. *osculor, -atus,* kiss; *osculum,* n. little mouth, pretty mouth, kiss: osculation, *Misocalius osculans* (black-eared cuckoo).

Gr. *philema, -tos,* n. kiss; *philemation,* n. dim.: *Philematium hottentottum* (a beetle).

L. *savior (suavior), -atus,* kiss; *savium,* n. a tender kiss:
See: **embrace**

kitchen < AS. *cycene,* < L. *coquina,* < *coquo, coctus,* cook; *coquinarius,* of the kitchen; see **cook**

L. *culina,* f. kitchen; *culinarius,* of the kitchen: culinary, kiln, Kilmer, *Lens culinarius* (lentil).

Gr. *ipnos,* kitchen; see **oven**

Gr. *optanion,* n. kitchen:

kite < AS. *cyta,* a bird belonging to the hawk family; see **hawk**

klepto, Gr. steal; see **steal**

knave < AS. *cnafa,* servant, boy; see **rogue**

knead < AS. *cnedan;* see **mix**

knee < AS. *cneo.*

L. *compernis,* knock-kneed; see **bend**

Gr. *epigonatis, -idos,* f. kneepan:

L. *genu, -us,* n.; *genus, -uis,* m. knee, joint, knot; *geniculum,* n.; *geniculus,* m. dim.; *geniculatus,* like the bent knee, knotty, jointed; *geniculosus,* knotty; *congenulatus,* fallen upon the knees; *ingeniculus,* kneeling: genuflection, geniculate, geniculum, *Drymocrinus geniculatus* (an Ordovician crinoid).

Gr. *gony, gonatos,* n. knee, joint, node; *gonation,* n. dim.; *gonykrotos,* knock-kneed; *gnyx; gonypetes; prochny,* with bent knee, kneeling; *mesogonation,* n. internode: gonagra, gonytheca, gonyocele, gonarthritis, pycnogonid, *Gonatorhynchus tumidus* (a crab), *Polygonatum latifolium* (a Solomon's-seal), *Gonyaulax polyedra* (a dinoflagellate), *Eriogonum saxatile* (a polygonacead), *Polygonum pennsylvanicum* (smartweed), *Pycnogonum littorale* (a sea-spider), *Prochnyanthes viridescens* (an amaryllidacead), *Gonatium cinctum* (a spider), *Antigonon leptopus* (coral-vine).

Gr. *ignya,* part behind the thigh and knee; see **leg**

Gr. *koleps, -epos,* f. hollow or bend of the knee; *kolepion,* n. dim. knuckle of beef or pork: *Coleps octospinus* (a protozoan).

Gr. *myle,* f. kneepan:

L. *patella,* kneepan; see *patina* under **pan**

L. *poples, -itis,* m. back part of the knee, knee: popliteus, popliteal.
See: **angle, knot, leg, joint**

knife < AS. *cnif.*

Gr. *arbelos,* m. shoemaker's rounded knife: *Arbelorrhina eximia* (a bird).

L. *cisorium,* n. a cutting instrument:

L. *copis, -idis* (Gr. *kopis, -idos*), f. cleaver, kitchen-knife; *koparion,* n. dim. small surgical knife: *Copidoglanus tandanus* (tandan), *Copidognathus glyptoderma* (a mite).

Sp. *cortadera,* f. knife, cutting instrument: *Cortaderia selloana* (pampas-grass).

L. *culter, -tri,* m. knife, plowshare; *cultellus,* m. dim.; *cultratus,* knife-shaped: cultrate, colter, cutlass, *Miliolina cultrata* (a foraminifer), *Spaniotoma cultriger* (a fly), *Acacia cultriformis* (an acacia), *Amphibelone cultellata* (a radiolarian), *Nereis cultrifera* (a worm), *Pterophyllum cultelliforme* (a fossil cycad), *Polypora cultellata* (a fossil bryozoan).

Gr. *dois, -idos,* f. sacrificial knife:

Gr. *glyphis, -idos,* f. penknife: *Glyphidodon chrysurus* (a fish), *Glyphidops obscurus* (a fly), *Glyphis cicatricosa* (a lichen).

Gr. *kainis, -idos,* f. knife: *Caenis lacustris* (a mayfly).

Gr. *kladeuterion; klasterion,* n. dim., pruning-knife; *kladeutes; klastes,* m. pruner: *Clasterosporium carpophilum* (a fungus).

Gr. *koiliskos,* m. a hollowed-out, surgical knife:

Gr. *kouris, -idos,* f. razor: *Curis aurifera* (a beetle).

Gr. *machairis, -idos,* razor; see *machaira* under **sword**

L. *novacula,* f. sharp knife, razor, dagger: *Novaculichthys kallosomus* (a damselfish), *Cantheliophorus novaculatus* (a fossil lepidophyte), *Xyrichthys novacula* (a fish), novaculite.

Gr. *rhamphe,* f. a curved knife:

L. *scalprum,* n. knife, chisel; *scalpellum,* n. dim. surgical knife, lancet; *scalptor, -is,* m. cutter, carver; *scalpratus,* having a sharp edge: scalpriform, *Scalpellum ornatum* (a cirriped), *Pholadomya scalprum* (a pelecypod).

Gr. *skalme,* f. knife, sword: *Scalmophorus reticulata* (a bug), *Syscalma prymnaea* (a moth).

L. *smila* (Gr. *smile*), f. carving-knife, chisel, graving tool; *smilion*, n. dim.: *Smilodon californicus* (a sabertooth), *Eusmilus dakotensis* (a sabertooth).

Gr. *sphagis, -idos,* f. sacrificial knife:

Gr. *tomeus,* m.; *tomis, -idos,* f. cutter, knife; *peritomeus,* m. shoemaker's knife; L. *phlebotomus,* m. (Gr. *phlebotomon,* n.), lancet: phlebotomy, *Phlebotomus africanus* (a sandfly).

Gr. *xyron,* n. razor; *xyrion,* n. dim.: *Xyrichthys psittacus* (razor-fish), *Xyrophoreus tonsor* (a bug).

See: **sword, dagger, cut**

knob < AS. *cnob;* see **projection, swell, head**

knock < AS. *cnocian;* see **strike**

knot < AS. *cnotta.*
L. *geniculatus,* knotted, jointed, with knees; *geniculosus,* knotty; see *genu* under **knee**

Gr. *gongros,* excrescence, swelling, knot; see **swell**

Gr. *gony,* knee, joint, node; see **knee**

L. *Gordius* (Gr. *Gordios*), m. mythical Phrygian king noted for the intricate knot on his chariot pole: Gordian knot, *Gordius aquaticus* (hairworm), *Glomospira gordialis* (a foraminifer).

Gr. *hamma, -tos,* knot; see *hapto* under **bind**

Gr. *krobylos,* roll or knot of hair on top of the head; see **hair**

L. *nodus,* m. knot; *nodulus,* m. dim.; *nodabilis; nodatus; nodosus,* knotty; *enodis,* without knots; *inenodabilis,* that cannot be freed of knots, inexplicable; *innodo, -atus,* fasten with a knot, entangle; *internodium,* n. space between knots or joints: node, nodule, internode, denouement, *Sagina nodosa* (knotted pearlwort), *Anicodes nodigera* (a moth), *Spirorbis nodulosus* (a fossil worm), *Nodosaria spinicosta* (a foraminifer), *Lingulina nodosaria* (a foraminifer), *Monograptus nodifer* (a graptolite), *Apium nodiflorum* (a celery), *Scaphites nodosus* (an ammonite).

Gr. *strangalis, -idos,* f. an intricate knot:

L. *torus,* knot, bulge; see **swell**

Gr. *tyle,* f.; *tylos,* m. knot, knob, callus; *tylarion; tyleion; tylion,* n. dim.; *tyligma, -tos,* n. weal, swelling; *tylosis,* f. a becoming callous; *tylotos,* knobby: tylus, tylosis, tylostyle, Tylopoda, *Tylosurus acus* (needle-fish), *Tylotriton verrucosus* (a newt), *Monotylus klingi* (a beetle), *Tritylodon longaevus* (a fossil mammal), *Protylopus annectens* (an Eocene camel).

See: **bind, complex, gland, joint, knee, lump, pillow, projection, secret, swell**

know < AS. *cnawan.*
L. *adagnitio, -onis,* f. knowledge:

L. *adeptus,* proficient: adept.

Gr. *alopekodes,* foxlike, sly; see *alopex* under **fox**

L. *argutia,* f. cunning, shrewdness, subtlety; *argutiola,* f. dim. piece of subtlety, quibble:

Gr. *Aristoteles,* m. Aristotle, distinguished natural scientist: Aristotelian, Aristotle's lantern, *Aristotelia racemosa* (a wineberry).

L. *astutus,* clever, cunning; *astus,* m. cleverness, cunning: astute, astuteness.

L. *Athena* (Gr. *Athene*), f. tutelary of Athens and goddess of wisdom; *Pallas, -ados,* f. another name for Athena: athenaeum, Athens, *Athena thetis* (a butterfly).

L. *auctoritas, -atis,* f. opinion, judgment, ability, power, record: authority, authoritarian.

Gr. *bathytropos,* crafty:

L. *bioneus,* witty, biting, < *Bion, -os,* m. witty Greek philosopher: *Bion brevis* (a spider), *Bionoblatta mastrucata* (a blattid).

L. *callidus; callens, -entis,* experienced, expert, shrewd, skillful, crafty, cunning; *calliditas, -atis,* f. skill, cunning, artfulness: *Aranea callida* (a spider).

L. *capax, -acis,* able, fit; see *capio* under **take**

L. *catus,* wise, intelligent, clever, cunning, sly:

Gr. *charientisma,* a witty remark; see **laugh**

L. *colubrinus,* serpentlike, cunning, wily; see *coluber* under **snake**

L. *compertus,* ascertained from personal knowledge or experience; see *comperio* under **learn**

L. *competens,* qualified, able: competent, incompetent.

L. *consiliosus,* full of prudence, wisdom: consilience, consiliary.

L. *consuesco, -etus,* become intimate with, know; see *suetus* under **use**

L. *cordatus,* wise, prudent; see *cor* under **heart**

L. *cultura,* care, polish, refinement; see *colo* under **till**

AS. *cunnan,* know, be able: cunning, can, could, uncouth.

Gr. *dexios*, clever, ready; see *dexia* under **hand**

L. *dicax, -acis*, sarcastic, satirical, witty; see **laugh**

Gr. *didaskalion*, science, art, lesson; see *didasko* under **teach**

L. *doctus*, learned, skilled, experienced; see *doceo* under **teach**

L. *dolosus*, crafty, cunning, deceitful; *subdolus*, somewhat crafty, sly; see *dolus* under **lie**

L. *dotatus*, gifted, endowed; see *do* under **give**

Gr. *eidemon; idmon, -os; eidylos*, expert, skilled; *eidesis; idmosyne*, f. knowledge; *eidetikos*, scientific: eidetic, *Idmon unicolor* (a butterfly).

L. *empiricus* (Gr. *empeirikos*); *empeiros*, experienced: empirical.

Gr. *entreches*, skillful:

Gr. *entribes*, practiced, skilled:

Gr. *episteme*, f. knowledge, science; *epistemos*, knowing, wise: epistemology, epistemic.

L. *eruditus*, learned, skilled, accomplished: erudite, erudition.

Gr. *eulabes*, cautious, discreet, wary: *Eulabeornis castaneoventris* (a bird).

Gr. *euphues*, shapely, clever; see **beauty**

Gr. *euthiktos*, to the point, clever; see **fit**

L. *experientia*, knowledge gained by trial; see *experimentum* under **try**

L. *faber, -bra, -brum*, skillful, ingenious; see **make**

L. *facultas, -atis*, f. capability, ability, skill: *facultatula*, f. dim.: faculty.

L. *familiaris*, friendly, knowing intimately; see *familia* under **kin**

L. *genius*, m. talent, wit, taste; *ingeniosus*, superior in intellect, gifted, talented; *ingenium*, n. genius: genius, ingenious, engine.

Gr. *glaphyros*, hollowed out, smoothed, polished, finished; see **hollow**

Gr. *hikanos*, competent, able, fit, sufficient: *Hicanodon cinerea* (a spider).

Gr. *histor, -os*, knowing, learned, informed; see *historia* under **story**

L. *humanitas, -atis*, f. cultivation, elegance, refinement: humanities, humanism, humanist.

Gr. *idris, -rios*, experienced, skillful, knowing: *Idris laeviceps* (a wasp).

L. *informatio, -onis*, f. knowledge, idea, < *informo, -atus*, instruct, teach: inform, information.

L. *intellego, -ectus*, perceive, comprehend; *intellectualis*, of understanding; *intellegibilis*, that can be understood: intellect, intellectual, intelligent, intelligible.

L. *intuitio, -onis*, f. instinctive knowledge, < *intueor, -uitus*, look at, contemplate: intuition, intuitive.

Gr. *kerdaleos*, wily, crafty, cunning, shrewd, < *kerdos, -eos*, n. craft, wile, trick; *kerdo, -oos*, f. the wily one, fox; L. *cerdo, -onis*, m. handicraftsman: *Cerdognathus greyi* (a fossil reptile), *Galeocerdo arcticus* (tiger-shark).

Gr. *kidaphos*, wily: *Cidaphus thuringiacus* (a wasp).

Gr. *-logy*, suffix denoting knowledge of, science of, study of, < *lego*, choose, gather, reason; *logikos*, rational, sensible: anthropology, bacteriology, biology, dendrology, etymology, geology, gynecology, histology, limnology, mineralogy, mycology, physiology, pseudology, teratology, logic, logical, syllogism.

Gr. *menyma*, information; see *menyo* under **treason**

Gr. *metis, -ios*, f. counsel, wisdom, skill, enterprise:

L. *Minerva*, f. goddess of wisdom, identified with the Greek Athena: Minerva.

Gr. *noëmon, noëros; noëtikos*, intellectual, intelligent, thoughtful; see *noëma* under **think**

Gr. *-nomy*, study of, knowledge of, science of, < *nomos*, m. usage, law, custom: astronomy, economy, agronomy, Deuteronomy, autonomy, gastronomic.

L. *nosco, notus* (Gr. *gignosko*), know; *cognosco, -nitus*, learn thoroughly, know; *cognoscibilis*, recognizable; *cognobilis*, understandable, intelligible; *gnarus*, knowing, skillful, expert; *ignotus*, unknown; *noscibilis*, knowable; *notorius*, widely but unfavorably known; *notitia*, f. knowledge; *gnome*, f. opinion, judgment, mind, maxim, mark; *gnomidion*, n. dim.; *gnomon, -os*, m. index, judge, rule; *gnomosyne*, f. judgment, prudence; *gnosis*, f. knowledge, wisdom; *gnostes*, m. knower; *gnostos*, known; *gnorimos; polygnotos*, well-known; *agnostos*, unknown, obscure; *apeirognostos*, omniscient; *diagnosis*, f. conclusion from understanding of appearances or symptoms; *micrognomon, -os*, narrow-minded; *micrognomosyne*, f. narrow-mindedness: notify, notion, notorious, notice, noble, recognize, cognition, reconnoiter, connoisseur, incognito, ignore, quaint, acquaintance, gnostic, agnosticism, gnome, gnomon, physiognomy, geognosy, diagnosis, prognosis, pathognomonic, *Limnobia ignota* (a fly), *Arachnocrinus ignotus* (a Devonian crinoid), *Gnorimocrinus cirrifer* (a Silurian crinoid), *Gnomonia ulmea* (a fungus), *Perna isognomon* (a pelecypod), *Agnostus granulatus* (a trilobite), *Polygnotus striolatus* (a wasp).

Gr. *nouneches*, discreet, sensible, wise: *Nuneches* (a wasp).

Gr. *paipalimos*, sly, subtle:

L. *peritus*, practiced, clever, skillful; *peritissimus*, very skillful; *peritia*, f. prac-

tical knowledge, skill; *expertus,* tried, proved, experienced: expert, experience.
L. *persibus,* extremely keen, knowing, acute:
L. *perspicax, -acis,* clear-seeing, keen, penetrating, acute: perspicacious, perspicacity, *Salifa perspicax* (a leech).
Gr. *phrade,* f. knowledge, understanding, hint; *phrades; phradmon, -os,* wise, shrewd; *phradmosyne,* f. understanding, cunning:
Gr. *phronimos,* sensible, wise: *Paraphronima gracilis* (an amphipod).
Gr. *pinytos,* wise, discreet, prudent:
L. *politus,* cultivated, accomplished; see *polio* under **smooth**
L. *prudens,* discreet, sagacious, foreseeing: prudent, prudence, prudential.
L. *sagax, -acis,* wise: sagacity, sagacious, *Sardina sagax* (a fish).
L. *salsus,* salty, keen, witty; see *sal* under **salt**
L. *sapio,* know, taste; *sapiens,* knowing, wise; *sapidus,* savory, wise, prudent; *consipio,* be of sound mind: sapient, savant, sage, insipid, presage, *Homo sapiens* (man).
L. *scientia,* f. knowledge, skill, < *scio, scitus,* know; *scientola,* f. dim. smattering; *scibilis,* knowable, discernible; *scitus; scius,* knowing, informed, experienced, skillful, wise; *scitum,* n. tenet, dogma; *conscientia,* f. consciousness of right and wrong; *conscius,* having knowledge in common with an accomplice or confidant: science, omniscience, conscience, subconscious, nescience, prescient, nice, unscientific, conscientious, scilicet, *Hypnum scitum* (a moss).
L. *sensatus,* intelligent:
L. *sollers, -ertis,* skillful, clever, expert:
Gr. *sophos,* wise; *sophia,* f. wisdom, skill; *sophikos,* of wisdom; *sophisma, -tos,* n. clever device, artifice, trick, fallacy; *sophistes,* m. expert, master; *sophistikos,* of a sophist; *philosophia,* f. love of wisdom: sophomore, sophist, Sophia, sophisticated, sophism, philosopher, philosophy, theosophy.
Gr. *sophron, -os,* sensible, prudent, wise; see *phren* under **mind**
Gr. *strategema(stratugema), -tos,* n. artifice, trick: stratagem, strategy.
L. *subidus,* knowing, sensible: *Homalota subida* (a beetle).
L. *subtilis,* acute, delicate, keen, elusive, sly, fine: subtile, subtle, subtleness, *Composita subtilita* (a fossil brachiopod), *Bacillus subtilis* (hay-bacillus).
Gr. *synetos,* intelligent, wise: *Synetocephalus autumnalis* (a beetle).
L. *trebax, -acis,* experienced, cunning, crafty, < Gr. *tribakos,* rubbed, worn, experienced:
L. *vafer, -fra, -frum,* artful, cunning, sly, subtle, crafty: *Carabus vafer* (a beetle), *Trochosa vufra* (a spider).
L. *versutus,* crafty, sly, dextrous, adroit: *Festuca versuta* (Texas fescue).
L. *veteranus,* experienced, skilled: veteran.
AS. *witan,* know: wit, witty, witness, wisdom, wise, wiseacre, wistful, wizard.
See: **remember, truth, think, art, feel, mind**

knuckle < uncertain origin; see **joint, projection, hand**

krypton < Gr. *kryptos,* hidden; see ELEMENTS under **thing**

L

la-, Gr. intensive suffix; see **very**
labarum, L. (Gr. *labaron*), Roman imperial standard; see **flag**
labe, Gr. handle, haft, hilt, hold, tool, instrument; *labion,* dim.; see **tool**
labecula, L. dim. of *labes,* blemish, blot, defect, spot, stain; see **spot**
label < OF. *label; lambel,* ribbon; see **word, mark**
labeo, L. one with large lips; see *labium* under **lip**
labes, L. a fall, subsidence, stroke of misfortune, first slip, see *labor* under **fall**; blemish, blot, defect, spot, stain, see **spot**

labido- < Gr. *labis, -idos,* forceps, tongs; *labidion,* dim. tweezers; see **tongs**

labilis, L. slipping, gliding, fleeting, transient; *labidus,* slippery; *labina,* a slippery place, landslide; see *labor* under **fall**

labium, L. lip; *labiatus,* lipped; *labiosus,* large-lipped; see **lip**

labo, *-atus,* L. shake, totter, sink, waver; see **shake**

labor, L. work; see **make, difficult**

labrax, *-akos,* Gr. a sea-bass; see **perch**

labros, Gr. furious, hasty, greedy; *labrosyne,* greed; see **greed**

labrum, L. lip, brim; *labellum,* dim., see **lip**; basin, tub, see **basin**

labrus, L. (Gr. *labros*), wrasse; see **wrasse**

labrusca, L. wild grapevine; see **grape**

labrys, Gr. two-bitted ax; see **ax**

laburnum, L. a genus of the bean family; see **bean**

labyrinthus, L. (Gr. *labyrinthos*), a structure with many winding passages; see **turn**

lac, *lactis,* L. milk, see **milk**; < Per. *lak,* a resinous substance, see **resin**

lacaro- < Gr. *lakara,* a kind of tree; see **tree**

lacca, L. an unknown plant; see **plant**

lacco- < Gr. *lakkos,* pond, cistern, reservoir; see **lake**

lace < OF. *laz,* < L. *laqueus,* snare, noose; see **net, weave**

lacer, L. torn, mangled, cut up, < *lacero, -atus,* tear, cut up; see **tear**

lacerna, L. a kind of cloak; *lacernula,* dim.; see **garment**

lacero- < Gr. *lakeros,* talkative, see **speak**

lacerta, L. lizard; see **lizard**

lacertus, L. upper arm from elbow to shoulder; *lacertosus,* muscular, strong; see **arm**

laceryzo- < Gr. *lakeryza,* screamer, howler; see **call**

lacesso, *-itus,* L. excite, provoke, irritate, stimulate; see **arouse**

lachano- < Gr. *lachanon,* potherb, vegetable, see **plant**; *lachania,* kitchen garden, see **field**

lachesis, Gr. lot, destiny; see **lot**

lachno- < Gr. *lachnos,* m.; *lachne,* f. soft, woolly hair, down; see **wool**

lacido- < Gr. *lakis, -idos,* a rent or tear; *lakistos,* torn; see **tear**

lacinia, L. lappet, fringe; *laciniosus,* full of flaps, fringed; see **fringe**

lacio, L. entice, charm; *lacto, -atus,* allure, wheedle, flatter, cajole; see **draw**

lacismato- < Gr. *lakisma, -tos,* tatted; see **rag**

lack < uncertain origin; see **empty, poor, not, little, free, fault, neglect**

lacrima (lacryma), L. tear; *lacrimosus,* tearful; *lacrimo, -atus,* shed tears, weep; see **drop**

lacteus, L. of milk, < *lac, lactis,* milk; *lactarius,* of milk, dairyman; *lacto, -atus,* yield milk, give suck; see **milk**

lactis, L. small intestine, see **intestine**; Gr. *laktis,* pestle, see **pestle**

lactisto- < Gr. *laktistes,* kicker; *laktisma,* a kick; *laktikos,* kicking; see **kick**

lactuca, L. lettuce; see **lettuce**

laculatus, L. four-cornered, checkered; see **four**

lacuna, L. cavity, hollow, cavern, pool, place where water collects, < *lacus,* lake; *lacunar,* a fluted or paneled ceiling; *lacunatus,* fluted, paneled; *lacunosus,* full of hollows, cavities, gaps, sloughs; see **hollow, lake**

lacus, L. lake; *lacusculus,* dim.; see **lake**

lacuturris, L. a kind of cabbage; see **cabbage**

ladanum, L. (Gr. *ladanon*), mastic; see **resin**

ladas, Gr. young stag; see **deer**

ladder < AS. *hlaeder.*
   Gr. *anabathra; diabathra; epibathra,* f. ladder, stairs, steps, gangway: *Epibathra crassa* (a trematode).
   L. *climax, -acis* (Gr. *klimax, -akos*), f. ladder, stairs; *klimakidion; klimakion,* n. dim.; *klimakter, -os,* m. rung of a ladder, critical period: climax, climactic, climacteric, *Climacteris scandens* (brown tree-creeper), *Climacium americanum* (a moss), *Fluvicola climacura* (a waterchat).

Gr. *katelips, -iphos,* f. ladder, stairway, upper story of a house:
L. *scala,* f. flight of stairs, ladder; *scalaris,* of a ladder: scale, scaler, scalariform, escalator, echelon, *Scala lineata* (a gastropod), *Scalaria lamellosa* (a fossil gastropod), *Dryobates scalaris* (a woodpecker), *Orthoceras scalariforme* (an Ordovician cephalopod).

ladle < AS. *hlaedel,* dipper; see **spoon**

lady < AS. *hlaefdie;* see **woman**

laedo, *laesus,* L. hurt by striking, injure, wound; see **strike**

laedos, Gr. a bird; see **bird**

laelaps, L. (Gr. *lailaps*), hurricane; see **storm**

laelia, NL. a genus of orchids; see **orchid**

laemargo- < Gr. *laimargos,* greedy; see **eat**

laemo- < Gr. *laimos,* throat; see **throat**

laena, L. (Gr. *laina*), cloak; see *chlaina* under **garment**

laeo- < Gr. *laios,* left, see **hand**; a kind of thrush, see **thrush**

laepho- < Gr. *laiphos,* shabby garment; piece of cloth, canvas, sail; see **sail**

laepsero- < Gr. *laipseros,* swift, nimble, light; see **swift**

laersino- < Gr. *laerkinon,* a kind of aromatic wood; see **wood**

laertes, Gr. a kind of ant; see **ant**

laesenio- < Gr. *laisenion,* shield; see **shield**

laetamen, L. dung, manure; see **dung**

laetmato- < Gr. *laitma, -tos,* the deep sea; see **sea**

laetus; laetabilis, L. joyful, glad, pleasant; *laetitia,* joy, pleasure; see **joy**

laeva, L. left hand; *laevus,* left; see **hand**

laevigatus (levigatus), L. smooth; see *levis* under **smooth**

laganum, L. (Gr. *laganon*), a thin, broad cake; see **cake**

lagaros, Gr. loose, slack, pliant, thin; see **free**

lagena, L. (Gr. *lagenos,* flagon), large jar or bottle with handles and a narrow neck; *laguncula,* dim.; see **bottle**

lagnos, Gr. lustful, < *lagneia,* coitus, lust; see **coitus, lewd**

lagodias, Gr. a horned owl; see **owl**

lagos, Gr. hare; *lagidion; lagodarion; lagodion; lagion,* dim.; see **hare**

lagynos, Gr. flagon, flask; *lagynion,* dim.; see *lagena* under **bottle**

lake < AS. *lacu,* < L. *lacus, -us,* m. body of standing water; *lacusculus,* m. dim.; *lacuna,* f. cavity, pool, pond; *lacustrinus,* of lakes; *lacunosus,* full of lakes, ponds, sloughs: lacuna, lagoon, lacunose, lacuscular, lacustrine, lacustral, Interlaken, Laguna del Inca, Beverly, *Scirpus lacustris* (tule-bulrush), *Nymphoides lacunosa* (floating-heart), *Drosophila lacicola* (a fly).
L. *cetarium,* n. fishpond:
Gr. *lakkos,* m. pond, cistern, reservoir, pit: laccolith, *Laccophilus decoratus* (a beetle), *Laccobius elongatus* (a fossil beetle).
Gr. *limne,* large pool of standing water, marshy lake; see **marsh**
C. *loch,* lake: Loch Katrine, Loch Lomond, Loch Linnhe, Carlow.
AS. *mere,* pool, lake, sea: mere, mermaid, marsh, meerschaum, Ellesmere, Edgemere.
L. *piscina,* f. fishpond:
See: **marsh, sea**

lalia, Gr. conversation, talk; *lalos,* talkative; *lalax, -agos,* babbler, croaker, frog; see *laleo* under **speak**

lallo, *-atus,* L. sing a lullaby; see **sing**

lama, L. slough, bog, fen; see **marsh**

lamb, AS. young sheep; see **sheep**

lambda, Gr. eleventh letter of the Greek alphabet; see **letter**

lambo, *-itus,* L. lick, wash, bathe; see **lick**

lame < AS. *lama;* see **hurt**

lamella, L. dim. of *lamina,* plate; see **plate**

lamentum, L. mourning, wailing; see **weep**

lamia, Gr. witch or monster said to suck blood and eat human flesh; a bugbear to frighten children; a sharklike fish; see **animal, shark**

lamina, L. a thin plate, blade, sheet; *lamella; lamnula,* dim.; see **plate**

laminaria, NL. a genus of seaweeds; see **alga**

lamium, L. dead-nettle; see **mint**

lamna, Gr. a voracious fish; see **shark**

lamp < Gr. *lampo,* shrine; *lampas, -ados; lampe; lampter, -os,* torch, lamp; see **light**

lamperos, Gr. scummy, slimy; see *lape* under **slime**

lamprey < ML. *lampetra,* an eellike fish; see **eel**

lampros, Gr. bright, brilliant, radiant; see *lanterna* under **light**

lampyris, Gr. glowworm; see **worm**

lamyros, Gr. greedy, wanton; see **greed**

lana, L. wool; *lanula,* dim.; *lanatus,* woolly; *laniger,* wool-bearing; see **wool**

lancea, L. a light spear; *lanceola,* dim.; *lanceolatus,* spearlike; see **spear**

lancino, *-atus,* L. tear to pieces, mangle; see **tear**

land < AS. *land;* see **earth, country, field**

language < OF. *langage,* < L. *lingua,* tongue; see *lingua* under **tongue, speak**

languidus, L. faint, weak, dull, weary; see **weak**

languria, L. a kind of lizard; see **lizard**

lanius, L. butcher, < *lanio, -iatus,* tear, rend, see **butcher;** the typical genus of shrikes, see **shrike**

lantana, L. old name for a viburnum, now a genus of the vervain family; see **vervain**

lantern < L. *lanterna* (Gr. *lampter*), lamp, torch; see **light**

lanthanum, NL. a chemical element, < Gr. *lanthano,* escape notice, forget; see ELEMENTS under **thing, forget**

lanugo, L. down on plants, first soft down of the beard; see *lana* under **wool**

lanx, *lancis,* L. plate, dish, charger, scale or pan of a balance; *lancicula; lancula,* dim.; see **balance**

lao- < Gr. *laos,* people, see **people;** < *las, laos,* stone, see **stone**

lap < AS. *laeppe,* front of the thighs.
    L. *gremium,* n. lap, bosom; *gremialis,* pertaining to a lap; growing in a pollard-like cluster: gremial.

lap < AS. *lapian,* lick up with the tongue; see **lick**

lapara, Gr. flank, loin; see **side**

lapathum, L. (Gr. *lapathon*), a kind of sorrel; see **sorrel**

lape, Gr. scum, slime; see **slime**

laphygmos, Gr. greediness, gluttony; *laphyktes,* gourmand; see *laphysso* under **eat**

laphyron, Gr. booty, plunder, spoils; see **plunder**

lapidarius, L. of stones; stonecutter, < *lapis, -idis,* stone; *lapillus,* dim. pebble; *lapidosus,* stony; see **stone**

lapistes, Gr. braggart; *lapisma,* a boasting, swaggering; see **brag**

laportea, NL. a genus of the nettle family; see **nettle**

lappa, L. bur; *lappaceus,* burlike; see **bur**

lapsana, L. (Gr. *lapsane*), a plant said to be the crucifer, charlock, but the name is now applied to a genus of the composite family; see **composite**

lapsus, L. a slipping, fall, error; see *labor* under **fall**

lapto, Gr. lap with the tongue, drink, suck; see **drink**

laquear, L. a pitted, paneled, fretted, or fluted ceiling; *laqueatus,* fluted, paneled; see *lacuna* under **hollow**

laqueus, L. snare, noose; *laqueo, -atus,* ensnare, entangle; see **trap**

larbason, Gr. antimony; see **antimony**

larch < L. *larix;* see **gymnosperm**

larco- < Gr. *larkos,* charcoal-basket; *larkidion,* dim.; see **basket**

lardum, L. lard, bacon; *lardarius,* butcher, pork-seller, pantry; see **fat**

Lares, L. household gods, hearth, home; see **house**

large < OF. *large,* < L. *largus,* abundant: largo, largess.   See **abundant**
    Gr. *ametros; ametretos,* beyond measure, immeasurable, immense, boundless; *apeirometros,* immeasurable, infinite: *Ametrosomus helmsi* (a katydid).
    Gr. *amphidochmos,* as large as can be grasped; *amphilaphes,* enormous, large, vast: *Amphilaphurus major* (a fossil fish).

L. *amplus*, large; *amplior; amplius*, larger; *amplissimus*, largest; *amplitudo, -inis*, f. bulk, breadth, size; *amplifico, -atus; amplio, -atus*, enlarge, expand, widen: ample, amplify, amplitude, amplification, ampliative, *Potamogeton amplifolius* (a pondweed).

Gr. *apletos*, immense:

Gr. *aspetos*, unspeakably great:

Gr. *athesphatos*, inexpressible, enormous, marvelously great:

Heb. *behemah*, beast, a large animal, probably the hippopotamus; hence, something large; see **animal**

ME. *big*, large, strong: big, bigger, biggest, bigwig, bigly, bighorn *(Ovis canadensis)*.

L. *bu-* (Gr. *bou-*), prefix meaning large, huge, great, monstrous: bulimia, *Bustomum phlebotomum* (a nematode), *Buphagoides erythrorhynchus* (a bird), *Bucephalomyia femorata* (a fly).

L. *capax, -acis*, roomy, spacious, able, fit; see *capio* under **take**

L. *capitaneus*, chief in size, large: *Conus capitaneus* (a gastropod).

Gr. *cheiroplethes*, as large as can be held in the hand:

L. *Cyclops, -opis* (Gr. *Kyklops, -opos*), m. mythical one-eyed giant; *Cyclopeus*, of the Cyclopes, large, massive, rough: Cyclops, Cyclopean, *Cyclops fimbriatus* (a copepod), *Hipposideros cyclops* (a bat).

L. *colossus* (Gr. *kolossos*), m. large statue; *kolossiaios; kolossikos*, gigantic: colossal, Colosseum, *Colossochelys atlas* (a fossil turtle), *Belostoma colossicum* (a waterbug).

L. *enormis*, immense, huge, vast: enormous, enormity.

F. *Gargantua*, m. a giant glutton: Gargantuan.

L. *gigas, -antis* (Gr. *-antos*), m. giant; *giganteus*, of giants, large: gigantic, giant, *Gigantoceras cancellatum* (a fossil cephalopod), *Gigantopithecus blacki* (a fossil primate), *Aristolochia gigas* (a birthwort), *Cereus giganteus* (sahuaro, a catus).

Heb. *Golyath*, Goliath, a Philistine giant: *Goliathinus aureosparsus* (a beetle), *Goliathocera antennalis* (a fly), *Rana goliath* (an African frog), *Hegemon goliathus* (a beetle).

L. *grandis*, great: grand, grandee, grandeur, grandiloquent, grandparent, *Abies grandis* (a fir), *Magnolia grandiflora* (southern magnolia), *Populus grandidentata* (large-toothed aspen).

L. *grossus*, big, coarse, thick: gross, grocer, engross, grosgrain, grosbeak, *Dasyllis grossa* (a robberfly), *Exaesiopus grossipes* (a beetle).

L. *Gyas*, m. a giant with a hundred arms: *Gyas annulatus* (a phalangid), *Gyascutus planicosta* (a beetle).

Gr. *hadros*, well-developed, bulky, stout, strong, great; *hadrotes*, f. thickness, fullness, vigor, abundance: *Hadrosaurus foulki* (a dinosaur), *Hadropithecus stenognathus* (a simian), *Hadronotus rugosus* (a wasp), *Hadrotes extensus* (a beetle).

Gr. *hamaxiaios*, large enough to fill a wagon:

Gr. *hyperphues*, enormous, excessive, immense, marvelous:

L. *immanis*, enormous, huge, monstrous, frightful, fierce: immane.

L. *immeasurabilis*, not measurable: immeasurable.

L. *immensus*, vast, boundless, immeasurable: immense, immensity.

L. *impensus*, ample, great, large, strong:

L. *indimensus*, immeasurable, innumerable:

L. *ingens*, vast, huge, great, immoderately large: *Aepyornis ingens* (an extinct bird).

L. *interminabilis; interminatus; interminis*, endless, boundless: interminable.

Gr. *keteios*, monstrous, large: *Cetiosaurus oxoniensis* (a sauropod).

Heb. *liwyathan*, a large aquatic animal; see **animal**

L. *magnus*, large, great; *major*, m., f.; *majus*, n. greater; *maximus*, greatest; *majusculus*, somewhat larger or greater; *majorinus*, of a larger kind of form: magnanimous, magnify, main, magnitude, magnificent, magnate, majority, majolica(Majorca), majestic, majuscule, master, mayor, merino, Myers, Charlemagne, maximum, *Monotrypa magna* (a fossil bryozoan), *Phacostaurus magnificus* (a radiolarian), *Vinca major* (a periwinkle), *Tropaeolum majus* (a nasturtium), *Spiropora majuscula* (a fossil bryozoan), *Busycon maximum* (a fossil gastropod).

Gr. *makros*, long, but often mistakenly used to mean large; see **long**

G. *mammuth* (Russ. *mammot*), a kind of elephant; something large; see **elephant**

Gr. *megas, megale, mega*, large, great, very; *meizon*, larger, greater; *megistos*, largest, greatest; *megethos*, n. greatness, size; *megalodynamos*, very powerful; *megaloploutos*, very rich: megaphone, megatherium, megalithic, acromegaly, megalops, megaspore, megalomania, megaloblast, *Megalosaurus oweni* (a dinosaur), *Megalobatrachus maximus* (giant salamander), *Megalonyx jeffersoni* (a fossil ground-sloth), *Megascops asio* (screech-owl), *Megarhyssa lunator* (thalessa, an ichneumon-wasp), *Meganthropus paleojavanicus* (a fossil primate), *Megistocrinus nodosus* (a Devonian crinoid), *Asaphus megistus* (a trilobite), *Otocyon megalotis* (an African mammal), *Platysternum megacephalum* (a tortoise).

Gr. *pelikos*, how large, how great, of what age: *pelikotes*, f. magnitude:

Gr. *peloros; pelorios*, huge, monstrous, prodigious, terrible; *pelor, -os*, n. monster, prodigy: peloria, *Pelorosaurus conybeari* (a fossil reptile), *Pelorus ambiguus* (a protozoan), *Nerita peloronta* (a gastropod).

Gr. *periosus*, immense:

L. *prodigiosus*, wonderful, large, vast; see **wonder**

L. *spatiosus*, roomy, large; see **space**

L. *stupendus*, astonishing, wonderful, large; see **wonder**

Gr. *Titan, -os*, m. son or daughter of Uranus and Gaea, symbolic of brute force and large size; *titanikos*, of the Titans: the *Titanic*, titanium, titaniferous, *Titanohyrax ultimus* (a fossil mammal), *Titanus giganteus* (a beetle), *Rubus titanus* (a blackberry), *Aepyornis titan* (an extinct bird), *Eotitanotherium osborni* (a fossil mammal).

L. *tremendus*, something to be trembled at, dreadful, terrible, large: tremendous.

L. *vastus*, waste, desolate, empty, enormous: vast, vastness, vastitude, devastate.

L. *voluminosus*, full of folds, of large bulk: voluminous.

See: **heavy, weight, strong, number, abundance, very.**   Derive: *Tipula brobdingnagia* (a Chinese cranefly).

largior, -itus, L. give bountifully; see **give**

larifuga, L. vagabond; see *fuga* under **leave**

larimnon, Gr. Arabic name for frankincense; see **resin**

larimos, Gr. a kind of fish; see **fish**

larinos, Gr. fatted, fat; see **fat**

larix, L. larch; see **gymnosperm**

lark < AS. *laferce.*
L. *alauda*, f. lark: *Alauda arvensis* (skylark).
Sp. *alondra*, f. lark:
L. *cassita*, f. crested lark:
Gr. *kalandros*, m. a kind of lark: *Melanocorypha calandra* (a lark).
Gr. *korydos; korydallis, -idos*, f.; *korydalos*, m. crested lark: *Corydalis(Capnoides) rosea* (a fumewort), *Corydalis cornutus* (hellgrammite).
Gr. *piphinx, -ingos*, m. a kind of lark:

larkspur; see *delphinium* under **buttercup**

larnaco- < Gr. *larnax, -akos*, box, chest, coffer; see **box**

laros, Gr. agreeable, pleasant; see **agreeable**

larus, L. (Gr. *laros*), gull, mew; see **gull**

larva, L. ghost, mask, early stage of some animals; *larvalis*, ghostly; see **young**

larynx, Gr. gullet; see **throat**

lasanum, L. (Gr. *lasanon*), stand, gridiron, nightstool; see **frame**

lascivus, L. wanton, sportive, lustful, licentious; *lascivulus*, dim.; see **lewd**

laserpitium, L. a genus of the carrot family; see **carrot**

lasios, Gr. hairy, woolly; see **wool**

lassus, L. weary; *lassitudo*, weariness; see **weak**

last < AS. *laet*, see **end, after, out, always;** < AS. *laest*, trace, track, see **shoe**

lastauros, Gr. lewd person; see **lewd**

latace, L. a magic herb; see **plant**

latax, Gr. a water quadruped, probably a beaver or otter; see **weasel**

late < AS. *laet;* see **after, old, slow, delay**

latens, L. hidden, secret, < *lateo*, lie hidden; *latebra*, hiding-place, retreat; *latebrosus*, full of hiding-places; see **cover**

later, -eris, L. brick; *laterculus*, dim.; *laterarius*, of bricks, bricklayer; see **brick**

lateralis, L. of the side; see *latus* under **side**

laterculum, L. list, register; see **list**

latex, L. fluid, liquid, sap, juice; see **juice**

lathargos, Gr. piece of leather; see **leather**

lathetico- < Gr. *lathetikos*, likely to escape notice; *lathiphron*, forgetful; see *lethe* under **forget**

lathraea, NL. a genus of the broomrape family; see **broomrape**

lathro- < Gr. *lathra,* secretly, covertly; *lathraios; lathrios,* secret, hidden, stealthy; see **secret**

lathyris, Gr. a kind of spurge; see **spurge**

lathyrus, L. (Gr. *lathyros*), a legume; see **bean**

latibulum, L. hiding-place, den, refuge; see *lateo* under **cover**

Latin < L. *Latinus,* pertaining to Latium, the province of Italy in which Rome is located: Latinist, latinization.

 L. *Etruscus,* of Etruria in Italy: *Crocus etruscus* (a crocus).

 L. *Italia,* f. Italy, supposedly < Gr. *italos,* bull: Italy, Italian, *Cryptochilus italicum* (a wasp).

 L. *Romanus,* Roman, Latin, < *Roma,* f. Rome; *Romulus,* m. fabled founder of Rome: Rome, Roman, romance, romantic, Romanesque, Romanism, *Sideritis romana* (Roman ironwort), *Musconiscus romanorum* (a sowbug).

latitudo, L. breadth, width; see *latus* under **broad**

latomium, L. (Gr. *latomion*), stone quarry, see **mine**; *latomus,* stone-cutter, see **mason**

lator, L. bearer, proposer; see *fero* under **carry**

latrans, L. barking, howling; *latrator,* barker; see *latro* under **bark**

latreus; latreutes; latris, Gr. servant; *latreia* (L. *latria*), service; see **servant**

latrina, L. bath, privy; see **room**

latro, *-onis,* L. hireling, robber, brigand; *latrunculus,* dim.; see **steal**

-latry < Gr. *latreia,* worship; see *latreus* under **servant**

lattice < F. *lattis;* see **screen, basket**

latypos, Gr. stone-cutter, mason; *latype,* chip of stone; see **mason**

laudabilis, L. praiseworthy; *laus, laudis,* praise; *laudo, -atus,* praise; see **honor**

laugh < AS. *hlehhan.*

 L. *balatro, -onis,* m. babbler, jester, buffon: *Balatro calvus* (a rotifer).

 L. *cachinno, -atus,* laugh immoderately; *cachinnus,* m. loud laugh: cachinnation, *Herpetotheres cachinnans* (laughing-falcon).

 L. *cavillor, -atus,* jest, joke, satirize, rail, scoff; *cavilla,* f. jeering, scoffing, irony; *cavillator,* m.; *cavillatrix,* f. jeerer: cavil.

 Gr. *charientisma, -tos,* n. a witty remark; *charitia,* f. jest, joke:

 Gr. *chleuastes,* m. mocker, scoffer; *chleue,* f. joke: *Chlevastes colubrinus* (a snake-eel).

 L. *comicus* (Gr. *komikos*), pertaining to comedy; *comoedia* (Gr. *komoidia*), f. a humorous drama; *komoidos,* m. comedian: comical, comedy, comedian, comic, encomium.

 L. *coprea* (Gr. *koprias*), m. obscene buffoon or jester:

 L. *dicax, -acis,* witty, satirical; *dicaulus,* facetious, witty: dicacity.

 L. *dossenus,* m. clown, jester:

 L. *facetia,* f. wit, jest, quip, drollery, humor; *facetus,* merry, witty: facetious.

 Gr. *gelao,* laugh; *gelanes,* laughing, cheerful; *gelasimos; gelastos,* laughable, ridiculous; *gelasinos; gelastes,* m. laugher, sneerer; *gelasis,* f.; *gelos, -otos,* m. laughter; *gelasma, -tos,* n. a laugh; *gelastikos,* inclined to laugh; *gelotopoios,* m. jester, buffoon; *ekgelos, -otos,* m. loud laughter; *perigelastos,* very ridiculous: gelastic, *Gelastocoris oculatus* (toad-bug), *Uca(Gelasimus) vocans* (a fiddler-crab).

 L. *gingilismus,* m. pealing laughter:

 L. *humor, -is,* m. fun, comicality: humor, humorous. Explain 'dry humor.'

 L. *jocus,* m. jest, game, sport; *joculus,* m. dim.; *jocabundus; jocundus(jucundus),* merry, pleasant, sporting; *jocosus; jocularis; jocularius,* facetious, droll, laughable; *joculator, -is,* m. jester, joker; *jocor, -atus,* jest: joke, jocose, jocularity, jocular, juggler, Joyce, jocundity, jeopardy, *Joculator caelata* (a gastropod), *Pycnonotus jocosus*(red-whiskered bulbul), *Macrocrinus jucundus* (a Mississippian crinoid), *Systolederus injucundus* (a grouse-locust).

 Gr. *kanchasmos,* m. loud laughter; *kanchas, -antos,* m. laughter, < *kanchazo; kanchalao,* laugh aloud:

 Gr. *katamokesis,* f. mockery:

 Gr. *kertomios,* heart-cutting, mocking; *kertomikos,* jeering:

 Gr. *kichlismos,* m. a tittering, giggling:

 L. *lepidus,* pleasant, witty; see **agreeable**

 L. *ludibrum,* derision, mockery, laughing-stock; *ludifico, -atus,* make a fool of, mock, ridicule; see *ludo* under **play**

 Gr. *meidao,* smile; *meidamon, -os,* smiling; *meidema, -tos,* n. smile:

 Gr. *mokos,* m. mocker:

L. *nugor, -atus,* jest, trifle, talk nonsense, play the fool; *nuga,* jest, joke, trifle; *nugator; nugo,* buffoo ɪ, trifler; see **trifle**
Gr. *phylax, -akos,* jester fool, buffoon; see **fool**
L. *rideo, risus,* laugh; *ridiculus,* exciting laughter, absurd, droll, funny; *ridibundus,* laughing, prone to laughter; *risibilis,* capable of laughing, causing laughter; *risor, -is,* m. laughter; *derisus,* mocked, scorned; *derisor; irrisor, -is,* m. derider, mocker, scoffer; *irridiculum,* n. laughing-stock; *irrisus,* scoffing, mocking, derision: ridiculous, ridicule, riant, risibility, arride, deride, derision, *Apobletes ridens* (a beetle), *Campsicnemus ridiculus* (a fly), *Larus ridibundus* (laughing gull), *Streptopelia risoria* (a dove), *Irrisor erythrorhynchus* (a wood-hoopoe).
L. *sannio, -onis,* m. a grimacer, buffoon:
Gr. *sarkasmos,* m. biting mockery: sarcasm, sarcastic.
L. *satira,* f. a poetic medley devoted to ridicule and sarcasm; *satiricus,* of ridicule: satire, satirical, *Neoclinus satiricus* (a blenny).
L. *scurra,* m. jester; *scurrilis,* buffooning, jesting: scurrilous, scurrility, *Acartauchenius scurrilis* (a spider), *Catasetum scurra* (an orchid).
Gr. *sillos,* m. lampoon, satire: sillograph.
Gr. *skomma, -tos,* n. jest, gibe, taunt; *skoptes,* m. scoffer; *skoptikos,* jesting, mocking, scoffing, < *skopto,* mock, jeer, scoff, jest, gibe; *aposkomma, -tos,* n. banter, raillery: *Scoptes crotopus* (a butterfly), *Scopticus herbeus* (a crab-spider).
Gr. *tothazo,* mock, scoff, scorn; *tothasmos,* m. a mocking; *tothastes,* m. scoffer; *tothastikos,* scornful:
See: **joy, play, fool, tease, agreeable, trifle**

laura, Gr. lane, alley; see **way**

laurel < L. *laurus,* f. bay-tree; *laureola,* f. dim.; *laurago, -inis,* f. a kind of laurel; *laureus; laurinus,* of laurel; *laureatus,* crowned or adorned with laurel: laureate, laurin, laurite, Laura, Lawrence, *Laurus nobilis* (Grecian laurel), *Smilax laurifolia* (a greenbrier), *Daphne laureola* (spurge-laurel), *Rhus laurina* (laurel-sumac), *Prunus laurocerasus* (laurel-cherry).
NL. *aniba,* f. a genus of the laurel family, < Ab.Am. *anhouibe,* a tropical plant: *Aniba coto* (a lauracead).
NL. *cassytha,* f. a genus of the laurel family, < L. *cadytas* (Gr. *kadytas; kasytas*), m. a parasitic plant: *Cassytha filiformis* (a woevine), *Rhipsalis cassytha* (a cactacead).
Gr. *daphne,* f. laurel, bay-tree; now a genus of the mezereon family, Thymeleaceae; *daphninos,* of laurel; *daphnis, -idos,* f. laurel berry; *daphnon, -os,* m. laurel grove; *daphnodes; daphnotos,* like laurel: daphnin, Daphne, *Daphne odora* (sweet daphne), *Chamaedaphne calyculata* (leatherleaf).
L. *euthalos,* f. a kind of laurel:
NL. *lindera,* f. a genus of the laurel family, < Johann Linder, Swedish botanist: *Lindera benzoin* (spicebush).
NL. *litsea,* f. a genus of the laurel family, < Chin. *li tsai,* little plum: *Litsea reticulata* (a litsea).
L. *mustax, -acis,* f. a kind of laurel:
NL. *ocotea,* f. a genus of the laurel family, < an Ab.Am. word: *Ocotea canaliculata* (a lauracead).
Gr. *persea,* f. an Egyptian tree; a genus of the laurel family: *Persea palustris* (swamp redbay).
NL. *sassafras,* n. a genus of the laurel family: *Sassafras albidum* (sassafras).
NL. *umbellularia,* f. a genus of the laurel family: *Umbellularia californica* (California laurel).

lautus, L. washed, neat, elegant; see *lavo* under **wash**

lavender < ML. *lavandula; lavendula;* see **mint**

lavo, L. wash; *lavabrum,* bathtub; *lavatorium,* place to wash; see **wash, basin**

law < AS. *lagu:* lawless, lawyer, unlawful, mother-in-law, scofflaw, inlaw, outlaw.
Gr. *ados,* n. decree:
Gr. *axioma, -tos,* n. self-evident principle: axiom, axiomatic.
L. *canon* (Gr. *kanon*), rule, standard; *kanonikos,* according to rule, regular; see **measure**
Gr. *dike,* f. right, order, law, manner; *dikaios,* observant of the right, decent, just, civilized; *dikaiosyne,* f. justice; *dikaspolia,* f. a judgment; *dikanikos,* lawyerlike; *dikasimos,* judicial; *dikastes,* m. judge; *dikastikos,* of law and trials; *dikasterion,* n. court of justice; *endikos,* right, just; *dosidikos,* letting the law take its course; *syndikos,* c. advocate: syndic, syndicate, Eurydice, *Dicastes endymion* (a beetle), *Dicasticus celatus* (a beetle), *Dicaiothrips[Dicaeothrips] denticollis* (a thysanopterid), *Dicaeum olivaceum* (a flowerpecker).
L. *formula,* f. rule, method: formula, formulate.

Gr. *horos,* rule, standard; see **border**

L. *jus, juris,* n. law, right; *juridicus,* pertaining to the administration of justice: just, justice, justify, justiciary, jurist, jurisdiction, jury, injury, unjustly, jurisprudence.

L. *jussum,* order, decree, law, < *jubeo, jussus,* bid, command, order; see **ask**

L. *lex, legis,* f. law; *legalis,* of law, according to law; *legitimus,* lawful; *legislator, -is,* m. lawmaker, lawgiver; *legirupa,* m. lawbreaker; *leguleius,* m. pettifogger, shyster; *exlex; illex, -egis,* without law, contrary to law, lawless: legal, legalize, legislature, legitimate, illegal, privilege, allege, loyal, leal, disloyalty, LL.D.

Gr. *Lykourgos,* m. Lycurgus, Spartan lawgiver: *Lycurgus subvitta* (a bug).

Gr. *nomos,* m. usage, law, rule, < *nemo,* distribute, regulate; *ennomos,* within the law, legal; *eknomos; paranomos,* illegal: nomology, nomography, antinomian, Deuteronomy, astronomy, economics, gastronomic, Nemesis.

L. *norma,* f. rule, pattern, carpenter's square; *normalis,* according to rule: normal, norm, abnormal, normalcy, enormous, *Carex normalis* (a sedge).

L. *praeceptum,* n. maxim, rule, injunction: precept.

L. *principium,* foundation, beginning; see **begin**

Gr. *psephisma, -tos,* n. an ordinance voted by the people:

L. *regimen, -inis,* n. guide, rule: regimen, regime.

L. *rogatio,* question, proposed law, bill; see *rogo* under **ask**

Gr. *Solon,* m. celebrated Greek legislator: Solon.

Gr. *themis, -mistos,* f.; *thesmos (tethmos),* m. law, decree, rule, justice; *themistos; themitos,* lawful; *athemistos; athesmos,* lawless, unlawful; *enthesmos,* lawful: Themis, Themistocles, *Themistonoe reticulata* (a beetle).

See: **justice, right, measure, govern, judge, true, forbidden**

lawyer

L. *advocatus,* m. counselor, attorney: advocate, advocacy, advocatress.

L. *causula,* f. petty lawsuit:

L. *cognitor, -is,* m. advocate, attorney, defender:

Gr. *dikorrhaphos,* m. pettifogger:

L. *leguleius,* m. pettifogging lawyer:

L. *rabula,* m. a brawling, wrangling advocate, pettifogger: *Rabula davisi* (a fish).

laxeutes, Gr. stone-cutter; see **mason**

laxus, L. loose, slack, unstrung; *laxo, -atus,* undo, unloose, slacken, lighten; see **free**

lay < AS. *lecgan,* put down, see **place**; layer, see **plate**

lazaros, Gr. corpse; see **death**

lazy < uncertain origin; see **rest**

lead < AS. *lead.*   See ELEMENTS under **thing**

L. *cerussa,* f. white-lead: cerussite.

L. *galena* (Gr. *galene*), f. an ore of lead: galena.

L. *liveo,* turn lead-colored, ashy; *lividus,* bluish; see **blue**

Gr. *miltos,* red earth, red-lead, minium; see **red**

L. *molybdus* (Gr. *molybdos*), m. lead; *molybdaena* (Gr. *molybdaina*), f. galena; *molybdion,* n.; *molybdis, -idos,* f. lead weight, sinker, plummet; *molybdotos,* leaded; *molybros,* lead-colored: molybdenum, molybdenite molybdic, *Molybdogompha biseriata* (a moth), *Molybdotus laxepunctatus* (a beetle), *Acarospora molybdina* (lichen).

Gr. *pelidnos; pelios,* black and blue, livid; see **blue**

L. *plumbum,* n. lead; *plumbago, -inis,* f. a leadlike ore, graphite, a plant; *plumbarius,* of lead; m. plumber; *plumbeus,* leaden; *plumbo, -atus,* solder with lead; *applumbo,* solder: plumb-bob, plumber, plummet, plumb, plunge, *Plumbago europaea* (leadwort), *Chimaera plumbea* (a sharklike fish), *Hippelates plumbellus* (a gnat).

Gr. *psimythion (psimmythion),* n. white-lead; *psimythistes,* m. painted: *Psimmythimas* (a bug).

lead < *laedan:* leader, leading, leadership, misleading.

Gr. *ago,* lead, train, guide, stimulate, promote; *agogos,* leading; m. guide; *agogeus; aktor, -os,* m. leader, leash; *agoge,* f. a carrying away, movement; *agogimos,* easy to lead; *agema, -tos,* n. anything led, division of an army: pedagogue, demagogue, synagogue, *Agoseris hirsuta* (a composite).

Gr. *apithynter; dieuthynter, -os,* m. director, pilot, guide:

Gr. *archi-,* chief, leader; see **govern**

L. *duco, ductus,* lead, guide, draw; *dux, ducis,* c.; *ductor, -is,* m.; *ductrix, -icis,* f. leader, guide; *ducalis; ducianus,* of a leader; *ducatus,* m. leadership; *ductilis,* that can be led, that can be drawn out thin; *seductor, -is,* m. misleader; *seductilis,* that may be misled, seducible: duct, aqueduct, viaduct, duke, ductile, duchy, duchess, ducat, douche, education, conducive, conduit, reduce, introduction, product, deduce, abductor, conduct, inductance, seduction, reproductive, traducer, *Campsiomorpha ducalis* (a beetle), *Naucrates ductor* (pilot-fish).

Gr. *hodegos*, guide, teacher; see **teach**

Gr. *hegemon, -os; hegeter, -os; hegetes; hyphegetes; kathegetes; periegetes,* m. leader, guide, teacher; *hegesis*, f. leadership: hegemony, *Hegeter politus* (a beetle), *Hegeterocasa arabica* (a beetle), *Hegetotherium mirabile* (a fossil mammal), *Hegesidesmus eliyanus* (a bug), *Hegemon regius* (a beetle), *Cathegesis tridentella* (a moth), *Archigetes neuropterorum* (a fossil neuropterid).

Gr. *karanos*, m. a chief:

Gr. *koryphaios*, leader, chief; see *koryphe* under **head**

Gr. *pompos,* m. leader, conductor, escort, guide; *pompinos*, conducting: *Pomponema mirabile* (a nematode).

L. *praetor, -is*, leader, chief, magistrate; see **govern**

L. *procer, -is*, m. leader, chief, noble:

Gr. *tagos*, m. commander: *Tagus productus* (a bug).

Gr. *xenagos*, m. commander of auxiliary troops:

See: **govern, teach, first, move, begin**

leadwort

NL. *armeria*, f. a genus of the leadwort family: *Armeria maritima* (thrift), *Silene armeria* (sweetwilliam-catchfly).

L. *limonium* (Gr. *leimonion*), n. sea-lavender: *Limonium angustatum* (a sea-lavender), *Phyteuma limonifolium* (a campanulacead).

L. *plumbago, inis*, f. a genus of the leadwort family: *Plumbago rosea* (a leadwort), *Ceratostigma plumbaginoides* (a plumbaginacead), *Nicotiana plumbaginifolia* (a tobacco), Plumbaginaceae.

L. *statice*, f. an astringent herb; a genus of the leadwort family: *Statice armeria* (now *Armeria maritima*), *Aster staticifolius* (an aster), *Chorizanthe staticoides* (a polygonacead).

L. *tripolium* (Gr. *tripolion*), n. leadwort: *Aster tripolium* (an aster).

leaf < AS. *leaf:* leaflet, leafy, loaf-mincr, goldleaf.

L. *bractea*, scale, small leaf; see **scale**

L. *folium*, n. leaf; *foliolum*, n. dim.; *foliaceus; foliatilis; foliatus; foliosus; folius*, leafy, leaved: foliage, foliole, folio, foliicolous, folic acid, cinquefoil, trefoil, defoliate, exfoliation, *Trifolium tomentosum* (woolly clover), *Kalmia latifolia* (mountain-laurel), *Argulus foliaceus* (carp-louse), *Lonicera caprifolium* (a honeysuckle), *Eupatorium perfoliatum* (boneset, thorowort), *Betula populifolia* (gray birch), *Acacia longifolia* (wattle), *Diplograptus foliaceus* (a graptolite), *Adenocarpus foliolosus* (a legume), *Clematis quinquefoliolata* (a clematis), *Hymenopappus filifolius* (a composite), *Coptis trifolia* (a goldthread), *Rhus quercifolia* (poison-oak), *Vaccinium parvifolium* (red huckleberry).

L. *frons, -ondis*, f. leafy branch, bough, foliage, leaf; *frondiculus*, m. dim.; *frondeus*, of leaves; *frondosus*, full of leaves, leafy: frondiferous, frondose, *Gaylussacia frondosa* (dangleberry), *Frondicularia compta* (a foraminifer), *Baetis frondalis* (a mayfly), *Rubus trifrons* (a dewberry).

L. *pagina*, f. leaf of a book, slab, tablet; *pagella; paginula*, f. dim.; *paginalis*, of a page: page, paginal, pagination.   Derive: pageant.

Gr. *petalon*, n. leaf: petal, petaloid, petalite, monopetalous, *Petalostemon albidus* (white prairie-clover), *Petaloproctus cirratus* (an annelid), *Zygopetalum rostratum* (an orchid), *Ceratopetalum gummiferum* (a cunoniacead), *Dryas octopetala* (a rosacead), *Phaenocalpis petalospyris* (a radiolarian), *Rollinia dolabripetala* (an annonacead), *Epiphyllum oxypetalum* (a cactus).

Gr. *phyllon*, n. leaf; *phyllarion; phyllion*, n. dim.; *phyllinos*, of leaves; *phyllas, -ados*, leafy; f. foliage, salad; *phyllodes*, leaflike: phyllode, phylloclad, phyllotaxy, cladophyll, cataphyll, sporophyll, chlorophyll, Phyllis, zygophyllaceous, Hymenophyllaceae, *Phyllopteryx foliatus* (a fish), *Phyllomorpha laciniata* (a bug), *Phylloxera vitifolia* (grape-aphid), *Phylladothrips karnyi* (a thysanopterid), *Phyllium frondosum* (a phasmid), *Epiphyllum strictum* (a cactus), *Buxbaumia aphylla* (a moss), *Podostemon ceratophyllum* (riverweed), *Castanopsis chrysophylla* (a chinquapin), *Oecophylla smaragdina* (red tree-ant), *Mitella diphylla* (miterwort).

L. *pinna*, f. leaflet, first division of a frond; *pinnula*, f. dim. division of a leaflet: pinna, pinnule.

Gr. *selis, -idos*, leaf, page or sheet of paper; *selidion*, dim.; see **paper**

NL. *sepalum*, n. a leafy division of the calyx: sepal, sepaloid, *Trichostema rubrisepalum* (a mint), *Chorisepalum ovatum* (a gentianacead).

Gr. *thrion*, n. fig-leaf:

L. *valva*, leaf of a folding door; see **door**

See: **plate, paper, tongue, flap, branch, scale**

leak < ON. *leka,* let water or anything come in or escape unintentionally; see **lose, flow, pour, run, let**

lean < AS. *hlinian,* deviate from a vertical position, see **slope;** < *hlaene,* deficient, wanting, see **thin**

leap < AS. *hleapan:* lope, elope, interloper, gallop, gantlet, orlop, lapwing.
    L. *assilio, -sultus,* leap upon; *desilio,* leap down, dismount; *desultor, -is,* m. leaper, inconstant person; *desultorius,* fickle, inconstant, superficial, skipping about; *exilio (exsilio),* bound forth, leap up, rejoice; *resilio,* leap back; *transilio,* leap across or over: desultory, exult, exultation, result, resultant, resilient.
    Gr. *elastikos,* rebounding, springing: elastic, elasticity, *Ficus elastica* (rubber-fig).
    Gr. *hallomai,* spring, leap, bound; *halma; exalma, -tos,* n. leap, bound; *halsis,* f. a leaping; *haltikos,* good at leaping; L. *halter, -is* (Gr. *-os*), m. weight used in jumping, balancer: halter (pl. halteres), *Allosaurus[Hallosaurus] agilis* (a dinosaur), *Halmaturus erubescens* (a fossil marsupial).
    Gr. *ixalos,* bounding, springing: *Ixalus punctatus* (a frog), *Ixalidium hematoscelis* (a grasshopper).
    Gr. *keletistes,* leaper, inconstant person, < *keletiza,* ride two or more horses at the same time; see *keles* under **ride**
    Gr. *metabates,* m. leaper:
    Gr. *pedao,* leap, spring; *pedema, -tos,* n. a leap; *pedesis,* f. a leaping; *pedetes,* m. leaper; *pedetikos,* good at leaping: diapedesis, *Pedetes caffer* (jumping hare), *Pedeticosaurus leviseuri* (a fossil crocodile), *Polypedetes pardalis* (a flying frog).
    Gr. *petauristes,* m. vaulter, tumbler; *petauron,* n. springboard: *Petauristes femorata* (a beetle), *Petaurus breviceps* (a gliding possum).
    L. *salio, saltus,* leap, jump, bound, dance, hop; *salto, -atus,* hop, dance; *saltito,* dance vigorously; *saltator, -is,* m.; *saltatrix, -icis,* f. dancer, hopper, leaper; *salax, -acis,* fond of leaping; *saliens,* leaping; *salticus,* of leaping: salient, salmon, sally, assail, salacious, somersault, assault, insult, *Salticus biguttatus* (a spider), *Talitrus saltator* (sandhopper), *Pomatomus saltatrix* (bluefish), *Neuroterus saltatorius* (jumping gall-wasp), *Thunnus saliens* (tuna), *Serrosalmo rhombeus* (a piranha), *Carpocapsa saltitans* (moth whose larva activates the Mexican jumping beans, which are seeds of the euphorbiacead, *Sebastiania pavoniana), Salius halticus* (a fish).
    Gr. *skarthmos,* m. a leaping, skipping, < *skairo,* leap, skip, dance, frisk; *skartes,* nimble, quick: *Scartichthys rubripunctatus* (a fish), *Rupiscartes atlanticus* (a blenny), *Poeciloscarta histrio* (a bug), *Euscarthmus agilis* (a bird).
    Gr. *skirtao,* leap, bound, spring; *skirtema, -tos,* n. bound, leap; *skirtetes,* m. leaper: *Petroscirtes atrodorsalis* (a blenny), *Alactaga(Scirtomys) tetradactylus* (a mouse), *Apteroscirtus denudatus* (a katydid).
    Gr. *throsko,* leap, spring, attack: *Throscinus crotchi* (a beetle).
    See: **run, goat, joy**

learn < AS. *leornian.*
    Gr. *aisthanomai,* perceive or apprehend by the senses, feel, learn; see **feel**
    L. *alumnus,* pupil; see **child**
    L. *cognosco, -nitus,* learn thoroughly, know; see *nosco* under **know**
    L. *comperio, -pertus,* ascertain, learn with certainty, discover: *Carex incomperta* (a sedge).
    L. *concipio, -ceptus,* take in, comprehend; *percipio,* observe, feel, comprehend: conceive, concept, conception, perceptible, perception.
    Gr. *didagma,* lesson; see *didasko* under **teach**
    L. *disco,* learn; *discipulus,* m. pupil, apprentice; *disciplina,* f. instruction: disciple, discipline, disciplinary.
    L. *docibilis,* learning easily, teachable; see *doceo* under **teach**
    Gr. *homiletes,* disciple, scholar; see *homiletikos* under **speak**
    Gr. *mathematikos,* disposed to learn, < *manthano,* learn; *mathetes,* m.; *mathetris, -idos,* f. *mathesis,* f. act of learning: mathematics, philomathean, polymathy, *Mathesis guttigera* (a beetle).
    L. *pupillus,* learner, dim. of *pupus,* child; see **child**
    Gr. *pystos,* learned, < *pynthanomai,* hear, learn:
    L. *scholaris,* of a school; see **school**
    L. *studens, -entis,* m. learner, pupil; *studiosus,* zealous, fond of study, < *studeo,* be diligent, zealous, learn: study, studious, student.
    See: **school, find**

leather < AS. *lether.*
    L. *aluta,* f. a kind of soft leather: *alutacius; alutarius,* of soft leather: *Aluta flexilis* (a Cambrian ostracode).
    Gr. *byrsa,* f. hide, leather; *byrsine,* f. strap, thong; *byrsinos,* leathern; *byrsodepses,* m. tanner: *Byrsonima crassifolia* (a malpighiacead).

L. *corium,* n. leather; *coriaceus; coriarius,* of leather; m. tanner: corium, coriace-
ous, excoriate, cuirass, *Coriaria thymifolia* (a sapindacead), *Ornithodorus
coriaceus* (pajaroello-tick), *Dermatochelys coriacea* (leatherback-turtle), *Rhus
coriaria* (Sicilian sumac).

Gr. *dermatinos,* of skin, leathern; see *derma* under **skin**

Gr. *diphthera,* f. leather; *diphtherias,* clad in a leather frock: diphtheria,
Diphtherogaster flava (a centiped).

Gr. *dora,* skin, hide, see **skin;** *doros,* leather bag, see **bag**

Gr. *kattys, -yos,* f. piece of leather:

Gr. *lathargos,* m. piece of leather:

Gr. *masthles, -etos,* m. leather whip, leather; *masthletinos,* of leather: *Masthletinus
abbreviatus* (a bug).

Gr. *spatos,* n. hide, leather:

L. *scorteus,* made of hides, leather, leathern; see *scortum* under **skin**

Gr. *sittybon,* small skin, piece of leather; see **skin**

Gr. *skytos,* hide, leather; *skytinos,* leathern; see **skin**

See: **skin, strap**

leave < AS. *laefan,* permit, abandon; see **let, abandon, depart**

leaven < OF. *levain,* < L. *levamen,* a lightening or raising; see **yeast**

leberis, Gr. sloughed skin; see **skin**

lebes, -etis, L. (Gr. *-etos*), bronze kettle; *lebetion,* dim.; see **kettle**

lebias, Gr. a kind of fish; see **fish**

lecano- < Gr. *lekane,* dish, pot, pan; *lekanidion; lekanion, lekanis; lekaniske,*
dim.; see *lekos* under **plate**

lechos, Gr. couch, bed, marriage bed, nest; *lechaios,* of a couch; see **bed**

lechrios, Gr. slanting, crosswise; see **slope**

lecitho- < Gr. *lekithos,* yolk of an egg; see **yolk**

leco- < Gr. *lekos,* plate, dish, pan; *lekis; lekiskion; lekiskos,* dim.; see **plate**

lectica, L. bier, litter, sedan; *lecticula,* dim.; see **vehicle**

lectro- < Gr. *lektron,* marriage-bed; see *lechos* under **bed**

lectus, L. bed, couch; *lectulus,* dim.; *lectuarius,* of a bed, see **bed;** < *lego, lectus*
(Gr. *lektos*), choose, gather, see *lego* under **choose, gather**

lecythus, L. (Gr. *lekythos*), flask, bottle, vase, jug; *lekythion,* dim.; see **bottle**

Leda, L., Gr. mythical mother of Castor and Pollux who came from two eggs
fathered by Jupiter in the form of a swan; see **mollusk**

ledger < uncertain origin; see **list**

ledos, Gr. a cheap dress; *ledion,* dim.; see **garment**

ledum, L. (Gr. *ledon,* a cistacead yielding ladanum), now a genus of the heath
family; see **heath**

leech < AS. *laece.*

Gr. *baityx, -ygos,* f. leech:

Gr. *bdella,* f. leech: bdellometer, *Bdellostoma forsteri* (a cyclostome), *Carcinobdella
tigrina* (a leech), *Bdelloura candida* (a flatworm), *Macrobdella decora* (a leech).

L. *hirudo, -inis,* f. leech: hirudin, *Hirudo medicinalis* (leech), *Macracanthorhynchus
hirudinaceus* (a worm).

L. *sanguisuga,* f. bloodsucker, leech: sanguisuge, *Sanguisuga decora* (a leech).

See: **suck**

leek < AS. *leac;* see **onion**

laes < OF. *lie,* sediment, dregs; see **dirt**

left < AS. *left,* weak (originally referring to the fact that in most persons muscular
action is weaker on that side of the body); see **hand**

leg < ON. *leggr.*

L. *artus,* joint, limb; see *articulus* under **joint**

L. *crus, cruris,* n. leg, shank; *crusculum,* n. dim.; *cruralis,* of the leg: crural, crus,
cruralium, *Panicum crus-galli* (cockspur-grass), *Melipona ruficrus* (a bee),
Phyllium crurifolium (a phasmid), *Brachyscelus crusculum* (an amphipod).

Gr. *dibamos,* on two legs: *Dibamus novaeguineae* (a lizard).

L. *femur, -moris,* n. thigh: femur, femoral, *Corixa femorata* (a water-boatman),
Paraclius setifemoratus (a fly), *Phrixocoma femoralis* (a trichopterid).

L. *gralla,* stilt; *grallator,* one who goes on stilts; see **pillar**

Gr. *gyion,* n. limb:

Gr. *ignya,* f. part behind the thigh and knee, ham:

Gr. *kneme,* f. leg between knee and ankle, spoke of a wheel; *euknemos,* with
beautiful legs; *gastroknemia,* f. calf of the leg: cnemidium, gastrocnemius,

Knema glomerata (a myristicacead), *Hoplocneme forcipata* (a beetle), *Chaetocnema punctifrons* (a beetle), *Oedicnemus crepitans* (thick-knee), *Pterocnemia pennata* (a rhea), *Polycnemum verrucosum* (a chenopodiacead).

Gr. *kolon,* n. limb, arm, leg, member of anything, clause; *kole; kolen, -os,* f. thigh, leg, ham; *kolenarion,* n. dim.; *dikolos,* with two legs; *brachykolos,* with short legs; *macrokolos,* long-legged: colon, *Brachycolus stellariae* (an aphid).

Gr. *melos,* n. limb; *melydrion,* n. dim.; *micromeles,* small-limbed:

Gr. *meros,* m. thigh, femur, ham: merocele, epimeron.

Gr. *parakerkis,* f. fibula:

L., Gr. *perna,* ham, haunch, leg; see **flesh**

Gr. *plichas, -ados,* f. the inside of the thigh:

Gr. *rhethos,* n. limb:

AS. *scina,* the part of the leg between knee and ankle: shin, shinbone. Derive: shindig.

Gr. *skelos,* n. leg; *skeliskos,* m.; *skelydrion,* n. dim.; *microskeles,* small-legged: scelalgia, sceloncus, isosceles, *Sceloporus undulatus* (fence-lizard), *Ancylosceles heterodoxa* (a bee), *Ascelichthys rhodorus* (a fish), *Cacosceles crassicornis* (a beetle).

L. *sura,* f. calf of the leg: *Brenthus suratus* (a beetle).

L. *tibia,* f. shinbone, pipe, flute; *tibialis,* of the tibia: tibia, tibial, *Striglina rufitibia* (a moth), *Gongrocnemis tibialis* (a katydid).

Gr. *trochanter, -os,* m. runner; process at end of femur; now a segment of an insect leg: trochanter.   See *trecho* under **run**

See: **pillar, knee, foot, branch, hip**

legalis, L. of law, according to law, < *lex, legis,* law; *legitimus,* lawful; see **law**

legatus, L. appointed, sent; ambassador; see *lego* under **send**

legio, -onis, L. body of soldiers, large number; *legiuncula,* dim.; see **number**

legitimus, L. legal, proper; see *lex* under **law**

legirupa, L. lawbreaker; see **fault**

legnon, Gr. border, edge; *legnotos,* with a colored border; see **border**

lego, -atus, L. send, appoint, see **send**; *lego, lectus* (Gr. *lego*), gather, choose, select, read, say, speak, see **choose, gather, read, speak**

legos, Gr. lewd; see **lewd**

legula, L. earlap; see **flap**

leguleius, L. pettifogging lawyer; see **lawyer**

legume < L. *legumen,* pulse; see **bean**

lelepris, L. a kind of wrasse; see **wrasse**

lema, L. (Gr. *leme*), pus in the corner of the eye; see **pus**

lembus, L. (Gr. *lembos*), small, swift vessel, cutter, fishing-boat; *lembulus,* dim.; see **ship**

lemma, Gr. premise, assumption, subject, see **hypothesis;** scale, husk, peel, skin, sheath, see *lepis* under **scale**

lemna, Gr. an aquatic plant; a genus of the duckweed family; see **duckweed**

lemniscus, L. (Gr. *lemniskos*), ribbon, band; *lemniscatus,* beribboned; see **ribbon**

lemo- < Gr. *laimos,* throat; see **throat**

lemon < F. *limon,* < Ar. *laymun;* see **citrus**

lemphos, Gr. mucus from the nose; see **slime**

lemur, L. shade, ghost of the departed; see **spirit, monkey**

lend < AS. *laenan;* see **trade**

length < AS. *length;* see **long**

lenis, L. soft, smooth, gentle, mild; *lenitas; lenitudo,* softness; see **soft**

leno, L. panderer, pimp, procurer; *lena,* bawd, procuress; see **prostitute**

lenos, Gr. trough, vat, anything with a troughlike cavity, see **basin;** wool, see *lana* under **wool**

lens < L. *lens, lentis,* f. lentil, anything shaped like the seed of the lentil; *lenticula,* f. dim. freckle; *lenticularis,* lentillike; *lentigo, -inis,* f. lentil-shaped spot, freckle; *lentiginosus,* full of freckles, spotted: lens, lenticle, lenticular, lentiginose, Lentibulariaceae, lentigo, lentiginous, *Lens culinaris* (lentil), *Lenticulina radiata* (a foraminifer), *Cristellaria lenticula* (a foraminifer), *Sphenodiscus lenticularis* (an ammonite), *Euphemites lentiformis* (a fossil gastropod), *Bufo lentiginosus* (a toad).

L. *conspicillum; perspicillum,* n. lens, spectacles: *Pelecanus conspicillatus* (spectacled pelican), *Artibeus perspicillatus* (leaf-nosed fruit-bat).

Gr. *phakos,* m. lentil, lens of the eye: phacolith, phacoid, phacocele, phacoscope, *Phacopteryx granulata* (a caddisfly), *Phacops cristata* (a trilobite).

lenticula; lentigo, L. freckle, lens; *lentiginosus,* freckled, spotted; see **lens**

lentiscus, L. mastic-tree; see **sumac**

lentus, L. flexible, slow, viscous; see **bend**

-lentus, L. full of, prone to; see **abundance**

leo, *-onis,* L. (Gr. *leon, -ontos*), lion; *leunculus,* dim.; *leoninus,* of lions; see **cat**

leontice, L. (Gr. *leontike*), a genus of the barberry family; see **barberry**

leontopodium, L. (Gr. *leontopodion*), lionsfoot; see **composite**

leonurus, NL. a genus of the mint family; see **mint**

leopard < L. *leopardus* (Gr. *leopardos*); see **cat**

lepas, Gr. limpet, see **mollusk;** a bare rock, see **stone**

lepaste, Gr. a limpet-shaped drinking-cup; see **cup**

lepidium, L. (Gr. *lepidion*), pepperwort; see **mustard**

lepido- < Gr. *lepis, -idos,* scale; *lepidion,* dim.; *lepisma, -tos,* peel, rind, scale; see **scale**

lepidotos, Gr. probably a carp; see **carp**

lepidus, L. pleasant, agreeable, neat, fine, witty; *lepor,* attractiveness, charm; see **agreeable**

leprosy < L. *leprosus,* scaly, scabby, < Gr. *lepra; lepros,* scaly; see **disease**

lepsis; -lepsy, Gr. seizure; *leptes,* seizer, taker, accepter; *leptikos,* disposed to take; see *lambano* under **take, disease**

leptos, Gr. peeled, cleaned of the husks, fine, small, thin, delicate; *leptaleos,* fine, delicate; see **thin**

lepus, *-oris,* L. hare; *lepusculus,* dim.; *leporarius; leporinus,* of hares; see **hare**

lepyron, Gr. rind, skin, husk; *lepyrion,* dim.; see **scale**

leros, Gr. silly, foolish, nonsensical; *lerema,* silly talk; see **fool**

Lesbius, L. (Gr. *Lesbios*), of the island of Lesbos; *lesbiazo,* be sensual like the Lesbians; see **lewd**

lescho- < Gr. *lesche,* gossip, talk; *lesches,* talker; see **speak**

lesmon, *-os,* Gr. forgetful, unmindful; *lesmosyne,* forgetfulness; see *lethe* under **forget**

lespedeza, NL. a genus of the bean family; see **bean**

-less < AS. *leas,* free from, destitution, without; see **not**

lessen < AS. *laessa,* less.

L. *aboleo, -litus,* check the growth of, retard, efface, decay; *abolesco,* decay gradually, cease, vanish: abolish, abolition.

Gr. *ampotis,* f. ebb of the tide:

Gr. *anaspastos; syspastos,* drawn up, contracted: *Anaspasis fasciolata* (a beetle), *Syspastos brevicornis* (an isopod).

Gr. *aniemi,* relax, slacken:

Gr. *apanthesis,* f. a fading, withering:

Gr. *apauxesis,* f. decrease, decline:

Gr. *aposklesis,* f. a drying up, withering:

Gr. *apoxys,* tapering off, becoming less; see *oxys* under **point**

Gr. *askeles,* dried up, withered: *Asceles margaritatus* (a phasmid).

L. *astringens,* shrinking, binding, contracting; see *stringo* under **bind**

Gr. *athales,* withered:

Gr. *atimaster, -os,* m. dishonorer:

L. *atrophus* (Gr. *atrophos*), wasting away, consumptive: atrophy.

L. *cado, casus,* fall, sink, subside, decline, diminish; see **fall**

Gr. *chalao,* loosen, relax, slacken; see *chalaros* under **free**

L. *consumo, -sumptus,* eat, destroy, waste, spend; se **eat**

L. *contraho, -tractus,* draw together, compress, abridge, narrow, restrict, wrinkle; *detraho,* withdraw, pull down, disparage: contract, contraction, contractor, contractile, detract, detraction, detractor, *Systenopora contracta* (a bryozoan).

L. *curto, -atus,* shorten, diminish; see **short**

L. *decrementum,* n. a diminution, lessening, < *decresco, -cretus,* grow less, diminish, vanish: decrement, decrease.

L. *densativus,* binding, astringent; see *densus* under **thick**

L. *derogo, -atus,* detract from, disparage, belittle: derogatory, derogation.

Gr. *elassoo (elattoo),* lessen, diminish; see *elachys* under **little**

L. *emacio, -atus,* waste away, make lean: emaciation, *Fenestrellina emaciata* (a fossil bryozoan).

Gr. *exitelos,* lessening, fading, weakening: *Exitelus exsanguis* (a fossil bug), *Exiteloceras jenneyi* (a Cretaceous ammonite).

Gr. *hyponostesis,* f. a sinking, subsiding, abatement:

Gr. *kachektikos,* consumptive, < *kachexia,* f. consumption, a wasting away: cachectic, cachexia.

Gr. *karpho,* dry up, wither; see **dry**

L. *levo, -atus,* raise, lighten, lessen; see **raise**

Gr. *lophao,* abate, lighten; *lophesis,* f. abatement:

Gr. *maraino,* waste away, wither; *maransis,* f.; *marasmos,* m. a dying off, a going out, a withering; *marantikos,* wasting away: marasmus, amaranth, *Marasmius oreades* (fairy-ring mushroom).

L. *marcidus,* withered, wasted, weak; *marcidulus,* dim.; *marcens,* drooping, languid; *marcor, -is,* m. faintness, decay: *Blatta marcida* (a cockroach).

Gr. *meiosis,* diminution, lessening; *miotikos,* lessening; see *mikros* under **little**

L. *minuo, -utus,* lessen, reduce, limit, lower; *deminuo (diminuo),* lessen, make smaller, abate; *minoro, -atus,* make smaller, lessen; see *parvus* under **little**

Gr. *minytho,* lessen, shorten, decay; *minythikos,* diminishing; see *minys* under **little**

L. *mitigo, -atus,* pacify, appease, soften, assuage; see *mitis* under **mild**

L. *palliatus,* cloaked, covered, mitigated, moderated; see *pallium* under **garment**

Gr. *paresis,* f. a letting go, slackening, paralysis: paresis.

L. *permities, -ei,* f. a wasting away, decay, ruin; *permitialis,* fatal, destructive, ruinous:

Gr. *phthio; phthino,* waste away, wane, decline, perish; *phthisis,* f. consumption, decay; *phthoë,* f. decline; *phthinodes; phthisikos,* consumptive: phthinoplasm, phthisis, phthisical, *Phthinomylacris cordiformis* (a fossil insect), *Phthoa graminea* (a phasmid).

L. *remitto, -missus,* loosen, relax, slacken:

Gr. *rhiknos,* shriveled, shrunk, withered, wrinkled; see **fold**

AS. *scrincan,* contract, lessen: shrink, shrinkage.

Gr. *skeletos,* dried up, withered, mummified; see *skeleton* under **frame**

Gr. *stalsis,* f. contraction; *staltikos,* contractive: peristalsis, peristaltic.

Gr. *stryphnos,* astringent, harsh; see **bind**

Gr. *stypho,* contract; *stypsis,* f. contraction, astringency; *styptikos,* astringent; *stymma, -tos,* n. an astringent: styptic, *Panus stypticus* (a fungus).

L. *sub-,* less than, < *sub,* under; see **under**

L. *succiduus,* sinking, failing, faltering:

Gr. *synitikos,* disposed to condensation:

Gr. *syspastos,* drawn up, contracted:

Gr. *systole,* f. contraction, < *systello,* draw together, contract: systole, *Systoloblatta ohioensis* (a fossil insect-wing), *Systolederus carli* (a grouse-locust), *Systellogaster ovivora* (a wasp).

L. *tabeo,* waste away, atrophy, melt; *tabes, -is,* f. a wasting away; *tabidus; tabificus,* wasting away, decaying, languishing; *contabesco,* waste away gradually: tabefy, tabetic, tabefaction, tabific, contabescence, *Cymindis tabida* (a beetle), *Psilota tabidosa* (a fossil fly), *Clitocybe tabescens* (a fungus).

Gr. *texis,* f. wasting, melting, dissolution; see *teko* under **flow**

L. *vietus,* shriveled, shrunken, withered; see *vieo* under **weave**

See: **little, thin, weak, lose, dry, press, narrow**

lesson < L. *lectio,* selection, reading; see **learn, read**

lessus, L. cry, lamentation, wail; see **weep**

lesto- < Gr. *lestes,* robber, plunderer, pirate; *lestikos; lestrikos,* piratical, predaceous; see **steal**

let < AS. *laetan,* allow, permit.

Gr. *anaschetikos,* enduring, patient; *anaschetos,* endurable:

Gr. *anektikos,* enduring, patient; *anektos,* bearable, tolerable: *Anectocleis indagatrix* (a wasp).

L. *concedo, -cessus,* give up; *concessio, -onis,* f. permission, leave: concession, concede.

Gr. *exousia,* f. license, authority, power:

L. *indulgeo, -dultus,* be forbearing, concede, grant; see **give**

L. *licet, -citus,* be allowed, permitted, lawful: license, licentiate, licentious, licitly, illicit, scilicet, videlicet, leisure.

Gr. *menetos,* patient, longsuffering, < *meno,* stand fast, wait patiently; see **stand**

L. *patior, passus,* suffer, bear, allow; see *patior* under **pain**

L. *permitto, -missus,* let go, grant, allow; *permissio, -onis,* f. leave: permission, permit, permissive, permissible.

L. *sino, situs,* place, permit; see *situs* under **place**
L. *venia,* indulgence, permission, forgiveness; see **free**
See: **abandon, free, law**

-let < F. *-lette,* diminutives; see *-et* under **little**
-letes, Gr. suffix meaning hidden, < *lethomai,* be unseen, forget; see **forget**
Lethaea, L. wife of Olenus; see *Olenus* under **stone**
lethalis, L. deadly, < *lethum (letum),* death; see **death**
lethe, Gr. a forgetting, oblivion; *lathetikos,* likely to be forgotten or overlooked; *lethargos,* drowsy, forgetful; see **forget**
lethusa, L. a poppy; see **poppy**
letter < L. *litera (littera),* f. alphabetical character, epistle, message; *literula,* f. dim.; *literalis,* of letters; *literatura,* f. writing; *literatus,* educated, learned: literal, literature, literate, literary, alliteration, belles-lettres, illiterate, LL.D., obliteration, *Oliva litterata* (a fossil gastropod).
L. *abecedarius,* of the alphabet, alphabetical: abecedarian.
L. *Cadmus* (Gr. *Kadmos*), m. fabled introducer of the Greek alphabet: Cadmus, Cadmean.
L. *character, -is* (Gr. *charakter, -os*), mark, letter, figure; see **mark**
L. *diploma, -tis* (Gr. *-tos*), n. folded letter, certificate: diploma, diplomacy, diplomat.
L. *elementum,* first principle, letter of the alphabet; see **thing**
L. *epistola* (Gr. *epistole*), f. letter: epistle, epistology.
Gr. *gramma, -tos,* n. written character, letter; *grammation,* n. dim.; *grammatikos,* knowing one's letters: gram, epigram, monogram, diagram, grammar, grammatical, glamor, *Grammatonotus laysanus* (a fish), *Grammatoptila striata* (a laughing-thrush).
Gr. *hieroglyphikos,* pertaining to the picture-characters used in writing by Egyptian priests: hieroglyphic.
L. *nota,* mark, cipher, character, sign, letter; see **mark**
AS. *run,* secret letter, character: runc, runic, aroon.   Derive: furthorc, ogam (ogham), thorn.

GREEK LETTERS

1. *alpha,* A, α: alphabet, alphol, alpha and omega(the first and the last), abraxas, *Alphapuffinus assimilis* (a bird), *Eutaxocrinus alpha* (a Devonian crinoid).
2. *beta,* B, β: betacism, alphabetical, *Betapsestis takeuchii* (a moth).
3. *gamma,* Γ, γ; *gammation,* n. dim.: digamma, gammadion, gamut, *Gammanema ferox* (a flatworm), *Crassatella gamma* (a Cretaceous pelecypod).
4. *delta,* Δ, δ; *deltoides; deltotos,* shaped like a delta: delta, deltoid, deltidium, *Deltocephalus semifuscus* (a bug), *Deltocyathus agassizi* (a coral), *Populus deltoides* (eastern cottonwood), *Euastrum didelta* (a desmid).
5. *epsilon,* E, ε: epsilon Lyrae, *Epsilonella gracilis* (a nematode).
6. *zeta,* Z, ζ: zeta Ceti, cedilla, *Polia zeta* (a butterfly).
7. *eta,* H, η: etacism, eta Leonis, *Etazeta aeneicolor* (a beetle).
8. *theta,* Θ, θ: theta Centauri.
9. *iota,* I, ι: iota, jot, *Iotaphora iridicolor* (a moth), *Iota cobbi* (a nematode), *Volvula iota* (a Miocene gastropod), *Plusia iota* (a moth).
10. *kappa,* K, κ: kappa Orionis.
11. *lambda,* Λ, λ: lambdacism, lambdoid, zalambdodont, *Lambdotherium magnum* (a titanothere), *Argyrolambda conidens* (a fossil mammal).
12. *mu,* M, μ: mu Scorpionis.
13. *nu,* N, ν: nu Andromedae.
14. *xi,* Ξ, ξ: *Xinema perfectum* (a flatworm).
15. *omicron,* O, o: *Omicronema litorium* (a flatworm), *Ephyra omicronaria* (a moth).
16. *pi,* Π, π: π(3.1416), pi Virginis.
17. *rho,* P, ρ: rhotacism.
18. *sigma,* Σ, σ, ς: sigmoid, *Sigmodon hispidus* (cotton-rat), *Sigmatosiphon gürichi* (a pedaliacead), *Gyrosigma acuminatum* (a diatom), *Dolichopus sigmatifer* (a fly), *Drosophila sigmoides* (a fruitfly), *Euglyphella sigmoidalis* (an ostracode), Sigma Xi *(spoudon xynones).*
19. *tau,* T, τ: tau cross, *Bombyx tau* (a butterfly).

20. *upsilon* (*ypsilon; hypsilon*), Υ, υ; *hyoeides*, like upsilon: ypsiliform, ypsiloid, hyoid, hyothyroid, hyolithid, *Agrotis upsilon* (a cutworm), *Rhopalastrum ypsilinum* (a radiolarian), *Ypsilonurus mutilatus* (an opilionid), *Eurydema ypsilon* (a bug), *Hyolithes quadricostatus* (a fossil mollusk).

21. *phi*, Φ, φ: Phi Beta Kappa *(philosophia biou kybernetes)*.

22. *chi*, X, χ; *chiazo*, mark with a cross like chi; *chiasmos; chiastos*, arranged diagonally, crosswise: chiasma, chiastolite, *Chiasmodon niger* (a fish), *Chirox[Chirhox] plicatus* (a Paleocene mammal).

23. *psi*, Ψ, ψ: *Acronycta psi* (a butterfly).

24. *omega*, Ω, ω: omegoid, *Omegasyrphus coarctatus* (a fly).

Relict Letters

koppa, a letter standing between pi and rho in the Doric alphabet. Kappa took its place in Ionic Greek but Latin borrowed it as *Q* and as such it survives in the English alphabet.

san, a Doric name for sigma. With pi it became *sampi*, the numeral 900.

vau, or *wau*, a letter standing between eta and zeta in the Doric alphabet. It was later called digamma because it resembled one gamma superposed on another and slightly offset upward. The Romans took it as *F* and as such it survives in the English alphabet. Although the letter disappeared from Greek words its sound was represented by *V* in Latin derivatives: *oFon* (*oon*), L. *ovum*, egg; *Foinos* (*oinos*), L. *vinum*, wine; *Foikos* (*oikos*), L. *vicus*, village; *Feido* (*eido*), L. *video*, see; *Fion* (*ion*), L. *viola*, violet; *Fis* (*is*), L. *vis*, strength; *kleFis* (*kleis*), L. *clavis*, key; *skaiFos* (*skaios*), L. *scaevus*, left; *Festhes* (*esthes*), L. *vestis*, garment; *diFos* (*dios*), L. *divus*, god; *Frhegnymi* (*rhegnymi*), L. *frango*, break; *Fear* (*ear*), L. *ver*, spring; *oFis* (*ois*), L. *ovis*, sheep; *boFus* (*bous*), L. *bovis*, ox; *Fergon* (*ergon*), G. *werk;* E. *work;* *Fesperos* (*hesperos*), L. *hesperus* (*vesperus*), evening.

The gender of all letters is neuter.

lettuce < L. *lactuca*, f. < *lac, lactis*, milk; *lactucula*, f. dim., young lettuce: *Lactuca sativa* (garden-lettuce), *Ulva lactuca* (sea-lettuce).
L. *ambubeia*, f. wild succory:
Gr. *apargia*, f. a kind of chicory: *Apargia(Leontodon) autumnalis* (fall-hawkbit).
Gr. *astytis, -idos*, f. a kind of lettuce:
L. *caesapon*, n. a kind of lettuce:
L. *chondrilla* (Gr. *chondrille*), f. a kind of endive or chicory: *Chondrilla juncea* (gum-chicory).
L. *cichorium* (Gr. *kichorion*), n. chicory: *Cichorium endivia* (endive), succory.
Gr. *hedypnois*, f. a kind of chicory: *Hedypnois cretica* (a chicory).
L. *hypochoeris, -idis* (Gr. *hypochoiris, -idos*), f. a kind of succory: *Hypochoeris glabra* (smooth catsear).
L. *intybus*, m. endive, succory: *Cichorium intybus* (chicory), endive.
Gr. *maroulion*, n. lettuce:
L. *picris, -idis* (Gr. *pikris, -idos*), f. bitter lettuce: *Picris hieracioides* (hawkweed-oxtongue).
Gr. *seirikon*, n. chicory:
L. *seris, -idis* (Gr. *-idos*), f. a kind of endive; *hyoseris*, f. pig-endive: *Agoseris arizonica* (a composite), *Hyoserites lingua* (a fossil composite), *Arnoseris minima* (lamb-succory).
Gr. *thridax, -akos; thridakine*, f. lettuce; *thridakinos*, of lettuce: thridacium.

letum, L. death; *letifer*, causing death; see *lethalis* under **death**

leucanthemum, L. (Gr. *leukanthemon*), a composite; see **composite**

leuciscus, L. (Gr. *leukiskos*), chub; see **carp**

leuco- < Gr. *leukos*, white; see **white**

Leucothoë, L. a princess said to have been changed into a sweet-scented shrub; a genus of the heath family; see **heath**

leucrium, L. a kind of houndstongue; see **borage**

leugaleos, Gr. sad, sorry, mournful; see **sad**

leuros, Gr. smooth, even; see **smooth**

leuster, Gr. stoner; *leustos*, stoned; see **throw**

levamentum, L. alleviation, comfort, consolation; see **soothe**

level < OF. *livel*, < L. *libella*, f. instrument for determining horizontality.
L. *aequor, -is*, n. an even, level surface; *aquoreus*, of any smooth, even surface, particularly the calm sea: *Acestra aequorea* (a fish).

L. *amussis, -is,* f. rule, level:
L. *campus,* field, plain; see **field**
Gr. *chorobates,* m. instrument for determining levels:
L. *degrumor, -atus,* level off:
Gr. *dioptra,* f. a sighting instrument for levelling: diopter.
Gr. *homalos,* level, even: *Homalonotus major* (a trilobite), *Homalium dolicho-phyllum* (a flacourtiacead), *Sauromalus ater* (chuckwalla).
F. *horizontal,* level; see *horizon* under **circle**
Gr. *hyptios,* supine; *hyptiotes,* flatness; see **lie**
Gr. *lagarodes,* sunken, flat, slack; see **free**
L. *libratus,* level, horizontal; see *libra* under **balance**
Gr. *listrotos,* leveled, smoothed; see *listron* under **shovel**
L. *oblatus,* flattened at the poles, depressed: oblate.
Sp. *pampa,* f. a plain, < Ab.Am.: pampa, pampean, pampas-grass *(Cortaderia argentea).*
Gr. *pedinos,* flat, level, < *pedion,* n. flat, open country, plain, < *pedon,* n. ground, earth; *pediakos; pediasios,* of a plain; *apedos,* even, flat, level; *arpedes,* flat, level; *dapedon; isopedon,* n. any level surface, floor, ground; *epipedos; homopedos,* on the ground, level, flat; *oropedion,* n. plateau, tableland: *Pedinopelta maculipennis* (a wasp), *Pedinopsis major* (a beetle), *Pedionomus micrurus* (a bird), *Dapedoglossus acutus* (an Eocene fish), *Dapedius politus* (a Triassic fish), *Epipedodema depressum* (a beetle), *Epipedosoma zanguebaricum* (a beetle), *Holopedium irregulare* (an alga).
L. *planus,* even, flat, level, smooth, clear; *planarius,* level; *applanatus; com-planatus,* flattened: plane, plain, explanation, planimeter, esplanade, plan, piano, pianissimo, *Sphaerium planum* (a fossil mollusk), *Planorbis albus* (a snail), *Vanilla planifolia* (Mexican vanilla), *Helicotoma planulata* (a fossil gastropod), *Lycopodium complanatum* (a clubmoss), *Agave applanata* (an agave), *Celeuthetes deplanatus* (a beetle).
Gr. *platys,* broad, wide, flat, level; see **broad**
Gr. *plax, -akos,* anything flat and wide; see **plate**
Sp. *sabana,* f. < Ab.Am. *zabana,* a treeless, grassy plain: savanna, *Ammodromus savannarum* (grasshopper-sparrow). Derive: Savannah(Ga.).
Gr. *sanidodes,* like a plank, flat; see **board**
See: **broad, board, lie, smooth**

lever < OF. *leveor,* < L. *levo,* raise.
Gr. *mochlos,* m. lever, crowbar; *hypomochlion,* n. fulcrum of a lever; *Mochlodon suessi* (a dinosaur).
L. *vectis, -is,* m. lever, crowbar, pole:

levis, L. light, mild, fickle; *leviculus,* dim.; *levitas,* lightness, see **light;** < *levis (laevis),* smooth, polished, bald; *levigatus,* smooth, slippery, see **smooth**

levo, *-atus,* lift up, lighten, see **raise;** < *laevus,* left, see **hand**

lewd < AS. *laewede,* of the laity, ignorant, lascivious, lustful, sensual, wanton.
Gr. *akolastos,* sensual, licentious:
Gr. *andromanes,* lusting after men; *hippomanes* (said of a mare in heat), lusting after horses; see *mania* under **mad**
Gr. *aselges,* dissolute, licentious, lascivious, lewd; *Aselgoides insularis* (a bug).
Gr. *asyres,* lewd, filthy; *anasyro,* pull up the the clothes, expose the body:
L. *catulio,* desire the male; *catulitio, -onis,* f. desire for the male:
L. *cinaedus* (Gr. *kinaidos*); *cinaedicus,* immodest, unchaste, lewd, shameless: *Labrus cinaedus* (a wrasse).
L. *concubinalis,* lascivious; see *concubina* under **prostitute**
L. *dissolutus,* lax, loose in morals, dissipated; see **bad**
Gr. *dysagnos,* unchaste:
L. *immodestus,* unrestrained, indecent: immodest.
L. *impudicus,* lewd, unchaste, unashamed; *impudens -entis,* shameless: impudent, *Ithyphallus impudicus* (a stinkhorn-fungus).
L. *incestus,* defiled, lewd, unchaste; *incestuosus,* lewd: incest, incestuous.
L. *indecens, -entis,* unseemly, improper, unchaste: indecent, indecency.
L. *inverecundus,* shameless, immodest:
L. *irrumator, -is,* m. one who practices beastly obscenity:
Gr. *kaprao* (said of sows), want the boar; *kaprios,* like a boar, lustful; *laiskapros,* very lustful:
Gr. *katapygon, -os,* given to unnatural lust, lecherous, lewd:
Gr. *lagnos,* lustful; see *lagneia* under **coitus**
Gr. *laiskarpos,* very lustful:
L. *lascivus,* sportive, frisky, lustful, wanton; *lascivulus,* dim.: lascivious.
Gr. *lastauros,* m. lewd person: *Lastaurus anthracinus* (a fly).
Gr. *legos,* lewd:

L. *Lesbius* (Gr. *Lesbios*), of the island of Lesbos; *lesbiazo*, be sensual like the Lesbians; *Lesbis, -idos*, f. Lesbian woman: lesbian, lesbianism, *Lesbia forficatus* (a beetle), *Cyanolesbia gorgo* (a hummingbird), *Melanesthes(Lesbidana) coriaria* (a beetle).

L. *libido, -inis*, f. desire, longing, passion, lust; *libidinosus*, lustful, licentious: libido, libidinous.

L. *Liburnus*, god of lustful enjoyment; see **joy**

L. *lupanus*, lewd; see *lupa* under **prostitute**

Gr. *machlos*, lustful, lewd; *machlosyne*, f. lewdness, lust: *Machlolophus xanthogenys* (a tit).

Gr. *misetia*, f. lust, lewdness:

L. *obscenus*, offensive, indecent; see **bad**

L. *oestrus* (Gr. *oistros*), gadfly, sexual heat, rut, desire; see **fly**

L. *orthophallicus*, obscene:

Gr. *pornikos*, of harlots; *pornographos*, writing about harlots; see *porne* under **prostitute**

L. *salax, -acis*, lustful, lecherous: salacity, salacious.

Gr. *skyza*, f. lust:

Gr. *spatalos*, wanton, lascivious; *spatalites*, m. profligate: *Spatalistes cyanoxantha* (a moth).

L. *subo, -atus*, be in heat, have sexual ardor; *subatio, -onis*, of oestrus:

L. *surio*, be in heat:

L. *tentigo, -inis*, f. tension, lust: tentigo, tentiginous.

L. *urigo, -inis*, f. lustful heat, desire:

L. *virosus*, fond of men, longing after men:

See: **bad, dirt, heat, coitus**, *mania* under **mad, prostitute, shame, wish**

lewisia, NL. a genus of the purslane family; see **purslane**

lex, *legis*, L. law; *legislator*, lawmaker; see **law**

lexico- < Gr. *lexikos*, pertaining to words, < *lexis*, speech, word; *lexidion*, dim.; see **word**

lexis, Gr. allotment, < *lanchano*, obtain by lot; see *lachesis* under **lot**

liatris, NL. a genus of the composite family; see **composite**

libanotos, Gr. frankincense, gum of the tree *libanos*; see **resin**

libas, -ados, Gr. anything that drops, trickles, or runs; *libadion*, dim.; see **stream**

libatio, L. a drink-offering, < *libo, -atus* (Gr. *leibo*), pour out, taste, take a little of a thing; see **drink**

libella, L. small silver coin, the tenth part of a denarius, see **money**; an instrument for determining levels, see **level**; *libellus*, dim. of *liber*, book, see **book**

liber, L. free; *libertas*, freedom, see **free**; *liber, -eri*, child, see **child**; *liber, -bri*, bark, book; *libellus*, dim., see **bark, book**

libethron, Gr. meadow; see **field**

libido, L. violent desire, longing, passion, lust; *libidinosus*, lustful; see **lewd**

libitinarius, L. undertaker, < *Libitina*, goddess of corpses; see **undertaker**

libitus, L. pleasing, agreeable; see **agreeable**

libo- < Gr. *libos*, tear, drop, see **drop**; < *lips, libos*, the southwest wind; *libonotos*, south southwest wind, see **wind**; < *leibo*, pour forth, see *libas* under **stream**

libra, L. balance; the Roman pound; see **balance**

librarius, L. pertaining to books; see *liber* under **book**

libum, L. cake, pancake; *libacunculus*, dim.; see **cake**

Liburnus, L. god of lustful enjoyment; see **joy**

Libya, L. (Gr. *Libye*), north Africa; *Libys, -yos*, a Libyan; *Libykos*, Libyan, hence foreign; see **Africa**

lichado- < Gr. *lichas, -ados*, space between thumb and forefinger; see **measure**

lichanos, Gr. forefinger; see **finger**

lichen < Gr. *leichen, -os*, m. a thallophyte composed of an alga and a fungus: lichenic, lichenography, lichenoid, *Gracilaria lichenoides* (a red alga, source of agar-agar).

ON. *litmose*, a kind of dye-yielding lichen: litmus.

NL. *usnea*, f. < Ar. *oshnah*, moss; now a genus of lichens: usnic, *Usnea barbata* beard-lichen), *Dendropogon usneoides* (Spanish 'moss').

lichnos, Gr. greedy, dainty; see **eat**

licinium, L. lint; see **thread**

licinus, L. bent or turned upward; see **bend**

licitus, L. permissible, lawful, < *licet,* be allowed, see **let;** < *liceo,* be on sale, be valued at; *liceor,* bid for, see **trade**

licium, L. thread of the web; see **thread**

lick < AS. *liccian.*
L. *catillo, -atus,* lick a plate:
L. *lambo, -itus,* lick: lambent, lamprey.
Gr. *lapto,* lap with the tongue; *lapsis,* f. a lapping:
Gr. *leicho,* lick up: eclegma, electuary, *Dermalichus mytilaspidis* (a mite).
Gr. *lichmao,* lick; *lichmas, -ados,* licked:
L. *lingo, linctus,* lick, lick up; *delingo,* lick off: linctus, lincture.
See: **suck, tongue, eat**

licmado- < Gr. *likmas, -ados,* a winnowing fan; see **fan**

licmetico- < Gr. *likmetikos,* pertaining to winnowing; *likmesis,* a winnowing; see **separate**

licmo- < Gr. *likmos,* a winnowing fan, basket, or shovel; see **fan**

licno- < Gr. *liknon,* basket or cradle of wickerwork used in winnowing; see **basket**

licorice < OF. *licoresse,* < Gr. *glykyrrhiza,* plant with a sweet root; see **bean**

licro- < Gr. *likros,* antler; see **horn**

lictor, L. attendant to a Roman magistrate; see **servant**

lictus, L. abandoned, left, forsaken; see *linquo* under **abandon**

lid < AS. *hlid;* see **cap**

lie < AS. *lyge,* falsehood, untruth, < *leogan,* prevaricate.   Derive: bogus, fake, quack.
Gr. *adokimos,* spurious, false, base: *Adocimus bellus* (a beetle).
L. *adultero, -atus,* commit adultery, falsify, corrupt; *adulterinus,* false, counterfeit, impure: adultery, adulteration, adulterate.
L. *adumbratus,* counterfeited, feigned, false, represented in outline, shaded imperfectly; see *umbra* under **shade**
Gr. *alazon, -os,* c. vagabond, imposter, quack, braggart: *Alazonia symbiblis* (a fish).
Gr. *analethes,* false, untrue:
Gr. *aniatros,* m. quack:
Gr. *apate,* f. deceit, fraud; *apatelos,* deceitful, deceptive, illusory, wily; *apatema, -tos,* n. stratagem; *apatetikos,* deceptive, fraudulent: apatite, apatetic, apatelite, *Apatosaurus amplus* (a dinosaur), *Apatemys bellus* (a fossil insectivore), *Apatela populi* (a moth).
Gr. *apistos,* faithless, untrustworthy; see **doubt**
Gr. *apoplanema, -tos,* n. deception; *apoplanos,* m. fallacy:
L. *bifax, -acis,* two-faced; see *facies* under **face**
L. *bisulcilingua,* double-tongued, hypocritical, deceitful:
L. *calumnior, -iatus,* misrepresent, depreciate, accuse falsely, practice trickery; *calumnia,* f. malicious, false accusation: calumny, calumniate, challenge.
L. *captio, -onis,* f. cheat, deception, quibble, fallacy, sophism; *captiosus,* deceptive; *captensula,* f. a fallacious argument: captious.
Gr. *chaunax, -akos,* m. liar, cheat: *Chaunax umbrinus* (a fish).
L. *cibdelus* (Gr. *kibdelos*), spurious, base, adulterated, counterfeit, false: *Cibdela poecilotricha* (a sawfly).
L. *circulator; circulatrix,* mountebank, quack, stroller; see **circle**
L. *decipio, -ceptus,* beguile, cheat; *deceptor, -is,* m.; *deceptrix, -icis,* f. deceiver: deception, deceive, deceit, deceptive, *Endomyces decipiens* (a fungus).
L. *deludo, -lusus,* play false, deceive: delude, *Echinomyia deludens* (a fly).
Gr. *diabole,* f. slander, false accusation; *diabletor, -os; diabolos,* m. slanderer; *diabolikos,* slanderous, devilish: diabolical, devil.
L. *dissimulo, -atus,* feign, disguise, hide: dissimulation, dissemble.
L. *dolus* (Gr. *dolos*), m. artifice, guile, deceit, bait, trap; *dolosus,* crafty, cunning, deceitful; *doleros; dolios; dolomedes,* crafty, deceitful, treacherous, wily; *doloma, -tos,* n. trick, deceit; *dolophrosyne,* f. craft, subtlety; *dolops, -opos,* m. ambusher; *dolosis,* f. trickery, fraud; *subdolus,* somewhat crafty, sly: dolerite, dolerophanite, *Dolophilus copiosus* (a trichopterid), *Dolops hastifer* (a wasp), *Dolomedes fimbriatus* (a diving-spider), *Archaeodolops clavulus* (a fossil mammal), *Neda subdola* (a beetle), *Dolophrosynella mirax* (a moth), *Dolerus nasutus* (a wasp).
Gr. *epamphoteristes,* m. a double-dealer; *epamphoteros,* ambiguous:
Gr. *eperopeus; eperopeutes,* m.; *eperopeis, -idos,* f. cheat, deceiver:
L. *fallo, falsus,* deceive; *fallax, -acis,* deceitful, false; *fallacia,* f. deceit, untruth; *fallaciosus,* deceitful; *falsarius,* m. forger; *falsidicus,* lying; *falsificus,* acting falsely; *infalsatus,* falsified: false, falsehood, falsify, fallacy, fallacious, infallible,

default, faultless, failure, faucet, falsetto, unfailing, *Bombylius fallax* (a bee-fly), *Paralina fallaciosa* (a beetle), *Falsifusus meyeri* (a fossil gastropod), *Falsicolus obrutus* (a fossil gastropod).

L. *fictus; fictitius,* artificial, counterfeit, not genuine: fiction, fictitious.

L. *fraus, fraudis,* f. deceit, cheating; *fraudabilis; fraudulentus; fraudulosus,* cheating, deceitful; *fraudator, -is,* m.; *fraudatrix, -icis,* f. cheat, deceiver; *fraudifer; fraudiger,* deceitful; *defraudo, -atus,* cheat: fraud, fraudulent, defraud, *Xanthophenax fraudator* (a wasp), *Rhyacophila fraudulenta* (a trichopterid).

L. *frustror, -atus,* deceive, thwart, disappoint, trick; see **hinder**

L. *fucosus,* painted, simulated, counterfeit; see *fuco* under **color**

L. *hyprocrita* (Gr. *hypocrites*), m. actor, pretender, dissembler: hypocrite, hypocritical, *Dendrotrogus hypocrita* (a beetle).   Derive: Pecksniffian.

Gr. *hypoxylos,* with wood underneath metal, that is, spurious, counterfeit:

L. *illicitator,* sham-bidder at an auction to raise others' bids; see *liceo* under **trade**

L. *illusio, -onis,* f. a mocking, deception: illusion, illusory.

L. *impostor, -is,* m. deceiver, cheat: impostor.

L. *infidelis; infidus; perfidus,* unreliable, faithless, false: infidel, infidelity, perfidy, perfidious.

L. *insidiosus,* cunningly deceitful; see **trap**

L. *insincerus,* hypocritical, dishonest: insincere, insincerity.

L. *ironia* (Gr. *eironeia*), f. dissimulation, the saying of one thing but meaning the opposite; *eiron, -os,* m. dissembler, pretender: *Ironomyia nigromaculata* (a fly), *Ironicus nothrus* (a fossil beetle), *Iron confusus* (a mayfly).

Gr. *katasophistes,* trickster; see *sophos* under **know**

Gr. *keratas,* m. cuckold:

Gr. *kopis, -eos,* m. prater, liar:

L. *malefidus,* faithless, unfaithful: *Otiorhynchus malefidus* (a beetle).

Gr. *manganeutes,* m. juggler, trickster, quack; *manganon,* n. means of charming or bewitching:

L. *mendax, -acis,* lying; *mendacium,* n. lie; *mentior, -titus,* lie, cheat, deceive, counterfeit; *ementitus,* forged, fabricated: mendacity, mendacious, *Bombylius mendax* (a bee-fly).

Gr. *metroxenos,* m. bastard:

OF. *mocquer,* imitate, counterfeit: mock, mock-orange, mock-turtle.

Gr. *nothos,* spurious, bastard, false, mongrel; m. illegitimate child: *Nothofagus antarctica* (a false-beech), *Nothoceras bohemicum* (a fossil cephalopod), *Nothosaurus mirabilis* (a fossil reptile).

Gr. *paleutes; paleutria; paleutris, -idos,* decoy-bird, < *paleuo,* decoy; see **draw**

Gr. *paracharaktes,* m. forger, falsifier, counterfeiter:

Gr. *paragogos,* misleading, deceitful: *Paragogus definitus* (a beetle).

Gr. *parakousma, -tos,* n. something misheard, false story:

Gr. *parasemos,* marked falsely, counterfeit: *Parasemus grouvellei* (a beetle).

L. *pellax, -acis,* seductive, deceitful; see *lacio* under **draw**

Gr. *penike,* false hair, wig; see **hair**

L. *perjurus,* lying under oath: perjury, *Acalles perjurus* (a beetle).

L. *perperus* (Gr. *perperos*), wrong, false:

Gr. *phelos,* deceitful; *pheletes,* m. cheat, thief: *Phelocalocera peregrina* (a beetle), *Phelochromus sexpunctatus* (a beetle), *Pheletes bructeri* (a beetle).

Gr. *phenax, -akos,* m. cheat, impostor: phenacite, *Phenacodus primaevus* (a fossil mammal), *Plectrophenax nivalis* (snow-bunting).

L. *praestigium,* n. illusion, trick, deception; *praestigiator, -is,* m. juggler, cheat, impostor, deceiver: prestigiator, prestige, prestigious.

L. *pretendo, -tus (-sus),* allege, simulate: pretend, pretender, pretense.

L. *pretextum,* n. pretense, excuse: pretext.

L. *prevaricor, -atus,* walk crookedly, deviate, transgress, lie: prevaricate, prevarication.

Gr. *psainythios,* false, vain: *Psaenythia picta* (a bee).

Gr. *pseudos,* n. fallacy, lie; *pseudes; psydros,* false, lying; *pseustes,* m. liar, cheat: pseudonym, pseudomorph, pseudaconitine, pseudograph, pseudopodium, pseudoscience, pseudology, *Pseudotsuga japonica* (a Douglas-fir), *Pseudaspis cana* (mole-snake), *Pseudomys auritus* (a rodent), *Pseudotriton montanus* (a salamander), *Pseudolmedia spuria* (milkwood), *Thiodina pseustes* (a spider), *Psydrus piceus* (a beetle).

Gr. *sophisma,* false conclusion, fallacy; see *sophos* under **know**

L. *spurius,* false, bastard: spurious, spurial, *Icterus spurius* (orchard-oriole), *Iris spuria* (seashore-iris), *Delphinium spurium* (a larkspur).

L. *subditicius; subditivus,* spurious, counterfeit; *subdo, -itus,* substitute, counterfeit:

L. *suppono, -positus,* substitute, forge; *suppositicius(supposititius),* substituted, false, counterfeit: suppose, supposition, supposititious.

lie 478

L. *surreptivus,* false, fraudulent:
Gr. *terthreus,* m. quack:
L. *tricor, -atus,* play tricks, make mischief; *tricosus,* full of tricks or wiles: trick, trickery, trickster.
L. *vanidicus,* m. liar:
See: **cover, secret, draw, two**

lie < AS. *licgan,* be prostrate, stretched out, situated.
L. *cubo (cumbo), cubitus,* lie down, recline, press upon; *cubans,* lying; *cubitorius,* of a lying posture; *Cuba,* f. goddess who protects sleeping children; *accumbens,* reclining; *decumbens,* lying down; *incumbens,* lying upon, filling an office; *procumbens,* lying forward, falling forward; *incubo, -itus,* lie upon, brood, hatch; *excubo,* lie out on guard, sleep out of doors; *concubo,* lie together: cubicle, covey, accubation, concubine, incubus, incumbent, incubator, procumbent, succubus, succumb, *Gaultheria procumbens* (wintergreen), *Petalostemon decumbens* (a prairie-clover).
Gr. *ektadios,* outstretched: *Ectadiophatnus tachardiae* (a wasp).
Gr. *epachnidios,* lie upon like dust:
L. *excussus,* stretched out, extended, stiff:
Gr. *hyptios,* supine; *hyptiotes, -etos,* f. flatness, supineness: *Hyptiogaster infumata* (a wasp), *Hyptiotes cavata* (triangle-spider), *Hyptis emoryi* (bushmint).
L. *jaceo, -citus,* lie, be situated: adjacent, circumjacent, interjacent, gist, joist, ease.
Gr. *klisis,* a bending, reclining, lying down; see **slope**
L. *nixor, -atus,* lean upon, rest upon:
Gr. *prenes,* drooping, face downward, prone; see **bend**
L. *prociduus,* prostrate; see **low**
L. *pronus,* lying on the face, face downward: prone, pronate, pronator, *Stenopilus pronus* (a trilobite).
L. *supinus,* lying on the back, back downward, outstretched; *resupinus,* bent back with the face upward, on the back: supine, supinator, resupinate, *Cytisus supinus* (a broom), *Linyphia resupina* (a spider), *Dicliptera resupinata* (an acanthacead).
See: **low, press, bed, creep, earth**

lienicus, L. of the spleen, < *lien, -is,* spleen; see **spleen**

life < AS. *lif:* lifelike, lifetime, lifelong, lifeless, lifeboat, lively, livelihood, living, live-oak, Livingston, alive.
L. *alacer, -cris, -cre,* lively; *alacritas, -atis,* f. eagerness, ardor, brashness: alacrity.
L. *anima,* f. air, breath of life, soul; *animus,* m. spirit, courage, mind; *animula,* f.; *animulus,* m. dim.; *animo, -atus,* quicken, give life to; *animalis; animatus,* lively; *animosus,* full of life, courageous, spirited: animation, animosity, animus, animal, inanimate, unanimity, longanimity, magnanimous, pusillanimous, animadversion.
Gr. *bios,* m. life; *biosis,* f. manner of life; *biotikos,* of life, lively; *bioteia,* f. livelihood; *biote,* f.; *biotos,* m. life, a living, means of living; *biothion,* n. dim. scant life; *biodoros,* life-giving; *biodotes,* m. life-giver; *biothalmios,* lively; *biothremmon; biotesios,* life-supporting; *amphibios,* living a double life; *anabiosis,* f. revival; *brachybios,* short-lived; *derobios; embios; makrobios,* long-lived; *symbiosis,* f. a living with: biology, biography, amphibian, biotin, symbiosis, biota, cenobite, antibiotic, biochemistry, biometric, *Biotus formicarius* (a beetle), *Dendrobium superbum* (an orchid), *Polybia flavitarsus* (a wasp), *Ochthebius aztecus* (a beetle), *Myrmecobius rufus* (a banded anteater), *Dermatobia cyaniventris* (a botfly), *Petrobius maritimus* (a thysanurid), *Polygonum amphibium* (water-polygonum), *Embiomyia australis* (a fly), *Streptomyces antibioticus* (a fungus), *Ophiactis symbiota* (a brittlestar).
L. *colo, cultus,* abide, dwell, inhabit, till; NL. *-colus, -a, -um,* dwelling in, living among, < *-cola,* dweller, inhabitant, tiller, as in *agricola,* m. farmer; *incola,* c. denizen, dweller, native; *limicola,* m. mud-dweller; *monticola,* c. dweller in the mountains; *paludicola,* c. marsh-dweller; *colonus,* m. inhabitant of a colony; *inquilinus,* inhabiting another's house temporarily, tenant, lodger: colony, colonist, colonial, Lincoln, Cologne, accolent, inquiline, Agricola, saxicolous, nidicolous, pratincole, arenicolous, aquicolous, deserticolous, *Petricola abbreviata* (a pelecypod), *Harpalus agricolus* (a beetle), *Parus monticolus* (green-backed tit), *Gecarcinus ruricola* (land-crab), *Cardium arenicolum* (a pelecypod), *Ceroxylon andicolum* (wax-palm).
L. *constitio, -onis,* f. a stay, an abiding:
L. *conversor, -atus,* abide, live, keep company with: conversant, conversation.
AS. *cwic,* alive: quick, quicksilver, quicksand, quickening, whitlow, "the quick and the dead", cushat.
Gr. *diaita,* f. life, mode of life, regimen: diet, dietetics.

Gr. *endemos*, living in, native; see **native**

Gr. *enthousiasmos*, m. inspiration, ardent zeal: enthusiasm, enthusiastic.

Gr. *entonos*, eager, vehement, intense:

L. *esse*, to be, < *sum*, I am, I be; *ens, entis*, n. being, thing, that which has existence; *essentia*, f. the substance of things; *futurus*, about to be; *adsum*, be present; *consum*, coexist: essence, essential, quintessence, absent, present, represent, interest, entity, nonentity, future, futurity, possible, power, *Entia non sunt multiplicanda praeter necessitatem* (Occam's razor).
Describing the death of Colonel Newcome, Thackeray said, "Just as the last bell struck, a peculiar sweet smile shone over his face, and, lifting up his head a little, he said quickly, 'Adsum!' and fell back."

Gr. *essymenos*, eager; see **swift**

L. *excitatus*, animated, lively, vigorous: excited, excitement, excitation.

L. *existo, exstitus*, appear, be manifest, be: exist, existence, existential.

L. *-gena*, born in, living in; see *gigno* under **born**

L. *habito, -atus*, dwell, abide, live; *habitabilis*, that may be lived in; *cohabito*, live together; *inhabito*, dwell in: habitation, habitat, cohabit, inhabitant.

Gr. *homaulos*, living together:

Gr. *homestios*, sharing the same fireside, dwelling with:

Gr. *metoikos*, m. emigrant, immigrant: *Metoecus paradoxus* (an amphipod).

Gr. *naio*, abide, dwell, inhabit; *naetes; nastes*, m. inhabitant: *Cryptonastes tersus* (a wasp).

Gr. *nomos*, m. place or condition for living; *gaionomos*, inhabiting: *Pedionomus torquatus* (a bird).

Gr. *oikeo*, inhabit, dwell; *oikizo*, build a house, settle, colonize; *oiketes*, dweller, servant; see *oikos* under **house**

Gr. *on, ontos*, n. being, thing, that which has existence, < *eimi*, I am, I be; *ousia*, f. essence, substance, property, reality: ontogeny, ontology, ontography, paleontology, homoousia, Homoousian.

Gr. *prothymos*, eager, zealous, active: *Prothymus novus* (a chalcid).

Gr. *psyche*, mind, spirit, life, soul; see **mind**

L. *sessor*, inhabitant; see *sedeo* under **sit**

Gr. *sphedanos*, eager, vehement; see **busy**

Gr. *sphriganos*, bursting with health and vigor, full, fresh, vigorous:

L. *vegetus*, lively, animated, brisk, < *vegeto, -atus*, quicken, arouse; *vegetabilis*, enlivening: vegetate, vegetation, vegetarian, vegetable, vegetivorous, *Crataegus vegeta* (a hawthorn).

L. *vehemens*, ardent, eager, vigorous; see **busy**

L. *vigeo*, thrive, be lively; *vigor, -is*, m. force, energy, liveliness: vigorous, invigorate.

L. *vita*, f. life; F. *vie*, life; *vitalis*, of life; *vivax, -acis*, lively, vigorous, brisk; *vivum*, n. that which is alive; *vividus*, full of life, animated; *vivus*, alive, lively; *viviparus*, bearing active, living young; *longivivax, -acis*, long-lived; *redivivus*, living again, renewed; *convictus*, living together, < *vivo, victus*, live: vivacity, vivid, viviparous, vivisection, viva voce, viand, victuals, Vivian, revive, survive, convivial, vitamine, vitalism, vitalism, viable, eau de vie, arbor-vitae, cevitamic acid, *Viviparus capillatus* (a snail), *Coryphanta vivipara* (a cactus), *Plasmodium vivax* (protozoan causing tertian malaria), *Sempervivum hirtum* (Italian house-leek), *Lunaria rediviva* (a satin-flower), *Lacerta vivipara* (common-lizard).

Gr. *zoë*, f. life; *zoös*, alive, living; *zosimos*, capable of life; *zotikos*, lively; *zoarkes*, life-supporting; *aeizoön; aeizoös*, ever-living; L. *aizoön*, n. an ever-green plant; *erizoös*, long-lived; *syzoös*, living together: Zoe, zoea, azote, Aizoaceae, *Zoarces viviparus* (an eelpout), *Azotobacter chroococcus* (a bacterium), *Androsace aizoon* (a rock-jasmine), *Aizoon canariense* (Canary carpet-weed), *Draba aizoides* (a whitlowwort).   See *zoon* under **animal**

See: **arouse, awake, breathe, busy, make, move, swift**

lift < ON. *lypta*, akin to Sw. *lyfta;* see **raise**

ligamentum; ligatura, L. band, tie; *ligatus*, bound; see *ligo* under **bind**

Liger, L. the Loire River; see **stream**

light < AS. *leoht*, radiant energy: lighting, lightning, lighthouse, lightship, light-year, sunlight, moonlight, starlight, twilight.

Gr. *aglaos*, splendid, bright, beautiful, noble; *aglaia*, f. splendor, beauty; *aigle*, f. radiance, brightness: *Aglaonema nitidum* (an aracead), *Aglaia odorata* (a meliacead), *Aegle marmelos* (bael or bel fruit), *Afraegle paniculata* (an African plant).

Gr. *amarysso*, sparkle; *amarygma, -tos*, n. sparkle, twinkle: *Amarygmus speciosus* (a beetle).

Gr. *anapsis*, f. a lighting up, kindling, appearance of the stars:

L. *ardeo, arsus*, burn, blaze, glow, sparkle, shine, glitter; see **burn**

Gr. *argos*, white, bright; see **white**

L. *argutus*, clear, bright, lively, shiny; see **clear**

Gr. *astrape*, f. flash of lightning: *Astrapephora romanovi* (a moth).

Gr. *auge*, f. luster, shine; *augasma, -tos*, n. brightness, radiance; *antauges*, reflecting light, sparkling: augite, augitite, *Leucauge venusta* (a spider).

L. *aura*, f. glow, emanation, exhalation: aura.

AS. *beorht; berht; briht*, bright: brightness, eyebright, Bertha, Bertram, Ethelbert, Albert, Robert.

L. *candeo*, shine, glow, be white; *candico*, shine, be white; *candidus*, shining white, bright, frank; *candidulus*, dim.; *candela*, f. candle; *candelabrum*, n. candlestick; *candelabrarius*, m. candlestick-maker: candid, candidate, chandelier, candle, cannel coal, incandescent, *Candide*, Candlemas, candelilla *(Euphorbia antisyphilitica)*, *Populus candicans* (a poplar), *Cypripedium candidum* (white moccasin-flower), *Calycophyllum candidissimum* (a rubiacead), *Pulvinulina candidula* (a foraminifer), *Aulographis candelabrum* (a radiolarian), *Pleurotus incandescens* (a mushroom).

L. *cereus*, wax-taper, candle; waxy; see *cera* under **wax**

L. *claresco*, begin to shine, become visible; *clarico*, glow, gleam, < *clarus*, clear; see **clear**

L. *corusco, -atus*, flash; *coruscus*, flashing, gleaming, glittering, tremulous: coruscate, *Crataegus corusca* (a hawthorn).

Gr. *das, dados (daïs, -ïdos)*, f. firebrand, pine-torch, pine-wood; *dadion*, n. dim., taper; *dadinos*, of torches, pine-wood; *dadodes; endados*, resinous; *dadouchos*, m. torch-bearer; *dalos*, m. firebrand: daduchus, *Daduchus flavocinctus* (a beetle), *Dadoxylon annulatum* (a fossil gymnosperm), *Pandalus annulicornis* (a decapod).

Gr. *deletron*, n. lantern: *Deletrocephalus dimidiatus* (a nematode).

Gr. *dete*, f. torch:

L. *fax, facis*, f. torch, firebrand, flame, light; *facula*, f. dim.: ultrafax, *Pleurotus facifer* (a mushroom).

L. *fulgeo*, flash; *fulgidus*, shining, gleaming; *fulgor, -is*, m. lightning, brightness; *fulgur, -is*, n. flash of lightning; *fulgetrum*, n. heat-lightning, lightning; *fulguralis*, of lightning; *fulguritus*, struck by lightning; *fulguro, -atus*, lighten; *effulgeo*, shine forth, gleam: fulgurite, effulgent, refulgent, *Fulgora lanternaria* (lantern-bug), *Opuntia fulgida* (cholla), *Oxydesmus effulgens* (a milleped), *Lychnis fulgens* (a campion), *Hoffmannia refulgens* (a rubiacead).

L. *fulmen, -inis*, n. flash of lightning, thunderbolt; *fulmineus*, of lightning; *fulmino, -atus*, lighten, thunder and lighten: fulminate, *Agrilus fulmineus* (a beetle).

Gr. *ganos, -eos*, n. brightness; *ganodes*, bright; *ganotos*, polished, brightened: ganoid, ganodont, ganomalite, ganophyllite, *Ganolytes aratus* (a fossil fish), *Ganoproctus longicornis* (a fly), *Ganorhynchus splendens* (a Devonian fish), *Ganoderma applanatum* (a fungus).

AS. *glom*, twilight; see **shade**

Gr. *grabion*, n. torch:

Gr. *helene*, f. torch, destroyer: Helen, Eleanor, Nellie, Nell, *Helenophorus collaris* (a beetle).

Gr. *ipnos*, oven, lantern; see **oven**

L. *jubar, -is*, n. radiance of the heavenly bodies, light, sunshine, splendor, glory:

L. *lanterna (laterna)*, f. lamp, torch, < Gr. *lampter, -os*, m. stand, grate, lantern, < *lampo*, shine; *lampas, -ados*, f. torch, beacon; *lampsis*, f. a shining; *lampabilis*, shining; *lampadarius*, m. lamp-bearer, torch-bearer; *aeilampes*, ever-shining; *lampros*, bright, brilliant, radiant; *lamprotes, -etos*, f. brilliance, splendor; *perilampros*, very bright, radiant: lamp, lantern, eclampsia, lamprophyre, *Laternaria[Lanternaria] phosphorea* (a lantern-bug), *Chrysolampis mosquitus* (a hummingbird), *Lampsilis ventricosus* (pocketbook-mussel), *Lampropeltis rhombomaculata* (brown king-snake), *Lampoxanthium pandora* (a radiolarian), *Thalassolampe maxima* (a radiolarian), *Nematolampas regalis* (a cephalopod), *Perilampus[Perilamprus] hyalinus* (a wasp), *Lampterocrinus inflatus* (a Silurian crinoid).

L. *lautus*, washed, splendid, brilliant, elegant; see *lavo* under **wash**

L. *liparos*, fatty, shiny, bright, brilliant; see *lipos* under **fat**

Gr. *lophnis, -idos*, f. a kind of torch; *lophnidion*, n. dim.:

L. *lumen, -inis*, n.; *luminosus*, full of light; *lumino, -atus*, light up; *illumino*, light up, embellish: lumen, luminous, luminary, luminiferous, luminescence, limn, luminal, luminosity, *Volucella lumina* (a fly).

L. *lustro, -atus*, illuminate, light up; *illustris*, lighted or lit up, bright: luster, lustring, illustrate, illustration, illustrious.

L. *lux, lucis*, f. light; *lucerna*, f. lamp; *lucernula*, f. dim.; *lucibilis*, bright, clear; *lucidus; luculentus*, full of light, clear, bright, splendid; *lucifer; lucinus*, light-bringing; *lucisator, -is*, m. light-maker; *lucificus*, light-making; *luceo*, beam,

glitter, shine; *lucubro, -atus,* work by lamplight, burn the midnight oil: lucid,
Lucifer, Lucius, Luke, Lucy, Lucile, de Luxe, lucubration, Lucina, luciferin,
translucent, pellucid, *Salix lucida* (shining willow), *Georgia pellucida* (a moss),
Lycopodium lucidulum (a clubmoss), *Nitella translucens* (a characead), *Ceratina
lucifera* (a bee), *Lucernaria quadricornis* (a jellyfish), *Lucilia cadaverina* (a
green-bottle fly), *Luxilius cornutus* (redfin).
Gr. *lychnos,* m. lamp, light; *lychnidion; lychnion,* n.; *lychniskos,* m. dim.;
lychnouchos, m. lampstand: lychnomancy, lychnoscope, *Lychnosphaera regina* (a
radiolarian), *Lychnidius scrobiculatus* (a Cretaceous crinoid), *Lychniscaulus
vannus* (a fossil sponge), *Lychnuchus olenus* (a butterfly).
Gr. *marmareos,* flashing, glistening, shining, < *marmairo,* flash, sparkle, gleam:
Per. *mazda,* god of light, < *Ahura Mazda,* supreme deity, spirit of good: Mazda,
Ormuzd.
L. *mico, -atus,* tremble, shine, twinkle; *micans, -antis,* sparkling, twinkling:
"*Mica, mica, parva stella*", *Buprestis micans* (a beetle), *Olethreutes micantana*
(a moth).   See *mica* under **little**
L. *niteo,* shine, glitter; *nitidus,* bright, shining, elegant; *nitidulus,* spruce, trim;
nitela (nitella), f. brightness, splendor; *nitido, -atus,* make bright, polish: nitid,
nitidous, *Nitella cernua* (a characead), *Nodosaria nitidula* (a foraminifer),
Turrilepas nitidulus (a fossil barnacle), *Phanocrinus nitidus* (a Mississippian
crinoid), *Diaphania nitidalis* (pickle-worm), *Phainopepla nitens* (a fly-snapper),
Chrysocoma nonnitens (a fly), *Chlorochrysa nitidissima* (a tanager).
Gr. *norops, -opos,* bright, flashing, gleaming: *Norops auratus* (a lizard).
Gr. *periglenes,* very bright:
Gr. *phaino; phao,* shine, appear, make manifest, < *phaos, phaeos,* n. light;
phaëthon, beaming, radiant, shining; *phaidimos,* shining; *phaidros,* bright,
beaming; *phanaios,* giving light; *phaneros; phantos,* evident, visible; *phanos,*
light, bright; m. torch; *phanion,* n. dim.; *phausis, -eos,* f. a lighting; *diaphanes,*
transparent, distinct; *emphanes; kataphanes; prophanes,* evident, manifest;
periphanes; telephanes, conspicuous; *pasiphaes,* shining on all; *phasis,* f. appearance, face, look, state; *symphasis,* f. coincidence, conjunction: Phaëthon,
phaeton, phenocryst, phenology, phenomenon, phenol, phenolate, phanerogam,
diaphanous, glaucophane, phantom, phanerosis, fancy, phase, *Phaethon rubicauda*
(a tropic-bird), *Phanaeus carnifex* (a dung-beetle), *Phaedrotettix angustipennis*
(a grasshopper), *Phaneropsolus sigmoideus* (a trematode), *Phanerobunus
armatus* (an opilionid), *Nyctiphanes australis* (a shrimp), *Phaneresthes
flavovariegata* (a beetle), *Phanerotis fletcheri* (an amphibian), *Phanostoma
senegalense* (a trichopterid), *Phaniomyia biguttata* (a fly), *Phaenotypus
palmarum* (a beetle), *Phaenobolus bicolor* (a beetle), *Eubranchipus diaphanus*
(a fairy-shrimp), *Adiantum diaphanum* (filmy maidenhair-fern), *Telephanus
squalidus* (a beetle), *Pasiphaea distincta* (a crab), *Phaedroctonus minutus* (a
wasp), *Melamphaes unicornis* (a fish), *Euphausia splendens* (a crustacean).
L. *pharus* (Gr. *pharos*), f. lighthouse, beacon: Pharos, *Phaenopharus struthioneus*
(a phasmatid).
Gr. *phengos,* n. light, splendor, luster, sunlight, < *phengo,* shine, gleam;
phengodes, bright, light, shining; *phengites,* m. a transparent mineral used
for windows: phengite, *Phengodes fusca* (a beetle).
Gr. *phiaros,* shining, bright: *Phiarus abdominalis* (a bee).
Gr. *phoibos,* bright, radiant: Phoebus, Phoebe.
Gr. *phos, photos,* m. light; *photeinos,* shining, bright, clear; *photismos,* m.
illumination; *photistes,* m. enlightener; *photistikos,* enlightening; *phosphoros,*
giving light; m. light-giver, Lucifer; *photodes,* lightlike, luminous; *photophanes,*
brilliant; *lykophos, -otos,* morning twilight: photograph, photosynthesis, photo-
chemical, photoelectric, photolysis, photometer, photomicrograph, photostat,
phototropism, photism, photistic, phosgene, phosphate, phosphorus, phospho-
rescence, phosphorylation, photophobic, sodium 2-methyl-4-dimethylaminophenyl-
phosphinite, adenosine-5-triphosphate, *Photocryptus photomorphus* (a wasp),
Photodotis prochalina (a moth), *Photoptera erythronota* (a wasp), *Photinopteris
speciosa* (a fern), *Photinia glabra* (Japanese toyon), *Photuris pennsylvanica*
(firefly).
Gr. *phryktos,* firebrand, torch, beacon; see *phrygo* under **cook**
L. *polimen, -inis,* polish, brightness; see *polio* under **smooth**
Gr. *prestikos,* flashing:
Gr. *pyrpolema, -tos,* beacon, signal-fire; *pyrsos,* beacon, torch, fire; see *pyr* under
fire
L. *radiatus,* rayed, beaming, shining; *irradio, -atus,* cast forth rays, illumine:
radiation, radiance, radiant, irradiate.
L. *scintillo, -atus,* sparkle, glitter; *scintilla,* f. spark; *scintillula,* f. dim.: scintilla,
scintillate, scintillation, *Scintilla stevensoni* (a pelecypod).

Gr. *selas, -aos,* n. light, flash, meteor; *selasma, -tos,* n. a shining, < *selasko,* shine: *Selaophora obscura* (a spider), *Selasphorus rufus* (a hummingbird).

Gr. *selenion,* moonlight; see *selene* under **moon**

Gr. *sigaloeis,* glossy, shiny:

Gr. *spinther, -os,* m.; *spintharis, -idos,* f. spark; see **fire**

L. *splendeo,* shine; *splendidus,* bright, shining; *splendor, -is,* m. luster, brilliance, glory: splendid, splendor, resplendent, *Haliotis splendens* (an abalone), *Calathea splendida* (a calathea), *Nodosaria splendidula* (a foraminifer).

Gr. *sterops, -opos,* flashing; *sterope,* f. flash of lightning, flash, gleam: *Steropus sigillatus* (a beetle), *Steromorpha lenis* (a beetle).

Gr. *stilbo,* glitter, shine; *stilbe,* f. lamp, mirror; *stilpnos,* glittering, glistening; *stilpsis,* f. a shining: stilbite, stilpnomelane, stilbene, diethylstilbestrol, *Stilbopteryx costalis* (a dragonfly), *Stilbella flavida* (coffee-disease fungus), *Stilpnothorax loricatus* (a katydid), *Stilpnotia salicis* (satin-moth), *Astilbe japonica* (a saxifrage), *Sphaerostilbe repens* (a fungus).

L. *taeda,* torch of pine-wood; see **pine**

Gr. *thystlon,* n. torch carried in the Bacchic festival:

Gr. *typhedon, -os,* f. a kindling, lighting, torch:

See: **fire, burn, clear, white, honor, sun, display**

light < AS. *liht,* having little weight: lightness, lightly, lighthearted, lightheaded, lighter, Lightfoot, lightsome.

Gr. *achne,* any light substance, foam, froth, chaff, down; see **scale**

Gr. *aesyros,* light as air, agile:

L. *agilis,* quick, light, nimble: agile, agility, *Oporornis agilis* (Connecticut warbler).

Gr. *chaunos,* empty, frivolous; see **hole**

Gr. *chnous (chnoos), -nou,* m. any light, porous substance, foam, fine down, bloom; see **foam**

Gr. *elaphros,* light in weight, nimble, < *elaphrizo,* lighten, alleviate; *elaphria,* f. lightness: *Elaphrus riparius* (a beetle), *Elaphrosaurus bambergi* (a fossil reptile), *Elaphrothrips nigripes* (a thrips), *Herminiera elaphroxylon* (a legume).

Gr. *kouphos,* light, airy, easy:

L. *levis,* light, mild, fickle; *leviculus,* dim., trivial, insignificant, lightheaded, vain; *levitas, -atis,* f. lightness, < *levo, -atus,* lift up, lighten, relieve, ease: levity, levitation, alleviate, leger, legerdemain.

See: **raise, weight, ease, laugh, lessen**

lignum, L. wood; *ligneus,* of wood, wooden; *ligneolus,* wooden; *lignarius,* carpenter; see **wood**

lignys, Gr. thick smoke and flame; *lignyodes,* smoky, sooty, murky; see **cloud**

ligo, -onis, L. grubbing-hoe, mattock; see **pick**

ligula; lingula, L. little tongue, strap, ladle, dim. of *lingua,* tongue; see **tongue**

ligurio, -itus, L. eat daintily, be fond of good things, lick; *liguritor,* epicure; see **eat**

ligusticum, L. lovage; see **carrot**

ligustrum, L. privet; see **privet**

ligys; ligyros, clear, sharp, distinct; see **clear**

like; likeness < AS. *gelic; gelicnes:* likely, likelihood, childlike, homelike, birdlike, manly, cowardly, saintly, swiftly, truly, freely, westerly, boldly, each, which, ilk, early, Lichfield.

Gr. *alinkios,* resembling, like:

Gr. *analogos,* conformable, resembling: analogous, analogy.

Gr. *anti,* equal to, like, in place of: *Antirrhinum glandulosum* (a snapdragon).

L. *citeria,* caricature, effigy, likeness; see **form**

L. *conformalis; conformis,* like, similar: conformity, conformable.

Gr. *empheres,* resembling, like; *emphereia,* f. likeness: *Empheremyia atra* (a fly).

Gr. *homoios,* like, resembling; *homologos,* agreeing, corresponding; see *homos* under **equal**

Gr. *ikelos,* like: *Icelus uncinalis* (a fish), iceloplasmic.

Gr. *instar,* form, image, likeness; see **form**

Gr. *isos,* equal, like; see **equal**

Gr. *paraplesios,* somewhat like: *Paraplesiobatis heinrichsi* (a Devonian fish).

L. *quasi,* appearing as if, simulating, < *quam,* as, *si,* if: quasi-contract, quasi-official, quasi-logical.

L. *sic,* thus, so, like, in the manner mentioned; a word within brackets inserted into a quoted passage to show that a statement, misspelling, etc., is exactly as in the original: sic.

L. *similis,* like; *simul,* at the same time; *simulo, -atus,* imitate, copy; *consimilis,* like in all respects; *simulatio, -onis,* f. affectation; *simulacrum,* n. image, likeness, portrait, effigy; *simulator, -is,* m.; *simulatrix, -icis,* f. imitator: similar, simulate, simile, assimilate, resemble, simple, simultaneous, dissemble, dissimulate, dissimilar, assembly, facsimile, simulacrum, simul-transit, *Simulium molestum* (blackfly), *Carex simulata* (a sedge), *Bombylius simulans* (a bee-fly), *Alampetis simulatrix* (a beetle), *Ceratorrhina simillima* (a beetle), *Hypericum dissimulatum* (a St. Johnswort), verisimilitude.

Gr. *symbletos*, comparable:

* * * * * * *

L. *-aster, -astra, -astrum*, diminutive suffix with derogatory implication; resemblance; see **little**

F. *-esque* (It. *-esco*), like in manner or style: arabesque, burlesque, romanesque, picturesque, grotesque.

Suf. *-ish* < AS. *-isc* (G. *-isch*), denoting origin, pertaining to, like; see **nature**

Suf. *-ite, -ites* < Gr. *-ites*, having the nature of, like; see **nature**

Suf. *-oid, -oides, -odes, -oideus, -a, -um*, like, resembling, having the form of,< Gr. *-eides*, like, < *eidos*, n. form, shape, likeness, as *dendroeides*, treelike; *kephaloeides*, headlike; *limnodes*, marshlike; *phyllodes*, leaflike. These are adjectives of two terminations, both of which in transliteration are rendered *-es*, and, therefore, give no clue to their gender. Generic terms ending with the noun *eidos* (as *-eidus* or *-idus*) are neuter; but such terms ending with the adjectival derivatives of *eidos* (as *-odes* and *-oides*) may, in practice, be masculine, feminine, or neuter, in accord with the genders of the governing nouns, as in *Cordyloides carbonarius* (a stegocephalian), < *Cordylus*, m.; *Cryptohalictoides spiniferus* (a bee), < *Halictus*, m.; *Dasychiroides obsoleta* (a moth), < *Dasychira*, f.; *Machiloides appendiculata* (a thysanurid), < *Machilis*, f.; *Centroptiloides bifasciatum* (an ephemerid), < *Centroptilum*, n.; *Isochromodes miniatum* (a moth), < *chroma*, n.; *Miltogrammoides maximum* (a fly), < *Miltogramma*, n. Perhaps the best recommendation here is: Do not create *-oides, -opsis*, and similar adjectival terms for use as nouns. The *oideus, -a, -um* terms have the gender indicated by the regular *-us, -a, -um* adjectival endings. Other examples: celluloid, cardioid, colloid, cycloid, crinoid, deltoid, ovoid, sigmoid, adenoid, geode, nematode, petalody, *Aplectoides calignea* (a moth), *Siculodes mellea* (a moth), *Limacodes argentifera* (a moth), *Aphobetoideus comperei* (a wasp), *Cycadeoidea marylandica* (a fossil cycadophyte).

Gr. *-opsis, -ops, -opos, -opis*, having the appearance of, like, < *opsis, -eos*, f. look, appearance, face, sight, eye, < *ops, opos*, f. eye, face, as *makropsis*, long-faced; *phakopsis*, freckled; *chrysops*, golden; *sterops*, flashing; *melanopos*, black-looking; *teratopos*, marvelous-looking; *dolopis*, artful-looking; *lithopis*, like stone. Generic terms ending with the noun *opsis* are feminine; but those in which *opsis* is adjectival may, in practice, be masculine, feminine, or neuter, in accord with the genders of the governing nouns, as in *Belemnopsis canaliculatus* (a fossil cephalopod), < *Belemnites*, m.; *Oryzopsis asperifolia* (a rice-grass), < *Oryza*, f.; *Trochonemopsis tricarinatum* (a Devonian gastropod), < *Trochonema*, n. The many current exceptions to this as well as to the similar rule for *-oides* terms emphasize the need for a guiding principle. Other examples: *Teneropsis sibuyanus* (a beetle), *Augochloropsis cuprea* (a wasp), *Rosa banksiopsis* (a rose), *Camelops kansanus* (a Pleistocene camel), *Chrysops vittatus* (a fly).

See: **equal, form, face**

lilac < Sp. *lilac*, < Per. *nilak*, bluish; see *nila* under **blue**

lily < AS. *lilie*, < L. *lilium*, n.; *liliaceus; lilinus*, of lilies: liliaceous, liliated, liliform, lilywort, lily-of-the-valley, wood-lily *(Lilium philadelphicum)*, *Lilium longiflorum* (Easter-lily), *Liparis liliifolia* (twayblade), *Frankliniella lilivora* (a bug), *Anthericum liliago* (St. Bernard lily).

L. *albucus*, m. bulb of the asphodel; asphodel: *Albuca major* (albuca).

L. *allium*, garlic; a genus of the lily family; see **onion**

Gr. *aloe*, f. aloe: aloes (from *Aloe vera*), aloin, lignaloes, *Yucca aloifolia* (Spanish dagger).

L. *anthericum* (Gr. *antherikon*), n. asphodel: *Anthericum ramosum* (an anthericum).

L. *asparagus* (Gr. *asparagos*), m. a genus of the lily family: *Asparagus officinalis* (asparagus).

L. *asphodelus* (Gr. *asphodelos*), m. a genus of the lily family: daffodil, *Asphodelus albus* (branching-asphodel), *Asphodeline lutea* (king's-spear), *Geranium asphodeloides* (a geranium).

NL. *camassia*, f. a genus of the lily family, < Ab.Am. *kamass (quamash)*: *Camassia quamash* (common camas), death camas *(Zygadenus venenosus)*.

NL. *clintonia,* f. a genus of the lily family, < DeWitt Clinton, governor of New York: *Clintonia umbellata* (speckled beadlily).

NL. *convallaria,* f. a genus of the lily family: *Convallaria majalis* (lily-of-the-valley).

L. *crinum* (Gr. *krinon*), n. lily; *krininos,* of lilies; *krinotos,* adorned with lilies: crinoid, *Crinum giganteum* (an amaryllis), *Leucocrinum montanum* (star-lily), *Neocrinus decorus* (a crinoid), *Dendrocrinus minutus* (an Ordovician crinoid), *Uintacrinus socialis* (a Cretaceous crinoid).

L. *erythronium* (Gr. *erythronion*), n. a genus of the lily family: *Erythronium citrinum* (lemon fawn-lily).

NL. *fritillaria,* f. a genus of the lily family, < *fritillus,* dice-box, spotted: *Fritillaria pudica* (yellow fritillary).

L. *hastula,* asphodel; see *hasta* under **spear**

Gr. *hemerokallis, -idos,* f. daylily: *Hemerocallis flava* (lemon-daylily).

L. *hyacinthus* (Gr. *hyakinthos*), m. hyacinth; *hyakinthinus,* like hyacinth: hyacinthine, *Hyacinthus orientalis* (a hyacinth), *Rhomborrhina hyacinthina* (a beetle).

Gr. *hypoglosson,* n. butcher's-broom: *Ruscus hypoglossum* (a butcher's-broom).

Gr. *leirion,* n. lily; *leirinos,* made of lilies; *leirios; leiros,* of or like lilies: *Liriodendron tulipifera* (tulip-tree), *Chamaelirium luteum* (fairywand), *Dasylirion serratifolium* (a liliacead, sotol), *Agave dasylirioides* (an agave).

L. *limnice* (Gr. *limnike*), f. a marsh-plant, pond-lily:

NL. *maianthemum,* n. mayflower, a genus of the lily family: *Maianthemum canadense* (beadruby).

L. *melanthium,* n. a genus of the lily family: *Melanthium virginicum* (bunch-flower).

L. *ornithogalum* (Gr. *ornithogalon*), n. star-of-Bethlehem: *Ornithogalum speciosum* (a star-of-Bethlehem).

Gr. *phasganion,* sword-lily, gladiolus; see *phasganon* under **sword**

L. *phormium* (Gr. *phormion*), n. a genus of the lily family: *Phormium tenax* (fiber-lily).

L. *polygonatum* (Gr. *polygonaton*), n. a genus of the lily family: *Polygonatum biflorum* (Solomon's-seal).

NL. *ruscus,* m. a genus of the lily family, < *ruscum (bruscum),* n. butcher's-broom; *ruscarius,* of broom: brusque, *Ruscus aculeatus* (a butcher's-broom), *Coriaria ruscifolia* (tutu), *Centroscyllium ruscosum* (a fish), *Elaphrus ruscarius* (a beetle).

L. *scilla; squilla* (Gr. *skilla*), f. sea-onion, sea-leek: red squill *(Urginea maritima),* *Scilla verna* (a sea-onion), *Camassia scilloides* (a camas).

L. *smilax, -acis* (Gr. *-akos*), f. a name given to several different plants but now restricted to a genus of the lily family: *Smilax rotundifolia* (a greenbrier), *Smilax hispida* (bristly greenbrier), *Smilacina racemosa* (a false Solomon's-seal), *Piper smilacifolium* (a pepper).

L. *susum* (Gr. *souson*), n. lily; *susinus* (Gr. *sousinos*), of lilies: *Susum anthelminticum* (a flagellariacead).

NL. *trillium,* n. a genus of the lily family: *Trillium cernuum* (nodding trillium).

NL. *tulipa,* f. tulip, < D. *tulp,* < F. *tulipe,* < Turk. *tulbend,* turban: *Tulipa rosea* (rosy tulip), *Siphonia tulipa* (a sponge), *Botrytis tulipae* (a fungus).

NL. *urginea,* f. a genus of the lily family: *Urginea indica* (Indian squill).

NL. *yucca,* f. a genus of the lily family, < Ab.Am. *yuca: Yucca filamentosa* (a yucca), *Agave yuccaefolia* (an agave).

lima, L. file; *limula,* dim.; *limatus,* polished, refined; see **scrape**

limaceus; limarius, L. of mud or slime; see *limus* under **earth**

limaco- < Gr. *leimax, -akos,* meadow; *leimakodes,* grassy, moist; see *leimon* under **field**

limax, -acis, L. slug, snail; see **mollusk**

limb < AS. *lim;* see **branch, arm, leg, part**

limbus, L. border, hem, fringe, edge, selvage; *limbatus,* bordered; see **border**

lime < AS. *lim,* a viscous substance; see **glue;** calcium oxide, calcium carbonate.

L. *calx, calcis,* f. lime; *calcarius (calcareus),* of lime, limy; m. lime-burner; *calcaria,* f. limekiln: calcite, calcium, calcimine, calcine, calcareous, chalk, *Gloeocapsa calcarea* (an alga), *Haplophragmium calcareum* (a foraminifer), *Delphinium calcicola* (a larkspur), *Cucumaria calcigera* (a holothurian), *Lecidea calcivora* (a lichen).

Gr. *chernites,* m. a kind of white marble:

L. *creta* (Gr. *Krete,* Crete), f. Cretan earth, white earth or clay, chalk; *cretula,* f. dim.; *cretaceus,* chalky; *cretarius,* of chalk; *cretatus,* marked with chalk;

cretosus, abounding in chalk: cretaceous, crayon, *Globigerina cretacea* (a foraminifer).

L. *eglecopala,* f. blue marl:

Gr. *gypsos,* chalk, gypsum; see **gypsum**

Gr. *kimolia,* a white, chalky clay from the island of Cimolus; see **earth**

Gr. *lygdos,* f. white marble; *lygdinos,* of white marble: *Lygdinus niveus* (a mite).

L. *marga,* f. marl: marl, marly, marlaceous.

L. *marmor, -is,* (Gr. *marmaros*), m. marble; *marmoreus; marmorosus,* made of marble, marblelike; *marmarinos,* of marble; *marmusculum,* n. small work in marble; *marmairo,* sparkle: marble, Marblehead, marbled, marmolite, marbrinus, marmoraceous, marmarosis, *Kalanchoë marmorata* (a houseleek).

Gr. *poros,* m. tufa, travertine; *poridion,* n. dim.; *porodes,* like tufa:

L. *tofus (tophus),* m. travertine: tufa.

See: **stone**

limen, *-inis,* L. threshold; *liminaris,* of a threshold; see **door**

limeno- < Gr. *limen, -os,* harbor, haven, refuge; see **harbor**

limeros, Gr. hungry; see *limos* under **hunger**

limeum, L. a Gallic herb yielding arrow poison; now a genus of the pokeweed family; see **pokeberry**

limit < L. *limes, -itis,* boundary; *limitaris,* on the border; see **border**

limma, L. (Gr. *leimma*), remnant; see *leipo* under **leave**

limnestis, Gr. a marsh-plant; see **plant**

limnice, L. a kind of water-plant; see **lily**

limno- < Gr. *lime,* marsh, lake, pool; *limnion,* dim.; *limnaios,* of a marsh; see **marsh**

limnobium, NL. a genus of waterweeds; see **waterweed**

limo- < Gr. *leimon,* meadow, see **field**; < Gr. *limos,* hunger, see **hunger**; < L. *limo, -atus,* file, see **scrape**

limodorum, L. (Gr. *limodoron*), a plant; a genus of orchids; see **orchid**

limonium, L. (Gr. *leimonion*), sea-lavender; see **leadwort**

limp < uncertain origin; see **walk, weak, hurt**

limpidus, L. clear, transparent, pure; *limpiditas; limpitudo,* clearness; see **clear**

limula, L. dim. of *lima,* file; see **scrape**

limulus, L. somewhat oblique, < *limus,* sidelong, sideways, askance; see **slope**

limus, L. mud, mire, slime; *limaceus,* of mud; *limosus,* muddy, miry, slimy, see **earth**; apron worn by priests, see **garment**

linaria, NL. a genus of the figwort family; see **figwort**

linctus, L. licked; see *lingo* under **lick**

linden < AS. *lind:* Hohenlinden, Unter den Linden, Carl von Linne, Lindley.

L. *corchorus* (Gr. *korchoros*), m. a genus of the linden family: *Corchorus hirsutus* (a jute plant), *Corchoropsis crenata* (a tiliacead).

Gr. *philyra,* f. linden:

L. *tilia,* f. linden: *Tilia americana* (basswood), *Agromyza tiliae* (linden gall-fly), *Hibiscus tiliaceus* (a hibiscus).

line < AS. *line,* < L. *linea,* f. linen thread, line; *lineola,* f. dim.; *linealis; linearis, lineatus,* of a line: linear, lineal, delineate, align, linotype, lineage, lineament, rectilinear, *Buteo lineatus* (red-shouldered hawk), *Schwagerina linearis* (a foraminifer), *Tabanus lineola* (green-headed fly), *Dyspnoetus lineellus* (a beetle), *Icogramma lineigera* (a beetle), *Lygus oblineatus* (a bug), *Dentalina multilineata* (a foraminifer), *Pnirsus linealiventris* (a bug), *Meromacrus lineascripta* (a fly), *Lymanopoda ignilineata* (a moth), *Cleptomita atriliniella* (a moth), *Hybanthus linearifolius* (a violet).

Gr. *alytos,* unbroken, continuous; see **one**

L. *continuus,* uninterrupted; see **one**

Gr. *eirmos,* m. series, train, < *eiro,* join in rows, string together: *Eirmocystis [Irmocystis] ventricosa* (a protozoan), *Epeira scutulata* (a spider, now *Aranea miniata*).

Gr. *epallelos,* in sequence, in close order, in unbroken succession, continuous:

L. *fasciatus,* banded, striped; see *fascia* under **belt**

Gr. *gramme,* f. line, stroke of the pen; *grammikos; grammodes,* linear, in lines: diagram, *Grammostola pulchripes* (a spider), *Grammostomum porosum* (a foraminifer), *Grammistes sexlineatus* (a soapfish), *Melanogrammus aeglefinus* (haddock), *Pityrogramma sulphurea* (a gold-fern), *Phanerogramma heeri* (a fossil beetle).

L. *laticlavius,* having a broad stripe, *laticlavus,* m., on the toga:
Gr. *oimos,* path, way, stripe; see **way**
Gr. *orchos,* m. row of trees; *orchatos,* m. orchard: *Orchoporus koehleri* (a Miocene echinoid).
L. *ordo, -inis,* methodical arrangement, line, series, row, order; see **class**
L. *phalanx, -angis* (Gr. *-angos*), f. line, battle-array, bone of the finger or toe; *phalangion,* n. dim. a spider: phalanx, phalanges, phalangitis, *Phalanges maculatus* (spotted phalanger), *Phalangium cinereum* (a daddylonglegs).
Gr. *rhabdotos,* striped, fluted: *Rhabdotorrhinus exaratus* (a bird).
Gr. *scheros,* in a line, successive:
L. *series,* f. row, succession, train: series, seriation, seriatim, *Psacadonotus seriatus* (a katydid), *Cicindela serieguttata* (a beetle).
Gr. *stathme,* f. carpenter's line or rule, < *stathmo,* measure; *stathmetikos,* of measuring: *Stathmophorus furcus* (a trichopterid), *Stathmepora gabbiana* (a fossil bryozoan), *Stathmonotus hemphilli* (a fish), *Stathmodera lineata* (a beetle).
Gr. *stichos; stoichos,* m. row, line, rank; *stichidion,* n. dim.: distich, hemistich, stichidium, stichomythia, stoichiometry, acrostic, *Stichopus fuscus* (a holothurian), *Stoecharthrum giardi* (a worm), *Acrostichum aureum* (golden-fern), *Orthostoechus maculicauda* (a grunt), *Distichlis dentata* (a saltgrass), *Systoechus oreas* (a bee-fly), *Atractosteus tristoechus* (an alligator-gar), *Hordeum hexastichon* (sixrow-barley).
Gr. *syrden,* in a long line: *Syrdenopsis grayi* (a beetle).
Gr. *taxis,* row, line, rank; see *tasso* under **class**
L. *versus,* furrow, line; *versiculus,* dim.; see **poem**
L. *virgulatus,* striped; see *virga* under **branch**
See: **furrow, thread, class, follow, ribbon, ridge, bruise**

linen < AS. *linen,* < L. *linum* (Gr. *linon*), n. flax; *linteum,* n. linen; *linteolum,* n. dim.; *linteus,* of linen; *linteatus,* clothed in linen; *linarius,* m. linen-weaver, dealer in linen: lint, *Cardium linteum* (a fossil pelecypod), *Tellina lintea* (a pelecypod). Derive: linsey-woolsey, lawn.
Gr. *motos,* m. shredded linen, lint:
Gr. *othone,* f. fine linen, sail-cloth, sail, membrane of the eye; *othonion,* n. dim.: *Othonops eos* (a fish), *Cynoscion othonopterus* (a weakfish), *Cyclothone elongata* (a fish).
See: **flax**

-ling, AS. diminutive; having the quality of: see DIMINUTIVES under **little**

linger < AS. *lengan,* prolong; see **stand, delay**

lingua, L. tongue; *lingula; ligula,* dim.; *lingulaca,* chatterbox, gossip; see **tongue**

linimentum, L. stuff to be rubbed on the skin, see **drug;** *linimen,* grease, see **fat;** *linio, -itus,* smear, spread, see **spread**

link < AS. *hlenc;* see **bind, chain**

linnaea, NL. a genus of the honeysuckle family, < Carolus Linnaeus, Carl von Linne, Swedish botanist; see **honeysuckle, plant**

lino- < Gr. *linon,* flax, anything made of flax, cord, net; *linarion; linidion,* dim.; see **flax, net**

lint < L. *linteus,* of linen; *linteum,* linen cloth; see **linen**

linter, L. boat, skiff, wherry, trough, tub, vat; *lintricula,* dim.; see **ship**

linum, L. (Gr. *linon*), flax, linen, thread; *linidion,* dim.; see **flax**

linurgo- < Gr. *linourgos,* weaver; see **weave**

linyphus, L. linen-weaver; see **weave**

lio- < Gr. *leios,* smooth; see **smooth**

lion < L. *leo, -onis* (Gr. *-ontos*); see **cat**

lip < AS. *lippa.*
Gr. *cheilos,* n. lip, rim; *cheilarion,* n. dim.: chiloplasty, chiloma, chilidium, Chilopoda, *Chilostomella ovoidea* (a foraminifer), *Chilomastix mesnili* (a flagellate), *Chilodonella uncinatus* (a protozoan), *Chilopsis linearis* (desert-willow), *Chilocorus bivulnerus* (two-spotted lady-beetle), *Isochilus lineare* (an orchid), *Ptychocheilus oregonense* (squawfish), *Prochilodus insignis* (a fish), *Cheilanthes gracillima* (a lipfern), *Megachile circumcincta* (a leaf-cutting bee), *Trichilia emetica* (mafura), *Plectrochilus machadoi* (a fish), *Lissochilus speciosum* (an orchid), *Brachychilus modestum* (a beetle).
Gr. *chelyne,* f. lip, jaw; *chelynion,* n. dim.: *Colochelyna basibifasciata* (a wasp).
L. *labium,* n. lip; *labeo, -onis,* m. one with large lips; *labiatus,* lipped; *labiosus,* large-lipped: labium, labial, Labiatae, bilabiate, *Labia minor* (an earwig),

Labeo niloticus (a fish), *Saccolabium guttatum* (an orchid), *Inoceramus labiatus* (a fossil pelecypod), *Labeotropheus curvirostris* (a fish).
L. *labrum,* n. lip, brim; *labellum,* n. dim.: labrum, labral, labret, *Cylichnostomum labratum* (a nematode).
Gr. *myllon,* n. lip:
Gr. *mystax,* upper lip, mustache; see **hair**
See: **border, mouth**

liparis, NL. a genus of orchids; see **orchid**

liparos, Gr. oily, greasy, sleek, shiny, comfortable; see *lipos* under **fat**

lipauges, Gr. dark, sunless; see *anaugetos* under **shade**

lipernes, Gr. homeless, outcast; see **alone**

liphaemo- < Gr. *liphaimos,* lacking blood, pale; see **gray**

lipo- < Gr. *leipo,* abandon, leave, be wanting or without, lack; *leipsanon,* remnant, see **abandon**; < *lipos,* fat, see **fat**

lipothrix, Gr. hairless; see **bare**

lippus, L. blear-eyed, dim-sighted, nearly blind; see **dim**

lipsano- < Gr. *leipsanon,* relic, remnant; see *leipo* under **abandon**

liquidambar, NL. a genus of the witchhazel family; see **witchhazel**

liquidus, L. fluid, clear; *liquidum; liquor,* fluid; *liqueo,* be fluid, melt; see *liqueo* under **flow, clear, water, juice, sauce, run, stream**

liquis, L. sloping; see *obliquus* under **slope**

liquor, L. fluid; see *liqueo* under **flow**

lira, L. earth or ridge thrown up by the plow, furrow-slice; *lirella,* dim.; see **ridge**

lirio- < Gr. *leirion,* lily; *leirios,* of lilies; see **lily**

liro- < Gr. *leiros,* like a lily, pale, see **gray**; *liros,* bold, shameless, see **bold**

lis, -itis, L. strife; *litigo, -atus,* dispute, sue at law; see **fight**

lisp < uncertain origin, probably imitative; see **falter**

lissos; lispos, Gr. smooth; see **smooth**

list < F. *liste,* roll, catalogue, register.
L. *album,* n. list, register, tablet: album.
Gr. *apographe,* f. register, list, census; *apographeus,* m. registrar:
L. *calendarium,* n. account-book, < *calendae,* first day of the month: calendar.
L. *catalogus* (Gr. *katalogos*), m. list, register: catalogue.
L. *census,* register of citizens and property; see *censeo* under **judge**
ML. *dictionarium,* n. alphabetical list of words with their definitions: dictionary.
"A dictionary is the universe in alphabetical order."—Anatole France.
L. *fastus,* m. calendar, almanac:
L. *glossarium,* n. list of words with definitions: glossary.
Gr. *hemerologion,* n. calendar:
L. *index, -icis,* c. catalogue, list, table; *indiculum,* n. dim. short list: index.
L. *inventarium,* n. list: inventory.
L. *laterculum,* n. register, list:
Gr. *lexikon,* dictionary, vocabulary; see *lexikos* under **word**
L. *matricula,* f. public list, register, roll: matriculation.
L. *nomenclatura,* f. list of names: nomenclature.
Gr. *onomastikon,* vocabulary, list of names; see *onoma* under **word**
Gr. *pandektes,* m. dictionary, encyclopedia, digest of laws: pandect.
L. *polyptychum,* n. account-book, register:
Gr. *programma, -tos,* n. proclamation, notice, catalogue of proceedings: program.
L. *prospectus,* m. preview, foresight: prospect, prospectus.
L. *recordatus,* m. recollection, remembrance: record, recorder, recordatory.
L. *regestum,* n. list, catalogue, record: register.
L. *repertorium,* n. catalogue, inventory: repertory, repertoire.
L. *schedula,* dim. of *scheda,* leaf of paper; see **paper**
L. *syllabus* (Gr. *syllabos*), m. list, register: syllabus.
L. *tabula,* board, writing-tablet, list; see **table**
ML. *vocabularium,* dictionary, lexicon; see *vocabulum* under **word**
See: **word, class, follow, number**

listen < AS. *hlystan;* see **hear**

listron, Gr. a leveling or smoothing tool, spade, shovel; *listrion,* dim.; see **shovel**

litamen, L. sacrifice; *litabilis,* fit for sacrifice; *lito, -atus,* offer a sacrifice; see **give**

litania, L. (Gr. *litaneia*), prayer, entreaty; see **ask**

litargos, Gr. swift, running; see **swift**

-lite < Gr. *lithos,* stone; see *lithos* under **stone**
litera (littera), L. letter; see **letter**
lithium < Gr. *lithos,* stone; see ELEMENTS under **thing**
litho- < Gr. *lithos,* stone; *lithidion,* dim. pebble; *lithikos; lithinos,* of stone; *lithax, -akos,* stony; see **stone**
lithospermum, L. (Gr. *lithospermon*), gromwell; see **borage**
lithostrotos, Gr. inlaid with stones, tesselated; see **mosaic**
liticen, L. trumpeter; see *lituus* under **rod**
litigator, L. disputant, < *litigo, -atus,* dispute, quarrel, sue at law; see *lis* under **fight**
litmus < ON. *litmose,* a kind of dye-yielding lichen; see **lichen**
litoralis, L. of the seashore, < *litus, -oris,* shore; see **shore**
litos, Gr. plain, simple, frugal; *litotes,* simplicity; see **simple,** FIGURES OF SPEECH under **form**
litsea, NL. a genus of the laurel family; see **laurel**
litter < L. *lectus,* bed; see **bed, carry, dirt**
little < AS. *lytel.*
  Gr. *acares,* short, small, tiny, momentary; *akariaios,* momentary: *Acaremys murinus* (a fossil porcupine).
  Gr. *akation,* little man, dwarf; see *acatus* under **ship**
  Gr. *akolos,* f. bit, morsel: *Acolus zabriskiei* (a serphid).
  L. *alpinus,* of high mountains, dwarfed; see **mountain**
  L. *aliquantulus,* little, small; see *aliquantus* under **few**
  Gr. *apoknisma, -tos,* n. snip, bit:
  Gr. *apoxesma, -tos,* a scraping, shred, shaving; see *xeo* under **scrape**
  L. *atomus* (Gr. *atomos,* indivisible, uncut), f. a small particle: atom, atomic, *Artemisia stomifera* (a sagebrush).
  Gr. *attaragos,* m. morsel, bit:
  Gr. *baios; ebaios,* little, small, scanty, few: *Baeus californicus* (black-widow killer), *Baeolophus bicolor* (tufted titmouse), *Baeodon alleni* (a bat), *Baeognatha turanica* (a wasp), *Baiostoma[Baeostoma] brachialis* (a fish), *Ebaeomorphus thalassinus* (a beetle), *Ceratobaeus cornutus* (a wasp).
  Gr. *batyle,* f. female dwarf: *Batyle sanguinicollis* (a beetle).
  AS. *bita,* bite, small piece: bit, titbit(tidbit), henbit.   Derive: rarebit.
  L. *buccella,* f. dim. of *buccea,* f. morsel, mouthful:
  Gr. *chamai,* on the ground, low, dwarf, creeping; see **low**
  Sp. *chico,* little, little one: Chico(Calif.)
  L. *derogo, -atus,* detract from, disparage, belittle; see **lessen**
  Gr. *elachys,* little, small, short, low; *elasson (elatton), -os,* smaller, less; *elachistos,* smallest, least, < *elassoo,* lessen, diminish: *Elassochiton albomarginata* (a beetle), *Elachista curta* (an alga), *Carex elachycarpa* (a sedge).
  L. *exiguus,* little, short, scanty, poor, meager; *exiguum,* n. trifle; *perexiguus,* very small: exiguous, *Pulvinulina exigua* (a foraminifer), *Agathidium exiguum* (a beetle), *Ochlia exigualis* (a butterfly).
  L. *frustum,* bit, piece, scrape; *frustillum; frustulum,* dim.; see **part**
  NL. *gnomus,* m.; *gnoma,* f. diminutive fabled being, dwarf: *Glaucidium gnoma* (American pygmy-owl), *Gnoma giraffa* (a beetle).
  Gr. *hypokoristikos,* use of baby talk, pet names, usually diminutives: hypocoristic English (as: Billy, snookums, palsie-walsie).
  Gr. *iota,* ninth letter of the Greek alphabet; anything very small; see **letter**
  Gr. *kerma, -atos,* n. small piece, mite; *kermation,* n. dim.: *Cermatobius martensi* (a centiped).
  Gr. *leptos,* fine, small; see **thin**
  L. *mica,* f. crumb, bit, morsel, grain; as to the mineral mica there may be some association with *mico,* shine, sparkle, twinkle; *micella; micula,* f. dim.; *micarius,* of crumbs, frugal, thrifty: micelle, mica, micaceous, *Coprinus micaceus* (a mushroom), *Liospira micula* (a fossil gastropod).
  Gr. *mikros (smikros),* small, little; *meion; hesson* (NL. *-essus, -a, -um,* dim. suffix), smaller, less; *hekistos,* least, worst; *mikrotes,* f. smallness; *hyphesson, -os,* somewhat smaller; *meiosis,* f. diminution, lessening; *meiotikos,* lessening: microscope, micropyle, microbe, micrometer, micron, microphone, microtome, meiobar, meiotherm, meionite, meiosis, essonite, meiotic, Miocene, miohippus, hekistotherm, hettocyrtosis, *Micromorphus perminutus* (a fly), *Micropogon undulatus* (croaker), *Microdrapetes bellus* (a wasp), *Microlestes exilis* (a beetle), *Smicroplectrus jucundus* (an ichneumon-wasp), *Smicronyx pygmaeus* (a beetle), *Nesophontes zamicrus* (an insectivore), *Gygis microrhyncha* (a fairy-

tern), *Pilocarpus microphyllus* (a rutacead, source of pilocarpine), *Miohippus acutidens* (a fossil horse), *Lophiomus miacanthus* (a fish), *Hessolestes ultimus* (a creodont), *Essonodontherium gervaisi* (a fossil edentate), *Meniscoëssus conquistus* (a fossil mammal), *Acoëssus siderolithicus* (a fossil ungulate), *Vampyressa pusilla* (a bat), *Xiphopteris hecistophylla* (a fern), *Hyphessobrycon rosaceus* (a characine fish), *Thysanoëssa gregaria* (a shrimp).

Gr. *minys*, little, small, short; *minyanthes*, blooming a short time; *minythesis*, f. a wasting, diminishing; *minythikos*, diminishing, lessening; *minytho*, make smaller, lessen, shorten: *Minyocerus angustus* (a decapod), *Menyanthes* [?*Minyanthes*] *trifoliata* (bogbean).

AS. *mite*, anything very small; a minute arachnid: mite.

L. *morsum*, n. bit, piece: morsel.

L. *nanus* (Gr. *nanos;* *nannos*), m. a dwarf; *nanion*, n. dim. puppet: nanosaur, nanocranous, nannoplankton, nanandrous, nanism, nanization, nanosomia, *Nanotragus tragulus* (steenbok), *Nannostomus trifasciatus* (a fish), *Amorpha nana* (a false-indigo), *Achyrodon nanus* (a fossil mammal), *Anthrax nanellus* (a fly), *Nannippus gratus* (a fossil horse).

L. *offa*, f. bit, morsel; *offella;* *offula*, f. dim.: *Offa barrandiana* (a fossil ostracode), *Offadesma angasi* (a pelecypod), *Offula patella* (a mite).

Gr. *oligos*, few, small, scanty; see **few**

L. *parcus*, frugal, thrifty, scanty, slight; see **stingy**

L. *parvus*, little; *minor*, m., f.; *minus*, n. less; *minimus*, least; *parvulus*, dim.; *parvitas, -atis*, f. littleness, insignificance; *parum*, too little, insufficient; *minutus*, little, small; *minusculus;* *minutulus*, dim.; *minutalis*, paltry, insignificant; *minuo, -utus*, diminish, lessen; *deminuo (diminuo)*, lessen: parvule, parvifolious, parvoline, parvirostrate, paraffin, minus, minuscule, minute, minuet, minor, Minorca, minority, Asia Minor, menu, miniver, minimize, mischief, minister, minimum, minim, diminish, diminutive, comminution, *Thysanoëssa parva* (a shrimp), *Schistocrinus parvus* (a Pennsylvanian crinoid), *Lepidocyclina parvula* (a foraminifer), *Ranunculus parviflorus* (a buttercup), *Ulmus parvifolia* (Chinese elm), *Vinca minor* (common periwinkle), *Arctium minus* (a burdock), *Primula minima* (a primrose), *Pollinula minuta* (a sponge), *Helix minutalis* (a snail), *Adjidaumo minutus* (a fossil squirrel), *Sorex minutus* (pygmy-shrew), *Monomorium minimum* (an ant), *Scaphosoma minutissimum* (a beetle), *Vulvulina minutissima* (a foraminifer), *Pisobia minutilla* (least sandpiper), *Triodontophorus diminutus* (a nematode), *Onthophagus diminutivus* (a dung-beetle).

Gr. *Pataikos*, m. an odd-shaped, dwarflike, Phoenician deity; *Pataikion*, n. dim. a mischief-maker: *Pataecus fronto* (a fish).

L. *paucus*, few, little; *pauculus;* *pauxillus* dim.; see **few**

L. *paulus*, little; *paululus*, dim.: Paul, *Cremastocheilus paulus* (a beetle), *Phidippus paululus* (a fossil spider).

Gr. *pauros*, little, few: pauropod.

Sp. *pequeno*, little, young: picaninny.

F. *petit*, small: petit jury, petite, petty.

F. *picaillon*, a coin, farthing; hence, something small or of little value: picayune, picayunish.

L. *pisinnus*, small, little:

Gr. *pollostos*, smallest, least:

Gr. *psakas (psekas) -ados*, bit, crumb, grain, small drop; see **drop**

Gr. *psix, -ichos*, m. bit, crumb, morsel; *psichion*, n. dim.: *Psichion miliaris* (a gastropod).

Gr. *psomos (blomos)*, m. morsel, bit; *psomion*, n. dim.; *psomizo*, feed in little bits: *Psomophilus capensis* (a bird).

L. *pumilus;* *pumilis*, dwarfish, little; *pumilio, -onis*, c.; *pumilus*, m. dwarf, pygmy: *Castanea pumila* (Allegheny chinquapin), *Loxodonta pumilio* (pygmy-elephant).

L. *pusillus*, very little, small, petty, puny, insignificant; *pusillulus*, dim.: pusillanimous, *Ranunculus pusillus* (a buttercup), *Lemna perpusilla* (a duckweed).

L. *pygmaeus* (Gr. *pygmaios*), m. dwarf: pygmy, *Pygmaeomorpha modesta* (a moth), *Pygmaeochelys michelobona* (a Cretaceous turtle), *Pygmaeodrilus quilimanensis* (an oligochaete), *Neotragus pygmaeus* (royal antelope), *Pandaka pygmaea* (the smallest fish).

L. *quantillus;* *quantulus*, how little, how small, dim. of *quantus*, how much:

L. *rarus*, thin, scanty, dispersed, sparse; see **few**

Gr. *rhathaminx*, drop, grain, bit; see **drop**

L. *salaputium*, manikin, dwarf; see **man**

L. *scintilla*, spark, glimmer, trace; see **fire**

Gr. *skopaios*, m. dwarf: *Scopaeothrips unicolor* (a thysanopterid), *Scopaeus delicatissimus* (a beetle).

ME. *tit*, a small thing: tit, titbit, titmouse, titlark, titling, tomtit.

L. *trica*, hindrance, trifle; see **trifle**
L. *trunculus*, small piece, bit; see *truncus* under **short**
Gr. *tytthos*, little, small, young: *Tytthus pygmaeus* (a bug), *Tytthocrinus comptus* (a Pennsylvanian crinoid), *Tytthonyx rubidus* (a beetle).
L. *vegrandis*, not large, little:
L. *vescus*, weak, poor, thin, little; *vesculus*, dim.; see **weak**

DIMINUTIVES

Besides literal littleness, as *metula*, small pyramid, noun diminutives may connote endearment, as *amicula*, dear feminine friend; and belittlement, as *regulus*, petty king. Adjective diminutives indicate rather or somewhat, as *acidulus*, rather or somewhat sour; *nigriculus*, blackish. The latter idea was effected in Greek by means of the prefix *hypo*, under, less than, somewhat, as *hypoloxos*, somewhat oblique; *hypoleukos*, whitish; and *hyperythros*, reddish. For intensiveness, diminutives may be doubled, as *area*, space; *areola*, small space; *areolula*, very small space; *liber*, book; *libellus*, small book; *libellulus*, very small book. Latin diminutives have the same gender as the parent nouns, but Greek diminutives, regardless of the gender of the parent nouns, are neuters in *-ion*, masculines in *-iskos*, or feminines in *-iske*. In this book diminutives have been listed with the parent nouns and adjectives, but some may have been omitted inadvertently. Many are also absent from the dictionaries. This, however, should not deter the wordcoiner from inventing them according to analogy with others of the same kind. Thus, the Latin dictionaries record no diminutive of *fusus*, spindle. On the analogy with *rivus*, stream, *rivulus*, streamlet, *fusulus*, little spindle, is indicated. Similarly, *cirratus*, curled, becomes *cirratulus*, somewhat curled. The Greek dictionaries record no diminutive of *odous, odontos*, tooth. On the analogy with *ous, otos*, ear, *otarion; otion*, little ear, the forms *odontarion, odontidion, odontion*, and *odontiskos*, little tooth, are indicated.

L. *-aster, -tra, -trum; -ister*, diminutive suffixes with derogatory implication; wild; resemblance: oleaster, poetaster, apiastrum, capistrum, raphanistrum, rapistrum, salicastrum, siliquastrum, sinapistrum, *Urticastrum divaricatum* (wood-nettle), *Trianthema portulacastrum* (horse-purslane), *Aspidistra elatior* (common aspidistra).

L. *-cle, -culus, -el, -ellus, -illus, -olus, -ulus, -unculus*, diminutive suffixes; when adjectival they mean rather or somewhat: animalcule, armadillo, bagatelle, canticle, Capella, carbuncle, caruncle, chapel, coccinellid, cockerel, codicil, corpuscle, cuticle, curriculum, damsel, disciple, domicile, fennel, flail, flotilla, fumarole, gladiolus, granule, homunculus, lapillus, lenticel, libel, lintel, male, miracle, modiolus, nucellus, nucleolus, operculum, oracle, ossicle, patella, peccadillo, pedicel, pedicle, peduncle, pencil, peril, petiole, pickerel, pinnacle, pistil, quadrille, radicle, receptaculum, siphuncle, spectacle, spicule, umbrella, uncle, vacuole, vehicle, *Carex cristatella* (a sedge), *Limosella subulata* (a mudwort), *Mimulus moschatus* (a monkey-flower), *Pennatula borealis* (a seapen), *Pipunculus cingulatus* (a fly), *Porcellio liliputanus* (a sowbug), *Ranunculus nemorosus* (a buttercup), *Setophaga ruticilla* (redstart), *Vorticella microstoma* (a protozoan).

The following examples illustrate the formation of nounal and adjectival diminutives by using these suffixes:

aqua, water, *aquula*, little water, small stream; *filia*, daughter, *filiola; lamina*, plate, *lamella; libra*, balance, *libella; mica*, crumb, *micula; porta*, gate, *portula; puella*, girl, *puellula; sica*, dagger, *sicula; solea*, slipper, *soleola; trua*, ladle, *trulla.*

ager, field, *agellus; alveus*, cavity, *alveolus; amicus*, friend, *amiculus; atrium*, hall, *atriolum; circus*, ring, *circulus; conus*, cone, *coniculus, conulus; filius*, son, *filiolus; gladius*, sword, *gladiolus; oleaster*, wild olive, *oleastellus; puer*, boy, *puellus; ramus*, branch, *ramulus; rivus*, stream, *rivulus; sparus*, bream, *sparulus; torus*, knot, *torulus.*

animal, beast, *animalculum; arbor*, tree, *arbuscula; auris*, ear, *auricula;*

calx, pebble, *calculus; caput*, head, *capitellum, capitulum; caro*, flesh, *caruncula; cliens*, customer, *clientulus; cor*, heart, *corcillum, corculum; corpus*, body, *corpusculum; crus*, leg, *crusculum; cutis*, skin, *cuticula; flos*, flower, *flosculus; flumen*, stream, *flumicellum; frater*, brother, *fraterculus; homo*, man, *homunculus; ignis*, fire, *igniculus; jus*, juice, *jusculum; lapis*, stone, *lapillus; leo*, lion, *leunculus; mas*, male, *masculus; mons*, mountain, *monticellus, monticulus; navis*, ship, *navicula; nubes*, cloud, *nubecula; nux*, nut, *nucella, nucula; ops*, work, *opusculum; os*, mouth, *osculum; pater*, father, *paterculus; pes*, foot, *pedicellus, pediculus; radix*, root, *radicula; rete*, net, *reticulum; rex*, king, *regulus; sedes*, seat, *sedecula; senex*, old man, *seniculus; sus*, hog, *suculus*, m., *sucula*, f.; *turris*, tower, *turricula; vertex*, whirl, *verticillus; virgo*, maid, *virguncula.*

cornu, horn, *corniculum; domus*, house, *domucula, domuncula; genu*, knee, *geniculum; lacus*, lake, *lacusculus; manus*, hand, *manicula, manciola.*

dies, day, *diecula; res*, thing, *recula; spes*, hope, *specula.*

aureus, golden, *aureolus; comatus*, long-haired, *comatulus; ferus*, wild, *fericulus; formosus*, beautiful, *formosulus; frigidus*, cold, *frigidulus; majus*, larger, *majusculus; minus*, smaller, *minusculus; morus*, dark like a mulberry, *morulus; parvus*, small, *parvulus; pronus*, inclined forward, *pronulus.*

glaber, smooth, *glabellus; niger*, black, *nigellus, nigriculus; pulcher*, beautiful, *pulchellus; ruber*, red, *rubellus; scaber*, rough, *scabellus; tener*, soft, *tenellus.*

acer, sharp, *acriculus; levis*, light, *leviculus; tristis*, sad, *tristiculus; vetus*, old, *vetulus, vetusculus.*

These adjectival diminutives take the regular *-us*, *-a*, *-um* gender endings.

NL. *-essus*, *-a*, *-um*, diminutives; see *mikros* under **little**
F. *-et*, *-ette*, *-let*, *-ot*, diminutives: armlet, ballot, banquet, bracelet, brooklet, bucket, budget, bullet, cabinet, chariot, cigarette, clarinet, cutlet, etiquette, eyelet, gauntlet, gazette, hamlet, kitchenette, leaflet, mignonette, minuet, Henrietta, palmetto, parrot, pellet, Pierrot, pullet, ringlet, rivulet, rosette, stiletto, streamlet, tablet, toilet, turret.   Derive: amulent, coverlet.
L. *-idium* (Gr. *-idion*); *-ium* (Gr. *-ion*); Gr. *arion; -ydrion; -yllion*, neuter diminutives: antheridium, bacterium, basidium, chromidium, cnemidium, coccidium, ctenidium, deltidium, idyl, sporidium, stephanion, *Cymbidium grandiflorum* (an orchid), *Dacrydium intermedium* (a conifer), *Hipparion antelopinum* (a fossil horse), *Lepidium campestre* (pepperweed), *Pteridium caudatum* (a bracken).
   The following examples illustrate the formation of diminutives by using these suffixes:
gephyra, bridge, *gephyrion*, little bridge; *glossa*, tongue, *glossarion, glossidion; oikia*, house, *oikidion, oikion; pelte*, shield, *peltarion; petra*, stone, *petridion; skene*, tent, *skenidion; skia*, shade, *skiadion; zone*, girdle, *zonarion, zonion.*

anthropos, man, *anthroparion, anthropion; hippos*, horse, *hipparion, hippidion; logos*, word, *logarion, logidion, logydrion; potamos*, river, *potamion; osteon*, bone, *ostarion.*

alopex, fox, *alopekion; anthos*, flower, *anthion, anthyllion; aspis*, shield, *aspidion; aster*, star, *astrion; bous*, ox, cow, *boidarion, boidion; dory*, spear, *doration, dorydion; drys*, oak, *dryarion; epos*, verse, *epyllion; gala*, milk, *galaktion; geron*, old man, *gerontion; gony*, knee, *gonation; hepar*, liver, *hepation; hydor*, water, *hydation; ichthys*, fish, *ichthydion; kreas*, flesh, *kreadion; lepis*, scale, *lepidion, lepion; mastix*, whip, *mastigion; melos*, song, *melydrion; onyx*, claw, *onychion; phleps*, vein, *phlebion; pyr*, fire, *pyridion; salpinx*, trumpet, *salpingion; skelos*, leg, *skelydrion; sphen*, wedge, *sphenarion; stoma*, mouth, *stomion.*
L. *-ina*, feminine suffix sometimes with diminutive implication; see **woman**
L. *iscus* (Gr. *-iskos*), masculine diminutives, as *anthropos*, man, *anthropiskos*, manikin; *aspis*, shield, *aspidiskos; aster*, star, *asteriskos; basileus*, king, *basiliskos; kreas*, flesh, *kreiskos; potamos*, river, *potamiskos; sphen*, wedge, *spheniskos*: asterisk, obelisk, basilisk, lychnisk, meniscus, *Oniscus ornatus* (a sowbug), *Spheniscus trifasciatus* (a penguin), *Trochiliscus bilineatus* (a fossil characead), *Pteronisculus cicatrosus* (a Triassic fish), *Basiliscus mitratus* (a lizard).
Sp. *-ita*, feminine diminutives: senorita, esquelita, estrellita, Paquita.
D. *-kin*, diminutives: napkin, lampkin, catkin, jerkin, pipkin, manikin, firkin, Kopkins, Jenkins, Tompkins.   Derive: pumpkin, welkin, bumpkin.
AS. *-ling*, diminutives; having the quality of, pertaining to: bitterling, changeling, darkling, darling, fingerling, footling, foundling, gosling, **groundling**, hireling, nestling, nurseling, sapling, scantling, spiderling, starveling, stripling, suckling, titling, twinkling, underling. Derive: cymling (simling), inkling.

AS. *-ock,* < *-oc, -uc,* diminutives: bullock, hillock, paddock, Pollock.   Derive: haddock, havoc, hammock, mattock, hemlock, shamrock, wedlock.

L. *-uncio,* diminutives, as *homuncio,* little man:

See: **few, thin, part, trifle, low, lessen, poor, stingy**

litura, L. erasure, blot, blur; see *lituro* under **cancel**

liturgus, L. (Gr. *leitourgos*), public servant, minister; see **servant**

litus; linitus, L. daubed, smeared, anointed, see *lino* under **spread**; *litus, -oris,* shore, see **shore**

lituus, L. curved staff or wand of an augur; curved trumpet; see **rod**

live < AS. *lifian;* see **life**

liver < AS. *lifer:* liverleaf, liverwort, liverwurst.

"Is life worth living? That depends upon the liver."

L. *ficatum,* n. liver of an animal fattened on figs:

L. *hepar, -atis* (Gr. *-atos*), n. liver; *hepation,* n. dim.; *hepaticus* (Gr. *hepatikos*), pertaining to the liver: hepatic, heparin, Hepaticae, hepatitis, hepatodynia, hepatolith, hepatectomy, *Hepatica triloba* (hepatica), *Hepaticola hepatica* (a nematode), *Cymbalaria hepaticaefolia* (a basket-ivy).

L. *jecur, -oris,* n. liver; *jecusculum,* n. dim.; *jecorosus,* with liver trouble: jecorin, jecoral.

lividus, L. bluish, black and blue; *livor,* bluish or lead color; see **blue**

livistona, NL. a genus of palms; see **palm**

lix, *licis,* L. ashes, lye; *lixivia,* lye; see **alkali**

lixa, L. sutler, camp-follower; see **follow**

lizard < OF. *laisard,* < L. *lacerta,* f.: lacertoid, Lacertilia, alligator, *Lacerta muralis* (wall-lizard), *Alligator mississippiensis* (American alligator), *Siren lacertina* (mud-eel).

Sp. *agama,* f. a genus of lizards: *Agama stellio* (starred lizard).

L. *amphisbaena* (Gr. *amphisbaina*), f. a fabulous serpent having a head at each end of the body; now applied to a genus of lizards: *Amphisbaena fuliginosa* (a legless lizard).

NL. *anolis,* m. < Ab.Am. *anoli,* a kind of lizard: *Anolis carolinensis* (American chameleon).

Gr. *askalabos; askalabotes,* m. spotted lizard: Ascalabota, *Ascalabotes ocellatus* (a lizard).

L. *basiliscus* (Gr. *basiliskos*), a lizard; see *basileus* under **govern**

L. *caecilia,* f. a kind of lizard; now applied to a genus of amphibians: *Caecilia tentaculata* (a çaecilian).

L. *chalcis, -idis* (Gr. *chalkis, -idos*), f. a lizard with copper-colored spots: *Chalcides ocellatus* (sand-skink).

L. *chamaeleon* (Gr. *chamaileon*), n. a kind of lizard: *Chameleon cristatus* (a chameleon), *Lopholatilus chamaeleonticeps* (tilefish).

Gr. *champsos,* m. crocodile: *Champsosaurus inflatus* (a fossil lizard), *Champsodon fimbriatus* (a fish).

L. *colotes* (Gr. *kolotes*), m. a kind of lizard: *Colotes maculatus* (a beetle).

L. *crocodilus* (Gr. *krokodeilos*), m. lizard: crocodilian, crocodility, crocodile (*Crocodilus niloticus*).

L. *draco, -onis; -ontis,* m. (Gr. *drakon, -ontos,* m.; *drakaina,* f.), a fabulous, lizardlike animal; *dracunculus,* m. dim.; *draconteus,* of a dragon: dragon, dracontic, dracontiasis, Draco, dracunculus, Dracula, firedrake, *Draco volans* (flying dragon), *Dracaena draco* (dragonsblood-tree), *Dracotettix monstrosus* dragon-grasshopper), *Draconetta xenica* (a fish), *Dracontomelum mangiferum* (a dragon-plum), *Dracunculus vulgaris* (an aracead), *Pegasus draconis* (a flying-sea-horse), *Dracocephalum speciosum* (Himalayan dragonhead), tarragon (*Artemisia dracunculus*).

Gr. *galeotes,* m. a kind of lizard: *Galeotes lophyrus* (a lizard).

NL. *gavialis,* m. a kind of crocodile: *Gavialis gangeticus* (gavial).

NL. *gecko,* m. a kind of lizard: *Gecko verticillatus* (a gecko).

Ab.Am. *iguana,* f. a kind of lizard: *Iguana iguana*(iguana), *Celtis iguanaea* (a hackberry), iguanodon.

L. *languria,* f. a kind of lizard; *langurium,* n. a kind of amber or probably some amber-colored stone said to have been formed from lizard urine:

Gr. *pipalis, -idos,* f. a kind of lizard:

Gr. *sauros,* m.; *saura,* f. lizard; *sauridion,* n. dim.: saurian, sauroid, saurichnite, sauropod, dinosaur, ichthyosaur, pterosaur, plesiosaur, *Saurocoris instans* (a bug), *Saurophthalmus oxyops* (a beetle), *Saururus cernuus* (lizardtail), *Scymnosaurus ferox* (a fossil reptile), *Scombresox saurus* (saury).   Derive: thesaurus.

L. *scincus* (Gr. *skinkos*), m. a kind of lizard: skink *(Scincus officinalis)*, scincoid, *Scincopus fasciatus* (a lizard), *Scincosaurus crassus* (a stegocephalian).

L. *seps, sepis* (Gr. *-os*), m. a kind of serpent, lizard: *Seps marmoratus* (a lizard), *Melanoseps ater* (a skink), *Batrachoseps attenuatus* (an amphibian).

Gr. *souchos*, m. Egyptian name for the crocodile: *Suchosaurus cultridens* (a lizard), *Rhamphosuchus crassidens* (a fossil reptile).

L. *stellio, -onis*, m. spotted lizard: *Stellio vulgaris* (a lizard).

Ab.Am. *teiu;* Pg. *teju*, a kind of lizard: teju *(Tupinambis teguixin)*, teguexin, teioid.

Gr. *typhlinos*, blindworm; see *typhlos* under **blind**

NL. *varanus*, m. a genus of lizards: *Varanus niloticus* (monitor-lizard), *Tejovaranus branicki* (a lizard).

Gr. *zignis, -idos*, f. a kind of lizard: *Cicigna ornata* (a lizard).

load < AS. *hlod;* see **weight**

loaf < AS. *hlaf;* see **bread**

loan < ON. *lan;* AS. *leon*, lend; see **trade**

loathe < AS. *lath*, hateful; see **hate, dirt, shame, bad**

lobe < L. *lobus* (Gr. *lobos*), a rounded projection or protuberance; *lobion*, dim., see **projection;** *lobe*, ill usage, outrage; *lobetos*, outraged, see **hurt**

lobelia, NL. f. type genus of the Lobeliaceae, < Matthias Lobel, Flemish botanist: *Lobelia cardinalis* (cardinal-flower).

localis, L. of a place; *localitas; locatio*, place; see *locus* under **place**

localo- < Gr. *lokalos*, probably an owl; see **owl**

lochia, L. (Gr. *locheia*), childbirth, delivery; *lochios*, of childbirth; see **birth**

lochme, Gr. bush, copse, thicket; *lochmaios; lichmios*, of thickets; *lochmodes, bushy;* see **bush**

lochos, Gr. a place for lying in wait, ambush; see **trap**

lock < AS. *loc;* see **bar, key, close, guard**

loculus; locellus, L. cell, compartment, box, purse, dim. of *locus*, place; see **box**

locuples, -etis, L. substantial, rich, opulent, wealthy; see **wealth**

locus, L. place; *locellus; loculus*, dim.; *loco, -atus*, quarter, place, put; see **place**

locust < L. *locusta*, lobster, crab, grasshopper; see **grasshopper, cicada, bean**

locutio, L. speech, utterance; *loquax, -acis*, talkative, garrulous, < *loquor, -cutus*, speak; see **speak**

lodix, -icis, L. coverlet, blanket; *lodicula*, dim.; see **blanket**

loedoro- < Gr. *loidoros*, abusive, railing; see *loedoria* under **curse**

loego- < Gr. *loigos*, ruin, havoc; *loigios*, ruinous; see **destroy**

loemo- < Gr. *loimos*, plague; see **disease**

loepo- < Gr. *loipos*, remaining, left; see *leipo* under **abandon**

loestho- < Gr. *loisthos*, last; see **after**

-log; -logue < Gr. *lego*, gather, choose, speak, reason; see **gather, speak, choose**

logania, NL. f. type genus of the Loganiaceae, < James Logan, secretary to William Penn: *Logania longifolia* (a logania).

NL. *gelsemium*, n. a genus of the logania family: *Gelsemium sempervirens* (yellow jessamine).

Gr. *strychnos*, m., f. nightshade; now a genus of the logania family: *Strychnos toxifera* (curare plant), strychnine.

loganion, Gr. dewlap of cattle; see **flap**

logicus, L. (Gr. *logikos*), rational, sensible; see *-logy* under **know**

logimos, Gr. notable, remarkable; see **worth**

logos, Gr. word, discourse; *logion*, saying; see **word**

-logy, knowledge of, science, study of, < Gr. *lego*, choose, speak; see **know**

loin < OF. *logne*, side between hip and ribs; see **side**

loiter < D. *leuteren*, linger, lag; see **stand, delay**

loligo, L. cuttlefish, squid; see **mollusk**

lolium, L. darnel, cockle, tares; a kind of grass; *loliaceus*, of darnel; see **grass**

loma, Gr. fringe, hem; *lomation*, dim.; see **border**

lomentum, L. bean meal; name for an indehiscent bean pod; see **bag**

lonchitis, Gr. a plant with spear-shaped leaves; see **orchid**

loncho- < Gr. *lonche*, spearhead, lance, spear, javelin; *loncharion; lonchidion*, dim.; *lonchaios*, of a spear; see **spear**

long < AS. *long* (G. *lang*): longbow, longhorn, longspur, Longfellow, Longman, length, linger, furlong.   Derive: headlong, sidelong.

L. *anfractus,* bending, winding, circuitous, prolix; see **bend**

L. *chronius* (Gr. *chronios*), lasting long, enduring, lingering; see *chronos* under **time**

Gr. *deros,* long (time), too long: *Deroderus vestitus* (a beetle), *Deropygus jocosus* (a beetle).

Gr. *detha,* for a long time:

L. *diutinus; diuturnus,* lasting long; *diutius,* longer:

Gr. *dolichos,* long: dolichocephalic, dolichurus, *Dolichopus hastatus* (a fly), *Dolichorhinus fluminalis* (a titanothere).

Gr. *ekteino,* stretch out, spread; see *teino* under **spread**

Gr. *erizoos,* long-lived; see *zoe* under **life**

Gr. *grammikos,* linear; see *gramme* under **line**

L. *longus,* long; *longior; longius,* longer; *longissimus,* longest; *longiusculus; longulus,* dim.; *longiquus,* long, extensive; *longitudo, -inis,* f. length; *longurio, -onis,* m. tall man; *elongatus,* prolonged; *oblongus,* longer than broad: longevity, longitude, elongated, oblong, prolonged, purloin, *Longicella mollis* (a moth), *Cyperus longus* (galingale), *Diabrotica longicornis* (corn-root worm), *Loxoconcha longipes* (an ostracode), *Bathynectes longispina* (a crab), *Dolichosoma longissimum* (a fossil amphibian), *Spinifex longifolius* (a grass), *Hypseloconus elongatus* (a fossil gastropod), *Stenopogon elongatissimus* (a fly), *Erimyzon oblongus* (a sucker), *Echinometra oblongata* (a sea-urchin), *Usnea longissima* (a lichen).   Derive: So long!

Gr. *makros,* long; often mistakenly used to mean large; *makroteros; masson, -os,* longer; *makrotatos; mekistos,* longest; *makraion, -os,* lasting long: macron, macrobian, macrostylous, macrocephalic, *Macropus rufus* (red kangaroo), *Zenaidura macrura* (mourning-dove), *Macrochelys temmincki* (alligator-snapper), *Macrobiotus schultzi* (a tardigrade), *Macrotis lagotis* (rabbit-bandicoot), *Mecistocephalus longichiliatus* (a centiped). Examples of misuse of *makros* to mean large: macrocosm, macrospore, *Cupressus macrocarpa* (Monterey cypress), *Magnolia macrophylla*(large-leaved magnolia), *Vaccinium macrocarpum* (cranberry).

Gr. *mekos,* n. length; *mekyno,* prolong; *eumekes,* of good length, tall, great; *katamekes,* lengthwise; *palimmekes,* as long again; *pammekes; perimekes; hypermekes,* very long; *paramekes,* oblong, oval: mecometer, Mecoptera, paramecium, *Mecynoceras rex* (a fossil cephalopod), *Catamekes thiophora* (a moth), *Eumeces fasciatus* (blue-tailed skink), *Palimmeces ithystica* (a moth), *Paramecocnemis erythrostigma* (a dragonfly), *Promeces(Metameces) suturalis* (a beetle).

L. *procerus,* tall, slender, long; see **high**

L. *prolatus,* extended, elongated: prolate, *Gentiana prolata* (a gentian).

L. *prolixus,* stretched out long: prolix, *Phthoa prolixa* (a phasmid).

L. *sesquipedalis,* excessively long; see *semi-* under **half**

Gr. *sychnos,* long (time), many; see **number**

Gr. *tany-,* long, < *tanyo,* stretch out; *tanaos,* outstretched, long, tall; *tanymetros,* of long measure: *Tanytarsus muticus* (a midge), *Tanyodes ochracea* (a moth), *Tanaognathus spinipes* (a mite), *Tanysiptera nympha* (a kingfisher).

See: **high, upright, spread, time**

longaevus, L. aged, ancient, old; see **old**

longano, L. straight gut, sausage; see **intestine**

longurius, L. long pole, rod; see **rod**

lonicera, NL. a genus of the honeysuckle family; see **honeysuckle**

look < AS. *locian;* see **see, face, like, ask, hunt, try**

loom < AS. *geloma;* frame for weaving; see **frame**

loon < a Scan. word.

L. *gavia,* f. a bird; now applied to loons: *Gavia stellata* (red-throated loon), *Gavia immer* (common loon).

loop < uncertain origin; see **bend, fold, wave, leap, trap**

loose < AS. *leas;* see **free, thin**

loosestrife

NL. *cuphea,* f. a genus of the loosestrife family: *Cuphea miniata* (a cuphea).

NL. *lawsonia,* f. a genus of the loosestrife family, < John Lawson, Scotch traveler: *Lawsonia inermis* (henna).

L., Gr. *lysimachia,* loosestrife; see **primrose**

L. *lythrum* (Gr. *lythron,* gore), n. a genus of the loosestrife family: Lythraceae, *Lythrum alatum* (winged loosestrife).

L. *peplis, -idis* (Gr. *-idos*), f.; *peplion,* n. name for several plants, but now a genus of the loosestrife family: *Peplis portula* (water-purslane), *Peplidium maritimum* (a figwort), *Arenaria peploides* (seabeach-sandwort).

lopas, *-ados,* Gr. flat vessel, plate; *lopadion,* dim.; see **plate**

lophema, Gr. rest; *lophao,* rest from toil, abate; see **rest, lessen**

lophido- < Gr. *lophis, -idos; lopheion,* a kind of case; see **box**

lophnido- < Gr. *lophnis, -idos,* a kind of torch; *lophnidion,* dim.; see **light**

lopho-; lophio- < Gr. *lophos; lophia,* mane, crest, comb, tuft; see **crest**

lopo- < Gr. *lope,* cloak, mantle, robe; *lopion,* dim., see **garment**; < *lopos,* husk, peel, see *lepis* under **scale**

lopodytes, Gr. thief; see **steal**

loquax, *-acis,* L. talkative; see *loquor* under **speak**

lordos, Gr. bent backward; see **bend**

lorica, L. leather cuirass or corselet; *loricula,* dim.; see **harness**

lorius, NL. < Malay *luri,* a kind of parrot; see **parrot**

lorum, L. strap, thong, scourge; see **strap**

lose < AS. *leosan:* lorn, forlorn, lovelorn, lost.
L. *amissus,* lost; *amissio, -onis,* f. loss; *amissibilis,* losable: *Stegotheca amissa* (a moth), amissible.
Gr. *apoktesis,* f. loss; *apoktetos,* lost:
L. *clades,* loss, damage, defeat; see **destroy**
Gr. *elattoma, -tos,* n. loss, defeat:
Gr. *hessa (hetta),* f. defeat:
L. *irreparabilis,* irrecoverable, irretrievable: irreparable.
L. *irrevocabilis,* that cannot be recalled, unchangeable; see **stand**
L. *orbo, -atus,* bereave; *orbus; orba,* orphan; see **alone**
Gr. *steresis,* f. deprivation, loss:
L. *strages, -is,* f. defeat, overthrow, ruin:
L. *viduus,* bereft; *vidua,* f. widow; *viduo, -atus,* deprive, cause to lose: void, avoidance, devoid, unavoidable, *Pipunculus viduus* (a fly), *Zodarion viduum* (a spider), *Schistoglossa viduata* (a beetle).
Gr. *zemia,* loss, damage; see **hurt**
See: **lessen, weak, death, fail**

lot < AS. *hlot:* allot, allotment, lottery, lotto.
L. *accidentia,* f. an unforeseen, unexpected event, chance: accident, accidental.
Gr. *aderkes,* unexpected; see **secret**
Gr. *adoketos,* unexpected, surprising: *Adocetus buprestoides* (a fossil beetle).
Gr. *aisios,* lucky, auspicious; *aisimos,* destined: *Aesiocrinus magnificus* (a Pennsylvanian crinoid).
L. *alea,* a game of chance with dice; *aleator,* dice-player, gamester; see **play**
Gr. *apotelesma, -tos,* n. influence of the stars on human destiny:
Gr. *aprophatos,* unforetold, unexpected: *Aprophata fausta* (a beetle).
L. *assigno, -atus,* allot, appoint: assign, assignation.
L. *auspicatus,* fortunate, favorable, lucky; see *auspex* under **prophecy**
L. *casus,* a falling down, event, accident, chance; *causalis,* fortuitous; see *cado* under **fall**
Gr. *diadexios,* presaging good luck: *Diadexia parodes* (a moth).
Gr. *dyspetes,* falling out badly, unfortunate: *Dyspetes praerogator* (a wasp).
Gr. *dyspraxia,* f. bad luck, misfortune:
Gr. *eikotos,* perhaps, probably, reasonably, naturally: *Icotopus arcurostris* (a decapod).
L. *emico, -atus,* appear suddenly, spring forth, become apparent; see **open**
Gr. *eulonchos,* fortunate, propitious: *Eulonchus smaragdinus* (a fly), *Eulonchopria psaenythioides* (a bee).
L. *eventum,* n. consequence, issue, result: event, eventually.
L. *fatum,* n. utterance of a god, prophecy, destiny, lot, death; *fatalis,* of fate, destined, fated; *fatidicus,* prophetic; m. prophet; *fatifer, -a, -um,* deadly, destructive: fate, fatal, fatality, fatalism, fairy, fay, Fayette, *Lycosa fatifera* (a spider).   The general term among the Romans for a Fate who presided over human destiny was *Parca,* f.; among the Greeks, *Moira,* f. Specifically, there were three Fates; L. *Nona;* Gr. *Clotho,* f. who spun life's thread; L. *Decuma;* Gr. *Lachesis,* f. who governed its length; and L. *Morta;* Gr. *Atropos,* f. who cut it off.
L. *faustus,* favorable, fortunate, lucky; *infaustus,* unfortunate, unlucky; *Rhaebothorax faustus* (a spider), *Cardamine infausta* (a bittercress), Faust.

L. *fors, fortis,* f. luck, chance; *fortuna,* f. fate, lot; *fortuitus,* by chance, accidental; *fortunatus,* happy, lucky, prosperous; *forsan; forsitan,* perchance, perhaps: fortune, fortunate, unfortunate, fortuitous, *Fortunella japonica* (round kumquat), *Anisonchus fortunatus* (a fossil mammal).

ON. *happ,* chance: hap, happen, happenstance, haphazard, happy, mayhap, mishap, perhaps.

Gr. *hermaion,* n. gift of Hermes, luck, windfall, godsend:

L. *improvidus,* not anticipating or foreseeing; *improvisus,* unexpected, unforeseen: improvident, improvise, improvisation, *Didelphys improvida* (an opossum).

L. *inominalis,* unlucky, inauspicious:

L. *inopinus,* unexpected: *Glyptopleura inopinata* (an ostracode).

L. *insperatus,* unhoped for, unexpected:

Gr. *kleros,* m. lot, chance; clergy; *klerion,* n. dim.; *klerotikos,* of casting lots; *klerotos,* appointed by lot; *kleroma, -tos,* n. the thing allotted; *dyskleros,* unlucky: cleromancy, cleruch, clerk, Clark, clergyman, clerical, *Clerodendron speciosissimum* (a verbenacead), *Synclerostola pampeana* (a moth).

Gr. *kybos,* cube, die; *kybeutes,* gambler; *kybeutikos,* of dicing; *kybeuterion,* gambling-house; see *cubus* under **three**

Gr. *lachesis,* f. lot, destiny, *lanchano,* obtain by lot; destiny, < *lanchano,* obtain by lot; *lachos,* n. allotted portion; *lexis,* f. allotment; *syllexis,* f. selection by lot: Lachesis, *Lachesilla pedicularia* (a psocid), *Lachesis atrox* (a snake), *Syllexis chartaria* (a moth).

Gr. *methektos,* participating, sharing, partaking:

Gr. *moira,* f. part, share, lot, destiny; *Moira,* f. goddess of destiny; *moros,* m. doom, fate, destiny, share, portion; *morsimos,* destined; *amoros,* without share, unlucky; *dysmoros,* ill-fated; *dysmoria,* f. a hard fate: *Dysmorodrepanis munroi* (a bird).

L. *necopinus,* unexpected:

L. *obventicius,* accidental, adventitious:

L. *occasio, -onis,* f. opportunity, favorable moment: occasion, occasionally.

Gr. *oitos,* m. fate, doom: *Oetophorus oculatus* (a wasp).

Gr. *perikakesis,* f. very bad luck:

L. *phenomenon* (Gr. *phainomenon*), an appearance, happening, event; see **display**

Gr. *potmos,* m. lot, destiny, luck; *apotmos; dyspotmos,* unlucky, unhappy:

L. *repentinus,* sudden, hasty, unexpected, new:

L. *reperticius,* accidental, random:

L. *scaevus,* on the left, unlucky; see **hand**

L. *sors, sortis,* f. lot, share, fate, kind; *sorticula,* f. dim.; *sortio, -itus,* draw lots: sort, sortition, sorcery, sortilege, assortment, consort, resort, unassorted.

Gr. *symphora,* f. event, circumstance, chance, mishap; *symphoros,* happening with, accompanying: *Symphora rugosa* (a beetle), *Symphorostola encomias* (a moth).

Gr. *symptoma,* anything that befalls one, chance, mishap; see **mark**

Gr. *synantema, -tos,* n. incident, happening, occurrence:

Gr. *synteuxis,* f. coincidence:

Gr. *tyche,* f. good fortune, luck, chance, accident; *tychadion,* n. dim.; *tycheros,* lucky; *tychaios; tychikos,* casual, fortuitous; *atychema, -tos,* n. misfortune; *dystyches,* unlucky; *eutychia,* f. good luck; *syntychikos,* accidental: *Tychostylops marculus* (a fossil mammal), *Tycherus* (a wasp), *Atychia notata* (a moth), *Ectyche erebea* (a beetle), *Eutychia caffra* (a butterfly).

Gr. *xenisma,* amazement, strangeness, surprise; see *xenos* under **guest**

See: **fall, choose, separate, prophecy, cut, break, swift, play**

lota, NL. a kind of fish; see **cod**

lotax, Gr. buffoon; see **fool**

lotium, L. urine; see **urine**

lotus, L. (Gr. *lotos*), name of several plants, see **bean, waterlily;** < *lotus,* washed, see *lavo* under **wash**

loud < AS. *hlud.*

Gr. *alalage; alalagmos,* a loud noise; see *alalazo* under **call**

L. *convicium,* n. loud noise, clamor, sound of wrangling: convicium.

Gr. *erigerys,* loud-speaking; see *gerys* under **speak**

L. *fragor, -is,* loud noise; see **sound**

Gr. *gegonos,* loud-sounding, sonorous:

Gr. *kelados,* sound of rushing waters, loud noise; see **sound**

Gr. *krambos,* loud, ringing; see **dry**

Gr. *otobus,* loud, startling noise, din; see **sound**

L. *sonorus,* loud, noisy, resounding; see *sono* under **sound**

Gr. *stombos,* deep or loud-sounding: *Stombus* (an amphibian).

Gr. *thorybos,* cheers, uproar; see **sound**

See: **sound, roar, ring, thunder**

louse < AS. *lus.*
L. *pediculus,* m. dim. of *pedis,* m. louse; *pedicularis; pedicularius,* of lice; *pedicosus; pediculosus,* lousy: *Pediculus vestimenti* (body-louse, cootie), *Pedicularis lanceolata* (swamp-lousewort), *Rickettsia pediculi* (microorganic cause of trench fever).
Gr. *phtheir, -os,* m. louse; *phtheirion,* n. dim.; *phtheirikos,* of lice; *phtheirodes,* lousy: *Phthirothrips pediculus* (a thysanopterid), *Phthirius pubis* (crab-louse), *Phthiria variegata* (a fly), *Phtheirichthys lineatus* (a remora), *Phthiridium pediculare* (a fly).
NL. *psocus,* m. a genus of book-lice or cabinet-mites: Psocoptera, *Psocus venosus* (a bark-louse), *Psocidium robustum* (a Permian psocid), *Metapsocidium loxoneurum* (a Permian psocid), *Archipsocus puber* (a psocid).
Gr. *sathrax, -akos,* m. louse:
See: **flea, tick, spider**

love < AS. *lufe:* lover, lovely, loveliness, loveless, Lovelace, Lovewell, Lowell, Lovejoy.
L. *affectio, -onis,* f. favorable disposition toward anyone, love, goodwill; *affectuosus,* affectionate: affection, affected, affectionate.
Gr. *agape,* f. love, charity; *agapesis,* f. affection; *agapetos,* beloved; *agapema, -tos,* n. darling; *agapetikos,* affectionate: agapetid, *Agapostemon radiatus* (a bee), *Agapornis cana* (a lovebird), *Agapema galbina* (a moth).
L. *allubentia,* f. liking or fondness for:
L. *amor, -is,* m. love, < *amo, -atus,* love physically or passionately; *amabilis,* lovely, lovable; *amasio, -onis; amasius; amator, -is,* m.; *amatrix, -icis,* f. lover; *amasiunculus,* m. dim. fond lover; *amatorculus,* m. dim. poor lover; *amatio, -onis,* f. a loving, fondling, caressing; *amatorius,* loving; *amorabundus,* amorous; *amorificus,* awakening or causing love; *peramans,* very fond: amative, amorous, amateur, Amadeus, Amanda, Amy, Mabel, enamor, inamorata, *Isoloma amabile* (a gesneriacead).
L. *amplexus,* an embrace, caress; see **embrace**
Gr. *anakampseros,* an herb, the touch of which was said to restore lost love; see **purslane**
L. *animula; animulus,* little soul, darling; see *anima* under **life**
Gr. *Aphrodite,* f. goddess of love; *aphrodisios,* of Aphrodite; *aphrodisiakos,* arousing sexual desire; *aphrodisiasmos,* m. sexual relations, lustfulness; *epaphroditos,* lovely, charming; L. *Cnidus* (Gr. *Knidos; Gnidos*), f. Doric city with Praxitelean statue of Aphrodite: Aphrodite, aphrodisiac, hermaphrodite, *Aphrodite aculeata* (a pelecypod), *Argynnis aphrodite* (a fritillary), *Gnidia penicillata* (a thymeleacead).
Gr. *appha,* f. dear one; *appharion; apphidion,* n. dim.:
Gr. *Astarte,* f. Phoenician goddess of love and fertility: *Astarte castanea* (a pelecypod), astartine.
L. *benevolentia,* f. friendliness, kindness: benevolent, benevolence.
L. *carus,* dear; *caritas, -atis,* f. regard, esteem, affection, love: caress, cherish, charity, charitable.
L. *comis,* courteous, affable, kind, friendly, loving; see **friend**
L. *Cupido, -inis,* m. instigator and god of love, < *cupio,* desire, long for: Cupid, *Tympanuchus cupido* (heath-hen).
L. *delicius,* m.; *delicia,* f. favorite, darling, pet, sweetheart, beloved; *delicola,* f. darling:
L. *dilectio, -onis,* f. love; *dilector, -is,* m. lover; *diligo, -lectus,* esteem highly, love; *diligibilis,* estimable, amiable:
Gr. *Dione,* f. Aphrodite; *Dionaios,* of Aphrodite: *Dionaea muscipula* (a flytrap), *Hysteroconcha dione* (a pelecypod).
Gr. *eros, -otos,* m. love; *Eros,* instigator and god of love, < *erao,* love physically or passionately; *eroeis, -essa, -en,* lovely, charming; *erannos; erasmios; eratos,* beloved, lovely; *Erato,* f. the Muse of lyric poetry; *erastes,* m.; *erastria,* f. lover; *erotylos,* m. darling, sweetheart; *anerastos,* unloved, unloving; *eperatos,* beloved, lovely: Eros, erotic, erotomania, erotopathy, Erasmus, Eratosthenes, erogenous, pederasty, *Eragrostis secundiflora* (a grass), *Erato perexigua* (a Miocene gastropod), *Eratocrinus ramosus* (a Mississippian crinoid), *Crypteronia paniculata* (bekoi).
L. *favor, -is,* m. goodwill, inclination toward: favor, favorite.
L. *foveo, fotus,* warm, cherish, pet:
AS. *freon,* love: friend, friendship, free, Friday.
Gr. *himertos,* desired, lovely, longed for; see **beauty**
L. *indulgens,* kind, loving, conceding; *indulgentia,* f. gentleness, tenderness, fondness, forbearance: indulgence, indulgent.
Gr. *kedeios,* cared for, beloved:

Gr. *korizomai*, fondle, caress: hypocoristic.

Father (to son who wanted spending money): "I want you to let those **wild** women alone."

Son: "Oh, dad, they're not wild. They're tame. Anybody can pet them!"

Dean of women reads notice to student body: "The president and I have decided to stop necking on the campus."

Gr. *melema, -tos,* n. object of care, something beloved, darling:

L. *melliculus; mellilla; mellitulus; mellitus,* darling, honey; see *mel* under **honey**

OHG. *minna,* love: minnesinger, mignonette.

L. *monnula,* f. darling:

L. *mulsa,* f. sweetheart, honey:

L. *oculissimus,* dearest:

Gr. *pallakos,* lover; *pallakion,* dim.; see *pallax* under **young**

Gr. *philia,* f. friendly love, affection, fondness, < *phileo,* love as a friend, regard with affection, esteem; *philos,* beloved, dear; *philteros,* dearer; *philtatos,* dearest; *philemon, -os; philios,* loving, friendly; *philetikos,* disposed to love; *philerastos,* amorous; *philadelphia,* f. brotherly love; *philtron,* n. love-charm; *philetes; philetor, -os,* m.; *philetria,* f. lover; *philesis,* f. a loving, feeling of affection; *aphilos,* friendless, unloved; *pasiphile,* f. one loved by all: philosophy, philology, philanthropist, philately, philter, philanderer, Philander, Philadelphia, Philip, Philippines, bibliophile, hydrophilous, spermophile, toxophily, *Philodendron verrucosum* (an aracead), *Philanthus gibbosus* (a bee-killer wasp), *Carpolithus hemipterus* (dried-fruit beetle), *Ammophila arenaria* (a beach-grass), *Drosophila guttifera* (a fruitfly).

L. *pietas,* duty, affection, loyalty; see *pio* under **holy**

L. *Salambo, -onis,* f. Babylonian goddess of love:

Gr. *storge,* f. affection, love, < *stergo,* love; *stergema, -tos; stergethron,* n. love-charm; *sterktikos,* affectionate:

L. *studium,* assiduity, eagerness, devotion; see *studeo* under **busy**

L. *Venus, -eris (Cypria; Cypris, -idis,* < Cyprus, supposed birthplace of Venus; *Cytherea),* f. goddess of love; *venereus (venerius),* of Venus; *venustus,* charming, lovely, beautiful, graceful: venereal, cypriphobia, *dies Veneris* (F. *vendredi,* Friday), *Venus magnifica* (a clam), *Venericardia borealis* (heart-cockle), *Hexoplon venus* (a beetle), *Simulium venustum* (a gnat), *Chiloneus veneriatus* (a beetle), *Cypris fusca* (an ostracode), *Cypraea tigris* (a gastropod), *Cypripedium arietinum* (a moccasin-flower), *Spondias cytherea* (a hog-plum), *Osmia cypricola* (a bee).

L. *voluntas, -atis,* f. goodwill, favor, affection:

See: **draw, friend, beauty, woo, marry**

low < ON. *lagr.*

LL. *bassus,* low, deep: base, debase, bass, basset, bassoon, bas-relief.

Gr. *chamai,* on the ground, dwarf; *chamelos,* on the ground, creeping: chameleon, chamomile, *Chamaecyparis thyoides* (white-cedar), *Chamaelirium luteum* (devilsbit), *Chamaerops humilis* (hair-palm), *Urtica chamaedrys* (a nettle), *Chamaea fasciata* (wren-tit).

Gr. *chthamalos,* on the ground, low: *Chthamalus fissus* (a barnacle), *Chthamalopteryx melbournensis* (a fish).

L. *contritus,* worn down, broken, humble: contrite.

L. *dejectus,* sunk down, low, dispirited: dejected, dejection.

L. *demissus,* low, hanging down, drooping, feeble; see **hang**

L. *depressus,* pressed down, low, flat: *Ampullina depressa* (a fossil brachiopod), *Megistocrinus depressus* (a Devonian crinoid).

Gr. *geleches,* sleeping on the ground; see **sleep**

L. *humilis,* low, on the ground, shallow, poor; *humifusus,* low, procumbent: humble, humility, humiliation, humifuse, *Agrostis humilis* (a grass), *Begonia humilicaulis* (a begonia), *Asperula humifusa* (a woodruff), *Astroloma humifusum* (an epacridacead).

L. *inferus; infer,* low; *inferior,* lower; *infimus; imus,* lowest, last; *infernalis,* nether, lower; *infernus,* lower, beneath; *infimo, -atus,* make low, lower: infernal, inferior, inferiority, infima, *Gerrhonotus infernalis* (alligator-lizard).

Gr. *nerthe,* beneath, below; *nerteros; enerteros,* lower; *nerterios,* underground: *Nerthomma stictica* (a beetle).

L. *prociduus,* fallen down, prostrate: *Formica procidua* (an ant).

L. *procumbo,* fall forward, become prostrate; see *cubo* under **lie**

L. *pronus,* lying on the face, face downward; see **lie**

L. *prostratus,* down flat, overthrown, laid low: prostrate, prostration, *Ceanothus prostratus* (squaw-carpet ceanothus).

L. *subterior,* lower; see *sub* under **under**

L. *supinus,* lying on the back; see **lie**

Gr. *tapeinos,* low, humble, abject, poor: *Tapinocomus subnudus* (a beetle), *Tapinotarsus maculatus* (a beetle).

L. *supplex, -icis,* beseeching, humble:

See: under, **lie, down, creep, poor, base, earth**

loxos, Gr. slanting; see **slope**

lozenge < F. *losange,* a diamond-shaped figure; see **diamond**

lubricus, L. slippery; *lubricitas,* slipperiness; see **slip**

lucanus, L. a kind of beetle; see **beetle**

lucaris, L. of a grove; see *lucus* under **forest**

lucerna, L. lamp; *lucernula,* dim.; *lucernaris,* of a lamp; see *lux* under **light**

lucidus, L. full of light, shining, bright; see *lux* under **light**

lucifugus, L. light-shunning; see *fuga* under **depart**

lucius, L. pike; see **pike**

luck < D. *geluk* (G. *glück*); see **lot**

lucrum, L. gain; *lucellum,* dim.; *lucrativus; lucrosus,* gainful, profitable; see **gain**

luctator, L. wrestler; *luctamen, -inis,* contest; see *luctor* under **fight**

luctus, L. grief, distress, lamentation; *luctuosus,* mournful; see *lugubris* under **sad**

lucubro, -atus, L. work by lamplight; see *lux* under **light**

luculentus, L. full of light, splendid; see *lux* under **light**

Lucullus, L. a Roman consul and general famous for his wealth, luxury, and banquets; see **wealth**

lucuma, Ab.Am. a genus of the sapodilla family; see **sapodilla**

lucunculus, L. dim. of *lucuns,* a kind of pastry; see **cake**

lucus, L. sacred grove; *luculus,* dim.; *lucaris,* of a grove; see **forest**

ludibundus, L. playful; *ludicer, -cra, -crum; ludicrus,* sportive; *ludifico, -atus,* ridicule; see *ludo* under **play**

Ludovicus, ML. Ludwig (Louis); *ludovicianus,* of Louis: Louisa, Aloysius, *Lanius ludovicianus* (a shrike), *Carludovica palmata* (a cyclanthacead, source of material for Panama hats), *Martynia louisiana* (a devilsclaw).

lues, L. pestilence, plague; see **disease**

lugubris, L. doleful, mournful; see **sad**

lukewarm

Gr. *chliaros (chlieros),* tepid, lukewarm; *chliodes,* slightly warm: *Chliarostoma relecta* (a moth).

L. *egelidus,* lukewarm, tepid, chilly:

Gr. *metakeras,* lukewarm: *Metaceradocus perdentatus* (a crustacean), *Metacerapterocerus fortunatus* (a wasp).

L. *tepidus,* lukewarm: tepid, tepefaction, tepidarium, tepidity, *Theridium tepidariorum* (domestic spider).

Gr. *thermodes,* lukewarm:

See: **heat, cold**

luma, L. thorn; *lumarius,* of thorns; *lumectum,* thorn-thickets; see **thorn**

lumbago, L. pain in the small of the back; see **disease**

lumber < uncertain origin; see **beam, board, rod, wood**

lumbricus, L. earthworm; see **worm**

lumbus, L. loin; *lumbellus,* dim.; see **side**

luminosus, L. full of light; see *lumen* under **light**

lump < uncertain origin, but probably from D. *lompe,* mass, piece: lumper, lumpy, lumpfish *(Cyclopterus lumpus).*

L. *blatta,* f. bloodclot:

L. *bolus* (Gr. *bolos*), m.; *bolax, -akos,* f. lump, clod; *bolarion; bolion,* n. dim.; *bolinos,* made of clay: *Bolacothrips jordani* (a thrips), *Actinobolina vorax* (a protozoan).

AS. *clott* (G. *klotz*), lump, piece, mass: clot.

L. *gleba,* f. clod, lump; *glebula,* f. dim.; *glebalis; glebarius,* of clods; *glebosus; glebulentus,* full of clods, lumpy: glebe, glebulose, glebous, gleba, *Unio(Glebula) rotundata* (a pelecypod), *Lecidea glebulosa* (a lichen).

L. *massa,* f. quantity, bulk, body, mass, lump; *massula,* f. dim.; *massalis,* of a mass: mass, massive, massif, massula, mass meeting.

Gr. *megethos,* greatness, size; see *magnus* under **large**

Gr. *onkos,* bulk, mass, weight, tumor; *onkeros; onkotos,* swollen; see **swell**

Gr. *platytes,* breadth, amplitude, bulk; see *platys* under **broad**

Gr. *solos,* a mass of iron; see **iron**

Gr. *thrombos,* m. lump, piece, bloodclot; *thrombion,* n. dim.: thrombus, thrombosis, *Trombidium[Thrombidium] tinctorium* (a harvest-mite).

L. *tumor,* a swelling, lump; see *tumeo* under **swell**

See: **ball, knot, heap, projection, swell, large**

luna, L. moon; *lunula,* dim.; *luno, -atus,* bend into a crescent or half-moon; see **moon, crescent**

lung < AS. *lungen.*

Gr. *pneumon, -os,* m. lung: pneumonia, pneumococccus, pneumectomy, pneumolith, pneumonic, pneumonocele, pneumonitis.

L. *pulmo, -onis,* m. lung; *pulmunculus,* m. dim.; *pulmonarius; pulmoneus,* of the lungs: pulmonary, pulmotor, pulmogastric, pulmonate, pulmoniferous, *Pulmonaria saccharata* (lungwort), *Lobaria pulmonaria* (a lichen).

See: **breathe**

lupanar, L. brothel; *lupanus,* lewd; see *lupa* under **prostitute**

lupinus, L. a genus of the bean family; see **bean**

lupulus, L. hop; see **hop**

lupus, L. wolf; *Luperca,* deified she-wolf that suckled Romulus and Remus; see **wolf**

lura, L. sack; see **bag**

lurco, L. gourmand, glutton; see **eat**

lure < F. *leurre;* see **draw, trap**

luridus, L. pale yellow, dull red, ghastly; see **yellow**

luscinia, L. nightingale; *lusciniola,* dim.; see **nightingale**

luscinus; luscus, L. one-eyed, with one eye shut, half-blind; *luscitiosus,* dim-sighted; see **dim**

lusor, L. player; *lusorius,* of a player; see *ludo* under **play**

lusso- < Gr. *lousson,* pith of a coniferous tree; see **pith**

lust < AS. *lust,* pleasure, longing; see **lewd, wish, heat, joy,** *mania* under **mad**

luster < F. *lustre,* < L. *lustro, -atus,* illuminate; see **light**

lusto- < Gr. *loustes,* one fond of bathing; see *loutron* under **wash**

lustrago, L. a kind of vervain; see **vervain**

lustralis, L. relating to purification from guilt; *lustrum,* a purificatory sacrifice; see **pure**

lusus, L. play, game, sport, fun; *lusor,* player; see *ludo* under **play**

luter, L. (Gr. *louter*), bathtub, basin; *louterion,* dim.; see *loutron* under **wash**

luteus, L. yellow; *luteolus,* dim. yellowish; see **yellow**

lutra, L. otter; *lutreola,* dim.; see **weasel**

lutro- < Gr. *loutron,* bath; see **wash**

lutum, L. mud, mire, clay; *lutulentus; lutosus,* muddy; see **earth**

lutus, L. washed; see *luo* under **wash**

lux, *lucis,* L. light; *lucifer,* light-bringing; *lucificus,* light-making; see **light**

luxus, L. dislocated, < *luxo, -atus,* put out of joint, dislocate; see **separate**

luxury < L. *luxuria,* profusion, rankness, excess, extravagance, delicacy; see **abundance**

luzula, NL. a genus of the Juncaceae; see **sedge**

lycaeno- < Gr. *lykaina,* she-wolf; see *lykos* under **wolf**

lyceum < Gr. *Lykeion,* n. a famous gymnasium at Athens; see **school**

lychnis, Gr. a plant with scarlet flowers; now a genus of the pink family; see **pink**

lychnos, Gr. lamp, light; *lychnidion; lychnion; lychniskos,* dim.; see **light**

lycium, L. a thorny plant of Lycia; see **nightshade**

lyco- < Gr. *lykos,* wolf; see **wolf**

lycoctonum, L. (Gr. *lykoktonon*), wolfbane; see **aconite**

lycopersicum, NL. (Gr. *lykopersion*), tomato; see **nightshade**

lycopodium, NL. clubmoss; see **clubmoss**

lycopsis, Gr. wild bugloss; see **borage**

lycopus, NL. a genus of the mint family; see **mint**

lycosa, NL. a genus of spiders; see **spider**

lyctus, NL. a genus of beetles; see **beetle**

Lycurgus, L. (Gr. *Lykourgos*), Spartan lawgiver; see **law**

lygdos, Gr. white marble; *lygdinos,* of white marble; see **lime**

lyge, Gr. twilight; *lygaios,* gloomy, shadowy; see **shade**

lygeros, Gr. pliant, flexible; *lygistos,* bent, pliable; see **bend**

lygistes, Gr. basket-maker; see *lygos* under **willow**

lygos, Gr. willow twig, tree with pliant twigs; *lyginos,* of willow twigs; *lygodes,* willowlike; see **willow**

lygros, Gr. baneful, harmful; see **hurt**

lyma, Gr. filth, dirt; see **dirt**

lymanterios, Gr. destructive; *lymanter,* destroyer; *lyme,* outrage, ruin; see **destroy**

lympha, L. clear water, see **water**; *lymphatus,* distracted, mad, see **mad**

lyncurium, L. (Gr. *lynkourion*), a gem; see **stone**

lyngo- < Gr. *lynx, lyngos,* hiccup; see **hiccup**

lynx, *lyncis,* L. (Gr. *lynx, -ynkos*), wildcat; *lynkion,* dim.; see **cat**

lype, Gr. pain, grief; *lyperos,* painful, distressing; see **pain**

lypros, Gr. poor, wretched; *lyprotes,* poverty; see **poor**

lyre < L., Gr. *lyra,* a stringed instrument; *lyricen,* harpist; see **harp**

lyron, L. a kind of alisma; see **waterweed**

lysichitum, NL. a genus of the arum family; see **arum**

lysimachia, L., Gr. loosestrife; see **primrose**

-lysis; -lytic < Gr. *lyo,* loose, dissolve, break up; see **free**

lyssa (lytta), Gr. madness, rage, fury; *lyseter,* madman; *lyssetikos,* furious; see **mad**

lyter, Gr. deliverer; see *lyo* under **free**

lythrum, L. (Gr. *lythron,* gore), a genus of the loosestrife family; see **loosestrife, blood**

lytron, Gr. ransom; see **pay**

-lyze, suffix meaning to do or make; see **make**

M

Mac; Mc < C. *mac,* son of; see **son**

macaca, NL. a genus of monkeys; see **monkey**

macario- < Gr. *makarios; makaros,* happy; see **joy**

maccus, L. buffoon, simpleton; see **fool**

macellarius, L. pertaining to the meat or provision market; meat-seller, butcher, victualler; *macellum,* market; see **trade**

macello- < Gr. *makella,* pickax with one point; see **pick**

macer, *-cra, -crum,* L. lean, thin, poor; *macellus; macriculus,* dim.; see **thin**

maceria, L. wall, enclosure; see **wall**

macero, *-atus,* L. soften; see **soft**

machaera, L. (Gr. *machaira*), bent sword, dirk, dagger, knife; see **sword**

machetico- < Gr. *machetikos; machinos,* warlike, quarrelsome; *machetes,* fighter, warrior; see **fight**

machine < L. *machina* (Gr. *mechane*), engine, device; *machilla; machinula,* dim.; see **tool**

machlos, Gr. lustful, lewd; see **lewd**

macia, L. pimpernel; see **primrose**

macilentus, L. thin, lean, poor, meager; see *macer* under **thin**

mackerel < OF. *maquerel*.
L. *colias* (Gr. *kolias*), m. a mackerel: *Scomber colias* (chub-mackerel).
NL. *gempylus*, m. a mackerellike fish: *Gempylus serpens* (snake-mackerel).
L. *scomber, -bri* (Gr. *skombros*), m. mackerel: *Scomber scombrus*(mackerel), *Scomberomorus maculatus* (Spanish mackerel), *Chloroscombrus chrysurus* (casabe).
Gr. *trachouros*, m. a kind of mackerel: *Trachurops crumenophthalmus* (a fish), *Caranx trachurus* (scad).

macraeo- < Gr. *makraion*, lasting long; see *makros* under **long**

macro- < Gr. *makros*, long; see **long**

mactans, L. killing, sacrificing; see *macto* under **kill**

mactro- < Gr. *maktra*, kneading-trough, see **basin**; < *maktron*, wiper, towel, see **napkin**

mactus, L. honored, worshipped; see **honor**

macula, L. spot, stain, mark; *maculosus*, dappled, spotted; see **spot**

mad < AS. *gemad*, crazy, insane, frantic, furious.
L. *alieno, -atus*, estrange, drive insane; see *alienus* under **strange**
L. *alogus* (Gr. *alogos*), irrational, absurd; see **fool**
L. *arrepticius*, raving, delirious:
L. *cerebrosus*, hare-brained:
L. *cerritus*, frantic, mad: *cerritulus*, dim.:
L. *delirus*, silly, doting, crazy, < *deliro*, get out of the furrow; *delirium*, n. madness, giddiness: delirium, delirious.
L. *demens, -entis*, insane; *dementia*, f. insanity; *amens, -entis*, senseless, mad, insane, foolish; *amentia*, f. madness, folly: demented, dementia praecox.
L. *deseps, -sipis*, insane:
Gr. *ekphron, -os*, out of one's mind, frenzied; *ekphrosyne*, f. madness, nonsense:
Gr. *ekthymos*, insane, demented:
Gr. *eleos (aleos)*, astray, distraught, crazed: *Aleomyia alpha* (a fly), *Aleodrilus keyesi* (a worm).
L. *fanaticus*, madly enthusiastic, over-zealous, unbalanced: fanatic, fanatical, fanaticism.
L. *furia*, f.; *furor, -is*, m. madness, rage, fierce passion; *Furia*, f. a goddess of vengeance causing madness; *furiatus*, enraged, maddened; *furiosus*, raving, mad; *furio*, be mad, rage: fury, furore, furious, infuriated, *Furia horrens* (a bat), *Arbutus furiens* (Chilean arbutus).
Gr. *hallucinor, -atus*, wander in mind, dream; see **dream**
L. *insanus; vesanus*, of unsound mind, crazy; *insania*, f. mental unsoundness, madness: insane, insanity.
L. *insensatus*, irrational: insensate.
L. *irrationalis*, without reason: irrational, irrationality.
Gr. *korybantikos*, of the Corybantes, < *Korybas, -antos*, m. priest of Cybele, whose religious rites were accompanied by frenzied music and dancing: corybantic.
Gr. *kybebos*, m. priest of Cybele *(Kybele; Kybebe)*; one in religious ecstasy or frenzy: *Cybebus dimidiatus* (a beetle).
L. *lymphatus*, distracted, crazy, mad; *lymphaticus*, frantic, panic-stricken: lymphatic.
Gr. *lyssa (lytta)*, f. madness, fury, rage; *lysseter, -os*, m. madman; *lyssaleos; lyssetikos*, furious, raging; *alyssos*, curing madness: lyssic, *Lytta vesicatoria* (Spanish 'fly', a beetle), *Alyssum atlanticum* (an alyssum), *Antholyza paniculata* (an iridacead), *Lyssomanes viridis* (a spider).
Gr. *mania*, f. madness, < *mainomai*, rage, be mad; *manikos*, mad, frenzied; *andromanes*, lusting for men; *ekmanes*, quite mad; *gynaikomanes*, mad for women; *hippomanes* (said of a mare in heat), lusting after horses, horse-mad; *perimanes*, furious: mania, maniacal, kleptomania, melomania, pyromaniac, gynecomania, maenad, *Hippomane mancinella* (manchineel), *Thryomanes bewicki* (a wren), *Trichomanes crispum* (a filmy-fern).
Gr. *margos*, raging mad, furious:
Gr. *oistros*, gadfly, frenzy, madness; see *oestrus* under **fly**
Gr. *paranoia*, f. derangement, madness: paranoia, paranoiac, paranoid, *Paranoea latescens* (a moth).
L. *phreneticus* (Gr. *phrenetikos*), suffering from phrenitis, delirium: frantic, frenzy, phrenesis, phrenitis, phrenetic (frenetic).
L. *rabidus*, mad, furious, < *rabio*, rave; *rabies*, f. madness, frenzy; *rabiosus*,

raving: rage, rave, rabies, rabid, enrage, outrageous, reverie, *Lycosa rabida* (a spider).

L. *saevus; saevis,* raging, savage, fierce, cruel; *saevitia,* rage; see **rough**

Gr. *thyas (thyias), -ados,* f. woman frantic for love, a Bacchante:

L. *vecors, -cordis,* senseless, silly, insane, foolish; see **fool**

See: **anger, wander, dream, wish, heat, arouse, love, strange**

madaros, Gr. bald; see **bare**

madder < AS. *maedere.*

Gr. *ampelokarpon,* n. a kind of bedstraw:

Gr. *aparine,* f. bedstraw: *Galium aparine* (catchweed-bedstraw), *Campanula aparinoides* (a bellflower).

NL. *asperula,* f. a genus of the madder family: *Asperula tinctoria* (dyer's woodruff).

NL. *cinchona,* f. a genus of the madder family, < Countess Chinchon: *Cinchona pallida* (a quinine-tree).

NL. *coffea,* f. a genus of the madder family, < Ar. *qahwah,* coffee: coffee, coffeehouse, cafe, cafeteria, caffeine, *Coffea arabica* (a coffee), *Drosophila coffeata* (a fruitfly), *Melampus coffea* (a gastropod).

Gr. *ereuthedanon,* n. madder:

Gr. *erithalis, -idos,* f. an unidentified plant; now a genus of the madder family: *Erithalis fruticosa* (a rubiacead).

L. *erythrodanum* (Gr. *erythrodanon*), n. madder:

L. *galium* (Gr. *galion*), n. bedstraw, cleavers: *Galium pilosum* (hairy bedstraw).

NL. *gardenia,* f. a genus of the madder family, < Alexander Garden, American physician: *Gardenia jasminoides* (Cape jasmine).

NL. *houstonia,* f. a genus of the madder family, < William Houston, English botanist: *Houstonia angustifolia* (a bluet).

NL. *manettia,* f. a genus of the madder family, < Xavier Manetti, Italian botanist: *Manettia inflata* (a rubiacead).

NL. *mitchella,* f. a genus of the madder family, < John Mitchell, American botanist: *Mitchella repens* (partridge-berry).

L. *rubia,* f. madder: Rubiaceae, *Rubia cordifolia* (a madder).

madidus, L. moist, soaked, sodden; *mador,* moisture; see **wet**

madonia, Gr. waterlily; see **waterlily**

madulsa, L. a drunken man; see *madidus* under **wet**

maena, L. (Gr. *maine*), a small, sparoid fish; see **bream**

Maenas, -adis, L. (Gr. *Mainas, -ados*), priestess of Bacchus; see **priest**

maestus, L. sad, dejected, gloomy; *maestificus,* saddening; see **sad**

magadis, Gr. a harplike instrument; see **harp**

magado- < Gr. *magas, -ados,* bridge of the cithara; *magadion,* dim.; see **bridge**

maganum, L. wine-vessel made of wood; see **barrel**

magdalium, L. cylindrical figure, pill; see **cylinder**

maggot < uncertain origin; see **worm**

magic < Gr. *magikos,* of the Magians, < *Magos,* m. priest or wise man of Persia, wizard: *mageia; mageutike,* f. magic; *mageutes,* m. wizard: Magi, magician, magical, *Magonympha chrysocoma* (a moth), *Magosphaera planula* (a protozoan), *Magilus antiquus* (a gastropod).

NL. *abraxas,* a magic word whose Greek letters as numerals amounted to 365: *Abraxas fasciaria* (a moth), *Abraximorpha davidi* (a butterfly), *Abraxaphantes perampla* (a moth).

L. *aeruscator, -is,* m. sleight-of-hand performer, juggler:

L. *amuletum,* n. a magic object worn as a charm against evil and disease: amulet.

L. *aquaelicium,* n. a sacrifice to produce rain:

Gr. *baskanos,* practicing sorcery, bewitching, slandering; *baskanion,* n. amulet, charm: *Bascanichthys scuticaris* (a fish), *Bascanium(Coluber) piceum* (a snake).

L. *cantamen, -inis,* n. charm, spell; *incanto, -atus,* charm: incantation.

L. *Circe* (Gr. *Kirke*), f. notorious enchantress: circean, *Circaea alpina* (dwarf enchanters-nightshade).

L. *fascino, -atus,* bewitch, charm, enchant; *fascinum,* n. spell, witchcraft: fascinate, fascination.

Gr. *goës, -ētos,* groaner, howler, wizard; see *goëros* under **weep**

L. *Hecate* (Gr. *Hekate*), f. goddess of magic and witchcraft: *Onthophagus hecate* (a dung-beetle).

Gr. *iyngikos,* magical; see *iynx* under **woodpecker**

Gr. *keletron,* charm, spell; see *keletikos* under **agreeable**

Gr. *lamia,* witch or monster said to suck blood and eat human flesh, a bugbear to frighten children; see **animal**

L. *larvo, -atus,* bewitch, enchant; see *larva* under **young**

Gr. *manganon,* n. means of charming or bewitching; see *manganeutes* under **lie**

Gr. *mantis,* diviner, soothsayer, prophet, seer; see **prophecy**

L. *Medea* (Gr. *Medeia*), f. a fabled enchantress: *Medeola virginiana*(cucumber-root).

L. *omen,* portent, sign, augury, foreshadowing; see **prophecy**

Gr. *pharmakis, -idos,* f. witch:

L. *philtrum* (Gr. *philtron*), n. love-charm, spell, potion: philter.

Gr. *phylakterion,* n. safeguard, amulet, charm: phylactery.

L. *praestigium,* illusion, trick; see **lie**

Gr. *sophis, -idos,* wise woman, witch:

Gr. *stergethron,* love-charm; see *storge* under **love**

Gr. *striglos,* m. wizard; *strigla,* f. witch: *Striglina lineola* (a moth).

Gr. *thelgo,* charm, enchant, spellbind, cheat; *thelkter, -os,* m. enchanter; *thelkterios,* charming, enchanting; *thelgema, -tos; thelgetron,* n. charm, spell; *thelxis,* f. an enchanting; *katathelxis,* f. enchantment: *Thelgetrum ampliatrum* (a beetle), *Thelcteria pupula* (a moth), *Thelxiope viridicyanea* (a beetle), *Thelxinoa epiphracta* (a moth).

L. *theurgus* (Gr. *theourgos*), m. magician:

L. *veneficus,* magical, poisonous; m. wizard; *venefica,* f. witch: *Depressaria veneficella* (a moth).

AS. *wicce,* female practitioner of the black art: witch, witchcraft.

See: **secret, draw, spirit, prophecy, rite, lie, lot, wonder**

magirus, L. (Gr. *mageiros*), cook; see **cook**

magis, Gr. dish, plate, platter, cake; *magidion,* dim.; see **plate, cake**

magister, L. master; *magistratus,* public administrator; see **govern**

magma, Gr. salve, kneaded mass; see *masso* under **mix**

magnet < L. *magnes, -etis,* lodestone, < Magnesia in Thessaly; see **draw,** ELEMENTS under **thing**

magnificus, L. noble, eminent, splendid; see **honor**

magnolia, NL., f. < Pierre Magnol, French botanist: Magnoliaceae, *Magnolia stellata* (star-magnolia), *Dendroica magnolia* (magnolia-warbler).

NL. *illicium,* n. a genus of the magnolia family: *Illicium verum* (star anise-tree).

NL. *liriodendron,* n. a genus of the magnolia family: *Liriodendron chinense* (Chinese tulip-tree).

magnus, L. large, great; *major; majus,* greater; *maximus,* greatest; see **large**

magpie; see **crow**

magulum, L. jaw, mouth; see **jaw**

mahogany < Ab.Am. *mahagoni: Swietenia mahagoni* (mahogany), *Anguillulina mahogani* (a nematode).

NL. *carapa,* f. a genus of the mahogany family, < Ab.Am.: *Carapa grandiflora* (Uganda crabwood).

NL. *cedrela,* f. a genus of the mahogany family: *Cedrela odorata* (cigarbox-cedrela).

Gr. *melia,* f. ash; now a genus of the mahogany family; *melinos,* ashen: Meliaceae, *Melia azedarach* (China berry).

NL. *swietenia,* f. a genus of the mahogany family, < Gerard van Swieten, Austrian botanist: *Swietenia macrophylla* (Honduras mahogany), *Chloroxylon swietenia* (satinwood).

NL. *toona,* f. a genus of the mahogany family: *Toona ciliata* (Burma toon).

NL. *trichilia,* f. a genus of the mahogany family: *Trichilia emetica* (a meliacead).

maia, Gr. a kind of crab, see **crab**; good mother, nurse, midwife, see **nurse**

maianthemum, NL. a genus of the lily family; see **lily**

maiden < AS. *maegden;* see **child**

Maius, L. May; *maialis,* of May, in May; see **month**

maize < Sp. *maiz,* < Ab.Am. *mahiz; mayz;* see **grain**

majalis, L. gelded boar, barrow, eunuch; see **sterile**

major; majus, L. greater; *majusculus,* dim. somewhat greater; see *magnus* under **large**

make < AS. *macian:* maker, makeshift, make-believe, peacemaker, remake, unmake.

L. *aedifico, -atus,* build, erect, establish; see *aedificium* under **house**

L. *affectus,* acted upon, influenced: affect, affection.

L. *ago, actus,* do, perform, drive; *activus,* in action, energetic; *actuosus,* full of activity; *actum,* n. deed, performance; *actor, -is,* m.; *actrix, -icis,* f. doer, performer: act, active, actual, actuary, actor, actress, agent, agitate, agile, transact, enactment, exact, reaction, counteract, activate, inactivity, redaction, essay, assay, cogent, cogitation, squat, ambiguity, navigation, castigate, chastize, purge, fumigate, litigation, mitigatory, prodigal, *Solidago speciosa* (a goldenrod), *Tussilago farfara* (coltsfoot).

Gr. *anysimos,* effectual; *anystos,* practicable, able; *anysis, -eos,* f. accomplishment, end, < *anyo,* effect, achieve, accomplish: *Anysis australiensis* (a wasp).

Gr. *askeo,* exercise, practice, fashion; *askema, -tos,* n. exercise; *askesis,* f. training, trade, profession; *asketes,* m. practician, athlete; *asketikos,* industrious, rigorous, austere; *asketos,* practiced, expert, curiously fashioned: ascetic, asceticism, *Ascetocrinus rusticellus* (a fossil crinoid).

L. *auctor, -is,* m.; *auctrix, -icis,* f. originator, causer, doer: author, authority, unauthorized.

Gr. *banausos,* working by the fire, mechanical; see *banausia* under **art**

Gr. *boupalis,* struggling hard:

AS. *byldan,* build: builder, building, rebuild.

L. *capax, -acis,* able, fit; see *capio* under **take**

L. *causa,* agent, reason, motive; see **begin**

L. *cerdo, -onis,* handicraftsman; see *kerdaleos* under **know**

Gr. *chernes, -etos,* m. laborer; poor, needy: *Chernes peninsularis* (a pseudoscorpion), *Adelphochernes aethiopicus* (a pseudoscorpion).

L. *coepio, coeptus,* begin, found, undertake; see **begin**

L. *commentus,* devised, fabricated:

L. *como, comptus,* form, construct, embellish; see **ornament**

L. *concoquo, -coctus,* boil together, prepare, devise; see **cook**

L. *condo, -ditus,* put together, form, found, establish, preserve; *conditor, -is,* m. founder, author: A.U.C.*(ab urbe condita), Leporillus conditor* (a stick-nest rat), *Bruchus inconditus* (a beetle).

L. *conor, -atus,* undertake, endeavor, try; *conabilis,* laborious, difficult; *conamen, -inis,* n. effort, struggle; *conatio, -onis,* f. undertaking, effort: conation, conative.

L. *constitutus,* established, founded, arranged: constitute, constitution, constituent.

L. *contigno, -atus,* join with beams, joists, or rafters; see *tignum* under **beam**

AS. *craeft,* strength, skill, art, cunning; see **art**

L. *creo, -atus,* make, produce; *creator, -is,* m.; *creatrix, -icis,* f. author, founder, maker; *creatio, -onis,* f. a making; *creatura,* f. the thing made; *creabilis,* makable: create, creator, creature, creation, Creole, recreation.

Gr. *demo; domeo,* build; *oikodomos,* architect; *lithodomos,* mason; see **carpenter, mason**

AS. *don,* act: do, doff, don, did, deed.

Gr. *drao,* do, accomplish, perform; *drama, -tos,* n. deed, play; *dranos,* n. deed; *draster (drester), -os; drestes,* m. laborer, worker; *drasterios,* active, efficacious, vigorous; *drastikos,* powerful, efficacious; *drasis,* f. strength, efficacy; *drastosyne,* f. service: drama, dramatic, drastic, dramamine(β-dimethylaminoethyl benzohydryl ether 8-chlorotheophyllinate).

Gr. *ergon,* n. work; *ergane,* f.; *ergaster; ergates; ourgates,* m. worker; *-ourgos,* suffix meaning work or worker; *ergasia,* f. work, business; *ergasimos,* workable; *ergastikos; ergatikos,* active, industrious, laboring; *ergodes,* irksome, troublesome; *ergocheiron,* n. manual labor; *energos,* active, busy; *periergos,* overcareful; *synergos,* working together; *chrysourgos,* m. goldsmith; *himatiourgos,* clothes-maker, tailor; L. *ergastulum,* n. workhouse, penitentiary: erg, energy, energetic, synergid, lethargy, liturgy, metallurgy, argon, surgeon, George, organ, *Ergates serrarius* (a beetle), *Linerges mercurius* (a jellyfish), *Polyergus lucidus* (an ant), *Periergates rodriguezi* (a beetle), *Periergopus sculptus* (a beetle), *Periurgus furcatus* (a fish), *Calliergon cordifolium* (a moss).

L. *exerceo, -citus,* practice, drill, work: exercise.

L. *exsecutio, -onis,* f. accomplishment, achievement: execute, execution, executive.

L. *faber, -bri,* m. craftsman, worker, smith; *faber; affaber, -bra, -brum,* workmanlike, skillful; *fabrica,* f. workshop; *fabrilis,* of a craftsman; *fabricor, -atus,* make, build: fabricate, fabric, forge, forgery, affabrous, *Baridius affaber* (a beetle).

L. *facio, factus,* do, make; *facesso, -itus,* do eagerly; *factum,* n. act, deed, achievement; *facetus,* well-made, fine; *facinus, -oris,* n. act, deed, misdeed, crime; *factor, -is,* m. doer, maker; *factura,* f. creation, work; *-fex,* m. suffix denoting doer, maker, agent, as in *artifex, -ficis,* artist, craftsman; *carnifex,* executioner, hangman; *munifex,* one on duty; *opifex,* workman; *pontifex,* chief priest; *conficio, -fectus,* make, prepare; *confector, -is,* m. maker, preparer, destroyer: fact, fiat, factory, factotum, facility, facsimile, faculty, facultative, feasible, feat, feature, fetish, forfeiture, faction, factitious, fashion, affect, effect, amplify, artifice, beatific, beautify, chafe, clarify, edifice, classification, coefficient, perfection, ossify, liquefaction, defective, olfactory, petrify, prolific, office, significant, affectionate, defeat, magnify, qualify, malfeasance, counterfeit, profit, imperfect, deficient, efface, beneficial, dissatisfaction, difficult, affair, discomfit, humidifier, hacienda, artificer, carnifex, opificer, pontifex, *Spinifex hirsutus* (an Australian grass), *Ichneumon confector* (a wasp).

L. *fingo, fictus,* form, fashion; *fictilis,* capable of being fashioned, especially clay; *fictio, -onis,* f. a making, forming; *figuro, -atus,* form, mold, shape: fiction, fictitious, figment, figure, figurative, configuration, fictile, feign, feint, faint, effigy.

L. *formo, -atus,* shape, prepare; see **form**

L. *fungor, functus,* perform, discharge, execute; *functio, -onis,* f. a doing, performance: function, functionary, perfunctory, defunct, fungible.

L. *genero, -atus,* beget, produce, create; see **begin**

L. *gestum,* n. act, deed, event; *gestuosus,* full of action: *Gesta Romanorum,* jest.

Gr. *gymnastikos,* pertaining to bodily exercise: gymnastics.

Gr. *heurisko,* invent, find out; *heuretes,* inventor; see **find**

Gr. *hidryo,* establish, found, settle; *hidryma, -tos,* n. an establishment, building, seat; *hidrysis,* f. a founding:

Gr. *hyphantes,* weaver; see **weave**

L. *impetro, -atus,* accomplish, effect, get:

L. *institutor, -is,* m. founder, creator: institute, institution, institutor.

L. *invenio, -ventus,* come or light upon a thing, devise, contrive, effect; *inventor, -is,* m.; *inventrix, -icis,* f. author, contriver, deviser: invent, inventor, invention.

Gr. *kamno,* work, suffer; *kamatos,* m. toil, labor, trouble; *kamateros; kamatodes,* toilsome, exhausting: *Camaterus dichaeta* (a fly), *Camatopsis rubida* (a crab).

Gr. *ktizo,* found, settle, create; *ktisis,* f. a founding, creation; *ktistes; ktistor, -os,* m. founder, creator; *ktistikos,* creative; *ktistos,* grounded, established: *Oreoctistes leucops* (a bird).

L. *labor, -is,* m. work, toil; *laboro, -atus,* work, strive, toil; *laboriosus,* toilsome, difficult, wearisome; *elaboro,* work out, develop: labor, laborer, laborious, Labrador, elaborate, collaborator, *Tetragnatha laboriosa* (a spider), laboratory, *Histiostoma laboratorium* (laboratory-mite).

Gr. *latypos,* stone-cutter, mason; see **stone**

L. *lucubro, -atus,* work by lamplight, burn the midnight oil; see *lux* under **light**

L. *machinor, -atus,* contrive skillfully, design, invent; *machinator,* architect, engineer; see *machina* under **tool**

L. *mimus* (Gr. *mimos*), imitator; see **equal**

Gr. *mogos,* toil, trouble; see *mochtheros* under **trouble**

L. *molior, -litus,* use power, strive, set in motion, build; *molimen, -inis,* n. attempt, effort, endeavor; *demolior,* destroy: demolish, demolition, emolument.

L. *munus, -eris,* n. service, office, performance of a duty; *munis, -e,* ready for service; *munium,* n. duty, function: immunity, municipal, community.

L. *nitor, nisus; nixus,* bring pressure to bear upon, labor, strain, strive; *nitabundus; nixabundus,* pressing, straining, striving; *nisus, -us,* m. effort, exertion; *connitor,* endeavor, struggle; *innitor,* press upon; *obnitor,* resist, strive, struggle: nisus, *Agulla adnixa* (a snakefly).

L. *officialis,* pertaining to duty, service, < *officium,* n. service, favor, duty; *officina,* f. workshop, laboratory; *officinalis,* of a laboratory, pertaining to drugs; *officinator, -is,* m. artisan, keeper of a workshop; *officiosus,* obliging, dutiful, ready to serve: office, official, officiousness, *Lithospermum officinale* (gromwell).

L. *opus, -eris,* n. work, literary or musical composition; *opusculum,* n. dim.; *opero, -atus,* work, be busy; *opera,* f. service, work, care; *opella; operula,* f. dim.; *operarius; operator, -is,* m.; *operatrix, -icis,* f. worker; *opifex, -icis,* m. worker, maker, artisan; *operosus,* active, busy, industrious, difficult; *operativus,* creative: opus, operate, operator, opera, operation, opificer, maneuver, inure, inoperative, cooperate, *Noctua operosa* (a moth), *Opifex fuscus* (a mosquito).

L. *paro, -atus,* prepare, make ready; see **fit**

L. *patro, -atus,* perform, accomplish, effect; *patrator, -is,* m. achiever; *perpetro, -atus,* effect; *perpetrator, -is,* m. doer: perpetrate, perpetrator, *Elpidophorus patratus* (a fossil mammal).

Gr. *phyteuo,* plant, engender, cause, produce; see *phyton* under **plant**

Gr. *plasso,* form, mold; *plasma, -tos,* n. the thing formed or molded, image, figure, fiction, substance; *plasis,* f. a molding; *plastes,* m.; *plastis, -idos,* f. molder, modeler; *plastikos,* suitable for molding, pliable; *plastos,* formed, molded: plasma, plastic, plastid, plasmolysis, protoplasm, plasmocyte, plasmodium, homoplasy, anaplasty, *Plasmatoceras plastum* (an ammonite), *Plasmodiophora brassicae* (a fungus), *Plasmodium ovale* (protozoan causing a kind of malaria), *Histoplasma capsulatum* (a fungus causing histoplasmosis), *Ceroplastes cerifera* (a scale-insect).

Gr. *poieo*, make; *poietes*, m. maker, poet; *poietikos*, creative; *ponema, -tos*, n. work; *poneros*, toilsome; *ponikos*, hard-working, laborious; *ponos*, m. hard work, toil; *ergoponos*, m. husbandman; *kalliponos*, beautifully made; *philoponos*, industrious; *chrysopoios*, m. goldsmith; *lithopoios*, turning to stone; *melopoios*, m. poet; *pelopoios*, m. potter; *toxopoios*, m. bow-maker; *trichopoios*, getting hair: geoponic, hydroponics, poet, poem, poetry, *Narcissus poeticus*(narcissus), *Melipona favosa* (a bee), *Philoponus pteropus* (a spider), *Termitopone marginata* (an ant), *Sphex pelopoeiformis* (a wasp), *Scenopoietes dentirostris* (a bower-bird).

Gr. *porizo*, cause, provide, invent; see **supply**

L. *possibilis*, that may exist or may be done, < *possum*, be able: possible, possibility, impossible.

Gr. *prasso (pratto)*, do, cause, effect; *pragma, -tos*, n. act, deed, fact; *praxis*, f. a doing, acting; *prakter; praktor, -os*, m. doer; *praktikos*, active, busy, able, effective; *pragmatikos*, pertaining to experience resulting from practice; *sympraktor, -os*, m. helper, assistant: pragmatism, pragmatic, practical, practice.

L. *procudo, -cusus*, fashion by hammering, forge; see *cudo* under **strike**

L. *produco, -ductus*, bring forth, make: produce, product, productivity, *Merluccius productus* (a hake).

L. *promo, -mptus*, put forth, produce, reveal: prompt, prompter.

Gr. *proxenetes*, m. agent, negotiator: *Proxenetes flabellifer* (a flatworm).

L. *resulto*, rebound, terminate as a consequence; see **end**

Gr. *rhekter, -os; rhektes*, m. worker, doer, < *rhezo*, do, act; *rhekterios*, active, busy: *Rhectes analogus* (a bird).

Gr. *skeuastes*, preparer; *skeuastos*, artificial, prepared; see *skeuos* under **tool**

AS. *smith*, a worker in metal, < *smitan*, strike: smith, blacksmith, silversmith, Goldsmith, Smith, smithcraft, jokesmith, wordsmith.

Gr. *speudo*, urge on, press, hasten; see **swift**

L. *struo, -uctus*, build; *structilis*, of building; *structor, -is*, m. builder, carpenter, mason; *structura*, f. something built, building: structure, structural, construct, construe, destroy, destruction, obstructive, instrument.

Gr. *synthesis*, a putting together, compounding, composition; *synthetos*, put together, compounded; see *tithemi* under **place**

Gr. *talaiporeo*, do hard work, suffer; *talaiporia*, f. hard work, hardship: *Talaeporia triquetrella* (a moth).

Gr. *techne*, art, craft, trade; see **art**

Gr. *tekton*, carpenter, joiner, builder, craftsman; *architekton*, master-builder, engineer; see **carpenter**

Gr. *thes; thessa*, hired laborer; see **servant**

Gr. *tryos*, n. labor, toil:

Gr. *tyktos*, created, wrought, < *teucho*, make, build, prepare; *teuxis(tyxis)*, f. a making, attainment:

AS. *wyrcan; weorcan*, labor; *wryhta*, workman, artisan: work, workshop, rework, handiwork, workmanship, wrought iron, wright, Wright, Cartwright, wheelwright, shipwright, millwright, bulwark.

*   *   *   *   *   *

L. *-arius*, masculine suffix denoting one who, agent, actor, as *aerarius*, coppersmith; *albarius*, plasterer, whitener; *argentarius*, silversmith; *aurarius*, goldsmith; *bractearius*, gold-beater, maker of goldleaf; *candelabrarius*, candlestick-maker; *carpentarius*, chariot, wagon, or carriage-maker; *cinerarius*, hair-curler; *clavicarius*, key-maker, locksmith; *claustrarius*, locksmith; *coactiliarius*, maker of thick cloth or felt; *coriarius*, tanner; *cornuarius*, maker of horns; *culcitarius*, cushion-maker; *cuparius*, cooper; *ferrarius*, blacksmith; *focarius*, cook; *furnarius*, baker; *gemmarius*, jeweler; *lactarius*, dairyman, milkman; *lapidarius*, gem-cutter, stone-cutter; *lignarius*, carpenter, turner, cabinet-maker; *monetarius*, coiner, minter; *pararius*, broker, agent; *piscarius*, fisherman; *plumbarius*, plumber; *scoparius*, sweeper; *scrutarius*, ragman, second-hand dealer; *scutarius*, shield-maker; *tignarius*, carpenter: antiquary, librarian, vicar, friar, secretary, sectarian, Unitarian, auxiliary, liar, auctioneer, cannoneer, engineer, privateer, charioteer, amateur, connoisseur, cashier, financier, gondolier, grenadier, brigadier, clothier, miller, baker, waiter, cutter, lawyer, sawyer, villager, Marylander, messenger, passenger, harbinger, scavenger, *Serrarius microcephala* (an acarinid).

F. *-ee*, denoting agent or object of an action: absentee, committee, drawee, grandee, grantee, guarantee, payee, referee, refugee, repartee, standee, trustee.

Gr. *-eus*, m.; *-eia*, f. suffixes meaning agent, doer, as *basileus*, king; *basileia*, queen; *goneus, goneia*, begetter; *grapheus, grapheia*, writer, painter; *halieus*, *halieia,* fisher, sailor; *koureus,* barber; *skyteus,* leather-worker; *tomeus,* cutter; *xyleus,* woodcutter: *Halieus leucogaster* (a cormorant), *Ironeus duplex* (a beetle), *Xyleus rosulentus* (a grasshopper), *Basilia ferruginea* (a fly).

L. *-fex,* denoting doer, maker, worker, < *facio,* do, make; see *facio* above.

Suf. *-ician,* < F. *-icien,* specialist, practitioner: physician, technician, mortician, musician, dietitian.

AS. *-ing,* < *-ende,* forms present participles and verbal nouns: building, counting, loving, making, reading, writing, meaning, feeling, clothing, leaving, roofing.

Suf. *-ist,* < F. *-iste,* < L. *-ista* (Gr. *-istes*), signifying agent, doer, maker, one who: anatomist, artist, botanist, Buddhist, Calvinist, geologist, florist, organist, linguist, pugilist, socialist, zoologist, *Bycanistes cristatus* (a bird), *Rhopalista ignifera* (a moth), *Hammamelistes spinosus* (a witchhazel gall-insect), *Glaphyristis marmorea* (a moth).

Gr. *-ite, -itis,* one who, descendant, doer; see **nature**

L. *-ivus,* suffix denoting quality, action, agent; see **nature**

Gr. *-ize, -lyze,* < *-izein,* verbal suffix meaning do, make, action: advertise, analyze, baptize, compromise, criticize, equalize, exercise, fertilize, fletcherize, ionize, legalize, macadamize, mineralize, minimize, ostracize, paralyze, sterilize, summarize.

L. *-mentum,* tool, means, action or result of action; see **tool**

L. *-monium; -monia,* action or result of action, as *testimonium,* evidence; Gr. *hegemonia,* leadership: ceremony, matrimony, patrimony, harmony, hegemony, parsimony, acrimony, vadimony, testimony, alimony, sanctimonious.

L. *-or, -sor, -tor,* m.; *-trix,* f. suffixes signifying one who, actor, agent, as *auctor, -is,* m.; *auctrix, -icis,* f. originator; *auditor, auditrix,* hearer; *doctor, doctrix,* teacher; *donor, donator, donatrix,* giver; *fusor,* pourer, caster, founder; *generator, generatrix; piscator,* fisherman; *sensor,* feeler; *sutor, sutrix,* sewer; *tonsor, tonstrix,* barber; *vexator, vexatrix,* trouble-maker; *victor, victrix,* conqueror: actor, author, doctor, donor, governor, inventor, operator, litigator, sponsor, Beatrice, Bellatrix, directrix, dictatrix, administratrix, executrix, testatrix.

Gr. *-sis, -sy, -cy,* denoting act or process of, as *likmesis,* a winnowing; *thesis,* a placing: analysis, ellipsis, emphasis, genesis, paralysis, synopsis, synthesis, hypocrisy, palsy, poesy, prophecy.

AS. *-ster, -stress,* suffixes meaning agency, condition, state, profession: barrister, Baxter, Brewster, Dempster, gamester, huckster, seamstress, songster, songstress, spinster, tapster, Webster, youngster.

L. *-sura, -tura, -ura,* suffixes denoting product or result of action, as *caesura,* a cutting, pause; *scriptura,* a writing: capture, creature, enclosure, figure, fissure, fixture, fracture, ligature, literature, measure, pasture, picture, pressure, rapture, scripture, signature, structure, tonsure.

Gr. *-ter, -tes, -tor,* m.; *-teira, -tis, -tria, -tris,* f. suffixes signifying agent, doer, as *aischynter,* dishonorer; *alexeter, alexetor,* averter; *basanistes,* examiner; *deleter,* destroyer; *dynastes, dynastis,* ruler; *erastes, erastris,* lover; *genetes, genetor, geneteira,* begetter; *halieutes,* fisher, sailor; *kolaster, kolastes,* punisher; *kyberneter, kybernetes, kyberneteira, kybernetis,* governor; *metanastes,* migrant; *misetes,* hater; *orchester, orchestes, orchestria, orchestris,* dancer; *orykter, oryktes,* digger; *penestes,* workman; *philetes, philetor,* lover; *plastes,* molder; *prophetes, prophetis,* prophet; *soter, soteira, sostria,* savior; *technites,* craftsman: *Deleter obscurus* (a wasp), *Heuretes picticornis* (a moth), *Metanastes australis* (a beetle), *Melletes papilio* (a fish), *Technites trifasciatus* (a beetle), *Hegetor herculeus* (a moth), *Barydotira hammeri* (a wasp), *Machetis aphroloba* (a moth), *Trichoclystis peregrina* (a moth), *Erastria venustula* (a moth), *Metanastria rubi* (a butterfly), *Trophocleptria variolosa* (a bee), *Damatris similis* (a beetle), *Parasectris tholaea* (a moth), *Philametris aethalopa* (a moth).

L. *-tion,* act of, result of, process of, state of; see *-ion* under **nature**

See: **begin, form, fit, art, edit, strike, think, tool, carpenter, cobbler, cook, mason, carry,** etc.

mala, Gr. very, much; *mallon,* more; *malista,* most, see **very;** L. *mala,* cheekbone, jaw, see **jaw**

malabathron, Gr. aromatic leaf of an Indian plant; see **smell**

malache, Gr. mallow; see **mallow**

malaco- < Gr. *malakos,* soft; *malakia,* softness; see **soft**

malagma, Gr. an emollient plaster, poultice; see **plaster**

malaria < It. *mala,* bad, *aria,* air; see **disease**

male < OF. *mascle,* < L. *masculus;* see **man**

maledicus, L. abusive, scurrilous, slanderous; see **curse**

malefactor, L. evildoer; *maleficus,* criminal, wicked; *malevolens,* ill-disposed, inimical; see *malus* under **bad**

maleros, Gr. fierce, fiery, mighty; see **rough**

malice < L. *malitia,* badness, spite; see *malus* under **bad**

malignus, L. bad, malevolent, fatal; see *malus* under **bad**

malinus, L. pertaining to apples, apple-green; see *malum* under **apple**

malleus, L. hammer; *malleolus,* dim.; *malleator,* hammerer; see **hammer**

mallos, Gr. wool; *mallotos,* fleecy, woolly; see **wool**

mallow < AS. *malwe,* < L. *malva* (Gr. *malache; moloche*), f.; *dendromalache,* f. tree-mallow: Malvaceae, malvaceous, malachite, mauve, marshallow*(Althaea officinalis),* Indian mallow *(Abutilon avicennae), Malva rotundifolia* (dwarf mallow), *Physocarpus malvaceus* (mallow-ninebark), *Anthophilax malachitus* (a beetle), *Malachiomimus latifrons* (a beetle).
> NL. *abutilon,* n. a genus of the mallow family: *Abutilon vitifolium* (grape-leaved abutilon).
> L. *alcea* (Gr. *alkea*), f. a kind of mallow: *Sphaeralcea rosea* (rose-globemallow).
> L. *althaea* (Gr. *althaia*), f. a mallow: *Althaea ficifolia* (a hollyhock).
> L. *hibiscus* (Gr. *hibiskos*), m. mallow: *Hibiscus syriacus* (shrubby althaea).
> L. *malope,* f. a kind of mallow: *Malope trifida* (a mallow).
> NL. *sida,* f. a genus of the mallow family: *Sida spinosa* (prickly sida), *Sidalcea malvaefolia* (a checker-mallow).
> NL. *urena,* f. a genus of the mallow family, < Malay *uren: Urena lobata* (cadillo).

malluvium, L. washbasin; see **basin**

maltha, L. (Gr. *malthe*), a kind of asphalt or wax; *malthodes,* pliant, soft; see **resin**

malum, L. apple; *malus,* apple-tree; *malinus,* of apples; see **apple**

malus, L. bad; *pejor; pejus,* worse; *pessimus,* worst; see **bad**

malva, L. mallow; see **mallow**

mamma, L. breast, teat; *mammilla; mammula,* dim.; see **udder**

mammoth < G. *mammuth;* see **elephant**

mamphula, L. a kind of Syrian bread; see **bread**

man < AS. *man:* mankind, gentleman, manly, alderman, clergyman, henchman, husbandman, nobleman, sportsman, statesman, superman, workman, manikin, men, women.   Derive: talisman.
> Gr. *aner, andros,* m. man, male; *andrarion; andrion,* n. dim.; *andreios,* of man; *andrikos,* of man, manly, masculine; *androdes,* like a man, manly; *andromeos,* of man, human; *andrias, -antos,* m. image of a man, statue; *androgynos,* m. hermaphrodite; *androsyne,* f. manliness; *epandros,* manly, masculine: android, androcyte, androcracy, androphore, androseme, androsin, androecium, androgynous, Andrew, Alexander, diandrous, polyandry, Philander, philanderer, Henderson, *Andropogon scoparius* (little bluestem), *Andrias scheuchzeri* (a fossil salamander), *Andrarion pumilio* (a gastropod), *Adinandra dumosa* (a tiup-tree), *Ceiba pentandra* (kapok-tree), *Heterandria formosa* (a fish), *Synandra hispidula* (a mint), *Philander laniger* (a woolly opossum).
> Gr. *anthropos,* m. man; *anthroparion; anthropion,* n.; *anthropiskos,* m. dim.; *anthropeios; anthropikos; anthropinos,* of man, human; *anthropismos,* m. humanity; *anthropodes,* manlike: anthropology, anthropoid, anthropomorphism, misanthrope, philanthropist, *Anthropithecus niger* (chimpanzee), *Sinanthropus pekinensis* (Peking man), *Anthropoides virgo* (Numidian crane).
> Gr. *arrhen (arsen), -os,* m. male, masculine; *arrhenikos; arrhenopos,* manly, masculine: arrhenoplasm, arrhenotoky, arsenoblast, *Arrhenatherum bulbosum* (an oatgrass), *Diplarrhena moraea* (an iridacead), *Diarrhena americana* (a grass).
> Gr. *brotos,* m. man, mortal, human; *broteios,* mortal, human; *brotosomos,* with human body:
> L. *casnar,* n. an old man:
> L. *concubinus,* male paramour; see *concubina* under **prostitute**
> L. *dominus,* master; see **govern**
> Gr. *enelix, -ikos,* of age, in prime of manhood; see **ripe**
> Gr. *geron, -tos,* old man; *gerontion,* dim.; *gerontikos,* like an old man, senile; see *geras* under **old**
> Gr. *gynnis, -idos,* m. a womanish man: *Gynnidomorpha mesoxutha* (a moth).
> L. *homo, -inis,* c. human being, man, mankind; *homullus; homunculus,* m. dim.; *humanus,* of man, human: hominal, hominoid, homage, Hominidae, homicide, human, humanitarian, inhumanity, hominivorous, humanism, omber, Bon Homme Richard, *Homo sapiens* (man), *Actinomyces hominis* (a fungus), *Anaptomorphus homunculus* (a fossil lemur).

L. *mas, maris,* m. a male; *masculus,* m. dim.; *masculinus,* manly, male; *maritus,* m. married man, husband; *marita,* f. married woman, wife: masculine, **male**, emasculate, mallard, marital, marry, marriage, *Cornus mas* (cornelian cherry), *Aspidium filix-mas* (male fern), *Orchis mascula* (satyrion), *Lithobotrys mascula* (a radiolarian).

L. *mortalis,* human being; see *mors* under **death**

L. *pappus* (Gr. *pappos*), old man, grandfather; see **kin**

Gr. *phos, photos,* m. man, hero: *Phos cancellatus* (a gastropod).

L. *salaputium,* n. manikin, dwarf: *Salaputium fulvidum* (a pelecypod).

L. *seniculus,* a little old man; see *senex* under **old**

Gr. *thoures,* m. male:

L. *vetulus,* old man; see *vetus* under **old**

L. *vir, -i,* m. man; *virilis,* manly, masculine, robust: virile, virility, virago, virtue, triumvirate.

manage < It. *maneggio,* < L. *manus,* hand, and F. *menage,* household, < L. *mansio,* dwelling; see **govern, guard**

manalis, L. flowing, < *mano, -atus,* flow, trickle; see **flow**

manciola, L. dim. of *manus,* hand; see **hand**

mancipatus, L. delivered, transferred; see *manceps* under **trade**

mancus, L. crippled, lame, maimed; see **hurt**

-mancy, pertaining to divination, < Gr. *mantis,* diviner, seer, prophet; see **prophecy**

mandalos, Gr. bolt, bar; *mandalotos,* with the bolt shot; see **bar**

mandatus, L. commanded, ordered; *mandativus; mandatorius,* of a command; see *mando* under **ask**

mandibula, L. jaw, < *mando, mansus,* chew; see **jaw, bite**

mandra, Gr. stall, stable; see **room**

mandrake < AS. *mandragora,* < Gr. *mandragoras;* see **nightshade**

manducus, L. glutton; see *mando* under **bite**

mandya, Gr. woolen cloak; see **garment**

mane < AS. *manu,* see **hair**; L. *mane,* morning, see **dawn**

manentia, L. stability; see *maneo* under **stand**

manes, Gr. a kind of cup, brazen figure, slave; see **cup,** *mania* under **mad,** *manis* under **spirit**

manettia, NL. a genus of the madder family; see **madder**

manganese < a corruption of Magnesia; see *magnes* under **draw,** ELEMENTS under **thing**

manganeutes, Gr. juggler, trickster, see **lie;** *manganeuma,* jugglery; *manganon,* charm, spell, see **magic**

manger < OF. *maingeure;* see **box**

mango < Malay *manga,* a fruit from a tree belonging to the sumac family, see **sumac;** < L. *mango, -onis,* dealer, see **trade**

mangrove < Sp. *mangle,* and E. *grove.*
NL. *anisophyllea,* f. a genus of the mangrove family: *Anisophyllea laurina* (a rhizophoracead).
NL. *crossostyles,* f. a genus of the mangrove family: *Crossostyles biflora* (a rhizophoracead).
NL. *gynotroches,* f. a genus of the mangrove family: *Gynotroches axillaris* (a rhizophoracead).
NL. *poga,* f. a genus of the mangrove family: *Poga oleosa* (a rhizophoracead).
NL. *rhizophora,* f. a genus of the mangrove family: Rhizophoraceae, *Rhizophora mangle* (common or red mangrove).

mania, Gr. madness, frenzy; *manikos,* mad, frenzied, see **mad;** bugbear to frighten children, see **fear**

manica, L. long sleeve of a tunic, glove; *manicarius,* of sleeves or gloves; see **sleeve**

manicula, L. dim. of *manus,* hand; see **hand**

manifestus, L. clear, distinct, evident; see **clear**

manifolium, L. a plant; see **plant**

maniola, L. dim. of *mania,* bugbear; see **fear**

manipulus, L. handful, bundle; see **bundle**

manis, L. ghost, shade, soul of the departed; pl. *manes;* see **spirit**

manna, Gr. morsel, grain, gum of a tree or exudate of an insect; see **resin**

manner < OF. *maniere*, < L. *manuarius*, of the hand, skillful.
L. *decorum*, propriety, fitness; see *decens* under **fit**
L. *ethicus* (Gr. *ethikos*), of habits, morals, and character; see *ethas* under **use**
L. *methodus* (Gr. *methodos*), mode of procedure; *eumethodos*, well-arranged; see **class**
L. *modus*, proper measure, custom, manner; see **measure**
L. *mos, moris*, m. manner, custom, practice, usage; *moralis*, of manners: moral, morale, mortality, immorality, immoral, demoralize, morose, demure.
L. *ratio*, understanding, procedure; see *reor* under **think**
L. *secta*, ´. way, mode, manner:
Gr. *tropos*, turn, manner, way; see *trepo* under **turn**
AS. *wise*, way, manner: edgewise, contrariwise, fanwise, lengthwise, likewise, otherwise, nowise, righteous.
See: **art, fit, beauty, ornament, rough, way**

mannos, Gr. collar, necklace; see **necklace**

mannus, L. a small Gallic horse, coach horse; *mannulus*, dim. pony; see **horse**

manon, Gr. sponge; see **sponge**

manos, Gr. thin, rare; see **thin**

mansio, L. abode, dwelling; *mansiuncula*, dim.; see **house**

mansor, L. guest, sojourner; see **guest**

mansuetus, L. tame, gentle; see **tame**

mansus, L. bitten, chewed, see *mando* under **bite**; < L. *maneo*, remain, stay, see **stand**

mantelum, L. cloak; *mantuelis*, cloaklike, see **garment**; *mantele*, towel, napkin, see **napkin**

mantica, L. handbag, wallet; *manticula*, dim.; see **bag**

mantichora, L. fabulous beast with human face, lion's body, and scorpion's tail; see **animal**

manticulator, L. pickpocket, purse-snatcher; see **steal**

mantis, Gr. diviner, soothsayer, seer, phophet; see **prophecy**

mantissa, L. a trifling addition; see **trifle**

manubrium, L. handle, haft; see **handle**

manus, L. hand; *manicola; manicula;* dim.; *manualis; manuarius*, of the hand; see **hand**

manutergium, L. towel; see **napkin**

many < AS. *manig*; see **number, abundance**

map < L. *mappa*, f. napkin, towel; *mappula*, f. dim.: mapping, mapmaker, mapland, napkin, napery.
L. *catagraphus* (Gr. *katagraphos*), drawn in profile, painted, depicted:
L. *charta* (Gr. *charte*), leaf of paper; see **paper**
L. *delineatio, -onis*, sketch, design, outline; see **form**
Gr. *diagramma, -tos*, n. outlined figure, form, plan: *Simochromis diagramma* (a fish), diagram.
Gr. *ichnographia*, f. ground plan, map: ichnography.
Gr. *pinax, -akos*, board, tablet, chart, register; see **board**
Gr. *schema*, plan, form, shape; see **aim**
L. *tabula*, board, tablet, map, chart; see **tablet**
See: **form, aim**

mapalium, L. African hut; see **house**

maple < AS. *mapol*.
L. *acer, -eris*, n. maple; *acerneus; acernus*, of maple: Aceraceae, acerin, acertannin, *Acer pennsylvanicum* (striped maple), *Aceranthus diphyllus* (maplewort), *Platanus acerifolia* (London plane-tree), *Phenacoccus acericola* (a mealybug), *Xylotrechus aceris* (maple-borer), *Viburnum acerifolium* (maple-leaved viburnum), *Phyllocoptes acericola* (a maple gall-insect), *Aegeria acerni* (a moth), *Rhytisma acerinum* (a fungus).
Gr. *glinos*, m. a kind of maple: *Glinus radiatus* (a carpet-weed).
Malay *negundo*, name for a chastetree; now a genus of the maple family; see **vervain**
L. *opulus*, f. a kind of maple: *Viburnum opulus* (a viburnum), *Physocarpus opulifolius* (common ninebark).
L. *rumpotinus*, f. a kind of maple:
Gr. *sphendamnos*, f. maple: *Sphendamnocarpus madagascariensis* (a malpighiacead).

mappa, L. napkin, towel; *mappula,* dim.; see **napkin, map**

maps, Gr. idly, vainly; *mapsidios,* vain, useless; see **worthless**

marantico- < Gr. *marantikos,* wasting away; *maransis; marasmos,* causing to decay, waste away; see *maraino* under **lessen**

marathrum, L. (Gr. *marathron*), fennel; see **carrot**

maraugia, Gr. loss of sight; see **blind**

marble < OF. *marbre,* < L. *marmor;* see **lime**

marcidus, L. withered, wasted, weak; *marcor,* faintness, decay; see **lessen**

marcus, L. a large hammer; *marculus,* dim.; see **hammer**

mare, *maris,* L. sea; *marinus; maritimus,* of the sea; see **sea**

mareca, NL. a genus of ducks; see **duck**

marga, L. a kind of earth, marl; see **earth**

margarita, L. (Gr. *margarites*), pearl; see **pearl**

margellion, Gr. a kind of palm; see **palm**

margin < L. *margo, -inis,* brink, border; *margino, -atus,* enclose with a border; see **border**

margos, Gr. raging mad, furious; see **mad**

marilo- < Gr. *marile,* embers of charcoal; see **coal**

marinus, L. of the sea; see *mare* under **sea**

mariposa, Sp. butterfly; see **butterfly**

mariscus, L. a kind of rush; see **sedge**

marita, L. wife; *maritus,* husband, see **spouse**; *maritalis,* pertaining to marriage, conjugal, nuptial, see **marry**

maritimus, L. of the sea; see *mare* under **sea**

marjoram < NL. *majorana;* see *amaracum* under **mint**

mark < AS. *mearc:* marker, marksman, Mark Twain, landmark, remarkable, hallmark.

L. *bulla,* seal, any round boss or knob, bubble; see **bubble**

L. *cauter* (Gr. *kauter*), branding-iron; *cauteroma,* brand; see *kaio* under **burn**

L. *character, -is* (Gr. *charakter, -os*), m. mark, figure, letter, habit, usage; *charagma, -tos,* n.; *charagmos,* m. any mark, cut, engraved, or stamped; *charaktos,* graven, cut in, < *charasso,* cut, engrave: character, characteristic, *Charagmophorus lineatus* (a beetle), *Charagmus gressorius* (a beetle), *Charactocnemus hintzi* (a beetle), *Charactophyllum radiculum* (a fossil coral), *Charassobates cavernosus* (a mite).

L. *cicatrix, -icis,* f. scar; *cicatricula,* f. dim.; *cicatricosus,* full of scars: cicatrice, cicatricle, cicatrize, cicatrization, *Ctenoloculina cicatricosa* (a fossil ostracode).

Gr. *emblema, -tos,* n. inlaid work, thing put in or on, mosaic, token: emblem, emblematic.

Gr. *eschara,* f. scar, scab; *escharion,* n. dim.; *escharotikos,* producing a scar: scar, scarry, eschar, escharotic, *Escharipora philomela* (a bryozoan).

Gr. *gnoma, -atos,* n.; *gnome,* f. mark, token, mind, judgment; *gnorisma, -atos,* n. mark of identification:

L. *hilum,* n. bit, scar: hilum.

Gr. *ichnos,* n. footprint, track, trail; *ichnion,* n. dim.; *ichneutes,* m. tracker, hunter; *ichneutikos,* good at tracking: ichnology, ichneumon, parichnus, *Ichniotherium cottae*(Permian amphibian tracks), *Ichneutes abdominalis* (a wasp), *Petalichnus multipartitus* (a fossil track), *Treptichnus bifurcus* (a fossil track), *Micrichnus scotti* (a Triassic footprint), *Gypsichnites pacensis* (a Cretaceous footprint).

L. *index, -icis,* c. sign, token, forefinger; *indicium,* n. sign, mark, discovery, proof; *indicivus,* indicative: index, indicator, indicium, indicial.

Gr. *klosma, -tos,* thread, line, clue; *klosmation,* dim.; see **thread**

L. *litera,* alphabetical character; see **letter**

L. *litura,* f. erasure, blot:

L. *menda,* bodily defect, blemish; see **fault**

Gr. *messoros,* boundary-stone; see **stone**

L. *naevus,* m. mole, birthmark, wart, blemish; *naevulus,* m. dim.; *naevius,* with a mole or birthmark: nevus, *Agalena naevia* (a spider), *Ixoreus naevius* (a thrush).

Gr. *noseros,* pertaining to symptoms, < *nosos,* disease; see **disease**

L. *nota,* f. mark, cipher, character, sign, letter; *notula,* f. dim.; *notaculum,* n. mark, sign; *notatio, -onis,* f. a marking, noting; *notarius,* m. shorthand writer; *notabilis,* noteworthy; *noto, -atus,* mark; *denotatus,* marked out, conspicuous:

note, notable, notary, annotation, connote, denote, prothonotary, *Penicillium notatum* (a mold, source of penicillin), *Euryscopa binotata* (a beetle), *Aphrophora quadrinotata* (a spittle-bug), *Nosema notabilis* (a protozoan).

Gr. *oule,* f. scar; *epoulosis,* f. formation of a scar: ulosis, ulotic, epulosis, epulotic, *Ulodendron minus* (a lepidodendron), *Lecanora epulotica* (a lichen).

L. *portentum,* sign, omen; see **prophecy**

Gr. *sema, -tos; semeion,* n. sign, token, flag, mark; *semation,* n. dim.; *semantikos,* significant, indicative; *semanterion; semantron,* n. seal: semaphore, sematic, semantics, semeiology, *Sematopoda elata* (a fly), *Semionotus elongatus* (a fossil fish), *Semiophora gothica* (a moth), *Semecarpus cuneifolius* (an anacardiacead), *Brachysema acuminatum* (pea-bush), *Semotilus atromaculatus* (horned dace), *Semantrum signarium* (a radiolarian).

L. *signum,* n. mark, flag, seal; *sigillum; sigilliolum,* n. dim.; *signaculum,* n. mark, sign; *signifer, -a, -um,* bearing marks, figures, signs; *sigillatus,* adorned with little figures or marks; *significans,* having meaning, important; *signifex, -icis,* m. sign or image maker; *signatorius,* of sealing; *signo, -atus,* mark, stamp, underwrite; *assigno,* allot, appoint; *designo,* mark out, appoint, *insignis,* marked, noted; *insignitus,* marked, plain, clear; *presignis,* distinguished, illustrious: sign, signal, signify, signet, signator, signature, seal, assignation, assign, consignment, countersign, designer, designation, ensign, insignificant, insignia, resigned, sigillate, *Sigillaria scutellata* (a fossil lycopod), *Combophora signata* (a bug), *Chrysobothris sexsignata* (a beetle), *Clytrasoma quadrisignata* (a beetle), *Thymallus signifer* (Alaska grayling), *Bumastus insignis* (a trilobite), *Cymbidium insigne* (an orchid).

Gr. *sphragis, -idos,* f. seal, signet; *sphragidion,* n. dim.; *sphragister, -os; sphragistes,* m. sealer, signet; *sphragistikos,* pertaining to sealing: *Sphragidiophorus violaceus* (a beetle).

Gr. *stigma, -tos,* n. mark, brand, spot, surface of the female floral structure that receives the pollen; *stigme,* f. prick, point, dot, spot, < *stizo,* mark, brand, prick; *stigmatidion; stigmation,* n. dim.; *stigmatias; stigon, -os,* m. a tattooed or branded person; *stigeus,* m. tattooer, needle used in tattooing; *stixis,* f. a marking, mark, spot; L. *stigmosus,* full of marks: stigma, stigmatize, stigmasterol, anastigmatic, astigmatism, physostigmine, neostigmine, tristigmatic, *Stigmatonotus australis* (a fish), *Stigmatophora micans* (a moth), *Stigmaphorus tesselatus* (a beetle), *Stigmatidium difficile* (a copepod), *Stigmatium dorsigerum* (a beetle), *Stizolobium giganteum* (oxeye bean), *Stizostedion[Stizostethium] vitreum* (wall-eyed pike), *Stixis punctata* (a beetle), *Stigonema ocellatum* (an alga), *Anisota stigma* (a moth), *Eucapnopsis stigmatica* (a plecopterid), *Buprestis chrysostigma* (a beetle), *Physostigma venenosum* (Calabar bean), *Thylogale stigmatica* (a wallaby), *Stigeoclonium flagelliferum* (an alga).

L. *symbolum* (Gr. *symbolon*), n. sign, mark, token; *symbolikos,* emblematic, figurative, significant: symbol, symbolic, symbolism.

Gr. *symptoma, -tos,* n. anything that befalls one, mishap, sign of disease: symptom, symptomatic.

Gr. *tekmar,* a fixed mark, boundary, goal; *tekmerion,* sure sign, positive proof; see **end, sure**

L. *titulus,* inscription, label; see **word**

L. *tropaeum* (Gr. *tropaion*), n. mark, token, shield or memorial of victory, that is, the enemy's turning in defeat: trophy, *Tropaeolum peregrinum* (canarynasturtium).

Gr. *typos,* impression, shape, figure, mark or effect of a blow; see **form, strike**

L. *vestigium,* n. footprint, track, < *vestigo, -atus,* trace, search out: vestige, vestigial, investigate, investigation.

L. *vibex, -icis,* f. mark, weal: *Nassa vibex* (a gastropod).

Derive: ampersand, breve, caret, cedilla, diaeresis, diesis, macron, obelus, serif, tilde, swastika.

See: **spot, flag, bruise, fault, color, end, way, letter, number, form**

marmor, L. (Gr. *marmaros*), marble; *marmareus,* of marble; see **lime**

marmot < F. *marmotte,* f.: *Marmota monax* (woodchuck), marmot *(Marmota flaviventris).*

marra L. hoe, weeding-hook; see **hoe**

marrubium, L. horehound; see **mint**

marry < OF. *marier,* < L. *marito, -atus,* wed; *maritalis,* pertaining to marriage, conjugal, nuptial: marital, marriage.

L. *confarreo, -atus,* marry:

L. *conjugalis,* pertaining to marriage, connubial; *conjux, -ugis,* c. consort: conjugal.

ON. *Frigg (Frigga),* wife of Odin and goddess of marriage and domestic life: *Frigga hamata* (a spider), Friday.

Gr. *gamos*, m. marriage, wedding, union; *gamelios*, of a wedding, nuptial; *gamikos*, of marriage; *artigamos*, just married: gamete, gametophyte, gamopetalous, cryptogam, monogamous, polygamous, *Thalictrum polygamum* (tall meadowrue).

Gr. *hednios*, nuptial; *hednon*, n. wedding-gift: *Hednophora pyritis* (a moth).

Gr. *Hymen, -os*, m. god of marriage; *hymenaios*, of a wedding: Hymen, hymeneal, *Hymenaea verrucosa* (a legume yielding copal).

Gr. *kouridios*, wedding, nuptial:

L. *matrimonium*, n. wedlock, marriage: matrimony, matrimonial.

Gr. *mnestos*, wooed, won, married; *mnestron*, n. betrothal, marriage; *mnester, -os*; *mnestes*, m. suitor, wooer: Clytemnestra.

L. *nubo, nuptus*, cover with a veil, marry; *nubilis*, marriageable; *nuptialis*, of marriage; *pronuba*, f. brideswoman; *pronubus*, m. groomsman; *connubium*, n. marriage; *connubialis*, of marriage; *denubo*, marry below one's rank: nubile, nuptial, connubial, *Panorpa nuptialis* (scorpion-fly), *Pronuba yuccasella* (yucca-moth).

Gr. *opyo*, marry:

L. *sponsa*, bride; *sponsus*, groom; see **spouse**

L. *Subigus*, m. god of the wedding-night:

L. *unio, -itus*, join together, make one; see *unus* under **one**

See: **bind, spouse**

Mars, L. god of war; *Martialis*, of Mars; see **fight**

marsh < AS. *mersc*.

Gr. *balte*, f. swamp; *baltodes*, swampy:

L. *carectum*, place covered with sedges or rushes; see *carex* under **sedge**

AS. *fen*, bog, marsh: fen, fenland, fenberry, fenfire.

Gr. *helos*, n. marsh, meadow; *heleios*, of a marsh, dwelling in a marsh; *helodes*, marshy; *helespis, -idos*, f. marshland, meadow; *heleionomos (helonomos)*, living in marshes: helobious, helophyte, *Eliomys[Heliomys] nitela* (a garden-dormouse), *Elodea[Helodea] densa* (a waterweed), *Eleocharis ovata* (a spike-rush), *Philohela minor* (woodcock).

L. *lacuna*, cavity, pond, pool, ditch; see **hollow**

L. *lama*, f. bog, fen, slough:

Gr. *limne*, f. marsh, lake, pool; *limnion*, n. dim.; *limnaios*, of a marsh; *limnodes*, marshy; *limnetes*, m. marsh-dweller; *Limnoreia*, f. a Nereid: limnology, limnophile, limnite, limnetic, epilimnion, hypolimnion, *Limnophilus combinatus* (a caddisfly), *Limnobium spongia* (American frogbit), *Limnanthes douglasi* (meadow-form), *Limnetis wahlbergi* (a phyllopod), *Lymnaea[Limnaea] stagnalis* (pond-snail), *Gammarus limnaeus* (an amphipod), *Limnodromus griseus* (dowitcher), *Limnoria terebrans* (a gribble).

L. *lustrum*, n. bog, morass, puddle, slough:

AS. *mor*, a peaty of heathery wasteland: moor, Dartmoor, Broadmoor.

ON. *myrr*, bog, swamp, deep mud: mire, miry, quagmire, quickmire.

L. *palus, -udis*, f. marsh, bog, fen, swamp, morass; *paludosus*; *paluster (palustris)*, *-tris, -tre*, marshy, boggy; *paludicola*, c. marsh-dweller: paludal, palustral, paludine, paludous, *Quercus palustris* (pin-oak), *Ledum palustre* (a ledum), *Crepis paludosa* (a composite), *Atilax paludinosus* (a mongoose), *Echinochloa paludigena* (a grass), *Paludicella articulata* (a fresh-water bryozoan).

L. *scaturiginosus*, full of springs, marshy, boggy; see **spring**

L. *stagnum*, n. pool, swamp, fen; *stagnalis*; *stagninus*, of pools; *stagnosus*, full of pools; *stagno, -atus*, form a pool of standing water: stagnate, stagnant, stagnicolous, *Callitriche stagnalis* (water-starwort), *Hydrometra stagnorum* (a water-strider).

Gr. *telma, -tos*, n. standing water, pool, marsh, mud of a pool: *Telmatoscopus albipunctatus* (a fly), *Telmatodytes griseus* (a marsh-wren).

Gr. *tiphos*, n. standing water, pool, marsh: *Tiphodytes gerriphagus* (a wasp).

L. *uliginosum*, marsh, swamp; see *udus* under **wet**

L. *volutabrum*, n. place where pigs roll, slough, puddle:

See: **field, wet**

marsilea, NL. a genus of aquatic ferns; see **fern**

marsupium, L. (Gr. *marsipion; marsypion*), pouch, bag, purse; see **bag**

marten < Teut. *martre*; L. *martes*, marten; see **weasel**

martin < F. *martin*, < L. *Martinus*, of Mars; see **swallow**

martyr, Gr. witness; see **witness**

marulium L. (Gr. *maroulion*), lettuce; see **lettuce**

marum, L. (Gr. *maron*), a kind of sage; see **mint**

marvel < OF. *marveille*, < L. *mirabilis*, wonderful; see *miror* under **wonder**

mas, *maris*, L. male, man; *masculus*, dim.; *masculinus*, manly, male; see **man**

-mas < AS. *maesse*, feast-day, festival, < L. *missa*, the Mass; see **joy**

maschale, Gr. armpit, hollow, bay; see **angle**

maseter, Gr. chewer; *masema*, quid; see **bite**

mash < AS. *masc;* see **plaster, break**

mask < F. *masque;* see **face, cover**

mason < OF. *maçon:* masonry, masonic.
  L. *caementarius*, m. mason, < *caementum*, n. rough stone chips, but now applied to a substance used as a binder: cement, *Sceliphron caementarium* (a mud-dauber).
  L. *latomus* (Gr. *latomos*), m. stone-cutter, mason:
  Gr. *latypos (laotypos)*, m. stone-cutter, mason; *latype*, f. chip of stone:
  Gr. *laxeutes*, m. stone-cutter; *laxeutos*, carved out of rock:
  Gr. *lithodomos*, m. mason: *Lithodomus plumula* (a pelecypod).

mass < L. *massa*, quantity, bulk, body, lump; *massula*, dim., see **lump, heap**; < AS. *maesse*, feast-day, festival, < L. *missa*, the Mass, see **joy, rite**

mast < AS. *maest;* see **rod**

mastaco- < Gr. *mastax, -akos*, mouth, jaw; see **jaw**

master < OF. *maistre*, < L. *magister*, see **govern**; < Gr. *master, -os; mastes*, searcher, seeker, see **try**

masthles, Gr. leather whip, leather; see **leather**

mastiche, Gr. resin from the mastic-tree; see **resin**

mastico, -atus, L. chew; see **bite**

mastigo- < Gr. *mastix, -igos*, whip; *mastigias*, one deserving whipping, knave, rogue; see **whip**

mastos, Gr. breast; see **udder**

mastrucatus, L. clothed in skins; see **skin**

masturbate < L. *masturbor, -atus*, practice self-orgasm without coitus, onanize: masturbation, masturbator.
  Gr. *anaphlasmos*, m. masturbation:
  L. *depso, deptus* (Gr. *depho; depso*), soften by working with the hand, masturbate; see **mix**

mat < AS. *matt*, < L. *matta;* see **rug**

mataeo- < Gr. *mataios*, vain, idle, foolish; see **fool**

mataris, L. a Celtic javelin; see **spear**

mataxa (metaxa), Gr. raw silk; see **silk**

mate < uncertain origin; see **spouse, companion**

mater, -tris, L. mother; *matercula*, dim.; *maternus*, of a mother; see **mother**

materia, L. matter; *materiola*, dim.; *materialis*, of matter; see **thing**

mathalis, Gr. a kind of cup; see **cup**

mathematicus, L. (Gr. *mathematikos*), disposed to learn, pertaining to mathematics; see **learn, number**

matricaria, NL. a genus of the composite family; see **composite**

matricula, L. public list, register, roll; see **list**

matrix, L. mother, womb, source, embedding or enclosing substance; *matricula*, dim.; *matricalis*, of the womb; see **womb**

matrimonium, L. wedlock, marriage; see **marry**

matrona, L. married woman; *matronalis*, of a married woman; see **spouse**

mattea, L. (Gr. *mattye*), dainty dish, delicacy; *matteola*, dim.; see **food**

matter < L. *materia*, stuff; see **thing**

mattock < AS. *mattuc*, instrument for digging and grubbing; see **pick**

matula, L. vessel, pot; *matella*, dim.; see **pot**

maturus, L. ripe; *maturitas*, ripeness; *maturo, -atus*, ripen; see **ripe**

matutinalis; matutinus, L. of the morning, early; see **dawn**

maulis, Gr. bawd; *maulistes*, pimp; see **prostitute**

mauros, Gr. dark; see **black**

mausoleum, L. (Gr. *mausoleion*), magnificent tomb erected for Mausolus at Halicarnassus; see **grave**

maw < AS. *maga*, stomach; see **stomach**

maxilla, L. jawbone; *maxillaris*, of the jaw; see **jaw**

maximus, L. greatest; see *magnus* under **large**

maytenus, NL. a genus of the staff-tree family; see **staff-tree**

maza, Gr. barley-cake; *mazion; maziske*, dim.; *mazinos*, of barley-meal; see **cake**

mazines, Gr. a kind of cod; see **cod**

mazo- < Gr. *maza*, barley-cake, placenta, see **cake**; < *mazos*, breast, see *mastos* under **udder**

me-, Gr. not; see **not**

meabilis, L. passable; see *meatus* under **way**

meadow < AS. *maedwe*, a grassy, treeless, usually moist, lowland; see **field**

meal < AS. *melu*, flour, see **flour**; < *mael*, measure, appointed time, repast, see **eat**

meander < L. *Maeander* (Gr. *Maiandros*), a river in Asia Minor noted for its winding course; see **turn**

meaning < AS. *maenan*, intend, signify; see **mark**

measles < D. *maselen;* see **disease**

measure < L. *mensura*, f. a measuring, < *metior, mensus*, measure; *metor, -atus*, measure off, mark out; *metator, -is*, m. a fixer of boundaries; *agrimensor, -is*, m. land-surveyor; *dimensio, -onis*, f. a measuring: mensuration, measurement, mete, commensurate, immense, dimension.
L. *amussis*, f. rule, level:
L. *canon, -onis* (Gr. *kanon, -os*), m. measuring-rod, rule, standard, model; *kanonion*, n. dim.; *kanonikos*, according to rule, regular: canon, canonical.
Gr. *choinix, -ikos*, a dry measure; see **box**
Gr. *gnomon*, carpenter's square, rule; see *nosco* under **know**
L. *groma*, f. surveyor's measuring-rod; *gromaticus*, pertaining to field surveying:
L. *iambus* (Gr. *iambos*), a metrical unit or foot consisting of an unaccented followed by an accented syllable: iamb, iambic, galliambic.
L. *jugerum*, n. acre:
Gr. *lichas, -ados*, f. space between thumb and forefinger, span: *Lichas ferrisi* (a trilobite), *Acrolichas cucullus* (a trilobite).
L. *medimnus* (Gr. *medimnos*), m. a corn measure corresponding to a bushel:
Gr. *metron*, n. measure, rule; *metrikos*, of measuring; *metrios*, within measure, moderate; *metretos*, measurable; *metretes*, m. measurer; *geometres*, m. surveyor, < *metreo*, measure, count: metrical, metric, meter, metronome, barometer, speedometer, thermometer, geometer, geometry, *Metrobates hesperius* (a bug), *Metriocharis atrocyanea* (a wasp), *Metriophilus discors* (a beetle), *Metretes inaequalis* (a beetle), *Metretopus norvegicus* (an ephemerid), *Lagena geometrica* (a foraminifer), *Isometrus maculatus* (a scorpion).
L. *modus*, m. proper measure, custom, manner; *modulus*, m. dim. norm, meter; *modius*, m. a dry measure, peck; *modiolus*, m. dim, nave or hub of a wheel, box; *modulor, -atus*, measure off properly, regulate; *modulator, -is*, m. regulator, director; *modifico, -atus*, qualify, regulate: mode, modish, model, moderator, modulator, modulation, modest, modify, commodious, modiolus, *Modiolus plicatulus* (a pelecypod), *Modiolaria nigra* (a pelecypod).
L. *norma*, rule, carpenter's square; see **law**
Gr. *orgyia*, f. length of the outstretched arms, fathom:
Gr. *plethron*, n. acre:
Gr. *podizo*, measure with the feet; see *pous* under **foot**
L. *rasta*, f. German mile:
L. *regula*, f. ruler, measure, pattern; *regularis*, according to rule: regula, regular, irregularity, *Spathognathodes regularis* (a conodont).
L. *spithama* (Gr. *spithame*), f. span: *Convolvulus spithamaeus* (a bindweed).
L. *stadium* (Gr. *stadion*), n. a measure of length, stade: stadium, stadia rod.
Gr. *stathme*, carpenter's rule or line, < *stathmo*, measure; *stathmikos*, of measuring; see **line**
L. *tornus* (Gr. *tornos*), lathe, compass; see **turn**
See: **long, broad, level, number, tool**

meat < AS. *mete;* see **flesh**

meatus, L. passage, course; *meabilis*, passable; see **way**

mecasmo- < Gr. *mekasmos*, a bleating; see **bleat**

mechano- < Gr. *mechane*, machine, contrivance; *mechanikos*, of machines; see *machina* under **tool**

mecisto- < Gr. *mekistos*, longest; see *makros* under **long**

meco- < Gr. *mekos*, length; *mekyno*, lengthen; see **long**

mecon, L. (Gr. *mekon*), poppy; *mekonidion; mekonion*, dim. poppy-juice, opium; see **poppy**

meconium, L. first excrement of new-born infants; see **dung**

medaminos, Gr. worthless, see **worthless**; *medamos,* none, see **not**

meddix, L. caretaker, curate; see **guard**

Medea, L. (Gr. *Medeia*), a fabled enchantress; see **magic**

mediastinus, L. drudge, menial; see **servant**

mediator, L. go-between, arbitrator, umpire; see **judge**

medica, L. lucerne, clover, < Gr. *Medike,* Mede, Persian; NL. *medicago,* alfalfa; see **clover**

medicatus, L. healing, curing, < *medico, -atus,* heal, cure; *medeor,* heal, cure; *medicabilis;* see **heal**

medicine < L. *medicina,* remedy, see **drug**; *medicus,* physician, see *medicatus* under **heal**

mediocris, L. middling, ordinary; see *medius* under **middle**

medion, Gr. a kind of plant; see **plant**

meditator, L. thinker; see *meditor* under **think**

medix, L. curator; see **guard**

medo- < Gr. *medos,* counsel, plan; pl. *medea,* see **aim**; < *medon,* guardian, see **guard**

medulla, L. marrow, inmost part, pith; *medullaris,* of pith; see **pith**

Medusa, L. (Gr. *Medousa*), a gorgon with snaky locks and capable of turning beholders to stone; see **fear**

meet < AS. *metan,* come together; see **gather**

mega-; megalo- < Gr. *megas,* large, great, very; *meizon,* larger, greater; *megistos,* largest, greatest; see **large**

megallium, L. an ointment; see **fat**

megaron, Gr. large chamber, room; see **room**

mel, *mellis,* L. (Gr. *meli, -itos*), honey; *melinus; melleus,* of honey; see **honey**

melampodium, L. (Gr. *melampodion*), blackfoot; see **composite**

melan-; melano-; mela- < Gr. *melas, melanos,* black; see **black**

melancholy < Gr. *melancholia,* black bile, depression, sadness; see **sad**

melanthemon, Gr. a composite; see **composite**

melanthium, L. a genus of the lily family; see **lily**

melathron, Gr. roof, ceiling; see **roof**

melea, Gr. apple-tree; see *malum* under **apple**

meleagris, Gr. guinea-fowl; see **peacock**

meledonos, Gr. caretaker, guardian; *meledone,* care, sorrow; see **guard**

meleos, Gr. idle, useless; see **worthless**

meles, L. marten, badger; *melinus,* of a marten; see **weasel**

meleteros, Gr. diligent; *meletikos,* of practice; see *meletao* under **busy**

melia, Gr. ash; now a genus of the mahogany family; see **mahogany**

melica, It. sorghum; see **grass**

meliceris, L. (Gr. *melikeris*), a kind of skin eruption; see **disease**

melicratum, L. (Gr. *melikraton*), a kind of mead; see **drink**

melicto- < Gr. *meliktes,* flute-player; see **music**

melicus, L. (Gr. *melikos*), musical, tuneful, melodious; see *melos* under **sing**

meligerys, Gr. sweet-voiced; see *gerys* under **speak**

melilotus, NL. a genus of the bean family; see *lotus* under **bean**

meline, Gr. millet; see **grain**

melinus, L. (Gr. *melinos*), of quinces, quince-yellow, see *malum* under **apple**; of honey, honey-colored, see *mel* under **honey**; of a marten, see *meles* under **weasel**; of ash, see *melia* under **mahogany**; of millet, see *meline* under **grain**

melior; melius, L. better; see *bonus* under **good**

melis, Gr. apple-tree; see *malum* under **apple**

melissa(melitta), Gr. honey-bee, see **bee**; a genus of the mint family, see **mint**

mellarium, L. beehive; see **hive**

mellitus, L. pertaining to honey, sweet; *melliculus; mellitulus,* honeyed; see **honey**

mello- < Gr. *mello,* be about to do, only think of doing; *melletes,* delayer; see **delay**

mellow < ME. *melwe,* soft, genial, mature; see **ripe**

mellum, L. dog-collar; see **collar**

melo- < Gr. *melos,* leg, limb, see **leg**; < *melos,* song, tune, air, see **sing**; < *melon,* apple, see **apple**; *melon,* cheek, see **cheek**; *melon,* sheep, see **sheep**; *melo,* care for, see *meledonos* under **guard**

melody < Gr. *melodia,* singing, song; *melodos,* musical; see *melos* under **sing**

melolonthe, Gr. cockchafer; see **beetle**

melon < L. *melo, -onis,* m. (Gr. *melon,* n.), an apple-shaped melon, a fruit of the gourd family; *melothron,* n. a bryonylike plant: meloniform, *Melo diadema* (melon-shell), *Cucumis melo* (muskmelon), *Melothria pendula* (creeping cucumber), *Melocactus caesius* (gray melon-cactus).
   Pg. *abobra,* f. a kind of gourd: *Abobra tenuifolia* (a gourd).
   L. *angurium* (Gr. *angourion*), n. watermelon: gherkin, *Cucumis anguria* (West Indian gherkin).
   L. *apronia,* f. bryony:
   L. *archezostis,* f. bryony:
   Gr. *boubalion,* n. a kind of gourd or wild cucumber:
   Gr. *bryonia,* f. a vine of the gourd family: *Bryonia alba* (white bryony).
   Sp. *calabaza,* f. gourd: calabash, calabazilla *(Cucurbita foetidissima).*
   L. *citrium,* n. a kind of gourd:
   NL. *citrullus,* f. a genus of the gourd family: *Citrullus vulgaris* (watermelon), *Solanum citrullifolium* (a nightshade).
   L. *colocynthis, -idis* (Gr. *kolokynthis, -idos*), f. a kind of gourd, pumpkin, or melon: *Citrullus colocynthis* (colocynth).
   NL. *crescentia,* f. a genus of the gourd family, < Pietro Crescenzi, Italian agriculturist: *Crescentia alcata* (a calabash tree).
   L. *cucumis, -eris,* m. cucumber: *Cucumis sativus* (cucumber), *Epitrix cucumeris* (a flea-beetle), *Trichosanthes cucumeroides* (a gourd), *Cucumaria miniata* (a holothurian), *Macrosporium cucumerinum* (a fungus).
   L. *cucurbita,* f. gourd; *cucurbitula,* f. dim.; *cucurbitinus,* of gourds: Cucurbitaceae, gourd *(Lagenaria vulgaris), Cucurbita maxima* (squash), *Cucurbitina cruciata* (a foraminifer), *Crescentia cucurbitina* (a calabash-tree), *Dacus cucurbitae* (melon-fly).
   L. *dinuptila,* f. a kind of bryony:
   Gr. *echetrosis,* f. bryony:
   Gr. *hedrostis,* f. bryony:
   NL. *luffa,* f. a genus of the gourd family: *Luffa cylindrica*(vegetable-sponge).
   L. *notion,* n. a wild cucumber:
   L. *pepo, -onis* (Gr. *pepon, -os*), m. melon; *pepunculus,* m. dim.: pepino, pumpkin, pompion, pepinella *(Sechium edule), Cucurbita pepo*(pumpkin), *Peponocyathus variabilis* (a coral).
   NL. *sechium,* n. a genus of the gourd family: *Sechium edule*(chayote).
   L. *sicyus* (Gr. *sikyos*), m. cucumber, gourd; *sikydion,* n. dim.: *Sicyos[Sicyus] angulatus* (bur-cucumber), *Sicyogaster concolor* (a fish), *Sicydium antillarum* (a fish), *Cissus sicyoides* (a treebine), *Acanthosicyos horridus* (a cucurbitacead).

melota, L. (Gr. *melote*), sheepskin with the wool; see **skin**

melothron, Gr. bryonylike plant; see **melon**

melt < AS. *meltan;* see **flow**

Melusina < F. *Melusine,* a legendary fairy; see **spirit**

melydrion, Gr. dim. of *melos,* leg, song; see **leg, sing**

mematos, Gr. desired; see *mao* under **wish**

membraco- < Gr. *membrax, -akos,* a kind of cicada; see **cicada**

membrado- < Gr. *membras, -ados,* a kind of anchovy; see *bembras* under **herring**

membrane < L. *membrana,* f. skin that covers special parts of the body, parchment; *membranaceus; membraneus,* of skin or parchment: membranaceous, *Vaccinium membranaceum* (big whortleberry).
   Gr. *amnion,* n. membrane around the fetus: amnion, amniotic, *Rhodamnia trinervia*(mempoyan).
   Gr. *chorion,* n. membrane enclosing the fetus: chorion, chorioid, allantochorion, *Rhus choriophylla* (a sumac).
   NL. *conjunctiva,* membrane lining the eyelid; see *jungo* under **bind**
   Gr. *dertron,* n. caul, omentum:
   Gr. *epiploon,* n. caul, omentum:
   Gr. *hymen, -os,* m. membrane; *hymenion,* n. dim.: hymenium, hymenectomy, Hymenoptera, Hymenomycetes, *Hymenophyllum recurvum* (a filmy-fern), *Hymenocallis tubiflora* (a spider-lily), *Hymenodictyon excelsum* (a rubiacead), *Hymenopus bicornis* (a mantid).

L. *meninga* (Gr. *meninx, -ingos*), f. membrane covering the brain and spinal cord: meningeal, meningitis.

Gr. *mesenterion*, n. membrane investing and supporting the intestines: mesentery, mesenteric.

L. *omentum*, n. caul, membrane that covers the bowels: omentum, omentitis, omentocele.

L. *zirbus*, m. caul, omentum:

See: **skin, cover**

membraphya, Gr. a kind of anchovy; see *aphya* under **herring**

membrum, L. part; see **part**

memecylon, L. (Gr. *mimaikylon*), fruit of the strawberry-tree; see **arbutus**

memnonius, L. black, dark, < Memnon, fabled king of the Egyptians; see **black**

memor, L. mindful; *memorabilis*, worthy of mention; *memorialis*, of remembrance; see **remember**

mempsis, Gr. blame, reproof, complaint; *memptos*, blameworthy; see **blame**

mend < OF. *amender*, < L. *emendo, -atus*, correct, improve, free from errors; see **sew, edit**

menda, L. fault, defect, blemish, error; *mendosus*, full of faults; see **fault**

mendicus, L. beggarly, poor; beggar; *mendiculus*, dim.; see **beggar**

mene, Gr. moon, < *men, -os*, month; *meniskos*, crescent; see **moon, month, crescent**

menetos, Gr. inclined to stand fast, patient; see **stand**

meninga, L. (Gr. *meninx, -ingos*), membrane covering the brain; see **membrane**

menis; **menithmos**, Gr. wrath; see **anger**

meniscus, L. (Gr. *meniskos*), crescent; see **crescent**

meno- < Gr. *menos*, force, courage, spirit, strength, see **strong**; < *mene*, moon, < *men, -os*, month, see **moon, month**; < *meno*, remain, stay, see **stand**

mens, *mentis*, L. mind; see **mind**

mensa, L. table; *mensula*, dim.; *mensalis*, of the table; see **table**

mensis, L. (Gr. *men*), month; see **month**

menstraulis; **menstruus**, L. happening monthly; see *mensis* under **month**

mensura, L. a measuring, < *metior, mensus*, measure; see **measure**

-ment < L. *-mentum*, a suffix denoting means, action, or result of action; see **tool**

mentalis, L. of the mind; see *mens* under **mind**

mentha, L. (Gr. *minthe*), mint; see **mint**

mentigo, L. an eruption on lambs; see **disease**

mentitus, L. counterfeit, feigned; see *mendax* under **lie**

mentula, L. penis; see **penis**

mentum, L. chin; see **chin**

menyo, Gr. disclose information, betray; see **treason**

mephitis, L. noxious exhalation, bad odor; skunk; see **smell**

meracus, L. pure, unmixed; *meraculus*, dim.; see *merus* under **pure**

mercator, L. merchant, < *mercor, -atus*, buy, trade; see *mercor* under **trade**

merces, L. pay, wages, salary; *mercedula*, dim.; *mercenarius*, doing merely for pay; see **pay**

merciless, without compassion, pitiless, unsparing.

Gr. *allistos*, inexorable:

Gr. *anaides*, shameless, ruthless: *Anaides[Anaedes] fossulatus* (a beetle).

Gr. *aneleos*; *neles, -eos*, pitiless, ruthless: *Neleopisthus elegans* (a wasp).

Gr. *anilastos*, merciless, cruel, pitiless: *Anilastus rapax* (a wasp).

Gr. *anoiktos*; *anoiktirmon, -os*, pitiless, ruthless, merciless:

Gr. *apenes*, rough, harsh, cruel; see **rough**

Gr. *astorgos*, heartless, cruel: *Astorga saccharicida* (a bug).

L. *atrox, -ocis*, cruel, harsh, horrible, severe: atrocious, atrocity, *Psammochares atrox* (a wasp).

L. *crudelis*, rude, severe, unmerciful; see *crudus* under **rough**

L. *durus*, hard, tough, unfeeling; see **hard**

Gr. *dysaxiotos*, inexorable:

L. *ferreus*, hard, firm, unfeeling; see *ferrum* under **iron**

L. *illacrimabilis*, unmoved by tears, pitiless, inexorable:

L. *immisericors, -ordis*, pitiless:

L. *immitigabilis,* that cannot be softened or allayed: immitigable.
L. *implacabilis,* unappeasable: implacable.
L. *inclemens,* rough, unmerciful; see **rough**
L. *inexorabilis,* that cannot be moved by entreaty, inflexible, relentless: inexorable.
L. *inhumanus,* cruel, unmerciful: inhuman, inhumanity.
L. *Nero, -onis,* m. Roman emperor, notorious for his cruelty and tyranny: Neronian.
L. *saevus,* cruel, harsh; see **rough**
See: **rough, hurt, stand**

mercury < L. *Mercurius,* m. messenger of the gods and fabled inventor of chemistry: mercuric, mercurous, mercurial, *Mercurialis perennis* (a euphorbiacead). See ELEMENTS under **thing**
L. *cinnabaris* (Gr. *kinnabari*), red sulfide of mercury; see **cinnabar**
Gr. *Hermes,* m. messenger of the gods: hermetical, hermeneutics, *Hermodactylus tuberosus* (snakeshead-iris), *Hermaea cruciata* (a sea-slug).
Gr. *hydrargyros,* m. quicksilver, mercury: $HgCl_2$ (mercury bichloride, corrosive sublimate), hydrargyrol, hydrargyriasis.

mercy < OF. *merci,* < L. *merces, -edis,* f. pay, reward: merciful, unmerciful, merciless.
L. *clementia,* kindness, sympathy, mercy; *clemens,* merciful; see **mild**
L. *eleemosyna* (Gr. *eleemosyne*), f. alms, < *eleos,* m. pity, mercy; *eleemon, -os,* compassionate, merciful: eleemosynary, alms, almoner.
L. *gratificus,* kind, obliging; see *gratia* under **give**
L. *misericordia,* f. pity, compassion, mercy; *misericors, -ordis,* pitiful, merciful, tender-hearted; *miseror, -atus,* lament, pity: misericorde, misericordious.
Gr. *oiktos,* m.; *oiktosyne,* f. pity, mercy, < *oiktyzo,* pity; *oiktikos,* of pity; *oiktros,* pitiable: *Oectoperia sincera* (a moth), *Oectropsis latifrons* (a beetle).
See: **mild, give, help**

merda, L. dung; see **dung**

-mere; -meri; -mero; -merous < Gr. *meros; meris,* part, portion, share; *meridion,* dim.; see **part**

merenda, L. afternoon luncheon; see **eat**

meretrix, L. harlot; *meretricula,* dim. public courtesan; see **prostitute**

merges, L. sheaf; see **bundle**

mergus, L. diver, a kind of waterfowl; *mergulus,* dim.; see *mergo* under **dip**

merico- < Gr. *merikos,* particular; see **choose**

meridies, L. midday, noon, south; *meridionalis,* southern; see *dies* under **day**

merimna, Gr. anxious care or thought; see **think**

meringo- < Gr. *merinx, -ingos,* bristle; see **hair**

merinthos, Gr. cord, line, string; see **rope**

meris, -idos; meros, Gr. part, portion, share; *meridion,* dim.; *merisma, -tos,* part; *merismos,* division; *meristos,* divided, divisible; see **part, separate**

merit < L. *mereo, -itus,* deserve, earn; see **honor**

merluccius, NL. a kind of hake; see **cod**

mermitho- < Gr. *mermis, -ithos,* cord, rope; see **rope**

mermnos, Gr. a kind of hawk; see **hawk**

merops, Gr. bee-eater; see **bird**

meros, Gr. thigh, ham, femur, see **leg**; part, portion, share, see **part**

merry < AS. *myrige,* gay, happy, mirthful; see **joy, laugh**

mertensia, NL. a genus of the borage family; see **borage**

merula, L. blackbird, thrush; see **thrush**

merulentus, L. drunk; see **drink**

merum, L. unmixed wine; see **wine**

merus, L. pure, unadulterated, genuine; see **pure**

meryco- < Gr. *meryx, -ykos,* a ruminant, < *merykizo,* chew the cud; see **bite**

merx, mercis, L. commodities, goods, wares; see *mercor* under **trade**

mesembria, Gr. midday, noon, south; see **day**

mesentery < Gr. *mesenterion,* membrane investing and supporting the intestines; see **membrane**

mesites, Gr. arbitrator; see **judge**

meso- < Gr. *mesos (messos),* middle; *messatos,* midmost; *mesotes,* mean, middle; see **middle**

mespilum, L. (Gr. *mespilon*), medlar; *mespilus,* medlar-tree; see **apple**

messenger < OF. *messagier,* < L. *mitto, missus,* send; see **send**

messoros, Gr. boundary-stone; see **stone**

messura, L. a reaping, < *meto, messus,* reap, gather; *messor,* reaper; see **reap**

mesto- < Gr. *mestos,* full, filled; *mestotes,* fullness; see **abundance**

mestor, Gr. adviser, counselor; see *medos* under **aim**

meta, L. conical or pyramidal column at ends of the Roman Circus, goal-post, boundary-stake; *metula,* dim.; see **pillar**

meta- < Gr. *meta,* between, among, near, after, over, reversely, implying change; see **over, change**

metal < L. *metallum* (Gr. *metallon*), mine, ore, mineral; see **stone**

metanastes, Gr. migrant; see *astatos* under **move**

metaphor < Gr. *metaphora,* a simile with the comparing word omitted; see FIGURES OF SPEECH under **form**

metaxy, Gr. between; see **middle**

metel, NL. < Ar. *mathil,* the fruit of a datura; see **nightshade**

metella, L. basket or crate filled with stones for hurling at besiegers; see **basket**

metelys, *-ydos,* Gr. emigrant, changing; see **move**

meteor < Gr. *meteoros,* high in the air, suspended, raised, pertaining to things or phenomena in the air: meteorite, meteoric, meteoritics, meteorology.
L. *baetulus; baetylus* (Gr. *baitylos*), m. meteorite, sacred stone; *baitylon,* n. dim.: baetylus (baetulus).
L. *ceraunus* (Gr. *keraunos*), a stone supposed to have fallen from the sky, meteorite; see **thunder**
Gr. *dokos; dokites,* m. a meteor:
L. *novendial, -is,* rites on the ninth day after, or lasting nine days, as on the occasion of a death or a shower of meteorites; *novendialis, -e,* of nine days; see **rite**
Gr. *selas, -aos,* light, flash, fiery meteor; see **light**

meter, Gr. mother, see **mother**; < *metron,* measure, see **measure**

methecto- < Gr. *methektos,* participating, sharing; see **lot**

methetico- < Gr. *methetikos,* letting go, relaxing; see **rest**

method < L. *methodus* (Gr. *methodos*), mode of procedure; see **class, manner, way, use**

methy, Gr. wine; *methysos,* drunk, see **wine**; < *methyl-,* pertaining to the hydrocarbon radical in wood alcohol, see **alcohol**

meticulosus, L. fearful, excessively careful; see *metus* under **fear**

metis, Gr. counsel, wisdom, skill, cunning, craft, plan; see **know**

metochos, Gr. sharing, partaking; see **companion**

metoeco- < Gr. *metoikos,* emigrating; see **move**

metonomy < Gr. *metonymia,* change of name; see FIGURES OF SPEECH under **form**

metopa, L. (Gr. *metope*), space between two triglyphs; see *ope* under **hole**

metopion, Gr. ointment, gum of an African tree; see **resin**

metopon, Gr. brow, forehead; *metopidios,* of the forehead; see **head**

metoporinos, Gr. autumnal; see *opora* under **year**

metra, Gr. womb, entrance to the womb, vulva, heart of a tree, matrix; see **womb**

metric < L. *metricus* (Gr. *metrikos*), of measuring; see *metron* under **measure**

metridios, Gr. fruitful, filled with seed; see **fertile**

metrios, Gr. within measure, moderate, temperate; see *metron* under **measure**

metro- < Gr. *meter, -tros,* mother; *metrarion,* dim., see **mother**; < *metra,* womb, heart of a tree, matrix, see **womb**; < *metron,* measure, rule, see **measure**

metula, L. dim. of *meta,* goal-post; see *meta* under **pillar**

metus, L. fear, anxiety; see **fear**

meum, L. (Gr. *meon*), spignel; see **carrot**

mezereon < Ar. *mazariyum,* a small shrub, *Daphne mezereum:* mezereum.
L. *chamelaea* (Gr. *chamelaia*), f. olive-daphne *(Daphne oleoides):*
L. *citocacium,* n. olive-daphne:
Gr. *daphne,* laurel, bay-tree; now a genus of the mezereon family, Thymeleaceae; see **laurel**

NL. *pimelea*, f. a genus of the mezereon family: *Pimelea graciliflora* (a pimelea).
L. *thymelaea* (Gr. *thymelaia*), f. a genus of the mezereon family: *Thymelaea hirsuta* (a thymelea).

mezo- < Gr. *meizon*, greater; see *megas* under **large**

miarolo, It. name for the Baveno granite that contains drusy cavities lined with crystals; see **hollow**

miasma, Gr. stain, defilement, taint; *miantos; miaros*, stained, defiled; see **dirt**

mica, L. crumb, bit, morsel, grain; *micella; micula*, dim.; see **little**

micans, L. shining, twinkling; see *mico* under **light**

micidus, L. thin, poor; see **thin**

mico, L. shine, twinkle; see **light**

micro- < Gr. *mikros*, small, little; *meion*, less; *hekistos*, least; see **little**

mictilis, L. worthless, bad; see **bad**

micto- < Gr. *miktos*, mixed; see *migma* under **mix**

micturio, L. make water, urinate; *mictualis*, urinary; see *mingo* under **urine**

micula, L. dim. of *mica*, bit, crumb, grain; see **little**

Midas, Gr. mythical king at whose touch everything turned to gold; see **gold**

middle < AS. *middel:* middling, middleman, Middletown, Middlesex, midland, midnight, midsummer, midwife, midshipman, midst.
  Gr. *adiaphoros*, indifferent, neutral: adiaphorous.
  L. *ascius* (Gr. *askios*, without shadow), under the equator: ascian.
  L. *centrum*, midpoint of a circle; see **point**
  L. *gremium*, lap, middle, center; see **lap**
  L. *inter*, between, among; *interim*, meantime, meanwhile: interfere, intermediate, intersect, intercollegiate, interrupt, intercept, interlude, interval, interloper, international, intermarry, interrogate, interpreter, entertain, intelligent, interim, *Epigaea intertexta* (a trailing arbutus).
  L. *medius*, middle; *mediocris, -cre*, middling, ordinary; *medium*, n. go-between or means; *medialis; medianus; mediatenus*, middle, in the middle; *intermedius*, that is between: median, mediator, medieval, intermediate, immediate, medulla, medium, mean, mediocrity, mediocre, mesne, Mediterranean, Metz, meridian, dimidiate, moiety, mezzanine, mizzen, mezzotint, *Stellaria media* (chickweed), *Loxoplocus medialis* (a fossil gastropod), *Pomolobus mediocris* (a herring), *Taxocrinus intermedius* (a fossil crinoid), *Rhyparida intermedia* (a beetle).
  Gr. *mesos (messos); mesidios*, middle; *mesatos (messotos)*, midmost; *meson*, n.; *mesotes*, f. middle or central position, mean; *mesomphalos*, central; *meseres; anamesos*, in the midst of; *mesodme*, f. something between: mesoderm, mesobar, mesoblast, mesochilium, mesolite, mesothelium, mesentery, mesoventral, meson, Mesopotamia, Mesozoic, *Mesosaurus brasiliensis* (a fossil reptile), *Messatoporus rufiventris* (an ichneumonid), *Mesomphalus hartleyi* (a Devonian ostracode), *Mesodma ambigua* (a fossil mammal), *Mesodmodus ornatus* (a fossil fish), *Stenomesson incarnatum* (an amaryllidacead), *Mesembryanthemum aureum* (a fig-marigold).
  Gr. *metaxy*, between: *Metaxyblata hadroptera* (a fossil cockroach), *Metaxyothrips expectatus* (a thysanopterid). See *meta* under **over**
  L. *modicus*, moderate, medium, middling, ordinary, average: modicum, *Lycosa modica* (a spider).
  L. *neuter, -tra, -trum*, neither one nor the other, middle, impartial: neuter, neutral, neutrality, neutron.
  See: *centrum* under **point, insert, bind**

midemato- < Gr. *meidema, -tos*, smile; see *meidao* under **laugh**

midge < AS. *mycge*, gnat, fly; see **fly**

migma, Gr. mixture, compound; *migas, -ados*, promiscuous; see **mix**

mignonette < F. *mignonnette*, dim. of *mignon*, darling, favorite.
  L. *reseda*, f. a plant with healing properties: resedaceous, Resedaceae, *Reseda odorata* (mignonette), *Primnoa resedaeformis* (a coral).

migrator, L. wanderer; see *migro* under **move**

mild < AS. *milde*.
  Gr. *acholos*, without gall, meek; *acholia*, f. meekness:
  Gr. *aganos*, mild, gentle, kind: *Aganosma acuminata* (an apocynacead).
  Gr. *akaskaios*, gentle:
  L. *blandus*, of a smooth tongue, flattering, friendly, kind, mild; see **agreeable**
  L. *cicur*, tame, mild; see **tame**
  L. *clemens, -entis*, calm, mild, merciful; *clementia*, f. kindness, sympathy, mercy: clement, clemency, inclement, Clementine.

Gr. *epios*, gentle, kind, soothing: *Epiopelmidea erythrogastra* (a wasp).
Gr. *euagogos*, docile, easily led:
Gr. *eukratos*, well-tempered, temperate, mild: *Eucratoscelus longiceps* (a spider).
L. *gentilis*, of the same race or clan: gentle, gentility.
L. *hapalus* (Gr. *hapalos*), soft, tender, gentle, delicate; see **soft**
Gr. *hilaros*, gracious, kind, gentle:
L. *levamen, -inis*, mitigation, comfort, solace; see **soothe**
L. *mansuetus*, mild, soft, gentle, tame, quiet: *Cygnus mansuetus* (a swan).
Gr. *meilichos*, gentle, kind, soothing; see *meilisso* under **soothe**
L. *mitis*, mild; *mitiusculus*, dim.; *mitigo, -atus*, soften, mellow, tame, appease, calm, assuage: mitigate, mitigation, unmitigated, immitigable, *Caryota mitis* (a fishtail-palm), *Andrena permitis* (a bee).
L. *placidus*, quiet, still, gentle; see **rest**
Gr. *praos*, mild, meek, gentle, tame; see **tame**
Gr. *prosenes*, gentle, soft, smooth:
L. *temperatus*, limited, moderate, sober; see **govern**
See: **soothe, smooth, soft, rest, mercy, tame**

mildew < AS. *mildeaw;* see **fungus**

miles, *-itis*, L. soldier; *militaris*, of soldiers and war; see **army**

milesium, L. a kind of kingfisher; see **kingfisher**

milicho- < Gr. *meilichos*, gentle, kind; *meiligma*, anything that soothes; see *meilisso* under **soothe**

milio, L. a kind of hawk, kite; see **hawk**

milium, L. millet; *miliaceus; miliarius*, of millet; see **grain**

milk < AS. *meolc*.
Gr. *amelktos*, milked, < *amelgo*, milk; *amelxis; amolge*, f. a milking; *amolgaios*, of milk:
Gr. *bdallo*, milk, suck; *bdalsis*, f. a milking: *Bdalsidothrips levis* (a thysanopterid).
L. *colostrum*, n. first milk after parturition, biesting: colostrum, colostral.
Gr. *gala, -aktos*, n. milk; *galaktion*, n. dim.; *galaktikos; galaxios*, milky, milk-white; *glageros*, full of milk; *galaxias*, m. milkyway, a kind of fish; *galax, -aktos*, f. a plant of the diapensia family; *agalax, -aktos*, without milk, dry; *eugalos*, giving milk; *euglages; periglages*, full of milk, with much milk; *aphrogala*, n. frothed milk; *oxygala*, n. sour milk, whey: galactic, galactose, galactan, galactite, galactophore, galaxy, *Galactopterus simillimus* (a wasp), *Galax aphylla* (galax), *Galanthus nivalis* (snowdrop), *Brosimum galactodendron* (cow-tree), *Galactia velutina* (a milkpea), *Galega officinalis* (goatsrue), *Shortia galacifolia* (a diapensiacead), *Glaux maritima* (sea-milkwort), *Euglages scripta* (a bee), *Ornithogalum speciosum* (a star-of-Bethlehem), *Polygala alba* (white polygala).
L. *lac, lactis*, n. milk; *lacteus*, of milk, milky; *lacteolus*, dim.; *lactarius*, of milk, milky; m. milkman, dairyman; *lactineus*, milklike, milk-colored; *lacto, -atus*, yield milk, give suck: lactic, lactation, lactose, lactase, lactant, lactamide, lactimide, lactarene, lactarium, lactary, lacteal, lacteous, lactescent, lactiferous, lactoglobulin, lactyl, ablactation, prolactin, lettuce, *Lactophrys tricornis* (cow-fish), *Lactarius deliciosus* (a mushroom), *Campanula lactiflora* (a bellflower), *Ornithogalum lacteum* (a star-of-Bethlehem), *Dendrocoelum lacteum* (a planarian), *Gymnema lactiferum* (a cow-tree).
L. *mulgeo, mulctus*, milk; *mulctra*, f. milk-pail: emulgent, emulsion, emulsin, *Caprimulgus pulchellus* (a goatsucker).
Gr. *opisma*, milky sap of plants; see *opos* under **juice**
Gr. *pyos*, m. first milk after parturition, colostrum:
See: **juice, udder, suck**

milkweed
NL. *acerates*, f. a genus of the milkweed family: *Acerates pumila* (an asclepiadacead).
L. *asclepias, -adis* (Gr. *asklepias, -ados*), f. a genus of the milkweed family, < *Aesculapius*, god of medicine: Asclepiadaceae, *Asclepias speciosa* (showy milkweed).
NL. *secamone*, f. a genus of the milkweed family: *Secamone emetica* (an asclepiadacead).
NL. *stapelia*, f. a genus of the milkweed family, < J. B. van Stapel, Dutch botanist: *Stapelia gigantea* (a carrion-flower).

milkwort
L. *polygala*, f. (Gr. *polygalon*, n.), milkwort: *Polygala sanguinea* (a milkwort).
NL. *securidaca*, f. a genus of the milkwort family: *Securidaca elliptica* (a polygalacead).

mill < AS. *myln,* < L. *molina,* mill; see **grind**

mille, L. thousand; *millenarius,* of a thousand; *millesimus,* thousandth; see **thousand**

millepeda, L. a myriapod with two pairs of legs to each segment; see **myriapod**

millet < L. *milium;* see **grain**

miltos, Gr. red lead, minium; red; see **red**

milvage, L. a fish, probably a gurnard; see **gurnard**

milvus, L. kite, hawk; *milvinus,* of a kite or hawk; see **hawk**

mimichmos, Gr. neigh; see **neigh**

mimicus, L. (Gr. *mimikos*), imitative; *mimus* (Gr. *mimos*), actor; *mimulus,* dim.; see *mimus* under **equal**

mimmulus, L. a plant; see **plant**

mimosa, NL. a genus of the bean family; see **bean**

mimulus, L. dim. of *mimus,* actor; see *mimus* under **equal, figwort**

mimusops, NL. a genus of the sapodilla family; see **sapodilla**

minatio, L. threat, menace; *minax, -acis,* jutting out, threatening; see *minor* under **projection**

mind < AS. *gemynd.*
    L. *animus,* mind, soul, spirit; see *anima* under **life**
    L. *dyscolus* (Gr. *dyskolos*), of bad temper, irritable, peevish, hard to please:
    Gr. *gnome,* mind, disposition; *gnomidion,* dim.; see *nosco* under **know**
    L. *halitus,* breath, spirit, soul; see *halo* under **breathe**
    L. *mens, mentis,* f. mind; *mentalis,* of the mind: mental, mention, dementia, memento, commentary, *mens sana in corpore sano.*
    Gr. *noös,* mind, intellect; *noëtikos,* intelligent; see *noëma* under **think**
    Gr. *phren (-phron), -os,* f. mind, brain; *agathophron, -os,* of good disposition; *artiphron,* of sound mind; *kenophron,* empty-minded; *megalophron,* high-minded, generous; *neophron,* childish; *sophron,* of sound mind, sensible, wise; *sophronikos,* temperate; *sophrosyne,* f. good sense, discretion, moderation; *tlesiphron,* staunch: phrenology, phrenic, phrenitis, frenzy, frantic, phrenoplegia, *Euphrasia officinalis* (eyebright), *Sophronorrhinus duvernoyi* (a beetle), *Sophrosyne murrayi* (an amphipod).
    Gr. *psyche,* f. breath, life, soul, spirit, mind, butterfly; *psychikos,* of the mind; *psychosis,* f. animation, quickening; *psychotria,* f. an enlivening; *apsychos,* lifeless, spiritless; *eupsychos,* courageous: psychology, psychoanalysis, psychosomatic, psychosis, psychial, psychopathic, psychiatry, metempsychosis, *Psychotria undata* (a rubiacead), *Helicopsyche borealis* (a caddisfly), *Apsychomyrmex myops* (an ant), *Hydropsychodes albocincta* (a trichopterid).
    L. *spiritus,* soul; see **spirit**
    Gr. *thymos,* m. mind, soul, spirit, temper, courage; *thymikos,* high-spirited, passionate; *athymos,* spiritless, fainthearted; *euthymos,* cheerful, kind; *megathymos,* high-minded; *mikrothymos,* narrow-minded; *tlethymos,* stouthearted, staunch: epithymetic, lypothymia.
    See: **think, know, spirit**

mindax, Gr. a Persian incense; see **resin**

mine < VL. *mina,* < a Celtic word like Ir. *mein,* W. *mwyn:* miner, mining, mineral.
    L. *arrugia,* shaft, pit; see **hole**
    Gr. *chryseion,* n. gold mine; *chrysoplysion,* n. placer:
    L. *columnar, -is,* n. stone quarry:
    L. *cotoria,* f. whetstone quarry:
    L. *cunicularius; cuniculator, -is,* m. miner; *cuniculum,* n. underground passage, hole, pit: *Furnarius cunicularius* (a bird), *Speotyto cunicularia* (burrowing-owl).
    L. *fodina,* f. pit, mine; *arenifodina,* f. sand-pit; *ferrifodina,* f. iron mine:
    Gr. *hyponome,* f. underground passage, tunnel, mine: hyponome, hyponomic.
    L. *lapidicina,* f. stone quarry; *lapidicida (lapicida),* m. quarryman, stone-cutter; *Raphistomina lapicida* (a fossil gastropod).
    L. *latomium* (Gr. *latomion*), n. stone quarry:
    L. *metallum* (Gr. *metallon*), mine, quarry, ore, mineral; *metallarius,* miner; see **stone**
    Gr. *orygma, -tos,* pit, mine, tunnel, excavation; see *orysso* under **dig**
    See: **dig, hole, stone**

mineral < F. *mineral;* see **stone, mine**

Minerva, L. goddess of wisdom, identified with the Greek Athena; see **know**

mingo, *-inctus; -ictus,* L. urinate; see **urine**

miniatus, L. bright red; see **red**

minimus, L. least; see *parvus* under **little**

minister, L. servant; see **servant**

minium, L. native cinnabar; red lead; see **cinnabar**

minnow < AS. *myne;* see **carp, fish**

minor, L. less; see *parvus* under **little**

Minotaurus, L. (Gr. *Minotauros*), monster with the head of a bull and body of a man; see **animal**

mint < AS. *minte,* < L. *mentha* (Gr. *minthe*), f.; *menthastrum,* n. wild mint; *calamintha* (Gr. *kalaminthe*), f. a mint: menthol, mintbush *(Prostanthera nivea),* calamint *(Satureia calamintha),* peppermint *(Mentha piperita),* spearmint *(Mentha spicata), Mentha citrata* (a mint), *Spilosoma menthastri* (a moth).
L. *abiga,* f. a species of mint said to produce abortion:
NL. *agastache,* f. a genus of mints: *Agastache urticifolia* (a giant-hyssop).
L. *ajuga,* f. a mint: *Ajuga pyramidalis* (a bugleweed).
Gr. *akinos,* m. basil-thyme: *Satureia acinos* (basil-balm), *Acinocoris cornutus* (a bug).
L. *amaracum* (Gr. *amarakon*), n. marjoram; ML. *majorana,* f.: sweet marjoram *(Majorana hortensis), Amaracus cordifolius* (a mint).
L. *ballota* (Gr. *ballote*), f. horehound: *Ballota nigra* (black horehound), *Salvia ballotaeflora* (a sage).
L. *betonica,* f. a genus of the mint family: betony *(Betonica officinalis).*
Sp. *bugula* (L. *bugillo, -onis,* m.), f. a kind of mint: bugleweed *(Ajuga reptans).*
L. *campsanema* (Gr. *kampsanema*), n. rosemary:
L. *chamaedrys* (Gr. *chamaidrys*), f. germander: *Teucrium chamaedrys* (germander).
L. *chamaepitys* (Gr. *chamaipitys*), f. ground-pine, a mint: *Ajuga chamaepitys* (a bugle).
Gr. *chondris, -idos,* a kind of mint:
L. *citrago, -inis,* f. lemon balm:
L. *cleonicum,* n. a kind of basil:
L. *clinopodium* (Gr. *klinopodion*), n. field-basil: *Clinopodium coccineum* (basil).
L. *coleus* (Gr. *koleos,* sheath), m. a genus of the mint family: *Coleus pumilus* (dwarf coleus).
NL. *collinsonia,* f. a genus of the mint family, < Peter Collinson, English botanist: *Collinsonia canadensis* (horse-balm).
L. *conula,* f. a kind of mint:
L. *corissum,* n. a kind of mint:
L. *crosmis,* f. a kind of sage:
L. *cunila* (Gr. *konile*), f. a mint: *Cunila origanoides* (dittany).
L. *dictamnus* (Gr. *diktamnos*), dittany; now applied to a genus of the Rutaceae; see **rue**
L. *dochlea,* f. a mint:
L. *galeopsis* (Gr. *galiopsis*), f. a kind of mint: *Galeopsis ladanum* (red hemp-nettle).
Gr. *glechon (blechon), -os,* f. pennyroyal: *Glechoma hederacea* (ground-ivy), *Aylax glechomae* (catnip gall-wasp).
NL. *gomphostemma, -tos,* n. a genus of the mint family: *Gomphostemma niveum* (a mint).
NL. *hedeoma,* f. a genus of the mint family: *Hedeoma hispida* (a pennyroyal).
Gr. *hedyosmos,* m. mint:
L. *holocyrum,* n. a mint:
L. *horminum* (Gr. *horminon*), n. a kind of sage: *Horminum pyrenaicum* (a sage), *Salvia horminum* (Joseph sage).
NL. *hyptis,* f. a genus of the mint family: *Hyptis spicigera* (a bushmint).
L. *hyssopus* (Gr. *hyssopos*), f. a genus of the mint family: *Hyssopus officinalis* (hyssop), *Sideritis hyssopifolia* (hyssop-ironwort), *Hedeoma hyssopifolium* (a pennyroyal).
Gr. *kamelopodion,* n. a kind of horehound:
Gr. *knekion,* n. marjoram:
L. *lamium,* n. dead-nettle: *Lamium maculatum* (spotted dead-nettle).
ML. *lavandula (lavendula),* f. a genus of the mint family; *Lavandula dentata* (toothed lavender), *Streptomyces lavendulae* (source of streptothricin), *Oenothera lavandulaefolia* (an onagracead).
NL. *leonurus,* m. a genus of the mint family: *Leonurus sibiricus* (Siberian mother-wort).
Gr. *linostrophon,* n. horehound:
NL. *lycopus,* m. a genus of the mint family: *Lycopus virginicus* (a mint).
L. *marrubium,* n. horehound: *Marrubium vulgare* (horehound).

L. *marum* (Gr. *maron*), n. a kind of sage: *Teucrium marum* (cat-thyme), *Eriogonum marifolium* (a polygonacead).

Gr. *melissa (melitta)*, f. honey-bee; a genus of the mint family; *melittaina*, f.; *melissophyllon*, n. a mint: *Melissa officinalis* (common balm), *Melittis melissophyllum* (a mint).

NL. *monarda*, f. a genus of the mint family, < Nicolas Monardes, Spanish botanist: *Monarda fistulosa* (wild bergamot), *Monardella odoratissima* (a mint).

L. *nepeta*, f. catnip: catnip(catnep), *Nepeta amethystina* (a catnip), *Agastache nepetoides* (a giant-hyssop), *Leonotis nepetaefolia* (lion's-ear), *Satureia nepeta* (catnip-savory).

L. *ocimum* (Gr. *okimon*), n. basil: *Ocimum sanctum* (holy basil), *Saponaria ocimoides* (rock-soapwort).

Gr. *onitis, -idos*, f. a kind of mint: *Onitis furcifer* (a beetle), *Oniticellus cinctus* (a beetle), *Majorana onitis* (pot-marjoram).

L. *origanum* (Gr. *origanon*), a mint: *Origanum vulgare* (wild marjoram), *Linaria origanifolia* (a toadflax).

L. *orozelum*, n. a kind of mint:

NL. *perilla*, f. a genus of the mint family: *Perilla frutescens* (perilla).

Gr. *phlomis*, mullein; now a genus of the mint family; see **mullein**

NL. *phyllostegia*, f. a genus of the mint family: *Phyllostegia vestita* (a mint).

L. *polium* (Gr. *polion*), n. an aromatic plant, probably a germander: *Teucrium polium* (golden germander), *Kalmia polifolia* (bog-laurel).

Gr. *prasion*, n. horehound: *Prasium majus* (a mint).

NL. *prunella*, f. a genus of the mint family: *Prunella[Brunella] vulgaris* (common selfheal).

L. *pulegium*, n. fleabane, pennyroyal: pulegol, pennyroyal *(Mentha pulegium)*, *Hedeoma pulegioides* (American pennyroyal).

NL. *pycnanthemum*, n. a genus of the mint family: *Pycnanthemum incanum* (hoary mountain-mint).

L. *rosmarinus*, m. rosemary: *Rosmarinus officinalis* (rosemary), *Haplostachys rosmarinifolia* (a mint), *Lithospermum rosmarinifolium* (a gromwell).

L. *salvia*, f. sage: sage, *Salvia splendens* (scarlet sage), *Salvia coccinea* (Texas sage).

Gr. *sampsychon*, n. marjoram:

L. *satureia*, f. savory: *Satureia hortensis* (savory).

L. *scordium* (Gr. *skordion*), n. a mint with a garliclike smell: *Teucrium scordium* water-germander).

NL. *scutellaria*, f. a genus of the mint family: *Scutellaria altissima* (a skullcap).

L. *serpyllum* (Gr. *herpyllion*, n.; *herpyllos*, m.), n. thyme: *Thymus serpyllum* (wild thyme), *Veronica serpyllifolia* (a speedwell).

L. *serratula*, betony; now a genus of the composite family; see **composite**

Gr. *sideritis, -idos*, f. ironwort; a genus of the mint family: *Sideritis libanotica* (Lebanon ironwort).

Gr. *sphakos*, m. sage; *sphakodes*, abounding in sage: *Sphacophyllum bojeri* (a composite), *Sphacele calycine* (wood-balm).

L. *stachys, -yos*, f. (Gr. *-yos*, m.), a mint: *Stachys sylvatica* (a betony), *Haplostachys truncata* (a mint).

L. *stoechas, -adis* (Gr. *stoichas, -ados*), f. a kind of mint: *Lavandula stoechas* (French lavender), *Arctotis stoechadifolia* (a composite).

NL. *synandra*, f. a genus of the mint family: *Synandra grandiflora* (a mint).

L. *teucrium* (Gr. *teukrion*), n. germander: *Teucrium montanum* (mountain-germander), *Teucridium parvifolium* (a verbenacead), *Verbena teucrioides* (fragrant verbena).

Gr. *thymbra*, f. savory: *Thymbra spicata* (a mint).

L. *thymum* (Gr. *thymon*), n. thyme: thymol, glycothymol, *Thymus vulgaris* (thyme), *Cuscuta epithymum* (a dodder), *Euphorbia thymifolia* (a spurge), *Hedeoma thymoides* (a pennyroyal).

NL. *trichostema*, n. a genus of the mint family: *Trichostema dichotomum* (blue curls).

L. *trixago, -inis*, f. germander:

NL. *zizyphora*, f. a genus of the mint family: *Zizyphora clinopodioides* (a mint).

Gr. *zygis, -idos*, f. wild thyme:

mint < AS. *mynet*, coin, money, < L. *moneta*, coin, place where money was coined; see **money**

minus, L. less, see *parvus* under **little**; bare, smooth, see **bare**

minute < L. *minuta*, see **time**; < *minutus*, little, small; *minusculus*; *minutulus*, dim., see *parvus* under **little**

minyas, L. a plant; see **plant**

minyorios, Gr. short-lived; see **short**

minyrisma, Gr. a warbling; *minyristria,* warbler; *minyros,* warbling, whimpering; see **sing**

minys, Gr. little, small, < *minytho,* lessen, shorten; see **little**

mio- < Gr. *meion,* less; see *mikros* under **little**

mirabilis, L. wonderful, strange; *mirificus,* causing wonder; see *miror* under **wonder**

miracle < L. *miraculum,* a strange, wonderful thing or occurrence; see *miror* under **wonder, strange**

miraco- < Gr. *meirax, -akos,* young girl, lass; *meirakidion; meirakion,* dim.; see **child**

mirandus, L. wonderful, strange, singular; see *miror* under **wonder**

mire < ON. *myrr,* bog, swamp, deep mud; see **marsh, earth**

mirmillo, L. a gladiator; see **fight**

mirror < OF. *mirour,* < L. *miror,* look at, wonder at.
   Gr. *enoptron; esoptron; katoptron,* n. mirror: catoptrics, catoptrite, catoptromancy, *Catoptrophorus semipalmatus* (willet), *Enoptroteuthis spinicauda* (a cephalopod).
   Gr. *hyalion,* n. mirror:
   Gr. *skaphion,* small bowl, basin, concave mirror; see **cup**
   L. *speculum,* n. mirror; *specularis,* of a mirror: speculum, *Specularia biflora* (a Venus' looking-glass), *Orpacophora speculifera* (a katydid), *Hirundichthys speculiger* (a fish), *Garrulus bispecularis* (a jay), *Distephanus speculum* (a flagellate).
   Gr. *stilbe,* lamp, mirror; see *stilbo* under **light**

mirth < AS. *myrth;* see *myrige* under **joy**

mirus, L. wonderful; see *miror* under **wonder**

mis- < AS. *mis-;* OF. *mes-,* < L. *minus,* less; wrong, ill, not; see **bad**

miscellaneus, L. mixed; see *misceo* under **mix**

mischief < OF. *meschief;* see **bad, hurt, trouble**

mischos, Gr. pedicel, stalk; see **stem**

miscix, -icis, L. changeable, inconstant; see **change**

miser, L. wretched; *misellus,* dim., see **sad;** *miseror, -atus,* lament, pity, see **mercy**

misettia, Gr. lewdness, lust; see **lewd**

miso- < Gr. *misos,* hate; *misetes,* hater; see **hate**

miss < AS. *missan,* fail; see **fault**

missile < L. *missile, -is,* n. anything cast, hurled, or thrown; *missilis,* that may be cast, hurled, or thrown:
   Gr. *bolis, -idos,* f. missile: bolide, *Dibolia timida* (a beetle).
   L. *falarica,* f. an incendiary missile of tow and pitch hurled by a catapult:
   See: **throw, ball, arrow, stone**

missus, L. sent; see *mitto* under **send**

mist < AS. *mist,* darkness; see **cloud, drop**

mistake < AS. *mis,* wrong, *tacan,* take; see **fault, falter, bad**

misthos, Gr. pay, wages; *misthoma,* price; see **pay**

misticius, L. hybrid, mongrel; see *misceo* under **mix**

mistletoe < AS. *misteltan.*
   Gr. *hyphear, -os,* n. a kind of mistletoe:
   Gr. *ixos,* mistletoe, birdlime; see **glue**
   NL. *loranthus,* m. a genus of the mistletoe family: Loranthaceae, *Loranthus coccineus* (a mistletoe).
   NL. *phoradendron,* n. a genus of the mistletoe family: *Phoradendron villosum* (Pacific mistletoe).
   Gr. *stelis, -idos,* f. a kind of mistletoe: *Stelis foliosa* (an orchid).
   L. *viscum,* n. mistletoe, birdlime: *Viscum album* (European mistletoe), *Turdus viscivorus* (missel-thrush).

mita, ML. mitten; see **glove**

mite < AS. *mite,* a minute arachnid, anything very small; see **tick, small**

mitella, L. turban; see *mitra* under **cap**

mithrax, L. a kind of Persian stone; see **stone**

mitis, L. mild; *mitigo, -atus,* soften, mellow, appease, soothe; see **mild**

mitos, Gr. thread; see **thread**

mitra, L., Gr. turban; *mitella; mitrula; mitrion,* dim.; see **cap**
mitten < OF. *mitaine;* see **glove**
mittendarius, L. provincial tax-collector; see **pay**
mitylos, Gr. hurt, shortened; see **hurt**
mitys, Gr. beeswax; see **wax**
miuro- < Gr. *meiouros,* curtailed, shortened; see **short**
mix < AS. *miscian.*
  Gr. *adiakritos,* undistinguishable, mixed:
  L. *adultero, -atus,* falsify, corrupt; see **lie**
  Gr. *androgynos,* m. a man-woman, hermaphrodite: androgynous, *Androgynoceras appressum* (an ammonite).
  L. *aquatus,* mixed with water, watery, thin; see **thin**
  L. *ardelio,* meddler, busybody; see **busy**
  L. *bigener, -a, -um,* hybrid, mongrel:
  L. *compono, -ositus,* put together, mix, unite: compose, composition, compound.
  L. *concoquo, -octus,* boil together, prepare, devise: concoct, concoction.
  L. *confusio, -onis,* mixture, disorder, < *confundo, -fusus,* pour together, mingle; see **confusion**
  L. *depso, depstus* (Gr. *depso; depho*), knead; *depsticius,* kneaded: *Dephomys defua* (a mouse).
  L. *dilutus,* mixed, thin, weak: dilute, *Chrysocoma dilutum* (a fly).
  L. *farrago, -inis,* f. medley, mixture: farrago.
  Gr. *hermaphroditos,* m. a person having both male and female characters, like Hermaphroditus, son of Hermes and Aphrodite: hermaphrodite, *Callitriche hermaphroditica* (a water-starwort).
  L. *hybrida,* f. mongrel: hybrid, hybridization, *Petunia hybrida* (common petunia).
  L. *infundo, -fusus,* pour into, mix: infuse, infusion.
  Gr. *kerannymi,* mix; *kerasma, -tos,* n. mixture; *kerastes,* m. mixer; *kerastikos,* of mixing; *kerastos,* mixed; *krama, -tos,* n. mixture; *kramation,* n. dim.; *krasis,* f. a mixing, compounding; *akrasia,* f. poor mixture; *synkerasma, -tos,* n. mixture, hash: crasis, idiosyncrasy, crater, grail, *Cerasma cornutum* (a trichopterid), *Achasia granulata* (a myxomycete), *Syncrasis fucicola* (a wasp).
  Gr. *kinetron,* ladle, stirring-rod; see **spoon**
  Gr. *kykao,* stir up; *kykeon, -os,* m. mixed drink; *kykesis,* f. a mixing; *kyketes,* m. mixer, agitator, stirrer; *kykethra,* f. mixture; *kykethron,* n. ladle, mixer, agitator: *Cycetes thyrsophoroides* (a psocopterid), *Cycethra simplex* (a starfish).
  Gr. *masso,* knead; *magma, -tos,* n. salve, kneaded mass; *maktra,* f. kneading-trough; *maktos,* kneaded: message, magma, magmatic, *Mactra solidissima* (a surf-clam).
  Gr. *migma, -tos,* n. mixture, compound, < *mignymi,* mix up, mingle; *miktos,* mixed, blended; *mixis,* f. a mixing, mingling; *mixias,* m. mixer; *migas, -ados,* promiscuous; *anamixis,* f. a mingling; *haimomixia,* f. incest; *symmigma, -tos,* n. a commixture: apomixis, migmatite, *Mictocrinus robustus* (a Devonian crinoid), *Polymixia japonica* (a fish), *Mixodectes crassiusculus* (a fossil mammal).
  L. *misceo, mixtus,* mix, mingle; *miscellaneus; miscellus,* mixed; *misticius; mixticius,* of mixed race, hybrid, mongrel; *mixtura,* f. a mingling, compound; *mixtarius,* m. a mixing-vessel; *permixtus,* mixed, confused; *promiscuus,* mixed, indiscriminate: mixture, admixture, miscible, miscellaneous, promiscuous, meddle, medley, melange, mestizo, mastiff, miscegenation, *Saprolegnia mixta* (a fungus). Derive: hodgepodge.
  Gr. *pelanos,* m. any thick liquid mixture:
  Gr. *phyrtos,* mixed, < *phyro,* mix, knead; *phyrama; phyrma, -tos,* n.; *phyrmos,* m. thing mixed or kneaded; *symphyrtos,* commingled:
  Gr. *stais, staitos,* n. dough; *staitinos,* of dough; *staitodes,* like dough, doughy:
  Gr. *synchlys, -ydos,* mixed, promiscuous, washed together by the waves:
  Gr. *synchysis,* f. commixture, confusion, < *syncheo,* pour together, confuse, destroy:
  Gr. *synthesis,* compounding, composition; *synthetos,* put together, compounded; see *tithemi* under **place**
  Gr. *syrphetodes,* jumbled together; see *syrphetos* under **dirt**
  Gr. *syrrhadios,* promiscuous; see *rhadios* under **rest**
  Gr. *tarasso,* stir up, confuse; see **confusion**
  L. *tempero, -atus,* blend or mix in due proportion; see **govern**
  Gr. *torynetos,* stirred; see *toryne* under **spoon**
  L. *tudiculo, -atus,* stir; see *tudicula* under **spoon**
  See: **confusion, complex, dishevel, different, turn, thin, spoon**

mnasium, L. (Gr. *mnasion*), a kind of sedge; see **sedge**

mnemo-; mnesi- < Gr. *mnemon, -os,* mindful, unforgetting; *mneme,* memory; *mnesios,* of memory; see *mnemon* under **remember**

mnester, Gr. wooer, suitor; *mnestos,* wooed, married; see **woo**

mnion, Gr. moss; *mniaros,* mossy; *mniodes,* mosslike; see **moss**

mnu- < Gr. *mnous (mnoos),* down; *mnoudion,* dim.; see **feather**

moan < AS. *maenan;* see **sad, weep**

mobilis, L. movable; *mobilitas,* movableness; see **move**

mochlos, Gr. crowbar, lever; see **lever**

mochtheros, Gr. suffering hardship, wretched, < *mochthos,* toil, trouble, distress; see **trouble**

mock < OF. *mocquer;* see **laugh**

moco- < Gr. *mokos,* mocker; see **laugh**

model < F. *modele,* < L. *modulus;* see **form**

moderator, L. director, governor, manager; see *modero* under **govern**

modernus, L. of the present; see **now**

modest < L. *modestus,* moderate, gentle, unassuming, chaste: modesty, *Hyrachyus modestus* (a fossil ungulate), *Lycosa modesta* (a spider).
  Gr. *aidemon, -os,* bashful, modest; *aidemosyne,* f. modesty; *aidos,* f. sense of modesty, respect, shame: *Aedemonophilus erirrhinoideus* (a beetle), *Aedophron aurorina* (a moth), *Aidemosyne[Aedemosyna] modesta* (a finch).
  Gr. *aischyntelos,* bashful, modest: *Aeschyntelus angularis* (a bug).
  Gr. *akompastos,* modest:
  Gr. *atyphos,* not puffed up, modest:
  Gr. *baukos,* affected, priggish, prudish:
  L. *modicus,* middling, temperate, modest; see **middle**
  L. *pudens,* modest, shamefaced; *pudicus,* bashful, chaste; see *pudeo* under **shame**
  L. *verecundus,* bashful, modest, shy, coy, demure: *Carpocanium verecundum* (a radiolarian), *Scirpus verecundus* (a sedge), *Zephyranthes verecunda* (a zephyr-lily).
  See: **shame, fear, govern**

modicus, L. moderate, medium, average, middling; see **middle**

modiolus, L. nave or hub of a wheel, dim. of *modius,* a dry measure; see *modus* under **measure**

modus, L. proper measure, custom, manner; *modulus,* dim. norm, meter; see **measure**

moecha, L. (Gr. *moiche*), adulteress; *moechus* (Gr. *moichos*), adulterer, paramour; see **prostitute**

moenium, L. rampart, bulwark, city wall; see **wall**

moero- < Gr. *moira,* part, share, lot, destiny; see **lot**

Moesiacus, L. of *Moesia,* f. Bulgaria and Serbia: *Crocus moesiacus* (a spring crocus).

mogos, Gr. distress, pain, trouble; *mogeros,* wretched, suffering; see *mochtheros* under **trouble**

moist < OF. *moiste,* < L. *mustus,* new, fresh; see **wet**

mola, L. millstone, mill; *molaris,* of a millstone or mill; see *molo* under **grind**

mold < *molde,* earth, soil, see **earth**; < ON. *mygla,* grow musty with mildew, see **fungus**; L. *modulus,* model, norm, see **form**

mole < uncertain origin.
  NL. *blarina,* f. shrew: *Blarina brevicauda* (short-tailed shrew).
  Gr. *hyrax, -akos,* m. shrew: hyracoid, hyracodon, hyraceum, hyrax *(Procavia ruficeps), Hyracotherium venticolum* (an eohippus).
  Gr. *siphneus,* m. mole: *Siphneus aspalax* (a mole).
  Gr. *skalops, -opos,* m. mole: *Scalopus aquaticus* (mole), *Scaloposaurus constrictus* (a fossil reptile).
  L. *sorex, -icis,* m. a shrew-mouse; *soricinus,* of a shrew: *Sorex personatus* (a long-tailed shrew), *Uropsilus soricipes* (uropsile), soricident, soricine, soricoid.
  Gr. *spalax, -akos,* f. mole: *Spalax typhlus* (mole-rat), *Spalacomys indicus* (a rodent).
  L. *talpa,* f. mole; *talpinus,* of moles, molelike: *Gryllotalpa hexadactyla* (a mole-cricket), Talpidae, *Talpa europaea* (mole).

moles, L. heap, mass; *molecula,* dim.; see **heap**

molestus, L. troublesome, irksome, annoying, disturbing; see **trouble**

molgos, Gr. hide, skin; *molginos,* of skin; NL. *molgula,* a tunicate; see **skin**

molina, L. mill; *molinus,* of a mill; *molinarius,* miller; see *molo* under **grind**

molitus, L. striven, built, see *molior* under **make;** ground, see *molo* under **grind**

mollis, L. soft; *molluscus,* soft; *mollio, mollitus,* soften; see **soft**

mollugo, L. carpet-weed; see *mollis* under **soft**

mollusk, an animal, with or without a shell, belonging to the group including the clam, oyster, slug, snail, nautilus, squid, and octopus, < L. *molluscus,* soft: Mollusca.

NL. *acmaea,* f. a genus of limpets: *Acmaea testudinalis* (a limpet).

Gr. *aporrhaïs, -idos,* f. a kind of mollusk: *Aporrhaïs calcarata* (a Cretaceous gastropod).

NL. *atrina,* f. a genus of mollusks: *Atrina serrata* (a pelecypod).

NL. *bittium,* n. a kind of snail: *Bittium tumidum* (a snail).

Gr. *bolbitis, -idos,* f. a kind of squid; *bolbidion; bolbition,* n. dim.; *bolitaina,* f. a squid: *Bolbitis serratifolia* (a fern), *Bolitaena pygmaea* (a squid).

L. *buccinum,* a horn-shaped mollusk; see **horn**

Sp. *cardita,* f. small vessel; F. *cardite,* a kind of mollusk: *Cardita floridana* (a pelecypod).

NL. *cerithium,* n. a genus of gastropods: *Cerithium ferrugineum* (horn-shell).

L. *chama* (Gr. *cheme*), f. a gaping mollusk, cockle: *Chama pulchella* (a rock-oyster).

Gr. *chamadytes,* m. a snail:

Gr. *cherambe,* f. a kind of pelecypod:

Gr. *cheramis, -idos,* f. a kind of scallop or mussel:

Gr. *Chione,* f. a mythological woman; a genus of pelecypods: *Chione cancellata* (a pelecypod).

Gr. *chiton,* garment; a genus of mollusks; see **garment**

Gr. *choirine,* f. a sea-mussel: *Choerina patula* (a gastropod).

L. *clacendix, -icis,* f. a kind of mollusk:

NL. *clione,* a genus of pteropods; see *Clio* under **story**

L. *cochlea,* f. (Gr. *kochlias; kochlos,* m.), snail with a spiral shell; *kochlidion; kochlion,* dim.: cochlea, *Cochliomyia macellaria* (a blowfly), *Cochlodesma leanum* (a pelecypod), *Cochlospermum religiosum* (cotton-shellseed), *Cochlearia officinalis* (scurvy-grass, a crucifer), *Hexastylus cochleatus* (a radiolarian), *Rotalia cochlea* (a foraminifer), *Actinoceras cochleatum* (a Silurian cephalopod).

L. *coluthium,* n. a kind of snail:

L. *concha,* f. (Gr. *konche,* f.; *konchos,* m.), snail, shell; *konchion,* n. dim.; L. *conchula* (Gr. *konchyle*), f.; L. *conchylium* (Gr. *konchylion*), n. dim.; Sp. *conchita; coquina,* f. shellfish; *coquillo,* m. dim.: conch, conk, conchology, conchoid, conchitis, conchylium, cockle, *Conhoderma virgatum* (a barnacle), *Conchidium laqueatum* (a fossil brachiopod), *Cryptoconchus porosus* (a chiton), *Pittoconcha concinna* (a gastropod), *Polyporus conchifer* (a bracket-fungus), Conchita, coquina *(Donax variabilis).*

L. *coryphium,* n. a snail yielding a purple dye:

NL. *cypraea,* f. a genus of mollusks, < L. *Cypria,* Venus: *Cypraea guttata* (spotted cowry).

L. *domiporta,* f. snail:

L. *donax, -acis,* m. a mollusk: *Donax denticulatus* (a wedge-shell), *Tellina donacina* (a pelecypod).

NL. *dosinia,* f. a genus of mollusks: *Dosinia subrosea* (a pelecypod).

L. *echinophora,* f. a kind of mollusk:

L. *glycymeris, -idis* (Gr. *glykymeris, -idos*), f. a mollusk: *Glycymeris laticostata* (a pelecypod).

NL. *haliotis,* f. a genus of mollusks: *Haliotis tuberculata* (an abalone), *Testacella haliotideae* (a snail), *Sinum haliotoideum* (a Miocene gastropod).

Gr. *heledone,* f. a mollusk: *Eledone[Heledone] verrucosa* (an octopus).

L. *helix, -icis* (Gr. *-ikos*), coil, spiral; snail; see **spiral**

Gr. *kalche,* f. the purple limpet, murex:

Gr. *kekibalos,* m. a kind of mollusk:

Gr. *keryx,* herald, shellfish; see **call**

Gr. *kokalion,* n. a land snail:

Gr. *krabyzos,* a shellfish; see **crab**

L., Gr. *Leda,* f. mythical mother of Castor and Pollux, who came from two eggs fathered by Jupiter in the form of a swan; a genus of mollusks: *Leda hamata* (hooked leda), *Loxonema leda* (a fossil gastropod).

L. *lepas, -adis* (Gr. *-ados*), f. a shellfish adhering closely to rocks; now applied to a genus of barnacles: *Lepas fascicularis* (goose-neck barnacle).

L. *limax, -acis* (Gr. *leimax, -akos*), f. slug, snail: limacel, limaçon, *Limax maximus* (gray slug), *Limacina elevata* (a fossil gastropod), *Agriolimax agrestis* (a slug).

L. *loligo, -inis,* f. cuttlefish, squid: *Loligo peali* (a squid).

Gr. *lopadion,* an oyster; see *lopas* under **plate**

NL. *mactra,* a genus of pelecypods; kneading-trough; see **basin**

Gr. *melainis, -idos,* f. a sea-shell: *Melaenis loveni* (an annelid).

L. *murex, -icis,* m. a shellfish yielding a purple dye; *muriculus,* m. dim.; *muricatus,* like a murex, spiny: murexide, muricate, muriculate, *Murex trunculus* (a gastropod yielding Tyrian purple), *Pedalium murex* (a pedaliacead), *Pinus muricata* (bishop-pine), *Cryptanthe muricatula* (a borage), *Gorgonocephalus muricatus* (a sandstar).

L. *mya,* f. a pelecypod; *myiscus,* m. dim.; *myax, -acis* (Gr. *-akos),* m. sea-mussel: *Mya arenaria* (soft clam).

L. *mytilus* (Gr. *mytilos),* m. mussel: *Mytilus edulis* (sea-mussel), *Stylonychia mytilus* (a protozoan), *Ostrea mytiloides* (an oyster), *Mytilaspis pomorum* (a scale-insect).

L. *nautilus* (Gr. *nautilos,* sailor), m. a cephalopod: *Nautilus pompilius* (pearly nautilus).

L. *nerita* (Gr. *nerites),* m. a sea-snail: *Nerita ustulata* (a snail), *Neritina zebra* (a snail).

NL. *octopus* (Gr. *oktopous,* eight-footed or armed), m. a cephalopod: *Octopus vulgaris* (common octopus).

Gr. *osmyle,* f.; *osmylos,* m. octopus; *osmylion,* n. dim.: *Osmylus maculatus* (a neuropterid), *Osmylopsis duplicata* (a fossil neuropterid), *Spilosmylus alticolus* (a neuropterid).

L. *ostrea,* f.; *ostreum* (Gr. *ostreon),* n. oyster; *ostrearius,* of oysters; *ostreosus,* abounding in oysters; *ostreatus,* pertaining to oyster shells, rough: *Ostrea virginica* (American oyster), *Pleurotus ostreatus* (a mushroom), *Navicula ostrearia* (a diatom).

L. *otia,* f. (Gr. *otion,* n. little ear), a mollusk:

Gr. *patelis, -idos,* f. a kind of limpet:

L. *pecten, -inis,* m. comb; a pelecypod: *Pecten gibbus* (a scallop), *Aviculopecten papyraceus* (a fossil pelecypod).

L. *pegris, -idis,* f. a sea-mussel:

Gr. *peloris, -idos,* f. giant mussel:

L. *perna,* f. a mussel: *Perna ephippium* (saddle-perna).

NL. *phalium,* n. a genus of mollusks: *Phalium granulatum* (a gastropod), *Xenophalium pyrum* (a gastropod).

Gr. *pholas, -ados,* f. a rock-boring mollusk: pholad, *Pholas orientalis* (piddock, angel-wing), *Pholadomya deltoidea* (a pelecypod).

NL. *physa,* a genus of snails; bellows; see **blow**

NL. *pinctada,* f. a genus of mollusks: *Pinctada radiata* (a pelecypod).

Gr. *pinna,* f. a pelecypod; *pinnikos,* of a pinna: *Pinna carnea* (a pelecypod).

Gr. *pomatias,* m. a snail:

Gr. *pontilos,* m. a nautilus:

Gr. *porphyra,* a mollusk that was the source of a purple dye; see **purple**

Gr. *rhomos,* m. woodworm, shipworm:

Gr. *selates,* m. a snail: *Selatosomus aeneus* (a beetle).

L., Gr. *sepia,* f. cuttlefish, squid; ink; *sepiola,* f.; *sepidarion; sepidion,* n. dim.; *sepion,* n. cuttlebone, proostracum: sepia, sepiolite, *Sepia officinalis* (a squid), *Sepiella japonica* (a squid), *Mimosa sepiaria* (a mimosa), *Periploca sepium* (Chinese silk-vine).

Gr. *sesilos,* m. a land snail:

Gr. *solen,* pipe, razor-clam; see **pipe**

NL. *spisula,* f. a genus of mollusks: *Spisula aequilateralis* (a surf-clam).

L. *spondylus* (Gr. *spondylos; sphondylos),* m. a pelecypod: *Spondylus regius* (thorny oyster).

Gr. *strabelos,* m. snail:

L. *striata,* f. a kind of mollusk:

L. *strombus* (Gr. *strombos),* a gastropod; cone; see **cone**

L. *tellina* (Gr. *telline),* f. a mollusk: *Tellina radiata* (a tellin-shell).

L. *teredo, -inis* (Gr. *teredon, -os),* f. woodworm, borer, shipworm: *Teredo navalis* (shipworm), *Xylocopa teredo* (a carpenter-bee).

Gr. *teuthis, -idos,* f. squid; *teuthidion,* n. dim.; *teuthidodes,* like a squid: *Teuthis media* (a squid), *Architeuthis dux* (a giant squid).

L. *tridacna,* f. in usage, but actually n. pl.; a large pelecypod: *Tridacna squamosa* (giant clam).

Gr. *xiphydrion (skiphydrion),* a mollusk; see *xiphos* under **sword**

See: **shell**

molo, *-itus,* L. grind; see **grind**

molobros, Gr. greedy, greedy fellow; see *bibrosko* under **eat**

Moloch < Heb. *Molech,* a deity worshipped by the sacrifice of firstborn children; see **spirit**

molops, *-opos,* Gr. weal, bruise; see **bruise**

molossus, L. (Gr. *Molossus,* famous for its hounds), a genus of bats; see **bat**

molpe, Gr. song; see *molpazo* under **sing**

molt (moult) < AS. *mutian,* < L. *muto,* change; see **change, bare**

molucrum, L. broom for sweeping out a mill; see **broom**

moluro- < Gr. *molouros,* a kind of snake; see **snake**

molva, NL. a fish, probably a kind of cod; see **cod**

moly, Gr. an herb with magic power, identified by some writers as garlic; see **onion**

molybdenum < L. *molybdaena,* galena; *molybdus,* L. (Gr. *molybdos*), lead; see **lead,** ELEMENTS under **thing**

molyno, Gr. stain, defile; *molysma,* spot, taint; see **dirt**

molys; molyx; molyros, Gr. soft, weak, feeble; see **weak**

molyxa, Gr. a kind of garlic; see *moly* under **onion**

momentum, L. motion, cause; see **move**

momos, Gr. blame, ridicule, carping criticism; see **blame**

monachos, Gr. single, solitary; see **alone**

monarda, NL. a genus of the mint family; see **mint**

monas, *-adis,* L. (Gr. *-ados*), a unit, alone, single; see *monos* under **one**

monedula, L. jackdaw; see **crow**

monela, L. admonition, reminder; see *moneo* under **warn**

moneses, NL. a genus of the heath family; see **heath**

moncy < L. *moneta,* f. the temple of Juno where money was coined, coin; *monetalis; monetarius,* of the mint, of money; m. minter: mint, monetary, mintage, demonetize, *Cypraea moneta* (cowry).
L. *argentum,* silver, money; *argentarius,* banker; see **silver**
Gr. *argyridion; argyrion,* a small coin, piece of money; see *argyros* under **silver**
L. *as, assis,* m. a copper coin; *assarium* (Gr. *assarion*), n. dim.:
L. *aurum,* gold, money; see **gold**
Gr. *chrema, -tos,* n. goods, money:
L. *collybus* (Gr. *kollybos*), m. a small coin, rate of exchange; *kollybistes,* m. money-changer:
L. *cusor,* coiner of money; *cusio, -onis,* the stamping of money, < *cudo,* strike, stamp; see **strike**
Gr. *danake,* f. a Persian coin:
L. *denarius,* m. a silver coin equivalent to ten asses: *Denariocrinus ferula* (a fossil crinoid).
L. *drachma* (Gr. *drachme*), f. a weight and coin: dram, drachma, *Collybus drachme* (a bramid fish).
L. *fiscalis,* relating to the treasury and money matters: fiscal, confiscate, confiscatory.
L. *libella,* a small coin, the tenth part of a denarius; see **balance**
L. *manubia,* f. money obtained by the sale of booty; *manubialis,* of booty:
L. *nummus,* m. (Gr. *nomisma, -tos,* n.), piece of money, coin; *nummulus,* m. dim.; *nummularius,* of money-changing: numismatics, nummulite, *Nummulina laevigata* (a foraminifer), *Nummulites reticulatus* (a foraminifer), *Cristellaria nummulitica* (a foraminifer), *Lysimachia nummularia* (moneywort), *Viola nummularifolia* (a violet), *Dadoxylon nummularium* (a fossil gymnosperm).
L. *obolus* (Gr. *obolos*), m. a small coin: obol, *Obolus dolatus* (a fossil brachiopod), *Obolaria virginica* (pennywort), *Obolella discoidea* (a fossil brachiopod).
L. *pecunia,* f. money; *pecunialis; pecuniarius,* of money; *pecuniosus,* moneyed, rich, wealthy: pecuniary, impecunious.
L. *raudus, -eris,* n. piece of bronze used as a coin; *raudusculum,* n. dim.:
L. *sembella,* f. a coin, half a *libella:*
L. *siclus* (Gr. *siklos*), m. < Heb. *sheqel,* skekel; *siklion,* n. dim.: shekel.
L. *talentum* (Gr. *talanton*), n. a weight and sum of money: talent, *Pleurotomaria (Talantodiscus) mirabilis* (a gastropod).
L. *temo, -onis,* m. money as commutation for recruits:
See: **wealth, pay, trade, gold, silver, copper**

-monger < AS. *mangian,* trade; see *mango* under **trade**

mongrel < a word akin to AS. *mengan,* mix; see **mix**

monile, L. necklace, collar; see **necklace**

monimos, Gr. steadfast, stable; see **stand**

monios, Gr. solitary, ferocious; see *monos* under **one**

monitor, L. admonisher, reminder; *monitorius,* serving to remind; see *moneo* under **warn**

monkey < uncertain origin.
> L. *callithrix, -trichis* (Gr. *kallithrix, -trichos*), f. a kind of monkey: Callitrichidae, *Callithrix personatus* (masked sagoin).
> L. *cebus* (Gr. *kebos*), m. a long-tailed monkey: *Cebus capucinus* (capuchin), *Cercocebus fuliginosus* (sooty mangabey), *Leontocebus rosalia* (a tamarin).
> L. *cercolips; corcholips,* m. a kind of monkey:
> L. *cercops, -opis* (Gr. *kerkops, -opos*), m. a long-tailed monkey: *Cercops holbolli* (an amphipod).
> L. *clura,* f. ape; *clurinus,* of apes:
> L. *cynocephalus* (Gr. *kynokephalos*), m. a dog-headed ape: *Cynocephalus volans* (flying lemur).
> African *gorilla,* a large ape: gorilloid, gorilla *(Gorilla savagei).*
> NL. *jacchus,* m. a kind of simian: *Jacchus penicillatus* (a simian), *Callithrix jacchus* (marmoset).
> Gr. *kallias,* m. a tame ape:
> Gr. *keblos,* m. a baboon:
> L. *lemur, -is,* m. ghost of the departed; now applied to a group of monkeylike mammals: *Lemur catta* (ring-tailed lemur), dwarf lemur *(Microcebus murinus), Phenacolemur citatus* (a fossil primate), lemuroid.
> NL. *loris,* m. a genus of lemurs: *Loris gracilis* (slender loris).
> NL. *macaca,* m. a genus of monkeys, < Ab.Am. *macaco: Macaca speciosus* (a macaque), *Lemur macaco* (black lemur).
> Gr. *mimo, -ous,* f. an ape:
> F. *papion,* baboon: *Papio mormon* (mandrill).
> Gr. *pithekos,* m. ape, monkey; *pithekion,* n.; *pithekis, -idos,* f. dim.; *pithekeios,* of apes; *pithekodes,* apelike: *pithekismos,* m. a playing the monkey; *choiropithekos,* m. hog-nosed monkey; *kerkopithekos,* m. a long-tailed monkey: pithecoid, pithecomorphic, *Pithecanthropus erectus* (Trinil, Java man), *Pithecoctenium echinatum* (Mexican monkey-comb), *Pithecellobium brevifolium* (a legume), *Cercopithecus callitrichus* (green monkey), *Spermatogonius cercopitheci* (a protozoan), *Choeropithecus leucophaeus* (a monkey), *Propithecus diadema* (a lemur).
> L. *satyrus* (Gr. *satyros*), a kind of ape; see **spirit**
> L. *simia,* f. ape; *simiolus,* m. dim.; *simininus,* of apes: simian, simioid, Simiidae, *Simia satyrus* (orangutan).
> Gr. *sphingia,* f. a kind of ape:
> NL. *tarsius,* m. a lemurlike primate: tarsioid, tarsier *(Tarsius spectrum).*

monnula, L. darling; see **love**

mono- < Gr. *monos,* one, single, alone; *moneres,* single, solitary; see **one, alone**

monopoly < L. *monopolium* (Gr. *monopolion*), exclusive control of a given trade; see *poleo* under **trade**

monotropa, NL. a genus of the heath family; see **heath**

monster < L. *monstrum,* an abnormal or supernatural wonder; see **animal, large, strange, wonder**

monstro, -atus, L. show, point out; see **teach**

montanus, L. pertaining to mountains, < *mons, montis,* mountain; see **mountain**

month < AS. *monath:* monthly.
> Months of the year:
> > Greek: Gamelion, Anthesterion, Elaphebolion, Mounychion, Thargelion, Skirophorion, Hekatombaion, Metageitnion, Boëdromion, Pyanepsion, Maimakterion, Poseideon.
> > Roman: Januarius, Februarius, Martius, Aprilis, Maius, Junius, Julius, Augustus, September, October, November, December.
> > Italian: Gennaio, Febbraio, Marzo, Aprile, Maggio, Giugno, Luglio, Agosto, Settembre, Ottobre, Novembre, Dicembre.
> > Spanish: Enero, Febrero, Marzo, Abril, Mayo, Junio, Julio, Agosto, Septiembre, Octubre, Noviembre, Diciembre.
> > French, Janvier, Fevrier, Mars, Avril, Mai, Juin, Juillet, Aout, Septembre, Octobre, Novembre, Decembre.
> > English: January, February, March, April, May, June, July, August, September, October, November, December.

Gr. *men, -os,* m. month; *emmenos; katamenios; menaios,* monthly: menolipsis, menopause, catamenia, emmenagogue.

L. *mensis,* m. month; *mensualis,* of a month; *mensurnus,* monthly; *menstrualis; menstruus,* happening monthly; *bimestris,* of two months; *semestris,* of six months: menses, menstrual, menstruation, semester.

See: **year, time**

monticellus; monticulus, L. dim. of *mons, montis,* mountain; see **mountain**

monumentum, L. memorial; see **remember, pillar**

-mony < L. *-monium,* action or result of action; see **make**

moon < AS. *mona:* moonlight, moonstruck, moony, moonseed, moonflower, moonwort, honeymoon, month, Monday (F. *lundi*).

L. *Cynthia* (Gr. *Kynthia*), f. goddess of the moon, emblem of Diana: Cynthia, *Cynthia pyriformis* (an ascidian), *Philosamia cynthia* (a moth).

Gr. *heliotis,* moon; see *helios* under **sun**

L. *luna,* f. moon; *lunula,* f. dim.; *lunaris,* of the moon; *lunatus,* shaped like a crescent moon; *lunaticus,* moonstruck, crazy; *interlunium; intermenstruum,* n. new moon, time of the new moon; *novilunium,* n. new moon; *plenilunium,* n. full moon; *illunis,* moonless, without moonlight: lunar, lunate, lunacy, lunula, lunule, semilunar, *Lunaria biennis* (honesty), *Phaseolus lunatus* (Sieva bean), *Tropaea luna* (luna moth), *Cristellaria semiluna* (a foraminifer), *Oreocincla lunulata* (a thrush), *Selenia bilunaria* (a moth), *Lunularia cruciata* (a liverwort), *Tetradella lunatifera* (a fossil ostracode).

Gr. *mene,* f. moon; *meniskos,* m. dim., crescent; *neomenia (noumenia),* f. new moon: meniscus, *Menispermum canadense* (moonseed), *Menura superba* (lyrebird), *Meniscotherium semicingulatum* (a Paleocene mammal), *Homalomena cordata* (an aracead).

L. *noctiluca,* that shines by night, the moon; see *nox* under **light**

L. *Phoebe* (Gr. *Phoibe*), f. goddess of the moon: Phoebe, *Phoebe porosa* (a lauracead), *Sayornis phoebe* (a flycatcher).

Gr. *selene,* f. moon; *selenion,* n. dim.; *selenis, -idos,* f. crescent; *seleniskos,* m. dim.: selenium, selenite, selenography, selenodont, *Selenicereus grandiflorus* (moonlight-cactus), *Selenophorus lucidulus* (a beetle), *Selenastrum bibraianum* (an alga).

Mopsus, L. (Gr. *Mopsos*), a celebrated soothsayer; see **prophecy**

mora, L. delay; *morula,* dim., < *moror, -atus,* linger, loiter, tarry; see **delay**

moral < L. *moralis,* belong to manners; see *mos* under **manner**

morbillus, ML. measles; see *morbus* under **disease**

morbus, L. sickness, disease; *morbidus; morbosus,* sickly, diseased; see **disease**

mordax, *-acis,* L. biting, corroding, pungent; *mordicus,* biting; see *mordeo* under **bite**

more < AS. *mara;* see **grow, abundance, other**

moretum, L. a kind of garlic salad; see **salad**

morigerus, L. complying, obedient; see **follow**

morio, L. fool; see **fool**

morion, Gr. piece, part, section, fraction, member; see *meros* under **part**

mormo, Gr. bugbear, hobgoblin, used to frighten children into good behavior; see **fear**

mormyr, L. (Gr. *mormyros*), a sparoid fish; see **bream**

morn < AS. *morgen,* morning; see **dawn**

morning-glory; see **bindweed**

morochthos; moroxos, Gr. a kind of pipe-clay; see **earth**

moros, Gr. stupid, foolish, see **fool;** fate, destiny, doom, share, see *moira* under **lot**

morosus, L. fretful, peevish, gloomy; see **fretful**

morphnos, Gr. a kind of eagle; see **eagle**

morpho- < Gr. *morphe,* form, shape; *Morpheus,* god of dreams; *Morpho,* Aphrodite, the shapely; see **form**

morrhua, NL. cod; see **cod**

mors, *mortis,* L. death; *moribundus,* dying; *morticinus,* dead; see **death**

morsel < OF. *morceau,* < L. *morsum,* bit, piece; see **little**

morsus, L. bitten; see *mordeo* under **bite**

mortal < L. *mortalis,* subject to death; see *mors* under **death**

mortar < L. *mortarium,* n. a basin or trough for mixing a plastic building material; plaster, see **plaster;** a vessel for grinding or triturating materials:
L. *coticula,* small mortar; see *cos* under **stone**
Gr. *holmos,* m. mortar; *holmiskos,* m. dim.: *Holmophyllum holmi* (a fossil coral).
Gr. *igdis,* f. a mortar; *igdion,* n. dim.:
L. *pila,* f. mortar:
L. *piso (pinso), -onis,* m. mortar:
Gr. *thyeia,* f. mortar; *thyeidion,* n. dim.:
See: **grind, pestle, plaster, strike**

mortise < F. *mortaise,* < Ar. *murtazz,* fastened in; see **bind**

mortuus, L. dead, < *morior, mortuus,* die; see *mors* under **death**

morulus, L. dark-colored, black; see *morus* under **black**

morus, L. (Gr. *morea*), mulberry-tree; *morum* (Gr. *moron*), mulberry; *morulus,* dim., see **mulberry;** < *morus* (Gr. *moros*), foolish, silly, see **fool**

mos, *moris,* L. manner, custom, way, practice; see **manner**

mosaic < F. *mosaique,* < L. *musaicus,* < Gr. *Mousaios,* of the Muses who presided over the fine arts; an inlay of small pieces of various colors and geometric patterns; *musivarius,* m. a worker in mosaic; *musivum,* n. mosaic work: mosaical, musive, potato mosaic.
L. *abaculus,* small, colored tile for inlay and mosaic work; see *abacus* under **number**
L. *asarotum* (Gr. *asaroton*), n. floor laid in mosaic:
Gr. *lithostrotos,* inlaid with stones, tesselated: *Lithostrotus caerulescens* (a beetle), *Lithostrotion striatum* (a coral).
L. *tesselatus,* inlaid with small, square stones, checkered, < *tesela,* f. small, square stone: tesselate, *Cryptolithus tesselatus* (a trilobite), *Eucalyptus tesselaris* (carbeen eucalyptus).

moschatus, L. perfumed with musk; see *moschus* under **smell**

moscho- < Gr. *moschos,* calf; *moschion,* dim.; see **cattle**

mosquito, Sp. little gnat, fly, < L. *musca,* fly; see **fly**

moss < AS. *mos,* bog.
L. *bryum* (Gr. *bryon*), n. moss: *Bryum argenteum* (silver moss), bryophyte, Bryozoa, Bryales.
L. *hypnum* (Gr. *hypnon*) n., a kind of moss: *Hypnum nitidulum* (a moss), *Cassiope hypnoides* (Arctic cassiope).
Gr. *mnion,* n. moss; *mniaros,* mossy: *Mnium undulatum* (a moss), *Mniotilta varia* (black and white warbler), *Ptilimnion capillaceum* (mock bishop-weed).
L. *muscus,* m. moss; *muscosus,* mossy: muscose, mushroom, Musci, *Nostoc muscorum* (an alga).
L. *polytrichum* (Gr. *polytrichon*), n. a kind of moss: *Polytrichum alpinum* (alpine haircap-moss).
NL. *sphagnum,* n. (Gr. *sphagnos,* m.), a moss: *Sphagnum cymbifolium* (a sphagnum-moss).
NL. *usnea,* < Ar. *oshnah,* moss; now a genus of lichens; see **lichen**

most < AS. *maest,* superlative of *much;* see **abundance**

motacilla, L. wagtail; see **wagtail**

motella, NL. a kind of rockling, < L. *mustela,* weasel, a fish; see **cod**

moth < AS. *moththe.*
NL. *ambulyx,* f. a kind of moth: *Ambulyx sexoculata* (a moth), *Oxyambulyx sericeipennis* (a moth).
L., Gr. *bombyx,* silkworm; see **silk**
Gr. *epiolos,* m. moth: *Epiolus lupulinus* (a moth).
L. *myloecus,* m. a moth that breeds in mills: *Colon(Myloecus) clavigerum* (a beetle).
L. *noctua,* owl, owlet moth; see **owl**
L. *phalaena* (Gr. *phalaina*), f. a kind of moth: *Phalaena typica* (a moth), *Phalaenopsis amabilis* (moth-orchid).
L. *pyralis, -idis* (Gr. *-idos*), f. a bird or insect; now a genus of moths: *Pyralis farinalis* (a meal-moth).
L. *pyrausta* (Gr. *pyraustes*), m. a moth that gets singed in the flame: *Pyrausta nubilalis* (European corn-borer).
Gr. *ses, seos; setos,* m. moth: Sesiidae, *Sesia muscaeformis* (a clearwing-moth), *Seopsis tricolor* (a psocopterid), *Setophaga ruticilla* (redstart), *Setolestes genalis* (a fly).

L. *tinea*, f. a gnawing worm, larva of a moth; *tineola*, f. dim.; *tineosus*, wormy: tineid, *tinea imbricata* (a form of ringworm), *Tinea pellionella* (case-making clothes-moth), *Tineola biselliella* (webbing clothes-moth).

L. *vappo*, *-onis*, m. a moth or butterfly: *Vappo ater* (a fly).

See: **butterfly**

mother < AS. *moder:* motherly, motherhood, motherwort, Mother Hubbard, mother-in-law, godmother.

Gr. *geneteira; goneia*, f. mother, parent:

Gr. *maia*, good mother, nurse, midwife; *maieutikos*, skilled in midwifery; see **nurse**

L. *mamma* (Gr. *mamme*), mother, breast; see **udder**

L. *mater*, *-tris* (Gr. *meter*, *-tros*), f. mother, origin, source; *matercula*, f. dim. a poor mother; *metrarion*, n. dim.; *maternus; matralis*, pertaining to mothers; It., Sp. *madre*, f. mother; *madrecita; madrecilla*, f. dim.; *matrix*, *-icis*, f. womb, enclosing substance, mother in the sense of original cause: alma mater, maternal, matron, matrix, matricide, matrimony, Marne, metropolitan, madrepore, *Metrodora amazonica* (a grasshopper), *Metrosideros lucida* (a myrtacead), *Azorella madreporica* (yareta), *Schidium matercula* (a bug), *Zoysia matrella* (Manila grass).

See: **begin, woman, nurse, womb, birth**

mothon, Gr. impudent fellow; see **rough**

mothos, Gr. battle; see **fight**

motivus, L. moving; *motiuncula*, dim. of *motio*, movement; see **move**

mountain < L. *mons, montis*, m.; *monticellus; monticulus*, m. dim.; *montanus*, of mountains; Sp. *montana*, f. mountain: mountaineer, montane, Monticello, Montello, Beaumont, Montana, mountebank, Mont Alto(Pa.), amount, demountable, remount, surmount, piedmont, tantamount, transmontane, marmot, paramount, seamount, promontory, catamount *(Felis concolor), Monticulipora arborea* (a fossil bryozoan), *Monticola rupestris* (Cape rock-thrush), *Pinus monticola* (western white-pine), *Rangifer montanus* (a caribou), *Aquilegia saximontana* (blue columbine), *Satureia montana* (a savory), *Erythronium montanum* (avalanche-lily).

Gr. *akrolophos; geolophos*, crest of a mountain, ridge; see *lophos* under **crest**

Gr. *akroreia*, f. mountain ridge: *Acroria baylei* (an Eocene gastropod).

L. *alpinus; alpestris*, *-e*, of high mountains, < *Alpis*, f. Alp: alpine, *Alpinibombus arcticus* (a bee), *Alpenoceras ulrichi* (a fossil cephalopod), *Salvelinus alpinus* (European char), *Arabis alpina* (a crucifer), *Dodecatheon alpinum* (a shooting-star), *Otocoris alpestris* (horned lark), *Blitophaga alpicola* (a beetle), *Poa alpigena* (a grass).

NL. *andinus*, pertaining to the Andes Mountains: Andean, Andine, andesite, andesine, *Begonia andina* (a begonia), *Hemixantha andorum* (a fly), *Ceroxylon andicola* (a wax-palm), *Solanum andigenum* (a potato).

Ab.Am. *apalachi*, a tribal name: Appalachian, Apalachee, Apalachicola, Appalachia.

AS. *beorg* (G. *berg*), a height, mountain, hill: bergschrund, iceberg, barrow, Königsberg.

Gr. *bounos*, m. hill, mound, knob; *bounis*, *-idos; bunodes*, hilly; *bounites*, m. dweller in the hills: bunodont, *Bunocephalus bicolor* (a catfish), *Astrobunus argentatus* (an opilionid), *Exochobunus pulcherrimus* (an opilionid).

Sp. *cerro*, m. hill; *cerillo*, m. dim.: Cerro de Pasco (Peru).

Gr. *choma*, *-tos*, bank, mound, dam; see **heap**

L. *clivus*, m. ascent, elevation, hill, sloping hillside; *clivulus*, m. dim. hillock; *clivosus*, hilly; *acclivis*, uphill, ascending, steep; *declivis*, downhill; *proclivis*, sloping, steep, downhill; *reclivis*, sloping backward: acclivity, declivity, *Clivicola riparia* (bank-swallow), *Rubus clivicola* (a blackberry), *Begonia clivalis* (a begonia), *Scobicia declivis* (cable-borer), *Neritina reclivata* (a gastropod).

L. *collis*, m. hill, high ground; *collicellus; colliculus*, m. dim.; *collinus*, hilly, on a hill: colline, colliculus, *Campanula collina* (a bellflower).

Sp. *cordillera*, f. chain of mountains: Cordilleras, Cordilleran, *Piper cordillerianum* (a pepper).

Sp. *cuesta*, f. hill or mountain with one side sloping upward to a steep escarpment: cuesta.

AS. *dun*, hill: dune, down, London, Verdun, Lyon, Dunkirk.

Gr. *Emodos*, m. Himalaya Mountains: emodin, *Rheum emodi* (Himalayan rhubarb), *Syringa emodi* (Himalayan lilac).

L. *grumus*, heap, mound, hillock; *grumulus*, dim.; see **heap**

Skt. *himalaya*, < *hima*, snow, *alaya*, abode: Himalaya, Himalayan black bear *(Ursus thibetanus), Aster himalaicus* (Himalayan aster).

L. *Hybla* (Gr. *Hyble*), f. mountain and town in Sicily famous for bees and honey: Hyblean.

AS. *hyll,* hill: hilly, hillock, Hillsdale, Foxhills, Churchill.

L. *Hymettus* (Gr. *Hymettos*), m. a mountain near Athens famous for its honey and marble: Hymettic, *Hymettus reticulatus* (a bug).

L. *Jura,* m. mountain range between France and Switzerland: Jura, Jurassic, Jurane.

Gr. *knemos,* shoulder of a mountain; see **shoulder**

Gr. *kolone,* f.; *kolonos,* m. hill, mound, barrow: *Colonosaurus mudgei* (a fossil reptile), *Colonocincla brevipes* (a bird).

Sp. *loma,* f. hill; *lomita,* f. dim.: lomita.

Ab.Am. *monadnock,* prominent mountain: Mt. Monadnock, monadnock, *Enophrys monadnock* (a spider).

Gr. *myrmekia,* f. anthill, wart: myrmekite, *Myrmekia karykina* (a milleped), *Myrmekiophila torreya* (a trapdoor-spider).

Gr. *ochthos,* m.; *ochthe,* f. any elevation, hill, bank, mound, tumor, wart; *ochtheros; ochthodes,* hilly; *epochthidios,* on or of the mountains: *Ochthobothrium merlangi* (a worm), *Ochthephilus ceylanicus* (a beetle), *Ochthebius deletus* (a beetle), *Ochthodocaryon wilkinsoni* (a fossil nut), *Calochthebius brevicollis* (a beetle).

L. *Olympus* (Gr. *Olympos*), m. a celebrated mountain in Thessaly, fabled to be the seat of the gods: Olympian, *Verbascum olympicum* (Olympic mullein).

Gr. *oros, -eos,* n. mountain, hill; *orion,* n. dim.; *oreinos,* mountainous, hilly, wild; *oreios,* of the mountains; *oresbios,* living in or on mountains; *oreites; orestes,* m. mountaineer; *oressinomos,* ranging the mountains; L. *oreas, -adis* (Gr. *oreias, -ados*), f. mountain-nymph: orogenic, orography, orohippus, oreodont, Orestes, oread, *Oreodoxa oleracea* (cabbage-palm), *Oreopteryx pictus* (mountain-quail), *Oreamnus montanus* (mountain-goat), *Oroxylum indicum* (Indian trumpet-flower), *Orosphaera serpentina* (a radiolarian), *Oresbius castaneus* (a wasp), *Abrotocrinus orestes* (a Mississippian crinoid), *Oreoscoptes montanus* (sage-thrasher), *Oreodaphne foetens* (a lauracead).

Gr. *pagos,* m. hill.: Areopagus, *Pagophila alba* (a gull).

Sp. *sierra,* f. saw; mountains with a craggy, irregular skyline: Sierra Nevada.

L. *teba,* f. hill:

L. *tumulus,* raised mound of earth, barrow, hillock; see **heap**

See: **heap, crest, projection, high, fold**

mourn < AS. *murnan;* see **sad, weep**

mouse < AS. *mus.*

NL. *cricetus,* m. hamster, a ratlike European rodent: *Cricetus frumentarius* (hamster), *Cricetomys gambianus* (a banana-rat), *Cricetulus griseus* (a rodent).

Gr. *eleios,* m. dormouse: *Eliomys melanurus* (a rodent).

NL. *gerbillus,* m. a genus of mouselike rodents: *Gerbillus aegyptius* (jerboa).

L. *glis, gliris,* m. dormouse: *Gliricidia sepium* (a legume), *Gliricola calcaratus* (a mallophagid), *Myoxus glis* (fat dormouse).

L. *mus, muris* (Gr. *mys, myos*), m. mouse; *musculus,* m.; *myidion,* n. dim.; *murinus,* of mice, mouse-gray; *muricus; myagros,* m. mouser; *muscipula,* f. mouse-trap: murine, muriform, Muridae, muriarium, *Mus musculus* (house-mouse), *Myosurus minimus* (a mousetail), *Myosotis arvensis* (a forget-menot), *Calomys americanus* (deer-mouse), *Oryzomys palustris* (a rice-rat), *Peromyscus californicus* (a deer-mouse), *Alsophila myosuroides* (a tree-fern), *Megamys patagonensis* (a fossil rodent), *Antechinomys laniger* (little jerboa), *Brachyderes murinus* (a beetle), *Hordeum murinum* (mouse-barley).

NL. *muscardinus,* m. < F. *muscardin,* dormouse: *Muscardinus avellanarius* (red dormouse).

Gr. *mygale,* f. a field-mouse: mygalid, mygaloid, *Mygale moschata* (a rodent), *Mygalina pyrenaica* (a rodent).

Gr. *myoxos,* m. dormouse: *Myoxus chrysurus* (a dormouse), *Myoxocephalus verrucosus* (a sculpin).

L. *nitedula,* f. small, red mouse, dormouse; *nitellinus,* of a dormouse: *Myoxus nitedula* (a dormouse).

ML. *rattus,* m. rat: *Rattus norvegicus* (house-rat).

Gr. *sminthos,* m. mouse: *Sminthurus aquaticus* (a springtail), *Sminthopsis psammophila* (a marsupial-mouse), *Acosminthus dimidiatus* (a mouse).

mouth < AS. *muth.*

L. *aestuarium,* n. channel subject to tidal action at the mouth of a river: estuary, estuarine, *Cerianthus estuari* (a sea-anemone).

Sp. *boca,* f. mouth, entrance; *boquilla,* f. dim.: Boca Chica, Bocagrande (Fla.).

Gr. *chanos,* the open mouth; *chasmema,* a yawn or gape; see *chaino* under **open**

Gr. *ekbole,* f. outlet, mouth of a river:

Gr. *exelysis,* f. outlet, mouth of a river:

L. *gingiva,* f. gum; *gingivula,* f. dim.: gingivitis.

Gr. *glottis, -idos,* f. mouth of the windpipe: glottid, epiglottis.

L. *os, oris,* n. mouth, opening; *oscillum; osculum,* n. dim.; *orificium,* n. opening: oscular, ósculate, osculum, oscule, oscitant, orifice, oral, orator, oracle, orotund, *Osculina polystomella* (a sponge), *Stomatodiscus osculatus* (a radiolarian).

Gr. *oulon,* n. gum: uloncus, ulorhagia, ulitis, parulis, *Haemulon macrostomum* (striped grunt).

L. *palatum,* n. roof of the mouth: palate, palatal, palatitis, palatonasal.

Gr. *prochoë,* f. mouth of a river:

L. *rictum; rictus,* open mouth; see *ringens* under **open**

Gr. *stoma, -tos,* n.; *stomion,* n. dim.; *stomias,* m. large or hard-mouthed animal; *stomalimne,* f. estuary; *mikrostomos,* small-mouthed: stoma (pl. stomata), stomodeum, stomach, peristome, anastomosis, *Stomoxys calcitrans* (stable-fly), *Polystomella crispa* (a foraminifer), *Ancistrodon rhodostoma* (a viper), *Batrachostomus auritus* (frogmouth, a goatsucker), *Oxystomina alpha* (a nematode), *Sclerostomum edentatum* (a nematode), *Grammostomum scabrum* (a foraminifer), *Stomias ferox* (a fish), *Atherestes stomias* (a halibut), *Ceratostomella pilifera* (blue-stain fungus), *Rhynchostoma cornigerum* (a fungus), *Stomatopora dichotoma* (a fossil bryozoan).

See: **jaw, bite, door, hole, enter**

move < L. *moveo, motus,* be in action, go, actuate; *mobilis,* movable; *momentum,* n. motion, cause; *motio, -onis,* f. motion; *motiuncula,* f. dim.; *motivus,* moving; *motor, -is,* m. mover; *motorius,* that has motion; *moto, -atus,* keep moving; *commotio, -onis,* f. agitation; *promotus,* advanced, pushed forward: movement, momentum, movable, motile, motive, mobility, motor, mob, commotion, emotion, immobile, automobile, remove, promoter, remote, unmoved, immovable, locomotive, mutiny, *Chironomus motilator* (a midge).

L. *agitatio, -onis,* motion, movement; see **arouse**, *ago* under **make**

Gr. *aiolos,* shifting, changing, variable; see *Aeolus* under **wind**

Gr. *akoimetos,* sleepless, unresting; see **busy**

Gr. *alysko,* wander uneasily; *alysmos,* m. disquiet, restlessness; *alyo,* be at a loss, be ill at ease:

Gr. *astatos,* unstable, restless; *apanastates; metanastes,* m. migrant: *Astatochroa fuscimargo* (a moth), *Metanastes australis* (a beetle).

Gr. *baino,* walk, go, pass; *basis,* a stepping; see **walk**

L. *cedo, cessus,* be in motion, go, yield; *decessus, -us,* m. departure; *incessus, -us,* m. a going: cede, cease, accede, secede, exceed, procedure, cession, excess, successor, decease, ancestor, incessant, precedence, antecedent, accessory, excessive, abscess, inaccessible, unsuccessful.

Gr. *cheironomia,* pantomimic movement, gesticulation; see **equal**

L. *cieo, citus,* actuate, excite, move, call; see **arouse**

L. *cillo,* move, put in motion:

L. *circo,* go about, traverse; see **circle**

Gr. *ektopizo,* migrate, wander; *ektopismos,* m. migration; *ektopistikos,* migratory: *Ectopistes migratorius* (passenger-pigeon).

L. *eo, ire, itus,* go; *itus, -us,* m. a going, departure; *iter, itineris,* n. journey; *itio, -onis,* f. a going, traveling; *abitus, -us,* m. departure, < *abeo,* go away; *ambio, -itus,* go around; *exeo, -itis,* go away, depart: transit, exit, itinerary, circuit, ambient, ambition, obituary, initial, sedition, perish, issue, coition, sudden, concomitant.

Gr. *eunomas,* mobile:

L. *expeditio, -onis,* f. journey, campaign: expedition.

AS. *faran,* go, travel: fare, welfare, thorofare, wayfarer, warfare, seafaring, eelfare, ferry, ford.

AS. *gan,* go: going, gone, begone, forego, gang, woebegone.

L. *gesticulor, -atus,* make an impressive motion of the body, pantomime; see **equal**

L. *grassor, -atus,* go about, loiter; *grassator,* vagabond, footpad; see *gradior* under **walk**

Gr. *hodos,* way, road, journey; *hodites,* traveler, wayfarer; *exodos,* departure, exit; see **way**

L. *impatientia,* f. restlessness, want of endurance; *impatiens, -entis,* desiring immediate action: *Impatiens biflora* (snapweed, touch-me-not), *Bombus impatiens* (a bumblebee), *Cecidomyia impatientis* (a gall-midge), impatience.

L. *inquietus,* restless; *irrequies,* always in motion:

Gr. *ion, -tos,* participle of *eimi,* go: ion, ionize, cations, anions, iontophoresis.

Gr. *kineo,* move; *kinema, -tos,* n.; *kinesis,* f. movement; *kinetikos,* setting in motion, exciting; *kinetos,* movable; *kinetes,* m. mover, agitator, author, initiator; *kinygma, -tos,* n. any moving object; *aeikinetos,* perpetually moving; *eukinetos,* easily moved, moving easily: kinetic, kinesis, cytokinesis, karyokinesis, kineto-scope, cinema, *Dyscinetus trachypygus* (rice-grub).

Gr. *klonos,* violent, confused motion, tumult, turmoil, rout; see **confusion**

Gr. *kykleo,* revolve, repeat; see *cyclus* under **circle**

L. *meo, -atus,* pass; see **passage**

Gr. *metelys, -ydos,* c. emigrant; changing:

Gr. *metoikos,* emigrating; c. emigrant: *Metoicoceras gibbosum* (an ammonite), *Metoecus paradoxus* (an amphipod), metic.

L. *migro, -atus,* move, change habitation; *migrator, -is,* m. wanderer; *emigro,* move out; *immigro,* move in: migrate, emigrate, immigrate, immigration, trans-migration, *Locusta migratoria* (a grasshopper), *Drosophila immigrans* (a fruit-fly).

Gr. *ochlizo,* move with a lever:

L. *peregrinor, -atus,* travel abroad or around; *peragro, -atus,* travel through; *peregrinus,* traveling about, foreign, strange; *pereger,* on a journey, abroad: peregrination, peregrine, *Tropaeolum peregrinum* (canary-bird flower), *Limnaea pereger* (a snail).

Gr. *phoitao,* come and go constantly; *phoitetes,* m. comer and goer; *phoitaleos,* going about wildly; *phoitos,* m. a constant coming or going: *Phoetaliotes nebrascensis* (a grasshopper), *Diaphoetes rugosus* (a beetle).

Gr. *poleo,* go about; *polesis,* f. movement:

Gr. *prokope,* f. advance, progress:

Gr. *stolos,* m. journey, voyage, expedition, army: *Stolotermes africanus* (a termite).

L. *turba,* disorder, hubbub; see **confusion**

L. *vado, vasus,* go, rush: invade, pervade, evasion, pervasive, vadose, wade, *Quo vadis?* (Henryk Sienkiewicz).

See: **depart, walk, wander, shake, confusion, turn, circle, spiral**

mow < AS. *mawan;* see **cut**

mu, Gr. twelfth letter of the Greek alphabet; see **letter**

mucedo, -inis, L. mold, mildew; *mucidus,* moldy; see **fungus**

much < AS. *micel;* see **abundance, wealth**

mucilago, L. a sticky, gelatinous juice; see **glue**

mucor, L. mold, mildew; see **fungus**

mucro, -onis, L. sharp point; see **point**

mucuna, NL. a genus of the bean family; see **bean**

mucus, L. secretion from the nose; *mucidus; mucosus,* slimy; see **slime**

mud < uncertain origin, but probably from MD. *modde;* see **earth**

mugil, L. a fish, probably a mullet; see **mullet**

mugitor, L. bellower, < *mugio, -itus,* bellow, low; see **roar**

mulberry < AS. *morbeam,* mulberry-tree.

NL. *artocarpus,* m. a genus of the mulberry family: *Artocarpus superbus* (a bread-fruit).

NL. *broussonetia,* f. a genus of the mulberry family, < Pierre Broussonet, French naturalist: *Broussonetia papyrifera* (paper-mulberry).

L. *Cecropia,* f. citadel of Athens, < Gr. Kekrops, mythical king of Athens; a genus of the mulberry family: *Cecropia pèltata* (trumpet-wood), *Samia cecropia* (a moth), *Thecla cecrops* (a hairstreak).

Gr. *habrynon,* n. mulberry:

L. *morus* (Gr. *morea*), f. mulberry-tree; *morum* (Gr. *moron*), n. black mulberry; *morulus,* m. dim.; dark-colored: morula, murrey, sycamore, *Morus rubra* (red mulberry), *Bombyx mori* (silkworm), *Rubus chamaemorus* (cloudberry), *Morinda citrifolia* (Indian mulberry), *Chrysanthemum morifolium* (a chrysanthemum), *Laportea moroides* (a nettle-tree).

L. *sycaminus* (Gr. *sykaminos*), f. mulberry-tree; *sykaminon,* n. mulberry: syca-mine (*Morus nigra*), *Sycamina nigrescens* (a flagellate).

mulceo, mulsus, L. stroke, touch lightly, charm, delight, soothe, soften; *Mulciber,* softener, a name for Vulcan; see **soft, fire**

mulcetra, L. heliotrope; see **borage**

mulco, -atus, L. beat, handle roughly, injure; see **hurt**

mulcta; multa, L. penalty, fine, < *multo, -atus,* punish by fining; see **pay, punish**

mule < L. *mulus;* see **horse**

mulgeo, mulctus, L. milk; *mulctra; mulgare,* milk-pail; see **milk, bucket**

mulier, L. woman, female; *muliercula,* dim.; *muliebris,* female, feminine; see **woman**

mullein < OF. *moleine*, < L. *mollis*, soft.
  L. *persolata*, f. a kind of mullein:
  Gr. *phlomis, -idos* (*phlomos*), f. mullein; *phlomodes*, like mullein: *Phlomis fruticosa* (Jerusalem sage), *Verbascum phlomoides* (clasping mullein).
  L. *verbascum*, n. mullein: *Verbascum crassifolium* (Portugal mullein), *Centaurea verbascifolia* (a centaurea), *Cucullia verbasci* (a moth).

mulleolus, L. reddish, < *mulleus*, a reddish shoe; see **red, shoe**

mullet < OF. *mulet*, prob. < L. *mullus*, m. red mullet; *mullulus*, m. dim.: surmullet, *Mullus barbatus* (red mullet), *Mullus surmulletus* (striped surmullet).
  Gr. *boreus*, m. a mullet; *boridion*, n. dim.: *Boreus brumalis* (a mecopterid).
  L. *cephalus* (Gr. *kephalos*), m. a kind of mullet: *Leuciscus cephalus* (a chub).
  Gr. *chellon* (*chelon*), *-os*, m. a mullet: *Mugil chelo* (a fish).
  Gr. *erythrinos*, m. a red mullet: *Pagellus erythrinus* (a fish).
  Gr. *gompharion*, n. a mullet:
  Gr. *kestreus; kestrinos*, m. mullet; *kestriniskos*, m. dim.: *Cestreus minimus* (a fish), *Cestrinus obscurus* (a beetle).
  Gr. *lineus*, m. a kind of mullet:
  L. *mugil, -is*, m. a fish, probably a mullet: *Mugil cephalus* (striped mullet).
  Gr. *myxinos*, a kind of mullet; see *myxa* under **slime**
  Gr. *peraias*, m. a kind of mullet:
  Gr. *plos, -otos*, mullet; see *pleo* under **swim**
  Gr. *spheneus*, m. a kind of mullet:
  Gr. *trigle*, red mullet; now a genus of gurnards; see **gurnard**

mulsus, L. mixed or sweetened with honey, see **honey**; softened, soothed, see **soft**

multicius, L. soft; see **soft**

multifarius, L. manifold, various; see **different**

multifidus, L. many-cleft; see *findo* under **cut**

multiplico, -atus, L. increase, < *multiplex*, manifold; see **grow**

multitudo, L. large number; see **number**

multus, L. much; *plus*, more; *plurimus*, most; see **abundance**

mulus, L. mule; *mulinus*, of mules; see **horse**

mumia, ML. mummy; see **body**

mundus, L. world, earth; *mundanus*, of the world, see **world**; nice, neat; *mundulus*, dim.; see **beauty**

muneralis, L. of gifts; *munerarius*, bestower of gifts; see *munus* under **give**

mungo, -unctus, L. blow the nose; *munctio, -onis*, a blowing of the nose; see **blow**

municipium, L. town; see **town**

munificus, L. bountiful, generous; see **abundance**

munio, -itus, L. fortify; see **strong**

munus, muneris, L. gift, duty, service; *munis*, ready for service; see **give, make**

muraena, L. (Gr. *myraina*), moray; see **eel**

muralis, L. of walls; *muratus*, walled, < *murus*, wall; see **wall**

murcidus, L. cowardly, slothful, < *murcus*, coward; see **coward**

murex, -icis, L. a mollusk yielding a purple dye; *muriculus*, dim.; *muricatus*, spiny like a *murex;* see **mollusk**

muria, L. brine; *muriaticus*, of brine; see **salt**

murinus, L. pertaining to mice, mouse-gray; see *mus* under **mouse**

murmur, -is, n. L. a low sound, hum, mutter, mumble, whisper: murmur, murmuring.
  Gr. *bembix, -ikos*, a buzzing sound; see **wasp**
  L. *bombus* (Gr. *bombos*), m. a booming, humming, buzzing; *bombito*, hum, buzz; *bombinator, -is*, m. buzzer, hummer: bomb, bombard, bombolo, *Bombus terrestris* (a bumblebee), *Bombinator igneus* (a toad).
  Gr. *enkelados*, buzzing; see *kelados* under **sound**
  Gr. *gryktos*, grumbling; *grysmos*, m. a grumbling, < *gryzo*, grumble:
  Gr. *laryno*, coo like a dove:
  Gr. *ligypterophonos*, buzzing with the wings:
  Gr. *minyros*, whimpering, whining; see *minyrisma* under **sing**
  L. *musso, -atus*, speak in an undertone, murmur; *mussitator, -is*, m. mutterer: mussitate.
  L. *mutio* (*muttio*), mutter, mumble, bleat; *muttum*, n. a mutter, mumble: motto, mot, bon mot.
  Gr. *mygmos*, moaning, muttering; see **sad**
  Gr. *myzo*, murmur, mutter:

Gr. *psithyros*, whispering; *psithyrisma*, *-tos*, n.; *psithyrismos*, m. whispering; *psithyristes*, m. whisperer, slanderer, < *psithyrizo*, whisper: *Psithyrus laboriosus* (a bee), *Psithyroedus locustella* (a bird).

L. *susurro*, *-atus*, whisper, murmur, hum, buzz; *susurrus*, m.; *susurramen*, *-inis*, n. whisper, mutter; *insusurro*, *-atus*, insinuate, suggest: susurrus, susurrant.

Gr. *tonthrystes*, m. mutterer, babbler; *tonthrys*, f. a muttering:

Gr. *tryzo*, murmur, mutter; *trysmos*, m. murmur:

See: **grunt, sing, speak, sound**

murrha, L. a kind of clay for making porcelain or china; *murrhinus*, of porcelain; see **earth**

murus, L. wall; see **wall**

mus, *muris*, L. mouse; *musculus*, dim.; *murinus*, of mice; *muricus*, mouser; see **mouse**

Musa, L. (Gr. *Mousa*), muse, one of the goddesses who presided over the fine arts, see **art**; NL. *musa*, a genus of the banana family, see **banana**

musca, L. fly; *muscula*, dim.; *muscarius*, of flies; see **fly**

muscardinus, NL. < F. *muscardin*, dormouse; see **mouse**

muscipula, L. mouse-trap; see **trap**

muscle < L. *musculus*, dim. of *mus*, mouse; *musculosus*, fleshy; see **flesh**

muscus, L. moss; *muscidus; muscosus*, mossy; see **moss**

museum, L. (Gr. *Mouseion*), temple of the Muses, art school, academy; see *Musa* under **art**

mushroom < OF. *mouscheron;* see **fungus**

music < L. *musica* (Gr. *mousike*), f. one of the arts presided over by the Muses: musical, musician, *Tubipora musica* (organ-pipe coral).

L. *aeneator*, *-is*, m. trumpeter:

Gr. *askaules*, bagpiper; see *aulos* under **pipe**

L. *buccino*, *-atus*, sound the trumpet; see **horn**

L. *campana*, bell; see **bell**

L. *canor*, *-oris*, tune, melody; *canorus*, melodious, harmonious, euphonious; *canticus*, musical; see *cano* under **sing**

L. *cicuticen*, *-inis*, m. player upon a pipe made of hemlock:

L. *cornicen*, hornblower; see *cornu* under **horn**

Gr. *krouma; krousma*, *-tos*, n. sound, note produced by stringed instruments; *kroumatikos*, of playing a stringed instrument:

L. *liticen*, clarinetist; see *lituus* under **rod**

L., Gr. *lyra*, lyre; see **harp**

Gr. *meliktes*, m. flute-player:

Gr. *nabla*, a stringed instrument; *nablistes*, the player of a *nabla;* see **harp**

Gr. *orchestra*, f. place where the chorus danced; see *orcheomai* under **dance**

L. *organum* (Gr. *organon*), n. pipe, church organ: organ, organist.

Gr. *Orpheus*, m. famous minstrel and husband of Eurydice: Orpheus, *Orpheus polyglottus* (a bird).

L. *pandura* (Gr. *pandoura*), f. a musical instrument of three strings: pandura, bandore, panduriform, pandurate, *Ficus pandurata* (fiddleleaf-fig).

Gr. *psallo*, play a stringed instrument; *psalmos*, m. sound of the harp, song; *psalterion*, n. a harplike instrument; *psaltikos*, of harp-playing: psalter, psalm, Psalmist, psalmody, psalterium.

L. *subulo*, *-onis*, m. flutist:

Gr. *symphonia*, f. concord in sound, unison: symphony.

Gr. *syristes*, m. piper: *Syristes sibilator* (a bird).

L. *tibicen*, *-inis*, m. piper, flutist, < *tibia*, f. shinbone, pipe, flute; *tibicino*, *-atus*, play the pipe, fife, flute: *Tibicen linnei* (dog-day cicada), *Gymnorhina tibicen* (a shrike).

Gr. *troparion*, n. a piece of ecclesiastical music:

L. *tympanum*, drum; see **drum**

L. *utricularius*, bagpiper; see *utriculus* under **bag**

See: **sound, harmony, harp, sing, pipe, drum, joy**

musimo, L. (Gr. *mousmon*), mouflon; see **sheep**

musk < L. *moschus* (Gr. *moschos*), < Per. *mushk*, a secretion of the male musk-deer; see **smell**

muskrat
Ab.Am. *ondatra*, f. muskrat: *Ondatra zibethica* (muskrat).

mussel < L. *musculus*, dim. of *mus*, mouse; see **mollusk**

musso, *-atus*, L. speak in an undertone, murmur; see **murmur**

must < AS. *moste,* obliged, required, necessary.
   L. *necessarius,* indispensable, requisite: necessary, necessarily, necessity.
   L. *requisitus,* demanded, necessary: requisite.

mustache < F. *moustache,* < Gr. *mystax;* see **hair**

mustard < OF. *moustarde,* < L. *mustum,* n. must, new wine with which the
      powdered mustard seed was mixed.
   L. *alyssum* (Gr. *alysson*), n. a genus of the mustard family: *Alyssum argenteum*
      (yellowtuft-alyssum), sweet alyssum (*Lobularia maritima*), *Oenothera alyssoides*
      (an evening-primrose), *Dasyneura alyssi* (a gall-midge).
   NL. *arabis* (Gr. *Arabis,* Arabian), f. a genus of the mustard family: *Arabis hirsuta*
      (hairy rockcress).
   L. *armoracia,* f. horseradish: *Armoracia lapathifolia* (horseradish).
   L. *berula,* f. a kind of watercress: *Berula erecta* (a water-parsnip).
   L. *brassica,* cabbage; a genus of the mustard family; see **cabbage**
   NL. *cakile,* f. a genus of the mustard family, < Ar. *qaqulla: Cakile maritima*
      (sea-rocket).
   L. *cardamum,* n. (Gr. *kardamon,* n.; *kardamine; kardamis, -idos,* f.), a kind of
      cress; *kardamomon,* n. a spice: *Cardamine bulbosa* (a bittercress), *Elettaria
      cardamomum* (cardamon), *Coreopsis cardaminefolia* (a tickseed), cardamon.
   L. *draba* (Gr. *drabe*), f. a plant of the mustard family: *Draba verna* (spring-
      draba), *Lepidium draba* (hoary cress), *Campanula drabifolia* (Greek bellflower).
   L. *eruca,* f. a colewort: *Eruca sativa* (rocket).
   L. *erysimum* (Gr. *erysimon*), n. hedge-mustard: *Erysimum pulchellum* (an erysi-
      mum).
   L. *glastum,* n. woad:
   L. *hesperis, -idis,* f. a genus of the mustard family: *Hesperis nivalis* (a rocket).
   L. *iberis, -idis* (Gr. *-idos*), f. a kind of pepperwort: *Iberis odorata* (candytuft),
      Brachycome iberidifolia (Swan River daisy), *Aethionema iberideum* (a cress).
   L. *irio, -onis,* f. a kind of cress:
   L. *isatis, -idis* (Gr. *-idos*), f. woad: *Isatis tinctoria* (dyer's woad).
   L. *lepidium* (Gr. *lepidion*), n. pepperwort: *Lepidium vesicarium* (a pepperwort).
   L. *nasturtium,* n. a kind of cress: nasturtium (*Tropaeolum speciosum*), *Rorippa
      nasturtium-aquaticum* (watercress).
   L. *raphanus* (Gr. *rhaphanos*), radish, a genus of the mustard family; see **radish**
   NL. *rorippa,* f. a genus of the mustard family: *Rorippa obtusa* (a cress).
   L. *siliquastrum,* n. pepperwort: *Cercis siliquastrum* (a redbud).
   L. *sinapis, -is,* f. (Gr. *sinapi,* n.), mustard: sinigrinase, sinigrin (from *Sinapis nigra*),
      Sinapis alba (white mustard), *Hebeloma sinapizans* (a mushroom).
   L. *sisymbrium* (Gr. *sisymbrion*), n. a fragrant plant; a genus of the mustard
      family: *Sisymbrium altissimum* (tall sisymbrium) *Solanum sisymbrifolium* (a
      nightshade).
   Gr. *thlaspis,* f. (*thlaspi,* n.), a kind of cress; *thlaspidion,* n. dim.; *Thlaspi arvense*
      (pennycress, fanweed), *Nothothlaspi rosulatum* (a cress), *Psylloides thlaspis* (a
      beetle).
   NL. *zilla,* f. a genus of the mustard family, < Ar. *sillah: Zilla spinosa* (a brassi-
      cacead).

mustax, L. a kind of laurel; see **laurel**

mustela, L. weasel; *mustecula,* dim.; *mustelinus,* of weasels; see **weasel**

mustum, L. new, unfermented wine; see **wine**

mustus, L. new, young, fresh; see **new**

mutabilis, L. changeable; *mutatus,* changed; see *muto* under **change**

mute < L. *mutus,* silent, dumb; see **silent**

muticus, L. shortened, curtailed; see *mutilus* under **hurt**

mutilago, L. a kind of spurge; see **spurge**

mutilla, NL. a genus of wasps; see **wasp**

mutilus, L. maimed, cut off, docked, shortened; see **hurt**

Mutinus, L. a name for Priapus; see **penis**

mutiny < OF. *muete,* riot; see **rebellion**

muto- -atus, L. change; *mutator,* changer; *mutatorius,* of changing; see **change**

mutton < F. *mouton,* sheep; see **flesh**

muttum, L. mutter, mumble; see *mutio* under **murmur**

mutual < F. *mutuel,* < L. *mutuus,* borrowed, loaned, reciprocal; see **common,
   alternate, trade, equal**

mutus, L. inarticulate, dumb, silent; see **silent**

muzzle < OF. *musel,* snout, jaws and face; see **nose, bind**

mya, L. a kind of mussel; see **mollusk**

myagra, Gr. mouse-trap; see **trap**

mycethmo- < Gr. *mykethmos,* a lowing, bellowing, roaring; *myketes,* bellower; see **roar**

myceto-; myco- < Gr. *mykes, -etos,* fungus; NL. *mycelium,* hyphal filaments; see **fungus**

mychos, Gr. inmost part, recess, closet, bay, sinus; see **room**

myclo- < Gr. *myklos,* ass; see **horse**

myctero- < Gr. *mykter, -os,* nose; see **nose**

mydion, Gr. boat; see **ship**

mydos, Gr. dampness; *mydaleos,* wet; see **wet**

mydriasis, Gr. prolonged dilatation of the pupil of the eye; see **disease**

mydros, Gr. mass of red-hot metal; see **heat**

myelo- < Gr. *myelos,* marrow, pith, spinal cord; see **pith**

mygale, Gr. a field-mouse; see **mouse**

mygmos, Gr. moaning, muttering; see **sad**

myio- < Gr. *myia,* fly; see **fly**

mylacris, L. (Gr. *mylakris*), a kind of cockroach; see **cockroach**

mylasia, L. a kind of hemp; see **hemp**

myllon, Gr. lip; see **lip**

myllos, Gr. bent, crooked; a fish; see **bend, fish**

mylo- < Gr. *myle,* mill, millstone, kneepan; *mylax, -akos,* millstone; *mylias; mylikos,* of a mill; see **grind, knee**

mymar, Gr. blame; see *momos* under **blame**

myndos, Gr. dumb; see **silent**

myo- < Gr. *mys, myos,* muscle, see **flesh;** mouse, see **mouse;** < *myo,* close, shut, wink, see **close**

myoctono- < Gr. *myoktonos,* mouse-killer, a kind of aconite; see **aconite**

myophonos, Gr. an umbelliferous plant; see **carrot**

myops, Gr. near-sighted, see *optikos* under **see;** horsefly, see **fly;** goad, spur, see **point**

myosotis, Gr. forgetmenot; see **borage**

myoxos, Gr. dormouse; see **mouse**

myraphion, Gr. dim. of *myron,* ointment; see **fat**

myrepsos, Gr. preparer of unguents; see *myron* under **fat**

myriapod < Gr. *myrios,* numberless, *pous,* foot: Myriapoda.
> Teacher: "What is a millennium?"
> Pupil: "It is something like a centennial, but it has more legs."

> L. *centipeda,* f. a myriapod with one pair of legs to each segment: centiped.
> L. *iulus* (Gr. *ioulos*), m. a multiped: *Julus terrestris* (a milliped), *Stemmatoiulus bioculatus* (a milleped), *Peripatus iuliformis* (a myriapod).
> L. *millepeda,* f. a myriapod with two pairs of legs to each segment: milleped.
> L. *scolopendra* (Gr. *skolopendra*), f. a multiped: *Scolopendra cingulata* (a centiped), *Scolopendrella immaculata* (a centiped).

myrica, L. (Gr. *myrike*), tamarisk, bayberry; see **bayberry**

myrio- < Gr. *myrios,* numberless; see **number**

myriophyllum, L. (Gr. *myriophyllon*), milfoil; see **water-milfoil**

myrismos, Gr. an anointing; see *myron* under **fat**

myristica, NL. nutmeg; see **nutmeg**

myrmeco- < Gr. *myrmex, -ekos,* ant; *myrmekion,* dim.; *myrmekia,* anthill, wart; *myrmedon,* ant-nest; see **ant, mountain**

myro- < Gr. *myron,* ointment; see **fat**

myrobalan < L. *myrobalanum,* n. (Gr. *myrobalanos,* f.) a kind of nut yielding an ointment; now applied to a family of trees, Combretaceae.
> L. *combretum,* n. a kind of rush; now applied to a genus of the myrobalan family: *Combretum coccineum* (scarlet combretum).
> L. *terminalia,* f. a genus of the myrobalan family: *Terminalia catappa* (Malabar almond).

myrrh < L., Gr. *myrrha;* see **resin**

myrrhis, Gr. sweet cicely; see **carrot**

myrsine, Gr. myrtle; see **myrtle**

myrsos, Gr. basket; see **basket**

myrtle < ML. *myrtillus,* < L. *myrtus* (Gr. *myrtos*), f. myrtle-tree; *myrtaceus; myrteus,* of myrtle; *myrtatus,* seasoned with myrtle; *myrton,* n. myrtleberry, clitoris: myrtol, Myrtle, Myrtaceae, *Myrtus communis* (myrtle), *Rhodomyrtus tomentosa* (downy myrtle), *Vaccinium myrtillus* (whortleberry), *Coriaria myrti-folia* (myrtle-coriaria).

NL. *eucalyptus,* f. a genus of the myrtle family: *Eucalyptus gigantea* (a eucalyptus).

NL. *eugenia,* f. a genus of the myrtle family, < Prince Eugene of Savoy: *Eugenia aromatica* (clove-tree).

Gr. *myrsine,* f. myrtle; now a genus of the Myrsinaceae: *Myrsine africana* (a myrsinacead), *Pachystima[Pachystigma] myrsinites* (Oregon boxwood).

NL. *pimenta,* f. a genus of the myrtle family, < Sp. *pimienta,* f. pepper: *Pimenta officinalis* (allspice).

NL. *psidium,* n. a genus of the myrtle family: *Psidium guajava* (guava).

NL. *syzygium,* n. a genus of the myrtle family: *Syzygium lineatum* (a myrtacead).

myrton, Gr. myrtleberry, clitoris; see **myrtle, vulva**

mys, Gr. mouse, muscle, mussel; see **mouse, flesh, mollusk**

mysaros, Gr. foul, impure, loathsome, < *mysos,* dirt, uncleanness, anything disgusting; see **dirt**

mysis, Gr. a closing of the eyes, lips, pores, etc.; see *myo* under **close**

mystaco- < Gr. *mystax, -akos,* upper lip, the hair upon it; *mystakion,* dim.; see **hair**

mystery < L. *mysterium* (Gr. *mysterion*), secret rite; *mystes.* priest; L. *mysticus,* secret; see **secret, magic, wonder, priest**

mystron, Gr. spoon; *mystrion,* dim.; see **spoon**

myth < Gr. *mythos,* fable, legend, speech; *mytharion; mythidion,* dim.; see **story**

mytilus, L. (Gr. *mytilos*), mussel; see **mollusk**

myxa, Gr. mucus. see **slime**; lamp-nozzle, see **nose**; a kind of plum, see **plum**

myzo, Gr. suck; see **suck**

N

nabis, L. giraffe, camelopard; see **giraffe**

nablium, L. (Gr. *nabla*), a stringed instrument; *nablistes,* player of a nabla; see **harp**

naco- < Gr. *nakos,* fleece; see **wool**

nacca; nacta, L. (Gr. *naktes*), fuller; see **thick**

nacto- < Gr. *naktos,* dense, pressed; see *nasso* under **press**

nactus; nanctus, L. found, got, stumbled upon; see *nanciscor* under **find**

naevus, L. mole, mark, spot; *naevulus,* dim.; *naevius,* having a mole or birthmark; see **mark**

nag < D. *negge,* horse, especially a poor or worn out horse, see **horse**; < the root of *gnaw,* irritate, pester, torment, see **tease**

nagma, Gr. anything piled up; see **heap**

Naias, *-adis,* L. (Gr. *-ados*), a water-nymph; see **water**

nail < AS. *naegel:* nailer, nailwort, agnail(hangnail), fingernail, toenail.

Gr. *balanos,* wooden peg, bolt; see **bar**

L. *chela* (Gr. *chele*), f. claw; *chelion; chelotion,* n. dim.: chela, cheliferous, chelicera, cheliped, chelate, cheloid(keloid), *Chelifer cancroides* (book-scorpion), *Chelodesmus marxi* (a milleped), *Electrochelifer mengei* (a pseudoscorpion).

L. *clavus,* m. nail; *clavulus,* m. dim.: *Clavulinoides compressa* (a foraminifer), *Nephila clavipes* (a spider).

L. *epigrus,* m. wooden pin, peg: *Epigrus ischnus* (a gastropod).

Gr. *epiouros,* m. wooden peg, pin: *Epiurus brevicornis* (a wasp).

L. *gamba,* hoof; see foot

Gr. *gomphos,* m. bolt, peg, nail; *gomphotikos,* fastened with nails: gomphodont, *Gomphus xanthemathus* (a dragonfly), *Gomphocarpus textilis* (a milkweed), *Gomphodesmus castaneus* (a milleped).

Gr. *helos,* m. nail, stud; *helarion,* n.; *heliskos,* m. dim.; *helotos,* nailed, nail-shaped; *ephelos,* nailed: *Heliscomys vetus* (a fossil rodent), *Heloderma horridum* (a Gila monster).

Gr. *katalabeus,* m. holder, nail:

Gr. *kyndalos,* m. wooden peg:

Gr. *onyx, -ychos,* m. fingernail, talon, claw, hoof; a gem stone; *onychion,* n. dim.; *onychinos,* nail-colored, resembling onyx; *aponychistikos,* pertaining to nail-polishing; *monyx, -ykos,* m. uncloven hoof: onyx, onychoid, onychophagist, onychia, paronychia, *Onychodus sigmoides* (a fossil fish), *Onychorhynchus regius* (king-tody), *Cyrtonyx montezumae* (a quail), *Platyonychus ocellatus* (a crab), *Acinonyx jubatus* (African cheetah), *Tetranychus bimaculatus* (a mite), *Paronychia serpyllifolia* (creeping nailwort), *Onygena[Onychogena] equina* (a fungus).

Gr. *passalos,* m. peg; *passaliskos,* m. dim.: *Passalus cornutus* (a beetle).

L. *paxillus,* m. peg, small stake: paxilliform, paxillose, *Paxillus involutus* (a mushroom).

L. *scalmus* (Gr. *skalmos*), m. peg or thole for fastening an oar:

L. *solipes,* one with an uncloven hoof, the horse; see *pes* under foot

Gr. *tylion,* pin. dim. of *tylos,* knot, bolt; see knot

L. *unguis, -is,* m. fingernail, claw, talon; *unguiculus,* m. dim.; *ungula,* f. hoof: unguiculate, unguligrade, Ungulata, triungulin, *Barbula unguiculata* (a moss), *Pithecellobium unguiscati* (catsclaw), *Manatus inunguis* (Amazon manatee).

See: bar, bind, close

naja, NL. < Skt. *naga,* snake; see snake

naked < AS. *nacod,* nude; see bare

nama, Gr. spring, stream; *namation,* dim.; *namatodes,* full of springs; see spring

name < AS. *nama;* see word

nannaris, Gr. prodigal, spendthrift; see *nannarion* under waste

nanus, L. (Gr. *nanos*), a dwarf; see little

naos, Gr. temple; *naiskos,* dim.; see temple

nap < MD. *noppe,* akin to AS. *ahneopan,* pluck off; see hair

napaeo- < Gr. *napaios,* of a dell, < *nape,* wooded vale, dell, glen; see valley

naphtha, Gr. a volatile rock oil; see fat

napkin < OF. *nappe,* < L. *mappa,* f. table napkin, towel; *mappula,* f. dim.: napery, map, apron, *Amanita mappa* (a fungus).

L. *facitergium,* n. towel:

Gr. *hemitybion,* n. towel, napkin:

Gr. *kapsidrotion,* n. napkin:

Gr. *maktron,* n. towel, wiper: *cheiromaktron,* n. napkin:

L. *mantele* (*mantile*), *-is,* n. towel, napkin:

L. *manutergium,* n. towel: manutergium.

L. *mucinium,* n. handkerchief:

L. *orarium,* n. napkin, handkerchief:

Gr. *phossonion,* coarse towel, dim. of *phosson,* sail, coarse linen; see sail

L. *sabanum* (Gr. *sabanon*), n. linen cloth, towel:

L. *sudarium,* n. handkerchief; *sudariolum,* n. dim.:

napus, L. turnip; *napellus; napulus,* dim.; see turnip

narcissus, L. (Gr. *narkissos*), a genus of the amaryllis family; see amaryllis

narcotic < Gr. *narkotikos,* benumbing, dulling, < *narke,* numbness, torpor; see dull

nardus, L. (Gr. *nardos*), a costly, perfumed oil; see fat

naris, L. nostril; *narinosus,* broad-nosed; see nose

naros, Gr. liquid, fluid, < *nao,* flow; see flow

narrativus, L. pertaining to reporting, < *narro, -atus,* relate; see story

narrow < AS. *nearu.*

L. *angustus,* narrow, tight, slender, thin; *angustia,* f. narrowness, strait, difficulty, distress; *angusto, -atus,* narrow, restrain; Sp. *angostura,* f. narrowness: angusti-

clave, angustirostrate, angostura(*Cusparia angostura*), *Crinum angustum* (an amaryllis), *Epilobium angustifolium* (fireweed), *Carex angustior* (a sedge).
L. *artus,* narrow, tight, close; see *arcto* under **press**
L. *astrictus,* drawn together, tight, narrow, close; *restrictus,* bound up, tight, limited, stingy; see **bind**
L. *collectus,* contracted, narrowed, concentrated; see *lego* under **gather**
L. *contractus,* drawn together, shortened, tightened; see **lessen**
Gr. *stenos* (*steinos*); *stenygros,* narrow, tight; *stenotes,* f. narrowness; *mesostenos,* narrow in the middle: stenographer, stenotype, stenosis, stenostomia, stenophyllous, *Stenochlaena palustris* (a fern), *Stenopogon indistinctus* (a fly), *Stenoproctus unipunctatus* (a fly), *Stenygrocercus silvicola* (a spider) *Stenanthium gramineum* (a melanthacead), *Mesostenoideus albomaculatus* (a wasp), *Hylastinus obscurus* (clover-root borer), *Metastenothorax punctatipennis* (a bug).
See: **lessen, bind, press, choke, close, thin**

narthecium, L. (Gr. *narthekion,* dim. of *narthex,* an umbellifer), ointment-box, medicine-chest; see **box, carrot**

nascens, L. arising, beginning, < *nascor, natus,* be born, spring forth; see **begin, birth**

nasiterna, L. a pot with a large spout; see **pot**

nasmos, Gr. spring, stream; see *nama* under **spring**

nassa; naxa, L. wicker-basket with a narrow neck; see **basket**

nastes, Gr. inhabitant; see *naio* under **life**

nastos, Gr. pressed close, firm, solid; a cake; see *nasso* under **press**

nasturtium, L. a kind of cress; see **mustard**

nasus, L. nose; *nasutus,* large-nosed; see **nose**

natalis, L. of birth; see *nascor* under **birth**

natator, L. swimmer; *natabulum,* swimming-place; see *nato* under **swim**

natis, L. rump, buttock; ML. *natica,* buttock; see **rump**

native < L. *nativus,* inherent or conferred by birth: naïve, naïvete, *Dolerus nativus* (a wasp).
L. *aborigineus,* ancestral, native, original: aborigines, aboriginal.
Gr. *autochthonos,* indigenous, native: autochthonous.
Gr. *chorites,* m. countryman, native:
Gr. *endemos,* living in, native: endemic, endemism.
Gr. *engaios,* of the land, native: *Astacus*(*Engaeus*) *cunicularius* (a crayfish).
L. *genuinus,* native, natural, authentic; see **true**
Gr. *hemedapos,* native, indigenous:
L. *indigenus,* born in a country, native; *indigena,* m. native: indigenous, *Phycita indiginella* (a moth).
L. *ingenuus,* native, indigenous, candid: ingenuous.
Gr. *palaichthon, -os,* c. an indigenous inhabitant:
Gr. *patriotes,* m. fellow-countryman:
L. *vernaculus; vernus,* native, < *verna,* home-born slave; see **servant**
See: **nature, birth, country, class**

natrix, L. water-snake; see **snake**

natro- < Ar. *natrum,* sodium carbonate; containing sodium; see **sodium**

nature < L. *natura,* f. birth, inherent quality or property of things; *naturalis,* innate, not artificial; *innatus,* inborn, inherent, natural: natural, *Systema naturae.*
Gr. *acheirotos,* untouched by the hand of man, untamed:
Gr. *adamastos; adamatos; admetos,* unconquered, unbroken, untamed, unwedded (females): *Admetus fuscimanus* (a pedipalpid).
L. *affectus, -us,* m. a state of mind, mood, disposition: affect, affectation, affectionate.
L. *agrestis,* of the land, rural, wild; Gr. *agrios,* living in the fields, wild; see *ager* under **field**
Gr. *anambatos,* unmounted, unbroken: *Anambates cleroides* (a beetle).
Gr. *anemeros,* wild, savage: *Anemerus peregrinus* (a beetle).
Gr. *apharotos,* unplowed, untilled; see **rough**
L. *attributum,* n. character, quality: attribute, attributable.
L. *character, -is,* mark, habit, usage; see **mark**
L. *Cybele* (Gr. *Kybele*), f. goddess of nature: *Argynnis cybele* (a butterfly).
AS. *dom,* jurisdiction, state, condition: kingdom, Christendom, pagandom, heathendom, serfdom, wisdom, freedom, fogydom, thraldom, doom, Domesday Book, deem, Dempster.
Gr. *emphytos,* implanted, inborn, natural:
Gr. *entheros,* savage, wild, rough:

Gr. *ethos,* character, usage, habit; see *ethas* under **use**

L. *ferox, -ocis,* wild, spirited, fierce, < *ferus,* wild, untamed; *feroculus,* dim.; *feritas, -atis,* f. wildness; *ferinus,* pertaining to wildness and wild animals; *ferus,* m.; *fera,* f. wild beast; *efferatus,* wild, fierce, savage: ferocity, feral, *Cryptoprocta ferox* (foussa), *Fierasfer acus* (a fish), *Ferocactus horridus* (a cactus), *Amaurobius ferox* (a spider), *Nyroca ferina* (pochard).

L. *habitus, -us,* m. condition, state, deportment: habit, habitual, habituate, habitat.

L. *hylaeus* (Gr. *hylaios*), of the wood or forest, savage, wild; see *hyle* under **wood**

L. *immansuetus,* untamed, wild, savage: *Acalles immansuetus* (a beetle).

L. *immeditatus,* artless, unstudied, natural:

L. *indoles, -is,* f. native quality, disposition, talent:

L. *indomitus,* untamed, wild: *Lycosa indomita* (a spider).

L. *infrenis; infrenus,* unbridled, untamed; see *frenum* under **hold**

L. *ingenuus,* native, natural, candid; see **native**

L. *instinctus, -us,* m. natural impulse; instigated, incited: instinct, instinctive.

L. *materia,* matter, stuff; see **thing**

Gr. *physis, -ios,* f. nature, condition; *physikos,* concerning nature and matter: physics, physical, physician, physicist, physiology, apophysis, metaphysics, symphysis, physiognomy, physiography.

Gr. *poios,* of a given nature, kind, quality; *poiotes, -etos,* f. quality:

L. *qualitas, -atis,* f. condition, nature, property: quality, qualification, *Quisqualis indica* (Rangoon creeper).

Gr. *schesis,* f. state, condition, habit, nature, quality, posture: *Anaroschesis chinensis* (a beetle).

L. *status, -us,* m. standing, condition, position: state, status.

AS. *wilde,* in a state of nature: wild, wildfire, wilderness, bewilder.

*   *   *   *   *   *

L. *-aceus, -icius,* pertaining to, belonging to, having the nature of, as *cretaceus,* chalky; *surrepticius,* clandestine: arenaceous, cretaceous, herbaceous, farinaceous, glumaceous, testaceous, latericeous, meretricious, surreptitious, *Smilax herbacea* (carrion-flower), *Carex paleacea* (a sedge).

L. *-acus, -icus* (Gr. *-akos, -ikos*), belonging to, like, of, pertaining to; denoting the greater valence in a chemical compound, as *cardiacus* (Gr. *kardiakos*), of the heart; *hypnoticus* (Gr. *hypnotikos*), soporific; *meracus,* pure; *posticus,* posterior; Gr. *physikos,* natural: cardiac, demoniac, elegiac, zodiac, ammoniacal, maniacal, athletic, aquatic, artistic, economic, erratic, fanatic, hypnotic, iconoclastic, gymnastic, lunatic, plastic, lithic, public, narcotic, prognostic, dynamics, mathematics, physics, technique, unique, oblique, Martinique, ferric, nitric, cupric, stannic, sulfuric, perchloric, *Lonicera japonica* (Japanese honeysuckle), *Nyssa aquatica* (water-tupelo), *Fagus sylvatica* (European beech), *Cyperus cylindricus* (a sedge).

Suf. *-ad, -ade,* pertaining to, relating to, akin to, made of, < F. *-ade,* < Gr. *-as, -ados,* as *monas, -ados,* unit; < F. *-ade,* Sp. *-ada,* < L. *-ata,* as arcade < *arco, -atus:* armada, ballad, blockade, bravado, brigade, brocade, bromeliad, cannonade, cavalcade, comrade, crusade, cycad, decade, desperado, dryad, dyad, esplanade, Iliad, lemonade, marmalade, masquerade, myriad, orangeade, palisade, parade, pomade, salad, stockade, tirade, tornado.

L. *-ae,* sing.; *-arum,* pl. first declension genitives; *-i,* sing.; *-orum,* pl. second declension genitives; *-is,* sing.; *-ium, -um,* pl. third declension genitives; *-us,* sing.; *-uum,* pl. fourth declension genitives; *-ei,* sing.; *-erum,* pl. fifth declension genitives; of, belonging to, pertaining to: *Ammobroma sonorae* (sand-food), *Puccinia gentianae* (a rust), *Puccinia malvacearum* (a rust), *Trametes pini* (pine-rust), *Exobasidium vaccinii* (a fungus), *Thermobia furnorum* (firebrat), *Pseudomonas juglandis* (walnut-blight), *Dermanyssus avium* (bird-mite).

The *-ae* and *-i* endings may be attached to proper names, both personal and geographic, to denote commemoration, dedication, and derivation. In the Augustan Age the genitive ending of masculine names in *-ius* was not *-ii,* as might be expected and as later developed, but *-i.* Modern New Latin follows the Augustan simplicity and the double *-ii* appears only in the genitives of names ending with *i,* as *Marconii,* < Marconi. This preserves the current spelling of modern names and avoids cumbersome manipulation of terminal letters. Usage indicates the following formulas for making such genitives, which, when used as specific terms of binomials, are, or should be, decapitalized.

Attach *-ae* to feminine proper names ending with any letter except *a,* to which attach *-e,* as: *Psen cheesemanae* (a wasp), *Osmia sandhouseae* (a bee), *Sorbus helenae* (a mountain-ash), *Astragalus thompsonae* (a vetch), *Begonia mexiae* (a begonia), *Calochortus catalinae* (Catalina mariposa).

Attach -*i* to masculine and neuter proper names ending with any letter except *a*, to which attach -*e*, as: *Quercus muhlenbergi* (an oak), *Astragalus blakei* (a vetch), *Pinus jeffreyi* (a pine), *Thymus przewalskii* (a thyme), *Triletes sahnii* (a Triassic plant), *Acer florissanti* (an Oligocene maple), *Rosa montezumae* (a rose).

F. -*age*, < L. -*aticum*, collection of, condition, state, as in *pontaticum*, bridge-toll; *viaticum*, travel-money: advantage, assemblage, beverage, courage, foliage, hostage, language, lineage, marriage, mileage, peerage, plumage, pontage, viaticum, voyage. Derive: carriage, dotage, mucilage, nonage, package.

L. -*ago*, -*igo*, -*ugo*, having the characteristics of, resemblance, as *plumbago*, a lead ore; *rubigo*, rust; *lanugo*, down: lumbago, plumbago, virago, impetigo, lentigo, vertigo, albugo, image, mucilage, cartilaginous, ferruginous, rubiginous, *Plantago lanceolata* (English plantain), *Medicago hispida* (bur-medic), *Viburnum lentago* (nannyberry).

Gr. -*aleos*, pertaining to, as *branchaleos*, hoarse; *rhomaleos*, strong:

L. -*alis*, -*elis*, -*ilis*, -*ulis*, pertaining to, condition, having the nature or quality of, as *naturalis*, natural; *fidelis*, faithful; *gracilis*, slender; *edulis*, edible: animal, political, mortal, musical, grammatical, personal, manual, social, adverbial, proverbial, ministerial, controversial, celestial, mental, removal, disposal, emphatically, hotel, crude, cruel, vowel, candle, civil, April, subtile, subtle, hostile, docile, fragile, juvenile, genteel, gentleman, frail, utensil, fidelity, *Digitalis lutea* (a foxglove).

L. -*and*, -*end*, -*und*, having the quality of: viand, multiplicand, brigand, memorandum, dividend, reverend, legend, tremendous, stupendous, jocund, rotund, second.

L. -*antia*, -*entia*; -*ans*, -*antis*; -*ens*, -*entis*, present participles; having the quality of, belonging to, as *substantia*, matter; *scientia*, knowledge; *clamans*, -*antis*, shouting; *audiens*, -*entis*, listening: defiant, militant, petulant, pliant, reluctant, stimulant, vagrant, competent, corpulent, pestilent, redolent, resident, student, acceptance, countenance, resemblance, constancy. audience, influence, indulgence, science, clemency, currency, potency.

L. -*anus*, -*enus*, -*inus*, -*unus*, belonging to, pertaining to; one who; chemical terms, as *urbanus*, of the city; *Romanus*, Roman; *serenus*, clear; *marinus*, of the sea; *jejunus*, hungry: artisan, African, veteran, Lutheran, captain, villain, chaplain, mountain, fountain, historian, Arabian, Italian, Christian, humane, mundane, arbutin, adrenalin, insulin, morphine, nyctanthin, pepsin, placentin, prolactin, precipitrin, pituitrin, luciferin, progestin, solanin, strychnine, saponin, methane, xylene, propane, nicotine, gasoline, theelin, thyroxin, benzene, chlorine, cocaine, coralline, saline, marine, medicine, doctrine, columbine, tamburine, basin, cousin, bulletin, fortune, opportunity, tribune, jejune, retina, canteen, intine, extine, perine, *Scirpus americanus* (a sedge), *Haeckeliana darwiniana* (a radiolarian).

F. -*ard*, -*art* (G. *hart*), quality in the highest degree, excessiveness, one who, that which; see **hard**

L. -*arius*, -*orius*, pertaining to, having the nature of, as *ordinarius*, customary; *desultorius*, jumping about: anniversary, ordinary, desultory, secondary, sanitary, stationary, January, February, introductory, *Utricularia inflata* (a bladderwort), *Carex vesicaria* (a sedge).

L. -*aster*, -*ister*, denoting wildness, diminutives; see **little**

L. -*atus*, -*etus*, -*itus*, -*otus*, -*utus*, -*tus*, perfect participles; provided with, having the nature of, pertaining to, as *ornatus*, adorned; *obsoletus*, worn out; *politus*, polished; *devotus*, attached, faithful; *acutus*, sharp; *victus*, conquered: desolate, magistrate, decorate, animate, elucidate, date, chordate, ornate, advocate, cantata, sonata, Vertebrata, Annulata, Edentata, complete, effete, obsolete, discreet, discrete, quiet, requisite, polite, deposit, digit, granite, depot, remote, devote, vote, debt, resolute, absolute, execute, institute, acute, minute, astute, tribute, statute, chlorate, stannate, chromate, nitrate, sulfate, phosphate, plumbate, carbonate, iodate, stibnate, acetate, citrate, tungstate, *Aster acuminatus* (an aster).

L. -*bilis*, tending to be, capable of, worthy of, having the quality of, as *domabilis*, tamable: curable, durable, favorable, legible, sensible, sociable, stable, soluble, terrible, visible, voluble, *Abies amabilis* (silver fir), *Abies nobilis* (noble fir).

L. -*bundus*, -*cundus*, -*undus*, increased quality or tendency toward; see **very**

L. -*bus*, having the quality of, as *acerbus*, sour: acerb, superb.

Gr. -*es*, of, pertaining to, as *akrates*, powerless; *eugenes*, well-born; *kerastes*, horned; *pseudes*, false: acraturesis, eugenic, cerastes, pseudomorph.

Suf. -*ese*, of, pertaining to, originating in, inhabitant of: Chinese, Japanese, Maltese, Portuguese, Vienese, journalese, Johnsonese, officialese.

L. -*eus*, -*ius*, made of, having the quality of, as *aureus*, golden; *corneus*, of horn, horny; *regius*, royal: arboreous, corneous, extraneous, ligneous, spontaneous,

Aster puniceus (swamp-aster), *Stegonotus plumbeus* (a snake), *Stachys coccinea* (Texas betony), *Juglans regia* (Persian walnut).

Suf. *-hood,* < AS. *had,* condition, state: motherhood, fatherhood, brotherhood, hardihood, likelihood, livelihood, neighborhood, falsehood, godhead, maidenhead.

L., Gr. *-ia, -cia, -tia,* pertaining to, state, condition, names of countries, diseases, biological, chemical, and medical terms, Greek and Latin plurals, as *audacia,* boldness; *tristitia,* sadness; Gr. *apoplexia,* stroke: bacchanalia, saturnalia, militia, regalia, justice, notice, malice, celibacy, confederacy, conspiracy, delicacy, fallacy, lunacy, privacy, aristocracy, democracy, Bulgaria, Australia, India, Virginia, diphtheria, dyspepsia, amnesia, dahlia, petunia, Mammalia, Reptilia, bacteria, morphia.

Suf. *-id, -ide,* < L. *-idus,* having the nature of; final part of a chemical term of which the first part is the name of a non-metal, as *lucidus,* clear: acid, florid, fluid, humid, lucid, morbid, rigid, solid, sordid, splendid, stupid, timid, tumid, turgid, viscid, vivid, chloride, bromide, oxide, sulfide, glucoside, nitride, iodide, boride, fluoride, cyanide, carbide.

L. *-imus* (Gr. *-imos*) pertaining to, related to, having the quality of, as *maritimus,* of the sea; *septimus,* seventh; Gr. *chresimos,* useful; *machimos,* warlike: maritime, *Festuca maritima* (sea-fescue).

Gr. *-inos,* of, made of, as *kyaminos,* of beans; *rhodinos,* made of roses; *xylinos,* of wood, wooden: *Apanteles xylinus* (a wasp).

L. *-ion, -sion, -tion,* act, process, result or state of, having the nature of, as *regio, -onis,* district; *percussio, -onis,* a striking; *actio, -onis,* performance: communion, complexion, contagion, rebellion, oblivion, emulsion, dissension, secession, succession, question, attention, education, preparation, objection, reason, lesson, poison, prison, venison, gudgeon, sturgeon, dungeon.

L. *-is,* m., f.; *-e,* n. adjectival suffix meaning with, having the nature of, as *bicornis,* two-horned; *conformis,* similar; *insomnis,* sleepless; *quadrijugis,* four-yoked; *multiforis,* many-doored: *Cladius pectinicornis* (a wasp), *Erianthus brevibarbis* (a grass), *Zonotrichia albicollis* (white-throated sparrow), *Coptotermes acinaciformis* (a termite), *Bromus latiglumis* (a grass), *Thomomys latirostris* (a pocket-gopher), *Toxostoma curvirostre* (a thrasher), *Carex laxiculmis* (a sedge), *Bibio albipennis* (a fly), *Empidonax flaviventris* (yellow-bellied flycatcher), *Cardiomanes reniforme* (a fern).

Suf. *-ish,* < AS. *-isc* (G. *-isch*), denoting origin, pertaining to, like: English, Irish, Moorish, whitish, reddish, foolish, fiendish, sluggish, modish, devilish.

Suf. *-ism,* < L. *-ismus* (Gr. *-ismos, -ysmos*), condition, quality, doctrine, result, sect, as *baptismos,* emersion; *kataklysmos,* deluge; *strabismos,* a squinting: strabismus, baptism, heroism, altruism, Buddhism, Judaism, paganism, fanaticism, humanism, idealism, mesmerism, isolationism, globalism, nepotism, favoritism, syllogism, prism, fantasm, paroxysm, enthusiasm, chasm, witticism, socialism, optimism, pessimism.

Suf. *-ite, -ites,* < L. *-ita;* Gr. *-ites,* m.; *-itis,* f., having the nature of, like, disciple, descendant, mineral terms, salts of *-ous* acids: The gender of Greek nouns in *-ites* is generally masculine, as *askites,* dropsy; *galaktites,* a kind of stone; *hoplites,* armed soldier; *molochites,* a kind of stone; *petasites,* coltsfoot; *phyllites,* wreath of leaves; *saurites,* a serpent; *siderites,* lodestone; *xylites,* woodcutter. To form the feminine, change *-ites* to *-itis,* as *hoplitis, -idos,* female warrior: Mennonite, Bryanite, Israelite, chlorite, plumbite, antimonite, sulfite, pyrites, *Phyllites perplexus* (a fossil leaf), *Hoplitis adunca* (a wasp), *Stephanitis mitrata* (a bug).

Suf. *-ity,* < L. *-itas,* denoting condition, degree, quality, state, as *celeritas, -atis,* swiftness; *humilitas, -atis,* lowness: ability, audacity, captivity, celerity, charity, electricity, geniality, gentility, nobility, quality, lucidity, plenty, sagacity, simplicity, utility, veracity.

L. *-ium,* quality or nature of, as *gaudium,* joy: cambium, delirium, equilibrium, medium, potassium, radium, sodium.

L. *-ius* (Gr. *-ios*), of, pertaining to, as *agrius* (Gr. *agrios*), of the fields, wild; *Daucius,* of Daucus; *fullonius,* of fullers; Gr. *ouranios,* of heaven; *pluvius,* rainy: Jupiter Pluvius, *Zapus hudsonius* (a jumping-mouse), *Microsciurus isthmius* (a squirrel).

L. *-ius,* m.; *-ia,* f.; *-ium,* n., suffixes attached to the stems of personal and other proper names to form nominative generic commemoratives.

For names ending with consonants, attach the suffixes unmodified, as *Dahlius, Dahlia, Dahlium* (< Dahl); *Carpenterius, Carpenteria, Carpenterium* (< Carpenter). Sometimes for those ending in *r,* omit the *i,* as *Gunnerus, Gunnera, Gunnerum* (< Gunner). Those ending in *ius* are considered already latinized, as *Confucius, Confucia, Confucium* (< Confucius).

For names ending with vowels, opinions and usage differ concerning the form of the suffix. Theoretical latinization may occasionally be waived in favor of the practical purpose of preserving the terminal letters of some modern names, as: *Danaius, Danaia, Danaium,* or *Danaeus, Danaea, Danaeum* (< Dana); *Saportaius, Saportaia, Saportaium,* or *Saportaeus, Saportaea, Saportaeum* (<Saporta); *Gilmoreus, Gilmorea, Gilmoreum* (< Gilmore); *Galvanius, Galvania, Galvanium* (< Galvani); *Serenous, Serenoa, Serenoum,* or *Serenonius, Serenonia, Serenonium* (< Sereno), as in *Petronius, Petronia, Petronium* (< Petro, *-onis,* Peter); *Canuus, Canua, Canuum* (< Canu); *Murraius, Murraia, Murraium,* or *Murrayus, Murraya, Murrayum* (< Murray); *Sperryus, Sperrya, Sperryum* (< Sperry).

F. *-ive,* < L. *-ivus,* nature or quality of, action, agent, as *festivus,* merry: active, negative, positive, effusive, affirmative, corrective, punitive, operative, sedative, motive, festive, lucrative, delusive, decisive, expensive, passive, fugitive, locomotive, relative, adjective, conclusive, motif, caitiff, plaintiff.

AS. *-ling,* diminutives; having the quality of; pertaining to; see DIMINUTIVES under little

L. *-mentum,* denoting action, means, result, condition, tool; see tool

Gr. *-mone,* denoting quality of, as *charmone,* joy: anemone, *Argemone grandiflora* (a prickly-poppy).

AS. *-ness,* condition, quality, state: brightness, dullness, selfishness, clearness, firmness, wilderness, witness, goodness, likeness, holiness, darkness, loneliness, forgetfulness, happiness, richness.

Suf. *-ome,* < Gr. *-oma,* signifying condition, having the nature of: caulome, coelom (coelome), mestome, phyllome, rhizome, telome, sarcoma.

L. *-or,* denoting condition, state, as *timor,* fear: error, favor, liquor, odor, splendor, tremor, vigor.

Gr. *-otes,* feminine suffix denoting quality, as *chlorotes,* greenness; *hippotes,* horsiness; *leptotes,* thinness; *mestotes,* fullness; *psathyrotes,* looseness:

Suf. *-ous, -ose, -osity,* < L. *-osus,* having the nature or quality of, usually in fullness or abundance; also the lesser valence in a chemical compound, as *verbosus,* wordy: ambitious, capacious, censorious, devious, disastrous, laborious, meritorious, malicious, mendacious, notorious, nutritious, nauseous, pugnacious, querulous, sedulous, tremulous, vivacious, virtuoso, wondrous, morose, globose, glucose, sucrose, varicose, animosity, monstrosity, pomposity, verbosity, ferrous, mercurous, nitrous, hypochlorous, stannous, sulfurous, *Artamus superciliosus* (a swallow-shrike), *Ledum glandulosum* (a ledum).

The successive stages of drunkenness are reputed to be: the jocose, verbose, bellicose, lachrymose, and comatose.

Suf. *-ship,* < AS. *scipe,* state, office, art: clerkship, courtship, fellowship, friendship, guardianship, hardship, kinship, marksmanship, stewardship, township, worship.   Derive: landscape, seascape, offscape.

Suf. *-som, -some,* < AS. *-sum,* having the quality of, like, full of: blithesome, burdensome, buxom, frolicsome, fulsome, handsome, lissome, lonesome, noisome, quarrelsome, tiresome.

L. *-syna* (Gr. *-syne*), feminine suffix denoting condition, state, quality, as *cherosyne,* widowhood; *eleemosyne,* pity; *gnomosyne,* prudence; *komporrhemosyne,* boastfulness; *leptosyne,* thinness; *orthosyne,* straightness; *sophrosyne,* moderation: *Charmosyna papuensis* (a parrot).

L. *-ter, -tris, -tre,* pertaining to, place where; see place

Suf. *-tude,* < L. *-tudo,* state, condition, pertaining to, as *pulchritudo, -inis,* beauty: aptitude, amplitude, exactitude, fortitude, gratitude, latitude, longitude, magnitude, multitude, rectitude, solitude, vicissitude.

L. *-us, -a, -um,* masculine, feminine, and neuter terminations of many adjectives, as *dubius, -a, -um,* doubtful; *glaucus, -a, -um,* grayish; *magnus, -a, -um,* large; *quadrimanus, -a, -um,* four-handed:

L. *-us* (Gr. *-os*), of, pertaining to, as *quintus,* fifth; *octavus,* eighth; *septimus,* seventh; Gr. *keros,* horned; *otos,* eared; *philos,* loving; *plinthotos,* brick-shaped; *polylogos,* talkative; *pterygotos,* winged; *skios,* shaded: quintus, octave, *Euryproctus macrocerus* (a wasp), *Hydrophilus ovatus* (a water-beetle).

See: thing, native, rough, field, earth, stone, stream, life, manner

natus, L. born; see *nascor* under birth

nau-; nausi-; nauti- < Gr. *naus,* ship; *nausiploos,* sailing in ships; *nautikos,* of ships; see *navis* under ship

naucrates, L. (Gr. *naukrates*), pilot, master; see *krateo* under govern

naucum, L. trifle; see trifle

naulum, L. (Gr. *naulon*), fare; see pay

nauplius, L. the larval form of many crustaceans; see young

nausea, L. (Gr. *nausia*), seasickness, retching; see disease

nauta, L. (Gr. *nautes*), sailor, seaman; see sail

nautilus, L. (Gr. *nautilos*), a cephalopod; see **mollusk**

nave < AS. *nafu,* hub of a wheel; see **box**

navel < AS. *nafela,* little nave.
 Gr. *omphalos,* m. navel, button, boss; *omphalion,* n. dim.; *omphalikos; omphalios,*
 with a boss; *omphalodes,* navellike: omphalism, omphalitis, *Omphalodes linifolia*
 (Venus navelwort), *Omphalia flavida* (a mushroom), *Chrysomphalus aurantii*
 (red orange-scab), *Mesomphalus submarginatus* (a Devonian ostracode).
 L. *umbilicus,* m. navel; *umbilicaris,* of the navel; *umbilicatus,* navellike: umbilical,
 nombril, *Umbilicaria arctica* (rock-tripe), *Umbilicus pendulinus* (navelwort),
 Neuroterus umbilicatus (a gall-wasp).

navis, L. (Gr. *naus*), ship; *navicula,* dim.; *navalis; nauticus,* of ships; *navigo, -atus,*
 sail; see **ship, sail**

navus, L. busy, active; see **busy**

naxium, L. a stone for polishing marble; see **stone**

ne- < L. *ne,* not; < AS. *ne,* not; see **not**

neanico- < Gr. *neanikos,* fresh, youthful; see *neanis* under **child**

near < AS. *neah,* nigh.
 L. *adibilis,* accessible; see **passage**
 L. *affinis,* neighboring, bordering, related; *confinis,* adjoining; see **kin**
 Gr. *anchi,* near; *anchion; anchoteros,* nearer; *anchistos,* nearest; *anchimolos,* near,
 close: *Anchisaurus colurus* (a dinosaur), *Anchitherium aurelianense* (a fossil
 mammal).
 L. *appono, -positus,* place near or by the side of: apposition, apposite.
 L. *assitus,* near: *Dusona assita* (a wasp).
 L. *attiguus; contiguus,* touching, bordering, neighboring: contiguous, contiguity,
 Callirhytis attigua (a wasp), *Polytribax contiguus* (a wasp).
 AS. *be; bi,* near: by, because, befall, behold, beyond, below, beneath, beseech,
 beware, berate, befitting, begin, beget, belie, believe, byword, bygone, byway,
 standby, about.
 L. *circa,* around, about, near by: circa.
 L. *circumjectus,* lying around, surrounding; *circumsitus,* situate around, neighbor-
 ing:
 L. *cis,* on this side; *citer, -tra, -trum,* lying near, near, close; *citerior,* nearer;
 citimus, nearest: cisalpine, cisatlantic, *Xenotipula cisatlantica* (a cranefly).
 L. *comminus,* near, at close quarters, hand to hand:
 L. *conterminus,* bordering, neighboring:
 L. *contiguus,* near, adjacent, bordering: contiguous, contiguity, *Polytribax contiguus*
 (a wasp).
 L. *continens,* holding together, lying near; see *teneo* under **hold**
 Skt. *eka,* one, next; see **one**
 Gr. *engys,* near, at hand, hard by; *engion,* nearer; *engistos,* nearest: *Engyophrys
 sanctilaurenti* (a fish), *Engistus brucki* (a bug), *Engistonura parallela* (a fly).
 L. *fere,* almost, nearly:
 L. *finitimus,* adjoining, bordering, near, related: *Trimorus finitimus* (a wasp).
 L. *juxta; juxtim,* near, close, next to, nigh: juxtaposition, juxtamarine, joust,
 jostle, *Osmia juxta* (a bee).
 Gr. *mello,* about to do; see **delay**
 Gr. *meta,* among, near, after, over, change; see **over, change**
 L. *mox,* soon, presently, anon:
 L. *pene (paene),* almost: peninsula, peneplain, penumbra, penultimate, *Andrena
 paenefulva* (a bee), *Paenepediculus simiae* (a louse).
 Gr. *para,* beside, near, by; *parergos,* subordinate, incidental: parasite, parable,
 paradise, paradox, parallel, parley, parenchyma, parody, paralysis, palsy, pare-
 goric, parabola, paragraph, paragenesis, paroxysm, parotid, parlor, paraphernalia,
 dyspareunia, *Paraspalangia annulipes* (a wasp).
 Gr. *paredros,* sitting beside, next to, near; m. associate: *Paredrocoris pectoralis*
 (a bug).
 Gr. *pareoros,* hitched beside: *Pareora incincta* (a gastropod).
 Gr. *pareunos,* lying beside or with; see **companion**
 Gr. *plesios,* near; *plesiaiteros,* nearer; *plesiaitatos,* nearest: plesiosaurus, plesio-
 morphous, plesiotype, *Plesiosorex germanicus* (a fossil mammal), *Zaleptopygus
 plesius* (a wasp).
 Gr. *procheiros,* at hand, near: *Procheirichthys ferox* (a Triassic fish).
 L. *prope,* near; *proprius,* n. nearer; *proximus,* nearest; *propinquus,* near, neighboring: propinquity, propinquous, approach, reproach, proximity, approximate,
 proximal, *Caryota propinqua* (a palm), *Cuterebra approximata* (a botfly).

Gr. *pros,* implying motion from, on, or to the side, beside, near, toward, in addition; *prosbletos,* added; *prosmeno,* wait longer: prosenchyma, prosody, proselyte.

Gr. *prosbatos,* accessible; see *baino* under **walk**

L. *relativus,* referring to, akin to, near to; see **kin**

Gr. *schedon,* near, close, almost: *Trapezium(Schedotrapezium) carinatum* (a fossil pelecypod).

L. *vicinus,* near, neighboring; see *vicus* under **town**

See: **kin, border, around, neighbor, companion, friend, touch, bind, accessible**

nearos, Gr. young, new, fresh; *neatos,* youngest, latest; see *neos* under **new**

neat < L. *nitidus,* shining; see **fit, beauty**

nebraco- < Gr. *nebrax, -akos,* a young animal; see **young**

nebrido- < Gr. *nebris, -idos,* fawnskin; *nebrites,* like a fawnskin; see **skin**

nebros, Gr. young deer, fawn; *nebrias,* spotted like a fawn; see **deer**

nebula, L. mist; *nebulosus,* misty, foggy, cloudy, dark; see **cloud**

necator; necatrix, L. killer, < *neco, -atus,* kill; see **kill**

necessarius, L. indispensable, requisite; see **must**

nechaleos, Gr. swimming; *necheion,* swimming-place; see *necho* under **swim**

neck < AS. *hnecca.*

Gr. *auchen, -os,* m. neck, throat: *Auchenophorus fulvicornis* (a wasp).

L. *cervix, -icis,* f. neck; *cervicula,* f. dim.; *cervicatus; cervicosus,* stiff-necked, stubborn: cervical, cerviculate, cervix.

L. *collum,* n. neck; *collaris,* pertaining to the neck; *decollo, -atus,* behead: collar, collet, accolade, decollation, decolleté, *Scarabaeus laticollis* (a beetle), *Eurymetopon convexicolle* (a beetle), *Naja nigricollis* (a cobra).

Gr. *deire; dere,* f. neck, throat: arthrodire, pleurodiran, *Oncideres cingulatus* (hickory twig-girdler), *Dirofilaria immitis* (a nematode), *Derichthys serpentinus* (long-necked eel), *Ectatoderus nigriventris* (a phasmid), *Chlamydera maculata* (spotted bowerbird), *Collyris lissodera* (a beetle), *Dolichoderus bispinosus* (an ant), *Exochoderes aurantiacus* (a grasshopper), *Brachyderes quercus* (a beetle).

Gr. *inion,* n. back of the head, nape of the neck: inion.

L. *isthmus* (Gr. *isthmos*), m. neck, narrow passage, neck of land; *isthmion,* n. dim.; *isthmios,* of a neck or isthmus: isthmus, paristhmitis, *Hydrochoerus isthmius* (a capybara).

ML. *nucha,* f. < Ar. *nukha,* nape of the neck: nuchal, nuchalgia, *Poecilogale albinucha* (a weasel), *Onthophagus nuchicornis* (a beetle), *Lophophanes rufonuchalis* (black tit).

Gr. *trachelos,* m. neck; *trachelion,* n. dim.; *tracheliaios,* of the neck; *trachelodes,* necklike: trachelitis, tracheloplasty, *Trachelomonas hispida* (a flagellate), *Conotrachelus nenuphar* (plum-curculio), *Helianthus trachelifolius* (a sunflower), *Campanula trachelium* (nettle-leaved campanula).

See: **throat, passage**

necklace

Gr. *deiropede,* f. collar, necklace; *deraion,* n. necklace, collar: *Diropeda clelia* (a snake), *Deraiophorus elegans* (a mite).

Gr. *hormos,* m. chain, necklace; *hormothion; hormidion,* n.; *hormiskos,* m. dim.; *hormathos,* m. chain of things, string, necklace; *kathormion,* n. necklace: hormion, hormogonium, *Hormocrinus anglicus* (a Silurian crinoid), *Hormidium flaccidum* (an alga), *Ormosia[Hormosia] monosperma* (necklace-tree), *Spatalistis hormota* (a moth), *Clastersporium hormiscioides* (a fungus), *Cathormiocerus vestitus* (a beetle), *Cathormiocerinus globulus* (a beetle), *Carex hormathodes* (a sedge), *Hormodendrum solani* (a fungus).

Gr. *kathema, -tos,* n. necklace, collar:

Gr. *mannos (manos)* m. necklace, collar; *mannakion,* n. dim.: *Mannophorus laetus* (a beetle).

L. *monile, -is,* n. necklace, string of beads: moniliform, monilethrix, *Monilia cinerea* (a fungus), *Monilothrips kempi* (a thysanopterid), *Armillifer moniliformis* (a tongue-worm), *Isoachlya monilifera* (a fungus), *Formicarius moniliger* (an ant-thrush), *Garrulax moniligera* (a laughing-thrush).

Gr. *periauchenion,* n.; *hypoderis; perideris, -idos,* f. necklace: *Stichopus (Perideris) chloronotus* (an echinoid).

L. *torques,* twisted neck-chain, necklace, collar; *torquatus,* adorned with a necklace or collar; see **collar**

See: **collar, chain**

necopinus, L. unexpected; see **lot**

necro- < Gr. *nekros,* a dead body, corpse; *nekrikos,* of the dead; see **death**

nectar, L. (Gr. *nektar*), drink of the gods; *nektareos,* sweet as nectar; see **drink**

necto- < Gr. *nektes,* swimmer; *nektos,* swimming, see *necho* under **swim**; < *necto, nexus,* bind, see **bind**

nectria, NL. a genus of fungi; see **fungus**

nectrido- < Gr. *nectris, -idos,* female swimmer; see *necho* under **swim**

necydalus, L. (Gr. *nekydalos*), larva of the silkworm; see **young**

necyo- < Gr. *nekys, -yos,* corpse; see **body**

nedymos, Gr. delightful; see **agreeable**

nedys, Gr. belly, paunch, stomach, womb; see **belly**

need < AS. *nead;* see **poor, empty, must**

needle < AS. *naedl.*

L. *acus,* f. needle, pin, bodkin; *acicula; acula,* f.; *aculeus; aculeolus,* m. dim.; *aculeatus,* prickly, sharp-pointed; *acicularis,* like a needle; Sp. *aguja,* f. needle: aculeate, aciculate, acicular, acutenaculum, aglet, aiguille, eglantine *(Rosa eglanteria),* Ruscus aculeatus (butcher's-broom), Nodosaria acicula (a foraminifer), Textularia aciculata (a foraminifer), Tylosurus acus (a needle-fish), Apeltes quadracus (a stickleback), Agulla bagnalli (a snakefly).

Gr. *akestra,* f. darning-needle; *akestria,* f. seamstress: *akesterion,* n. tailor's shop: Acestra aequorea (a fish), Acestrocephalus anomalus (a fish), Acestridium discus (a fish).

Gr. *akis, -idos,* point, barb, pointed instrument; see **point**

Sp. *alfiler,* m. pin: alfileria *(Erodium cicutarium).*

Gr. *belone,* f. needle, arrowhead, dart; gar; *belonis, -idos,* f.; *belonion,* n. dim.: belonite, Belone pacifica (a gar), Belonidium gracile (a gastropod), Beloniscus simalurus (an opilionid), Belonisculus jacobroni (an opilionid), Beloniscellus floresianus (an opilionid), Belinurus[Belonurus] reginae (a eurypterid).

Gr. *chelotion,* n. netting-needle:

Gr. *enete,* f. pin, brooch:

Gr. *epetrion,* n. needle: Epetrium griguonense (a Tertiary gastropod).

Gr. *kenteterion,* n. awl, needle:

Gr. *kobele,* f. needle; *kobeline,* f. needle-woman: Cobelura prolixa (a beetle).

Gr. *perone,* f. pin, anything pointed for piercing, small bone of arm or leg (radius or fibula), buckle, brooch, clasp; *peronion,* n. dim.: Peronospora parasitica (a downy-mildew), Beloperone guttata (shrimp-plant).

Gr. *rhaphis, -idos,* f. needle; *rhaphidotheke,* f. needle-box: Rhaphis pauciflorus (a grass), Rhaphidograptus tornquisti (a graptolite), Raphidia[Rhaphidia] oblita (a neuropterid), Rhaphiolepis umbellata (Japanese hawthorn), Rhaphithamnus venustus (a verbenacead), Tylosurus rhaphidoma (a needle-fish).

Gr. *stigeus,* tattooer, needle used in tattooing; see *stigma* under **mark**

See: **awl, point, sew, bind, bore, thorn**

nefandus; nefarius; nefastus, L. heinous, abominable, criminal, unlawful; see **bad**

nefrens, L. toothless, that cannot bite; see **tooth**

negativus, L. denying; *negator; negatrix,* denier, < *nego, -atus,* say no; see **no**

neglect < L. *neglego, -lectus,* pay no attention, disregard, slight, overlook, forget; *negligens, -entis,* heedless, careless, indifferent: neglection, negligent, Epicaerus neglectus (a beetle), Sturnella neglecta (western meadowlark), Aesculus neglecta (a horsechestnut), Ribes neglectum (a currant).

Gr. *aboulos,* unconsidered, ill-advised, without plan:

Gr. *akedes; apokedes,* careless, heedless, negligent: Acedes lapella (a butterfly).

Gr. *ameles,* careless, heedless, negligent; *ameletes,* m. a neglecter; *ameletos,* unworthy of care; *atemeles,* neglected, neglectful, careless: Ameletus lineatus (a mayfly), Atemeles strumosus (a beetle), Atemelia torquatella (a moth), Ameles nana (dwarf mantis).

Gr. *aneleges,* unconcerned, reckless:

Gr. *anepibletos,* heedless, inattentive:

Gr. *aprosektos,* careless, heedless:

Gr. *atheristos,* unheeded:

L. *desum,* be absent, fail, neglect:

Gr. *epimethes,* afterthinking, careless; see *promethes* under **think**

L. *ignoro, -atus,* pay no attention to, disregard: ignore.

L. *immemor, -is,* unmindful, negligent, forgetful:

L. *improvidens, -entis,* not foreseeing or anticipating; see *improvidus* under **lot**

L. *imprudens, -entis,* not foreseeing or anticipating: imprudent, imprudence.

L. *incautus,* heedless, improvident: incautious.

L. *incogitatus,* thoughtless, inconsiderate, unstudied:

L. *inconsideratus,* thoughtless, heedless: inconsiderate.
L. *inconspectus,* indiscreet, imprudent: *Ceutorhynchus inconspectus* (a beetle).
L. *incultus,* uncultivated, neglected: *Buprestis inculta* (a beetle).
L. *incuria,* f. carelessness, neglect; *incuriosus,* careless, indifferent:
L. *incustoditus,* unguarded, heedless, imprudent:
L. *indeliberatus,* not reflected on, unconsidered:
L. *indifferens, -entis,* neutral, uninterested: indifferent, indifference.
Gr. *methemon, -os,* careless, remiss; *methemosyne,* f. carelessness:
Gr. *oligoria,* f. a slighting, negligence; *oligoros,* neglectful:
L. *omissus,* negligent, remiss: omit, omission, *Polygonum omissum* (a knotweed).
L. *perfunctorius,* superficial, careless, negligent: perfunctory.
L. *remissus,* slack, negligent, loose: remiss.
See: **forget, abandon, free**

negotiator, L. agent, trader; see *negotior* under **trade**

negretos, Gr. unwaking, in deep sleep; see **sleep**

negundo < a Malay word for a chastetree; see **vervain**

neigh < AS. *hnaegan,* sound off like a horse.
Tommy: "The horse doth neigh."
Scotty: "Does the cock neigh too?"
Gr. *chremetisma, -tos,* n.; *chremetismos,* m. neigh; *chremetistikos,* fond of neigh-
ing, < *chremetizo,* neigh, whinny:
L. *hinnio, -itus,* neigh, whinny; *hinnitus, -us,* m. a neigh; *hinnibilis,* neighing;
hinnibundus, constantly neighing: hinnible.
Gr. *mimichmos,* m. neigh:
L. *onco, -atus,* bray; Gr. *onkema, -tos,* n. a braying; *onketes,* m. brayer, ass;
onketikos, given to braying:
L. *rudo, -itus,* bray; *rudor, -is,* m. a braying:

neighbor < AS. *neahgebur.*
L. *accola,* c. neighbor, < *accolo, -cultus,* dwell by or near: accolent, *Gongrocnemis
accola* (a katydid).
Gr. *agyiates,* m.; *agyiatis, -idos,* f. neighbor:
Gr. *amphiktyon, -os,* m. neighbor: amphictyonic.
Gr. *anchitermon, -os,* near the borders, neighboring; *anchouros,* neighboring:
Anchura carinata (a gastropod).
L. *circumsocius,* neighborly:
L. *confinis,* neighboring, adjoining; see *affinis* under **kin**
L. *contiguus,* lying near, neighboring; see **near**
Gr. *etes,* m. neighbor:
Gr. *geiton, -os,* m.; *geitaina,* f. neighbor; *geitonia,* f. neighborhood: *Geitonichthys
ornatus* (a Triassic fish), *Potamogeton crispus* (riverweed), *Aponogeton dis-
tachyus* (Cape lattice-leaf), *Halogeton glomeratus* (a saltwort).
Gr. *homoros,* having the same borders, neighboring; m. neighbor:
Gr. *kometes,* m.; *kometis, -idos,* f. neighbor: *Cometes pretiosus* (a beetle).
Gr. *pelates (pelastes),* m. neighbor, approacher, hireling; *pelatikos,* of neighbors:
Pelates sexlineatus (a fish), *Pelastoneurus emasculatus* (a fly), *Dasypelates
fasciatus* (a beetle).
Gr. *prosoikos,* dwelling near, neighboring:
L. *rivalis,* user of the same brook, neighbor, competitor; see *rivus* under **stream**
L. *vicinus,* near, neighboring; see *vicus* under **town**
See: **near, companion, border, friend**

neleo- < Gr. *neles,* pitiless, ruthless; see *aneleos* under **merciless**

nelipos, Gr. barefooted; see **bare**

nelites, Gr. guiltless, harmless; see **innocent**

nelumbo, NL. a genus of the waterlily family; see **waterlily**

nemato- < Gr. *nema, -tos,* thread; see **thread**

nemertes, Gr. infallible, certain; see **sure**

Nemesis, Gr. goddess of retributive justice; see **punish**

nemo- < Gr. *nemo,* graze, dispense, see **eat, give**; < L. *nemo,* no one, nobody,
see *ne* under **not**; < Gr. *nemos,* forest, grove, see **forest**

nemoralis, L. woody, sylvan, < L. *nemus, -oris* (Gr. *nemos*), forest or wood with
pasture for cattle, grove, glade; see **forest**

nenia, L. funeral song, dirge; see **sing**

neo- < Gr. *neos,* new, young, recent; *neoteros,* newer, younger; *neatos; neistos,*
newest, youngest, latest, last; *neotes,* youth, see **new, young**; < *neo,* pile,
heap up, see **heap**; < L. *neo, netus* (Gr. *neo*), spin, see **spin**; < Gr. *neo,*
swim, see **swim**

neon < Gr. *neos,* new; see ELEMENTS under **thing**

neorion, Gr. dockyard; *neoros,* superintendent of a dockyard; see *navis* under **ship**

neossia (neottia), Gr. nest, brood; *neossis, -idos; neossos,* young bird, nestling, chick; see **nest, bird**

nepa, L. scorpion; see **scorpion**

nepenthes, Gr. banishing pain and sorrow, free from sorrow; see **painless, pitcher-plant**

nepeta, L. catnip; see **mint**

nephalios, Gr. drinking no wine, sober; see **govern**

nephelo- < Gr. *nephele; nephos,* cloud; *nephelion,* dim., cloudlike spot, white speck; see **cloud, spot**

nephew < OF. *neveu,* < L. *nepos,* grandson; see *nepos* under **kin**

nephon, -os, Gr. abstemious, sober; see **govern**

nephos, Gr. cloud; *nephion,* dim.; see **cloud**

nephro- < Gr. *nephros,* kidney; see **kidney**

Nephthys, Gr. an Egyptian household goddess; see **house**

nepios, Gr. infant; see **child**

nepos, -otis, L. nephew; see **kin**

nepotalis; nepotinus, L. extravagant, prodigal, profuse; see **abundance**

neptis, L. granddaughter; *nepticula,* dim.; see **kin**

Neptune < L. *Neptunus,* god of the sea; see **sea**

nequitia, L. badness, worthlessness; see **bad**

Nereis, Gr. a sea-nymph, daughter of Nereus; see **sea**

nerita, L. (Gr. *nerites*), a sea-snail; see **mollusk**

nerithmos; neritos, Gr. countless, immense; see *arithmos* under **number**

nerium, L. (Gr. *nerion*), oleander; see **oleander**

nerius, NL. a genus of flies; see **fly**

neros, Gr. flowing, liquid; see *naros* under **flow**

nerteros, Gr. lower; see *nerthe* under **low**

nertos, Gr. a vulture; see **vulture**

nerve < L. *nervus,* m. (Gr. *neuron,* n.), sinew, tendon; *nervalis; nervinus,* of nerves; *nervicus,* nervous; *nervosus,* full of nerves or sinews: nervous, enervate, neuritis, polyneuritis, neuralgia, neuropath, Neuroptera, *Schizoneura paradoxa* (a fossil plant), *Calliandra trinervis* (silkflower), *Mahonia nervosa* (Cascade mahonia). Gr. *ganglion,* n. knot or plexus of nerves: ganglion, ganglionic, paraganglion.

nescius, L. unaware, ignorant; see **dull**

nesos, Gr. island; *nesidion; nesion; nesydrion,* dim.; *nesiotes,* islander; see **island**

-ness, AS. condition, quality, state; see **nature**

nessa (netta), Gr. duck; *nessarion (nettarion); nettion,* dim.; see **duck**

nest < AS. *nest.*
    Gr. *anthrenion,* n. wasp-nest:
    Gr. *gype,* f. vulture-nest; *gyparion,* n. dim.:
    Gr. *kalia,* hut, bird-nest; see **house**
    Gr. *lechos,* bed, bird-nest; see **bed**
    Gr. *myrmedon,* ant-nest; see **ant**
    Gr. *neossia (neottia),* f. nest, brood: neossin, *Neottia nidus-avis* (bird-nest orchid).
    L. *nidus,* m. nest; *nidulus,* m. dim.; *nidamentum,* n. material for a nest; *nidifico; nidulor,* build a nest: nidify, nidification, nidifugous, nidulant, nidamental, eyas, *Nidularia australis* (a bird-nest fungus), *Eriogonum nidularium* (a polygonacead).
    Gr. *sphekia,* f. wasp-nest; *sphekion,* n. dim. cell in a wasp-nest: *Sphecia apiformis* (a butterfly), *Sphecidion* (a fossil sponge).
    Gr. *tenthrenion,* n. subterranean nest of a wasp:
    See: **house, den, hive, bed**

nestico- < Gr. *nestikos,* of spinning; see *neo* under **spin**

nestis, L., Gr. jejunum, see **intestine**; fasting, hungry, see **hunger**

Nestor, Gr. a famous, aged warrior and wise counselor; see **teach**

nestoris, Gr. a kind of cup; see **cup**

net < AS. *net,* an open meshwork.
    Gr. *agrenon,* n. net:
    Gr. *amphiblestron,* n. a casting-net: *Amphiblestrum membranaceum* (a bryozoan).

Gr. *anastomosis,* f. formation of a network: anastomosis.
Gr. *arachnion,* cobweb; see **weave**
Gr. *arkys, -yos,* f. net, toils: *Pteronarcys regalis* (a stonefly).
Gr. *brochos,* noose, snare; *brochotos,* ensnared; in meshes, netted; see **rope**
L. *cassiculus,* m. dim. of *cassis,* m. hunting-net, snare: *Cassiculus coronatus* (a bird).
Gr. *chelinos; cheleutos,* netted, plaited; *cheleusis,* f. a netting:
L. *conopium* (Gr. *konopeion*), n. mosquito-net: canopy.
Gr. *diktyon,* n. net; *diktydion,* n. dim.; *diktyotos,* reticulate: dictyogen, dictyo-stele, *Dictyophora indusiata* (a fungus), *Dictyonema flabelliforme*(a graptolite), *Dictydium cancellatum* (a slime-mold), *Dictymna laeta* (a wasp), *Dictyna sublata* (a spider), *Eriodictyon tomentosum* (a hydrophyllacead), *Ileodictyon cibarium* (a mushroom), *Hydrodictyon reticulatum* (water-net alga), *Mahonia dictyota* (a mahonia), *Nephrospyris paradictyum* (a radiolarian).
Gr. *gangamon,* n. net: *Gangamopteris cyclopteroides* (a fossil fern).
Gr. *hapsis, -idos,* f. mesh, network, loop: *Hapsidolema lichensis* (a beetle).
Gr. *histos,* web, tissue; see **weave**
Gr. *kekryphalos,* m. hairnet, net; *kekryphalion,* n. dim.: *Cecryphalium lamprodiscus* (a radiolarian).
Gr. *linon,* n. flax, anything made of flax, linen, cord, net; *linarion; linidion,* n. dim.; *linoplokos,* net-making; *linoptes,* m. net-watcher; *linotheras,* m. net-user; *linostates,* m. net-layer; *linouchos,* using nets; *lineuo,* take with a net; *epilineutes,* m. one who catches with a net: *Linophryne arborifera* (a fish), *Linopteris obliqua* (a pteridosperm), *Linuche unguiculata* (a jellyfish).
L. *rete, -is,* n.; *retia,* f. net; *reticulum; retiolum,* n. dim.; *reticulatus,* netted, net-like; *irretitus,* caught in a net: retina, reticulate, reticule, *Reticularia lycoperdon* (a myxomycete), *Fenestella retiformis* (a bryozoan), *Esox reticulatus* (pickerel), *Lima retifera* (a fossil pelecypod), *Conus retifer* (a gastropod), *Reticulotermes hesperus* (a termite), *Panaeolus retirugis* (a mushroom), *Cratis retiaria* (a pelecypod), *Cyclodesmus irretitus* (a milleped).
L. *sagena* (Gr. *sagene*), f. fish-net, snare; *saganeutes,* m. one who fishes with a net: scine, sagenate, sagenite, sagenitic, *Sagenopteris elliptica* (a fossil fern), *Sagenocrinus expansus* (a Silurian crinoid).
Gr. *tarsos,* woven mat or grate; see **foot**
L. *verriculum* (*everriculum*), n. dragnet, seine: *Conus verriculum* (a gastropod).
See: **weave, screen, trap, fold, rope**

nethis, Gr. spinster; see *neo* under **spin**

netopon, Gr. oil of bitter almonds; see **fat**

netos, Gr. heaped up, see **heap**; spun, see *neo* under **spin**

netron, Gr. spindle; *netrion,* dim.; see **spindle**

netta, Gr. duck; see *nessa* under **duck**

nettle < AS. *netle.*
L. *acalepha* (Gr. *akalephe*), f. nettle: acalephoid, acaleph, *Acalepha deperdita* (a jellyfish).
Gr. *adike,* f. nettle:
L. *cania,* f. a kind of nettle:
L. *cneorum* (Gr. *kneoron; knestron*), n. a stinging plant, nettle: *Cneorum tricoccum* (spurge-olive), *Cneoridium dumosum* (a rue), *Cneorhinus illibatus* (a beetle), *Daphne cneorum* (rose-daphne).
Gr. *knide,* f. nettle: cnidarian, cnidoblast, cnidocil, cnidophore, *Acnida cannabina* (water-hemp), *Cnidoscolus urens* (drug-treadsoftly), *Eucnide urens* (stingbush), *Cnidocampa flavescens* (a moth), *Limnocnida tanganyikae* (a fresh-water jellyfish), *Cnidium suffruticosum* (an umbellifer).
NL. *laportea,* f. a genus of the nettle family, < Francois Laporte, entomologist: *Laportea gigas* (a nettle-tree).
NL. *pilea,* f. a genus of the nettle family: *Pilea microphylla* (artillery-clearweed).
L. *urtica,* f. nettle; Sp. *ortiga,* f. nettle: urticate, urticaria, ortiga, *Urtica gracilis* (a nettle), *Salvia urticifolia* (a sage).
See: **point, itch, burn, arouse, tease**

neuma, Gr. nod; *neusis,* inclination; see **yes**

neuro- < Gr. *neuron,* nerve, sinew, tendon; see **nerve**

neuster, Gr. swimmer; *neustikos,* able to swim; see *neo* under **swim**

neuter, L. neither the one nor the other, middle, impartial, sexless; see **middle, sterile**

nevus < L. *naevus,* birthmark, mole; see **mark**

new < AS. *niwe:* news, newness, newspaper, Newton, Newcastle, Newfoundland, New York, renewal.
Gr. *akopos,* without weariness, untired, fresh; *akopon,* n. restorative drug: acopon.

Gr. *anaplasis,* f. a remodeling, renewing:
L. *inauditus,* unheard of, new:
L. *instauro; restauro, -atus,* renew, repair, refresh: store, instauration, restore, restaurant.
L. *integro, -atus,* renew, make entire; see **all**
Gr. *kainos,* new, recent; *kainistes,* m. innovator; *kainotes,* f. newness; *kainolektos,* newfangled; *kainoprepes,* looking new, novel; *kainophanes,* appearing new; *kainophilos,* desiring new things; *kainourgos,* producing new things; *kainoschemon, -os,* new or strangely formed: Cenozoic, Paleocene, Eocene, kainite, *Caenophanes incompletus* (a wasp), *Caenopimpla ruficollis* (a wasp), *Caenophthalmus tridentatus* (a crab), *Caenoprymnus spinosus* (a wasp).
L. *mustus,* new, young, fresh: must, musty, mustard.
Gr. *neos,* new, recent; young; *neoteros,* newer, younger; *neatos; neistos,* newest, youngest, last; *neophytos,* newly planted, beginner; *neoterikos,* new, modern; *nearos,* young: neolithic, neophyte, Neocene, neo-Darwinian, Neapolitan, Naples, neologism, neoplasm, *Neosorex aquaticus* (a shrew), *Neatus tenebrioides* (a beetle), *Nearoblatta parvula* (a fossil blattid), *Nearomyia flavovaria* (a fly), *Neatocnemis punctata* (a beetle).
L. *novus,* new, young, fresh, recent; *novellus; noviciolus,* dim.; *novalis,* newly plowed; *novissimus,* newest, last, youngest; *novitas, -atis,* f. newness; *innovo, -atus,* introduce something new in an established setting, alter; *renovo,* renew, refresh, restore: novel, novice, novitiate, innovation, renovate, novelty, Nova Scotia, Renova (Pa.), novocaine, *Vernonia noveboracensis* (ironweed).
L. *nuperus,* new, fresh, recent, late: *Cheirurus nuperus* (a trilobite).
Gr. *potainios,* fresh, new:
Gr. *prosphatos,* new, fresh:
L. *recens, -entis,* new, fresh, young: recent, recency.
L. *recreo, -atus,* remake, restore, renew: recreation, recreate.
L. *redivivus,* living again, renewed: *Lewisia rediviva* (bitterroot).
L. *reficio, -ectus,* renew, restore, refresh: refection, refectory.
L. *refocillo, -atus,* warm into life again, revive:
L. *refoveo, -otus,* warm again, refresh, restore, revive:
L. *repentinus,* sudden, hasty, unexpected, new; see **lot**
L. *repetitus,* anew, again: repetition, repeat.
See: **young, begin, change, now, return**

newt < AS. *efete,* salamander: crested newt (*Triturus cristatus*), red eft (*Triturus viridescens*).
Gr. *kordylos,* m. a kind of newt: *Heterocordylus malinus* (a bug), *Cordylus giganteus* (a salamander).
Gr. *salamandra,* f. a lizardlike animal, newt: salamander, *Salamandra maculosa* (a salamander). Derive: gerrymander.

nex, *necis,* L. violent death; *neco, -atus,* kill, slay; see **death, kill**

nexilis, L. tied together; *nexus; nexuosus,* tied, interlaced, coiled, complicated; see *necto* under **bind**

nexis, Gr. a swimming; see *nektos* under **swim**

nice < OF. *nice,* foolish, < L. *nescius,* ignorant; see **beauty, agreeable, good**

nick < uncertain origin; see **cut**

nickel < G. *nickel,* demon; see **spirit,** ELEMENTS under **thing**

nico- < Gr. *nike,* victory, see **victory;** < *neikos,* quarrel, wrangle, dispute, see **fight**

nicotine < NL. *nicotiana,* a genus of the nightshade family that includes tobacco; see **nightshade**

nicto, *-atus,* L. wink; *nictus,* a winking; *nictitans,* winking, see **wink;** < *nico,* beckon, see **call**

nidor, L. vapor, steam, smell from cooking; see **cloud**

nidus, L. nest; *nidulus,* dim.; *nidifico, -atus,* build a nest; see **nest**

niger, *-gra, -grum,* L. black, dark, dusky; *nigellus,* dim.; see **black**

night < AS. *niht:* nightly, nightmare, nightingale, nightshade, benighted, fortnight.
Gr. *achlyo,* grow dark; see **black**
Gr. *amolgos,* m. dead of night:
L. *concubium,* n. that part of night covered by the first sleep:
L. *conticinium,* n. first part of the night, evening:
Gr. *deile,* f. afternoon, evening; *deielos,* of the evening: *Deilemera evergista* (a moth), *Deilephila lineata* (a moth), *Dilobates platycephalus* (an amphibian), *Chordeiles acutipennis* (a nighthawk), *Dielocerus ellisi* (a wasp).
Gr. *dnophos* (*gnophos*), m. darkness, gloom; *dnopheros,* dark, gloomy, murky: *Dnopherula callosa* (a grasshopper), *Dnopheropsis scotaea* (a moth), *Gnophothrips megaceps* (a thysanopterid).

Gr. *dorpestos,* m. suppertime, evening:

Gr. *euphrone,* f. the kindly time, night:

L. *hesperus* (Gr. *hesperos*), m. the evening star, evening, west; *hesperis; hesperius,* of the evening, western: Hesperian, Hesperides, hesperidium, hesperidin, *Hesperornis regalis* (a Cretaceous bird), *Hesperocallis undulata* (desert-lily), *Hesperis fragrans* (a rocket), *Hesperethusa crenulata* (a rutacead), *Hesperonoë complanata* (an annelid).

L. *nox, noctis,* f. night; *nocturnalis; nocturnus,* of the night; *nocticola,* c. nightlover; *noctiluca,* f. that shines by night, the moon, lantern: nocturnal, nocturne, noctambulist, equinox, equinoctial, *Noctiluca scintillans* (a protozoan), *Echidna nocturna* (an eel), *Cestrum nocturnum* (night-cestrum), *Pleurotus noctilucens* (a mushroom), *Nyctalus noctula* (noctule), *Smilax bona-nox* (a greenbrier) *Nicotiana noctiflora* (a tobacco), *Zenilla nox* (a fly), *Pyrophorus noctilucus* (a firefly).

Gr. *nyx, nyktos,* f. night; *nychios; nykterinos; nykteros,* at night, of night; *nyktelios; nyktios,* nightly; *akronychos,* at nightfall; *ennychios,* in the night, at night; *mesonyktion,* n. midnight; *pannychios(pannychos),* lasting all night: nyctanthous, nyctophobia, acronical, Nyctaginaceae, *Nyctotherus africanus* (an infusorian), *Nycteroleter ineptus* (a fossil reptile), *Nyctanthes arbortristis* (hursinghar), *Nycticorax hoactli* (a night-heron), *Nyctibius jamaicensis* (a nightjar), *Acronycta americana* (dagger-moth), *Nyctea nyctea* (snowy owl), *Calonyction aculeatum* (moonflower), *Gennaeus nycthemerus* (silver pheasant), *Nycticircus trimaculatus* (a bird), *Meganyctiphanes norvegica* (a shrimp).

Gr. *opsia,* f. latter part of the day, evening:

Gr. *orphne,* the darkness of night, night; see *orphnos* under **black**

L. *tenebrae,* shades or darkness of night, gloom, obscurity; see *tenebrosus* under **black**

L. *vesper, -i,* m. evening, west; *vesperalis; vespertinus; vesperus,* of the evening, western: vesper, vespers, vespertine, vesperian, *Lychnis vespertina* (evening-lychnis), *Speciospongia vespera* (a sponge).

Gr. *zophos,* m. darkness, dusk, gloom; *zopheros; zophodes,* dusky, gloomy: *Zophocrinus globosus* (a Silurian crinoid), *Zopheromantis trimaculata* (a mantid).

See: **black, shade, west**

nightingale < AS. *nihtegale,* night-singer.

Gr. *aedon, -os,* f. nightingale: *Aedon meridionalis* (a bird).

Gr. *boutalis,* f. nightingale: *Boutalis cerealella* (a bird).

Gr. *Daulias,* f. a name for Philomela, nightingale: *Daulias hafzi* (a bird).

L. *luscinia,* f. nightingale; *lusciniola,* f. dim.: *Luscinia megarhyncha* (English nightingale), *Acrocephalus luscinia* (Guam reed-warbler).

Gr. *Philomela,* f. daughter of Pandion, changed into a nightingale: *Philomela major* (a bird).

nightmare < AS. *nihtmara;* see **dream, spirit**

nightshade < AS. *nihtscada.*

L. *altercum,* n. henbane:

L. *apollinaria; apollinaris,* f. a kind of nightshade, henbane:

NL. *atropa,* f. a genus of the nightshade family: *Atropa belladonna* (belladonna).

L. *auginus,* f. henbane:

L. *baccina,* f. henbane:

NL. *capsicum,* n. a genus of the nightshade family: *Capsicum frutescens* (red pepper, chili).

L. *cestrum* (Gr. *kestron,* a plant), n. a genus of the nightshade family: *Cestrum laurifolium* (a cestrum).

L. *cucubalus,* f. probably a kind of nightshade: *Cucubalus baccifer* (a pink).

NL. *datura,* f. a genus of the nightshade family, < Hind. *dhatura: Datura quercifolia* (oak-leaved datura).

L. *dercea,* f. a kind of nightshade:

NL. *duboisia,* f. a genus of the nightshade family, < F. N. Dubois, French botanist: *Duboisia hopwoodi* (pituri).

L. *hyoscyamus* (Gr. *hyoskyamos*), m. henbane: *Hyoscyamus niger* (henbane).

L. *lycium* (Gr. *lykion*), n. a thorny plant of Lycia; a genus of the nightshade family: *Lycium pallidum* (tomatilla), *Bumelia lycioides* (buckthorn-bumelia).

NL. *lycopersicum* (Gr. *lykopersion,* an Egyptian plant), n. tomato: *Lycopersicum esculentum* (tomato).

Gr. *mandragoras,* m. mandrake: *Mandragora officinarum* (mandrake).

NL. *metel* (*methel*), < Ar. *mathil,* the fruit of a datura: metel, *Datura metel* (Hindu metel), *Datura meteloides* (sacred datura).

Gr. *morion,* n. a kind of mandrake:

NL. *nicotiana,* f. a genus of the nightshade family, < Jean Nicot, introducer of tobacco to France in 1560: nicotine, isonicotinic acid hydrazide, *Nicotiana tomentosa* (giant tobacco).

Sp. *patata,* f. < Ab.Am. *batata,* sweet potato: potato (*Solanum tuberosum*), *Ipomoea batatas* (sweet potato), *Sclerotium bataticolum* (a fungus).

NL. *petunia,* f. a genus of the nightshade family, < Ab.Am. *petum: Petunia violacea* (violet petunia).

NL. *physalis* (Gr. *physallis, -idos,* a plant with bladderlike fruits), f. a genus of the nightshade family: *Physalis pruinosa* (hairy groundcherry).

NL. *scopolia,* f. a genus of the nightshade family, < J. A. Scopoli, Italian botanist: *Scopolia carniolica* (scopolia), scopolamine.

L. *solanum,* n. nightshade: *Solanum umbelliferum* (a nightshade), *Ardisia solanacea* (an ardisia), *Fusarium solani* (a fungus).

NL. *stramonium,* n. a plant of the nightshade family: *Datura stramonium* (jimsonweed).

Gr. *strychnos,* nightshade; now a genus of the logania family; see **logania**

NL. *tabacum,* n. < Ab.Am. *tabaco,* tobacco: tobacco (*Nicotiana tabacum*), *Fistularia tabacaria* (tobacco-pipe fish), *Peronospora tabacina* (downy mildew of tobacco), *Telephanus tabaciphilus* (a beetle), *Thrips tabaci* (onion-thrips).

Ab.Am. *tomatl,* tomato: tomato (*Lycopersicum esculentum*), *Macrosporium tomato* (a fungus).

niglaros, Gr. whistle; see **whistle**

nihil, L. nothing; see **not**

Nile < L. *Nilus* (Gr. *Neilos*), a river in northern Africa; *Niloticus,* of the Nile; see **stream**

nimble < uncertain origin; see **swift**

nimbus, L. rain-cloud; see **cloud**

nimius, L. too much, excessive, superfluous; *nimietas,* superfluity; see **abundance**

nimma, Gr. water for washing; see *nipto* under **wash**

nine < AS. *nigon.*

L. *dodrans,* nine-twelfths or three-fourths of anything:

Gr. *ennea,* nine; *enneas, -ados,* f. the number nine; *enneadikos,* of nine; *enatos,* ninth: ennead, enneatic, enneasyllabic, *Enneadesmus forficula* (a beetle), *Oxalis enneaphylla* (a wood-sorrel).

L. *novem,* nine; *novenarius,* consisting of nine; *novenus,* nine; *nonus,* ninth; *nonarius,* of the ninth; *novemdecim (novendecim),* nineteen; *nonaginta,* ninety; *nonagenarius,* of ninety; *nonagesimus,* ninetieth: November, novennial, nonadecane, nonagesimal, nones, noon, nonagenarian, *Dasypus novemcinctus* (an armadillo).

ninguis, L. snow; *ninguidus,* snowy; *ningor,* snowfall; see **ice**

ningulus, L. nobody; see **not**

Nioba, L. (Gr. *Niobe*), daughter of Tantalus, changed by Zeus into stone; see **stone**

nipho- < Gr. *nipha,* snow; *niphas,* snowflake; *niphetos,* snowstorm; see **ice**

nipple < uncertain origin; see **udder, bud, projection**

nipto (nizo), Gr. bathe, wash; *nipter,* basin, laver; *niptron,* water for washing; see **wash, basin**

nisus; nixus, L. labored, striven; see *nitor* under **make**

nit < AS. *hnitu,* egg of a louse; see **egg**

nitedula, L. small, red mouse, dormouse; *nitellinus,* of a dormouse; see **mouse**

nitella, L. brightness, splendor; see *niteo* under **light**

nitibundus, L. straining, pressing forward with effort; see *nitor* under **make**

nitidus, L. shining, neat, elegant; *nitidulus,* dim.; *nitido, -atus,* make bright, polish; see *niteo* under **light**

nitrogen < L. *nitrum* (Gr. *nitron*), n. natron, sodium carbonate; nitrogen compounds: natron, niter, nitrate, nitroglycerin, nitrogenous, nitrocellulose, TNT, nitramine, nitric, nitride, nitrification, nitrile, nitrite, nitrous, nitrobenzene, nitrosochloride, nitrometer, diethylparanitrophenylthiophosphate, *Nitrosomonas europaea* (a bacterium), *Flavobacterium denitrificans* (a bacterium).

Gr. *azo-* < *azoos,* lifeless; pertaining to nitrogen compounds: azote, azotic, azolitmin, azotemia, azoturia, azoxybenzene, diazobenzene, *Azotobacter chroococcus* (a nitrogen-fixing bacterium), hydrazide, hydrazine, isoniazid.

See: ELEMENTS under **thing**

nivalis; nivarius; niveus, L. of snow, snowy, < *nix, nivis,* snow; see **ice**

nivens, L. winking; see *conniveo* under **wink**

no < AS. *na,* < *ne,* not, *a,* ever: not, nill, willy-nilly.

L. *abdico, -atus,* deny, disown, renounce: abdicate, abdication.

L. *abjuro, -atus,* deny on oath: abjure, abjuration.

L. *abnuo, -utus,* deny, refuse, decline, reject; *abnuto, -atus,* deny often, refuse; *abnutivus,* denying; *renuo,* deny:

Gr. *aparnos,* denying utterly:

Gr. *apomosia; exomosia,* denial under oath; see *omnymi* under **swear**

Gr. *aporrhema, -tos,* n.; *aporrhesis.* f. prohibition, renunciation; *aporrhetos,* forbidden:

Gr. *arnesis,* f. denial; *arnesimos; arnetikos,* of denial, negative; *arneomai,* deny, disown:

L. *confuto; refuto, -atus,* disprove, repel, put to silence: confute, refutation, irrefutable.

L. *contradico, -ictus,* reply in the negative: contradict, contradiction.

L. *dedignor, -gnatus,* reject, refuse, scorn:

L. *infiteor, -tiatus,* contradict, deny, disown; *infitialis,* negative; *infitia,* f. denial:

L. *nego, -atus,* say no, deny, refuse, decline; *negativus,* that denies; *negator, -is,* m.; *negatrix,* f. denier; *denego,* say it is not, refuse: negative, negation, negator, abnegation, denial, deny, renegade, renege, undeniable.

L. *nolo,* do not wish: nolition, nolleity, *nolo contendere.*

See: **not, lie, against, forbidden**

noble < L. *nobilis,* well-known, excellent; see **honor**

nocaro- < Gr. *nokar, -os,* sleep, sloth; *nokarodes,* sleepy, slothful; see **sleep**

noceo, -citus, L. hurt, harm; *nocivus; noxius,* harmful, injurious; see **hurt**

nocheles, Gr. moving slowly, sluggish; see **slow**

noctua, L. night-owl; see **owl**

nocturnus, L. of the night; *nocticola,* night-lover; *noctiluca,* night-shiner, moon, lantern; see *nox* under **night**

nod < uncertain origin; see **bend, yes**

nodosaria, NL. a genus of foraminifers; see **protozoan**

nodus, L. knot, swelling; *nodulus,* dim.; *nodosus,* full of knots, knotty; see **knot**

noëo, Gr. perceive, think; *noëma,* thought; *noëtikos,* intelligent; see **think**

nogalon, Gr. sweetmeat, dessert; see **food**

noise < uncertain origin; see **sound**

nola, L. a little bell; see **bell**

nolens, L. unwilling; see **against**

noma, L. (Gr. *nome*), an eating ulcer; see **sore**

nomado- < Gr. *nomas, -ados,* roaming about for pasture, roving; see **wander**

nomen, L. name; *nominalis,* of a name; *nominator,* namer; see **word**

nomo-; nomy- < Gr. *nomos,* usage, law, pasture, place or condition for living; knowledge or science of; see **know, law,** *nemo* under **eat**

non- < L. *noenum,* < *ne,* not, *unus,* one; see *ne-* under **not**

nonnus, L. monk; *nonna,* nun; see **alone**

nonus, L. ninth, < *novem,* nine; see **nine**

noon < AS. *non,* < L. *nona,* the ninth hour; see **day**

noös, Gr. mind, intellect; see *noëma* under **think**

noose < L. *nodus,* knot; see **trap**

nops, -opos, Gr. blind; see *amblyopos* under **blind**

norma, L. rule, pattern, carpenter's square; *normalis,* according to rule; see **law**

norops, Gr. bright, flashing, gleaming; see **light**

north < AS. *north.*

L. *aquilonaris; aquilonius,* north, northern, northerly: *Aquilonaria turneri* (a gastropod), *Coccygomimus aquilonius* (a wasp).

Gr. *arktos,* m. bear, north; *arktikos; arktios; arktous,* northern: arctic, *Arctophila fulva* (a grass), *Arcticoceras ishmae* (an ammonite), *Arctioblepsis rubida* (a moth), *Arctous alpinus* (ptarmiganberry), *Salix arctica* (arctic willow), *Lepus arcticus* (polar hare), *Erigone palaearctica* (a spider).

Gr. *boreas, -ou,* m. north wind, north; *boreios,* northern; L. *borealis; hyperboreus,* northern: boreal, hyperborean, aurora borealis, *Boreodromia bicolor* (a fly), *Boreobdella verrucata* (a leech), *Boriomyia disjuncta* (a neuropterid), *Clintonia borealis* (yellow beadlily), *Sparganium hypoboreum* (northern burreed), *Tanacetum boreale* (arctic tansy).

L. *septentrionalis,* north, northern; *septentrio, -onis,* m. north wind: *Hyla septentrionalis* (a frog).

L. *Thule* (Gr. *Thoule*), f. farthest north: thulium, thulia, thulite, ultima Thule,
Ergetta thula (snowy egret).

nosco, *notus,* L. (Gr. *gignosko*), know; see **know**

nose < AS. *nosu.*
L. *beccus,* m. beak, bill:
Gr. *mykter, -os,* m. nose, nostril; *mykterismos,* m. a turning up the nose, sneering:
mycterism, *Mycteroperca venenosa* (yellow-fin grouper), *Mycterosaurus
longiceps* (a fossil reptile), *Mycteria americana* (wood-ibis, jabiru), *Metopo-
mycter denticulatus* (a fish).
Gr. *myxa,* f. lamp-nozzle:
L. *naris, -is,* f. nostril; *narinosus,* broad-nosed: nares, naricorn, *Nasua narica*
(brown coati).
L. *nasus,* m. nose; *nasicus; nasutus,* large-nosed: nasal, nasality, nasicorn,
nasopharynx, nasturtium, *Nasalis larvatus* (proboscis-monkey), *Nasutitermes
corniger* (a termite), *Nasua rufa* (red coati), *Heterodon nasicus* (western hog-
nosed snake), *Tryxalis nasuta* (a grasshopper), *Ligyrus ruginasus* (a beetle).
AS. *nebb,* beak, nose: neb, nib, nipple.
L. *proboscis, -idis* (Gr. *proboskis, -idos*), f. trunk of an elephant, snout:
Proboscidea, *Proboscidea(Martynia) parviflora* (a devil's-claw), *Cactocrinus
proboscidialis* (a Mississippian crinoid), *Halosauropsis proboscidea* (a fish).
Gr. *rhamphos,* n. curving beak, bill; *rhamphion,* n. dim.; *rhamphodes,* beaklike:
rhamphoid, *Rhamphorhynchus phyllurus* (a pterodactyl), *Rhamphastes toco*
(toucan), *Hemirhamphus unifasciatus* (a halfbeak).
Gr. *rhis, rhinos,* f. nose, snout, beak, bill; *rhinion,* n. dim.; *epirrhinos;
makrorrhinos,* long-nosed; *mikrorrhis, -inos,* small-nosed: rhinitis, rhinion, rhin-
encephalon, rhinolith, rhinology, rhinoplasty, rhinorrhea, rhinoscope, rhinotheca,
Rhinoceros unicornis (rhinoceros), *Rhina nigra* (a beetle), *Rhinichthys atronasus*
(black-nosed dace), *Rhinanthus cristagalli* (cockscomb-rattleweed), *Rhinacan-
thus nasutus* (a fish), *Rhiniopora aviculosa* (a bryozoan), *Rhinobatus spinosus*
(a guitar-fish), *Callorhinus ursinus* (a fur-seal), *Siphonorhis americana* (a goatsucker), *Bothrorrhina ochreata* (a beetle), *Ceratorrhina guttata* (a beetle),
Choerorrhinus squalidus (a beetle), *Nothorrhina muricata* (a beetle), *Rhomborrhina setulosa* (a beetle), *Platyrrhinus marmoratus* (a beetle), *Antirrhinum
majus* (snapdragon), *Heterodon platyrhinus*[now *contortrix*] (hog-nosed snake).
Gr. *rhothon, -os,* m. nose, beak, nostril; *rhothonion,* n. dim.:
Gr. *rhynchos,* n. snout, muzzle; *rhynchion,* n. dim.: ornithorhynchus, rhynchodont,
Rhynchonella aptycha (a fossil brachiopod), *Rhynchium nitidulum* (a wasp),
Rhynchops nigra (skimmer), *Rhynchites betulae* (a weevil), *Amblyrhynchus
cristatum* (a Galapagos lizard), *Leiorhynchus castaneum* (a fossil brachiopod),
Echinorhynchus moniliforme (a nematode), *Coregonus oxyrhynchus* (a white-
fish), *Callorhynchus capense* (a chimaera).
L. *rostrum,* n. bill, beak, snout, muzzle; *rostellum,* n. dim.; *rostratus,* beaked,
curved: rostrum, rostrate, rostral, rostellum, *Aptenodytes longirostris* (a
penguin), *Acarna rostrifera* (a bug), *Rostellaria fusus* (spindle-shell), *Nuculana
rostellata* (a fossil pelecypod), *Rostricellula rostrata* (a fossil brachiopod).
L. *silus,* pugnosed: *Crotaphytus silus* (a leopard-lizard).
L. *simus* (Gr. *simos*), pugnosed; *simulus,* dim.; *anasimos,* with turned-up nose;
resimus, with turned-up nose, snub-nosed; *simo, -atus,* flatten: *Simocephalus
vetulus* (a water-flea), *Simorhynchus pusillum* (an auklet), *Anasimus fugax*
(a crab). *Rhipicephalus simus* (a tick), *Ceratotherium simum* (a rhinoceros).
See: **projection, face, before**

nosos, Gr. disease, sickness; *noseros,* diseased; *nosema,* sickness; see **disease**

nosphidios, Gr. clandestine, stealthy; see **secret**

nosphistes, Gr. embezzler, peculator; *nosphimos,* peculation; see **steal**

nossax, Gr. chick; see **chicken**

noster, *-tra, -trum,* L. our, ours, our own; see **hold, drug**

nostos, Gr. return home; see **return**

not < AS. *nawiht,* < *ne,* not, *wiht,* jot, particle: nought, naught, naughty, nothing,
neither, nor, never, none, nonesuch, whatnot.
Gr. *a-, an-, ar-,* < *an,* not, without, privative, as in *abotanos,* without plants;
anoplos, unarmed; *arrhektos,* unbroken: agnostic, atheist, aphasia, asymmetric,
amethyst, atom, Azoic, amorphous, aphanitic, anarchy, anomalous, anesthetic,
anhydrous, anonymous, anhedral, anisotropic, Anura, arrhythmical, arrhinia,
arrhizal, *Apteryx australis* (kiwi), *Aphyllon uniflorum* (naked broomrape),
Pyrola aphylla (a pyrola), *Atrypa gibbosa* (a fossil brachiopod), *Agnostus
acadicus* (a trilobite), *Achromaticus vesperuginis* (a protozoan).
L. *careo, -itus,* be without, want, be deprived of; see **poor**

L. *de*, from; see **from**

L. *dis*, without, not, as in *discalceatus*, unshod; *discordia*, dissension; *dissocialis*, unfriendly, selfish: discord, disunion, dissatisfaction, discontent, disease.

L. *e, ex*, without; see **out**

L. *expers, -tis*, having no part in, not privy to, free from, without:

AS. *for-*, prefix with privative force: forbearance, forbidden, forfend, forgetful, forgo, forlorn, forsaken, forswear.

L. *i-, il-, im-, in-, ir-*, < *in*, not, without, as in *ignarus*, not knowing; *illimitatus*, unbounded; *immature*, unripe; *infrequens*, not often; *irrasus*, unshaved: ignore, ignoble, ignominy, illicit, illegitimate, illiterate, immaterial, immovable, immodest, immortal, impure, impossible, indefinite, insane, infallible, innocent, inimitable, insensible, insoluble, intolerant, inimical, enemy, irregular, irrational, irresistible, irreverent, irrevocable, *Tiphia inornata* (a wasp).

L. *imparatus*, unprepared, not ready: *Pipunculus imparatus* (a fly).

Gr. *leipo*, abandon, be wanting, be without; see **abandon**

AS. *-less*, < *leas*, free from, without: careless, faultless, harmless, sunless, worthless, headless, helpless, useless, nameless, nevertheless, spotless, smokeless.

Gr. *me*, not; *medamos*, none:

L. *ne* (Gr. *ne-*); *non*, not; *nemo, -inis*, c. no one, nobody: neuter, neutralize, nefarious, necessary, neglect, negotiate, nepenthe, nepenthic, nestitherapy, denial, non-alcoholic, non-Christian, non-excusable, nonsense, nondescript, nonentity, nonconformity, nonage, nonchalant, nonpareil, noncombatant, noncommittal, umpire.

L. *nihil*, nothing; *annihilo, -atus*, bring to nothing, destroy: nihilism, annihilate, hilum, nil.

L. *ningulus*, nobody:

L. *nullus*, not any, none, nobody, no: annul, nullification, nulliparous, nullipore, null and void.

Gr. *ou, ouk, ouch*, no, not; *oudamos*, no one, none, in no wise; *oudenakis*, not once; *oudenizo*, bring to nothing; *oudeterous*, neither; *ouketi*, not now, no longer; *oupos*, not at all; *outis*, no one, nobody: Utopia, *Udamomitra campestris* (a fly).

F. *sans*, without, < L. *absentia*, absence: sans, Sans Souci, sansculotte, Sanka.

L. *sine*, without: sinecure, sine qua non, sine die.

AS. *un*, not, back, off: uncertain, unnatural, unavoidable, unconscious, unexpected, unfaithful, unloose, unhorse, untruth, unconformity, unfold, unyoke, undo.

L. *ve-*, without, as in *vecors*, without reason, silly:

AS. *wan*, deficient, lacking: want, wane, wanton.

See: **poor, cancel, empty, no, bare, from, abandon**

nota, L. mark; *notula*, dim.; *noto, -atus*, mark; *notabilis*, remarkable; see **mark, honor**

notarius, L. stenographer, secretary, clerk; see **write**

notch < uncertain origin.

Gr. *charaktos*, notched, toothed; see *character* under **mark**

Gr. *cheloma, -tos*, n. notch: *Acheloma cumminsi* (a fossil amphibian).

L. *crena*, f. notch, rounded projection; *crenula*, f. dim.; *crenulatus*, minutely crenate: crenate, crenulate, crenelated, cranny(?), *Crenilabrus melops* (cunner), *Calceolaria crenatiflora* (a slipper-flower), *Bryophyllum crenatum* (a houseleek), *Catoryctis tricrena* (a moth), *Rhamnus crenata* (Oriental buckthorn).

L. *emarginatus*, without margin, notched at the apex: emarginate, *Viola emarginata* (a violet).

Gr. *entmema; entome*, incision, notch; see *temno* under **cut**

Gr. *glyphis, -idos*, f. notch or groove on an arrow: *Glyphidoptera polymita* (a moth).

Sp. *mella*, f. notch, gap:

L. *retusus*, blunted, rounded, notched at the apex: retuse, *Crotalaria retusa* (a rattlebox).

See: **cut, dig**

nothing < AS. *na*, not, *thing*; see **not, no, empty, cancel**

notho- < Gr. *nothos*, spurious, bastard; see **lie**

nothro- < Gr. *nothros*, sluggish, torpid, dull; see **slow**

notidanos, Gr. with pointed dorsal fin; see **fin**

notido- < Gr. *notis, -idos*, dampness, moisture, wetness; see **wet**

noto- < Gr. *notos*, back, see **back**; south; *notios*, southern, wet, rainy, see **south**

notorius, L. widely but unfavorably known; see *nosco* under **know**

notula, L. dim. of *nota*, mark; see **mark**

novacula, L. sharp knife, razor; see **knife**

novellus, L. dim. of *novus,* new; see **new**

novem, L. nine; *novenarius,* of nine; *novenus,* nine; see **nine**

noverca, L. stepmother; see **kin**

novus, L. new; *novellus,* dim.; *novicius,* rather new; *novitas,* newness; see **new**

now < AS. *nu,* the present moment.

Gr. *arti,* now, at this time, just, as in *artibaphes,* newly dyed; *artichytos,* just poured; *artidomos,* just built; *artigamos,* just married; *artigenes,* just made; *artiteles,* just finished; *artitokos,* newborn:

Gr. *epautika,* immediately:

L. *instans, -antis,* present, near, standing by, < *insto, -atus,* stand in or by: instant, instantly, instantaneity, instantaneous.

L. *modernus,* of the present: modern.

L. *nunc,* now, at this time: Nunc Dimittis.

Gr. *parakrema,* forthwith, immediate:

L. *presens, -entis,* in sight, at hand, now: present, presence.

Gr. *proka,* forthwith, straightway:

See: **new, life**

nox, noctis, L. night; *nocturnus,* of the night; see **night**

noxius, L. hurtful, harmful, injurious; see *noceo* under **hurt**

nu, Gr. thirteenth letter of the Greek alphabet; see **letter**

nubecula, L. dim. of *nubes,* cloud; see **cloud**

nubilis, L. marriageable; see *nubo* under **marry**

nubilus, L. cloudy, gloomy; *nubilosus,* cloudy; see *nubes* under **cloud**

nucha, ML. < Ar. *nukha,* nape of the neck; see **neck**

nucleus, L. kernel, < *nux, nucis,* nut; *nucella; nucula,* dim.; see **nut**

nudus, L. bare, naked; *nudulus,* dim.; *nuditas,* nakedness; see **bare**

nugalis, L. trifling, frivolous, worthless, < *nugor, -atus,* trifle, be frivolous, talk nonsense; *nugator,* trifler; *nugax, -acis,* jesting, trifling; see *nugor* under **trifle**

nullus, L. nothing, nobody; see **not**

numb < AS. *numen,* taken, deprived of sensation; see **painless**

number < OF. *nombre,* < L. *numerus,* m. a quantity; *numeralis; numerius,* of number; *numerarius; numerator, -is,* m. counter; *numerabilis,* that can be counted; *numerosus,* many: numeral, enumerate, numerical, numerator, innumerable, numerous, numberless, supernumerary.

L. *abacus; abax, -acis* (Gr. *-akos*), m. counting-board, gaming-board divided into square compartments; *abaculus,* m. dim.; *abakiskos,* m.; *abakion,* n. dim. small, colored tile for inlay and mosaic work: abacus, abacist, *Abacocrinus tesselatus* (a Silurian crinoid), *Abaciscus tristis* (a moth).

L. *agmen, -inis,* army on the march, multitude, crowd, train; see **army**

Gr. *anchistinos,* close, crowded, in heaps; see **thick**

Gr. *apeiros,* infinite, boundless, countless; see **infinite**

Gr. *arithmos,* m. number; *arithmetes,* m. calculator; *arithmetikos,* of reckoning; *anarithmos; nerithmos; neritos,* countless: arithmetic, arithmetical, logarithm.

Gr. *artios,* complete, perfect, even (as applied to numbers); see **all**

Gr. *athroos,* assembled in crowds; see **gather**

L. *calculo, -atus,* compute, reckon, < *calx, calcis,* stone, pebble; see **stone**

L. *caterva,* f. crowd, troop, gang:

L. *census,* register, list, assessment; see *censeo* under **judge**

L. *chydaeus* (Gr. *chydaios*), poured out in streams, common, numerous; see *adaios* under **abundance**

OF. *cifre,* < Ar. *sifr,* empty: cipher, zero.

L. *cohors,* pen, company; see **pen**

L. *complures,* several, many; *complusculi,* several:

L. *Euclides* (Gr. *Eukleides*), m. Greek geometer: Euclid, Euclidean.

L. *examen, -inis,* n. swarm, throng, crowd, shoal:

L. *frequens, -entis,* taking place repeatedly; *frequentia,* f. assembly, crowd, throng; *infrequens,* seldom: frequent, frequency, infrequent.

Gr. *hales; halis,* in heaps, crowds, abundance; see **abundance**

Gr. *hesmos,* swarm, flock; *aphesmos,* swarm of bees; see **herd**

Gr. *homilos,* m. crowd, throng: *Homilostola taeniata* (a moth).

Gr. *ile (eile),* f. crowd, band, troop, company; *iladon,* in troops, bodies:

L. *increbro, -atus,* do frequently or repeatedly:

L. *legio, -onis,* f. body of soldiers, large number; *legiuncula,* f. dim.: legion, American Legion, legionary.

L. *mathematicus* (Gr. *mathematikos*), disposed to learn, pertaining to mathematics: mathematical.

L. *multiplex, -icis,* manifold, numerous; *multitudo, -inis,* f. large number: multiplex, multiplication, multitude, multitudinous. See *multus* under **abundance**
Gr. *myrios,* numberless; *myrias, -ados,* f. ten thousand, countless number: myriad, Myriapoda, *Myriophyllum tenellum* (slender water-milfoil).
Gr. *ochlos,* m. crowd, mob, multitude; *ochlikos,* of the mob: ochlocracy.
Gr. *perissos,* beyond the regular number or size, extraordinary, odd (as applied to numbers); see **different**
Gr. *plethos; plethys,* great number, mass, throng, crowd, fullness; see *pleres* under **abundance**
Gr. *pollache,* often, diverse:
Gr. *polys,* many; *pleion (pleon),* more; *pleistos,* most: polychrome, polygon, polygamy, polyglot, polypetalous, polytechnic, Polynesia, polyp, polysyllabic, polyzoan, hoi polloi, pleochroic, pleonasm, Pliocene, pliohippus, Pleistocene, *Polycotyle ornata* (a trematode), *Polyporus schweinitzi* (a fungus), *Gnaphalium polycephalum* (common everlasting), *Epipolycystus asilus* (a wasp), *Plioplarchus sexspinosus* (an Oligocene fish), *Polystichum setiferum* (a fern), *Polynoë squamata* (an annelid), *Pleospora graminea* (a fungus causing barley-stripe).
L. *populosus,* abounding in people, numerous; see **people**
Gr. *posos,* how many, how much; *posotes,* f. quantity: posology.
Gr. *psephistes,* accountant, < *psephos,* pebble; see **stone**
Gr. *Pythagoras,* m. celebrated Greek mathematician and philosopher: Pythagorean, *Pythagoraea categorica* (a moth).
L. *quantus,* how many, how much: quantity, quantitative, quantic, quantum.
L. *quot; quotus,* how many: quote, quotation, misquote, quota, quotient, aliquot.
Gr. *sychnos,* many, great, far, long(time); *episychnos,* often: *Sychnomerus hirticornis* (a beetle), *Paepalanthus sychnophyllus* (a pipewort).
Gr. *thama; thameios; thaminos,* often, crowded: *Thamiocolus viduatus* (a beetle).
L. *turba,* disorderly crowd; see **turn**
See: **abundance, heap, herd, thick, few, gather, mark, infinite, army**

numella, L. fetter, shackle; see **bind**

numen, L. nod, will, sway, majesty, deity, godhead; see **spirit**

numenius, L. (Gr. *noumenios*), a curlew; see **snipe**

numerus, L. number; *numeralis,* of number; *numerosus,* many; see **number**

numida, NL. the genus of birds including the guinea-fowl; see **peacock**

nummus, L. (Gr. *nomisma*), coin, piece of money; *nummulus,* dim.; see **money**

nuncupo, *-atus,* L. call by name, name, appoint; see **call**

nundinor, *-atus,* L. attend market, trade, traffic; see **trade**

nuntio, *-atus,* L. proclaim, call out, announce; *nuntius,* reporter, messenger; see **call**

nuperus, L. new, fresh, recent, late; see **new**

nuphar, L. (Gr. *nouphar*), a kind of waterlily; see **waterlily**

nupta, L. wife; see **spouse**

nuptialis, L. pertaining to marriage; see *nubo* under **marry**

nura, L. daughter-in-law; see **kin**

nurse < L. *nutrix, -icis,* f. nourisher; *nutricula,* f. dim.; *nutricius,* nursing, nourishing: nursery, nurseling, nutritious, *Blastotrochus nutrix* (a coral).
Gr. *altrix, -icis,* f. nourisher, wet-nurse: altricial, altricious.
L. *assa,* f. dry-nurse:
L. *cura,* care, attention; see **guard**
Gr. *maia,* f. good mother, nurse, midwife; *maias, -ados; maieutikos,* of midwifery; *maieutria,* f. midwife: maieutic.
L. *nosocomus* (Gr. *nosokomos*), m. nurse:
L. *obstetrix, -icis,* f. midwife: obstetrics, obstetrician, *Alytes obstetricans* (a toad).
Gr. *thelastria,* nurse; see *thelazo* under **suck**
Gr. *tithene,* f.; *tithenos,* m.; *titthe,* f. nurse, rearer:
Gr. *trophos,* feeder, nurse; see *trepho* under **eat**
See: **guard, eat, suck, mother**

nut < AS. *hnutu.*
Gr. *akrodryon,* n. hard-shelled fruit:
Gr. *karyon,* n. nut, kernel; *karydion,* n.; *karyiskos,* m. dim.; *karya,* f. walnut-tree; *karyarion,* n. dim.; *karyinos,* of nuts: caryopsis, caryopilite, karyokinesis, karyolysis, karyotin. *Carya tomentosa* (mockernut-hickory), *Caryomyia persicoides* (hickory gall-insect), *Caryothraustes poliogaster* (a finch), *Caryocar nuciferum* (a souari-nut), *Caryota urens* (wine-palm), *Hedycarya dentata* (kaiwhiria), *Pterocarya fraxinifolia* (a tree), *Curculio caryae* (pecan-weevil), *Astrocaryum aureum* (a palm), *Nucifraga caryocatactes* (European nutcracker).

L. *nux, nucis,* f. nut; *nucella; nucula,* f. dim.; *nucleus,* m. kernel; *nucleolus,* m. dim.; *nucalis; nuceus,* of a nut; *nucetum,* n. nut orchard; *enucleo, -atus,* remove the kernels, clear from the husk: nucleus, nucleolus, nucellus, newel (nowel), nux-vomica, nucleic, nuclein, nucleotide, desoxyribonuclease, nougat, *Nucifraga columbiana* (western nutcracker), *Nucula proxima* (a pelecypod), *Nuculana bellula* (a pelecypod), *Torreya nucifera* (Japanese torreya), *Cromyocrinus nuciformis* (a fossil crinoid), *Thalassicola nucleata* (a protozoan), *Pinicola enucleator* (pine-grosbeak), *Balaninus nucum* (nut-weevil).

Gr. *pryen, -os,* m. pit or hard seed; *pyrenion,* n. dim.: pyrenoid, pyrenocarp, pyrenematous, pyrenin, *Ilex dipyrena* (Himalayan holly).

See: **seed, fruit, grain**

nuthatch
L. *sitta* (Gr. *sitte*), f. nuthatch: *Sitta caesia* (European nuthatch), *Sittella chrysoptera* (a nuthatch), *Geositta cunicularia* (minera).

nutmeg < OF. *nois,* nut, *muguete* (L. *moschus*), musk: muscade.
L. *macir* (Gr. *maker*), an Indian spice: mace.
F. *muscade,* nutmeg: muscade.
NL. *myristica,* f. nutmeg: *Myristica fragrans* (nutmeg), Myristicaceae, *Myristicivora melanura* (a nutmeg-pigeon), *Carya myristicaeformis* (nutmeg-hickory), *Monodora myristica* (African nutmeg).

nuto, *-atus,* L. nod, droop; *nutabilis; nutabundus,* tottering; see **bend**

nutrio, *-itus,* L. nourish; *nutritorius,* nourishing; *nutrimentum,* food; see **eat**

nutrix, L. nurse; *nutricula,* dim.; see **nurse**

nux, *nucis,* L. nut; see **nut**

nychos, Gr. night; see *nyx* under **night**

nycterido- < Gr. *nykteris, -idos,* bat; see **bat**

nycti-; nycto- < Gr. *nyx, nyktos,* night; *nyktelios,* nightly; *nykteros; nychios,* at night; see **night**

nygma, Gr. puncture, sting; see *nysso* under **bore**

nymph < L. *nympha* (Gr. *nymphe*), bride, young woman, a goddess of the sea, rivers, springs, woods, trees, and mountains; *nymphalis,* nymphlike; see **woman, vulva**

nymphaea, L. (Gr. *nymphaia*), waterlily; see **waterlily**

nyos, Gr. daughter-in-law; see **kin**

nyroca, NL. (Russ. *nyrok,* diver), a genus of ducks; see **duck**

nysium, L. ivy; see **vine**

nyssa, Gr. a turning-post, see **pillar**; a genus of plants, see **dogwood**

nysso (nytto), Gr. prick, spur, pierce, puncture; *nyxis,* a pricking; see **bore**

nystalos, Gr. drowsy, sleepy; *nystagmos,* drowsiness; *nystaktes,* nodder; see **sleep**

nythos, Gr. dumb; see **silent**

nyxis, Gr. a pricking; see *nysso* under **bore**

O

O', Ir. son, < *ua,* descendant; see **son**

o- < L. *ob,* toward; see **to**

oa, Gr. hem, border; see **border**

oak < AS. *ac;* G. *eiche:* oakling, oak-apple, oak-fern, Oakland (Md.), Oakley, Oakham, Ackley, Acton, Acworth, Eichstadt.   Derive: Roanoke, acorn.
L. *aesculus,* f. ancient name of an Italian oak, now applied to the horsechestnut; *aesculeus; aesculinus,* of oak; *aesculetum,* n. an oak forest: *Aesculus carnea* (red horsechestnut), *Ceiba aesculifolia* (a silk-cotton tree), esculin.

Gr. *aria*, f. a kind of oak: *Sorbus aria* (whitebeam).

Gr. *aspris, -idos*, f. a kind of oak:

L. *cerrus*, f. a kind of oak; *cerreus; cerrinus*, of the Turkey oak: *Quercus cerris* (Turkey oak).

Sp. *chaparro*, m. evergreen oak, bush: chaparral.

Gr. *drys*, tree, oak; *phellodrys*, an evergreen oak; see **tree**

Sp. *encina*, f. evergreen or live oak; *encinilla*, f. dim.: encina, Las Encinas, encinillo (*Drypetes ilicifolia*).

Gr. *haliphloios*, m. a kind of oak:

L. *ilex, -icis*, holm-oak; name now applied to holly; see **holly**

Gr. *phegos*, oak, in some dictionaries, but now applied to beech; see **beech**

Gr. *phellos*, cork-oak *(Quercus suber)*; see **bark**

L. *prinus* (Gr. *prinos*), f. a kind of oak: *Quercus prinus* (chestnut-oak), *Quercus prinoides* (chinquapin-oak), *Prinobius scutellarius* (a beetle).

L. *quercus*, f. oak; *querceus; quercinus; querneus; quernus*, of oaks, oaken: quercitrin, quercitol, quercitron, quercitannic, *Quercus alba* (white oak), *Cerococcus quercus* (oak wax-scale), *Rhus quercifolia* (poison-oak), *Bufo quercicus* (oak-toad), *Delphinium quercetorum* (oakwoods-larkspur), *Daedalea quercina* (a fungus), *Agrilus quercicola* (a beetle), *Lecidea quernea* (a lichen).

L. *robur (robor), -is*, n. oak, strength; *robustus*, oaken, strong like oak: robust, roble, roborean, roborant, corroborate, *Quercus robur* (English oak), *Euphorbia robusta* (Rocky Mountain spurge).

Gr. *saronis, -idos*, f. an old hollow oak:

L. *suber*, cork-oak, cork; see **bark**

C. *tann*, oak: tan, tanner, tannin, tannic acid, tannate, tawny, tanbark-oak (*Lithocarpus edulis*).

oar < AS. *ar.*

Gr. *elate*, oar made of pine or fir; see **fir**

Gr. *eretmon*, n. oar; *eretmion*, n. dim.; *eretes*, m. rower; *eretikos*, of rowing, *eresso*, row: *Eretmocaris remipes* (a decapod), *Eretmochelys imbricata* (hawks-bill-turtle), *Xeneretmus leiops* (a fish).

Gr. *kope*, f. oar; *kopion*, n. dim.; *kopelates*, m. rower; *kopelatos*, like an oar; *kopelatikos*, of a rower; *epikopos*, furnished with oars: copepod, *Copecrypta ruficauda* (a fly), *Copiophora brevicornis* (a katydid), *Copelatus posticatus* (a beetle).

L. *palmula*, blade of an oar, oar; see *palma* under **hand**

Gr. *pedos(pedon)*, oar, rudder; *pedalion*, rudder; see **rudder**

Gr. *plate*, f. blade of an oar:

L. *remus*, m. oar; *remulus*, m. dim.; *remex, -igis*, m. oarsman, rower; *remigulus*, m. dim.; *remigo, -atus*, row: remigate, remiges, trireme, *Remigulus tridens* (a copepod), *Eurypterus remipes* (a Silurian arthropod), *Hygrotrechus remigis* (a water-strider), *Eretmocrinus remibrachiatus* (a Mississippian crinoid), *Oestocephalus* [*Oïstocephalus*] *remex* (a fossil amphibian).

L. *tonsa*, f. oar:

Gr. *tropex, -ekos*, m. oar:

See: **rudder, drive**

oaros, Gr. chat, talk, see **speak;** < *oar, -os*, wife, see **spouse**

oasis, Gr. fertile spot in a desert; see **field**

oat; oats < AS. *ate;* see **grain**

oath < AS. *ath*, an affirmation by appeal to deity or revered persons or things; see **swear**

ob-; o-; oc-; of-; og-; op-; os- < L. *ob*, toward, to, upon; see **to**

obba, L. beaker, decanter; see **bottle**

obelos, Gr. spit; *obeliskos*, dim.; see **spear**

Oberon, OHG. king of the fairies; see **govern**

obesus, L. fat; see **fat**

obex, L. bar, bolt, boom; see **bar**

obey < OF. *obeir*, < L. *oboedio*, listen to; see **follow**

obisium, NL. a genus of pseudoscorpions; see **scorpion**

obitus, L. a going down, downfall, destruction, death; see **destroy**

object < L. *objectum*, something thrown before, a concrete entity; see **thing**

oblativus, L. freely given; see *offero* under **give**

oblatus, L. flattened at the poles; see **level**

oblectator, L. charmer, delighter, pleaser; see **joy**

oblido, -isus, L. squeeze together; see **press**

obligatus < L. *obligo,* bind, pledge, engage; see *ligo* under **bind**

oblinitus; oblitus, L. smeared, bedaubed; see *lino* under **spread**

obliquus, L. slanting, sideways, indirect; see **slope**

oblitero, *-atus,* L. blot out, efface, erase; see **cancel**

oblivius, L. forgotten; *oblivialis,* causing forgetfulness; see *obliviscor* under **forget**

oblongus, L. longer than broad; see *longus* under **long**

obnisus < L. *obnitor,* resist, strive, struggle; see *nitor* under **make**

obnixus, L. steadfast, firm, resolute; see **stand**

obolus, L. (Gr. *obolos*), a small Greek coin; see **money**

obrimos, Gr. strong, mighty; see **strong**

obrussa, L. (Gr. *obryzon*), test or assay of gold; see **try**

obrutus, L. buried, hidden; see **cover**

obscenus, L. offensive, indecent, filthy; see **bad**

obscurus, L. dark, indistinct; *obscuritas,* darkness, dimness; see **dim**

obsequialis, L. complying, yielding; see *sequor* under **follow**

observo, *-atus,* L. pay attention to, notice, watch; see **see**

obses, *-idis,* L. hostage, pledge, surety; *obsidatus,* of a hostage; see **promise**

obsessio, L. blockade, siege; see *sedeo* under **sit**

obsidatus, L. condition of a hostage; see *obses* under **promise**

obsidian < L. *Obsidianus,* pertaining to the discoverer of the rock in Ethiopia; see **stone**

obsitus, L. sown, covered, filled; see *sativus* under **sow**

obsoletus, L. worn out, decayed; see **old**

obstacle < L. *obstaculum,* hindrance; see **hinder**

obstetrix, L. midwife; see **nurse**

obstinatus, L. firmly set, steadfast, stubborn; see *sto* under **stand**

obstipus, L. inclined to one side, oblique; see **slope**

obstragulum, L. strap, lace, latchet; see **strap**

obstreperus, L. clamorous, noisy; see *strepo* under **sound**

obstruction < L. *obstructio, -onis,* barrier; see *obstruo* under **hinder, wall, bar**

obtectus, L. covered over, concealed; see *tego* under **cover**

obturatus, L. closed, stopped up; see *obturo* under **close**

obtusus, L. blunt, dull; see **dull**

obtutus, L. look, gaze, stare; see **see**

obuncus, L. bent in, hooked; see *uncus* under **hook**

obvius, L. at hand, lying open, easily seen; see **open**

oc- < L. *ob,* toward; see **to**

occa, L. harrow; *occo, -atus,* harrow; *occamen, -inis,* a harrowing; see **till**

occidentalis, L. of the west, < *occidens; occiduus,* L. setting (of the sun), western, westerly; see **west**

occiput, L. back part of the head; see **head**

occisus, L. lost, ruined, undone; see **destroy**

occlusus, L. closed, shut; see *claudo* under **close**

occo- < Gr. *okkos,* eye; see **eye**

occulco, *-atus,* L. tread, trample; see **step**

occultus, L. hidden, concealed; *occultatio, -onis,* concealment; see **secret**

oce, L. a small bird; see **bird**

ocean < L. *oceanus* (Gr. *okeanos*), the sea; see **sea**

ocellatus, L. having little eyes, marked with spots; see *oculus* under **eye**

ochema, Gr. any kind of carrier, vehicle, horse, ship; see **vehicle**

ochetos, Gr. aqueduct, conduit, pike; see **pipe**

ocheuma, Gr. embryo; *ocheutos,* impregnated; see *ocheuo* under **coitus**

ocheus, Gr. anything for fastening or holding; see *echo* under **hold**

ochlos, Gr. multitude, crowd, mob; see **number**

ochleros, Gr. troublesome; see **trouble**

ochma, Gr. tie, band; see *echo* under **hold**

ochne, Gr. a pear-tree; see **pear**

ochos, Gr. holding, securing, bearing; see *echo* under **hold**

ochotona, NL. a rabbitlike rodent, pika; see **hare**

ochro- < Gr. *ochros,* pale-yellow, sallow; *ochroma,* paleness; *ochra,* earthy oxide of iron; see **yellow**

ochthos, Gr. any elevation, hill, bank, mound; *ochthodes,* hilly; see **mountain**

ochyros, Gr. strong, stout, firm; see **strong**

ocido- < Gr. *okis, -idos,* earring; see **earring**

ocimum, L. (Gr. *okimon*), basil; see **mint**

ocior, L. swifter; *ocissimus,* swiftest; see *okys* under **swift**

-ock, AS. diminutive suffix; see DIMINUTIVES under **little**

ocno- < Gr. *oknos,* hesitation, sluggishness, reluctance; *okneros,* hesitating, shrinking from; see **slow**

ocotea, NL. a genus of the laurel family; see **laurel**

ocrea, L. greave, legging, sheath; see **sheath**

ocrio- < Gr. *okris, -ios,* projection, roughness; see **projection**

octo, L. (Gr. *okto*), eight; *octavus,* eighth; *octonus,* consisting of eight; see **eight**

oculus, L. eye; *ocellus,* dim.; *ocularis,* of the eyes; see **eye**

ocy- < Gr. *okys,* swift, quick, sharp; see **swift**

ocymoro- < Gr. *okymoros,* dying early, short-lived; see **short**

odagmos, Gr. itching, irritation; see **itch**

odax, Gr. biting; see *dakno* under **bite**

odd < ON. *oddi,* point, triangle; see **different, strange**

ode, Gr. song; see *aeido* under **sing**

-ode < Gr. *hodos,* way; see *hodos* under **way**

-odes, Gr. suffix denoting likeness, fullness; see *-oid* under **like, abundance**

odino- < Gr. *odis, -inos,* childbirth pains; see **pain**

odinolytes, Gr. a fish; see **fish**

odium, L. hatred; *odibilis,* hateful; *odiosus,* hateful; see **hate**

odode, Gr. odor, scent; *odmeros,* smelling; see *ozo* under **smell**

odonto-; -odon < Gr. *odous (odon), odontos,* tooth; see **tooth**

odor, L. smell; see **smell**

-ody < Gr. *ode,* song, see **sing**; < *-odes,* likeness, see *-oid* under **form**

odyno- < Gr. *odyne,* pain; *odyneros,* painful; see **pain**

odyrtos, Gr. lamentable; *odyrtes,* complainer, wailer; *odyrtikos,* querulous; see **weep**

odysis, Gr. anger, wrath; *odyssomai,* hate; see **anger**

oeaco- < Gr. *oiax, -akos,* rudder, tiller, helm; see **rudder**

oeceto- < Gr. *oiketes,* dweller, servant; see *oikos* under **house**

oeco-; -oeci < Gr. *oikos,* house, home, dwelling; *oikarion; oikidion; oikion,* dim.; see **house**

oecodomo- < Gr. *oikodomos,* architect, builder; see **carpenter**

oecto- < Gr. *oiktos,* compassion, mercy, pity; *oiktros,* pitiable; see **mercy**

oedema (edema) < Gr. *oidema,* swelling, tumor; *oidos,* swelling, tumor; see **swell**

Oedipus, L. (*Oidipous*), reputed solver of the riddle of the Sphinx; see **ask**

oego- < Gr. *oigo,* open; see **open**

oemo- < Gr. *oimos,* path, road, stripe; see **way**

oemocto- < Gr. *oimoktos,* pitiable; *oimoktikos,* given to wailing; *oimogma; oimoxia,* lamentation, wail; see *oimozo* under **weep**

oenado- < Gr. *oinas, -ados,* a wild pigeon; see **dove**

oenanthe, L. (Gr. *oinanthe*), blossom of the vine; wheatear; see *oine* under **grape**

oeno- < Gr. *oinos,* wine, see **wine**; < *oine,* die, ace, see **play**

oenothera, L. (Gr. *oinotheras*), a genus of the evening-primrose family; see **evening-primrose**

oeo- < Gr. *oios,* alone; see **alone**

oeono- < Gr. *oionos,* omen, large bird; *oionistes,* augur; see **prophecy**

oeso- < Gr. *oisos,* a kind of willow; see **willow**

oestrus, L. (Gr. *oistros*), gadfly, horsefly, frenzy, sexual heat, rut, desire; see **fly**

oesypo- < Gr. *oisypos,* grease from sheep's wool; see **fat**

oeto- < Gr. *oitos,* fate, doom; see **lot**

of < AS. *of,* from, derived, made; see **nature**

of- < L. *ob,* toward; see **to**

off < AS. *of,* from; see AS. *a* under **from**

offa, L. morsel, bit; *offella; offula,* dim.; see **little**

offendo, *-fensus,* L. hit or strike against something, do or suffer damage or injury, stumble, blunder, displease; see **hurt**

offer < L. *offero,* present for acceptance, give; see **give**

officialis, L. pertaining to duty, office, or service, < *officium,* service; see **make**

officinalis, L. pertaining to drugs and medicines, < *officina,* laboratory; see **drug, make**

officio, *-fectus,* L. hinder, obstruct, oppose; see **against**

often < AS. *oft;* see **number**

og- < L. *ob,* toward; see **to**

ogdoos, Gr. eighth; see *octo* under **eight**

ogmos, Gr. furrow, swath, row, path, orbit; see **furrow**

ogygius, L. (Gr. *ogygios*), pertaining to Ogyges, mythical king of Thebes, and suggesting primeval or ancient; see **old**

-oid; -oides, like, resembling, having the form of, < Gr. *-eides,* like < *eidos,* form, shape, likeness; see **like, form**

-oideae, L. suffix denoting a tribe of plants; see **class**

oil < L. *oleum;* see **fat**

ointment < OF. *oignement,* < L. *unguentum,* perfume, ointment; see **fat, heal, smell, drug**

oïs, Gr. sheep; see **sheep**

oïstos, Gr. arrow; see **arrow**

oïzyo- < Gr. *oïzys, -yos,* distress, suffering, misery, woe; *oïzyros,* distressing; see **trouble**

-ol, suffix signifying an alcohol, an oil, or a benzene derivative; see **alcohol,** *oleum* under **fat, benzene**

olax, *-acis,* L. odorous; see *oleo* under **smell**

olbio- < Gr. *olbios,* blessed, happy, < *olbos,* happiness; see **joy**

old < AS. *ald (eald):* older, oldest, old-fashioned, eld, elder, Oldham, world, alderman.   How old was Methuselah?

L. *annosus,* full of years, long-lived, old, aged: *Fomes annosus* (a shelf-fungus).

L. *antiquus,* old, ancient, former, < *ante,* before; *antiquitas, -atis,* f. ancient times; *antiquarius,* pertaining to ancient times and things: antique, antiquity, antiquary, antiquated, *Notolophus antiqua* (a tussock-moth), *Chrysodomus antiquus* (red whelk), *Colocasia antiquorum* (an elephant-ear).

Gr. *aphebos; exebos,* past one's youth:

Gr. *aphelix, -ikos,* past youth, elderly:

Gr. *archaios,* ancient, old; see **begin**

L. *avitus,* grandfatherly, ancestral, old, ancient; see *avus* under **kin**

L. *cascus,* old: *Anthrax casca* (a fly).

L. *decrepitus,* worn down and feeble with the infirmities of old age; see **weak**

Gr. *denaios,* long-lived, old:

Gr. *ekkairos,* out of date, antiquated:

L. *emeritus,* honorably discharged or retired, veteran, worn out; see **honor**

Gr. *geras, -aos; geratos; geros,* n. old age; *geraios,* old, aged; *geron, -tos,* m. old man; *gerontion,* n. dim.; *gerontikos,* like an old man, senile; *erigeron, -tos,* early old: geratic, geratology, gerontology, geriatrics, gerontal, gerontic, gerocomia, *Ageratum houstonianum* (Mexican ageratum), *Gerontochoerus scotti* (a fossil mammal), *Amictogeron fuscipes* (a fly), *Erigeron canus* (hoary fleabane), *Thalassogeron chlororhynchus* (yellow-nosed albatross).

L. *grandaevus; longaevus,* old, ancient: longevity.

Gr. *henos,* old, former:

Gr. *Kronos,* m. Cronos, a nickname for old dotards, old fools: *Kronosaurus queenslandicus* (a fossil reptile).

L. *longivivax, -acis,* long-lived:

Gr. *messogenes,* middle-aged:

L. *obsoletus,* worn out, decayed: obsolete, *Heliothis obsoleta* (corn ear-worm), *Rallus obsoletus* (California clapper-rail).

L. *ogygius* (Gr. *ogygios*), pertaining to Ogyges, mythical king of Thebes, and suggesting primeval or ancient: ogygian.

Gr. *palaios*, ancient, old: paleolithic, paleontology, paleology, Paleozoic, Paleocene, *Palaeotherium magnum* (a fossil mammal), *Palaeaster eucharis* (a fossil starfish).

Gr. *presbys*, m. old man, elder, old; *presbyteros*, older; *presbytatos; presbistos*, oldest; *presbytikos*, elderly; *presbyterion*, n. an assembly or council of elders: presbyter, presbytery, Presbyterian, presbyopia, presbyophrenia, priest.

L. *primaevus*, early, young, original; *primitivus*, early, primeval; see **before**, *primus* under **first**

L. *priscus*, ancient, former, aboriginal: priscan, Priscilla, *Ambloctonus priscus* (a creodont), *Monomerella prisca* (a fossil brachiopod).

L. *pristinus*, early, primitive; see **before**

L. *senex; senicus*, old; *senior*, older; *senex, senis*, m. old man; *seniculus*, m. little old man; *senecio, -onis*, m. old man, groundsel; *senectus*, very old, aged; *senectus, -utis*, f. old age, senility; *senilis*, of old people, aged: senescent, senate, seniority, senile, Seneca, sir, sire, senectuous, senectitude, senicide, senior, surly, monsignor, *Senecio vulgaris* (groundsel). *Onthophagus seniculus* (a dung-beetle), *Cephalocereus senilis* (a cactus), *Mormoops senicula* (a bat), *Caunaca seniella* (a moth).

Gr. *sorelle*, f. nickname for an old man with one foot in the grave: *Sorella eminibey* (a weaver-finch).

L. *Tithonus* (Gr. *Tithonos*), m. mythical consort of Aurora, symbolic of decrepit old age, changed into a cicada: tithonic, *Tithonoceras zittlei* (an ammonite).

L. *tritus*, well-worn, commonplace; see *tero* under **rub**

L. *vetus, -eris*, old; *vetulus*, dim. somewhat old; m. little old man; *vetula*, f. little old woman; *vetusculus*, dim. oldish; *veteranus; veterarius; veternus; vetustus*, old, experienced, aged: *inveteratus*, chronic: veteran, inveterate, *Veterna decorata* (a bug), *Campeloma vetulum* (a fossil gastropod), *Balistes vetula* (oldwife, triggerfish).

L. *vietus*, shriveled, shrunken, withered; see *vieo* under **weave**

See: **begin, before, first, time, age, pappus**

olea, L. olive; *oleaceus*, of olive, oily; *olivarius*, of olive; *olivetum*, olive-grove; see **olive**

oleander < uncertain origin: oleandrin, *Oleandra neriiformis* (a fern).

L. *nerium* (Gr. *nerion*), n. oleander: *Nerium oleander* (oleander), *Acacia neriifolia* (an acacia).

oleaster, *-tri*, L., m. wild olive-tree, < *olea*, olive, *-aster*, wild.

NL. *elaeagnus* (Gr. *elaiagnos*), m. a marsh plant; now a genus of the oleaster family: *Elaeagnus commutatus* (silverberry).

NL. *hippophaë* (Gr. *hippophaës*, n.), f. a spiny plant; now a genus of the oleaster family: *Hippophaë salicifolia* (a sea-buckthorn).

NL. *shepherdia*, f. a genus of the oleaster family, < John Shepherd, English botanist: *Shepherdia argentea* (buffalo-berry).

olene, Gr. elbow, forearm; see **arm**

olens, L. odorous; *olidus*, smelling, odorous; see *oleo* under **smell**

Olenus, L. (Gr. *Olenos*), husband of Lethaea; see **stone**

oleraceus, L. herbaceous, vegetable; see *olus* under **plant**

olesco, L. grow; see **grow**

olethros, Gr. ruin, destruction, death; *olethrios*, destructive, deadly; see **destroy**

oleum, L. oil; *oleosus*, oily; see **fat**

olfactus, < L. *olfacio*, smell; see *oleo* under **smell**

olibanum, ML. frankincense; see **resin**

olibros, Gr. slippery; see *olistheros* under **slip**

olidus, L. emitting an odor; see *oleo* under **smell**

oligo- < Gr. *oligos*, few, scanty; see **few**

oligobios, Gr. short-lived; see **short**

olingos, Gr. a kind of locust; see **grasshopper**

olisbos, Gr. penis; see **penis**

olisthos, Gr. slipperiness; *olisbros; olistheros*, slippery; see **slip**

olitorius, L. pertaining to potherbs, vegetables; see *olus* under **plant**

olive, L. *olea; oliva* (Gr. *elaia*), f. olive-tree; *oleaceus; oleaginus; olearius; olearis; olivarius*, of the olive, of oil, oily; *oleum* (Gr. *elaion*), n. olive oil, any oil; *oleaster, -tri*, m. wild olive-tree; *oleastellus*, m. dim.: Oleaceae, olivaceous, olivine, Oliver, Olivia, elaeolite, elaeoplast, Thymeleaceae, *Olea europaea* (common olive), *Olearia virgata* (a daisy-tree), *Oliva porphyria* (a gastropod), oleaster (*Elaeagnus angustifolius*), *Heterodendron oleaefolium* (a sapindacead), *Elaeis guineensis* (African oil-palm), *Chrysophyllum oliviforme* (satinleaf starapple). *Cercococcyx olivinus* (a cuckoo).

Gr. *agrippos*, m. wild olive:

L. *cercitis, -idis,* f. a kind of olive-tree:

L. *drupa* (Gr. *dryppa*), an overripe, wrinkled olive; see **fruit**

NL. *forsythia,* f. a genus of the olive family, < William Forsyth, English horticulturist: *Forsythia suspensa* (weeping forsythia).

NL. *jasminum,* n. a genus of the olive family, < Ar. *yasmin: Jasminum nudiflorum* (winter jasmine), jessamine, *Solanum jasminoides* (jasmine-nightshade).

Gr. *kotinos,* the wild olive-tree; now a genus of the sumac family; see **sumac**

L. *pausea,* f. a kind of olive:

NL. *phillyrea* (Gr. *philyrea,* a kind of shrub), f. a genus of the olive family: *Phillyrea latifolia* (a phillyrea).

NL. *syringa,* f. a genus of the olive family: *Syringa persica* (Persian lilac).

olla, L. earthen pot, jar; *ollicula; ollula,* dim.; *ollaris,* of a pot; see **pot**

olma, L. another name for the plant *ebulus;* see **elder**

-ology; see *-logy* under **know**

ololygon, Gr. croaking of the frog; see **croak**

olor, L. swan; *olorinus,* of swans; see **goose**

olos, Gr. mud, dirt, sepia; see *tholos* under **earth**

olpe; olpis, Gr. leather flask, vessel; see **bottle**

-olus, L. diminutives; see *-cle* under **little**

Olympus, L. (Gr. *Olympos*), a celebrated mountain in Thessaly, fabled to be the seat of the gods; see **mountain**

olynthos, Gr. an untimely fig; see **fig**

olyra, Gr. a kind of grain; see **grain**

-oma, Gr. denoting a tumor or morbid condition; see **disease**

omasum, L. psalterium or third division of the ruminant stomach; see **stomach**

ombros, Gr. rainstorm; see **drop**

-ome, NL. suffix denoting condition, having the nature of; see **nature**

omega, Gr. last letter of the Greek alphabet; see **letter**

omen, L. sign, augury, portent, prophecy; *ominosus,* portentous; see **prophecy**

omentum, L. caul; see **membrane**

omichma, Gr. urine; see **urine**

omicron, Gr. fifteenth letter of the Greek alphabet; see **letter**

omilla, Gr. a circle for playing a game; see **circle**

omissus, L. negligent, heedless; see **neglect**

omistes, Gr. porter; see **carry**

omma, Gr. eye; *ommation,* dim.; see **eye**

omnis, L. all; see **all**

omo- < Gr. *omos,* raw, rough, unripe; *omotes,* rawness, see **rough**; shoulder, upper arm, see **shoulder**

omorgma, Gr. spot; see **spot**

omphaco- < Gr. *omphax, -akos,* an unripe grape, sour grape; see **grape**

omphalos, Gr. navel; see **navel**

ompheter, Gr. soothsayer; see **prophecy**

ompne, Gr. food; see **food**

on < AS. *on, an,* upon, in, at: abroad, about, afield, aloft, anon, anvil, again, asleep, athwart, away, onlooker, onslaught, onward, thereon, upon, unless.

Gr. *epi,* upon, beside, over, after: epidemic, epiphyte, epididymis, epitaph, epithet, epitome, epicycloid, epidermis, epiglottis, episode, Episcopalian, ephemeral, epoch, ephectic, *Epidendrum odoratissimum* (an orchid), *Epilachna sanscrita* (a beetle), *Ephydra hians* (a brinefly).

L. *im-, in-* < *in,* on: impinge, impose, impress, impugn, incubate, incumbent, insistent.

L. *inevectus,* borne or mounted upon:

See: **over, top**

-on; -oon < L. *-onus,* suffix that augments the force or meaning; see **very**

onager, L. (Gr. *onagros*), wild ass; see **horse**

onagra, Gr. a plant; see **evening-primrose**

oncero- < Gr. *onkeros,* bulky, swollen; see *onkos* under **swell**

onceto- < Gr. *onketes,* brayer, ass; see *onco* under **neigh**

onchno- < Gr. *onchne,* pear; see *ochne* under **pear**

oncidium, NL. a genus of the orchid family; see **orchid**

onco- < Gr. *onkos,* hook, barb, see *uncus* under **hook;** mass, bulk, swelling, see **swell**

ondatra, Ab.Am. muskrat; see **muskrat**

one < AS. *an;* G. *ein:* a, an, once, only, oneness, atone, any, adder, auger, orange, nickname, none, einkorn.
Gr. *alytos,* indissoluble, continuous: *Alytopistis tortricitella* (a moth).
Gr. *arrhaphos,* of one piece, without seam: *Arrhaphipterus olivetorum* (a beetle).
Gr. *arrhektos; arrhox, -ogos,* unbroken:
Gr. *artios,* complete, perfect, even; see **all**
L. *as, assis,* m. unit: ace, decussate.
Gr. *aschides; aschistos,* not split, entire:
Gr. *athraustos,* unbroken, unhurt, sound:
Gr. *atmetos,* uncut, undivided: *Atmetonychus peregrinus* (a beetle).
L. *continuus,* unbroken, uninterrupted, successive: continuity, continuous, discontinuous.
Skt. *eka,* one, next: eka-caesium, cka-manganese.
L. *ens, entis,* being, thing, that which has existence; see *esse* under **life**
Gr. *hapax,* once only, once for all: hapaxanthous.
Gr. *haploos (haplos),* single, simple; *haplosyne; haplotes,* f. singleness, simplicity: haploid, haplodont, haplosis, haploma, aplite(haplite), *Haplococcus reticulatus* (a protozoan), *Haplochromis multicolor* (a fish), *Aplopappus[Haplopappus] squarrosus* (a goldenweed), *Platyxantha(Haplotes) curvicornis* (a beetle), *Haploceras elimatum* (a Jurassic ammonite).
Gr. *heis, henos,* m.; *mia, mias,* f.; *hen, henos,* n. one; *henas, -ados,* f. a unit; *henikos,* single; *henotes,* f. unity: hendiadys, hendecandrous, henogeny, henotic, hendecyl, hyphen, henotheism, *Henicophaps albifrons* (a bird), *Henicops maculata* (a centiped), *Enikocephalus[Henicocephalus] formicina* (a bug), *Enicurus[Henicurus] maculatus* (a bird).
L. *impertilis,* indivisible:
L. *inconsutus,* seamless:
L. *indiscretus,* unseparated, united, not distinguishing, closely connected: indiscrete, indiscreet, indiscretion, *Clavulina indiscreta* (a foraminifer).
L. *insecabilis; indissecabilis,* indivisible:
L. *integer,* whole, complete, unhurt, a whole number; see **all**
L. *item,* also, part of a series: item, itemize.
Gr. *monos,* one, single, alone; *moneres,* single; *monios,* solitary; *monas, -ados,* f. a unit, alone, single: monochrome, monotone, monogram, monolith, monopetalous, monocarpic, monoecious, monad, monarchy, monopoly, monk, monastery, monogamy, monologue, monoplane, monotreme, monoxide, moneran, *Monotropa uniflora* (Indian-pipe), *Monodon monoceros* (narwhal), *Monocephalus fuscipes* (a spider), *Monoxenus spinator* (a beetle), *Monas amyli* (a protozoan), *Monadopsis vampyrelloides* (a protozoan), *Pseudomonas rosea* (a bacterium), *Pinus monophylla* (a pinyon-pine).
L. *primus,* first; see **first**
L. *semel,* once, a single time, once for all: *Macrocrinus semelfurcatus* (a Devonian crinoid).
L. *simplex, -icis,* onefold, unmixed, single; see **simple**
L. *singulus,* one; *singularis,* one at a time, alone: single, singular, singly, singleton, *Polyblepharides singularis* (an alga).
Gr. *tis, tinos,* any, anyone, someone:
L. *ullus,* any, anyone:
L. *unus,* one, whole; *unicus,* sole, only, singular; *universalis,* pertaining to the whole; *universus,* one, entire, whole; *unio, -onis,* f. the number one, one, oneness; *unio, -itus,* join together, make one; *aduno; coaduno, -atus,* make one, unite: uniform, universal, unicorn, Unitarian, universe, unicellular, unique, unity, unanimous, onion, triune, unison, union, United States, *E pluribus unum,* Inadunata, *Unio complanatus* (fresh-water mussel), *Acanthurus unicornis* (unicorn-fish), *Lycostomus univittatus* (a beetle), *Abrotocrinus unicus* (a Mississippian crinoid), *Campanula uniflora* (arctic bluebell), *Orbulina universa* (a foraminifer), *Eleocharis uniglumis* (a spikerush).
See: **simple, all, alone**

-one, suffix signifying a ketone; see **ketone**

onear, -*atis,* L. an evening-primrose; see **evening-primrose**

onerosus, L. burdensome, heavy, < *onus, -eris,* burden; *oneratus,* filled, loaded; see **weight**

onesimos, Gr. beneficial, useful; see **use**

onetes, Gr. buyer; *onetos,* bought; *onios,* for sale; see *oneomai* under **trade**

onias, Gr. a sea-fish; see **wrasse**

onido- < Gr. *oneidos,* blame, reproach; see **blame**

onion < F. *oignon,* < *unio, -onis,* f. one, a kind of onion.
  L. *acrocorium,* n. a kind of onion:
  Gr. *aglis, -ithos,* f. head of garlic:
  L. *allium,* n. garlic: alliaceous, allyl, diallyl disulfide, thioallylaldehyde, *Allium sativum* (garlic), *Allium ascalonicum* (shallot), *Alliaria officinalis* (hedge-garlic), *Cordia alliodora* (a boraginacead), *Adenocalymma alliaceum* (a bignonia-cead), *Campanula alliariaefolia* (spurred bellflower), *Sisymbrium alliaria* (garlic-mustard).
  L. *alsidena,* f. a kind of onion:
  L. *cepa,* f. onion; *cepula,* f. dim.; Sp. *cebolla,* f. onion; *cepina,* f. bed of onions: cepaceous, Cebolleta (N. Mex.), cebollite, *Allium cepa* (onion), cibol *(Allium fistulosum),* chive *(Allium schoenoprasum),* *Phorbia ceparum* (onion-fly), *Sclerotium cepivorum* (neck-rot fungus), *Lepiota cepaestipes* (a mushroom), *Ovula cepula* (a gastropod), *Succinea cepulla* (a gastropod), *Succinea cepulla* (a gastropod), *Oncidium cebolleta* (an orchid), *Cromyatractus ceparius* (a fungus), *Urocystis cepulae* (a fungus).
  Gr. *detis, -idos,* f. head of garlic:
  Gr. *gelgis, -ithos,* f. head or clove of garlic:
  L. *gethyum* (Gr. *gethyon*), n. leek; *gethyllis, -idos,* f. dim.; *epithyllis, -idos,* f. a leek: *Gethyum atropurpureum* (a liliacead), *Gethyllis afra* (an amaryllidacead).
  Gr. *kidalon,* n. an onion:
  Gr. *kromyon* (*krommyon*), n. onion; *krommydion,* n. dim.: *Cromyocrinus ornatus* (a fossil crinoid), *Stylocromyum amphiconus* (a radiolarian)
  Gr. *moly,* n. an herb with magic powers, identified by some as garlic; *molyxa,* f. a kind of garlic with a single clove: *Allium moly* (lily-leek).
  L. *opition, -onis,* m. a kind of onion:
  L. *pallacana,* f. a kind of onion:
  Gr. *physinx, -ingos,* f. a kind of garlic:
  L. *porrum,* n. leek; *porraceus,* of leeks; *porrina,* f. bed of leeks: porraceous, *Allium porrum* (leek), *Tilia porracea* (Okaloosa linden), *Phytophthora porri* (a fungus).
  Gr. *prason,* n. leek; *prasinos,* of leeks, leek-green: prasophagous, praseodymium, prase, praseolite, prasochrome, chrysoprase, *Dryophis prasinus* (a snake), *Scarus prasiognathus* (a parrot-fish), *Carex prasina* (a sedge).
  L. *scilla; squilla* (Gr. *skilla*), sea-onion, sea-leek; see **lily**
  Gr. *skorodon* (*skordon*), n. garlic; *skordonion,* n. dim.: scorodite, *Allium scorodo-prasum* (rocambole), *Nothoscordum fragans* (a false-garlic), *Teucrium scorodonia* (a germander).
  L. *ulpicum,* n. a kind of leek:

oniro- < Gr. *oneiros,* dream; see **dream**

oniscus, L. (Gr. *oniskos,* dim. of *onos,* ass), sowbug; see **sowbug,** *onos* under **horse**

onisto- < Gr. *oneistos,* most useful, < *oneios,* useful; see **use**

onitis, Gr. a kind of mint; see **mint**

onobrychis, Gr. a legume; see **bean**

onoclea, NL. (Gr. *onokleia*), a genus of ferns; see **fern**

onocrotalus, L. (Gr. *onokrotalos*), pelican; see **pelican**

onomato- < *onoma, -tos,* name; *onomation,* dim.; see **word**

onomatopoeia, L. (Gr. *onomatpoiia*), making of words to imitate natural sounds; see FIGURES OF SPEECH under **form**

ononis, Gr. restharrow; see **bean**

onophyllon, Gr. a kind of borage; see **borage**

onopordum, L. (Gr. *onopordon*), a kind of thistle; see **thistle**

onopyxos, Gr. an unidentified plant; see **plant**

onorhynchus, L. (Gr. *onorhynchos*), a plant; see **plant**

onos, Gr. ass, beaker, fish, see **horse**; price, see *oneomai* under **trade**

onthos, Gr. dung; see **dung**

onto- < Gr. *on, ontos,* being, thing, that which has existence; see **life**

onustus, L. laden, burdensome; see *onus* under **weight**

onycho- < Gr. *onyx, -ychos,* fingernail, talon, claw, hoof, see **nail**; a yellowish gem stone, see **stone**

-onym < Gr. *onoma (onyma),* name; see **word**

onyx, Gr. a yellowish gem stone, see **stone**; fingernail, talon, claw, hoof, see **nail**

oön, Gr. egg; *oarion,* dim.; see **egg**

op- < L. *ob,* toward; see **to**

opacus, L. shady, dark, obscure, dim; *opacitas,* shadiness, dimness; see **shade**

opados, Gr. accompanying, attending; *opadeter,* attendant; see **servant**

opaeo- < Gr. *opaios,* with a hole, < *ope,* hole; see **hole**

opalus, L. (Gr. *opallios*), opal; see **stone**

ope, Gr. hole; see **hole**

opeato- < Gr. *opeas, -atos,* awl; see **awl**

open < AS. *open.*
Gr. *aerktos,* unfenced, open:
Gr. *akalyptos; akalyphos,* uncovered, unveiled; *apokalypsis,* f. a manifestation, revelation: *Acalyptonotus violaceus* (a water-mite), apocalypse.
Gr. *akleistos,* open, unfastened, unclosed: *Acleistoceras olla* (a fossil cephalopod), *Aclistoides retractus* (a wasp).
Gr. *amphadios,* open, public:
Gr. *anapetes,* wide open, expanded; *diapetes,* unfolded, open: *Agrion(Anapetes) virgatum* (an odonatid).
Gr. *anochyros,* open, clear:
Gr. *anoiktos,* opened; *anoige,* f. act of opening; *anoigeus,* m. opener; *anoigma, -tos,* n.; *anoxis,* f. an opening, < *anoignymi,* open, unfold: *Anoectochilus regalis* (an orchid), *Anoectomychus pudens* (a moth), *Anoiganthus breviflorus* (an amaryllidacead), *Anigozanthus flavida* (an amaryllidacead).
L. *aperio, -ertus,* open; *apertibilis,* causing to open; *apertio, -onis,* f. an opening; *comperio,* lay open, disclose wholly: aperient, aperture, aperitive, April, pert, malapert, overt, overture, *Bulimina aperta* (a foraminifer).
L. *apricus,* lying open, uncovered, exposed to the sun, sunny; *apricatio, -onis,* f. a basking in the sun; *apricum,* n. a sunny spot; *apricarius,* of the open: *Antennaria aprica* (Rocky Mountain pussytoes), *Apion apricans* (clover-weevil), *Pluvalis apricaria* (European golden-plover).
L. *candor,* openness, sincerity; see **true**
Gr. *chaino; chasko,* open, yawn, gape; *chanos; chasmema, -tos,* n. the open mouth, a wide yawn or gape; *chaskax, -akos,* m. gaper; *chaos,* n. yawning abyss, condition of formlessness, utter disorder and confusion; *chasma, -tos,* n. yawning abyss, gulf; *chasme,* f. a yawn; *amphichasko,* gape around: chaos, chaotic, casemate, achene, *Chaenolobus virgatus* (goldenlocks), *Chaenomeles japonica* (a flowering-quince), *Chaenactis artemisiaefolia* (a composite), *Chaograptis crystallodes* (a moth), *Chaolaimus[Chaolaemus] pellucidus* (a nematode), *Chaos carolinense* (a protozoan), *Myiochanes pertinax* (a pewee), *Chasmatosaurus yuani* (a fossil reptile), *Chasmorhynchus nudicollis* (a bellbird), *Chasmichthys gulosus* (a goby), *Macrochasma crenulata* (a gastropod), *Chaenostoma hispidum* (a figwort), *Chanos cyprinella* (a milkfish), *Chama squamosa* (an Eocene pelecypod).
L. *coram; palam,* in the presence of, openly:
L. *detego, -ectus,* unroof, disclose, reveal: detect, detective, detection.
L. *dissutus,* unstitched, ripped, opened: *Xystocheir dissuta* (a myriapod).
L. *emico, -atus,* appear suddenly, spring forth, become apparent:
Gr. *endios,* at midday, in the open:
Gr. *enopadios,* to one's face, openly:
Gr. *epipolaios,* superficial, manifest; *epipole,* f. surface: epipolic, epipolism, *Epipolaeus caliginosus* (a beetle), *Epipolops quadrispinus* (a bug).
L. *expositus,* open, accessible, free: *Bombyx exposita* (a moth).
L. *fatisco,* crack open, fall apart:
L. *foras; foris; forus,* out of doors, in the open, abroad; *forasticus,* public; *forensis,* of the forum, public; *forum,* n. an open, public place, market-place; *forinsecus,* on the outside: forest, forum, forensic, foreign, foreclosure, forfeiture, forisfamiliate, *Calpodes forulus* (a butterfly).
L. *hio, -atus; hisco,* open, gape; *hiatus,* m. opening, aperture, gap; *hiulcus,* gaping, split open, < *hiulco, -atus,* cause to gape or split open; *dehisco,* split open, yawn; *inhio, -atus,* gape at: hiatus, dehisce, dehiscence, dehiscent, *Argonauta hians* (a nautilus), *Orectochilus dehiscens* (a beetle).
Gr. *horatos,* visible; see *horao* under **see**

L. *hyaethrus* (Gr. *hyaithros*), under the sky, open, uncovered:

L. *improtectus,* uncovered, undefended:

L. *inhumatus,* unburied; see *humus* under **earth**

L. *inobseptus,* not hedged up, not closed, open:

Gr. *kechenos,* gaping, yawning; *kechenotos,* open-mouthed: *Cechenosternum nigromaculatum* (a beetle).

Gr. *menyo,* reveal, disclose, betray; see **treason**

L. *obvius,* at hand, lying open, easily seen: obvious.

L. *oscito, -atus,* open the mouth wide, gape, yawn; *oscitans,* yawning, listless, drowsy: oscitate, *Micrasterias oscitans* (a desmid).

L. *pansus; passus,* spread out, extended, open; see **spread**

L. *pateo,* lie open; *patulus,* open, spread out, broad; *pator, -is,* m. an opening; *patefacio, -factus,* lay open, disclose; *dispatens,* opening in different directions: patent, patentee, patulin (from *Penicillium patulum*), *Spartina patens* (a grass), *Didymograptus patulus* (a graptolite), *Campanula patula* (a bellflower), *Dodecatheon patulum* (a shooting-star).

L. *percernis,* easily visible:

L. *pervius,* affording a passage, open, penetrable:

L. *promo, -omptus,* bring out, produce; see **make**

L. *propalo, -atus,* make public, manifest, divulge:

L. *publicus,* common, open; see *populus* under **people**

L. *resero, -atus,* unlock, open:

L. *revelo, -atus,* unveil, lay bare: reveal, revelation.

L. *rimor, -atus,* lay open, turn over; *rima,* cleft, fissure; see **break**

L. *ringens,* gaping, < *ringor, rictus,* gape, grin; *rictum,* n.; *rictus, -us,* m. open mouth: rictus, rictal, ringent, *Mimulus ringens* (monkey-flower), *Geminiricta virgata* (a foraminifer).

Gr. *sairo,* draw back the lips and show the teeth, grin, snarl:

See: **clear, bare, see, spread, hole, display, space, cut, broad, free**

opera, L. work; *operula,* dim.; *operor, -atus,* be busy, work; *operator,* worker; see *opus* under **make**

operculina, NL. a genus of foraminifers; see **protozoan**

operculum, L. cover, lid, < *operio, -ertus,* cover, hide; see **cap, cover**

opes, pl. of *ops, opis,* L. riches, wealth, property, power; *opicillum,* dim.; see **wealth**

ophelma, Gr. broom; see **broom**

ophelos, Gr. useful, helpful; see **use**

ophileto- < Gr. *opheiletes,* debtor; see *ophelio* under **pay**

ophio- < Gr. *ophis, -ios,* serpent, reptile; *ophidion,* dim.; see **snake**

ophioglossum, NL. a genus of ferns; see **fern**

ophrys, Gr. brow, eyebrow; *ophrydion,* dim.; see **face**

ophthalmos, Gr. eye; see *ops* under **eye**

-opia < Gr. *ops,* eye; see *optikos* under **see**

opicus, L. rude, stupid, foolish; see **fool**

opidnos, Gr. awful, terrifying; see **fear**

opifer; opitulus, L. helper; see *ops* under **wealth**

opifex, L. worker, artisan; see *opus* under **make**

opilio, L. shepherd; see **guard**

opimus, L. fruitful, fertile, well-fed, fat, wealthy, rich; see *ops* under **wealth**

opinor, -atus, L. suppose, think; *opinio, -onis,* conjecture, view; see **think**

opipes; opipeuter, Gr. gaper, starer; see **see**

opisma, Gr. milky sap of plants; see *opos* under **juice**

opistho- < Gr. *opisthen,* behind; see **after**

opium, L. (Gr. *opion*), poppy-juice; see *opos* under **juice**

opo- < Gr. *ops, opos,* voice, see **speak**; eye, face, sight, see **eye**; *opos,* juice, see **juice**

opora, Gr. autumn; *oporinos,* autumnal; *metoporon,* late autumn; see **year**

opos, Gr. juice; *opion,* dim., poppy-juice; see **juice**

oppidum, L. town; *oppidulum,* dim.; *oppidanus,* of a town; see **town**

oppilo, -atus, L. close up, shut up; see **close**

opportunus, L. fit, suitable, appropriate; *opportunitas,* timely advantage; see **fit**

opposite < L. *oppositus,* setting against; see **against**

oppressus, L. pressed together, pressed down; see *primo* under **press**

opprobriosus, L. abusive, taunting, shameful; see *probrum* under **shame**

ops, *opis,* L. riches, means, strength, help, see **wealth;** < Gr. *ops, opos,* eye, face, voice, see *ophthalmos* under **eye,** *ops* under **speak,** *-opis* under **like**

opsios, Gr. late; *opsiotes,* lateness; see **after**

-opsis; -opy, relating to sight and appearance, < Gr. *opsis,* sight, appearance, view, face, likeness, < *ops, opos,* eye, face; see *optikos* under **see,** *ops* under **face,** *-opsis* under **like**

opso- < Gr. *opson; opsonion,* meat, rich fare; *opsaridion; opsarion,* dim.; see **food**

optico- < Gr. *optikos,* pertaining to sight; *opter, -os,* scout, spy; see **see**

optimus, L. best; see *bonus* under **good**

opto, *-atus,* L. choose, wish, desire; see **choose**

opulentus, L. rich, wealthy; see *ops* under **wealth**

opulus, L. a kind of maple; see **maple**

opuntia, L. a genus of cacti; see **cactus**

opus, *-eris,* L. work; *opusculum,* dim.; *operosus,* busy; see **make**

-or, L. suffix meaning one who, agent, actor, condition, state; see **make, nature**

ora, L. edge, border, margin, coast, zone, region; *orarius,* of the coast, see **border**

oraculum, L. prophetic saying; see **prophecy**

orange < OF. *orenge,* < Ar. *naranj;* see *aurantium* under **citrus**

orator, L. speaker; see *oro* under **speak**

orbis, L. circle; *orbiculus,* dim.; *orbitus,* circular; *orbita,* wheel-rut, circuit; see **circle**

orbo, *-atus,* L. bereave; *orbus,* bereft of parents; *orba,* orphan; see **alone**

orca, L. a kind of whale, earthenware pot with a large belly; *orcula,* dim.; see **whale**

orchamos, Gr. first; see **first**

orchard < AS. *orceard; ortgeard;* see **tree**

orchestra, Gr. place where the chorus danced; *orchestes,* dancer; see *orcheomai* under **dance**

orchid < L. *orchis, -is,* f. (Gr. *orchis, -ios,* m. testicle, orchid), a plant so named from the shape of its tuberous root; *orchidion,* n. dim.: Orchidaceae, *Orchis rotundifolia* (an orchid), *Iris orchioides* (orchid-iris), *Veronica orchidea* (a speedwell), *Gladiolus orchidiflorus* (a gladiolus).
 NL. *cattleya,* f. a genus of orchids, < William Cattley, English botanist: *Cattleya labiata* (an orchid).
 NL. *coelogyne,* f. a genus of orchids: *Coelogyne speciosa* (an orchid).
 NL. *corallorrhiza,* f. a genus of orchids: *Corallorrhiza trifida* (early coral-root).
 NL. *cymbidium,* n. a genus of orchids: *Cymbidium eburneum* (an orchid).
 NL. *cypripedium,* n. a genus of orchids: *Cypripedium passerinum* (Canada moccasin-flower).
 NL. *dendrobium,* n. a genus of orchids: *Dendrobium formosum* (an orchid).
 L. *entaticus* (Gr. *entatikon*), a name for *satyrion;* see *entatikos* under **arouse**
 NL. *epidendrum,* n. a genus of orchids: *Epidendrum gracile* (an orchid).
 L. *epipactis, -idis* (Gr. *epipaktis, -idos*), f. a plant; now a genus of orchids: *Epipactis gigantea* (an orchid).
 NL. *goodyera,* f. a genus of orchids, < John Goodyer, English botanist: *Goodyera tesselata* (a rattlesnake-plantain).
 NL. *habenaria,* f. a genus of orchids: *Habenaria fimbriata* (purple fringed-orchid).
 NL. *laelia,* f. a genus of orchids, < L. *Laelia,* a proper name: *Laelia crispa* (an orchid).
 L. *limodorum* (Gr. *limodoron,* gift of the meadow), n. a plant; now a genus of orchids: *Limodorum(Calopogon) pulchellum* (grass-pink).
 NL. *liparis,* f. a genus of orchids: *Liparis ramosa* (an orchid).
 Gr. *lonchitis, -idos,* f. a kind of orchid: *Alethopteris lonchitica* (a fossil fern), *Polystichum lonchitis* (a fern).
 NL. *odontoglossum,* n. a genus of orchids: *Odontoglossum maculatum* (an orchid).
 NL. *oncidium,* n. a genus of orchids: *Oncidium pulchellum* (an orchid).
 NL. *pholidota,* f. a genus of orchids: *Pholidota articulata* (an orchid).
 NL *pogonia,* f. a genus of orchids: *Pogonia divaricata* (southern pogonia).
 Gr. *satyrion,* n. an orchid with aphrodisiac properties: *Satyrium carneum* (an orchid), *Satyridium rostratum* (an orchid).

NL. *spiranthes,* f. a genus of orchids: *Spiranthes porrifolia* (a ladies-tresses).

NL. *vanilla,* f. a genus of orchids, < Sp. *vainilla,* dim. of *vaina,* sheath, scabbard: vanilla, vanillin, *Vanilla planifolia* (Mexican vanilla).

orchilos, Gr. wren; see **wren**

orchis, Gr. testicle; *orchidion,* dim., see **testicle**; a plant, see **orchid**

orchos, Gr. row of trees; see **line**

Orcus, L. the nether world, abode of the dead; *Orcinus,* of the nether world; see **hell**

orcynus, L. (Gr. *orkys, -ynos*), a kind of tunny; see **tunny**

order < L. *òrdo, -inis,* methodical arrangement, line, row, series; *ordinarius,* of regular or usual manner; see **class, govern, line, law, send, ask**

ordior, *orsus,* L. begin a web; see **weave**

ore < AS. *ar,* copper, bronze; a mineral with recoverable metallic or non-metallic constituents of economic value; see **stone, mine**

oreo- < Gr. *oros, -eos,* mountain; *orion,* dim.; *oreios,* of mountains; *oreinos,* mountainous, hilly; *oresbios,* living on mountains; L. *oreas* (Gr. *oreias*), a mountain-nymph; see **mountain**

orecto- < Gr. *orektos,* stretched out, held out; *oregma,* a stretching out; see *orego* under **spread**

orestes, Gr. mountaineer; see *oros* under **mountain**

orestium, L. (Gr. *orestion*), elecampane; see **composite**

oreus, Gr. mule; *orikos,* of a mule; see **horse**

orexis, Gr. appetite, desire, longing; *orektikos,* of desire, appetite; see **wish**

orgado- < Gr. *orgas, -ados,* meadow, field; see **field**

organ < L. *organum* (Gr. *organon*), instrument; see **tool**

orgasm < Gr. *orgasmos,* excitement, swelling, kneading; especially the turgidity, pulsation, and discharge at the culmination of sexual intercourse; see **coitus**

orgilos, Gr. easily angered, irritable; see *orge* under **anger**

orgy < Gr. *orgion,* secret rite; see **rite**

orgyia, Gr. length of the outstretched arms, fathom; see **measure**

orient < L. *oriens, -entis,* the rising sun, east, < *orior, ortus,* rise, grow, appear; see **east, begin, raise**

orificium, L. opening; see *os* under **mouth**

origanum, L. (Gr. *origanon*), a mint; see **mint**

origin < L. *origo, -inis,* source, birth; *originalis,* of the beginning; see **begin**

orino, Gr. excite, stir; *orintes,* exciter; see **arouse**

oriole < OF. *oriol,* < L. *aureolus,* golden: *Oriolus kundoo* (a golden oriole), *Haemoproteus orioli* (a protozoan).
Gr. *chlorion, -os,* m. probably the golden oriole: *Chlorion azureum* (a wasp).
L. *icterus* (Gr. *ikteros*), m. a yellow bird: *Icterus cucullatus* (hooded oriole), *Hypolais icterina* (a warbler), *Icteria longicauda* (long-tailed chat).

Orion, L., Gr. fabled hunter transformed into a constellation; see **hunt**

-orium, L. place where, place for; see *-arium* under **place**

oriundus, L. born in, descended or risen from; see *origo* under **begin**

ormenos, Gr. shoot, sprout, stalk; see **stem**

ornament < L. *ornamentum,* n. equipment, decoration, honor, < *orno, -atus,* decorate, furnish; *adorno, -atus,* decorate, embellish: adorn, ornamental, ornamentation, ornate, adornment, suborn, *Lophornus ornatus* (a hummingbird), *Hemixantha ornamentata* (a fly), *Aphis ornata* (an aphid), *Goniobasis ornatella* (a snail), *Cosmoceras ornatum* (a Jurassic ammonite).
L. *antefixum,* n. little ornament or image for roofs of houses and temples:
Gr. *asketos,* curiously wrought, ornamented; see *askeo* under **make**
L. *basilium* (Gr. *basileion*), n. royal ornament, diadem:
L. *bulla,* bubble, ornament; see **bubble**
Gr. *chlidon, -os,* m. ornament; *chlidosis,* f. ornamentation: *Terebratula*(*Chlidonophora*) *incerta* (a brachiopod).
L. *comptus; comtus,* ornamented, adorned; *comptulus,* dim.: *Gennaeocrinus comptus* (a Devonian crinoid), *Cordylocrinus comtus* (a Silurian crinoid).
L. *decoro, -atus,* grace, ornament; *decoris,* adorned, elegant, beautiful; *decus, -oris,* n. ornament, splendor; *decorum,* n. propriety, fitness; *decorus,* becoming, fitting, proper, beautiful: decorate, decoration, decorous, decorum, *Sorbus decora* (showy mountain-ash), *Gennaeocrinus decorus* (a Devonian crinoid), *Calliurich-*

thys decoratus (a fish), *Ribes indecorum* (a currant), *Aeonium decorum* (a crassulacead).

L. *elimatus*, finished, elaborated, adorned; see *lima* under **scrape**

Gr. *emblema*, inlaid work, thing put in or on, ornament; see **mark**

L. *excultus*, adorned, polished, refined; *percultus*, highly adorned: *Cassidula inexculta* (a foraminifer).

L. *frontalium*, n. an ornament for the forehead of horses:

Gr. *kommos*, m. decoration, embellishment; *kommosis*, f. ornamentation; *kommotes*, m. embellisher, beautifier; *kommotikos*, of embellishment: *Commophila macrocarpana* (a moth), *Commolenda deusta* (a bug), *Commoptera solenopsidis* (a fly).

Gr. *kosmos*, m. ornament, decoration, dress; *kosmarion; kosmion*, n. dim.; *kosmetikos*, pertaining to the art of dress, skilled in the art of arraying, decorating, and beautifying; *kosmetos*, trim, neat; *kosmios*, well-ordered, neat, decent; *eukosmos*, orderly, decorous, graceful: cosmetic, *Cosmos bipinnatus* (common cosmos), *Cosmocrinus holzapfeli* (a Devonian crinoid), *Cosmetocrinus gracilis* (a Mississippian crinoid), *Cosmioconcha parvula* (a gastropod), *Dryocosmus deciduus* (an oak gall-wasp), *Cosmarium botrytis* (a desmid), *Cyclocosmia truncata* (a trapdoor-spider), *Cleptocosmia mutabilis* (a moth).

L. *phalera*, f. military decoration worn on the breast, a trapping worn on horse's head and breast: *Phalera bucephala* (a butterfly), *Cratosomus phaleratus* (a beetle).

L. *polymitarius*, highly wrought or finished:

L. *topium*, n. ornamental gardening; *topiarius*, pertaining to ornamental gardening: topiary.

See: **beauty, bracelet, necklace, earring, agreeable, bind, needle**

ornitho- < Gr. *ornis, -ithos; orneon*, bird; *ornithion*, dim.; see **bird**

ornus, L. mountain-ash; *orneus*, of the mountain-ash; see **rose**

oro-; oreo-; oressi- < Gr. *oros, -eos*, mountain; see **mountain**; < *oros*, watery or serous part of anything, see **juice**

orobanche, L., Gr. broomrape; see **broomrape**

orobax, Gr. peony; see **peony**

orobus, L. (Gr. *orobos*), vetch; see **bean**

orodamnos, Gr. bough, branch; see **branch**

orontium, NL. a genus of the arum family; see **arum**

oropedion, Gr. plateau; see **level**

orophe, Gr. roof, ceiling; see **roof**

orospizos, Gr. mountain-finch; see *spiza* under **finch**

orozelum, L. a kind of mint; see **mint**

orpeco- < Gr. *orpex, -ekos*, sapling; see **tree**

orphan < L. *orphanus* (Gr. *orphanos*), bereft; *orphanotropheion*, orphanage; see **alone**

Orpheus, Gr. famous minstrel and husband of Eurydice; see **music**

orphnos, Gr. dark, dusky; *orphne*, the darkness of night, night; see **black**

orphus, L. (Gr. *orphos*), sea-perch; see **perch**

orpine < L. *auripigmentum*, orpiment; see **stonecrop**

orrhodia, Gr. dread, terror; see **fear**

orrhos, Gr. rump, see **rump**; serum, see **juice**

orsi- < Gr. *ornymi*, stir up, excite, raise; see *orino* under **arouse**

orsinus, L. a kind of crocus; see **crocus**

orsites, Gr. a Cretan dance; see **dance**

ortalis; ortalichos, Gr. young bird, chick, fowl; see **bird**

ortho- < Gr. *orthos*, straight, correct, normal, right, direct; see **straight, right**

orthros, Gr. at daybreak, dawn, early in the morning; *orthrios*, early; see **dawn**

ortus, L. risen; a rising of the heavenly bodies; see *orior* under **raise**

ortygo- < Gr. *ortyx, -ygos*, quail; see **quail**

orycto- < Gr. *oryktos*, dug, mined; *oryktes*, digger; *oryxis*, a digging; *orygma*, pit, trench, moat, tunnel; see *orysso* under **dig**

orymagdos, Gr. loud noise, din, roar; see **roar**

orythmos, Gr. a howling; see **sound**

oryx, L., Gr. pick for digging; *orygion*, dim., see **pick**; gazelle, antelope, see **antelope**

oryza, Gr. rice; see **grain**

os- < L. *ob*, toward; see **to**

os, *oris*, L. mouth, opening; *oscillum; osculum*, dim., see **mouth**; < *os, ossis*, bone; *ossiculum*, dim., see **bone**

oscen, *-inis*, L. a singing bird; see **bird**

osche, Gr. scrotum; see **bag**

oschos, Gr. short, young branch; see **branch**

oscillo, *-atus*, L. swing; *oscillum*, swing; see **swing**

oscito, L. open the mouth wide, gape, yawn; *oscitans*, yawning, listless, drowsy; see **open**

osculor, *-atus*, L. kiss; *osculum*, little mouth, kiss; see **kiss**

-ose, suffix indicating a carbohydrate; the product of: dulcose, galactose, glucose, inose, levulose, maltose, proteose, sucrose.

-ose; -osity < L. *-osus*, having the nature or quality of, usually denoting fullness or abundance; see *-ous* under **nature**

-osis, Gr. a suffix denoting a condition, usually morbid; see **disease**

osme, Gr. smell, odor; *osmeros*, odorous; *osphradion*, strong smell; see *ozo* under **smell, smelt**

osmo- < Gr. *osmos*, Gr. a pushing; see *otheo* under **push**

osmunda, ML. a kind of fern; see **fern**

osmyle; osmylos, L. an octopus; see **mollusk**

osor, L. hater; see *odium* under **haet**

osphrantos, Gr. smellable; *osphresis*, smell; see *ozo* under **smell**

osphys, Gr. loin, lower part of the back; *osphydion*, dim.; see **side**

osprion, Gr. pulse; see **bean**

ossa (otta), Gr. rumor; see **speak**

osseus, L. bone, bony; *ossiculum*, dim. of *os, ossis*, bone; see **bone**

ossifragus, L. sea-eagle, osprey; see **eagle**

ostentatio, L. display, parade; *ostentus*, spread out, displayed; see *ostendo* under **display**

osteo-; osto- < Gr. *osteon*, bone; see **bone**

ostigo, L. an eruption on lambs; see **disease**

ostinon, Gr. bone-pipe; see **pipe**

ostium, L. door, entrance; *ostiolum*, dim.; see **door**

ostlingo- < Gr. *ostlinx, -ingos*, curled hair, lock of hair, tendril; see **hair**

ostocatacto- < Gr. *ostokataktes*, osprey; see **eagle**

ostraco- < Gr. *ostrakon*, shell, potsherd; *ostrakion*, dim.; see **shell**

ostrea, L. (Gr. *ostreon*), oyster; *ostrearius*, of oysters; *ostreosus*, abounding in oysters; see **mollusk**

ostrich < OF. *ostruche*, < L. *avis struthio*, the bird *struthio*, < L. *struthio, -onis* (Gr. *strouthion, -os*), m. < Gr. *strouthokamelos*, m. camel-bird: *Struthio camelus* (ostrich), *Struthiopteris nodulosa* (ostrich-fern), *Peucedanum ostruthium* (masterwort), *Struthiomimus ingens* (a dinosaur), *Struthiola ciliata* (a thymeleacead).

    NL. *casuarius*, m. a genus of ostrichlike birds, < Malay *kasuari: Casuarius papuanus* (a cassowary), *Casuarina stricta* (a beefwood), *Corythosaurus casuarius* (a dinosaur).

    NL. *dromiceius*, m. a genus of ostrichlike birds: *Dromiceius irroratus* (an emu).

    NL. *rhea*, f. a genus of ostrichlike birds, < *Rhea*, f. wife of Cronus: *Rhea darwini* (Patagonian rhea).

ostrimon, Gr. stable; see **house**

ostrinus, L. purple, < *ostrum*, purple dye from a mollusk; see **purple**

ostrya, L. (Gr. *ostrye*), hop-hornbeam, ironwood; see **hornbeam**

-osus, L. suffix signifying nature or quality of, usually fullness or abundance; see *-ous* under **nature**

osyris, Gr. an unknown plant; now a genus of the sandalwood family; see **sandalwood**

-otes, Gr. suffix denoting quality; see **nature**

other < AS. *other:* otherwise, another, others, neither, either.

L. *alibi* (*aliubi*), elsewhere, at another place, with some other person: alibi.

L. *alius,* another, other; *alienus,* pertaining to another: alias, alien, aliquot, else, *Ofcookogona alia* (a milleped).

Gr. *allos,* other: allotropic, allomorph, allograph, allegorical, allergy, parallel, allochroite, allophane, allopathy, *Allomyces arbuscula* (a fungus), *Allacodon pumilus* (a fossil mammal).

L. *alter, -a, -um,* the other: alternate, altercation, alterative, altruism, adulterate, adultery, subaltern.

L. *ceterus,* other: et cetera(etc.).

L. *subdo, -itus,* substitute, counterfeit; see **lie**

L. *uter, -tra, -trum,* other: neuter, neutrality.

See: **change, lie, alternate, different, companion, over**

othnio- < Gr. *othneios,* strange, alien; see **strange**

othono- < Gr. *othone,* fine linen, sail-cloth, sail; *othonion,* dim.; see **linen**

otilo- < Gr. *oteile,* wound; see **sore**

otiosus, L. at leisure, idle, < *otium,* leisure; see **rest**

otis, L., Gr. bustard; see **plover**

otlos, Gr. distress, suffering; see **trouble**

oto- < Gr. *ous, otos,* ear; *otion,* dim.; see **ear**

otobos, Gr. loud noise, din; see **sound**

otreros, Gr. nimble, quick, busy; see **swift**

otter < AS. *oter;* see **weasel**

otus, L. (Gr. *otos*), a kind of owl; see **owl**

-otus, L. having the nature of, pertaining to; see *-atus* under **nature**

ounce < L. *uncia,* a twelfth part; see **twelve**

-ous; -ose; -osity < L. *-osus,* having the nature or quality of, denoting usually fullness or abundance; see **nature**

ousia, Gr. essence, substance, property, reality; see *on* under **life**

out < AS. *ut,* from, beyond: outrun, outplay, outcome, outfield, outside, outlandish, outmost, utmost, utter, lookout, about.

L. *e-, ef-, ex-,* < *ex, out,* out of, from, by reason of, as in *emergo,* come out, rise up; *effero,* carry out; *expurgo,* cleanse out: educate, emit, elongate, election, evolve, effect, effloresce, effluent, effusion, escape, essay, escort, escheat, estrange, exclude, exaggerate, excuse, exacerbate, exasperate, excel, express, except, export, expose, exhibit, explosion, extend, exquisite, exuviate, ex-President, ex officio, enormous, amend, award, issue, scamper, scarcity, spend, sample, abash, *Gallinula ecaudata* (a bird).

Gr. *ek-, ekto-, ex-, exo-,* < *ek; ex,* out, out of, from, without, as in *ekbaino,* step out; *ektos,* without; *exantheo,* flower out, blossom; *exoteros,* outer; *exotatos,* outermost; *exoterikos,* outside, external: eclipse, eclectic, eccentric, eclogue, ecstasy, ectoderm, ectoplasm, ectogenous, ectotrophic, exoteric, exoskeleton, exogenous, exotic, exodus, exarch, exipodite, ectal, exorcise, anecdote, ellipse, *Ectocarpus siliculosus* (an alga), *Exogyra costata* (a fossil pelecypod).

Gr. *epipole,* surface; see **open**

L. *exterus* (*exter*), out; *exterior,* outer; *extimus; extremus,* outermost, utmost; *externus,* outward, outside; *extra,* outside, besides, beyond; *extraneus,* outside, unrelated; *extrinsecus,* on the outside, outwardly; *extrorsus,* in an outward direction: external, exterior, extra, extreme, extremity, extrinsic, extrorse, extraneous, extra-curricular, extrapolate, extravagance, strange, estrange, exine(extine), *Alampetis extrema* (a beetle), *Extra extra* (a gastropod), *Extracrinus briareus* (a crinoid).

L. *foras; foris,* out of doors; see **open**

L. *superficies, -ei,* f. upper side, outside: surface, superficial.

Gr. *thyraios,* outside the door, abroad:

See: **from, open, not, banish**

ovary < NL. *ovarium,* n. the egg-producing organ of the female, < L. *ovarius,* m. egg-keeper: ovarian, ovaritis, ovarotomy.

F. *pistil,* female or seed-producing organ of a flower, < L. *pistillum,* n. pestle: pistil, pistillate, pistillidium.

ovatus, L. egg-shaped, see *ovum* under **egg;** a rejoicing, see *ovo* under **joy**

oven < AS. *ofen,* furnace, stove.

Gr. *baunos,* m. furnace, forge:

L. *calcaria,* f. limekiln:

L. *caminus,* m. (Gr. *kaminos,* f.), fireplace, furnace, oven; *kaminion,* n. dim.: chimney.

AS. *cyln,* furnace, < L. *culina,* kitchen: kiln, brickkiln, limekiln.

Gr. *eschara,* f. hearth, fireplace; *escharion,* n. dim.:

L. *focus,* m. hearth, fireplace, center, central point; *foculus,* m. dim.; *focacius,* of the hearth: Sp. *fuego,* m. fire, hearth, fireplace: focus(pl. foci), focimeter, epifocal, fuel, fusillade, Tierra del Fuego, Fuegian.

L. *fornax, -acis,* f. oven; *furnus,* m. oven, kiln; *fornacula,* f. dim.; *furnaceus,* of ovens; *furnarius,* of ovens; m. baker; *Fornax,* f. goddess of ovens: furnace, Fornacalia, *Furnarius rufus* (red oven-bird), *Thermobia furnorum* (firebrat), *Fornax obrutus* (a gastropod).

Gr. *hestia,* f. hearth, fireside: see house

Gr. *ipnos,* m. oven, furnace, lantern; *ipnios,* of ovens: *Ipnops murrayi* (a fish), *Argyripnus ephippiatus* (a fish).

Gr. *kribanos* (*klibanos*), vessel used for baking bread, small oven; see pot

L. *laterina,* f. brickkiln:

Gr. *plintheion,* n. brickkiln:

Gr. *pnigeus,* m. oven, cover or damper to smother a fire:

Gr. *trasia,* grate, kiln; see screen

See: fire, house

over < AS. *ofer,* above, beyond, higher, more than: overhead, overawe, hangover, overlook, Passover.

Gr. *ana,* up, again; see return

L. *anatonus* (Gr. *anatonos*), stretching or extending upward:

Gr. *anekas,* upward; *aneko,* reach up to, come up to, amount to:

Gr. *ano-,* up, upward, above; *anothen,* above, on high: *Anostoma depressum* (a snail), *Anobium striatum* (death-watch beetle), *Opsanus pardus* (leopard-toadfish).

L. *antistatus,* m. superiority in rank:

Gr. *epi,* upon, beside, over, after; see on

Gr. *exaisios,* beyond demand, extraordinary: *Exaesiognatha ivanovi* (a beetle), *Exaesiopus torvus* (a beetle).

L. *extraordinarius,* above or beyond the common or ordinary; see wonder

Gr. *hyper; hyperos,* above, over, beyond, very; *hyperteros,* upper, better; *hypatos; hypertatos,* uppermost, highest, topmost, best: hyperbole, hyperbola, hypertrophy, hypercritical, hyperemia, hypersthene, hyperbaton, hyperthyroidism, *Hyperchoristus tanneri* (a fish), *Hyperothrix orophura* (a milliped), *Plectrophenax hyperboreus* (snowflake).

Gr. *meta,* between, among, beyond, after, over, reversely, implying change: metacarpal, metaphor, metamere, metaborate, metabolism, metaplasm, metaphyte, metatarsal, metachrome, metaphysics, metempsychosis, meteor, *Metatitan relictus* (a fossil mammal).

Gr. *pera,* beyond, across, further, very; *peraiteros,* further; *peraios,* on the other side, opposite: *Peraiocynodon*[*Peraeocynodon*] *inexpectans* (a fossil mammal), *Peraphyllum ramosissimum* (a squaw-apple).

L. *praeter,* beyond, past, more than; *praeteritus,* gone by, past: pretergress, preterhuman, preternatural, preterit.

L. *redundans,* superfluous: redundant.

Gr. *simoo,* turn up the nose; *simoma, -tos,* n. anything turned up; see bend

L. *super-, supra-, sur-,* < *super,* above, over, beyond; *supernus; superus,* over, above, on high, high; *superior,* higher; *supremus; summus,* highest: superb, superficial, superfluous, superhuman, supercilious, supervisor, superintendent, superlative, superstition, supersede, supraorbital, suprarenal, suprascapular, surcease, survey, surtax, surplus, surcingle, surface, surmise, surname, surfeit, surrender, survivor, surround, supernal, superior, summary, summit, somersault, sirloin, sovereign, soprano, insuperable, *Anomalophrys superciliosus* (watted plover), *Pomatostomus superciliosa* (a babbler).

L. *sursus,* upward:

L. *trans,* across, over, beyond, through: transport, transmit, transverse, transfusion, transatlantic, Transylvania, transfigure, transmutation, transfer, transom, traduce, trajectory, travesty, treason, tramontane(transmontane), tradition, trespass, trestle, *Cristellaria translucida* (a foraminifer).

L. *ultra-, ulter,* beyond, far; see far

AS. *up,* toward a higher point or place: upheave, uphold, upset, upon, upper, upward, upland, upstart, upright, flareup, upholster.

See: on, high, top, far, and

ovis, L. sheep; *ovicula,* dim.; *ovilis; ovillus; ovinus,* of sheep; see sheep

ovo, -atus, L. exult, rejoice; *ovatio,* acclaim, public homage; see joy

ovum, L. egg; *ovulum,* dim.; *ovatus,* egg-shaped; see egg

owe < AS. *agan,* have; see **pay**

owl < AS. *ule.*

Londoner (at night in an American countryside): "What was that sound?"
Host: "That was an owl."
Londoner: "Yes, I know; but what did the 'owling?"

Gr. *aigolios,* m. probably an owl: *Aegolius montanus* (an owl).

NL. *aluco,* f. screech-owl: *Aluco(Strix) flammea* (an owl).

L. *asio, -onis,* m. horned owl: *Asio flammeus* (short-eared owl).

Gr. *askalaphos,* m. probably an owl: *Ascalaphus ictericus* (an owlfly).

L. *bubo, -onis,* m. owl: *Bubo arcticus* (an owl).

Gr. *byas,* m. an owl: *Byas nobilis* (an owl).

L. *cavannus,* m. night-owl:

L. *cicuma* (Gr. *kikymis, -idos*), f. screech-owl:

Gr. *eleos* (*eleas*), m. a kind of owl: *Eleopicus caboti* (a woodpecker).

Gr. *epolios,* m. probably an owl:

Gr. *glaux, glaukos,* f. owl; *glaukidion,* n. dim.: *Glaucidium passerinum* (European pigmy-owl), *Cryptoglaux acadica* (sawwhet-owl).

Gr. *hybris, -idos,* f. a kind of owl:

Gr. *kikkabe,* f. screech-owl: *Ciccaba albogularis* (an owl).

Gr. *lagodias,* m. a horned owl: *Lagodias albidipennis* (a fly).

Gr. *lokalos,* m. probably an owl:

NL. *ninox,* f. a genus of owls: *Ninox odiosa* (boobook-owl).

L. *noctua,* f. night-owl, owl: Noctuidae, *Noctua clandestina* (an owlet-moth).

NL. *nyctea,* f. snowy owl: *Nyctea nyctea* (snowy owl).

L. *otus* (Gr. *otos*), m. horned owl: *Otus asio* (screech-owl).

Gr. *ptynx, ptyngos,* f. eagle-owl: *Rhinoptynx clamator* (an owl).

L. *scops, -opis* (Gr. *skops, -opos*), m. a small kind of owl: *Otus scops* (a screech-owl).

L. *strix, strigis* (Gr. *strix, strigos*), f. screech-owl: *Strix varia* (barred owl), *Epeira strix* (a spider), *Stringocephalus[Strigocephalus] obesus* (a fossil brachiopod).

NL. *surnia,* f. a kind of owl: *Surnia ulula* (hawk-owl), *Surniculus lugubris* (a cuckoo), *Surniculoides clamosus* (a cuckoo).

Gr. *tyto, -ous,* f. night-owl: *Tyto pratincola* (barn-owl), *Speotyto cunicularia hypogaea* (burrowing owl).

L. *ulula,* f. an owl: *Peloropus ulula* (a beetle).

own < AS. *agan,* have; see **hold, alone**

ox < AS. *oxa;* see **cattle**

oxalis, -idis, L. (Gr. *-idos*), f. woodsorrel: *Oxalis violacea* (violet woodsorrel). NL. *averrhoa,* f. a genus of the woodsorrel family, < Averrhoes, Arabian philosopher: *Averrhoa carambola* (carambola).

oxalme, Gr. a sour sauce; see **sauce**

oxina, Gr. harrow; see **rake**

oxya, Gr. beech; see **beech**

oxybaphon, Gr. saucer; see **plate**

oxygen < Gr. *oxys,* sour, acid, *-gen,* forming; see ELEMENTS under **thing**

oxys, Gr. sharp, acute, keen, quick, sour, acid, oxygen compounds; *oxynter,* sharpener; *oxyno,* sharpen; see **sour, point**

oyster < OF. *oistre,* < L. *ostrea* (Gr. *ostreon*); see *ostrea* under **mollusk**

ozaena, L. (Gr. *ozaina*), a fetid polyp in the nose; see *ozo* under **smell**

ozo, Gr. smell; see **smell**

ozos, Gr. branch, bough, twig, offshoot; *ozaleos; ozotos,* branched; see **branch**

P

pabo, L. wheelbarrow; *pabillus,* dim.; see **vehicle**

pabulum, L. food, fodder; *pabulor, -atus,* forage; *pabularis,* of fodder; see **food**

pachne, Gr. frost; *pachnodes,* chilly, cold, frosty; see **ice**

pachys, Gr. thick; *pachylos,* thickish; *pachos,* thickness; see **thick**

pacificus, L. peacemaking, peaceful, < *pax, pacis,* peace; *pacalis,* of peace; *paco, -atus,* quiet, soothe; see *pax* under **rest**

pack, prob. < the same source as D. *pak,* bundle; see **press, weight, bag**

pactus, L. agreed upon, settled, covenanted, bound; *pactilis,* entwined, plaited; *pactum,* agreement, covenant; see *pango* under **bind**

paddle < uncertain origin; see **spoon, oar**

pados, Gr. a tree, probably a cherry; see **plum**

paean, L. (Gr. *paian*), choral song; see **sing**

paecto- < Gr. *paiktos,* played with; *paiktes,* player; see *paizo* under **play**

paedidus, L. nasty, filthy, stinky; see **smell**

paedo- < Gr. *pais, paidos,* child; *paidarion,* dim.; *paidia,* childhood; see **child**

paegnio- < Gr. *paignion,* plaything, toy; *paignios,* sportive; see *paizo* under **play**

paeminosus, L. uneven, rough; see **rough**

Paeon, L. (Gr. *Paion*), physician to the gods; see **heal**

paepalimo- < Gr. *paipalimos,* sly, subtle; see **know**

paepalo- < Gr. *paipale,* fine flour, meal, pollen; see *pale* under **flour**

paetus, L. with a wink or blink of the eyes; *paetulus,* dim.; see **wink**

paganus, L. of the country, rustic; see *pagus* under **country**

page < L. *pagina,* leaf of a book; see **leaf**

pageto- < Gr. *pagetos,* frost; *pageros,* frosty, cold; see **ice**

pagina, L. page; *pagella; paginula,* dim.; *paginalis,* of a page; see **leaf**

pagios, Gr. firm, solid, steadfast; see **stand**

pagis, Gr. trap, snare; see **trap**

pagos, Gr. hill; see **mountain**

pagrus, L. (Gr. *pagros*), sea-bream; see **bream**

pagurus, L. (Gr. *pagourus*), a kind of crab; see **crab**

pagus, L. country, district; *paganus,* of the country, rustic; see **country**

pail < AS. *paegel;* see **bucket**

pain < OF. *peine,* < L. *poena* (Gr. *poine*), f. fine, penalty, punishment.
AS. *acan,* pain: ache, headache, backache, toothache, bellyache.
L. *acerbus,* painful; *acerbissimus,* excruciating; see *acidus* under **sour**
Gr. *achos,* n. pain, distress, grief, sorrow:
L. *afflictio, -onis,* f. pain, suffering, torment: affliction.
Gr. *aganaktesis,* f. pain, irritation; *aganaktetos,* vexatious; *aganaktikos,* fretful, irritable, peevish: *Aganactesis indecora* (a moth).
Gr. *agonia,* f. contest, pain, anguish, intense suffering: agony, agonize.
Gr. *alastos,* insufferable:
Gr. *algos,* n. pain; *algeros,* painful; *algesis,* f. sense of pain; *algeinos,* painful; *algion,* more painful; *algistos,* most painful; *algeo,* feel pain, suffer; *-algia,* suffix denoting pain: algometer, algolagnia, analgesic, cardialgia, cephalalgia, neuralgia, odontalgia, otalgia.
Gr. *alysis,* f. distress, anguish:
L. *ango, anctus; anxus,* cause to feel or suffer pain, vex, distress, choke; see **choke**
Gr. *chalepos,* painful, grievous; see **difficult**
L. *cruciabilis,* painful, tormenting; see **cross**
L. *dolor, -is,* m. pain, ache, distress, grief, sorrow; *dolidus,* painful; *dolorosus,* painful, full of sorrow; *dolentia,* f. pain; *cordolium,* n. sorrow at heart; *doleo,* feel pain, grieve: dolorous, doleful, condolence, indolent, dole, dolorific, Via Dolorosa, *Pardosa condolens* (a spider).
Gr. *dye,* f. pain, misery, anguish; *dyeros,* miserable: *Dyerocera gravida* (a beetle).
L. *exanclo, -atus,* suffer, endure to the end:
Gr. *gomphiasis,* toothache; see *gomphos* under **tooth**
Gr. *karebareia,* f. headache:
Gr. *kolikos,* pain in the colon; see *colon* under **intestine**

Gr. *lype*, f. pain, grief; *lyperos*, painful, distressing; *barylypos*, very sad; *ellypos*, grieving, mournful: lypemania, lypothymia, alypin, *Lyperosia irritans* (hornfly), *Lyperanthus antarcticus* (an orchid), *Lyperosomus aterrimus* (a beetle).

L. *morbeo*, be sick; see *morbus* under **disease**

Gr. *odaxo*, feel a biting, stinging pain; see *dakno* under **bite**

Gr. *odis, -inos*, f. childbirth or labor pains: parodinia.

Gr. *odyne*, f. pain; *odyneros*, painful; *polyodynos*, very painful: osteodynia, odontodynia, *Odynerus tempiferus* (a wasp).

Gr. *pascho*, suffer, feel; *pathos*, n. suffering, tender emotion; *-pathy*, denoting suffering, disease, treatment; *pathetikos*, capable of feeling, sensuous; *aeipathes*, ever-suffering; *tlepathes*, wretched: pathos, pathetic, pathology, pathogenic, apathy, sympathy, antipathy, aeipathy, psychopathic, osteopath, homeopathy. Derive: paschal.

L. *patior, passus*, suffer, endure, feel; *patiens, -entis*, suffering, enduring; *passio, -onis*, f. suffering, strong emotion; *passibilis*; *passivus*, capable of feeling or suffering; *patibilis*, endurable, sensitive; *perpetior, -pessus*, endure, suffer with patience; *perpessicius*, patient: passion, passive, patient, compatible, compassion, impatiently, *Passiflora incarnata*(passion-flower), *Impatiens biflora* (touch-me-not, jewelweed).

Gr. *ponos*, labor, pain, suffering; see *poieo* under **make**

L. *tormentum*, n. instrument for hurling stones, windlass, rack, anguish, pain; *tortor, -is*, m. twister, tormentor; *tortura*, f. torment, pain, < *torqueo, tortus*, twist: torment, tormentor, torture, tormentil *(Potentilla tormentilla)*.

Gr. *talas, -anos*, suffering, wretched; *tlemon, -os*, patient, suffering; *tlemosyne*, endurance, < *tlao*, bear, suffer; see *atlas* under **carry**

See: **disease, hurt, trouble, sad, feel**

painless

Gr. *alypos*, without pain; *alypon*, n. a plant with pain-killing properties; *pausilypos*, ending pain: alypin, *Globularia alypum* (gutwort).

Gr. *anaisthesia*, f. insensibility, loss of sensation; *anaisthetos*, insensible, without feeling: anesthesia, anesthetic.

Gr. *analgesia*, f. insensibility; *analgetos*, insensible, without pain: analgesia, analgesic.

Gr. *ananios*, without pain or grief:

Gr. *anodynos*; *pausodynos*, allaying pain; *anodynia*, f. freedom from pain; *odynephatos*, killing pain: anodyne, anodynia, anodynous.

L. *indolorius*, painless: *indoloris, -e*, free of pain; *indolentia*, f. insensibility, inactivity: indolent.

Gr. *lysipemon, -os*, ending sorrow or pain:

Gr. *nepenthes*, banishing pain and sorrow: nepenthe, *Nepenthes sanguinea* (a pitcher-plant).

AS. *numen*, taken, seized, deprived of sensation: numb, numbness, benumb, numbskull(numskull).

L. *sopio, -itus*, deprive of feeling, lull to sleep, quiet; see *sopor* under **sleep**

See: **dull, forget**

paint < OF. *peint*, < L. *pingo, pictus*, color, stain; see **color**

pair < OF. *paire*, < L. *par*, equal; see **two, equal**

pal- < Gr. *pas*, all; see **all**

pala, L. shovel, see **shovel**; socket of a ring for a jewel, bezel, see **hollow**

palabundus, L. wandering about; see *palor* under **wander**

palace < OF. *palais*, < L. *Palatium*, one of the seven hills of Rome; see **house**

palaestes, L. (Gr. *palaistes*), wrestler; see **fight**

palaga, L. ingot of gold; see **gold**

palame, Gr. palm of the hand; see *palma* under **hand**

palasia, L. buttock or rump of an ox; see **rump**

palasso, Gr. sprinkle, spot, bespatter; see **spot**

palatha, L. (Gr. *palathe*), a cake of dried fruit; see **fruit**

palatum, L. roof of the mouth; see **mouth**

palatus, L. wandered, straggled; see *palor* under **wander**

pale < L. *pallidus*, ashen, wan, see **gray**; < *palus*, stake, see **pillar**; < Gr. *pale*, wrestling, fight, see **fight**; < Gr. *pale*, fine meal, flour, pollen, see **flour**

palea, L. chaff; *palealis*, of chaff; see **scale**

palear, L. dewlap; see **flap**

palema, Gr. fine meal; see *pale* under **flour**

paleo-; **palaeo-** < Gr. *palaios*, ancient, old; see **old**

Pales, L. god of shepherds and cattle; *Palalis,* of Pales; see **cattle**

paleuma, Gr. allurement; *paleutes; paleutria,* decoy-bird; see *paleutes* under **draw**

palin-; palim- < Gr. *palin,* again, back, repetition; see **return**

Palinurus, L. (Gr. *Palinouros*), pilot of Aeneas; see **govern**

palitans, L. wandering about; see *palor* under **wander**

paliurus, L. (Gr. *paliouros*), a genus of the buckthorn family; see **buckthorn**

palla, Gr. ball, see **ball**; robe, mantle, see *pallium* under **garment**

pallacana, L. a kind of onion; see **onion**

pallacido- < Gr. *pallakis, -idos,* concubine, mistress; *pallakidion,* dim.; see *pellex* under **prostitute**

pallaco- < Gr. *pallax, -akos,* youth; *pallakos,* lover; see **young**

pallidus, L. ashen, pale, wan; *palleo,* be pale; *pallor,* paleness; see **gray**

pallium, L. mantle, robe, cloak, coverlet; *palliolum,* dim.; *palliatus,* cloaked, protected; see **garment**

pallo, Gr. shake, poise, leap; see **shake**

palm < L. *palma,* f. hand, inside of the hand, palm-tree; *palmula,* f. dim.; *palmaris; palmarius; palmeus; palmicius,* of palms; *palmosus,* full of palms; *palmatus,* like the palm of the hand or a palm leaf: palmate, palmetum, palmification, palmitate, palmitin, palmitoleic, palmitic, palmetto(*Sabal palmetto*), *Carludovica palmata* (jipijapa, source of leaves for Panama hats), *Setaria palmifolia* (palm-grass), *Rhynchophorus palmarum* (palm-weevil), *Phytophthora palmivora* (a fungus), *Palmella miniata* (an alga), *Pachyrhizus palmatilobus* (a legume).

NL. *areca,* f. a kind of palm: *Areca catechu* (betel-nut palm), arecoline.

NL. *arenga,* f. a genus of palms: *Arenga pinnata* (a sugar-palm).

NL. *attalea,* f. a genus of palms, < L. Attalus (Gr. Attalos), king of Pergamum: *Attalea funifera* (piasaba-palm), *Attalea cohune* (cohune-palm).

NL. *bactris,* f. a genus of palms: *Bactris pallidispina* (a palm).

L. *baia* (Gr. *bais*), f. palm branch:

Gr. *borassos,* m. palm fruit: *Borassus flabellifer* (palmyra-palm).

L. *caryota,* f. a palm with walnutlike fruit: *Caryota mitis* (a fishtail-palm), *Cyrtomium caryotideum* (a fern).

NL. *chamaedorea,* f. a genus of palms: *Chamaedorea pumila* (a palm).

L. *chamaerops, -opis* (Gr. *chamairops, -opos*), f. a genus of palms: *Chamaerops humilis,* var. *argentea* (a palm).

NL. *cocos,* f. a genus of palms, < Pg. *coco: Cocos nucifera* (coconut-palm), coconut, *Poria cocos* (tuckahoe), *Aphelenchus cocophilus* (a nematode).

NL. *copernicia,* f. a genus of palms, < Nikolaus Copernicus, Polish astronomer: *Copernicia cerifera* (carnauba-palm).

NL. *corypha,* f. a genus of palms: *Corypha umbraculifera* (talipot-palm).

L. *cuci* (Gr. *kouki*), n. coconut-palm: *Hyphaene*(formerly *Cucifera*) *crinita* (doum-palm).

L. *dabla,* f. a kind of Arabian palm:

NL. *daemonorops,* m. a genus of palms: *Daemonorops fissus* (a rattan-palm).

NL. *desmoncus,* m. a genus of palms: *Desmoncus horridus* (a bramble-palm).

NL. *geonoma,* f. a genus of palms: *Geonoma gracilis* (a shadow-palm).

NL. *licuala,* f. a genus of palms: *Licuala acutifida* (penang-lawyer).

NL. *livistona,* f. a genus of palms, < Livistone, Scotland: *Livistona rotundifolia* (Java palm).

NL. *lodoicea,* f. a genus of palms: *Lodoicea maldivica* (a coconut).

NL. *manicaria,* a genus of palms; see **sleeve**

Gr. *margellion,* n. a kind of palm:

NL. *nipa,* f. a kind of palm: *Nipa fruticans* (a palm).

Gr. *petele,* f. a small kind of palm:

L. *phoenix, -icis* (Gr. *phoinix, -ikos*), f. date-palm, date: *Phoenix dactylifera* (date-palm), *Phoenicophilus palmarum* (a tanager), *Pseudophoenix vinifera* (cherry-palm).

NL. *phytelephas,* a genus of palms; see **elephant**

NL. *raphia,* f. a genus of palms: *Raphia vinifera* (wine-palm).

NL. *roystonea,* f. a genus of palms, < Roy Stone, American engineer: *Roystonea (Oreodoxa) regia* (royal palm).

NL. *sabal,* f. a genus of palms, prob. < an Ab.Am. word: *Sabal parviflora* (Cuban palm).

L. *sandalis, -idis,* f. a kind of palm:

NL. *serenoa*, f. a genus of palms, < Sereno Watson, American botanist: *Serenoa repens* (saw-palmetto).

L. *syagrus*, f. a kind of palm: *Syagrus comosa* (a palm).

NL. *thrinax*, *-acis*, f. a genus of palms: *Thrinax punctulata* (a thatch-palm), *Coccothrinax radiata* (a silver palm).

NL. *washingtonia*, f. a genus of palms, < George Washington, first President of the United States: *Washingtonia robusta* (a palm).

palma, L. (Gr. *palame*), hand, palm of the hand, palm-tree; *palmula*, dim.; *palmatus*, palmlike; see **hand, palm**

palme, Gr. shield; see **shield**

palmes, *-itis*, L. young branch or shoot; see **branch**

palmos, Gr. pulsation, palpitation, quivering, throbbing; see *pallo* under **shake**

palor, *-atus*, L. wander about, straggle, stray; see **wander**

palpebra, L. eyelid, see **eyelid**; *palpebro*, *-atus*, wink, see **wink**

palpito, *-atus*, L. tremble, throb, beat, pant; see **strike**

palpo, *-atus*, L. touch, stroke, feel; *palpabilis*, touchable, that may be felt; *palpator*, stroker; *palpus*, soft palm of the hand, feeler; see **touch**

palton, Gr. dart, javelin; see **spear**

paludamentum, L. military cloak; see **garment**

paludosus, L. boggy, marshy; see *palus* under **marsh**

palumbus, L. wood-pigeon, ring-dove; see **dove**

palus, L. stake, see **pillar**; *palus*, *-udis*, marsh, swamp, see **marsh**

paluster, *-tris*, *-tre*, L. marshy, swampy; see *palus* under **marsh**

palyno, Gr. sprinkle; see **wet**

pam- < Gr. *pas*, all; see **all**

pampinus, L. tendril; *pampineus*, of tendrils; *pampinosus*, full of tendrils; see **curl**

pan < AS. *panne*, prob. < L. *patina*, f. dish, wide shallow vessel; *patella*, f. dim.; *patellarius*; *patinarius*, of a dish or pan: paten, patella, *Patella vulgata* (a limpet), *Patelloidea corticata* (a limpet), *Anisomyon patelliformis* (a fossil gastropod).
Gr. *andrachle*, f. a kind of pan or chafing-dish:
L. *batillum*, fire-pan, incense-pan; see **shovel**
Gr. *escharis*, *-idos*, f. pan of coals:
L. *fretale*, n. frying-pan:
L. *frixorium*, n.; *frixura*, f. frying-pan:
Gr. *kelebe*, cup, jar, pan; *kelebeion*, dim.; see **cup**
L. *lanx*, *lancis*, pan of a balance; see **balance**
Gr. *lekane*, dish, pot, pan; see **plate**
L. *sartago*, *-inis*, f. frying-pan:
Gr. *taganon (tegenon)*, n. frying-pan: *Tagenostola turkestanica* (a beetle), *Tagenodes mouffleti* (a beetle).
See: **plate**

Pan, Gr. god of hills, forests, pastures, flocks, sportsmen; *Paniskos*, dim.; *Panikos*, of Pan; see **herd**

pan-; **panto-** < Gr. *pas*, *pantos*, all, the whole, every; see **all**

panaca, L. a drinking-vessel; see **cup**

panacea, L. (Gr. *panakeia*), a universal remedy, heal-all; see **drug**

panax, L., Gr. a plant with cure-all porperties; see **ginseng**

pancratium, L. (Gr. *pankration*), a genus of the amaryllis family; see **amaryllis**

pancreas < Gr. *pankreas*, sweetbread; see **gland**

panctus, L. fastened; see *pango* under **bind**

pandanus, NL. a genus of monocotyledons, < Malay *pandan:* Pandanaceae, *Pandanus utilis* (common screwpine), *Eryngium pandanifolium* (an umbellifer).

Pandion, Gr. a king of Athens; now a genus of hawks; see **hawk**

pando, *pansus; passus*, L. stretch, spread, lay open; see **spread**

Pandora, Gr. fabled recipient of presents from all the gods; see *do* under **give**

pandura, L. musical instrument with three strings; see **music**

pandus, L. bent, crooked, curved; see **bend**

panel < OF. *panel;* see **hollow**

panegyris, Gr. national or festal assembly; see *agora* under **gather**

paniceus, L. of bread; see *panis* under **bread**

paniculus, L. tuft, dim. of *panus,* ear of millet; see **bush**

panicum, L. a grass; see **grass**

panis, L. bread, loaf; *panifex; panifica,* bread-maker; *panarium,* bread-basket; see **bread**

pannus, L. piece of cloth, garment, rag, patch; *panniculus,* dim.; *pannosus,* ragged, tattered; *pannuceus,* wrinkled, shriveled; see **rag, fold**

Panope, Gr. a sea-nymph; see **sea**

panoros, Gr. produced in every season; see **fertile**

pansus, L. spread out, extended; see *pando* under **spread**

panteles, Gr. complete, entire; see *telos* under **end**

pantex, L. paunch, bowels; see **belly**

panther < L. *panthera* (Gr. *panther, -os*); see **cat**

panto- < Gr. *pas, pantos,* all; see **all**

pants, dim. of pantaloon, < It. *Pantalone,* a Venetian buffoon; see **trousers**

panus, L. ear of millet; *paniculus,* dim.; see **bush**

papas, Gr. father; see **father**

papaver, L. poppy; *papaverculum,* dim.; *papavereus,* of poppies; see **poppy**

paper < OF. *papier,* < L. *papyrus* (Gr. *papyros*), m. the sedge, *Cyperus papyrus,* from whose pith paper was made; *papyrion,* n. dim.; *papyraceus; papyrius; papyrinus(papyrinos),* of papyrus; *papyrodes,* like papyrus: newspaper, papyrus, papyrine, *Betula papyrifera* (paper-birch), *Papyridea hiatus* (a gastropod).
Gr. *byblos,* papyrus; see *biblion* under **book**
L. *charta* (Gr. *charte*), f. leaf of paper, thin plate; *chartula,* f. dim.; *chartaceus; charteus,* of paper: card, charter, cartoon, chartaceous, cartridge, Magna Charta, cartographer, *Ascotricha chartarum* (a fungus).
ML. *pergamentum,* n. writing material prepared from the skins of animals, < Gr. *Pergamos,* Pergamum, a city of Asia Minor: parchment, pergameneous, *Chaetopterus pergamentaceus* (parchment-worm).
L. *plagula,* f. sheet of paper:
L. *scheda* (Gr. *schede*), f. strip of papyrus, sheet of paper; *schedula,* f.; *schedarion,* n. dim.: schedule.
Gr. *selis, -idos,* f. plank, leaf of papyrus, page of a book, sheet of paper; *selidion,* n. dim.: *Selidosemia plumaria* (a butterfly).

paphlasma, Gr. a boiling, blustering; see **boil**

papilio, L. butterfly, tent; *papiliunculus,* dim.; see **butterfly, tent**

papilla, L. nipple, teat, bud; *papillatus,* budlike; see **bud**

papio < F. *papion,* baboon; see **monkey**

pappus, L. (Gr. *pappos*), old man, grandfather; the wool, hair, bristles, setae, teeth or awns that surmount the achene in some Compositae; see **hair, kin**

paprax, Gr. probably a perch; see **perch**

papula, L. pustule, pimple; see **pimple**

papyrus, L. (Gr. *papyros*), an Egyptian sedge, paper made from it; see **paper, sedge**

par, L. equal; *parilis,* even, like; see **equal**

para, Gr. beside, near, by; see **near**

parabilis, L. easily procured; see **rest**

parabola, L. (Gr. *parabole*), comparison, parallel, allegory, analogy, tale, curve; see **story, bend,** FIGURES OF SPEECH under **form**

parabolos, Gr. bold, reckless, venturesome; see **bold**

paradigma, L. (Gr. *paradeigma*), pattern, model, plan; see **form**

paradise < L. *paradisus* (Gr. *paradeisos*), park, pleasure grounds, delightful spot; see **park, heaven**

paradoxus, L. (Gr. *paradoxos*), strange, contrary to expectation; see **strange**

paragio- < Gr. *parageios,* pertaining to shallow water; see **shallow**

paragogos, Gr. misleading, deceitful; see **lie**

paragraph < Gr. *paragraphe,* part of a composition, originally distinguished by a special character in the margin; see **part**

paraitios, Gr. accessory to; see **help**

paralios; paralos, Gr. by or near the sea; see **sea**

parallel < L. *parallelus* (Gr. *parallelos*), side by side equidistantly; see **equal**

paralysis, Gr. palsy; see **disease**

paramecium, NL. a genus of protozoans; see **protozoan**

parameco- < Gr. *paramekes,* longish, oblong; see *mekos* under **long**

paranoea < Gr. *paranoia,* derangement, madness; see **mad**

paraplesios, Gr. somewhat like; see **like**

parasite < L. *parasitus* (Gr. *parasitos*), one who eats at the table of another, guest, sponger; see **guest**

parastas, Gr. square pillar, doorpost; see **pillar**

paratus, L. prepared, equipped; *paratura,* preparation; see *paro* under **supply**

Parca, L. goddess of destiny; see **lot**

parcus, L. frugal, scanty, thrifty, penurious; *parcitas,* scantiness; see **stingy**

pardaco- < Gr. *pardakos,* damp, wet; see **wet**

pardon < OF. *pardoner,* < ML. *perdono,* spare; see **free**

pardos; pardalis, Gr. leopard, ounce, panther; *pardalotos,* spotted like a leopard; see **cat**

paredros, Gr. sitting beside, near, assistant; see **near**

paregoricus, L. (Gr. *paregorikos*), consoling, assuaging; *paregoria,* consolation; see **soothe**

parent < L. *parens, -entis,* procreator, progenitor; see **begin**

parenthesis, Gr. putting in extra, insertion; see *enthesis* under **insert**

paresis, Gr. a letting go, slackening, paralysis; see **lessen**

pareunos, Gr. lying beside or with, bedfellow; see *euneter* under **companion**

parias, L. (Gr. *pareias*), a kind of snake; see **snake**

parietalis; parietarius, L. of walls, < *paries,* wall; *parietaria,* wall-plant, pellitory; see *paries* under **wall, pellitory**

parilis, L. equal, like; see *par* under **equal**

pario- < Gr. *pareion,* cheek; see **cheek**

paritus, L. present, visible, see *pareo* under **see**; < *pario,* beget, create, produce, see **birth**

park < OF. *parc.*
L. *paradisus* (Gr. *paradeisos*), m. delightful spot, park, pleasure grounds: paradise, paradisiacal, *Paradisea apoda* (bird-of-paradise), *Paradisia liliastrum* (a lily), *Paradigalla carunculata* (a bird), *Citrus paradisi* (grapefruit), *Ptiloris paradisea* (rifle-bird), *Musa paradisiaca* (plantain-banana).
L. *vivarium,* n. preserve, park: vivarium.
See: **field, forest**

parma, L. (Gr. *parme*), a small shield; *parmula,* dim.; see **shield**

parnassia, NL. a genus of the saxifrage family; see **saxifrage**

parnops, Gr. a kind of locust; see **grasshopper**

paro, L. (Gr. *paron*), a small, light ship; *parunculus,* dim.; see **ship**

parocha, L. (Gr. *paroche*), a supplying or furnishing; see **supply**

parodos, Gr. byway; see *hodos* under **way**

paronomasia, Gr. pun; see FIGURES OF SPEECH under **form**

paronychia, Gr. whitlow, nailwort; see **sore**

paropsis, Gr. dessert, dessert-dish; see *opson* under **food**

paroros, Gr. untimely, unseasonable; see *aoros* under **unfit**

-parous, denoting production, giving birth to; see *pario* under **birth**

paroxysm < Gr. *paroxysmos,* attack, fit, spasm, throe; see **disease**

parra, L. a bird of ill omen; see **bird**

parrot < F. *perrot* (*pierrot; perruche*); Sp. *perico,* < L. *Petrus,* Peter: parakeet.
NL. *lorius,* m. < Malay *luri,* a kind of parrot: *Lorius roratus* (eclectus-parrot), lory.
Sp. *papagayo,* m. parrot:
L. *psittacus* (Gr. *psittakos; bittakos*), m. parrot; *psittacinus,* of parrots: psittacine, psittaceous, psittacinite, psittacosis, *Psittacus erithacus* (gray parrot), *Psittacula cyanocephala* (a parakeet), *Bittacus chlorostigma* (a scorpion-fly), *Eupsittula aurea* (a parrot), *Melopsittacus undulatus* (budgereegah, a parakeet), *Bittacomorpha clavipes* (a cranefly), *Apterobittacus apterus* (a mecopterid), *Hemithyris psittacea* (a brachiopod), *Rhynchopsitta pachyrhyncha* (thick-billed parrot), *Gladiolus psittacinus* (a gladiolus), *Heliconia psittacorum* (a musacead), *Cyclorrhynchus psittaculus* (parrot-auklet).

parsimonia, L. frugality; see *parcus* under **stingy**

parsley < Gr. *petroselinon;* see **celery**

parsnip < OF. *pasnaie,* < L. *pastinaca;* see **carrot**

part < L. *pars, partis,* f. piece, portion, share; *particula,* f. dim.; *partilis,* in part, single: particle, particular, partake, parcel, parse, participate, party, partition, jeopardy, impartial, departure, counterpart, *Cyclanthus bipartitus* (a cyclanthacead).

Gr. *agma; age,* fragment, splinter; see *agmos* under **break**

Gr. *aposmileuma, -tos,* n. chip, shaving:

Gr. *apospasma, -tos,* n. piece torn or broken off: *Apospasmica fasciata* (a fly).

L. *appendix,* appendage, addition; see **and**

L. *articulus,* m. member, part, division: article, articulation.

L. *assula,* f. shaving, chip, splinter; *assulosus,* in splinters: *Enallocrinus assulosus* (a Silurian crinoid).

L. *buccella,* dim. of *buccea,* morsel, mouthful; see **little**

Gr. *chnauma, -tos,* n. piece, slice:

L. *clausula,* close or end of a part of a sentence, clause; see **close**

L. *comma, -tis* (Gr. *komma, -tos*), n. part cut off, chip, piece, part of a sentence, clause, mark of punctuation; *kommation,* n. dim.; *kommatikos,* of clauses; *apokomma, -tos,* n. chip, splinter; *perikomma, -tos,* n. trimmings: comma, *Commatocerus subnitidus* (a beetle), *Polygonia comma* (a butterfly), *Vibrio comma* (bacterium of Asiatic cholera).

Gr. *daitron,* portion; see *daio* under **cut**

L. *fragmen, -inis; fragmentum,* n. bit, piece: fragment, fragmentary, fragmental.

L. *frustum,* n. bit, piece, morsel, scrap; *frustillum; frustulum,* n. dim.: frustum, frustule, frustulent, *Stereum frustulatum* (a fungus), *Lecanora frustulosa* (a lichen).

L. *incohatus (inchoatus),* only begun, incipient, incomplete: inchoate.

L. *ineffectus,* incomplete, unfinished:

L. *item,* part of a series; see **one**

Gr. *klasma,* fragment, piece, chip; see *klao* under **break**

Gr. *klerion,* small portion; see *kleros* under **lot**

Gr. *kolon,* member of anything, part of a sentence, clause; see **leg**

Gr. *kopaion; kopadion,* piece; see *kopto* under **cut**

L. *membrum,* n. part: member, dismember, membrane, *Crioceris trimembris* (a beetle).

Gr. *meros; merisma, -tos,* n.; *meris, -idos,* f. part, portion, share; *meridion,* n. dim.; *meristes,* m. divider; *merismos,* m. division; *meristos,* divided, divisible, < *merizo,* divide, split, distribute; *morion,* n. piece, portion, section, sexual member: meristem, merosome, blastomere, metamerous, americtic, *Meristobelus forcipatus* (a beetle), *Meristocarpus fuscus* (a tunicate), *Meristocrinus tuberosus* (a Silurian crinoid), *Merismopedia punctata* (an alga), *Meridion circulare* (a diatom), *Catamerus revoili* (a beetle), *Trimerus delphinocephalus* (a trilobite), *Cryptomeria japonica* (a conifer), *Moriodema paramattensis* (a beetle), *Cynomorium coccineum* (a parasitic plant), *Phaeomeria magnifica* (a zingiberacead).

Gr. *moira,* part, share, destiny; see **lot**

Gr. *paragraphe,* f. part of a composition, originally distinguished by a special character in the margin: paragraph, paragrapher.

Gr. *pelekema, -tos,* n. chip:

Gr. *pharsos,* n. part, piece, portion: *Pharsophorus lacerans* (a fossil mammal).

L. *portio, -onis,* f. part of a whole; *portiuncula,* f. dim.: portion, apportion, disproportionate, proportional, *El pueblo de nuestra senora la reina de los angeles de porciuncula* (Los Angeles, Calif.).

Gr. *psacas, -ados,* bit, crumb, morsel; see **little**

L. *recissamentum,* n. chip, shaving:

L. *reduvia,* hangnail, remnant, fragment; see **sore**

L. *resigminum,* clipping, paring; see *seco* under **cut**

Gr. *rhox, -ogos; aporrhegma,* fragment; see *rhegnymi* under **break**

Gr. *schiza,* f.; *schidax, -akos,* m. splinter; *schidion; schizion,* n. dim.: *Schidium lemur* (a bug), *Yucca schidigera* (Mohave yucca).

L. *segmen, -inis; segmentum,* n. piece, cutting, shred: segment, segmentation.

Gr. *skarphion,* fragment, splinter; see *karphos* under **stem**

Gr. *sparagma, -tos,* n. piece, shred:

Gr. *tmema,* part cut off, portion, piece; *ektmema,* section, segment; see *temno* under **cut**

Gr. *thrausma,* fragment, piece; see *thrauo,* under **break**

Gr. *thryligma, -tos,* n. fragment:

Gr. *thrymma, -tos,* n. piece, bit: *Dyschimus thrymmifer* (a trichopterid).

L. *truncus,* piece cut off, tip, end; *trunculus,* dim.; see **short**

Gr. *tryphos,* n. piece, morsel, lump: *Tryphomys adustus* (a murid).
Gr. *xysma,* shaving, scraping, particle; see *xeo* under **scrape**
See: **break, separate, cut, tear, little, trifle, lot**
parthenium, L. (Gr. *parthenion*), a plant of the composite family; see **composite**
parthenos, Gr. virgin; see **woman**
particeps, L. comrade, partner, sharer; sharing, see **companion**
particular < L. *particularis,* of parts, partial; see **choose**
partition < L. *partitio,* a dividing, that which separates, wall; *partitus,* divided; see **wall, separate**
partner < OF. *parconier,* < L. *partitio,* sharing, division; see **companion**
partridge < OF. *pertris,* < L. *perdix;* see **quail**
partus; paritus, L. given birth to, produced; *partualis,* of birth; *parturio,* bear, give birth to; see *pario* under **birth**
parulido- < Gr. *paroulis, -idos,* gumboil; see **sore**
parus, L. titmouse; see **titmouse**
parvus, L. little; *minor; minus,* less; *minimus,* least; *parvulus,* dim.; see **little**
paryphe, Gr. border woven along a robe; see **border**
pascuus, L. of pasture; *pascuum,* pasture; see *pasco* under **eat**
pasi- < Gr. *pas,* all; see **all**
pasis, Gr. gain, possession; see **gain**
pasma, Gr. a sprinkling; see *pastos* under **wet**
paspalum, NL. (Gr. *paspalos*), a kind of millet; see **grain**
passage < OF. *passer,* < L. *passus,* step; see **way, accessible, door**
passalos, Gr. peg; *passaliskos,* dim.; see **nail**
passer, L. sparrow; *passerculus,* dim.; *passerinus,* of sparrows; see **finch**
passernix, L. whetstone; see **stone**
passion < L. *passio, -onis,* a feeling, enduring, suffering; *passibilis; passivus,* capable of feeling or suffering; see *patior* under **pain**
passula, It. raisin, dried grape; see **grape**
passus, L. step, stride, pace; see **step**
past < OF. *passer,* < L. *passus,* step; see **before, old**
paste < L. *pasta* (Gr. *paste*), barley broth; see **glue, plaster**
pastillus, L. dim. of *panis,* loaf of bread; see **bread**
pastinaca, L. parsnip, carrot; see **carrot**
pastinum, L. two-pronged instrument for digging and trenching ground; see **hoe**
pastor, L. herdsman, shepherd; *pastoralis,* of shepherds, rustic; see **guard**
pastos, Gr. sprinkled; see **wet**
pastus, L. fodder, food, pasture; see *pasco* under **eat**
Pataeco- < Gr. *Pataikos,* a dwarflike Phoenician deity; *Pataikion,* dim.; see **little**
patagium, L. a gold edging or border; see **border**
patagos, Gr. clatter, rattle, crash; *patagema,* rattle; *patagetikos,* clattering; see **rattle**
patane, Gr. flat dish; *patanion,* dim. see **plate**
patasso, Gr. beat, throb; see **strike**
patch < uncertain origin.
  L. *assumentum,* n. patch:
  L. *centunculus,* small patch; see *cento* under **rag**
  L. *pannus,* piece of cloth, rag, patch; see **rag**
  Gr. *rhytisma, -tos,* n. patch: *Rhytisma acerinum* (a fungus).
  L. *splenium* (Gr. *splenion*), plaster or patch with a medicinal preparation; see **spleen**
  See: **sew, part, rag**
patella, L. small pan, dish, kneepan; *patellarius,* of a pan; see **pan**
patens, L. open, exposed; see *pateo* under **open**
pater, L., Gr. father; *paterculus,* dim.; *paternus; patrius,* of a father; see **father**
patera, L. saucer; see **plate**
patetes, Gr. treader; see *peripatos* under **walk**
path < AS. *path;* see **way**
pathetic < Gr. *pathetikos,* capable of feeling, sensuous; see **feel**

patho-; pathy, denoting disease, suffering, < Gr. *pathos,* suffering, tender emotion; see *pascho* under **pain, disease**

patibilis, L. endurable; see *patior* under **pain**

patibulum, L. fork-shaped yoke; see **bind**

patiens, L. suffering, enduring; see *patior* under **pain**

patina, L. a wide, shallow vessel, dish; *patella,* dim.; *patinarius,* of a pan; see **pan**

patos, Gr. trodden path, way; see **way**

patratus, L. accomplished; see *patro* under **make**

patria, L. fatherland, native country; *patrioticus,* of one's country; see **country**

patrimonium, L. inheritance; see **receive**

patro- < Gr. *pater, patros,* father; *patrios,* of a father; see **father**

patronus, L. protector, supporter; *patronalis,* of a supporter; see **guard**

patruus, L. uncle; *patruelis,* of an uncle, cousin; see **kin**

pattern < OF. *patron;* see **form, equal**

patulus, L. open, spread out, broad; see *pateo* under **open**

paucus, L. few, little; *pauculus; pauxillus; pauxillulus,* dim.; see **few**

paula, Gr. rest, pause; see *pausa* under **rest**

paulus, L. little; *paululus,* dim.; see **little**

pauper, L. poor; *pauperculus,* dim.; *paupertas,* poverty; see **poor**

pauros, Gr. little, few; see **little**

pause < L. *pausa* (Gr. *pausis*), halt, rest; see **rest**

pausilypos, Gr. ending pain; see **painless**

pavicula, L. rammer; see *pavio* under **strike**

pavidus, L. trembling, quaking, fearful; *pavibundus,* anxious, fearful; see **fear**

pavimentum, L. floor of stones; see **base**

pavo, L. peacock; *pavonaceus; pavoninus,* of peacocks; see **peacock**

pawpaw (papaw) < Sp. *papaya,* < Ab.Am. origin: *Carica papaya* (papaya).
  NL. *annona,* f. a genus of the custard-apple family, < Ab.Am. *anon,* custard-apple: Annonaceae, anonol, anoncillo, *Annona muricata* (soursop).
  NL. *asimina,* f. a genus of the custard-apple family, < Ab.Am. *rassimina,* pawpaw: *Asimina triloba* (pawpaw).

pax, *pacis,* L. peace, tranquillity, rest; see **rest**

paxillus, L. peg, small stake; see **nail**

pay < OF. *paier,* < L. *paco, -atus,* appease, pacify, < *pax, pacis,* peace.
  Gr. *agertes,* m. collector of dues:
  L. *agraticum,* n. land tax:
  L. *alimonium,* sustenance, allowance for support; see **receive**
  Gr. *allagma,* that which is given in exchange, price; see **trade**
  Gr. *amoibe,* exchange, recompense, payment; *amoibaios,* retributive, interchanging; see **change**
  Gr. *analoma, -tos,* n. cost, expense; *analotikos,* expensive:
  Gr. *anaprasso,* exact, levy, collect:
  Gr. *antellogos,* m. compensation:
  L. *apocha* (Gr. *apoche*), f. receipt:
  L. *arrha,* f.; *arrhabo, -onis* (Gr. *arrhabon, -os*), m. down-payment, pledge; *arrhalis,* of a pledge:
  L. *capitatio, -onis,* f.; *capitulare,* n. poll tax:
  L. *caritas,* dearness, costliness, high price; see *carus* under **love**
  L. *chartula,* bill; see *charta* under **paper**
  Gr. *chreos,* n. debt, obligation; *chreostes,* m. debtor: *Chreonoma venusta* (a beetle), *Chreostes ephippiatus* (a beetle).
  L. *collybus* (Gr. *kollybos*), a small coin, rate of exchange; *kollybistes,* money-changer; see **money**
  L. *columnarium,* n. pillar tax:
  Gr. *dapanos,* extravagant, lavish, expensive; *dapane,* f. expense, cost: *Dapanus cinctorius* (a wasp), *Dapanoptera plenipennis* (a tipulid).
  Gr. *dasmos,* m. impost, tax, tribute: *Dasmophora xeropila* (a moth).
  L. *debeo, debitus,* owe; *debitor, -is,* m. one bound to fulfil his obligation; *debitum,* n. what is owed: debt, debtor, indebtedness, due, duty, debenture, endeavor, dyvour.
  L. *decima,* tenth part, tithe; *decimanus; decumanus,* tithe-gatherer; see *decem* under **ten**
  L. *dego,* spend, pass, continue:

Gr. *ellimenistes,* m. collector of harbor dues: *Ellimenistes rusticus* (a beetle).
L. *emolumentum,* n. reward, gain, profit, pay: emolument.
Gr. *epibathron,* n. fare:
Gr. *epicheiron,* n. wages, reward:
Gr. *eponion,* duty, tax; see *oneomai* under **trade**
Gr. *eranos,* m. contribution to a common fund; *eranikos,* of a contribution:
L. *erogo, -atus,* pay out, expend, disburse; *erogator, -is,* m. disburser: erogate, supererogatory.
Gr. *euteles,* cheap, worthless: *Euteles corollana* (a moth), *Philobota eutelopsis* (a moth).
L. *exactor,* demander, tax-collector; see **push**
Gr. *exilasis,* f. atonement:
Sp. *finta,* f. tax: finta, *Alosa finta* (thwaite-shad).
Gr. *habrotimos,* delicate and costly; see *habros* under **beauty**
L. *honorarium,* n. reward, fee: honorarium.
L. *imputo, -atus,* reckon, charge, ascribe: impute, imputation.
L. *interest,* premium paid for the use of money; see **gain**
Gr. *karkadon, -os,* f. Charon's ferry fee:
Gr. *katatheke; katathesis,* f. deposit, down-payment:
Gr. *komistron,* n. pay, reward:
Gr. *latron,* n. pay, hire:
L. *libripens, -endis,* m. paymaster:
L. *luitio, -onis,* f. payment:
Gr. *lytron,* n. ransom; *lytrosis,* f. redemption; *lytrotes,* m. ransomer, redeemer; *lytroo,* hold for ransom: *Lytrophila phaulopa* (a moth), *Lytrosis unitaria* (a moth).
L. *merces, -edis,* f. wages, hire, salary, fee, reward; *mercedula,* f. dim.; *mercenarius,* doing merely for pay: mercenary, amerce, mercy, unmerciful, *Venus mercenaria* (quahog).
Gr. *misthos,* m. pay, wages; *mistharion,* dim.; *mistharnetikos,* of hired work, mercenary; *misthotikos,* of hiring; *misthoma, -tos,* n. contract price; *misthotos,* hired; *misthosis,* f. a hiring; *misthodotes,* m. hirer, payer: *Misthodotes obtusus* (a fossil ephemerid).
L. *mittendarius,* m. provincial tax-collector:
L. *mulcta; multa,* f. fine, penalty; *multaticius,* of fines: mulct.
L. *multinummus,* costly, expensive:
L. *naulum* (Gr. *naulon*), n. fare:
Gr. *opheilo,* owe; *opheiletes,* m. debtor; *opheilema, -tos,* n. debt.
L. *pario, -atus,* balance an account, pay in full:
Gr. *paragogion,* n. toll paid by ships:
L. *pendo, pensus,* weigh, pay; *pensio, -onis,* f. payment; *dispendium,* n. expense, cost, loss; *dispenso, -atus,* disburse, pay out; *expensum,* n. payment; *impendium,* n.; *impensa,* f. outlay, cost, charge, interest; *stipendium,* n. tax, tribute, pay: pension, compensation, dispensation, expend, expenditure, stipend, Spenser.
L. *portorium,* n. customs, duty, excise, toll:
L. *prebenda; prebita,* f. allowance for support: prebend, prebendal, prebendary.
L. *pretium,* n. worth, value, price, money, pay; *pretiosus,* valuable, costly, expensive; *impretiabilis,* priceless, invaluable: price, prize, precious, appreciate, depreciation, priceless, appraise, apprise, praise, *Bryobia pretiosa* (almond-mite), *Scala pretiosa* (a gastropod), *Ruvettus pretiosus* (escolar).
Gr. *prodosis,* f. advance money:
L. *remunero, -atus,* pay, reward, recompense: remunerate.
L. *salarium,* n. salt-money, stipend, pension: salary.
L. *satisfacio,* content, pay, make amends; see *satis* under **abundance**
L. *solvo,* loosen, free; *exsolvo; persolvo,* pay debts; see **free**
L. *sumptus,* m. cost; *sumptuosus,* costly; *insumptum,* n. expense: sumptuous, *Gymnopleurus sumptuosus* (a beetle).
L. *susceptor, -is,* m. receiver, collector of taxes:
Gr. *synkoition,* n. harlot's fee:
Gr. *syntimesis,* f. estimate of value, price:
L. *taxo, -atus,* rate, appraise, estimate, reproach: tax, taxation, task.
Gr. *telos,* n. tax, duty, toll; *telesma, -tos,* n. payment; *telones* (L. *telonarius*), m. collector of tolls; *telonion,* n. toll-house, custom-house; *telonikos,* of excise; *atelia,* f. tax exemption; *polyteles,* very costly: talisman, philately, philatelist.
Gr. *tino,* pay; *tisis,* f. payment; *ektisma, -tos,* n. penalty, < *ektino,* pay in full: *Tinosaurus stenodon* (a fossil reptile), *Tisiphone rhodostoma* (a snake).
L. *toculio, -onis,* m. usurer:
L. *tributum,* n. tax contribution, < *tribuo, -utus,* allot, bestow assign, give, divide: tribute, contribution, distribute, attribute, contributory, tributary, retribution.
L. *usura,* interest on money lent; see **gain**

L. *vadimonium*, bail, security; see *vador* under **bind**
L. *vectigal*, *-is*, n. tax, impost, duty, revenue, tribute; *vectigalis*, of taxes:
L. *viaticum*, n. travel-money, fare; *viaticulum*, n. dim.: viaticum.
See: **give, trade, punish, pain, justice, honor, atone**

pea < AS. *pise*, < L. *pisum*; see **bean**

peace < OF. *pais*, < L. *pax, pacis*; see **rest, harmony, painless**

peach < OF. *pesche*, < L. *persica*, f. < *Persicus*, Persian: *Amygdalus persica* (peach),
Campanula persicifolia (peachleaf-bellflower), *Fusanus persicarius* (a santalacead), *Lecanium persicae* (peach-scale), *Dianthidium persicolum* (a bee), *Polygonum persicaria* (a polygonacead).
L. *amygdalus* (Gr. *amygdalos*), f. almond-tree; *amygdala* (Gr. *amygdale*), f.
almond; *amygdalion*, n. dim.: amygdaloid, amygdule, almond (*Amygdalus communis*), *Amygdalophyllum inopinatum* (a coral), *Ambonychia amygdalina* (a
fossil pelecypod), *Caryocar amygdaliferum* (a souari-nut), *Salix amygdaloides*
(a willow).
Gr. *barbilos*, f. wild peach-tree:
Gr. *mykeros*, m. almond:
See: **plum**

peacock < AS. *pea*, < L. *pavo*, *-onis*, m.; *pavoninus*, of peacocks: pavonine, *Pavo
cristatus* (peacock), *Pavoncella pugnax* (ruff), *Pedilanthus pavonis* (a slipper-
flower), *Papaver pavoninum* (peacock-poppy), *Crenilabrus pavo* (peacock-fish),
Saturnia pavonia (emperor-moth).
L. *Argus* (Gr. *Argos*), m. hundred-eyed guardian of Io, whose eyes became the
spots in the peacock's tail: *Cypraea argus* (a gastropod), *Argusianus bipunctatus*
(an argus-pheasant), *Argulus foliaceus* (carp-louse).
Gr. *katreus*, m. an Indian peacock: *Catreus ornaticollis* (a grasshopper).
Gr. *meleagris*, *-idos*, f. guinea-fowl: *Meleagris gallopavo* (turkey), *Meleagrina
margaritifera* (pearl-oyster), *Fritillaria meleagris* (checkered lily).
NL. *numida*, f. the genus of birds including the guinea-fowl, < Numidia, in
northern Africa: *Numida meleagris* (guinea-fowl).
Gr. *taos*, m. peacock: *Taonurus caudagalli* (a problematical fossil).

peak < AS. *pic*, point; see **top**

pear < AS. *peru*, < L. *pirum* (*pyrum*), n.; *pirus* (*pyrus*), f. pear-tree; Sp. *pera*, f.
pear; *perilla*, f. dim.: pyriform, *Pyrus communis* (pear), *Pyrola rotundifolia*
(European pyrola), perry, *Acrocrinus pirum* (a Pennsylvanian crinoid), *Corbula
pyriformis* (a fossil pelecypod), *Macrocystis pyrifera* (giant kelp), *Contarinia
pyrivora* (pear-midge), *Pyrula ficus* (a fig-shell), *Psylla pyricola* (pear-psylla),
Venturia pyrina (pear-scab), *Erythrospermum pyrifolium* (a flacourtiacead).
Gr. *achras*, *-ados*, f. wild pear: *Achras sapota* (sapodilla), *Achradocrinus ventrosus*
(a Devonian crinoid).
Gr. *apios*, f. pear-tree; *apion*, n. pear; *apidion*, n. dim.: *Apios tuberosa* (groundnut,
wild bean), *Apiocephalus punctipennis* (a beetle).
Gr. *ochne* (*onchne*), f. pear-tree, pear: *Ochna arborea* (Cape ochna).
L. *phocis*, *-idis*, f. a kind of pear-tree:

pearl < OF. *perle*.
L. *bacatus*, set or adorned with pearls:
L. *margarita*, f. (Gr. *margarites*, m.), pearl; *margaritatus*, adorned with pearls;
margarodes, pearllike: margarite, oleomargarine, Margaret, Madge, Padge, Peg,
Margarita(*Margarites*) *helicina* (a gastropod), *Margaroptilon woodwardi* (a
Jurassic bug), *Margarodes formicarum* (a bug), *Anaphalis margaritacea* (pearly
everlasting), *Meleagrina margaritifera* (pearl-oyster).

peat < uncertain origin; see **earth**

pebble < AS. *papol*; see **stone**

peccary < Sp. *pecari*, < Ab.Am. *pakira*, a piglike animal; see **hog**

pecco, *-atus*, L. err, sin, commit a crime; *peccamen*, fault, sin; see **fault**

pechys, Gr. forearm; see **arm**

peco- < Gr. *pekos*, hide, rind, skin; see **skin**

pecoralis, L. of cattle; *pecorarius*, farmer; *pecorosus*, rich in cattle; see *pecus*
under **cattle**

pecten, L. comb; *pectinatus*, comblike, see **comb**; scallop, see **mollusk**

pectis, L. a kind of plant; now a genus of the composite family; see **composite**

pecto- < Gr. *pektos*, fixed, compacted, congealed; see *pegma* under **thick**

pectus, *-oris*, L. breast, chest; *pectoralis*, of the breast; see **breast**

pecualis; **pecudalis**, L. of cattle; see *pecus* under **cattle**

peculator, L. defrauder, embezzler; see *peculor* under **steal**

peculiar < L. *peculiaris*, one's own, singular; see **alone, different, strange, new**

pecunia, L. money, property; *pecuniarius*, of money; *pecuniosus*, rich; see **money**

pecus, *-oris*, L. cattle collectively; *pecus*, *-udis*, cattle singly; herd, flock, animal; *pecusculum*, dim.; see **cattle**

ped-; **peda-**; **pedi-**; **pedo-** < Gr. *pais, paidos*, child, see **child**; < L. *pes, pedis*, foot; *pedalis*, of the foot, see **foot**

pedalion, Gr. rudder; see **rudder**

pedamen, L. stake, prop; see **pillar**

pedamos, Gr. short; see **short**

pedema, Gr. leap, spring; *pedesis*, a leaping; see *pedao* under **leap**

pedestal < L. *pes, pedis*, foot, and G. *stal*, a stable or standing place; see **base**

pedetes, Gr. leaper; *pedetikos*, good at leaping, see *pedao* under **leap**; fetterer, prisoner, < *pede*, fetter, see **pen**

pedica, L. fetter, shackle, anklet, bangle; see **bind**

pedicel < L. *pedicellus*, dim. of *pes*, foot; see **stem**

pediculus, L. dim. of *pedis*, louse; *pedicularis*, of lice; *pedicosus*; *pediculosus*, full of lice, lousy, see **louse**; dim. of *pes*, foot, see **foot**, *pedunculus* under **stem**

pedilon, Gr. sandal; see **shoe**

pedinos, Gr. flat, level; see **level**

pedio- < Gr. *pedion*, flat, open country, dim. of *pedon*, earth, ground, see *pedinos* under **level**; *pedion*, dim. of *pede*, anklet, shackle, see *pedica* under **bind**; vulva, see **vulva**

pedo, *-itus*, L. break wind; see **smell**

pedon, Gr. ground, earth, see **earth**; *pedon*; *pedos*, blade of an oar, oar, see *pedalion* under **rudder**

pedum, L. shepherd's crook; see **rod**

peduncle < L. *pedunculus*, dim. of *pes*, foot; see **stem, foot**

peel, prob. < AS. *peolian*, and OF. *peler*; see **bark, scale, shell, skin**

peg, prob. < D. *pegge*; see **bind, close, nail**

peganum, L. (Gr. *peganon*), rue; see **rue**

Pegasus, L. (Gr. *Pegasos*), fabled winged horse of the Muses; see **horse**

pege, Gr. water, stream, spring, tear; see **spring**

pegma, Gr. anything congealed, fastened, fixed; see **thick**

pegos, Gr. strong, solid; see **strong**

pegris, L. a sea-mussel; see **mollusk**

pejor, L. worse; see *malus* under **bad**

pel- < L. *per*, through; see **through**

pelagos, Gr. sea; *pelagios*, of the sea; see **sea**

pelamis, L. (Gr. *pelamys*), tunny; see **tunny**

pelamos, Gr. any liquid of thick consistency; see **flow**

pelargonium, NL. a genus of the geranium family; see **geranium**

pelargos, Gr. stork; see **stork**

pelates (pelastes), Gr. approacher, neighbor, hireling; see **neighbor**

pelecanto- < Gr. *pelekas, -antos*, woodpecker; see **woodpecker**

pelecemato- < Gr. *pelekema, -tos*, chip; see **part**

peleco- < Gr. *pelex, -ekos*, helmet; see **cap**

pelecy- < Gr. *pelekys*, ax, hatchet; *pelekion*, dim.; see **ax**

pelethos, Gr. human dung; see **dung**

pelia < Gr. *peleia*, dove; see **dove**

pelica, L. (Gr. *pelika*; *pelike*), bowl, basin, cup; see *pella* under **cup**

pelican < L. *pelecanus* (Gr. *pelekan, -os*), m. a water-bird: *Pelecanus erythrorhynchus* (American white pelican), *Pelecanoides exsul* (a diving-petrel), *Eurypharynx pelecanoides* (pelican-fish), *Aporrhais pes-pelecani* (a gastropod).
   Sp. *alcatraz*, m. pelican: Alcatraz.
   L. *cofanus*, m. pelican:
   L. *onocrotalus* (Gr. *onokrotalos*), m. pelican: *Pelecanus onocrotalus* (a pelican).

pelichne, Gr. cup, bowl; see *pella* under **cup**

pelico- < Gr. *pelikos*, how large, how great, of what age; see **large**

pelidnos; **pelios**, Gr. black and blue, livid; see **blue**

pelinos, Gr. pertaining to clay; see *pelos* under **earth**

pella; pellis, Gr. bowl, milk-pail; see **cup**

pellaea, NL. a genus of ferns; see **fern**

pellatus, L. called; see *pello* under **call**

pellax, -acis, L. seductive, deceitful; *pellicio, -ectus,* allure, coax, decoy, entice, inveigle, wheedle; see *lacio* under **draw**

pellex, L. concubine, mistress; see **prostitute**

pellis, L. skin; *pellicula,* dim.; *pelliceus; pellinus,* of skins; *pellio, -onis,* furrier; *pellitus,* covered with skins; see **skin**

pellitory < OF. *paritoire,* < L. *parietaria,* f. wall plant: *Parietaria officinalis* (wall-pellitory).
L. *amnacum,* n. pellitory:
L. *clybatis* (Gr. *klybatis*), f. a kind of pellitory:
Gr. *helxine,* f. probably a kind of pellitory: *Helxine soleiroli* (babytears).
Gr. *kissanthemon,* n. probably a kind of pellitory:
L. *perdicium* (Gr. *perdikion*), n. pellitory:

pello, *pulsus,* L. beat, drive, push; see **push**

pellos, Gr. dark-colored, dusky; see **black**

pellucidus, L. clear, transparent; see *lucidus* under **clear**

pelma, Gr. sole of the foot; see **sole**

pelo- < Gr. *pelos,* clay, mud; *pelinos,* of clay; *pelopios,* potter; see **earth**

peloris, Gr. giant mussel; see **mollusk**

peloros, Gr. huge, monstrous, prodigious, < *pelor,* prodigy, monster; see **large**

pelta, L. (Gr. *pelte*), small shield; *peltarion,* dim.; see **shield**

peltastes, Gr. soldier armed with a small shield; see **army**

pelvis, L. basin; *pelvicula,* dim.; see **basin**

pelyco- < Gr. *pelyx, -ykos,* bowl, basin, pelvis; see **basin**

pema, Gr. suffering, woe; see **trouble**

pemma, Gr. pastry; see **cake**

pempheris, Gr. a kind of fish; see **fish**

pemphigo- < Gr. *pemphix, -igos (pemphis, -idos),* bubble; see **bubble**

pemphredon, Gr. a kind of wasp; see **wasp**

pempobolon, Gr. five-pronged fork; see **fork**

pemptos, Gr. sent; *pempsis,* mission, see *pempo* under **send**; fifth, see *pente* under **five**

pen < OF. *penne,* < L. *penna,* feather; an instrument for writing.
L. *calamus* (Gr. *kalamos*), reed, pen; see **reed**
Gr. *graphis, -idos,* f. stilus, pen; *grapheion,* n. pencil, pen; *grapheidion,* n. dim.: *Graphidium rudicaudatum* (a nematode), *Aulographis serrulata* (a radiolarian), *Graphis scripta* (a lichen).
Gr. *skariphos,* m. pencil, stilus:
L. *stilus* (incorrectly *stylus*), m. pointed instrument used for writing on waxen tablets: stilus (stylus), stilletto, stylet, *Styliola striatula* (a fossil pteropod), *Styliolina fissurella* (a fossil pteropod), *Stylatula elongata* (a sea-pen), *Stylonurus excelsior* (a eurypterid).
See: **write, reed, feather**

pen < AS. *penn,* inclosure, cage: pigpen, penknife, penstock.   Derive: penitentiary.
Gr. *anankaion,* n. prison:
Gr. *apokleisma, -tos,* n. guardhouse; *apokleistos,* shut off, enclosed:
L. *cancellarius,* kept behind bars; m. doorkeeper:
L. *capsus,* m. pen or enclosure for animals:
L. *captivus,* taken prisoner; m. a prisoner: captive, captivity.
L. *carcer, -eris,* m. (Gr. *karkaron,* n.), jail, prison; *carceralis,* of prison; *carcerarius,* m. jailer: incarcerate.
L. *cavea,* f. cage, stall, den, coop, hive, enclosure, cavity, < *cavus,* hollow; *caveatus,* encaged, cooped up: jail, gabion, gabionade.
L. *clathrum,* bar or grate making a cage for animals; see *clathratus* under **screen**
L. *cohors, -ortis,* f. enclosed space, pen, company: cohort, court, courtier, cortege, courteous, Curtis, courtmartial.
Sp. *corral,* yard, pen: corral.
L. *custodia,* guard, guardhouse; see *custos* under **guard**
Gr. *desmotes,* m. prisoner; *desmoterion,* n. prison:

L. *ergastulum,* n. workhouse, house of correction, penitentiary:
L. *hara,* f. pen, coop, sty:
Gr. *heirkte,* f. enclosure, prison; *heirgmos,* m. cage, prison:
Gr. *herkane (horkane),* fence, enclosure; see *herkos* under **wall**
Gr. *hyollos,* m. pigsty:
Gr. *klobos,* m. bird-cage; *klobion,* n. dim.:
L. *obses,* hostage, pledge, security; see **promise**
Gr. *pedetes,* m. fetterer, prisoner:
L. *phylaca* (Gr. *phylake*), prison; *phylakistes,* jailer; *phylakterion,* safeguard; see *phylax* under **guard**
Gr. *sekos,* m. pen, fold, shrine; *sekites,* m. one kept in the fold:
Gr. *zogrion,* n. menagerie, zoo:
See: **bind, guard, close, box, hold, house**

penalty < L. *poena,* punishment, pain; see **punish, pain, pay**

penarius, L. storehouse, granary; see **store**

Penates, L. household gods, hearth, home; see **house**

pencil < L. *penicillum; penicillus,* dim. of *penis,* tail; see **pen, bush**

pendens; pendulus, L. hanging, < *pendo, pensus,* hang, weigh; see **hang**

pendigo, L. internal tumor; see **swell**

pene (paene), L. almost; see **near**

Penelope, Gr. faithful wife of Ulysses; see *odysis* under **anger**

penelops, Gr. a kind of duck; see **duck**

peneto- < Gr. *penes, -etos; penestes,* poor man, day-laborer; see **poor**

penetralis, L. entering; see *penetro* under **enter**

penguin < uncertain origin.
NL. *aptenodytes,* m. a genus of penguins: *Aptenodytes patagonicus* (a penguin).
NL. *spheniscus,* m. a genus of penguins: *Spheniscus magellanicus* (a jackass-penguin).

penichros, Gr. needy, poor; *penia,* poverty; see **poor**

penicillum; penicillus, L. painter's brush, pencil, tuft, < *peniculus,* dim. of *penis,* tail; see **bush**

penico- < Gr. *penike,* false hair, wig; see **hair**

peniculus, L. dim. of *penis,* tail; see **penis**

peninsula, L., f. a relatively long, narrow body of land almost surrounded by water: peninsular, *Thanatus peninsularis* (a spider).
Gr. *chersonesos,* f. peninsula: chersonese.

penion, Gr. bobbin, spool, thread on the bobbin; see *penos* under **weave**

penis, L., m. tail, male copulatory organ, *membrum virile; peniculus,* m. dim.; *penirus,* tailed: penis, penitis, *Peniculus clavatus* (a copepod), *Phaneropsolus longipenis* (a trematode).
Gr. *akrobystia,* f. foreskin: acrobystitis.
Gr. *akroposthia,* f. foreskin:
Gr. *balanos,* f. acorn, head or glans of the penis: balanitis.
L. *colis, -is,* m. penis; *coleatus,* of the penis:
Gr. *drilos,* m. penis: *Ocnerodrilus occidentalis* (a worm), *Eudriloides parvus* (a worm).
Gr. *epagogion,* n. foreskin:
L. *fascinum,* n. penis; *Fascinus,* m. the penis as a deity: *Fascinus typicus* (a gastropod).
Gr. *kontilos,* penis, dim. of *kontos,* long pole; see **rod**
Gr. *leko,* f. penis:
Skt. *linga; lingam,* a phallic symbol: Lingayat, *Linga columbella* (a pelecypod), *Phoma lingam* (fungus causing cabbage-blackleg). Derive: yoni. Explain 'pillar' and 'grove' in Genesis 28:18; 2 Kings 17:10; Isaiah 17:8.
L. *mentula,* f. penis: *Mentula marina* (a holothurian), *Ascidiella mentula* (a tunicate), *Camelostrongylus mentulatus* (a nematode).
L. *Mutinus,* m. a name for Priapus, hence penis: *Mutinus caninus* (a stinkhorn-fungus).
Gr. *olisbos,* m. penis:
Gr. *peos,* n. penis: peotomy, peophobia.
Gr. *phallos,* m. penis; *phalletarion,* n. dim.; *phallikos,* of the penis: phallic, phallism, phallitis, *Phallocerus caudomaculatus* (a fish), *Phallogaster saccata* (a fungus), *Amorphophallus titanum* (a Sumatran arum), *Gymnophallus deliciosus* (a trematode), *Solenophallus ctenophorus* (a fish), *Amanita phalloides* (deadly amanita), *Capparis cynophallophora* (Jamaica caper).

AS. *pintel,* penis: pintle, cuckoopint *(Arum maculatum).*

L. *pipinna,* f. small penis:

Gr. *posthe,* f. penis; *posthion,* n. dim.; *posthon, -os,* m. one with a large penis: posthetomy, posthitis, *Posthon gracilis* (a fossil fly), *Actinoposthia caudata* (a worm).

L. *praeputium,* n. foreskin: prepuce.

L. *Priapus* (Gr. *Priapos),* m. god of reproduction, represented with a large penis, hence a name for that organ: *priapiskotos,* like a penis; *priapismos,* m. morbid erection: priapic, priapism, *Priapulus caudatus* (a worm).

Gr. *psolos,* m. one circumcised or with the prepuce drawn back, penis: *Psolus operculatus* (a holothurian), *Thyonepsolus nutricans* (a holothurian).

Gr. *sathe,* f. penis; *sathon, -os,* m. one with a large penis:

L. *stamen, -inis* (Gr. *stemon, -os),* thread, fiber, male organ of a flower; Gr. *stema, -tos,* penis; see **thread**

L. *verpa,* f. penis; *verpus,* m. a circumcised man: *Verpa bohemica* (a fungus).

L. *vomer,* plowshare, penis; see **plow**

penitent < L. *poenitens, -entis,* sorry, contrite; see **sad**

penitus, L. internal, within; see **in**

penna; pinna, L. feather, wing, arrow, pen; *pennula; pinnula,* dim.; see **feather**

penos, Gr. web; *Penelope,* weaver; see **weave**

pensilis, L. hanging; *pensus,* hung, weighed, < *pendo, pensus,* hang, consider; see **hang**

pensio, L. payment; see *pendo* under **pay**

penta- < Gr. *pente,* five; see **five**

pentheros, Gr. father-in-law; see **kin**

penthos, Gr. sorrow; *penthesis,* mourning; *pentheter,* mourner; see **sad**

penula, L. mantle, cloak; see **garment**

penuria, L. want; see **poor**

penus, L. inmost part, sanctuary, storeroom, provisions; see *penarius* under **store**

peony < OF. *pione,* < L. *paeonia* (Gr. *paionia),* f.: *Paeonia officinalis* (peony).

Gr. *glykyside,* f. peony:

Gr. *orobax,* f. peony:

Gr. *pentorobon,* n. peony:

Gr. *selenion,* n. peony:

Gr. *theodonion,* n. peony:

people < OF. *poeple,* < L. *populus,* m. the people; *popellus,* m. dim., rabble, mob; *populatio, -onis,* f. a people; *popularis,* of the people; *populosus,* many peopled, numerous; *publicus,* of the people, common, open; *publico, -atus,* make common or known to all, disclose, impart; *respublica,* f. commonwealth, state: popular, populace, population, public, John Q. Public, pueblo, S.P.Q.R., publish, populous, publication, publisher, publicist, depopulate, republican, *Uloborus republicanus* (a spider).

L. *civis,* c. citizen; *civicus; civilis,* of citizens; *civitas, -atis,* f. citizenship, the body-politic; *concivis,* m. fellow-citizen: civil, civilian, civilization, citizen, city, citadel, incivility, uncivilized.

L. *compatriota,* m. fellow-citizen: compatriot.

Gr. *demos,* m. the people; *demios; demotikos,* of the people, common, popular; *demokratikos,* popular; *demosios,* belonging to the people: democracy, democratic, demotic, demagogue, deme, Demosthenes, epidemic, *Papilio demodocus* (a butterfly), *Eudemoticus soror* (a fly).

AS. *folc,* people: folk, folkland, folklore, folk-etymology, kinsfolk, Norfolk, Suffolk.

Gr. *laos,* m. the people; *laikos,* of the people; *laodikos,* tried by the people: lay, laity, laic, layman, liturgy, Laodice, Menelaus, *Laomedea angulata* (a hydrozoan), *Sosilaus spiniger* (a spider).

L. *municeps, -ipis,* c. inhabitant of a town, burgher, citizen:

L. *persona,* f. mask, person; *personalis,* of a person: person, personal, personnel, personify, impersonate, parson, parsonage.

L. *plebs, plebis,* f. the common people; *plebeius,* of the people, common: *plebicola,* m., one who courts the favor of the common people, a friend of the people: plebeian, plebiscite, *Rhyparochromus plebeius* (a bug), *Gentiana plebeia* (a gentian).

Gr. *polites,* citizen; *politikos,* of citizens; *sympolites,* fellow-citizen; see *polis* under **town**

Gr. *polloi,* the many; see *polys* under **number**

L. *turba,* disorderly crowd; see **turn**

L. *vulgus,* the people, public, rabble; see *vulgo* under **common**

peos, Gr. penis; see **penis**

pepiro- < Gr. *pepeiros,* ripe; see **ripe**

peplis, Gr. name for several plants; now a genus of the loosestrife family; see **loosestrife**

peplum; peplus, L. (Gr. *peplos*), robe, tunic; see **garment**

pepo, L. (Gr. *pepon*), melon, gourd; ripe, mellow, tender; see **melon, ripe**

pepper < AS. *pipor,* < L. *piper, -eris* (Gr. *peperi*), n.; *piperatus,* peppered: piperine, peperino, piperitone, piperonal, piperonyl cyclohexenone, piperylene, piperidine, pimpernel, peppermint, peppercorn, pebrine, paprika, pepper *(Piper nigrum), Piper betle* (betel-pepper, the leaf of which with lime and the betel-nut, *Areca catechu,* is chewed), *Peperomia pellucida* (shiny peperomia), *Polygonum hydropiper* (a smartweed), *Eucalyptus piperita* (peppermint-eucalyptus), *Xanthoxylum piperitum* (Japanese prickly-ash).
   NL. *capsicum,* n. chili pepper: capsicin, capsaicin, *Capsicum frutescens* (chili, red pepper).
   Sp. *pimienta,* f. pepper; *pimiento,* m. capsicum: pimienta, pimiento, pimento (from *Capsicum tetragonum,* paprika).

peptos, Gr. cooked; *pepsis,* digestion; *peptikos,* of digestion; see *pepto* under **digest**

per- < L. *per,* through, by, very; see **through, very**

pera, Gr. pouch, wallet; *perula; peridion,* dim., see **bag;** beyond, across, more than, see **over**

peraeo- < Gr. *peraios,* beyond, across, apposite; see *pera* under **over**

perasimos, Gr. passable; see **accessible**

perates, Gr. wanderer; see **wander**

peratico- < Gr. *peratikos,* alien, foreign; see **strange**

peratos, Gr. on the opposite side; *perate,* west; *peras, -atos,* end, goal, limit; see **end**

perch < L. *perca* (Gr. *perke; perkis, -idos*), f.; *perkidion,* n. dim.: perciform, *Perca flavescens* (yellow perch), *Percopsis guttatus* (sand-roller), *Percina rex* (a darter), *Amphiprion percula* (a fish), *Helioperca incisor* (bluegill).
   Gr. *akarnax,* f. a basslike fish:
   AS. *baers,* perch: bass *(Micropterus salmoides).*
   Gr. *beryx, -ykos,* m. a perchlike fish: *Beryx splendens* (a fish), berycoid, berycine, *Scopeloberyx opercularis* (a fish).
   Gr. *channe,* f.; *channos,* m. a sea-perch: *Neochanna apoda* (a New Zealand mud-fish).
   L. *coracinus* (Gr. *korakinos*), m. a perchlike fish: *Coracinus chalcis* (a coracine).
   L. *gerres, -is,* m. a basslike, marine fish: *Eugerres plumieri* (mojarra), *Gerres rhombeus* (a fish).
   Gr. *labrax, -akos,* m. a sea-bass; *labrakion,* n. dim.: *Labrax notatus* (a fish), *Paralabrax clathratus* (cabrilla), *Labracoglossa argenteiventris* (a fish).
   Gr. *latos,* m. Nile perch: *Lates niloticus* (a fish), *Percalates colonorum* (an Australian fish).
   L. *orphus* (Gr. *orphos*), m. sea-perch: *Pagrus orphus* (a fish).
   Gr. *paprax, -akos,* m. probably a perch:
   Gr. *platax, -akos,* m. a perchlike fish: *Platax pinnatus* (a fish).
   L. *sciaena* (Gr. *skiaina*), f.; *skiadeus,* m. a perchlike, marine fish: *Sciaena deliciosa* (a fish), *Ctenosciaena dubia* (a fish), *Sciaenops ocellata* (red drum), *Rhinoscion saturnus* (a roncador).
   NL. *serranus,* m. a genus of perchlike fishes: *Serranus hepatus* (a fish).
   L. *umbra,* f. a sciaenid fish: *Umbrina cirrosa* (umbra), *Umbra limi* (mud-minnow).

percitus, L. greatly agitated; see *cieo* under **arouse**

percno- < Gr. *perknos,* dark-colored; see **black**

percnoptero- < Gr. *perknopteros,* a vulture; see **vulture**

perco- < Gr. *perkos,* a kind of hawk; see **hawk**

perculsus, L. struck, smitten, upset; see *percello* under **strike**

percussus, L. struck, beaten, shocked; see *concutio* under **strike**

perdicium, L. (Gr. *perdikion*), pellitory; see **pellitory**

perditus, L. ruined, lost, corrupt; *perditor; perditrix,* destroyer; see *perdo* under **destroy**

perdix, L., Gr. partridge; see **quail**

perdomai, Gr. break wind; see **smell**

peregrinus, L. traveling about, foreign, exotic, strange; see *peregrinor* under **move**

perennis, L. through the years, perpetual, everlasting; see **always**

peresus, L. eaten up, consumed, wasted; see *edo* under **eat**

perfect < L. *perfectus,* complete, finished; see **all, right, pure, innocent**

perfidus, L. faithless, false; see *infidelis* under **lie**

perfume < F. *parfum,* < L. *per,* through, *fumus,* smoke; see **smell**

pergamentum, ML. writing material prepared from the skins of animals; see **paper**

pergula, L. booth, stall, shop; see **store**

pergulo- < Gr. *pergoulos,* a bird; see **bird**

perhaps < L. *per,* by, and ON. *happ,* chance; see **lot**

peri, Gr. around, near, very; see **around, very**

periclymenum, L. (Gr. *periklymenon*), woodbine; see **honeysuckle**

periculum, L. danger, trial, risk; see **danger**

perideris, Gr. necklace; see **necklace**

peridineo, Gr. whirl around; see *dino* under **turn**

peridinos, Gr. rover, pirate; see **steal**

peridion, Gr. dim. of *pera,* pouch; see **bag**

periergos, Gr. over-careful, meddlesome, busybody; see *ergon* under **make**

perilla, NL. a genus of the mint family, see **mint**; dim. of Sp. *pera,* pear, see **pear**

perillus, L. a genus of bugs; see **bug**

perimesos, Gr. edged with purple; see **purple**

perimeter < Gr. *perimetron,* circumference; see **circle**

perineum, L. (Gr. *perinaion*), area between the anus and pudenda; see **rump**

period < L. *periodus* (Gr. *periodos*), a completed course; see **time, age, end, spot, point**

periosios, Gr. immense; see **large**

peripatos, Gr. a walking about; see **walk**

periphery < Gr. *periphereia,* circumference; see **circle, border**

periscelis, L. (Gr. *periskelis*), garter, anklet; see **garter**

perish < OF. *perir,* < L. *pereo, -itus,* pass away, die; see **death**

perisso- < Gr. *perissos (perittos),* beyond the regular number or size, extraordinary, odd as applied to numbers, see **different**; *perisseia,* abundance, surplus, see **abundance**

peristera, Gr. pigeon; *peristerion,* dim.; see **dove**

peristicto- < Gr. *peristiktos,* dappled; see **spot**

peritus, L. practiced, clever, skillful, see **know**; < *pereo,* pass away, perish, see **death**

perjurus, L. lying under oath; see **lie**

permanent < L. *permanens,* remaining forever, enduring; see **always, stand**

permit < L. *permitto, -missus,* let go, grant; see **let**

perna, L., Gr. leg of pork, ham, haunch, see **flesh**; a shellfish, see **mollusk**

perniciosus, L. baneful, injurious, destructive; see **destroy**

pernio, L. chilblain; see **sore**

pernix, *-icis,* L. nimble, agile, swift; see **swift**

pero- < Gr. *peros,* disabled, maimed; see **hurt**

peronatus, L. booted, < *pero, -onis,* a kind of boot; see **shoe**

perone, Gr. pin, anything pointed for piercing, small bone of arm or leg, buckle, clasp; see **needle**

perpendicular < L. *perpendicularis,* upright, at right angles; *perpendiculum,* plummet, plumbing; see **upright**

perperus, L. (Gr. *perperos*), wrong, false; see **lie**

perpetrator, L. doer; see *patro* under **make**

perpetuus, L. continuous, constant, forever; see **always**

perplexus, L. intricate, involved, puzzling, tangled; see **complex**

persea, Gr. an Egyptian tree; a genus of the laurel family; see **laurel**

persecutor, L. one who pursues with intent to injure; see **hurt**

Persephone, Gr. wife of Hades; see *Haides* under **hell**

persibus, L. extremely keen, knowing, acute; see **know**

persica, L. peach; see **peach**

persimmon < an Ab.Am. word.
Gr. *diospyros,* m. a kind of plant; a genus of the ebony family: *Diospyros virginiana*[*Diospyrus virginianus*] (persimmon).
L. *ebenus* (Gr. *ebenos*), f. ebony-tree, a species of persimmon: Ceylon ebony *(Diospyros ebenum), Ebenus cretica* (a legume).

persolata, L. a kind of mullein; see **mullein**

persona, L. mask, person; *persolla,* dim.; *personalis,* of a person; *personatus,* masked; see **face, people**

perspicax, -*acis*, L. clear-seeing, keen, acute; see **know**

perspicillum, L. lens, spectacles; see **lens**

perspicuus, L. clear, transparent, evident; see *conspectus* under **clear**

persuadeo, L. prevail upon, induce, win over; see **victory**

pertaining to; relating to; of; see **nature**

pertho, Gr. destroy, kill, plunder; see **destroy**

pertica, L. long pole, flail; *perticalis,* of a pole; see **rod**

pertinax, -*acis*, L. firm, persistent, obstinate; see **stubborn**

perturbatus, L. agitated, troubled; see **trouble**

pertusus, L. perforated; see **bore**

perula, L. dim. of *pera,* bag, pouch; see **bag**

perversus, L. wrong, evil; *perversio; perversitas,* evil; see **bad**

pervicax; pervicus, L. firm, stubborn; see **stubborn**

pervius, L. affording a passage, open, see **open;** *pervium,* passage, thorofare, see *via* under **way**

perysimos, Gr. of last year; see **year**

pes, *pedis,* L. foot; *pediculus; petiolus,* dim.; *pedalis,* of the foot; see **foot**

pesco- < Gr. *peskos,* skin, hide, rind; see **skin**

pessary < L. *pessum* (Gr. *pesson*), plug, tampon; *pessulum,* dim.; see **drug**

pessimus, L. worst; see *malus* under **bad**

pessos, Gr. a round stone used in playing checkers; see **stone**

pessulus, L. bolt; see **bar**

pessum, L. (Gr. *pesson*), plug, tampon; see **drug**

pestis, L. plague; *pestilentus,* unhealthful; see **disease**

pestle < L. *pistillum,* n. a club-shaped pounder used in a mortar: pistil, pistillate, piston, *Stylophora pistillata* (a coral).
Gr. *doidyx, -ykos,* m. pestle: *Daedicurus*[*Doedycurus*] *clavicaudatus* (a fossil edentate).
L. *fundulus,* m. piston:
Gr. *hyperos,* m. pestle, club: *Hyperophora brasiliensis* (a katydid), *Hyperomorpha squamosa* (a beetle).
Gr. *kopunon; igdokopanon,* n. pestle: *Copanognathus crassus* (a fossil fish).
Gr. *kotalis,* f. pestle:
Gr. *laktis, -ios,* f. pestle:
Gr. *leanter, -os,* m. pestle:
L. *pilum,* n. pounder, pestle: pilum, Pilumnus.
Gr. *thyestes,* m. pestle: *Thyestes verrucosus* (a Devonian fish).
Gr. *tripter, -os,* m. pestle; *tripterion,* n. dim.: *Tripteroceras hastatum* (a fossil cephalopod).
See: **mortar, strike, grind**

pet < uncertain origin; see **love**

petachnon, Gr. a broad, flat cup; see **cup**

petal < Gr. *petalon,* leaf, thin metal plate; see **leaf**

petalos, Gr. broad, flat, outspread; see **broad**

petasites, Gr. a broad-leaved plant; a genus of the composite family; see **composite**

petasma, Gr. carpet, rug; see **rug**

petasus, L. (Gr. *petasos*), a broad-brimmed hat; see **cap**

petauristes, Gr. tumbler, vaulter; *petauron,* springboard; see **leap**

petax, -*acis*, L. greedy; see **greed**

petelis, Gr. a locust; see **grasshopper**

petes, Gr. flyer; see *petomai* under **fly**

petigo, L. scab, eruption; *petiginosus,* scabby; see **scale**

petilium, L. a plant; see **plant**

petilus, L. thin, slender; see **thin**

petimen, L. sore on the shoulder of beasts of burden; see **sore**

petiolus, L. little foot, stalk, stem; see *pedunculus* under **stem, foot**

peto, *-itus,* L. seek, ask, desire, strive after, attack; see **ask**

petrel < L. *Petrellus,* dim. of *Petrus,* Peter; see **albatross**

petro- < Gr. *petra,* rock; *petridion,* dim.; see **stone**

petulans, L. pert, saucy, impudent, peevish, irritable; see **fretful**

petulcus, L. butting, frisky; see **play**

petunia, NL. a genus of the nightshade family; see **nightshade**

peucedanum, L. (Gr. *peukedanon*), a genus of the carrot family; see **carrot**

peuco- < Gr. *peuke,* pine; *peukinos,* of pine; see **pine**

peusis, Gr. inquiry, question; *peuthen,* inquirer, spy; see **try**

pexatus, L. covered with a napped garment; see **garment**

pexis, Gr. fixation, solidification, coagulation; see *pegma* under **thick**

pexus, L. combed, carded, < *pecto,* comb; see *pecten* under **comb**

peza, Gr. edge, border, foot; *pezidion,* dim.; see **border, foot**

pezica, L. (Gr. *pezis*), a kind of mushroom; see **fungus**

pezo- < Gr. *pezos,* on foot, walking; see *peza* under **foot**

phabo- < Gr. *phaps, phabos,* a wild pigeon; see **dove**

phacelo- < Gr. *phakelos,* bundle, cluster; see **cluster**

phaco- < Gr. *phakos,* lentil, anything shaped like the seed of a lentil; see **bean, lens**

phaecado- < Gr. *phaikas, -ados,* a white shoe; see **shoe**

phaedro- < Gr. *phaidros,* bright, beaming; see *phaino* under **light**

phaedrynto- < Gr. *phaidryntes,* shine, appear; see **light**

phaeo- < Gr. *phaios,* dusky; see **brown**

phaethon, Gr. radiant, shining; see *phaino* under **light**

-phage; phago- < Gr. *phagein,* to eat; see **eat**

phagedaena, L. (Gr. *phagedaina*), an eating ulcer; see **sore**

phalacro- < Gr. *phalakros,* baldheaded, bald, smooth; see **bare**

phalacrocorax, L. cormorant; see **coot**

phalaena, L. (Gr. *phalaina*), a kind of moth, see **moth**; whale, see *balaena* under **whale**

phalanga, L. carrying-pole, roller; see **rod**

phalangium, L. (Gr. *phalangion*), a kind of spider; see **spider**

phalanx, L., Gr. line, battle-array, round piece of wood, bone of the finger or toe; see **line**

phalaris, Gr. name of a grass, see **grass**; coot, see **coot**

phalaros, Gr. having a white patch, white-spotted; see **spot**

phalco- < Gr. *phalke,* a kind of bat, see **bat**; < *phalkes,* beam, rib of a ship, see **beam**

phalera, L. (Gr. *phalaron*), disk or boss of metal used as an ornament; see **ornament**

phallos, Gr. penis; see **penis**

phalos, Gr. bright, shining, white; see *phaino* under **light**

phanaeo- < Gr. *phanaios,* giving light; see *phaino* under **light**

phaneros, Gr. visible, evident; see *phaino* under **light**

phano- < Gr. *phanos,* light, bright, torch; *phanion,* dim.; *phanotes,* brightness, clearness; see *phaino* under **light**

phantasia; phantasma, Gr. image, appearance, show, apparition; see **fancy, spirit**

phantos, Gr. visible; see *phaino* under **light**

phaps, *phabos,* Gr. a wild pigeon; see **dove**

phar, Gr. spelt; see *far* under **grain**

pharango- < Gr. *pharanx, -angos,* chasm, gully, ravine; see **valley**

pharcido- < Gr. *pharkis, -idos,* wrinkle; see **fold**

pharetra, Gr. quiver or case for arrows; see **bag**

pharmaco- < Gr. *pharmakon,* drug, medicine, poison; *pharmakion,* dim.; see **drug**

pharnaceon, L. a kind of ginseng; see **ginseng**

pharos, Gr. cloak, mantle, shroud, see **garment**; plow, see **plow**; lighthouse, see **light**

pharsos, Gr. part, portion; see **part**

pharynx, Gr. throat; see **throat**

phascolo- < Gr. *phaskolos,* leather bag; see **bag**

phase < Gr. *phasis,* appearance, look, state; see *phaino* under **light**

phaseolus, L. dim. of *phaselus* (Gr. *phaselos*), kidney-bean; see *phaselus* under **bean**

phasganon, Gr. sword; *phasganion; phasganis,* dim.; see **sword**

phasianus, L. (Gr. *phasianos*), pheasant; see **pheasant**

phasis, Gr. appearance, look, state, see *phaino* under **light**; a saying, statement, see *phemi* under **speak**

phasma, Gr. apparition, specter; see **spirit**

phassa (phatta), Gr. ringdove; *phattion,* dim.; see **dove**

phatne, Gr. manger, crib, socket; *phatnion,* dim.; see **basin**

phatnoma, Gr. panel, lacunar; see **hollow**

phatos, Gr. spoken, that may be spoken; see *phemi* under **speak**

phattages, Gr. probably the pangolin; see **anteater**

phaulos; phlauros, Gr. paltry, petty, trivial, abject, poor; see **trifle**

phausis, Gr. a lighting; see *phaino* under **light**

pheasant < OF. *faisant,* < L. *phasianus* (Gr. *phasianos*), m., < Phasis, a river in Colchis: *Phasianus colchicus* (common pheasant).
 Gr. *tatyras,* m. pheasant:
 Gr. *tetaros,* m. pheasant:
 Gr. *tetraon, -os,* m. pheasant: *Tetraonoperdix nivicola* (a bird).

phegos, Gr. beech; see **beech**

phellandrium, L. a plant; see **plant**

phelleus, Gr. stony ground; see **stone**

phellos, Gr. cork-tree, cork; see **bark**

phelos, Gr. deceitful; *pheletes,* cheat, knave; see **lie**

pheme, Gr. voice, speech; see *phemi* under **speak**

phenaco- < Gr. *phenax, -akos,* cheat, imposter, see **lie**; < *phenake,* false hair, wig, see **hair**

phene, Gr. bearded vulture; see **eagle**

phengos, Gr. light, splendor, luster, sunlight; *phengodes,* bright, shining; *phengites,* a mineral used for windows; see **light**

phenico- < Gr. *phoinix, -ikos,* purple-red, see **purple**; date-palm, see **palm**; a fabulous bird, see **bird**

pheno- < Gr. *phaino,* appear, shine; see **light**

phenol; phenyl, denoting benzene and its compounds; see **benzene**

phenomenon < L. *phenomenon* (Gr. *phainomenon*), appearance, happening, event; see **display**

pheos, Gr. a spiny plant; see **thorn**

phepsalos, Gr. spark, ember; see **fire**

-pher; -phor < Gr. *phero,* bear, carry; see **carry**

pheretron, Gr. litter, bier; see *phero* under **carry**

pherma, Gr. burden, load, produce; see *phero* under **carry**

pherne, Gr. dowry; see **give**

phernion, Gr. fish-basket; see **basket**

phertos, Gr. brave; *pherteros,* braver; *phertatos,* bravest; see **bold**

phi, Gr. twenty-first letter of the Greek alphabet; see **letter**

phiala, L. (Gr. *phiale*), a broad, flat vessel, saucer, bowl; see **cup**

phiaros, Gr. shining, bright; see **light**

phibaleos, Gr. a kind of fig; see **fig**

phido- < Gr. *pheidos,* thrifty, stingy; *pheidole,* thrift; *pheidolos,* niggard, miser; see **stingy**

phidono- < Gr. *pheidon, -os,* oil-can; see **vessel**

phil-; philo- < Gr. *phileo,* love as a friend, regard with affection; see *philia* under **love**

philadelphus, L. (Gr. *philadelphon*), a genus of the saxifrage family; see **saxifrage**

philautos, Gr. selfish; see **selfish**

philax, Gr. tree; see **tree**

philemon, Gr. loving, friendly; see *philia* under **love**

philetes; philetor, Gr. lover; see *philia* under **love**

phillyrea, NL. a genus of the olive family; see **olive**

Philomela, Gr. daughter of Pandion changed into a nightingale; see **nightingale**

philosophia, Gr. love of wisdom; see *sophos* under **know**

philtrum, L. (Gr. *philtron*), love-charm, spell, potion; see **magic**

philyca, L. (Gr. *phylike*), a kind of buckthorn; see **buckthorn**

philyra, Gr. linden; see **linden**

phimos; phimotron, Gr. muzzle; *phimosis,* muzzling or stopping an orifice, see **close**; cup used as a dice-box, see **cup**

phityo- < Gr. *phitys, -yos,* begetter, father; see **begin**

phlasco- < Gr. *phlaske,* wine-flask; *phlaskon,* flagon; see **bottle**

phlauros, Gr. paltry, trivial; see **trifle**

phlebo- < Gr. *phleps, phlebos,* vein; *phlebion,* dim.; see **pipe**

phledon, Gr. babbler, idle talker; *phlenaphos,* idle talk; see **speak**

phlegma, Gr. inflammation, heat, morbid humor; see *phlego* under **burn**

phleum, NL. a kind of reed; see **grass**

phlexis, Gr. a kind of bird; see **bird**

phlia, Gr. doorpost; see **pillar**

phloeo- < Gr. *phloios,* bark; see **bark**

phlogmos; phlogos, Gr. flame, inflammation; see *phlego* under **burn**

phlomis, Gr. mullein; see **mullein**

phlonitis, Gr. a kind of borage; see **borage**

phlox, -ogis, L. (Gr. *-ogos*), f. flame, a kind of plant: *Phlox divaricata* (a phlox). See *phlego* under **burn**
　　NL. *gilia,* f. a genus of the phlox family, < Philip S. Gil, Spanish botanist: *Gilia aggregata* (scarlet gilia).
　　L. *polemonium* (Gr. *polemonion*), n. a genus of the phlox family: *Polemonium caeruleum* (Greek valerian).

phlyaco- < Gr. *phlyax, -akos,* jester, fool, buffoon; *phlyaros,* nonsense, silly talk, babbling; see *phledon* under **speak**

phlyctaena, L. (Gr. *phlyktaina; phlyktis*), blister, pustule; *phlyzakion,* dim.; see **bubble**

phlydaros, Gr. soft, flabby; see **soft**

phlyzacium, L. (Gr. *phlyzakion*), dim. of *phlyktaina,* blister, vesicle; see **bubble**

phoba, L. (Gr. *phobe*), corymb, curl, tuft; see **cluster**

phobia < Gr. *phobos,* fear; *phoberos,* fearful, formidable, terrible; *phobetron,* bugbear, scarecrow; see **fear**

phoca, L. (Gr. *phoke*), seal; see **seal**

phocaena, L. (Gr. *phokaina*), porpoise; see **whale**

phocio- < Gr. *phokion,* a kind of bird; see **bird**

phocis, L. a kind of pear-tree; see **pear**

phocto- < Gr. *phoktos,* roasted, baked; see **cook**

phodo- < Gr. *phos, -odos,* blister, burn; see *phois* under **bubble**

Phoebe, L. (Gr. *Phoibe*), goddess of the moon, < *phoibos,* bright, radiant; see **moon, light**

phoebeto- < Gr. *phoibetes,* prophet; see **prophecy**

phoebetria, L. (Gr. *phoibetria*), purifier; see **pure**

phoenicuro- < Gr. *phoinikouros,* redstart; see **bird**

phoenix, L. (Gr. *phoinix*), Phoenician, purple-red, see **purple**; date-palm, see palm; a fabulous bird symbolic of resurrection and immortality, see **bird**

phoeteto- < Gr. *phoitetes,* comer and goer; see **move**

phois, Gr. blister, burn; see **bubble**

pholado- < Gr. *pholas, -ados,* a rock-boring mollusk; see **mollusk**

pholco- < Gr. *pholkos,* bowlegged; see **bend**

pholeos, Gr. hole, cave, den; *pholeter,* one who lurks in a hole; *pholion,* dim.; see **hole**

pholido- < Gr. *pholis, -idos,* scale, spot, fleck; *pholidotos,* clad in scales; see **scale**

phollico- < Gr. *phollix, -ikos,* scab, sore; *phollikodes,* scabby; see **sore**

phoma, NL. a kind of fungus; see **fungus**

phonaco- < Gr. *phonax, -akos,* blood-thirsty; see *phonos* under **kill**

phono- < Gr. *phone,* sound, voice; *phonion,* dim.; *phonetes,* sounder, speaker; *phonetikos,* of sound, vocal, see **sound**; < *phonos,* slaughter, murder; *phoneus;* *phonergates; phontes,* murderer, slayer, see **kill**

phor, Gr. thief; *phorios,* stolen, see **steal**; *phorao,* search for a thief, see **try**

phoras, -ados, Gr. fruitful, bearing; see *phero* under **carry**

phorbe, Gr. pasture, food; see *pherbo* under **eat**

phorco- < Gr. *phorkos,* gray, white; see **gray**

-phore; phoro- < Gr. *phoreus,* bearer, carrier; *phoretos,* borne; *phorimos,* bearing, fruitful; see *phero* under **carry**

phoringes, Gr. truffle; see **fungus**

phorino- < Gr. *phorine,* thick skin, hide; see **skin**

phorminx, Gr. a kind of cithara or lyre; *phormiktes,* harper; see **harp**

phormio, L. mat; see **rug**

phormium, L. (Gr. *phormion*), a genus of the lily family; see **lily**

phormos, Gr. basket, plaited mat; *phormion,* dim.; *phormis, -idos,* basket; see **basket**

phortax, Gr. carrier, porter; see *phero* under **carry**

phortis, Gr. merchantman; see **ship**

phortos, Gr. load, cargo; *phortion,* weight, burden, embryo; see **weight**

phorycto- < Gr. *phoryktos,* stained, defiled; see **dirt**

phorytos, Gr. sweepings, rubbish, chaff; see **dirt**

phosphorus, L. (Gr. *phosphoros*), light-bringer, Lucifer, a chemical element; see *phos* under **light,** ELEMENTS under **thing**

phosson, Gr. coarse linen, sail; see **sail**

phoster, Gr. illuminator, window; see **window**

photino- < Gr. *photeinos,* shining, bright; see *phos* under **light**

photinx, Gr. a kind of flute; *photingion,* dim.; *photingistes,* flutist, fifer; see **pipe**

photo- < Gr. *phos, photos,* light, see **light**; man, hero, see **man**

phoxinos, Gr. probably a minnow; see **carp**

phoxos, Gr. pointed, peaked; see **point**

phoyx, Gr. a kind of heron; see **heron**

phrade, Gr. knowledge, understanding, hint; *phrades,* wise, shrewd; *phradmosyne,* cunning; see **know**

phragma; phragmos, Gr. fence, partition, screen, hedge, wall; *phraktos,* fenced; see **wall**

phragmites, L. a reed, now a genus of grasses; see **grass**

phrase < Gr. *phrasis,* speech; *phraster,* expounder, guide; see **speak**

phrater, Gr. clansman, one of a brotherhood, *phratra;* see **kin**

phreato- < Gr. *phrear, -atos,* well, reservoir; see **spring**

phreno- < Gr. *phren, -os,* mind, heart, diaphragm; see **mind, wall**

phrico- < Gr. *phrix, -ikos,* ruffling of a smooth surface, ripple, shivering; *phrikaleos; phriknos,* shivering with cold, goose-pimply, rough; *phriktos,* horrible, terrible; *phrixos,* standing on end, bristling, shuddering, shivering; see **shake**

phron, Gr. mind; see *phren* under **mind**

phronimos, Gr. sensible, wise; see **know**

phrontido- < Gr. *phrontis, -idos,* thought, attention; see **think**

phrudo- < Gr. *phroudos,* fled, departed, gone, ruined; see **depart**

phruro- < Gr. *phrouros,* guard; *phrourion,* fort; see **guard**

phrycto- < Gr. *phryktos,* roasted, firebrand, torch, beacon, see **cook**; < *phrykte,* a kind of resin, see **resin**

phrygano- < Gr. *phryganon,* dry stick, brushwood, firewood; see **wood**

phrygilos, Gr. a kind of finch; see **finch**

phrygios, Gr. dry; see **dry**

phryno- < Gr. *phryne,* toad; see **toad**

phthalo- < naphthalene; see *naphtha* under **fat**

phthano, Gr. anticipate, do first; see **before**

phthartos, Gr. perishable; *phtharsis,* corruption; *phthartikos,* destructive; see *phtheiro* under **destroy**

phthegma, Gr. sound of the voice; *phthegmatikos,* sounding, vocal; *phthenxis,* speech, utterance; *agaphthenktos,* loud-sounding; see *phthenxis* under **speak**

phthino- < Gr. *phthio,* waste away, decline, decay, wane; *phthinodes,* consumptive; *phthisis,* consumption, decline, decay; see **lessen**

phthirio-; phthiro- < Gr. *phtheir, -os,* louse; *phtheirion,* dim.; see **louse**

phthoë, Gr. decline; see *phthio* under **lessen**

phthoïs, Gr. a kind of cake; see **cake**

phthongus, L. (Gr. *phthongos*), sound, tone; see **sound**

phthonos, Gr. ill-will, envy, jealousy; *phthoneros,* envious, jealous; see **jealous**

phthoro- < Gr. *phthora,* corruption, decay, destruction; *phthorimos,* destructive; see **destroy**

phyas, Gr. shoot, sucker; see **branch**

phycis, L. (Gr. *phykis*), a fish living among seaweeds; see **wrasse**

phyco- < Gr. *phykos,* seaweed, alga; *phykion; phykarion;* dim.; see **alga**

phygo- < Gr. *phyge,* flight, escape, avoidance; *phygas, -ados,* fugitive; *phygadeion,* asylum, refuge; see *fuga* under **depart, safe**

phylaco- < Gr. *phylax, -akos,* guard; *phylake,* prison; *phylakterion,* safeguard, amulet; see **guard**

phyllanthus, NL. a genus of the spurge family; see **spurge**

phyllitis, Gr. hartstongue-fern; see **fern**

phyllo- < Gr. *phyllon,* leaf; *phyllarion; phyllion,* dim.; *phyllikos; phyllinos,* of leaves; see **leaf**

Phyllodoce, L. (Gr. *Phyllodoke*), a sea-nymph; see **sea**

phylo- < Gr. *phyle,* tribe, race; *phyletikos,* pertaining to a tribe; see **class**

phyma, Gr. tumor; *phymation,* dim.; see **swell**

phyo; phyteuo, Gr. produce, beget, make grow; *phye,* full growth, prime; see *phyton* under **plant, grow**

phyrtos, Gr. mixed; *phyrama; phyrmos,* mixture; see **mix**

physa, Gr. bellows, bubble, win; see **blow, bag**

physalis, NL. a genus of the nightshade family, < Gr. *physallis, -idos,* a plant with bladderlike fruits; see **nightshade**

physalos, Gr. a kind of toad, see **toad**; whale, see **whale**

physco- < Gr. *physke,* blister, sausage; *physema,* something inflated, bubble; see *physema* under **bubble**

physeter, Gr. blower; *physetos,* blown; see *physa* under **blow**

physician < OF. *physicien;* see **heal**

physicus, L. (Gr. *physikos*), concerning nature and matter; *physis,* nature, condition; see **nature,** *phyo* under **grow**

physinx, Gr. bladder, bubble; garlic; see *physema* under **bubble, onion**

physo- < Gr. *physa,* bellows; see **blow**

phyto- < Gr. *phyton,* plant; *phytarion,* dim.; *phytikos,* of plants; *phytodes,* plantlike; see **plant**

phytolacca, NL. pokeweed; see **pokeberry**

phyxelis; phyzelos, Gr. cowardly, shy; see **coward**

phyxion; phyxis, Gr. asylum, refuge, escape; *phyximos; phyxios,* of flight and refuge; see *fuga* under **depart**

pi, Gr. sixteenth letter of the Greek alphabet; see **letter**

piacularis, L. atoning; *piaculum*, propitiatory sacrifice, < *pio, -atus,* sacrifice, appease, perform sacred rites; see *pietas* under **rite**

piar, Gr. fat, cream; *piaros*, fat, rich; see **fat**

pica, L. magpie; see **crow**

picea, L. spruce, < *pix, picis,* pitch; *piceus*, pitchy, pitch-black; see **gymnosperm, resin**

pick < AS. *pic*, pike, a handled tool for grubbing or digging, mattock.
   L. *dentiscalpium*, n. toothpick:
   L. *dolabra*, mattock, pickax; see **ax**
   L. *ligo, -onis*, m. grubbing-hoe, mattock: *Ligonipes illustris* (a spider).
   Gr. *makella*, f. pick with one point: *Macellodus brodiei* (a Jurassic reptile).
   L. *oryx, -ygis* (Gr. *-ygos*), m. pick for digging; *orygion*, n. dim.:
   Gr. *skalis, -idos,* hoe, mattock; see **hoe**
   Gr. *skapane*, shovel, mattock; *skapanion,* dim.; see **shovel**

pick < OF. *piquer*, pierce, separate; see **choose**

pickle, prob. < D. *pekel;* see **salt, sour, save, melon**

picris, L. bitter lettuce; see **lettuce**

picro- < Gr. *pikros*, bitter; see **bitter**

pictus, L. painted, colored; *pictilis*, painted, embroidered; *pictura*, a painting; see *pingo* under **color**

Picumnus; Pilumnus, L. brother tutelary deities of wedlock and fertility; see **fertile**

picus, L. woodpecker; see **woodpecker**

pidaco- < Gr. *pidax, -akos,* spring, fountain; see **spring**

pie < uncertain origin; see **cake**

piece < OF. *pece;* see **part**

pierce < OF. *percer;* see **bore**

Pieris, L., Gr. one of the Pierides; see **art**

piestos, Gr. compressible; *piester,* press, squeezer; see *piezo* under **press**

pietas, L. sense of duty, loyalty, kindness, < *pius,* dutiful, conscientious, devout; see **rite**

piezo, Gr. press; see **press**

pig < uncertain origin; see **hog**

pigeo, *-itus*, L. feel annoyance, be irked; see **fretful**

pigeon < F. *pigeon*, < L. *pipio,* a chirping bird; see **dove**

piger, *-gra, -grum,* L. slow, lazy, reluctant, dilatory, slothful; see **slow**

pigment < L. *pigmentum*, color, paint; see **color**

pignus, L. pledge, pawn, security; *pignero, -atus,* pledge, pawn; see **promise**

pike < AS. *pic:* pickerel *(Esox niger).*
   Gr. *belone*, needle, gar; see **needle**
   L. *esox, -ocis,* m. pike: *Esox masquinongy* (muskellunge), *Scombresox saurus* (saury, a billfish).
   Gr. *kestra*, hammer, probably with one pointed end, a pikelike fish; see **hammer**
   L. *lucius,* m. pike: *Esox lucius* (a pike).
   Gr. *rhamphestes*, m. probably a pike:
   Gr. *sargios*, m. a gar: *Sarginites pygmaeus* (a fish).
   L. *sphyraena* (Gr. *sphyraina*), f. a pikelike fish: *Sphyraena barracuda* (barracuda).
   L. *sudis*, a fish, probably a pike; see **pillar**

pila, L. ball; *pilula,* dim., see **ball**; pillar, see **pillar**; mortar, see **mortar**

pilatus, L. thick, dense, see **thick**; < *pilo, -atus,* grow hairy, plunder, see **hair, steal**

pile < L. *pila*, pier, column, mole, see **pillar, heap**; < AS. *pil,* arrow, < L. *pilum,* javelin, see **arrow, spear**

pilema, Gr. felt; see *pileo* under **press**

piles < L. *pila*, ball; see *exochadium* and *haemorrhoida* under **disease**

pileus, L. (Gr. *pileos; pilos*), cap; *pileolus; pilidion; pilion,* dim.; *pileatus,* capped; see **cap**

pilatus, L. grown hairy, see **hair**; plundered, robbed, see **steal**; armed with a javelin, see *pilum* under **spear**

pilentum, L. carriage, coach; see **vehicle**

pill < L. *pilula*, dim. of *pila,* ball; see **ball, drug**

pillage < F. *piller,* < L. *pilo,* deprive of hair, plunder; see **destroy, steal, plunder**
pillar < L. *pila,* f. column, pier: pilaster, pile.
  L. *adminiculum,* n. prop, support, stay; *adminiculo, -atus,* prop up, support:
    adminicle, adminiculate.
  L. *ames, -itis,* m. forked pole:
  Gr. *anteris, -idos,* f. prop, stay, support, buttress; *anteridion,* n. dim.; L. *anta,* f.
    pillar at the side of a door or corner of a building: *Anteris simulans* (a wasp),
    Anteriscus abyssinicus (a beetle).
  L. *asser,* beam, post, pole; see **beam**
  L. *canteriatus,* supported by props:
  Gr. *charax, -akos,* m. pointed stake, prop, pole; *charakion,* n. dim.; L. *charactus,*
    provided with stakes, propped up: *Characopygus moricei* (a wasp), *Characium*
    marinum (an alga).
  L. *cippus,* m. post, pillar, gravestone, palisade:
  L. *columna,* f. pillar; *columella,* f. dim.; *columellaris; columnaris,* pillarlike;
    columnarius, of pillars; *columnatus,* supported by pillars: column, columnar,
    columniferous, columella, colonel, colonnade, *Columella edentula* (a snail),
    Idria columnaris (cirio).
  L. *conamen, -inis,* n. support, stay, prop:
  L. *destina,* f. support, prop:
  L. *erisma, -tis* (Gr. *eirisma, -tos*), n. buttress, stay, support: *Erismatopterus*
    levatus (an Eocene fish).
  L. *firmamentum,* support, prop, sky; see **sky**
  L. *fulcimen, -inis,* n. prop, pillar, < *fulcio, fultus,* prop up, support; *fultor, -is,*
    m.; *fultura,* f. prop, supporter; *fulcimentum (fulmentum),* n. prop, bedpost;
    effultus; suffultus, propped up: *Phyllactinia suffulta* (a fungus).
  L. *fulcrum,* n. bedpost, prop, support: fulcrum, *Solenella fulcrata* (a conodont).
  L. *furcula,* f. forked prop:
  L. *gralla,* f. stilt; *grallator, -is,* m. one who walks on stilts: grallatorial, *Grallina*
    cyanoleuca (magpie-lark), *Eocathartes grallator* (an Eocene vulture).
  Gr. *hermis; hermin, -os,* m. bedpost:
  Gr. *hypomochlion,* n. fulcrum of a lever:
  L. *impedatio, -onis,* f. a supporting with props:
  Gr. *kalobamon, -os,* walking on stilts; *kalobates,* m. walker on stilts: *Calobamon*
    (name for a fossil fly), *Calobates radiatus* (a bird).
  Gr. *kamax, -akos,* f. pole, prop, shaft: *Camacolaimus tardus* (a nematode),
    Camacopselaphus fulvus (a beetle).
  Gr. *keleon,* upright beam of a loom; see **rod**
  Gr. *kion, -os,* m. column, pillar, uvula; *kionion; kioniskos,* dim.; *perikion, -os,*
    surrounded by pillars: cionocranial, *Cionus scrophulariae* (a beetle), *Ectocion*
    collinus (a Paleocene mammal).
  Gr. *kolobathron,* n. stilt; *kolobathristes,* m. walker on stilts: *Colobathristes*
    chalcocephalus (a bug).
  L. *meta,* f. conical or pyramidal column at ends of the Roma circus, goal-post,
    boundary-stake; *metula,* f. dim.: metes and bounds, *Meta menardi* (a spider),
    Eucidaris metularia (an echinoid).
  L. *monumentum,* n. memorial: monument, monumental.
  Gr. *nyssa,* f. turning-post, goal-post: *Nyssa sylvatica* (black gum), *Gennaeocrinus*
    nyssa (a Devonian crinoid).
  Gr. *obeliskos,* dim. of *obelos,* spit; pointed pillar; see **spear**
  Gr. *oche,* f. prop, support:
  L. *palus,* m. stake, stick; *palaris,* of stakes: pale, palo verde *(Cercidium*
    torreyanum).
  Gr. *parastas, -ados,* f. square pillar, doorpost, pilaster:
  L. *pedamen, -inis,* n. stake, prop:
  Gr. *phlia,* f. doorpost:
  L. *postis, -is,* m. doorpost, door: post.
  Gr. *pyramis, -idos,* f. an angular structure or figure with pointed apex: pyramidal,
    Pyramidotettix citri (a bug), *Triactiscus tripyramis* (a radiolarian), *Actinostro-*
    bus pyramidalis (a conifer), *Goniophyllum pyramidale* (a Silurian coral).
  L. *ridica,* f. stake, prop:
  Gr. *schalis, -idos,* f.; *schalidoma, -tos,* n. forked stick used as a prop: *Schalidomitra*
    ambages (a moth).
  Gr. *stalix, -ikos,* f. stake for fastening nets: *Stalix histrio* (a fish), *Sigmatostalix*
    radicans (an orchid).
  Gr. *stathmos,* pillar, post; *stathmidion; stathmion,* dim.; see **house**
  L. *statumen, -inis,* n. support, stay, prop:
  Gr. *stauros,* upright pale, stake; see **cross**

L. *stela* (Gr. *stele*), f. column, pillar; *stelidion,* n. dim.: *stelechos,* m. trunk, log: stele, stelar, meristele, *Stelorrhinus carinirostris* (a beetle), *Stelidiocrinus argutus* (an Ordovician crinoid), *Stelechopus hyocrini* (a worm).

Gr. *sterigma, -tos,* n.; *sterinx, -ingos,* f. prop, stalk, base: sterigma, *Steringophorus furciger* (a worm).

L. *stilus* (incorrectly *stylus*), pen; see **pen**

Gr. *stylos,* m.; *styloma, -tos,* n. pillar, post; *stylarion; stylidion,* n.; *stylis, -idos,* f.; *styliskos,* m. dim.; *stylotos,* with pillars: style, stylite, stylewort, stylolite, stylobate, *Stylocrinus tabulatus* (a Devonian crinoid), *Stylophora subreticulata* (a fossil coral), *Stylops aterrima* (a strepsipterid), *Stylochus speciosus* (a worm), *Stylidium graminifolium* (a stylewort), *Stylidodon politum* (a wasp), *Osmorrhiza brevistylis* (sweet cicely), *Stylosanthes biflora* (a legume), *Rhopalostylis sapidus* (Nikau palm), *Microstylum nigrisetosum* (a fly).

L. *sublica,* f. stake, pile; *sublicius,* resting on piles:

L. *sudis,* f. stake, pile, a fish: *Sudis borealis* (a fish).

L. *sustentaculum,* n. support, prop:

L. *vacerra,* f. log, stock, post: *Vacerra litania* (a butterfly).

L. *vallus,* stake, pale; see **wall**

See: **axis, rod, stem, stone**

pillow < AS. *pylu,* < L. *pulvinus,* m. cushion, pad, pillow; *pulvillus; pulvinulus,* m. dim.; *pulvinar, -is,* n. a cushioned couch, seat; *pulvinaris,* of cushions; *pulvinatus,* cushioned: pulvinar, pulvillus, pulvinus, pulvinate, *Pulvinaria innumerabilis* (cottony maple-scale), *Pulvinulina scitula* (a foraminifer), *Paepalanthus pulvinatus* (a pipewort), *Paronychia pulvinata* (a whitlowwort).

L. *anaclinterium* (Gr. *anaklinterion*), n. cushion:

L. *cervical, -is,* n. pillow, bolster:

L. *cubital, -is,* n. cushion for leaning on:

L. *culcita,* f. cushion, pillow, bolster, mattress; *culcitella; culcitula,* f. dim.; *culcitarius,* m. cushion-maker: *Culcita coriacea* (a starfish), *Acteocina culcitella* (a fossil gastropod).

Gr. *hypokephalaion,* n. pillow, cushion:

L. *plumacium,* n. feather-pillow:

L. *sedularium,* n. cushion of a carriage:

Gr. *stoibe,* f. cushion, pad:

Gr. *tyleion,* n. cushion, bolster; see *tyle* under **knot**

pilo- < Gr. *pilos,* hair, felt, felt cap, ball; see **hair,** *pila* under **ball,** *pileus* under **cap**

pilot < It. *pilota,* steersman; see **govern**

pilum, L. javelin, pestle; see **spear, pestle**

pilus, L. (Gr. *pilos*), hair, cap, ball; *pilidion; pilion,* dim.; *pilosus,* hairy; see **hair, cap, ball**

pimele, Gr. fat; see **fat**

pimelea, NL. a genus of the mezereon family; see **mezereon**

pimenta, NL. a genus of the myrtle family, < Sp. *pimienta,* pepper; see **myrtle**

pimpinella, It. pimpernel; a genus of the carrot family; see **carrot**

pimple, prob. < AS. *piplian,* form pustules.

Gr. *aitholix, -ikos,* f. pimple, pustule: *Aetholix flavibasalis* (a moth).

L. *bullula,* a watery vesicle; see *bulla* under **bubble**

Gr. *chalaza,* f. hailstone, sleet, pimple, tubercle; *chalazion,* n. dim.; *chalazios,* of or like hail or sleet: chalaza, chalazion, chalazotomy.

Gr. *ekphyma, -tos,* n. eruption of pimples: *Ecphymatotes torquatus* (a lizard).

Gr. *ekthyma, -tos,* n. pimple: ecthyma.

Gr. *ionthos,* young hair, pimple; see *ionthas* under **hair**

L. *papula,* f. pustule, pimple: *Papula sceptrifera* (an echinoid), *Umbilicaria papulosa* (a lichen).

L. *pustula,* blister, pimple; see **bubble**

L. *tuberculum,* small swelling, protuberance, pimple; see *tuber* under **swell**

L. *varus,* m. blotch, pimple; LL. *variola,* f. smallpox: variola, varioloid.

L. *vesicula,* pustule, dim. of *vesica,* bladder; see **bag**

See: **bubble, bag, swell, wart, spot, disease**

pin < AS. *pinn,* peg; see **needle, nail, bind**

pinaco- < Gr. *pinax, -akos,* board, tablet, chart, register; *pinakion,* dim.; see **board**

pinaleo- < Gr. *peinaleos,* hungry; see *peina* under **hunger**

pinaros, Gr. dirty; see **dirt**

pincerna, L. cupbearer, butler; see **carry**

pinch, prob. < OF. *pincier;* see **tongs, lessen**

pindalos, Gr. an Indian bird; see **bird**

pine < L. *pinus,* f.; *pinea,* f. pine-cone; *pineus,* of pine: pinetum, pineapple, pinite, pineal, pinyon, pignon(nut of *Pinus pinea*), *Pinus radiata* (Monterey pine), *Pinicola enucleator* (pine-grosbeak), *Spinus pinus* (pine-siskin), *Fomes pinicola* (a bracket-fungus), *Parharmonia pini* (pine-pitch moth), *Chionaspis pinifolii* (pine-leaf scale), *Neodiprion pinetum* (pine-sawfly), *Aquilegia pinetorum* (a columbine), *Callistemon pinifolius* (a bottlebrush), *Hyla pinorum* (a frog), *Lophodermium pinastri* (a fungus).
  Gr. *das, dados,* pine torch, pine wood; *dadinos,* of pine wood; see **light**
  Gr. *peuke,* f. pine; *peukinos,* of pine: *Peucoglyphus corporalis* (a beetle), *Peucephila essoni* (a moth), *Pinus peuce* (Balkan pine), *Peucestes coronatus* (a grasshopper).
  Gr. *pitys, -yos,* f. pine; *pitydion,* n. dim.: pityline, *Pitylus grossus* (a grosbeak-tanager), *Pityophthorus pulicarius* (a beetle), *Pityophis melanoleucus* (pine- or bull-snake), *Hypopitys lanuginosa* (a pinesap), *Monotropa hypopitys* (false beechdrops).
  L. *sapinus,* f. a kind of pine or fir; *sapineus,* of pine:
  L. *strobus,* m. a tree yielding an aromatic resin; name applied to a pine by Linnaeus: *Pinus strobus* (white pine).
  L. *taeda,* f. pine, torch: *Pinus taeda* (loblolly-pine).
  L. *tibulus,* f. a kind of pine:

pinea, L. pine-cone; see **cone**

pineapple
  NL. *ananas,* m. pineapple, < an Ab.Am. word: *Ananas comosus* (pineapple), *Fragaria ananassa* (a strawberry).
  NL. *bromelia,* f. a genus of the pineapple family, < Olaf Bromel, Swedish botanist: *Bromelia pinguin* (pinguin), Bromeliaceae, bromelin, *Diaspis bromeliae* (a scale-insect).
  NL. *tillandsia,* f. a genus of the pineapple family, < Elias Tillands, Swedish physician: *Tillandsia fasciculata* (a bromeliacead).

pinguis, L. fat; *pinguiculus,* dim.; see **fat**

pink < uncertain origin; a family of plants including the carnation.   See **red**
  Gr. *alsine,* f. probably a chickweed: *Alsine graminea* (little starwort).
  L. *condurdum,* n. a kind of pink:
  NL. *dianthus,* m. carnation, pink: *Dianthus plumarius* (grass-pink), *Silene dianthifolia* (a catchfly), *Metridium dianthus* (a sea-anemone).
  Gr. *drypis, -idos,* f. a thorny plant; a genus of the pink family: *Drypis spinosa* (a pink).
  Gr. *lychnis, -idos,* f. a plant with scarlet flowers; now a genus of the pink family: *Lychnis coronaria* (mullein-pink, rose-campion).
  NL. *saponaria,* f. a genus of the pink family: *Saponaria bellidifolia* (a soapwort).
  NL. *silene,* f. a genus of the pink family: *Silene pendula*(drooping-silene).
  NL. *spergula,* f. a genus of the pink family: *Spergula sativa* (field-spurry), *Spergularia marina* (sand-spurry).
  NL. *stellaria,* f. a genus of the pink family: *Stellaria media* (chickweed).

pinna; penna, L. feather, wing, fin, pen, leaflet; *pinnula,* dim.; *pinniger,* bearing feathers, wings; *pinnatus,* feathered, plumed, winged, see **feather**; < Gr. *pinna,* a pelecypod, see **mollusk**

pinnaculum, L. peak; see **top**

pino, Gr. drink; see *posis* under **drink**

pinon, Gr. beer; see **beer**

pinos, Gr. dirt, filth, squalor; *pinaros, pinodes,* dirty; *pinodia,* dirt; see **dirt**

pinso, *pinsus; pistus,* L. pound, stamp, crush, grind; see **grind**

pinus, L. pine; see **pine**

pinytos, Gr. wise, discreet; see **know**

pion, Gr. fat, plump, rich; see **fat**

pipalis, Gr. a kind of lizard; see **lizard**

pipe < AS. *pipe,* tube: piping, pipette, pipkin, pipefish, bagpipe, *Pipetta tuba* (a radiolarian).
  Gr. *aorte,* f. the great artery: aorta.
  L. *aquagium,* n. aqueduct:
  Gr. *arteria,* f. windpipe, aorta: artery, arterial, arteriosclerosis, arteriomalacosis.
  Gr. *aulon, -os,* m. channel, pass, pipe; *aulos,* m. flute, pipe, tube; *aulidion,* n.; *auliskos,* m. dim.; *auleter, -os; auletes,* m. flute-player; *auletris, -idos,* f. flute-girl; *auletikos,* of the flute; *askaules,* m. bagpiper; *keraules,* m. hornblower: aulophyte, hydraulic, carol. *Aulostomus maculatus* (a fish), *Aulonocephalus*

lindquisti (a nematode), *Auletes rhynchitoides* (a beetle), *Cystauletes mam-millosus* (a fossil sponge), *Pseudauletes modestus* (a beetle).
Gr. *bronchos*, windpipe; see **throat**
L. *calamus*, a reed-pipe; see **reed**
L. *canalis*, water-pipe, channel; see **ditch**
L. *canna*, a reed-pipe, flute; see **reed**
L. *catheter*, *-is* (Gr. *katheter*, *-os*), m. a pipelike instrument for emptying the bladder: catheter, catheterize.
L. *cloaca*, f. canal, sewer, drain; *cloacula*, f. dim.; *cloacalis*, of a canal: cloacal, Cloaca Maxima.
L. *cuniculum*, rabbit burrow, underground passage; see *cunicularius* under **mine**
Gr. *diabetes*, m. siphon:
L. *ductus*, m. a leading; tube, pipe, canal; *aquaeductus*, m. conduit: duct, ductless, aqueduct, *Fusarium aquaeductum* (a fungus).
Gr. *exagogos*, m. waste-pipe, drain:
L. *fistula*, f. pipe, tube, ulcer; *fistella*, f. dim.; *fistularis*, of a pipe; *fistulatus*, with pipes; *fistulosus*, full of pipes, holes, porous: fistula, fistulous, fester, *Fistulipora carbonaria* (a fossil bryozoan), *Fistulina hepatica* (beefsteak-fungus), *Fistulana elongata* (a fossil pelecypod), *Cassia fistula* (a senna).
Gr. *ginglaros*, m. flute, fife:
Gr. *gingras*, m. a Phoenician flute or fife; L. *gingrina*, f. small flute:
Gr. *glottis*, *-idos*, f. mouth of the windpipe: glottis, epiglottis.
Gr. *gorgyra*, f. sewer, drain: *Gorgyra minima* (a butterfly).
L. *gula*; *gurgulio*, gullet, weasand, throat; see **throat**
Gr. *kapne*, smoke-hole, chimney; see *kapnos* under **cloud**
Gr. *niglaros*, m. small pipe, whistle:
Gr. *ochetos*, m. water-pipe, conduit, aqueduct: *Ochetorhynchus ruficaudus* (a bird), *Parochetus communis* (a legume), *Ochetoceras flexuosum* (a Jurassic ammonite).
Gr. *oisophagos*, gullet, weasand; see **throat**
Gr. *ostinon*, n. bone-pipe, tibia: *Ostinops decumanus* (crested cacique).
Gr. *phleps*, *phlebos*, f. vein; *phlebion*, n. dim.; *phlebodes*, full of veins: phlebitis, phlebotomy, aphlebia, *Phlebodium aureum* (a fern), *Phlebotomus minutus* (a sandfly), *Hyla phlebodes* (a frog), *Flacourtia euphlebia* (a flacourtiacead), *Habrophlebia vibrans* (a mayfly).
Gr. *photinx*, *-ingos*, f. a kind of flute; *photingion*, n. dim.; *photingistes*, m. flutist, fifer:
Gr. *physeter*, blowpipe, blower; see *physa* under **blow**
L. *plumbarius*, plumber; see *plumbum* under **lead**
Gr. *rhapate*, f. shepherd's pipe:
Gr. *salpinx*, trumpet, tube; see **horn**
ML. *saphena*, f. a vein in the leg: saphena, saphenous.
L. *sipho*, *-onis* (Gr. *siphon*, *-os*), m. pipe, bent tube; *siphunculus*, m.; *siphonion*, n. dim.: siphon, siphonal, siphonostome, siphonostele, siphonogamous, siphuncle, Siphonaptera, *Siphonops annulatus* (a caecilian), *Siphonocladus pusillus* (an alga), *Siphonorhinus angustus* (a milleped), *Siphunculus nudus* (a worm), *Macrosiphum pisi* (pea-aphid), *Orthosiphon rubicundus* (a mint), *Stenosiphon linifolius* (an onagracead), *Glossiphonia complanata* (a leech).
Gr. *solen*, *-os*, m. pipe, channel, razor-clam; *solenarion*; *solenidion*; *solenion*, n.; *soleniskos*, m. dim.; *solenotos*, channeled: solenoid, typhlosole, *Solenodon paradoxus* (agouta), *Solenostomus cyanopterus* (a fish), *Solenidium nitidum* (a sponge), *Solen ensis* (a razor-clam), *Ostrea soleniscus* (a fossil oyster).
Gr. *syrinx*, *-ingos*, f. pine, tube; *syringion*, n. dim.; *syringodes*, pipelike: syrinx, syringe, *Syringopora ramulosa* (a fossil bryozoan), *Syringa vulgaris* (lilac), *Ancistrosyrinx radiata* (a gastropod).
L. *tibia*, shinbone, pipe, flute; *tibicen*, piper, flutist; see **leg, music**
Gr. *tityros*, m. shepherd's pipe; *tityristes*, m. piper: *Dynastes tityrus* (a beetle).
L. *trachea* (Gr. *tracheia*), windpipe; see **throat**
L. *tubus*, m. pipe; *tubulus*, m. dim.: tube, tubular, tubulate, tubuliferous, tubiparous, tubicolous, *Tubifex multisetosus* (a worm), *Tubulipora serpens* (a bryozoan), *Caryomyia tubicola* (hickory tube-gall), *Tuscarora tubulosa* (a radiolarian), *Sceliphron tubifex* (a mud-dauber).
L. *urethra* (Gr. *ourethra*), f. urinary canal leading from the bladder; *oureter*, *-os*, m. duct from the kidney to the bladder: urethra, urethral, ureter, urethroscope.
L. *vas*, vessel, duct; see **vessel**
L. *vena*, f. bloodvessel, pipe; *venula*, f. dim.; *venosus*, full of veins: vein, venule, venose, venous, venation, venesection, *Delphinium venulosum* (a larkspur).
See: **music, horn, cylinder, reed**

piper, L. pepper; *piperatus*, peppered; *piperatorium*, pepper-shaker; see **pepper, vessel**

piphinx, Gr. a kind of lark; see **lark**

pipilo; pipio; pipo, L. chirp, peep, twitter; *pipulus,* a chirping; see **chirp**

pipistrellus, NL. (It. *pipistrello*), a kind of bat; see *vespertilio* under **bat**

pipos, Gr. young bird, chick; see **bird**

pipra, Gr. woodpecker; see **woodpecker**

pipto, Gr. fall; see **fall**

pipunculus, NL. a genus of flies; see **fly**

pira < Gr. *peira,* attempt, experiment, trial; *peirasis,* attempt; see **try**

pirate < L. *pirata* (Gr. *peirates*), sea-robber, corsair; see **steal**

pirus, L. pear-tree; *pirum,* pear; see *pyrus* under **pear**

piscis, L. fish; *pisciculus,* dim.; *piscator,* fisherman; see **fish, fisherman**

pisinnus, L. small, little; see **little**

pismato- < Gr. *peisma, -tos,* ship's cable; see **rope**

piso- < Gr. *pison,* pea, see *pisum* under **bean**; < *peisa,* obedience, see **follow**; < *piso, -onis,* mortar, see **mortar**

pissa (pitta), Gr. pitch, gum; see **resin**

pissocero- < Gr. *pissokeros,* beeswax; see *cera* under **wax**

pistacia, L. (Gr. *pistake*), pistachio-tree; see **sumac**

pistana, L. arrowleaf; see **waterweed**

pistia, NL. a genus of the arum family; see **arum**

pistil, F. female or seed-bearing part of a flower; see **ovary**

pistillum, L. a club-shaped pounder used in a mortar; see **pestle**

pisto- < Gr. *pistos,* faithful, genuine, < *pistis,* faith; sce **believe**

piston < It. *pistone,* < L. *pinso, pistus,* pound, stamp, grind; see **pestle, pump**

pistor, L. miller, baker; *pistrinum,* mill; *pistrina,* bakery; see **grind, cook**

pistra, Gr. water-trough; see **basin**

pisum, L. (Gr. *pison*), pea; see **bean**

pisyngos, Gr. shoemaker; see **cobbler**

pisynos, Gr. relying on, trusting in; see **believe**

pit < AS. *pytt,* hole; see **hole, hollow, grave, ditch**

pitch < AS. *pic,* < L. *pix, picis;* see *pix* under **resin, glue**

pitcher < OF. *picher;* see **pot**

pitcher-plant
   NL. *cephalotus,* m. a genus of pitcher-plants (Cephalotaceae): *Cephalotus follicularis* (a pitcher-plant).
   NL. *darlingtonia,* f. a genus of pitcher-plants (Sarraceniaceae), < William Darlington, American botanist: *Darlingtonia californica* (California pitcher-plant).
   NL. *nepenthes,* f. a genus of pitcher-plants (Nepenthaceae): *Nepenthes maxima* (a pitcher-plant).
   NL. *sarracenia,* f. a genus of pitcher-plants (Sarraceniaceae), < Michel Sarrasin, Canadian physician and naturalist: *Sarracenia purpurea* (common pitcher-plant), *Sarcophaga sarraceniae* (a fly).

pith < AS. *pith.*
   Gr. *enkephalos,* brain, pith of a palm; see **brain**
   Gr. *enterione,* f. inmost part, pith:
   Gr. *lousson,* n. pith of a coniferous tree:
   L. *medulla,* f. marrow, inmost part, pith; *medullosus,* pithy; *medullaris,* of pith: medulla oblongata, medullary, medullose, medullation.
   Gr. *metra,* pith or heart of a tree; see **womb**
   Gr. *myelos,* m. marrow, pith, spinal cord; *myelodes,* marrowlike: myelon, myelitis, osteomyelitis, poliomyelitis, myelalgia, myeloblast.
   See: **middle, heart, point**

pithanos, Gr. persuasive, plausible, probable, obedient; see *peitho* under **victory**

pithecus, L. (Gr. *pithekos*), ape; *pithekion,* dim.; *pithekideus,* young ape; see **monkey**

pithon, Gr. cellar, see **room**

pithos, Gr. earthen wine-jar; *pithiskos,* dim.; see **pot**

pitiless; see **merciless**

pittacium, L. (Gr. *pittakion*), tablet, parchment, label; see **book**

pituita, L. phlegm, rheum; *pituitosus,* full of phlegm; see **slime**

pity < OF. *pitie*, < L. *pietas*, sense of duty; see **mercy**

pitylos, Gr. splash of oars; see **sound**

pityusa, NL. (Gr. *pityousa*), a kind of spurge; see **spurge**

pityro- < Gr. *pityron*, bran, scale; *pityrodes*, scalelike; see **scale**

pitys, Gr. pine; see **pine**

pius, L. devout, dutiful, kind, tender; see *pietas* under **rite**

pivot, F.; see **axis, point**

pix, *picis*, L. pitch; see **resin**

placatus, L. appeased, calmed; see *placo* under **soothe**

place < OF. *place*, < L. *platea* (Gr. *plateia*), f.; Sp. *plaza*, f.; It. *piazza*, f. square, market-place: placement, placer, emplace, displace, replace, plaza, piazza.
Gr. *arthrembolesis*, f. the setting of a broken limb:
Gr. *chora; chorema*, room, place, space; *choraphion*, dim.; see **space**
Gr. *choros*, place, land, country; see **country**
L. *committo, -missus*, intrust to, resign to: commit, commission, committee.
L. *epibole*, f. a throwing or placing upon, penalty, purpose:
L. *forum*, an open, public place, market-place; see **open**
Gr. *hapante*, everywhere, on all sides:
L. *indo, -itus*, put, set, place into or upon, apply to, impose:
AS. *lecgan*, put down: lay, lie, laid.
L. *locus*, m. place, spot; *locellus; loculus*, m. dim.; *localis*, of a place; *loco, -atus*, put, set, arrange, dispose, quarter, place: locus(pl. loci), locality, location, locomotive, locular, locule, loculicidal, allowance, disallow, allocate, couch, lieu.
Gr. *oötokos*, laying eggs, oviparous:
L. *oviparus*, laying eggs: oviparous.
Gr. *pantothen*, from all sides: pantothenic.
L. *passim*, everywhere, promiscuously:
L. *pono, positus*, place, put, lay; *depositus*, laid, put, or set down; *dispositus*, distributed, arranged; *impositus*, placed upon, assigned to, taken advantage of: post, position, positive, postilion, poster, postage, posture, apropos, appositeness, composition, compost, deposit, dispose, depot, exponent, impostor, expounder, opponent, ovipositor, postpone, propound, proposition, preposition, provost, proposal, repository, supposition, suppository.
L. *scena* (Gr. *skene*), tent, stage, decorative setting or place; see **play**
L. *situs, -us*, m. location, place; *situatus*, located, < *sino, situs*, put, set down: site, situated, situation, *in situ*.
AS. *stede* (G. *stadt*), place, town: stead, steady, steadfast, instead, bedstead, homestead, East Grimstead, Hampstead, Freistadt, Neustadt.
Gr. *stello*, place, arrange, send; see **send**
Gr. *tithemi*, put, place; *thema, -tos*, n. thing placed, laid down, deposited; *thesis*, f. a placing, proposition; *thetos*, placed; *diathesis*, f. arrangement, disposition; *epithetos*, added to; *euthemon, -os*, well-arranged, ordered, neat; *euthemosyne*, f. good management, tidiness; *metathesis*, f. change of position; *synthesis*, f. a putting together; *synthetos*, put together, compounded: thesis, diathesis, parenthesis, hypothesis, antithesis, synthesis, synthetic, epithet, apothecary, anathema, theme, thesaurus, treasury, cyclothem, thesmothete, Themistocles, metathesis, *Epithetica typhoscia* (a moth), *Epithetosoma norvegicum* (a gephyrean), *Diathetes ruficollis* (a beetle).
Gr. *topos*, m. place, position, spot; *topion*, n. dim.; *topikos*, of a place, local; *ektopos*, out of place, displaced, strange: topectomy, topophobia, topotype, Utopia, isotope.

* * * * * * *

L. *-arium, -orium, -eria*, denoting place where, place for: sanitarium(sanatorium), aquarium, emporium, moratorium, auditorium, dormitory, rectory, territory, oratorio, scenario, aviary, granary, library, glossary, commentary, dictionary, cafeteria, sudsateria (laundry), bakery, confectionery, grocery, colliery, vestry, ambry.

L. *-ensis, -e*, denoting place, locality, country. For euphony and smoothness some place-name endings may need adjustment, such as elision or addition of a vowel, keeping a *y* or changing it to *i*, etc., before attaching *-ensis*. Examples of current practice are: *Bromus arvensis* (a grass), *Helix hortensis* (a snail), *Aster yunnanensis* (Yunnan aster), *Epidendrum kewense* (an orchid), *Ulocrinus kansasensis* (a Pennsylvanian crinoid), *Viola missouriensis* (a violet), *Solidago ohioensis* (a goldenrod), *Prunus yedoensis* (an Oriental cherry), *Sisyrinchium idahoense* (a blue-eyed grass), *Poteriocrinites illinoisensis* (a Mississippian crinoid), *Vaccinium canadense* (Canada blueberry), *Acnida alabamensis* (a water-hemp), *Crepicephalus iowensis* (a trilobite), *Begonia paraguayensis* (a begonia), *Rubus allegheniensis* (a blackberry), *Phoenix canariensis* (a palm), *Zygospira kentuckiensis* (a Silurian brachiopod), *Aristolochia brasiliensis* (a birthwort), *Stanhopea devoniensis* (an orchid).

AS. *-ern*, direction toward: eastern, northern, southern, western.

L. *-etum, -tum*, place where or for; grove: arboretum, pinetum, arbustum, rosetum.

L. *-ile*, place for, as *caprile*, n. goat-stall; *ovile*, sheepfold: campanile.

L. *-ter, -tris, -tre; -ester, -estris, -estre*, belonging to, place where: *Rosa palustris* (swamp-rose), *Alyssum alpestre* (alpine alyssum).

Gr. *-terion*, place where: *deipneterion*, n. dining-room; *gymnasterion*, gymnasium; *hestiaterion*, banquet-hall; *koimeterion*, sleeping-room, cemetery; *lesterion*, nest of robbers; *louterion*, bathtub:

See: **house, country, nest, space, class, fort, stand, town**

placenta, L. cake; *placentarius*, confectioner; see **cake**

placidus, L. quiet, still, gentle; *placidulus*, dim.; *placiditas*, calmness; see **rest**

placitus; placibilis, L. pleasing; see *placeo* under **agreeable**

placo- < Gr. *plax, -akos*, anything flat and wide, tablet; see **plate**

pladaros, Gr. wet, damp, insipid; see **wet**

plaesio- < Gr. *plaision*, an oblong body, figure, or form; see **form**

plaga, L. district, zone, tract, region, see **country**; *plege*, stroke, blow, wound, see **strike**; snare, net, see **trap**

plagiarius, L. kidnapper, literary thief; *plagium*, kidnapping; see **steal**

plagios, Gr. oblique; see **slope**

plagula, L. bed-curtain, sheet of paper; see **curtain, paper**

plagusia, L. probably a crustacean; see **crab**

plain < L. *planus*, level, flat; see **level, simple**

plan < L. *planus*, flat, level; see **aim, map, form**

plancto- < Gr. *planktos*, wandering, roaming; see *planes* under **wander**

plancus, L. flat-footed, see **foot**; < Gr. *plangos*, a kind of eagle, see **eagle**

plane < L. *planus*, even, flat, level, see **level, face, side, smooth, sit**

planet < L. *planeta* (Gr. *planes, -etos*, wanderer), wandering star; see **star, wander**

plane-tree

L. *platanus* (Gr. *platanos*), f. plane-tree, sycamore: *Platanus occidentalis* (western sycamore), *Sterculia platanifolia* (Japanese varnish-tree), *Acer platanoides* (Norway maple), *Acer pseudoplatanus* (sycamore-maple).

plango, planctus, L. strike; see **strike**

planguncula, L. wax doll; see **doll**

plank < OF. *planche*, < L. *planca*, board, slab; see **board**

plant < L. *planta*, f. vegetation; *plantarium*, n. nursery: plantlet, plantule, planting, plantation, implant, transplant, Plantagenet.

Gr. *botane*, f. herb, grass, pasture; *botanion*, n. dim.; *botanikos*, of herbs; *botanodes*, herbaceous, full of herbs: botany, botanical, botanist, paleobotany.

Gr. *Dioskorides*, m. Dioscorides, Greek herbalist and physician: *Dioscorea sativa* (a yam).

L. *herba*, f. soft, usually green vegetation as contrasted with woody plants; *herbula*, f. dim.; *herbaceus*, of herbs; *herbarius*, m. botanist; *herbosus*, full of herbs; *herbilis*, fed with herbs: herb, herbage, herbal, herbarium, herbaceous, herbivorous, arbor, *Dorycnium herbaceum* (a canary-clover), *Aspergillus herbariorum* (herbarium-mold), *Phoma herbarum* (a fungus), *Salix herbacea* (dwarf willow),

Gr. *lachanon*, n. potherb, vegetable; *lachanidion; lachanion*, n. dim.; *lachaneros; lachanios*, of vegetables; *lachanites*, m. vegetable-gardener:

L. *Linnaeus, Carolus*, Carl von Linne, Swedish botanist, establisher of binomial nomenclature; see *linnaea* under **honeysuckle**

L. *olus, oleris*, n. potherbs, vegetables, greens; *oleraceus*, herbaceous, vegetable; *olitor, -is*, m. gardener; *olitorius*, of potherbs: oleraceous, olitory, olericulture, *Brassica oleracea* (cabbage), *Corchorus olitorius* (a jute-plant).

Gr. *phyton*, n. plant; *phytarion*, n. dim.; *phytikos*, of plants; *phytourgos*, m. gardener; *phyteuma, -tos*, n. that which is planted; *phyteutos*, planted; *phytorion*, n. nursery; *phyteuo*, plant: phytoma, phytophagous, phytosis, phytopathology, phytobezoar, dermatophyte, hydrophyte, neophyte, saprophyte, epiphyte, Schizophyta, emphyteusis, *Phyteuma spicatum* (a campanulacead), *Hydrophytum formicarum* (a plant), *Sarcophyte sanguinea* (a balanophoracead), *Trichophyton rosaceum* (a ringworm fungus), *Phytophaga destructor* (Hessian fly).

L. *Theophrastus* (Gr. *Theophrastos*), m. a renowned botanist and philosopher: *Abutilon theophrasti* (Indian mallow), *Theophrasta americana* (a primulalad).
AS. *weod*, weed: weedy, bindweed, duckweed, driftweed, knapweed, milkweed, seaweed, smartweed.
AS. *wyrt*, plant: figwort, liverwort, lungwort, milkwort, soapwort, spleenwort, swallowwort, whortleberry.

CLASSICAL PLANTS OF UNCERTAIN IDENTITY

L. *achaemenis, -idis*, f. an Indian plant: *Achimenes grandiflora* (a gesneriacead).
L. *acrifolium*, n. an unknown tree:
Gr. *agallis, -idos*, f. a plant, probably an iris:
Gr. *akeanos*, m. a plant:
Gr. *akidoton*, n. a plant:
Gr. *apharke*, f. a plant:
L. *apiastellum*, n. a plant:
L. *aproxis*, f. a plant:
L. *arianis, -idis*, f. a Persian plant:
Gr. *arkeion*, n. a plant, probably burdock:
L. *assefolium*, n. a plant:
L. *asyla*, f. a plant:
L. *auletica*, f. a plant:
Gr. *ballis, -eos*, f. a plant:
Gr. *belenion*, n. a kind of plant:
L. *burrhinum* (Gr. *bourrhinon*), n. oxnose, a plant:
L. *callicia*, f. a plant:
L. *caragogus*, f. a plant:
L. *chalcetum*, n. a plant:
L. *conula*, f. a plant:
L. *cotonea*, f. a name given to several plants: *Cotoneaster bullata* (a cotoneaster), melocoton.
L. *corysidia*, f. a plant:
L. *culix, -icis*, m. a plant:
L. *diachetum*, n. a plant:
Gr. *echetrosis*, f. a plant: *Echetrosis pentasperma* (a composite).
L. *echinopus, -podis* (Gr. *echinopous, -podos*), m. a prickly plant:
L. *euplia*, f. a plant:
L. *exedum*, n. a plant:
L. *gabalium*, n. an Arabian plant:
L. *gith (git)*, n. a plant: *Githopsis specularioides* (a campanulacead), *Agrostemma githago* (corn-cockle).
L. *halicacabus*, f. (Gr. *halikakabon*, n.), a plant with inflated fruits: *Cardiospermun halicacabum* (heartseed).
Gr. *halimon*, n. a plant growing near the sea: *Baccharis halimifolia* (groundsel).
L. *helvella*, f. a small potherb: *Helvella elastica* (a fungus).
Gr. *ikme*, f. a plant growing in wet places:
Gr. *iphyon*, n. a plant:
Gr. *ipnon*, n. a marsh-plant: *Ipnum mendocinum* (a grass, now called *Leptochloa dubia*).
Gr. *karpyke*, f. an Indian plant:
Gr. *kolymbatos*, f. a plant:
Gr. *konnaros*, m. a thorny, evergreen plant; *konnaron*, n. the fruit of *konnaros*: Connaraceae, *Connarus guianensis* (zebrawood).
Gr. *kopethron*, n. a plant:
Gr. *kynops, -opos*, m. a plant:
Gr. *kyoura*, f. a plant used to produce abortion:
Gr. *labyzos*, f. a plant:
L. *lacca*, f. a plant:
L. *lactilago, -inis*, f. a plant:
L. *lago, -inis*, f. a plant:
L. *lappago, -inis*, f. a plant with burs: *Lappago racemosa* (a grass, now *Tragus racemosus*).
L. *latace*, f. a magic herb:
L. *laver, -is*, f. a water-plant:
Gr. *limnestis*, f. a marsh-plant:
Gr. *malinathalle*, f. an Egyptian plant:
L. *manifolium*, f. a plant:
L. *maurella*, f. a plant:
Gr. *medion*, n. a plant: *Campanula medium*(Canterbury bells), *Epimedium sagittatum* (a berberidacead).

L. *mimmulus*, m. a plant:
L. *minyas*, *-adis*, f. a plant:
Gr. *mnasion*, n. an Egyptian water-plant:
L. *mustellago*, *-inis*, f. a plant:
Gr. *myagros*, m. a plant: *Myagrum perfoliatum* (a crucifer).
L. *nyma*, f. a plant:
Gr. *onopyxos*, m. a plant:
Gr. *onorhynchos*, f. a plant:
Gr. *othonna*, f. a Syrian plant: *Othonna graveolens* (a composite).
L. *pesoluta*, f. an Egyptian plant:
L. *petilium*, n. a plant:
L. *phaunus*, m. a plant:
L. *phellandrion*, n. a plant: *Oenanthe phellandrium* (water-fennel).
Gr. *philistion*, n. a plant:
Gr. *phrynion*, a plant; see *phryne* under **toad**
L. *pistolochia*, f. a plant that facilitates childbirth:
Gr. *polyknemon*, n. a plant:
Gr. *probataia*, f. a plant:
Gr. *protogonon*, n. a plant:
L. *pythonium*, n. dragonwort:
L. *rhexia*, f. a plant: *Rhexia virginica* (meadow-beauty).
L. *scardia*, f. a plant:
L. *semnium*, n. a plant:
L. *sirpe*, *-is*, n. a plant:
L. *staphis*, *-idis*, f. licebane: stavesacre *(Delphinium staphisagria)*.
L. *statumaria*, f. a plant:
L. *teuthalis*, *-idis*, f. a plant:
Gr. *theangelis*, *-idos*, f. a plant:
Gr. *thelygonon*, n. a plant:
Gr. *theobrotion*, n. a plant:
Gr. *thorybethron*, n. a plant:
L. *tinus*, m. a plant: *Viburnum tinus* (laurustine), *Clethra tinifolia* (a white-alder).
See: **tree, flower, leaf, alga, fungus, gymnosperm**

planta, L. vegetation, sprout, see **plant**; sole of the foot, see **sole**

plantain < L. *plantago*, *-inis*, f.: *Plantago aristata* (a plantain), *Antennaria plantaginifolia* (an everlasting), *Schizocapsa plantaginea* (a tacca).
L. *arnoglossa*, f. (Gr. *arnoglosson*, n.), sheep's-tongue, a kind of plantain:
L. *psyllium* (Gr. *psyllion*), n. fleawort, a plantain: *Plantago psyllium* (fleawort).
L. *sicelicum* (Gr. *sikelikon*), n. a kind of fleawort:

planus, L. even, flat, level, plain, clear, smooth; *planarius*, level; see **level, clear**

plasma, Gr. that which is formed or molded, image, figure, substance; *plasis*, a molding; *plastes*, molder, modeler; *plastikos*, fit for molding, pliable; *plastos*, formed, molded; see *plasso* under **make**

plaster < L. *emplastrum* (Gr. *emplastron*), n. a medicinal preparation of pasty consistency for external application, salve, coating for walls, mortar; *emplastros*, daubed over; *epiplasma*; *kataplasma*, *-tos*, n. poultice, plaster, salve; *epiplastos*, plastered; *kataplastes*, m. plasterer: plasterer, plasterwork, plasterboard, plaster-of-Paris, emplastrum, cataplasm.
L. *albarium*, n. white plaster, stucco; *albarius*; *albinus*; *dealbator*, *-is*, m. whitener, plasterer:
L. *arenatum*, n. mortar made with sand:
L. *caementum*, a substance used as a binder; see *caementarius* under **mason**
L. *cycnarium*, n. a kind of eye-salve:
L. *dropax*, *-acis* (Gr. *-akos*), m. a pitch plaster used as a depilatory: dropax.
L. *gypsum* (Gr. *gypsos*), chalk, lime-plaster; see **gypsum**
L. *intrita*, f. paste, mash, < *intero*, *-tritus*, rub, bruise, or crumble to pieces:
Gr. *koniao*, plaster, daub, stucco; *koniatos*, plastered; *koniama*, *-tos*, n. plaster, stucco; *koniates*, m. plasterer: *Coniatus[Coniates] minusculus* (a fossil beetle).
L. *lipara*, an emollient plaster; see *lipos* under **fat**
Gr. *lithokolla*, cement; see *kolla* under **glue**
Gr. *malagma*, *-tos*, n. an emollient plaster, poultice: malagma, amalgam.
L. *mortarium*, n. a plastic building material, plaster: mortar.
L. *pasta* (Gr. *paste*), f. barley broth, porridge: paste, pastel, pastry, pasteboard.
Gr. *plangonion*, n. a kind of ointment:
L. *tector* *-is*, m. plasterer; *tectorium*, n. plaster, stucco:
L. *trullisso*, *-atus*, plaster, trowel: *Gaudryina trullissata* (a foraminifer).
L. *unguentum*, ointment, perfume; see *unguen* under **fat**
See: **spread, form, earth, fat, glue, mason**

plasticus, L. (Gr. *plastikos*), pertaining to molding, pliable; see *plasso* under **make**

platagema, Gr. clap, slap; see **strike**

platalea, L. spoonbill; see **heron**

platamon, Gr. any broad, flat surface; see *platys* under **broad**

platanus, L. (Gr. *platanos*), plane-tree, sycamore; see **plane-tree**

platax, Gr. a perchlike fish; see **perch**

plate < OF. *plate*, < ML. *platus*, prob. < Gr. *platys,* flat.

L. *bractea,* thin metal plate, gold-leaf, scale; see **scale**

L. *catillus,* small bowl, dish, plate; see **cup**

L. *charta* (Gr. *charte*), thin leaf of paper, lamina; see **paper**

L. *cymbalum* (Gr. *kymbalon*), n. a musical instrument consisting of two brass plates; *kymbalion,* n. dim.: cymbals, *Cymbalopus bigibbosus* (a bug).

L. *discus* (Gr. *diskos*), m. quoit, flat, circular plate: discal, discoid, disk, discobolus, discus, desk, dish, *Discosira sulcata* (a diatom), *Discina convexa* (a fossil brachiopod), *Discinisca lamellosa* (a brachiopod), *Holodiscus dumosus* (a rosacead), *Porodiscus elegans* (a diatom), *Chrysops discalis* (a deerfly), *Acromyrmex disciger* (an ant).

Gr. *elasma, -tos,* n.; *elasmos,* m. metal plate: elasmobranch, *Elasmophyllum attenuatum* (a fossil coral), *Elasmosaurus platyurus* (a fossil reptile), *Streptelasma profundum* (a fossil coral).

L. *gabata,* f. dish, platter: *Gabata semipunctata* (a beetle).

Gr. *kernos,* n. large earthen dish with wells in the bottom for fruits:

L. *lamina,* f. thin plate, veneer, blade; *lamella; lamnula,* f. dim.: laminal, lamination, lamellal, lamellibranch, omelet, laminarin, *Laminaria bulbosa* (kelp), *Anastomus lamelligerus* (a stork), *Paradoxides lamellatus* (a trilobite), *Conopeum lamellosum* (a fossil bryozoan).

L. *lanx, lancis,* plate, dish, scale of a balance; see **balance**

Gr. *lekos,* n. plate, platter, dish, pan; *lekis, -idos,* f.; *lekiskos,* m.; *lekarion; lekiskion,* n. dim.; *lekane,* f. dish, pot, pan; *lekanis, -idos; lekaniske,* f.; *lekanidion; lekanion,* n. dim.: lecotropal, lecanomancy, *Lecanium viride* (coffeescale), *Lecanidium atratum* (a fungus), *Lecanora esculenta* (a lichen), *Lecidea granulosa* (a lichen).

Gr. *lopas, -ados,* f. flat dish, plate; *lopadion,* n. dim.: *Lopadiocrinus granulatus* (a Permian crinoid).

Gr. *magis, -idos,* f. dish, platter, kneading trough; *magidion,* n. dim.:

L. *mazonomus* (Gr. *mazonomos*), m. large dish, charger, trencher:

L. *missorium,* n. dish, charger:

Gr. *oxybaphon,* n. saucer: *Oxybaphus himalaicus* (a nyctaginacead), *Clypeaster oxybaphon* (a fossil echinoid).

L. *pagina,* f. leaf of a book, slab, tablet: page, pagination.

Gr. *paropsis,* a dessert-dish; see *opson* under **food**

Gr. *patane,* f. a flat dish; *patanion,* n. dim.: *Patanocnema ovata* (a bug).

L. *patera,* f. saucer, a broad, flat dish: *Cinachyrella paterifera* (a sponge).

L. *patina,* a wide, shallow vessel, pan; see **pan**

L. *petalum* (Gr. *petalon*), a thin, metal plate, leaf; see **leaf**

L. *phalera* (Gr. *phalaron*), disk or boss of metal used as an ornament; see **ornament**

L. *phiala* (Gr. *phiale*), saucer, bowl; see **cup**

Gr. *plathanon,* n. platter, dish: *Plathanocera uniformis* (a beetle).

Gr. *platysma, -atos,* n. flat piece, plate; *platysmation,* n. dim.:

Gr. *plax, plakos,* f. anything flat and wide, plate, tablet; *plakodes,* flat; *plakinos,* made of boards or flat parts: placoid, placoderm, placosaurus, siphonoplax, *Placobdella rugosa* (a leech), *Placonotus longicornis* (a beetle), *Placodes amethystina* (a beetle), *Scleroplax granulata* (pea-crab), *Notoplax violacea* (a chiton), *Monomma triplacinum* (a beetle), *Camptoplax coppingeri* (a crab).

Gr. *ptyx; ptyche,* plate; see **fold**

L. *scheda* (Gr. *schede*), strip of papyrus, leaf of paper; see **paper**

L. *scutella; scutula,* f. little, flat dish, platter, plate, diamond-shaped figure, dim. of *scuta (scutra),* f. flat tray, platter: *Scutellaria lateriflora* (a skullcap), *Catactygnus scutellaris* (a beetle). See *scutum* under **shield**

Gr. *sphallos,* m. round, metal plate for throwing: *Sphalloglandulus unicus* (a fly), *Sphallonycha roseicollis* (a beetle).

L. *tabula,* board, flat piece of anything on which to write; *tabularis,* of boards or plates; see **table**

See: **pan, vessel, scale, board, leaf**

platea, L. (Gr. *plateia*), street; see **way**

platessa, L. flounder; see **flatfish**

plathanon, Gr. dish, platter; see **plate**

platinum < Sp. *platina,* a silverlike substance; see *plata* under **silver, ELEMENTS** under **thing**

platys, Gr. broad, wide, level, flat; *platanon,* anything broad and flat; see **broad**

plaudo, *plausus,* L. strike, clap the hands in praise; *plausibilis,* praiseworthy; see **strike**

plaustrum, L. wagon, cart; *plostellum,* dim.; see **vehicle**

plautus, L. broad, flat, flat-footed; see **broad**

plax, Gr. plate, tablet; see **plate**

play < AS. *plegan:* player, playful, playmate, plaything, playwright, Playfair, misplay.

Gr. *abax, -akos,* gaming-board divided into square compartments; see *abacus* under **number**

L. *actor, -is,* m. performer, player: actor, actress.

Gr. *akroama, -tos,* n. entertainment with reading or music:

L. *Aeschylus* (Gr. *Aeschylos*), m. Greek dramatist: *Aeschylus clathratus* (a beetle).

L. *alea,* f. game of chance with dice; *aleator, -is,* m. dice-player, gamester; *alearis,* of dicing; *aleatorius,* of a gamester: aleatory, *Misumena aleatoria* (crab-spider).

L. *arena,* f. place for games and contests: arena.

Gr. *athlos,* m. contest for a prize; *athletes,* m. contestant; *athletikos,* of athletes: athlete, athletic, Pentathlon.

Gr. *athyrma, -tos,* n. toy, plaything, < *athyro,* play, sport: *Athyrma interpuncta* (a butterfly).

L. *avocatio, -onis,* f. side interest: avocation.

Gr. *cheironomia,* f. pantomimic movement, gesticulation: *Chironomus plumosus* (a midge).

L. *circus,* circular enclosure for entertainment purposes; see **circle**

L. *comoedia* (Gr. *komoidia*), a light and amusing play, humorous drama; see *comicus* under **laugh**

L. *cottabus* (Gr. *kottabos*), m. a game in which wine was thrown into cups or saucers: cottabus.

L. *cubus* (Gr. *kybos*), a solid with six square sides, cube, die; *kybeutes,* gambler; *kybeutikos,* of dicing; *kybeuterion,* gambling-house; see *cubus* under **three**

L. *delecto, -atus,* cause pleasure; see **joy**

L. *diverto, -versus,* turn aside, entertain: divert, diversion.

Gr. *drama, -tos,* n. act, play; *dramation,* n. dim.; *dramatikos,* of drama: drama, dramatic.

Gr. *ereschelia,* f. sport, raillery:

OF. *eschec,* check, stop, < Per. *shah,* king; *eschequier,* chessboard: check, checkers, chess.

L. *fritillus,* dice-box, spotted; see **spot**

L. *histrio, -onis; hister,* m. actor; *histriculus,* m. dim.; *histricus; histrionalis; histrionicus,* of actors: histrionic, *Hister stercorarius* (a beetle), *Histriophoca fasciata* (ribbon-seal), *Histrionicus pacificus* (harlequin-duck), *Pterophryne histrio* (a sargassum-fish).

L. *jocus,* jest, game, sport; see **laugh**

Gr. *kondax, -akos,* m. a kind of game:

L. *lascivus,* sportive, frolicsome, frisky, wanton; see **lewd**

L. *ludo, lusus,* play; *lusus, -us,* m. play, sport, jest, fun; *lusor, -is,* m. player; *lusorius,* of a player; *ludibundus,* playful, frolicsome; *ludicer, -cra, -crum; ludicrus,* sportive; *ludibrum,* n. derision, mockery, sport; *ludibriosus,* mocking, scornful; *ludimentum,* n. plaything; *collusor, -is,* m. playmate; *illusorius,* ironical, mocking; *ludifico, -atus,* make sport of, make a fool of, tease, banter, mock, ridicule: ludicrous, allusion, collusion, delusion, elude, illusive, prelude, interlude, postlude, *Anastrepha ludens* (orange-maggot), *Clytocybe illudens* (a mushroom), *Chrysobothris ludificata* (a beetle).

Gr. *manganeutes,* juggler, trickster, quack; see **lie**

L. *mimus* (Gr. *mimos*), imitator, actor; *mimulus,* dim.; see **equal**

Gr. *oine,* f. die, ace: *Oene plana* (a pelecypod).

Gr. *paizo,* play like a child, sport; *paidia,* f. childish play; *paidodes,* playful; *paigma, -tos,* n. play, sport; *paignidion,* n. farce; *paignion,* n. plaything, toy; *paignios,* sportive; *paiktos,* played with; *paiktes,* m. player; *sympaistes,* m. playmate: *Paectophyllum escherichi* (a milleped), *Paegniodes cupulatus* (an ephemerid).

Gr. *parodia,* f. burlesque: parody.

L. *petulcus,* butting, frisky: *Melissodes petulca* (a bee).

Gr. *plethrion,* n. circus:

L. *sannio, -onis,* buffoon, harlequin; see **fool**

L. *scena* (Gr. *skene*), f.; *skenos*, m. tent, stage, decorative setting or place; *skenidion*, n. dim.; *skenikos*, of the stage, theatrical; *episkenos*, on the stage: scene, scenery, scenic, scenical, proscenium, *Salticus scenicus* (a spider), *Scenedesmus bijugatus* (an alga), *Rhaphidoscene conica* (a foraminifer), *Pteroscenium arcuatum* (a radiolarian).

L. *stadium*, n. arena, race-course: stadium.

L. *talus*, anklebone, knucklebone, die; *taxillus*, dim.; see **foot**

L. *tessera*, die; see **four**

L. *theatrum* (Gr. *theatron*), n. playhouse: theatre, theatrical, amphitheatre.

Gr. *Thespis*, founder of Greek drama; see *thespesios* under **wonder**

L. *tragoedia* (Gr. *tragoidia*), a serious or heroic drama; see *tragicus* under **serious**

Gr. *zatrikion*, n. the game of chess; *zatrikizo*, play chess:

See: **rest, joy, laugh, tease, art, music, magic**

please; pleasure < OF. *plaisir*, < L. *placeo*, satisfy, be agreeable; see **agreeable, joy**

plebs, L. the common people; *plebeius*, of the people; see **people**

plecto- < Gr. *plektos*, plaited, twisted; < L. *plecto, plexus*, braid, plait, interweave; *plectilis*, plaited, complicated, see **weave**; < Gr. *plektes*, striker, see *plaga* under **strike**

plectrum, L. (Gr. *plektron*), a tool for plucking or striking a stringed instrument, spur; see **point**

pledge < OF. *plege;* see **promise**

pleganon, Gr. rod, stick; see **rod**

plegas, Gr. sickle; see **sickle**

-plegia < Gr. *plege,* stroke; see *plaga* under **strike**

plegma, Gr. anything twined or plaited, net, wreath, chaplet; see *plecto* under **weave**

Pleiades, Gr. the constellation of the seven stars; see **star**

pleistos, Gr. most; see *polys* under **number**

plemmyris, Gr. flood-tide; see **wave**

plemno- < Gr. *plemne,* nave of a wheel; see **box**

plenty < L. *plenitas*, fullness, repletion; *plenus*, full; see **abundance**

pleo- < Gr. *pleo,* swim, sail; *pleustes; ploter,* sailor, see **swim;** < *pleon,* more, comp. of *polys*, many, see **number;** < *pleos; pleres,* full, see **abundance**

pleonasmos, Gr. more than enough, < *pleonazo,* claim too much, exaggerate; see **exceed**

plesios, Gr. near; see **near**

plesmone, Gr. satiety, fullness, plenty; see **abundance**

pletho- < Gr. *plethos; plethys,* crowd, multitude, throng, fullness; *plethora,* fullness, satiety; *plethysmos,* increase, enlargement; see *pleos* under **abundance**

plethrion, Gr. circus; see **play**

pletura, L. fullness; *pletus,* filled; see *pleo* under **abundance**

pleumon, Gr. probably a kind of jellyfish; see **animal**

pleuro- < Gr. *pleura,* side, see **side;** < *pleuron,* rib; *pleurikos,* of the ribs and side, see **rib**

plexus, L. braided, plaited, interwoven; see *plecto* under **weave**

plicatus; plicitus, L. folded; see *plico* under **fold**

plicio- < Gr. *plikion,* a kind of cake; see **cake**

plinthus, L. (Gr. *plinthos*), brick; see **brick**

plio- < Gr. *pleion,* more; see *polys* under **number**

plisto- < Gr. *pleistos,* most; see *polys* under **number**

ploco- < Gr. *plokos; plokamos,* lock of hair, braid, curl, chaplet, wreath; *plokion,* dim.; *plokios,* twined, twisted; *plokeus,* braider, plaiter, weaver; see *plecto* under **weave**

plodo, *plosus,* L. clap, strike; see *plaudo* under **strike**

ploeo- < Gr. *ploion,* a floating vessel, any kind of ship; see **ship**

plorabilis, L. lamentable; see *ploro* under **weep**

plot, prob. < AS. *plot;* see **map, aim**

plotos, Gr. floating; see *pleo* under **swim**

plover < OF. *plovier*, the rain-bird, < *pluvia*, rain: *Pluvialis apricaria* (European golden plover).
L. *charadrius* (Gr. *charadrios*), m. a yellowish bird; a genus of plovers: *Charadrius melodus* (piping plover).
L. *otis, -idis* (Gr. *-idos*), f. bustard, a ploverlike bird: *Otis tarda* (a great bustard), *Otidiphaps insularis* (a pigeon).

plow; plough < AS. *ploh* (G. *pflug*): plowman, plowshare, plowfish, *Plowrightia ribesia* (a fungus).
L. *aratrum* (Gr. *arotron*), n. plow; *arator, -is* (Gr. *aroter, -os*); *arotes; arotreus*, m. plowman, husbandman; *arabilis*, tillable; *arosimos; arotos*, arable, fruitful; *arotrios*, of husbandry; *aro, -atus*, plow, till: arable, aratory, *Coelorhynchus aratrum* (a fish), *Arotrophora lividana* (a moth), *Saxifraga exarata* (a saxifrage).
L. *auris*, ear, moldboard; see **ear**
Gr. *bootes*, ox-driver, plowman; see *bos* under **cattle**
ML. *caruca*, f. plow: carucate, caruage.
L. *culter*, plowshare; see **knife**
Gr. *hynis (hynnis); hynne*, f. plowshare: *Hynnis cubensis* (a fish), *Metynnis luna* (a fish), *Hynnodus atherinoides* (a fish).
L. *liro, -atus*, plow; see *lira* under **ridge**
Gr. *pharos*, n. plow; *apharotos*, unplowed:
L. *stiva*, plow-handle; see **handle**
L. *vomer, -is*, m. plowshare: vomer, vomerine, *Vomer dorsalis* (a fish), *Selene vomer* (horsehead-fish), *Cristivomer namaycush* (Mackinaw trout).
See: **till, hoe**

pluck < AS. *pluccian;* see **tear**

plug < D. *plug;* see **close**

plum < AS. *plume*, < L. *prunum* (Gr. *prounon; proumnon*), n.; *prunus*, f. plum-tree: plum pudding, plumcot, prune, prunitrin, *Prunus domestica* (plum), *Prunus ilicifolia* (islay), *Viburnum prunifolium* (black haw), *Clitopilus prunulus* (a mushroom), *Taphrina pruni* (a fungus), *Evernia prunastri* (a lichen).
L. *armeniacum*, n. apricot: *Prunus armeniaca* (apricot).
Gr. *atalymnos*, m. a plum-tree:
Gr. *berikokkon*, n. apricot:
Gr. *brabylos*, f. wild plum-tree; *brabylon*, n. sloe:
L. *cerasus* (Gr. *kerasos*), f. cherry-tree; *cerasum* (Gr. *kerasion*), n. cherry; *cerasinus*, of cherry: *Prunus cerasus* (sour cherry), cerasein, cerasin, *Pholiota cerasina* (a mushroom).
Sp. *ciruela*, f. plum:
L. *coccymelum* (Gr. *kokkymelon*), n. a plum: *Prunulum coccymelium* (a radiolarian).
Gr. *dorakinon*, n. apricot:
Ab.Am. *icaco*, coco-plum: *Chrysobalanus icaco* (coco-plum), *Icacorea paniculata* (marlberry).
L. *myxa*, f. a kind of plum-tree: *Cordia myxa* (sebesten-plum).
Gr. *pados*, f. a tree, probably a cherry: *Prunus padus* (bird-cherry), *Eriophyes padi* (cherryleaf-gall), *Ficus padifolia* (a fig).
NL. *parinarium*, n. a genus of the plum group, < Ab.Am. *parinari: Parinarium macrophyllum* (gingerbread-tree).
Gr. *spodias; spondias*, a kind of plum-tree; now a genus of the sumac family; see **sumac**

pluma, L. soft feather, down; *plumella; plumula*, dim.; *plumosus*, feathery, downy; see **feather**

plumbago, L. a leadlike ore, graphite, a plant; see **leadwort**, *plumbum* under **lead**

plumbum, L. lead; *plumbeus*, of lead; *plumbarius*, plumber; see **lead, pipe**

plunder < G. *plunder*, household goods, lumber, trash.
Gr. *anairema*, booty, spoil; see *anaireo* under **destroy**
Gr. *enaron*, n. spoil, booty:
Gr. *harpagma*, booty, plunder; see *harpago* under **take**
Gr. *helor*, n. booty, spoil, prey:
Gr. *kyrma, -tos*, n. booty, prey, plunder:
Gr. *laphyron*, n. booty, plunder: *Laphyroscopus capitatus* (a wasp).
Gr. *leis*, booty, spoil; see *lestes* under **steal**
L. *pilo, -atus*, rob, plunder; see **steal**
L. *praeda*, f. booty, plunder; *praedator, -is*, m.; *praedatrix, -icis*, f. plunderer; *praedo, -atus*, pillage, plunder, rob; *praedonulus*, m. petty robber: prey, predatory, predaceous, predacity, depredation.
L. *raptum*, plunder; *raptor*, robber, plunderer; see *rapio* under **take**
AS. *reaf*, spoil: reave, bereave, bereavement.

Gr. *rhysion,* n. booty, prey: *Rhysium bimaculatum* (a beetle).
Gr. *sinis,* plunderer, destroyer; see *sinos* under **hurt**
Gr. *skylon,* n. spoil, booty:
L. *spolium,* n. booty; *spoliator, -is,* m. plunderer; *spolio, -atus,* strip, rob, plunder: spoil, despoil, spoliary, spoliation, *Henicoptera spoliata* (a fly).
Gr. *sylema, -tos,* n. booty, plunder; *syletes; syletor, -os,* m. robber; *aposylesis,* f. a plundering: *Syletor imerinae* (a beetle), *Syletria angulata* (a grasshopper), *Charadrodromia syletor* (a fly).
See: **steal, take, hurt**

plunge < OF. *plungier;* see **dip**

plus, L. more; *plurimus,* most; *pluralis,* pertaining to more than one; see *multus* under **abundance**

pluteus, L. shed, parapet, breastwork, backboard; see **wall**

pluto- < Gr. *ploutos,* wealth, riches, < *Ploutos,* god of wealth; *plousios,* rich, wealthy, ample, abundant, see **wealth;** < L. *Pluto* (Gr. *Pluton*), god of the nether world, see **hell**

pluvius, L. rainy; *pluvialis,* of rain; *pluviosus,* rainy; see *pluo* under **drop**

plyno-; plysi-; plyto- < Gr. *plynos; plynter,* washtub, basin; *plysis,* a washing; *plytos,* washed; see *plyno* under **wash, basin**

plysma, Gr. dish-water; see **water**

pneumo- < Gr. *pneuma,* wind, air, breath; *pneumatikos,* of the wind, breath; *pneumon,* lung; see **breathe, lung**

pnicto- < Gr. *pniktos,* strangled, suffocated; see *pnigo* under **choke**

pnigalion, Gr. nightmare; see **dream**

pnigeus, Gr. oven; see **oven**

pnigma; pnigmos, Gr. a choking, stifling; see *pnigo* under **choke**

pnyx, Gr. place for public assembly in Athens; see **gather**

poa, Gr. grass; see **grass**

pocket < OF. *poke; poque,* pouch; see **bag, hollow**

poco- < Gr. *pokos,* fleece, wool; *pokas, -ados,* hair, wool; *pokarion,* dim.; see **wool**

poculum, L. cup, goblet; *pocillum,* dim.; *pocillator,* cupbearer; see **cup**

pod < uncertain origin; see **bag**

podagra, L., Gr. gout; *podagricus,* gouty; see **disease**

podex, L. fundament, anus; see **rump**

podistra, Gr. foot-trap; see **trap**

podium, L. platform, balcony; see **projection**

podo-; podi-; -poda < Gr. *pous, podos,* foot; *podarion; podion,* dim.; *podister,* one who does anything with the feet; see **foot**

poecilido- < Gr. *poikilis, -idos,* a kind of finch; see **finch**

poecilo- < Gr. *poikilos,* varicolored, pied, mottled, spotted; see **color**

poem < L. *poema* (Gr. *poiema*), n. a composition in verse; *poietes,* m. maker, creator of verse: poematic, poesy, poet, poetry, poetical, *Dodecatheon poeticum* (a shooting-star).
Gr. *elegos,* m. mourning song or poem: elegy, elegiac.
Gr. *epos,* n. word, tale, song; *epikos,* heroic: epic.
L. *idyllium* (Gr. *eidyllion*), n. a poem describing pleasant rustic conditions: idyl, idyllic.
Gr. *rhapsodia,* f. recitation of an epic poem: rhapsody.
L. *versus, -us,* m. line of poetry, furrow; *versiculus,* m. dim.: verse, versicle.

poemeno- < Gr. *poimen, -os,* shepherd; *poimne,* flock; see **herd**

poet < Gr. *poietes,* maker, creator of verse; see **poem, make**

-poeus < Gr. *poios,* of a given nature, kind, quality, see **nature;** < *poieo,* make, see **make**

pogon, Gr. beard; see **hair**

pogonia, NL. a genus of orchids; see **orchid**

poinciana, NL. a genus of the bean family; see **bean**

point < OF. *pointe,* < L. *pungo, punctus,* prick, sting, punch; *pungens,* sharp, acrid, biting, piercing; *punctum,* n. point, hole, spot; *puncta,* f. prick, puncture: punctum, puncture, pungent, punctual, punctilious, punctuate, puncheon, poignant, punt, appointment, disappointing, compunction, contrapuntal, expunge, spontoon, *Hypericum punctatum* (a St. Johnswort), *Picea pungens* (Colorado spruce), *Pygosteus pungitius* (a stickleback).

L. *acuo, acutus,* make pointed, sharpen, whet; *acer, acris, acre,* pointed, pungent, piercing, sharp, stinging, sour; *acerbus,* sharp, harsh, sour; *acies, -ei,* f. sharp edge or point, keenness of vision, acuteness of mind; *acriculus,* testy; *aculeus,* m. sting, spur; *acumen, -inis,* n. sharp point of anything; *acuminatus,* pointed, sharpened; *acutalis; acutus,* sharpened, pointed, keen; *acutulus,* dim.: acute, ague, acerbity, acrid, acrimony, acrolein, exacerbate, eager, vinegar, aculeate, acumen, acuminate, *Erigeron acris* (bitter fleabane), *Sedum acre* (wall-pepper), *Parkinsonia aculeata* (Jerusalem thorn), *Solanum aculeatissimum* (a nightshade), *Arabis aculeolata* (a rockcress), *Viverravus acutus* (a creodont), *Quercus acutissima* (sawtooth-oak), *Aster acuminatus* (an aster), *Luffa acutangula* (towel-gourd), *Magnolia acuminata* (cucumber-magnolia).   Derive: cute.

Gr. *akis, -idos,* f. point, beak, barb; *ake; akoke,* f. point, sharp edge; *akidodes; akidotos,* pointed; suf. *-eces (-ekes),* pointed, sharp, as in *amphekes,* two-edged; *euekes,* well-pointed; *leptekes,* fine-pointed; *proekes,* pointed in front; *tanekes,* long-pointed; *xyrekes,* sharp as a razor: *Acidaspis fimbriata* (a trilobite), *Acinonyx jubatus* (cheetah), *Acocanthera venenata* (bushman's-poison), *Pentace burmanica* (thitka), *Isacis migrans* (a nematode), *Spermacoce glabra* (a rubiacead), *Anampses(Ampheces) geographicus* (a fish), *Leptecodon rectus* (a fossil fish), *Amphecostephanus rex* (a mantid).

Gr. *akme; akros,* highest point, end, tip, top; *akmaios,* in full bloom, prime; *epakros,* pointed at the end; see **top**

Gr. *akonesis,* f. a sharpening:

Gr. *amphigyos,* double-pointed, pointed at both ends: *Amphigyus piceus* (a beetle).

Gr. *amphitomos,* cutting on both sides, two-edged:

L. *amycticus* (Gr. *amyktikos*), sharp, biting, pungent, strong:

Gr. *antherix, -ikos,* m. ear of wheat, beard of the ear: *Anthericomma barberi* (a beetle).

L. *apiculus,* point, tip, dim. of *apex,* tip, top; see **top**

L. *aquifolius,* having pointed leaves; see **holly**

Gr. *ardis,* f. point of an arrow: *Ardisia crenulata* (a myrsinacead).

L. *arista,* f. beard of grain, awn, ear; *aristatus,* with ears, awns; *aristosus,* full of ears or awns: aristate, arete, arris, *Aristida longiseta* (a grass), *Pinus aristata* (bristlecone-pine), *Juncus aristulatus* (a rush).

Gr. *ather, -os,* m. ear of wheat, spike, awn; *atherodes,* like ears of wheat: atheroid, atherure, *Atherosperma moschatum* (a monimiacead), *Atherura africana* (a porcupine), *Arrhenatherum elatius* (a grass), *Optonurus atherodon* (a fish).

L. *barba,* beard, barb; *barbula,* dim.; see **hair**

Gr. *belos,* dart, arrow, bolt, sting; *acrobeles; oxybeles,* pointed at the end, sharp-pointed; *tribeles; tribolos,* three-pointed; see **arrow**

L. *cacumen, -inis,* extreme point, end; see **end**

L. *calcar, -is,* n. spur: calcarate, *Calcarius pictus* (a longspur), *Calcaritermes nearcticus* (a termite), *Robulus calcar* (a foraminifer), *Phaseolus calcaratus* (rice-bean), *Hapsiphyllum calcariforme* (a fossil coral), *Impatiens ecalcarata* (a jewelweed).

ML. *calcitrapa,* f. a four-pointed weapon placed upon the ground to impede the movements of the enemy: caltrop, *Centranthus calcitrapa* (a valerianacead), *Trapa natans* (water-chestnut), *Hemitrapa trapelloides* (a Pliocene water-chestnut), *Trapella sinensis* (a pedaliacead).

L. *centrum* (Gr. *kentron*), n. point, prickle, spur, sting, midpoint; *kentrion,* n. dim.; *kentriskos,* m. a fish; *centralis; centratus,* in the middle; *centrosus,* inward, internal: center, central, centrifuge, centrifugal, centrosome, centripetal, eccentricity, concentrated, geocentric, heliocentric, concentric, *Centrosema virginianum* (butterfly-pea), *Centranthus ruber* (a valerianacead), *Centella asiatica* (marsh-pennywort), *Polycentropsis abbreviata* (a fish), *Schizolobus concentricus* (a fossil brachiopod), *Dicentra spectabilis* (bleeding-heart), *Tetracentron sinense* (a tree), *Microcentron retinerve* (a katydid), *Macrocentrus bicolor* (a wasp), *Isoachlya eccentrica* (a fungus), *Heterocentrotus mammillatus* (a sea-urchin), *Macrorhamphosus* (formerly *Centriscus*) *scolopax* (a fish), *Philander centralis* (a woolly opossum).

L. *cuspis, -idis,* f. point, pointed end of anything; *cuspido, -atus,* make pointed: cusp, cuspidate, cuspidal, cuspate, cuspule, cuspidine, bicuspid, *Fraxinus cuspidata* (fragrant ash), *Cheiropleuria bicuspis* (a fern), *Jacaranda cuspidifolia* (a bignoniacead).

Gr. *drimys,* piercing, sharp, keen, pungent: *Drimys aromatica* (a tree).

AS. *ear* (G. *ähre*), the fruiting spike of a cereal: ear.

AS. *ecg,* point, sharp border or margin: edge, edging, edgestone, edgeways, edgewise.

Gr. *echinos,* hedgehog, sea-urchin, spiny; see **hedgehog**

Gr. *epakros*, pointed at the end; see *akros* under **top**

F. *ergot*, cockspur: ergot, ergotinine, ergotism.

Gr. *exoche*, projection, point; see **projection**

L. *fastigo*, *-atus*, bring to a point, sharpen: *Cassiope fastigiata* (Himalaya cassiope).

L. *focus*, hearth, a central point; see **oven**

ON. *gaddr*, sting, goad: gad, gadfly.

Gr. *genys*, jaw, edge of an ax; *amphigenys*, double-edged; see **jaw**

Gr. *glochin*, *-os*; *glochis*, *-idos*, f. point of an arrow; *glox*, *-ochos*, f. beard of wheat, awn: glochidium, *Triglochin maritima* (arrow-grass), *Glochidion coronatum* (a pinflower-tree), *Echinocereus triglochidiatus* (a cactus), *Halicapsa triglochin* (a radiolarian).

L. *intentus*, eager, attentive; see *attentus* under **busy**

Gr. *karcharos*, sharp-pointed, jagged: carcharodont, *Carcharodon megalodon* (a Miocene shark).

Gr. *kenteo*, goad, prick, spur, sting; *kentetes;* *kentor*, *-os*, m. goader, driver, pricker, piercer; *kentetos*, pierced: Centetidae, *Centetes ecaudatus* (a tenrec), *Centeter cinereus* (a fly), *Dermacentor venustus* (tick that transmits Rocky Mountain spotted-fever).

Gr. *koryphe*, head, crown, top, highest point; see **head**

L. *locus*, place, spot; see **place**

L. *mucro*, *-onis*, m. sharp point; *mucronatus*, pointed: mucronate, *Scirpus mucronatus* (a bulrush), *Proctonotus mucroniferus* (a nudibranch).

L. *muricatus*, pointed or spiny like a murex; see *murex* under **mollusk**

Gr. *myops*, *-opos*, m. goad, spur, incentive: *Myopocera basalis* (a beetle).

Gr. *obeliskos*, dim. of *obelos*, spit; see **spear**

Gr. *okys*, swift, quick, sharp; see **swift**

Gr. *orpex*, sapling, goad; see **tree**

Gr. *oxys*, sharp, acute, keen, quick; *oxyno*, sharpen, point, goad, provoke; *oxynter*, *-os*, m. sharpener; *oxypages*, sharp-pointed; *apoxys*, tapering to a point, becoming less; *katoxys*, very sharp: amphioxus, oxymoron, paroxysm, *Oxyechus vociferus* (killdeer-plover), *Oxynoblatta alutacea* (a fossil blattid), *Pomoxys annularis* (crappie), *Paroxyna aequalis* (a fly), *Crataegus oxyacantha* (English hawthorn), *Acacia oxycedrus* (an acacia).

Gr. *perone*, pin, anything pointed for piercing; see **needle**

Gr. *phoxos*, pointed, peaked: *Phoxichilidium maxillare* (a pycnogonid), *Ethesostoma phoxocephalum* (a darter).

AS. *pic*, point: pike, pickerel, picket, pick, peck, peak, picador.

Sp. *picudo*, acuminate, beaked: *Sphyraena picudillo* (a barricuda).

L. *plectrum* (Gr. *plectron*), n. a tool for plucking or striking a stringed instrument, spur: plectrum, plectridium, *Aplectrum hyemale* (puttyroot), *Buccinum plectrum* (a fossil gastropod), *Plectrogenium nanum* (a fish), *Plectronia spinosa* (a rubiacead), *Smicroplectrus robustus* (an inchneumonid), *Polyplectrum bicalcaratum* (peacock-pheasant), *Plectranthus saccatus* (a mint), *Comanthus plectrophorum* (a crinoid).

Gr. *sauroter*, *-os*, m. spike at the butt end of a spear; *saurotos*, spiked:

Gr. *skolops*, anything pointed, thorn, pale; see **thorn**

L. *spica*, f.; *spicus*, m.; *spicum*, n. ear of grain, point; *spicula*, f.; *spiculus*, m.; *spiculum*, n. dim.; *spiceus*, of ears of grain; *spico*, *-atus*, put forth ears; *spiculo*, *-atus*, sharpen to a point: spike, spicate, spicule, spikenard, spikelet, spiciferous, spicigerous, spiciform, Spica *(alpha Virginis)*, *Distichlis spicata* (saltgrass), *Eragrostis sessilispica* (a grass), *Ergates spiculatus* (a beetle), *Aristida spiciformis* (a grass), *Straboscopus spiculosus* (a beetle), *Jacobinia spicigera* (an acanthacead), *Calcigorgia spiculifera* (a coral).

L. *spina*, thorn; see **thorn**

AS. *spir*, young shoot, tapering point: spire.

AS. *spitu* (G. *spitz*), point, pointed rod: spit, spet *(Sphyraena spet).*

Gr. *stachys*, *-yos*, m. ear of grain, spike, a mint; *stachyodes*, like ears of grain; *agastachys*, with many ears or spikes: *Stachys lanata* (woolly betony), *Stachyocrinus zea* (a Permian crinoid), *Agastache rugosa* (a giant-hyssop), *Phyllostachys nigra* (black bamboo), *Aechmaea polystachya* (a bromeliacead), *Ephedra distachya* (an ephedra), *Pterocaulon pycnostachyum* (blackroot).

Gr. *stigme*, point, dot, prick, spot; see *stigma* under **mark**

Gr. *stonyx*, *-ychos*, m. any sharp point: *Stonychophora fulva* (a gryllacridid), *Dicrostonyx rubricatus* (a lemming).

Gr. *storthynx*, *-yngos*, m. point, spike: *Storthyngocrinus fritillus* (a Devonian crinoid).

Gr. *styrax*, *-akos*, m. spike at the butt end of a spear: *Styracosaurus ovatus* (a Cretaceous dinosaur).

L. *subula*, awl; *subulatus*, awl-shaped, pointed; see **awl**

Gr. *systenos,* running or tapering to a point: *Systenognathus porosus* (a beetle), *Systenocentrus quinquedentatus* (a phalangid).

L. *tersus,* cleansed, neat, succinct, pithy, pointed; see *tergo* under **wash**

Gr. *thegaleos,* pointed, sharp; *thektos,* sharpened; *thexis,* f. a sharpening, < *thego,* sharpen; *amphithektos; dithektos,* sharp on both sides, two-edged: *Thegalea haemorrhanta* (a moth), *Amphthectus dahlbomi* (a gallfly).

See: **thorn, hawthorn, cactus, nettle, hedgehog, porcupine, thistle, spear, tooth, cone, sword, hook, arrow, mark, aim, needle, awl, horn, dagger, top, bore, cut, knife, sour, arouse, spot, projection**

poison < OF. *poison,* < L. *potio,* drink.

Gr. *ios,* m. poison; *iotikos,* poisonous, venomous: *Iophoroxenus exilimanus* (a spider).

L. *noxius,* harmful, sometimes in the sense of poisonous; see *noceo* under **hurt**

Gr. *pharmakon,* drug, poison; *pharmakos,* poisoner; see **drug**

Gr. *sepsis,* decay, putrefaction, poisoning from bacterial action; see *sapros* under **rot**

L. *toxicum* (Gr. *toxikon*), n. poison for arrows: toxicology, toxemia, hematoxin, neurotoxin, antitoxin, intoxication, toxiferin, *Rhus toxicodendron*(poison-ivy), *Strychnos toxifera* (curare-plant), *Antiaris toxicaria* (upas-tree), *Vincetoxicum hirsutum* (an asclepiadacead), *Cynanchum vincetoxicum* (white swallowwort), *Toxicodendron capense* (a euphorbiacead).

L. *venenum,* n. poison; *veneficus,* poisonous, magical; *venenosus,* very poisonous: venom, venomous, *Zygadenus venenosus* (death camas), *Physostigma venenosum* (Calabar bean), *Hydnocarpus venenata* (a flacourtiacead).

L. *virus,* n. slime, poison; *virosus,* poisonous, foul, slimy; *virulentus,* full of poison: virus, virulent, virulence, *Cicuta virosa* (water-hemlock).

pokeberry

L. *limeum,* n. a Gallic herb yielding arrow-poison; now a genus of the pokeweed family: *Limeum africanum* (a phytolaccacead).

NL. *phytolacca,* f. a genus of the pokeweed family: Phytolaccaceae, *Phytolacca americana* (pokeweed), phytolaccin.

pole < AS. *pal,* < L. *palus,* stake; < L. *polus* (Gr. *polos*), pivot, axis, pole of the earth; see **axis, rod, stem, pillar, tree**

polecat < OF. *poule,* fowl, hen, and *cat;* see **weasel**

polemonium, L. (Gr. *polemonion*), a genus of the phlox family; see **phlox**

polemos, Gr. war; *polemios,* of war; see **fight**

polenta, L. pearl-barley, barley-meal; *polentarius,* of pearl-barley; see **flour**

polesis, Gr. movement; see *poleo* under **move**

police < L. *politia* (Gr. *politeia*), administration, condition of a state, business of government; see **guard**

polimen, L. polish, brightness; see *polio* under **smooth**

polimentum, L. testicle; see **testicle**

polios, Gr. gray; see **gray**

polis, Gr. city; *polidion,* dim.; *polistes,* founder of a city; *polites,* citizen; see **town**

polish < L. *polio, -itus,* make smooth, adorn, finish, furbish; see **smooth, rub, scrape, manner, know, light**

polium, L. (Gr. *polion*), an aromatic plant; see **mint**

pollache, Gr. often, diverse; see **number**

pollachius, NL. a kind of cod; see **cod**

pollen, L. fine flour, mill-dust, dust; *pollinarius,* of flour; see **flour**

pollex, L. thumb; *pollicaris,* of a thumb; see **finger**

pollicitum, L. a promise; *pollicitator; pollicitatrix,* promiser; see **promise**

pollinctor, L. undertaker; see **undertaker**

pollostos, Gr. smallest, least; see **little**

pollubrum, L. washbasin; see **basin**

pollucibilis, L. sumptuous, rich, magnificent; see **wealth**

pollutus, L. defiled, unchaste; *pollutio,* defilement; see *polluo* under **dirt**

polos, Gr. foal, colt, filly, pony; *polion,* dim., see **horse**; axis, pivot, see **axis**

polphos, Gr. a kind of starchy food; see **food**

poltos, Gr. porridge; see *puls* under **food**

polus, L. pivot, axis; see **axis**

poly- < Gr. *polys,* many, very; *pleion,* more; *pleistos,* most; see **number, very**

polygala, L. (Gr. *polygalon*), milkwort; see *gala* under **milk**

polygonatum, L. (Gr. *polygonaton*), a genus of the lily family; see **lily**

polygonum, L. (Gr. *polygonon*), knotweed; see **buckwheat**

polyp < L. *polypus* (Gr. *polypous*); see **coral**

polypodium, L. a fern; see **fern**

polytrichum, L. (Gr. *polytrichon*), a kind of moss; see **moss**

pomarius, L. of applelike fruits; see *pomum* under **apple**

poma, Gr. cover, lid, operculum; *pomation,* dim., see **cap**; *poma,* drink, drinking-cup, see **drink**

pomegranate < L. *pomum,* apple, *granatum,* with many seeds.
  L. *balaustium* (Gr. *balaustion*), n. flower of the wild pomegranate: balausta, balaustine, *Balaustion pulcherrimum* (solitary balaustium).
  NL. *punica,* f. a genus of the pomegranate family, < L. *Punicus,* Carthaginian: *Punica granatum* (pomegranate), punicine, Punic, Punicaceae.
  Gr. *rhoa,* f. pomegranate; *rhoiskos,* m. dim.:
  Gr. *side,* f. pomegranate-tree and fruit; *sidion,* n. pomegranate peel:

pomeridianus, L. in the afternoon; see *dies* under **day**

pomerium, L. space free of buildings along the inside and outside of the city wall; see **space**

pompa, L. (Gr. *pompe*), solemn procession, parade, display; see **display**

pomphos, Gr. blister; *pompholyx,* bubble, slag; see **bubble, slag**

pompilus, L. (Gr. *pompilos*), a fish that follows ships; see **fish**

pompos, Gr. conductor, guide; see **lead**

pomum, L. apple, fruit of any kind; *pomulum,* dim.; *pomosus,* full of fruit; see **apple**

pond < AS. *pund,* enclosure, body of water; see **lake, marsh**

ponderosus, L. heavy, weighty; *pondero, -atus,* weigh, consider; see *pondus* under **weight, think**

ponema, Gr. work; *ponikos,* hardworking, laborious; *ponos,* hard work, toil, labor; see **make**

poneros, Gr. bad, worthless, useless; see **bad**

pons, *pontis,* L. bridge; *ponticulus,* dim.; see **bridge**

pontifex, L. high priest; see **priest**

pontus, L. (Gr. *pontos*), the open sea, the high sea; *pontios,* of the sea; see **sea**

pool < AS. *pol,* hole, basin; see **marsh, lake**

poor < OF. *povre,* < L. *pauper, -a, -um,* without means; m. a poor man; *pauperculus,* dim.; *paupertinus,* poor, sorry; *paupero, -atus,* make poor, deprive: poorly, poorhouse, pauper, pauperdom, pauperitic, poverty, impoverish, *Vestula paupera* (a bug), *Gerardia paupercula* (a scrophulariacead), *Empusa pauperata* (a mantid).
  L. *abjectus,* cast away, low, mean: abject, *Libellula abjecta* (an odonatid).
  Gr. *achen, -os,* poor, needy: *Achenomorphus columbicus* (a beetle).
  Gr. *achrematos,* poor, needy:
  Gr. *agenes,* of low family, mean: *Agenocimbex maculata* (a sawfly), *Ageneotettix deorum* (a grasshopper).
  Gr. *akleros,* without lot, poor, needy: *Aclerus leucopyge* (a butterfly).
  Gr. *aktemon (akten), -os,* poor; *aktemosyne,* f. poverty: *Actenobius saginatus* (a beetle) *Actenochroma farinosa* (a moth).
  Gr. *amechanos,* want of resources, helpless, impotent, impossible: *Amechanus fossatus* (a beetle).
  Gr. *anerges,* ineffectual, inefficient:
  L. *angustia,* f. want, poverty, scarcity, brevity:
  Gr. *anolbos,* wretched, luckless, poor:
  L. *caritus,* lacking, devoid of, free from, < *careo,* be without, lack: caret.
  Gr. *chernes,* poor, needy; laborer; see **make**
  Gr. *chetos,* n. need, want; *chetosyne,* f. need, destitution: *Ichniochetus stigma* (a beetle).
  L. *deficiens,* lack; see *defectivus* under **fault**
  Gr. *deilos,* wretched, miserable, paltry, weak, cowardly: *Dilodendron bipinnatum* (a sapindacead), *Deilotherium simplex* (a fossil mammal).
  L. *destitutus,* lacking, without possessions: destitute, *Stathmophorus destitutus* (a trichopterid).
  L. *deter,* poor, bad; *deterior,* poorer, worse, less; *deterrimus,* poorest, worst; see **bad**

L. *egenus; egens, -entis,* poor, needy; *egestas, -atis,* f. poverty, need; *indigens, -entis; indigus; indiguus,* needy, in want: indigent, indigence, *Potamopyrgus egenus* (a gastropod), *Zodarion egens* (a spider), *Aphodius indigens* (a beetle), *Sigodesmus indigus* (a centiped).

L. *exilis,* thin, slender, meager, poor; see **thin**

L. *indotatus,* poor, portionless: *Andrena indotata* (a bee).

L. *inops, -opis,* without means, poor, helpless, weak; *inopia,* f. want, lack, scarcity: *Platycephalus inops* (a fish), *Pyrgops inops* (a beetle).

L. *Irus* (Gr. *Iros*), proverbial name for a beggar, a poor man; see **beggar**

Gr. *kakobios,* living poorly:

Gr. *lipokteanos,* without property, poor:

Gr. *lypros,* poor, wretched; *lyprotes,* f. poverty; *paralypros,* rather poor: *Lyprocorrhe anceps* (a beetle), *Lyprosodes quadricostata* (a beetle).

L. *macer,* lean, thin, poor; see **thin**

L. *mendicus,* poor, beggarly; see **beggar**

L. *miser,* wretched; see **sad**

L. *pannosus,* ragged, tattered, poor, shriveled; see *pannus* under **rag**

Gr. *penes, -etos; penichros,* poor; *penestes,* m. poor man, day-laborer; *penia,* f. poverty, need: *Penestes tigris* (a beetle), *Penichrolucanus copricephalus* (a beetle), *Peniagone horrifer* (a holothurian).

L. *penuria,* f. want, need: penury, penurious.

Gr. *phaulos,* trivial, bad, poor; see **trifle**

Gr. *ptochos,* beggar, a poor man; *ptocheion,* poorhouse; see **beggar**

Gr. *rhakos,* rag, remnant; see **rag**

Gr. *skolythros,* mean, shabby, poor: *Scolyphrus*[*Scolythrus*] *obesus* (a beetle).

Gr. *tapeinos,* low, humble, poor; see **low**

Gr. *tlepathes,* wretched; *tlepathema,* wretchedness; see *pascho* under **pain**

AS. *wan,* deficient: wan, wane, want, wanton.

See: **beggar, not, empty, hollow, thin, weak, pain, worthless, bad, few, little, sad, low, fault**

popillia, NL. a genus of beetles; see **beetle**

popina, L. cook-house, eating-house; *popinalis,* of a cook-shop; *popinarius,* cook; see **eat**

poplar < OF. *poplier,* < L. *populus,* f. aspen; *populeus; populneus,* of poplar: *Populus grandidentata* (large-toothed aspen), *Betula populifolia* (gray birch), *Saperda populnea* (a beetle), *Melampsora populina* (a rust), *Ranunculus populago* (a buttercup).

Gr. *acherois, -idos,* f. white poplar:

Gr. *aigeiros,* f. black poplar:

Sp. *alamo,* m. poplar, cottonwood; *alameda,* f. poplar grove: Ojo Alamo, Alameda, Alamogordo, Alamosa.

poples, L. knee, back part of the knee; see **knee**

poppy < AS. *popig,* < L. *papaver, -is,* n.; *papaverculum,* n. dim.; *papavereus,* of poppies: papaveraceous, papaverin, *Papaver setigerum* (hairy poppy), *Callirrhoe papaver* (a poppy-mallow), *Cistanthera papaverifera* (a tiliacead). Derive: poppycock.

Gr. *argemone,* f. a kind of poppy: *Argemone mexicana* (Mexican prickly-poppy).

L. *arsella,* f. probably a kind of poppy:

L. *chelidonium* (Gr. *chelidonion*), n. swallowwort: chelidonine, chelidonic, celandine (*Chelidonium majus*).

NL. *eschscholtzia,* f. a genus of the poppy family, < J. F. Eschscholtz, German botanist: *Eschscholtzia californica* (a poppy).

L. *glaucium* (Gr. *glaukion*), n. a genus of the poppy family: *Glaucium corniculatum* (horn-poppy).

Gr. *hypekoon,* n. a narcotic plant; a genus of the poppy family: *Hypecoum grandiflorum* (a papaveracead).

Gr. *kodeia,* poppy-head; see **head**

L. *lethusa,* f. white poppy:

L. *mecon, -is* (Gr. *mekon, -os*), f. poppy; *mekonion,* n. poppy-juice, opium; *mekonikos,* of poppies: meconin, meconic, meconidium, meconium, *Dendromecon rigida* (bush-poppy), *Meconopsis cambrica* (Welsh poppy).

L. *rhoeas, -adis* (Gr. *rhoias, -ados*), f. a kind of poppy: *Papaver rhoeas* (corn-poppy).

populus, L. people; *popularis,* of the people, see **people**; aspen, poplar; *populeus; populneus,* of poplar, see **poplar**

por- < L. *pro,* before; see **before**

porca, L. the ridge between two furrows, see **ridge**; sow; *porcula,* dim., see **hog**

porcellio, L. sowbug; see **sowbug**

porch < OF. *porche,* < L. *porticus,* f. piazza, gallery, colonnade, walk; *porticula,*
  f. dim.: portico.
  Gr. *aithousa,* f. sun-porch: *Aethusa cynapium* (fools-parsley).
  ML. *lobia* (*laubia*), f. porch, gallery: lodge.
  Gr. *pastas, -ados,* f. porch, colonnade:
  L. *podium,* platform, balcony; see **projection**
  Gr. *pronaos,* m. porch, vestibule: *Pronaonota fornicata* (a blattid).
  L. *proscenium* (Gr. *proskenion*), stage; see **frame**
  Gr. *prostas, -ados,* f. porch, vestibule:
  Gr. *prothyron,* n. porch, veranda:
  L. *solarium,* sun-porch, balcony, sundial; see **dial**
  Gr. *stoa,* f. porch, colonnade; *stoidion,* n. dim.; *stoikos,* of a colonnade: Stoic,
  stoical, *Stoastoma pisum* (a gastropod).
  L. *xystus* (Gr. *xystos*), m. a covered colonnade, portico, or gallery; *xysticus,* per-
  taining to exercise in a *xystus: Xysticus elegans* (a spider).
porcinus, L. of hogs, < *porcus,* hog, pig, swine; *porculus; porcula,* dim.; *porcina,*
  pork; see **hog, flesh**
porcupine < L. *porcus,* hog, *spina,* thorn.
  NL. *erethizon,* m. porcupine: *Erethizon dorsatus* (Canada porcupine).
  L. *hystrix, -icis* (Gr. *hystrix, -trichos*), f. porcupine; *hystricosus,* prickly, thorny:
  hystricine, hystricoid, hystriciasis, *Hystrix cristata* (European porcupine), *Dio-
  don hystrix* (a porcupine-fish), *Hystrichopsylla gigas* (a siphonapterid), *Syzygops hystrix* (a beetle), *Carex hystricina* (a sedge).
pordon, Gr. stinkard; see *perdomai* under **smell**
poristos, Gr. provided; *poristes,* provider; see **supply**
pork < L. *porcus,* hog, swine; see **flesh**
porno- < Gr. *porne,* prostitute, harlot; see **prostitute**
poro- < Gr. *poros,* hole, passage; see **hole**
poroma, Gr. callus; see **skin**
porpaco- < Gr. *porpax, -akos,* handle of a shield; see **handle**
porpe, Gr. buckle, clasp, brooch; see **bind**
porphyrio, L. (Gr. *porphyrion*), a gallinule; see **crane**
porphyro- < Gr. *porphyra,* purple, a mollusk; see **purple, mollusk**
porrectus, L. spread out, stretched out, extended; see *porrigo* under **spread**
porrho- < Gr. *proso,* forward; see *pro* under **before**
porrigo, L. scurf, dandruff; *porriginosus,* scurfy; see **scale**
porrum, L. leek; *porraceus,* of leeks; *porrina,* bed of leeks; see **onion**
porta, L. gate, door; *portula,* dim.; see **door**
portabilis, L. that may be carried; see *porto* under **carry**
portaco- < Gr. *portax, -akos,* calf; see *portis* under **cattle**
portentum, L. sign, omen; see **prophecy**
portheo, Gr. destroy, plunder, ravage; see *pertho* under **destroy**
porthmos, Gr. ferry, passage, strait; see **way**
porticula, L. dim. of *porticus,* porch; see **porch**
portio, L. part of a whole; *portiuncula,* dim.; see **part**
portis, Gr. calf, heifer; see **cattle**
portisculus, L. hammer for beating time for rowers; see **hammer**
portorium, L. customs, duty, excise, toll; see **pay**
portulaca, L. purslane; see **purslane**
portus, L. harbor, haven, entrance; *portuosus,* with many harbors; see **harbor**
porus, L. (Gr. *poros*), hole, passage; see **hole**
Posidon, L. (Gr. *Poseidon*), god of the sea; see **sea**
posis, Gr. drink, beverage, see **drink;** husband, spouse, see **spouse**
positivus, L. settled, definite, explicit, certain; see **sure**
positus, L. placed, put, situated; see *pono* under **place**
posos, Gr. how many, how much; see **number**
possessor, L. holder, owner; see *possideo* under **hold**
possibilis, L. that may exist or may be done; see **make**
post < AS. *post,* < L. *postis,* doorpost; see **pillar**

post- < L. *post,* after, behind; *posterus,* following; *posterior,* next, later, hinder; *postremus; postumus,* hindmost, last; see **after**

posthe, Gr. penis; *posthon,* one with a large penis; see **penis**

posticus, L. that is behind; *posticum,* back-door; see *post* under **after, door**

postilena, L. crupper; see **crupper**

postpone < L. *post,* after, *pono,* place; see **delay**

postulo, *-atus,* L. request, ask; see **ask**

pot < AS. *pott,* a round, deep vessel: pottery, potash, potassium, potbelly, potluck, potsherd, potpie, poteen, pottle, jackpot.　　　Derive: potlatch, potato, Potomac.
Gr. *akratophoros,* m.; *akratophoron,* n. a vessel for wine:
Gr. *amis, -idos,* f. chamber-pot; *amidion,* n. dim.: *Amidostomum anseris* (a round-worm).
L. *amphora,* f. (Gr. *amphoreus,* m.), two-handled vase, pitcher, jar, jug, cinerary urn; *amphoriskos,* m. dim.: *Amphoroides polydesmi* (a protozoan), *Amphoridium viride* (a sponge), *Amphoriscus testiparus* (a sponge), *Heliamphora nutans* (a pitcher-plant), *Acrocrinus amphora* (a Mississippian crinoid), *Nepenthes phyllamphora* (a pitcher-plant).
L. *aulula,* f. small pipkin or pot:
L. *authepsa,* f. (Gr. *authepses,* m.), self-cooker, teapot; *panthepses,* m. a vessel for cooking:
L. *auxilla,* f. small pots:
L. *bria (hebria),* f. a wine-vessel:
L. *burranicum,* n. vessel for milk and must:
L. *caccabus* (Gr. *kakkabos*), m. a cooking-pot; *caccabulus,* m. dim.:
L. *cadus* (Gr. *kados*), m. jar, jug; *kadion,* n.; *kadiskos,* m. dim.; *cadialis,* of a jar: *Cadoderus bellus* (a beetle), *Cadium pomum* (a beetle), *Cadiscocrinus south-worthi* (a Devonian crinoid).
L. *caldaria,* f. pot or vessel for heating; *caldariola,* f. dim.: caldron.
L. *cantharus* (Gr. *kantharos*), m. wide-bellied vessel with handles; *kantharion,* n. dim.: cantharus, *Cantharus globularis* (a fossil gastropod), *Cantharellus cibarius* (chanterelle, a mushroom).
L. *caprunculum,* n. an earthen vessel:
L. *catinus,* m. deep vessel, pot, bowl, dish, crucible, cup; *catillus; catinulus,* m. dim.: *Peziza catinus* (a fungus), *Catinulus quadrificus* (a radiolarian), *Catillo-crinus turbinatus* (a Mississippian crinoid), *Laudonocrinus catillus* (a Pennsyl-vanian crinoid), *Cylichnostomum catinatum* (a nematode).
Gr. *chernibion,* n. chamber-pot:
Gr. *choanos,* melting-pot; crucible; *chonion,* dim.; see *choane* under **funnel**
Gr. *chytra,* f.; *chytros,* m. earthen pot, pipkin; *chytridion; chytrion,* n. dim.; *chytreus,* m. potter; *chytrinos,* of pottery; L. *chytropus* (Gr. *chytropous*), m. pot or caldron with feet: chytridiosis, *Chytrocrinus laevis* (a crinoid), *Chytridium olla* (a fungus), *Cladochytrium hyalinum* (a fungus), *Sethochytris triconiscus* (a radiolarian).
ML. *crucibulum,* n. earthen pot: crucible, *Crucibulum vulgare* (a birdsnest-fungus).
L. *diota,* f. a two-handled jar: *Diota rostrata* (a moth).
L. *fictiliarius,* potter; see *fictilis* under **earth**
L. *fidelia,* f. earthen vessel, pot:
L. *figulinus (figlinus),* pertaining to pottery; *figulus,* m. potter: figulate, figuline.
L. *gemellar, -is,* n. vessel for holding oil:
L. *hirnea,* jug; *hirnula,* dim.; see **bottle**
Gr. *hydria,* f. water-pot, jug, ewer, urn, bucket; *hydrion,* n.; *hydriske,* f. dim.: *Hydria undulata* (a butterfly), *Nymphydrium delicatum* (a neuropterid).
Gr. *hyrcha (yrcha),* f. pickle-jar:
L. *infusorium,* n. can, reservoir for a lamp:
Gr. *kalpis, -idos; kalpe,* f. vessel for drawing water, pitcher, urn; *kalpion,* n. dim.: *Calpis approximaria* (a moth), *Calpidopora auritulus* (a bryozoan), *Calpionella elliptica* (a fossil infusorian), *Coenocalpa testaceata* (a moth), *Cyrtocalpis urceola* (a radiolarian).
Gr. *katachytlon,* n. watering-pot:
Gr. *keramion,* n. pot, jar; *keramidion,* n. dim.; *kerameus,* m. potter; *keramos,* m. anything made of clay, pot, vessel, tile; *keramikos,* of pottery: ceramics, ceramidium, *Ceramium roseum* (a red alga), *Inoceramus labiatus* (a fossil pelecypod), *Astragalus ceramicus* (a vetch).
Gr. *klerotris, -idos,* f. vase for receiving ballots:
Gr. *kribanos (klibanos),* m. pot or pan wider at the bottom than at the top:
Gr. *krossos,* m. pitcher, pail, urn; *krossion,* n. dim.:
L. *lasanum* (Gr. *lasanon*), chamber-pot; see **frame**

L. *matula*, f. vessel, pot, chamber-pot; *matella*, f. dim.:

L. *nasiterna (nassiterna)*, f. watering-pot with a large spout: *Choenicosphaera nassiterna* (a radiolarian).

L. *olla*, f. earthen pot, jar; *ollicula; ollula*, f. dim.; *ollaris; ollarius*, of pots: olla, *Lecythis ollaria* (cauchillo), *Lecythiocrinus olliculaeformis* (a Pennsylvanian crinoid).

L. *orca*, whale, earthenware pot or jar with a large belly, tun, butt; *orcula*, dim.; *orcularis*, of a tun; see **whale**

Gr. *ourana; ouretris, -idos*, f. chamber-pot:

Gr. *pelopoios; peloplathos; pelourgos*, worker in clay, potter; see *pelos* under **earth**

Gr. *pithos*, m. large earthen wine-jar; *pithiskos*, m. dim.: *Pithophorus tetraglobus* (a worm).

Gr. *prochoos (prochous)*, f. vessel for pouring, pitcher, ewer:

Gr. *psykter*, wine-cooler; see *psychros* under **cold**

L. *seria*, f. large earthen jar; *seriola*, f. dim.:   Derive: *Seriola zonata* (an amber-fish).

Gr. *sipya (sipye); sipydnos*, f. meal-jar:

Gr. *stamnos*, m. earthen jar or bottle; *stammarion; stamnion*, n. dim.: *Stamnoctenis volucer* (a moth), *Stamnaria equiseti* (a fungus).

L. *tina*, f. a wine-vessel:

L. *urceus*, m. pitcher, urn; *urceolus*, m. dim.: urceolate, urceiform, *Urceola esculenta* (an apocynacead), *Urceolaria scruposa* (a lichen), *Vaccinium urceolatum* (a blueberry), *Urceolina pendula* (urn-flower), *Urceolabrum tuberculatum* (a fossil gastropod), *Tetraplodon urceolatus* (a moss).

L. *urna*, f. pitcher for drawing water, vase, container for the ashes of the dead; *urnula*, f. dim.; *urnalis*, of an urn: urn, *Urnula geaster* (a puffball), *Sphex urnaria* (a wasp), *Uvigerina urnula* (a foraminifer).   Traditional last words of Thomas Hood: "I am dying out of charity for the undertaker who wants to earn a livelihood."

L. *vas, vasis*, vessel of any kind, jar; see **vessel**

See: **vessel, kettle, barrel, bottle, cup, plate, bucket**

potaenio- < Gr. *potainios*, fresh, new; see **new**

potamo- < Gr. *potamos*, river; *potamion; potamiskos*, dim.; see **stream**

potamogeton (potamogiton), L. (Gr. *potamogeiton*), pondweed; see **waterweed**

potanos; potetos, Gr. flying, winged; see *petomai* under **fly**

potassium < D. *potasch*, < *pot* and *ash:* potash, potassium nitrate, potassamide. See ELEMENTS under **thing**

potato < Sp. *patata*, < Ab.Am. *batata*, sweet potato; see **nightshade**

potens, L. powerful; *potestativus*, denoting power; see **strong**

potentilla, NL. a genus of the rose family; see **rose**

poterium, L. (Gr. *poterion*), a drinking-cup, see **cup**; a genus of the rose family, see **rose**

potes, Gr. drinker; see *posis* under **drink**

pothos, Gr. fond desire, see **wish**; NL. a genus of the arum family, see **arum**

potio, L. drink, draught; *potiuncula*, dim.; see *poto* under **drink**

potis, L. able; *potior*, abler; *potissimus*, ablest; see *potens* under **strong**

potitor, L. master, possessor; see **govern**

potmos, Gr. lot, destiny, luck; see **lot**

potnia, Gr. mistress, queen; see **govern**

poto, -atus; potus, L. drink; *potor; potrix*, drinker; see **drink**

pouch < OF. *poche*; see **bag**

pound < AS. *punian*, bruise, strike, see **strike, mortar**; < L. *pondus*, weight, see **weight**; < AS. *pund*, enclosure, see **pen**

pour < uncertain origin.

L. *capulo, -atus*, pour forth:

Gr. *cheo*, pour; *cheuma, -tos; chyma, -tos*, n. anything poured, stream; *choë*, f. a pouring, stream; *chytlon*, n. anything that can be poured, fluid, liquid; *chytos*, poured; *anachoë*, f. eruption; *anachysis*, f. effusion; *ekchymosis*, f. extravasation of blood; *enchyma, -tos*, n. infusion; *enchytos*, poured in; *synchysis*, f. commixture: chyme, chyle, parenchyma, oenochoë, prochoös, ecchymoma, ecchymosis, *Eucheuma speciosum* (jelly-plant), *Chytolita fulicosa* (a moth).

L. *fundo, fusus*, pour; *fusilis*, molten, fluid, pourable; *fusio, -onis*, f. a melting; *fusor, -is*, m. pourer, founder; *effusus*, poured out or forth; *infusus*, poured in, mixed: fusion, diffuse, effusive, infusion, fusile, infusorial, confuse, confound,

found, foundry, infundibulum, profusion, fusibility, refund, transfuse, **funnel,** *Juncus effusus* (a rush), *Griphoneura suffusa* (a fly), *Peronospora effusa* (a mildew), *Lembus infusionum* (a protozoan).
L. *libo, -atus* (Gr. *leibo*), pour out; see *libatio* under **drink,** *libas* under **stream**
L. *scateo,* gush forth, bubble out; see **spring**
See: **flow, stream, belch, vomit, drink**

powder < OF. *poudre,* < L. *pulvis,* dust; see **dust**

power < F. *poer,* < L. *possum,* be able; see *potens* under **strong**

practico- < Gr. *praktikos,* active, busy, able, effective; *prakter; praktor,* doer; *praxis,* action, deed; see *prasso* under **make**

prae, L. before; see *pre-* under **before**

praecia, L. public crier, herald; see *cieo* under **call**

praedium, L. farm, estate, manor; *praediolum,* dim.; see **field**

praenum, L. hatchel; see **comb**

praeses, L. guard, protector; see **guard**

praestans, L. preeminent, distinguished, superior, excellent; see **honor**

praeter, L. beyond, past, more than; see **over**

pragma, Gr. act, deed, fact, matter; see *prasso* under **make**

praise < OF. *preisier,* < L. *pretium,* price; see **honor**

prandium, L. lunch; see **eat**

praos, Gr. mild, meek, gentle, tame; see **tame**

prapedilum, L. lionsfoot; see **composite**

prason, Gr. leek; *prasinos; prasios,* of leeks; see **onion**

pratum, L. meadow; *pratens,* meadow-green, grassy; *pratensis,* found in meadows; see **field**

pravus, L. crooked, deformed, perverse, bad; see **bad**

pray < OF. *preier,* < L. *precor, -atus,* entreat, ask; see **ask**

pre- < L. *prae,* before, very; see **before, very**

prebenda; prebita, L. allowance for support; see **pay**

precarius, L. obtained by prayer, doubtful, uncertain, transient; see **doubt**

precator, L. one who prays; see *precor* under **ask**

precept < L. *praeceptum,* maxim, injunction; see **law**

precipice < L. *praecipitium,* steep place, cliff; *praeceps,* headlong; see **cliff**

precipuus, L. particular, peculiar, special; see **different**

precocious < L. *praecox, -ocis,* too early ripe, premature; see *pre-* under **before**

preda, L. prey, plunder; *predator, -is,* plunderer; see **plunder**

predict < L. *praedico, -ictus,* foretell, prophesy; see **prophecy**

preface < L. *praefatio,* introduction, prologue; see *pre-* under **before**

prefectus, L. chief, overseer; see **govern**

pregnant < L. *praegnans,* with child; see **fertile**

prehendo, -hensus, L. seize; see **take**

prejudicial < L. *praejudicialis,* of judgment beforehand, damaging; see **hurt**

prelum, L. a press of any kind; see **press**

premature < L. *praematurus,* too early, untimely; see *immaturus* under **before**

premiosus, L. rich; see **wealth**

premium, L. profit, reward; see **gain**

premnas, Gr. a kind of tunny; see **tunny**

premon, Gr. stump of a tree, stem; see **stem**

prenes, Gr. hanging forward, drooping, prone; see **bend**

prenuntius, L. harbinger, omen; see **stem**

prepare < L. *praeparo, -atus,* make ready; see **fit**

prepes, L. flying swiftly, fleet, quick; see **swift**

prepon, Gr. a sea-fish; see **fish**

preposterus, L. absurd; see **fool**

preptos, Gr. distinguished, renowned; see **honor**

preputium, L. foreskin; see **penis**

prerogativus, L. asked before others, privileged; see *pre-* under **before**

presagus, L. foreboding, foretelling; see **prophecy**

presbys, Gr. old man, elder, old; *presbyteros,* older; *presbytatos,* oldest; see **old**

present < L. *praesens, -entis,* at hand, in sight, now; see **now**

preserve < L. *praeservo, -atus,* save; see **save**

president < L. *praesidens, -entis,* director, governor, ruler; see **govern**

press < OF. *presser,* < L. *premo, pressus,* bear down upon, crowd; *prelum,* n. a press; *pressulus,* hemmed in, compressed; *compressus,* pressed together, squeezed; *repressus,* restrained, curbed: compress, depressive, expression, irrepressible, pressure, suppress, repression, impressionable, print, imprint, sprain, reprimand, *Corynophorus compressus* (a fish), *Ephippigera compressicollis* (a grasshopper), *Polygale appressipilis* (a polygala), *Ceanothus impressus* (Santa Barbara ceanothus.)

Gr. *amorgeus; amorgos,* m. squeezer, presser, drainer: *Amorgos indicus* (a bug).

L. *arcto (arto), -atus,* compress, confine, contract, press together; *arctus(artus),* confined, close, strait, narrow; *coarctatus,* pressed together, shortened: arctation, coarctate, *Andropogon arctatus* (a beardgrass), *Platophrys coarctatus* (a fish).

Gr. *byo,* stuff, bung up; *symbo,* cram, huddle together; see *bysma* under **close**

L. *calcatorium,* n. wine-press:

L. *coangusto, -atus,* confine, compress, enclose; see **close**

L. *consuo, -utus,* stuff, stop up, fill up, sew up: *Pachymerus consutus* (a beetle).

L. *convaso, -atus,* pack up:

AS. *crammian,* stuff, crowd, pack, press: cram.

Gr. *epikrotos,* beaten hard, trod down:

L. *farcio, fartus (farctus),* stuff, cram; *fartor, -is,* m. stuffer; *farsilis; fartilis,* stuffed, crammed; *confercio, -ertus,* stuff, press, cram together; *infarctus,* crammed, stuffed, copious; *refertus,* stuffed, crammed, filled: farce, farcical, farcy, forcemeat, *Rhactorhynchia farcta* (a fossil brachiopod), *Marginulina infarcta* (a foraminifer), infarct, infarctation.

L. *inculco, -atus,* tread down, stuff, cram: inculcate.

L. *incunctabilis,* admitting no delay, pressing:

L. *infulcio, -fultus,* cram in, put in, foist:

L. *ingero, -estus,* carry into, press upon, obtrude: ingest.

L. *insicium,* stuffing, forcemeat; see **food**

L. *insisto, -institus,* press upon, urge: insist, insistence, *Megachile instita* (a bee).

Gr. *ipoo,* press down, weigh down; *iptomai,* press down hard, oppress, hurt, harm; *ipos,* m. a press, the piece of wood that catches the mouse in a trap; *ipoterion,* n. a pressing-place, press; *ipsis,* f. a pressing:

Gr. *kandytalis, -idos,* f. clothespress:

Gr. *nasso (natto),* press, squeeze, stuff; *naktos; nastos,* pressed close, solid, a kind of cake:

L. *nitor, nisus; nixus,* bring pressure to bear upon, strive, labor; see **make**

L. *oblido, -isus,* squeeze together:

L. *pavio, -itus,* beat, ram; see **strike**

Gr. *piezo,* press; *piesis,* f. a pressing; *piesma; sympiesma, -tos;* n. anything pressed; *piester, -os,* m. press, squeezer; *piestos,* pressed, compressible: piezometer, piezoelectric, *Piestocystis rugosa* (a worm), *Piestometopon luteiceps* (a fly), *Sympiezocnemis gigantea* (a beetle).

Gr. *pileo,* compress wool or other material into felt; *pilema,* felt; see **wool**

Gr. *satto,* pack, load; *saktor, -os,* m. packer; *saktos,* packed, stuffed: *Sactogaster mucronata* (a proctotropid).

Gr. *stalsis,* compression, contraction; restriction; *staltikos,* contractable; see **lessen**

Gr. *steinos,* confined, narrow, crowded; see **narrow**

L. *stipo, -atus,* press, cram, crowd, fill up, stuff; *constipo, -atus,* press together; Gr. *stiptos,* crammed, packed, trodden down: stipation, costive, constipation, obstipation, stevedore, *Stipocrinus spinosus* (a fossil crinoid).

Gr. *stoibazo,* pack, stuff, heap up; *stoibasimos,* packed; *stoibastes,* m. packer:

Gr. *sympyknos,* pressed together, tight; see *pyknos* under **thick**

Gr. *thlibo; thlipso,* press, rub, gall; *thlipsis,* f. pressure, pinching, constriction; *ekthlibe,* f. oppression; *ekthlimma, -tos,* n. pressure; *ekthlipsis,* f. a squeezing out: thlipsis, *Thlipsura furca* (a fossil ostracode), *Thliptocnemis barbipes* (a moth), *Synthliborhamphus antiquus* (a murrelet).

L. *torcular, -is,* n. wine-press; *torcularius,* of a press; *tortivus,* pressed out:

L. *urgeo,* press: urge, urgent, urgency.

See: **close, thick, push, weight, lie, arouse**

pressus, L. slow, measured, precise; see **slow**

prester, Gr. hurricane; see **wind**

prestigium, L. illusion, trick, deception; *prestigiator,* deceiver; see **lie**

presul, L. public dancer, presider; see **dance**

pretendo, L. allege, simulate; see **equal**

preter- < L. *praeter,* beyond, past, more than; see **over**

pretext < L. *praetextum,* allegation, excuse; see **excuse**

pretho, Gr. swell, blow up; see **swell**

pretium, L. worth, value, price, money, pay; *pretiosus,* valuable, precious, costly; see **pay**

pretor < L. *praetor,* leader, chief, magistrate; see **govern**

pretty < AS. *praettig,* clever, crafty; see **beauty**

previus, L. going before, leading; see **before**

prex, L. prayer; see *precor* under **ask**

Priapus, L. (Gr. *Priapos*), god of reproduction, penis; see **penis**

price < OF. *pris,* < L. *pretium,* worth, value; see **pay**

prick < AS. *prica,* point, puncture; see **point, hole**

pridianus, L. of yesterday; see *dies* under **day**

priest < AS. *preost,* < Gr. *presbyteros,* elder.
L. *antistes,* c. priest, overseer:
L. *druis, -idis,* m. a Celtic priest or wiseman: druid.
L. *episcopus* (Gr. *episkopos*), overseer, bishop; see **govern**
L. *fetialis,* pertaining to the *fetiales* or priests who sanctioned war and approved treaties: fetial.
L. *flamen, -inis,* n. priest:
Gr. *hiereus,* priest; *hieris, -idos,* priestess; *hierosynos,* priestly; see *hieros* under **holy**
L. *Maenas, -adis* (Gr. *Mainas, -ados*), f. priestess of Bacchus, a Bacchante: maenad.
Gr. *mystes,* m. one initiated, priest: *Sebastodes mystinus* (a rockfish).
L. *pontifex, -ficis,* m., chief or high priest: pontiff, pontifical, Pontifex Maximus, *Pelastoneurus pontifex* (a fly).
L. *popa,* m. an inferior priest or priest's assistant with a fat paunch: *Extatosoma popa* (a phasmid).
L. *sacerdos, -otis,* m. priest; *sacerdotalis,* of priests: sacerdotal.

primaevus, L. early, young; see **before**

primitivus, L. first, early; *primotinus,* happening early; see *primus* under **first**

primordius, L. original; *primoticus; primotinus,* early; see *primus* under **first**

primrose < OF. *primerole,* < L. *primula,* f.: *Primula rosea* (a primrose), *Viola primulifolia* (a violet), *Androsace primuloides* (a rock-jasmine), *Campanula primulaefolia* (a bellflower), *Gladiolus primulinus* (a gladiolus).
Gr. *anagallis, -idos,* f. pimpernel: *Anagallis arvensis* (scarlet pimpernel), *Hypericum anagalloides* (trailing St. Johnswort).
L. *androsaces* (Gr. *androsakes*), n. a genus of the primrose family: *Androsace villosa* (a rock-jasmine), *Arabis androsacea* (a rockcress).
Gr. *dodekatheon,* n. a genus of the primrose family: *Dodecatheon dentatum* (a shooting-star).
Gr. *glaux (glax),* f. milk-vetch: *Glaux maritima* (sea-milkwort).
Gr. *kyklaminos,* f. cyclamen: *Cyclamen europaeum* (European cyclamen).
L. *lysimachia,* f. (Gr. *lysimachion,* n.), loosestrife: *Lysimachia punctata* (a loosestrife).
L. *macia,* f. pimpernel:
L. *samolus,* m. a plant; now a genus of the primrose family: *Samolus floribundus* (brookweed), *Plectritis samolifolia* (a valerianacead).
It. *soldanella,* f. a genus of the primrose family: *Soldanella alpina* (glacier-alpenclock), *Schizodon soldanelloides* (a diapensiacead).

primus, L. first; *primulus,* dim.; *primarius,* of the first; see **first**

princeps, L. first man, chief; *principalis,* first, chief; see **govern**

principle < L. *principium,* foundation, beginning; see **begin**

print < L. *premo,* press; see **write, type**

prinus, L. (Gr. *prinos*), an evergreen oak; see **oak**

priono- < Gr. *prion, -os,* saw; *prionotos,* jagged, serrated; see **saw**

prior; prius, L. earlier, former; see *primus* under **first**

priscus, L. of former times, ancient; see **old**

prism < Gr. *prisma,* anything sawed, sawdust; see **form, dust**

prison < L. *prehensio,* a seizing; see **pen**

pristinus, L. early, original, primitive; see **before**

pristis, Gr. shark, sawfish; see **shark**

pristo- < Gr. *pristes*, sawyer; *pristos*, sawed; see *prion* under **saw**

privet < uncertain origin.
  L. *ligustrum*, n. privet: *Ligustrum vulgare* (common privet), *Lyonia ligustrina* (maleberry).

privignus, L. stepson; *privigna*, stepdaughter; see **kin**

privilegium, L. law in favor of an individual, special right; see *privus* under **alone**

privo, -*atus*, L. rob, strip, separate; see **separate**

privus, L. alone, each, single; *privatus*, individual; see **alone**

prize < OF. *pris*, < L. *pretium*, value; see **pay**

pro- < L. *pro*, before, forward, in front of; *prior*, earlier; *primus*, first; < Gr. *pro*, before; *proteros*, earlier; *protos*, first; see **before**

proales, Gr. springing forward, abrupt, overhanging; see **projection**

probabilis, L. likely, credible; see **believe**

probato- < Gr. *probaton*, sheep; *probation*, dim.; see **sheep**

problem < Gr. *problema*, question propounded for solution, puzzle, riddle; see **ask, try, hypothesis, make**

probleto- < Gr. *probles*, -*etos*, thrown forward, projected; see *ballo* under **throw**

probo, -*atus*, L. test; *probatio*, test, trial, examination; see **try**

probolos, Gr. any projecting or jutting object or prominence; see **projection**

proboscis, L. (Gr. *proboskis*), trunk of an elephant, snout; see **nose**

probrum, L. shame, disgrace; *probrosus*, shameful; see **shame**

probus, L. good, excellent, upright; see **good**

proca < Gr. *proka*, forthwith, straightway; see **now**

procax, -*acis*, L. bold, forward, impudent; see **bold**

procella, L. storm, hurricane, tempest; *procellosus*, stormy; see **storm**

procer, L. leader, chief, noble; see **lead**

procerus, L. tall, slender, long; *procerulus*, dim.; see **high**

process < L. *procedo*, -*essus*, go forward, progress; see *cedo* under **move**

prochny, Gr. with bent knee, kneeling; see **knee**

prociduus, L. fallen down, prostrate; see **low**

proclivis, L. sloping, going downward; see *clivus* under **slope**

procne (progne), L. (Gr. *Prokne*, fabled woman changed into a swallow); see **swallow**

proco- < Gr. *prox*, -*okos*, a kind of deer, see **deer**; dewdrop, see **drop**

procrastino, -*atus*, L. put off till tomorrow; see *cras* under **day**

Procrustes, L. (Gr. *Prokroustes*), a legendary highwayman who conformed his captives to a celebrated bed by stretching or amputation as required; see **fit**

procto- < Gr. *proktos*, anus, rectum, fundament, tail; see **intestine**

procumbens, L. prostrate, face downward; see *cubo* under **lie**

procurator, L. agent, manager, steward; see **govern**

procus, L. wooer, suitor; see **woo**

prodigiosus, L. strange, wonderful, vast, extraordinary; see **wonder**

prodigus, L. wasteful, lavish, extravagant; see **waste**

proditor; proditrix, L. betrayer, traitor; Gr. *prodosia*, betrayal, treason; *prodotes*, traitor; see *prodo* under **treason**

prodromos, Gr. running before, precursor; see *pro* under **before**

produce < L. *produco*, -*uctus*, bring out, make, beget; see **make, birth**

proeco- < Gr. *proix*, -*ikos*, gift, present, dowry; see **give**

proelium, L. battle; see **fight**

proeo- < Gr. *proios*, early, early morn; *proiotes*, earliness; see *proimos* under **dawn**

profane < L. *profanus*, outside the temple, unholy, esoteric, irreverent, wicked: profanity, profanation.
  Mother: "Where did you learn those terrible words?"
  Son: "From Santa Claus, when he stubbed his toe against my bed on Christmas Eve."
  Gr. *amystos*, profane:
  Gr. *anosios*, unholy, profane: *Anosius angustulus* (a beetle).
  Gr. *asebema*, -*tos*, n.; *asebia*, f. sacrilege; *asebes; dysebes*, profane, sacrilegious: *Asebeomyia epira* (a mosquito).

Gr. *bebelos*, profane: *Bebelothrips latus* (a thysanopterid).
Gr. *blasphemia*, f. evil, defamatory speech, profanity: blaspheme, blasphemy, blasphemous, blasphemer, blame.
Gr. *dysages*, impious:
Gr. *dystheos*, ungodly:
Gr. *hierosyleo*, rob a temple, commit sacrilege; *hierosylesis*, f. sacrilege; *hierosylos*, m. a sacrilegious person; *anieros*, unholy:
L. *impius*, irreverent, wicked: impious, impiety.
L. *irreverens, -entis*, lack of veneration, disrespectful: irreverent, irreverence.
L. *sacrilegus*, that violates sacred things, impious, profane: sacrilegious, sacrilege.
L. *temero, -atus*, violate, dishonor, disgrace, profane: temerity.
Gr. *theosylia*, f. sacrilege:
L. *violo, -atus*, dishonor, profane, injure: violate, violation, violator.
See: **common**

professor, L. teacher; see **teach**

profligatus, L. corrupt, dissolute; see **bad**

profundus, L. deep; see **deep**

profusus, L. lavish, extravagant; see **abundance**

progeny < L. *progenies*, descent, lineage, offspring; see *gigno* under **birth, young**

progne (procne), L. swallow; see **swallow**

prognosis, Gr. forecast; see **prophecy**

program < Gr. *programma*, public notice; see **list**

progress < L. *progredior, -gressus*, go forward; see **grow**

prohibitio, L. a forbidding, preventing; see **forbid**

projection < L. *projicio, -jectus*, throw at, push forth.
L. *acta* (Gr. *akte*), headland, foreland, promontory, shore; see **shore**
Gr. *akra*, f. headland, cape; *akrokolion*, n. any extremity of the body; *akroterion*, n. extremity of anything, cape, promontory:
Gr. *anteris*, buttress; see **pillar**
Gr. *apophysis*, f. offshoot, process, prominence; *ekphysis*, f. outgrowth, projection: apophysis.
Gr. *bounos*, hill, mound, knob; see **mountain**
L. *broccus; brochus; broncus*, projecting, particularly projection of teeth: broach, brooch.
L. *bulla*, knob, boss, stud, bubble; see **bubble**
L. *caput*, head, end, point; see **head**
Gr. *chaulios*, outstanding, prominent; *chauliodous*, with prominent teeth, tusky: *Chauliognathus pennsylvanica* (soldier-beetle), *Chauliodus sloani* (a viper-fish), *Chaulelasmus streperus* (gadwall).
L. *condylus* (Gr. *kondylos*), m. knuckle, knob, prominence, enlarged end of a bone, bony knob; *kondylion*, n. dim.: condyle, condyloma, Condylarthra, *Condylocrinus verrucosus* (a fossil crinoid), *Condylura cristata* (star-nosed mole), *Condylostylus nigrosetosus* (a fly).
L. *conspicuus*, manifest, visible, standing out, prominent; *prospicuus*, that may be seen from afar; see *conspectus* under **clear**
L. *convexus*, with curved surface arched toward the observer, protuberant; see **arc**
L. *crena*, notch, rounded projection; see **notch**
L. *denotatus*, marked out, conspicuous; see *nota* under **mark**
L. *ecphora* (Gr. *ekphora*), f. a projection in buildings: *Ecphora quadricostata* (a fossil gastropod).
Gr. *ektatos*, capable of extension or prolongation; see *teino* under **spread**
Gr. *eperephes*, overhanging, beetling:
Gr. *exochos; exeches*, jutting out, projecting, prominent, eminent; *exoche; hyperoche*, f. projection, prominence: *Exochoblatta hastata* (a fossil cockroach), *Exechocentrus lancearius* (a spider), *Hyperoche lütkeni* (an amphipod).
L. *exsero* (*exero*), *exsertus* (*exertus*), project, thrust forth: exserted, exert, exertion, *Polygonum exsertum* (knotweed), *Cephalocereus exerens* (a cactus).
L. *gibber, -a, -um; gibbus*, humpbacked, humped, bent, protuberant, swollen; *gibberosus; gibbosus; gibbosus*, very humped, crooked: gibbous, *Holopedium gibberum* (a cladoceran), *Lemna gibba* (a duckweed), *Eupomotis gibbosus* (sunfish), *Murex gibbulus* (a fossil gastropod), *Kloedenella gibberosa* (an ostracode), *Ctenodonta gibberula* (a fossil pelecypod), *Strombus gibberulus* (a gastropod), *Gibberella zeae* (a fungus), *Articulina gibbosula* (a foraminifer), *Echeveria gibbiflora* (a crassulacead), *Hybogaster gibberosa* (a wasp), *Gibbium psylloides* (a beetle).
Gr. *hybos*, m.; *hyboma, -tos*, n. hump: hybodont, *Hybocrinus tumidus* (an Ordovician crinoid), *Hybodus plicatilis* (a Triassic fish), *Hybanthus verticillatus*

(nodding violet), *Hybosa insculpta* (a beetle), *Megalybus obesus* (a fly), *Acrohybus argutus* (a trilobite), *Schizolobium parahybum* (a legume).

L. *impendeo*, overhang, be imminent, threaten: impend, impending.

Gr. *knemos*, shoulder of a mountain; see **leg**

Gr. *kremnos*, overhanging wall or bank; see **high**

Gr. *kyphoma*, hump; see *kyphos* under **bend**

Gr. *kyrtos*, curved, humped; *dikyrtos*, two-humped; see **bend**

L. *lobus* (Gr. *lobos*), m. an elongated projection or protuberance, capsule, pod; *lobion*, n. dim.; *ellobos*, in a pod: lobe, lobar, lobule, *Epilobium hirsutum* (hairy willowherb), *Bilobites bilobus* (a fossil brachiopod), *Coccolobis diversifolia* (a sea-grape), *Trachylobium verrucosum* (a legume, source of copal), *Allolobophora foetida* (an oligochaete), *Delphinium trilobatum* (a larkspur), *Dennstaedtia punctilobula* (a fern), *Nucula biloba* (a pelecypod), *Physetocrinus lobatus* (a Mississippian crinoid).

Gr. *mastos*, breast, nipple, teat; see **udder**

L. *minor, -atus; mineo*, project, overhang, threaten, menace; *mina*, f. projecting point, threat, menace; *minatio, -onis*, f. threat, menace; *minator, -is*, m. threatener; *minax, -acis*, jutting out, threatening; *eminens; prominens, -entis*, projecting, standing out; *prominulus*, projecting; *promontorium*, n. headland: menace, minatory, eminent, eminence, preeminence, imminent, prominent, prominence, amenable, demeanor, *Natica eminula* (a fossil gastropod), *Salebius minax* (a beetle), *Sisyphus prominens* (a beetle).

Gr. *okris, -ios*, f. any projection, roughness:

L. *papilla*, nipple, teat, bud; see **bud**

L. *podium*, n. elevated place, platform, balcony, parapet: podium.

L. *polypus* (Gr. *polypous*), a growth in the nose, the coral animal; see **coral**

Gr. *proales*, overhanging, abrupt: *Proales gibba* (a rotifer).

Gr. *probolos*, m. any projecting or jutting object or prominence; *probolion*, n. dim.: *Probolomyrmex petiolatus* (an ant).

Gr. *pron, -os*, m. headland, promontory: *Pronopyge ocreata* (a trematode), *Pronopharynx nematoides* (a trematode).

L. *prosto, -atus*, stand out, project:

L. *protrudo, -usus*, push forward, thrust out: protrude, protrusion.

L. *protubero, -atus*, swell or bulge out: protuberance, *Barycrinus protuberans* (a Mississippian crinoid).

L. *protumidus*, protuberant, gibbous:

Gr. *rhion*, n. a jutting part of a mountain, peak, headland: *Rhion pallidum* (a spider).

L. *scopulus* (Gr. *skopelos*), observation point, projecting rock, crag; *scopulosus*, rocky, craggy; see **stone**

L. *solarium*, sun-porch, balcony; see **dial**

AS. *stubb*, stump: stub. See **stem**

L. *torus*, m. round elevation, protuberance, bulge; *torulus*, m. dim.; *torosus*, bulging, muscular, fleshy, lusty: torus, torose, torosity, torulosis, *Torula utilis* (a yeast), *Torilis nodosa* (hedge-parsley), *Casuarina torulosa* (a beefwood), *Stigmatium torulentum* (a beetle), *Triturus torosus* (a newt), *Rhodotorula rubra* (red yeast).

L. *umbo, -onis*, m. boss, rounded protuberance, knob, shield: umbo, umbonal, *Ambocoelia umbonatus* (a fossil brachiopod).

See: **point, tooth, mountain, horn, heap, club, cliff, udder, face, swell, upright, wart, bud**

prolatus L. extended, elongated; see **long**

proles, L. offspring, youth; see **young**

proletarius, L. a citizen of the lowest class, low, common; see **common**

prolifer; prolificus, NL. fruitful, productive; see **fertile**

prolixus, L. stretched out, long; see **long**

prolobos, Gr. crop; see **stomach**

promethes, Gr. cautious, forethinking, wary; see **think**

Prometheus, Gr. fabled Titan who stole fire from heaven and gave it with its derivative arts to man; see **fire**

prominent < L. *prominens*, projecting, standing out; see *minor* under **projection**

promiscuus, L. mixed, indiscriminate; see *misceo* under **mix**

promise < OF. *promisse*, < L. *promissum*, n. pledge, engagement: promissory.

Gr. *apotimema, -tos*, n. mortgage, security:

L. *arrha*, pledge; *arrhalis*, of a pledge; see **pay**

Gr. *asphalisma*, guarantee, security; see *asphales* under **stand**

Gr. *engyos*, giving surety, bail, pledge, promise: *Engyophlebus obesus* (a moth).

Gr. *homeros*, m.; *homereia*, f. hostage, security, surety, pledge: *Homeria pallida* (an iridacead).

Gr. *hypotheke*, f. pledge, security, mortgage:

L. *obligo, -atus*, bind, engage, pledge; see *ligo* under **bind**

L. *obses, -idis*, c. hostage, pledge, surety; *obsidatus*, m. condition of a hostage:

L. *pactio, -onis*, f. agreement, covenant, contract, bargain: pact.

L. *pignus, -eris*, n. pledge, pawn, security; *pignerator, -is*, m. pledger; *pignero, -atus*, pledge, pawn: oppignerate.

L. *pollicitum*, n. promise; *pollicitor, -atus,* promise: pollicitation.

L. *spondeo, sponsus,* promise; *sponsor, -is,* m. bondsman, surety: sponsor, respond, spouse, espousal, despondent, correspondence, irresponsible, R.S.V.P.

L. *stipulatio, -onis,* f. agreement, bargain, promise; *stipulator, -is,* m. one who demands a promise; *stipulor, -atus,* exact a promise, bargain, covenant: stipulate, stipulation.

L. *vas, vadis,* bail, security; *vadimonium,* bail; see *vador* under **bind**

L. *voveo, votus,* promise solemnly; *votivus,* of a promise: vow, vote, votary, votive, avow, devout, devotional.

Goth. *wadi* (AS. *wed*), pledge: wed, wedlock, wedding, wager, wages, gage, engagement, mortgage.

See: **swear, bind, speak**

promontorium, L. headland; see *minor* under **projection**

promotus, L. advanced, pushed forward; see *moveo* under **move**

promptus, L. at hand, ready, on time; see **time**

promulgo, -atus, L. make known, publish; *promulgator,* publisher; see **display**

promus, L. pertaining to giving or receiving; *promum,* storeroom; see **give, store**

prono- < Gr. *pron, -os,* headland, promontory; see **projection**

pronomos, Gr. grazing forward; see *nemos* under **eat**

pronuba, L. brideswoman; see *nubo* under **marry**

pronus, L. inclined forward, lying face downward; see **lie**

proof < L. *proba,* test; see *probo* under **try**

prop < MD. *proppe;* see **pillar**

propago, L. layer, set, shoot, slip, see **branch**; *propago, -atus,* set, generate, see **begin**

prope, L. near; *propior,* nearer; *proximus,* nearest; *propinquus,* neighboring; see **near**

proper < L. *proprius,* one's own, special, particular; see **fit**

properus, L. quick, speedy; see **swift**

prophecy < OF. *profecie,* < L. *prophetia* (Gr. *propheteia*), f. a foretelling or predicting the future; *propheta* (Gr. *prophetes*), m. foreteller, soothsayer; *propheticus,* predicting, oracular: prophet, prophetic, prophesy, *Cucumis prophetarum* (a cucumber).

L. *augur, -is,* c. diviner, soothsayer: augury, inaugural, inaugurate, *Pterolestes augur* (a bird).

L. *auspex, -icis,* c. bird inspector, augur, diviner; *auspicalis,* pertaining to divination; *auspicatus,* fortunate, favorable, lucky: auspicious.

L. *Cassandra* (Gr. *Kassandra*), f. legendary prophetess of impending evil but believed by no one: Cassandra.

Gr. *chresmos,* m. oracle; *chrester, -os; chrestes,* m. prophet, soothsayer; *chresterios,* prophetic; *chresmodes,* like an oracle: *Chresmodes obscura* (a phasmid).

L. *conjector,* seer, soothsayer, diviner; see *conjectura* under **hypothesis**

L. *Delphi* (Gr. *Delphoi*), m. city of the famous oracle of Apollo: Delphic, Delphian.

L. *divino, -atus,* foresee, foretell, prophesy; *divinator, -is,* m. soothsayer: divine, divination, *Troctes divinatorius* (a book-louse).

L. *fatum,* utterance of a god, prophecy, lot; see **lot**

L. *hariolus,* m.; *hariola,* f. soothsayer, prophet: hariolate, *Hariola tiarata* (a bug).

L. *haruspex, -icis,* m.; *haruspica,* f. soothsayer, diviner: *Haruspex ornatus* (a beetle).

Gr. *kledon, -os,* f.; *kledonisma, -tos,* n. omen, portent, sign; *kledonistes,* m. observer of omens: cledonism.

Gr. *mantis; manteutes,* m. diviner, seer, soothsayer, prophet; *-mancy,* pertaining to divination; *mantikos,* of soothsaying; *mantosyne,* f. divination; *nekromantis,* c. spiritualist: crystallomancy, chiromancy, necromancy, *Mantis religiosa* (praying mantis), *Mantispa brunnea* (a neuropterid), *Cheiromantis rufescens* (a frog), *Mantisia saltatoria* (a zingiberacead), *Tachysphex mantiraptor* (a wasp).

L. *Mopsus* (Gr. *Mopsos*), m. a celebrated soothsayer: *Onthophagus mopsus* (a dung-beetle).

Gr. *oionos*, m. omen, a large bird; *oionistes*, m. augur: *Oeonoscopus striatissimus* (a Jurassic fish).

L. *omen, -inis*, n. augury, sign, portent, prophecy; *ominosus*, portentous; *ominor, -atus*, forebode, forecast, predict, prophesy: omen, ominous, abomination.

Gr. *ompheter, -os*, m. soothsayer; *omphe*, f. voice of a god:

L. *oraculum*, n. divine announcement, prophetic declaration; *oracularis*, prophetic: oracle, oracular.

L. *portentum*, n. sign, omen, < *portendo, -itus*, foretell, presage: portent, portentous, *Trophithauma portentus* (a fly).

Gr. *phoibetes*, m. prophet; *phoibastria*, f. prophetess; *phoibastikos*, inspired; *phoibasma, -tos*, n. prophecy: *Phoebastria nigripes* (a bird).

L. *praedico, -ictus*, foretell, forecast, prophesy: predict, prediction.

L. *praemoneo, -itus*, forewarn, foretell: premonition, premonitory.

L. *praenuntio, -atus*, foretell, foreshadow, predict; *praenuntius*, that foretells or forebodes; *praenuntium*, n. harbinger, omen: prenuncial, *Dissacus prenuntius* (a creodont).

L. *praesagatus*, perceived beforehand; *praesagus*, foretelling, foreboding, < *sagus*, m.; *saga; sagana*, f. diviner, fortune-teller: presage.

L. *praescius*, foreknowing: prescient, prescience.

L. *praesentio, -sensus*, feel or perceive beforehand, divine: presentiment.

Gr. *proagoreuo*, foretell; *proagoreuma, -tos*, n. prophecy; *proagoreutikos*, prophetic:

L. *profor, -atus*, speak out, foretell, predict:

Gr. *prognosis*, f. foreknowledge, forecast: prognosis, prognostication.

L. *pronoea* (Gr. *pronoia*), f. foresight, providence:

Gr. *Sibylla*, f. a female soothsayer: sibyl, Sibylline, Sibylla.

L. *strena*, f. sign, omen:

L. *Telemus* (Gr. *Telemos*), m. a soothsayer: *Monoculus telemus* (a gastropod).

Gr. *thespizo*, prophesy, foretell, divine; *thespistes*, m. prophet:

Gr. *Thria*, f. nurse of Apollo and inventor of divination by drawing pebbles from an urn; the pebble so drawn; *thriobolos*, m. soothsayer; *thriazo*, prophesy: *Thria robusta* (a moth).

L. *vates, -is*, c. seer, soothsayer, prophet; *vaticinus*, prophetical, < *vaticinor, -atus*, foretell, prophesy: vatic, vaticinate.

See: **magic, lot**

propinquus, L. near; see *prope* under **near**

propitius, L. favorable, gracious, kind; see **agreeable**

propolis, Gr. bee-glue; see **glue**

propositio, L. plan, theme, statement; see **hypothesis**

proprius, L. one's own, special, particular, peculiar; see **alone**

propylaeum, L. (Gr. *propylaion*), gateway, entrance; see *pyle* under **door**

prora, Gr. prow, bow; see **face**

prorsus, L. onward, straight, direct, forward; see **face**

pros, Gr. implying motion from, on, or to the side, beside, near, toward; see **near**

prosator, L. ancestor; see **kin**

prosbatos, Gr. accessible; see *baino* under **walk**

proscenium, L. (Gr. *proskenion*), stage; see **frame**

proselytus, L. (Gr. *proselytos*), convert; see **change**

prosenes, Gr. gentle, soft, smooth; see **mild**

Proserpina, L. wife of Pluto and queen of Hades; see *Pluto* under **hell**

proserpinaca, L. a genus of the water-milfoil family; see **water-milfoil**

prosilo- < Gr. *proseilos*, toward the sun, sunny, warm; see **heat**

prositos, Gr. approachable, accessible; see **accessible**

proso- < Gr. *proso*, forward, onward, in front; see *pro* under **before**

prosodia, Gr. song, tone or accent of a syllable; see *aeido* under **sing**

prosodos, Gr. approach, advance, onset, attack; see **come**

prosopis, Gr. a plant, now a genus of the bean family; see **bean**

prosopon, Gr. face, front; see **face**

prosperus, L. agreeable, favorable; see **agreeable**

prospicuus, L. that may be seen from afar; see *conspectus* under **clear**

prostato- < Gr. *prostates*, one who stands first, chief, protector; see **guard, gland**

prostatus, L. projecting; see *prosto* under **projection**

prostethion, Gr. girdle; see **belt**

prosthe (prosthen), Gr. before, in front; *prosthios,* foremost; see *pro* under **before**

prostheco- < Gr. *prostheke,* addition, appendage, supplement; see **and**

prosthetos, Gr. applied; see **give**

prostitute < L. *prostituta,* f. whore, harlot, < *prostituo, -utus,* expose publicly, dishonor, sully: prostitution.
L. *adulter, -i,* m.; adulterer; *adultera,* f. adultress: adulterer, adulteress(adultress).
L. *amicarius,* m. procurer of a mistress, *amica,* f.:
Gr. *anasyrtolis,* f. a lewd woman:
Gr. *blitas, -ados,* f. a worthless woman:
L. *concubina,* f. kept mistress; *concubinus,* m. male paramour; *concubinalis,* lascivious: concubine, concubinage, concubinary.
Gr. *epimisthis, -idos,* f. courtesan:
L. *fornicarius; fornicator, -is,* m. male who commits illicit intercourse; *fornicaria; fornicatrix, -icis,* f. prostitute: fornication, fornicator.
Gr. *hetaira,* companion, concubine, courtesan; see *hetairos* under **companion**
AS. *hore,* harlot: whore, whoredom, whoremonger.
L. *intermuculus,* m. one who devotes himself to prostitution:
Gr. *kasalbas, -ados,* f. strumpet, whore; *kasorion,* n. brothel:
Gr. *laikastes,* m. wencher; *laikastria,* f. strumpet; *lekema, -tos,* n. wenching:
L. *leno, -onis,* m. pimp, panderer, procurer; *lena; vitilena,* f. bawd, procuress; *lenullus; lenunculus,* m. young go-between; *lenonius,* of pimping or pandering; *lenocinium,* n. trade of a procurer; *leno, -atus,* pimp, pander, procure:
L. *lupa,* f. she-wolf, prostitute; *lupula,* f. dim.; *lupanus,* lewd; *lupanar, -is,* n. brothel: lupanarian.
Gr. *mastropos,* m. panderer, pimp, procurer:
Gr. *maulis, -idos,* f. bawd; *maulistes,* m. pimp:
L. *meretrix, -icis,* f. harlot, prostitute, < *mereo,* sell for pay; *meretricula,* f. dim.; *meretricius,* pertaining to harlots: meretricious.
Gr. *misete,* f. prostitute:
L. *moecha* (Gr. *moiche*), f. adulteress; *moechus* (Gr. *moichos*), m. adulterer; *moikichos,* adulterous; *moechor, -atus,* commit adultery, fornicate: *Moecha molitor* (a beetle).
Gr. *myllas, -ados,* prostitute; see *myllo* under **coitus**
Gr. *myonia,* f. mouse-hole, figurative name for a lewd woman:
Gr. *mysachne,* f. prostitute:
Gr. *myzouris,* f. tail-sucker, fellatrix:
L. *nonaria,* f. a public prostitute:
L. *Pandarus* (Gr. *Pandaros*), m. son of Lycaon; synonym for procurer: pander, panderer.
Gr. *parthenopipes,* m. seducer:
L. *pellex, -icis* (Gr. *pallakis, -idos*), f. kept mistress, concubine; *pallakidion,* n. dim.; *pellicatus,* m. concubinage: *Cecidomyia pellex* (an ash gall-insect).
Gr. *phorbas, -ados,* giving pasture, prostitution; see *pherbo* under **eat**
Gr. *porne,* f. prostitute, harlot; *pornos,* m. male prostitute; *pornidion,* n. dim.; *pornikos,* of harlots; *porneusis,* f. prostitution; *porneion,* n. brothel: pornographic, Pornerastic, pornocracy, *Pornothemis serrata* (a dragonfly).
L. *prostibula,* f. harlot; *prostibulum,* n. brothel:
Gr. *saperdion,* n. nickname for a courtesan, dim. of *saperdes,* a kind of fish:
L. *scortator, -is,* m. whoremonger, < *scortor, -atus,* whore; *scortum,* n. harlot; *scortillum,* n. dim.; *scortulum,* n. young harlot: scortation.
L. *scratta,* f. an unchaste woman:
L. *seductor, -is,* m.; *seductrix, -icis,* f. misleader, enticer: seduce, seduction, seducer, seductor, seductress, *Oncopsia seductrix* (a fly).
L. *sellarius,* m. male prostitute: sellary.
Gr. *stegitis, -idos,* f. prostitute:
L. *strictivilla,* f. a vile woman:
L. *succuba,* lecher, strumpet; a female demon supposed to prostitute men in their sleep; see *incubus* under **spirit**
Gr. *tribas, -ados,* f. a woman who practices lewdness with herself or other women,< *tribo,* rub:
See: **coitus, shame**

prostratus, L. down flat, laid low; see **low**

protect < L. *protego, -tectus,* cover, defend, guard; see **guard**

protein < Gr. *proteion,* holding first place, primary; a nitrogenous, essential constituent of all living cells: proteinaceous, proteogenous, proteose, proteolytic.

L. *albumen, -inis,* n. white of an egg: albumen, albumin, albuminoid, albumose, albuminous, albuminosis, albuminate.

proteles, Gr. perfect before; see *telos* under **end**

protelum, L. team of oxen; see **team**

protensus (protentus), L. stretched out, extended; see *tendo* under **spread**

protenus, L. before, forward; see *pro* under **before**

proteros, Gr. earlier; see *pro* under **before**

protervus, L. bold, impudent; see **bold**

Proteus, Gr. a sea-god capable of changing his form; see **change**

protos, Gr. first; *protistos,* the very first; see **first**

protozoan < Gr. *protos,* first, *zoon,* animal: Protozoa.
　　NL. *amoeba,* f. (Gr. *amoibe,* f. change), a genus of protozoans: amebicidal, *Amoeba dubia* (an ameba), *Endamoeba histolytica* (cause of amebic dysentery).
　　NL. *fusulina,* f. a genus of foraminifers: *Fusulina cylindrica* (a foraminifer).
　　NL. *globigerina,* f. a genus of foraminifers: *Globigerina cretacea* (a foraminifer).
　　NL. *nodosaria,* f. a genus of foraminifers: *Nodosaria radicula* (a foraminifer).
　　NL. *operculina,* f. a genus of foraminifers: *Operculina complanata* (a foraminifer).
　　NL. *paramecium,* n. a genus of protozoans: *Paramecium caudatum* (a paramecium).
　　NL. *robulus,* m. a genus of foraminifers: *Robulus inornatus* (a foraminifer).
　　NL. *textularia,* f. a genus of foraminifers: *Textularia sagittula* (a foraminifer).
　　NL. *triticites,* m. (< L. *triticum,* n. wheat), a genus of foraminifers: *Triticites tumidus* (a foraminifer).
　　NL. *trypanosoma,* n. a genus of protozoans: *Trypanosoma gambiense* (cause of African sleeping sickness), trypanosome, trypanosomiasis.

proud < AS. *prud,* pride.
　　Gr. *agerochos,* haughty, arrogant:
　　Gr. *anenios,* insolent:
　　L. *arrogans,* assuming, haughty; *arrogantia,* f. presumption, conceit, < *arrogo, -atus,* appropriate, assume, take: arrogant, arrogance, arrogate.
　　Gr. *brenthos,* m. haughtiness, arrogance; a kind of bird: *Brenthus armillatus* (a beetle).
　　L. *fastus, -us,* m. pride, haughtiness, arrogance; *fastosus; fastuosus,* proud, haughty: *Bromelia fastuosa* (a bromelia), fastuous, *Deima fastosum* (a holothurian).
　　Gr. *gauros,* haughty, arrogant, disdainful: *Gaura coccinea* (scarlet gaura), *Gaurocrinus magnificus* (a fossil crinoid).
　　Gr. *hyperphialos,* arrogant, overbearing:
　　Gr. *hyperthymos,* overweening:
　　Gr. *hypsiphron, -os,* haughty:
　　L. *insolens, -entis,* proud, haughty, arrogant: insolent, insolence, *Cucumaria insolens* (a holothurian).
　　Gr. *kenodoxia,* f. conceit, vanity:
　　L. *ostentatio, -onis,* display, parade; see **display**
　　Gr. *sobaros,* haughty, pompous: *Sobarus anomalus* (a fly), *Sobarocephalus rubsaameni* (a fly).
　　L. *sublatus,* raised aloft, proud, haughty; see *tollo* under **raise**
　　L. *superbus,* proud, haughty, distinguished, splendid; see **good**
　　L. *superciliosus,* haughty, arrogant: supercilious, *Poecilodryas superciliosa* (a bird).
　　L. *tarquinius,* proud, haughty, < *Tarquinius,* a proud Tarquin of ancient Rome: tarquinish.
　　L. *vanus,* empty, idle, proud; see **empty**

prove < L. *probo,* test; see **try**

proverb < L. *proverbium,* adage, maxim; see **adage**

provide < L. *provideo,* foresee, prepare for; *providus,* foreseeing; see **thing, supply**

provincia, L. a division of a country or empire; see **country**

prow < Gr. *prora,* bow; see **face**

proximus, L. nearest; see *prope* under **near**

prudent < L. *prudens, -entis,* foreseeing, experienced; see *providus* under **think**

pruina, L. hoarfrost, rime; *pruinosus,* covered with frost; see **ice**

pruna, L. a live coal; see **coal**

prune < L. *prunum* (Gr. *prounon*), plum; *prunus,* plum-tree, see **plum**; < OF. *proignier,* lop off, see **cut**

prurio, L. itch or long for; *prurigo,* the itch; see **itch**

prymna, Gr. stern of a ship, see **stern**; *prymnos,* hindmost, see **after**

prytanis, Gr. a chief magistrate; see **govern**

psacado- < Gr. *psakas, -ados,* small drop, bit, morsel; *psakadion; psakion,* dim.; *psakastos,* dripping; see **drop**

psaenythio- < Gr. *psainythios,* false, vain; see **lie**

psagdan, Gr. an Egyptian unguent; see **fat**

psalido- < Gr. *psalis, -idos,* clipper, scissors; *psalidion,* dim.; *psalistos,* clipped, see **scissors**; low building with a vaulted roof, crypt, see **house**

psalion, Gr. curb-chain, part of a bridle; see **chain**

psalm < L. *psalmus* (Gr. *psalmos*); *psalterium,* a harplike instrument; see *psallo* under **music, harp**

psamathis, Gr. a fish; see **fish**

psammos, Gr. sand; *psammion,* dim.; *psamathos,* shore sand; *psamathion,* dim.; see **sand**

psaro- < Gr. *psaros,* speckled, dappled, like a starling, < *psar, -os,* starling; see **starling**

psapharos; psathyros, Gr. easily powdered, crumbling, friable; see **brittle**

psecado- < Gr. *psekas, -ados,* bit, crumb, small drop; *psekadion,* dim.; see *psakas* under **little**

psecto- < Gr. *psektes,* blamer; *psektos,* blamable; see *psogos* under **blame**

psectro- < Gr. *psektra,* scraper, currycomb; *psegma,* scrapings; see **scrape**

psedno- < Gr. *psednos,* thin, scanty, bald; see **bare**

pselaphetos, Gr. palpable, < *pselaphao,* grope about, feel, touch; see **touch**

pselio- < Gr. *pselion (psellion),* armlet; see **bracelet**

psellos, Gr. faltering in speech; see **falter**

pseno- < Gr. *psen, -os,* a gall-insect, see **wasp**; < *psenos,* smooth, bald, see **bare**

psephos, Gr. pebble, see **stone**; < *psephisma,* an ordinance voted by the people, see **law**; < *psephenos,* dark, obscure; *psephos,* darkness, see **black**

psesto- < Gr. *psestos,* scraped, rubbed; *psexis,* a rubbing down, currying; see *psao* under **rub**

psetta, Gr. flatfish, sole, turbot; see **flatfish**

pseudo- < Gr. *pseudos,* lie; *pseudes,* false; *pseustes,* liar; see **lie**

psi, Gr. twenty-third letter of the Greek alphabet; see **letter**

psiado- < Gr. *psias, -ados,* drop; see **drop**

psiathos, Gr. rush-mat; see **rug**

psicho- < Gr. *psix, -ichos,* bit, crumb, morsel; *psichion,* dim.; see **little**

psidium, NL. a genus of the myrtle family; see **myrtle**

psilo- < Gr. *psilos,* bare, smooth; see **bare**

psilothrum, L. (Gr. *psilothron*), depilatory; see *psilos* under **bare**

psimythion, Gr. white-lead; see **lead**

psinathos, Gr. wild goat; see **goat**

psithyros, Gr. whispering, twittering, rustling; *psithyristes,* whisperer; see **murmur**

psittacus, L. (Gr. *psittakos*), parrot; see **parrot**

psoa, Gr. a muscle of the loin; see **side**

psochos, Gr. rubbed small; see *psao* under **rub**

psocus, NL. a genus of book-lice; see **louse**

psogo- < Gr. *psogos,* blame; see **blame**

psolos, Gr. soot, smoke, see **dirt**; one circumcised, penis, see **penis**

psomos, Gr. morsel, bit; *psomion,* dim.; see **little**

psophos, Gr. sound, noise; see **sound**

psoralea, NL. a genus of the bean family; see **bean**

psoro- < Gr. *psora,* itch, scurvy, scab, mange; *psoraleos,* itchy, scabby, mangy; *psoriasis,* a skin disease; see **itch**

psycho- < Gr. *psyche,* breath, life, soul, spirit, mind, butterfly; see **mind, butterfly**

psychros, Gr. cold, frigid; see **cold**

psycto- < Gr. *psyktos,* cool; *psyktikos,* cooling; see *psychros* under **cold**

psydrax, Gr. blister; see **bubble**

psydros, Gr. false, lying, untrue; see *pseudos* under **lie**

psygma, Gr. fan, a means of cooling; *psygmos,* chilliness; see **fan,** *psychros* under **cold**

psylla, Gr. flea; see **flea**

psyllium, L. (Gr. *psyllion*), fleawort; see **plantain**

psyllo- < Gr. *psylla; psyllos,* flea; see **flea**

psyros, Gr. a fish; see **fish**

psytton, Gr. spit; see **spit**

ptacismo- < Gr. *ptakismos,* shyness, timidity; see **fear**

ptaesmato- < Gr. *ptaisma, -tos,* n. mistake, false step, stumble; see **falter**

ptarmicus, L. (Gr. *ptarmikos*), causing to sneeze; *ptarmos,* sneeze, < *ptairo,* sneeze, see **sneeze**; *ptarmike,* yarrow, see **yarrow**

ptelas, Gr. a wild boar; see **hog**

ptelea, Gr. elm; see **elm**

ptenos, Gr. feathered, winged; see **feather**

pteris, L., Gr. fern; see **fern**

pternis, Gr. bottom of a dish, see **base**; a kind of hawk, see **hawk**

pternix, Gr. stem of a plant; see **stem**

ptero- < Gr. *pteron,* feather, wing, fin; *pteridion; pteriskos,* dim.; *pterinos; pterotos,* feathered, winged; see **feather, wing**

pterygion, Gr. dim. of *pteryx, -ygos,* wing; see *pteron* under **wing**

ptesis, Gr. flight; *ptesimos,* able to fly; see *petomai* under **fly**

ptexis, Gr. terror, < *ptesso,* cower, crouch; see **fear**

ptilo- < Gr. *ptilon,* down, feather, wing, leaf; *ptilion,* dim.; *ptilos,* plumage; *ptilotos,* feathered, winged; see **feather**

ptinus, NL. a genus of beetles; see **beetle**

ptisane, Gr. hulled barley; see **grain**

ptisso, Gr. winnow grain; *ptistes,* winnower; *ptismos,* a winnowing; see **separate**

ptochos, Gr. beggar, a poor person, see **beggar**; < *ptox, -ochos,* hare, see **hare**

ptoma, Gr. that which has fallen, corpse, see **body**; *ptosis,* a falling, see *pipto* under **fall**

ptolemos, Gr. variant of *polemos,* war; see **fight**

ptomatis, Gr. cup that must be emptied at once because it will not stand upright, tumbler; see **cup**

ptorthos, Gr. young branch, shoot, sucker; see **branch**

ptox, -ochos, Gr. the cowering one, hare; see **hare**

ptyalon, Gr. saliva, spit; see **spit**

ptyas, Gr. spitter, a kind of snake; see **spit**

ptycho- < Gr. *ptyx, -ychos,* fold, leaf, layer, plate; *ptygma,* anything folded; *ptychios; ptyktos,* folded; see **fold**

ptyngo- < Gr. *ptynx, -ngos,* eagle-owl; see **owl**

ptyon, Gr. winnowing fan; *ptyarion,* dim.; see **fan**

ptyrmos, Gr. consternation; *ptyrtikos,* timorous; see **fear**

ptysma, Gr. spit; *ptysis,* a spitting; see *ptyalon* under **spit**

ptyx, Gr. fold, layer, plate; *ptyktion,* dim.; see **fold**

puber, L. downy, ripe; see *pubes* under **hair**

pubertas, L. period of transition from youth to manhood and womanhood; see **young**

pubes; pubis, L. down of adulthood and maturity, the signs of manhood, grown up, adult, the private parts; *pubesco,* develop the hair of puberty, ripen; see **hair**

publico, -atus, L. make known, disclose, impart to all; *publicus,* belonging to the people, open; see *populus* under **people**

puddle < AS. *pudd,* ditch, pool; see **marsh**

pudenda, L. external genitals; see **sex organs**

pudendus, L. shameful; *pudicus,* bashful, modest; see *pudeo* under **shame**

puella, L. girl; *puellula,* dim.; see **child**

puer, L. boy; *puera,* girl; *puerculus,* dim.; *puerilis,* youthful, childish, silly; *puerperus,* parturient; see **child**

pueraria, NL. a genus of the bean family; see **bean**

puffinus, NL. shearwater; see **albatross, auk**

pugilator, L. one who fights with his fists; see *pugilo* under **fight**

pugillar, L. writing tablet; see **book**

pugillus, L. handful, dim. of *pugnus,* fist; *pugillaris,* of the fist; see **hand**

pugio, L. dagger, dirk, poniard; see **dagger**

pugnax, *-acis,* L. combative, contentious; see *pugno* under **fight**

pugnus, L. fist; *pugneus,* of the fist; see **hand**

pulcher, L. beautiful; *pulchellus,* dim.; *puchritudo; pulchrum,* beauty; see **beauty**

pulegium, L. pennyroyal; see **mint**

pulex, L. flea; *pulicarius,* of fleas; *pulicosus,* full of fleas; see **flea**

pull < AS. *pullian;* see **draw, tear, suck**

pullatus, L. clothed in dark garments, < *pullus,* dark, dusky, blackish; *puligo, -inis,* a dark color; see **black**

pullus, L. young animal, particularly a fowl; *pullulus,* dim.; *pullarius,* of young animals; *pullulo, -atus,* produce young, put forth, see **chicken;** dark, dusky, blackish, see **black**

pulmentum, L. sauce, relish; see **sauce**

pulmo, L. lung; *pulmunculus,* dim.; *pulmonarius,* of the lungs; see **lung**

pulpa, L. flesh of animals and fruits; *pulposus,* fleshy; see **flesh**

pulpitum, L. frame, stage, desk; see **table**

puls, *pullis,* L. porridge, pottage; *pulticula,* dim.; see **food**

pulse < L. *puls, pultis,* porridge, see **bean;** < *pulsus,* beating of the pulse; *pulsator,* striker, beater, see *pello* under **push**

pulvereus, L. dusty, < *pulvis, -eris,* dust, powder; see **dust**

pulvinus, L. cushion, pad, pillow; *pulvillus; pulvinulus,* dim.; see **pillow**

pulvis, L. dust, powder; *pulvisculus,* dim.; see **dust**

pumex, *-icis,* L. pumice; *pumiceus,* of pumice; *pumicosus,* like pumice, porous; see **stone**

pumilus; pumilis, L. dwarfish, diminutive, little; *pumilio,* a dwarf, pygmy; see **little**

pump < MD. *pompe.*
> Gr. *antlia,* f. bilge-water in the hold of a ship, pump; *exantleo,* pump out: *Antliarrhinus signatus* (a beetle).
> L. *clyster, -is* (Gr. *klyster, -os*), m. syringe; *klysis,* f. a drenching; *klysma, -tos,* n. injection, drench: clyster, clysis, clysma.
> L. *embolus* (Gr. *embolos*), piston or sucker of a pump; see **close**
> Gr. *eneter, -os,* m. syringe; *enesis,* f. injection; *enetos,* injected; *enema, -tos,* n. clyster, injection: enema, *Enetia spinosissima* (a phasmid).
> L. *fundulus,* piston, sucker; see **pestle**
> L. *haustrum,* n. a machine for drawing water, pump: see *haurio* under **suck**
> See: **suck**

pumpkin < OF. *pompon,* < Gr. *pepon,* a large melon; see **melon**

punctum, L. small hole, dot, spot, < *pungo, punctus,* prick, punch, sting; see **spot, point**

punctura, L. hole, prick; see **hole**

pungens, L. sharp, acrid, piercing, biting; see **point**

punica, NL. a genus of the pomegranate family; see **pomegranate**

puniceus, L. reddish, purplish-red; see **red**

punish < OF. *punir,* < L. *punio, -itus,* inflict pain, take revenge: punitive, impunity, punishment.
> Gr. *alastos,* unforgetting, avenging; *alastor, -os,* m. avenger: *Asilus alastor* (a gadfly), *Alastor atropos* (a wasp).
> Gr. *antektisis,* f. retribution:
> Gr. *anteros,* m. an avenger of slighted love: *Anterus formosus* (a butterfly), *Anteromorpha australica* (a wasp).
> Gr. *antitos,* requited, revenged:
> L. *castigo, -atus,* punish, censure, correct: castigate, chasten, chastize.
> L. *damno, -atus,* sentence to punishment, doom; *damnabilis,* worthy of condemnation; *condemno, -atus,* charge with, convict, disapprove: damn, damned, damnation, indemnity, condemn, condemnation, damnable.
> Gr. *ekdikastes; ekdiketes,* m. avenger; *ekdikesis,* f. vengeance:

Gr. *Erinys, -yos,* f. an avenging goddess, a Fury (Alecto, Megaera, Tisiphone): Erinys.

Gr. *euthynter, -os,* straightener, corrector, chastiser; see *euthys* under **straight**

L. *Furia,* f. an avenging goddess: Fury.

Gr. *kolasis,* f.; *kolasma, -tos,* n. punishment, correction; *kolastes,* m. punisher; *kolastikos,* corrective: *Colastes braconis* (a wasp).

L. *multo, -atus,* punish by fining; *multa; mulcta,* f. fine, penalty: mulct.

Gr. *Nemesis,* f. goddess of retributive justice: Nemesis, *Nemesia cellicola* (a spider), *Nemesis damna* (a copepod), *Nemesiellus montanus* (a spider).

L. *poena* (Gr. *poine*), f. punishment, pain; *poenalis,* of punishment: pain, penalty, penance, subpena, penalize, penology, pine, penal, penitentiary, impenitent, repentance.

AS. *scrifan,* impose penance or punishment: shrive, sheriff, shrievalty (sheriffwick, sheriffship, sheriffalty).

L. *talio, -onis,* f. punishment in kind for the injury sustained: retaliate, talion, retaliatory, *lex talionis.*

Gr. *thoë,* f. penalty: *Thoë erosa* (a crab), *Parathoë rotundata* (a crab).

Gr. *timoros,* upholding honor, avenging:

Gr. *Tityos,* m. a mythical giant punished in Hades for attempted rape by having a vulture feed constantly on his liver: *Tityus infamatus* (a scorpion), *Dynastes tityus* (a rhinoceros-beetle).

L. *ulciscor, ultus,* avenge, punish; *ultio, -onis,* f. revenge; *ultor, -is,* m.; *ultrix, -icis,* f. punisher; *ultorius,* of vengeance:

L. *vindico, -atus,* avenge, deliver, justify; *vindex, -icis,* c. avenger, deliverer; *revindico,* exact retribution: vindicate, vindictive, avenger, revenge, vengeance.

Gr. *zemiotes,* punisher; see *zemia* under **hurt**

See: **justice, hurt, strike, whip, hell, tease**

pupa, L. girl, doll; *pupilla; pupula,* dim.; see **child, doll**

pupil < F. *pupille,* < L. *pupillus; pupilla,* dim. of *pupus,* boy; *pupa,* girl; see **learn, eye**

puppis, L. poop, stern; see **stern**

pupula, L. pupil of the eye, dim. of *pupa,* girl, child; see **eye, child**

pupus, L. boy; *pupillus,* dim.; see **child**

pure < L. *purus,* clean, unadulterated: purity, purify, purification, purine, Puritan, impurity.

Gr. *abatos,* untrodden, inviolate, pure, chaste: *Aster abatus* (Mohave aster).

Gr. *achrantos,* undefiled, immaculate:

Gr. *achrostos,* untouched, uncolored:

Gr. *adiaphthoros,* pure, chaste: *aphthoros,* uncorrupt:

Gr. *adoros,* taking no gifts, incorruptible:

Gr. *aeiparthenos,* ever a virgin:

Gr. *akerastos,* pure, unmixed:

Gr. *akibdelos,* unadulterated, genuine: *Acibdela alba* (a moth).

Gr. *akratos,* pure, unmixed:

Gr. *amiantos,* undefiled: *Amiantus rusticus* (a beetle).

Gr. *amiges; amiktos,* unmixed, pure, unsociable: *Amictogeron anomalus* (a fly), *Amictoides breviventris* (a fly).

Gr. *amyschros,* undefiled:

Gr. *anothos,* genuine:

Gr. *apines,* without dirt, clean: *Apinoglossa comburana* (a moth).

Gr. *aspilos,* unspotted, spotless: *Aspilocoryphus mendicus* (a bug), *Aspilomyia alba* (a fly).

Gr. *athiktos,* untouched, chaste:

Gr. *atrygos,* without lees, clear:

Gr. *azymos,* unleavened, uncorrupted:

L. *auricoctor, -is,* m. refiner of gold:

L. *castus,* pure; *castificus,* purifying: caste, chaste, chastity, incest, incestuous.

Gr. *chrysepsetes,* m. refiner of gold:

L. *colatus,* cleansed, purified, strained; see *colo* under **sieve**

L. *denicalis,* purifying from death:

L. *effaecatus,* purified from dregs, refined:

Gr. *eilikrines,* pure, unmixed:

L. *enucleatus,* clear of husks, unadulterated:

L. *expurgatus,* clean, pure; see *purgo* under **wash**

L. *februo, -atus,* purify, expiate: February.

Gr. *hagnos,* pure, innocent, chaste, sacred; *hagneia,* f. purity, chastity; *aphagnismos,* m. purification: Agnes, *Hagnometopias pater* (a beetle), *Elaeagnus umbellata* (an oleaster).

L. *immaculatus,* unstained, unspotted: immaculate.
L. *impeccabilis,* faultless, sinless: impeccable.
L. *impromiscus,* unmixed:
L. *inalienatus,* unspoiled, uncorrupted:
L. *incoinquinatus,* undefiled, unpolluted:
L. *incontaminatus,* undefiled: uncontaminated.
L. *incorruptus,* unspoiled: incorrupt, incorruptible.
L. *ingenuilis,* sincere:
L. *intactus,* untouched, undefiled: intact.
L. *intaminatus,* unpolluted:
L. *intemeratus,* inviolate, virgin:
L. *intentatus,* untouched, unattempted:
Gr. *katharos,* clean, pure; *katharios,* cleanly, tidy; *katharmos,* m.; *katharsis,* f. a
    cleansing, purification; *kathartes,* m. cleanser, purifier; *kathariotes,* f. cleanliness:
    cathartic, catharsis, acatharsis, Cathartidae, *Catharista atrata* (black vulture),
    Cathartes aura (turkey-buzzard), *Rhamnus cathartica* (a buckthorn), *Linum
    catharticum* (purging flax), *Bromus catharticus* (a grass).   Derive: Catharine,
    Kate.
L. *lustrum,* n. a purificatory sacrifice; *lustralis,* of purification from guilt; *lustro,*
    -atus, purify by propitiation: lustrum, lustral, lustration.
L. *merus,* pure, unadulterated, genuine; *meracus,* pure; *meraculus,* dim.: *Eumolpus
    merus* (a beetle).
Gr. *phoibetria,* f. purifier, < *phoibao,* cleanse, purify: *Phoebetria fuliginosa* (sooty
    albatross).
L. *putus,* cleansed, purified, clear: *Noctua puta* (a butterfly).
L. *sincerus,* pure, unmixed, genuine: sincere, sincerity.
Gr. *zoros,* pure, sheer: *Zorotypus longicercatus* (a zorapteran), *Zoropsocus
    delicatulus* (a fossil psocopterid).
See: **wash, clear, innocent, white, light**

purgativus, L. cathartic, cleansing; see *purgo* under **wash**
purple < OF. *purpre,* < L. *purpura* (Gr. *porphyra*), f. the mollusk yielding a
    purple dye, the dye itself; *purpureus,* purple; *porphyrites,* like purple, a purple
    stone: purpura, purpurate, purpurescent, porphyry, porphyritic, lamprophyre,
    vitrophyre, *Purpura persica* (a gastropod), *Eupatorium purpureum* (a joepye-
    weed), *Acanthephyra purpurea* (a shrimp), *Laelia purpurata* (an orchid),
    Scabiosa atropurpurea (sweet scabious), *Phoebe porphyria* (a lauracead),
    Peltogyne porphyrocardia (a legume).
L. *blatteus,* purple:
L. *conchyliatus,* of a purple color, clothed in purple; *conchyliarius,* m. a purple
    dyer:
Gr. *halourgos,* sea-purple; *paralourgos,* edged with purple; *halourgis, -idos,* f. a
    purple robe; *halourgidion,* n. dim.:
L. *ostrum,* n. purple dye from a mollusk; *ostrinus,* purple:
Gr. *perinesos,* edged with purple:
Gr. *phoinix, -ikos,* purple-red: phenicite, phoeniceous, Phoenician, *Phoenicopsis
    roseus* (a flamingo).
L. *puniceus,* purplish-red; see **red**
Gr. *tyrianthinos,* of Tyrian purple:
purpose < L. *propositum,* design; see **aim**
purpureus, L. purple; see **purple**
purse < AS. *purs,* < L. *bursa* (Gr. *byrsa*); see **bag**
purslane < OF. *porcelaine,* < L. *porcilaca,* a corruption of *portulaca,* f.: Por-
    tulacaceae, *Portulaca oleracea* (purslane), *Sesuvium portulacastrum* (sea-
    purslane).
L. *anacampseros* (Gr. *anakampseros*), m. a plant the touch of which was said to
    restore lost love: *Anacampseros albissima* (a portulacacead), *Sedum anacamp-
    seros* (shy stonecrop).
Gr. *andrachne,* f. purslane: *Andrachne colchica* (andrachne).
NL. *lewisia,* f. a genus of the purslane family, < Capt. Meriwether Lewis: *Lewisia
    rediviva* (bitterroot).
NL. *talinum,* n. a genus of the purslane family: *Talinum calycinum* (rock-pink).
pursue < OF. *porsivre,* < L. *prosequor,* follow; see **follow, hunt**
purus, L. clean; see **pure**
pus, *puris,* L., n. corrupt matter, discharge from a sore; *purulentus,* festering;
    suppuro, -atus, form pus: pus, pustule, purulent, suppurate, suppuration,
    seropurulent.
Gr. *glamyros,* with pus in the eyes, blear-eyed, watery: *Glamyromyrmex beebei*
    (an ant).

L. *gramia*, f. pus or matter in the corners of the eyes; *gramiosus*, full of pus, watery, blear-eyed:

L. *lema* (Gr. *leme*; *gleme*), f. pus in the corners of the eyes; *glemion*, n. dim.:

Gr. *pyon*, n. pus; *pyodes*, like pus: pyorrhea, empyema, pyoid, pyuria.

L. *sanies*, f. bloody matter, pus; *saniosus*, full of pus: sanies, sanious.

See: **sore, slime**

push < OF. *pousser*, < L. *pello, pulsus*, beat, drive, impel; *pulsator, -is*, m. beater, striker; *appello*, drive forward; *compello*, drive together; *expello*, drive out; *impello*, drive against: pulse, pulsation, impulse, compulsory, expulsion, repellent, dispel, propeller, pelt, *Pulsatrix perspicilluatus* (spectacled owl), *Anemone pulsatilla* (pasque-flower), *Cercyon pulsatus* (a beetle).

Gr. *ananche*, force, compulsion; see **strong**

Gr. *apokroustikos*, able to repel; see *krouma* under **strike**

L. *arieto, -atus*, butt like a ram: *Bitis arietans* (a puff-adder).

L. *coactus*, a forcing, constraint; *exactor*, demander, enforcer, tax-collector; see *abigo* under **drive**

L. *coerceo, -itus*, compel, restrain: coerce, coercion.

L. *cornupeta*, f. a goring with the horns:

L. *fistuca*, f. rammer; *fistuco, -atus*, ram:

L. *fundulus*, piston; see **pestle**

L. *impetus*, impulse, attack, charge; see **strike**

Gr. *koryptilos*, m. one that butts with the head: *Coryptilomyia armigera* (a fly).

Gr. *kydebazo*, butt with head and horns:

Gr. *otheo*, push; *osmos*; *othismos*, m.; *othesis*, f. a pushing: osmosis, osmotic, endosmose, osmotaxis, exosmotic, *Exothea paniculata* (inkwood).

L. *petulcus*, butting, frisky:

AS. *scufan*, push: shove, shuffle, scuffle, shovel, scoop.

L. *trudo, trusus*, thrust; *trusatilis*, that may be pushed; *contrudo*, thrust together; *intrudo*, thrust in; *obtrudo*, thrust in, press upon; *retrudo*, thrust back: abstruse, extrusion, intrude, obtrusive, protrude, *Nannippus retrusus* (a fossil horse).

See: **press, strike, drive, banish, pump, draw**

pusillus, L. very small, little, petty; *pusillulus*, dim.; see **little**

pusio, L. little boy; *pusiola*, little girl; see *pusus* under **child**

pustula, L. blister, pimple, bubble; *pustulosus*, full of blisters; see **bubble**

pusus, L. a little boy; *pusa*, girl; see **child**

put < AS. *putian*, set, place; see **place**

putamen, L. cutting, paring, shred, shell; see **shell**

putativus, L. imaginary, suppositional; see *puto* under **think**

puteus, L. well, cistern, pit; *puteolus*; *puticulus*, dim.; *putealis*; *puteanus*, of wells; see **hole**

putidus, L. stinking, fetid; *putidulus*, dim.; see **smell**

putillus, L. child; see *pusus* under **child**

putor, L. stench, rottenness; see *putidus* under **smell**

putridus, L. decayed, rotten; *putridulus*, dim.; *putribilis*, corruptible; *putruosus*, very rotten; see *putreo* under **rot**

putus, L. cleansed, purified, clear; see **pure**

puzzle < E. *opposal*, pose a problem; see **ask, secret, try**

pycno- < Gr. *pyknos*, dense, thick; see **thick**

pycta, L. (Gr. *pyktes*), boxer, pugilist; see **fight**

pyctido- < Gr. *pyktis, -idos*, probably a beaver; see **beaver**

pyelis, Gr. socket, setting; see *pyelos* under **basin**

pyelos, Gr. trough, tub, bathtub, pelvis; see **basin**

pygargus, L. (Gr. *pygargos*), a kind of eagle; see **eagle**

pygmo- < Gr. *pygme*, fist; see **hand**

pygmy < L. *pygmaeus* (Gr. *pygmaios*), dwarf; see **little**

pygo- < Gr. *pyge*; *pyx, -ygos*, rump, buttocks; *pygidion*, dim.; see **rump**

pylo- < Gr. *pyle*, gate, orifice; *pylis, -idos*, dim.; *pyloros*, outlet of the stomach, gatekeeper; see **door, guard**

pymatos, Gr. hindmost, last; see **after**

pyndaco- < Gr. *pyndax, -akos*, bottom of a cup or other vessel; see **base**

pyon, Gr. pus; see **pus**

pyos, Gr. first milk after parturition; see **milk**

pyralis, Gr. a kind of insect said to live in fire; see **moth**

pyramid < Gr. *pyramis, -idos;* see **pillar**

pyrausta, L. (Gr. *pyraustes*), a moth that gets singed in the flame; see **moth**

pyrdalon, Gr. firewood, brushwood; see **wood**

pyreno- < Gr. *pyren, -os,* pit or hard seed; see **nut**

pyrethrum, L. (Gr. *pyrethron*), a plant of the composite family; see **composite**

pyretos, Gr. burning heat, fever; see *pyr* under **fire, disease**

pyrgos, Gr. tower; *pyrgidion,* dim.; *pyrgotos,* like a tower; see **tower**

pyria, Gr. vapor bath; *pyriaterion,* sudatorium; see **wash**

pyrites, Gr. a mineral that strikes fire; see **stone**

pyro- < Gr. *pyr, -os,* fire; *pyridion,* dim. spark; *pyrios,* of fire; see **fire**

pyrola, NL. a genus of the heath family; see **heath**

pyros, Gr. wheat, grain; *pyraminos; pyrinos,* of wheat; see **grain**

pyrrho- < Gr. *pyrrhos,* flame-colored, red, yellowish-red, tawny; *pyrrhias,* redhead;
see **red**

pyrrhocorax, L. (Gr. *pyrrhokorax*), a kind of crow; see *corax* under **crow**

pyrrhula, NL. (Gr. *pyrrhoulas*), a kind of finch; see **finch**

pyrsos, Gr. fire, torch, beacon; see *pyr* under **fire**

pyrum, L. pear; *pyrus,* pear-tree; see **pear**

pysma, Gr. question; *pysmatikos,* interrogative; see **ask**

pystos, Gr. learned; see **learn**

Pythagoras, Gr. celebrated Greek mathematician; see **number**

pythmeno- < Gr. *pythmen, -os,* bottom, foundation, stock; see **base**

pythedon, Gr. putrefaction; see **rot**

Python, Gr. mythical serpent slain by Apollo near Delphi; see **snake**

pythonium, L. dragonwort; see **plant**

pytine, Gr. flask covered with plaited work; see **bottle**

pytinos, Gr. a fish; see **fish**

pyxis, L., Gr. box; *pyxidicula; pyxidion; pyxion,* dim.; *pyxos,* boxwood; see **box**

Q

quack < imitative origin; see **cackle, lie**

quadri- < L. *quattuor,* four; *quadrigeminus,* fourfold; see **four**

quadrus, L. square; *quadrula,* dim.; see *quattuor* under **four**

quaesitus, L. sought; see *quaero* under **ask, question**

quaestor, L. a Roman magistrate; see **govern**

quail < OF. *quaille.*
   Gr. *attagas,* m. a kind of partridge, francolin: *Attagis[Attagas] falklandica* (a
     bird).
   Gr. *chennion,* n. a kind of quail: *Chenniopsis madescassa* (a beetle), *Chennium
     bituberculatum* (a beetle).
   NL. *colinus,* m. < Ab.Am. *colin,* partridge: *Colinus virginianus* (bobwhite, quail).
   L. *coturnix, -icis,* f. quail: *Coturnix communis* (European quail).
   Gr. *kakkabis (kakkabe), -idos,* f. partridge: *Caccabis saxatilis* (a bird).
   Gr. *ortyx, -ygos,* m. quail; *ortygion,* n. dim.; *Ortyx plumifera* (a quail), *Lophortyx
     californica* (California quail), *Oreortyx picta* (mountain-quail), *Collyris ortygia*
     (a beetle).

L. *perdix, -icis,* (Gr. *-ikos*), c. partridge; *perdikion,* n. dim.: *Perdix cinerea* (gray partridge), *Galloperdix lunatulus* (a spur-partridge), partridge.

quake < AS. *cwacian;* see **shake**

quality < L. *qualitas,* condition, nature, property; see **nature**

qualum, L. wicker-basket; *quasillum,* dim.; see **basket**

quantus, L. how many, how much; see **number**

quarrel < L. *querella,* complaint; see **fight**

quarry < OF. *quarriere,* < L. *quadrum,* something square, a squared stone; place for squaring stone; see **mine**

quartus, L. fourth; *quaternarius,* fourfold; see *quattuor* under **four**

quasi, L. appearing as if, simulating; see **like**

quasillum, L. dim. of *qualum,* wicker-basket; see **basket**

quassia, NL., f. a genus of the Simarubaceae, < Quassi, a Surinam negro: quassin, *Quassia amara* (quassia).
    NL. *ailanthus,* f. < East Indian *aylanto,* tree of heaven: *Ailanthus altissima* (ailanthus).
    NL. *simaruba,* f. < Ab.Am. *simarouba,* a tropical tree: Simarubaceae, *Simaruba amara* (Orinoco simaruba), *Bursera simaruba* (gumbolimbo).

quassus, L. shaken, broken, shattered; see *quatio* under **shake**

quattuor, L. four; see **four**

queen < AS. *cwen,* wife; see **govern**

quench < AS. *cwencan,* extinguish, slake, wet; see **wet**

quercus, L. oak; *querceus; quercinus; querneus,* of oak; see **oak**

querquedula, L. a kind of duck; see **duck**

querquerus, L. shivering; see **shake**

querulus, L. plaintive, complaining, < *queror, questus,* complain, find fault; see **fault**

question < L. *quaestio, -onis,* inquiry, interrogation, investigation, problem; see **ask, try, doubt, secret**

-quetrus, L. suffix meaning angled, cornered, sided; see **angle**

qui; quod; quis; quid, L. who, which, what; see **ask**

quick < AS. *cwic,* living; see **life, swift**

quiet < L. *quiesco, quietus,* repose, rest; see *quies* under **rest**

quillwort
    Gr. *isoëtes,* n. an evergreen plant, now a genus of quillworts: *Isoëtes riparia* [*riparium*] (a quillwort), *Sagittaria isoëtiformis* (an arrowleaf).

quinarius, L. of five; *quini,* five; see *quinque* under **five**

quince < L. *cydonium* (Gr. *kydonion*), n.; *cydonia* (Gr. *kydonia*), f. quince-tree, < Kydonia, ancient town on the north coast of Crete: cydonium, *Cydonia oblonga* (quince), *Cydonocrinus turbinatus* (a Permian crinoid), *Cydonium mulleri* (a sponge), *Adhatoda cydoniaefolia* (Brazil bower-plant).
    Gr. *chrysomelon,* n. a kind of quince:
    Pg. *marmelos,* quince; see *malum* under **apple**

quincunx, L. five-twelfths; arrangement in diagonal rows; see **twelve, slope**

quinine < Ab.Am. *quinquina,* bark of barks: quinaldine, quinamine, quinene, quinoline, anthraquinone, quinone, quinidine, hydroquinone, quinovin, *Strychnos pseudoquina* (copalchi).
    NL. *cinchona,* f. a genus of the madder family, species of which yield quinine and other alkaloids, < Countess Chinchon: cinchonamine, cinchonine, cinchonism, *Cinchona officinalis* (a quinine-tree).

quinque, L. five; *quintus,* fifth; *quinqueginta,* fifty; *quinquagesimus,* fiftieth; see **five**

quiris, L. spear; see **spear**

quirito, L. cry out, shriek, wail; see **call**

quirrito, L. grunt; see **grunt**

quiscalus, NL. a genus of grackles; see **starling**

quisquilium, L. refuse, trash, rubbish; see **dirt**

quitus, L. enabled, strong; see *queo* under **strong**

quot, L. how many; *quotus,* how many; see **number**

quotidianus, L. daily, recurring; see *dies* under **day**

R

rabbit < OF. *rabot, prob.* < D. *robbe;* see **hare**

rabidus, L. mad, furious; *rabies,* madness, frenzy; *rabiosus,* raving; see **mad**

rabula, L. wrangler, pettifogger; see **fight**

race < ON. *ras,* a running, see **run;** F. *race,* member of the same stock or lineage, see **class**

racemus, L. bunch or cluster of grapes; *racemosus,* full of clusters; see **grape**

rachis < Gr. *rhachis,* spine, backbone, ridge, axis, stem or stalk bearing flowers or leaflets; see **back**

rack, prob. < D. *rec,* framework; see **frame, screen, trouble, pain**

radiatus, L. rayed, beaming, shining; see *radius* under **rod, light**

radicle < L. *radicula,* dim. of *radix, -icis,* root; *radicalis,* of roots; see **root**

radish < L. *radix,* root.
  L. *armoracia,* horseradish; see **mustard**
  L. *raphanus,* m. (Gr. *rhaphanos; rhaphanis, -idos,* f.), radish; *rhaphanidion,* n. dim.; *rhaphaninos,* of radishes: *Raphanus sativus* (radish), *Raphanus raphanistrum* (jointed charlock), *Dentalina raphanistriformis* (a foraminifer).

radium < L. *radius,* ray; see ELEMENTS under **thing**

radius, L. rod, ray, spoke; *radiolus,* dim.; see **rod**

radix, L. root; *radicula,* dim.; *radicosus,* full of roots; see **root**

radula, L. scraper; see *rado* under **scrape**

rafflesia, NL. a genus of the order Aristolochiales; see **birthwort**

raft < ON. *raptr,* rafter; see **ship**

rag < AS, *ragg:* ragged, ragfish, ragamuffin, ragman, ragpicker.
  Gr. *berberion,* a shabby garment; see **garment**
  L. *cento, -onis,* m. patchwork, covering of rags; *centunculus,* m. dim.: *Centunculus minimus* (chaffweed), *Megachile centuncularis* (a bee).
  Gr. *lakisma, -tos,* n. tatter:
  L. *pannus,* m. piece of cloth, rag, patch; *panniculus,* m. dim.; *pannosus,* ragged, tattered, poor: panniculus, *Cotoneaster pannosa* (a cotoneaster).
  Gr. *rhakos,* n. rag, remnant; *rhakion,* n. dim.; *rhakodes,* ragged: *Rhacophorus marmoratus* (a frog), *Rhacochlaena permagna* (a fly), *Phororhacus platygnathus* (an Eocene bird).
  L. *scrutum,* trash, rags, rubbish; see **worthless**
  Gr. *trychos,* n. worn garment, rag, shred; *trychion,* n. dim.; *trycheros,* ragged, tattered, worn; *trychinos,* of rags; *trychosis,* f. exhaustion, < *trycho,* wear out: *Trycherodon megalops* (a fish), *Trychioplectus geminatus* (a beetle), *Trychopeplus multilobatus* (a phasmid), *Trycherus elongatus* (a beetle), *Brachytrycherus concolor* (a beetle), *Microtrycherus luteosignatus* (a beetle).
  See: **garment, worthless**

rage < L. *rabio,* be mad; see **anger**

raia, L. ray, skate; see **shark**

raid < AS. *rad,* road, a riding; see **destroy, plunder, steal**

rail < OF. *ralle,* a bird, see **crane;** < L. *regula,* a straight piece of wood, see **rod**

rain < AS. *regn;* see **drop, storm**

rainbow; see **arc, color**

raise < ON. *risa,* rise.
  Gr. *airo,* raise, lift; *arsis,* f. a raising; *epartikos,* causing to rise or swell: arsis, *Arsilonche venosa* (a moth), *Argyroploce arsiptera* (a moth), *Arsinoë quadriguttata* (a beetle).
  Gr. *anaphora,* f. a rising, ascension:
  Gr. *anastasis,* f. a raising up again, erection, convalescence: anastasis, anastatic.
  Gr. *anatole,* a rising, rise, sunrise, east; see **east**
  Gr. *anecho,* lift up, raise, hold up:
  L. *antarius,* for raising or hoisting:
  L. *arrigo, -rectus,* set up, raise, erect; *erigo,* raise up, set up: erection, erect.
  L. *ascendo, -ensus,* mount, climb, rise, grow: ascend, ascension, *Taxodium ascendens* (a bald-cypress), *Epinephelus adscensionis* (a spotted grouper).
  Gr. *bibazo,* cause to rise, lift up, mount; *bibastes,* m. mounter, stallion: *Bradybibastes costulatus* (a beetle).
  L. *cretus,* arisen, sprung, descended from, born of; see *cresco* under **grow**
  L. *emergo, -ersus,* come forth, rise: emerge, emergency.

Gr. *epembas, -ados,* f. a rising, elevation, increase:

Gr. *epitole,* f. rising of anything, rise, appearance: *Epitola conjuncta* (a butterfly).

L. *excello, -celsus,* rise: excel, excellent, excellency.

AS. *hebban,* raise: heave, heavy.

Gr. *hypsosis,* f. a raising high, an exalting:

L. *levo, -atus,* lift up, raise; *elevator, -is,* m. a raiser, an uplifter: lever, levy, levee, Levant, leaven, elevate, elevator, elevation, relieve, relevant, unleavened, *Erisocrinus elevatus* (a Pennsylvanian crinoid).

L. *orior, ortus,* rise, grow, appear; *oriundus,* risen from, descended; *ortivus,* of rising; *exortivus,* rising, ascendant: oriental, origin, aborigines, abortive.

Gr. *ornymi,* stir up, excite, raise; see **arouse**

AS. *stigan,* rise: sty, stile.

L. *sublimo, -atus,* elevate, lift up; see *sublimis* under **high**

L. *sucula,* windlass, winch, capstan; see **drum**

L. *supero, -atus,* rise above, surmount, exceed:

L. *surgo, -rectus,* rise; *surgens; assurgens,* ascending, rising: surge, source, insurgent, resurrection, resources, resourceful, resurgence, assurgent, *Lavatera assurgentiflora* (a tree-mallow).

L. *suscito,* raise, stir up, arouse; see *cieo* under **arouse**

L. *tollo, sublatus,* raise; *attollens,* raising; *sublatus,* raised aloft, proud, haughty: extol, maletolt, *Cardium sublatum* (a fossil pelecypod).

Gr. *zale,* surge of the sea; see **storm**

See: **climb, ladder, upright, exceed, swell, light, birth, grow**

raisin < OF. *raisin,* < L. *racemus,* cluster of grapes; see **grape**

rake < AS. *raca;* G. *rechen.*

Gr. *agrophe,* f. harrow, rake:

L. *cratis,* f. harrow:

L. *hirpex, -icis,* m. harrow, rake: hearse, rehearse.

L. *occa,* harrow; see *occo* under **till**

Gr. *oxina,* f. harrow: *Oxina maculata* (a fly).

L. *pecten,* comb, rake; see **comb**

L. *rastrum,* hoe, rake; see **hoe**

L. *rutabulum,* n. fire-rake, stirrer:

See: **comb, till, hoe**

rallum, L. a kind of scraper; see **scrape**

rallus, NL. rail, see **crane;** thin, see **thin**

ram < AS. *ramm;* see **sheep, push**

ramentum, L. chaffy, loose scales, shavings, scrapings; see **scale**

ramex, L. hernia, rupture; see **break**

ramula, L. hoof; see **foot**

ramus, L. branch, antler; *ramulus; ramusculus,* dim.; *ramosus,* full of branches; see **branch**

rana, L. frog; *ranula,* dim.; *ranunculus,* dim. tadpole; see **frog**

rancidus, L. stinking, offensive, see **smell;** *rancor,* stink, grudge, hate, see **hate**

rangifer, L. reindeer; see **deer**

ranunculus, L. tadpole, dim. of *rana,* frog, see **frog;** buttercup, crowfoot, see **buttercup**

rapax, -acis, L. grasping, greedy; see *rapio* under **take**

raphanus, L. (Gr. *rhaphanos; rhaphanis, -idos*), radish; *rhaphanidion,* dim.; see **radish**

raphia, NL. a genus of palms; see **palm**

rapid < L. *rapidus,* quick, swift; see **swift**

rapina, L. robbery, pillage; *raptor,* robber, plunderer; see *rapio* under **take**

rapum, L. turnip; *rapulum; rapunculus,* dim.; *rapistrum,* wild turnip; see **turnip**

rarus, L. scarce, scattered, thin; *raritas; raritudo,* scarceness; see **few, thin**

rascal < OF. *rascaille,* rabble, rubbish; see **rogue**

rasor, L. scraper; *rasilis,* scraped, smooth; *rasus,* scraped, shaved; see *rado* under **scrape**

raspberry. See **blackberry**

rastrum, L. hoe, rake, mattock; *rastellum,* dim.; see **hoe**

rat < AS. *raet;* see **mouse**

ratio, L. reckoning, transaction, relation; *rationalis,* reasonable; see *reor* under **think**

ratis, L. raft; *ratitus,* marked with the figure of a raft; see **ship**

rattle < AS. *hraetel.*
> Gr. *arabos,* m. chattering, rattling, or gnashing of the teeth, < *arabeo,* rattle:
> Gr. *aragmos,* m.; *aragma, -tos,* n. rattle, clatter, clash:
> Gr. *bambaino,* chatter with the teeth:
> L. *crepo, -itus,* rattle, crackle, creak, clatter; *crepax, -acis,* rattling; *crepans; crepitans,* rattling; *crepitaculum,* n. rattle; *crepundia,* f. a child's rattle; *crepulus,* rattling: crepitate, decrepit, discrepancy, decrepitude, crepitant, crepitaculum, craven, crevasse, crevice, *Hura crepitans* (sandbox-tree).
> Gr. *krotalon,* n. rattle, castanet, clapper; *krotos,* m. a rattling noise, < *kroteo,* rattle; *eukrotos,* well-sounding; *katakrotos,* noisy: crotalum, crotalin, *Crotalus horridus* (rattlesnake), *Crotalaria mucronata* (rattlebox).
> Gr. *patagos,* m. clatter, rattle, crash; *pategma, -tos,* n. rattle; *patagetikos,* clattering:
> L. *sistrum* (Gr. *seistron*), n. rattle: *Sistrum nodulosum* (a gastropod), *Sistrurus miliaris* (pygmy-rattlesnake).
> L. *strepo, -itus,* make a noise, rattle; see **sound**
> See: **sound**

rattus, ML. rat; see **mouse**

ratus, L. calculated, fixed, settled, certain; see *reor* under **think**

rauca, L. worm in oak roots; see **worm**

raucus, L. hoarse; *raucidulus,* dim.; *raucedo; raucitas,* hoarseness; see **hoarse**

raudus, L. piece of bronze used as a coin; *raudusculum,* dim.; see **money**

ravidus, L. somewhat gray, grayish; see *ravus* under **gray**

ravine < OF. *ravine,* < L. *rapina,* robbery; see **valley**

ravus, L. grayish-yellow, gray, see **gray**; hoarse, see **hoarse**

raw < AS. *hreaw;* see **rough**

ray < L. *radius,* rod, see **rod**; < *raia,* ray, see **shark**

razor < OF. *rasor,* < L. *rado,* scrape; see **knife**

re-; red-; retro- < L. *re,* back, again, down, very; see **return, very**

reach < AS. *raecan,* extend, stretch out; see **spread, come, take**

read < AS. *raed,* counsel; see **speak, teach**

ready < AS. *raede,* finished, arranged; see **fit**

real < OF. *real,* < ML. *realis,* of a thing, < L. *res,* thing; see **thing, true**

reap < AS. *reopan,* gather, harvest.
> Gr. *ametos,* m. harvest; *ametikos,* of harvest; *ameter, -os,* m.; *ametris, -idos,* f. reaper:
> Gr. *analego,* glean, pick up crumbs:
> L. *annona,* f. yearly produce, crop, grain:
> L. *carpo, carptus,* pick, pluck, gather; *carptura,* a sucking or gathering by bees from flowers; see **tear**
> L. *decerpo, -ptus,* pluck off, break off, gather:
> AS. *haerfest,* crop, gathering of crops: harvest.
> Gr. *leion,* n. crop:
> L. *meto, messus,* reap; *messis; messura,* f. a reaping, harvest, crop; *messor, -is,* m. reaper; *messorius,* of a reaper; *demeto,* cut off, gather, reap: *Messor arenatus* (an ant).
> L. *spicilegium,* n. gleaning:
> Gr. *synkomide,* harvest; see *synkomizo* under **gather**
> Gr. *therizo,* mow, reap; *therisis,* f.; *therismos; theristos,* m. harvest; *therister, -os; theristes,* m. reaper; *theristikos,* of reaping, harvesting: *Theristes cultrella* (a moth), *Theristicus brevirostris* (a bird), *Missourium theristocaulodon* (a fossil mammal).
> Gr. *tryge,* f.; *trygema, -tos,* n.; *trygetos,* m. crop, harvest, vintage; *trygeter, -os; trygetes,* m. harvester: *Trygetes nitidissimus* (a spider).
> See: **gather, take**

rear < OF. *ariere,* hindmost part of, < L. *retro,* back, backward; see **after, back, rump, tail, end**

reason < OF. *raison,* < L. *ratio,* reckoning; see **think, begin**

rebellion < L. *rebellio, -onis,* f. renewal of war, revolt, insurrection: rebel, rebellious.
> Gr. *apostasis,* defection, revolt; see **abandon**
> L. *seditio, -onis,* f. insurrection, mutiny, revolt: sedition, seditious.

rebuke < OF. *rebuchier,* repel; see **scold**

reburrus, L. one with bristling hair; see **hair**

receive < OF. *receivre,* < L. *recipio, -ceptus,* take back, regain, take in; *receptor, -is,* m.; *receptrix, -icis,* f. receiver: receiver, receptacle, reception, receptive, receptor.

L. *acquiro, -isitus,* get: acquire, acquisition, acquisitive.

L. *adscitus,* accepted, received:

L. *alimonium,* n. sustenance, allowance for support: alimony.

Gr. *dekter, -os; dektes; apodektes,* m. receiver, beggar; *dektikos,* acceptable, capacious; *dechomai,* take, accept, receive; *apodektos,* welcome; *diadektor, -os,* m. inheritor; *diadochos,* succeeding, receiving: diadoche, diadochite, Polydectes, *Decticogaster zonularis* (a moth), *Apodecter stromeri* (a Miocene murid).

L. *heres, -edis,* c. heir; *hereditas, -atis,* f.; *herctum; heredium,* n. inheritance; *herediolum,* n. dim.: heredity, heir, heiress, heritage, heirloom, inherit, disinherit.

Gr. *paraleptor, -os,* m. inheritor:

L. *patrimonium,* n. inheritance: patrimony.

See: **take, gain**

recens, *-entis,* L. new, fresh, young; see **new, young**

receptacle < L. *receptaculum,* place or vessel for keeping things; see **vessel**

receptor; receptrix, L. receiver; see **receive**

recessus, L. corner, nook, retreat; see **hollow**

recidivus, L. recurring, returning; see **return**

reciproco, *-atus,* L. return, move back and forth, alternate; see **alternate**

recisamentum, L. chip, shaving; see **part**

reclinis, L. leaning back; see *clino* under **slope**

reclusus, L. shut up, separated, removed; see **separate**

reconditus, L. hidden, concealed; see *condo* under **cover**

record < L. *recordor, -atus,* remember; see **book, list, story, remember**

rectum, L. end of the large intestine; see **intestine**

rectus, L. straight, upright, proper, right; *rector,* leader, director; see **right, *rex*** under **govern**

recula, L. trifle, dim. of *res,* thing; see **thing**

red- < L. *re,* back, again; see **return**

red < AS. *read:* redbird, redfin, redstart, redtop, redwood, Redfield, Red River.

Gr. *aithos,* burnt, fiery, reddish-brown; see *aitho* under **burn**

Ab.Am. *annatto,* a tree yielding a yellowish-red or salmon-colored dye: annatto (from *Bixa orellana*).

Ab.Am. *attamusco,* stained with red: *Zephyranthes atamasco* (zephyr-lily).

L. *cardinalis,* red (by transfer of meaning); see *cardo* under **hinge**

L. *cerasinus,* cherry-colored; see *cerasus* under **plum**

L. *cinnabarinus* (Gr. *kinnabarinos*), red like cinnabar; see **cinnabar**

L. *coccineus; coccinus,* red like a berry, scarlet; *coccinatus,* clad in scarlet: *Hygrophorus coccineus* (a mushroom), *Quercus coccinea* (scarlet oak).

L. *corallinus,* coral-red; see **coral**

L. *cruentus,* blood-red; see *cruor* under **blood**

Gr. *daphoinos,* blood-red, gory: *Daphoenositta miranda* (a bird).

Gr. *erythros; erythraios,* red; *erythrotes, -etos,* f. redness, ruddiness; *erythema, -tos,* n.; *erythriasis,* f. blush; *erythrokomos,* red-haired; *dierythros,* spotted or variegated with red; *exerythros,* very red; *katerythros,* deep red: erythrin, Erythrean, erysipelas, erythrocyte, haemoerythrin, *Erythroxylum coca* (coca tree), *Erythronium americanum* (a fawn-lily), *Erythrina speciosa* (a coral-bean), *Erysibe*(now *Erysiphe*) *communis* (a mildew), *Melanerpes erythrocephalus* red-headed woodpecker), *Ichthyocampus erythraeus* (a fish), *Eperythrozoon suis* (a blood-parasite of the pig).

L. *ferrugineus; ferruginus,* rust-colored, dark-red, dusky, rusty; see *ferrugo* under **rust**

L. *flammeus,* fiery-red, flame-colored; see *flamma* under **fire**

L. *fuco, -atus,* rouge, dye, color; see **color**

Gr. *haimaleos; haimatinos; haimatodes; haimonios,* blood-red, bloody; see *haima* under **blood**

Gr. *hysginon,* n. crimson or scarlet dye; *hysginoeis,* scarlet: *Hysginum nivale* (a protozoan).

L. *ignicolor,* fire-colored; see *ignis* under **fire**

Gr. *kokkyginos,* purple-red; see *coccygia* under **sumac**

Gr. *miltos,* red, red earth; *miltokarenos,* red-headed; *emmiltos,* tinged with red: *Miltogramma oculata* (a fly).

L. *miniatus; miniaceus; minianus; mineus; minius,* bright red, cinnabar-red, vermilion; *miniatulus,* dim.; *minio, -atus,* paint red with cinnabar; *minium,* n.

cinnabar, red-lead: miniature, carmine, *Aechmea miniata* (a bromeliacead), *Polytrema miniaceum* (a foraminifer).

L. *mulleolus*, reddish, < *mulleus*, a reddish shoe; see **shoe**

Gr. *phlogeos; phlogeros*, fiery-red; see *phlego* under **burn**

Gr. *phoinix*, purple-red; see **purple**

L. *puniceus*, reddish, purplish-red, < *Punicus*, Carthaginian, < *Poenus*, Phoenician: puniceous, punicin, Punicaceae, *Punica granatum* (pomegranate), *Mimulus puniceus* (red monkey-flower).

Gr. *pyrodes*, like fire, fiery-red; see *pyr* under **fire**

Gr. *pyrrhos* (L. *burrus; byrrus*), red, flame-colored, yellowish-red; *pyrrhias*, m. redhead; *empyrrhos*, ruddy; *epipyrrhos*, reddish; *katapyrrhos*, very red: pyrrole, pyrrhotite, pyrrhotism, Pyrrhocoridae, *Pyrrhocoris apterus* (a bug).

Gr. *pyrsokomos*, red-haired; see *pyr* under **fire**

Ar. *qirmiz*, deep red; F. *kermes*, a coccid insect: kermes (on *Quercus coccifera*), kermesite, crimson, carmine, *Kermes galliformis* (a coccid), *Chermes*[*Kermes*] *pinicorticis* (pine-bark aphid), *Kermesia albida* (a bug), *Callistochermes rubrovariegata* (a bug), *Gladiolus carmineus* (a gladiolus), *Kermococcus vermilio* (a bug).

Gr. *rhodon*, n. rose; red; *rhodochroos*, rose-colored; *rhodopos*, rosy: rhodium, rhodonite, rhodophyll, rhodizite, rhodamine, rhodochrosite, Rhoda, Rhode Island, Rhodophyceae, *Rhodococcus cinnabareus* (a bacterium), *Rhodocyathus ceylonensis* (a coral) *Rhodopteryx pulchripennis* (a katydid).   See **rose**

Gr. *rhousios*, reddish:

L. *roseus*, rose-colored; see *rosa* under **rose**

L. *ruber, -bra, -brum; rubellus; rubellulus*, dim.; *rubens; rubeus; rubicundus; rubidus; surruber; surrubeus*, reddish, ruddy, red; *rubor, -is*, m. redness, blush; *rubesco; erubesco; irrubesco*, grow red, redden, blush; *rubrico, -atus*, color red: ruby, rubeola, rubella, rubellite, rubefaction, rubefacient, erubescent, rubidium, rubiginous, rubicund, Rubicon, rouge, rubric, *Phoenicopterus ruber* (a flamingo), *Myrica rubra* (red bayberry), *Acer rubrum* (red maple), *Callicarpa rubella* (a beautyberry), *Picea rubens* (red spruce), *Crataegus rubicunda* (a hawthorn), *Mastostethus rubricollis* (a beetle), *Mesembryanthemum rubricaule* (a figmarigold), *Vermivora rubricapilla* (Nashville warbler), *Pyrocephalus rubineus* (vermilion flycatcher), *Astacus rubricatus* (a lobster), *Glymmatophora rubripes* (a bug), *Sterculia rubiginosa* (rusty sterculia), *Crinum erubescens* (an amaryllis), *Apoderus rubidus* (a beetle), *Arisaema atrorubens* (a jack-in-the-pulpit).

L. *rufus*, red, reddish; *rufulus*, dim.; *surrufus*, somewhat reddish: Rufus, rufescent, ruficaudate, rufofulvous, *Macropus rufus* (red kangaroo), *Toxostoma rufum* (brown thrasher), *Crataegus rufula* (a hawthorn), *Xylocopa rufipes* (a carpenterbee), *Begonia rufipila* (a begonia), *Rhynchotus rufescens* (tinamou), *Macropus ruficollis* (a wallaby), *Cephalophus rufilatus* (duiker), *Viburnum rufidulum* (rusty viburnum), *Terpsiphone unirufa* (a flycatcher).

L. *russus*, red; *russulus*, dim.; *russeus*, reddish; *russatus*, clothed in red: russet, Barbarossa, *Russula emetica* (a mushroom), *Tricholoma russula* (a mushroom), *Cydnocoris russulatus* (a bug).

L. *rutilus*, red, < *rutilo, -atus*, color red; *irrutilo*, grow red, be ruddy: rutile, *Rutilaria epsilon* (a diatom), *Phalonia rutilana* (a moth), *Tytthonyx rutilis* (a beetle).

Gr. *sandix, -ikos*, f. vermilion: sandix.

L. *sanguineus*, bloody, blood-red; see *sanguis* under **blood**

ML. *scarlatum*, n. cloth of a red color: scarlet, scarlatina.

L. *sinopis, -idis* (Gr. *-idos*), a red ocher from Sinope on the Black Sea; see **earth**

OF. *sorel*, dim. of *sor*, reddish-yellow, brown: sorrel, surmullet.

Gr. *tingabarinos*, vermilion:

L. *usta*, a kind of red color; see *uro* under **burn**

OF. *vermeil*, bright red, the kermes insect, < L. *vermiculus*, little worm: vermilion.

redactus, L. edited, reduced; see **edit, short**

redditus, L. given back, returned; see *reddo* under **return**

redimiculum, L. band, fillet, girdle; see **belt**

redivivus, L. living again, revived, renewed; see *vita* under **life**

reductus, L. withdrawn, remote; see **separate**

reduncus, L. bent or curved backward; see *uncus* under **hook**

redundans, L. superfluous; see **over**

reduvia, L. hangnail, remnant, fragment; see **sore**

reed < AS. *hreod*; see **grass, sedge, cattail, waterweed**

reef < ON. *rif*, rocks just below the surface of the water; see **stone**

refertus, L. crammed, stuffed; see **press**

reflexus < L. bent or turned back; see *flecto* under **bend**

refractarius, L. obstinate, stubborn; see **stubborn**

refrain < OF. *refrener*, < L. *refreno, -atus,* bridle, check; see **hold**

refuge < L. *refugium,* safety, shelter; see *confugium* under **safe**

refuse < OF. *refuser,* < L. *refundo, -fusus,* pour back; see **no, dirt**

refute < L. *refuto, -atus,* disprove, repel; see **no**

regalis; regius, L. royal, < *rex, regis,* king; *regina,* queen; see *rex* under **govern**

regelo, *-atus,* L. thaw; see **flow**

regimen, L. guide, rule; *regimentum,* government; see **law**

region < L. *regio, -onis,* district, quarter, tract; see **country**

regret < OF. *regreter;* see **sad**

regula, L. ruler, measure, pattern; *regularis,* according to rule; see **measure**

regulus, L. dim. of *rex, regis,* king; see **govern**

rein < OF. *rene,* < L. *retineo,* hold back; see **hold**

rejectus, L. cast back, refused, worthless; see *jacio* under **throw**

relativus, L. referring to, akin to, near to; see **kin**

relax < L. *relaxo, -atus,* loosen, slacken, unbend; see *laxus* under **free**

relectus, L. gathered, recollected; see *lego* under **gather**

relic < L. *reliquia,* leaving, remnant; see *linquo* under **abandon**

relieve < OF. *relever,* < L. *relevo,* lift up, mitigate, lessen, comfort; see *levis* under **light, lessen, soothe**

religion < L. *religio, -onis,* sense of piety, duty, taboo; *religiosus,* devout, pious; see **rite**

relish < OF. *relais,* < L. *relaxo,* unloose, open; see **sauce, lick, joy**

reluctans, L. struggling against, resisting; see *luctor* under **fight**

remain < L. *remaneo,* stay; see **stand**

remedy < L. *remedium,* medicine, cure; see **drug, help**

remember < OF. *remembrer,* < L. *memor,* mindful; *memorabilis,* worthy of being remembered; *memorialis,* of memory: memorable, memory, immemorial, memoir, memorandum, commemorate, memorial, reminiscence.

Gr. *alastos,* not to be forgotten, unforgettable, avenging; see **punish**

Gr. *mnemon, -os,* mindful, unforgetting; *mnemonikos; mnesios,* of memory; *mneme; mnemosyne; mnestis,* f. memory, remembrance; *mnema, -tos,* n. memorial, record; *mnemation,* n. dim.; *mnesikakos,* remembering wrongs, revengeful; *hypomnema, -tos,* n. written reminder, memorandum, note, < *mnaomai,* remember: mnemonic, Mnemosyne, amnesty, amnesia, *Mnemosyne cubana* (a bug), *Mnesipenthe obliquisignata* (a moth), *Parnassius mnemosyne* (a butterfly).

L. *monumentum,* n. memorial: monument, monumental.

remigo, L. row; *remex, -igis,* rower; *remigulus,* dim.; see *remus* under **oar**

remissus, L. slack, relaxed, negligent; see **neglect**

remora, L. delay, hindrance; a sucking fish; *remorator,* delayer; see *mora* under **delay**

remorse < L. *remorsus,* a biting back, regret; see **sad**

remotus, L. distant; see **far**

remulcum, L. tow-rope; see **rope**

remus, L. oar; *remulus,* dim.; see **oar**

ren, *renis,* L. kidney; *reniculus; renunculus,* dim.; see **kidney**

renodis, L. unbound, loose; see **free**

repagulum, L. barrier, bar, bolt; see **bar**

repair < L. *reparo,* renew, restore; see **heal, sew, cobbler**

repandus, L bent backward, turned up, undulate; see *pandus* under **bend**

repeal < OF. *rapeler,* < L. *re,* back, *appello,* call; see **cancel**

repeat < OF. *repeter,* < L. *repeto,* seek or attack again; see **return**

repens, L. creeping; see *repo* under **creep**

repent < OF. *repentir,* < L. *re,* back, *poeniteo,* make satisfaction; see *poenitens* under **sad**

repentinus, L. sudden, hasty, unexpected, new; see **lot**

reperticius, L. accidental, random; see **lot**

repertor, L. inventor; *repertorium,* catalogue, inventory; see *reperio* under **find, list**

repetitus, L. again, anew; see *repeto* under **return**

repigratus, L. retarded, slow; see *piger* under **slow**

repletus, L. filled, full; see *pleo* under **abundance**

replum, L. bolt for closing folding doors, dissepiment in fruits; see **bar, wall**

reply < OF. *replier,* < L. *replico,* fold back; see **answer**

reprove < OF. *reprover,* < L. *reprobo,* condemn; see **scold**

reptile < L. *reptile,* snake, serpent; *reptilis,* creeping, crawling; see **snake, lizard, turtle, creep**

republic < L. *respublica,* commonwealth, state; see *publicus* under **people**

repudio, -atus, L. cast off, reject, divorce; see **separate**

requisitus, L. demanded, necessary; see **must**

res, L. thing; *recula,* dim.; see **thing**

rescindo, L. abrogate, repeal; see **cancel**

reseda, L. a kind of plant, mignonette; see **mignonette**

resero, -atus, L. unlock, open; see **open**

reses, L. remain sitting, idle, inactive; see **rest**

resex, L. stub left after pruning; see **stem**

residuus, L. remaining, left over; see **abandon**

resigminum, L. clipping, paring; see *seco* under **cut**

resigno, -atus, L. cancel, give up, relinquish; see **abandon**

resimus, L. turned up, bent back; see **bend**

resin < L. *resina* (Gr. *rhetine*), f. pitch, gum; *resinula,* f. dim.; *resinaceus; resinalis,* of resin; *resinosus,* full of resin, gummy: resinous, resinol, rosin, *Eucalyptus resinifera* (kino-eucalyptus), *Pinus resinosa* (red pine), *Rhetinangium arberi* (a fossil plant), *Retinodiplosis resinicola* (resin-gnat).

Gr. *agallochon,* n. an East Indian wood yielding a dark, aromatic resin: agallochum, *Aquilaria agallocha* (lignaloes, agallochum, or eaglewood).

Gr. *aloe,* a plant yielding a resinous juice; see *aloe* under **lily**

OF. *ambre,* < Ar. *anbar,* a fossilized resin, amber; see **amber**

L. *ammoniacum* (Gr. *ammoniakon*), n. a gum-resin: ammoniac (from the plant sumbul, *Dorema ammoniacum*).

Javanese *antjar,* a poisonous resin: antiar, *Antiaris toxicaria* (upas-tree).

L. *arabicus,* Arabic, < Arabia: gum arabic (derived from *Acacia senegal* and *A. arabica*).

Gr. *asphaltos,* f. bitumen: asphalt.

Per. *aza,* mastic: asafetida (from *Ferula assafoetida*), asadulcis.

Sp. *balata,* f. a gum, < Ab.Am.: balata gum (from *Mimusops globosa*).

L. *balsamum* (Gr. *balsamon*), n. a fragrant resin; *balsameus; balsaminus,* of balsam: balsam, balm, embalm, balsamiferous, *Balsamorrhiza helianthemoides* (balsam-root), *Abies balsamea* (balsam-fir), *Myroxylon balsamum* (tolu-balsam tree), *Commiphora opobalsamum* (Mecca balsam), *Euphorbia balsamifera* (afernan), *Momordica balsamina* (a gourd), *Cecidomyia balsamicola* (a gall-midge), *Chrysanthemum balsamita* (costmary).

L. *bdellium* (Gr. *bdellion*), n. a gum-resin: bdellium (from *Commiphora africana*).

NL. *benzoin,* an aromatic resin; see **benzene**

L. *bitumen, -inis,* n. asphalt, pitch; *bitumineus,* of asphalt; *bituminosus,* full of asphalt or pitch: bitumen, bituminous, bitulithic, bitumastic, *Psoralea bituminosa* (a scurfpea).

ML. *camphora* (Gr. *kaphoura*), f. < Ar. *kapur,* camphor: camphor, camphene, campholic, camphoric, *Camphorosma monspeliaca* (a chenopodiacead), *Cinnamomum camphora* (camphor-tree), *Pluchea camphorata* (a composite).

L. *cancamum* (Gr. *kankamon*), n. an Arabian gum:

Sp. *carana,* f. a kind of resin, < Ab.Am.: caranna gum (from *Protium carana*).

L. *carfiathum,* n. a kind of incense:

Ab.Am. *cauchu,* rubber: caoutchouc (from *Hevea brasiliensis* and *Ficus elastica*), caucho (from *Castilla elastica*).

NL. *cerasina,* f. cherry gum, < L. *cerasus,* cherry-tree: cerasin.

Ab.Am. *chictli,* a gum derived from *Achras sapota* (sapodilla): chicle.

Ab.Am. *copaiba,* an aromatic oleoresin: copaiba, *Copaifera bracteata* (a tree yielding phenin), *Prioria copaifera* (cativo).

Ab.Am. *copalli,* a resin from tropical leguminous trees: copal (from the courbaril, *Hymenaea courbaril,* and *Trachyphyllum mosambicense*), copalcocote (*Crypto-*

carpa procera), copalite, *Copaifera copallifera* (a copal-tree), *Rhus copallina* (a sumac).

Gr. *dadodes,* resinous, < *das, dados,* torch, pine wood; *endados,* resinous; see *das* under **light**

Malay *damar,* a resin: dammar, *Dammara acicularis* (a fossil conifer).

NL. *dracina,* f. dragon's-blood, a red resin from *Dracaena draco* (dragon-tree): dracine.

L. *electrum* (Gr. *elektron*), amber; see **amber**

Sp. *elemi* (Ar. *al lami*), m. a fragrant oleoresin: elemin, *Amyris elemifera* (torchwood).

NL. *euphorbium,* n. a resin derived from *Euphorbia resinifera* (gum-euphorbia): euphorbium.

L. *galbanum* (Gr. *chalbanon*), n. resinous sap of an umbellifer in Syria: galbanum (from *Ferula galbaniflua*).

Malay *gambir,* a yellowish catechu: gambier (from *Uncaria gambir*).

NL. *gambogium,* n. a gum-resin, < Cambodia: gamboge (from *Garcinia hanburyi*), *Garcinia cambogia* (a gamboge-tree).

Malay *getah,* gum, *perca,* tree: guttapercha (from *Palaquium gutta*).

Ab.Am. *guayacan,* a tree and the resin derived therefrom: guaiacum, guaiacol, guaiol, guaiasanol, *Guaiacum officinale* (lignum-vitae).

L. *gummi* (Gr. *kommi*), n. an exudate from plants, resin, sticky substance; *gummatus; gummeus,* containing gum, gummy; *gummosus,* full of gum: gum, gummy, gummite, gum arabic, *Commiphora myrrha* (Arabian myrrh-tree), *Eucommia ulmoides* (eucommia), *Commidendron gummiferum* (gumwood), *Acacia gummifera* (mogador acacia), *Machaerocereus gummosus* (a cactus).

L. *incensum,* n. resinous material that yields a fragrant odor or smoke when burned: incense, cense, frankincense.

Sp. *jalapa,* f. a resin from *Exogonium purga,* < Ab.Am. Xalapa, a town in Mexico: jalap, *Mirabilis jalapa* (common fouro'clock).

Ar. *kababah,* dried fruit of *Piper cubeba:* cubeb.

Malay *kachu,* a kind of resin: catechu, protocatechuic, cutch, *Acacio catechu* (catechu-tree).

Maori *kauri,* a tree and the resin therefrom: kauri gum (from *Agathis australis*), *Synagathis kauricola* (a beetle).

African *kino,* a dark-red gum from *Lingoum marsupium* (Amboina kino): kino, kinoin, kinofluous.

L. *ladanum* (Gr. *ladanon*), n. mastic: ladanum (labdanum), laudanum, *Cistus ladaniferus* (a rock-rose).

Per. *lak,* a resinous substance exuded from the scale-insect, *Tachardia lacca:* lacquer, shellac, lake, laccase, *Phytolacca dioica* (umbra-tree), *Cyrtostachys lakka* (sealing-wax palm).

Gr. *larimnon,* n. Arabic name for frankincense:

L. *laser, -is,* n. a gum-resin similar to asafetida: laser, laserwort *(Laserpitium latifolium).*

L. *latex, -icis,* fluid, liquid, sap; see **juice**

Gr. *libanotos,* m. frankincense, resin of the tree *libanos:* libaniferous, *Seseli libanotis* (an umbellifer).

L. *maltha* (Gr. *malthe*), f. a kind of asphalt or wax; *malthodes,* pliant, soft: *Malthopterus pallidus* (a beetle), *Malthodes costatipennis* (a beetle), *Micromalthus debilis* (a beetle).

Gr. *manna,* f. morsel, grain, gum of a tree or exudate of an insect: manna, mannitol, mannoheptose, *Gossyparia mannifera* (a scale-insect).

Gr. *mastiche,* resin from the mastic-tree: mastic (from *Pistacia lentiscus*), *Sideroxylon mastichodendron* (mastic-ironwood).

Gr. *metopion,* n. ointment, gum of an African tree: *Metopium toxiferum* (Florida poisonwood).

Gr. *mindax, -akos,* f. a Persian incense:

L., Gr. *myrrha,* f. an aromatic gum-resin; *myrrhatus,* spiced with myrrh; *myrrheus; myrrhinus,* of myrrh: myrrh (from *Commiphora abyssinica*), myrrhophore, myrrhol.

ML. *olibanum,* n. frankincense: olibanum (from *Boswellia carteri, B. frereana, B. papyrifera, B. serrata*).

L. *opium* (Gr. *opion*), poppy-juice; see *opos* under **juice**

Gr. *phrykte,* f. a kind of resin:

Gr. *pissa (pitta),* f. pitch; *pissinos (pittinos),* of pitch; *pissodes,* like pitch, yielding pitch; *pissosis,* f. a pitching: pittacal, pittizite, *Pittosporum crassifolium* (karo), *Pissodes strobi* (white-pine weevil).

L. *pix, picis,* f. pitch; *picula,* f. dim.; *piceus,* pitchy, pitch-black, of pitch; *piceatus,* bedaubed with pitch, pitchy; *pico, -atus,* bedaub with pitch, tar: pitch, pitchblende, picoline, *Melichthys piceus* (a trigger-fish), *Picea engelmanni* (Engelmann spruce), *Melanolestes picipes* (a bug).

Gr. *propolis,* bee-glue, a resinous material collected by bees and used as a cement; see **glue**

L. *sacopenium* (Gr. *sagapenon*), n. gum or resin from an umbellifer, *Ferula persica:* sagapenum.

L. *sandaraca* (Gr. *sandarake*), realgar; now applied to a resin; see **arsenic**

L. *sarcocolla* (Gr. *sarkokolla*), f. a Persian gum: sarcocolla, sarcocollin, *Sarcocolla squamosa* (a shrub), *Astragalus sarcocolla* (a vetch).

Gr. *smyrna,* f. a kind of resin from an Arabian tree:

L. *storax; styrax, -acis* (Gr. *-akos*), m. a tree or the resin therefrom; *styracinus,* of storax: styracin, styracol, storax, *Styrax officinalis* (storax-tree), *Liquidambar styraciflua* (sweetgum), *Pterostyrax hispida* (epaulette-tree).

L. *succinum,* amber; see **amber**

Ab.Am. *tecomahiyac,* stinking-pot tree: tacamahac, *Populus tacamahaca* (balsam poplar), *Calophyllum tacamahaca* (a clusiacead).

L. *terebinthus* (Gr. *terebinthos*), f. terebinth-tree; *terebinthinus,* of the terebinth-tree: turpentine, sesquiterpene, terpinol, terpene, *Pistacia terebinthus* (terebinth), *Silphium terebinthinaceum* (dock-rosinweed), *Rhus terebinthifolia* (a sumac).

L. *thus, thuris (tus, turis),* n. incense; *thuralis,* of incense; *thurifer,* incense-yielding; *thurifico, -atus,* burn incense: thurible, *Juniperus thurifera* (incense-juniper).

Sp. *tolu,* a balsam, < Santiago de Tolu: tolu (from *Myroxylon balsamum*), toluene, toluic, toluidine, toluol, toluyl, tolyl, tolylene.

L. *tragacanthum,* n. a gum from *tragacantha* (Gr. *tragakantha*), f.: tragacanth (from *Astragalus gummifer*), *Astragalus tragacantha* (a vetch).

Ab.Am. *umiri,* a kind of balsam: umiri (from *Humiria balsamifera*).

NL. *vernix, -icis,* < OF. *vernis,* varnish: varnish, vernicose, *Rhus vernix* (poison-sumac), *Rhus verniciflua* (lacquer-tree), *Acacia vernicosa* (an acacia).

See: **glue, amber, juice, jelly**

resolutus, L. bold, firm; see **bold**

responsum, L. reply, answer; *responsivus,* answering; see **answer**

rest < AS. *rest.*

Gr. *aderitos,* without strife, uncontested:

Gr. *aergos (argos); apergos; dysergos,* not working, idle; *argikos,* indolent: argic.

L. *aequanimus,* calm, composed: equanimity.

Gr. *akinetos,* not moving, motionless, idle, steadfast: acinetan, *Acineta tuberosa* (a protozoan), *Acinetopsis rara* (a protozoan).

Gr. *akonitos,* without the dust of the arena, without contest or struggle:

Gr. *akymatos,* waveless, calm:

L. *alcyon* (Gr. *alkyon*), kingfisher, supposed period of calm and quiet; see **kingfisher**

Gr. *amerimnos,* free from care, unconcerned:

Gr. *anenemos,* without wind, calm:

Gr. *anetos,* relaxed, slack; see *aneticus* under **free**

Gr. *anoche; epoche,* f. stop, pause, armistice: epoch, epochal.

Gr. *aochletos,* undisturbed, calm: *Aochletus cinctus* (a fly).

Gr. *aphrontis, -idos,* carefree:

Gr. *aponos,* idle, untroubled, lazy, easy: *Aponogeton fenestralis* (lattice-leaf).

Gr. *ataraktos,* undisturbed, calm, cool, steady:

Gr. *blakikos,* lazy, stupid; see *blax* under **dull**

L. *caesura,* a pause or break in a verse; see **break**

L. *cesso, -atus,* rest, loiter, idle; *cessator, -is,* m. dawdler, idler, a dilatory person: cessation, cease, incessant, *Arotrophora incessana* (a moth).

L. *comma* (Gr. *komma*), part cut off, clause, a punctuation mark denoting a pause; see **part**

L. *contentus,* satisfied; see **agreeable**

L. *deses, -idis,* idle, lazy, slothful; *desidiosus,* slothful; *reses, -idis,* remaining sitting, inactive, idle: *Euglena deses* (a flagellate).

Gr. *eirene,* peace; *eirenikos,* of peace: irenical, irenarch, Irene, *Irenesauripus* [*Irenosauropus*] *occidentalis* (a Cretaceous footprint).

Gr. *ekecheiria,* f. rest, holiday, vacation, holding of hands, armistice:

Gr. *ekklites,* m. one who shuns work, drone: *Ecclites clypeatus* (a braconid).

Gr. *elinyo,* rest:

Gr. *enauros,* windless, calm, still:

Gr. *eremaios,* quiet, still, gentle: *Eremaeus modestus* (a mite), *Eremaeozetes tuberculatus* (a mite).

Gr. *eudios,* calm, fine, clear; see **clear**

Gr. *eukalos (eukelos),* carefree; *eukalia (eukelia),* f. quiet:

Gr. *eukolos,* easily satisfied, contented, at peace:
Gr. *eumareia,* f. ease, comfort; *eumares,* easy:
Gr. *eupetes,* easy; *eupeteia,* f. ease: *Eupetodromus exhibitus* (a beetle).
L. *facilis,* easy, without difficulty: facile, facility.
L. *feria,* day of rest, holiday; *ferio, -atus,* rest from work, keep holiday; see **day**
L. *flustrum,* n. calm, quietness:
AS. *frith* (OHG. *fridu;* G. *friede*), peace: affray, fray, afraid, frith, Friedensville (Pa.), belfry.
L. *fucus,* drone; see **bee**
Gr. *galenos,* n. calm: *Galenomys garleppi* (a mouse).
L. *grassator, -is,* idler, vagabond; see *gradior* under **walk**
Gr. *hekelos (hekalos; akalos),* at rest, at ease, at will; *eukelos,* free from care: *Hecalus pallescens* (a bug), *Acalodegma marginipennis* (a beetle).
Gr. *hesychios; hesychos,* still, quiet, at ease; *hesychia,* f. stillness, quiet, rest: *Hesychotypa miniata* (a beetle).
L. *hiberno, -atus,* spend the winter, keep in winter quarters, be inactive; see *hibernum* under **year**
Gr. *iachros,* softened, calm, tranquil:
L. *ignavus,* idle, slothful, listless, sluggish; see **dull**
L. *imbellis,* peaceful:
L. *imperturbatus,* unruffled, calm: imperturbation.
L. *impetrabilis,* easily obtainable, attainable:
L. *indutia,* f. temporary cessation of hostilities, truce, armistice:
L. *iners, -ertis,* inactive, idle, lazy; *inerticulus,* dim.; *inertia,* f. idleness: inert, inertia, *Cinnamomum iners* (a lauracead).
L. *inexcitus,* unmoved, quiet, calm:
L. *inficiens, -entis,* inactive:
L. *intercapedo,* interval, pause, respite; see **break**
Gr. *kephen,* drone, lazy fellow; see **bee**
Gr. *lophao,* rest from toil, abate; *lophema, -tos,* n. rest:
Gr. *malakia,* calm at sea; see *malakos* under **soft**
Gr. *methetikos,* letting go, relaxing:
L. *nequitia,* f. laziness, idleness, inactivity:
L. *nixor,* lean upon, rest upon; see **lie**
L. *oscito, -atus,* gape, yawn, be lazy, idle; see **open**
L. *otium,* n. leisure, rest; *otiosus,* idle, useless: otium, otiose, negotiate, *Agrilus otiosus* (a beetle).
L. *parabilis,* easily procured, easily attained:
L. *pausa* (Gr. *pausis; pausole; paula*), f. temporary stop, halt, rest; *pauso, -atus,* halt, cease, rest; *anapauma, -tos,* n. repose, rest: pause, pose, paulospore, menopause.
L. *pax, pacis,* f. peace; *pacalis,* of peace; *pacator; pacificator, -is,* m. peacemaker; *pacatorius; pacifer, -a, -um; pacificus,* peacebringing, peaceful; *paco, -atus,* quiet, soothe: peace, pacify, Pacific Ocean, peaceably, appease, pay, unpaid, La Paz, *Rhinochimaera pacifica* (a fish).
L. *piger, -gra, -grum,* disinclined, lazy, slow; see **slow**
L. *placatus,* appeased, calmed, quiet, still; see **soothe**
L. *placidus,* quiet, still, gentle; *placidulus,* dim.: placid, *Rhabdospilus placidus* (a wasp).
L. *quies, -etis,* f. rest; *quiesco, -etus,* repose, rest; *requietus,* rested, refreshed: quiet, quiescent, quit, acquit, acquiesce, disquietude, requiem, R.I.P., requital, unrequited, coy.
L. *resupinus,* lying on the back, lazy; see *supinus* under **lie**
Gr. *rhadios,* easy, ready, light; *rhaon,* easier; *rhastos,* easiest; *syrrhadios,* very easy, promiscuous:
Gr. *rhathymos,* carefree, inactive, indifferent, lazy; *rhathymia,* f. lightheartedness: rhathymia, *Rhathymodes lechrioides* (a moth).
L. *schola* (Gr. *schole*), leisure, given to learning, school; see **school**
L. *securus,* free from care, tranquil; see **safe**
L. *sedatus,* composed, calm, quiet, sober, tranquil; see *sedo* under **soothe**
Heb. *shabbath,* day of rest: Sabbath, sabbatical.
Gr. *sponde,* f. truce:
L. *tranquillis,* quiet; *tranquillitas, -atis,* f. quietness, stillness: tranquil, tranquillity.
Gr. *triodites,* m. frequenter of crossroads, loafer: *Triodites mus* (a fly).
L. *umbraticus; umbratilis,* retired, private; see *umbra* under **shade**
See: **lie, health, forget, painless, harbor, safe, soothe, silent, smooth, free, sit**

restis, L. rope, cord; *resticula,* dim.; *restio,* rope-maker; see **rope**
restless; see **move**

restore < L. *restauro, -atus,* repair, renew; see **return, new**

restrain < OF. *restraindre,* < L. *restringo,* draw back, tighten; see **hold**

result < L. *resulto, -ultus,* leap back, rebound; see **end**

resupinus, L. bent back; see *supinus* under **lie**

reticulatus, L. netlike, netted, < *rete, retis,* net; *reticulum,* dim.; see **net**

retinaculum, L. holdfast, tether, cable, bound; see *teneo* under **hold**

retro- < L. *retro,* back, backward; see *re* under **return**

retrorsus, L. turned or bent backward; see *verto* under **turn**

return < OF. *retourner,* < L. *re,* back, *torno,* turn.
  Gr. *ampotis, -eos,* f. ebb of the tide:
  Gr. *ana,* up, back, again; *anabiosis,* f. revival; *anakope,* f. recoil, repulsion; *ananeusis,* f. return, revival; *ananeotes,* m. renewer, reviver; *anaepsis,* f. recovery; *ananoseo,* be sick again, have a relapse; *anapodisis,* f. a going back: analysis, anatomy, anachronism, anathema, analects, analogous, Xenophon's *Anabasis, Anastomus oscitans* (a stork).
  Gr. *antechema, -tos,* n. echo; *antechesis,* f. a reechoing:
  Gr. *antektisis,* retribution; see **punish**
  Gr. *antimisthia,* f. requital, recompense:
  Gr. *apechema, -tos,* n. echo; *apechetikos,* resounding:
  Gr. *apodisis,* restitution, return; see *do* under **give**
  Gr. *apokatastatikos,* recurring in a cycle:
  Gr. *aps,* backward, back again:
  AS. *baec,* to the rear; see **after**
  L. *denuo,* anew, again, a second time:
  L. *domuitio, -onis,* f. a returning home:
  L. *duplico, -atus,* double, repeat; see *duo* under **two**
  Gr. *echo,* a returned sound; see **sound**
  Gr. *enkyklios,* periodical: encyclical.
  L. *frequens,* taking place repeatedly; see **number**
  L. *increbro, -atus,* do frequently or repeatedly; see **number**
  L. *itero, -atus,* repeat: iterate, iteration, reiterate, *Crataegus iterata* (a hawthorn).
  Gr. *mapsylakas,* idly barking, vainly repetitive:
  Gr. *nostos,* m. a return home; *anostos,* unreturning: nostalgia, nostomania.
  Gr. *palin-, palim-* < *palin,* again, back, repetition; *anapalin,* back again, over again, reversely; *palirrhytos,* refluent: palinal, palindrome, palilogy, palingenesis, palinode, palimpsest, *Palintropa hippica* (a moth), *Palimbolus mirandus* (a beetle), *Palimmecomyia calaenospila* (a fly).
  L. *re-, red-, retro-* < *re,* again, back, backward; *recidivus,* recurring, returning; *reciproco, -atus,* return, move back and forth, alternate; *reddo, -itus,* give back, restore; *reditus,* returned; *redivivus,* living again, revived, renewed; *repercussus, -us,* m. reverberation, echo; *repetitus,* anew, again; *resonus,* reechoing, ringing; *resulto, -atus,* resound, reecho; *reverbero, -atus,* whip back and forth, rebound; *reverto, -versus,* turn back, return: reaction, recall, recede, recreation, redditive, redundant, reflect, report, repeat, repetitive, redemption, respect, residue, reprobate, repercussion, revenue, revert, reversion, rendezvous, rampant, rally, Renaissance, rest(remainder), relic, rent, ragout, retrograde, retroversion, retroactive, arrear, *Cryptocoryne retrospiralis* (an aracead), *Formica microgyna recidiva* (an ant).
  L. *rursus,* turned back, backward, reverse:
  L. *talio, -onis,* punishment in kind for the injury sustained; see **punish**
  See: **alternate, swing, new, equal, answer, atone, punish, like, after, circle**

retusus, L. blunted, rounded, notched at the apex; see **notch**

reus, L. accused person, defendant; *reatus,* charge; see **fault**

reveal < L. *revelo, -atus,* unveil; see **display, speak**

revenge < OF. *revengier,* < L. *revindico,* exact retribution; see **punish**

reverens, L. respectful, regardful, filled with awe; see *vereor* under **honor**

revolutus, L. turned over, rolled back; see *volvo* under **turn**

reward < OF. *rewarder;* see **pay, justice, give, take**

rex, *regis,* L. king; *regulus,* dim.; see **govern**

rhabdo- < Gr. *rhabdos,* rod, stick, staff; *rhabdotos,* striped, fluted; *rhabdouchos,* staff-bearer, judge, umpire; see **rod, line, judge**

rhachio- < Gr. *rhachis, -ios,* spine, backbone, ridge, stem, see **back**; < *rhachia,* shore, beach, surf, see **shore**

rhacho- < Gr. *rhachos,* thorn-bush, brier; see **thorn**

rhaco- < Gr. *rhakos,* rag, remnant; *rhakion,* dim.; *rhakinos,* ragged; see **rag**

rhadico- < Gr. *rhadix, -ikos,* branch, frond; see **branch**

rhadinos, Gr. slender, tapering, lithe; see **thin**

rhadion, Gr. a kind of easy shoe; see **shoe**

rhadios, Gr. easy, ready, light; *rhaon,* easier; *rhastos,* easiest; see **rest**

rhaebo- < Gr. *rhaibos,* bent, bowlegged; see **bend**

rhaestero- < Gr. *rhaister, -os,* hammer; see **hammer**

rhagado- < Gr. *rhagas, -ados,* crack, chink, rent; see *rhegnymi* under **break**

rhago- < Gr. *rhax, -agos,* grape, berry; *rhagion,* dim., see **grape;** < *rhage,* burst, crack, rent, see *rhegnymi* under **break**

rhamma, Gr. anything sewn, seam, thread; see *rhapto* under **sew**

rhamnus, L. (Gr. *rhamnos*), buckthorn; see **buckthorn**

rhamphestes, Gr. probably a pike; see **pike**

rhamphis, Gr. hook; see **hook**

rhampho- < Gr. *rhamphos,* curving beak, bill; *rhamphion,* dim., see **nose;** < *rhamphe,* a curved knife, see **knife**

rhanido- < Gr. *rhanis, -idos,* drop, rain, spot; see **drop**

rhantos, Gr. sprinkled; *rhanter,* sprinkler, wetter; see **wet**

rhaphanis, Gr. radish; see **radish**

rhaphanos, Gr. cabbage, but applied to the radish by the Romans; see **radish**

rhaphido- < Gr. *rhaphis, -idos,* needle; *rhaphe,* seam; *rhapheus; rhaphideus,* sewer; see **needle, sew**

rhapido- < Gr. *rhapis, -idos,* rod, stick; see **rod**

rhapisma, Gr. blow with the palm of the hand, slap; see **strike**

rhaponticum, NL. rhubarb; see *rheum* under **buckwheat**

rhapsody < Gr. *rhapsodia,* recitation of an epic poem; see **poem**

rhapto- < Gr. *rhaptos,* stitched, sewn; see *rhapto* under **sew**

rhasma, Gr. shower; see *rhanis* under **drop**

rhastos, Gr. superlative of *rhadios,* easy; see **rest**

rhatane, Gr. ladle, stirrer; see **spoon**

rhathymos, Gr. indifferent, lightly, easy; see **rest**

rhea, NL. a genus of ostrichlike birds; see **ostrich**

rhectero- < Gr. *rhekter, -os,* worker, doer; *rhekterios,* active, busy; see **make**

rhecto- < Gr. *rhektos,* breakable, brittle; *rhektes,* breaker, earthquake; see *rhegnymi* under **break**

rheda, L. (Gr. *rhede*), four-wheeled wagon or carriage; see **vehicle**

rhegeus, Gr. dyer; see **color**

rhegma, Gr. break, fracture, tear; see *rhegnymi* under **break**

rhegmin, Gr. line of breakers, surf; see **shore**

rhegos, Gr. rug, blanket; see **rug**

rhema, Gr. word, saying, verb; *rhemation,* dim.; see **word**

rhembos, Gr. a wandering, roving; see *rhembo* under **turn**

rhen, Gr. sheep; *rhenikos,* of sheep; see **sheep**

rhenco- < Gr. *rhenkos (rhenchos),* a snore; see *rhonchus* under **snore**

rheos, Gr. stream, current; see *rheo* under **flow, stream**

rhethos, Gr. limb (arm, leg); see **arm**

rhetine, Gr. pitch, resin; see **resin**

rhetor, Gr. orator, teacher of oratory; see **speak**

rheum, L. (Gr. *rheon; rha*), rhubarb, a genus of the buckwheat family; see **buckwheat**

rheuma, Gr. flow, current, see *rheos* under **stream;** *rheumatismos,* a disease, see **disease**

rhexia, L. a kind of plant; see **plant**

rhexis, Gr. a breaking; see *rhegnymi* under **break**

rhicno- < Gr. *rhiknos,* shriveled, shrunk, withered; see **fold**

rhigos, Gr. frost, cold; *rhigelos,* chilling; see **ice**

rhimma, Gr. cast, throw; see *rhiptos* under **throw**

rhimpha, Gr. swiftly; see **swift**

rhine, Gr. file, rasp, shark; *rhinion,* dim.; see **scrape, shark**

rhino- < Gr. *rhis, rhinos,* nose, snout, beak, bill; *rhinion,* dim.; see **nose**

rhinoceros, *-otis,* L. (Gr. *rhinokeros, -otos*), m. rhino: *Rhinoceros unicornis* (Indian rhino).

rhion, Gr. peak, headland; see **projection**

rhipido- < Gr. *rhipis, -idos,* fan; *rhipidion,* dim.; see **fan**

rhipo- < Gr. *rhips, -ipos,* wickerwork, mat, screen; see **screen**

rhipsalis, NL. a genus of the cactus family; see **cactus**

rhiptos, Gr. thrown; *rhipsis,* a casting; see *rhipo* under **throw**

rhisco- < Gr. *rhiskos,* box, chest; see **box**

rhithro- < Gr. *rheithron,* stream; see **stream**

rhizo- < Gr. *rhiza,* root; *rhizion,* dim.; see **root**

rho, Gr. seventeenth letter of the Greek alphabet; see **letter**

rhoa, Gr. pomegranate; see **pomegranate**

rhochmos, Gr. cleft, gully, see *rhegnymi* under **break**; a wheeze, see **sound**

rhochthos, Gr. roaring of the sea; see **roar**

rhodane, Gr. the spun thread; see **thread**

rhodanos, Gr. waving, flickering; see **wave**

rhodo- < Gr. *rhodon,* rose, red; see **rose, red**

rhododendron, Gr. rosebay, a genus of the heath family; see **heath**

rhodora, L. a kind of plant; see **heath**

rhoë, Gr. stream, a flowing; *rhoedion,* dim.; see *rheo* under **flow**

rhoeas, L. (Gr. *rhoias*), a kind of poppy; see **poppy**

rhoeco- < Gr. *rhoikos,* crooked; see **bend**

rhoezo- < Gr. *rhoizos,* whistle or whizz of an arrow; *rhoizetor,* whizzer; see **whistle**

rhogo- < Gr. *rhox, -ogos,* cleft, break; *rhogaleos,* broken; *rhogme,* fracture; see *rhegnymi* under **break**

rhoi- < L. *rhus, rhois* (Gr. *rhous*), sumac; see **sumac**

rhomaleos, Gr. able-bodied, strong; *rhome,* bodily strength, might; see **strong**

rhombus, L. (Gr. *rhombos*), spinning-top, magic wheel, see *rhembo* under **turn**; lozenge, see **diamond**; turbot, see **flatfish**

rhomphaeo- < Gr. *rhomphaia,* a Thracian sword; see **sword**

rhomphus, Gr. shoemaker's waxed thread; see **thread**

rhonchus, L. (Gr. *rhenchos*), snore; see **snore**

rhopalon, Gr. club; *rhopalion,* dim.; see **club**

rhope, Gr. downward inclination, bias; see *rhepo* under **slope**

rhophema, Gr. gruel, soup, *rhopheo,* sip, gulp; see **soup**

rhopo- < Gr. *rhops, -opos,* shrub, bush; *rhopion,* dim., see **bush**; *rhopos,* small, petty, or trifling wares, see **trifle**

rhoptron, Gr. spring bar or rod of a trap; see **rod**

rhoros, Gr. strong; *rhosis,* strength, strengthening; see **strong**

rhothon, Gr. nose, beak; *rhothonion,* dim.; see **nose**

rhothos, Gr. rushing or dashing noise; *rhothios,* rushing, dashing, as of breakers and surf; see **sound**

rhubarb < ML. *rhabarbarum,* < *Rha,* Volga River, *barbarus,* foreign; see *rheum* under **buckwheat**

rhus, L. (Gr. *rhous*), sumac; see **sumac**

rhyaco- < Gr. *rhyax, -akos,* stream, torrent; see **stream**

rhydon, Gr. abundantly; see **abundance**

rhymos, Gr. pole of a vehicle; see **rod**

rhyncho- < Gr. *rhynchos,* nose, snout, muzzle; *rhynchion,* dim.; see **nose**

rhyndaco- < Gr. *rhyndake,* probably a bird-of-Paradise; see **bird**

rhypos, Gr. dirt, filth; *rhyparos,* dirty, filthy; see **dirt**

rhyptico- < Gr. *rhyptikos,* cleaning; see **wash**

rhysi- < Gr. *rhysis,* deliverance, see **save**; *rhysion,* booty, see **plunder**

rhysos (rhyssos), Gr. wrinkled; *rhysema,* wrinkle; see *rhytis* under **fold**

rhyter, Gr. rein, trace; see **hold**

rhythm < L. *rhythmus* (Gr. *rhythmos*), regular recurring motion, measure, proportion; see **swing**

rhytido- < Gr. *rhytis, -idos,* wrinkle, fold, pucker; *rhysos,* drawn up, wrinkled; see **fold**

rhytisma, Gr. patch; see **patch**

rhytium, L. (Gr. *rhytion,* dim. of *rhyton*), drinking-horn; see **horn**

rhytos, Gr. flowing, fluid, liquid; see *rheo* under **flow**

rib < AS. *rib.*
    L. *costa,* f. rib, side, ridge; *costula,* f. dim.; *costabilis,* riblike; *costatus,* ribbed: costal, costalgia, intercostal, coast, accost, cutlet, Costa Rica, *Exogyra costata* (a fossil pelecypod), *Dyera costulata* (jelutong), *Cristellaria bicostata* (a foraminifer), *Polystomella costifera* (a foraminifer), *Cinyra costulifera* (a beetle).
    Gr. *phalkes,* beam, rib of a ship; see **beam**
    Gr. *pleuron,* n. rib, side; *pleurion,* n. dim.; *pleurikos,* of the ribs: pleuron, pleurapophysis, *Pleuracanthus dilatatus* (a fossil fish), *Bupleurum rotundifolium* (a thorowax).
    Gr. *skelis (schelis), -idos,* f. rib, side, sometimes leg; *skelidion,* n. dim.: scelidosaur, ornithoscelidan, *Scelidotherium leptocephalum* (a fossil mammal), *Acanthoscelides obtectus* (bean-weevil), *Euscelis striatulus* (a leafhopper).
    See **side, beam, frame**

ribbon < OF. *riban.*
    Gr. *ampyx, -ykos,* m. headdress, fillet: ampyx, *Ampyx nasutus* (a trilobite), *Leucampyx newberryi* (wild cosmos), *Cercopithecus leucampyx* (a monkey).
    Gr. *diadema,* headband, fillet, crown; see **crown**
    L. *fascia,* band, fillet; see **belt**
    L. *galbeum,* n. armband, fillet:
    L. *infula,* f. band, fillet; *infulatus,* adorned with a fillet:
    L. *lemniscus* (Gr. *lemniskos*), m. ribbon, fillet, band; *lemniscatus,* adorned with ribbons: lemniscate, lemniscus, *Lemniscomys pulchellus* (an African rodent), *Rhacochlaena lemniscata* (a fly).
    Gr. *metopis, -idos,* f. headband:
    L. *redimiculum,* band, fillet, girdle; see **belt**
    Gr. *sparganon,* n. band, swaddling cloth; *sparganion,* n. dim.: *Sparganum proliferum* (a cestode), *Sparganophasma concitata* (a beetle), *Sparganium minimum* (little burreed).
    L. *taenia* (Gr. *tainia*), f. ribbon, fillet, band, stripe, tapeworm; *taeniola,* f.; *tainidion; tainion,* n. dim.: taenidium, taeniodont, taenioglossate, taeniolite, *Taenia solium* (tapeworm), *Taeniopteris vittata* (a fossil fern), *Taenidia integerrima* (yellow pimpernel), *Cryptotaenia canadensis* (honewort), *Linotaenia crassipes* (a centiped), *Spirotaenia bryophila* (a desmid), *Cepola taenia* (ribbonfish), *Hemiodus semitaeniatus* (a fish).
    L. *vitta,* f. ribbon, band, fillet, plait, stripe; *vittatus,* decorated or bound with a ribbon: vittate, *Vittaria lineata* (ribbon-fern), *Agave univittata* (an amaryllidacead), *Amblyopus vittatus* (a beetle), *Pedicia albivitta* (a cranefly), *Zaprionus vittiger* (a fly), *Hyla evittata* (a frog).
    See: **line, collar, strap, bind, belt**

ribes, NL. currant, gooseberry, < Ar. *ribas,* a berry with an acid flavor; see **currant**

ribo-, a chemical affix derived from a transposition of letters in *arabinose;* see **arabicus**

rica, L. veil; *ricula,* dim.; see **curtain**

rice < Gr. *oryza;* see **grain**

rich < AS. *rice,* powerful, wealthy; see **wealth, abundance, money**

ricinus, L. castor-oil plant, tick, see **spurge**; veiled, see *rica* under **curtain**

riddle < AS. *hriddel;* see **question**

ride < AS. *ridan.*
    L. *angarius* (Gr. *angaros*), m. mounted courier, messenger: *Angarodon kumsassiensis* (a pelecypod).
    L. *caballarius,* m. rider, horseman:
    Gr. *epibates; anabates; ambates,* m. rider, passenger: *Epibates funestus* (a fly), *Ambates pictus* (a beetle).
    L. *equito, -atus,* ride on horseback; *eques, -uitis,* m. horseman, rider; *equester, -tris, -tre,* pertaining to riding and cavalry: equitation, equitant, *Eques punctatus* (a ribbon-fish).
    Gr. *hippeuo,* be a horseman, ride; *hippeus; hippeutes; hippotes,* m. horseman, rider, driver; *hippasia,* f. a riding; *hipposyne,* f. horsemanship, driving:
    Gr. *kathistes,* m. rider:

Gr. *keles, -etos,* m. courser, riding-horse; *keletion,* n. dim.; *keletistes,* m. leaper, inconstant person, < *keletizo,* ride two or more horses at the same time:

ridge < AS. *hrycg,* back of an animal, any elongated elevation.

Gr. *akrolophos,* mountain crest, ridge; see *lophos* under **crest**

Gr. *ambon,* ridge, crest, rim; see **border**

Sp. *amelga,* f. ridge between two furrows:

AS. *balca,* ridge: balk.

L. *carina,* f. keel; *carinatus,* keeled: carinal, carinate, careen, *Cariniana excelsa* (a lecythidacead), *Ostrea carinata* (a fossil oyster), *Clavulina tricarinata* (a foraminifer).

L. *costa,* rib, side, ridge; see **rib**

L. *crista,* crest, ridge; see **crest**

L. *culmen,* summit, ridge of a roof; see **top**

Gr. *deiras, -ados,* f. ridge of a hill: *Deiradognathus* [*Diradognathus*] *fasciatus* (a beetle).

L. *dorsum,* back, ridge of a hill; see **back**

L. *fastigium,* top of a gable, ridge of a roof; see **gable**

L. *jugum,* n. ridge, summit of a mountain:

L. *lira,* f. earth or ridge thrown up by the plow, furrow-slice; *lirella,* f. dim.; *liro, -atus,* plow: lirate, delirium, delirious, delirifacient, *Venus latilirata* (a fossil pelecypod), *Dichocrinus liratus* (a Mississippian crinoid).

Gr. *lophos,* nape, crest, ridge; see **crest**

L. *monticellus; monticulus,* little mountain, mount; see *mons* under **mountain**

L. *porca,* f. earth or ridge between two furrows; *imporco, -atus,* put into furrows; *Imporcitor, -is,* m. god of plowing: *Scaurus porcatus* (a beetle).

L. *pulvinus,* cushion, ridge, bank; see **pillow**

Gr. *rhachis,* ridge, midrib; see **back**

L. *scamnum,* balk or ridge of earth left unplowed; see **sit**

L. *striga,* f. row of grain or hay, swath, windrow, bristle; *strigilla; strigula,* f. dim.; *strigosus,* full of strigae, thin, lean: strigose, strigillose, *Strigilla carnaria* (a pelecypod), *Strigula complanata* (a lichen), *Lythrodes discistriga* (a moth), *Ditropia strigata* (a wasp), *Mallochia strigosa* (a wasp).

L. *tropis, -eos, -idos,* f. keel: *Tropidorhynchus corniculatus* (friar-bird), *Calotropis gigantea* (Akund calotrope), *Notropsis cornutus* (redfin), *Molossus tropido-rhynchus* (a bat).

AS. *walu,* ridge, stripe, or mark caused by the blow of a whip or stick; see **bruise**

See: **back, fold, furrow, crest, top, stem, line, bar, rib**

ridibundus, L. laughing, prone to laughter; see *rideo* under **laugh**

ridica, L. stake, prop; see **pillar**

ridiculus, L. exciting laughter, absurd; see *rideo* under **laugh**

rigator, L. waterer; *riguus,* wetting; see *rigo* under **wet**

right < AS. *riht,* straight, direct, correct: rightness, rightful, righteous, rightward, upright, downright.

L. *accuratus,* careful, exact: accurate, inaccuracy.

L. *adamussim; examussim,* according to rule, exactly, accurately, <*amussis,* rule, level:

Gr. *akribes,* exact, precise, accurate, perfect; *akribos,* exactly, precisely:

Gr. *artios,* complete, perfect, exact; see **all**

Gr. *atrekeia,* f. accuracy, strict truth, reality; *atrekes,* exact, true, sure, real:

Gr. *dikaios,* observant of the right, decent, just, < *dike,* right, custom, law; see **law**

L. *emendo, -atus,* free from errors, improve, correct; see **edit**

Gr. *euthynter,* straightener, corrector; see *euthys* under **straight**

L. *exactus,* accurate, precise: exact, exactly, exactness.

L. *fandus,* right, correct, lawful; see **justice**

L. *justus,* right, equitable, fair; see **justice**

L. *legitimus,* lawful, right, proper; see *lex* under **law**

Gr. *orthos,* straight, right; see **straight**

L. *polio, -itus,* make smooth, adorn, finish; see **smooth**

L. *pressus,* close, exact, accurate:

L. *rectus,* straight, upright, proper, right; *correctus,* improved, amended; *directus,* straight, arranged: rectify, rectangle, correct, direct, incorrigible, *Volsella recta* (a fossil pelecypod), *Paltorhynchus rectirostris* (a fossil beetle).

See: **straight, justice, edit, law, hand, true, sure**

rigidus, L. stiff, hard, inflexible; *rigor,* stiffness; see **stiff**

rim < AS. *rima,* border, edge, margin; see **border**

rima, L. cleft, fissure; *rimula,* dim.; *rimator,* investigator, < *rimor, -atus,* lay open, turn over, examine; see **break, try**

ring < AS. *hringan,* sound like a bell; < AS. *hring,* circle, see **circle, collar, earring, belt, bind**
L. *clangor, -is,* m. loud, ringing sound, noise: clang, clangor, clangorous, *Glaucionetta clangula* (golden-eye).
Gr. *kanache,* f. sharp sound, ring, clang; *kanachos,* noisy: *Canachus crocodilus* (a phasmid).
Gr. *krambos,* loud, ringing; see **loud**
L. *resonus,* reechoing, ringing; see *sono* under **sound**
L. *tinnio (tintinnio), -itus; tintinno, -atus,* ring, clink, jingle, tinkle; *tinnitus, -us,* m. a ringing, tinkling; *tinnulus,* ringing, tinkling; *tintinnaculus,* tinkling, clanging: tintinnabulation.   See **bell**

ringens, L. gaping, < *ringor, rictus,* open the mouth wide, gape; see **open**

rio, Sp. river, < L. *rivus,* stream; see **stream**

riot < OF. *riote,* quarrel, dispute; see **fight**

ripa, L. bank, shore; *ripula,* dim.; *riparius,* of stream banks; see **shore**

ripe < AS. *ripe,* mature.
L. *adultus,* grown up: adult.
Gr. *akmaios,* in full bloom, prime; *akmenos,* full grown; *epakmos,* in full bloom, mature; see *akme* under **top**
Gr. *aphelix, -ikos,* beyond youth, elderly; *enelix, -ikos,* of age, in prime of manhood:
Gr. *drypetes,* ripened on the tree, ready to fall, mature: *Drypetes diversifolia* (a Guiana plum).
Gr. *ephebos,* arrived at puberty; see *hebe* under **young**
L. *exoletus,* grown up, mature: *Polystomella exoleta* (a foraminifer).
L. *fracidus,* overripe, soft, mellow; *fracesco,* soften, rot, spoil: fracid, fracedinous.
Gr. *hadrynis,* f. a coming to maturity, a ripening; *hadryntikos,* of ripening, < *hadryno,* ripen:
L. *horaios; horimos,* seasonable, timely, ripe: *Horaeomorphus eumicroides* (a beetle).
L. *imago, -inis,* the adult or mature form of an insect; see **form**
L. *maturus,* ripe: mature, maturity, maturation, immature, premature, *Wellerella immatura* (a fossil brachiopod).
L. *parectatus* (Gr. *parektatos*), grown up, marriageable: *Parectatosoma hystrix* (a phasmid).
Gr. *pepeiros,* mature, ripe; *pepon, -os,* ripe, mellow, tender; *drypepes,* ripened on the tree:
L. *praecox, -ocis; precoquus,* too early ripe, premature; see **before**
L. *puber,* downy, ripe; *pubes,* grown up, adult; see *pubes* under **hair**
Gr. *spargao,* swell to ripeness; see **swell**
See: **all, abundance, strong, fit**

ripple < uncertain origin; see **wave, fold, furrow, curl, rough**

riscus, L. (Gr. *riskos*), trunk, chest; see **box**

rise < AS. *risan;* see **raise**

risibilis, L. capable of laughing, causing laughter; *risor,* laughter; see *rideo* under **laugh**

risk < F. *risque,* < It. *risco;* see **danger**

rissa, NL. a kind of gull; see **gull**

rite < L. *ritus, -us,* m. ceremony; *ritualis,* of ceremonies: ritual, ritualistic, ritualism.
L. *adoro, -atus,* honor, esteem, worship; see **honor**
L. *amburbium,* n. expiatory sacrificial procession around the city of Rome:
L. *caeremonia,* f. rite; *caeremonialis,* of rites; *caeremoniosus,* devoted to rites: ceremony, ceremonial.
Gr. *eulabes,* reverent, pious, religious: *Eulabes religiosa* (hill-myna).
Gr. *eusebes,* dutiful, pious, religious, holy; see **holy**
Gr. *hosia,* sacred rite, worship; see *hosios* under **holy**
Gr. *latreia,* worship, service; see *latreus* under **servant**
LL. *missa,* f. celebration of the Lord's Supper: Mass.
L. *novendial, -is,* n. rite or ceremony on the ninth day after or lasting nine days, as on the occasion of a death or a shower of meteorites; *novendialis, -e,* of nine days:
Gr. *orgion,* n. secret rite; *orgiastes,* m.; *orgiastis, -idos,* f. celebrator of rites; *orgiophantes,* m. priest, teacher of orgies: orgy, orgiast, orgiastic.
L. *pietas, -atis,* f. sense of duty, loyalty, kindness, < *pius,* dutiful, conscientious, devout, tender; *piaculum,* m. propitiatory sacrifice; *piacularis,* atoning, < *pio, -atus,* sacrifice, appease, perform sacred rites: piety, pious, pity, pitiful, piteous, pia mater, pittance, expiate, impious, *Scarabaeus impius* (a dung-beetle).

L. *religio, -onis,* f. sense of piety, duty, taboo; *religiosus,* pious, devout: religion, religious, *Ficus religiosa* (pipal, sacred fig).

L. *sacramentum,* n. a sacred rite: sacrament.

Gr. *threskia,* f. religious worship; *threskos,* religious; *threskeutes,* m. worshipper: *Ibis*(or *Thresciornis) aethiopica* (sacred ibis).

Gr. *thyos; thysia,* offering, sacrifice, sacred rite; see **give**

See: **display, honor, servant, holy, magic, law, feast**

rival < L. *rivalis,* user of the same brook; see *rivus* under **stream, enemy**

rivea, NL. a genus of the mahogany family; see **bindweed**

rivus, L. stream, brook; *rivulus,* dim.; *rivalis,* of a stream; see **stream**

rixa, L. quarrel, brawl, strife, scuffle; *rixosus,* quarrelsome; see **fight**

road < AS. *rad;* see **way**

roar < AS. *rarian.*

Gr. *adinos,* thick, loud; see **thick**

Gr. *agaphthenktos,* loud-sounding; see *phthegma* under **speak**

L. *boö, -atus* (Gr. *boao*), roar, shout, bellow; *boama, -tos,* n.; *boë; boësis,* f. loud cry, bellow, roar; *boëtes,* m. clamorer, clamorous; *amphiboëtos,* resounding; *hedyboës,* sounding sweet; *kalliboas,* sounding beautiful:

Gr. *brychema, -tos,* n. roar, bellow; *brychetes,* m. roarer, bellower; *brychetikos,* roaring, bellowing:

AS. *bylgan,* sound like a bull: bellow.

Gr. *erygmelos,* loud-bellowing:

L. *fremo, -itus,* roar, howl, growl, grumble, snort; *fremor, -is,* m. roar, growl; *fremitus, -us,* m. loud noise, roaring sound; *confremo,* sound aloud; *infremo,* growl, roar: fremitus.

L. *mugio, -itus,* bellow, low; *mugitor, -is,* m. bellower: mugient.

Gr. *mykethmos,* m. a lowing, bellowing, roaring; *myketes,* m. bellower: *Mycetes seniculus* (howling monkey).

Gr. *orymagdos,* m. loud noise, din, roar:

Gr. *rhochthos,* m. the roaring of the sea:

L. *ululo, -atus,* howl, make a mournful outcry; see **call**

See: **call, thunder, loud**

roast < OF. *rostir;* see **cook, heat**

rob < OF. *rober,* < OHG. *roubon;* see **steal**

robin < F. Robin, dim. of Robert; see **thrush**

robulus, NL. a genus of foraminifers; see **protozoan**

robur, L. oak, strength; see **oak**

robustus, L. hard and strong like oak; see **strong**

rock < AS. *rocc* (F. *roche*); see **stone**

rockrose

L. *cistus* (Gr. *kistos; kisthos*), m. rockrose: Cistaceae, *Cistus hirsutus* (hairy rockrose), *Cisticola robusta* (a grass-warbler), *Rhodothamnus chamaecistus* (an ericacead).

NL. *helianthemum,* n. a genus of the rockrose family: *Helianthemum canadense* (a sunrose).

NL. *hudsonia,* f. a genus of the rockrose family, < William Hudson, English botanist: *Hudsonia tomentosa* (beach-heather).

NL. *lechea,* f. a genus of the rockrose family,< Johan Leche, Swedish botanist: *Lechea villosa* (hairy pinweed).

rod < AS. *rod.*

L. *aerumnula,* f. traveler's stick for carrying a bundle:

L. *agolum,* n. shepherd's staff or crook:

Gr. *aktis, -inos,* f. ray, beam; *aktinotos,* with rays; *aktinodes,* like rays: actinolite, actinic, Actinozoa, actinomycosis(caused by *Actinomyces* bovis), *Actinicyclus ovalis* (a diatom), *Actinidia arguta* (a dilleniacead), *Actinea scaposa* (a composite), *Asteractis expansa* (a sand-anemone), *Heliactin cornutus* (a hummingbird), *Psilactis coulteri* (a composite), *Triactiscus tricuspis* (a radiolarian), *Hydractinia echinata* (a hydroid).

L. *ames, -itis,* m. pole:

L. *asser, -eris,* beam, stake, pole; see **beam**

L. *axis,* pole, line, hinge; see **axis**

L. *baculum,* n.; *baculus,* m. stick, rod, cane, staff; *bacillum,* n.; *bacillus,* m. dim.: bacillus, bacillary, bacillosis, baguet, *Baculipalpus rarus* (a beetle), *Baculites compressus* (an ammonite), *Bacillus ellenbachensis* (a bacterium), *Nephila baculigera* (a spider), *Prionocidaris baculosa* (an echinoid), *Fuchsia bacillaris* (a fuchsia).

Gr. *baktron*, n. staff, cane, stick, rod; *bakterion*, n. dim.: bacterium, *Bactrognathus hamata* (a conodont), *Bactrites elegans* (an ammonite), *Bactris horrida* (a palm), *Corynebacterium diphtheriae* (diphtheria-bacillus), *Nitrobacter winogradskyi* (a bacterium), *Abactrus championi* (a beetle).

OF. *baston* (F. *baton*), rod, staff, stick: baton, Baton Rouge (La.).

L. *caduceus*, m. wand of Mercury, herald's staff: caduceus, *Isograptus caduceus* (a graptolite).

L. *canon* (Gr. *kanon*), measuring-rod, rule, model; see **measure**

L. *cerycium* (Gr. *kerykion*), n. caduceus, herald's staff: *Cerycium paradoxum* (a gastropod).

Gr. *chaios*, m. herald's staff:

L. *colus*, distaff; see **axis**

L. *contus* (Gr. *kontos*), m. long pole for pushing a boat, pike; *kontarion*, n.; *kontilos*, m. dim.; L. *contiger*, m. spear-bearer: chondriocont.

ML. *croca*, f. shepherd's crook:

Gr. *dokis, -idos*, small beam, rod; see *dokos* under **beam**

Gr. *dryphaktos*, m. handrail:

L. *ferula*, f. rod, whip, staff: ferule.

Gr. *garka*, f. rod:

AS. *gierd; gyrd*, rod, stick: yard, yardstick, halyard, sailyard.

L. *groma*, f. surveyor's measuring-rod:

L. *hastile, -is*, n. shaft of a spear: *Xanthorrhea hastilis* (a liliacead).

Gr. *kalamis, -idos*, a reed fishing-rod; *calamus* under **reed**

Gr. *kalaurops, -opos*, f. shepherd's staff: *Calaurops lituiformis* (a fossil gastropod).

Gr. *kampyle*, f. a crooked staff:

Gr. *kerkis, -idos*, f. rod for striking the woof in a loom, shuttle; *kerkidion*, n. dim.: *Cercidocerus niger* (a beetle), *Cercidochela lankesteri* (a sponge).

Gr. *kneme*, leg, spoke of a wheel; see **leg**

L. *lituus*, m. curved staff or wand of an augur, curved trumpet; *liticen, -inis*, m. trumpeter: lituuus, *Lituola subglobosa* (a foraminifer), *Spirula* (formerly *Lituus*) *vulgaris* (a cephalopod).

L. *longurius*, m. long pole, rod:

L. *pedum*, n. shepherd's staff or crook:

L. *pertica*, f. long pole, flail; *perticalis*, of a pole: perch.

L. *phalanga*, f. carrying-pole, roller:

Gr. *phryganon*, dry stick, brushwood; see **wood**

Gr. *pleganon*, n. rod, stick: *Pleganophorus bispinosus* (a beetle).

L. *radius*, m. ray, rod, spoke, shuttle; *radiolus*, m. dim.: ray, radiant, radio, radium, radiator, irradiate, interradial, radiolarian, *Radius volva* (shuttle-shell), *Dendrosoma radians* (a protozoan), *Cheilanthes radiata* (a lip-fern), *Theridiosoma radiosa* (a spider), *Actinopterella radialis* (a fossil pelecypod), *Pecten irradians* (a scallop), *Entomaspis radiatus* (a trilobite), *Anisocrinus interradiatus* (a Silurian crinoid).

Gr. *rhabdos*, f. rod; *rhabdion*, n. dim.; *rhabdouchos*, m. staff-bearer: rhabdosome, rhabdite, rhabdocyst, rhabdosphere, rhabdomancy, *Rhabdophaga strobiloides* (a willow-cone gall-gnat), *Magorrhabda elytrata* (a moth), *Hephessobrycon heterorhabdus* (a fish).

Gr. *rhapis, -idos*, f. rod: *Rhapis excelsa* (a palm), *Rhapidophyllum hystrix* (needle-palm), *Rhapidostreptus virgator* (a milleped).

Gr. *rhoptron*, n. spring bar or rod of a trap: *Rhoptrocerus xylophagorum* (a wasp).

Gr. *rhymos*, m. pole of a vehicle: *Rhymodus oblongus* (a fossil fish).

L. *rudis*, f. stirring-rod, staff; *rudicula*, f. dim. wooden spoon, spatula:

L. *sceptrum* (Gr. *skeptron*), n. staff, walking-stick, baton, wand: scepter, *Sceptrophorus sceptriger* (a chalcid), *Hexacontium sceptrum* (a radiolarian).

L. *scipio, -onis* (Gr. *skipon, -os*), m. staff, wand: *Scipio tripedatus* (a louse), *Sciponoceras baculoides* (a fossil cephalopod), *Scipopus sexguttatus* (a fly).

L. *scytala* (Gr. *skytale*), f. staff, cudgel; *skytalion*, n.; *skytalis, -idos*, f. dim.: *Scytale scytale* (a snake), *Sctyalocrinus robustus* (a Mississippian crinoid), *Scytalium tentaculatum* (a coral).

L. *talea*, f. slender staff, rod, stick, bar, set, layer; *taleola*, f. dim.: tally, tallage, tail, detail, tailor.

L. *trabs*, beam; see **beam**

L. *vectis*, lever, bar, bolt, spar; see **lever**

L. *virga*, twig, rod; see **branch**

See: **beam, bar, lever, axis, club, pillar, reed, spear, whip**

rodent < L. *rodo*, gnaw, nibble at; see **beaver, hare, mouse, squirrel, bite**

rogo, *-atus*, L. ask, beg; see **ask**

rogue < uncertain origin.

L. *furcifer*, yoke-bearer, jail-bird; see **fork**

Gr. *kobalos*, m. rogue, knave: *Cobalus triangularis* (a butterfly), *Cobaloblatta simulans* (a fossil cockroach), goblin(?).

Gr. *kyphon*, one pilloried, knave; see *kyphos* under **bend**

L. *mastigia* (Gr. *mastigias*), one deserving whipping, knave, rogue; see **whip**

Gr. *palaiomolops, -opos,* m. old rogue:

Gr. *panourgos,* m. knave, rogue, villain: Panurge, panurgy.

Gr. *phrynondeios,* m. cheat, rogue, swindler:

L. *scelio, -onis,* m. scoundrel: *Scelio ater* (a wasp), *Sceliotrachelus braunsi* (a wasp).

Gr. *strophis, -ios,* f. a twisting, slippery person, rogue:

Gr. *tribon, -os,* m. crafty fellow, rogue:

L. *verbero, -onis,* one worthy of stripes, rascal, scoundrel; see **whip**

L. *veterator, -is,* m. old fox, cunning fellow:

rogus, L. funeral pile; *rogalis,* of a funeral pile; see **heap**

roll < L. *rotula,* dim. of *rota, wheel;* see **wheel, cylinder, turn**

Romanus, L. Roman, Latin, < *Roma,* Rome; *Romulus,* fabled founder of Rome; see **Latin**

roof < AS. *hroof,* top.

L. *caelum,* arched covering, sky, heaven; see **heaven**

L. *constratum,* n. covering, deck:

Gr. *erepsis,* f. roof; *erepsimos,* of roofing: *Erepsimus setiferus* (a beetle).

Gr. *melathron,* n. roof, ceiling:

Gr. *orophe,* f. roof of a house, ceiling; *orophos,* m. material for thatching; *orophikos,* of a roof: *Orophus mexicanus* (a katydid), *Orophorhynchus spinoculatus* (a crab), *Orophicus antelmi* (a beetle).

Gr. *ouraniskos,* roof, ceiling, dim. of *ouranos,* sky; see **sky**

L. *palatum,* roof of the mouth; see **mouth**

Gr. *stege,* f.; *stegos; tegos, -eos,* n. roof; *stegion,* n. dim.: stegodont, *Stegosaurus undulatus* (a dinosaur), *Tegonotus serratus* (a mite), *Cryptostegia grandiflora* (a rubber-vine), *Schistostega pennata* (luminous moss), *Rhynchostegium serrulatum* (a moss).

L. *tectum; tegulum,* n. roof; *tectulum; tegillum,* n. dim.: tegulum, protegulum, *Crepis tectorum* (a composite).

See: **cover, cap, heaven**

room < AS. *rum:* roomy, roomful, roommate, barroom, sickroom, schoolroom. Derive: mushroom.

L. *andron, -onis* (Gr. *-os*), m. men's apartment:

Gr. *apodyterion,* n. undressing room at a bathhouse:

L. *atrium,* n. vestibule, hall, entry, passage; *atriolum,* n. dim.: atrium, atrial.

L. *auditorium,* n. a room for hearing: auditorium.

L. *aula* (Gr. *aule*), court, courtyard, hall; see **space**

L. *bubile (bovile), -is,* n. stall for oxen:

L. *caliclarium,* n. cupboard:

L. *camera* (Gr. *kamara*), f. chamber or vault with arched roof; *camella,* f. dim.: camerate, cameritelous, bicameral, camarilla, chamber, Camerata, comrade, *Camerella plicata* (a fossil brachiopod).

L. *carnarium,* n. larder, pantry:

L. *cella,* f. storeroom, chamber; *cellula,* f. dim.; *cellarium,* n. storeroom; *cellarius,* m. storekeeper, steward, butler; *cellulosus,* full of cells: cell, cellar, cellular, celluloid, cellulose, cellulosic, methylcellulose, intracellular, multicellular, *Orbicella annularis* (star-coral), *Cyperus cellulosa* (a sedge), *Clathrodictyon cellulosum* (a fossil bryozoan), *Talaeporia lapidicella* (a moth), *Celliforma nuda* (a fossil bee-chamber).

L. *cenaculum,* dining-room, upper story; see *cena* under **eat**

Gr. *chora; chorema,* room, place, space; *choraphion,* dim.; see **space**

L. *coillum,* n. inmost part of the house where the Lares were worshipped:

L. *conclave, -is,* n. room, private chamber: conclave.

L. *consistorium,* n. a place of assembly, council chamber where the emperor's cabinet met: consistory.

L. *cubiculum,* n. sleeping-room, bedchamber: cubicle.

L. *dolarium,* n. place for casks, cellar:

L. *dormitorium,* n. sleeping-room: dormitory.

Gr. *eunaterion,* n. bedroom:

Gr. *exedra,* f. hall, arcade:

L. *favissa,* f. cellar or reservoir of a temple:

L. *fornix,* arch, vault; see **arc**

L. *frigidarium,* n. cooling-room; *frigidaria,* f. cool larder, provision-room: Frigidaire.

L. *fumarium,* smoke-chamber; see *fumus* under **cloud**

AS. *heall* (G. *halle*), large room or apartment: hall, hallmark, Valhalla.
Gr. *histon*, a weaving-room; see *histos* under **weave**
L. *hypogeum*, n. cellar:
Gr. *koimesterion; koiton*, n. sleeping-room, bedchamber:
Gr. *koureion*, n. barbershop:
Gr. *kytos, -eos*, n. any hollow place, container, vessel, cell: cytoplasm, leucocyte, cytula, cytatoxin, *Cytocrinus laevis* (a Silurian crinoid), *Melicytus ramiflorus* (mahoe).
Gr. *kyttaros*, m. cell of a comb, any cell; *kyttarion*, n. dim.: poecilocyttarous, *Cyttarocrinus eriensis* (a Devonian crinoid).
L. *latrina*, f. bath, privy: latrine.
L. *loculus*, dim. of *locus*, place, cell, room; see **box**
Gr. *mandra*, f. stall, stable: madrigal, mandrel.
Gr. *megaron*, n. large room, chamber, hall:
Gr. *mellition*, n. cell of a honeycomb:
Gr. *mychos*, m. innermost part, recess, bay, alcove, sinus: *Mychodesmus macramma* (a milleped), *Endomychus biguttatus* (a beetle).
L. *oecus* (Gr. *oikos*), room, house; see **house**
L. *officina*, workshop, laboratory; see *officialis* under **make**
Gr. *pastos*, m. bridal chamber, bridal bed:
L. *penus*, inmost part, sanctuary, storeroom; see *penarius* under **store**
Gr. *pithon, -os*, m. cellar:
Gr. *prytaneion*, town hall; see *prytanis* under **govern**
Gr. *schadon*, cell in a honeycomb; the larva in the cell; see **young**
L. *sirus* (Gr. *siros*), m. underground pit for grain, cellar:
L. *spatium*, room, distance; see **space**
Gr. *sphekion*, n. cell in a wasp-nest:
Gr. *stathmos*, stall, stable, abode, quarters; see *stasis* under **stand**
L. *thalamus* (Gr. *thalamos*), m. room, chamber, bedroom, bridal chamber; *thalame*, f. den, hole; *thalamios*, of a room: thalamus, thalamite, thalamifloral, epithalamium, *Thalamocrinus globosus* (a Silurian crinoid).
Gr. *thyron, -os*, m. hall, vestibule:
L. *vestibulum*, n. entrance court, hall: vestibule.
L. *zothecula*, f. dim. of *zotheca* (Gr. *zotheke*), f. chamber, closet: *Zotheca meridionalis* (a moth).
See: **space, house, hollow, den, vessel, bag, box, pen**

root < AS. *rot*.
L. *radix, -icis*, f. root; *radicula*, f. dim.; *radicalis*, of roots; *radicitus*, by the roots; *radicor, -atus*, take root; *radicosus*, full of roots: radix, radical, radicle, radish, eradicate, ineradicably, *Nodosaria radicula* (a foraminifer), *Diastrophus radicum* (raspberry root-gall), *Bouteloua radicosa* (a grama-grass), *Hypochaeris radicata* (spotted catsear), *Fusarium radicicola* (a fungus), *Begonia radicans* (a begonia), *Polyporus radiciperda* (a fungus).
Gr. *rhiza*, f. root; *rhizion*, n. dim.; *rhizikos*, of roots; *rhizoma, -tos*, n. mass of roots; *rhizosis*, f. a taking root: rhizoid, rhizome, rhizophilous, rhizanthous, rhizopod, rhizotomy, mycorrhiza, licorice, *Rhizophora mucronata* (a mangrove), *Rhizopus elegans* (a mold), *Rhizobium radicicola* (a nitrogen-fixing bacterium), *Rhizoctonia lanuginosa* (a fungus), *Rhizoctonus ampelinus* (a bug), *Corallorhiza multiflora* (coralroot).

rope < AS. *rap*.
Gr. *anchone*, a cord for hanging, halter; see *ango* under **choke**
L. *ancorale, -is*, n. cable:
L. *anquina*, f. rope binding the sailyard to the mast:
Gr. *artane*, f. rope, noose, halter:
Gr. *balbis, -idos*, f. the rope across the race-course, the start and finish line: *Balbis assumptana* (a moth).
Gr. *boeus, -eos*, m. rope of ox-hide:
Gr. *brochos*, m. noose, halter, snare; *brochotos*, ensnared: *Brochocephalus paradoxus* (a cestode), *Brochopeltis mjöbergi* (a milleped).
Gr. *cheleuma, -tos*, n. cord, bond:
L. *chorda* (Gr. *chorde*), f. gut, string of a musical instrument, twine, rope; *chordarion*, n. dim.; Sp. *cuerda*, f. cord, rope; *cordilla*, f. dim.: cord, chord, cordage, cordwood, cordon, cordonette, corduroy, cordillera, Chordata, *Chordeiles virginianus* (nighthawk), *Exochorda racemosa* (pearlbush).
L. *corrigia*, f. shoelace:
Gr. *desmos*, cable; see *deo* under **bind**
L. *fidicula*, cord, line, string, dim. of *fides*, harp, lyre; see **harp**
L. *funis, -is*, m. rope, line, cord; *funiculus*, m. dim.; *funalis; funarius*, of a rope

or cord: funicle, funambulist, funicular, funiform, *Funaria hygrometrica* (cord-moss), *Funisciurus auriculatus* (a squirrel), *Attalea funifera* (a palm).

Gr. *harpedon*, f. cord, thread:

Gr. *himonia*, f. rope of a well:

Gr. *hormia*, f. fishing-line:

Gr. *illas, -ados*, f. rope, band:

Gr. *kalos*, m. rope, line, cable; *kalodion*, n. dim.: *Calodium annulosum* (a nematode).

Gr. *kamilos*, m. rope, cable:

L. *linea*, linen thread, line, string; see **line**

Gr. *linon*, flaxen cord; see **flax**

L. *medipontus*, m. a kind of thick rope:

Gr. *merinthos (smerinthos)*, f. cord, line, string: *Smerinthus ocellatus* (a hawk-moth), *Merinthichthys sanchezi* (an eel).

Gr. *mermis, -ithos*, f. cord, rope, string:

Gr. *parolkos*, m. tow-rope:

Gr. *peisma, -tos*, n. ship's cable:

L. *pendiculus*, m. cord, noose:

Gr. *plekte*, rope, cord; see *plekto* under **weave**

Sp. *reata*, f. rope: lariat, riata.

L. *remulcum* (Gr. *rhymoulkeo*, tow); *promulcum*, n. tow-rope; *Astragulus remulcum* (towline-vetch).

L. *restis*, f. rope, cord; *resticula*, f. dim.; *restio, -onis*, m. ropemaker: *Resticula gelida* (a rotifer), *Restio complanatus* (rope-grass), restiform.

L. *rudens, -entis*, m. rope, line, cord:

L. *saphon, -onis*, m. ship's cable:

Gr. *sardon, -os*, f. a kind of rope; *sardonion*, n. dim.:

L. *schoenus* (Gr. *schoinos*), sedge, rush-rope, rope; *schoinion*, dim.; *schoinos, -idos*, rope, cord; see **sedge**

Gr. *seira*, f. cord, rope, band, string; *seiraios*, of a cord, tied by a cord: sirogonim-ium, *Melosira hyperborea* (a diatom), *Cytoseira ericoides* (a seaweed).

L. *serilium*, n. rope, cord:

Gr. *sparte; spartine*, f.; *sparton*, n. cord, rope, cable; *spartion*, n. dim.: *Spartina cynosuroides* (a cord-grass), *Spartium ochroleucum* (a broom).

Gr. *strophos*, m. a twisted cord, rope, band; *strophion* (L. *strophium; strophio-lum*), n. dim.: *Strophanthus hispidus* (an apocynacead yielding strophanthin), *Acacia estrophiolata* (an acacia).

Gr. *syrtes*, m. cord, rein:

Gr. *tenon*, sinew, thread; see **thread**

Gr. *terthrios*, m. rope or brace at end of sailyard:

Gr. *thominx, -ingos (thomix, -ikos)*, m. cord, string, line, thread: *Thominx manica* (a worm).

Gr. *throsis*, f. cord, line:

Gr. *tonos*, cord, rope; see **draw**

See: **thread, bind, chain**

ros, *roris*, L. dew; *roridus; rorulentus; roscidus*, dewy, bedewed; see **drop**

rose < L. *rosa*, f.; *rosula*, f. dim.; *rosaceus; rosarius; roseus*, of roses; *rosulentus*, full of roses, rosy; *rosetum*, n. rose garden: rosy, roseate, roseola, rosette, rosary, rosulate, *sub rosa*, Rosicrucian, Rosaceae, rosewood, rosebud, roselle (*Hibiscus sabdariffa*), *Rosa cinnamomea* (cinnamon-rose), *Rosalia alpina* (a beetle), *Althea rosea* (hollyhock), *Crassula rosularis* (a stonecrop), *Aspidospira rosula* (a foraminifer), *Archips rosana* (rose-worm), *Hylotoma rosarum* (rose-sawfly), *Typhlocyba rosae* (rose-leafhopper), *Discorbis rosacea* (a foraminifer), *Leontocebus rosalia* (silky tamarin), *Sisyrinchium rosulatum* (a blue-eyed grass), *Erythropteryx roseotincta* (a moth), *Potentilla rosulata* (a cinquefoil), *Robulina rosetta* (a foraminifer), *Hibiscus oculiroseus* (a rose-mallow). Derive: primrose, tuberose.

L. *agrimonia*, f. a genus of the rose family: agrimony, *Agrimonia hirsuta* (hairy agrimony).

NL. *alchemilla*, f. a genus of the rose family, < an Ar. name: *Alchemilla sericea* (a lady's-mantle).

NL. *amelanchier*, f. serviceberry, a genus of the rose family: *Amelanchier alni-folia* (Saskatoon serviceberry).

L. *amygdalus*, almond-tree; see **peach**

NL. *aronia*, f. a genus of the rose family: *Aronia arbutifolia* (red chokeberry).

L. *aruncus*, m. a genus of the rose family: *Aruncus sylvester* (goatsbeard).

NL. *cercocarpus*, m. a genus of the rose family: *Cercocarpus argenteus* (a moun-tain-mahogany), *Aphis cercocarpi* (an aphid).

L. *coroniola*, f. an autumnal rose:

L. *crataegus* (Gr. *krataigos*), m. (but f. in use), flowering thorn, a genus of the rose family: *Crataegus prunifolia* (a hawthorn).

L. *cydonia* (Gr. *kydonia*), quince; see **quince**

NL. *filipendula*, f. a genus of the rose family: *Filipendula purpurea* (Japanese meadowsweet), *Sargassum filipendula* (a gulfweed).

NL. *fragaria*, f. a genus of the rose family; see *fragum* under **strawberry**

L. *geum*, n. avens, bennet: *Geum coccineum* (an avens), *Saxifraga geum* (a saxifrage).

Gr. *kynosbatos*, f. dogthorn, dogrose:

L. *malus*, apple-tree; see *malum* under **apple**

Gr. *oa* (*oia*), f. mountain-ash, equivalent to *sorbus*:

L. *ornus*, f. mountain-ash, equivalent to *sorbus*; *orneus*, of mountain-ash: *Fraxinus ornus* (flowering ash).

Gr. *paideros*, *-otos*, m. a plant with rosy flowers, rouge:

L. *persica*, peach; see **peach**

NL. *photinia*, f. a genus of the rose family: *Photinia arbutifolia* (toyon).

NL. *potentilla*, f. a genus of the rose family: *Potentilla villosa* (hairy cinquefoil).

L. *poterium* (Gr. *poterion*), n. a plant, now a genus of the rose family: *Poterium spinosum* (spiny burnet).

L. *prunus*, plum-tree; see **plum**

L. *pyrus*, pear-tree; see **pear**

Gr. *rhodon*, n. rose, red; *rhodarion*, n. dim.; *rhodeon*, *-os*, m.; *rhodonia*, f. rose garden; *rhodeos*; *rhodinos*, of roses; *rhodochroos*, rose-colored; *rhodopos*, rosy; *kynorhodon*, dogrose: rhodora, Rhodesia, cynorrhodon, Rhoda, *Rhododendron macrophyllum* (coast-rhododendron), *Chamaerhodus erecta* (a rosacead).

L. *rubus*, bramble, blackberry; see **blackberry**

NL. *sanguisorba*, f. a genus of the rose family: *Sanguisorba obtusa* (Japanese burnet).

L. *sentix*, *-icis*, m. dogrose:

L. *sorbus*, f. mountain-ash; AS. *syrfe*, service, sorb; a genus of the rose family: sorbose, *Sorbus domestica* (a mountain-ash), *Sorbaria arborea* (a false-spiraea), *Xanthoceras sorbifolium* (a sapindacead).

L. *spiraea* (Gr. *speiraia*), f. a genus of the rose family: *Spiraea salicifolia* (a spirea), aspirin.

rosemary < L. *rosmarinus;* see **mint**

rostrum, L. beak, bill, snout, muzzle; *rostellum*, dim.; *rostratus*, beaked, curved; see **nose**

rosus, L. gnawed; see *rodo* under **bite**

rot < AS. *rotian*, decay.

L. *caries*, f. decay; *cariosus*, decayed, rotten, withered: caries, carious, carrion, *Tilletia caries* (a fungus).

L. *corruptus*, spoiled; see **bad**

Gr. *euros*, *-otos*, m. mold, decay: *Eurotia lanata* (common winterfat), *Eurotium herbariorum* (a mold), *Cephaleuros virescens* (an alga).

L. *fracesco*, soften, rot, spoil; see *fracidus* under **ripe**

Gr. *phtharsis*, corruption; see *phtheiro* under **destroy**

Gr. *phthisis*, wasting, decay; *phthoë*, infection; see *phthio* under **lessen**

L. *putreo*, rot; *putrilago*, *-inis*, f. rottenness; *putridus*, decayed, rotten, corrupt; *puter*, *-tris*, *-tre*, rotten: putrid, putrefy, putrefaction, potpourri, putrescin.

Gr. *pythedon*, *-os*, f. putrefaction, < *pytho*, decay, rot:

Gr. *sabakos*, rotten, enervated:

Gr. *sapros*, decayed, rotten, putrid, < *sepo*, rot, fester, mortify; *sepsis*, f. putrefaction, decay, poisoning from bacterial action; *sepedon*, *-os*, f. rottenness, decay; *sepedonikos*, causing decay; *seps*, *sepos*, f. a putrefying sore; *septikos*, putrefactive; *septos*, rotten; *aseptos*, not subject to decay: saprophyte, saprolite, saprophagous, saprine, sapremia, sepsis, urosepsis, antiseptic, septic, aseptic, septicemia, Sepsidae, sepedonium, *Saprolegnia ferax* (a fish-mold), *Saproglyphus neglectus* (a mite), *Saprinus cruciatus* (a beetle), *Fulgio septica* (a slime-mold), *Chaoborus antisepticus* (a fly).

Gr. *sathros*, decayed, rotten, unsound: *Sathrocercus cinnamomeus* (a bird).

Gr. *sphakelos*, m. gangrene, mortification, necrosis; *sphakelodes*, like gangrene: sphacelus, sphacelous, sphacelate, *Sphacelotheca hydropiperis* (a fungus), *Brachychaeta sphacelata* (false goldenrod).

Gr. *tangos*, rancid; *tange*, rancidity, a putrid abscess; see **smell**

See: **dirt, poison, slime, disease, smell, bad, destroy, death**

rota, L. wheel; *rotula*, dim.; *rotalis*, of wheels; see **wheel**

rotatio, L. a turning around; see *roto* under **turn**

rotundus, L. circular, round, spherical; see **circle**

rough < AS. *ruh.*

Gr. *adiazetos,* unpolished:

L. *agrestis,* of the fields or country, wild, rough, boorish; Gr. *agriotes,* savageness, wildness; *agroikos,* rustic, boorish, coarse; see *ager* under **field**

Gr. *akompsos,* rude, boorish: *Acompsosaurus wingatensis* (a fossil crocodile).

Gr. *amotos,* furious, insatiate, savage: *Amotus longisternus* (a beetle).

Gr. *anasteios,* unmannerly:

Gr. *anergastos,* not thoroughly wrought, raw:

Gr. *anerotos,* unplowed, untilled: *Anerota falcata* (a grasshopper).

Gr. *apenes,* rough, harsh, cruel: *Apenes lucidula* (a beetle), *Apenophyseter patagonicus* (a fossil cetacean).

Gr. *apeptos,* uncooked, raw:

Gr. *apharotos,* unplowed, untilled:

Gr. *arrhenes,* fierce, savage: *Arrhenophagus chionaspidis* (a wasp).

Gr. *aschetos,* irrepressible, ungovernable: see **strong**

Gr. *askalos,* unhoed, unweeded:

L. *asper, -a, -um,* rough, harsh, uneven; *asperulus,* dim.; *aspratilis,* rough, scaly; *aspredo, -inis; aspritudo, -inis,* f. roughness; *aspretum,* n. a rough place; *aspero, -atus,* roughen: asperity, exasperate, *per aspera ad astra,* asperate, asperite, asperulous, *Asperula odorata* (sweet woodruff), *Asperugo procumbens* (German madwort), *Xylergastes asper* (a beetle), *Eurynotus asperatus* (a beetle), *Schismatoglottis asperata* (an aracead), *Ficus asperrima* (a fig), *Galium asprellum* (rough bedstraw), *Cristellaria asperula* (a foraminifer), *Cornus asperifolia* (rough-leaved dogwood).

L. *atrox, -ocis,* cruel, harsh, severe; see **merciless**

L. *barbarus* (Gr. *barbaros*), foreign, therefore outlandish, crude, rude; *barbarikos,* foreign, crude; see **strange**

Gr. *barycholos,* savage:

L. *beluinus,* animal, brutal, bestial; see *belua* under **animal**

Gr. *biastos; biastikos; biaios,* violent, forcible; *biastes,* m. one who uses force; *biasmos,* m. violence: *Biastoides punctata* (a bee), *Biasmia guttata* (a beetle), *Biastes schotti* (a bee), *Biasticus impiger* (a bug).

L. *brutus,* stupid, rough, heavy: brute, brutish, brutality, brutalize, imbrute, *Eriboea bruta* (a butterfly), *Sphenophorus brutus* (a billbug).

L. *censorius,* rigid, severe: censorious.

L. *crudus,* bloody, raw, rough, rude; *crudelis,* rude, unmerciful, severe; *cruditas, -atis,* f. rawness: crude, crudity, cruel, recrudescence.

Gr. *dasys,* hairy, shaggy, rough; *dasytes,* roughness, hairiness; see **hair**

L. *ferus; ferox, -ocis,* wild, untamed, rough, savage; see **nature**

L. *fragosus; confragus,* broken, rough, uneven: *Daedalea confragosa* (a fungus).

L. *hirtus,* shaggy, rough, hairy, prickly, uncultivated; see **hair**

L. *hispidus,* rough, bristly, hairy; see **hair**

L. *horridus,* rough, dreadful; see **fear**

L. *illepidus,* impolite, rude, disagreeable:

L. *immitis,* harsh, rough, rude, inexorable: *Coccidioides immitis* (a fungus).

L. *impetuosus,* violent; see *impetus* under **strike**

L. *impexus,* uncombed, disheveled: *Agrilus impexus* (a beetle).

L. *implacidus,* rough, fierce: *Lycosa implàcida* (a spider).

L. *implanus,* uneven: *Tentyria implana* (a beetle).

L. *impolitus,* unpolished, rough: *Anolotichia impolita* (a fossil bryozoan).

L. *importunus,* unmannerly, rude, churlish: importunate, importunity.

L. *imputatus,* unpruned, untrimmed: *Ostrea imputata* (a pelecypod).

L. *inaequalis; iniquus,* uneven, unequal, unjust, unlike, rough: inequality, iniquity.

L. *inaratus,* unplowed, fallow: *Pleurotoma inarata* (a gastropod).

L. *incicur,* untamed, wild:

L. *inclemens,* rigorous, harsh, rough, severe, unmerciful: inclement.

L. *incoctus,* uncooked, raw:

L. *incomptus,* unadorned, rough: *Cerithium incomptum* (a gastropod).

L. *inconditus,* unformed, rough, rude: *Begonia incondita* (a begonia), *Lachnopus inconditus* (a beetle).

L. *inconsitus,* unsown, untilled;

L. *incultus,* untilled, unpolished, rude: *Tachina inculta* (a fly), *Deretaphrus incultus* (a beetle).

L. *indolatus,* unhewn:

L. *ineffigiatus,* formless, shapeless:

L. *infabricatus,* unfinished, unwrought:

L. *infacetus,* coarse, rude, blunt, unmannerly: *Lordops infacetus* (a beetle).

L. *informis,* shapeless, unformed, hideous: *Cryptorhynchus informis* (a beetle).

L. *inhumanus,* uncivil, rude, savage: inhuman, inhumane.
L. *intonsus,* unshorn, bearded, rude: *Lamia intonsa* (a beetle).
L. *intractabilis,* unmanageable, rough: intractable.
L. *inurbanus,* rustic, boorish, unpolished: *Brachycerus inurbanus* (a beetle).
L. *irrasus,* unshaved, unpolished: *Batostoma irrasum* (a fossil bryozoan).
Gr. *kerchaleos; kerchnos,* rough, fierce, hoarse; see **hoarse**
Gr. *kestros,* m. roughness on the tongue:
Gr. *kranaos,* rugged, rocky:
Gr. *ktenodes,* like a beast, brutish; see *ktenos* under **cattle**
Gr. *labros,* furious, hasty, greedy; see **greed**
Gr. *lepros,* scaly, rough; see *lepsis* under **scale**
Gr. *maleros,* fierce, fiery, mighty:
Gr. *mothon, -os,* m. impudent fellow; *mothonikos,* impudent: *Mothonodes obtusata* (a moth), *Aphodius(Mothon) sarmaticus* (a beetle).
Gr. *okris, -ios,* f. any roughness:
Gr. *omos,* raw, uncooked, unripe; *omotes, -etos,* f. rawness; *enomos,* rather **raw:** omophagous, omotocia, *Omomys minutus* (a fossil mammal), *Omophagus artus* (a beetle), *Omotrachelus difformis* (a beetle).
L. *ostreatus,* pertaining to oyster shells, rough; see *ostrea* under **mollusk**
L. *paeminosus,* uneven, rough:
Gr. *phaulios,* coarse:
Gr. *phonax, -akos,* bloodthirsty; see *phonos* under **kill**
Gr. *phrix, -ikos,* a ruffling of a smooth surface, ripple, shivering; see **shake**
Gr. *psoros,* itchy, scabby, mangy, rough; see **itch**
L. *rudis,* rough, unpolished, raw, wild: rude, rudiment, erudite, erudition, *Littorina rudis* (a gastropod), *Cortaderia rudiuscula* (pampas-grass).
L. *ruidus,* rough: *Dionychus ruidus* (a beetle).
L. *rusticus,* of the country, plain, homely, awkward, boorish; see *rus* under **country**
L. *saevus; saevis,* raging, fierce, cruel, harsh: *Acanthoclonia saevissima* (a phasmid).
L. *salebra,* f. roughness in a road; *salebrosus,* rough, rugged, uneven: *Talaurinus salebrosus* (a beetle).
L. *scaber, -bra, -brum,* rough, scurfy, scabby, mangy, itchy; *scabiosus,* rough, scurfy; *scabridus,* rough, rugged; *scabies,* f. roughness, scurf, mange, itch: scabies, scabrous, scabious, scabrites, scaberulous, scabrescent, scabridulous, scabrin (from *Heliopsis scabra*), *Scabiosa stellata* (star-scabious), *Leptospermum scabrum* (ti-tree), *Centaurea scabiosa* (a dusty-miller), *Calceolaria scabiosaefolia* (a slipper-flower), *Coreus scabrator* (a bug), *Physocalymma scaberrimum* (a lythracead), *Festuca scabrella* (rough fescue), *Actinomyces scabies* (potato-scab fungus), *Sarcoptes scabiei* (an itch-mite), *Impatiens scabrida* (a snapweed), *Carex scabrata* (a sedge).
L. *scrupeus; scruposus; scrupulosus,* rough, rugged; see *scrupus* under **stone**
L. *sentus,* thorny, rough; see *sentis* under **thorn**
L. *severus,* harsh, rough, sharp, stern, rigorous; *assevero, -atus,* pursue earnestly, assert strongly: severe, severity, perseverance, asseverate, *Heliocausta severa* (a moth).
L. *solox, -ocis,* coarse, rough, bristly:
Gr. *sphedanos,* eager, vehement; see **busy**
Gr. *sphodros,* violent, impetuous, strong; see **strong**
L. *squalidus,* rough from want of attention, filthy, dirty; see **dirt**
L. *squarrosus,* rough with stiff scales, bracts, leaves, or processes: squarrose, *Rubus squarrosus* (bush-lawyer), *Grindelia squarrosa* (gumweed).
Gr. *strenes,* harsh, hard, rough:
Gr. *styphelos; styphlos,* hard, harsh, rough, sour: *Styphelia viridiflora* (an epacridacead), *Styphloderes exsculpatus* (a beetle).
L. *tetricus,* forbidding, harsh; see *teter* under **bad**
Gr. *theriodes,* wild, brutal, savage; see *therion* under **animal**
L. *torvidus,* wild, fierce, < *torvus,* wild, severe, grim: *Exaesiopus torvus* (a beetle).
Gr. *trachys,* rough; *trachodes,* of rough nature; *trachoma, -tos,* n. roughness: trachyte, trachoma, trachea, tracheid, trachyspermous, trachyphonia, *Trachys nomas* (a beetle), *Trachystoma multidens* (a fish), *Trachinus draco* (weever), *Trachypterus iris* (a fish), *Trachysycon nodosum* (a Cretaceous sponge), *Trachodon annectens* (a dinosaur), *Trachymene caerulea* (laceflower-didiscus), *Disporum trachycarpum* (fairy-bells).
L. *trux, -ucis,* rough, savage, fierce, ferocious, grim; *truculentus,* savage, fierce, harsh: truculence, truculent, *Trucifelis fatalis* (a fossil carnivore).
L. *verrucosus,* full of warts; see *verruca* under **wart**

AS. *wilde,* in a state of nature; see **nature**
See: **scale, wart, wave, hoarse, nature, hair, scrape, rub**

round < L. *rotundus;* see **circle, ball, wheel, berry, turn, cylinder**

row < AS. *raw,* line, rank, see **line**; < *rowan,* propel with oars, see **oar**

roystonea, NL. a genus of palms; see **palm**

rub < uncertain origin.
L. *frico, -ictus, -catus,* rub, rub down; *confrico,* rub vigorously; *infrico,* rub in: friction, fricative, fray.
L. *frio, -atus,* rub or crumble into small pieces; *friabilis,* easily broken; see **break**
L. *polio, -itus,* smooth, furbish, rub; *politor, -is,* polisher; see **smooth**
Gr. *psao; psecho,* rub; *psestos,* rubbed, scraped; *psexis,* f. a rubbing down, currying; *psokos,* rubbed small: palimpsest, *Psestophleps neobisignata* (a bug), *Psochodesmus crescentis* (a milleped).
L. *scabo,* scratch, rub; see *scaber* under **rough**
L. *tero, tritus,* rub, grind, wear, waste; *teres, -etis,* rubbed off, rounded; *trituro,* thresh; *attritus,* rubbed off, wasted; *contritus,* worn out, broken, penitent; *detritus,* worn away; *intertriginosus,* chafed, galled; *pertritus,* very worn; *protritus,* worn out, trite: trite, trituration, contrite, detritus, attrition, detrimental, terete, *Triticum turgidum* (a wheat), *Talinum teretifolium* (fameflower), *Catostomus teres* (sucker).
Gr. *tribo,* rub; *tribakos,* rubbed, worn; *trimma, -tos,* n. anything rubbed or crushed; *tripsis; anatripsis,* f. a rubbing, friction; *triptos,* rubbed; *diatribe,* f. a rubbing away, waste, pastime, delay, bitter discussion; *epitrimma, -tos,* n. anything rubbed in or on: diatribe, tribade, tribadism, anatripsis, triboluminescence, *Tripsacum dactyloides* (gama-grass), *Trimma caesiura* (a fish).
Gr. *trycho,* wear, consume; see **rag**
Gr. *tryo,* rub; *trysis,* f. a wearing away: trypsin, chymotrypsin, trypsinogen.
See: **scrape, smooth, itch, grind, wash, wipe**

rubber < rub; see **resin**

rubbish < uncertain origin; see **dirt, worthless, trifle**

ruber, -bra, -brum, L. red; *rubellus,* dim.; *rubicundus; rubidus,* reddish; see **red**

rubeta, L. a kind of toad; see **toad**

rubetum, L. bramble thicket; see *rubus* under **blackberry**

rubia, L. madder; see **madder**

rubidium < L. *rubidus,* reddish; see ELEMENTS under **thing**

rubigo, L. rust (iron), blight; *rubiginosus,* rusty; see **rust**

rubrica, L. red earth; *rubricosus,* full of red earth; see **earth**

rubus, L. bramble, blackberry, raspberry; see **blackberry**

ructo, -atus, L. belch; *ructabundus,* belching repeatedly; see **belch**

rudbeckia, NL. a genus of the composite family; see **composite**

rudder < AS. *rother.*
L. *gubernum; gubernaculum,* n. helm, rudder, guidance: gubernaculum.
Gr. *holkeion; holkion,* n. rudder: *Holciophorus ater* (a beetle).
Gr. *oiax, -akos,* m. rudder, tiller, helm; *oiakion,* n. dim.; *oiakistes,* m. steersman, pilot: *Oeax triangularis* (a beetle).
Gr. *pedalion,* n. rudder; *pedaliotos,* with a rudder: Pedaliaceae, *Pedalium murex* (a pedalium).
L. *serraculum,* n. rudder:
See: **govern**

rudens, L. rope, line, cord; see **rope**

rudero, -atus, L. pave with crushed stones or rubbish, < *rudus, -eris,* crushed stone, rubbish, debris; see **stone**

rudicula, L. wooden spoon, spatula; see *rudis* under **rod**

rudimentum, L. first principle, beginning; see **begin**

rudis, L. rough, raw, wild, see **rough**; stirring rod, staff; *rudicula,* dim., see **rod**

rudor, L. a braying; see *rudo* under **neigh**

rudus, -eris, L. crushed stone, rubbish, debris; see **stone**

rue < L. *ruta* (Gr. *rhyte*), f. a bitter herb; *rutaceus,* made of rue; *rutatus,* flavored with rue: *Ruta graveolens* (common rue), Rutaceae, rue, rueful, rutin.
NL. *amyris,* f. a genus of the rue family: *Amyris balsamifera* (an amyris).
L. *citrus,* citron tree; a genus of the rue family; see **citrus**
NL. *cusparia,* f. a genus of the rue family: *Cusparia odoratissima* (a rutacead).

L. *dictamnus* (Gr. *diktamnos*), f. dittany, a mint; now a genus of the rue family: *Dictamnus alba* (gasplant-dittany), *Origanum dictamnus* (Cretan dittany), *Ballota pseudodictamnus* (bastard dittany).

L. *Feronia*, f. a Roman goddess; a genus of the rue family: *Feronia limonia* (wood-apple), *Feroniella oblata* (a rutacead).

Gr. *harmala*, f. wild rue: *Peganon harmala* (harmel), harmaline.

L. *peganum* (Gr. *peganon*), n. rue: *Peganum mexicanum* (Mexican harmel).

NL. *xanthoxylum*, n. a genus of the rue family: *Xanthoxylum americanum* (prickly-ash).

ruffle < D. *ruifelen*, rumple, wrinkle; see **fold**, **wave**, **rough**

rufus, L. red, reddish; *rufulus*, dim.; see **red**

rug < uncertain origin.

Gr. *dapis, -idos*, f. carpet, rug; *dapidion*, n. dim.: *Dapidodigma liger* (a moth).

Gr. *epeuchion*, n. prayer-rug:

Gr. *kanes, -etos*, m. reed-mat:

L. *matta*, f. mat made of rushes: mat.

Gr. *peristoma, -tos*, n. coverlet, tapestry, carpet:

Gr. *petasma, -tos*, n. carpet, rug:

L. *phormio, -onis* (Gr. *phormos*), m. mat:

Gr. *psiathos*, f. rush-mat: *psiathion*, n. dim.: *Psiatholasius bombyliformis* (a fly).

Gr. *rhegos*, n. rug, blanket: regolith [rhegolith].

Gr. *rhips, -ipos*, wickerwork, mat; see **screen**

Gr. *sumax, -akos*, m. rush-mat:

L. *storea*, f. mat:

L. *stragulum*, spread, rug, blanket; see **blanket**

L. *tapete, -is*, n. (Gr. *tapes, -etos*, m.), carpet, drapery, coverlet; *psilotapis, -idos*, f. carpet without nap: tapestry, *Trichophaga tapetzella*[*tapetizella*] (tapestry-moth), tapetum.

L. *teges, -etis*, f. covering, mat; *tegeticula*, f. dim.: *Tegeticula alba* (a moth), *Cyperus tegetiformis* (mat-sedge).

ruga, L. wrinkle, crease; *rugosus*, wrinkled, shriveled; see **fold**

ruidus, L. rough; see **rough**

ruin < L. *ruina*; *ruinosus*, falling to pieces; see *ruo* under **destroy**

rule < OF. *ruiler*, < L. *regulo*, direct, guide; see **govern**, **law**, **right**, **straight**, **measure**

rumble < imitative origin; see **thunder**

rumen, L. throat, gullet, paunch; see **stomach**

rumex, L. dock, sorrel, see **sorrel**; a missile weapon, see **spear**

ruminator, L. one who chews again; see *rumino* under **bite**

rumis, L. breast, teat; see **udder**

rumor, L. report, gossip, hearsay; *rumusculus*, dim.; *rumigerulus*, newsmonger; see **speak**

rump, prob. < ON. *rumpr*.

Sp. *anca*, f. buttock:

L. *anus*, rectal opening, fundament; see **intestine**

Gr. *batalos*, m. anus, hinder parts:

Gr. *chodanos*, m. breech, buttocks:

L. *clunis, -is* (Gr. *klonis, -ios*), f. buttocks, rump; *cluniculus*, m.; *klonion*, n. dim.; *clunalis*, of the hind parts: *Nucleolites*(*Cluniculus*) *kimmeridgensis* (a Jurassic echinoid).

NL. *crissum*, n. hind parts of a bird around the cloacal opening, < *crisso*, move the haunches: *Pipilo crissalis* (California towhee), *Toxostoma crissale* (a thrasher).

L. *culus*, m. fundament, buttocks:

L. *fundamentum*, n. bottom, buttocks: fundament.

Gr. *gloutos*, m. rump, buttocks; *apogloutos*, with small rump; *exechegloutos*, with prominent buttocks: gluteal, gluteus, *Glutoxys elegans* (a fly).

AS. *ham*, thigh and buttock: ham.

Gr. *hedra*, seat, breech, rump; see **sit**

Gr. *ignya*, f. ham:

Gr. *kathisma*, seat, buttocks; see **sit**

Gr. *kochone*, f. area between the anus and pudenda:

L. *natis*, f. rump, buttocks; ML. *natica*, f. buttock: nates, natiform, aitchbone, *Natica immaculata* (a gastropod).

Gr. *orrhos*, m. rump: *Onopygia* (formerly *Orrhopygia*) *myrmecophila* (a psela-phid).

L. *palasea*; *plasea*, f. buttock or rump of an ox: *Palasea albimacula* (a butterfly).

L. *perineum* (Gr. *perinaion*), n. area between the anus and pudenda: perineum, perineal.

L. *podex, -icis,* m. fundament, anus: *Podiceps ruficollis* (dabchick, a grebe), *Podilymbus podiceps* (pied-billed grebe).

L. *posticum,* n. fundament, buttocks:

Gr. *proktos,* anus, fundament, tail; see **intestine**

Gr. *pyge; pyx, -ygos,* f. rump, buttocks; *pygidion,* n. dim.; *pygaios,* of the rump; *mesopygion,* n. part between the buttocks; *orrhopygion,* n. rump or tail of a bird: pygidium, isopygous, heteropygous, pygarg, callipygous, callipygian, pygope, pygopod, pygopodoid, *Pygocryptus grandis* (a wasp), *Pygorchis affixus* (a trematode), *Pygeum vulgare* (a rosacead), *Megalopyge crispata* (a moth), *Deropygus histrio* (a beetle), *Thyropygus poseidon* (a milleped), *Xanthopygia bella* (a flycatcher), *Molpastes pygaeus* (black bulbul).

L. *strebula,* f. flesh about the haunches:

L. *suffrago, -inis,* f. ham:

Gr. *tramis,* f. perineum:

See: **stern**

rumpotinus, L. a kind of maple; see **maple**

run < AS. *rinnan.*

Gr. *apostates,* deserter, renegade; see **abandon**

Gr. *astandes,* m. courier: *Astandes densatus* (a Cretaceous gastropod).

L. *Atalanta* (Gr. *Atalante*), f. a legendary maiden noted for her fleetness of foot: *Atalantia ceylonica* (a rutacead), *Pyrameis atalanta* (red admiral).

L. *commeator,* messenger; see **send**

L. *curro, cursus, cursus,* run; *cursus, -us,* m. march, course, journey; *cursor, -is,* m. runner; *cursorius; cursualis; decursorius,* of running, of a course; *curriculum,* n. race, course, career; *cursilitas, -atis,* f. a running about; *cursito, -atus,* run hither and thither; *discursus,* running about; *occursus,* running together; *recursus,* running back, returning: current, currency, curriculum, cursorial, cursory, course, coarse, concurrently, concourse, corral, courier, corridor, corsair, discourse, excursion, discursive, incur, intercourse, incursive, occurrence, occur, precursor, recourse, recurrent, scour, succor, *Trichopteryx cursitans* (a beetle), *Cursorius bicinctus* (a courser), *Libocedrus decurrens* (California incense-cedar), *Orbitopsella praecursus* (a foraminifer).

Gr. *dromos,* m. a running, race; *dromas, -ados,* running: *dromaios,* running fast, fleet, swift; *dromeus,* m. runner; *dromikos,* fleet, swift; *drasmos,* m. flight; L. *drapeta* (Gr. *drapetes*), m. runaway slave, fugitive; *dromon, -os,* m. light, fast vessel; *apodrome,* f. a running from, divergence, error; *didrasko,* run: dromedary, hippodrome, anadromous, syndrome, aerodrome, drapetomania, *Apodroma subcaerulea* (a moth), *Ammodramus bimaculatus* (a grasshopper-sparrow), *Paradrapetes villosus* (a beetle), *Dromomeryx americanus* (a fossil mammal), *Dromicodryas bernieri* (a snake), *Dromicosaurus gracilis* (a fossil reptile).

L. *fugio, -itus,* fly from, flee, run away; see **depart**

Gr. *iochmos,* m.; *ioke,* f. rout, < *dioko,* cause to run; see *dioktos* under **hunt**

Gr. *kele,* riding-horse, racer; see **ride**

ON. *ras,* a running: race, racer, horserace, tailrace.

L. *stadium,* arena, race-course; see **play**

L. *tolutarius; tolutilis,* on a trot, trotting:

Gr. *trecho,* run: *trochalos,* running; *trochanter, -os,* m. runner, segment of an insect leg; *threktikos,* able to run; *trochis,* m. runner, messenger; *epitrochos,* running, flowing: trochanter, trechometer, *Trechomys bonduelli* (a fossil rodent), *Hygrotrechus conformis* (a water-strider), *Trochalopteron cachinnans* (a laughing-thrush).

See: **flow, walk, swift, fly, leap, send, move, depart, pour**

runa, L. javelin, dart; see **spear**

runcator, L. weeder; see *runco* under **till**

runcina, L. a plane; see **smooth**

rupes, L. rock; *rupecula,* dim.; *rupestris,* of rocks, rocky; see **stone**

rupex, L. uncultivated man, boor; see **dull**

rupina, L. rocky chasm; see **valley**

ruppia, NL. a genus of aquatic plants; see **waterweed**

rupture < L. *ruptura,* fracture, break; see *rumpo* under **break**

ruralis, L. of the country, < *rus, ruris,* country, farm, land; see **country**

rursus, L. turned back, backward, reverse; see **return**

ruscum, L. butcher's-broom; *ruscarius,* of broom; see *ruscus* under **lily**

rush < OF. *ruser,* see **run, swift;** < AS. *rysc,* see **grass, reed, sedge, waterweed**

russus, L. red; *russatus,* clothed in red; *russeus; russulus,* reddish; see **red**

rust < AS. *rust.*

    L. *aerugo, -inis,* copper rust; *aeruginosus,* rusty; see *aes* under **copper**

    L. *ferrugo, -inis,* f. iron rust; *ferrugineus; ferruginus,* rusty, rust-colored: ferruginous, *Digitalis ferruginea* (a foxglove), *Cyperus ferruginescens* (a sedge).

    Gr. *ios,* m. rust: *Iospilus phalacroides* (an annulate).

    L. *rubigo (robigo), -inis,* f. rust, blight, mildew, mold; *rubiginosus,* rusty: rubiginous, Robigalia, *Erioglossum rubiginosum* (kelatlayu).

    See: **fungus, red, rot, lessen**

rusticus, L. of the country, rural; see *rus* under **country**

ruta, L. rue; *rutaceus,* of rue; see **rue**

rutabulum, L. instrument for raking or stirring; see **rake**

ruthless < AS. *hreow,* sorrow, mercy, *-less,* without; see **rough, merciless**

rutilus, L. red, < *rutilo, -atus,* color red; *rutilesco,* grow reddish; see **red**

rutrum, L. shovel, spade; *rutellum,* dim.; see **shovel**

rutus, L. everything dug up, ruined; see *ruo* under **destroy**

rye < AS. *ryge;* see **grain**

S

sabaco- < Gr. *sabakos,* rotten, enervated; see **rot**

sabaia, L. drink made of barley; see **beer**

sabal, NL. a genus of palms; see **palm**

sabanum, L. (Gr. *sabanon*), linen cloth, towel; see **napkin**

sabatenum, L. a kind of slipper; see **shoe**

Sabbath < Heb. *shabbath,* day of rest; see **rest**

sabbatio, NL. a genus of the gentian family; see **gentian**

sabellum, L. dim. of *sabulum,* sand; see **sand**

sabina, L. savin, the Sabine juniper; see **juniper**

Sabrina, L. Severn River; a river-nymph; see **stream**

sabulum, L. gravel, sand; *sabellum,* dim.; *sabulosus,* sandy; see **sand**

saburra, L. sand used as ballast; see **sand**

sabyttos, Gr. female pudenda; see **vulva**

saccatum, L. urine; see **urine**

saccharum, L. (Gr. *sakcharon*), sugar; see **sugar**

saccus, L. (Gr. *sakkos*), bag; *sacculus,* dim.; *saccarius,* of bags; *sakkinos,* of sackcloth; *sacco, -atus,* strain or filter through a bag; see **bag**

sacellum, L. small shrine, dim. of *sacrum,* temple; see *sacer* under **holy**

sacer, *-cra, -crum,* L. holy; *sacerdos,* priest; *sacrum,* holy thing, temple; see **holy, priest**

sack < AS. *sacc,* < L. *saccus;* see **bag**

saco- < Gr. *sakos,* shield; see **shield**

sacoma, L. (Gr. *sekoma*), counterpoise; see **balance**

sacrificium, L. offering to the gods; see **give**

sacrilegus, L. that steals or violates sacred things; see **profane**

sacrum, L. holy thing, temple; *sacellum,* dim.; see **holy, temple**

sacto- < Gr. *saktos,* crammed, stuffed; see *satto* under **press**

sacuto- < Gr. *sakoutos,* a kind of fish; see **fish**

sad < AS. *saed.*
Gr. *achoreutos,* joyless, melancholy, banished from the dance; see *choros* under **dance**
Gr. *achos,* pain, distress, grief, sorrow; see **pain**
Gr. *achthos,* burden, sorrow, grief; see **weight**
L. *aegritudo,* sickness, sorrow; see *aeger* under **disease**
Gr. *ailinos,* mournful, plaintive; see **weep**
Gr. *ania,* trouble, grief; see **trouble**
Gr. *apotmos; dyspotmos,* unlucky, unhappy; see *potmos* under **lot**
Gr. *aterpes,* joyless: *Aterpia anderreggana* (a moth).
Gr. *baryphron; dysphron, -os,* sad, melancholy, sorrowful; *baryphrosyne,* f. melancholia:
Gr. *barypsychos,* dejected:
Gr. *barystonos,* groaning heavily:
L. *dejectus,* cast down, sunk down, dispirited: dejection, *Artemon dejectus* (a gastropod).
L. *dolor,* pain, grief, sorrow; *dolorosus,* painful, sorrowful; see **pain**
Gr. *dysdaimon, -os,* unhappy; *dysdaimonia,* f. misery: *Dysdaemonia boreas* (a moth).
Gr. *dystenos,* unhappy, wretched:
Gr. *dysthymia,* f. despair, melancholy; *dysthymos,* desponding: *Dysthymus brunneus* (a milleped).
Gr. *elegos,* mourning song; see **poem**
Gr. *ellypos,* in grief, mournful; see *lype* under **pain**
L. *flebilis,* tearful, doleful, lamentable; see *fleo* under **weep**
L. *funebris,* pertaining to a funeral; see *funero* under **cover**
L. *gemo, -itus,* sigh, groan, moan, bewail, lament; *gemulus,* moaning; *gemebundus,* groaning, sighing; *congemo,* sigh deeply; *ingemo,* mourn over, lament: gemitorial.
Gr. *ialemos,* sad, melancholy:
L. *illaetabilis,* cheerless, gloomy, sad: *Geonemus illaetabilis* (a beetle).
L. *infelix, -icis,* unhappy, miserable: infelicity.
Gr. *katastygnos,* of sad countenance: *Catastygnus rivulosus* (a beetle).
Gr. *katephes,* downcast, mute; *katepheia,* f. dejection, sorrow: *Catephia alchymista* (a moth).
Gr. *kedos,* sorrow, trouble, care; see *kednos* under **guard**
Gr. *leugaleos,* sad, sorry, mournful:
L. *lugubris,* mournful, doleful, < *lugeo, luctus,* mourn, lament; *luctuosus,* sorrowful, mournful, doleful: lugubrious, luctiferous, *Discinisca lugubris* (a Miocene brachiopod), *Dolichoderus lugens* (an ant), *Dorcopsis luctuosa* (a kangaroo).
Gr. *lygros,* baneful, mournful; see **hurt**
L. *maestus,* sad, mournful, dejected, melancholy; *maestitia,* f. sadness, sorrow; *maestifico, -atus,* sadden, cause sorrow: *Chthonius maestus* (a pseudoscorpion).
Gr. *melancholia,* f. black bile, depression, sadness: melancholy, *Catamonus melancholicus* (a beetle).
Gr. *meledone,* care, sorrow; see *meledonos* under **guard**
L. *miser,* wretched; *misellus; miserulus,* dim.; *miserabilis,* pitiable, wretched; *miserandus,* lamentable: misery, miserable, miser, misericorde, commiserable.
L. *morosus,* fretful, peevish, gloomy; see **fretful**
Gr. *mygmos,* m. moaning, muttering:
Gr. *odyrtos,* lamentable; see **weep**
Gr. *oimoktos,* pitiable; see *oimozo* under **weep**
Gr. *penthos,* n. grief, sadness, sorrow; *penthema, -tos,* n.; *penthesis,* f. mourning; *pentheros,* of mourning; *pentheter, -os,* c. mourner; *penthikos; penthimos,* mournful: *Penthocrates bigenita* (a moth), *Penthestes atricapillus* (black-capped chickadee), *Nepenthes villosa* (a pitcher-plant).
L. *poenitens, -entis,* contrite, sorry: penitent, penitential, repentant.
L. *remorsus, -us,* m. a biting back, regret: remorse, remorseless, remorseful.
Gr. *stenagma, -tos,* n. groan, moan, sigh; *stonos,* m. groan, sigh, < *stenazo,* groan sigh:
L. *suspirium,* sigh; see *spiro* under **breathe**
Gr. *talas, -anos,* wretched, sad, sorry; *dystalas,* very unhappy:
Gr. *threnos,* funeral song, dirge; see **sing**
L. *tristis,* sad; *tristiculus,* dim.; *tristitia,* f. sadness, gloominess, grief; *tristificus,* causing sadness; *contristo, -atus,* sadden, darken: tristful, tristisonous, tristiloquy, tristemania, *Astragalinus tristis* (goldfinch).
See: **weep, trouble, pain, shade, rue, discontent**

saddle < AS. *sadol.*

Gr. *astrabe*, f. mule's saddle: *Astrabe lactisella* (a fish), *Astrabodus expansus* (a fossil fish).

L. *clitella*, f. pack-saddle: clitellum.

L. *ephippium* (Gr. *ephippion*), n. saddle, saddle-cloth: ephippium, *Ephippigera carinata* (a grasshopper), *Anomia ephippium* (saddle-shell), *Nepenthes ephippiata* (a pitcher-plant), *Micrurus ephippifer* (a snake), *Psolus ephippiger* (a holothurian), *Zetes ephippiatus* (a pycnogonid).

Gr. *kanthelion*, n. pack-saddle, pannier: *Cantheliophorus robustus* (a fossil lepidophyte).

Gr. *sagma, -tos*, n. pack-saddle; *sagmation*, n. dim.; *sagmarion*, n. pack-horse; *sagmarius*, of a saddle: *Sagmatorrhina lathami* (a bird), *Sagmidium crucicorne* (a radiolarian).

L. *sella*, seat, saddle; see **sit**

saeno- < Gr. *saino*, wag the tail, fawn upon; see **flatter**

saevus; saevis, L. raging, wild, violent, fierce; *saevitia*, rage; see **rough**

safe < OF. *sauf*, < L. *salvus*, preserved, unhurt, well, sound: Safety First!, vouchsafe, unsafe, sage, safeguard, save, saving, savior.

Gr. *ablabes*, harmless, innocent: *Ablabesmyia fastuosa* (a midge), *Ablabus ornatus* (a beetle), *Ablabophis rufula* (a snake).

Gr. *abletos*, unhit: *Abletobium pallescens* (a beetle).

Gr. *adeia*, f. freedom from fear, safety, security: *Nipponysson adiaphilus* (a wasp).

Gr. *adeletos*, unhurt:

Gr. *akindynos*, without danger, safe:

Gr. *aklystos*, sheltered:

Gr. *anatos*, harmless: *Anatochilus glabratus* (a beetle).

Gr. *anepaphos*, unharmed, untouched:

Gr. *anipsalos*, unhurt:

Gr. *anoutatos*, unwounded:

Gr. *apemon, -os*, harmless; *apemantos*, unharmed, unhurt; *apemosyne*, f. safety: *Apemon nigriventris* (a fly).

Gr. *asphales*, steadfast, safe; see **stand**

Gr. *asylon*, n. sanctuary, refuge; *asylos*, safe: asylum.

Gr. *atrotos*, unwounded, invulnerable: *Atrotus forcipatus* (a beetle).

L. *columis*, unhurt, safe; *incolumis*, unimpaired, safe, sound, entire, whole: *Oxytelus incolumis* (a beetle).

L. *confugium; perfugium; refugium; suffugium*, n. safety, shelter; see *fuga* under **depart**

Gr. *diaphyge*, f. refuge:

Gr. *echyros*, strong, secure; see **strong**

Gr. *euerkes*, well-fenced, well-walled, safe:

L. *impunitus*, free from danger, safe, secure; see **free**

L. *indelibatus*, untouched, uninjured:

L. *indemnis*, unhurt, uninjured: indemnify, indemnity, *Ectopimorpha indemnis* (a wasp).

L. *innocuus; innoxius*, safe, harmless: innocuous, *Hapalocrinus innoxius* (a Devonian crinoid).

L. *inviolatus*, unhurt; *inviolabilis*, invulnerable: inviolate, inviolable.

L. *invulnerabilis*, unwoundable: invulnerable, invulnerability.

Gr. *phygadeion; phyxion*, n.; *phyxis*, f. asylum, escape, refuge: *Phyxium ignarum* (a beetle).

L. *securus*, free from care, safe: secure, security.

Gr. *skepe*, f.; *skepas, -aos; skepasma, -tos*, n. shelter, protection, cover; *skepanon*, n. a covering; *skepanos*, sheltering; *skepastos*, covered; *skepastron*, n. veil: sepal, sepaloid, *Scepasma gigas* (a fossil insect), *Scepanotrocha rubra* (a worm), *Hedyscepe canterburyana* (an umbrella-palm).

Gr. *sos*, safe, sound; see *soter* under **save**

L. *tutus*, safe, secure; *tutum*, refuge; see *tueor* under **guard**

See: **save, innocent, sure, all**

saffron < OF. *safran*, < ML. *safranum*, < Ar. *safra;* see **yellow, crocus**

sagana, L. fortune-teller, soothsayer; see *praesagatus* under **prophecy**

sagapenon, Gr. a plant of the carrot family; see **carrot**

sagaris, Gr. a kind of sword; see **sword**

sagax, -acis, L. wise; *sagacitas*, intellectual keenness; see **know**

sage < L. *salvia*, see **mint**; < *sapidus*, savory, wise, see *sapio* under **know**

sage, Gr. pack, baggage; see **weight**

sagebrush; see **wormwood**

sagena, L. (Gr. *sagene*), fish-net; *sageneutes,* one who fishes with a net; see **net**

sagido- < Gr. *sagis, -idos,* wallet; see **bag**

sagina, L. a feasting, stuffing, fattening; *saginatus,* fattened; see **eat**

sagitta, L. arrow; *sagittula,* dim.; *sagittarius,* of an arrow, archer; see **arrow**

sagittaria, NL. a genus of aquatic plants; see **waterweed**

sagma, Gr. pack-saddle; *sagmation,* dim.; *sagmarion,* pack-horse; see **saddle, horse**

sagmen, L. tuft of sacred herbs rendering the bearer inviolable; see **cluster**

sagum; sagus, L. (Gr. *sagos*), cloak, mantle; *sagulum,* dim.; see **garment**

sagus, L. presaging, predicting, prophetic; see *praesagatus* under **prophecy**

sail < AS. *segel.*

Gr. *artemon, -os,* m. foresail, pulley: *Artemon candidus* (a snail).

L. *eremigo, -atus,* navigate, row, or sail:

Gr. *histion,* n. sail: *Istiophorus*[*Histiophorus*] *volador* (sailfish), *Histiurus amboiensis* (a lizard).   See *histos* under **weave**

Gr. *laiphos,* n. shabby garment, canvas, sail: *Laephotis wintoni* (a bat).

L. *nauta* (Gr. *nautes*); *nautilos,* m. sailor, seaman: Argonaut, nautilus, *Nautilus pompilius* (pearly nautilus), *Argonauta nodosa* (a cephalopod), *Aeronautes melanoleucus* (white-throated swift), *Exonautes rondeleti* (a flying-fish).

L. *navigator, -is,* m. sailor, mariner; *navigo, -atus,* sail: navigate, navigator, *Striglina navigatorum* (a moth).

Gr. *othone,* fine linen, sail-cloth, sail; *othonion,* dim.; see **linen**

Gr. *paraseion,* n. topsail:

Gr. *phosson, -os,* m. coarse linen garment, sail; *phossonion,* n. dim. coarse towel:

Gr. *ploter,* sailor; see *plotos* under **swim**

Gr. *sipharos,* m. topsail:

L. *supparum,* n. topsail:

L. *velum,* n. sail, covering, curtain, veil; *velifer,* sail-bearing; *velivolus,* with flying sails; *velificor, -atus,* set sail, sail: veliferous, veliger, velic, *Plecoglossus altivelis* (ayu, sweetfish).   See *velo* under **cover**

Gr. *Zetes,* m. son of Boreas and one of the Argonauts: *Zetes hispidus* (a pycnogonid), *Ceratozetes pusillus* (a mite).

See: **ship, swim**

St. Johnswort

L. *androsaemon* (Gr. *androsaimon*), n. a kind of St. Johnswort: *Hypericum androsaemum* (tutsan St. Johnswort), *Apocynum androsaemifolium* (spreading dogbane).

L. *ascyrum* (Gr. *askyron*), n. a kind of St. Johnswort: *Ascyrum hypericoides* (St. Andrew's cross), *Hypericum ascyron* (giant St. Johnswort).

L. *hypericum* (Gr. *hyperikon*), n. St. Johnswort: *Hypericum calycinum* (a St. Johnswort), *Chrysolina hyperici* (a beetle).

sal, *salis,* L. salt; *salinus,* of salt, salty; salted; see **salt**

Salacia, L. wife of Neptune and goddess of the sea; see **Neptune**

salaco, L. (Gr. *salakon*), braggart, see **brag**; < *salax, -akos,* sieve, see **sieve**

salad < OF. *salade,* < L. *sal,* salt.

Gr. *abyrtake,* f. salad of leeks, cresses, with sour sauce:

L. *acetaria,* f. salad prepared with vinegar and oil:

L. *moretum,* n. a kind of garlic salad:

Gr. *phyllas, -ados,* litter of leaves, salad; see *phyllon* under **leaf**

Gr. *tybaris,* m. a kind of salad:

salamander < Gr. *salamandra,* a lizardlike animal, an amphibian; see **newt**

salambe, Gr. vent, chimney, door; see **hole**

salanx, Gr. a kind of fish; see **fish**

salapitta, L. box on the ear; see **strike**

salaputium, NL. manikin; see **man**

salar, L. a kind of trout; see **trout**

salax, -acis, L. lustful, see **lewd**; < Gr. *salax, -akos,* sieve, see **sieve**

sale, Gr. road, see **way**; < ON. *sala,* sale, see **trade**

salebra, L. roughness in a road; *salebrosus,* rough, rugged, uneven; see **rough**

salgamum, L. pickle; see **sour**

salicornia, NL. a genus of the goosefoot family; see **goosefoot**

salinum, L. saltcellar; *salillum,* dim.; see **vessel**

salio, *saltus,* L. leap, jump, bound, dance; *saliens,* leaping; see **leap**

saliunca, L. a kind of valerian; see **valerian**

saliva, L. spit, spittle; *salivarius,* of spit, slimy; see **spit**

salix, L. willow; *salictum,* willow-thicket; *salignus,* of willow; see **willow**

salmon < L. *salmo, -onis;* see **trout**

salos, Gr. a shaking, tossing motion; see **shake**

salpa, L. (Gr. *salpe*), a kind of gilthead; see **bream**

salpingo- < Gr. *salpinx, -ingos,* trumpet; *salpinktes,* trumpeter; see **horn**

salsola, NL. a genus of the goosefoot family; see **goosefoot**

salt < AS. *sealt.*

L. *adarca* (Gr. *adarke*), f. salt coating on marsh vegetation; *adarkion,* n. dim.:
G. *ester,* an alkyl salt: ester, glyceryl ester.

Gr. *hals, halos,* m. salt, sea; *halimos,* of salt; *halistos,* salted, pickled; *halimon,* n.
a plant growing on the seashore, saltwort; *halme,* f. salt water, brine; *halmyros,*
salty, briny; *ephalmos,* pickled, salted; *kathalos,* full of salt: halophyte,
halmyrolysis, halophilous, halogen, halite, halide, haloid, *Halimodendron halo-
denron* (Siberian salt-tree), *Halogeton sativus* (a saltwort), *Halosaurus oweni* (a
fossil fish), *Lycium halimifolium* (matrimony-vine), *Aster halophilus* (an aster).

L. *muria,* f. brine; *muriaticus,* of brine: muriate, muriatic.

L. *sal, salis,* m. salt; *salarius; salinus,* of salt, salty; *salsura,* f. a salting; *salsus,*
salted, salty, witty; *salsiusculus,* dim. brackish; *salsitudo; salsugo, -inis,* f. salt-
ness, brackishness; *salio, -itus,* salt, salt down; *insulsus,* unsalted, insipid: saline,
salinity, salary, salad, sauce, *cum grano salis,* sausage, saucy, insulse, *Salsola
kali* (common saltwort), *Salicornia europaea* (samphire, glasswort), *Artemia
salina* (a brine-shrimp), *Erigeron salsuginosus* (a fleabane), *Carex salina* (salt-
sedge), *Sphaerophysa salsula* (globe-pea).

Gr. *tarichos,* pickled meat, fish, etc.; see **flesh**

See: **sea, taste**

saltator, L. hopper, dancer; *salticus,* of dancing; see *salio* under **leap**

saltus, L. woodland, forest; *saltuarius,* of forests, forester, ranger; see **forest**

salum, L. the open sea; see **sea**

salus, L. health; *salubris; salutaris,* healthful, wholesome, beneficial; see **health**

saluto, -atus, L. greet, address; *salutatorius,* of greeting; see **speak**

salvator, L. deliverer, preserver; see **save**

salve < AS. *sealf;* see **plaster**

salvelinus, NL. char, trout; see **trout**

salvia, L. sage; see **mint**

salvus, L. preserved, unhurt, well, sound; see **safe**

salyx, Gr. a kind of borage; see **borage**

samamithion, Gr. a kind of worm; see **worm**

samara, L. seed of the elm, any dry, indehiscent, winged fruit, key; see **fruit**

samardacus, L. (Gr. *samardakos*), buffoon; see **fool**

sambucus, L. elder; *sambuceus,* of elder; see **elder**

same < ON. *samr;* see **equal, like**

samio, -atus, L. polish; *samiator,* polisher; see **smooth**

samolus, L. a plant, now a genus of the primrose family; see **primrose**

sample < OF. *example,* < L. *exemplum,* imitation, copy; see **form, thing**

sampsychon, Gr. marjoram; see **mint**

sanctify < L. *sanctifico, -atus,* make holy; see *sancio* under **holy**

sand < AS. *sand:* sandy, sandwort, sandworm, sandstone, sandpiper, sandman,
quicksand, greensand.   Derive: sandwich.

Gr. *ammos (psammos; amathos),* f. sand; *psammion,* n. dim.; *psamathos,* f. sand
of the seashore; *psamathion,* n. dim.; *psamminos,* of sand: ammophilous
(psammophilous), ammotherapy, ammodytid, psammitic, psammophyte, *Ammophila breviligulata* (a beach-grass), *Ammobia ichneumonea* (a wasp), *Ammo-
spermophilus leucurus* (antelope-squirrel), *Amathomyia persiana* (a fly),
Psammobia vespertina (sunset-shell), *Psammichthys nudus* (a fish), *Psammato-
dendron[Psammodendron] arborescens* (a foraminifer), *Psamathiomyia pectinata*
(a fly), *Psamathocrita osseella* (a moth), *Georyssus*(or *Cathammistes*) *pygmaeus*
(a beetle).

L. *arena,* f. sand, sandy place; *arenula,* f. dim.; *arenaceus; arenarius,* of sand,
sandy; *arenatus,* with sand; *arenosus,* full of sand: arena, arenaceous,
arenicolous, arenite, *Arenaria serpyllifolia* (a sandwort), *Arenicola marina* (lug-
worm), *Ammophila arenaria* (a beach-grass), *Hyla arenicolor* (a frog), *Elymus
arenicola* (sand wild-rye), *Helichrysum arenarium* (yellow everlasting).

L. *ballux, -ucis; balluca,* gold-sand, gold-dust; see **gold**

Gr. *konia,* dust, sand, powder; see **dust**

L. *sabulum,* n.; *sabulo, -onis,* m. coarse sand, gravel; *sabellum,* n. dim.; *sabulosus,* full of sand, sandy: *Sabella penicillus* (a worm), *Sabellaria vulgaria* (a worm), *Sphex sabulosa* (a wasp).

L. *saburra,* f. sand used as ballast: saburrate, *Halichondria saburrata* (a sponge).

Gr. *Syrtis, -idos,* f. name of a sandbar off the coast of Africa, quicksand; *Syrtikos,* of the sandbar: *Salix syrticola* (a willow).

See: **grain, stone**

sandal < L. *sandalium* (Gr. *sandalion*), slipper; see **shoe**

sandalis, L. a kind of palm; see **palm**

sandalwood

NL. *comandra,* f. a genus of the sandalwood family: *Comandra umbellata* (common comandra).

Gr. *osyris,* f. a plant, now a genus of the sandalwood family: *Osyris alba* (a santalacead).

L. *santalum* (Gr. *santalon*), n. sandalwood-tree: Santalaceae, santalol, santalin, *Santalum album* (sandalwood), *Pterocarpus santalinus* (padauk).

L. *thesium* (Gr. *theseion*), n. a genus of the sandalwood family: *Thesium alpinum* (a santalacead), *Thesidium fruticulosum* (a santalacead).

sandapila, L. bier for the lower classes; see **bed**

sandaraca, L. (Gr. *sandarake*), realgar, see **arsenic**; beebread, see **bread**

sandix, Gr. vermilion; see **red**

sandpiper; see **snipe**

sanguis, L. blood; *sanguinarius,* of blood, bloody; see **blood**

sanguisorba, NL. a genus of the rose family; see **rose**

sanguisuga, L. bloodsucker, leech; see **leech**

sanicula, NL. a genus of the carrot family; see **carrot**

sanido- < Gr. *sanis, -idos,* board, plank, tablet; see **board**

sanies, L. bloody matter, pus; *saniosus,* full of pus; see **pus**

sanitas, L. health, soundness of mind; see *sanus* under **health**

sannio, L. one who makes faces, grimacer, buffoon, harlequin; see **fool**

santalum, L. (Gr. *santalon*), sandalwood-tree; see **sandalwood**

santerna, L. borax; see **borax**

santonium, L. (Gr. *santonion; santonikon*), a kind of wormwood; see **wormwood**

sanus, L. sound, healthy; *sano, -atus,* heal, cure; see **health**

sap < AS. *saep;* see **juice**

saperda, L. (Gr. *saperdes*), a kind of fish from the Black Sea, herring or sardine; see **herring**

saperdion, Gr. courtesan; see **prostitute**

saperion, Gr. an animal; see **animal**

saphena, ML. a vein in the leg; see **pipe**

saphes, Gr. clear, distinct, certain; see **clear**

sapidus, L. savory, tasty, witty, < *sapio,* taste, know; see **taste, know**

sapiens, L. prudent, wise; see *sapio* under **know**

sapindus, NL. soapberry; see **soapberry**

sapinus, L. a kind of pine or fir; see **pine**

sapium, NL. a genus of the spurge family; see **spurge**

sapo, L. soap; see **soap**

sapodilla < Sp. *sapotillo,* dim. of *sapote,* < Ab.Am. *tzapotl,* a tree that yields chicle (*Achras sapota,* formerly *Sapota achras*): Sapotaceae, sapote (*Calocarpum sapota*).

L. *bumelia* (Gr. *boumelia*), f. a kind of ash, now a genus of the sapodilla family: *Bumelia monticola* (a bumelia).

NL. *chrysophyllum,* n. a genus of the sapodilla family: *Chrysophyllum africanum* (African star-apple).

Ab.Am. *lucuma,* f. a genus of the sapodilla family: *Lucuma salicifolia* (a lucuma).

NL. *mimusops,* f. a genus of the sapodilla family: *Mimusops caffra* (a bulletwood).

NL. *sideroxylon,* n. a genus of the sapodilla family: *Sideroxylon foetidissimum* (a jungle-plum).

saponaria, NL. a genus of the pink family; see **pink**

sapor, L. taste, flavor; see *sapidus* under **taste**

sapphire < L. *sapphirus* (Gr. *sappheiros*); *sapphirinus,* of sapphire; see **stone**

sapros, Gr. rotten; see **rot**

sarabos, Gr. female pudenda; see **vulva**

sarcasm < Gr. *sarkasmos,* biting mockery; see **laugh, form**

sarcina, L. package, bundle, burden, load; *sarcinula,* dim.; see **bundle**

sarcio, *sartus,* L. sew, patch, mend, repair; see *sartor* under **sew**

sarco- < Gr. *sarx, sarkos,* flesh; *sarkion,* dim.; *sarkodes,* fleshy; see **flesh**

sarcocolla, L. (Gr. *sarkokolla*), a Persian gum; see **resin**

sarcodes, NL. a genus of the heath family; see **heath**

sarcoma < Gr. *sarkoma,* tumor, cancer; see **disease**

sarculum, L. hoe; see **hoe**

sardina, L. a kind of fish caught near Sardinia and pickled; see **herring**

sardius, L. a kind of carnelian; see **stone**

sardon, Gr. a kind of rope; *sardonion,* dim.; see **rope**

sargane, Gr. wickerwork, plait, braid; see **weave**

sargasso, Pg. gulfweed; see **alga**

sarginos, Gr. a gar; see **pike**

sargus, L. (Gr. *sargos*), a sparoid fish; see **bream**

sarisa (sarissa), Gr. a long, Macedonian lance; see **spear**

sarma, Gr. chasm in the earth; see **hole**

sarmentum, L. twig, light branch, brush; *sarmentosus,* full of twigs; see **branch**

sarmos, Gr. heap of earth or sand; see **heap**

saron; sarotron, Gr. broom; *saroma,* sweepings; see **broom**

saronis, Gr. an old, hollow oak; see **oak**

saros, Gr. Babylonian cycle for the recurrence of lunar and solar eclipses; see **circle**

sarpo, *sarptus,* L. trim, prune; see **cut**

sarracenia, NL. a genus of pitcher-plants; see **pitcher-plant**

sarracum, L. a kind of wagon or cart; see **vehicle**

sarrio, *-itus,* L. hoe, weed; *sarritor,* hoer; *sarritorius,* of hoeing; see **till**

sartago, L. frying-pan; see **pan**

sartor; sartrix, L. tailor, patcher; *sartura,* a mending; see **sew**

sassafras, NL. a genus of the laurel family; see **laurel**

satagius, L. anxious, troubled, worried; see **trouble**

Satan < Heb. *satan,* an enemy, the devil; see **spirit**

satelles, L. guard, companion, lackey; see **companion**

sathe, Gr. penis; see **penis**

satherion, Gr. probably a kind of beaver; see **beaver**

sathrax, Gr. louse; see **louse**

sathros, Gr. decayed, rotten, unsound; *sathroma,* something unsound, flaw; see **rot**

satine, Gr. chariot; see **vehicle**

satire < L. *satira,* a poetic medley devoted to ridicule and sarcasm; see **laugh, form**

satis, L. enough; *satio, -atus,* fill, satisfy; see **abundance**

sativus, L. sown, planted, cultivated, < *sero, satus,* sow; see **sow**

satrapa, L. (Gr. *satrapes*), governor of a province; see **govern**

satur, L. full, rich; *satullus,* filled, satisfied; *saturo, -atus,* fill, glut; see **abundance**

satureia, L. savory; see **mint**

Saturnus, L. god of agriculture; see **till**

satyrion, Gr. an orchid; see **orchid**

satyros, Gr. a woodland deity typifying lasciviousness, a kind of ape; see **spirit**

sauce < OF. *sauce,* < L. *salsus,* salted.
    Gr. *abyrtake,* a sour sauce; see **salad**
    L. *alec, -is,* f. sauce prepared from small fish; *alecula,* f. dim.: *Halecodon denticulatus* (a fossil fish), *Regalecus glesne* (oar-fish).

L. *cepolindrum*, n. a kind of condiment:

L. *condimentum*, n. a pungent or spicy substance added to food: see **spice**

L. *cremor, -is*, m. thick juice, broth, **gravy**:

Gr. *embamma, -tos*, n. sauce, soup:

L. *garum* (Gr. *garon*), n. sauce made of small fish:

Gr. *hedysma*, seasoning, sauce, spice, perfume; see *hedys* under **sweet**

L. *jus*, broth, soup, sauce; see **juice**

Gr. *katachysma, -tos*, n. sauce:

Malay *kechap*, a kind of sauce: catchup (ketchup).

L. *liquamen, -inis*, n. sauce, gravy; *liquaminosus*, full of gravy:

Gr. *opson*, meat, rich fare, sauce; see **food**

Gr. *oxalme*, f. pickle or sauce made of vinegar and brine:

L. *pulmentum*, n. sauce, relish:

Gr. *zomos*, soup, sauce; see **soup**

See: **juice, soup, spice, jelly**

saucer < OF. *saussier*, a dish for sauce; see **plate**

saucius, L. wounded, hurt; see **hurt**

saucro- < Gr. *saukros*, graceful, pretty; see **beauty**

saulos, Gr. a waddling, swaying, or swaggering gait; see **walk**

saunion, Gr. javelin; see **spear**

sauros, Gr. lizard, reptile; *sauridion*, dim.; see **lizard**

sauroter, Gr. spike at the butt end of a spear; see **point**

sausage < OF. *saucisse*, < L. *salsus*, salted.

Gr. *allas, -antos*, m. sausage; *allantion*, n. dim.: allantoid, allantois, allantoic.

L. *apexado, -onis*, m. a kind of sausage:

L. *botulus*, m. sausage; *botellus*, m. dim.: botuline, botulism, botuliform, bowel, *Clostridium botulinum* (bacterium causing botulism), *Dentalina botuliformis* (a foraminifer).

Gr. *chordeum, -tos*, n. sausage:

L. *farcimen, -inis*, n. sausage: *Farciminaria uncinata* (a cheilostome).

L. *fundulus*, a kind of sausage; see *fundus* under **base**

L. *hilla*, small intestine, a kind of sausage; see *hira* under **intestine**

L. *longano*, straight gut, sausage; see **intestine**

Gr. *physke*, sausage, blister; see **bubble**

L. *scrutillus*, a pork-sausage:

L. *tomacina*, f.; *tomaculum*, n. a kind of sausage:

L. *tucetum*, n. a kind of sausage:

See: **intestine, press**

save < OF. *salver*, < L. *salvo, -atus*, rescue, preserve, retain; *salvator, -is*, m. deliverer, preserver; *salvus*, preserved, unhurt: salvation, salvage, savior, salvarsan (606), safe, salver, salvo, sage (*Salvia officinalis*), *Hydrosaurus salvator* (a monitor-lizard), San Salvador, *Salvadora persica* (toothbrush-tree).

Gr. *asphalizo*, make safe, secure; see *asphales* under **stand**

L. *condio, -itus*, pickle, preserve; *condo, -itus*, lay up, treasure, preserve; *conditum*, n. the thing preserved, store; *conditaneus; conditivus*, suitable for preserving:

L. *frugalis*, economical, sparing, thrifty: frugal, frugality.

L. *micarius*, frugal, economical:

L. *parco, parsus*, spare; *parcus*, frugal, thrifty, scanty: see **stingy**

Gr. *pheidos*, saving, sparing, thrifty, stingy; *pheidolos*, niggard, miser; see **stingy**

Gr. *rhysis*, f. deliverance; *rhysios*, saving: *Rhysipolis meditator* (a wasp).

L. *servo, -atus*, keep preserve; *conservo, -atus*, retain, keep in existence: conserve, conservation, conservatory, observant, observatory, preservative, reservoir, reservation.

L. *sospes, -pitis*, saving, delivering; *sospitalis*, salutary: Juno Sospita.

Gr. *soter, -os*, m.; *soteira; sostria*, f. savior, deliverer, preserver; *soterios*, saving, preserving; *soteriodes*, wholesome; *sos*, safe, sound; *sosis*, f. salvation; *sostikos*, able to save; *sostos*, saved; *sozo*, save: soteriology, Socrates, sozobranchiate, sozolic, Sozura, Sozodont (a dentifrice). *Hyposoter fugitivus* (a wasp), *Sycosoter lavagnei* (a wasp), *Sosippus floridanus* (a spider), *Sosthenes dyschirioides* (a beetle).

See: **safe, hold, guard, cover, heal, stingy, store**

saw < AS. *saga* (G. *säge*): sawbuck, sawdust, sawmill, sawyer, jig-saw, seesaw.

Gr. *prion, -os*, m. saw; *prio; prizo*, saw; *prionotos*, jagged, serrate; *prisma, -tos*, n. anything sawed, sawdust; *pristes*, m. sawyer; *pristos*, sawed; *pristis*, f. sawfish: prionodont, Prionodesmaceae, *Prionosaurus regularis* (a fossil lizard), *Prionotus birostratus* (a gurnard), *Prionus laticollis* (a beetle), *Priononyx atratus* (a wasp), *Pristomyrmex pungens* (an ant), *Helicoprion nevadensis* (a fossil fish), *Polyprion americanus* (a stobe-bass), *Zaprionus spinosus* (a fly), *Centropristes striatus* (sea-bass), *Dasyscopelus pristilepsis* (a fish), *Diprion polytomum* (spruce-sawfly).

L. *serra*, f. saw; Sp. *sierra*, f. saw; *serrula*, f. dim.; *serrator, -is*, m. sawyer; *serratus*, toothed like a saw: serrate, serration, serrulate, Sierra, serried, serriferous, *Serrasalmo piraya* (a piranha), *Serranus cabrilla* (a Mediterranean fish), *Serratula tinctoria* (a sawwort), *Sierra cavalla* (cero), *Hemipristis serra* (a fossil shark), *Barosma serratifolia* (a buchu), *Eratocrinus serratus* (a Mississippian crinoid), *Amphidolops serrifer* (a fossil mammal), *Mergus serrator* (red-breasted merganser), *Brontes serricollis* (a beetle), *Cleome serrulata* (bee spider-flower), *Lactuca serriola* (prickly lettuce), *Lasioderma serricorne* (cigarette-beetle).

See: **tooth, notch, scrape, cut**

saxifrage < L. *saxifraga*, f.: Saxifragaceae, sassafras, *Saxifraga aquatica* (water-saxifrage).

L. *apruco, -onis*, f. a kind of saxifrage:

NL. *astilbe*, f. a genus of the saxifrage family: *Astilbe simplicifolia* (star-astilbe).

NL. *hydrangea*, f. a genus of the saxifrage family: *Hydrangea quercifolia* (oak-leaved hydrangea), *Schizophragma hydrangeoides* (a saxifragacead).

Gr. *itea*, f. willow, now a genus of the saxifrage family; *iteinos*, of willow: *Itea virginica* (Virginia sweetspire).

NL. *parnassia*, f. a genus of the saxifrage family, < Mt. Parnassus in Greece: *Parnassia fimbriata* (Rocky Mountain parnassia), *Parnassius glacialis* (a butterfly), *Ranunculus parnassifolium* (a buttercup), *Proctilopha parnassiella* (a moth).

L. *philadelphus*, m. (Gr. *philadelphon*, n.), a shrub, a genus of the saxifrage family: *Philadelphus floribundus* (a mockorange).

NL. *tellima*, f. a genus of the saxifrage family: *Tellima odorata* (a fringecup), *Boykinia tellimoides* (Japanese boykinia).

Saxon < OF. *Saxon*, perhaps ultimately from the same root as AS. *seax*, short sword; see **sword**

saxum, L. rock; *saxulum*, dim.; *saxeus*, of stone; *saxatilis*, among rocks; *saxosus*, rocky, stony; see **stone**

say < AS. *secgan*; see **speak**

sbestos, Gr. quenched; *sbester*, extinguisher; see *sbennymi* under **wet**

scab < AS. *scaeb*, crust over a sore; see **scale, disease, fungus**

scabellum, L. footstool; see **sit**

scaber, *-bra, -brum*, L. rough, scabby, mangy; *scabies*, roughness, scurf, mange, itch; *scabiosus*, scabby, mangy; see **rough**

scaeo- < Gr. *skaios*, left, awkward, ill; see **hand**

scaero- < Gr. *skairo*, dance, skip; see **leap**

scaevus, L. (Gr. *skaios*), left, toward the left, west; see **hand, west**

scaffold < OF. *escafaut*; see **frame**

scala, L. flight of stairs, staircase, ladder; *scalaris*, of a ladder; see **ladder**

scale < ON. *skal*, bowl, dish, balance; see **balance**

scale < OF. escale, husk, shell.

Gr. *achne*, f. any light substance, foam, down, chaff: *Achna phalaenoides* (a moth), *Achnocampa ilicis* (a moth), *Neurachne mitchelliana* (a grass), *Trichachne hitchcocki* (a grass).

Gr. *achor, -os*, m. scurf, dandruff: *Achorocephalus cinctipes* (a wasp), *Achorion schoenleini* (fungus causing favus).

Gr. *achyron*, n. chaff, bran, husk; *achyrinos*, of chaff: *Achyrospermum densiflorum* (a mint), *Achyranthes australis* (an amaranth).

L. *acus, aceris*, n. husk, chaff; *aceratus*, with chaff; *acerosus*, chaffy: acerate, acerose, *Acalles acerosus* (a beetle).

L. *apluda*, f. chaff, bran: *Apluda varia* (a grass).

Gr. *apoptisma, -tos*, n. chaff, husk:

L. *bractea*, f. thin, metal plate, scale, small leaf, gold-leaf; *bracteatus*, gilt; *bractearius*, m. gold-beater, maker of gold-leaf; *imbracteo, -atus*, overlay with leaf-metal: bract, bracteate, *Plusia bractea* (a moth), *Helichrysum bracteatum* (strawflower), *Pedicularis bracteosa* (a lousewort), *Samolus ebracteatus* (a brookweed).

L. *canica*, f. a kind of bran:

L. *cantabrum*, n. a kind of bran:

L. *cretura*, f. bran, chaff, siftings:

L. *derbiosus,* scabby:

L. *furfur, -uris,* m. bran, scurf, scale, dandruff; *furfuricula,* f. dim.; *furfuraceus; furfurarius; furfureus; furfurosus,* of bran, like bran: furfur, furfuraceous, furfurol, furfuramide, furan, furoin, *Begonia furfuracea* (a begonia), *Chionaspis furfura* (scurfy scale).

L. *gluma,* f. hull, husk, bract: glume, glumiferous, glumaceous, *Bromus latiglumis* (a grass), *Puccinia glumarum* (a rust), *Dendrochilum glumaceum* (an orchid).

Gr. *ichthyema, -tos,* n. fish-scale:

Gr. *karphos,* n. chaff, chip, straw; *karphion,* n. dim.: *Carphophis amoena* (a worm-snake), *Psilocarphus tenellum* (a composite), *Hemicarpha aristulata* (a sedge).

Gr. *kyrebion,* n. husk, bran: *Cyrebion*[*Cyrebium*] *laticornis* (a beetle).

Gr. *lepis, -idos,* f. scale; *lepidion; lepion,* n. dim.; *lepidotos,* scaly; *lemma; lepisma, -tos; lepyron,* n. peel, husk, sheath, shell, scale; *lepyrion,* n. dim.; *lepros,* scaly, rough; *leptos,* like a scale or peel, thin, fine, small, delicate; *lopos,* m.; *lopisma, -tos,* n. peel, scale: leper, leprosy, lepidolite, lopolith, lepidocrocite, lemma, telolemma, Lepidoptera, *Lepidodiscus squamosus* (a fossil echinoderm), *Lepidosaphes fici* (fig-scale), *Lepidodendron selaginoides* (a fossil clubmoss), *Lepidium sativum* (a pepperweed), *Lepisma saccharina* (silverfish), *Lepiota americana* (a mushroom), *Lepionurus sylvestris* (catberry), *Lepomis auritus* (red-breasted bream), *Lepyrolobus recurvatus* (a gastropod), *Lopomorphus dolabratus* (a bug), *Lopodytes dolichomerus* (a bug), *Lemmatophila atomella* (a moth), *Lemmatophoropsis sibirica* (a fossil stonefly), *Selaginella lepidophylla* (resurrection-plant), *Nephrolepsis pendula* (a fern), *Salix lasiolepis* (arroyo-willow), *Stephanolepis hispidus* (a filefish), *Xenophora leprosa* (a fossil gastropod), *Repipta depidula* (a bug), *Sclerolepis uniflora* (a composite), *Monolopia major* (a composite).

L. *palea,* f. chaff; *palealis; palearis,* of chaff: palet, paleaceous, paleola, *Carex paleacea* (a sedge), *Calamogrostis brevipaleata* (a grass).

L. *petigo, -inis,* f. scab, eruption: petechia, petechial.

Gr. *Pholis, -idos,* f. scale, spot, fleck; *pholidotos,* scaly: pholidolite, *Pholidosaurus schaumbergensis* (a lizard), *Pholidota imbricata* (an orchid), *Cephalopholis fulvus* (a fish), *Pholiota squarrosa* (a mushroom).

Gr. *pityron,* n. bran, scale; *pityrodes,* branlike: pityriasis, pityroid, *Pityrogramma chrysophylla* (gold-fern).

L. *porrigo, -inis,* f. dandruff, scurf: porrigo.

L. *ramentum,* n. chaffy, loose scales, shavings, scrapings: ramentum.

L. *scabies,* roughness, scurf, mange, itch; see **itch**

NL. *scariosus,* thin, papery, or scaly like a bract: scarious.

L. *squama,* f. scale; *squamula,* f. dim.; *squamatus; squameus; squamosus,* scaly; *desquamo, -atus,* scale off, peel off: squamoid, squamosal, squamella, Squamata, desquamate, *Annona squamosa* (sweetsop), *Macromischa squamifera* (an ant), *Barynotus squamosus* (a beetle), *Hypopitys latisquama* (a pinesap), *Chironema squamiceps* (a fish), *Silene squamigera* (a catchfly).

NL. *squarrosus,* rough with stiff scales, bracts, leaves or prominences; see **rough**

See: **plate, leaf, scrape, rough, bug**

scalenus, L. (Gr. *skalenos*), unequal; see **different**

scalido- < Gr. *skalis, -idos; skaleuthron,* hoe; *skalidion,* dim.; see **hoe**

scalidris, L. (Gr. *skalidris*), a speckled shore bird; see *calidris* under **snipe**

scalmo- < Gr. *skalme,* knife, sword, see **knife**; < *skalmos,* peg or thole for fastening an oar, see **nail**

scalopo- < Gr. *skalops, -opos,* mole; see **mole**

scalpel < L. *scalpellum,* surgical knife, lancet, dim. of *scalprum,* chisel, knife; see **knife**

scambus, L. (Gr. *skambos*), bent, crooked, bowlegged; see **bend**

scamellum; scammillus, L. dim. of *scamnum,* bench, stool, balk; see **sit**

scammato- < Gr. *skamma, -tos,* that which has been dug, trench, pit; see **ditch**

scammony < L. *scammonia* (Gr. *skammonia*); see **bindweed**

scamnum, L. (Gr. *skamnon*), bench, stool, step, balk; *scamellum; scamillus,* dim.; see **sit**

scandal < Gr. *skandalon,* snare; see **trap**

scandix, L. (Gr. *skandix*), an umbellifer; see **carrot**

scando, *scansus,* L. climb; see **climb**

scapano- < Gr. *skapane,* a digging tool, spade; *skapanion,* dim.; see **shovel**

scapha, L. (Gr. *skaphe*), light boat, skiff; *scaphula,* dim.; see **ship**

scaphio- < Gr. *skaphion; skaphidion,* dim. of *skaphe; skaphis,* bowl, basin, small boat, shovel; see **cup, ship, shovel**

scapho- < Gr. *skaphe;* *skaphos,* anything dug or hollowed out, scoop, trough, vessel; see **hollow, cup, ship, shovel**

scapto- < Gr. *skaptos,* dug, < *skapto,* dig; *skapter,* digger; see **dig**

scapula, L. shoulder-blade; see **shoulder**

scapus, L. (Gr. *skapos*), stem, staff; see **stem**

scar < Gr. *eschara;* see **mark**

scarabaeus, L. a kind of beetle; see **beetle**

scarce < OF. *escars,* few, rare, sparse; see **few, thin, stingy, little**

scardamycto- < Gr. *skardamyktes,* blinker, winker; see **wink**

scardia, L. a plant; see **plant**

scardinius, NL. name for a cyprinoid fish; see **carp**

scare < ON. *skirren,* frighten; see **fear**

sacrifico, -atus, L. scratch open; see **scrape**

scariosus, NL. thin, papery, scaly; see **scale**

scarlet < ML. *scarlatum,* cloth of a red color; see **red**

scarto- < Gr. *skartes,* nimble, quick; *skarthmos,* leap, skip; see *skairo* under **leap**

scarus, L. (Gr. *skaros*), a kind of fish; see **wrasse**

scato- < Gr. *skor, skatos,* dung; see **dung**

scatter < from the same root as *shatter;* see **spread**

scatula, ML. a kind of pill-box; see **box**

scaturigo, L. bubbling spring; see **spring**

scaurus, L. (Gr. *skauros*), having large or swollen ankles; see **swell**

scedasto- < Gr. *skedastos,* scatterable; *skedastes,* scatterer; see *skedannymi* under **spread**

scelestus, L. wicked, infamous, abominable; *scelus, -eris,* evil deed, crime; see **bad**

sceleto- < Gr. *skeletos,* dried up, withered; see **frame**

scelido- < Gr. *skelis, -idos,* rib, side, leg; *skelidion,* dim.; see **rib, leg**

scelio, L. scoundrel; see **rogue**

sceliphro- < Gr. *skeliphros,* dry-looking, lean, thin; see **thin**

scello- < Gr. *skellos,* bowlegged; see **bend**

scelo- < Gr. *skelos,* leg; *skeliskos; skelydrion,* dim.; see **leg**

scelus, -eris, L. evil deed, crime; see **bad**

scemmato- < Gr. *skemma, -tos,* question, speculation; see *skeptikos* under **think**

scene < L. *scena* (Gr. *skene*), tent, stage, decorative setting or place; see **play**

scent < OF. *sentir,* perceive, especially by smell; see **smell**

scepano- < Gr. *skepanon,* a covering; *skepanos,* sheltered; see *skepe* under **safe**

sceparno- < Gr. *skeparnon,* ax; see **ax**

scepasto- < Gr. *skepastos,* covered; see *skepe* under **safe**

scepastro- < Gr. *skepastron,* veil; see **curtain**

scepto- < Gr. *skeptos,* thunderbolt; see **arrow**

sceptrum, L. (Gr. *skeptron*), wand, baton, staff; see **rod**

scerbolo- < Gr. *skerbolos,* scolding, abusive; see **scold**

sceuo- < Gr. *skeuos,* vessel, implement, utensil, gear; *skeuastos,* artificial; see **tool**

schadon, Gr. cell in a honeycomb, the larva in the cell; see **room**

schalidoma, Gr. forked prop; see **pillar**

scheda, L. (Gr. *schede*), strip of papyrus, sheet of paper; *schedula; schedarion,* dim.; see **paper**

schedius, L. (Gr. *schedios*), hastily made; see **swift**

schedo- < Gr. *schedon,* near, almost; see **near**

schelis, Gr. rib; *schelidion,* dim.; see **rib**

schema, Gr. form, shape, plan; *schemation,* dim.; see **aim**

scheros, Gr. in a line, successive; see **line**

schesis, Gr. state, condition, habit, nature; see **nature**

schidaco- < Gr. *schidax, -akos,* splinter; *schidion,* dim.; see **part**

schinus, L. (Gr. *schinos*), mastic-tree; see **sumac**

schistos, Gr. split, divided; *schisma,* a split; see *schizo* under **cut**

schiza, Gr. splinter or chip of wood; *schidion; schizion,* dim.; see **part**

schoeniclo- < Gr. *schoiniklos,* a reed-bunting; see *schoenus* under **sedge**

schoenus, L. (Gr. *schoinos*), a rush, rush-rope; *schoinion,* dim.; see **sedge**

school < AS. *scolu,* < L. *scola; schola* (Gr. *schole*), f. leisure, learned disputation, a place for such activity; *scholaris; scholasticus,* of schools and scholars: scholar, scholastic, scholium, schoolcraft, scholarly, escolar (*Ruvettus pretiosus*), *Alstonia scholaris* (an apocynacead).
  L. *academia* (Gr. *akademia*), f. an association of learned men: academy, academical, academic.
  L. *collegium,* n. an association of colleagues: college, collegiate.
  Gr. *didaskaleion,* school; see *didasko* under **teach**
  L. *gymnasium* (Gr. *gymnasion*), n. school for bodily exercises; *gymnikos,* of gymnastic exercises: gymnasium, *Sciurus gymnicus* (northern red-squirrel).
  Gr. *Lykeion,* n. famous gymnasium at Athens: Lyceum.
  L. *universitas, -atis,* f. body of scholars, guild: university.
  See: **learn, teach, think**

schyr, Gr. hedgehog; see **hedgehog**

sciado- < Gr. *skias, -ados; skiadeion,* canopy, umbrella; *skia,* shade, shadow; *skiaros,* shady; see *skia* under **shade**

sciaena, L. (Gr. *skiaina*), a perchlike marine fish; see **perch**

sciaro- < Gr. *skiaros,* shady; see *skia* under **shade**

science < L. *scientia,* knowledge, skill, < *scio, scitus,* know; see **know, think, try, question, hypothesis**

scilla; squilla, L. (Gr. *skilla*), squill, shrimp; see **lily, crab**

scimpodium, L. (Gr. *skimpodion*), low bed, couch; see **bed**

scinaco- < Gr. *skinax, -akos,* quick, nimble; see **swift**

scincus, L. (Gr. *skinkos*), a kind of lizard; see **lizard**

scindapho- < Gr. *skindaphos,* vixen; see **fox**

scindapso- < Gr. *skindapsos,* a plant, now a genus of the arum family; see **arum**

scindo, *scissus,* L. cut, cleave; see **cut**

scindula, L. shingle; see **shingle**

scintho- < Gr. *skinthos,* diving; see **dip**

scintilla, L. spark; *scintillula,* dim.; see *scintillo* under **light**

scio- < Gr. *skia,* shade, shadow; *skiotos,* shady; see **shade**

scion < F. *scion,* branch, shoot, slip; see **branch**

scipio, L. staff; see **rod**

sciridio- < Gr. *skiridion,* a kind of fish; see **fish**

sciro- < Gr. *skiros,* gypsum, stucco; see **gypsum**

scirpus, L. rush; *scirpeus,* of rushes; see **sedge**

scirrho- < Gr. *schirrhos,* hard, hardening, gypsum, tumor; see **hard**

scirto- < Gr. *skirtao,* bound, leap; *skirtetes,* leaper; see **leap**

scisco, *scitus,* L. seek to know, inquire, search; *scitor, -atus,* ask, inquire; see **try**

scissors < OF *cisoires,* < L. *caedo, caesus,* cut.
  L. *axicia,* f. shears:
  L. *forfex, -icis,* f. scissors, shears; *forficula,* f. dim.; *forficatus,* scissors-shaped, forked: forficate, *Forficula auriculata* (earwig), *Muscivora forficata* (a flycatcher), *Ophiogomphus forficula* (a dragonfly).
  Gr. *psalis, -idos,* f. clipper, scissors; *psalidion,* n. dim.; *psalistos,* clipped; *psalizo,* cut with a scissors: *Psalidognathus modestus* (a beetle), *Psalidomyrmex foveolatus* (an ant), *Idopsalis riveti* (an earwig), *Eupsalis minuta* (a beetle), *Psalidium forcipatum* (a beetle), *Psalistus sordidus* (a beetle).

scissus, L. cut, rent, split; *scissura,* rent, cleft; *scissilis,* that may be split readily; see *scindo* under **cut**

scitulus, L. beautiful, elegant, handsome, pretty; see **beauty**

scitus, L. experienced, informed, knowing, skillful, wise, fine; *scitulus,* dim.; see *scio* under **know, beauty**

sciurus, L. (Gr. *skiouros*), the shadow-tail or squirrel; see **squirrel**

sclero- < Gr. *skleros,* tough, hard; see **hard**

scnipo- < Gr. *sknipos,* niggardly, stingy; see knipos under **stingy**

scobina, L. rasp; see **scrape**

scobis, L. scrapings, powder, dust; see **dust**

scoff < uncertain origin; see **tease**
scold < uncertain origin.
  L. *accuso, -atus,* blame, charge; see **blame**
  L. *compellatio, -onis,* f. reprimand, reproof, rebuke; *compello, -atus,* accost reproachfully: compellative.
  L. *concerpo, -cerptus,* tear to pieces, abuse, revile:
  L. *convicior, -atus,* revile, reproach, taunt; *conviciator, -is,* m. reviler: convicium.
  L. *denuntio, -atus,* upbraid, threaten: denounce, denunciation, denunciatory.
  Gr. *enipe,* f. reproof, rebuke, < *enipto; enisso,* reprove, reproach:
  Gr. *epiplexis,* f. reproof: epiplexis, epiplectic.
  Gr. *epitimesis,* f. reproof: epitimesis.
  L. *incilo, -atus,* blame, scold, rebuke:
  L. *increpo, -itus,* make a noise, upbraid loudly, chide, rebuke, accuse: increpation.
  L. *invectus,* scolding, reproachful, abusive: invective.
  L. *jurgo, -atus,* contend in words, quarrel, scold; *jurgium,* n. verbal quarrel, brawl; *objurgator, -is,* m. chider, scolder: objurgate, objurgator, objurgation.
  Gr. *loidoros,* abusive, railing; *loidoresis,* abuse, reproach; see **curse**
  L. *reprehendo, -hensus,* seize, blame, censure, reprove: reprehensive, reprehensible.
  Gr. *skerbolos,* scolding, abusive:
  Gr. *stobos,* m. abuse, scolding:
  L. *vitupero, -atus,* censure, scold, disparage: vituperation, vituperative.
  See: **curse, blame, fight, fault, hurt**
scoleco- < Gr. *skolex, -ekos,* worm; *skolekion,* dim.; see **worm**
scolibrochum, L. a kind of fern; see **fern**
scolio- < Gr. *skolios,* curved, bent, oblique; see **bend**
scolo- < Gr. *skolos,* pointed object, thorn, see **thorn;** < *skolon,* stumbling-block, hindrance, see **hinder**
scolopax, L. (Gr. *skolopax*), snipe, woodcock; see **snipe**
scolopendra, L. (Gr. *skolopendra*), a multiped; see **myriapod**
scolopo- < Gr. *skolops, -opos,* anything pointed, thorn, pale; see *skolos* under **thorn**
scolymus, L. (Gr. *skolymos*), a kind of artichoke; see **thistle**
scolythro- < Gr. *skolythros,* mean, shabby, poor; see **poor**
scomber, L. (Gr. *skombros*), mackerel; see **mackerel**
scomma, L. (Gr. *skomma*), taunt, jeer, jest; see **laugh**
scoop < D. *schope;* see **shovel, cup**
scopa, L. twigs, broom; *scopula,* dim.; *scoparius,* sweeper; see **broom**
scopaeo- < Gr. *skopaios,* dwarf; see **little**
scope < Gr. *skopos,* watcher, mark, range of view; *skopelos,* observation point, peak; see **see, stone**
scopimo- < Gr. *skopimos,* suitable for a purpose; see **fit**
scopio, L. stem, stalk; see **stem**
scopolia, NL. a genus of the nightshade family; see **nightshade**
scops, L. (Gr. *skops*), a small kind of owl; see **owl**
scopto- < Gr. *skoptikos,* jesting, mocking; *skoptes,* scoffer; see *skomma* under **laugh**
scopulus, L. (Gr. *skopelos*), projecting rock, shelf, ledge, cliff; see **stone**
scopus, L. mark at which to shoot; see **aim**
scorbutus, ML. scurvy; see **disease**
scordalus, L. wrangler, quarreler, brawler; see **fight**
scordium, L. (Gr. *skordion*), a mint with a garliclike smell; see **mint**
scordylo- < Gr. *skordyle,* a young tunny; see **tunny**
scoria, L. dross, slag; see **slag**
scorn < OF. *escarn;* see **hate, laugh**
scorobylo- < Gr. *skorobylos,* a kind of beetle; see **beetle**
scorodo- < Gr. *skorodon (skordon),* garlic; see **onion**
scorpaena, L. (Gr. *skorpaina*), sculpin; see **sculpin**
scorpion < L. *scorpio, -onis* (Gr. *skorpion, -os; skorpios*), m.: *skorpidion,* n. dim.: scorpioid, Scorpio, Scorpio afer (a scorpion), Scorpiurus vermiculata (a legume), Myosotis scorpioides (forgetmenot), Ophiorrhiza scorpioidea (a madder), Myxocephalus scorpius (a sculpin).

L. *nepa*, f. scorpion: *Nepa apiculata* (water-scorpion).

NL. *obisium*, n. a genus of pseudoscorpions: *Obisium lubricum* (a pseudoscorpion).

scortator, L. whoremonger; *scortor*, *-atus*, whore; see **prostitute**

scortum, L. skin, hide; *scorteus*, of hide or leather; see **skin**

scorzonera, NL. a genus of the composite family; see **composite**

scoto- < Gr. *skotos*, darkness; *skoteinos*, dark; *skotios*, dark, obscure; see **shade**

Scotus, L. an individual of a Celtic people in northern Britain; see **Celt**

scoundrel < uncertain origin; see **rogue**

scrape < AS. *scrapian.*

Gr. *apomaktron*, n. strickle:

Gr. *knetho; knao*, scrape, scratch, tickle, itch, irritate, sting; *knesis*, f.; *knesmos*, m. a scratching, tickling, itching, irritation; *knestis*, *-idos*, f.; *knestron*, n. scraper; *knestos*, scraped, rasped: *Cnethocampa processionea* (oak-procession-ary), *Cnestis ferruginea* (a connaracead), *Cnestidium rufescens* (a connaracead), *Cnestrostoma polyodon* (a fish), *Cneorhinus fossulatus* (a beetle).

Gr. *knizo*, scrape, tickle, tease, nettle; see **tease**

L. *lima*, f. file; *limula*, f. dim.; *limo*, *-atus*, file, polish; *limatulus*, dim.; *delimator*, *-is*, m. filer; *elimatus*, filed, polished: *Lima reticulata* (a fossil pelecypod), *Limatula maoria* (a pelecypod), *Aphelecrinus limatus* (a fossil crinoid), *Acmaea limatula* (a limpet), *Ammonites elimatus* (an ammonite).

Gr. *psektra*, f. scraper, currycomb; *psektrion*, n. dim.; *psegma*, *-tos*, n. scrapings; *psegmation*, n. dim.: *Psectrocephalus caecus* (a bug), *Psegmatopterus uncho-menoides* (a beetle).

L. *rado*, *rasus*, shave, scrape; *radula*, f. scraper; *rasilis*, scraped, shaved, smoothed, polished; *interrasilis*, scraped to give the appearance of low relief, embossed: radula, abrade, abrasion, erase, corrasion, razor, raze, ramentum, rastrum, rastellum, rasorial, *Taeniopsetta radula* (a fish), *Ficus radulina* (a fig), *Vaginulina raduliformis* (a foraminifer).

L. *rallum*, n. a kind of scraper:

Gr. *rhine*, f. file, rasp, a kind of shark; *rhinion*, n. dim.; *rhinema*, *-tos*, n. fillings:

L. *runcino*, *-atus*, plane off; see *runcina* under **smooth**

L. *scabo*, scrape, scratch: scab, scabies, scabious, scabrous.

L. *scalpo*, *scalptus*, scrape, scratch, engrave: sculpture.

L. *scarifico*, *-atus*, scratch open: scarify, scarification.

L. *scobina*, f. rasp; *scobis*, f. scrapings, filings, sawdust; *discobino*, *-atus*, file away, scrap, scratch: *Scobicia declivis* (lead-cable worm).

Gr. *stelgis* (*stlengis*), *-idos*, f.; *stelgisma*, *-tos; stelgistron*, n. scraper; *stlengidion*, n. dim.: *Stelgidopteryx serripennis* (a swallow), *Stelgistrum stejnegeri* (a fish), *Cyttomimus stelgis* (a fish), *Stlengis osensis* (a fish).

L. *strigilis*, f. scraper; *strigilecula*, f. dim.: strigil, *Formicivora strigillata* (an ant-wren), *Madarus bistrigellus* (a beetle), *Delima(Strigilodelima) platystoma* (a gastropod).

Gr. *xeo; xyo*, scrape, scratch; *xesis*, f. a scraping; *xester*, *-os; xestes*, scraper, polisher; *xestos*, scraped, polished; *xesma*, *-tos*, n. scrapings; *xyele*, f. tool for scraping wood, rasp, dagger; *xyron*, n. razor; *xysilos*, shaved, smoothed; *xysma*, *-tos*, n. shaving/scraping, particle; *xysmation*, n. dim.; *xyster*, *-os*, m.; *xystra*, f.; *xystron*, n. scraper, rasp, file; *xystikos*, of or for scraping; *xystos*, scraped, polished: xyster, *Xestobium affine* (a beetle), *Xestoderma atrum* (a beetle), *Xestocephalus pulicarius* (a bug), *Xestoleberis punctata* (an ostracode), *Euxestina spoliata* (a fly), *Xesurus laticlavus* (a surgeon-fish), *Xystocheir obtusa* (a myriapod), *Xysticus gulosus* (a spider), *Xyster fuscus* (a chopa), *Xyela minor* (a sawfly), *Prionoxystus robiniae* (a moth), *Xysmalobium undulatum* (an asclepiadacead).

See: **rub, itch, smooth, bite**

scream < ON. *skraema;* see **call**

screatus, L. a hawking, coughing; *screator*, hawker; see *screo* under **cough**

screech < imitative origin; see **squeak**

screen < OF. *escren.*

L. *cancellus*, m. lattice, grating, grill; *cancellatus*, cross-barred, latticed: cancel, chancel, *Exogyra cancellata* (a fossil pelecypod).

L. *canteriolus*, m. trellis for plants:

L. *clathratus*, latticed, grated, screened; *clathrum* (Gr. *kleithron*), n. lattice, grate; *kleithrion*, n. dim.: clathrate, *Clathrus cancellatus* (a fungus), *Clathrocystis aeruginosa* (an alga), *Clathrella pusilla* (a fungus), *Clathrocanium diadema* (a radiolarian), *Cleithrolepis granulatus* (a fossil fish), *Fenestella clathrata* (a fossil bryozoan).

L. *cratis*, f. wickerwork, hurdle; *craticula*, f. dim.; *craticius; craticulus*, of wickerwork, latticed, wattled: grate, crate, craticular, grill, *Cratis delicatula* (a pelecypod), *Craticularia paradoxa* (a fossil sponge), *Aphrodes craticula* (a bug), *Polystomella craticulata* (a foraminifer).

Gr. *dryphaktos*, m. latticed partition, railing:

Gr. *gerrhon*, n. screen, wickerwork: *Gerrhosaurus major* (African skink), *Gerrhonotus caeruleus* (a lizard).

Gr. *kinklis, -idos*, f. latticed gate or bar, hence any network: *Cincliderma quadratum* (a fossil sponge), *Cinclidoceramus pinniformis* (a fossil pelecypod), *Cinclidotus riparius* (a lattice-moss).

Gr. *plegma*, plaited work, wickerwork; see *plecto* under **weave**

Gr. *prokalymma, -tos*, n. covering, screen, veil:

Gr. *rhips, rhipos*, f. wickerwork, mat, screen: *Rhipocarabus alysidotus* (a beetle), *Rhipsalis rosea* (a cactus).

Gr. *tarsos*, woven mat or grate; see **foot**

Gr. *trasia*, f. grate, kiln:

L. *umbraculum*, shaded spot, screen; see *umbra* under **shade**

See: **sieve, weave, net, cover, garment, shield**

screw < OF. *escroue*, < L. *scrofa*, breeding-sow, and *scrobis*, ditch, hole, vulva; see **turn**

scriblita, L. tart; see **cake**

scribo, *scriptus*, L. write; *scriptor*, writer; *scriptorius*, of writing; see **write**

scrinium, L. cylindrical case, portfolio; see **box**

scriptura; scriptio, L. a writing; *scriptilis*, that can be written; see *scribo* under **write**

scrobis, L. ditch, dike, trench; *scrobiculus*, dim.; see **ditch**

scrofa, L. breeding-sow; *scrofula*, dim.; *scrofinus*, of sows; see **hog**

scrofula, L. dim. of *scrofa*, sow; a swelling, like little pigs, of the glands in the neck; see **disease**

scrophularia, NL. a genus of the figwort family; see **figwort**

scrotum, L. pouch containing the testicles; see **bag**

scrupus, L. sharp stone, anxiety; *scrupulus*, dim.; *scruposus*, rough stony; see **stone**

scrutor, -atus, L. examine thoroly; *scrutator*, examiner; see **try**

scrutum, L. rubbish, trash; *scrutarius*, second-hand dealer, ragman; see **worthless**

sculna, L. arbiter, mediator, umpire; see **judge**

sculpin < uncertain origin.

L. *cottus* (Gr. *kottos*), m. sculpin: *Cottus gracilis* (miller's-thumb), *Cottunculus microps* (a fish), *Acanthocottus groenlandicus* (sculpin), *Pterygiocottus macouni* (a fish).

L *scorpoena* (Gr. *skorpaina*), f. sculpin: *Scorpaena guttata* (a scorpion-fish), *Scorpaenobdella elegans* (a leech), *Scorpaenichthys marmoratus* (cabezon).

sculpo, *sculptus*, L. carve; *sculptor*, graver, cutter; *sculptilis*, carved; see **cut**

sculponea, L. wooden shoe; see **shoe**

scum < D. *schum*; see **slime, dirt**

scurf < AS. *scurf; sceorf*; see **scale**

scurra, L. jester, buffoon; *scurrilis*, jesting; *scurrilitas*, abusive buffoonery; see **laugh**

scutella, L. dim. of *scuta(scutra)*, tray, platter see **plate**; *scutellum*, dim. of *scutum*, shield, see **shield**

scutellaria, NL. a genus of the mint family; see **mint**

scutica, L. whip; see **whip**

scutula, L. small plate, diamond or lozenge-shaped figure; see *scutella* under **plate**

scutum, L. shield; *scutellum; scutulum*, dim.; see **shield**

scybalo- < Gr. *skybalon*, dung, refuse; see **dung**

scydmaeno- < Gr. *skydmainos*, angry; see **anger**

scylaco- < Gr. *skylax, -akos*, young dog, whelp, puppy; see **dog**

scylio- < Gr. *skylion*, dogfish; see **shark**

Scylla, L. (Gr. *Skylla*), a sea-monster; famous rock between Italy and Sicily, opposite the dangerous whirlpool, Charybdis; see **animal**

scyllaro- < Gr. *skyllaros (kyllaros)*, hermit-crab; see **crab**

scylmo- < Gr. *skylmos*, a tearing, rending, plucking; see *skyllo* under **tear**

scylo- < Gr. *skylon*, spoil, booty, see **plunder**; < *skylos*, skin, hide, see **skin**

scymnus, L. (Gr. *skymnos*), a young animal, cub, whelp; see **young**

scynium, L. (Gr. *skynion*), the skin above the eyes; see **face**

scyphus, L. (Gr. *skyphos*), cup; *skyphion*, dim.; see **cup**

scytala, L. (Gr. *skytale; skytalon*), staff, cudgel; *skytalion; skytalis*, dim.; see **rod**

scytanum, L. mordant; see **bind**

scythe < AS. *sithe;* see **sickle**

scythro- < Gr. *skythros*, angry, sullen; see **anger**

scyto- < Gr. *skytos*, hide, skin, leather; *skytinos*, of leather; see **skin**

scyzinum, L. a kind of wine; see **wine**

scyzo- < Gr. *skyza*, lust; see **lewd**

se- < L. *se*, apart, aside, separation; see **separate**

sea < AS. *sae*.
  Gr. *abyssos*, the deep sea; see **deep**
  ON. *Aegir*, god of the sea: aegerite, aegirinolite.
  L. *aequor*, the calm, smooth sea; see **level**
  L. *aestuarium*, n. firth, inlet, creek: estuary.
  Gr. *anachysis*, f. estuary:
  L. *angulus*, m. bay, gulf:
  Gr. *bythos*, the deep, the depths of the sea; see **deep**
  L. *Cydippe* (Gr. *Kydippe*), f. a Nereid: *Cydippe dimidiata* (a ctenophore),
    Alazonia cydippe (a butterfly).
  L. *Cymodoce* (Gr. *Kymodoke*), f. a Nereid: *Cymodocea manatorum* (manatee-
    grass).
  L. *Doris, -idis* (Gr. *-idos*), f. daughter of Oceanus and wife of Nereus: *Doris
    tuberculata* (a sea-slug), Dorididae, doridoid.
  Sp. *ensenada*, f. creek, bay, cove: Ensenada.
  ON. *fjördhr*, an arm of the sea: fjord, frith, Firth of Forth, Firth of Clyde.
  L. *Galatea* (Gr. *Galateia*), f. a sea-nymph: *Galatea rugosa* (a crab), *Papilio
    galatea* (a butterfly).
  Gr. *hals, halos*, f. sea, salt; *halme*, f. salt water, brine, the sea; *halimos; halios*, of
    the sea, marine; *haliplous*, sailing the sea, seaman; *halirrhytos*, washed by the
    sea; *Halosydne*, f. a name for Amphitrite; *amphialos*, seagirt; *anchialos*, near
    the sea, maritime; *enalios*, in, on, or of the sea; *ephalos*, on the sea; *exalos*,
    out of the sea; *hyphalos*, under the sea, submerged; *paralios; paralos*, by or
    near the sea: halophilous, paralic, *Halobates sobrinus* (a water-strider),
    Halosydna brevisetosa (a scale-worm), *Halimeda gracilis* (an alga), *Haliotis
    rufescens* (red abalone), *Haliplus fasciatus* (a beetle), *Hyphalonedrus chalybeius*
    (a fish), *Anchialus spinosus* (a water-beetle), *Enaliosuchus macrospondylus*
    (a fossil crocodile), *Paralichthys dentatus* (summer-flounder), *Atriplex halimus*
    (orach).
  Gr. *kolpos*, bay, gulf; *kolpodes*, full of bays, sinuous; see **womb**
  Gr. *laitma, -tos*, n. the deep sea: *Laetmatophilus tuberculatus* (an amphipod),
    Laetmogone[*Laetmatogone*] *violacea* (a holothurian).
  L., Gr. *Leucothea*, f. a sea-goddess, sometimes called Ino: *Leucothea formosa* (a
    coelenterate).
  L. *mare, maris*, n. sea, ocean; *marinus; maritimus*, of the sea: marine, marinate,
    maritime, submarine, ultramarine, maricolous, rosemary, *Hordeum marinum*
    (seaside-barley), *Thalarctus maritimus* (polar bear), *Alnus maritima* (seaside-
    alder).   Derive: meerschaum, mermaid, marigold.
  AS. *mere*, pool, lake, sea; see **lake**
  Gr. *mesaktos*, in mid-ocean:
  Gr. *mychos*, bay or creek running far inland; see **room**
  L. *Neptunus*, m. Neptune, god of the sea; *Salacia*, f. wife of Neptune and goddess
    of the sea: neptunium, *Neptunea tabulata* (a gastropod), *Virgula neptunia* (a
    fossil sponge).
  L. *Nereis, -idis* (Gr. *-idos*), f. a sea-nymph, daughter of Nereus, a sea-god;
    Nerinus, of Nereus; *Polynoë*, f. daughter of Nereus: Nereid, neritic, *Nereis
    pelagica* (a sea-worm), *Nerinea dilatata* (a fossil gastropod), *Lumbriconereis
    flabellicola* (an annelid), *Sterna nereis* (a tern), *Polynoë squamata* (a worm).
  L. *oceanus* (Gr. *okeanos*), m. the sea: ocean, oceanid, oceanography, Oceanus,
    Oceanops lativittata (a fish), *Oceanodroma furcata* (a petrel), *Gobionellus
    oceanicus* (emerald-fish).
  L., Gr. *Panope*, f. a sea-nymph: *Panope generosa* (gweduc), *Panopaea americana*
    (a Miocene pelecypod).
  Gr. *pelagos*, n. sea; *pelagikos; pelagios*, of the sea: pelagic, pelagosaur, archipelago,
    Pelagodroma marina (a storm-petrel), *Pelagobia longiciviata* (an annelid),
    Phalacrocorax pelagicus (a cormorant).

L. *Phyllodoce* (Gr. *Phyllodoke*), f. a sea-nymph: *Phyllodoce caerulea* (blue mountain-heath).

L. *pontus* (Gr. *pontos*), m. the open sea, the high sea; *pontios,* of the sea, the Black sea: Hellespont, *Pontobdella muricata* (a leech), *Pontioceramus grandis* (a starfish), *Rhododendron ponticum* (Pontic rhododendron).

L. *Posidon, -is* (Gr. *Poseidon, -os*), m. god of the sea, Neptune; *Amphitrite,* f. wife of Poseidon: *Posidonia oceanica* (a marine plant), *Ornithoptera(Papilio) poseidon* (a butterfly), *Amphitrite ornata* (a marine worm).

L. *salum,* n. the open sea, high sea, main, deep:

L. *sinus,* bay, gulf; see **hollow**

L. *Spio, -us* (Gr. *Speio, -ous*), f. a sea-nymph, daughter of Nereus: Spionidae, *Spio filicornis* (an annelid), *Prionospio cirrifera* (an annelid).

L., Gr. *Tethys, -yos,* f. a sea-goddess, wife of Oceanus: Tethys, *Tethys fimbriata* (a gastropod), *Tethyceramus emigrans* (a fossil pelecypod), *Procellaria tethys* (Galapagos petrel), *Tethya aurantia* (a sponge), *Tethyopsis steinmanni* (a fossil sponge).

Gr. *thalassa (thalatta),* f. sea; *thalassikos; thalassinos; thalassios,* of the sea, like the sea, sea-green; *limnothalassa,* f. salt-marsh estuary: thalassic, thalassophilous, thalassophobia, thalassiophyte, *Thalassophryne maculosa* (a toadfish), *Thalassiosolen atlanticus* (a radiolarian), *Thalattosaurus alexandrae* (a fossil reptile), *Circotettix thalassinus* (a grasshopper).

L. *Thetis, -idis* (Gr. *-idos*), f. a sea-nymph, the sea: *Macropus thetidis* (a wallaby), *Thetidicrinus piriformis* (a Permian crinoid).

L. *Triton, -is* (Gr. *-os*), m. a sea-god who used a conch to control the waves: *Triton variegatum* (a sea-snail), *Tritonia crocata* (saffron tritonia), *Tritonofusus roseus* (a gastropod), *Typhlotriton speleus* (a cave-salamander), *Dolomedes triton* (a diving-spider).

Gr. *zale,* surge of the sea; see **wave**

See: **wave, water, deep**

seal < AS. *seolh,* an aquatic animal; < L. *sigillum,* dim. of *signum,* mark, sign, see **mark**

L. *phoca* (Gr. *phoke*), f. seal: phocacean, phocoid, phocomelus, *Phoca hispida* (a seal).

ON. *rosmhvalr,* walrus: rosmarine, walrus (*Odobenus rosmarus*).

seam < AS. *seam,* suture; see **joint, border, sew, bind**

search < OF. *cerchier,* < L. *circum,* around; see **try, hunt, ask**

season < OF. *seson,* sowing time, < L. *sero, satus,* sow; see **year, salt, spice**

seat < AS. *sittan,* sit; see **sit**

sebastos, Gr. venerable, august; *sebasmos,* reverence; see **honor**

sebenion, Gr. sheath of the palm flower; see **sheath**

sebum, L. tallow, suet, grease, oil; *sebaceus,* fatty, greasy; see **fat**

secale, L. a kind of grain, rye; see **grain**

secamone, NL. a genus of the milkweed family; see **milkweed**

sechium, NL. a genus of the gourd family; see **melon**

seclusus, L. separated, removed; see **separate**

seco- < Gr. *sekos,* pen, fold, shrine; see **pen**

second < L. *secunda,* see **time;** < *secundus,* following, see *sequor* under **follow**

secotero- < Gr. *sekoter, -os,* beam of a balance; see **balance**

secret < OF. *secret,* < L. *secretum,* n. mystery, something hidden, < *secerno, secretus,* separate, set aside: secretive. Explain: *sub rosa.*

L. *abditus,* hidden, concealed, < *abdo, -itus,* put away: abditive, abditory, *Microtus abditus* (a vole).

L. *abstrusus,* hidden, concealed: abstruse.

Gr. *adaëtos,* unknown, unknowing: *Adaëtontherium[Adaëtotherium] incognitum* (a fossil mammal).

Gr. *adelos,* unseen, unknown, obscure: adelite, adelopod, *Adelosina pulchella* (a foraminifer), *Adeloposthia elegans* (a worm).

Gr. *aderkes,* unseen, invisible, unexpected: *Aderces suturalis* (a beetle).

Gr. *adespotos,* anonymous:

L. *aenigma, -atis* (Gr. *ainigma, -tos*), n. something obscure, riddle, mystery; *ainiktos,* baffling, obscure: enigma, enigmatic, aenigmatite, *Oedaleonotus enigma* (valley-grasshopper), *Aenigmatomyia unipuncta* (a fly), *Aenigmatophyllum bipinnatum* (a fossil plant), *Enigmatichthys attenuata* (a fossil fish), *Aenictomorpha variipes* (a beetle), *Aenictomyia chapmani* (a fly), *Araeostoma aenicta* (a moth), *Icosteus enigmaticus* (a ragfish).

Gr. *agnostos,* unknown, unknowing: agnostic, agnosticism, *Agnostus bolivianus* (a trilobite).
Gr. *aïdelos,* unseen, unknown, obscure:
Gr. *aïstos,* unseen: aistopod.
Gr. *alektos,* not to be told, indescribable:
Gr. *anekdotos,* unpublished, kept secret: anecdote.
Gr. *aneuretos,* undiscovered: *Aneuretus simoni* (a wasp).
Gr. *anonymos; nonymos,* nameless, unknown: anonymous, anonym, *Nonyma egregia* (a beetle), *Carpoglyphus anonymus* (a mite).
Gr. *anoptos,* unseen:
Gr. *apeuthes,* unknown, unknowing:
Gr. *aphanes; aphantos,* unseen, invisible, secret, obscure: aphanite, aphanophyre, *Aphananthe aspera* (an ulmacead), *Aphanes arvensis* (a rosacead), *Aphanopteryx bonasia* (a bird), *Aphanopetalum resinosum* (a cunoniacead).
Gr. *aporrhetos,* not to be spoken, secret:
L. *arcanus,* shut up, secret, private, mysterious: arcanum, arcanite, Royal Arcanum.
Gr. *atheatos,* unseen, invisible, secret, blind:
ML. *cabala,* < Heb. *qabbalah,* tradition, secret: cabal, cabalistic, cabala.
L. *clancularius,* secret, anonymous, unknown, < *clanculum,* secretly: *Clanculus pharaonicus* (a gastropod).
L. *clandestinus,* secret, concealed, hidden, < *clam,* secretly, private: clandestine, *Utricularia clandestina* (a bladderwort).
L. *colludo, -lusus,* play together, cooperate with a secret understanding: collusion.
L. *conditus,* hidden, secret; *absconditus,* concealed, secret; see *condo* under **cover**
Gr. *esoterikos,* inside, abstruse: esoteric.
L. *furtivus,* stolen, secret, concealed, clandestine; see *fur* under **steal**
Gr. *hyphalos,* under the sea, secret; see *hals* under **sea**
L. *ignotus,* unknown: *Chirotonetes ignota* (an ephemerid).
L. *incognitus,* unknown; *incognoscibilis,* not to be known, incomprehensible: incognoscible, incognizable, *Aphis incognita* (an aphid).
L. *incompertus,* unknown:
L. *incomprehensibilis,* that cannot be grasped, inconceivable: incomprehensible.
L. *indemonstrabilis,* that cannot be proved: indemonstrable.
L. *indeprensus,* unobserved, undiscovered; *indeprensibilis,* undiscoverable: indeprensible.
L. *ineditus,* not made known, unpublished: *Buprestis inedita* (a beetle).
L. *inenodabilis,* that cannot be freed from knots, inexplicable:
L. *inexcogitabilis,* inconceivable:
L. *inexplicabilis,* unexplainable; *inexplicitus,* unexplained, obscure: inexplicable.
L. *inexploratus,* unexamined, unknown:
L. *innominatus,* nameless; *innominabilis,* unnameable: innominate.
L. *inscrutabilis,* incomprehensible, indemonstrable: inscrutable.
L. *intenibilis,* not to be grasped, intangible: intenible.
L. *invisibilis,* unseen: invisible, invisibility.
Gr. *kryphios,* hidden, secret; *apokryphos,* concealed, obscure, spurious; see *krypto* under **cover**
L. *latens, -entis,* concealed, secret, hidden: latent.
Gr. *lathra* (*lathre*), secretly, covertly; *lathraios; lathridios; lathrios,* secret, hidden, stealthy: *Lathraea squamaria* (toothwort), *Lathrolestes nasoni* (a wasp).
L. *mysterium* (Gr. *mysterion*), n. secret rite; *mystes,* m. one initiated; *mystikos,* secret: mystery, mysterious, mystic, *Sphenoptera mystica* (a beetle).
Gr. *nosphidios,* stealthy, clandestine:
L. *obscurus,* dark, indistinct; see **dim**
L. *occultus,* secret, hidden, reserved: occult, occultation, *Phrixocoma occulta* (a trichopterid).
L. *opertaneus,* concealed, secret; see *operio* under **cover**
L. *silendum,* n. mystery, secret:
L. *surreptitius,* concealed, clandestine: surreptitious.
L. *tacendus,* to be kept silent, secret; see *taceo* under **silent**
L. *tectus,* covered, concealed, secret, disguised; see *tego* under **cover**
See: **cover, magic, wonder, strange, ineffable, ask, silent**

secretio, L. a separation; see **separate**

secta, L. way, mode, manner; see **manner**

sectator; sectatrix, L. follower; see *sequor* under **follow**

sectilis, L. cut; see *seco* under **cut**

secula, L. sickle; see **sickle**

seculum, L. age, race, times, world; see **age**

secundus, L. following, ranking below, unilateral; see *sequor* under **follow**

secure < L. *securus,* free from care, safe; see **safe**

securidaca, L. a genus of the milkwort family; see **milkwort**

securis, L. ax, hatchet; *securicula,* dim.; *securiger,* ax-bearing; see **ax**

secus, L. otherwise, different; see **different**

secutus, L. followed; *secutor; secutrix,* follower; see **follow**

sedatus, L. calm, sober, tranquil; see *sedo* under **soothe**

sedentarius, L. sitting; *sedes,* seat, place, abode; *sedecula,* dim.; see *sedeo* under **sit**

sedge < AS. *secg.*
  L. *anthalium,* n. a kind of sedge:
  Gr. *byblos,* papyrus, a sedge; see *biblion* under **book**
  L. *carex, -icis,* f. sedge; *carectum,* n. a sedgy place: *Carex comosa* (bristly sedge).
  L. *cyperus* (Gr. *kyperos; kypeiros*), m. a sedge: Cyperaceae, *Cyperus alterni-folius* (umbrella-sedge), *Scirpus cyperinus* (a bulrush).
  L. *dulichium,* n. a kind of sedge: *Dulichium arundinaceum* (a sedge).
  L. *euripice* (Gr. *euripike*), f. a kind of rush:
  L. *juncus,* m. rush, bogrush, woodrush; *junceus; juncinus,* of rushes, rushlike; *juncosus,* full of rushes; *ejuncidus,* rushlike, slender: Juncaceae, jonquil, junk, junket, *Juncus nodosus* (jointed rush), *Junco hiemalis* (a snowbird), *Acacia juncifolia* (an acacia), *Spartium junceum* (Spanish broom), *Aster junceus* (swamp-aster).
  NL. *luzula,* f. a genus of the Juncaceae: *Luzula spicata* (a woodrush), *Hemizonia luzulaefolia* (a tarweed).
  L. *mariscus,* m. a kind of rush: *Mariscus californicus* (a saw-sedge), *Cladium mariscoides* (a saw-sedge).
  Gr. *mnasion,* n. a kind of sedge:
  L. *papyrus* (Gr. *papyros*), an Egyptian sedge from which paper was made; see **paper**
  L. *schoenus* (Gr. *schoinos*), m. a rush, rope, cord: *schoinion,* n. dim.; *schoininos,* made of rushes; *schoinis, -idos,* f. cord, rope; *schoiniklos,* m. reed-bunting: *Schoenus ferrugineus* (a sedge), *Schoenocaulon officinale* (sabadilla), *Schoeni-cola brevirostris* (a bird), *Cymbopogon schoenanthus* (a grass), *Emberiza schoeniclus* (reed-bunting), *Acrocephalus schoenobaenus* (sedge-warbler), *Schoeniclus brunneus* (a bird).
  L. *scirpus,* m. rush, bulrush; *scirpeus,* of rushes; *scirpiculus,* made of rushes; *scirpea,* f. basket made of rushes: scirpean, *Scirpus paludosus* (alkali-bulrush), *Eunicea scirpea* (a coral), *Acrocephalus scirpaceus* (a reed-warbler), *Pleospora scirpicola* (a fungus), *Carex scirpoidea* (a sedge).
  Gr. *thryon,* n. rush: *Thryocrex rubra* (a rail), *Thryonomys[Thryomys] semi-palmatus* (a rodent).
  L. *ulva,* sedge; now a genus of seaweeds; see **alga**

sedibilis, L. that can be sat upon; see *sedeo* under **sit**

sediment < L. *sedimentum,* a settling; see **stone, wash**

seditio, L. insurrection, mutiny; *seditiosus,* full of discord; see **rebellion**

seductus, L. remote, apart, see **separate**; *seductilis,* that may be misled, see *duco* under **lead**

sedulus, L. busy, diligent; see **busy**

sedum, L. houseleek; see **stonecrop**

see < AS. *seon.*
  L. *additus,* observing or watching in hostile manner; see *addo* under **and**
  Gr. *atenismos,* m. intent observation:
  Gr. *blepsis,* f. seeing, sight; *blemma, -tos; blepos,* n. look, glance; *bleptikos,* of sight; *bleptos,* worth seeing; *epiblepsis,* f. view; *blepo,* look, see: ablepsia, *Bleptonema paraguayensis* (a fish).
  L. *cernentia,* f. sight, seeing, < *cerno,* distinguish:
  L. *contueor, -utus,* gaze upon, behold, look at attentively; *obtutus, -us,* m. look, stare:
  Gr. *delos,* evident, visible; see **clear**
  Gr. *derkomai,* see clearly; *dergma, -tos,* n. look, glance; *derxis,* f. the sense of sight; *aderktos,* not seeing, sightless; *dysderkes,* seeing badly; *oxyderkes,* sharp-sighted: *Oxyderces coelestinus* (a beetle), *Dysdercus cingulatus* (red cotton-bug).
  L. *emissarius,* scout, spy; see *mitto* under **send**
  Gr. *horao,* see; *horama, -tos,* n. thing seen, object, scene; *horasis,* f. a seeing;

horatikos, able to see; *horatos,* visible; *dioratikos,* clear-sighted: panorama, diorama, *Horatopyga caligata* (a beetle).
L. *inconnivus,* that does not close the eyes, sleepless; see **awake**
Gr. *korykaios,* m. spy: *Corycaeus crassiusculus* (a cyclops).
Gr. *mantis,* seer, diviner, prophet; see **prophecy**
L. *observo, -atus,* pay attention to, notice, watch, regard: observe, observation, observatory.
Gr. *opipes; opipeuter, -os,* m. gaper, starer; *arrhenopipes,* m. one who looks lewdly on males; *gynaikopipes,* m. one who looks lewdly on females:
Gr. *optikos; -opsia; -opsis; -opsy; -opy; -opia,* pertaining to sight, vision, < *opsis, -eos,* f. sight, appearance, look, face, aspect, likeness; *apopsis, -eos,* f. outlook, view; *synopsis, -eos,* f. general view, summary; *diopter; katopter; opter, -os,* m. eyewitness, scout, optical instrument; *euops, -opos,* of good appearance; *myops, -opos,* near-sighted: optical, optics, optometrist, myopia, synopsis, autopsy, dioptrics, diopside, dioptase, *Dioptrornis fischeri* (a bird), *Coreopsis palmata* (a tickseed), *Lycopsis arvensis* (small bugloss), *Amblyopsis spelaeus* (a Mammoth Cave blindfish), *Myopias amblyops* (an ant), *Myopopone maculata* (an ant), *Telopea speciosissima* (a waratah).
Gr. *parakypto,* stoop for the purpose of looking, peep:
L. *pareo, -itus,* appear, become visible, evident, manifest; *appareo,* come in sight: appear, appearance, disappear, apparent, transparent, peer, apparition.
Gr. *phantos,* visible; see **clear**
Gr. *skopos,* m. watcher, mark, range of view, < *skopeo,* look, behold, contemplate; *skope; skopia,* f.; *skopelos,* m. observation point, watch-tower, peak: scope, telescope, microscope, stereoscope, electroscope, scopograph, bishop, San Luis Obispo, Episcopal, Gillespie, *Astroscopus guttatus* (stargazer fish), *Euryscopa pulchella* (a beetle), *Dasyscopelus spinosus* (a fish).
L. *specio; spicio, spectus,* behold, look; *specto, -atus,* look at, behold, observe; *spectator, -is,* m.; *spectatrix, -icis,* f. beholder, observer; *specula,* f. observation point; *speculator, -is,* m. explorer, searcher; *speciosus,* showy, beautiful; *spectabilis,* visible, remarkable; *aspectus, -us,* m. look, appearance; *circumspectus,* looking cautiously; *conspicabilis; conspicuus; conspectus,* visible, striking; *inspector, -is,* m. observer, examiner; *perspicax, -acis,* sharp-sighted, keen; *prospector, -is,* m. foreseer, investigator: spectator, spectacle, species, specialist, specialty, specimen, specific, specter, speculation, spice, spy, spectrum, aspect, auspicious, especially, espionage, frontispiece, despise, expectancy, circumspect, conspicuous, suspicion, inspector, despicable, respectable, respite, perspicacity, *Campanula speciosa* (a bellflower), *Sedum spectabile* (showy stonecrop), *Rubus spectabilis* (salmon-berry), *Specularia perfoliata* (Venus looking-glass).
Gr. *theaomai,* behold, view; *theates,* m. spectator; *theatikos,* for seeing: *Theatops postica* (a centiped), theater.
Gr. *theoreo,* look at, view; *theorema, -tos,* n. sight, spectacle, speculation; *theoretes,* m. spectator: theorem, theory.
L. *video, visus,* see; *visibilis,* that may be seen; *visio, -onis,* f.; *visum,* n. something seen, sight; *visualis,* of sight: vision, visual, visibility, visage, visit, vidette, vizor, q.v., viz., advisable, Buena Vista, video, Montevideo, review, invisible, envy, clairvoyant, supervisor, invidious, survey, evident, providential, purveyor, improvise, Belvidere.
AS. *waeccan,* watch: watch, watchful, watchman, watchword.
See: **open, clear, light, form, witness**

seed < AS. *saed:* seeder, seedling, seedtime, tickseed.
Gr. *gigarton,* n. grape seed; *agigartos,* seedless: *Gigartina mammillosa* (an alga).
Gr. *gone,* f.; *gonos,* m. that which is begotten, that which produces seed, seed, young; *gonion; gonidion,* n. dim.; *gonikos,* of seed; *polygonos,* prolific: gonad, gonophore, gonangium, gonadectomy, gonidial, gonorrhea, archegonium, epigonium, *Mycogone rosea* (a fungus), *Gonactinia prolifera* (a sea-anemone).
L. *granum,* small kernel, seed; see **grain**
Gr. *kenchramis, -idos,* f. fig seed:
Gr. *kokkalos,* seed from a cone; *kokkos,* seed, grain, berry; *kokkion,* dim.; see **berry**
Gr. *pyren,* pit or hard seed; see **nut**
L. *samara,* seed of the elm; any dry, indehiscent, winged fruit, key; see **fruit**
L. *semen, -inis,* n. seed; *seminulum,* n. dim.; *seminalis; seminarius,* of seed; *seminarium,* n. seed-plot, nursery; *semino, -atus,* sow, seed, beget; *sementis,* f. a seeding, sowing; *sementivus,* of sowing; *consemineus,* sown with several kinds of seed; *dissemino,* sow, scatter; *insemino,* implant, impregnate: semen, seminule, seminal, seminary, disseminate, inseminate, seminiferous, *Seminula argentea* (a fossil brachiopod), *Miliolina seminulum* (a foraminifer), *Serpula seminulum* (a marine worm).

Gr. *sperma, -tos,* n. seed; *spermation,* n. dim.; *spartos,* grown from seed, cultivated; *spora,* f.; *sporos,* m. a sowing, the seed sown: sperm, spermatozoon, spermophile, spermatic, spermatium, spermatogenesis, gymnosperm, spore, sporophyte, sporadic, diaspore, sporiferous, teleutospore, teliospore, *Spermatolonchaea minuscula* (a fly), *Spermatozopsis exsultans* (a protozoan), *Lithospermum arvense* (corn-gromwell), *Aspidosperma quebracho* (quebracho), *Atherosperma moschatum* (an Australian tree), *Spermestes cucullata* (a finch), *Dolichos sphaerospermus* (a bean called black-eyed pea), *Cylindrosporium acicolum* (a fungus), *Disporum maculatum* (spotted fairy-bells), *Rhynchospora stipitata* (a sedge), *Pityrosporum ovale* (bottle-bacillus), *Sporobolus microspermus* (a dropseed), *Cladosporium herbarum* (a fungus), *Helminthosporium sativum* (a fungus).

Gr. *thoros,* m. seed of the male, semen; *thorikos,* of semen: *Cistothorus stellaris* (marsh-wren).

See: **fruit, egg, young, grain, sand, sow, spread, begin**

seek < AS. *secan;* see **hunt, try, ask, woo**

seem < uncertain origin; see **like, lot, see, dream**

seges, *-etis,* L. grainfield, standing grain, crop; *segetalis,* of standing grain; see **field**

segmentum, L. piece, cutting, shred; see **part**

segnis, L. slow, slothful, dilatory; see **slow**

segrego, *-atus,* L. set apart, separate; *segrex; segregus,* apart; see **separate**

segutilum (segullum), L. earth supposed to indicate the presence of gold; see **earth**

seismos, Gr. earthquake; see **shake**

seize < OF. *seisir;* see **take, hold**

sejunctus, L. disjoined, separated; see **separate**

selachos, Gr. a cartilaginous fish, shark, ray; see **shark**

selago, L. a kind of clubmoss; see **clubmoss**

selas, Gr. light, flash, meteor; *selasma,* a shining; see **light**

selates, Gr. snail; see **mollusk**

seldom < AS. *seldum;* see **few, little, number**

select < L. *seligo, selectus,* choose; see **choose, sieve, judge, separate**

selenium < Gr. *selene,* moon; see ELEMENTS under **thing**

seleno- Gr. *selene,* moon; *selenion,* dim. moonlight, peony; *selenites,* moonstone, gypsum; see **moon, peony, gypsum**

seleucido- < Gr. *seleukis, -idos,* a kind of woman's shoe from Seleucia; a bird; see **shoe**

self < AS. *self;* see **alone**

selfish < AS. *self,* and *-ish.*
   L. *cupiditas,* avarice, lust; see *cupido* under **wish**
   L. *dissocialis,* selfish, unsocial; see **alone**
   Gr. *dysareskos,* unaccommodating:
   Gr. *philautos,* loving oneself, selfish:
   See: **stingy, alone**

selia, Gr. Attic for *telia,* gaming table with raised border; see **table**

selido- < Gr. *selis, -idos,* plank, leaf of papyrus, page of a book, sheet of paper; see **paper**

selinum, L. (Gr. *selinon*), parsley; see **carrot**

selion, Gr. small vessel used by bakers; see **vessel**

sell < AS. *sellan;* see **trade**

sella, L. seat, chair, stool, saddle; *sellula,* dim.; *sellaris; sellularius,* of a chair; see **sit**

selma, Gr. deck, floor; see **base**

selmido- < Gr. *selmis, -idos,* noose; see **trap**

sema, Gr. sign, token, seal, mark; *semation,* dim.; *semantikos,* significant, indicative; see **mark**

sembella, L. a coin; see **money**

semel, L. once; see **one**

semen, L. seed; *seminulum,* dim.; *seminalis; seminarius,* of seed; see **seed**

semestris, L. semiannual; see **six**

semi- < L. *semis,* a half; see **half**

seminulum, L. dim. of *semen,* seed; see **seed**

semio- < Gr. *semeion,* sign, signal, flag, standard; see *sema* under **mark**

semita, L. footpath; *semitalis,* of footpaths; see **way**

semnos, Gr. grave, solemn, august, sacred; see **serious**

semotus, L. distant, removed; see **far**

semper, L. always; see **always**

semyda, Gr. probably a birch; see **birch**

senatus, L. senate; see **gather**

send < AS. *sendan.*
> L. *angarius* (Gr. *angaros*), mounted courier, messenger; see **ride**
> L. *angelus* (Gr. *angelos*), m. messenger, envoy; *angeleia,* f. message, tidings, news; *angelikos,* of messengers: angel, angelic, evangelist, Evangeline, evangelical, Angelina, Los Angeles, *Angelica pinnata* (an umbellifer), *Heliangelus micraster* (a hummingbird).
> L. *commeator, -is,* m. messenger:
> Gr. *diadekter, -os; diadektes,* m. transmitter: *Diadectes latibuccatus* (a fossil reptile).
> Gr. *hiemi,* send, let go; *hesis,* f. a sending forth, impulse; *hesia,* f. mission: aphesis, aphetize, enema, catheter, paresis.
> Gr. *ialtos,* sent forth:
> L., Gr. *Iris,* female messenger of the gods; see **color, iris**
> L. *lego, -atus,* send, appoint, bequeath: legatee, legacy, allegation, delegate, relegate, college, colleague.
> L. *mitto, missus,* send, quit; *amitto,* send away, dismiss, lose; *emissarius,* m. messenger, spy: missile, remiss, commission, remittance, omit, transmit, mission, missal, missionary, emissary, commissary, demise, dismissal, promise, compromise, surmise, message, mass, mess, committee, premise, permit, emit, inadmissable, promissory, recommittal, submissive, intermittently.
> Gr. *pempo,* send; *pempsis,* f. mission; *pemptos,* sent: *Apopempsis pulchra* (a beetle).
> Gr. *stello,* place, arrange, equip, send; *apostolos,* m. messenger, ambassador: apostle, epistle, apostolic.
> See: **throw**

senecio, L. groundsel; see **composite**

senex; senectus; senilis, senicus, L. old; *senior,* older; see **old**

senna < Ar. *sana,* a species of cassia; see *cassia* under **bean**

sense < L. *sentio, sensus,* perceive, feel, experience, think; *sensibilis,* perceptible; see **feel**

sentence < L. *sententia,* f. opinion, judgment; *sententiosus,* full of meaning: sententious.
> Gr. *apokrima, -tos,* n. judicial sentence, condemnation:
> L. *decretum,* decision, ordinance; see *cerno* under **separate**
> Gr. *gnome,* opinion, judgment; see *nosco* under **know**
> L. *irrogo, -atus,* impose, ordain, inflict:
> L. *judicium,* decision, opinion, sentence; see **judge**
> Gr. *krisis,* decision, issue, judgment; see *crisis* under **turn**

sentina, L. bilge-water; see **water**

sentis, L. thorn, brier, bramble; *sentus; senticosus; sentosus,* thorny; see **thorn**

sentix, L. dogrose; see **rose**

seorsus, L. apart, separate, several; see **separate**

sepal < NL. *sepalum,* < Gr. *skepe,* covering; see **leaf**

separate < L. *separo, -atus,* sever, divide, part: separation, separatist, *Hydropsyche separata* (a trichopterid).
> L. *abjunctus; disjunctus; sejunctus,* unyoked, disunited, separated; *sejugis,* unyoked: disjunctive, disjoined.
> Gr. *aloao,* thresh; *aloesis,* f.; *aloetos,* m. a threshing.
> Gr. *analysis,* f. resolution of a whole into its parts; *analytikos,* of analysis; *dialysis,* f. dissolution, separation: analysis, analytical, dialysis.   See *lyo* under **free**
> L. *anatomia* (Gr. *anatome*), f. a cutting up, dissection: anatomy, anatomical.
> Gr. *apahrizo,* skim off froth; *exaphrismos,* m. a foaming off:
> Gr. *apidiastikos,* retired, secluded:

Gr. *apo*, from, separate; *apodasmios*, separated; *apodasmos*, m. a division; *apodastos*, separated, apportioned; *apospastos*, separated: *Apodasmia rufonigraria* (a moth).

L. *articulatus*, jointed, distinct; see *articulus* under **joint**

Gr. *asynaptos*, unconnected; *asynchytos*, not confused, unmixed; *asyndetos*, unconnected, loose;

L. *avello, avulsus*, tear away, separate: avulsion.

Gr. *azyx, -ygos*, unyoked, unpaired:

Gr. *brasso*, shake violently, winnow grain; see **shake**

L. *cerno, cretus*, sift, separate, distinguish; *decerno*, decide, judge; *decretorius*, decisive; *decretum*, n. decision; *discretus*, separated; *secretus*, separated, isolated: decree, decretal, decretory, discern, discretion, excrement, excretory, indiscreet, secretary, secretion, secret, *Metriophilus discretus* (a beetle).

Gr. *choris*, apart, asunder; *chorisis*, f.; *chorismos*, m. separation; *choristes*, m. separator; *choristikos*, separative; *choristos*, separated, < *chorizo*, separate, distinguish: choripetalous, chorisis, chorism, *Chorizema ilicifolium* (holly-flamepea), *Chorisepalum carnosum* (a gentianacead), *Chorizomma sequoiae* (a spider).

AS. *daelan*, apportion, distribute, divide, parcel out; *dael*, part, portion, share: deal, ordeal.

Gr. *daithmos*, division, boundary; see *daio* under **cut**

L. *despumo, -atus*, skim; see *spuma* under **foam**

Gr. *dia*, through, between, asunder, as in *diadelos*, distinguishable; *diaireo*, divide, separate; *diairesis*, f. division, separation; *diairetes*, m. divider; *diairetos*, divided, divisible; *diastatikos*, separable, distinctive; *diastatos*, divided; *diedros*, sitting apart, separated: *Diadelia biplagiata* (a beetle), diaeresis, diastatic, diastasis, diastase, *Diedrus areolatus* (a wasp), *Diedronotus rosulentus* (a grasshopper).

Gr. *dicha*, in two, apart, asunder; see *di-* under **two**

L. *dis (di-)*, asunder, separate, as in *diduco, -uctus*, separate; *digero, -estus*, separate, classify, render food assimilable; *dimitto, -missus*, send apart, dissolve; *diribeo, -itus*, lay apart, separate, sort; *diribitor, -is*, m. separator, sorter; *dirimo, -emptus*, take apart, separate, divide; *discapedino*, hold the hands apart; *discidium*, n. separation, disagreement, divorce; *discrimino, -atus*, separate, divide; *disruptus*, broken up, separated; *dissitus*, apart, remote; *dissolvo, -solutus*, disunite, separate; *dissuo, -utus*, rip open; *disterminus*, separated, divided; *distinguo, -stinctus*, separate, discriminate; *disto, -atus*, stand apart, be separate, differ; *divarico, -atus*, spread apart, spread out, fork; *dividus*, separated; *divortium*, n. separation, dissolution of marriage: diductor, diduction, digest, dimission, dimissary, diremption, discrimination, indiscriminate, disjunct, disparate, disruption, dissolve, dissolution, distinguish, distinct, distinction, distant, distal, dividual, dividuity, dividuous, divide, divorce, *Ceramoporella distincta* (a fossil bryozoan), *Deltocephalus distinguendus* (a bug), *Lacuna divaricata* (a gastropod).

Gr. *ekkrima, -tos*, n.; *ekkrisis*, f. separation, secretion: eccresis, eccritic.

Gr. *ekpales*, out of joint; *ekpalesis*, f. dislocation:

L. *enucleo, -atus*, remove the kernels, clear of husks; see *nux* under **nut**

Gr. *eukrines*, clear, distinct, well-separated; see **clear**

L. *excerptum*, extract, selection; see **choose**

L. *extractus*, drawn out; see *traho* under **draw**

L. *factiosus*, forming parties, usually given to contention: factious.

Gr. *hyphen*, f. a punctuation mark: hypen, hyphenate.

L. *incommiscibilis*, that cannot be mixed: incommiscible.

L. *inconnexus*, unjoined:

L. *interstinctus*, marked off, separated: *Heliolites interstinctus* (a Silurian coral).

NL. *isolatus*, detached, separate; see *insula* under **island**

Gr. *krino*, judge, estimate, discern, distinguish; *kritos*, separated; *diakritikos*, separating, distinguishing; see *criticus* under **judge**

Gr. *likmao*, winnow; *likmesis*, f. a winnowing; *likmeter, -os*; *likmetes*, m. winnower; *likmetikos*, of winnowing:

L. *luxo, -atus*, put out of joint, dislocate; *luxatura*, f. dislocation; *luxus*, dislocated: luxate, luxation.

Gr. *merismos*, division; *meristos*, divided, divisible; see *meros* under **part**

L. *particularis*, of parts, partial; see **choose, part**

L. *partitus*, divided: partitive, partition.

L. *privo, -atus*, rob, strip, separate: deprive, privation.

Gr. *ptisso*, winnow grain; *ptismos*, m. a winnowing; *ptistes*, m. winnower; *ptistikos*, of winnowing: *Ptistes coccineopterus* (a bird).

L. *reclusus; seclusus,* shut up, separated, removed: recluse, seclude, seclusion, *Batyle seclusa* (a beetle).

L. *remotus,* distant; see **far**

L. *repudio, -atus,* cast off, reject, divorce: repudiate, repudiation.

AS. *sundrig,* separate, several; *sundrian,* separate, sever: sunder, asunder, sundry.

L. *se,* apart, aside, separation: seclude, separate, secure, seduce, sedition, segregate, secrete, solution, sober, saunter, sure.

L. *secretio, -onis,* f. separation: secretion.

L. *segrego, -atus,* set apart, separate; *segrex, -egis,* apart, separate: segregate, segregation, *Paryphostomum segregatum* (a worm).

L. *seorsus,* severed, apart, separate:

L. *sepono, -positus,* set aside, separate, select, reserve, hoard; *sepositus,* apart, distinct, special, remote: sepose, seposition, *Echinaster sepositus* (a starfish).

L. *sequestro, -atus,* set apart, remove, separate; *sequester, -tris,* m. a trustee: sequester, sequestrate, sequestrum.

L. *tribulo, -atus,* thresh, afflict, oppress; *tribulum,* n. threshing flail: tribulate, tribulation.

See: **banish, break, choose, cut, different, free, from, screen, sieve, steal, wall, wean, alone,** *di-* under **two**

sepedon, Gr. rottenness, decay, see *sapros* under **rot**; a kind of snake, see **snake**

sepia, L., Gr. cuttlefish, squid, ink; *sepiola,* dim.; *sepion,* cuttlebone; see **mollusk**

sepicula, L. dim. of *sepes,* hedge, fence; see *septum* under **wall**

sepimentum, L. fence, hedge; see **wall**

sepion, Gr. internal shell or proostracum of the squid; see *sepia* under **mollusk**

sepo- < Gr. *seps, sepos,* a serpent, lizard, see **lizard**; a putrefying sore, see *sapros* under **rot**

sepositus, L. apart, special, distinct, remote; see **separate**

sepsis, Gr. decay, putrefaction, poisoning from bacterial action; *septikos,* putrefactive; *septos,* rotten; see *sapros* under **rot**

septem, L. seven; *septimus,* seventh; see **seven**

septentrionalis, L. north, northern; see **north**

septum, L. hedge, fence, partition; see **wall**

septuosus, L. obscure; see **dim**

sepulcrum, L. grave, tomb; *sepultus,* buried; see **grave**

sequax, -acis, L. following, follower, attendant; see **follow**

sequence < L. *sequentia,* consecutive continuity, orderly succession; *sequela,* that which follows; see *sequor* under **follow**

sequestro, -atus, L. set apart, remove, separate; see **separate**

sequoia, NL. a genus of conifers; see **gymnosperm**

sera, L. bar for fastening doors; see **bar**

seranx, Gr. cave, hollow; *serangion,* dim.; see **hollow**

serenus, L. clear, bright; *serenitas,* clearness, calmness; see **clear**

seria, L. large, earthen jar; *seriola,* dim.; see **pot**

sericum, L. (Gr. *serikon*), silk; *sericus; sericeus,* silken; see **silk**

series, L. row, succession, train; see **line**

serilium, L. rope, cordage; see **rope**

serinus, NL. a finch; see **finch**

serious < L. *serius,* grave, earnest.

Gr. *agelastos,* grave, gloomy, not laughing: *Agelastus meleagrides* (a bird).

Gr. *asketikos,* given to rigorous discipline, austere; see *askeo* under **make**

L. *augustus,* venerable, majestic; see **honor**

L. *austerus* (Gr. *austeros*), stern, harsh, gloomy, hard: austere, austerity.

L. *gravis,* heavy, severe, serious; see **heavy**

Gr. *semnos,* grave, solemn, august, sacred: *Semnopithecus entellus* (hanuman or sacred monkey).

L. *severus,* harsh, stern, strict; see **rough**

L. *sobrius,* not drunk, moderate, temperate, cautious; see **sober**

L. *solemnis,* stated, religious, serious: solemn, solemnity.

L. *sonticus,* serious, dangerous, critical, < *sons, sontis,* guilty; m. criminal:

L. *strictus,* drawn together, tight, close, severe; see *stringo* under **bind**

L. *tetricus,* forbidding, gloomy, stern; see *teter* under **bad**

L. *torvus,* wild, gloomy, severe, grim; see *torvidus* under **rough**

L. *tragicus* (Gr. *tragikos*), pertaining to tragedy; *tragoedia* (Gr. *tragoidia*), f. a serious or heroic drama: tragedy, tragic.
See: **think, sad**

seriphos, Gr. a kind of wormwood; see **wormwood**

seris, L. a kind of endive; see **lettuce**

serius, L. grave, earnest; see **serious**

sermo, L. talk, conversation, discourse; see **speak**

sero, *sertus*, L. join, knit, connect, see **bind**; *sero, satus*, sow, see *sativus* under **sow**

serotinus, L. happening late; see *serus* under **after**

serpent < L. *serpens, -entis*, snake, < *serpo, serptus*, creep; see **snake**

serphos, Gr. a small, winged insect; see **wasp**

serpula, L. little snake; see *serpens* under **snake**

serpyllum, L. thyme; see **mint**

serra, L. saw; *serrula*, dim.; *serrago*, sawdust; see **saw, dust**

serraculum, L. rudder; see **rudder**

serranus, NL. a genus of perchlike fishes; see **perch**

serratula, L. betony, now a genus of the composite family; see **composite**

sertula, L. dim. of *serta*, garland, wreath; see **branch**

serum, L. a thin, watery fluid, whey of curdled milk; see **juice**

serus, L. late; *senior*, later; *serotinus*, happening late; see **after**

servant < L. *servio, -itus*, serve, assist; *servus*, slavish, subject; m. male slave; *serva*, f. female slave or servant; *servulus*, m.; *servula*, f. dim.; *servitor, -is*, m. servant; *servilis*, slavish; *servitudo, -inis*, f. slavery: serve, service, serf, deserve, dessert, sergeant, servitude, subservient.

L. *accensus*, m. attendant, servant:

L. *agaso, -onis*, m. groom, stable-boy, driver, jockey, lackey:

Gr. *aletris*, female slave who ground grain; see *aleo* under **grind**

L. *alipilus*, m. slave who plucked hair from the armpits of bathers:

L. *amanuensis*, m. clerk, secretary: amanuensis.

L. *ambactus*, m. vassal, dependent: ambassador, embassy.

Gr. *amorbos*, m. follower, attendant: *Amorbus alternatus* (a bug).

L. *ancus*, m. servant; *anculus*, m. manservant; *ancula*, f. maidservant; *ancilla*; *ancillula*, f. dim.; *ancillaris*, of maidservants, subservient; *ancillor, -atus*, serve, attend upon: *Ancilla subulata* (a fossil gastropod).

Gr. *andrapodon*, n. slave; *andrapodistes*, m. slave-dealer; *andrapodes*, slavish, servile:

L. *apothecarius*, m. warehouse-man, clerk: apothecary.

L. *apparitor, -is*, m. public servant; *apparitura*, f. service: apparitor.

L. *aquariolus*, m. an attendant of lewd women:

L. *assecla*, c. attendant, sycophant; *assectator, -is*, follower, attendant; *sectator, -is*, m. follower, attendant, adherent:

Gr. *atmen, -os*, m. slave, servant:

Gr. *aulites*, m. farm-servant:

Gr. *bolize*, f. female slave:

L. *cacula*, m. servant, particularly servant of a soldier:

L. *calator, -is*, m. servant, attendant, especially of priests; see *calo* under **call**

L. *clericus* (Gr. *klerikos*), m. priest: clerk, clergyman, Clark, *Scarabaeus clericus* (a dung-beetle).

L. *cliens, -entis*, m. dependent upon a patron, protege, vassal, retainer, customer: client, clientele.

L. *conservus*, m.; *conserva*, f. fellow-slave:

L. *cubicularius*, m. valet:

L. *dapifer*, waiter, steward; see *daps* under **eat**

L. *diaconus* (Gr. *diakonos*), m. servant, minister of the church; *diaconicus*, of a deacon: deacon, deaconess.

Gr. *dmos*; *dmoë*, slave taken in war; *dmios*, servile; see *dmetos* under **tame**

L. *domiseda*, f. a domestic:

Gr. *doulos*, m. bondman, slave; *doule*, f. bondwoman; *doularion*, n. dim.; *doulikos*; *doulios*, servile, slavish; *douleia*, f. servitude: dulosis, dulia.

Gr. *erithos*, m. hired servant: *Erithophilus neopus* (a centiped).

L. *famulus*, serving, servile; m. servant; *famula*, f. servant; *famularis*, of servants; *famulatorius*, slavish; *famulor, -atus*, serve, attend:

L. *focaria*, f. kitchen-maid, housekeeper:

Gr. *habra*, f. favorite slave: *Habra securifera* (a grasshopper).

Gr. *Harpalus*, m. one of Cicero's slaves: *Harpalus hirsutus* (a beetle).

L. *Helotes;* *Hilota* (Gr. *Heilotes*), m. Spartan serf: helot.
Gr. *hiereus,* priest; *hiereia,* priestess; see *hieros* under **holy**
Gr. *hippokomos,* m. groom, squire, attendant to a horseman:
Gr. *hyperetes,* m.; *-etis,* f. underling, servant, menial:
Gr. *hypochos,* subject to, liable to, subordinate to:
L. *infertor, -is,* m. waiter, steward:
L. *latreus; latreutes,* m.; *latris, -ios,* c. hired servant; *latria* (Gr. *latreia*), f. service, worship: latria, *Latreutes ensiformis* (a crustacean), *Latris ciliaris* (a fish).
L. *lictor, -is,* m. attendant to a Roman magistrate; *lictorius,* of a lictor: lictor, *Salebius lictor* (a beetle).
L. *liturgus* (Gr. *litourgos*), m. public servant, minister: liturgy, liturgical.
L. *mancipium,* n. slave by purchase or prisoner of war: mancipium, mancipate, *Mancipium hellica* (a butterfly).
L. *mediastinus,* m. servant, drudge, menial, intermediary: mediastinum.
L. *minister, -tri,* m. attendant, servant, waiter; *minister, -tra, -trum,* serving, helping; *ministro, -atus,* attend, serve: minister, ministerial, minstrel, administer.
L. *munifex, -icis,* m. one on duty; *munium,* n. duty, function: mean, municipal, immunity.
L. *officialis,* pertaining to duty, office, or service; *officium,* n. service, duty, favor; *officiosus,* obliging; see **make**
Gr. *oiketes; oiketis,* house-servant, menial, housewife; *oiketikos,* pertaining to servants; see *oikos* under **house**
Gr. *opados,* accompanying, attending; *opadeter, -os,* m. attendant: *Opadothrips fritschianus* (a fossil thysanoterid).
Gr. *paratiltria,* f. female slave who depilated her mistress:
L. *pedisequus,* m. lackey, footman:
Gr. *pelates,* approacher, hireling; see *pelazo* under **come**
Gr. *penes, -etos,* servant, poor man; see *poieo* under **make**
L. *pincerna,* f. cupbearer, butler: *Pincerna liratula* (a gastropod).
L. *proriga,* f. stable-boy:
L. *satelles,* guard, companion, attendant; see **companion**
L. *stipator, -is,* m.; *stipatrix, -icis,* f. attendant:
L. *strator, -is,* m. groom, hostler:
L. *subjectus,* brought under: subject, subjective, subjection.
AS. *thegn,* servant to a lord, warrior: thane, thaneland, thaneship.
Gr. *therapeuo,* attend, serve, give medical treatment; *therapon, -os; theraps, -apos; therapeutes,* m. male attendant; *therapne,* f. female attendant, hand-maid: therapy, therapeutic, *Theraps irregularis* (a fish), Therapsida.
Gr. *thes, -etos,* hired man; *thessa,* f. hired girl; *thetikos,* of a hireling, menial: *Thetomys nanus* (a mouse).
ON. *thraell,* serf: thrall, thraldom, enthrall.
Gr. *threptos,* servant; see *trepho* under **eat**
LL. *vassus,* < C. *gwaz,* servant: vassal, vassalage, valet, varlet.
L. *verna,* c. home-born slave; *vernula,* f. dim.; *vernaculus,* of home-born slaves; *vernilis,* cringing, fawning, servile: vernacular.
See: **make, guard, help, rite**

servo, *-atus,* L. keep, preserve; *servator; servatrix,* preserver; see **save**

ses, *seos; setos,* Gr. moth; see **moth**

sesame < Gr. *sesame,* f. a plant with oily seeds; L. *sesamum* (Gr. *sesamon*), n. the seed of sesame; *sesamion,* n. dim.; *sesaminos,* of sesame: *Sesamum indicum* (sesame), sesamoid, *Sesamothamnus benguellensis* (a pedaliacead).

sesbania, NL. a genus of the bean family; see **bean**

seselis, Gr. an unknown plant; now a genus of the carrot family; see **carrot**

seserinos, Gr. a sea-fish; see **fish**

sesilos, Gr. a snail; see **mollusk**

sesqui, L. one and one-half; *sesquipedalis,* one and one-half feet long; see *semi-* under **half**

sessilis, L. sitting; see *sedeo* under **sit**

sestertius, L. two and one-half; see *semi-* under **half**

sestron, Gr. sieve; see **sieve**

set < AS. *settan,* put, place, cause to sit, go down; see **place, sit, down, west**

seta, L. bristle; *setula,* dim.; *setiger; setosus,* bristly; see **hair**

setania, L. a kind of medlar; see **apple**

setanios, Gr. this year's; see **year**

setho, Gr. sift, sieve; see **sieve**

seto- < Gr. *ses, seos; setos,* moth; see **moth**

setodocido- < Gr. *setodokis, -idos,* a butterfly; see **butterfly**

settle < AS. *setlan;* see **sit, down, agreeable, yes**

seven < AS. *seofon.*
>   Gr. *hepta,* n. seven; *hebdomas, -ados; heptas, -ados,* f. the number seven, a group of seven, a week; *hebdomatos; hebdomos*(for *heptomos*), seventh; *heptakis,* seven times; *heptaplasios,* sevenfold: heptagonal, heptane, heptene, heptahydric, Heptateuch, hebdomad, heptadecane, heptarchy, heptamerous, *Heptathela kimurai* (a spider), *Heptasogaster stultus* (a mallophagid), *Protium heptaphyllum* (a burseracead).
>   L. *septem,* seven; *septenarius,* consisting of seven; *septimus,* seventh; *septifarius; septiformis; septemplex,* sevenfold; *septeni,* seven each; *septendecim,* seventeen; *septuaginta,* seventy; *septigenti,* seven hundred: septamerous, septentrional, September, Septuagint, *Tibicen septendecim* (seventeen-year cicada, 'locust'), *Eriocaulon septangulare* (a pipewort).

severus, L. harsh, sharp, stern, rough, rigorous; *severitas,* harshness; see **rough**

sew < AS. *seowian.*
>   Gr. *akestria,* seamstress; see *akestra* under **needle**
>   Gr. *epetes,* m. mender; *epesis,* f. a mending:
>   Gr. *rhapto,* sew, stitch, seam; *rhaphe,* f.; *rhamma, -tos,* n. seam, suture; *rhapheus; rhaphideus; rhaphideutes; rhaptes,* m. sewer, stitcher, cobbler; *rhaptos,* sewn, stitched; *syrrhaptos,* sewn or stitched together: rhaphe, rhapsody, *Rhammatopoda opilionoides* (a katydid), *Syrrhaptes tibetanus* (a bird).
>   L. *sartor, -is,* m.; *sartrix, -icis,* f. tailor, mender ,< *sarcio, sartus,* mend; *sarcinator, -is,* m.; *sarcinatrix, -icis,* f. mender; *sarcimen, -inis,* n. seam, suture; *sarsurius,* mending, patching; *consarcino, -atus,* patch together: sartorial, sartorius, *Sartor resartus* (by Thomas Carlyle), *Hilara sartor* (a fly), *Hyracotherium resartum* (an eohippus).
>   L. *suo, sutus,* sew; *sutor, -is,* m.; *sutrix, -icis,* f. sewer; *sutura,* f. seam; *sutilis,* sewed together; *consutilis,* sewed together: suture, sutural, *Sutoria longicauda* (tailorbird), *Dysdercus suturellus* (cotton-stainer), *Peliocypas suturalis* (a beetle), *Hemirhopalum suturale* (a beetle).
>   OF. *tailleor,* < *taillier,* cut: tailor, tailorbird, tailor-made, Taylor.
>   L. *vestitor, -is,* m. tailor:
>   See: **bind, needle**

sewer < OF. *sewiere;* see **ditch, pipe**

sex, L. six; *senarius,* of six; *sextus,* sixth; *sedecim,* sixteen; see **six**

sex organs < L. *sexus, -us,* m.; *secus,* n. male or female sex, < *seco,* cut, divide; *sexualis,* of sex: sex, sexual, bisexual, sexuality, *Cylloepus sexualis* (a beetle).
>   Gr. *aidoion,* n. private part, genital: aedeagus, edeology.
>   L. *genitalis,* pertaining to procreation: genital, genitalia.
>   Gr. *medea,* pl. of *medos,* n. genitals:
>   Gr. *morion,* piece, part, member; see *meros* under **part**
>   L. *muliebria* (pl.), n. female pudenda:
>   L. *obscenum,* n. private part:
>   L. *pubes, -is,* f. private parts: pubic.
>   L. *pudenda* (pl.), n. external genitals, particularly female: pudenda, pudendal.
>   L. *veretrum,* n. private parts; *veretillum,* n. dim.: *Veretillum cynomorum* (an alcyonarian).
>   L. *virilia* (pl.), n. male genitals:
>   See: **ovary, penis, testicle, vulva, womb**

shackle < AS. *scacul;* see **bracelet, collar, bind**

shad < AS. *sceadd.*
>   L. *alosa,* f. shad: *Alosa sapidissima* (common shad).

shade < AS. *sceadu.*
>   Gr. *akronychos; akronyktos,* at nightfall: *Acronyches fenestratulus* (a fly), *Acronyctodes insignata* (a moth).
>   Gr. *alampetos,* dark, obscure: *Alampetis ambigua* (a beetle).
>   Gr. *anaugetos,* rayless, sunless; *lipauges,* dark, sunless:
>   Gr. *anelios,* sunless:
>   Gr. *aphos, -otos,* without light; *aphotistos,* dark, obscure: *Aphotismus niger* (a wasp).
>   L. *caeco, -atus,* make blind, darken; see **blind**
>   L. *caliginosus,* foggy, dark; see **black**
>   L. *cnephosus,* dark, < Gr. *knephas, -tos,* n. darkness, dusk, twilight; *knephaios,* dark, dusky, at twilight: *Cnephaotachina* [*Cnephaeotachina*] *crepusculi* (a fly).

L. *creper, -a, -um,* dark, obscure; *crepusculum,* n. twilight, dusk: crepuscle, crepuscular.

Gr. *epelyx, -ygos,* overshadowing: *Epelyx latus* (a beetle).

L. *Erebus* (Gr. *Erebos*), m. a place of darkness in the nether world, god of darkness; *erebennos,* dark, gloomy: Erebus.

AS. *glom,* twilight: gloaming, gloom.

Gr. *lyge,* f. twilight; *lygaios,* gloomy, shadowy; *elyge,* f. shadow, shade, darkness: *Lygaeospilus tripunctatus* (a bug).

Gr. *lykophos, -otos,* n. twilight, gloaming:

L. *obscurus,* dark, dim; see **dim**

L. *opacus,* shady, dark, obscure, dim: opaque, opacity, opacite, *Ilex opaca* (American holly), *Pipunculus subopacus* (a fly), *Ponera opaciceps* (an ant).

Gr. *paropion,* n. eyeshade, blinker: *Paropioxys opulentus* (a bug).

Gr. *skia, -as,* f. shade, shadow; *skiadion,* n.; *skiadiske,* f. sunshade, umbrella, parasol; *skias, -ados,* f. anything providing shade; *skiasma, -tos,* n. shadow; *skiaros; skiodes,* giving shade, shady; *skiastikos,* shading; *skiotos,* shaded; *skiophos, -otos,* n. twilight; *daskios,* very shady, bushy; *hyposkios,* shaded; *syskios,* densely shaded: sciomancy, sciophilous, squirrel, antiscians, *Sciobius muricatus* (a beetle), *Sciurus hudsonicus* (red squirrel), *Sciadopitys verticillata* (umbrella-fir), *Sciara hirtella* (a fungus-gnat), *Dasciopteryx polymenes* (a moth), *Episcia fulgida* (a gesneriacead), *Polyscias fruticosa* (an araliacead), *Xyloscia subpersata* (a moth).

Gr. *skotos,* m. darkness, gloom, obscurity; *skotia,* f. darkness, gloom; *skoteinos; skotios,* dark, obscure; *skotoma, -tos,* n. dimness of vision: scotophobia, scotophilous, scotoma, scotomatous, *Scotiotrichia ocreata* (a trichopterid), *Scoteinus balstoni* (a bat).

Gr. *synerephes,* densely shaded:

L. *tenebrae,* shades of night, darkness; see *tenebrosus* under **black**

L. *umbra,* f. shade, shadow; *umbella,* f. dim.; *umbraculum,* n. shady place, arbor; *umbraticus; umbratilis,* of shade, retired, private; *umbrosus,* shady; *umbro, -atus,* cast a shadow; *adumbratus,* vaguely outlined, hazy, foreshadowed, feigned: umbrageous, umber, umbral, umbel, adumbration, penumbra, somber, Umbelliferae, *Umbra limi* (mud-minnow), *Umbellula pellucida* (a coral), *Umbellularia californica* (California laurel), *Umbellulifera planoregularis* (a coral), *Umbraculum flabellatum* (a liverwort), *Anthomyia umbratica* (a fly), *Pteris umbrosa* (forest-brake), *Pelastoneurus umbricola* (a fly), *Sabal umbraculifera* (a palmetto), *Tanacetum umbelliferum* (a tansy), *Ornithogalum umbellatum* (star-of-Bethlehem), *Laccophilus umbrinus* (a beetle), *Baridius obumbratus* (a beetle).

See: **black, brown, night, dim, secret**

shaggy < AS. *sceacga,* thick mat of hair or wool; see **thick**

shake < AS. *sceacan.*

Gr. *abebaios,* uncertain, unsteady, wavering, fickle; see **change**

Gr. *astatos,* unstable, unsteady: *Astatochroa fuscimargo* (a moth).

Gr. *asteriktos,* unstable:

Gr. *brasso,* shake violently; *brastes,* m. shaker, earthquake:

L. *casabundus,* ready to fall, tottering; *caso, -atus,* shake, waver:

L. *ceveo, -etus,* move the haunches, wag the tail:

L. *convulsus,* shaking spasmodically, wrenching, tearing: convulsive, convulsion, convulse.

L. *crisso, -atus,* move the haunches:

AS. *cwacian,* shake, tremble, quiver, wag: quake, quagmire, Quaker, quakegrass (*Briza maxima*).

L. *desultorius,* skipping about, fickle, wavering, superficial; see **change**

Gr. *diadoneo,* shake to pieces:

Gr. *elelichthon,* earth-shaking:

Gr. *enosis,* f. a shaking, quake: *Enosis quadrinotata* (a moth).

Gr. *epiklintes,* moving sideways, shaking:

Gr. *gakinos,* m. earthquake:

L. *inconstans,* changeable, fickle, unstable: inconstancy, inconstant, *Eucalia inconstans* (brook-stickleback).

L. *instabilis,* unsteady, changeable, tottering: unstable: instability, *Pheidole instabilis* (an ant).

Gr. *klonos,* m. confused motion: clonic spasm.

Gr. *kradalos,* quivering:

L. *labo, -atus,* totter, shake, waver: labefy, labefaction.

L. *mico,* tremble, twinkle; see **light**

L. *nutabilis; nutabundus,* tottering, staggering, swaying:

L. *oscillo, -atus,* swing; see **swing**

Gr. *pallo*, shake, sway, brandish; *palmos*, m. pulsation, quivering, throbbing; *palmikos*, of palpitation: pallograph, pallometric, palmoscopy.

L. *palpito*, *-atus*, tremble, throb, beat; see **strike**

L. *pavidus*, trembling, fearful; see **fear**

Gr. *phrix*, *-ikos*, f. ruffling of a smooth surface, ripple, shivering; *phrike*, f. a shuddering or shivering from fear or cold; *phrikaleos; phriknos*, shivering from cold, goose-pimply, rough; *phrixos*, standing on end, bristling, shuddering, shivering; *phriktos*, horrible, terrible; *phrixothrix*, *-trichos*, with bristling hair; *phrisso (phritto)*, bristle up, roughen, shiver; *metaphrisso*, shiver: *Phricocarabus glabratus* (a beetle), *Phrixothrix pallida* (a beetle), *Phrixocoma forcipata* (a trichopterid), *Misophrice hispida* (a beetle), *Cacophrissus pauper* (a beetle).

L. *quatio*, *quassus*, shake, agitate; *quasso*, *-atus*, shake severely: quash, cask, fracas. See *concutio* under **strike.**

L. *querqerus*, shivering:

Gr. *rhektes*, breaker, render, earthquake; see **break**

Gr. *rhigosis*, a shivering from cold; see *rhigos* under **ice**

Gr. *saino*, wag the tail; see **flatter**

Gr. *salos*, m. a tossing by the sea; *episalos*, be tossed on the sea, unstable: *Salobrama tecomae* (a moth).

L. *scintillo*, *-atus*, sparkle, flash; see **light**

Gr. *seismos*, m. a shaking, shock, earthquake, < *seio*, shake; *seismatias*, shaking, tremulous; *anaseisma*, *-tos*, n. a shaking up and down, a threatening gesture: seismic, seismograph, seismogram, seismology, *Seisura inquieta* (restless flycatcher), *Seiurus auricapillus* (oven-bird).

Gr. *spasmos*, convulsion, fit; see **disease**

Gr. *sphygmos*, pulse, throbbing, vibration; see **strike**

Gr. *stembo*, shake, agitate:

AS. *swingan*, flutter, sway: swing, swingletree (singletree).

Gr. *tinagma*, *-tos*, n. a quake, shake, < *tinasso*, shake; *tinaktor*, *-os*, m. shaker: *Tinagma dentella* (a moth), *Tinactor fuscus* (a bird).

L. *titubo*, *-atus*, stagger, totter, reel: titubate, titubancy, titubation.

L. *tremo*, shake, quiver, tremble; *tremor*, *-is*, m. a shaking, quivering; *tremebundus; tremulus*, shaking, quivering; *contremulus*, trembling violently: tremble, tremor, delirium tremens, tremolo, tremulous, *Populus tremula* (European aspen), *Populus tremuloides* (trembling aspen).

L. *trepido*, *-atus*, tremble, be agitated; *trepidus*, agitated, alarmed, disturbed; *trepidulus*, dim.: trepidation, intrepid.

Gr. *tromos*, m. a trembling, quivering, < *tromeo*, tremble, quiver; *tromeros; trometos; tromikos; tromodes*, trembling: *Tromosternus haagi* (a beetle), *Tromicosoma koehleri* (an echinoid).

L. *vacillo*, *-atus*, waver: vacillate, vacillation, vacillatory, *Vaccinium vacillans* (late low-blueberry), *Rhyssomatus vacillatus* (a beetle).

L. *vexo*, *-atus*, shake, agitate, annoy; see **tease**

L. *vibro*, *-atus*, shake, quiver; *vibrabilis; vibrabundus*, quivering: vibrate, vibration, vibraculum, vibrissa, *Vibrio rugola* (a bacterium).

See: **move, dizzy, whip, strike, wander, change**

shallow < AS. *sceolh*, not deep: shoal.

Gr. *abathes*, shallow: *Abathoceramus percostatus* (a fossil pelecypod).

Gr. *brachos*, n. shallows:

Gr. *parageios*, pertaining to shallow water:

L. *superficialis*, on the surface, shallow: superficial, superficiality.

Gr. *tenagos*, n. shoal, shallows: *Tenagocrinus sulcatus* (a Permian crinoid).

L. *vadosus*, full of shoals, shallow, < *vadum*, n. shoal, ford: vadose.

sham < AS. *scamu*, shame; see **lie**

shame < AS. *scamu.*

Gr. *aidos*, sense of modesty, respect, shame; see **modest**

Gr. *aischyne*, f. shame, disgrace; *aischyntos*, shameful: *Aeschynomene viscidula* (sticky joint-vetch), *Aeschynanthus marmoratus* (marbled basket-vine), aeschynite.

Gr. *atimia*, f. disgrace; *atimos*, dishonored:

L. *dedecus*, *-oris*, n. shame, disgrace, dishonor; *dedecorus*, disgraceful, shameful:

L. *defamis*, shameful; *infamis*, disreputable; *infamia*, f. shame, disgrace: defame, infamy, infamous.

L. *flagitium*, n. shame, disgrace, infamy; *flagitiosus*, shameful: flagitious.

Gr. *elenchos*, disgrace, reproach; see **try**

L. *ignominia*, f. disgrace, dishonor: ignominy, ignominious.

L. *indecorus*, unseemly, shameful: indecorous.

L. *inglorius*, without honor, shameful, ignominious: inglorious.

L. *inhonestus*, dishonorable, shameful:

L. *probrum,* n. disgrace, shame; *probrosus,* shameful, infamous; *opprobriosus,* shameful, abusive, taunting: opprobrium, opprobrious.

L. *pudeo, -itus,* be ashamed; *pudendus,* shameful; *pudicus,* modest, bashful; *pudor, -is,* m. shame, shyness; *pudoratus,* shamefaced, modest; *propudiosus; pudibundus,* ashamed, modest; *impudicus,* lewd, unchaste, unashamed, shameful, infamous: pudency, pudenda, pudic, impudent, repudiate, *Mimosa pudica* (sensitive-plant), *Dasychira pudibunda* (pale tussock-moth).

L. *teter,* foul, shameful, abominable; see **bad**

L. *turpitudo, -inis,* f. disgrace, infamy, < *turpis,* ugly, foul, shameful: turpitude. See: **bad, lewd, cover**

shape < AS. *sceap,* form; see **form**

share < AS. *sceran;* see **lot, cut, part, give**

shark < uncertain origin.
Gr. *akanthias,* m. a prickly thing, a kind of shark: *Squalus acanthias* (spiny dogfish, formerly called *Acanthias vulgaris*).
Gr. *batis, -idos,* f.; *batos,* m. a flatfish, skate, ray: *Raia batis* (skate), *Aetobatis aquila* (an eagle-ray), *Rhinobatus productus* (a guitar-fish).
L. *canicula,* f. dogfish, dim. of *canis,* dog:
Gr. *galeos,* m. shark: *Galeorhinus galeus* (tope).
Gr. *karcharias,* m. shark: *Carcharias obscurus* (dusky shark), *Agave carchariodonta* (an agave).
Gr. *kentrines,* m. a kind of shark: *Centrina centrina* (a fish).
Gr. *lamia,* f. a sharklike fish: *Carcharias lamia* (cub-shark).
Gr. *lamna,* f. a kind of shark: *Lamna nasus* (porbeagle).
Gr. *pristis,* f. shark, sawfish: *Pristis pectinata* (sawfish), *Orthopristis chrysoptera* (a pigfish), *Diplograptus pristis* (a graptolite).
L. *raia,* f. ray: *Raja maculata* (homelyn), *Raja laevis* (smooth skate), *Squaloraja polyspondyla* (a fossil fish).
Gr. *rhine,* f. shark with a rough skin like a file: *Cetorhinus maximum* (a baskingshark, largest fish), *Rhineodon* (originally *Rhincodon*) *typus* (whale-shark).
Gr. *selachos,* n. a cartilaginous fish, shark, ray: selachian, *Selachochthonius serridentatus* (a pseudoscorpion).
Gr. *skylion,* n. dogfish: *Scyliorhinus canicula* (spotted dogfish), *Chiloscyllium* [*Chiloscylium*] *punctatum* (a shark).
NL. *sphyrna,* f. a genus of sharks: *Sphyrna prisca* (a hammerhead-shark).
L. *squalus,* m. a kind of sea-fish: *Squalus fernandinus* (a dogfish), squaliform, *Squalodon atlanticus* (a fossil aquatic mammal).
L. *squatina,* f. skate: *Squatina squatina* (angelfish).
L. *torpedo, -inis,* f. numbness; a ray: *Torpedo occidentalis* (a ray), *Narcobatus torpedo* (a ray), torpedo.
Gr. *trygon, -os,* f. sting-ray: Trygonidae, *Trygon tuberculata* (a fish), *Trygonobatus torpedinus* (a ray).
Gr. *zygaina,* f. a kind of shark: *Sphyrna zygaena* (a hammerhead-shark).

sharp < AS. *scearp;* see **point, cut, sour**

shatter < ME. *schateren;* see **break**

sheaf < AS. *sceaf;* see **bundle**

shears < AS. *sceran,* cut; see **scissors**

sheath < AS. *sceath,* case, cover.
L. *coleus* (Gr. *koleos*), m. sheath, scabbard, scrotum: coleorhiza, coleoptile, Coleoptera, cullion, *Coleus thyrsoides* (a mint), *Coleonyx variegatus* (a lizard), *Lepidocoleus jamesi* (a Silurian cirriped), *Coleanthus subtilis* (a grass), *Scutigera coleoptrata* (a centiped).
Gr. *eilema (eilyma), -tos,* n. wrapper, envelope: *Ilemodes heterogyna* (a moth), neurilemma.
Gr. *elytron,* n. cover, sheath, husk, shell; *elymos,* m. case, quiver, < *elyo,* wrap up: elytron, *Elytridium rugulosum* (a fossil beetle), *Atriplex hymenelytra* (desert-holly saltbush), *Elytranthe flavida* (yellow mistletoe), *Elymus glaucus* (a wild-rye), *Elymocaris siliqua* (a Devonian crustacean), *Brachyelytrum erectum* (a grass).
Gr. *epitelis,* with a husk or shell:
Gr. *gorytos,* m. quiver, bow-case: *Gorytodes personaria* (a moth).
L. *involucrum,* n. wrapper, case, envelope: involucre, involucrate, involucel, *Lonicera involucrata* (a honeysuckle).
Gr. *kalypter,* covering. wrapper, sheath; see *kalypto* under **cover**
Gr. *kolpos,* bosom, fold, bay, uterus, vagina; see **womb**
Gr. *lemma,* scale, husk, sheath; see *lepis* under **scale**
L. *ocrea,* f. greave, husk, sheath: *Scotiotrichia ocreata* (a trichopterid).
Gr. *sebenion,* n. sheath of the palm flower:

L. *vagina*, f. sheath, scabbard, case; *vaginula*, f. dim.; Sp. *vaina*, f. sheath, scabbard, pod; *vainilla*, f. dim.; *evagino*, *-atus*, unsheath: vagina, vaginal, evaginate, invagination, *Vanilla aromatica* (an orchid), *Vaginula daudini* (a pteropod), *Vaginella depressa* (a fossil gastropod), *Vaginulina denudata* (a foraminifer), *Vaginicola cristalina* (a protozoan), *Sporobolus vaginiflorus* (a grass), *Salpa vagina* (a tunicate), *Eriophorum vaginatum* (a cotton-sedge), *Trichomonas vaginalis* (a protozoan).

L. *vulva (volva)*, f. cover, wrapper; a cup-shaped sheath or envelope at the base of some mushrooms; see **vulva**

See: **cover, bag, sleeve**

shed < AS. *sceadan*, part, take off; see **abandon, bare, cancel, drop, throw**

sheep < AS. *sceap*.

L. *agnus*, m. lamb; *agnicellus*; *agniculus*, dim.; *agninus*, of a lamb:

Gr. *amnos*, m. lamb; *amnion*, n. dim.; *amneios*, of a lamb: amnion, *Oreamnus americanus* (a mountain-goat).

L. *aries*, *-etis*, m. ram; *arietinus*, of a ram: Aries, *Ovis aries* (sheep), *Cicer arietinum* (chickpea).

Gr. *arnos*, m. sheep, lamb; *arnion*, n. dim.; *arneios*, of a lamb or sheep: *Arnoseris minima* (lamb-succory).

L. *arvix*, *-igis*, f. a ram for sacrifice:

Gr. *krios*, m. ram; *kridion*, n. dim.: *Crioceris smaragdina* (a beetle).

Gr. *ktilos*, m. ram: *Ctilocephala pellucens* (a beetle).

AS. *lamb*, young sheep: lamb, lambiness, lambkin, lambskin.

Gr. *melon*, n. sheep, goat: *Melophagus ovinus* (sheep-tickfly).

L. *musimo*, *-onis*, m. mouflon: *Ovis musimon* (mouflon).

Gr. *oïs*, *oïos*, c. sheep: *Oephronistus[Oïphronistus] australicus* (a beetle).

L. *ovis*, f. sheep; *ovicula*, f. dim.; *oviarius*; *ovilis*; *ovillinus*; *ovillus*; *ovinus*, of sheep: ovine, Ovillus, *Ovis poli* (Marco Polo sheep), *Ovibos moschatus* (musk-ox), *Festuca ovina* (sheep-fescue), *Eperythrozoon ovis* (a blood-parasite).

Gr. *phagilos*, m. lamb:

Gr. *probaton*, n. sheep; *probation*, n. dim.: *Archosargus probatocephalus* (sheep-head fish).

Gr. *rhen*, *-os*, f. sheep, lamb; *rhenikos*, of sheep:

L. *vervex*, *-ecis*, m. wether, stupid fellow; *vervecinus*, of a wether: vervecine.

AS. *wether*, a castrated ram: wether, bellwether.

sheet < AS. *sciete;* see **cover, paper, plate, leaf, scale**

shelf < AS. *scielfe*.

L. *caliclarium*, n. cupboard:

L. *forulus*, bookcase; see **box**

Gr. *herma*, sunken rock, reef; see **stone**

Gr. *kylikeion*, n. sideboard, buffet:

L. *pegma*, fixture, bookcase, shelf; see **thick**

L. *pluteus*, backboard, breastwork, shelf; see **wall**

See: **frame, table**

shell < AS. *scell*.

Gr. *chelonion; chelyon*, tortoise-shell; see **turtle**

L. *cochlea* (Gr. *kochlias; kochlos*), snail with a spiral shell; see **mollusk**

L. *concha* (Gr. *konche*), snail, shell; see **mollusk**

L. *crusta*, hard outer surface of a body, rind, shell; *crustula*, dim.; see **skin**

Gr. *lepyron*, husk, rind, shell; see **lepis** under **scale**

Gr. *ostrakon*, n. shell, tile, potsherd; *ostrakion*, n.; *ostrakis*, *-idos*, f. dim.: ostracoderm, ostracism, *Ostracion[Ostracium] trigonum* (a trunk-fish).

L. *putamen*, *-inis*, n. shell, shred, paring: putamen.

Gr. *sepion*, internal shell or proostracum of the squid; see **sepia** under **mollusk**

L. *testa*, f. piece of burned clay, potsherd, shell, skull; *testula*, f. dim.; *testaceus*, with a shell, of earthenware; *testeus*, earthen: test, testaceous, tester, tete a tete, Testudinata, *Testacella haliotoides* (a slug), *Lilium testaceum* (Nankeen lily), *Lysiphlebus testaceipes* (a wasp).

L. *valva*, leaf of a folding door; *valvola*, dim. pod, shell; see **door**

See: **mollusk, cover, skin, bark**

shelter < AS. *scildtruma*, a troop with shields; see **cover, house, harbor, save, guard, fort**

shield < AS. *scild* (ON. *skjöldr;* G. *schild*); shield-fern, shieldtail, windshield, Nordenskjold, Rothschild.

L. *aegis*, *-idis* (Gr. *aigis*, *-idos*), f. shield of Jupiter: aegis.

L. *ancile*, n. sacred shield that fell from heaven: ancile.

Gr. *aspis*, *-idos*, f. shield; *aspidion*, n.; *aspidiskos*, f. dim.; *aspidiotes*; *aspister*, *-os; aspistes*, m. warrior: aspidate, aspidium, aspidinol, *Aspidosperma tomento-*

sum (woolly quebracho), *Aspidiotes bromeliae* (a beetle), *Aspidiotus hederae* (white scale-insect), *Acidaspis mira* (a trilobite), *Acraspis erinacei* (hedgehog gall-wasp), *Aspidistra typica* (a lily).

L. *cetra*, f. a short Spanish shield; *cetratus,* armed with a shield: *Cetraria glauca* (a lichen).

L. *clypeus,* m. shield; *clypeolus,* m. dim.; *clypeo, -atus,* provide with a shield: clypeus, clypeate, clypeole, *Clypeaster reticulatus* (a fossil echinoid), *Cycloclypeus neglectus* (a foraminifer), *Achillea clypeolata* (a yarrow), *Begonia clypeifolia* (a begonia), *Cerceris clypeata* (a wasp), *Pithecellobium clypeatum* (a legume), *Phacodiscus clypeus* (a radiolarian).

Gr. *hoplon,* armor, heavy shield used by foot-soldiers; see **tool**

Gr. *laisenion,* n. a kind of shield:

Gr. *palme,* f. shield:

L. *parma* (Gr. *parme*), f. small light shield; *parmula,* f. dim.; *parmatus,* armed with a shield: parmelin, *Parmelia stellaris* (a lichen).

L. *pelta* (Gr. *pelte*), f. small shield; *peltarion,* n. dim.; *peltatus,* shield-shaped, armed with a shield; *peltates,* m. warrior armed with a shield: pelta, *Peltigera canina* (a lichen), *Peltophorum pterocarpum* (yellow-flame), *Peltogyne paniculata* (a legume), *Peltandra virginica* (arrow-arum), *Dipelta floribunda* (a caprifoliacead), *Apeltes quadrans* (a stickleback), *Peltastes hastatus* (a grasshopper), *Eriopeltastes leucoprymnus* (a beetle), *Dasypeltis unicolor* (an African snake), *Mallotus apelta* (a euphorbiacead), *Tropaeolum peltophorum* (a nasturtium), *Peltigera scutata* (a lichen).

Gr. *sakos,* n. shield of wickerwork or wood covered with oxhide; *sakellion,* n. dim.; *sakesphoros,* shield-bearing: *Sacodiscus fasciatus* (a copepod), *Sacesphorus maculatus* (an opilionid), *Pomatosace filicula* (a primulacead).

L. *scutum,* n. shield; *scutellum; scutulum,* n. dim.; *scutatus,* shield-shaped, armed with a shield: scutum, scutate, scutellum, scutellate, escutcheon, esquire, escuage, *Scutellum niagarense* (a Silurian trilobite), *Scutigera forceps* (a centiped), *Scutacarus tlazolteotli* (a mite), *Anthonomus scutellaris* (a weevil), *Testacella scutulum* (a snail), *Coleochaete scutata* (an alga), *Rumex scutatus* (French sorrel), *Begonia scutulum* (a begonia), *Conops aureoscutellatus* (a fly), *Cyclops scutifer* (a copepod).

Gr. *talaurinos,* m. a shield of tough hide: *Talaurinus foveatus* (a beetle).

Gr. *thorax,* breastplate, chest; see **chest**

Gr. *thorektes,* m. warrior armed with a breastplate: *Thorectes rotundatus* (a beetle), *Thorectospamma irregularis* (a sponge).

Gr. *thyreos,* m. large, oblong shield; *thyreoeides,* like a shield: thyroid, thyroxin, thyreoitis, *Thyreocephalus lynceus* (a beetle), *Sphyrapicus thyroideus* (a sapsucker).

L. *tropaeum* (Gr. *tropaion*), shield or other trophy from the battlefield, a token of the enemy's defeat; see **mark**

L. *umbo, -onis,* boss of a shield, shield; see **projection**

See: **cover, tool, guard, mark, ornament**

shine < AS. *scinan;* see **light**

shingle < L. *scindula (scandula),* f.:

L. *assula,* shingle, dim. of *assis,* board; see **board**

L. *imbrex, -icis,* f. roofing-tile, gutter-tile; *imbricatus,* overlapping like roofing-tiles and shingles; *imbrico, -atus,* cover with tiles: imbricate, imbrication, *Araucaria imbricata* (monkey-puzzle), *Quercus imbricaria* (shingle-oak), *Prionus imbricornis* (a beetle).

Gr. *keramis, -idos,* f. roofing-tile:

L. *tegula,* f. roofing-tile: tile, *Tegulaplax matthewsi* (a chiton).

See: **roof, brick**

ship < AS. *scip:* shipper, shipwreck, shipshape, shipping, flagship, midshipman.

L. *acatus* (Gr. *akatos*), f. a light boat, sail, woman's shoe; *akation,* n. dim.: *Acatus rhombicus* (a beetle).

L. *actuariolum,* n. rowboat, barge, dim. of *actuarium,* n. a swift sailer:

L. *barca,* f. a small boat: bark, embark, barge, barcarole, *Ophiocephalus barca* (a fish).

Gr. *baris, -idos,* f. a flat-bottomed boat, bateau: *Limnobaris rectirostris* (a beetle).

L. *batalaria,* f. a kind of warship:

L. *carabus* (Gr. *karabos*), a small boat, beetle; see **beetle**

L. *caupulus,* m. a kind of small ship:

L. *celox, -ocis,* f. cutter, yacht: *Celox dima* (a beetle).

L. *cercurus* (Gr. *kerkouros*), m. a light vessel, boat:

L. *classis,* f. navy, fleet, marines; *classiarius,* of the navy:

L. *corbita,* f. a slow, freight ship: *Corbitella speciosa* (a sponge).

L. *cydarum* (Gr. *kydaron*), n. a kind of small ship:

L. *cymba* (Gr. *kymbe*), f. small boat, skiff; *cymbella*; *cymbula*, f.; *kymbion*, n.
dim.: *Cymbella lanceolata* (a diatom), *Cymbium proboscidale* (a gastropod),
Cymbopogon flexuosus (a grass), *Derbyia cymbula* (a fossil brachiopod).
Gr. *dromon*, light, fast vessel; see *dromos* under **run**
Gr. *epaktris*, *-idos*, f. skiff: *Epactris melanchata* (a moth).
L. *geseoreta*, f. a kind of boat:
L. *hexeris* (Gr. *hexeres*), f. vessel with six banks of oars:
Gr. *holkas*, *-ados*, f. a towed ship; *epholkion*, n. a small, towed boat; *epholkis*,
-idos, f. burdensome appendage: *Epholcis divergens* (a beetle), *Laphria
(Epholciolaphria) gilva* (a fly).
L. *horia (oria)*, f. fishing-smack; *horiola*, f. dim.: *Horia maculata* (a beetle).
L. *lembus* (Gr. *lembos*), m. small, swift vessel, cutter; *lembulus*, m. dim.:
Lembodes solitarius (a beetle), *Lembus pusillus* (a protozoan).
L. *linter*, *-tris*, f. boat, skiff, wherry; *lintricula*, f. dim.; *lintrarius*, m. boatman:
Linter acutata (a Cretaceous pelecypod), *Olivancillaria(Lintricula) vesica* (a
gastropod).
Gr. *mydion*, n. boat:
L. *navis* (Gr. *naus, neos*), f. ship; *naucella*; *navicella*; *navicula*, f. dim.; *navalis*;
navicularis; *navicularius*; *nauticus* (Gr. *nautikos*), of ships; *navigator*, *-is*;
navita; *nauta* (Gr. *nautes*); *nautilus* (Gr. *nautilos*); *naubates*, m. seaman, sailor;
nausiphoretos, carried by ship, seafaring; *nausiploos*, sailing in ships; *neorion*,
n. dock, shipyard: navigate, navigator, navigation, naval, nave, navy, circum-
navigate, Argonaut, nausea, nautical, nautilus, nauplius, aeronautics, *Navicula
crucigera* (a diatom), *Nauclea esculenta* (burflower), *Naucrates ductor*(pilot-fish),
Nausigaster punctulata (a fly), *Argonauta argo* (paper-nautilus).
L. *paro*, *-onis* (Gr. *paron, -os*), m. small, light ship; *parunculus*, m. dim.:
Gr. *phortis*, *-idos*, f. merchantman: *Phortis gibbosa* (a hydrozoan).
Gr. *ploion*, n. a floating vessel, any kind of ship:
L. *prosumia*, f. small boat for spying:
L. *ratis*, f. raft, a float made of logs fastened together; *ratitus*, marked with the
figure of a raft: ratite.
Gr. *sangaron*, n. a kind of boat:
L. *scapha* (Gr. *skaphe*); *scaphos*, m. light boat, skiff; *skaphidion*; *skaphion*, n.;
skaphis, *-idos*; *scaphula*, f. dim.; *skaphites*, m. steersman, pilot, boatlike:
Scaphites ventricosus (an ammonite), *Hydroscapha natans* (a beetle).
L. *trabica*, f. vessel made of beams fastened together, raft:
Gr. *trampis*, *-idos*, f. ship:
See: **hollow, cup, sail, swim**

-ship < AS. *scipe*, state, condition; see **nature**

shirk, prob. < a German word; see **shun**

shirt < AS. *scyrte*; see **garment**

shiver < uncertain origin; see **shake**

shoal < AS. *sceolh*; see **shallow**

shoe < AS. *sceoh*: shoemaker, shoecraft, shoestring, shoehorn, shoebill, snowshoe,
horseshoe.
L. *aluta*, shoe made of soft leather; see **leather**
Gr. *arbyle*; *arbylis*, *-idos*, f. half-boot:
Gr. *arter*, *-os*, m. felt shoe:
Gr. *askera*, f. shoe lined with fur; *askeriskos*, m. dim.: *Ascerodes prochlora* (a
moth).
L. *baxa*, f. a kind of woven shoe:
Gr. *blaute*, f. a kind of slipper; *blaution*, n. dim.; *ablautos*, unslippered: *Ablautus
trifarius* (a fly).
L. *calceus*, m. shoe; *calceolus*, dim. half-boot; *calceolarius*, m. shoemaker; *calceo*,
-atus, provide with shoes: calceolate, Chaucer, *Calceola sandalina* (slipper-coral),
Calceolaria pinnata (a slipper-flower), *Cypripedium calceolus* (European
moccasin-flower).
L. *caliga*, f. soldier's shoe, boot; *caligula*, f. dim.; *caligaris*; *caligarius*, of shoes or
boots; *caligatus*, booted, shod: Caligula, *Marmota caligata* (a marmot),
Cristellaria caligula (a foraminifer).
L. *campagus*, m. boot worn by military officers: *Campages furcifera* (a
brachiopod).
L. *carbatina* (Gr. *karbatine*), f. shoe of undressed leather, brogue: *Carbatina
levigata* (a moth).
L. *carpisculus*, m. a kind of shoe:
L. *cernuus*, m. a kind of shoe:
L. *cothurnus* (Gr. *kothornos*), m. high boot, hunting-boot, buskin: *Cothurnocystis
curvata* (a cystid), *Stibus kothornostibus* (an ostracode).

L. *crepida* (Gr. *krepis, -idos*), f. boot, sandal, shoe, base; *crepido, -onis,* f. base, foundation; *crepidula,* f. dim.; *crepidarius,* of shoes; *crepidatus,* wearing sandals; *eukrepis, -idos,* well-shod, well-based: crepidoma, *Crepidula lirata* (a fossil slipper-limpet), *Crepidomenus decoratus* (a beetle), *Hippocrepis comosa* (horse-shoe-vetch), *Cristellaria crepidulaeformis* (a foraminifer), *Scaphites hippocrepis* (an ammonite).

L. *diabathrum* (Gr. *diabathron*), n. a kind of slipper:

Gr. *elips, -ipos,* m. a Dorian shoe:

Gr. *embas, -ados,* f. a felt shoe; *embadion,* n. dim.: *Embadomonas agilis* (a protozoan).

Gr. *eumaris, -idos,* f. an Asiatic shoe or slipper:

L. *gallica,* f. a Gallic shoe; *gallicula,* f. dim.: galosh.

Gr. *harpis, -idos,* f. a kind of shoe or sandal:

Gr. *hipposideros,* m. horseshoe: *Hipposideros cervinus* (a horseshoe-bat).

Gr. *hypodema, -tos,* n. sandal; *hypodemation,* n. dim., horseshoe:

L. *impilium,* n. a felt shoe:

Gr. *kalopous, -odos,* m. shoemaker's last; *kalopodion,* n. dim.:

Gr. *kroupeza,* f. high, wooden shoe; *kroupezion,* n. dim.:

L. *mulleus,* m. a reddish or purplish shoe; *mulleolus,* dim. reddish:

L. *mustricola,* f. shoemaker's last:

Gr. *pedilon,* n. sandal; *kalopedilon,* n. wooden shoe: *Pedilanthus macrocarpus* (a slipper-flower), *Paphiopedilum insigne* (an orchid).

L. *pero, -onis,* m. a kind of rawhide boot; *peronatus,* booted:

Gr. *phaikas, -ados,* f. a kind of white shoe; *phaikasion,* n. dim.: *Phaecadophora fimbriata* (a moth).

Gr. *rhadion,* n. a kind of easy shoe: *Calorhadia[Calorhadium] pharcida* (a fossil mollusk).

L. *sabatenum,* n. a kind of slipper:

F. *sabot,* wooden shoe, clog: sabot, sabotage.

L. *sandalium* (Gr. *sandalion,* dim. of *sandalon,* n. wooden sole), n. slipper: sandal.

L. *scrupeda,* f. high, wooden shoe:

L. *sculponea,* f. wooden shoe:

Gr. *seleukis, -idos,* f. a kind of woman's shoe from Seleucia; a bird: *Seleucides ignotus* (a bird).

L. *soccus,* m. a low-heeled, light shoe, slipper; *socculus,* m. dim.: sock.

L. *solea,* sandal, sole of the foot, see **sole**

Gr. *synchis, -idos,* f. a kind of shoe:

Gr. *trochas, -ados,* f. a light shoe for running:

L. *udo, -onis* (Gr. *oudon, -os*), a felt shoe or sock; see **stocking**

See: **sole, base, iron, leather, cobbler**

shoot < AS. *sceotan,* move rapidly, dart, rush; see **swift, throw, pour, branch**

shop < AS. *sceoppa;* see **make, store, room**

shore < AS. *score,* edge, border.

L. *acta* (Gr. *akte*), f. seashore, beach, strand, promontory; *aktaios; aktios; epaktios,* coastal; *aktites,* m. shore or coast dweller; *paraktios,* on the coast: Actaeus(Attica), *Actitis macularia* (spotted sandpiper), *Epactiothynnus opaciventris* (a wasp).

Gr. *aigialos,* m. seashore, beach; *aigialeios,* of the shore; *aigialeus; aigialites,* m.; *aigialitis,* f. one on the shore: *Aegialitis hiaticula* (ringed plover).

L. *costa,* rib, side, coast; see **rib**

Gr. *eïon, -os,* f. shore, beach: *Eïoneus bilineatus* (a bug).

Gr. *krokale,* f. beach, seashore; rounded pebble on the shore; *halikrokalos,* shingly, pebbly: *Crocalia aglossalis* (a moth).

L. *litus (littus), litoris,* n. shore; *litoralis; litoreus,* of the seashore: littoral, *Littorella uniflora* (a plantain), *Littorina littorea* (a periwinkle), *Grammostomum littorale* (a foraminifer).

L. *ora,* edge, border, coast; *orarius,* of the coast; see **border**

Sp. *playa,* f. shore, strand, beach: playa.

Gr. *rhachia,* f. shore, beach, surf:

Gr. *rhegmin, -os,* f. line of breakers, surf:

Gr. *rhothion,* n. breakers, surf:

L. *ripa,* f. bank, shore; *ripula,* f. dim.; *riparius,* of stream banks: riparian, riparious, river, Riviera, *Roripa obtusa* (a marsh-cress), *Lycosa riparia* (a spider), *Begonia ripicola* (a begonia).

Gr. *this, thinos,* m. beach, shore, strand, sand-heap: thinolite, *Thinopus antiquus* (a Devonian footprint), *Thinolestes anceps* (a fossil mammal).

short < AS. *sceort:* shorten, shortcoming, shortcake, shorthand, shorthorn.

L. *abortivus,* prematurely terminated, shortened; see **birth, abortion**

L. *actuarius,* shorthand writer; see **write**

Gr. *akares,* short, small, brief, momentary; see **little**

L. *amputo, -atus,* cut away, lop off, shorten: amputate, amputation.

L. *apocope* (Gr. *apokope*), f. a cutting off, dropping, stoppage: apocope, apocopate, *Apocope oscula* (a fish).

Gr. *aporrhox,* broken off, abrupt, steep; see *rhegnymi* under **break**

Gr. *apotomos,* cut off, abrupt, precipitous; *syntomos,* concise, abridged, abbreviated: *Apotorhamphus splendens* (a beetle), *Syntomomelus rossicus* (a wasp).

Gr. *brachys,* short; *brachyteros,* shorter; *brachystatos; brachistos,* shortest; *brachytes, -etos,* f. shortness; *brachyno,* shorten: brachycephalic, brachydont, brachypterous, brachistochrone, brachypinacoid, amphibrach, *Brachyrhinus ovatus* (strawberry-root weevil), *Brachycome iberidifolia* (Swan River daisy), *Brachyderes signatus* (a beetle).

L. *brevis,* short; *brevior,* shorter; *brevissimus,* shortest; *breviculus,* dim.; *breviarius,* shortened; *brevitas, -atis,* f. shortness, narrowness; *brevio, -atus,* shorten; *abbreviatus,* shortened: brevity, brief, brevet, abridge, abbreviate, unabridged, breviary, *Breviceps gibbosum* (an amphibian), *Brevipes melanogaster* (a bird), *Aphis abbreviata* (an aphid), *Gryllus abbreviatus* (a cricket), *Yucca brevifolia* (Joshua tree).

L. *cohibilis,* short, abridged:

L. *compendium,* n. abridgment, abstract, summary; *compendiarius; compendiosus,* short, brief; *compendio, -atus,* shorten, abridge: compendium, compendious.

L. *concisus,* brief, short: concise.

L. *contractus,* drawn together, abridged, shortened; see *traho* under **draw**

L. *curtus,* short; *curto, -atus,* shorten: curt, curtail, curtate, curtness, *Lucina curta* (a fossil pelecypod), *Mulhlenbergia curtifolia* (a grass), *Melipona curtula* (a bee), *Corycaeus decurtatus* (a cyclops).

Gr. *ephemeros,* living only a day, short-lived, temporary; see **day**

Gr. *epitome,* f. abstract, abbreviation: epitome, epitomize.

L. *evanesco,* disappear, pass away; see *vanesco* under **depart**

L. *improcerus,* short, undersized:

Gr. *kartos,* shorn close, shortened: *Cartocometes io* (a fly).

Gr. *kolos; kolobos; kolobikos; anakolos,* docked, curtailed, shortened, stunted, incomplete; *kolobodes,* stunted; *kolobotes,* f. stuntedness; *kolouros,* bobtailed; *kolouo,* shorten, dock: coloboma, colobium, colure, colocephalous, Colobognatha, *Colobus guereza* (a monkey), *Colobodes magicus* (a beetle).

Gr. *kontos,* short: *Contoderus acanthocinoides* (a beetle), *Contopteryx illustris* (a fly).

Gr. *meiouros,* curtailed, shortened: *Ameiurus melas* (bullhead).

Gr. *mikroschemos,* of small stature, short; *mikrophyes,* short, small:

Gr. *minyorios,* short-lived: *Minyoriolus pusillus* (a spider).

L. *mutilus,* maimed, cut off, docked; see **hurt**

Gr. *okymoros,* dying early, short-lived: *Hemiteles(Ocymorus) cingulator* (a wasp).

Gr. *oligobios,* short-lived: *Oligobiotherium divisum* (an Eocene marsupial).

Gr. *pedanos,* short: *Pedanus quadrilunatus* (a beetle), *Pedanostethus lividus* (a spider).

L. *premordicus,* bitten off in front, at the end, or at the top:

L. *redactus,* edited, reduced, abridged; see **edit**

L. *siglum,* n. a sign of character signifying an abbreviation: *Siglophora bella* (a moth).

L. *simus,* flat-nosed, pug-nosed; *simulus,* dim.; see **nose**

L. *succinctus,* brief, short, concise: succinct, *Cyrtacanthacris succincta* (Bombay locust).

L. *summarium,* n. abstract, epitome: summary.

L. *syncope* (Gr. *synkope*), f. abbreviation by elision of letters; *synkoptos,* chopped, shortened: syncope, *Syncoptoblatta thoracica* (a fossil blattid).

Gr. *synopsis,* f. a general view, sketch, outline, abstract, summary: synopsis, synoptic.

Gr. *syntomos,* abridged, shortened:

L. *transitorius,* passing, evanescent, fleeting, temporary; see **depart**

L. *truncus,* maimed, cut off; *truncus,* m. piece cut off, tip, end; *trunculus,* m. dim.; *trunco, -atus,* maim or shorten by cutting off: truncate, trunk, truncheon, trunnion, detruncate, *Truncatella truncatula* (a gastropod), *Truncatulina tenella* (a foraminifer), *Zygocactus truncatus*(Christmas cactus).

See: **little, cut**

shoulder < AS. *sculdor.*

L. *armus* (Gr. *harmos*), m. shoulder, shoulder-joint:

L. *humerus,* m. shoulder, bone of the upper arm; *humerulus,* m. dim.: humerus, humeral, *Falsophrixothrix humeralis* (a beetle), *Lycopolemius inhumeralis* (a beetle), *Turritella humerosa* (a fossil gastropod).

Gr. *knemos*, m. shoulder or limb of a mountain:

Gr. *omos*, m. shoulder, upper arm; *akromion*, n. point of the shoulder-blade;
L. *catomus*, m. shoulder; *epomis*, *-idos*, f. point of the shoulder, acromion;
epiomidios, on the shoulder: acromion, omophore, omodynia, *Epomidiostomum anatinum* (a nematode).

L. *scapula*, f. shoulder-blade: scapular, scapulated, scapulohumeral, *Pteropus scapulatus* (a fruit-bat). '

shout < uncertain origin; see **call**

shovel < AS. *scofl*.

Gr. *ame*, shovel, water-bucket; see **bucket**
L. *batillum*, n.; *batillus*, m. shovel, fire-pan, chafing-dish: *Batillipes mirus* (a tardigrade).
Gr. *listron*, n. a leveling or smoothing tool, spade, shovel; *listrion*, n. dim.; *listrotos*, leveled, smoothed: *Listroderes obliquus* (a weevil), *Listrognathus cornutus* (a wasp), *Listriotherium patagonicum* (a fossil mammal).
L. *pala*, f. shovel: pala, palette, palaceous, pallet.
L. *rutrum*, n. shovel, spade; *rutellum*, n. dim.: *Rutripalpus limicola* (an arachnid).
L. *spatula*, a broad blade for stirring liquids, a small spade; see **spoon**
Gr. *skapane*, f. spade, hoe; *skapanion*; *skapheion*; *skapheidion*, n. dim.: *Scapanorhynchus elongatus* (a fish), *Scaphiopus holbrooki* (an amphibian), *Scapania nemorosa* (a liverwort), *Scapanonyx palpatus* (a cricket).
L. *vanga*, f. spade or mattock: *Vanga flaviventris* (a bird), *Vanganella taylori* (a pelecypod).

show < AS. *scaewian*; see **display, lead, teach, bare, treason, open**

shower < AS. *scour*; see **storm, drop**

shred < AS. *screade*, thin, cut, or torn strip; see **part, tear, fringe, flap**

shrew < AS. *screawa*; see **mole**

shriek, prob. < ON. *shraekja*; see **call**

shrike < AS. *scric*, butcher-bird.
L. *lanius*, m. butcher, the typical genus of shrikes: *Lanius ludovicianus* (a shrike).
Gr. *malakokraneus*, m. a shrike:

shrimp < uncertain origin; see **crab**

shrine < AS. *scrin*; see **temple, holy**

shrink < AS. *scrincan*; see **lessen**

shrivel < uncertain origin; see **dry, fold, lessen**

shrub < AS. *scrob*; see **bush**

shun < AS. *scunian*.
Gr. *alyktos*, to be shunned, < *alysko*, shun, flee from: *Alyctus rusci* (a beetle), *Alyxia buxifolia* (an apocynacead).
Gr. *apotropos*, turned away, averted, banished:
L. *averrunco*, avert, remove; see **guard**
L. *averto, aversus*, turn away from: avert, aversion.
Gr. *diadidrasko*, run away from, avoid, shirk, shun; *diadrasis*, f. a shirking:
L. *eludo, elusus*, avoid, evade, frustrate, baffle: elude, elusive.
Gr. *pheuktos*, to be avoided or shunned; see *fuga* under **depart**
L. *vito, -atus*, shun, avoid; *vitabilis*, avoidable, that ought to be shunned; *vitabundus*, shunning; *evito*, shun; *evitabilis*, avoidable: evitable, inevitable.
See: **depart, run, abandon, hate**

shut < AS. *scyttan*; see **close, bar**

shuttle < AS. *scyttel*, bolt; see *kerkis* and *radius* under **rod, swing**

shy < AS. *sceoh*; see **modest, fear**

siagon, Gr. jaw; *siagonion*, dim.; see **jaw**

sialis, Gr. a kind of bird; see **bird**

sialon, Gr. saliva; see **spit**

sialos, Gr. fat; see **fat**

sibilo, *-atus*, L. hiss, whistle; *sibilatrix, -icis*, hissing; see **whistle**

sibyl < Gr. *Sibylla*, a female soothsayer; see **prophecy**

sibyna, L. (Gr. *sibyne*), a kind of spear; see **spear**

sic, L. thus, so, like; see **like**

sica, L. dagger; *sicula*, dim.; *sicarius*, assassin; see **dagger, kill**

siccus, L. dry; *siccativus*, drying; see **dry**

sicelicum, L. a kind of fleawort; see **plant**

sicera, L. (Gr. *sikera*), a spirituous drink; see **drink**

sicilis, L. sickle; *sicilicula,* dim.; see **sickle**

sicinnis, L. (Gr. *sikinnis*), a satyric dance; see **dance**

sick < AS. *seoc;* see **disease, pain**

sickle < AS. *sicel.*
  Gr. *ameterion,* n. sickle:
  Gr. *drepane,* f.; *drepanon,* n. sickle, scimitar; *drepanion,* n. dim.: drepanium, drepaniform, *Drepane punctata* (sickle-fish), *Drepanophorus modestus* (a nemertean), *Drepanorhynchus schistaceus* (a bird), *Drepanidium ranorum* (a protozoan), *Asparagus drepanophyllus* (an asparagus), *Acacia drepanolobium* (an acacia).
  L. *falx, falcis,* f. sickle, scythe; *falcicula; falcula,* f. dim.; *falcatus,* sickle-shaped, curved; *falcipedius,* bowlegged: falciform, falcate, falchion, defalcation, *Amphicoryne falx* (a foraminifer), *Habrodesmus falx* (a milleped), *Chrysopsis falcata* (a golden aster), *Anchura falciformis* (a fossil gastropod), *Plasmodium falciparum* (protozoan causing a kind of malaria).
  Gr. *harpe,* f. sickle, hook, a kind of hawk: *Harpethrix plana* (a milleped), *Harpidium rotundum* (a fossil brachiopod), *Harpephyllum caffrum* (an anacardiacead).
  Gr. *kropion,* n. scythe:
  Gr. *plegas, -ados,* f. sickle: *Plegadis falcinellus* (glossy ibis).
  Gr. *poastrion,* n. sickle for cutting grass:
  L. *sicilis,* f. sickle; *sicilicula,* f. dim.:
  Gr. *zanklon,* n. sickle: *Zanclus cornutus* (a fish), *Zanclopalpus rasalis* (a moth), *Zanclorhynchus spinifer* (a fish), *Dermatolepis zanclus* (a fish).
  See: **sword, cut, knife**

siclus, L. (Gr. *siklos*), shekel; *siklion,* dim.; see **money**

sicula, L. dim. of *sica,* dagger; see **dagger**

sicyus, L. (Gr. *sikyos*), cucumber, gourd; *sikydion,* dim.; see **melon**

sida, NL. a genus of the mallow family; see **mallow**

side < AS. *side.*
  L. *costa,* rib, side; see **rib**
  Gr. *eurax,* on one side, sideways:
  Gr. *hedra,* base, plane, side; see **sit**
  L. *ilium,* groin, flank; see **groin**
  Gr. *ixys, -yos,* f. waist, loins:
  Gr. *keneon, -os,* m. flank:
  Gr. *lagon, -os,* f. flank: *Lagonosticta nitidula* (a bird).
  Gr. *lapara,* f. flank, loin: laparocele, laparotome.
  L. *latus, -eris,* n. side, flank; *latusculum,* n. dim.; *lateralis,* of the side: lateral, bilateral, collateral, *Spirifer lateralis* (a fossil brachiopod), *Scutellaria lateriflora* (a skullcap), *Tmesisternus laterimaculatus* (a beetle), *Poa unilateralis* (seacliff-bluegrass).
  L. *lumbus,* m. loin; *lumbulus,* m. dim.; *delumbis,* lamed in the loins, weak: lumbar, lumbago, lumbarization, lumbosacral, iliolumbar.
  Gr. *osphys, -yos,* f. lower part of the back, loin; *osphydion,* n. dim.: osphyalgia, *Osphyolax pellucidus* (a fish), *Osphyoplesius anophthalmus* (a beetle).
  Gr. *para,* beside, alongside, near; see **near**
  Gr. *pleura,* f. side, rib; *pleurikos,* of the side and ribs: pleural, pleurisy, pleuropneumonia, pleurodont, pleuropodium, *Pleuronectes platessa* (plaice), *Pleurotoma babylonica* (a snail), *Pleurotus ulmarius* (a mushroom), *Pleuricospora fimbriolata* (an ericacead), *Coscinopleura digitata* (a fossil bryozoan), *Empleurum serrulatum* (a rutacead), *Gymnopleurus azureus* (a beetle).
  Gr. *psoa,* f. a muscle of the loin: psoas, psoitis.
  L. *secundus,* ranking below, unilateral; see **follow**
  L. *skelis, -idos,* rib, side; see **rib**
  See: **border, hip, wall, rib**

sidereal < L. *sidereus,* of the stars, < *sidus, -eris,* star, constellation; see **star**

sideritis, Gr. ironwort, a genus of the mint family; see **mint**

sideros, Gr. iron; see **iron**

sieve < AS. *sife.*
  Gr. *athelbo,* filter:
  L. *colum,* n. sieve, strainer, < *colo, -atus,* filter, strain: colander, cullet, colation, colature, colatorium, culvert, percolator.
  L. *cribrum,* n. sieve; *cribellum,* n. dim.; *cribrarius,* of a sieve, < *cribro, -atus,* sift: cribrate, cribrose, cribellum, cribble, *Cribrostomum pyriforme* (a foraminifer), *Coscinium cribriforme* (a fossil bryozoan), *Crabro cribrarius* (a wasp),

Dorcus cribellatus (a beetle), *Pronaonota cribrosa* (a blattid), *Grammostomum cribrum* (a foraminifer), *Pagiocerus cribricollis* (a beetle).

Gr. *dierama, -tos,* n. strainer: *Dierama pendulum* (elfin-wand).

L. *eliquo, -atus,* clarify, strain, sift: eliquate.

Gr. *ethmos; etheter, -os,* m. strainer, colander, < *etheo,* filter, strain; *apetheo,* strain off; *dietheo,* strain through, percolate: ethmoid, *Ethmocyathus lineatus* (a sponge), *Ethmosphaera polysiphonia* (a radiolarian), *Etheostoma maculatum* (a darter).

ML. *filtrum (feltrum),* n. fulled wool for straining: felt, filter, infiltration.

Gr. *haino,* sift, winnow: *Haenoblattina tenuis* (a fossil blattid).

Gr. *hylister, -os,* m. colander, filter, strainer; *hylistos,* strained, filtered; *hylizo,* filter, strain:

L. *incerniculum,* n. sifter, sieve, < *cerno, cretus,* separate, sift, distinguish: discern.

Gr. *kinachyra,* f. bag or sieve for bolting flour: *Cinachyra barbata* (a sponge), *Cinachyrella crustata* (a sponge).

Gr. *koskinon,* n. sieve; *koskinion,* n. dim.: *Coscinium latum* (a fossil bryozoan), *Coscinodiscus perforatus* (a diatom), *Metacoscinus reteseptatus* (a coral).

Gr. *kresera,* f. flour sieve; *kreserion,* n. dim.: *Cresera annulata* (a moth).

Gr. *ptisso,* winnow grain; *ptistes,* winnower; see **separate**

L. *sacco, -atus,* strain or filter through a bag; see *saccus* under **bag**

Gr. *salax, -akos,* m. sieve: *Salax lacordairei* (a beetle).

Gr. *sestron,* n. sieve, < *setho,* sieve, sift: *Stylatractus sethoporus* (a radiolarian).

Gr. *sinion,* n. sieve: *Sinion hageni* (a trichopterid), *Siniopelta costazi* (a bryozoan).

Gr. *trypeter, -os,* m. colander, sieve, strainer:

See: **choose, judge, separate, screen, fan**

sift < AS. *siftan;* see **sieve, separate, choose**

sigaloma, Gr. a smoothing, polishing instrument; see **smooth**

sigelos (sigalos), Gr. mute, silent; *sige,* silence; *Sigalion,* Egyptian god of silence; see **silent**

sigh < AS. *sican;* see **sad, breathe**

sigillum; sigilliolum, L. dim. of *signum,* mark; see **mark**

siglos, Gr. shekel, earring; see **earring**

sigma, Gr. eighteenth letter of the Greek alphabet; see **letter**

sign < L. *signum,* mark, flag, seal; see **mark**

sigras, Gr. a wild fig; see **fig**

sigyno- < Gr. *sigynes (sigynos),* spear; see **spear**

sil, L. a kind of yellowish earth, yellow ocher; see **earth**

silanus, L. (Gr. *silenos*), a fountain or jet of water from a head of Silenus; see **spring**

silaus, L. a genus of the carrot family; see **carrot**

silene, NL. a genus of the pink family; see **pink**

silenus, L. (Gr. *seilenos*), a bearded, bald, woodland deity; see **spirit**

silent < L. *silens, -entis; silentium,* n. stillness, quietness; *silentus,* still, quiet; *consilesco,* become still, hushed, dumb: silence.

Gr. *abakes,* speechless, childlike, innocent: *Abacetus antiquus* (a beetle), *Metabacetus immarginatus* (a beetle).

Gr. *achanes,* mute with astonishment: *Achanodes sympathetica* (a moth).

Gr. *aglossos,* without tongue, dumb:

Gr. *agryktos,* not to be spoken of:

Gr. *aklytos,* unheard:

Gr. *alalos,* speechless, dumb: alalia, *Alalomantis muta* (a mantid).

L. *allapsus, -us,* m. silent or stealthy approach:

Gr. *alogos,* speechless, absurd, irrational; see **fool**

Gr. *anaudia,* f. loss of speech; *anaudos,* speechless: *Anaudia felderi* (a moth).

Gr. *anepes,* without a word, speechless:

Gr. *aphasia,* f. speechlessness: aphasia.

Gr. *aphonos,* dumb, silent: *Aphonogryllus apteryx* (a cricket), *Aphonopelma seemanni* (a tarantula).

Gr. *aphthongos; aphthenktos,* speechless: aphthong, aphthongia.

Gr. *arrhemon, -os,* silent, dumb: *Arrhemon atropileus* (a bird).

L. *elinguis,* speechless:

Gr. *ellops, -opos,* mute (generally said of fish): *Ellopostoma megalomycter* (a fish).

Gr. *eneos,* dumb, stupid; *eneotes, -etos,* f. dumbness: *Eneodes hirsuta* (a beetle), *Eneoptera fasciata* (a cricket).

L. *inarticulatus,* dumb, indistinct, not fluent: inarticulate, inarticulateness.
L. *infans, -antis,* speechless: infant, infantile, infantilism.
L. *insonus,* silent: *Sylvilagus insonus* (a rabbit).
Gr. *katephas,* downcast, mute; see **sad**
Gr. *kophos,* dumb; see **dull**
L. *musso, -atus,* keep relatively silent, murmur; see **murmur**
L. *mutus,* speechless, inarticulate, silent, dumb; m. a mute person: mute, *Lachesis mutus* (bushmaster).
Gr. *myndos,* dumb: *Myndus pictifrons* (a bug).
Gr. *nythos,* dumb: *Nythobia pusio* (a wasp), *Nythosaurus larvatus* (a fossil reptile).
Gr. *sigelos (sigalos); sigodes,* silent; *sige,* f. silence; *Sigalion,* m. god of silence: *Sigelotroxis parvus* (a fly), *Sigaloceras calloviense* (an ammonite), *Sigodesmus innotatus* (a centiped), *Sigalon arenosum* (a polychaete), *Codosiga botrytis* (a protozoan).
Gr. *siopelos,* silent, still; *siope,* f. silence; *siopesis,* f. taciturnity: *Siopelus calabaricus* (a beetle).
L. *taceo, -itus,* be silent, say nothing; *tacendus,* to be kept silent, secret; *taciturnus,* not talkative, of few words; *tacitus,* implied, unmentioned; *tacitulus,* dim.; *reticio,* keep silent: tacit, taciturn, reticent.
See: **rest, deaf, sound, speak, ineffable**

siler, L. a kind of willow; see **willow**

silica; silicon < L. *silex, -icis,* any hard stone, a flint; *siliceus,* flinty; see **stone,** ELEMENTS under **thing**

siligo, L. a white wheat; see **grain**

siliqua, L. pod; *silicula,* dim.; see **bag**

silk < AS. *seolc,* < L. *sericum* (Gr. *serikon*), < *Serikos,* of the Seres (Chinese); *sericeus; sericus,* silken, silky; *sericarius,* of silks: silken, silky, silkworm, sericulture, serific, sericeous, serge, seric, serine, *Sericotachina vulpecula* (a fly), *Sericulus chrysocephalus* (a bower-bird), *Xylopia sericea* (a annonacead), *Aranea sericata* (a spider).
L. *bombyx, -ycis* (Gr. *-ykos*), m. silk, silkworm; *bombycinus,* silken: bombycine, *Bombyx mori* (silkworm), *Volvaria bombycina* (a mushroom), *Nosema bombycis* (a protozoan).
NL. *ceiba,* f. a genus of the silk-cotton family, < Ab.Am. *zeiba: Ceiba pentandra* (kapok-tree).
Gr. *mataxa (metaxa),* f. silk; *metaxion,* n. dim.; *metaxoton,* n. silk cloth: *Mataxa elegans* (a gastropod).

sillos, Gr. lampoon, satire; see **laugh**

silly < AS. *saelig;* see **fool, simple**

silphe, Gr. a kind of insect, now applied to a genus of beetles; see **beetle**

silphium, L. (Gr. *silphion*), a genus of the composite family; see **composite**

silurus, L. (Gr. *silouros*), sheatfish; see **catfish**

silus, L. pugnosed; see **nose**

silva (sylva), L. forest; *silvula,* dim.; see **forest**

silver < AS. *seolfor:* silversmith, silverware, silverfish, quicksilver, silverweed (*Argentina anserina*). See ELEMENTS under **thing**
L. *argentum,* n. silver; *argentatus; argenteus; argentinus,* of silver, silvery; *argentarius,* of silver; m. silversmith: argent, Argentina, argentite, argentiferous, argentol, argentate, argentous, argentic, *Leucodendron argenteum* (silvertree), *Parthenium argentatum* (guayule), *Ganoproctus argentifer* (a fly), *Bryophila argentacea* (a moth), *Sphyraena argentea* (a barracuda), *Begonia argentinensis* (a begonia).
Gr. *argyros,* m. silver; *argyridion; argyrion,* n. dim. a small coin; *argyreos; argyrikos; argyrites,* of silver, silvery; *argyphos,* silver-white, bright: argyrocephalous, argyrosis, argyria, argyrol, pyrargyite, argyrodite, *Argyrochlamys impudicus* (a fly), *Argyroneta palustris* (a spider).
Sp. *plata,* f. silver; *platina,* f. a substance like silver: platinum, platinoid, platan-iridium, La Plata, hexachlorplatinate.

silybum, L. (Gr. *silybon*), a kind of thistle; see **thistle**

simaruba, NL. a genus of the quassia family; see **quassia**

simblos (simblon), Gr. beehive; *simblios,* of a hive; see **hive**

simia, L. ape; see **monkey**

simila; similago, L. finest wheat flour; see **flour**

similis, L. like, resembling; see **like,** FIGURES OF SPEECH under **form**

simoma, Gr. anything turned up; see **over**
simple < L. *simplex, -icis,* onefold, unmixed, single: simplicity, *Hastula simplex* (a fossil gastropod), *Thorectes simplicidens* (a beetle), *Jasminum simplicifolium* (a jasmine), simpleton.
Gr. *apheles,* even, smooth, simple; see **smooth**
L. *elementarius,* pertaining to rudiments and first principles: elementary.
L. *inaffectus,* natural, simple:
L. *inalienatus,* unspoiled:
L. *incomptus,* unadorned: *Miltogramma incompta* (a fly).
L. *inornatus,* unadorned: *Amblyornis inornata* (gardener-bird).
Gr. *litos,* plain, simple, frugal; *litotes,* f. plainness, simplicity: litotes, *Litotherium complicatum* (a fossil mammal), *Litotethrips pasaniae* (a bug), *Harmolita tritici* (wheat joint-worm).
See: **one, clear, bare, letter, begin**

simpulum, L. small ladle; see **spoon**
simpuvium, L. a kind of bowl; see **cup**
simulo, -atus, L. imitate, copy; *simul,* at the same time; *simulacrum,* image, likeness; see *similis* under **like**
simus, L. (Gr. *simos*), flat-nosed; *simulus,* somewhat pug-nosed; see **nose**
sin < AS. *syn;* see **fault, falter, wander, bad, hurt**
sinado- < Gr. *sinas, -ados,* destructive; see *sinos* under **hurt**
sinamoros, Gr. hurtful, mischievous; see *sinos* under **hurt**
sinapis, L. (Gr. *sinapi*), mustard; see **mustard**
sincerus, L. pure, unmixed, genuine, see **pure, true, free, open**
sinciput, L. front part of the head; see *caput* under **head**
sindon, Gr. muslin; see **cloth**
sindron, Gr. mischievous; see **bad**
sine- < L. *sine,* without; see **not**
sinea, NL. a genus of bugs; see **bug**
sinew < AS. *sinu;* see **thread, flesh**
sing < AS. *singan:* singer, singing, singsong, song, songster.
Gr. *aeido,* sing; *asma, -tos,* n. song; *asmation,* n. dim.; *astes,* m. singer; *ode,* f. song; *odikos,* musical, of singing; *prosodia,* f. song, tone or accent of a syllable: ode, melody, comedy, rhapsody, parody, prosody, tragedy, threnody.
Gr. *ailinos,* song of lament, dirge; see **weep**
L. *axamentum,* n. religious hymn:
L. *bardus,* m. minstrel, singer; *barditus,* m. war song of the Germans: bard.
L. *cano, cantus,* sing; *cantillo, -atus,* sing low, hum; *cantito, -atus,* sing often; *canturio,* chirp; *canor, -is,* m. song, tune, sound; *canticum,* n. song in comedy; *canticulum,* n. dim.; *cantus, -us,* m. song; *cantulus,* m. dim.; *cantabilis,* worthy to be sung; *cantator; cantor, -is,* m. singer, minstrel, poet; *cantilena,* f. same old song, ditty, twaddle, chatter; *canorus; canticus,* singing, melodious, musical; *cantilenosus,* of song, poetic; *accentor, -is,* m. song-leader; *sincinium,* n. solo: cant, chant, canto, canticle, cantillate, cantata, incantation, enchant, accent, recant, incentive, descant, accentor, oscine, chanticleer, *Accentor modularis* (hedge-sparrow), *Melierax canorus* (a hawk), *Calamodyta cantillans* (a bird).
L. *carmen, -inis,* n. song, poetry, ballad: charm.
Gr. *elegos,* mourning song or poem; see **poem**
L. *epicedium* (Gr. *epikedeion*), n. funeral song, dirge: epicedium, epicedial, *Epicedia bigemmata* (a beetle).
AS. *galan,* sing: gale, nightingale.
Gr. *hymnos,* m. song of praise: hymn, hymnody, hymnology, Polymnia (Polyhymnia), *Polymnia uvedalia* (leafcup).
Gr. *ibykter, -os,* m. leader in a war song: *Ibycter ater* (a caracara).
Gr. *katalegma, -tos,* n. dirge:
L. *lallo, -atus,* sing a lullaby:
Gr. *melos,* n. tune, air, song; *melydrion,* n. dim.; *melodia,* f. singing; *melodos,* musical; L. *melicus* (Gr. *melikos*), musical, tuneful; *melopoios,* tuneful; *eumeles,* musical, euphonious: melody, melodrama, melodious, melodeon, Melpomene, dulcimer, Philomela, *Melospiza melodia* (song-sparrow), *Zamelodia ludoviciana* (rose-breasted grosbeak), *Melierax cantans* (a hawk).
Gr. *minyrisma, -tos,* n. a warbling; *minyristria,* f. warbler; *minyros,* warbling, whining; *minyrizo,* warble, hum: *Minyrus exaratus* (a beetle).
Gr. *molpazo,* sing; *molpe,* f. song; *molpastes,* m. minstrel; *molpetis, -idos,* f. singer and dancer; *eumolpos,* singing sweetly, singing well: *Molpemyia purpurea* (a fly), *Molpastes atricapillus* (a bird), *Eumolpus obscurus* (a beetle).

L. *nenia*, f. funeral song, dirge:

L. *oscen*, a singing bird; see **bird**

Gr. *paian, -os*, m. a hymn to Apollo to whom was transferred the office of Paian, physician to the gods: paean.

L. *psalmus* (Gr. *psalmos*), sacred song; see *psallo* under **music**

Gr. *skolion*, n. a kind of banquet song:

Gr. *Thamyris, -idos*, m. a boastful singer: *Thamyris antipodes* (an amphipod).

Gr. *threnos*, m. funeral song, dirge, lament; *threnodia*, f. lamentation: threnody, *Threnopipo borealis* (a bird).

See: **music, call**

singultus, L. sobbing, hiccup; see **hiccup**

singulus, L. one; *singularis*, one at a time, alone; see **one, alone**

sinido- < Gr. *sinis, -idos*, ravager, destroyer; see *sinos* under **hurt**

sinister, -tra, -trum, L. left, on the left hand; see **hand**

sink < AS. *sincan*; see **dip, sit, down**

sino- < Gr. *sinos*, harm, hurt, damage, injury, see **hurt**; < *Sino-*, pertaining to China and the Chinese, see **word**

sinodon, Gr. a fish; see **bream**

sinopis, Gr. a red ocher from Sinope on the Black Sea; see **earth**

sintor, Gr. ravenous, tearing; see **greed**

sinum, L. large cup or bowl; see **cup**

sinuosus, L. full of bendings, windings; see *sinuo* under **bend**

sinus, L. pocket, recess, bay; see **hollow**

-sion, L. suffix meaning act of, process of, having the nature of; see *-ion* under **nature**

siopelos, Gr. silent, still; see **silent**

sipalos, Gr. crippled; see *siphlos* under **hurt**

siparium, L. a theater curtain; see **curtain**

siphlos, Gr. crippled, maimed; see **hurt**

siphneus, Gr. mole; see **mole**

siphon < L. *sipho* (Gr. *siphon*), pipe, bent tube; *siphunculus*, dim.; see **pipe**

siptachoras, Gr. an Indian tree; see **plant**

sipya; sipydnos, Gr. meal-jar; see **pot**

sircula, L. a kind of grape; see **grape**

siren < *Siren* (Gr. *Seiren*), a sea-nymph who lured mariners to destruction; see **draw**

sirico- < Gr. *seirikon*, chicory; see **lettuce**

sirium, L. a plant, mugwort; see **wormwood**

Sirius, L. (Gr. *seirios; seiros*, the scorcher), dog-star; see **heat**

siro- < Gr. *seira*, cord, rope, band, string, see **rope**; < *siros*, pit, pitfall, see **hole**

sirpe, L. an asafetida plant; see **carrot**

sirus, L. (Gr. *siros*), underground pit for grain, cellar; see **room**

-sis; -sy; -cy, Gr. act or process of; see **make**

sisal < Ab.Am. *Sisal*, a seaport of Yucatan; see **thread**

sisarum, L. (Gr. *sisaron*), skirret; see **carrot**

sismos; sisilismos, Gr. a hissing; see **whistle**

sisopygido- < Gr. *seisopygis, -idos*, a wagtail; see **wagtail**

sister < AS. *sweoster*.

Gr. *adelphe*, f. sister:

Gr. *kasignete*, f. sister; *kasis, -ios*, f. sister: *Casigneta cochleata* (a grasshopper), *Casignetella argentula* (a moth).

L. *soror, -is*, f. sister; *sororcula*, f. dim.: *sororius*, of a sister, sisterly: sororal, sororiation, sorority, soroptimist, *Diabrotica soror* (a cucumber-beetle), *Typhlocyba sororcula* (a bug), *Festuca sororia* (ravine-fescue).

See: **woman, brother**

sistrum, L. (Gr. *seistron*), a rattle; see **rattle**

sisum, L. (Gr. *sison*), a genus of the carrot family; see **carrot**

sisymbrium, L. (Gr. *sisymbrion*), a fragrant plant; see **mustard**

Sisyphus, L. (Gr. *Sisyphos*), a mythological king whose perpetual punishment in Hades was to roll to the top of a hill a large stone that always rolled down again; typical of unavailing effort; see **hopeless**

sisyra, Gr. shaggy garment; see **garment**

sisyrinchium, L. (Gr. *sisyrinchion*), a genus of the iris family; see **iris**

sit < AS. *sittan*.
  Gr. *amischos*, without stalk, sessile: *Amiscogaster[Amischogaster] ruskini* (a wasp).
  L. *basterna*, f. sedan chair, litter:
  L. *clitella*, pack-saddle; see **saddle**
  Gr. *diphros*, seat, chariot; see **vehicle**
  Gr. *enolmos*, sitting on a tripod:
  L. *ephippium*, saddle; see **saddle**
  Gr. *epoazo*, sit on eggs, brood; *epoasis*, f. brooding; *epoastikos*, fond of sitting:
  Gr. *epochos*, sitting upon, mounted:
  L. *fertorium*, n. sedan chair:
  Gr. *hedos*, n. seat, abode, base, pedestal; *hedolion*, n. seat of the rowers, bench in a theater: *Hedotettix rusticus* (a grasshopper).
  Gr. *hedra*, f. seat, chair, base, plane, side; *hedrion*, n. dim.; *hedraios*, sitting, sessile, sedentary, stable, steadfast; *hedranon*, n. abode, dwelling, seat; *ephedra*, f. a sitting by, siege; *ephedros*, sitting upon; *kathedra*, f. seat of a bishop, abode, fundament, rump: hedrocele, tetrahedral, dodecahedron, octahedrite, cathedral, Sanhedrin, chair, chaise, *Edrioaster[Hedrioaster] saratogensis* (a cystoid), *Ephedra distachys* (a joint-fir), *Gonyaulax polyhedra* (a flagellate).
  Gr. *hisma, -tos*, n. foundation, seat; *hizema, -tos*, n. a settling down, sinking, < *hizo*, seat, place, settle, sink; *enizema, -tos*, n. seat; *kathisma -tos*, n. that on which one sits, seat, buttock:
  Gr. *hypopodion*, n. footstool:
  L. *lasanum* (Gr. *lasanon*), n. nightstool, closestool:
  L. *pulvinar*, cushioned couch, seat; see **pillow**
  L. *scamnum* (Gr. *skamnon*), n. bench, stool, seat, balk or ridge of earth left unplowed; *scabellum; scamellum*, n.; *scamillus*, m. dim.: shamble, *Scamnoceras angulatum* (an ammonite), *Heptium scamillatum* (a milleped).
  L. *sedeo, sessus*, sit; *sessilis*, sitting, seated; *sessibulum*, n. seat, chair; *sessio, -onis*, f. a sitting; *sessor, -is*, m. sitter, inhabitant; *sessorium*, n. seat, chair, abode; *sedes, -is*, f. seat, place, abode, home; *sedecula*, f. dim.; *sedile, -is*, n. seat, bench, stool; *sedentarius*, of sitting; *sedibilis*, that can be sat upon; *sedimentum*, n. a settling; *assideo*, sit beside; *insideo*, sit upon; *obsideo*, stay; *sessito*, sit long; *subsido*, sit down, settle, sink: session, sessile, sediment, sedentary, see, seance, siege, assessor, assize, size, residence, residue, residual, dissident, president, assiduous, possessive, supersede, subsidy, insidious, surcease, incertae sedis, *Cornus sessilis* (a dogwood), *Trillium sessile* (a trillium), *Chaetodon sedentarius* (butterfly-fish), *Smilacina sessilifolia* (a false Solomon's-seal), *Acanthopanax sessiliflorus* (an araliacead), *Sisyphus subsidens* (a beetle).
  L. *seliquastrum*, n. seat, stool:
  L. *sella*, f. seat, stool, settle, saddle; *sellula*, f. dim.; *sellaris*, of a seat; *bisellium*, n. seat of honor; *subsellium*, n. bench, seat: selliform, sellary, angustisellate, *Mnuphorus sellatus* (a beetle), *Coprophilus sellula* (a beetle), *Dascyllus albisella* (a fish), *Iridio sellifer* (a fish), *Cypripedium selligerum* (a moccasin-flower), *Amia sellicauda* (a fish), *Tineola biselliella* (a clothes-moth).
  Gr. *selma*, deck, rowing-bench, seat; see **base**
  L. *sido*, sit down, settle, alight, sink:
  Gr. *skolythron*, n. stool:
  Gr. *skoramis, -idos*, f. night-stool:
  L. *solium*, n. royal seat, throne, chair of state: *Taenia solium* (a tapeworm).
  Gr. *sphelas, -tos*, n. footstool, pedestal: *Sphelatus lehoni* (an Eocene echinoid).
  Gr. *sympselion*, n. bench, seat:
  Gr. *thakema, -tos*, n. a sitting, seat; *thakesis*, f. a sitting; *thakos*, m. chair, seat, office; *thasso (thatto)*, sit, sit idle: *Coprothassa sordida* (a beetle).
  Gr. *thranos*, m. bench, form, stool; *thranion*, n. dim.: *Thranium tenuipes* (a dino-flagellate).
  L. *thronus* (Gr. *thronos*), m. an elevated seat; *thronion*, n. dim.: throne.
  L. *transtrum*, n. bench for rowers, bank of oars:
  L. *tripus*, three-legged chair or stool; see **frame**
  See: **rest, place, house, base, frame**

sitanion, NL. a genus of grasses; see **grass**

sitis, L. thirst, dryness; *sitiens*, thirsty; *siticulosus*, thirsty; see **dry**

sitos, Gr. food, grain; *sitarion; sition*, dim.; *sitikos*, of food; see **food**

sitta, L. (Gr. *sitte*), nuthatch; see **nuthatch**

sittybus, L. strip of parchment; Gr. *sittybon*, small skin, piece of leather; see **skin**

situla, L. bucket or urn for drawing water; *sitella*, dim.; see **bucket**

situs, L. place, site, location; *situatus*, located; see **place**

sium, L. (Gr. *sion*), water-parsnip; see **carrot**

six < AS. *six*.

Gr. *hex*, six; *hexas, -ados*, f. the number six; *hexadikos*, of six; *hektos*, sixth; *hexakis*, six times; *hexekonta*, sixty; *hexekostos*, sixtieth; *hexakosioi*, six hundred: hexameter, hexagram, hexose, hexylamine, hexachlorocyclohexane (gammexane), *Hexagenia bilineata* (a mayfly), *Hexamitus intestinalis* (a protozoan), *Hexathele hochstetteri* (a tarantula).

L. *sex*, six; *sextus; sextarius*, sixth; *seni*, six, six each; *senarius*, consisting of six; *sexdecim(sedecim)*, sixteen; *sexaginta*, sixty; *sexagesimus*, sixtieth; *sescenti*, six hundred; *semestris*, of six months, semi-annual; *sexvir*, m. board or committee of six: sextet, sextant, sextile, Sistine, semester, sexagesimal, sexangular, sexennial, sexenary, sextuplet, sextan, *Cnemidophorus sexlineatus* (a lizard), *Eucapnodes sexmaculata* (a moth), *Atta sexdens* (an ant).

sixis, Gr. a hissing; see *sismos* under **whistle**

skate < D. *schaats*, see **slip**; < ON. *skata*, see **shark**

skeleton, Gr. dried body, mummy; see **bone, frame**

skeptic < Gr. *skeptikos*, thoughtful, critical, doubtful; see **doubt, think**

sketch < D. *schets*, < It. *schizzo*, < L. *schedius* (Gr. *schedios*), made suddenly, offhand, on the spur of the moment, hence a rough outline of a design; see **form, map**

skill < ON. *skil;* see **art, know**

skim < OF. *escumer,* < OHG. *scum;* see **separate**

skin < ON. *skinn:* skinny, skinflint, redskin, foreskin, Skinner.

L. *arnacis, -idis* (Gr. *arnakis, -idos*), f. sheepskin:

L. *brisa*, f. grape-skin, refuse after pressing: *Brisa lucida* (a crab).

Gr. *byrsa*, hide, skin; see **leather**

L. *callum*, n.; *callus*, m. hard skin; *callosus*, with a hard skin; *occallatus*, indurated: callus, callous, callosity, corpus callosum, *Catasetum callosum* (an orchid).

Gr. *chroa; chroia; chroma; chros*, surface of the body, skin, color of the skin; see *chroma* under **color**

L. *crusta*, f. hard outer surface of a body, coat, rind, bark, shell; *crustula*, f. dim.; *crustosus*, covered with a rind; *crusto, -atus*, cover with a coating: crust, crustate, crustose, crustaceous, crustacean, crustification, incrustation, crusty, *Penicillium crustaceum* (green cheese-mold), *Conularia crustula* (a fossil coelenterate), *Sisyrinchium incrustum* (a blue-eyed grass), *Chamaesiphon incrustans* (an alga).

L. *cutis*, f. skin; *cuticula*, f. dim.; *intercus*, under the skin: cuticle, cutis, cutin, subcutaneous, cuticular, cuticolor, cutigeral, *Neascus cuticola* (a worm).

L. *deglubo, -uptus*, peel off, husk, shell:

Gr. *derma, -tos; deros*, n.; *dora*, f. skin, hide; *dermation*, n.; *dermatis, -idos*, f. dim.; *dermatikos; dermatinos*, of skin, leathern; *dorikos*, of skin or hide; *dartos*, skinned, flayed, < *dero*, skin, flay; *doreus*, m. flayer: *apodora*, f. a peeling of the skin; *epidermis, -idos*, f. outer skin, cuticle: dermis, epidermis, dermatitis, dermatogen, dermatophyte, dermatosis, dermatoglyphics, *Dermatolepis inermis* (a fish), *Dermatinus lugens* (a bug), *Dermacentor venustus* (a tick that transmits Rocky Mountain spotted fever), *Dermanysus avium* (a mite), *Dermestes vulpinus* (leather-beetle), *Lasioderma serricorne* (cigarette-beetle), *Scleroderma vulgare* (a fungus), *Chalcodermus segnis* (a beetle), *Cryptococcus epidermidis* (a fungus), *Apoderus biguttatus* (a beetle), *Lasiodora rosea* (a tarantula), *Leptodora kindti* (a cladoceran), *Acanthodoras spinosissima* (a catfish).

Gr. *derris*, leather covering, coat, skin; see **garment**

Gr. *eklemma, -tos*, n. peel, rind:

Gr. *episkynion*, n. skin over the eyebrows: *Episcynia inornata* (a gastropod).

L. *exuvia*, f. cast skin, slough: exuvial, exuviate.

Gr. *ixale*, f. goatskin worn by actors:

Gr. *koas*, fleece, sheepskin; *kodion*, dim.; see **wool**

Gr. *leberis, -idos*, f. sloughed skin: *Scapholeberis mucronata* (a water-flea).

L. *liber*, bark of a tree; see **bark**

Gr. *lopos*, m.; *lopisma, -tos*, n. bark, husk, peel, shell, skin; see *lepis* under **scale**

L. *mastruca*, f. sheepskin, skin; *mastrucatus*, clad in skins: *Bion mastrucatus* (a blattid).

L. *melota* (Gr. *melote*), f. sheepskin with the wool: melote.

L. *membrana,* skin that covers parts of the body; see **membrane**

Gr. *molgos,* m. hide, skin; *molginos,* of hide or skin: *Molgosporidium ellipticum* (a protozoan), *Molgula tubulosa* (a tunicate).

Gr. *nebris, -idos,* f. fawnskin; *nebridion,* n. dim.; *nebrites,* like a fawnskin: nebris, nebrismus.

L. *omentum,* apron of skin, caul; see **membrane**

Gr. *pekos (peskos),* n. hide, rind, skin: *Pecomyia maculata* (a fly).

L. *pellis,* f. skin; *pellicula,* f. dim.; *pelliceus; pellinus; pellitus,* of skins, made of skins, covered with skin; *pellio, -onis,* m. furrier: pellicle, pelliculate, pelt, peltry, pelisse, pellagra, erysipelas, surplice, *Baridius pelliceus* (a beetle), *Eucalyptus pellita* (a eucalyptus), *Tinea pellionella* (clothes case-moth), *Trachonurus sentipellis* (a fish).

Gr. *phorine,* f. thick skin, hide; *phorinion,* n. dim.:*Phoriniophylax phoeda* (a fly).

Gr. *poroma, -tos,* n. callus; *porosis,* f. formation of a callus:

L. *scortum,* n. skin, hide; *scorteus,* of hide or leather:

L. *sittybus,* m. strip of parchment; Gr. *sittybon,* n. small skin, piece of leather:

Gr. *skytos,* n. hide, skin, leather; *skytinos,* of leather, leathern: scytodepsic, *Scytonema mirabile* (an alga), *Scytonotus nodulosus* (a milleped).

Gr. *sterphos (terphos),* n. hide, skin: *Sterphus antennalis* (a fly), *Terphothrix lanaria* (a fly).

AS. *sweard,* skin, covering, turf; see **sod**

Gr. *syphar, -os,* n. slough of skin, old or wrinkled skin: *Sypharochiton pelliserpentis* (a chiton).

L. *talla,* f. onion-skin: *Talla quercus* (a coccid).

L. *tergilla,* f. skin or sward of pork; *terginum,* n. rawhide: *Tergilla familiaris* (a mite).

L. *tunica,* garment, skin, husk, peel; see **garment**

Gr. *tyle,* callus, pad, bolster; see **knot**

L. *vellus,* fleece, pelt; see **wool**

Gr. *zalmos,* m. Thracian word for a skin:

See: **membrane, cover, leather, shell, bark, bag, sheath**

skull < ON. *skal;* see **head**

skunk < Ab.Am. *segonku;* see **weasel**

sky < ON. *sky,* cloud; see **heaven**

-sky; -off; -vitch, Slav. relative of, son of, residence at; see **son**

slack < AS. *slaec;* see **free, rest, weak**

slag < G. *slagge; schlacke,* dross, scoria.

Gr. *diphryges,* n. slag from copper-smelting:

Gr. *ekbolas, -ados,* f. anything thrown out, dross:

Gr. *pompholyx,* bubble, slag; see **bubble**

L. *scoria* (Gr. *skoria*), f. dross, slag: scoria, scoriaceous.

See: **ash, dirt**

slander < OF. *esclandre,* < L. *scandalum* (Gr. *skandalon*), trap; see **curse, lie**

slant < uncertain origin; see **slope**

slate < OF. *esclat,* splinter, chip; see **stone, part**

slave < ML. *sclavus,* one taken in war, a Slavonian; see **servant, make, bind**

sleep < AS. *slaep:* sleeper, sleepless, sleeping sickness, Sleepy Hollow, oversleep.

Gr. *brizo,* be sleepy, nod:

L. *concubius,* lying in sleep; *concubium,* n. time of first sleep; *Cuba,* goddess who protected sleeping children; see *cubo* under **lie**

Gr. *darthano,* sleep:

Gr. *derkeunes,* sleeping with the eyes open:

L. *dormio, -itus,* sleep; *dormito, -atus,* be drowsy, begin to sleep; *dormitatio, -onis,* f. sleep; *dormitator; dormitor, -is,* m. sleeper; *dormitorium,* n. sleeping-room; *dormitorius,* of sleeping: dormant, dormitory, dormer, dormouse, obdormition, *Dormitator maculatus* (a fish).

Gr. *dysegertos,* hard to waken:

Gr. *geleches,* sleeping on the ground: *Gelechia notatella* (a butterfly).

L. *hiberno, -atus,* keep inactive in winter quarters; see *hibernum* under **year**

Gr. *hypnos,* m. sleep, slumber; *hypnikos,* of sleep; *hypnaleos; hypnelos; hypneros; hypnidios; hypnodes,* drowsy; *hypnotikos,* putting to sleep; *kathypnos,* fast asleep; *philypnos,* sleep-loving: hypnotism, hypnocyst, hypnophobia, hypnosis, hypnobate, hypnotoxin, agrypnia:

Gr. *iauo,* sleep; *iauthmos,* m. sleeping-place, den, lair:

Gr. *karos,* m. heavy sleep, torpor; *karodes,* drowsy; *karosis,* f. drowsiness: *karotikos,* soporific: carotid.

Gr. *katanyxis,* f. stupefaction, slumber:

Gr. *katheudo,* lie down to sleep, lie still, sleep:

Gr. *knosso,* slumber, sleep:

Gr. *koma, -tos,* n. deep sleep, slumber, < *koimao,* put to sleep; *koimesis,* f. a lying down to sleep; *koimethra,* f. a place to sleep; *koimeterion,* n. sleeping-room, burial-place; *koimismos,* m. a putting to sleep; *koimistikos,* of putting to sleep: coma, comatose, cemetery, *Comatopselaphus parcepunctatus* (a beetle), *Cephalocoema lineata* (a grasshopper).

Gr. *lethargos; lethargikos,* drowsy, forgetful; see *lethe* under **forget**

Gr. *Morpheus,* god of dreams; see *morphe* under **form**

Gr. *negretos,* unwaking, in deep sleep: *Negretus ertli* (a neuropterid).

Gr. *nokar, -os,* n. sleep, sloth; *nokarodes,* sleepy, slothful:

Gr. *nystalos,* drowsy, sleepy, < *nystazo,* nod, nap; *nystagmos,* m. drowsiness; *nystaktes,* m. nodder: nystagmus, *Nystactes tamatia* (a bird).

L. *oscitans,* listless, drowsy, yawning; see **dull**

L. *somnus,* m. sleep; *somnialis,* of sleep and dreams; *somniator, -is,* m. dreamer; *somniculosus; somniosus; somniolentus,* drowsy, sleepy; *somnifer; somnificus;* bringing or causing sleep, soporific, narcotic; *somnium,* n. dream: somnambulist, somnolent, somnifacient, somniferous, somnific, somnifugous, somniloquent, insomnia, *Ellopia somniaria* (a beetle), *Papaver somniferum* (opium-poppy).

L. *sopor, -is,* m. deep sleep, lethargy, stupor; *soporus,* sleep-bringing; *sopio, -itus,* lull to sleep, calm, quiet: soporific, sopient, *Bulla sopita* (a gastropod), *Terebellum sopitum* (a gastropod).

See: **rest, dull, forget, soothe**

sleet < AS. *slete* (G. *schlosse,* hail); see *grando* under **ice**

sleeve < AS. *slief.*

Gr. *cheiris, -idos,* sleeve, glove; *cheiridion,* dim.; see **glove**

Gr. *knemis, -idos,* f. greave, legging; *knemidotos,* with leggings: *Cnemidophorus tesselatus* (a lizard), *Cnemidothrix protensus* (a beetle), *Gongrocnemis mutica* (a katydid), *Pelorocnemis puntigera* (a beetle), *Priocnemis flavicornis* (a wasp).

L. *manica,* f. long sleeve of a tunic, glove; *manicarius,* of sleeves; *manicatus,* with sleeves; *manulea,* f. a long sleeve; *manuleatus,* with long sleeves; *manucium,* n. glove, muff: *Manicaria saccifera* (sleeve-palm), *Gunnera manicata* (a halorhagidacead), *Anthidium manicatum* (a carder-bee).

L. *ocrea,* greave, legging, sheath; see **sheath**

See: **sheath, garment, cover, stocking**

slender < uncertain origin; see **thin**

slide < AS. *slidan;* see **slip**

slime < AS. *slim.*

Gr. *asis,* mud, slime; see **earth**

Gr. *blennos,* n. mucus: blennorrhea, antiblennorrhagic, *Blennosperma californicum* (a composite).

Gr. *blichodes,* running at the nose; *blichanodes,* clammy:

Gr. *ekbrasma, -tos,* n. scum:

Gr. *koryza,* a running at the nose, catarrh; see **disease**

Gr. *lape (lampe),* f. scum, mold, phlegm; *lamperos,* scummy, slimy: *Lamperos micans* (a beetle), *Lamperostoma maculatum* (a gastropod).

Gr. *lemphos,* m. mucus from the nose, snotty: *Lemphus mancus* (a beetle).

L. *limus,* mud, mire, slime; see **earth**

L. *mucus,* m. secretion from the nose; *mucidus,* slimy: mucus, mucin, mucilage, mucosity, pyromucic, *Treponema mucosum* (spirochaete of pyorrhea), *Ptyas mucosa* (rat-snake).

Gr. *myxa,* f. mucus; *myxarion,* n. dim.; *myodes,* mucid; *myxinos,* m. slime-fish, a kind of mullet: myxedema, myxadenitis, myxoma, myxomatosis, Myxomycetes, myxamoeba, polymyxin, *Myxine glutinosa* (hagfish), *Lithomyxa calcigena* (an alga), *Myxosoma cerebralis* (a protozoan), *Myxidium folium* (a protozoan).

L. *pituita,* f. phlegm, rheum; *pituitosus,* full of phlegm: pituite, pituitary, pituitous.

L. *virus,* slimy liquid, poison; see **poison**

See: **glue, earth, smooth**

sling < AS. *slingan;* see **throw**

slip < AS. *slippan.*

L. *labor, lapsus,* slide, glide, slip, fall; *labidus,* slippery; see **fall**

L. *lubricus,* slippery, < *lubrico, -atus,* make smooth or slippery: lubricate, lubricator, *Nemalion lubricum* (a red alga), *Holothuria lubrica* (a sea-cucumber).

L. *olisthos,* m. slipperiness; *olistheros; olibros,* slippery; *olisthesis,* f. a slipping: *Diapterus olisthostomus* (mojarra), *Olibrosoma testacea* (a beetle).

AS. *slidan,* slip: slide, landslide, slide-rule.

Gr. *sphaleros,* slippery, treacherous: sphalerite, *Sphalerostoma layardi* (a gastropod).

See: **smooth, fall, fat**

slit < AS. *slitan;* see **cut, hole**
slope < AS. *aslopen.*
 Gr. *anantes,* uphill, steep; *katantes,* downhill: *Catantostoma clathratum* (a fossil
  gastropod).
 L. *arduus,* lofty, steep: arduous.
 F. *biais,* slant, slope: bias.
 L. *clima* (Gr. *klime*), f. supposed inclination of the earth from the equator to the
  poles; region, zone: clime, climate, acclimated.
 L. *clino, -atus* (Gr. *klino*), bend, slope, slant, tend: *klisis,* f. inclination, bending;
  klitos, n.; *klitys, -yos,* f. slope, hillside; *acclinis; declinis; inclinis; reclinis,*
  leaning, sloping (in the direction indicated by the preposition); *apoklines;*
  enklines; epiklines; periklines, leaning, sloping; *epiklintes,* moving sideways:
  incline, declination, anticline, syncline, diclinic, microcline, pericline, recline,
  declension, enclitic, *Clinopodium coccineum* (basil), *Apoclisis rupestris* (a moth),
  Colubrina reclinata (a rhamnacead), *Clitocybe dealbata* (a mushroom), *Fundulus
  heteroclitus* (a killifish), *Trillium declinatum* (a trillium), *Epiclintes radiosa* (a
  protozoan).
 L. *clivus,* sloping hillside; *clivulus,* dim.; *acclivis,* uphill, steep; *declivis,* down-
  hill; see **mountain**
 L. *descensus,* m. downward course, inclined way, declivity: descent.
 L. *devexus,* inclined, descending, sloping, steep; *subvexus,* sloping upward:
 L. *diagonalis,* < Gr. *diagonos,* from angle to angle, oblique: diagonal.
 Gr. *dochmos; dochmios,* slanting, oblique: *Dochmiogramma filamentosa* (a moth).
 Gr. *eurax,* on one side, sideways; see **side**
 L. *fastigium,* slope, declivity, pitch, gable-end; see **gable**
 L. *gradus,* degree, stage, pitch; see **step**
 Gr. *illos; illis; illis, -idos; illodes,* looking obliquely, squinting; *illosis,* f. distortion, <
  illaino, look obliquely, squint: *Illops corniculata* (a beetle), *Illis medana* (a
  beetle).
 Gr. *katapheres,* inclined, downhill:
 Gr. *lechrios,* slanting, crosswise: lechriodont, *Lechriopyla mystax* (a protozoan),
  Lechriolepis anomala (a moth).
 L. *limus,* sidelong, sideways, oblique; *limulus,* dim.; *Limulus polyphemus* (king-
  crab), *Pteria limula* (a fossil pelecypod), *Hemiospis limuloides* (a Silurian
  crustacean), *Paleolimulus avitus* (a eurypterid).
 Gr. *loxos,* slanting, crosswise: loxoclase, loxodromic, loxic, *Loxocrinus globulus* (a
  Permian crinoid), *Loxorhynchus crispatus* (a crustacean), *Loxoconcha granulata*
  (an ostracode), *Loxops virens* (a bird).
 Gr. *neusis,* inclination, < *neuo,* incline, nod; see **yes**
 L. *obliquus,* slanting, sideways, indirect, < *liquis,* oblique: oblique, obliquity,
  Ctenodonta obliqua (a fossil pelecypod), *Euxestina obliquistriata* (a fly),
  Ptaeroxylon obliquum (a sneezewood).
 L. *obstipus,* inclined to one side, oblique: *Marginulina obstipa* (a foraminifer).
 Gr. *parablops, -opos,* looking askance, squinting: *Parablops pauper* (a beetle).
 L. *pentoncium* (Gr. *pentonkion*), n. quincunx:
 Gr. *pronopes,* stooping forward:
 Gr. *rhepo,* incline, slope; *rhepsis; rhope,* f. inclination, downward bias; *anarrhopos,*
  tilted up; *anorhepes,* inclined upward; *eurhopos,* inclining or sliding easily;
  katarrhopos, sloping downward: *Dirrhope rufa* (a wasp).
 L. *rhombus* (Gr. *rhombos*), a figure with sloping sides; see **diamond**
 Gr. *skolios,* curved, bent, oblique; see **bend**
 L. *strabus* (Gr. *strabos*), squinting, oblique; *strabo, -onis,* m. squinter: strabismus,
  Straboscopus aculeatus (a beetle).
 F. *zigzag,* alternately changing direction by sharp angles: zigzag, *Zigzagiceras
  postpollubrum* (a Jurassic ammonite), *Quinqueloculina zigzag* (a foraminifer),
  Pecten zigzag (a scallop), *Aturia ziczac* (a fossil cephalopod).
 See: **bend, cliff, fall, mountain**

slough < AS. *sloh;* see **marsh**

slow < AS. *slaw:* slowly, slowness, sloth.   Derive: slowworm.
 Gr. *bradys,* slow, late; *bradyteros,* slower; *bradistos (bardistos),* slowest; *bradytes,*
  f. slowness, < *bradyno,* slow up, delay; *bradykarpos,* late-fruiting; *bradytokos,*
  late-bearing: bradycrotic, bradypepsia, bradysuria, *Bradypus tridactylus* (three-
  toed sloth), *Bradynobaenus gayi* (a wasp), *Bardistopus papuanum* (a fly).
 L. *desidiosus,* slothful, idle; see *deses* under **rest**
 L. *dilatorius,* delaying; see **delay**
 L. *impromptus,* not ready, not quick:
 L. *instrenuus,* inactive, sluggish:
 L. *lentus,* viscous, slow; see **bend**
 Gr. *mocheles,* moving slowly, sluggish:

Gr. *nothros; nothes,* sluggish, slothful, torpid: *Nothrotherium shastense* (a fossil mammal).

Gr. *oknos,* m. hesitation, sluggishness, reluctance; *okneros,* hesitating, shrinking: *Ocnotherium gigas* (a fossil sloth), *Ocnerostoma piniarella* (a moth).

L. *piger, -gra, -grum,* reluctant, slow, dull, lazy; *pigredo, -inis,* f. slothfulness, < *pigro, -atus,* be indolent, slow, dilatory; *repigratus,* retarded, slow: *Didelphys pigra* (Florida opossum).

L. *pressus,* moderate, slow, suppressed:

L. *segnis,* slow, slothful, dilatory; *segnitia,* f. slowness, dilatoriness: *Thinobadistes segnis* (a fossil ground-sloth).

L. *spissus,* thick, slow, tardy, late; see **thick**

L. *tardus,* slow; *tardo, -atus,* hinder, delay; *tardabilis,* that renders slow: *retardatus,* delayed, hindered: tardy, retard, retardation, tardigrade, bustard *(Otis tarda).*

See: **delay, stiff, thick, against, lessen, hold, rest, stand, dull**

sly < ON. *slaegr;* see **know, art**

small < AS. *smael;* see **little**

smaragdus, L. (Gr. *smaragdos*), a precious stone of green color, emerald; *smaragdinus,* emerald-green; see **stone**

smaris, L., Gr. a sparoid fish; see **bream**

smart < AS. *smeortan;* see **bite, burn, pain, nettle, arouse**

smear < AS. *smerian;* see **spread, plaster, fat, dirt, spot**

smeche, Gr. a kind of beet; see **beet**

smecticus, L. (Gr. *smektikos*), cleansing; *smegma,* any cleansing agent, soap; see **wash**

smell < uncertain origin.

Gr. *aroma, -tos,* n. smell, spice; *aromatikos,* fragrant, spicy: aromatic, aromatite, *Aromia moschata* (musk-beetle), *Aromachelys odorata* (musk-turtle), *Rhus aromatica* (fragrant sumac).

L. *aspalathus* (Gr. *aspalathos*), a thorny shrub yielding a fragrant oil; see **bean**

L. *balsamum* (Gr. *balsamon*), a fragrant gum; see **resin**

Gr. *bdeo,* break wind, stink; *bdesma, -tos,* n.; *bdolos,* m. stench: *Bdeogale crassicauda* (a civet), *Lamium galeobdolon* (yellow dead-nettle).

Gr. *bromos,* m. stench, smell; *bromodes; bromosus,* stinking, fetid: bromine, bromide, hydrobromic acid, tribromethanol (avertin).

Gr. *eurhis, -inos,* keen-scented: *Leistus(Eurhinophorus) depressus* (a beetle).

L. *feteo (foeteo),* stink; *fetidus; fetulentus,* stinking; *fetor, -is,* m. bad or offensive smell; *fetutina,* f. a stinking place: fetid, asafetida, fetor, *Symplocarpus foetidus* (skunk-cabbage), *Synodus foetens* (galliwasp), *Ferula foetida* (devil-dung), *Sideroxylon foetidissimum* (an ironwood).

L. *fragro, -atus,* smell sweetly: fragrance, *Osmanthus fragrans* (yellow olive), *Gaultheria fragrantissima* (fragrant wintergreen).

Gr. *grasos,* m. odor of a goat; *grason, -os,* smelling like a goat:

L. *halatus,* m. odor, fragrance, perfume:

Gr. *hedypnoos,* sweet-smelling, fragrant:

L. *incensum,* incense; see **resin**

Gr. *keodes,* fragrant like incense:

Gr. *kinabra,* f. smell of a he-goat: *Cinabra hyperbia* (a moth).

Gr. *knisa (knissa),* f. steam and smell from cooking fat meat; *knisarion,* n. dim.; *knisaleos,* filled with the smell of cooking; *knisodes; knisotos,* steaming, smelling: *Cnissocnema neuhaussi* (a moth).

Gr. *libanotos,* frankincense; see **resin**

Gr. *malabathron,* n. aromatic leaf of an Indian plant: malabathrum, *Cinnamomum malabathrum* (Malabar leaf), *Melastoma malabathricum* (a melastoma).

L. *mephitis, -is,* f. noxious exhalation, bad odor, stench: mephitic, *Mephitis occidentalis* (skunk), *Psoralea mephitica* (a scurfpea).

L. *moschus* (Gr. *moschos*), m. < Per. *mushk,* musk (secretion of the musk-deer, *Moschus moschiferus*), musky, moschiferous, muscatel, nutmeg, muscat, moschatel *(Adoxa moschatellina),* muskrat, muskmallow *(Hibiscus abelmoschus), Rosa moschata* (musk-rose), *Abelmoschus esculentus* (okra, gumbo).

L. *nidor,* steam and smell from cooking; see **cloud**

L. *odor, -is,* m. smell, scent; *odoratus; odorus,* having a smell, fragrant; *inodorus,* without smell: odor, odoriferous, *Viola odorata* (sweet violet), *Artocarpus odoratissimus* (marang), *Solidago odora* (sweet goldenrod), *Monarda citriodora* (lemon bee-balm), *Myrica inodora* (a wax-myrtle).

L. *oleo,* emit an odor, smell; *olax, -acis; olens; olidus,* smelling, odorous; *olfacio, -actus.* smell, scent; *olfactorium,* n. smelling-bottle; *beneolens,* fragrant; *graveolens,* strong-smelling, noisome; *odoriferous; suaveolens,* fragrant: olfactory, redolent, *Olax imbricata* (stinkwood), *Staphylinus olens* (a beetle), *Pelargonium graveolens* (a geraniacead), *Artemisia redolens* (Chihuahua sagebrush), *Callichroma suaveolens* (a beetle).

Gr. *ozo*, smell; *odme; odode; osme*, f. odor, scent; *odneros; osmeros*, odorous; *osphrantos*, smellable; *osphrantikos*, capable of smelling; *osphradion*, n. strong smell; *osphresis*, f. sense of smell; *odmaleos*, stinking; *anosmos*, without smell; *dysodia*, f. stench; *enodes*, fragrant; *dysodes; kakodes*, stinking, foul; *epodes; kakosmos*, ill-smelling; *ozaina*, f. fetid polyp in the nose: ozone, ozocerite, ozena, osmium, anosmia, anosmatic, cacodyl oxide, osphradium, osmaterium, *Osmanthus ilicifolius* (holly-leaved olive), *Barosma ovata* (a buchu), *Diosma ericoides* (a rutacead), *Osphromenus olfax* (a fish), *Osphranter robustus* (wallaroo), *Cacosmus gracilicornis* (a bug), *Osmerus mordax* (smelt), *Osmaronia cerasiformis* (osoberry), *Dysodia acerosa* (prickly dogweed), *Xylosma heterophylla* (hollyxylosma), *Melinosma pannosa* (a sabiacead), *Evodia[Euodia] fraxinifolia* (a rutacead), *Klebsiella ozaenae* (a bacterium).

L. *paedidus*, nasty, filthy, stinking; *paedor, -is*, m. filth, stench: *Paederia foetida* (Chinese fever-vine).

L. *pedo, -itus*, break wind; *peditum*, n. a breaking wind:

Gr. *perdomai*, break wind; *pepradile (pradile); porde*, f. a breaking of wind; *pordon, -os*, m. stinkard: *Lycoperdon lilacinum* (a puffball).

L. *putidus*, stinking, fetid, < *puteo*, stink; *putor, -is*, m. foul smell, stench; *oriputidus*, having a stinking mouth, with bad breath: *Mustela putorius* (European polecat, fitchew).

L. *rancidus*, stinking, offensive; *rancidulus*, dim.; *rancor, -is*, m. bad smell or flavor: rancid, rancidity, rancescent, rancor.

AS. *stincan*, smell offensively: stink, stinker, stinkard, stench.

L. *suffio, -itus*, fumigate, perfume, scent; *suffimentum*, n. incense; *suffitor, -is*, m. fumigator; *suffitus, -us*, m. a fumigation, perfuming:

Gr. *tangos*, rancid; *tange*, f. rancidity, a putrid abscess:

L. *teter*, foul, noisome, offensive, hideous; see **bad**

L. *thus, thuris (tus, turis)*, incense; *thuralis*, of incense; see **resin**

L. *unguentum*, ointment, perfume; see **fat**

L. *virosus*, slimy, fetid, foul, poisonous; see *virus* under **poison**

L. *visium*, n. stench:

ML. *zibethum*, < Ar. *zabad*, civet; see *civetta* under **weasel**

smelt < AS. *smelt*.

Gr. *atherine*, f. a kind of smelt: *Atherina presbyter* (a fish), *Atherinopsis californiensis* (a silversides).

Gr. *hepsetos*, m. a small fish: *Atherina hepsetus* (sand-smelt).

NL. *osmerus* (Gr. *osmeros*), odorous; m. smelt: *Osmerus mordax* (smelt).

smenos, Gr. beehive, swarm of bees; see **hive**

smerdaleos, Gr. terrible; see **fear**

smerdos, Gr. a fish; see **fish**

smeringo- < Gr. *smerinx, -ingos*, bristle; see *merinx* under **hair**

smicro- < Gr. *smikros*, Ionic for *mikros*, little; see *mikros* under **little**

smilax, L., Gr. a name for several different plants but now restricted to a genus of the lily family; see **lily**

smile < uncertain origin; see **laugh, joy**

smilo- < Gr. *smile*, knife, chisel, graving tool; *smilion*, dim.; see **knife**

sminthos, Gr. mouse; see **mouse**

sminye, Gr. two-pronged hoe; *sminydion*, dim.; see **hoe**

smith < AS. *smith*, a worker in metal; see **make**

smodix, *-ingos*, Gr. bruise, weal; see **bruise**

smoke < AS. *smoca*; see **cloud**

smooth < AS. *smoth*.

L. *aequor*, any level, even surface; see **level**

Gr. *akymatos*, waveless, calm:

L. *allapsus*, silent or stealthy approach; see **silent**

Gr. *anaxeo*, hew smooth, polish:

Gr. *apheles*, even, smooth, simple: *Apheloglossa rufipennis* (a beetle), *Aphelops megalodus* (a fossil rhinoceros), *Aphelandra squarrosa* (an acanthacead), *Metapheles dinora* (a butterfly), *Aphelocoma insularis* (a jay).

L. *calvus*, bald; see *calvatus* under **bare**

L. *enodis*, without knots, smooth; see *nodus* under **knot**

L. *erudio, -itus*, polish, educate: erudite, erudition.

L. *erugo, -atus*, clear of wrinkles, smooth:

L. *fastidiosus,* full of loathing, nice, dainty; see **hate**

L. *glaber,* without hair, smooth; see **bare**

Gr. *glaphyros,* hollowed out, smoothed, polished; see **hollow**

L. *imberbis; imberbus,* beardless; see *barba* under **hair**

L. *inermis,* unarmed; see **bare**

Gr. *leios,* smooth, bald: liodermatous, *Liosternus vittatus* (a grasshopper), *Liostethus gladius* (a grasshopper), *Leiothrix callipyga* (a hilltit), *Leiophyllum buxifolium* (sand-myrtle), *Sambucus leiosperma* (an elderberry), *Liopeltis vernalis* (grass-snake).

L. *lenis,* soft, smooth, mild; see **soft**

Gr. *leuros,* smooth, level, polished, even: *Leurognathus marmorata* (a salamander), *Leuresthes tenuis* (grunion).

L. *levis (laevis),* smooth, polished, bald; *levigatus,* smooth, slippery; *levigo, -atus,* make smooth, polish: *Aster laevis* (smooth aster), *Equisetum laevigatum* (smooth horsetail), *Pentstemon laevigatus* (a pentstemon), *Frondicularia laevissima* (a foraminifer).

L. *limatulus,* somewhat filed, polished, smoothed; see *lima* under **scrape**

Gr. *lissos; lispos,* smooth, polished: lissotrichous, lissoflagellate, *Lispothrips wasastjernae* (a thysanopterid), *Perilissus lucidulus* (a wasp), *Lissanthe sapida* (an epacridacead).

L. *lubricus,* slippery; see **slip**

L. *planus,* smooth, even; see **level**

L. *polio, -itus,* make smooth, adorn, finish; *polimen, -inis,* n. polish; *politor, -is,* m. polisher: polish, politeness, politesse, impolite, unpolished, interpolate, *Fusimitra polita* (a fossil gastropod), *Cirrhoscyllium expolitum* (a dogfish), *Ceratina politula* (a bee).

Gr. *psilos,* bare, smooth; see **bare**

L. *rasilis,* scraped, shaved, smoothed, polished; see *rado* under **scrape**

L. *runcina,* f. plane; *runcino, -atus,* plane off: *Crepis runcinata* (hawksbeard).

L. *samio, -atus,* polish; *samiator, -is,* m. polisher:

Gr. *sigaloma, -tos,* n. a smoothing, polishing instrument:

Gr. *sphaleros,* slippery; see **slip**

Gr. *teren,* smooth, delicate; see *teramnos* under **soft**

L. *teres, -etis,* rubbed off, rounded, smooth; see **circle**

Gr. *xestos; xysilos; xystos,* scraped, shaved, planed, polished, smoothed; see *xeo* under **scrape**

See: **bare, slip, level, mild, scrape, spread, rest**

smut, prob. from the same root as G. *schmutz,* dirt, filth, stain; see **dirt, fungus, slime**

smylla, Gr. a fish; see **fish**

smyrna, Gr. myrrh; see **resin**

smyrnium, L. (Gr. *smrynion*), a genus of the carrot family; see **carrot**

smyrus, L. (Gr. *smyros*), a kind of eel; see **eel**

snail < AS. *snaegel;* see **mollusk**

snake < AS. *snaca.*

Sp. *aboma,* f. any large snake: *Aboma etheostoma* (a fish).

Gr. *ammodytes,* m. a kind of serpent: *Ammodytes tobianus* (sand-launce).

L. *anguis,* m. snake; *anguiculus,* m. dim.; *anguineus; anguinus,* of snakes, snaky: anguine, *Anguis fragilis* (a lizard), *Chlamydoselachus anguineus* (frilled shark), *Trichosanthes anguina* (snake-gourd), *Anguispira alternata* (a snail), *Chiococca anguifuga* (a Brazilian shrub).

Gr. *argetes,* m. a kind of snake:

Gr. *argolas,* m. a kind of serpent:

L. *aspis, -idis* (Gr. *-idos*), f. viper: asp *(Naja haje), Aspidoboa curta* (a snake), *Dendraspis angusticeps* (mamba).

NL. *bitis,* f. a genus of snakes: *Bitis nasicornis* (a puffadder).

L. *boa,* f. a large serpent: boa *(Constrictor imperator), Sthenelais boa* (a worm).

NL. *bungarus,* m. a genus of snakes: *Bungarus caeruleus* (krait), *Naja bungarus* (a cobra).

L. *cerastes* (Gr. *kerastes*), m. a horned serpent: *Cerastes cornutus* (horned viper), *Crotalus cerastes* (sidewinder).

L. *chamaedracon, -ontis,* m. an African snake:

Gr. *chelydros,* a kind of snake; see *chelys* under **turtle**

Gr. *chersydros,* m. a kind of snake: *Chersydrus fasciatus* (a snake).

L. *coluber, -bri,* m. snake, serpent; Sp. *culebra,* f. snake; *colubrinus,* snakelike, wily; *Coluber constrictor* (blacksnake), *Colubrina glabra* (snakebark), *Strychnos colubrina* (snakewood), *Trochilus colubris* (ruby-throated hummingbird).

Gr. *diphas,* f. a kind of serpent:

Gr. *dipsas, -ados,* f. a venomous snake: *Dipsas nebulata* (a snake), *Dipsadophidium weileri* (a snake).

Gr. *dryinas,* m. a kind of snake:

Gr. *echis, -eos,* m. viper; *echidion,* n. dim.; *echidna,* f. viper, adder: *Echis carinatus* (Indian viper), *Echidna catenata* (an eel), *Echium fastuosum* (a viper's-bugloss), *Echites umbellata* (an apocynacead), *Pseudechis porphyriacus* (a blacksnake), *Notechis scutatus* (tiger-snake).

NL. *elaps, -pis,* m. a kind of snake: *Elaps corallinus* (coral-snake), elapine, *Aspidelaps lubricus* (an African snake).

L. *excetra,* f. snake, serpent:

Gr. *herpes, -etos,* m. a creeping thing, shingles; *herpeton,* n.; *herpestes,* m. a creeping animal, snake, reptile; *herpilla (herpele),* f. a marine animal; *herpestikos,* creeping, < *herpo,* creep: herpetology, herpes, *Herpetomonas muscae* (a protozoan), *Herpestes nyula* (a mongoose), *Herpobdella punctata* (a leech), *Herpele squalostoma* (a caecilian), *Caulerpa crassifolia* (a seaweed).

Gr. *hiera,* f. a kind of serpent, an antidote against poison:

Gr. *hydra,* f. a fabulous many-headed serpent, whose slaying was difficult because no sooner was a head lopped off than two others came up in its place; *hydros,* m. a water-snake; *hyllos,* m. dim.: *Hydra viridis* (a fresh-water polyp), *Hydrurus foetida* (a flagellate).

L. *hypnale,* f. a kind of adder: *Hypnale hypnale* (a snake).

Gr. *kausos,* m. a kind of serpent: *Causus rhombeatus* (a snake).

Gr. *kinopeton,* n. serpent, a venomous beast; *knops, -opos,* m. a short form of *kinopeton: Cnopus flohri* (a beetle).

Gr. *kophias,* m. an adder: *Cophias atrox* (a snake).

Gr. *libys, -yos,* a kind of snake; see **Libya** under **Africa**

L. *Medusa* (Gr. *Medousa*), a gorgon with snaky locks; see **fear**

Gr. *molouros,* m. a kind of serpent:

Gr. *myagros,* m. mouser, a kind of snake: *Myagrostoma plexum* (a fossil gastropod).

NL. *naja,* f. a genus of snakes, < Skt. *naga,* snake: *Naja tripudians* (cobra).

L. *natrix, -icis,* f. water-snake: *Natrix rhombifera* (a snake), *Ononis natrix* (snake-root restharrow).

Gr. *ophis, -ios,* m. serpent, reptile, snake; *ophidion,* n. dim.; *ophiakos,* of snakes: ophicleide, ophite, ophitic, Ophidia, ophiolite, Ophiuchus, *Ophioglossum engelmanni* (an adderstongue-fern), *Ophiosaurus ventralis* (glass-snake), *Ophiodon elongatus* (a greenling, buffalo-cod), *Ophileta complanata* (a fossil gastropod), *Ophiura multispina* (a bitterling), *Ophidion barbatum* (a cusk-eel), *Eremochloa ophiuroides* (centiped-grass), *Cristatella ophidea* (a bryozoan), *Ophidiaster confertus* (a starfish), *Ophiopholis aculeata* (a brittle-star), *Acanthophis antarcticus* (a snake), *Ichthyophis glutinosus* (a caecilian), *Ophiopsis dorsalis* (a fossil fish).

L. *parias, -ae* (Gr. *pareias, -ou*), m. a kind of snake: *Parias variegatus* (a snake).

Gr. *prester,* hurricane, a kind of serpent; see **wind**

Gr. *ptyas, -ados,* spitter, a kind of snake; see *ptyalon* under **spit**

Gr. *Python, -os,* m. mythical serpent slain by Apollo near Delphi; *pythonion,* n. a plant: *Python reticulatus* (Malay python).

L. *reptile, -is,* n. snake, serpent, < *reptilis,* creeping, crawling: reptile, reptilian.

Gr. *sepedon, -os,* f. a serpent whose bite causes mortification; *seps, sepos,* m. a kind of serpent: *Sepedon pusilla* (a fly), *Natrix sipedon[sepedon]* (water-snake).

L. *serpens, -entis,* f. snake, < *serpo, serptus,* creep, crawl; *serpula,* f. dim.; *serpentinus,* of or like a serpent, twisting, winding; *serpentaria,* f. snakeweed: serpent, serpentine, serpentarium, serpigo, serpiginous, *Serpula arenicola* (a worm), *Aristolochia serpentaria* (Virginia snakeroot), *Proserpinaca palustris* (mermaidweed), *Gempylus serpens* (snake-mackerel), *Rauwolfia serpentina* (an apocynacead).

L. *vipera,* f. adder, snake, serpent; *viperalis; viperinus,* of adders; *viperina,* f. snakeweed: viperine, wivern, viperiform, viper *(Vipera berus),* *Cerastes vipera* (an Egyptian snake), *Trachinus vipera* (lesser weever).

snare < AS. *snear;* see **trap**

sneer < uncertain origin; see **laugh**

sneeze < AS. *fneosan.*

Gr. *errhinon,* n. a sternutative medicine: errhine.

Gr. *ptarmos,* m. a sneeze, < *ptairo,* sneeze; *ptarmikos,* causing sneezing: *Achillea ptarmica* (sneezewort-yarrow), *Ptaeroxylon utile* (sneezeweed).

L. *sternuto, -atus,* sneeze; *sternutatio, -onis,* f. a sneeze: sternutation, sternutatory.

snipe, prob. < ON. *snipe.*

Gr. *aktites,* m. shore-dweller: *Actitis hypoleuca* (European sandpiper).

Gr. *askalopas,* m. probably a woodcock:

It. *avocetta,* f. a shore bird: avocet *(Recurvirostra avocetta).*

Gr. *erythropous*, m. redshank: *Erythropus vespertinus* (a hawk), *Totanus erythropus* (spotted redshank).

L. *himantopus* (Gr. *himantopous*), m. a stilt: *Himantopus mexicanus* (black-necked stilt).

Gr. *kalidris; skalidris, -idos,* f. a speckled shore bird: *Calidris leucophaea* (sanderling).

Gr. *noumenios,* m. a kind of curlew: *Numenius americanus* (long-billed curlew).

NL. *phaeopus,* m. a kind of curlew: *Phaeopus borealis* (Eskimo curlew), *Numenius phaeopus* (whimbrel).

L. *scolopax, -acis* (Gr. *skolopax, -akos*), m. snipe, woodcock: *Scolopax rusticola* (European woodcock), *Macrorhamphosus scolopax* (a snipe-fish), *Nemichthys scolopaceus* (snipe-eel).

NL. *totanus,* m. < It. *totano,* moor hen: *Totanus flavipes* (yellowlegs, tattler).

Gr. *tryngas,* m. a kind of sandpiper: *Tringa[Tryngas] canutus* (red sandpiper, knot), *Palaeotringa vetus* (a Mesozoic bird).

snore < uncertain origin.

L. *rhonchus* (Gr. *rhenchos; rhenkos*), m. a snore; *rhenxis, -eos,* f. stertorous snore, < *rhenko,* snore: rhonchus, rhonchial, Serra do Roncador, *Roncador stearnsi* (a fish), *Pomadasis(Rhenchus) panamensis* (a fish).

L. *sterto,* snore; *stertens, -entis,* snoring, snorer: stertor, stertorous.

snow < AS. *snaw;* see **ice**

soak < AS. *socian,* saturate, steep; see **wet**

soap < AS. *sape.*

Gr. *rhymma, -tos,* n. any cleansing agent, soap:

L. *sapo, -onis* (Gr. *sapon, -os*), m. soap: saponify, saponin, saponite, saponule, saponaceous, *Saponaria officinalis* (soapwort), *Sapindus saponaria* (a soapberry), *Rhypticus saponaceus* (a soapfish).

Gr. *smegma (smema), -tos,* n. any cleansing agent, soap: smegma, *Mycobacterium smegmatis* (a bacterium).

See: **wash**

soapberry

NL. *cardiospermum,* n. a genus of the soapberry family: *Cardiospermum dissectum* (a heartseed).

NL. *dodonaea,* f. a genus of the soapberry family, < Rambert Dodoens, Dutch botanist: *Dodonaea viscosa* (a sapindacead).

Gr. *hypelate,* f. a shrub, now a genus of the soapberry family: *Hypelate trifoliata* (inkwood).

NL. *koelreuteria,* f. a genus of the soapberry family, < Joseph Koelreuter, German botanist: *Koelreuteria paniculata* (a Chinese tree).

NL. *paullinia,* f. a genus of the soapberry family, < C. F. Paullini, German botanist: *Paullinia thalictrifolia* (a sapindacead).

NL. *sapindus* (f. in part, by Linnaeus; m. in usage; but cf. *tamarindus,* f.), soapberry: Sapindaceae, Sapindales, *Sapindus marginatus* (a soapberry).

NL. *serjania,* f. a genus of the soapberry family: *Serjania fuscifolia* (a serjania).

NL. *ungnadia,* f. a genus of the soapberry family, < Ferdinand Ungnad, who introduced the horsechestnut to western Europe: *Ungnadia speciosa* (Mexico buckeye).

sob < AS. *seofian;* see **sad**

sobaros, Gr. frightening, pompous; see **fear**

sobe, Gr. horse's tail; see **tail**

sober < L. *sobrius,* not drunk, temperate, serious: sobriety, *Lepthyphantes sobrius* (a spider).

Gr. *amethystos,* not drunk, sober; see **stone**

sobetes, Gr. frightener; see *sobaros* under **fear**

soboles (suboles), L. sprout, shoot, twig; see **branch**

sobrinus, L. cousin on the mother's side; see **kin**

sobrius, L. not intoxicated, serious; see **sober**

soccus, L. a low-heeled light shoe, slipper; *socculus,* dim.; see **shoe**

socer; socrus, L. father-in-law; *socra,* mother-in-law; see **kin**

socius, L. companion; *socialis,* of companionship; *sociabilis,* companionable; see **companion**

sock < AS. *socc,* < L. *soccus,* a kind of light shoe; see **stocking**

socket < OF. *soket,* < C. *soc,* plowshare; see **hollow, joint, cup**

soco- < Gr. *sokos,* stout, strong; see **strong**

socors, L. stupid, silly, negligent, inactive; see *excors* under **fool**

Socrates, L. (Gr. *Sokrates*), a celebrated Greek philosopher; see **ask**

sod < D. *sode*.

L. *caespes, -pitis,* m. sod, turf, mat, tuft, clump, hovel; *caespiticius,* made of turf; *caespitosus; caesposus,* turfy, matted: cespitose, *Cymbella caespitosa* (a diatom).

AS. *sweard,* skin, turf, covering: sward, greensward.

AS. *turf,* sod, peat: turf, *Campylopus turficola* (a moss from Rana Kao, Easter Island).

L. *viretum,* n. greensward, sod, turf:

See: **grass**

sodalis, L. companion, comrade, crony; *sodalitas,* club; see **companion**

sodium < ML. *sodanum,* headache remedy, < Ar. *suda,* headache: sodium bicarbonate, sodalite, soda, *Salsola soda* (a saltwort).   See ELEMENTS under **thing** Ar. *natrun,* sodium carbonate; *natro-,* containing sodium: natron, natrolite, trona.

soft < AS. *softe.*

Gr. *amalos,* soft, tender, weak: *Amalocichla brevicauda* (a thrush), *Amalopteryx maritima* (a fly).

Gr. *apages,* not firm, flabby, soft: *Apagomera suturella* (a beetle).

Gr. *atalos,* tender, delicate: *Atalotriccus pilaris* (a flycatcher).

Gr. *chalaros,* loose, slack, supple, languid; see **free**

Gr. *chaunosis,* a slackening, lightening; see *chaunos* under **hole**

Gr. *chlidanos,* delicate, luxurious, < *chlide,* f. delicacy, luxury: *Chlidanthus fragrans* (an amaryllidacead).

Gr. *chnodes,* soft, porous, downy; see *chnous* under **foam**

L. *delicatus,* delightful, soft, tender; *delicatulus,* dim.: delicate, *Thuidium delicatulum* (a fern-moss).

Gr. *depho; depso,* soften by working with the hand, knead; see **mix**

Gr. *eiktikos,* yielding readily, pliable; see **bend**

L. *flaccidus,* weak, drooping, flabby; see **weak**

L. *fracidus,* overripe, soft, mellow; see **ripe**

Gr. *habros,* delicate, soft, dainty, graceful; see **beauty**

L. *hapalus* (Gr. *hapalos*), soft to the touch, tender, gentle, delicate; *hapalia,* f. softness, tenderness: *Hapalodectes compressus* (a creodont), *Hapale pygmaeus* (pygmy marmoset), *Hapalemur griseus* (gray lemur), *Hapalocrinus tuberculatus* (a Silurian crinoid).

Gr. *hygros,* moist, loose, soft; see **wet**

L. *lanatus,* soft like wool, velvety, downy; see *lana* under **wool**

L. *laxativus,* loosening, mitigating: laxative.

L. *lenis,* soft, smooth, mild, gentle, easy, calm; *lenimentum,* n. a soothing remedy; *lenitas, -atis; lenitudo, -inis,* f. softness, smoothness, gentleness, mildness; *lenio, -itus,* soften, mollify, alleviate, soothe, calm; *delenitorius,* serving to soften: leniency, lenity, lenitive, lenitude, lenigallol, lenitic.

Gr. *leptaleos,* fine, delicate, feeble; see *leptos* under **thin**

L. *macero, -atus,* soften: macerate, macaroni, macaroon.

Gr. *malakos,* soft, supple, pliant; *malakion,* dim.; *malakia,* f. softness; *malaxo (malasso),* soften: malacoid, malaxation, osteomalacia, Malacostraca, amalgam, *Malacosoma americanum* (tent-caterpillar), *Malacobdella grossa* (a nemertean), *Malacothrix saxatilis* (a composite).

Gr. *malthodes,* pliant, soft like resin; see *maltha* under **resin**

L. *mitigo, -atus,* soften, mellow, tame, appease, assuage; see *mitis* under **mild**

L. *mollis,* soft; *molliculus,* dim.; *molluscus,* soft; *mollio, -itus,* soften: mollify, mollient, mollitious, mollusk, mullein, Mollusca, *Mollugo verticillata* (carpetweed), *Castanea mollissima* (Chinese chestnut), *Astragalus mollissimus* (woolly locoweed), *Geranium molle* (a geranium), *Thermopsis mollis* (a legume).

Gr. *molys; molyros; molyx,* soft, weak, feeble; see **weak**

L. *mulceo, mulsus,* touch lightly, allay, soften, soothe; *Mulciber, -is,* m. softener, a name for Vulcan: demulcent, *Mulciber biguttatus* (a beetle), *Micromulciber notatus* (a beetle).

L. *multicius,* soft:

Gr. *pepon, -os,* ripe, mellow, tender; see *pepeiros* under **ripe**

Gr. *phlydaros,* soft, flabby:

L. *plasticus* (Gr. *plastikos*), capable of being molded or modeled, pliable; see *plasso* under **make**

L. *summissus,* lowered, subdued, soft, calm:

Gr. *takeros,* melting, tender, dissolving; see *teko* under **flow**

L. *tener, -a, -um,* soft, delicate; *tenellus; tenellulus,* dim.: tender, *Adiantum tenerum* (a maidenhair-fern), *Viola tenella* (a violet).

Gr. *teramnos; teramon, -os; teren, -os,* soft, tender, smooth: *Teramnus uncinatus* (a legume), *Lepidopleurus(Terenochiton) subtropicalis* (a chiton).

Gr. *thelyno,* make womanish, soften; *ekthelynsis,* f. a becoming soft:

Gr. *tryphe*, f. softness, delicacy, luxury; *trypheros; tryssos*, delicate, dainty, soft: *Tryphomys adustus* (a mouse), *Trypheromimum nankineus* (a beetle), *Tryssothele pisii* (an arachnid).

L. *velutinus*, velvety; see *villus* under **hair**

See: **weak, ripe, dainty, smooth, soothe**

soil < L. *solum*, ground, bottom; see **earth**

sol, L. sun; *solaris*, of the sun; *solatus*, sunburned; see **sun**

solago, L. a plant; see **plant**

solanum, L. nightshade; see **nightshade**

solarium, L. sundial, balcony; see **dial**

solatus, L. sunburned, see *sol* under **sun**; comforted, soothed, see *solor* under **soothe**

soldanella, NL. a genus of the primrose family; see **primrose**

solder < OF. *soulder*, < L. *solido, -atus*, fasten together; see **bind, glue**

soldier < OF. *soldier*, < L. *solidus*, a coin, the pay of a soldier; see **army**

sole < L. *solea*, f. sandal, bottom of the foot; *solea ferrea*, horseshoe: insole, galosh, soleiform, *Diactis soleata* (a milleped).

Gr. *hylia*, f. shoe sole: *Hylia prasina* (a bird).

Gr. *kassyma, -tos*, n. shoe sole: *Cassyma quadrinata* (a moth).

Gr. *pelma, -tos*, n. sole of the foot, shoe sole: pelmatogram, Pelmatozoa, *Chaetopelma olivaceum* (a spider), *Gongropelma formosana* (a wasp), *Stenopelmatus longispira* (sand-cricket).

L. *planta*, f. sole of the foot: plantigrade, plantar, plantain (*Plantago major*).

See: **base**

solea, L. bottom of the foot, see **sole**; a fish, see **flatfish**

solemnis, L. stated, religious, serious; see **serious**

solen, Gr. pipe, channel, a mollusk; *solenarion; solenidion; solenion; soleniskos*, dim.; see **pipe, mollusk**

solicito, -atus, L. disturb, tempt, vex; see **trouble**

solidago, NL. goldenrod; see **composite**

solidus, L. dense, hard, firm, thick, entire enduring; see **thick, hard**

solitarius, L. alone; *solitudo*, aloneness, isolation; see *solus* under **alone**

solitus, L. customary, habitual, usual; see **use**

solium, L. royal seat, throne, chair of state; see **sit**

sollers, -tis, L. skilled, clever, adroit; see **know**

sollicitus, L. agitated, anxious, uneasy; *sollicitudo*, uneasiness; see **fear**

sollus, L. entire, complete; see **all**

soloeco- < Gr. *soloikos*, making a grammatical error; see **fault**

Solon, Gr. celebrated Greek legislator; see **law**

solor, -atus, L. comfort, soothe; *solator*, comforter; see **soothe**

solos, Gr. a mass of iron; see **iron**

solox, -ocis, L. coarse, rough, bristly; see **rough**

solpuga, L. probably a kind of spider; see **spider**

solstitium, L. longest day of the year, summer solstice; see **day**

solubilis, L. that can be freed, loosed, explained; see *solvo* under **free**

solum, L. bottom of anything, ground, floor, earth, soil; see **earth**

solus, L. alone, single; see **alone**

solvo, -utus, L. set free, loosen; see **free**

-som; -some < AS. *-sum*, like, full of, having the quality of; see **nature**

soma, Gr. body, flesh; *somation*, dim.; *somatikos*, of the body; see **body**

some < AS. *sum*, an indefinite quantity or number; see **few**

somnium, L. dream; see *somnus* under **sleep**

somnus, L. sleep; see **sleep**

somphos, Gr. porous, spongy; see **hole**

son < AS. *sunu;* G. *sohn*: sonship, grandson, stepson, Thompson, Dixon, Harrison, Mendelssohn, Edwards, Roberts, Jones, Davis.

W. *ap*, son of: Apjohn, Powell (ap Howell), Price (ap Rice), Pugh (ap Hugh), Bevan (ab Evan), Blake (ab Lake).

L. *filius*, m. son; *filiolus*, m. dim.; *filialis*, of a son or daughter; ME. *fitz*, < OF.

filz, son of, < L. *filius*, son: filial, affiliate, affiliation, hidalgo, Fitzgerald, Fitz-
hugh, Fitzpatrick, Fitzjames, Fitzsimmons.
Gr. *hyios*, m. son; *hyionos*, m. grandson:
Gr. *-ides; -ades*, m. patronymic suffixes denoting son of, descendant of, as
Priamides, m. son of Priam; *Perseides* or *Perseiades*, m. son of Perseus:
Aristides, Euripides, Simonides, Atlantiades.
AS. *-ing*, belonging to, son of: Browning, Spalding, Manning, Reading, Harding,
Billing, Nutting, Nottingham.
Gr. *inis*, m., f. son or daughter:
Gr. *kelor, -os*, m. son: *Celor semenowi* (a wasp).
C. *mac*, son of: Macdonald, Macmillan, MacGinitie, McDowell, McCarthy,
McKeesport (Pa.), mackintosh, macadamize.   Derive: mackinaw.
L. *natus*, m. son; *natulus*, m. dim.:
Ir. *O'* son of: O'Brien, O'Donnell, O'Toole, O'Callaghan.
Scan. *sen*, son of: Johannsen, Jespersen, Amundsen, Olsson, Nilsson.
Slav. *-sky, -off, -vitch*, relative of, son of, inhabitant of: Rojestvensky, Dos-
toyefsky, Stokowski, Sobieski, Paderewski, Metchnikoff, Petrovitch, Petrunke-
vitch.
See: **child, kin**

sonax, -acis, L. noisy; see *sono* under **sound**

sonchus, L. (Gr. *sonchos*), sowthistle; see **thistle**

song < AS. *singan;* see **sing**

sono, -itus, L. sound; *sonivius*, noisy; *sonorus*, resounding; see **sound**

sonticus, L. serious, dangerous, critical, < *sons, sontis*, a guilty person, criminal;
see **serious**

soon < AS. *sona*, shortly, presently; see **near**

soot < AS. *sot;* see **dust, dirt**

soothe < AS. *soth*.
Gr. *acholos*, allaying anger, soothing:
L. *chalasticus (chalastikos)*, alleviating, soothing:
L. *condoleo*, suffer with another, sympathize with; see *dolor* under **pain**
L. *delenificus*, soothing, caressing; *delenitorius*, serving to soften, soothe:
L. *levamen, -inis; levamentum*, n. alleviation, comfort, consolation:
Gr. *meilisso*, soothe, appease; *meilichos*, gentle, kind, soothing; *meiligma, -tos*,
n. anything that soothes: *Milichilinus decorus* (a beetle).
L. *mitigo, -atus*, soften, mellow, tame, appease, soothe; see *mitis* under **mild**
L. *mulceo, mulsus*, stroke, soothe, appease; see **soft**
L. *paco, -atus*, pacify, soothe; see *pax* under **rest**
L. *paregoricus* (Gr. *paregorikos*), consoling, assuaging, < *paregoria*, f. consolation,
easement: paregoric.
L. *placo, -atus*, pacify, reconcile, soothe, appease: placate, implacable.
L. *sedo, -atus*, allay, still, calm, assuage, settle, quiet, stay; *sedamen, -inis*, n.
sedative: sedate, sedative, *Reseda alba* (white mignonette).
L. *solor, -atus*, comfort, soothe; *solatium*, n. comfort, relief; *solator, -is*, m. com-
forter; *consolabilis*, that may be comforted; *consolativus*, comforting:
See: **rest, heal, soft**

sophora, NL. a genus of the bean family; see **bean**

sophos, Gr. wise; *sophistes*, expert, master; see **know**

sophron, -os, Gr. of sound mind, sensible, wise; *sophrosyne*, good sense, discretion;
see *phren* under **mind**

sopor, L. sleep, stupor, lethargy; *sopio, -itus*, deprive of feeling, lull to sleep, quiet;
see **sleep**

-sor, L. agent; see *-or* under **make**

soracum, L. (Gr. *sorakos*), basket, hamper, pannier; see **basket**

sorbeo, L. suck in; *sorbilis*, that may be sucked in; *sorbitio*, drink, portion; see
suck, drink

sorbus, L. mountain-ash, a genus of the rose family; see **rose**

sordidus, L. dirty, filthy; *sordes*, dirt, filth; *sordicula*, dim.; see **dirt**

sore < AS. *sar*, aching, pain, wound.
L. *abscessus, -us*, m. a sore with accumulation of pus: abscess.
L. *aegilops, -opis*, f. (Gr. *aigilops, -os*, m.), ulcer at the inner corner of the eye:
egilops.
Gr. *amyche*, cut, laceration, tear; see *amysso* under **tear**
Gr. *anchilops, -opos*, m. sore at the inner corner of the eye, sty: *Anchilops
nodulosa* (a decapod).
Gr. *anthrax, -akos*, m. ulcer, carbuncle: *anthrax* (caused by *Bacillus anthracis*).

Gr. *apostema, -tos,* n. abscess, ulcer: aposteme, apostemate.

L. *argema, -tis,* n. a small, white ulcer on the eyeball:

AS. *blegen,* inflammatory swelling, kibe: blain, chilblain.

L. *cancer, -cri,* m. a benign or malignant growth: cancer, cancerous, canker, chancre, chancroid.

L. *carbunculus,* m. a severe boil: carbuncle.

L. *carcinoma, -tis,* n. (Gr. *karkinoma, -tos,* n.; *karkinos,* m.), cancerous ulcer: carcinoma.

Gr. *chimetlon,* n. chilblain, kibe:

L. *contusio,* bruise; see *contundo* under **bruise**

Gr. *dothien, -os,* m. abscess, boil: *Dothidella ulmi* (a fungus).

Gr. *ektrimma, -tos,* n. sore caused by rubbing:

Gr. *empyema, -tos,* n. a suppurating internal abscess: empyema.

Gr. *epoulis, -idos,* f. gumboil: epulis.

L. *faredo, -inis,* f. a kind of abscess:

L. *fistula,* f. a kind of ulcer: fistulous, fistula, fester.

L. *fleminum,* n. a kind of boil:

L. *furunculus,* petty thief, boil; see *fur* under **steal**

L. *gangrena* (Gr. *gangraina*), f. an eating sore, mortification: gangrene.

Gr. *helkos,* n. wound, sore, ulcer; *helkydrion,* n. dim.; *helkoma, -tos,* n. sore, ulcer; *helkodes,* sore, ulcerous; *helkosis,* f. ulceration: helcoid, helcosis.

L. *hordeolus,* m. sty in the eye:

Gr. *hypopyon,* n. ulcer:

L. *laesio, -onis,* wound, injury; see *laedo* under **strike**

L. *noma* (Gr. *nome*), f. an eating ulcer: noma.

Gr. *oteile,* f. a fresh wound:

Gr. *paronychia,* f. whitlow, felon; nailwort: paronychia, *Paronychia argentea* (whitlowwort).

Gr. *paroulis, -idos,* f. gumboil: parulis.

L. *pernio, -onis,* m. chilblain: pernio.

L. *petimen, -inis,* n. sore on the shoulder of beasts of burden:

L. *phagedaena* (Gr. *phagedaina*), f. an eating ulcer: phagedena.

Gr. *phlegmone,* fiery heat, inflammation, boil; see *phlego* under **burn**

Gr. *phollix, -ikos,* f. scab, sore; *phollikodes,* scabby: *Phollicodes plebeius* (a beetle).

L. *reduvia,* f. hangnail, agnail, whitlow, remainder: *Reduvius personatus* (masked bedbug-hunter).

Gr. *seps, sepos,* a putrefying sore; see *sapros* under **rot**

L. *suppuratio, -onis,* f. abscess, gathering: suppuration.

Gr. *tange,* a putrid abscess; see **rot**

Gr. *trauma (troma), -tos,* n. wound; *traumatikos,* of wounds; *trosis,* f. a wounding; *trotos,* wounded, vulnerable: trauma, traumatic, *Traumatocampa pinivora* (a moth), *Trotothyris abnormalis* (a moth).

L. *ulcus, -eris,* n. sore, abscess; *ulcusculum,* n. dim.; *ulcerosus,* full of sores; *ulcero, -atus,* make or become sore: ulcer, ulcerous, ulcuscule, ulceration, *Adesma ulcerosa* (a beetle), *Libertella ulcerata* (a fungus).

L. *vomica,* f. sore, boil, ulcer; *vomicosus,* full of sores; *vomicus,* ulcerous: nux vomica (seed of *Strychnos nux-vomica*).

L. *vulnus, -eris,* n. wound; *vulnusculum,* n. dim.; *vulnerabilis,* susceptible to wounding; *vulnerarius,* of wounds; *vulnero, -atus,* hurt, wound: vulnerable, vulnerary, *Anthyllis vulneraria* (kidney-vetch), *Godetia quadrivulnera* (an onagracead), *Anemone quinquevulnera* (an anemone).

See: **bruise, hurt, pain, disease**

sorex, L. a shrew-mouse; *soricinus,* of the shrew-mouse; see **mole**

sorghum, NL. (It. *sorgo*), a genus of grasses; see **grass**

soror, L. sister; *sororcula,* dim.; *sororius,* of a sister, sisterly; see **sister**

soros, Gr. coffin, cinerary urn; *soridion,* dim., see **box**; heap, see **heap**

sorrel < OF. *surele.*

L. *lapathum* (Gr. *lapathon*), n. a kind of sorrel: *Rumex hydrolapathum* (water-dock), *Piper lapathifolium* (a pepper).

L. *rumex, -icis,* f. dock, sorrel: *Rumex verticillatus* (swamp-sorrel), *Aphis rumicis* (an aphid).

sorrow < AS. *sorg;* see **sad**

sort < L. *sors, sortis,* lot, share, kind, fate; *sorticula,* dim.; see **lot, class, part, separate**

sorus, L. (Gr. *soros*), heap; *soredion,* dim.; see **heap**

sory, Gr. probably an ore of sulfur; see **sulfur**

sos; sosi-, Gr. safe, sound; *sosis,* salvation; see *soter* under **save**

sospes, -*itis,* L. saving, delivering; *sospitalis,* salutary; see **save**

sostron, Gr. reward, fee; see **pay**

soter, Gr. savior, deliverer, preserver; *soterios,* saving, preserving; see **save**

sotron, Gr. felly or rim of a wheel; see **border**

soul < AS. *sawl;* see **spirit, mind, life**

sound < L. *sonus,* m. noise, tone; *sonabilis; sonax, -acis; sonivius; sonorus,* noisy, sounding; *sono, -itus,* make a noise, sound; *absonus,* discordant, inharmonious; *resonus,* reechoing, ringing: sonorous, sonata, sonnet, absonant, consonant, dissonant, person, resonance, resounding, unison, Sonora, *Solatisonax injussa* (a gastropod), *Hernandia sonora* (jack-in-a-box).

L. *accentus, -us,* m. tone, signal: accent, accentuation, *Cymindis accentifera* (a beetle).

Gr. *alalage,* a loud noise; see *alalazo* under **call**

L. *assensus, -us,* approval, echo; see **yes**

Gr. *babazo,* chatter, chirp; *babax,* chatterer; see **speak**

L. *balo, -atus,* bleat; see **bleat**

L. *barritus,* the trumpet cry of the elephant; see **call**

Gr. *bex, bechos,* cough; see **cough**

L. *bilbo, -itus,* swash, as a liquid in a vessel:

Gr. *bombos,* a booming or humming sound; see **murmur**

Gr. *borborygmos,* a rumbling in the bowels; see **thunder**

Gr. *bromios,* noisy, boisterous, < *bromos,* m. any loud noise: *Bromiodes indicus* (a beetle).

Gr. *bronte,* thunder; see **thunder**

L. *bubulo, -atus,* hoot like an owl; see **hoot**

L. *buccino, -atus,* sound the trumpet; see *buccina* under **horn**

L. *cano, cantus,* sing; see **sing**

Gr. *chromados,* m. a clashing noise:

L. *clamosus,* noisy; see *clamo* under **call**

L. *clangor, -is,* a loud, ringing sound, noise; see **ring**

L. *coaxo,* croak; see **croak**

L. *convicium,* n. loud noise, clamor, outcry, sound of wrangling: convicium.

L. *crepo, -itus,* rattle, clatter, crackle; see **rattle**

L. *destico, -atus,* squeak like a mouse; see **squeak**

Gr. *doupos,* m. dead, heavy sound, thud; *doupema, -tos,* n. crash, peal; *doupetor, -os,* m. clatterer: *Dupetor flavicollis* (a bird).

Gr. *echo,* f. a returned sound; *eche,* f.; *echos,* m. sound, noise; *echetes,* m. sounder, noise-maker, musical, shrill; *echetikos,* ringing; *glykeches,* sweet-sounding; *echeo,* sound, ring: echo, echoic, echolalia, anechoic, catechism.

Gr. *epye,* voice, sound, < *epyo,* call; see **call**

L. *fragor, -is,* m. loud noise, crash, din: fragor.

L. *fremo, -itus,* roar, growl, snarl, grumble; see **roar**

L. *gemo, -itus,* groan, moan, sigh; see **sad**

L. *gingritus,* cackling of geese; see **cackle**

L. *grunnio, -itus,* grunt; see **grunt**

L. *hinnio, -itus,* neigh, whinny; *hinnibilis,* neighing; see **neigh**

Gr. *homados,* m. din, noise:

Gr. *iache,* cry, shout, sound; see **call**

Gr. *ioë,* f. any loud sound:

Gr. *kelados,* m. sound of rushing waters, loud noise: *dyskelados,* ill-sounding, shrieking; *enkelados,* buzzing; *eukelados,* melodious: *Dysceladus tuberculatus* (a beetle), *Enceladus gigas* (a beetle).   Derive: celadonite.

Gr. *kompos,* din, noise, boast; see **brag**

Gr. *konabos,* m. a clashing, din:

Gr. *kraktikos,* noisy, < *krazo,* cry, scream, shriek; see **call**

Gr. *krektos,* struck, sounded (of stringed instruments), < *kreko,* sound an instrument:

Gr. *krotalon,* rattle; see **rattle**

Gr. *krouma, -tos,* n. beat, note, sound; *kroumation,* n. dim:

Gr. *ktypos,* m. any loud noise; *ktypodes,* noisy; *kataktypos,* making a loud noise:

L. *latro, -atus,* bark, howl; see **bark**

Gr. *minyros,* warbling, whining; see **sing**

L. *mugio, -itus,* bellow, low; see **roar**

L. *murmur,* a low sound, hum, mutter; see **murmur**

L. *musica* (Gr. *mousike*), music, one of the arts presided over by the Muses; see **music**

L. *mutio, -itus,* mutter; see **murmur**

Gr. *otobos,* m. loud, startling noise, din:

Gr. *phone*, f. sound; *phonion*, n. dim.; *phonetes*, m. sounder, speaker; *phonetikos*, of sound, vocal; *phoneo*, produce sound, speak; *diaphonos*, discordant; *euphonos*, sounding well; *kakophonos*, sounding ill: phonetics, phonograph, phonolite, saxophone, telephone, anthem, euphony, cacophonous, symphony, *Hesperiphona vespertina* (evening-grosbeak), *Terpsiphone perspicillata* (a fly-catcher), *Charadrodromia microphona* (a fly).

Gr. *phthegma*, sound of the voice, voice; see **speak**

L. *phthongus* (Gr. *phthongos*), m. sound, tone; *phthonge*, f. voice; *phthongarion*, n. dim.: phthongal, phthongometer, diphthong.

L. *pipilo*, *-atus*, chirp, twitter; see **chirp**

Gr. *pitylos*, m. plash of oars: *Pitylus violaceus* (a bird).

L. *plausus*, a clapping sound; see *plaudo* under **strike**

Gr. *poppyzo*, smack or cluck with the tongue and lips; see **cackle**

Gr. *psithyos*, whispering; see **murmur**

Gr. *psophos*, m.; *psophema*, *-tos*, n. sound, noise; *psophax*, *-akos*, m. noisy fellow; *psophesis*, f. a sounding; *psophetikos*, able to make a sound; *psophodes*, noisy: *Psophocarpus tetragonolobus* (Goa bean), *Psophia crepitans* (trumpeter), *Psophodes olivaceus* (coachwhip-bird).

Gr. *rhochmos*, m. a wheezing, < *rhocho*, wheeze:

Gr. *rhochthos*, roaring of the sea; see **roar**

Gr. *rhothos*, m. rushing or dashing noise; *rhothios*, rushing, dashing, as of waves and surf:

L. *rideo*, *risus*, laugh; see **laugh**

L. *sibilus*, hissing, whistling, < *sibilo*, hiss, whistle; see **whistle**

L. *singulto*, *-atus*, sob, hiccup; *singultus*, sobbing, hiccup; see **hiccup**

Gr. *spharagos*, m. noise: *Spharagemon cristatum* (a grasshopper).

Gr. *stenagma*, moan, groan, sigh; see **sad**

L. *sterto*, snore; see **snore**

L. *strepo*, *-itus*, make a noise, rattle; *strepitus*, *-us*, m. loud noise, < *strepito*, *-atus*, make a loud noise; *obstrepo*, roar, clamor, oppose: strepitant, streperous, obstreperous, *Strepera fuliginosa* (a crow-shrike).

L. *stridulus*, creaking, grating; see **squeak**

L. *susurro*, *-atus*, murmur, whisper, hum, buzz; see **murmur**

Gr. *symphonia*, concord in sound, unison; see **harmony**

Gr. *syrbeneus*, m. a noisy person:

Gr. *syrizo* (*syritto*), pipe, whistle, hiss; see **whistle**

Gr. *thorybos*, m. noise of a crowded assembly, applause, cheers, uproar: *Thorybothrips graminis* (a thysanopterid).

Gr. *thrylos*, noise of many voices; see *thryleo* under **speak**

L. *tonus* (Gr. *tonos*), m. sound; *tono*, *-itus*, sound, resound, thunder; *tonitrus*, *-us*, m. thunder: tone, tune, tonic, monotonous, baritone, attune, detonate, astonish, stun, overtone, astounding, tonitruous.

L. *tumultus*, noise, confusion, uproar; see *tumeo* under **swell**

L. *ululo*, *-atus*, howl, shriek, wail; see **call**

L. *vagitus*, crying, squalling, whimpering of children; see **sad**

L. *voco*, *-atus*, call; *vocifero*, *-atus*, cry out, clamor; see **call**

See: **call**, **speak**, **sing**, **music**, **ring**, **thunder**, **squeak**, **chirp**, **roar**, **weep**, **cackle**, **murmur**, **whistle**, **grunt**

soup < F. *soupe*.

Gr. *embamma*, *-tos*, n. sauce, soup;

Gr. *etnos*, n. pea soup: *Etnodona phalacropis* (a moth).

L. *jus*, juice, soup, broth; see **juice**

L. *puls*, porridge, pottage; see **food**

Gr. *rhophema* (*rhomma*), *-tos*, n. gruel, soup; *rhophetikos*, pertaining to sipping; *rhophetos*, that can be sipped; *rhoptos*, sipped, < *rhopheo*, sip, sup, gulp:

Gr. *zomos*, m. soup, sauce; *zomidion*, n. dim.: zomotherapy.

sour < AS. *sur*.

L. *acidus*, sour, tart, < *aceo*, be sour; *acescens*, *-entis*, turning sour; *acerbus*, sour, sharp, harsh; *acerbissimus*; *acetosus*, very sour, excruciating; *acetum*, n. sour wine, vinegar; *acor*, *-is*, m. sour taste, sourness; *coacesco*, become sour, deteriorate: acid, acetic, acetate, acetone, acetylsalicylate, acescent, ketone, acidosis, acerbity, acetylene, acetylglycine, phenylacetamide, acetylmethylcarbinol, *Acetobacter acetus* (vinegar-bacterium), *Rumex acetosa* (garden-sorrel), *Rumex acetosella* (sheep-sorrel), *Anguillula aceti* (vinegar-eel), *Phyllanthus acidus* (Malay gooseberry), *Lactobacillus acidophilus* (a bacterium), *Polyporus subacidus* (a fungus).

Gr. *agleukes*, not sweet, sour:

Gr. *omphax*, *-akos*, unripe grape, sour grape; see **grape**

Gr. *oxys*, sour, acid, sharp, oxygen compounds; *oxos*, n. sour wine, vinegar;

oxalis, -idos, f. sorrel; *oxalme,* f. vinegar and brine, pickle; *oxotos,* pickled; *enoxio,* turn sour; *hypoxys,* subacid; *katoxos,* very sour: oxygen, oxidize, peroxide, oxides, oxyhydrogen, oxalic acid, oxymyoglobin, *Oxydendrum arboreum* (sourwood), *Oxalis acetosella* (wood-sorrel), *Oxyria digyna* (mountain-sorrel), *Oxylobium cordifolium* (a legume), *Oxypetalum caeruleum* (an asclepiadacead), *Hypoxys hirsuta* (star-grass), *Vaccinium oxycoccus* (small cranberry).

L. *salgamum,* n. a pickled cucumber or other object; *salmacidus,* salty, sour: See: **point, cut, sad**

south < AS. *suth.*
Gr. *antarktikos,* southern: antarctic, *Nothofagus antarctica* (a false-beech).
Gr. *antichthon, -os,* f. southern hemisphere:
L. *auster, -tri,* m. south wind, south; *austellus,* m. dim., gentle south wind; *australis; austrinus,* southern: austral, Australia, *Australopithecus africanus* (a fossil primate), *Microcitrus australis* (a wild lime).
L. *meridies, -ei,* midday, noon, south; *meridionalis,* southern; see *dies* under **day**
Gr. *mesembria,* midday, noon, south; see *hemera* under **day**
Gr. *notos,* m. south, south wind; *notios,* southern, wet, rainy; L. *notialis,* southern: notalian, Notogaea, *Notomys megalotis* (a hopping-mouse), *Notornis hochstetteri* (takahe), *Notiotitanops mississippiensis* (a titanothere), *Notiocrinus timoricus* (a Permian crinoid), *Notiocichla baeticata* (a bird).

sow < AS. *sugu,* female hog; see **hog**
sow < AS. *sawan.*
L. *jacio, jactus,* throw, scatter, sow; see **throw**
Gr. *phityo; phyteuo,* sow, plant, beget, engender, produce; see *phyton* under **plant**
L. *sativus,* sown, planted, cultivated, < *sero, satus,* sow, plant; *satio, -onis,* f. a sowing; *sator, -is,* m. sower; *satorius,* of sowing; *consatus; consitus,* planted with something; *consitor, -is,* m. planter, sower; *dissitus,* scattered, sown; *obsitus,* sown, covered, filled: Saturn, *Sator angustus* (a lizard), *Lactuca sativa* (garden-lettuce), *Otiorhynchus obsitus* (a beetle).
L. *semino, -atus,* sow, seed; *sementivus,* of sowing; see *semen* under **seed**
L. *spargo, sparsus,* strew, scatter; see **spread**
Gr. *speiro,* sow; *spartos,* grown from seed, cultivated; *spora,* f. a sowing, seed; *sporadikos,* scattered; *sporeus,* m. sower; *sporimos,* sown: sporadic.
See: **spread, seed, till, begin**

sowbug, an isopod crustacean.
L. *asellus,* m. dim. of *asinus,* ass: *Asellus communis* (an isopod).
L. *oniscus* (Gr. *oniskos*), m. dim. of *onos,* ass: *Oniscus armatus* (a sowbug), *Paranotoniscus tuberculatus* (a sowbug) *Parastenoniscus elbanus* (a sowbug), *Phylloniscus braunsi* (a sowbug), *Psiloniscus sticticus* (a beetle), *Corbula oniscus* (a fossil pelecypod).
L. *porcellio, -onis,* m. sowbug: *Porcellio limbatus* (a sowbug).

sozo, Gr. keep, save; see *soter* under **save**

sozusa, L. a kind of artemisia; see **wormwod**

space < L. *spatium,* n. room, distance; *spatiolum,* n. dim.; *spatiosus,* roomy, large; *spatior, -atus,* walk about, promenade: spatial, spacious, expatiate.
Gr. *anachyma, -tos,* n. an expanse:
L. *area,* f. an open or vacant space, court; *areola,* f. dim.; *areolatus,* with small spaces: area, areal, areole, areola, areolate, areolar, *Woodwardia areolata* (a chain-fern).
L. *aula* (Gr. *aule*), f. court, courtyard, hall: aula, aularian.
L. *capax, -acis,* roomy, spacious, able, fit; see *capio* under **take**
Gr. *chora,* f.; *chorema, -atos,* n. room, space; *choraphion,* n. dim.; *choretos,* containing; *choreo,* make room, withdraw, retire:
L. *compluvium,* n. open space in the middle of a Roman house for collecting rain water from the roof: compluvium.
L. *desertum,* a waste place, wilderness; see **desert**
Gr. *dialeima, -tos,* n.; *dialeipsis,* f. interval; *diastema, -tos,* n. space between, interval; *diarrhoge,* f. gap, interstice: diastem.
L. *internodium,* n. space between knots; *interstitium,* n. space between, interval; *intervallum,* n. open space between two palisades, pause, intermission: internode, interstice, interstitial, interval, *Psalidium interstitiale* (a beetle).
Gr. *kochone,* area between the anus and pudenda; see **rump**
Gr. *kosmos,* universe; see **world**
Gr. *makrorhysmos,* m. long interval:
Gr. *mesodactylon,* n. space between fingers or toes; *mesogonation,* n. internode; *mesomazion,* n. space between the breasts; *mesomerion,* n. space between the

hips; *mesophrynon,* n. space between the eyebrows; *mesopleurios,* between the ribs, intercostal; *mesostylon,* n. space between pillars:

Gr. *metazytes, -etos,* f. interval:

L. *metopa* (Gr. *metope*), space between two triglyphs; see *ope* under **hole**

Gr. *parateichion,* n. pomerium:

L. *perineum,* area between the anus and pudenda; see **rump**

L. *plenum,* n. space filled with matter: plenum.

L. *pomerium,* n. free space along the inside and outside of the city wall:

Gr. *prostasia,* f. courtyard, area:

L. *vacuum,* empty space, void; see *vaco* under **empty**

L. *vastus,* empty, immense; see **large**

See: **open, hole, hollow, empty, world, country, spread, alone**

spade < AS. *spadu;* see **shovel**

spadix, L., Gr. palm branch, spike of flowers enclosed in a spathe; see **branch**

spado, L. (Gr. *spadon*), an impotent person, eunuch; *spadonius,* sterile, unfruitful; see **sterile**

Spain < L. *Hispania,* f.; *Hispanicus; Hispanus,* Spanish: Hispanic.

Gr. *Iberia,* f. Spain: Iberian, *Iberis sempervirens* (evergreen candytuft).

L. *Lusitania,* f. western Spain, Portugal: Lusitania, *Erica lusitanica* (Spanish heath).

spalaco- < Gr. *spalax, -akos,* mole; see **mole**

span < AS. *spann,* spread of space between end of thumb and end of little finger; see **measure, bridge**

spanios; spanos; sparnos, Gr. rare, scarce; see **few**

sparacto- < Gr. *sparaktes,* render, tearer; *sparagmos; sparaxis,* a tearing; see *sparasso* under **tear**

sparganium, L. (Gr. *sparganion*), burreed; see **waterweed**

sparganon, Gr. band, swaddling cloth; see **ribbon**

spargosis, Gr. distention, swelling; see **swell**

spark < AS. *spaerca;* see **fire**

sparrow < AS. *spearwa,* a kind of finch; see **finch**

sparse < L. *spargo, sparsus,* strew, scatter, sprinkle; see **few, spread**

sparteus, L. of broom; see *spartum* under **bean**

spartina, NL. a genus of grasses; see **grass**

sparto- < Gr. *spartos; sparton* (L. *spartum*), Spanish broom; *spartion,* dim., see *spartum* under **bean;** < *sparte; spartine; sparton,* rope, cable, cord; *spartion,* dim., see **rope;** < *spartos,* grown from seed, cultivated, see *speiro* under **sow**

sparus, L. (Gr. *sparos*), gilthead; *sparulus,* dim.; see **bream**

spasmos, Gr. convulsion, fit; *spastikos,* afflicted with spasms; see *spasma* under **disease**

spatalium, L. (Gr. *spatalion*), bracelet; see **bracelet**

spatalos, Gr. wanton, lascivious; *spatalistes,* profligate; see **lewd**

spatangius, L. (Gr. *spatanges*), a sea-urchin; see **echinoderm**

spathe < L. *spatha* (Gr. *spathe*), any broad blade, paddle for stirring and mixing, stem of a palm leaf, sheath enclosing a spadix; *spathula, spatula; spathion,* dim.; see **spoon**

spatium, L. room, distance; *spatiolum,* dim.; *spatiosus,* roomy, large; see **space**

spatos, Gr. hide, leather; see **leather**

spatula, L. dim. of *spatha* (Gr. *spathe*), a broad, flat tool for stirring or mixing, broad-sword; see **spoon, shovel, sword**

spay < OF. *espee,* cut with a sword, remove the ovaries; see **geld**

speak < AS. *specan; spreccan;* G. *sprechen.*

L. *affania,* f. chatter, idle talk:

Gr. *agoretes,* m. speaker; *agoretys, -yos,* f. eloquence; *agoreutos,* utterable, to be spoken of:

L. *aio,* say yes, affirm, assent, assert; see **yes**

Gr. *akroasis,* f. public lecture; *akroama, -tos,* n. oral teaching, something heard: acroama, acroamatic.

L. *anfractus,* bending, winding, circuitous, prolix; see **bend**

L. *arguo, -utus,* debate, contend, attempt to prove; *argutator, -is,* m.; *argutatrix, -icis,* f. subtle disputant; *argutulus,* noisy, talkative; *argutor, -atus,* prattle; *argumentum,* n. evidence, contention: argue, argument, argumentative.

L. *arilator, -is,* m. haggler:

L. *assero, -tus,* affirm, declare; *dissertatio, -onis,* f. discourse: assert, assertion, dissertation.

L. *assevero, -atus,* assert strongly, affirm: asseveration.

Gr. *asteismos,* m. clever talk, wit:

Gr. *aude,* f. voice, speech, < *audao,* speak:

L. *babulus,* m. babbler, fool: babble, babbler, babbling.   Derive: gibberish.

Gr. *baxis,* f. saying, report, rumor, < *bazo,* speak, say; *babazo,* chatter; *babax, -akos; babaktes,* m. chatterer, loud talker; *bagma, -tos,* n. speech: *Babax lanceolatus* (a bird).

L. *blatero, -atus,* talk idly, babble, prate; *blatero (balatro), -onis,* m. babbler, buffoon, fool: blaterate.

L. *bucco, -onis,* m. babbler, fool: Bucconidae, *Bucco collaris* (a puffbird).

L. *causidicus,* m. pleader, advocate:

L. *Cicero, -onis,* m. Roman orator and writer: Ciceronian, cicerone, *Cicerocrinus elegans* (a Silurian crinoid).

L. *contionor, -atus,* address an assembly; *contiuncula,* f. short harangue, trifling speech:

AS. *cwethan,* say: quoth, bequeath.

L. *declamo, -atus,* speak, orate: declaim, declamation.

L. *declaro, -atus,* show, manifest, proclaim: declare, declaration, declaratory.

L. *Delphi* (Gr. *Delphoi*), city of the famous oracle of Apollo; see **prophecy**

Gr. *Demosthenes,* m. Greek orator: Demosthenes, Demosthenic.

L. *dico, dictus,* speak, say, declare; *dicabula,* f. chatter, idle talk; *dicax, -acis,* talking sharply, sarcastic; *dicaculus,* loquacious; *dictio, -onis,* f. saying, style or mode of expression; *dictum,* n. saying, word; *edictum,* n. ordinance, proclamation; *dico, -atus,* proclaim, devote, dedicate; *indico, -atus,* point out, disclose, declare, reveal, betray; *indicivus,* indicative: diction, dictionary, edict, addict, dictation, dictator, dictum, benediction, valedictory, contradict, benison, verdict, ditto, ditty, prediction, preach, indictment, *ipse dixit,* abdicate, dedication, predicament, vindicate, avenge, revengeful, indicate, indicative, *Potnia indicator* (a bug).

L. *effutio, -itus,* blab, babble, prate, chatter:

Gr. *eipon,* speak, name, call:

Gr. *enteuxis; entychia,* f. conversation; *enteuxidion,* n. dim.:

Gr. *epesbolia,* f. hasty or scurrilous language:

L. *for, fatus,* speak, say, predict; *fateor, fassus,* admit, avow, own; *fabulor, -atus,* speak, talk; *facundus,* eloquent; *confabulor, -atus,* discuss with, converse; *effabilis,* utterable; *profiteor, -fessus,* declare publicly, avow, teach: fable, fabulous, fate, fatal, nefarious, preface, prefatory, fairy, fame, profession, confess, *Farrea facunda* (a sponge).

L. *garrio, -itus,* chatter, prate, babble; *garrulus,* talkative: garrulous, garrulity, *Garrulax leucolophus* (a jay-thrush), *Garrulus atricapillus* (a jay), *Myxantha garrula* (a honey-eater).

Gr. *gerys, -yos,* f. voice, speech, < *geryo,* speak, shout; *Geryon, -os,* m. a shouting, three-bodied monster; *geryma, -tos,* n. voice, sound; *gerygone,* f. born of sound, echo; *meligerys, -yos,* sweet-voiced, melodious; *erigerys, -yos,* loudvoiced: *Geryon tridens* (a decapod), *Geryonia rosacea* (a medusa), *Gerygone fusca* (a babbler), *Pseudogerygone igata* (a warbler).

L. *gestor, -is,* m. tattler:

Gr. *gnomikos,* dealing in maxims or pithy sayings:

L. *grammatica* (Gr. *grammatike*), f. grammar: grammatical.

Gr. *homiletikos,* affable, conversable; *homilema, -tos,* n.; *homilia,* f. conversation; *homiletes,* m. disciple, scholar: homiletics.

Gr. *hythlos,* m. idle talk, gossip, nonsense:

L. *intimo, -atus,* announce, publish: intimate, intimation.

Gr. *kleo,* tell of, celebrate, rumor; see *kleos* under **honor**

Gr. *kolax, -akos,* flatterer, fawner; see **flatter**

Gr. *kopis,* prater, wrangler, liar; see **lie**

Gr. *kotilos,* babbling, chattering, prattling, < *kotillo,* babble, chatter, prattle:

Gr. *lakeros,* talkative:

Gr. *laleo,* talk, babble, prattle; *lalia,* f. conversation, talk; *lalos,* talkative; *lalax, -agos,* m. babbler, croaker; *lalage,* f. a babbling; *laletikos,* given to babbling; *katalalos,* m. slanderer: lalopathy, laloplegia, laliatry, alalia, bradylalia, Eulalia, *Eulalia exigua* (a fly), *Catalalus obscurus* (a beetle).

L. *Larunda,* f. a nymph whose tongue was cut out by Jupiter for her talkativeness: *Larunda ceti* (an amphipod).

L., Gr. *lego,* say, speak, read; *lektikos,* pertaining to speaking or reading; *lexis,* f. a speaking or reading; *lexidion,* n. dim.; *logos,* m. word, speech, talk; *lectio, -onis,* f. a reading, perusal; *legibilis,* distinct enough to be read; *aeilogia,* f. continual talking; *amphilektos,* discussed on all sides, disputed; *dialektos,* f.

common language; *kenologos*, prattling, chattering; *leptologia*, f. subtle discourse, detailed discussion of trifles, quibbling: eulogy, homology, syllogism, dialogue, genealogy, dialect, dialectic, dyslexia, lecture, legend, legible, lectern, lesson, intellectual, intelligible, *Amphiletus papillatus* (a sponge).

Gr. *lerema*, silly talk; see *leros* under **fool**

Gr. *lesche*, f. gossip, talk; *leschema*, *-tos*, n. idle talk, chatter; *lesches*, m. talker; *adoleschos*, prating, garrulous; *adolesches*, m. idle talker:

L. *lingua*, tongue, language; *lingulaca*, chatterbox, gossip; *lingulatus*, eloquent; *lingulus; linguosus*, talkative, loquacious; see **tongue**

L. *loquor, locutus*, speak; *loquax, -acis*, talkative, garrulous; *loquela*, f. speech, language, words; *loquelaris*, of speech; *eloquor, -cutus*, speak out, declare: loquacious, colloquial, eloquent, elocution, circumlocution, obloquy, soliloquy, ventriloquist, *Sciurus loquax* (southern red-squirrel).

Gr. *menyo*, reveal, disclose, betray; *menytes*, informer; see **treason**

L. *murmur*, a low sound, mutter, mumble; see **murmur**

Gr. *mytheomai*, speak, recount; see *mythos* under **story**

L. *narro, -atus*, relate; see **story**

L. *nuntio, -atus*, report, make known; *enuntio*, speak out; see **call**

Gr. *oaros*, m.; *oarisma, -tos*, n. familiar talk, chat, < *oarizo*, chat with, daily: *Oarisma powesheik* (a butterfly).

Gr. *ops, opos*, f. voice: Calliope, *Calliope pectoralis* (Himalayan rubythroat), *Opopsitta diophthalma* (a parrot), *Trochilus calliope* (a hummingbird).

L. *oro, -atus*, speak, plead, pray; *oratio, -onis*, f. speech; *orator, -is*, m. speaker; *exoro*, persuade by entreaty; *exorabilis*, easily moved or persuaded: orator, oration, orison, oracle, oracular, oratorio, oratrix, inexorable, *Amazona oratrix* (a parrot), *Ponera inexorata* (an ant).

Gr. *ossa (otta)*, f. rumor: *Ossa dimidiata* (a bug).

Gr. *panegyrikos*, a festal oration; see *agora* under **gather**

Gr. *periglossos*, eloquent; *glossalgos*, talking until the tongue aches, very talkative:

Gr. *phemi*, speak; *pheme*, f. voice, speech; *phatos*, spoken, that may be spoken; *aphasia*, f. speechlessness; *aphatos*, ineffable, unutterable; *euphemia*, f. good speech, use of good names for bad things, praise: blasphemy, blame, euphemism, aphasia, prophet, Polyphemus, *Telea polyphemus* (a moth).

Gr. *phledon, -os*, m. idle talker, babbler; *phlenaphos*, m. idle talk; *phlyaria*, f.; *phlyaros*, m. silly talk, nonsense; *phlyax, -akos*, m. jester, fool, buffoon: *Phlyarus basalis* (a beetle).

Gr. *phoneo*, produce sound, speak; see *phone* under **sound**

Gr. *phrasis*, f. speech, < *phrazo*, speak, tell, indicate; *phraster, -os*, m. expounder; *phrastor, -os*, m. guide; *phrastikos*, expressive, eloquent: phrase, phaseology, paraphrase, periphrastic, *Phrasterothrips conducans* (a thysanopterid).

Gr. *phthenxis*, f. speech, utterance, < *phthengomai*, speak, utter; *phthegma, -tos*, n. sound of the voice; *phthegmatikos*, sounding, vocal; *phthengodes*, like a voice; *agathenktos*, loud-sounding; *apothegma, -tos*, n. a terse saying: apothegm.

L. *prodo, -itus*, produce, publish, report, disclose, betray; see **treason**

L. *promulgo, -atus*, make known, publish; see **display**

L. *publico, -atus*, make known, impart, disclose; see *populus* under **people**

L. *recito, -atus*, read aloud: recite, recitation.

Gr. *rhesis*, f., saying, speech, tale; *rheter; rhetor, -os*, m. orator, public speaker; *rhetorikos*, oratorical: rhetoric, rhetorician, rhetorical.

L. *rumor, -is*, m. common talk, hearsay, gossip; *rumusculus*, m. dim.; *rumigerulus*, m. newsmonger, gossiper: rumor, rumorous.

L. *saluto, -atus*, greet, address: salute, salutation.

AS. *secgan*, utter in words: say, gainsay, aforesaid, soothsayer, saw, saga, wiseacre.

L. *sermo, -onis*, m. talk, discourse; *sermunculus*, m. dim.: sermon, sermonize.

Gr. *stomphax, -akos*, a bombastic talker, < *stomphazo*, speak mouthfuls; see **brag**

Gr. *stomylos*, talkative, wordy; *stomylma, -tos*, n. chatterbox: *Stomylomyia leonina* (a fly).

L. *susurro, -atus*, whisper, murmur; see **murmur**

AS. *tellan*, speak, number: tell, tale, talk.

Gr. *thema, -tos*, n. subject of discourse: theme, thematic.

Gr. *thryleo*, babble, chatter, keep talking; *thrylos*, m. noise of many voices; *thrylema, -tos*, n. common talk:

L. *vates, -is*, prophet, soothsayer, seer, poet; see **prophecy**

L. *verbigero, -atus*, talk, chat, dispute; see *verbum* under **word**

L. *voco, -atus*, call; *vocifero, -atus*, cry out, clamor; *vox, vocis*, voice, sound, call; see **call**

L. *volubilis*, turning, spinning, fluent; see *volvo* under **turn**

See: **call, ask, word, sound, scold, story, adage, swear**

spear < AS. *spere:* spearman, spearmint, spearwood, speargrass, Shakespear.
  L. *aclys, -ydis,* f. a small javelin:
  Gr. *aichme,* f. point of a spear, spear; *aichmetes,* m. spearman, warrior;
    Aechmophorus occidentalis (western grebe), *Aechmea bracetata* (a bromeliacead).
  Gr. *akon, -ontos,* m. javelin, dart; *akontion,* n. dim.; *akontistes,* m. darter;
    akontizo, hurl a javelin: *Acontosceles hydroporoides* (a beetle), *Acontiostoma*
    magellanicum (an amphipod), *Acontistella amazonica* (a mantid), *Acontistoptera*
    melanderi (a fly), *Acontias punctatus* (a lizard), *Hexacontium floridum* (a
    radiolarian), *Acontistes gladiator* (a bird).
  Gr. *angon, -os,* m. a Frankish javelin:
  Gr. *ankyleton,* n. javelin:
  Gr. *belemnon,* n. dart, javelin: belemnite, *Belemnoteuthis antiqua* (a fossil squid).
  L. *cateia,* f. a kind of spear: *Catia druryi* (a butterfly).
  Gr. *dibolia,* f. a kind of lance; *dibolos,* two-pointed: *Dibolia decorata* (a beetle),
    Dibolosoma quadripustulatum (a beetle).
  Gr. *dory, doratos,* n. spear; *doration; dorydion,* n. dim.; *dorybolos,* hurling spears;
    doryphoros, spear-bearing; *doryssos,* wielding the lance; *doryxos,* m. spearmaker;
    dorikranos, spear-headed; *dorimachos,* fighting with the spear; *doritmetos,* spear-
    pierced: doryphorus, *Doratonotus megalepis* (a fish), *Dorydrilus mirabilis* (an
    annelid), *Dorylus helvolus* (a legionary ant), *Dorydium lanceolatum* (a bug),
    Acacia doratoxylon (currawang-acacia), *Coelorhynchus doryssus* (a fish),
    Phytodietus (Doratistes) absyrtinus (a wasp), *Doryphora sassafras* (a mon-
    imiacead), *Doryopteris arifolia* (a fern).
  Gr. *embolion,* n. something thrown, a javelin:
  Gr. *enchos,* n. spear, lance: *Enchocrates glaucopsis* (a moth).
  L. *framea,* f. spear, javelin:
  L. *fuscina,* three-pronged fork, trident; see **fork**
  L. *gaesum* (Gr. *gaison*), n. a long, heavy javelin: *Gaesomyrmex corniger* (a fossil
    ant).
  AS. *gar,* spear, lance: gar *(Belone vulgaris),* garlic *(Allium sativum),* garpike
    (Lepidosteus osseus).
  L. *grosphus* (Gr. *grosphos*), m. javelin: *Grosphus madagascariensis* (a scorpion).
  L. *hasta,* f. spear; *hastula,* f. dim.; *hastatus,* spear-shaped, armed with a spear:
    hastate, hastilude, *Hastigerina pelagica* (a foraminifer), *Viola hastata* (a violet),
    Phyllostoma hastatum (a vampire), *Xanthosoma hastifolium* (calalu), *Frigga*
    hastigera (a spider), *Xanthorrhoea hastilis* (a grass-tree).
  Gr. *hema, -tos,* n. dart, javelin:
  Gr. *hyssos,* m. javelin: *Hyssura producta* (an isopod).
  L. *jaculum,* n. dart, javelin: jaculiferous.
  Gr. *kontax, -akos,* m. shaft, spear; *kontakion,* n. dim.:
  L. *lancea,* f. a light spear; *lanceola,* f. dim.; *lanceatus; lanceolatus,* lancelike;
    lanciarius, m. lancer: lance, lancet, lancelet, lanceolate, Lancelot, *Solanum*
    lanceolatum (a nightshade).
  Gr. *lonche,* f. spear, lance, javelin; *loncharion; lonchidion,* n. dim.; *lonchites,* m.
    spearman; *lonchaios,* of spears: *Lonchocarpus latifolius* (a lancepod), *Hexalonche*
    octocolpa (a radiolarian), *Spermatolonchaea magnicornis* (a fly).
  L. *mataris,* f. a Celtic javelin: *Mataris grouvellei* (a beetle).
  Gr. *obelos,* m. spit; *obeliskos,* m. dim.: obelus, obelion, obelisk, obeliscal,
    Obelopteryx angusta (a moth), *Obelophorus terebratus* (a fly), *Obeliscus cal-*
    careus (a gastropod), *Pelurga obeliscata* (a butterfly).
  Gr. *palton,* n. light spear, dart; *paltos,* brandished, hurled: *Paltorhynchus narwhal*
    (a fossil beetle).
  L. *pilum,* n. javelin; *pilatus,* armed with a javelin: pile.
  L. *quiris, -itis,* f. a Sabine spear:
  L. *runa,* f. javelin, dart:
  Gr. *sarisa (sarissa),* f. a long, Macedonian pike: *Echinocereus sarissophorus* (a
    cactus).
  Gr. *saunion,* n. javelin:
  L. *sibyna* (Gr. *sibyne*), f. a kind of spear: *Sibynomorphus mikani* (a snake).
  Gr. *sigynes (sigynos),* m. spear:
  L. *spica,* ear of grain, point, spear; see **point**
  Gr. *stochasma,* the thing aimed, arrow, spear; see **arrow**
  Gr. *styrax, -akos,* spike at the butt end of a spear; see **point**
  L. *telum,* n. dart, javelin, spear, missile:
  L. *tragula,* f. javelin; *tragularius,* m. javelin-thrower:
  L. *trifax, -acis,* f. a kind of spear:
  L. *trudis, -is,* f. pointed pole, pike: *Trudis caeruleopunctatus* (a fish).
  L. *venabulum,* n. hunting spear:

L. *verutus,* armed with a dart or javelin, < *veru; verum; verutum,* n. spit for roasting meat, dart, javelin; *veruculum(vericulum),* n. dim.; *veruina,* f. a small javelin: *Lactistes vericulatus* (a bug).

Gr. *xyston,* n. shaft of a spear, spear: *Xystophora lucidella* (a moth), *Xystosomus turgidus* (a beetle), *Xystodesmus martensi* (a milleped), *Exyston cinctulus* (a wasp).

See: **arrow, point, rod, throw**

special < L. *specialis,* individual, particular; see **alone**

species, L. look, appearance, form, model, kind; see **class**

specillum, L. probe; see **tool**

specimen, L. example, token, evidence, sign; see **thing**

speciosus, L. beautiful, splendid, showy; *spectabilis,* notable, showy, < *specio; spicio, spectus,* behold, look; see **beauty, see**

speckled < AS. *specca,* bit, spot, stain; see **spot**

spectabilis, L. notable, showy; see *speciosus* under **beauty**

spectacle < L. *spectaculum,* show, sight, view, eyeglass; see *specio* under **see, lens**

specter < L. *spectrum,* vision, appearance, apparition, image; see **spirit**

specula, L. watch-tower, see *specio* under **see**; dim. of *spes,* hope, see *spero* under **believe**

speculator, L. explorer, scout, investigator; see *inspector* under **try**

speculum, L. mirror; see **mirror**

specus, L. cave, den; see **hollow**

speechless; see **silent, ineffable, wonder**

speed < AS. *sped,* swiftness; see **swift**

speleum, L. (Gr. *spelaion*); *spelunca* (Gr. *spelynx*), < Gr. *speos,* cave, den, grotto; see **hollow**

spelt < AS. *spelt,* < L. *spelta,* a kind of wheat; see **grain**

spend < AS. *spendan,* < L. *expendo, -pensus,* weigh out; see **pay**

speos, Gr. cave, den, grotto; see **hollow**

speratus, L. expected, hoped; see *spero* under **believe**

sperchnos, Gr. hasty, hurried, rapid; see **swift**

spergula, NL. a genus of the pink family; see **pink**

sperma, Gr. seed; *spermation,* dim.; *spermatikos,* of seed; see **seed**

spernax, L. despising, contemptuous; see *sperno* under **hate**

spes, L. hope; see *spero* under **believe**

speudo, Gr. hasten, quicken, press, urge on; *speustikos,* to be done or pushed eagerly; see **swift**

sphacelo- < Gr. *sphakelos,* gangrene, mortification; see **rot**

sphaco- < Gr. *sphakos,* sage; see **mint**

sphacto- < Gr. *sphaktos,* slain; *sphagion,* victim; see *sphazo* under **kill**

sphaero- < Gr. *sphaira,* ball; see **ball**

sphagido- < Gr. *sphagis, -idos,* sacrificial knife; *sphagidion,* dim.; see **knife**

sphagnum, NL. (Gr. *sphagnos*), a kind of moss; see **moss**

sphago- < Gr. *sphax, -agos,* throat; see **throat**

sphaleros, Gr. slippery, treacherous; see **slip**

sphallos, Gr. round, metal plate for throwing; see **plate**

sphalma, Gr. false step, stumble, trip, fault; see **falter**

spharagos, Gr. noise; see **sound**

spheco- < Gr. *sphex, -ekos,* wasp; *sphekion,* dim.; *sphekia,* wasp-nest; see **wasp, nest**

sphedanos, Gr. eager, vehement; see **busy**

sphelas, Gr. footstool, pedestal; see **base**

sphendamnos, Gr. maple; see **maple**

sphendone, Gr. sling; *sphendonestes,* slinger; see **throw**

spheneus, Gr. a kind of mullet; see **mullet**

spheno- < Gr. *sphen, -os,* wedge; *sphenarion; spheniskos,* dim.; see **wedge**

sphere < L. *sphaera* (Gr. *sphaira*), ball; *spherula,* dim.; see **ball**

sphex, Gr. wasp; see **wasp**

sphincter, L. (Gr. *sphinkter*), binder, a closing muscle; *sphigmos,* bound tight; see *sphingo* under **bind**

sphingia, Gr. a kind of ape; see **monkey**

sphingion, Gr. bracelet, necklace; see **bracelet**

Sphinx, Gr. female monster of Thebes who asked riddles; see **ask**

sphodros, Gr. strong, violent; see **strong**

sphondyle, Gr. an insect that lives on the roots of plants; see **insect**

sphondylium, L. (Gr. *sphondylion*), cow-parsnip; see **carrot**

sphondylos, Gr. vertebra; see **back**

sphragis, Gr. seal, signet; see **mark**

sphriganos, Gr. bursting with health and vigor, full, fresh, vigorous; see **life**

sphygmos, Gr. pulse, throb, beat; *sphygmikos,* of the pulse; see **strike**

sphyra, Gr. hammer; *sphyrion,* dim.; see **hammer**

sphyraena, L. (Gr. *sphyraina*), a pikelike fish; see **pike**

sphyrna, NL. a genus of sharks; see **shark**

sphyron, Gr. ankle; see **foot**

spica, L. ear of grain, point, spear; *spicula,* dim.; *spiculus,* pointed; see **point**

spice < OF. *espice,* < L. *species,* kind, sort.
    Gr. *artytos,* flavored, seasoned; *artyter, -os,* m. seasoner; *artytikos,* of seasoning:
    Gr. *asyphe,* f. a kind of spice:
    L. *carum* (Gr. *karon*), caraway; see **carrot**
    L. *cicilendrum,* n. spice:
    L. *condimentum,* n. a pungent or spicy substance added to food; *conditus,* savory, seasoned: condiment, condite.
    L. *cuminum* (Gr. *kyminon*), a plant whose seeds are used for seasoning; see **carrot**
    L. *hapalopsis, -idis,* f. a spice:
    See: **smell, taste, cassia, cinnamon, clove, cuminum, dill, ginger, mustard, myrtle, nutmeg, pepper, salt, sauce**

spicula, L. dim. of *spica,* ear of grain, point; see **point**

spider < AS. *spither,* < *spinnan,* spin.
    NL. *agalena,* f. a genus of spiders: *Agalena labyrinthica* (a spider).
    Gr. *arachne,* f.; *arachnes,* m. spider; *arachnidion; arachnion,* n. dim.; *arachnaios,* of spiders: arachnology, arachnid, arachnoid, Arachne, arachnidium, arachnitis, *Echinarachnius parma* (a sand-dollar), *Ilyarachna longicornis* (an isopod), *Anacampseros arachnoides* (a portulacacead),
    L. *aranea,* f. spider; *araneola,* f. dim.; *araneus,* of spiders: araneous, *Aranea diadema* (a spider), *Araneaster granulosus* (a bug), *Galeades araneoides* (a solpugid), *Ophrys aranifera* (an orchid).
    Gr. *Argiope,* f. a nymph; a genus of spiders: *Argiope argentata* (a spider), *Metargiope trigasciata* (banded garden-spider).
    Gr. *asterion,* n. a kind of spider:
    NL. *epeira,* f. a genus of spiders: *Epeira foliata*(now *Aranea frondosa*) (a spider), *Metepeira labyrinthea* (white-cross spider).
    Gr. *hypodromos,* m. a kind of spider:
    NL. *lycosa,* f. a genus of spiders: *Lycosa riparia* (a wolf-spider), *Geolycosa turricola* (a spider).
    Gr. *nethis,* spinster; see *neo* under **spin**
    L. *phalangium* (Gr. *phalangion*), n. a kind of spider: *Phalangium dorsatum* (daddylonglegs), *Pholcus phalangioides* (a spider), *Stenorhynchus falangium* (a spider-crab).
    Gr. *rhagion,* a kind of spider; see *rhax* under **grape**
    L. *solpuga (solipuga),* f. probably a kind of spider: solpugid, *Metasolpuga picta* (a solpugid).
    It. *tarantola,* f. a large spider, < L. *Tarentum,* now Taranto: tarantella, tarantism, *Tarantula aculeata* (a spider), *Lycosa tarantula* (tarantula), *Tarentola mauretanica* (a gecko).
    NL. *uloborus,* m. a genus of spiders: *Uloborus geniculatus* (a spider).
    See: **spin, weave, mite**

spike < L. *spica,* ear of grain, tuft, point, see **point, nail**

spilas, Gr. rock, crag, cave; see **stone**

spill < AS. *spillan,* let fall, let out, spread; see **fall, empty, drop, spread, speak**

spilos, Gr. spot, speck, stain, blemish; *spilotos,* stained, soiled; see **spot**

spin < AS. *spinnan:* spinner, spinneret, spinning-wheel, spinster, spider, spindle, spun.

L. *colus*, distaff; see **axis**

Gr. *kataktria*, f. spinner:

Gr. *klotho*, twist, spin; *klostes*, m. spinner; *klostos*, spun: Clotho, *Clostes priscus* (a spider).

L. *neo, netus* (Gr. *neo; netho*), spin, weave; *nethis, -idos*, f. spinster; *nestikos*, of spinning; *netos*, spun, twisted; *reneo*, unravel: *Nephila clavipes* (silk-spider), *Nesticus cellulanus* (a spider), *Argyroneta aquatica* (water-spider), *Callinethis elegans* (a spider), *Hilara aeronetha* (a fly).

L. *quasillaria*, f. spinning-girl:

Gr. *talasios*, pertaining to wool-spinning: *Talasius quadricornis* (a beetle).

See: **weave, spindle**

spinach < ML. *spinacia*, a plant of the goosefoot family; see **goosefoot**

spindle < AS. *spinel*.

Gr. *ataktos*, m. spindle, arrow; *ataktion*, n. dim.: *Atractosteus spatula* (an alligator-gar), *Atractocerus luteolus* (a beetle), *Atractaspis irregularis* (an African viper), *Megalatractus aruanus* (a gastropod), *Paratractus caballus* (jurel).

L. *colus, -us*, f. distaff: colulus, *Colus hirundinosus* (a fungus).

Gr. *elakate*, f. distaff, spindle: *Elacate atlantica* (a sergeant-fish), *Elacatinus oceanops* (a fish).

L. *fusus*, m. spindle: fusil, fusiform, fusee, fusain, fuselage, *Fusus salebrosus* (a gastropod), *Fusicladium dendriticum* (now *Venturia inaequalis*) (apple-scab), *Fusulina robusta* (a foraminifer), *Fusarium viride* (a fungus), *Fusanus acuminatus* (quandong nut), *Fusulinella sphaerica* (a foraminifer), *Salpa fusiformis* (a tunicate), *Tritonofusus brunneus* (a gastropod).

Gr. *kloster, -os*, m. spindle; *klosterion*, n. dim.: *Closterocrinus elongatus* (a Silurian crinoid), *Closterium moniliferum* (a desmid), *Clostridium histolyticum* (a proteolytic bacterium).

Gr. *netron*, n. spindle; *netrion*, n. dim.: *Netroneurus carolinus* (a neuropterid), *Netrocoryne repanda* (a moth), *Gymnetron thapsicola* (a weevil), *Netrium lamellosum* (a desmid).

L. *turbo, -inis*, top, spindle; see **cone**

spine < L. *spina*, thorn, backbone; *spinula*, dim.; *spinosus*, thorny; see **thorn, back**

spingos, Gr. a fish; see **fish**

spinter, L. a kind of bracelet; see **bracelet**

spinther, Gr. spark; see **fire**

spinus, L. (Gr. *spinos*), finch; see **finch**

Spio, L. (Gr. *Speio*), a sea-nymph; see **sea**

spionia, L. a kind of grapevine; see **grape**

spiraculum, L. air-hole, breathing-pore, vent; see *spiro* under **breathe**

spiraea, L. (Gr. *speiraia*), a genus of the rose family; see **rose**

spiral < L. *spiralis*, < *spira* (Gr. *speira*), f. coil, twist; *spirula*, f. dim.: spire, spiral, spirifer, spirochete, *Spirogyra crassa* (a green alga), *Spiroceras bifurcatum* (a Jurassic ammonite), *Spirorbis nipponica* (a worm), *Spirula australis* (a cuttlefish), *Spiranthes odorata* (an orchid), *Spiraea bella* (a spirea), *Spirillum rufum* (a bacterium), *Vallisneria spiralis* (tapegrass), *Cornuspira archimedes* (a foraminifer), *Coelospira hemispherica* (a fossil brachiopod), *Sceliphron spirifex* (a mud-dauber), *Potamogeton spirillus* (a pondweed).

L. *helix, -icis* (Gr. *-ikos*), f. whorl, spiral; *helikter, -os*, m. anything twisted; *heliktos*, rolled, twisted; *anthelix, -ikos*, f. inner curve of the ear: helix, helicoid, helictite, demonelix, *Helix albolabris* (forest-snail), *Helicopsyche limnella* (a caddisfly), *Helicella caperata* (a snail), *Helicodiceros muscivorus* (an arum), *Globigerina helicina* (a foraminifer), *Helicteres spicata* (a screw-tree).

L. *voluta*, f. spiral: volute, *Voluta musica* (music-shell), *Volutopsius castaneus* (a gastropod).

See: **turn, circle**

spiratus, L. breath; see *spiro* under **breathe**

spire < L. *spira* (Gr. *speira*), coil, see **spiral**; < AS. *spir*, young shoot, tapering point, see **point**

spirio- < Gr. *speirion*, light garment, dim. of *speiron*, piece of cloth; see **garment**

spirit < L. *spiritus, -us*, m. air, breath of life, energy, apparition, god, ghost, soul, < *spiro, -atus*, breathe: spiritual, spirited, inspiration, Holy Spirit, *spiritus frumenti*, spirituous.

AS. *aelf*, a diminutive, sprightly, sometimes mischievous, fairylike being: elf, elfin, elfish, Alfred, oaf.

Gr. *akko*, a bugbear; see **fear**

Ar. *Allah,* the supreme being of the Mohammedans: *La ilaha illa Allah.*

Egypt. *Amun (Amen),* the sun god; see **amido-**

Gr. *anousios,* immaterial, without substance; see *ousia* under **thing**

L. *apparitio, -onis,* f. unusual appearance, phantom, ghost: apparition.

Heb., Gr. *ariel,* m. an airy sprite: Ariel, *Lagenoplastes ariel* (fairy-martin).

L. *asomatus* (Gr. *asomatos),* incorporeal:

Heb. *Baal,* lord, deity; Babylonian *Bel;* Heb. *Baalzebub,* lord of flies: Baal, Beelzebub, Bel, Belshazzar, *Ateles belzebuth* (a spider-monkey).

Skt. *brahman,* soul or essence of the universe: Brahma, Brahman, Brahmaputra, Brahmanism, *Cremastocheilus brahma* (a beetle).

Skt. *buddha,* enlightened; name of a deified religionist, Gautama Buddha: Buddha, Buddhism, *Hagiastrum buddhae* (a radiolarian).

L. *caelicola,* m. dweller in heaven, god:

L. *Cronus* (Gr. *Kronos),* m. former ruler of heaven and earth, equivalent to Saturn: Cronian.

Gr. *daimon, -os,* c. god, spirit, evil spirit; *daimonion,* n. dim.; *daimonios,* of demons; *daimonikos,* possessed of a demon, demonlike: demon, demoniac, demoniast, demonifuge, demonism, demonology, eudemonic, pandemonium, demonelix, *Daemonorops draco* (a rattan-palm).   Henry Thoreau in *Walden:* "What demon possessed me that I behaved so well?"

L. *deus; divus (dius),* m. god; *dea; diva,* f. goddess; Skt. *deva,* god; *divus; divinus; divalis,* of a god, godlike: deity, deist, deify, deism, *deus ex machina,* divine, divination, 'dear me!', adieu, joss, Amadeus, devadasi, devaloka, deodar *(Cedrus deodara), Primula deorum* (Olympian primrose).

L. *diabolus* (Gr. *diabolos),* m. devil; *diabolicus* (Gr. *diabolikos),* slanderous, devil-ish: devil, devilish, diabolical, bedevil, *Vespa diabolica* (yellow-jacket), *Mobula diabolus* (a devilfish).

Gr. *Dis, Dios,* m. Zeus, chief of the Greek gods; *Hera,* f. wife of Zeus; *Zen, -os,* m. a form of *Zeus;* L. *Dialis,* pertaining to Dis (Jupiter), ethereal: Diogenes, Dioscurian, Zenobia, Diana, *Diospyros virginiana* (persimmon), *Diomedea exulans* (wandering-albatross), *Dialiptila guinea* (a bird), *Hera mikii* (a fly), *Heranice miltoglypta* (a bug), *Zenodoxus maculipes* (a moth).

L. *Empusa* (Gr. *Empousa),* f. hobgoblin, specter: *Empusa muscae* (a fungus), *Squilla empusa* (a shrimp).

L. *Fornax,* goddess of ovens; see **oven**

AS. *gast,* soul, spirit, apparition: ghost, ghostly, Holy Ghost, ghastly, aghast.

L. *genius,* m. protecting attendant spirit of a person or place: genius.

AS. *god,* a spiritual being, deity: god, goddess, godly, godlike, godless, godfather, godsend, Goddard, Godwin, Godfrey, gossip, demigod, ungodly.

L. *inconcretus,* bodiless:

L. *incorporalis; incorporeus,* bodiless, immaterial, intangible: incorporeal.

L. *incubus,* m. nightmare, male demon supposed to lie upon persons, especially women, in their sleep; *succuba,* f. female demon supposed to prostitute men in their sleep: incubus, succubus.

L. *Isis, -idis* (Gr. *-idos),* f. the Egyptian goddess of fertility: Isis, isidium, isidiferous, Isidore, *Isis hippuris* (king-coral), *Isidium corallium* (a lichen), *Isidella neapolitana* (a gorgonid coral).

L. *Jupiter, Jovis,* m. chief of the Roman gods, equivalent to the Greek Zeus; *Juno, -onis,* f. wife of Jupiter: jovial, jovite, by Jove!, joubarb, apojove, *Jovis die* (F. *jeudi,* Thursday), jovilabe, *Juglans nigra* (black walnut), *Scaphella junonis* (a gastropod), *Lychnis flosjovis* (a campion).

Gr. *kobold,* goblin, earth spirit: kobold, cobalt, cobaltite, Godbold.

L. *larva,* ghost, mask; the young stage of some animals; see **young**

L. *Lateranus,* god of the hearth, < *later,* brick; see **brick**

L. *lemur, -is,* m. shade, ghost of the departed: lemur, Lemuria, lemuroid, *Lemur varius* (ruffed lemur).

L. *Liburnus,* god of lustful enjoyment; see **joy**

L. *mania,* a bugbear to frighten children; *maniola,* dim.; see **fear**

L. *manis,* m. ghost, shade, soul of the departed: Manes, *Manis tricuspis* (pangolin).

Ab.Am. *manito,* spirit, god: manito, Manitou (Colo.), Manitowoc (Wis.), Manitoba.

F. *Melusine,* f. a legendary fairy: Melusina, *Gazeletta melusina* (a radiolarian).

Heb. *Molekh,* a deity worshipped by sacrifice of firstborn children: Moloch, *Moloch horridus* (a lizard).

Gr. *Mormo (Mormon),* bugbear, hobgoblin; see **fear**

G. *nickel,* demon: nickel, nickelodeon, kupfernickel, niccolite.   Derive: Belsnickel.

Gr. *noar, -os,* n. phantom, specter:

L. *numen, -inis,* n. nod, will, sway, majesty, deity, godhead: numen, numenism, numinous.

Gr. *opsis,* face, likeness, vision, apparition; see *ops* under **face**

Gr. *Pan,* god of hills, forests, pastures, flocks, sportsmen; see **herd**

Gr. *phasma, -tos; phantasma, -tos,* n. apparition: phantom, *Phasmatocoris spectrum* (a bug), *Leucophasma phantasmella* (a moth).

Gr. *psyche,* breath, mind, soul, spirit; see **mind**

Heb. *satan,* an enemy, the devil: Satan, satanic, *Satanoperca crassilabris* (a fish), *Satanellus hallucatus* (a marsupial-cat), *Boletus satanus* (a mushroom).
Professor: "What is the origin of the word Satan?"
Student: "I think it is just an Old Nick name."

L. *Saturnus,* god of agriculture; see **till**

Gr. *satyros,* m. a woodland deity typifying lasciviousness; *satyridion,* n. dim.; *satyrikos,* of or like a satyr: satyric, satyriasis, *Tragopan satyrus* (a pheasant).

AS. *sawl,* essence of life: soul, soulful.

Jap. *shinto,* way of the gods: Shintoism.

L. *silenus* (Gr. *seilenos*), m. a bearded, bald, woodland deity, similar to but older than a satyr: *Scarabaeus silenus* (a dung-beetle).

L. *simulacrum,* n. image, phantom: simulacrum.

Skt. *Siva,* supreme deity of destruction and restoration: *Sivapithecus indicus* (a fossil simian).

Gr. *skia,* shade, shadow, phantom; see **shade**

L. *species,* apparition, vision, kind; see **class**

L. *spectrum,* n. appearance, form, image, apparition, ghost: specter, spectrum, *Vampyrus spectrum* (a bat).

F. *sylphe,* f. a fairylike spirit of the air: sylph, sylphlike, sylphid.

Gr. *theos,* m. god; *thea,* f. goddess; *theios,* divine; *theiotes,* f. divinity, religion; *theskelos,* godlike; *thespis, -ios; thespesios,* inspired, divine; *thesphatos,* spoken by a god, ordained; *apotheosis,* f. deification; *antitheos,* equal to the gods, godlike, impious; *entheos,* divinely inspired: theology, theocratic, Pantheon, apotheosis, Theodore, Theophilus (Gottlieb), polytheistic, Thespian, enthusiasm, Dorothea, Dorothy, Dot, timothy *(Phleum pratense), Dodecatheon salinum* (a shooting-star), *Theocapsa aristotelis* (a radiolarian), *Thespesia populnea* (a malvacead), *Thescelocichla leucopleurus* (a bird).

L. *Venus,* goddess of love; see **love**

Slav. *wampir,* a blood-sucking ghost or demon; see *vampyrus* under **bat**

AS. *Woden* (ON. *Odhinn*), chief of the gods: Woden, Wotan, Odin, Odinism, Wednesday, Winsborough, Wanstead.

Heb. *Yehowah (Yahweh),* the supreme being of the Hebrews: Yahweh, Jehovah.

Identify: afrit(afreet), angel, apparition, Ariel, banshee, bogey, brownie, demon, devil, dryad, dwarf, elf, fairy, Fate, faun, fay, fenodryee, fiend, fifinella, fomorian, genie, genius (genii), ghost, ghoul, giant, gnome, goblin, god, goddess, gremlin, griffin, hamadryad, harpy, hex, hobgoblin, hoodoo (voodoo), hopschneider, imp, incubus, jinnee (jinn, djinn), jinx, kobold, lar (lares), larva, lemur, leprechaun, manes, mermaid, merman, muse, naiad, nemesis, nereid, nix, nixie, nymph, oaf, oceanid, ogre, oread, phantom, pixy, poltergeist, Puck, salamander, Santa Claus, saskwatch, satyr, shade, silenus, sorcerer, spandule, specter, spirit, spook, sprite, succubus, sylph, troll, undine, valkyrie, vampire, wee folk, widget, witch, wizard, wraith, zombie.

See: **wind, life, form, fancy, magic, alcohol**

spiro- < Gr. *speira,* coil, twist, see **spiral**; < *speiro,* sow, see **sow**; < *speiron,* piece of cloth, see **garment**; < L. *spiro, -atus,* breathe, see **breathe**

spissus, L. thick, crowded, dense, slow; *spissitudo,* thickness; see **thick**

spisula, NL. a genus of mollusks; see **mollusk**

spit < AS. *spittan.*

Gr. *aperao,* spit out, disgorge; *aperasis,* f. a spitting out:

Gr. *chremma, -tos; chrempton,* n. spit, saliva; *chrempsis,* f. a hawking, spitting:

L. *expectoro, -atus,* drive from the breast, banish, spit: expectorate.

Gr. *psytton,* n. spit, < *psytto,* spit:

Gr. *ptyalon,* n. saliva, spit; *ptysma, -tos,* n. spit; *ptyas, -ados,* f. spitter, < *ptyo,* spit: ptyalin, ptyalogogue, ptyalism, ptyalose, ptyalith, *Ptyoprocne fuligula* (a bird), *Ptyas blumenbachi* (a snake), *Ptyelus lineatus* (spittle-insect).

L. *saliva,* f. spit; *salivosus,* full of spit, slavering; *salivo, -atus,* spit out, drivel, slaver: salivary, salivation.

Gr. *sialon,* n. spit: sialogogue, sialosyrinx.

L. *sputum,* n. spit; *spuo, sputus,* spit; *sputamen, -inis,* n. spit; *conspuo,* spit upon in contempt: sputum, cuspidor, *Melanophila conspula* (a beetle).

spite < OF. *despit,* < L. *despectus,* contempt; see **hate**

spithama, L. (Gr. *spithame*), span; see **measure**

spiza, Gr. finch; see **finch**

spizias, Gr. sparrow-hawk; see **hawk**

splanchnon, Gr. entrail; *splanchnidion,* dim.; see **intestine**

spleen < L. *splen, -is* (Gr. *-os*), m. milt; *splenicus* (Gr. *splenikos*), of the spleen; *spleneticus,* depressed, melancholy; *splenium* (Gr. *splenion*), n. a fern, spleen-wort, plaster or patch with a medicinal preparation: splenic, splenetic, splenitis, splenalgia, splenomegaly, splenectomy, splenculus, splenitive, splenoid, splenule, splenical, green spleenwort *(Asplenium viride), Chrysosplenium alternifolium* (a golden-saxifrage).
L. *lien, -is,* m. spleen; *lienicus; lienosus,* splenetic: lienitis, lienocele.

splendid < L. *spendidus,* bright, shining; splendor, brightness, brilliance; see *splendeo* under **light, beauty**

splenium, L. (Gr. *splenion*), plaster or patch with a medicinal preparation, spleen-wort; see **spleen**

split < MD. *splitten;* see **cut, separate**

splinter < D. *splinter;* see **part**

spodos, Gr. ashes, embers, dross, slag; *spodios,* ash-colored, gray; see **ashes, gray**

spoil < L. *spolium,* booty; see **plunder, rot, destroy**

spoke < AS. *spaca;* see **rod**

spolium, L. booty, plunder; *spoliator; spoliatrix,* pillager; see **plunder**

sponda, L. bedstead, bed; see **bed**

sponde, Gr. drink-offering, libation, truce; *spondeios* (L. *spondeus*), of libations; *spondeion* (L. *spondeum*), cup or vessel used in libations; see **drink, rest**

spondeo, *sponsus,* L. promise; *sponsor,* bondsman; see **promise**

spondias, Gr. a kind of plum-tree, now a genus of the sumac family; see **sumac**

spondylus, L. (Gr. *spondylos; sphondylos; sphondylios*), vertebra, joint, see **back**; a kind of mussel, see **mollusk;** cow-parsnip, see **carrot**

sponge < L. *spongia* (Gr. *spongia; sphongia*), f.; *spongos (sphongos),* m. any spongy substance; *spongion (sphongion),* n. dim.: spongin, spongiose, spongy, spongioblast, *Spongospora subterranea* (a slime-mold), *Spongostylum perpusillum* (a fly), *Spongophora bipunctata* (an earwig), *Spongilla lacustris* (a fresh-water sponge), *Catulus spongiceps* (a fish), *Laetmogone spongiosa* (a holothurian).
NL. *ectyon,* m. a kind of sponge: *Ectyon sparsus* (a sponge), *Ectyodoryx loyningi* (a sponge).
Gr. *epikoilos,* spongy, porous:
Gr. *manon,* n. a kind of soft sponge: *Manon favosum* (a sponge).
Gr. *somphos,* porous, spongy; see **hole**
L., Gr. *tethea,* f. a kind of sponge: *Tethea mammillosa* (a sponge), *Tetheodus pephredo* (a fossil fish).

sponsa, L. bride; *sponsus,* groom; see **spouse**

spool < D. *spoel;* see **axis**

spoon < AS. *spon,* chip.
Gr. *aryter, -os,* m. ladle, dipper; *arytaina,* f. dipper: arytenoid.
L. *bacrio, -onis,* m. vessel with a long handle, dipper, ladle:
L. *cochlear, -is* (Gr. *kochliarion*), n. spoon, ladle: *Cochlearia officinalis* (scurvy-weed).
Sp. *cuchara,* f. spoon; *cucharita,* f. dim. teaspoon:
L. *cyathus* (Gr. *kyathos*), cup, ladle; see **cup**
Gr. *kinetron,* n. ladle, stirring-rod:
Gr. *kykethron,* ladle, agitator; see *kykao* under **mix**
L. *ligula,* ladle, skimmer; see *lingua* under **tongue**
L. *misisula,* f. piece of bread used as a spoon:
Gr. *mystron,* n. spoon; *mystrion,* n. dim.; *mystile,* f. piece of bread used as a spoon: *Mystropomus subcostatus* (a beetle).
Gr. *rhatane,* f. ladle:
L. *rudicula,* wooden spoon, spatula; see *rudis* under **rod**
L. *simpulum,* n. small ladle: *Simpulum trapezium* (a gastropod).
L. *spatha* (Gr. *spathe*); *spathis, -idos,* f. paddle for stirring or mixing, broad two-edged sword, any broad blade, stem of a palm leaf, sheath enclosing a spadix; *spathula; spatula,* f.; *spathion,* n. dim.: spathe, spathose, spatula, spatulate, spade, epaulet, *Spatula clypeata* (shoveler-duck), *Spathognathodus primus* (a conodont), *Spathiopsilopus papuasinus* (a fly), *Spathidium caudatum* (a protozoan), *Spathodea campanulata* (a bignoniacead), *Circospathis furcata* (a radiolarian), *Anthurium spathiphyllum* (an aroid), *Polyodon spathula* (paddle-fish), *Salvia spathacea* (crimson sage), *Urtica spathulata* (a nettle), *Hoplogyron*

spatulifer (a wasp), *Thamnocalamus spathiflorus* (a bamboo), *Micromorphus spatulifer* (a fly), *Listrocryptus spatulatus* (a wasp).
Gr. *toryne*, f. ladle, stirrer; *torynetos*, stirred, < *toryno*, stir: *Toryna vespoides* (a wasp), *Torynobelodon barnumbrowni* (a fossil mammal), *Torynesis mintha* (a butterfly).
L. *trua*, f. stirring spoon, skimmer; *trulla*, f. dim.: *Trullula nitidula* (a fungus), *Plagusia trulla* (a crustacean), *Clematis trullifera* (a clematis), *Pterostylis trullifolia* (an orchid).
Gr. *tryelis, -idos*, f. stirrer, ladle:
L. *tudiculo, -atus*, stir; see *tudes* under **hammer**
Gr. *zomerysis*, f. soup ladle:
See: **mix, dip, cup**

spore < Gr. *spora*, a sowing, seed; *sporadikos*, scattered; see *sperma* under **seed, sow**

sporta, L. basket, hamper; *sportella; sportula*, dim., gift; see **basket, give**
spot < ME. *spotte:* spotted, spotty, spotless, spot (*Leiostomus xanthurus*).
L. *albugo*, white spot on the eye; see *albus* under **white**
L. *alea*, a game of chance with dice, spot; see **lot**
L. *alphus* (Gr. *alphos*), white spot on the skin, a kind of leprosy; see **disease**
Gr. *argema; argemos*, a white fleck on the eyeballs; see *argos* under **white**
Gr. *astralos*, spotted with stars, speckled; see *aster* under **star**
Gr. *balios*, spotted, dappled, piebald: *Balionycteris maculatus* (a bat).
L. *cruentus*, spotted or stained with blood, bloody; see *cruor* under **blood**
Gr. *daidaleos*, dappled, spotted; see *daedalus* under **art**
Gr. *ephelis (epelis), -idos*, f. freckle: ephelis.   See *epelis* under **cap**
L. *fritillus*, m. dice-box, spotted: fritillary (*Argynnis diana*), *Fritillaria lanceolata* (riceroot-fritillary).
L. *gutta*, drop, spot; *guttula*, dim.; *guttatus*, dappled, spotted, speckled; see **drop**
Gr. *kelis, -idos*, f. stain, spot, blemish: celidography, *Celidodacus apicalis* (a fly).
Gr. *knekis, -idos*, f. pale spot, dim cloud:
L. *labes, -is*, f. spot, stain, blemish, disgrace; *labecula*, f. dim.:
L. *lenticula; lentigo*, a lentil-shaped spot, freckle; *lentiginosus*, full of freckles, spotted; see **lens**
L. *lino, litus*, daub, smear, anoint, spread over; see **spread**
L. *litura*, blot, blur; see *lituro* under **cancel**
L. *macula*, f. spot, stain, mark; *maculosus*, spotted, dappled, pied, < *maculo, -atus*, spot, stain, speckle; *immaculatus*, unspotted, unstained: macula, mackle, mascle, maqui, immaculate, *Maculifer japonicus* (a trematode), *Maculopeplum junonium* (a gastropod), *Euphorbia maculata* (spotted spurge), *Anopheles maculipennis* (a mosquito), *Eucalyptus maculosa* (a eucalyptus), *Silphomorpha maculigera* (a beetle), *Clematostigma maculiceps* (a psocid), *Aphrodes unimacula* (a bug), *Mochtherus immaculatus* (a beetle).
Gr. *melasma*, a black spot; see *melas* under **black**
Gr. *miasma*, stain, defilement, taint; *miantos*, stained; *miaros*, stained with blood; see *miaros* under **dirt**
Gr. *molysma, -tos*, n. spot, taint:
ML. *morbillus*, measle or measles; see *morbus* under **disease**
Gr. *nebrias*, spotted like a fawn; see *nebros* under **deer**
Gr. *nephelion*, cloudlike spot, white fleck, dim. of *nephele*, cloud; see **cloud**
L. *nota*, mark, cipher, character, sign, letter; see **mark**
L. *ocellatus*, having little eyelike spots; see *oculus* under **eye**
Gr. *omorgma, -tos*, n. spot, stain, < *omorgnymi*, wipe:
Gr. *palasso*, sprinkle, spot, bespatter: *Palassopora eocenica* (a fossil coral).
Gr. *pardalotos*, spotted like a leopard; see *pardos* under **cat**
L. *petigo*, scab, eruption, spot; see **scale**
Gr. *phalaros*, having a white patch, white-spotted: phalarope.
Gr. *pholis, -idos*, scale, spot, fleck; see **scale**
Gr. *poikilos*, variegated, pied, mottled, spotted; see **color**
Gr. *psaros*, speckled, dappled, like a starling, < *psar*, starling; see **starling**
L. *punctum*, n. little hole, dot, point; *puncta*, f. prick, puncture; *punctillum; punctulum*, n. dim.: puncture, punctate, *Tabanus punctifer* (a horsefly), *Ichthyobdella punctata* (a leech), *Microperca punctulata* (a darter), *Capelopterum punctatellum* (a bug), *Subcoccinella 24-punctata* (a beetle), *Eurymetopon punctulatum* (a beetle), *Rhyparochromus puncticollis* (a bug). *Selidosema punctularia* (a butterfly), *Hemerobius sexpunctatus* (a lacewing).
Gr. *rhantos*, sprinkled, spotted; see **wet**
Gr. *rhytisma*, patch; see **patch**
L. *sparsus*, strewn, sprinkled, flecked, spotted; see *spargo* under **spread**
AS. *specca*, bit, spot, stain: speck, speckled.

Gr. *spilos,* m. spot, fleck, speck, stain, blemish; *spilotos,* spotted, stained: *Spilophora lyra* (a beetle), *Spilogale pygmaea* (a spotted shark), *Zaspilota cupricollis* (a beetle), *Cinyra spilota* (a beetle), *Platophrys chlorospilus* (a fish), *Ctenobrycon spilurus* (a fish), *Aspilogeton nubicola* (a bug).

L. *stellatus,* starred, spotted; see *stella* under **star**

Gr. *stiktos,* punctured, dappled, spotted, < *stizo,* prick; *pantostiktos,* spotted all over; *peristiktos,* dappled: *Sticta crocata* (a lichen), *Stictoporella cribrosa* (a fossil bryozoan), *Polystictus versicolor* (a bracket-fungus), *Scaurus sticticus* (a beetle), *Lophornis stictolophus* (a hummingbird).   See *stigma* under **mark**

L. *sugillatum,* black and blue spot, bruise; *sugillo, -atus,* beat black and blue; see **bruise**

LL. *variola,* f. smallpox, spotted: variola, variolate, variolite, varioloid, *Scarabaeus variolosus* (a dung-beetle).

L. *vitiligo,* a skin disease manifested by milk-white spots; see **disease**

L. *vulnus,* wound, sore, spot; see **sore**

See: **mark, point, drop, eye, peacock, tiger, color, trout, pimple, starling**

spouse < OF. *espous,* < L. *sponsus,* betrothed, engaged, promised; *sponsus,* m. groom; *sponsa,* f. bride: *Aix sponsa* (wood-duck).

Gr. *akoites,* bedfellow, spouse; see **companion**

Gr. *alochos,* bedfellow, spouse; see **companion**

L. *conjunx (conjux), -jugis,* c. consort; *conjuga,* f. wife: conjugal.

L. *consors, -tis,* partner, spouse; see **companion**

AS. *bryd,* bride: bride, bridal, bridesmaid.

Gr. *damar, -tos,* f. one tamed or yoked, wife, spouse, < *damao,* subdue, tame; *dysdamar,* ill-wedded:

L. *deserta,* the abandoned wife; see *desero* under **abandon**

L. *destinata,* f. betrothed female, bride:

Gr. *gamete,* f. married woman, wife; *gametes,* m. husband: gamete, gametophyte.

AS. *husbonda,* master of a house, < *hus,* house, *bonda,* dwelling in: husband, husbandman, husbandry.

L. *marita,* f. wife; *maritus,* m. husband: marital.

L. *matrona,* f. married woman; *matronalis,* of a married woman: matron, *Hesperis matronalis* (dames-rocket).

L. *nupta,* f. wife; *nuptula,* f. dim. young wife:

L. *nympha* (Gr. *nymphe*), bride, young woman; *nymphikos,* of a bride; *nymphios,* bridegroom; see **woman**

Gr. *oar, -os,* f. wife:

Gr. *posis,* m. husband, spouse:

L. *uxor, -is,* f. wife; *uxorcula,* f. dim.; *uxorius,* of a wife: uxorious, uxoricide, uxorial.

AS. *wif,* wife: wife, wifely, woman, fishwife.

See: **companion, marry**

spout; see **nose, spring, vomit**

spread < AS. *spraedan,* expand, extend, stretch, open.

Gr. *anapetes,* expanded, wide open; see **open**

AS. *brad,* extended, wide; see **broad**

L. *delibuo, -utus,* besmear, anoint:

Gr. *diaspora,* f. a scattering, dispersion: diaspore, Diaspora.

Gr. *diastasis,* f. separation, distention: diastase, diastatic.

Gr. *diastole,* f. dilatation: diastole, diastolic.

L. *differo, dilatus,* spread, scatter; *dilato, -atus,* spread out, enlarge, extend; *collatatus,* extended, diffuse; *prolatus,* extended, elongate: dilate, dilation, prolate, *Acheilops dilatus* (a trilobite), *Loxocrinus dilatatus* (a Permian crinoid).

L. *diffusus; effusus; profusus,* spread out, extended, dispersed: diffuse, diffusion, effusive, profuse, profusion, *Aristida diffusa* (a grass).

L. *discutio, -cussus,* break up, disperse, dissipate: discuss, discussion.

L. *dispello, -pulsus,* scatter, disperse, drive away: dispel.

L. *displico, -atus,* scatter: *Aranea displicata* (a spider).

L. *displodo, -plosus,* spread out, dilate, extend:

L. *dissemino, -atus,* sow, scatter, spread; see *semen* under **seed**

L. *dissipo, -atus,* scatter, squander, waste: dissipate, dissipation.

L. *distribuo, -butus,* divide, apportion: distribute, distribution, distributive.

L. *divaricatus,* spread apart: divaricate, *Aster divaricatus* (an aster).

L. *divulgo, -atus,* spread, publish: divulge.

Gr. *elastikos,* rebounding, springing, stretching; see **leap**

Gr. *epidemos,* prevalent, spreading through a large number: epidemic.

Gr. *eurys,* broad, widespread; *aneurysmos,* a dilatation; see **broad, disease**

L. *explicatus,* unfolded, unrolled, spread out, plain, clear; see **clear**

L. *interverto, -versus,* turn aside, squander:

Gr. *lichas, -ados,* space between thumb and forefinger, span; see **measure**
L. *lino, litus (linio, linitus),* daub, smear, anoint, spread over; *allitus; collitus; illitus; oblitus; perlitus,* smeared, bedaubed, polluted: illinition, liniment, *Escallonia illinita* (a saxifrage), *Apatella oblinita* (a moth), *Ptomatophagus oblitus* (a beetle).
Gr. *nemo,* graze, dispense; see **eat, give**
Gr. *orego,* stretch out; *oregma, -tos,* n. a stretching out; *orektos,* stretched out, held out: *Orectochilus costatus* (a beetle), *Orectoderus obliquus* (a bug), *Oregma bambusae* (an aphid).
L. *pando, passus; pansus,* stretch, spread, extend; *pandiculor, -atus,* stretch oneself: expand, expansion, pandiculation, pass, trespass, passenger, impassable, passage, spawn, *Tachina pansa* (a fly), *Lycophris dispansus* (a foraminifer), *Tetragonia expansa* (an aizoacead), *Penicillium expansum* (a mold).
Gr. *pastos,* sprinkled, < *passo,* sprinkled; see **wet**
L. *patulus,* spread out, open; see *pateo* under **open**
Gr. *petasma, -tos,* n. anything spread out:
L. *porrigo, -rectus,* stretch, spread, extend; *exporrectus,* spread out, smoothed: *Cyrtia exporrecta* (a fossil brachiopod).
L. *prolixus,* stretched out, long; see **long**
L. *promulgo, -atus,* make known, publish; see **display**
L. *propago, -atus,* extend, increase, spread: propagate, propagation.
Gr. *skedannymi,* disperse, scatter; *skedasis,* f. a scattering; *skedastos,* scatterable; *skedastes,* m. scatterer:
Gr. *skordinaomai,* stretch, yawn:
L. *spargo, sparsus,* strew, scatter, sprinkle; *sparsus,* scattered, few, rare, sprinkled, spotted; *aspersus,* scattered, sprinkled; *conspersus,* strewn, sprinkled; *dispersio, -onis,* f. a scattering; *interspersus,* strewn or sprinkled upon or between: sparse, aspersion, disperse, intersperse, *Chrysopa aspersa* (a lacewing), *Pupa conspersa* (a gastropod), *Spergula arvensis* (corn-spurry), *Phrixocoma sparsa* (a trichopterid), *Adenostoma sparsifolium* (a rosacead).
L. *spithama* (Gr. *spithame*), span; see **measure**
Gr. *sporadikos,* scattered; see *speiro* under **sow**
L. *sterno, stratus,* spread, pave; *stratura,* f. pavement; *stragulus,* spreading, covering: stratum, stratification, street, consternation, prostrate, Stratford, Streatham, estrade, stray, *Eryngium prostratum* (creeping eryngo).
Gr. *strotos,* spread, laid; *strotes,* m. spreader, < *stronnymi,* spread, scatter, strew: *Strotocrinus venustus* (a Mississippian crinoid).
Gr. *syrden,* dragging, in a long line: *Syrdenopsis grayi* (a beetle).
Gr. *teino,* stretch, strain, extend, spread; *tasis,* f.; *tonos,* m. stretching or tightening; *anastasis,* f. a stretching out, extension; *ektadios,* outstretched, spread; *ektasis,* f. extension; *ektatos,* capable of extension; *entanysis; katatasis,* f. a stretching or straining: catatonic, ectasin, anectasin, peritoneum, *Tinognathus parviceps* (a beetle), *Teinotrachelus ruficollis* (a wasp), *Ectenocrinus elongatus* (a Silurian crinoid), *Ectinogramma collare* (a beetle), *Ectenostoma nigriventris* (a beetle), *Ectatocera longicornis* (a beetle), *Ectatomma tuberculatum* (an ant).
L. *tendo, tensus; tentus,* stretch, extend, spread out; *contentus,* stretched, strained, tight; *distentus,* inflated, full; *extensus,* stretched out, spread out; *intensus,* stretched tight, strained; *ostentus,* stretched out, exposed; *protensus (protentus),* stretched forth, extended: tense, tensile, tent, tendon, tendency, tender (offer), temptation, attention, attendant, attempt, contend, extensive, intention, intensity, ostensible, ostentation, pretense, portentous, standard, superintend, *Paradelocrinus protensus* (a Pennsylvanian crinoid).
See: **broad, level, sow, open, space, waste**

spretus, L. despised, scorned; see **hate**

spring < AS. *spring,* fountain.
L. *Arethusa* (Gr. *Arethousa*), f. famous spring at Syracuse, named for a woodnymph: *Arethusa bulbosa* (an orchid), *Plumatella arethusa* (a bryozoan).
L. *Artesium,* n. Artois, a French province where deep water wells were bored: artesian.
Gr. *Byblis, -idos,* f. a nymph changed into a spring: *Byblis gigantea* (a sundewlike plant).
L. *Callirrhoë* (Gr. *Kallirrhoë*), f. a celebrated spring at Athens: *Callirrhoë involucrata* (a poppy-mallow).
L. *Castalia* (Gr. *Kastalia*), f. famous spring on Mount Parnassus, sacred to the Muses: Castalian, *Castalia* (now *Nymphaea*) *alba* (European white waterlily).
L. *Dirca* (Gr. *Dirke*), f. a spring in Thebes: *Dirca palustris* (leatherwood).
L. *fons, fontis,* m. spring; *fonticulus,* m. dim.; *fontaneus; fontanus; fontinalis (fontanalis),* of a spring: font, fountain, fontanel, Bellefonte (Pa.), Fontainbleau,

Philonotis fontana (a moss), *Salvelinus fontinalis* (brook-trout), *Ribes fontinale* (a currant).

L. *Juturna*, f. a spring in Latium; a nymph: *Euemera juturnaria* (a moth).

Gr. *krene*, f. spring; *krenidion*, n. dim.; *eukrenos*, well-watered: Hippocrene, *Crenothrix polyspora* (an iron-depositing bacterium), *Phytocrene macrophylla* (a water-vinewort).

Gr. *krounos*, m. spring: *Crunobia schineri* (a fly).

Gr. *libas*, anything that drops, trickles, or runs, spring, stream; see **stream**

Gr. *nama*, *-tos*, n.; *nasmos*, m. spring, stream; *namation*, n. dim.; *namatodes*; *nasmodes*, full of springs: *Nama lobbi* (a hydrophyllacead).

Sp. *ojo*, m. eye, spring: Ojo Alamo.

Gr. *pege*, f. water, spring; *pegidion*, n. dim.: pegology, Pegasus, *Ceropegia gemmifera* (an asclepiadacead).

Gr. *phrear*, *-atos*, n. well, reservoir; *phreation*, n. dim.; *phreoryktes*, m. well-digger: *Phreatogammarus fragilis* (an amphipod), *Phreatichthys andruzzi* (a fish), phreatic, phreatophyte.

Gr. *pidax*, *-akos*, f. spring, fountain, < *piduo*, gush forth; *pidylis*, *-idos*, f. a gushing: *Dipidax ciliata* (a liliacead).

L. *Pirene* (Gr. *Peirene*), f. a spring at Corinth:

L. *puteus*, well, cistern, pit; *puteolus*; *puticulus*, dim.; *putealis*; *puteanus*, of a well; see **hole**

L. *scaturigo*, *-inis*, f. spring of bubbling water; *scaturiginosus*, full of springs, marshy, boggy, < *scateo*, gush forth, bubble out: scaturient, scaturiginous.

L. *silanus* (Gr. *silenos*), m. a fountain or jet of water from a head of Silenus: *Silanus praefectus* (a bug).

Gr. *thermydron*, n. hot spring:

See: **stream, water, flow, begin, hypothesis,** *ear* under **year**

spring < AS. *springan*, leap; see **leap**

sprinkle; see **wet, spread**

sprout < AS. *spreatan*, shoot; see **branch, bud, young, begin**

spruce < ML. *Prussia*, Prussia; see *picea* under **gymnosperm**

spudo- < Gr. *spoude*, zeal, exertion, see **busy**; *spoudastes*, partisan, see **help**

spuma, L. foam, froth; *spumeus*; *spumidus*, foamy, frothy; see **foam**

spur < AS. *spura*; see **point, arouse**

spurcus, L. dirty, filthy, base, foul; *spurcamen*, dirt, filth; see **dirt**

spurge < OF. *espurge*, < L. *expurgo*, *-atus*, purge, cleanse, purify.

NL. *acalypha*, f. a genus of the spurge family: *Acalypha virginica* (Virginia copper-leaf).

NL. *codiaeum*, n. a genus of the spurge family: *Codiaeum variegatum* (a spurge).

L. *cyparissias* (Gr. *kyparissias*), m. a kind of spurge: *Euphorbia cyparissias* (a spurge).

L. *euphorbea*, f.; *euphorbeum*; NL. *euphorbium* (Gr. *euphorbion*), n. an African plant with acrid juice, < Euphorbus, Greek physician: euphorbium, Euphorbiaceae, *Euphorbia esculenta* (edible spurge).

NL. *hevea*, f. a genus of the spurge family: *Hevea brasiliensis* (Para rubber-tree).

NL. *jatropha*, f. a genus of the spurge family: *Jatropha hastata* (a euphorbiacead).

Gr. *kroton*, *-os*, m. a euphorbiaceous shrub with seeds resembling ticks: *Croton glandulosus* (a croton).

Gr. *lathyris*, *-idos*, f. a kind of spurge: *Euphorbia lathyrus[lathyris]* (caper-spurge).

NL. *mallotus*, m. a genus of the spurge family: *Mallotus tenuifolius* (a euphorbiacead).

NL. *manihot*, f. a genus of the spurge family, < Ab.Am.: manioc, *Manihot utilissima* (bitter cassava, source of tapioca).

L. *mutilago*, *-inis*, f. a kind of spurge:

NL. *phyllanthus*, m. (Gr. *phyllanthes*, n.), a genus of the spurge family: *Phyllanthus calycinus* (a euphorbiacead).

Gr. *pityousa*, f. a kind of spurge: *Euphorbia pithyusa[pityusa]* (a spurge).

L. *ricinus*, m. castor-oil plant, tick; *cici* (Gr. *kiki*), n. castor-oil plant; *cicinus*, of the castor-oil plant: ricinolein, ricin, *Ricinus communis* (a castor-oil plant), *Mallotus ricinoides* (alem), *Acanthopanax ricinifoius* (an araliacead).

NL. *sapium*, n. a genus of the spurge family: *Sapium japonicum* (Japanese sapium).

NL. *tiglium*, n. croton: tiglic, *Croton tiglium* (croton-oil plant).

L. *tithymalus* (Gr. *tithymalos*), m. spurge: *Pedilanthus tithymaloides* (a slipper-flower).

spurius, L. false, bastard; see **lie**

spurn < AS. *spurnan;* see **kick, hate**

sputum, L. spit; see **spit**

spy < OF. *espier;* see **see, hear, secret**

spyras (sphyras), Gr. ball of dung; see **dung**

spyrido- < Gr. *spyris, -idos,* basket, creel; *spyridion; spyrichnion,* dim.; see **basket**

squalidus; squalus, L. neglected, filthy, dirty; *squalor,* dirtiness; see **dirt**

squalus, L. a kind of sea-fish; see **shark**

squama, L. scale; *squamula,* dim.; *squamosus,* scaly; see **scale**

squander < uncertain origin; see **spread, waste**

square < OF. *esquarre,* < L. *quadratus,* four-cornered; see **four, right**

squarrosus, L. rough with stiff scales, bracts, leaves, or processes; see **rough**

squatina, L. skate; see **shark**

squeak < uncertain origin.
 L. *destico, -atus,* squeak like a mouse:
 Gr. *hyizo,* squeal like a pig:
 Gr. *koizo,* squeal like a pig:
 Gr. *krizo,* creak:
 L. *mintrio, -itus,* squeak like a mouse:
 L. *murrio, -itus,* squeak like a mouse:
 L. *stridulus,* creaking, grating, < *strido,* make a harsh, shrill sound, creak, grate, *stridor, -is,* m. any harsh, shrill sound: strident, stridulous, stridulation, stridulatory, stridulator.
 Gr. *stringizo,* screech, scream:
 Gr. *trigmos,* m. a strident sound, cry, < *trizo,* chirp, creak, squeak:

squeeze < AS. *cwesan;* see **press, tongs, lessen**

squid < uncertain origin; see *teuthis, loligo,* and *sepia* under **mollusk**

squill < L. *scilla; squilla* (Gr. *skilla*), a sea-onion, shrimp; see **lily, crab**

squint < uncertain origin; see **slope**

squirm < uncertain origin; see **turn**

squirrel < L. *sciurus* (Gr. *skiouros*), m. the shadow-tail: *Sciurus carolinensis* (gray squirrel), *Haemulon sciurus* (yellow grunt).
 Sp. *ardilla,* f. squirrel:
 NL. *citellus,* m. a genus of ground-squirrels: *Citellus tridecemlineatus* (striped spermophile).
 Gr. *tamias,* m. treasurer; the genus including the chipmunk: *Tamias striatus* (eastern chipmunk), *Eutamias quadrivittatus* (western chipmunk).

stable < L. *stabilis,* firm, steadfast, see *sto* under **stand;** < *stabulum,* stall, shed, abode, see **house**

stachyo- < Gr. *stachys, -yos,* ear of grain, spike, see **point;** a mint, see **mint**

stacto- < Gr. *staktos,* in drops, trickling; see *stazo* under **drop**

stacula, L. a kind of vine; see **vine**

stadium, L. arena, race-course; see **play**

staeto- < Gr. *stais, staitos,* dough; *staitinos,* of dough; *staitodes,* doughy; see **mix**

staff < AS. *steaf,* stick; see **rod**

staff-tree
 NL. *catha,* f. a genus of the staff-tree family: cathine, cathidine, cathinine, *Catha edulis* (kat or hkat).
 L. *celastrus* (Gr. *kelastros*), f. an evergreen tree; a genus of the staff-tree family: Celastraceae, *Celastrus orbiculata* (a bittersweet), *Celastrina xanthippe* (a butterfly).
 L. *euonymus* (Gr. *euonymos*), f. spindle-tree: *Euonymus europaeus[europaea]* (a spindle-tree).
 NL. *maytenus,* f. a genus of the staff-tree family, < Ab.Am.: *Maytenus ilicifolia* (holly-leaved mayten).

stage < OF. *estage,* story of a dwelling, platform in a theater; see **play, frame**

stagma; stagon, Gr. drop; see **drop**

stagnum, L. pool; *stagninus,* of standing water; *stagnosus,* full of pools; *stagno, -atus,* form a pool of standing water; see **marsh, dull**

stain < OF. *desteign,* dye; see **spot, color, mark, dirt**

stair < AS. *staeger;* see **step**

stake < AS. *staca,* post; see **pillar**

stalagma, Gr. drop; *stalagmion,* dim.; *stalagmos,* a dropping, dripping; *stalaktos,* dropping, dripping; see **drop**

stale < uncertain origin; see **dull, old**

stalico- < Gr. *stalix, -ikos,* stake; see **pillar**

stalk < AS. *stealc;* see **stem**

stall < AS. *steall,* place, seat, station, stable; see **room**

stalsis, Gr. compression, contraction, restriction; *staltikos,* contractile; see **lessen**

stamen, L. (Gr. *stemon*), warp, thread, fiber, male organ of a flower; *stamineus,* thready, fibrous; see **thread, penis**

stammer < AS. *stammerian;* see **falter**

stamnos, Gr. earthen jar or bottle; *stammarion; stamnion,* dim.; see **pot**

stamp < OHG. *stampfen,* impress, strike, thrust the foot upon; see **strike, step, mark, mortar**

stand < AS. *standan,* get or place in an upright position, endure, last, tolerate.
Gr. *adonetos,* unshaken: *Adoneta bicaudata* (a moth).
Gr. *adrastos,* not running away: *Adrastotherium dimotum* (a fossil mammal).
Gr. *akinetos,* not moving; see **rest**
Gr. *ametabletos; ametagnostos; ametathetos; ametaptotos; ametastatos; ametastrophos,* unchangeable:
Gr. *amiktos,* immiscible, inaccessible, unsociable; see **pure**
Gr. *anallaktos,* unchangeable: *Anallacta methanoides* (a blattid).
Gr. *aniketos,* unconquered, invincible: *Anicetus ceylonensis* (a wasp).
Gr. *antereisis,* f. resistance; *antereistikos,* of resistance:
Gr. *aplanes,* steady, unerring: *Aplanetes fasciatus* (a beetle).
Gr. *apolos,* immovable: *Apolopsyche minusculus* (a trichopterid).
Gr. *asphales,* steadfast, stable; *asphalisma, -tos,* n. guarantee, security; *asphalistos,* made secure: *Asphalidesmus leae* (a milleped).
Gr. *atinaktos,* unshaken, immovable:
Gr. *bebaios,* firm, steadfast, sure, constant; *bebaiotes, -etos,* f. firmness, stability, certainty: *Bebaiotes[Bebaeotes] nigrigaster* (a bug).
Gr. *chronios,* lasting long, lingering; see *chronos* under **time**
L. *contumax, -acis,* defiant, insolent, obstinate, stubborn; see **stubborn**
L. *duro, -atus,* harden, last, remain; *durabilis,* lasting, enduring; see *durus* under **hard**
Gr. *dyslytos,* indissoluble:
Gr. *empedos,* firm, steadfast, lasting: Empedocles, *Empedomorpha empedoides* (a fly).
L. *firmus,* strong, stout, durable; see **strong**
L. *haesito, -atus,* stick fast, remain fixed, be undecided: hesitate, hesitation.
Gr. *hedraios,* sitting, stable, steadfast; see *hedra* under **sit**
Gr. *hektor, -os,* holding fast; see *echo* under **hold**
L. *immobilis,* not movable; *immotus,* unmovable, steadfast, firm: immobile, immotile, immotive.
L. *immutabilis,* unchangeable; *immutatus,* unchanged, unaltered: immutable.
L. *imperturbabilis,* that cannot be ruffled: imperturbable.
L. *inagitabilis,* immovable; *inagitatus,* unmoved:
L. *incommiscibilis,* that cannot be mixed: incommiscible.
L. *inconcussus,* unshaken, firm, constant: inconcussible.
L. *indeclinabilis,* inflexible, unchangeable; *indeclinatus,* unchanged, constant: indeclinable.
L. *indemutabilis,* unchangeable:
L. *ineluibilis,* that cannot be washed out, indelible:
L. *inexorabilis,* that cannot be moved by entreaty, inflexible, relentless: inexorable.
L. *intitutabilis,* firm, unswerving:
L. *irrevocabilis,* that cannot be recalled, unchangeable: irrevocable.
L. *maneo, mansus,* remain, stay in place; *manentia,* f. stability; *emansor, -is,* m. one who overstays his furlough: permanent, mansion, manor, menial, messuage, remain, remnant, menagerie, immanence, immament.
Gr. *meno,* stand fast, abide, remain, continue; *menetos; menetikos,* inclined to stand fast, patient; *menemachos,* steadfast in battle; *menestheus,* m. abider; *emmenes,* abiding, persistent, steadfast; *mone,* f. a tarrying, stay, delay: menotyphlic, Menestheus, Menelaus, *Emmenanthe penduliflora* (yellow whispering-bells).
Gr. *monimos,* steadfast, lasting: monimolite, monimostylic.
L. *obnixus,* steadfast, firm, resolute:
Gr. *pagios,* firm, steadfast, solid: *Pagiocerus rimosus* (a beetle).
L. *patiens, -entis,* enduring, suffering; see *patior* under **pain**
L. *pertinax, -acis,* firm, persistent, obstinate; see **stubborn**
OHG. *stal,* stand, place: stall, stallion, stalwart, stale, stalemate.

Gr. *stasis*, f. a placing, a standing, a position; *statikos*, causing to stand, of standing; *statos*, fixed, placed, standing; *stadios*; *statheros*, standing firm; *stasimos*, stationary, steady; *ekstasis*, f. a standing aside; *enstatikos*, stubborn, opposing, resisting; *enstates*, m. adversary; *epistema, -tos*, n. anything set up; *eustathes*, steady, well-based; *systas, -ados*, standing together, < *histemi*, stand, make firm: stasis, stamen, statoblast, statolith, statocyst, stele, ecstasy, rheostat, cryostat, thermostat, anastasis, apostate, system, diastem, prostate, hemostat, enstatite, *Anastatica hierochuntica* (a crucifer).

Gr. *steriktos*, fixed, firmly set:

L. *stipulus*, firm:

L. *sto, status*, stand; *sisto*, cause to stand; *stabilis*, standing firm, steady; *statarius*, standing fast; *stativus*, standing still; *statuo, -utus*, cause to stand, set up; *statura*, f. upright posture, height of the body; *status, -us*, m. position, attitude, condition; *consisto, -stitus*, stand still, assume an attitude; *consto, -atus*, stand firm; *constituo, -stitutus*, cause to stand, set up, establish; *obsisto, -stitus*; *obsto, -atus*, stand against, oppose; *obstino, -atus*, persist firmly and inflexibly, be stubborn; *praesto, -atus*, stand out, be superior; *resto, -atus*, stand firm: stable, stability, stage, state, statue, stature, station, stationary, stationery, stanchion, status, statistics, stance, stanza, stay, armistice, arrest, assistance, circumstance, coexistent, constabulary, constant, Constantine, consistency, constitute, contrast, cost, distance, destiny, destitution, establish, estate, extant, insist, interstice, instantaneous, irresistible, obstacle, obstetrical, obstinate, persist, restive, solstice, substantial, unconstitutional, *Clematis stans* (a clematis), *Acanthocheilonema perstans* (a nematode), *Rhizotrogus solstitialis* (a cockchafer).

See: **always, delay, indelible, long, pillar, place, slow, strong, stubborn, sure**

stannum, L. tin; see **tin**

stapelia, NL. a genus of the milkweed family; see **milkweed**

stapes, L. stirrup; see **stirrup**

staphis, L. licebane; see **plant**

staphylea, NL. bladdernut; see *staphyle* under **grape**

staphylinos, Gr. a kind of carrot or parsnip, see **carrot**; an insect, now applied to the rove-beetle, see **beetle**

staphylo- < Gr. *staphyle*, bunch or cluster of grapes, uvula; see **grape, uvula**

star < AS. *steorra*: starry, starlit, starlight, starflower, starfish, stargazer, dog-star.

L. *aster, -is* (Gr. *-os*), m.; *astrum* (Gr. *astron*), n. star; a genus of the composite family; Sp. *estrella*, f. star; *estrellita*, f. dim.; *asteriscus* (Gr. *asteriskos*), m.; *astrion*, n. dim.; *asterikos*; *astrikos*, of the stars; *asterias*; *asterios*, starry; *asterismos*, m. constellation; *astralos*, spotted with stars, speckled; *astroplethes*, full of stars; *astrifer*; *astriger*, starred; *astrologos*; *astronomos*, m. students of the stars but for radically different purposes; *katasteros*, set with stars: aster, astral, asteroid, asterisk, astronomy, astrologer, astronomer, astrosphere, chlorastrolite, Estrellita, *Aster paniculatus* (an aster), *Asterina atyphoida* (a starfish), *Asterocoris australis* (a bug), *Asteromyia asterifolia* (a midge), *Asterionella formosa* (a diatom), *Asterias vulgaris* (a starfish), *Astropyga pulvinata* (a sea-urchin), *Astronotus ocellatus* (a fish), *Astrodendrum pustulatum* (a brittlestar), *Aspidistra lurida* (a liliacead), *Geaster formicatus* (a puffball), *Astrocaryum giganteum* (a palm), *Polyporus tuberaster* (a fungus), *Micrasterias rotata* (a desmid), *Calliaster spinosus* (a starfish), *Euastrum verrucosum* (a desmid), *Diospyros ebenaster* (Indian persimmon), *Stapelia asterias* (an asclepiadacead), *Pentacrinus asteriscus* (a Jurassic crinoid).

L. *cometa*; *cometes* (Gr. *kometes*), m. comet: comet, cometary.

L. *planeta* (Gr. *planetes*), m. planet: planet, planetary, planetarium, planetesimal.

Gr. *Pleiades*, f. the constellation of the seven stars representing the seven daughters (Celaeno, Electra, Halcyone, Maia, Merope, Sterope, Taygete) of Atlas and Pleione; *Pleias, -ados*, f. a Pleiad: Pleiades.

L. *sidus, sideris*, n. star, constellation; *sideralis*; *sidereus*, of the stars: sidereal, desire, desideratum.

L. *stella*, f. star; *stellula*, f. dim.; *stellaris*, of the stars; *stellatus*, starred, starry; *constellatio, -onis*, f. a group of stars; *constellatus*, studded with stars, starry: stellar, stelliferous, Stella, Estelle, interstellar, constellation, *Stellaria pubera* (starwort), *Stellula calliope* (a hummingbird), *Stellaster inspinosus* (a starfish), *Quercus stellata* (post-oak), *Botaurus stellaris* (bittern), *Dibranchus stellulatus* (a fish), *Xenisma stellifer* (a studfish), *Astrononion stelligera* (a foraminifer).

starch < AS. *stearc*; see **flour**

stare < AS. *starian*, gaze fixedly; see **see**

starling < AS. *staerlinc*.

NL. *cassidix, -icis*, m. a kind of grackle: *Cassidix mexicanus* (a grackle).

Gr. *kerkion, -os,* f. probably the myna:

Gr. *psar, -os,* m. starling; *psaros,* speckled, dappled, like a starling: psarolite, *Psaronius asterolithus* (a fossil fern).

NL. *quiscalus,* m. a kind of grackle: *Quiscalus quiscula* (purple grackle).

L. *sturnus,* m. starling: *Sturnus vulgaris* (starling), *Sturnella magna* (meadowlark).

start; see **begin**

starve < AS. *steorfan;* see **hunger**

stasis, Gr. a standing, a position; *statikos,* causing to stand, standing still; see **stand**

state < OF. *estat,* < L. *status,* condition, position, standing; see **nature, country**

stater, Gr. a weight and coin; see **weight**

stathme, Gr. carpenter's rule or line; *stathmetikos,* of measuring; see **line**

stathmo, Gr. stall, stable, abode, post; *stathmidion,* dim., see **house;** *stathmion,* dim. balance, weight, see **balance**

statice, L. an astringent herb, a genus of the leadwort family; see **leadwort**

stato- < Gr. *statos,* fixed, placed, standing; *statikos,* standing, resting; see *stasis* under **stand**

statua, L. image in marble, bronze, or other material; see **form**

statumaria, L. a plant; see **plant**

statumen, L. support, prop; see **pillar**

status, L. a standing, a position; see *sto* under **stand**

stauro- < Gr. *stauros,* cross, upright stake; *stauridion; staurion,* dim.; see **cross**

stauroma, Gr. palisade, stockade; see **wall**

staxis, Gr. a dropping; see *stagma* under **drop**

stay < OF. *ester,* < L. *sto,* stand; see **stand, delay**

stead < AS. *stede,* place, town; see **place**

steal < AS. *stelan.*

L. *abductio, -onis,* f. forcible carrying off, kidnapping: abduction, abductor.

L. *abigeator, -is; abigeus; abactor, -is,* m. cattle-stealer, rustler; *abigeatus,* m. cattle-stealing, < *abigo, abactus,* drive away:

Gr. *andrapodistes,* m. slave-dealer, kidnapper:

Gr. *blisso (blitto),* take honey from the hive, rob, steal: *Blissus leucopterus* (chinchbug).

L. *contrectator, -is,* m. thief: contrectation.

L. *depopulor, -atus,* ravage, plunder, pillage: depopulate.

L. *direptor; ereptor; raptor,* robber, ravager, plunderer; see *rapio* under **take**

L. *effractor, -is,* m. house-beaker, burglar:

L. *fur, -is,* c. thief; *furunculus,* m. dim., petty thief, boil; *furax, -acis,* inclined to steal, thievish; *furtivus,* stolen, secret, concealed; *furtum,* n. petty theft: furtive, furunculus, ferret, *Ptinus fur* (museum-pest), *Cricetulus furunculus* (a rodent), *Herpetomonas furunculosa* (a protozoan), *Abacetus furax* (a beetle), *Scydmaenus furtivus* (a beetle).

L. *grassans, -antis,* m. robber, thief:

Gr. *haliphthoros,* m. pirate:

L. *harpago, -atus* (Gr. *harpazo*), seize, rob, plunder; *harpakter,* robber; see **take**

Gr. *keraistes,* m. ravager, robber:

Gr. *Kerkyon, -os,* m. a notorious Attic robber: *Cercyon fimbriatus* (a beetle).

Gr. *kixalles,* m. highway robber:

Gr. *klepto* (L. *clepo, cleptus*), steal; *kleptes; klops, -opos,* m.; *kleptis, -idos,* f. thief; *kleptides,* m. son of a thief; *kleptikos; klopikos,* thievish; *klepsia; klopeia,* f. theft; *klemma, -tos,* n. thing stolen; *klemmatistes,* m. thievish person; *epiklopos,* thievish; *mikrokleptes,* m. petty thief: clepsydra, cleptobiosis, kleptomaniac, *Phonergates (Clopophora) basilicus* (a bug), *Cleptophasia scissa* (a moth), *Cleptodiscus reticulatus* (a trematode).

L. *latro, -onis,* m. hireling, robber; *latrunculus,* m. dim.; *latrocinium,* n. robbery, piracy: larceny, latrocinium, ladrone, *Birgus latro* (coconut-crab), *Latrodectus formidabilis* (a black-widow spider).

Gr. *lestes,* m. robber, plunderer, pirate; *lestikos; lestrikos,* piratical; *leia; leis, -idos,* f. booty, plunder; *leidios; leistos,* of booty: *Lestes uncatus* (a damselfly), *Ornitholestes hermanni* (a bird-catching dinosaur), *Melanolestes morio* (a bug), *Perilestes castor* (a damselfly), *Leistus punctatissimus* (a beetle), *Lestidium nudum* (a fish), *Lestricothynnus subtilis* (a wasp), *Archilestidium cinnabarium* (a bug).

L. *levator, -is,* m. lifter, thief:

Gr. *lopodytes,* m. clothes-stealer, thief; *lopodysia,* f. highway robbery:

L. *manticulor, -atus,* act slyly, steal; *manticulator, -is,* m. pickpocket, purse-snatcher:

Gr. *nekrosylia,* f. robbery of the dead:

Gr. *nosphistes,* m. embezzler, peculator; *nosphismos,* m. peculation: *Nosphistia perplexa* (a butterfly).

L. *peculor, -atus,* defraud, embezzler; *peculator, -is,* m. embezzler: peculate, peculator, peculation, *Phraedrotoma depeculator* (a wasp).

Gr. *peridinos,* c. rover, pirate:

Gr. *perikoptes,* m. thief, robber:

Gr. *phor, -os,* m. thief; *phorios,* stolen: *Phororhacus affinis* (an Eocene bird), *Phoradendron flavescens* (American mistletoe).

L. *pilo, -atus,* rob, plunder; *compilo; expilo,* gather, rob: pillage, compile, compilation, expilate.

L. *pirate* (Gr. *peirates*), m. sea-robber, corsair; *piraticus* (Gr. *peiratikos*), of pirates: pirate, piratical, piracy, *Piratosaurus plicatus* (a fossil reptile), *Lycosa piratica* (a spider).

L. *plagium,* n. kidnapping, plundering; *plagiarius; plagiator, -is,* m. plunderer, kidnapper, literary thief; *plagio, -atus,* steal: plagiarism, plagiarist, *Euporus plagiatus* (a beetle).

L. *praedor, -atus,* plunder, pillage, rob; *praedonulus,* petty robber; see *praeda* under **plunder**

L. *raptor; rapinator,* robber, ravisher; see *rapio* under **take**

L. *saccularius,* m. pickpocket:

Gr. *syletes; syletor, -os,* robber; see *sylema* under **plunder**

L. *tagax, -acis,* light-fingered, thievish:

See: **take, plunder, waste**

stealthy < AS. *stelan,* steal; see **secret, silent, slip, steal**

steam < AS. *steam,* vapor; see **cloud**

steato- < Gr. *stear, -atos,* suet, fat; *steation,* dim.; see **fat**

steel < AS. *stel* (G. *stahl*); see **iron**

steep < AS. *steap;* see **high, upright, cliff, dip**

steer < AS. *stieran,* guide, govern; see **lead, govern, rudder**

stegnus, L. (Gr. *stegnos*), closed, constricted; see **close**

stego- < Gr. *stege; stegos; tegos,* roof; *stegion,* dim., see **roof**; < *stego,* cover; *steganos; stegastos,* covered, enclosed, sheathed; *stegaster,* coverer, see **cover**

stela, L. (Gr. *stele*), pillar, column; *stelidion,* dim.; *stelechos,* trunk, log; see **pillar**

steleon, Gr. handle; *steilarion,* dim.; see **handle**

stelgido- < Gr. *stelgis, -idos; stelgisma; stelgistron,* scraper; see **scrape**

stelis, Gr. a kind of mistletoe; see **mistletoe**

stella, L. star; *stellula,* dim.; *stellaris,* of the stars; *stellatus,* starry; see **star**

stellaria, NL. a genus of chickweeds; see **pink**

stellio, L. a spotted lizard; see **lizard**

stello, Gr. arrange, array, dispatch; see **send**

stem < AS. *stemm.*

L. *caudex, -icis,* m. stem, trunk; *caudiculus,* m.; *caudeus; caudicalis; caudicarius; caudiceus,* of stems or trunks: caudex, caudicle.

L. *caulis* (Gr. *kaulos*), m. stem, stalk; *cauliculus; kauliskos,* m.; *kaulion,* n. dim.; *kaulikos; kaulinos,* of or like a stalk: cauline, cauline, cauliculus, cauliflower, cauliflory, cole, kohlrabi, caulescent, acaulescent, *Caulerpa prolifera* (a seaweed), *Caulophyllum thalictroides* (blue cohosh), *Eriocaulon compressum* (pipewort), *Papaver nudicaule* (Iceland poppy), *Cypripedium acaule* (moccasin-flower), *Pinus albicaulis* (white-bark pine), *Cynometra cauliflora* (a bean), *Polygonum amplexicaule* (a knotweed), caulicle.

L. *culmus,* m. stalk, halm, straw; *culmulus,* m. dim.; *culmeus,* of a stalk or straw; *culmosus,* full of stalks, stalklike: culm, *Carex laxiculmis* (a sedge), *Fusarium culmorum* (a fungus).

L. *fabalium,* beanstalk; see *faba* under **bean**

L. *festuca,* f. stalk, stem, straw; *festucula,* f. dim.: festucine, *Festuca subulata* (bearded fescue), *Fluminea festucacea* (river-grass), *Noctua festucae* (a moth).

Gr. *kalame,* stalk, stem; see *calamus* under **reed**

Gr. *karphos,* chaff, chip, straw; see **scale**

Gr. *kormos,* m. trunk of a tree; *kormion,* n. dim.: corm, cormus, cormophyte, *Cormobates leucophaea* (a tree-creeper), *Trachycormocephalus mirabilis* (a centiped), *Cormonema ovalifolium* (a rhamnacead), *Pachycormus discolor* (an anacardiacead).

Gr. *kremaster,* suspender; *kremastra,* pedicel, peduncle; see **hang**

Gr. *mischos,* m. stalk, petiole, peduncle: *Mischocephalus spinicollis* (a beetle), *Platymischus dilatatus* (a wasp), *Macromischa fuscata* (an ant).

Gr. *ormenos,* m. shoot, sprout, stalk:

L. *pedunculus; pedicellus; pediculus; petiolus,* m. small, slender stalk, dim. of *pes, pedis,* foot: peduncle, pedunculate, pedicel, pedicellar, pedicellate, petiole, petiolar, *Quercus pedunculiflora* (an oak), *Raphia pedunculata* (a raffia-palm), *Pedicellina cernua* (a bryozoan), *Pedicellaster reticulatus* (a starfish), *Thelephora pedicellata* (a fungus), *Cuphea petiolata* (waxweed), *Asteromyia petiolicola* (a grape gall-insect), *Helianthus petiolaris* (a sunflower).

Gr. *phitros,* m. bole, trunk:

Gr. *premnon, n.; premnos,* m. stump of a tree, stem: *Premnotrypes solani* (a beetle), *Epipremnum pinnatum* (an aroid).

Gr. *pternix, -ikos,* m. stem of a plant:

Gr. *pythmen,* bottom, stock, stalk; see **base**

L. *resex, -ecis,* m. stub left after pruning:

Gr. *rhachis,* spine, axis, stem, stalk; see **back**

L. *scapus* (Gr. *skapos*), m. stem, staff: scape, scapose, scapolite, *Kalanchoë scapigera* (a crassulacead), *Psathyrotes scaposa* (a composite), *Arctotis breviscapa* (blue-eyed daisy).

L. *scopio, -onis,* m. stalk or stem of grapes:

Gr. *stelechos,* trunk, log; see *stela* under **pillar**

L. *stipes, -pitis,* m. stalk, stem, straw, trunk; *stipula,* f. dim.; *stipitatus,* borne on a stipe; *stipidiosus,* branchy, woody: stipe, stipule, etiolate, etiolin, *Gillenia stipulata* (an Indian-physic), *Mitragyna stipulosa* (abura), *Begonia latistipula* (a begonia), *Isopyrum stipitatum* (a ranunculacead), *Colobopsis etiolatus* (an ant).

L. *stirps, -pis,* f. stalk, stem, stock, race, family: stirps, stirpiculture, extirpate, *Puposyrnola stirps* (a fossil gastropod).

AS. *stocc,* stick, stem, trunk, race: stock, livestock.

Gr. *stypos,* n. stem, stump: *Stypodon signifer* (a fish).

L. *truncus,* m. stem of a tree; *trunculus,* m. dim.: trunk.

See: **pillar, axis, rod, reed, branch**

stema, Gr. penis, stamen; see *stamen* under **thread**

stemma, Gr. garland, wreath, crown; *stemmation,* dim.; see **crown**

stemon, Gr. warp, thread; see *stamen* under **thread**

stemphylon, Gr. pomace of pressed olives or grapes; see **dirt**

stenagma, Gr. groan, moan, sigh; see **sad**

stenos; stenygros, Gr. narrow; see **narrow**

Stentor, Gr. herald with a loud voice; see **call**

step < AS. *staepe.*

L. *ascensus, -us,* m. a flight of stairs:

Gr. *bathmos (basmos); anabathmos,* m. step, stair; *bathron,* n. step, rung of a ladder; *bema, -tos,* n. step, pace, < *baino,* walk, go; *bematistes,* m. pacer: bema, *Bathmodon semicinctus* (an Eocene mammal), *Bematistes umbra* (a butterfly).

Gr. *elysis,* f. step:

L. *gradus, -us,* m. step, pace, stair, degree, stage, pitch, < *gradior, gressus,* step, walk, go; *gradilis,* of steps; *gradalis; gradatus,* step by step, by degrees: grade, gradate, graduation, gradually, centigrade, degree.

L. *insilium; insubulum,* n. treadle of a loom:

Gr. *ithma, -tos,* n. step:

L. *occulco, -atus,* tread, trample; *proculco,* trample upon:

L. *passus, -us,* m. step, stride: pace, pass.

Gr. *plix, -ikos,* f. step:

L. *scamnum,* bench, stool, step; see **sit**

L. *vestigium,* footstep, step; see **mark**

See: **walk, ladder**

step- < AS. *steop-,* not a blood relative; see **different**

stephanos, Gr. crown, wreath, diadem; *stephanion; stephaniskos,* dim.; *steptos,* crowned; see **crown**

-ster; -stress, AS. suffixes denoting agency; see **make**

sterctico- < Gr. *sterktikos,* affection; see *storge* under **love**

sterculia, NL. a genus of the chocolate family; see **chocolate**

stercus, L. dung; *stercoratus,* dunged, manured; *Sterculius (Stercutius),* god of manuring; see **dung**

stereos, Gr. solid, firm, hard, three-dimensional; see **thick**

stereum, NL. a genus of fungi; see **fungus**

stergema; **stergethron**, Gr. love-charm; see *storage* under **love**

stericto- < Gr. *steriktos,* fixed, firmly set; see **stand**

sterigma, Gr. support, prop; see **pillar**

sterile < L. *sterilis,* unfruitful, barren, empty: sterility, sterilize, *Bromus sterilis* (a grass).
Gr. *agalax, -aktos,* without milk, giving none: agalactia.
Gr. *agonos,* seedless, barren, impotent:
Gr. *akarpos,* fruitless, barren: acarpous.
Gr. *akouros,* childless, without an heir:
Gr. *akymon, -os,* barren:
Gr. *akythos,* unfruitful:
Gr. *aphoros,* not bearing, barren: aphoria, *Aphorogrammus coleoptratus* (a bug).
Gr. *apoios,* neutral:
Gr. *asporos,* unseeded, seedless: asporous, asporogenic.
Gr. *asyllepsia,* f. inability to conceive, barrenness:
Gr. *ateknos,* childless, barren:
Gr. *atokos,* never having brought forth; *atokia,* f. barrenness: atocia, atoke.
Gr. *atrygetos,* yielding no harvest, barren:
Gr. *bagoas,* m. a Persian eunuch:
Gr. *bakelos,* m. a eunuch in the service of Cybele:
L. *eunuchus* (Gr. *eunouchos*), m. a castrated male person: eunuch.
Gr. *hemiandros,* m. eunuch:
L. *illiberis,* childless: *infecundus,* unfruitful; *inferax, -acis,* unfruitful; *infructuosus,* unfruitful:
Gr. *ithris, -idos,* m. eunuch:
L. *majalis,* m. gelded boar, barrow, eunuch: *Majalis indigena* (a beetle), *Fundulus majalis* (a killifish).
L. *neuter,* sexless: neuter.
L. *spado, -onis* (Gr. *spadon, -os*), m. an impotent person, eunuch; *spadoninus; spadonius,* sterile, unfruitful: *Pentaspadon velutinus* (a pelong-tree).
Gr. *steiros,* barren, sterile: *Stirocnemia flavicosa* (a fly), *Steironema*[*Stironema*] *ciliatum* (a primulacead).
Gr. *thladias; thlibias,* m. eunuch: *Thladiantha dubia* (a cucurbitacead).
See: **not, geld**

sterilicula, L. womb of an unfarrowed sow; see **womb**

steringo- < Gr. *sterinx, -ingos,* support; see **pillar**

steriphos, Gr. firm, solid; see *stereos* under **thick**

stern < AS. *styrne;* see **serious**

stern, the steering or rear end of a ship, < the root of AS. *steor,* steer.
L. *aplustre, -is,* n. stern of a ship: *Aplustrum aplustre* (a gastropod).
Gr. *cheniskos,* end of a ship's stern shaped like a goose; see **goose**
Gr. *prymna,* f. stern of a ship; *prymnesios,* of the stern: *Psalidoprymna gracilis* (a bird).
L. *puppis,* f. stern of a ship, poop, aft:
See: **after, rump**

sterna, NL. tern; see **tern**

sternax, *-acis,* L. that throws the rider; see **throw**

sterno- < Gr. *sternon,* chest, breast; see **breast**

sternutatio, L. a sneeze; see *sternuto* under **sneeze**

sterops, *-opos,* Gr. flashing; see **light**

sterphos, Gr. hide, skin; see **skin**

sterquilinum, L. dung-pit; see *stercus* under **dung**

sterrhos, Gr. firm, stiff, solid; see *stereos* under **thick**

sterto, L. snore; *stertens,* snorer; see **snore**

stethos, Gr. breast, chest; *stethion,* dim.; *stethikos,* of the breast; see **breast**

stew < OF. *estuver;* see **soup, boil, cook, mix**

sthenos, Gr. strength; *sthenaros,* strong; see **strong**

stia, Gr. pebble; *stion,* dim.; see **stone**

stibado- < Gr. *stibas, -ados,* bed, mattress; see **bed**

stibaros, Gr. strong, sturdy; see **strong**

stibium, L. antimony; see **antimony**

stibos, Gr. path, track, way; *stibeus; stibeutes,* walker, treader; see **way,** *stibeo* under **walk**

stichos, Gr. line, row; *stichidion,* dim.; see **line**

stick < AS. *sticca; stician;* see **rod, axis, stem, glue, bind**

sticto- < Gr. *stiktos,* punctured, spotted, dappled; see **spot**

sticula, L. a kind of grape; see **grape**

stiff < AS. *stif.*
Gr. *ankylosis,* f. stiffening of the joints: ancylosis.
Gr. *astreptos,* inflexible:
L. *incurvabilis,* inflexible:
L. *inflexibilis,* that cannot be bent: inflexible, inflexibility.
L. *rigidus,* stiff, hard, inflexible; *rigor, -is,* m. stiffness, < *rigeo,* be stiff: rigid, rigidity, rigescent, rigor, rigorous, *Pinus rigida* (pitch-pine), *Muhlenbergia rigens* (deergrass), *Echinocereus rigidissimus* (a cactus), *Gilia rigidula* (a gilia), *Gladiolus rigescens* (a gladiolus), *Oxypolis rigidior* (cowbane).
AS. *stearc,* stiff, strong: stark, starch.
Gr. *stereos; sterrhos,* stiff, firm, three-dimensional; see **thick**
Gr. *styo,* erect, stiffen; see **upright**
L. *tetanus* (Gr. *tetanos),* stiff, rigid; lockjaw: tetanus (caused by the bacterium, *Clostridium tetani), Tetanocephalus rugosus* (a beetle), *Tetanonema clarum* (a trichopterid).
See: **hard, upright, strong, thick, slow**

stigeus, Gr. tattooer, needle used in tattooing; *stigon,* a tattooed or branded person; see *stigma* under **mark**

stigma, Gr. mark, brand, spot, surface of the female floral structure that receives the pollen; see **mark**

stilbo- < Gr. *stilbe,* lamp, mirror; *stilpnos,* glittering, < *stilbo,* glitter, shine; see **light**

still < AS. *stille;* see **rest, silent**

stilla, L. drop; *stillo, -atus,* drop, trickle, distill; see **drop**

stilpnos, Gr. glittering, shining; see *stilbo* under **light**

stilt < Sw. *stylta;* see **pillar**

stilus, L. stake, pen; see **pen**

stimulus, L. goad; see **arouse**

stinctus, L. quenched; see *stinguo* under **wet**

sting < AS. *stingan;* see **arouse, burn, bite, strike, nettle, bore, point**

stingy < uncertain origin.
Gr. *adapanos,* costing nothing, not spending:
L. *frugalis,* economical, sparing, thrifty; see **save**
Gr. *glischron, -os,* niggard; *glischros,* sticky, greedy, niggardly; see **glue**
L. *illiberalis,* ungenerous, niggardly: illiberal.
L. *immunificus,* stingy:
Gr. *kimbix, -ikos,* n. niggard, miser; *kimbeia,* f. stinginess:
Gr. *knipos (sknipos),* niggardly, miserly, stingy; *sknipotes, -etos,* f. stinginess; *Gniphon, -os,* m. a niggard: *Cnipodectes subbrunneus* (a bird).
Gr. *kyminopristes; kyminopristokardamoglyphos,* m. cumin-splitter, hence skin-flint:
Gr. *mikrodosia,* f. stinginess:
L. *parcus,* frugal, thrifty, scanty, niggardly, penurious, < *parco, parsus,* spare, save; *parcipromus,* m. niggard, curmudgeon; *parsimonia,* f. frugality: parsimony, parsimonious, *Begonia parcifolia* (a begonia).
Gr. *pheidos,* thrifty, niggardly, stingy; *pheidole,* f. thrift; *pheidolia,* f. stinginess; *pheidolos,* m. niggard, miser, < *pheidomai,* spare, save; *apheides,* lavish, unsparing; *biopheides,* penurious: *Phidodonta modesta* (a beetle), *Phidippus frenatus* (a fossil spider), *Phidologeton ocelliferus* (an ant), *Pheidole[Phidola] opaca* (an ant), *Aphis aceris* (maple-aphid).
L. *tenax,* holding fast, stingy; see *teneo* under **hold**
See: **selfish, few, little, hold**

stink < AS. *stincan,* smell offensively; see **smell**

stino- < Gr. *steinos,* confined, narrow, close, see **narrow**

stipa, NL. a genus of grasses; see **grass**

stipator, L. attendant; see **servant**

stipatus, L. crammed, pressed, crowded, packed; see *stipo* under **press**

stipe < L. *stipes, -pitis,* stock, stem, trunk; *stipula,* dim.; *stipitatus,* borne on a stipe; see **stem**

stipendium, L. tax, tribute, pay; see *pendo* under **pay**

stiphros, Gr. firm, solid, stout; see **strong**

stips, *-ipis,* L. gift, contribution; see **give**

stiptos, Gr. crammed, packed, trodden down; see *stipo* under **press**

stipulatio, L. promise, bargain; *stipulor, -atus,* bargain, covenant; see **promise**

stipule < L. *stipula,* dim. of *stipes,* stalk; see **stem**

stipulus, L. firm; see **stand**

stir < AS. *styrian;* see **mix, turn, spoon**

stiria, L. a frozen drop, icicle; see **ice**

stiro- < Gr. *steiros,* sterile, barren; see **sterile**

stirps, L. stalk, stem, stock; see **stem**

stirrup < AS. *stigrap.*
L. *stapes, -edis,* m. stirrup: stapes, stapedius, stapediform, *Dictyocha stapedia* (a radiolarian).

stiva, L. plow-handle; see **handle**

stizo, Gr. prick, puncture, mark, brand, tattoo; *stixis,* a pricking, mark, spot; see *stigma* under **mark**

stlengido- < Gr. *stlengis, -idos,* scraper; see *stelgis* under **scrape**

stoa, Gr. porch, colonnade; *stoidion,* dim.; *stoikos,* of a colonnade; see **porch**

stochos, Gr. aim, shot, guess; *stochasma,* the thing aimed at; see **aim**

stock < AS. *stocc,* stick, stem, trunk, race; see **stem**

stocking < AS. *stocc,* stem, support.
L. *ocrea,* greave, legging, sheath; see **sheath**
Gr. *pellytron,* n. sock, band around the ankle:
Gr. *podeion,* n. sock:
L. *soccus,* a kind of light shoe, slipper; see **shoe**
L. *tibiale, -is,* n. legging, stocking:
L. *udo, -onis* (Gr. *oudon, -os*), m. felt sock or shoe: *Udonedus diabolicus* (a beetle).
See: **sleeve, sheath**

stoebe, L. (Gr. *stoibe*), a composite, see **composite**; cushion, pad, see **pillow**

stoechas, *-adis,* L. (Gr. *stoichas, -ados*), a kind of mint; see **mint**

stola, L. (Gr. *stole*); *stolis, -idos,* a long garment, robe; *stolidion; stolion,* dim.; *stolidotos,* in folds, folded; see **garment**

stolidus, L. unmovable, dull, stupid; *stoliditas,* dullness; see **dull**

stolismos, Gr. equipment, dress; see *stolizo* under **supply**

stolmus, Gr. raiment; see *stola* under **garment**

stolo, L. branch, shoot, runner, sucker; see **branch**

stolos, Gr. journey, voyage, expedition; see **move**

stoma, Gr. mouth; *stomion,* dim.; *stomatikos,* of the mouth; see **mouth**

stomach < L. *stomachus* (Gr. *stomachos*), m.; *stomachicus* (Gr. *stomachikos*), of the stomach: stomachic, stomacher, *Stomachobia pecorum* (a fly), *Stomachicola secunda* (a trematode). The stomach of ruminants consists of four chambers—rumen, reticulum, omasum, and abomasum.
Gr. *enystron,* n. fourth chamber of the ruminant stomach:
Gr. *gaster, -os; -tros,* f. stomach, belly, womb, paunch; *gastridion; gastrion,* n. dim.; *gastris, -idos; gastrodes; megalogaster, -os; pithogastros,* potbellied; *hypogastrion,* n. abdomen: gastronomy, gastritis, gastric, hypogastric, gastrectomy, gastroenteritis, gastrolith, gastropod, epigastric, gastrophile, gastrula, *Gastrophilus equi* (botfly), *Gasterosteus aculeatus* (a stickleback), *Gastridium ventricosum* (a grass), *Eugaster spinulosa* (a tettigonid), *Atractogaster semisculpta* (a wasp), *Pithogaster inflata* (a fly), *Chelidon erythrogaster* (barnswallow).
L. *ingluvies, -ei,* f. crop, maw; *ingluviosus,* voracious, gluttonous: ingluvies.
Gr. *koilia,* belly, abdomen, stomach; see *coeliacus* under **belly**
L. *omasum,* n. psalterium, manyplies, or third division of the ruminant stomach: omasum, abomasum.
Gr. *pregoreon; pregoron, -os,* m. crop:
Gr. *prolobos,* m. crop:
L. *rumen, -inis,* n. throat, gullet, paunch: rumen, rumenitis, ruminate, ruminator.
See: **belly, womb, bag**

stomachosus, L. angry, irritable, choleric; see **anger**

stombos, Gr. deep or loud-sounding; see **loud**

stomico- < Gr. *stomix, -ikos,* wooden beam; see **beam**

stomotos, Gr. hardened; see *stomoo* under **hard**

stomphos, Gr. bombastic; *stomphastes; stomphax,* braggart; see **brag**

stomylos, Gr. talkative, wordy; see **speak**

stone < AS. *stan;* G. *stein:* stony, stonebreak, stonecrop, stonewort, Stonewall
  Jackson, Stonehenge, Stanfield, Stanford, Stanton, keystone, lodestone, millstone,
  brimstone, grindstone, whetstone, madstone, bloodstone, cobblestone, curbstone,
  gallstone, flagstone, capstone, ironstone, limestone, mudstone, Rosetta Stone,
  Blarney Stone, stein, Goldstein.
  L. *achates* (Gr. *Achates,* a river in Sicily), m. agate; *kerachates,* m. wax-agate:
    agate, agatize, agatiform, agatiferous, *Achatina variegata* (a gastropod),
    Achatocarpus oaxacanus (a phytolaccacead), *Crotalaria agatiflora* (a rattlebox).
  L. *adamas, -antis* (Gr. *-antos*), the diamond; see **diamond**
  L. *aerizusa* (Gr. *aerizousa*), f. a kind of precious stone, probably turquoise:
  Gr. *akone,* f. whetstone, hone; *akonion,* n. dim.: *Aconeceras nisum* (an am-
    monite), paragon.
  L. *alabaster* (Gr. *alabastros*), a perfume box carved out of white gypsum or
    calcite; see **gypsum**
  L. *amethystus* (Gr. *amethystos*), f. a preventive of drunkenness; a bluish precious
    stone; *amethystinus,* of amethyst, like amethyst: amethyst, amethystine, *Nepeta
    amethystina* (a catnip), *Python amethystinus* (a snake).
  L. *amphitane,* f. a kind of precious stone:
  Gr. *anthrax, -akos,* m. a dark-red, precious stone, carbuncle: anthrax.
  L. *ariste,* f. a precious stone:
  L. *aspilates,* m. a precious stone of Arabia:
  L. *asteria,* f. a precious stone:
  L. *baetulus; baetylus* (Gr. *baitylos*), a meteorite, sacred stone; see **meteor**
  Gr. *balanites,* m. a precious stone:
  L. *baroptenus; barippe,* f. a precious stone:
  L. *basaltes,* m. a dark, African rock: basalt, basaltic, *Rana basaltica* (a fossil frog).
  Gr. *basanos,* touchstone; see *basanistes* under **try**
  Gr. *batrachites,* m. a stone of frog-green color:
  L. *beryllus* (Gr. *beryllos*), m. a precious stone; *beryllion,* n. dim.: beryl, berylline,
    beryllium, beryllonite, brilliant(?), *Cryptotomus beryllinus* (a parrot-fish).
  L. *boria,* f. a kind of jasper:
  Gr. *bostrychites,* m. a precious stone:
  L. *brontea(brontia),* f. thunder-stone:
  L. *bucardia,* f. a precious stone:
  L. *cadmitis,* f. a precious stone:
  L. *caementum,* rough, unhewn stone, chips; now applied to a substance used as a
    binder, mortar; see *caementarius* under **mason**
  L. *callaïs, -idis,* f. a greenish-blue, precious stone, probably the turquoise:
  L. *calx, calcis* (Gr. *chalix, -ikos*), f. pebble, gravel, stone, lime; *calculus,* m. dim.;
    calculosus, pebbly, gravelly; *calculo, -atus,* compute, reckon: calculus, calcula-
    tion, chalicosis, *Chalicotherium goldfussi* (a fossil mammal), *Mesembryan-
    themum calculus* (a fig-marigold).   See **lime**
  L. *carbunculus,* m. a reddish, precious stone: carbuncle.
  L. *cautes, -is,* f. a rough, pointed rock:
  L. *cepitis; cepalotitis,* f. a precious stone:
  L. *ceraunus,* a stone supposed to have fallen from the sky, meteorite; see **meteor**
  L. *chalcedonius,* of Chalcedon (Gr. *Chalkedon,* f.), a town in Asia Minor:
    chalcedony, *Lychnis chalcedonica* (Maltese cross).
  Gr. *cherados,* n. silt, mud, debris, sand, gravel:
  Gr. *chermas, -ados,* f. large pebble, stone; *chermadion,* n. dim.:
  Gr. *cholas, -ados,* f. a kind of emerald:
  Gr. *chrysolithos,* m. topaz: chrysolite.
  L. *cos, cotis,* f. any hard stone, flintstone, whetstone, hone; *coticula,* f. dim.;
    cotoria, f. whetstone quarry:
  L. *crystallum,* n. (Gr. *krystallos,* m.), ice, rock-crystal; *crystallinus* (Gr.
    krystallinos), of crystal: crystal, crystalline, phenocryst, *Crystallonema fusci-
    cephalum* (a nematode), *Mesembryanthemum crystallinum* (ice-plant).
  L. *cyamias,* f. bean-stone, a precious stone:
  Gr. *epalxites,* m. a coping-stone:
  L. *eristalis,* a precious stone, now applied to a genus of flies; see **fly**
  L. *erythallis,* f. a kind of gem:
  L. *eureos,* m. a precious stone:

L. *gemma*, bud, jewel, precious stone; *gemmarius*, of gems, jeweler; see **bud**

L. *glarea*, f. gravel; *glareosus*, gravelly: glareal, glareose, *Glareola pratincola* (pratincole), *Paepalanthus glareosus* (a pipewort).

L. *haematites* (Gr. *haimatites*), m. an ore of iron: hematite.

Gr. *herma, -tos*, n. prop, ballast, mound, sunken rock, reef; *hermax, -akos*, f. heap of stones: hermatypic, bioherm, *Herma insignis* (a beetle), *Hermatostroma guelphica* (a coelenterate).

Gr. *iaspis, -idos*, f. jasper; *iaspachates*, m. a kind of agate; *iasponyx, -ychos*, m. a kind onyx; L. *iaspideus; iaspius*, of jasper: jasper, jasperoid, jaspideus, jaspachate, jasponyx, jaspilite.

Gr. *ikterias*, m. a yellowish, precious stone:

L. *-ite, -ites* (Gr. *-ites*), suffixes signifying mineral, rock, rocklike, petrified, fossilized. In some instances *-ite* is a contraction of *-lite*, < *lithos*, stone: apatite, barite, calcite, chromite, cuprite, fluorite, graphite, greenalite, molybdenite, cryolite, nephelite, pegmatite, pyrite(pyrites), stibnite, syenite, tridymite, vanadinite, wolframite, xanthophyllite, yttrialite, zincite, *Betulites populifolius* (a fossil leaf-species), *Ceratites nodosus* (an ammonite). Terms derived from proper names: andesite(Andes Mountains), colemanite(W. T. Coleman), danalite(J. D. Dana), dolomite (Deodat Dolomieu), franklinite(Franklin, N. J.), glauberite(J. R. Glauber), labradorite(Labrador), muscovite(Moscow), powellite (J. W. Powell), rosenbuschite(K. H. F. Rosenbusch), scheelite(K. W. Scheele), sillimanite(Benjamin Silliman), vivianite(J. G. Vivian), wernerite(Abraham G. Werner).

Gr. *kachlex, -ekos*, m. pebble on a stream bed, gravel:

Gr. *kiseris, -eos*, f. pumice; *kiserion*, n. dim.; *kiserodes*, like pumice:

Gr. *klomax (kromax), -akos*, m. heap of stones, stony place:

Gr. *krokale*, shore, pebble on the beach; *halikrokalos*, shingly, pebbly; see **shore**

L. *lapis, -idis*, m. stone, rock; *lapillus; lapillulus*, m. dim.; *lapideus*, of stone; *lapidosus*, stony; *lapicida (lapidicida); lapidarius*, m. stone-cutter, gem-cutter; *lapido, -atus*, throw stones, stone: lapis lazuli, lapidary, lapillus, dilapidated, *Bombus lapidaria* (orange-tailed bee), *Ceuthophilus lapidicola* (a grasshopper), *Purpura lapillus* (a gastropod).

Gr. *las, laos*, m. stone, rock, crag; *lainos*, of stone, stony: *Laopithecus robustus* (a fossil primate), *Laopteryx priscus* (a fossil bird).

Gr. *latype*, chip of stone; *latypos*, stone-cutter, mason; see **mason**

Gr. *lepas, -tos*, n. bare rock, crag:

Gr. *leuster*, stoner; *leustos*, stoned; see **throw**

Gr. *lithos*, m. stone, rock; *litharion; lithidion; lithion*, n. dim.; *lithax, -akos*, stony; *lithikos; lithinos*, of stone; *lithosis*, f. turning to stone, petrification; *-lite*, a suffix derived from *lithos* and, like *-ite* (q.v.), signifying mineral, rock: lithology, lithic, lithify, lithium, lithogenesis, lithograph, lithotomy, paleolithic, rhabdolith, coccolith, monolithic, statolith, chrysolite, melilite, lepidolite, *Cryptolithus tesselatus* (a trilobite), *Lithobius forficatus* (a centiped), *Lithospermum asperum* (a gromwell), *Lithops olivacea* (an aizoacead), *Dialithus magnificus* (a beetle), *Heliolites porosus* (a Devonian coral).

L. *lyncurium* (Gr. *lynkourion*), n. a gem thought to have been formed from lynx urine:

L. *marga*, a kind of earth, marl; *acaunumarga*, marl; see **earth**

L. *Medusa* (Gr. *Medousa*), a gorgon with snaky locks and capable of turning beholders to stone; see **fear**

Gr. *messoros*, m. boundary-stone: *Messorus brunneus* (a bug).

L. *metallum* (Gr. *metallon*), n. mine, ore, mineral; *metalleus; metalleutes*, m. prospector, miner: metal, metallic, metallurgy, bimetallism, *Ctenostoma metallicum* (a beetle), *Cryptorhopalium submetallicum* (a beetle), *Coptocephala metalliconotata* (a beetle).

F. *mineral*, < VL. *mina*, mine: mine, mineral, mining, mineralogy, permineralization.

Gr. *misy, -yos*, n. an ore, perhaps a sulfide or sulfate of copper:

L. *mithrax, -acis*, m. a Persian precious stone:

Gr. *molochites*, a precious stone, probably malachite; see **mallow**

L. *muchula*, f. a precious stone also called *telicardios:*

Gr. *mylax*, millstone; see **grind**

L. *naxium*, n. a stone for polishing marble, Naxian whetstone:

L. *nebritis, -idis*, f. a precious stone sacred to Bacchus:

L. *nilius* (Gr. *neilios*), f. a precious stone:

L. *Nioba* (Gr. *Niobe*), f. daughter of Tantalus, changed by Zeus into stone: niobium, niobate.

L. *nipparene*, f. a kind of gem:

L. *Obsidianus* (for *Obsianus*), of Obsius, discoverer of volcanic glass in Ethiopia: obsidian.

L. *Olenus* (Gr. *Olenos*), m. husband of Lethaea, for whose pride both were turned into stone: *Olenus truncatus* (a Cambrian trilobite), *Lethaea geognostica* (a book by H. G. Bronn).

Gr. *onyx*, *-ychos*, m. a yellowish gem stone; *onychinos*, of onyx: onyx, sardonyx, *Crepidula onyx* (a slipper-shell).

L. *opalus* (Gr. *opallios*), m. a precious stone: opal, opaline, opalescence, *Opalina ranarum* (a protozoan), *Conopia opalecsens* (a peach-tree borer).

L. *passernix*, *-icis*, m. whetstone:

Gr. *pessos (pettos)*, m. a round stone used in playing checkers:

L., Gr. *petra*, f. rock, shelf or ledge of rock; *petridion*, n. dim.; Sp. *piedra*, f. rock; *petrinus*; *petronius*, of stone; *petralis*; *petrodes*; *petrosus*, rocky, stony; *petraeus*; *petrensis*, among rocks: petrify, petroleum, petrography, petrology, petrofabric, petrophilous, petrous, Peter, Pierre, pierrot, parrot, petrel, samphire, saltpeter, petricolous, *Petrodromus tetradactylus* (a shrew), *Petrobium arboreum* (whitewood), *Petricola lithophaga* (a pelecypod), *Empetrum atropurpureum* (a crowberry), *Hymenolobium petraeum* (a legume), *Pteria petrosa* (a pelecypod), *Petrea arborea* (a verbenacead).

Gr. *phelleus*, m. stony ground:

Gr. *poroo*, petrify, harden, become calloused:

Gr. *porphytrites*, like purple, a purple stone; see **purple**

Gr. *psephos*, f. pebble; *psephidion; psephion*, n.; *psephis*, *-idos*, f. dim.; *psephistes*, m. calculator: psephite, psephism, psephomancy, *Psephurus gladius* (a fish), *Psephophorus polygonus* (a turtle), *Psephosaurus suevicus* (a fossil reptile).

L. *pumex*, *-icis*, m. a light, porous, volcanic glass; *pumiceus*, of pumice; *pumicosus*, porous like pumice: pumice, *Dactylocalyx pumiceus* (a sponge).

Gr. *pyrites*, m. a mineral that strikes fire: pyrite, chalcopyrite.

AS. *rocc*, stone; F. *roche*: rock, rocky, Rocky Mountains, Rockford (Ill.), rococo, roche moutonnee, Rochelle, traprock.

L. *rudus*, *-deris*, n. crushed stone, rubbish, debris; *rudero*, *-atus*, pave with crushed stone or rubbish; *ruderalis; ruderarius*, pertaining to rubbish: ruderal, rudaceous, rudite, *Lithospermum ruderale* (wayside-gromwell).

L. *rupes*, *-is*, f. rock; *rupecula*, f. dim.; *rupestris*, *-e*, of rocks, rocky: rupestral, rupestrine, rupicoline, *Rupicapra tragus* (chamois), *Rupicola crocea* (cock-of-the-rock), *Draba rupestris* (rockcress), *Lagopus rupestris* (rock-ptarmigan), *Sedum rupestre* (a stonecrop), *Fendlera rupicola* (a saxifragacead).

L. *sacondios*, m. a kind of amethyst:

L. *sagda*, f. a leek-green stone:

L. *sandaresus*, f. a kind of onyx:

L. *santerna*, f. material for soldering gold:

L. *sapenos*, m. a kind of amethyst:

L. *sapphirus* (Gr. *sappheiros*), f. sapphire; *sapphirinus*, of sapphire, sapphire-blue: sapphire, sapphirine, *Elonchus sapphirinus* (a fly), *Callithea sapphira* (a blue butterfly).

L. *sardius* (Gr. *sardios*), m. a kind of chalcedony, carnelian: sard, sardonyx, sardachate.

L. *saxum*, n. rock; *saxulum*, n. dim.; *saxialis*, of rock, rocky; *saxatilis*, found among rocks; *saxificus*, turning into stone, petrifying; *saxosus*, full of rocks, rocky: saxicoline, saxicavous, saxifrage, sassafras, *Saxicava arctica* (a pelecypod), *Saxicola rubetra* (whinchat), *Parmelia saxatilis* (a lichen), *Gentiana saxosa* (a gentian), *Juncus saximontanus* (Rocky Mountain rush), *Pardosa saxatilis* (a spider), *Aethionema saxatile* (a stonecress).

L. *scopulus* (Gr. *skopelos*), m. observation point, projecting rock, crag, cliff, ledge; *scopulosus*, rocky, craggy: *Gentiana scopulorum* (a gentian), *Rhinoscopelus rarus* (a fish).

L. *scrupus*, m. sharp stone, anxiety; *scrupulus*, m. dim.; *scrupeus*, of sharp stones; *scruposus; scrupulosus*, full of sharp stones, rough, rugged: scruple, unscrupulous, *Urceolaria scruposa* (a lichen).

L. *sedimentum*, n. a settling: sediment, sedimentary, sedimentation.

L. *segullum; segutilum*, n. a kind of earth or ore supposed to indicate the presence of gold.

Gr. *siderites*, m. an ore of iron: siderite.

L. *sil*, *-is*, n. ocher; *silaceus*, ocherous; see **earth**

L. *silex*, *-icis*, m. any hard stone, flint; *siliceus*, flinty: silica, silicic acid, silicate, siliceous, silicosis, silicon, silicone, sial, sima, *Juniperus silicicola* (West Indian juniper), *Nonionina silicea* (a foraminifer), *Phacoides silicatus* (a Miocene pelecypod).

Gr. *skyros*, m. chips of stone from hewing:

L. *smaragdus* (Gr. *smaragdos*), f. a green-colored stone; *smaragdinus*, like an emerald, emerald-green: emerald, emeraude, Emerald Isle, Esmeralda, esmeraldite, *Gobius smaragdus* (a fish), *Oecophylla smaragdina* (an ant), *Eulonchus smaragdinus* (a fly), *Halictus emeraldensis* (a bee), *Cleistocactus smaragdiflorus* (a cactus).

Gr. *sory*, n. a kind ore:

Gr. *spilas, -ados,* f. rock, slab, cliff, crag; stony: *Spiladomyia simplex* (a fossil fly).

Gr. *stia,* f. pebble; *stion,* n. dim.; *stizo,* throw pebbles at: *Stiodesmus stratus* (a milleped).

L. *Syenites,* pertaining to Syene in Egypt: syenite.

Gr. *synagma, -tos,* n. concretion:

L. *tephritis, -idis,* f. an ash-colored stone: *Tephritis maculata* (a fly).

L. *tessera,* paving tile, square piece of stone; *tessella,* dim.; see **four**

Gr. *thegane,* f. whetstone, < *thego,* sharpen: *Theganopteryx jucunda* (a blattid).

Gr. *thria,* a pebble drawn from a divining urn; see **prophecy**

L. *tofus,* m. a soft, porous rock: tufa, tuff, tufaceous, tuffaceous, tophus, *Prioria tufficola* (a fossil legume).

Gr. *topazos,* m. a yellowish mineral: topaz, *Topaza pella* (a bird).

OF. *turquois,* Turkish: turquoise.

See: **earth, sand, ashes, lime, glass, coral, brick, coal, diamond, plaster, grind, pearl**

stonecrop < AS. *stancropp,* houseleek, orpine.

L. *amerimnum* (Gr. *amerimnon*), n. a kind of houseleek:

NL. *bryophyllum,* n. a genus of the stonecrop family: *Bryophyllum uniflorum* (a houseleek).

L. *cepaea* (Gr. *kepaia*), f. a succulent salad herb: *Sedum cepaea* (a stonecrop).

Gr. *chrysothales; erithales; trithales,* n. a kind of houseleek:

L. *cotyledon, -is* (Gr. *kotyledon, -os*), f. a genus of the stonecrop family: *Cotyledon oppositifolia* (a crassulacead).

NL. *crassula,* f. a genus of the stonecrop family, < L. *crassus,* thick: Crassulaceae, *Crassula argentea* (silver crassula).

L. *digitellum,* n. houseleek:

NL. *echeveria,* f. a genus of the stonecrop family, < Atanasio Echeverra, Mexican botanical artist: *Echeveria secunda* (a crassulacead).

Gr. *epipetron,* n. a kind of sedum:

Gr. *isoëtes,* a kind of houseleek, now applied to the quillworts; see **quillwort**

NL. *kalanchoë,* f. a genus of the stonecrop family: *Kalanchoë flammea* (a crassulacead).

Gr. *krinanthemon,* n. houseleek:

L. *sedum,* n. houseleek, stonecrop, orpine: *Sedum acre* (a stonecrop), *Crassula sedifolia* (a crassula), *Penthorum sedoides* (Virginia stonecrop).

L. *sempervivum,* n. houseleek: *Sedum sempervivum* (a stonecrop).

L. *telephium* (Gr. *telephion*), n. a kind of stonecrop: *Telephium imperati* (orpine), *Sedum telephioides* (Allegheny stonecrop), *Sedum telephium* (liveforever).

stonos, Gr. groan, sigh; see *stenagma* under **sound**

stonycho- < Gr. *stonyx, -ychos,* any sharp point; see **point**

stool < AS. *stol,* seat; see **sit, dung**

stoop < AS. *stupian;* see **bend**

stop < AS. *stoppian;* see **end, stand, rest, delay, destroy, cancel, bar, close, key**

stopper; see **close**

storage, Gr. affection, love; see **love**

storax, L. resin from the styrax tree; see **resin**

store < OF. *estorer,* < L. *instauro, -atus,* renew, repair: storage, restore, instauration.

L. *aerarium* (Gr. *airarion*), n. treasury:

Gr. *apodocheion,* n. receptacle, storehouse:

L. *apotheca* (Gr. *apotheke*), f. storehouse, barn, magazine; *apothecarius,* m. warehouseman, clerk; *apothesis,* f. a storing up; *apothetos,* stored up; *enapothesis,* f. deposit, depot; *entheke,* f. store: apothecary, apothecium.

L. *arcarius,* m. treasurer:

L. *archivum* (Gr. *archeion*), n. depository of original and historic records; the documents themselves: archives.

L. *argentaria,* f. bank:

L. *armamentarium,* n. arsenal, armory:

Gr. *chryson, -os,* m. treasury:

L. *conditum,* n. thing laid up, store:

L. *corbona,* f. treasure-chamber:

L. *fenilium*, n. hayloft:

L. *fiscalis*, relating to the treasury; see *fiscus* under **basket**

L. *granarium*, n. a place for storing grain; see **grain**

L. *horreum*, n. barn, storehouse, granary; *horreaticus*, of a storehouse:

Gr. *keimelion*, n. treasurer; *keimelios*, treasured; *keimeliorches*, m. treasurer: cimelium, cimeliarch.

L. *nubilarium*, n. barn, shed:

L. *penarius*, m. storehouse, granary, < *penus*, c. store of provisions:

L. *pergula*, f. bower, booth, stall, shop: pergola, *Pergula turneri* (a wasp).

L. *promum*, n. storeroom.

L. *repositorium*, n. cabinet, place where a thing is stored:

L. *taberna*, booth, shop, inn; see **house**

L. *tabularium*, n. archives:

Gr. *tamias*, m. dispenser, treasurer, steward, paymaster; *tamiakon*, n. treasury; *tamiakos*, of the treasury: *Tamias striatus* (eastern chipmunk).

L. *thesaurus* (Gr. *thesauros*), m. treasure, storehouse, magazine, chest: thesaurus, treasury, treasure, treasurer.

See: **house, trade, abundance**

storea, L. mat; see **rug**

stork < AS. *storc*.

L. *ciconia*, f. stork: *Ciconia alba* (stork), *Erodium ciconium* (stork-heronbill).

Gr. *kelas*, m. a stork, adjutant:

Gr. *pelargos*, m. stork: *Pelargonium odoratissimum* (nutmeg-geranium).

storm < AS. *storm*.

Gr. *cheima*, *-tos*, winter, storm; see *cheimon* under **year**

L. *imber*, *-bris*, rain, shower, storm; see **drop**

L. *jactabundus*, tossed about, stormy; see *jacio* under **throw**

Gr. *kataigis*, *-idos*, f. squall, hurricane:

Gr. *keraunobolia*, f. thunderstorm:

L. *laelaps*, *-apis*, m. (Gr. *lailaps*, *-apos*, f.), hurricane: *Proctolaelaps productus* (a mite).

Gr. *niphatos*, snowstorm; see *nipha* under **ice**

Gr. *ombros*, rainstorm; see *imber* under **drop**

Gr. *prester*, hurricane; see **wind**

L. *procella*, f. storm, tempest, hurricane; *procellosus*, stormy: procellarian.

Gr. *rhasma*, shower; see *rhanis* under **drop**

L. *tempestas*, *-atis*, f. storm; *tempestuosus*, stormy: tempest, tempestuous.

Gr. *thyella*, f. storm, hurricane, whirlwind: *Puffinus(Thyellodroma) sphenurus* (a shearwater).

Gr. *zale*, surge of the sea; see **wave**

Gr. *zaps*, f. storm:

See: **wind, drop, weather, turn**

storthynx, Gr. point, spike; see **point**

story < OF. *estoire*, < L., Gr. *historia*, f. narrative, account; *histor*, *-os*, informed, learned: history, historical, historiettes.

L. *Aesopus* (Gr. *Aisopos*), m. Aesop, a famous Greek fabulist: Aesop's *Fables*, *Oromys aesopi* (a fossil rodent), *Aesopus japonicus* (a gastropod).

Gr. *ainos*, m. tale, story, fable:

Gr. *akousma*, thing heard, rumor, tale; see *akouo* under **hear**

L., Gr. *allegoria*, f. a figurative story: allegory, allegorical.

Gr. *anekdotos*, unpublished, secret: anecdote, anecdotal.

Gr. *aphegema*; *diegema*, *-tos*, n. narrative, tale; *diegemation*, n. dim.:

Gr. *apomnemoneuma*, *-tos*, n. memoir, narrative:

L. *Clio*, *-us* (Gr. *Kleio*, *-ous*), f. Muse of history; a sea-nymph: Clio, *Clio pyramidata* (a pteropod), *Cliona sulphurea* (a boring sponge), *Clione papilionacea* (a pteropod).

Gr. *epos*, word, tale, speech, song; see **word**

L. *fabula*, f. narrative, story; *fabella*, f. dim.; *fabulosus*, celebrated in fable, fictitious; *fabularis*, of fables: fable, fabulous, fabulary.

L. *fama*, rumor, report, common talk; *famigeratus*, tale-bearing; see **honor**

Gr. *logos*, word, story; see **word**

Gr. *mythos*, m. fable, fiction, legend; *mytharion*; *mythidion*, n. dim.; *mythikos*, of fables and legends: myth, mythology, mythopoeic, mythoclast.

L. *narratio*, *-onis*, f. story of events; *narratiuncula*, f. dim.; *narrativus*, pertaining to reporting; *narro*, *-atus*, relate: narration, narrative, narrate, narrator.

L. *parabola* (Gr. *parabole*), f. a comparison, allegory, analogy, tale: parable, palaver, parabola, parabolic.

See: **speak, write, word**

stout < OF. *estout,* bold, strong; see **strong, bold, thick**

stove < AS. *stofa;* see **oven**

strabelos, Gr. a kind of snail; see **mollusk**

strabus, L. (Gr. *strabos*), squinting; see **slope**

stragulum, L. blanket, pall, spread; *stragulus,* spreading or covering; see **blanket**

straight < AS. *streht.*
    Gr. *astraphes (astrabes); astreptos,* straight:
    L. *cathetus* (Gr. *kathetos*), f. a perpendicular line: *Cathetostoma laeve* (a fish).
    L. *dirigo, directus,* set straight, arrange in a staight line, order, guide; see **govern**
    Gr. *euthys; ithys,* straight, upright, direct, ruled; *euthynter, -os; euthyntes; euthynos,* m. straightener, chastiser, auditor; *euthyntos,* drawn straight; *euthyno; ithyno,* make straight, guide, rule: *Euthynotus micropodius* (a fossil fish), *Ithyphallus impudicus* (a stinkhorn-fungus), *Paepalanthus ithyphallus* (a pipewort).
    Gr. *itamos,* hasty, headlong, impetuous, reckless; see **swift**
    Gr. *kymbachos,* head foremost, headlong: *Cymbachus pulchellus* (a beetle).
    Gr. *orthos,* straight, normal, right, correct; *orthotenes,* stretched out, straight; *epanorthosis,* f. revision, correction: orthoclase, orthodox, orthography, orthophonic, Orthoptera, orthodontia, anorthite, *Orthocarpus luteus* (a figwort), *Orthoceras strigatum* (an Ordovician cephalopod).
    L. *praeceps,* headlong, steep; see **cliff**
    L. *rectus,* straight, upright, proper, right; *corrigo, correctus,* make straight, set right; see **right**
    L. *strictus,* straight, tight: strict, stricture, *Dendrocalamus strictus* (a bamboo), *Oxalis stricta* (yellow woodsorrel), *Potamogeton strictifolius* (a pondweed).
    See: **right, upright, line**

strain < OF. *estraindre,* < L. *stringo, strictus,* bind; see **bind, draw, sieve**

strait < OF. *estreit,* < L. *strictus,* tight, narrow; see **way, narrow**

stramen, L. straw, litter; *stramineus,* of straw; see **grass**

stramonium, NL. jimsonweed; see **nightshade**

strangale, Gr. halter; see **hold**

strangalis, Gr. knot; see **knot**

strange < OF. *estrange,* < L. *extraneus,* outside, foreign: strangeness, stranger.
    L. *advena,* c. stranger, foreigner, alien: *Nuphar advena* (spatterdock).
    Gr. *aethes,* unusual, strange: *Aetherrhinus aurichalceus* (a beetle).
    Gr. *agnostos,* unknown, unknowing; see *nosco* under **know**
    L. *alienus,* foreign, strange, not belonging, not related; *alieno, -atus,* estrange, drive insane: alien, alienate, alienation.
    Gr. *allodapos,* belonging to another kind, foreign, strange; *allophylos; aprosphylos; ekphylos,* of another tribe, foreign; *allotrios,* of another, alien, strange; *allokotos,* strange, unusual: *Allodaposuchus precedens* (a fossil crocodile), *Allophylus occidentalis* (a sapindacead), *Aprosphylus hybridus* (a grasshopper), *Allotriozoon prodigiosum* (a wasp), *Allocotops calvus* (a bladder).
    Gr. *anarsios,* incongruous, strange, hostile; see **against**
    L. *ascitus,* alien, foreign, derived:
    Gr. *atopos; ektopios; ektopos,* out of place, odd, strange, unnatural: ectopic, ectopia, *Atopognathus collaris* (a wasp), *Atopomyrmex mocquersyi* (an ant), *Ectopiocephalus vanduzeei* (a bug), *Ectopocerus verticornis* (a beetle).
    L. *barbarus* (Gr. *barbaros*), foreign, strange, therefore outlandish, crude, rude; *barbarikos,* foreign, crude: barbarian, barbaric, barbarism, barbarity, Barbara, rhubarb.
    L. *curiosus,* odd, strange, inquisitive; see **wish**
    Gr. *eknomios,* unususal, marvelous; see **wonder**
    Gr. *ektrapelos,* devious, strange:
    Gr. *epelys, -ydos,* c. stranger, foreigner:
    Gr. *exhomilos,* strange, unfamiliar, alien: *Exomilus lutarius* (a gastropod).
    Gr. *exotikos,* from the outside, alien, foreign: exotic, *Venerupis exotica* (a pelecypod).
    L. *extraordinarius,* above or beyond the common or ordinary; see **wonder**
    L. *glossa,* an obsolete or foreign word; see **word**
    L. *hostis,* stranger, enemy; see **enemy**
    L. *ignotus,* unknown, strange: *Diplolepis ignota* (a wasp).
    L. *incognitus,* unknown, unidentified: incognito, *Iliosuchus incognitus* (a fossil reptile).
    L. *insolitus,* unusual, uncommon, strange, odd, queer: *Trigonocelia insolita* (a pelecypod).

L. *insuetus,* inexperienced in, unused to, unusual: *Callistomimus insuetus* (a beetle).

L. *inusitatus,* unusual, extraordinary, strange: *Achatina inusitata* (a gastropod).

L. *invecticius; invectus,* brought in, imported, exotic:

L. *invisitatus,* strange, new, uncommon:

L. *metoecus* (Gr. *metoikos*), c. stranger, resident alien: *Metoicoceras[Metoecoceras] gibbosum* (an ammonite).

Gr. *othneios,* strange, alien: *Othnius guttulatus* (a beetle), *Othniocryptus variegatus* (a beetle).

L. *paradoxus* (Gr. *paradoxos*), strange, contrary to expectation, marvelous: paradox, paradoxical, *Spinifex paradoxus* (an Australian grass), *Paradoxides lamellatus* (a trilobite), *Paradoxurus hermaphroditus* (musang), *Aleuroparadoxus iridescens* (a whitefly).

Gr. *peratikos,* alien, strange, foreign:

L. *peregrinus,* traveling about, foreign, exotic, strange; see *peregrinor* under **move**

Gr. *perissos,* odd, strange; see **different**

L. *praecipuus,* peculiar, special, extraordinary: *Sciara praecipua* (a fly).

AS. *wealh,* strange, foreign: Welsh, walnut, Walloon, Cornwall.

Gr. *xenos,* stranger, guest; *xenikos,* strange, foreign; see **guest**

See: **other, different, wonder, far, secret, over, and, new**

strangle < *strangulo, -atus,* choke, stifle; see **choke**

strangos, Gr. twisted, crooked; see **bend**

stranx, Gr. something squeezed out, drop; see **drop**

strap < AS. *stropp,* < L. *stroppus,* thong, band.
L. *amentum,* strap, thong, catkin; see **catkin**
Gr. *anamaschalister, -os,* m. shoulder-strap:
L. *antilena,* f. strap on a yoke:
Gr. *aorter, -os,* m. strap, sword-belt:
Gr. *boeus,* m. strap of ox-hide:
Gr. *byrsine,* thong, strap; see *byrsa* under **leather**
L. *cohum,* n. strap for fastening plowbeam to yoke:
L. *corrigia,* f. shoe-latchet, thong, rein: *Corrigiola littoralis* (strapwort).
L. *epiredium,* n. strap for hitching horse to cart:
L. *habena,* f. thong, strap, rein; *habenula,* f. dim.: *Habenaria grandiflora* (a fringed orchid), *Sphex habena* (a wasp).
Gr. *himas, -antos,* m. leather strap, leash, thong; *himantidion; himantion,* n. dim.: *Himantopus albus* (a stilt), *Himantoglossum hircinum* (an orchid).
Gr. *holkos,* strap, rein; see **draw**
Gr. *lepadnon,* n. strap on a yoke:
L. *ligula,* strap, latchet, dim. of *lingua,* tongue; see **tongue**
L. *lorum* (Gr. *loron*), n. strap, thong, rein; *loramentum,* n. thong; *loratus,* bound with thongs: lorate, *Loranthus europaeus* (a mistletoe), *Corythoderus loripes* (a beetle), *Sagittaria lorata* (strap-leaved arrowleaf), *Cobelura lorigera* (a beetle).
L. *obstragulum,* n. strap, lace, latchet:
Gr. *rhyter,* rein, trace, strap; see **hold**
Gr. *telamon, -os,* m. strap, belt:
Gr. *tropos,* m. twisted leather thong:
See: **belt, ribbon, tongue, bind**

stratagem < Gr. *strategema(stratagema),* artifice, trick; see **know**

stratos, Gr. army; *strategos,* general; *stratiotes,* soldier; see **army, govern**

stratum, L. cover, blanket, bed, layer, see **bed**; *stratus,* stretched out, prostrate, see *sterno* under **spread**; *strata,* a paved road, see **way**

straw < AS. *streaw,* ripened stalk of grasses; see **grass, stem**

strawberry
L. *fragum,* n. strawberry: *Fragaria virginiana* (a strawberry), *Trifolium fragiferum* (strawberry-clover), *Waldsteinia fragarioides* (barren strawberry).

stream < AS. *stream.*
Gr. *Acheron, -tos,* m. a river of the nether world, the nether world itself: *Acherontemys heckmani* (a fossil turtle), *Acherontia atropos* (death'shead-hawkmoth).
L. *amnis, -is,* m. stream of water, river; *amniculus,* m. dim., rivulet, brook; *amnicola,* c. inhabitant of or by a river; *amnicus,* of a stream: amnigenous, amnicolist, *Amnicola limosa* (a gastropod), *Pisidium amnicus* (a pelecypod).
L. *aquula,* small stream of water; see *aqua* under **water**
Sp. *arroyo,* m. creek, dry channel: arroyo.

L. *Baetis, -is,* m. Guadalquivir River; *baeticus,* pertaining to the Baetis and the excellent wool from that region: *Baetis ephippiatus* (a mayfly), *Baetisca obesa* (a mayfly).

AS. *burna; brunna* (G. *brunnen*), spring, brook, rivulet: burn(bourne), Bannockburn, Blackburn, Burnham, Eastbourne, Kilbourne, Bournemouth, Osborn.

L. *cataclysmus* (Gr. *kataklysmos*), m. deluge, flood, inundation: cataclysm.

Gr. *charadra,* mountain stream, torrent; see **valley**

Gr. *cheuma,* stream, flow; see *cheo* under **pour**

Gr. *enaulos,* m. bed of a stream, torrent:

L. *Eridanus* (Gr. *Eridanos*), m. name of a legendary river in western Europe, either the Po, Rhine, or Rhone: Eridanus, Eridanid, *Eridanosaurus brambillae* (a fossil reptile).

L. *flumen, -inis; fluentum,* n.; *fluvius,* m. stream, river, flow; *flumicellum,* n. dim.; *fluminalis; flumineus; fluvialis; fluviatilis,* of a stream: fluminose, Fiume, fluent, fluviatile, fluvicolous, *Fluminea festucacea* (river-grass), *Ochthebius flumineus* (a beetle), *Perca fluviatilis* (European perch), *Fluvicola pica* (a waterchat).

L. *Inachus* (Gr. *Inachos*), m. a river and river-god; *Io, -us,* f. daughter of Inachus: *Inachus muricatus* (a decapod), *Automeris io* (a moth), *Vanessa io* (peacockbutterfly), *Io spinosa* (a gastropod).

D. *kill,* stream, creek, channel: Catskill, Kaaterskill, Bushkill, Schuylkill, Kill Van Kull.

ON. *kriki* (AS. *crecca*), a small stream: creek, Cottonwood Creek, Sage Creek, Pohopoco Creek, Creccanford, creek-nettle *(Urtica holoserica).*

Gr. *libas, -ados,* f. anything that drops, trickles, or runs; *libadion,* n. dim., < *leibo,* pour forth, run; *aeilibes,* even-flowing; L. *Libethra* (Gr. *Leibethra*), f. a spring sacred to the Muses:

L. *Liger, -is,* m. the Loire River: Ligerian.

Gr. *nama,* spring, stream; see **spring**

L. *Nilus* (Gr. *Neilos*), m. the Nile River; *Niloticus,* of the Nile: *Ceratina nilotica* (a bee).

Gr. *Phlegethon, -tos,* m. a river of fire in Hades: Phlegethontal.

Gr. *potamos,* m. river, stream; *potamion,* n.; *potamiskos,* m. dim.; *potamios,* of a river; *potamites,* m. water-finder: potamic, potamometer, potamology, potamophilous, hippopotamus, Mesopotamia, *Cervus mesopotamicus* (a deer), *Helioctamenus hippopotamus* (a beetle), *Abutilon megapotamicum* (Brazilian abutilon).

Gr. *rheos; rheithron; rheuma, -tos,* n. current, stream; *rhyax, -akos,* m. rushing stream, torrent, < *rheo,* flow; *rheumation,* n. dim.; *cheimarrhos,* m. a stream swollen by the melting of winter snows; *hyporrhysis,* f. underground stream or channel: *Rheumatobates rileyi* (a bug), *Rhithrum rivulatum* (a fly), *Rhithrogena semicolorata* (an ephemerid), *Rheithronycteris aphylla* (a bat), *Rhyacophila relicta* (a trichopterid).

Pg. *Rio das Amazonas,* Amazon River; see *Amazon* under **woman**

L. *rivus,* m. stream, brook, creek; Sp. *rio,* m. river, stream; *rivulus,* m. dim., rill; *rivalis; rivularis,* of a brook; *rivalis, -is,* m. user of the same brook, neighbor, competitor; *corrivatio, -onis,* f.; *corrivium,* n. confluence of streams: rivulet, rivose, rivulose, rival, arrive, derive, derivative, Rio Grande, Rio de la Plata, Rialto, *Rivularia bullata* (an alga), *Rivulus cylindraceus* (a fish), *Ursirivus pyriformis* (a fossil pelecypod), *Metriophilus rivalis* (a beetle), *Geum rivale* (water-avens), *Bidessus rivulorum* (a beetle), *Laccophilus rivulosus* (a beetle), *Cyperus rivularis* (brook-sedge).

L. *Sabrina,* f. Severn River; a river-nymph: *Sabrina tesselata* (a worm-snake).

L. *Styx, -ygis* (Gr. *-ygos*), f. a river in the nether world; *Stygius* (Gr. *Stygios*), of the Styx, of the nether world: Stygian, *Stygicola dentata* (a blind brotula), *Juncus stygius* (a rush).

L. *torrens, -entis,* m. swift or violent stream: torrent, torrential, *Charadromyia torrenticola* (a fly).

See: **flow, pour, spring, water, valley**

streblos, Gr. twisted, crooked, wrinkled; see **bend**

strebula, L. flesh about the haunches; see **rump**

street < AS. *straet,* < L. *strata,* a paved road; see **way**

stremma, Gr. a twist, wrench, roll, thread; see *streptos* under **turn**

strena, L. sign, omen; see **prophecy**

strenes, Gr. harsh, hard, rough; see **rough**

strength < AS. *strengthu;* see **strong**

strenuus, L. brisk, prompt, active, vigorous; see **busy**

strepitus, L. din, noise; see **sound**

streptos, Gr. twisted; *strepsis,* a turning, twisting; see **turn**

stretch < AS. *streccan;* see **spread, draw**

stria, L. furrow, channel, hollow, line; *striola,* dim.; see **furrow**

stribligo, L. error in language, violation of propriety, solecism; see **fault**

strictus, L. drawn together, tight, straight; see *stringo* under **bind, straight**

stridulus, L. creaking, grating; see **squeak**

striga, L. swath, windrow, bristle; *strigula,* dim., see **ridge;** hag, witch, see **woman**

strigilis, L. scraper; *strigilecula,* dim.; see **scrape**

strigo- < Gr. *strix, -igos,* screech-owl; see **owl**

strigosus, L. thin, lean, meager; see **thin**

strike < AS. *strican.*

L. *aggressor, -is,* m. attacker, assailant; *aggredior, -essus,* attack, assault: aggressor, aggression, aggressive.

L. *alapa; salapitta,* f. box or cuff on the ear:

Gr. *apolakema, -tos,* n. snapping of the fingers:

Gr. *arasso,* strike hard, dash to pieces; *katarasso,* dash down: cataract.

L. *arieto, -atus,* butt like a ram; see **push**

L. *assultus, -us,* m. attack, < *assulto, -atus,* leap toward, attack: assault.

L. *attonitus,* thunderstruck, stunned, confounded:

L. *battuo,* beat, hit, strike: battle, battalion, batter, battery, abate, combat, debate, embattled, battleship.

Gr. *bletos,* stricken, smitten:

L. *cancello, -atus,* make like a lattice, cross out, strike out; see **cancel**

L. *colaphus* (Gr. *kolaphos*), m. blow, cuff, box on the ear; F. *coup,* master stroke; Gr. *kope,* f. stroke, cutting; *kolaptes,* m. pecker, < *kolapto,* strike, peck, chisel: cope, coup, coup d'etat, coppice, coupon, *Colaphomegra rufifrons* (a beetle), *Colaptes auratus* (flicker), *Chrysocolaptes sultaneus* (a bird).

L. *concutio, -cussus,* strike together, shake; *decutio,* beat off; *discutio,* shatter; *excutio,* drive out; *incutio,* strike upon, dash against; *percutio,* strike, beat, shock; *succutio,* toss up, shake: concussion, discuss, incuss, percussive, succussion, succussatory, successive, rescue, repercussion.

L. *convitio, -atus,* attack or injure at the same time:

L. *cudo, cusus,* strike, beat, pound, stamp money; *cusio, -onis,* f. a stamping of money; *cusor, -is,* m. coiner of money; *incudo,* forge with a hammer; *procudo,* fashion by hammering, forge: incuse, incusate.

L. *depalmo, -atus,* strike with the open hand, box the ear:

Gr. *eisdrome,* f. attack, onslaught:

Gr. *elauno,* strike, beat, drive: *elatos,* ductile; *exelatos,* beaten out; *sphyrelatos,* wrought with a hammer:

Gr. *enelysios,* struck by lightning:

Gr. *ereiko,* pound, bruise, shiver; *ereiktos,* bruised, pounded; *ereixis,* f. a pounding, grinding:

L. *expungo, -unctus,* strike out, blot out, erase; see **cancel**

L. *fendo, fensus,* strike: fend, fender, fence.   See *defendo* under **guard**

L. *ferio,* strike, beat: ferule, interfere, *Ferula communis* (giant fennel).

L. *fligo, flictus,* strike down, beat; *affligo,* strike down; *confligo,* strike against, contend; *infligo,* strike upon, impose on: affliction, conflict, inflict.

L. *fulguritus,* struck by lightning; see *fulgeo* under **light**

L. *fustigo,* cudgel to death; see *fustis* under **club**

L. *ictus, -us,* m. blow, stroke, stab, sting, < *ico, ictus,* strike, hit, stab, sting: ictus.

L. *impetus, -us,* m. attack, assault; *impetibilis,* assailing; *impetuosus,* violent: impetus, impetuous, impetuosity.

L. *impingo, -pactus,* push, strike, drive, thrust: impinge, impact.

L. *incurro, -cursus,* run into, rush at, assail, attack: incursion.

Gr. *keraunos,* thunderbolt; *keraunios,* of a thunderbolt; see **thunder**

Gr. *kopto,* cut, strike, chop; see **cut**

Gr. *kossos,* m. box or cuff on the ear:

Gr. *kroteo,* beat, strike, clap, rattle; *krotema, -tos,* n. a piece of hammered work; *krotetos,* sounding with blows; *akrotos,* unclapped, unapplauded; *apokrotema, -tos,* n. a snap of the fingers; *dikrotos,* double-beating: anacrotism, dicrotic, *Acrotus willoughbyi* (a fish).

Gr. *krouo,* strike, smite, slap; *kroma, -tos,* n. beat, stroke; *krousis,* f. a striking; *kroustikos,* striking, impressive; *apokroustikos,* able to repel: *Crusimetra verecunda* (a moth).

L. *laedo, laesus,* hurt by striking, attack, injure; *laesio, -onis,* f. wound, injury; *allido, -lisus,* dash one thing against another; *collido,* strike together, clash;

elido, strike out, eliminate; *illido,* strike into, beat upon: lesion, collide, collision, elide, elision, lese majesty.

L. *mulco, -atus,* beat, handle roughly; see **hurt**

Gr. *oimao,* pounce or swoop upon:

Gr. *paio,* strike, smite:

Gr. *palmos,* pulsation, palpitation, vibration; *palmikos,* of palpitation; see *pallo* under **shake**

L. *palpito, -atus,* throb, beat, pant, tremble: palpitate, palpitation.

Gr. *paroxysmos,* spasm, attack, throe, fit; see **disease**

Gr. *patasso,* beat, strike; *pataktikos,* striking:

L. *pavio, -atus,* beat, ram, tread down; *pavicula,* f. rammer: pave, paver, pavement.

L. *pello, pulsus,* drive, strike, beat; *pulso (pulto), -atus,* push, strike, beat, knock; see **push**

L. *percello, -culsus,* beat down, strike down, shatter:

L. *pinso, pinsus; pistus,* pound, stamp, crush, grind; see **grind**

L. *plaga* (Gr. *plege*), f. blow, stroke, wound; *plesso,* strike; *plektes,* m. striker; *plektron; apoplexia,* f. stroke; *astrapoplektos,* struck by lightning; *plagigerulus,* stripe-bearing: plectrum, apoplexy, apoplectic, hemiplegia, plague, *Plectrophenax nivalis* (snow-bunting).

L. *plango, -anctus,* strike; *plangor, -is,* m. a striking: plaint, plaintive, plaintiff, complain, complaint, plangent.

Gr. *platagema, -tos,* n. clap, slap, < *platasso,* clap or slap two flat surfaces together:

L. *plaudo, plausus (plodo, plosus),* clap, acclaim, strike: applause, plaudit, plausible, explode, explosion, implosion.

Gr. *rhapisma, -tos,* n. blow with the palm of the hand, slap, stroke, < *rhapizo,* strike: *Rhapisma viridipennis* (a neuropterid).

L. *sileratus,* planet-struck, sunstruck:

AS. *slaegen,* strike: slay, slaughter, sledge, slug, onslaught.

AS. *smitan,* strike: smite, smith, blacksmith.

Gr. *sphygmos,* m. throbbing, pulsation, vibration, beating of the heart: sphygmograph, asphyxiate.

Gr. *stochos,* aim, shot, guess; see **aim**

L. *sugillo, -atus,* beat black and blue; *sugillatum,* black and blue spot, bruise; see **bruise**

L. *talitrum,* n. a rap with the finger: *Talitrus locusta* (an amphipod), *Taliatriator eastwoodae* (an amphipod).

L. *tundo, tusus,* beat, bruise; *contundo,* beat, crush, grind; *obtundo; retundo,* beat, dull: contusion, obtund, obtuse, retuse, *Rumex obtusifolius* (bitter-dock).

Gr. *tychano,* hit, meet, happen:

Gr. *typto,* beat, strike; *typsis,* f. a beating; *tymma, -tos,* n. blow; *typos,* m. blow or its result: type, typography, linotype, typtology, *Philotymma* (a wasp).

L. *vapulo, -atus,* cudgel, flog: vapulation.

L. *verbero, -atus,* beat, strike; see *verber* under **whip**

See: **cut, hammer, whip, club, sword, enter, hurt, ax, push, bruise, mortar**

string < AS. *strenge;* see **rope, thread, nerve, line**

stringizo, Gr. screech, see **squeak;** *stringo, strictus,* bind; see **bind**

strip < AS. *strypan,* deprive, divest; a relatively long narrow piece of anything; see **bare, skin, destroy, strap, ribbon, tongue, belt**

stripe < D. *stripe;* see **line, belt, furrow, ribbon, ridge, bruise, zebra**

striphnos, Gr. firm, solid; see **hard**

strix, L., Gr. screech-owl, see **owl;** furrow, channel, groove, flute, see **furrow**

strobilus, L. (Gr. *strobilos*), anything twisted or that turns, top, pine-cone; see **cone**

strobus, L. a tree yielding an aromatic resin; applied to a pine by Linnaeus, see **pine;** < Gr. *strobos,* anything twisted, cone, see *strobilus* under **cone**

stroma, Gr. bed, mattress; *stromation,* dim.; *stromne,* couch, mattress; see **bed**

stromateus, Gr. patchwork coverlet; see **blanket**

strombos, Gr. a top, a spiral shell; *strombeion,* dim.; see **cone**

strong < AS. *strang:* strongly, stronghold, strongbox, strength, headstrong.

Gr. *aaptos,* invincible: *Aaptus chopi* (a bird).

L. *Achilles, -is* (Gr. *Achilleus*), m. a strong and handsome Trojan warrior: Achillean, *Achillea argentea* (silver yarrow).

L. *agens, -entis,* effective, powerful: agent, agency.

Gr. *aizenos,* strong, active, vigorous:

Gr. *akamas, -antos; akamatos,* untiring. unresting; *akmes, -etos,* untiring, unwearied: *Acamatoxenus suavis* (a beetle).

Gr. *akopos,* unwearied, without fatigue:

Gr. *akrotonos,* muscular:

Gr. *alke,* f. strength, courage; *alkaios,* strong; *alkimos,* strong, stout, brave: *Alcaeorrhynchus grandis* (a bug), *Alcimonotus gounellei* (a spider), *Alcinoë rosea* (a coelenterate).

Gr. *amogetos,* untiring, unwearied:

Gr. *ananke,* f. force, compulsion, necessity; *katananke,* f. force: *Catananche* [*Catananca*] *caerulea* (a composite used for making a love-philter).

Gr. *aniketos,* unconquered, invincible; see **stand**

Gr. *anolotos,* invincible, impregnable, unattainable:

L. *Antaeus* (Gr. *Antaios*), m. giant Libyan wrestler whose strength was renewed when he touched the earth: Antean, *Dorcus antaeus* (a beetle).

Gr. *arrhen; arsen,* male, masculine, strong; see **man**

Gr. *aschetos,* ungovernable, resistless:

Gr. *athletikos,* pertaining to an athlete: athletic.

Gr. *atrytos,* not worn, unwearied: Atrytone.

L. *auctoritas, -atis,* f. power, will, influence: authority.

Gr. *bia,* bodily strength, force, might; *biastes,* one who uses force, violence; see *biastos* under **rough**

Gr. *briaros,* strong; *Briareos,* m. a hundred-armed giant: Briareus, *Briaromys trouessartianus* (a rodent), *Pentacrinus briareus* (a fossil crinoid).

Gr. *brime,* f. strength, bulk; see **weight**

L. *competens,* qualified, able; see **know**

Gr. *drastikos,* efficacious, powerful; *drasis,* strength, efficacy; see *drao* under **make**

Gr. *dynamis,* f. power, strength; *dynamikos,* powerful, efficacious; *dynatos,* strong, powerful; *dynastes,* m. ruler, master, lord: dynamic, dynamite, dyne, dynasty, hydrodynamics, dynamo, dynamometer, *Dynastes neptunus* (a beetle).

Gr. *dyspsyktos,* tolerant of cold, hardy:

Gr. *echyros; ochyros,* firm, strong, secure, stout: *Ochyrocoris electrina* (a fossil bug).

L. *efficax, -acis,* effectual, powerful: efficacious.

Gr. *emphasis,* f. intensive stress: emphasis, emphatic, emphasize.

Gr. *epizaphelos,* violent, furious, vehement:

Gr. *eri-,* intensive particle, very; see *ari-* under **very**

Gr. *eridmatos,* strongly built, immovable:

Gr. *erymnos,* fortified, strong; see *eryma* under **wall**

Gr. *exousia,* f. power, authority, means:

L. *factiosus,* powerful, eager for power: factious.

L. *facultas, -atis,* f. capability, power, skill: faculty, facultative.

AS. *faest,* firm, strong: fast, fasten, fastness, steadfast, bedfast, shamefaced.

L. *firmus,* fast, powerful, strong; *confirmo, -atus,* establish, strengthen: firm, firmness, affirmation, confirm, firmament, infirmity, comfrey.

L. *fortis,* powerful, strong, brave; *forticulus,* dim.: fortify, fortitude, fortress, force, forcible, forceful, comfort, effort, enforce, pianoforte, reenforcement, uncomfortably.

L. *fulcio, fultus,* prop up, support, strengthen, sustain; *effultus,* propped up; see *fulcimen* under **pillar**

L. *fundatus,* firm, grounded, established; see *fundus* under **base**

Gr. *hadros,* thick, bulky, stout, strong; see **large**

L. *Hercules, -is* (Gr. *Herakles, -os*), m. mythical hero celebrated for his strength and twelve great labors: Herculean, Herculaneum, Heracleopolis, Heraclea, *Heracleum villosum* (a cow-parsnip), *Camponotus herculeanus* (a carpenter-ant), *Xanthoxylum clavaherculis* (Hercules-club), *Polypodium heracleum* (a fern), *Dynastes hercules* (a beetle), *Cerebratulus herculeus* (a nemertean).

Gr. *ikanos,* competent, sufficient: *Icanodus convexus* (a fossil fish).

L. *indefatigabilis,* that cannot be tired; *indefatigatus,* untiring: indefatigable.

L. *indefectus,* undiminished, unweakened:

L. *indefessus,* unwearied, indefatigable: *Formica indefessa* (an ant).

L. *indomitabilis,* that cannot be conquered or subdued: indomitable.

L. *inexcoctus,* unexhausted:

L. *inexpugnabilis,* impregnable; *inexpugnatus,* unconquered: inexpugnable.

L. *infatigatus,* unwearied:

L. *infragilis,* unbreakable, strong:

L. *invictus,* unconquered, strong; *invincibilis,* unconquerable: invincible, *Opuntia invicta* (a cactus).

Gr. *iphi; iphios,* strongly, mightily; *iphthimos,* strong, stout, stalwart; *iphigeneia,* f. one born strong: Iphigenia, *Iphthimus serratus* (a beetle), *Iphiaulax testaceus* (a braconid).

Gr. *is, inos,* muscle, fiber, strength; see **thread**

Gr. *ischys, -yos*, f. strength, force, might; *ischyros*, strong, mighty, stubborn, excessive: *Ischyodus emarginatus* (a fossil fish), *Ischyroplectron isolatum* (a grasshopper), *Inopsetta ischyra* (a fish).

Gr. *kikys, -yos*, f. strength, vigor: *Cicynethus acanthopous* (a spider).

Gr. *kratys*, strong, sturdy; *krateros (karteros); kreisson, -os*, stronger; *kratistos*, strongest; *kratos*, n. strength; *kratesis*, f. might, dominion: cratometer, democrat, autocracy, autocrat, *Cratypedes neglectus* (a grasshopper), *Craterocercus phytophagicus* (a sawfly), *Carterorhinus major* (a beetle), *Cratoxylon formosum* (a hypericacead), *Pancratium maritimum* (a sea-daffodil). See *krateo* under **govern**

L. *lacertosus*, muscular, strong; see *lacertus* under **arm**

AS. *maeg*, be able; *meaht*, power, force: may, main, might, mighty, dismay.

Gr. *menos*, n. force, strength, courage, spirit; *erromenos* stout, strong, vigorous: *Menodora scoparia* (broom-menodora), *Menopteryx binocula* (a neuropterid), *Erromenosteus lucifer* (a placoderm), *Cecentromenus marmoratus* (a grasshopper).

Gr. *monimos*, steadfast, permanent: monimolite, monimostylic.

L. *munio, -itus*, fortify; *munimentus*, n. fortification; *munitor, -is*, m. fortifier, engineer: munitions, ammunition, *Polystichum munitum* (a fern).

Gr. *obrimos*, strong, mighty: *Obrimus bufo* (a phasmid).

L. *ops*, might, power, resources; see **wealth**

Gr. *pegos*, strong, solid: *Pegohylemyia spinosa* (a fly).

L. *potens, -entis*, powerful, < *possum, posse, potui*, be able; *potestativus*, denoting power; *altipotens*, mighty, high-powered: potent, potential, potency, potentate, power, puissant, possibility, impossible, impotent, omnipotent, *Potentilla tridentata* (a cinquefoil).

L. *potis*, able, capable; *potior*, abler, better, stronger, preferable; *potissimus*, ablest, strongest, chief; *compos, -otis*, having the mastery or control: *non compos mentis*.

L. *queo, quitus*, be able, can:

Gr. *rhome*, f. bodily strength, force, might; *rhomaleos*, ablebodied, strong: *Rhomaleosaurus cramptoni* (a Jurassic reptile).

Gr. *rhoros*, strong; *rhosis*, f. strength; *rhoster, -os*, m. strengthener; *rhostikos*, strengthening, < *rhonnymi*, strengthen, confirm; *eurhostos*, stout, strong: *Rhorus mesoxanthus* (a wasp), *Eurosta solidaginis* (a gall-fly), *Eurostopus guttatus* (a bird).

L. *robustus*, hard and strong like oak; *corroboro, -atus*, strengthen, confirm; robust, corroborate, *Cryptodrilus robustus* (an oligochaete), *Grevillea robusta* (silk-oak tree), *Cyathophyllum robustum* (a fossil coral), *Acanthiza robustirostris* (a thornbill). See *robur* under **oak**

Gr. *sokos*, stout, strong:

Gr. *sphodros*, strong, violent: *Sphodroschema crampeli* (a beetle), *Sphodrolestes vittaticollis* (a bug).

AS. *stearc*, stiff, strong; see **stiff**

Gr. *steriphos*, firm, solid; see *stereos* under **thick**

Gr. *sthenos*, n. strength; *sthenaros*, strong, mighty: *agasthenes*, very strong, powerful: sthenic, sthenochire, asthenia, hypersthene, neurasthenia, Demosthenes, Callisthenes, *Sthenocephalus indicus* (an ophiuroid), *Sthenarosaurus dawkinsi* (a fossil reptile), *Sthenelais picta* (a worm), *Peristhenes adustus* (a beetle).

Gr. *stibaros*, strong, sturdy: *Stibarobdella superba* (a leech).

Gr. *stiphros*, firm, solid, stout: *Stiphromyrmex robustus* (an ant).

Gr. *strenes*, strong, hard, harsh; see **rough**

Gr. *striphnos*, firm, hard, solid; see **hard**

L. *torosus*, brawny, muscular, bulging; see *torus* under **projection**

L. *validus*, strong, powerful, sound, < *valeo, -itus*, be strong, have strength and worth; *valens, -entis*, strong, powerful; *valentulus*, dim.; *valentia*, f. strength, vigor: valid, valiant, valence, value, valor, Valentine, valedictory, available, *ad valorem*, convalescence, equivalent, invalid, invalidate, invaluable, prevailing, *Valeriana supina* (Austrian valerian), *Valerianella olitoria* (corn-salad), *Dichelonyx valida* (a beetle), *Scirpus validus* (a bulrush), valeric, pivalic.

L. *vehemens, -entis*, eager, ardent, vigorous; see **busy**

L. *viriosus*, robust, strong, < *vis*, f. force, strength: virial, *vis vitae*, vim, violent, violate.

See: **health, make, guard, large, hard, oak, know, bold, govern, honor, stand, pillar**

strongylos, Gr. round, rounded; see **circle**

strontium < Strontian, Scotland; see ELEMENTS under **thing**

strophe, Gr. a turning point; *stropheus*, a vertebra, socket; see *strephos* under **turn**

strophinx, Gr. axle, pivot; see **axis**

strophis, Gr. a twisting person, rogue; see **rogue**

strophos, Gr. a twisting, a twisted cord; *strophion,* dim.; see *streptos* under **turn, rope**

stroppus, L. strap, thong; see **strap**

stroter, Gr. crossbeam; *stroterion,* dim.; see **beam**

strotos, Gr. spread, laid; *strotes,* spreader; see **spread**

structilis, L. of building, < *struo, structus,* build, arrange; see **build**

strues; struix, L. heap, pile; see **heap**

struggle < uncertain origin; see **fight, make**

struma, L., scrofulous tumor; *strumella,* dim.; *strumosus,* scrofulous; see **swell**

struthio, L. (Gr. *strouthion*), ostrich; see **ostrich**

strutho- < Gr. *strouthos,* sparrow; *strouthion; strouthis,* dim.; see **finch**

strychnos, Gr. nightshade; now a genus of the logania family; see **logania**

stryphnos, Gr. astringent; see **bind**

stubborn < uncertain origin. Derive: bigot.
  Gr. *anelatos,* not malleable or ductile, stubborn; see **hard**
  Gr. *authades; authadikos,* stubborn, willful: *Anthades dominus* (a beetle).
  L. *cervicatus; cervicosus,* stiffnecked, stubborn, obstinate; see *cervix* under **neck**
  L. *contumax, -acis,* defiant, insolent, obstinate, stubborn: contumacy, contumacious.
  Gr. *dyslytos,* indissoluble:
  Gr. *dyspeistos; dyspeithes,* hard to persuade, obstinate:
  Gr. *ischyros,* strong, obstinate, stubborn; see *ischys* under **strong**
  L. *obduro, -atus,* harden, hold out; see *durus* under **hard**
  L. *obstinatus,* firm, stubborn: obstinate, obstinacy.
  L. *pertinax, -acis,* firm, obstinate, persistent: pertinacious, *Acerastes pertinax* (a wasp).
  L. *pervicax, -acis; pervicus,* firm, stubborn, obstinate:
  L. *refractarius,* obstinate, stubborn; *refractariolus,* dim.: refractory.
  See: **stand, against, hard, slow**

study < OF. *estudier,* < L. *studeo,* be diligent, eager, zealous; *studiosus,* eager, assiduous; see **busy, think, try, learn**

stuff < OF. *estoffer,* cram, see **press**; material, see **thing**

stultus, L. foolish; see **fool**

stumble < uncertain origin; see **falter**

stump < D. *stomp;* see **stem, base, short, projection**

stun < OF. *estoner;* see **dull, strike**

stupa, L. (Gr. *styphe*), coarse fiber of flax or hemp, tow, oakum; see **flax**

stupendus, L. astonishing, wonderful, large; see **wonder**

stupidus, L. senseless, dull; *stupor,* dullness, insensibility, numbness; see **dull**

stupro, -atus, L. ravish, debauch, defile; *stuprum,* adultery, illicit intercourse; see **coitus**

sturgeon < OF. *esturgeon,* < ML. *sturio* (OHG. *sturjo*).
  L. *acipenser, -is* (Gr. *akkipesios*), m. sturgeon: *Acipenser sturio* (sturgeon), *Acipenser rubicundus* (lake-sturgeon), *Podothecus acipenserinus* (a fish).
  Gr. *antakaios,* m. a kind of sturgeon: *Antaceus guldenstadti* (a fish).
  L. *attilus,* m. a large fish in the Po River, probably a sturgeon: *Attilus cirrhosus* (a fish).
  Gr. *elops; ellops, -opos,* m. a large fish, probably a sturgeon: *Elops saurus* (chiro).
  ML., OHG. *huso,* m. sturgeon: *Acipenser huso* (beluga), isinglass.
  Gr. *ichthyokolla,* f. a kind of sturgeon from which glue is made:
  Gr. *saperdis,* f. a kind of sturgeon:
  Russ. *sterlyadi,* a small sturgeon: sterlet (*Acipenser ruthenus*), *Averruncus sterletus* (a fish).

sturnus, L. starling; *sturninus,* of a starling, speckled; see **starling**

stutter < uncertain origin; see **falter**

sty < AS. *sti; stig;* see **pen**

stygetos; stygeros; stygnos; styganos, Gr. hated, hateful; see *stygeo* under **hate**

stylos, Gr. pillar, column; *stylarion; stylidion; stylis; styliskos,* dim.; see **pillar**

styma, Gr. erection, stiffening; see *styo* under **upright**

stymma, Gr. an astringent; see *stypho* under **lessen**

stymphalis, Gr. man-eating bird in Arcadia; see **bird**

styphelos; styphlos, Gr. hard, harsh, rough, sour; see **rough**

styptico- < Gr. *styptikos,* astringent; see *stypho* under **lessen**

stypos, Gr. stem, stump; see **stem**

styrax, L., Gr. a tree yielding storax, see **resin;** spike at the butt end of a spear, see **point**

stytico- < Gr. *stytikos,* causing erection, priapism; see *styo* under **upright**

Styx, L., Gr. a river in the nether world; *Stygius,* of the Styx; see **stream**

suadeo, *suasus,* L. advise, impel; see **teach**

suavis, L. sweet; *suavior,* sweeter; *suavissimus,* sweetest; see **sweet**

sub- < L. *sub,* under, from, somewhat, less than; see **under**

subatus, L. oestrual; see *subo* under **lewd**

subditivus, L. spurious, counterfeit; see **lie**

subdolus, L. somewhat crafty, sly; see **know**

subdue < OF. *soduire,* < L. *subdo, -itus,* put under; see **victory**

suber, L. cork-oak, cork; *suberinus,* of cork; *suberosus,* corky; see **bark**

subex, L. basal layer, substratum, support, underlayer; *subicula,* dim.; see **base**

subidus, L. knowing, sensible; see **know**

subis, L. an unidentified bird; see **bird**

subitaneus; subitus, L. sudden; see **swift**

subjectus, L. brought under; see **servant**

sublatus, L. raised aloft, proud, haughty; see *tollo* under **raise**

sublestus, L. slight, trifling, trivial; see **trifle**

sublica, L. pile, palisade; *sublicius,* resting on piles; see **pillar**

subligaculum, L. apron; see **garment**

sublimis, L. uplifted, exalted, lofty, distinguished; see **high**

subnuba, L. rival; see **enemy**

subscus, L. tongue or tenon of a dovetail; see **tongue**

subsellium, L. bench, seat; see *sella* under **sit**

subsessa, L. ambush; see **trap**

subsicivus, L. that remains over and above, extra, odd, spare; see **and**

subsidiarius, L. belonging to a reserve; see **and**

substantia, L. the material of which a thing consists; see **thing**

substitute < L. *substitutus,* replaced; see **equal**

subter, L. below, beneath; see *sub* under **under**

subtilis, L. thin, fine, slender, sly, acute; *subtilitas,* keenness; see **know**

subucula, L. shirt; see **garment**

subula, L. awl; *subulatus,* awl-shaped, pointed; see **awl**

subulcus, L. swineherd; see **guard**

subulo, L. flute-player; see **music**

suc- < L. *sub,* under; see **under**

succedaneus, L. following after, substitute; see *sequor* under **follow**

success < L. *successus,* happy issue, good result; see **victory**

succidia, L. flitch or side of bacon; see **flesh**

succiduus, L. sinking, failing, faltering; see **lessen**

succinctus, L. short, concise; see **short**

succinum, L. amber; see **amber**

succuba, L. female demon supposed to prostitute men in their sleep; see *incubus* under **spirit**

succus, L. juice, sap; *succulentus; succidus,* juicy; see **juice**

succussus, L. tossed up, flung, shaken; see *concutio* under **strike**

sucerda, L. swine dung; see **dung**

such < AS. *swilc;* see **equal**

sucho- < Gr. *souchos,* Egyptian name for a crocodile; see **lizard**

suck < AS. *sucan.*
    Gr. *daptes,* m.; *daptria,* f. eater of blood by sucking, as a gnat, bloodsucker; see **eat**
    L. *fello, -atus,* suck; *fellator, -is,* m.; *fellatrix, -icis,* f. sucker; *fellebris,* sucking:

L. *haurio, haustus,* draw out, drain, suck; *haustor, is,* m. drawer; *haustrum,* n. a machine for drawing water, pump; *haustus, -us,* m. drink, draught: haurient, haustorial, haustorium, haustrum, haustellum, haustellate, exhaust, inexhaustible.

L. *irrumo, -atus,* give suck to; *surrumus,* sucking:

Gr. *lapto,* lap with the tongue, drink, suck; *haimolaptis,* blood-sucking; leech:

Gr. *myzo,* suck; *myzouris, -idos,* f. sucker: myzostomous, *Myzodendron brachystachyum* (a myzodendracead), *Agromyza pusilla* (a leaf-miner), *Petromyzon marinus* (lamprey).

Gr. *rhophetos,* that can be sucked up or absorbed:

L. *sorbeo,* suck in, sip; *sorbilis,* that may be sucked or sipped; *sorbitio, -onis,* f. sip, drink; *sorbitiuncula,* f. dim.: absorb, absorption, absorbent, absorbefacient, *Sanguisorba officinalis* (burnet).

L. *sugo, suctus,* suck; *exsuctus,* sucked out, dried up: suction, suctorial, sanguisuge, *Mellisuga minima* (a hummingbird), *Ligyrocoris exsuctus* (a fossil bug).

Gr. *thelazo,* suckle, nurse; *thelasmos,* m. a giving suck; *thelaminos,* m. a suckling; *thelastria,* f. nurse: *Thelazomenus poecilocerus* (a bird).

See: **draw, drink, sponge, leech, milk**

sucula, L. piglet, dim. of *sus,* hog, see **hog**; winch, windlass, see **turn**

sudatorius, L. of sweating, < *sudo, -atus,* sweat, see **drop**; *sudarium,* handkerchief, see **napkin**

sudden < OF. *sudain,* < L. *subitaneus;* see **swift**

sudis, L. stake, pile, a fish; see **pillar**

sudor, -is, L. sweat; *sudorus,* sweating; see *sudo* under **drop**

sudus, L. without moisture, dry, cloudless, bright, clear; see **clear**

suetus, L. customary, habitual, usual; see **use**

suf- < L. *sub,* under; see **under**

suffer < OF. *sufrir,* < L. *suffero,* endure, bear; see **pain, disease, let, carry**

suffitus, L. a fumigation, perfuming; see *suffio* under **smell**

sufflamen, L. brake, clog, check, impediment; see **hold**

suffoco, -atus, L. choke, stifle, strangle; see **choke**

suffragium, L. ballot; see **vote**

suffultus, L. propped up; see *fulcimen* under **pillar**

sug- < L. *sub,* under; see **under**

sugar < F. *sucre,* < Ar. *sukkar:* sucrose, sucaryl.

L. *saccharum* (Gr. *sakcharon*), n. sugar: saccharine, saccharate, saccharic, saccharide, sacchariferous, saccharimeter, saccharose, saccharoid, *Saccharum officinarum* (sugarcane), *Saccharomyces ellipsoideus* (wine-yeast), *Acer saccharum* (sugar-maple), *Lepisma saccharina* (silver-fish), *Beta saccharifera* (sugar-beet), *Glyciphagus sacchari* (sugar-mite), *Diatraea saccharalis* (sugarcane-borer).

See: **sweet, -ose, honey**

suggestus, L. elevation, height, platform; see **high**

suggrunda, L. eaves of a house; see **border**

sugillo, -atus, L. beat black and blue; *sugillatum,* black and blue spot, bruise; see **bruise**

sugo, suctus, L. suck; see **suck**

suillus; suinus, L. of swine; see *sus* under **hog**

suitable < OF. *siute;* see **fit**

sulcus, L. furrow, groove; *sulculus,* dim.; see **furrow**

sulfur < L. *sulfur, -is,* n. brimstone; *sulfureus,* of sulfur, sulfury: sulfuric, sulfate, sulfide, sulfurous, sulfuretted, sulfonate, sulfadiazine, sulfonaphthol, phthalylsulfacetamide, *Polyporus sulfureus* (a fungus), *Frankliniella sulfuripes* (a bug), *Eremobates sulfurea* (a solpugid).   See ELEMENTS under **thing**

L. *egula,* f. a kind of sulfur:

Gr. *sory,* n. probably an ore of sulfur:

Gr. *theion; theaphion,* n. brimstone, sulfur; *theiodes,* sulfurlike: thio acid, thio ether, thiosulfate, thiophene, thionyl, thiocarbonyl, thiol, thiourea, thiocyanate, thiamin, thiadiazole, sulfathiazole, dithiopropanol, *Thiopsyche pryeri* (a moth), *Thiothrix nivea* (a bacterium).

sum- < L. *sub,* under; see **under**

sumac < Ar. *summaq.*

NL. *anacardium,* n. a genus of the sumac family: Anacardiaceae, *Anacardium excelsum* (espave), *Semecarpus anacardium* (a marking-nut), *Cupania anacardioides* (a sapindacead).

NL. *astronium*, n. a genus of the sumac family: *Astronium fraxinifolium* (a star-tree).

L. *coccygia* (Gr. *kokkygia*), f. a kind of sumac; *kokkyginos*, purple-red:

L. *cotinus* (Gr. *kotinos*), m. an olivelike shrub; a genus of the sumac family; *Cotinus coggygria* (European smoke-tree), *Ficus cotinifolia* (a fig), *Viburnum cotinifolium* (a viburnum).

L. *lentiscus*, f. mastic-tree: lentiscine, *Pistacia lentiscus* (source of mastic).

Malay *manga*, a fruit from a tree belonging to the sumac family: mango *(Mangifera indica)*.

L. *pistacia* (Gr. *pistake*), f. pistachio-tree: *Pistacia vera* (common pistachio), fustic *(Chlorophora tinctoria)*.

L. *rhus, rhois* (Gr. *rhous*), f. sumac: *Rhus trilobata* (skunk-bush sumac), *Pterocarya rhoifolia* (a wingnut), *Blepharida rhois* (a beetle).

L. *schinus* (Gr. *schinos*), f. mastic-tree: *Schinus latifolia* (Chile pepper-tree), *Euroschinus foliata* (an anacardiacead), *Xanthoxylum schinifolium* (a prickly-ash).

Gr. *spodias; spondias, -ados*, f. a tree belonging to the sumac family: *Spondias pinnata* (Andaman mombin).

sumen, L. breast, sow's udder; see **udder**

summa, L. the entire quantity, total; see **all**

summary < L. *summarium*, abstract, epitome; see **short**

summer < AS. *sumer;* see **year**

summissum, L. substitute; see **equal**

summus, L. highest, superlative of *superus*, high; see *super* under **over**

sumptus, L. cost; *sumptiosus*, costly; see **pay,** *sumo* under **take**

sun < AS. *sunne:* sunny, sunbeam, sunshine, sunlight, sunrise, sunset, sunburn, sunfish, sunflower, Sunday.

L. *apricus*, exposed to the sun, sunny; *apricum*, a sunny spot; see **open**

Gr. *elektor, -os*, m. the beaming sun:

Gr. *helios*, m. sun; *heliakos*, of the sun; *Helias, -ados*, f. a daughter of Helios who wept tears of amber on being changed into a poplar-tree; *heliotis, -idos*, f. a name for the moon: helium, heliocentric, heliograph, helioscope, heliostat, heliolithic, heliotrope, heliophilous, heliophobia, aphelion, perihelion, *Helianthus tomentosus* (woolly sunflower), *Helianthemum pilosum* (a cistacead), *Heliopsis scabra* (a composite), *Heliodrymus viminalis* (a radiolarian), *Heliornis fulica* (finfoot), *Helipterum roseum* (an everlasting), *Euphorbia helioscopia* (sun-spurge), *Cryptohelia pudica* (a coral), *Lophohelia prolifera* (a coral), *Heliothis armigera* (corn-ear worm, formerly *Chloridea obsoleta*).

L. *oriens, -entis*, the rising sun, east; see **east**

L. *sol, -is*, m. sun; *solaris*, of the sun; *solatus*, sunburned, sunstruck; *insolatus*, exposed to the sun: solar, solstice, solarium, parasol, insolation, girasol (Jerusalem artichoke, *Helianthus tuberosus*), *Astrophacus solaris* (a radiolarian), *Phalaena solata* (a moth).

sup- < L. *sub*, under; see **under**

supellex, L. household goods, furniture, utensils; see **tool**

super, L. over, above; *supernus; superus,* above, over, high; *superior,* higher; *supremus; summus,* highest, latest; see **over**

superbus, L. excellent, superior, splendid; see **good**

supercilium, L. eyebrow; see **face**

superficialis, L. on the surface, shallow; see **shallow**

superior, L. higher; see *super* under **over**

superlativus, L. in the highest or utmost degree; see **very,** COMPARISON OF ADJECTIVES

supernus, L. on high, celestial; see *super* under **over**

superstes, L. bystander, witness; see **witness**

superstition < L. *superstitio*, unreasoning belief; see **believe, rite, spirit, magic, prophecy**

supinus, L. lying on the back, outstretched; see **lie**

supo, L. throw; see **throw**

supparum, L. topsail; see **sail**

suppedito, *-atus,* L. have in abundance, supply abundantly; see **abundance**

suppes, L. with feet turned under, with twisted feet; see *pes* under **foot**

suppetia, L. aid, help; see **help**

supplementum, L. addition, completion; see **and**

supplico, *-atus,* L. beseech, beg humbly; *supplex, -icis,* humble; see **ask**

supply < OF. *supplier,* < L. *suppleo, -etus,* fill up, furnish: suppletory, supplement.
 Gr. *biarkes,* m. provisioner, supplier:
 Gr. *ephodion,* n. provisions for traveling:
 Gr. *kateres,* furnished, fitted:
 Gr. *komide,* f. care, provision, carriage, recovery:
 Gr. *korysso,* equip, arm: *Coryssocnemis uncata* (a spider).
 F. *loger,* transport, quarter, and supply troops: logistics.
 Gr. *paraskeuo,* prepare; *paraskeue,* f. preparation; *parasyeuastes,* m. provider,
  preparator:
 L. *paro, -atus,* prepare, make ready, furnish, equip, fit out, procure: pare, parry,
  parapet, parade, parachute, parasol, prepare, preparation, apparatus, emperor,
  empire, imperative, imperial, repair, reparation, rampart, dissever, irreparable,
  sever, several, separate, unprepared, vituperation.
 L. *parocha* (Gr. *paroche*), f. a supplying or furnishing; *parochus* (Gr. *parochos*),
  m. purveyor:
 Gr. *poristos,* provided; *poristes,* m. provider; *porismos,* m. a getting, procuring,
  < *porizo,* supply, provide: *Porismus strigatus* (a neuropterid).
 Gr. *stello,* prepare, equip, make ready, dispatch; see **send**
 Gr. *stolizo,* make ready, trim, equip, adorn; *stolismos,* m. an equipping, dress,
  equipment:
 L. *viaticum,* n. provisions for traveling:
 See: **fit, make**

suppose < OF. *supposer,* < L. *suppono, -positus,* put under, substitute; see **think**

supposititius, L. false, counterfeit; see **lie**

suppuro, *-atus,* L. form pus; *suppuratio,* abscess, gathering; see **pus**

supra, L. above, over; see *super* under **over**

supremus, L. highest; see *super* under **over**

sur- < L. *sub,* under, see **under;** < *super,* over, see **over**

sura, L. calf of the leg; see **leg**

-sura; -tura, L. suffixes denoting result of action; see **make**

surculus, L. young branch, twig, shoot, sprig; *surculosus,* branchy; see **branch**

surdus, L. deaf, silent, mute, faint; see **deaf**

sure < OF. *sur,* < L. *securus,* free from care, anxiety, and danger, safe: secure,
  security, insecure, surely, surety.
 Gr. *adiaptotos,* not likely to err, faultless, infallible; *anekdromos,* inevitable:
 L. *appromissor, -is,* m. one who goes bail for another:
 Gr. *arkios,* sure, certain, enough:
 Gr. *asphalistos,* made secure, safe; see *asphales* under **stand**
 Gr. *athestos,* inexorable: *Athestia elongata* (a bug).
 Gr. *atrekes,* certain, sure, true:
 Gr. *bebaios,* firm, steadfast, constant, sure; see **stand**
 L. *certus,* fixed, settled, definite, sure: certain, certificate, certify, certitude, as-
  certain, concert, disconcert, uncertainty.
 Gr. *diengya,* f. bail, surety:
 L. *fidelis; fidus,* trustworthy, faithful, sure, true; see *fido* under **believe**
 L. *indubitatus,* certain, sure; *indubius,* certain; *ineffugibilis,* unavoidable, in-
  evitable; *ineluctabilis,* unavoidable, inevitable; *inerrabilis,* unerring; *inevitabilis,*
  certain to happen; *inexpedibilis,* unavoidable, inevitable; *infallibilis,* not making
  mistakes, inerrant: indubitable, indubitably, ineluctable, inevitable, infallible,
  infallibility.
 Gr. *nemertes,* infallible, certain; a nereid: nemertine, Nemertinea, *Paranemertes
  peregrina* (a nemertean).
 L. *positivus,* settled, definite, explicit, certain: positive, positively.
 L. *quippe,* certainly, surely, forsooth: quip, quipful.
 L. *ratus,* calculated, fixed, settled, certain; see *reor* under **think**
 Gr. *tekmerion,* n. sure sign, positive proof: *Tecmerium anthophagum* (a moth).
 L. *vador, -atus,* bind by requiring bail or surety; see **bind**
 See: **safe, guard, true, right, stand, strong**

surema, L. a kind of fish; see **fish**

surface < L. *superfacies,* upper side, outside; see **out, area, face**

surgeon < Gr. *cheirourgos,* working with the hand; an operating physician; see
  heal

surgo, *surrectus,* L. rise; see **raise**

surnia, NL. a kind of owl; see **owl**

surprise < OF. *surpris,* < L. *super,* above, *prehendo,* grasp; see **lot, strange**

surreptitius, L. concealed, clandestine; see **secret**

surrogatus, L. substituted; see *surrogo* under **equal**

surus, L. branch; see *surculus* under **branch**

sus, L. pig, hog; *suilla,* pork; *suillus; suinus,* of swine, swinish; see **hog, flesh**

sus- < L. *sub,* under; see **under**

suscito, L. raise, arouse, stir up; see *cieo* under **arouse**

susinus, L. of lilies; see *susum* under **lily**

suspect < L. *suspicio, -ectus,* regard with mistrust; *suspicax, -acis,* distrustful; see **doubt**

sustain < OF. *sustenir,* < L. *sustineo, -entus,* uphold, support; see **help**

susum, L. (Gr. *souson*), lily; see **lily**

susurro, -atus, L. whisper, murmur, buzz; *susurratio,* a whispering; see **murmur**

sutor, L. sewer, cobbler; *sutorius,* of cobbling; see *suo* under **sew**

sutura, L. seam; *sutilis,* sewed together; see *suo* under **sew**

swallow < AS. *swalewe.*
  L. *apus, apodis* (Gr. *apous, apodos*), m. a kind of swallow, swift: *Motacilla apus* (a wagtail).
  Gr. *chelidon, -os,* f. swallow; *chelidonios,* of the swallow: *chelidonion,* n. swallow-wort: chelidonian, celandine, *Chelidon urbica* (house-martin), *Petrochelidon albifrons* (cliff-swallow).
  Gr. *drepanis, -idos,* f. a bird, perhaps the swift: *Drepanis pacifica* (mamo).
  Sp. *golondrina,* f. swallow: *La Golondrina.*
  L. *hirundo, -inis,* f. swallow; *hirundineus; hirundininus,* of swallows: hirundine, *Hirundo erythrogaster* (barn-swallow), *Sterna hirundo* (common tern), *Dicaeum hirundinaceum* (a flowerpecker).
  Gr. *kypselos,* m. swift, swallow: cypseline, *Cypselurus furcatus* (a flying-fish), *Cypseloides niger* (black swift), *Psalidoprocne cypselina* (a bird).
  L. *procne (progne),* f. swallow, < Gr. *Prokne,* fabled woman changed into a swallow: *Progne subis* (purple martin), *Liospira progne* (a fossil gastropod), *Iridoprocne bicolor* (tree-swallow).

swallow < AS. *swelgan,* gulp, eat, drink, take in; see **eat**

swamp < uncertain origin; see **marsh**

swan < AS. *swan;* see **goose**

sward < AS. *sweard;* see **skin, sod**

swarm < AS. *swearm;* see **number**

sway < ON. *sveigja,* bend, swing; see **swing, wave, bend, turn, govern**

swear < AS. *swerian,* affirm, declare or promise solemnly, curse: swearer, swear-word, forswear, answer.
  L. *assevero, -atus,* assert firmly, declare positively: asseverate, asseveration.
  AS. *ath,* an affirmation by appeal to deity, revered persons or things, curse: oath.
  Gr. *horkos,* m.; *horkion,* n. oath; *horkios,* of an oath; *horkotos,* bound by an oath, *horkotes,* m. officer who administered the oath; *enorkos; epiorkos; exorkos,* sworn falsely, perjured; *horkizo,* swear an oath, abjure: exorcise, *Horcotes quadricostatus* (a spider).
  L. *juro, -atus,* take an oath, swear; *juramentum; jurandum,* n.; *juratio, -onis,* f. oath; *jurativus,* of an oath; *jurator, -is,* m. swearer; *adjuro,* swear to; *conjuro,* swear together; *expejuro,* swear falsely: jury, adjuration, conjure, perjury.
  Gr. *omnymi,* swear, affirm by oath; *apomosia; exomosia,* f. denial under oath: enomoty, enomotarch.
  See: **curse, promise, yes**

sweat < AS. *swat;* see **drop**

sweep < AS. *sweop.*
  Gr. *kallyno,* sweep clean, beautify; see *kallyntron* under **broom**
  Gr. *koreo,* sweep out; see *korema* under **broom**
  Gr. *saroo,* sweep; *saron* under **broom**
  L. *scopa,* twigs, broom; *scoparius,* sweeper; see **broom**
  Gr. *syrtos,* swept or carried along by a stream; *syro,* sweep, drag, drift, or trail along:
  L. *verro, versus,* sweep, brush, scour, scrape; *converro,* sweep along, clear away; *converritor, -is,* m. sweeper; *averro,* sweep out: *Deverra,* f. goddess of sweeping: *Deverra tortuosa* (an umbellifer).
  See: **broom, carry, scrape, clear**

sweet < AS. *swete:* sweetly, sweeten, sweetness, sweetheart, sweetmeat, sweetbrier, sweetbread, sweetroot, sweetsop, meadowsweet, bittersweet.
L. *dulcis, -e,* sweet; *dulciculus,* dim.; *dulcedo, -inis; dulcitas,* f. sweetness; *dulcacidus,* sour-sweet; *dulcator, -is,* m. sweetener; *dulciarius,* m. confectioner, pastry-cook; *dulcoratus,* sweetened; *dulcifer,* bringing sweetness; *indulco, -atus,* sweeten: dulcet, dulcifluous, dulcimer, dulcitol, dulcamara, dulcify, Dulcinea, *Rhinichthys dulcis* (black-nosed dace), *Synsepalum dulcificum* (a sapotacead).
Gr. *glykys; glykeros,* sweet; *glykasma; glykysma, -tos,* n.; *glykismos,* m.; *glykytes, -etos,* f. sweetness; *glykantikos,* of sweetening; *glykypikros,* sweetly bitter: glycerine, glycogen, glycerol, glycosuria, glycine, glycocholic, glyceride, *Glycobius speciosus* (maple-borer), *Glycyrrhiza lepidota* (American licorice), *Glyceria pallida* (a manna-grass), *Glycine gracilis* (a bean), *Smilax glyciphylla* (green-brier).
Gr. *hedys; hedymos,* sweet, pleasant, dear; *hedion,* sweeter; *hedistos,* sweetest; *hedysma, -tos,* n.; *hedysmos; hedytes,* m. sweetness; *hedynter, -os,* m. sweetener; *hedyntos,* sweetened: hedyphane, *Hedysarum coronarium* (a vetch), *Hedychium flavum* (yellow ginger-lily), *Hedycarya arborea* (pigeon-wood), *Hedyosmum scabrum* (a chloranthacead).
Skt. *madhu,* sweet, honey; see **honey**
L. *mellitus,* sweetened with honey; see *mel* under **honey**
L. *mulsus,* mixed or sweetened with honey; see **honey**
L. *suavis, -e,* sweet; *suaveolens, -entis,* sweet-smelling; *suavidicus; suaviloquens, -entis,* sweet-spoken: suave, suavity, assuage, *Trichosma suavis* (an orchid), *Ocimum suave* (a basil), *Melilotus suaveolens* (a sweet-clover), *Trigonella suavissima* (a legume).
See: **sugar, honey, -ose, smooth**
swell < AS. *swellan.*
L. *albucus,* m. bulb of the asphodel:
L. *bova,* f. a swelling of the legs:
Gr. *bryo,* swell, teem: embryo, bryophyte, *Bryophyllum pinnatum* (a houseleek), *Bryonia dioica* (a bryony).
L. *bubo, -onis,* m. a swelling in the groin, < Gr. *boubon, -os,* m. groin: bubo (buboes), bubonic, bubonocele.
L. *bulbus* (Gr. *bolbos*), m. a fleshy, usually underground, stem or bud; *bulbillus; bulbulus; bolbiskos,* m.; *bolbarion; bolbidion; bolbion,* n. dim.; *bulbosus,* swollen; *bolbodes,* bulblike: bulb, bulbil, bulbilla, bulbous, bulbiferous, bulbodium, *Bulbophyllum comosum* (an orchid), *Bolboceras mobilicorne* (a beetle), *Ranunculus bulbosus* (a buttercup), *Cystopteris bulbifera* (a bladder-fern), *Sclerotinia bulborum* (a fungus), *Narcissus bulbocodium* (petticoat-daffodil).
L. *bulla,* bubble, knob, boss, stud; see **bubble**
L. *condylus* (Gr. *kondylos*), knob, knuckle, enlarged end of a bone; see **projection**
L. *dilato, -atus,* spread out, enlarge; see *differo* under **spread**
L. *distendo, -entus (-ensus),* swell out, extend: distend, distention.
Gr. *ektyphos,* puffed up: *Ectyphus pinguis* (a fly).
Gr. *emphysema,* something inflated, swollen; see **bubble**
Gr. *epartikos,* causing to rise or swell; *eparma, -tos,* n. a swelling: *Eparmatostethus madecassus* (a wasp).
L. *galla,* a pathologic swelling or excrescence on plants; see **gall**
L. *gambosus,* having a swelling near the hoof:
Gr. *ganglion,* a swelling, a knot or plexus of nerves; see **nerve**
Gr. *gastris, -idos,* potbellied; see *gaster* under **stomach**
L. *gemursa,* f. a swelling between the toes:
L. *gibber,* humped, protuberant, swollen; see **projection**
Gr. *gongros,* m. excrescence, swelling, knot: *Gongrocnemis bivittata* (a katydid), *Gongroneura brevicornis* (a bug).
L. *inflatus,* puffed up, swollen; see *flo* under **blow**
Gr. *kanthyle,* f. swelling, tumor: *Canthyloscelis antennata* (a fly).
Gr. *kele,* f. tumor, hernia: hydrocele, varicocele, lymphocele, celotomy.
Gr. *kordyle,* club, bump, swelling; see **club**
Gr. *kydonios,* swell like a quince, become round and plump; see **quince**
Gr. *kyo (kyeo),* swell, be pregnant; *kyma,* anything swollen, sprout, wave; see **fertile, branch, wave**
L. *nodus,* knot, swelling; see **knot**
L. *obesus,* fat, corpulent; see **fat**
Gr. *oidema, -tos,* n. a swelling, tumor, < *oidea,* swell; *oidesis,* f.; *oidos,* n. a swelling; *oidaleos,* swollen: edema, myxedema, *Oedogonium ciliatum* (an alga), Oedipus, *Ectoedemia populella* (a petiole gall-insect on poplar), *Bryum oediloma* (a moss), *Oedaleonotus enigma* (valley-grasshopper).

Gr. *-oma*, suffix signifying tumor or morbid growth; see **disease**

Gr. *onkos*, m. bulk, mass, weight, tumor; *onkosis*, f. a swelling; *onkeros; onkodes; onkotos*, swollen, rounded, bulky: oncometer, oncology, oncograph, oncosimeter, oncosphere, oncotomy, deradenoncus.

Gr. *orgao*, swell; see *orgasmos* under **coitus**

L. *pendigo, -inis,* f. internal tumor:

Gr. *phyma, -tos,* n. tumor, growth; *phymation,* n. dim.; *phymatodes,* tumorlike: phymatic, phymatoid, phymatosis, phymatorhysin, *Phymatothynnus monilicornis* (a wasp), *Phymatopsis nigrirostris* (a tipulid), *Chonophyma perforatum* (a fossil sponge), *Taxophyma lyonsi* (a fossil echinoid), *Phymatodes vulgaris* (a fern).

Gr. *physao*, puff, distend, inflate; see *physa* under **blow**

Gr. *pneumatikos,* inflated; see *pneuma* under **breathe**

Gr. *pretho*, swell, blow up, inflate: *Buprestis gigantea* (a beetle).

L. *scaurus* (Gr. *skauros*), having large and swollen ankles: *Scaurus atratus* (a beetle).

Gr. *spargosis*, f. distention, swelling, < *spargao*, swell to ripeness:

L. *struma*, f. any glandular swelling, a scrofulous tumor; *strumella,* f. dim.; *strumosus,* scrofulous: struma, strumatic, strumose, strumitis, *Strumella coryneoidea* (a fungus), *Horatopyga strumifera* (a beetle), *Ocotea strumosa* (an octea), *Xanthium strumarium* (a cocklebur).

Gr. *thrombos*, lump, blood-clot; see **lump**

L. *torus*, round elevation, protuberance, bulge; see **projection**

L. *tuber, -is,* n. swelling, lump, bulb; *tuberculum,* n. dim.; *tuberosus,* full of lumps: tuber, tubercle, protuberant, tubercular, tuberculosis, tuberous, tuberose (*Polianthes tuberosa*), truffle (*Tuber melanosporum*), *Tubercularia fici* (a fungus), *Iguana tuberculata* (a lizard), *Oxalis tuberosa* (a wood-sorrel, source of oca), *Ullucus tuberosus* (ulluco), *Tropaeolum tuberosum* (a nasturtium), *Phyllodactylus tuberculosus* (warty gecko), *Diplodia tubericola* (a fungus), *Scirpus etuberculatus* (a sedge).

L. *tumeo*, swell; *tumidus,* swollen; *tumidulus,* dim.; *tumor, -is,* m. a swelling; *tumidosus; tumorosus,* bloated, inflated: tumid, tumor, tumefacient, tumefy, tumidity, tumorous, intumescence, *Trophocrinus tumidus* (a Mississippian crinoid), *Ophthalmidium tumidulum* (a foraminifer), *Phytomonas tumefaciens* (a bacterium), *Carex intumescens* (a sedge).

L. *turgeo*, swell, be inflated; *turgidus,* distended, inflated, swollen; *turgidulus,* dim.; *turgor, -is,* m. a swelling: turgid, turgor, turgidity, turgescence, *Quinqueloculina turgida* (a foraminifer).

Gr. *tyligma, -tos,* weal, swelling; see *tyle* under **knot**

L. *umbo*, boss, rounded protuberance, knob; see **projection**

L. *ventricosus*, potbellied, bulging; see *venter* under **belly**

See: **blow, bubble, bag, bud, lump, knot, rump, fat, womb, projection, belly, gall, wave, spread**

swift < AS. *swift*.   Derive: jiffy.

L. *abruptus*, sudden, hasty: abrupt.

L. *agilis*, quick, light, nimble; see **light**

Gr. *aïke*, f. flight, rapid motion:

Gr. *aiolos*, rapid, shifting, wily; see *Aeolus* under **wind**

Gr. *aiphnidios*, abrupt, sudden, unexpected: *Aephnidiogenes barbarus* (a trematode).

Gr. *aipseros*, quick, sudden, swift: *Aepsera ferruginea* (a beetle).

Gr. *aithykter, -os,* m. darter: *Aethyctera electa* (a moth).

L. *alacer*, quick, lively; see **life**

L. *alipes, -pedis,* wing-footed, fleet, quick, swift; *levipes,* light-footed: aliped.

Gr. *argos*, swift; *chelargos,* with fleet hoofs; *litargos,* running fast; *podargos,* swift-footed: Argonaut, *Podargus cuvieri* (a goatsucker), *Palinurus argus* (a spiny lobster), *Litargosomus maculatus* (a beetle).

Gr. *artipous, -podos,* swift-footed: *Artipus psittacus* (a beetle).

L. *celer, -eris, -e,* swift; *celero, -atus,* quicken, hasten; *celox, -ocis,* quick, swift: celerity, accelerate, accelerator, acceleration, *Scaphula celox* (a pelecypod).

L. *citatus; citus,* quick, rapid, speedy, swift; *citior,* swifter; *citellus,* dim.; *citipes, -pedis,* swift-footed, fleet: *Citellus tridecimlineatus* (a spermophile), *Spermophilus citillus* (suslik).

L. *currax, -acis,* running fast, quick, swift:

Gr. *dieros*, active, nimble: *Dierobia juncorum* (a beetle).

Gr. *dromikos*, swift; see *dromos* under **run**

Gr. *elaphros*, light, nimble; see **light**

Gr. *epeixis*, f. haste, hurry:

Gr. *essymenos*, hurrying, eager:

L. *festino, -atus,* move rapidly, hasten, hurry; *festinus,* hasty: *Stictocephala festina* (a leaf-hopper).

L. *fugax, -acis,* flying swiftly, fleet, fleeting, transitory: fugacious, *Melica fugax* (little onion-grass).

L. *improvisus,* unforeseen, unexpected, sudden; see **lot**

Gr. *itamos,* hasty, headlong, impetuous, reckless: *Itamus castaneus* (a beetle).

Gr. *karpalimos,* swift: *Carpalimus arcuatus* (a beetle).

Gr. *keles, -etos,* racer, courser; see **ride**

Gr. *kraipnos,* swift, rushing:

Gr. *labrosytos,* rushing furiously; see *labros* under **greed**

Gr. *laipseros,* swift, nimble, light: *Laepserus* (a wasp).

L. *necopinus,* unexpected, sudden:

Gr. *okys,* swift, quick, sharp; *okion,* swifter; *okistos,* swiftest; L. *ocior,* swifter; *ocissimus,* swiftest: ocypodian, *Ocypoda arenaria* (a crab), *Ocydromus australis* (weka), *Ocypterosoma politum* (a fly), *Ocythoë tuberculata* (a cephalopod).

Gr. *otreros,* nimble, quick, busy:

L. *pernix, -icis,* swift, nimble, agile, active: *Noctua pernix* (a butterfly).

L. *praeceps,* headlong, hasty; see *praecipitium* under **cliff**

L. *praepes, -etis,* flying swiftly, fleet, quick: *Halictus praepes* (a bee).

It. *presto,* < *praesto,* ready, quick: presto!

L. *promptus,* ready, quick: prompt.

L. *properus,* quick, rapid, hasty; *propero, -atus,* hasten, hurry, quicken, speed: *Aleochara propera* (a beetle).

L. *rapidus,* quick, swift: rapid, rapidity, Grand Rapids (Mich.).

L. *repentinus,* sudden, unexpected; see **lot**

Gr. *rhimpha,* swiftly; *rhimphaleos,* swift, light: *Rhimphoctona fulvipes* (a wasp).

Gr. *rhothios,* rushing, dashing, as of waves and surf; see *rhothos* under **sound**

L. *schedius* (Gr. *schedios*), hastily made: *Schedius kuvanae* (a wasp).

Gr. *skartes,* agile, nimble, < *skairo,* skip, frisk; see **leap**

Gr. *skinax, -akos,* quick, nimble: *Scinacopus cnemargus* (a wasp).

Gr. *sperchnos,* hasty, hurried, rapid, < *spercho,* move rapidly: *Spercheus stasimus* (a beetle).

Gr. *speustikos,* hasty; *speustos,* to be done or pursued eagerly; *kataspeusis,* f. haste; *speudo,* urge on, hasten, quicken, press: *Speudoteloclerus cyanipennis* (a beetle).

L. *spissus,* thick (in rapid succession), fast; see **thick**

L. *strenuus,* brisk, nimble, quick, active: strenuous, *Anax strenuus* (a dragonfly).

L. *subitaneus,* sudden, < *subeo, -itus,* come on stealthily, spring up:

Gr. *tachys,* swift, quick, fast; *tachos,* n. speed; *tachinos,* swift; *tachytes, -etos,* f. swiftness: tachygraph, tachylite, tachometer, tachyphrasia, *Tachysphex mantiraptor* (a wasp), *Tachinus fimbriatus* (a beetle), *Tachinoderus fulvipes* (a beetle), *Tachytes nigra* (a wasp).

Gr. *thoös, thoë, thoön,* quick, nimble, swift, < *thoäzo,* move quickly, hurry; *thoösa,* f. speed; *amphithoäzo,* rush around: *Eurythoë paupera* (a worm), *Actinothoë lacerta* (a coral), *Amphithoë femorata* (an amphipod), *Nicothoë astaci* (a copepod), *Cymothoa ovalis* (a fish-louse), *Hippothoa*(now *Corynotrypa*) *inflata* (a fossil bryozoan), *Baccha amphithoë* (a fly), *Thoatherium crepidatum* (a fossil mammal).

Gr. *thouros,* rushing, impetuous:

Gr. *thyo; thyno,* rush, dart:

L. *torrens, -entis,* rushing, rapid; a swift stream; see **stream**

L. *velox, -ocis,* swift, speedy: velocity, velocipede, velodrome, *Potamogale velox* (giant water-shrew).

L. *volucer,* flying, swift; see *volo* under **fly**

See: **run, fly, make, bold, life, short**

swim < AS. *swimman.*

AS. *flotian,* swim: float, flotsam, afloat, fleet.

L. *fluito, -atus,* float, swim: *Glyceria fluitans* (a manna-grass).

Gr. *kolymbethra,* f. swimming-pool:

L. *nato, -atus; no, natus,* swim, float; *natans; nans,* swimming, floating; *natabilis; natatilis,* that can swim or float; *natator, -is,* m. swimmer; *natabulum; natatorium,* n. swimming-place: natation, natatorium, natatorial, natant, supernatant, *Salvinia natans* (a water-fern), *Oryzomys natator* (a rice-rat).

Gr. *necho,* swim; *nektes,* m.; *nektris, -idos,* f. swimmer; *nechaleos; nektos,* swimming; *necheion,* n. a swimming-place; *nexis,* f. a swimming: necton, nectophore, nectopod, nectrianin, *Nectonema agile* (a worm), *Nectophryne afra* (a toad), *Necturus maculosus* (mud-puppy), *Nectria cinnabarina* (a fungus), *Callinectes hastatus* (a blue crab), *Lacconectes fulvescens* (a beetle), *Eunectes murinus* (anaconda), *Metronectes auboei* (a beetle).

Gr. *neo*, swim, float; *neuster, -os*, m. swimmer; *neustos*, swimming; *neustikos*, able
to swim: neuston, *Neusterophis laevissima* (a snake), *Neusticosaurus pusillus*
(a Triassic reptile).

Gr. *pleo*, swim, sail, float; *pleustes*; *plos, -otos*; *ploter, -os*, m. sailor, swimmer;
plotos, floating, swimming; *plotikos*, skilled in seamanship; *acroploos*, swimming
at the surface: pleopod, Pleiades, *Pleustes tuberculata* (an amphipod), *Plotocnide
borealis* (a jellyfish), *Plotornis deltortrii* (a bird), *Plotobia simplex* (an annelid),
Euplotes patella (a protozoan).

See: **dip, sail**

swine < AS. *swin;* see **hog**

swing < AS. *swingan.*

L. *aestus*, tide, surge; see **heat**

Gr. *aiora*, f. swing, hammock: *Aeora cretacea* (a fossil pelecypod).

L. *oscillo, -atus*, swing; *oscillum*, n. swing: oscillate, oscillator, oscillograph,
Oscillatoria limosa (an alga).

Gr. *palirrhoia*, f. ebb and flow of the tide:

L. *rhythmus* (Gr. *rhythmos*), m. regular recurring motion, measure, proportion:
rhythm, rhythmic, arrhythmia.

Gr. *saulos*, a waddling, swaying, or swaggering gait; see **walk**

L. *vibro, -atus*, shake, oscillate; see **shake**

See: **wave, shake, return, oestrus, strike, whip**

switch < uncertain origin; see **branch, whip**

sword < AS. *sweord.*

L. *acinaces, -is* (Gr. *akinakes*), m. scimitar: acinaciform, acinacifolius, *Mesembryanthemum acinaciforme* (a fig-marigold).

Gr. *aor*, sword, any tool or weapon; see **tool**

L. *copis* (Gr. *kopis*), short sword, kitchen-knife, cleaver; see **knife**

Gr. *drepanon*, curved sword, scimitar; see **sickle**

L. *ensis, -is*, m. sword; *ensiculus*, m. dim.; *ensifer; ensiger*, sword-bearing: ensi-
form, *Ensis directus* (a razor-clam), *Cymbidium ensifolium* (an orchid), *Canavalia ensiformis* (a jack-bean), *Docimastes ensifera* (a hummingbird), *Sisyrinchium ensigerum* (a blue-eyed grass), *Iris ensata* (Russian iris).

L. *gladius*, m. sword; *gladiolus*, m. dim.; *gladiator, -is*, m. swordsman: gladiator,
glaive, *Gladiolus coccineus* (a gladiolus), *Orca gladiator*(*Orcinus orca*) (a killer-
whale), *Lepidosperma gladiatum* (sword-edge).

Gr. *knodon, -ontos*, m. sword: *Cnodostethus seminudus* (a beetle).

L. *machaera* (Gr. *machaira*), f. bent sword, dirk, dagger, knife; *machairidion;
machairion*, n.; *machairis, -idos*, f. dim.: machaerodont, *Machaerodus cultridens*
(a sabertooth), *Machaeromeryx tragulus* (a fossil ruminant), *Makaira*[*Machaera*]
ampla (marlin).

Gr. *phasganon*, n. sword; *phasganion*, n.; *phasganis, -idos*, f. dim.: *Phasganodus
dirus* (a fossil fish), *Phasganocnema melanianthe* (a beetle), *Notacanthus phas-
ganorus* (an eel).

Gr. *rhomphaia*, f. a Thracian sword: *Rhomphaea cometes* (a spider).

Gr. *sagaris, -eos*, f. a kind of sword: *Sagariphora heliochlaena* (a moth).

AS. *seax*, short-sword, knife: Saxon, Essex, Middlesex, Wessex, Sussex, Saxony.

Gr. *skalme*, knife, sword; see **knife**

L. *spatha*, a broad sword without a point; see **spoon**

Gr. *xiphos*, n. sword; *xiphidion; xiphion*, n. dim.; *xiphourgos*, m. sword-cutler:
xiphoid, xiphodont, xiphopagus, Xiphosura, *Xiphorhynchus pusillus* (a bird),
Xiphocera asina (a grasshopper), *Xiphidium fasciatum* (a grasshopper), *Xiphi-
diopicus percussus* (a bird), *Iris xiphium* (Spanish iris), *Argyroxiphium virescens*
(a composite).

See: **knife, dagger, ax, sickle, cut**

swordfish

Gr. *thranis, -idos*, f. swordfish:

Gr. *xiphias*, m. swordfish: *Xiphias gladius* (swordfish).

syaco- < Gr. *syax, -akos*, a fish; *syakion*, dim.; see **fish**

syaena, L. (Gr. *syaina*), sow; see *sus* under **hog**

syagrus, L. a kind of palm; see **palm**

sybaco- < Gr. *sybax, -akos*, swinish; see *sus* under **hog**

Sybarite < L. *Sybarita* (Gr. *Sybarites*), a citizen of Sybaris, a town noted for its
luxury and debauchery; see **joy**

sycalis, L. (Gr. *sykalis*), a kind of warbler; see **warbler**

sycaminus, L. (Gr. *sykaminos*), mulberry-tree; see **mulberry**

sycamore < L. *sycomorus* (Gr. *sykomoros*), a kind of fig, see *sykon* under **fig;**
plane-tree, see **plane-tree**

sychneon, Gr. thicket; see **bush**

sychnos, Gr. many, great, long; see **number**

syco- < Gr. *sykon,* fig; *sykarion; sykidion,* dim.; see **fig**

sycophant < Gr. *sykophantes,* slanderer, flatterer; see **flatter**

syenite < L. *Syenites,* pertaining to Syene in Egypt; see **stone**

syl- < Gr. *syn,* together, with; see **with**

syletes; syletor; Gr. robber; see *sylegma* under **plunder**

syllable < L. *syllaba* (Gr. *syllabe*); see **word**

syllabus, L. (Gr. *syllabos*), list, register; see **list**

sylleptor, Gr. partner, assistant; see **help**

syllexis, Gr. contribution, selection by lot; see *lachesis* under **lot**

syllis, NL. a worm; see **worm**

sylph < *sylphe,* a fairylike spirit of the air; see **spirit**

sylva (silva), L. forest; see **forest**

sym- < Gr. *syn,* together, with; see **with**

symbol < L. *symbolum* (Gr. *symbolon*); see **mark, form**

symmetros, Gr. in measure with, proportional, corresponding part for part; see **equal**

sympathy < L. *sympathia* (Gr. *sympatheia*), fellow-feeling; see **feel**

sympheron, Gr. useful; see **use**

symphony < Gr. *symphonia,* concord in sound; see **music**

symphora, Gr. event, circumstance, chance, luck; *symphoros,* happening with, accompanying; see **lot**

symphoricarpus, NL. a genus of the honeysuckle family; see **honeysuckle**

symphysis, Gr. junction, seam; see **joint**

symphytum, L. (Gr. *symphyton*), a genus of the borage family; see **borage**

symplocarpus, NL. a genus of the arum family; see **arum**

symploco- < Gr. *symplokos,* intertwined, interwoven; see *plecto* under **weave**

symptom < Gr. *symptoma,* anything that has befallen one, a sign of disease; see **mark**

syn- < Gr. *syn,* together, with; see **with**

-syna, L. (Gr. *-syne*), a suffix denoting condition, state, quality; see **nature**

synactero- < Gr. *synakter, -os,* collector; see **gather**

synagma, Gr. accumulation, concretion; see **stone**

synagoga, L. (Gr. *synagoge*), a bringing together, gathering, place of meeting; see **gather**

synagris, Gr. a sparoid fish; see **bream**

synancia, NL. < Gr. *synankeia,* confluence of glens or valleys; see **bind**

synaptos, Gr. joined together, united; *synaphe; synapsia,* connection, union, junction; see *hapto* under **bind**

synchis, Gr. shoe, sock; see **shoe**

synchysis, Gr. commixture, confusion; see **mix**

synclydo- < Gr. *synklys, -ydos,* mixed, promiscuous; see **mix**

syncope, Gr. elision, abbreviation, faint, swoon; see **short, weak**

syndetos, Gr. bound together; see *deo* under **bind**

synecdoche, Gr. figurative substitution; see FIGURES OF SPEECH under **form**

synechos, Gr. holding together, continuous; see *echo* under **hold**

synemo- < Gr. *synaimos,* of common blood, kindred, see **kin**; < *synemon,* united, comrade, see **companion**

synergos, Gr. associate, partner; see **help**

synetos, Gr. intelligent, wise; see **know**

syngnome, Gr. pardon, forgiveness; see **free**

synod < L. *synodus* (Gr. *synodos*), assembly, meeting; *synodia,* a journey in company, caravan; see **gather**

synodontis, Gr. a kind of tunny; see **tunny**

synodus, L. (Gr. *synodous*), a sparoid fish; see **bream**

synoeco- < Gr. *synoikos,* living in the same house, associated with; see **companion**

synonym < Gr. *synonymos,* of like meaning; see **like**

synopsis, Gr. a general view, sketch, outline, abstract, summary; see **short**

synoria, Gr. borderland; see **country**

syntegma; syntexis, Gr. a melting or fusing together; see *teko* under **flow**

synthesis, Gr. compounding, composition; *synthetos,* put together, compounded; see *tithemi* under **place**

syntresis, Gr. channel, passage, strait; see **way**

synyphe, Gr. web; see **weave**

syphar, Gr. old or wrinkled skin; see **skin**

syphax, Gr. sweet new wine; see **wine**

syphilis, NL. said to be from Syphilus, a swineherd in a Latin poem by Fracastoro; see **disease**

syr- < Gr. *syn,* together, with; see **with**

syrictero- < Gr. *syrikter, -os,* male crane; see **crane**

syringa, NL. a genus of the olive family; see **olive**

syringe < Gr. *syrinx, -ingos,* pipe; see **pump**

syrinx, Gr. pipe; *syristes,* piper; see **pipe, music**

syrizo; syritto, Gr. pipe, whistle, hiss; see **whistle**

syrma, Gr. robe with a train, anything trailed along, sweepings, litter; *syrmas,* drift; **heap**

syrmaeo- < Gr. *syrmaia,* purge; *syrmos,* a vomiting or purging; see **wash**

syro, Gr. drag, draw, sweep, trail; see *syrtos* under **carry**

syrphetos, Gr. sweepings, litter, rubbish; *syrphetodes,* jumbled together; see **dirt**

syrphos, Gr. a kind of fly; see **fly**

syrrho- < Gr. *syrrhoos,* flowing together, confluent; see *rheo* under **flow**

syrtes, Gr. cord, rein; see **rope**

Syrtis, Gr. name of a sandbar off the north coast of Africa, quicksand; see **sand**

syrtos, Gr. washed along by a stream, as alluvial material; see **carry**

syrus, L. a broom; see **broom**

sys- < Gr. *syn,* together, with; see **with**

syspastos, Gr. drawn up, contracted; see **lessen**

systello, Gr. draw together, contract; see *systole* under **lessen**

system < Gr. *systema,* an ordered arrangement of things; see **class**

systenos, Gr. tapering to a point; see **point**

systole, Gr. contraction; see **lessen**

systremma, Gr. anything aggregated, consolidated, or twisted together, generally a ball or round object; see **ball**

syzygium, NL. a genus of the myrtle family; see **myrtle**

syzygos, Gr. yoked, united; see *zygon* under **bind**

T

tabacum, NL. tobacco, a plant of the nightshade family; see **nightshade**

tabanus, L. gadfly; see **fly**

tabebuia, NL. a genus of the bignonia family; see **bignonia**

tabella, L. dim. of *tabula,* board, table; see **table**

taberna, L., hut, store, shop, inn; *tabernula,* dim., see **house;** *tabernaculum,* tent, see **tent**

tabidus; tabificus, L. wasting away, melting, pining, languishing; *tabes,* a dwindling, a wasting disease; see *tabeo* under **lessen**

table < L. *tabula,* f. board, plank, flat piece of anything on which to write; *tabella,* f. dim.; *tabellarius,* of tablets; *tabularis,* of boards or plates; *tabulatus,* boarded, plated: tablet, tablature, tableau, tabulate, tabella, tabellary, table d'hote, tabloid, tableland, *Tabellaria fenestrata* (a diatom), *Syringopora tabulata* (a fossil coral), *Catasetum tabulare* (an orchid).

Gr. *ambon, -os,* m. pulpit, reading-desk:

L. *cartibulum,* n. an oblong table of stone on a pedestal:

L. *cilliba,* f. (Gr. *killibas, -antos,* m.), round dining-table, three-legged stand:

Gr. *eleos,* m. kitchen-table, dresser; *eleastros,* m. manager of the table:

L. *mensa,* f. table; *mensula,* f. dim.; *mensalis; mensarius,* of a table or counter; Sp. *mesa,* f. table; *mesilla,* f. dim.: mensal, commensalism, mesa, Mesa Verde.

L. *pulpitum,* n. frame, stage, desk: pulpit.

L. *scamnum,* n. bench, stool, step; *scamillus,* m. dim.: shamble.

Gr. *telia (selia),* f. gaming table with raised border, stage, board: *Megaselia castanea* (a fly).

Gr. *trapeza,* f. table, dining-table; *trapezion,* n. dim.: trapezium, trapezoid. See **form**

See: **board, level, plate, side**

tablet < L. *tabula,* board, a flat piece of any material on which to write; see **book, table**

tacero- < Gr. *takeros,* melting, tender, dissolving; see *teko* under **flow**

tachina, NL. a kind of fly; see **fly**

tachys; tachinos, Gr. swift; *tachos,* speed; see **swift**

taciturnus, L. quiet, of few words; *tacitus,* implied, unmentioned; *tacitulus,* dim.; see *taceo* under **silent**

tactilis, L. that may be touched, concrete, < *tango, tactus,* touch; see **touch**

tadpole < uncertain origin; see **young, frog**

taeda, L. pine-tree, torch; see **pine**

taenia, L. (Gr. *tainia*), ribbon, fillet, tapeworm; *taeniola,* dim.; see **ribbon**

tagax, -acis, L. light-fingered, thievish; see **steal**

tagenias, Gr. pancake; see **cake**

tagenon, Gr. frying-pan; see **pan**

tagetes, NL. a genus of the composite family; see **composite**

tagma, Gr. something ordered or arranged; see *tasso* under **class**

tagos, Gr. commander, chief, ruler; see **govern**

tail < AS. *taegl.*

Gr. *alkaia,* f. lion's tail:

L. *cauda,* f. tail, appendage; *caudicula,* f. dim.: caudal, caudate, caudicle, caudiform, coward, queue, cue, *Pinguicula caudata* (a butterwort), *Perla bicaudata* (a stonefly), *Bulimina caudigera* (a foraminifer), *Stercorarius longicaudus* (long-tailed jaeger), *Hemigrammus caudivittatus* (a fish).

L. *equisetum,* horsetail; see **horsetail**

Gr. *kerkos,* f. tail; *kerkion,* n. dim.: cercal, cercaria, diphycercal, heterocercal, homocercal, platycercine, *Cercocarpus montanus* (a mountain-mahogany), *Cercotmetus asiaticus* (a bug), *Cercion linderi* (a dragonfly), *Dicerca divaricata* (peach-borer), *Cryptocercus punctulatus* (a cockroach), *Eupetes macrocercus* (a bird), *Schistocerca americana* (a grasshopper).

Gr. *kolouros,* bobtailed; see *kolos* under **short**

Gr. *oura,* f. tail; *ouradion,* n. dim.; *ouraios,* of the tail; *exouros,* ending in a tail or point: urochord, urohyal, uropatagium, uropod, urosthenic, hippurid, platyurous, Anura, Urodela, *Uroctonus mordax* (a scorpion), *Uropterygius marmoratus* (a fish), *Uromyces caryophyllinus* (carnation-rust), *Urophlyctis alfalfa* (a fungus), *Urocystis tritici* (wheat flag-smut), *Cynosurus cristatus* (a grass), *Crocidura russula* (a shrew), *Sciurus niger* (fox-squirrel), *Myosurus minimus* (a mousetail), *Anthurium magnificum* (tail-flower), *Xanthoura luxuosus* (green jay), *Xiphosurus bimaculatus* (a lizard), *Coryphaena hippurus* (dolphin).

L. *penis,* tail, male copulatory organ; *peniculus,* dim.; see **penis**

L. *puppis,* poop, stern; see **stern**

Gr. *pyge,* rump, buttocks; see **rump**

Gr. *sobe,* f. a horse's tail, hence any rough hair:

See: **end, rump, penis, catkin**

tailor < OF. *tailleor,* < *tallier,* cut; see **sew**

taint < F. *teint,* < L. *tingo, tinctus,* dye, paint, stain; see **color, spot**

take < AS. *tacan,* lay hold of, seize.
L. *adopto, -atus,* take or receive as one's own: adopt, adoption, adoptive.
Gr. *amerdo,* deprive of, take away; *amersis,* f. deprivation:
L. *apiscor, aptus,* pursue, take, reach, attain; *indipiscor, indeptus,* obtain, attain, reach:
L. *arrogo, -atus,* appropriate, assume; see *arrogans* under **proud**
L. *ascisco, -itus,* take, receive, adopt:
L. *ausceps,* bird-catcher; see *avis* under **bird**
Gr. *blisso (blitto),* take honey from the hive, rob, steal; see **steal**
L. *capio, captus,* take, receive; *captivus,* taken prisoner; m. a prisoner; *captor, -is,* m.; *captrix, -icis,* f. catcher, hunter; *captura,* f. the thing caught or taken; *captabilis,* that can take; *capax, -acis,* that can take or hold much, roomy, spacious, able, fit; *accipio, -ceptus,* take, receive; *concipio,* lay hold of, seize; *decipio,* catch, deceive; *incipio,* take in, begin; *percipio,* take, comprehend; *recipio,* take back, recover; *suscipio,* take up, assume, beget: captive, capture, captor, captation, captivate, captious, capstan, capsule, case, capable, capacity, capacious, catch, chase, caitiff, accept, anticipate, conceive, concept, conceit, deception, imperceptible, incipient, intercept, forceps, occupation, perception, precept, prince, principle, municipal, participate, receptive, receipt, recipe, receptacle, receiver, recover, recuperate, susceptible, mercaptan, *Muscicapa striata* (a flycatcher), *Noctua receptricula* (a butterfly), *Rhynchotrema capax* (a fossil brachiopod).
L. *carpo, carptus,* pluck, pick, gather; see **tear**
L. *confisco, -atus,* seize for the public treasury, cause to forfeit, appropriate: confiscate.
L. *complector, -plexus,* clasp, embrace, seize; see **complex**
Gr. *dechomai,* take, accept, receive; see **receive**
L. *demo, demptus,* take away, subtract, withdraw; *demptio, -onis,* f. a taking away; *adimo,* take away, seize; *ademptor, -is,* m. seizer; *eximo,* take out; *redimo,* take back, ransom: exempt, redeem.
L. *detraho, -tractus,* take away, remove, disparage: detract, detractor, detraction.
L. *devirgino, -atus,* deprive of virginity, deflower:
Gr. *diakoresis,* f. rape:
Gr. *drassomai,* grasp, take a handful; *dragmos,* grasping; *peridraxis,* f. a grasping: drassid, *Drassodes rubidus* (a spider).
L. *exuo, -utus,* take off, divest, strip; see *exutus* under **bare**
AS. *gripan,* seize: grip, gripe, grasp, grope.
Gr. *haireo,* take, adopt an opinion; *hairesis,* f. a taking; *hairetos,* eligible; *hairetikos,* able to choose or take; *anaireo,* take away, destroy; *aphairesis,* f. removal; *aphairema, -tos,* n. the thing removed; *aphairetos,* taken away; *diaireo,* distinguish, divide; *paraireo,* withdraw, remove; *periaireo,* take off: aphaeresis(apheresis), diaeresis(dieresis), heresy, heretical, *Aphaerema spicatum* (a flacourtiacead).
Gr. *halosimos,* easy to take, catch, conquer, attain; *halosis,* f. capture; *halotos,* attainable; *dysalotos,* hard to catch: *Halosimus noticollis* (a beetle), *Dysalotus alcocki* (a fish), *Dysalotosaurus lettwoorbecki* (a dinosaur).
Gr. *hapto,* lay hold of, grasp, seize; see **bind**
L. *harpago, -atus* (Gr. *harpazo*), seize, snatch, plunder; *harpago, -onis,* m. grappling hook, drag; *harpage,* f. seizure, robbery, rapine, booty; *harpagimos,* ravished; *harpaktos,* obtained by rapine, stolen; *harpax, -agos,* rapacious; *harpakter, -os,* m. robber; *harpagma, -tos,* n. booty, plunder: harpy, *Harpagomyia splendens* (a mosquito), *Harpagosaurus silberlingi* (a fossil reptile), *Harpactor cinctus* (a bug), *Harpagophora diplocrada* (a millepede), *Harpagophytum procumbens* (a pedaliacead), *Hyparpax aurora* (a moth), *Onthophagus harpax* (a beetle), *Harpagornis moorei* (extinct New Zealand bird).
L. *irretitus,* caught in a net; see *rete* under **net**
Gr. *ixeutes,* bird-catcher, fowler; see *ixos* under **glue**
Gr. *karkineutes,* crab-catcher; see *karkinos* under **crab**
Gr. *lambano (labein),* take, seize, receive; *eulabes,* take hold well, holding fast; *leptikos,* disposed to take or accept; *leptes,* m. taker, accepter; *lepsis,* f. a taking hold, seizing: astrolabe, prolepsis, catalepsy, androlepsia, syllepsis, epileptic, epilepsy, *Cercolabes prehensilis* (a porcupine), *Eulabes intermedia* (a bird), *Cercoleptes caudivolvulus* (kinkajou).
Gr. *lineuo,* take with a net; *epilineutes,* one who catches with nets; see *linon* under **net**
Gr. *methektos,* participating, sharing, partaking; see *metochos* under **companion**
Gr. *myiagros,* m. fly-catcher:
Gr. *nekroryktes,* m. body-snatcher:
L. *occupo, -atus,* take possession of: occupy, occupant, occupation.
Gr. *ornitheutes,* bird-catcher; see *ornithos* under **bird**
L. *potior, -itus,* get, obtain, become master of; see **govern**

L. *praedo, -atus,* pillage, plunder, rob; see *praeda* under **plunder**

L. *prehendo, -hensus,* seize; *comprehensibilis,* seizable, intelligible: apprehend, comprehensible, prehensile, prison, prisoner, apprentice, enterprise, comprise, reprisal, prize, surprise, *Capromys prehensilis* (hutia).

L. *psilothrum* (Gr. *psilothron*), depilatory; see *psilos* under **bare**

L. *rapio, raptus,* seize, snatch; *raptor, -is,* m. robber, plunderer; *rapax, -acis,* grasping, greedy, violent; *rapina,* f.; *raptum,* n. robbery, plunder; *abripio, -reptus,* take away by violence, drag off; *arripio,* seize, appropriate; *corripio,* snatch up, seize; *diripio,* take apart, rape; *direptor, -is,* m. plunderer; *eripio,* take away; *erepticius,* taken away; *ereptor, -is,* m. robber, plunderer: ravish, ravine, surreptitious, usurp, rapacity, raptorial, *Lycosa raptoria* (a spider), *Aquila rapax* (tawny eagle), *Eciton rapax* (army-ant), *Tachysphex mantiraptor* (a wasp).

L. *removeo, -motus,* take away, withdraw: remove, removable, remote.   See **far, move**

L. *sumo, sumptus,* take, apply: assume, presume, comsume, resumption, sumptuous, 'mumpsimus'.

Gr. *symmeristes,* m. partaker, sharer: *Symmerista albicosta* (a butterfly).

Gr. *therao,* hunt, catch; see **hunt**

L. *usurpo, -atus,* take unlawfully, seize; *usurpativus,* wrongly taken, improper; *usurpator, -is,* m.; *usurpatrix, -icis,* f. one who takes possession unlawfully: usurp, usurper, usurpative, usurpatory, usurpation.

Gr. *zogreus,* m. catcher of animals:

See: **steal, receive, trap, hunt, hold, plunder**

talano- < Gr. *talas, -anos,* wretched, sad, sorry; see **sad**

talanton, Gr. balance, scales; the thing weighed; a piece of money; see **balance, money**

talaris, L. of the ankle; see *talus* under **foot**

talaros, Gr. basket; *talarion; talariskos,* dim.; see **basket**

talasios, Gr. pertaining to wool-spinning; see **spin**

tale < AS. *talu,* speech, narrative; see **story**

talea, L. slender staff, rod, stick, cutting, set, layer; *taleola,* dim.; see **rod**

talentum, L. (Gr. *talanton*), an ancient weight and sum of money; see **money**

talido- < Gr. *talis, -idos,* a marriageable maiden; see **woman**

talinum, NL. a genus of the purslane family; see **purslane**

talio, L. punishment in kind for the injury sustained; see **punish**

talipes, L. clubfoot; see *talus* under **foot**

talis, L. such, the like, in kind; see **equal**

talitrum, L. a rap with the finger; see **strike**

talk < AS. *talian,* speak; see **speak**

tall < W. *tal,* high; see **high**

talla, L. onion-skin; see **skin**

talpa, L. mole; *talpinus,* of moles; see **mole**

talpona, L. a kind of grapevine; see **grape**

talus, L. ankle, heel, anklebone; see **foot**

tamarindus, NL. a genus of the bean family; see **bean**

tamarix, -icis, L. f. tamarisk: *Tamarix gallica* (French tamarisk).

Gr. *akakalis, -idos,* f. white tamarisk:

tame < AS. *tamian,* render docile, domesticate, civilize.

Gr. *cheirotos,* tamable; *cheirotikos,* good at taming: *Chirotica inermis* (a wasp), *Chirotonetes manca* (a mayfly).

L. *cicur, -uris,* tame, mild: *Colymbetes cicur* (a beetle), *Philoica cicurea* (a spider).

Gr. *dmetos,* tamed; *dmesis,* f. a subjecting, taming, < *damao,* break, subdue, tame; *damanter, -os; damastes; dmeter, -os,* m. a tamer; *dmoë,* f.; *dmos,* m. slave taken in war; *dmoios,* servile; *adamantos; adamastos; adamatos,* unconquerable, inflexible, untamable: adamant.

L. *domesticus,* of the house, house-broken; see *domus* under **house**

L. *domo, -itus,* tame, break, subdue; *domabilis,* tamable; *domitor, -is,* m.; *domitrix, -icis,* f. tamer: domitable, daunt, indomitable, undaunted.

Gr. *euagogos,* easily lead, docile:

Gr. *hemeros,* tame, cultivated; *hemerotes,* m. tamer:

Gr. *ktilos,* docile, obedient, tame, < *ktiloo,* tame:

L. *mansuetus,* tame, gentle; *mansuetudo, -inis,* f. tameness, mildness: mastiff(?), mansuetude.

Gr. *praos*, mild, meek, gentle, tame: *Praopsylla powelli* (a siphonapterid).
L. *sativus*, sown, cultivated; see **sow**
Gr. *tithasos*, tame, domesticated, < *tithaseuo*, tame:

tamias, Gr. dispenser, steward, treasurer; see **store, squirrel**

tamnus, L. wild grapevine; see **grape**

tan < AS. *tannian*, prob. < C. *tann*, oak; see **brown, oak**

tanacetum, NL. a genus of the composite family; see **composite**

tanaos, Gr. outstretched, long, tall; see *tany-* under **long**

tangens, L. touching; *tangibilis*, touchable; see *tango* under **touch**

tangle < uncertain origin; see **complex, mix, confusion, dishevel, trap, turn, weave, bind**

tangos, Gr. rancid; *tange*, rancidity, a putrid abscess; see **rot**

Tantalus, L. (Gr. *Tantalos*), a mythological character symbolic of eternal torment; see **tease**

tantillus; tantulus, L. so little, such a trifle; see **trifle**

tantus, L. so much; see **abundance**

tany- < Gr. *tanyo*, stretch out; see **long**

taos, Gr. peacock; see **peacock**

taper < AS. *taper*, a slender candle; narrow to a point like a candle flame; see **point, lessen**

tapete, -is, L. (Gr. *tapes, -etos*), carpet, drapery, coverlet; see **rug**

taphos, Gr. grave, tomb; see **grave**

taphros, Gr. trench, ditch; see **ditch**

tapino- < Gr. *tapeinos*, low, humble, poor; see **low**

tapir < Ab. Am. *tapira*: *Tapirus terrestris* (Brazilian tapir), *Tapirella bairdi* (a tapir).

taracho- < Gr. *tarache*; *taraktes*, disturber; *taragmos*, disturbance; see **confusion**

tarandus, L. (Gr. *tarandos*), animal of northern latitudes, reindeer; see **deer**

tarantula < L. *Tarentum*, now Taranto, a city in Italy; see **spider**

taraxacum, NL. a genus of the composite family; see **composite**

taraxis, Gr. confusion, disorder; see *tarache* under **confusion**

tarbos, Gr. alarm, fear, terror; *tarbaleos*, fearful, frightful, terrible; see **fear**

tardus, L. slow; *tardiusculus*, dim.; see **slow**

tarichos, Gr. mummy, pickled meat, fish, etc.; *tarichion*, dim.; see **flesh**

tarpe, Gr. large wicker-basket; see **basket**

tarphys, Gr. close, thick; *tarphos*, thicket; see **thick**

tarpon < uncertain origin: *Tarpon atlanticus* (tarpon).

tarquinius, L. proud, haughty; see **proud**

tarrhos, Gr. basket, crate, mat; *tarrhion*, dim.; see **basket**

tarry < AS. *tergan*, delay, hinder; see **stand, delay**

tarsius, NL. a lemurlike primate; see **monkey**

tarsos, Gr. woven mat or grate, any flat surface, flat of the foot between toes and heel, ankle; see **foot**

tartar < OF. *tartre*, < an Ar. word: cream of tartar, tartar emetic, tartaric acid, tartareous, tartrate, tartronic, tartronate, tartronyl, tartramide, tartarous.
L. *faex, faecis*, lees, sediment, tarar; *faecula*, dim.; see **dirt**

Tartarus, L. (Gr. *Tartaros*), the infernal regions; see **hell**

tasconium, L. a white, clayey earth; see **earth**

tasis, Gr. a stretching, extension; see *teino* under **spread**

tassel < OF. *tassel*; *taisel*, a tuft of loose cords, fibers, etc.; see **fringe, catkin, thread, bush, border**

taste < OF. *taster*, feel, touch, try.
L. *acredo, -inis*, f. a sharp or pungent taste:
Gr. *artytos*, flavored, seasoned; *artytikos*, of seasoning; see **spice**
L. *conditus*, seasoned, savory; see *condimentum* under **spice**
L. *delibo, -atus*, sip, taste: *delibamentum*, n. libation: delibate.
L. *deliciosus*, delightful, tasty; see *delecto* under **joy**
Gr. *geuo*, taste, try; *geuma, -tos*, n. a taste; *geusis*, f. sense of taste; *geustes*, m. taster; *geustikos*, of taste; *geustos*, tasted: ageusia, cacogeusia, *Telegeusis debilis* (a beetle).

L. *gustus, -us,* m. taste; *gustulum,* n. dim. relish; *gustabilis,* tasty, appetizing; *gustator, -is,* m. taster, < *gusto, -atus,* taste; *degusto,* taste, try; *praegustator, -is,* m. foretaster: gusto, gustatory, ragout, disgust, pregustator, *De gustibus non est disputandum.*

Gr. *poppyzo,* smack or cluck with the tongue and lips; see **cackle**

L. *sapidus,* savory, tasty; *sapor, -is,* m. taste, flavor; *saporatus,* seasoned, savory; *saporosus,* of good flavor, < *sapio,* taste, know: sapid, savor, savory, insipid, *Callinectes sapidus* (blue crab), *Alosa sapidissima* (shad), *Musa sapientum* (banana).

See: **spice, bitter, sweet, sour, salt**

tasteless

Gr. *achylos,* without juice, inspid:

Gr. *anaphes,* tasteless, insipid, not to be touched: *Anaphothrips obscurus* (a thysanopterid).

Gr. *anedyntos,* unseasoned:

L. *bliteus,* tasteless, insipid, foolish; see **fool**

L. *insipidus,* tasteless: insipid.

L. *insulsus,* unsalted, insipid, tasteless, silly: insulse, insulsity, *Sitona insulsus* (a beetle).

tata, L. daddy; see **father**

tathrision, Gr. a fish; see **fish**

tau, Gr. nineteenth letter of the Greek alphabet; see **letter**

taunt < OF. *tanter,* tempt, try; see **tease, laugh**

taurus, L. (Gr. *tauros*), bull; *taurulus,* dim.; *taureus; taurinus,* of bulls; see **cattle**

tautos, Gr. the same; see **equal**

tax < L. *taxo, -atus,* rate, appraise the value of, reproach; see **pay**

tax-; taxia-; taxis; taxo- < Gr. *tasso,* arrange, classify, place; see **class**

taxillus, L. dim. of *talus,* ankle; see **foot**

taxodium, NL. bald cypress; see *taxus* under **yew**

taxus, L. yew, see **yew**; badger, see **weasel**

tea < a Chinese word; NL. *thea,* f. a genus of the tea family: thein, theiform, *Thea sinensis* or *Camellia thea* (tea), *Pestalozzia theae* (gray blight), *Malus theifera* (an apple), *Neumannia theiformis* (a flacourtiacead).

NL. *camellia,* f. a genus of the tea family, < George J. Kamel, Moravian traveler: *Camellia japonica* (camellia), *Pterophylla camellifolia* (a katydid), camellin.

NL. *franklinia,* f. a genus of the tea family, < Benjamin Franklin, American philosopher, scientist, and statesman: *Franklinia alatamaha[altamaha]* (franklinia).

NL. *gordonia,* f. a genus of the tea family, < James Gordon, English nurseryman: *Gordonia axillaris* (a gordonia).

NL. *stuartia,* f. a genus of the tea family, < John Stuart, Scotch patron of botany: *Stuartia pentagyna* (mountain-stuartia).

teach < AS. *taecan.*

L. *alipta* (Gr. *aleiptes*), m. trainer and teacher in gymnastic schools:

L. *alumno, -atus,* nourish, bring up, educate; see *alumnus* under **child**

Gr. *deiktes,* an exhibitor, < *deiknymi,* show; see *deiktos* under **try**

Gr. *delotikos,* indicative; *delotos,* demonstrable; *delosis,* explanation, < *deloo,* explain, reveal, show; see **clear**

Gr. *didasko,* teach; *didagma, -tos,* n. lesson; *didaktikos,* apt at teaching; *didaktos,* taught; *didaskalos,* m. teacher; *didaxis,* f. instruction; *didaskalion,* n. science, art, lesson; *didaskaleion,* n. school: didactics.

L. *disciplina,* training, instruction; see *disco* under **learn**

L. *doceo, doctus,* teach, instruct; *docibilis; docilis,* easily taught; *doctor, -is,* m.; *doctrix, -icis,* f. teacher, instructor; *doctrina,* f. teaching, instruction; *documentum,* n. lesson, example, paper; *doctus,* learned, skilled, experienced: doctor, doctrine, docile, docility, document.

Gr. *dogma,* opinion, doctrine, principle; see **think**

L. *educo, -atus,* bring up a child, rear, train; *educator, -is,* m.; *educatrix, -icis,* f. rearer, tutor: educate, education, educator, educative.

L. *erudio, -itus,* instruct, educate; see *eruditus* under **know**

L. *exerceo, -itus,* practice, train: exercise.

Gr. *hodegos; hodegeter, -os,* m. guide, teacher; *hodegia,* f. teaching: *Hodegia apatela* (a moth).

L. *inculco, -atus,* tread in, tread down, stuff in, impress on, educate: inculcate.

L. *instillo, -atus,* drop in, inspire: instill.

L. *instruo, -uctus,* build in, furnish, teach; *instructor, -is,* m. teacher, preparer: instruct, instructor, instruction, instructive.

Gr. *katechetikos,* of instruction; *katechetes,* m. teacher; *katechesis,* f. instruction: catechism, catechetical.

Gr. *kathegetes; perigetes,* guide, showman, teacher; see *hegemon* under **lead**

Gr. *lakedon, -os,* f. doctrine, saying:

L. *lanista,* m. trainer of gladiators:

Gr. *mestor; mestris,* adviser, counselor; see *medos* under **aim**

L. *moneo, -itus,* advise, remind, warn; *monstro, -atus,* show, point out, teach; see **warn**

Gr. *Nestor, -os,* m. an aged warrior and wise counselor: Nestorian, *Nestor notabilis* (kea), *Pseudonestor xanthophrys* (a bird).

Gr. *paidagogos,* m. the slave who took the children to school, tutor, teacher: pedagogue, pedagogical, pedant, pedantic.

L. *professor, -is,* m. teacher: professor, professorial, professorship.

L. *promulgo, -atus,* make known, publish; see **destroy**

L. *scitum,* tenet, dogma; see *scientia* under **know**

L. *suadeo, suasus,* advise, convince; *suadibilis,* that may be convinced; *suadus; suasorius,* persuasive; *suasor, -is,* m. adviser; *persuasus,* convinced, established, settled: suasion, dissuade, persuade, persuasion, persuasive, persuader.

Gr. *symboulos,* m. adviser, counselor:

See: **school, learn, think, aim, display, lead**

team < AS *team,* animals hitched together.

L. *biga,* f. a pair of horses, team: *quadriga,* four-horse team; *triga,* three-horse team:

L. *bijugis,* m. two-horse team; *quadrijugis,* four-horse team; *sejugis,* six-horse team; *trijugis,* three-horse team:

L. *protelum,* n. team of oxen:

Gr. *zeugos,* yoke, pair, team; see *zygon* under **bind**

tear < AS. *tear,* drop of fluid from the eye; see **drop**

tear < AS. *teran,* pull apart, rend, separate.

Gr. *amergo,* pull, pluck:

Gr. *amysso,* tear; *amyche,* f. a scratch, tear, wound; *amyxis,* f. a tearing, rending; *amyktikos,* tearing: amyctic, *Amyxodon sivalensis* (a mustelid).

Gr. *anaxaino,* tear open:

L. *artuatus,* torn to pieces:

L. *carpo, carptus,* pluck, tear, slander; *carptura,* f. a sucking or gathering (by bees) from flowers; *concerpo, -erptus,* tear to pieces, abuse; *decerpo,* pluck off, break off, gather; *discerpo,* tear to pieces; *excerpo,* extract, select: carpet, excerpt, scarce, scarcity.

L. *conscindo, -scissus,* tear to pieces; see *scindo* under **cut**

Gr. *dreptos,* plucked, < *drepo,* pluck:

Gr. *drypto,* tear; *drypsos,* torn, worn: *Dryptosaurus incrassatus* (a dinosaur).

Gr. *eryo,* drag, draw, pluck; see **draw**

L. *lacero, -atus,* tear to pieces, mutilate, cut up; *lacer, -a, -um,* mangled, torn: lacerate, laceration, *Ctenostoma laceratum* (a beetle), *Philodendron lacerum* (an aracead).

Gr. *lakis, -idos,* f. rent, tear; *lakistos,* torn: *Lacistorhynchus benedeni* (a flatworm), *Lacistodes tauropis* (a moth).

L. *lancino, -atus,* tear to pieces, rend, mangle, destroy:

L. *lanio, -atus,* tear, rend; see *lanius* under **butcher**

Gr. *skyllo,* rend, tear, pluck; *skylmos,* m. a rending, tearing, plucking; *skylma, -tos,* n. the thing plucked:

Gr. *sparasso,* rend, tear; *sparagma, -tos,* n. thing torn, piece, shred; *sparagmos,* m.; *sparaxis,* f. a tearing; *sparaktes,* m. render, tearer: Sparassodonta, *Sparassis radicata* (a fungus), *Sparagmites lacertinus* (a stegocephalian), *Sparaxis grandiflora* (an iridacead), *Catosparactes eburneus* (a gull), *Sparactolambda looki* (a fossil mammal).

Gr. *spasma; spasmos,* convulsion, fit, throe; see **disease**

Gr. *tillo,* pluck, tear; *tilma, -tos,* n. anything plucked, pulled, or shredded, lint; *tilmation,* n. dim.; *tilsis,* f. a plucking; *tiltos,* plucked, shredded: tillodont, *Tillotherium hyracoides* (a fossil mammal), *Tilmatura lepida* (a bird).

Gr. *trycheros,* ragged, tattered, torn; see **worthless**

L. *vello, vulsus,* pluck, tear, pull, twitch; *vulsio, -onis,* f. a plucking; *avulsus,* torn away, separated: avulsion, revulsion, convulse.

See: **break**

tease < AS. *taesan,* pull, pluck, and separate wool or other fibers: teaser, teasel (*Dipsacus sylvestris*).

L. *illudo, -lusus,* make sport of, mock; see *ludo* under **play**

Gr. *skomma,* jeer, jest, taunt; see **laugh**

L. *Tantalus* (Gr. *Tantalos*), m. a mythological character symbolic of eternal torment. For revealing the secrets of the gods he was compelled to stand in water up to his chin under branches laden with fruit, but both receded just beyond his reach whenever he attempted to satisfy his hunger and thirst: tantalize, tantalum, tantalite, *Mycteria*(or *Tantalus*) *loculator* (wood-ibis).

L. *tormentum,* anguish, torture; see **pain**

L. *vellico, -atus,* pinch, twitch, nip, twist:

See: **arouse, trouble, laugh**

teasel < AS. *taesel;* see **thistle**

teat < AS. *tit;* see **udder**

teba, L. hill; see **mountain**

tebenna, Gr. a robe of state; see **garment**

tecedono- < Gr. *tekedon, -os,* a melting, wasting, decline; see *teko* under **flow**

techno- < Gr. *techne,* art, craft; see **art**

tecmarto- < Gr. *tekmar, -tos,* a fixed mark, boundary, goal; see **end**

tecmerio- < Gr. *tekmerion,* sure sign, positive proof; see **sure**

tecno- < Gr. *teknon,* that which is born, child, young; *teknidion; teknion,* dim.; see **child**

tecoma, NL. a genus of the bignonia family; see **bignonia**

tecto- < Gr. *tektos,* soluble, molten, fluid; *tektikos,* of solubility; see *teko* under **flow**

tectono < Gr. *tekton, -os,* carpenter, joiner, craftsman, builder; *tektonikos,* of building; see **carpenter**

tector, L. plasterer; *tectorium,* plaster; see **plaster**

tectum, L. roof; *tectulum,* dim., see **roof**; *tectus,* covered, see **tego** under **cover**

tediosus, L. wearisome, < *taedium,* weariness, disgust; see **weak**

teges, L. covering, mat; *tegeticula,* dim.; see **rug**

tegmen, L. cover; see *tego* under **cover**

tegos, Gr. roof; see *stego* under **cover**

tegula, L. roofing-tile; see **shingle**

tegulum, L. roof, ceiling; *tegillum,* dim.; see *tectum* under **roof**

tegumentum, L. cover; see *tego* under **cover**

tela, L. web; see **weave**

telamon, Gr. strap, belt; see **strap**

telchin, Gr. a person with malicious or spiteful disposition; see **bad**

tele, Gr. far; *teloteros,* farther; *telistos,* farthest; see **far**

Telemus, L. a soothsayer; see **prophecy**

teleos; teleios, Gr. having reached its end, finished, complete, perfect; see *telos* under **end**

telephanes, Gr. visible from afar, conspicuous; see *phaino* under **light**

telephium, L. (Gr. *telephion*), a kind of stonecrop; see **stonecrop**

telesma, Gr. payment; see *telos* under **pay**

teleta, L. (Gr. *telete*), initiation; see **begin**

teleuto- < Gr. *teleute,* completion; see *telos* under **end**

telia, Gr. gaming table, stage, board; see **table**

telis, Gr. fenugreek; see **bean**

tell < AS. *tellan,* speak, number; see **speak**

tellima, NL. a genus of the saxifrage family; see **saxifrage**

tellina, L. (Gr. *telline*), a kind of mollusk; see **mollusk**

tellurium < L. *tellus,* earth; see ELEMENTS under **thing**

tellus, *-uris,* L. earth; *telluster, -tris, -e,* of the earth; see **earth**

telma, Gr. standing water, pool, marsh; see **marsh**

telos, Gr. end, completion, see **end**; tax, duty, toll, see **pay**

telson, Gr. end, boundary; see **end**

telum, L. dart, javelin, spear, missile; see **spear**

temachos, Gr. slice; *temachion,* dim.; see *temno* under **cut**

temenos, Gr. piece of land marked off as a reservation; see **earth**

temeritas, L. venturesome boldness, rashness; see **bold, profane**

temetum, L. any intoxicating drink; see *temulentus* under **drink**

temna, Gr. cut, slice; see *temno* under **cut**

temnibilis, L. despicable, hateful; see *temno* under **hate**

temno, Gr. cut, divide, sever; see **cut**

temo, L. beam, pole, tongue; see **beam**

temperatus, L. limited, moderate, sober; see *tempero* under **govern**

tempest < L. *tempestas,* storm; see **storm**

tempestivus, L. happening at the right time; see **fit**

temple < L. *templum,* n. sanctuary: templet, Knights Templars, contemplation.
L. *adytum (adyton),* inmost portion of a temple, shrine: adytum.
L. *aedicula,* f. small temple, shrine: aedicula.
L. *Capitolium,* n. temple of Jupiter, statehouse: Capitol, Capitoline.
AS. *cyrice,* temple, < Gr. *kyriakon,* n. the Lord's house: church, kirk, Kirk, Kirkdale, Dunkirk, Selkirk.
L. *delubrum,* n. temple, shrine: delubrum.
L. *ecclesia* (Gr. *ekklesia*), f. the church; *ecclesiola,* f. dim.; *ecclesiasticus,* of the church; *ecclesiastes* (Gr. *ekklesiastes*), m. preacher, clergyman, priest: ecclesiastic, *Ecclesiastes.*
L. *fanum,* n. temple; *fanulum,* n. dim.; *fanitalis,* of a temple: fane, fanatic, profanity.
Gr. *naos,* m. temple, nave; *naiskos,* m. dim.: naosaurus, naology.
Pg. *pagode,* a temple or memorial, < a Skt. word; see **tower**
L. *sacellum,* n. small shrine, temple: sacellum.
L. *sacrum,* holy thing, temple; see *sacer* under **holy**
L. *sanctuarium,* n. place for sacred things, shrine: sanctuary.

tempt < L. *tento (tempto), -atus,* try; see **try**

tempus, L. time; *temporalis; temporarius,* of time; see **time**

temulentus, L. drunken, intoxicated; see **drink**

ten < AS. *ten:* tenth, thirteen, nineteen, twenty, sixty, tithe, fortnight.
L. *decem* (Gr. *deka*), ten; *decimus; decumanus; decumus,* tenth; *decuplatus; decuplus,* tenfold; *decanus,* m. chief of ten, dean; *decuria,* f. division of ten; *decurialis,* of a decuria; *decussis,* m. the number X; *denarius,* containing ten; *deni,* ten at a time; *dekas, -ados,* f. the number ten, company of ten; *dekatos,* tenth: December, decimal, decemfid, decimate, decussate, denarius, dime, dean, decade, decalogue, decapod, Boccaccio's *Decameron,* decane, decylic, decylene, *Decodon verticillatus* (swamp-loosestrife), *Decumaria barbara* (a saxifragacead), *Decadocrinus crassidactylus* (a Devonian crinoid), *Melastoma decemfidum* (a shrub), *Melanophila decastigma* (a beetle), *Amygdala decussata* (a pelecypod), *Plectaster decanus* (a starfish), *Anemone decapetala* (an anemone).
Gr. *eikosi* (eikosa-), twenty; *eikostos,* twentieth: icosian, icosahedron, icosatetrahedron, *Icosidactylocrinus reticulatus* (a fossil crinoid), *Polydesmus(Icosidesmus) hochstetteri* (a milleped).
L. *viginti,* twenty:

tenagos, Gr. shoal, shallows; see **shallow**

tenax, *-acis,* L. holding fast; see **hold**

tencto- < Gr. *tenktos,* that may be wetted; *tenxis,* a wetting; see **wet**

tender < OF. *tendre,* < *tener,* delicate, soft; see **soft**

tendo, tensus; tentus, L. stretch; *tendicula,* dim. a little stretcher; see **spread**

tendon < L. *tendo,* stretch; see **thread**

tendril < uncertain origin; see **curl**

tenebricus; tenebrosus, L. dark, gloomy, < *tenebra,* darkness (*tenebrae,* shades of night); see **black**

tener, L. soft, delicate; *tenellus,* dim.; see **soft**

tenon, Gr. sinew; see **thread**

tensus, L. stretched, extended; see *tendo* under **spread**

tent < F. *tente,* < L. *tendo, tentus,* stretch; *tentorium,* n. tent: tentorium, tentorial, tentiform, tentmaker, tentwort, *Tentorium semisuberites* (a sponge).
L. *attegia,* f. tent:
Gr. *aulis, -idos,* f. tent, stall:
L. *baeta* (Gr. *baite*), coat of skin, tent of skin; see **garment**
L. *papilio, -onis,* m. tent: pavilion.
Gr. *skene,* tent, booth, stage, decorative setting; see **play**
L. *tabernaculum,* n. tent: tabernacle.

tentacle < L. *tentaculum,* feeler, holdfast; see *teneo* under **hold**

tentamen, L. trial, test, attempt; see *tento* under **try**

tenthes, Gr. epicure; see **eat**

tenthredon, Gr. a kind of wasp; see **wasp**

tentigo, L. tension, lust; see **lewd**

tentus, L. holding, see *teneo* under **hold**; stretched, see *tendo* under **spread**

tenuis, L. thin; see **thin**

tephra, Gr. ashes; *tephros; tephrinos,* ash-colored; see **ashes**

tephras, Gr. a cicada; see **cicada**

tephritis, L. an ash-colored precious stone; see **stone**

tepidus, L. lukewarm; see **lukewarm**

ter, L. thrice; see **three**

-ter; -tes; -tor; -teira; -tis; -tria; -tris, Gr. suffixes signifying agent, doer, see **make**; **-ter, -tra, -trum,** L. signifying tool, instrument, see **tool**; **-ter, -tris, -tre,** L. pertaining to, place where, see **place**

teramnos; teramon, Gr. soft; see **soft**

terato- < Gr. *teras, -atos,* monster, sign, marvel, wonder; *terastios,* marvelous; see **wonder**

terchnos, Gr. twig; see **branch**

terebinthus, L. (Gr. *terebinthos*), terebinth-tree; see **resin**

terebro, *-atus,* L. perforate, pierce; *terebra,* gimlet, auger, borer; see **bore**

teredo, L. (Gr. *teredon*), woodworm, borer, shipworm; see **mollusk**

tereno- < Gr. *teren, -os,* soft, tender; see *teramnos* under **soft**

teres, *-etis,* L. rubbed off, rounded, cylindrical; see **cylinder**

teretico- < Gr. *teretikos,* watchful, observant; *teretes,* keeper, observer; see **guard**

teretron, Gr. borer; *teretrion,* dim.; see *teiro* under **bore**

tergens, L. cleansing; see *tergeo* under **wash**

tergilla, L. skin or sward of pork; *terginum,* rawhide; see **skin**

tergum, L. back; see **back**

-terion, Gr. place where, see **place**; tool, instrument, see **tool**

termes, L. an insect that eats wood; see **termite**

terminalia, NL. a genus of the myrobalan family; see **myrobalan**

terminus, L. (Gr. *termon*), boundary, limit, end, goal; *terminalis,* of boundaries; Gr. *termios,* at the end, last; see **end**

termite < L. *termes (tarmes), -itis,* m. an insect that eats wood: termitarium, termitid, termitophile, *Termes bellicosus* (a termite), *Termitophagus synterminus* (a beetle), *Calotermes(Electrotermes) affinis* (a fossil termite), *Reticulotermes arenicola* (a termite), *Agaricus termitigena* (a fungus).

tern < uncertain origin.
  L. *ceyx, -ycis* (Gr. *keyx, -ykos; kex, -ekos; kauax, -akos*), m. a sea-bird that dives for its prey: *Ceyx tridactyla* (three-toed kingfisher), *Clytoceyx rex* (a kingfisher), *Celastrina ceyx* (a butterfly).
  Gr. *gyges,* m. a water-bird, tern: *Gygis alba* (a fairy-tern).
  NL. *sterna,* f. tern: *Sterna paradisea* (arctic tern).

ternarius; ternatus, L. consisting of three, < *terni,* in three's; see *tres* under **three**

tero, *tritus,* L. rub; see **rub**

terphos, Gr. hide, skin; see *sterphos* under **skin**

terpnos, Gr. delightful, agreeable; *terpsis,* delight, enjoyment, pleasure, < *terpo,* delight, cheer, make merry; see **joy**

Terpsichore, Gr. the Muse of dancing; see *choros* under **dance**

terra, L. earth; *terrula,* dim.; *terrestris, -e,* of the earth; see **earth**

terrace < L. terra, earth; see **shelf, mountain**

terribilis, L. dreadful, frightful; *terriculum,* bugbear, scarecrow; see **fear**

territorium, L. district, domain; see **country**

tersus, L. cleansed, clean, neat, succinct; see *tergeo* under **wash**

terthreus, Gr. quack; see **lie**

terthrios, Gr. rope or brace at end of sailyard; see **rope**

terthron, Gr. end, crisis; see **end**

tertius, L. third; *tertianus; tertiarius,* of the third part; see *tres* under **three**

tescum, L. waste place, desert; see **desert**

tessellatus, L. inlaid with small square stones, mosaic, < *tessella,* small square stone, < *tessera* (Gr. *tessares*), square piece of anything, tablet, ticket, tile, die; see **mosaic,** *tessares* under **four**

test < L. *testum,* earthen vessel or cupel for examining, reducing, or refining ores and metals, trial; see **try**

testa, L. potsherd, urn, brick, shell, skull; *testula,* dim.; *testaceus,* having a shell, shelly; see **shell**

testamentum, L. will; see *testor* under **wish**

testicle < L. *testiculus,* m. dim. of *testis,* m. male genital gland: testes, testicular, testiculate, testicond, testosterone.
L. *coleus* (Gr. *koleos*), scrotum; see *coleus* under **sheath**
Gr. *didymos,* m. testicle: didymitis, epididymis.
Gr. *kolythros,* m. testicle:
Gr. *orchis, -ios (-eos),* m. testicle, orchid (from the shape of the tuberous roots of some species); *orchidion,* n. dim.; *enorches,* with testicles, uncastrated; *triorches,* with three testicles, very lecherous: orchitis, orcheotomy(orchid-ectomy), cryptorchidism, *Orchis mascula* (male orchid), *Lithobotrys orchidea* (a radiolarian), *Cynorchis villosa* (an orchid), *Prosthogonimus macrorchis* (a fluke).

testis, L. witness; *testificor, -atus,* bear witness; *testor, -atus,* declare, make a will, see **witness, wish;** testicle, see **testicle**

testudo, *-inis,* L. tortoise; see **turtle**

teta, L. a kind of dove; see **dove**

tetanus, L. (Gr. *tetanos*), stiff, rigid; lockjaw; see **stiff**

tetaros, Gr. pheasant; see **pheasant**

teter, *-tra, -trum,* L. foul, hideous, offensive, shameful; see **bad**

tethe, Gr. grandmother; *tethis, -idos,* aunt; see **kin**

tethea, L., Gr. a kind of sponge; see **sponge**

tethyon, Gr. a tunicate; *tethynakion,* dim.; see **animal**

Tethys, L., Gr. a sea-goddess, wife of Oceanus; see **sea**

tetra < Gr. *tessares (tettares),* four; *tetartos,* fourth; see **four**

tetralix, L. a heath; see **heath**

tetrax, Gr. a kind of grouse; see **grouse**

tetricus, L. forbidding, harsh, stern; see *teter* under **bad**

tetrinnio, L. quack like a duck; see **cackle**

tetrix, Gr. a kind of bird; see **bird**

tettix, *-igos,* Gr. cicada; *tettiganion,* dim.; see **cicada**

teuchos, Gr. tool, gear, armor, vessel, book; see **tool**

teucrium, L. (Gr. *teukrion*), germander; see **mint**

teusios, Gr. vain, futile, idle; see **worthless**

teuthalis, L. a plant; see **plant**

teuthido- < Gr. *teuthis, -idos,* squid; see **mollusk**

teutlum, L. (Gr. *teutlon*), beet; *teutlion,* dim.; see **beet**

texis, Gr. a wasting, melting, dissolution; *teximeles,* wasting of limbs; see *teko* under **flow**

textilis, L. woven; *textrinus,* of weaving; *textum,* web, cloth; see *texo* under **weave**

textularia, NL. a genus of foraminifers; see **protozoan**

thacemato- < Gr. *thakema, -tos,* a sitting, seat; *thakesis,* a sitting, seat; *thakos,* chair, seat, office; see **sit**

thaero- < Gr. *thairos,* hinge of a door or gate, axle of a chariot; see **hinge**

thalamus, L. (Gr. *thalamos*), chamber, room, bedroom, bridal chamber; *thalame,* den, hole; *thalamios,* of a room; see **room, den**

thalasso- < Gr. *thalassa (thalatta),* sea; *thalassikos,* like the sea; *thalassinos,* sea-colored, green; *thalassios,* of the sea; see **sea**

thaleros, Gr. blooming, vigorous, fresh, active; *thaleia,* blooming, luxuriance; *thalia,* abundance, wealth, < *thallo,* abound, grow, flourish; see **grow, abundance**

thalictrum, L. (Gr. *thaliktron*), meadow-rue; see **buttercup**

thallus, L. (Gr. *thallos*), young shoot, sprout, green branch, plant body of algae and fungi; *thallion,* dim.; see **branch**

thalos, Gr. young (of persons), child; see **child**

thalpos, Gr. warmth, heat; *thalporos,* warm; *thalpsis,* a warming; see **heat**

thalycro- < Gr. *thalykros,* hot, glowing; see **heat**

thambos, Gr. astonishment; *thambema,* monster; see **wonder**

thamino-; thamio- < Gr. *thama; thaminos; thameios,* often, crowded; see **number**

thamnos, Gr. shrub, bush; *thamnion,* dim.; see **bush**

thanatos, Gr. death; see **death**

thank < AS. *thanc,* express gratitude; see **give**

thapsia, L. (Gr. *thapsia; thapsos*), a poisonous umbellifer from Thapsos that was the source of a yellow dye; *thapsinos,* yellow; see **carrot**

tharsos, Gr. courage, boldness; *tharsaleos (tharrhaleos),* bold; see **bold**

thasso (thatto), Gr. sit, sit idle; see *thakema* under **sit**

thauma, Gr. wonder, marvel; *thaumastos,* wonderful; see **wonder**

thaw < AS. *thawian;* see **flow, heat**

thea, NL. a genus of the tea family; see **tea**

theangelis, Gr. a kind of plant; see **plant**

theaphion, Gr. sulfur; see *theion* under **sulfur**

theater < OF. *theatre,* < L. *theatrum* (Gr. *theatron*), playhouse; *theates,* spectator; see **play,** *theaomai* under **see**

theca, L. (Gr. *theke*), case, container, envelope, sheath; *thekion,* dim.; see **bag**

thecla, NL. a genus of butterflies; see **butterfly**

thegane, Gr. whetstone, < *thego,* sharpen; *thegaleos,* pointed, sharp; *thektos,* sharpened; see **stone, point**

thele, Gr. teat, nipple; *thelion,* dim.; *thelazo,* suckle, suck; see **udder, suck**

thelema, Gr. will; *theleos,* willing; *theletes,* willer; see *ethelo* under **wish, free**

thelgo, Gr. enchant, charm, spellbound; *thelgema; thelgetron; thelxis,* charm, spell; *thelkter,* charmer; *thelkterion,* charm, enchantment; see **magic**

thelymnon, Gr. element, foundation of things; see **thing**

thelypteris, Gr. a fern; see **fern**

thelys, Gr. female; *thelykos,* feminine; see **woman**

theme < Gr. *thema,* subject of discourse, something laid down or deposited; see **speak**

themelio- < Gr. *themelios,* of a foundation, < *themethlon,* foundation, bottom, root; see **base**

themis, Gr. law, decree, order, justice; *themistos,* lawful; see **law**

themon, Gr. heap; see **heap**

thenar, Gr. palm of the hand; see **hand**

theobroma, NL. a genus of the chocolate family; see **chocolate**

theobrotion, Gr. an evergreen plant; see **plant**

Theophrastus, L. (Gr. *Theophrastos*), renowned botanist and philosopher; see **plant**

theory < Gr. *theoria,* something seen in the mind, plan, scheme, guess; *theorama,* spectacle, < *theoreo,* look at, view; see **hypothesis, see**

theos, Gr. god; see **spirit**

therapeutico- < Gr. *therapeutikos,* of medical treatment, < *therapeuo,* serve, care for, give medical treatment; *therapidion,* means of cure; *therapon,* attendant; see **heal, drug, servant**

therio-; thero- < Gr. *therion; theridion; theraphion,* beast, dim. of *ther, theros,* wild animal; *therao,* hunt; *theratron,* trap; see **animal, hunt, trap**

therismos, Gr. harvest; *therister,* reaper, < *therizo,* reap; see **reap**

theristrion, Gr. a light, summer dress; see **garment**

thermastris, Gr. tongs; see **tongs**

thermo- < Gr. *therme,* heat; *thermos,* hot, see **heat;** < *thermos,* lupine, see **bean**

theros, Gr. summer; *thereios; therinos,* of summer; see **year**

thersigenes, Gr. race-destroying; see *phthora* under **destroy**

thesaurus, L. (Gr. *thesauros*), treasure, storehouse, chest; see **store**

thescelo- < Gr. *theskelos,* marvelous; see **wonder**

thesis, Gr. an arranging, position, proposition; see **hypothesis**

thesium, L. (Gr. *thesion*), a plant; see **sandalwood**

thesmos, Gr. law, ordinance, rule; see *themis* under **law**

thespesios; thespios, Gr. ineffable, divine, wondrous; see **wonder**

theta, Gr. eighth letter of the Greek alphabet; see **letter**

Thetis, L. a sea-nymph; see **sea**

theto- < Gr. *thes, thetos; thessa,* serf, servant; see **servant**

thiasos, Gr. band of street-singers; *thiasodes,* festive; see **joy**

thibe, Gr. basket, ark; see **basket**

thick < AS. *thicce.*

Gr. *adinos,* thick, crowded, loud: *Adina cordifolia* (adina), *Adinandra maculosa* (a tiup-tree).

Gr. *anchistinos,* close, crowded, in heaps:

Gr. *athroos,* crowded, assembled, collected; see **gather**

L. *caespes,* turf, sod, clump of plants; see **sod**

L. *coactilis,* made thick, < *cogo, coactus,* compress, assemble, collect: cogent, *Salix coactilis* (a willow).

L. *coagulo, -atus,* curdle; *coagulum,* n. rennet, curdler: coagulate, coagulation.

L. *compactus,* thick, firm: compact, *Lecanospira compacta* (a fossil gastropod).

L. *compressus,* pressed together, close, squeezed; see **press**

L. *concretus,* thick, hard, stiff: concrete:

L. *confertus,* compressed, dense, crowded, thick: *Juniperus conferta* (shore-juniper), *Atriplex confertifolia* (shadscale-saltbush), *Arctostaphylos confertiflora* (a manzanita).

L. *congestus,* dense, heaped up, thick: congested, congestion, *Ostrea congesta* (a fossil oyster).

L. *crassus,* thick, fat, stout; *crassedo, -inis; crassitas, -atis; crassitudo, -inis,* f. thickness; *crasso, -atus,* thicken; *incrasso, -atus* thicken, make stout: crass, Crassus, crassitude, crassamentum, crassilingual, grease, Mardi Gras, *Crassula cornuta* (a houseleek), *Crassatella pulchra* (a pelecypod), *Ulmus crassifolia* (cedar-elm), *Eriodictyon crassifolium* (a hydrophyllacead), *Acacia crassiuscula* (an acacia), *Chaetophora incrassata* (an alga).

L. *creber, -bra, -brum,* thick, close, numerous; *crebratus,* thick, close: crebritude, crebrous, *Dosinula crebra* (a pelecypod), *Dentalina crebricosta* (a foraminifer), *Betulodes crebraria* (a moth), *Carex crebriflora* (a sedge).

Gr. *dasys,* thick with hair, hairy, shaggy; see **hair**

Gr. *daulos,* thick, shaggy, dark: *Daulotypus picticornis* (a beetle).

L. *densus,* thick, close, compact; *densabilis, densativus,* thickening, binding; *densitas, -atis,* f. thickness; *denso, -atus,* thicken: dense, density, condensation, condenser, *Pugnellus densatus* (a fossil gastropod), *Elymus condensatus* (a wild-rye), *Lithocarpus densiflorus* (tanbark-oak).

Gr. *epetrimos,* closely woven, dense, thronged:

L. *fullo, -onis,* m. fuller: full, fuller, *Dipsacus fullonum* (fuller's-teasel), *Melolontha fullo* (mottled chafer).

L. *gelo, -atus,* freeze, solidify; see *gelidus* under **cold**

Gr. *hadros,* thick, bulky, stout, strong; see **large**

Gr. *lasios,* hairy, woolly; see **wool**

L. *nacca; nacta* (Gr. *naktes*), m. fuller; *naccinus* (*nactinus*), of a fuller:

Gr. *nasso* (*natto*), press close, thicken; see **press**

Gr. *pachys,* thick; *pachos,* n.; *pachytes,* f. thickness; *pachylos,* thickish: pachyderm, pachycephalous, pachydactyly, pachymeter, pachytene, *Pachysandra procumbens* (Allegheny pachysandra), *Pachylocrinus globosus* (a Mississippian crinoid), *Pachyscelus purpureus* (a beetle), *Pachyrhizus angulatus* (a legume).

Gr. *pegma, -tos,* n. anything congealed, thickened, fastened, < *pegnymi,* fasten; *pektos,* fixed, compacted, congealed, fastened; *pexis,* f. fixation, solidification: pegmatite, parapegm, pectolite, pectin, pectose, amylopectin, calcipexy, colpopexy, galactopexy, hepatopexy, nephropexy, desmopexia.

L. *pilatus,* thick, dense, < *pilo, -atus,* ram down, thrust home:

Gr. *pyknos,* dense, thick; *pyknotes,* f. denseness, thickness; *pyknosis,* f. condensation: pycnium, pycnidium, pycnosis, pycnogonid, *Pycnogonum littorale* (a sea-spider), *Pycnonotus plumosus* (a bulbul), *Pycnobracon niger* (a wasp), *Pycnothrix monocystoides* (a protozoan), *Pycnanthemum virginianum* (a mint), *Acacia pycnantha* (golden wattle), *Sympycnus minor* (a fly), *Sympycnodes trigonocosma* (a moth).

AS. *sceacga,* thick mat of hair or wool: shag, shaggy.

L. *solidus,* dense, firm, hard, thick, entire; *solido, -atus,* make firm, establish: solid, solidarity, solidification, solder, consolidate, *Solidago arguta* (a goldenrod), *Soldanella alpina* (a primrose), *Spisula solidissima* (surf-clam), *Panaeolus solidipes* (a mushroom), *Ephebe solida* (a lichen).

L. *spissus*, close, dense, thick, slow, < *spisso, -atus,* thicken, condense; *conspissatus; inspissatus,* thickened: spissitude, inspissate, inspissosis, *Ancillaria spissa* (a gastropod), *Eriophorum spissum* (a sedge), *Orthoceras spississeptum* (an Ordovician cephalopod).

Gr. *stereos; steriphos; sterrhos,* solid, hard, three-dimensional; *stereoma, -tos,* n. skeleton: stereogram, stereopticon, stereotype, stereochemistry, stereome, stereoplasm, stereoscopic, sterol, cholesterol, corticosterone, *Stereospermum fimbriatum* (a bignoniacead), *Stereum purpureum* (a fungus), *Steriphus solidus* (a beetle), *Steriphon bedeli* (a beetle), *Sterrhoptilus capitalis* (a bird), *Polyipnus sterope* (a fish), *Sterrhurus musculus* (a trematode).

Gr. *tarphys* (*tarpheios*), thick, close; *tarphos,* n. thicket: *Tarphypygus ellipticus* (a holothurian), *Tarphiomimus indentatus* (a beetle), *Stachytarpheta coccinea* (a false valerian).

Gr. *thrombosis,* a clotting, curdling; see *thrombos* under **lump**

See: **broad, large, fat, flesh, herd, number, press, dull, hard, slow**

thief < AS. *thiof;* see **steal**

thigh < AS. *thioh;* see **leg**

thigma, Gr. touch; see **touch**

thimble < AS. *thymel;* see **cap**

thin < AS. *thynne.*

Gr. *aknisos,* without fat, meager, spare; *anousios,* without substance, immaterial; *apaches,* without thickness, thin; *apimelos,* without fat, lean; *asarkos; kenosarkos; mikrosarkos,* not fleshy, lean: *Apachemyia pallida* (a fly), *Asarcomyia cadaver* (a fossil fly), *Asarcopus palmarum* (a bug).

L. *aquatus,* mixed with water, watery, thin, dilute:

Gr. *araios,* thin, lean, narrow, loose, porous, few: *Araeomorpha atmota* (a moth), *Araiospora[Araeospora] spinosa* (a fungus).

L. *collatus,* extended, diffuse:

L. *colluco, -atus,* let light into the forest, clear, thin:

L. *cracens, -entis,* slender, graceful:

L. *dilutus,* mixed, weak, thin; see **mix**

L. *ejuncidus,* rushlike, lean, slender: *Dentalina ejuncida* (a Jurassic foraminifer), *Opilio ejuncidus* (a harvestman).

Gr. *enischnos,* thin, slight, slender:

Gr. *eukteanos; ithykteanos,* tall, slender: *Eucteanus caelestinus* (a beetle).

L. *exilis,* thin, slender, meager, weak: *Anobium exile* (a beetle), *Ixobrychus exilis* (least bittern), *Leiobunum[Liobunum] exilipes* (a phalangid).

L. *gracilis,* slender: gracile, gracilescent, *Acalypha gracilens* (a euphorbiacead), *Andropogon gracilis* (a grass), *Epidendrum gracile* (an orchid), *Cenchrus gracillimus* (a grass), *Rhynchospora gracilenta* (a sedge), *Strombus gracilior* (a conch).

Gr. *ischnos,* dry, withered, thin, lean, weak; *ischnotes,* f. thinness, leanness: *Ischnoptera pennsylvanica* (a cockroach), *Ischnogaster albibucca* (a wasp), *Podischnus agenor* (a beetle).

Gr. *kanabinos,* lean, slender; *kanabos,* m. lean person, mere skeleton:

Gr. *kolekanos,* m. a long, thin person:

Gr. *lagaros,* loose, slack, thin; see **free**

Gr. *leptos,* thin, fine, small, slender, subtle, delicate; *leptosyne; leptotes,* f. thinness, slenderness; *leptysmos,* m. a thinning; *leptaleos,* fine, delicate; *leptyno,* thin out: leptocephalus, leptosporangiate, leptology, *Leptocoris trivittatus* (a bug), *Leptodactylus albilabris* (a bug), *Leptarrhena amplexifolia* (a saxifragacead), *Leptotes bicolor* (an orchid), *Leptysma obscura* (a grasshopper), *Leptaleoceras leptum* (an ammonite), *Leptospermum scoparium* (manuka), *Leptynopterus femoratus* (a beetle), *Leptinotarsa decemlineata* (potato-beetle), *Leptinus testaceus* (a beetle), *Asterina leptalacantha* (a starfish), *Tropidoleptus carinatus* (a fossil brachiopod).

L. *macer, -cra, -crum; macilentus,* lean, thin, poor; *macellus; macriculus,* dim.; *macies,* f. leanness: meager, emaciated, *Hylastes macer* (a beetle), *Elphidium macellum* (a foraminifer), *Rubus macilentus* (a blackberry), *Cladonia macilenta* (a lichen).

Gr. *manos,* thin, rare, scanty, sometimes implying the gaseous state: manometer, manograph, manoxylic, *Trichomanes punctatum* (a filmy-fern).

L. *micidus,* thin, poor:

L. *petilus,* thin, slender:

L. *procerus,* slender, tall, long; see **high**

Gr. *psednos,* thin, scanty, bald; see **bare**

L. *rallus,* thin:

L. *rarus,* scarce, thin; see **few**

Gr. *rhadinos,* slender, tapering, lithe: *Rhadinocrinus dactylus* (a Devonian crinoid).

Gr. *skeliphros* (*sklephros*), dry-looking, thin, slender: *Sceliphron femoratum* (a mud-dauber).

L. *strigosus*, lean, thin, meager: *Erigeron strigosus* (a fleabane).

L. *subtilis*, thin, fine, slender, acute; see **know**

L. *tenuis*, thin; *tenuo, -atus*, make thin, slenderize, rarefy, dilute; *attenuatus*, drawn out, weakened, thin: tenuity, tenuous, tenuate, tenuirostral, attenuated, extenuation, *Alternaria tenuis* (a mold), *Aster tenuifolius* (marsh-aster), *Murex tenuispina* (a gastropod), *Pinus attenuata* (knobcone-pine).

See: **narrow, little, poor, weak**

thing < AS. *thing*, cause, deed, entity, fact, idea, material object, judicial assembly: anything, nothing, something, husting, Althing, Storthing.

Gr. *chrema, -tos*, needful or useful thing, goods, money; see **money**

Gr. *deigma; deixis*, sample, specimen, pattern; see **form**

L. *elementum*, n. first principle, rudiment: element, elemental, elementary.

L. *ens, entis*, being, thing, that which has existence; *essentia*, the substance of things; see *esse* under **life**

L. *examplum*, pattern, model, copy, specimen; see **form**

L. *factum*, n. act, certainty, reality: fact, factual.

Gr. *hyle*, raw material, matter, stuff, wood; *hylikos*, of matter, material, worldly; see **wood, alcohol**

Gr. *hypar, -tos*, n. actual appearance, reality; *hyparktos*, real:

L. *materia*, f. stuff, constituent of things; *materiola*, f. dim.; *materialis*, of matter: matter, material, materialism, materia medica, immaterial.

L. *matrix*, womb, embedding or enclosing material; see **womb**

Gr. *metra*, pith or heart of a tree, matrix; see **womb**

L. *objectum*, n. a concrete entity: object, objective.

Gr. *on, ontos*, being, thing, that which has existence; *ousia*, essence, substance, property, reality; see **life**

Gr. *pragma, -tos*, deed, act, thing; see *prasso* under **make**

L. *quidvis*, anything:

L. *res, rei*, f. thing, circumstance; *recula*, f. dim.; ML. *realis*, of a thing: reality, realism, realize, real, rebus, *res publica* (republic), Lucretius' *De rerum natura*, *Recula parva* (a fossil insect).

L. *specimen, -inis*, n. example, token, evidence, sign: specimen.

L. *subjectum*, n. basis or substance for thought, theme, topic: subject, subjective.

L. *substantia*, f. the material of which a thing consists: substance, substantial.

Gr. *thelymon*, n. element, foundation of things:

AS. *waru*, article, commodity: ware, warehouse, hardware, tinware, earthenware, chinaware, glassware.

See: **nature, form, tool**

CHEMICAL ELEMENTS

actinium (Ac), < Gr. *aktis, -inos*, ray.

actinon (An), < actinium.

aluminum (Al), < L. *alumen, -inis*, alum.

antimony (Sb, < L. *stibium*), < uncertain origin.

argon (A), < Gr. *argon* (*argos*), inactive.

arsenic (As), < L. *arsenicum* (Gr. *arsenikon*), orpiment.

barium (Ba), < Gr. *barys*, heavy.

beryllium (Be), < Gr. *beryllion*, dim. of *beryllos*, beryl.

bismuth (Bi), < G. *bismuth* (*wismuth*).

boron (B), < Per. *burah*, borax.

bromine (Br), < Gr. *bromos*, stench.

cadmium (Cd), < L. *cadmia* (Gr. *kadmia*), calamine.

calcium (Ca), < L. *calx, calcis*, lime.

carbon (C), < *carbo, -onis*, charcoal, coal.

cerium (Ce), < L. *Ceres*, goddess of agriculture, whose name was given to the first planetoid discovered.

cesium (Cs), < L. *caesius*, bluish-gray.

chlorine (Cl), < Gr. *chloros*, green.

chromium (Cr), < Gr. *chroma*, color.

cobalt (Co), < G. *kobalt*, < *kobold*, goblin, earth spirit.

copper (Cu), < AS. *coper*, < L. *cuprum*, < Gr. *Kypros*, Cyprus, ancient source of copper.

dysprosium (Dy), < Gr. *dysprositos,* hard to get at.

erbium (Er), < Ytterby, Sweden.

europium (Eu), < Europe.

fluorine (F), < L. *fluor, -is,* flow, flux.

gadolinium (Gd), < J. Gadolin, Finnish chemist.

gallium (Ga), < *Gallia,* Gaul, France.

germanium (Ge), < L. *Germania,* Germany.

gold (Au, < L. *aurum*), < AS. *gold.*

hafnium (Hf), < L. *Hafnia,* Copenhagen.

helium (He), < Gr. *helios,* sun; first observed in the spectrum of the sun.

holmium (Ho), < L. *Holmia,* Stockholm.

hydrogen (H), < Gr. *hydor, hydro-,* water, *-gen,* forming.

indium (In), < L. *indicum,* indigo.

iodine (I), < Gr. *iodes,* violet-colored.

iridium (Ir), < L. *iris, -idis,* rainbow.

iron (Fe, < L. *ferrum*), < AS. *iren.*

krypton (Kr), < Gr. *kryptos,* hidden.

lanthanum (La), < Gr. *lanthano,* escape notice.

lead (Pb, < L. *plumbum*), < AS. *lead.*

lithium (Li), < Gr. *lithos,* stone.

lutecium (Lu), < L. *Lutetia,* former name of Paris.

magnesium (Mg), < L. *Magnesia,* a district in Thessaly.

manganese (Mn), < It. *manganese,* corruption of Magnesia.

mercury (Hg, < L. *hydrargyrus*), < L. *Mercurius,* Mercury, messenger of the gods.

molybdenum (Mo), < L. *molybdaena,* galena.

neodymium (Nd), < Gr. *neos,* new, and didymium (< *didymos,* twin).

neon (Ne), < Gr. *neos,* new.

nickel (Ni), < G. *nickel,* demon.

niobium (Nb), < Gr. *Niobe,* daughter of Tantalus. Formerly columbium.

nitrogen (N), < L. *nitrum* (Gr. *nitron*), niter, native soda, *-gen,* forming.

osmium (Os), < Gr. *osme,* smell.

oxygen (O), < Gr. *oxys,* sour, acid, *-gen,* forming.

palladium (Pd), < L. *Palladium,* a statue of Gr. *Pallas,* a name for Athena, goddess of wisdom.

phosphorus (P), < Gr. *phosphoros,* bringing light.

platinum (Pt), < Sp. *platina,* < *plata,* silver.

polonium (Po), < ML. *Polonia,* Poland.

potassium (K, < NL. *kalium*), < D. *potasch,* < pot, and ash.

praseodymium (Pr), < Gr. *prasios,* leek-green, and didymium.

protoactinium (Pa), < Gr. *protos,* first, and actinium.

radium (Ra), < L. *radius,* ray.

radon (Rn), < *rad,* in radium, and *on,* as in argon.

rhenium (Re), < L. *Rhenus,* Rhine.

rhodium (Rh), < Gr. *rhodon,* rose, red.

rubidium (Rb), < L. *rubidus,* reddish.

ruthenium (Ru), < ML. *Ruthenia,* Russia.

samarium (Sm), < samarskite, < *Samarski,* a Russian mine officer.

scandium (Sc), < L. *Scandia,* Scandinavia.

selenium (Se), < Gr. *selene,* moon.

silicon (Si), < L. *silex, -icis,* flint.

silver (Ag, < L. *argentum*), < AS. *seolfor.*

sodium (Na, < Ar. *natrun,* sodium carbonate), < ML. *sodanum,* headache remedy, < Ar. *suda,* headache.

strontium (Sr), < Strontian, Scotland.

sulfur (S), < L. *sulfur,* brimstone.

tantalum (Ta), < Gr. *Tantalos,* Tantalus, symbol of eternal torment.

technetium (Tc), < Gr. *techne,* craft.

tellurium (Te), < L. *tellus, -uris,* earth.

terbium (Tb), < Ytterby, Sweden.

thallium (Tl), < Gr. *thallion,* dim. of *thallos,* a young, green shoot. The spectral line is green.

thorium (Th), < thorite, < ON. *Thor,* god of thunder.

thoron (Tn), < thorium.

thulium (Tm), < L. *Thule,* farthest north.

tin (Sn, < L. *stannum*), < AS. *tin.*

titanium (Ti), < Gr. *Titan,* one of a mythologic race symbolic of brute force and large size.

uranium (U), < L. *Uranus* (Gr. *Ouranos*), god of heaven, for whom the planet was named.

vanadium (V), < ON. *Vanadis,* a name of Freya, goddess of love.

wolfram (W), < G. *wolfram,* < uncertain origin. Formerly tungsten.

xenon (Xe), < Gr. *xenos,* stranger.

ytterbium (YB), < Ytterby, Sweden.

yttrium (Y), < Ytterby, Sweden.

zinc (Zn), < G. *zink.*

zirconium (Zr), < F. *zircon,* < Ar. *zarqun,* gold-colored.

Isotopic or Short-lived Elements

americium (Am), < America.

astartine (At), < Gr. *Astarte,* Phoenician goddess of love.

berkelium (Bk), < Berkeley, California.

californium (Cf), < California.

curium (Cm), < Marie and Pierre Curie.

deuterium (H^2, D), < Gr. *deuteros,* second.

francium (Fr), < France.

neptunium (Np), < L. *Neptunus,* Neptune, god of the sea.

plutonium (Pu), < L. *Pluto,* god of the nether world.

promethium (Pm), < Gr. *Prometheus,* a Titan who gave fire and the arts to man.

tritium (H^3, T), < Gr. *tritos,* third.

think < AS. *thencan.*

L. *assumo, -umptus,* take for granted, suppose; see **hypothesis**

L. *circumspectus,* considered, guarded, cautious; see **guard**

Gr. *boule,* will, determination, counsel, advice; *bouleutes,* adviser; *aboulos,* thoughtless; see *bouletos* under **aim**

L. *coctus,* well-considered, well-digested; see *coquo* under **cook**

L. *cogito, -atus,* pursue mentally, think, consider, ponder; *cogitabilis,* conceivable, imaginable; *cogitabundus,* thoughtful; *cogitamen, -inis,* n. thought: cogitate, "*Cogito, ergo sum*" (Rene Descartes).

L. *conceptivus,* considered, thought out; *conceptus, -us,* m. thought: conceive, concept, conceptible, conceptual.

L. *conjectura,* opinion, guess; see **hypothesis**

L. *considero, -atus,* contemplate, examine, reflect upon: consider, consideration, inconsiderate.

L. *consulo, -ultus,* consider, reflect, deliberate; *consilium,* n. a considering together, deliberation; *consiliarius,* suitable for counsel; *consiliator, -is,* m. counselor; *consilor, -atus,* take counsel: consult, consultation, consul, counsel, counselor.

L. *contemplo, -atus,* consider, survey, observe: contemplate, contemplation.

L. *cura,* care, attention; see **guard**

L. *delibero, -atus,* consider maturely, advise upon, weigh: deliberate, deliberation, deliberator.

L. *digero, -estus,* separate, classify, consider; see **digest**

L. *diligens, -entis,* assiduous, attentive, careful; see **busy**

L. *discussio, -onis,* f. a shaking down, examination, consideration: discuss, discussion.

L. *dispecto, -atus,* consider, contemplate, examine; *dispector, -is,* m. examiner, searcher:

Gr. *dokeo,* think, suppose, have an opinion; *dogma, -tos,* n. opinion, doctrine, principle; *doxa,* f. opinion, glory, praise; *doxarion,* n. dim.; *doxastos,* conjectural;

dogmatikos, of opinions; *endoxos*, esteemed, notable; *eudokesis; eudokia*, f. satisfaction, approval: dogma, dogmatic, orthodox, heterodox, doxology, *Chionodoxa luciliae* (a liliacead), *Endoxocrinus alternicirrus* (a crinoid), *Photinia amphidoxa* (a rosacead), *Phyllodoce empetriformis* (red heather).

Gr. *echephron, -os*, prudent, sensible; *echephrosyne*, f. prudence, good sense:

Gr. *enthymema, -tos*, n. thought, argument; *enthymistos*, deeply considered, taken to heart: enthymeme.

L. *examino, -atus*, weigh, consider; see try

Gr. *gnome*, f. opinion, maxim; *gnomidion*, n. dim.; *gnomikos*, of maxims; *gnomosyne*, f. prudence, judgment:

Gr. *idea*, mental image, form, appearance; see *eidos* under form

L. *illativus*, inferring, concluding, < *infero, illatus*, bring in, conclude: *Melissodes illata* (a bee).

Gr. *kedos*, care, concern, anxiety; see guard

Gr. *lemma*, assumption, premise, subject; see hypothesis

L. *logicus* (Gr. *logikos*), reasonable: logical, illogical.

L. *meditor, -atus*, ponder; *meditator, -is*, m. thinker: meditate, meditator, meditation, premeditated, unpremeditation.

Gr. *medos*, counsel, plan; see aim

Gr. *melo*, care for; *melema*, object of care, darling, duty; see *meledonos* under guard

Gr. *merimna*, f. anxious care or thought, solicitude; *merimnema, -tos*, n. anxiety: *Merimna atrata* (a beetle).

Gr. *noëma, -tos*, n. perception, thought; *noëmation*, n. dim.; *noëmon, -os; noëros, noëtikos*, intellectual, intelligent, thoughtful; *noësis*, f. intelligence, thought; *noëtos*, thinkable, reasonable; *noös*, m. mind, intellect; *noötes, -etos*, f. intellectuality, < *noëo*, perceive, think; *anchinoia*, f. sagacity; *anchinoös*, sagacious; *autonoös*, self-minded, independent; *ennoia*, f. idea, concept, thought; *ennoös*, thoughtful, shrewd: noetic, noematics, dianoetic, paranoiac, noumenon, *Ennoia briareus* (a gastropod), *Autonoë longipes* (an amphipod).

L. *opinor, -atus*, suppose, imagine, think; *opinio, -onis*, f. conjecture, view, fancy, belief: opinion, opine, opinionated.

Gr. *phrontis, -idos*, f. thought, attention, < *phroneo*, think; *phronimos*, thoughtful, prudent:

L. *pondero, -atus*, consider, meditate upon, < *pendo, pensus*, weigh: ponder, pensive, prepense, Milton's *Il penseroso*, pansy *(Viola tricolor)*.

Gr. *promethes*, forethinking, cautious; *epimethes*, afterthinking, careless: Prometheus, Epimetheus, *Promethes pulchellus* (a wasp).

L. *providus; provisus; prudens, -entis*, foreseeing, cautious: provident, provision, improvidence, provide, prudent, prudential, jurisprudence.

L. *puto, -atus*, clear up, think, consider, imagine; *putativus*, imaginary, suppositional; *deputo*, consider, esteem, allot: putative, computation, dispute, imputation, amputate, discount, deputy, disreputable, unaccountable.

L. *reor, ratus*, reckon, think; *ratio, -onis*, f. reckoning, procedure, understanding; *ratiuncula*, f. dim.; *rationalis*, reasonable: rate, ratify, ratification, rational, ratio, reason, ratiocination, irrational, arraign.

L. *sentio, sensus*, feel, think; see feel

Gr. *skeptikos*, thoughtful, critically doubtful, < *skeptomai*, look carefully, examine, consider; *skemma, -tos*, n. question, speculation; *syskemma*, joint consideration: skeptic, skepticism, *Skemmatopyge*[*Scemmatopyge*] *tietzei* (a trilobite).

Gr. *sophron*, sensible, wise; see *phren* under mind

L. *suspicor, -atus*, mistrust, surmise; see *suspicio* under doubt

See: **judge, separate, try, hypothesis, ask, read, know, learn, mind**

thino- < Gr. *this, thinos*, beach, shore, strand, sand-heap; see shore

thio- < Gr. *theion*, sulfur; see sulfur

thirst < AS. *thurst;* see dry, drink

thistle < AS. *thistel*.

L. *acanthium* (Gr. *akanthion*), n. a kind of thistle: *Onopordum acanthium* (Scotch thistle).

L. *acorna* (Gr. *akarna; akorna*), f. a kind of thistle, prickly plant: *Acarna ustulata* (a bug).

Gr. *akanos*, m. a kind of thistle: acanaceous, *Akania hilli* (an Australian tree).

L. *ascalia*, f. edible part of an artichoke:

L. *atractylis, -idis* (Gr. *atraktylis, -idos*), f. a thistlelike plant: *Atractylis gummifera* (a composite).

L. *cactus* (Gr. *kaktos*), a prickly plant; name now applied to cactus: see cactus

L. *carduus*, m. thistle: card, chard *(Beta cicla)*, *Carduus crispus* (a thistle), *Cynara cardunculus* (cardoon), *Aphis cardui* (an aphid), *Salvia carduacea* (thistle-sage).

NL. *carthamus*, m. a genus of thistles, < Ar. *qartam*, safflower: *Carthamus tinctorius* (safflower).

Gr. *chalkeios*, f. a thistlelike plant:

L. *cirsium* (Gr. *kirsion*), n. thistle: *Cirsium arvense* (Canada thistle).

L. *cnicus (cnecus)*, m. < Gr. *knekos*, f. a kind of thistle: cnicin, *Cnicus benedictus* (blessed thistle).

L. *cynara* (Gr. *kynara*), f. a kind of artichoke: *Cynara scolymus* (artichoke).

L. *dipsacus* (Gr. *dipsakos*), m. teasel: *Dipsacus pilosus* (hairy teasel), *Cucumis dipsacus* (teasel-gourd).

L. *eryngium* (Gr. *eryngion*), n. probably a thistle, but name now given to a genus of the carrot family; see **carrot**

L. *gallidraga*, f. hairy teasel:

Gr. *ixine*, f. a kind of thistle:

Gr. *knaphos (gnaphos)*, m. teasel, carding-comb; *knapheus*, m. fuller: *Cnaphoscapus bisignatus* (a beetle), *Gnaphocercus adela* (a bird).

L. *onopordum* (Gr. *onopordon*), n. a kind of thistle: *Onopordum polycephalum* (a cotton-thistle).

L. *scolymus* (Gr. *skolymos*), m. a kind of thistle, artichoke: *Scolymus hispanicus* (Spanish oyster-plant), *Anthocharis scolymus* (a butterfly).

L. *silybum*, n. (Gr. *silybon*, n.; *silybos*, m.), a kind of thistle: *Silybum eburneum* milk-thistle).

L. *sonchus* (Gr. *sonchos*), m. sow-thistle: *Sonchus oleraceus* (a sow-thistle), *Francoa sonchifolia* (a saxifragacead).

thixis, Gr. touch; see *thigma* under **touch**

thladias; thlibias, Gr. eunuch; see **sterile**

thlaspis (thlaspi), Gr. a kind of cress; see **mustard**

thlastos, Gr. bruised, crushed; *thlasma*, a bruise; see *thlao* under **bruise**

thlipsis, Gr. pressure; see *thlibo* under **press**

thnetos, Gr. mortal; see *thanatos* under **death**

thoë, Gr. penalty, see **punish**; quick, nimble, swift, see *thoös* under **swift**

thoeno- < Gr. *thoine*, meal, banquet, feast; *thoinater*, giver of a feast; *thoinatikos*, of a feast; see **eat**

tholia, Gr. conical hat with a broad brim; see **cap**

tholos, Gr. mud, dirt; *tholeros*, muddy, see **earth**; dome, cupola, see **arc**

tholus, L. (Gr. *tholos*), dome, cupola; *tholion*, dim.; see **arc**

thominx, Gr. cord, string, line, thread; see **rope**

thomisso, Gr. whip, scourge; see **whip**

thomos, Gr. heap; see **heap**

thoös, thoë, thoön, Gr. quick, nimble, swift; see **swift**

thong < AS. *thwang, thwong;* see **strap**

thops, Gr. flatterer, fawner; see **flatter**

thorax, Gr. breastplate, cuirass; the part of the body covered by this armor, chest; see **breast**

thorecto- < Gr. *thorektes*, warrior armed with a breastplate; see **shield**

thorium < ON. *Thor*, god of thunder; see ELEMENTS under **thing**

thorn < AS. *thorn*.

Gr. *akaina*, f. thorn, spine: *Acaena microphylla* (redspine sheep-bur), *Acinonyx jubatus* (African cheetah), *Acinopterus acuminatus* (a bug).

Gr. *akantha*, f. thorn, prickle; *akanthos*, m. a prickly plant; *akanthikos; akanthinos; akanthodes*, thorny: acanthus *(Acanthus mollis)*, *Acanthophorus serraticornis* (a beetle), *Acanthopanax senticosus* (an araliacead), *Acanthoproctus cervinus* (a tettigonid), *Pyracantha coccinea* (firethorn), *Gleditschia triacanthos* (honey-locust), *Cephalacanthus volitans* (gurnard), *Glymmatacanthus irishi* (a fossil fish), *Cyrtacanthacris rubella* (a grasshopper).

L. *batus* (Gr. *batos*), bramble, blackerry, thorn-bush; see **blackberry**

AS. *brer*, thorn, prickle: briar, brier, greenbrier, sweetbrier.

Gr. *chaskanon*, cocklebur; see **composite**

L. *dumus*, thorn-bush, bramble; see **bush**

L. *echinatus*, spiny, prickly; see *echinos* under **hedgehog**

AS. *hagathorn*, hedgethorn, < *haga (hecg)*, hedge, *thorn*, thorn: hawthorn, haw, hag, haggard, Hawthorne.

L. *hystricosus*, prickly, thorny, < *hystrix*, porcupine; see **porcupine**
L. *luma*, f. thorn; *lumarius*, of thorns; *lumectum*, n. thorn-thicket:
L. *muricatus*, pointed or spiny like a murex; see *murex* under **mollusk**
Gr. *pheos*, m. a spiny plant: *Hippophae salicifolia* (a sea-buckthorn).
Gr. *rhachos*, f. thorn-bush, brier:
L. *sentis, -is*, m. thorn, brier, bramble; *senticosus; sentosus; sentus*, thorny, rough;
  senticetum, n. thorn-thicket: *Xylosma senticosa* (shiny xylosma), *Zaphrentis
  sentosa* (a fossil coral), *Hydatina senta* (a rotifer), *Coelosternus sentus* (a
  beetle).
Gr. *skolos; skolops, -opos*, m. anything pointed, thorn, pale; *skolopion*, n. dim.:
  Scolochloa festucacea (a grass), *Scologaster fuscipennis* (a fly), *Scolocephalus
  mirabilis* (a grasshopper), *Scolopia crenata* (a flacourtiacead), *Scolopocerus
  secundarius* (a bug), *Cnidoscolus texanus* (a treadsoftly).
L. *spina*, f. thorn; *spineola; spinula*, f. dim.; *spineus*, of thorns; *spinosus*, thorny;
  spinetum, n. thorn-hedge: spine, spinal, spinate, spinel, spinescent, porcupine,
  Acanthus spinosus (spiny acanthus), *Prunus spinosa* (sloe), *Rosa spinosissima*
  (Scotch rose), *Ancyloceras spinigerum* (a fossil cephalopod), *Pycnogonum
  spinipes* (a sea-spider), *Carissa spinarum* (Ceylon carissa), *Systenocerus spinifer* (a beetle), *Ophiacantha spinulosa* (a brittlestar), *Androsace spinulifera* (a
  rock-jasmine), *Opilio spinulatus* (a harvestman), *Spinifex hirsutus* (an Australian
  grass), spinulosin (from *Penicillium spinulosum*), *Opuntia spinosior* (a cane-
  cactus), *Glossopetalon spinescens* (a celastracead).
AS. *sticel* (G. *stachel*), prickle, sting, thorn: stick, stickle, stickleback *(Gasterosteus
  aculeatus)*.
L. *tribulus* (Gr. *tribolos*), m. caltrop; *tribulosus*, thorny: *Tribulus terrestris*
  (puncture-vine), *Cenchrus tribuloides* (dune-sandbur), *Tribolium ferrugineum*
  (a flour-beetle).
L. *vepres (vepris), -is*, m.; *veprecula*, f. dim.: *Vepris lanceolata* (a rutacead).
See: **point, thistle**, *crataegus* under **rose**, *rubus* under **blackberry, bush, rough,
  cactus**

thoros, Gr. male seed, semen; *thorikos*, of seed; see **seed**

thorp < AS. *thorp*, village; see **town**

thorybos, Gr. noise of a crowded assembly, applause, cheers, uproar; see **sound**

thos, *thoos*, Gr. jackal; see **wolf**

thoughtless; see **neglect**

thousand < AS. *thusend*.
Gr. *chilios* (pl. *chilioi*), thousand; *chilias, -ados*, f. a thousand; *chiliostos*,
  thousandth: chiliad, chiliarch, chiliasm, chiliomb, kilogram, kilometer, kilowatt,
  kiloliter, *Chilianthus arboreus* (a loganiacead), *Chiliophyllum densifolium* (a
  composite).
L. *mille*, a thousand; *millenarius*, of a thousand; *millesimus*, thousandth; *milli-
  modus*, thousandfold: milleped, millennium, millimeter, mile, mileage, million,
  millionaire, *Achillea millefolium* (yarrow, milfoil), *Millepora alcicornis* (a coral).

thoysmos, Gr. a barking, baying; see **bark**

thranis, Gr. swordfish; see **swordfish**

thranos, Gr. bench, stool; *thranion*, dim.; see **sit**

thrassa (thratta), Gr. a small, herringlike fish; *thrattidion*, dim.; see **herring**

thrasys, Gr. bold, courageous, bragging; *thrasos*, boldness, courage; see **bold**

thraulos, Gr. brittle; see *thrauo* under **break**

thraupalos, Gr. a shrub; see **honeysuckle**

thraupis, Gr. a kind of finch; see **finch**

thraustos, Gr. brittle; see *thrauo* under **break**

thread < AS. *thraed*.
L. *acia*, f. thread for sewing:
Gr. *agathis -idos*, f. ball of thread; see **ball**
L. *byssus* (Gr. *byssos*), fine flax, cotton, thread; see **flax**
L. *cirritus*, having fine filaments; see *cirrus* under **curl**
L. *fibra*, f. thread, filament, sinew: fiber, fibrilla, fibril, fibrillar, fibrillose, fibroid,
  fibrous, fibroma, fibrovascular, fibrolite, fibrin, *Sisyrinchium fibrosum* (a blue-
  eyed grass).
L. *filum*, n. thread: filiform, filigree, filament, fillet, filoplume, filopodium, filar,
  filoselle, bifilar, filarial, filariasis, defile, enfilade, *Filipendula hexapetala* (dropwort), *Filistata incerta* (a spider), *Filago californica* (fluffweed), *Gifola germanica*
  (cotton-rose), *Filaria bancrofti* (a nematode causing elephantiasis), *Yucca filamentosa* (a yucca), *Pigafetta filaris* (a palm), *Uromitus filiferus* (a swallow),

Schizocrania filosa (a fossil brachiopod), *Microfilaria columbae* (a nematode), *Perilissus filicornis* (a wasp), *Vittaria filifolia* (a grass-fern), *Stylocline filaginea* (a composite), *Washingtonia filifera* (a palm), *Oxypolis filiformis* (cowbane).

L. *fimbria*, fiber, thread, fringe, shred; *fimbriatus*, fringed, fibrous; see **fringe**

Gr. *harpedon*, cord, thread; see **rope**

Ab.Am. *henequen*, a kind of fiber: henequen (from *Agave fourcroydes*).

Ab.Am. *ichtle*; Sp. *ixtle*, a kind of fiber: istle.

Gr. *is, inos*, f. fiber, sinew, muscle; L. *ina*, f. fiber; *inodes*, fibrous: inoma, inosin, inositol, inosite, hinoid, amacrine, *Inoceramus umbonatus* (a fossil pelecypod), *Inodes yapa* (yapa-palm), *Calophyllum inophyllum* (a guttifer).

Gr. *kairos*, m. thrum or ends of the threads in a loom: *Caerosternus laevissimus* (a beetle).

Gr. *klosma, -tos*, n. thread, line, clue; *klostron*, n. clue; *klosmation*, n. dim.:

Gr. *kroke*, thread, woof in a piece of cloth; see **wool**

Gr. *ktedon, -os*, f. fiber; *euktedon, -os*, with straight fibers: *Ctedonia bipunctata* (a fly), *Ctedoctema acanthocrypta* (a protozoan).

L. *licinium*, n. lint:

L. *licium*, n. thrum or ends of the threads in a loom; *bilix, -icis*, with a double thread; *trilix*, triple-twilled:

L. *linteus*, of linen, linty; see **linen**

Gr. *mitos*, m. thread; *mitodes*, threadlike; *mitolinon*, n. linen thread: mitosis, mitotic, dimity, samite, *Leptomitus epidermidis* (a fungus), *Dicromita agassizi* (a fish), *Mitobates conspersus* (a spider).

L. *motarium*, n. lint:

Gr. *nema, -tos*, n. thread; *nemation*, n. dim.; *nematodes*, threadlike, filamentous: nematocyst, nemaline, nemalite, nematode, nematocide, Nemathelminthes, *Nematospira turgida* (a nematode), *Nemichthys avocetta* (thread-eel), *Nemalion multifidum* (a red alga), *Gymnema sylvestre* (an asclepiadacead), *Treponema pallidum* (the spirochaete of syphilis), *Nemopanthus mucronata* (mountainholly).

Gr. *pene*, thread on the bobbin; see *penos* under **weave**

Gr. *rhamma*, suture, seam, thread; see *rhapto* under **sew**

Gr. *rhodane*, f. spun thread:

Gr. *rhompheus*, m. shoemaker's waxed thread:

Ab.Am. *Sisal*, a seaport of Yucatan: sisal (from *Agave sisalana*).

L. *stamen, -inis*, n. (Gr. *stemon, -os*, m.), thread, fiber, warp, male organ of a flower; *stemonarion*; *stemonion*, n. dim.; *stema, -tos*, n. penis; *staminatus*; *stamineus*, consisting of threads, thready, fibrous: stamen, staminate, stamina, *Stemonidium hypomelas* (a fish), *Callistemon coccineus* (scarlet bottle-brush), *Pentstemon hirsutus* (hairy pentstemon), *Petalostemon villosus* (silky prairie-clover), *Xanthostemon pubescens* (a myrtacead), *Distemonanthus benthamianus* (a legume), *Exostema industum* (a rubiacead), *Trichostema dichotomum* (blue curls), *Conomitra staminea* (a fossil gastropod), *Vaccineum stamineum* (deerberry), *Stemonitis fusca* (a myxomycete).

Gr. *stremma*, that which is twisted, thread; see *steptos* under **turn**

L. *stupa* (Gr. *styppe*), coarse fiber of flax or hemp, tow, oakum; *stupeus*, of tow; see **flax**

Gr. *tenon, -ontos*, m. sinew: *Tenontomyia gracilipes* (a fly).

Gr. *thysanos*, fringe, tassel; see **fringe**

Gr. *tilma, -tos*, n. lint:

Gr. *tilos*, m. shred, fiber:

Gr. *tolype*, ball of yarn; see **ball**

See: **rope, cotton, flax, hair, silk, wool, fringe, nerve, sew, weave**

threat < AS. *threat;* see **danger**

three < AS. *threo; thri:* threefold, thrice, third, thirty, thirteen.

L. *cubus* (Gr. *kybos*), m. a solid with six equal square sides, die; *cubicus*, three-dimensional; *kybeutes*, m. gambler; *kybeuterion*, n. gambling-house; *kybeutikos*, of dicing: cube, cubic, cubical, cuboid, *Acrocubus arcuatus* (a radiolarian), *Cybocephalus pulchellus* (a beetle).

L. *tres* (Gr. *treis, trion*); *trinalis*; *trinus*, three; *ter-; tris-; tri-*, thrice; *ternarius; ternatus; ternus;* Gr. *trias, -ados*, of three, in three's; *tertiarius; tertius; triens, -entis;* Gr. *tritos*, third; *tertianus*, of a third; *trientalis*, third of a foot; *trifarius; triplaris; triplex, -icis;* Gr. *tricha; trichtha; tridymos; triplax, -akos; triptychos; trissos; trixos*, threefold; *tredecim*, thirteen; *triceni; triginta;* Gr. *triakonta*, thirty; *tricesimus;* Gr. *triakostos*, thirtieth; *tricenti*, three hundred: triangle, tripod, trio, trident, trinity, triple, trilobite, Triassic, triquetrous, triad, Trinidad, trivial, trefoil, treble, trey, ternate, Tertiary, tercentenary, ternary, tierce, testis, trilling, sesterce, trammel, tridymite, triplite, triploid, *Trifolium incarnatum* (crimson clover), *Trithrinax campestris* (a palm), *Triadocidaris canaliculata* (a sea-urchin), *Trixis californica* (a composite), *Triplaris americana* (an ant-tree), *Tritochaeta prosopoides* (a fly), *Trichthacerus peristedi* (a copepod), *Trissocyclus stauroporus* (a radiolarian), *Trientalis borealis* (star-flower), *Clitoria ternatea* (a legume), *Macadamia ternifolia* (Queensland nut), *Magnolia tripetala* (umbrella-magnolia), *Ambrosia trifida* (tall ragweed), *Nassa trivittata* (a gastropod), *Viola tricolor* (pansy), *Trillium grandiflorum* (snow-trillium), *Arisaema triphyllum* (jack-in-the-pulpit), *Chasmorhynchus tricarunculatus* (a bell-bird).

thremma, Gr. creature, slave; see **animal**

threnos, Gr. funeral song, dirge, lament; see **sing**

threpsis, Gr. nourishment; *threpter,* feeder, rearer; see *trepho* under **eat**

thresco- < Gr. *threskos,* religious; see **rite**

thresh < AS. *threscan,* remove grain from the stalks; see **strike, separate**

Thria, Gr. nurse of Apollo and inventor of divination by drawing pebbles from an urn; *thriobolos,* soothsayer; see **prophecy**

thriambos, Gr. triumph; see **victory**

thridaco- < Gr. *thridax, -akos,* lettuce; *thridakinos,* of lettuce; see **lettuce**

thrift < ON. *thrift; thrif;* see **save, grow**

thrill < AS. *thyrel,* < *thyrlian,* perforate, < *thurh,* through; see **joy, move**

thrinaco- < Gr. *thrinax, -akos,* trident, see **fork**; a genus of palms, see **palm**

thrinco- < Gr. *thrinkos,* eaves, cornice, coping; *thrinkion,* dim.; see **border**

thrion, Gr. fig-leaf, leaf; see **leaf**

thrips, L., Gr. a genus of Thysanoptera; see **bug**

thrissa, Gr. a kind of anchovy; see **herring**

thrix, *trichos,* Gr. hair; *trichion,* dim.; see **hair**

throat < AS. *throte.*
  Gr. *brochthos,* m. throat:
  Gr. *bronchos,* m. windpipe, trachea: bronchial, bronchitis, bronchotherapy, bronchotomy, *Saxifraga bronchialis* (a saxifrage).
  L. *faux, faucis,* f. throat, pharynx; pl. *fauces:* faucal, fauces, suffocate.
  Gr. *glottis, -idos,* f. mouth of the windpipe: glottis, epiglottis.
  L. *gula,* f.; *gurgulio, -onis,* m. gullet, weasand, throat: gullet, gular, gargoyle, gules, gully, *gularis (Fondulopanchax caeruleus), Cnemidophorus gularis* (a lizard), *Anas fulvigula* (a duck), *Chrysococcyx flavigularis* (a cuckoo), *Neotoma albigula* (a wood-rat).
  L. *guttur, -uris,* n. throat; *gutturosus,* with enlarged throat, goitered: guttural, goiter.
  L. *jugulum,* n. collarbone, throat: *jugulator, -is,* m. cutthroat, murderer: jugular, jugulate, jugulator.
  Gr. *laimos,* m. throat, gullet: brachylaemid, laemoparalysis, laemostenosis, *Laemobothrium nigrum* (a bird-louse), *Cryptolaemus subviolaceus* (a beetle), *Pyrrholaemus brunneus* (a wren), *Xantholaema haematocephala* (a barbet).
  Gr. *larynx, -yngos,* m. upper part of the windpipe, gullet: larynx, laryngeal, laryngitis, laryngoscope, laryngectomy.
  Gr. *laukanie (leukanie),* f. throat:
  Gr. *oisophagos,* m. gullet, weasand: esophagus, esophageal.
  Gr. *pharynx, -yngos,* f. throat: pharynx, pharyngeal, pharyngitis, *Saccopharynx ampullaceus* (a fish).
  L. *rumen,* throat, gullet, paunch; see **stomach**
  Gr. *sphax, -agos,* f. throat: *Sphagolobus atratus* (a hornbill).
  L. *trachia* (Gr. *tracheia*), f. windpipe: trachea, Tracheata, tracheid, tracheary, tracheal, tracheitis, tracheole, *Syngamus trachealis* (gapeworm), *Pseudopeziza tracheiphila* (a fungus).
  See: **neck, pipe, mouth, hollow**

throb < uncertain origin; see **strike**

thrombos, Gr. lump, bloodclot; *thrombion,* dim.; see **lump**

throne < L. *thronus* (Gr. *thronos*), an elevated seat; see **sit**

throsco- < Gr. *throsko,* leap; see **leap**

throsis, Gr. cord, line; see **rope**

through < AS. *thurh.*
  Gr. *dia,* through, between, during, as in *diametreo,* measure through: diameter, dialogue, diagnosis, diabetes, diagonal, diabase, dialect, dialysis, diaphanous, diaphragm, diarrhea, diastole, diatom, diatribe, diallage, diastem, deacon, *Diabrotica vittata* (striped cucumber-beetle).

L. *pel-, per-,* < *per,* through, by, by means of, as in *pellego,* read through; *permeo,* pass through: perfect, perforation, perjure, perfume, perception, permission, perspiration, perplex, perversion, persecute, persuasive, permutation, impervious, perish, parboil, pardon, paramount, paramour, pellucid, pilgrim, purlieu, ampersand, *Specularia perfoliata* (a campanulacead).

throw < AS. *thrawan.*

Gr. *akontizo,* hurl a javelin:

L. *allido, allisus,* dash, hurl, break:

Gr. *anochlizo,* heave upward:

Gr. *ballo,* throw, cast, change; *ballistes,* m. thrower; L. *ballista,* f. machine for throwing missiles; *blema, -tos,* n. throw, cast; *bole,* f.; *bolos,* m. a throw, cast, stroke; *bolis, -idos,* f. missile; *ekballo,* throw out; *ekbletos,* thrown away, rejected; *emballo,* throw in, insert; *lithobletos; lithobolos,* throwing stones; *probles, -etos,* thrown forward, projected: ballistics, parabola, parable, hyperbole, hyperbola, amphibole, symbol, bolide, bolometer, embolism, diabolical, emblem, problem, parley, parliament, catabolism, *Balistes[Ballistes] carolinensis* (a triggerfish), *Boleosoma nigrum* (a darter), *Ecballium elaterium* (squirting cucumber), *Ecbletus simplex* (a beetle), *Ecbletodes psephenias* (a moth), *Emballonura nigrescens* (a bat), *Sporobolus ramulosus* (red dropseed).

L. *catapulta,* f. a machine for hurling weapons: catapult.

Gr. *chermaster, -os,* m. slinger:

L. *conicio, conjectus,* throw together, cast, project: conjecture, conjectural.

L. *decutio, decussus,* shake off, cast off: decussion.

L. *demolior, demolitus,* throw down, tear down, destroy; see **destroy**

Gr. *exhoristos,* banished, expelled; see **banish**

L. *funda,* f. sling; *fundalis,* of a sling; *fundibularius; fundibulor, -is,* m. slinger: *Dolichopus funditor* (a fly).

L. *fustibalus,* m. a slingstaff:

Gr. *hemon, -os,* m. thrower, slinger; *hemosyne,* f. skill in throwing: *Synemosyna formica* (antlike spider).

L. *jacio, jactus; jacto, -atus; jaculor, -atus,* hurl, throw; *jactabilis,* that can be thrown; *jactabundus,* tossing to and fro, stormy; *jaculabilis,* that may be thrown; *jaculus,* thrown; m. thing thrown; *disjicio, -jectus,* throw around, scatter; *ejectus,* cast out, expelled; *ejaculor, -atus,* throw out, shoot out; *injicio,* cast in, put in; *rejectus,* thrown away: jactation, jaculatory, jaculative, ejaculate, conjecture, dejection, subjective, abjectly, eject, rejection, injector, adjective, projectile, gist, jet, jut, jettison, jetsam, jetty, *Alactaga jaculus* (jerboa), *Apate rejecta* (a beetle).

ON. *kasta,* throw: cast, castaway, casting, broadcast, forecast, telecast, outcast.

Gr. *leuster, -os,* m. stoner; *leustos,* stoned:

Gr. *paltos,* brandished, hurled; *palton,* anything hurled, spear, dart; see **spear**

L. *petrabulum,* n. catapult:

Gr. *rhipto,* throw; *rhimma, -tos,* n. cast, throw; *rhipsis,* f. a throwing, casting; *rhiptos,* thrown: rhiptoglossate.

Gr. *sphendone,* f. sling; *sphendonetes,* m. slinger; *sphendonesis,* f. a slinging: *Sphendononema camerunense* (a centiped).

L. *sternax, -acis,* throwing to the ground, unhorsing:

L. *supo (sipo), -atus,* throw about, scatter: dissipate, dissipation.

See: **banish, spear, ball, vomit**

thrush < AS. *thrysce:* song-thrush or throstle *(Turdus musicus).*

L. *cichla* (Gr. *kichle*), f. thrush: *Cichlasoma nebuliferum* (a fish), *Hylocichla guttata* (hermit-thrush).

L. *erithacus* (Gr. *eritkakos*), m. probably a robin: *Erithacus rubecula* (robin-redbreast), *Psittacus erithacus* (gray parrot).

Gr. *illas (ilias), -ados,* f. a kind of thrush: *Illadopsis fulvescens* (a babbler), *Diaphorillas striata* (a bird).

Gr. *ixoboros,* f. missel-thrush:

Gr. *kollyrion, -os,* m. a kind of thrush: *Lanius collurio* (red-backed shrike).

Gr. *kossyphos (kopsichos),* m. probably a kind of thrush: *Copsychus saularis* (magpie-robin), *Cossyphopsis reevei* (a thrush), *Cossypha caffra* (a robin-chat), *Plesius cossyphus* (a beetle).

Gr. *laedos,* m. probably a thrush:

Gr. *laios,* m. a kind of thrush: *Urolais mariae* (a bird).

L. *merula,* f. a blackbird, thrush: *Merula torquata* (ring-ouzel), *Merulinus tremellosus* (a fungus), *Turdus merula* (European blackbird, a thrush).

NL. *sialia,* a genus of bluebirds; see *sialis* under **bird**

Gr. *trichas, -ados,* f. a kind of thrush: *Trichas personata* (a bird).

L. *turdus,* m. thrush: *Turdus migratorius* (robin).

Gr. *tylas, -ados,* f. a kind of thrush: *Tylas eduardi* (a bird).

thrust < ON. *thrysta,* press, force, drive; see **push, press, projection**

thryallis, Gr. wick, a plant used for making wicks; see **wick**

thryligma, Gr. fragment; see **part**

thrylos, Gr. noise of many voices; see *thryleo* under **speak**

thrymma, Gr. piece, bit; see **part**

thryon, Gr. rush; see **sedge**

thrypsis, Gr. a breaking; *thryptikos,* breakable, brittle; see **break**

thuja, NL. < L., Gr. *thya; thyia,* a resinous sweet-scented tree, arborvitae; see **gymnosperm**

Thule, L. (Gr. *Thoule*), farthest north; see **north**

thumb < AS. *thumba;* see **finger**

thunder < AS. *thunor,* akin to ON. *Thor,* god of thunder: thunderbird, thunderbolt, thunderhead, thunderous, Thursday, thorium, *Thor floridanus* (a decapod).
　　Gr. *arados,* m. rumbling in the bowels: *Aradosyrtis ghiliani* (a bug).
　　Gr. *borborygmos,* m. rumbing in the bowels: borborygmus.
　　Gr. *bronte,* f. thunder; *brontema, -tos,* n. clap of thunder; *brontaios,* thundering; *brontes,* m. thunderer; *brontodes,* like thunder: *Brontosaurus excelsus* (a dinosaur), *Brontes ceylonicus* (a beetle), *Bronteopsis gregaria* (a trilobite).
　　L. *ceraunus* (Gr. *keraunos*), m. thunderbolt, a stone supposed to have fallen from the sky, meteorite; *keraunios,* of a thunderbolt: ceraunics, ceraunite, ceraunia.
　　L. *tonitrus,* thunder; see *tonus* under **sound**

thunnus, L. (Gr. *thynnos*), tunny; see **tunny**

thuro- < Gr. *thouros,* rushing, impetuous, see **swift**; < *thoures,* male, see **man**

thus, *thuris,* L. incense, see **smell**; *thuribulum,* censer, see **cup**

thwaite < ON. *thveit,* a tract of tilled land, clearing, meadow; see **field**

thwart < ON. *thvert,* transverse, across; see **against, cross**

thyado- < Gr. *thyas (thyias), -ados,* woman frantic for love, a Bacchante; see **mad**

thyaros, Gr. darnel, a kind of grass; see **grass**

thyella, Gr. storm, hurricane, whirlwind; see **storm**

thyestes, Gr. pestle; see **pestle**

thygater, Gr. daughter; see **daughter**

thyio- < Gr. *thyeia,* mortar; *thyeidion,* dim.; see **mortar**

thylaco- < Gr. *thylakos; thylax, -akos,* bag, sack, pouch; see **bag**

thymallos, Gr. a fish of the trout family, grayling; see **trout**

thymbra, Gr. savory; see **mint**

thyme < OF. *tym,* < L. *thymum* (Gr. *thymon*); see **mint**

thymelaea, L. (Gr. *thymelaia*), a genus of the mezereon family; see **mezereon**

thymele, Gr. place for sacrifice, altar; *thyma,* victim, sacrifice; see *thyos* under **give**

thymos, Gr. mind, soul, spirit, temper, see **mind**; warty excrescence, see **wart**

Thyone (Semele), Gr. mother of Bacchus; see *Bacchus* under **wine**

thyos, Gr. sacrifice, offering; see **give**

thyra, Gr. door; *thyrion; thyris, -idos,* dim. window, see **door**; *thyron,* vestibule, see **room**

thyreos, Gr. a large, oblong, door-shaped shield; see **shield**

thyrsus, L. (Gr. *thyrsos*), a close-branched cluster or panicle; see **cluster**

thysanos, Gr. fringe, tassel; see **fringe**

thysio- < Gr. *thysia,* rite, sacrifice; see *thyos* under **give**

thysthlon, Gr. torch carried in the Bacchic festival; see **light**

-tia, L. suffix denoting condition, pertaining to; see *-ia* under **nature**

tiara, L., Gr. turban; see **cap**

tibia, L. shinbone, pipe, flute; see **leg**

tibicen, L. flutist, piper, fifer; see *tibia* under **leg**

tibulus, L. a kind of pine; see **pine**

ticho- < Gr. *teichos,* wall around a city; *teichion; teichydrion,* dim.; see **wall**

tick < AS. *tica;* G. *zecke.*
　　NL. *acarus,* m. (Gr. *akari,* n.), mite: acarina, acarinosis, acariasis, *Acarus serotinae* (a mite), *Abacarus hystrix* (a mite), *Acarinicola claviger* (a nematode), *Acariscus multisetosa* (a mite).
　　NL. *argas,* m. a genus of ticks: *Argas persicus* (chicken-tick).

Gr. *kroton, -os,* m. tick, a euphorbiaceous plant with seeds resembling ticks: croton oil (from *Croton tiglium*), *Crotophaga sulcirostris* (groove-billed ani), *Crotonus variegatus* (a mite).

Gr. *kynorhaistes,* m. dog-tick:

AS. *mite,* a minute arachnid: mite, miticidal.

L. *ricinus,* tick, castor-oil plant; see **spurge**

See: **flea**

tickle < uncertain origin.

Gr. *gargalizo,* tickle; *gargalismos,* m. a tickling; *dysgargalis,* very ticklish, skittish:

Gr. *ginglismos,* m. tickling, laughter:

L. *titillo, -atus,* tickle, touch lightly: titillate, titillation, *Monochamus titillator* (a beetle).

See: **itch, tease, touch, laugh**

tide < AS. *tid,* time; see **time, swing, wave**

tidy < AS. *tid,* time; see **class, fit**

tie < AS. *tigan;* see **bind**

tiger < L. *tigris; tigrinus,* of tigers; see **cat**

tight < uncertain origin; see **close, narrow, bind, press, draw, lessen**

tignum, L. beam of wood, log; *tigillum,* dim.; *tignarius,* carpenter; see **beam**

tile < AS. *tigle,* < L. *tegula;* see **shingle, brick**

tilia, L. linden; see **linden**

till < AS. *tilian,* cultivate.

Gr. *aneastos,* unplowed:

L. *aro, -atus,* plow; see **plow**

Gr. *aulakizo,* plow; see *aulax* under **furrow**

L. *colo, cultus,* till, farm, inhabit; *colonia,* f. farm, estate, settlement; *colonus,* m. farmer, settler; *cultum,* n. cultivated land, farm, garden, plantation; *cultor, -is,* m. husbandman, planter, tiller; *cultura,* f. care, polish, refinement: cultivate, culture, agriculture, horticulture, colonist, inquiline.

Gr. *Demeter,* goddess of agriculture, equivalent to the Roman Ceres; see **grain**

Gr. *georgyos,* farming, husbandman; *georgikos,* agricultural; see **farmer**

L. *imporco, -atus,* put into furrows; see *porca* under **ridge**

L. *liro, -atus,* plow or harrow; see *lira* under **ridge**

L. *occo, -atus,* harrow; *occa,* f. harrow; *occator, -is,* m. harrower: *Farrea occa* (a sponge).

L. *pastino, -atus,* prepare the ground by digging and trenching; see *pastinum* under **hoe**

Gr. *phyteuo,* plant; see *phyton* under **plant**

L. *runco, -atus,* remove weeds; *runcator, -is,* m. weeder; *runco, -onis,* m. hoe:

L. *sarculo, -atus,* hoe; see *sarculum* under **hoe**

L. *sarrio, -itus,* hoe, weed; *sarritor, -is,* m. hoer, weeder; *sarritorius,* of hoeing or wedding; *consarrio,* hoe or rake to pieces:

L. *sativus,* sown, planted, cultivated, < *sero, satus,* sow; see **sow**

L. *Saturnus,* m. god of agriculture; *Saturnalis,* of Saturn: Saturn, Saturnalia, saturnine, Saturday, *Saturnia atlantica* (a moth), *Saturnulus planetes* (a radiolarian), *Leptogium saturninum* (a lichen).

L. *subigo, subactus,* plow, cultivate, conquer, subdue:

L. *sulco, -atus,* plow, harrow; *desulco,* plow up, furrow through; see *sulcus* under **furrow**

L. *vervago,* plow a fallow field; see *vervactum* under **field**

See: **dig, hoe, pick, plow, rake, shovel, plant, field, farmer**

tillandsia, NL. a genus of the pineapple family; see **pineapple**

tillo, Gr. pluck, tear; see **tear**

tilma, Gr. anything plucked, pulled, or shredded; *tilmation,* dim.; *tiltos,* shredded; see *tillo* under **tear**

tilon, Gr. a fresh-water fish; see **fish**

tilos, Gr. shred, fiber, see **thread**; a thin stool as in diarrhea, see **dung**

time < AS. *tima:* timely, timeless, timekeeper, timepiece, time-signal, meantime, mealtime, bedtime, springtime.

L. *aliquantisper,* for a time:

Gr. *chronos,* m. time; *chronikos,* of time; *chronios,* lasting, enduring, lingering; *chronizo,* spend time, pass time, continue, delay, linger: chronometer, chronicle, chronic, chronology, anachronism, synchronize, synchronous, synchrotron.

L. *clepsydra* (Gr. *klepsydra*), f. water-clock: clepsydra, *Ganoëssus clepsydra* (a fossil fish).

Gr. *deipnestos,* mealtime; see *deipnon* under **eat**

L. *-ernus, -urnus,* pertaining to time, as *hibernus,* of winter; *diurnus,* of the day: hibernal, diurnal, sempiternal, nocturnal.

Gr. *hama,* at the same time with; see **with**

L., Gr. *hora,* f. hour; *horalis,* of an hour; *horarium,* n. dial, clock; *horologium* (Gr. *horologion*), n. clock: horal, hour, hourglass, horologue, horoscope, horology.

L. *interim,* meantime, meanwhile: interim.

Gr. *kairos,* fit, opportune, seasonable, of the right or critical time; see **fit**

L. *minuta,* f. minute: minute.

L. *mox,* soon, directly:

L. *promptus,* at hand, on time, ready: prompt.

L. *secunda,* f. second: second.

L. *semel,* once, a single time:

L. *simul,* at the same time: simultaneous.

L. *solarium,* sundial; see **dial**

L. *tempus, temporis,* n. time; *temporalis; temporarius,* of time, lasting only a time; *tempestivus; temporaneus,* timely, opportune, seasonable: temporal, contemporary, extemporaneous, *tempus fugit,* pro tem, tempest, tempestive, tense, *Rana temporaria* (a frog).

AS. *tid,* time: tide, betide, eventide, Whitsuntide, Christmastide, Yuletide, tidy, "Time and tide wait for no man".

See: **day, week, month, year, age, young, old**

timidus, L. fearful; see **fear**

timo- < Gr. *time,* honor; *timetos,* honorable; *timios,* worthy; *timoros,* upholding honor; see **honor**

tin < AS. *tin.*   Derive: tinsel, tinker.   See ELEMENTS under **thing**
Gr. *kassiteros,* m. tin: cassiterite.

L. *stannum,* n. tin; *stanneus,* of tin: stannic, stannous, stanniferous, stannate, stannary, stannite, *Mycena stannea* (a fungus).

tina, L. a wine-vessel; see **pot**

tinagma, Gr. a shake, quake; *tinaktor,* shaker; see **shake**

tinca, L. tench; see **carp**

tinctorius, L. of dyeing; *tinctura,* a dyeing; see *tingo* under **color**

tinder < AS. *tynder,* kindling, punk; see **fire**

tinea, L. larva of a moth; *tineola,* dim.; see **moth**

tingabarinos, Gr. vermilion; see **red**

tingens, L. dyer; see *tingo* under **color**

tinnitus, L. a ringing; *tinnulus,* ringing, tinkling; see *tinnio* under **ring**

tinnunculus, L. kestrel; see **hawk**

tino- < Gr. *teino,* stretch, see **spread**; < *tino,* pay a penalty, see **pay**

tintinnabulum, L. bell; see **bell**

tinus, L. a plant; see **plant**

-tion, L. suffix denoting act, process, result, or state of; see *-ion* under **nature**

tiphia, NL. a genus of wasps; see **wasp**

tiphos, Gr. standing water, pool, marsh; see **marsh**

tiphys, Gr. nightmare; see **dream**

tipula, L. water-spider, now applied to the cranefly; see **fly**

tire < AS. *teorian,* be tired, weakened, see **weak**; cover, rim, tread of a wheel, < OF. *atirer,* dress, cover, see **border**

tiro, L. recruit, beginner; *tirunculus,* dim.; see **begin**

tissue < OF. *tissu,* < L. *texo,* weave; see **weave, skin, flesh, net, cloth**

Titan, Gr. a son or daughter of Uranus and Gaea, symbolic of brute force and large size; see **large**

tithasos, Gr. tame, domesticated; see **tame**

tithene; tithenos; titthe, Gr. nurse, rearer; see **nurse**

Tithonus, L. (Gr. *Tithonos*), consort of Aurora, symbolic of decrepit old **age**; see **old**

tithymalus, L. (Gr. *tithymalos*), spurge, euphorbia; see **spurge**

titillo, *-atus,* L. tickle; see **tickle**

titio, L. firebrand; see **fire**

titivillitium, L. very small trifle, bagatelle; see **trifle**

title < L. *titulus,* inscription, label; see **word**

titmouse < AS. *titmase,* a bird.

 L. *acanthyllis, -idis* (Gr. *akanthyllis, -idos*), f. a small bird, probably a titmouse or goldfinch: *Acanthyllis spinicauda* (a bird).

 Gr. *aigithalos,* m. titmouse: *Aegithalus caudata* (bottle-tit), *Aegithaliscus concinnus* (red-headed tit).

 L. *parus,* m. titmouse; *parula,* f. dim.: Paridae, *Parus caeruleus* (blue titmouse), parula warbler, *Auriparus flaviceps* (verdin).

 NL. *penthestes,* m. chickadee: *Penthestes atricapillus* (black-capped chickadee).

 Gr. *spizites,* m. a titmouse:

titter < uncertain origin; see **laugh**

titthos, Gr. nipple, teat; *titthion,* dim.; see **udder**

titubo, *-atus,* L. stagger, totter, reel; see **shake**

titulus, L. inscription, label; see **word**

Tityos, Gr. a giant, punished in Hades for attempted rape by having a vulture feed constantly on his liver; see **punish**

tityros, Gr. shepherd's pipe; *tityristes,* piper; see **pipe**

tlemono-; tlesi-; tleto-, Gr. patient, enduring, suffering, < *tlao,* bear, carry, suffer; see *Atlas* under **carry**

tmema, Gr. portion, piece, section; *tmesis,* a cutting, separation; *tmetos,* cut; see *temno* under **cut**

to < AS. *to,* toward: today, tonight, tomorrow, together, heretofore, hitherto.

 L. *a-, ab-, ac-, ad-, af-, ag-, al-, an-, ap-, ar-, as-, at-,* < *ad,* to, as in *ascendo,* climb up, rise; *abbrevio,* abridge, shorten; *accedo,* come to, assent; *adduco,* lead to, bring; *affingo,* fasten to; *agglutino,* cement to; *alloquor,* speak to; *annuto,* nod to; *applico,* put to; *arrepo,* creep to; *assumo,* take to; *attraho,* draw to: abet, abeyance, amount, agreeable, abbreviate, abridge, ascend, aspiration, avalanche, accumulate, accept, adhere, admiration, add, affect, affirmative, aggregate, alleviation, allude, alluvium, announcement, annul, apprehend, applaud, arrest, arrange, arrogant, arbiter, assiduous, assimilate, attract, attach, caudad, dorsad, proximad, ventrad, *Nomada adducta* (a bee).

 Gr. *eis,* in, into, to; see **in**

 Gr. *epi,* on, motion toward; see **on**

 L. *o-, ob-, oc-, of-, og-, op-, os-,* < *ob,* to, toward, against, opposite, upon, reverse, as in *omitto,* leave off, neglect; *objecto,* throw against; *occludo,* close up; *offendo,* strike against; *oggero,* proffer; *opprobo,* reproach; *oscillo,* swing: omit, objection, obedience, obesity, obligation, oblong, obovate, obscure, observation, obsolete, obtuse, occlude, occupy, occurrence, offensive, offer, ogganition, opponent, opportunity, opposite, oppression, ostensible, ostentation, oscillation, *Quercus oblongifolia* (an oak).

 Gr. *para,* beside, motion to the side of; see **near**

 Gr. *pros,* beside, near, toward; see **near**

 AS. *weard,* in the direction of: eastward, westward, homeward, toward, forward, windward, backward, awkward, froward, wayward.

 See: **against, near, on**

toad < AS. *tadie.*

 L. *bufo, -onis,* m. toad: Bufonidae, bufonite, bufagin, bufotalin, bufonin, *Bufo superciliaris* (a toad), *Juncus bufonius* (toad-rush), *Phyxium bufonium* (a beetle), *Trachypetra bufo* (a grasshopper).

 Gr. *phryne,* f.; *phrynos,* m. toad: phrynin, *Phrynus coronatus* (a spider), *Phrynosoma cornutum* (horned 'toad'), *Pterophryne histrio* (a sargassum-fish), *Phrynium capitatum* (a marantacead).

 Gr. *physalos,* m. a kind of toad:

 L. *rubeta,* f. a kind of toad:

tobacco < Ab.Am. *tabaco;* see *tabacum* under **nightshade**

toco- < Gr. *tokos,* birth, delivery, offspring; *tokas, -ados,* of birth; *toketos,* that which is brought forth, see **birth**; interest, usury; *tokarion,* dim., see **gain**

toculio, L. usurer; see **pay**

todus, L. a small bird; see **bird**

toe < AS. *ta;* see **finger, nail**

toecho- < Gr. *toichos,* wall of a house, side of a ship; *toichidion; toichion,* dim.; see **wall**

tofus, L. *tufa,* tuff; see **stone**

toga, L. the outer garment of a Roman citizen; *togula,* dim.; see **garment**

together < AS. *togaedere;* see **with, gather, all**

tolero, *-atus,* L. bear, endure; *tolerabilis,* bearable; see **carry**

tolleno, L. swing-beam; see beam

tolma, Gr. courage, boldness; *tolmeros,* daring; see bold

tolu < Sp. balsam from Santiago de Tolu; *tolyl,* denoting toluene; see resin

tolutarius, L. trotting; see run

tolype, Gr. ball of yarn; see ball

tomaculum; tomacina, L. a kind of sausage; see sausage

tomato < Ab.Am. *tomatl;* see nightshade

tomb < OF. *tombe,* < LL. *tumba* (Gr. *tymbos*), sepulchral mound, grave; see
    grave

tome < L. *tomus* (Gr. *tomos*), part, book, volume; *tomarion,* dim.; see book

tomentum, L. woolly hairs; see hair

tomido- < Gr. *tomis, -idos,* knife; *tomeus,* cutter, knife; see knife

tomorrow; see day

-tomy; -tome, dissection, excision; < Gr. *temno,* cut; *tomikos,* of cutting; see cut

-ton < AS. *tun,* town; see town

tone < L. *tonus* (Gr. *tonos*), sound, pitch; see sound

tongs < AS. *tang.*
    L. *chela,* claw; see nail
    L. *forceps, -cipis,* f. nippers, tongs: forcipiform, forcipulate, *Forcipula quadrispinosa*
        (an earwig), *Forcipiger longirostris* (a butterfly-fish), *Orodesmus forceps* (a
        milleped), *Oxyaena forcipata* (a creodont).
    Gr. *helkyster, -os,* instrument for drawing, midwife's forceps; see *helko* under
        draw
    Gr. *labis, -idos,* f. forceps, tongs; *labidion,* n. dim. tweezers; *pyrolabis, -idos,* f.;
        xylabion, n. fire-tongs: *Labidomera circumpuncta* (a beetle), *Labidoplax digitata*
        (a holothurian), *Labidura riparia* (an earwig), *Anisolabis annulipes* (an earwig).
    Gr. *thermastris, -idos,* f. tongs: *Thermastris brasiliensis* (an earwig).
    L. *volsella* (*vulsella*), f. tweezers, pincers: *Volsella multilinigera* (a fossil pelecy-
        pod).

tongue < AS. *tunge.*
    Gr. *glossa* (*glotta*), f. tongue; *glossarion; glossidion* (*glottidion*), n. dim.; *glottikos,*
        of the tongue: glossary, epiglottis, polyglot, proglottid, bugloss, *Glossopteris
        retifera* (a fossil fern), *Glottidium vesicarium* (bagpod), *Diglossa pectoralis* (a
        honey-sucker), *Glossina morsitans* (a tsetse-fly), *Elaphoglossum crinitum* (a
        fern), *Saccoglottis amazonica* (a humiriacead), *Tachyglossus aculeatus* (echidna
        or spiny anteater), *Pogonia ophioglossoides* (an orchid).   According to Pan-
        gloss in Voltaire's *Candide,* "All is for the best in this best of all possible
        worlds".
    L. *lingua,* f. tongue, speech; *ligula; lingula,* f. dim.; *lingulaca,* f. chatterbox, gossip;
        linguatus, eloquent; *linguax, -acis; lingulus; linguosus, loquacious,* talkative;
        lingulatus, tonguelike: linguist, linguistics, trilingual, lingo, lingulate, ligulate,
        language, lingula, ligula, ligule, liguliferous, *Linguatula serrata* (a tongue-worm),
        Lingulina gracillima (a foraminifer), *Ligularia japonica* (golden-ray), *Ranun-
        culus lingua* (tongue-buttercup), *Atrypa linguifera* (a fossil brachiopod), *Exo-
        glossum maxillingua* (stone-toter), *Dichomeris liguellus* (palmer-worm), *Agrostis
        longiligula* (a bent-grass), *Ammophila breviligulata* (American beach-grass),
        Passiflora ligularis (sweet granadilla).
    L. *subscus, -udis,* f. tongue or tenon of a dovetail:
    See: strap, flap, fringe, leaf, projection

tonitrus, L. thunder; see *tonus* under sound

tonos, Gr. a stretching, tightening, bracing; *tonikos,* of tension; see draw

tonsa, L. oar; see oar

tonsil < L. *tonsilla,* f.: tonsillectomy, tonsillitis.
    Gr. *antias, -ados,* f. tonsil: antiaditis.
    L. *glandula,* f. gland of the throat, tonsil:
    Gr. *paristhmion,* n. tonsil:

tonsor, L. barber; *tonsorius,* of barbering; *tonsura,* a clipping, shearing; see barber

tonthrystes, Gr. mutterer; see murmur

tonus, L. (Gr. *tonos*), sound, pitch; see sound

tool < AS. *tol.*
    L. *accinctus,* well-girded, armed, equipped; see *cingo* under bind
    Gr. *aor, -os,* n. any tool or weapon, particularly the sword: *Aorus ferrugineus*
        (a beetle), *Chrysaor hercinius* (a cephalopod).
    L. *apparatus, -us,* m. equipment, machine, instrument, trappings: apparatus.

L. *arma* (pl.), n. weapons; *armatura*, f. armor, equipment; *armiger, -a, -um,* bearing weapons, armed; *armo, -atus,* furnish with weapons: army, armory, armada, armature, armadillo, armistice, armament, alarm, *Armadillidium nasutum* (a pillbug), *Opilio armatus* (a harvestman), *Cremastochilus armatura* (a beetle), *Euphoberia armigera* (a fossil centiped), *Otiorhynchus armadillo* (a beetle).

L. *ballista,* a machine for throwing missiles; see *ballo* under **throw**

L. *catapulta,* a machine for hurling weapons; see **throw**

Gr. *enteon,* n. fighting gear, armor, appliance:

Gr. *ergaleion,* n. tool, instrument:

L. *fabrilium,* n. mechanical tool:

Gr. *graphis, -idos,* stylus, graving tool; see **pen, write**

Gr. *hoplon,* n. any tool or implement, armor, shield; *hoplites,* m. man in armor, soldier; *hoplitikos,* of arms; *enoplos,* armed; *epiplon,* n. implement, utensil: hoplite, hoplology, panoply, Hoplophora, anoplotherium, Hoplocarida, *Hoplocephalus curtus* (tiger-snake), *Hoplochelys paludosa* (a fossil turtle), *Holothrips quercinus* (a bug), *Hoplia mucorea* (a beetle), *Hoplichthys fasciatus* (a fish), *Hoplitotrachelus spinifer* (a beetle), *Hoplitosaurus marshi* (a dinosaur), *Enoplosus armatus* (a zebra-fish), *Archoplites interruptus* (Sacramento perch), *Melanoplus spretus* (Rocky Mountain grasshopper), *Sternoplistes temmincki* (a beetle), *Propalaeohoplophorus incisivus* (a glyptodont), *Cratosomus hoplites* (a beetle).

Gr. *katias, -ados,* f. a surgical instrument:

Gr. *labe,* f. handle, haft, hold, hilt, tool or instrument: astrolabe, mesolabe, trilabe.

L. *machina* (Gr. *mechane*), f. apparatus, contrivance; *machilla; machinula,* f. dim.; *machinator, -is,* m. architect, engineer; *machinor, -atus,* contrive skillfully, design, invent; *mechanikos,* of machines: machine, machinery, machination, mechanism, mechanic.

Gr. *mele; melotris, -idos,* f. instrument for probing:

L. *organum* (Gr. *organon*), n. instrument; *organikos,* of implements: organ, organic, organization, organism, Francis Bacon's *Novum organum,* inorganic, disorganize, *Sarcinula organum* (a fossil coral).

Gr. *skapane,* a digging tool, mattock, shovel; see **shovel**

Gr. *skeuos,* n. implement of any kind, gear, utensil; *skeuarion; skeuyphion,* n. dim.; *skeuastes,* m. preparer; *skeuastos,* artificial; *skeuazo,* equip, supply, prepare:

Gr. *sosanion,* n. shoulder-piece of a coat of mail:

L. *specillum,* n. probe:

L. *supellex, -lectilis,* f. apparatus, utensil, furniture: *Supella supellectilium* (a cockroach), supellectile.

Gr. *teuchos,* n. tool, armor, gear, book; *teucheres,* armed; *teuchester, -os,* m. armed man, soldier; *ateuches,* unarmed: Pentateuch, *Teuchocnemis bacuntius* (a fly).

L. *utensilium,* n. a domestic implement: utensil.

* * * * * * *

L. *-brum, -crum,* n. suffixes signifying instrument, tool, means, as *candelabrum,* candlestick; *cribrum,* sieve; *flabrum,* blast; *fulcrum,* support; *sacrum,* holy thing:

L. *-bulum, -culum,* n. suffixes signifying instrument, tool, means, as *tribulum,* flail; *sarculum,* hoe; *vehiculum,* wagon:

L. *-mentum.* n. suffix signifying tool, means, action or result of action, condition, as *ferramentum,* iron tool; *implementum,* tool; *instrumentum,* tool; *tormentum,* windlass, rack: implement, instrument, instrumental, achievement, arrangement, astonishment, development, document, entertainment, experiment, fragment, fulfillment, government, integument, judgment, merriment, momentum, monument, movement, nourishment, ornament, regiment, segment, torment, treatment.

L. *-ter,* m., *-tra,* f., *-trum,* n. suffixes denoting tool, instrument, means, apparatus, as *culter,* knife; *mulctra,* pail; *capistrum,* halter: colter.

Gr. *-tros,* m., *-tra,* f., *-tron,* n.; *-terion,* n. tool, instrument, means, as *daitros,* carver; *iatros,* physician; *rhetra,* agreement; *sostra,* reward; *kestron,* a graving tool; *plektron,* tool for plucking a stringed instrument; *sestron,* sieve; *skaleuthron,* hoe; *teretron,* auger; *kauterion,* branding-iron; *poterion,* drinking-cup; *skepterion,* proof: cestrum, plectrum, chytra, electron, cyclotron, cosmotron, betatron, bevatron, infinitron (a waggish invention for helping anyone to neatly and effectively split an infinitive).

See: **make, ax, hoe, knife, chisel, spear, hammer, plow, sickle, harp, iron, bore**

toona, NL. a genus of the mahogany family; see **mahogany**

tooth < AS. *toth:* toothache, toothpick, toothwort, teeth, teething, bucktooth.
L. *brochus,* with projecting teeth, projecing; see **projection**
L. *crena,* notch, rounded projection; see **notch**
L. *cuspis, -idis,* pointed end of anything; see **point**
L. *dens, dentis,* m. tooth; *denticulus,* m. dim.; *dentarius,* of teeth; *dentatus,* toothed, pointed; *denticulatus,* with small teeth; *dentio; dentitio, -onis,* f. a teething; *edentatus; edentulus,* toothless: dentate, dentition, dentist, dentifrice, indent, indenture, edentate, trident, dandelion, denticulate, dentiform, dentine, dental, dentilingual, *Dentaria diphylla* (a toothwort), *Dentella repens* (a rubiacead), *Dentalium pretiosum* (a tooth-shell), *Bidens frondosa* (beggarticks), *Tridens melanops* (a fish), *Tridentiger obscurus* (a fish), *Fissidens hyalinus* (a moss), *Diplograptus dentatus* (a graptolite), *Pelargonium denticulatum* (a pelargonium), *Ichthyurus dentatipes* (a beetle), *Didymictis altidens* (a creodont), *Acer grandidentatum* (a maple), *Cakile edentula* (a sea-rocket), *Caulepis longidens* (a fish), *Linyphia multidenta* (a spider), *Salebius 6-dentatus* (a beetle).
Gr. *gomphios,* m. molar tooth; *gomphiasis,* f. toothache: *Gomphiocephalus hodgsoni* (a collembolid).
L. *laciniosus,* full of flaps, indentations, jagged; see **fringe**
L. *molaris,* grinder; see *molo* under **grind**
Gr. *mylos,* grinder, molar; see *myle* under **grind**
L. *nefrens, -dis,* toothless, that cannot bite:
Gr. *odous (odon), odontos,* m. tooth; *odontidion; odontion,* n. dim.; *odontikos,* of teeth; *odontotos,* with teeth; *chauliodous,* having prominent teeth; *kenodontis, -idos; nodos,* toothless; *odontokeras,* n. tusk: odontalgia, odontitis, odontiasis, odontoclast, odontography, odontoid, odontolite, odontology, odontophore, bunodont, creodont, labyrinthodont, orthodontist, oreodon, solenodon, Odonata, *Odontoglossum pulchellum* (an orchid), *Odontomachus rixosus* (an ant), *Odontidium rugulosum* (a gastropod), *Odontium* (now *Bembicidium*) *nitidulum* (a beetle), *Odocoileus virginianus* (white-tailed deer), *Achyrodon pusillus* (a fossil mammal), *Anodonta cygnea* (swan-mussel), *Chauliodus macouni* (a viper-fish), *Cynodon incompletus* (a grass), *Hybodus reticulatus* (a Jurassic fish), *Leontodon autumnalis* (fall-hawbit), *Loxodonta africana* (African elephant), *Platybelodon grangeri* (a mastodon), *Odontospermum aquaticum* (a composite), *Solenodella lacerata* (a conodont), *Triodia flava* (purpletop), *Desmodus rufus* (a vampire-bat), *Mastodon americanus* (a mastodon), *Tetraplodon paradoxus* (a moss).
Gr. *tomeus,* cutting tooth, incisor; see *temno* under **cut**
See: **point, saw, comb, jaw, bite**

top < AS. *top:* topple, topper, topmost, topsail, topsoil, tiptop, topsy-turvy, redtop.
See **cone**
Gr. *akme,* f. highest point, prime, best time; *akmaios,* in prime, vigorous; *akmenos,* full grown; *epakmos,* in full bloom, mature: acme, acmite, *Acmopyle pancheri* (a gymnosperm), *Acmaea persona* (a limpet), *Acmaeoblatta lanceolata* (a fossil cockroach), *Acmaedera pulchella* (a beetle), *Pentacme contorta* (a dipterocarpacead), *Epacmus modestus* (a fly).
Gr. *akron,* n. top, tip, end, summit, peak; *akros,* at the top, highest; *akris, -ios,* f. hilltop, peak; *epakros,* pointed at the end: acrobat, acrodont, acrodromous, acrogenous, acromegaly, acromion, acrocarpous, acropodium, acropolis, acrostic, acronym, acrospore, Akron (Ohio), *Acrobates pygmaeus* (flying phalanger), *Acropogon barbatipes* (a fly), *Acrioceras tabarelli* (an ammonite), *Acrophylax vernalis* (a trichopterid), Epacridaceae, *Epacris impressa* (an Australian shrub), *Polystichum acrostichoides* (a fern).
L. *apex, -icis,* m. top; *apiculus,* m. dim.: apex, apical, apicular, apiculate, *Apicotermes angustus* (a termite), *Calopteryx apicalis* (a dragonfly), *Coscinodiscus apiculatus* (a diatom), *Spongostylum candidapex* (a fly).
L. *cacumen,* extreme point or end; see **end**
L. *climax* (Gr. *klimax*), ladder, peak, top, crisis; *climacter,* critical period; see **ladder**
L. *columen, -inis,* n. top, crown, summit:
L. *crisis* (Gr. *krisis*), issue, decision, turning-point; see **turn**
L. *culmen (columen), -inis,* n. top, crown, summit: culmen, culminate, culmination.
L. *fastigium,* a slope, up or down, to point, gable; see **gable**
Gr. *hyperoche,* projection, prominence, summit; see **projection**
Gr. *kolophon, -os,* m. summit, top, climax: colophon, *Colophonodon holmesi* (a fossil cetacean).
Gr. *korymbos,* peak, summit; see **cluster**
Gr. *koryphe,* top, crown, head; *koryphaios,* leader, chief; see **head**
Gr. *mesokranon,* crown of the head; see *kranion* under **head**

L. *pinnaculum,* n. top, peak: pinnacle.

L. *summum,* n. highest place, top: summit.

L. *superficies, -ei,* f. top, surface: superficial.

L. *trunculus,* tip, end; see *truncus* under **short**

L. *vertex, -icis,* m. height, elevation, peak, pole of the heavens: vertex, vertical, *Chaetoctesius albovertex* (a beetle).

See: **end, head, point, crest, ridge, over, honor**

topaz, < Gr. *topazos;* see **stone**

topium, L. ornamental gardening; see **ornament**

topos, Gr. place; *topion,* dim.; see **place**

-tor, L. suffix signifying agent, usually masculine, in contrast with *-trix,* feminine; see *-or* under **make**

toral, L. valance, border; see **border**

torch < L. *torqueo,* twist; see **light**

torcular, L. wine-press; see **press**

tordylium, L. (Gr. *tordylion*), an umbelliferous plant; see **carrot**

toretos, Gr. bored, pierced; see *toreo* under **bore**

toreuma, Gr. work in relief, embossed work; *toreutos,* turned on a lathe; see **cut**

torgos, Gr. vulture; see **vulture**

tormentum, L. instrument for hurling stones, windlass, rack, torture; see **pain**

torminum, L. gripes, colic; *torminalis,* of or for colic; see **disease**

tormos, Gr. hole, socket; *tormion,* dim.; see **hole**

torno, -atus, L. turn in a lathe, make round; *tornus* (Gr. *tornos*), turner's wheel, lathe; *turnatilis,* turned, finished; see **turn**

toros, Gr. borer, see *toreo* under **bore**; piercing, sharp, clear, see **clear**

torosus, L. muscular, bulging, lusty; see *torus* under **projection**

torpidus, L. numb, stiff; *torpedo,* numbness, see **dull**; a ray, see **shark**

torquatus, L. adorned with a necklace or collar, < *torques,* necklace, collar; see **collar**

torrens, -entis, L. swift stream; see **stream**

torridus, L. dry, parched, hot, scorched; see **dry**

torris, L. firebrand; see **fire**

tortilis; tortuosus, L. twisted, winding; *tortor,* twister; see *torqueo* under **turn**

tortura, L. torment, pain; see *tormentum* under **pain**

torus, L. round elevation, protuberance, bulge; *torulus,* dim.; see **projection**

torvus, L. wild, gloomy, severe, grim; *torvidus,* wild, fierce; see **rough**

toryne, Gr. ladle, stirrer; *torynetos,* stirred; see **spoon**

tosos, Gr. so much, so very; see **abundance**

tostus, L. burned, parched; see *torreo* under **burn**

totanus, NL. a limicoline bird; see **snipe**

tothastes, Gr. scoffer; *tothastikos,* mocking, scornful; see **laugh**

totietas, L. the whole; see *totus* under **all**

totter < uncertain origin; see **shake**

totus, L. whole, all; see **all**

touch < OF. *toucher:* touching, touchstone, touchdown, retouch.   Derive: touchy.

L. *attrecto; contrecto, -atus,* touch, handle, feel for, grope for: contrectation.

Gr. *blimazo,* feel hens for the presence of eggs; *blimasis,* f. a lewd handling:

Gr. *haptos,* touchable, approachable, < *hapto,* touch, fasten to, lay hold of, grasp; *haphe; epaphe,* f. touch, contact, grip; *aaptos,* untouchable, unapproachable: haphephobia, *Haphospatha scolopax* (a fly).

L. *manutigium,* n. a touching or feeling with the hand:

L. *mulceo, mulsus,* stroke, touch lightly, charm, soothe, soften; see **soft**

L. *palpo, -atus,* touch, stroke, feel; *palpabilis,* touchable, that may be felt; *palpator, -is,* m. stroker; *palpus,* m. soft palm of the hand, feeler: palpable, impalpable, palpus, palpulus, palpiferous, pedipalp, *Glossina palpalis* (a tsetse-fly), *Thelxiope palpigera* (a beetle), *Calopteron palpale* (a beetle).

Gr. *pselaphao,* grope about, feel, examine, touch; *pselaphema, -tos,* n. a touch; *pselaphesis; pselaphia,* f. a feeling, handling, touching; *pselaphetikos,* palpatory; *pselaphetos,* palpable: *Pselaphus fustifer* (a beetle), *Pselaphostomus stussineri* (a beetle), *Araeopselaphus myrmecophilus* (a beetle).

L. *tango, tactus,* touch; *tactilis; tangibilis,* touchable, concrete; *tagax, -acis,* apt to touch, light-fingered: tactile, tactful, tangent, tangible, attainment, contagious, contact, contaminate, contiguous, contingent, intact, intangible, entire, *Impatiens noli-me-tangere* (English touch-me-not).

Gr. *thigma, -tos,* n. touch; *thixis,* f. a touching, < *thigano,* touch, handle: thigmotropism, thigmotaxis, *Thigmophrya bivalviorum* (a protozoan).

See: **feel, rub, near**

tough < AS. *toh;* see **hard, stiff, strong, thick**

tour < OF. *tor,* < L. *torno,* turn; see **turn**

towel < OF. *toaille;* see **napkin**

tower < AS. *torr,* < L. *turris, -is* (Gr. *tyrris; tyrsis*), f.; *turricula; turritella,* f.; *tyrridion,* n. dim.; *turritus,* with towers, castellated: turret, towering, towerwort, towerman, turret-spider *(Lycosa arenicola), Turritella imbricata* (a gastropod), *Turrilites catenatus* (an ammonite), *Pachylomeris turris* (a trapdoor-spider).

Pg. *pagode,* storied tower, temple, memorial, < a Skt. word: pagoda, *Zeacolpus pagodus* (a gastropod), *Quercus pagodaefolia* (an oak), *Tectaria pagoda* (a gastropod).

Gr. *pyrgos,* m. tower; *pyrgidion; pyrgion,* n.; *pyrgiskos,* m. dim.; *pyrgodes,* tower-like; *pyrgoma, -tos,* n. something with towers; *pyrgotos,* having towers: pyrgoidal, *Pyrgula tesselata* (a fossil gastropod), *Pyrgidium nodotianum* (a gastropod), *Pyrgodiscus armatus* (a diatom), *Pyrgoma cancellatum* (a barnacle), *Pyrgulopsis coronatus* (a gastropod), *Potamopyrgus spelaeus* (a gastropod), *Pyrgota undata* (a fly), *Leptopyrgota minuta* (a fly).

L. *specula,* watch-tower; see *specio* under **see**

town < AS. *tun:* town hall, townish, townling, township, townsman, Townsend, Washington, Germantown, Easton, Allentown, Northampton, Palmerton, Lehighton, Hazleton, Scranton.

Gr. *asty, -eos,* n. town, city; *asteios; astikos (astykos),* urban, urbane, refined, nice; *astos,* m. citizen with civil rights only; *astyanax, -aktos,* m. mayor: *Astycophobus cretaceus* (a beetle), *Astyanax bimaculatus* (a fish).

AS. *burg,* < ML. *burgus,* fort, fortified town; *Pergamos,* f. citadel of Troy, any citadel: burgess, burgher, burglar, bourgeois, barrow, borough, Edinboro, Edinburgh, Cherbourg, Waynesboro, Chambersburg, Harrisburg, Canterbury, Hamburg, hamburger, limburgite.

Dan. *by,* town, village: Buxby, Derby, Kirby, Selby, Whitby, bylaws, Willoughby.

AS. *ceaster,* walled town, chester, < L. *castrum,* camp, settlement; see **fort**

L. *civitas, -atis,* f. citizenship, the body politic, city; *civitula,* f. dim.; *civis,* c. citizen; *civilis,* of citizens: city, citizen, civil, civilization.

AS. *ham,* home, a group of houses, village, town; see **house**

Gr. *kome,* f. village, country town; *komidion; komydrion,* n. dim.; *komestes,* m. a countryman; *kometikos,* pagan:

L. *municipium,* n. town, < *municeps, -cipis,* c. citizen; *municipialis,* of a town: municipal, *Municeps redempta* (a gastropod).

L. *oppidum,* n. town; *oppidulum,* n. dim.; *oppidaneus,* of a town; *oppidanus,* m. townsman:

Gr. *polis, -ios (-eos),* f. city; *polichne,* f.; *polidion,* n. dim.; *polites,* m. citizen; *polistes,* m. founder of a city; *politikos,* of citizens, civil: police, policy, political, propolis, acropolis, metropolis, metropolitan, Annapolis, Constantinople, Grenoble, Naples, Neapolitan, Napoleon, Minneapolis, Tripoli, poliorcetics, *Polistes pallipes* (a wasp), *Memythrus polistiformis* (a moth).

Sp. *pueblo,* village, < L. *populus,* people; see **people**

AS. *stead,* place, town; see **place**

AS. *thorp* (G. *dorf*), village: Althorpe, Halethorpe, Oglethorpe, Northrop, Düsseldorf.

L. *urbs, urbis,* m. city; *urbiculus,* m. dim.; *urbanus,* of the city; *urbicus,* civic: urban, urbane, urbicolous, interurban, suburb, *Chenopodium urbicum* (a goosefoot), *Geum urbanum* (wood-avens).

L. *vicus,* m. village; *viculus,* m., dim. hamlet; *villa,* f. dim., farmhouse, country-seat, farm; *villula,* f. dim.; *vicanus; villaticus,* of a village; *vicinalis; vicinarius; vicinus,* near, neighboring: vicinage, vicinal, vicinity, villa, village, villain, villainous, Villanova, Newville, Oakville, Fogelsville, Warwick, Berwick, New Brunswick, Norwich, Greenwich, Sandwich, *Villanova dissecta* (a composite), *Crabro vicinus* (a wasp).

See: **place, fort**

toxetesia, Gr. a plant; see **plant**

toxicum, L. (Gr. *toxikon*), poison for arrows; see **poison**

toxo- < Gr. *toxon,* bow; *toxarion,* dim.; *toxeutes; toxotes,* archer, see **arc;** *toxeuma,* arrow, see **arrow**

toy < D. *tuig,* tool, plaything; see **play**

trabea, L. a robe of state, toga with border of colored stripes; *trabealis,* of a toga; *trabeatus,* dressed in a toga; see **garment**

trabs, L. beam; *trabecula,* dim.; *trabalis,* of beams; see **beam**

trace; track < OF. *trace,* track of animals, mark, path; see **mark, way**

trachea < L. *trachia* (Gr. *tracheia*), windpipe; see **throat**

trachelos, Gr. neck; *trachelion,* dim.; see **neck**

trachinus, NL. a genus of fishes; see **fish**

trachurus, L. (Gr. *trachouros*), a mackerellike fish; see **mackerel**

trachys, Gr. rough; *trachoma,* roughness; see **rough**

track < OF. *trak,* trace, trail, mark; see **mark, way**

tract < L. *tractus,* a space drawn out, district, region; see **country**

tractable < L. *tractabilis,* manageable, < *tracto, -atus,* handle, manage, haul, < *traho, tractus,* draw, haul; see **govern, draw**

tractuosus, L. clammy, gluey; see **glue**

trade < AS. *trod,* track, path, activity.
Gr. *agora,* f. assembly, market; *agorasia,* f.; *agorasmos,* m. a buying, purchase; *agorasma, -tos,* n. merchandise, article bought or sold; *agorastes,* m. buyer; *agorastikos,* of trade; *agorastos,* buyable, purchasable:
Gr. *allaxis,* f. exchange, barter, trade; *allagma, -tos,* n. that which is given in exchange, price; *antallaktos,* taken as equivalent in exchange, < *alasso,* change, alter: *Enallagma cucurbitina* (a bignoniacead).
L. *arillator, -is,* m. haggler, broker:
L. *aromatarius,* m. dealer in spices:
L. *auctio, -onis,* f. public sale: auction, auctioneer.
L. *cambio,* exchange, barter; see **change**
L. *caupo, -onis,* m. tradesman, huckster; *caupona; copa,* f. landlady; tavern; *cauponula,* f. dim.: *Urechis caupo* (an echiuroid).
AS. *ceapian,* buy; *ceap,* bargain, sale: cheap, chap, Chapman, chaffer.
L. *chartopola* (Gr. *chartopoles*); *chartoprates,* m. paper merchant; *chartula,* f. bill:
L. *circitor; circuitor, -is,* m. peddler:
L. *cocio; cotio, -onis,* m. broker, factor:
L. *collybista* (Gr. *kollybistes*), m. money-changer, banker:
L. *conducticius,* pertaining to hire or rent; *conductor, -is,* m. lessee, tenant, contractor:
L. *conforaneus,* selling at the same market-place:
L. *copiarius,* m. purveyor:
L. *creditum,* n. loan: credit.
L. *danista* (Gr. *daneistes*), m. money-lender, usurer; *danisticus,* money-lending, usurious:
L. *dardanarius,* m. grain-speculator:
L. *emo, emptus,* buy; *emax, -acis,* eager to buy; *emptor, -is,* m. buyer; *empticius,* bought; *emptio, -onis,* f. purchase; *inemptus,* unbought: redemption, peremptory, redeem, ransom, premium, consume, *caveat emptor.*
L. *emporium* (Gr. *emporion*), n. market: emporium.
L. *forum,* market, public place; see *foras* under **open**
L. *institor, -is,* m. peddler, huckster:
Gr. *kapelos,* m.; *kapelis, -idos,* f. huckster, retailer, peddler; *theokapelos,* dealing in sacred things: *Capelopterum sellatum* (a bug).
Gr. *kichremi,* lend:
L. *largitio, -onis,* f. bribe:
L. *liceo, licitus,* be on sale, be valued at; *liceor; licitor,* offer a price for, bid for; *illicitator, -is,* m. sham-bidder at an auction to raise others' bids:
L. *lixa,* m. sutler, peddler:
L. *macellum,* n. market; *macellarius,* of the meat or provision market, meat-seller, butcher: *Macellaria compressa* (a fly), *Ephydra macellarius* (a brinefly).
L. *manceps, -cipis,* m. purchaser, contractor, speculator; *mancipium,* n. a legal form of conveyance of property; *mancipo, -atus,* deliver, transfer: mancipium.
L. *mango, -onis,* m. dealer who furbishes his wares to give them an appearance of greater value; *mangonicus,* of a dealer; AS. *mangian,* trade: fishmonger, gossipmonger, newsmonger, whoremonger.
L. *mercor, -atus,* buy, trade, < *merx, mercis,* f. goods, wares; *mercabilis; mercalis,* purchasable; *mercator, -is,* m. trader, dealer; *mercimonium,* n. goods, wares; *Mercurius,* Mercury, god of traders and thieves: merchant, merchandise, mer-

cantile, market, Markham, commerce, commercial, Mercer, Mercury, mart, mercurial, F. *mercredi* (Wednesday).

Gr. *metabolos*, m. trader, merchant:

Gr. *metadosis*, f. exchange, barter:

L. *mutuum*, n. loan; *mutuus*, borrowed, loaned; *mutuor, -atus*, borrow: mutual.

L. *negotior, -atus*, engage in business, deal, trade; *negotialis*, of business; *negotiator, -is*, m. trader, agent; *negotiosus*, full of business, busy; *negotium*, n. business, affair, thing: negotiate, negotiable, negotiator, negotiation.

L. *nundinor, -atus*, attend market, trade, traffic; *nundina*, f. market-day; *nundinalis*, of market-day; *nundinarius*, of a market; *nundinator, -is*, m. trader:

Gr. *oneomai*, buy; *onema, -tos*, n. purchase; *onetes*, m. buyer; *onetos*, bought; *onios*, of or for sale; *onos*, m. price; *eponion*, n. duty, tax: *Onetes sanguinolentus* (a grasshopper).

L. *opsono, -atus* (Gr. *opsoneo*), buy provisions, cater, purvey:

Gr. *piprasko*, sell:

L. *piscarius*, m. fishmonger:

Gr. *poleo*, sell, barter; *empole*, f. merchandise, profit; *empoleus*, m. merchant; *monopolion*, n. exclusive control of a given trade; *propoles*, m. middleman: monopoly.

Gr. *prater, -os*, m. dealer; *praterion*, n. market:

L. *precia*, m. public crier, herald, auctioneer; *preconius*, of a public crier:

L. *proxeneta* (Gr. *proxenetes*), m. agent, broker, negotiator:

L. *scrutarius*, m. dealer in secondhand goods:

AS. *sellan*, sell; ON. *sala*, sale: sell, sale.

L. *stipulor, -atus*, bargain, state a requirement: stipulate.

L. *trapezita* (Gr. *trapezites*), m. banker:

L. *vendo, -itus*, sell; *venditor, -is*, m.; *venditrix, -icis*, f. seller; *venditum; venum*, n. sale; *vendax, -acis*, fond of selling; *vendibilis*, salable; *venalis*, of selling; *invendibilis*, unsalable; *invenditus*, unsold: venal, vend, vendor, vendue.

L. *Vertumnus*, god of change and trade; see **change**

AS. *waru*, article, commodity; see **thing**

See: **change, money**

traditio, L. a surrender, a saying handed down, < *trado, -itus*, give up, deliver, transmit; see *do* under **give**

tragacanthum, L. a gum; see **resin**

traganon, Gr. cartilage, gristle; see **gristle**

tragema, Gr. sweetmeat; see **food**

tragicus, L. (Gr. *tragikos*), of tragedy; *tragoedia* (Gr. *tragoidia*), a serious or heroic drama; see **serious**

tragopan, Gr. a fabulous bird; see **bird**

tragopogon, Gr. goatsbeard; see **composite**

tragula, L. javelin; see **spear**

tragus, L. (Gr. *tragos*), goat; *tragion; tragiskos*, dim.; *tragicus* (Gr. *tragikos*), of a goat, pertaining to tragedy; see **goat**

traha, L. drag, sled; see *traho* under **draw**

trahax, -acis, L. greedy, grasping; see **greed**

train < OF. *trainer*, < L. *traho, tractus*, draw, drag; see **teach, make**

trajectio, L. a crossing over, pathway of passage; see **way**

trama, L. warp, weft; see **weave**

trames, L. byway, path; see **way**

tramis, Gr. perineum; see **rump**

trampis, Gr. ship; see **ship**

trample < ME. *trampen*, tread heavily; see **step, walk**

tranos, Gr. clear, distinct; see **clear**

tranquillus, L. quiet; *tranquillitas*, quietness; see **rest**

trans- < L. *trans*, across, beyond, through; see **over**

transenna, L. snare, trap; see **trap**

transitorius, L. passing, evanescent, fleeting; see **depart**

translator, L. transferrer, one who renders from one language into another; see **change**

transtrum, L. crossbeam; see **beam**

transversus, L. crosswise; see **cross**

trap < AS. *traeppe,* snare.
Gr. *arachnion,* cobweb; see **weave**
L. *araneum,* cobweb; *araneosus,* full of webs; see **weave**
Gr. *arkys,* net, toils; see **net**
L. *auceps,* fowler; *aucupor, -atus,* go bird-catching; see *avis* under **bird**
Gr. *brochos; embroche,* noose, snare; *brochotos,* ensnared; see **rope**
L. *cassis,* trap, snare, hunter's net, spider's web; *cassiculus,* dim.; see **net**
L. *decipula,* f. snare, trap:
Gr. *deleastron,* baited trap; see *delear* under **draw**
Gr. *dilemma, -tos,* n. a double proposition consisting of unwelcome alternatives: dilemma.
L. *dolus* (Gr. *dolos*), bait, trap, trick, treachery; *dolops,* ambusher; see *dolus* under **lie**
L. *fovea,* pit, pitfall:
L. *immendo, -atus,* implicate:
L. *implico, -atus,* involve, entangle: implicate, implicit.
L. *inexpeditus,* not free, entangled:
L. *inextricabilis,* that cannot be disentangled: inextricable.
L. *insidia,* f. ambush; *insidiosus,* cunningly deceitful; treacherous, dangerous; *insidiator, -is,* m. ambusher, lurker: *Liognathus insidiator* (a fish), *Chrysarachnion insidians* (a flagellate).
L. *intrico, -atus,* entangle, perplex, embarrass: intricate, intricacy.
L. *laqueus,* m. noose, snare, trap; *laqueo, -atus,* ensnare, entangle: laqueary, lasso, lace, *Cuna laqueus* (a pelecypod).
Gr. *lochos,* m. a place for lying in wait, ambush; *lochites,* m. ambusher, comrade: *Archilochus colubris* (a hummingbird).
L. *muscarium,* n. fly-trap, fly-brush; see *musca* under **fly**
L. *muscipula,* f. mouse-trap: *Dionaea muscipula* (Venus' fly-trap).
Gr. *myagra,* f. mouse-trap: *Zilla myagroides* (a brassicacead).
Gr. *myiosobe,* f. fly-trap; *myiosobion,* dim.:
L. *nassa,* wicker-basket used as a fish-trap; see **basket**
Gr. *pagis, -idos; page,* f. snare, trap; *pagideuma, -tos,* n. snare; *pagideutikos,* ensnaring:
L. *pendiculus,* m. cord, noose, snare:
L. *plaga,* f. snare, net:
Gr. *podistra,* f. foot-trap: *Podistra alpina* (a beetle), *Malthodes (Podistrella) meloiformis* (a beetle).
Gr. *selmis, -idos,* f. a kind of noose: *Zygoselmis nebulosa* (a protozoan).
Gr. *skandalon,* n. snare, trap, offense; *skandalodes,* offensive: scandal, scandalize, scandalous, *Scandalon ridiculum* (a grasshopper).
L. *subsessa,* f. ambush; *subsessor, -is,* m. waylayer:
L. *tenus, -oris,* n. snare, noose:
Gr. *theratron,* n. trap, noose:
L. *transenna,* f. snare, trap:
L. *verriculum,* dragnet; see **net**
See: **hunt, net, take, pen, draw, bind**

trapa, NL. contraction of ML. *calcitrapa,* caltrop; the water-chestnut; see *calcitrapa* under **point**

trapeza, Gr. table; *trapezion,* dim.; see **table, form**

trapheco- < Gr. *traphex, -ekos,* beam, piece of timber; see **beam**

trash < uncertain origin; see **dirt, trifle, worthless**

trasia, Gr. grate, kiln; see **screen**

traulos, Gr. lisping; see **falter**

trauma, Gr. wound; *traumation,* dim.; *traumatikos,* of wounds; see **sore**

travel < OF. *travaillier,* labor, toil; see **walk, move, make**

tray < AS. *treg;* see **plate, vessel**

tread < AS. *tredan;* see **walk, step**

treason < OF. *traison,* < L. *traditio, -onis,* f. surrender, delivery: tradition, traditor, traitor, traitorous, treasonable, betray, betrayal, extradite.
L. *detego, -ectus,* uncover, disclose, reveal, betray; see **bare**
Gr. *ekodotos,* given up, betrayed:
L. *infidelis,* faithless, false; see **lie**
Gr. *menyo,* reveal, disclose, betray; *menytes,* m. informer; *menytikos,* traitorous; *menyma, -tos,* n. information: *Menyanthes trifoliata* (bogbean).
L. *prodo, -itus,* disclose, betray; *proditor, -is,* m.; *proditrix, -icis,* f. betrayer, traitor, traitress; *prodicius,* treacherous; Gr. *prodosia; prodosis,* f. a giving up,

betrayal, treason; *prodotes,* m. betrayer, traitor; *prodotos,* abandoned: prodition, proditorious, *Prodosia mycha* (a moth), *Prodotiscus regulus* (a bird).

See: **bare, open, speak, lie**

treasure < OF. *tresor,* < *thesaurus* (Gr. *thesauros*), treasury; see **store**

trebax, *-acis,* L. experienced, cunning, crafty; see **know**

trecho, Gr. run, move quickly; *threktikos,* able to run; see **run**

tree < AS. *treo:* treebine, tree-fern, treefish, singletree, treelet, treetop, Trowbridge, L. *arbor, -is,* f. tree; *arbuscula,* f. dim.; *arborarius; arboreus,* of trees; *arboretum,* n. a tree garden; *arbustum,* n. orchard; *arbustivus; arbustus,* with trees: arboreal, arboretum, arboriculture, arborescent, arboricolous, arbuscule, arbuscular, Arbor Day, *Vaccinium arboreum* (farkleberry), *Hydrangea arborescens* (smooth hydrangea), *Artemisia arbuscula* (low sagebrush), *Salix arbusculoides* (a willow).

Gr. *dendron,* n. tree; *dendrion; dendryphion,* n. dim.; *dendras, -ados,* woody; *dendrikos,* of trees; *dendrodes,* treelike; *syndendros,* thickly wooded: dendrology, dendrite, dendritic, dendroid, dendrolite, lepidodendron, *Dendrobium speciosum* (an orchid), *Dendroica coronata* (myrtle-warbler), *Dendrocalamus strictus* (a bamboo), *Leiophyllum* (formerly *Dendrium*) *buxifolium* (sand-myrtle), *Eudendrium cochleatum* (a hydroid), *Philodendron spectabile* (an aracead), *Epidendrum roseum* (an orchid), *Dendrocygna arborea* (a tree-duck), *Sedum dendroideum* (a houseleek), *Boiga dendrophila* (golden tree-snake).

Gr. *drys, dryos,* f. tree, oak; *dryarion,* n. dim.; *Dryas, -ados,* f. a wood-nymph whose life was that of her tree; *dryotomos,* m. woodcutter; *daryllos,* f. Macedonian for *drys; phellodrys,* f. an evergreen oak: dryad, hamadryad, *Dryobates pubescens* (hairy woodpecker), *Dryopteris cristata* (a fern), *Dryopithecus germanicus* (a fossil primate), *Dryas drummondi* (a rosacead), *Halidrys siliquosa* (a marine alga), *Primula dryadifolia* (a primrose), *Drypetes lateriflora* (a Guiana plum), *Veronica chamaedrys* (a speedwell), *Polyporus dryadeus* (a fungus).

Gr. *ide,* f. tree; L. *Idaeus* (Gr. *Idaios*), pertaining to wooded Mount Ida: *Rubus idaeus* (red raspberry).

Gr. *lakara,* f. a kind of tree:

Gr. *orpex (orpax), -ekos,* m. sapling, pike: *Orpacophora coronata* (a katydid).

Gr. *philax, -akos,* f. tree:

L. *pomarium; pometum,* n. orchard:

Gr. *siptachoras,* m. an Indian tree:

See: **forest, bush, wood**

trellis < OF. *treliz,* < L. *trilix,* triple-twilled; see **screen**

trema, Gr. hole; *tremation,* dim.; see **hole**

tremella, NL. a kind of fungus; see **fungus**

tremendus, L. something to be trembled at, dreadful, terrible, large; see **large**

tremulus, L. shaking, quivering; *tremor,* a shaking; see *tremo* under **shake**

trench < OF. *trenche,* < L. *trunco, -atus,* shorten by cutting off; see **ditch, furrow, valley**

trepho, Gr. feed, nourish; see **eat**

trepidus, L. agitated, alarmed; see *trepido* under **shake, fear**

trepo, Gr. turn; *treptos,* turned, changed; see **turn**

treron, Gr. wild dove; see **dove**

tresis, Gr. perforation; see *teiro* under **bore**

trestes, Gr. coward; see **coward**

tretos, Gr. perforated; see *teiro* under **bore**

tri- < L. *tres* (Gr. *treis, trion*), three; *trias, -ados,* in three's; *tris,* thrice; see **three**

triaeno- < Gr. *triaina,* trident; see **fork**

triangle < L. *triangulus,* having three angles; see **angle**

tribaco- < Gr. *tribax, -akos,* rubbed, worn, experienced; see *tribo* under **rub**

tribe < L. *tribus,* a division of the people; see **class**

tribelo- < Gr. *tribeles (tribolos),* three-pointed; see *belos* under **arrow**

triben, Gr. tripod; see **frame**

tribon, Gr. cloak; *tribonarion; tribonion,* dim., see **garment**; crafty fellow, see **rogue**

tribos, Gr. worn path, track; see **way**

tribulum, L. threshing flail; see *tribulo* under **separate**

tribulus, L. (Gr. *tribolos*), caltrop; *tribulosus,* thorny; see **thorn**

tribune < L. *tribunus,* chieftain, representative; see **govern**
tribute < L. *tributum,* tax, contribution; see **pay**
trica, L. trifle, hindrance; *tricor, -atus,* dally, trifle; see **trifle**
triccus, NL. (Gr. *trikkos*), a small bird; see **bird**
trichas, Gr. a kind of thrush; see **thrush**
trichila, L. bower, arbor, summer-house; see **house**
trichilia, NL. a genus of the mahogany family; see **mahogany**
trichis, Gr. a kind of anchovy; see **herring**
tricho- < Gr. *thrix, trichos,* hair; *trichidion; trichion,* dim.; *trichinos,* of hair; see **hair**
trichomanes, Gr. a kind of fern; see **fern**
trick < OF. *trichier,* deceit; L. *tricor, -atus,* play tricks, make mischief; see **lie**
tridacna, L. (Gr. *tridakna*), a large pelecypod; see **mollusk**
tridens, L. fork with three tines; see **fork**
trientalis, L. third of a foot; see *tres* under **three**
trifax, L. a kind of spear; see **spear**
trifidus, L. three-cleft; see *findo* under **cut**
trifle < OF. *trufle,* jest.
    L. *bulla,* bubble, trifle, vanity; see **bubble**
    L. *butubattum,* n. trifle, worthless thing:
    L. *ciccus,* m. trifle, bagatelle: *Ciccus marmoratus* (a bug).
    Gr. *elematos,* vain, trifling:
    L. *exiguum,* trifle; see *exiguus* under **little**
    L. *gerra,* f. trifle, nonsense: *Gerra radicalis* (a moth).
    Gr. *gry,* something insignificant, trifle; see *gryte* under **worthless**
    L. *hilum,* bit, trifle, scar; see **mark**
    Gr. *leros,* silly talk, nonsense, trumpery, trifle; see **fool**
    L. *leviculus,* trivial, insignificant; see *levis* under **light**
    L. *mantissa,* f. a trifling addition: mantissa.
    Gr. *mikrologos,* gathering trifles, careful about trifles; *mikrolypos,* vexed at trifles; *mikrothaumastos,* admiring trifles:
    L. *minutalis,* paltry, insignificant; see *parvus* under **little**
    L. *naucum,* n.; *naucus,* m. trifle: *Lepiota naucina* (a mushroom).
    L. *nugor, -atus,* trifle, jest, talk nonsense, be frivolous; *nuga,* f. trifle, joke; *nugula,* f. dim.; *nugamentum,* n. trifle, trash; *nugalis; nugatorius; nugax, -acis,* trifling, futile, worthless; *nugator, -is,* m.; *nugatrix, -icis,* f. trifler, jester: nugatory, nugation, *Bryostemma nugator* (a fish), *Anonyx nugax* (an amphipod).
    L. *paulum; paululum,* bit, trifle; see *paulus* under **little**
    Gr. *phaulos; phauros,* trivial, paltry, petty, mean, poor; *pamphaulos,* all bad: *Phauloblatta clathrata* (a fossil cockroach), *Phaulothamnus spinescens* (a phytolaccacead).
    Gr. *phlyaros,* foolery, nonsense, silly talk; see *phledon* under **speak**
    L. *recula,* trifle, dim. of *res,* thing; see **thing**
    Gr. *rhopos,* m. petty wares, knickknacks; *rhopikos,* of trumpery:
    L. *sublestus,* slight, trifling, trivial:
    L. *tantillus; tantulus,* so little, such a trifle: *Cardiophorus tantillus* (a beetle).
    L. *titivillitium,* n. very small trifle, bagatelle: *Spermophagus titivillitius* (a beetle).
    L. *trica,* f. trifle, hindrance; *tricula,* f. dim.; *trico, -onis,* m. trickster, mischief-maker; *tricor, -atus,* play tricks, trifle, dally: intricate, intrigue, extricate, inextricable.
    L. *trivialis,* belonging to the crossroads, commonplace, vulgar: trivial, triviality.
    L. *vesculus,* little, trifling; see **weak**
    See: **little, worthless, empty, laugh**
trifolium, L. trefoil, clover; see **clover**
trigle, Gr. red mullet; now applied to a genus of gurnards; see **gurnard**
triglochin, NL. a genus of bog plants; see **waterweed**
trigmos, Gr. a strident sound, cry; see **squeak**
trigonos, Gr. triangular; see *gonia* under **angle**
trillium, NL. a genus of the lily family; see **lily**
trimma, Gr. anything rubbed or crushed; see *tribo* under **rub**
trimulus, L. of three years; see **year**
trinus, L. in three's, three each; see *tres* under **three**
triorchis, Gr. a kind of hawk: see **hawk**

triplaris, L. threefold; a genus of the buckwheat family; see *tres* under **three, buckwheat**

triplex, L. threefold; see *tres* under **three**

tripod < L. *tripus* (Gr. *tripous*), a three-legged stand; see **frame**

tripolium, L. (Gr. *tripolion*), a plant; see **leadwort**

tripsacum, NL. a genus of grasses; see **grass**

tripter, Gr. pestle; see **pestle**

tripudium, L. a measured stamping, dancing; see **dance**

triquetrus, L. three-cornered, triangular; see *-quetrus* under **angle**

-tris; -tria, Gr. suffixes signifying agent, doer; see *-ter* under **make**

trismos, Gr. a gnashing, grating, grinding, rasping; see **grind**

trissos, Gr. threefold; see *tres* under **three**

tristis, L. sad; *tristiculus,* dim.; *tristificus,* saddening; see **sad**

triticum, L. wheat; see **grain**

Triton, L., Gr. a sea-god who used a conch to control the waves; see **sea**

tritos, Gr. third; see *tres* under **three**

tritus, L. well-worn, familiar, commonplace; *tritor,* rubber, grinder; see *tero* under **rub**

triumphus, L. victory; *triumphalis,* of victory; see **victory**

trivialis, L. belonging to the crossroads, commonplace, vulgar; *trivius,* three-wayed; see *via* under **way, trifle, common**

-trix, L. suffix signifying agent, usually feminine, in contrast with *-tor,* masculine; see *-or* under **make**

trixago, L. germander; see **mint**

trochado- < Gr. *trochas, -ados,* a light shoe for running; see **shoe**

trochalos, Gr. running, round; see *trecho* under **run**

trochanter, Gr. segment of an insect leg, runner; see *trecho* under **run**

trochilus, L. (Gr. *trochilos*), a small bird, probably a wren; see **wren**

trochlea, L. (Gr. *trochilia*), pulley; see **wheel**

trocho- < Gr. *trochos,* anything round or circular, wheel, ball; *trochion; trochiskos,* dim.; see **wheel, ball**

trocto- < Gr. *troktos,* gnawed, eatable; *troktes; trox, trogos,* gnawer, nibbler; see *trogo* under **bite**

troglo- < Gr. *trogle,* hole; *troglodytes,* a hole-dweller; see **hole**

trollius, NL. a genus of the buttercup family; see **butercup**

tromos, Gr. a trembling, quivering; *tromeros; tromikos,* trembling; see **shake**

-tron, Gr. suffix denoting tool, instrument; see *-tros* under **tool**

tropaeolum, NL. nasturtium in the garden sense; see *tropaeum* under **mark**

trope; tropos, Gr. a turn, turning, direction, way; see *trepo* under **turn**

tropalis, Gr. bundle, bunch; see **bundle**

tropeco- < Gr. *tropex, -ekos,* oar; see **oar**

trophe, Gr. food, nourishment; *trophimos,* nutritious; *trophos,* feeder, nurse; see *trepho* under **eat**

trophy < L. *tropaeum* (Gr. *tropaion*), mark, token, or memorial of victory, that is, the enemy's turning in defeat; see **mark**

tropicus, L. (Gr. *tropikos*), a turning, solstice; see *trepo* under **turn**

tropis, Gr. keel; see **ridge**

tropos, Gr. turn, direction, way, see *trepo* under **turn**; a twisted, leather thong, see **strap**

trossulus, L. fop, coxcomb, dandy; see **fool**

trotos, Gr. wounded, vulnerable; *trosis,* a wounding; see *trauma* under **sore**

trouble < OF. *turbler,* < L. *turbula,* f. dim. of *turba,* f. turmoil, disorder, tumult; *perturbatus,* troubled: troublesome, troublous, perturbation, perturbed, turbulence.
    Gr. *achthos,* weight, burden, distress; *achtheros (achtheres),* annoying, distressing; see **weight**
    Gr. *ademon, -os,* troubled, distressed; *ademonia,* f. trouble, distress: *Ademon decrescens* (a wasp).
    L. *aerumna,* f. trouble, distress, need; *aerumnabilis,* wretched, miserable; *aerumnosus,* full of trouble, suffering:

L. *angor, -is,* m. anguish, trouble: *anxius,* uneasy, troubled: anxious.
Gr. *ania,* f. trouble, distress, grief; *aniaros,* troublesome, annoying, grievous: *Aniaropsis latifrons* (a beetle).
Gr. *aphertos,* insufferable, intolerable:
Gr. *aremenos,* distressed, harassed:
Gr. *argaleos,* troublesome, vexatious: *Argaleus attenuatus* (a beetle).
It. *briga,* f. trouble, strife: brigand, brigade, brigadier, brigantine.
Gr. *chalepos,* difficult, harsh, troublesome; see **difficult**
L. *displicentia,* f. discontent, dissatisfaction:
Gr. *dyscheres,* annoying, vexatious: *dysdaimonia,* f. misery; *dysdaimon, -os,* unhappy; *dysphrosyne,* f. anxiety, care; *dystonos,* grievous, lamentable: *Dyscheres griseus* (a beetle).
Gr. *epistropresis,* f. vicissitude:
Gr. *ergodes,* irksome, troublesome:
L. *fatigo, -atus,* tire, plague, vex; see **weak**
L. *gravamen, -inis,* trouble, inconvenience; see *gravis* under **heavy**
Gr. *kamatos,* toil, labor, distress, trouble, weariness; see *kamno* under **make**
Gr. *kedos; kedosyne,* trouble, affliction, care; see *kednos* under **guard**
Gr. *meledone,* care, sorrow; see *meledonos* under **guard**
Gr. *mochtheros (mogeros),* toiling, suffering, wretched; *mochthos (mogos),* m. toil, distress, trouble; *emmochthos,* toilsome, fatiguing: *Mochtherus tetraspliotus* (a beetle), *Mogoplistes brunneus* (a cricket), *Barymochtha entherastis* (a moth).
L. *molestus,* troublesome, annoying: molest, *Simulium molestum* (blackfly).
Gr. *ochleros; ochletikos,* troublesome, turbulent; *ochlesis; diochlesis; enochlesis,* f. annoyance, disturbance: ochlesis, *Ochleroptera oblita* (a wasp), *Enochletica ostentatrix* (a grasshopper).
Gr. *oïzys, -yos,* f. distress, suffering, misery, woe; *oïzyros,* distressing, miserable, woeful; *oïzyo,* be miserable, suffer:
Gr. *otlos,* m. distress, suffering: *Otlophorus vepretorum* (a wasp).
Gr. *pema, -tos,* n. suffering, woe:
L. *satagius,* anxious, troubled, worried:
L. *solicito, -atus,* disturb, tempt, vex: solicit, solicitation, solicitor, solicitous, solicitude.
L. *tortura,* torment; see **pain**
L. *tribulo, -atus,* afflict, oppress; see *tribulum* under **separate**
L. *vexo, -atus,* agitate, harass, annoy: vex, vexation, vexatious, *Aëdes vexans* (a mosquito).
See: **pain, sad, difficult, hurt, arouse, shake, move, confusion**

trough < AS. *trog;* see **basin**

trousers < uncertain origin.
Gr. *anaxyris, -idos,* f. trousers:
L. *braca,* f. breeches, trousers; *bracatus,* wearing trousers: bracate, *Chalcis bracata* (a wasp).
L. *sarabarum (saraballum),* n. wide trousers:

trout < AS. *truht,* < L. *tructa* (ML. *trutta*), f. (Gr. *troktes,* m.), a fish with sharp teeth: truttaceous, troctolite (forellenstein), *Salmo trutta* (sea-trout), *Galaxias truttaceus* (a fish, kokopu), *Trichodina truttae* (a protozoan).
L. *fario, -onis,* m. a sea-trout: *Salmo fario* (brown trout).
L. *salar, -is,* m. a kind of trout: *Salmo salar* (Atlantic salmon).
L. *salmo, -onis,* m. salmon: salmon, *Salmo irideus* (rainbow-trout), *Salmonella psittacosis* (bacterium causing parrot-fever), *Serrasalmo scapularis* (a caribe, piranha), *Corticium salmonicolor* (a fungus), *Micropterus salmoides* (large-mouthed black-bass), *Basanistes salmoneus* (a copepod).
NL. *salvelinus,* m. char, trout, < G. *salmling,* < L. *salmo,* salmon: *Salvelinus fontinalis* (brook-trout).
Gr. *thymallos,* m. a fish of the trout family, grayling: *Thymallus tricolor* (Michigan grayling).

trowel < OF. *truele,* < L. *trulla,* small ladle; see **spoon**

trox, *trogos,* Gr. nibbler, gnawer; see **bite**

troxalis, L. grasshopper, cricket; see **grasshopper**

troxanon, Gr. twig; see **branch**

trua, L. stirring spoon, skimmer; *trulla,* dim.; see **spoon**

trucido, -atus, L. cut to pieces, kill, butcher; see **kill**

tructa, L. (Gr. *troktes*), trout; see **trout**

truculentus, L. savage, fierce, harsh, cruel; see *trux* under **rough**

trudis, L. pointed pole, pike; see **spear**

trudo, *trusus,* L. thrust; see **push**

true < AS. *treowe,* faithful, trusty: truth, troth, truce, betrothal, untrue, untruthful.
Gr. *adolos,* without fraud, honest: *Adolopus helmsi* (a beetle).
Gr. *alethes,* true; *aletheia,* f. truth; *aletheutikos,* truthful; *alethinos,* genuine; *philalethes,* loving truth: Alethea, Alice, *Alethopteris robusta* (a fossil fern).
Heb. *amen,* true, truly, verily, so be it: amen.
Gr. *anamilletos,* undisputed:
Gr. *anamphilogos,* undisputed, undoubted:
Gr. *aneristos,* undisputed, uncontested: *Aneristus ceroplastae* (a wasp).
Gr. *apodeixis,* f. conclusive proof, demonstration: apodixis.
Gr. *apseudes,* truthful, sincere:
Gr. *atrekes,* exact, sure, true, real; see **right**
L. *auctoritas, -atis,* validity, security; see *auctor* under **make**
L. *authenticus* (Gr. *authentikos*), genuine, true, original: authentic, authenticity.
Gr. *axioma,* principle; see **law**
L. *candor, -is,* m. openness, sincerity, frankness, brightness, clearness: candor.
L. *certus,* definite, fixed, sure, true; see **sure**
Gr. *chaios,* genuine, true, good: *Chaeomyias incompta* (a bird).
L. *convictio, -onis,* f. demonstration, proof: convincing, conviction.
L. *cordicitus,* from the heart, sincere:
L. *devotus,* attached, dedicated, faithful: devoted, devout.
L. *effatum,* n. axiom:
Gr. *eikos,* likely, probable, reasonable; *eikotos,* reasonably, suitably, fairly: *Icotopus arcurostris* (a decapod), *Icogramma obscura* (a beetle).
Gr. *eteos,* true, real, genuine: *Eteophilus rufulus* (a beetle).
Gr. *etymos,* true; *etymon,* n. the true, original, literal meaning of a word: etymon, etymology, *Quercus etymodrys* (a fossil oak).
L. *fidelis; fidus,* trustworthy, true, sure, faithful; see *fido* under **believe**
L. *genuinus,* natural, authentic: genuine, *Lasioglossum genuinum* (a bee).
L. *germanus,* having the same parents, genuine, real; see **kin**
Gr. *gnesios,* genuine, true, legitimate: *Gnesiomyia crassiseta* (a fly).
L. *honestus,* honored, free from fraud: honest, honesty.
L. *indubitabilis,* that cannot be doubted: indubitable, indubiety.
L. *insubditivus,* genuine:
Gr. *ithagenes,* born in wedlock, legitimate, genuine: *Ithagenes* (a wasp).
L. *pisticus* (Gr. *pistikos*), true, genuine, faithful; see *pistis* under **believe**
L. *principium,* foundation, beginning; see **begin**
L. *sincerus,* pure, unmixed, genuine: sincere, sincerity, insincere.
AS. *soth,* true: sooth, soothe, soothsayer, forsooth.
L. *validus,* strong, sound; see **strong**
L. *verus,* true; *verax, -acis,* speaking truly; *veritas, -atis,* f. truth; *veridicus; veriloquus,* truthful, true: veracity, verify, veritable, verdict, verily, very, aver, Lake Itasca (Minn.), *Galium verum* (yellow bedstraw). Asked Pontius Pilate: *"Quid est veritas (Ti estin aletheia)?"*
See: **right, sure, believe, know, good, try**

trulla, L. dipper, ladle, scoop, see **spoon**; *trullisso, -atus,* plaster, trowel, see **plaster**

trulleum, L. basin; see **basin**

-trum, L. (Gr. *-tron*), tool, instrument; see *-tros* under **tool**

trumpet < OF. *trompe;* see **horn**

truncus, L. maimed, cut off; *trunculus,* dim.; see **short, stem**

trust < ON. *traust,* confidence; see **believe**

truth < AS. *treowth;* see **true**

trutina, L. balance, scales; see **balance**

trutta, ML. trout; see **trout**

trux, -ucis, f. fierce; *truculentus,* savage, fierce, cruel; see **rough**

try < OF. *trier,* cull, pick out, sift: tryout, trial, mistrial.
L. *anquiro, -isitus,* seek on all sides, inquire into, examine carefully:
L. *aquilex, -egis,* m. water-inspector:
L. *audeo, ausus,* venture, dare, attempt, hazard; see *audax* under **bold**
Gr. *autopsia,* f. a seeing for oneself, a post-mortem examination: autopsy.
Gr. *basanistes,* m. examiner, investigator; *basanos,* f. touchstone, trial or test by touchstone, < *basanizo,* test, examine, prove: basanite, *Basanus forticornis* (a beetle), *Basanistes huchonis* (a copepod).
Gr. *chrysakonion,* n. touchstone:
L. *conor, -atus,* try, attempt, endeavor; see **make**
L. *coticula,* touchstone, dim. of *cos,* stone; see **stone**

Gr. *deiktos*, capable of proof; *deiktikos*, able to prove; *deiktes*, m. exhibitor; *deiknymi*, bring to light, make known, inform; *apodeiktos*, demonstrable, demonstrated; *dysdeiktos*, hard to prove: apodictic.

L. *demonstratio, -onis*, f. designation, description: demonstration.

Gr. *diphetor, -os*, m. searcher:

L. *dispunctio, -onis*, f. examination, test:

Gr. *dokimos*, assayed, tested, approved, < *dokimazo*, examine, test, prove; *dokimastes*, m. assayer, examiner: *Docimocephalus gregori* (a trilobite), *Docimastes ensiferus* (a hummingbird).

Gr. *elenchos*, m. trial, test, examination; *elenktikos*, fond of cross-questioning or examining: *Elenchus walkeri* (a strepsipterid).

Gr. *ereuna*, f. inquiry, search, examination; *ereuneter, -os; ereunetes*, m. inquirer, searcher: *Ereunetes pusillus* (semipalmated sandpiper).

Gr. *erotema*, question; see **ask**

L. *examino, -atus*, scrutinize, test, consider: examine, examination, examiner.

Gr. *exetastes*, m. examiner, auditor, inspector; *exetasis*, f. examination: *Exetastes maculatus* (a wasp), *Exetastica ignobilis* (a beetle).

L. *experimentum*, n. test, trial, < *experior, -ertus*, test, try; *experientia*, f. knowledge gained by trial; *perior, peritus*, try, risk: experiment, experience, expert, inexpertness.

L. *expiscor, -atus*, fish out, search out:

L. *exploro, -atus*, search out, investigate, try; *explorator, -is*, m. searcher, investigator: explore, exploration, explorer, exploratory.

Gr. *geuo*, taste, try; see **taste**

Gr. *ichneuo*, trace, seek out; see *ichnos* under **mark**

L. *inceptum*, beginning, attempt, undertaking; see **begin**

L. *indago, -atus*, trace out, track, hunt for, search out, investigate, explore; *indagabilis*, investigating, inquiring; *indagator, -is*, m.; *indagatrix, -icis*, f. hunter, searcher: *Ichneumon indagator* (a wasp), *Lycosa indagatrix* (a spider).

L. *inspector; dispector, -is*, m. examiner, searcher; *introspicio, -ectus*, look into, examine; *speculator, -is*, m. explorer, scout, investigator, < *speculor, -atus*, explore, examine; *specillum*, n. surgeon's probe, < *specio*, look at: inspect, inspector, introspection, speculate, speculator, speculation, *Telostylinus speculator* (a fly).

L. *investigo, -atus*, search after, seek out, trace; *investigator, -is*, m. inquirer, searcher: investigator, *Microdon investigator* (a fly).

Gr. *kriterion*, means of judging, standard; see *criticus* under **judge**

Gr. *martyrion*, proof, testimony; see *martyr* under **witness**

Gr. *master, -os; masteutes*, m. searcher, seeker: *Masteutes saxifer* (a fossil beetle), *Heliomaster squamosus* (a hummingbird).

L. *obrussa*, f. (Gr. *obryzon*, n.), test or assay of gold: *Obrussa ochrifasciella* (a moth).

Gr. *peira*, f. attempt, experiment, trial, attack; *peirasis*, f. attempt; *peirastes*, m. experimenter; *empeirikos*, experienced: peirameter, peirastic, pirate, piratical, empirical, empiricism.

L. *periclitor, -atus*, put to the test, try, expose to danger; *periculum*, n. trial, risk, danger: periclitate, peril.

L. *peto, -itus*, seek, ask, strive after; see **ask**

Gr. *peuthen, -os; peustes*, m. inquirer, spy; *peusis*, f. inquiry, question, < *peuthomai*, inquire:

Gr. *phorao*, search for a thief, detect, discover; *phoratikos*, of searching; *phoratos*, discoverable:

L. *probo, -atus*, test; *proba*, f. test; *probatio, -onis*, f. test, trial, examination: probe, probate, probative, probable, proof, probity, approbation, approve, approval, disprove, disapproval, reprove, reprobate.

Gr. *problema*, question propounded for solution, puzzle, riddle; see **ask**

Gr. *pselapho*, grope about, feel, touch, examine; *pselaphetes*, searcher; see **touch**

L. *quaero, quaesitus*, seek to learn, ask; *quaestio, -onis*, f. inquiry, interrogation, investigation, problem; *quaesitor, -is*, m. seeker, searcher; *conquiro, -isitus*, seek or search out eagerly; *disquiro*, investigate; *exquiro*, search out; *inquiro*, search into; *requiro*, seek again: question, query, inquest, request, requisition, acquire, conquer, conquest, conquistador, disquisition, exquisite, inquiry, inquisition, inquisitive, inquisitorial, require, requisite, *Meniscoëssus conquisitus* (a Cretaceous mammal), *Pimpla inquisitor* (an ichneumonid wasp).

L. *rimor, -atus*, lay open, turn over, examine; *rimator, -is*, m. investigator: *Bathystoma rimator* (tomtate).

L. *ruspor, -atus*, search through, examine, explore:

L. *scisco, scitus; scitor, -atus*, seek to know, search, inquire; *sciscitator, -is*, m. investigator: adscititious.

L. *scrutor, -atus*, examine thoroly; *scrutator, -is*, m.; *scrutatrix, -icis*, f. examiner, investigator: scrutinize, scrutiny, scrutineer, inscrutable, *Anthrax scrutata* (a fly), *Calosoma scrutator* (a beetle).

Gr. *skepterion*, n. proof:

L. *solicito, -atus,* disturb, tempt; see **trouble**

Gr. *tekmerion,* sure sign, positive proof; see **sure**

L. *tento (tempto), -atus,* try, prove, put to the test; *tentamen, -inis,* n. trial, attempt: tempt, attempt, temptation, tentative, taunt.

L. *testum,* n. earthen vessel or cupel for examining, reducing, or refining ores or metals, trial: test, tester, test-tube.

Gr. *zetetes,* m. seeker; *zetesis,* f. a seeking, searching; *zetetikos,* disposed to search; *ekzetetes,* m. searcher; *perizetesis,* f. diligent search: *Zetetes niger* (a bird), *Zeteticontus abilis*[*habilis*] (a wasp), *Eczetesis paniscoides* (a wasp).

See: **judge, hunt, ask, science, true, think, make**

tryblium, L. (Gr. *tryblion*), cup, bowl; see **cup**

trycheros, Gr. ragged, tattered, worn; *trychosis,* exhaustion; *trychos,* worn garment; *trychion,* dim.; see **rag**

trygetos, Gr. harvest, vintage; see **reap**

trygon, Gr. sting-ray, see **shark**; turtledove, see **dove**

tryma, Gr. hole; *trymation,* dim.; see **hole**

tryngas, Gr. a kind of sandpiper; see **snipe**

tryo, Gr. rub, wear out; *trysis,* a wearing away; see **rub**

tryos, Gr. labor, toil; see **make**

trypa; trypema, Gr. hole; *trypemation,* dim.; see *tryma* under **hole**

trypanosoma, NL. a genus of protozoans; see **protozoan**

tryphe, Gr. softness, delicacy, luxury; *trypheros,* dainty, delicate; see **soft**

tryphos, Gr. piece, morsel, lump; see **part**

trysmos, Gr. murmur; see *tryzo* under **murmur**

tryssos, Gr. dainty, delicate; see *tryphe* under **soft**

tryx, -ygos, Gr. new wine, lees of wine, dregs; see **dirt**

tsuga, Jap. hemlock; see **gymnosperm**

tub < D. *tobbe;* see **basin, barrel**

tuba, L. war-trumphet; *tubula,* dim.; see **horn**

tuber, L. a swelling, bulb; *tuberculum,* dim.; *tuberosus,* full of lumps or protuberances; see **swell**

tubicen, L. trumpeter; see *tuba* under **horn**

tuburcinor, -atus, L. eat greedily, devour; see **eat**

tubus, L. pipe; *tubulus,* dim.; see **pipe**

tucetum, L. a kind of sausage; see **sausage**

-tude < L. *-tudo,* state, condition, pertaining to; see **nature**

tudes, L. hammer, mallet; *tudicula,* dim.; see **hammer**

tufa < L. *tofus,* travertine; see **lime, stone**

tuft < OF. *toffe;* see **bush**

tugurium, L. hut, cottage; see **house**

tulip < D. *tulp,* < F. *tulipe,* < Turk. *tulbend,* turban; see *tulipa* under **lily**

tumba, L. sepulchral mound, tomb; *tumbula,* dim.; see **grave**

tumble < AS. *tumbian;* see **turn, fall, dip**

tumidus, L. swollen; *tumor,* a swelling; see *tumeo* under **swell**

tumultus, L. commotion, confusion, disturbance, uproar; see **confusion**

tumulus, L. mound, barrow, hillock; see **heap**

tungsten < Sw. *tung,* heavy; *sten,* stone; wolfram; see ELEMENTS under **thing**

tunica, L. garment, skin, husk; *tunicula,* dim.; see **garment**

tunnel < OF. *tonnel;* see **hole, mine**

tunny < L. *thunnus* (Gr. *thynnos*), m. a large, mackerellike fish: tuna, *Thunnus thynnus* (tunny).

Gr. *amia,* f. a kind of tunny; now applied to the bowfin: *Amia calva* (bowfin), *Paramia quinquevittata* (cardinal-fish).

Gr. *attageinos,* m. a kind of tunny:

Gr. *auxis, -idos,* f. a young tunny: *Auxis mediterranea* (a tunny).

L. *cybium* (Gr. *kybion*), n. a tunny: *Cybium guttatum* (a mackerel).

L. *elacaten* (Gr. *elakaten*), probably a tunny; see *elakate* under **spindle**

Gr. *melandrys, -yos,* m. a large kind of tunny:

L. *orcynus* (Gr. *orkynos; orkys, -ynos*), m. a kind of tunny: *Orcynus subulatus* (a fish).

L. *pelamis, -idis* (Gr. *pelamys, -ydos*), f. a young tunny: pelamyd, *Pelamis bicolor* (a sea-snake), *Pelamycybium vindobonensis* (a fish), *Pelamichthys unicolor* (a fish).

Gr. *premnas, -ados,* f. a kind of tunny: *Premnas unicolor* (a fish).

Gr. *skepanos,* m. a kind of tunny:

Gr. *skordyle,* f. a young tunny:

Gr. *synodontis, -idos,* f. a kind of tunny: *Synodontis nigrita* (a fish).

turba, L. (Gr. *tyrbe*), tumult, disorder; *turbula,* dim.; *turbidus,* confused, disordered; see **confusion**

turbinatus, L. top-shaped, conical, < *turbo,* top; see **cone**

turbulentus, L. agitated, stormy, troubled; see *turba* under **confusion**

turdus, L. thrush; see **thrush**

-ture < L. *-tura,* a suffix denoting result of action; see *-sura* under **make**

turf < AS. *turf;* see **sod**

turgidus, L. inflated, swollen, distended; *turgor,* swelling; see *turgeo* under **swell**

turifer, L. incense-bearing, < *tus, turis,* incense; see *thus* under **resin**

turio, L. shoot, sprout; see **branch**

turkey. Erroneously so named because the bird was confused with another supposed to have come from Turkey; see *meleagris* under **peacock**

turma, L. troop of cavalry; *turmalis,* of a troop; see **army**

turn < AS. *turnian,* < L. *torno, -atus,* revolve, fashion in a lathe, round off; *tornus* (Gr. *tornos*), m. lathe; *tornatilis;* Gr. *tornotos,* turned or rounded in a lathe; *entornos,* turned: tornado, turnpike, return, tour, tourist, tourniquet, tournament, attorney, contour, *Tornoceras simplex* (a fossil cephalopod).

L. *ambitus,* a going round, revolution, circuit; see **around**

L. *amplector, -exus,* wind, twine, embrace; see **embrace**

L. *anfractus,* bending, winding, crooked, circuitous; see **bend**

L. *chamulcus* (Gr. *chamoulkos*), m. windlass for hauling ships on land:

L. *circuitus,* a going round, circling; see **circle**

L. *colubrosus,* serpentine, winding; see *coluber* under **snake**

L. *crisis* (Gr. *krisis*), f. turning point; *krisimos,* critical, decisive: crisis, *Pterostichus (Crisimus) placidus* (a beetle).

Gr. *diaptyxis,* f. evolution, explication:

Gr. *dine,* f.; *dinos,* m. whirl, whirlpool, eddy; *dinesis,* f. rotation; *dinetos,* whirled round; *dinodes,* eddying; *dinotos,* turned, rounded; *peridineo,* whirl, spin: dinical, dinoflagellate, oticodinia, peridinian, *Dinenympha gracilis* (a protozoan), *Peridinium ovatum* (a dinoflagellate), *Aplodinotus grunniens* (croaker).

Gr. *eilo,* roll up, wind, twist; *eiletos,* wound; *eiletikos,* wriggling: ileus, eileton, *Cephaëlis acuminata* (an ipecac).

L. *fidicularis,* like a cord, twisted; see *fides* under **harp**

L. *flexuosus,* full of turns, tortuous, crooked, winding: flexuous.

L. *gurges, -itis,* m. whirlpool, eddy, abyss; *gurgito, -atus,* surge, boil, rage, toss about, engulf: gurgitation, gorge, disgorge, gorgeous, regurgitate.

L. *gyro, -atus,* turn around, < L. *gyrus* (Gr. *gyros*), m. circle, round; *gyraleos; gyrinos,* rounded, curved; *perigyris, -idos,* f. circumference: gyrate, gyration, gyroscope, *Gyrocarpus americanus* (a hernandiacead), *Gyrinus borealis* (whirligig-beetle), *Anagyris foetida* (bean-trefoil), *Exogyra arietina* (a fossil gastropod), *Cornuspira polygyra* (a foraminifer).

Gr. *helisso,* turn round; *heligmos,* m. a winding, convolution; *helix, -ikos,* f. spiral, whirl, eddy, tendril, coil; *heliktos,* rolled, spiral; *anatylisso; anelisso,* unroll; *exelixis,* f. evolution: helix, helicoid, helicopter, helicon.   See **spiral**

ON. *hverfa,* turn: whirl, whorl, wharf, warble.

Gr. *ilinx, -ingos,* f. a whirling, agitation, dizziness: *Ilingocerus alexandrae* (a fossil mammal).

Gr. *kampe,* a turn, bend, winding; see *kamptos* under **bend**

Gr. *kollops, -opos,* m. peg, screw for tightening the lyre-strings: *Collops cribrosus* (a beetle).

Gr. *kybistema, -tos,* n. somersault, < *kybistao,* tumble headfirst, somersault; *kybistesis,* f. a tumbling; *kybisteter, -os,* m. tumbler, diver: *Cybister*[*Cybisteter*] *explanatus* (a water-beetle), *Cybistetes longifolia* (an amaryllidacead).

Gr. *kylindo; kylio,* roll, roll along, roll up; *kylindros,* m. roller; *kylisis,* f.; *kylisma, -tos,* n. a rolling; *kylistikos,* expert at rolling; *kylistos,* rolled, turned; *ekkyliomai,* unroll, spread out: cylinder, cylindrical, *Cylisticus convexus* (a sowbug), *Eccyliopterus alatus* (a fossil gastropod).

L. *labyrinthus* (Gr. *labyrinthos*), m. structure with many winding passages; *labyrinthicus*, of a labyrinth, intricate: labyrinth, labyrinthine, *Labyrinthula macrocystis* (a slime-mold).

L. *Maeander* (Gr. *Maiandros*), m. Meander, a river in Asia Minor noted for its winding course: meander, *Maeandrina cerebriformis* (a brain-coral), *Caularchus meandricus* (a clingfish).

Gr. *periagogeus*, m. capstan:

Gr. *periamphis, -idos*, f. a turning around:

Gr. *perikochlion*, n. female screw:

L. *petaminarius*, m. tumbler, rope-dancer:

L. *petaurista* (Gr. *petauristes*), tumbler, vaulter; see **leap**

L. *postomis, -idis*, f. twitch for controlling horses:

Gr. *rhembo*, turn, revolve, roll, roam; *rhombos*, spinning top, magic wheel: *Rhembobius quadrispinosus* (a wasp), *Rhomboceras welchi* (a fossil cephalopod).

L. *roto, -atus*, turn around: rotation, rotate, *Hymenocallis rotata* (a spider-lily).

L. *serpentinus*, serpentlike, twisting, winding; see *serpens* under **snake**

L. *spira* (Gr. *speira*), coil, twist; see **spiral**

Gr. *streptos*, twisted; m. a twisted or linked collar or chain, < *strepho*, turn; *strepsis*, f. a turning, twisting; *stremma, -tos*, n. twist, wrench, roll; *strophe*, f. a turning, twist; *stropheus, -eos*, m. vertebra, socket; *strophos*, m. a twisting, twisted band, cord, rope; *aeistrephes*, ever-turning; *apostrophos*, turned away; *diastremma, -tos*, n. twist, wrench: streptoneurous, Strepsiptera, strophe, apostrophe, catastrophe, strophulus, *Strophostyles helvola* (a legume), *Strophanthus sarmentosus* (an apocynacead), *Streptococcus pyogenes* (a bacterium), *Strepsicerus kudu* (koodoo), *Streptorhynchus vetustum* (a fossil brachiopod), *Stremmatognathus catesbyi* (a snake), *Platystrophia lynx* (a fossil brachiopod), *Anastrepha fraterculus* (a fruitfly), *Diastremma marmoratum* (a grasshopper).

Gr. *strobos*, anything twisted, a twisting, turning; *strobilos*, top, cone; see **cone**

L. *sucula*, f. winch, windlass, capstan:

AS. *thrawan*, twist: throw, overthrow, thread.

L. *torqueo, tortus*, twist, turn, wind, writhe; *tortura*, f. a twisting, torment; *torta*, f. a twist, turn; *tortula*, f. dim.; *tortilis; tortuosus*, twisted, winding; *tortor, -is*, m. twister, torturer; *contortus; intortus*, twisted, curled, complicated, intricate: torque, tort, torch, torture, torsion, torment, tortoise, torticollis, contortion, distortion, extortion, retort, tart, tortuous, truss, nasturtium, bistort *(Polygonum bistortoides)*, *Pinus contorta* (lodgepole-pine), *Helix tortula* (a snail), *Cerinthe retorta* (honeywort), *Clidastes tortor* (a mosasaur), *Molgula retortiformis* (a tunicate), *Oenothera bistorta* (an evening-primrose), *Eleocharis tortilis* (a spikerush), *Melocactus intortus* (Turkscap-cactus), *Heterodon contortrix* (hog-nosed snake).

Gr. *trepo*, turn; *treptos*, turned, changed; *trope*, f. a turn, a turning; *tropikos*, of turning, of the solstice, figurative; *tropos*, m. turn, manner, way; *anatrope*, f. upset, overturn: trope, tropic, tropical, heliotrope, trophy, anatropous, Atropos, Tropic of Capricorn, Tropic of Cancer, *Tropocyclops prasinus* (a copepod), *Tropicoperdix chloropus* (a bird), *Xenopus tropicalis* (a toad), *Treponema pertenue* (spirochaete of yaws), *Monotropa uniflora* (Indian-pipe), *Monotropsis odorata* (sweet pinesap), *Monotropamyces nigrescens* (a fungus).

L. *turba*, turmoil, uproar, hubbub; *turbula*, dim.; see **confusion**

L. *turbo, -inis*, anything that turns in a circle, top, spindle, eddy; *turbinatus*, conical; see **cone**

L. *vergo*, tend or turn toward: verge, converge, divergence, *Andropogon divergens* (a grass).

L. *verto, versus*, turn; *vertex (vortex), -icis*, m. whirl, whirlpool, eddy, pole of the heavens; *verticulus; verticillus*, m. dim.; *versabilis; versatilis*, turning with ease, changeable, adaptable; *conversus*, turned around; *inversus*, turned upside down; *obversus*, turned toward; *reversus*, turned back; *retrorsus*, turned or bent backward: vertex, vortex, vertebra, verse, versatile, verticillate, Vertumnus, adversary, advertise, aversion, anniversary, conversation, controversy, conversely, diversity, divorce, divers, dextrorse, extrorse, introrse, inadvertent, inversion, obverse, prose, perversely, traverse, subversive, university, reverse, retrorse, *Vorticella microstoma* (a protozoan), *Telostylus inversus* (a fly), *Schizocosa retrorsa* (a spider), *Ilex verticillata* (winterberry), *Gephyrometra versicolor* (a crinoid), *Cryptochilus vorticosus* (a wasp), *Megaptera versabilis* (a whale), *Zea mays everta* (popcorn).

L. *volvo, volutus*, turn round; *volubilis*, rolling, twisting around, rapid, fluent; *voluta*, f. spiral; *voluto, -atus*, roll, turn or tumble about; *convolvo*, roll together, roll up, interlace; *evolvo*, roll forth, unroll, unfold; *revolutus*, turned over, rolled back: volute, revolve, devolve, voluble, volume, vault, involved, revolution, revolt, evolution, revolting, involucre, *Volvox aureus* (a flagellate), *Convoluta roscoffensis* (a flatworm), *Convolvulus pubescens* (bindweed), *Petrea*

volubilis (purple wreath), *Cycas revoluta* (sago-cycad), *Rhyacophila evoluta* (a trichopterid), *Myelodactylus convolutus* (a Silurian crinoid).
AS. *wrigian,* turn, twist: wry, wryneck, wriggle.
AS. *writhan,* twist: wreath, wrist, writhe.
See: **circle, spiral, around, change, dizzy, axis, spindle, fold, bend, storm, worm**

turnip < AS. *turnian,* turn, *naep,* < L. *napus,* m. turnip; *napellus; napulus,* m. dim.: *Brassica napobrassica* (rutabaga), *Geranium napuligerum* (a geranium), *Aconitum napellus* (monkshood).
Gr. *gongylis, -idos,* f. turnip:
L. *rapum,* n.; *rapa,* f. (Gr. *rhapys, -yos,* f.), turnip; *rapulum,* n.; *rapunculus,* m. dim.; *rapistrum,* n. wild rape; *rapicius,* of rape or turnips: rape *(Brassica napus),* *Brassica rapa* (turnip), *Campanula rapunculus* (rampion), *Ammophilactis rapiformis* (a sea-anemone).

turpentine < OF. *turbentine,* < L. *terebinthinus* (Gr. *terebinthinos*), of the terebinth-tree; see **resin**

turpis, L. ugly, foul, filthy, base; *turpiculus,* dim.; see **dirt**

turquoise < OF. *turquois,* Turkish; see **stone**

turret < L. *turris,* tower; *turricula; turritella,* dim.; see **tower**

tursio, L. an animal like the dolphin, porpoise; see **whale**

turtle < F. *tortue* (Sp. *tortuga*): turtleback, turtledom, turtlehead, snapping-turtle, tortoise, Tortuga Island.
Gr. *chelone,* f. tortoise, turtle; *chelonarion,* n. dim.; *chelonion,* n. tortoise-shell: chelonian, chelonin, *Chelone glabra* (turtlehead), *Chelonarium atrum* (a beetle), *Chelonidium punctatissimum* (an isopod), *Chelonia mydas* (a green turtle), *Archelon ischyros* (a Cretaceous turtle).
Gr. *chelys, -yos,* f.; *chelyros,* m. tortoise, turtle, water-serpent; *chelyon,* n. tortoise-shell: *Chelyocephalus varicolor* (a beetle), *Chelyosoma macleayanum* (a tunicate), *Chelydra serpentina* (snapping-turtle), *Aromachelys odorata* (musk-turtle), *Ophiacantha chelys* (a brittlestar), *Stereospermum chelonoides* (a bignoniacead).
Gr. *emys, -ydos,* f. fresh-water tortoise, turtle: *Emys orbicularis* (a tortoise), *Pseudemys rubriventris* (red-bellied terrapin), *Chrysemys picta* (painted turtle), *Ptychogaster emydoides* (a turtle).
Sp. *galapago,* m. tortoise, turtle: Galapagos Islands.
Gr. *klemmys, -yos,* f. tortoise, turtle: *Clemmys guttata* (spotted turtle).
L. *testudo, -inis,* f. tortoise, turtle; *testudineus,* of turtles; *testudinatus,* like a turtle-shell: Testudinidae, testudinal, testudinate, testudineous, *Testudo gigantea* (giant tortoise), *Testudinaria paniculata* (a dioscoriacead), *Cypraea testudinaria* (turtle-cowry), *Euxestoxenus testudo* (a beetle), *Thalassia testudinum* (turtle-grass).

turunda, L. a ball of paste for fattening geese; see **food**

tus, *turis,* L. incense; see *thus* under **resin**

-tus, L. having the nature of, pertaining to; see *-atus* under **nature**

tussilago, L. coltsfoot; see **composite**

tussis, L. cough; *tussicula,* dim.; *tussicularis,* of a cough; *tussiculosus,* coughing much; see **cough**

tusus, L. struck, beaten; see *tundo* under **strike**

tutor, L. watcher, protector; *tutelaris,* of guardianship; see *tueor* under **guard**

tutubo, L. hoot like an owl; see **hoot**

tutulus, L. hair dressed in a high cone over the forehead; see **crest**

tutus, L. safe, secure; see *tueor* under **guard**

tweezer < F. *etui,* case, sheath; see **tongs**

twelve < AS. *twelf.*
Gr. *dodeka; dyodeka,* twelve; *dodekatos,* twelfth: dodecahedron, *Dodecatheon dentatum* (a shooting-star), *Dodecactenus standingeri* (a beetle), *Dizygocrinus dodecadactylus* (a Mississippian crinoid).
L. *duodecim,* twelve; *duodecimus,* twelfth; *duodeni,* twelve at a time: dozen, duodenum, 12 mo., *Crioceris duodecimpunctata* (an asparagus-beetle).
L. *uncia,* f. twelfth part; *unciola,* f. dim.; *uncialis,* of a twelfth part; *quincunx,* five-twelfths: uncial, inch, ounce, quincuncial, quincunx, *Aegilops triuncialis* (a goat-grass), *Abyla quincunx* (a siphonophore).

twi- < AS. *twi; twa,* two; see **two**

twig < AS. *twig,* shoot, branch, spray; see **branch**

twilight < AS. *twileoht;* see **shade**

twin < AS. *twinn,* two; see **two**

twine < AS. *twin;* see **turn, weave, circle, curl, vine, rope, complex, climb**

twinkle < AS. *twinclian;* see **light, wink**

twist < AS. *twist;* see **turn, weave, fold, complex**

twitter < imitative origin; see **chirp**

two < AS. *twa; twi:* twofold, two-faced, two-handed, twopence, twosome, twice, twin, twilight, Mark Twain, twenty, twill, betwixt, between. Identify: Castor and Pollux.

Gr. *adelphos,* twin, brother; see **brother**

Gr. *agastor,* from the same womb, twin, brother; see **brother**

L. *ambiguus,* of double meaning, doubtful, uncertain; *ambifarius,* of double meaning, ambiguous: ambiguous, ambiguity, *Digitalis ambigua* (a foxglove).

Gr. *amphi,* around, on both sides, double; *amphibios,* living a double life (in water and on land); *amphibolos,* ambiguous, doubtful; *amphidymos,* double, twofold: amphidetic, amphibious, amphibole, *Hippopotamus amphibius* (hippo).

Gr. *amphoteros,* each, both; *epamphoteros,* ambiguous: *Amphoteromorphus peniculus* (a worm).

L. *anceps, -pitis,* two-headed, two-sided, double, ambiguous, uncertain: ancipital, ancipitous, *Baculites anceps* (an ammonite).

Gr. *androgynos,* m. man-woman, hermaphrodite: androgynous.

L. *bi-,* two, double, < *bis,* twice; *binus,* two by two, couple, pair; *binarius,* of two; *bifarius,* double; *bijugus,* yoked together, double; *bisulcus,* cloven, forked: bisect, biped, bissextile, bireme, biduous, bifid, bipinnatifid, binocular, biscuit, balance, bivalve, binary, biceps, biennial, bigamy, bifurcation, bilateral, binomial, combine, combination, *Bidens cernua* (nodding beggarticks), *Capelopterum bimaculatum* (a bug), *Dendrobium bigibbum* (an orchid), *Discodon bisbinotatum* (a beetle), *Drosera binata* (a sundew), *Corbula bisulcata* (a fossil pelecypod), *Lepidopsetta bilineata* (a fish), *Quercus bicolor* (swamp white-oak), *Verbena bipinnatifida* (a verbena), *Trapa bispinosa* (singhara nut). Derive: billion.

L. *comparo, -atus,* couple or pair for the purpose of discovering resemblances or differences; *comparabilis,* that may be compared; *comparativus,* of comparison: compare, comparative, comparison, comparable, incomparable.

Gr. *deuteros,* second; *deuterios,* secondary, of inferior quality: deuterocone, deuteromorphic, Deuteronomy, deuterium, deutonymph, deutoplasm.

Gr. *di-,* two, double, < *dis,* twice; *dicha,* in two, asunder, apart; *dichastes,* m. divider; *dichotomos,* cut in two; *dichthadios,* twofold, double; *didymos,* double, twin; *dieres,* double; *diklis, -idos,* double-folding, double-swinging; *dikros,* forked, cloven; *diphasios,* double; *diphyes,* of double nature, twofold; *diplax, -akos; diplasios,* double; *diploos,* twofold, double; *dissos (dittos; dixos),* twofold, double; *distichos,* in two rows; *distolos,* in pairs; *dizyx, -ygos,* double-yoked, double: digraph, dilemma, Diptera, dioecious, diadelphous, dichromatic, diazomethane, dicrotic, dipyrenous, diethyl, diacetate, dicotyledon, diphthong, diselenide, disomus, disdiaclast, disdodecahedroid, dissyllabic, dichogamy, dichotic, dichotomy, didymous, didymolite, didymium (neodymium and praseodymium), epididymis, epididymite, diphase, diphycercal, diphyodont, diploma, diploid, diplococcus, amphodiplopia, Diplopoda, dissoconch, dissogeny, dittograph, distichous, dizygotic, *Dicentra eximia* (fringed bleeding-heart), *Didiplis diandra* (water-purslane), *Diopsis apicalis* (a fly), *Dipterocarpus turbinatus* (gurjun-tree), *Disanthus cercidifolius* (a hamamelidacead), *Dichognathus typica* (a conodont), *Dichapetalum toxicarium* (an African shrub), *Dichastops subaeneus* (a beetle), *Dichthadia glaberrima* (an ant), *Didymium difforme* (a protozoan), *Dieropsis quadriplagiata* (a beetle), *Diphyodactylus singularis* (a beetle), *Diplacodon alatus* (a fossil mammal), *Diplasiotherium robustum* (a fossil mammal), *Diplodocus carnegiei* (a dinosaur), *Diplodicyton heteromorpha* (a sponge), *Dissotis grandiflora* (a melastomacead), *Dixippus nodosus* (a phasmid), *Distoloceras hystrix* (an ammonite), *Dizygocrinus biturbinatus* (a Mississippian crinoid), *Fraxinus dipetala* (an ash), *Jeffersonia diphylla* (American twinleaf), *Coniothyrium diplodiella* (a fungus), *Macrodipteryx longipennis* (a goatsucker), *Monarda didyma* (bee-balm), *Lychnis dioica* (red campion).

L. *duo* (Gr. *dyo*), two; Gr. *dyas, -ados,* two; *dualis,* of two; *duplaris; duplex, -icis; duplus,* double, twice; *duplico, -atus,* repeat: dual, deuce, double, doubloon, duplex, duplicity, dubious, doubt, indubitable, reduplicate, dyad, dyarchy, dyotheism, dypnone, *Ceratina dupla* (a carpenter-bee), *Pugnoides duplicata* (a fossil brachiopod), *Megaselia dupliciseta* (a fly), *Stylidium reduplicatum* (a stylewort), *Syndyograptus pecten* (a graptolite).

L. *geminus,* twin-born; m. twin; *gemellus,* dim. born at the same time; m. twin; *gemino, -atus,* double, pair, unite; *congemino,* reduplicate: geminate, gemellus, gemel, gimbal, trigeminal, Gemini, by jimminy!, *Solenopsis geminata* (fire-ant), *Agave geminiflora* (an agave), *Chrysolina gemellata* (a beetle).

Gr. *hermaphroditos*, m. man-woman: hermaphrodite.

L. *par, -is*, n. a pair; *pararius*, of a pair:

Gr. *properysi*, two years ago; *properysinos*, of the year before last; see *perysinos* under **year**

L. *secundus*, following, second; see *sequor* under **follow**

Gr. *zeugos*; *zygas*, team, pair; see *zygon* under **bind**

See: **half, team, separate,** *amphi* under **around**

tycho- < Gr. *tyche*, good fortune, luck, chance, accident; *tycheros*, lucky; see **lot**

tyco- < Gr. *tykos*, mason's hammer; *tykion*, dim.; see **hammer**

tycto- < Gr. *tyktos*, created, wrought; see **make**

tylo- < Gr. *tyle*; *tylos*, knot, knob, callus, lump, bolt; *tylarion*; *tylion*, dim.; see **knot**

tymbos, Gr. sepulchral mound, tomb; see *tumba* under **grave**

tymma, Gr. blow; see *typto* under **strike**

tympanum, L. (Gr. *tympanon*), drum; *tympanion*, dim.; see **drum**

tyntlos, Gr. mud; *tyntlodes*, muddy; see **earth**

typado-; typido- < Gr. *typas, -ados*; *typis, -idos*, hammer; see **hammer**

type < L. *typus* (Gr. *typos*), impression, shape, figure, model, example; *typarion*, dim.; *typicalis*, conformable; see **form**

typha, L. (Gr. *typhe*), cattail; see **waterweed**

typhle, Gr. a Nile fish; see **fish**

typhlos; typhlops, Gr. blind; *typhlinos*, blindworm; see **blind**

typhon, L., Gr. whirlwind, tempest; see **wind**

typhos, Gr. smoke, vapor; see **cloud**

typicalis; typicus, L. of the type, conformable; see *typos* under **form**

typis, Gr. hammer, mallet; see **hammer**

typo- < Gr. *typos*, impression, shape, figure, mark or effect of a blow; see **form**, *typto* under **strike**

tyrannus, L. (Gr. *tyrannos*), master, despot; see **govern**

tyrbastes, Gr. agitator, disturber; see *turba* under **confusion**

tyrianthinos, Gr. of Tyrian purple; see **purple**

tyros, Gr. cheese; *tyridion*; *tyrion*, dim.; see **cheese**

tyrris (tyrsis), Gr. tower; *tyrridion*, dim.; see **tower**

tyto, Gr. a kind of owl; see **owl**

tytthos, Gr. little, small, young; see **little**

tyxis, Gr. a making; see *tyktos* under **make**

U

u- < Gr. *ou*, no, not; see **not**

uber, L. udder, teat, breast, see **udder**; fruitful, fertile, full, copious, see **abundance**

ubi, L. where; *ubique*, everywhere; see **place**

uca, Ab.Am. fiddler-crab; see **crab**

-uchus, NL. < Gr. *echo*, hold, bear; see **carry, hold**

udaeo- < Gr. *oudaios*, on the earth, earthly, < *oudas*, surface of the earth, ground; see **earth**

udamino- < Gr. *oudaminos*, good for nothing; see **worthless**

udamo- < Gr. *oudamos*, no one, none; see **not**

udder < AS. *uder,* mammary or milk gland: udderful, udderless, udderlike.
L. *mamma,* f. breast, teat; *mammicula; mammilla; mammula,* f. dim.; *mammillaris,* of the breast or nipple; *mammillatus,* breast-shaped, nipple-bearing; *mammosus; bumammus,* with large breasts: mammal, mammary, mammilla, mammiferous, mammillary, mammillate, mammilliform, *Mammillaria vivipara* (a cactus), *Mammillopora tuberosa* (a fossil bryozoan), *Euphorbia mammillaris* (a spurge), *Raoulia mammillaria* (sheep-plant), *Cycloclypeus mammillatus* (a foraminifer), *Coryphanta bumamma* (a cactus), *Erica mammosa* (a heath).
Gr. *mastos (mazos),* m. breast, nipple, any mammillate object; *mastarion,* n. dim.; *boumastos,* with large breasts: mastoid, mastitis, mastodon, mastotomy, Amazon, *Mastophora cornigera* (a spider), *Mastotermes wheeleri* (an Eocene termite), *Mastostethus tricinctus* (a beetle), *Mazodus kepleri* (a fossil fish), *Bumastus globosus* (a trilobite).
Gr. *outhar, -atos,* n. udder, breast; *outhatios,* of the udder:
L. *papilla,* nipple, teat, bud; *papillatus,* bud-shaped; see **bud**
L. *rumis, -is,* f. breast, teat; *surrumus,* sucking:
L. *sumen, -inis,* n. breast, sow's udder: *Tectisumen compressa* (a gastropod).
Sp. *teta,* f. teat; *tetilla,* f. dim.: *Tetilla polyura* (a sponge).
Gr. *thele,* f. nipple, teat; *thelion,* n. dim.: thelalgia, thelitis, thelium, epithelium, endothelium, *Thelodiscus indivisus* (a fly), *Acrothele coriacea* (a fossil brachiopod), *Blastothela rosea* (a medusa), *Thelesperma ambiguum* (a composite), *Thelephora anthocephala* (a fungus).
AS. *tit,* nipple: tit, teat.   Derive: Grand Teton.
Gr. *titthos,* m. nipple, teat, nurse, rearer; *titthidion; titthion,* n. dim.; *titthe,* f. nurse: *Titthodomus koeneni* (a fossil gastropod).
See: **breast, suck, milk**

udo, L. (Gr. *oudon*), sock, shoe, see **stocking**; < *oudos,* threshold, see **door**

udus, L. wet, damp, humid; see **wet**

ugly < ON. *uggligr,* dreadful.
Gr. *aischros,* ugly, base; see **bad**
Gr. *akalles,* ugly: *Acalles nubilosus* (a beetle), *Acallophilus scrobicollis* (a beetle), *Laccoproctus acalloides* (a beetle).
Gr. *aschemon, -os,* misshapen, ugly: *Aschemonella scabra* (a protozoan).
L. *deformis; informis,* misshapen, hideous, ugly: *deformo, -atus,* disfigure, cause to be out of shape: deform, deformed, deformity, *Inoceramus deformis* (a fossil pelecypod), *Taphrina deformans* (a fungus).
Gr. *dyscides,* ugly, unshapely; *dyseidia,* f. ugliness: *Dyseidopus*[*Disidopus*] *sericeus* (a wasp).
Gr. *dysmorphos; kakomorphos,* misshapen, ugly; *dysmorphia,* f. ugliness: *Dysmorphorhynchus amabilis* (a beetle), *Dysmorphia astyocha* (a butterfly).
Gr. *eidechthes,* of repulsive appearance, ugly:
L. *inhonestus,* ugly, unseemly, shameful: *Melania inhonesta* (a gastropod).
L. *invenustus,* unattractive:
L. *turpis,* ugly, filthy, unseemly; see **dirt**
See: **bad, rough, dirt**

ulamo- < Gr. *oulamos,* crowd of warriors; see **army**

ulcer < L. *ulcus, -eris,* abscess, sore; *ulcusculum,* dim.; *ulcerosus,* full of sores; see **sore**

-ule, L. diminutive suffix; see *-cle* under **little**

ulex, L. a genus of the bean family; see **bean**

uliginosus, L. full of moisture; see *udus* under **wet**

ulio- < Gr. *oulios,* baneful, deadly; see **kill**

ullus, L. any, anyone; see **one**

ulmus, L. elm; *ulmeus,* of elm; see **elm**

ulna, L. elbow, forearm; see **arm**

ulo- < Gr. *oulon,* gum, see **mouth**; *oule,* scar, see **mark**; *oulos,* woolly, curly, see **wool, curl**

uloborus, NL. a genus of spiders; see **spider**

ulpicum, L. a kind of leek; see **onion**

ultimus, L. farthest, last; see *ultra* under **far**

ultra- < L. *ultra,* on the other side, beyond, < *ulter,* beyond, far; *ulterior,* farther; *ultimus,* farthest, last; see **far**

ultroneus, L. voluntary; see **free**

ultus, L. avenged, punished; *ultio,* revenge, punishment; see *ulciscor* under **punish**

ulula, L. a screech-owl, < *ululo, -atus,* howl, shriek, wail; see **owl, call**

-ulus, L. diminutive suffix; see *-cle* under **little**

ulva, L. sedge; now a genus of seaweeds; see **alga**

umbella, L. parasol, sunshade; see *umbra* under **shade**

umbilicus, L. navel; see **navel**

umbo, L. boss, knob, shield; see **projection**

umbra, L. shade, shadow; *umbella,* dim.; *umbraculum,* shady place, arbor; *umbraticus; umbratilis,* of the shade, retired, private, see **shade;** a sciaenid fish, see **perch**

umpire < OF. *nonper,* uneven; see **judge**

un- < AS. *un,* not, back, off; see **not**

unarmed; see **bare**

uncatus; uncinatus, L. hooked, barbed; see *uncus* under **hook**

uncertain; see **doubt, wander, dizzy**

uncia, L. the twelfth part; *uncialis,* of a twelfth part; see **twelve**

uncinus, L. hook, barb; *uncinatus,* hooked, barbed; see *uncus* under **hook**

unctus, L. anointed; see *unguen* under **fat**

-unculus, L. diminutive suffix; see *-cle* under **little**

uncus, L. hook, barb; see **hook**

unda, L. wave; *undatus; undulatus,* wavy; see **wave**

under < AS. *under,* below, beneath: understanding, underbrush, undergo, undertaker, underling, underestimate, underwriter, underneath, underline, undertone.
Gr. *enerthe,* below, under; see *nerthe* under **low**
Gr. *hypo,* under, beneath, less than: hypodermic, hypocrisy, hypothesis, hypogenous, hyposulfite, hyposulfurous, hypochlorous, hyphen, hypogeous, *Hypoderma lineatum* (a heel or oxwarble fly), *Hypericum perforatum* (a St. Johnswort), *Arachis hypogaea* (peanut), *Hypoxis hirsuta* (a stargrass).
L. *infra,* underneath, below: infrared, infrahuman, *Colletes infracognitus* (a bee).
Gr. *kata,* down, below; see **down**
L. *sub-, suc-, suf-, sug-, sum-, sup-, sur-, sus-,* < *sub,* under, from, somewhat, less than; *subter,* below, beneath; *subternus,* underneath, lower; *subtus,* below, beneath; *succido,* undercut; *suffigo,* fasten below; *suggero,* advise, subjoin; *summergo (submergo),* dip under; *suppono,* place under, substitute; *surrepo,* creep under; *sustineo,* support: subjugate, subterranean, subtract, subaqueous, sublet, subsoil, submarine, submit, subscribe, subsequent, suburb, subway, substance, subconscious, subnormal, suboval, succumb, succeed, succinct, suffer, suffocate, sufficient, suffruticose, suffuse, suffrage, suggest, summon, sumptuous, support, supplement, supplicate, supply, suppose, suppurate, surrogate, surreptitious, susceptible, suspicion, suspension, sustain, sojourn, somber, souvenir, supine, subterfuge, *Batocrinus subovatus* (a Mississippian crinoid), *Macrodactylus subspinosus* (rose-beetle).
See: **low, base, deep, down**

undertaker
Gr. *entaphiopoles,* m. undertaker:
Gr. *kteristes,* m. undertaker, < *kterizo,* bury:
L. *libitinarius,* m. undertaker, < *Libitina,* f. goddess of corpses, in whose temple the register of the dead was kept and funeral apparatus was hired out or sold:
L. *pollinctor, -is,* m. undertaker, < *pollingo, -inctus,* prepare a corpse for the funeral:
See: **death, dig, cover**

undulatus, L. wavy; see *unda* under **wave**

-undus, L. suffix denoting continuance or augmentation; see *-bundus* under **very**

unedo, L. strawberry-tree; see **arbutus**

unequal; see **different**

unfit
Gr. *abrotos,* not eating, not fit to eat; see **hunger**
Gr. *akairos,* inopportune, unseasonable: *Porcellio(Acaeroplastes) areolatus* (an isopod).
Gr. *anarmostos,* unfit, incongruous, unsuitable: *Anarmostodera crassicornis* (a beetle).
Gr. *aneuthetos,* inconvenient:
Gr. *aoros; exoros; paroros,* untimely, unfitting:
Gr. *apaxios,* unworthy:
Gr. *athetos,* set aside, invalid, useless, unfit: *Athetocephus maculatus* (a cephid).
Gr. *dysedros,* awry, not fitting:

L. *improprius,* unfit, unseemly, unsuitable: improper.

L. *incommodus,* inconvenient, unsuitable, unfit: *Sclerosomus incommodus* (a beetle).

L. *incongruus,* unsuitable, unfit: incongruous, *Arca incongrua* (a pelecypod).

L. *inconsantaneus,* unfit, unsuitable:

L. *inconveniens, -entis,* unsuitable, inexpedient: inconvenience, inconvenient.

L. *indecens,* unbecoming, unseemly, improper, unfit: indecent, *Pholiota indecens* (a mushroom).

L. *ingustabilis,* not fit to eat or drink:

L. *inhabilis,* unfit, incapable, unmanageable: *Siphonostoma inhabile* (an annelid), *Conops inhabilis* (a fly).

See: **worthless**

unfruitable; unfruitful; see **sterile**

unguentum, L. ointment, perfume; *unguilla,* ointment-box; see *unguen* under **fat**

unguis, L. nail, claw, talon; *unguiculus,* dim.; *ungula,* hoof; see **nail**

unhappy; see **sad, discontent, trouble, disease, pain**

unicus, L. only, sole, singular; see *unus* under **one**

uniola, L. a plant; now a genus of grasses; see **grass**

unite < L. *unio, -itus,* join together; see **bind, marry, one**

universus, L. one, entire, whole; see *unus* under **one**

unknown; see **secret**

unstable; see **shake**

untidy; see **dirt, confusion**

unus, L. one, the whole; *unio, -onis; unitas,* one, oneness; see **one**

-unus, L. pertaining to; see *-anus* under **nature**

unusual < L. *un-,* not, *usualis,* common; see **different, strange, wonder**

up < AS. *up,* toward a higher point or place; see **over**

upright < AS. *upriht,* erect in position, vertical.

L. *erectus,* upright; *arrectus,* upright, steep: erect, erection, *Saccelatia arrectus* (a fossil ostracode), *Cellepora arrecta* (a bryozoan), *Trillium erectum* (purple trillium).

Gr. *histos,* anything set upright, loom, web; see **weave**

Gr. *ithys,* straight, direct, upright; see **straight**

Gr. *kathetos,* perpendicular: *Cathetostoma laeve* (a fish).

L. *perpendicularis,* upright, at right angles; *perpendiculum,* n. plummet, plumb-line: perpendicular.

Gr. *styo,* erect, stiffen; *stytikos,* causing erection; *stysis,* f. erection; *styma, -tos,* n. erection, stiffening:

ML. *verticalis,* at the vertex, directly overhead: vertical, *Ischnura verticalis* (a damselfly), *Tyrannus verticalis* (Arkansas kingbird).

See: **straight, right, stiff, high, cliff**

upsilon, Gr. twentieth letter of the Greek alphabet; see **letter**

upupa, L. hoopoe; see **hoopoe**

uraeo- < Gr. *ouraios,* of the tail, hindmost; see *oura* under **tail**

urago- < Gr. *ourax, -agos,* a kind of bird; see **bird**

uranium < L. *Uranus* (Gr. *Ouranos*), god of heaven, for whom the planet was named; see ELEMENTS under **thing**

urano- < Gr. *ouranos,* heaven, sky; see **heaven**

urbs, L. city; *urbanus; urbicus,* of the city; see **town**

urceus, L. pitcher, urn; *urceolus,* dim.; see **pot**

uredo, L. blight; see **fungus**

urens, L. burning, stinging; see *uro* under **burn**

urethra, L. (Gr. *ourethra*), urinary canal leading from the bladder; *oureter,* duct from the kidney to the bladder; see **pipe**

uretrido- < Gr. *ouretris, -idos,* chamber-pot; see *ourana* under **pot**

-urge; -urgo-; -urgy < Gr. *-ourgos,* denoting work; see *ergon* under **make**

urgent < L. *urgens, -entis,* pressing, demanding immediate attention; see *urgeo* under **press**

urginea, NL. a genus of the lily family; see **lily**

uria, L. (Gr. *ouria*), a water-bird; see **bird**

urigo, L. lustful heat, desire; see **lewd**

urinator, L. diver; see *urinor* under **dip**

urine < L. *urina,* f. (Gr. *ouron,* n.), an excretion from the kidneys; *urinalis,* of urine; *ouresis,* f. urination; *ouretikos,* inclined to pass water; *oureter,* m. duct from the kidney to the bladder: urea, uremia, urologist, urinary, ureter, urethra, uric acid, urinalysis, urease, uroscopy, urorubin, urodynia, urotoxin, diuretic, adiposuria, thiourea, melituria, glucuronic, hyaluronidase, purine, ischuria, anischuria, enuresis.
L. *effusum,* n. urine:
L. *lotium,* n. urine:
L. *mingo, minctus; mictus,* urinate; *micturio,* urinate; *mictualis,* of urination, urinary: micturate, micturition, miction, retromingent.
Gr. *omichma, -tos,* n. urine:
ME. *pissen,* < OF. *pissier,* urinate: piss, pismire.
L. *saccatum,* n. urine:

urio- < Gr. *ourios,* fair, prospering, happy; see **agreeable**

urn < L. *urna,* pitcher for drawing water, container for the ashes of the dead; *urnula,* dim.; see **pot**

uro- < Gr. *oura,* tail, see **tail;** < *ouron,* urine, see **urine;** < *ouros,* a fair wind, see **wind;** guard, watcher, see **guard;** < L. *uro, -ustus,* burn, see **burn**

ursus, L. bear; *ursa,* female bear; *ursula,* dim. female cub; *ursinus,* of bears; see **bear**

urtica, L. nettle; see **nettle**

urus, L. wild ox, aurochs; see **cattle**

-us, -a, -um, L. masculine, feminine, and neuter terminations of many adjectives; see **nature**

use < OF. *user,* < L. *utor, usus; usuarius; usurarius,* of use, fit for use; *usitor, -atus,* use habitually; *usualis,* common, ordinary; *usura,* f. use, interest; *usurpo, -atus,* take or use unlawfully; *utensilis; utilis; utibilis,* useful, usable; *abusus, -us,* m. misuse: usage, usury, useful, useless, abuse, misuse, utility, utensil, usurpation, unusual, usual, usufruct, *Linum usitatissimum* (flax), *Melanorrhoea usitata* (a varnish-tree), *Brosimum utile* (a breadnut), *Elaeagia utilis* (a rubiacead), *Manihot utilissima* (cassava, topioca), *Eridantes utibilis* (a spider).
L. *adhibitus,* applied, employed, used:
Gr. *chrao,* use; *chresis,* f. use; *chrestikos,* knowing how to use; *chresimos; chrestos,* useful, wholesome, good; *katachresis,* f. misuse: chrestomathy, catrachresis, *Typhlochrestus bifurcatus* (a spider).
Gr. *dike,* right, custom; see **law**
Gr. *ethas, -ados,* customary, ordinary; *ethos,* n. custom, habit, character; *ethikos,* of habits, morals, and character; *ethistos,* acquired by habit, < *ethizo,* accustom, use: ethics, ethical, ethology, *Ethas carbonarius* (a beetle), *Ethadopselaphus cicatricosus* (a beetle), *Ethophallus cervantes* (a beetle).
Gr. *onesimos,* beneficial, useful; *oneios,* useful; *oneistos,* most useful; *onesis,* f. use, advantage: *Onesia bisetosa* (a fly).
Gr. *ophelema, -tos,* n.; *ophelia,* f. aid, benefit, help, use; *ophelimos; ophelos,* useful, helpful; *anopheles,* useless, hurtful; *epopheles,* helpful, useful: Ophelia, *Anopheles maculipennis* (a mosquito).
Gr. *prosphoros,* suitable, serviceable:
L. *receptus,* usual, customary:
L. *soleo, -itus,* be used to, accustomed; *solitus,* customary, habitual, usual: obsolete, obsolescent, insolent, insolence.
L. *sueo, suetus,* be used to; *suetus,* customary, habitual, usual; *assuetus; consuetus,* usual, customary; *consuetudo, -inis,* f. habit, usage; *desuetudo, -inis,* f. disuse, discontinuance; *insuetus,* unaccustomed, inexperienced, unusual: consuetude, desuetude, customary, custom, costume, accustom, customer, unaccustomed, insuetude.
Gr. *sympheron, -os,* useful; *symphoros,* useful, good: *Sympherobius umbratus* (a neuropterid), *Symphoromyia melaena* (a fly).
See: **good, make, help**

useless; see **worthless, unfit**

usio- < Gr. *ousia,* property, substance, reality, essence; see **thing**

usitatus, L. common, customary, familiar, usual; see **use, common**

usnea, NL. a genus of lichens; see **lichen**

usticius, L. color produced by burning, scorched, brown; *usta,* burnt color; see **uro** under **burn**

ustilago, L. plant name now applied to a smut; see **fungus**

ustricula, L. hair-curler; see **curl**

ustulatus, L. scorched, brown; see *uro* under **burn**

usualis, L. common, ordinary; see **use**

usurpo, *-atus,* L. take unlawfully, seize; see **take**

usury < OF. *usure,* < L. *usura,* use of money lent, interest paid on a loan; *usurula,* dim.; see **gain**

utensilis, L. useful, see **use**; *utensilium,* a domestic implement, see **tool**

uter, L. one or the other, see **other**; *uter, -utris,* leather bag, see **bag**

uterus, L. womb; *uterculus,* dim.; see **womb**

utesi- < Gr. *outesis,* a wounding; *outetes,* wounder; see *outao* under **hurt**

uthato- < Gr. *outhar, -atos,* udder, breast; *outhatios,* of the udder; see **udder**

utidano- < Gr. *outidanos,* worthless; see **worthless**

utilis, L. useful, beneficial; see **use**

utriculus, L. dim. of *uter, utris,* leather bag or bottle; see **bag**

-utus, L. having the nature of, pertaining to; see *-atus* under **nature**

uva, L. grape; *uvula,* dim.; see **grape**

uvidus, L. wet, moist, damp; see **wet**

uvula, L. f. pendant, fleshy lobe of the soft palate, dim. of *uva,* grape: uvula, uvular, uvulitis, *Uvularia sessilifolia* (a bellwort), *Pylodexia uvula* (a foraminifer).

    L. *columella,* uvula; see *columna* under **pillar**

    Gr. *kion, -os,* m. uvula: cionitis, kiotome.

    Gr. *staphyle,* f. uvula: staphylitis.

    Gr. *gargareon, -os,* m. uvula:

uxor, L. wife; *uxorcula,* dim.; *uxorius,* of a wife; see **spouse**

V

vacans, L. empty, unoccupied; *vacivus; vacuus,* empty, void; see *vaco* under **empty**

vacca, L. cow; *vaccula,* dim. heifer; *vaccinus,* of cows; see **cattle**

vaccinium, L. blueberry; see **heath**

vacerra, L. log, stock, post; see **pillar**

vacillo, *-atus,* L. waver; see **shake**

vacuum, L. empty space, void; see *vaco* under **empty**

vadimonium, L. bail, security, < *vas, vadis,* bail, surety; see *vador* under **bind**

vado, *vasus,* L. go, rush; see **move**

vador, *-atus,* L. bind by requiring bail or surety; see **bind**

vadum, L. shallow place, shoal, ford; *vadosus,* full of shoals, shallow; see **shallow**

vafer, *-fra, -frum,* L. artful, crafty, cunning, sly, subtle; see **know**

vagabundus, L. strolling about, roaming; see *vagus* under **wander**

vagina, L. sheath, scabbard, case; *vaginula,* dim.; see **sheath**

vagitus, L. crying, squalling, whimpering; see **weep**

vague < L. *vagus,* wandering, indefinite; see **dim**

vagus, L. wandering; *vagor, -atus,* stroll about, roam; see **wander**

valde, L. exceedingly, very; see **very**

valens, L. strong, vigorous, significant; see *validus* under **strong**

valeria, L. a kind of eagle; see **eagle**

valerian < OF. *valeriane,* < L. *valeo,* be strong: *Valeriana officinalis* (common valerian), *Valerianella dentata* (toothed corn-salad), valeric acid.

    L. *saliunca,* f. a kind of valerian *(Valeriana celtica); saliuncula,* f. dim.:

valgus, L. bowlegged; see **bend**

validus, L. strong, sound, powerful; see **strong**

valley < L. *valles (vallis), -is,* f.; *vallecula (vallicula),* f. dim. dell, glen; *convallis,* f. a valley enclosed on all sides: vale, vallecular, Rosevale, Valparaiso, *Convallaria japonica* (a lily-of-the-valley), *Ramularia vallisumbrosae* (a fungus).
Gr. *ankos,* mountain glen, valley; see **bend**
Sp. *arroyo,* creek, dry channel; see **stream**
Gr. *bessa,* f. wooded glen: *Bessaphilus cephalotes* (a beetle), *Bessobia monticola* (a beetle).
Sp. *cañon,* gorge: canyon, Grand Canyon.
Gr. *charadra,* f.; *charadros,* m. deep gully, rift, ravine, bed of a mountain torrent: *Chardromyia abnormis* (a fly), *Charadronota quadrisignata* (a beetle).
AS. *dael* (G. *thal),* valley: dale, Carbondale, Glendale, Rosedale, Springdale, Blumenthal, Joachimsthal, Siebenthal, Dalberg, dollar, *Homo neanderthalensis* (a fossil man).
Gr. *diasphax, -agos,* f. cleft, rocky gorge:
ON. *gil,* narrow, wooded ravine with a brook: gill.
C. *gleann,* valley, ravine: glen, Glenwood, Glengary, Glen Echo, Glen Onoko.
Gr. *nape,* f. a wooded vale, ell, glen; L. *napaeus* (Gr. *napaios),* of a wooded vale; *Napaea,* f. a wood-nymph: *Napaea dioica* (glade-mallow), Napoleon.
Gr. *pharanx, -angos,* f. chasm, gully, ravine; *pharangites,* of a gully: *Pharangitis spathias* (a moth).
Sp. *quebrada,* f. ravine, gully; *quebradilla,* f. dim.: Quebradillas (Porto Rico).
L. *rupina,* f. a rocky chasm:
See: **furrow, ditch, fold, stream**

vallisneria, NL. a genus of aquatic plants; see **waterweed**

vallum, L. wall; *vallaris,* of walls; see **wall**

value < L. *valeo,* be strong, have worth; see **worth**

valva, L. leaf of a folding door; *valvola (valvula),* dim.; see **door**

vampyrus, NL. a genus of bats; see **bat**

vanadium < ON. *Vanadis,* a name of Freya, goddess of love: see ELEMENTS under **thing**

vanga, L. spade or mattock; see **shovel**

vandicus, L. liar; see **lie**

vanilla, NL. a genus of orchids; see **orchid**

vanish < L. *vanesco,* pass away, disappear; see **move**

vanity < L. *vanus,* empty; see **proud**

vannus, L. fan; see **fan**

vanus, L. empty, fruitless, idle; *vanitas,* emptiness; see **empty**

vapidus, L. flat, lifeless, stale; see **dull**

vapor, L. steam; *vaporalis,* of steam; *vaporarium,* steam bath; see **cloud**

vappo, L. moth; see **moth**

vapulo, -atus, L. flog, cudgel; see **strike**

vara, L. trestle, horse; see **frame**

varanus, NL. a genus of lizards; see **lizard**

vargus, L. vagabond; see **wander**

variegatus, L. of different sorts, particularly colors; see *varius* under **different**

variola, LL. smallpox, spotted; see **disease, spot, pimple**

varius, L. different; *variabilis,* changeable; *varietas,* difference; see **different**

varix, L. a dilated, twisted vein; *varicula,* dim.; *varicosus,* with dilated veins; see *varus* under **bend**

varnish < OF. *vernis;* see *vernix* under **resin**

varus, L. blotch, pimple, see **pimple;** bent, distorted, spread, see **bend**

vas, *vadis,* L. bail, surety, see *vador* under **bind;** < *vas, vasis,* vessel, duct, utensil; *vasculum,* dim., see **vessel**

vase < L. *vas, vasis,* vessel; see **vessel**

vassus, LL. servant; see **servant**

vastus, L. empty, waste, see **large,** *devasto* under **destroy**

vat < AS. *faet;* see **barrel, vessel**

vates, L. seer, soothsayer, prophet; *vaticinus,* prophetical; see **prophecy**

vatius, L. bent outward, bowlegged; see **bend**

vatrax; vatricosus, L. with crooked feet, clubfooted; see **foot**

ve-, L. without, intensive; see **not, very**

vecors, -ordis, L. senseless, silly, insane, foolish; see *excors* under **fool**

vectigal, L. tax, impost, duty, revenue; see **pay**

vectis, L. spar, lever, bar; see **lever**

vector, L. carrier, passenger, rider; see *veho* under **carry**

vegetus, L. lively, animated, brisk; *vegetabilis,* enlivening; see **life**

vegrandis, L. not large, see **little**; very great, see **ve-** under **very**

vehemens, L. eager, ardent, vigorous; see **busy**

vehicle < L. *vehiculum,* n. conveyance, < *veho,* carry: vehicular.
  Gr. *apene,* f. a four-wheeled wagon:
  L. *arcera,* f. covered carriage for the sick:
  L. *benna,* f. carriage or wagon made of wickerwork:
  L. *birota,* f. a two-wheeled cab:
  L. *carpentum,* n. a two-wheeled carriage, coach; *carpentarius,* of a wagon; m.
    carriage-maker: carpenter.
  L. *carrus,* m. a two-wheeled wagon, cart; *carrulus,* m. dim.; *carruca,* f. four-
    wheeled coach; *carracutium,* n. two-wheeled carriage: car, carriage, cargo, carrier,
    chariot, charge, cark, caricature, career, carryall, surcharge.
  L. *cisium,* n. gig, cabriolet; *cisarius,* m. cab-driver:
  L. *clabulare, -is,* n. large, open wagon; *clabularis; clabularius,* of transport
    wagons:
  L. *covinus,* m. war-chariot of the Britons:
  AS. *craet* (from a probable *caert*), two-wheeled vehicle, perhaps originally made
    of wickerwork: cart, cartwheel.
  L. *currus, -us,* m. chariot, car; *currulis; currilis (curilis),* of a chariot: curule.
  Gr. *diphros,* m. chariot, seat; *diphrion,* n.; *diphriskos,* m. dim.; *diphrios,* of a
    chariot:
  L. *esseda,* f.; *essedum,* n. a two-wheeled war-chariot:
  L. *ferculum,* litter, bier, tray; see *fero* under **carry**
  Gr. *hamaxa,* f. wagon, carriage; *hamaxis, -idos,* f. dim.; *hamaxikos,* of a wagon;
    cheiramaxa, f. barrow, handcart; *cheiramaxion,* n. dim.:
  Gr. *harma, -tos,* n. chariot; *harmation,* n. dim.; *harmelates,* m. charioteer:
  L. *lectica,* f. bier, litter, sedan; *lecticula,* f. dim.; *lecticarius,* m. litter-bearer:
    lectica, litter.
  Gr. *ochema, -tos,* n. any kind of carrier, vehicle, horse, ship:
  L. *pabo, -onis,* m. wheelbarrow; *pabillus,* m. dim.:
  L. *petoritum (petorritum),* n. an open, four-wheeled carriage:
  Gr. *pheretron,* litter, bier; see *phero* under **carry**
  L. *pilentum,* n. carriage, coach:
  L. *plaustrum,* n. wagon, cart; *plostellum,* n. dim.:
  L. *rheda* (Gr. *rhede*), f. a four-wheeled wagon or carriage; *rhedion,* n. dim.:
  L. *sarracum,* n. a kind of wagon or cart:
  Gr. *satine,* f. chariot:
  L. *traha,* drag, sled; see *traho* under **draw**
  D. *wagen* (AS. *waegen; waen*), four-wheeled vehicle: wagon, Wagon Mound (N.
    Mex.), bandwagon, wain, Wainwright.
  See: **carry, horse, ship**

veil < L. *velum,* awning, curtain; *velaris,* of a veil; see **curtain**

vein < L. *vena,* bloodvessel; see **pipe**

velamen, L. cover, garment, < *velo, -atus,* cover, conceal; see **garment, cover**

velitor, -atus, L. skirmish; see **fight**

vellico, -atus, L. pinch, twitch, nip, twit; see **tease**

vellus, -eris, L. fleece, pelt; see **wool**

velo, -atus, L. cover, conceal; see **cover**

velocity < L. *velox, -ocis,* swift; see **swift**

velum, L. veil, curtain, sail; *velifer; veliger,* sail-bearing; see **sail**

velutinus, NL. velvety, < L. *villus,* shaggy hair; see *villus* under **hair**

vena, L. bloodvessel; *venula,* dim.; *venosus,* veiny; see **pipe**

venabulum, L. hunting-spear; see **spear**

venator, L. hunter; see *venor* under **hunt**

vendo, -itus, L. sell; *venum,* sale; *venalis,* of selling; *venditor,* seller; see **trade**

venenum, L. poison; *veneficus,* poisonous, magical; *venenosus,* very poisonous; see
  poison

venereus (venerius), L. pertaining to Venus; see *Venus* under **love**

veneror, *-atus,* L. honor, respect; see **honor**

venetus, L. sea-colored, blue; see **blue**

venia, L. grace, pardon, forgiveness, indulgence; *venialis,* pardonable; see **free**

venio, *ventus,* L. come; see **come**

venosus, L. veiny; see *vena* under **pipe**

venter, L. belly; *ventriculus,* dim.; *ventralis,* of the belly; see **belly**

venturia, NL. a genus of fungi; see **fungus**

ventus, L. wind; *ventulus,* dim.; *ventosus,* windy; *ventilo, -atus,* winnow, fan; see **wind, fan**

Venus, L. goddess of love; *venereus,* of love; *venustus,* charming, lovely, beautiful, graceful; *venustulus,* dim.; see **love, beauty**

vepres, L. brier, bramble; *veprecula,* dim.; see **bush**

ver, L. spring, springtime; *verculum,* dim.; see **year**

veratrum, L. hellebore; see **hellebore**

verax, *-acis,* L. speaking truly; see *verus* under **true**

verbascum, L. mullein; see **mullein**

verbena, L. vervain; see **vervain**

verber, L. whip; *verbero, -atus,* whip, lash, flog; see **whip**

verbum, L. word; *verbosus,* wordy; see **word**

verdant < OF. *verdoier,* < L. *viridis,* green; see **green**

verecundus, L. bashful, shy, coy; see **modest**

veredus, L. post-horse, swift horse; *veredarius,* post-boy, courier; see **horse**

vereor, *veritus,* L. fear, respect, honor; *verendus,* awful, fearful, terrible; see **honor, fear**

veretrum, L. the private parts; *veretillum,* dim.; see **sex organs**

vergo, L. bend, incline, or tend toward; see **turn**

vericulum, L. a small javelin, dim. of *verum,* spit, dart; see **spear**

vermilion < OF. *vermeil,* bright red, the kermes insect, < L. *vermiculus,* little worm; see **red**

verminum, L. bellyache; see **disease**

vermis, L. worm; *vermiculus,* dim.; *vermiculatus,* wormy, wormlike, worm-eaten; see **worm**

verna, L. home-born slave; *vernula,* dim.; *vernaculus,* of a home-born slave, domestic, indigenous, common; see **servant**

vernalis; vernus, L. of springtime; see *ver* under **year**

vernilis, L. cringing, fawning, servile; see **flatter**

vernix, NL. varnish; see **resin**

vernonia, NL. a genus of the composite family; see **composite**

veronica, NL. a genus of the figwort family; see **figwort**

verpa, L. penis; see **penis**

verres, L. boar; *verrinus,* of pigs; see **hog**

verriculum, L. dragnet; see **net**

verro, *versus,* L. sweep, clear away; see **sweep**

verruca, L. wart; *verrucula,* dim.; *verrucosus,* full of warts; see **wart**

versabilis; versatilis, L. changeable, mobile, turning with ease, adaptable; see *verto* under **turn**

verse < L. *versus,* line of poetry; *versiculus,* dim.; see **poem**

versicolor, L. of various colors, variegated; see **color**

versus, L. furrow, line of poetry, see **furrow, poem;** *verto,* turn, see **turn**

versutus, L. crafty, sly, dexterous, adroit; see **know**

vertebra, L. one of the bones of the backbone; see **back**

vertex, L. top, peak, pole of the heavens, whirl, eddy; *verticillus; verticulus,* dim.; see **top,** *verto* under **turn**

verticalis, L. directly overhead, upright; see **upright**

verto, *versus,* L. turn; see **turn**

verus, L. true; *veridicus,* truthful, true; see **true**

L. *impossibilis,* incapable, unable, hopeless: impossible, impossibility.
L. *incapabilis; incapax, -acis,* incompetent: incapable, incapability, incapacitated.
L. *incompetens, -entis,* insufficient, incapable: incompetent, incompetence.
L. *inefficax, -acis,* weak, ineffectual: inefficient, inefficacious.
L. *inermis,* unarmed, defenseless; see **bare**
L. *infirmus,* weak, feeble; *infirmo, -atus,* weaken, enfeeble: infirm, infirmity, *Pipunculus infirmatus* (a fly).
L. *inops, -opis,* poor, weak, helpless; see **poor**
L. *invalidus,* infirm, weak: invalid, *Drasodes invalidus* (a spider).
Gr. *ischnos,* thin, weak; see **thin**
Gr. *kamatos,* result of toil and trouble, weariness; see **trouble**
Gr. *kladaros,* easily broken, frail: *Cladarodes peloptera* (a moth), *Cladara atroliturata* (a moth).
Gr. *kopos,* m. weariness, fatigue; *kopasis,* f. weariness; *kopodes,* wearisome; *akopos,* untiring, unwearying, easy: copiopsia, *Copostigma fumatum* (a psocid).
L. *languidus,* faint, weak, weary, dull; *languor, -is,* m. weariness, lassitude: languid, languish, languor, *Caranistes languidus* (a beetle), *Poa languida* (a grass).
L. *lassus,* weary; *lassulus,* dim.; *lassitudo, -inis,* f. weariness; *delassabilis,* that can be tired out: lassitude, alas.
L. *laxus,* loose, slack, unstrung; see **free**
Gr. *leptos,* small, fine, weak, thin; *leptaleos,* delicate, feeble; see **thin**
L. *marcor,* faintness, decay; *marcidus,* wasted, weak; see **lessen**
Gr. *molys, -yos; molyros; molyx, -ykos,* soft, weak, feeble:
L. *nequeo, -itus,* be unable, be impossible; *nequitia,* f. worthlessness:
L. *succiduus,* sinking, failing, faltering; see **lessen**
L. *syncope* (Gr. *synkope*), f. faint, swoon: syncope.
L. *taedium,* n. weariness, disgust; *taediosus,* wearisome; *extaediatus,* worn out, utterly wearied: tedium, tedious.
Gr. *trycho,* wear out; *trychosis,* exhaustion; *trycheros,* worn; see **rag**
L. *vescus,* weak, poor, thin, little; *vesculus,* dim.: *Fragaria vesca* (European strawberry).
See: **soft, bend, thin, poor, dull, lessen, coward**

wealth < AS. *wela:* wealthy, weal, commonwealth.
Gr. *aphenos,* n. riches, wealth; *aphneios,* rich, wealthy, abundant: *Aphenoserica fallax* (a beetle), *Aphnaeus vulcanus* (a butterfly), *Aphniolaus pallene* (a butterfly).
L. *bonusculum,* n. a small estate:
L. *censum,* n. wealth, riches:
L. *Croesus* (Gr. *Kroisos*), m. a king of Lydia renowned for his riches: a Croesus, *Papilio(Ornithoptera) croesus* (a butterfly).
L. *dis, ditis,* rich, abundant; *ditior,* richer; *ditissimus,* richest; *ditesco,* grow rich: *Nectria ditissima* (a fungus).
L. *dives, -itis,* rich, wealthy; *divitia,* f. richness, riches; *divitior,* richer; *divitissimus,* richest: Dives (see Luke 16:19-31), *Eucalyptus dives* (a eucalyptus).
Gr. *enchalkos,* rich; see **chalkos** under **copper**
Gr. *enolbos,* wealthy, prosperous:
Gr. *euporia,* f. means, resources, plenty; *euporos,* well-provided: *Euporus torquatus* (a beetle), *Euporodesmus solitarius* (a centiped).
AS. *feoh,* cattle, property, money: fee.
Gr. *gaza,* f. royal treasure of Persia, wealth, riches: *Gazacrinus stellatus* (a Silurian crinoid).
Gr. *ktema, -tos,* n. property, possession; *ktesis,* f. acquisition, possession; *kteseidion,* n. dim.; *ktesios,* of property; *ktetikos,* acquisitive; *ktesibios,* possessing property; *ktetos,* that may be had; *ktetor, -os,* m. owner; *periktetos,* rich; *ktaomai,* get: Ctesiphon, Epictetus, *Ctedoctema acanthocrypta* (a protozoan), *Ctesibius eumolpoides* (a beetle), *Ctesicles maritimus* (a beetle), *Termitoctesis gridelli* (a beetle).
L. *locuples, -etis,* substantial, rich, wealthy, opulent: *Sabethes locuples* (a fly), *Dapsilia locupletana* (a butterfly).
L. *Lucullus,* m. a Roman consul and general, famous for his wealth, luxury, and banquets: Lucullan, lucullite.
Heb. *mamona,* riches: mammon, mammonish.
Gr. *Midas,* mythical king at whose touch everything turned to gold; see **gold**
L. *ops, opis,* f. riches, wealth, property, power, strength, help; *opicillum,* n. dim.; *opimus,* fat, rich, sumptuous, abundant; *opiparus,* richly furnished, splendid, sumptuous; *opifer, -a, -um,* aid-bringing, helping; *opitulus,* m. helper; *opitulor, -atus,* help, aid; *opimo, -atus,* fatten enrich, fertilize; *opulens; opulentus,* rich: opulence, opulent, *Ctenomys opimus* (a rodent).

L. *pecuniosus,* having much money, rich; see *pecunia* under **money**
Gr. *ploutos,* m. wealth, riches; *plousios,* rich, wealthy, abundant: Plutus, Pluto, plutocracy, plutarchy, plutocrat, plutonic, Plutarch, *Plusia moneta* (a moth), *Plutella cruciferarum* (a moth), *Zaplutus madagascariensis* (a bug).
L. *pollucibilis,* sumptuous, rich, magnificent:
L. *premiosus,* rich:
L. *proprium,* n. property, possession:
L. *satur, -a, -um,* full, rich; see **abundance**
See **abundance, money**

wean < AS. *wenian,* accustom young to do without mother's milk.
L. *ablacto, -atus,* wean: ablactate.
Gr. *apotithos,* weaned:
Gr. *athelos,* weaned:
L. *delicus,* weaned:
See: **separate**

weapon < AS. *waepen;* see **tool**

wear < AS. *werian,* carry, use, waste; see **carry, use, rub**

weary < AS. *werig,* tired; see **weak**

weasel < AS. *wesle.*
NL. *civetta,* f. < It. *zibetto,* < Ar. *zabad,* civet: civet, civetone, *Civittictis civetta* (civet-cat), zibet *(Viverra zibetha).*
Gr. *enhydris, -idos,* f. otter; water-snake:
LL. *furo,* ferret, fur, thief: *Mustela furo* (ferret).
Gr. *gale,* f. polecat, weasel; *galideus,* m. a young weasel: *Spilogale putorius* (a spotted skunk), *Phascogale penicillata* (Australian pouched-mouse), *Onychogalea unguifer* (nail-tailed wallaby), *Galemys pyrenaicus* (an insectivore), *Galeriscus nigripes* (white mongoose).
NL. *genetta,* f. a kind of civet: *Genetta genetta* (genet).
AS. *hearma,* a weasel; NL. *ermineus,* white like ermine: ermine *(Mustela erminea), Begonia erminea* (a begonia).
Gr. *hyllos,* m. ichneumon: *Hyllus aeruginosus* (a bug).
Gr. *ichneumon, -os,* m. the tracker, mongoose: *Herpestes ichneumon* (an Egyptian mongoose).
L. *ictis, -idis* (Gr. *iktis, -idos),* f. marten, weasel: *Ictidomys mexicanus* (a squirrel), *Ictocyon[Ictidocyon] venaticus* (bush-dog), *Ictonyx striata* (zoril), *Arctitis binturong* (a civet).
Gr. *latax, -agos,* f. probably a beaver or otter: *Latax lutris* (sea-otter), *Latagognoma dacryodes* (a bug).
L. *lutra,* f. otter; *lutreola,* f. dim.: *Lutra vulgaris* (European otter), *Mustela lutreola* (a European mink).
L. *martes, -is,* f. marten: *Martes americana* (American sable, marten).
L. *meles, -is,* f. badger, marten; *melinus,* of a marten: *Meles taxus* (a badger), *Perameles nasuta* (long-nosed bandicoot).
L. *mephitis,* bad odor; skunk; see **smell**
Skt. *mungus,* a kind of civet: mongoose *(Herpestes mungo), Lemur mongoz* (a lemur), *Ophiorrhiza mungos* (mongoose-plant).
L. *mustela,* f. weasel; *mustecula,* f. dim.; *mustelinus,* of weasels: mustelid, musteline, *Mustela arctica* (a weasel), *Hylocichla mustelina* (wood-thrush).
NL. *taxus,* m. badger; *taxoninus,* of badgers: *Taxidea taxus* (a badger).
NL. *vison,* m. mink: *Mustela·vison* (mink).
L. *viverra,* f. ferret: viverrine, *Viverra megaspila* (a civet-cat), *Dasyurus viverrinus* (dasyure).

weather < AS. *weder.*
Gr. *aithria,* clear weather, open sky; *mixaithria,* changeable weather; see *aithrios* under **clear**
L. *caelum,* air, sky, climate, weather; see **heaven**
Gr. *cheimon, -os,* winter weather; see **year**
L. *clima,* region, zone; see **slope**
Gr. *dysaeria,* bad weather; see *aer* under **wind**
Gr. *eudia,* fair weather, calm; see **clear**
Gr. *euemeria,* f. fine day, good weather: *Euemera viridirufaria* (a moth).
L. *sudum,* bright, clear weather; see **clear**
See: **heat, cold, rain, snow, wind, storm, clear**

weave < AS. *wefan:* web, web-footed, weaver, Webster, weft, woof.
Gr. *antion,* n. loom:
Gr. *arachnion,* n. cobweb: *Litharachnium araneosum* (a radiolarian), *Sempervivum arachnoideum* (a houseleek), *Saxifraga arachnoidea* (a saxifrage).

L. *araneum*, n. cobweb; *araneosus*, full of webs:

Gr. *arkane*, f. bar to which the threads of the warp were fastened:

L. *conopeum* (Gr. *konopeion*), curtain of fine gauze, mosquito net; see **net**

Gr. *etrion*, n. warp in a piece of cloth:

L. *gerdius*, m. weaver:

Gr. *gerrhon*, anything made of wickerwork; see **screen**

Gr. *histos*, m. the upright web-beam of a loom, loom, frame, web, tissue; *histion*, n. dim. sail: histoblast, histology, histolysis, histamine, histidine, histiocyte, histone, histogram, histoplasmosis, *Histoplasma capsulatum* (a fungus), *Histoderma appendiculatum* (a sponge), *Histiophorus pulchellus* (a sailfish), *Histoderma hibernicum* (a Cambrian worm-tube), *Histiophryne bougainvillei* (a fish), *Endamoeba histolytica* (a protozoan), *Torula histolytica* (a fungus).

L. *hypha* (Gr. *hyphe*), f. web; *hyphadion*, n. dim.; *hyphantes*, m.; *hyphantria*, f. weaver; *hyphantos*, woven; *hyphaino*, weave; *synyphes*, woven together: hypha, hyphal, paryphodrome, *Hypholoma incertum* (a fungus), *Hyphantes euonymella* (a spider), *Hyphaene thebaica* (doum-palm), *Lepthyphantes sabulosus* (a spider), *Synyphocrinus cornutus* (a fossil crinoid).

Gr. *kroke*, woof in a piece of cloth; *krokismos*, a weaving, web; see *krokis* under **wool**

L. *linarius*; *linificus*; *linyphus*, m.; Gr. *linourgos*, m. linen-weaver: linyphid, *Linyphia signata* (a spider), *Linyphoides typus* (a spider), *Linurgus keniensis* (a bird).

Gr. *lygizo*, bend, twist; *lygistes*, basket-maker; see **bend**

Gr. *nexipous*, web-footed:

L. *ordior*, *orsus*, begin a web, lay the warp; *orsorius*, of a loom: exordium.

L. *pactilis*, plaited together, wreathed: *Mitra pactilis* (a fossil gastropod).

Gr. *penos*, m. web; *pene*, f. thread on the bobbin, web; *penion*, n. dim. bobbin, spool; *penisma*, *-tos*, n. woof on the spool; *penitis*, *-idos*, f. weaver: Penelope, *Penelope cristata* (guan), *Penium interruptum* (a desmid).

Gr. *phormos*, basket, plaited mat; see **basket**

L. *plecto*, *plexus* (Gr. *pleko*), twine, twist, weave), braid, interweave, plait; *plectilis*, plaited, complicated; *plektos*, plaited, twisted; *plegma*, *-tos*, n. anything twined or plaited, net, wreath, wickerwork; *plektane*, f. anything coiled, twined, wreathed; *plektanion*, n. dim.; *plochmos*; *plokamos*, m.; *plokanon*, n.; *ploke*, f.; *plokos*, m. anything twisted or twined, lock of hair, curl, tendril; *plokion*, n. dim.; *plokios*, twined; *plokeus*, m. braider, plaiter, weaver; *emplektos*; *symplokos*, intertwined, interwoven, inwoven; *euplokamos*, having beautiful locks; L. *implexus*, entwined, interwoven: Plectognathi, Plecoptera, emplectite, plectospondylous, haploid, diploid, *Plegmatograptus nebula* (a graptolite), *Ploceus nigerrimus* (black weaver), *Plocionus pallens* (a beetle), *Plociopterus comptus* (a beetle), *Plocopsylla achilles* (a siphonapterid), *Plocamophorus ocellatus* (a slug), *Anaplectes rubriceps* (a bird), *Emplectocladus fasciculatus* (a rosacead), *Filograna implexa* (a tubeworm), *Euplectella aspergillum* (a sponge, Venus' flower-basket), *Alloplectus sparsiflorus* (a gesneriacead), *Biblioplectus ambiguus* (a beetle), *Periploca graeca* (Grecian silk-vine), *Ophioplocus imbricatus* (a brittlestar), *Rhizoplegma radicatum* (a radiolarian), *Symplocos* [*Symplocus*] *tinctoria* (sweetleaf).

L. *rete*, net; see **net**

Gr. *rhips*, wickerwork, mat, screen; see **screen**

Gr. *sargane*, f. braid, plait, basket: *Sargana stantoni* (a fossil gastropod), *Sarganura maculipes* (a bird).

L. *stamen*, warp, thread; see **thread**

Gr. *steganopous*, web-footed:

L. *subtemen (subtegmen)*, *-inis*, n. weft or woof:

Gr. *tarrhos*, crate, basket, wickerwork, hurdle, mat; *tarrhion*, dim.; see **basket**

Gr. *tarsos*, woven mat or grate; see **foot**

L. *tela*, f. web, loom, warp: tela, telary, aulatela, cameritelous, toil, toilet, mantle, subtle, subtile, subtlety, *Tetranychus telarius* (red spider).

L. *texo*, *textus*, weave; *textilis*, woven; *textrinus*, of weaving; *textor*, *-is*, m.; *textrix*, *-icis*, f. weaver; *textum*, n. that which is woven, web, cloth; *textura*, f. web, structure; *contextus*; *intextus*, interwoven, interlaced: textile, text, textbook, texture, tissue, toil, context, pretext, textrine, textorial, *Textor albirostris* (a weaver-bird), *Textularia gibbosa* (a foraminifer), *Cisticola textrix* (pincpinc, a warbler), *Musa textilis* (abaca-palm, source of manila fiber), *Frondicularia texta* (a foraminifer), *Liopistha protexta* (a fossil pelecypod).

L. *trama*, f. woof, weft: trama, *Trama troglodytes* (a bug), *Trametes pini* (a fungus), *Lepidotrama arrosa* (a moth).

L. *trilix*, *-icis*, triple twilled:

L. *vieo, vietus,* twist together, plait, weave, shrink, wrinkle; *vitilis,* plaited, interwoven; *vietor(vitor), -is,* m. basket-maker: *Helix vieta* (a gastropod), *Mycena vitilis* (a fungus).

See: **spin, fold, turn, curl, basket, complex, net, cloth**

web < AS. *webb;* see **weave, net, trap**

wed < AS. *weddian,* marry; see **marry**

wedge < AS. *wecg.*

L. *cuneus,* m. wedge; *cuneolus,* m. dim.; *cuneatus,* wedge-shaped: cuneate, cuneiform, obcuneate, coin, coinage, *Cuneolina conica* (a foraminifer), *Primula cuneifolia* (a primrose), *Flabellum cuneiforme* (a fossil coral), *Henicoptera cuneilineata* (a fly), *Adiantum cuneatum* (a maidenhair-fern), *Acarinicola cuneifer* (a nematode), *Longicilium flexicuneus* (a flagellate).

Gr. *embolos,* wedge, plug;·see **close**

Gr. *sphen, -os,* m. wedge; *sphenarion,* n.; *spheniskos,* m. dim.: sphenoid, sphenic, sphenion, sphenodont, alisphenoid, *Sphenodon punctatus* (tuatara), *Sphenophyllum cuneifolium* (a fossil plant), *Sphenophorus pulchellus* (a billbug), *Spheniscus demersus* (jackass-penguin), *Spheniscocrinus spinosus* (a Permian crinoid), *Sphenarium carinatum* (a grasshopper), *Entosphenus tridentatus* (a lamprey).

weed < AS. *weod;* see **plant**

week < AS. *wice.*

Gr. *hebdomas, -ados,* f. seven, week; L. *hebdomadalis,* of a week, weekly: hebdomad, hedomadary, hebdomadal.

L. *septimana,* f. week:

See: **day, seven**

weep < AS. *wepan.* Derive: jeremiad.

Gr. *ailinos,* mournful, plaintive:

Gr. *dakryo,* shed tears, weep; *dakrytos,* tearful; see *dakryon* under **drop**

L. *ejulo, -atus,* lament, wail: ejulate.

Gr. *elegos,* mourning song or poem; see **poem**

L. *fleo, fletus,* weep; *flebilis,* tearful, doleful, lamentable: flebile, feeble, foible.

Gr. *goëros,* mournful, distressing, lamentable; *goës, -ëtos; goëtes,* m. howler, wailer, wizard; *goëtikos,* of howling, witchcraft: goetic, goety, *Goës tigrinus* (a beetle), *Goërodes cornigera* (a trichopterid), *Goëtothrips terrestris* (a thsanopterid).

Gr. *ialemos,* m. dirge, lament, wail; melancholy, mournful:

Gr. *kinyros,* plaintive, wailing, < *kinyrizo,* lament, wail:

Gr. *knyzema, -tos,* n.; *knyzethmos,* m. a whimpering, whining:

Gr. *klaio,* weep; *apoklaio,* weep aloud, mourn deeply; *klausma, -tos,* n.; *klauthmos,* m. a weeping, wailing; *klauster, -os,* m. weeper; *klauthmeros,* tearful; *klausimos,* plaintive; *klaustos,* mournful; *apoklausis,* f. lamentation:

Gr. *kokytos,* m. a wailing, lamentation, shrieking, < *kokyo,* cry, lament, wail: *Cocytotettix linearis* (a grasshopper).

Gr. *kommos,* m. a striking or beating of the breast in lamentation: kommos.

L. *lacrimo, -atus,* shed tears, weep; see *lacrima* under **drop**

L. *lamentor, -atus,* wail, weep; *lamentum,* n. wailing, mourning; *lamentabilis,* causing sorrow; *illamentatus,* unmourned: lamentation, lamentable, lamented.

L. *lessus,* m. lamentation, wailing:

L. *lugeo, luctus,* lament, mourn; see *lugubris* under **sad**

Gr. *Magdalene,* of or from Magdala: Mary Magdalene (who wept in repentance), Madeline, maudlin.

Gr. *minyros,* warbling, whimpering, whining; see *minyrisma* under **sing**

Gr. *odyrmos,* m.; *odyrma, -tos,* n. complaint, wailing, lamentation; *odyrtes,* m. complainer, wailer; *odyrtikos,* querulous; *odyrtos,* lamentable, < *odyrtomai,* lament, mourn:

Gr. *oiktros,* lamentable, pitiable; see *oiktos* under **mercy**

Gr. *oimozo,* lament, wail;ʻ *oimogma, -tos,* n.; *oimoge; oimoxia,* f. lamentation, wail; *oimoktos,* pitiable; *oimoktikos,* given to wailing:

L. *plangimonium,* n. lamentation:

L. *ploro, -atus,* weep aloud, lament; *plorabilis,* lamentable; *deploro,* bemoan, lament: implore, deplorable, *Chrysopa plorabunda* (aphis-lion).

L. *prefica,* f. a woman hired to lament at the head of a funeral procession:

L. *singulto, -atus,* sob, hiccup; *singultus,* sobbing, hiccup; see **hiccup**

Gr. *threneo,* lament, sing a dirge; *threnodia,* lamentation; see **sing**

L. *vagio, -itus,* cry, whimper, whine, squall; *vagitus, -us,* m. whimpering, squalling of children:

See: **drop, sad, pain**

weevil < AS. *wifel;* see **beetle**

weight < AS. *gewiht* (G. *gewicht*).

Gr. *achthos, -eos,* n. weight, burden, load, grief, distress, trouble; *achthedon, -os,* f. weight, burden, annoyance, distress; *achtheinos,* burdensome, oppressive, wearisome; *achtheros* (*achtheres*), distressing, grievous; *achthophoros,* bearing burdens; m. porter; *homachthes,* heavy to the shoulders; *seisachtheia,* f. an unburdening: *Achthophora tristis* (a beetle), *Achtheinus oblongus* (a copepod), *Homachthes histrio* (a bee), *Achtheres extensus* (a copepod).

Gr. *baros,* weight, pressure; *barema, -tos,* burden, load; *baryllion,* dim.; see *barys* under **heavy**

Gr. *bastagma, -tos,* n. burden, baggage; *bastaktes,* m. porter, < *bastazo,* lift, carry; L. *bastagarius,* m. baggage-master: *Bastactes bituberculatus* (a beetle).

Gr. *brime,* f. bulk, strength: *Brimosaurus grandis* (a dinosaur).

Gr. *brithos,* weight; see *brithys* under **heavy**

L. *drachma* (Gr. *drachme*), f. a weight and coin: dram.

Gr. *emphasis,* suggested importance, stress; see **strong**

Gr. *epholkion,* n.; *epholkis, -idos,* f. burdensome appendage; *epholkos,* lagging, laggard: *Epholcis divergens* (a beetle).

L. *exagium,* n. weight, balance:

Gr. *gemos,* n.; *gomos,* m. load, freight, cargo; *gemistos,* laden: *Gemophaga rufa* (a wasp).

L. *gestamen, -inis,* n. burden, load:

L. *gravo, -atus,* load, weigh down, oppress; *gravidus,* laden; see *gravis* under **heavy**

L. *halter, -is* (Gr. *-os*), m. weight used in jumping, balancer: halter (pl. halteres), *Bibioides halteralis* (a fly), *Delphinium halteratum* (a larkspur).

L. *impedimentum,* n. encumbrance, baggage, hindrance: impediment, impedimenta.

L. *libra,* f. the Roman pound of twelve ounces; *libralis,* of a pound: £.

Gr. *molybdis, -idos,* lead sinker, plummet; see **lead**

L. *obesus,* fat, corpulent; see **fat**

L. *onus, oneris,* n. burden, load; *oneratus,* filled, loaded; *onerosus,* burdensome, heavy; *onustus,* laden, burdened; *deonero, -atus,* unload, lighten: onerous, exonerate, *Polistes oneratus* (a wasp), *Lupinus onustus* (a lupine).

Gr. *phortos,* load, cargo; *phortion,* dim.; *phorema, -tos,* that which is carried, burden; *phortikos,* of a load, < *phero,* carry; see **carry**

L. *pondus, ponderis,* n. weight; *pondiculum; pondusculum,* n. dim.; *ponderosus,* heavy, weighty; *pondero, -atus,* weigh, consider; *appensor, -is,* m. weigher: ponderous, preponderate, ponder, poise, avoirdupois, imponderable, *Pinus ponderosa* (western yellow-pine), *Dendroctonus ponderosae* (a bark-beetle).

L. *saburra,* sand used as ballast; see **sand**

Gr. *sage,* f. baggage, knapsack, equipment: *Saga ephippigera* (a katydid).

L. *sarcina,* package, bundle, burden, load; *sarcinula,* dim.; *sarcinalis; sarcinarius,* of burdens; *sarcinatus,* laden; see **bundle**

Gr. *satto,* load, pack:

Gr. *sekoma,* counterpoise; see **balance**

Gr. *stater, -os,* m. a weight and coin: stater.

Gr. *stathmos,* weight of a balance; *stathmion,* dim.; see **balance**

L. *talentum* (Gr. *talenton*), n. an ancient weight, sum of money: talent.

See: **heavy, bundle, fat, balance, think**

welcome < AS. *wilcuma,* a guest bringing pleasure; see **agreeable**

well < AS. *wel,* good, suitable, healthy, see **good, health**; < *wella,* a spring or pit, see **spring, hole**

Welsh < AS. *Waelisc,* a Celtic people in southwestern Britain; see **Celt**

welwitschia, NL. a genus usually referred to the jointfir family; see **gnetum**

west < AS. *west.*

Gr. *dysmikos; dytikos,* western, < *dyo,* set, sink: *Dysmicus sentus* (a flea).

L. *hesperus* (Gr. *hesperos*), the evening star, evening, west; *hesperius,* western; see **night**

L. *occidens, -entis,* m. direction of the setting sun, evening, west; *occidaneus; occidentalis; occidualis,* western; *occiduus,* going down, setting, western: occident, occidental, *Blissus occiduus* (western chinchbug), *Dermacentron occidentale* (Pacific Coast tick), *Poa occidentalis* (New Mexico bluegrass).

Gr. *perate,* west; see *peras* under **end**

Gr. *skaios,* left, on the left, western, westward; see **hand**

L. *vesper,* the evening star, evening, west; *vespertinus,* of the evening, western; see **night**

Gr. *zophos,* west; see **night**

wet < AS. *waet.*

Gr. *apobamma, -tos,* n. infusion, tincture:

Gr. *ardeutos,* watered, < *ardo,* irrigate; *ardeusis,* f.; *ardeuma, -tos,* n. a watering; *ardeutes,* m. waterer; *ardmos,* m. watering-place: *Ardeutica spumosa* (a moth).

Gr. *brecho,* wet, moisten, steep, rain; *brexis; broche,* f.; *brochetos,* m. a wetting, rain; *apobregma, -tos,* n. an infusion; *diabrecho,* wet through, soak; *diabrochos,* soaked, sodden; *embregma; epibregma, -tos,* n. a wet application; *epibroche,* f. a wetting: embrocate, *Brechites vaginiferus* (watering-pot shell).

L. *conspergo, -ersus,* sprinkle, moisten, strew; see *spergo* under **spread**

Gr. *deuo,* drench, wet; *deuma, -tos,* n. anything wetted:

Gr. *diaino,* wet; *diantos,* wettable: *Adiantum pedatum* (a maidenhair-fern).

Gr. *enchymos,* moistened; *enchyma, -tos,* n. infusion: *Enchymus punctonotatus* (a beetle).

Gr. *ephydros,* wet, moist, rainy, abounding in water, living on water; see *hydor* under **water**

Gr. *epipastos,* sprinkled:

Gr. *eukrenos,* well-watered; see *krene* under **spring**

L. *humidus,* moist, damp, wet; *humidulus,* dim.; *humectus,* moist, damp; *humor, -is,* m. liquid, moisture; *humorosus,* moist, wet: humid, humidity, humidor, humectant, humectation, humor, humorous.

L. *Hyas, -adis* (Gr. *-ados*), f. one of the seven stars in the constellation Taurus, whose rising with the sun presaged rainy weather; *hyetios,* rainy, < *hyo,* rain: Hyad, Hyades, *Hyemoschus aquaticus* (water-chevrotain).

Gr. *hydatinos,* watery, wet; see *hydor* under **water**

Gr. *hygros,* wet; *enhygros,* damp, wet; *kathygros,* very wet: hygrometer, hygroscopic, hygrology, *Hygrobates galbinus* (a spider).

Gr. *ikmas, -ados,* f. moisture; *ikmaios,* m. the rainer; *ikmaleos,* damp, wet; *enikmos,* humid; *exikmazo,* exude: *Enicmoderes apfelbecki* (a beetle), *Icmadophila ericetorum* (a lichen).

L. *imbuo, -utus,* wet, dip, infect, inspire, instruct: imbue.

L. *infusio, -onis,* f. a pouring into, a steeping: infusion.

L. *insiccatus,* not dried up, wet:

L. *irroratus,* bedewed, moistened; see *ros* under **drop**

L. *madidus,* soaked, drenched, sodden, drunk; *madido, -atus,* moisten, make drunk; *madulsa,* f. a drunken man: madid, madescent, *Oniscus madidus* (a sowbug).

Gr. *mydos,* m. dampness; *mydaleos,* wet, damp: mydaleine, mydatoxin, *Mydaus meliceps* (teledu).

Gr. *naros (neros),* liquid, flowing; see *naros* under **flow**

Gr. *notis, -idos,* f. dampness, moisture, wetness; *noteros,* wet, damp; *notios,* wet, rainy, southern; *notizo,* moisten; *ennotios,* wet, moist: *Notidobia pallipes* (a caddisfly), *Noterophagus politus* (a beetle), *Noterus imbricatus* (a beetle), *Philonotis calcarea* (a moss). See *notos* under **south**

Gr. *palyno,* sprinkle, strew upon: palynology.

Gr. *parakos,* damp, wet:

Gr. *pastos,* sprinkled; *pasma, -tos,* n. a sprinkling, < *passo,* sprinkle: *Catapastus diffusus* (a beetle).

Gr. *pladaros,* wet, damp, insipid, < *plados,* m. abundance of liquid:

Gr. *rhantos,* sprinkled, spotted, < *rhaino,* sprinkle; *rhanter, -os,* m.; *rhantistron,* n. sprinkler, wetter; *rhantismos,* m. a sprinkling: *Rhantus aberratus* (a beetle), *Rhantistes raoulensis* (a bird).

L. *rigo, -atus,* water, moisten; *rigator, -is,* m. waterer; *riguus,* wetting; *irrigo, -atus,* supply with water; *irriguus,* watered, wet: irrigate.

Gr. *sbennymi,* quench, extinguish; *sbester, -os,* m. extinguisher; *sbestos,* quenched: asbestus, *Sbesterium variouncinatum* (a worm).

L. *stinguo, stinctus,* quench, wet: extinguish.

Gr. *tenktos,* wettable, < *tengo,* moisten; *tenxis,* f. a wetting:

L. *udus; uvidus,* damp, moist, humid, wet; *uvidulus,* dim.; *uviditas, -atis; uligo (uviligo), -inis,* f.; *uvor, -is,* m. moisture; *uliginosus,* full of moisture, wet, marshy: uliginal, uliginose, *Solidago uliginosa* (bog-goldenrod), *Begonia udisilvestris* (a begonia).

See: **water, drop, juice, sweat, marsh, flow**

whale < AS. *hwael* (ON. *hvalr*): whaler, whaleback, whalebird, whalebone, narwhal, rorqual, walrus.

L. *balaena* (Gr. *phalaina*), f. whale: baleen, *Balaeniceps rex* (shoebill), *Balaenoptera borealis* (a rorqual), *Balaena mysticetus* (bowhead), *Eubalaena glacialis* (southern right-whale).

L. *cetus* (Gr. *ketos*), m. whale: cetacean, cetotolite, cetyl alcohol, spermaceti, *Diaphorocetus poucheti* (a fossil cetacean), *Cetorhinus maximus* (basking shark), *Basilosaurus cetoides* (a fossil mammal).

L. *delphinus* (Gr. *delphis, -inos*), m. dolphin, porpoise: dolphin, dauphin, *Cyamus delphini* (whale-louse).

NL. *grampus*, m. a kind of whale: *Grampus griseus* (grampus).

L. *orca*, f. whale, large-bellied vessel, butt, tun, cask; *orcula*, f. dim.; *orcularis*, of a tun or cask: *Orca destructor* (a killer-whale).

L. *phocaena* (Gr. *phokaina*), f. porpoise: *Phocaena phocaena* (harbor-porpoise).

Gr. *physalos*, m. a kind of whale: *Balaenoptera physalus* (finback).

Gr. *physeter*, blower, a kind of whale; see **blow**

L. *tursio, -onis*, m. a dolphinlike animal, porpoise: *Tursiops truncatus* (bottlenosed dolphin, porpoise).

wheat < AS. *hwaet;* see **grain**

wheel < AS. *hweol.*

L. *canthus* (Gr. *kanthos*), tire of a wheel, rim, wheel; see **border**

Gr. *rhombos*, spinning-top, magic wheel; see *rhembo* under **turn**

L. *rota*, f. wheel; *rotula*, f. dim.; *rotalis*, of wheels; *rotatilis*, wheellike; *birotus*, two-wheeled: rota, rotal, rotula, rotuliform, roll, enroll, roulette, control, comptroller, barouche, Rotifera, rotiform, *Rotala serpiculoides* (a lythracead), *Rotalia formosa* (a foraminifer), *Lenticulina rotulata* (a foraminifer), *Lomatogonium rotatum* (a gentianacead).

L. *sucula*, windlass, winch, capstan; see **turn**

Gr. *trochos*, m. wheel; *trochion; trochiskion*, n.; *trochiskos*, m. dim.; *trochilodes*, like a pulley or wheel; L. *trochlea* (Gr. *trochalia; trochilia*), f. pulley: trochee, trochlea, trochiferous, *Trochoceras costatum* (an ammonite), *Trochodendron aralioides* (a tree), *Trochocarpa laurina* (an epacridacead), *Trochiscocoris hemipterus* (a bug), *Trochiliscus bellatulus* (a fossil characead), *Ceratotrochus nobilis* (a coral).

L. *tympanum*, drum for raising weights; see **drum**

whine < AS. *hwinan*, make a plaintive sound, whimper; see **sad**

whip < uncertain origin.

L. *ferula*, rod, staff, whip; see **rod**

L. *flagellum*, n. dim. of *flagrum*, n. whip, lash; *flagello, -atus*, scourge, whip: flagellum, flagellula, flagellate, flagellation, *Achlya flagellata* (a fungus), *Prothyma flagellicornis* (a beetle), *Masticophis flagellum* (coachwhip-snake).

Gr. *himasthle*, f. whip: *Himasthla alincia* (a flatworm), *Himasthlephora glauca* (a holothurian).

L. *lorum*, strap, thong, scourge, whip; see **strap**

Gr. *maragna (smaragna)*, f. whip, scourge: *Maragnicrinus portlandicus* (a Devonian crinoid).

Gr. *mastix, -igos*, f. whip; *mastigion*, n. dim.; *mastikter, -os*, m. scourger; L. *mastigia* (Gr. *mastigias*), m. one deserving whipping, knave, rogue: mastigium, mastigure, mastigate, Mastigophora, *Mastigamoeba minuta* (a protozoan), *Mastigias ocellata* (a medusa), *Mastigia epitusalis* (a moth), *Chilomastix mediterranea* (a protozoan), *Mastigoproctus giganteus* (whip-scorpion, vinegaroon).

L. *scutica*, f. whip, lash: *Bascanichthys scuticaris* (a fish).

Gr. *thomisso*, whip, scourge: Thomisidae, thomisid, *Thomisus onustus* (a crab-spider).

L. *verber, -is*, n. whip; *verbero, -atus*, lash, whip, flog, beat; *verbero, -onis*, m. one worthy of whipping; *verbereus*, of whipping: reverberate.

See: **rod, club, strap, punish, strike, swing**

whirl < ON. *hvirfla;* see **turn**

whisky < C. *uisge*, water, *beatha*, life; usquebaugh; see **water**

whisper < AS. *hwisprian;* see **murmur**

whistle < AS. *hwistle.*

Gr. *lapizo*, whistle, brag; see **brag**

Gr. *ligypterophonos*, buzzing with the wings; see **murmur**

Gr. *niglaros*, m. whistle:

Gr. *rhoizos*, m. whistle or whiz of an arrow; *rhoizema, -tos*, n. whiz; *rhoizetikos*, whizzing; *rhoizetor, -os*, m. whizzer:

L. *sibilus*, m. a hissing, whistling, < *sibilo, -atus*, hiss, whistle; *sibilatrix, -icis*, hissing: sibilant, sibilate, sibilator, persiflage.

Gr. *sismos; sisilismos*, m.; *sixis, -eos*, f. hiss, < *sizo*, hiss: *Sixeonotus insignis* (a bug).

Gr. *syrizo (syritto)*, pipe, whistle, hiss; *syrigmos*, m. whistle, hiss: *Syrittomyia syrphoides* (a fly), *Syrigma sibilatrix* (a heron).

Gr. *teretistes*, m. whistler, < *teretizo*, whistle; *teretisma, -tos*, n. a whistling:

white < AS. *hwit:* whiteness, whiten, whitecap, whitewash, whitefish, White, Whitefield. Whitehall, Whitehead, Whitman, Whitney, Whitsunday, snowwhite.

L. *albus,* white; *albidulus; albulus,* dim. whitish; *albidus; albineus,* white; *albedo, -inis; albitudo, -inis,* f.; *albor, -is,* m. whiteness; *albarius; albinus,* m. whitener, plasterer; *albesco,* become white; *albico, -atus,* whiten; *albugo, -inis,* f. white spot on the eye; *albatus,* clothed in white; *dealbo, -atus,* whitewash: albescent, albino, Alba Longa, Elvira, albite, albumen, album, auburn, albugo, Albuquerque, abele *(Populus alba), Pinus albicaulis* (white-bark pine), *Ixodes albipictus* (a tick), *Clitocybe dealbata* (a mushroom), *Artemisia albula* (a sagebrush), *Saccharomyces albicans* (yeast that causes thrush), *Mergus albellus* (smew), *Thalia dealbata* (a marantacead), *Spiraea albiflora* (Japanese spirea). Derive: Albion.

L. *alphus* (Gr. *alphos*), white spot on the skin, a kind of leprosy; see **disease**

Gr. *aphyo,* become white or bleached; *aphyodes,* whitish: *Aphyocharax rubripinnis* (a fish).

Gr. *argos; argennos; argetos,* white, bright; *argestes,* m. whitener; *argema, -tos,* n.; *argemos,* m. a white fleck on the eye; *diargemos,* flecked with white; *lepargos,* with white skin: argil, pygarg, *Argema mimosea* (a moth), *Scatophagus argus* (a fish), *Argestes mollis* (a copepod), *Olearia argophylla* (a composite).

Gr. *aspros,* white: diaper, *Alphitobius diaperinus* (a beetle), *Asprocottus herzensteini* (a fish), *Asprotilapia leptura* (a fish).

F. *blanc,* white, < OHG. *blanch,* shining: blanch, blank, blanket, Mont Blanc, *vin blanc,* Casablanca, Blanche, *Gymnognathus blanca* (a beetle), blanquillo *(Caulolatilus chrysops).*

L. *candidus,* shining, white, snowwhite, bright, dazzling; see *candeo* under **light**

L. *cerussatus,* colored or painted with white-lead; see **lead**

Gr. *chioneos; chionodes; chionopos,* like snow, snowwhite; see *chion* under **ice**

L. *cretatus,* marked with chalk, whitened; see *creta* under **lime**

NL. *ermineus,* white like the fur of the European ermine; see **weasel**

Gr. *exauges,* dazzling white:

Gr. *galaktikos,* like milk, milk-white; see *gala* under **milk**

L. *lacteus,* milk-colored, milk-white; see *lac* under **milk**

Gr. *leukos,* white; *leukaino,* whiten, turn white; *dialeukos,* marked with white; *epileukos,* superficially white, whitish; *hololeukos; palleukos,* all white; *kataleukos,* very white; *paraleukos,* partly white; *perileukos,* edged with white: leucite, leucocyte, leucodermia, leucocratic, leucoma, leucorrhea, leucine, leucoxene, leucyl, *Leucochlaena hispanica* (a moth), *Leucojum aestivum* (summer snowflake), *Leucothrix barbata* (a fly), *Leucobryum glaucum* (a moss), *Leucichthys lucidus* (a fish), *Leucaena glauca* (a legume), *Leucosigma uncifera* (a moth), *Hemerocampa leucostigma* (a tussock-moth), *Melaleuca leucadendron* (cajuput-tree).

L. *niveus,* snowy, snowwhite; see *nix* under **ice**

Gr. *phalaros,* having a white patch, white-spotted; see **spot**

Gr. *titanotos,* whitened; see *titanos* under **gypsum**

See: **light, pure, ivory, snow, milk**

whole < AS. *hal,* well, entire; see **health, one, all**

whore < AS. *hore;* see **prostitute**

whorl < ON. *hvirfla;* see **turn, circle**

-wich; -wick < AS. *wic,* town, village, camp, < L. *vicus,* village; see **town**

wick < AS. *weoc.*

L. *ellychnium* (Gr. *ellychnion*), n. lampwick: *Ellychnia minuta* (a beetle).

Gr. *haptra,* f. wick; *haptrion,* n. dim.:

Gr. *thryallis, -idos,* f. wick, a plant used for making wicks: *Thryallis glauca* (a malpighiacead).

wicked < uncertain origin; see **bad, fault, profane**

wicker < AS. *wican,* bend, yield; see **weave, basket, screen, net**

wide < AS. *wid;* see **broad**

widow < AS. *weodwe;* see **woman**

wife < AS. *wif;* see **spouse**

wild < AS. *wilde;* see **nature, rough, wander**

willing < AS. *willan,* wish; see **wish, agreeable, free**

willow < AS. *welig.*

ML. *ausaria,* f. a kind of willow: osier, red osier *(Salix purpurea).*

Gr. *itea,* willow; now a genus of the saxifrage family; *iteinos,* of willow; see **saxifrage**

Gr. *lygos,* f. willow or other tree with pliant twigs; *lyginos,* of willow; *lygistes,* m. basket-maker; *lygodes,* willowlike: *Lygus pratensis* (a bug), *Lygodesmia juncea* (a composite), *Lygodium japonicum* (a climbing-fern).

Gr. *oisos*, m. willow, osier: *Oesocerus murrayi* (a beetle).

L. *salix, -icis*, f. willow; *salictarius; salignus*, of willow; *salicetum; salictum*, n. willow plantation or thicket; AS. *sealh*, willow: Salicaceae, salicin, salicylic acid, acetyl salicylate (aspirin), methyl salicylate (wintergreen oil), sallee, sallow *(Salix caprea), Salix bablylonica* (weeping willow), *Dipholis salicifolia* (bustic), *Aphis salicicola* (an aphid), *Lythrum salicaria* (purple loosestrife), *Fusicladium saliciperdum* (a fungus), *Callistemon salignus* (willow bottle-brush), *Acacia salicina* (cooba, a wattle), *Bacterium salicis* (a bacterium).

L. *siler, -is*, n. a kind of willow: *Siler trilobum* (an umbellifer).

L. *vimen, -inis*, n. pliant twig, osier, withe; *viminalis; vimineus*, of osiers, pliant, willowy; *viminetum*, n. willow copse: *Salix viminalis* (basket or osier-willow), *Viminaria denudata* (a legume), vimineous, *Baccharis viminea* (mulefat).

win < AS. *winnan;* see **victory**

winch < AS. *wince*, reel; see **turn**

wind < AS. *wind:* windbreak, windfall, windiness, windrow, windward, windmill, windy, windpipe, windlass, windflower.

Gr. *aëlla*, f. stormy wind, whirlwind; *aëllodes*, stormy:

Gr. *aema, -tos*, n. blast, wind; *aetes*, m. blast, gale:

L. *Aeolus* (Gr. *Aiolos*), m. god of the winds; *aiolos*, shifting, changeable, variable, light: eolian, eolipile, *Aeolothrips fasciata* (a thysanopterid), *Aeoliscus strigatus* (a fish), *Panaeolus campanulatus* (a mushroom), *Corythaeola cristata* (a plantain-eater).

L. *aer, -is* (Gr. *-os*), m. air, the lower atmosphere; *aerios*, of the air, airy; *aerinos; aerodes*, like the sky or air, sky-blue; *dysaeria*, f. bad weather: aerial, aeronautics, aeroplane, aerobic, air, airy, debonair, malaria, aerosporin (from *Bacillus aerosporus*), *Aerobacter aerogenes* (a bacterium), *Crocus aerius* (a crocus), *Aerides maculosum* (an orchid).

L. *aether, -is* (Gr. *aither, -os*), m. the upper atmosphere; *aetherios*, of the upper air: ether, ethereal, ethyl.

L. *africus*, m. southwest wind:

L. *altanus*, m. south-southwest wind between *africus* and *libonotus:*

Gr. *anemos*, m. wind; *anemios; anemodes*, windy; *anemia*, f. flatulency; *dienemos*, windswept; *epanemos; hypenemos*, windy, vain: anemometer, anemology, anemophilous, anemotaxis, *Anemone ludoviciana* (American pasque-flower).

L. *anima*, air, breath of life; see **life**

Gr. *aparktias*, m. a north wind:

Gr. *apeliotes*, m. east wind:

L. *aquilo, -onis*, m. north wind:

Gr. *atmos; atmis*, steam, vapor; see **cloud**

L., Gr. *aura*, f. gentle breeze, air, wind, glow, exhalation; *aurula*, f. dim., whiff, puff: aura, aural.

L. *auster, -tri*, south wind, south; *austellus*, dim. gentle south wind; see **south**

Gr. *boreas*, m. north wind, north; see **north**

Gr. *byktes*, swelling, blustering; m. a hurricane: *Bycticus populi* (a beetle).

L. *caecias* (Gr. *kaikias*), m. northeast wind:

L. *carbas*, m. east-northeast wind:

L. *cataegis, -idis* (Gr. *kataigis, -idos*), f. hurricane, whirlwind:

L. *caurus (corus)*, m. northwest wind; *caurinus*, of the northwest wind:

L. *circius*, m. west-northwest wind: cers.

Gr. *dine*, whirlwind; see **turn**

Gr. *eriole*, f. hurricane, whirlwind: *Eriolus caraibeus* (a grasshopper).

L. *etesia*, f. tradewind:

Gr. *euros*, m. east wind: Euroclydon.

L. *favonius*, m. west wind: *Thecla favonius* (a hairstreak), *Flabellum favonium* (a fossil coral).

L. *flabrum*, n. blast of wind, breeze, fan; *flabilis*, airy; *flamen, -inis*, n. blast, wind, gale; *flatus, -us*, m. breath, breeze, < *flo, flatus*, blow: flatus, flatulent.

L. *halitus*, breath, exhalation; see **breathe**

Gr. *iayx, -ygos*, m. northwest wind, south Italian: *Iapyx solifugus* (an iapygid), *Anajapyx vesiculosus* (a proiapygid), *Catajapyx confusus* (a thysanurid).

Gr. *kaikias*, m. northeast wind:

Gr. *lips, libos*, m. southwest wind; *libonotos*, m. south-southwest wind:

Gr. *notos*, south, southwest wind; *notios*, southerly; see **south**

Gr. *ouros*, m. a fair wind:

Gr. *pneuma*, wind, air, breath, spirit; see **breathe**

Gr. *prester, -os*, m. hurricane with lightning:

L. *procella*, a violent wind, hurricane, tempest; *procellosus*, stormy; see **storm**

L. *septentrio, -onis*, north wind; see *septentrionalis* under **north**

L. *spiritus*, air in gentle motion, draft, breath of life, ghost; see **spirit**

L. *subsolanus,* m. east wind:

L. *typhon, -is* (Gr. *-os*), m. whirlwind; *typhonicus,* of a tempest or whirlwind: typhonic.   Derive: typhoon.

L. *ventus,* m. wind; *ventulus,* m. dim. breeze; *ventosus,* windy; *ventilo, -atus,* air, fan, winnow: ventose, ventiduct, ventifact, ventilagin, ventilate, ventometer, *Ventilago calyculata* (a rhamnacead), *Notharctus venticolus* (a fossil primate).

L. *vulturnus,* m. a south-southeast wind:

L. *zephyrus* (Gr. *zephyros*), m. west wind: zephyr, *Zephyrus hisamatsusanus* (a butterfly), *Zephyranthes rosea* (Cuban zephyr-lily).

See: **blow, breathe, fan, storm**

wind < AS. *windan,* twist; see **turn**

window < ON. *vindauga,* wind eye.

L. *fenestra,* f. window; *fenestella,* f. dim.: fenestral, *Fenestralia compacta* (a fossil bryozoan), *Fenestella cribrosa* (a fossil bryozoan), *Fenestrellina stellata* (a fossil bryozoan), *Protospongia fenestrata* (a fossil sponge), *Limeum fenestratum* (a phytolaccacead).

L. *impluvium,* the skylight in the roof of the atrium of a Roman house; see **hole**

Gr. *phoster, -os,* m. illuminator, window; *phosterikos,* of illumination:

Gr. *thyris,* small door, window; see **door**

wine < L. *vinum,* n. wine, juice; *vinalis,* of wine; *vinarius,* m. vintner; *vinosus; vinolentus,* full of wine, drunk: vine, vinegar, vineyard, vintage, *vin rouge,* winesap, wineberry, brandywine, *Vitis vinifera* (European grape), *Cathegesis vinitincta* (a moth), *Blastodacna vinolentella* (a moth).

L. *Bacchus* (Gr. *Bakchos; Iakchos*), m. god of wine, shouting, and revelry; L. *Dionysus* (Gr. *Dionysos*), m. earlier Greek equivalent of Bacchus; *Bacchanalis,* of Bacchus and the carousing festivals in his honor; *Thyone (Semele),* f. mother of Bacchus: Bacchus, bacchanalian, Dionysian, bacchante, *Physiculus bacchus* (a red cod), *Hapale jacchus* (a marmoset), *Thyone rosacea* (a holothurian), *Thyonella gemmata* (a holothurian), *Thyonidium pellucidum* (a holothurian), *Trachythyone muricata* (a holothurian).

L. *bublum,* n. a kind of wine:

L. *calidum,* n. warm wine and water, punch:

L. *capnius* (Gr. *kapnios*), f. a kind of wine from smoke-colored grapes:

L. *carenum* (Gr. *karoinon*), n. a sweet wine boiled down:

Gr. *gleukos,* n. sweet new wine, must; *gleukinos,* of new wine: gluconic acid, glucose, glucoside.   See *glykys* under **sweet**

Gr. *herpis,* m. an Egyptian word for wine:

L. *horconia,* f. a kind of wine:

L. *Icarus* (Gr. *Ikaros*), m. protege of Dionysus, who taught him viticulture:

L. *melina,* f. honey-wine, mead:

L. *merum,* n. unmixed wine; *meribibulus,* wine-bibbing:

Gr. *methy,* n. wine; *methysis,* f. drunkenness; *methysos,* drunk with wine; *methystikos,* intoxicating: methyl alcohol, methane, methanol, amethyst, *Piper methysticum* (kava-pepper):

Gr. *molax, -akos,* m. Lydian name for wine:

L. *mulsum,* n. wine sweetened with honey, mead:

L. *mustum,* n. new unfermented wine: *musteus,* of new wine; *mustulentus,* abounding in new wine; *mustuosus,* full of new wine: must, mustard.

Gr. *oinos,* m. wine; *oinidion,* n.; *oiniskos,* m. dim. poor wine; *oineros; oinikos; oininos,* of wine; *oinophlyx, -ygos,* drunk; *oinosis,* f. drunkenness; *exoinos,* drunken: enology (oenology), oenomancy, oenomel, oenophilist, *Oenothera speciosa* (an evening-primrose).

L. *passum,* n. raisin-wine:

L. *rosatum* (Gr. *rhosaton*), n. rose-wine:

L. *sapa,* f. new wine boiled thick:

L. *scyzinum,* n. a kind of wine:

Gr. *syphax, -akos,* m. sweet new wine: *Syphax fuliginosus* (a fossil spider).

Gr. *tryx, -ygos,* new, unfermented wine, must; see **dirt**

L. *vappa,* f. insipid wine:

L. *visula,* f. a kind of wine:

Gr. *zela,* n. Thracian word for wine:

See: **drink, grape, juice, berry**

wing < ON. *vaengr.*

L. *ala,* f. wing, upper part of arm that joins the shoulder; *alicula; alula,* f. dim.; *alaris; alarius,* of wings; *alatus; aliger, -a, -um; ales, -itis,* winged: alar, alate, alula, aisle, ailette, aileron, alinasal, alisphenoid, aliped, alalonga *(Germo alalunga), Alicula cylindrica* (a gastropod), *Bulla alicula* (a gastropod), *Dipterocarpus alatus* (a tree), *Rhus semialata* (a sumac), *Idiastes alaticollis* (a beetle), *Apteroscirtus inalatus* (a katydid), *Dioscorea alata* (a yam).

L. *penna (pinna)*, feather, wing, fin; see **feather**
Gr. *potanos*, winged, flying:
Gr. *pteron*, n. feather, wing; *pteryx, -ygos*, f. wing; *pteridion*; *pterygion*, n. dim.
fin; *ptesimos*, winged; *aptenos*, wingless, unable to fly: pteropod, pterygoid,
pterocarpous, pterospermous, pterotic, Diptera, Orthoptera, pterygium, pterygio-
phore, eurypterid, pterodactyl *(Pterodactylus spectabilis)*, *Pterocarpus erinaceus*
(a legume), *Pteropus conspicillatus* (spectacled fruit-bat), *Pterygogramma
acuminata* (a wasp), *Pterygioteuthis giardi* (a squid), *Pteridium aquilinum*
(bracken-fern), *Pteris vittata* (Chinese brake), *Archaeopteryx lithographica* (a
Jurassic bird), *Dipteryx oppositifolia* (British tonka-bean), *Rhagopteryx brahma*
(a beetle), *Ornithoptera victoriae* (a butterfly), *Calouteron reticulatum* (a
beetle), *Tetrapturus imperator* (spearfish), *Aptericula villosa* (a beetle),
Loxopterygium sagoti (an anacardiacead), *Valgus hemipterus* (a beetle).
Gr. *ptilotos*, winged; see *ptilon* under **feather**
See: **feather, fin, arm, fly, bird**

wink < AS. *wincian*.
Gr. *blepharizo*, wink:
L. *conniveo*, blink, half close the eyes, overlook; *nivens*, winking: connive, con-
nivent.
Gr. *epillos*, leering, squinting, < *epillizo*, wink:
L. *mico*, shine, twinkle, wink; see **light**
Gr. *myo*, close the eyes, wink:
L. *nicto, -atus*, wink, blink; *nictitans*, winking: nictitate, *Cercopithecus nictitans*
(guenon).
L. *paetus*, with a wink or blink in the eyes; *paetulus*, dim.: *Rhadinocerus paetulus*
(a beetle) *Rhaebothorax paetutus* (a spider).
L. *palpebro, -atus*, wink: palpebrate.
Gr. *skardamysso*, blink, wink; *skardamyktikos*, winking, blinking: *Scardamyctes
rufus* (a beetle).

winnow < AS. *windwian*, fan out chaff; see **separate**

winter < AS. *winter;* see **year**

wipe < AS. *wipian*.
Gr. *apomasso*, wipe off, wipe clean; *apomaxis*, f. a wiping off:
Gr. *apomysso*, wipe the nose:
Gr. *omorgnymi*, wipe; see **spot**
Gr. *smao; smecho*, wipe clean; *smektikos*, cleansing; *smexis*, a wiping off,
cleansing; see *smecticus* under **wash**
L. *tergeo, tersus*, cleanse, wipe, scour; see **wash**
See: **rub, wash**

wire < AS. *wir;* see **rope, bind**

wise < AS. *wis;* see **know, manner**

wish < AS. *wyscan*.
Gr. *agaios*, enviable:
L. *ambitio, -onis*, f. going about of candidates in canvassing and soliciting votes,
desire for honor, power, success; *ambitiosus*, desiring excessively: ambition,
ambitious.
Gr. *andromanes*, lusting after men; *hippomanes* (said of a mare in heat), lusting
after horses; see *mania* under **mad**
L. *appeto, -itus*, desire eagerly, long for; *appetibilis*; *expetibilis*, desirable:
appetite, appetizing.
L. *aspiro, -atus*, desire to reach or obtain: aspire, aspiration.
L. *aveo*, crave; *avarus; avidus*, eagerly desirous, covetous, greedy: avid, avidity,
avaricious, *Scellus avidus* (a fly).
L. *catulio*, desire the male; see **lewd**
L. *cupio, -itus*, desire; *cupidus*, desiring, longing, eager; *cupiditas, -atis*, f. avarice,
lust; *cupitor, -is*, m. wisher; *concupio*, be very desirous: cupidity, Cupid,
Kewpie, concupiscence, covet, covetousness.
L. *cuppes, -edis*, fond of delicacies, dainty:
L. *curiosus*, inquisitive, odd, strange: curious, curiosity.
L. *desidero, -atus*, long for, desire; *desiderabilis*, desirable; *desiderium*, n. a long-
ing, ardent wish: *Schuchertella desiderata* (a fossil brachiopod).
Gr. *ethelo (thelo)*, wish, be disposed to; *etheletos*, voluntary; *thelema, -tos*, n.
will; *theleos*, willing; *theletes*, m. willer; *theletikos*, of the will: *Ethelomorus
parallelus* (an isopod).
Gr. *himeros*, yearning for, desiring, desirable; see **beauty**
Gr. *ichar*, n. strong desire:
L. *invidus; invidiosus*, envious, jealous, hostile; see **jealous**
L. *libido*, desire, longing, passion, lust; see **lewd**

Gr. *mao*, desire, wish for eagerly, yearn for; *mematos*, desired:

L. *numen*, divine will, command; see **spirit**

L. *oestrus* (Gr. *oistros*), gadfly, sting, madness, sexual heat, rut, desire; see **fly**

L. *opto*, *-atus*, choose, wish; *exopto*, desire greatly, long for; see **choose**

Gr. *orexis*, f. appetite, desire, longing; *orektikos*, of desire, appetite; *orektos*, longed for, desired; *anorektos*, undesired, without appetite for; *anorexia*, f. lack of appetite: orexin, anorexia, *Orexita subgibbosa* (a beetle).

Gr. *pothos*, m. fond desire, < *potheo*, long for, yearn after:

L. *subo*, *-atus*, be in heat, have sexual ardor; see **lewd**

Gr. *taurao*, want the bull; see **lewd**

L. *testor*, *-atus*, declare, assert, make a will; *testamentum*, n. will; *intestatus*, without a will: testator, testament, intestate.

L. *volo*, wish, desire, will: volition, benevolent, voluntary.   See *voluntarius* under **free**

See: **hope, ask, believe, greed, lewd, agreeable, jealous**

wistaria, NL. a genus of the bean family; see **bean**

wit < AS. *witan*, know; see **know, laugh**

witch < AS. *wicce*, sorceress; see **magic**

witchhazel

NL. *fothergilla*, f. a genus of the witchhazel family, < John Fothergill, English physician: *Fothergilla monticola* (a hamamelidacead).

Gr. *hamamelis*, *-idos*, f. a tree with pomelike fruit; now applied to the witchhazel: hamamelin, Hamamelidaceae, *Hamamelis virginiana* (witchhazel), *Hormaphis hamamelidis* (an aphid).

NL. *liquidambar*, f. a genus of the witchhazel family: *Liquidambar formosana* (Formosa sweetgum).

NL. *parrotia*, f. a genus of the witchhazel family, < F. W. Parrot, German naturalist: *Parrotia persica* (a hamamelidacead).

with < AS. *with:* within, without, notwithstanding, withstand, withdraw, withhold, wherewithal, forthwith, drawing-room.

Gr. *a-*, with, union, sameness, copulative: Adelphia.

Gr. *amydis*, together, at the same time:

L. *apud*, at, near, in, with, among:

L. *axitiosus*, acting together, in combination; see **harmony**

L. *co-*, *col-*, *com-*, *con-*, *cor-*, < *cum*, together, with, as in *cooperor*, work with; *collaboro*, work with; *communico*, share with; *concurro*, run with; *corrivo*, run streams together; *cumprimis*, with the foremost, especially: coeducation, coincidence, cooperate, cogitate, cognate, cognomen, recognize, collection, collapse, collateral, commission, committee, combustion, companion, comfort, compete, complication, conserve, congregate, confusion, concomitant, conductor, constable, conclude, corrupt, corrode, correlation, correspond, vademecum, curry, cuspidor, cost, couch, council, viscount, custom, abscond, *Elampus connexus* (a wasp).

Gr. *hama*, together, at the same time, mutual: hamadryad, hamarchy, *Hamamelis mollis* (Chinese witchhazel).

Gr. *syl-*, *sym-*, *syn-*, *syr-*, *sys-*, < *syn*, together, with, as in *syllektos*, gathered together; *symbiosis*, a living together; *syndetikos*, binding together; *syrrhizos*, rooted together; *syssition*, a meal in common: syllable, syllepsis, syllogism, symbol, sympathetic, symposium, synagogue, synchronous, syncopation, syncline, syntax, syndicate, synonym, system, systole, syzygy, *Syllophodus fraternus* (a fossil rodent), *Symplocos chinensis* (Chinese sweetleaf), *Synedra superba* (a diatom), *Syrrhizus delusorius* (a wasp).

See: **companion, harmony, other, bind**

wither < AS. *wedrian;* see **lessen**

without; see **not, empty, poor, out**

witness < AS. *gewitness.*

L. *cognitor*, *-is*, m. advocate, protector, witness:

Gr. *martyr*, *-os*, c. witness; *martyria*, f. evidence, testimony; *martyrion*, n. proof: martyr, martyrdom, martyrize, protomartyr, Montmartre.

L. *notor*, *-is*, m. witness:

L. *superstes*, *-itis*, m. bystander, witness: *Superstes innominandus* (a fossil gastropod), *Ectocion superstes* (a condylarth).

Gr. *symphetor*, *-os*, m. witness:

Gr. *synistor*, *-os*, m. witness:

L. *testis*, *-is*, c. witness; *testificor*, *-atus*, bear witness; *testor*, *-atus*, declare, assert, bear witness, make a will; *testimonium*, n. evidence, proof, witness: testify, testimony, testimonial, testator, testament, contest, attestation, Protestant, detest, incontestable.

See: **try, judge, see**

wizard < AS. *wis,* know, *-ard,* in a high degree; see **magic**

woe < AS. *wa;* see **sad, trouble**

wolf < AS. *wulf.*

L. *irpus (hirpus),* m. wolf:

L. *lupus,* m. wolf; *lupulus,* m. dim.; *lupa,* f. she-wolf; *lupula,* f. dim.; *lupinus,* of wolves; *Luperca,* f. deified she-wolf that suckled Romulus and Remus: lupus, Lupercalia, lupine, *Canis lupus* (European wolf), *Oxyaena lupina* (a creodont), *Lupercalia ignita* (a moth).   Derive: Guadelupe.

Gr. *lykos,* m. wolf; *lykaina,* f. she-wolf; *lykideus,* m. dim.: *Lycogala epidendrum* (a myxomycete), *Lycopus uniflorus* (bugleweed), *Lycoperdon pyriforme* (a puffball), *Lycopodium complanatum* (a clubmoss), *Lycurus phleoides* (wolftail), *Lycosa arenicola* (a spider), *Lycaena marcida* (a butterfly), *Lycaon pictus* (Cape hunting-dog), *Lycaenopsis argiolus* (common blue butterfly), *Lycaenesthes bengalensis* (a butterfly), *Lycaenognathus angusticeps* (a fossil reptile), *Lycoteuthis diadema* (a squid).

Gr. *thos, thoos,* m. jackal: thooid, *Thos vulgaris* (former name of *Canis aureus*).

wolfram < G. *wolfram,* of uncertain derivation; see ELEMENTS under **thing**

woman < AS. *wifman:* womanly, womanhood, womanish, women, charwoman.

Gr. *Amazon, -os,* f. one of a race of warrior women; a strong, masculine woman: Amazon River, *Amazona aestiva* (a parrot), *Amazonetta brasiliensis* (a parrot).

L. *anus, -us,* f. old woman; *anicilla; anicula,* f. dim.; *anicularis; anilis,* of an old woman: anile.

L. *carisa,* f. an artful woman:

Gr. *chera,* f. widow; *chereios; cherikos,* of widows; *chereia; cherosyne,* f. widowhood; *cherosis,* f. bereavement: *Chera serratilinea* (a butterfly).

L. *concubina,* kept mistress; see **prostitute**

L. *domina;* Sp. *dona; duenna;* It. *donna,* lady; see *dominus* under **govern**

L. *Dryas, -adis* (Gr. *-ados*), a wood or tree nymph; see *drys* under **tree**

Gr. *Erigone,* f. daughter of Icarus, translated to the sky as the constellation Virgo: *Erigone tolucana* (a spider), *Eperigone tlaxcalana* (a spider).

L. *femina,* f. female, woman; *femella,* f. dim. girl; *femineus; femininus,* womanly, of women; *effeminatus,* womanish: feminine, femininity, female, effeminate, *Athyrium filixfemina* (ladyfern).

L. *formosa,* beautiful woman, belle; see **form**

Gr. *graus, -aos; graia,* f. old woman, old; *graidion,* n. dim. hag:

Gr. *gyne, gynaikos,* f. woman, female; *gynaikarion; gynaikon; gynaion,* n. dim.; *gynaikeios; gynaios,* of women; *gynaikikos; gynaikodes,* womanish: gynecology, gynaecocoenic, gynecomastia, gynecophore, gynandrous, gyniatrics, gynaeceum, misogynous, gyneocracy, epigynum, hypogynous, *Coelogyne cristata* (an orchid), *Coleogyne ramosissima* (a rosacead), *Glossogyne tenuifolia* (a composite), *Zygogynum pomiferum* (a magnoliacead), *Gynocardia odorata* (a flacourtiacead), *Gynopogon stellatus* (an apocynacead), *Gynaecomeloë opaca* (a beetle), *Mitragyna africana* (josswood), *Hymenoclea monogyna* (jacate).

L. *hera,* f. mistress of a house, lady, queen:

L. *innupta,* f. virgin, damsel, spinster:

Gr. *kymas, -ados,* a pregnant woman; see **fertile**

Gr. *lecho, lechoos,* f. woman in childbed, one who has just given birth:

L. *matrona,* married woman; see **spouse**

L. *mulier, -is,* f. woman; *muliercula,* f. dim.; *muliebris; mulierarius,* of a woman, womanly, female, feminine; *muliebrosus; mulierosus,* fond of women: mulier, muliebral, muliebrile, muliebrity, mulierose, muliebria.

Gr. *nethis, -idos,* spinster; see *neo* under **spin**

L. *nympha* (Gr. *nymphe*), f. bride, young woman, a maiden-spirit supposed to inhabit trees, springs, mountains, rivers, and the sea; *nymphalis,* of a nymph; *nymphaios,* of nymphs; *nymphidios; nymphikos,* of a bride, bridal; *nymphios,* m. bridegroom: nymph, nymphal, nymphosis, *Nymphaea caerulea* (Egyptian lotus), *Nymphalis antiopa* (mourning-cloak), *Nymphoides peltatum* (floating-heart), *Nymphula maculalis* (a moth), *Trichonympha agilis* (a protozoan). *Oreonympha nobilis* (a hummingbird).

L. *obstetrix, -icis,* midwife; see **nurse**

Gr. *parthenos,* f. maid, virgin; *partheniskarion; partheniske,* f. dim.; *parthenikos; parthenios,* of maidens, maidenly; *parthenon, -os,* m. maiden's apartment; *aeiparthenos,* ever a virgin: parthenic, Parthenon, Parthenope, parthenogenesis, *Parthenium integrifolium* (feverfew), *Parthenocissus tricuspidata* (Japanese creeper), *Cinnamomum parthenoxylon* (a lauracead).

L. *pellex, -icis* (Gr. *pallakis, -idos*), concubine, mistress; see **prostitute**
Gr. *phthinylla*, f. a thin or delicate woman: *Phthinylla fracta* (a Miocene rodent).
Gr. *potnia*, mistress, queen; see **govern**
L. *puerpera*, f. woman in childbed:
L. *quasillaria*, spinning-girl; see **spin**
L. *rapta*, f. the ravished or seduced one:
L. *striga*, f. hag:
Gr. *talis, -idos*, f. a marriageable maiden:
Gr. *thelys*, female, of or belonging to women; *thelykos*, feminine; *thelyphron*,
    effeminate: thelyblast, thelyotopy, theelin, *Thelymitra longifolia* (an orchid),
    Thelyphassa diaphana (a beetle).
L. *vetula*, old woman, crone; see *vetus* under **old**
L. *vidua*, f. widow; *vidualis*, of a widow: vidual, viduity, widow, *Vidua hypo-
    cherina* (whidah-bird).
L. *vira*, f. woman: *virago, -inis*, f. female warrior, quarrelsome woman: virago,
    viraginous.
L. *virgo, -inis*, f. maid, maiden; *virginalis*; *virgineus*, of maidens, maidenly;
    devirgino, -atus, deflower: virgin, Virginia, virginal, virginity, Virgo, *Anthro-
    poides virgo* (demoiselle), *Dasyscypha virginea* (a fungus), *Chionanthus virginica*
    (fringetree), *Philadelphus virginalis* (a mockorange).

* * * * * *

L. *-aena* (Gr. *-aina*), feminine suffix: Hyaenidae, hyena, *Leptaena plicatella* (a
    fossil brachiopod), *Oxyaena lupina* (a creodont).
OF. *-esse*, < L., Gr. *-essa, -issa*, feminine suffix, sometimes with diminutive signifi-
    cance: duchess, countess, princess, empress, Clarissa, lioness, laundress, goddess,
    heiress, waitress, *Auletrissa proletaria* (a bug). Derive: burgess.
Hen: "You say Caesar was killed by a woman?"
Peck: "Yes, for at the fatal moment he said, 'Oh, you brutess!'"
F. *-et, -ette*, diminutives, feminines: Harriet, Henrietta, suffragette, yeomanette.
L. *-ina*, feminines, sometimes with diminutive significance: Christina, Katrina,
    Czarina, Josephine, heroine, comedienne, *Dentalina acicula* (a foraminifer).
AS. *-ster, -stress*, originally denoting a female agent; see **make**
Gr. *-teira, -tis, -tria, -tris*, denoting female agent or doer; see *-ter* under **make**
L. *-trix*, signifying feminine agent, in contrast with *-tor*, masculine; see *-or* under
    make
See: **mother, sister**

womb < AS. *wamb*, belly.
L. *alvus*, belly, womb; see **belly**
Gr. *angos*, vessel, container, womb; see *angeion* under **bag**
Gr. *delphys, -yos*, f. womb: monadelphous, *Didelphys virginiana* (opossum).
Gr. *gaster*, stomach, belly, womb; see **stomach**
Gr. *hystera*, f. womb; *hysterikos*, of the womb: hysteria, hysteralgia, hysterectomy,
    hysterical, hysterocele, hysteropexy, hysteromania, *Hysteromorpha trilobum*
    (a trematode).
Gr. *kolops*, m. bosom, fold, uterus, but now applied particularly to the vagina;
    kolpidion, n. dim.; *kolpotos*, folded, sinuous; *kolpodes*, full of bays, folded:
    colpalgia, colpitis, colpoplasty, colpenchyma, gulf, *Colposcelis longicollis* (a
    beetle), *Colpothrinax wrighti* (barrel-palm), *Hexacolpus infundibulum* (a radio-
    larian), *Zeacolpus vittatus* (a gastropod).
L. *matrix, -icis*, f. mother, womb, embedding or enclosing substance: matrix,
    matriculate, *Matricaria grandiflora* (a composite), *Botrychium matricariaefolium*
    (a grape-fern).
Gr. *metra*, f. womb, entrance to the womb, vulva, matrix, pith or heart of a tree:
    Metrosideros tomentosa (pohutukawa), metrectomy, metrocele, metralgia,
    endometrium, endometriosis, *Cynometra floribunda* (a legume), *Stephanometra
    spicata* (a crinoid).
L. *stericula; sterilicula*, f. womb of an unfarrowed sow:
L. *uterus*, m. womb; *uterculus*, m. dim.: uterus, uterine, uterotomy.
L. *vulva*, wrapper, womb, external female genitals; see **vulva**

wonder < AS. *wundor*.
Gr. *athesphatos*, inexpressible, marvelous:
Gr. *eknomos; eknomios*, wonderful, marvelous, unusual, monstrous: *Ecnomogna-
    thus sericus* (a beetle).
Gr. *ekpaglos*, exceedingly, wondrous, marvelous, terrible; *ekplektikos*, astounding:
    Ecpaglus brevicornis (a wasp).
Gr. *exaisios*, beyond demand, extraordinary; see **over**
L. *extraordinarius*, above, beyond the common or ordinary, unusual, singular:
    extraordinary, *Neopromachus extraordinarius* (a grasshopper).

Gr. *glenos,* thing to stare at, show, wonder; see *glene* under **eye**

L. *incredibilis,* unbelievable, extraordinary: incredible.

L. *inusitatus,* unusual, extraordinary:

L. *miror, -atus,* wonder at, be astonished at; *mirabilis; mirandus; mirus,* wonderful, strange; *mirabundus,* full of wonder; *miraculum,* n. marvel; *mirificus,* causing wonder; *admirabilis,* producing wonder: Miranda, miracle, mirage, mirror, marvel, marvelous, *mirabile dictu!,* admirable, admire, *Mirabilis jalapa* (a fouro'clock), *Mirax insularis* (a braconid), *Hydrichthys mirus* (a hydroid), *Mesembryanthemum mirabile* (a fig-marigold), *Citropsis mirabilis* (a cherry-orange), *Anthophilax mirificus* (a beetle).

L. *monstrum,* an abnormal or supernatural wonder; see **animal**

Gr. *paragrapsimos,* exceptional:

Gr. *pelor,* monster, prodigy; see **large**

L. *prodigiosus,* strange, wonderful, extraordinary; *prodigium,* n. omen, wonder, monster: prodigy, *Scarabaeus prodigiosus* (a beetle).

L. *singularis,* unique, extraordinary; see **alone**

L. *stupendus,* astounding, wonderful, large: stupendous.

Gr. *teras, -atos,* n. monster, sign, marvel, wonder; *terastios,* marvelous, monstrous: teratology, teratogenic, teratoma, teratosis, teramorphous, *Teratophyllum aculeatum* (a fern), *Teratosaurus suevicus* (a fossil reptile), *Terataspis grandis* (a Devonian trilobite), *Terastiomyia lobifera* (a fly), *Teratornis merriami* (a Pleistocene bird).

Gr. *tethepa,* be amazed, astonished:

Gr. *thambema, -tos,* n. monster; *thambetos,* astonishing; *thambos,* n. astonishment; *ekthambos,* amazed, astonished: *Thamboceras mirum* (an ammonite), *Thambema emicorum* (an isopod).

Gr. *thauma, -tos,* n. wonder, marvel; *thaumasios; thaumastos,* wonderful, marvelous; *thaumastikos,* inclined to wonder: *Thaumasiocyclops insulans* (a copepod), *Bathothauma lyromma* (a squid), *Thaumastocoris australicus* (a bug), *Trophithauma dissitum* (a fly), *Agathaumas sylvestris* (a dinosaur).

Gr. *theskelos,* marvelous, wonderful: *Thescelosaurus neglectus* (a fossil reptile).

Gr. *thespesios; thespios,* ineffable, divine, wondrous: Thespis, Thespian, *Thespesia populnea* (banago), *Thespesiopsyllus paradoxus* (a copepod).

See: **fear, animal, secret, ineffable, magic**

woo < AS. *wogian.*

Gr. *mnester, -os,* m. wooer, suitor, < *mnaomai,* court, solicit; *mnestos,* wooed and won; *mnestys, -yos,* f. courtship:

L. *procus,* m. wooer, suitor: *Procus latrunculus* (a moth), *Scopaeus proculus* (a beetle).

See: **love**

wood < AS. *wudu:* wooden, woody, woodland, woodman, woodpecker, woodcraft, Woodlawn, firewood, plywood, cordwood, redwood.   Derive: woodchuck.

L. *alburnum,* n. sapwood of a tree:

L. *cala,* f. (Gr. *kalon,* n.), billet of wood, log for burning; *calamentum,* n. dry wood on the vine; *kalinos,* wooden; *kalopeo,* cut wood:

L. *caudeus,* of wood, wooden; see *caudex* under **stem**

NL. *cembra,* f. < G. *zember; zimmer,* timber: *Pinus cembra* (Swiss nut-pine), *Pinus cembroides* (Mexican pinyon).

L. *coctilis,* very dry wood that burns without smoke; *coctilicius,* of dried wood:

L. *cremium,* n. dry firewood, brushwood:

Gr. *dete,* f. fagot:

L. *duramen, -inis,* n. heartwood of a tree: duramen.

L. *fomes,* tinder, kindling wood; see **fungus**

Gr. *grynos,* m. fagot: *Grynobius castaneus* (a beetle).

Gr. *hyle,* f. wood, stuff, material; *hylikos,* of wood; *hylodes,* woody, wooded; *hylaios,* of the wood, forest, savage, wild; *hylekoites,* m. dweller or lodger in wood; *hyloros,* m. forester; *hylourgos,* m. carpenter: hyle, hylic, hylism, hylophagous, hylephobia, methyl, ethyl, *Hyla versicolor* (a tree-frog), *Hylobates agilis* (gibbon), *Hylocichla fuscescens* (veery), *Hylastes salebrosus* (a beetle), *Gulo hylaeus* (a wolverine), *Hylecoetus lugubris* (a beetle).

Gr. *laerkinon,* n. a kind of aromatic wood:

L. *lignum,* n. wood; *ligneolus; ligneus,* of wood, wooden; *lignosus,* woody; *lignarius,* m. carpenter: lignification, lignite, lignum-vitae, *Calceolaria lignosa* (a slipper-flower), *Selenops lignicolus* (a spider), *Limnoria lignorum* (a gribble).

Sp. *madera,* f. wood, timber:

Gr. *phryganon,* n. dry stick, firewood, brushwood; *phryganion,* n. dim.: *Phryganea cinerea* (a caddisfly).

Gr. *pyrdalon,* n. small wood for burning, brushwood:

L. *sarmentum,* fagot, brushwood; see **branch**

L. *tignum*, beam of wood, log of timber; *tigillum*, dim.; *tignarius*, carpenter; see **beam**

Gr. *xylon*, n. wood; *xylarion*; *xylephion*, n. dim.; *xylikos*; *xylinos*; *xylodes*, of wood, wooden, woody; *xyleboros*; *xylophagos*, eating wood; *xyleus*, m. wood-cutter: xylograph, xylophone, xylocarpous, xylem, xyloid, xylose, xylophilous, xylotile, xylolyl, *Xyleborus xylographus* (ambrosia-beetle), *Xylopia ferruginea* (an annonacead), *Xylion cylindricum* (a beetle), *Xylosoma japonica* (tung-ching-tree), *Xylaria hypopoxylon* (a fungus), *Xylocopa violacea* (a carpenter-bee), *Xylergates lacteus* (a beetle), *Xylinophorus scobinatus* (a beetle), *Xylobosca bispinosa* (a beetle), *Xyleutes maculatus* (a moth), *Brothylus conspersus* (a beetle), *Erythroxylum ovatum* (a coca-tree), *Nyctomyces entoxylinus* (a fungus), *Xanthoxylum alatum* (a prickly-ash), *Citharexylum ilicifolium* (a fiddlewood), *Sideroxylum foetidissimum* (mastic), *Trypoxylon rugifrons* (a wasp), *Aster xylorrhiza* (woody aster).

See: **tree, forest, beam, board, rod, pillar, stem, branch**

woodcock; see **snipe**

woodpecker

Gr. *dryokolaptes*, m. woodpecker: *Picus(Dryocolaptes) tridactylus* (a wood-pecker).

Gr. *dryops*, *-opos*, m. a woodpecker: *Dryops musgravei* (a beetle), *Dryopomera indica* (a beetle), *Ceradryops punctatus* (a beetle).

Gr. *ipne*, f. a woodpecker:

L. *iynx*, *-yngis* (Gr. *-os*), f. wryneck; *iyngikos*, magical, as the bird was used in magic: *Jynx[Iynx] torquilla* (wryneck), iyngine, *Iyngipicus kizuki* (a wood-pecker).

Gr. *kalotypos*, m. woodpecker:

Gr. *keleos*, f. a woodpecker: celeomorphic, *Celeopicus smaragdinicollis* (a wood-pecker).

Gr. *knipologos*, m. woodpecker: *Cnipologus anthracinus* (a bird).

Gr. *kolios*, m. a woodpecker: *Colius macrourus* (a coly).

Gr. *kraugos*, m. woodpecker: *Craugus auratus* (a bird).

Gr. *pelekas*, *-antos*, m. woodpecker:

L. *picus*, m. woodpecker: *Picus martius* (black woodpecker), *Picoides arcticus* (arctic woodpecker), piculet *(Picumnus lepidotus)*, *Ceocephalus picipes* (a beetle), *Odontopteryx toliapicus* (a fossil bird), *Sphyrapicus ruber* (red-breasted sapsucker).

Gr. *pipo*; *pipra*, f. woodpecker: *Pipra aureola* (a bird).

Gr. *typanos*, m. a kind of woodpecker: *Typanus aegyptius* (a bird).

wool < AS. *wull*.

L. *arnacis* (Gr. *arnakis*), sheepskin; see **skin**

Gr. *bathyrrhenos*, with thick wool:

Gr. *eiros*, n. wool; *eiresione*, f. garland or wreath wound with wool; *eiropokos*, woolly:

Gr. *erion*, n. wool; *eridion*, n. dim.; *erineos*, of wool, woolly; *eriodes*, like wool, woolly: erineum, erinose, erionite, eriophyllous, *Eriophorum angustifolium* (a cotton-sedge), *Eriodictyon lanatum* (yerba santa), *Eriobotrya japonica* (loquat), *Eriogonum nodosum* (a polygonacead), *Eriophyllum jepsoni* (a composite), *Fusicladium eriobotryae* (a fungus), *Gynerium sagittatum* (a grass), *Somateria spectabilis* (king-eider).

L. *floccus*, m. tuft or lock of wool; *floccosus*, tufty, woolly: floccose, floccillation, flocculable, flocculant, flocculence, flocculus, floccule, flock, floss, *Persea floccosa* (Oaxaca avocado), *Tabellaria flocculosa* (a diatom), *Begonia floccifera* (a begonia).

Gr. *gnaphalon* (*knaphalon*), n. wool; *gnaphalodes* (*knaphalodes*), like wool; *gnaphalion*, n. a downy plant: *Gnaphalium obtusifolium* (common everlasting), *Cnaphalocrocis rutilalis* (a moth).

Gr. *katagma*, *-tos*, n. flock of wool: *Catagma porcatum* (a fossil sponge), *Catagmatus japonicus* (a beetle).

Gr. *koas*, n. fleece; *kodarion*; *kodion*, n. dim.: *Bulbocodium vernum* (meadow-saffron).

Gr. *koleros*, short-wooled:

Gr. *krokis*, *-idos* (*krokys*, *-ydos*), f. nap or downy fibers on woolen cloth; *kroke*, f. thread, woof in a piece of cloth; *krokidismos*, m. a plucking at blankets by persons in delirium; *krokismos*, m. a weaving, web: crocidolite, *Crocidura occidentalis* (a shrew), *Croce filipennis* (a neuropterid), *Crocidium multicaule* (a composite), *Crocydoscelus ferrugineum* (a moth), *Crocethia alba* (sanderling).

Gr. *lachnos*, m.; *lachne*, f. soft, woolly hair, down: *Lachnosterna fusca* (June-bug), *Lachnanthes tinctoria* (redroot), *Epilachna ocellata* (a beetle).

L. *lana*, f. (Gr. *lenos*, n.), wool, woolly hair, fleece, down on leaves or fruit; *lanula*, f.; *lanugo, -inis*, f. down on plants and cheeks; *laniger, -a, -um*, bearing wool; *lanatus, laneus; lanicius*, of wool, woolly; *lanosus; lanugineus; lanuginosus*, woolly; *lanarius*, m. fuller, wool-worker; of wool; *lanificus*, spinning wool; *altilaneus*, of deep, thick wool: lanugo, laniferous, lanigerous, lanolin, lanoresin, lanose, delaine, *Chinchilla laniger* (chinchilla), *Eriosma lanigerum* (woolly aphid), *Heracleum lanatum* (a cow-parsnip), *Cheilanthes lanosa* (hairy lip-fern), *Cladothrix lanuginosa* (an amaranth), *Disporum lanuginosum* (fairy-bells), *Thymus lanicaulis* (a thyme), *Lenothrix canus* (a rat), *Digitalis lanata* (Grecian foxglove).

Gr. *lasios*, hairy, woolly: lasiocarpous, *Lasius niger* (an ant), *Lasioderma serricorne* (cigarette-beetle), *Lasiorhinus latifrons* (a wombat), *Lasiurus cinereus* (a bat), *Abies lasiocarpa* (subalpine fir), *Gordonia lasianthus* (loblolly-bay).

Gr. *mallos*, m. wool; *mallotos*, fleecy, woolly; *amphimallos*, woolly on both sides: mallophagous, Mallophaga, *Hypermallus villosus* (a twig-pruner), *Mallotus villosus* (capelin), *Rhododendron mallotum* (a rhododendron).

Gr. *nakos*, n. fleece; *nakyrion*, n. dim.: *Nacospatangus gracilis* (a sea-urchin).

Gr. *oulos*, woolly, curly; *oulotes, -etos*, f. woolliness, curliness; *ioulos*, m. down, wool, catkin: ulotrichan, oulopholite, Julius, July, *Ulothrix zonata* (an alga), *Ulochlaena hirta* (a moth), *Parmelia ulophylla* (a lichen), *Acrulocercus niger* (a bird).

L. *pappus* (Gr. *pappos*), wool or down on some seeds; see **hair**

Gr. *pilema, -tos*, n. felt; *pileo*, compress wool or other material into **felt**; *sympilema, -tos*, n. anything pressed or matted together: *Pilematechinus rathbuni* (an echinoid).

Gr. *pokos*, m. fleece, wool; *pokas, -ados*, f. wool, hair; *pokarion*, n. dim.; *epipokos*, woolly: *Pocobletus coroniger* (a spider), *Pocadius niger* (a beetle).

L. *tomentum*, woolly hairs; see **hair**

L. *vellus, -eris*, n. fleece, pelt; *velumen, -inis*, n. fleece; *Umbilicaria vellea* (a lichen), *Ateles vellerosus* (a spider-monkey).

Gr. *xasma, -tos*, n. carded wool:

See: **hair, thick**

word < AS. *word:* wordy, wording, wordcraft, Wordsworth, password, reword.

L. *appositum*, n. an epithet, adjective: apposition.

L. *dictum*, n. saying, word: dictum.

Gr. *epiklesis*, f. surname:

Gr. *epitheton*, n. term: epithet.

Gr. *epos*, n. word, tale, speech, song; *epyllion*, n. dim.; *epikos*, heroic: epic, cacoepy, orthoepy.

Gr. *etymon*, the ultimate or original meaning of a word; see *etymos* under **true**

Gr. *glossa*, f. an obsolete, foreign, or other word needing explanation: gloss, glossary.

L. *inscriptio, -onis*, f. title, epitaph: inscription.

Gr. *lexikos*, pertaining to words, < *lexis, -eos*, f. speech, word; *lexidion*, n. dim.; *lexikon*, n. dictionary, vocabulary: lexicon, lexicography.

Gr. *logos*, m. word, reason, discourse, story; *logarion; logidion; logydrion*, n. dim.; *logotechnes*, m. artificer in words; *philologos*, fond of words, studious of words: logomachy, philology, dialogue, apology, doxology, eulogy, logotype.   See *-logy* under **know**, *lego* under **speak**

AS. *nama*, term, title, word: name, nameless, namely, namesake, nickname, surname, Nome (Alaska).

L. *nomen, -inis*, n. name; *nominalis*, of a name; *nomino, -atus*, call by name; *praenomen*, first or personal name; *nomen*, clan name; *cognomen*, family name, surname; *agnomen*, additional surname or nickname, as in Quintus Fabius Maximus Cunctator: nominal, nomination, noun, nom de plume, nomenclature, binomial, cognomen, denominator, ignominious, innominate, misnomer, pronoun, renown.

L. *nuncupo, -atus*, call by name, name; see **call**

Gr. *onoma; onyma, -tos*, n. name; *onomation*, n. dim.; *onomasia*, f. a naming; *onomastes*, m. namer; *onomatikos*, of naming; *onomastikon*, n. vocabulary, list of names; *onomastos*, named, noted, famous; *anonymos; nonymos*, nameless, unknown: onomatopeia, onomasticon, onomastic, onomancy, onomatology, anonymous, Jerome, antonym, synonym, pseudonym, metonymy, eponym, antonomasia, paronomasia, *Onomastus quinquenotus* (a spider), *Stathmonyma homomorpha* (a moth), *Nonyma egregia* (a beetle).

Sp. *palabra*, f. word; *palabrita*, f. dim.:

Gr. *paroimia*, f. byword, maxim, proverb:

Gr. *rhema, -tos*, n. word, verb, phrase, saying; *rhemation*, n. dim.:

L. *syllaba* (Gr. *syllabe*), f. a sound-division of a word: syllable, syllabication.

Gr. *synthema, -tos,* n. watchword, sign, signal; *synthemation,* n. dim.:

L. *titulus,* m. inscription, label, sign: title, titled, titular, entitle, tilde, titrate.

L. *verbum,* n. word; *verbalis,* of words; *verbialis,* of a word; *verbosus,* wordy, prolix; *verbigero, -atus,* chat, talk, dispute: verb, adverb, proverbial, verbal, verbosity, verbatim, verbigerate, verbiage.

L. *vocabulum,* n. name, appellation; *vocamen, -inis,* n. name, title: vocabulary.

See: **speak, write, letter**

SOME GEOGRAPHIC NAMES

Africa: *Asparagus africanus* (an asparagus).
Allegheny: *Rubus allegheniensis* (blackberry).
America: *Ulmus americana* (American elm).
Andes: *Oenothera andina* (Andean sundrops).
Antilles: *Mymar antillarum* (a wasp), *Leptomastidea antillicola* (a wasp).
Arabia: *Acacia arabica* (babul-acacia).
Arizona: *Cupressus arizonica* (Arizona cypress).
Asia: *Ranunculus asiaticus* (Persian buttercup).
Atlantic: *Pipunculus atlanticus* (a fly).
Australia: *Stolotermes australicus* (a termite), *Microcitrus australasica* (a wild-lime).
Austria: *Arenaria austriaca* (Austrian sandwort).
Bengal: *Ficus bengalensis* (banyan-fig).
Brazil: *Aristolochia brasiliensis* (a birthwort).
California: *Eremobates californica* (a solpugid), *Helianthus californicus* (a sun-flower).
Canada: *Amelanchier canadensis* (serviceberry), *Asarum canadense* (Canada wild-ginger).
Chile: *Fragaria chiloensis* (Chilean strawberry).
China: *Aesculus chinensis* (Chinese horsechestnut), *Primula sinensis* (Chinese primrose).
Europe: *Trientalis europaea* (European starflower), *Asarum europaeum* (a wild-ginger).
France: *Rosa gallica* (French rose).
Germany: *Mespilus germanica* (medlar).
Great Britain: *Inula britannica* (a composite), *Pyrgoma anglicum* (a barnacle).
Greece: *Periploca graeca* (Grecian silk-vine).
Hudson: *Circus hudsonius* (marsh-hawk).
Iceland: *Pecten islandica* (a scallop).
India: *Ficus indica* (Indian fig), *Bos indicus* (zebu).
Ireland: *Megaceros hibernicus* (extinct Irish elk).
Italy: *Scilla italica* (Italian squill).
Japan: *Sorbus japonica* (Japanese mountain-ash), *Chrysanthemum nipponicum* (a Japanese daisy), *Japanopsychrolutes dentatus* (a fish).
Lapland: *Limosa lapponica* (godwit).
Lima: *Phaseolus limensis* (lima bean).
Louisiana: *Artemisia ludoviciana* (Louisiana sagebrush).
Macedonia: *Anthemis macedonica* (a chamomile).
Madagascar: *Asparagus madagascariensis* (an asparagus).
Mexico: *Salazaria mexicana* (bladder-sage).
Nevada: *Arnica nevadensis* (a composite).
New York: *Dryopteris noveboracensis* (New York fern).
Norway: *Aricia norvegica* (a worm).
Panama: *Sweetia panamensis* (a legume).
Paris: *Circaea lutetiana* (Paris circaea).
Pennsylvania: *Polygonum pennsylvanicum* (a knotweed), *Prunus pennsylvanica* (fire-cherry).
Persia: *Rosa persica* (one-leaved rose).
Portugal: *Prunus lusitanica* (Portuguese laurel-cherry).
Pyrenees: *Aquilegia pyrenaica* (a columbine).
Russia: *Prosthorhynchus rossicus* (a worm).
Siberia: *Larix sibirica* (Siberian larch)
Sitka: *Picea sitchensis* (Sitka spruce).
Spain: *Scilla hispanica* (Spanish squill), *Iberis gibraltarica* (a candytuft).
Sweden: *Luscinia suecica* (bluethroat).
Switzerland: *Androsace helvetica* (Swiss androsace).
Turkey: *Helix turcica* (a snail).
Utah: *Juniperus utahensis* (Utah juniper).
Yosemite: *Lewisia yosemitana* (Yosemite bitterroot).
Yukon: *Polytrichum yukonense* (a haircap-moss).

work < AS. *weorc;* see **make, busy**

world < AS. *weorold.*
    Gr. *kosmos,* m. world, universe; *kosmikos,* of the world; *kosmopolites,* m. citizen of the world: cosmic, cosmopolite, cosmopolitan, microcosm.
    L. *mundus,* m. world, earth, heavens; *mundanus,* of the world; *mundialis,* worldly: mundane, supramundane.
    Gr. *oikoumene,* f. the world: *oikoumenikos,* worldwide, general: ecumenical.
    L. *seculum,* age, lifetime, world; see **age**

worm < AS. *wyrm.*
    Gr. *askaris, -idos,* f. an intestinal worm: *Ascaris megalocephala* (a nematode), *Scaridium longicaudum* (a rotifer).
    L. *cicindela,* glowworm; a genus of beetles; see **beetle**
    Gr. *dex, dekos,* m. a worm in wood: *Demodex folliculorum* (follicle-mite).
    Gr. *eule,* f. worm, maggot:
    L. *galba,* f. a kind of larva: *Galba hieroglyphica* (a mite).
    Gr. *helmins, -inthos,* f. worm; *helminthion,* n. dim.: helminthology, anthelmintic, platyhelminth, *Helminthostachys zeylanica* (a fern), *Helmintherus vermivorus* (a warbler), *Chenopodium anthelminticum* (wormseed), *Hydnocarpus anthelminticus* (a flacourtiacead).
    Gr. *ips, ipos,* m. a kind of worm: *Ips radiatae* (Monterey-pine engraver), *Ipoctonus gabonensis* (a woodpecker), *Ipobracon mundella* (a braconid), *Ipomoea batatas* (sweet potato), *Amphibolips caelebs* (a gall-insect), *Ceratostomella ips* (a fungus).
    Gr. *ix, ikos,* f. a kind of worm or grub: *Ix porrecta* (a bug).
    Gr. *keraïs, -idos,* f. a worm that eats horn:
    Gr. *lampyris, -idos,* f. glowworm: *Lampyris noctiluca* (European glowworm).
    L. *lumbricus,* m. earthworm: lumbricine, *Lumbricus terrestris* (an earthworm), *Ascaris lumbricoides* (a roundworm), *Monocystis lumbrici* (a sporozoan).
    Gr. *pygolampis, -idos,* f. glowworm:
    L. *rauca,* f. worm in oak roots:
    Gr. *rhomos,* woodworm, shipworm; see **mollusk**
    NL. *sabella,* f. a genus of marine annelids: *Sabella pavonina* (an annelid).
    Gr. *samamithion,* n. a kind of worm:
    L. *scolex, -ecis* (Gr. *skolex, -ekos),* m. worm; *skolekion,* n. dim.; *skolekodes,* wormlike: scoleciform, scoleciasis, scolecite, scolecodont, scolecoid, scolecophagous, *Scolithus linearis* (a Cambrian fossil), *Scolecoseps boulengeri* (a lizard), *Mytilidon scolecosporum* (a fungus), *Bothrioscolex prussicus* (a cestode), *Aspidodera scoleciformis* (a nematode).
    NL. *syllis,* f. a genus of annelids: *Syllis pusilla* (an annelid).
    L. *taenia* (Gr. *tainia),* ribbon, tapeworm; see **ribbon**
    L. *thrips, -ipis* (Gr. *-ipos),* a woodworm; a genus of Thysanoptera; see **bug**
    L. *tinea,* a gnawing worm, larva of a moth; see **moth**
    L. *vermis, -is,* m. worm; *vermiculus,* m. dim.; *vermiculatus,* like worms, of worms; *verminosus,* wormy: vermin, verminal, verminous, vermicular, vermicelli, vermifuge, vermilion, vermiform, *Vermetus lumbricalis* (a gastropod wormshell), *Vermileo degeeri* (a worm-lion) *Vermivora peregrina* (Tennessee warbler), *Vermicularia spirata* (a worm-shell), *Sarcobatus vermiculatus* (greasewood), *Eosentomon vermiforme* (a proturan), *Myzocytium vermicolum* (a fungus), *Oxyuris vermicularis* (pinworm).
    See: *larva* under **young**

wormwood < AS. *wermod* (G. *wermuth):* vermouth.
    L. *absinthium* (Gr. *apsinthion),* n. wormwood: *Artemisia absinthium* (wormwood), absinthe, absinthial, absinthium, absinthin.
    L. *artemisia,* f. mugwort, sagebrush: *Artemisia tridentata* (sagebrush).
    Gr. *santonion; santonikon,* n. a kind of wormwood: santonica, santonin, *Artemisia santonica* (a wormwood).
    Gr. *seriphos,* f. a kind of wormwood:
    L. *sirium,* n. mugwort:
    L. *sozusa,* f. a kind of artemisia:
    Gr. *toxetesia,* f. another name for *artemisia:*

worry < AS. *wyrgan;* see **trouble**

worship < AS. *weorthscipe;* see **honor, rite**

wort < AS. *wyrt,* plant; see **plant**

worth < AS. *weorth,* value: worthy, worthily, untrustworthy, seaworthiness, Wordsworth, Lapworth, Ellsworth, worship, worthless.
    L. *aestimabilis,* valuable; see *aestimo* under **judge**
    Gr. *ainetos,* praiseworthy: *Aenetus prasinus* (a moth).

Gr. *axios,* worthy, fit; *axia,* f. worth, value; *axiosis,* f. worth, excellence; *axiologos,* worthy of mention, notable, remarkable; *antaxios,* worth as much as; *epaxios,* worthy, deserving: chronaxy, *Axiocrita cataphanes* (a moth), *Axiologus thoracicus* (a fossil insect), *Axiologa pura* (a moth), *Axiagastus rosmarus* (a bug).

L. *caritas,* dearness, costliness, high price; see *carus* under **love**

L. *dignus,* worthy, deserving, fitting; see **honor**

L. *diligo, -lectus,* value highly, esteem, love; see **love**

Gr. *epikairios,* important:

Gr. *epitimion,* n. value or assessment of a thing; *epitimetes,* m. estimator, valuer; *eritimos,* precious: *Epitimetes lutosus* (a beetle), *Coeranica eritima* (a moth).

Gr. *keimelios,* treasured; see *keimelion* under **store**

Gr. *logimos,* notable, remarkable: *Logimus erminea* (a psychodid).

L. *oculitus,* as one's own eyes, dearly:

L. *pretiosus,* valuable, dear: precious.

L. *valeo,* be strong, have value; see **strong**

See: **wealth, honor, money**

worthless

L. *abjectus; rejectus,* cast off, worthless; see *jacio* under **throw**

Gr. *achrestos,* useless: *Achrestocoris cinerarius* (a fossil bug).

Gr. *akrantos,* futile, idle, unaccomplished: *Acrantus teyou* (a lizard), *Acrantophis dumerili* (a snake).

Gr. *anaxios; apaxios,* unworthy, worthless: *Anaxius obesus* (a beetle).

Gr. *anopheles,* useless, hurtful: *Anopheles punctipennis* (a mosquito).

Gr. *apobletos,* rejectable, worthless: *Apobletes amphibius* (a beetle).

Gr. *apodokimos,* worthless:

Gr. *asymphoros,* useless, inconvenient: *Asymphorodes sphenocopa* (a moth).

Gr. *asynergos,* not helpful:

Gr. *asynteles,* useless:

Gr. *asyphelos,* mean, paltry, worthless:

L. *brisa,* grape-skin, refuse after pressing; see **skin**

Gr. *deilos,* wretched, miserable, paltry, weak; see **poor**

Gr. *etosios,* fruitless, useless, in vain:

L. *futilis,* useless, vain, worthless: futile, futility.

Gr. *gryte,* f. trash, frippery; *grytopoles,* m. seller of knickknacks:

L. *inanis,* empty, useless, vain; see **empty**

L. *inutilis,* useless, worthless, injurious: *Bulimus inutilis* (a gastropod).

L. *irritus,* invalid, vain, useless, infertile: *Centrinus irritus* (a beetle).

Gr. *maps,* idly, vainly; *mapsidios,* vain, useless; *mapsilogos,* talking idly: *Mapsidius auspicata* (a moth).

Gr. *medaminos; oudaminos,* good for nothing:

Gr. *meleos,* in vain, useless: *Meleoma signoreti* (a neuropterid), *Meleonoma heterota* (a moth).

L. *mictilis,* worthless, bad; see **bad**

L. *nequam,* worthless, bad; *nequitia,* badness, worthlessness; see **bad**

L. *otiosus,* idle, useless; see *otium* under **rest**

Gr. *oudeneia,* f. worthlessness:

Gr. *outidanos,* worthless, insignificant: *Utidana pleurostigma* (a moth).

Gr. *rhopos,* m. petty wares, trumpery; *rhopikos,* of petty wares, worthless: *Rhopophilus pekinensis* (a bird), *Rhopica cornigera* (a fly), *Rhopobota naevana* (fireworm).

L. *scrutum,* n. trash, rags, trumpery, rubbish; *scrutarius,* m. second-hand dealer, ragman: *Trochus scrutarius* (a wasp).

Gr. *teusios,* vain, futile, idle:

Gr. *tycheros,* ragged, tattered, worn; see **rag**

L. *vilis,* cheap, base, worthless: vile, vilify, vilification, revile.

See: **bad, poor, fool, unfit, empty, weak, trifle, dirt, rag, rest**

wound < AS. *wund;* see **sore, bruise, cut, tear**

wrapper < uncertain origin; see **bag, cover, fold, garment, sheath, skin**

wrasse < C. *urach.*

Gr. *alphestes,* m. a wrasse: *Alphestes afer* (a fish).

Gr. *chromis, -ios,* m. a sea-fish: *Chromis caeruleus* (a damsel-fish), *Haplochromis obliquidens* (a fish), *Hemichromis bimaculatus* (a fish), Chromidae.

L. *iulis, -idis* (Gr. *ioulis, -idos*), f. rainbow-wrasse: *Julis bifasciata* (a fish), *Julidochromis ornatus* (a fish), *Stethojulis trilineata* (a fish), *Labrus julis* (a fish).

Gr. *kichle,* f. a kind of wrasse: *Cichla ocellaris* (a fish), *Cichlasoma fenestrum* (a fish).

L. *labrus* (Gr. *labros*), m. wrasse: *Labrus vulgaris* (rainbow-wrasse), *Labrichthys psittacula* (parrot-fish).
L. *lelepris*, f. a kind of wrasse:
Gr. *onias*, m. a sea-fish:
L. *phycis*, *-idis* (Gr. *phykis*, *-idos*), m. a fish living among seaweeds, probably a kind of wrasse; *phykidion*, n. dim.: *Phycis mediterraneus* (figo, Mediterranean hake), *Phycidimorpha rosea* (a moth).
L. *scarus* (Gr. *skaros*), m. a kind of fish, probably a wrasse: scar, *Scarus strongylocephalus* (Indian parrot-fish), *Pseudoscarus caelestinus* (loro).

wrath < AS. *wrath;* see **anger**

wreath < AS. *wraeth;* see **crown, branch, turn**

wreck < AS. *wrec;* see **destroy**

wren < AS. *wrenna.*
Gr. *basileus*, king; name for a wren; see **govern**
Gr. *orchilos*, m. a kind of wren: *Acrorchilus erythrops* (an ovenbird), *Thryorchilus browni* (a wren).
L. *regulus*, dim. of *rex*, king; name for a wren or other small bird; see *rex* under **govern**
L. *trochilus* (Gr. *trochilos*), m. a small bird, probably a wren: *Trochilus polytmus* (a hummingbird), *Phylloscopus trochilus* (willow-wren).
Gr. *troglodytes*, m. cave man; name for a wren: *Troglodytes aedon* (house-wren).

wrestle < AS. *wrestlian;* see **fight**

wretch < AS. *wrecca*, an unhappy person; see **sad**

wright < AS. *wryhta;* see *wyrcan* under **make**

wrinkle < AS. *wrincle;* see **fold**

wrist < AS. *wrist;* see **hand**

write < AS. *writan:* writing, writer, written, rewrite, wrote, underwriter.
L. *actuarius*, m. shorthand writer, copyist, clerk: actuary, actuarial.
L. *amanuensis*, m. clerk, secretary: amanuensis.
L. *codicillus*, letter, note, appendix to a will, dim. of *codex*, *-icis*, tablet, writing; see *codex* under **book**
Gr. *diploma*, *-tos*, n. a letter folded double, letter of recommendation, passport, certificate: diploma, diplomacy, diplomat.
L. *epistola* (Gr. *epistole*), f. letter: epistle, epistolary.
Gr. *grapho*, write; *grapheus*, m. writer, painter; *graphe*, f. a writing; *graphikos*, of writing, drawing, painting; *graptos*, written, marked, inscribed; *gramma*, *-tos; graphos*, *-eos*, n. letter, line, mark, picture; *grammateus*, m. scribe, secretary; *kalligraphos*, writing beautifully: graphite, graptolite, graft, gram, grammar, biography, diagram, chirography, epigraph, lithographic, photograph, geographer, geography, telegraph, epigram, monogram, *Graphocephala coccinea* (a leafhopper), *Phyllograptus angustifolius* (a graptolite), *Graptophyllum pictum* (morado), *Grammatophyllum speciosum* (an orchid), *Ips calligraphus* (a beetle), *Opegrapha betulina* (a lichen).
L. *notarius*, m. stenographer, secretary, clerk: notary, prothonotary, *Protonotaria citrea* (prothonotary warbler).
L. *scribo, scriptus*, write; *scriptilis*, writable; *scriptura; scriptio*, *-onis*, f. a writing; *scriba*, m. public writer, clerk, secretary; *scriptor*, *-is*, m. writer; *conscribillo*, *-atus*, scrawl; *describo*, copy, sketch, delineate; *inscriptio*, *-onis*, f. a writing upon something: scribe, scribble, scripture, scrivener, shrive, conscription, describe, description, escritoire, inscription, manuscript, nondescript, prescription, postscript, subscribe, *Tragelaphus scriptus* (guib), *Sphaerophoria scripta* (a fly), *Arion circumscriptus* (a slug), *Etheostoma inscriptum* (a darter), *Scilla non-scripta* (blue squill).
Gr. *synthetes*, m. author, composer, writer:
L. *tabellio*, *-onis*, m. notary, scrivener: tabellion.
See: **pen, letter, line, word, book, list**

wrong < AS. *wrang*, contrary to best experience; see **bad, fault, lie, hurt, wander**

wry < AS. *wrigian*, turn, twist; see **turn**

X

xandaros, Gr. a fabulous sea-monster; see **animal**

xanion, Gr. comb for carding wool; *xantes,* carder; see **comb**

xanthium, L. (Gr. *xanthion*), a genus of the composite family; see **composite**

xanthos, Gr. yellow; see **yellow**

xanthoxylum, NL. a genus of the rue family; see **rue**

xasma, Gr. carded wool; see **wool**

xenagos, Gr. commander of auxiliary troops; see **lead**

xenium, L. (Gr. *xenion*), gift to a guest; *xeniolum,* dim.; see **give**

xenodochium, L. (Gr. *xenodocheion*), inn, lodge; *xenodochos,* innkeeper; see **house**

xenos, Gr. stranger, guest; *xenyllion,* dim.; *xenikos,* strange, foreign; see **guest**

xeros, Gr. dry; see **dry**

xestos, Gr. scraped, planed, smoothed, polished; *xesis,* a scraping; *xesma,* scrapings; see *xeo* under **scrape**

xi, Gr. fourteenth letter of the Greek alphabet; see **letter**

xiphias, Gr. swordfish; see **swordfish**

xiphos, Gr. sword, saber; *xiphidion; xiphion,* n. dim.; see **sword**

xoanon, Gr. wooden image; *xoanoglyphos,* sculptor; see **form**

xutho- < Gr. *xouthos,* yellowish, yellowish-brown; see **yellow**

xyele, Gr. a tool for scraping wood, rasp, dagger; see *xeo* under **scrape**

xylabion, Gr. fire-tongs; see *labis* under **tongs**

xyleus, Gr. woodcutter; see *xylon* under **wood**

xylinum, L. (Gr. *xylinon*), cotton; see **cotton**

xylo- < Gr. *xylon,* wood; *xylarion; xylephion,* dim.; *xylikos; xylinos,* wooden; see **wood**

xylochos, Gr. copse, thicket; see **bush**

xynos, Gr. common; *xynon,* companion, partner; see **common**

xyris, L., Gr. wild iris; see **iris**

xyron, Gr. razor; *xyrion,* dim.; see **knife**

xysilos, Gr. shaven, smooth; *xysma,* a shaving, scraping; see *xeo* under **scrape**

xysticus, L. (Gr. *xystikos*), pertaining to scraping or polishing; *xyster; xystra,* scraper; see *xeo* under **scrape**

xyston, Gr. shaft of a spear; see **spear**

xystus, L. (Gr. *xystos*), a covered colonnade, portico, gallery, walk; *xystikos,* of a gallery; see **porch**

Y

yam < African *nyami,* eat.
NL. *dioscorea,* f. a genus of the yam family, < Dioscorides, Greek botanist: *Dioscorea esculenta* (Chinese yam), Dioscoreaceae.
Gr. *epipetron,* n. a plant; now a genus of the yam family: *Epipetrum humile* (a dioscoreacead).

yard < AS. *gierd; gyrd,* rod, stick, see **rod, measure**; < AS. *geard,* garden, see **field**

yarrow < AS. *gearwe.*
L. *achillea,* f. yarrow, < Achilles, hero of the Trojan war: *Achillea argentea* (silver yarrow), *Gilia achilleaefolia* (a gilia).

Gr. *ptarmike*, f. yarrow: *Achillea ptarmica* (sneezewort-yarrow), *Aster ptarmicoides* (upland-aster), *Chrysanthemum ptarmicaeflorum* (a chrysanthemum), *Dodonaea ptarmicaefolia* (a sapindacead).

yawn < AS. *ganian;* see **open**

year < AS. *gear:* yearly, yearling, Goodyear, leap year, yore.
L. *annus,* m. year; *anniculus,* m. dim. yearling, of a year; *annalis; annarius,* of a year; *anniversarius; annualis; annuus,* yearly; *annosus,* full of years; *annotinus,* of last year; *biennis,* lasting two years; *perennis,* lasting through the years, everlasting, perpetual; *sollennis,* annual: annuity, annals, anniversary, annual, centennial, biennial, perennial, decennial, millennium, superannuated, 1893 A.D., annalist, solemn, *Erigeron annuus* (fleabane), *Fomes annosus* (a bracket-fungus), *Lycopodium annotinum* (a clubmoss), *Oenothera biennis* (an evening-primrose), *Bellis perennis* (European daisy), *Linum perenne* (a flax), *Scleranthus annuus* (knawel), *Agrostis perennans* (a grass).
L. *bimus,* of two years; *bimulus,* dim. only two years; *trimulus,* of three years:
Gr. *etos,* n. year; *etesios,* yearly; *amphietes,* yearly; *epeteios,* annual; *dietesios,* lasting through the year: eteostic, etesian, *Isoëtes lacustris* (a quillwort).
Gr. *hora,* f. any time or season of the year; *horaios,* seasonable, timely, ripe:
L. *hornotinus; hornus,* of this year, this year's:
Gr. *perysinos,* of last year; *properysi,* two years ago; *properysinos,* of year before last:
Gr. *setanios,* this year's:
L. *Vertumnus,* god of the changing year, the seasons, and trade; see **change**

SEASONS

L. *aestas, -atis,* f. summer; *aestivus,* of summer: estival, estivation, *Dendroica aestiva* (yellow warbler), *Vitis aestivalis* (summer-grape).
L. *autumnus(auctumnus),* m. season of increase and abundance, fall; *autumnalis,* of autumn: autumn, autumnal, *Trombicula autumnalis* (harvest-mite).
L. *brumalis,* of the winter solstice, wintry; see *bruma* under **day**
Gr. *cheimon, -os,* m.; *cheima, -tos,* n. winter, winter weather, storm; *cheimas, -ados,* f. winter, a winter garment; *cheimatikos; cheimerinos; cheimerios,* of winter, wintry, bleak, cold: *Chimonophila signella* (a moth), *Chimonobambusa falcata* (a bamboo), *Chimonanthus praecox* (wintersweet), *Chimatophila castaneana* (a moth), *Chimaphila maculata* (spotted pipsissewa), *Achimenes longiflora* (a gesneriacead).
Gr. *ear, -os (er, -os),* n. spring; *earinos,* of spring, the color of spring, green: *Earomyia lonchaeoides* (a fly), *Earinus jezoensis* (a braconid), *Earina autumnalis* (an orchid), *Eranthis hyemalis* (winter-aconite).
AS. *haerfest,* crop, season for gathering crops; see **reap**
L. *hibernum,* n. winter; *hibernalis; hibernus,* of winter, wintry; *hibernaculum,* n. winter quarters; *hiberno, -atus,* spend the winter, keep in winter quarters, be inactive: hibernal, hibernation, hibernaculum, *Filistata hibernalis* (a house-spider).
L. *hiems, -mis,* f. winter; *hiemalis (hyemalis),* of winter, wintry: hiemal, hiemation, *Troglodytes hiemalis* (winter-wren), *Equisetum hyemale* (a horsetail), *Junco hyemalis* (snowbird).
Gr. *opora,* f. autumn; *oporinos,* autumnal; *metoporon,* n. late autumn; *metoporinos; phthinoporinos,* autumnal: *Oporornis formosus* (Kentucky warbler).
L. *solstitium,* longest day of the year, summer solstice, summer; see **day**
Gr. *thereia,* f.; *theros,* m. summer; *thereios; therinos,* of summer; *theretron,* n. summer home; *zatheres,* scorching: therophyte, *Theroscopus pedestris* (a wasp), *Theriophila miara* (a moth), *Theriobius rugatus* (a beetle), *Therinopsilus fuscicornis* (a wasp).
L. *ver, -is,* n. spring; *verculum,* n. dim.; *vernalis; vernus,* of spring: vernal, vernation, primaveral, *Primula veris* (cowslip), *Crocus vernus* (spring-crocus), *Eubranchipus vernalis* (fairy-shrimp), *Amanita verna* (a mushroom).
See: **time, month, day**

yeast < AS. *gist.*
NL. *-ase,* a suffix denoting a ferment or enzyme: alcoholase, aldehydase, amidase, amylase, apozymase, butyrinase, carbohydrase, carotenase, casease, cellulase, cozymase, cytase, desoxyribonuclease, diastase, hyaluronidase, hydrogenase, hydrolase, invertase, lactase, ligninase, lipase, luciferase, maltase, peptidase, phytase, reductase, sucrase, tyrosinase, urease, zymase, zymohexase.
AS. *breowan,* ferment: brew, Brewster, brewmaster, broth.
L. *fermentum,* n. yeast; *fermento, -atus,* leaven with yeast; *infermentatus,* un-leavened: ferment, fermentation.
L. *levamen, -inis,* n. a lightening or raising: leaven.

Gr. *zyme*, f. yeast, leaven; *zymosis*, f. fermentation; *zymotikos*, causing fermentation; *azymos*, unleavened: zymogen, zymology, zymolysis, zymoplastic, zymotic, zymohexase, zymosterol, enzyme, enzymatic, azymic.

yellow < AS. *geolu:* yellowish, yellowhead, yellowwood, yellowwort, Yellowstone.
Sp. *amarillo*, yellow: Amarillo (Texas), amarillo *(Chlorophora tinctoria).*
L. *aurigineus*, jaundiced, yellow; see *aurigo* under **disease**
L. *cerinus* (Gr. *kerinos*), wax-colored, yellowish; see *cera* under **wax**
L. *crocatus; croceus*, saffron-yellow; see **crocus**
L. *flavus*, golden-yellow, yellow; *flavidus*, yellowish: flavescent, flavedo, flavaniline, flavine, flavone, flavopurpurin, riboflavin, *Sarracenia flava* (a pitcher-plant), *Aurelia flavidula* (a jellyfish), *Reticulotermes flavipes* (a termite), *Gentiana flavida* (a gentian), *Cnidocampa flavescens* (a moth), *Iris flavissima* (an iris), *Chamaecrista flavicoma* (a partridge-pea), *Eristalis flavoscutellata* (a fly), *Pipistrellus subflavus* (a bat), *Halictus flavipes* (a bee).
L. *galbus*, yellow; *galbinus*, yellowish: jaundice.
L. *gilvus (gilbus)*, pale-yellow; *albogilvus*, whitish-yellow: *Vireo gilvus* (warbling vireo).
L. *helvus*, bay, yellow; *helvenacus; helvinus; helvolus (helveolus)*, yellowish, pale yellow: helvite, helvolous, *Cyperus helvus* (a sedge), *Strophostyles helvola* (a legume).
Gr. *ikterikos*, jaundiced, yellow; see *ikteros* under **disease**
Gr. *kirrhos*, yellow, tawny: cirrhosis, cirrhonosus, cirrolite[cirrholite], *Cirrhosoma translucida* (a moth), *Cirrhopetalum appendiculatum* (an orchid).
Gr. *kitrinos*, of citron, citron-yellow; see **citrus**
Gr. *knekos*, pale-yellow: *Brachynus (Cnecostolus) limbellus* (a beetle).
L. *luridus*, pale-yellow, dull-red, ghastly; *luror, -is*, m. a yellowish color, sallowness: lurid, *Aspidistra lurida* (a liliacead), *Brachychiton luridus* (a bottle-tree).
L. *luteus*, yellow; *luteolus*, dim. yellowish: corpus luteum, lutein, luteolin, luteous, luteofuscous, luteolous, lutescent, *Betula lutea* (yellow birch), *Reseda luteola* (a mignonette), *Chrysalidocarpus lutescens* (a palm), *Microtrycherus luteosignatus* (a beetle).
L. *melinus*, honey-colored, see *mel* under **honey;** < Gr. *melinos*, quince-yellow, see *malum* under **apple**
Gr. *ochros*, pale-yellow, sallow, wan; *ochra*, f. earthy oxide of iron; *ochriasis*, f.; *ochroma, -tos*, n. paleness, wanness; *enochros*, somewhat pale; *exochros*, very pale: ocher(ochre), ocherous, ochrolite, ochraceous, *Ochrocarpus longifolius* (nagkassar), *Ochroma pyramidale* (balsa), *Ochrosia elliptica* (an apocynacead), *Leptothrix ochracea* (a bacterium), *Dillenia ochreata* (a dilleniacead), *Anthidiellum polyochrum* (a bee).
L. *ravus*, grayish-yellow, gray; see **gray**
Ar. *safra*, yellow: saffron, safranine, safranol, safflower *(Carthamus tinctorius).*
L. *sandaraca* (Gr. *sandarake*), a reddish-yellow color; see **resin**
L. *silaceus*, like yellow ocher; see *sil* under **earth**
L. *sulfureus*, like sulfur, sulfur-yellow; see **sulfur**
Gr. *thapsinos*, yellow, < *thapsia*, a plant from Thapsos; see **carrot**
L. *vitellinus*, yellow like the yolk of an egg; see *vitellus* under **yolk**
Gr. *xanthos*, yellow, yellowish-red, orange, golden; *xanthotes*, f. yellowness; *epixanthos*, yellowish: xanthic, xanthin, xanthite, xanthophyll, xanthopterin, xanthosis, xanthous, *Xanthorrhiza simplicissima* (yellowroot), *Xanthoxylum americanum* (prickly-ash), *Xanthostemon oppositifolius* (a myrtacead), *Xanthium italicum* (Italian cocklebur), *Anthoxanthum odoratum* (sweet vernalgrass), *Liostomus xanthurus* (a spot), *Microbracon xanthonotus* (a braconid).
Gr. *xouthos*, yellowish, yellowish-brown: *Xuthotrichia ochracea* (a trichopterid).
See: **gold, citrus, crocus, carrot, amber,** *gambogium* under **resin**

yes < AS. *gese.*
L. *accedo, -essus*, assent to, approve, nod to: *Senecio accedens* (a groundsel).
L. *acquiesco, -etus*, rest, assent to: acquiesce, acquiescence.
Gr. *agapao*, be content, acquiesce:
Gr. *aineo*, praise, recommend; *ainesis*, f. praise; *synainos*, agreeing with:
L. *aio*, say yes, affirm, assert: adage.
L. *annuo, -utus*, nod to, assent, approve, affirm; *annuto, -atus*, nod to: *Annuit coeptis.* See *nuto* under **bend**
L. *approbo, -atus*, assent to, favor: approbation, approve, approval.
L. *assentor, -atus*, agree constantly, flatter; *assensus, -us*, m. approval, echo: assent, assentator.
L. *autumo, -atus*, say aye, affirm, aver:
Gr. *deta*, of course, to be sure:
L. *ita*, so, yes:
Gr. *nai*, yes, yea verily:

Gr. *neuma, -tos,* n. nod, approval; *neusis,* f. a nodding, inclination; *aponeuma, -tos,* n. slope: *Neumatoceras gibberosum* (a fossil cephalopod).

Gr. *symphemos,* agreeing with, < *symphemi,* assent, approve:

Gr. *synchoreo,* agree, assent; *synchoretes,* m. conceder, forgiver; *synchorema, -tos,* n. concession: *Synchorema zygoneura* (a trichopterid).

Gr. *synemptosis,* f. concurrence:

Gr. *stergo,* be satisfied, acquiesce:

See: **agreeable, harmony**

yew < AS. *iw.*

Gr. *milos,* f. yew:

L. *taxus,* f. yew; *taxeus; taxicus,* of yew: *Taxus cuspidata* (a yew), *Taxodium distichum* (bald cypress), *Cephalotaxus drupacea* (plum-yew), *Torreya taxifolia* (stinking-cedar).

yield < AS. *gieldan,* give, pay; see **give, pay, let**

-yl, a suffix signifying stuff, matter, generally an alcohol, < Gr. *hyle,* wood, raw material; see **alcohol, wood**

yoke < AS. *geoc;* see **bind, bridge**

yolk < AS. *geolca.*

Gr. *lekithos,* f. yolk of an egg: lecithin, alecithal, lecithinase, *Lecithobotrys putrescens* (a worm), *Sterrhurus monolecithus* (a trematode).

L. *modiolum,* n. yolk:

L. *vitellus,* m. yolk of an egg; *vitellinus,* of or like the yolk of an egg: vitelline, vitellicle, vitellose, *Crocus vitellinus* (Syrian crocus).

young < AS. *geong:* younger, youngest, youngster, youngling, Youngstown, youth.

L. *adolescens, -entis,* young, growing up: adolescent.

Gr. *agouros,* m. a youth:

Gr. *anathremma, -tos,* n. nurseling:

Gr. *anebos,* immature, not yet come to manhood, beardless: *Anebocaris quadroculus* (a crab).

Gr. *anelikos,* immature, not yet come to manhood:

L. *Apollo, -inis* or *-onis* (Gr. *Apollon, -os*), m. god of youth, music, healing: Apolline, Apollonian, Apollinaris, *Parnassius apollo* (a butterfly), *Apollocrinus geometricus* (a fossil crinoid).

Gr. *apten, -os,* unable to fly, unfledged, callow; see **fly**

Gr. *brephos,* n. fetus, embryo, babe; *brephion; brephyllion,* n. dim.; *brephikos; brephodes,* childish: *Brephoctonus californicus* (a wasp), *Brephidium exile* (a butterfly).

L. *catellus; catulus,* young of an animal, whelp, puppy; see **dog, cat**

NL. *cercaria,* f. tailed, tadpolelike larval stage of trematodes: cercaria, *Bilharzia cercariae* (a blood-fluke).

Gr. *chrysallis, -idos,* f. gold-colored pupa of a butterfly: chrysalis, *Chrysalidocarpus lutescens* (a butterfly-palm), *Goniobasis chrysalis* (a fossil gastropod), *Cyrtocapsa chrysalidium* (a radiolarian).

L. *conceptum,* n. fetus:

L. *cossus,* m. a kind of larva under the bark of trees: *Cossus ligniperda* (goat-moth).

Gr. *embryon,* n. fetus, the unborn young (in humans, the first two months): embryo, embryonic, *Diospyros embryopteris* (gaub-persimmon), *Embryopteris glutinifera* (wild mangosteen), polyembryony.

Gr. *eule,* worm, maggot, larva; see **worm**

L. *fetus (foetus), -us,* m. offspring, fruit, young, the unborn young (in humans, the last seven months): fetus, fetal, feticide, effete.

Gr. *galathenos,* sucking, young, tender:

L. *gyrinus* (Gr. *gyrinos*), tadpole; *gyrinodes,* tadpolelike; see **frog**

Gr. *hebe,* f. youth, time of pubescence, youthful prime; *hebetikos,* of youth, youthful; *hebos,* at the threshold of manhood; *ephebeia,* f. puberty; *ephebos,* arrived at puberty; *ephebikos,* of youth: Hebe, hebetic, hebeanthous, ephebic, ephebian, *Hebe buxifolia* (a speedwell), *Hebeloma crustuliniforme* (a fungus), *Epheboblatta attenuata* (a fossil blattid), *Delphinium hebegynum* (a larkspur), *Indigofera hebepetala* (an indigo-plant).

AS. *hwelp,* young of a dog or beast of prey, puppy, cub: whelp.

L. *immaturus,* unripe, green, untimely: immature, immaturity.

L. *implumis,* unfledged, callow:

L. *infans, -antis,* speechless, babe: infant.

L. *instar,* form, figure, image, likeness; stage of an insect between molts; see **form**

L. *juvenis,* young; *junior,* younger; *juvencus,* young; *juvenilis,* of youth: juvenile, junior, rejuvenate, rejuvenescence, *Curculio juvencus* (a beetle).

Gr. *koreios*, youthful; see *kore* under **child**

Gr. *kourios*, youthful; see *koros* under **child**

Gr. *kyema, -tos*, n. embryo, fetus: *Cyema atrum* (a fish), *Dicyema paradoxum* (a mesozoan).

L. *larva*, f. ghost, mask, early stage of some animals: larva, larval, larvicidal, *Bacillus(Achromobacter) larvae* (bacterium of foulbrood), *Mutilla larvata* (a wasp), *Xenopicus albolarvatus* (white-headed woodpecker).

L. *nauplius*, m. the larval form of many crustaceans: nauplius.

Gr. *neanikos*, fresh, youthful; see *neanis* under **child**

Gr. *nearos*, young; *neoteros*, younger; *neatos*, youngest; see *neos* under **new**

Gr. *nebrax, -akos*, m. young deer, young animal: *Lagonebrax javanicus* (a mammal).

L. *necydalus* (Gr. *nekydalos*), m. larva of the silkworm: *Necydalopsis trigonata* (a beetle).

Gr. *neossos (neottos)*, m. any young animal, nestling, chick: neossology, neossoptile, *Neossus marylandica* (a fly).

L. *nympha* (Gr. *nymphe*), f. pupa of an insect: nymph, nymphal.

Gr. *obrikala*, n. pl. the young of animals: *Obricala foveicollis* (a beetle).

Gr. *ocheuma*, embryo, the result of coitus; see *ocheuo* under **coitus**

Gr. *pallax, -akos (pallex, -ekos)*, c. youth; *pallakos*, m. lover; *pallakion*, n. dim.: palikar, *Pallacocris suavis* (a bug).

L. *partus, -us*, m. that which is born, embryo: parturient.

Gr. *phortion*, burden, weight, embryo; see *phortos* under **weight**

L. *primaevus*, early, young; see *primus* under **first**

L. *proles, -is*, f. offspring, youth: proletarian, proletariat, prolific, proliferate, *Filago prolifera* (a composite).

Gr. *psakalos*, new born, young:

L. *pubertas, -atis*, f. period of transition from youth to manhood and womanhood; *impubes, -eris*, not having attained manhood: puberty, pubertal.

L. *puerilis*, youthful, childish, silly; see *puer* under **child**

L. *pullus*, a young animal, young fowl; see **chicken**

L. *pupa*, girl, doll; in some insects a stage intermediate between larva and adult; see **child**

L. *ranunculus*, tadpole; see **frog**

L. *recens, -entis*, new, fresh, young: recent, recently.

Gr. *schadon, -os*, f. larva of a bee or wasp; the breeding cell itself: *Chaoborus (Schadonophasma) trivittatus* (a culicid).

L. *scymnus* (Gr. *skymnos*), m. a young animal, cub, whelp; *skymnion*, n. dim.: *Scymnus punctatus* (a beetle), *Scymnophagus townsendi* (a chalcid-fly), *Scymnodon ringens* (a fish), *Scymnognathus angusticeps* (a fossil reptile).

Gr. *teknon*, that which is born, child, young; *teknidion; teknion*, dim.; see **child**

L. *tiro (tyro), -onis*, m. young soldier, recruit, beginner: tyro.

See: **first, new, begin, before, dawn, caterpillar, birth, bird**

ypsilo- < Gr. *ypsilon (hypsilon)*, upsilon, twentieth letter of the Greek alphabet; see **letter**

yrcha, Gr. jar; see *hyrcha* under **jar**

yucca, NL. a genus of the lily family; see **lily**

Z

za-, Gr. intensive particle; see **very**

zabros, Gr. gluttonous; see *bibrosko* under **eat**

zacholos, Gr. very angry; see *cholos* under **anger**

zacoto- < Gr. *zakotos*, very angry; see *kotos* under **anger**

zale, Gr. surging sea, storm; see **wave**

zalmos, Gr. a skin; see **skin**

zamia, L. (Gr. *zemia*), hurt, damage, see **hurt**; a genus of cycads, see **gymnosperm**

zanclo- < Gr. *zanklon*, sickle; see **sickle**

zannichellia, NL. a genus of aquatic plants; see **waterweed**

zarza, Sp. bramble; see **blackberry**

zatricio- < Gr. *zatrikion*, the game of chess; see **play**

zea, L. (Gr. *zea; zeia*), a kind of grain, spelt; now applied to maize; see **grain**

zeal < L. *zelus* (Gr. *zelos*), ardor, enthusiasm; see **busy**

zebra, the Abyssinian name for the striped equine of Africa; see **horse**

zela, Gr. wine; see **wine**

zelkova, NL. a genus of the elm family; see **elm**

zelos, Gr. emulation, rivalry, ardor; *zelotes*, a devoted enthusiast, zealous follower; see *zelus* under **busy**

zema, Gr. a decoction; see **drink**

zemia, Gr. damage, loss; *zemiodes*, ruinous; see **hurt**

zene, Gr. a goldfinch; see **finch**

Zeno- < Gr. *Zen, -os*, a form of *Zeus*; see *Dis* under **spirit**

zeo, Gr. boil; *zestos*, boiled; see **boil**

zephyrus, L. (Gr. *zephyros*), west wind; see **wind**

zero < It. *zero*, < Ar. *sifr*, empty; see **empty, number**

zestos, Gr. boiled; see **boil**

zeta, Gr. sixth letter of the Greek alphabet; see **letter**

Zetes, Gr. son of Boreas and one of the Argonauts; see **sail**

zetetes, Gr. searcher, seeker; see **try**

zeuglo- < Gr. *zeugle*, loop of a yoke; see *zygon* under **bind**

zeugos, Gr. pair, team; *zeuktos*, yoked; *zeugma*, anything used for joining; see *zygon* under **bind**

Zeus, Gr. chief of the gods, see *Dis* under **spirit**; < L. *zeus*, a kind of fish, see **dory**

zibetha, ML. civet; see *civetta* under **weasel**

zignis, Gr. a kind of lizard; see **lizard**

zigzag, F. alternately changing direction by sharp angles; see **slope**

zilla, NL. a genus of the mustard family; see **mustard**

zinc < G. *zink*.   See ELEMENTS under **thing**
 L. *cadmia* (Gr. *kadmia*), f. calamine, an ore of zinc: cadmium, calamine.

zingiber, L. (Gr. *zingiberis*), ginger; see **ginger**

zinnia, NL. a genus of the composite family; see **composite**

zinzala, L. a kind of gnat; see **fly**

zinzilulo, L. chirp; see **chirp**

zirbus, L. caul, omentum; see **membrane**

zirconium < Ar. *zarqun*, gold-colored; see ELEMENTS under **thing**

ziro- < Gr. *zeira*, a wide garment, robe; see **garment**

zizanium, L. (Gr. *zizanion*), a kind of weed; see *zizania* under **grass**

zizia, NL. a genus of the carrot family; see **carrot**

ziziphora, NL. a genus of the mint family; see **mint**

zizyphus, L. (Gr. *zizyphos*), jujube-tree; see **buckthorn**

zoarco- < Gr. *zoarkes*, life-supporting; see *zoë* under **life**

zoarion, Gr. dim. of *zoön*, animal; see **animal**

zodiac < Gr. *zodiakos*, of animals; *zodarion; zodion*, dim. of *zoön*, animal; see *zoön* under **animal**

zoë, Gr. life; see **life**

zogrion, Gr. menagerie, see **pen**; *zogreus*, catcher of animals, see **take**

zoma, Gr. a girded garment, the position of the girdle, girdle; see **belt**

zomerysis, Gr. soup ladle; see **spoon**

zomile, Gr. a kind of dill; see **carrot**

zomos, Gr. soup, sauce; *zomidion*, dim.; see **soup**

zona, L. (Gr. *zone*), belt, girdle; *zonula*, dim.; *zonalis; zonarius*, of a belt; see **belt**

zoön, Gr. animal; *zoarion; zodarion; zodion; zoyphion,* dim.; see **animal**

zophos, Gr. darkness, dusk, gloom, nether world; *zopheros,* dusky, gloomy; see **night**

zopyron, Gr. spark; see *pyr* under **fire**

zoros, Gr. pure, sheer; see **pure**

zoster, Gr. belt, girdle; see **belt**

zostera, NL. a genus of aquatic plants; see **waterweed**

zothecula, L. dim. of *zotheca* (Gr. *zotheke*), chamber, chest; see **room**

zoyphion, Gr. dim. of *zoon,* animal; see **animal**

zygaena, L. (Gr. *zygaina*), a kind of shark; see **shark**

zygastron, Gr. box, chest, archives; see **box**

zygia, Gr. probably the hornbeam; see **hornbeam**

zygis, Gr. wild thyme; see **mint**

zygoma, Gr. bar, bolt; see **bar**

zygon; zygos, Gr. yoke, pair, balance; *zygion,* dim.; *zygios,* of a yoke; see **bind**

zyme, Gr. leaven, yeast; *zymotikos,* causing fermentation; see **yeast**

zythum, L. (Gr. *zythos*), a kind of Egyptian beer; *zythion,* dim.; see **beer**

BIBLIOGRAPHY

Agassiz, Louis. *Nomenclator zoologicus.* Soloduri, 1842-1846.

American Speech. Quarterly. Columbia University Press, 1925-

Atkinson, B. F. C. *The Greek language.* Faber & Faber, 1931.

Bailey, Dorothy and K. C. *Etymological dictionary of chemistry and mineralogy.* E. Arnold & Co., 1929.

Bardsley, Charles W. *A dictionary of English and Welsh surnames.* Frowde, 1901; *English surnames, their sources and significations.* Chatto & Windus, 1906.

Barfield, Owen. *History in English words.* Methuen & Co., 1926.

Baugh, Albert C. *A history of the English language.* D. Appleton-Century Co., 1935.

Baxter, J. H., and Johnson, Charles. *Medieval Latin word-list.* Oxford University Press, 1934.

Bell, Ralcy H. *The worth of words.* 1903; *The changing values of English speech.* 1909. Hinds.

Bentley, Harold W. *A dictionary of Spanish terms in English.* Colorado University Press, 1932.

Berrey, Lester V., and Van den Bark, Melvin. *The American thesaurus of slang.* Thomas Y. Crowell, 1947.

Bierce, Ambrose. *The Devil's dictionary.* World Publishing Co., 1943.

Blackie, Miss C. *A dictionary of place-names.* J. Murray, 1887.

Boisacq, Emile. *Dictionnaire etymologique de langue Grecque.* Winter, 1938.

Bowman, William D. *The story of surnames.* G. Routledge & Sons, 1931.

Brachet, Auguste. *Dictionnaire etymologique de langue Francaise.* Paris, 1870.

Bradley, Henry. *The making of English.* Macmillan Co., 1904.

Brown, Roland W. *Materials for word-study.* Van Dyck & Co., 1927.

Buriss, E. E., and Casson, L. *Latin and Greek in current use.* Prentice-Hall, 1939.

Charnock, Richard S. *Praenomina, or the etymology of the principal Christian names of Great Britain and Ireland.* Trübner & Co., 1882.

Chemical dictionaries.
  Hackh, Ingo W. D. *Chemical dictionary* (Grant revision). Blakiston Co., 1944.
  Rose, Arthur and Elizabeth. *The condensed chemical dictionary* (Turner, editor). Reinhold Pub. Corp., 1950.

Classical dictionaries.
  Carey, M., et al. *Oxford classical dictionary.* Oxford University Press, 1950.
  Lewis, Charlton T., and Short, Charles. *Latin dictionary* (revision of E. A. Andrews' translation of Freund). American Book Co., 1907.
  Liddell, H. G., and Scott, Robert. *A Greek-English lexicon.* Oxford University Press, 1940.
  Peck, Harry T. *Harper's dictionary of classical literature and antiquities.* Harper & Brothers, 1898.
  Smith, William, and Hall, T. D. *English-Latin dictionary.* American Book Co., 1871.
  Yonge, Charles D. *English-Greek lexicon* (Drisler revision). Harper & Brothers, 1870.

Clements, Frederick E. *Greek and Latin in biological nomenclature.* University of Nebraska Studies, vol. 3, no. 1, 1902.

Curtius, George. *Principles of Greek etymology.* J. Murray, 1886.

Dalla Torre, C. G. de, and Harms, H. *Genera siphonogamarum.* Engelmann, 1900-1907.

Dellquest, A. W. *These names of ours.* Thomas Y. Crowell, 1938.

Diringer, David. *The alphabet: a key to the history of mankind* (English translation). Hutchinson's Sci. and Tech. Pub., 1948.

Engeln, O. D. von, and Urquhart, Jane M. *The story-key to geographic names.* D. Appleton & Co., 1924.

English dictionaries.
    Century. Century Co., 1911.
    New English. Oxford University Press, 1888-1933.
    Standard. Funk & Wagnalls, 1953.
    Webster. New International. G. & C. Merriam Co., 1954; *New World.* World Publishing Co., 1951.

English Place-name Society. *English place-names.* Guardian, 1923.

Ernst, Margaret S. *Words.* A. A. Knopf, 1937.

Fernald, J. C. *Historic English.* Funk & Wagnalls, 1921.

Fowler, H. W. *A dictionary of modern English usage.* H. Milford, 1937.

Freeman, William. *Plain English.* D. Appleton-Century Co., 1939.

Funk, Wilfred J. *Word origins and their romantic stories.* Wilfred Funk, 1950.

Gannett, Henry S. *Origin of certain place-names in the United States.* U. S. Geol. Survey Bull. 258, 1905.

Gover, J. E., et al. *Place-names in Devon,* etc. Cambridge University Press, 1931-

Greenough, J. B., and Kittredge, G. L. *Words and their ways in English speech.* Macmillan Co., 1922.

Gudde, Erwin G. *California place-names.* University of California Press, 1949.

Hargrave, Basil. *Origins and meanings of popular phrases and names.* J. B. Lippincott Co., 1911.

Hellquist, Elof. *Svensk etymologisk ordbok.* Gleerup, 1939.

Henderson, I. F. and W. D. *A dictionary of scientific terms* (Kenneth revision). D. Van Nostrand Co., 1952.

Horwill, H. W. *A dictionary of modern American usage.* Oxford University Press, 1935.

Hough, John N. *Scientific terminology.* Rinehart & Co., 1953.

Index Kewensis of flowering plants (with supplements). Oxford University Press, 1886-

Jackson, Benjamin D. *A glossary of botanic terms.* Gerald Duckworth & Co., 1949.

Jaeger, Edmund C. *A source-book of biological names and terms.* Charles C. Thomas, 1944.

Jespersen, O. *Growth and structure of the English language.* Blackwell, 1923; *Language, its nature, development, and origin.* Henry Holt & Co., 1934.

Johnson, Burges. *The lost art of profanity.* Bobbs-Merrill, 1948.

Jones, W. Paul. *Writing scientific papers and reports.* W. C. Brown & Co., 1946.

Kelly, James P. *Workmanship in words.* Little, Brown & Co., 1917.

Kelsey, Harlan P., and Dayton, William A. *Standardized plant names.* J. Horace McFarland Co., 1942.

Kent, Roland G. *Language and philology.* Marshall Jones Co., 1923.

Kluge, Friedrich. *An etymological dictionary of the German language* (Davis translation). G. Bell & Sons, 1891.

Körting, Gustav. *Lateinisch-Romanisches Wörterbuch*. Paderborn, 1907.

Krapp, George P. *The English language in America*. Century Co., 1925.

Laird, Charlton. *The miracle of language*. World Publishing Co., 1953.

Link, J. T. *The origin of the place-names of Nebraska*. Nebraska Geol. Survey Bull. 7, 1933.

McArthur, Lewis A. *Oregon geographic names*. Binfords, 1944.

McDonald, P. B. *Scientific terms in American speech*. American Speech, vol. 2, pp. 67-70, Nov., 1926.

McKnight, George H. *English words and their background*. 1923; *Modern English in the making*. 1928. D. Appleton & Co.

McNicoll, David H. *Dictionary of natural history terms*. Lovell, Reeve & Co., 1863.

Mason, W. A. *A history of the art of writing*. Macmillan Co., 1920.

Mayne, R. G. *An expository lexicon*. John Churchill, 1860.

Mead, Leon. *How words grow*. Thomas Y. Crowell, 1907.

Medical dictionaries.
   Dorland, W. A. N. *The American illustrated medical dictionary*. W. B. Saunders Co., 1951.
   New Gould medical dictionary. Blakiston Co., 1949.

Melander, A. L. *Source-book of biological terms*. The Comet Press, 1937.

Mencken, H. L. *The American language*. 4th ed., 1936; Supplement 1, 1945; 2, 1948. Alfred Knopf.

Meyer-Lübke, W. *Romanisches etymologisches Wörterbuch*. Winter, 1924.

Miller, Walter. *Scientific names of Latin and Greek derivation*. California Acad. Sci. Proc., 3rd ser., Zoology, vol. 1, no. 3, pp. 115-143, 1897.

Moorhouse, Alfred C. *The triumph of the alphabet*. Schuman, 1953.

Müller, F. M. *Biographies of words*. Longmans, Green & Co., 1888.

Muller, Henri F., and Taylor, Pauline. *A chrestomathy of Vulgar Latin*. D. C. Heath & Co., 1932.

Names. Quarterly. American Names Society. University of California Press, 1953-

Neave, S. A. *Nomenclator zoologicus*. Zoological Soc. of London, 1939-1950.

Northrop, Stuart A. *A glossary of scientific terms*. Albuquerque, N. Mex., 1949.

O'Neill, Elizabeth. *Stories that words tell us*. T. C. and E. C. Jack, 1918.

Palmer, A. Smythe. *Folk-etymology*. G. Bell & Sons, 1882.

Palmer, T. S. *Index generum mammalium*. U. S. Dept. Agri., Biological Survey. North American Fauna 23. Government Printing Office, 1904.

Partridge, Eric. *A dictionary of slang and unconventional English*. Macmillan Co., 1938; *Name into word*. Seeker & Marbury, 1949.

Pei, Mario. *The story of language*. 1949; *The story of English*. 1952. J. B. Lippincott Co.

Pfeiffer, Ludovicus. *Nomenclator botanicus*. Casellis, 1874.

Pyles, Thomas. *Words and ways of American English*. Random House, 1952.

Rice, Clara M. *Dictionary of geological terms*. Edwards Brothers Inc., 1945.

Roget, T. M. *Thesaurus of English words and phrases* (Mawson revision). T. Y. Crowell, 1922.

Roller, Duane. *The terminology of physical science*. University of Oklahoma Press, 1929.

Salmon, Lucy. *Place-names and personal names as records of history.* American Speech, vol. 2, pp. 228-232, Feb., 1927.

Savory, Theodore H. *Latin and Greek for biologists.* University of London Press, 1946; *The language of science.* Andre Deutsch, 1953.

Schenk, Edward T., and McMasters, John H. *Procedure in taxonomy.* Stanford University Press, 1936.

Schenkling, Sigmund. *Nomenclator coleopterologicus.* Gustav Fischer, 1922.

Schlauch, Margaret. *The gift of tongues.* Modern Age Books, 1942.

Scudder, Samuel H. *Nomenclator zoologicus.* U. S. Nat. Mus. Bull. 19, 1882.

Sherborne, Charles D. *Index animalium, 1758-1800.* 1902; *1801-1850.* 1922-1932. Cambridge University Press.

Shimer, Hervey W. *Origin and significance of plant names.* Hingham, Mass, 1943.

Shipley, J. T. *Dictionary of word origins.* The Philosophical Library, 1945.

Skeat, Walter W. *Etymological dictionary of the English language.* Clarendon Press, 1910.

Skillin, Marjorie E., and Gay, Robert M. *Words into type.* Appleton-Century-Crofts, 1948.

Skinner, Henry A. *The origin of medical terms.* Williams & Wilkins Co., 1905.

Skinner, Hubert M. *The story of letters and figures.* O. Brewer Co., 1905.

Smith, Elsdon C. *The story of our names.* Harper & Brothers, 1950.

Smith, Logan P. *Words and idioms; The English language.* Constable & Co., 1950.

Smith, S. S. *The command of words.* T. Y. Crowell, 1935.

Spilman, Mignonette. *Medical Latin and Greek.* Edwards Brothers Inc., 1949.

Steinmetz, E. F. *Vocabularium botanicum.* Amsterdam, 1947.

Stewart, George R. *Names on the land.* Random House, 1945.

Taylor, Isaac. *Names and their histories.* Macmillan Co., 1896; *Words and places.* E. P. Dutton, 1927.

Thompson, D'Arcy W. *A glossary of Greek birds.* 1936; *A glossary of Greek fishes.* 1947. Oxford University Press.

Trench, R. C. *On the study of words.* George Routledge & Sons, 1905; *English past and present.* Charles Scribner, 1871.

Vizetelly, F. H. *Essentials of English speech and literature.* Funk & Wagnalls, 1915.

Voss, Andreas. *Botanisches Hilfs und Wörterbuch* (Martin Tessenow revision). P. Parrey, 1929.

Walde, Alois. *Lateinisches etymologisches Wörterbuch.* Winter, 1938.

Weekley, Ernest. *The romance of names.* 1914; *Surnames.* 1916; *The romance of words.* 1917; *An etymological dictionary of modern English.* 1921; *Words ancient and modern.* 1926; *More words ancient and modern.* 1927; *Words and names.* 1932. J. Murray.

Wells, Evelyn. *What to name the baby.* Garden City Books, 1953.

Woods, Robert S. *The naturalist's lexicon.* Abbey Garden Press, 1944.

Wyld, Henry C. *A short history of English.* J. Murray, 1927.

Yancey, P. H. *Origins from mythology of biological names and terms.* Bios, vol. 16, 1945.

Yonge, Charlotte M. *History of Christian names.* Macmillan Co., 1878.

Zoological Record, volumes 1864-1951. Zoological Soc. of London.

INDEX